Genomics, Circuits, and Pathways in Clinical Neuropsychiatry

Genomics, Circuits, and Pathways in Clinical Neuropsychiatry

Edited by

Thomas Lehner
Office of Genomics Research Coordination
National Institute of Mental Health
Bethesda, MD, United States

Bruce L. Miller
Department of Neurology, School of Medicine
University of California San Francisco
San Francisco, CA, United States

Matthew W. State
Department of Psychiatry, School of Medicine
University of California San Francisco
San Francisco, CA, United States

ELSEVIER

AMSTERDAM • BOSTON • HEIDELBERG • LONDON • NEW YORK • OXFORD • PARIS
SAN DIEGO • SAN FRANCISCO • SINGAPORE • SYDNEY • TOKYO

Academic Press is an imprint of Elsevier

Academic Press is an imprint of Elsevier
125 London Wall, London EC2Y 5AS, UK
525 B Street, Suite 1800, San Diego, CA 92101-4495, USA
50 Hampshire Street, 5th Floor, Cambridge, MA 02139, USA
The Boulevard, Langford Lane, Kidlington, Oxford OX5 1GB, UK

Notices
Knowledge and best practice in this field are constantly changing. As new research and experience broaden our understanding, changes in research methods, professional practices, or medical treatment may become necessary.

Practitioners and researchers may always rely on their own experience and knowledge in evaluating and using any information, methods, compounds, or experiments described herein. In using such information or methods they should be mindful of their own safety and the safety of others, including parties for whom they have a professional responsibility.

To the fullest extent of the law, neither the Publisher nor the authors, contributors, or editors, assume any liability for any injury and/or damage to persons or property as a matter of products liability, negligence or otherwise, or from any use or operation of any methods, products, instructions, or ideas contained in the material herein.

British Library Cataloguing-in-Publication Data
A catalogue record for this book is available from the British Library

Library of Congress Cataloging-in-Publication Data
A catalog record for this book is available from the Library of Congress

ISBN: 978-0-12-800105-9

For information on all Academic Press publications
visit our website at https://www.elsevier.com/

Publisher: Nikki Levy
Acquisition Editor: Nikki Levy
Editorial Project Manager: Barbara Makinster
Production Project Manager: Caroline Johnson
Designer: Matthew Limbert

Typeset by TNQ Books and Journals
www.tnq.co.in

Contents

Section IV
Clinical Translation and Drug Development

List of Contributors

Nii A. Addy, Yale University School of Medicine, New Haven, CT, United States

Laura Almasy, University of Texas Rio Grand Valley, Brownsville, TX, United States

Lynn M. Almli, Emory University, Atlanta, GA, United States; Harvard Medical School, Belmont, MA, United States

Nicholas AuYong, University of California Los Angeles, Los Angeles, CA, United States

Hanwen Bai, Yale University School of Medicine, New Haven, CT, United States

Daniel J. Barrera, University of California San Diego, San Diego, CA, United States

Carrie E. Bearden, University of California Los Angeles, Los Angeles, CA, United States

James D. Berry, Neurological Clinical Research Institute (NCRI), Massachusetts General Hospital, Boston, MA, United States

Brianne M. Bettcher, University of Colorado School of Medicine, Aurora, CO, United States

Carolyn Bevan, University of California San Francisco, San Francisco, CA, United States

John Blangero, University of Texas Rio Grand Valley, Brownsville, TX, United States

Yvette M. Bordelon, University of California Los Angeles, Los Angeles, CA, United States

Jesse A. Brown, University of California San Francisco, San Francisco, CA, United States

Cynthia M. Bulik, University of North Carolina Center of Excellence for Eating Disorders, Chapel Hill, NC, United States; Karolinska Institutet, Stockholm, Sweden

Victoria E. Clark, Yale University School of Medicine, New Haven, CT, United States

Aiden Corvin, Trinity College Dublin, Dublin, Ireland

Joanne E. Curran, University of Texas Rio Grand Valley, Brownsville, TX, United States

Bruce N. Cuthbert, National Institute of Mental Health, National Institutes of Health, Bethesda, MD, United States

Alissa M. D'Gama, Howard Hughes Medical Institute, Boston, MA, United States; Harvard Medical School, Boston, MA, United States

Mark J. Daly, Massachusetts General Hospital, Boston, MA, United States; Broad Institute of MIT and Harvard, Cambridge, MA, United States

Ryan S. Dhindsa, Columbia University Medical Center, New York, NY, United States

Gül Dölen, Johns Hopkins University, Baltimore, MD, United States

Andrew Evans, Royal Melbourne Hospital, Melbourne, VIC, Australia

Nelson B. Freimer, University of California Los Angeles, Los Angeles, CA, United States

Daniel H. Geschwind, University of California Los Angeles, Los Angeles, CA, United States

Michael D. Geschwind, University of California San Francisco, San Francisco, CA, United States

David C. Glahn, Hartford Hospital, Hartford, CT, United States; Yale University School of Medicine, New Haven, CT, United States

David B. Goldstein, Columbia University Medical Center, New York, NY, United States

John M. Greally, Albert Einstein College of Medicine, Bronx, NY, United States

Michael Greicius, Stanford University School of Medicine, Stanford, CA, United States

Lea T. Grinberg, University of California San Francisco, San Francisco, CA, United States; University of São Paulo Medical School, São Paulo, Brazil

Jennifer M. Günel, Yale University School of Medicine, New Haven, CT, United States

Murat Günel, Yale University School of Medicine, New Haven, CT, United States

Raquel E. Gur, University of Pennsylvania, Philadelphia, PA, United States

Ruben C. Gur, University of Pennsylvania, Philadelphia, PA, United States

Claudia M. Haase, Northwestern University, Evanston, IL, United States

Helmut Heinsen, University of São Paulo Medical School, São Paulo, Brazil; University Hospital Würzburg, Würzburg, Germany

Derrek P. Hibar, University of Southern California, Los Angeles, CA, United States

Karen Hodgson, Yale University School of Medicine, New Haven, CT, United States

Basavaraj Hooli, Massachusetts General Hospital, Boston, MA, United States; Harvard Medical School, Boston, MA, United States

Branka Hrvoj-Mihic, University of California San Diego, La Jolla, CA, United States

Louis Jacob, Stanford University Center for Narcolepsy, Palo Alto, CA, United States

Neda Jahanshad, University of Southern California, Los Angeles, CA, United States

Saumya S. Jamuar, KK Women's and Children's Hospital, Singapore; Howard Hughes Medical Institute, Boston, MA, United States; Harvard Medical School, Boston, MA, United States

Adrienne M. Keener, University of California Los Angeles, Los Angeles, CA, United States

Emma E.M. Knowles, Yale University School of Medicine, New Haven, CT, United States

Stephan Lammel, University of California Berkeley, Berkeley, CA, United States

Suzee E. Lee, University of California San Francisco, San Francisco, CA, United States

Robert W. Levenson, University of California Berkeley, Berkeley, CA, United States

Mingfeng Li, Yale School of Medicine, New Haven, CT, United States

Mauro F.C. Lins, Clínica-Escola de Autismo, Itaboraí, Rio de Janeiro, Brazil

Daniel H. Lowenstein, University of California San Francisco, San Francisco, CA, United States

Joshua Mahlios, Stanford University Center for Narcolepsy, Palo Alto, CA, United States

Robert C. Malenka, Stanford University School of Medicine, Stanford, CA, United States

Christopher E. Mason, Cornell University, New York, NY, United States

Samuel R. Mathias, Yale University School of Medicine, New Haven, CT, United States

Mario F. Mendez, University of California Los Angeles, Los Angeles, CA, United States; Greater Los Angeles VA Healthcare System, Los Angeles, CA, United States

Kathleen R. Merikangas, National Institutes of Health, Bethesda, MD, United States

Alison K. Merikangas, University of Pennsylvania, Philadelphia, PA, United States

Emmanuel Mignot, Stanford University Center for Narcolepsy, Palo Alto, CA, United States

Ramon Mocellin, Royal Melbourne Hospital, Melbourne, VIC, Australia; University of Melbourne and Melbourne Health, Melbourne, VIC, Australia

Rebecca A. Muhle, Yale University School of Medicine, New Haven, CT, United States

Alysson R. Muotri, University of California San Diego, La Jolla, CA, United States

Georges Naasan, University of California San Francisco, San Francisco, CA, United States

Srikantan S. Nagarajan, University of California San Francisco, San Francisco, CA, United States

Benjamin M. Neale, Massachusetts General Hospital, Boston, MA, United States; Broad Institute of MIT and Harvard, Cambridge, MA, United States

Priscilla D. Negraes, University of California San Diego, La Jolla, CA, United States

Eric J. Nestler, Icahn School of Medicine at Mount Sinai, New York, NY, United States

Hanna M. Ollila, Stanford University Center for Narcolepsy, Palo Alto, CA, United States

Rene L. Olvera, University of Texas Health Science Center San Antonio, San Antonio, TX, United States

Pongsatorn Paholpak, Khon Kaen University, Khon Kaen, Thailand; University of California Los Angeles, Los Angeles, CA, United States

Neelroop N. Parikshak, University of California Los Angeles, Los Angeles, CA, United States

Roy H. Perlis, Harvard Medical School, Boston, MA, United States

Sirisha Pochareddy, Yale School of Medicine, New Haven, CT, United States

Carlos Portera-Cailliau, University of California Los Angeles, Los Angeles, CA, United States

Nader Pouratian, University of California Los Angeles, Los Angeles, CA, United States

Shaun M. Purcell, Harvard Medical School, Boston, MA, United States; Icahn School of Medicine at Mount Sinai, New York City, NY, United States

Gil D. Rabinovici, University of California San Francisco, San Francisco, CA, United States

Kamalini G. Ranasinghe, University of California San Francisco, San Francisco, CA, United States

Elena Ratti, Neurological Clinical Research Institute (NCRI), Massachusetts General Hospital, Boston, MA, United States

Hannah E. Reed, Yale University School of Medicine, New Haven, CT, United States

Kerry J. Ressler, Emory University, Atlanta, GA, United States; Harvard Medical School, Belmont, MA, United States

Elise B. Robinson, Massachusetts General Hospital, Boston, MA, United States; Broad Institute of MIT and Harvard, Cambridge, MA, United States

Stephan J. Sanders, University of California San Francisco, San Francisco, CA, United States

Eric E. Schadt, Icahn School of Medicine at Mount Sinai, New York City, NY, United States

Daniel R. Schonhaut, University of California San Francisco, San Francisco, CA, United States

Jonathan Sebat, University of California San Diego, San Diego, CA, United States

William W. Seeley, University of California San Francisco, San Francisco, CA, United States

Katerina Semendeferi, University of California San Diego, La Jolla, CA, United States

Nenad Sestan, Yale School of Medicine, New Haven, CT, United States

Kevin A. Shapiro, University of California San Francisco, San Francisco, CA, United States

Elliot H. Sherr, University of California San Francisco, San Francisco, CA, United States

John C. Silbereis, Yale School of Medicine, New Haven, CT, United States

Wojciech B. Solecki, Institute of Pharmacology PAS, Krakow, Poland; Jagiellonian University, Krakow, Poland

Edoardo G. Spinelli, University of California San Francisco, San Francisco, CA, United States; Vita-Salute San Raffaele University, Milan, Italy

Matthew W. State, University of California San Francisco, San Francisco, CA, United States

Jason L. Stein, University of North Carolina, Chapel Hill, NC, United States

Garret D. Stuber, University of North Carolina at Chapel Hill, Chapel Hill, NC, United States

Virginia E. Sturm, University of California San Francisco, San Francisco, CA, United States

Patrick F. Sullivan, University of North Carolina at Chapel Hill, Chapel Hill, NC, United States; Karolinska Institutet, Stockholm, Sweden

Rudolph E. Tanzi, Massachusetts General Hospital, Boston, MA, United States; Harvard Medical School, Boston, MA, United States

Maria L.G. Tempini, University of California San Francisco, San Francisco, CA, United States

Panos Theofilas, University of California San Francisco, San Francisco, CA, United States

Paul M. Thompson, University of Southern California, Los Angeles, CA, United States

Cleber A. Trujillo, University of California San Diego, La Jolla, CA, United States

Dennis Velakoulis, Royal Melbourne Hospital, Melbourne, VIC, Australia; University of Melbourne, Melbourne, VIC, Australia

Keith A. Vossel, University of California San Francisco, San Francisco, CA, United States

Christopher A. Walsh, Howard Hughes Medical Institute, Boston, MA, United States; Harvard Medical School, Boston, MA, United States; Broad Institute of MIT and Harvard, Cambridge, MA, United States

Mark Walterfang, Royal Melbourne Hospital, Melbourne, VIC, Australia; University of Melbourne, Melbourne, VIC, Australia

Aliza P. Wingo, Emory University, Atlanta, GA, United States; Harvard Medical School, Belmont, MA, United States

Nailin Yao, Yale University School of Medicine, New Haven, CT, United States

Jennifer S. Yokoyama, University of California San Francisco, San Francisco, CA, United States

Mark W. Youngblood, Yale University School of Medicine, New Haven, CT, United States

Peter Zandi, Johns Hopkins School of Medicine, Baltimore, MD, United States

Ying Zhu, Yale School of Medicine, New Haven, CT, United States

Preface

Clinical neuroscience is in the midst of a profound transformation. Parallel advances in genomics and neurobiology are revealing the pathophysiological mechanisms underlying both common and rare disorders and empowering the search for novel and more effective therapies. Over the last few years, a revolution in gene discovery has changed the landscape for many neurological and psychiatric conditions. The path to reliable and systematic gene discovery has been firmly established, providing a solid molecular foundation upon which to build a deeper understanding of the pathophysiology of multiple clinical syndromes. The next wave of the "-omics" revolution is already upon us, providing the opportunity to study germline and somatic variation, gene expression and, soon, the proteome, at scale, offering the ability to readily edit the genome and thereby illuminate function, and increasingly leveraging powerful informatics and systems biological approaches. Recent advances in neuroscience have been no less dramatic: the field is steadily characterizing brain in a wide range of species at cellular resolution and across development; leveraging an array of tools to assess and manipulate circuits in unprecedented ways; integrating multiple levels of analysis from developmental and molecular to computational and systems neuroscience; and developing stem cell technologies promising to provide the opportunity to study previously inaccessible biological processes restricted to the human central nervous system.

These advances are driving the reintegration of the clinical neurosciences most dramatically neurology and psychiatry. The demands on the contemporary clinician, regardless of discipline, to understand a wide range of fields from genomics to molecular, cellular, and circuit level neuroanatomy has never been greater and will undoubtedly increase over the coming years. For example, genetic testing is becoming a standard of care for multiple diagnoses across psychiatry and neurology, and recent findings promise to make these assays even more commonplace in clinical and clinical research settings. And recent efforts to transform the psychiatric diagnostic nosology are pushing investigators and clinicians to prioritize the understanding of CNS anatomy and function, and their links to psychopathology, in an unprecedented fashion.

This is the motivation for this volume: to provide relevant foundational scientific knowledge to clinicians, researchers, and trainees in a wide range of neuroscience disciplines as a resource for understanding the relevance of recent advances for clinical syndromes.

The first section provides a comprehensive review of genetics and genomics relevant to clinical neuroscience. Chapters address the contemporary view of the human genome, provide an overview of the state-of-the-art methods to detect and analyze common and rare variation both in clinical and epidemiological samples, describe emerging tools to examine and manipulate the genome, and offer an introduction to a range of related topics including epigenetics, bioinformatics, stem cell models, imaging genomics, pathway analysis, and systems biology.

The second section of the book focuses on the neuroanatomy and neurocircuitry associated with psychiatric and neurological conditions. Chapters in this section address molecular, cellular, and circuit level analyses, describe state-of-the-art methods to study and manipulate circuits, and review specific functional and behavioral domains and their anatomical underpinnings—including mood and emotion, apathy, delusions, hallucinations, and movement.

The third and fourth sections endeavor to bring this foundational knowledge to bear on a discussion of clinical syndromes and therapeutics development. This begins with a discussion of the overlap of genetic risk for a wide-range disorders and the challenges of and emerging alternatives to categorical diagnosis—particularly in psychiatry. Disorders and syndromes are then covered individually and range from those that have been traditionally considered the domain of psychiatry, to neurological and neurodegenerative and neurooncological disorders. As the title of the volume suggests, all of the authors address what is currently known about the genomic, pathway, and circuit level aspects of pathology. Not surprisingly, there is considerable variation in the state-of-the-science for various conditions. As a general proposition, disorders that have traditionally been the remit of neurology and neurosurgery have already accumulated a stronger knowledge base with regard to genetics, molecular pathology, and anatomy. Of course, there are several syndromes traditionally tied to psychiatry in which tremendous recent progress has been made in elaborating molecular, cellular, and circuit level mechanism.

The use of the term neuropsychiatry in the title of this volume is a prediction. As the basic science of brain and behavior continues to advance rapidly, easy distinctions between neurological and psychiatric disorders will continue to erode and a new synthesis—and new professional pathways—will emerge. It is not, however, meant to reify the term versus other similar ones. Indeed, it is likely that any current distinctions between, for instance, neuropsychiatry, clinical neuroscience, and behavioral neurology will also become less and less clear with time. Moreover, our view of the future celebrates the differences in emphasis and culture that currently characterize distinct subfields focused on mind and brain. Recent advances do not necessitate a reductionist view. The astonishing scientific progress that is challenging the conventional wisdom leaves plenty of room for the types of differences that drive aspiring surgeons versus psychotherapists, neuroimmunologist, neurogeneticists, and basic neuroscientists. Our hope is that this text will serve as a useful resource for anyone interested in a core set of principles, practices, and methods relevant to the contemporary understanding of disorders of the human brain.

Thomas Lehner, Bruce Miller and Matthew State

Section I

The Genome Tools and Methods

Chapter 1

The Newly Emerging View of the Genome

Stephan J. Sanders
University of California San Francisco, San Francisco, CA, United States

Christopher E. Mason
Cornell University, New York, NY, United States

INTRODUCTION

A genomic revolution has been underway since the beginning of this century. A series of transformative advances, starting with the sequencing of the human genome (Lander et al., 2001; Venter et al., 2001), have enabled high-resolution, genome-wide analysis to become a routine technique in both research and clinical practice. After generation of the human reference genome by the Human Genome Project, the Haplotype Mapping (HapMap) Consortium began the process of cataloging human genetic variation. The combination of these HapMap variants and microarray technology enabled genome-wide association studies (GWAS) to find common variants associated with human disease.

The subsequent invention of high-throughput sequencing reduced the price of DNA sequencing by 50% every 5 months between 2006 and 2014: data generation required for the 10-year, $3 billion human genome project thus could be achieved in 3 days for $2000. This advance has enabled gene discovery in complex disorders such as intellectual disability and autism spectrum disorder (ASD), provided high-resolution maps of human variation (1000 Genomes Project and Exome Aggregation Consortium), and enabled tissue-specific maps of gene expression (Genotype-Tissue Expression Project [GTEx] and BrainSpan).

However, rather than yielding a working model of the genome, these advances have simply begun to map the sheer complexity of the dynamic role played by DNA in cells. On top of the DNA genome lies an epigenome composed of myriad interacting factors that regulate how, when, and where DNA is expressed as RNA. The extent of this noncoding regulatory architecture correlates with organismal complexity (Taft, Pheasant, & Mattick, 2007) to a far greater degree than the protein-coding regions; furthermore, many of the genetic loci identified in human disease through GWAS and sequencing analyses target regulatory regions or the genes involved in DNA regulation. The Encyclopedia of DNA Elements (ENCODE) Project has generated tissue-specific maps of chromatin, the first-level of this regulation, yet the list of regulatory processes is expanding faster than the understanding of how these processes function or interact.

Therefore, the genomic revolution has transformed our understanding of biology, human diversity, and human disease, laying a foundation for a future era of genome-guided precision medicine. To achieve this goal the modern clinician will require familiarity not just with the 1.5% of protein-coding DNA but also the 98.5% of noncoding DNA that describes how the resulting proteins create a human and a human brain.

CONCEPTUALIZING DNA

Early descriptions of DNA describe a passive molecule that stores information to be read by the cell as required, analogous to a book. Contemporary molecular analysis has revealed a vastly more dynamic and active molecule capable of both storing and processing information, analogous to a computer. To extend this computer analogy, the genome contains "hardware" in the form of nucleotides, "software" in the form of epigenetic modifications, such as methylation, and

Genomics, Circuits, and Pathways in Clinical Neuropsychiatry. http://dx.doi.org/10.1016/B978-0-12-800105-9.00001-9

information processing capabilities in the form of factors that integrate internal and external signals to regulate mRNA transcription. Furthermore, this genomic "computer" can be reconfigured to run different operating systems (eg, Windows, Mac OS, Linux) in the form of hundreds or possibly thousands of cell types (eg, stem cells, neurons, or lymphocytes). This entire process takes place in concert with the other molecular and structural components of the cell, in a manner similar to the bidirectional feedback between the central nervous system and the body.

However, several aspects of the genome go beyond a computer analogy. Foremost among these is that the genome contains the information to reproduce itself, including replication of the information and development from a single copy (ie, zygote) to a fully developed adult. In addition, the components of a computer are usually discrete, with a hard drive for storage and a processor for data integration, whereas in the genome these processes are fully integrated with the regulatory machinery acting directly on the nucleotides and their surroundings. Finally, whereas computers are based on the reliable transmission of bits of information (ie, 0 or 1), the genome acts through inherently stochastic chemical interactions which are essentially continuous (0−1 rather than 0 or 1). Therefore, whereas computers act by enforcing conformity, biological systems act by manipulating stochastic processes, constraining them to the degree that is required to achieve the necessary outcome reliably.

INFORMATION FLOW IN BIOLOGICAL SYSTEMS

To make sense of the data generated by genomics, it is necessary to understand how the information encoded in DNA leads to an observable human phenotype. This chapter will describe the first four molecular steps in this process in detail (Fig. 1.1A), identifying specific examples relevant to neuropsychiatry where possible. However, the information from the DNA continues to exert an influence through the actions of proteins across further levels of organization, ultimately resulting in an observable phenotype (Fig. 1.1B). This includes the formation of circuits and pathways in the brain, in tandem with environmental and regulatory influences.

KEY DEFINITIONS

Cis and *Trans*

These terms describe whether a DNA sequence acts on a nearby region, eg, a promoter acting in *cis* to initiate transcription of a gene, or a distant region, eg, a gene encoding a transcription factor that acts in *trans* at multiple binding sites across the genome.

Gene

The word "gene" was used initially to describe a particle that is inherited to form a specific trait ("pangene," 1889, Hugo de Vries) or an independent determinant of a hereditary characteristic ("gene," 1909, Wilhelm Johannsen). After the discovery of transcription, translation, and the triplet code, the word "gene" became synonymous with "protein-coding gene," and this is still the most common usage of the term. However, one gene can have multiple isoforms, a noncoding enhancer could be considered part of the gene unit, and noncoding transcripts can have an influence on a hereditary characteristic. Two contemporary definitions have been proposed: (1) "A gene is a union of genomic sequences encoding a coherent set of potentially overlapping functional products" (Gerstein et al., 2007), and (2) "A gene is a discrete genomic region whose transcription is regulated by one or more promoters and distal regulatory elements and which contains the information for the synthesis of functional proteins or noncoding RNAs, related by the sharing of a portion of genetic information at the level of the ultimate products (proteins or RNAs)" (Pesole, 2008).

Transcript

This is a catchall phrase used to describe lengths of RNA that are produced from DNA. Many, but not all of these transcripts will be the mRNAs transcribed from genes.

High-Throughput Sequencing

This is also called next-generation sequencing and it has revolutionized genomics. The previous generation, called Sanger sequencing or dideoxy sequencing, involved amplifying a specific region of DNA using the polymerase chain

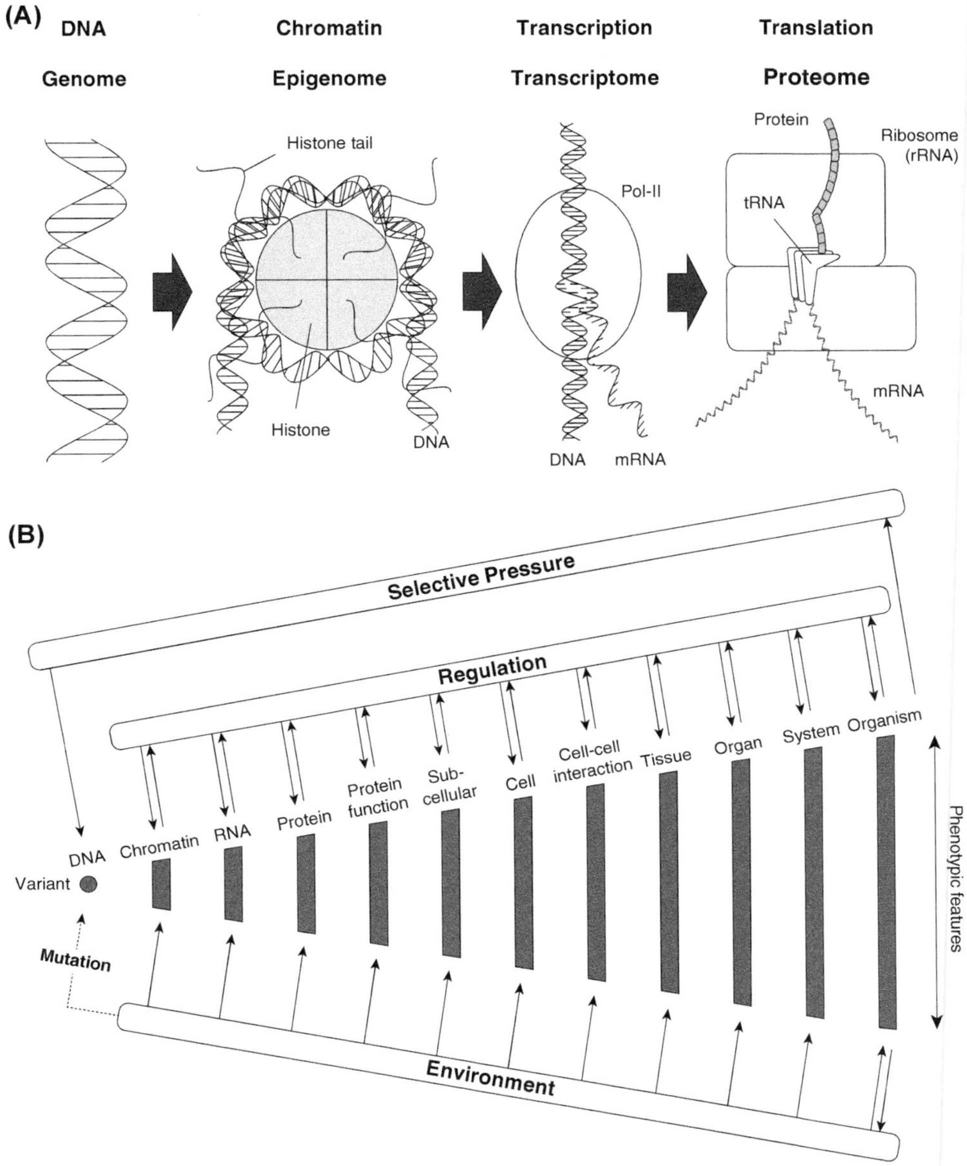

FIGURE 1.1 Information flow in biological systems. (A) The genome describes the totality of information encoded in DNA nucleotides. Proteins and RNAs surround the DNA to form chromatin; the complete set of these reversible chemical changes in a cell or groups of cells is called the epigenome. Chromatin regulates the transcription of DNA into RNA; the complete set of RNA transcripts in a cell or groups of cells is called the transcriptome. Protein-coding mRNA is translated into proteins through the action of ribosomes and tRNAs; the complete set of proteins in the cell or groups of cells is called the proteome. (B) Information encoded by DNA nucleotides is amplified through multiple molecular and structural stages, eventually resulting in an observable phenotype in the organism. Throughout this amplification process the information flow can be regulated by upstream or downstream processes and further influenced by environmental factors. Finally, the phenotype of the organism governs its ability to reproduce, leading to selective pressure that determines whether the DNA is passed on to the next generation.

reaction (PCR), inserting one of four fluorescently labeled nucleotides with chain terminators that prevent further DNA replication, and then comparing the fluorescent color with the length of the DNA to "read" the output. This method was used to map the human genome in 2001 (Lander et al., 2001). In contrast, high-throughput methods read multiple strands of DNA simultaneously: for example, using the Illumina sequencing by synthesis method that involves amplifying a strand of DNA to form a cluster of identical DNA strands and then replicating all of the strands in the cluster using fluorescently labeled nucleotides with reversible chain terminators. An image of the cluster is taken to record the fluorescent color present, and then the chain termination is removed, allowing the next nucleotide to be added and imaged in a similar manner.

These high-throughput methods produce numerous "reads" of sequencing that have to be aligned to the genome for interpretation. High-throughput sequencing has been applied to many different research methods, including:

- **Whole exome sequence (WES)**: The 1.5% of DNA in protein-coding exons is captured then sequenced to identify DNA variants in at least 80% of protein-coding nucleotides.
- **Whole genome sequence (WGS)**: All of the DNA in the genome is sequenced to identify DNA variants in at least 95% of nucleotides.
- **RNA sequencing (RNA-Seq)**: RNA transcripts are captured (eg, by selecting for a poly(A) tail) and converted to complementary DNA, which can be sequenced. This is used to assess the quantity and type of RNAs in the cell.
- **Chromatin immunoprecipitation sequencing (ChIP-Seq)**: Chromatin immunoprecipitation creates cross-links between proteins and DNA, and then uses an antibody to recognize a specific marker (eg, a histone mark or transcription factor). The antibody and bound protein/DNA are isolated. Then the DNA is released and can be sequenced. This allows identification of the location of epigenetic markers in the genome.
- **Methyl sequencing**: This technique identifies regions of DNA that are methylated. This can be achieved by either bisulfite treatment, which converts unmethylated cytosine (C) to uracil (U) and comparing treated to untreated DNA, or immunoprecipitation (methylated DNA immunoprecipitation) with an antibody specific to methylated DNA.
- **DNase sequencing**: DNase I is an enzyme that digests DNA. Active regions of DNA that have been exposed through chromatin remodeling are hypersensitive to this enzyme so comparing treated to untreated DNA identifies these regions.
- **Assay for Transposase-Accessible Chromatin sequencing (ATAC-Seq)**: This technique uses a highly active transposase enzyme to insert sequencing adapters into exposed (open) DNA. The technique is complementary to DNase sequencing, but requires less starting DNA and provides higher resolution of the open DNA (Buenrostro, Wu, Chang, & Greenleaf, 2015).

THE GENOME (DNA)

The term "genome" describes the complete set of DNA found in the cells of an organism. DNA is made up of two polymer chains arranged in a double helix and bound together by hydrogen bonds. Each polymer chain is composed of multiple copies of four nucleotides, each of which is composed of a phosphate group, a deoxyribose sugar, and one of four nucleobases: adenine (A), cytosine (C), guanine (G), and thymine (T). Both polymer chains start with a phosphate group on the fifth carbon in the deoxyribose sugar (called the 5′ end) and ends with a hydroxyl group on the third carbon in the deoxyribose sugar (3′); this distinction gives directionality to the DNA which influences function, because both DNA and RNA polymerases act only in a 5′ to 3′ direction. The two polymer chains are arranged in opposite directions, so that the 5′ end of one chain opposes the 3′ end of the other. The nucleotides on each chain are paired, so that A binds to T with two hydrogen bonds, whereas C binds to G with three hydrogen bonds. Along with creating redundancy (ie, each nucleotide can be imputed from the opposite strand) the differing number of hydrogen bonds makes CG pairing slightly stronger than AT pairing. This GC bias has an influence on numerous molecular techniques, eg, PCR and whole-genome sequencing.

Each chromosome is composed of a single DNA molecule; after the Human Genome Project (Lander et al., 2001; Venter et al., 2001), most of the human genome was mapped. Every nucleotide pair is numbered starting at the 5′ end of the short arm of the chromosome (*p* arm) (Fig. 1.2) and continuing to the 3′ end at the end of the long arm of the chromosome (*q* arm). The polymer chain (that is, 5′ to 3′ in this arrangement) is called the positive strand, whereas the opposing chain (3′ to 5′) is the negative strand. By convention, the nucleotide at each position is expressed as that on the positive strand.

In humans, the genome is composed of 46 chromosomes (two pairs of 22 autosomes and two sex chromosomes: XX or XY) and the mitochondrial chromosome, of which there may be numerous copies per cell. In total, this equates to about

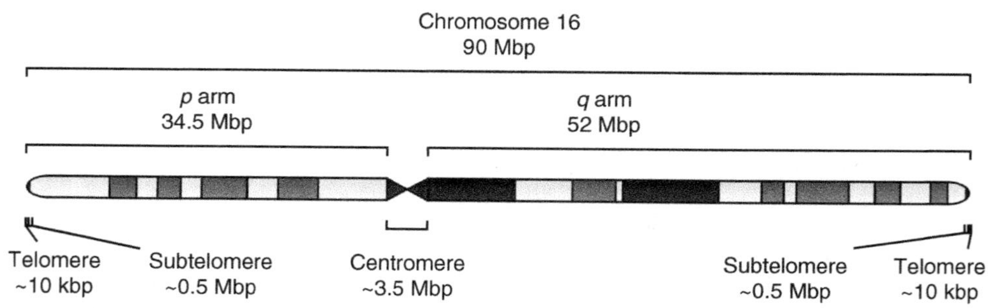

FIGURE 1.2 Chromosomal structure. An ideogram of human chromosome 16 with Giemsa banding is shown as an example of chromosomal structure. The centromere splits the chromosome into a short *p* arm and a longer *q* arm. Each arm is capped by a subtelomeric region and telomere.

12.8 billion nucleotides (including both strands and both chromosome copies), 6.4 billion nucleotide pairs (the diploid genome, including both chromosome copies), or 3.2 billion numbered nucleotide pairs (the haploid genome, including only one chromosome copy). The haploid genome measure is commonly used to measure DNA or describe a position in the chromosome, with a single nucleotide pair being described as 1 base pair (bp); 1 kbp is 1000 nucleotide pairs, and 1 Mbp is 1 million nucleotide pairs.

VISUALIZING THE GENOME

Three of the most frequently used tools for visualizing the genome are UCSC Genome Browser (genome.ucsc.edu), the Ensembl Genome Browser (www.ensembl.org), and the Integrative Genome Viewer (www.broadinstitute.org/igv/). All of these are publicly accessible and have a wealth of information about the genome of humans and those of other species, including known genes and common variants.

LARGE-SCALE STRUCTURES IN THE GENOME

To visualize chromosomes, cell division is arrested during metaphase, when the chromosomes that have been replicated are at their most condensed, forming the familiar X shape. Giemsa banding is frequently used to stain the condensed chromosomes; this solution binds to the phosphate in DNA with a preference for AT nucleotides rather than GC nucleotides (Fig. 1.2). This leads to a series of discrete bands, each of a few million nucleotides that can be observed through a microscope. Broadly, the AT-rich darkly staining bands tend to be heterochromatin with fewer genes and more repetitive DNA, whereas the GC-rich, lightly staining bands tend to be euchromatin; however, this distinction rarely has practical applications. The human chromosomes are numbered from 1 to 22 based on their length in this condensed state, with 1 being the largest (247 million nucleotide pairs) and 21 being the smallest (50 million nucleotide pairs).

Telomeres

Much like the hard cover on the front and back of a book, at each end of the chromosome there is a region of repetitive DNA called the telomere. In all vertebrates, including humans, the telomere is composed of 1000 to 3000 six-nucleotide "TTAGGG" repeats (Fig. 1.2). At the end of the telomere there is an overhanging 5′ strand that forms a T-loop, shaped like a paperclip and protecting the end of the DNA polymer, and preventing chromosomes from joining together. Of note, the length of telomeres decreases with each cell division in differentiated cells, preventing further cell division once a critical length is reached. This limitation of 50–70 cell divisions is called the "Hayflick limit." The length of telomeres varies among chromosomes, tissues, individuals, and species, and according to exposure to environmental factors (Gomez et al., 2012; Riethman, 2008).

Subtelomeres

Immediately after the telomeres lie the subtelomeric regions. These are 100 to 500 kbp long and composed of numerous segmental duplications (long stretches of DNA that are nearly identical to other stretches of DNA on the same or other chromosomes), subtelomeric repeats, and stretches of the TTAGGG telomeric sequence (Fig. 1.2) (Riethman, 2008). These regions are prone to mutation, particularly translocation between chromosomes, and vary widely between individuals and species. Despite the repetitive sequence, the subtelomeres initiate transcription of both the subtelomeric and telomeric regions, resulting in numerous transcripts including *Telomeric Repeat-containing RNA*, a long noncoding RNA (lncRNA) that has an important role in regulating the telomere, and *Wiskott-Aldrich syndrome protein and SCAR Homolog* genes which have an important role in the actin cytoskeleton. These subtelomeric genes are related to the *Wiskott–Aldrich Syndrome Protein* (*WASP*) that is mutated in Wiskott–Aldrich X-linked immunodeficiency, though *WASP* itself is not subtelomeric (Linardopoulou et al., 2007).

Centromeres

The "middle" of the chromosome is called the centromere and it is the point at which the kinetochore protein complex forms to allow the attachment of spindle fibers that pull sister chromatids apart during cell division (Fig. 1.2). This region is largely composed of repetitive DNA, particularly α-satellite repeats, but it is the histone markers, rather than the DNA sequence, that are recognized by the kinetochore. Specifically, the usual H3 histone protein is replaced by centromere protein A (Fukagawa & Earnshaw, 2014).

SMALL-SCALE STRUCTURES IN THE GENOME

Understanding the information content and functional role of specific DNA regions requires integrating numerous types of data. Comparisons between the genomes of different organisms allow regions of conserved DNA to be detected, indicating key information content. Analysis of transcribed RNA through expressed sequence tags and RNA-Seq has identified active transcribed regions of DNA and numerous types of RNA (Table 1.5). The GTEx project has brought tissue-level resolution to this endeavor (www.gtexportal.org). Bioinformatic analysis of DNA sequences has identified protein-coding genes through open reading frames, transcription factor motifs, and binding sites, and the ENCODE project (www.encodeproject.org) has begun to form a comprehensive map of epigenetic markers in various tissues. This endeavor is proceeding rapidly, and undoubtedly there are further insights to be found that will fundamentally alter our perception of how DNA functions, such as an investigation of three-dimensional (3D-structure). The following discussion summarizes some key components that have been identified to date.

Protein-Coding Genes

Regions of DNA are transcribed to form messenger RNAs (mRNAs), which are translated by ribosomes to form protein products. Several gene standard groups, including the HUGO Gene Nomenclature Committee (www.genenames.org), the Consensus Coding Sequence project (www.ncbi.nlm.nih.gov/CCDS/), and the Genes of ENCODE (www.gencodegenes.org), identify 18,826–19,797 distinct protein-coding genes. The ambiguity is due to the multiple isoforms that each gene can have, with differing transcription start sites (TSS), exons, and ends, and these names and annotations are continuously being updated. Protein-coding genes are made up of the following components (Fig. 1.3).

Promoters

These are regions of DNA that are not transcribed, in which the preinitiation complex (PIC) can form. The "classic" core promoter sequence is the TATA-box, an eight-nucleotide sequence that begins with TATA and has three variable sites

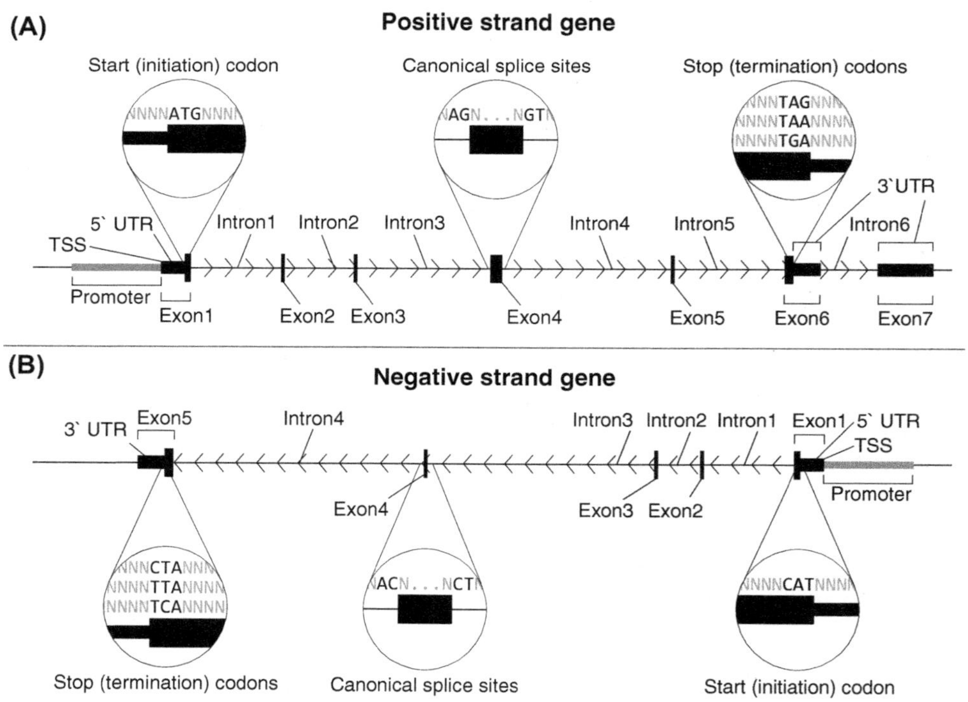

FIGURE 1.3 Features of a protein-coding gene. (A) A positive-strand gene is read from left to right on the positive strand. The promoter region is immediately upstream of the transcription start site (TSS). The first exon begins at the TSS, usually with a 5′ UTR, which may contain an intron (not shown). The 5′ UTR is followed a protein-coding sequence that begins with the ATG initiation codon. Exons and introns alternate along the transcript with the canonical splice sites marking the start and end of each intron. The protein-coding sequence ends with one of three stop codons, which are usually followed by a 3′ UTR, which may contain an intron as shown here. (B) The negative-strand gene is read from right to left on the positive strand. It contains the same elements and start/splicing/stop sequences as a positive-strand gene; however, because DNA is represented on the positive strand, the reverse complement of these sequences is shown, as would be observed in a genome browser.

shown by single-nucleotide polymorphism (SNP) codes (Table 1.3): TATAWAWR. However, only 10% of human genes have this exact sequence whereas a further 14% of genes have a TATA-like element (Yang, Bolotin, Jiang, Sladek, & Martinez, 2007). A higher proportion of human genes (46%) contain the eight-nucleotide initiator (INR) element (INR: YYANWYY), although 46% lack either TATA or INR elements. Along with the core promoter sequences, the surrounding nucleotides can modulate PIC binding (Sandelin et al., 2007). Although variation in the promoter region can have significant impacts on gene expression, it is rarely possible to predict what this effect will be without experimental testing. The promoter is commonly defined as 1000 bp upstream of the transcription start site (TSS).

Untranslated Regions

The PIC binds to the promoter and brings the RNA polymerase II (Pol II) enzyme to the TSS. DNA is transcribed into mRNA from the TSS; however, the first protein-coding sequence is usually some distance downstream. The transcribed region before the first protein-coding sequence is called the 5′ UTR, and DNA variants in this region have the potential to influence the resulting protein through transcriptional pausing, the formation of loops of RNA that affect ribosomal binding, and as a binding for site for regulators of transcription, eg, micro RNAs (miRNAs). Similarly, at the end of the gene, the transcribed region after the last first protein-coding sequence is called the 3′ UTR, and DNA variants in this region can also influence the resulting protein through variation in polyadenylation and translational efficiency. Notably, the 3′ UTRs of genes are significantly longer in neuronal and brain tissues than in other tissue types (Miura, Shenker, Andreu-Agullo, Westholm, & Lai, 2013).

Exons

These are regions of DNA that are not removed from the pre-mRNA when it is spliced to form the mature mRNA. They may be either protein-coding sequence or UTRs.

Introns

These are regions of DNA that are removed from the pre-mRNA when it is spliced to form the mature mRNA. They alternate with exons and similarly may be present between either protein-coding sequence or untranslated regions. While introns are removed before translation, they can still perform a wide range of regulatory functions, including variation in exon splicing and the presence of other types of RNA, eg, lncRNA (Chorev & Carmel, 2012).

Splice Sites

These are the boundaries between exons and introns. Most splice sites (about 99%) are marked by the two-nucleotide canonical splice site sequences: the GT "donor" at the start of each intron and the AG "acceptor" at the end of each intron. Variants in these canonical sites frequently lead to exon skipping and an altered reading frame. The intronic nucleotides near the canonical splice site are also highly conserved and can contribute to splicing behavior.

Protein-Coding Sequence

This includes all DNA in exons that are translated into proteins and it accounts for about 1.5% of the genome. The coding sequence is made up of three nucleotide blocks, each of which encodes one amino acid in the ribosome through the binding of tRNA (Tables 1.1 and 1.2). DNA variants in protein-coding sequence frequently change the resulting protein (a nonsynonymous variant) by changing one amino acid (a missense variant) or creating a premature stop codon (a nonsense variant). If the DNA change does not alter the resulting amino acid, it is described as synonymous or silent; however, these variants can still alter RNA stability, ribosomal pausing, and protein folding (Sauna & Kimchi-Sarfaty, 2011).

Noncoding Transcripts/Genes

Most transcripts and genes do not encode proteins; to date, over 40,000 noncoding transcripts and genes have been described, compared with about 20,000 protein-coding genes (www.gencodegenes.org). These noncoding genes are split into four categories of approximately similar size.

TABLE 1.1 Triplet Code of Amino Acids From Positive Strand Genes as Observed in the DNA on the Positive Strand of a Genome Browser

Base 1	Base 2: T		C		A		G		Base 3
	T		C		A		G		
T	TTT	(F/Phe) Phenylalanine	TCT	(S/Ser) Serine	TAT	(Y/Tyr) Tyrosine	TGT	(C/Cys) Cysteine	T
	TTC		TCC		TAC		TGC		C
	TTA	(L/Leu) Leucine	TCA		TAA	Stop (Ochre)	TGA	Stop (Opal)	A
	TTG		TCG		TAG	Stop (Amber)	TGG	(W/Trp) Tryptophan	G
C	CTT		CCT	(P/Pro) Proline	CAT	(H/His) Histidine	CGT	(R/Arg) Arginine	T
	CTC		CCC		CAC		CGC		C
	CTA		CCA		CAA	(Q/Gln) Glutamine	CGA		A
	CTG		CCG		CAG		CGG		G
A	ATT	(I/Ile) Isoleucine	ACT	(T/Thr) Threonine	AAT	(N/Asn) Asparagine	AGT	(S/Ser) Serine	T
	ATC		ACC		AAC		AGC		C
	ATA		ACA		AAA	(K/Lys) Lysine	AGA	(R/Arg) Arginine	A
	ATG	(M/Met) Methionine	ACG		AAG		AGG		G
G	GTT	(V/Val) Valine	GCT	(A/Ala) Alanine	GAT	(D/Asp) Aspartic acid	GGT	(G/Gly) Glycine	T
	GTC		GCC		GAC		GGC		C
	GTA		GCA		GAA	(E/Glu) Glutamic acid	GGA		A
	GTG		GCG		GAG		GGG		G

TABLE 1.2 Triplet Code of Amino Acids From Negative Strand Genes as Observed in the DNA on the Positive Strand of a Genome Browser

Base 1	Base 2: A		G		T		C		Base 3
	A		G		T		C		
A	AAA	(F/Phe) Phenylalanine	AGA	(S/Ser) Serine	ATA	(Y/Tyr) Tyrosine	ACA	(C/Cys) Cysteine	A
	AAG	(L/Leu) Leucine	AGG	(P/Pro) Proline	ATG	(H/His) Histidine	ACG	(R/Arg) Arginine	G
	AAT	(I/Ile) Isoleucine	AGT	(T/Thr) Threonine	ATT	(N/Asn) Asparagine	ACT	(S/Ser) Serine	T
	AAC	(V/Val) Valine	AGC	(A/Ala) Alanine	ATC	(D/Asp) Aspartic acid	ACC	(G/Gly) Glycine	C
G	GAA	(F/Phe) Phenylalanine	GGA	(S/Ser) Serine	GTA	(Y/Tyr) Tyrosine	GCA	(C/Cys) Cysteine	A
	GAG	(L/Leu) Leucine	GGG	(P/Pro) Proline	GTG	(H/His) Histidine	GCG	(R/Arg) Arginine	G
	GAT	(I/Ile) Isoleucine	GGT	(T/Thr) Threonine	GTT	(N/Asn) Asparagine	GCT	(S/Ser) Serine	T
	GAC	(V/Val) Valine	GGC	(A/Ala) Alanine	GTC	(D/Asp) Aspartic acid	GCC	(G/Gly) Glycine	C
T	TAA	(L/Leu) Leucine	TGA	(S/Ser) Serine	TTA	Stop (Ochre)	TCA	Stop (Opal)	A
	TAG		TGG	(P/Pro) Proline	TTG	(Q/Gln) Glutamine	TCG	(R/Arg) Arginine	G
	TAT	(I/Ile) Isoleucine	TGT	(T/Thr) Threonine	TTT	(K/Lys) Lysine	TCT		T
	TAC	(V/Val) Valine	TGC	(A/Ala) Alanine	TTC	(E/Glu) Glutamic acid	TCC	(G/Gly) Glycine	C
C	CAA	(L/Leu) Leucine	CGA	(S/Ser) Serine	CTA	Stop (Amber)	CCA	(W/Trp) Tryptophan	A
	CAG		CGG	(P/Pro) Proline	CTG	(Q/Gln) Glutamine	CCG	(R/Arg) Arginine	G
	CAT	(M/Met) Methionine	CGT	(T/Thr) Threonine	CTT	(K/Lys) Lysine	CCT		T
	CAC	(V/Val) Valine	CGC	(A/Ala) Alanine	CTC	(E/Glu) Glutamic acid	CCC	(G/Gly) Glycine	C

Pseudogenes

These genes are similar to functional genes but lack coding potential. There are three main classes: (1) processed, mature mRNA that has been reintegrated into the genome as DNA and lacks introns; (2) unprocessed, duplicated versions of existing genes; and (3) unitary, nonfunctional genes that are the only copy of previously functional genes (eg, premature stop codons in olfactory receptor genes). Although these genes do not become translated into proteins, they can still be transcribed and have regulatory roles (Ding, Qu, & Fang, 2014). Despite their degenerate state, DNA variants in pseudogenes may influence human disorders, possibly by acting as a sponge to suppress the actions of miRNA. A small number of pseudogenes have been identified as being differentially expressed in neurodegenerative disorders, including Alzheimer disease (Costa, Esposito, Aprile, & Ciccodicola, 2012).

Long Noncoding RNAs

These are defined as transcripts greater than 200 bp that have been processed and polyadenylated, but which lack any suggestion of an open reading frame that defines a protein-coding gene or that is degenerate in a pseudogene. Many lncRNAs represent antisense transcripts of protein-coding genes and therefore bind to the complementary mRNAs. Understanding of the role of lncRNAs continues to advance; however, they appear to have an important regulatory role, especially in fine-tuning mRNA transcription and protein translation. This regulatory role is facilitated by their ability to bind to DNA/RNA, bind to proteins, and form complex 3D structures. They are frequently observed to act as intermediaries guiding chromatin regulators to specific genomic loci (eg, methyltransferase genes). They can also act to edit mRNA, alter mRNA stability, influence protein folding, and influence posttranslational modification of proteins. Finally, lncRNAs have a critical role in X chromosome inactivation (*X Inactive Specific Transcript*) and genomic imprinting (eg, *Antisense of IGF2R Non-protein-coding RNA*) (Fatica & Bozzoni, 2014). There are several databases of lncRNAs, with some transcript annotations varying between databases (Fritah, Niclou, & Azuaje, 2014). Because of their regulatory roles, it is likely that DNA variants in lncRNAs have important roles in human disorders. One such example is an expansion of a triplet repeat in the lncRNA, *ATXN8 opposite strand*, that is the cause of spinocerebellar ataxia 8, possibly through the toxic effects of excess RNA.

Small Noncoding RNAs

These are RNAs smaller than 200 bp with varying degrees of posttranslational processing. Numerous classes have been described (see Table 1.5 for a complete list). They frequently act by targeting mRNA to regulate stability and translation. For example, miRNAs form part of the RNA-induced silencing complex that binds to mRNA to prevent translation. Like lncRNAs, it is likely that DNA variants in small noncoding RNAs have key roles in human disorders. Examples of this include both the miRNA *MIR137* and its regulatory targets that are associated with schizophrenia through two large GWAS (Ripke et al., 2011; Schizophrenia Working Group of the Psychiatric Genomics Consortium, 2014). In addition, absence of the paternal copy of the small nucleolar RNA, *SNORD116*, is the cause of Prader–Willi syndrome, which is associated with developmental delay and overeating (Duker et al., 2010). There are several databases of small noncoding RNAs, including miRNAs (www.mirbase.org, mirdb.org/miRDB/), noncoding elements (www.noncode.org), and genome browsers.

Circular RNAs

These RNAs form a circle without exposed 5′ and 3′ ends, are consequently resistant to RNA degradation, and therefore are highly stable. They are usually formed from mRNA by a process of "back-splicing" and are often composed of one or more exons joined at the canonical splice sites. Their expression pattern is similar to that of mRNAs, including the degree of cell type specificity and abundance. It has been suggested that they act as miRNA "sponges," serving to prevent mRNA degradation as a form of posttranscriptional regulation (Guo, Agarwal, Guo, & Bartel, 2014). The role of circRNA DNA variants in human disorders remains to be evaluated; however, it is likely to be difficult to distinguish the contribution of the variant through a circRNA from that of the parent protein-coding gene. CircBase is a database of circRNAs (www.circbase.org).

Enhancers

These are regions of DNA that act as binding sites for transcription factors that increase (activate) the transcription of nearby genes (ie, acting in *cis*). Whereas there are about 20,000 protein-coding genes in the human genome, there are estimated to be over 100,000 enhancers (E. P. Consortium, 2012) and possibly millions (Arnold et al., 2013). The location of many of these has been predicted (www.encodeproject.org), although only a small number of these have been experimentally validated (enhancer.lbl.gov). Enhancers are thought to be one of the main determinants of cell type by determining which genes are expressed. Most enhancers are within 100 kbp of the transcription start site of a gene; however, there are well-documented examples over 1000 kbp away: for example, the zone of polarizing activity regulatory sequence (ZRS), which is an enhancer of the *Sonic Hedgehog* (*SHH*) and in which DNA variants can lead to polydactyly (Lettice et al., 2003). A 28-bp deletion of the enhancer of the gene *proteolipid protein 1* leads to X-linked recessive hypomyelinative leukodystrophy, characterized by nystagmus, spastic quadriplegia, ataxia, and developmental delay (Hobson et al., 2002).

Silencers

These are the opposite of enhancers; they act as transcription factor binding sites that repress the transcription of nearby genes; like enhancers, they are usually found in close proximity to the genes they regulate in *cis*. Silencers are considerably less well characterized than enhancers. However, some enhancers may double as silencers in different cell types (Kim et al., 2008). One common example of silencers is the neuron-restrictive silencer elements (NRSEs; also called Repressor Element-1 [RE1]) that silence neuronal genes in nonneuronal cell types. The *RE1-Silencing Transcription Factor* (*REST*; also called *Neural-Restrictive Silencer Factor*) gene encodes a transcription factor that acts as a repressor when bound to these NRSEs. Altered expression of *REST* in neurons has been implicated in Down syndrome (Canzonetta et al., 2008), Huntington disease (Zuccato et al., 2003), and Alzheimer disease (Lu et al., 2014).

Insulators

These are regions that block the activity of enhancers and silencers on nearby genes. A common mechanism of this insulation is through the formation of DNA loops that physically prevent interaction between neighboring regions of DNA, potentially creating regulatory domains in the genome. These loops are held in place through the action of cohesins that form a ringlike structure over two strands of DNA; these are often associated with the protein CTCF.

Repetitive Elements

Although most biological investigation focuses on nonrepetitive DNA, about 50% of the genome is duplicated to some extent (Treangen & Salzberg, 2012). Regions of dense repetitive DNA remain as about 160 gaps in the human reference genome, because they cannot be mapped with short DNA reads. Their repetitive nature interferes with DNA replication, making these regions hypermutable and therefore highly polymorphic between individuals.

About 3% of these regions are variable number tandem repeats (VNTRs), ie, stretches of DNA repeated multiple times next to each other. VNTRs are divided into satellites (>60 bp, eg, the 171-bp α-satellite seen in centromeres), minisatellites (10–60 bp, eg, short tandem repeats used for DNA fingerprinting), or microsatellites (<10 bp, eg, the 6-bp repeat in telomeres). Segmental duplications are near-identical (≥90% homology) stretches of DNA over 1000 bp long. Although segmental duplications make up only 0.2% of the genome, they have a disproportionate contribution to human disease because they are the mechanism behind nonallelic homologous recombination (NAHR) that leads to the copy number variants (CNVs) responsible for numerous syndromes (eg, DiGeorge syndrome, deletion of 22q11.2). The remaining 46.8% of the genome is composed of interspersed repeat elements (ie, occurring in multiple locations throughout the genome rather than next to each other, in contrast to tandem repeats) (Treangen & Salzberg, 2012). There are four main groups of interspersed repeat elements.

DNA Transposons

Transposons are regions of DNA (200–2000 bp, 3% of the genome) capable of cutting and pasting themselves into another region of the genome by recruiting transposase enzymes that cut the target site and the original transposon donor site followed by the action of DNA polymerase, which mends the cuts, leaving the transposon in the new position. Occasionally the transposon is replicated, leading to a copy at the donor site and target site.

Short Interspersed Nuclear Elements

Short interspersed nuclear elements (SINEs) (100–300 bp, 15% of the genome) are retrotransposons, which are regions of DNA capable of copying and pasting themselves into another region of the genome via an RNA intermediate and the action of reverse transcriptase to convert the RNA to DNA. SINE elements are copied by RNA polymerase III and rely on the presence of reverse transcriptase from other sources. The *Alu* element is an approximately 350-bp SINE and makes up 11% of the human genome, which makes it the most abundant feature.

Long Interspersed Nuclear Elements

Long interspersed nuclear elements (500–8000 bp, 21% of the genome) are retrotransposons that are transcribed by RNA polymerase II. They contain a promoter, three genes (RNA binding protein, endonuclease, and reverse transcriptase), and a poly(A) tail.

Long Terminal Repeat Retrotransposons

Long terminal repeat retrotransposons (200–5000 bp, 9% of the genome) are endogenous retroviruses that have lost the ability to make a viral envelope.

The interspersed repetitive elements are often considered to be genomic parasites and the existence of Piwi-interacting RNAs specifically to silence retrotransposons would support this view. However, they have been coopted to create the V(D)J recombination system that underlies immunoglobulin diversity in the adaptive immune system (Jones & Gellert, 2004).

MULTIPLE ROLES

Conceptually it is simple to imagine each section of DNA having a single function (eg, exon, promoter, enhancer), much like words and spaces in a sentence. However, this is not always the case and a single DNA locus may perform multiple different roles. For example, it may act as an exon for a protein-coding gene or as an enhancer, or, in reverse, as a long noncoding RNA (Birnbaum et al., 2012; E. P. Consortium, 2012).

VARIATION IN THE GENOME

Approximately 0.1% (about 3.5 million nucleotides) of the human genome varies between any two individuals. An individual's genome is usually expressed as the variation from the human reference genome, often in a file called Variant Call Format. Of note, the human reference genome is European biased, therefore individuals with non-European ancestry, especially African ancestry, have comparatively more variation. These DNA variations are usually named according to their size (Fig. 1.4).

Entire Chromosomes

The presence of an extra chromosome is called trisomy whereas the absence of one copy of a chromosome is called monosomy (Fig. 1.4A). This level of variation is poorly tolerated; most trisomies or monosomies are incompatible with life. The exceptions to this are trisomy of chromosomes 13 (Patau syndrome), 18 (Edward syndrome), 21 (Down syndrome), X (XXY/Klinefelter syndrome or XXX), and Y (XYY); the only monosomy exception is the sex chromosomes (X0/Turner syndrome). Of note, this implies that deletion of a section of DNA generally has more functionally consequences than duplication, and this trend is observed for smaller variants, too.

Structural Variation

Large-scale genomic variation that involves only a part of a chromosome is called structural variation and it is split into two subcategories: (1) Copy number neutral variation, in which no DNA is gained or lost but the structure has been rearranged. Examples include inversions, in which a section of a chromosome is back to front, and translocations, in which a section of a chromosome has been moved, often to a different chromosome (Fig. 1.4B). These variants can cause pathology by disrupting a gene or removing a section of DNA from its regulatory context. They remain difficult to detect even with modern sequencing and their overall contribution to human disease is unknown (Steinberg et al., 2014). (2) CNVs, in which DNA is either lost (deletion) or gained (duplication) (Fig. 1.4C). De novo CNVs have been associated with intellectual disability (Deciphering Developmental Disorders Study, 2015), ASD (Pinto et al., 2014; Sanders et al., 2011; Sebat et al., 2007), and schizophrenia (ISC, 2008; Szatkiewicz et al., 2014). Recurrent CNVs frequently occur where regions of DNA in close proximity are very similar (segmental duplications) through NAHR; this mechanism is responsible for several well-recognized disorders including DiGeorge syndrome (22q11.2 deletion), William–Bueren syndrome (7q11.23 deletion), and 16p11.2 deletions and duplications. Large CNVs (>20 kbp) are readily detected using SNP genotyping arrays or comparative genomic hybridization arrays; detecting smaller CNVs with sequencing remains a challenge.

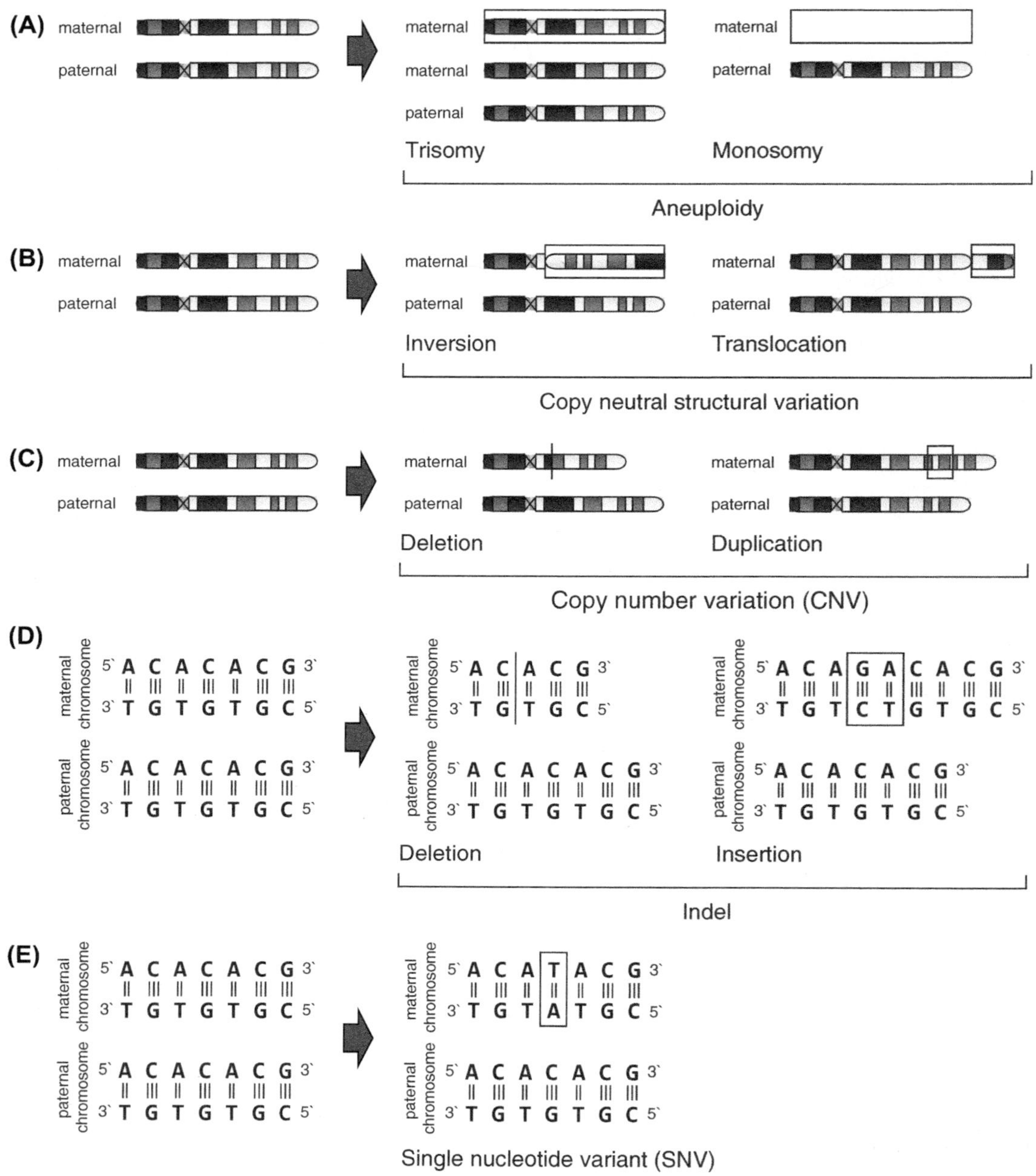

FIGURE 1.4 Categories of genomic variation. (A) Aneuploidies are whole chromosome duplications (trisomy) or deletions (monosomy). (B) Copy neutral structural variations are large-scale abnormalities that do not result in the duplication or deletion of DNA: for example, inversions of a portion of a chromosome or a translocation of one portion of a chromosome to another. (C) CNVs are duplications or deletions of DNA >1000 bp in length. (D) Indels are duplications (insertions) or deletions of DNA <1000 bp in length. (E) SNVs are the substitution of one nucleotide for another with no duplication or deletion of DNA. All variants are shown on the maternal chromosome but could equally occur on the paternal chromosome.

Insertions and Deletions

Insertions and deletions (indels) are when DNA is lost (deletion) or gained (insertion) on a smaller scale, defined as <1000 bp (Fig. 1.4D). If the indel occurs within a coding region, it is described as being "in-frame" if the amount of DNA lost or gained is divisible by 3; alternately, it is called a "frameshift" because the triplet reading code is altered for all subsequent nucleotides. Frameshift indels frequently lead to premature stop codons (Tables 1.1 and 1.2) and usually have more functional impact than in-frame indels. Indels ≤40 bp are readily discovered using sequencing technology, whereas the detection of indels >40 bp remains a challenge. Triplet repeats represent a subtype of indels in which a repetitive block of three nucleotides can be prone to further insertions of the same three nucleotides as the number of repeats increases. Fragile X syndrome is an example of one such triplet repeat disorder.

Single-Nucleotide Variants

If a nucleotide is altered without the gain or loss of DNA, it is a single-nucleotide variant (SNV) (Fig. 1.4E). These represent the most frequently observed variation in the human genome and account for at least 80% of known variants. They are readily detected using sequencing technology. If the SNV is frequently observed in human populations ($\geq$1% of a population), it is called an SNP. A catalog of human SNPs is recorded by the HapMap project (hapmap.ncbi.nlm.nih.gov) and the variable nucleotides are often recorded using standardized SNP codes (Table 1.3).

Loss-of-Function Variants

These variants lead to "complete" disruption of that copy of a gene. This class of variants is also referred to as likely gene-disrupting. This term is usually used to describe three types of variant: (1) nonsense/premature stop codon variants; (2) variants in the canonical splice site; and (3) insertions/deletions that alter the reading frame of the gene (frameshift indels). In all three cases the transcript is likely to contain an abnormal stop codon and to be targeted by nonsense-mediated decay (NMD) if it occurs more than 50 nucleotides from the last exon–intron boundary of the gene (Nagy & Maquat, 1998). However, this assumption of NMD has rarely been tested and genomic analysis suggests that the efficiency of the NMD process is variable (MacArthur et al., 2012). A number of other variants could arguably be included in this definition, including deletion CNVs, inversions/translocations that disrupt a transcript, variants at the initiation site or transcription start site in the promoter, and disruption of the AUG start codon.

ETHNICITY AND POPULATION FREQUENCY OF VARIANTS IN THE GENOME

In any individual there are approximately 3.5 million variants that differ from the reference genome (G. P. Consortium, 2010). This number is somewhat higher in populations that are ancestrally diverse from Europeans, eg, Africans, because the human reference is largely based on European individuals (Lander et al., 2001; Venter et al., 2001).

The vast majority of these variants are observed at a high population frequency (over 1% of individuals) (Table 1.4) across all ancestries. Based on the principle of natural selection, we would expect these to have minimal impact on human reproduction, and therefore limited contribution to severe early-onset human disease. A more nuanced view would state that any detrimental effects on reproductive fitness in specific situations are balanced by beneficial effects on reproductive fitness in others, leading to a minimal overall effect on reproductive fitness.

About 5% of variants are rare (<1% of individuals) (Table 1.4) and these rare variants are more likely to differ between populations of different ancestry. Rare variants may represent new mutations in a specific population or longstanding

TABLE 1.3 SNP Codes

SNP Code	Nucleotides
M	A/C
R	A/G
W	A/T
S	C/G
Y	C/T
K	G/T
V	A/C/G
H	A/C/T
D	A/G/T
B	C/G/T
N	A/C/G/T

TABLE 1.4 Approximate Number of Genomic Variants in an Average Individual

		Population Frequency		
		De Novo (≤0.001%)	Rare Inherited (≤1%)	Common Inherited (>1%)
Variant type	SNV (0 bp)	65	150,000	3,000,000
	Indel (1–1000 bp)	5	15,000	250,000
	CNV (>1000 bp)	<1	150	2000

variation that has not become more common owing to chance (genetic drift) or potentially because of a more complex and potentially detrimental effect on reproductive fitness and human disorders.

Finally, a tiny proportion of variants (about 0.002%) (Table 1.4) are de novo, meaning that the variant was not present in either parent but arose as a spontaneous mutation in the parental germline, during meiosis, or very early in the zygote. Because these mutations have not been subjected to natural selection, beyond survival of the individual to birth, they have the potential to mediate significant impairments of reproductive fitness and therefore human disorders.

This relationship between common and rare variation with human disorders is explored at greater length in chapter "Contribution of Genetic Epidemiology to Our Understanding of Psychiatric Disorders". There are several databases of human variation including a catalog of all known single nucleotide variants (www.ncbi.nlm.nih.gov/SNP/), a summary of all variants identified in whole genome sequencing of over 2500 individuals (www.1000genomes.org), and the variants identified in whole exome sequencing of over 60,000 unrelated individuals (exac.broadinstitute.org).

VARIATION IN COMPLEX AND MENDELIAN DISORDERS

Mendelian disorders are those in which a small number of genomic variants at a small number of loci are responsible for the vast majority of cases of a disorder with high penetrance. This results in a clear pattern of inheritance within a family: for example, autosomal dominant inheritance in which all individuals with the variant are affected (eg, variants in the *Neurofibromin 1* gene causing neurofibromatosis type 1) or autosomal recessive inheritance in which only individuals with two copies of the disease-associated allele are affected (eg, variants in the *Phenylalanine hydroxylase* gene that cause phenylketonuria with intellectual disability). Of note, the variants that lead to severe early-onset Mendelian disorders are almost always rare in the population (≤1% population frequency); in fact, the variants that lead to autosomal dominant disorders are often de novo, whereas those that lead to autosomal recessive inheritance usually occur in families with a history of consanguinity. Many of the variants that cause Mendelian disorders are known through linkage analysis, and this progress has accelerated considerably with the availability of high-throughput sequencing (www.omim.org). The variants identified frequently result in missense or loss-of-function of protein-coding sequences. However, a rigorous search for noncoding mutations has not been performed for most Mendelian disorders, and specific examples exist (eg, variants in the ZRS enhancer of the *SHH* gene, described in the section on "Enhancers").

In contrast, complex disorders are caused by a large number of variants at a large number of genomic loci (eg, about 1000 genes in ASD) (He et al., 2013; Sanders et al., 2012), often in parallel with environmental risk factors. These disorders often represent extremes of a quantitative trait in the population. For example, blood pressure varies across the population, with the individuals with higher pressures being diagnosed with hypertension. Most of these genetic risk factors tend to be common (>1% of the population) (Gaugler et al., 2014; Lee et al., 2013) and are often detected in regulatory, noncoding regions using GWAS (Schizophrenia Working Group of the Psychiatric Genomics Consortium, 2014). Rare variants can still have a substantial contribution in complex disorders, especially those that are severe and occur early in life, such as intellectual disability (Deciphering Developmental Disorders Study, 2015) and ASD (De Rubeis et al., 2014; Iossifov et al., 2014; Sanders et al., 2015). Many of the rare variants discovered in these disorders affect genes involved in chromatin and transcription regulation. In combination with the known common variants, this suggests that understanding gene regulation and noncoding variation may be critical for a complete understanding of neuropsychiatric disorders in humans.

CHROMATIN (DNA, RNA, AND PROTEIN) AND EPIGENETICS

The nucleotides of DNA encode information, but generating proteins from this information requires a complex interplay between DNA, RNAs, and proteins. These RNAs and proteins regulate the DNA, determining when, how, and to what extent regions of DNA are opened, regulated, transcribed into RNA and, for protein-coding mRNAs, translated into a protein. Whereas the DNA nucleotides remain almost entirely constant throughout the life of the organism and between different cells, the other components of chromatin are dynamic and vary with time and location, and between cell types. The complete set of these reversible chemical modifications to chromatin is called the epigenome, and the epigenetic changes allow different cell types to work from a common genome. This dynamism allows the cell to respond to its environment, including responding to cell signaling, metabolic demands, and circadian variation.

Chromatin State

Examples of known classes of epigenetic factors are described next. These factors act in concert to regulate the extent of DNA transcription across the genome. By considering the degree of binding of several epigenetic factors, an estimation of the type of region (eg, enhancer, promoter) and the degree of activity (eg, active, weakly active, inactive) can be made. These combinations of epigenetic factors are called the chromatin state (Ernst et al., 2011).

This complexity limits the ability to describe specific epigenetic markers as being "activators" or "repressors" in isolation. There are certainly markers usually associated with activation (eg, H3K27ac) or repression (eg, DNA methylation); however, there will always be exceptions. Furthermore, it is unclear which markers cause the activity and which are the consequence. It is better to conceptualize these epigenetic markers as the machinery for integrating multiple variables (eg, signaling pathways, metabolic state) to produce a suitable transcriptional response for a particular gene in a particular cell type.

Similarly, it is tempting to think of genes as being "on" or "off" in specific cells, but again this misses the complexity of the system. Rather, the multiple epigenetic markers each influence the degree of transcription leading to a dynamic equilibrium around a set point of transcription for a given gene that can rapidly change in response to the cellular environment. Moreover, even a set of purified cells of the same type, each cell may have their own transcriptional landscape, cell cycle, and regulatory states that vary from neighboring cells of the same type.

Heterochromatin

This is DNA that is tightly coiled. This densely packed DNA prevents access by transcriptional machinery leading to genes that are not transcribed. Heterochromatin is frequently observed in regions of repetitive DNA (eg, telomeres and centromeres) and is often associated with the nuclear envelope.

Euchromatin

This is the opposite of heterochromatin. The DNA is less tightly coiled, allowing access to transcription factors and chromatin remodelers so that transcription can occur. This DNA tends to lie away from the nuclear envelope.

DNA Methylation

Cytosine (C) can be methylated to form 5-methylcytosine (5mC) without affecting its ability to form hydrogen bonds with the guanine (G) on the opposing strand. In humans this methylation occurs when the cytosine is followed by a guanine, This arrangement is called CpG, where the "p" stands for the phosphate in the DNA backbone. Regions that are enriched for CpG nucleotides are called CpG islands, and these are frequently observed in the promoter region at the start of a gene (Fig. 1.3). The methylation and hydroxyl-methylation marks can be modified by the action of ten-eleven translocation and Wilms tumor 1 enzymes, which convert 5mC to 5-hydroxymethylcytosine (5hmC), 5-formylcytosine, and 5-carboxylcytosine. These three chemicals can revert to unmethylated cytosine spontaneously or through the action of thymine-DNA glycosylase. Methylation (5mC) of the CpG island usually reduces expression of the gene, whereas 5hmC of a gene usually increases its expression. DNA methylation is generally preserved between cell divisions but is mostly erased in the gametes and between generations (Schübeler, 2015). Rett syndrome, associated with encephalopathy and severe mental retardation in females, is caused by loss-of-function mutations in the gene *methyl CpG binding protein 2* on chromosome X. This gene is highly expressed in neurons and binds to methylated DNA to regulate gene expression (Pohodich & Zoghbi, 2015).

In addition to silencing specific genes, changes in DNA methylation and packaging also underlie X inactivation in females, in which one copy of chromosome X in each cell is largely untranscribed. Finally, some regions of the genome are "imprinted" with different methylation patterns observed on the paternal and maternal chromosomes leading to altered gene expression (Schübeler, 2015). Methylation patterns in chromosome region 15q12 are used to diagnose Angelman syndrome, in which only the paternal copy is present, and Prada–Willi syndrome, in which only the maternal copy is present.

Histone Proteins

DNA exists in a highly ordered and condensed state in the nucleus of cells. The first level of this ordered state is the nucleosome, in which DNA is wrapped twice around a complex of eight histone proteins (Fig. 1.1A). Each of these histone proteins has an amino acid tail; chemical modifications of the amino acids in this tail are called histone markers. For example, acetylation (ac) of the lysine (K) at position 27 of the amino acid tail of histone 3 (H3) is described as H3K27ac and is a marker of active transcription. The histone marks are created, removed, and recognized by numerous chromatin and transcription regulating proteins, many of which are associated with risk for ASD including *lysine (K)-specific methyltransferase 2E* (Dong et al., 2014).

Chromatin Remodelers

For replication or transcription to occur, the DNA needs to be exposed to the polymerase proteins and transcription factors by moving the histone proteins, a process called chromatin remodeling. Five major classes of chromatin remodeling complexes are known: SWI/SNF, ISWI, CHD, INO80, and ATRX (Bartholomew, 2014). Numerous neurological disorders have been associated with the gene in these complexes. Coffin–Siris syndrome, characterized by mental retardation and dysmorphic features, is associated with mutations in several genes in the SWI–SNF complex (Tsurusaki et al., 2012), including *AT-rich interactive domain 1B*, which is also one of the genes most strongly associated with ASD (Iossifov et al., 2014). Mutations in *Chromodomain Helicase DNA Binding Protein 7* (*CHD7*) are a frequent cause of CHARGE syndrome, characterized by coloboma of the eye, heart anomaly, atresia of the choanae, retardation of mental and somatic development, genital abnormalities, and ear abnormalities syndrome (Bajpai et al., 2010), whereas loss-of-function mutations in the chromodomains *CHD8* and *CHD2* are strongly associated with ASD (Iossifov et al., 2014).

Transcription Factors

Transcription factors are proteins that bind to specific sequences of DNA to modify transcription. This modification may be direct (for example, an activator that binds to a promoter region of DNA and to Pol II to start transcription) or indirect (for example, acting through one or many cofactors). Furthermore, the extent and nature of this transcriptional modification varies among genes, transcription factors, and cell types (Lee & Young, 2013). Transcription factors bind to specific motifs of DNA that frequently range from 5 to 15 bp in length. Variation of the nucleotides within this motif alters the degree of binding and can have functional implications. Numerous genes act as transcription factors (including histone modifiers and chromatin remodelers), many of which are associated with human disorders including cancer and neurological conditions. Specific examples are described in the section on regulatory complexes.

Cofactors

These are proteins involved in the transcription regulatory process that do not bind directly to DNA, in contrast to transcription factors. Cofactors frequently form complexes (eg, the PIC and the mediator complex), often involving transcription factors as well. As with transcription factors, many genes act as cofactors, leading to an association with numerous disorders. Specific examples are described in the section on regulatory complexes.

Elongation Factors

These are factors that promote continued transcription of RNA, as opposed to the initiation of transcription. It has become clear that this is a major component of the regulation of transcription in humans, but less so in yeast (Adelman & Lis, 2012).

Cohesins

In the nucleus, DNA forms loops that allow distant regions of DNA to be brought into apposition; the ring-shaped cohesins maintain these loop structures (Fig. 1.5). These loops are thought to create regulatory regions in which chromatin state may

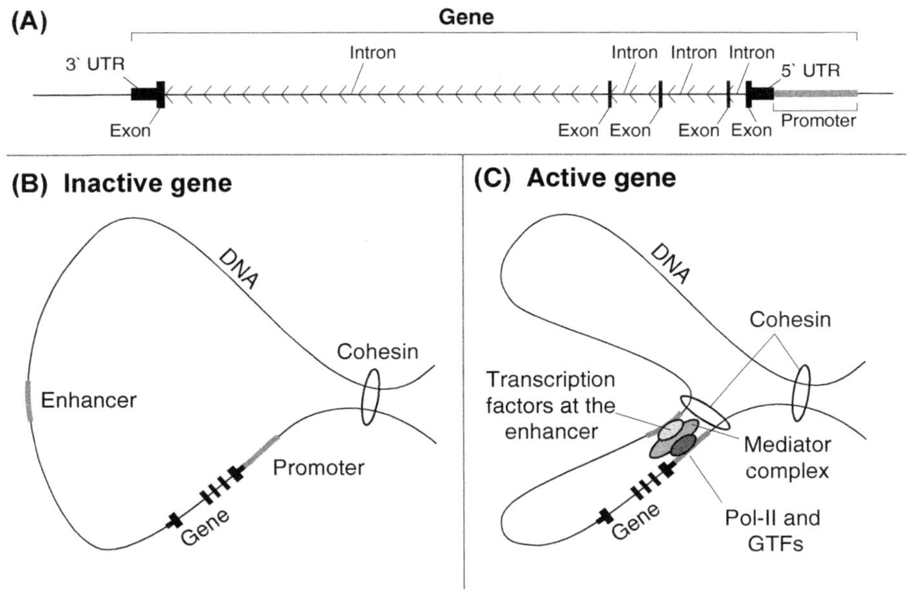

FIGURE 1.5 Gene regulation by enhancers. (A) Key elements of a gene, including the promoter, UTRs, exons, and introns. (B) An example of the same gene in a regulatory loop of DNA maintained by a cohesin complex acting as an insulator. The enhancer is not associated with the promoter inhibiting transcription of the gene. (C) The enhancer is brought into proximity of the promoter through the binding of transcription factors and the PIC formed by mediator, Pol II (Pol-II), and GTFs leading to transcription of the gene. A new cohesin complex maintains this arrangement.

be shared, with the cohesins in combination with the protein CTCF acting as insulators to prevent the regulatory signals extending beyond the loop. In addition, cohesins can bind enhancers to promoters, leading to active transcription of the downstream gene (Fig. 1.5). Finally, cohesins have important roles in DNA replication and the binding of sister chromatids. The cohesin complex is composed of the proteins made of four genes: *NIPBL*, *SMC1A*, *SMC3*, and *RAD21*. Mutations in any of these can lead to Cornelia de Lange syndrome, which is characterized by growth retardation, developmental delay, a distinctive facial appearance, and upper limb anomalies. These symptoms probably result from the role of cohesions in transcriptional regulation rather than in replication (Lee & Young, 2013).

REGULATORY COMPLEXES

Preinitiation Complex (PIC)

The main "output" of the regulatory epigenetic factors described earlier (eg, methylation, histones, chromatin remodelers) is to determine the extent to which a section of DNA is transcribed to RNA. For transcription to occur, the RNA Pol II enzyme must be brought into contact with the promoter (a section of DNA immediately before the start of the gene) and released to start transcription. This process involves a large number of transcription factors and cofactors forming a PIC around Pol II and the promoter region of DNA. The PIC includes three families of proteins.

General Transcription Factors

General transcription factors (GTFs) form six key subunits that make up most of the PIC (Luse, 2014; Sainsbury, Bernecky, & Cramer, 2015):

- TFIID binds to the promoter through the action of *TATA-box binding protein* (*TBP*); triplet repeat expansions in this gene lead to the neurodegenerative disorder spinocerebellar ataxia 17. It also contains numerous TBP-associated factors.
- TFIIB also has a role in binding to DNA, specifically B recognition elements near the promoter and selection of the TSS.
- TFIIH contains helicases that unwind DNA, including *excision repair cross-complementing rodent repair deficiency, complementation group 3*, mutations of which cause extreme skin sensitivity to sunlight in xeroderma pigmentosum B.
- TFIIE recruits TFIIH to the PIC.

- TFIIF recruits Pol II to the PIC.
- TFIIA is not essential for PIC function, but it serves to stabilize the binding of TFIID to the DNA.

These are the main subunits of the PIC but other factors may be involved in specific cell types, eg, *general transcription factor IIi* (*GTF2I*) forms part of the TFII–I complex and is one of the many genes deleted in Williams Syndrome, which is characterized by cardiac abnormalities, developmental delay, and a hypersociable personality. *GTF2I* may have a role in the cognitive features of this disorder (Borralleras, Sahun, Pérez-Jurado, & Campuzano, 2015).

RNA Polymerase II

This is an enzyme that transcribes DNA into mRNA. It is recruited to the transcription start site by the PIC and is bound by TFIIF and the mediator complex. It is composed of 12 subunits named *POLR2A* to *POLR2L* in humans (Sainsbury et al., 2015).

Mediator (Coactivator)

Whereas the GTFs recruit the Pol II to the promoter, mediator allows transcription factors at other DNA sites (eg, enhancers) and other cofactors to influence this process. The components of the GTF subunits listed previously are relatively constant, but those of mediator vary with time, cell type, and species. Furthermore, different subunits bind to different transcription factors (Poss, Ebmeier, & Taatjes, 2013), and this forms the basis for tissue-specific associations. For example, both *Mediator complex subunit 23* and *Mediator complex subunit 12* have been associated with intellectual disability (Lee & Young, 2013). The mediator complex binds to Pol II and TFIIF; however, the presence of the *cyclin-dependent kinase (CDK) 8* subunit blocks this binding behavior (Poss et al., 2013). In addition, mediator modulates PIC binding at the promoter and the stability of the PIC. Finally, noncoding RNA can also interact with the mediator complex to govern its function (Luse, 2014).

The vast majority of research on the function and members of the PIC has been performed on genes with TATA-box promoters, which account for fewer than 25% of protein-coding genes in humans. Although extensive progress has been made in understanding the PIC, much remains unknown.

Proximal Promoter-Pausing Complexes

In yeast the PIC recruits Pol II to the TSS, after which Pol II is released and transcribes the DNA. In animals, including humans, an extra level of regulation exists, with Pol II coming to a pause 20–60 nucleotides downstream of the TSS owing to the action of the negative elongation factor (NELF) (composed of the proteins from four genes: *NELFA*, *NELFB*, *NELFC*, and *NELFD*) and DRB sensitivity–inducing factor (DSIF) (composed of the proteins from two genes: *SUPT4H1* and *SUPT5H*) (Adelman & Lis, 2012; Jonkers & Lis, 2015). Of note, *NELFA* is one of the genes deleted in Wolf–Hirschhorn syndrome caused by a distal deletion of the short arm of chromosome 4 (4p16.3) and associated with growth retardation, developmental delay, seizures, and a distinctive facial appearance.

Elongation Complex

Releasing the paused Pol II at the proximal promoter to allow transcription requires the protein complex positive transcription elongation factor b (P-TEFb), which is composed of the protein encoded by *CDK9* and one of several other cyclin proteins (eg, *cyclin T1*). P-TEFb phosphorylates NELF, which dissociates, and DSIF, which then promotes transcription as an elongation factor throughout the rest of transcription. P-TEFb can form a large "super elongation complex" in combination with other elongation factors and chromatin regulators that strongly promotes further transcription (Jonkers & Lis, 2015).

TRANSCRIPTION (RNA)

To act on the cell, the double-stranded DNA must be transcribed into RNA. RNA differs from DNA in that it is single-stranded, has a ribose sugar instead of deoxyribose, and uses the nucleobase uracil (U) instead of thymine (T). There are three broad categories of RNA.

1. mRNA: Pre-mRNA is transcribed from DNA by RNA Pol II and undergoes processing to form mature mRNA, which is translated by ribosomes to form proteins. Over 20,000 genes form mRNA in humans, and variants in the genes that code these mRNAs are frequently responsible for human disorders.

2. Transfer RNA (tRNA) is small molecules that act as a bridge between amino acids and mRNA. Each tRNA binds a single amino acid and has a distinct pattern of three nucleotides that complements the triplet code in mRNA (Tables 1.1 and 1.2). There are about 500 tRNA genes throughout the human genome. They are transcribed by RNA Pol III.
3. Ribosomal RNA (rRNA) forms ribosomes for making proteins from mRNA. Excluding mitochondria, there are four rRNA genes in humans, although multiple copies of each are often arranged in clusters. rRNA is transcribed by RNA Pol I and is one of the most abundant transcripts in the cell. The absence of a poly(A) tail excludes it from RNA-Seq analysis.

Along with these three "classical" types of RNA, more than two dozen other types of RNA have been discovered (Table 1.5) (Morris & Mattick, 2014).

TABLE 1.5 Types of RNA

Type	Abbreviation	Function	Organisms
7SK RNA	7SK	Negatively regulating CDK9/cyclin T complex	Metazoans
Signal recognition particle RNA	7SRNA	Membrane integration	All organisms
Antisense RNA	aRNA	Regulatory	All organisms
Circular RNAs	circRNA	Regulatory	All organisms
CRISPR RNA	crRNA	Resistance to parasites	Bacteria and archaea
Guide RNA	gRNA	mRNA nucleotide modification	Kinetoplastid mitochondria
Long noncoding RNA	lncRNA	XIST (dosage compensation), HOTAIR (cancer)	Eukaryotes
MicroRNA	miRNA	Gene regulation	Most eukaryotes
Messenger RNA	mRNA	Codes for protein	All organisms
Piwi-interacting RNA	piRNA	Transposon defense, maybe other functions	Most animals
Repeat associated siRNA	rasiRNA	Type of piRNA; transposon defense	*Drosophila*
Retrotransposon	retroRNA	Self-propagation	Eukaryotes and some bacteria
Ribonuclease MRP	RNase MRP	rRNA maturation, DNA replication	Eukaryotes
Ribonuclease P	RNase P	tRNA maturation	All organisms
Ribosomal RNA	rRNA	Translation	All organisms
Small Cajal body-specific RNA	scaRNA	Guide RNA to telomere in active cells	Metazoans
Small interfering RNA	siRNA	Gene regulation	Most eukaryotes
SmY RNA	SmY	mRNA *trans*-splicing	Nematodes
Small nucleolar RNA	snoRNA	Nucleotide modification of RNAs	Eukaryotes and archaea
Small nuclear RNA	snRNA	Splicing and other functions	Eukaryotes and archaea
Trans-acting siRNA	tasiRNA	Gene regulation	Land plants
Telomerase RNA	telRNA	Telomere synthesis	Most eukaryotes
Transfer-messenger RNA	tmRNA	Rescuing stalled ribosomes	Bacteria
Transfer RNA	tRNA	Translation	All organisms
Viral response RNA	viRNA	Antiviral immunity	*Caenorhabditis elegans*
Vault RNA	vRNA	Self-propagation	Expulsion of xenobiotics
Y RNA	yRNA	RNA processing, DNA replication	Animals

POSTTRANSCRIPTIONAL mRNA MODIFICATIONS

Regulation continues to have a major role after transcription as the pre-mRNA is processed to form a mature mRNA. This occurs through several mechanisms.

Epitranscriptome

mRNA nucleotides can be chemically modified in a manner similar to DNA nucleotides (Saletore et al., 2012). For example, the sixth nitrogen in adenosine (A) may be reversibly methylated to methyl-6-adenosine, particularly at stop codons, where they have a destabilizing effect on the mRNA, and at the transcriptional start site (Meyer & Jaffrey, 2014). Over 100 such modifications have been recognized (modomics.genesilico.pl).

RNA Editing

The nucleotides within the mRNA can be edited by the action of proteins encoded by the genes *Adenosine Deaminase, RNA-specific* (*ADAR*) and *ADAR1B*, the latter of which is highly expressed in the human brain. The *ADAR* enzymes deaminate adenosine (A) into inosine (I), which is then recognized as guanosine (G) by both the spliceosome and the ribosome (Fig. 1.6). This editing may modify the protein-coding sequence, miRNA binding sites, or splice sites. Whereas RNA editing occurs in numerous tissues, it is most frequently observed in the cells of the central nervous system, including numerous receptors (eg, glutamate, $GABA_A$, serotonin 2c) and ion channels (voltage-gated potassium and calcium) (Tariq & Jantsch, 2012). It is especially critical in the processing of the gene *Glutamate Receptor, Ionotropic, AMPA 2*, with a glutamine (Q) being converted to an arginine (R) in all transcripts (Nishikura, 2010).

RNA Processing

Introns are removed from the pre-mRNA by the spliceosome, a complex of five small nuclear ribonucleic proteins, named U1, U2, U4, U5, and U6, that are composed of a combination of proteins and RNA. In addition, the ends of the mRNA are

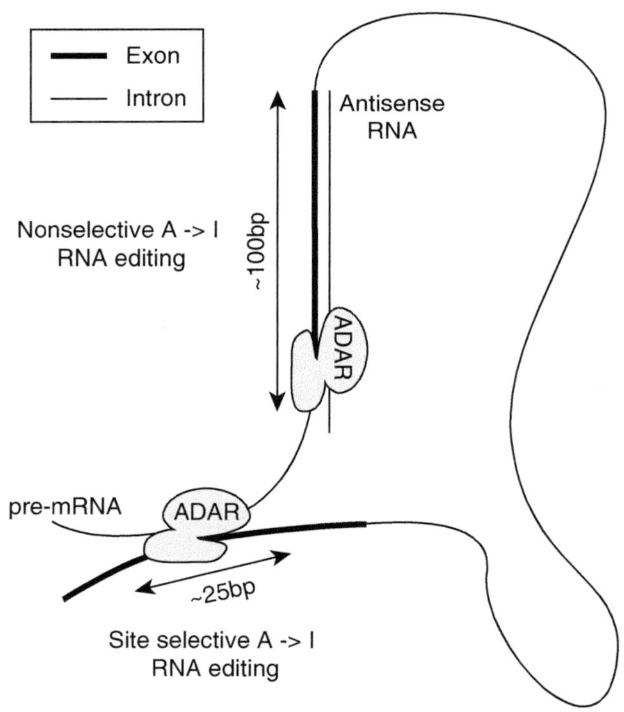

FIGURE 1.6 RNA editing. Adenosine (A) nucleotides are converted to inosine (I) nucleotides by the action of the *ADAR* enzyme on double-stranded RNA (dsRNA). This can be highly selective in short regions of dsRNA, often between a coding exon and an intron, or promiscuous, with most adenosines being edited in long regions of dsRNA with perfect homology, eg, in the presence of an antisense RNA (Nishikura, 2010).

stabilized, often by the addition of a 7-methylguanosine cap at the 5′ end and polyadenylation (the addition of 150–200 adenosine nucleotides, a poly[A] tail) at the 3′ end. The mature mRNA is exported from the nucleus by the nuclear RNA export factors (eg, *Nuclear RNA Export Factor 1*), which bind to the poly(A) tail.

TRANSLATION

In the cytoplasm, ribosomes can form at the AUG initiation codon of the mature mRNA and form a protein through binding of tRNAs with the specific amino acids to which they are bound (Tables 1.1 and 1.2). As the ribosome moves along the mRNA molecule, further ribosomes can bind so that one mRNA molecule can be translated into multiple proteins. The rate of translation varies between mRNA molecules, possibly owing to the relative abundance of specific tRNAs. This variability is called ribosomal pausing and it may have a role in protein folding and the number of proteins made. When the ribosome encounters a stop codon (Tables 1.1 and 1.2), a release factor (eg, *eukaryotic translation termination factor 1*) binds instead of a tRNA, which prompts the dissociation of the mRNA, ribosome, and protein.

In the cytoplasm, the mRNA is constantly attacked by exonucleases which degrade the poly(A) tail. Once the tail is removed, degradation proteins are attracted, which rapidly degrade the mRNA from both ends. This process is called the deadenylation-dependent mRNA decay pathway.

POSTTRANSLATIONAL MODIFICATIONS

The protein products of translation can also be modified by adding chemical groups (eg, phosphorylation), by adding large molecules (eg, glycosylation, palmitoylation, ubiquitination, SUMOylation), and by cleavage. dbPTM is a database of known posttranslational modifications (PTMs) at specific sites (dbptm.mbc.nctu.edu.tw). Although DNA and RNA sequences can be read rapid, accurately, and cheaply through high-throughput sequencing, the technology for reading protein sequences and PTMs is less developed. However, it is likely that developments under way in proteomics will yield transformative into human biology, specifically in genomics, epigenomics, and transcriptomics. Whereas there are about 20,000 genes in the human genome, these may yield up to two million distinct proteins. Efforts to map these proteins are ongoing in, for example, the Universal Protein Resource (UniProt, www.uniprot.org) and the Human Proteome Project (HPP, www.thehpp.org).

SUMMARY

Sequencing the human genome was a landmark achievement that ushered in an era of genomic research. In addition to refining the raw DNA sequence, substantial progress has been made in annotating the genome, including locating common variants, protein-coding genes, noncoding regulatory elements, and epigenetic marks. The availability of the genome and high-throughput sequencing has helped reveal complex regulatory mechanisms at multiple levels, including epigenetic (eg, DNA methylation and histone markers), pretranscription (eg, enhancers and transcription factors), posttranscription (eg, miRNAs and RNA editing), and posttranslation (eg, chemical modification and cleavage).

Progress has already revealed insights into the causes of human neurological disorders. The genes responsible for many, possibly most, Mendelian disorders are known and there is strong evidence of association between numerous genetic loci and some complex neuropsychiatric disorders including congenital brain malformations, intellectual disability, ASD, schizophrenia, multiple sclerosis, and Alzheimer disease. Many of these genetic loci implicate the regulatory machinery across multiple levels, including the epigenome (eg, *CHD8* in ASD), pretranscription (eg, mediator genes in intellectual disability), and posttranscription (eg, *MIR137* miRNA and its targets in schizophrenia). This progress presents the opportunity and challenge of applying these discoveries to understand the pathology of neuropsychiatric disorders with the view to transforming patient diagnosis and treatment.

REFERENCES

Adelman, K., & Lis, J. T. (2012). Promoter-proximal pausing of RNA polymerase II: emerging roles in metazoans. *Nature Reviews Genetics, 13*(10), 720–731. http://dx.doi.org/10.1038/nrg3293.

Arnold, C. D., Gerlach, D., Stelzer, C., Boryń, Ł. M., Rath, M., & Stark, A. (2013). Genome-wide quantitative enhancer activity maps identified by STARR-seq. *Science, 339*(6123), 1074–1077. http://dx.doi.org/10.1126/science.1232542.

Bajpai, R., Chen, D. A., Rada-Iglesias, A., Zhang, J., Xiong, Y., Helms, J., ... Wysocka, J. (2010). CHD7 cooperates with PBAF to control multipotent neural crest formation. *Nature, 463*(7283), 958–962. http://dx.doi.org/10.1038/nature08733.

Bartholomew, B. (2014). ISWI chromatin remodeling: one primary actor or a coordinated effort? *Current Opinion in Structural Biology, 24*, 150–155. http://dx.doi.org/10.1016/j.sbi.2014.01.010.

Birnbaum, R. Y., Clowney, E. J., Agamy, O., Kim, M. J., Zhao, J., Yamanaka, T., ... Ahituv, N. (2012). Coding exons function as tissue-specific enhancers of nearby genes. *Genome Research, 22*(6), 1059–1068. http://dx.doi.org/10.1101/gr.133546.111.

Borralleras, C., Sahun, I., Pérez-Jurado, L. A., & Campuzano, V. (2015). Intracisternal Gtf2i gene therapy ameliorates deficits in cognition and synaptic plasticity of a mouse model of Williams-Beuren syndrome. *Molecular Therapy*. http://dx.doi.org/10.1038/mt.2015.130.

Buenrostro, J. D., Wu, B., Chang, H. Y., & Greenleaf, W. J. (2015). ATAC-seq: A Method for Assaying Chromatin Accessibility Genome-Wide. *Current Protocols in Molecular Biology, 109*. http://dx.doi.org/10.1002/0471142727.mb2129s109.

Canzonetta, C., Mulligan, C., Deutsch, S., Ruf, S., O'Doherty, A., Lyle, R., ... Nizetic, D. (2008). DYRK1A-dosage imbalance perturbs NRSF/REST levels, deregulating pluripotency and embryonic stem cell fate in Down syndrome. *American Journal of Human Genetics, 83*(3), 388–400. http://dx.doi.org/10.1016/j.ajhg.2008.08.012.

Chorev, M., & Carmel, L. (2012). The function of introns. *Frontiers in Genetics, 3*, 55. http://dx.doi.org/10.3389/fgene.2012.00055.

Consortium, E. P. (2012). An integrated encyclopedia of DNA elements in the human genome. *Nature, 489*(7414), 57–74. http://dx.doi.org/10.1038/nature11247.

Consortium, G. P. (2010). A map of human genome variation from population-scale sequencing. *Nature, 467*(7319), 1061–1073. http://dx.doi.org/10.1038/nature09534. pii:nature09534.

Costa, V., Esposito, R., Aprile, M., & Ciccodicola, A. (2012). Non-coding RNA and pseudogenes in neurodegenerative diseases: "The (un)Usual Suspects". *Frontiers in Genetics, 3*, 231. http://dx.doi.org/10.3389/fgene.2012.00231.

Deciphering Developmental Disorders Study. (2015). Large-scale discovery of novel genetic causes of developmental disorders. *Nature, 519*(7542), 223–228. http://dx.doi.org/10.1038/nature14135.

De Rubeis, S., He, X., Goldberg, A. P., Poultney, C. S., Samocha, K., Cicek, A. E., ... Consortium, U. K. (2014). Synaptic, transcriptional and chromatin genes disrupted in autism. *Nature, 515*(7526), 209–215. http://dx.doi.org/10.1038/nature13772.

Ding, N., Qu, H., & Fang, X. (2014). The ENCODE project and functional genomics studies. *Yi Chuan, 36*(3), 237–247.

Dong, S., Walker, M. F., Carriero, N. J., DiCola, M., Willsey, A. J., Ye, A. Y., ... Sanders, S. J. (2014). De novo insertions and deletions of predominantly paternal origin are associated with autism spectrum disorder. *Cell Reports, 9*(1), 16–23. http://dx.doi.org/10.1016/j.celrep.2014.08.068.

Duker, A. L., Ballif, B. C., Bawle, E. V., Person, R. E., Mahadevan, S., Alliman, S., ... Sahoo, T. (2010). Paternally inherited microdeletion at 15q11.2 confirms a significant role for the SNORD116 C/D box snoRNA cluster in Prader-Willi syndrome. *European Journal of Human Genetics, 18*(11), 1196–1201. http://dx.doi.org/10.1038/ejhg.2010.102.

Ernst, J., Kheradpour, P., Mikkelsen, T. S., Shoresh, N., Ward, L. D., Epstein, C. B., ... Bernstein, B. E. (2011). Mapping and analysis of chromatin state dynamics in nine human cell types. *Nature, 473*(7345), 43–49. http://dx.doi.org/10.1038/nature09906.

Fatica, A., & Bozzoni, I. (2014). Long non-coding RNAs: new players in cell differentiation and development. *Nature Reviews Genetics, 15*(1), 7–21. http://dx.doi.org/10.1038/nrg3606.

Fritah, S., Niclou, S. P., & Azuaje, F. (2014). Databases for lncRNAs: a comparative evaluation of emerging tools. *RNA, 20*(11), 1655–1665. http://dx.doi.org/10.1261/rna.044040.113.

Fukagawa, T., & Earnshaw, W. C. (2014). The centromere: chromatin foundation for the kinetochore machinery. *Developmental Cell, 30*(5), 496–508. http://dx.doi.org/10.1016/j.devcel.2014.08.016.

Gaugler, T., Klei, L., Sanders, S. J., Bodea, C. A., Goldberg, A. P., Lee, A. B., ... Buxbaum, J. D. (2014). Most genetic risk for autism resides with common variation. *Nature Genetics, 46*(8), 881–885. http://dx.doi.org/10.1038/ng.3039.

Gerstein, M. B., Bruce, C., Rozowsky, J. S., Zheng, D., Du, J., Korbel, J. O., ... Snyder, M. (2007). What is a gene, post-ENCODE? History and updated definition. *Genome Research, 17*(6), 669–681. http://dx.doi.org/10.1101/gr.6339607.

Gomez, D. E., Armando, R. G., Farina, H. G., Menna, P. L., Cerrudo, C. S., Ghiringhelli, P. D., & Alonso, D. F. (2012). Telomere structure and telomerase in health and disease (review). *Internet Journal of Oncology, 41*(5), 1561–1569. http://dx.doi.org/10.3892/ijo.2012.1611.

Guo, J. U., Agarwal, V., Guo, H., & Bartel, D. P. (2014). Expanded identification and characterization of mammalian circular RNAs. *Genome Biology, 15*(7), 409. http://dx.doi.org/10.1186/s13059-014-0409-z.

He, X., Sanders, S. J., Liu, L., De Rubeis, S., Lim, E. T., Sutcliffe, J. S., ... Roeder, K. (2013). Integrated model of de novo and inherited genetic variants yields greater power to identify risk genes. *PLoS Genetics, 9*(8), e1003671. http://dx.doi.org/10.1371/journal.pgen.1003671.

Hobson, G. M., Huang, Z., Sperle, K., Stabley, D. L., Marks, H. G., & Cambi, F. (2002). A PLP splicing abnormality is associated with an unusual presentation of PMD. *Annals of Neurology, 52*(4), 477–488. http://dx.doi.org/10.1002/ana.10320.

Iossifov, I., O'Roak, B. J., Sanders, S. J., Ronemus, M., Krumm, N., Levy, D., ... Wigler, M. (2014). The contribution of de novo coding mutations to autism spectrum disorder. *Nature, 515*(7526), 216–221. http://dx.doi.org/10.1038/nature13908.

ISC. (2008). Rare chromosomal deletions and duplications increase risk of schizophrenia. *Nature, 455*(7210), 237–241. http://dx.doi.org/10.1038/nature07239. pii:nature07239.

Jones, J. M., & Gellert, M. (2004). The taming of a transposon: V(D)J recombination and the immune system. *Immunological Reviews, 200*, 233–248. http://dx.doi.org/10.1111/j.0105-2896.2004.00168.x.

Jonkers, I., & Lis, J. T. (2015). Getting up to speed with transcription elongation by RNA polymerase II. *Nature Reviews Molecular Cell Biology, 16*(3), 167–177. http://dx.doi.org/10.1038/nrm3953.

Kim, C. S., Hwang, C. K., Song, K. Y., Choi, H. S., Kim, D. K., Law, P. Y., ... Loh, H. H. (2008). Novel function of neuron-restrictive silencer factor (NRSF) for posttranscriptional regulation. *Biochimica et Biophysica Acta, 1783*(10), 1835–1846. http://dx.doi.org/10.1016/j.bbamcr.2008.06.019.

Lander, E. S., Linton, L. M., Birren, B., Nusbaum, C., Zody, M. C., Baldwin, J., … Consortium, I. H. G. S. (2001). Initial sequencing and analysis of the human genome. *Nature, 409*(6822), 860–921. http://dx.doi.org/10.1038/35057062.

Lee, S. H., Ripke, S., Neale, B. M., Faraone, S. V., Purcell, S. M., Perlis, R. H., … I. I. B. D. G. C. (2013). Genetic relationship between five psychiatric disorders estimated from genome-wide SNPs. *Nature Genetics, 45*(9), 984–994. http://dx.doi.org/10.1038/ng.2711.

Lee, T. I., & Young, R. A. (2013). Transcriptional regulation and its misregulation in disease. *Cell, 152*(6), 1237–1251. http://dx.doi.org/10.1016/j.cell.2013.02.014.

Lettice, L. A., Heaney, S. J., Purdie, L. A., Li, L., de Beer, P., Oostra, B. A., … de Graaff, E. (2003). A long-range Shh enhancer regulates expression in the developing limb and fin and is associated with preaxial polydactyly. *Human Molecular Genetics, 12*(14), 1725–1735.

Linardopoulou, E. V., Parghi, S. S., Friedman, C., Osborn, G. E., Parkhurst, S. M., & Trask, B. J. (2007). Human subtelomeric WASH genes encode a new subclass of the WASP family. *PLoS Genetics, 3*(12), e237. http://dx.doi.org/10.1371/journal.pgen.0030237.

Lu, T., Aron, L., Zullo, J., Pan, Y., Kim, H., Chen, Y., … Yankner, B. A. (2014). REST and stress resistance in ageing and Alzheimer's disease. *Nature, 507*(7493), 448–454. http://dx.doi.org/10.1038/nature13163.

Luse, D. S. (2014). The RNA polymerase II preinitiation complex. Through what pathway is the complex assembled? *Transcription, 5*(1), e27050. http://dx.doi.org/10.4161/trns.27050.

MacArthur, D. G., Balasubramanian, S., Frankish, A., Huang, N., Morris, J., Walter, K., … Tyler-Smith, C. (2012). A systematic survey of loss-of-function variants in human protein-coding genes. *Science, 335*(6070), 823–828. http://dx.doi.org/10.1126/science.1215040.

Meyer, K. D., & Jaffrey, S. R. (2014). The dynamic epitranscriptome: N6-methyladenosine and gene expression control. *Nature Reviews Molecular Cell Biology, 15*(5), 313–326. http://dx.doi.org/10.1038/nrm3785.

Miura, P., Shenker, S., Andreu-Agullo, C., Westholm, J. O., & Lai, E. C. (2013). Widespread and extensive lengthening of 3′ UTRs in the mammalian brain. *Genome Research, 23*(5), 812–825. http://dx.doi.org/10.1101/gr.146886.112.

Morris, K. V., & Mattick, J. S. (2014). The rise of regulatory RNA. *Nature Reviews Genetics, 15*(6), 423–437. http://dx.doi.org/10.1038/nrg3722.

Nagy, E., & Maquat, L. E. (1998). A rule for termination-codon position within intron-containing genes: when nonsense affects RNA abundance. *Trends in Biochemical Sciences, 23*(6), 198–199.

Nishikura, K. (2010). Functions and regulation of RNA editing by ADAR deaminases. *Annual Review of Biochemistry, 79*, 321–349. http://dx.doi.org/10.1146/annurev-biochem-060208-105251.

Pesole, G. (2008). What is a gene? An updated operational definition. *Gene, 417*(1–2), 1–4. http://dx.doi.org/10.1016/j.gene.2008.03.010.

Pinto, D., Delaby, E., Merico, D., Barbosa, M., Merikangas, A., Klei, L., … Scherer, S. W. (2014). Convergence of genes and cellular pathways dysregulated in autism spectrum disorders. *American Journal of Human Genetics, 94*(5), 677–694. http://dx.doi.org/10.1016/j.ajhg.2014.03.018.

Pohodich, A. E., & Zoghbi, H. Y. (2015). Rett syndrome: disruption of epigenetic control of postnatal neurological functions. *Human Molecular Genetics*. http://dx.doi.org/10.1093/hmg/ddv217.

Poss, Z. C., Ebmeier, C. C., & Taatjes, D. J. (2013). The mediator complex and transcription regulation. *Critical Reviews in Biochemistry and Molecular Biology, 48*(6), 575–608. http://dx.doi.org/10.3109/10409238.2013.840259.

Riethman, H. (2008). Human subtelomeric copy number variations. *Cytogenetic and Genome Research, 123*(1–4), 244–252. http://dx.doi.org/10.1159/000184714.

Ripke, S., Sanders, A. R., Kendler, K. S., Levinson, D. F., Sklar, P., Holmans, P. A., … Consortium, S. P. G.-W. A. S. (2011). Genome-wide association study identifies five new schizophrenia loci. *Nature Genetics, 43*(10), 969–976. http://dx.doi.org/10.1038/ng.940.

Sainsbury, S., Bernecky, C., & Cramer, P. (2015). Structural basis of transcription initiation by RNA polymerase II. *Nature Reviews Molecular Cell Biology, 16*(3), 129–143. http://dx.doi.org/10.1038/nrm3952.

Saletore, Y., Meyer, K., Korlach, J., Vilfan, I. D., Jaffrey, S., & Mason, C. E. (2012). The birth of the Epitranscriptome: deciphering the function of RNA modifications. *Genome Biology, 13*(10), 175. http://dx.doi.org/10.1186/gb-2012-13-10-175.

Sandelin, A., Carninci, P., Lenhard, B., Ponjavic, J., Hayashizaki, Y., & Hume, D. A. (2007). Mammalian RNA polymerase II core promoters: insights from genome-wide studies. *Nature Reviews Genetics, 8*(6), 424–436. http://dx.doi.org/10.1038/nrg2026.

Sanders, S. J., Ercan-Sencicek, A. G., Hus, V., Luo, R., Murtha, M. T., Moreno-De-Luca, D., … State, M. W. (2011). Multiple recurrent de novo CNVs, including duplications of the 7q11.23 Williams syndrome region, are strongly associated with autism. *Neuron, 70*(5), 863–885. http://dx.doi.org/10.1016/j.neuron.2011.05.002.

Sanders, S. J., He, X., Willsey, A. J., Ercan-Sencicek, A. G., Samocha, K. E., Cicek, A. E., … State, M. W. (2015). Insights into Autism Spectrum Disorder Genomic Architecture and Biology from 71 Risk Loci. *Neuron, 87*(6), 1215–1233. http://dx.doi.org/10.1016/j.neuron.2015.09.016.

Sanders, S. J., Murtha, M. T., Gupta, A. R., Murdoch, J. D., Raubeson, M. J., Willsey, A. J., … State, M. W. (2012). De novo mutations revealed by whole-exome sequencing are strongly associated with autism. *Nature, 485*(7397), 237–241. http://dx.doi.org/10.1038/nature10945.

Sauna, Z. E., & Kimchi-Sarfaty, C. (2011). Understanding the contribution of synonymous mutations to human disease. *Nature Reviews Genetics, 12*(10), 683–691. http://dx.doi.org/10.1038/nrg3051.

Schizophrenia Working Group of the Psychiatric Genomics Consortium. (2014). Biological insights from 108 schizophrenia-associated genetic loci. *Nature, 511*(7510), 421–427. http://dx.doi.org/10.1038/nature13595.

Schübeler, D. (2015). Function and information content of DNA methylation. *Nature, 517*(7534), 321–326. http://dx.doi.org/10.1038/nature14192.

Sebat, J., Lakshmi, B., Malhotra, D., Troge, J., Lese-Martin, C., Walsh, T., … Wigler, M. (2007). Strong association of de novo copy number mutations with autism. *Science, 316*(5823), 445–449. http://dx.doi.org/10.1126/science.1138659.

Steinberg, K. M., Schneider, V. A., Graves-Lindsay, T. A., Fulton, R. S., Agarwala, R., Huddleston, J., … Wilson, R. K. (2014). Single haplotype assembly of the human genome from a hydatidiform mole. *Genome Research, 24*(12), 2066–2076. http://dx.doi.org/10.1101/gr.180893.114.

Szatkiewicz, J. P., O'Dushlaine, C., Chen, G., Chambert, K., Moran, J. L., Neale, B. M., ... Sullivan, P. F. (2014). Copy number variation in schizophrenia in Sweden. *Molecular Psychiatry*. http://dx.doi.org/10.1038/mp.2014.40.

Taft, R. J., Pheasant, M., & Mattick, J. S. (2007). The relationship between non-protein-coding DNA and eukaryotic complexity. *BioEssays, 29*(3), 288–299. http://dx.doi.org/10.1002/bies.20544.

Tariq, A., & Jantsch, M. F. (2012). Transcript diversification in the nervous system: a to I RNA editing in CNS function and disease development. *Frontiers in Neuroscience, 6*, 99. http://dx.doi.org/10.3389/fnins.2012.00099.

Treangen, T. J., & Salzberg, S. L. (2012). Repetitive DNA and next-generation sequencing: computational challenges and solutions. *Nature Reviews Genetics, 13*(1), 36–46. http://dx.doi.org/10.1038/nrg3117.

Tsurusaki, Y., Okamoto, N., Ohashi, H., Kosho, T., Imai, Y., Hibi-Ko, Y., ... Matsumoto, N. (2012). Mutations affecting components of the SWI/SNF complex cause Coffin-Siris syndrome. *Nature Genetics, 44*(4), 376–378. http://dx.doi.org/10.1038/ng.2219.

Venter, J. C., Adams, M. D., Myers, E. W., Li, P. W., Mural, R. J., Sutton, G. G., ... Zhu, X. (2001). The sequence of the human genome. *Science, 291*(5507), 1304–1351. http://dx.doi.org/10.1126/science.1058040.

Yang, C., Bolotin, E., Jiang, T., Sladek, F. M., & Martinez, E. (2007). Prevalence of the initiator over the TATA box in human and yeast genes and identification of DNA motifs enriched in human TATA-less core promoters. *Gene, 389*(1), 52–65. http://dx.doi.org/10.1016/j.gene.2006.09.029.

Zuccato, C., Tartari, M., Crotti, A., Goffredo, D., Valenza, M., Conti, L., ... Cattaneo, E. (2003). Huntingtin interacts with REST/NRSF to modulate the transcription of NRSE-controlled neuronal genes. *Nature Genetics, 35*(1), 76–83. http://dx.doi.org/10.1038/ng1219.

Chapter 2

Contribution of Genetic Epidemiology to Our Understanding of Psychiatric Disorders

Kathleen R. Merikangas
National Institutes of Health, Bethesda, MD, United States

Alison K. Merikangas
University of Pennsylvania, Philadelphia, PA, United States

INTRODUCTION

Advances in molecular genetics have led to a major transformation in the field of genetic epidemiology. With its roots in population genetics and epidemiology, the first stage of research relied on inference of the contribution of genetic risk factors to disease etiology based on the adherence of patterns of transmission of disease in families to Mendelian principles. Diseases that followed the classic modes of Mendelian transmission were the first to profit from our ability to identify specific genetic markers of disease transmission, and the molecular genetic basis of nearly all of these diseases has now been established. Identification of genetic risk factors underlying complex, multifactorial diseases such as neuropsychiatric conditions has been far more challenging.

This chapter will: (1) provide background on the discipline of genetic epidemiology; (2) summarize evidence regarding the genetic epidemiology of psychiatric disorders; and (3) describe challenges and future strategies for application of the tools of genetic epidemiology of psychiatric disorders in the molecular era.

BACKGROUND: EPIDEMIOLOGY

Epidemiology is defined as the study of the distribution and determinants of diseases in human populations. Epidemiologic studies are concerned with the extent and types of illnesses in groups of people and with the factors that influence their distribution. Epidemiologists investigate the interactions that may occur among the host, agent, and environment (the classic epidemiologic triangle) to produce a disease state. The important goal of epidemiologic studies is to identify the etiology of a disease in order to prevent or intervene in the progression of the disorder. To achieve this goal, epidemiologic research generally proceeds from studies that specify the distribution of a disease within a population by person, place, and time (that is, descriptive epidemiology), to more focused studies of the determinants of disease in specific groups (that is, analytic epidemiology).

Differences in prevalence rates by particular characteristics (eg, sex, age, ethnicity, urban vs. rural) provide clues regarding potential underlying etiologic factors. Potential risk factors that are identified at the descriptive level may then be tested systematically with systematic case—control designs that compare the association between a particular risk factor or disease correlate and the presence or absence of a given disease, after controlling for relevant confounding variables. Case—control designs generally employ either cross-sectional or retrospective studies of associations between risk factors among cases compared with controls. Significant associations in case—control studies can then be tested in prospective cohort studies, which can formally test the temporal direction of such associations by observing large populations of unaffected individuals over time, monitoring risk factors and the onset of new cases of a disease.

Genomics, Circuits, and Pathways in Clinical Neuropsychiatry. http://dx.doi.org/10.1016/B978-0-12-800105-9.00002-0

The core measure in epidemiology is estimation of morbidity or illness in a defined population. Rates of morbidity include incidence, the number of new cases of disease that occur during a specified period in a defined population at risk, and prevalence, the number of affected persons in the population at a particular time divided by the number of persons in the population at that period. Prevalence estimates must be anchored in time ranging from lifetime to the current period. These approaches yield a range of risk estimates, including relative risk, the magnitude of the association between an exposure and disease incidence; absolute risk, the overall probability of developing a disease in an individual or in a particular population; attributable risk, the difference in the risk of the disease in those exposed to a particular risk factor compared with the background risk of a disease in a population (ie, in the unexposed); and population attributable risk, the risk of a disease in a total population (exposed and unexposed) indicating the amount the disease can be reduced in a population if an exposure is eliminated (Gordis, 2000). The odds ratio, defined as the ratio of the odds that cases were exposed to a risk factor to the odds that the control subjects were exposed, is the chief measure of association derived from case—control studies. The odds ratio is an approximation of the relative risk, which is based on incidence rates from prospective studies.

There are several criteria for assessing the extent to which a risk factor is causally associated with a trait or disease. These include strength of the association, a dose—response effect, and a lack of temporal ambiguity. Broader criteria that can be applied to a set of studies on a putative etiologic risk factor include consistency of the findings across studies, biologic plausibility of the hypothesis, and a specificity of association (Hill, 1953). The field of epidemiology, which is intricately tied to biostatistics, has employed increasingly sophisticated analytic methods to investigate risk factors for disease while simultaneously accounting for confounders, effect mediation, interactions, and varying time periods of observation of both disease and covariates to test associations to build causal models.

Contributions to Psychiatry

Application of the tools of epidemiology to psychiatry over the past 50 years has led to advances in our understanding of psychiatric disorders in the general population across the world. These contributions include: (1) development of standardized tools that operationalize diagnostic criteria to obtain reliable estimates of psychiatric disorders; (2) estimation of the magnitude and correlates of disorders across the world; (3) characterization of the full spectrum of manifestations of disorders and their underlying dimensions; (4) documentation of patterns of both mental and physical comorbidity with index psychiatric disorders; (5) identification of the gaps between clinical and population samples; (6) estimation of current and lifetime disability attributable to psychiatric disorders; and (7) identification of the demographic, biologic, and environmental risk and protective factors for psychiatric disorders.

Genetic Epidemiology

With advances in molecular biology, there has been a flood of information about the contribution of genetic risk factors to complex diseases. This has transformed the subdiscipline of genetic epidemiology that previously relied solely on inferences based on phenotypic disease manifestations in relatives. Commonly, genome-wide association studies (GWAS) are completed to compare common genetic variants between patients and control subjects to examine their association with a trait or disease. GWAS typically focus on associations between single-nucleotide polymorphisms (SNPs) (common DNA variants with $>1\%$ population frequency), and owing to the number of SNPs under investigation, a statistical significance threshold of 5×10^{-8} is employed. Association studies can also investigate structural variation such as segmental duplications and deletions, or copy number variants (CNVs).

The subdiscipline of genetic epidemiology focuses on identifying the role of genetic factors and their joint influence with environmental factors in disease etiology. Genetic epidemiology employs traditional epidemiologic study designs including case—control and cohort studies to evaluate the aggregation in groups as closely related as twins or as loosely related as migrant cohorts. Before the molecular genetic era, study designs in genetic epidemiology were designed to infer genetic causation by controlling for genetic background while letting the environment vary (eg, migrant studies, half siblings, separated twins) or conversely, controlling for the environment while allowing variance in the genetic background (eg, siblings, twins, adoptees, nonbiologic siblings). Investigations in genetic epidemiology are typically based on a combination of study designs including family, twin, and adoption studies. Measures of risk in genetic epidemiology include familial relative risk (disease risk in relatives of patients versus control subjects), and genetic attributable risk (the proportion of a particular disease that would be eliminated if a particular gene or genes were not involved in the disease). As described subsequently, sophisticated methods have been developed to compare combinations of genetic markers between patients and control subjects (eg, polygenic scores), and genome-wide complex trait analysis (GCTA), that estimates the proportion of phenotypic variance explained by genetic variants (typically SNPs) for complex traits (Yang, Lee, Goddard, & Visscher, 2011).

Family Studies

Familial aggregation is generally the first source of evidence that genetic factors may have a role in a disorder. The most common indicator of familial aggregation is the relative risk ratio, computed as the rate of a disorder in families of affected persons divided by the corresponding rate in families of controls. The patterns of genetic factors underlying a disorder can be inferred from the extent to which patterns of familial resemblance adhere to the expectations of Mendelian laws of inheritance. The degree of genetic relatedness among relatives is based on the proportion of shared genes between a particular relative and an index family member or proband. First-degree relatives share 50% of their genes in common, second-degree relatives share 25% of their genes in common, and third-degree relatives share 12.5% of their genes in common. If familial resemblance is wholly attributable to genes, there should be a 50% decrease in disease risk, with each successive increase in degree of relatedness, from first to second to third, and so forth. This information can be used to derive estimates of familial recurrence risk within and across generations as a function of population prevalence (λ) (Risch, 1990). Whereas λ tends to exceed 20 for most autosomal dominant diseases, values of λ derived from family studies of many complex disorders tend to range from 2 to 5. Diseases with strong genetic contributions tend to be characterized by a 50% decrease in risk across successive generations. Decrease in risk according to the degree of genetic relatedness can also be examined to detect interactions between several loci. If the risk to second- and third-degree relatives decreases by more than 50%, this implies that more than a single locus must contribute to disease risk and that no single locus can largely predominate.

The major advantage of studying diseases within families is that etiologic factors are more likely to be homogeneous than those studied between families that may result from heterogeneous etiology. In fact, the family study approach is particularly valuable when sources of heterogeneity are unknown. Evidence for specificity of familial aggregation of diagnostic subgroups and core features of disorders can inform both phenotypic and etiologic heterogeneity. Phenotypic heterogeneity is suggested by variable expressivity of symptoms and diseases within families, whereas etiologic heterogeneity is demonstrated by similar phenotypes manifesting between families. Moreover, the family study method permits assessment of associations between disorders by evaluating specific patterns of cosegregation of two or more disorders within families (Merikangas, 1990).

Twin Studies

The twin study design is one of the most powerful study designs in genetic epidemiology because it yields estimates of heritability and also permits evaluation of multigenerational patterns of expression of genetic and environmental risk factors (Dick, Riley, & Kendler, 2010). Twin studies that compare concordance rates for monozygotic twins (who share the same genotype) with those of dizygotic twins (who share an average of 50% of their genes) provide estimates of the degree to which genetic factors contribute to the etiology of a disease phenotype. Path analytic approaches that estimate the proportion of variance attributable to additive genes and common and unique environment have been the standard method of analysis of data from large twin studies. Twin studies can also provide information on the genetic and environmental sources of sex differences in a disease by investigating sex-specific concordance rates. Environmental exposures may also be identified by comparing differential exposures among discordant monozygotic twins. Twin studies may also inform the genetic mode of transmission of a disease by inspection of the degree of adherence of the difference in risk between monozygotic and dizygotic twins to the Mendelian ratio of 50%. Finally, twin studies may inform the spectrum of expression of diseases and disease subtypes through identifying the components of phenotypes that are most heritable (van Beijsterveldt et al., 2013; Lichtenstein, Tuvblad, Larsson, & Carlstrom, 2007; McGuffin et al., 2003; Verhulst, Neale, & Kendler, 2015).

Adoption Studies

Adoption studies have been the major source of evidence regarding the joint contribution of genetic and environmental factors to disease etiology. Adoption studies compare the similarity between an adoptee and his or her biological versus adoptive relatives, or the similarity between biological relatives of affected adoptees with those of unaffected or control adoptees. The latter approach is more powerful because it eliminates the potentially confounding effect of environmental factors. Similar to the familial recurrence risk, the genetic contribution in adoption studies is estimated by comparing the risk of disease to biological versus adoptive relatives, or the risk of disease in biological relatives of affected versus control adoptees. These estimates of risk are often adjusted for sex, age, ethnicity, and other factors that may confound the links between adoption status and an index disease.

With trends toward selective adoption and the diminishing frequency of adoptions in the United States, adoption studies are becoming less feasible methods for identifying genetic and environmental sources of disease etiology (National Adoption Information Clearinghouse, 2007). However, the increased rate of reconstituted families (families composed of both siblings and half siblings) may offer a new way to evaluate the role of genetic factors in the transmission of complex disorders (Risch et al., 2014). Genetic models predict that half siblings should have a 50% reduction in disease risk compared with that of full siblings. Deviations from this risk provide evidence for polygenic transmission, gene × environment (G × E), or other complex modes of transmission.

GENETIC EPIDEMIOLOGY OF PSYCHIATRIC DISORDERS

There has been a substantial body of research on the genetic epidemiology of psychiatric disorders, as summarized in numerous reviews of aggregate evidence (Merikangas & Karayiorgou, 2015), as well as of specific disorders (Boraska et al., 2014; Hinney & Volckmar, 2013; Kim & Leventhal, 2015; Merikangas & Yu, 2002; Sullivan, 2005; Taylor, 2013; Wetherill et al., 2014; Wilde et al., 2014). Increased recognition of the role of biologic and genetic vulnerability factors for psychiatric disorders has led to research with increasing methodological sophistication over the past two decades. Next, we review the current evidence regarding the genetic epidemiology of psychiatric disorders.

Neurodevelopmental Disorders

Attention-Deficit Hyperactivity Disorder

Attention-deficit hyperactivity disorder (ADHD) is both familial and heritable (Wallis, Russell, & Muenke, 2008). The average twin concordance rate of 0.76 from numerous twin studies indicates that 70–80% of the etiology of ADHD can be attributed to genetics (Biederman & Faraone, 2005). In addition, a higher rate of ADHD is reported in biological parents of ADHD probands than in their adoptive parents (18% vs. 6%), and in only 3% of the biological parents of control probands (Sprich, Biederman, Crawford, Mundy, & Faraone, 2000).

Autism Spectrum Disorder

Both the twin and family studies converge in implicating genetic risk factors in the etiology of autism spectrum disorder (ASD). Concordance rates of ASD in siblings range from about 3% to 14% (Bolton et al., 1994; Constantino, Zhang, Frazier, Abbacchi, & Law, 2010; Muhle, Trentacoste, & Rapin, 2004; Sumi, Taniai, Miyachi, & Tanemura, 2006). Studies of ASD and related phenotypes in infant siblings of patients with autism have also shown that there is a range of manifestations of social and communication as well as cognitive deficits in these siblings compared with control subjects (Georgiades et al., 2012; Szatmari et al., 1996). The results of these studies suggest that approximately 20% of infant siblings of autism probands proceed to develop ASDs (Georgiades et al., 2012; Ozonoff et al., 2011), and there are sex differences in the development of autism phenotypes (Zwaigenbaum et al., 2012). These familial influences have been investigated as sources of heterogeneity of autism (Szatmari et al., 2008) as well as a source of nongenetic influence on functioning among children with autism (MacLean et al., 1999).

Estimates of the heritability of ASD from twin studies of clinical samples range from approximately 36% to greater than 95% (Colvert et al., 2015; Hallmayer et al., 2011). The largest twin study with direct clinical assessment of a systematic sample of cases of autism from northern California yielded concordance rates of 0.58 (male) and 0.60 (female) for monozygotic pairs, and 0.21 (male) and 0.27 (female) for dizygotic pairs, yet the heritability (37% for strict autism and 38% for ASD) remained highly significant (Hallmayer et al., 2011).

Schizophrenia

There is an extensive body of research on the genetic epidemiology of schizophrenia (Doherty, O'Donovan, & Owen, 2012; McGrath, Mortensen, Visscher, & Wray, 2013; Sullivan, 2005). Despite wide differences in methods, samples, and geographic locations, controlled family studies yield a remarkably similar average relative risk of 8.9 to first-degree relatives. The fourfold greater proband-wise concordance rate of schizophrenia in monozygotic compared with dizygotic twins found in 12 studies to date demonstrates the role of genetic factors in the familial aggregation of schizophrenia. The average heritability in liability to schizophrenia across the studies is 0.81 (Sullivan, Kendler, & Neale, 2003). Similarly, adoption studies using traditional paradigms and modern diagnostic criteria (if available) demonstrate that the average risk to first-degree relatives was 15.5% compared with 3.6% for control subjects, yielding a relative risk of 4.3.

Despite evidence regarding the importance of genetic risk factors for schizophrenia, the lack of expected Mendelian risk ratios in the difference in risk of schizophrenia as a function of genetic similarity suggests that schizophrenia is a genetically complex disorder (Risch, 1990). Despite its high heritability, schizophrenia has been shown to result from multifactorial etiologic influences (Agerbo et al., 2011; Fazel, Wolf, Palm, & Lichtenstein, 2014; Hallmayer, 2000; Kendler, Ohlsson, Sundquist, & Sundquist, 2015; McGuffin, 2004; Portin & Alanen, 1997; Sariaslan et al., 2015; Sullivan et al., 2003; Tsuang, 2001). As described subsequently, the neurocognitive correlates of schizophrenia and increased neurodevelopmental abnormalities among children at high risk for schizophrenia have led to the widespread recognition of schizophrenia as a neurodevelopmental disorder (Dealberto, 2007).

Eating Disorders

Eating disorders (EDs) are a group of disorders involving disturbed or distorted body image coupled with eating and/or weight loss behaviors, and include anorexia nervosa (AN), bulimia nervosa (BN), binge eating disorder (BED), and EDs not otherwise specified. The prevalence is low; eg, the average prevalence rate for AN in young females is 0.3%, and 1% for BN, whereas higher prevalence rates have been shown for partial syndromes of AN (Hoek, 2006). Prevalence rates are even lower in males (Hudson, Hiripi, Pope, & Kessler, 2007). Because of the low prevalence rates, EDs have often been excluded from examination in national surveys, and data are primarily derived from clinical samples. Because most people with EDs do not seek treatment, those in research studies may comprise a different group from that of the general population.

Relatives of probands with AN are 11 times more likely to have AN than are relatives of control subjects and there is a higher prevalence of AN and other subthreshold EDs in first-degree relatives of probands than in relatives of control subjects (Strober, Freeman, Lampert, Diamond, & Kaye, 2000). Increased prevalence of BN is also found in relatives of BN probands (Thornton, Mazzeo, & Bulik, 2011). Multiple twin studies have found high heritability for AN, and one clinically ascertained study reports that additive genetic effects accounted for 88% of the liability to AN (Bulik, Sullivan, Wade, & Kendler, 2000), whereas a population-based study reports a lower rate (58%) (Wade, Bulik, Sullivan, Neale, & Kendler, 2000). Heritability estimates on ED symptoms from an adoption study ranged from 59% to 82% (Klump, Suisman, Burt, McGue, & Iacono, 2009), which indicated the importance of underlying genetic risk factors. Other studies of the heritability of EDs yielded slightly lower estimates: 22% for AN, 52% for BN, and 57% for BED (Mitchison & Hay, 2014).

Mood Disorders

Genetic factors have long been implicated in the etiology of bipolar (BP) disorder. The bulk of evidence is based on controlled family studies that have yielded risk ratios averaging from 8.0 to 12.0 (Merikangas & Yu, 2002; Sullivan, Neale, & Kendler, 2000). In a meta-analysis of the familial loading of mood disorders, Wilde et al. (2014) reported close to an eightfold increase in odds of BP among first-degree relatives of probands with the disorder. In two population-based studies, a sevenfold and a 6.5-fold increased risk of BP disorder among first-degree relatives of probands with BP disorder were reported from the Swedish (Lichtenstein et al., 2009) and Dutch (Aukes et al., 2012) registries. Furthermore, a meta-analysis of BP disorder in children found that the odds of BP-I were almost sevenfold in first-degree relatives of probands with pediatric BP disorder compared with control probands (Wozniak, Faraone, Martelon, McKillop, & Biederman, 2012).

Twin studies of BP disorder reported concordance rates in monozygotic twins that are about eight times those of dizygotic twins and yielded high heritability estimates (Smoller & Gardner-Schuster, 2007). For example, a study of twins from the Maudsley Twin Registry reported a heritability estimate of 85% when employing a narrow definition of concordance and 89% when employing a broad definition (McGuffin et al., 2003). Such high heritability estimates indicate that a substantial proportion of phenotypic variation in BP disorder is attributable to genes.

The familial aggregation of major depression has also been examined in numerous controlled studies. Aggregate estimates of the familial associations for major depression based on reviews and meta-analyses of controlled family study (Sullivan et al., 2000) and registry data (Wilde et al., 2014) are substantially lower than those of BP disorder, with familial risk ratios averaging about 2.5–2.8, familial heritability of 0.32 (Wray & Gottesman, 2012), and average twin heritability of 0.3–0.4.

The results of two large controlled family studies demonstrated that the familial transmission of mania and major depression may be independent (Merikangas et al., 2013; Vandeleur, Merikangas, Strippoli, Castelao, & Preisig, 2013). This suggests that mania and depression do not comprise opposite ends of a common underlying diathesis (Hickie, 2014). Aside from the BP—major depression distinction, there is substantial ongoing effort to distinguish subtypes of both BP disorder and major depression, which may increase our ability to identify their genetic and environmental correlates. The most discriminating subtypes of BP disorder are early age at onset (Kendler, Gatz, Gardner, & Pedersen, 2005) and psychotic subtypes (Ostergaard, Waltoft, Mortensen, & Mors, 2013), whereas the atypical subtype of major depression appears to differ from that of nonatypical, particularly melancholic depression (Lamers et al., 2013). Therefore, future studies that investigate mania and major depression as well as early age at onset and the atypical subtype may yield results different from those cited earlier.

Aggregate adoption study data on mood disorders reveal a moderate increase in rates of mood disorders among biologic compared with adoptive relatives of adoptees with mood disorders (Faraone, Kremen, & Tsuang, 1990). With respect to BP disorder, there is little evidence for differential risk among biologic compared with adoptive relatives of adoptees with BP disorder. However, the small numbers of BP adoptees who have been studied (ie, fewer than 50) do not provide an adequate test of genetic and environmental influences. The most compelling finding from adoption studies, however, is the dramatic increase in completed suicide among biologic, as opposed to adoptive, relatives of mood disorder probands (Mendlewicz & Rainer, 1977; Wender et al., 1986).

Anxiety Disorders: Panic Disorder

The lifetime prevalence of panic disorder is reported to be 4.8%, typically with a young adult age at onset, and it occurs more frequently in females (Kessler et al., 2006; Roy-Byrne, Craske, & Stein, 2006). Panic disorder has the strongest degree of familial aggregation of any of the anxiety disorder subtypes. A review of 13 family studies of panic disorder by Gorwood, Feingold, and Ades (1999) shows a sevenfold relative risk of panic among relatives of panic probands compared with control subjects. In addition, early-onset panic, panic associated with childhood separation anxiety, and panic associated with respiratory symptoms each have been shown to have a higher familial loading than other varieties of panic disorder (Goldstein, Wickramaratne, Horwath, & Weissman, 1997). Although there has been some inconsistency reported among twin studies of panic disorder, studies show that panic disorder has the highest heritability of all anxiety disorders (44%) (Kendler, Gardner, & Prescott, 2001). Comorbidity with other anxiety disorders, such as social anxiety disorder, generalized anxiety disorder, major depressive disorder, and substance use disorders, often occurs with panic disorder and may be the result of phenotypic overlap within the internalizing disorder spectrum as well as shared genetic and nongenetic risk factors (Hettema, Neale, Myers, Prescott, & Kendler, 2006). There is also a strong association between panic disorder and suicidal ideation and suicide attempts, even after adjustment for affective comorbidity and other suicide risk factors (Sareen et al., 2005).

Obsessive—Compulsive Disorder

Controlled family studies of obsessive—compulsive disorder (OCD) reveal an elevated familial risk in probands with OCD compared with control subjects. There is greater familial aggregation associated with early age at onset and obsessions rather than compulsions (Grabe et al., 2006; Nestadt et al., 2000; Pauls, Alsobrook, Goodman, Rasmussen, & Leckman, 1995). Other studies show that the risk for OCD among relatives of OCD probands increases proportionally to the degree of genetic relatedness, and relatives at similar genetic distances have similar risks for OCD despite different degrees of shared environment. The proportion of familial risk for OCD attributable to additive genetic factors is 47%. There are no significant sex differences in the familial pattern or heritability estimates (Mataix-Cols et al., 2013). Heritability of OCD in twins across multiple studies ranges from 0.48 to 0.54; however, heritability of obsessive—compulsive trait is substantially greater than the disorder (Bellodi, Sciuto, Diaferia, Ronchi, & Smeraldi, 1992; Browne, Gair, Scharf, & Grice, 2014; Carey & Gottesman, 1981; Lenane et al., 1990; Taylor, 2013). First-degree relatives of OCD probands are more likely to have mood disorders than those of control relatives (Black et al., 2013). Furthermore, relatives of OCD probands with unaffected-OCD first-, second-, and third-degree relatives have a significantly increased risk for schizophrenia, BP disorder, and schizoaffective disorder (Cederlöf et al., 2015). Interestingly, nonbiologic relatives (spouses or partners who have at least one child together) also have an elevated risk for OCD (Mataix-Cols et al., 2013). The rate of OCD among the relatives of children with OCD (onset before age 12/prepubertal) is significantly higher than the rates in families of adults with OCD, which indicates that childhood-onset OCD may have a different etiology from adult-onset OCD (Pauls, Abramovitch, Rauch, & Geller, 2014).

Substance Use Disorders

A positive family history of a substance use disorder is a consistent and robust risk factor for substance use in first-degree relatives (for comprehensive reviews of alcoholism, see Heath et al. (1997, 2001), Merikangas (2002), and Tsuang, Bar, Harley, and Lyons (2001)). Twin studies have demonstrated the contribution of both genetic and environmental risk factors to both alcoholism and drug abuse (Kendler et al., 2012; Kendler, Neale, Heath, Kessler, & Eaves, 1994). Heritability of alcoholism (narrowly defined) has been estimated at 59% by some researchers (Kendler, Karkowski, Neale, & Prescott, 2000), whereas the heritability of problem drinking (using broad definitions) has been estimated at 8–44% in females and 10–50% in males (Heath et al., 1997). Several adoption studies conducted in the 1980s and 1990s in Scandinavia demonstrated the importance of genetic factors underlying alcoholism (Cloninger, Bohman, Sigvardsson, & von Knorring, 1985; Goodwin, 1985; Sigvardsson, Bohman, & Cloninger, 1996).

Although there has been less systematic research on the familial aggregation of drug use disorders, numerous family history studies and uncontrolled and controlled family studies have demonstrated that rates of substance use disorders are elevated among relatives of drug abusers compared with those of control subjects and compared with population expectations (Bierut et al., 1998; Merikangas, Stevens, et al., 1998). One controlled family study of drug use disorders using contemporary family study data (Merikangas, Stolar, et al., 1998) showed an eightfold increased risk of substance use disorders (opioids, cocaine, cannabis, and alcohol) among relatives of probands with drug disorders compared with relatives of people with psychiatric disorders and normal control subjects. Family, twin, and adoption studies have also demonstrated common genetic and environmental factors that contribute to cannabis use disorders and other drug use disorders (Agrawal & Lynskey, 2006). The results of numerous twin studies of substance use disorders in general, as well as those of specific drugs, have shown that genetic factors underlie drug abuse in general (Kendler, Karkowski, & Prescott, 1999) in addition to unique genetic factors associated with specific drugs of abuse including nicotine and cannabis (Kendler et al., 2012, 2000; Kendler & Prescott, 1998a, 1998b; Tsuang et al., 2001).

Summary

This review demonstrates that all of the major mental disorders are familial with the highest recurrence risks for BP disorder, schizophrenia, and ASD, intermediate for AN, OCD, and substance abuse, and lowest for major depression and ADHD. Heritability in twin studies parallel the familial risks, demonstrating the contribution of genetic factors contributing to their etiology (Table 2.1). There has been a shift from small clinical studies to those with larger samples from registries and population samples that have increased our ability to identify shared familial risk across these conditions, as well as

TABLE 2.1 Summary of Estimates of Familial Risk, Phenotypic, and Genotypic Heritability

Disorder	Family-Based Risk Ratios	Phenotypic Heritability	SNP-Based Heritability (h^2_{SNP})
Neurodevelopmental disorders			
Attention-deficit hyperactivity disorder	2–8	0.30–0.76	0.281
Autism spectrum disorder	7–10	–	0.130
Schizophrenia	8–10	0.80–0.84	0.235
Eating disorder			
Anorexia nervosa	7–10	0.48–0.88	–
Mood disorders			
Bipolar disorder	7–10	0.60–0.80	0.250
Major depression	2–3	0.28–0.40	0.210
Obsessive–compulsive disorder	5–6	0.38–0.51	0.370
Panic disorder	3–8	0.40–0.60	–
Substance dependence	4–8	0.30–0.50	–

patterns of familial specificity. The newer generation of family studies has also begun to expand phenotypic assessments to include dimensional measures of phenotypes and biologic measures that may be closer manifestations of the underlying genetic factors (Glahn et al., 2012; Gur et al., 2007).

APPLICATIONS OF GENETIC EPIDEMIOLOGY IN MOLECULAR ERA

The next section describes future directions in genetic epidemiology in the molecular era. We first summarize the results of GWAS and sequencing studies of neuropsychiatric disorders using case−control designs. We then consider future study designs and methods that can incorporate advances in molecular genetics and neuroscience, as well as the role of environmental exposures that influence the timing and manifestation of psychiatric disorders.

Study Designs and Samples

Genome-Wide Association Studies: Case−Control Studies

There has been dramatic growth in large-scale case−control studies to identify genetic markers associated with psychiatric disorders, especially with the advent of large international data-sharing initiatives such as the Psychiatric Genomics Consortium (PGC) (www.med.unc.edu/pgc/). This initiative has been most successful for schizophrenia, in which, in the largest study to date (36,989 patients and 113,075 control subjects) the PGC reported 128 independent associations spanning 108 loci that meet genome-wide significance (Schizophrenia Working Group of the Psychiatric Genomics Consortium, 2014). The PGC has also completed the largest GWAS to date of major depressive disorder and found no statistically significant hits (Ripke et al., 2013). However, a meta-analysis combining major depression and BP disorder did demonstrate genome-wide significant associations for six SNPs on chromosome 3p21 ($P < 7.2 \times 10^{-8}$) (McMahon et al., 2010). More robust results have been found in GWAS of BP disorder, in which researchers identified statistically significant associations at *CACNA1C* and *ODZ4* (Psychiatric GWAS Consortium Bipolar Disorder Working Group, 2011).

Despite the high heritability of ADHD, to date no SNPs have reached the genome-wide significance threshold in GWAS (Neale et al., 2010); however, genome-wide CNV studies have demonstrated some promising findings (Hawi et al., 2015). Given heritability estimates of 60−80%, ADHD is assumed to be a complex, polygenic disorder with substantial genetic contributions (Froehlich et al., 2011). In ASD, GWAS completed by the Autism Genome Project yielded a statistically significant association for an SNP in *MACROD2* (Anney et al., 2010), but this association was not replicated in the second stage of the analysis (Anney et al., 2012). More promising findings have been reported in CNV studies of ASD, in which rare CNV is now considered an important risk factor in the development of ASD (Pinto et al., 2010). Pinto et al. (2014) demonstrated an excess of genic deletions and duplications in affected versus control groups.

Studies of AN and BN have not yielded genome-wide associations (Boraska et al., 2014; Wade et al., 2013), nor have examining general eating disorders been successful (Boraska et al., 2012). However, Boraska and colleagues stated that intronic variants in *SOX2OT* and *PPP3CA* were suggestively associated in their analyses, and signals specific to Europeans were found between *CUL3* and *FAM124B* and near *SPATA13* (Boraska et al., 2014).

Although twin and family studies have consistently shown that genetic factors explain almost half the variance of panic disorder (Maron, Hettema, & Shlik, 2010), GWAS have not yet led to the identification of potential underlying genetic variants (Schumacher et al., 2011). The largest GWAS and meta-analysis (718 patients and 1717 control subjects) of panic disorder detected no genome-wide significant SNPs, although the authors reported suggestive associations for several loci (eg, *BDKRB2*) (Otowa et al., 2012).

Likewise, a GWAS of OCD by the International OCD Foundation Genetics Collaborative did not report genome-wide significant associations in the combined case−control and trio meta-analysis of this sample (Stewart et al., 2013). The OCD Collaborative Genetics Association Study, composed of comprehensively assessed OCD patients with an early age of OCD onset, did not report genome-wide significant associations; however, follow-up analyses of signals from previously published data highlighted *DLGAP1* and *GRIK2*, as well as another set of genes including *NEUROD6*, *SV2A*, *GRIA4*, *SLC1A2*, and *PTPRD* (Mattheisen et al., 2015).

A GWAS of aggregate substance dependence categories using multiple case−control studies yielded one significant hit in *PKNOX2* for European women; however, the association signal was not statistically significant in males or African-origin population (Chen, Cho, Singer, & Zhang, 2011). Binary analysis of data from the Collaborative Study on the Genetics of Alcoholism sample (2322 subjects from 118 European American families) revealed a significant SNP in the uncharacterized gene LOC151121 on chromosome 2, whereas quantitative analysis implicated another SNP in *ARHGAP28*.

The result of the binary analysis was replicated in the Study of Addiction: Genetics and Environment sample, whereas the results of the quantitative analysis were not (Wetherill et al., 2015).

Despite the enthusiasm generated by the positive findings for many of these disorders, the total proportion of variance explained by even the largest international collaborative studies with hundreds of thousands of cases is still small (17–29% in the cross-disorder analyses and much lower in others) (Lee et al., 2013). This has led to substantial discussion regarding the so-called "missing heritability" in GWAS. However, the finding that common genetic variants explain only a limited proportion of the variance is not surprising in light of growing evidence regarding the role of undetected rare variants, environmental factors, and sources of misclassification of patients and control subjects in GWAS, including etiologic and clinical heterogeneity within patients, misclassification of control subjects, and other factors that might reduce the power of these studies. The clinical samples from these studies have been highly heterogeneous in terms of sampling source and diagnostic characteristics, and few of the control samples would meet traditional criteria for controls in epidemiology (Wacholder, Garcia-Closas, & Rothman, 2002). Future studies will require large systematic samples that are directly recruited for a cohort study or existing registries and/or biobanks that have sufficiently large and well-characterized samples of cases, as well as built-in controls without the index conditions.

Whole Genome/Exome Sequencing of Case–Control Studies

International collaborations have focused on genome-wide or exome (protein-coding) sequencing techniques that have been proposed as a tool that can be used in risk prediction and may lead to greater understanding of the etiology, prognosis, and treatment response in psychiatric disorders (Biesecker & Peay, 2013). Moreover, the analytic challenges will be complex, and novel techniques are under development (Pabinger et al., 2014). Success of the sequencing approach has been shown in schizophrenia, in which an exome sequencing study of 2536 schizophrenia patients and 2543 control subjects showed polygenic burden from rare (less than one in 10,000), disruptive mutations distributed across many genes (Purcell et al., 2014). Although no individual gene-based test achieves significance after correcting for multiple testing, enriched gene sets included those previously implicated by GWAS and CNV studies, specifically the voltage-gated calcium ion channel, and the signaling complex formed by the activity-regulated cytoskeleton-associated scaffold protein of the postsynaptic density.

Population Registries and Biobanks

One of the most important sources of future data for genetic epidemiology will be systematic data from population samples, disease or medical registries, and biobanks that have sufficiently large samples of diseases coupled with comprehensive information on phenotypes, genotypes, treatment patterns, laboratory measures, environmental exposures, and health behaviors. Population-based registry data have made a growing contribution to epidemiologic research because of their large sample size, the representativeness of their target populations, and the reduced likelihood of bias owing to recall, or nonresponse. Registries such as the Scandinavian administrative registries that are inclusive of the entire population (Olsen, Basso, & Sorensen, 1999) can be combined with other ones (eg, population, multigeneration, hospital discharge, outpatient care, etc.) because they contain common identifying information for the entire population. The ability to link health data with family information, social factors, and morbidity and mortality information is a bonus of registry-based research that can provide a rich series of correlates of disease, especially in registries that include biologic data. In fact, some registries even include genomic data for the entire population (ie, deCODE) (Gudbjartsson et al., 2015). The National Health and Nutrition Examination Survey collected DNA and disease information on a nationally representative sample of US adults that is available for analysis for research (http://www.cdc.gov/nchs/nhanes/genetics/genetic_types.htm). Analysis of registry data as a retrospective cohort study can provide a cost-effective way to gain insight into important associations that can later be tested in more expensive prospective studies.

Similar to all secondary data, the quality of diagnosis and risk factors is not under the control of the researcher, and often there are missing data that preclude comprehensive analysis of risk. There is often a lack of clinical information beyond categorical diagnosis (eg, patient discharge registers) which precludes the examination of subclinical features of disorders and can be subject to greater misclassification of cases and controls. The registers also represent treated cases that may reflect the "tip of the iceberg" phenomenon for common disorders. Finally, the retrospective nature of a registry limits causal inference.

Another type of population sample that has been of growing interest is biobanks, or repositories for biologic samples (typically including blood and extracted DNA, but also other tissues and cells) with associated phenotype information. Biobanks can be disease-specific, such as the EuroBioBank (http://www.eurobiobank.org/), which focuses on a particular rare disease, or population-specific, such as the United Kingdom Biobank (http://www.ukbiobank.ac.uk/about-biobank-uk/; de la Paz et al., 2010). Biobank data are also available in the United States from large health care networks such as Kaiser-Permanente (http://www.dor.kaiser.org/external/DORExternal/rpgeh/index.aspx) and the Geisinger My Code Project (https://c.ymcdn.com/sites/www.isber.org/resource/resmgr/2014_Presentations/D.Ledbetter-GeisingersByCode.pdf).

There are strengths and limitations to both of these approaches, as well as specific biases that should be considered when interpreting findings. The observational and retrospective nature of opt-in biobanks and registries is perhaps the most serious impediment to the generalizability of the findings (Krumholz, 2009). Finally, ethical concerns such as informed consent, reporting of incidental findings, and confidentiality will continue to be an important concern in biobanks and registry settings (de la Paz et al., 2010). The most powerful studies of the future will require large numbers of participants with in-depth assessment of both phenotypes and potential risk factors for these diseases, particularly cohort studies that can detect incident patient who are observed prospectively over time.

Statistical Approaches to Estimate Genetic Influence

Several statistical approaches that take advantage of markers identified in GWAS have also advanced our understanding of the genetic architecture of psychiatric disorders, notably genomic profile (or polygenic) risk scores and GCTA. Polygenic scores summarize genetic effects in a GWAS by computing a weighted sum of associated "risk" alleles within each subject. Initially, markers (typically SNPs) are selected based on their evidence for association, typically their *P* values, using a training sample, and the weighted score is then constructed in an independent replication sample. The association threshold used in the score is determined by researchers, but is typically examined at multiple thresholds (eg, 0.5, 0.1, 0.01, 0.001, 0.0001, <0.0001) (Wray et al., 2014). If an association is found between a trait/disorder and the polygenic score, one assumes that a genetic signal is present among the selected markers. Later, this score can be used to predict individual trait values (Dudbridge, 2013). The original use of polygenic scores has been extended to include detecting shared genetic etiology among traits, to infer the genetic architecture of a trait, and to establish the presence of a genetic signal in underpowered studies, and can act as a biomarker for a phenotype (Euesden, Lewis, & O'Reilly, 2014).

Polygenic score analysis was first applied successfully to a GWAS of 3322 European individuals with schizophrenia and 3587 control subjects. Findings provided molecular genetic evidence for a substantial polygenic component to the risk of schizophrenia involving thousands of common alleles of very small effect (Purcell et al., 2009). This approach has also been used to examine the shared genetic effect between purportedly distinct disorders, notably schizophrenia and BP disorder. For example, in a study of 1829 patients with BP disorder and a replication sample of 506 patients, researchers found that patients with schizoaffective bipolar disorder were significantly discriminated from the remaining participants with BP disorder in both the primary and replication data sets, whereas those with psychotic BP disorder were not significantly different from those with nonpsychotic BP disorder in either data set (Hamshere et al., 2011). Subsequently, polygenic score methodology has been applied to a host of complex disorders beyond psychiatric phenotypes, including Parkinson disease (Escott-Price et al., 2015), breast cancer (Vachon et al., 2015), Alzheimer disease (Martiskainen et al., 2015), and obesity (Domingue et al., 2014), among others, and has examined differences in the genetic architecture of behavioral traits and psychiatric disorders (Stergiakouli et al., 2015).

GCTA, which estimates the proportion of phenotypic variance explained by genetic variants (typically SNPs) for complex traits, is another method that can be used to characterize the genetic architecture of complex disorders (Yang et al., 2011). Genomic-relatedness matrix-restricted maximum likelihood, either chromosome- or genome-wide, provides an estimate of narrow heritability that does not rely on the assumptions defined in standard twin studies. Instead, it assumes that environmental factors are not correlated with differences in the degree of genetic similarity for individuals who are not in the same extended families. Contrary to standard behavioral genetics studies that define relatedness by pedigree structure, it estimates genetic relatedness directly from the SNP data (Benjamin et al., 2012).

In addition to defining the genetic relationship from genome-wide SNPs, GCTA can also be used to predict the genome-wide additive genetic effects for individual subjects and for individual SNPs, to estimate the linkage disequilibrium structure encompassing a list of target SNPs and the genetic correlation between two traits or diseases using a bivariate SNP-based model that estimates the average genome-wide relationship between two disorders (Lee et al., 2012). This method was first applied successfully to analyze the genetic contribution to human height (Yang et al., 2010), but its use has expanded to psychiatric disorders such as ADHD (Yang et al., 2013), anxiety-related behaviors (Trzaskowski et al., 2013), ASD (Klei et al., 2012) and others, to neurological disorders such as Parkinson disease (Keller et al., 2012) and multiple sclerosis (Watson, Disanto, Breden, Giovannoni, & Ramagopalan, 2012), and to childhood obesity (Llewellyn, Trzaskowski, Plomin, & Wardle, 2013).

However, polygenic scoring and GCTA assume additive genetic variance, which does not take into account potential multiplicative gene–gene interactions nor do they consider G × E interactions. Furthermore, polygenic scoring and GCTA estimates that are typically derived from SNPs do not include other types of genetic variants (CNVs, segmental duplications, etc.) that may also underlie disease etiology.

As shown in Table 2.1, there are striking differences in the discrepancy between SNP-based heritability estimates for GWAS "hits" compared with all genome-wide SNPs. In general, the estimates of SNP based heritability $\left(h^2_{SNP}\right)$ are substantially lower than those for phenotypic heritability. For example, the SNP-based heritability for BP disorder and schizophrenia are about one-third that of phenotypic heritability $\left(\text{eg, } h^2_{phenotype} = 0.70 \text{ vs. } h^2_{SNP} = 0.20, \text{ and } h^2_{phenotype} = 0.80 \text{ vs. } h^2_{SNP} = 0.30, \text{ respectively}\right)$. Moreover, in contrast to the much greater heritability of BP disorder compared with major depression, the estimates of SNP-based heritability for BP disorder and major depression are nearly equal (ie, 0.25 for BP and 0.21 for major depression) (Wray & Gottesman, 2012). Whereas the SNP-based estimate for major depression is nearly equal to the phenotypic heritability, there is a large discrepancy between these estimates for BP disorder. However, phenotypic blurring between manic episodes, the core feature of BP disorder, and major depressive episodes may have led to misclassification of patients in current GWAS studies on which these heritability estimates are based.

Differences between phenotypic and SNP-based heritability have important implications for risk prediction. There has been increasing concern regarding the low attributable risk of GWAS "hits" because few of the molecular genetic findings have improved our ability to predict risk. For example, the variance explained by GWAS for BP disorder is 0.02 and for schizophrenia is 0.01 (Visscher, Brown, McCarthy, & Yang, 2012). By contrast, the phenotypic variance explained for nonpsychiatric disorders has been much greater (eg, 0.60 for type 1 diabetes, 0.24 for Alzheimer disease (Ridge, Mukherjee, Crane, & Kauwe, 2013), and 0.10 each for multiple sclerosis (Sawcer et al., 2011) and high-density lipoprotein cholesterol (Teslovich et al., 2010). This disparity can be attributed to the fact that GWAS do not examine the full range of genetic variation such as rare or structural variants (Wray, Goddard, & Visscher, 2008). Furthermore, they do not reflect sources of complexity of the genetic architecture of diseases such as pleiotropy (multiple phenotypic effects of single variants) (Visscher et al., 2012) or the dichotomous classification of disease, which may lead to misclassification of patients with subthreshold manifestations of disease as control subjects. In addition, phenotypic heritability estimates from family studies may reflect environmental factors that may influence familial resemblance. In a comparison of the use of family history and SNPs to predict risk of complex diseases, Do, Hinds, Francke, and Eriksson (2012) found that family history surpasses SNPs for highly common heritable conditions such as cardiovascular disease and type 2 diabetes, whereas SNP-based estimates have superior prediction of risk for diseases of low frequency such as multiple sclerosis and Crohn's disease. Ultimately, combinations of family history and genome-wide SNPs may have increased prognostic ability for complex diseases.

Phenotypes: Moving Beyond Dichotomous Classification

The most important impediment to progress in psychiatric genetics is the lack of validity in the classification of psychiatric disorders. The development of structured interviews to ascertain diagnostic criteria for psychiatric disorders has enhanced the comparability of diagnostic methods within the United States and worldwide, as reflected in the World Mental Health Initiative, the largest international study of psychiatric disorders. Despite the enhancement of reliability conferred by operational diagnostic criteria, however, there is still a lack of validity for these broad categories (Kendell, 1989). There is substantial evidence that all of the major mental disorders are reflections of continuous underlying liabilities as represented by the multifactorial polygenic threshold model of complex diseases introduced by Falconer (1960). The spectrum concept of schizophrenia (Kraepelin, 1919), mood disorders (Angst, 2007), and autism (Wing, 1981) has provided alternatives to dichotomous classification of these conditions. Future studies that consider the full range of expression of these conditions as well as more homogeneous subtypes may improve our ability to define psychiatric phenotypes. For example, a study of ADHD symptoms and clinical disorder demonstrated that the same genetic factors influenced ADHD as both a disorder and trait (Stergiakouli et al., 2015). Therefore, studies that impose high disease thresholds may misclassify those with less severe manifestations as control subjects. The Research Domain Criteria initiative established by the National Institutes of Mental Health, that was designed to develop new ways of classifying psychiatric disorders based on behavioral dimensions and neurobiological measures for research purposes, has led to widespread research that extends dichotomous phenotypic classification beyond dichotomous diagnostic entities (http://www.nimh.nih.gov/research-priorities/rdoc/index.shtml).

The boundaries between disorders are also blurred, and most individuals with one disorder tend to have others as well. For example, 70% of people with BP disorder have a history of three or more other psychiatric disorders

(Merikangas et al., 2007). Despite these well-established patterns of overlap among disorders and their dimensional underpinnings, genetic studies to date have been based on dichotomous categories that are certainly replete with comorbidity and characterized by heterogeneous etiologic factors. Twin and family studies will have major value as we move from dichotomous diagnostic categories to identify the underlying components and subtypes of these categories, as described next.

Phenotypic Boundaries

One finding regarding psychiatric phenotypes that has generated substantial attention is the overlap in genetic factors underlying schizophrenia, BP disorder, ASD, and ADHD from analyses of registries and large GWAS (Lee et al., 2013; Smoller et al., 2013). Using data from the Swedish treatment registry, Lichtenstein et al. (2009) reported heritability coefficients of 0.59 for BP disorder and 0.64 for schizophrenia. However, they also found substantial familial overlap in the two conditions, most of which was attributable to common genetic influences. Wray and Gottesman (2012) reported remarkably similar heritability of these conditions in the Danish Registry (0.62 for BP disorder and 0.67 for schizophrenia, respectively).

Cross-disorder studies of SNP-based overlap have also yielded suggestive evidence of shared genetic etiology across a range of psychiatric disorders. In a combined case sample of 33,332 cases of European ancestry which included ASD, ADHD, BP disorder, major depressive disorder, and schizophrenia compared with 27,888 control subjects, there was shared heritability between schizophrenia and BP disorder; moderate heritability between schizophrenia and major depressive disorder, and BP disorder and major depressive disorder; and limited overlap between schizophrenia and ASD. This overlap also occurred at the genome level. For example, four loci surpassed the genome-wide significance cutoff for common genetic risk, with a particularly intriguing finding in the calcium channel subunits (Maier et al., 2015).

There is also substantial evidence for specificity of familial factors underlying schizophrenia and BP disorder. Analyses of the Danish registry data by Aukes et al. (2012) showed strong evidence for familial specificity of schizophrenia, BP disorder, and major depressive disorder, but also a less potent but significant degree of shared familial risk of these conditions. Likewise, Goldstein, Buka, Seidman, and Tsuang (2010) found substantial specificity of familial transmission of affective psychosis and schizophrenia in the New England Family Study. Several large studies of the mood disorder spectrum demonstrated specificity of familial aggregation of BP disorder, major depression, and schizophrenia, as have studies of registries that demonstrate substantial specificity (Merikangas et al., 2013; Steinhausen, Foldager, Perto, & Munk-Jorgensen, 2009; Vandeleur et al., 2013). Another meta-analysis of 33 high-risk studies based on parents with severe psychiatric disorders such as schizophrenia, BP disorder, and major depression yielded evidence for specificity of manifestation of the index parental disorder in youth (Rasic, Hajek, Alda, & Uher, 2013).

Discrepancies in findings regarding familial specificity may be partly attributable to diagnostic methods across studies. Specificity has been much greater in direct family interview studies than in registry studies that are based solely on diagnostic codes. A review by Cosgrove and Suppes (2013) documented the importance of distinctions between the major classes of psychiatric disorders based on evidence regarding the specificity of course, clinical features, and treatment response that continues to discriminate these two major disorders. These aggregate findings suggest that there may be common underlying factors as well as risk factors that are specific to particular disorders.

Use of Endophenotypes for Classification

Numerous studies have begun to deconstruct psychiatric phenotypes by their component features or subtypes, including BP disorder (Benazzi, 2007), generalized anxiety disorder (Angst et al., 2006), OCD (Eapen, Pauls, & Robertson, 2006), schizophrenia (Braff, Freedman, Schork, & Gottesman, 2007), and panic disorder (Smoller & Tsuang, 1998). Identification of phenotypic traits or markers, which are themselves heritable and which may represent intermediate forms of expression between the output of underlying genes and the broader disease phenotype, have been termed "endophenotypes" (Gottesman & Gould, 2003). Studies of the role of genetic factors involved in these systems may be more informative than studies of the aggregate psychiatric phenotypes because they may more closely represent the expression of underlying biologic systems. To the extent that particular endophenotypes more clearly represent the expression of genotypes, they may help to unravel the complexity of transmission of the psychiatric disorders. For example, some of the endophenotypes that may underlie mood disorders include circadian rhythm, stress reactivity, and mood, sleep, and appetite regulation (Lenox, Gould, & Manji, 2002). Cognitive, neurophysiological, and structural measures continue to be tested as potential schizophrenia endophenotypes (Calkins et al., 2007; Gur et al., 2007; Horan et al., 2008). However, before applying endophenotypes in gene identification studies, there should be evidence that the endophenotype has a stronger genetic signal than the broader phenotype. A meta-analysis of psychiatric endophenotypes (Flint & Munafo, 2007) and a review of the genetic architecture of traits in model organisms are not yet evidence that endophenotypes are superior to current phenotypic disease definitions (Valdar et al., 2006).

Continued reliance on the descriptive approach as the sole basis for diagnosis in psychiatry is attributable to the lack of identification of biologic markers for these conditions. Although there has been some progress in identifying neuroimaging and cognitive tests that may enhance the diagnosis of schizophrenia (Smucny, Wylie, & Tregellas, 2014; Targum & Keefe, 2008), there is still a lack of such markers for most other psychiatric disorders. The difficulty in classifying disorders of human cognition, behavior, and emotion is understandable in light of the complex psychological and physiological states underlying mental function, which is the culmination of human adaptation to the environment to date. Progress in neuroscience that reveals information about the biologic pathways underlying psychiatric disorders should inform our understanding of the classification of psychiatric phenotypes.

Computational Phenotype Analysis

There are also a growing number of methods for refinement of phenotypes that move beyond the current dichotomous disease classification approach. First, there are increasingly sophisticated methods for classification analysis to identify disease subgroups that may be more direct manifestations of genetic liability, such as machine learning, regression trees and random forests, support vector machines, artificial neural networks, and latent variable analysis. Second, there have also been numerous developments in strategies that examine multiple phenotypes simultaneously, such as the Human Phenotype Ontology Project (Kohler et al., 2014), which creates a structured vocabulary for phenotypic features and can be used in computational phenotypic analysis that ranks variants based on their pathogenicity and the semantic similarity of patients' phenotypes (Zemojtel et al., 2014). The most critical step will involve linking biologic markers and systems to these more refined indices of phenotype subtypes and their underlying dimensions.

Incorporation of Environmental Factors

To date, most molecular genetic studies in psychiatry have not incorporated environmental factors as a source of variance in etiologic models. Most evidence for the role of the common and unique environment has been based on residuals from path analytic models employed in twin studies. Yet there are now numerous well-established environmental correlates of most psychiatric disorders, particularly schizophrenia, ASD, and other neurodevelopmental disorders. Aggregate evidence clearly demonstrates that multiple common exposures across all phases of development are associated with a broad range of neurodevelopmental disorders in three chief domains of risk, including prenatal and perinatal exposures and infections, parental age at birth of the affected individual, and a range of social factors such as urban environment, social class, and immigration status.

There is accumulating evidence that prenatal exposures to potential teratogens including maternal infections, autoimmune disease, gestational diabetes (Buka, Tsuang, Torrey, Klebanoff, Bernstein, et al., 2001; Buka, Tsuang, Torrey, Klebanoff, Wagner, et al., 2001; Lyall, Pauls, Santangelo, Spiegelman, & Ascherio, 2011), and other prenatal factors (Meli, Ottl, Paladini, & Cataldi, 2012) including preeclampsia (Mann, McDermott, Bao, Hardin, & Gregg, 2010), maternal psychotropic (Croen, Grether, Yoshida, Odouli, & Hendrick, 2011) and antiseizure medication use (Williams et al., 2001) or pollution exposure (Volk, Hertz-Picciotto, Delwiche, Lurmann, & McConnell, 2011), birth difficulties including obstetric complications and preterm birth (Clarke, Harley, & Cannon, 2006; Johnson et al., 2010), maternal stress (Froehlich et al., 2011; Li, Olsen, Vestergaard, & Obel, 2010), and family interactions (McGuffin, 2004) are associated with an increased risk of both ASD and schizophrenia, whereas only reduced birth weight has been significantly associated with an increase in ADHD symptoms (Pettersson et al., 2015). Urbanicity has been reported as a risk factor for schizophrenia as well (Mortensen et al., 1999), but this finding is complicated by socioeconomic status (Matheson, Shepherd, Laurens, & Carr, 2011) and associated stressors, which have also been identified as risk factors for the later development of schizophrenia. Finally, of specific interest in schizophrenia is cannabis use as a precursor to psychotic symptoms (Arseneault et al., 2002; Dean & Murray, 2005; Giordano, Ohlsson, Sundquist, Sundquist, & Kendler, 2014; Wilkinson, Radhakrishnan, & D'Souza, 2014).

Parental age at birth has been linked with the subsequent development of psychiatric disorders (McGrath et al., 2014), specifically autism (Reichenberg et al., 2006), schizophrenia (Malaspina et al., 2002), and ADHD (Chang et al., 2014). Interestingly, whereas ADHD is associated with younger parental age, later parental age is associated with both ASD and schizophrenia. Data also show that both maternal and paternal age effects contribute to increased risk of both schizophrenia and ASD (Risch et al., 2014; Shelton, Tancredi, & Hertz-Picciotto, 2010).

Other environmental correlates of schizophrenia and other psychotic disorders are the degree of urbanization of the place of birth, and urban rearing, presumably attributable to social isolation or disorganization in cities, and the course is poorer in developed countries. An excess of stressful life events and other social stresses seems to trigger relapse. Low socioeconomic status also increases the risk for schizophrenia, but it remains unclear whether this is a causal factor or a consequence of delayed diagnosis and treatment, because lack of access to intervention could lead to more severe disease in lower-socioeconomic status patients (McDonald & Murray, 2000; Mortensen et al., 1999). The incidence of schizophrenia and ASD is also increased among immigrants in several different countries (Hultman, Sparen, & Cnattingius, 2002; Selten, Cantor-Graae, & Kahn, 2007). Environmental factors have also been shown to increase the risk of mood disorders. Infections (Benros et al., 2013) and inflammation or immune response (Maes, 1995) are purported environmental factors associated with subsequent depression, though this association may be related to comorbid conditions, rather than depression per se influencing the development of inflammation or inflammation instigating depression (Glaus et al., 2014). Whereas season of birth, urban residence, and some other risk factors for schizophrenia have also been found among those with bipolar disorder, the overlap appears to be attributable to psychosis that is common to both conditions (Demjaha, MacCabe, & Murray, 2012; Lichtenstein et al., 2009; Ostergaard et al., 2013).

The most widely studied factors for mood disorders are stressful live events and low social supports; however, their bi-directional association with mood disorders blurs causal inferences, and prospective research suggests that life stressors may have a provocative rather than causal influence on depression (Hammen, 2005; Kessler, 1997). However, many forms of childhood adversity appear to be nonspecific risk factors for adult mental disorders. For example, several studies have shown that childhood physical abuse increased risk for bipolar disorder after adjusting for socioeconomic characteristics and other psychiatric comorbidity, but it also increased risk for subsequent onset of drug abuse, nicotine dependence, posttraumatic stress disorder, ADHD, generalized anxiety disorder, panic, MDD, and suicide attempt (Nierenberg et al., 2010; Sugaya et al., 2012). Moreover, retrospective reporting bias of child abuse had been demonstrated by comparisons with prospective data in the same sample (Scott, Smith, & Ellis, 2010).

Retrospective studies, particularly those of dietary exposures, have been notoriously unreliable, and have produced erroneous conclusions regarding etiologic links between cancer and dietary patterns that have not been replicated in prospective research (Taubes, 2007). Prospective cohort studies have been shown to be the only valid approach to identify causal environmental factors for some exposures such as environmental toxicants, dietary factors, or infectious agents because of the necessity of demonstrating that the environmental factor predates the onset of disease (Wareham, Young, & Loos, 2008).

Another major challenge to the incorporation of environmental risk factors in genetic studies is that the clustering of multiple risk factors limits inferences regarding the attributable risk of specific factors. Patel and Ioannidis (2014) proposed environment-wide association studies as a method to characterize environmental factors and their temporal relationships to health conditions across several studies that appropriately adjust for confounders of their associations. Such methods are appropriate for environmental toxicology and other studies that have numerous correlated biomarkers, but these methods have not been applied to diseases for which measured environmental exposures are less reliable.

Environmental risk factors for many psychiatric disorders appear to have impact on early neural development (perinatal exposures to toxins, malnutrition, and infections) or later physical or emotional trauma, whereas others may influence the context of expression of psychiatric disorders (eg, social class, urban environment, immigration). The lack of specificity and clustering of risk factors challenge our ability to identify specific environmental factors that may interact with underlying genetic susceptibility in the etiology of any of the psychiatric disorders.

Combining Genetic and Environmental Factors

The next phase of research in genetic epidemiology will require integrating research on these genetic and environmental risk factors. Traditional study designs in genetic epidemiology that can be used to study the joint influence of genetic and environmental factors include case-only studies, cross-sectional case–control studies, as well as cohort studies of G × E interaction (Beaty, 1997; Ottman, 1990; Yang & Khoury, 1997). These study designs can be extended to conditions with broad or specific genotypic similarity based on GWAS to identify environmental exposures that influence gene expression.

A growing number of studies have incorporated findings from genetic studies to identify the effect of environmental factors for diseases that have well-established genetic risk factors. The large scope of studies that will be required to detect the joint impact of genetic with environmental factors is daunting, as illustrated by the InterAct study, an eight-country prospective cohort study of the incidence of type 2 diabetes in a subcohort of 16,154 European individuals that includes 3.99 million person-years of follow-up (Langenberg et al., 2014). Although G × E interaction is generally assumed to be a key mechanism for links between genetic susceptibility and environmental exposures in complex diseases, analyses of several of these cohort studies have yielded additive rather than interactive influences of genetic and environmental risk factors. For example, no interactions were found between the genetic risk score based on the 49 established SNP loci for type 2 diabetes and environmental/lifestyle variables including family history, obesity, waist circumference, physical activity, and dietary factors in the InterAct prospective cohort study. In fact, genetic risk factors were found to have greater influence on incidental diabetes in young, thin people than in older individuals with elevated body mass index and other risk factors. Likewise, studies of environmental factors among those with the apolipoprotein-E_4 genotype that confers increased risk of cognitive decline and Alzheimer disease have found additive rather than interactive influences of other environmental risk factors (Andrews, Das, Anstey, & Easteal, 2015; Wirth, Villeneuve, & La Joie, 2014).

The bulk of evidence for the joint influence of genetic and environmental factors for psychiatric disorders has been based on indirect estimates from twin studies and other variations of this design such as the twin-family study (Posthuma & Boomsma, 2000). Although numerous studies have reported G × E interaction between several genes that interact with nonspecific environmental exposures such as life stress or childhood adversity and a range of outcomes including depression, cannabis dependence, and conduct disorder (Caspi et al., 2003), replication of these findings has not been consistent (Risch et al., 2009; Zammit & Owen, 2006). Duncan and Keller (2011) concluded that low power along with low prior probability that a G × E hypothesis is true suggests that most or even all positive candidate-gene G × E findings represent type I errors.

Over the next decades, it will be important to identify and evaluate the effects of specific environmental factors on disease outcomes and to refine measurement of environmental exposures to evaluate the specificity of effects. There have been a few efforts to integrate findings from GWAS with well-established risk factors for psychiatric disorders such as schizophrenia. Using data from the Danish-population registry, Agerbo et al. (2015) demonstrated the complexity of analysis of the joint effect of the polygenic risk score, parental socioeconomic status, and family history for schizophrenia risk. The attributable risk of socioeconomic status was substantially greater than either family history of or polygenic risk score for schizophrenia. This work will be facilitated by growing efforts to enhance the quality of measurement of environmental exposures and to include environmental measures in large-scale population registries and collaborative cohort studies.

Resurrection of the extended family study, particularly as an extension of case–control GWAS, is a promising direction for the next generation of genetic studies of psychiatric disorders (Benyamin, Visscher, & McRae, 2009; Cordell, 2009; Ott, Kamatani, & Lathrop, 2011). Family studies can minimize the false-positive risk induced by population stratification, increase the rate of genetic susceptibility variants, reduce heterogeneity that plagues current GWAS case–control studies, and inform GWAS findings by discriminating positive findings and elucidating patterns of transmission and biologic pathways underlying SNPs or genetic variants identified in GWAS. The polygenicity and heterogeneity of the psychiatric disorders as well as the indisputable role of environmental influences will also require study designs that are conditioned on either genetic or environmental factors as described previously to hone in on specific factors underlying these sources of risk. Incorporation of phenotyping that taps the domains underlying broad diagnostic categories based on knowledge of underlying biologic pathways, coupled with built-in hypothesis-based environmental exposures, will facilitate the integration of advances in molecular genetics and environmental science.

Team Science

This chapter summarizes the contributions of epidemiology to our current and future understanding of the complex risk factors underlying psychiatric disorders (Table 2.2). When taken together with other chapters by molecular geneticists, bioinformaticians, statistical geneticists, neuroscientists, psychiatrists, and clinical geneticists, it is increasingly evident that future progress in genomics will require a team effort that moves beyond molecular biology and bioinformatics. Epidemiology is critical to refining phenotypes and identifying appropriate samples, study designs, and analytic methods to evaluate the role of genetic and environmental risk factors in complex diseases (Lehner, Senthil, & Addington, 2015). Therefore, it will be essential to include epidemiologists as key members of teams that seek to identify the role of genetic risk factors and environmental influences on psychiatric disorders. Furthermore, the tools and methods of epidemiology will be extremely valuable in translating the advances in molecular genetics to the general population and public health.

Effective translational research teams require mutual respect across multiple disciplines, as well as at least familiarity with the key goals and methods of each discipline to facilitate communication. Epidemiology has generally been misconstrued solely as a descriptive discipline that provides estimates of morbidity rather than a field that seeks to identify causal mechanisms for diseases. Familiarity with the goals, methods, and analytic tools of epidemiology will be necessary for other members of collaborative groups engaged in psychiatric genetics. However, it will be equally important for epidemiologists to have greater understanding of molecular genetics and disease pathology to contribute to research teams that seek to discover disease etiology (Kuller, 2012).

Clinicians will also be essential members of teams involved in translating genomics to the public. An understanding of the significance of genetic risk factors and proper interpretations of their meaning for patients and their families will also become part of clinical practice. Epidemiologists should also devote effort to educating clinicians and patients in interpreting genetic findings. Clinicians will become increasingly involved in helping patients to comprehend the meaning and potential impact of genetic risk for both psychiatric and nonpsychiatric disorders. As our knowledge of the role of genetic risk factors in psychiatric disorders advances, it will be incumbent upon clinicians to become familiar with knowledge gleaned from genetic epidemiologic and genomics research. In the meanwhile, use of recurrence risk estimates from family studies best predicts the risk of the development of psychiatric disorders.

TABLE 2.2 Future Applications of Epidemiology in Molecular Genetics Era

Study designs	• Extended prospective family studies with deep phenotyping • Prospective cohort studies based on genetic variants or environmental exposures • Extend case–control studies to include relatives • Incorporation of families in registries and biobanks
Statistical methods	• Genes as independent variables • Classification models to refine phenotypes • Models that incorporate clustering and specificity of genes and environmental exposures
Phenotypes	• Moving beyond dichotomous classification • Incorporate endophenotypes in genome-wide association studies • Include multiple disorders in case–control studies
Environmental factors	• Stratify genetic case–control studies by environmental exposures • Studies of specific versus general effects of environmental risk factors • Identify "critical" timing of environmental exposures
Team science	• Epidemiologists as part of study teams • Team education regarding concepts of epidemiology • Incorporate study designs informed by genetic epidemiology • Training in molecular genetics and pathology in epidemiology training programs

REFERENCES

Agerbo, E., Mortensen, P. B., Wiuf, C., Pedersen, M. S., McGrath, J., Hollegaard, M. V., ... Pedersen, C. B. (2011). Modelling the contribution of family history and variation in single nucleotide polymorphisms to risk of schizophrenia: a Danish national birth cohort-based study. *Schizophrenia Research*. http://dx.doi.org/10.1016/j.schres.2011.10.025. pii:S0920-9964(11)00560-3.

Agerbo, E., Sullivan, P. F., Vilhjalmsson, B. J., Pedersen, C. B., Mors, O., Borglum, A. D., ... Mortensen, P. B. (2015). Polygenic risk score, parental socioeconomic status, family history of psychiatric disorders, and the risk for schizophrenia: a Danish population-based study and meta-analysis. *JAMA Psychiatry*. http://dx.doi.org/10.1001/jamapsychiatry.2015.0346.

Agrawal, A., & Lynskey, M. T. (2006). The genetic epidemiology of cannabis use, abuse and dependence. *Addiction, 101*(6), 801–812. http://dx.doi.org/10.1111/j.1360-0443.2006.01399.x. pii:ADD1399.

Andrews, S., Das, D., Anstey, K. J., & Easteal, S. (2015). Interactive effect of APOE genotype and blood pressure on cognitive decline: the PATH through life study. *Journal of Alzheimer's Disease, 44*(4), 1087–1098. http://dx.doi.org/10.3233/jad-140630.

Angst, J. (2007). The bipolar spectrum. *British Journal of Psychiatry, 190*, 189–191. http://dx.doi.org/10.1192/bjp.bp.106.030957. pii:190/3/189.

Angst, J., Gamma, A., Joseph Bienvenu, O., Eaton, W. W., Ajdacic, V., Eich, D., & Rossler, W. (2006). Varying temporal criteria for generalized anxiety disorder: prevalence and clinical characteristics in a young age cohort. *Psychologie Medicale, 36*(9), 1283–1292.

Anney, R., Klei, L., Pinto, D., Almeida, J., Bacchelli, E., Baird, G., ... Devlin, B. (2012). Individual common variants exert weak effects on the risk for autism spectrum disorders. *Human Molecular Genetics, 21*(21), 4781–4792. http://dx.doi.org/10.1093/hmg/dds301.

Anney, R., Klei, L., Pinto, D., Regan, R., Conroy, J., Magalhaes, T. R., ... Hallmayer, J. (2010). A genome-wide scan for common alleles affecting risk for autism. *Human Molecular Genetics, 19*(20), 4072–4082. http://dx.doi.org/10.1093/hmg/ddq307.

Arseneault, L., Cannon, M., Poulton, R., Murray, R., Caspi, A., & Moffitt, T. E. (2002). Cannabis use in adolescence and risk for adult psychosis: longitudinal prospective study. *BMJ, 325*(7374), 1212–1213.

Aukes, M. F., Laan, W., Termorshuizen, F., Buizer-Voskamp, J. E., Hennekam, E. A., Smeets, H. M., ... Kahn, R. S. (2012). Familial clustering of schizophrenia, bipolar disorder, and major depressive disorder. *Genetics in Medicine, 14*(3), 338–341. http://dx.doi.org/10.1016/gim.2011.16 gim201116.

Beaty, T. H. (1997). Evolving methods in genetic epidemiology. I. Analysis of genetic and environmental factors in family studies. *Epidemlotogic Reviews, 19*(1), 14–23.

van Beijsterveldt, C. E., Groen-Blokhuis, M., Hottenga, J. J., Franic, S., Hudziak, J. J., Lamb, D., ... Boomsma, D. I. (2013). The young Netherlands twin register (YNTR): longitudinal twin and family studies in over 70,000 children. *Twin Research and Human Genetics, 16*(1), 252–267. http://dx.doi.org/10.1017/thg.2012.118. pii:S1832427412001181.

Bellodi, L., Sciuto, G., Diaferia, G., Ronchi, P., & Smeraldi, E. (1992). Psychiatric disorders in the families of patients with obsessive-compulsive disorder. *Psychiatry Research, 42*(2), 111–120.

Benazzi, F. (2007). Bipolar disorder—focus on bipolar II disorder and mixed depression. *Lancet, 369*(9565), 935–945. http://dx.doi.org/10.1016/S0140-6736(07)60453-X.

Benjamin, D. J., Cesarini, D., van der Loos, M. J., Dawes, C. T., Koellinger, P. D., Magnusson, P. K., ... Visscher, P. M. (2012). The genetic architecture of economic and political preferences. *Proceedings of the National Academy of Sciences of the United States of America, 109*(21), 8026–8031. http://dx.doi.org/10.1073/pnas.1120666109.

Benros, M. E., Waltoft, B. L., Nordentoft, M., Ostergaard, S. D., Eaton, W. W., Krogh, J., & Mortensen, P. B. (2013). Autoimmune diseases and severe infections as risk factors for mood disorders: a nationwide study. *JAMA Psychiatry, 70*(8), 812–820. http://dx.doi.org/10.1001/jamapsychiatry.2013.1111.

Benyamin, B., Visscher, P. M., & McRae, A. F. (2009). Family-based genome-wide association studies. *Pharmacogenomics, 10*(2), 181–190. http://dx.doi.org/10.2217/14622416.10.2.181.

Biederman, J., & Faraone, S. V. (2005). Attention-deficit hyperactivity disorder. *Lancet, 366*(9481), 237–248. http://dx.doi.org/10.1016/s0140-6736(05)66915-2.

Bierut, L. J., Dinwiddie, S. H., Begleiter, H., Crowe, R. R., Hesselbrock, V., Nurnberger, J. I., Jr., … Reich, T. (1998). Familial transmission of substance dependence: alcohol, marijuana, cocaine, and habitual smoking: a report from the collaborative study on the genetics of alcoholism. *Archives of General Psychiatry, 55*(11), 982–988.

Biesecker, B. B., & Peay, H. L. (2013). Genomic sequencing for psychiatric disorders: promise and challenge. *International Journal of Neuropsychopharmacology, 16*(7), 1667–1672. http://dx.doi.org/10.1017/s146114571300014x.

Black, D. W., Stumpf, A., McCormick, B., Allen, J., Blum, N., & Noyes, R. (2013). A blind re-analysis of the Iowa family study of obsessive-compulsive disorder. *Psychiatry Research, 209*(2), 202–206. http://dx.doi.org/10.1016/j.psychres.2013.04.013.

Bolton, P., Macdonald, H., Pickles, A., Rios, P., Goode, S., Crowson, M., … Rutter, M. (1994). A case-control family history study of autism. *Journal of Child Psychology and Psychiatry, 35*(5), 877–900.

Boraska, V., Davis, O. S., Cherkas, L. F., Helder, S. G., Harris, J., Krug, I., … Zeggini, E. (2012). Genome-wide association analysis of eating disorder-related symptoms, behaviors, and personality traits. *American Journal of Medical Genetics, Part B: Neuropsychiatric Genetics, 159b*(7), 803–811. http://dx.doi.org/10.1002/ajmg.b.32087.

Boraska, V., Franklin, C. S., Floyd, J. A., Thornton, L. M., Huckins, L. M., Southam, L., … Bulik, C. M. (2014). A genome-wide association study of anorexia nervosa. *Molecular Psychiatry, 19*(10), 1085–1094. http://dx.doi.org/10.1038/mp.2013.187.

Braff, D. L., Freedman, R., Schork, N. J., & Gottesman, I. I. (2007). Deconstructing schizophrenia: an overview of the use of endophenotypes in order to understand a complex disorder. *Schizophrenia Bulletin, 33*(1), 21–32.

Browne, H. A., Gair, S. L., Scharf, J. M., & Grice, D. E. (2014). Genetics of obsessive-compulsive disorder and related disorders. *The Psychiatric Clinics of North America, 37*(3), 319–335. http://dx.doi.org/10.1016/j.psc.2014.06.002.

Buka, S. L., Tsuang, M. T., Torrey, E. F., Klebanoff, M. A., Bernstein, D., & Yolken, R. H. (2001). Maternal infections and subsequent psychosis among offspring. *Archives of General Psychiatry, 58*(11), 1032–1037.

Buka, S. L., Tsuang, M. T., Torrey, E. F., Klebanoff, M. A., Wagner, R. L., & Yolken, R. H. (2001). Maternal cytokine levels during pregnancy and adult psychosis. *Brain, Behavior, and Immunity, 15*(4), 411–420.

Bulik, C. M., Sullivan, P. F., Wade, T. D., & Kendler, K. S. (2000). Twin studies of eating disorders: a review. *The International Journal of Eating Disorders, 27*(1), 1–20.

Calkins, M. E., Dobie, D. J., Cadenhead, K. S., Olincy, A., Freedman, R., Green, M. F., … Braff, D. L. (2007). The consortium on the genetics of endophenotypes in schizophrenia: model recruitment, assessment, and endophenotyping methods for a multisite collaboration. *Schizophrenia Bulletin, 33*(1), 33–48. http://dx.doi.org/10.1093/schbul/sbl044.

Carey, G., & Gottesman, I. (1981). Twin and family studies of anxiety, phobic and obsessive disorders. In J. Rabkin (Ed.), *Anxiety: New research and changing concepts* (Vol. 117). New York: Raven Press.

Caspi, A., Sugden, K., Moffitt, T. E., Taylor, A., Craig, I. W., Harrington, H., … Poulton, R. (2003). Influence of life stress on depression: moderation by a polymorphism in the 5-HTT gene. *Science, 301*(5631), 386–389.

Cederlöf, M., Lichtenstein, P., Larsson, H., Boman, M., Rück, C., Landén, M., & Mataix-Cols, D. (2015). Obsessive-compulsive disorder, psychosis, and bipolarity: a longitudinal cohort and multigenerational family study. *Schizophrenia Bulletin, 41*(5), 1076–1083. http://dx.doi.org/10.1093/schbul/sbu169.

Chang, Z., Lichtenstein, P., D'Onofrio, B. M., Almqvist, C., Kuja-Halkola, R., Sjolander, A., & Larsson, H. (2014). Maternal age at childbirth and risk for ADHD in offspring: a population-based cohort study. *International Journal of Epidemiology, 43*(6), 1815–1824. http://dx.doi.org/10.1093/ije/dyu204.

Chen, X., Cho, K., Singer, B. H., & Zhang, H. (2011). The nuclear transcription factor PKNOX2 is a candidate gene for substance dependence in European-origin women. *PLoS One, 6*(1), e16002. http://dx.doi.org/10.1371/journal.pone.0016002.

Clarke, M. C., Harley, M., & Cannon, M. (2006). The role of obstetric events in schizophrenia. *Schizophrenia Bulletin, 32*(1), 3–8.

Cloninger, C. R., Bohman, M., Sigvardsson, S., & von Knorring, A. L. (1985). Psychopathology in adopted-out children of alcoholics. The Stockholm Adoption Study. *Recent Developments in Alcoholism, 3*, 37–51.

Colvert, E., Tick, B., McEwen, F., Stewart, C., Curran, S. R., Woodhouse, E., … Bolton, P. (2015). Heritability of autism spectrum disorder in a UK population-based twin sample. *JAMA Psychiatry*. http://dx.doi.org/10.1001/jamapsychiatry.2014.3028.

Constantino, J. N., Zhang, Y., Frazier, T., Abbacchi, A. M., & Law, P. (2010). Sibling recurrence and the genetic epidemiology of autism. *American Journal of Psychiatry, 167*(11), 1349–1356. http://dx.doi.org/10.1176/appi.ajp.2010.09101470.

Cordell, H. J. (2009). Estimation and testing of gene-environment interactions in family-based association studies. *Genomics, 93*(1), 5–9. http://dx.doi.org/10.1016/j.ygeno.2008.05.002.

Cosgrove, V. E., & Suppes, T. (2013). Informing DSM-5: biological boundaries between bipolar I disorder, schizoaffective disorder, and schizophrenia. *BMC Medicine, 11*, 127. http://dx.doi.org/10.1186/1741-7015-11-127.

Croen, L. A., Grether, J. K., Yoshida, C. K., Odouli, R., & Hendrick, V. (2011). Antidepressant use during pregnancy and childhood autism spectrum disorders. *Archives of General Psychiatry, 68*(11), 1104–1112. http://dx.doi.org/10.1001/archgenpsychiatry.2011.73.

Dealberto, M. J. (2007). Why are immigrants at increased risk for psychosis? Vitamin D insufficiency, epigenetic mechanisms, or both? *Medical Hypotheses, 68*(2), 259–267.

Dean, K., & Murray, R. M. (2005). Environmental risk factors for psychosis. *Dialogues in Clinical Neuroscience, 7*(1), 69–80.

Demjaha, A., MacCabe, J. H., & Murray, R. M. (2012). How genes and environmental factors determine the different neurodevelopmental trajectories of schizophrenia and bipolar disorder. *Schizophrenia Bulletin, 38*(2), 209–214. http://dx.doi.org/10.1093/schbul/sbr100.

Dick, D. M., Riley, B., & Kendler, K. S. (2010). Nature and nurture in neuropsychiatric genetics: where do we stand? *Dialogues in Clinical Neuroscience, 12*(1), 7–23.

Doherty, J. L., O'Donovan, M. C., & Owen, M. J. (2012). Recent genomic advances in schizophrenia. *Clinical Genetics, 81*(2), 103–109. http://dx.doi.org/10.1111/j.1399-0004.2011.01773.x.

Do, C. B., Hinds, D. A., Francke, U., & Eriksson, N. (2012). Comparison of family history and SNPs for predicting risk of complex disease. *PLoS Genetics, 8*(10), e1002973. http://dx.doi.org/10.1371/journal.pgen.1002973.

Domingue, B. W., Belsky, D. W., Harris, K. M., Smolen, A., McQueen, M. B., & Boardman, J. D. (2014). Polygenic risk predicts obesity in both white and black young adults. *PLoS One, 9*(7), e101596. http://dx.doi.org/10.1371/journal.pone.0101596.

Dudbridge, F. (2013). Power and predictive accuracy of polygenic risk scores. *PLoS Genetics, 9*(3), e1003348. http://dx.doi.org/10.1371/journal.pgen.1003348.

Duncan, L. E., & Keller, M. C. (2011). A critical review of the first 10 years of candidate gene-by-environment interaction research in psychiatry. *American Journal of Psychiatry, 168*(10), 1041–1049. http://dx.doi.org/10.1176/appi.ajp.2011.11020191 appi.ajp.2011.11020191.

Eapen, V., Pauls, D. L., & Robertson, M. M. (2006). The role of clinical phenotypes in understanding the genetics of obsessive-compulsive disorder. *Journal of Psychosomatic Research, 61*(3), 359–364.

Escott-Price, V., Nalls, M. A., Morris, H. R., Lubbe, S., Brice, A., Gasser, T., ... Williams, N. M. (2015). Polygenic risk of Parkinson disease is correlated with disease age at onset. *Annals of Neurology, 77*(4), 582–591. http://dx.doi.org/10.1002/ana.24335.

Euesden, J., Lewis, C. M., & O'Reilly, P. F. (2014). PRSice: polygenic risk score software. *Bioinformatics*. http://dx.doi.org/10.1093/bioinformatics/btu848.

Falconer, D. S. (1960). *Introduction to quantitative genetics*. New York: Ronald Press Company.

Faraone, S. V., Kremen, W. S., & Tsuang, M. T. (1990). Genetic transmission of major affective disorders: quantitative models and linkage analyses. *Psychological Bulletin, 108*(1), 109–127.

Fazel, S., Wolf, A., Palm, C., & Lichtenstein, P. (2014). Violent crime, suicide, and premature mortality in patients with schizophrenia and related disorders: a 38-year total population study in Sweden. *Lancet Psychiatry, 1*(1), 44–54. http://dx.doi.org/10.1016/s2215-0366(14)70223-8.

Flint, J., & Munafo, M. R. (2007). The endophenotype concept in psychiatric genetics. *Psychologie Medicale, 37*(2), 163–180.

Froehlich, T. E., Anixt, J. S., Loe, I. M., Chirdkiatgumchai, V., Kuan, L., & Gilman, R. C. (2011). Update on environmental risk factors for attention-deficit/hyperactivity disorder. *Current Psychiatry Report, 13*(5), 333–344. http://dx.doi.org/10.1007/s11920-011-0221-3.

Georgiades, S., Szatmari, P., Zwaigenbaum, L., Bryson, S., Brian, J., Roberts, W., ... Garon, N. (2012). A prospective study of autistic-like traits in unaffected siblings of probands with autism spectrum disorder. *Archives of General Psychiatry*, 1–7. http://dx.doi.org/10.1001/2013.jamapsychiatry.1.

Giordano, G. N., Ohlsson, H., Sundquist, K., Sundquist, J., & Kendler, K. S. (2014). The association between cannabis abuse and subsequent schizophrenia: a Swedish national co-relative control study. *Psychologie Medicale*, 1–8. http://dx.doi.org/10.1017/s0033291714001524.

Glahn, D. C., Curran, J. E., Winkler, A. M., Carless, M. A., Kent, J. W., Jr., Charlesworth, J. C., ... Blangero, J. (2012). High dimensional endophenotype ranking in the search for major depression risk genes. *Biological Psychiatry, 71*(1), 6–14. http://dx.doi.org/10.1016/j.biopsych.2011.08.022. pii:S0006-3223(11)00857-2.

Glaus, J., Vandeleur, C. L., von Kanel, R., Lasserre, A. M., Strippoli, M. P., Gholam-Rezaee, M., ... Preisig, M. (2014). Associations between mood, anxiety or substance use disorders and inflammatory markers after adjustment for multiple covariates in a population-based study. *Journal of Psychiatric Research, 58*, 36–45. http://dx.doi.org/10.1016/j.jpsychires.2014.07.012.

Goldstein, J. M., Buka, S. L., Seidman, L. J., & Tsuang, M. T. (2010). Specificity of familial transmission of schizophrenia psychosis spectrum and affective psychoses in the New England family study's high-risk design. *Archives of General Psychiatry, 67*(5), 458–467. http://dx.doi.org/10.1001/archgenpsychiatry.2010.38 67/5/458.

Goldstein, R. B., Wickramaratne, P. J., Horwath, E., & Weissman, M. M. (1997). Familial aggregation and phenomenology of 'early'-onset (at or before age 20 years) panic disorder. *Archives of General Psychiatry, 54*(3), 271–278.

Goodwin, D. W. (1985). Alcoholism and genetics. The sins of the fathers. *Archives of General Psychiatry, 42*(2), 171–174.

Gordis, L. (2000). *Epidemiology*. Philadelphia: W.B. Saunders.

Gorwood, P., Feingold, J., & Ades, J. (1999). [Genetic epidemiology and psychiatry (I): scope and limitations of familial studies. Case of panic disorder]. *Encephale, 25*(1), 21–29.

Gottesman, I. I., & Gould, T. D. (2003). The endophenotype concept in psychiatry: etymology and strategic intentions. *American Journal of Psychiatry, 160*(4), 636–645.

Grabe, H. J., Ruhrmann, S., Ettelt, S., Buhtz, F., Hochrein, A., Schulze-Rauschenbach, S., ... Wagner, M. (2006). Familiality of obsessive-compulsive disorder in nonclinical and clinical subjects. *American Journal of Psychiatry, 163*(11), 1986–1992. http://dx.doi.org/10.1176/appi.ajp.163.11.1986.

Gudbjartsson, D. F., Helgason, H., Gudjonsson, S. A., Zink, F., Oddson, A., Gylfason, A., ... Stefansson, K. (2015). Large-scale whole-genome sequencing of the Icelandic population. *Nature Genetics*. http://dx.doi.org/10.1038/ng.3247.

Gur, R. E., Calkins, M. E., Gur, R. C., Horan, W. P., Nuechterlein, K. H., Seidman, L. J., & Stone, W. S. (2007). The consortium on the genetics of schizophrenia: neurocognitive endophenotypes. *Schizophrenia Bulletin, 33*(1), 49–68.

Hallmayer, J. (2000). The epidemiology of the genetic liability for schizophrenia. *Australian and New Zealand Journal of Psychiatry, 34*(Suppl.), S47–S55. discussion S56–S47.

Hallmayer, J., Cleveland, S., Torres, A., Phillips, J., Cohen, B., Torigoe, T., ... Risch, N. (2011). Genetic heritability and shared environmental factors among twin pairs with autism. *Archives of General Psychiatry*. http://dx.doi.org/10.1001/archgenpsychiatry.2011.76.

Hammen, C. (2005). Stress and depression. *Annual Review of Clinical Psychology, 1*, 293–319. http://dx.doi.org/10.1146/annurev.clinpsy.1.102803.143938.

Hamshere, M. L., O'Donovan, M. C., Jones, I. R., Jones, L., Kirov, G., Green, E. K., … Craddock, N. (2011). Polygenic dissection of the bipolar phenotype. *British Journal of Psychiatry, 198*(4), 284–288. http://dx.doi.org/10.1192/bjp.bp.110.087866.

Hawi, Z., Cummins, T. D., Tong, J., Johnson, B., Lau, R., Samarrai, W., & Bellgrove, M. A. (2015). The molecular genetic architecture of attention deficit hyperactivity disorder. *Molecular Psychiatry, 20*(3), 289–297. http://dx.doi.org/10.1038/mp.2014.183.

Heath, A. C., Bucholz, K. K., Madden, P. A., Dinwiddie, S. H., Slutske, W. S., Bierut, L. J., … Martin, N. G. (1997). Genetic and environmental contributions to alcohol dependence risk in a national twin sample: consistency of findings in women and men. *Psychologie Medicale, 27*(6), 1381–1396.

Heath, A. C., Whitfield, J. B., Madden, P. A., Bucholz, K. K., Dinwiddie, S. H., Slutske, W. S., … Martin, N. G. (2001). Towards a molecular epidemiology of alcohol dependence: analysing the interplay of genetic and environmental risk factors. *The British Journal of Psychiatry. Supplement, 40*, s33–40.

Hettema, J. M., Neale, M. C., Myers, J. M., Prescott, C. A., & Kendler, K. S. (2006). A population-based twin study of the relationship between neuroticism and internalizing disorders. *American Journal of Psychiatry, 163*(5), 857–864. http://dx.doi.org/10.1176/ajp.2006.163.5.857.

Hickie, I. B. (2014). Evidence for separate inheritance of mania and depression challenges current concepts of bipolar mood disorder. *Molecular Psychiatry, 19*(2), 153–155. http://dx.doi.org/10.1038/mp.2013.173.

Hill, A. B. (1953). Observation and experiment. *New England Journal of Medicine, 248*(24), 995–1001.

Hinney, A., & Volckmar, A. L. (2013). Genetics of eating disorders. *Current Psychiatry Report, 15*(12), 423. http://dx.doi.org/10.1007/s11920-013-0423-y.

Hoek, H. W. (2006). Incidence, prevalence and mortality of anorexia nervosa and other eating disorders. *Current Opinion in Psychiatry, 19*(4), 389–394. http://dx.doi.org/10.1097/01.yco.0000228759.95237.78.

Horan, W. P., Braff, D. L., Nuechterlein, K. H., Sugar, C. A., Cadenhead, K. S., Calkins, M. E., … Green, M. F. (2008). Verbal working memory impairments in individuals with schizophrenia and their first-degree relatives: findings from the consortium on the genetics of schizophrenia. *Schizophrenia Research, 103*(1–3), 218–228. http://dx.doi.org/10.1016/j.schres.2008.02.014.

Hudson, J. I., Hiripi, E., Pope, H. G., Jr., & Kessler, R. C. (2007). The prevalence and correlates of eating disorders in the National Comorbidity Survey Replication. *Biological Psychiatry, 61*(3), 348–358.

Hultman, C. M., Sparen, P., & Cnattingius, S. (2002). Perinatal risk factors for infantile autism. *Epidemiology, 13*(4), 417–423.

Johnson, S., Hollis, C., Kochhar, P., Hennessy, E., Wolke, D., & Marlow, N. (2010). Psychiatric disorders in extremely preterm children: longitudinal finding at age 11 years in the EPICure study. *Journal of the American Academy of Child and Adolescent Psychiatry, 49*(5), 453–463, e451.

Keller, M. F., Saad, M., Bras, J., Bettella, F., Nicolaou, N., Simon-Sanchez, J., … Nalls, M. A. (2012). Using genome-wide complex trait analysis to quantify 'missing heritability' in Parkinson's disease. *Human Molecular Genetics, 21*(22), 4996–5009. http://dx.doi.org/10.1093/hmg/dds335.

Kendell, R. E. (1989). Clinical validity. *Psychologie Medicale, 19*(1), 45–55.

Kendler, K. S., Chen, X., Dick, D., Maes, H., Gillespie, N., Neale, M. C., & Riley, B. (2012). Recent advances in the genetic epidemiology and molecular genetics of substance use disorders. *Nature Neuroscience, 15*(2), 181–189. http://dx.doi.org/10.1038/nn.3018.

Kendler, K. S., Gardner, C. O., & Prescott, C. A. (2001). Panic syndromes in a population-based sample of male and female twins. *Psychologie Medicale, 31*(6), 989–1000.

Kendler, K. S., Gatz, M., Gardner, C. O., & Pedersen, N. L. (2005). Age at onset and familial risk for major depression in a Swedish national twin sample. *Psychologie Medicale, 35*(11), 1573–1579. http://dx.doi.org/10.1017/s0033291705005714.

Kendler, K. S., Karkowski, L. M., Neale, M. C., & Prescott, C. A. (2000). Illicit psychoactive substance use, heavy use, abuse, and dependence in a US population-based sample of male twins. *Archives of General Psychiatry, 57*(3), 261–269.

Kendler, K. S., Karkowski, L., & Prescott, C. A. (1999). Hallucinogen, opiate, sedative and stimulant use and abuse in a population-based sample of female twins. *Acta Psychiatrica Scandinavica, 99*(5), 368–376.

Kendler, K. S., Neale, M. C., Heath, A. C., Kessler, R. C., & Eaves, L. J. (1994). A twin-family study of alcoholism in women. *American Journal of Psychiatry, 151*(5), 707–715.

Kendler, K. S., Ohlsson, H., Sundquist, J., & Sundquist, K. (2015). IQ and schizophrenia in a Swedish national sample: their causal relationship and the interaction of IQ with genetic risk. *American Journal of Psychiatry, 172*(3), 259–265. http://dx.doi.org/10.1176/appi.ajp.2014.14040516.

Kendler, K. S., & Prescott, C. A. (1998a). Cannabis use, abuse, and dependence in a population-based sample of female twins. *American Journal of Psychiatry, 155*(8), 1016–1022.

Kendler, K. S., & Prescott, C. A. (1998b). Cocaine use, abuse and dependence in a population-based sample of female twins. *British Journal of Psychiatry, 173*, 345–350.

Kessler, R. C. (1997). The effects of stressful life events on depression. *Annual Review of Psychology, 48*, 191–214. http://dx.doi.org/10.1146/annurev.psych.48.1.191.

Kessler, R. C., Chiu, W. T., Jin, R., Ruscio, A. M., Shear, K., & Walters, E. E. (2006). The epidemiology of panic attacks, panic disorder, and agoraphobia in the National Comorbidity Survey Replication. *Archives of General Psychiatry, 63*(4), 415–424. http://dx.doi.org/10.1001/archpsyc.63.4.415.

Kim, Y. S., & Leventhal, B. L. (2015). Genetic epidemiology and insights into interactive genetic and environmental effects in autism spectrum disorders. *Biological Psychiatry, 77*(1), 66–74. http://dx.doi.org/10.1016/j.biopsych.2014.11.001.

Klei, L., Sanders, S. J., Murtha, M. T., Hus, V., Lowe, J. K., Willsey, A. J., … Devlin, B. (2012). Common genetic variants, acting additively, are a major source of risk for autism. *Molecular Autism, 3*(1), 9. http://dx.doi.org/10.1186/2040-2392-3-9.

Klump, K. L., Suisman, J. L., Burt, S. A., McGue, M., & Iacono, W. G. (2009). Genetic and environmental influences on disordered eating: an adoption study. *Journal of Abnormal Psychology, 118*(4), 797–805. http://dx.doi.org/10.1037/a0017204.

Kohler, S., Doelken, S. C., Mungall, C. J., Bauer, S., Firth, H. V., Bailleul-Forestier, I., … Robinson, P. N. (2014). The human phenotype ontology project: linking molecular biology and disease through phenotype data. *Nucleic Acids Research, 42*(Database issue), D966–D974. http://dx.doi.org/10.1093/nar/gkt1026.

Kraepelin, E. (Ed.). (1919). *Dementia praecox and paraphrenia* (Robert E. Krieger).

Krumholz, H. M. (2009). Registries and selection bias: the need for accountability. *Circulation: Cardiovascular Quality and Outcomes, 2*(6), 517–518. http://dx.doi.org/10.1161/circoutcomes.109.916601.

Kuller, L. H. (2012). Invited commentary: the 21st century epidemiologist—a need for different training? *American Journal of Epidemiology, 176*(8), 668–671. http://dx.doi.org/10.1093/aje/kws227.

Lamers, F., Vogelzangs, N., Merikangas, K. R., de Jonge, P., Beekman, A. T., & Penninx, B. W. (2013). Evidence for a differential role of HPA-axis function, inflammation and metabolic syndrome in melancholic versus atypical depression. *Molecular Psychiatry, 18*(6), 692–699. http://dx.doi.org/10.1038/mp.2012.144.

Langenberg, C., Sharp, S. J., Franks, P. W., Scott, R. A., Deloukas, P., Forouhi, N. G., … Wareham, N. J. (2014). Gene-lifestyle interaction and type 2 diabetes: the EPIC interact case-cohort study. *PLoS Medicine, 11*(5), e1001647. http://dx.doi.org/10.1371/journal.pmed.1001647.

Lee, S. H., DeCandia, T. R., Ripke, S., Yang, J., Sullivan, P. F., Goddard, M. E., … Wray, N. R. (2012). Estimating the proportion of variation in susceptibility to schizophrenia captured by common SNPs. *Nature Genetics, 44*(3), 247–250. http://dx.doi.org/10.1038/ng.1108.

Lee, S. H., Ripke, S., Neale, B. M., Faraone, S. V., Purcell, S. M., Perlis, R. H., … Wray, N. R. (2013). Genetic relationship between five psychiatric disorders estimated from genome-wide SNPs. *Nature Genetics, 45*(9), 984–994. http://dx.doi.org/10.1038/ng.2711 ng.2711.

Lehner, T., Senthil, G., & Addington, A. M. (2015). Convergence of advances in genomics, team science, and repositories as drivers of progress in psychiatric genomics. *Biological Psychiatry, 77*(1), 6–14. http://dx.doi.org/10.1016/j.biopsych.2014.01.003.

Lenane, M. C., Swedo, S. E., Leonard, H., Pauls, D. L., Sceery, W., & Rapoport, J. L. (1990). Psychiatric disorders in first degree relatives of children and adolescents with obsessive compulsive disorder. *Journal of the American Academy of Child and Adolescent Psychiatry, 29*(3), 407–412.

Lenox, R. H., Gould, T. D., & Manji, H. K. (2002). Endophenotypes in bipolar disorder. *American Journal of Medical Genetics, 114*(4), 391–406.

Lichtenstein, P., Tuvblad, C., Larsson, H., & Carlstrom, E. (2007). The Swedish twin study of child and adolescent development: the TCHAD-study. *Twin Research and Human Genetics, 10*(1), 67–73.

Lichtenstein, P., Yip, B. H., Bjork, C., Pawitan, Y., Cannon, T. D., Sullivan, P. F., & Hultman, C. M. (2009). Common genetic determinants of schizophrenia and bipolar disorder in Swedish families: a population-based study. *Lancet, 373*(9659), 234–239. http://dx.doi.org/10.1016/S0140-6736(09)60072-6.

Li, J., Olsen, J., Vestergaard, M., & Obel, C. (2010). Attention-deficit/hyperactivity disorder in the offspring following prenatal maternal bereavement: a nationwide follow-up study in Denmark. *European Child and Adolescent Psychiatry, 19*(10), 747–753. http://dx.doi.org/10.1007/s00787-010-0113-9.

Llewellyn, C. H., Trzaskowski, M., Plomin, R., & Wardle, J. (2013). Finding the missing heritability in pediatric obesity: the contribution of genome-wide complex trait analysis. *The International Journal of Obesity (London), 37*(11), 1506–1509. http://dx.doi.org/10.1038/ijo.2013.30.

Lyall, K., Pauls, D. L., Santangelo, S. L., Spiegelman, D., & Ascherio, A. (2011). Maternal early life factors associated with hormone levels and the risk of having a child with an autism spectrum disorder in the nurses health study II. *Journal of Autism and Developmental Disorders, 41*(5), 618–627. http://dx.doi.org/10.1007/s10803-010-1079-7.

MacLean, J. E., Szatmari, P., Jones, M. B., Bryson, S. E., Mahoney, W. J., Bartolucci, G., & Tuff, L. (1999). Familial factors influence level of functioning in pervasive developmental disorder. *Journal of the American Academy of Child and Adolescent Psychiatry, 38*(6), 746–753. http://dx.doi.org/10.1097/00004583-199906000-00023.

Maes, M. (1995). Evidence for an immune response in major depression: a review and hypothesis. *Progress in Neuro-Psychopharmacology and Biological Psychiatry, 19*(1), 11–38.

Maier, R., Moser, G., Chen, G. B., Ripke, S., Coryell, W., Potash, J. B., … Lee, S. H. (2015). Joint analysis of psychiatric disorders increases accuracy of risk prediction for schizophrenia, bipolar disorder, and major depressive disorder. *American Journal of Human Genetics, 96*(2), 283–294. http://dx.doi.org/10.1016/j.ajhg.2014.12.006.

Malaspina, D., Brown, A., Goetz, D., Alia-Klein, N., Harkavy-Friedman, J., Harlap, S., & Fennig, S. (2002). Schizophrenia risk and paternal age: a potential role for de novo mutations in schizophrenia vulnerability genes. *CNS Spectrums, 7*(1), 26–29.

Mann, J. R., McDermott, S., Bao, H., Hardin, J., & Gregg, A. (2010). Pre-eclampsia, birth weight, and autism spectrum disorders. *Journal of Autism and Developmental Disorders, 40*(5), 548–554. http://dx.doi.org/10.1007/s10803-009-0903-4.

Maron, E., Hettema, J. M., & Shlik, J. (2010). Advances in molecular genetics of panic disorder. *Molecular Psychiatry, 15*(7), 681–701. http://dx.doi.org/10.1038/mp.2009.145.

Martiskainen, H., Helisalmi, S., Viswanathan, J., Kurki, M., Hall, A., Herukka, S. K., … Hiltunen, M. (2015). Effects of Alzheimer's disease-associated risk loci on cerebrospinal fluid biomarkers and disease progression: a polygenic risk score approach. *Journal of Alzheimer's Disease, 43*(2), 565–573. http://dx.doi.org/10.3233/jad-140777.

Mataix-Cols, D., Boman, M., Monzani, B., Ruck, C., Serlachius, E., Langstrom, N., & Lichtenstein, P. (2013). Population-based, multigenerational family clustering study of obsessive-compulsive disorder. *JAMA Psychiatry, 70*(7), 709–717. http://dx.doi.org/10.1001/jamapsychiatry.2013.3.

Matheson, S. L., Shepherd, A. M., Laurens, K. R., & Carr, V. J. (2011). A systematic meta-review grading the evidence for non-genetic risk factors and putative antecedents of schizophrenia. *Schizophrenia Research, 133*(1–3), 133–142. http://dx.doi.org/10.1016/j.schres.2011.09.020.

Mattheisen, M., Samuels, J. F., Wang, Y., Greenberg, B. D., Fyer, A. J., McCracken, J. T., … Nestadt, G. (2015). Genome-wide association study in obsessive-compulsive disorder: results from the OCGAS. *Molecular Psychiatry, 20*(3), 337–344. http://dx.doi.org/10.1038/mp.2014.43.

McDonald, C., & Murray, R. M. (2000). Early and late environmental risk factors for schizophrenia. *Brain Research. Brain Research Reviews, 31*(2–3), 130–137.

McGrath, J. J., Mortensen, P. B., Visscher, P. M., & Wray, N. R. (2013). Where GWAS and epidemiology meet: opportunities for the simultaneous study of genetic and environmental risk factors in schizophrenia. *Schizophrenia Bulletin, 39*(5), 955–959. http://dx.doi.org/10.1093/schbul/sbt108.

McGrath, J. J., Petersen, L., Agerbo, E., Mors, O., Mortensen, P. B., & Pedersen, C. B. (2014). A comprehensive assessment of parental age and psychiatric disorders. *JAMA Psychiatry, 71*(3), 301–309. http://dx.doi.org/10.1001/jamapsychiatry.2013.4081.

McGuffin, P. (2004). Nature and nurture interplay: schizophrenia. *Psychiatrische Praxis, 31*(Suppl. 2), S189–S193.

McGuffin, P., Rijsdijk, F., Andrew, M., Sham, P., Katz, R., & Cardno, A. (2003). The heritability of bipolar affective disorder and the genetic relationship to unipolar depression. *Archives of General Psychiatry, 60*(5), 497–502. http://dx.doi.org/10.1001/archpsyc.60.5.497 60/5/497.

McMahon, F. J., Akula, N., Schulze, T. G., Muglia, P., Tozzi, F., Detera-Wadleigh, S. D., ... Rietschel, M. (2010). Meta-analysis of genome-wide association data identifies a risk locus for major mood disorders on 3p21.1. *Nature Genetics, 42*(2), 128–131. http://dx.doi.org/10.1038/ng.523.

Meli, G., Ottl, B., Paladini, A., & Cataldi, L. (2012). Prenatal and perinatal risk factors of schizophrenia. *The Journal of Maternal-Fetal & Neonatal Medicine, 25*(12), 2559–2563. http://dx.doi.org/10.3109/14767058.2012.699118.

Mendlewicz, J., & Rainer, J. D. (1977). Adoption study supporting genetic transmission in manic–depressive illness. *Nature, 268*(5618), 327–329.

Merikangas, K. R. (1990). *Comorbity for anxiety and depression: Review of family and genetic studies*. Washington, DC: American Psychiatric Press.

Merikangas, K. R. (2002). *Genetic epidemiolgy of substance-use disorders. Textbook of biological psychiatry* (p. 537).

Merikangas, K. R., Akiskal, H. S., Angst, J., Greenberg, P. E., Hirschfeld, R. M., Petukhova, M., & Kessler, R. (2007). Lifetime and 12-month prevalence of bipolar spectrum disorder in the national comorbidity survey replication. *Archives of General Psychiatry, 64*(5), 543–552.

Merikangas, K. R., Cui, L., Heaton, L., Nakamura, E., Roca, C., Ding, J., ... Angst, J. (2013). Independence of familial transmission of mania and depression: results of the NIMH family study of affective spectrum disorders. *Molecular Psychiatry*. http://dx.doi.org/10.1038/mp.2013.116 mp2013116.

Merikangas, K. R., & Karayiorgou, M. (2015). Genetics of psychiatric disorders: advances in genetic epidemiology and molecular genetics. In A. Tasman, J. Kay, J. A. Lieberman, M. B. First, & M. Riba (Eds.), *Psychiatry* (4th ed.). Hoboken, NJ: Wiley-Blackwell.

Merikangas, K. R., Stevens, D. E., Fenton, B., Stolar, M., O'Malley, S., Woods, S. W., & Risch, N. (1998). Co-morbidity and familial aggregation of alcoholism and anxiety disorders. *Psychologie Medicale, 28*(4), 773–788.

Merikangas, K. R., Stolar, M., Stevens, D. E., Goulet, J., Preisig, M. A., Fenton, B., ... Rounsaville, B. J. (1998). Familial transmission of substance use disorders. *Archives of General Psychiatry, 55*(11), 973–979.

Merikangas, K. R., & Yu, K. (2002). Genetic epidemiology of bipolar disorder. *Clinical Neuroscience Research, 2*, 127.

Mitchison, D., & Hay, P. J. (2014). The epidemiology of eating disorders: genetic, environmental, and societal factors. *Clinical Epidemiology, 6*, 89–97. http://dx.doi.org/10.2147/clep.s40841.

Mortensen, P. B., Pedersen, C. B., Westergaard, T., Wohlfahrt, J., Ewald, H., Mors, O., ... Melbye, M. (1999). Effects of family history and place and season of birth on the risk of schizophrenia. *New England Journal of Medicine, 340*(8), 603–608. http://dx.doi.org/10.1056/nejm199902253400803.

Muhle, R., Trentacoste, S. V., & Rapin, I. (2004). The genetics of autism. *Pediatrics, 113*(5), e472–486.

National Adoption Information Clearinghouse. (2007).

Neale, B. M., Medland, S. E., Ripke, S., Asherson, P., Franke, B., Lesch, K. P., ... Nelson, S. (2010). Meta-analysis of genome-wide association studies of attention-deficit/hyperactivity disorder. *Journal of the American Academy of Child and Adolescent Psychiatry, 49*(9), 884–897. http://dx.doi.org/10.1016/j.jaac.2010.06.008.

Nestadt, G., Samuels, J., Riddle, M., Bienvenu, O. J., 3rd, Liang, K. Y., LaBuda, M., ... Hoehn-Saric, R. (2000). A family study of obsessive-compulsive disorder. *Archives of General Psychiatry, 57*(4), 358–363.

Nierenberg, A. A., Akiskal, H. S., Angst, J., Hirschfeld, R. M., Merikangas, K. R., Petukhova, M., & Kessler, R. C. (2010). Bipolar disorder with frequent mood episodes in the national comorbidity survey replication (NCS-R). *Molecular Psychiatry, 15*(11), 1075–1087. http://dx.doi.org/10.1038/mp.2009.61.

Olsen, J., Basso, O., & Sorensen, H. T. (1999). What is a population-based registry? *Scandinavian Journal of Public Health, 27*(1), 78.

Ostergaard, S. D., Waltoft, B. L., Mortensen, P. B., & Mors, O. (2013). Environmental and familial risk factors for psychotic and non-psychotic severe depression. *Journal of Affective Disorders, 147*(1–3), 232–240. http://dx.doi.org/10.1016/j.jad.2012.11.009.

Otowa, T., Kawamura, Y., Nishida, N., Sugaya, N., Koike, A., Yoshida, E., ... Sasaki, T. (2012). Meta-analysis of genome-wide association studies for panic disorder in the Japanese population. *Translational Psychiatry, 2*, e186. http://dx.doi.org/10.1038/tp.2012.89.

Ott, J., Kamatani, Y., & Lathrop, M. (2011). Family-based designs for genome-wide association studies. *Nature Reviews Genetics, 12*(7), 465–474. http://dx.doi.org/10.1038/nrg2989.

Ottman, R. (1990). An epidemiologic approach to gene-environment interaction. *Genetic Epidemiology, 7*(3), 177–185.

Ozonoff, S., Young, G. S., Carter, A., Messinger, D., Yirmiya, N., Zwaigenbaum, L., ... Stone, W. L. (2011). Recurrence risk for autism spectrum disorders: a baby siblings research consortium study. *Pediatrics, 128*(3), e488–495. http://dx.doi.org/10.1542/peds.2010-2825.

Pabinger, S., Dander, A., Fischer, M., Snajder, R., Sperk, M., Efremova, M., ... Trajanoski, Z. (2014). A survey of tools for variant analysis of next-generation genome sequencing data. *Briefings in Bioinformatics, 15*(2), 256–278. http://dx.doi.org/10.1093/bib/bbs086.

Patel, C. J., & Ioannidis, J. P. (2014). Studying the elusive environment in large scale. *JAMA, 311*(21), 2173–2174. http://dx.doi.org/10.1001/jama.2014.4129.

Pauls, D. L., Abramovitch, A., Rauch, S. L., & Geller, D. A. (2014). Obsessive-compulsive disorder: an integrative genetic and neurobiological perspective. *Nature Reviews Neuroscience, 15*(6), 410–424. http://dx.doi.org/10.1038/nrn3746.

Pauls, D. L., Alsobrook, J. P., 2nd, Goodman, W., Rasmussen, S., & Leckman, J. F. (1995). A family study of obsessive-compulsive disorder. *American Journal of Psychiatry, 152*(1), 76–84.

de la Paz, M. P., Villaverde-Hueso, A., Alonso, V., Janos, S., Zurriaga, O., Pollan, M., & Abaitua-Borda, I. (2010). Rare diseases epidemiology research. *Advances in Experimental Medicine and Biology, 686*, 17–39. http://dx.doi.org/10.1007/978-90-481-9485-8_2.

Pettersson, E., Sjolander, A., Almqvist, C., Anckarsater, H., D'Onofrio, B. M., Lichtenstein, P., & Larsson, H. (2015). Birth weight as an independent predictor of ADHD symptoms: a within-twin pair analysis. *Journal of Child Psychology and Psychiatry, 56*(4), 453–459. http://dx.doi.org/10.1111/jcpp.12299.

Pinto, D., Delaby, E., Merico, D., Barbosa, M., Merikangas, A., Klei, L., ... Scherer, S. W. (2014). Convergence of genes and cellular pathways dysregulated in autism spectrum disorders. *American Journal of Human Genetics, 94*(5), 677–694. http://dx.doi.org/10.1016/j.ajhg.2014.03.018.

Pinto, D., Pagnamenta, A. T., Klei, L., Anney, R., Merico, D., Regan, R., … Betancur, C. (2010). Functional impact of global rare copy number variation in autism spectrum disorders. *Nature, 466*(7304), 368–372. http://dx.doi.org/10.1038/nature09146.

Portin, P., & Alanen, Y. O. (1997). A critical review of genetic studies of schizophrenia. I. Epidemiological and brain studies. *Acta Psychiatrica Scandinavica, 95*(1), 1–5.

Posthuma, D., & Boomsma, D. I. (2000). A note on the statistical power in extended twin designs. *Behavior Genetics, 30*(2), 147–158.

Psychiatric GWAS Consortium Bipolar Disorder Working Group. (2011). Large-scale genome-wide association analysis of bipolar disorder identifies a new susceptibility locus near ODZ4. *Nature Genetics, 43*(10), 977–983. http://dx.doi.org/10.1038/ng.943.

Purcell, S. M., Moran, J. L., Fromer, M., Ruderfer, D., Solovieff, N., Roussos, P., … Sklar, P. (2014). A polygenic burden of rare disruptive mutations in schizophrenia. *Nature.* http://dx.doi.org/10.1038/nature12975.

Purcell, S. M., Wray, N. R., Stone, J. L., Visscher, P. M., O'Donovan, M. C., Sullivan, P. F., & Sklar, P. (2009). Common polygenic variation contributes to risk of schizophrenia and bipolar disorder. *Nature, 460*(7256), 748–752. http://dx.doi.org/10.1038/nature08185.

Rasic, D., Hajek, T., Alda, M., & Uher, R. (2013). Risk of mental illness in offspring of parents with schizophrenia, bipolar disorder, and major depressive disorder: a meta-analysis of family high-risk studies. *Schizophrenia Bulletin.* http://dx.doi.org/10.1093/schbul/sbt114.

Reichenberg, A., Gross, R., Weiser, M., Bresnahan, M., Silverman, J., Harlap, S., … Susser, E. (2006). Advancing paternal age and autism. *Archives of General Psychiatry, 63*(9), 1026–1032. http://dx.doi.org/10.1001/archpsyc.63.9.1026.

Ridge, P. G., Mukherjee, S., Crane, P. K., & Kauwe, J. S. (2013). Alzheimer's disease: analyzing the missing heritability. *PLoS One, 8*(11), e79771. http://dx.doi.org/10.1371/journal.pone.0079771.

Ripke, S., Wray, N. R., Lewis, C. M., Hamilton, S. P., Weissman, M. M., Breen, G., … Sullivan, P. F. (2013). A mega-analysis of genome-wide association studies for major depressive disorder. *Molecular Psychiatry, 18*(4), 497–511. http://dx.doi.org/10.1038/mp.2012.21 mp201221.

Risch, N. (1990). Linkage strategies for genetically complex traits. I. Multilocus models. *American Journal of Human Genetics, 46*(2), 222–228.

Risch, N., Herrell, R., Lehner, T., Liang, K. Y., Eaves, L., Hoh, J., … Merikangas, K. R. (2009). Interaction between the serotonin transporter gene (5-HTTLPR), stressful life events, and risk of depression: a meta-analysis. *JAMA, 301*(23), 2462–2471. http://dx.doi.org/10.1001/jama.2009.878 301/23/2462.

Risch, N., Hoffmann, T. J., Anderson, M., Croen, L. A., Grether, J. K., & Windham, G. C. (2014). Familial recurrence of autism spectrum disorder: evaluating genetic and environmental contributions. *American Journal of Psychiatry, 171*(11), 1206–1213. http://dx.doi.org/10.1176/appi.ajp.2014.13101359.

Roy-Byrne, P. P., Craske, M. G., & Stein, M. B. (2006). Panic disorder. *Lancet, 368*(9540), 1023–1032. http://dx.doi.org/10.1016/s0140-6736(06)69418-x.

Sareen, J., Cox, B. J., Afifi, T. O., de Graaf, R., Asmundson, G. J., ten Have, M., & Stein, M. B. (2005). Anxiety disorders and risk for suicidal ideation and suicide attempts: a population-based longitudinal study of adults. *Archives of General Psychiatry, 62*(11), 1249–1257. http://dx.doi.org/10.1001/archpsyc.62.11.1249.

Sariaslan, A., Larsson, H., D'Onofrio, B., Langstrom, N., Fazel, S., & Lichtenstein, P. (2015). Does population density and neighborhood deprivation predict schizophrenia? A nationwide Swedish family-based study of 2.4 million individuals. *Schizophrenia Bulletin, 41*(2), 494–502. http://dx.doi.org/10.1093/schbul/sbu105.

Sawcer, S., Hellenthal, G., Pirinen, M., Spencer, C. C., Patsopoulos, N. A., Moutsianas, L., … Compston, A. (2011). Genetic risk and a primary role for cell-mediated immune mechanisms in multiple sclerosis. *Nature, 476*(7359), 214–219. http://dx.doi.org/10.1038/nature10251.

Schizophrenia Working Group of the Psychiatric Genomics Consortium. (2014). Biological insights from 108 schizophrenia-associated genetic loci. *Nature, 511*(7510), 421–427. http://dx.doi.org/10.1038/nature13595.

Schumacher, J., Kristensen, A. S., Wendland, J. R., Nothen, M. M., Mors, O., & McMahon, F. J. (2011). The genetics of panic disorder. *Journal of Medical Genetics, 48*(6), 361–368. http://dx.doi.org/10.1136/jmg.2010.086876.

Scott, K. M., Smith, D. R., & Ellis, P. M. (2010). Prospectively ascertained child maltreatment and its association with DSM-IV mental disorders in young adults. *Archives of General Psychiatry, 67*(7), 712–719. http://dx.doi.org/10.1001/archgenpsychiatry.2010.71.

Selten, J. P., Cantor-Graae, E., & Kahn, R. S. (2007). Migration and schizophrenia. *Current Opinion in Psychiatry, 20*(2), 111–115. http://dx.doi.org/10.1097/YCO.0b013e328017f68e.

Shelton, J. F., Tancredi, D. J., & Hertz-Picciotto, I. (2010). Independent and dependent contributions of advanced maternal and paternal ages to autism risk. *Autism Research, 3*(1), 30–39. http://dx.doi.org/10.1002/aur.116.

Sigvardsson, S., Bohman, M., & Cloninger, C. R. (1996). Replication of the stockholm adoption study of alcoholism. Confirmatory cross-fostering analysis. *Archives of General Psychiatry, 53*(8), 681–687.

Smoller, J. W., Craddock, N., Kendler, K., Lee, P. H., Neale, B. M., Nurnberger, J. I., … Sullivan, P. F. (2013). Identification of risk loci with shared effects on five major psychiatric disorders: a genome-wide analysis. *Lancet, 381*(9875), 1371–1379. http://dx.doi.org/10.1016/s0140-6736(12)62129-1.

Smoller, J. W., & Gardner-Schuster, E. (2007). Genetics of bipolar disorder. *Current Psychiatry Report, 9*(6), 504–511.

Smoller, J. W., & Tsuang, M. T. (1998). Panic and phobic anxiety: defining phenotypes for genetic studies. *American Journal of Psychiatry, 155*(9), 1152–1162.

Smucny, J., Wylie, K. P., & Tregellas, J. R. (2014). Functional magnetic resonance imaging of intrinsic brain networks for translational drug discovery. *Trends in Pharmacological Sciences, 35*(8), 397–403. http://dx.doi.org/10.1016/j.tips.2014.05.001.

Sprich, S., Biederman, J., Crawford, M. H., Mundy, E., & Faraone, S. V. (2000). Adoptive and biological families of children and adolescents with ADHD. *Journal of the American Academy of Child and Adolescent Psychiatry, 39*(11), 1432–1437. http://dx.doi.org/10.1097/00004583-200011000-00018.

Steinhausen, H. C., Foldager, L., Perto, G., & Munk-Jorgensen, P. (2009). Family aggregation of mental disorders in the nationwide Danish three generation study. *European Archives of Psychiatry and Clinical Neuroscience, 259*(5), 270–277. http://dx.doi.org/10.1007/s00406-008-0865-0.

Stergiakouli, E., Martin, J., Hamshere, M. L., Langley, K., Evans, D. M., St Pourcain, B., ... Davey Smith, G. (2015). Shared genetic influences between attention-deficit/hyperactivity disorder (ADHD) traits in children and clinical ADHD. *Journal of the American Academy of Child and Adolescent Psychiatry, 54*(4), 322–327. http://dx.doi.org/10.1016/j.jaac.2015.01.010.

Stewart, S. E., Yu, D., Scharf, J. M., Neale, B. M., Fagerness, J. A., Mathews, C. A., ... Pauls, D. L. (2013). Genome-wide association study of obsessive-compulsive disorder. *Molecular Psychiatry, 18*(7), 788–798. http://dx.doi.org/10.1038/mp.2012.85 mp201285.

Strober, M., Freeman, R., Lampert, C., Diamond, J., & Kaye, W. (2000). Controlled family study of anorexia nervosa and bulimia nervosa: evidence of shared liability and transmission of partial syndromes. *American Journal of Psychiatry, 157*(3), 393–401.

Sugaya, L., Hasin, D. S., Olfson, M., Lin, K. H., Grant, B. F., & Blanco, C. (2012). Child physical abuse and adult mental health: a national study. *Journal of Traumatic Stress, 25*(4), 384–392. http://dx.doi.org/10.1002/jts.21719.

Sullivan, P. F. (2005). The genetics of schizophrenia. *PLoS Medicine, 2*(7), e212. http://dx.doi.org/10.1371/journal.pmed.0020212.

Sullivan, P. F., Kendler, K. S., & Neale, M. C. (2003). Schizophrenia as a complex trait: evidence from a meta-analysis of twin studies. *Archives of General Psychiatry, 60*(12), 1187–1192. http://dx.doi.org/10.1001/archpsyc.60.12.1187 60/12/1187.

Sullivan, P. F., Neale, M. C., & Kendler, K. S. (2000). Genetic epidemiology of major depression: review and meta-analysis. *American Journal of Psychiatry, 157*(10), 1552–1562.

Sumi, S., Taniai, H., Miyachi, T., & Tanemura, M. (2006). Sibling risk of pervasive developmental disorder estimated by means of an epidemiologic survey in Nagoya, Japan. *Journal of Human Genetics, 51*(6), 518–522. http://dx.doi.org/10.1007/s10038-006-0392-7.

Szatmari, P., Jones, M. B., Holden, J., Bryson, S., Mahoney, W., Tuff, L., ... Hoult, L. (1996). High phenotypic correlations among siblings with autism and pervasive developmental disorders. *American Journal of Medical Genetics, 67*(4), 354–360. http://dx.doi.org/10.1002/(sici)1096-8628(19960726)67, 4<354::aid-ajmg7>3.0.co;2-m.

Szatmari, P., Merette, C., Emond, C., Zwaigenbaum, L., Jones, M. B., Maziade, M., ... Palmour, R. (2008). Decomposing the autism phenotype into familial dimensions. *American Journal of Medical Genetics, Part B: Neuropsychiatric Genetics, 147B*(1), 3–9. http://dx.doi.org/10.1002/ajmg.b.30561.

Targum, S. D., & Keefe, R. S. (2008). Cognition and schizophrenia: is there a role for cognitive assessments in diagnosis and treatment? *Psychiatry (Edgmont), 5*(12), 55–59.

Taubes, G. (September 16, 2007). Do we really know what makes us healthy? *New York Times*. Retrieved from http://www.nytimes.com/2007/09/16/magazine/16epidemiology-t.html?pagewanted=all&_r=0.

Taylor, S. (2013). Molecular genetics of obsessive-compulsive disorder: a comprehensive meta-analysis of genetic association studies. *Molecular Psychiatry, 18*(7), 799–805. http://dx.doi.org/10.1038/mp.2012.76.

Teslovich, T. M., Musunuru, K., Smith, A. V., Edmondson, A. C., Stylianou, I. M., Koseki, M., ... Kathiresan, S. (2010). Biological, clinical and population relevance of 95 loci for blood lipids. *Nature, 466*(7307), 707–713. http://dx.doi.org/10.1038/nature09270.

Thornton, L. M., Mazzeo, S. E., & Bulik, C. M. (2011). The heritability of eating disorders: methods and current findings. In R. A. Adan, & W. H. Kaye (Eds.), *Behavioral neurobiology of eating disorders*. Berlin: Springer-Verlag.

Trzaskowski, M., Eley, T. C., Davis, O. S., Doherty, S. J., Hanscombe, K. B., Meaburn, E. L., ... Plomin, R. (2013). First genome-wide association study on anxiety-related behaviours in childhood. *PLoS One, 8*(4), e58676. http://dx.doi.org/10.1371/journal.pone.0058676.

Tsuang, M. T. (2001). Defining alternative phenotypes for genetic studies: what can we learn from studies of schizophrenia? *American Journal of Medical Genetics, 105*(1), 8–10.

Tsuang, M. T., Bar, J. L., Harley, R. M., & Lyons, M. J. (2001). The Harvard twin study of substance abuse: what we have learned. *Harvard Review of Psychiatry, 9*(6), 267–279.

Vachon, C. M., Schaid, D. J., Ingle, J. N., Wickerham, D. L., Kubo, M., Mushiroda, T., ... Couch, F. J. (2015). A polygenic risk score for breast cancer in women receiving tamoxifen or raloxifene on NSABP P-1 and P-2. *Breast Cancer Research and Treatment, 149*(2), 517–523. http://dx.doi.org/10.1007/s10549-014-3175-4.

Valdar, W., Solberg, L. C., Gauguier, D., Burnett, S., Klenerman, P., Cookson, W. O., ... Flint, J. (2006). Genome-wide genetic association of complex traits in heterogeneous stock mice. *Nature Genetics, 38*(8), 879–887.

Vandeleur, C. L., Merikangas, K. R., Strippoli, M. P., Castelao, E., & Preisig, M. (2013). Specificity of psychosis, mania and major depression in a contemporary family study. *Molecular Psychiatry*. http://dx.doi.org/10.1038/mp.2013.132 mp2013132.

Verhulst, B., Neale, M. C., & Kendler, K. S. (2015). The heritability of alcohol use disorders: a meta-analysis of twin and adoption studies. *Psychologie Medicale, 45*(5), 1061–1072. http://dx.doi.org/10.1017/s0033291714002165.

Visscher, P. M., Brown, M. A., McCarthy, M. I., & Yang, J. (2012). Five years of GWAS discovery. *American Journal of Human Genetics, 90*(1), 7–24. http://dx.doi.org/10.1016/j.ajhg.2011.11.029.

Volk, H. E., Hertz-Picciotto, I., Delwiche, L., Lurmann, F., & McConnell, R. (2011). Residential proximity to freeways and autism in the CHARGE study. *Environmental Health Perspectives, 119*(6), 873–877. http://dx.doi.org/10.1289/ehp.1002835.

Wacholder, S., Garcia-Closas, M., & Rothman, N. (2002). Study of genes and environmental factors in complex diseases. *Lancet, 359*(9312), 1155. author reply 1157.

Wade, T. D., Bulik, C. M., Sullivan, P. F., Neale, M. C., & Kendler, K. S. (2000). The relation between risk factors for binge eating and bulimia nervosa: a population-based female twin study. *Health Psychology, 19*(2), 115–123.

Wade, T. D., Gordon, S., Medland, S., Bulik, C. M., Heath, A. C., Montgomery, G. W., & Martin, N. G. (2013). Genetic variants associated with disordered eating. *The International Journal of Eating Disorders, 46*(6), 594–608. http://dx.doi.org/10.1002/eat.22133.

Wallis, D., Russell, H. F., & Muenke, M. (2008). Review: genetics of attention deficit/hyperactivity disorder. *The Journal of Pediatric Psychology, 33*(10), 1085–1099. http://dx.doi.org/10.1093/jpepsy/jsn049.

Wareham, N. J., Young, E. H., & Loos, R. J. (2008). Epidemiological study designs to investigate gene-behavior interactions in the context of human obesity. *Obesity (Silver Spring), 16*(Suppl. 3), S66–S71. http://dx.doi.org/10.1038/oby.2008.521.

Watson, C. T., Disanto, G., Breden, F., Giovannoni, G., & Ramagopalan, S. V. (2012). Estimating the proportion of variation in susceptibility to multiple sclerosis captured by common SNPs. *Scientific Reports, 2*, 770. http://dx.doi.org/10.1038/srep00770.

Wender, P. H., Kety, S. S., Rosenthal, D., Schulsinger, F., Ortmann, J., & Lunde, I. (1986). Psychiatric disorders in the biological and adoptive families of adopted individuals with affective disorders. *Archives of General Psychiatry, 43*(10), 923–929.

Wetherill, L., Agrawal, A., Kapoor, M., Bertelsen, S., Bierut, L. J., Brooks, A., … Foroud, T. (2015). Association of substance dependence phenotypes in the COGA sample. *Addiction Biology, 20*(3), 617–627. http://dx.doi.org/10.1111/adb.12153.

Wetherill, L., Kapoor, M., Agrawal, A., Bucholz, K., Koller, D., Bertelsen, S. E., … Foroud, T. (2014). Family-based association analysis of alcohol dependence criteria and severity. *Alcoholism Clinical and Experimental Research, 38*(2), 354–366. http://dx.doi.org/10.1111/acer.12251.

Wilde, A., Chan, H. N., Rahman, B., Meiser, B., Mitchell, P. B., Schofield, P. R., & Green, M. J. (2014). A meta-analysis of the risk of major affective disorder in relatives of individuals affected by major depressive disorder or bipolar disorder. *Journal of Affective Disorders, 158*, 37–47. http://dx.doi.org/10.1016/j.jad.2014.01.014.

Wilkinson, S. T., Radhakrishnan, R., & D'Souza, D. C. (2014). Impact of cannabis use on the development of psychotic disorders. *Current Addiction Reports, 1*(2), 115–128. http://dx.doi.org/10.1007/s40429-014-0018-7.

Williams, G., King, J., Cunningham, M., Stephan, M., Kerr, B., & Hersh, J. H. (2001). Fetal valproate syndrome and autism: additional evidence of an association. *Developmental Medicine and Child Neurology, 43*(3), 202–206.

Wing, L. (1981). Language, social, and cognitive impairments in autism and severe mental retardation. *Journal of Autism and Developmental Disorders, 11*(1), 31–44.

Wirth, M., Villeneuve, S., & La Joie, R. (2014). Gene-environment interactions: lifetime cognitive activity, APOE genotype, and beta-amyloid burden. *Journal of Nanoneuroscience, 34*(25), 8612–8617. http://dx.doi.org/10.1523/jneurosci.4612-13.2014.

Wozniak, J., Faraone, S. V., Martelon, M., McKillop, H. N., & Biederman, J. (2012). Further evidence for robust familiality of pediatric bipolar I disorder: results from a very large controlled family study of pediatric bipolar I disorder and a meta-analysis. *Journal of Clinical Psychiatry, 73*(10), 1328–1334. http://dx.doi.org/10.4088/JCP.12m07770.

Wray, N. R., Goddard, M. E., & Visscher, P. M. (2008). Prediction of individual genetic risk of complex disease. *Current Opinion in Genetics and Development, 18*(3), 257–263. http://dx.doi.org/10.1016/j.gde.2008.07.006.

Wray, N. R., & Gottesman, I. I. (2012). Using summary data from the danish national registers to estimate heritabilities for schizophrenia, bipolar disorder, and major depressive disorder. *Frontiers in Genetics, 3*, 118. http://dx.doi.org/10.3389/fgene.2012.00118.

Wray, N. R., Lee, S. H., Mehta, D., Vinkhuyzen, A. A., Dudbridge, F., & Middeldorp, C. M. (2014). Research review: polygenic methods and their application to psychiatric traits. *Journal of Child Psychology and Psychiatry, 55*(10), 1068–1087. http://dx.doi.org/10.1111/jcpp.12295.

Yang, J., Benyamin, B., McEvoy, B. P., Gordon, S., Henders, A. K., Nyholt, D. R., … Visscher, P. M. (2010). Common SNPs explain a large proportion of the heritability for human height. *Nature Genetics, 42*(7), 565–569. http://dx.doi.org/10.1038/ng.608.

Yang, Q., & Khoury, M. J. (1997). Evolving methods in genetic epidemiology. III. Gene-environment interaction in epidemiologic research. *Epidemiologic Reviews, 19*(1), 33–43.

Yang, J., Lee, S. H., Goddard, M. E., & Visscher, P. M. (2011). GCTA: a tool for genome-wide complex trait analysis. *American Journal of Human Genetics, 88*(1), 76–82. http://dx.doi.org/10.1016/j.ajhg.2010.11.011.

Yang, L., Neale, B. M., Liu, L., Lee, S. H., Wray, N. R., Ji, N., … Hu, X. (2013). Polygenic transmission and complex neuro developmental network for attention deficit hyperactivity disorder: genome-wide association study of both common and rare variants. *American Journal of Medical Genetics, Part B: Neuropsychiatric Genetics, 162b*(5), 419–430. http://dx.doi.org/10.1002/ajmg.b.32169.

Zammit, S., & Owen, M. J. (2006). Stressful life events, 5-HTT genotype and risk of depression. *British Journal of Psychiatry, 188*, 199–201.

Zemojtel, T., Kohler, S., Mackenroth, L., Jager, M., Hecht, J., Krawitz, P., … Robinson, P. N. (2014). Effective diagnosis of genetic disease by computational phenotype analysis of the disease-associated genome. *Science Translational Medicine, 6*(252), 252ra123. http://dx.doi.org/10.1126/scitranslmed.3009262.

Zwaigenbaum, L., Bryson, S. E., Szatmari, P., Brian, J., Smith, I. M., Roberts, W., … Roncadin, C. (2012). Sex differences in children with autism spectrum disorder identified within a high-risk infant cohort. *Journal of Autism and Developmental Disorders*. http://dx.doi.org/10.1007/s10803-012-1515-y.

Chapter 3

Natural Selection and Neuropsychiatric Disease: Theory, Observation, and Emerging Genetic Findings

Mark J. Daly, Elise B. Robinson and Benjamin M. Neale
Massachusetts General Hospital, Boston, MA, United States; Broad Institute of MIT and Harvard, Cambridge, MA, United States

INTRODUCTION

Natural selection is almost universally effective at eliminating or at the least dramatically suppressing the frequency of alleles that predispose to diseases that substantially affect reproductive fitness. In practice, and as a direct result, most genetic conditions that substantially reduce or eliminate the opportunity to reproduce are indeed rare. Neuropsychiatric disease, particularly severe mental illness, stands in contrast to this picture because:

1. Neuropsychiatric diseases are common, consistently observed across all populations, and are actually rising in the frequency with which they are diagnosed.
2. Neuropsychiatric diseases have been established as being among the most heritable of common human diseases.
3. Neuropsychiatric diseases [most substantially autism spectrum disorders (ASDs) and schizophrenia (SCZ), and many others to a smaller degree] have been reliably observed to result in significantly reduced fecundity.

Such seemingly strong selective pressure suggests that only rare and de novo variation should contribute to neuropsychiatric diseases because strong-acting alleles must be rapidly removed from the population via natural selection. However, this assertion is inconsistent with the extremely high frequency, high heritability, and apparently non-Mendelian genetic epidemiology of severe mental illnesses.

Numerous theories have been proposed by which this conundrum might be resolved. In this section we cover basic epidemiologic observations and theoretical proposals for resolving them, and examine how emerging genetic data begin to inform the ultimate resolution of these questions.

EPIDEMIOLOGY OF NEUROPSYCHIATRIC DISEASE

Neuropsychiatric diseases are devastating conditions not only individually but societally, given their high prevalence (Kessler et al., 2005) and early age of onset compared with most other serious common chronic diseases. Even without considering the burden of substance abuse disorders, 2012 National Institutes of Mental Health estimates suggest that 18.6% of American adults have a mental illness and nearly a quarter of those have a serious mental illness that severely limits major life activities. Unlike many diseases which exhibit substantial heterogeneity across different regions of the world, the impact of neuropsychiatric disease is felt everywhere (Demyttenaere et al., 2004; Whiteford et al., 2013).

The 2010 Global Burden of Disease (GBD) study (Whiteford et al., 2013) is particularly informative on several key issues. This study estimated that 7.4% of disability-adjusted life-years, an integrated estimate of the value of future years lost to disease mortality and disability, result from mental illness as a category. Despite the elevation of mortality of all causes (Harris & Barraclough, 1998; Hiroeh, Appleby, Mortensen, & Dunn, 2010) (including estimates indicating that mental illness has a role in 90% of suicides (Nock et al., 2008)), the major contribution to this number comes from years

Genomics, Circuits, and Pathways in Clinical Neuropsychiatry. http://dx.doi.org/10.1016/B978-0-12-800105-9.00003-2

lived with disability. Because mental illness frequently appears in young adulthood (if not childhood, as in the case of ASDs) and responds to therapy only in some dimensions, the lifetime burden of disease per case is enormous. Moreover, neuropsychiatric disorders have been described as "disproportionately disabling" because they substantially impair a range of faculties including cognition, emotional regulation and perception, executive function, and motivation. In addition, the impairments often arise at a time that disrupts higher education, career development, and employment opportunity as well as the formation of personal relationships that might lead to long-term commitment and marriage. On top of this, mental illness leads to increased substance abuse rates and lower access to medical treatment for other nonmental illnesses, which further exacerbates their impact.

Owing to the dramatically increased disability and mortality attributable to serious mental illness at the earliest stages of adulthood, it is perhaps unsurprising that for decades epidemiologic studies have demonstrated substantially reduced reproductive fitness. Strong and consistent reports of reduced marriage and fertility rates among schizophrenic patients in particular have existed for decades; an overview of many of these is contained in Saugstad (1989). Epidemiological studies have left little question as to the magnitude and veracity of these effects. A population-wide study of all Danes born from 1950 through 1990 (Laursen & Munk-Olsen, 2010) provides a striking view. With cohort follow-up of 2.8 million individuals from 1970 through 2006 using the Civil Registration System and the Central Psychiatric Registry, the authors described strikingly reduced rates at which patients hospitalized for a psychiatric disorder subsequently had offspring, compared with their population counterparts who had not been hospitalized. Most dramatic was the small incidence rate ratio (IRR), the rate relative to the reference nonhospitalized cohort, with which schizophrenic patients subsequently had a first child (men, 0.10; and women, 0.18). Similar but less extreme IRRs were seen for the categories of bipolar (men, 0.32; and women, 0.36) and unipolar depression (men, 0.46; and women, 0.57) and even for the catchall category of all other psychiatric admissions (men, 0.51; and women, 0.70). The rates of having a second or third child were also significantly reduced although not as dramatically as the rates of having a first child.

A similarly powerful study examined a broader set of diagnoses using the entire 2.3-million-person birth cohort of Sweden between 1950 and 1970 (Power et al., 2013).

In that study, six major psychiatric diagnoses were examined individually (schizophrenia, autism, bipolar disorder, depression, anorexia nervosa, and substance abuse) using a fertility ratio indicating the fold increase or decrease in the number of offspring, correcting for year of birth. Schizophrenia and autism showed the most dramatic and almost identical reduced fecundity (schizophrenia: men, 0.23; women, 0.47; autism: men, 0.25; women, 0.48). This confirmed the previously noted stronger effect in male than in female patients. All other diagnoses except depression showed a similar but markedly less substantial pattern of reduced fecundity, which was greater in males than in females (Fig. 3.1).

Another important observation from the 2010 GBD and previous studies is the recognition of strikingly consistent and substantial rates of mental illnesses in almost every region of the world despite great variations in ancestry, culture, and level of development (Demyttenaere et al., 2004; Whiteford et al., 2013). Whereas phenotypic consistency across cultural backgrounds just beginning to be studied is for some disorders, diseases with strong forces of selection at play, as well as most chronic diseases that have an environmental contribution, often show substantial variation in rates across the world.

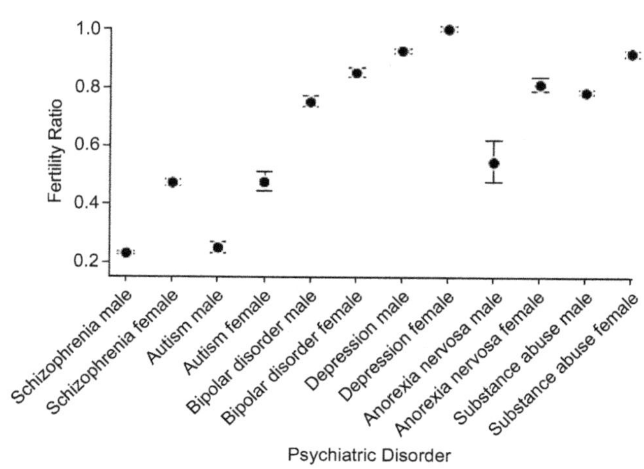

FIGURE 3.1 Fertility ratios for individuals with schizophrenia, autism, bipolar disorder, depression, anorexia nervosa, and substance abuse. A fertility ratio of 1 represents that of the general population. *Reproduced from Power, R. A., Kyaga, S., Uher, R., MacCabe, J. H., Langstrom, N., Mikael, L., … Svensson, A. C. (2013). Fecundity of patients with schizophrenia, autism, bipolar disorder, depression anorexia nervosa, or substance abuse vs their unaffected siblings.* JAMA Psychiatry, *70(1), 22–30.*

Thus it is perhaps informative to some of the considerations below that marked rate differences are not systematically noted on geographic or national development lines.

In addition to their individual and collective commonality, severe mental illnesses share a feature that is an important element of this conundrum and ultimately a key to disentangling it: extremely high heritability. Despite the lack of a pathology with which to make definitive diagnosis, there is a prevailing sense that better characterization of patients is needed to improve the genetic (and therefore molecular) homogeneity of patient groups. This lack of biological insight means that most mental illnesses are simply referred to as disorders; despite this, mental illnesses, particularly schizophrenia, ASDs, and bipolar disorder, are among the most heritable common diseases known. With a heritability of 60–90% documented in numerous twin and family studies over many decades for each of these major diagnoses, the genetic component of severe mental illness is as large or larger than every other common disease and among the most heritable of human traits generally. Such consistent population measures are not only striking in the presence of this diagnostic uncertainty and lack of pathology, the familial recurrence in the presence of the drastic reduction in fitness is all the more puzzling.

For extremely rare severe diseases such as cystic fibrosis, twin and family studies established the importance of genetic risk factors. Building on the observed heritability, family-based linkage studies were pursued to map the causal locus. These linkage studies succeeded because the inheritance pattern for these rare severe diseases was effectively Mendelian or single-gene, in which highly penetrant mutations in a single gene explain most of all cases in the population. Unlike these Mendelian diseases, severe and common mental illnesses were not amenable to these approaches. Indeed, despite some sizable efforts, linkage studies by and large failed to elucidate regions reliably and have led to inconclusive genetic findings. Although this is unsurprising in retrospect (indeed common human diseases outside the brain have also rarely benefited from this approach), these early efforts strongly suggest that high-penetrance variants in one or a small number of genes is not the general mechanism by which the heritability of mental illnesses arises. For example, a single gene in which high- or even intermediate-penetrance (eg, odds ratio = 20), recessive-acting variation explained 10–20% of schizophrenia would have had highly significant linkage results in studies with far fewer than 100 sibling pairs. A thought experiment proposed by Keller and Miller (2006) also points out that such models make little sense because selection against such alleles suggests that schizophrenia would have been substantially and inconsistently more common only tens of generations ago.

Furthermore, a role for truly high-penetrance Mendelian mutations may be limited overall: Many genes for early-onset and severe forms of cancers, neurodegenerative disease, diabetes, and cardiovascular disease have all been discovered by the presence of unusual, densely affected families that were individually large enough to map genetic variants unique to those families. Such families have been far less forthcoming in mental illnesses and to date have not provided the same early foothold into the molecular underpinnings of disease as in these other medical areas.

THEORETICAL CONSIDERATIONS

How then can psychiatric diseases that are so evidently devastating to the individual and his or her probability of reproducing be so common in today's population and be a growing, rather than shrinking, public health concern? Unsurprisingly this observation has not escaped notice, but rather been the topic of active hypothesis generation among epidemiologists, psychiatrists, and geneticists for many years. We first articulate these hypotheses and then address each in turn, integrating results from large-scale genetic studies to bolster traditional epidemiologic analyses to evaluate their potential validity.

Such reproductive disadvantages as seen in severe mental illness must exert a significant influence on the genetic architecture of disease. For example, in the case of autosomal recessive diseases that are lethal before maturity, disease-causing mutations must generally be at very low frequencies in the general population, because in each generation any homozygous individuals are eliminated from the breeding pool, and along with them, the two disease-causing alleles. Specifically, if in generation N the recessive lethal allele a is at frequency f, then of the fraction of the population that is aa (f^2), Aa ($2f[1-f]$), and AA ($[1-f]^2$), only these latter two groups survive and reproduce, and as a result the expected allele frequency of a in generation $N+1$ is $f/(1+f)$, demonstrating the inexorable cleansing of natural selection. Scenarios exist, however, by which such mutations can be unexpectedly common. A heterozygote advantage may exist: that is, a survival advantage conferred upon heterozygous carriers of the mutation as in the canonical example of hemoglobin-S, which when it is homozygous causes sickle-cell anemia, but in the heterozygous form affords the carrier protection from malaria. Such an advantage is a specific instance of **balancing selection**, in which the forces of natural selection may act upon an allele in different directions owing to the impact of the allele on diverse favorable and unfavorable phenotypes. Other potential instances of balancing selection include cases in which the same phenotype was previously favorable in a different point in evolution or alternate environmental setting, such as the white-bodied versus black-bodied peppered moth in Manchester before and after the industrial revolution.

Against a certain and strong force of selection against alleles predisposing to schizophrenia, and ASDs in particular, commonly proposed theories have often focused on balancing selection as a potential for how such heritable disorders can maintain high frequency. These include:

1. Benefits are conferred upon nondiseased individuals with high genetic risk.
2. Sexual antagonism exists: alleles damaging to one sex are favorable to the other.
3. Genetic risk to disease was ancestrally neutral or under positive selection.

Each of these forms of balancing selection makes specific predictions about the epidemiology and genetic architecture of disease.

The major competing hypothesis explaining the high frequency of mental illness is that of **mutation–selection balance**. Specifically, deleterious disease-causing variants are restored to the population via mutation as rapidly as they are removed by natural selection. This hypothesis, too, can be evaluated afresh with emerging genetic data, particularly from exome and genome-sequencing efforts possible owing to dramatic technical improvements in DNA sequencing in the past few years.

HYPOTHESIS TESTING: EPIDEMIOLOGY AND EMERGING GENETIC DATA

Hypothesis 1 proposes there might be a reproductive advantage to individuals carrying genetic risk factors but not manifesting disease. The simplest way in which this might be recognized is through increased reproductive success among healthy siblings of cases. Without knowledge of the specific genes involved in mental illness, we can rely on the obligatory fact that siblings of cases must be genetically enriched (because they share half of their genomes) for risk of disease compared with the rest of the population at large. Continuing the earlier example of hemoglobin-S, two-thirds of the healthy siblings of homozygotes with hemoglobin-S will themselves be carriers and protected from malaria.

Epidemiologically this question has been addressed in numerous studies across recent decades. For example, Haukka, Suvisaari, and Lonnqvist (2003) explored siblings of patients of schizophrenia born between 1950 and 1959 (drawn from the total population cohort of more than 870,000 nationally). While observing the same reduced fecundity as later observed in the Sweden and Denmark studies noted earlier among 10,000 patients with schizophrenia, their more than 24,000 siblings showed no significant differences in reproductive age or number of children, which suggested no evidence of compensation for the reduced fitness of patients via more favorable outcomes in high-risk but healthy relatives.

The more recent Sweden study (Power et al., 2013) described also addressed this hypothesis and similarly concluded that no significant increase in fecundity was consistently observed after correcting for potential confounders, again with the conclusion that perhaps hypotheses other than direct balancing selection to increase the fitness of healthy relatives should be considered more likely.

One limitation of the sibling construct is that it might casually be taken to suggest that the increased fitness needs be strong. As articulated in Haukka et al. (2003), because 2% of the population consists of siblings of schizophrenic people, these individuals would have to have dramatically increased fitness to make up for the more than 50% deficit in offspring produced by their (1% population rate) schizophrenic siblings. Such a comparison is inherently flawed, however, particularly under a polygenic model. Most individuals who carry a greater than average number of risk alleles for schizophrenia will not in fact have schizophrenia. Consequently, those allele carriers in the population may well benefit from the positive side of the balancing selection.

Take our simple recessive lethal/heterozygote advantage model once again: If we imagine a population in which the allele is at a frequency of 0.10, 1% of the population is affected and the 1–2% of the population that is their healthy carrier siblings would indeed need to reproduce at a dramatically higher rate to make up for the loss of the homozygotes. However, a full 18% of the population at random enjoys the benefit of carrier status, and each carrier therefore needs to have an advantage roughly comparable to $1 + f$ to make up for the lost heterozygotes. In the case of recessive lethal alleles, most of which exist at frequencies less than 0.001, this indeed constitutes an almost imperceptible advantage as far as most epidemiological studies are concerned. Of course new mutation does arise and contribute to risk, which suggests some contribution from mutation–selection balance, but just as the failure to exclude balancing selection does not constitute proof of its existence, neither does the existence of mutation–selection balance preclude a contribution from balancing selection.

In addition, advantages that could provide the balance in the balancing–selection equation with respect to mental illness may not themselves be obvious in direct fecundity of siblings or high-risk individuals in the first place. If, for example, siblings/carriers were advantaged in their career and life trajectory, they might not themselves have more children but their children might be greatly advantaged, more likely themselves to have children, and so forth. For example, if genetic risk for mental illness also conferred increased creativity, risk taking that increased odds of career success, or perhaps even greater cognitive performance in general, one might not expect a greater number of children but perhaps greater achievement, filial success rates, and ultimately greater likelihood of having grandchildren.

Historically, such conjecture has rested entirely in the realm of the hypothetical (a fantastical possible connection between mental illnesses and human behavioral, emotional, and cognitive traits) and the epidemiological (in which correlational relationships can be found but establishing causation is vexingly elusive because potential confounding explanations simply cannot be excluded). As the unfortunately complicated genetic architecture of mental illnesses and human traits has begun to come into focus in the past decade, a compelling opportunity to take more concrete steps in this direction has emerged in an exciting way.

MENDELIAN RANDOMIZATION AND POLYGENIC RISK

Genetic associations are unique among the myriad correlations between measurable risk factors and common disease outcomes. Because diseases do not modify the germline DNA carried in each of our cells, when a correlation is observed between a genetic variant and disease, one can conclude (assuming the genetic study is free from flaws in the study design and uses technically and ancestry-matched cases and controls) that a causal link exists between DNA variation and outcome.

As pertains to modifiable risk factors for common disease, a powerful concept has emerged from this simple observation for determining causation in the absence of a definitive randomized controlled trial. Originally articulated in the evaluation of bone marrow transplantation (BMT) versus chemotherapy, Gray and Wheatley (1991) pointed out that the existence of a human leukocyte antigen—matched sibling suitable for BMT was a matter of chance and that because there would be an intent to perform such in all cases in which a donor was available, the comparison of two groups was therefore justified since essentially, individuals were assigned at random to the two groups. If there a common genetic variant exists with a strong impact on the risk factor, such a genetic variant provides a powerful influence on the problem. If the risk factor is causally related to the disease, any genetic variant influencing the risk factor must have a predictable and defined effect on the risk of disease (assuming there are no pleiotropic effects of the allele on other traits which countermand or attenuate the influence on disease). By contrast, if the variant strongly influences the risk factor but does not change disease risk, a hypothesized causal link can be rejected confidently.

Because they enable robust inferences from cross-sectional and observational data, such approaches (Davey Smith & Ebrahim, 2003; Youngman et al., 2000) have gained great popularity and have, for example, been widely used to evaluate cardiovascular risk factors, affirming causal roles of low-density lipoprotein, triglycerides, and lipoprotein(a) levels but significantly challenging others which many had presumed likely causal (eg, high-density lipoprotein, C-reactive protein). At face value, nothing could seem less relevant to psychiatry and mental illness as strong-acting common variants theoretically cannot and empirically do not exist. Additional developments (in both genetic discovery and analytic methods) have now enabled this concept to be carried much further than expected.

By 2005 both the heritability of mental illness and the failure of linkage methods to define regions in which higher penetrance variants explained meaningful amounts of that heritability had been well established. Dramatic technical advances, spurred by the Human Genome Project and follow-on projects such as the HapMap to characterize common human genome variation, were suddenly making new study designs possible. Along with catalogs of DNA variants and an understanding of their correlation patterns (linkage disequilibrium [LD]) emerged array-based genotyping technology that enabled simultaneous assessment (genotyping) of hundreds of thousands of sites across the genome. Rather than linkage analysis, which evaluated cosegregation of long genomic regions within a family, the new (but conceptually simpler) approach of genome-wide association suddenly became possible. Instead of a series of within-family tests, it became possible to evaluate the contribution of individual common DNA variants across the entire population. The attractiveness of the approach was considerable: As emerging genome variation catalogs (International HapMap Consortium, 2005) reinforced the theoretical principles that humans have limited genetic variation (Kimura & Ota, 1973) and that most human heterozygosity is due to common DNA variants with a population distribution suggesting that they have existed for thousands of generations (Lewontin, 1972), new technologies enabled the potential for an association study that could potentially comprehensively assess the vast majority of DNA variation in the human genome for a role in disease. In theory and practice, the approach was dramatically more sensitive: Instead of an indirect assessment of the role of a DNA variant (an effect on disease risk indirectly leading to an increased likelihood that related family members sharing a trait would share the chromosomal region carrying that variant), one could now directly compare large numbers of cases and controls at the genotype level and assess those variants which even only modestly attenuate risk of disease.

TABLE 3.1 Theoretical Selection and Power in Schizophrenia/Autism

OR	s	MedianAF	SS_5e^{-8}
20	0.12	<0.00001	
10	0.06	<0.00001	
5	0.025	0.00001	700,000
3	0.013	0.00004	500,000
2	0.006	0.0001	500,000
1.5	0.003	0.0005	400,000
1.2	0.001	0.01	108,000
1.1	0.0006	0.05	86,000
1.05	0.0003	0.11	163,000

Estimated impact of reduced fitness of autism/schizophrenia. MedianAF, median allele frequency extrapolated from simulations provided in Zuk et al., 2014; SS_5e^{-8}, sample size required in a conventional case–control design for such a median allele; *s*, the selection coefficient.

Early genome-wide association studies (GWASs) in mental illness, particularly in schizophrenia and bipolar illness, where they have been most fervently pursued to date, were only modestly more successful than their linkage-based predecessors. Most studies turned up evidence of few if any genuine risk factors. Viewed retrospectively, this is expected in light of nothing more than the reduced fitness of severe mental illness presented previously and the consequent selective pressure against variants conferring strong risk. Table 3.1, derived from mathematical formulation in Zuk et al. (2014), using the selection against schizophrenia/ASDs as an example, shows a range of effect sizes, the average allele frequencies at which such an effect should be found, and the sample size required to achieve "genome-wide significance" (a widely used figure of $P < 5 \times 10^{-8}$ is most often used, reflecting a classical $P = 0.05$ corrected for what effectively corresponds to roughly 1 million independent statistical trials (eg, univariate association tests across the genome)).

Although few significantly associated loci were identified in these early studies, a particularly insightful analysis was introduced by Purcell and colleagues (International Schizophrenia Consortium et al., 2009) in one that radically altered the course of GWAS interpretation afterward. An early large-scale schizophrenia GWAS effort demonstrated that a large number of variants must exist across the genome with individual effects too small to detect, but when combined these effects could explain and predict a meaningful amount of schizophrenia risk. Essentially by weighting each single nucleotide polymorphism across the genome by its contribution to risk in one study, it was possible to define a risk score that predicted case–control status in a completely independent study. These scores demonstrated conclusively that schizophrenia was a highly polygenic disease and a large slice of heritability could be explained by the combination of very weak-acting common variants.

A simple extension of this approach was then to create polygenic risk scores (PRSs) which, in their simplest form, are log(odds ratio) weighted sums of nominally associated alleles from the largest GWAS. PRSs can be standardized across a population and evaluated in new studies and populations as a measure of risk. As the GWAS efforts in schizophrenia have become larger and more powerful through a series of collaborations forged by the Psychiatric Genomics Consortium (PGC), such scores have increased in value. In an instance of schizophrenia GWAS, nearly 37,000 cases of schizophrenia were evaluated and 108 distinct genomic loci were robustly associated to schizophrenia at the cited genome-wide significance threshold (PGC-SCZ, 2014). When these loci, and many others not reaching certain significance, were combined into a PRS, roughly 7% of the population-wide variance in liability to schizophrenia is accounted for by this PRS, and the ability to predict risk in other population samples was demonstrated, with the top decile of PRS showing between 7 and 20 times increased risk over the lowest decile in independent studies. Capturing even this much genetic risk in a single variable opens up entirely novel and impactful avenues for genetic and epidemiological work.

A few things are noteworthy considering this theoretical approximation in light of the empirical findings. Whereas from Table 3.1 it would appear that no study to date should have defined risk factors, a great many risk factors will be substantially more frequent than the median and have more attainable power given the sample numbers achieved by the PGC in schizophrenia. If many such alleles are present in the population, some of these alleles will become lucky and achieve a better than average strength of association from the initial 30,000–40,000 cases in GWAS (a corollary of this is the often discussed winners curse—in genetic association terms, the first study to detect a true effect has almost certainly substantially overestimated its effect because in practice, all of our current studies are badly underpowered to find the alleles that exist). Given the success of GWAS in schizophrenia (with the most substantial reduced fitness), it would seem likely that the study of other highly heritable mental illnesses on a similar scale should provide substantial genetic insights.

Fig. 3.2 expands on these points. The plot shows the risk allele frequency plotted against the odds ratio for the 128 common variants conclusively associated with schizophrenia risk in the PGC-SCZ findings in 2014. As described, odds ratios are consistently low and their combination with allele frequency places them in the range where the reduced fitness of schizophrenia exerts little pressure to eliminate them. Furthermore, these alleles exist in a thin odds ratio band bounded on top by selection and on the bottom by power to discover. At the same time, theory suggests that some modest pressure will work against common, low-penetrance alleles; however, little if any signature of this is seen. In fact, Fig. 3.2 seems to show that the risk allele is almost as often the more common rather than the less common allele and even common (>90%) alleles that exert stronger risk (ie, new, low-frequency variants that are protective) seem not to drift to higher frequencies. Although not affirmative proof of any specific hypothesis, these last effects are consistent with an impact of additional forces, potentially balancing selection, beyond direct selection based on reduced fitness.

One additional, unexpected observation was noted early on in the first schizophrenia study to introduce this concept (International Schizophrenia Consortium et al., 2009): Not only did a schizophrenia score predict schizophrenia in a new study, it was also demonstrated to predict bipolar case or control status in independent studies (albeit with not as strong a correlation). However, it did not predict any of a range of autoimmune or cardiometabolic traits. This indicated that many of the weak polygenic risk factors were shared between schizophrenia and bipolar illnesses. At the same time, rare copy number variants (CNVs) were observed to be shared between autism and schizophrenia more often than not (Sebat et al., 2009) and these CNVs were often associated with a more diverse range of neurodevelopmental phenotypes. Thus, it became the exception rather than the rule that genetic risk factors might often be shared across related neuropsychiatric diseases and that underlying genetic risk factors, often acting at a fundamental molecular level, would be unlikely to respect traditional diagnostic boundaries or have intimately specific influences on complex adult behavioral, emotional, and cognitive traits.

To estimate the genetic overlap better between related traits such as schizophrenia and bipolar disorder, genome-wide complex trait analysis (GCTA) was developed (Yang et al., 2011). Building from the conceptual PRS observations and from foundational approaches to sibling pair linkage analysis years earlier, the basic principle behind GCTA is that if a diagnosis or quantitative trait has a genetic component, people who share that diagnosis or are more phenotypically similar will tend to be more genetically similar. This approach and others that have followed provided for the first time a mechanism to estimate heritability arising from common DNA variation from population GWAS data directly, complementing estimates from twin and family studies. Importantly and satisfyingly, these new approaches have strongly confirmed the epidemiological determination of high heritability for nearly every severe mental illness. Perhaps more importantly, the approach also enables the joint analysis of multiple traits and diseases and can provide an estimation of the proportion of heritability shared between distinct diseases, and in an initial application of such approaches, an estimated genetic correlation of 0.7 between schizophrenia and bipolar was detected (Cross-Disorder Group of the Psychiatric Genomics et al., 2013) as well as significant correlations between schizophrenia and autism and both schizophrenia and bipolar with major depression.

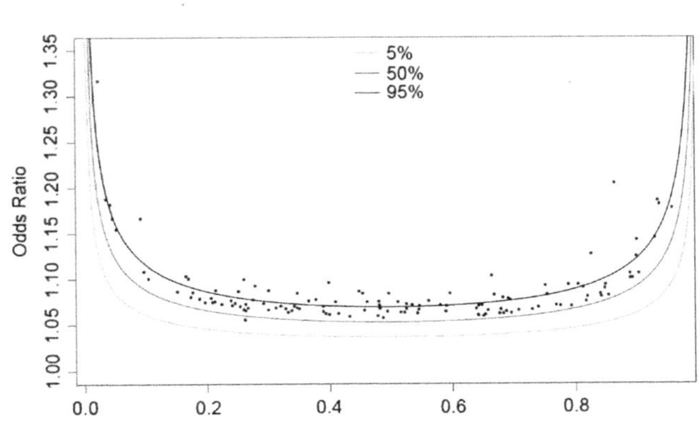

FIGURE 3.2 Schizophrenia risk allele frequency versus odds-ratio (PGC-SCZ, 2014). Colored lines represent power of the study to detect variation at specific (frequency, odds ratio) pairings.

An alternate approach to estimating heritability based on genome-wide association results was developed based solely on summary statistics. This approach, LD score regression (Bulik-Sullivan et al., 2015; PGC-SCZ, 2014), capitalizes on the premise that the more genetic variation an individual variant tags, the more likely it is to tag causal allele when the trait is polygenic. This approach allows for a rapid evaluation of genetic correlation across large-scale meta-analyses of complex traits. These methods have recapitulated the same correlations among neuropsychiatric diseases and then seamlessly extended them in a broad cross-disease study of GWAS summary statistics, confirming genetic correlations in distinct diseases with prior epidemiological support (eg, schizophrenia and inflammatory bowel disease).

This rapid evolution of findings and methodology from global efforts in psychiatric genetics now brings us the potential to bring our discussions of balancing selection from the realm of untestable philosophical conjecture to scientifically testable hypotheses. If we imagine there could be advantages in creativity, cognition, or other traits leading to success that might be conferred by genetic risk to schizophrenia or autism, we now have a direct means of inquiry, rather than the nonspecific and indirect assessment of sibling fecundity. We need only perform a powerful GWAS on the trait in the general population to explore the nature of the relationship between these traits and mental illness. These approaches including genetic risk scores, Mendelian randomization, and powerful genetic correlation techniques such as are implemented in GCTA to enable direct evaluation of these relationships. Intriguingly, although the concept is extremely new, two independent studies have suggested that the genetic risk for autism is positively correlated with genetic influences that raise IQ and, using LD–score regression correlation, increase educational attainment (Bulik-Sullivan et al., 2015). This almost paradoxical observation (given the substantial average IQ deficit in patients with autism) is nonetheless supported by some historical epidemiological and clinical observations (Baron-Cohen, Wheelwright, Skinner, Martin, & Clubley, 2001; Kanner, 1943) and is at the leading edge of what will likely be countless efforts in upcoming years to take a fresh and much more powerful look at the potential role of balancing selection in psychiatric and neurodevelopmental traits.

Other hypotheses by which balancing selection might play a role have not been similarly supported by emerging genetics. Balancing selection via sexual antagonism has become similarly evaluable across large-scale GWAS efforts of the PGC and others. Little evidence has emerged to support the idea that individual implicated loci show different let alone opposite effects on disease risk, and PRSs in total have not been shown to predict outcomes in males and females with any meaningful difference. As data sharing efforts of the PGC expand, this hypothesis will become conclusively evaluable over all psychiatric diseases, even though early analyses do not suggest that this is likely to be a major contributor.

That traits, particularly complex polygenic ones, might have been neutral or even positively selected for in ancient times, accounting for their modern prevalence, is a particularly challenging conjecture to test. To the extent that correlations between traits and diseases as described previously are now discoverable, it could certainly help focus hypotheses in this space as well. Such correlations may not confer significant discernible reproductive advantage in our modern environment and society, but they could in some cases be hypothesized to have been much more important for survival evolutionarily.

DE NOVO MUTATION AND MUTATION–SELECTION BALANCE

With decades of epidemiological observations indicating reduced fitness, a common, informal line of thinking took hold in much of psychiatric genetics. Specifically, because alleles contributing to neuropsychiatric disease have strong selection acting against them, common variation is unlikely to contribute because risk variants will never rise in frequency to become common, and if they happen to be common they are rapidly purged. As noted earlier, by example and in Table 3.1, selection against the disease does not equate with selection against the alleles predisposing to disease. Even in the absence of pleiotropic effects that might balance the effect on a reproductively deleterious disease, different penetrances and modes of inheritance attract the attention of natural selection with markedly different ardor. Regarding the unexpectedly large number of common alleles that make up a substantial portion of heritability, their individual contributions to risk are so modest that it would potentially take thousands of generations to move many of these to the ranks of the uncommon (consider a 0.50 frequency allele with an odds ratio of 1.05; after 1000 generations this is still a respectable common allele in the range of 0.20, assuming no other balancing effects or linked loci with opposite effects on risk).

As pertains to higher-penetrance alleles, however, the informal assumption that selection would not permit these to achieve common frequencies is consistent with theory. Among the earliest recognized genetic risk factors for neuropsychiatric disease were large and generally spontaneously arising (so-called de novo) microdeletions or microduplications (referred to collectively as CNVs). An unexpectedly common form of DNA variation, a global excess of such events in ASDs and schizophrenia was demonstrated in several early studies (Sebat et al., 2007), predating the polygenic risk findings described earlier and fueling further focus on rare and de novo mutation as a potential source of risk. Such major alleles have indeed been shown by themselves to result in reduced fecundity (Stefansson et al., 2014)), although importantly this study also points out a systematic flaw in our simplistic construction of penetrance–selection, as in Table 3.1. Specifically, whereas carriers of particular CNVs such as those recurrently observed at 22q11.21 and 16p11.2 are at high risk of developing ASDs or schizophrenia (perhaps odds ratios in the 10–20 range), those individuals who escape such diagnoses likely have other substantial impairments that are not selectively neutral, and thus even those high odds ratios may underestimate the force selection may bring to bear on them (Moreno-De-Luca, Moreno-De-Luca, Cubells, & Sanders, 2014).

Although only a small fraction of patients with neuropsychiatric disease carry a major genomic perturbation such as a CNV (Gaugler et al., 2014), the idea that more commonly observed de novo point mutations might have a similar but more substantial role was a logical next consideration. Given the emergence of remarkable advances in genome sequencing technology between 2005 and 2010, it became possible to sequence the exomes (the roughly 1% of the genome that encodes for proteins) efficiently in parents and affected children, and zero in on spontaneously arising mutations. Conveniently, the mutation rate of human genomes is such that each of us carries on the order of 100 new mutations, with the expectation that one will land in a protein-coding sequence. A wave of such studies in ASDs (De Rubeis et al., 2014; Iossifov et al., 2014) in particular have indicated that anywhere from 5% to 20% of cases of ASD may have a contribution to risk from a de novo point mutation. However, such mutations appear to have a larger role in intellectual disability and epilepsy and a more muted contribution to schizophrenia and other neuropsychiatric disease (Fromer et al., 2014; Samocha et al., 2014). Indeed when the ASD data are further dissected, it is clear that the vast majority of the excess resides in patients with concomitant substantial or severe intellectual disability, whereas higher-functioning patients demonstrate little if any excess of mutations but by contrast have a stronger family history of schizophrenia and other neuropsychiatric diseases (Robinson et al., 2014).

Regarding selection, the excess of de novo mutations appears distributed, like the polygenic risk, over many hundreds of genes. Strong-acting mutations in one or a small number of genes could not explain common diseases with reduced fitness simply because new mutation will not occur frequently enough to replace the celerity with which selection removes higher-penetrance mutations.

If we accept that the odds ratio = 20 allele could survive only 5 to 10 generations, mutation rates are now reasonably well established, and for an average gene we would estimate a rate of damaging mutations to be roughly 1 in 100,000 births and new mutation in that gene would therefore add a new case for every 500,000 individuals. If, however, we postulate 100–1000 such genes, mutation–selection balance could readily accommodate a contribution to severe mental illness similar to what is seen in autism.

The actual model by which the mutation–selection balance exerts effect will be more complex and vary from disease to disease. Even within ASDs, it is becoming clear that likely only a substantially smaller set of genes has high odds ratios of greater than 20 and most of these have a strong role in the far less heritable trait of severe intellectual disability, whereas the bulk of the mutational excess has a smaller impact on risk and will more often be inherited than de novo but will still (as in the estimates in Table 3.1) be held to very low frequencies. Nonetheless, this segment of the genomic architecture of disease, as with the polygenic common variation, is beginning to come into a more concrete focus, and with it a better picture of the forces of selection acting on it.

CONCLUSIONS

The compelling paradox of the high frequency of clearly reproductively deleterious mental illnesses continues to be a great focus of debate and study in behavioral epidemiology and psychiatric genetics. Although the complete etiology of neuropsychiatric disease is still in its early stages, substantial insights have been gained into genetic architecture that are advancing this discussion. Specifically, significant roles for both polygenic common variation (distributed over hundreds to thousands of common variants) and much rarer and de novo variants of intermediate or higher risk (distributed over many hundreds of genes) are now well established. More than the balance of the two categories, the sheer number in each case helps clarify some of the challenges in this debate, which began in a much earlier time when we considered single-gene models, or single genes making substantial individual contributions to common disease plausible.

In the case of rare variation, we can now extend the traditional mutation–selection balance hypotheses to a number of genes whereby mutation can plausibly keep up with the obvious swift scythe of selection. Given the equally if not more important parallel role of common, low-penetrance variation, this may not likely be the entire story. Consistent with the observed reduced fitness in schizophrenia, GWAS has begun to have dramatic success only when sample sizes began to permit detection of alleles with effects modest enough that they may rise to common frequencies. At the same time, the apparent uniformity of risk allele frequency and early hints that the polygenic risk in autism may correspond to genetic predisposition to higher cognitive performance suggest that balancing selection may well have a role in maintaining the polygenic signature that is also keeping mental illness common. Of importance to this hypothesis is that the polygenic combination of so many low-impact common alleles, individually and collectively of little concern to selection, does not demand that balancing selection come with increased fecundity among siblings as many prior studies have postulated.

We have much more to learn regarding the genetic architecture of neuropsychiatric disease, but even in advance of the articulation of all genetic risk factors, the patterns we learn from polygenic risk from GWAS, shared polygenic risk with other traits and diseases, and the specific role of rare variation derived from genome sequencing are greatly advancing the discourse on selection and mental illness. It seems at this point there is no definitive and simple answer (just as there is no one simple model by which genetic variation influences disease risk) but the emerging genetic architecture suggests that many of the proposed rejections of both mutation–selection balance and balancing selection were likely premature, and furthermore that the considerable difference between selection against disease and selection against alleles contributing to disease, particularly in extremely polygenic models, suggests that there may be less of a conundrum than once imagined.

REFERENCES

Baron-Cohen, S., Wheelwright, S., Skinner, R., Martin, J., & Clubley, E. (2001). The autism-spectrum quotient (AQ): evidence from Asperger syndrome/high-functioning autism, males and females, scientists and mathematicians. *Journal of Autism and Developmental Disorders, 31*, 5–17.

Bulik-Sullivan, B. K., Loh, P. R., Finucane, H. K., Ripke, S., Yang, J., Schizophrenia Working Group of the Psychiatric Genomics Consortium, … Neale, B. M. (March 2015). LD-score regression distinguishes confounding from polygenicity in genome-wide association studies. *Nature Genetics, 47*(3), 291–295. http://dx.doi.org/10.1038/ng. 3211. Epub February 2, 2015.

Cross-Disorder Group of the Psychiatric Genomics Consortium, Lee, S. H., Ripke, S., Neale, B. M., Faraone, S. V., Purcell, S. M., … Wray, N. R. (2013). Genetic relationship between five psychiatric disorders estimated from genome-wide SNPs. *Nature Genetics, 45*, 984–994.

Davey Smith, G., & Ebrahim, S. (2003). 'Mendelian randomization': can genetic epidemiology contribute to understanding environmental determinants of disease? *International Journal of Epidemiology, 32*, 1–22.

De Rubeis, S., He, X., Goldberg, A. P., Poultney, C. S., Samocha, K., Cicek, A. E., … Buxbaum, J. D. (November 13, 2014). Synaptic, transcriptional and chromatin genes disrupted in autism. *Nature, 515*(7526), 209–215.

Demyttenaere, K., Bruffaerts, R., Posada-Villa, J., Gasquet, I., Kovess, V., Lepine, J. P., … WHO World Mental Health Survey Consortium. (2004). Prevalence, severity, and unmet need for treatment of mental disorders in the World Health Organization World Mental Health Surveys. *Journal of the American Medical Association, 291*, 2581–2590.

Fromer, M., Pocklington, A. J., Kavanagh, D. H., Williams, H. J., Dwyer, S., Gormley, P., … O'Donovan, M. C. (2014). De novo mutations in schizophrenia implicate synaptic networks. *Nature, 506*(7487), 179–184.

Gaugler, T., Klei, L., Sanders, S. J., Bodea, C. A., Goldberg, A. P., Lee, A. B., … Buxbaum, J. D. (2014). Most genetic risk for autism resides with common variation. *Nature Genetics, 46*(8), 881–885.

Gray, R. G., & Wheatley, K. (1991). How to avoid bias when comparing bone marrow transplant with chemotherapy. *Bone Marrow Transplantation, 7*(Suppl. 3), 9–12.

Harris, E. C., & Barraclough, B. (1998). Excess mortality of mental disorder. *British Journal of Psychiatry, 173*, 11–53.

Haukka, J., Suvisaari, J., & Lonnqvist, J. (2003). Fertility of patients with schizophrenia, their siblings, and the general population: a cohort study from 1950 to 1959 in Finland. *American Journal of Psychiatry, 160*, 460–463.

Hiroeh, U., Appleby, L., Mortensen, P. B., & Dunn, G. (2001). Death by homicide, suicide, and other unnatural causes in people with mental illness: a population-based study. *Lancet, 358*, 2110–2112.

International HapMap Consortium. (2005). A haplotype map of the human genome. *Nature, 437*(7063), 1299–1320. http://www.ncbi.nlm.nih.gov/pubmed/16255080.

International Schizophrenia Consortium, Purcell, S. M., Wray, N. R., Stone, J. L., Visscher, P. M., O'Donovan, M. C., … Sklar, P. (2009). Common polygenic variation contributes to risk of schizophrenia and bipolar disorder. *Nature, 460*, 748–752.

Iossifov, I., O'Roak, B. J., Sanders, S. J., Ronemus, M., Krumm, N., Levy, D., … Wigler, M. (2014). The contribution of de novo coding mutations to autism spectrum disorder. *Nature, 515*(7526), 216–221.

Kanner, L. (1943). Autistic disturbances of affective contact. *Nervous Child, 2*, 217–250.

Keller, M. C., & Miller, G. (August 2006). Resolving the paradox of common, harmful, heritable mental disorders: which evolutionary genetic models work best. *Behavioral and Brain Sciences, 29*(4), 385–404.

Kessler, R. C., Berglund, P., Demler, O., Jin, R., Merikangas, K. R., & Walters, E. E. (2005). Lifetime prevalence and age-of-onset distributions of DSM-IV disorders in the National Comorbidity Survey Replication. *Archives of General Psychiatry, 62*, 593–602.

Kimura, M., & Ota, T. (September 1973). The age of a neutral mutant persisting in a finite population. *Genetics, 75*, 199–212.

Laursen, T. M., & Munk-Olsen, T. (2010). Reproductive patterns in psychotic patients. *Schizophrenia Research, 121*, 234–240.

Lewontin, R. (1972). The apportionment of human diversity. *Evolutionary Biology, 6*, 381–398.

Moreno-De-Luca, D., Moreno-De-Luca, A., Cubells, J. F., & Sanders, S. J. (2014). Cross-disorder comparison of four neuropsychiatric CNV loci. *Current Genetic Medicine Reports, 2*(3), 151–161.

Nock, M. K., Borges, G., Bromet, E. J., Alonso, J., Angermeyer, M., Beautrais, A., … Williams, D. (2008). Cross-national prevalence and risk factors for suicidal ideation, plans and attempts. *British Journal of Psychiatry, 192*, 98–105.

Power, R. A., Kyaga, S., Uher, R., MacCabe, J. H., Langstrom, N., Mikael, L., … Svensson, A. C. (2013). Fecundity of patients with schizophrenia, autism, bipolar disorder, depression anorexia nervosa, or substance abuse vs their unaffected siblings. *JAMA Psychiatry, 70*(1), 22–30.

Robinson, E. B., Samocha, K. E., Kosmicki, J. A., McGrath, L., Neale, B. M., Perlis, R. H., & Daly, M. J. (October 21, 2014). Autism spectrum disorder severity reflects the average contribution of de novo and familial influences. *Proceedings of the National Academy of Sciences of the United States of America, 111*(42), 15161–15165.

Samocha, K. E., Robinson, E. B., Sanders, S. J., Stevens, C., Sabo, A., McGrath, L. M., … Daly, M. J. (2014). A framework for the interpretation of de novo mutation in human disease. *Nature Genetics, 46*, 944–950.

Saugstad, L. F. (1989). Social class, marriage, and fertility in schizophrenia. *Schizophrenia Bulletin, 15*(1), 9–43.

Schizophrenia Working Group of the Psychiatric Genomics Consortium. (July 24, 2014). Biological insights from 108 schizophrenia associated loci. *Nature, 511*(7510), 421–427. http://dx.doi.org/10.1038/nature13595. Epub July 22, 2014.

Sebat, J., Lakshmi, B., Malhotra, D., Troge, J., Lese-Martin, C., Walsh, T., … Wigler, M. (2007). Strong association of de novo copy number mutations with autism. *Science, 316*(5823), 445–449.

Sebat, J., Levy, D. L., & McCarthy, S. E. (2009). Rare structural variants in schizophrenia: one disorder, multiple mutations; one mutation, multiple disorders. *Trends in Genetics, 25*(12), 528–535.

Stefansson, H., Meyer-Lindenberg, A., Steinberg, S., Magnusdottir, B., Morgen, K., Arnarsdottir, S., … Stefansson, K. (2014). CNVs conferring risk of autism or schizophrenia affect cognition in controls. *Nature, 505*(7483), 361–366.

Whiteford, H. A., Degenhardt, L., Rehm, J., Baxter, A. J., Ferrari, A. J., Erskine, H. E., … Vos, T. (2013). Global burden of disease attributable to mental and substance use disorders: findings from the Global Burden of Disease Study 2010. *Lancet, 382*, 1575–1586.

Yang, J., Manolio, T. A., Pasquale, L. R., Boerwinkle, E., Caporaso, N., Cunningham, J. M., … Visscher, P. M. (2011). Genome partitioning of genetic variation for complex traits using common SNPs. *Nature Genetics, 43*, 519–525.

Youngman, L. D., Keavney, B. D., Palmer, A., Parish, S., Clark, S., Danesh, J., … Collins, R. (2000). Plasma fibrinogen and fibrinogen genotypes in 4685 cases of myocardial infarction and in 6002 controls: test of causality by "Mendelian randomization". *Circulation, 102*(Suppl. II), 31–32.

Zuk, O., Schaffner, S. F., Samocha, K., Do, R., Hechter, E., Kathiresan, S., … Lander, E. S. (2014). Searching for missing heritability: designing rare variant association studies. *Proc Natl Acad Sci USA, 111*(4), E455–E464.

Chapter 4

Genome Tools and Methods: Rare Genetic Variation

Daniel J. Barrera and Jonathan Sebat
University of California San Diego, San Diego, CA, United States

RARE VARIANTS IN PSYCHIATRIC DISEASE

Rare variants have a storied history in the field of neurogenetics. Mapping of the first human disease gene was achieved by linkage analysis of large families with the autosomal dominant neurodegenerative disorder Huntington's disease (HD) (Gusella et al., 1983) and similar breakthroughs soon followed for a variety of other neurodegenerative disorders (Goate et al., 1991; Levy-Lahad et al., 1995; Sherrington et al., 1995) and movement disorders (Schöls, Bauer, Schmidt, Schulte, & Riess, 2004). Linkage analysis of large pedigrees of autosomal dominant Alzheimer's disease (AD), for example, led to the identification of amyloid precursor protein (Goate et al., 1991) and presenilins 1 (Schellenberg et al., 1992) and 2 (Levy-Lahad et al., 1995). Since their discovery, most autosomal dominant familial AD cases can be attributed to one of these three genes. Whereas such families are rare in AD, the identification of these genetic factors has contributed significantly to our understanding of AD pathogenesis.

Until recently, the rare genetic causes of mental disorders were not known. In contrast to dominant neurological disorders (Fig. 4.1A and B), behavioral and psychiatric traits do not segregate in families in a Mendelian fashion and therefore have been less amenable to linkage analysis. The familial segregation of severe psychiatric disorders differs from that of autosomal dominant HD and AD for variety of reasons. Chief among them is the early age at onset of disease and

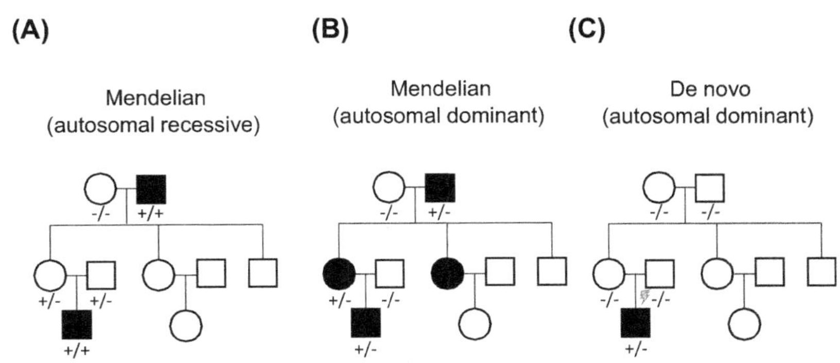

FIGURE 4.1 Modes of disease inheritance. (A) Family tree of the inheritance pattern of a recessive autosomal disease in which both deleterious alleles are required for disease presentation. Disease alleles are represented as plus signs and wild-type alleles are represented as minus signs. An affected individual (+/+) who produces offspring with an unaffected individual of genotype (−/−) will produce unaffected carriers with genotype (+/−). Offspring from (+/−) individuals will produce either affected (+/+), unaffected carriers (+/−), or disease allele-free (−/−) individuals depending on the genotype of their partner and the alleles contributed by each person. (B) Family tree of the inheritance pattern of an autosomal dominant disease. There are no unaffected carriers in autosomal dominant diseases because the deleterious allele is sufficient to cause disease even in the presence of the functional, wild-type allele. Affected individuals can still produce unaffected offspring if they contribute their wild-type allele and their partner is unaffected. (C) De novo inheritance is a form of autosomal dominant inheritance except that the disease allele is derived sporadically, often through a paternal germline mutation (mutation density increases with age). A lightning bolt indicates a de novo mutation not present in previous generations.

Genomics, Circuits, and Pathways in Clinical Neuropsychiatry. http://dx.doi.org/10.1016/B978-0-12-800105-9.00004-4

the effect of the disease on reproduction. For example, individuals with schizophrenia have far fewer children (Laursen and Munk-Olsen, 2010) than their healthy family members. Consequently, a severe dominant form of schizophrenia is less likely to segregate over multiple generations in a family.

For psychiatric disorders such as autism, schizophrenia, and bipolar disorder, genetic architectures have proven to be complex, spawning a lively debate as to the nature of this complexity (Klein, Xing, Mukherjee, Willis, & Hayes, 2010; McClellan & King, 2010). According to a common variant model, genetic risk can be explained by the additive effects of many common alleles (>5% in the population), and the evidence for this model is explored in detail elsewhere in this book. According to a rare variant model, many rare variants (<1% frequency) each may confer significant risk to an individual and in aggregate could explain a substantial proportion of risk in the population. The modern genomic methods for identifying rare variants and the study designs for investigating their impact on neuropsychiatric disease are detailed here.

As knowledge of the genetic basis of psychiatric disorders has emerged, we have begun to understand that rare functional variants of large effect are significant contributors to disease risk (Malhotra & Sebat, 2012) (Fig. 4.1C). This chapter will address the new genetic approaches for identifying causal rare variants and the emerging new understanding of the genetic basis of mental illness.

GENETIC VARIATION

Variation in the human genome comes in many flavors, including multiple forms of nucleotide sequence variation and structural variation. Fig. 4.2 provides a concise summary of the most prevalent categories of variation and the nomenclature that is used elsewhere in this book.

MODERN TECHNOLOGIES FOR DISCOVERY OF RARE VARIANTS

New genomic technologies have replaced linkage analysis for disease gene discovery. In contrast to traditional methods, modern approaches rely on the direct detection of causal variants in the genome. The characterization of common genetic variation based on single nucleotide polymorphism (SNP) genotypes is discussed in detail in chapter "Natural Selection and Neuropsychiatric Disease: Theory, Observation, and Emerging Genetic Findings." This chapter will focus on knowledge that has been gained from studies of rare genetic variation.

MICROARRAY TECHNOLOGY AND LARGE-SCALE COPY NUMBER VARIANTS IN NEURODEVELOPMENTAL AND NEUROPSYCHIATRIC DISEASE

Microarrays are a reliable method for high-throughput genotyping of common SNPs and detection of large-scale copy number variants (CNVs). A microarray consists of an array of many thousands or millions of short (<100-base pair) oligonucleotide probes. Each probe is complementary to a specific sequence in the human genome. DNA copy number can be determined by hybridizing the microarray with a fluorescently labeled genomic DNA sample and by measuring the probe signal intensities using optical techniques (Fig. 4.3A). A duplication or deletion of a genomic region in a given sample is evident as a segment of multiple probes with intensities significantly increased or decreased, respectively, relative to the average intensities of reference samples (where the typical copy number is 2) (Fig. 4.3B).

Computational methods for the normalization of probe intensities and CNV calling (Xu, Hou, Bickhart, Song, & Liu, 2013) have become increasingly reliable, thus enabling the analysis of CNV across large samples.

COPY NUMBER VARIANT STUDIES OF DISEASE

Current genotyping platforms and CNV discovery algorithms enable the genotyping of 1000 common copy number polymorphisms and the discovery of additional new CNVs, including mutations that are rare or unique to an individual (Alkan, Coe, & Eichler, 2011). These rare CNVs have provided the first glimpse into the many rare mutations that contribute to common psychiatric disease. New findings have begun to emerge from genome-wide studies of CNV in three major psychiatric disorders: autism spectrum disorders (ASDs), schizophrenia, and bipolar disorder.

Two study designs in particular have been effective in elucidating the rare genetic variation that contributes to disease. These approaches were originally pioneered within the context of microarray-based studies of CNV. Similar strategies are now being applied in the context of high-throughput sequencing.

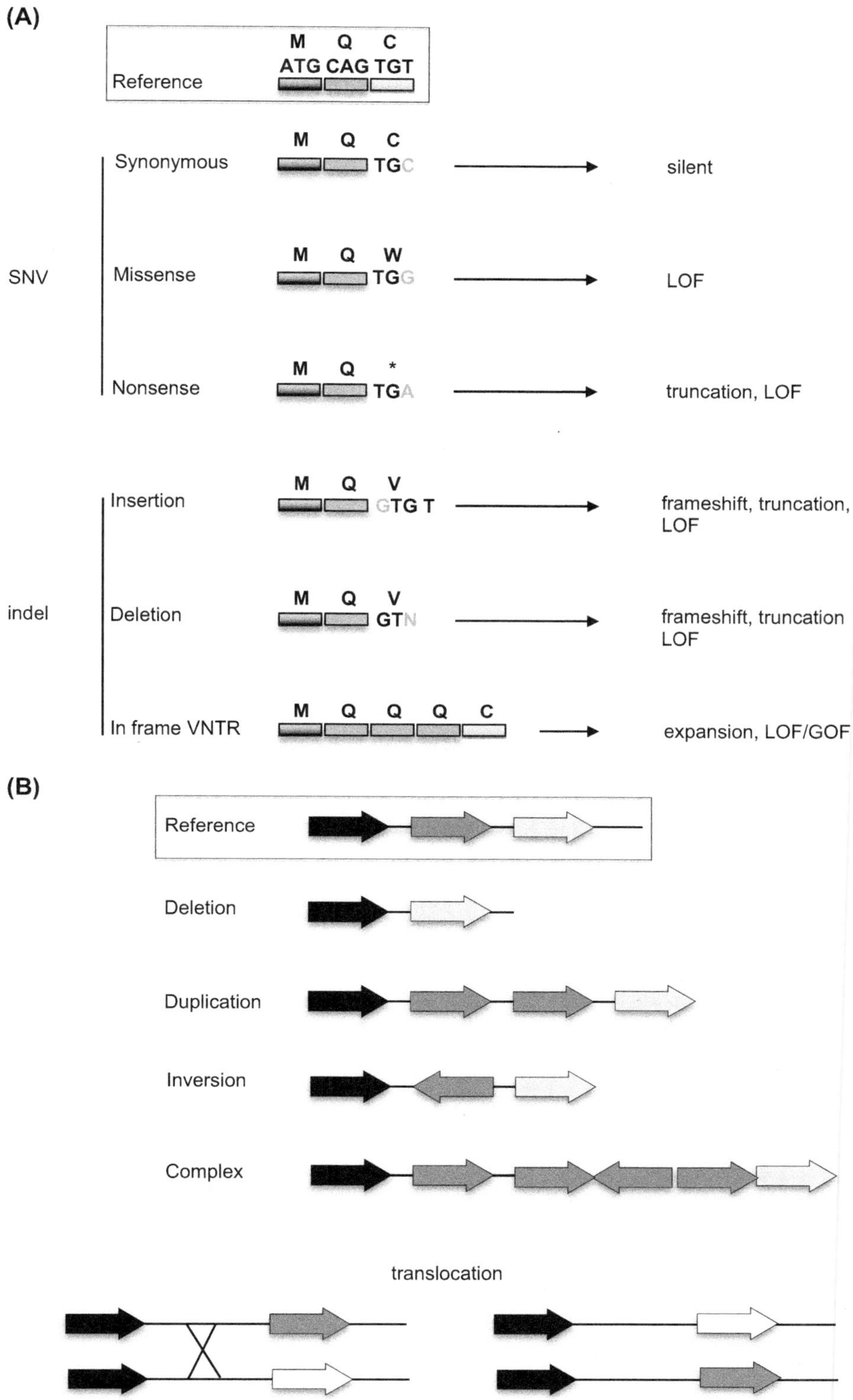

FIGURE 4.2 Classes of mutation which lead to genetic variation. (A) Reference sequence with three codons encoding the M, Q, and C amino acids are subject to point mutations (also known as single-nucleotide variants [SNVs]) or small insertions or deletions (indels) which create genetic variation. SNVs can produce the same amino acid if a degenerate base of the codon is mutated (synonymous), a different amino acid if a critical base is mutated (missense), or a truncated protein if mutated into a stop codon (nonsense). Indels generally produce truncated, nonfunctional proteins by inserting or removing base pairs out of frame with the start codon. If multiples of three bases are inserted or deleted, an in-frame amino acid addition or deletion can occur, altering the peptide sequence but still producing nontruncated protein. Variable nucleotide tandem repeats (VNTRs) can produce in-frame expansions of amino acid sequences; these are at the core of several neurodegenerative diseases. (B) Structural variation is larger scale sequence diversity than SNVs or indels and can cover multiple genes. Several types of structural variation are presented in relation to a reference chromosome segment with a *black*, *a dark gray*, and *a light gray gene*. The *dark gray gene* is deleted, duplicated, inverted, or subjected to complex rearrangement with multiple types of structural variation of the *dark gray gene*. Translocation is presented with a set of reference sequences because this type of structural variation occurs between different chromosomes. *LOF*, loss-of-function; *GOF*, gain-of-function.

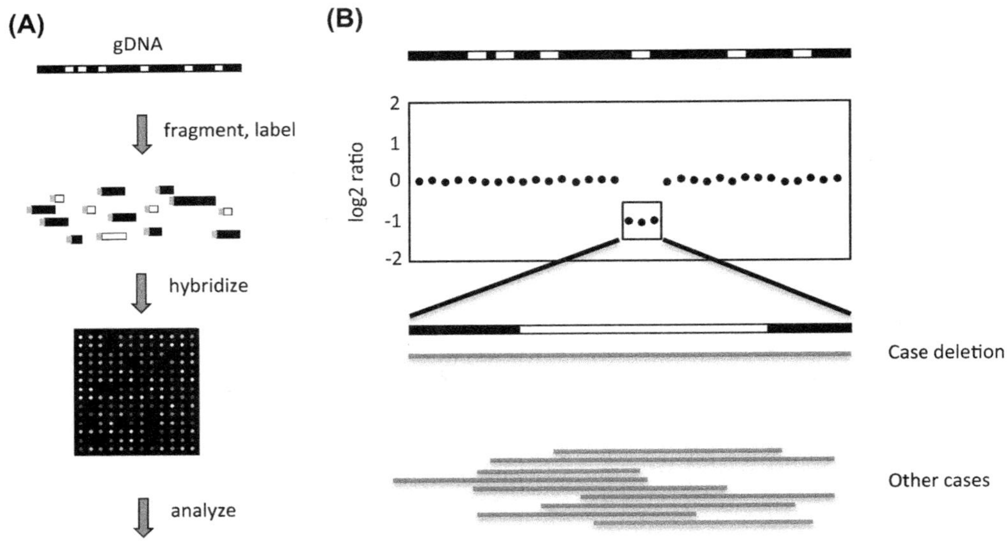

FIGURE 4.3 Microarray technology can detect enrichments of copy number variation in disease. (A) Technical workflow process for data generation from microarrays. Genomic DNA (gDNA) is fragmented and labeled with fluorescent tags before being applied to an array with complementary probe sequences covalently attached to a surface. The arrays are scanned with a fluorescence scanner to generate signals corresponding to relative presences of target sequences (*dot matrix*). Probes that anneal to a target sequence will produce an optical signal when interrogated with an excitation light source (*gray dots*), whereas the limited presence or absence of a target sequence will produce a weak signal or none at all (*black or dark gray dots*). A high relative optical signal indicates that more of a target sequence is present (*white dots*). Computational methods are used to normalize signal intensities and infer target abundance. (B) Illustration of a case deletion from a patient with a neuropsychiatric disease. Log2 ratios are plotted and a deletion of a gene (*white box*) is evident from a reduced signal originating from three contiguous probes that fall within the same region. The prevalence of similar deletions from other affected individuals covering the same region in a cohort implicates that region in the respective disease.

STUDIES OF DE NOVO MUTATION IN TRIO FAMILIES

Neurodevelopmental disorders and severe psychiatric illness are characterized by high heritability, early age at onset, and significantly reduced fecundity. As we have come to recognize, such characteristics are consistent with an etiology composed of de novo (spontaneous) mutations. A series of studies by our group and others have demonstrated a significant association of de novo CNVs with schizophrenia (Karayiorgou, Simon, & Gogos, 2010; Malhotra et al., 2011), early-onset bipolar disorder (Malhotra et al., 2011), autism (Sebat et al., 2007), and intellectual disability (Hehir-Kwa et al., 2011). The aggregate frequency of de novo CNVs in patients ranges from 5% to 10% and is several times higher than the frequency in control subjects (1–2%). Thus, although this class of variants does not explain a large fraction of cases, the large effect that we observe is consistent with de novo CNVs contributing significant risk to individual patients.

CASE–CONTROL STUDIES

We have also shown that independent of the mode of inheritance, there is a genome-wide enrichment of rare CNVs in patient cohorts with schizophrenia compared with the general population (Walsh et al., 2008). Thus a similar approach can be applied to the analysis of large patient and control cohorts, including large array datasets that have been collected in conjunction with genome-wide association studies of several psychiatric disorders, including schizophrenia (Schizophrenia Working Group of the Psychiatric Genomics, 2014), bipolar disorder (Green et al., 2015), and major depression (O'Dushlaine et al., 2014).

ANALYSIS STRATEGIES

Using either of these designs, the association of rare CNVs with disease traits can be interrogated at multiple levels: for example, across the genome (Fig. 4.4A), pathways, (Fig. 4.4B), or genes (Fig. 4.4C) or at an individual SNV (Fig. 4.4D). Analysis of CNV has thus has yielded key insights into the genetics of disease at all levels.

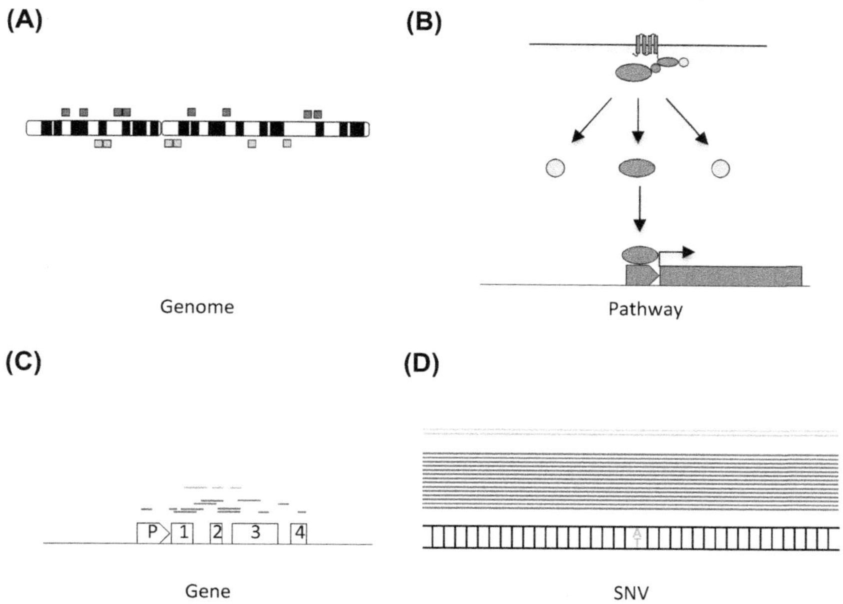

FIGURE 4.4 Analysis strategies for case control or family design studies of disease genetics. (A) A chromosome with *dark gray boxes* marking case CNV loci against *light gray boxes* marking control CNV loci. (B) Pathway analysis attempts to identify enrichments of CNVs in genetic pathways for cases versus controls. *Dark gray shapes* represent gene products affected by significantly more CNVs in cases than in controls. *Light gray shapes* are gene products without an enrichment of CNVs for cases. Enrichments are detected by assessing CNV burden in a randomized pathway of similar size. (C) An enrichment in unique CNVs detected from many case samples (*dark gray lines*) helps implicate a single gene when there are significantly fewer CNVs detected in the same gene for control samples (*light gray lines*). (D) In contrast to the previous example, when there are statistically enough case CNVs across an SNV (*dark gray lines*) compared with case CNVs (*light gray lines*), a single polymorphism can be implicated in the disease.

Genome

The contribution of risk CNVs to disease can be examined by comparing the aggregate burden of deletion or duplication in cases and controls. An enrichment of large (>100 kb) CNVs in patients has been reported in schizophrenia (I.S. Consortium, 2008; Walsh et al., 2008), autism (Pinto et al., 2010), and bipolar disorder (Zhang et al., 2008). Measuring a genetic association based on CNV burden has obvious limitations. When all deletions across the genome are collapsed into a single category, the genetic effect can be measured only in terms of the excess of risk alleles in patients. By contrast, "protective" alleles (ie, variants associated with control subjects) will not be captured and their presence in the dataset will tend to diminish the overall effect. Rare protective alleles appear to influence case–control status to some extent (Rees et al., 2014). Accurate estimation of the combined genetic effect of rare CNVs will require more complete knowledge of the directionality and effect size of the individual variants.

Pathways

A pathway-based analysis of CNV seeks to quantify the enrichment of deletions and duplications accurately across specific gene set by comparing the observed CNV count in a pathway with a null model that accurately reflects the case–control difference in CNV burden that would be expected for a random pathway with the same number and sizes of genes. Findings indicate that risk CNVs are not randomly distributed in the genome and that CNV burden is concentrated within pathways related to neurodevelopment (Kirov et al., 2012; Pinto et al., 2014; Walsh et al., 2008). For instance, pathways implicated in autism include genes involved in synapse function, chromatin regulation, and transcripts that are bound by Fragile X mental retardation protein (FMRP) (Pinto et al., 2014). Pathways implicated in schizophrenia include proteins located at the postsynaptic density and components of the activity-regulated cytoskeleton complex (Kirov et al., 2012).

Genes

Because of low frequencies of the individual variants, statistical power to detect the association for a specific allele may be limited. In some cases additional power can be gained by testing association across individual genes.

SNVs

For some genomic loci, risk CNVs occur at sufficient frequencies to test the association at individual sites. In autism and schizophrenia, several loci have been specifically implicated (for review see Malhotra & Sebat, 2012).

Although these approaches were pioneered in the context of CNV studies, the same principles apply to any mutation discovery platform including exome and whole genome sequencing (WGS).

DETECTION OF RARE GENETIC VARIATION BY HIGH-THROUGHPUT SEQUENCING

High-throughput sequencing technologies carry out massively parallel sequencing of shotgun libraries. In contrast to traditional Sanger sequencing, the technical workflow for next-generation sequencing exchanges bacterial cloning for a library preparation step which involves ligating universal amplification primers onto target DNA that has been mechanically or enzymatically sheared to a length appropriate for the platform. Generally the target library is then annealed to immobilized primers that are complementary to the universal sequencing adapters. A variety of different chemistries are applied to generate signals that represent nucleotide addition during elongation, and these signals are interpreted as sequence. Unlike Sanger sequencing, the nucleotide addition and detection steps are incorporated into a single process, and the physical separation of immobilized templates allows upward of billions of unique, elongating DNA molecules to be recorded at the same time.

High-throughput sequencing has allowed researchers to examine sequence diversity that is not limited to predefined arrays. This makes the technology particularly useful for identifying rare variants that are unaccounted for through array technologies. As the cost per accurate base call lowers, the ability to perform WGS of very large sample sizes will enable association studies with rare variants. Currently, read length and depth required to confidently call a base are the biggest technological limiting factors hindering these studies. A combination of exome sequencing (Fig. 4.5A), targeted sequencing (Fig. 4.5B), and WGS (Fig. 4.5C) is currently being employed to elucidate the genetic underpinnings behind neuropsychiatric diseases.

Exome

It is currently more cost-effective to sequence an informed subset of the genome selectively. For this reason, researchers have focused on whole exome sequencing, which reduces complexity to 1% of the entire genome. Genetic variation within protein-coding regions of the genome is of high interest because mutations within these regions are more likely to have functional consequences. Exon capture and sequencing is economical and sequencing depth is not sacrificed as quickly when sample size increases compared with WGS.

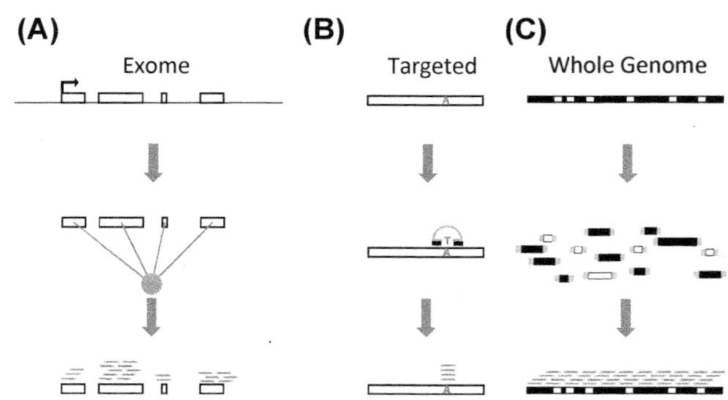

FIGURE 4.5 High-throughput DNA sequencing applications. (A) Whole exome sequencing attempts to selectively sequence the 1% of the genome that encodes mature transcripts. Genomic DNA is fragmented and exons (*white boxes*) enriched through a combination of hybridization to bead probe conjugates (*gray circle*) and amplification by polymerase chain reaction, leave intergenic and intronic DNA. The resulting sequence information after high-throughput sequencing exclusively covers the exons (*gray lines* above exons). (B) Targeted sequencing with molecular inversion probes (MIPs) selectively sequences regions inherent in the design of the MIPs. In this simplified example, an MIP is designed against an SNV, and the A allele is confirmed by reading the T that is incorporated through a gap-filling process. It is now possible to design MIPs with targeting arms that are farther apart, and capture a larger region through polymerization and ligation (circularization of MIP). (C) Whole genome sequencing is able to assess variation within the accessible genome. Genomic DNA is fragmented (*black and white boxes* representing intergenic and genic space, respectively) and ligated to adapter sequences (*gray flanking boxes*) specific to the high-throughput sequencing platform used. After high-throughput sequencing is performed, the reads can be assembled into larger contigs and matched to a reference sequence.

Using the analytical approaches similar to those described previously, whole exome sequencing has been successfully applied to relatively large trio and case–control samples of ASD and schizophrenia. In this context, however, the relevant genes can be elucidated with much greater precision.

In particular, trio-based analysis of de novo mutation has proven to be a highly effective approach to gene discovery. The observation of recurrent de novo mutations in individual genes can provide strong statistical evidence for association. In this context, the association of recurrent gene mutations with disease is assessed by estimating the probability of observing a given number of mutations per gene by chance, taking into consideration the per-base mutation rate, gene length, guanine and cytosine content, and sample size (Sanders et al., 2012). In a sample of 1000 trios, for example, there is a low (<0.05) probability of observing four mutations in any one gene throughout the genome (Sanders et al., 2012). Thus, any mutation observed at this frequency in cases would achieve genome-wide significance. Recurrent de novo protein-disrupting mutations in genes have provided credible evidence implicating specific genes in ASD (Iossifov et al., 2012; Neale et al., 2012; Sanders et al., 2012) and have greatly strengthened the evidence implicating synaptic networks, genes involved in the regulation of chromatin, and targets of FMRP (Iossifov et al., 2014). Exome sequencing of trio samples in schizophrenia have similarly strengthened the evidence for synaptic proteins (Fromer et al., 2014).

Gene Panels

Highly economical sequencing of small panels of genes (typically fewer than 50) is possible using specialized target capture procedures such as "padlock" or molecular inversion probes (MIPs). This has proven useful for resequencing candidate disease genes that have been implicated from exome and CNV studies. In ASD, there are hundreds of implicated genes with a small number of recurrent mutations. Thus, whereas more exomes may help expand the candidate gene list, targeted gene sequencing can extend coverage of specific gene sets and help identify recurrent mutations.

One example of this comes from a study that used MIPs to resequence a prioritized list of 44 ASD candidate risk genes in 2446 probands (O'Roak et al., 2012). Using a family-based trio design looking for de novo variation, recurrent sporadic mutations in six genes were found. This study also finds new exonic de novo mutations that were not found in exome sequencing studies as a result of the increased coverage inherent in sequencing smaller, focused target sets.

Whole Genome Sequencing

Sequencing of the complete human genome at greater than 30 times coverage provides accurate detection of nucleotide variants across 85% of the reference genome (G.P. Consortium, 2010). This portion is referred to as the "accessible" genome. The inaccessible portion consists mainly of highly repetitive sequences, and a significant fraction of reads in these regions have a low mapping quality. WGS enables a far more complete ascertainment of common and rare variation throughout the genome, including structural variation and variation in nucleotide sequence (G.P. Consortium, 2010). Furthermore, the cost of high-throughput sequencing has fallen to a point that enables WGS of thousands of samples (Gudbjartsson et al., 2015; Michaelson et al., 2012; Genome of the Netherlands Consortium, 2014).

To date, multiple small studies of autism have been published (Michaelson et al., 2012; Yuen et al., 2015). The focus of these small studies has been to explore new genetic questions that can be addressed using complete genomic information on families. For instance, our team examined global patterns of de novo germline mutation in autism by WGS of identical twins concordant for ASD autism. We showed that variation in global mutation rates are strongly influenced by paternal age. We further showed that local mutation rates are influenced by intrinsic properties of DNA sequencing and chromatin structure. Two studies by Scherer and colleagues sought to apply a clinical genetics perspective to the interpretation of genome sequence data. The first, a study of 32 families (Jiang et al., 2013), performed expert curation of variants in coding regions and yielded clinically interpretable variants in about 50% of cases. However, a lack of strong statistical support for most of these variants makes it difficult to assess the accuracy of the interpretations. The second study expanded this approach to 85 "multiplex" families with multiple affected offspring. The results of this second study were fraught with similar challenges. Expert curation of genetic variants yielded a variety of putative causal variants, most which were not shared among multiple affected family members, thus raising doubt about their causal role. The results of this study highlight the difficulties in trying to infer pathogenicity of genetic variants (Yuen et al., 2015). Every genome sequence possesses remarkable narrative potential because every human genome contains a variety of CNVs and protein-altering variants in genes that can plausibly be associated with disease.

Although only a limited number of WGS studies have been published, numerous studies are ongoing in autism, schizophrenia, and bipolar disorder, and in the general population (Gudbjartsson et al., 2015). We are likely to see a flood of new data emerging in the coming years.

RARE GENETIC VARIATION IN NONCODING DNA: A NEW FRONTIER

The strategies described here have been successful. Knowledge of rare genetic contributors to risk is growing rapidly along with an improved understanding of the molecular pathways underlying mental illness. However, large gaps remain in our knowledge.

Studies have focused on coding sequence variation, largely for economic reasons. A focus on protein-coding sequences is also justified because it is possible to infer with some degree of accuracy whether changes in protein sequences are deleterious. However, protein coding sequence represents only 1% of the human genome. It has been has been more difficult to explore and interpret noncoding DNA, which represents 99% of the genome. The DNA sequences that comprise this "dark matter" perform a variety of regulatory functions pertaining to transcriptional regulation and chromatin accessibility (E.P. Consortium, 2007). We are only beginning to understand the underlying genetic mechanisms and how they are influenced by genetic variation. Analysis of genetic regulatory elements is becoming increasingly tractable. Genes are governed by promoters, enhancers, untranslated regions, and noncoding RNAs; these can be prioritized based on what genes have been previously implicated and then targeted for sequencing in case–control or family design studies. Then, variant loci that are enriched in affected individuals can be tested for functionality using reporter assays.

RARE GENETIC VARIANTS AND THE POTENTIAL TO AFFECT CLINICAL CARE

Rare and de novo mutations at specific genetic loci may confer substantial risk to an individual patient. In addition, specific clinical features may be associated with a given genotype (Sebat, Levy, & McCarthy, 2009). Thus rare variants of large effect could have a significant impact on the diagnoses and treatment of mental disorders. Future progress will continue to depend heavily on new genomic technologies and their application in the clinical arena.

The development of cell-based and animal models of genetic disorders has been greatly accelerated by the development of highly efficient gene editing techniques, in particular the CRISPR/Cas9 system (Cong et al., 2013; Mali et al., 2013). The extraordinary efficiency of this system has made it possible to introduce a variety of specific mutations into mammalian genomes, including nucleotide substitutions, indels, and structural variants (Kraft et al., 2015). Model systems that faithfully recapitulate human mutations will enable the design of novel therapeutics based on the underlying molecular pathology of disease.

By design, personalized treatments that emerge from such studies will be effective for specific patient groups. Therefore, from the early phases of clinical trials to the eventual application of such treatments, diagnostic testing is needed to identify genetic subgroups (Brandler & Sebat, 2015). Diagnostic methods will require more accurate prediction algorithms for inferring the functional impact of mutations in both coding and noncoding regions. In addition, if the set of genotypes that defines a diagnostic group is rare, genetic screening must be applied to large populations to identify them. Hence routine clinical genetic testing of patient populations ultimately will be required to make good on the promise of precision medicine.

REFERENCES

Alkan, C., Coe, B. P., & Eichler, E. E. (2011). Genome structural variation discovery and genotyping. *Nature Reviews Genetics, 12*(5), 363–376.

Brandler, W. M., & Sebat, J. (2015). From de novo mutations to personalized therapeutic interventions in autism. *Annual Review of Medicine, 66*(1), 487–507.

Cong, L., Ran, F. A., Cox, D., Lin, S., Barretto, R., Habib, N., … Zhang, F. (2013). Multiplex genome engineering using CRISPR/Cas systems. *Science, 339*(6121), 819–823.

Consortium, E. P. (2007). Identification and analysis of functional elements in 1% of the human genome by the ENCODE pilot project. *Nature, 447*(7146), 799–816.

Consortium, G. P. (2010). A map of human genome variation from population-scale sequencing. *Nature, 467*(7319), 1061–1073.

Consortium, I. S. (2008). Rare chromosomal deletions and duplications increase risk of schizophrenia. *Nature, 455*(7210), 237–241.

Fromer, M., Pocklington, A. J., Kavanagh, D. H., Williams, H. J., Dwyer, S., Gormley, P., … O'Donovan, M. C. (2014). De novo mutations in schizophrenia implicate synaptic networks. *Nature, 506*(7487), 179–184.

Genome of the Netherlands Consortium. (2014). Whole-genome sequence variation, population structure and demographic history of the Dutch population. *Nature Genetics, 46*(8), 818–825.

Goate, A., Chartier-Harlin, M.-C., Mullan, M., Brown, J., Crawford, F., Fidani, L., … Hardy, J. (1991). Segregation of a missense mutation in the amyloid precursor protein gene with familial Alzheimer's disease. *Nature, 349*(6311), 704–706.

Green, E. K., Rees, E., Walters, J. T. R., Smith, K. G., Forty, L., Grozeva, D., … Kirov, G. (2015). Copy number variation in bipolar disorder. *Molecular Psychiatry, 21*(1), 89–93.

Gudbjartsson, D. F., Helgason, H., Gudjonsson, S. A., Zink, F., Oddson, A., Gylfason, A., … Stefansson, K. (2015). Large-scale whole-genome sequencing of the Icelandic population. *Nature Genetics, 47*(5).

Gusella, J. F., Wexler, N. S., Conneally, P. M., Naylor, S. L., Anderson, M. A., Tanzi, R. E., … Martin, J. B. (1983). A polymorphic DNA marker genetically linked to Huntington's disease. *Nature, 306*(5940), 234–238.

Hehir-Kwa, J. Y., Rodríguez-Santiago, B., Vissers, L. E., de Leeuw, N., Pfundt, R., Buitelaar, J. K., … Veltman, J. A. (2011). De novo copy number variants associated with intellectual disability have a paternal origin and age bias. *Journal of Medical Genetics, 48*(11), 776–778.

Iossifov, I., O'Roak, B. J., Sanders, S. J., Ronemus, M., Krumm, N., Levy, D., … Wigler, M. (2014). The contribution of de novo coding mutations to autism spectrum disorder. *Nature, 515*(7526), 216–221.

Iossifov, I., Ronemus, M., Levy, D., Wang, Z., Hakker, I., Rosenbaum, J., … Wigler, M. (2012). De novo gene disruptions in children on the autistic spectrum. *Neuron, 74*(2), 285–299.

Jiang, Y. H., Yuen, R. K. C., Jin, X., Wang, M., Chen, N., Wu, X., … Scherer, S. W. (2013). Detection of clinically relevant genetic variants in autism spectrum disorder by whole-genome sequencing. *American Journal of Human Genetics, 93*(2), 249–263.

Karayiorgou, M., Simon, T. J., & Gogos, J. A. (2010). 22q11.2 microdeletions: linking DNA structural variation to brain dysfunction and schizophrenia. *Nature Reviews. Neuroscience, 11*(6), 402–416.

Kirov, G., Pocklington, A. J., Holmans, P., Ivanov, D., Ikeda, M., Ruderfer, D., … Owen, M. J. (2012). De novo CNV analysis implicates specific abnormalities of postsynaptic signalling complexes in the pathogenesis of schizophrenia. *Molecular Psychiatry, 17*(2), 142–153.

Klein, R. J., Xing, X., Mukherjee, S., Willis, J., & Hayes, J. (2010). Successes of genome-wide association studies. *Cell, 142*(3), 350–351.

Kraft, K., Geuer, S., Will, A. J., Chan, W. L., Paliou, C., Borschiwer, M., … Andrey, G. (2015). Deletions, inversions, duplications: Engineering of structural variants using CRISPR/Cas in mice. *Cell Reports, 10*(5), 833–839.

Laursen, T. M., & Munk-Olsen, T. (2010). Reproductive patterns in psychotic patients. *Schizophrenia Research, 121*(1–3), 234–240.

Levy-Lahad, E., Wasco, W., Poorkaj, P., Romano, D., Oshima, J., Pettingell, W. H., … Wang, K. (1995). Candidate gene for the chromosome 1 familial Alzheimer's disease locus. *Science, 269*(5226), 973–977.

Malhotra, D., McCarthy, S., Michaelson, J. J., Vacic, V., Burdick, K. E., Yoon, S., … Sebat, J. (2011). High frequencies of de novo CNVs in bipolar disorder and schizophrenia. *Neuron, 72*(6), 951–963.

Malhotra, D., & Sebat, J. (2012). CNVs: harbingers of a rare variant revolution in psychiatric genetics. *Cell, 148*(6), 1223–1241.

Mali, P., Yang, L., Esvelt, K. M., Aach, J., Guell, M., DiCarlo, J. E., … Church, G. M. (2013). RNA-guided human genome engineering via Cas9. *Science, 339*(6121), 823–826.

McClellan, J., & King, M. C. (2010). Genetic Heterogeneity in human disease. *Cell, 141*(2), 210–217.

Michaelson, J. J., Shi, Y., Gujral, M., Zheng, H., Malhotra, D., Jin, X., … Sebat, J. (2012). Whole genome sequencing in autism identifies hotspots for de novo germline mutation. *Cell, 151*(7), 1431–1442.

Neale, B. M., Kou, Y., Liu, L., Ma'ayan, A., Samocha, K. E., Sabo, A., … Daly, M. J. (2012). Patterns and rates of exonic de novo mutations in autism spectrum disorders. *Nature, 485*(7397), 242–245.

O'Dushlaine, C., Ripke, S., Ruderfer, D. M., Hamilton, S. P., Fava, M., Iosifescu, D. V., … Perlis, R. H. (2014). Rare copy number variation in treatment-resistant major depressive disorder. *Biological Psychiatry, 76*(7), 536–541.

O'Roak, B. J., Vives, L., Fu, W., Egertson, J. D., Stanaway, I. B., Phelps, I. G., … Shendure, J. (2012). Multiplex targeted sequencing identifies recurrently mutated genes in autism spectrum disorders. *Science, 338*(6114), 1619–1622.

Pinto, D., Delaby, E., Merico, D., Barbosa, M., Merikangas, A., Klei, L., … Scherer, S. W. (2014). Convergence of genes and cellular pathways dysregulated in autism spectrum disorders. *American Journal of Human Genetics, 94*(5), 677–694.

Pinto, D., Pagnamenta, A. T., Klei, L., Anney, R., Merico, D., Regan, R., … Betancur, C. (2010). Functional impact of global rare copy number variation in autism spectrum disorders. *Nature, 466*(7304), 368–372.

Rees, E., Kirov, G., Sanders, A., Walters, J. T. R., Chambert, K. D., Shi, J., … Owen, M. J. (2014). Evidence that duplications of 22q11.2 protect against schizophrenia. *Molecular Psychiatry, 19*(1), 37–40.

Sanders, S. J., Murtha, M. T., Gupta, A. R., Murdoch, J. D., Raubeson, M. J., Willsey, A. J., … State, M. W. (2012). De novo mutations revealed by whole-exome sequencing are strongly associated with autism. *Nature, 485*(7397), 237–241.

Schellenberg, G., Bird, T., Wijsman, E., Orr, H., Anderson, L., Nemens, E., … Alonso, M. E. (1992). Genetic linkage evidence for a familial Alzheimer's disease locus on chromosome 14. *Science, 258*(5082), 668–671.

Schizophrenia Working Group of the Psychiatric Genomics Consortium. (2014). Biological insights from 108 schizophrenia-associated genetic loci. *Nature, 511*(7510), 421–427.

Schöls, L., Bauer, P., Schmidt, T., Schulte, T., & Riess, O. (2004). Autosomal dominant cerebellar ataxias: clinical features, genetics, and pathogenesis. *The Lancet Neurology, 3*(5), 291–304.

Sebat, J., Lakshmi, B., Malhotra, D., Troge, J., Lese-Martin, C., Walsh, T., … Wigler, M. (2007). Strong association of de novo copy number mutations with autism. *Science, 316*(5823), 445–449.

Sebat, J., Levy, D. L., & McCarthy, S. E. (2009). Rare structural variants in schizophrenia: one disorder, multiple mutations; one mutation, multiple disorders. *Trends in Genetics, 25*(12), 528–535.

Sherrington, R., Rogaev, E. I., Liang, Y., Rogaeva, E. A., Levesque, G., Ikeda, M., … St. George-Hyslop, P. H. (1995). Cloning of a gene bearing missense mutations in early-onset familial Alzheimer's disease. *Nature, 375*(6534), 754–760.

Walsh, T., McClellan, J. M., McCarthy, S. E., Addington, A. M., Pierce, S. B., Cooper, G. M., … Sebat, J. (2008). Rare structural variants disrupt multiple genes in neurodevelopmental pathways in schizophrenia. *Science, 320*(5875), 539–543.

Xu, L., Hou, Y., Bickhart, D., Song, J., & Liu, G. (2013). Comparative analysis of CNV calling algorithms: literature survey and a case study using bovine high-density SNP data. *Microarrays, 2*(3), 171–185.

Yuen, R. K. C., Thiruvahindrapuram, B., Merico, D., Walker, S., Tammimies, K., Hoang, N., … Scherer, S. W. (2015). Whole-genome sequencing of quartet families with autism spectrum disorder. *Nature Medicine, 21*(2), 185–191.

Zhang, D., Cheng, L., Qian, Y., Alliey-Rodriguez, N., Kelsoe, J. R., Greenwood, T., … Gershon, E. S. (2008). Singleton deletions throughout the genome increase risk of bipolar disorder. *Molecular Psychiatry, 14*(4), 376–380.

Chapter 5

Neuroepigenomics and Human Disease

John M. Greally
Albert Einstein College of Medicine, Bronx, NY, United States

Eric J. Nestler
Icahn School of Medicine at Mount Sinai, New York, NY, United States

EPIGENETICS

Most of the time when the word "epigenetics" has been defined, it has nothing to do with the concept originally proposed by Waddington. This issue was excellently reviewed (Gilbert, 2012) but bears repeating. Waddington's training in developmental biology and appreciation of the potential for genetic mutations to cause disease put him at the intersection of two fields of research: embryology (which he referred to as epigenesis) and the emerging field of genetics. The conflict between the groups was based on a simple question: If genes are involved in embryogenesis and the DNA content of each cell is the same, how can the DNA cause different cell types to be created? Waddington's goal was to reconcile the fields of epigenesis and genetics with what he called an "epigenetic" model of genes influencing cell lineage commitment but without proposing how such lineage commitments could be mediated by variability in function of the same DNA.

In the interim, we have learned that the fixed information content of DNA is used in different ways in different cell types, associated with a large number of distinctive nuclear events, including those involving chromatin structure and nucleosomal organization, posttranslational modifications of chromatin proteins, covalent modifications of DNA, and transcription factor (TF) binding to DNA with recruitment of enzymatic complexes influencing local chromatin organization, all resulting in cell type–specific transcriptional regulatory patterns with subsequent feedback from small and long noncoding RNA species to influence these regulatory events (Choukrallah & Matthias, 2014; Ciabrelli & Cavalli, 2014; Delatte, Deplus, & Fuks, 2014; Henikoff & Smith, 2015; Hiragami-Hamada & Fischle, 2014; Liebers, Rassoulzadegan, & Lyko, 2014; Messerschmidt, Knowles, & Solter, 2014; Morris & Mattick, 2014; Peschansky & Wahlestedt, 2014; Pikaard & Mittelsten Scheid, 2014; Sarkies & Miska, 2014; Tee & Reinberg, 2014; Vogel-Ciernia & Wood, 2014; Weick & Miska, 2014). With a focus on DNA methylation, Holliday revisited the original Waddingtonian concept of an epigenetic landscape and proposed that the maintenance of a cell commitment in the canals (or "creodes," as Waddington (1957) described them) required information to be propagated from parent to daughter cells (Holliday, 1979), updating the definition of epigenetics accordingly to encompass all heritable information that is not encoded in the DNA sequence directly, and resulting in one currently accepted definition.

However, a further complication arises when the definition of epigenetics becomes conflated with all transcriptional regulatory processes, which is commonplace at present. For example, whereas posttranslational modifications of histones are frequently referred to as epigenetic, and there is certainly some evidence in model organisms that these modifications can self-propagate at particular loci through cell division (Ragunathan, Jih, & Moazed, 2014), there are few mechanistic insights into how this takes place. This has prompted the alternative viewpoint that the maintenance of cell lineage commitment is in fact primarily influenced by TFs (Ptashne, 2013a, 2013b). This is a reasonable model because it only requires the presence of the TF at the same locus in parent and daughter cells to maintain the cell state, a "bookmarking" of mitotic chromosomes for which there is some evidence (Caravaca et al., 2013). Because TFs recruit enzyme complexes that influence chromatin characteristics (Benner et al., 2013; Cheng, Shou, Yip, & Gerstein, 2011; Luu, Scholer, & Arauzo-Bravo, 2013), as noted earlier, the presence of many regulatory processes described as epigenetic may in fact result

Genomics, Circuits, and Pathways in Clinical Neuropsychiatry. http://dx.doi.org/10.1016/B978-0-12-800105-9.00005-6

from TF actions. As a counterpoint, it could be argued that the activity of a TF is at least in part determined by preexisting chromatin organization and DNA methylation states (Iwafuchi-Doi & Zaret, 2014; Magnani, Eeckhoute, & Lupien, 2011), so the system is more likely to reflect an interplay of these mediators than a model in which all TFs are completely deterministic.

We should therefore use care and perspective when using the term "epigenetic," recognizing that not all transcriptional regulation is epigenetic, because not all transcriptional regulatory processes have clear mechanisms of self-propagating heritability, and that the study of one component of the transcriptional regulatory process ignores or may be confounded by influences from many other components that are not being tested.

So where does this leave us when we want to discuss the field of epigenetics as it relates to neurobiological processes? The question becomes one of how the identification of transcriptional regulatory changes associated with specific phenotypes, including diseases, allows insights into pathogenesis and the potential for novel avenues of therapy. In this review, we describe the experimental approaches that have been used to suggest links between transcriptional regulatory mechanisms and nervous system phenotypes, but maintain the perspective that these links have less certain mechanistic interpretability than might have been supposed, because of the considerations described earlier. However, even applying this cautious perspective, it is clear that observations linking DNA methylation and chromatin states with psychiatric or neurological disease phenotypes show promise for gaining insights into pathogenetic mechanisms.

EPIGENETICS IN NEUROBIOLOGICAL RESEARCH

There has been a surge of epigenetic studies of the nervous system over the past decade (Epigenetic Regulation in the nervous system: Basic mechanisms and clinical impact, 2013). The vast majority of this work has used the broader definition of epigenetics noted previously: namely, denoting stable mechanisms of transcriptional regulation and not denoting transgenerational inheritance. One of the primary interests of the field is to explain lifelong changes in disease vulnerability (Robison & Nestler, 2011; Zhang, Labonte, Wen, Turecki, & Meaney, 2013). Exposure to severe stress early in life can increase an individual's vulnerability to a range of mental disorders in adulthood, such as depression and drug addiction. Likewise, repeated exposure to a drug of abuse during adolescence can permanently enhance that individual's risk for addiction for a lifetime. Conversely, a several-month course of treatment with an antidepressant medication may not only alleviate the symptoms of depression while a person is being treated but can often render many individuals illness-free for the rest of their lives. The ability of these and other discrete environmental exposures to produce such long-lasting changes in behavior means that they are altering the brain in stable ways, for which epigenetic control over gene expression is one plausible mechanism.

A second imperative to study epigenetic mechanisms in neuropsychiatric diseases is to gain a more complete appreciation of how specific genes or nongenic loci are affected. Studies focused solely on measures of RNA expression levels (whether polymerase chain reaction for individual genes or sequencing or microarray platforms for all RNAs) are limited by the fact that they quantify steady-state RNA levels with no indication of the dynamic nature of an RNA's regulation. By contrast, assays of DNA methylation or chromatin states reveal many layers of regulation of a genomic locus that might not be evident from measures of that locus's transcription at a given time point (Robison & Nestler, 2011). Thus, epigenetic modifications can represent latent forms of regulation that, although not reflected in altered steady-state levels of RNA expression, modify that gene's inducibility in response to a subsequent stimulus. For this reason, epigenetic studies can uniquely identify genes whose future transcription is sensitized (primed) or desensitized. Epigenetic investigations also provide information for the first time about the molecular mechanisms associated with an altered state of gene transcription. Before such approaches, any effort to understand mechanisms of gene transcription have relied on studies in cell culture, even though we know that what occurs in a cultured cell, even a cultured neuron, is not always an accurate reflection of what occurs in the brain in vivo (Robison & Nestler, 2011). Finally, epigenetic studies reveal regulation occurring at nongenic regions of the genome, which might have profound influences on normal cell function and perhaps disease states.

There has also been increasing attention given to true, heritable epigenetic regulation in the nervous system and its contribution to neuropsychiatric disease (Bale, 2014; Dias & Ressler, 2014; Dietz et al., 2011; Franklin, Saab, & Mansuy, 2012; Szutorisz et al., 2014; Vassoler, White, Schmidt, Sadri-Vakili, & Pierce, 2013). Such interest is based partly on observations that whereas most nervous system disorders are highly heritable, it has been difficult to identify individual risk genes of strong effect. However, whether epigenetic inheritance contributes to disease risk remains uncertain, largely owing to a lack of appreciation of plausible mechanisms that might be involved.

Lack of access to the human brain represents a unique obstacle for epigenetic investigations of the nervous system. Unlike every other organ system, it is not possible to obtain brain tissue from living patients during the course of an illness. For this reason, the vast majority of epigenetic studies of the nervous system have relied on animal models. Genetic tools in mice that enable the isolation of individual cell types from a microdissected brain region offer another driver of the use of animal models. However, there is growing interest in epigenetic analyses of human *postmortem* brain tissue as well as peripheral tissues from living patients as surrogate indicators of the brain. Each of these approaches has strengths and limitations; the integrated use of multiple approaches is currently the most effective way to elucidate epigenetic mechanisms of brain diseases.

THE GENOMIC CONTEXT OF TRANSCRIPTIONAL AND EPIGENETIC MECHANISMS

The process of gene transcription involves numerous mediators. Ultimately the RNA message is produced by the movement of RNA polymerase through a region of the genome, generating the primary transcript that should be a perfect copy of the transcribed DNA strand.

The process by which this transcription is established at a specific locus is generally the focus of studies trying to understand how phenotypic variability is established, but some of the same mediators appear to act as a result of, or associated with, transcription. For example, specific histone modifications are present almost exclusively in transcribed regions (histone H3 lysine 36 trimethylation (H3K36me3) and H3K79me3 (Vakoc, Sachdeva, Wang, & Blobel, 2006), H3K9me3, and HP1 (Vakoc, Mandat, Olenchock, & Blobel, 2005)). DNA methylation is also generally increased in transcribed regions (Suzuki et al., 2011), where it may have roles in transcriptional (Kulis, Queiros, Beekman, & Martin-Subero, 2013) and splicing regulation (Maunakea, Chepelev, Cui, & Zhao, 2013). The association of DNA methylation with transcription is sufficiently strong that on a macro-organizational level, there is increased DNA methylation in euchromatic regions of the mammalian genome compared with heterochromatic regions (Suzuki et al., 2011). This suggests that there are two types of heterochromatin in the mammalian genome: one that is associated with highly repetitive, pericentromeric DNA and is highly DNA methylated, probably involving the effects of DNA methyltransferase (DNMT) 3B, and one in interstitial regions of chromosomes, which has low levels of DNA methylation.

When performing a study of transcriptional and epigenetic regulators associated with a phenotype, there may therefore be changes identified that reflect the response to transcription rather than processes that determine transcription. The transcriptional differences may be the primary distinctive characteristic distinguishing the groups being studied, reflected by the transcriptional regulatory and epigenetic differences. The observation remains valuable because it indicates a transcriptional change associated with the phenotype being studied. However, the observation does not support a model in which transcriptional regulatory and epigenetic processes are primary events in determining the phenotype. Advances which enable studies of the causal role of epigenetic mechanisms in inducing disease-like abnormalities in animal models (Heller et al., 2008) are discussed subsequently.

However, the focus of phenotypic association studies involving the epigenome is to understand how transcriptional changes represent the consequences of altered epigenetic regulation. The paradigm established for the field is from cancer epigenetics, in which there have been both global and local changes observed for DNA methylation (Torano, Petrus, Fernandez, & Fraga, 2012), with some of the changes associated causally with transcriptional changes. It becomes a reasonable question whether similar events take place in nonneoplastic phenotypic or disease states, including brain disorders.

The most direct mediator of transcription is the RNA polymerase holoenzyme, which is recruited by a group of TFs forming the preinitiation complex at a nucleosome-free region of a gene (Thomas & Chiang, 2006). Events at the carboxyl terminal domain of RNA polymerase II are required to start the polymerase moving along the DNA of protein-coding genes, including the binding of transcriptional coactivators such as the mediator complex (Myers & Kornberg, 2000). This places our focus directly on the promoter as the locus of critical importance for gene transcriptional regulation. The most conclusive finding from cancer epigenetics studies is the demonstration of DNA methylation acquisition in cancer cells at promoters of genes that were transcriptionally silenced (Kulis & Esteller, 2010). These promoters are generally rich in the CG dinucleotides typical of CpG islands, and the DNA methylation events occurred consistently enough to allow a subset of colorectal carcinomas to be described as having a distinctive 'CpG island methylator phenotype' (Issa, 2004). The associated silencing of transcription of known tumor-suppressor genes completes the chain of causality in this CpG island methylation contributing to the neoplastic process (Lopez-Serra & Esteller, 2012; Sarkar, Goldgar, Byler, Rosenthal, & Heerboth, 2013; Taberlay & Jones, 2011).

Whereas this model has provided the foundation for exploring epigenetic dysregulation in human diseases other than cancer, even in the cancer epigenome it is being recognized that a substantial proportion of changes do not occur at canonical promoters but are located more distally at regulatory sequences generally classified as enhancers (Aran, Sabato, & Hellman, 2013; Blair et al., 2013; Hu et al., 2014; Ko et al., 2013; Taberlay, Statham, Kelly, Clark, & Jones, 2014; Zhang et al., 2014). How the physically distant enhancers interact with promoters is being tested using studies that reveal the three-dimensional architecture of chromatin in the cell nucleus (Dostie et al., 2006; Kalhor, Tjong, Jayathilaka, Alber, & Chen, 2012; Lieberman-Aiden et al., 2009). That these physical interactions occur is apparent; what is less clear is how these interactions are controlled and how they affect the polymerase complex located at the promoter.

The properties of enhancers have been reviewed (Plank & Dean, 2014), including their transcription to form enhancer RNAs, a type of noncoding RNA that may be involved in their function (Kim, Hemberg, & Gray, 2015). Although the property of transcription is in common with promoters, a distinction is that the histone posttranslational modifications that occur at enhancers differ from those at promoters, in which H3K4me3 and histone H4 lysine 16 acetylation (H4K16ac) are enriched compared with the enrichment of H3K4me1 and H3K27ac at enhancers, for example. Another difference is that enhancer locations vary among cell types to a much greater extent than do promoter locations (Won et al., 2013). Promoters tend to be "poised" and ready for transcription constitutively (Guenther, Levine, Boyer, Jaenisch, & Young, 2007) but require additional signals from transcription factors such as MYC (Rahl et al., 2010) to allow their transition to an active state. Because enhancers can be defined by their property of TF binding, a simple model for their activation of promoters may be that they bring into proximity the TFs needed to convert promoters from a poised to an actively transcribing state.

There are several reasons for our placing emphasis on the complementary roles of promoters and enhancers. One is to stress the need for enhancer actions in normal transcriptional regulation; a second is in terms of the information content from the assays used to study the epigenome. A problem with promoter studies reflects their constitutive nature: They tend to have too little normal variability for phenotypic or even transcriptional associations to be made. For example, DNA methylation at CG dinucleotide-rich promoters tends to be almost completely absent whether the associated gene is transcriptionally active or silent (Weber et al., 2007).

This finding was highlighted in a study of DNA methylation at and flanking CpG islands in which it was observed that the variability in DNA methylation within the CpG island was poorly correlated with both local gene transcription and changes associated with cancer (Irizarry et al., 2009). However, the 2-kb regions flanking these promoter-associated CpG islands were much more predictive of these changes, which led the authors to propose that the "shores" of CpG islands are more valuable for study than the CpG islands themselves (Irizarry et al., 2009). Further extension of this notion is to consider so-called "shelves" (beyond the shores) and the remaining "open seas" of the genome, which have also proven popular as ways of describing more or less informative contexts for DNA methylation (Sandoval et al., 2011). In a study of CD34+ hematopoietic stem and progenitor cells, we showed that the CpG island shore is in fact highly enriched with histones with posttranslational modifications typical of enhancer sequences (Wijetunga et al., 2014). This observation ties together the emerging role of enhancers in transcriptional regulation and its dysregulation in disease states, with studies that indicate CpG island shores to be DNA methylation-dependent enhancers (Perisic, Holsboer, Rein, & Zschocke, 2012) and informative in phenotypic and disease associations (Aran et al., 2013; Blair et al., 2013; Hu et al., 2014; Ko et al., 2013; Taberlay et al., 2014; Zhang et al., 2014). It is likely that CpG island shores represent a subset of the broader group of enhancers genome-wide which would be the optimal targets of DNA methylation studies in disease states.

A problem with studying DNA methylation at enhancers is that they are located in distinct locations in different cell types, which makes a single analytical platform surveying fixed locations in the genome unlikely to be universally informative. For example, the popular Illumina Infinium microarray, which tests about 485,000 sites for DNA methylation throughout the human genome, is focused on annotated promoters (Bibikova et al., 2011). Although it also covers some CpG island shores and other nonpromoter locations, even if it performs well in testing enhancer locations in one cell type, it is unlikely to be comprehensive in enhancer coverage in other cell types. The reduced representation bisulphite sequencing assay (Meissner et al., 2008) is targeted to very CG-rich regions of the genome and likewise targets promoters preferentially. The more comprehensive approach of shotgun bisulphite sequencing (Lister et al., 2008) is more likely to report informative loci as a result, but it suffers in comparison by being substantially more costly and generates a sizable proportion of sequence from uninformative regions of the genome. We discuss choices of genome-wide epigenetic assays subsequently, taking some of these considerations into account.

TRANSCRIPTIONAL AND EPIGENETIC REGULATORY MECHANISMS

DNA methylation is the most studied regulator in epigenetic association studies of mammalian phenotypes and diseases. It warrants further discussion as a result. DNA methylation refers to the addition of a methyl group to cytosine in the context of a CG dinucleotide. A CG on one strand of DNA is mirrored on the other strand by a complementary CG, which makes this the smallest possible unit of DNA in which a cytosine can be present on both strands. This becomes interesting when the biochemistry of DNA methylation is considered, because it is now recognized that cytosine methylation is symmetrical on both strands and that this state is transmitted from parent to daughter cells. The process of DNA replication creates complements to each parental DNA strand using unmodified cytosines, creating a hemimethylated state in which the cytosine on one strand is methylated, but not that on the other strand. This hemimethylated state is recognized by the UHRF1 protein (Bostick et al., 2007), which recruits DNMT1 that then restores symmetrical methylation on both daughter strands. This mechanism of heritability makes DNA methylation a genuinely heritable process, whereas chromatin states lack similar mechanisms for self-propagation unless mediated by a primary TF binding event, as described earlier.

5-Methylcytosine (5mC) is not the only modification of nucleotides in the mammalian genome. Oxidases exist that convert 5mC to 5-hydroxymethylcytosine (5hmC), which is then further oxidized to 5-formylcytosine and to 5-carboxylcytosine, reactions catalyzed by ten-eleven translocases (TETs), of which there are three homologues in mammals (Delatte et al., 2014). These enzymatic steps create substrates for glycosylases, which remove the modified cytosine to create an abasic site that is then filled with an unmodified cytosine (Maiti & Drohat, 2011). This provides a means of reversing DNA methylation that appears to be active in early embryogenesis (Guo et al., 2014) and may persist in adult brain (Kriaucionis & Heintz, 2009; Lister et al., 2013). The other means of removing DNA methylation is for DNMT1 not to maintain the mark through cell division, allowing the mark to be diluted passively. Nondividing cells do not have this option, which makes the increased amount of 5hmC in nondividing cells of the brain (Kriaucionis & Heintz, 2009; Lister et al., 2013) an intriguing observation that could be consistent with the preferential use of this alternative mechanism. A related question is whether there is a cell cycle–independent means of establishing DNA methylation that may be especially important in terminally differentiated nondividing cells. The maintenance DNMTs, DNMT3A, and DNMT3B do not require DNA replication but can add methyl groups to cytosines de novo (Jurkowska, Jurkowski, & Jeltsch, 2011). A report suggests that DNMT3A not only establishes DNA methylation at CG dinucleotides but also establishes and maintains DNA methylation at a substantial number of non-CG cytosines in postmitotic neurons (Guo et al., 2014), which indicates a degree of complexity in the brain beyond what we have come to expect in mammalian genomes. Interestingly, levels of 5hmC are particularly enriched in the nervous system (Kriaucionis & Heintz, 2009; Lister et al., 2013), but most studies of DNA methylation in the brain to date have not distinguished between these two forms.

This discussion of DNA methylation can be framed in terms of the regulatory mechanisms that establish the mark in the first place: the chromatin components that recognize the mark and the enzymes that remove the mark subsequently. This has led to the common description of epigenetic writers, readers, and erasers. For DNA methylation, the writers are DNMTs, the erasers are the TET and glycosylase enzymes, whereas the readers are those DNA binding proteins that are positively or negatively influenced by the presence of the methyl group at the intended site of binding. Of the DNA binding proteins whose binding is enhanced by 5mC, the paradigm is established by the proteins containing methyl-binding domains (MBDs) but also include some of the zinc finger domain transcription factors (Liu et al., 2014; Liu, Zhang, Blumenthal, & Cheng, 2013). The methyl group protrudes into the major groove of the DNA where sequence-specific recognition of DNA by binding proteins can occur. Interestingly, the conversion of 5mC to 5hmC by TET oxidases changes the binding characteristics of some MBDs for the DNA, including methyl CpG binding protein 2 (MECP2), which is abundantly expressed in the brain (Nan, Campoy, & Bird, 1997). Because *MECP2* mutations are responsible for Rett syndrome (Amir et al., 1999) and a subset of individuals with autism spectrum disorders (ASDs) (Cukier et al., 2010), the relative degrees of interaction of MECP2 with 5mC versus 5hmC need to be understood to gain insights into the pathogenic effects of the mutations. The methyl group can also interfere with binding of transcription factors, as exemplified by the CTCF protein (Bell & Felsenfeld, 2000; Wang et al., 2012), which acts to prevent enhancers from binding to promoters, referred to as an insulator function that partitions chromatin into functional domains (Bell & Felsenfeld, 2000). The influence of DNA methylation has been especially well-characterized for CTCF binding within imprinted domains, leading to imprinted insulator function and enhancer-dependent transcription of the *Igf2* locus in mouse (Bell & Felsenfeld, 2000). Genomic imprinting represents an example of an epigenetic mark being written distinctively in male compared with female gametes, proving resistant to subsequent erasure and leading to differences in how the mark is read on each parental chromosome.

The same structure of writers, readers, and erasers can be used when considering the vastly more diverse world of posttranslational modifications of chromatin components. The number of posttranslational modifications of the canonical H2A, H2B, H3, and H4 histones in eukaryotic chromatin is in the hundreds (Nardelli et al., 2013), an indication of the complexity and potential diversity of the information encoded in these marks. This complexity is further expanded when one considers that a substantial proportion of nucleosomes contain alternative types of histones other than the four listed (Maze, Noh, Soshnev, & Allis, 2014) (so-called histone variants) which can themselves undergo their own repertoire of posttranslational modifications, whereas TFs themselves also represent targets of frequently poorly understood posttranslational modifications (Filtz, Vogel, & Leid, 2014).

The enzymes that write and erase the posttranslational modifications of histones and other chromatin constituents are diverse and incompletely understood, as might be expected from the enormous range of modifications, sites, and biological effects. Readers of the posttranslational modifications can even recognize combinations of different marks occurring in close proximity (Musselman, Lalonde, Cote, & Kutateladze, 2012). This complex and emerging field has been the subject of several excellent reviews (Du & Patel, 2014; Fischle, 2014; Fuchs & Oren, 2014; Gayatri & Bedford, 2014; Musselman, Khorasanizadeh, & Kutateladze, 2014; Rothbart & Strahl, 2014; Sawicka & Seiser, 2014).

EPIGENETIC PERTURBATION AND EPIGENOME-WIDE ASSAYS

This all leads to the question that is central to epigenetic studies of human disease phenotypes: How do epigenetic changes take place causing those associated with a normal cellular function to modify patterns associated with disease? This depends on the researcher's perspective. In a TF-centered model, the primary signal would come from a change in the quantitative or qualitative characteristics of the TF, resulting in an altered pattern of association with chromatin in the genome. The differential recruitment of enzyme complexes to loci would then result in distinctive patterns of posttranslational modifications of chromatin components and of DNA methylation. Someone with a more chromatin-centered perspective may invoke the need for nucleosomal remodeling as a primary event to allow TF binding to DNA, whereas those interested in DNA methylation may invoke the need for DNA methylation patterns to change to allow TF binding to take place.

To date, most epigenome-wide assays of human disease tissue have been based on the study of DNA methylation. This partly reflects the success of such approaches in cancer epigenomics, partly the relative maturity of the assays and associated analytical approaches, and partly the perception that we understand how to interpret DNA methylation changes in the genome. It also assumes that even if DNA methylation is not the primary event in changing the cell's state to that associated with disease, the interdependence of the transcriptional and epigenetic regulatory mechanisms is such that DNA methylation will reflect changes to any component of the process. No DNA methylation assay is genuinely comprehensive in testing the entire genome, so the assay will need to interrogate the loci where dysregulation is occurring if it is to be informative.

Chromatin mapping approaches, on the other hand, should be comprehensive in identifying the location of a chromatin component genome-wide (apart from highly repetitive loci), but have the problem of choice: Can you predict in advance which of at least hundreds of different chromatin states is going to be informative for your phenotype of interest? Although chromatin studies of human diseases are relatively rare, the general approach has been to study some of the better understood posttranslational modifications of histones, usually both activating and repressive marks such as H3K4me3 and H3K27me3, respectively. The alternative approach of profiling loci of open (nucleosome-free) chromatin has been applied successfully in a mouse model of dietary effects (Leung et al., 2014), which may be a sensible compromise in combining a relatively generic chromatin characteristic with genome-wide mapping potential.

To study DNA methylation, three broad approaches are available. The reference standard is based on bisulphite mutagenesis of unmethylated cytosines to uracil, which is then amplified as thymine. Quantifying the ratio of unconverted cytosine to converted thymine reflects the proportion of 5mC to C at that nucleotide before conversion. To measure 5mC at specific loci in the genome, microarrays based on C or T primer extension have been employed (Bibikova et al., 2011), whereas massively parallel sequencing (MPS) of the entire genome (Lister et al., 2008) or subsets enriched by restriction enzyme representations (Meissner et al., 2008) or hybridization-based capture (Hodges et al., 2009) have been employed. A second approach involves the use of DNA methylation-sensitive restriction enzymes, also using microarrays or MPS to map and quantify 5mC (Khulan et al., 2006; Suzuki et al., 2010). A disadvantage of these methods is that they do not distinguish between 5mC and 5hmC. To do so, oxidized bisulphite approaches are available that either protect (TET-assisted bisulfite sequencing) (Yu et al., 2012) or selectively oxidize (oxidative bisulfite sequencing) (Booth et al., 2012) 5hmC. The third approach uses antibodies or other reagents with selective affinity for 5mC, 5hmC, or unmethylated CG dinucleotides followed by deep sequencing (Long et al., 2013; Song et al., 2011; Weber et al., 2005). Whereas the last category is most informative in CG-dense regions of the genome (Down et al., 2008), these assays have the characteristic of being quantitative, allowing discrimination of modest, as opposed to extreme, changes of DNA methylation. This is potentially the most valuable characteristic of these assays, as will be described subsequently.

Other chromatin components of different types can be identified and mapped throughout the genome as well, using some means of preserving the physical interactions of chromatin with DNA, most commonly using formaldehyde to cross-link chromatin proteins to "cage" the associated DNA (Solomon & Varshavsky, 1985), or by using nucleases to digest DNA exposed between chromatin proteins (O'Neill & Turner, 2003). The individual units of chromatin can then be exposed to a reagent with binding properties that allow selective enrichment of one of the chromatin components; the reagent is usually an antibody, which leads to the description of the enrichment as chromatin immunoprecipitation (ChIP) (Carey, Peterson, & Smale, 2009). Reagents other than antibodies are also used (eg, biotin (Furuyama & Henikoff, 2006),

SNAP-tagging (Dunleavy, Almouzni, & Karpen, 2011)), but the general principle of affinity-based enrichment remains the same.

Once the affinity-based enrichment has been performed, the DNA sequences that are simultaneously captured can be identified to determine where the chromatin component was located in the genome. In previous years this was performed using microarrays (ChIP on chip); it currently uses MPS instead (ChIP sequencing [ChIP-seq]). The result of ChIP-seq is a binary state defining the presence or absence of a chromatin constituent at a locus, and not the more quantitative output of DNA methylation assays. This is an important consideration because most DNA methylation changes in human diseases have been of modest magnitude. DNA methylation at a diploid locus can be present only at 0%, 50%, or 100%, depending on whether neither, one, or both alleles are methylated at that single site. A change of 20% DNA methylation at a given site has to reflect a subset of alleles or cells in the population studied changing their DNA methylation state. ChIP-seq obviously does not have the capacity to discriminate between samples in which one has 10% of alleles containing a chromatin component and the other 30% of alleles. This lack of quantitative resolution is a potential drawback for the application of ChIP-based studies of the epigenome.

The frequently limited cell numbers in human disease studies also favor the choice of DNA methylation assays, which generally require less material for robust performance (Plongthongkum, Diep, & Zhang, 2014), but there have been significant advances to allow ChIP-seq to be performed with limited sample amounts (Gilfillan et al., 2012). ChIP critically depends on antibody quality and needs samples to be processed carefully and early after biopsy or necropsy if high-quality chromatin is to be obtained.

Some potentially valuable assays are emerging. The assay for transposase-accessible chromatin using sequencing (ATAC-seq) allows nucleosome-free regions to be mapped genome-wide in tens of thousands of cells (Buenrostro, Giresi, Zaba, Chang, & Greenleaf, 2013), allowing a representation of the repertoire of regulatory loci to be predicted empirically. How these regulatory loci interact *in cis* and *in trans* with promoters will be valuable information that can potentially be generated from new assays that study three-dimensional chromatin interactions (Dostie et al., 2006; Kalhor et al., 2012; Lieberman-Aiden et al., 2009). As described earlier, the ideal DNA methylation assay would target enhancers and other regulatory loci, which may be possible using new assays that select specific regions of the genome for targeted bisulphite sequencing (Hodges et al., 2009; Lee et al., 2011). This field of assay development is fast-moving and is likely to progress significantly in the next few years, in part aided by new MPS technologies.

CELL TYPE CHOICES IN NEUROEPIGENOMIC STUDIES

For studies of neurobiological phenotypes, a major problem is that cells mediating the phenotype are not accessible from living human subjects. This has led to three avenues of research to overcome this problem. One is to use postmortem brains, the second is to use animal models, and the third is to use surrogate cells from human subjects. Each approach has strengths and weaknesses. The concern with postmortem material is that the way in which one dies and how quickly the brain is isolated from the deceased individual can dramatically affect how well the biomolecules in the cell reflect their state when the individual was alive, at least for studies of RNA (Cummings, Strum, Yoon, Szymanski, & Hulette, 2001; Li et al., 2007, 2004; Preece & Cairns, 2003; Tomita et al., 2004), although DNA methylation does not appear to be as susceptible to the brain tissue pH marker of agonal variability (Ernst et al., 2008). A second problem with the use of *postmortem* tissue is the inherent heterogeneity of the cell types composing the tissue, which is discussed further later. Another problem is that the numbers of brains available to study through repositories are frequently limited, which compromises the statistical power of studies. However, a major advantage in epigenomic studies is the ability to be able to examine the cells mediating the phenotype, because there is always uncertainty about the accuracy of animal models or surrogate human cells (see subsequent discussion). Animal models have numerous advantages, including the ability to control exposures, diet, and genotype, while allowing sampling of discrete regions of brain in highly controlled necropsies designed to minimize sample degradation. Disadvantages include how well the animal phenotype represents that in humans, and the relatively limited amounts of tissue available from the smaller brains of rodents, and whether rodent brains contain certain frontal cortical regions homologous to those in humans. Accessible surrogate cell use allows humans with the well-characterized phenotype to be studied and can allow large cohorts to be assembled to power studies looking for limited or heterogeneous epigenetic changes in affected individuals. The critical question with surrogate cells is always whether epigenetic changes found in those cells reliably represent those in the brain. To some extent, that depends on the biological hypothesis being tested. We performed a study of ASD using buccal epithelial cells with the rationale that we were looking for epigenetic dysregulation occurring in gametes or very early embryogenesis, so that this accessible ectodermal cell type should have the same epigenetic changes as those in ectodermally derived neurons in the brain (Berko et al., 2014). On the other hand, most epigenetic studies of rodent tissues have revealed abnormalities in histone modifications or DNA methylation associated with a given disease model to be highly region specific: that is, expressed in one brain region (even one cell type) but not many others (Dias et al., 2014). It thus appears unlikely that environmentally induced epigenetic markers in peripheral cells would reflect patterns of epigenetic regulation that occur in a given cell type within a narrowly defined brain region.

The use of mixed cell types in tissue samples is recognized to carry the risk of misinterpretation of the epigenetic results obtained. If there is a systematic bias toward the altered representation of a specific cell subtype in the disease compared with the control group of samples, the epigenetic differences that characterize that cell type will begin to emerge as distinctive in the disease group. There is obvious value to identifying cell subtype differences characterizing a disease state, but if the starting hypothesis were limited to the need to discover changes in the epigenome within a specific cell type, and the alternative hypothesis of cell subtype changes is not embraced or tested, there is potential to misinterpret the results obtained. Purifying cells is one way to overcome this problem, with ingenious fluorescence-based solutions applied to the sorting of specific types of cells or nuclei from brain (Jordi et al., 2013; Okada et al., 2011). Statistical modeling based on prior knowledge of epigenetic patterns of the cell subtypes is another strategy primarily applied in peripheral blood leukocytes (Houseman et al., 2012), but also in neuronal/nonneuronal cells from brain tissue samples (Montano et al., 2013).

As described earlier, the changes in an epigenetic end point associated with disease phenotypes are generally small in magnitude and can be interpreted to reflect changes in the levels of mosaicism for epigenetic states between control and disease states. Clearly the same pattern will occur when a cell subcomposition difference exists between disease and controls, leading to the potential that this could be a confounding influence if unrecognized. As such, cell subcomposition represents a broader group of potential confounding influences upon epigenomic studies, which include various technical factors (Michels et al., 2013) and the effect of variable transcription of that locus (Suzuki et al., 2011; Zilberman, Gehring, Tran, Ballinger, & Henikoff, 2007). In performing epigenomic studies, it is essential for as many possible sources of variability as possible to be catalogued and quantified so that the results of downstream analyses will be of as high confidence as possible. Systematic metadata capture is an essential component of experimental design in epigenome-wide studies and needs to be planned for implementation at the outset of a project.

INTERACTION OF THE GENOME WITH THE EPIGENOME

A major source of variability on the epigenome is from the DNA sequence of the genome itself. It has become apparent that DNA sequence variants are associated nonrandomly with DNA methylation patterns *in cis*, with some alleles associated with increased and others decreased DNA methylation. These methylation quantitative trait loci (mQTLs) can act over at least tens of kilobase pairs (Bell et al., 2011; Gertz et al., 2011; Grundberg et al., 2013), are generally located in and target *cis*-regulatory loci (Shi et al., 2014), and may overlap with expression quantitative trait loci (eQTLs) physically and functionally (Zhang et al., 2014). Of concern for studies of genetically diverse individuals is the proportion of DNA methylation variability for which these mQTLs appear to account, which is estimated to be between 22% and 80% (Bell et al., 2011; Gertz et al., 2011; Grundberg et al., 2013). This is a profound influence that is almost never addressed in epigenomic association studies, and which renders most such studies of questionable interpretability. Our use of microarray genotyping in a study of individuals with ASD revealed such an influence of ancestry upon DNA methylation (Berko et al., 2014), allowing its removal in the analytical approach used.

It is not clear mechanistically how mQTLs exert their effects; a tenable model is that the DNA sequence variant influences local binding of one or more TFs, with effects on DNA methylation as a consequence. In yeast, DNA sequence variants with effects on chromatin accessibility (Lee et al., 2013) and transcription factor binding (Zheng, Zhao, Mancera, Steinmetz, & Snyder, 2010) have been identified, which suggests that there may also be chromatin QTLs that potentially overlap with mQTLs and eQTLs and supports the TF-based hypothesis. To overcome this strong source of variability in epigenomic studies of human disease, a cross-sectional study design involving cohorts of genetically diverse subjects will need to capture genotypic information from those same individuals, allowing potential mQTLs to be identified and accounted for analytically. Studies of monozygotic twins discordant for the phenotype offer advantages in this regard, as do longitudinal studies in which the phenotype changes within the individual over time, whereas isogenic animal models represent another means of overcoming this issue. The effects of DNA sequence variability are looming as the biggest challenge to date to the interpretability of epigenomic studies of human diseases, and have to be accounted for in designing new studies prospectively. The ideal epigenomic study of human disease would therefore include a means of addressing mQTL effects and would incorporate transcriptional studies to account for alterations of the epigenome induced by transcription, while testing the purified cell type mediating the phenotype in a cohort sufficiently large for statistical power, and while capturing metadata systematically using one or more epigenomic assays targeted to the most informative regions of the genome for the cell type being studied. These are nontrivial but critically important issues that will allow us to gain higher-confidence insights into the possible role of epigenetic dysregulation in human disease.

EPIGENETICS MECHANISMS OF BRAIN DISEASES: GENERAL PRINCIPLES AND FUTURE DIRECTIONS

Most studies of epigenetic mechanisms of brain diseases to date have focused on a range of histone modifications (acetylation, methylation, phosphorylation, adenosine diphosphate ribosylation), with many fewer studies examining DNA methylation, in part based on the technical hurdles discussed earlier. A large majority of studies have used animal models, with increasing attention given to human tissue. Available models of epigenetic contributions to a range of neuropsychiatric syndromes are reviewed elsewhere (Fischer, 2014; Jakovcevski & Akbarian, 2012; Pena, Bagot, Labonte, & Nestler, 2014; Robison & Nestler, 2011; Rudenko & Tsai, 2014; Zhang et al., 2013). The current discussion summarizes some general themes and future directions.

Surprising findings in the field are alterations in total cellular levels of a given histone modification or DNA methylation in an animal model or human tissue. For example, depression and drug addiction are associated with reduced H3K9me2 (a form of repressive histone methylation) in nucleus accumbens (a brain reward region), an effect seen in mouse models and human postmortem brain (Covington et al., 2011; Maze et al., 2010), whereas many diseases are associated with altered levels of acetylated histones in several brain regions (Covington et al., 2009; Fischer, 2014; Kennedy et al., 2013; Sadri-Vakili & Cha, 2006). Such global changes are usually associated with altered expression levels of an enzyme that controls regulated modification: for example, a histone methyltransferase or histone deacetylase. Altered global levels of DNA methylation and DNMTs have also been reported (LaPlant et al., 2010). Interestingly, when ChIP-seq has been used to map such modifications genome-wide, far more complex patterns are seen: Only a subset of genes and nongenic regions reflect the global change, whereas many loci show changes in the opposite direction (Feng et al., 2014; Ferguson et al., 2015; Scobie et al., 2014). It is possible that transcription factors drive the deposition of marks at specific sites, although this has not yet been demonstrated.

There are still limited numbers of genome-wide studies of epigenetic mechanisms performed to date on mouse or human brain tissue (Table 5.1) (reviewed in Maze, Shen, et al. (2014)). Such studies of brain offer particular challenges, such as the heterogeneity of cell types, as noted in earlier sections of this review. Nevertheless, this evolving literature is providing new insight into the range of genes and nongenic regions as well of epigenetic mechanisms that contribute to disease pathophysiology. As one example, the mapping of β-catenin (a type of transcription factor) in nucleus accumbens of mouse depression models, in concert with genome-wide changes in H3K4me3, implicated numerous previously unexplored genes in mediating resilience (resistance) to chronic stress (Dias et al., 2014). As another example, the genome-wide mapping of 5mc and 5hmc, at single-base resolution, in frontal cortical neurons and glia of mice and humans as a function of age revealed shifting cell type–specific patterns of gene regulation over time and identified sets of genes possibly involved in age-related cognitive decline (Lister et al., 2013).

Genome-wide exploration of epigenetic regulation in neuropsychiatric disease models has also begun to identify "chromatin signatures"—combinations of histone modifications, other chromatin state changes, and DNA methylation alterations—that mark genes for altered levels of inducibility (Feng et al., 2014; Ferguson et al., 2015; Guo et al., 2014; Lister et al., 2013; Peleg et al., 2010; Scobie et al., 2014; Zhou, Yuan, Mash, & Goldman, 2011; Zhu et al., 2013). For example, in the case of reduced H3K9me2 in nucleus accumbens of depression and addiction models, decreased H3K9me2 deposition is associated with the priming of large numbers of genes and their increased induction in response to a subsequent stress or drug exposure (Feng et al., 2014; Maze, Shen, et al., 2014). Nevertheless, despite the power of these genome-wide approaches, results have demonstrated clearly that no single epigenetic modification and not even a combination of several are perfectively predictive of a gene's transcriptional activity (Feng et al., 2014; Zhu et al., 2013). Presumably, this reflects the extraordinary complexity of epigenetic regulation and the likely involvement of perhaps hundreds of epigenetic modifications in concert with the altered transcription of a given gene.

A glaring weakness in the epigenetics field, as stated previously, is the limited ability to demonstrate whether an epigenetic modification causes a change in transcription or reflects it. As an example, we have proposed that cocaine induction of the transcription factor ΔFosB is mediated in part through downregulation of the histone methyltransferase, G9a, in nucleus accumbens. We demonstrated that knocking out G9a in this brain region of adult mice promotes ΔFosB induction by cocaine, whereas overexpression G9a blocks cocaine action (Maze, Shen, et al., 2014). However, knocking down or overexpressing G9a influences H3K9me2, the mark catalyzed by G9a, at hundreds of genes, which makes it impossible to connect a change in H3K9me2 at the *FosB* gene directly to altered expression of the gene and to altered

TABLE 5.1 Progress Report of Epigenomic Data From Brain

Year	Species/Brain Region(s) Examined	Modification(s)/DNA Binding Protein(s) Examined	Platform(s)	Variable(s)	Key Finding(s)	References
2009	Mouse/embryonic forebrain and midbrain	p300	ChIP-seq	Basal state	Genome-wide map of p300 identifying tissue-specific enhancers	Visel et al. (2009)
2010	Human/prefrontal cortex, neurons vs. nonneuronal cells	H3K4me3	ChIP-seq	Age	Age-correlated reorganization of H3K4me3: Cell type- and subject-specific regulation	Cheung et al. (2010)
2010	Mouse/adult hippocampus	H4K12ac	ChIP-seq/microarray	Fear conditioned learning	Dysregulated H4K12ac and gene expression in aging	Peleg et al. (2010)
2011	Mouse/adult hippocampal dentate granule cells	5mC and 5hmC	MethylC-seq/BS-seq/microarray	Electroconvulsive stimulation	Genome-wide single base resolution maps of 5mC and 5hmC	Guo, Su, Zhong, Ming, and Song (2011)
2011	Mouse/adult nucleus accumbens	H3K9me3	ChIP-seq	Chronic cocaine	Reduced H3K9me3 at heterochromatic loci, and induction of retrotransposable elements, after chronic cocaine	Maze et al. (2011)
2011	Mouse/early postnatal and adult hippocampus and cerebellum human/adult cerebellum	5hmC	Chemical labeling and immunoprecipitation/ChIP-seq	Age/*Mecp2* overexpression and knockout	Genome-wide maps of 5hmC during development and aging, including mouse models of Rett syndrome	Szulwach et al. (2011)
2011	Human/adult hippocampus	H3K4me3	ChIP-seq/RNA-seq	Cocaine and alcohol addiction	Transcriptional and chromatin changes after cocaine or alcohol exposure	Zhou et al. (2011)
2012	Rat/adult hippocampus	H3K9me3	ChIP-seq	Acute restraint stress	Increased H3K9me3 at heterochromatic loci, and repression of retrotransposable elements, after acute stress	Hunter et al. (2012)
2012	Mouse/adult cerebellar Purkinje, granule and Bergmann glial cells	5mC, 5hmC and non-CpG methylation	MeDIP-seq/TRAP-seq	Cell type, *Mecp2* knockout	Cell type–specific relationships between 5hmC, 5mC and gene expression, and evidence that MeCP2 is major 5hmC-binding protein in brain	Mellén, Ayata, Dewell, Kriaucionis, and Heintz (2012)

2012	Human, chimpanzee and macaque/adult prefrontal cortex, neurons vs. nonneuronal cells	H3K4me3	ChIP-seq	Species	Insights into human-specific modifications of neuronal epigenome, with evidence for coordinated regulation across distant sites	Shulha et al. (2012)
2012	Mouse/adult nucleus accumbens	H3K9me2	ChIP-seq/RNA-seq	Chronic morphine	Genome-wide map of H3K9me2 and identification of regulated target genes, after chronic morphine	Sun et al. (2012)
2013	Human and mouse/frontal cortex, neurons vs. nonneuronal cells	5mC and 5hmC	MethylC-seq/RNA-seq	Cell type, age	Genome-wide single-base resolution maps of 5mC throughout lifespan showing increased non-CpG methylation during development	Lister et al. (2013)
2013	Mouse/adult hippocampus	H4K5ac	ChIP-seq/microarray	Fear-conditioned learning	Insights into mechanisms of gene priming and "bookmarking" by histone acetylation during memory activation	Park, Rehrauer, and Mansuy (2013)
2013	Human/adult	H3K4me1, H3K4me3, H3K9me3, H3K27me3, H3K36me3, H3K9ac, and H3K27ac	ChIP-seq/microarray	Cell/tissue types	Global chromatin state transitions accompanying cell specification during development as well as age-related changes	Zhu et al. (2013)
2014	Mouse/adult nucleus accumbens	H3K4me3, H3K4me1, H3K27me3, H3K9me2, H3K9me3, H3K36me3, and RNA Pol II	ChIP-seq/RNA-seq	Chronic cocaine	Identification of combinations of chromatin changes (ie, signatures) that predict regulation of pre-mRNA splicing by chronic cocaine	Feng et al. (2014)
2014	Mouse/adult hippocampal dentate granule cells human brain	5mC (CpG) and non-CpG methylation (CpH)	BS-seq/ChIP-seq/RNA-seq	Age/triple *Dnmt1*, *Dnmt3a*, and *Dnmt3b* knockout	Genome-wide single-base resolution maps of neuronal DNA methylome, identifying high levels of both CpG and CpH methylation	Guo, Li, et al. (2014), Guo, Su, et al. (2014)
2014	Mouse/adult nucleus accumbens	PARP-1 and H3K4me3	ChIP-seq/RNA-seq	Chronic cocaine	Genome-wide map of PARP1, and identification of regulated target genes, after chronic cocaine	Scobie et al. (2014)

Continued

TABLE 5.1 Progress Report of Epigenomic Data From Brain—cont'd

Year	Species/Brain Region(s) Examined	Modification(s)/DNA Binding Protein(s) Examined	Platform(s)	Variable(s)	Key Finding(s)	References
2014	Mouse/developing forebrain	ISL1-LHX8	ChIP-seq	Development	Genome-wide map of ISL1-LHX8 reveals crucial role in expression of genes involved in cholinergic neurotransmission	Cho et al. (2014)
2014	Baboon/adult hippocampal progenitor cells	H3K4me3 and H3K9me3	ChIP-seq/RNA-seq	Cell type	Insight into cell differentiation	Sandstrom et al. (2014), Sandstrom et al. (2014)
2014	Mouse/developing cerebellum	Several histone marks	ChIP-seq/RNA-seq	Cell type, *NuRD* knockout	Identification of chromatin changes at genes controlled by *NuRD* chromatin remodeling complex	Yamada et al. (2014)
2014	Mouse/developing forebrain	AUTS2	ChIP-seq/RNA-seq	Development	Genome-wide map of AUTS2 and identification of downstream regulatory networks	Oksenberg et al. (2014)
2014	Mouse/adult striatum	H3K27me3S28p	ChIP-seq/RNA-seq	L-DOPA, Parkinson models	Genome-wide map of H3K27me3S28p and identification of regulated target genes after L-DOPA	Södersten et al. (2014)
2014	Mouse/developing forebrain	AUTS2	ChIP-seq/RNA-seq	Development	Genome-wide map of AUTS2 and identification of downstream regulatory networks	Gao et al. (2014)
2014	Mouse/adult nucleus accumbens	β-Catenin	ChIP-seq/RNA-seq	Chronic stress	Genome-wide map of β-catenin and identification of regulated target genes after chronic social stress	Dias et al. (2014)
2014	Mouse/adult nucleus accumbens	SIRT1 and H4K16ac	ChIP-seq	Chronic cocaine	Genome-wide map of SIRT1 and H3K16ac and identification of regulated target genes after chronic cocaine	Ferguson et al. (2015)

Modified from Maze, I., Noh, K.M., Soshnev, A.A., & Allis, C.D. (2014). Every amino acid matters: essential contributions of histone variants to mammalian development and disease. *Nature Reviews Genetics, 15*(4), 259–271. doi: 10.1038/nrg3673 and Maze, I., Shen, L., Zhang, B., Garcia, B.A., Shao, N., Mitchell, A., ... Nestler, E.J. (2014). Analytical tools and current challenges in the modern era of neuroepigenomics. *Nature Neuroscience, 17*(11), 1476–1490. doi: 10.1038/nn.3816 with permission.

neural function. To overcome this obstacle and better establish a causal relationship between ΔFosB and G9a, we generated synthetic zinc finger proteins (ZFPs) that target different regions of the *FosB* gene. The ZFPs were incorporated into Herpes simplex virus vectors and injected into nucleus accumbens. We identified one ZFP which when fused to G9a suppresses *FosB* expression in NAc and reduces behavioral responses to cocaine, but when fused to a histone acetylating moiety induces *FosB* expression and increases cocaine's behavioral effects (Heller et al., 2008). These changes are associated with selective histone methylation or acetylation, respectively, at the *FosB* gene promoter, effects not seen for other histone modifications. Moreover, no such changes are seen at about 30 of the most homologous genomic loci, and expression of those loci is not altered. This method therefore targets a single type of histone modification to a single gene within a given brain region in vivo. Interestingly, we have found that the ZFP fused to G9a completely blocks the ability of cocaine to induce ΔFosB, and this is associated with blockade of the activation of the transcription factor cAMP response element binding protein at the *FosB* promoter, which is required for *FosB* induction by cocaine (Heller et al., 2008). In contrast, the ZFP-G9a fusion did not affect activation of another transcription factor at *FosB*, which is also required for *FosB* induction. These findings suggest that a single histone modification can regulate transcription by interfering with activation of specific transcription factors and induce downstream behavioral changes. Application of this approach should help establish the causal role of epigenetic modifications in controlling gene expression in the nervous system and help decipher the extent to which transcriptional compared with epigenetic modes of regulation dominate under various conditions.

Because the brain tissue of living patients will remain off-limits for epigenetic investigations, it is imperative that alternative approaches be validated. As noted earlier, current efforts focus on peripheral tissues, with increasing attention being given to neuron-like cells induced from those tissues. Although epigenetic modifications induced in certain cell types in brain by life experiences (eg, stress or drug exposure) are likely not reflected in peripheral tissues or induced neurons, subsets of heritable epigenetic changes might be. In addition, even if such epigenetic end points in peripheral tissues do not perfectly reflect changes that occur within a given brain region, they may still be useful as biomarkers: that is, as predictors of disease vulnerability or treatment response. Genome-wide analyses of candidate epigenetic mechanisms, such as DNA methylation and microRNA levels, in peripheral tissues and induced neurons are needed to evaluate these possibilities empirically.

Finally, we return to where we started: to the notion of transgenerational epigenetic inheritance of disease vulnerability. Several studies in rodents have demonstrated the ability of lifetime exposure of a male animal to stress or a drug of abuse to alter the stress or drug sensitivity of offspring, in some cases for several generations (Bale, 2014; Dias & Ressler, 2014; Dietz et al., 2011; Franklin et al., 2012; Szutorisz et al., 2014; Vassoler et al., 2013). The patterns of inheritance are sometimes complicated: for example, being restricted to male compared with female offspring, and for this sex balance to shift across generations. The major challenge in the field is to rule out the involvement of behavioral modes of transmission, because it is known that bird and mammalian mothers allocate different levels of nutrients and care to their offspring based on the robustness of the male with which they procreated (Dietz et al., 2011). An even greater challenge is to establish the mechanism by which transgenerational inheritance might proceed. Current efforts focus on DNA methylation and microRNA regulation based on studies in other systems (Bale, 2014; Dias & Ressler, 2014; Franklin et al., 2012). As causal mechanisms are established, it will be interesting to compare the magnitude of their contribution to offspring phenotype with that of genetic and environmental factors.

Together, epigenetic investigations of the nervous system promise major advances in understanding the pathophysiology of neurological and psychiatric disorders. Major progress is expected as tools evolve to make it possible to implicate precise epigenetic mechanisms directly in molecular and neural phenomena and as improving genome-wide assays make it possible to map hosts of epigenetic modifications in single types of cells obtained from discrete regions of brain from both animal models and humans *postmortem*. It is hoped that this knowledge will lead to the development of improved therapeutics for a range of brain disorders.

REFERENCES

Amir, R. E., Van den Veyver, I. B., Wan, M., Tran, C. Q., Francke, U., & Zoghbi, H. Y. (1999). Rett syndrome is caused by mutations in X-linked MECP2, encoding methyl-CpG-binding protein 2. *Nature Genetics, 23*(2), 185–188. http://dx.doi.org/10.1038/13810.

Aran, D., Sabato, S., & Hellman, A. (2013). DNA methylation of distal regulatory sites characterizes dysregulation of cancer genes. *Genome Biology, 14*(3), R21. http://dx.doi.org/10.1186/gb-2013-14-3-r21.

Bale, T. L. (2014). Lifetime stress experience: transgenerational epigenetics and germ cell programming. *Dialogues in Clinical Neuroscience, 16*(3), 297–305.

Bell, A. C., & Felsenfeld, G. (2000). Methylation of a CTCF-dependent boundary controls imprinted expression of the Igf2 gene. *Nature, 405*(6785), 482–485. http://dx.doi.org/10.1038/35013100.

Bell, J. T., Pai, A. A., Pickrell, J. K., Gaffney, D. J., Pique-Regi, R., Degner, J. F., ... Pritchard, J. K. (2011). DNA methylation patterns associate with genetic and gene expression variation in HapMap cell lines. *Genome Biology, 12*(1), R10. http://dx.doi.org/10.1186/gb-2011-12-1-r10.

Benner, C., Konovalov, S., Mackintosh, C., Hutt, K. R., Stunnenberg, R., & Garcia-Bassets, I. (2013). Decoding a signature-based model of transcription cofactor recruitment dictated by cardinal cis-regulatory elements in proximal promoter regions. *PLoS Genetics, 9*(11), e1003906. http://dx.doi.org/10.1371/journal.pgen.1003906.

Berko, E. R., Suzuki, M., Beren, F., Lemetre, C., Alaimo, C. M., Calder, R. B., ... Greally, J. M. (2014). Mosaic epigenetic dysregulation of ectodermal cells in autism spectrum disorder. *PLoS Genetics, 10*(5), e1004402. http://dx.doi.org/10.1371/journal.pgen.1004402.

Bibikova, M., Barnes, B., Tsan, C., Ho, V., Klotzle, B., Le, J. M., ... Shen, R. (2011). High density DNA methylation array with single CpG site resolution. *Genomics, 98*(4), 288–295. http://dx.doi.org/10.1016/j.ygeno.2011.07.007.

Blair, J. D., Yuen, R. K., Lim, B. K., McFadden, D. E., von Dadelszen, P., & Robinson, W. P. (2013). Widespread DNA hypomethylation at gene enhancer regions in placentas associated with early-onset pre-eclampsia. *Molecular Human Reproduction, 19*(10), 697–708. http://dx.doi.org/10.1093/molehr/gat044.

Booth, M. J., Branco, M. R., Ficz, G., Oxley, D., Krueger, F., Reik, W., & Balasubramanian, S. (2012). Quantitative sequencing of 5-methylcytosine and 5-hydroxymethylcytosine at single-base resolution. *Science, 336*(6083), 934–937. http://dx.doi.org/10.1126/science.1220671.

Bostick, M., Kim, J. K., Esteve, P. O., Clark, A., Pradhan, S., & Jacobsen, S. E. (2007). UHRF1 plays a role in maintaining DNA methylation in mammalian cells. *Science, 317*(5845), 1760–1764. http://dx.doi.org/10.1126/science.1147939. pii:1147939.

Buenrostro, J. D., Giresi, P. G., Zaba, L. C., Chang, H. Y., & Greenleaf, W. J. (2013). Transposition of native chromatin for fast and sensitive epigenomic profiling of open chromatin, DNA-binding proteins and nucleosome position. *Nature Methods, 10*(12), 1213–1218. http://dx.doi.org/10.1038/nmeth.2688.

Caravaca, J. M., Donahue, G., Becker, J. S., He, X., Vinson, C., & Zaret, K. S. (2013). Bookmarking by specific and nonspecific binding of FoxA1 pioneer factor to mitotic chromosomes. *Genes & Development, 27*(3), 251–260. http://dx.doi.org/10.1101/gad.206458.112.

Carey, M. F., Peterson, C. L., & Smale, S. T. (2009). Chromatin immunoprecipitation (ChIP). *Cold Spring Harbor Protocols, 2009*(9). http://dx.doi.org/10.1101/pdb.prot5279. pdb.prot5279.

Cheng, C., Shou, C., Yip, K. Y., & Gerstein, M. B. (2011). Genome-wide analysis of chromatin features identifies histone modification sensitive and insensitive yeast transcription factors. *Genome Biology, 12*(11), R111. http://dx.doi.org/10.1186/gb-2011-12-11-r111.

Cheung, I., Shulha, H. P., Jiang, Y., Matevossian, A., Wang, J., ... Akbarian, S. (2010). Developmental regulation and individual differences of neuronal H3K4me3 epigenomes in the prefrontal cortex. *Proceedings of the National Academy of Sciences of the United States of America, 107*(19), 8824–8829.

Cho, H. H., Cargnin, F., Kim, Y., Lee, B., Kwon, R. J., Nam, H., ... Lee, S. K. (2014). Isl1 directly controls a cholinergic neuronal identity in the developing forebrain and spinal cord by forming cell type-specific complexes. *Public Library of Science Genetics, 10*(4), e1004280.

Choukrallah, M. A., & Matthias, P. (2014). The interplay between chromatin and transcription factor networks during B cell development: who pulls the trigger first? *Frontiers in Immunology, 5*, 156. http://dx.doi.org/10.3389/fimmu.2014.00156.

Ciabrelli, F., & Cavalli, G. (2014). Chromatin-driven behavior of topologically associating domains. *Journal of Molecular Biology*. http://dx.doi.org/10.1016/j.jmb.2014.09.013.

Covington, H. E., 3rd, Maze, I., LaPlant, Q. C., Vialou, V. F., Ohnishi, Y. N., Berton, O., ... Nestler, E. J. (2009). Antidepressant actions of histone deacetylase inhibitors. *Journal of Neuroscience, 29*(37), 11451–11460. http://dx.doi.org/10.1523/JNEUROSCI.1758-09.2009.

Covington, H. E., 3rd, Maze, I., Sun, H., Bomze, H. M., DeMaio, K. D., Wu, E. Y., ... Nestler, E. J. (2011). A role for repressive histone methylation in cocaine-induced vulnerability to stress. *Neuron, 71*(4), 656–670. http://dx.doi.org/10.1016/j.neuron.2011.06.007.

Cukier, H. N., Rabionet, R., Konidari, I., Rayner-Evans, M. Y., Baltos, M. L., Wright, H. H., ... Gilbert, J. R. (2010). Novel variants identified in methyl-CpG-binding domain genes in autistic individuals. *Neurogenetics, 11*(3), 291–303. http://dx.doi.org/10.1007/s10048-009-0228-7.

Cummings, T. J., Strum, J. C., Yoon, L. W., Szymanski, M. H., & Hulette, C. M. (2001). Recovery and expression of messenger RNA from postmortem human brain tissue. *Modern Pathology, 14*(11), 1157–1161. http://dx.doi.org/10.1038/modpathol.3880451.

Delatte, B., Deplus, R., & Fuks, F. (2014). Playing TETris with DNA modifications. *EMBO Journal, 33*(11), 1198–1211. http://dx.doi.org/10.15252/embj.201488290.

Dias, B. G., & Ressler, K. J. (2014). Parental olfactory experience influences behavior and neural structure in subsequent generations. *Nature Neuroscience, 17*(1), 89–96. http://dx.doi.org/10.1038/nn.3594.

Dias, C., Feng, J., Sun, H., Shao, N. Y., Mazei-Robison, M. S., Damez-Werno, D., ... Nestler, E. J. (2014). beta-catenin mediates stress resilience through Dicer1/microRNA regulation. *Nature, 516*(7529), 51–55. http://dx.doi.org/10.1038/nature13976.

Dietz, D. M., Laplant, Q., Watts, E. L., Hodes, G. E., Russo, S. J., Feng, J., ... Nestler, E. J. (2011). Paternal transmission of stress-induced pathologies. *Biological Psychiatry, 70*(5), 408–414. http://dx.doi.org/10.1016/j.biopsych.2011.05.005.

Dostie, J., Richmond, T. A., Arnaout, R. A., Selzer, R. R., Lee, W. L., Honan, T. A., ... Dekker, J. (2006). Chromosome Conformation Capture Carbon Copy (5C): a massively parallel solution for mapping interactions between genomic elements. *Genome Research, 16*(10), 1299–1309. http://dx.doi.org/10.1101/gr.5571506. pii:gr.5571506.

Down, T. A., Rakyan, V. K., Turner, D. J., Flicek, P., Li, H., Kulesha, E., ... Beck, S. (2008). A Bayesian deconvolution strategy for immunoprecipitation-based DNA methylome analysis. *Nature Biotechnology, 26*(7), 779–785. http://dx.doi.org/10.1038/nbt1414. pii:nbt1414.

Du, J., & Patel, D. J. (2014). Structural biology-based insights into combinatorial readout and crosstalk among epigenetic marks. *Biochimica et Biophysica Acta (BBA) – Gene Regulatory Mechanisms, 1839*(8), 719–727. http://dx.doi.org/10.1016/j.bbagrm.2014.04.011.

Dunleavy, E. M., Almouzni, G., & Karpen, G. H. (2011). H3.3 is deposited at centromeres in S phase as a placeholder for newly assembled CENP-A in G(1) phase. *Nucleus, 2*(2), 146–157. http://dx.doi.org/10.4161/nucl.2.2.15211.

Ernst, C., McGowan, P. O., Deleva, V., Meaney, M. J., Szyf, M., & Turecki, G. (2008). The effects of pH on DNA methylation state: in vitro and post-mortem brain studies. *Journal of Neuroscience Methods, 174*(1), 123–125. http://dx.doi.org/10.1016/j.jneumeth.2008.06.027.

Feng, J., Wilkinson, M., Liu, X., Purushothaman, I., Ferguson, D., Vialou, V., … Shen, L. (2014). Chronic cocaine-regulated epigenomic changes in mouse nucleus accumbens. *Genome Biology, 15*(4), R65. http://dx.doi.org/10.1186/gb-2014-15-4-r65.

Ferguson, D., Shao, N. Y., Heller, E., Feng, J., Neve, R., & Kim, H.-D. (2015). SIRT1-FOXO3a regulate cocaine actions in the nucleus accumbens. *Journal of Neuroscience, 35*(7), 3100–3111.

Filtz, T. M., Vogel, W. K., & Leid, M. (2014). Regulation of transcription factor activity by interconnected post-translational modifications. *Trends in Pharmacological Sciences, 35*(2), 76–85. http://dx.doi.org/10.1016/j.tips.2013.11.005.

Fischer, A. (2014). Epigenetic memory: the Lamarckian brain. *EMBO Journal, 33*(9), 945–967. http://dx.doi.org/10.1002/embj.201387637.

Fischle, W. (2014). Molecular mechanisms of histone modification function. *Biochimica et Biophysica Acta (BBA) – Gene Regulatory Mechanisms, 1839*(8), 621–622. http://dx.doi.org/10.1016/j.bbagrm.2014.06.003.

Franklin, T. B., Saab, B. J., & Mansuy, I. M. (2012). Neural mechanisms of stress resilience and vulnerability. *Neuron, 75*(5), 747–761. http://dx.doi.org/10.1016/j.neuron.2012.08.016.

Fuchs, G., & Oren, M. (2014). Writing and reading H2B monoubiquitylation. *Biochimica et Biophysica Acta (BBA) – Gene Regulatory Mechanisms, 1839*(8), 694–701. http://dx.doi.org/10.1016/j.bbagrm.2014.01.002.

Furuyama, T., & Henikoff, S. (2006). Biotin-tag affinity purification of a centromeric nucleosome assembly complex. *Cell Cycle, 5*(12), 1269–1274.

Gao, Z., Lee, P., Stafford, J. M., von Schimmelmann, M., Schaefer, A., & Reinberg, D. (2014). An AUTS2-Polycomb complex activates gene expression in the CNS. *Nature, 516*(7531), 349–354.

Gayatri, S., & Bedford, M. T. (2014). Readers of histone methylarginine marks. *Biochimica et Biophysica Acta (BBA) – Gene Regulatory Mechanisms, 1839*(8), 702–710. http://dx.doi.org/10.1016/j.bbagrm.2014.02.015.

Gertz, J., Varley, K. E., Reddy, T. E., Bowling, K. M., Pauli, F., Parker, S. L., … Myers, R. M. (2011). Analysis of DNA methylation in a three-generation family reveals widespread genetic influence on epigenetic regulation. *PLoS Genetics, 7*(8), e1002228. http://dx.doi.org/10.1371/journal.pgen.1002228.

Gilbert, S. F. (2012). Commentary: 'The epigenotype' by C.H. Waddington. *International Journal of Epidemiology, 41*(1), 20–23. http://dx.doi.org/10.1093/ije/dyr186.

Gilfillan, G. D., Hughes, T., Sheng, Y., Hjorthaug, H. S., Straub, T., Gervin, K., … Lyle, R. (2012). Limitations and possibilities of low cell number ChIP-seq. *BMC Genomics, 13*, 645. http://dx.doi.org/10.1186/1471-2164-13-645.

Grundberg, E., Meduri, E., Sandling, J. K., Hedman, A. K., Keildson, S., Buil, A., … Deloukas, P. (2013). Global analysis of DNA methylation variation in adipose tissue from twins reveals links to disease-associated variants in distal regulatory elements. *American Journal of Human Genetics, 93*(5), 876–890. http://dx.doi.org/10.1016/j.ajhg.2013.10.004.

Guenther, M. G., Levine, S. S., Boyer, L. A., Jaenisch, R., & Young, R. A. (2007). A chromatin landmark and transcription initiation at most promoters in human cells. *Cell, 130*(1), 77–88. http://dx.doi.org/10.1016/j.cell.2007.05.042.

Guo, F., Li, X., Liang, D., Li, T., Zhu, P., Guo, H., … Xu, G. L. (2014). Active and passive demethylation of male and female pronuclear DNA in the Mammalian zygote. *Cell Stem Cell, 15*(4), 447–458. http://dx.doi.org/10.1016/j.stem.2014.08.003.

Guo, J. U., Su, Y., Shin, J. H., Shin, J., Li, H., Xie, B., … Song, H. (2014). Distribution, recognition and regulation of non-CpG methylation in the adult mammalian brain. *Nature Neuroscience, 17*(2), 215–222. http://dx.doi.org/10.1038/nn.3607.

Guo, J. U., Su, Y., Zhong, C., Ming, G. L., & Song, H. (2011). Hydroxylation of 5-methylcytosine by TET1 promotes active DNA demethylation in the adult brain. *Cell, 145*(3), 423–434.

Heller, G., Schmidt, W. M., Ziegler, B., Holzer, S., Mullauer, L., Bilban, M., … Zochbauer-Muller, S. (2008). Genome-wide transcriptional response to 5-aza-2′-deoxycytidine and trichostatin a in multiple myeloma cells. *Cancer Research, 68*(1), 44–54. http://dx.doi.org/10.1158/0008-5472.CAN-07-2531.

Henikoff, S., & Smith, M. M. (2015). Histone variants and epigenetics. *Cold Spring Harbor Perspectives in Biology, 7*(1). http://dx.doi.org/10.1101/cshperspect.a019364.

Hiragami-Hamada, K., & Fischle, W. (2014). RNAs – physical and functional modulators of chromatin reader proteins. *Biochimica et Biophysica Acta, 1839*(8), 737–742. http://dx.doi.org/10.1016/j.bbagrm.2014.03.015.

Hodges, E., Smith, A. D., Kendall, J., Xuan, Z., Ravi, K., Rooks, M., … Hicks, J. B. (2009). High definition profiling of mammalian DNA methylation by array capture and single molecule bisulfite sequencing. *Genome Research, 19*(9), 1593–1605. http://dx.doi.org/10.1101/gr.095190.109.

Holliday, R. (1979). A new theory of carcinogenesis. *British Journal of Cancer, 40*(4), 513–522.

Houseman, E. A., Accomando, W. P., Koestler, D. C., Christensen, B. C., Marsit, C. J., Nelson, H. H., … Kelsey, K. T. (2012). DNA methylation arrays as surrogate measures of cell mixture distribution. *BMC Bioinformatics, 13*, 86. http://dx.doi.org/10.1186/1471-2105-13-86.

Hu, C. Y., Mohtat, D., Yu, Y., Ko, Y. A., Shenoy, N., Bhattacharya, S., … Verma, A. (2014). Kidney cancer is characterized by aberrant methylation of tissue-specific enhancers that are prognostic for overall survival. *Clinical Cancer Research, 20*(16), 4349–4360. http://dx.doi.org/10.1158/1078-0432.CCR-14-0494.

Hunter, R. G., Murakami, G., Dewell, S., Seligsohn, M., Baker, M. E., Datson, N. A., … Pfaff, D. W. (2012). Acute stress and hippocampal histone H3 lysine 9 trimethylation, a retrotransposon silencing response. *Proceedings of the National Academy of Sciences of the United States of America, 109*(43), 17657–17662.

Irizarry, R. A., Ladd-Acosta, C., Wen, B., Wu, Z., Montano, C., Onyango, P., … Feinberg, A. P. (2009). The human colon cancer methylome shows similar hypo- and hypermethylation at conserved tissue-specific CpG island shores. *Nature Genetics, 41*(2), 178–186. http://dx.doi.org/10.1038/ng.298. pii:ng.298.

Issa, J. P. (2004). CpG island methylator phenotype in cancer. *Nature Reviews Cancer, 4*(12), 988–993. http://dx.doi.org/10.1038/nrc1507. pii:nrc1507.

Iwafuchi-Doi, M., & Zaret, K. S. (2014). Pioneer transcription factors in cell reprogramming. *Genes & Development, 28*(24), 2679–2692. http://dx.doi.org/10.1101/gad.253443.114.

Jakovcevski, M., & Akbarian, S. (2012). Epigenetic mechanisms in neurological disease. *Nature Medicine, 18*(8), 1194–1204. http://dx.doi.org/10.1038/nm.2828.

Jordi, E., Heiman, M., Marion-Poll, L., Guermonprez, P., Cheng, S. K., Nairn, A. C., … Girault, J. A. (2013). Differential effects of cocaine on histone posttranslational modifications in identified populations of striatal neurons. *Proceedings of the National Academy of Sciences of the United States of America, 110*(23), 9511–9516. http://dx.doi.org/10.1073/pnas.1307116110.

Jurkowska, R. Z., Jurkowski, T. P., & Jeltsch, A. (2011). Structure and function of mammalian DNA methyltransferases. *ChemBioChem, 12*(2), 206–222. http://dx.doi.org/10.1002/cbic.201000195.

Kalhor, R., Tjong, H., Jayathilaka, N., Alber, F., & Chen, L. (2012). Genome architectures revealed by tethered chromosome conformation capture and population-based modeling. *Nature Biotechnology, 30*(1), 90–98. http://dx.doi.org/10.1038/nbt.2057.

Kennedy, P. J., Feng, J., Robison, A. J., Maze, I., Badimon, A., Mouzon, E., … Nestler, E. J. (2013). Class I HDAC inhibition blocks cocaine-induced plasticity by targeted changes in histone methylation. *Nature Neuroscience, 16*(4), 434–440. http://dx.doi.org/10.1038/nn.3354.

Khulan, B., Thompson, R. F., Ye, K., Fazzari, M. J., Suzuki, M., Stasiek, E., … Greally, J. M. (2006). Comparative isoschizomer profiling of cytosine methylation: the HELP assay. *Genome Research, 16*(8), 1046–1055. http://dx.doi.org/10.1101/gr.5273806.

Kim, T., Hemberg, M., & Gray, J. M. (2015). Enhancer RNAs: a class of long noncoding RNAs synthesized at enhancers. *Cold Spring Harbor Perspectives in Biology, 7*(1). http://dx.doi.org/10.1101/cshperspect.a018622.

Ko, Y. A., Mohtat, D., Suzuki, M., Park, A. S., Izquierdo, M. C., Han, S. Y., … Susztak, K. (2013). Cytosine methylation changes in enhancer regions of core pro-fibrotic genes characterize kidney fibrosis development. *Genome Biology, 14*(10), R108. http://dx.doi.org/10.1186/gb-2013-14-10-r108.

Kriaucionis, S., & Heintz, N. (2009). The nuclear DNA base 5-hydroxymethylcytosine is present in Purkinje neurons and the brain. *Science, 324*(5929), 929–930. http://dx.doi.org/10.1126/science.1169786.

Kulis, M., & Esteller, M. (2010). DNA methylation and cancer. *Advances in Genetics, 70*, 27–56. http://dx.doi.org/10.1016/B978-0-12-380866-0.60002-2.

Kulis, M., Queiros, A. C., Beekman, R., & Martin-Subero, J. I. (2013). Intragenic DNA methylation in transcriptional regulation, normal differentiation and cancer. *Biochimica et Biophysica Acta, 1829*(11), 1161–1174. http://dx.doi.org/10.1016/j.bbagrm.2013.08.001.

LaPlant, Q., Vialou, V., Covington, H. E., 3rd, Dumitriu, D., Feng, J., Warren, B. L., … Nestler, E. J. (2010). Dnmt3a regulates emotional behavior and spine plasticity in the nucleus accumbens. *Nature Neuroscience, 13*(9), 1137–1143. http://dx.doi.org/10.1038/nn.2619.

Lee, E. J., Pei, L., Srivastava, G., Joshi, T., Kushwaha, G., Choi, J. H., … Shi, H. (2011). Targeted bisulfite sequencing by solution hybrid selection and massively parallel sequencing. *Nucleic Acids Research, 39*(19), e127. http://dx.doi.org/10.1093/nar/gkr598.

Lee, K., Kim, S. C., Jung, I., Kim, K., Seo, J., Lee, H. S., … Choi, J. K. (2013). Genetic landscape of open chromatin in yeast. *PLoS Genetics, 9*(2), e1003229. http://dx.doi.org/10.1371/journal.pgen.1003229.

Leung, A., Parks, B. W., Du, J., Trac, C., Setten, R., Chen, Y., … Schones, D. E. (2014). Open chromatin profiling in mice livers reveals unique chromatin variations induced by high fat diet. *Journal of Biological Chemistry, 289*(34), 23557–23567. http://dx.doi.org/10.1074/jbc.M114.581439.

Li, J. Z., Meng, F., Tsavaler, L., Evans, S. J., Choudary, P. V., Tomita, H., … Myers, R. M. (2007). Sample matching by inferred agonal stress in gene expression analyses of the brain. *BMC Genomics, 8*, 336. http://dx.doi.org/10.1186/1471-2164-8-336.

Li, J. Z., Vawter, M. P., Walsh, D. M., Tomita, H., Evans, S. J., Choudary, P. V., … Myers, R. M. (2004). Systematic changes in gene expression in postmortem human brains associated with tissue pH and terminal medical conditions. *Human Molecular Genetics, 13*(6), 609–616. http://dx.doi.org/10.1093/hmg/ddh065.

Lieberman-Aiden, E., van Berkum, N. L., Williams, L., Imakaev, M., Ragoczy, T., Telling, A., … Dekker, J. (2009). Comprehensive mapping of long-range interactions reveals folding principles of the human genome. *Science, 326*(5950), 289–293. http://dx.doi.org/10.1126/science.1181369.

Liebers, R., Rassoulzadegan, M., & Lyko, F. (2014). Epigenetic regulation by heritable RNA. *PLoS Genetics, 10*(4), e1004296. http://dx.doi.org/10.1371/journal.pgen.1004296.

Lister, R., Mukamel, E. A., Nery, J. R., Urich, M., Puddifoot, C. A., Johnson, N. D., … Ecker, J. R. (2013). Global epigenomic reconfiguration during mammalian brain development. *Science, 341*(6146), 1237905. http://dx.doi.org/10.1126/science.1237905.

Lister, R., O'Malley, R. C., Tonti-Filippini, J., Gregory, B. D., Berry, C. C., Millar, A. H., & Ecker, J. R. (2008). Highly integrated single-base resolution maps of the epigenome in *Arabidopsis*. *Cell, 133*(3), 523–536. http://dx.doi.org/10.1016/j.cell.2008.03.029. pii:S0092-8674(08)00448-0.

Liu, Y., Olanrewaju, Y. O., Zheng, Y., Hashimoto, H., Blumenthal, R. M., Zhang, X., & Cheng, X. (2014). Structural basis for Klf4 recognition of methylated DNA. *Nucleic Acids Research, 42*(8), 4859–4867. http://dx.doi.org/10.1093/nar/gku134.

Liu, Y., Zhang, X., Blumenthal, R. M., & Cheng, X. (2013). A common mode of recognition for methylated CpG. *Trends in Biochemical Sciences, 38*(4), 177–183. http://dx.doi.org/10.1016/j.tibs.2012.12.005.

Long, H. K., Sims, D., Heger, A., Blackledge, N. P., Kutter, C., Wright, M. L., … Klose, R. J. (2013). Epigenetic conservation at gene regulatory elements revealed by non-methylated DNA profiling in seven vertebrates. *eLife, 2*, e00348. http://dx.doi.org/10.7554/eLife.00348.

Lopez-Serra, P., & Esteller, M. (2012). DNA methylation-associated silencing of tumor-suppressor microRNAs in cancer. *Oncogene, 31*(13), 1609–1622. http://dx.doi.org/10.1038/onc.2011.354.

Luu, P. L., Scholer, H. R., & Arauzo-Bravo, M. J. (2013). Disclosing the crosstalk among DNA methylation, transcription factors, and histone marks in human pluripotent cells through discovery of DNA methylation motifs. *Genome Research, 23*(12), 2013–2029. http://dx.doi.org/10.1101/gr.155960.113.

Magnani, L., Eeckhoute, J., & Lupien, M. (2011). Pioneer factors: directing transcriptional regulators within the chromatin environment. *Trends in Genetics, 27*(11), 465–474. http://dx.doi.org/10.1016/j.tig.2011.07.002.

Maiti, A., & Drohat, A. C. (2011). Thymine DNA glycosylase can rapidly excise 5-formylcytosine and 5-carboxylcytosine: potential implications for active demethylation of CpG sites. *Journal of Biological Chemistry, 286*(41), 35334–35338. http://dx.doi.org/10.1074/jbc.C111.284620.

Maunakea, A. K., Chepelev, I., Cui, K., & Zhao, K. (2013). Intragenic DNA methylation modulates alternative splicing by recruiting MeCP2 to promote exon recognition. *Cell Research, 23*(11), 1256–1269. http://dx.doi.org/10.1038/cr.2013.110.

Maze, I., Covington, H. E., 3rd, Dietz, D. M., LaPlant, Q., Renthal, W., Russo, S. J., … Nestler, E. J. (2010). Essential role of the histone methyltransferase G9a in cocaine-induced plasticity. *Science, 327*(5962), 213–216. http://dx.doi.org/10.1126/science.1179438.

Maze, I., Feng, J., Wilkinson, M. B., Sun, H., Shen, L., & Nestler, E. J. (2011). Cocaine dynamically regulates heterochromatin and repetitive element unsilencing in nucleus accumbens. *Proceedings of the National Academy of Sciences of the United States of America, 108*(7), 3035–3040.

Maze, I., Noh, K. M., Soshnev, A. A., & Allis, C. D. (2014). Every amino acid matters: essential contributions of histone variants to mammalian development and disease. *Nature Reviews Genetics, 15*(4), 259–271. http://dx.doi.org/10.1038/nrg3673.

Maze, I., Shen, L., Zhang, B., Garcia, B. A., Shao, N., Mitchell, A., … Nestler, E. J. (2014). Analytical tools and current challenges in the modern era of neuroepigenomics. *Nature Neuroscience, 17*(11), 1476–1490. http://dx.doi.org/10.1038/nn.3816.

Meissner, A., Mikkelsen, T. S., Gu, H., Wernig, M., Hanna, J., Sivachenko, A., … Lander, E. S. (2008). Genome-scale DNA methylation maps of pluripotent and differentiated cells. *Nature, 454*(7205), 766–770. http://dx.doi.org/10.1038/nature07107. pii:nature07107.

Mellén, M., Ayata, P., Dewell, S., Kriaucionis, S., & Heintz, N. (2012). MeCP2 binds to 5hmC enriched within active genes and accessible chromatin in the nervous system. *Cell, 151*(7), 1417–1430.

Messerschmidt, D. M., Knowles, B. B., & Solter, D. (2014). DNA methylation dynamics during epigenetic reprogramming in the germline and preimplantation embryos. *Genes & Development, 28*(8), 812–828. http://dx.doi.org/10.1101/gad.234294.113.

Michels, K. B., Binder, A. M., Dedeurwaerder, S., Epstein, C. B., Greally, J. M., Gut, I., … Irizarry, R. A. (2013). Recommendations for the design and analysis of epigenome-wide association studies. *Nature Methods, 10*(10), 949–955. http://dx.doi.org/10.1038/nmeth.2632.

Montano, C. M., Irizarry, R. A., Kaufmann, W. E., Talbot, K., Gur, R. E., Feinberg, A. P., & Taub, M. A. (2013). Measuring cell-type specific differential methylation in human brain tissue. *Genome Biology, 14*(8), R94. http://dx.doi.org/10.1186/gb-2013-14-8-r94.

Morris, K. V., & Mattick, J. S. (2014). The rise of regulatory RNA. *Nature Reviews Genetics, 15*(6), 423–437. http://dx.doi.org/10.1038/nrg3722.

Musselman, C. A., Khorasanizadeh, S., & Kutateladze, T. G. (2014). Towards understanding methyllysine readout. *Biochimica et Biophysica Acta (BBA) – Gene Regulatory Mechanisms, 1839*(8), 686–693. http://dx.doi.org/10.1016/j.bbagrm.2014.04.001.

Musselman, C. A., Lalonde, M. E., Cote, J., & Kutateladze, T. G. (2012). Perceiving the epigenetic landscape through histone readers. *Nature Structural & Molecular Biology, 19*(12), 1218–1227. http://dx.doi.org/10.1038/nsmb.2436.

Myers, L. C., & Kornberg, R. D. (2000). Mediator of transcriptional regulation. *Annual Review of Biochemistry, 69*, 729–749. http://dx.doi.org/10.1146/annurev.biochem.69.1.729.

Nan, X., Campoy, F. J., & Bird, A. (1997). MeCP2 is a transcriptional repressor with abundant binding sites in genomic chromatin. *Cell, 88*(4), 471–481. pii:S0092-8674(00)81887-5.

Nardelli, S. C., Che, F. Y., Silmon de Monerri, N. C., Xiao, H., Nieves, E., Madrid-Aliste, C., … Weiss, L. M. (2013). The histone code of *Toxoplasma gondii* comprises conserved and unique posttranslational modifications. *mBio, 4*(6), e00922–13. http://dx.doi.org/10.1128/mBio.00922-13.

Okada, S., Saiwai, H., Kumamaru, H., Kubota, K., Harada, A., Yamaguchi, M., … Ohkawa, Y. (2011). Flow cytometric sorting of neuronal and glial nuclei from central nervous system tissue. *Journal of Cellular Physiology, 226*(2), 552–558. http://dx.doi.org/10.1002/jcp.22365.

Oksenberg, N., Haliburton, G. D., Eckalbar, W. L., Oren, I., Nishizaki, S., Murphy, K., … Ahituv, N. (2014). Genome-wide distribution of Auts2 binding localizes with active neurodevelopmental genes. *Translational Psychiatry, 4*, e431.

O'Neill, L. P., & Turner, B. M. (2003). Immunoprecipitation of native chromatin: NChIP. *Methods, 31*(1), 76–82.

Park, C. S., Rehrauer, H., & Mansuy, I. M. (2013). Genome-wide analysis of H4K5 acetylation associated with fear memory in mice. *BMC Genomics, 14*, 539.

Peleg, S., Sananbenesi, F., Zovoilis, A., Burkhardt, S., Bahari-Javan, S., Agis-Balboa, R. C., … Fischer, A. (2010). Altered histone acetylation is associated with age-dependent memory impairment in mice. *Science, 328*(5979), 753–756. http://dx.doi.org/10.1126/science.1186088.

Pena, C. J., Bagot, R. C., Labonte, B., & Nestler, E. J. (2014). Epigenetic signaling in psychiatric disorders. *Journal of Molecular Biology, 426*(20), 3389–3412. http://dx.doi.org/10.1016/j.jmb.2014.03.016.

Perisic, T., Holsboer, F., Rein, T., & Zschocke, J. (2012). The CpG island shore of the GLT-1 gene acts as a methylation-sensitive enhancer. *Glia, 60*(9), 1345–1355. http://dx.doi.org/10.1002/glia.22353.

Peschansky, V. J., & Wahlestedt, C. (2014). Non-coding RNAs as direct and indirect modulators of epigenetic regulation. *Epigenetics, 9*(1), 3–12. http://dx.doi.org/10.4161/epi.27473.

Pikaard, C. S., & Mittelsten Scheid, O. (2014). Epigenetic regulation in plants. *Cold Spring Harbor Perspectives in Biology, 6*(12). http://dx.doi.org/10.1101/cshperspect.a019315.

Plank, J. L., & Dean, A. (2014). Enhancer function: mechanistic and genome-wide insights come together. *Molecular Cell, 55*(1), 5–14. http://dx.doi.org/10.1016/j.molcel.2014.06.015.

Plongthongkum, N., Diep, D. H., & Zhang, K. (2014). Advances in the profiling of DNA modifications: cytosine methylation and beyond. *Nature Reviews Genetics*. http://dx.doi.org/10.1038/nrg3772.

Preece, P., & Cairns, N. J. (2003). Quantifying mRNA in postmortem human brain: influence of gender, age at death, postmortem interval, brain pH, agonal state and inter-lobe mRNA variance. *Brain Research Molecular Brain Research, 118*(1–2), 60–71.

Ptashne, M. (2013a). Epigenetics: core misconcept. *Proceedings of the National Academy of Sciences of the United States of America, 110*(18), 7101–7103. http://dx.doi.org/10.1073/pnas.1305399110.

Ptashne, M. (2013b). Faddish stuff: epigenetics and the inheritance of acquired characteristics. *FASEB Journal, 27*(1), 1–2. http://dx.doi.org/10.1096/fj.13-0101ufm.

Ragunathan, K., Jih, G., & Moazed, D. (2014). Epigenetic inheritance uncoupled from sequence-specific recruitment. *Science*. http://dx.doi.org/10.1126/science.1258699.

Rahl, P. B., Lin, C. Y., Seila, A. C., Flynn, R. A., McCuine, S., Burge, C. B., … Young, R. A. (2010). c-Myc regulates transcriptional pause release. *Cell, 141*(3), 432–445. http://dx.doi.org/10.1016/j.cell.2010.03.030.

Robison, A. J., & Nestler, E. J. (2011). Transcriptional and epigenetic mechanisms of addiction. *Nature Reviews Neuroscience, 12*(11), 623–637. http://dx.doi.org/10.1038/nrn3111.

Rothbart, S. B., & Strahl, B. D. (2014). Interpreting the language of histone and DNA modifications. *Biochimica et Biophysica Acta (BBA) — Gene Regulatory Mechanisms, 1839*(8), 627–643. http://dx.doi.org/10.1016/j.bbagrm.2014.03.001.

Rudenko, A., & Tsai, L. H. (2014). Epigenetic modifications in the nervous system and their impact upon cognitive impairments. *Neuropharmacology, 80*, 70–82. http://dx.doi.org/10.1016/j.neuropharm.2014.01.043.

Sadri-Vakili, G., & Cha, J. H. (2006). Mechanisms of disease: histone modifications in Huntington's disease. *Nature Clinical Practice Neurology, 2*(6), 330–338. http://dx.doi.org/10.1038/ncpneuro0199.

Sandoval, J., Heyn, H., Moran, S., Serra-Musach, J., Pujana, M. A., Bibikova, M., & Esteller, M. (2011). Validation of a DNA methylation microarray for 450,000 CpG sites in the human genome. *Epigenetics, 6*(6), 692–702.

Sandstrom, R. S., Foret, M. R., Grow, D. A., Haugen, E., Rhodes, C. T., Cardona, A. E., … Lin, C. H. (2014). Epigenetic regulation by chromatin activation mark H3K4me3 in primate progenitor cells within adult neurogenic niche. *Science Reports, 4*, 5371.

Sarkar, S., Goldgar, S., Byler, S., Rosenthal, S., & Heerboth, S. (2013). Demethylation and re-expression of epigenetically silenced tumor suppressor genes: sensitization of cancer cells by combination therapy. *Epigenomics, 5*(1), 87–94. http://dx.doi.org/10.2217/epi.12.68.

Sarkies, P., & Miska, E. A. (2014). Small RNAs break out: the molecular cell biology of mobile small RNAs. *Nature Reviews Molecular Cell Biology, 15*(8), 525–535. http://dx.doi.org/10.1038/nrm3840.

Sawicka, A., & Seiser, C. (2014). Sensing core histone phosphorylation — A matter of perfect timing. *Biochimica et Biophysica Acta (BBA) — Gene Regulatory Mechanisms, 1839*(8), 711–718. http://dx.doi.org/10.1016/j.bbagrm.2014.04.013.

Scobie, K. N., Damez-Werno, D., Sun, H., Shao, N., Gancarz, A., Panganiban, C. H., … Nestler, E. J. (2014). Essential role of poly(ADP-ribosyl)ation in cocaine action. *Proceedings of the National Academy of Sciences of the United States of America, 111*(5), 2005–2010. http://dx.doi.org/10.1073/pnas.1319703111.

Shi, J., Marconett, C. N., Duan, J., Hyland, P. L., Li, P., Wang, Z., … Landi, M. T. (2014). Characterizing the genetic basis of methylome diversity in histologically normal human lung tissue. *Nature Communications, 5*, 3365. http://dx.doi.org/10.1038/ncomms4365.

Shulha, H. P., Crisci, J. L., Reshetov, D., Tushir, J. S., Cheung, I., Bharadwaj, R., … Akbarian, S. (2012). Human-specific histone methylation signatures at transcription start sites in prefrontal neurons. *Public Library of Science One, 10*(11), e1001427.

Södersten, E., Feyder, M., Lerdrup, M., Gomes, A. L., Kryh, H., Spigolon, G., … Hansen, K. (2014). Dopamine signaling leads to loss of Polycomb repression and aberrant gene activation in experimental parkinsonism. *Public Library of Science Genetics, 10*(9), e1004574.

Solomon, M. J., & Varshavsky, A. (1985). Formaldehyde-mediated DNA-protein crosslinking: a probe for in vivo chromatin structures. *Proceedings of the National Academy of Sciences of the United States of America, 82*(19), 6470–6474.

Song, C. X., Szulwach, K. E., Fu, Y., Dai, Q., Yi, C., Li, X., … He, C. (2011). Selective chemical labeling reveals the genome-wide distribution of 5-hydroxymethylcytosine. *Nature Biotechnology, 29*(1), 68–72. http://dx.doi.org/10.1038/nbt.1732.

Sun, H., Maze, I., Dietz, D. M., Scobie, K. N., Kennedy, P. J., Damez-Werno, D., … Nestler, E. J. (2012). Morphine epigenomically regulates behavior through alterations in histone H3 lysine 9 dimethylation in the nucleus accumbens. *Journal of Neuroscience, 32*(48), 17454–17464.

Suzuki, M., Jing, Q., Lia, D., Pascual, M., McLellan, A., & Greally, J. M. (2010). Optimized design and data analysis of tag-based cytosine methylation assays. *Genome Biology, 11*(4), R36. http://dx.doi.org/10.1186/gb-2010-11-4-r36.

Suzuki, M., Oda, M., Ramos, M. P., Pascual, M., Lau, K., Stasiek, E., … Greally, J. M. (2011). Late-replicating heterochromatin is characterized by decreased cytosine methylation in the human genome. *Genome Research, 21*(11), 1833–1840. http://dx.doi.org/10.1101/gr.116509.110.

Sweatt, J. D., Meaney, M. J., Nestler, E. J., & Akbarian, S. (Eds.). (2013). *Epigenetic regulation in the nervous system: Basic mechanisms and clinical impact*. New York: Academic Press.

Szulwach, K. E., Li, X., Li, Y., Song, C. X., Wu, H., Dai, Q., … Jin, P. (2011). 5-hmC-mediated epigenetic dynamics during postnatal neurodevelopment and aging. *Nature Neuroscience, 14*(12), 1607–1616.

Szutorisz, H., DiNieri, J. A., Sweet, E., Egervari, G., Michaelides, M., Carter, J. M., … Hurd, Y. L. (2014). Parental THC exposure leads to compulsive heroin-seeking and altered striatal synaptic plasticity in the subsequent generation. *Neuropsychopharmacology, 39*(6), 1315–1323. http://dx.doi.org/10.1038/npp.2013.352.

Taberlay, P. C., & Jones, P. A. (2011). DNA methylation and cancer. *Progress in Drug Research, 67*, 1–23.

Taberlay, P. C., Statham, A. L., Kelly, T. K., Clark, S. J., & Jones, P. A. (2014). Reconfiguration of nucleosome-depleted regions at distal regulatory elements accompanies DNA methylation of enhancers and insulators in cancer. *Genome Research, 24*(9), 1421–1432. http://dx.doi.org/10.1101/gr.163485.113.

Tee, W. W., & Reinberg, D. (2014). Chromatin features and the epigenetic regulation of pluripotency states in ESCs. *Development, 141*(12), 2376–2390. http://dx.doi.org/10.1242/dev.096982.

Thomas, M. C., & Chiang, C. M. (2006). The general transcription machinery and general cofactors. *Critical Reviews in Biochemistry and Molecular Biology, 41*(3), 105–178. http://dx.doi.org/10.1080/10409230600648736.

Tomita, H., Vawter, M. P., Walsh, D. M., Evans, S. J., Choudary, P. V., Li, J., … Bunney, W. E., Jr. (2004). Effect of agonal and postmortem factors on gene expression profile: quality control in microarray analyses of postmortem human brain. *Biological Psychiatry, 55*(4), 346–352. http://dx.doi.org/10.1016/j.biopsych.2003.10.013.

Torano, E. G., Petrus, S., Fernandez, A. F., & Fraga, M. F. (2012). Global DNA hypomethylation in cancer: review of validated methods and clinical significance. *Clinical Chemistry and Laboratory Medicine, 50*(10), 1733–1742. http://dx.doi.org/10.1515/cclm-2011-0902.

Vakoc, C. R., Mandat, S. A., Olenchock, B. A., & Blobel, G. A. (2005). Histone H3 lysine 9 methylation and HP1gamma are associated with transcription elongation through mammalian chromatin. *Molecular Cell, 19*(3), 381–391. http://dx.doi.org/10.1016/j.molcel.2005.06.011. pii:S1097-2765(05)01390-0.

Vakoc, C. R., Sachdeva, M. M., Wang, H., & Blobel, G. A. (2006). Profile of histone lysine methylation across transcribed mammalian chromatin. *Molecular and Cellular Biology, 26*(24), 9185–9195. http://dx.doi.org/10.1128/MCB.01529-06.

Vassoler, F. M., White, S. L., Schmidt, H. D., Sadri-Vakili, G., & Pierce, R. C. (2013). Epigenetic inheritance of a cocaine-resistance phenotype. *Nature Neuroscience, 16*(1), 42–47. http://dx.doi.org/10.1038/nn.3280.

Visel, A., Blow, M. J., Li, Z., Zhang, T., Akiyama, J. A., Holt, A., … Pennacchio, L. A. (2009). ChIP-seq accurately predicts tissue-specific activity of enhancers. *Nature, 457*(7231), 854–858.

Vogel-Ciernia, A., & Wood, M. A. (2014). Neuron-specific chromatin remodeling: a missing link in epigenetic mechanisms underlying synaptic plasticity, memory, and intellectual disability disorders. *Neuropharmacology, 80*, 18–27. http://dx.doi.org/10.1016/j.neuropharm.2013.10.002.

Waddington, C. H. (1957). *The strategy of the genes: A discussion of some aspects of theoretical biology*. London, UK: Allen & Unwin.

Wang, H., Maurano, M. T., Qu, H., Varley, K. E., Gertz, J., Pauli, F., … Stamatoyannopoulos, J. A. (2012). Widespread plasticity in CTCF occupancy linked to DNA methylation. *Genome Research, 22*(9), 1680–1688. http://dx.doi.org/10.1101/gr.136101.111.

Weber, M., Davies, J. J., Wittig, D., Oakeley, E. J., Haase, M., Lam, W. L., & Schubeler, D. (2005). Chromosome-wide and promoter-specific analyses identify sites of differential DNA methylation in normal and transformed human cells. *Nature Genetics, 37*(8), 853–862. http://dx.doi.org/10.1038/ng1598. pii:ng1598.

Weber, M., Hellmann, I., Stadler, M. B., Ramos, L., Paabo, S., Rebhan, M., & Schubeler, D. (2007). Distribution, silencing potential and evolutionary impact of promoter DNA methylation in the human genome. *Nature Genetics, 39*(4), 457–466. http://dx.doi.org/10.1038/ng1990.

Weick, E. M., & Miska, E. A. (2014). piRNAs: from biogenesis to function. *Development, 141*(18), 3458–3471. http://dx.doi.org/10.1242/dev.094037.

Wijetunga, N. A., Delahaye, F., Zhao, Y. M., Golden, A., Mar, J. C., Einstein, F. H., & Greally, J. M. (2014). The meta-epigenomic structure of purified human stem cell populations is defined at cis-regulatory sequences. *Nature Communications, 5*, 5195. http://dx.doi.org/10.1038/ncomms6195.

Won, K. J., Zhang, X., Wang, T., Ding, B., Raha, D., Snyder, M., … Wang, W. (2013). Comparative annotation of functional regions in the human genome using epigenomic data. *Nucleic Acids Research, 41*(8), 4423–4432. http://dx.doi.org/10.1093/nar/gkt143.

Yamada, T., Yang, Y., Hemberg, M., Yoshida, T., Cho, H. Y., Murphy, J. P., … Bonni, A. (2014). Promoter decommissioning by the NuRD chromatin remodeling complex triggers synaptic connectivity in the mammalian brain. *Neuron, 83*(1), 122–134.

Yu, M., Hon, G. C., Szulwach, K. E., Song, C. X., Jin, P., Ren, B., & He, C. (2012). Tet-assisted bisulfite sequencing of 5-hydroxymethylcytosine. *Nature Protocols, 7*(12), 2159–2170. http://dx.doi.org/10.1038/nprot.2012.137.

Zhang, B., Xing, X., Li, J., Lowdon, R. F., Zhou, Y., Lin, N., … Wang, T. (2014). Comparative DNA methylome analysis of endometrial carcinoma reveals complex and distinct deregulation of cancer promoters and enhancers. *BMC Genomics, 15*(1), 868. http://dx.doi.org/10.1186/1471-2164-15-868.

Zhang, T. Y., Labonte, B., Wen, X. L., Turecki, G., & Meaney, M. J. (2013). Epigenetic mechanisms for the early environmental regulation of hippocampal glucocorticoid receptor gene expression in rodents and humans. *Neuropsychopharmacology, 38*(1), 111–123. http://dx.doi.org/10.1038/npp.2012.149.

Zhang, X., Moen, E. L., Liu, C., Mu, W., Gamazon, E. R., Delaney, S. M., … Zhang, W. (2014). Linking the genetic architecture of cytosine modifications with human complex traits. *Human Molecular Genetics, 23*(22), 5893–5905. http://dx.doi.org/10.1093/hmg/ddu313.

Zheng, W., Zhao, H., Mancera, E., Steinmetz, L. M., & Snyder, M. (2010). Genetic analysis of variation in transcription factor binding in yeast. *Nature, 464*(7292), 1187–1191. http://dx.doi.org/10.1038/nature08934.

Zhou, Z., Yuan, Q., Mash, D. C., & Goldman, D. (2011). Substance-specific and shared transcription and epigenetic changes in the human hippocampus chronically exposed to cocaine and alcohol. *Proceedings of the National Academy of Sciences of the United States of America, 108*(16), 6626–6631. http://dx.doi.org/10.1073/pnas.1018514108.

Zhu, J., Adli, M., Zou, J. Y., Verstappen, G., Coyne, M., Zhang, X., … Bernstein, B. E. (2013). Genome-wide chromatin state transitions associated with developmental and environmental cues. *Cell, 152*(3), 642–654. http://dx.doi.org/10.1016/j.cell.2012.12.033.

Zilberman, D., Gehring, M., Tran, R. K., Ballinger, T., & Henikoff, S. (2007). Genome-wide analysis of *Arabidopsis thaliana* DNA methylation uncovers an interdependence between methylation and transcription. *Nature Genetics, 39*(1), 61–69. http://dx.doi.org/10.1038/ng1929. pii:ng1929.

Chapter 6

Bioinformatics in Neuropsychiatric Genomics

Shaun M. Purcell
Harvard Medical School, Boston, MA, United States; Icahn School of Medicine at Mount Sinai, New York City, NY, United States

Progress in bioinformatics has facilitated dramatic advances in the scope and efficiency of the biological sciences. As a powerful and transformative tool in the service of genomics and allied disciplines, it has also brought up entirely new research questions. In this chapter, we review some of the core components in this relatively new field and illustrate how bioinformatics has been and will continue to be crucial to continued progress in our understanding of the genetic basis of neuropsychiatric disease.

WHAT IS BIOINFORMATICS?

Bioinformatics represents a broad class of approaches that use concepts and tools from computer science and information technology to make sense of biological data. On one level, the core focus of bioinformatics (essentially, working with data to make it manageable and interpretable) is something that life scientists have been doing for many decades: performing assays, collating observations, and testing hypotheses. What has changed in more recent years to promote bioinformatics to the status of being a discipline in itself is primarily the sheer size and scale of biological datasets, which has necessitated an increasingly sophisticated suite of bioinformatics tools (Luscombe, Greenbaum, & Gerstein, 2001). Indeed, genomics and brain imaging are among the scientific fields that have been at the forefront of generating massive datasets, and many molecular biologists and neuroscientists would have struggled if not for work in bioinformatics.

The critical role of bioinformatics is evidenced by a compilation by the journal *Nature* of the 100 most highly cited scientific articles of all time (Van Noorden, Maher, & Nuzzo, 2014). Six of these articles are categorized as "bioinformatics" and have collectively been cited 167,502 times as of 2014, which illustrates how central these approaches have become to a wide range of life sciences. Some of the top-cited articles relate to two widely used sequence alignment algorithms and their associated software implementations, BLAST (Altschul et al., 1997) and CLUSTAL (Thompson, Higgins, & Gibson, 1994). In the future, bioinformatics is likely to become so ubiquitous and implicitly such a central part of all modern biology that it may even cease to be recognized as its own identifiable discipline as a consequence of its success (Ouzounis, 2012).

Bioinformatics can be conceptualized as composed of four interrelated areas: databases, methods, tools, and standards. In most cases, the entry point for bioinformatics is in the accumulation of data in **databases**: for example, the human reference genome, catalogs of allelic variation, and tissue-specific maps of gene expression. The next step is to develop **methods** (algorithms and statistics) that can make these data interpretable: for example, how to find protein-coding genes within a genome sequence, and how to quantify and use patterns of linkage disequilibrium between nearby allelic variants. Third, **tools** in the form of computer software or Web services need to be developed to make these methods efficient and accessible. Finally, as this work evolves with ever more tools being applied to growing numbers of datasets, the establishment of **standards** (from file formats to ontologies) becomes increasingly important so that the tools can "speak to each other" (interoperability) and work across diverse types of databases. Next, we present some examples of databases, methods, tools, and standards from applications in genomics.

Genomics, Circuits, and Pathways in Clinical Neuropsychiatry. http://dx.doi.org/10.1016/B978-0-12-800105-9.00006-8

The growth of bioinformatics has also been related to other trends in the quantitative life sciences, namely systems biology, machine learning, and "big data" approaches. The systems biology perspective stresses the importance of simultaneously considering all possible levels of an organism (from gene to protein, cell, circuit, physiology, behavior, symptom, and finally disease) to understand complex phenotypes and bioinformatics that have a central role in this effort. Technological advances have enabled substantially greater collections of diverse types of cellular "omic" data: Genomic, transcriptomic, proteomic, metabolomic, and connectomic data can be brought to bear for any given disease or trait. An important role of bioinformatics is to provide tools that can collate diverse but interconnected datasets: for example, to be able to study whether the impact of DNA variation on a disease is mediated by changes in gene expression, or to use knowledge of protein–protein interactions to guide the search for disease genes, or to find functionally related genes across different species. Computational biology is another field closely related to bioinformatics, although whereas bioinformatics is primarily concerned with organizing and manipulating large and complex biological datasets, computational biology places a greater emphasis on building theoretical models based on such datasets.

DATABASES

Work in bioinformatics has led to the generation of many forms of primary database that are now readily accessible and widely used. Here we focus on some of the ones most directly relevant for genomic research in complex traits. In many respects the starting point is the reference genome sequence from DNA sequences that have been assembled by bioinformatics software to produce the three billion bases of the human reference genome. Databases such as GenBank now contain raw sequence information on over 300,000 organisms. GenBank continues to grow at an exponential rate: from 1982 until now, the number of bases stored has doubled every 18 months.

The sequences in repositories such as GenBank have been assembled into nonredundant and richly annotated sequences. For example, the RefSeq database from the US National Center for Biotechnology Information (NCBI) connects sequence, genetic, expression, and functional information in a unified framework and allows searches based on genomic location. Web-based "genome browsers" have been developed to provide an easy means to view and search these integrated genomic databases, notably the UCSC Genome Browser and Ensembl. As well as human genomes, these aggregated resources encompass a wide variety of nonhuman data for comparative genomics research.

Perhaps most crucially for genomic studies of disease, information on the genetic variation among individuals (variants or alleles that differ from the reference sequence) have been collated in various databases by different projects, for common alleles [eg, single nucleotide polymorphisms (SNPs)] as well as for rare variants such as point mutations, short insertions and deletions (indels), and larger structural variants. Centralized databases such as dbSNP catalog and characterize all discovered SNPs and indels. Projects such as the International HapMap Project subsequently study the genome-wide patterns of variations within and across human populations and provide a basis for designing genome-wide association studies (GWASs) of disease. The 1000 Genomes Project has provided a map of very low-frequency variants through the application of whole genome sequencing to large populations. As well as providing primary data, these projects have served to coordinate the efforts of many research groups internationally to spur theoretical model development and have encouraged tools and common standards and protocols.

Higher levels of information have subsequently been coordinated and mapped on top of the basic reference sequence. For example, databases such as Online Mendelian Inheritance in Man and ClinVar relate particular genetic mutations to risk for disease and other traits. Although such genotype–phenotype databases are well-suited to Mendelian disease, appropriately quantifying the genetic basis of complex, polygenic diseases will prove more challenging.

Using experimental data and *in silico* prediction applied to the reference sequence, protein coding versus noncoding regions of the genome have been identified. For specific genes, bioinformatic approaches have been used to identify exons, introns, and promoter regions. The ENCODE project aimed to identify all functional elements of the human genome for a variety of tissue and cell types. These types of projects continue to provide windows into many areas of genome biology that are currently not well understood: the role of noncoding RNAs, different types of regulatory elements and epigenetic marks, and the determinants of chromosome structure and function. These databases and the broader mechanistic insights they generate are likely to be critical in interpreting the results of genome-wide association and sequencing studies of disease, in which many associated disease alleles do not straightforwardly affect gene function in an obvious manner.

Other types of database, many of which are linked to genome databases, focus on gene expression (the transcriptome) and protein sequence, structure, and interaction (the proteome). Increasingly, these resources are critical in guiding genetic studies of disease. For example, when studying a psychiatric or neurological disease, it may be important to know whether a given gene is expressed in the brain, and if so, where exactly in the brain, when during development, and which other genes tend to be coexpressed with it (indicative of functional biological modules). Resources such as the Allen Brain Atlas provide information on the extent of mRNA expression within the cell for all gene transcripts, stratified by tissue type, brain region, and developmental period, primarily in mouse and human. Other resources focus on collating genome-wide expression profiles on individuals with disease versus healthy individuals; for example, the CommonMind Consortium includes patients with schizophrenia and bipolar disorder as well as healthy controls.

Another ubiquitous application of informatics is in the systematic curation of scientific manuscripts in databases such as PubMed. Information in manuscripts is increasingly being connected directly to genomic databases. For example, the NCBI Gene database integrates information from a variety of sources on all known genes, including manuscripts which report on the gene as well as known mutations, associated diseases, and the biological pathways to which the gene is believed to belong. The NCBI Entrez portal provides a single search engine to query a huge amount of information on the literature, health, genomes, genes, proteins, and chemicals, aggregating information from these resources and many others.

Finally, although many of the databases mentioned are the end points of large national and international consortia, increasingly the Internet is being leveraged to allow data sharing for much smaller projects and individual laboratories. Platforms such as Sage Bionetworks' Synapse in genomics, or in the imaging domain, neuGRID, aim to act as "virtual laboratories" that let individual investigators share data and reuse analytic pipelines with the aim of transparent and reproducible research. The dream of the digital age, that we are all connected, is rapidly becoming true in modern biology. This then raises the question as to what we can actually do with these data. This topic is discussed next and concerns the methods and tools that allow us to learn from these data.

METHODS

Making sense of the vast amounts of data collected at different levels is a key goal of bioinformatics (Hogeweg, 2011). To this end, the methods of bioinformatics are primarily algorithms and statistical approaches, many of which have revolutionized areas of scientific enquiry. Naturally, this is a vast and open-ended topic that cannot be comprehensively reviewed here. Below we offer a few examples of important algorithmic or statistical approaches that facilitated new avenues of research.

Perhaps the most influential and in some ways fundamental application of bioinformatics methods in genomics arises from the problem of sequence alignment. Sequences that share similar functions or have close evolutionary relationships will tend to be more similar to each other. Alignment is the task of identifying these regions of similarity that are embedded in otherwise divergent sequences. A large number of algorithms have been developed to make alignment efficient, such as the Needleman–Wunsch algorithm for global alignment (of entire sequences) and the Smith–Waterman algorithm for local alignment (of regions within larger sequences), both of which are based on dynamic programming methods. By knowing which regions correspond to one another (eg, across the genomes of different but evolutionarily related species), we can determine which bases represent point mutations by finding the mismatches within those aligned sequences. Alignment can also be applied at the protein level, which can be important for identifying groups of functionally related proteins and identifying which proteins, or parts of proteins, are likely to be under evolutionary constraint (namely, regions that do not show high levels of divergence among species). Beyond pairwise alignment (of two sequences), multiple sequence alignment compares multiple sequences simultaneously, which is important in phylogenetic analysis. Ultimately, sequence alignment methods led to genome assembly methods and tools, the stitching together of huge numbers of short reads from sequencing machines to construct contiguous maps of entire chromosomes. Similar methods underlie modern exome and whole genome sequencing, in which reads are aligned against the reference genome to find mutations in individuals, some of which could be related to disease.

Algorithms have profoundly affected other areas of genetics, including classical disease-mapping approaches. For example, although it is no longer believed to be widely applicable to common, complex traits, the development of linkage mapping in pedigrees was greatly advanced by the application of the Elston–Stewart and Lander–Green algorithms, both of which take different approaches to computing the likelihood of the locus being linked to disease given the observed marker data.

Bioinformatic approaches are currently particularly important in predicting the likely impact of rare variants that arise from sequencing studies of disease (Ritchie & Flicek, 2014). Even healthy individuals carry rare mutations that nominally alter the protein-coding sequence of many genes; the key is to determine which variants are more likely to be functional and "deleterious": that is, that affect the function of that gene and ultimately the biological processes that, when disrupted, lead to disease.

Based on a set of known gene transcripts, it is relatively easy to use "rule-based" approaches to determine whether a given variant is a nonsense, frameshift, missense, or silent mutation with tools such as Ensembl's Variant Effect Predictor, based on the reference sequence and the known structure of gene transcripts. However, even here one can falsely annotate variants, for example, if the gene model is incorrect or if the mutation is not a simple SNP. Thus careful algorithms are called for that try to catch known sources of bias. Predicting whether a variant affects gene splicing can be more involved; methods that consider overlap with known sequence motifs can make predictions for variants that are near intron–exon boundaries but beyond the "canonical" essential splice-site nucleotides right at the boundaries. On top of this, not all genes are created equal or are equally likely to be implicated in a given disease. A loss-of-function mutation may be more likely to be related to disease if it occurs in a gene that is under heavy selective constraint, meaning that it does not tolerate severe mutations in a manner compatible with life, as inferred by the relative absence of such alleles in the population compared with the expected rate given the gene size and sequence context (Samocha et al., 2014). Similarly, for particular diseases, whether a mutation impacts transcripts of a gene that are expressed in the relevant tissue or cell type may also be important to consider; this type of information is not routinely captured by standard annotation pipelines, however.

Beyond the annotation of nonsense, splice, and frameshift mutations, it is even harder to predict whether mutations belonging to the much larger class of missense mutations are likely to have an effect (and a deleterious one) on gene function. Many missense mutations will be effectively neutral and equivalent to silent mutations, whereas others may be functionally equivalent to gene-disruptive ("loss-of-function") mutations. One approach to determine *a priori* which missenses are more likely to be deleterious is to look at nucleotide-by-nucleotide evolutionary constraint across species by multiple sequence alignment [eg, genomic evolutionary rate profiling (GERP)]. The hypothesis is that bases that are more conserved are more likely to be functional and thus intolerant of mutation. A second broad class of methods uses different types of classification algorithms to compare the properties of mutations that are known to be implicated in disease against the background of presumably neutral variants across the genome. For example, PolyPhen2 (Adzhubei et al., 2010) uses information about protein structure as well as multiple sequence alignment conservation to determine whether a novel variant looks more similar to the set of known disease mutations (in which case it is labeled "deleterious") versus randomly selected variants (in which case it is labeled "benign"). The Combined Annotation Dependent Depletion (CADD) method takes a related approach in that it assigns a score to all possible mutations by comparing the properties of two groups of mutations (Kircher et al., 2014). Unlike other approaches, CADD contrasts all known mutations (ie, those that have survived natural selection) against simulated sets of mutations (ie, a group that has not passed this benchmark and that we therefore expect to be more deleterious, on average). Like GERP, this approach is able to estimate the likely impact of all possible single nucleotide variants, not just missense mutations. Filtering or statistical weighting of mutations based on these metrics is critical in genetic association studies based on sequence data as well as in interpretation of sequence data in clinical contexts.

Currently, bioinformatic algorithms are widely used in network models of biological data, which have a central role in integrative, multimodal models of disease. For example, networks based on gene coexpression or protein–protein interaction can be intersected with genetic association data to ask whether putative risk genes appear to colocalize in the network above chance expectation. Network models have been applied to studies of somatic mutation in cancer (Leiserson et al., 2015) and *de novo* gene mutations in autism (O'Roak et al., 2012). Such analyses can yield biological insight, pointing to the types of processes involved in pathogenesis, and can be used to predict new candidate genes.

TOOLS

Often methodological work in bioinformatics goes hand in hand with implementation and distribution of software tools that enable research groups around the world to apply (and extend) these methods. Typically, these tools are "open source," meaning that others can freely view, modify, and redistribute the software. Bioinformatic tools come in a variety of flavors from Web services to general-purpose software suites and individual scripts focused on specific problems.

One widely used bioinformatics resource is Bioconductor (Huber et al., 2015), which is based on the R statistical programming language. Although largely centered on gene expression analysis, the framework extends to cover microarray, sequence, SNP, and other data types. A primary aim of the project is to provide a robust and comprehensive set of well-documented analytic tools. Because it has a common set of tools (as opposed to every laboratory evolving its own set of brittle, "home-grown" scripts for analysis), it massively promotes the transparency and reproducibility of research, which is also a central goal of the Bioconductor project. Just as best-practice standards (of reproducibility, quality control measures, etc.) have evolved for wet-laboratory research, it is increasingly being recognized that "dry-laboratory" computational research also needs such a framework to allow disparate groups to replicate and understand bioinformatic methods rather than keeping communications relegated only to "black box" descriptions in methods sections or appendices of manuscripts.

Oriented more toward technically advanced users, community-based projects such as BioPerl, BioJava, and Biopython aim to provide core sets of functions that allow different researchers to avoid "reinventing the wheel": for example, by having to develop the basic software infrastructure to deal with sequence alignment data or accessing functional annotations for genes. At the other end of the spectrum, projects such as Galaxy aim to provide accessible Web-based platforms to allow scientists of any background to be able to leverage powerful bioinformatics tools. Most bioinformatics research is still carried out on high-performance computers, typically running operating systems such as Linux, although the rise of cloud computing and more sophisticated Web interfaces will likely mean that tools such as Galaxy will become more popular over time.

In addition to general-purpose tools and frameworks that can be used in many contexts, specific research areas will typically have focused tools for particular commonly performed tasks. For example, within the field of genetic association analysis, a wide number of tools perform steps from linkage analysis (eg, Merlin) to genome-wide association analysis (eg, PLINK) and to the analysis of next-generation sequence data (eg, GATK). Although general-purpose tools (such as R, Perl, and other generic programming languages) could theoretically be used to perform all of these steps, in practice it is often desirable to adopt tools more closely matched to the task at hand. Even within genetic association analysis, some tools will be ideally suited to family-based datasets, to samples from admixed populations, or to datasets with large numbers of measured phenotypes, and so on, which means that a "one-size-fits-all" strategy is typically suboptimal.

Visualization is another key component of the bioinformatics toolset. Powerful tools have been developed to view different types of data, such as genomic data via browsers (eg, UCSC genome browser) or standalone software (eg, Integrative Genome Viewer) and network data (eg, CytoScape). Alternatively, packages such as R offer flexible graphical environments that can produce complex, publication-quality images.

STANDARDS

Many databases and tools have been developed over the past decade, but there is concern that many of them will fall into a state of dysfunction or irrelevance owing to a lack of maintenance or interoperability with other tools (Tan et al., 2010). One important strategy to future-proof many resources is to adopt standards now in terms of both minimum requirements that must be met to produce robust tools (eg, adoption of open-source software practices and public repositories for code such as GitHub) and a lingua franca that can connect different projects, databases, and tools.

Standardization can involve relatively prosaic initiatives such as the adoption of common file formats. For example, the Sequence Alignment/Map format, which grew out of the 1000 Genomes Project, provides a common yet flexible way to represent data from a variety of sequencing platforms (Li et al., 2009). Numerous tools (for tasks such as variant calling and viewing alignments) have adopted this format along with its binary implementation. As datasets become larger and more complex, a proliferation of incompatible file formats can lead to more than mere inconvenience; it can effectively determine the types of questions that researchers are realistically able to ask of their data. For large projects, conversion between file formats incurs significant costs not just in terms of computation and disk space but also in the risk of errors creeping in. Another focus is on producing application program interfaces rather than file formats, with the idea that rather than copying huge files, researchers point Web-enabled tools to perform computations on data that reside "in the cloud."

Another route to standardization is the adoption of universally accepted and stable identifiers. Many of these ideas grew out of the widespread use of URLs in the Internet, terms that uniquely identify and locate a given resource. Schemes exist for data (life science identifiers), for publications (digital object identifiers), and increasingly for authors [open researcher and contributor identification (ORCID) IDs]. Nonetheless it is often challenging to unify resources and adopt common identification schemes because different resources have been developed for different goals and so have different needs. Even for the central questions regarding information on gene and proteins, there can be significant bottlenecks in connecting databases when different (and not always consistent) identification schemes have been used. Indicative of this problem, researchers have been motivated to create automated tools to perform mappings across these terms (Huang et al., 2008).

Beyond simple identifiers, there has been considerable effort to formalize biological knowledge into ontologies (the formal naming of objects and the definition of their types and relationships). For example, the Gene Ontology (GO) Project attempts to describe the properties of gene products systematically in terms of their cellular component, molecular function, and biological processes (The Gene Ontology project in 2008, 2008). This ontology is hierarchical, which means that genes can be grouped by GO terms at different levels, for example, from genes involved in the biological process of "pigmentation" to "pigmentation during development," "regulation of pigmentation during development," "negative regulation of eye pigmentation," and so on. The grouping of genes by "GO terms" or by other sources of functional annotation (eg, the Kyoto Encyclopedia of Genes and Genomes a collection of manually created pathways based on experimental data) enables different forms of "gene set enrichment analysis," testing whether groups of genes are over- or underrepresented in particular predefined functional groups.

Finally, whereas genomics and proteomics have made considerable progress in standardization, developing shared semantics and syntaxes for interpreting and processing data, the effectively unbounded universe of phenotypic data is also in need of greater standardization (Deans et al., 2015). So-called "phenomics"—that is, taking a systems biology approach to the measurement of phenotypes and environmental factors—is an emerging field that aims to adopt some of the standardization in genomics and other sciences. For example, there are efforts to produce more standardized approaches to measuring complex traits efficiently using paradigms that different researchers can adopt rapidly (eg, the PhenX project). There are also efforts to use ontologies for traits and tools to process and extract "computable phenotypes" automatically from imaging data or the supplemental data that accompany manuscripts. Once common standards are in place, it is easier to apply powerful analytic tools to mine those data and connect them to the genomic and proteomic knowledge bases.

SUMMARY

There has been considerable progress in the application of computer science and information technology to the management and analysis of biological data. Neuropsychiatric genetics is being driven by large-scale studies that critically depend on bioinformatics methods for both their data generation and the analysis and interpretation of those datasets. For example, the Psychiatric Genomics Consortium GWAS of schizophrenia (Schizophrenia Working Group of the Psychiatric Genomics Consortium, 2014) leveraged previous work in bioinformatics, including: (1) previous SNP detection efforts including the HapMap Project, (2) a variety of statistical genetic analysis tools and methods, (3) integration with brain gene expression datasets and the expression quantitative trait loci derived from them, (4) integration with data on epigenetic markers indicative of active enhancers in a variety of different tissues and cell lines, and (5) gene set enrichment analysis based on functionally related groups of genes. The combination of these different approaches allowed for greater interpretation of the statistical association signals observed in this study: for example, highlighting the role of genes involved in glutamatergic neurotransmission and providing evidence for links between the immune system and schizophrenia.

Similar large-scale projects are also emerging in neuroscience which rely heavily on "neuroinformatics." For example, the Human Connectome is applying diffusion tensor imaging and resting-state functional magnetic resonance imaging to map the physical and functional connections among all brain regions, which requires massive and sophisticated computing resources. Imaging datasets are also being augmented with GWAS data to begin to map the genetic basis of brain structure and function. For example, the ENIGMA consortium's GWAS of brain structure (Hibar et al., 2015) featured over 30,000 individuals from 50 cohorts and found common gene variants that influence the volume of human subcortical brain structures.

Taken together, all of these efforts should provide a solid base upon which our understanding of neuropsychiatric disease can develop when bioinformatics analysis is coupled with the necessary sample collection efforts and functional screening or follow-up studies.

LIST OF URLs

Core Genomic Databases and Portals

NCBI: http://www.ncbi.nlm.nih.gov/
Entrez: http://www.ncbi.nlm.nih.gov/sites/gquery
RefSeq: http://www.ncbi.nlm.nih.gov/refseq/
NCBI Gene: http://www.ncbi.nlm.nih.gov/gene/
GenBank: http://www.ncbi.nlm.nih.gov/genbank/
PubMed: http://www.ncbi.nlm.nih.gov/pubmed

Genome Browsers

UCSC Genome Browser: https://genome.ucsc.edu/cgi-bin/hgGateway
Ensembl: http://www.ensembl.org/index.html
Integrative Genome Viewer: http://www.broadinstitute.org/igv/

Genetic Variation

dbSNP: http://www.ncbi.nlm.nih.gov/SNP/
HapMap: http://hapmap.ncbi.nlm.nih.gov/
1000 Genomes: http://www.1000genomes.org/

ClinVar: http://www.ncbi.nlm.nih.gov/clinvar/
Online Mendelian Inheritance in Man: http://omim.org/

Gene Expression Resources

Allen Brain Atlas: http://www.brain-map.org/
CommonMind Consortium: http://commonmind.org/

Programming Environments for Bioinformatics

BioPerl: http://bioperl.org/
R: http://www.r-project.org/
Bioconductor: http://bioconductor.org/
Galaxy: http://galaxyproject.org/
neuGRID: https://neugrid4you.eu/
Synapse: http://www.sagebase.org/synapse/
GitHub: https://github.com/

Genetic Linkage, Association, and Variation Detection

Merlin: http://csg.sph.umich.edu/abecasis/Merlin/
PLINK: http://pngu.mgh.harvard.edu/purcell/plink/

Network and Pathway Analysis

HotNet: http://compbio.cs.brown.edu/projects/hotnet/
Cytoscape: http://cytoscape.org/
Kyoto Encyclopedia of Genes and Genomes: http://www.genome.jp/kegg/

Rare Variant Detection and Annotation

GATK: https://www.broadinstitute.org/gatk/
PolyPhen2: http://genetics.bwh.harvard.edu/pph2/
CADD: http://cadd.gs.washington.edu/

Identifiers, Ontologies, and Standards

Digital Object Identifier System: http://www.doi.org/
ORCID: http://www.orcid.org/
Sequence Ontology: http://www.sequenceontology.org/
Gene Ontology: http://geneontology.org/
PhenX: https://www.phenx.org/

Psychiatric/Neuroscience Projects

Psychiatric Genomics Consortium: http://www.med.unc.edu/pgc/
ENIGMA: http://enigma.ini.usc.edu/
Human Connectome Project: http://humanconnectome.org/

REFERENCES

Adzhubei, I. A., Schmidt, S., Peshkin, L., Ramensky, V. E., Gerasimova, A., Bork, P., ... Sunyaev, S. R. (2010). A method and server for predicting damaging missense mutations. *Nature Methods, 7*(4), 248–249. http://dx.doi.org/10.1038/nmeth0410-248.

Altschul, S. F., Madden, T. L., Schäffer, A. A., Zhang, J., Zhang, Z., Miller, W., & Lipman, D. J. (1997). Gapped BLAST and PSI-BLAST: a new generation of protein database search programs. *Nucleic Acids Research, 25*(17), 3389–3402.

Deans, A. R., Lewis, S. E., Huala, E., Anzaldo, S. S., Ashburner, M., Balhoff, J. P., ... Mabee, P. (2015). Finding our way through phenotypes. *PLoS Biology, 13*(1), e1002033. http://dx.doi.org/10.1371/journal.pbio.1002033.

Hibar, D. P., Stein, J. L., Renteria, M. E., Arias-Vasquez, A., Desrivières, S., Jahanshad, N., ... Medland, S. E. (2015). Common genetic variants influence human subcortical brain structures. *Nature.* http://dx.doi.org/10.1038/nature14101.

Hogeweg, P. (2011). The roots of bioinformatics in theoretical biology. *PLoS Computational Biology, 7*(3). http://dx.doi.org/10.1371/journal.pcbi.1002021.

Huang, D. W., Sherman, B. T., Stephens, R., Baseler, M. W., Lane, H. C., & Lempicki, R. A. (2008). DAVID gene ID conversion tool. *Bioinformation, 2*(10), 428–430.

Huber, W., Carey, V. J., Gentleman, R., Anders, S., Carlson, M., Carvalho, B. S., ... Morgan, M. (2015). Orchestrating high-throughput genomic analysis with Bioconductor. *Nature Methods, 12*(2), 115–121. http://dx.doi.org/10.1038/nmeth.3252.

Kircher, M., Witten, D. M., Jain, P., O'Roak, B. J., Cooper, G. M., & Shendure, J. (2014). A general framework for estimating the relative pathogenicity of human genetic variants. *Nature Genetics, 46*(3), 310–315. http://dx.doi.org/10.1038/ng.2892.

Leiserson, M. D. M., Vandin, F., Wu, H.-T., Dobson, J. R., Eldridge, J. V., Thomas, J. L., ... Raphael, B. J. (2015). Pan-cancer network analysis identifies combinations of rare somatic mutations across pathways and protein complexes. *Nature Genetics, 47*(2), 106–114. http://dx.doi.org/10.1038/ng.3168.

Li, H., Handsaker, B., Wysoker, A., Fennell, T., Ruan, J., Homer, N., ... 1000 Genome Project Data Processing Subgroup. (2009). The Sequence Alignment/Map format and SAMtools. *Bioinformatics (Oxford, England), 25*(16), 2078–2079. http://dx.doi.org/10.1093/bioinformatics/btp352.

Luscombe, N. M., Greenbaum, D., & Gerstein, M. (2001). What is bioinformatics? An introduction and overview. *Yearbook of Medical Informatics.* Retrieved from http://www.genopomii.unina.it/genopom/files/bioinformatics_review.pdf.

O'Roak, B. J., Vives, L., Girirajan, S., Karakoc, E., Krumm, N., Coe, B. P., ... Eichler, E. E. (2012). Sporadic autism exomes reveal a highly interconnected protein network of de novo mutations. *Nature, 485*(7397), 246–250. http://dx.doi.org/10.1038/nature10989.

Ouzounis, C. A. (2012). Rise and demise of bioinformatics? Promise and progress. *PLoS Computational Biology, 8*(4), e1002487. http://dx.doi.org/10.1371/journal.pcbi.1002487.

Ritchie, G. R., & Flicek, P. (2014). Computational approaches to interpreting genomic sequence variation. *Genome Medicine, 6*(10), 87. http://dx.doi.org/10.1186/s13073-014-0087-1.

Samocha, K. E., Robinson, E. B., Sanders, S. J., Stevens, C., Sabo, A., McGrath, L. M., ... Daly, M. J. (2014). A framework for the interpretation of de novo mutation in human disease. *Nature Genetics, 46*(9), 944–950. http://dx.doi.org/10.1038/ng.3050.

Schizophrenia Working Group of the Psychiatric Genomics Consortium. (2014). Biological insights from 108 schizophrenia-associated genetic loci. *Nature, 511*(7510), 421–427. http://dx.doi.org/10.1038/nature13595.

Tan, T. W., Tong, J. C., Khan, A. M., de Silva, M., Lim, K. S., & Ranganathan, S. (2010). Advancing standards for bioinformatics activities: persistence, reproducibility, disambiguation and Minimum Information about a Bioinformatics investigation (MIABi). *BMC Genomics, 11*(Suppl. 4), S27. http://dx.doi.org/10.1186/1471-2164-11-S4-S27.

The Gene Ontology project in 2008. (2008). *Nucleic Acids Research, 36*(Database issue), D440–D444. http://dx.doi.org/10.1093/nar/gkm883.

Thompson, J. D., Higgins, D. G., & Gibson, T. J. (1994). CLUSTAL W: improving the sensitivity of progressive multiple sequence alignment through sequence weighting, position-specific gap penalties and weight matrix choice. *Nucleic Acids Research, 22*(22), 4673–4680.

Van Noorden, R., Maher, B., & Nuzzo, R. (2014). The top 100 papers. *Nature, 514*(7524), 550–553. http://dx.doi.org/10.1038/514550a.

Chapter 7

Imaging Genomics and ENIGMA

Paul M. Thompson and Derrek P. Hibar
University of Southern California, Los Angeles, CA, United States

Jason L. Stein
University of North Carolina, Chapel Hill, NC, United States

Neda Jahanshad
University of Southern California, Los Angeles, CA, United States

INTRODUCTION

The quest to identify genetic factors that change the structure or function of the living brain has been greatly accelerated by advances in genetic sequencing technology. Rapid genotyping methods can assess millions of variants in the genomes of thousands of subjects based on saliva or other biological samples; whole genome sequences are being collected for studies of Alzheimer disease (AD) and major psychiatric illnesses. Using these technologies, often in hundreds of thousands of people, genome-wide association studies (GWASs) can pinpoint within the human genome significant and unexpected sources of disease risk for illnesses such as schizophrenia and bipolar disorder. The mechanism of these disorders has long been an enigma, but one whose mystery is close to being cracked.

Entire genomes can now be sequenced and related to clinical and other biomarker data. This has ushered in a new level of enthusiasm for biobanks.[1] Many international projects are under way or just beginning to relate genetic variation to disease risk and treatment outcomes. Psychiatric genetics consortia worldwide have amassed genetic data relevant to AD, schizophrenia, bipolar illness, major depression, and attention-deficit hyperactivity disorder (ADHD); this led to significant breakthroughs and identified new molecular and physiological mechanisms that contribute to these illnesses, as well as potential new drug targets. These discoveries are changing and advancing our understanding of the causes of devastating brain diseases, particularly those in need of novel therapeutic development.

POWER OF MRI

At the same time, neuroimaging methods have rapidly advanced for studying the structure and function of the living brain, and uncovering patterns of brain connectivity and neural circuits. Magnetic resonance imaging (MRI) is now commonly used in major medical centers worldwide. MRI is the mainstay of radiologic diagnosis for many major brain diseases. These include neurodegenerative dementias, vascular diseases and stroke, malignancies of the nervous system such as glioma, immune diseases such as multiple sclerosis, and disorders of myelin such as amyotrophic lateral sclerosis and motor neuron disease. Imaging is often used for differential diagnosis of conditions whose symptoms somewhat overlap but whose treatments differ, such as frontotemporal dementia, stroke, and AD.

Aside from its wide use for surgical planning, diagnostic radiology, and drug trials, numerous large-scale research initiatives assess patient populations with brain imaging and include in-depth clinical, behavioral, and genetic assessments. Some of these efforts are now scanning thousands of subjects (Ikram et al., 2011; Jack et al., 2010; Schumann et al., 2010; de Zubicaray et al., 2008), offering a new source of power to discover factors that help or harm the brain. Analytic methods to handle the sheer volume of clinical, imaging and genetic data, and extract information from it, still

1. A biobank is a type of repository for biological specimens to be used in medical research.

Genomics, Circuits, and Pathways in Clinical Neuropsychiatry. http://dx.doi.org/10.1016/B978-0-12-800105-9.00007-X

lag behind the pace of data collection. As such, the mathematics and statistics used to distil patterns from biomedical data are quickly advancing, leading to efforts in big data analytics. Data are being collected on an unprecedented scale and at a previously unimaginable rate. Many patterns can be discovered and verified that no human radiologist could ever identify. These include genetic predictors of brain structure and decline, inferred from computerized analyses of thousands of brain scans.

WHAT IS IMAGING GENOMICS?

Worldwide experts in imaging and genetics have joined forces to create a new field called **imaging genomics**. Using methods from both fields, scientists are tackling two key sets of questions:

1. **Imaging of Disease.** How do genetically influenced major neurological and psychiatric diseases affect the brain? Which imaging-derived brain measures best differentiate patients from controls?
2. **Genetic Variation.** How does individual genetic variation affect brain structure and function? What kinds of genetic variations contribute to differences in brain structure and function? Do these same genetic variants affect disease risk, and how?

Whereas imaging genomics can refer to many types of image modalities, MRI is less invasive than some other types of brain scans, such as positron emission tomography, which requires a radiotracer injection. In addition, MRI's high-contrast scans and detailed neuroanatomic resolution provide large amounts of information on brain structure and function; this information is further enhanced by accumulating numerous scans collected daily for decades at medical centers worldwide.

The Endophenotype Concept and Biomarkers

To discover underlying the genetic promoters of disease, many researchers in psychiatry have sought endophenotypes for brain disorders, biological measures related to a disease process that may be closer to its molecular causes than clinical or behavioral assessments. Endophenotypes are sought to uncover genetic and mechanistic pathways that pave the route to disease.

In parallel, much neurological research is devoted to efficiently identifying and refining measures of disease burden and progression from clinical brain images: for example, rates of brain tissue loss, markers of abnormal metabolism, blood flow, vascular disease, or brain amyloid load. A disease process may have many complementary biomarkers; even for relatively well-understood conditions such as AD or Parkinson disease, national research initiatives collect a variety of structural and functional brain scans, blood analytes, behavioral and cognitive data, and genomic assays. There is great interest in understanding the added value, for diagnosis and prognosis, of newer imaging techniques versus older ones (Demirhan et al., 2015; Madsen et al., 2015a,b). These include resting-state functional MRI, arterial spin labeling (an MRI-based measure of brain perfusion), and diffusion imaging, which can map anatomic connections in the brain in exquisite detail (Daianu, Jacobs, et al., 2015; Daianu, Mezher, et al., 2015).

In this chapter, we discuss how global alliances, new analysis methods, and ever-increasing computing power are used to answer the two main questions of imaging genetics: How do the major neurological and psychiatric diseases affect the brain? and How do genetic variations relate to brain measures and disease risk?

A decade ago there was some confidence that imaging studies would quickly and definitively answer both questions as scanning technology improved. Here, we review some of the major findings in the field and some surprises encountered along the way.

MAPPING BRAIN DISEASES IN NEUROLOGY

As soon as MRI became commonplace in the 1980s, image-derived patterns of brain differences were discovered that distinguish patients from healthy individuals; these patterns now form the basis of diagnostic neuroradiology.

Patients with AD, for example, differ from cognitively normal elderly controls in their overall rate of brain tissue loss. They also differ in the level of atrophy in key learning and memory systems such as the hippocampus, and in the level of brain amyloid, which can be measured using positron emission tomography. Repeating the scanning of patients at different stages of their disease also led to time-lapse maps of AD progression (Thompson et al., 2001). In AD, tissue is lost in the brain's memory systems, particularly regions of the temporal lobes and hippocampus. As the disease progresses, cell loss and atrophy spread to the association cortices and limbic areas involved in emotion and affect. Much later, cell loss proceeds into frontal areas involved in planning and executive function. Perhaps remarkably, the primary sensory areas of the brain which are involved in vision and touch are spared, even late in the illness; disturbances in touch and visual function are not typical in dementia, in line with the anatomy of the disease.

The pattern of atrophy on MRI does not proceed identically in every patient, but there is a characteristic sequence of disease effects. Some brain regions are more vulnerable than others. There are other dementia subtypes, including **vascular dementia,** which is usually evident as small strokes or scattered white matter lesions on brain MRI images. Parkinson and Huntington disease, by contrast, involve subcortical atrophy. Even infectious diseases such as HIV and AIDS are sometimes associated with a selective pattern of brain atrophy distinct from that seen in AD. Brain tissue loss in people who are HIV-positive is generally much milder, especially in patients receiving antiretroviral therapy, but it is also linked with measures of disease severity, such as T-cell counts and with cognitive impairment (Chiang et al., 2007; Thompson et al., 2005)

Neurological diseases differ in their effects on the brain, and this is borne out by histological studies. In diseases whose pathological basis is known, such as amyloid and tau pathology in the case of AD, and white matter disruption in the case of multiple sclerosis, some of the molecular substrates of the disease are found in many brain regions that show measurable signal changes on imaging.

MRI VERSUS CELLULAR MEASURES

It is vital to realize that neuroimaging measures are at best proxies for biochemical and cellular changes that we cannot readily measure in living patients. The structure of the brain is extraordinarily complex, and neuroimaging provides a unique window into the structure and function of the brain that cannot otherwise be measured in living humans, which makes it indispensable for the study of gross brain structure and function. Some disease biomarkers are derivable from human neuroimaging that no cellular pathologist could identify. The functional synchrony of brain activation visible with resting-state MRI is not observable postmortem and visualization of the three-dimensional structure of the postmortem human brain at a cellular level is in its infancy. Whole-brain imaging also allows for systems-level insights into the functioning of the human brain. Connectomics, for example, is a relatively new field of brain imaging that attempts to understand the anatomical wiring or circuitry of the living brain, allowing discovery of the principles of communication among brain regions, and how they break down in a variety of brain diseases (Daianu et al., 2013; Jahanshad et al., 2012). Such a holistic view of brain function would not be easy to appreciate via molecular analysis of brain tissue from any one brain region via postmortem evaluations.

MAPPING BRAIN DISORDERS: PSYCHIATRY

The whole field of biological psychiatry is predicated on a vast and growing body of evidence that psychiatric disorders, from schizophrenia to bipolar disorders and to major depression and ADHD, have biological correlates that can be measured and analyzed.

Clinical diagnosis of these disorders still relies largely on cognitive and behavioral evaluations. Even today, neuroimaging tests are not routinely used to diagnose psychiatric illness, but instead are used to determine whether symptoms have a clear organic cause such as a neoplastic change or congenital anomaly in the brain.

Brain imaging has yielded over 2 decades of discoveries of brain abnormalities that correlate with symptom profiles, diagnosis, prognosis, and outcomes in psychiatry. However, most major psychiatric disorders have no standard radiologic diagnosis. In a patient diagnosed with AD, for example, the brain's temporal lobes may have already lost around 10–15% of their volume compared with age-matched healthy controls. In major depression, many people with protracted illness show brain differences in the same region, the hippocampus. But by all accounts, these differences have a much smaller effect size, which makes them hard to distinguish from normal variations when individual scans are reviewed.

PSYCHIATRIC NEUROIMAGING EXPANDS WORLDWIDE

To assess systematic differences in brain structure and function across many psychiatric illnesses, international consortia were formed, such as the Enhancing Neuro Imaging Genetics through Meta-Analysis (ENIGMA) Consortium, to aggregate evidence of brain abnormalities from multiple cohorts worldwide (Thompson et al., 2014).

Meta-Analysis

Meta-analysis is an approach to assess the strength of statistical associations based on their effect sizes across numerous studies. Commonly, this is based on a retrospective review of the published literature. However, when the original researchers are not involved in the meta-analysis, little can be done to standardize the traits being analyzed. In imaging, for example, vastly different approaches are used to measure seemingly similar aspects of brain structure, leaving unresolved discrepancies among studies that may be pooled together unjustly.

Global alliances have formed to analyze imaging data sets in a coordinated way. Prior agreements are made, along with defined protocols on how to compute regional brain volumes on MRI, and measures of tracts and connectivity on diffusion tensor imaging (DTI). Similar protocols may be used to standardize behavioral and clinical measures collected using different scales, tests, and references.

ENIGMA STUDIES OF BRAIN DISEASE

The ENIGMA Consortium is one such effort that aggregates brain imaging, clinical, and genome-wide genetic data from cohorts in 35 countries worldwide, and studies 12 major brain diseases. ENIGMA's members meta-analyzed the evidence for structural brain differences on MRI in schizophrenia (van Erp et al., 2015; Okada et al., 2015), bipolar illness (Hibar et al., 2016), major depression (Renteria et al., 2015; Schmaal et al., 2015), addictive disorders (Mackey and the ENIGMA Addictions Working Group, 2015), obsessive–compulsive disorder (Boedhoe et al., 2016), and ADHD (Franke, Hoogman, Zwiers, Mennes, for the Enigma-ADHD Working Group, 2014) by measuring the volumes of many brain regions repeatedly implicated in these illnesses.

These MRI studies of thousands of patients reveal clear, systematic, and unquestionable brain differences in all of these illnesses. This is already an advance. Not long ago, skeptics claimed that evidence of brain differences in psychiatry was limited (Szasz, 2006). The patterns differ across illnesses, and the magnitude of the differences depends on numerous factors. In schizophrenia, for example, patients show volume deficits on MRI in several brain regions including the hippocampus in the temporal lobes. The cellular basis of these differences has long been debated (Weinberger & McClure, 2002). Initial claims of cellular disarray in schizophrenia were not replicated, and the disease has been termed "the graveyard of neuropathologists." It is possible that a disorder of brain function may not have a correlate that is readily identifiable postmortem; if the symptoms relate to the functional interaction or synchrony of brain systems, these abnormalities may be observable only through functional assessments.

Brain abnormalities in schizophrenia depend on medication exposure, among other things; psychotic patients taking second-generation antipsychotics (eg, olanzapine) may have greater regional volumes in the basal ganglia than do control subjects, and apparently preserved gray matter in the cortex (van Erp et al., 2015; Thompson et al., 2009). Some researchers believe that this is due to a trophic effect of the medication. Bipolar patients who take lithium tend to have greater gray matter volumes in some brain regions than those who do not (Bearden et al., 2008; Hibar et al., 2016). It is not yet known whether the enlargement is beneficial or related to functional outcomes. Brain differences are often confounded with factors that affect medication dose, such as disease severity. These factors can be disentangled only with a randomized design of the kind commonly used in a drug trial.

Patients with major depression may also have hippocampal abnormalities that depend on how long they have been ill (Schmaal et al., 2015). Ideally, these evolving brain differences could be resisted, or the rate of change slowed, with targeted medications. This would have an enormous impact on society and public health, especially because increased deviations from normal brain structure and function are generally a sign of poorer patient outcomes.

DEFUSING CONTROVERSY WITH META-ANALYSIS

One of the many reasons meta-analysis is valuable in all fields of science is brought to light in biological psychiatry. Few researchers would disagree that effect sizes for case–control brain differences in psychiatry are typically smaller than those for neurological illnesses such as AD. Larger samples are needed to detect them and assess whether they generalize to other populations. A major benefit of global consortia such as ENIGMA is the greater power to test hypotheses about factors that influence disease burden and progression without limiting the analysis to any particular population. A screen of medication or even dietary or environmental effects on the brain worldwide is within reach. Effects must be interpreted cautiously, though; there are many unmodeled factors and confounds, known and unknown. These can affect any epidemiological study.

IMAGING GENOMICS AND GENOME-WIDE ASSOCIATION STUDIES

Neuroimaging can identify typical signatures of disease, but disease effects are not uniform across patients. Also, because brain structure and function vary widely even among healthy people, these characteristic signatures have varying uses for diagnosis.

A still more challenging project is to try to discover genetic factors that shape or structure the living brain and lead to individual differences in brain structure, function, and behavior. Some of these factors may affect the way information in the brain is processed in health and disease. Many of these factors lie hidden in the genome; one such method to identify them is a **GWAS**. GWAS is covered more fully in a separate chapter in this book, but we briefly summarize its goals here.

GENETIC INFLUENCES ON THE BRAIN

Many brain measures are heritable: that is to say, the variability of the traits across the population is influenced by individual genetic differences. Heritable brain measures obtained from MRI are numerous and include the overall volume of the brain, the volume of gray or white matter, and the volumes of specific brain substructures. Even aspects of brain shape (Gutman et al., 2015) and structural (Jahanshad, Kochunov, et al., 2013; Jahanshad, Rajagopalan, et al., 2013) and functional connectivity (Glahn et al., 2010) have been found to be under strong genetic influences.

Quantitative genetic studies of twins or large family pedigrees reveal that over half of the variance in brain volume, for example, is explainable by genetic factors. Many additional nongenetic factors explain variance in brain measures: a person's age, sex, some environmental factors (such as drug use and toxin exposure), and measurement errors. These family studies of genetic influences have been conducted without needing to examine any individual person's DNA; they are based on knowing the kinship or the degree of relatedness between different family members, and adjusting for common household or environmental upbringing, another source of resemblance among family members. Heritability studies of imaged traits show that brain structures can be measured reliably enough to identify genetic effects, among all the other factors that affect the brain throughout life.

GENES AND DISEASE RISK

Genotyping chips can now test a person's genetic code at over a million common variants in the genome. Single-nucleotide polymorphisms (SNPs) are locations in which the genetic sequence differs among different individuals. Many other kinds of genetic variation affect the brain and disease risk, such as insertions and deletions of genetic material, and rare or even private variants. These genetic variants may greatly affect disease risk but they may also be hard to identify in a screen of a few thousand individuals if they are uncommon. One such rare genetic variant is *TREM2*; the adverse variant of this gene is found in only around 1% of healthy individuals. However, it appears to boost a person's lifetime risk of AD by a factor of two to four times and may be associated with faster brain tissue loss in the elderly (Jonsson et al., 2013; Rajagopalan, Hibar, & Thompson, 2013).

Some common genetic variants greatly affect a person's lifetime risk for disease; a variant within the *APOE* gene is a major risk factor for late-onset AD, the more common form of the disease. Over 20% of Caucasians have at least one copy of the APOE4 form of the gene, which is associated with a threefold increased risk of AD; if other factors are equal, two copies of APOE4 may increase AD risk by as much as 10–15 times compared to noncarriers. *APOE* was discovered via linkage analysis of affected families as long ago as 1993 (Strittmatter et al., 1993). Unbiased searches of the genome still find it to be consistently associated with AD (Fig. 7.1); patient DNA differs statistically from that of control subjects at multiple locations in the *APOE* gene, including the two SNPs that are classically tested to determine APOE4 status.

MANHATTAN PLOTS

Fig. 7.1 shows a Manhattan plot from the first stage of a large-scale **GWAS** of AD risk loci (Lambert et al., 2013). The plot shows on the y-axis the negative log-base-10 of the *P* value for each of the polymorphisms in the genome (along the x-axis), when tested for differences in frequency between 17,008 cases and 37,154 controls. The line shows the threshold for genome-wide significance ($P < 5 \times 10^{-8}$). Diamonds represent SNPs with the smallest *P* values in the overall analysis from that article.

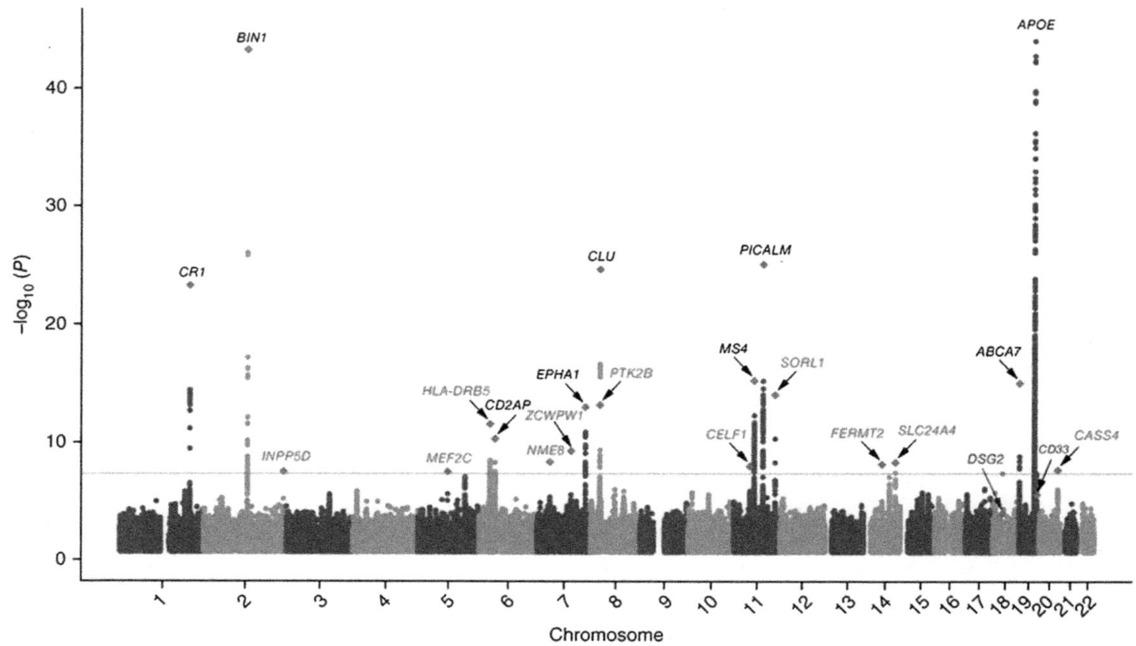

FIGURE 7.1 Genomic variants associated with AD. This Manhattan plot shows the evidence of association with AD for over a million common variants in the genome. Peaks of association are found in well-known AD risk genes, such as *APOE*, but also in over 20 genes. For many of these, worldwide efforts are under way to understand their function. *Reproduced with permission from Lambert, J. C., Ibrahim-Verbaas, C. A., Harold, D., Naj, A. C., Sims R., Bellenguez, C., … Amouyel, P. (December 2013). Meta-analysis of 74,046 individuals identifies 11 new susceptibility loci for Alzheimer's disease.* Nature Genetics, *45(12), 1452–1458. http://dx.doi.org/10.1038/ng.2802. Epub Oct 27, 2013.*

An up-to-date list of these risk loci is curated online at www.alzgene.org; based on current evidence, over 20 SNPs have a consistently replicated effect on AD risk. Carriers of APOE4 have detectable differences in brain images, as well; even young adults who carry APOE4 have differences in cortical development (Shaw et al., 2007). Cognitively intact young adults and older APOE4 carriers show differences in anatomical brain networks that decline in dementia (Brown et al., 2011; Daianu, Jacobs, Zlokovic, Montagne & Thompson, 2015; Daianu, Mezher, et al., 2015). APOE4 may also interact with other conditions such as HIV infection or traumatic brain injury and may intensify patterns of brain differences (Jahanshad et al., 2012).

A trickier question is whether AD risk genes other than *APOE* are consistently associated with differences in brain structure; some small imaging studies have computed polygenic risk scores from the discovered AD risk loci and used them to predict brain measures on MRI.

In polygenic prediction models of AD, the effect of *APOE* is the major contributor to risk. Evidence for the predictive value of other SNPs is much lower. A major goal of AD genetics is to supplement the classical *APOE* SNPs with other markers of AD risk, perhaps from whole genome sequence data. Other small studies suggest that the AD risk gene *CLU* may affect white matter microstructure in young adults (Braskie et al., 2011) and even resting-state functional MRI activation (Erk et al., 2013). These results await replication in larger samples.

CANDIDATE GENES AND GENOME-WIDE ASSOCIATION STUDIES

By 2009, a large number of imaging genetics studies claimed to have found associations between specific candidate genes and brain measures, often in studies of fewer than 100 subjects. Unlike many journals focused on genetics, few neuroimaging journals required a study to show a replication in an independent sample of subjects when an effect was reported. As such, many reports of candidate gene associations with imaging-derived brain traits remain uncontested and unverified, partly because of the large cost of an imaging study, and the pressure to not simply repeat studies, but to design new ones or use novel methods.

In the words of a prominent behavioral geneticist, "It does not make the genetic associations any more reliable, just because imaging data is more expensive to collect." Around the same time, complaints began to emerge in the neuroimaging literature about some clearly underpowered studies (Button et al., 2013; Ioannidis, 2011; Dorothy Bishop's blog: http://deevybee.blogspot.co.uk/2012/03/time-for-neuroimaging-to-clean-up-its.html). Brain structure differences in bipolar disorder, for example, are consistent but typically have small effect sizes, requiring samples of many hundreds to detect with a reasonable level of power (Hibar et al., 2016). Large multisite studies such as those in ENIGMA are only beginning to reevaluate the effect of candidate genes on brain measures in the general population (Hibar, Stein, et al., 2015). Candidate gene studies have largely fallen out of favor in biological psychiatry because such approaches have proven generally unreliable in psychiatric illness (Farrell et al., 2015). Previously suggested candidate gene effects on brain structure have not been replicated in much larger consortium sample sizes, except for APOE4 (Stein et al., 2012; Hibar, hundreds of authors, CHARGE ENIGMA, Thompson, & Ikram, 2015). As well as assessing candidate gene effects, there have been efforts to test for the enrichment (overrepresentation) of specific biological pathways in genome-wide association results (Becker et al., 2015). If specific biological pathways, networks, or features were known that were generally implicated in brain–gene association studies, true associations could be discovered more quickly and efficiently in smaller samples. Such gene annotation, filtering, or preselection methods are being tested to see whether they allow more efficient or powerful gene discovery. ENIGMA's genomic screens of brain measures are enriched in known risk genes for Parkinson disease, obsessive–compulsive disorder, and AD (Hibar, Stein, et al., 2015). There are also reports assessing the relationship of ENIGMA's brain-related loci and schizophrenia risk loci discovered by the Psychiatric Genomics Consortium (Franke et al., 2016).

APOE is perhaps the only well-known disease susceptibility gene with unquestionable evidence of association with brain measures. For hippocampal volume, the strength of *APOE's* effect depends on the cohort assessed, and older cohorts show the strongest effects. If disease risk genes' effects depend on the age or demographics of the cohort, it is important not to dismiss them as spurious if a meta-analysis of diverse cohorts does not find them.

GENOME-WIDE SIGNIFICANCE

A related criticism in psychiatric genetics was that many candidate gene studies in psychiatry were hard to replicate, perhaps because researchers had been conducting too many statistical tests of many different candidate genes on multiple traits and not correcting adequately for multiple comparisons. GWASs conventionally set the bar very high for reporting a significant finding, at around 20 million to 1. This widely used statistical threshold, also called the **genome-wide significance threshold,** divides the standard $P = 0.05$ significance threshold by 1 million to adjust for the risk of reporting false-positive findings when testing 1 million independent genomic loci. In fact, in a GWAS in which the trait assessed is completely unrelated to any of the SNPs, just by chance one in every million SNPs should still have a P value smaller than 10^{-6}, so we need an even stricter threshold at $0.05/10^{-6}$ or 5×10^{-8}.

With this in mind, GWASs now employ truly vast samples to detect effects that achieve genome-wide significance, that is, they correct for all possible association tests at independent genetic loci. In addition, they then carry forward results from an initial discovery phase to test them in independent (replication) data sets. This approach has led to a flurry of findings that are widely agreed to be consistent and robust; over 100 genetic loci were associated with schizophrenia (Schizophrenia Working Group of the Psychiatric Genomics Consortium, 2014).

Several other successful GWASs have been published for a range of psychiatric disorders, and even for neurological conditions such as multiple sclerosis (IMSGC et al., 2013), Parkinson disease (Nalls et al., 2014), and epilepsy (ILAE et al., 2014). In schizophrenia, some disease-associated SNPs are in dopamine-related genes long implicated in psychosis and its treatment. Other schizophrenia-associated SNPs lie in regions of the genome related to immunity and histocompatibility; these genomic screens are the first steps toward new studies that assess enrichment of those same SNPs and their biological pathways in relation to other biological traits.

FORMATION OF ENIGMA IN 2009

As explained previously, many imagers believed that some SNPs would be more strongly associated with endophenotypes, including neuroimaging traits, than with clinical measures such as a diagnosis of AD or schizophrenia. In fact, one reason to perform an imaging genetics study is to identify genetic predictors of brain measures that also relate to disease.

To perform a GWAS analysis of brain imaging measures, the ENIGMA Consortium was formed in 2009 and reported the first large-scale genetic analysis of brain volumes, intracranial volumes, and hippocampal volumes (Stein et al., 2012).

A growing partnership with the Cohorts for Heart and Aging Research in Genomic Epidemiology Consortium (CHARGE) and involvement of other large imaging genetics consortia such as IMAGEN and BIG allowed the discovery and replication of genetic hits in 21,151 individuals scanned with brain MRI and assessed with GWAS (Bis et al., 2012; Ikram et al., 2012; Stein et al., 2012). Not all samples had been genotyped or scanned in the same way or with the same goals in mind, but ENIGMA developed standardized protocols for image analysis and quality control. Genomic imputation protocols were standardized to map different people's genetic data to a common reference panel, such as HapMap 3 or the 1000 Genomes reference data sets. As a result, all groups would be testing the same set of SNPs, regardless of how the genotyping was performed.

A more extensive ENIGMA GWAS analysis (Hibar, hundreds of authors, et al., 2015) evaluated more structures and discovered six SNPs reliably associated with volumes of crucial brain regions including the putamen and hippocampus. The study also strengthened the evidence for SNPs that ENIGMA had previously discovered. Because of the typical upswing in rates of genomic associations as consortia expand, many groups joined ENIGMA. By 2014, ENIGMA expanded to include over 500 scientists in 35 countries (Thompson et al., 2014, 2015) (Fig. 7.2 and 7.3).

SURPRISES FROM IMAGING GENOMICS CONSORTIA

The global analysis of imaging and genetic data yielded several surprises. The first surprise was that consistent SNP effects were detectable at all, and were replicated worldwide in cohorts of widely varying geography and demographics.

Until ENIGMA and CHARGE, the largest imaging studies focused on harmonizing image acquisition across sites. This is not completely achievable with multiple vendors of MRI scanners and multiple magnetic field strengths in use today (1.5- and 3-T scanners are most commonly used). ADNI, for example, developed a standard brain imaging protocol that older people and patients with dementia could tolerate, collecting anatomical MRI, FLAIR (a scan to assess white matter), and resting-state functional MRI, diffusion, or perfusion MRI in around 30-min scan time.

Scan harmonization is no doubt valuable, but ENIGMA's studies found that brain measures can be extracted reliably enough to detect effects of genetic variants that explain only 0.5–1% of the variance (Hibar, hundreds of authors, et al., 2015), even in cohorts whose scanning protocols were not harmonized.

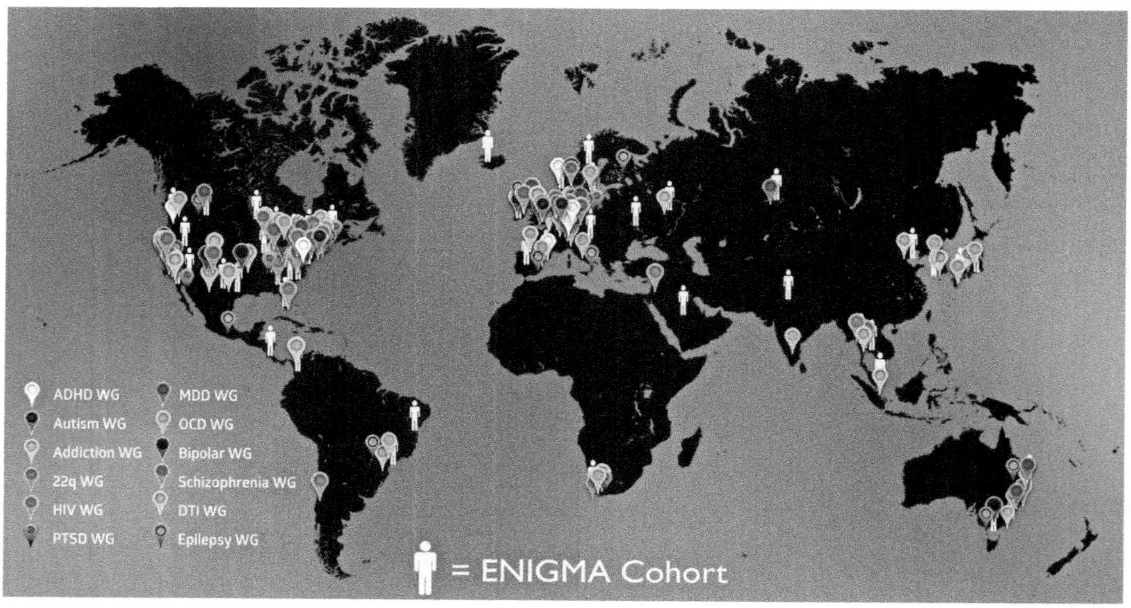

FIGURE 7.2 ENIGMA as a global partnership. In 2009, ENIGMA started off as three groups working together: one in the United States, one in the Netherlands, and one in Australia, with an interest in replicating genetic associations with hippocampal volume and total brain volume. In 6 years (by 2015), ENIGMA had expanded to include over 500 scientists from over 35 countries, focusing on many brain traits and diseases (see Thompson et al., 2015 for an updated review).

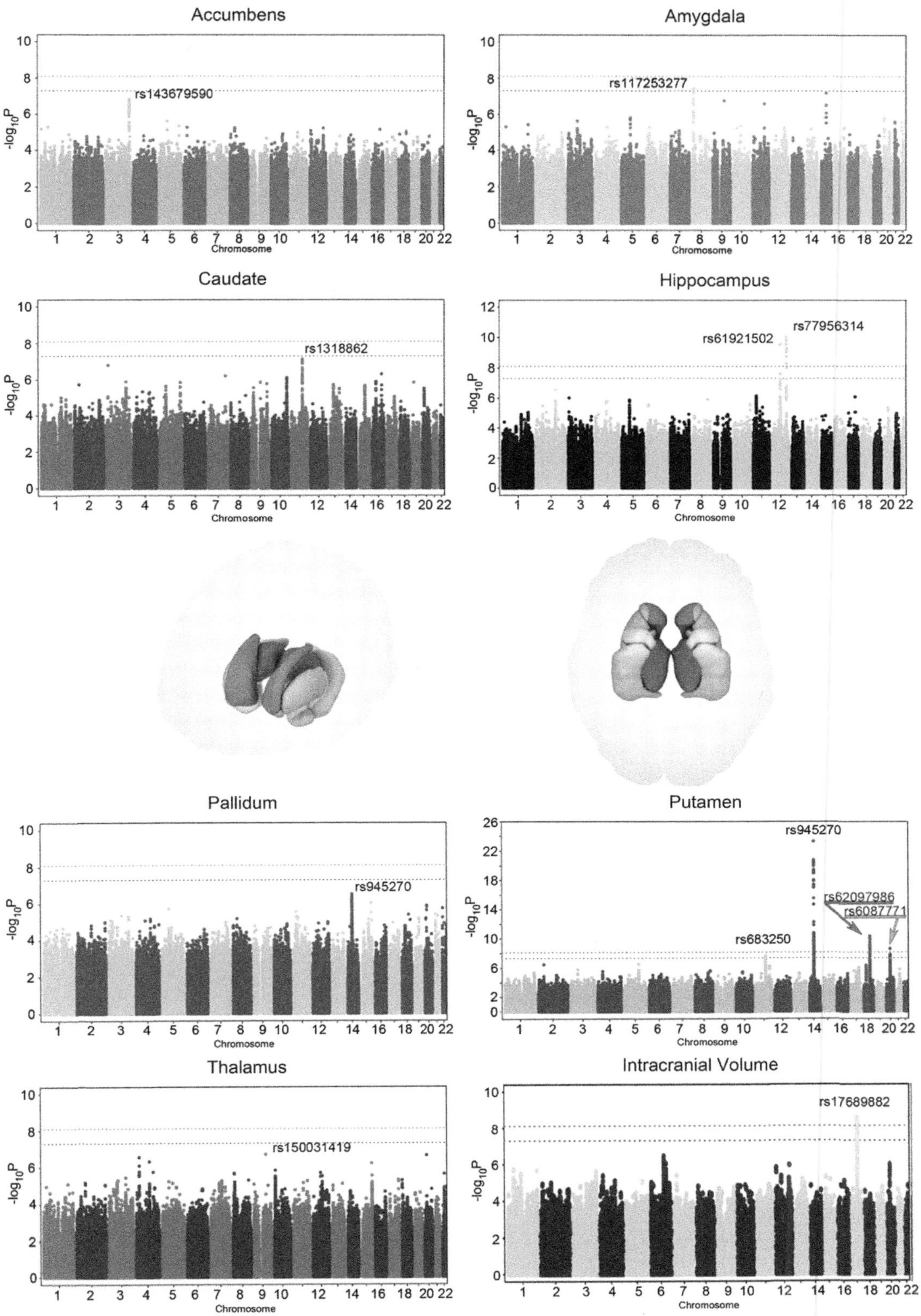

FIGURE 7.3 ENIGMA2: Worldwide genomic screening discovered multiple genetic variants that affect the volumes of subcortical brain regions that are implicated in psychiatric and neurological illness. Color-coded plots show evidence (on a $-\log_{10}$ scale) that common variants in the genome affect the volume of the structure shown. Consistent effects were found in 48 cohorts across the world, showing that single-nucleotide variants in DNA can be discovered that change brain structure, by screening MRI data worldwide. *Reproduced with permission from Hibar, D., hundreds of authors, CHARGE, ENIGMA, Thompson, P. M., & Ikram, M. A. (2015). Novel genetic loci associated with hippocampal volume are relevant to aging and dementia (submitted to* Nature Genetics, *August 14, 2015).*

Second, if such a global alliance in neuroimaging can successfully detect genetic effects, many other effects should be detectable with the same collaborative approach. ENIGMA's lifespan working group has reconstructed growth and aging trajectories for different brain structures in over 10,000 people scanned with MRI (Dima et al., 2015). ENIGMA has also begun to study 12 major brain diseases without requiring GWAS data collection from the cohorts. The intent was to identify brain measures that differed most between patients and healthy matched control subjects in each cohort, and best differentiated disorders. If we rank brain measures in terms of their heritability and ability to detect disease effects, we would also have a rational basis to prioritize brain measures for genetic studies (Glahn et al., 2012). Collaborative efforts between disease GWAS consortia and ENIGMA are well under way to test for enrichment of disease risk genes among genes with known effects on the brain. In many cases, a disease risk SNP is discovered with no known effects on any biological pathway. Thus the joint analysis of multiple traits for genetic overlap is likely to generate new insights with increasing power as brain and biological databanks expand.

Third, and perhaps most surprising, is that many of the major candidate genes that had been thought to affect hippocampal volume (or brain volume) showed no evidence of an effect. There are common variants in many genes that encode key growth factors in the brain, such as the *BDNF* gene. There are also well-studied polymorphisms in neurotransmitter receptors, transporters, and enzymes; some of these are already the targets of drug therapy in psychiatry. This had led to hypotheses that polymorphisms in these genes might predict brain morphology to some extent. A very small effect cannot be ruled out, but it is interesting that their effect was undetected in studies vast enough to detect variants that account for as little as 0.5–1% of the observed variance in the population.

There is still vigorous debate in imaging genetics over **candidate gene** studies, relative to unbiased searches of the genome and the need to verify each kind of effect, and the stringency of verification. These issues are discussed in Medland, Jahanshad, Neale, and Thompson (2014).

Finally, evidence is still coming in regarding whether imaging genetics studies can find true associations more efficiently, that is, with smaller samples, or with higher effect sizes, than other psychiatric GWAS. The efficiency of a GWAS in terms of the sample sizes needed to detect a true association signal depends on the genetic architecture of the trait and how reproducibly it can be measured. Holland et al. (2015) used empirical association statistics from ENIGMA and from the Psychiatric Genomics Consortium to argue that imaging measures may be less polygenic than the major mental illnesses, in the sense that a smaller fraction of the genome may explain almost all of their heritability. If that is true, they argue, ENIGMA may be especially efficient in the sense that SNPs explaining almost all of the heritability may be discoverable in smaller samples. If the heritability of a trait, measure, or illness is spread out over a very large part of the genome, with a very small percentage of the heritability explained per SNP, samples exceeding a million individuals may be needed to detect true associations; this may be the case for disorders such as major depression, in which genome-wide screens have unearthed no true hits despite samples well over a 100,000 individuals. Franke et al. (2015) also presented a preliminary comparison of SNP effect sizes between imaging traits and psychiatric GWAS and found slightly higher effect sizes for top imaging-associated SNPs, along with caveats of interpretation which related to the selection of measures, winner's curse[2], and the relative value of the results. Clearly, more data will be useful to resolve the statistical efficiency of different GWAS and the sample sizes needed to detect any or almost all true associations. Also, the usefulness of the results depends on factors other than the efficiency.

OTHER NEUROIMAGING METHODS

DTI is a variant of brain MRI that maps the diffusion of water along neural pathways in the living brain. Because the MRI signal is attenuated by water diffusion in a mathematically well-understood way (according to the Stejskal–Tanner equation), DTI scans can be used for whole-brain tractography, recovering entire tracts and tracking neural pathways as they weave and intermix in the living brain. Even more advanced methods of network analysis, such as graph theory, have been used to show genetic effects on the topology and organization of the brain's networks (Daianu, Jacobs, et al., 2015; Daianu, Mezher, et al., 2015; Jahanshad, Kochunov, et al., 2013; Jahanshad, Rajagopalan, et al., 2013, Zhan et al., 2015).

ENIGMA is conducting the first global GWAS of DTI measures, focusing first on standard measures of white matter integrity and microstructure. These include the fractional anisotropy of major white matter bundles identified automatically in the images. This GWAS was preceded by a major effort to show that DTI measures are heritable and reproducible across multiple cohorts worldwide (Jahanshad, Kochunov, et al., 2013; Jahanshad, Rajagopalan, et al., 2013; Kochunov et al., 2014).

2. "winner's curse": the winner's curse is the phenomenon that the initial reported finding is a false positive and is not replicated in follow-up.

A GWAS of EEG-derived measures is also being conducted by the ENIGMA Consortium. EEG has been used in neurological research, specifically epilepsy and sleep research, for decades and in twin and family genetics studies (Malone et al., 2014; Smit et al., 2010). In addition, the growing popularity of resting-state functional MRI makes it a natural candidate for imaging genomics studies in the future.

MULTIVARIATE IMAGING GENOMICS AND BIG DATA

One arguable limitation of most imaging genomics studies to date is that they take one or at most a handful of imaging measures from the many millions of measures that could in principle be analyzed or derived from a brain scan.

If there were some brain measure more amenable to genetic analysis, more reliably computed in large data sets, and proven to show larger SNP effects, it would be used instead of the measures employed so far. One reason why MRI and DTI have been used for the largest studies to date is that the methods to analyze them are relatively mature. Also, ENIGMA was able to develop standardized protocols that accelerated multisite analysis.

SEARCHING THE BRAIN FOR GENE EFFECTS

Voxelwise GWAS (Stein et al., 2010) and voxelwise GWAS (Hibar, Kohannim, et al., 2011; Hibar, Stein, et al., 2011) are extensions of GWAS in which the signal at each location in the brain is subjected to genome-wide association testing. In other words, mass univariate genetic association tests are used to assess on a very fine spatial scale whether the gray matter volume, fiber integrity, or activation of the brain is associated with any SNP without having to predefine a specific region of the image. In a standard three-dimensional image, this involves testing each of the possibly millions of locations in the image against the millions of variants within the genome, a true big data problem with some uniquely interesting challenges. Efforts to launch this computation and make it more efficient are under way. Computations are made scalable by distributing them across a massive computational infrastructure available at sites worldwide (Jahanshad et al., 2015).

The P value threshold needed to correct for 10^{12} tests at the strict Bonferroni level is $0.05/10^{12}$, or 5×10^{-14}. This seems impossible to achieve without a vast sample size. Even so, the ENIGMA studies have already found robust SNP effects at $P < 10^{-23}$ (Hibar, hundreds of authors, et al., 2015), so such a correction is feasible. To make such an analysis work, the sheer volume of data to process is hard to transfer. Substantial benefits can be gained from distributed computation at multiple sites globally (Jahanshad et al., 2015). Genetic analysis of brain connectivity can also screen thousands of detected connections and millions of SNPs at the same time using GWAS, leading to a connectome-wide genome-wide search (Jahanshad, Kochunov, et al., 2013; Jahanshad, Rajagopalan, et al., 2013). In these searches, we can sort and prioritize network connections if they show significant heritability determined from family or twin studies (Fig. 7.4).

There is also interest in developing and testing **multivariate and machine learning methods** to reduce the number of statistical tests in imaging genomics (Ge, Schumann, & Feng, 2013; Hibar, Kohannim, et al., 2011; Hibar, Stein, et al., 2011; Thompson, Ge, Glahn, Jahanshad, & Nichols, 2013). This allows informative data reduction and modeling. Imaging data shows a spatial coherence, allowing specialized statistical tests that can search for clusters, features, and higher-order topology when entire images are tested. Methods such as **seemingly unrelated regression** (Jahanshad et al., 2015) can boost the power to test for associations by taking into account the correlation between measures. Because of this coherence, the effective dimensionality of the imaging data is often much less than the number of voxels, and simulations can be conducted to estimate it (Medland et al., 2014). Other machine learning approaches such as LASSO can be used for dimension reduction in the genome (Yang et al., 2015). So-called "sparse" methods have been developed to lock onto weak but robust genomic predictors in large data sets. In these cases, most of the predictors are not useful and need to be screened out.

Cooperative Machine Learning

Cooperative machine learning or statistical estimation methods can use multiple data sets at the same time, even ones in which different cohorts were assessed with partially overlapping measures or completely independent measures (Xiang et al., 2014).

Empirical data will be useful to determine which kinds of imaging features and which genetic pathways might be prioritized in such a search. Apart from the ENIGMA results, there is little empirical information so far to rank brain features or even genomic regions that might show strongest associations.

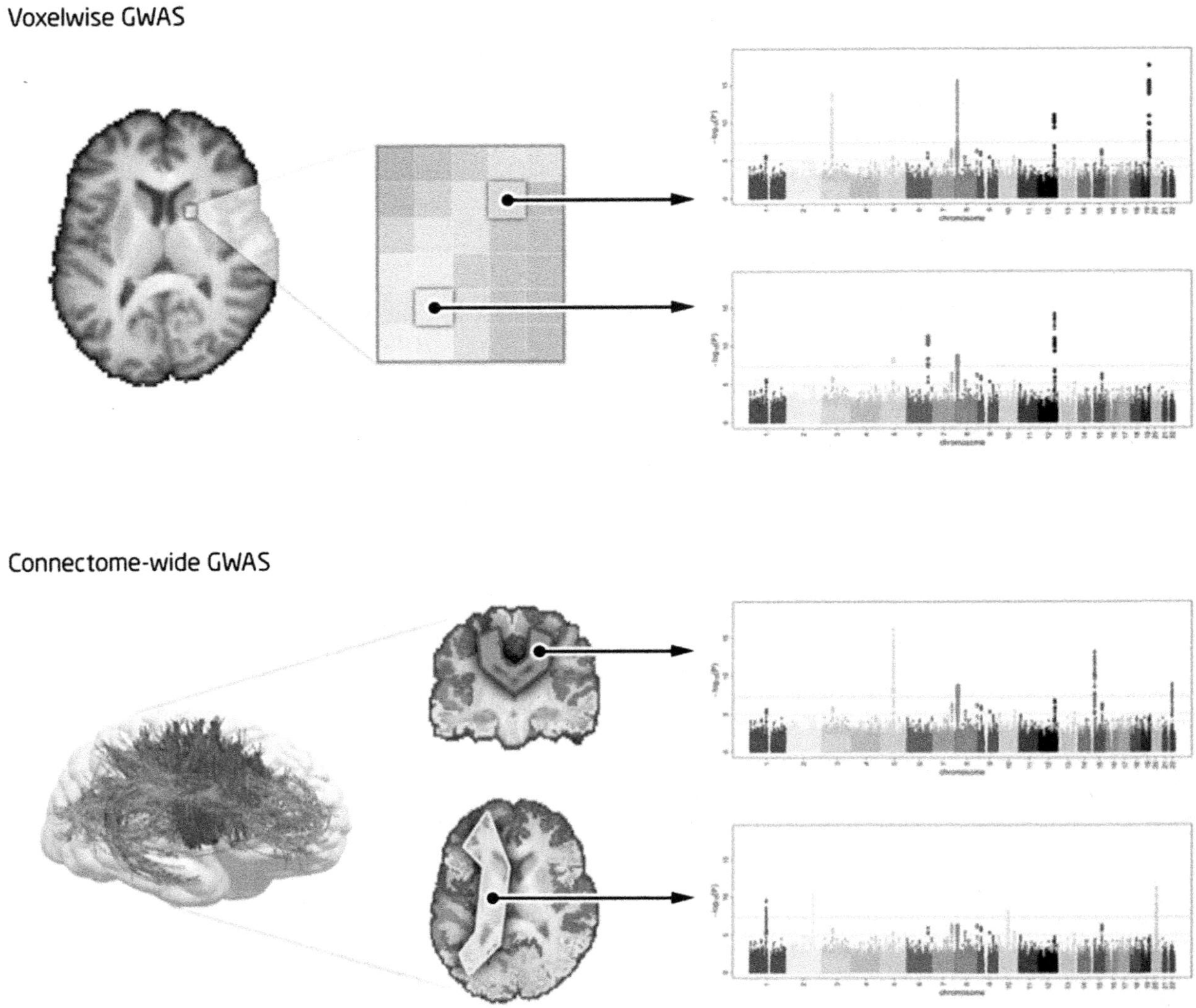

FIGURE 7.4 Screening images and connectomes for gene effects. Genome-wide tests of association can be performed at each pixel in a brain image (voxel-based GWAS) and at each connection in a brain network. In these diagrams, the signal at each location could be a measure of functional activation, metabolism, or fiber integrity, and the genome is screened for evidence of common variants that predict this signal. The free search of the image and genome involves billions of tests, but alliances such as ENIGMA are finding statistical effects that replicate across cohorts worldwide. *Figure reproduced with permission from Medland, S. E., Jahanshad, N., Neale B. M., & Thompson P. M. (2014). Whole-genome analyses of whole-brain data: working within an expanded search space.* Nature Neuroscience, 17*(6), 791–800 and adapted from original studies by Stein, J. L., Hua, X., Lee, S., Ho, A. J., Leow, A. D., Toga, A. W., ... Alzheimer's Disease Neuroimaging Initiative (2010). Voxelwise genome-wide association study (vGWAS).* Neuroimage, 53*(3) 1160–1174, Jahanshad, N., Kochunov, P. V., Sprooten, E., Mandl, R. C., Nichols, T. E., Almasy, L. & Glahn D. C. (2013). Multi-site genetic analysis of diffusion images and voxelwise heritability analysis: a pilot project of the ENIGMA-DTI working group.* Neuroimage, 81, *455–469 and Jahanshad, N., Rajagopalan, P., Hua, X., Hibar, D. P., Nir, T. M., Toga, A. W., ... Alzheimer's Disease Neuroimaging Initiative (2013). Genome-wide scan of healthy human connectome discovers SPON1 gene variant influencing dementia severity.* Proceedings of the National Academy of Sciences of the United States of America, 110*(12), 4768–4773.*

CONCLUSIONS

Global alliances such as ENIGMA are beginning to crack the code that governs how genetic variation shapes the brain. Imaging genetics is an emerging field. It is already clear that consistent effects of genetic variants on the brain are detectable worldwide. This is true even when the data pooled were collected for unrelated goals.

The ongoing pace of discoveries that link genetics with imaging is spurring initiatives to relate brain-related genes found in ENIGMA to those implicated in psychiatric and neurological illness. Tests of genetic overlap are under way to identify common genetic pathways and understand the scope of gene effects.

This previously unimaginable cascade of global scientific activity will likely identify key genomic determinants of brain circuitry, function, and risk for disease.

The united effort will also yield a deeper and increasingly mechanistic understanding of the most enigmatic illnesses of all time.

ACKNOWLEDGMENTS

Funding for the ENIGMA Center for Worldwide Medicine Imaging and Genomics is provided as part of the 2014 NIH Big Data to Knowledge (BD2K) Initiative under grant number U54 EB 020403 (PI: Thompson) to support big data analytics, management, and distribution of programs. J.L.S. is supported by NIH grant K99 MH 102357.

REFERENCES

Bearden, C. E., Thompson, P. M., Dutton, R. A., Frey, B. N., Peluso, M. A., Nicoletti, M., … Soares, J. C. (2008). Three-dimensional mapping of hippocampal anatomy in unmedicated and lithium-treated patients with bipolar disorder. *Neuropsychopharmacology, 33*(6), 1229–1238.

Becker, M., Guadalupe, T., Franke, B., Hibar, D. P., Thompson, P. M., ENIGMA Consortium, … Fisher, S. E. (2015). *A hypothesis-driven approach to identify genetic factors affecting brain subcortical volumes* (submitted for publication, May 2015).

Bis, J. C., DeCarli, C., Smith, A. V., van der Lijn, F., Crivello, F., Fornage, M., … Cohorts for Heart and Aging Research in Genomic Epidemiology Consortium. (2012). Common variants at 12q14 and 12q24 are associated with hippocampal volume. *Nature Genetics, 44*(5), 545–551.

Boedhoe, P. S. W., Schmaal, L., Abe, Y., Ameis, S., Arnold, P., Batistuzzo, M. C., … van den Heuvel, O. A. (2016). Age-specific subcortical volumetric abnormalities in Obsessive-Compulsive Disorder (OCD): findings from the ENIGMA-OCD Working Group. Presented at the Organization of Human Brain Mapping: Geneva, Switzerland.

Braskie, M. N., Jahanshad, N., Stein, J. L., Barysheva, M., McMahon, K. L., de Zubicaray, G. I., … Thompson, P. M. (2011). Common Alzheimer's disease risk variant within the *CLU* gene affects white matter microstructure in young adults. *Journal of Neuroscience, 31*(18), 6764–6770.

Brown, J. A., Terashima, K. H., Burggren, A. C., Ercoli, L. M., Miller, K. J., Small, G. W., & Bookheimer, S. Y. (2011). Brain network local interconnectivity loss in aging APOE-4 allele carriers. *Proceedings of the National Academy of Sciences of the United States of America, 108*(51).

Button, K. S., Ioannidis, J. P., Mokrysz, C., Nosek, B. A., Flint, J., Robinson, E. S., & Munafo, M. R. (2013). Power failure: why small sample size undermines the reliability of neuroscience. *Nature Reviews Neuroscience, 14*(5), 365–376.

Chiang, M. C., Dutton, R. A., Hayashi, K. M., Lopez, O. L., Aizenstein, H. J., Toga, A. W., … Thompson, P. M. (2007). 3D pattern of brain atrophy in HIV/AIDS visualized using tensor-based morphometry. *Neuroimage, 34*, 44–60.

Daianu, M., Jahanshad, N., Nir, T. M., Toga, A. W., Jack, C. R., Jr., Weiner, M. W., … Alzheimer's Disease Neuroimaging Initiative. (2013). Breakdown of brain connectivity between normal aging and Alzheimer's disease: a structural k-core network analysis. *Brain Connectivity, 3*(4), 407–422.

Daianu, M., Jacobs, R. E., Zlokovic, B. V., Montagne, A., & Thompson, P. M. (2015). Reconstruction of major fibers using 7T multi-shell Hybrid Diffusion Imaging in mice. In *International symposium for biomedical imaging*. NY, USA: Brooklyn (submitted for publication).

Daianu, M., Mezher, A., Jahanshad, N., Hibar, D. P., Nir, T. M., Clifford, J., … Thompson, P. M. (2015). Spectral graph theory and graph energy metrics show evidence for the Alzheimer's disease disconnection syndrome in APOE-4 risk gene carriers. In *International symposium for biomedical imaging*. NY, USA: Brooklyn (submitted for publication).

Demirhan, A., Nir, T. M., Zavaliangos-Petropulu, A., Clifford, J., Jack, R., Weiner, M. W., … Jahanshad, N. (2015). Effect of feature selection to improve the accuracy of classifying Alzheimer disease using diffusion tensor images. In *International symposium for biomedical imaging*. NY, USA: Brooklyn.

Dima, D., Papachristou, E., Turner, J., Glahn, D. C., Hibar, D. P., van Erp, T. G. M., … Frangou, S. (June 14–18, 2015). *ENIGMA Lifespan Working Group. Subcortical brain volumes across the lifespan based on 10,722 people aged 2 to 92*. Honolulu, Hawaii, USA: Organization for Human Brain Mapping annual meeting (OHBM).

Erk, S., Meyer-Lindenberg, A., Schmierer, P., Grimm, O., Tost, H., Muhleisen, T., … Walter, H. (2013). Functional impact of a recently identified quantitative trait locus for hippocampal volume with genome-wide support. *Translational Psychiatry, 3*, e287.

van Erp, T. G. M., Hibar, D. P., Rasmussen, J. M., Glahn, D. C., Pearlson, G. D., Andreassen, O. A., … for the ENIGMA – Schizophrenia Working Group. (August 18, 2015). Subcortical brain volume abnormalities in 2,028 patients with schizophrenia and 2,540 healthy controls via the ENIGMA consortium. *Molecular Psychiatry*. http://dx.doi.org/10.1038/mp.2015.118.

Farrell, M. S., Werge, T., Sklar, P., Owen, M. J., Ophoff, R. A., O'Donovan, M. C., … Sullivan, P. F. (2015). Evaluating historical candidate genes for schizophrenia. *Molecular Psychiatry, 20*, 555–562.

Franke, B., Stein, J., Ripke, S., Anttila, V., Hibar, D., van Hulzen, K., … Sullivan, P. F. (2015). Is there overlap between common genetic influences on schizophrenia and subcortical brain volumes? (submitted to *Nature Neuroscience*, August 2015).

Franke, B., Stein, J. L., Ripke, S., Anttila, V., Hibar, D. P., van Hulzen, K. J., … Sullivan, P. F. (2016). Genetic influences on schizophrenia and subcortical brain volumes: large-scale proof of concept. *Nature Neuroscience, 19*, 420–431.

Franke, B., Hoogman, M., Zwiers, M., Mennes, M., & For the Enigma-ADHD Working Group. (2014). *Brain structure in ADHD across the life span: The ENIGMA ADHD working group* (submitted to the Society for Neuroscience (SFN) 2014).

Ge, T., Schumann, G., & Feng, J. (2013). Imaging genetics—towards discovery neuroscience. *Quantitative Biology, 1.4*, 227–245.

Glahn, D. C., Curran, J. E., Winkler, A. M., Carless, M. A., Kent, J. W., Jr., Charlesworth, J. C., … Blangero, J. (2012). High dimensional endophenotype ranking in the search for major depression risk genes. *Biological Psychiatry, 71*, 6–14.

Glahn, D. C., Winkler, A. M., Kochunov, P., Almasy, L., Duggirala, R., Carless, M. A., ... Blangero, J. (2010). Genetic control over the resting brain. *Proceedings of the National Academy of Sciences of the United States of America, 107*(3), 1223–1228.

Gutman, B. A., Jahanshad, N., Wang, Y., Kochunov, P., Nichols, T. E., & Thompson, P. M. (2015). Medial demons registration localizes the degree of genetic influence over subcortical shape variability: An N= 1480 meta-analysis. In *International symposium in biomedical imaging*. NY, USA: Brooklyn (submitted for publication).

Hibar, D. P., Kohannim, O., Stein, J. L., Chiang, M. C., & Thompson, P. M. (2011). Multilocus genetic analysis of brain images. *Frontiers in Genetics, 2*, 73.

Hibar, D. P., Stein, J. L., Kohannim, O., Jahanshad, N., Saykin, A. J., Shen, L., ... Alzheimer's Disease Neuroimaging Initiative. (2011). Voxelwise gene-wide association study (vGeneWAS): multivariate gene-based association testing in 731 elderly subjects. *Neuroimage, 56*(4), 1875–1891.

Hibar, D. P., Stein, J. L., Renteria, M. E., Arias-Vasquez, A., Desrivières, S., Jahanshad, N., ... Medland, S. E. (2015). Common genetic variants influence human subcortical brain structures. *Nature, 520*, 224–229.

Hibar, D. P., Westlye, L. T., van Erp, T. G., Rasmussen, J., Leonardo, C. D., Faskowitz, J., ... Andreassen, O. A. (2016). Subcortical volumetric abnormalities in bipolar disorder. *Molecular Psychiatry* (in press).

Hibar, D., hundreds of authors, CHARGE, ENIGMA, Thompson, P. M., & Ikram, M. A. (2015). Novel genetic loci associated with hippocampal volume are relevant to aging and dementia (submitted to *Nature Genetics*, August 14, 2015).

Holland, D., Wang, Y., Thompson, W. K., Schork, A., Chen, C.-H., Lo, M.-T., ... Dale, A. M. (2015). Estimating effect sizes and expected replication probability from GWAS summary statistics (submitted).

Ikram, M. A., van der Lugt, A., Niessen, W. J., Krestin, G. P., Koudstaal, P. J., Hofman, A., ... Vernooij, M. W. (2011). The Rotterdam Scan Study: design and update up to 2012. *European Journal of Epidemiology, 26*(10), 811–824.

Ikram, M. A., Fornage, M., Smith, A. V., Seshadri, S., Schmidt, R., Debette, S., ... Cohorts for Heart and Aging Research in Genomic Epidemiology Consortium. (2012). Common variants at 6q22 and 17q21 are associated with intracranial volume. *Nature Genetics, 44*(5), 539–544.

International League Against Epilepsy Consortium on Complex Epilepsies. (2014). Genetic determinants of common epilepsies: a meta-analysis of genome-wide association studies. *Lancet Neurology, 13*(9), 893–903.

International Multiple Sclerosis Genetics Consortium, Beecham, A. H., Patsopoulos, N. A., Xifara, D. K., Davis, M. F., Kemppinen, A., ... McCauley, J. L. (2013). Analysis of immune-related loci identifies 48 new susceptibility variants for multiple sclerosis. *Nature Genetics, 45*(11), 1353–1360.

Ioannidis, J. P. (2011). Excess significance bias in the literature on brain volume abnormalities. *Archives of General Psychiatry, 68*(8), 773–780.

Jack, C. R., Jr., Bernstein, M. A., Borowski, B. J., Gunter, J. L., Fox, N. C., Thompson, P. M., ... Weiner, M. W. (2010). Update on the magnetic resonance imaging core of the Alzheimer's disease neuroimaging initiative. *Alzheimer's & Dementia, 6*(3), 212–220.

Jahanshad, N., Valcour, V. G., Nir, T. M., Kohannim, O., Busovaca, E., Nicolas, K., & Thompson, P. M. (2012). Disrupted brain networks in the aging HIV+ population. *Brain Connectivity, 2*(6), 335–344.

Jahanshad, N., Kochunov, P. V., Sprooten, E., Mandl, R. C., Nichols, T. E., Almasy, L., ... Glahn, D. C. (2013). Multi-site genetic analysis of diffusion images and voxelwise heritability analysis: a pilot project of the ENIGMA-DTI working group. *Neuroimage, 81*, 455–469.

Jahanshad, N., Rajagopalan, P., Hua, X., Hibar, D. P., Nir, T. M., Toga, A. W., ... Alzheimer's Disease Neuroimaging Initiative. (2013). Genome-wide scan of healthy human connectome discovers *SPON1* gene variant influencing dementia severity. *Proceedings of the National Academy of Sciences of the United States of America, 110*(12), 4768–4773.

Jahanshad, N., Faskowitz, J., Roshchupkin, G., Hibar, D. P., Gutman, B. A., Tustison, N. J., ... Thompson, P. M. (2015). Multi-site meta-analysis of morphometry. In *Bio-KDD workshop, Sydney, Australia; 14th international workshop on data mining in bioinformatics, August 10, 2015*.

Jonsson, T., Stefansson, H., Steinberg, S., Jonsdottir, I., Jonsson, P. V., Snaedal, J., ... Stefansson, K. (2013). Variant of TREM2 associated with the risk of Alzheimer's disease. *New England Journal of Medicine, 368*(2), 107–116.

Kochunov, P., Jahanshad, N., Sprooten, E., Nichols, T. E., Mandl, R. C., Almasy, L., ... Glahn, D. C. (2014). Multi-site study of additive genetic effects on fractional anisotropy of cerebral white matter: comparing meta and mega-analytical approaches for data pooling. *Neuroimage, 95*, 136–150.

Lambert, J. C., Ibrahim-Verbaas, C. A., Harold, D., Naj, A. C., Sims, R., Bellenguez, C., ... Amouyel, P. (December 2013). Meta-analysis of 74,046 individuals identifies 11 new susceptibility loci for Alzheimer's disease. *Nature Genetics, 45*(12), 1452–1458. http://dx.doi.org/10.1038/ng.2802. Epub Oct 27, 2013.

Mackey, S., & the ENIGMA-Addictions Working Group. (2015). Genetic imaging consortium for addiction medicine; from neuroimaging to genes. *Progress in Brain Research, 224*, 203–223.

Madsen, S. K., Steeg, G. V., Mezher, A., Jahanshad, N., Nir, T. M., Hua, X., ... Thompson, P. M. (2015a). Information-theoretic characterization of blood panel predictors for brain atrophy and cognitive decline in the elderly. In *International symposium for biomedical imaging*. NY, USA: Brooklyn.

Madsen, S. K., Steeg, G. V., Mezher, A., Jahanshad, N., Nir, T. M., Hua, X., ... Thompson, P. M. (2015b). Relative value of diverse brain MRI and blood-based measures for predicting cognitive decline in the elderly. In *International symposium for biomedical imaging*. NY, USA: Brooklyn.

Malone, S. M., Burwell, S. J., Vaidyanathan, U., Miller, M. B., McGue, M., & Iacono, W. G. (2014). Heritability and molecular-genetic basis of resting EEG activity: a genome-wide association study. *Psychophysiology, 51*(12), 1225–1245.

Medland, S. E., Jahanshad, N., Neale, B. M., & Thompson, P. M. (2014). Whole-genome analyses of whole-brain data: working within an expanded search space. *Nature Neuroscience, 17*(6), 791–800.

Nalls, M. A., Pankratz, N., Lill, C. M., Do, C. B., Hernandez, D. G., Saad, M., ... Singleton, A. B. (2014). Large-scale meta-analysis of genome-wide association data identifies six new risk loci for Parkinson's disease. *Nature Genetics, 46*(9), 989–993.

Okada, N., Fukunaga, M., Yamashita, F., Koshiyama, D., Yamamori, H., Ohi, K., ... COCORO. (2015). Abnormal asymmetries in subcortical brain volume in schizophrenia (submitted to *Molecular Psychiatry*, July 2015).

Rajagopalan, P., Hibar, D., & Thompson, P. (2013). TREM2 and neurodegenerative disease. *New England Journal of Medicine, 369*(16), 1565.

Rentería, M. E., Schmaal, L., Hibar, D. P., Couvy-Duchesne, B., Strike, L. T., Mills, N. T., … for the ENIGMA-Major Depressive Disorder Working Group. (2015). Subcortical brain structure and suicidal behaviour in major depressive disorder: A meta-analysis from the ENIGMA-MDD working group (submitted to *Translational Psychiatry*, July 28, 2015).

Schizophrenia Working Group of the Psychiatric Genomics Consortium. (2014). Biological insights from 108 schizophrenia-associated genetic loci. *Nature, 511*(7510), 421–427.

Schmaal, L., Veltman, D. J., van Erp, T. G. M., Sämann, P. G., Frodl, T., Jahanshad, N., … for the ENIGMA-Major Depressive Disorder Working Group. (June 30, 2015). Subcortical brain alterations in major depressive disorder: findings from the ENIGMA major depressive disorder working group. *Molecular Psychiatry*. http://dx.doi.org/10.1038/mp.2015.69.

Schumann, G., Loth, E., Banaschewski, T., Barbot, A., Barker, G., Buchel, C., … IMAGEN Consortium. (2010). The IMAGEN study: reinforcement-related behaviour in normal brain function and psychopathology. *Molecular Psychiatry, 15*(12), 1128–1139.

Shaw, P., Lerch, J. P., Pruessner, J. C., Taylor, K. N., Rose, A. B., Greenstein, D., … Giedd, J. N. (2007). Cortical morphology in children and adolescents with different apolipoprotein E gene polymorphisms: an observational study. *Lancet Neurology, 6*(6), 494–500.

Smit, D. J., Boersma, M., van Beijsterveldt, C. E., Posthuma, D., Boomsma, D. I., Stam, C. J., & de Geus, E. J. (2010). Endophenotypes in a dynamically connected brain. *Behavior Genetics, 40*(2), 167–177.

Stein, J. L., Hua, X., Lee, S., Ho, A. J., Leow, A. D., Toga, A. W., … Alzheimer's Disease Neuroimaging Initiative. (2010). Voxelwise genome-wide association study (vGWAS). *Neuroimage, 53*(3), 1160–1174.

Stein, J. L., Medland, S. E., Vasquez, A. A., Hibar, D. P., Senstad, R. E., Winkler, A. M., … Enhancing Neuro Imaging Genetics through Meta-Analysis Consortium. (2012). Identification of common variants associated with human hippocampal and intracranial volumes. *Nature Genetics, 44*(5), 552–561.

Strittmatter, W. J., Saunders, A. M., Schmechel, D., Pericak-Vance, M., Enghild, J., Salvesen, G. S., & Roses, A. D. (1993). Apolipoprotein E: high-avidity binding to beta-amyloid and increased frequency of type 4 allele in late-onset familial Alzheimer disease. *Proceedings of the National Academy of Sciences of the United States of America, 90*(5), 1977–1981.

Szasz, T. (2006). The pretense of psychology as science: the myth of mental illness in Statu Nascendi. *Current Psychology, 25*(1), 42–49.

Thompson, P. M., Andreassen, O. A., Arias-Vasquez, A., Bearden, C. E., Boedhoe, P. S., Brouwer, R. M., … for the ENIGMA Consortium. (2015). ENIGMA and the Individual: Predicting factors that affect the brain in 35 countries worldwide (submitted to *Neuroimage* (special issue on individual prediction, Ed. Calhoun V., et al.), August 14, 2015).

Thompson, P. M., Bartzokis, G., Hayashi, K. M., Klunder, A. D., Lu, P. H., Edwards, N., … Lieberman, J. A. (2009). Time-lapse mapping of cortical changes in schizophrenia with different treatments. *Cerebral Cortex, 19*(5), 1107–1123.

Thompson, P. M., Dutton, R. A., Hayashi, K. M., Toga, A. W., Lopez, O. L., Aizenstein, H. J., & Becker, J. T. (2005). Thinning of the cerebral cortex visualized in HIV/AIDS reflects CD4+ T lymphocyte decline. *Proceedings of the National Academy of Sciences of the United States of America, 102*(43), 15647–15652.

Thompson, P. M., Ge, T., Glahn, D. C., Jahanshad, N., & Nichols, T. E. (2013). Genetics of the connectome. *Neuroimage, 80*, 475–488.

Thompson, P. M., Mega, M. S., Woods, R. P., Zoumalan, C. I., Lindshield, C. J., Blanton, R. E., … Toga, A. W. (2001). Cortical change in Alzheimer's disease detected with a disease-specific population-based brain atlas. *Cerebral Cortex, 11*(1), 1–16.

Thompson, P. M., Stein, J. L., Medland, S. E., Hibar, D. P., Vasquez, A. A., Renteria, M. E., … the Alzheimer's Disease Neuroimaging Initiative. (2014). The ENIGMA Consortium: large-scale collaborative analyses of neuroimaging and genetic data. *Brain Imaging and Behavior, 8*(2), 153–182.

Weinberger, D. R., & McClure, R. K. (2002). Neurotoxicity, neuroplasticity, and magnetic resonance imaging morphometry: what is happening in the schizophrenic brain? *Archives of General Psychiatry, 59*(6), 553–558 (review).

Xiang, S., Yuan, L., Fan, W., Wang, Y., Thompson, P. M., Ye, J., & for the Alzheimer's Disease Neuroimaging Initiative. (2014). Bi-level multi-source learning for heterogeneous block-wise missing data. *Neuroimage, 102P1*, 192–206.

Yang, T., Wang, J., Sun, Q., Hibar, D. P., Jahanshad, N., Zhan, L., … Ye, J. (2015). Detecting genetic risk factors for Alzheimer's disease in whole genome sequence data via Lasso screening. *International Symposium for Biomedical Imaging*, 985–989.

Zhan, L., Jahanshad, N., Faskowitz, J., Prasad, G., Martin, N. G., Zubicaray, G. I. D., … Thompson, P. M. (2015). Heritability of brain network topology in 853 twins and siblings. In *International symposium for biomedical imaging*. NY, USA: Brooklyn.

de Zubicaray, G. I., Chiang, M. C., McMahon, K. L., Shattuck, D. W., Toga, A. W., Martin, N. G., … Thompson, P. M. (2008). Meeting the challenges of neuroimaging genetics. *Brain Imaging and Behavior, 2*, 258–263.

Chapter 8

Brain in a Dish: Stem Cell Technologies to Study Disorders of the Central Nervous System

Cleber A. Trujillo, Priscilla D. Negraes and Branka Hrvoj-Mihic
University of California San Diego, La Jolla, CA, United States

Mauro F.C. Lins
Clínica-Escola de Autismo, Itaboraí, Rio de Janeiro, Brazil

Katerina Semendeferi and Alysson R. Muotri
University of California San Diego, La Jolla, CA, United States

Gaining access to the organ (or cells) of interest in human neurodevelopmental, psychiatric, and neurological syndromes presents a fundamental challenge. The difficulty in accessing living human brain places significant limits on our ability to capture the molecular-, cellular-, and circuit-level mechanisms underlying these disorders. Whereas the power of the current technical and methodological armamentarium for studying the central nervous system (CNS) is self-evident, it has some marked limitations. In particular, with regard to humans, studying neurons and glia in primary culture is technically challenging and has limited ability to capture the complexity of the functioning human CNS. Most studies of human brain and neuronal function in research subjects have been performed on postmortem tissue, with all the associated and well-known confounds. Peripheral tissues such as blood are clearly not ideal for relevant biological experiments. The tremendous anatomical diversity and developmental dynamism of human brain, coupled with what has been learned about the cellular specificity of gene expression, regulation, and splice variation, highlight the fundamental limitations of relying on peripheral assays. Mouse and other model systems certainly have an important role in elaborating conserved pathophysiological mechanisms, but have obvious constraints in illuminating human pathology that involves higher-order brain functions including emotion, social interaction, and language.

All of these considerations point to the value of being able to leverage the programming of human somatic cells to recapitulate the development and ongoing function of the CNS. These reprogrammed cell types, induced pluripotent stem cells (iPSCs), can be derived from cells isolated from tissues of normal individuals or those affected with a variety of conditions. Rapidly evolving technologies that allow for coculture of neurons and glia as well as the recapitulation of aspects of brain development through the generation of organoids are quickly making the iPSC model a critical complement to more traditional approaches to accessing human brain.

GENERATING INDUCED PLURIPOTENT STEM CELLS

In 2006, Shinya Yamanaka (Takahashi & Yamanaka, 2006) used retroviral-mediated vectors to introduce transcriptional factors into a mouse fibroblast to reprogram them to behave as an embryonic stem (ES) cell. After eliminating irrelevant factors, a minimum of four factors proved necessary to generate mouse iPSCs, including OCT4, SOX2, KLF4, and c-MYC. One year later, using the same method and the four "Yamanaka" factors, human dermal fibroblasts were successfully reprogrammed into iPSC with similar ES properties such as proliferation, morphology, differentiation

Genomics, Circuits, and Pathways in Clinical Neuropsychiatry. http://dx.doi.org/10.1016/B978-0-12-800105-9.00008-1

potential in both in vitro and teratoma assays, gene expression, and cell surface markers (Takahashi et al., 2007; Yu et al., 2007).

In principle, any somatic cell line can generate iPSC with a specific reprogramming method, typically classified as integrative (viral vector are integrated into the host cell genome) and nonintegrative systems (DNA plasmid, recombinant proteins, and messenger RNA (mRNA)) (Rao & Malik, 2012). The virus-based reprogramming system is still widely used owing to its efficient transduction of the target cells. The induced modification of the host cell genome is the main disadvantage of this method. The integration has the potential to cause unpredictable expression changes, which can increase risks of cancer formation and potentially affect the differentiation potential. Although there are strategies to remove the transgenes after reprogramming (Cre-lox technology) (Soldner et al., 2009), it is still necessary to screen all clones. For all these reasons, nonintegrative systems are becoming more attractive.

The most popular nonintegrative method is Sendai virus. Sendai virus is an RNA virus that does not enter the nucleus, can produce large number of proteins, and decreases transgene expression over cell division. Other approaches, such as episomal plasmids or mRNA-containing reprogramming factors, have been used for reprogramming, but the expression period is short with low efficiency (Malik & Rao, 2013). Therefore, the balance of reprogramming efficiency and transgene genomic footprint is the key starting point in iPSC generation. Therefore, the research aims and clinical perspectives must drive the decision of the compatible reprogramming system.

PROGRESS IN MODELING HUMAN CENTRAL NERVOUS SYSTEM PATHOLOGY: NEURODEVELOPMENTAL DISORDERS

iPSC technology has already been used to model multiple disorders, including a range of neurodevelopmental syndromes that include intellectual and social disability. In general, as described in detail subsequently, these studies have supported the assertion that disease-specific iPSC-derived neurons can recapitulate relevant cellular and molecular phenotypes previously observed using different approaches and offer new avenues to elaborate relevant pathophysiological mechanisms.

RETT SYNDROME

Rett syndrome (RTT) is a progressive neurological disorder that affects mainly females (Chahrour & Zoghbi, 2007) and is primarily caused by mutations in the X-linked gene encoding methyl CpG-binding protein 2 (*MECP2*) (Amir et al., 1999; Francke, 2006). Patients with RTT apparently develop normally until 6–18 months of age, when they enter a stage of developmental regression with major symptoms including autistic-like behavior, loss of speech, hypotonia, abnormal breathing, seizures, microcephaly, and mental retardation (Amir et al., 1999; Chahrour & Zoghbi, 2007). Because females with RTT are genetically MECP2 mosaics owing to X inactivation (XCI), a spectrum of phenotypic severity is observed (Amir et al., 2000; Ishii et al., 2001; Young & Zoghbi, 2004).

Several works have independently described the generation and characterization of iPSC-derived neurons from patients with RTT (Ananiev, Williams, Li, & Chang, 2011; Cheung et al., 2011; Hotta et al., 2009; Kim, Hysolli, & Park, 2011; Marchetto et al., 2010). Together, these reports represent the most comprehensive set of iPSC research to date conducted on a neurological disorder by studying neurons derived from RTT; many aspects of the disorder could be reproduced in vitro. It was observed that the neuronal phenotype arises owing to neuronal maturation defects (K.-Y. Kim et al., 2011), which supports the assumption that RTT symptoms appear during establishment and refinement of neural networks in early postnatal development. Compared with nonaffected cells, RTT neurons exhibit reduced dendritic arborization and spine density, smaller soma and nuclei sizes, and less excitatory synapses (Cheung et al., 2011; Marchetto et al., 2010). The compromised calcium signaling and electrophysiology profile confirmed the decreased number of glutamatergic synapses. Although many of these characteristics could be rescued after the cells were exposed to insulin-like growth factor 1 (IGF1) or gentamicin (Marchetto et al., 2010), the time and duration of treatments need to be tuned to avoid side effects.

iPSC lines were also generated by taking advantage of alternative parental nonrandom XCI. Neurons derived from these cells showed that *MECP2* expression follows XCI pattern, allowing the obtainment of isogenic mutant and control cell lines (Ananiev et al., 2011; Cheung et al., 2011; K.-Y. Kim et al., 2011). In addition, there are atypical forms of RTT in which different genes are compromised, such as the Hanefeld variant caused by mutations in the *CDKL5* gene (Evans et al., 2005; Scala et al., 2005). Neurons derived from CDKL5-mutant iPSC showed an increased number of aberrant dendritic spines and reduced functional synaptic contacts, which suggests that *CDKL5* gene absence compromises neuronal activity (Ricciardi et al., 2012).

FRAGILE X SYNDROME

Fragile X syndrome (FXS) is the most common cause of inherited mental retardation (Crawford, Acuña, & Sherman, 2001). Affected individuals may exhibit a spectrum of cognitive abnormalities ranging from learning problems and a normal IQ to severe mental retardation. Autistic-like behaviors such as hand flapping, hand biting, poor eye contact, hyperactivity, anxiety, and impaired social skills can also be present (Brown et al., 1986; Hagerman et al., 2009; Penagarikano, Mulle, & Warren, 2007). In the vast majority of patients, the fragile X mental retardation 1 (*FMR1*) gene is not expressed owing to an expansion of a CGG repeat in its 5′-untranslated region (Pearson, Edamura, & Cleary, 2005; Verkerk et al., 1991). Normal subjects have 5 to 55 CGG repeats, whereas carriers of permutations have 50 to 200 repeats, and affected patients have more than 200 copies (Crawford et al., 2001; O'Donnell & Warren, 2002).

Fragile X mental retardation protein (FMRP) is a cytoplasmic mRNA-binding protein (Ashley, Wilkinson, Reines, & Warren, 1993) expressed in many tissues, but is most abundant in neurons and in the testes (Devys, Lutz, Rouyer, Bellocq, & Mandel, 1993; Feng et al., 1997). FMRP participates in mRNA transport from the nucleus to the dendrites in neurons, where it is involved in the synthesis of proteins important for synaptic development and plasticity in an activity-dependent manner (Antar, Afroz, Dictenberg, Carroll, & Bassell, 2004; Sidorov, Auerbach, & Bear, 2013; Siomi, Siomi, Nussbaum, & Dreyfuss, 1993). Gross abnormalities were not observed in brains of individuals with FXS after autopsy (Hinton, Brown, Wisniewski, & Rudelli, 1991); nonetheless, in some areas their neurons have immature and filopodia-shaped dendritic spines (Irwin et al., 2001), which suggests a defect in the development and maturation of excitatory synapses. In this context, a potential role for FMRP at the synapse is to inhibit the protein synthesis stimulated by metabotropic glutamate receptor (mGluR) activation (Bear, Huber, & Warren, 2004).

The lack of FMRP is associated with an increase in α-amino-3-hydroxy-5-methyl-4-isoxazolepropionic acid (AMPA) receptor internalization that could ultimately affect synaptic activity. In this context, treatment of FMRP-deficient neurons with mGluR antagonist was able to rescue the defect in AMPA receptor trafficking (Nakamoto et al., 2007), and also some of the defects observed in *Drosophila* and mouse models (McBride et al., 2005; Yan, Rammal, Tranfaglia, & Bauchwitz, 2005). In addition, it has been shown that the silencing of *FMR1* gene by CGG expansion is mainly a consequence of an epigenetic transcriptional regulation. Full expansion of the CGG repeat usually coincides with hypermethylation of the repeat region and its upstream CpG island promoter (Oberlé et al., 1991), together with other chromatin modifications (Coffee, Zhang, Ceman, Warren, & Reines, 2002; Coffee, Zhang, Warren, & Reines, 1999; Tabolacci et al., 2005). These epigenetic events seem to be critical for the shift from active to inactive chromatin. Increasing evidence suggests that FXS could be an epigenetic disorder in which the epigenetic state of *FMR1* gene is determinant to the pathogenesis and treatment response more than the CGG repeat itself (Jacquemont et al., 2011; Kumari & Usdin, 2009).

Human ES cells carrying a full CGG mutation were derived through an FXS-affected embryo preimplantation and revealed that downregulation of the *FMR1* gene is triggered by differentiation (Eiges et al., 2007). In other words, the gene is expressed in FXS-ES cells and then silenced during differentiation. Interestingly, the *FMR1* gene of FXS-iPSCs remains inactive and exhibits the same DNA methylation and histone modifications of somatic cells (Urbach, Bar-Nur, Daley, & Benvenisty, 2010). However, Sheridan et al. (2011) reprogrammed cells from a mosaic patient with both normal and premutation-length CGG repeats generating isogenic clones and reported that an altered neuronal differentiation correlated with epigenetic modification of the *FMR1* gene and a lack of FMRP.

In another study, differentiation of FXS-iPSCs showed that these cells have an abnormal neurogenesis with fewer and smaller neurites accompanied by reduced spontaneous synaptic activity (Doers et al., 2014; Telias, Segal, & Ben-Yosef, 2013). These findings are similar to what was observed in *FMR1* knockout mouse models (Braun & Segal, 2000; Luo et al., 2010) and postmortem fetal FXS brain tissue (Castrén et al., 2005). The glial population was altered (Sheridan et al., 2011), which corroborated the enriched glia/neuron ratio already described (Luo et al., 2010). In addition, iPSC derived from mutated heterozygous individuals resulted in active neurons that also have defective synapses and neurite outgrowth, besides functional defects revealed by aberrant calcium signaling (Liu et al., 2012).

ANGELMAN SYNDROME

Patients with Angelman syndrome (AS) exhibit severe developmental delay, cognitive disability, absent speech, ataxia, and a characteristic happy and excitable personality with frequent laughing. Microcephaly, seizures, and sleeping problems can also be present (Angelman, 1965; Clayton-Smith & Laan, 2003; Williams, Driscoll, & Dagli, 2010). AS is a complex genetic disorder associated with genomic imprinting and a reduced expression of the maternally inherited ubiquitin-protein ligase E3A gene (*UBE3A*) on chromosome 15 (Angelman, 1965). Mutations in the *UBE3A* gene or deletion of region

15q11.2-q13 on maternal chromosome or uniparental disomy of chromosome 15 can also result in this condition (Bird, 2014; Williams et al., 2010).

There are different mouse models already established which mimic major features of AS patients (Jana, 2012), such as motor and cognitive impairments, ataxia, sleep disturbance, and inducible seizures. These models have provided significant information about the disease mechanism, including the involvement of *UBE3A* in synapse development and plasticity (Greer et al., 2010; Yashiro et al., 2009). However, Cavaillé et al. (2000) showed that the timing and mechanism of *UBE3A* repression may be different between humans and mice, and pointed out the necessity of a human model. To address this issue, iPSCs were generated and it was observed that the in vitro model recapitulates the species tissue-specific pattern of *UBE3A* imprinting (Chamberlain et al., 2010). The evaluation of functional AS-derived human neurons revealed that *UBE3A* expression was significantly repressed in neuronal cells over time, recapitulating the main epigenetic characteristic of the disease in vitro.

TIMOTHY SYNDROME

Timothy syndrome (TS) is a rare autosomal dominant disorder in which patients show a spectrum of physical manifestations characterized by cardiac arrhythmias, heart malformations, syndactyly, hypoglycemia, developmental delay, and autism. It is caused by mutations in an alternatively spliced exon of the *CACNA1C* gene, which encodes the α subunit of the calcium channel Cav1.2 (Cohen-Kutner, Yahalom, Trus, & Atlas, 2012; Splawski et al., 2004), leading to decreased calcium- and voltage-dependent inactivation of the channel (Barrett & Tsien, 2008). The *CACNA1C* gene mutations lead to longer opening conformation time of the channel, resulting in an abnormal accumulation of intracellular calcium.

Neuronal dendrites are crucial to the process and integrate into neuronal networks. In this context, CaV1.2 is a major player in the regulation of dendritic response to electrical activity (McAllister, Katz, & Lo, 1996; Wong & Ghosh, 2002). The mutation that causes TS leads to a dendrite retraction in rodent neurons after electrical stimuli (Krey et al., 2013). By generating iPSC from patients with TS, it was observed that these cells also exhibited activity-dependent dendrite retraction (Krey et al., 2013; Pasca et al., 2011). Moreover, calcium signaling is compromised in human TS-derived neurons, which show a decreased expression of lower cortical layer and callosal projection neuron markers (Krey et al., 2013; Pasca et al., 2011). On the other hand, increased expression of tyrosine hydroxylase (*TH*), along with norepinephrine and dopamine-producing cells, were observed (Dulcis & Spitzer, 2008).

Although mouse models of the disease recapitulate some behavioral abnormalities, partially mimicking the human condition, they lack the postnatal lethality common in patients (Bader et al., 2011), probably because mouse cardiomyocytes have different electrical properties compared with their human counterparts. In this context, Yazawa and Dolmetsch (2013) derived cardiac cells and observed that the ventricular-like cardiomyocytes exhibit irregular contraction and prolonged action potentials, besides an excess of calcium influx. These human cardiomyocytes displayed electrical properties similar to those found in patients, and rescue their defects after in vitro treatment with roscovitine, which prevents the occurrence of arrhythmias (Yazawa et al., 2011). This same L-type channel blocker reduced the number of TH-positive TS neurons, showing that calcium channel modulators could be important for future clinical trials.

CONTRIBUTION OF DIFFERENT CELL TYPES TO NEURODEVELOPMENTAL DISORDERS

Neurons have been studied extensively and considered the primary players in neurodevelopmental and neurodegenerative disorders. Therefore, different protocols have been developed aiming at specific neuronal subtypes. Neural progenitor cells become cortical neurons upon SMAD signaling inhibition (Marchetto et al. 2010). However, for dopaminergic neurons generation, it is necessary to stimulate midbrain differentiation by adding Sonic Hedgehog (*Shh*) and fibroblast growth factor 8 into the culture media, in association with ascorbic acid, brain-derived neurotrophic factor, glial cell line–derived neurotrophic factor, transforming growth factor-β3, and cyclic adenosine monophosphate for cell maturation (Chambers et al., 2009; Swistowski et al., 2009, 2010; Kriks et al., 2011; Nishimura & Takahashi, 2013). Similarly, motor neurons can be obtained in the presence of Shh, in addition to retinoic acid and some trophic factors that potentiate cell differentiation and survival (Chipman, Toma, & Rafuse, 2012).

Increasing attention is being given to the contribution of glial cells to different pathologies. In the CNS, glial cells include astrocytes, oligodendrocytes, microglia, and chondroitin sulfate proteoglycan NG2–positive cells. Previously considered support cells, astrocytes interact with neurons and blood vessels, and can detect and modulate neuronal networks by actively participating in the maintenance of homeostasis and cellular cross-talk (Yamamuro, Kimoto, Rosen, Kishimoto, & Makinodan, 2015). On the other hand, microglia constitute the immune cells of the CNS and are derived from hematopoietic progenitors (Frick, Williams, & Pittenger, 2013). They participate in tissue homeostasis, synaptic regulation, and neural plasticity by responding to changes in the environment (Yamamuro et al., 2015), whereas the primary function of oligodendrocytes is to form myelin, an insulator of axonal segments that is critical for neuronal signaling propagation. All these features highlight the importance of these cells for the neural network (Baumann & Pham-Dinh, 2001).

Studies using coculture experiments or astrocyte-conditioned medium (ACM) have been conducted to understand the contribution of such cells in different neuropathies. However, protocols using iPSC and ES cells are being developed and optimized to generate specific glial cell types (Duan, Peng, Pan, & Kessler, 2015; Hu, Du, Li, Ayala, & Zhang, 2009; Hu et al., 2010; Krencik, Weick, Liu, Zhang, & Zhang, 2011; Shaltouki, Peng, Liu, Rao, & Zeng, 2013). Most studies include a step in which aggregates of progenitor cells are formed in suspension and exposed to a chemically defined medium capable of direct astrocyte differentiation (Chen et al., 2014; Krencik et al., 2011; Serio et al., 2013; Shaltouki et al., 2013; Williams et al., 2014). Specific molecules used for this purpose include bone morphogenetic proteins (BMPs) (Chen et al., 2014; Shaltouki et al., 2013) and/or ciliary neurotrophic factor—related proteins (Krencik et al., 2015; Serio et al., 2013; Shaltouki et al., 2013).

Increased expression of astrocyte-specific markers was found in multiple regions in postmortem brains from autistic patients (Fatemi, Folsom, Reutiman, & Lee, 2008; Laurence & Fatemi, 2005), accompanied by decreased cell body size and reduced complexity and branch length in the cortex (Cao et al., 2012). Through co-culture experiments using isogenic cells lines derived from patients with RTT, Williams et al. (2014) did not find significant changes in proliferation rate and differentiation efficiency between the control and iPSC-*MECP2*—mutant astrocyte progenitor cells. However, the authors observed a negative effect of mutant-derived astrocytes in control neurons from human and mouse, which exhibited smaller soma size, shorter total neurite length, and a smaller number of terminal ends under these conditions. Similar morphological effects were mimicked by ACM cultures, which suggests that factors secreted by glial cells contribute to the phenotype displayed by neurons. The opposite was also true, that is, there was an increase in neurite length and in the number of terminal ends of mutated neurons in the presence of control astrocytes. Similar results were observed in mouse models of autism spectrum disorder (Ballas, Lioy, Grunseich, & Mandel, 2009; Jacobs & Doering, 2010; Lioy et al., 2011), which indicates that whereas neurons could control the initiation of the pathology, astrocytes might be responsible for its progression (McGann, Lioy, & Mandel, 2012). A study developed by Krencik et al. corroborates this assumption. Functional astrocytes derived from iPSCs from patients with Costello syndrome, a neurodevelopmental disorder caused by alterations in Ras pathway signaling, revealed the influence of these glial cells on the disease phenotype (Krencik et al., 2015).

In addition, astrocytes derived from Down syndrome (DS) patient iPSCs produce increased levels of nitric oxide and reactive oxygen species. Moreover, DS-iPSC—derived ACM compromised neurogenesis in vitro, failing to promote neural maturation and synapse formation besides being toxic to neurons (Chen et al., 2014); suggesting that such glial cells are relevant to the pathogenesis of the disease. In the same way, coculture experiments using mouse-derived FXS cells showed that FMRP1-deficient astrocytes reduce the survival rates, but also affect the morphology of wild-type neurons (McGann et al., 2012). These findings point out that defects in astrocyte differentiation might contribute to neuronal defects later observed in the syndrome.

Although protocols for the effective generation of microglia using stem cells has not been established, iPSC-derived macrophages can be generated successfully. First, it is necessary to drive the differentiation to mesoderm, usually by adding BMP4 to the media culture. Next, after exposure to different molecules and/or cytokines, the cells mature in the presence of macrophage colony-stimulating factor (Brault et al., 2014; Van Wilgenburg, Browne, Vowles, & Cowley, 2013; Yanagimachi et al., 2013). In 2005, Vargas et al. observed activation of astroglial and microglial cells in postmortem brains of autistic patients, characteristic of a neuroinflammatory phenotype (Vargas, Nascimbene, Krishnan, Zimmerman, & Pardo, 2005). Increasing evidence suggests that there is immune dysfunction/inflammation, not just in patients with autism spectrum disorder (Onore, Careaga, & Ashwood, 2012; Suzuki et al., 2013; Theoharides et al., 2012), but also in those with schizophrenia and obsessive-compulsive disorder (Frick et al., 2013). In addition, by investigating Gaucher disease, Aflaki et al. (2014) and Panicker et al. (2014) reproduced some of the pathological phenotypes of the disorder using iPSCs, including macrophages with reduced glucocerebrosidase activity and increased storage of glucocerebroside and glucosylsphingosine in lysosomes. These results show the potential to use monocytic-derived cells for the comprehension of complex diseases.

Little is known about the contribution of oligodendrocytes to the phenotype of neurodevelopmental disorders. However, studies are clarifying the relevance of these cells for proper neural network functioning. Using an in vivo approach to RTT, Nguyen et al. (2013) observed that the selective removal or reexpression of *MECP2* in oligodendrocytes had a negative or positive impact, respectively, on autistic features/behavior exhibited by animals. Myelination impairments were also observed in a mouse model for FXS (Pacey et al., 2013). Different groups started to generate oligodendrocytes using stem cells. The common feature of this specific differentiation is the presence of triiodothyronine and insulin, among other small molecules (Hu et al., 2009, 2010; Izrael et al., 2007; Nistor, Totoiu, Haque, Carpenter, & Keirstead, 2005). The myelination properties of the generated cells were also observed in vivo after mouse transplantation (Izrael et al., 2007; Nistor et al., 2005), which shows the potential of further investigating the effect of these glial cells to specific disease initiation and progression.

Increasing evidence supports the view that neurodevelopmental disorders are not exclusively neuronal pathologies, but also conditions that evolve due to glial impairment. Although most results in the literature still rely on animal models, human ES- and iPSC-based studies have been performed and gradually are enlightening and strengthening the relevance of interactions between neurons and glia to the progress of a neural phenotype. The initial idea that neurons are the major and/or only cells responsible for neural deficits is being replaced by a more dynamic and intricate interpretation of the function and effects of the neural network.

THE NEXT STEP OF INDUCED PLURIPOTENT STEM CELLS: THREE-DIMENSIONAL CULTURES AND MINI-BRAINS

The idea that iPSC-derived neurons can mimic in-the-dish neurodevelopmental programs has gradually become stronger over the past decade, lighting the way to build tissues and organs in vitro. Biomedical researchers and tissue engineers have successfully combined neural stem cells and different types of polymer and matrix scaffolds to generate functional neurons and astrocytes that emulate the mammalian brain (Ma et al., 2004, 2008; Martinez-Ramos et al., 2008; Park, Teng, & Snyder, 2002).

Three-dimensional (3D) cell cultures can represent an environment closer to in vivo model and could ultimately serve as a more predictive preclinical translational model for studying neurodevelopmental disease mechanisms. A classical 2D cell culture model is an important tool to understand cell behavior under controlled conditions, however, cells cultured in 3D scaffolds have been found to exhibit a better distribution of cell—cell and cell—extracellular matrix interactions, which alter cell morphology, signaling mechanisms, and subsequent cell function (Cukierman, Pankov, & Yamada, 2002). Presynthesized scaffolds such as Matrigel, collagen, and hydrogel, which are designed to recreate an extracellular matrix environment, often do not support neurite outgrowth and do not allow the long-term survival of cells (Dewitt, Kaszuba, Thompson, & Stege-Mann, 2009; Dutta & Dutta, 2009; Subramanian, Krishnan, & Sethuraman, 2009; Tibbitt & Anseth, 2009;). However, to overcome these difficulties and the elevated costs of classical 2D culture maintenance, interesting progress is being achieved by growing cells in a microfluidic chip, that is, lab-in-a-chip. These chips allow more controlled spatiotemporal conditions in a versatile cellular microenvironment setup (Moreno et al., 2015). An alternative to presynthesized scaffolds is the generation of an extracellular matrix by the cells themselves in 3D tissue culture, as cell aggregates.

In general, human pluripotent stem cells require biological signals from the matrix and from the adjacent ones for cell survival, proliferation, and normal physiological process (Amit et al., 2011; Ohgushi & Sasai, 2011; Pamies, Hartung, & Hogberg, 2014; Peerani et al., 2007; Steiner et al., 2010; Zweigerdt, Olmer, Singh, Haverich, & Martin, 2011). In this context, Eiraku et al. (2008) reported self-organized formation of human ES cell—derived cortical tissues using a 3D aggregation culture. With this method, it was demonstrated that the in vitro development of human stem cells mimic in some aspects early embryonic corticogenesis with regional specification and layer-specific neurogenesis. Muguruma et al. also showed the formation of a cerebellar structure of ES cell culture comparable to an early human cerebellum in the first trimester of gestation. This self-organized cerebellar neuroepithelium also was able to differentiate into functional Purkinje cells (Muguruma, Nishiyama, Kawakami, Hashimoto, & Sasai, 2015). Therefore, the cellular 3D aggregated culture allows us to access a new set of information regarding tissue self-organization and long-range neuronal networks, and can be established as a new tool for drug discovery and regenerative medicine. A future challenge is to develop a long-term culture system that enables human pluripotent stem cells to generate organs or organoids to study disease-relevant phenotypes in a macroscopic, structural, and multifactorial view.

We have a limited understanding of the brain's function under healthy and pathological conditions. Architecture, cellular number and diversity, and neural migration patterns are some examples of limitations that are difficult to reproduce with traditional in vitro methods. The organoid model developed by Lancaster et al. exhibited the typical organization of cortical and ventricular regions and the same type of cellular migration found in vivo during development (Lancaster & Knoblich, 2014a,b; Lancaster et al., 2013; Paşca et al., 2015). The authors observed that when iPSCs are cultured as aggregates under appropriate conditions for a couple of weeks, the cells self-organize into cerebral organoids, grow to a half centimeter, and can live for several months in a bioreactor. These "mini-brains" resemble early cortical development, forming ventricles and other structures. The tissue surrounding the ventricles differentiates into cortical layers in which the innermost layer filled with radial glial cells divides and becomes neurons in the same pattern observed in human embryos. This type of culture platform could represent a bridge between classical neural culture and in vivo studies, representing a great potential for a new step toward disease-modeling studies. To demonstrate this, Lancaster et al. used iPSC from patients with human microcephaly to generated organoids. The organoids derived from iPSC from patients with microcephaly were smaller than those obtained from normal individuals and had reduced numbers of progenitors (Lancaster et al., 2013). Although the organoid model does not fully recapture the brain, such a model will likely be particularly valuable for studies of neurodevelopmental and disease mechanisms (Lancaster & Knoblich, 2014a).

All of the strategies and methodologies mentioned share the cell−cell interaction and maintenance of the microenvironment as the main advantages that favor neurogenesis. However, the in vitro neurons are still far from the human brain in terms of maturation. Compared with other species, our neurodevelopment is slow and needs an intricate network of soluble and genetic factors and niche clues to mature properly. In 2005, Muotri et al. transplanted human ES cells into the brain ventricles of embryonic mice to study the differentiation and migration potential of human ES cells in vivo. The transplanted human neurons integrated successfully into the adult mouse forebrain and displayed mature morphology and functionality (Muotri, Nakashima, Toni, Sandler, & Gage, 2005). In 2013, Espuny-Camacho et al. transplanted embryonic and iPSC-derived cells into newborn mouse brains. The differentiation and connectivity of the transplanted human neurons increased over 6 months in vivo, culminating in the formation of functional synapses (Espuny-Camacho et al., 2013).

In this context, human chimeric mice brains could ultimately serve as a more predictive preclinical translational model for studying disease mechanisms and brain repair (Steinbeck & Studer, 2015; Thompson & Björklund, 2015). As a key chimeric translational model, researchers have derived inhibitory neurons from iPSC in vivo and demonstrated that these interneurons migrate and integrate into the circuitry of the epileptic mouse brain. More important, the transplanted human neurons were able to diminish the number of seizures and ameliorate behavioral abnormalities (Cunningham et al., 2014).

LIMITATIONS OF INDUCED PLURIPOTENT STEM CELL MODEL

As described earlier, iPSC technology is a breakthrough in developmental biology and translational medicine. However, several limitations remain regarding reprogramming efficiency, clone-to-clone variation, and heterogeneity between iPSC and the capability of generating mature functional cell types. Therefore, genetic and epigenetic backgrounds, iPSC derivation protocol, and culture conditions are key components to cellular variability.

It is becoming clear that the presence of such variability among iPSC lines highlights the importance of validating results with multiple cell lines and clones from the same genotype (Hu et al., 2010; Marchetto et al., 2010). Inconsistent outcomes obtained from different clones from the same cell lines can result from the number and sites of viral integrations and heterogeneity in the starting somatic cell populations (Gore et al., 2011). In addition, there is growing evidence that epigenetic changes among iPSC lines or in the same line may contribute to cellular heterogeneity (Mekhoubad et al., 2012) Therefore, multiple clones from multiple patients must be generated to truly establish in vitro iPSC disease modeling. Finally, variability is also caused by undefined and less than optimal culture conditions. The use of chemically defined formulations and optimized conditions was achieved for iPSC using mathematical and statistical methods (Marinho, Chailangkarn, & Muotri, 2015). Such an approach could be applied to all steps of neural differentiation, standardizing protocols across different laboratories worldwide.

Another concern is the difficulty of obtaining mature functional cell types that reflect our lack of understanding of developmental signaling for the maturation step. For instance, several studies suggest that neurons derived from iPSCs show immature morphology, filopodia synaptic protrusions, and an electrophysiological profile with a low number of cells that fire action potential (Marchetto et al., 2010). In 2014, an important study highlighted how immature the iPSC-derived neurons are. Using a transcriptomic analysis to compare stem cell−derived neurons with developing human fetal brain, Stein et al. (2014) developed a tool (CoNTExT) that classified the developmental stage and regionally identified the derived neurons. The authors reported that the derived neurons could be compared with a 3-month fetal stage of maturity. Thus, optimization of culture conditions to allow better maturation in vitro is an active and promising research topic in this field.

GENOME EDITING: CREATING DISEASED EMBRYONIC STEM CELLS

iPSC technology allows us to obtain stem cells from skin biopsies, blood, dental pulp, and other tissue samples (Dimos et al., 2008; Park et al., 2008; Seki et al., 2010; Staerk et al., 2010), making it possible to generate differentiated cells predisposed to diseases. However, as noted earlier, iPSC-based studies have some limitations. Because any observed phenotype in an iPSC model must be interpreted by comparison with unaffected cells, the quality of iPSC studies relies on the availability or quality of those related controls. Genetic/epigenetic background differences could affect the phenotypic characterization of the disease, even when siblings are used as controls (Bock et al., 2011; Kim, Choi, Choi, & Do, 2011; Kim et al., 2010; Nishino et al., 2011; Ruiz et al., 2012). Furthermore, differences in reprogramming methodology, less than optimal culture conditions, and expansion and passaging of iPSC lines also can lead to the accumulation of a variety of genetic alterations (Gore et al., 2011; Hussein et al., 2011; Ji et al., 2012; Laurent et al., 2011; Mayshar et al., 2010; Musunuru, 2013).

Perhaps the most rigorous evaluation would be between cell lines with the same background (isogenic), differing only by a relevant disease mutation. This approach would also exclude other variables, enabling investigators to establish causal relationships among genotypes and phenotypes. Several studies took advantage of XCI in female iPSCs to generate isogenic pairs of control and mutant cells for neurodevelopmental disorders (Ananiev et al., 2011; Liu et al., 2012). Alternatively, genome editing technology such as zinc finger nuclease (ZFN), transcription activator-like effector nucleases (TALENs), or CRISPR/Cas9 systems are being used to manipulate genomes and introduce or correct locus-specific mutations via homologous recombination (Capecchi, 1989; Hockemeyer & Jaenisch, 2010; Lombardo et al., 2007; Zou et al., 2009).

ZFNs are artificial modular restriction enzymes designed to recognize a target DNA sequence locus by combining zinc finger proteins and an endonuclease to induce double-strand breaks (Bibikova, Beumer, Trautman, & Carroll, 2003). By taking advantage of cellular DNA machinery, the induced break can be repaired by homologous recombination and nonhomologous end-joining. In 2009, Zou et al. demonstrated that ZFNs are able to induce target gene modifications in human ES cell and iPSC lines, providing a solid basis for genomic editing in pluripotent cells. In parallel, different research groups have proven the impact of ZFN as a powerful tool to target genes that are not even expressed (Hockemeyer et al., 2009) and to create specific chromosomal translocation (Brunet et al., 2009). Although several studies have reported no effect on the cellular karyotype and pluripotency of stem cells after prolonged culture, zinc fingers may be further refined for higher specificity.

Activator-like effector transcription factors were discovered in plant pathogens to bypass host immunity machinery. TAL effector proteins consist of highly conserved repetitive amino acid sequences that determine DNA binding specificity. Similar to ZFN, TALENs are associated with endonuclease to create a double-strand break at a particular genomic site (Miller et al., 2011). The relatively fast and more flexible protocol to clone and generate TALENs over ZFN is the major advantage of the method. The authors stated that the specific site was produced with similar efficiency and precision as ZFN but with fewer off-target effects. The power of this technology is being proved by the increasing number of studies applying and generating new TALENs, encouraging the development of more efficient genomic editing tools (Cermak et al., 2011; Mussolino et al., 2011; Scholze & Boch, 2010).

Clustered regularly interspaced short palindromic repeats (CRISPRs) are the latest technology for gene editing of a specific locus. In bacteria, the CRISPR system provides acquired "immunity" against viral DNA (Wiedenheft, Sternberg, & Doudna, 2012). Short segments of viral DNA are integrated within the CRISPR genomic loci, transcribed, and processed into short guide RNAs (gRNAs). These gRNAs specifically bind to the target pathogenic DNA to be cleaved by Cas9 endonucleases. Using this idea, the CRISPR/Cas system can edit (silence, enhance, or change specific genes) any DNA sequence with custom-designed gRNAs. Therefore, the CRISPR/Cas9 system requires only a specific gRNA to bind and cut the DNA. The simplicity, high efficiency, and versatility of the system favor rapid progress in developing CRISPR/Cas9 into a set of tools for cellular and molecular biology research, revealing that its potential is limitless. Together, these technologies promise to expand our ability to explore and alter any genome to screen for disease-relevant phenotypes. In addition, genome-editing technologies open new paths for stem cell—corrective cellular therapies.

DRUG DISCOVERY WITH INDUCED PLURIPOTENT STEM CELLS

For many years, pharmacological studies have been an important part of the preclinical drug evaluation before clinical trials and human therapy. The use of chemical compound library screening and different cellular or animal models allows researchers to investigate specific pathogenic mechanisms and identify relevant disease targets. In this context, the impact of human iPSC as a well-defined model to screen and test lead compounds is enormous, making patient-specific therapy a real possibility (Bosnjak, 2012). Moreover, iPSC-based disease models are transforming translational science by reducing the need to use animal approaches (Fig. 8.1).

The evaluation of synthetic small molecules and natural compounds will bring new discoveries to the stem cell biology field, and may also contribute to the establishment of more effective regenerative medicine (Ding & Schultz, 2005; Ding et al., 2003). With a library of about 50,000 compounds, Chen et al. (2006) used a multiwell, high-throughput platform to

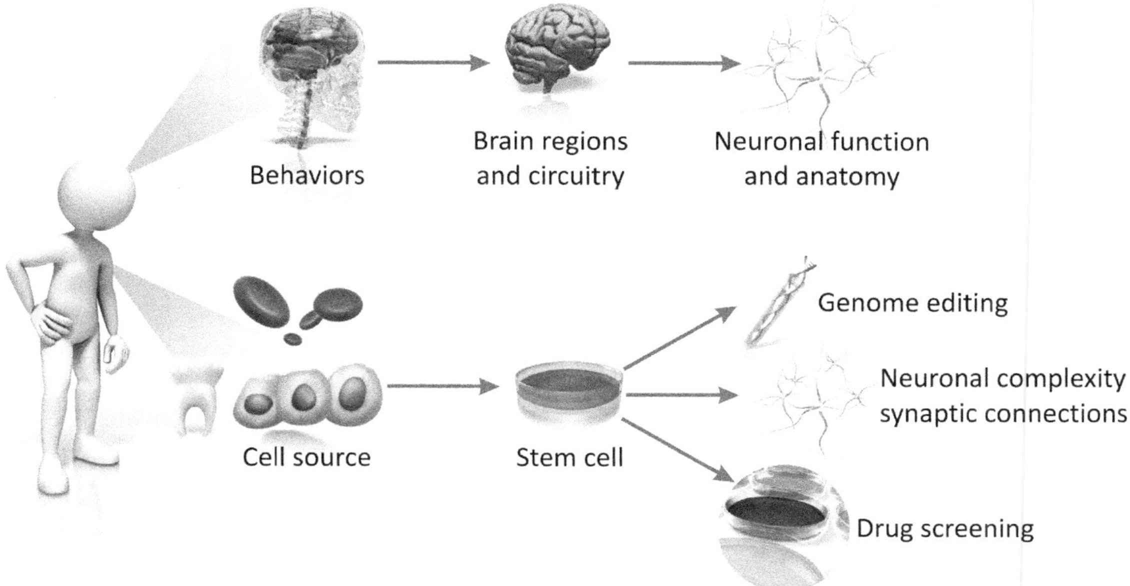

FIGURE 8.1 Modeling neurodevelopmental disorders with human iPSCs. iPSC technology can be used in disease modeling, pathogenesis research, drug screening, and cell replacement therapy. Genetic mutations can be corrected by genome-editing approaches after reprogramming. Neural cells derived from patient-specific iPSCs are being used for disease modeling and transplantation purposes. Cellular phenotypes are assessed by measuring neuronal morphology, complexity, and connectivity. Once the phenotype is identified, drug-screening platforms can be used to rescue the phenotype. New therapies and drugs are emerging from this approach, benefiting numerous neurological patients.

select those able to keep mouse stem cells in the pluripotent stage. In the same way, Desbordes et al. (2008) developed a high-throughput assay (over 2500 compounds) to screen compounds to promote cellular proliferation and affect human ES cell self-renewal and differentiation fate. These studies highlight the impact of high-throughput screening in the study of ES biology, drug discovery, and therapy (Underhill & Bhatia, 2007).

Until recently, no studies have used a drug-screening platform to select drugs that can rescue abnormal phenotypes of neurological diseases. For instance, using RTT-derived neurons as a proof-of-principle, our group demonstrated the potential of iPSC models in drug screening and toxicology, accelerating the search for new therapeutic agents and anticipating results from translational medicine (Marchetto et al., 2010). Moreover, the use of spinal muscular atrophy (SMA) patient-specific iPSCs by Ebert et al. (2009) provided the basis of iPSC technology in drug screening. In that study, the authors treated iPSC-derived motor neurons with selected drugs to increase the expression of a specific cellular survival protein. The authors observed that fibroblast- and iPS-SMA cells respond similarly to drug treatment and suggested that iPSC-derived motor neurons could be useful for new drug screening approaches. In another study, Egawa et al. (2012) used a low-throughput assay with amyotrophic lateral sclerosis (ALS) patient-specific neurons derived from iPSC to evaluate only four histone acetyltransferase inhibitors to rescue the ALS motor neuron phenotype. These exciting observations show the great potential of patient-specific iPSCs in drug screening to test the efficacy of candidate drugs for the treatment of human neurological diseases.

CONCLUSIONS

The iPSC field has moved with astonishing rapidity from being able to induce a single neuron to producing organoids or mini-brains that recapitulate the early steps of development of the human brain and the interconnected functioning of living neurons and glia. These advances provide an unprecedented degree of access to human neural development. Although important limitations and technical challenges remain, these advances mark a profound inflection point in the ability to study neurological and psychiatric disorders. Although this technology is still in its early stage, it has demonstrated the ability to recapitulate relevant neuronal defects of neurodevelopmental diseases. New protocols for the generation of subtypes of neurons and glia will help researchers to identify the contribution of these cell types to a wide range of pathologies. Previous studies established that astrocytes secrete signaling molecules that stimulate the formation, pruning, and function of synapses throughout the brain (Chung et al., 2013; Diniz et al., 2012). By mixing different cell types in coculture experiments, one can identify eventual non—cell autonomous molecular and cellular mechanisms that almost certainly have a role in both neurodevelopmental and neurodegenerative disorders.

Genome-editing technologies will help to determine whether specific mutations correlate to gene expression differences or whether a clinical pharmacological response is predictable by drug treatments. Incorporating genomic data into stem cell work may allow us to characterize alterations in the genome of thousands of patients, linking these variants to specific cellular phenotypes. Nonetheless, prospective work should take advantage of larger, better-characterized patient cohorts with well-defined clinical endophenotypes, pharmacological history, and genetic predisposition. Contrary to "one-gene at the time," the study of idiopathic conditions will likely bring novel insights for neurodevelopmental disorders. For example, our group has focused on reprogramming human dental pulp stem cells from children with autism, using the deciduous tooth as a source of somatic cells. The "tooth-fairy project" has dramatically increased in the past years, probably owing to the use of social networks to connect with families (Beltrão-Braga et al., 2011). By generating human iPSC neurons from large cohorts of pediatric disorders, one can test whether the clinical outcome is predictive of the magnitude of cellular phenotype.

REFERENCES

Aflaki, E., Stubblefield, B. K., Maniwang, E., Lopez, G., Moaven, N., Goldin, E. ... Sidransky, E. (2014). Macrophage models of Gaucher disease for evaluating disease pathogenesis and candidate drugs. *Science Translational Medicine, 6*(240), 240ra73.

Amir, R. E., Van den Veyver, I. B., Schultz, R., Malicki, D. M., Tran, C. Q., Dahle, E. J. ... Zoghbi, H. Y. (2000). Influence of mutation type and X chromosome inactivation on Rett syndrome phenotypes. *Annals of Neurology, 47*(5), 670—679.

Amir, R. E., Van den Veyver, I. B., Wan, M., Tran, C. Q., Francke, U., & Zoghbi, H. Y. (1999). Rett syndrome is caused by mutations in X-linked MECP2, encoding methyl-CpG-binding protein 2. *Nature Genetics, 23*(2), 185—188.

Amit, M., Laevsky, I., Miropolsky, Y., Shariki, K., Peri, M., & Itskovitz-Eldor, J. (2011). Dynamic suspension culture for scalable expansion of undifferentiated human pluripotent stem cells. *Nature Protocols, 6*(5), 572—579.

Ananiev, G., Williams, E. C., Li, H., & Chang, Q. (2011). Isogenic pairs of wild type and mutant induced pluripotent stem cell (iPSC) lines from Rett syndrome patients as in vitro disease model. *PLoS One, 6*(9), e25255.

Angelman, H. (1965). "Puppet" children a report on three cases. *Developmental Medicine and Child Neurology, 7*(6), 681—688.

Antar, L. N., Afroz, R., Dictenberg, J. B., Carroll, R. C., & Bassell, G. J. (2004). Metabotropic glutamate receptor activation regulates fragile x mental retardation protein and FMR1 mRNA localization differentially in dendrites and at synapses. *The Journal of Neuroscience : The Official Journal of the Society for Neuroscience, 24*(11), 2648—2655.

Ashley, C. T., Wilkinson, K. D., Reines, D., & Warren, S. T. (1993). FMR1 protein: conserved RNP family domains and selective RNA binding. *Science (New York, NY), 262*(5133), 563—566.

Bader, P. L., Faizi, M., Kim, L. H., Owen, S. F., Tadross, M. R., Alfa, R. W. ... Shamloo, M. (2011). Mouse model of Timothy syndrome recapitulates triad of autistic traits. *Proceedings of the National Academy of Sciences of the United States of America, 108*(37), 15432—15437.

Ballas, N., Lioy, D. T., Grunseich, C., & Mandel, G. (2009). Non-cell autonomous influence of MeCP2-deficient glia on neuronal dendritic morphology. *Nature Neuroscience, 12*, 311—317.

Barrett, C. F., & Tsien, R. W. (2008). The Timothy syndrome mutation differentially affects voltage- and calcium-dependent inactivation of CaV1.2 L-type calcium channels. *Proceedings of the National Academy of Sciences of the United States of America, 105*(6), 2157—2162.

Baumann, N., & Pham-Dinh, D. (2001). Biology of oligodendrocyte and myelin in the mammalian central nervous system. *Physiological Reviews, 81*, 871—927.

Bear, M. F., Huber, K. M., & Warren, S. T. (2004). The mGluR theory of fragile X mental retardation. *Trends in Neurosciences, 27*(7), 370—377.

Beltrão-Braga, P. C. B., Pignatari, G. C., Maiorka, P. C., Oliveira, N. A. J., Lizier, N. F., Wenceslau, C. V. ... Kerkis, I. (2011). Feeder-free derivation of induced pluripotent stem cells from human immature dental pulp stem cells. *Cell Transplantation, 20*(11—12), 1707—1719.

Bibikova, M., Beumer, K., Trautman, J. K., & Carroll, D. (2003). Enhancing gene targeting with designed zinc finger nucleases. *Science (New York, NY), 300*(5620), 764.

Bird, L. M. (2014). Angelman syndrome: review of clinical and molecular aspects. *The Application of Clinical Genetics, 7*, 93—104.

Bock, C., Kiskinis, E., Verstappen, G., Gu, H., Boulting, G., Smith, Z. D. ... Meissner, A. (2011). Reference Maps of human ES and iPS cell variation enable high-throughput characterization of pluripotent cell lines. *Cell, 144*(3), 439—452.

Bosnjak, Z. J. (2012). Developmental neurotoxicity screening using human embryonic stem cells. *Experimental Neurology, 237*(1), 207—210.

Brault, J., Goutagny, E., Telugu, N., Shao, K., Baquié, M., Satre, V. ... Stasia, M. J. (2014). Optimized generation of functional neutrophils and macrophages from patient-specific induced pluripotent stem cells: ex vivo models of X(0)-linked, AR22(0)- and AR47(0)- chronic granulomatous diseases. *BioResearch Open Access, 3*(6), 311—326.

Braun, K., & Segal, M. (2000). FMRP involvement in formation of synapses among cultured hippocampal neurons. *Cerebral Cortex (New York, NY : 1991), 10*(10), 1045—1052.

Brown, W. T., Jenkins, E. C., Cohen, I. L., Fisch, G. S., Wolf-Schein, E. G., Gross, A. ... Ritvo, E. (1986). Fragile X and autism: a multicenter survey. *American Journal of Medical Genetics, 23*(1—2), 341—352.

Brunet, E., Simsek, D., Tomishima, M., DeKelver, R., Choi, V. M., Gregory, P. … Jasin, M. (2009). Chromosomal translocations induced at specified loci in human stem cells. *Proceedings of the National Academy of Sciences of the United States of America, 106*(26), 10620–10625.

Cao, F., Yin, A., Wen, G., Sheikh, A. M., Tauqeer, Z., Malik, M. … Li, X. (2012). Alteration of astrocytes and Wnt/β-catenin signaling in the frontal cortex of autistic subjects. *Journal of Neuroinflammation, 9*, 223.

Capecchi, M. R. (1989). Altering the genome homologous recombination. *Science, 244*(4910), 1288–1292.

Castrén, M., Tervonen, T., Kärkkäinen, V., Heinonen, S., Castrén, E., Larsson, K. … Akerman, K. (2005). Altered differentiation of neural stem cells in fragile X syndrome. *Proceedings of the National Academy of Sciences of the United States of America, 102*(49), 17834–17839.

Cavaillé, J., Buiting, K., Kiefmann, M., Lalande, M., Brannan, C. I., Horsthemke, B. … Hüttenhofer, A. (2000). Identification of brain-specific and imprinted small nucleolar RNA genes exhibiting an unusual genomic organization. *Proceedings of the National Academy of Sciences of the United States of America, 97*(26), 14311–14316.

Cermak, T., Doyle, E. L., Christian, M., Wang, L., Zhang, Y., Schmidt, C. … Voytas, D. F. (2011). Efficient design and assembly of custom TALEN and other TAL effector-based constructs for DNA targeting. *Nucleic Acids Research, 39*(12), e82.

Chahrour, M., & Zoghbi, H. Y. (2007). The story of Rett syndrome: from clinic to neurobiology. *Neuron, 56*(3), 422–437.

Chamberlain, S. J., Chen, P.-F., Ng, K. Y., Bourgois-rocha, F., Lemtiri-chlieh, F., Levine, E. S., & Lalande, M. (2010). Induced pluripotent stem cell models of the genomic imprinting disorders Angelman and Prader-Willi syndromes. *Proceedings of the National Academy of Sciences of the United States of America, 107*(41), 17668–17673.

Chambers, S. M., Fasano, C. A., Papapetrou, E. P., Tomishima, M., Sadelain, M., & Studer, L. (2009). Highly efficient neural conversion of human ES and iPS cells by dual inhibition of SMAD signaling. *Nature Biotechnology, 27*(3), 275–280.

Chen, C., Jiang, P., Xue, H., Peterson, S. E., Tran, H. T., McCann, A. E. … Deng, W. (2014). Role of astroglia in Down's syndrome revealed by patient-derived human-induced pluripotent stem cells. *Nature Communications, 5*, 4430.

Chen, S., Do, J. T., Zhang, Q., Yao, S., Yan, F., Peters, E. C. … Ding, S. (2006). Self-renewal of embryonic stem cells by a small molecule. *Proceedings of the National Academy of Sciences of the United States of America, 103*(46), 17266–17271.

Cheung, A. Y. L., Horvath, L. M., Grafodatskaya, D., Pasceri, P., Weksberg, R., Hotta, A. … Ellis, J. (2011). Isolation of MECP2-null Rett Syndrome patient hiPS cells and isogenic controls through X-chromosome inactivation. *Human Molecular Genetics, 20*(11), 2103–2115.

Chipman, P. H., Toma, J. S., & Rafuse, V. F. (2012). Generation of motor neurons from pluripotent stem cells. *Progress in Brain Research, 201*, 313–331.

Chung, W.-S., Clarke, L. E., Wang, G. X., Stafford, B. K., Sher, A., Chakraborty, C. … Barres, B. A. (2013). Astrocytes mediate synapse elimination through MEGF10 and MERTK pathways. *Nature, 504*(7480), 394–400.

Clayton-Smith, J., & Laan, L. (2003). Angelman syndrome: a review of the clinical and genetic aspects. *Journal of Medical Genetics, 40*(2), 87–95.

Coffee, B., Zhang, F., Ceman, S., Warren, S. T., & Reines, D. (2002). Histone modifications depict an aberrantly heterochromatinized FMR1 gene in fragile X syndrome. *American Journal of Human Genetics, 71*(4), 923–932.

Coffee, B., Zhang, F., Warren, S. T., & Reines, D. (1999). Acetylated histones are associated with FMR1 in normal but not fragile X-syndrome cells. *Nature Genetics, 22*(1), 98–101.

Cohen-Kutner, M., Yahalom, Y., Trus, M., & Atlas, D. (2012). Calcineurin controls voltage-dependent-inactivation (VDI) of the normal and Timothy cardiac channels. *Scientific Reports, 2*, 366.

Crawford, D. C., Acuña, J. M., & Sherman, S. L. (2001). FMR1 and the fragile X syndrome: human genome epidemiology review. *Genetics in Medicine : Official Journal of the American College of Medical Genetics, 3*(5), 359–371.

Cukierman, E., Pankov, R., & Yamada, K. M. (2002). Cell interactions with three-dimensional matrices. *Current Opinion in Cell Biology, 14*, 633–639.

Cunningham, M., Cho, J.-H., Leung, A., Savvidis, G., Ahn, S., Moon, M. … Chung, S. (2014). hPSC-derived maturing GABAergic interneurons ameliorate seizures and abnormal behavior in epileptic mice. *Cell Stem Cell, 15*(5), 559–573.

Desbordes, S. C., Placantonakis, D. G., Ciro, A., Socci, N. D., Lee, G., Djaballah, H., & Studer, L. (2008). High-throughput screening assay for the identification of compounds regulating self-renewal and differentiation in human embryonic stem cells. *Cell Stem Cell, 2*(6), 602–612.

Devys, D., Lutz, Y., Rouyer, N., Bellocq, J. P., & Mandel, J. L. (1993). The FMR-1 protein is cytoplasmic, most abundant in neurons and appears normal in carriers of a fragile X premutation. *Nature Genetics, 4*(4), 335–340.

Dewitt, D. D., Kaszuba, S. N., Thompson, D. M., & Stege-Mann, J. P. (2009). Collagen I-matrigel scaffolds for enhanced Schwann cell survival and control of three-dimensional cell morphology. *Tissue Engineering. Part A, 15*(10), 2785–2793.

Dimos, J. T., Rodolfa, K. T., Niakan, K. K., Weisenthal, L. M., Mitsumoto, H., Chung, W. … Eggan, K. (2008). Induced pluripotent stem cells generated from patients with ALS can be differentiated into motor neurons. *Science (New York, NY), 321*(5893), 1218–1221.

Ding, S., & Schultz, P. G. (2005). Small molecules and future regenerative medicine. *Current Topics in Medicinal Chemistry, 5*(4), 383–395.

Ding, S., Wu, T. Y. H., Brinker, A., Peters, E. C., Hur, W., Gray, N. S., & Schultz, P. G. (2003). Synthetic small molecules that control stem cell fate. *Proceedings of the National Academy of Sciences of the United States of America, 100*(13), 7632–7637.

Diniz, L. P., Almeida, J. C., Tortelli, V., Vargas Lopes, C., Setti-Perdigão, P., Stipursky, J. … Gomes, F. C. A. (2012). Astrocyte-induced synaptogenesis is mediated by transforming growth factor β signaling through modulation of D-serine levels in cerebral cortex neurons. *The Journal of Biological Chemistry, 287*(49), 41432–41445.

Doers, M. E., Musser, M. T., Nichol, R., Berndt, E. R., Baker, M., Gomez, T. M. … Bhattacharyya, A. (2014). iPSC-derived forebrain neurons from FXS individuals show defects in initial neurite outgrowth. *Stem Cells and Development, 23*(15), 1777–1787.

Duan, L., Peng, C. Y., Pan, L., & Kessler, J. A. (2015). Human pluripotent stem cell-derived radial glia recapitulate developmental events and provide real-time access to cortical neurons and astrocytes. *Stem Cells Translational Medicine, 4*(5), 437–447.

Dulcis, D., & Spitzer, N. C. (2008). Illumination controls differentiation of dopamine neurons regulating behaviour. *Nature, 456*(7219), 195–201.

Dutta, R. C., & Dutta, A. K. (2009). Cell-interactive 3D-scaffold; advances and applications. *Biotechnology Advances, 27*(4), 334–339.

Ebert, A. D., Yu, J., Rose, F. F., Mattis, V. B., Lorson, C. L., Thomson, J. A., & Svendsen, C. N. (2009). Induced pluripotent stem cells from a spinal muscular atrophy patient. *Nature, 457*(7227), 277–280.

Egawa, N., Kitaoka, S., Tsukita, K., Naitoh, M., Takahashi, K., Yamamoto, T. … Inoue, H. (2012). Drug screening for ALS using patient-specific induced pluripotent stem cells. *Science Translational Medicine, 4*(145), 145ra104.

Eiges, R., Urbach, A., Malcov, M., Frumkin, T., Schwartz, T., Amit, A. … Ben-Yosef, D. (2007). Developmental study of fragile X syndrome using human embryonic stem cells derived from preimplantation genetically diagnosed embryos. *Cell Stem Cell, 1*(5), 568–577.

Eiraku, M., Watanabe, K., Matsuo-Takasaki, M., Kawada, M., Yonemura, S., Matsumura, M. … Sasai, Y. (2008). Self-organized formation of polarized cortical tissues from ESCs and its active manipulation by extrinsic signals. *Cell Stem Cell, 3*(5), 519–532.

Espuny-Camacho, I., Michelsen, K. A., Gall, D., Linaro, D., Hasche, A., Bonnefont, J. … Vanderhaeghen, P. (2013). Pyramidal neurons derived from human pluripotent stem cells integrate efficiently into mouse brain circuits in vivo. *Neuron, 77*(3), 440–456.

Evans, J. C., Archer, H. L., Colley, J. P., Ravn, K., Nielsen, J. B., Kerr, A. … Clarke, A. J. (2005). Early onset seizures and Rett-like features associated with mutations in CDKL5. *European Journal of Human Genetics : EJHG, 13*(10), 1113–1120.

Fatem, I. S. H., Folsom, T. D., Reutiman, T. J., & Lee, S. (2008). Expression of astrocytic markers aquaporin 4 and connexin 43 is altered in brains of subjects with autism. *Synapse, 62*(7), 501–507.

Feng, Y., Gutekunst, C. A., Eberhart, D. E., Yi, H., Warren, S. T., & Hersch, S. M. (1997). Fragile X mental retardation protein: nucleocytoplasmic shuttling and association with somatodendritic ribosomes. *The Journal of Neuroscience : The Official Journal of the Society for Neuroscience, 17*(5), 1539–1547.

Francke, U. (2006). Mechanisms of disease: neurogenetics of MeCP2 deficiency. *Nature Clinical Practice Neurology, 2*(4), 212–221.

Frick, L. R., Williams, K., & Pittenger, C. (2013). Microglial dysregulation in psychiatric disease. *Clinical and Developmental Immunology, 2013*, 608654.

Gore, A., Li, Z., Fung, H.-L., Young, J. E., Agarwal, S., Antosiewicz-Bourget, J. … Zhang, K. (2011). Somatic coding mutations in human induced pluripotent stem cells. *Nature, 471*(7336), 63–67.

Greer, P. L., Hanayama, R., Bloodgood, B. L., Mardinly, A. R., Lipton, D. M., Flavell, S. W. … Greenberg, M. E. (2010). The Angelman Syndrome protein Ube3A regulates synapse development by ubiquitinating arc. *Cell, 140*(5), 704–716.

Hagerman, R. J., Berry-Kravis, E., Kaufmann, W. E., Ono, M. Y., Tartaglia, N., Lachiewicz, A. … Tranfaglia, M. (2009). Advances in the treatment of fragile X syndrome. *Pediatrics, 123*(1), 378–390.

Hinton, V. J., Brown, W. T., Wisniewski, K., & Rudelli, R. D. (1991). Analysis of neocortex in three males with the fragile X syndrome. *American Journal of Medical Genetics, 41*(3), 289–294.

Hockemeyer, D., & Jaenisch, R. (2010). Gene targeting in human pluripotent cells. *Cold Spring Harbor Symposia on Quantitative Biology, 75*, 201–209.

Hockemeyer, D., Soldner, F., Beard, C., Gao, Q., Mitalipova, M., DeKelver, R. C. … Jaenisch, R. (2009). Efficient targeting of expressed and silent genes in human ESCs and iPSCs using zinc-finger nucleases. *Nature Biotechnology, 27*(9), 851–857.

Hotta, A., Cheung, A. Y. L., Farra, N., Vijayaragavan, K., Séguin, C. A., Draper, J. S. … Ellis, J. (2009). Isolation of human iPS cells using EOS lentiviral vectors to select for pluripotency. *Nature Methods, 6*(5), 370–376.

Hu, B. Y., Du, Z. W., Li, X. J., Ayala, M., & Zhang, S. C. (2009). Human oligodendrocytes from embryonic stem cells: conserved SHH signaling networks and divergent FGF effects. *Development, 136*(9), 1443–1452.

Hu, B. Y., Weick, J. P., Yu, J., Ma, L. X., Zhang, X. Q., Thomson, J. A., & Zhang, S. C. (2010). Neural differentiation of human induced pluripotent stem cells follows developmental principles but with variable potency. *Proceedings of the National Academy of Sciences of the United States of America, 107*(9), 4335–4340.

Hussein, S. M., Batada, N. N., Vuoristo, S., Ching, R. W., Autio, R. … Otonkoski, T. (2011). Copy number variation and selection during reprogramming to pluripotency. *Nature, 471*(7336), 58–62.

Irwin, S. A., Patel, B., Idupulapati, M., Harris, J. B., Crisostomo, R. A., Larsen, B. P. … Greenough, W. T. (2001). Abnormal dendritic spine characteristics in the temporal and visual cortices of patients with fragile-X syndrome: a quantitative examination. *American Journal of Medical Genetics, 98*(2), 161–167.

Ishii, T., Makita, Y., Ogawa, A., Amamiya, S., Yamamoto, M., Miyamoto, A., & Oki, J. (2001). The role of different X-inactivation pattern on the variable clinical phenotype with Rett syndrome. *Brain and Development, 23*(Suppl. 1), S161–S164.

Izrael, M., Zhang, P., Kaufman, R., Shinder, V., Ella, R., Amit, M. … Revel, M. (2007). Human oligodendrocytes derived from embryonic stem cells: effect of noggin on phenotypic differentiation in vitro and on myelination in vivo. *Molecular and Cellular Neuroscience, 34*(3), 310–323.

Jacobs, S., & Doering, L. C. (2010). Astrocytes prevent abnormal neuronal development in the fragile x mouse. *Journal of Neuroscience, 30*, 4508–4514.

Jacquemont, S., Curie, A., des Portes, V., Torrioli, M. G., Berry-Kravis, E., Hagerman, R. J. … Gomez-Mancilla, B. (2011). Epigenetic modification of the FMR1 gene in fragile X syndrome is associated with differential response to the mGluR5 antagonist AFQ056. *Science Translational Medicine, 3*(64), 64ra1.

Jana, N. R. (2012). Understanding the pathogenesis of Angelman syndrome through animal models. *Neural Plasticity, 2012*, 710943.

Ji, J., Ng, S. H., Sharma, V., Neculai, D., Hussein, S., Sam, M. … Batada, N. N. (2012). Elevated coding mutation rate during the reprogramming of human somatic cells into induced pluripotent stem cells. *Stem Cells (Dayton, Ohio), 30*(3), 435–440.

Kim, J. S., Choi, H. W., Choi, S., & Do, J. T. (2011). Reprogrammed pluripotent stem cells from somatic cells. *International Journal of Stem Cells, 4*(1), 1–8.

Kim, K., Doi, A., Wen, B., Ng, K., Zhao, R., Cahan, P. … Daley, G. Q. (2010). Epigenetic memory in induced pluripotent stem cells. *Nature, 467*(7313), 285–290.

Kim, K.-Y., Hysolli, E., & Park, I.-H. (2011). Neuronal maturation defect in induced pluripotent stem cells from patients with Rett syndrome. *Proceedings of the National Academy of Sciences of the United States of America, 108*(34), 14169–14174.

Krencik, R., Hokanson, K. C., Narayan, A. R., Dvornik, J., Rooney, G. E., Rauen, K. A. … Ullian, E. M. (2015). Dysregulation of astrocyte extracellular signaling in Costello syndrome. *Science Translational Medicine, 7*(286), 286ra66.

Krencik, R., Weick, J. P., Liu, Y., Zhang, Z. J., & Zhang, S. C. (2011). Specification of transplantable astroglial subtypes from human pluripotent stem cells. *Nature Biotechnology, 29*(6), 528–534.

Krey, J. F., Pasca, S. P., Shcheglovitov, A., Yazawa, M., Schwemberger, R., Rasmusson, R., & Dolmetsch, R. E. (2013). Timothy syndrome is associated with activity-dependent dendritic retraction in rodent and human neurons. *Nature Neuroscience, 16*(2), 201–209.

Kriks, S., Shim, J. W., Piao, J., Ganat, Y. M., Wakeman, D. R., Xie, Z. … Studer, L. (2011). Dopamine neurons derived from human ES cells efficiently engraft in animal models of Parkinson's disease. *Nature, 480*(7378), 547–551.

Kumari, D., & Usdin, K. (2009). Chromatin remodeling in the noncoding repeat expansion diseases. *The Journal of Biological Chemistry, 284*(12), 7413–7417.

Lancaster, M. A., & Knoblich, J. A. (2014a). Generation of cerebral organoids from human pluripotent stem cells. *Nature Protocols, 9*(10), 2329–2340.

Lancaster, M. A., & Knoblich, J. A. (2014b). Organogenesis in a dish: modeling development and disease using organoid technologies. *Science (New York, NY), 345*(6194), 1247125.

Lancaster, M. A., Renner, M., Martin, C.-A., Wenzel, D., Bicknell, L. S., Hurles, M. E. … Knoblich, J. (2013). Cerebral organoids model human brain development and microcephaly. *Nature, 501*(7467), 373–379.

Laurence, J. A., & Fatemi, S. H. (2005). Glial fibrillary acidic protein is elevated in the superior, frontal, parietal and cerebellar cortices of patients with autism. *Cerebellum, 4*, 206–210.

Laurent, L. C., Ulitsky, I., Slavin, I., Tran, H., Schork, A., Morey, R. … Loring, J. F. (2011). Dynamic changes in the copy number of pluripotency and cell proliferation genes in human ESCs and iPSCs during reprogramming and time in culture. *Cell Stem Cell, 8*(1), 106–118.

Lioy, D. T., Garg, S. K., Monaghan, C. E., Raber, J., Foust, K. D., Kaspar, B. K. … Mandel, G. (2011). A role for glia in the progression of Rett's syndrome. *Nature, 475*, 497–500.

Liu, J., Koscielska, K. A., Cao, Z., Hulsizer, S., Grace, N., Mitchell, G. … Hagerman, P. J. (2012). Signaling defects in iPSC-derived fragile X premutation neurons. *Human Molecular Genetics, 21*(17), 3795–3805.

Lombardo, A., Genovese, P., Beausejour, C. M., Colleoni, S., Lee, Y.-L., Kim, K. A. … Naldini, L. (2007). Gene editing in human stem cells using zinc finger nucleases and integrase-defective lentiviral vector delivery. *Nature Biotechnology, 25*(11), 1298–1306.

Luo, Y., Shan, G., Guo, W., Smrt, R. D., Johnson, E. B., Li, X. … Zhao, X. (2010). Fragile x mental retardation protein regulates proliferation and differentiation of adult neural stem/progenitor cells. *PLoS Genetics, 6*(4), e1000898.

Ma, W., Fitzgerald, W., Liu, Q. Y., O'Shaughnessy, T. J., Maric, D., Lin, H. J. … Barker, J. L. (2004). CNS stem and progenitor cell differentiation into functional neuronal circuits in three-dimensional collagen gels. *Experimental Neurology, 190*(92), 276–288.

Ma, W., Tavakoli, T., Chen, S., Maric, D., Liu, J. L., O'Shaughnessy, T. J., & Barker, J. L. (2008). Reconstruction of functional cortical-like tissues from neural stem and progenitor cells. *Tissue Engineering. Part A, 14*(10), 1673–1686.

Malik, N., & Rao, M. S. (2013). A review of the methods for human iPSC derivation. *Methods in Molecular Biology, 997*, 23–33.

Marchetto, M. C. N., Carromeu, C., Acab, A., Yu, D., Yeo, G. W., Mu, Y. … Muotri, A. R. (2010). A model for neural development and treatment of Rett syndrome using human induced pluripotent stem cells. *Cell, 143*(4), 527–539.

Marinho, P. A., Chailangkarn, T., & Muotri, A. R. (2015). Systematic optimization of human pluripotent stem cells media using design of experiments. *Scientific Reports, 5*, 9834.

Martinez-Ramos, C., Lainez, S., Sancho, F., García Esparza, M. A., Planells-Cases, R., García Verdugo, J. M. … Soria, J. M. (2008). Differentiation of postnatal neural stem cells into glia and functional neurons on laminin-coated polymeric substrates. *Tissue Engineering. Part A, 14*(8), 1365–1375.

Mayshar, Y., Ben-David, U., Lavon, N., Biancotti, J.-C., Yakir, B., Clark, A. T. … Benvenisty, N. (2010). Identification and classification of chromosomal aberrations in human induced pluripotent stem cells. *Cell Stem Cell, 7*(4), 521–531.

McAllister, A. K., Katz, L. C., & Lo, D. C. (1996). Neurotrophin regulation of cortical dendritic growth requires activity. *Neuron, 17*(6), 1057–1064.

McBride, S. M. J., Choi, C. H., Wang, Y., Liebelt, D., Braunstein, E., Ferreiro, D. … Jongens, T. A. (2005). Pharmacological rescue of synaptic plasticity, courtship behavior, and mushroom body defects in a Drosophila model of fragile X syndrome. *Neuron, 45*(5), 753–764.

McGann, J. C., Lioy, D. T., & Mandel, G. (2012). Astrocytes conspire with neurons during progression of neurological disease. *Current Opinion in Neurobiology, 22*(5), 850–858.

Mekhoubad, S., Bock, C., de Boer, A. S., Kiskinis, E., Meissner, A., & Eggan, K. (2012). Erosion of dosage compensation impacts human iPSC disease modeling. *Cell Stem Cell, 10*(5), 595–609.

Miller, J. C., Tan, S., Qiao, G., Barlow, K. A., Wang, J., Xia, D. F. … Rebar, E. J. (2011). A TALE nuclease architecture for efficient genome editing. *Nature Biotechnology, 29*(2), 143–148.

Moreno, E. L., Hachi, S., Hemmer, K., Trietsch, S. J., Baumuratov, A. S., Hankemeier, T. … Fleming, R. M. (2015). Differentiation of neuroepithelial stem cells into functional dopaminergic neurons in 3D microfluidic cell culture. *Lab on a Chip, 15*, 2419–2428.

Muguruma, K., Nishiyama, A., Kawakami, H., Hashimoto, K., & Sasai, Y. (2015). Self-organization of polarized cerebellar tissue in 3D culture of human pluripotent stem cells. *Cell Reports, 10*(4), 537–550.

Muotri, A. R., Nakashima, K., Toni, N., Sandler, V. M., & Gage, F. H. (2005). Development of functional human embryonic stem cell-derived neurons in mouse brain. *Proceedings of the National Academy of Sciences of the United States of America, 102*(51), 18644–18648.

Mussolino, C., Morbitzer, R., Lütge, F., Dannemann, N., Lahaye, T., & Cathomen, T. (2011). A novel TALE nuclease scaffold enables high genome editing activity in combination with low toxicity. *Nucleic Acids Research, 39*(21), 9283–9293.

Musunuru, K. (2013). Genome editing of human pluripotent stem cells to generate human cellular disease models. *Disease Models and Mechanisms, 6*(4), 896–904.

Nakamoto, M., Nalavadi, V., Epstein, M. P., Narayanan, U., Bassell, G. J., & Warren, S. T. (2007). Fragile X mental retardation protein deficiency leads to excessive mGluR5-dependent internalization of AMPA receptors. *Proceedings of the National Academy of Sciences of the United States of America, 104*(39), 15537–15542.

Nguyen, M. V., Felice, C. A., Du, F., Covey, M. V., Robinson, J. K., Mandel, G., & Ballas, N. (2013). Oligodendrocyte lineage cells contribute unique features to Rett syndrome neuropathology. *Journal of Neuroscience, 33*(48), 18764–18774.

Nishimura, K., & Takahashi, J. (2013). Therapeutic application of stem cell technology toward the treatment of Parkinson's disease. *Biological and Pharmaceutical Bulletin, 36*(2), 171–175.

Nishino, K., Toyoda, M., Yamazaki-Inoue, M., Fukawatase, Y., Chikazawa, E., Sakaguchi, H. ... Umezawa, A. (2011). DNA methylation dynamics in human induced pluripotent stem cells over time. *PLoS Genetics, 7*(5), e1002085.

Nistor, G. I., Totoiu, M. O., Haque, N., Carpenter, M. K., & Keirstead, H. S. (2005). Human embryonic stem cells differentiate into oligodendrocytes in high purity and myelinate after spinal cord transplantation. *Glia, 49*(3), 385–396.

Oberlé, I., Rousseau, F., Heitz, D., Kretz, C., Devys, D., Hanauer, A. ... Mandel, J. L. (1991). Instability of a 550-base pair DNA segment and abnormal methylation in fragile X syndrome. *Science, 252*(5009), 1097–1102.

O'Donnell, W. T., & Warren, S. T. (2002). A decade of molecular studies of fragile X syndrome. *Annual Review of Neuroscience, 25*, 315–338.

Ohgushi, M., & Sasai, Y. (2011). Lonely death dance of human pluripotent stem cells: ROCKing between metastable cell states. *Trends in Cell Biology, 21*(5), 274–282.

Onore, C., Careaga, M., & Ashwood, P. (2012). The role of immune dysfunction in the pathophysiology of autism. *Brain Behavior and Immunity, 26*, 383–392.

Pacey, L. K., Xuan, I. C., Guan, S., Sussman, D., Henkelman, R. M., Chen, Y. ... Hampson, D. R. (2013). Delayed myelination in a mouse model of fragile X syndrome. *Human Molecular Genetics, 22*(19), 3920–3930.

Pamies, D., Hartung, T., & Hogberg, H. T. (2014). Biological and medical applications of a brain-in-a-chip. *Experimental Biology and Medicine (Maywood, NJ)*, 1096–1107.

Panicker, L. M., Miller, D., Awad, O., Bose, V., Lun, Y., Park, T. S. ... Feldman, R. A. (2014). Gaucher iPSC-derived macrophages produce elevated levels of inflammatory mediators and serve as a new platform for therapeutic development. *Stem Cells, 32*(9), 2338–2349.

Park, I.-H., Zhao, R., West, J. A., Yabuuchi, A., Huo, H., Ince, T. A. ... Daley, G. Q. (2008). Reprogramming of human somatic cells to pluripotency with defined factors. *Nature, 451*(7175), 141–146.

Park, K. I., Teng, Y. D., & Snyder, E. Y. (2002). The injured brain interacts reciprocally with neural stem cells supported by scaffolds to reconstitute lost tissue. *Nature Biotechnology, 20*, 1111–1117.

Pasca, A. M., Sloan, S. A., Clarke, L. E., Tian, Y., Makinson, C. D., Huber, N. ... Pasca, S. P. (2015). Functional cortical neurons and astrocytes from human pluripotent stem cells in 3D culture. *Nature Methods, 12*(7), 671–678.

Pasca, S. P., Portmann, T., Voineagu, I., Yazawa, M., Shcheglovitov, A., Pasca, A. M. ... Dolmetsch, R. E. (2011). Using iPSC-derived neurons to uncover cellular phenotypes associated with Timothy syndrome. *Nature Medicine, 17*(12), 1657–1662.

Pearson, C. E., Edamura, K. N., & Cleary, J. D. (2005). Repeat instability: mechanisms of dynamic mutations. *Nature Reviews Genetics, 6*(10), 729–742.

Peerani, R., Rao, B. M., Bauwens, C., Yin, T., Wood, G. A., Nagy, A. ... Zandstra, P. W. (2007). Niche-mediated control of human embryonic stem cell self-renewal and differentiation. *The EMBO Journal, 26*(22), 4744–4755.

Penagarikano, O., Mulle, J. G., & Warren, S. T. (2007). The pathophysiology of fragile X syndrome. *Annual Review of Genomics and Human Genetics, 8*, 109–129.

Rao, M. S., & Malik, N. (2012). Assessing iPSC reprogramming methods for their suitability in translational medicine. *Journal of Cellular Biochemistry, 113*(10), 3061–3068.

Ricciardi, S., Ungaro, F., Hambrock, M., Rademacher, N., Stefanelli, G., Brambilla, D. ... Broccoli, V. (2012). CDKL5 ensures excitatory synapse stability by reinforcing NGL-1-PSD95 interaction in the postsynaptic compartment and is impaired in patient iPSC-derived neurons. *Nature Cell Biology, 14*(9), 911–923.

Ruiz, S., Diep, D., Gore, A., Panopoulos, A. D., Montserrat, N., Plongthongkum, N. ... Izpisua Belmonte, J. C. (2012). Identification of a specific reprogramming-associated epigenetic signature in human induced pluripotent stem cells. *Proceedings of the National Academy of Sciences of the United States of America, 109*(40), 16196–16201.

Scala, E., Ariani, F., Mari, F., Caselli, R., Pescucci, C., Longo, I. ... Renieri, A. (2005). CDKL5/STK9 is mutated in Rett syndrome variant with infantile spasms. *Journal of Medical Genetics, 42*(2), 103–107.

Scholze, H., & Boch, J. (2010). TAL effector-DNA specificity. *Virulence, 1*(5), 428–432.

Seki, T., Yuasa, S., Oda, M., Egashira, T., Yae, K., Kusumoto, D. ... Fukuda, K. (2010). Generation of induced pluripotent stem cells from human terminally differentiated circulating T cells. *Cell Stem Cell, 7*(1), 11–14.

Serio, A., Bilican, B., Barmada, S. J., Ando, D. M., Zhao, C., Siller, R. ... Chandran, S. (2013). Astrocyte pathology and the absence of non-cell autonomy in an induced pluripotent stem cell model of TDP-43proteinopathy. *Proceedings of the National Academy of Sciences of the United States of America, 110*(12), 4697–4702.

Shaltouk, I. A., Peng, J., Liu, Q., Rao, M. S., & Zeng, X. (2013). Efficient generation of astrocytes from human pluripotent stem cells in defined conditions. *Stem Cells, 31*(5), 941–952.

Sheridan, S. D., Theriault, K. M., Reis, S. A., Zhou, F., Madison, J. M., Daheron, L. … Haggarty, S. J. (2011). Epigenetic characterization of the FMR1 gene and aberrant neurodevelopment in human induced pluripotent stem cell models of fragile X syndrome. *PLoS One, 6*(10), e26203.

Sidorov, M. S., Auerbach, B. D., & Bear, M. F. (2013). Fragile X mental retardation protein and synaptic plasticity. *Molecular Brain, 6*(1), 15.

Siomi, H., Siomi, M. C., Nussbaum, R. L., & Dreyfuss, G. (1993). The protein product of the fragile X gene, FMRI, has characteristics of an RNA-binding protein. *Cell, 74*, 291–298.

Soldner, F., Hockemeyer, D., Beard, C., Gao, Q., Bell, G. W., Cook, E. G. … Jaenisch, R. (2009). Parkinson's disease patient-derived induced pluripotent stem cells free of viral reprogramming factors. *Cell, 136*(5), 964–977.

Splawski, I., Timothy, K. W., Sharpe, L. M., Decher, N., Kumar, P., Bloise, R. … Keating, M. T. (2004). Ca(V)1.2 calcium channel dysfunction causes a multisystem disorder including arrhythmia and autism. *Cell, 119*(1), 19–31.

Staerk, J., Dawlaty, M. M., Gao, Q., Maetzel, D., Hanna, J., Sommer, C. A. … Jaenisch, R. (2010). Reprogramming of human peripheral blood cells to induced pluripotent stem cells. *Cell Stem Cell, 7*(1), 20–24.

Stein, J. L., de la Torre-Ubieta, L., Tian, Y., Parikshak, N. N., Hernández, I. A., Marchetto, M. C. … Geschwind, D. H. (2014). A quantitative framework to evaluate modeling of cortical development by neural stem cells. *Neuron, 83*(1), 69–86.

Steinbeck, J. A., & Studer, L. (2015). Moving stem cells to the clinic: potential and limitations for brain repair. *Neuron, 86*(1), 187–206.

Steiner, D., Khaner, H., Cohen, M., Even-Ram, S., Gil, Y., Itsykson, P. … Reubinoff, B. (2010). Derivation, propagation and controlled differentiation of human embryonic stem cells in suspension. *Nature Biotechnology, 28*(4), 361–364.

Subramanian, A., Krishnan, U. M., & Sethuraman, S. (2009). Development of biomaterial scaffold for nerve tissue engineering: biomaterial mediated neural regeneration. *Journal of Biomedical Science, 16*, 108.

Suzuki, K., Sugihara, G., Ouchi, Y., Nakamura, K., Futatsubashi, M., Takebayashi, K. … Mori, N. (2013). Microglial activation in young adults with autism spectrum disorder. *JAMA Psychiatry, 70*(1), 49–58.

Swistowski, A., Peng, J., Han, Y., Swistowska, A. M., Rao, M. S., & Zeng, X. (2009). Xeno-free defined conditions for culture of human embryonic stem cells, neural stem cells and dopaminergic neurons derived from them. *PLoS One, 4*(7), e6233.

Swistowski, A., Peng, J., Liu, Q., Mali, P., Rao, M. S., Cheng, L., & Zeng, X. (2010). Efficient generation of functional dopaminergic neurons from human induced pluripotent stem cells under defined conditions. *Stem Cells, 28*(10), 1893–1904.

Tabolacci, E., Pietrobono, R., Moscato, U., Oostra, B. A., Chiurazzi, P., & Neri, G. (2005). Differential epigenetic modifications in the FMR1 gene of the fragile X syndrome after reactivating pharmacological treatments. *European Journal of Human Genetics, 13*(5), 641–648.

Takahashi, K., Tanabe, K., Ohnuki, M., Narita, M., Ichisaka, T., Tomoda, K., & Yamanaka, S. (2007). Induction of pluripotent stem cells from adult human fibroblasts by defined factors. *Cell, 131*(5), 861–872.

Takahashi, K., & Yamanaka, S. (2006). Induction of pluripotent stem cells from mouse embryonic and adult fibroblast cultures by defined factors. *Cell, 126*(4), 663–676.

Telias, M., Segal, M., & Ben-Yosef, D. (2013). Neural differentiation of fragile X human embryonic stem cells reveals abnormal patterns of development despite successful neurogenesis. *Developmental Biology, 374*(1), 32–45.

Theoharides, T. C., Angelidou, A., Alysandratos, K. D., Zhang, B., Asadi, S., Francis, K. … Kalogeromitros, D. (2012). Mast cell activation and autism. *Biochimica et Biophysica Acta, 1822*(1), 34–41.

Thompson, L. H., & Björklund, A. (2015). Reconstruction of brain circuitry by neural transplants generated from pluripotent stem cells. *Neurobiology of Disease, 79*, 28–40.

Tibbitt, M. W., & Anseth, K. S. (2009). Hydrogels as extra- cellular matrix mimics for 3D cell culture. *Biotechnology and Bioengineering, 103*, 655–663.

Underhill, G. H., & Bhatia, S. N. (2007). High-throughput analysis of signals regulating stem cell fate and function. *Current Opinion in Chemical Biology, 11*(4), 357–366.

Urbach, A., Bar-Nur, O., Daley, G. Q., & Benvenisty, N. (2010). Differential modeling of fragile X syndrome by human embryonic stem cells and induced pluripotent stem cells. *Cell Stem Cell, 6*(5), 407–411.

Van Wilgenburg, B., Browne, C., Vowles, J., & Cowley, S. A. (2013). Efficient, long term production of monocyte- derived macrophages from human pluripotent stem cells under partly-defined and fully-defined conditions. *PLoS One, 8*(8), e71098.

Vargas, D. L., Nascimbene, C., Krishnan, C., Zimmerman, A. W., & Pardo, C. A. (2005). Neuroglial activation and neuroinflammation in the brain of patients with autism. *Annals of Neurology, 57*(1), 67–81.

Verkerk, A. J., Pieretti, M., Sutcliffe, J. S., Fu, Y. H., Kuhl, D. P., Pizzuti, A. … Zhang, F. P. (1991). Identification of a gene (FMR-1) containing a CGG repeat coincident with a breakpoint cluster region exhibiting length variation in fragile X syndrome. *Cell, 65*(5), 905–914.

Wiedenheft, B., Sternberg, S. H., & Doudna, J. A. (2012). RNA-guided genetic silencing systems in bacteria and archaea. *Nature, 482*(7385), 331–338.

Williams, C. A., Driscoll, D. J., & Dagli, A. I. (2010). Clinical and genetic aspects of Angelman syndrome. *Genetics in Medicine : Official Journal of the American College of Medical Genetics, 12*(7), 385–395.

Williams, E. C., Zhong, X., Mohamed, A., Li, R., Liu, Y., Dong, Q. … Chang, Q. (2014). Mutant astrocytes differentiated from Rett syndrome patients-specific iPSCs have adverse effects on wild-type neurons. *Human Molecular Genetics, 23*(11), 2968–2980.

Wong, R. O. L., & Ghosh, A. (2002). Activity-dependent regulation of dendritic growth and patterning. *Nature Reviews Neuroscience, 3*(10), 803–812.

Yamamuro, K., Kimoto, S., Rosen, K. M., Kishimoto, T., & Makinodan, M. (2015). Potential primary roles of glial cells in the mechanisms of psychiatric disorders. *Frontiers in Cellular Neuroscience, 9*, 154.

Yan, Q. J., Rammal, M., Tranfaglia, M., & Bauchwitz, R. P. (2005). Suppression of two major Fragile X Syndrome mouse model phenotypes by the mGluR5 antagonist MPEP. *Neuropharmacology, 49*(7), 1053–1066.

Yanagimachi, M. D., Niwa, A., Tanaka, T., Honda-Ozaki, F., Nishimoto, S., Murata, Y. … Saito, M. K. (2013). Robust and highly-efficient differentiation of functional monocytic cells from human pluripotent stem cells under serum- and feeder cell-free conditions. *PLoS One, 8*(4), e59243.

Yashiro, K., Riday, T. T., Condon, K. H., Roberts, A. C., Bernardo, D. R., Prakash, R. … Philpot, B. D. (2009). Ube3a is required for experience-dependent maturation of the neocortex. *Nature Neuroscience, 12*(6), 777–783.

Yazawa, M., & Dolmetsch, R. E. (2013). Modeling Timothy syndrome with iPS cells. *Journal of Cardiovascular Translational Research, 6*(1), 1–9.

Yazawa, M., Hsueh, B., Jia, X., Pasca, A. M., Bernstein, J. A., Hallmayer, J., & Dolmetsch, R. E. (2011). Using induced pluripotent stem cells to investigate cardiac phenotypes in Timothy syndrome. *Nature, 471*(7337), 230–234.

Young, J. I., & Zoghbi, H. Y. (2004). X-chromosome inactivation patterns are unbalanced and affect the phenotypic outcome in a mouse model of Rett syndrome. *American Journal of Human Genetics, 74*(3), 511–520.

Yu, J., Vodyanik, M. A., Smuga-Otto, K., Antosiewicz-Bourget, J., Frane, J. L., Tian, S. … Thomson, J. A. (2007). Induced pluripotent stem cell lines derived from human somatic cells. *Science (New York, NY), 318*(5858), 1917–1920.

Zou, J., Maeder, M. L., Mali, P., Pruett-Miller, S. M., Thibodeau-Beganny, S., Chou, B.-K. … Cheng, L. (2009). Gene targeting of a disease-related gene in human induced pluripotent stem and embryonic stem cells. *Cell Stem Cell, 5*(1), 97–110.

Zweigerdt, R., Olmer, R., Singh, H., Haverich, A., & Martin, U. (2011). Scalable expansion of human pluripotent stem cells in suspension culture. *Nature Protocols, 6*(5), 689–700.

Chapter 9

Association Strategies

Benjamin M. Neale
Massachusetts General Hospital, Boston, MA, United States; Broad Institute of MIT and Harvard, Cambridge, MA, United States

In 1918, Ronald Fisher published the groundbreaking work, "The Correlation Between Relatives on the Supposition of Mendelian Inheritance" in the *Proceedings of the Royal Society of Edinburgh*. Here, Fisher reconciled Mendel's Laws of Inheritance with Galton's observation of the apparent similarity that relatives show in continuous phenotypes. This seminal work not only laid the foundation for much of current genetic analytic thought but also defined the critical statistic variance which is the focus of most genetic analysis. The following century saw dramatic technological progress in measuring the human genome and with it largely the validation of Fisher's model for how genes influence complex traits. Nevertheless biology is complex and necessitates diverse methods for both common and rare variations, which are detailed in this chapter.

COMMON VARIANTS

At the most basic level, association analysis is a test of mean differences in phenotype conditional on genotype. Any test of group differences such as the chi-square, the Armitage trend test, or the Cochran—Mantel—Haenszel can be applied. However, regression methods are generally preferred because they allow the inclusion of additional variables as covariates to protect against biases. As with variance, regression also finds its roots in genetic analysis, first coined by Galton in examining the similarity between parents and children in height, which Pearson and Yule went on to generalize. Typically the phenotype is the dependent variable (Y) and the genotype is the independent variable (X) or regressor. The regression can be flipped to perform the "genotype—conditional association test" which models the genotype as the dependent or outcome variable (Song, Hao, & Storey, 2015).

In 1996, Risch and Merikangas published a prophetic work envisioning the direct testing of genetic variation, genome-wide (Risch & Merikangas, 1996). Roughly a decade later, the development of genome-wide genotyping arrays began to enable precisely this, with the ability to assess hundreds of thousands of common genetic variants and thus perform genome-wide association studies (GWAS). With this wealth of data comes a range of challenges and opportunities. One of the primary challenges is technical artifact. For example, if 500,000 markers are genotyped and 1% of markers are error prone, 5000 markers will be severely biased. Consequently, much early GWAS work focused on the development and refinement of quality control measures. These techniques are largely mature and are essential for a robust analysis (see Anderson et al., 2010 for a nice review).

The typical GWAS genotypes hundreds of thousands to millions of markers across the genome. With these variants it is possible to impute most untyped common genetic variations. Genotypic imputation leverages the correlational structure within the genome to estimate the probability of each possible genotype at an untyped locus. The correlation between genetic markers is termed linkage disequilibrium (LD). Large-scale genetics projects such as the International HapMap Project (International HapMap, 2003, 2005; International HapMap et al., 2007, 2010) and the 1000 Genomes Project (Genomes Project et al., 2010, 2012) have generated references of LD patterns in human populations. With these reference samples, it is possible to take genotype array data for an individual and estimate what the probability is of each possible genotype at each untyped locus (Li, Willer, Sanna, & Abecasis, 2009, 2010). Using these dense sets of imputed and directly genotyped variants, GWAS perform single locus association at each of these markers. Although haplotype methods have been proposed, these tend to gain power only in circumstances where there is cis interaction

Genomics, Circuits, and Pathways in Clinical Neuropsychiatry. http://dx.doi.org/10.1016/B978-0-12-800105-9.00009-3

TABLE 9.1 Example of False-Positive Association Induced by Population Stratification

	Population 1	Population 2
Frequency of allele 1	0.3	0.5
Number of cases	300	700
Number of controls	700	300
	Combined Sample	
	Allele 1	**Allele 2**
Cases	0.44	0.56
Controls	0.36	0.64

(Chapman, Cooper, Todd, & Clayton, 2003). As a result, millions of single-marker tests are performed for the typical GWAS; this results in a substantial multiple testing problem, but the patterns of LD within the genome greatly reduce the number of independent tests being performed. Although this was a substantial concern in the nascent GWAS literature, this issue was quickly resolved. Working independently, two groups estimated the effective number of tests being performed, assuming whole genome analysis and arrived at 5×10^{-8} as the genome-wide significant threshold, which equates to approximately 1 million independent tests (Dudbridge & Gusnanto, 2008; Pe'er, Yelensky, Altshuler, & Daly, 2008).

Another major challenge within GWAS is the control of confounding biases. Population stratification arises when cases have a different distribution of ancestries than do controls. Consequently, any allele that differs in frequencies between populations will tend to show association to the disease in the study if individuals of any given ancestry are over-represented among the cases compared with the controls. For example, Table 9.1 shows how mixing together cases and controls from two different populations induces association. When performing any genetic association test, it is absolutely essential to control for population stratification.

One of the earliest approaches to detecting population stratification was proposed by Pritchard and Rosenberg (1999) based on the clustering of participants using independent markers. The concept behind this approach was to consider what would happen to allele frequencies if the sample were actually a mixture of individuals with different population ancestries. These populations were modeled as discrete ancestral groups and subsets of the sample could be assigned to different population clusters based on the genetic data.

However, this clustering approach does not accommodate the continuous nature of population ancestry such as is observed across European populations. This continuous variability in the frequency across continental populations can be detected by applying principal component analysis (PCA) to genetic variation (Patterson, Price, & Reich, 2006; Price et al., 2006). Conceptually, PCA capitalizes on the emergent correlation structure that arises when you mix individuals of different population ancestries in a single sample. As an intuitive example of why this correlation arises, consider two populations (A and B) and two unlinked single nucleotide polymorphisms (rs1 and rs2). In population A, the frequency of the ancestral allele at rs1 is 0.1, and at rs2 is 0.1. In population B, the frequency of the ancestral allele at rs1 is 0.9, and at rs2 is 0.9. When we have an equal mixture of population A and B, the frequency of the ancestral allele is 0.5 in both populations. However, there is now a strong correlation between rs1 and rs2 in the mixed sample: If you are from population A, you are unlikely to have the ancestral allele at both rs1 and rs2, whereas if you are from population B, you are highly likely to have the ancestral allele at both rs1 and rs2. This emergent correlation structure that comes from mixing the frequencies of alleles across the genome dominates the variance covariance structure of unlinked loci across the genome. As a technique, PCA identifies linear combinations that explain the most variant covariance structure and thus captures this kind of information.

Principal components defined based on genome-wide common genetic variants have been shown to correlate with geographic ancestry (Novembre et al., 2008). Consequently such PCs are effective at correcting the impact of population stratification in the analysis of common variants.

The other major development in the control of population stratification is linear mixed models (LMMs) (sometimes referred to as mixed linear models). When using mixed models in regression, the main or fixed effect of the variant being tested is still the primary test, but a random effect, which is a term that specifies the variance covariance structure of the individuals genome-wide, is included. These LMMs can also be used to control for population stratification (Yang, Zaitlen, Goddard, Visscher, & Price, 2014).

This random effect term, however, also captures genome-wide heritability, both stratification and polygenic inheritance. For most complex traits, polygenic inheritance has been conclusively demonstrated, which means that a large number of variants scattered genome-wide confer risk to disease. As a consequence, using LMMs with principal components can provide an estimate of heritability of the trait. Conceptually, this method is the basis of genome-wide complex trait analysis (GCTA), which was developed to quantify the amount of variability in the population that common variation explains, if we assume a model about the distribution of effect sizes (Yang, Lee et al., 2011; Yang et al., 2012; Yang, Weedon et al., 2011). These LMMs thus serve two purposes: control of stratification and estimation of heritability. When analyzing continuous traits such as height or body mass index, controlling for both stratification and heritability can boost the power to detect association, because the PCAs and the random effect component effectively "de-noise" the phenotype. For binary traits such as disease status, LMMs introduce additional challenges. If the cases are oversampled (eg, a 50/50 split of cases and controls for schizophrenia, in which the actual disease prevalence in the population is about 1%), LMMs can lead to a loss in power rather than gains because of the correlation structure between causal alleles in the case sample (Yang et al., 2014). These approaches are targeted at ensuring the robustness of single locus association for complex traits.

Beyond single locus association analysis, GWAS data can be used to estimate the contribution of common variation to heritability (ie, the proportion of trait variability that results from genetic influences). Three main approaches to estimating the genome-wide contribution have been developed: polygenic risk scores (PRS) (International Schizophrenia Consortium et al., 2009), GCTA (Yang, Lee et al., 2011), and methods derived from the summary LD score (Bulik-Sullivan, Loh et al., 2015). Each of these methods is constructed differently and has a slightly different interpretation, but in general they all attempt to capture the impact of common variation genome-wide.

Typically, the GWAS for many complex traits has followed the following arc: Initial small sample sizes fail to identify genome-wide significant loci, leading to collaborative meta-analyses; as sample size increases, loci eventually emerge; and then the pace of discovery accelerates, with additional samples yielding greater numbers of loci. Thus, many truly associated loci lurk beneath the genome-wide significance threshold of 5×10^{-8} but remain undetected because of a lack of power. Another way to conceptualize this is that the variants yielding P values in the range of 0.05 to 5×10^{-8} are a mixture of true effects and variants that are not associated. However, it is possible to construct a PRS that captures this information. The simplest form of PRS is to sum the number of risk alleles that each individual carries. This score can then be used to predict the complex trait in an independent sample.

The PRS can be improved in a number of ways. The first is by removing variants from the predictor (sometimes termed feature selection) that are likely to have no effect on the phenotype in question. The earliest approach to performing this feature selection was to apply a threshold on P values for variant selection. This P value threshold was tuned based on out-of-sample prediction performance. A second approach to improve performance is by applying weights for the variants included in the score. The most common weighting scheme is to use the estimated effect size from a meta-analysis. However, the estimation of effects genome-wide is noisy and often overestimated. To improve prediction, a number of statistical approaches have been proposed to regularize these effect sizes, which provide some gains in predictive power (Meuwissen, Hayes, & Goddard, 2001). Finally, annotation information can be used to further improve the performance of these scores and variant prioritization (Schork et al., 2013), as can modeling the LD, as demonstrated in LDpred (Vilhjalmsson et al., 2015).

Beyond prediction, the genome-wide contribution to heritability can also be estimated using GWAS data. The primary approach, GCTA, operates on comparing overall genetic similarity between a pair of individuals and phenotypic similarity. For quantitative traits, this approach naturally lends itself to a direct measure of the heritability. For binary traits such as schizophrenia, it is necessary to assume a model, typically the liability threshold model, which assumes that the binary classification is simply the extreme end of a normal distribution. However, such an approach may be biased when cases are oversampled (Golan, Lander, & Rosset, 2014).

An alternate approach to estimating heritability is LD score regression (LDSC) (Bulik-Sullivan, Finucane et al., 2015; Bulik-Sullivan, Loh et al., 2015). LDSC works on the premise that the more genetic variation an individual variant tags, the higher the probability is that the variant tags a causal variant (ie, a variant that has an effect on the trait in question). Initially LDSC was proposed to quantify how much population stratification or polygenic inheritance inflates the distribution of test statistics from large-scale GWAS (Bulik-Sullivan, Loh et al., 2015). This quantification of polygenic inheritance can provide an estimate of heritability based on the relationship between the LD score (ie, the sum of the r^2 between an index variant and all surrounding variants) and the chi-square.

Both GCTA and LD score have been applied across a wide range of traits and have shown clear evidence of a polygenic contribution to most complex traits. These methods also allow for two key extensions: genetic correlation and partitioning (Finucane et al., 2015; Gusev et al., 2014). Both of these extensions enable further interpretation of the GWAS results (Lee et al., 2013).

A genetic correlation is defined as the proportion of the heritability that is shared between two traits divided by the square root of the product of the heritability for each trait. Genetic correlation can be interpreted in a variety of ways. For example, it may mean that certain upstream risk factors are shared for both traits, such as obesity being a risk factor for both myocardial infarction and type 2 diabetes. Under this scenario, all genetic influences on obesity will be part of the heritability of both diseases and these heritable influences will be shared as part of the genetic correlation. Another possible interpretation is that genetic risk factors influence specific biological mechanisms which are shared for both traits (eg, calcium channel dysfunction as a risk factor for multiple psychiatric disorders). Finally, the presence of a genetic correlation may reflect pleiotropy, in which the same genes act across multiple tissues and are subject to similar genetic influences, meaning that there is emergent correlation between these different traits, but the specific biological mechanism is not same. The application of genetic correlation in psychiatric disorders has revealed widespread sharing of genetic risk across schizophrenia, bipolar disorder, and major depression, which underscores that genetic risk factors do not respect traditional diagnostic boundaries.

Partitioning heritability aims to assess the impact of different functional classes on the genome-wide contribution to heritability. For example, do coding variants, on average, explain more heritability per variant than intergenic variants? Significant enrichments in heritability per variant (calculated as class h^2/n variants) can mean that a larger fraction of variants in the class influence trait variation, that the effect size for variants in the class is on average larger, or some combination of both. Such partitioning analyses can be applied across functional classes, such as enhancer marks or other classes derived from ENCODE and Roadmap (Bernstein et al., 2010; Consortium et al., 2007) or across the minor allele frequency spectrum, or to determine whether specific cell types are more relevant by aggregating across functional marks across cell types. Partitioning analyses have been applied using both GCTA and LDSC for psychiatric traits, which demonstrates the importance of neuronal cell types and conserved and coding regions (Finucane et al., 2015; Gusev et al., 2014).

An alternate method to extract biological insights from GWAS is to apply pathway or network analysis. The basic principle behind these approaches is to take external data sources as a basis for grouping genes or regions. These groups are then tested for enrichment. Group definition can be derived from coexpression networks, protein–protein interaction, or other external biological data sets. A diverse array of methods has been proposed for these kinds of analyses (see Network, Pathway Analysis Subgroup of the Psychiatric Genomics, 2015) for a comparison of methods and application to psychiatric disorders.

RARE VARIANTS

Sequencing technologies enable a more comprehensive capture of genetic variation, although in practice, sequence data for a given individual within a study will still contain regions which are poorly covered or untyped. The earliest studies focused on exome sequencing, capturing the coding regions of genes. Subsequently, genome sequencing became feasible with decreasing costs of data generation. As a consequence, analysis of rare variations is now possible.

When analyzing rare variations, it is ill-advised to test individual variants because such tests are underpowered; as such, it is necessary to employ omnibus tests that group variants together. How best to group variants is a complex question, with different considerations for exome and genome sequencing.

The most natural unit for grouping rare variants together is the gene and, in particular, the coding region (ie, the exons). Genetic variation in coding regions is more readily annotated for function, with four primary classes for point mutations: synonymous, which maintains the same amino acid; missense, which changes the amino acid; nonsense, which generates a premature stop codon; and splice site, which disrupts the essential splice site at the intron–exon boundary. The value of these annotations is threefold: Synonymous mutations show no real evidence of evolutionary conservation, and in most instances are unlikely to be functional, meaning that they can be removed from consideration (this equates to the removal of 30–40% of variants for a typical gene); nonsense and splice mutations are likely to have the same functional consequence, knocking down the level of expression; and significant association to coding variation is readily interpretable with respect to which gene is affected. Typical tests for genes will focus on grouping together all nonsense mutations; nonsense and splice site mutations; all missense mutations; and all nonsense, splice site, and missense mutations.

For noncoding variations, the challenges of grouping rare variants are substantial. No annotation class equivalent to synonymous mutations exists, which means that vast numbers of neutral variants are included in the analyses. Second, the annotations that exist, such as transcription factor binding sites, are small, typically fewer than 50 base pairs, which means that the amount of genetic variation is insufficient for rare variant testing (Zuk et al., 2014). Consequently, it will be important to design strategies for grouping across these regulatory elements, such as all elements that influence a given gene or all elements within a given class. Alternately, sliding windows across the genome may provide an unbiased approach to testing for the impact of rare variations.

Two main classes of approaches have been proposed to testing rare variations: burden tests (eg, the collapsing and combining method), in which variants are grouped together and tested for differences in the number between cases and controls (Li & Leal, 2008; Neale & Sham, 2004) and variance or kernel-based tests that assess whether individuals who carry the same rare variants tend to have more similar phenotypes (Lee et al., 2012; Neale et al., 2011; Wu et al., 2010, 2011). The primary distinction between these two approaches is that burden tests operate under the assumption that all rare variants influence the phenotype in the same direction, whereas kernel-based tests are robust to scenarios in which some rare variants increase risk and others decrease it.

A key approach to improving the power to detect rare variant association is weighting the contribution of each variant to the overall test. A natural starting point is to increase the weight of rarer variations, because the minor allele frequency is one of the best indicators of selective pressure, and by extension the probability of functionality (Madsen & Browning, 2009). Further weights can also developed to reflect regional functional annotations, such as DNAse I hypersensitivity sites or mammalian conservation. Another way to incorporate additional information is by selecting variants for inclusion in the test (which is equivalent to setting the weight to 0) as implemented in the variable threshold (VT) test (Price et al., 2010). The way VT works is to maximize the strength of the association analysis across multiple different thresholds for a given external data type, such as frequency. However, doing so will bias the test because of overfitting. To control for this, the phenotypes are permuted and then the entire process is run again thousands of times to obtain the empirical distribution of test statistics when applying such selection considerations.

With sequencing, it is possible to identify de novo mutations, variants that newly arise in the individual. The identification of de novo mutations requires sequencing of the offspring and both parents and determining what mutations are new in the child. Analyzing these de novo mutations poses both a tremendous opportunity to evaluate variants that are subject to little selective pressure (ie, viability is the sole selective force) and the substantial challenge of how best to test these mutations, because the expected number of mutations for any gene is extremely small (Zuk et al., 2014). Further complicating matters is that de novo mutations occur at a background rate of about 1 per exome and about 60 per genome. By extension, as more trios are sequenced, some genes will have more than one de novo functional mutation, simply by chance. To analyze these data, it is necessary to build a model of the rate of mutations at a base-level resolution. Then the probability of each type of mutation per gene can be calculated. With these probabilities, it is then possible to test whether the observed rate of de novo mutations in each gene is overrepresented. This model framework has been applied to de novo mutation analysis of intellectual disability (ID), autism, and schizophrenia to identify genes for ID and autism and demonstrate enrichment of such mutations across these three disorders (Samocha et al., 2014).

This model for testing de novo mutations described has also been used to identify genes for which there is a reduction in genetic diversity, a hallmark of purifying selection. The basic premise is that the mutation rate serves as a guide for the expected rate of variants in a gene, and that genes for which missense or nonsense mutations are reduced are subject to these selective pressures (Samocha et al., 2014). Thus large-scale sequencing studies such as the Exome Sequencing Project (Fu et al., 2013; Tennessen et al., 2012) and the Exome Aggregation Consortium (exac.broadinstitute.org) can be used to identify such genes under mutational constraint. An alternate approach to this, Residual Variation Intolerance Score, compares the rate of genetic variation in the gene with the rate of common functional variation (eg, 0.1% frequency) (Petrovski, Wang, Heinzen, Allen, & Goldstein, 2013). Another signature of purifying selection is a deficiency of common missense and nonsense mutations. These two approaches to identifying genes intolerant of mutation clearly strengthen the rare variant signals emerging from de novo screens of autism and ID.

CONCLUSIONS

Methods for analyzing genetic data have matured with the technological abilities to assay the genome. With increasing sample sizes and sophistication of analysis, robust findings are emerging from genetic data. These findings have wide applicability across a range of central questions in disease pathogenesis, including forming the foundation for discovering novel biology, understanding how to classify different disorders, and the identifying intermediate traits relevant to outcomes.

ACKNOWLEDGMENTS

I would like to thank Drs. Sarah Medland and Mark Daly for carefully reading the chapter and providing helpful feedback on structure and content.

REFERENCES

Anderson, C. A., Pettersson, F. H., Clarke, G. M., Cardon, L. R., Morris, A. P., & Zondervan, K. T. (2010). Data quality control in genetic case-control association studies. *Nature Protocols, 5*(9), 1564–1573. http://dx.doi.org/10.1038/nprot.2010.116. PubMed PMID:21085122; PMCID: 3025522.

Bernstein, B. E., Stamatoyannopoulos, J. A., Costello, J. F., Ren, B., Milosavljevic, A., Meissner, A., … Thomson, J. A. (2010). The NIH roadmap epigenomics mapping consortium. *Nature Biotechnology, 28*(10), 1045–1048. http://dx.doi.org/10.1038/nbt1010-1045. PubMed PMID:20944595; PMCID: 3607281.

Bulik-Sullivan, B., Finucane, H., Anttila, V., Gusev, A., Day, F., ReproGen Consortium, … Neale, B. (2015). An atlas of genetic correlations across human diseases and traits. *Nature Genetics, 47*(11), 1236–1241. http://dx.doi.org/10.1038/ng.3406. PubMed PMID:26414676.

Bulik-Sullivan, B. K., Loh, P. R., Finucane, H. K., Ripke, S., Yang, J., Schizophrenia Working Group of the Psychiatric Genomics Consortium, … Neale, B. M. (2015). LD score regression distinguishes confounding from polygenicity in genome-wide association studies. *Nature Genetics, 47*(3), 291–295. http://dx.doi.org/10.1038/ng.3211. PubMed PMID:25642630.

Chapman, J. M., Cooper, J. D., Todd, J. A., & Clayton, D. G. (2003). Detecting disease associations due to linkage disequilibrium using haplotype tags: a class of tests and the determinants of statistical power. *Human Heredity, 56*(1–3), 18–31. http://dx.doi.org/10.1159/000073729. PubMed PMID:14614235.

Dudbridge, F., & Gusnanto, A. (2008). Estimation of significance thresholds for genomewide association scans. *Genetic Epidemiology, 32*(3), 227–234. http://dx.doi.org/10.1002/gepi.20297. PubMed PMID:18300295; PMCID: 2573032.

ENCODE Project Consortium, Birney, E., Stamatoyannopoulos, J. A., Dutta, A., Guigo, R., Gingeras, T. R., … de Jong, P. J. (2007). Identification and analysis of functional elements in 1% of the human genome by the ENCODE pilot project. *Nature, 447*(7146), 799–816. http://dx.doi.org/10.1038/nature05874. PubMed PMID:17571346; PMCID: 2212820.

Finucane, H. K., Bulik-Sullivan, B., Gusev, A., Trynka, G., Reshef, Y., Loh, P., … Price, A. L. (2015). Partitioning heritability by functional category using genome-wide association summary statistics. *Nature Genetics, 47*(11), 1228–1235. http://dx.doi.org/10.1038/ng. 3404. PubMed PMID:26414678.

Fu, W., O'Connor, T. D., Jun, G., Kang, H. M., Abecasis, G., Leal, S. M., … Akey, J. M. (2013). Analysis of 6,515 exomes reveals the recent origin of most human protein-coding variants. *Nature, 493*(7431), 216–220. http://dx.doi.org/10.1038/nature11690. Epub 2012/12/04. PubMed PMID:23201682; PMCID: PMC3676746.

Genomes Project Consortium, Abecasis, G. R., Altshuler, D., Auton, A., Brooks, L. D., Durbin, R. M., … McVean, G. A. (2010). A map of human genome variation from population-scale sequencing. *Nature, 467*(7319), 1061–1073. http://dx.doi.org/10.1038/nature09534. PubMed PMID:20981092; PMCID: 3042601.

Genomes Project Consortium, Abecasis, G. R., Auton, A., Brooks, L. D., DePristo, M. A., Durbin, R. M., … McVean, G. A. (2012). An integrated map of genetic variation from 1,092 human genomes. *Nature, 491*(7422), 56–65. http://dx.doi.org/10.1038/nature11632. PubMed PMID:23128226; PMCID: 3498066.

Golan, D., Lander, E. S., & Rosset, S. (2014). Measuring missing heritability: inferring the contribution of common variants. *Proceedings of the National Academy of Sciences of the United States of America, 111*(49), E5272–E5281. http://dx.doi.org/10.1073/pnas.1419064111. PubMed PMID:25422463; PMCID: 4267399.

Gusev, A., Lee, S. H., Trynka, G., Finucane, H., Vilhjalmsson, B. J., Xu, H., … SWE-SCZ Consortium. (2014). Partitioning heritability of regulatory and cell-type-specific variants across 11 common diseases. *American Journal of Human Genetics, 95*(5), 535–552. http://dx.doi.org/10.1016/j.ajhg.2014.10.004. PubMed PMID:25439723; PMCID: 4225595.

International HapMap Consortium. (2003). The international HapMap project. *Nature, 426*(6968), 789–796. http://dx.doi.org/10.1038/nature02168. PubMed PMID:14685227.

International HapMap Consortium. (2005). A haplotype map of the human genome. *Nature, 437*(7063), 1299–1320. http://dx.doi.org/10.1038/nature04226. PubMed PMID:16255080; PMCID: 1880871.

International HapMap Consortium, Altshuler, D. M., Gibbs, R. A., Peltonen, L., Altshuler, D. M., Gibbs, R. A., … McEwen, J. E. (2010). Integrating common and rare genetic variation in diverse human populations. *Nature, 467*(7311), 52–58. http://dx.doi.org/10.1038/nature09298. PubMed PMID:20811451; PMCID: 3173859.

International HapMap Consortium, Frazer, K. A., Ballinger, D. G., Cox, D. R., Hinds, D. A., Stuve, L. L., … Stewart, J. (2007). A second generation human haplotype map of over 3.1 million SNPs. *Nature, 449*(7164), 851–861. http://dx.doi.org/10.1038/nature06258. PubMed PMID:17943122; PMCID: 2689609.

International Schizophrenia Consortium, Purcell, S. M., Wray, N. R., Stone, J. L., Visscher, P. M., O'Donovan, M. C., … Sklar, P. (2009). Common polygenic variation contributes to risk of schizophrenia and bipolar disorder. *Nature, 460*(7256), 748–752. http://dx.doi.org/10.1038/nature08185. PubMed PMID:19571811; PMCID: 3912837.

Lee, S., Emond, M. J., Bamshad, M. J., Barnes, K. C., Rieder, M. J., Nickerson, D. A., … Lin, X. (2012). Optimal unified approach for rare-variant association testing with application to small-sample case-control whole-exome sequencing studies. *American Journal of Human Genetics, 91*(2), 224–237. http://dx.doi.org/10.1016/j.ajhg.2012.06.007. Epub 2012/08/07. PubMed PMID:22863193; PMCID: PMC3415556.

Lee, S. H., Ripke, S., Neale, B. M., Faraone, S. V., Purcell, S. M., Perlis, R. H., … Wray, N. R. (2013). Genetic relationship between five psychiatric disorders estimated from genome-wide SNPs. *Nature Genetics, 45*(9), 984–994. http://dx.doi.org/10.1038/ng.2711. Epub 2013/08/13. PubMed PMID:23933821; PMCID: PMC3800159.

Li, B., & Leal, S. M. (2008). Methods for detecting associations with rare variants for common diseases: application to analysis of sequence data. *American Journal of Human Genetics, 83*(3), 311–321. http://dx.doi.org/10.1016/j.ajhg.2008.06.024. Epub 2008/08/12. PubMed PMID:18691683; PMCID: PMC2842185.

Li, Y., Willer, C., Sanna, S., & Abecasis, G. (2009). Genotype imputation. *Annual Review of Genomics and Human Genetics, 10*, 387–406. http://dx.doi.org/10.1146/annurev.genom.9.081307.164242. PubMed PMID:19715440; PMCID: 2925172.

Li, Y., Willer, C. J., Ding, J., Scheet, P., & Abecasis, G. R. (2010). MaCH: using sequence and genotype data to estimate haplotypes and unobserved genotypes. *Genetic Epidemiology, 34*(8), 816–834. http://dx.doi.org/10.1002/gepi.20533. PubMed PMID:21058334; PMCID: 3175618.

Madsen, B. E., & Browning, S. R. (2009). A groupwise association test for rare mutations using a weighted sum statistic. *PLoS Genetics, 5*(2), e1000384. http://dx.doi.org/10.1371/journal.pgen.1000384. Epub 2009/02/14. PubMed PMID:19214210; PMCID: PMC2633048.

Meuwissen, T. H., Hayes, B. J., & Goddard, M. E. (2001). Prediction of total genetic value using genome-wide dense marker maps. *Genetics, 157*(4), 1819–1829. PubMed PMID:11290733; PMCID: 1461589.

Neale, B. M., Rivas, M. A., Voight, B. F., Altshuler, D., Devlin, B., Orho-Melander, M., … Daly, M. J. (2011). Testing for an unusual distribution of rare variants. *PLoS Genetics, 7*(3), e1001322. http://dx.doi.org/10.1371/journal.pgen.1001322. PubMed PMID:21408211; PMCID: 3048375.

Neale, B. M., & Sham, P. C. (2004). The future of association studies: gene-based analysis and replication. *American Journal of Human Genetics, 75*(3), 353–362. http://dx.doi.org/10.1086/423901. Epub 2004/07/24. PubMed PMID:15272419; PMCID: Pmc1182015.

Network, Pathway Analysis Subgroup of the Psychiatric Genomics Consortium. (2015). Corrigendum: psychiatric genome-wide association study analyses implicate neuronal, immune and histone pathways. *Nature Neuroscience, 18*(6), 926. http://dx.doi.org/10.1038/nn0615-926c. PubMed PMID:26007216.

Novembre, J., Johnson, T., Bryc, K., Kutalik, Z., Boyko, A. R., Auton, A., … Bustamante, C. D. (2008). Genes mirror geography within Europe. *Nature, 456*(7218), 98–101. http://dx.doi.org/10.1038/nature07331. PubMed PMID:18758442; PMCID: 2735096.

Patterson, N., Price, A. L., & Reich, D. (2006). Population structure and eigenanalysis. *PLoS Genetics, 2*(12), e190. http://dx.doi.org/10.1371/journal.pgen.0020190. Epub 2006/12/30. PubMed PMID:17194218; PMCID: PMC1713260.

Pe'er, I., Yelensky, R., Altshuler, D., & Daly, M. J. (2008). Estimation of the multiple testing burden for genomewide association studies of nearly all common variants. *Genetic Epidemiology, 32*(4), 381–385. http://dx.doi.org/10.1002/gepi.20303. PubMed PMID:18348202.

Petrovski, S., Wang, Q., Heinzen, E. L., Allen, A. S., & Goldstein, D. B. (2013). Genic intolerance to functional variation and the interpretation of personal genomes. *PLoS Genetics, 9*(8), e1003709. http://dx.doi.org/10.1371/journal.pgen.1003709. PubMed PMID:23990802; PMCID: 3749936.

Price, A. L., Kryukov, G. V., de Bakker, P. I., Purcell, S. M., Staples, J., Wei, L. J., & Sunyaev, S. R. (2010). Pooled association tests for rare variants in exon-resequencing studies. *American Journal of Human Genetics, 86*(6), 832–838. http://dx.doi.org/10.1016/j.ajhg.2010.04.005. Epub 2010/05/18. PubMed PMID:20471002; PMCID: PMC3032073.

Price, A. L., Patterson, N. J., Plenge, R. M., Weinblatt, M. E., Shadick, N. A., & Reich, D. (2006). Principal components analysis corrects for stratification in genome-wide association studies. *Nature Genetics, 38*(8), 904–909. http://dx.doi.org/10.1038/ng1847. Epub 2006/07/25. PubMed PMID:16862161.

Pritchard, J. K., & Rosenberg, N. A. (1999). Use of unlinked genetic markers to detect population stratification in association studies. *American Journal of Human Genetics, 65*(1), 220–228. http://dx.doi.org/10.1086/302449. PubMed PMID:10364535; PMCID: 1378093.

Risch, N., & Merikangas, K. (1996). The future of genetic studies of complex human diseases. *Science, 273*(5281), 1516–1517. PubMed PMID:8801636.

Samocha, K. E., Robinson, E. B., Sanders, S. J., Stevens, C., Sabo, A., McGrath, L. M., … Daly, M. J. (2014). A framework for the interpretation of de novo mutation in human disease. *Nature Genetics, 46*(9), 944–950. http://dx.doi.org/10.1038/ng.3050. PubMed PMID:25086666; PMCID: 4222185.

Schork, A. J., Thompson, W. K., Pham, P., Torkamani, A., Roddey, J. C., Sullivan, P. F., … Dale, A. M. (2013). All SNPs are not created equal: genome-wide association studies reveal a consistent pattern of enrichment among functionally annotated SNPs. *PLoS Genetics, 9*(4), e1003449. http://dx.doi.org/10.1371/journal.pgen.1003449. PubMed PMID:23637621; PMCID: 3636284.

Song, M., Hao, W., & Storey, J. D. (2015). Testing for genetic associations in arbitrarily structured populations. *Nature Genetics, 47*(5), 550–554. http://dx.doi.org/10.1038/ng.3244. PubMed PMID:25822090; PMCID: 4464830.

Tennessen, J. A., Bigham, A. W., O'Connor, T. D., Fu, W., Kenny, E. E., Gravel, S., … NHLBI Exome Sequencing Project. (2012). Evolution and functional impact of rare coding variation from deep sequencing of human exomes. *Science, 337*(6090), 64–69. http://dx.doi.org/10.1126/science.1219240. PubMed PMID:22604720; PMCID: 3708544.

Vilhjalmsson, B. J., Yang, J., Finucane, H. K., Gusev, A., Lindstrom, S., Ripke, S., … Price, A. L. (2015). Modeling linkage disequilibrium increases accuracy of polygenic risk scores. *American Journal of Human Genetics, 97*(4), 576–592. http://dx.doi.org/10.1016/j.ajhg.2015.09.001. PubMed PMID:26430803.

Wu, M. C., Kraft, P., Epstein, M. P., Taylor, D. M., Chanock, S. J., Hunter, D. J., & Lin, X. (2010). Powerful SNP-set analysis for case-control genome-wide association studies. *American Journal of Human Genetics, 86*(6), 929–942. http://dx.doi.org/10.1016/j.ajhg.2010.05.002. Epub 2010/06/22. PubMed PMID:20560208; PMCID: PMC3032061.

Wu, M. C., Lee, S., Cai, T., Li, Y., Boehnke, M., & Lin, X. (2011). Rare-variant association testing for sequencing data with the sequence kernel association test. *American Journal of Human Genetics, 89*(1), 82–93. http://dx.doi.org/10.1016/j.ajhg.2011.05.029. Epub 2011/07/09. PubMed PMID:21737059; PMCID: PMC3135811.

Yang, J., Lee, S. H., Goddard, M. E., & Visscher, P. M. (2011). GCTA: a tool for genome-wide complex trait analysis. *American Journal of Human Genetics, 88*(1), 76–82. http://dx.doi.org/10.1016/j.ajhg.2010.11.011. PubMed PMID:21167468; PMCID: 3014363.

Yang, J., Loos, R. J., Powell, J. E., Medland, S. E., Speliotes, E. K., Chasman, D. I., … Visscher, P. M. (2012). FTO genotype is associated with phenotypic variability of body mass index. *Nature, 490*(7419), 267–272. http://dx.doi.org/10.1038/nature11401. PubMed PMID:22982992; PMCID: 3564953.

Yang, J., Weedon, M. N., Purcell, S., Lettre, G., Estrada, K., Willer, C. J., … Visscher, P. M. (2011). Genomic inflation factors under polygenic inheritance. *European Journal of Human Genetics, 19*(7), 807–812. http://dx.doi.org/10.1038/ejhg.2011.39. Epub 2011/03/17. PubMed PMID:21407268; PMCID: 3137506.

Yang, J., Zaitlen, N. A., Goddard, M. E., Visscher, P. M., & Price, A. L. (2014). Advantages and pitfalls in the application of mixed-model association methods. *Nature Genetics, 46*(2), 100–106. http://dx.doi.org/10.1038/ng.2876. PubMed PMID:24473328; PMCID: 3989144.

Zuk, O., Schaffner, S. F., Samocha, K., Do, R., Hechter, E., Kathiresan, S., … Lander, E. S. (2014). Searching for missing heritability: designing rare variant association studies. *Proceedings of the National Academy of Sciences of the United States of America, 111*(4), E455–E464. http://dx.doi.org/10.1073/pnas.1322563111. Epub 2014/01/21. PubMed PMID:24443550; PMCID: Pmc3910587.

Chapter 10

Reconstructing Causal Network Models of Human Disease

Eric E. Schadt
Icahn School of Medicine at Mount Sinai, New York City, NY, United States

INTRODUCTION

We live in a big data universe composed of zettabytes of data that holds promise in improving human well-being across a broad range of areas, from more accurate weather predictions to more accurate modeling of financial markets, to achieving the vision of precision medicine by enabling a truly personalized understanding of an individual's health. In 2011, the digital universe of information surpassed 1 zettabyte of data (a zettabyte is a 1 followed by 21 zeros), an astonishing number by any measure, equivalent to the amount of data you could store on 30 billion 32-gigabyte iPhones. The digital universe of information continues to grow at roughly 2 zettabytes a year, and that growth continues to increase at an exponential pace. The life and biomedical sciences are no longer strangers to this big data revolution, given dramatic advances in technologies such as next-generation DNA sequencing in which the cost of sequencing entire cancer and germ-line genomes has dropped at a super Moore's law pace (few if any other technologies have evolved at a similar pace). Today we can routinely generate whole genomes, transcriptomes, metabolomes, proteomes, glycomes, lipidomes, and phenomes at population scales given the relatively low (and dropping) cost of running such assays, and the hope in generating large-scales of such data in human and experimental populations is that these deeper and more relevant forms of big data will lead to a better understanding not only of disease but of wellness. However, to move effectively from data to understanding, we must employ advanced computational frameworks capable of simultaneously organizing and modeling big data, to learn from the data how to classify individuals accurately along the wellness and disease spectrum, model their health course trajectories, and assist in either maintaining their course along trajectories associated with wellness, or to identify the most appropriate interventions accurately to change their trajectories to treat or provide protection against disease. Once we can accurately model health course trajectories at the level of individuals, like a global positioning satellite navigator we can help guide individuals through changes in behavior, changes in diet, or small molecule compounds that can favorably maintain or alter their trajectories with respect to treating or preventing disease. This is the hope of precision medicine operating at a highly personalized level.

But how do we integrate the digital universe of data to model complex diseases more accurately, such as schizophrenia and autism, so that we can better diagnose, treat, or prevent disease altogether? To organize very large-scale, high-dimensional data from the digital universe of data, we must employ mathematical algorithms that can establish how the vast space of features associated with, say, a given disease or set of physiologic traits, relate to one another. For example, in biological systems, a complete characterization of genetic and epigenetic variations in germ line and somatic genomes can provide the necessary blueprint from which to understand many (if not most) of the causal relationships between genes and higher-order phenotypic variations such as pathophysiological states associated with disease. Detecting small nucleotide variations, moderate-to-large structural variations, and a whole range of chemical modifications to bases in nucleic acid sequences can accelerate our understanding of genetic and epigenetic variants that underlie our risk of disease, disease severity, and response to treatments targeting specific diseases. However, on their own these changes are not sufficient for understanding how a given variant or constellation of variants increases (or decreases) our risk of disease.

Genomics, Circuits, and Pathways in Clinical Neuropsychiatry. http://dx.doi.org/10.1016/B978-0-12-800105-9.00010-X

Even in diseases such as cancer, a disease of the genome, that have been studied extensively at the genomic level, identification of the variations in DNA will not necessarily lead to an understanding of the disease. For instance, today the focus in cancer genomes is on identifying obvious mutations in protein-coding sequences that activate oncogenes or inactivating tumor suppressor genes. However, not only are variations in DNA in the germ line and somatic genomes that affect regulation largely unexplored with respect to their role in cancer, but even with knowledge of such variations we would not necessarily understand the genes that are affected, the pathways in which those genes operate, the larger networks in which the pathways operate, knowledge of whether you activate or suppress such genes, and pathways for treatment, and so on. The same is perhaps truer for complex diseases such as autism and schizophrenia, in which knowledge of a causal variant for these diseases does not necessarily directly implicate a gene, does not define the context under which the variant will have its adverse impact, does not define the pathways and networks perturbed by the variant, and, importantly, does not necessarily inform on whether you want to activate, partially activate, or inactivate the gene to treat disease.

With roughly 3 billion nucleotides making up the human genome, the number of nucleotide changes that can affect the activities of a moderate to large number of genes is effectively infinite with respect to our ability to experimentally determine the effects of combinations of such changes. Whereas the focus in the past regarding DNA variation and its association with disease had focused on protein-coding sequences, because of declarations that intergenic DNA is composed mainly of "junk" (Smith, Brookhaven National Laboratory, & U.S. Atomic Energy Commission, 1972), today we know that greater than 80% of the human genome is actively bound by proteins that regulate the expression of genes (Ecker et al., 2012), providing a vast array of knobs and switches to modulate not only the activity of genes but of whole gene networks. Therefore, leveraging naturally occurring DNA variation in human populations can be considered among the most attractive approaches to inferring the constellation of genes that affect disease risk. For most noncancer human diseases such as Alzheimer disease, autism, and schizophrenia, changes in DNA that correlate with changes in disease can be inferred as tagging or directly representing causal components of disease. In this way, the DNA variation directly elucidates disease etiology, and so is extremely useful. Genome-wide association studies are well proven to uncover genetic loci that affect disease risk or disease progression (Stranger, Stahl, & Raj, 2011).

Of course, genetics is but one dimension in a big sea of data dimensions that we can now leverage to better understand human conditions such as psychiatric disorders. Models of disease that consider a greater diversity of data that inform on disease will necessarily deliver more accurate diagnoses. In fact, we are in the midst of a big data revolution that permeates nearly every aspect of our lives. Electronic devices that consume much of our attention on a daily basis enable rapid transactions among individuals on unprecedented scales, in which all of the information involved in these daily transactions can be seamlessly stored in digital form, whether the transactions involve monitoring of activity levels using Fitbit-like devices, cell phone calls, text messages, credit card purchases, e-mails, or visits to the doctor's office in which all tests carried out are digitized and entered into your electronic medical record.

If we want to achieve understanding from these big data, organize it, compute on it, and build predictive models from it, we must employ statistical reasoning beyond the more classic hypothesis testing of yesteryear. We have moved well beyond the idea that we can simply repeat experiments to validate findings generated in populations. In fact, whereas first instances of the Central Dogma of Biology looked something like the simple graph depicted in Fig. 10.1, today, because the complex interplay of multiple dimensions of data (DNA, RNA, protein, metabolite, cellular, physiologic, ecologic, and social structures more generally) demands that a more holistic view be taken in which we embrace complexity in its entirety, the central dogma is evolving to look something more like the more complex graph depicted in Fig. 10.1. Our emerging view of complex biological systems is one of a dynamic, fluid system that is able to reconfigure itself as conditions require, as depicted in Fig. 10.2 (Califano, Butte, Friend, Ideker, & Schadt, 2012; Chang, Karr, & Schadt, 2015; Schadt, 2009; Schadt, Buchanan, Brennand, & Merchant, 2014; Schadt, Friend, & Shaywitz, 2009). Despite these transformative advances in technology and the need to embrace complexity, it remains difficult to assess where we are with respect to our understanding of living systems, relative to a complete comprehension of such systems. One of the primary difficulties in our making such an assessment is that the suite of research tools available to us seldom provides insights into aspects of the overall picture of the system that are not directly measured.

In this chapter, I discuss a particular class of integrative modeling that can use panomic data (DNA, RNA, metabolites, proteins, DNA—protein interactions, DNA—small molecule interactions, protein—protein interactions, and so on) along with other types of data such as cellular and physiologic data to derive causal relationships among thousands of intermediate molecular traits and between molecular and higher-order physiological traits associated with disease (Barabasi & Oltvai, 2004; Zhu et al., 2012). Although a very broad set of modeling techniques are applicable across a diversity of problems in disease biology for which mathematical modeling would be appropriate (Fig. 10.3), I will focus primarily on causal probabilistic modeling as a convenient way of incorporating highly diverse data in ways that

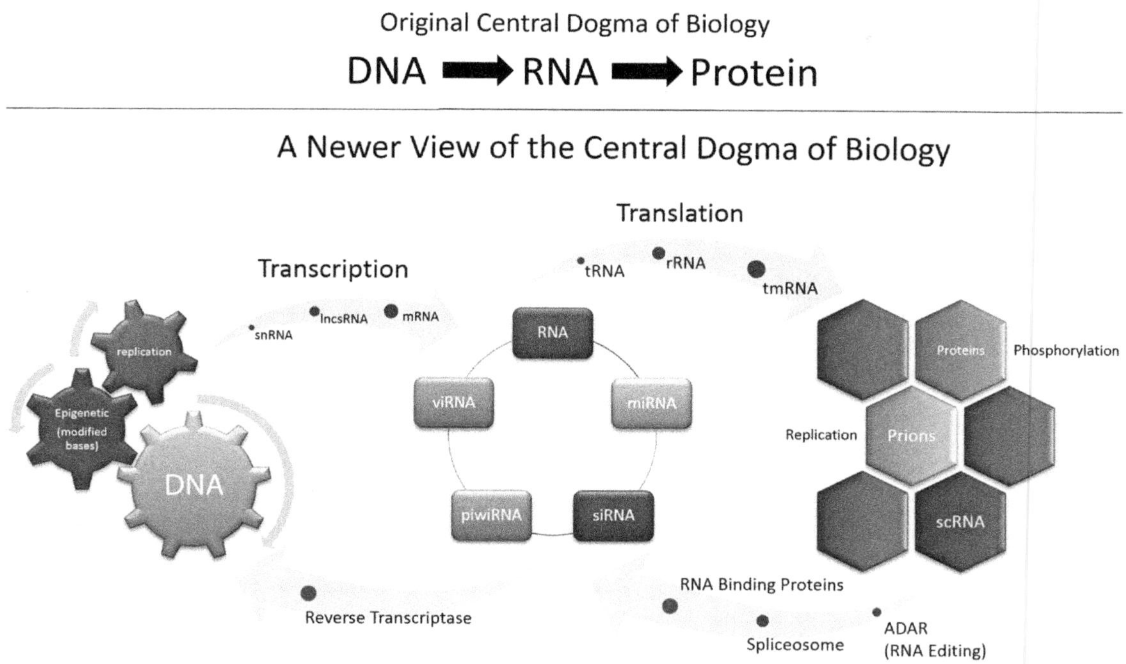

FIGURE 10.1 Evolving central dogma of biology. The *upper area* represents the original central dogma of biology, a simple view driven by early observations with low-resolution tools that uncovered a central relationship among DNA, RNA, and proteins: namely that RNA is transcribed from DNA, and RNA in turn is translated into proteins. New higher-resolution technologies have enabled a far more complex view of the central dogma to emerge (*bottom area*), with epigenetic changes to DNA that are transgenerational leading to non-Mendelian patterns of inheritance, a complex array of RNA molecules such as microRNA (mRNA), virus-induced RNA (viRNA), piwiRNA, and synthetic RNA (siRNA) that do not code for proteins but carry out complex regulatory functions, and sophisticated protein complexes involved in splicing, RNA editing, and RNA binding, all feeding back on transcription, leading to a more network-oriented view of the central dogma. *ADAR*, adenosine deaminases acting on RNA; *scRNA*, small conditional RNA; *lncsRNA*, long noncoding RNA; *snRNA*, small nuclear RNA; *tRNA*, transfer RNA; *rRNA*, ribosomal RNA; *tmRNA*, transfer-messenger RNA.

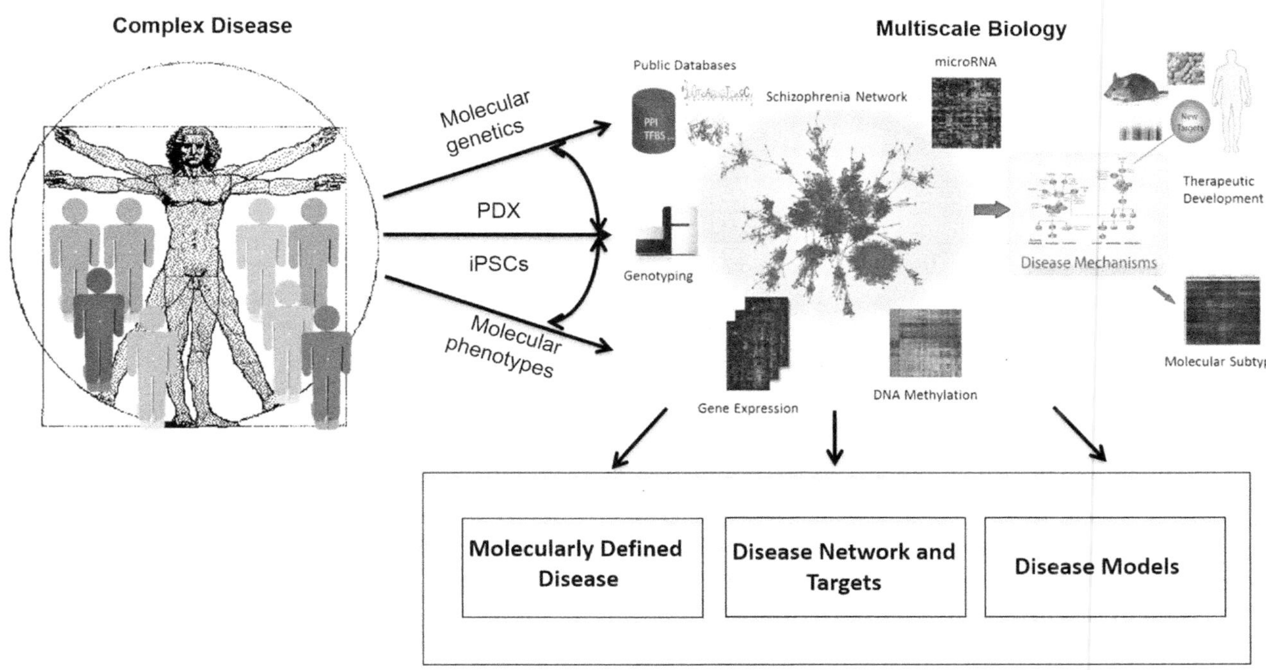

FIGURE 10.2 Integrative biology view of achieving a better understanding of disease and how best to diagnose, treat, and prevent it. Because of the vast stores of data, we can generate and store for populations of individuals, we must integrate the diversity of these data to construct predictive models of disease. The network-driven framework discussed in the main text can lead to multiscale models of disease that in turn can be used to elucidate disease mechanisms, stratify patient populations, and develop novel therapeutics. *PDX*, patient-derived xenograft; *iPSCs*, induced pluripotent stem cell.

Provides Strong Mechanistic Insights	Provides Direct Mechanistic Insights	Potential to Provide Mechanistic Insights	Can Learn Novel Causal Relationships	Informs on Relationships but Does Not Implicitly Infer Causality	Association Based, Little Ability to Provide Mechanistic Insights
Decreasing Mechanistic Insights/Increased Abilty to Learn From Data					
Small, Focused Data to Fit the Model	Small-to-Moderate Data to Fit the Model	Moderate-to-Larger Data to Fit the Model	Larger Data Sizes Needed to Fit Model	Large Datasets Required to Fit Model	Very Large Datasets Required to Fit Model
Increasing Scale of Data Required to Fit Model					
Requires Extensive Prior Knowledge	Requires Strong Prior Knowledge	Requires Less Prior Knowledge	Can Leverage Prior Knowledge When Available, but Not Required	No Prior Information Required, Ability to Model Some Prior Data	No Prior Information Required, Limited Ability to Incorporate Prior Knoweldge
Decreasing Dependence on Prior Knowledge					
Models Limited to Small Number of Nodes	Models Limited to Small-to-Moderate Number of Nodes	Models Can Have Moderate Number of Nodes	Models Can Have Moderate-to-Large Number of Nodes	Models Can Have Large Number of Nodes	Models Can Have Very Large Number of Nodes
Increasing Model Size					
Kinetic Models	**Fuzzy Logic Models**	**Boolean Network Models**	**Bayesian Network Models**	**Partial Least Squares Regr Models**	**PCA Multi-Regression Models**
Bottom-Up Modeling Begins with Prior Knowledge of Mechanism		Top-Down Modeling Seeks to Learn Causal Relationships Given Big Data		Correlation-Based Modeling Explores Associations Among Variables	

FIGURE 10.3 Table highlighting the different classes of mathematical modeling that can be applied to biological data to uncover relationships in the data that may help predict phenotypes of interest, elucidate causal relationships among traits and biological processes, and derive mechanistic insights into the causes of disease, wellness, drug response, and other phenotypes of interest. A more detailed description of the different modeling approaches is given in the text.

can lead to robust causal inference, a necessity if we hope to understand cause-and-effect relationships between high-dimensional data scored across a number of diverse modalities such as molecular, cellular, imaging, physiologic, and disease.

MODELING BIOLOGICAL DATA

A true understanding of complex systems and the complex behaviors they exhibit can be achieved only if we understand the causal relationships among the hierarchy of constituent components comprising the system. For example, in living systems we seek to understand the relationships in and between the molecular, cellular, tissue, organ, organism, and community dimensions assayed in all relevant contexts. However, inferring causality between variables, especially recovering causal networks from observational data, is a particularly challenging task. Because of the complexity of biological data and the complexity of methods that can be applied to deriving meaning from such data, an awareness of the different classes of models is warranted, although I will focus primarily on probabilistic causal reasoning. The different types of modeling that can be applied to biological data can be broken down into a number of different classes, and the selection of modeling approach to employ depends on a number of factors such as the extent of prior knowledge, the dimensionality of the data to be modeling, the scale of data available to model, and, of course, what you hope to derive from the data and the model (Fig. 10.3).

In the spectrum of modeling classes ranging from those assuming the most complete knowledge of pathways and networks to those assuming no knowledge, preferring instead to learn the network structures directly from the data, the kinetic models are at the most extreme end of the distribution with respect to requiring extensive prior knowledge. Kinetic models are typically represented as systems of ordinary differential equations (ODEs), which require extensive prior knowledge as the ODEs fix the connectivity structure among the variables being modeled (eg, the pathway is assumed to be known). The model is then defined by a series of parameters that are fit from the data, and with these parameter estimates the behavior of the system can be directly explored via simulations run on the model. Via these simulations, kinetic models provide for greater mechanistic insights. These models can also be fit from smaller, more focused data sets, although typically this modeling approach is restricted to smaller network structures and the models can be difficult to calibrate (Azeloglu & Iyengar, 2015). Modeling of the dynamics of physiologic glucose-insulin levels, metabolic flux, and drug response are just a few of many examples that have been effectively modeled using this approach. Logic models represent another class of models that require significant prior knowledge, but that also have an adaptive component that can be learned from the data and thus serves to reduce dependence on the extent of knowledge required to model the biological system of interest. Logic models also maintain a simple and intuitive framework for understanding complex signaling networks (Morris, Saez-Rodriguez, Sorger, & Lauffenburger, 2010). In addition, this type of modeling approach still provides for direct mechanistic insights to be derived from simulations on these models. Kinetic and logic models are more representative of what I refer to as bottom-up modeling approaches that begin with strong prior knowledge regarding how pathways are put together, but then define the kinetic parameters on those pathways that describe the flow of information through the system.

Boolean network modeling is another class of approaches that provide an even more flexible framework for modeling biomolecules as binary variables that directly relate to state information that is relevant to downstream biological processes. However, regulation of the different states represented is described in a parameter-free way (in contrast to kinetic models that are defined by kinetic parameters), providing for an approach that enables a more exploratory characterization of the dynamics of a complex system (Albert & Thakar, 2014). Whereas these types of models can represent many more variables than kinetic models, they provide less mechanistic insight. Bayesian network models, the approach discussed in depth in this chapter, are an even more flexible framework for modeling complex biological processes, requiring no prior knowledge, but still providing for a natural and mathematically elegant way to incorporate prior knowledge. Bayesian networks provide a way to learn regulatory relationships directly from the data. With the use of heuristic searching, networks composed of many thousands of variables can be constructed, although equally large sets of data are required to construct this type of model effectively. The causal relationships represented in these models are statistically inferred and so it is more difficult to derive mechanistic insights. Boolean and Bayesian network modeling approaches are examples of what I refer to as top-down modeling approaches that seek to learn relationships directly from the data (structure-based learning).

The final classes of modeling approaches are correlation based and are more exploratory in nature, seeking to elucidate the correlation structures in extensive data sets to begin to understand the relationships that may be well reflected in them and that may aid in understanding key processes involved in complex processes associated with phenotypes of interest such as disease. Partial least-squares regression and principal component analysis multiregression are examples of two classes of such modeling approaches. They do not require prior knowledge to fit the models; they can operate on extremely large data sets, scaling to any number of variables that give rise to very large-scale networks; and they are easy to calibrate. However, such models do not explicitly infer causality but rather reflect connections and influences on those connections, a first step for learning important relationships that are involved in complex processes such as disease.

In this broad spectrum of methods, Bayesian networks strike a nice balance between resolving mechanisms and structure and more broadly reflecting connections and their influences, thereby providing an efficient path for understanding information flow. Whereas ODEs are hypothesis driven, in which the relationships among variables is assumed to be known, Bayesian methods operate in a hypothesis-free context in which we attempt to infer the relationships among variables given the data. As a result, Bayesian networks have emerged as a state-of-the-art approach for understanding complex systems in which the relationships among the constituent components of the system are not generally known, because they can seamlessly incorporate existing knowledge as structural and parameter priors and then infer directed relationships among the nodes in the network using conditional dependency arguments (Chang et al., 2015; Zhu et al., 2008, 2012). However, there are also limitations with this modeling approach that relate to the ability of Bayesian networks to distinguish causal structures that have equivalent joint probability and conditional independence structures (Markov equivalence). The severity of this problem cannot be understated because of statistically indistinguishable structures that may reflect completely contradictory causal relationships. I will explore next how appropriate prior information can be incorporated to help resolve these and related issues.

CAUSALITY AS A STATISTICAL INFERENCE

How molecular traits and disease traits causally relate to each other can be modeled using pairwise causality tests (Millstein, Zhang, Zhu, & Schadt, 2009; Schadt et al., 2005) or probabilistic graphical models such as RIMBANet (Zhu et al., 2004, 2007, 2008, 2010, 2012), in which all available traits are considered simultaneously. A number of studies across a variety of species have demonstrated that predictive networks such as Bayesian networks can capture fundamental properties of complex systems in states that result in complex phenotypes (Jansen et al., 2003; Lee, Date, Adai, & Marcotte, 2004; Schadt et al., 2008; Zhu et al., 2004, 2007, 2008, 2012, 2013). Available molecular data that informs on disease, derived from different tissues in different states, provides the necessary ingredients to reconstruct causal network models of disease. Understanding how the constellation of genes identified for diseases such as schizophrenia or autism (Neale et al., 2012; Ripke et al., 2011) are actually related to each other in probabilistic causal ways can lead to an understanding of how perturbing a given gene or genes (say, for treatment) will affect the corresponding molecular networks and ultimately the pathophysiology of the diseases they affect. The key to constructing predictive models is to elucidate causal relationships between traits of interest. Resolving causal relationships requires a systematic source of perturbation; here, I discuss the use of DNA variation as a systematic perturbation source to infer causal relationships among molecular traits and between molecular traits and higher-order traits such as disease (Chen et al., 2008; Emilsson et al., 2008; Mehrabian et al., 2005; Millstein et al., 2009; Schadt et al., 2005; Yang et al., 2009; Zhu et al., 2004, 2008, 2012).

In the life sciences, most researchers are accustomed to thinking about causality from the standpoint of physical interactions. In the setting of molecular biology or biochemistry, when two molecular traits are indicated as causally related, we typically mean that one of the molecular entities (eg, a small molecule compound) has been determined experimentally to physically interact with or induce processes that directly affect the other molecular entity (eg, the target protein of the small molecule) and consequently lead to a phenotypic change of interest (eg, lower-density lipoprotein cholesterol levels). In this case we have an understanding of the causal factors relevant to the activity of interest, so that careful experimental manipulation of these factors allows for an identification of genuine causal relationships. However, in the context of many thousands of variables related in unknown ways, the aim is to examine the behavior of those variables across populations in ways that facilitate statistically inferring causal relationships. For example, statistical associations among changes in DNA, changes in molecular phenotypes, and changes in higher-order phenotypes such as functional MRI readouts or disease can be examined for patterns of conditional dependency among the variables that allow directionality to be inferred. In this case we can employ indirect measures of processes that mediate changes in one trait conditional on another to make a statistically inferred causal link. This is like the types of statistical inferences that are leveraged in other disciplines to make new discoveries. For example, less than 5% of known extrasolar planets have been directly observed, so that most are observed indirectly. One method for detecting planets that cannot be directly observed considers that when a planet is orbiting a star, the gravitational pull of the planet on the star will place the star into a subtle orbit which from our vantage point will appear as the star moving closer to and farther away from the earth in a cyclical fashion. Such movement can be measured as displacements in the star's spectral lines as a result of the Doppler effect (Erskine, Edelstein, Harbeck, & Lloyd, 2005), and so the presence of the planet acting on the star can be statistically inferred.

Similarly, consider genetic variants associated with, say, schizophrenia or autism (many such loci have now been identified) (Elia et al., 2012; Glessner et al., 2009; Neale et al., 2012; Pinto et al., 2010; Ripke et al., 2011; Wang et al., 2009). Furthermore, suppose the expression of some number of genes assayed in relevant regions of the brain relating to these disorders was also associated with these same genetic variants. By examining the changes in levels of expression of these genes in response to changes in genotype at any of the genetic loci of interest, one can directly assess the extent to which these expression changes induced by the genetic loci will explain the degree of association between the locus genotypes and disease trait (Fig. 10.5). In this way, just as the characteristic wobble of a star induced by an orbiting planet

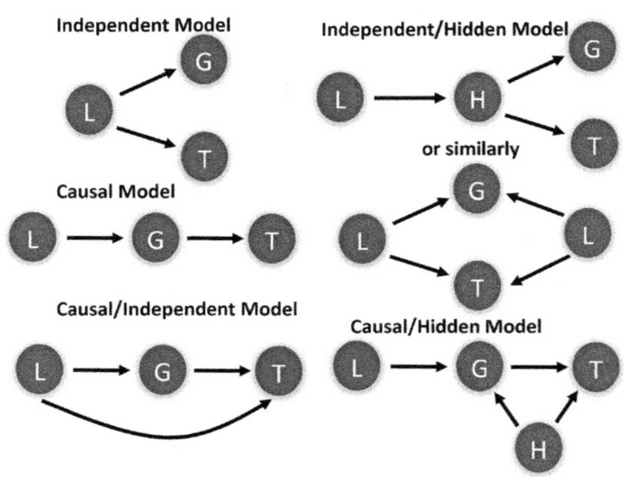

FIGURE 10.4 Because two traits, *G* and *T*, are correlated in a population with changes in DNA at locus *L*, there are five basic causal models to consider in testing the hypothesis that variations in trait *G* cause variations in trait *T*. Here, *H* denotes an unmeasured molecular or higher-order trait.

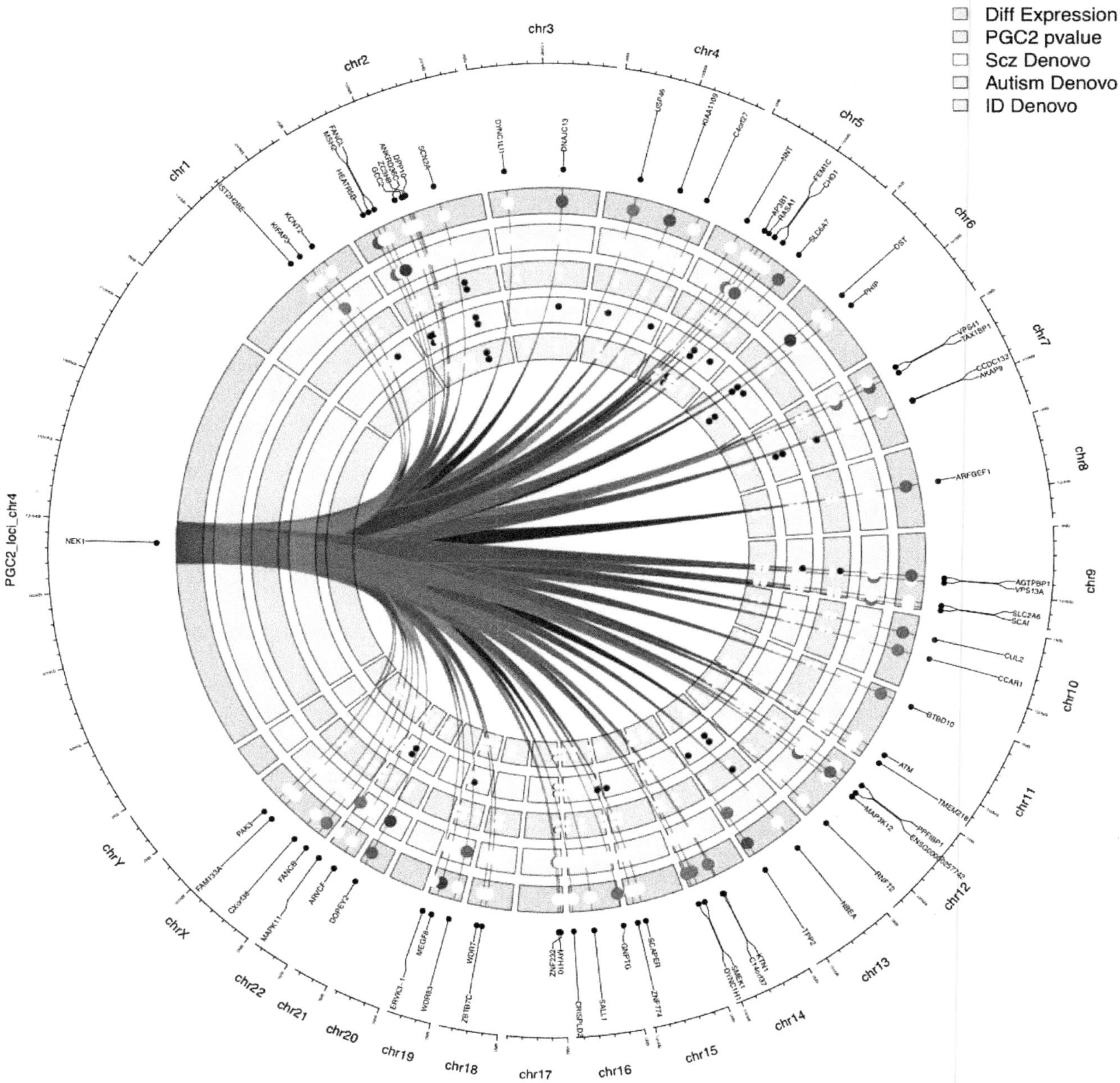

FIGURE 10.5 *Circos plot* representing inferred causal links between NEK1, a gene coincident with a genome-wide significant schizophrenia locus on chromosome 4, and gene expression traits from prefrontal cortex whose expression levels also associate with the NEK1 schizophrenia locus. The location and gene symbol for the *trans* genes are given on the outer portion of the plot. The causal links depicted by the *blue edges* (dark gray in print versions) were inferred via the causal inference test described in the text. The different colored bands in the circos plot serve to annotate the *trans*-acting expression traits linked to NEK1 to further enhance the relevance of this network structure to schizophrenia. The *purple band* (darker gray in print versions) indicates *trans* genes that were significantly differentially expressed between patients with schizophrenia and control subjects in prefrontal cortex. *White* indicates genes that were not differentially expressed; *red dots* (darkest gray in print versions) indicate significance. *Green band* (gray in print versions) indicates genes that have been identified in genome-wide association studies for schizophrenia; *red* (darkest gray in print versions) and *yellow bands* (dark gray in print versions) indicate genes identified in whole-exome sequencing studies for autism and schizophrenia, respectively. In the *green* (gray in print versions), *red* (darkest gray in print versions), and *yellow bands* (dark gray in print versions), a *black dot* indicates that the gene was identified as genome-wide significant in the respective study, whereas a *white dot* indicates nonsignificance. *PGC2*, Psychiatric Genomics Consortium 2; *Scz*, schizophrenia; *ID*, intellectual disability.

predicts the presence of the planet, the characteristic "wobble" of the expression levels of a gene and its association to the disease state predicts a causal path between the gene and disease state, as described in more detail subsequently.

Critical to identifying causal relationships is distinguishing between correlation and causation. The old adage, "correlation does not imply causation," is familiar to most. This is among the first fallacies one learns about in beginning logic courses: *Post hoc ergo propter hoc* (Latin for "After this, therefore because of this"). Measurements taken over time on independent variables can be correlated because trends reflected by such variables are coincidentally similar, or changes in each variable are independently caused by a common source, in addition to being correlated as a result of a cause—effect relationship. Also, whereas correlation and causation are related, our intuitive notation that causation implies correlation is not always correct. For example, suppose U and V are random variables with the same distribution, and suppose $X = U + V$ and $Y = U - V$. In this case, the covariance between X and Y (defined as $E(XY) - E(X)E(Y)$, where E represents the expectation function) is 0, and so the correlation is 0, even though there is a direct functional dependence between the variables (Feller, 1967). Only when two variables are linearly dependent (which is often the case in research) does our intuitive notion of functional dependence imply that perfect correlation has been met.

Structure learning approaches that seek to infer causal relationships among correlated variables often employ conditional dependency arguments or mutual information measures to resolve causality by introducing a third correlated variable. By conditioning each of the variables on the third and examining the residual correlation between them in each case, a decision can be made as to the direction of the flow of information between the variables. However, this type of reasoning has generally failed to result in predictive causal inference, because in the absence of systematic perturbations, the number of graphs that can be represented among just three traits is large (125 graphs representing directed and undirected relationships among three correlated variables are possible), and many of these possible relationships between the traits are not statistically distinguishable (Sieberts & Schadt, 2007). For example, if variables X, Y, and Z are observed in a population to be correlated (eg, suppose X, Y, and Z represent the expression levels of three genes assayed in a given region of the brain in a population of individuals with schizophrenia) and the true relationship between the variables is $X \rightarrow Z \leftarrow Y$, this relationship cannot be statistically distinguished from $X \rightarrow Y \leftarrow Z$ and $Z \rightarrow X \leftarrow Y$, even though these relationships result in to contradictory causal relationships.

To break this type of statistical symmetry, a source of perturbation is required. The classic approach in the experimental sciences is to introduce an artificial perturbation by knocking a gene out, overexpressing a gene, or chemically perturbing a given protein to assess the consequences on a given trait of interest. More recently, in the neurosciences, optogenetics methods have provided novel ways to perturb genes on the short time scales needed to elucidate the complexity of networks at play in neurons in living mammals (Boyden, Zhang, Bamberg, Nagel, & Deisseroth, 2005) and genome-editing technologies are coming into their own as well (Pelletier, Gingras, & Green, 2015). If experimentally controlled artificial perturbations on a given gene cause a change in a trait of interest, we infer a causal relationship between that gene and trait. However, DNA variation in the germ line provides an excellent systematic perturbation source that can also be used to resolve causal relationships in biological systems. Because variations in DNA cause variations in RNA, proteins, metabolites, and subsequently higher-order phenotypes, this source of variation can be leveraged to infer causality. Unlike artificial perturbations such as gene knockouts, transgenics, chemical, or optogenetic perturbations that may induce artificial correlations that are not observed in more natural settings, naturally occurring genetic variation defines those perturbations that give rise to the broad array of phenotypic variations (such as disease and drug response) that we are precisely interested in elucidating. Research has demonstrated that causal links between DNA variations and molecular and higher-order phenotypes can provide information on causal relationships between those traits (Chen, Emmert-Streib, & Storey, 2007; Chen et al., 2008; Davey Smith & Ebrahim, 2003; Didelez & Sheehan, 2007; Emilsson et al., 2008; Kulp & Jagalur, 2006; Millstein et al., 2009; Schadt et al., 2003, 2005; Yang et al., 2009; Zhu et al., 2004, 2008). Causality in this instance can be inferred because there is random segregation of the chromosomes during gametogenesis, thus providing the appropriate randomization mechanism to protect against confounding similar to what is achieved in randomized clinical trials by randomly assigning patients to treatments to test the causal effects of a drug of interest (Lawlor, Harbord, Sterne, Timpson, & Davey Smith, 2008; Nitsch et al., 2006). However, it has been challenging to quantify the uncertainty in making such causal calls. For example, causal effect estimates often considered in Mendelian randomization approaches can be confounded by pleiotropic effects and reverse causation, which limiting the use of such approaches for problems that involve reconstructing regulatory networks in which pleiotropy is common and there may be little a priori information regarding the structure of the causal relationships between the traits of interest (Millstein et al., 2009).

However, formal statistical tests for inferring causal relationships between quantitative traits mediated by a common genetic locus have been developed (Millstein et al., 2009). To understand how such a test works, consider marker genotypes at a given DNA locus L that are correlated with two quantitative phenotypes of interest, G, and T (Fig. 10.4).

There is no constraint on the types of traits G and T can represent, as long as they are scored in the same population. The causal relationship $G \rightarrow T$ is implied if three conditions are satisfied under the assumption that L is sufficiently randomized: (1) L and G are associated; (2) L and T are associated; and (3) L is independent of T given G (ie, L and $T|G$ are not associated) (Chen et al., 2007). If a given locus L is independent of G given T ($G|T$), this is consistent with T being causal for G ($T \rightarrow G$), and if L is associated with $G|T$, this is consistent with G being causal for T ($G \rightarrow T$). We can boil all of these observations down to four conditions from which a statistical test can be formed to test for causality: (1) L and T are associated; (2) L is associated with $G|T$; (3) G is associated with $T|L$; and (4) L is independent of $T|G$. Each of these conditions can be assessed with a corresponding statistical test. For example, if we assume the marker corresponding to locus L is biallelic, where L_1 and L_2 represent indicator variables for the two alleles in a codominant coding scheme, these four conditions can be tested in the parameters of the following three regression models (Eqs. [10.1]–[10.3]):

$$T_i = \alpha_1 + \beta_1 L_{1i} + \beta_2 L_{2i} + \varepsilon_{1i} \quad [10.1]$$

$$G_i = \alpha_2 + \beta_3 T_i + \beta_4 L_{1i} + \beta_5 L_{2i} + \varepsilon_{2i} \quad [10.2]$$

$$T_i = \alpha_3 + \beta_6 G_i + \beta_7 L_{1i} + \beta_8 L_{2i} + \varepsilon_{3i}, \quad [10.3]$$

where G_i and T_i represent the trait levels for the respective traits for individual i in a population of interest, and ε_{ij} represents independently distributed random noise variables with variance σ_j^2 (Chen et al., 2007). Given these models, the four component tests of interest are (Eqs. [10.4]–[10.7]):

$$H_0: \{\beta_1, \beta_2 = 0\},\ H_1: \{\beta_1, \beta_2\} \neq 0 \quad [10.4]$$

$$H_0: \{\beta_4, \beta_5 = 0\},\ H_1: \{\beta_4, \beta_5\} \neq 0 \quad [10.5]$$

$$H_0: \beta_6 = 0,\ H_1: \beta_6 \neq 0 \quad [10.6]$$

$$H_0: \{\beta_7, \beta_8 \neq 0\},\ H_1: \{\beta_7, \beta_8\} = 0. \quad [10.7]$$

The four conditions of interest can be tested using standard F tests for linear model coefficients (conditions 1–3) and a slightly more involved test for the last condition because it is an equivalence testing problem (Millstein et al., 2009). With these individual statistical tests on the different regression parameters, a causal inference test can then be carried out by testing the strength of the chain of mathematical conditions that collectively are consistent with causal mediation (ie, the strength of the chain is only as strong as its weakest link, so that the intersection of the rejection regions of the component tests provides for the causality test we seek). For a series of statistical tests of size α_γ and rejection region R_γ, the "intersection-union" test with rejection region $\cap R_\gamma$ is a level sup(α_γ) test, so that the P value for the causal inference test corresponds to the P value for an intersection-union test, or simply, the supremum of the four P values for the component tests (Chen et al., 2007). This test has been implemented as the CIT package in the R statistical programming language and is freely available.

Applications of this type of test can be applied to resolve the types of causal relationships depicted in Fig. 10.4. As an example, I examined genetic risk factors for schizophrenia, which include both rare variants conferring large relative risks (eg, copy number variants) as well as common single-nucleotide polymorphism (SNP) variants, the latter with modest individual effect sizes (Purcell et al., 2009), for association with prefrontal cortex gene expression levels from a previously published study (Zhang et al., 2013). Among the many findings, I identified that the gene expression levels of NIMA (never in mitosis gene a)-related kinase 1 (NEK1) in this cohort had a strong association with a chromosome 4 genetic locus identified as significantly associated with schizophrenia (Schizophrenia Working Group of the Psychiatric Genomics, 2014). We refer to this association as a *cis*-acting association because of NEK1 gene and the schizophrenia locus are coincident in the human genome. However, in addition to NEK1 associated with this schizophrenia locus, there were many other gene expression traits in brain associated with this locus in *trans* (ie, the gene corresponding to the associated expression trait was distal to the chromosome 4 genetic locus). This is exactly the setup for which the CIT test can be applied to resolve whether the NEK1 gene serves as an intermediate phenotype for the *trans*-acting gene expression traits associated with the same schizophrenia locus. After applying the CIT test to the set of genes linked to the schizophrenia locus in *trans*, considering whether NEK1 was supported as causal for these *trans*-acting expression traits, a significant number of the *trans*-linked genes were supported as reacting to changes in NEK1 expression (Fig. 10.5). In fact, not only was NEK1 supported as causal for these *trans*-acting expression traits, the trans genes themselves were enriched for genes harboring variants that were associated with schizophrenia and autism (see the *yellow, red, and green bands* in Fig. 10.5) and for genes that are differentially expressed in the prefrontal cortex between patients with schizophrenia and control subjects (*purple band* in Fig. 10.5). Together, these data and the results of the CIT begin to illustrate the complex web of causal interactions that are at play in diseases such as schizophrenia. This type of approach has previously been applied to a diversity of other diseases including obesity, diabetes, heart disease, inflammatory bowel disease (IBD), Alzheimer disease, and cancer (Chen et al., 2008; Emilsson et al., 2008; Lamb et al., 2011; Schadt et al., 2005, 2008; Su, Kleinhanz, & Schadt, 2011; Tu et al., 2012; Wang et al., 2012, 2007; Yang et al., 2009; Zhang et al., 2013; Zhong et al., 2010; Zhong, Yang, Kaplan, Molony, & Schadt, 2010).

FROM ASSESSING CAUSAL RELATIONSHIPS AMONG TRAIT PAIRS TO PREDICTIVE GENE NETWORKS

Leveraging DNA variation as a systematic perturbation source to resolve the causal relationships among traits is necessary but not sufficient for understanding the complexity of living systems. Cells are composed of many tens of thousands of proteins, metabolites, RNA, and DNA, all interacting in complex ways. Complex biological systems are composed of many different types of cells operating within and between many different types of tissues that make up different organ systems, all of which interact in complex ways to create a vast array of phenotypes that manifest themselves in living systems. Modeling the extent of such relationships among molecular entities, between cells, and between organ systems is a daunting task. Networks are a convenient framework for representing the relationships among these different variables. In the context of biological systems, a network can be viewed as a graphical model that represents relationships among DNA, RNA, protein, metabolite, and higher-order phenotypes such as disease state. In this way, networks provide a way to represent extremely large-scale and complex relationships among molecular and higher-order phenotypes such as disease in any given context.

Building From the Bottom Up or Top Down?

Two fundamental approaches to the reconstruction of molecular networks dominate computational biology today. The first is what is referred to as the bottom-up approach, in which fundamental relationships between small sets of genes that may comprise a given pathway are established, thus providing the fundamental building blocks of higher-order processes that are then constructed from the bottom up. This approach typically assumes that we have more complete knowledge regarding the fundamental topology (connectivity structure) of pathways, and given this knowledge, models are constructed that precisely detail how changes to any component of the pathway affect other components as well as the known functions carried out by the pathway (ie, bottom-up approaches are hypothesis driven). The second approach is referred to as a top-down approach, in which we take into account all data and our existing understanding of systems and construct a model that reflects whole system behavior, and from there tease apart the fundamental components from the top down. This approach typically assumes our understanding of how the network is actually wired is sufficiently incomplete, that our knowledge is sufficiently incomplete, and that we must objectively infer the relationships by considering large-scale, high-dimensional data that inform on all relationships of interest (ie, top-down approaches are data driven).

With our incomplete understanding of more general networks and pathways in living systems, in this chapter I focus on a top-down approach to reconstructing predictive networks, because this type of structure learning from data is critical to derive hypotheses that cannot otherwise be efficiently proposed in the context of what is known (from the literature, pathway databases, or other such sources). However, top-down and bottom-up approaches are complementary to each other, although these approaches have largely been pursued as separate disciplines with little cross-talk occurring between them. One of the future directions I discuss in the conclusion is the need to unify these two classes of predictive modeling mathematically to produce probabilistic causal networks that more maximally leverage all available data and knowledge.

In the context of integrating genetic, molecular profiling, and higher-order phenotypic data, biological networks are composed of nodes that represent molecular entities that are observed to vary in a given population under study (eg, DNA variations, RNA levels, protein states, or metabolite levels). Edges between the nodes represent relationships between the molecular entities, and these edges can either be directed, indicating a cause—effect relationship, or undirected, indicating an association or interaction. For example, a DNA node in the network representing a locus that varies in a population of interest may be connected to a transcript abundance trait, indicating that changes at the particular DNA locus induce changes in the levels of the transcript. The potentially millions of such relationships represented in a network define the overall connectivity structure of the network, or what is otherwise known as the topology of the network. Any realistic network topology will be necessarily complicated and nonlinear from the standpoint of the more classic biochemical pathway diagrams represented in textbooks and pathway databases such as KEGG (Kanehisa, 2002). The more classic pathway view represents molecular processes on an individual level, whereas networks represent global (population-level) metrics that describe variations between individuals in a population of interest that in turn define coherent biological processes in the tissue or cells associated with the network. One way to manage the complexity of network structures that can be obtained is to impose constraints on network structures to make them more computationally tractable. For example, it is common when learning network structures to disallow loops or cycles in the network structure (otherwise known as the network topology, the connectivity structure of the network), in which case we refer to the network as acyclic.

The neurosciences have a rich history of employing network-based approaches to understanding the complexity of the human brain and the causes of psychiatric illnesses. Resources such as the Allen Brain Atlas (http://www.alleninstitute.org) provide an anatomically comprehensive map of gene expression of the human brain that can facilitate network-based analyses (Hawrylycz et al., 2012). Others have employed techniques developed for constructing gene coexpression networks to construct interaction networks on functional MRI data (Mumford et al., 2010), and still others have generated protein interaction networks to reflect features of the network architecture in brains of those with illnesses such as Huntington disease (Shirasaki et al., 2012). Larger-scale efforts have also been undertaken to integrate larger-scale transcriptomic data in the context of diseases such as autism to understand how changes in these networks may cause autism or reflect the types of pathways or biological processes involved in such a disease (Voineagu et al., 2011). These efforts are important not only for better understanding psychiatric diseases but also for elucidating novel drug targets or biomarkers that better assess disease risk or severity. However, most of these current efforts do not lead to predictive models of disease, but rather provide a descriptive framework within which to uncover associations among a myriad of molecular, cellular, imaging, and clinical traits and disease.

An Integrative Genomics Approach to Constructive Predictive Network Models

Systematically integrating different types of data into probabilistic networks using Bayesian networks has been proposed and applied for the purpose of predicting protein–protein interactions (Jansen et al., 2003) and protein function (Lee et al., 2004). However, these Bayesian networks are still based on associations between nodes in the network as opposed to causal relationships. As discussed earlier for the simple case of two traits, from these types of networks we cannot infer whether a specific perturbation will affect a complex disease trait. To make such predictions, we need networks capable of representing causal relationships. Probabilistic causal networks are one way to model such relationships from the top down, in which causality again in this context reflects a probabilistic belief that one node in the network affects the behavior of another. Bayesian networks (Pearl, 1988) are one type of probabilistic causal network that provide a natural framework for integrating highly dissimilar types of data.

Bayesian networks are directed acyclic graphs in which the edges of the graph are defined by conditional probabilities that characterize the distribution of states of each node given the state of its parents (Pearl, 1988). The network topology defines a partitioned joint probability distribution over all nodes in a network, such that the probability distribution of states of a node depends only on the states of its parent nodes: formally, a joint probability distribution $p(X)$ on a set of nodes X can be decomposed as $p(X) = \prod_i p(X^i|\mathrm{Pa}(X^i))$, where $\mathrm{Pa}(X^i)$ represents the parent set of X^i. The biological networks of interest we wish to construct are composed of nodes that represent a quantitative trait such as the transcript abundance of a given gene or levels of a given metabolite. The conditional probabilities reflect not only relationships between genes but also the stochastic nature of these relationships as well as noise in the data used to reconstruct the network.

The aim in any network reconstruction such as this is to find the best model, the model that best reflects the relationships among all of the variables under consideration, given a set of data that informs on the variables of interest. In a probabilistic sense, we want to search the space of all possible networks (or models) for that network that gives the highest likelihood of occurring given the data. Bayes' formula allows us to determine the likelihood of a network model M given observed data D as a function of our prior belief that the model is correct and the probability of the observed data given the model: $P(M|D) \sim P(D|M)P(M)$. The number of possible network structures grows superexponentially with the number of nodes, so an exhaustive search of all possible structures to find the one best supported by the data is not feasible, even for a relatively small number of nodes. A number of algorithms exist to find the optimal network without searching exhaustively, such as Markov chain Monte Carlo (MCMC) simulation (Madigan, York, & Allard, 1995). With the MCMC algorithm, optimal networks are constructed from a set of starting conditions. This algorithm is run thousands of times to identify different plausible networks, each time beginning with different starting conditions. These most plausible networks can then be combined to obtain a consensus network. For each of the reconstructions using the MCMC algorithm, the starting point is a null network. Small random changes are made to the network by flipping, adding, or deleting individual edges, ultimately accepting those changes that lead to an overall improvement in the fit of the network to the data. To assess whether a change improves the network model, information measures such as the Bayesian information criterion (BIC) (Schwarz, 1978) are employed, which reduces overfitting by imposing a cost on the addition of new parameters. This is equivalent to imposing a lower prior probability *P*(*M*) on models with larger numbers of parameters.

Although edges in Bayesian networks are directed, we cannot in general infer causal relationships from the structure directly, just as I discussed earlier in relation to the causal inference test. For a network with three nodes, X_1, X_2, and X_3, there are multiple groups of structures that are mathematically equivalent. For example, the three models, M1: $X_1 \rightarrow X_2$, $X_2 \rightarrow X_3$; M2: $X_2 \rightarrow X_1$, $X_2 \rightarrow X_3$; and M3: $X_2 \rightarrow X_1$, $X_3 \rightarrow X_2$, are all Markov equivalent, meaning that they all encode for the same conditional independence relationship: $X_1 \perp X_3 | X_2$, X_1 and X_3 are independent conditional on X_2. In addition, these models are mathematically equivalent:

$$\begin{aligned} p(X) &= p(\mathrm{M1}|D) = p(X_2|X_1)p(X_1)p(X_3|X_2) \\ &= p(M2|D) = p(X_1|X_2)p(X_2)p(X_3|X_2) \\ &= p(M3|D) = p(X_2|X_3)p(X_3)p(X_1|X_2) \end{aligned}$$

Thus, from correlation data alone we cannot infer whether X_1 is causal for X_2 or vice versa from these types of structures. However, V-shape structures (eg, Mv: $X_1 \rightarrow X_2$, $X_3 \rightarrow X_2$) have no Markov equivalent structure. In such cases it is not possible based on correlation data alone to infer causal relationships. Because there are more parameters to estimate in the Mv model than in the M1, M2, or M3 models, there is a large penalty in the BIC score for the Mv model. Therefore, in practice a large sample size is needed to differentiate the Mv model from the M1, M2, or M3 models.

Integrating Genetic Data as a Structure Before Enhancing Causal Inference in the Bayesian Network Reconstruction Process

In general, Bayesian networks can only be solved to Markov equivalent structures, so that it is often not possible to determine the causal direction of a link between two nodes even though Bayesian networks are directed graphs. However, the Bayesian network reconstruction algorithm can take advantage of genetic data to break the symmetry among nodes in the network that lead to Markov equivalent structures, thereby providing a way to infer causal directions in the network in an unambiguous fashion (Zhu et al., 2004). The reconstruction algorithm can be modified to incorporate genetic data as prior evidence that two quantitative traits may be causally related based on a previously described causality test (Zhu et al., 2004). The genetic priors can be constructed from three basic sources. First, gene expression traits associated with DNA variants that are coincident with the gene's physical location (referred to as *cis*-acting expression quantitative trait loci [*cis*-eQTL]) (Doss, Schadt, Drake, & Lusis, 2005) are allowed to be parent nodes of genes with coincident *trans*-eQTLs (the gene in this case does not physically reside at the genetic locus of interest), $p(cis \rightarrow trans) = 1$, but genes with *trans*-eQTLs are not allowed to be parents of genes with *cis*-eQTLs, $p(trans \rightarrow cis) = 0$. Second, after identifying all associations between different genetic loci and expression traits at some reasonable significance threshold, genes from this analysis with *cis*- or *trans*-eQTL can be tested individually for pleiotropic effects at each of their eQTLs to determine whether any other genes in the set are driven by common eQTLs (Jiang & Zeng, 1995; Lum et al., 2006). If such pleiotropic effects are detected, the corresponding gene pair and locus causing the pleiotropic effect can then be used to infer a causal/reactive or independent relationship based on the causality test described. If an independent relationship is inferred, the prior probability that gene *A* is a parent of gene *B* can be scaled as $p(A \rightarrow B) = 1 - \frac{\sum_i p(A \perp B|A,B,l_i)}{\sum_i 1}$, where the sums are taken over all loci used to infer the relationship. If a causal or reactive relationship is inferred, the prior probability is scaled as $p(A \rightarrow B) = \frac{2\sum_i p(A \rightarrow B|A,B,l_i)}{\sum_i p(A \rightarrow B|A,B,l_i) + p(B \rightarrow A|A,B,l_i)}$. Finally, if the causal/reactive relationship between genes *A* and *B* cannot be determined from the first two sources, the complexity of the eQTL signature for each gene can be taken into consideration. Genes with a simpler, albeit stronger eQTL signature (ie, a small number of eQTLs that explain the genetic variance component for the gene, with a significant proportion of the overall variance explained by the genetic effects) can be considered as more likely to be causal compared with genes with more complex and possibly weaker eQTL signatures (ie, a larger number of eQTLs explaining the genetic variance component for the gene, with less of the overall variance explained by the genetic effects). The structure prior in which gene *A* is a parent of gene *B* can then be taken to be $p(A \rightarrow B) = 2\frac{1+n(B)}{2+n(A)+n(B)}$, where $n(A)$ and $n(B)$ are the number of eQTLs at some predetermined significance level for genes *A* and *B*, respectively.

Incorporating Other -Omics Data as Network Priors in the Bayesian Network Reconstruction Process

Just as genetic data can be incorporated as a network prior in the Bayesian network reconstruction algorithm, so can other types of data such as transcription factor binding site (TFBS) data, protein–protein interaction (PPI) data, and protein–small molecular interaction data. PPI data can be used to infer protein complexes to enhance the set of manually curated protein complexes (Guldener et al., 2006). PPI-inferred protein complexes can be combined with manually curated sets, and each protein complex can then be examined for common transcription factor binding sites at the corresponding genes. If some proportion of the genes in a protein complex (eg, half) carry a given TFBS, all genes in the complex can be included in the TFBS gene set as being under the control of the corresponding transcription factor.

Because the scale-free property is a general property of biological networks (ie, most nodes in the network are linked to a small number of nodes whereas a smaller number of nodes are linked to many nodes) (Albert, Jeong, & Barabasi, 2000), inferred and experimentally determined TFBS data can be incorporated into the network reconstruction process by constructing scale-free priors, in a manner similar to the scale-free priors others have constructed to integrate expression and genetic data (Lee, Pe'er, Dudley, Church, & Koller, 2006). Given a transcription factor T, and a set of genes, G, that contain the binding site of T, the TF prior, p_{tf}, can be defined so that it is proportional to the number of expression traits correlated with the TF expression levels, for genes carrying the corresponding TFBS:

$$\log(p_{tf}(T \rightarrow g)) \propto \log\left(\sum_{g_i \in G} p_{\mathrm{qtl}}(T \rightarrow g_i)\delta\right),$$

where $p_{\mathrm{qtl}}(T \rightarrow g)$ is the prior for the QTL and $\delta = \begin{cases} 1, \text{ if } \operatorname{corr}(T, g_i) \geq r_{\mathrm{cutoff}} \\ 0, \text{ if } \operatorname{corr}(T, g_i) < r_{\mathrm{cutoff}} \end{cases}$. The correlation cutoff r_{cutoff} can be determined by permuting the data and then selecting the maximum correlation values in the permuted data sets (corresponding to some predetermined, reasonable false discovery rate). This form of the structure prior favors transcription factors that have a large number of correlated responding genes. From the set of priors computed from the inferred and experimentally determined TFBS set, only nonnegative priors should be used to reconstruct the Bayesian network. For those protein complexes that could not be integrated into the network reconstruction process using scale-free priors, uniform priors were used for pairs of genes in these complexes (ie, $p_{pc}(g_i \rightarrow g_j) = p_{pc}(g_j \rightarrow g_i) = c$).

Small molecule–protein interactions can also be incorporated into the Bayesian network reconstruction process. Chemical reactions reflected in biochemical pathways and the associated catalyzing enzymes can be identified as metabolite–enzyme pairs from existing pathway databases such as KEGG. These relationships can then be stored in an adjacency matrix in which a 1 in a cell represents a direct connection between the metabolite and the enzyme. The shortest distance $d_{m,e}$ from an enzyme e to a metabolite m can then calculated using the repeated matrix multiplication algorithm. The structure prior for the gene expression of an enzyme e affecting the metabolite concentration is related to their shortest distance $d_{m,e}$ as $p(m \rightarrow e) \propto e^{-\lambda d_{m,e}}$. The shorter the distance, the stronger the prior.

Elucidating the Complexity of Human Disease: From the Metabolic to the Psychiatric

We have carried out studies using the modeling described in detail for human and mouse populations segregating a number of different diseases such as obesity, diabetes, and heart disease. For example, in a segregating mouse population in which an extensive suite of disease traits associated with metabolic syndrome was manifested, including obesity, diabetes, and atherosclerosis (Chen et al., 2008), we carried out the type of network analysis discussed earlier using genetic data typed in all animals and gene expression data generated from the liver and adipose tissues of all animals in the population. With this approach we found that of the many functional units (subnetworks) identified in the networks that reflected core biological processes specific to the liver and adipose tissues, only a handful were strongly causally associated with the metabolic syndrome traits. One module (referred to here as the inflammatome module) in particular stood out not only because it was conserved across the liver and adipose tissues, between the sexes, and between species (Emilsson et al., 2008), but because it was supported as strongly causal for nearly all of the metabolic traits scored in the cross (fat mass, weight, plasma glucose, insulin, and lipid levels, and aortic lesions) (Chen et al., 2008). Again, the causal relationship between the inflammatome module and the disease traits was established by leveraging changes in DNA in this population that were simultaneously associated with disease and expression traits. The entire subnetwork was shown to be under the control of genomic loci associated with the metabolic traits, whereas the predictive network modeling strongly indicated that the module was causal for the disease traits, and was not simply reacting to or acting independently of these traits.

Of the more than 100 genes supported in the inflammatome module as causal for metabolic disease traits such as obesity and diabetes, many genes such as *Zfp90*, *Alox5*, *C3ar1*, and *Tgfbr2* had been previously identified and validated as causal for metabolic traits (Mehrabian et al., 2005; Schadt et al., 2005). In addition, three other genes were selected for validation because they were independently supported as causal for metabolic traits in other studies (*Lpl* and *Lactb*) or because they were supported as causal for such a wide variety of metabolic traits (*Ppm1l*) (Chen et al., 2008). Interestingly, the degree of connectivity in this causal metabolic subnetwork was extreme. Perturbations to genes in this module that were previously validated as causal for the metabolic traits caused expression changes in many other genes validated as causal for metabolic traits. For example, overexpression of *Zfp90* in mouse not only generated an expression response that was significantly overlapping with the causal metabolic module, it caused changes in other genes such as *Pparg* that are known to have an impact on metabolic traits (Chen et al., 2008).

This same type of procedure also can be applied to schizophrenia. By integrating the genetic data for schizophrenia, gene expression data from brain regions relevant to schizophrenia, and clinical data relating to this disease, we can generalize the type of network depicted in Fig. 10.5 to provide a richer context within which to examine the causes of schizophrenia. The network depicted in Fig. 10.6 was constructed by integrating the DNA variation information for schizophrenia, brain gene expression data, expression quantitative trait loci from these expression data (our perturbation source), along with protein—protein interaction data and protein—DNA binding data, as described previously. From this probabilistic causal network framework, we can interpret genetic findings in a more informed context, identify master regulators of the network, and better understand the flow of information among biological processes that are defined by this network.

APPLICATION OF PREDICTIVE NETWORK MODELS TO HIGH-THROUGHPUT SCREENING

Reconstructing reliable network models of disease positions you to simultaneously consider the ensemble of data for a given disease to make more informed decisions on putative disease drivers, biomarkers, and therapeutic intervention points. Network models can be directly queried to identify key drivers of disease or assess how a set of perturbations (eg, drug, environmental factor, genetic variant) may affect molecular states associated with disease, which in turn may help provide direct links to therapeutics that may positively affect disease states or maintain wellness. More generally, we can leverage network models for a neurological disease to construct molecular assays for high-throughput screening (Fig. 10.7). The effect of any given perturbagen, whether a small molecule compound, natural product, RNA interference—based construct, and so on, on a specific network of interest can be assayed directly in cell-based systems (such as human induced pluripotent cell—derived neurons derived from schizophrenia patients), which more accurately reflect the states of networks underlying disease. Complementing these types of network-based screens that use molecular network state as a readout are cellular phenotyping assays that also aid in linking molecular states of disease to pathophysiological states. Screening carried out in this way can lead to the rapid identification of compounds that affect disease networks in favorable ways while simultaneously identifying compounds that hit networks associated with toxicity or other adverse events (Fig. 10.7). In this way, compounds can be identified that target specific subtypes of disease without targeting networks that can lead to toxicity or adverse events.

Network constructs can be used to inform on molecular responses to perturbations with small molecules or other perturbagens, in which the networks enable a direct link between molecular biology and pathophysiology. In a high-throughput screening context, in which transcription or other molecular features are the readout, in addition to cell-based phenotypes, the aim is to identify molecular responses to the perturbagens that are predicted to associate with physiological changes in favorable directions while simultaneously being predicted to have a minimal adverse event profile. Networks can be integrated with molecular screening data to identify those perturbagens from the screen that have similar mechanisms of action, that affect key disease-related processes, or that affect key driver genes of diseases of interest. We and others have previously used network models to inform on perturbagen-induced molecular signatures as a means of predicting and validating the impact a gene or genes had on molecular states and the pathophysiology of disease associated with those states (Chen et al., 2008; Mehrabian et al., 2005; Schadt et al., 2005; Zhu et al., 2008, 2012).

This type of approach has been more generally applied to repurpose existing drugs for novel indications. For example, IBD signatures were derived from surgical specimens and intersected with Connectivity Map data representing transcriptional readouts across a number of cell lines in response to treatment with many hundreds of drugs using a novel

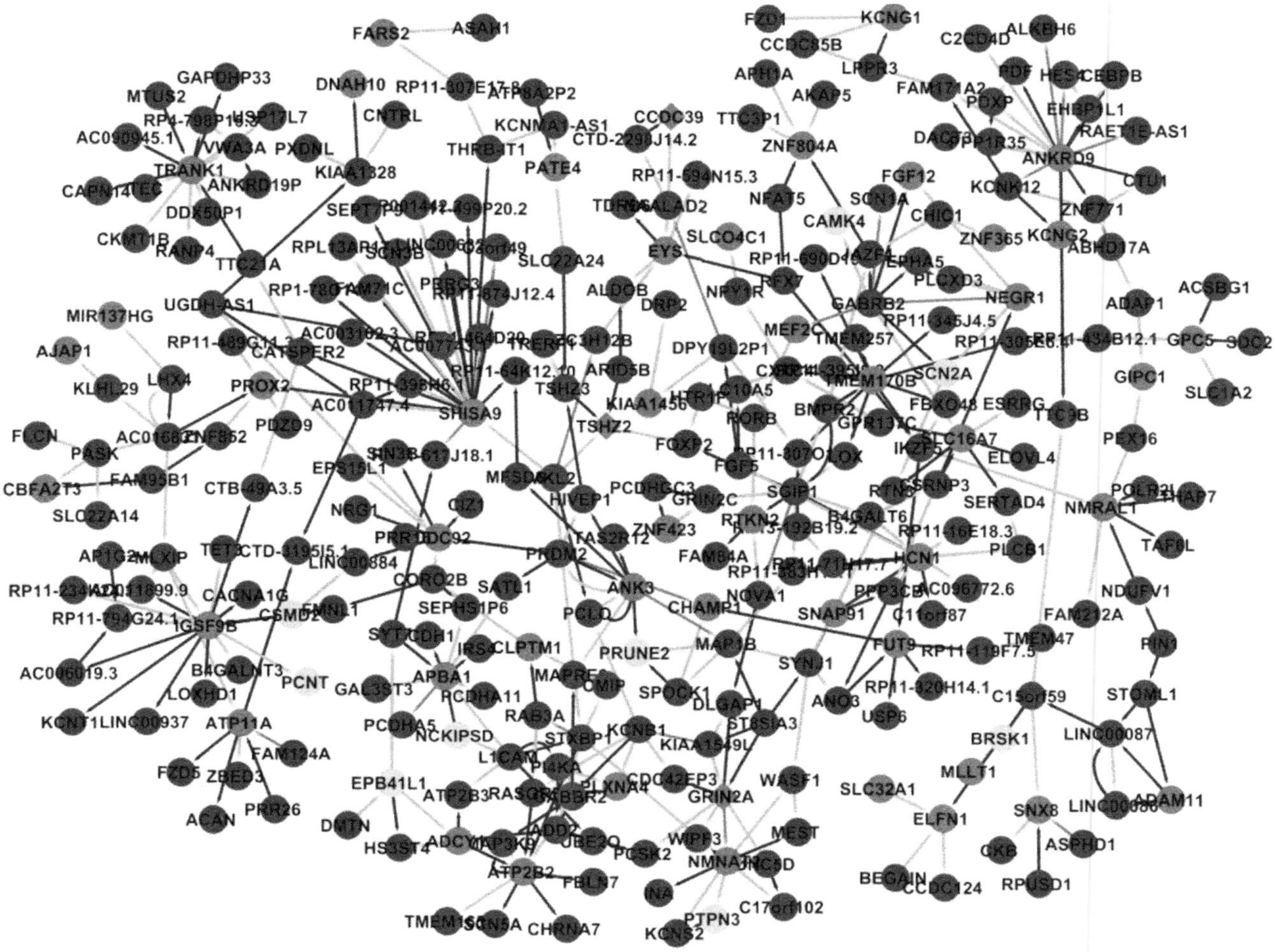

FIGURE 10.6 Probabilistic causal network model of schizophrenia constructed from controls (nondisease samples), bipolar individuals, and individuals with schizophrenia. *Blue* (darkest gray in print versions), *green* (light gray in print versions), and *red edges* (dark gray in print versions) represent links identified in control subjects, individuals with schizophrenia, and bipolar individuals. *Red nodes* (lighter gray in print versions) indicate genes identified from genome-wide association studies in schizophrenia and the *orange nodes* (lightest gray in print versions) represent genes with loss of function variants identified in whole-exome sequencing studies carried out for schizophrenia. This network was derived from three different Bayesian networks constructed using the Bayesian network reconstruction algorithm described in the text: control, bipolar, and schizophrenia networks were constructed from prefrontal cortex samples collected from control subjects and subjects with bipolar disorder and schizophrenia, respectively. This network represents the largest connected subgraph identified by projected schizophrenia genes from genome-wide association studies onto the merged network. Causal links among the genes provide a richer context within which to interpret the genes identified in the genetic studies, potentially implicating genes (*blue nodes* (darkest gray in print versions)) and biological processes in schizophrenia that have not yet been identified in genetic studies.

pattern-matching algorithm (Dudley et al., 2011). From this search, the anticonvulsant drug topiramate was identified and experimentally validated as a novel treatment for IBD (Dudley et al., 2011). Topiramate has primary indications for seizure disorders and no history of efficacious use for IBD or other inflammatory diseases. Using a chemically induced (2,4,6-trinitrobenzenesulfonic acid) rodent model of IBD to evaluate the activity of topiramate administered in the presence of an IBD phenotype, a statistically significant reduction was observed in gross pathophysiological and histopathological measures of severity of the induced IBD phenotype in the population of animals receiving topiramate compared with untreated vehicle controls.

Whereas low-cost sequencing assays have provided an unprecedented amount of data on genetic loci and variants associated with common syndromic disorders, these loci by themselves are insufficient to garner the most informative insights into disease mechanisms upon which new drug discovery efforts may be founded. As already demonstrated in the oncology field, integrative, predictive network models now provide a computational platform that has the potential to improve the success rate of neurological drug discovery significantly if integrated appropriately.

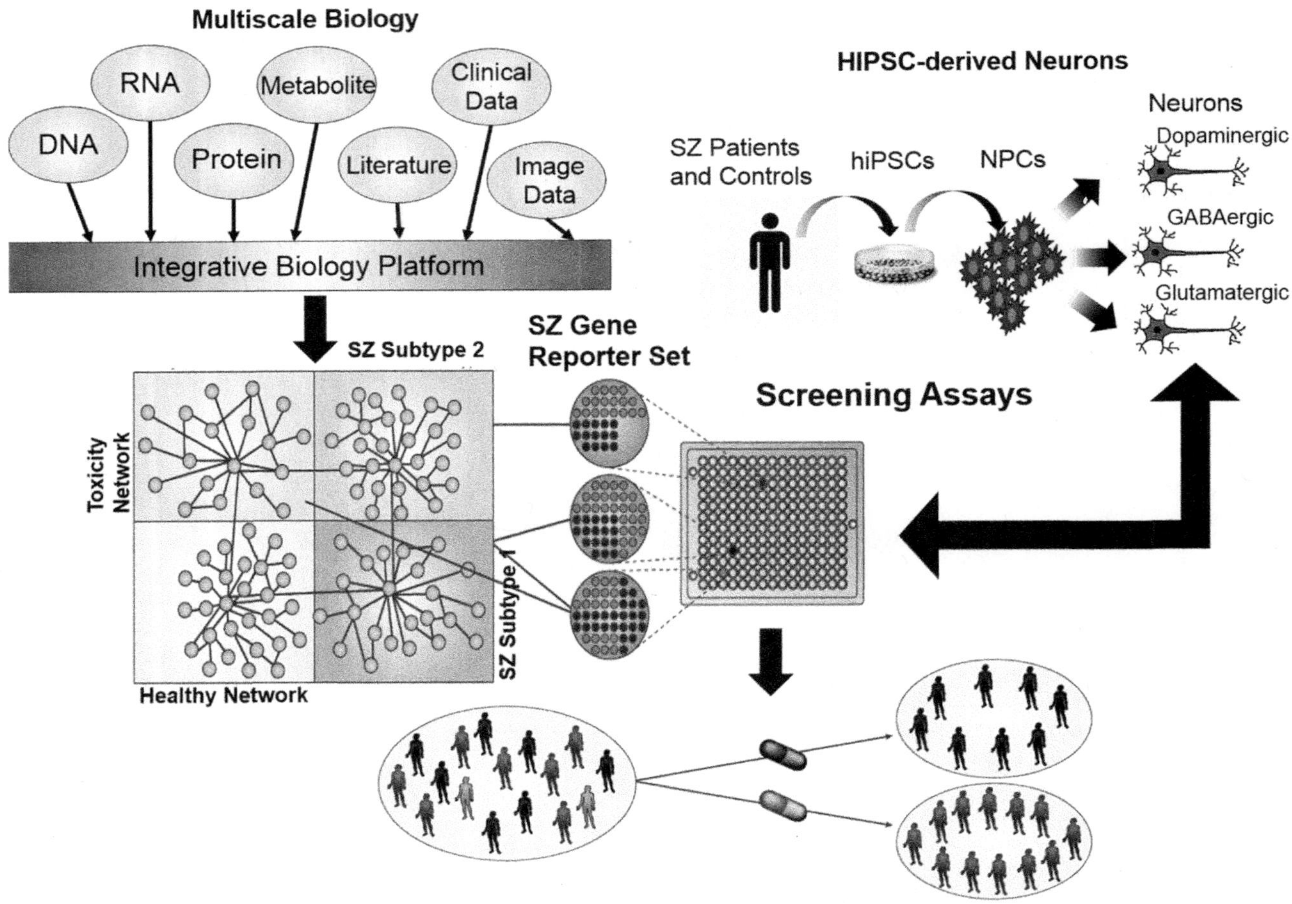

FIGURE 10.7 Integrating predictive network models for disease, wellness, and toxicity with high-throughput screen approaches to drug discovery. First, predictive networks of disease can be reduced to a minimal gene reporter set specific to a disease of interest: in this case, schizophrenia. Second, expression of this gene reporter set can be used in high-throughput screening platforms of cell-based systems to identify perturbagens (eg, small molecules) that favorably affect disease-causing networks but that also do not cause toxicity or other molecular changes that may lead to adverse events. For example, network reporter sets could be used to screen neurons derived from induced pluripotent stem cells isolated from patients with schizophrenia and control subjects to identify compounds capable of ameliorating the disease-associated gene expression signature in patient-derived neurons. *SZ*, schizophrenia; *HIPSC*, human induced pluripotent stem cells; *NPC*, neural progenitor cells.

CONCLUSION AND FUTURE DIRECTIONS

The generation of ever higher dimensional data (DNA sequencing, RNA sequencing, epigenomic profiling, proteomic profiling, metabolomic profiling, and so on) at ever higher scales demands sophisticated mathematical approaches to integrate these data in more holistic ways to uncover patterns of molecular, cellular, and higher-order activities that underlie the biological processes that define physiological states of interest and to uncover causal relationships among molecular and cellular phenotypes and between these phenotypes and clinical traits such as disease or drug response. One of the more successful frameworks for representing large-scale, high-dimensional data is networks. Here I have detailed one particular approach to reconstructing predictive network models of living systems that leverages DNA variation as a systematic variation source and Bayesian network reconstruction algorithms to take a top-down approach to modeling complex systems. Because state-of-the-art therapies in the future will be based on targeting combinations of genes (Schadt, 2009; Schadt et al., 2009), and for such applications it is important not just to infer the direction of each interaction (ie, do you antagonize or activate a given target) but also to be able to predict the degree to which each gene should be knocked down or activated (in a quantitative sense), it is only by generating accurate predictive models of complex phenotypes that we can search most efficiently for such combinations to pursue for experimental proof of concept.

In the future the success of modeling complex systems will depend on constructing networks that are predictive of complex behavior, not merely descriptive. To achieve these more predictive models in complex systems such as humans, we must expand existing networks so that they reflect relationships between cell types and tissues, not just within a single cell type or tissue; capture a greater range of molecular phenotypes to enhance understanding of relevant functional units that define biological processes of interest; and improve modeling capabilities, ideally drawing on the expertise of other fields that have pioneered causality-type reasoning. The complex phenotype-associated molecular networks we can construct today are necessarily based on grossly incomplete sets of data. Even given the ability to assay DNA and RNA variation in whole populations in a comprehensive manner, the information is not complete because rare variations, DNA variations other than SNP/copy number, variations in noncoding RNA levels, and variations in the different isoforms of genes are far from being completely characterized in any sample, let alone in entire populations. Beyond DNA and RNA, measuring all protein-associated traits, interactions between proteins and DNA/RNA, metabolite levels, epigenetic changes, as well as other molecular entities important to the functioning of living systems is not yet possible with existing technologies. Furthermore, the types of high-dimensional data we are able to generate routinely today in populations represent only a snapshot at a single time point, which may enable the identification of the functional units of the system under study and how these units relate to each other, but they do not enable a complete understanding of how the functional units are put together, the mechanistic underpinnings of the complex set of functions carried out by individual cells and by entire organs and whole systems composed of multiple organs.

One future development expected to be most important in this context is the unification of bottom-up and top-down modeling approaches that maximally leverage the strengths of each approach while minimizing the weaknesses. Integrating models derived from bottom-up approaches into top-down approaches is currently hampered by the fact that existing approaches do not typically fully parameterize the network structure in ways that match the intrinsic quantitative nature of top-down approaches. In bottom-up approaches, the structural information detailing how different molecular entities are connected is typically derived from the literature or pathway databases, but such structural information is only qualitative and fails to define quantitatively how one node responds to another. On the other hand, in existing top-down approaches, unless a tremendous amount of training data are available to cover all of the categories represented in the conditional probability distribution defining how nodes are connected in the network (such as with Bayesian network reconstruction approaches), it is generally not possible to estimate accurately the full set of parameters associated with the reconstructed network structure. Worse, carrying out parameter estimation on a network structure that is not correct can be misleading, given false-positive and false-negative predictions. In cases in which heuristic searches are used to orient the edges in a network structure, the end result is that model parameters have not been fitted accurately if the network itself is not correct. Without proper parameterization of network structures from these conventional systems biology approaches, the networks serve only as descriptive models that are not generally capable of generating in silico predictions.

The limitations of bottom-up and top-down approaches can be addressed by devising bottom-up modeling approaches that deliver structures that can serve as prior information for top-down approaches, thereby providing a direct path for parameterizing bottom-up models in the context of a richer set of -omics data and network architectures, while simultaneously reducing the size of the search space for top-down approaches. Such bottom-up approaches are beginning to emerge (Chang, Shoemaker, & Wang, 2011). By automatically parameterizing large networks given a particular network structure and corresponding interaction functions (eg, activation or repression of gene activity) associated with all node pairs by either leveraging prior information or performing a heuristic search, bottom-up approaches will be capable of generating direct quantitative predictions that are compatible with top-down approaches. Central to the success of this approach is the observation that the complexity of the structure of biological networks leads to robust parameter estimates in a constrained parameter space (Blanchini & Franco, 2011; Wilhelm, Behre, & Schuster, 2004; Wu, Zhang, Yu, & Ouyang, 2009) and that a statistical model's parameters are in fact constrained to a cubic space (eg, the conditional probabilities that represent parameters in our modeling approach are constrained to fall between 0 and 1). This stands in contrast to current bottom-up modeling approaches such as continuous ODE modeling, in which the parameter space is generally unconstrained (infinitely large).

Current systems biology approaches relating to network learning and modeling have exclusively used a top-down (reverse-engineering) approach to learn network structure based on association scores (Carro et al., 2010; Fiedler et al., 2009; Margolin et al., 2006; Stuart, Segal, Koller, & Kim, 2003; Zhu et al., 2008). Association scores are designed to uncover the best correlations between variables. Bayesian networks are among the most popular models for this purpose. In theory, it is known that learning the optimal (global maximum) Bayesian network structure from the data is a nondeterministic polynomial-time—hard problem; furthermore, because many substructures that must be considered during the reconstruction process are from classes of structures that are equivalent (the Markov equivalence issue noted previously), the statistical scores for all of the structures in an equivalence class are equal, so that completely contradictory causal relationships are indistinguishable from each other. The integration of bottom-up and top-down approaches in a more holistic mathematical framework has the potential to address these issues further, potentially enhancing the power to uncover true causal relationships.

REFERENCES

Albert, R., Jeong, H., & Barabasi, A. L. (2000). Error and attack tolerance of complex networks. *Nature, 406*, 378–382.

Albert, R., & Thakar, J. (2014). Boolean modeling: a logic-based dynamic approach for understanding signaling and regulatory networks and for making useful predictions. *Wiley Interdisciplinary Reviews: Systems Biology and Medicine, 6*, 353–369.

Azeloglu, E. U., & Iyengar, R. (2015). Good practices for building dynamical models in systems biology. *Science Signaling, 8*, fs8. http://dx.doi.org/10.1126/scisignal.aab0880.

Barabasi, A. L., & Oltvai, Z. N. (2004). Network biology: understanding the cell's functional organization. *Nature Reviews Genetics, 5*, 101–113.

Blanchini, F., & Franco, E. (2011). Structurally robust biological networks. *BMC Systems Biology, 5*, 74. http://dx.doi.org/10.1186/1752-0509-5-74.

Boyden, E. S., Zhang, F., Bamberg, E., Nagel, G., & Deisseroth, K. (2005). Millisecond-timescale, genetically targeted optical control of neural activity. *Nature Neuroscience, 8*, 1263–1268. http://dx.doi.org/10.1038/nn1525.

Califano, A., Butte, A. J., Friend, S., Ideker, T., & Schadt, E. (2012). Leveraging models of cell regulation and GWAS data in integrative network-based association studies. *Nature Genetics, 44*, 841–847. http://dx.doi.org/10.1038/ng.2355.

Carro, M. S., et al. (2010). The transcriptional network for mesenchymal transformation of brain tumours. *Nature, 463*, 318–325. http://dx.doi.org/10.1038/nature08712.

Chang, R., Karr, J. R., & Schadt, E. E. (2015). Causal inference in biology networks with integrated belief propagation. *Pacific Symposium on Biocomputing. Pacific Symposium on Biocomputing, 20*, 359–370.

Chang, R., Shoemaker, R., & Wang, W. (2011). Systematic search for recipes to generate induced pluripotent stem cells. *PLoS Computational Biology, 7*, e1002300. http://dx.doi.org/10.1371/journal.pcbi.1002300.

Chen, L. S., Emmert-Streib, F., & Storey, J. D. (2007). Harnessing naturally randomized transcription to infer regulatory relationships among genes. *Genome Biology, 8*, R219. http://dx.doi.org/10.1186/gb-2007-8-10-r219.

Chen, Y., et al. (2008). Variations in DNA elucidate molecular networks that cause disease. *Nature, 452*, 429–435. http://dx.doi.org/10.1038/nature06757.

Davey Smith, G., & Ebrahim, S. (2003). 'Mendelian randomization': can genetic epidemiology contribute to understanding environmental determinants of disease? *International Journal of Epidemiology, 32*, 1–22.

Didelez, V., & Sheehan, N. (2007). Mendelian randomization as an instrumental variable approach to causal inference. *Statistical Methods in Medical Research, 16*, 309–330. http://dx.doi.org/10.1177/0962280206077743.

Doss, S., Schadt, E. E., Drake, T. A., & Lusis, A. J. (2005). Cis-acting expression quantitative trait loci in mice. *Genome Research, 15*, 681–691. http://dx.doi.org/10.1101/gr.3216905.

Dudley, J. T., et al. (2011). Computational repositioning of the anticonvulsant topiramate for inflammatory bowel disease. *Science Translational Medicine, 3*, 96ra76. http://dx.doi.org/10.1126/scitranslmed.3002648.

Ecker, J. R., et al. (2012). Genomics: ENCODE explained. *Nature, 489*, 52–55. http://dx.doi.org/10.1038/489052a.

Elia, J., et al. (2012). Genome-wide copy number variation study associates metabotropic glutamate receptor gene networks with attention deficit hyperactivity disorder. *Nature Genetics, 44*, 78–84. http://dx.doi.org/10.1038/ng.1013.

Emilsson, V., et al. (2008). Genetics of gene expression and its effect on disease. *Nature, 452*, 423–428. http://dx.doi.org/10.1038/nature06758.

Erskine, D. J., Edelstein, J., Harbeck, D., & Lloyd, J. (2005). Externally dispersed interferometry for planetary studies. *Proceedings of SPIE, 5905*, 249–260.

Feller, W. (1967). *An introduction to probability theory and its applications* (3rd ed.). Wiley.

Fiedler, D., et al. (2009). Functional organization of the *S. cerevisiae* phosphorylation network. *Cell, 136*, 952–963. http://dx.doi.org/10.1016/j.cell.2008.12.039.

Glessner, J. T., et al. (2009). Autism genome-wide copy number variation reveals ubiquitin and neuronal genes. *Nature, 459*, 569–573. http://dx.doi.org/10.1038/nature07953.

Guldener, U., et al. (2006). MPact: the MIPS protein interaction resource on yeast. *Nucleic Acids Research, 34*, D436–D441.

Hawrylycz, M. J., et al. (2012). An anatomically comprehensive atlas of the adult human brain transcriptome. *Nature, 489*, 391–399. http://dx.doi.org/10.1038/nature11405.

Jansen, R., et al. (2003). A Bayesian networks approach for predicting protein-protein interactions from genomic data. *Science, 302*, 449–453.

Jiang, C., & Zeng, Z. B. (1995). Multiple trait analysis of genetic mapping for quantitative trait loci. *Genetics, 140*, 1111–1127.

Kanehisa, M. (2002). The KEGG database. *Novartis Foundation Symposium, 247*, 91–101. discussion 101–103, 119–128, 244–252.

Kulp, D. C., & Jagalur, M. (2006). Causal inference of regulator-target pairs by gene mapping of expression phenotypes. *BMC Genomics, 7*, 125.

Lamb, J. R., et al. (2011). Predictive genes in adjacent normal tissue are preferentially altered by sCNV during tumorigenesis in liver cancer and may rate limiting. *PloS One, 6*, e20090. http://dx.doi.org/10.1371/journal.pone.0020090.

Lawlor, D. A., Harbord, R. M., Sterne, J. A., Timpson, N., & Davey Smith, G. (2008). Mendelian randomization: using genes as instruments for making causal inferences in epidemiology. *Statistics in Medicine, 27*, 1133–1163. http://dx.doi.org/10.1002/sim.3034.

Lee, I., Date, S. V., Adai, A. T., & Marcotte, E. M. (2004). A probabilistic functional network of yeast genes. *Science, 306*, 1555–1558.

Lee, S. I., Pe'er, D., Dudley, A. M., Church, G. M., & Koller, D. (2006). Identifying regulatory mechanisms using individual variation reveals key role for chromatin modification. *Proceedings of the National Academy of Sciences of the United States of America, 103*, 14062–14067. http://dx.doi.org/10.1073/pnas.0601852103.

Lum, P. Y., et al. (2006). Elucidating the murine brain transcriptional network in a segregating mouse population to identify core functional modules for obesity and diabetes. *Journal of Neurochemistry, 97*(Suppl. 1), 50–62. http://dx.doi.org/10.1111/j.1471-4159.2006.03661.x.

Madigan, D., York, J., & Allard, D. (1995). Bayesian graphical models for discrete data. *International Statistical Review, 63*, 215–232.

Margolin, A. A., et al. (2006). Reverse engineering cellular networks. *Nature Protocols, 1*, 662–671. http://dx.doi.org/10.1038/nprot.2006.106.

Mehrabian, M., et al. (2005). Integrating genotypic and expression data in a segregating mouse population to identify 5-lipoxygenase as a susceptibility gene for obesity and bone traits. *Nature Genetics, 37*, 1224–1233. http://dx.doi.org/10.1038/ng1619.

Millstein, J., Zhang, B., Zhu, J., & Schadt, E. E. (2009). Disentangling molecular relationships with a causal inference test. *BMC Genetics, 10*, 23. http://dx.doi.org/10.1186/1471-2156-10-23.

Morris, M. K., Saez-Rodriguez, J., Sorger, P. K., & Lauffenburger, D. A. (2010). Logic-based models for the analysis of cell signaling networks. *Biochemistry, 49*, 3216–3224. http://dx.doi.org/10.1021/bi902202q.

Mumford, J. A., et al. (2010). Detecting network modules in fMRI time series: a weighted network analysis approach. *NeuroImage, 52*, 1465–1476. http://dx.doi.org/10.1016/j.neuroimage.2010.05.047.

Neale, B. M., et al. (2012). Patterns and rates of exonic de novo mutations in autism spectrum disorders. *Nature, 485*, 242–245. http://dx.doi.org/10.1038/nature11011.

Nitsch, D., et al. (2006). Limits to causal inference based on Mendelian randomization: a comparison with randomized controlled trials. *American Journal of Epidemiology, 163*, 397–403. http://dx.doi.org/10.1093/aje/kwj062.

Pearl, J. (1988). *Probabalistic reasoning in intelligent systems: Networks of plausible inference*. Morgan Kaufmann Publishers.

Pelletier, S., Gingras, S., & Green, D. R. (2015). Mouse genome engineering via CRISPR-Cas9 for study of immune function. *Immunity, 42*, 18–27. http://dx.doi.org/10.1016/j.immuni.2015.01.004.

Pinto, D., et al. (2010). Functional impact of global rare copy number variation in autism spectrum disorders. *Nature, 466*, 368–372. http://dx.doi.org/10.1038/nature09146.

Purcell, S. M., et al. (2009). Common polygenic variation contributes to risk of schizophrenia and bipolar disorder. *Nature, 460*, 748–752.

Ripke, S., et al. (2011). Genome-wide association study identifies five new schizophrenia loci. *Nature Genetics, 43*, 969–976. http://dx.doi.org/10.1038/ng.940.

Schadt, E. E. (2009). Molecular networks as sensors and drivers of common human diseases. *Nature, 461*, 218–223. http://dx.doi.org/10.1038/nature08454.

Schadt, E. E., Buchanan, S., Brennand, K. J., & Merchant, K. M. (2014). Evolving toward a human-cell based and multiscale approach to drug discovery for CNS disorders. *Frontiers in Pharmacology, 5*, 252. http://dx.doi.org/10.3389/fphar.2014.00252.

Schadt, E. E., et al. (2003). Genetics of gene expression surveyed in maize, mouse and man. *Nature, 422*, 297–302. http://dx.doi.org/10.1038/nature01434.

Schadt, E. E., et al. (2005). An integrative genomics approach to infer causal associations between gene expression and disease. *Nature Genetics, 37*, 710–717. http://dx.doi.org/10.1038/ng1589.

Schadt, E. E., et al. (2008). Mapping the genetic architecture of gene expression in human liver. *PLoS Biology, 6*, e107. http://dx.doi.org/10.1371/journal.pbio.0060107.

Schadt, E. E., Friend, S. H., & Shaywitz, D. A. (2009). A network view of disease and compound screening. *Nature Reviews. Drug Discovery, 8*, 286–295. http://dx.doi.org/10.1038/nrd2826.

Schizophrenia Working Group of the Psychiatric Genomics, C. (2014). Biological insights from 108 schizophrenia-associated genetic loci. *Nature, 511*, 421–427. http://dx.doi.org/10.1038/nature13595.

Schwarz, G. (1978). Estimating the dimension of a model. *Annals of Statistics, 6*, 461–464.

Shirasaki, D. I., et al. (2012). Network organization of the huntingtin proteomic interactome in mammalian brain. *Neuron, 75*, 41–57. http://dx.doi.org/10.1016/j.neuron.2012.05.024.

Sieberts, S. K., & Schadt, E. E. (2007). Moving toward a system genetics view of disease. *Mammalian Genome : Official Journal of the International Mammalian Genome Society, 18*, 389–401. http://dx.doi.org/10.1007/s00335-007-9040-6.

Smith, H. H., Brookhaven National Laboratory, & U.S. Atomic Energy Commission. (1972). *Evolution of genetic systems* (pp. 366–370). Gordon and Breach.

Stranger, B. E., Stahl, E. A., & Raj, T. (2011). Progress and promise of genome-wide association studies for human complex trait genetics. *Genetics, 187*, 367–383. http://dx.doi.org/10.1534/genetics.110.120907.

Stuart, J. M., Segal, E., Koller, D., & Kim, S. K. (2003). A gene-coexpression network for global discovery of conserved genetic modules. *Science, 302*, 249–255.

Su, W. L., Kleinhanz, R. R., & Schadt, E. E. (2011). Characterizing the role of miRNAs within gene regulatory networks using integrative genomics techniques. *Molecular Systems Biology, 7*, 490. http://dx.doi.org/10.1038/msb.2011.23.

Tu, Z., et al. (2012). Integrative analysis of a cross-loci regulation network identifies App as a gene regulating insulin secretion from pancreatic islets. *PLoS Genetics, 8*, e1003107. http://dx.doi.org/10.1371/journal.pgen.1003107.

Voineagu, I., et al. (2011). Transcriptomic analysis of autistic brain reveals convergent molecular pathology. *Nature, 474*, 380–384. http://dx.doi.org/10.1038/nature10110.

Wang, I. M., et al. (2012). Systems analysis of eleven rodent disease models reveals an inflammatome signature and key drivers. *Molecular Systems Biology, 8*, 594. http://dx.doi.org/10.1038/msb.2012.24.

Wang, K., et al. (2009). Common genetic variants on 5p14.1 associate with autism spectrum disorders. *Nature, 459*, 528–533. http://dx.doi.org/10.1038/nature07999.

Wang, S. S., et al. (2007). Identification of pathways for atherosclerosis in mice: integration of quantitative trait locus analysis and global gene expression data. *Circulation Research, 101*, e11–30. http://dx.doi.org/10.1161/CIRCRESAHA.107.152975.

Wilhelm, T., Behre, J., & Schuster, S. (2004). Analysis of structural robustness of metabolic networks. *Systems Biology, 1*, 114–120.

Wu, Y., Zhang, X., Yu, J., & Ouyang, Q. (2009). Identification of a topological characteristic responsible for the biological robustness of regulatory networks. *PLoS Computational Biology, 5*, e1000442. http://dx.doi.org/10.1371/journal.pcbi.1000442.

Yang, X., et al. (2009). Validation of candidate causal genes for obesity that affect shared metabolic pathways and networks. *Nature Genetics, 41*, 415–423. http://dx.doi.org/10.1038/ng.325.

Zhang, B., et al. (2013). Integrated systems approach identifies genetic nodes and networks in late-onset Alzheimer's disease. *Cell, 153*, 707–720. http://dx.doi.org/10.1016/j.cell.2013.03.030.

Zhong, H., et al. (2010). Liver and adipose expression associated SNPs are enriched for association to type 2 diabetes. *PLoS Genetics, 6*, e1000932. http://dx.doi.org/10.1371/journal.pgen.1000932.

Zhong, H., Yang, X., Kaplan, L. M., Molony, C., & Schadt, E. E. (2010). Integrating pathway analysis and genetics of gene expression for genome-wide association studies. *American Journal of Human Genetics, 86*, 581–591. http://dx.doi.org/10.1016/j.ajhg.2010.02.020.

Zhu, J., et al. (2004). An integrative genomics approach to the reconstruction of gene networks in segregating populations. *Cytogenetic and Genome Research, 105*, 363–374. http://dx.doi.org/10.1159/000078209.

Zhu, J., et al. (2007). Increasing the power to detect causal associations by combining genotypic and expression data in segregating populations. *PLoS Computational Biology, 3*, e69. http://dx.doi.org/10.1371/journal.pcbi.0030069.

Zhu, J., et al. (2008). Integrating large-scale functional genomic data to dissect the complexity of yeast regulatory networks. *Nature Genetics, 40*, 854–861. http://dx.doi.org/10.1038/ng.167.

Zhu, J., et al. (2010). Characterizing dynamic changes in the human blood transcriptional network. *PLoS Computational Biology, 6*, e1000671. http://dx.doi.org/10.1371/journal.pcbi.1000671.

Zhu, J., et al. (2012). Stitching together multiple data dimensions reveals interacting metabolomic and transcriptomic networks that modulate cell regulation. *PLoS Biology, 10*, e1001301. http://dx.doi.org/10.1371/journal.pbio.1001301.

Chapter 11

Gene Networks in Neuropsychiatric Disease

Neelroop N. Parikshak and Daniel H. Geschwind
University of California Los Angeles, Los Angeles, CA, United States

INTRODUCTION

Genetic studies of neuropsychiatric disorders have led to a new appreciation of their remarkable etiological heterogeneity (Gratten, Wray, Keller, & Visscher, 2014). However, because neuropsychiatric diagnoses are defined by shared alterations in behavior and cognition, there is hope that disparate genetic alterations will affect common molecular pathways that mediate disease pathophysiology. For example, despite heterogeneity across the autism spectrum, individuals with autism spectrum disorder (ASD) share core deficits in social communication and mental inflexibility, and have similar onsets and developmental trajectories of their behavioral changes (Berg & Geschwind, 2012). Moreover, despite heterogeneity across neuropsychiatric disorders as a whole, there is considerable sharing of both rare (Fromer et al., 2014; Iossifov et al., 2014) and common (Cross-Disorder Group of the Psychiatric Genomics Consortium et al., 2013) genetic variation both within and across disorders, as well as evidence for convergence at the molecular pathway level (Gilman et al., 2012; Network and Consortium, 2015; Parikshak et al., 2013). Many advances are still necessary to understand how a large number of risk variants and distinct molecular pathways converge to cause disorders sharing many phenotypic features. A major approach enabling such investigation is the application of large-scale assays of molecular function (Geschwind & Konopka, 2009; Maze et al., 2014) followed by network-based methods to organize, analyze, and understand biological functions affected in neuropsychiatric disease (Barabási & Oltvai, 2004; Carter, Hofree, & Ideker, 2013).

In this chapter, we discuss how high-throughput molecular biology has connected disparate genetic findings to neurobiological function (Geschwind & Konopka, 2009; Maze et al., 2014; Nord, Pattabiraman, Visel, & Rubenstein, 2015; Parikshak, Gandal, & Geschwind, 2015). Specifically, we describe how gene networks can organize large-scale functional molecular data and how these networks in turn provide insight into neurobiological function (Barabási & Oltvai, 2004; Beyer, Bandyopadhyay, & Ideker, 2007). We discuss how methods have used transcriptomic and other high-throughput molecular measurements to place genes in a network context, providing a rubric for creating a more synthesized understanding of the brain in health and disease. Rather than concentrating on methodological issues revolving around network construction and details of the technical differences (eg, between weighted and binary networks, or between Bayesian or non-Bayesian approaches) (Horvath, 2011; Parikshak et al., 2015), we focus on issues relevant to the neurobiological interpretation of gene networks and their relationship to disease biology.

Our goal is to convey how gene network studies identify insights about neurobiological function and disease. To this end, we start by covering important considerations for evaluating large-scale functional genomic methods, including the major challenges faced in neurobiology. We then describe the general principles of network analysis, discuss different network approaches used to understand neuropsychiatric disorders, compare representative methods, and identify overarching themes and future directions.

RNA, PROTEIN, AND EPIGENETIC MOLECULAR LEVELS IN NEUROBIOLOGY

The fundamental question for molecular biology investigations of the nervous system is what to measure. Currently, many high-throughput methods exist for evaluating multiple levels of molecular organization (Consortium, 2012; Geschwind &

Genomics, Circuits, and Pathways in Clinical Neuropsychiatry. http://dx.doi.org/10.1016/B978-0-12-800105-9.00011-1

Konopka, 2009), including the profiling of RNA (Wang, Gerstein, & Snyder, 2009), protein (Bayés & Grant, 2009), and epigenetic changes (Consortium et al., 2015). Methods have also been rapidly developed for measuring molecular interactions, including protein—protein interactions (PPIs) (Beyer et al., 2007), protein binding to DNA (eg, the profiling of histone modifications or transcription factor binding sites) (Consortium, 2012), protein binding to RNA (Zhang & Darnell, 2011), and chromosome—chromosome interactions (Lieberman-Aiden et al., 2009). These methods measure different levels of molecular changes or interactions within or between levels that provide distinct and overlapping insights about the functional consequence of a mutation.

Notably, epigenetic methods are just reaching maturity (Consortium et al., 2015; Maze et al., 2014), whereas RNA and proteins have been measured in a high-throughput manner for over a decade. We therefore focus discussion of the considerations taken when choosing the molecular level for study between the RNA and protein levels. Some changes in RNA levels reflect protein levels accurately, but many changes at the protein level will be missed by evaluating RNA alone (Battle et al., 2015; Schwanhäusser et al., 2011). In addition, many alterations in regulatory RNA (eg, microRNAs and other regulatory noncoding RNAs) and the competitive interactions among regulatory RNAs cannot be detected at the protein level (Salmena, Poliseno, Tay, Kats, & Pandolfi, 2011). It would therefore be ideal to measure both levels of organization because they are complementary, but many studies in neurobiology must make a choice, particularly given cost constraints and tissue availability (Geschwind & Konopka, 2009).

Major considerations in studies include the cost and quality of the technology. For example, nucleotide-based assays are relatively inexpensive owing to hybridization-based technologies such a microarrays and sequencing. Moreover, for any given method, high coverage (the extent to which all the possible entities are measured) and high resolution (the precision to which the entity is measured) are important for reducing biases (Beyer et al., 2007). For example, methods that measure genome-wide levels of molecules may require unrealistic labor or cost to measure the entire distribution of molecular levels accurately, leading to incomplete measurement of low-expressed RNAs or proteins. Furthermore, transcripts and proteins have multiple isoforms, and some molecular assays may not be able to resolve these important structural changes.

Because of these considerations, most studies to date have relied on genome-wide RNA measurement, as these methods are less inexpensive, have better coverage and resolution, and are more unbiased than most widely available protein quantification methods. There are several important studies that evaluate large number of proteins (proteomic studies) in brain (Bayés & Grant, 2009; Bayés et al., 2010), but most development-, aging-, and disease-relevant studies are at the RNA level. These studies measure hundreds to thousands of RNA transcripts, together called the transcriptome, with cDNA microarray technology (Geschwind & Gregg, 2002; Oldham & Geschwind, 2006) and with massively parallel high-throughput sequencing of cDNA libraries from RNA (RNA-seq) (Wang et al., 2009). In general, how a given gene is expressed at the RNA level is an important intermediate step in bridging the genome to function relationship. RNA transcript levels can serve as a valuable marker for understanding how genes and environment affect cell types, tissue composition, and ultimately neural circuits and phenotypes (Carter, Brechbühler, Griffin, & Bond, 2004; Geschwind & Konopka, 2009; Parikshak et al., 2015). Because of the preponderance of transcriptomics in neurobiology, we focus on these studies but also discuss protein-level studies where applicable.

THE CHALLENGE OF SPATIAL AND TEMPORAL HETEROGENEITY IN THE CENTRAL NERVOUS SYSTEM

Understanding transcriptomic changes in the context of spatial and temporal complexity in human central nervous system disorders necessitates a thorough knowledge of the molecules that distinguish different brain regions, circuits, and cells in normal brain development and aging. Extensive work by the Allen Brain Institute and others has mapped the molecular heterogeneity across space and time in mice, nonhuman primates, and humans using genetically targeted cell identification, microdissection coupled with transcriptomics, large-scale in situ hybridization, and other complementary methods (Bernard et al., 2012; Gong et al., 2003; Lein et al., 2006; Miller et al., 2014; Sunkin et al., 2013; Zeng et al., 2012).

Comparison of the transcriptome across neurodevelopmental stages reveals striking changes in gene expression and alternative splicing of most genes in the human genome (Colantuoni et al., 2011; Johnson et al., 2009; Kang et al., 2011), with a key inflection point at birth that is marked by large-scale changes in gene expression programs. Importantly, this dynamic regulation is not restricted to prenatal and childhood periods. Relative to rodent or primate brains, the human brain exhibits prolonged development, with biological processes such as synaptic pruning and stabilization extending into the third decade of life (Changeux & Danchin, 1976; Geschwind & Rakic, 2013).

Research shows how the transcriptome changes at an even higher spatial and temporal resolution during the development of three major populations of pyramidal neurons in mouse cortex (callosal projection neurons, corticothalamic projection neurons, and subcerebral projection neurons) (Molyneaux et al., 2015), which highlights that future work in constructing transcriptomic atlases must delve into greater spatial and temporal specificity. It is apparent that cell markers at one stage of development may not apply to another stage (Bystron, Blakemore, & Rakic, 2008; Molyneaux, Arlotta, Menezes, & Macklis, 2007), and some major cell-type features such as neurotransmitter phenotype may require maintained expression of specific transcription factors (Deneris & Hobert, 2014).

GENE NETWORKS PROVIDE A FRAMEWORK FOR NEUROBIOLOGICAL INTERPRETATION

Because of the massive changes in gene expression that occur over development and between cell types and different cell states, network approaches have an increasingly important role in organizing transcriptomic data (Hawrylycz et al., 2012; Miller et al., 2014; Oldham et al., 2008; Parikshak et al., 2013; Willsey et al., 2013; Winden et al., 2009). These methods allow investigators to move beyond long gene lists ordered alphabetically, by genomic location, or by significance of differential expression changes, and allow functionally related genes to be subdivided and associated with pathways, neuroanatomical regions, critical time points, and other measurements that are relevant to neurodevelopmental disorders (Geschwind & Konopka, 2009).

The architecture of gene expression networks was initially investigated in yeast and across species in an evolutionary context (Barabási & Oltvai, 2004; Stuart, 2003). Multiple metabolic and protein interaction network studies demonstrated that biological network architecture can be approximated by underlying scale-free topology, which was applied to characterize gene coexpression networks (Barabasi, 2009; Horvath et al., 2006; Zhang & Horvath, 2005). From a practical perspective, network methods reduce the dimensionality of genome-wide RNA expression patterns (Carter et al., 2004; Langfelder & Horvath, 2007) using correlation, mutual information, or other metrics to organize thousands of genes corresponding to millions of relationships into a relatively small collection of modules (also referred to as clusters, cliques, or communities), which in turn can be related to each other and to relevant phenotypes (Langfelder & Horvath, 2007; Oldham et al., 2008; Parikshak et al., 2013). Box 11.1 provides a basic outline of the essential steps in network analysis.

BOX 11.1 Four Components of Gene Network Approaches

Functional genomic data can be modeled as a network in which molecules or genes are nodes and their functional relationships with each other are edges. Gene network analysis can be summarized in five basic steps:

1. Node specification:
   a. Seeded (prior-based): nodes are selected using prior knowledge, eg, disease-associated genes from genome-wide association studies (GWASs)
   b. Unseeded (genome-wide or whole-genome): all usable measurements from the genome are included in an unsupervised analysis
2. Edge specification:
   a. Experimentally observed pairwise statistical relationships (Butte & Kohane, 2000; Carter et al., 2004; Horvath, 2011) evaluating shared patterns of molecular levels across experiments: eg, coexpression
   b. Experimentally observed or literature-curated physical interactions: eg, protein interactions from immunohistochemistry and yeast two-hybrid experiments
   c. Computationally predicted relationships: eg, transcription factor binding based on DNA motifs

   Notably, edges are susceptible to ascertainment biases (Hakes et al., 2008; Lee & Marcotte, 2009) and confounding factors that can induce spurious relationships (Leek et al., 2010).
3. Modules are identified from an adjacency matrix to simplify biological relationships at a higher-order level. Assessing connectivity can identify hubs and enables comparison of changes between health and disease at the module level.
4. Annotation of modules and gene connectivity by:
   a. Relating external measures of gene importance (such as cell type specificity or GWAS signal) with module membership, intramodular connectivity, or whole-network connectivity of genes
   b. Associating module summary or hub gene measurements, such as eigengenes or average expression levels, to biological traits
   c. Assessing preservation or changes in network connectivity for specific genes or modules between health and disease
   d. Integrating data at the edge level or module level across biological levels, such as different cell types or brain regions, or different regulatory levels, such as gene expression and chromatin immunoprecipitation signal from binding proteins
   e. Addressing the crucial issue of reproducibility by validating network observations in independent data or experiments (see Table 11.1 for examples)

Adapted from Parikshak, N. N., Gandal, M. J., & Geschwind, D. H. (2015). Systems biology and gene networks in neurodevelopmental and neurodegenerative disorders. Nature Reviews Genetics, 16*(8), 441–458.*

Coexpression network modules represent genes that show similar patterns of expression, which may result from co-occurrence in a particular cell type (Oldham, Horvath, & Geschwind, 2006; Oldham et al., 2008; Winden et al., 2009), response to a stimulus (Tian et al., 2014; Wexler et al., 2011), or co-involvement in any number of cellular processes, such as translation, energy metabolism, or synaptic function (Ben-David & Shifman, 2012; Gaiteri, Ding, French, Tseng, & Sibille, 2013; Kang et al., 2011; Parikshak et al., 2013; Winden et al., 2009). Modules thus identify key levels of organization whose specificity depends on the experimental design. For example, networks derived from whole tissue profiling across brain regions will often define modules that are specific to cell types or subcellular compartments, because these are the biological features varying across samples in such a study design (Johnson et al., 2009; Oldham et al., 2008; Winden et al., 2009). Higher-resolution anatomical profiling, for example, by laser capture microdissection, can yield greater cell type and biological process spatial specificity for modules (Bernard et al., 2012; Hawrylycz et al., 2012; Miller et al., 2014; Miller, Woltjer, Goodenbour, Horvath, Geschwind, 2013).

Because network edges reflect shared biological functions, a gene of unknown function can be annotated based on genes of known function that are highly connected to it, so-called, "guilt by association" (Langfelder & Horvath, 2008; Lui et al., 2014; Oldham et al., 2008). Similarly, if genes within a particular module have been implicated in disease, it suggests that other genes in the module may be associated with disease at a higher probability. Because modules are highly interconnected, knowing the function of some genes in a module may inform the functions of others in the same module. Genes with the highest levels of intramodular connectivity (also known as module membership) are known as hubs and are particularly informative for guilt by association reasoning (Barabási & Oltvai, 2004; Horvath & Dong, 2008).

In addition, multiple studies have demonstrated that the human brain transcriptome has a fundamental coexpression structure that can be reproducibly identified across diverse technologies and different individuals (Hawrylycz et al., 2012; Oldham et al., 2008). This permits comparative network analyses across different disease states (Torkamani, Dean, Schork, & Thomas, 2010; Voineagu et al., 2011), experimental perturbations (Tian et al., 2014; Wexler et al., 2011), or even species (Konopka et al., 2012; Oldham et al., 2006), in which module preservation or gene network position is assessed to identify specific biological processes or genes that are the most different between conditions (Langfelder, Luo, Oldham, & Horvath, 2011). Similarly, how a specific gene's ranking in a module or in the whole network changes in health and disease can be evaluated to identify substantial alterations in the normal coexpression structure of gene networks (Choi, Yu, Yoo, & Kim, 2005; Gaiteri et al., 2013; Geschwind & Konopka, 2009; Hudson, Reverter, & Dalrymple, 2009; Zhang et al., 2013).

We focus here on the interpretation of gene coexpression networks in neurobiology because microarray and RNA-seq experiments cover the spatial and temporal heterogeneity in the brain. PPI networks are also commonly used and offer unique insights. However, although tissue-specific (Bayés et al., 2010; Kim et al., 2014; Uhlén et al., 2015; Wilhelm et al., 2014) and cell compartment−specific (Bayés et al., 2010) maps of the proteome are becoming available, there are no large-scale, neuronal tissue-specific PPI databases. We discuss this issue subsequently, but Table 11.1 provides a comparison between coexpression and PPI networks.

GENE NETWORKS IN NEUROPSYCHIATRIC DISORDERS

Because of the genetic heterogeneity in neuropsychiatric disorders (Gratten et al., 2014), it is particularly useful to evaluate individual genes in the context of their interactions with genetic background and other developmental processes, a task for which network methods are particularly well-suited. Here we focus on ASD and schizophrenia (SCZ), which represent the complexity at hand, and the insights that have been gained in neuropsychiatric disease. We first discuss some examples of both genome-wide and seed-based (Box 11.1) approaches for constructing networks to understand ASD and SCZ. We then cover network studies that use coexpression and describe the use of PPIs and integrative networks. We summarize the network studies on neuropsychiatric disorders that we cover in the main text, as well as several other representative studies, in Table 11.2.

Genome-Wide Approaches

Microarray-based analyses of postmortem brain have identified what appear to be consistent transcriptional alterations in over half of individuals with ASD (Chow et al., 2012; Garbett et al., 2008; Ginsberg, Rubin, Falcone, Ting, & Natowicz, 2012; Purcell, Jeon, Zimmerman, Blue, & Pevsner, 2001). Most of these studies contained fewer than 10 individuals and were underpowered to identify many significant pathway-level changes, but consistent patterns have emerged. The only study to use a network approach identified two modules, one upregulated and one downregulated in ASD. Analysis of these modules showed that the downregulated module, referred to here as asdM12, was enriched in genes involved in

TABLE 11.1 Coexpression Versus Physical Interaction Networks: Practical and Theoretical Considerations Comparison of the Steps of Network Analysis From Box 11.1 for Coexpression and Protein–Protein Interaction (PPI) Networks, Highlighting the Strengths of Each Approach

	Gene Coexpression	Protein–Protein Interaction
Nodes	Microarray and RNA sequencing offer genome-wide coverage across nodes. • Genome-wide data can easily be generated for a specific biological question. • Data may also be acquired from previous experiments via the Gene Expression Omnibus.	Most PPI studies have low coverage across the genome for the tissue or cell of interest. • Obtaining high-coverage data for a specific biological question requires direct experiments (yeast two-hybrid, co-immunoprecipitation, etc.). • Genome-wide coverage requires using literature-curated PPIs (BioGRID, DAPPLE, GeneMANIA).
Edges	Relationships between genes depend on how they covary in expression level across samples. • Edge strength between two genes depends on spatial, temporal, or disease-related expression patterns in input data. • Edge accuracy depends on sample size. • Edge strength can be spuriously affected by factors such as RNA quality, batch effects, or other biases which introduce false correlations. • Edges may represent co-regulation of genes in different cells, cellular compartments, or within cells that do not interact physically, thus revealing co-regulatory networks not observable otherwise.	Edges reflect physical interactions and thus identified pathways are concrete rather than inferred. • For most experiments and literature-curated interactions, edges are not tissue-specific or isoform-aware. • In humans, the false-positive rate of experimental PPIs has been estimated at >80% (Hart, Ramani, & Marcotte, 2006). • The positive predictive value of literature-curated PPIs ranges from 10% to 40% for experimental yeast two-hybrid PPIs (data not shown). • Edges will not reveal co-regulation that does not involve physical interactions.
Connectivity and modularity	Global network structure is modular and is: • predictive of co-regulation, physical interaction, shared pathway membership, and tissue, cell type, or cell compartment specificity • this structure can also be driven by confounding factors as discussed, so confounders should be carefully evaluated • edges or modules can be evaluated for agreement with PPIs or other -omic data	Network structure reflects structural, enzymatic, and signaling protein complexes. • These can be related to coexpression and shared pathway membership. • Ascertainment biases may be an issue, particularly when using literature-curated databases. • The most studied genes (eg, those involved in cancer biology) are overrepresented in literature-curated data and databases and can form false "modules"
Annotation	Modules can be annotated as follows: • Pathway enrichment, eg, gene ontology and *Kyoto Encyclopedia of Genes and Genomes* • Both modules and hubs can be assessed for relationships with traits, including potential biasing factors. • Module preservation can be evaluated across tissue or diseases.	Modules can be annotated. • Pathway enrichment and PPI modules can be evaluated for agreement with PPIs or other -omic data. • Connectivity metrics such as path length can be evaluated to predict the effect of disrupting physical interactions mediated by highly interconnected genes on the network.
Reproducibility	Reproducibility can be evaluated: • Resampling the genes in modules can ensure that a few genes are not driving the network structure. • Evaluating independent coexpression data for edge structure can ensure that biases from an outlier dataset or sample are not driving network structure.	Reproducibility can be evaluated: • Resampling can ensure reproducibility and assess significance of modular interconnectivity; such analyses should control for the ascertainment bias when using literature-curated data. • Independent validation can be performed with PPI data from a different method or literature database (with nonoverlapping studies). • Relevance to in vivo tissue can be supported by assessing coexpression of interacting proteins from in vivo data.

Coexpression networks represent pairwise statistical relationships, or correlational networks (Box 11.1), which are often less intuitive to biologists compared with PPI networks, which are a type of direct physical interaction network.

TABLE 11.2 Selected Transcriptomic, Protein Interaction, and Integrative Network Studies in Neuropsychiatric Disorders

References	Nodes	Edges	Description of Network	Network-Level Findings	Validation	Major Finding
How Are Transcriptional Patterns Perturbed in Brains of Those With Disease?						
Voineagu et al. (2011)	Genome-wide microarray of about 10,000 genes	Coexpression using 58 samples from 19 people with ASD, 17 CTL subjects in frontal and temporal cortex	• WGCNA with unsigned weighted network, but genes were assigned into modules using signed criteria; • Modules annotated with pathways, disease status, age, sex, and potential confounders.	• Networks structure preserved between ASD and controls; • In the ASD versus CTL network, one module up with ASD, one down with ASD; • Module upregulated in ASD enriched for immune function, microglial markers; • Module downregulated in ASD enriched for synaptic function, known ASD genes, and ASD genome-wide association signal.	• A splicing regulator hub in the downregulated module was shown to alter splicing of genes in the same module using RNA-seq; • Modules correlated to confounders were excluded from the analysis.	Autism involves convergent cellular changes between normal aging involving synaptic function and cellular metabolic processes.
Torkamani et al. (2010)	Genome-wide profiling of about 4000 genes	Coexpression using 47 patients with SCZ, 54 control subjects from frontal cortex	• Coexpression based on combining ARACNE and WGCNA methods; • Modules annotated with pathway/cell-type enrichment, association to disease and age.	• Network structure preserved between SCZ and control subjects; • In SCZ versus CTL network, five modules were enriched for genes differentially expressed in SCZ; • One module related to dopamine signaling and neurogenesis that is downregulated with age in controls but not SCZ.	• Used differential expression from previous studies to show enrichment in four of five SCZ modules; • Multiple regression used to control postmortem factor covariates.	SCZ involves failure of normal developmental downregulation of genes related to neuronal development and dopamine signaling.
Chen et al. (2012)	Genome-wide	Coexpression using SCZ, bipolar, and similarly aged control brains	• Coexpression using WGCNA; • Modules in a similar manner as above, but also compared across disorders.	• Network structure preserved between SCZ and control subjects; • Identified two coexpression modules related to SCZ status, one enriched for neuron differentiation and neuronal development, another enriched for neuron projection; • Modules were also associated with disease status in bipolar disorder, which has genetic overlap with SCZ.	• Multiple regression used to reduce effect of postmortem interval, sex, and pH; • Two coexpression modules were seen across brain regions and brains from different brain banks.	SCZ and bipolar disorder have similar transcriptomic perturbations.

Are Specific Pathways, Cell Types, or Circuits Defined by Genome-Wide Transcriptional Networks Affected by Disease Risk Genes or Mutations?						
Ben-David and Shifman (2012)	Genome-wide microarray profiling of about 16,000 genes	Coexpression using 1340 samples from two adults, spanning hundreds of anatomic regions	• WGCNA with unsigned weighted network; • Modules annotated with pathway/cell-type enrichment, rare and common variant enrichment, regional specificity.	• Three modules were enriched for candidate ASD genes, two were enriched for ASD genome-wide association signal; • All three of these modules were enriched for neuronal cell markers.	• Three ASD risk gene-enriched modules showed increasing expression throughout brain development; • Genome-wide association signal confirmed using replication data.	Autism risk genes affected by inherited rare and common variants are coexpressed in adult brain and affect neuronal function.
Parikshak et al. (2013)	Genome-wide RNA-seq of about 15,000 genes	Coexpression using 146 cortical samples spanning fetal cortical development to early life from BrainSpan	• WGCNA with signed weighted network; • Modules annotated with enrichment for ASD risk genes, pathways/cell types, transcription factor binding, and cortical layer specificity.	• Two transcriptional/chromatin regulation-related modules enriched for risk genes with rare de novo mutation; • Three neuronal/synaptic development modules enriched for candidate ASD genes and ASD gene coexpression; • ASD risk modules enriched for glutamatergic neuron markers, expression in cortical layers two to three in adult cortex; • Enrichment in modules for common transcription factor and Fragile X mental retardation protein binding.	• Network module robustness assessed via bootstrapping analysis; • Validated modules via PPI enrichment and independent expression data; • Replicated risk gene enrichments across multiple studies; • Used multiple null sets as controls.	ASD genetic risk converges on neurodevelopmental transcriptional and chromatin regulation, neuronal and synaptic development, particularly L2–4 glutamatergic neurons.
Are Selected Disease Risk Genes Associated With Each Other at the Transcriptomic Network Level, and Do They Affect Specific Time Points, Brain Regions, or Cell Types?						
Willsey et al. (2013)	Seeded using nine statistically supported risk genes	Search for enriched seeded coexpression modules across four brain region groups and 13 time windows	• Search by constructing 52 modules with about 180 genes and assessing enrichment for the remaining held-out 120 risk genes.	• Modules identify multiple converging points for ASD de novo hits: frontal cortex in midfetal development, parietal/temporal/visual cortex and thalamus/cerebellum around birth; • Characterized midfetal frontal cortex modules, showing marker enrichment for neurons, inner cortical plate, and developing L5/6 glutamatergic neurons.	• Used leave-one-out cross-validation for seed genes; • Showed module enrichment with integrated de novo and inherited mutation signal; • Validated expression of seeds in inner/outer cortical plate in midfetal development.	ASD risk genes affect gene set most coexpressed in fetal dorsolateral prefrontal cortex.

Continued

TABLE 11.2 Selected Transcriptomic, Protein Interaction, and Integrative Network Studies in Neuropsychiatric Disorders—cont'd

References	Nodes	Edges	Description of Network	Network-Level Findings	Validation	Major Finding
Gulsuner et al. (2013)	Seeded RNA-seq using 54 risk genes	Search for maximal coexpression across four brain region groups, three time epochs in 578 samples	• Search for high coexpression of 54 risk genes using binary network edges thresholded at $\|r\| > 0.8$; • Compared 54 risk genes with permuted set of 54 similar genes from controls.	• Noted enrichment identified that SCZ de novo hits affect gene sets highly interconnected by PPI and coexpression during in fetal prefrontal cortex; • Ensured network more interconnected than random de novo hits from unaffected siblings across neuropsychiatric disorders.	• Confirmed 54 genes enriched for PPIs using GeneMANIA.	SCZ risk genes affect gene set most coexpressed in fetal dorsolateral prefrontal cortex.
Do Disease Risk Genes Interact With Each Other at the Protein Level?						
Sakai et al. (2011)	Seeded with 35 genes (192 baits) implicated in monogenetic autism syndromes	Y2H used to create PPI network from 539 genes (7933 preys)	• 848 direct interactions, 24 genes were part of one interconnected module; • This module was expanded using curated PPIs (from BioGRID and HPRD); • Expanding the initial network yielded 3507 proteins and 6881 interactions.	Gene ontology enrichment suggested enrichment of "cellular compartment." In addition, the PPI set had high coexpression in independent mouse brain gene expression data, particularly in hypothalamus, cerebellum, and amygdala.	• Network edge robustness was validated by assessing 44 of 52 randomly selected PPIs from yeast in human cells (nonneuronal).	ASD genes affect common, interacting pathways at protein level.
Corominas et al. (2014)	Seeded with 191 candidate autism risk genes (422 baits) with varying evidence; fragments related to brain RNA splice isoforms	Y2H used to create PPI network from human ORFeome 5.1 (about 15,000 ORFs)	• 71 baits and 291 preys formed a module of 506 interactions (629 isoform level PPIs).	• Showed interacting proteins shared function by other, independent methods including GO pathways, co-regulation by common transcription factors; • Demonstrated 1.5-fold enrichment of preys for genes hit by 198 de novo copy number variants (2267 genes), noted additional importance of isoforms in PPIs and cross-disorder PPI relationships.	• Edge robustness was supported by four Y2H experiments performed for each interaction; three of four had to confirm interaction; • Validation of edges with mammalian PPI trap assay, precision of about 90%.	ASD genes affect common, interacting pathways and assessment of splice isoforms is necessary for their complete characterization.

Li et al. (2014) (Snyder PPI paper)	Started with all proteins covered by physical interactions in humans in BioGRID	PPI edges from BioGRID	• Topological overlap based clustering of genome-wide PPIs.	• Identified two modules, one related to transcriptional regulation and another to synaptic function; • Found synaptic function module was enriched for mutations found in a small set of ASD whole genomes; this module was also enriched for gene expression in oligodendrocytes and corpus callosum.	• Validation of specific interactions or network as a whole was not performed.	ASD-associated genes may disrupt protein interactions between genes highly expressed in oligodendrocytes and particularly in corpus callosum, leading to disruption of cortical connectivity.
Do Disease Risk Genes Affect Molecular Networks From Multiple Data Sources?						
Gilman et al. (2011)	Seeded with 47 rare de novo CNVs spanning 433 genes (Levy et al., 2011)	Genome-wide phenotypic background network of likelihood ratios primarily reflecting probability of shared PPI and pathway	• Greedy search for clusters connecting seed genes, allowing up to one or two genes per CNV; • Modules annotated by network connectivity to GO and KEGG pathways using background network.	• Identified interconnected gene set enriched for axonogenesis and morphogenesis of dendritic spines; • No clusters were found with ultrarare inherited CNVs; • CNVs found in females affected genes that were more interconnected in networks than male genes, suggesting disruption of these genes leads to more severe consequences.	• Network edges were robust, weighted using independent data on molecular phenotypes shared in humans (Feldman, Rzhetsky, & Vitkup, 2008); • No network-level validation was performed.	Genes affected by de novo CNVs in ASD affect a convergent molecular network related to neuronal functioning.
Gilman et al. (2012)	Seeded with 1044 genes from de novo CNVs and SNVs, as well as common variants	Similar to NETBAG in Gilman et al. (2011)	Similar to NETBAG in Gilman et al. (2011), with special considerations for SNVs and CNVs when computing the significance of modules.	• Identified two clusters, both of which had CNVs and SNVs; • Both clusters have higher median expression during brain development; • One cluster enriched for synaptic function, related to ASD and ID genes; • Other cluster enriched for chromatin regulators; • Literature mining suggested that genes in ASD and SCZ clusters result in different phenotypes despite overlap.	• Pathways enriched in one cluster were also found to be differentially expressed in SCZ cell culture model.	Genes affected by diverse types of genetic variation in SCZ affect similar molecular pathways as ASD and ID, but may have different functional consequences.

Continued

TABLE 11.2 Selected Transcriptomic, Protein Interaction, and Integrative Network Studies in Neuropsychiatric Disorders—cont'd

References	Nodes	Edges	Description of Network	Network-Level Findings	Validation	Major Finding
Chang, Gilman, Chiang, Sanders, and Vitkup (2015) (NETBAG+ in ASD)	Seeded with 580 genes affected by de novo SNVs, 434 affected by de novo CNVs	Similar to NETBAG in Gilman et al. (2011)	Similar to NETBAG in Gilman et al. (2012)	• Identified one major cluster, divided it into four subclusters; • Annotated clusters based on gene expression in brain development, IQ of individuals harboring mutation in genes in cluster, and cell type–specific gene expression bias.	• No validation of network was performed.	Genes affected by diverse types of genetic variation in ASD affect pathways related to specific cell types. More severe mutations are found in females and low-IQ individuals, in genes with higher expression during fetal brain development.
Hormozdiari, Penn, Borenstein, and Eichler (2015) (MAGI)	Seeded with 877 genes implicated by de novo SNVs in ASD and ID	Coexpression from gene expression in brain and PPI from BioGRID	• Started with "seed pathways," annotated pathways with many mutations; • Expanded pathways via an objective function maximizing intramodular connectivity relative to intermodular connectivity but also based on maximizing mutations found in patients compared with control subjects.	• Identified different but mostly overlapping modules across diseases ranging in size from about 50 to 100 genes; • Main finding was two modules, one comprising earlier expressed genes involved in transcriptional regulation and chromatin modification, another with genes involved in synaptic development; • All modules identified were highly constrained: they had few mutations in controls, which suggests they identify genes under purifying selection.	• Authors showed that identification of modules was similar using different brain regions and different PPI databases.	Genes affected by de novo SNVs in ASD, ID, and other disorders affect convergent sets of gene expressed during brain development that are under considerable purifying selection.

ARACNE, Algorithm for the Reconstruction of Accurate Cellular Networks; *ASD*, autism spectrum disorder; *CTL*, control; *GO*, gene ontology; *IQ*, intelligence quotient; *KEGG*, *Kyoto Encyclopedia of Genes and Genomes*; *ORFs*, open reading frames; *PPI*, protein–protein interaction; *RNA-seq*, RNA sequencing; *SCZ*, schizophrenia; *WGCNA*, weighted gene coexpression network analysis; *Y2H*, yeast two-hybrid.

neurotransmitter signaling and release and vesicle transport. This module also contained a significant signal from rare and common variants, which suggests that it is the primary causal module. The upregulated module, referred to here as asdM16, corresponded to genes expressed in microglia and, to a lesser extent, astrocytes, which is consistent with upregulation of microglia observed in small studies of ASD postmortem brain (Morgan et al., 2012; Tetreault et al., 2012). Inspection of the hub genes in the downregulated asdM12 suggested alterations in splicing connected to downregulation of the hub gene RBFOX1, which was experimentally validated by RNA sequencing and alternative splicing analysis. In addition to the identification of biologically coherent processes, the use of weighted gene coexpression network analysis (WGCNA) (Zhang & Horvath, 2005) allowed the authors to determine the strength of the relationship of asdM12 and asdM16 with postmortem changes and other potential confounders including postmortem interval, RNA quality, age, seizure status, and neuropsychiatric medication status. Thus, transcriptomic networks provide an essential organizational framework that demonstrates a shared molecular pathology across two-thirds of ASD cases profiled. Importantly, similar synaptic and microglial pathways have been identified in other brain regions with RNA-seq and network analysis (Gupta et al., 2014).

Consistent results have also emerged from transcriptomic studies of postmortem brain from schizophrenic subjects, highlighting downregulation of GABAergic markers and mitochondrial genes, as well as neural immune changes (Hashimoto et al., 2007; Mirnics, Middleton, Marquez, Lewis, & Levitt, 2000). Network analysis has identified SCZ-specific modules by applying a modification of the WGCNA approach (Torkamani et al., 2010). Interestingly, overlap between SCZ and bipolar disorder has been supported by cross-disorder transcriptome analyses (Akula et al., 2014; Chen et al., 2012; Shao & Vawter, 2008), which is consistent with genetic studies revealing significant overlap in risk alleles between the disorders (Cross-Disorder Group of the Psychiatric Genomics Consortium et al., 2013; Network and Consortium, 2015). As a next step, transcriptomic analysis with WGCNA could provide a framework for cross-disease comparison at a functional genomic level. This may reveal disorder overlap and specificity at a molecular pathway level (Parikshak et al., 2015).

Because of the influx of neurobiologically relevant genomic and transcriptomic data, it has become possible to integrate risk genes identified by whole-exome sequencing (WES) (Iossifov et al., 2012; Neale et al., 2012; O'Roak et al., 2012; Sanders et al., 2012) with transcriptome data spanning multiple brain regions and developmental stages (Colantuoni et al., 2011; Kang et al., 2011). One study used WGCNA to construct genome-wide coexpression networks (Box 11.1) from 8 weeks postconception to 1 year of age in human cortex and assessed how both ASD and intellectual disability (ID) risk genes are involved in cortical development, cell types, and circuits (Parikshak et al., 2013). Robust, independently reproducible coexpression modules that were supported by PPIs were identified. Module hubs were shown to reflect the known timing of molecular processes closely during human cortical development. Enrichment analysis with multiple ASD risk gene sets identified five developmentally regulated coexpression modules enriched for ASD-associated risk genes but not ID or multiple control genes. Two of these risk modules were enriched for de novo mutations from WES and were found to be related to pathways involved in transcriptional and chromatin regulation, including the BAF (SWI/SNF) complex (Ronan, Wu, & Crabtree, 2013). The three other modules were later upregulated in cortical development and were enriched for heterogeneously defined ASD candidate genes (Basu, Kollu, & Banerjee-Basu, 2009) that harbored mostly inherited mutations. These modules were also enriched for genes in the module downregulated in ASD cortex found by Voineagu et al. (2011), which was consistent with observations by the coexpression study of adult brain (Ben-David & Shifman, 2012).

Seed-Based Approaches

Another study applied a seed-based approach (Box 11.1) to ask a similar question about how genes identified by WES affect brain development. Willsey and colleagues searched spatial and temporal combinations of microarray data from brain tissue across neurodevelopment to identify coexpression networks enriched for ASD risk that were seeded around nine high-confidence ASD risk genes identified by WES. These authors asked if, when, and where this subset of genes affected by rare variation in ASD converges during brain development. Using the seed genes, the authors constructed binary coexpression networks based on the top 20 correlations for each seed for various spatial and temporal combinations, resulting in 85 networks. Each of these networks was evaluated for enrichment from 122 additional genes supported by WES evidence, which identified four spatiotemporal combinations that passed stringent correction for multiple testing. The authors then focused on convergence in the prefrontal cortex subnetwork that spanned PCW 10–24, and supported ASD risk enrichment by demonstrating increased genetic risk based on WES data that included inherited variation (He et al., 2013). Notably, the authors did not find enrichment for known PPIs or molecular pathways. This was likely because of the relatively small size of their coexpression modules, as well as the use of unsigned correlation analysis, which does not account for the directionality of correlations when defining modules.

Both genome-wide and the seeded approaches identified enrichment for ASD risk genes during fetal development and further demonstrated enrichment for cortical glutamatergic projection neurons. This suggests that despite the locus heterogeneity of ASD, there is a specific cell-type disruption in many cases. Notably, the genome-wide network study (Parikshak et al., 2013), which assessed both ASD and ID genes, suggested that upper cortical layer disruption would result in more ASD-like phenotypes, whereas lower layer disruption would result in ID. Other studies have also identified fetal cortical development as affected by mutations in ASD (Miller et al., 2014; Stein et al., 2014; Steinberg & Webber, 2013).

In SCZ, a seeded spatial and temporal search of coexpression networks identified fetal development of the PFC as the point of molecular convergence among risk genes (Gulsuner et al., 2013). These authors took genes hit by rare de novo single-nucleotide variants (SNVs) from a small WES screen as seeds and demonstrated that the genes formed a network that is enriched for PPI. They next screened binary coexpression networks formed by these genes across brain regions and time points and identified significant coexpression among risk genes in fetal prefrontal cortex relative to similarly sized gene sets from mutations in unaffected individuals. However, this study did not evaluate cell type, laminar, or co-regulatory relationships or assess whether genes outside the seed set might be involved in SCZ during brain development. Because larger sets of risk genes are now available in SCZ (Fromer et al., 2014; Purcell et al., 2014), it will be valuable to replicate and extend this finding.

Protein–Protein Interaction Networks

As mentioned in the "Introduction" section, although PPI networks have the advantage of representing strong interactions, there are also several disadvantages, the most important of which are incomplete coverage of the interactome and reliance on nonneural or in vitro experiments (Hakes, Pinney, Robertson, & Lovell, 2008; Hart et al., 2006). Despite this, PPI networks provide an important and useful alternative to gene coexpression, because PPIs reflect true molecular interactions. Three important examples of the use of PPI in ASD are the work of Sakai et al. (2011), Corominas et al. (2014), and O'Roak et al. (2012) (Table 11.2). Sakai et al. (2011) performed the first PPI analysis of neurodevelopmental disease genes using a yeast two-hybrid system to perform an in vitro screen of 35 syndromic or major effect candidate ASD genes. They found that these seed genes were highly interconnected at a protein level, and identified many previously unknown interactions involving major risk genes for neurodevelopmental disorders. Corominas et al. (2014) analyzed interactions between the protein products of several hundred ASD risk genes. Importantly, they used neuronal splice isoforms of these genes and found that these isoform-level PPIs identified distinct interactions and corresponded with known ASD risk genes and copy number variants (CNVs).

O'Roak et al. employed seed-based networks with literature-curated PPIs compiled from many studies (Mostafavi, Ray, Warde-Farley, Grouios, & Morris, 2008) to ask whether ASD risk genes identified by exome sequencing interacted more than expected by chance. Based on the hypothesis that proteins interacting with ASD risk genes are more likely to harbor risk mutations as well, these investigators performed targeted sequencing of proteins within a core subnetwork across thousands of additional individuals and identified significantly more mutations affecting the genes chosen by PPI compared with chance. Remarkably, this work demonstrates that despite immense locus heterogeneity in rare mutations, convergence exists at the level of protein interactions among some mutations.

However, because PPIs in these studies are not from neural-specific tissue or cell types, like most current approaches based solely on PPI, these approaches cannot define neural circuits or developmental stages without integrating additional data (Table 11.1). To circumvent this problem, we suggest studies start with tissue- or cell-specific transcriptional networks that then combine known PPIs (Parikshak et al., 2013) or evaluate coexpression in PPI-defined modules (Li et al., 2014; Sakai et al., 2011). In addition, proteomic approaches can provide compartmental specificity, so filtering coexpression relationships by experimentally identified postsynaptic density proteins (Bayés et al., 2010) may provide a compartment-level specificity to networks.

Integrative Network Approaches

Multiple levels of -omic data may contribute unique functional insights and increase power to detect molecular convergence. This has motivated the integration of genetic and functional evidence to identify genes or pathways in ASD. The network-based analysis of genes (or NETBAG) approach (Gilman et al., 2011) defines networks in which highly connected genes are more likely to participate in the same phenotype. Edges in the network are primarily based on shared PPI edges, shared gene ontology (GO) (Ashburner et al., 2000), or *Kyoto Encyclopedia of Genes and Genomes* (Ogata et al., 1999) term membership. Each data type's contribution to edges is weighted based on relationships among a known set of disease genes (Feldman et al., 2008), and these different weights are combined to construct a phenotypic "background network" in which gene lists of interest are assessed for high connectivity using a permutation analysis.

When seeded with 75 disparate de novo CNV deletions spanning 746 genes found in ASD, NETBAG found a cluster related to synaptic function, whereas a similarly sized set of genes affected by inherited rare CNVs did not cluster (Levy et al., 2011). Furthermore, genes in CNVs from females contributed more to the cluster connectivity than did those from males, which suggests that females are affected by more severe genetic hits, which has been replicated in exome sequencing studies (Ronemus, Iossifov, Levy, & Wigler, 2014). In addition, a later study (Noh et al., 2013) evaluated a larger number of cases and demonstrated that in addition to CNV deletions, duplications affect an interconnected PPI network. This network was enriched for multiple knockout mouse phenotypes and synaptic transmission, which is consistent with the initial NETBAG findings (Gilman et al., 2011).

NETBAG has also been applied to assess the effect of rare de novo CNVs and SNVs in SCZ (Gilman et al., 2012) in a method called NETBAG+. NETBAG+ identified two clusters: one related to axonal guidance, neuronal migration, and synaptic function, and another enriched for chromatin modifiers. Both clusters were expressed more highly during fetal brain development, and the first cluster was shown to be highly connected in the phenotypic background network with candidate lists of ASD and ID genes as well as with the previously identified cluster that was defined by CNVs in ASD. The authors disentangled the overlap between ASD and SCZ genes by mining the literature to suggest that mutations in ASD increase spine or dendrite growth whereas those in SCZ decrease it (Gilman et al., 2012). Taken together, it is clear that integrative network analysis can identify shared pathways among disorders. However, the ultimate reliance on transcriptional profiles to place identified clusters in a neurobiological context suggests that it is necessary to add more brain- and disorder-relevant data into these approaches to derive more refined insights.

Themes From Cross-Method Comparisons

We have reviewed how network approaches provide unique insights for understanding biological pathways and points of convergence in neurodevelopmental disorders such as ASD. Each of the analyses that have been used in ASD has its strengths and weaknesses. The comparisons that we provide in this chapter demonstrate a degree of agreement among methods, and several informative conclusions can be made:

1. Existing methods are limited in resolution for offering new neurobiological insights based on enrichment analyses with existing laminar and cell type–specific gene sets. One reason for this could be that current coexpression and PPI data sets are of limited resolution, and better data are needed to construct these networks. Another reason could be that the set of biological pathways affected by genetic risk in ASD is actually large, and it will be necessary to subdivide ASD mutations based on phenotypic heterogeneity.
2. Seeded approaches identify smaller modules but do not necessarily offer greater specificity for neurobiological pathways compared with genome-wide approaches. Therefore, either more informative seeds will need to be chosen or additional data will need to be integrated into these approaches to identify more specific pathways.
3. There is a large overlap among seeded and unseeded methods, it is apparent that severe ASD risk mutations implicated by many methods disrupt common genes and biological processes that occur during brain development. This may reflect a disruption of core neurobiological pathways that have been subjected to canalization (Masel & Siegal, 2009; Suliman, Ben-David, & Shifman, 2014; Waddington, 1942) and have been under purifying selection in humans (Samocha et al., 2014; Uddin et al., 2014). The consequence of canalization would be that these genes and biological processes are buffered against small perturbations, but large-effect mutations can disrupt them. Thus, the most severe (rare de novo variant) mutations lead to variable phenotypic effects and severe neurodevelopmental disorders including ASD, SCZ, epilepsy, and ID (Fromer et al., 2014; Krumm, O'Roak, Shendure, & Eichler, 2014), whereas milder mutations with additive effects may cause more specific phenotypes (Cross-Disorder Group of the Psychiatric Genomics Consortium et al., 2013).

CONCLUSIONS AND FUTURE DIRECTIONS

To identify more specific pathways using network methods, it will be important to evaluate genome-wide data derived from neurobiological studies at high spatial and temporal resolution, particularly during early brain development. Not every gene in these early neurobiological processes, which include transcriptional regulation and chromatin modification, will be observed as mutated with an observable high-impact effect. Many mutations could be lethal and cause spontaneous abortion, whereas others may not yield a phenotype because a gene performs a redundant role in a biological pathway.

Lack of appropriate prior evidence is likely to limit the predictive value of the seed-based approach for identifying novel genes or novel neurobiological pathways in ASD, whereas a lack of ideal spatial and temporal sampling of human brain development limits the resolution of both seed-based and genome-wide methods. Based on the distinct biological processes implicated by coexpression modules in brain development (Parikshak et al., 2013), we predict that disruption of genes that peak at different points of development yields different phenotypic outcomes. To evaluate this properly, very high temporal resolution transcriptomes will be necessary—many individuals' worth of data per week of fetal development and postnatal development owing to the rapid changes occurring at these developmental epochs. This will benefit both genome-wide coexpression networks and seeded coexpression networks to achieve greater specificity. Current data sets lack this temporal resolution because only one to three individuals are available for most time points, and most fetal time points are weeks apart from each other. In addition, the appropriate data may never be available for the third trimester of fetal development, so it may be necessary to evaluate these stages in nonhuman primates (Bernard et al., 2012; Sunkin et al., 2013) or in vitro, once viable comparisons between in vitro and in vivo cell types can be established (Stein et al., 2014).

To gain additional biological insights about the affected cell types, it will be necessary to pursue temporal trajectories in individual cell types or a lineage tree of cortical cell types. Molyneaux et al. (2015) demonstrated how this could be done using mice, and claimed a similar approach could work in humans. Because currently available PPIs contain no cell type–specific information, a promising avenue is to use cell type–specific transcriptomes and epigenomes to elucidate the important regulatory networks during cell fate determination in the cortex, as has been done for the development of blood cells (Lara-Astiaso et al., 2014). Such a method, which would likely track cells from neural progenitor status to differentiated neuronal subtypes and profile transcriptomes, histone markers, and open chromatin for homogeneous populations defined by combinations of cellular markers, could identify the important regulatory changes at each lineage branch point and identify which steps of cortical development might specifically be affected in ASD by different mutational processes.

Here we have reviewed and summarized multiple approaches to gene network analysis in neuropsychiatric disorders (Table 11.2). We find that the general framework in Box 11.1 aids in understanding many network biology studies, and we hope the comparisons provided in this chapter highlight the benefits and challenges inherent to different approaches (Allen et al., 2012). Molecular biology, neuroscience, and psychiatry are on an inexorable path to sequence more genomes and measure neurobiological function at ever-increasing depth and resolution. With massive amounts of data inundating the field, it will be essential to apply network approaches to organize this information into knowledge, and systematically link genetic variants to neurobiological pathways and phenotypes.

ACKNOWLEDGMENTS

Box 11.1 and some descriptions of network studies in the neuropsychiatric disorders section are adapted from Parikshak et al. (2015). Funding for this work came from an NRSA fellowship to NNP (F30MH099886) and the following NIH grants to DHG: NIMH MERIT award for asymmetrically expressed genes in the brain (5R37MH060233), Epigenetics of autism (5R01MH094714), and ACE Network2, autism heterogeneity (5R01MH100027). We thank Lauren Kawaguchi for helping assemble this chapter.

REFERENCES

Akula, N., Barb, J., Jiang, X., Wendland, J. R., Choi, K. H., Sen, S. K., ... McMahon, F. J. (2014). RNA-sequencing of the brain transcriptome implicates dysregulation of neuroplasticity, circadian rhythms and GTPase binding in bipolar disorder. *Molecular Psychiatry, 19*, 1179–1185.

Allen, J. D., Xie, Y., Chen, M., Girard, L., & Xiao, G. (2012). Comparing statistical methods for constructing large scale gene networks. *PLoS One, 7*, e29348.

Ashburner, M., Ball, C. A., Blake, J. A., Botstein, D., Butler, H., Cherry, J. M., ... Sherlock, G. (2000). Gene ontology: tool for the unification of biology. *Nature Genetics, 25*, 25–29.

Barabasi, A. L. (2009). Scale-free networks: a decade and beyond. *Science, 325*, 412–413.

Barabási, A.-L., & Oltvai, Z. N. (2004). Network biology: understanding the cell's functional organization. *Nature Reviews Genetics, 5*, 101–113.

Basu, S. N., Kollu, R., & Banerjee-Basu, S. (2009). AutDB: a gene reference resource for autism research. *Nucleic Acids Research, 37*, D832–D836.

Battle, A., Khan, Z., Wang, S. H., Mitrano, A., Ford, M. J., Pritchard, J. K., & Gilad, Y. (2015). Genomic variation. Impact of regulatory variation from RNA to protein. *Science, 347*, 664–667.

Bayés, À., & Grant, S. G. N. (2009). Neuroproteomics: understanding the molecular organization and complexity of the brain. *Nature Reviews Neuroscience, 10*, 635–646.

Bayés, À., van de Lagemaat, L. N., Collins, M. O., Croning, M. D. R., Whittle, I. R., Choudhary, J. S., & Grant, S. G. N. (2010). Characterization of the proteome, diseases and evolution of the human postsynaptic density. *Nature Neuroscience, 14*, 19–21.

Ben-David, E., & Shifman, S. (2012). Networks of neuronal genes affected by common and rare variants in autism spectrum disorders. *PLoS Genetics, 8*, e1002556.

Berg, J. M., & Geschwind, D. H. (2012). Autism genetics: searching for specificity and convergence. *Genome Biology, 13*, 247.

Bernard, A., Lubbers, L. S., Tanis, K. Q., Luo, R., Podtelezhnikov, A. A., Finney, E. M., ... Lein, E. S. (2012). Transcriptional architecture of the primate neocortex. *Neuron, 73*, 1083–1099.

Beyer, A., Bandyopadhyay, S., & Ideker, T. (2007). Integrating physical and genetic maps: from genomes to interaction networks. *Nature Reviews Genetics, 8*, 699–710.

Butte, A. J., & Kohane, I. S. (2000). Mutual information relevance networks: functional genomic clustering using pairwise entropy measurements. *Pacific Symposium on Biocomputing*, 418–429.

Bystron, I., Blakemore, C., & Rakic, P. (2008). Development of the human cerebral cortex: Boulder Committee revisited. *Nature Reviews Neuroscience, 9*, 110–122.

Carter, H., Hofree, M., & Ideker, T. (2013). Genotype to phenotype via network analysis. *Current Opinion in Genetics & Development, 23*, 611–621.

Carter, S. L., Brechbühler, C. M., Griffin, M., & Bond, A. T. (2004). Gene co-expression network topology provides a framework for molecular characterization of cellular state. *Bioinformatics, 20*, 2242–2250.

Chang, J., Gilman, S. R., Chiang, A. H., Sanders, S. J., & Vitkup, D. (2015). Genotype to phenotype relationships in autism spectrum disorders. *Nature Neuroscience, 18*, 191–198.

Changeux, J.-P., & Danchin, A. (1976). Selective stabilisation of developing synapses as a mechanism for the specification of neuronal networks. *Nature, 264*, 705–712.

Chen, C., Cheng, L., Grennan, K., Pibiri, F., Zhang, C., Badner, J. A., ... Liu, C. (2012). Two gene co-expression modules differentiate psychotics and controls. *Molecular Psychiatry, 18*, 1308–1314.

Choi, J. K., Yu, U., Yoo, O. J., & Kim, S. (2005). Differential coexpression analysis using microarray data and its application to human cancer. *Bioinformatics, 21*, 4348–4355.

Chow, M. L., Pramparo, T., Winn, M. E., Barnes, C. C., Li, H.-R., Weiss, L., ... Courchesne, E. (2012). Age-dependent brain gene expression and copy number anomalies in autism suggest distinct pathological processes at young versus mature ages. *PLoS Genetics, 8*, e1002592.

Colantuoni, C., Lipska, B. K., Ye, T., Hyde, T. M., Tao, R., Leek, J. T., ... Kleinman, J. E. (2011). Temporal dynamics and genetic control of transcription in the human prefrontal cortex. *Nature, 478*, 519–523.

Corominas, R., Yang, X., Lin, G. N., Kang, S., Shen, Y., Ghamsari, L., ... Iakoucheva, L. M. (2014). Protein interaction network of alternatively spliced isoforms from brain links genetic risk factors for autism. *Nature Communications, 5*.

Cross-Disorder Group of the Psychiatric Genomics Consortium, Lee, S. H., Ripke, S., Neale, B. M., Faraone, S. V., Purcell, S. M., ... Wray, N. R. (2013). Genetic relationship between five psychiatric disorders estimated from genome-wide SNPs. *Nature Genetics, 45*, 984–994.

Deneris, E. S., & Hobert, O. (2014). Maintenance of postmitotic neuronal cell identity. *Nature Neuroscience, 17*, 899–907.

ENCODE Project Consortium. (2012). An integrated encyclopedia of DNA elements in the human genome. *Nature, 489*(7414), 57–74.

Feldman, I., Rzhetsky, A., & Vitkup, D. (2008). Network properties of genes harboring inherited disease mutations. *Proceedings of the National Academy of Sciences of the United States of America, 105*, 4323–4328.

Fromer, M., Pocklington, A. J., Kavanagh, D. H., Williams, H. J., Dwyer, S., Gormley, P., ... O'Donovan, M. C. (2014). De novo mutations in schizophrenia implicate synaptic networks. *Nature, 506*, 179–184.

Gaiteri, C., Ding, Y., French, B., Tseng, G. C., & Sibille, E. (2013). Beyond modules and hubs: the potential of gene coexpression networks for investigating molecular mechanisms of complex brain disorders. *Genes, Brain and Behavior, 13*, 13–24.

Garbett, K., Ebert, P. J., Mitchell, A., Lintas, C., Manzi, B., Mirnics, K., & Persico, A. M. (2008). Immune transcriptome alterations in the temporal cortex of subjects with autism. *Neurobiology of Disease, 30*, 303–311.

Geschwind, D. H., & Gregg, J. P. (2002). *Microarrays for the Neurosciences: An Essential Guide*. Cambridge, MA: MIT Press.

Geschwind, D. H., & Konopka, G. (2009). Neuroscience in the era of functional genomics and systems biology. *Nature, 461*, 908–915.

Geschwind, D. H., & Rakic, P. (2013). Cortical evolution: judge the brain by its cover. *Neuron, 80*, 633–647.

Gilman, S. R., Chang, J., Xu, B., Bawa, T. S., Gogos, J. A., Karayiorgou, M., & Vitkup, D. (2012). Diverse types of genetic variation converge on functional gene networks involved in schizophrenia. *Nature Neuroscience, 15*, 1723–1728.

Gilman, S. R., Iossifov, I., Levy, D., Ronemus, M., Wigler, M., & Vitkup, D. (2011). Rare de novo variants associated with autism implicate a large functional network of genes involved in formation and function of synapses. *Neuron, 70*, 898–907.

Ginsberg, M. R., Rubin, R. A., Falcone, T., Ting, A. H., & Natowicz, M. R. (2012). Brain transcriptional and epigenetic associations with autism. *PLoS One, 7*, e44736.

Gong, S., Zheng, C., Doughty, M. L., Losos, K., Didkovsky, N., Schambra, U. B., ... Heintz, N. (2003). A gene expression atlas of the central nervous system based on bacterial artificial chromosomes. *Nature, 425*, 917–925.

Gratten, J., Wray, N. R., Keller, M. C., & Visscher, P. M. (2014). Large-scale genomics unveils the genetic architecture of psychiatric disorders. *Nature Neuroscience, 17*, 782–790.

Gulsuner, S., Walsh, T., Watts, A. C., Lee, M. K., Thornton, A. M., Casadei, S., ... McClellan, J. M. (2013). Spatial and temporal mapping of de novo mutations in schizophrenia to a fetal prefrontal cortical network. *Cell, 154*, 518–529.

Gupta, S., Ellis, S. E., Ashar, F. N., Moes, A., Bader, J. S., Zhan, J., ... Arking, D. E. (2014). Transcriptome analysis reveals dysregulation of innate immune response genes and neuronal activity-dependent genes in autism. *Nature Communications, 5*, 5748.

Hakes, L., Pinney, J. W., Robertson, D. L., & Lovell, S. C. (2008). Protein-protein interaction networks and biology—what's the connection? *Nature Biotechnology, 26*, 69–72.

Hart, G. T., Ramani, A. K., & Marcotte, E. M. (2006). How complete are current yeast and human protein-interaction networks? *Genome Biology, 7*, 120.

Hashimoto, T., Arion, D., Unger, T., Maldonado-Avilés, J. G., Morris, H. M., Volk, D. W., … Lewis, D. A. (2007). Alterations in GABA-related transcriptome in the dorsolateral prefrontal cortex of subjects with schizophrenia. *Molecular Psychiatry, 13*, 147–161.

Hawrylycz, M. J., Lein, E. S., Guillozet-Bongaarts, A. L., Shen, E. H., Ng, L., Miller, J. A., … Jones, A. R. (2012). An anatomically comprehensive atlas of the adult human brain transcriptome. *Nature, 489*, 391–399.

He, X., Sanders, S. J., Liu, L., De Rubeis, S., Lim, E. T., Sutcliffe, J. S., … Roeder, K. (2013). Integrated model of de novo and inherited genetic variants yields greater power to identify risk genes. *PLoS Genetics, 9*, e1003671.

Hormozdiari, F., Penn, O., Borenstein, E., & Eichler, E. E. (2015). The discovery of integrated gene networks for autism and related disorders. *Genome Research, 25*, 142–154.

Horvath, S. (2011). *Weighted Network Analysis: Applications in Genomics and Systems Biology*. New York: Springer.

Horvath, S., & Dong, J. (2008). Geometric interpretation of gene coexpression network analysis. *PLoS Computational Biology, 4*, e1000117.

Horvath, S., Zhang, B., Carlson, M., Lu, K. V., Zhu, S., Felciano, R. M., … Mischel, P. S. (2006). Analysis of oncogenic signaling networks in glioblastoma identifies ASPM as a molecular target. *Proceedings of the National Academy Sciences of the United States of America, 103*, 17402–17407.

Hudson, N. J., Reverter, A., & Dalrymple, B. P. (2009). A differential wiring analysis of expression data correctly identifies the gene containing the causal mutation. *PLoS Computational Biology, 5*, e1000382.

Iossifov, I., O'Roak, B. J., Sanders, S. J., Ronemus, M., Krumm, N., Levy, D., … Wigler, M. (2014). The contribution of de novo coding mutations to autism spectrum disorder. *Nature, 515*, 216–221.

Iossifov, I., Ronemus, M., Levy, D., Wang, Z., Hakker, I., Rosenbaum, J., … Wigler, M. (2012). De novo gene disruptions in children on the autistic spectrum. *Neuron, 74*, 285–299.

Johnson, M. B., Kawasawa, Y. I., Mason, C. E., Krsnik, Ž., Coppola, G., Bogdanović, D., … Šestan, N. (2009). Functional and evolutionary insights into human brain development through global transcriptome analysis. *Neuron, 62*, 494–509.

Kang, H. J., Kawasawa, Y. I., Cheng, F., Zhu, Y., Xu, X., Li, M., … Šestan, N. (2011). Spatio-temporal transcriptome of the human brain. *Nature, 478*, 483–489.

Kim, M.-S., Pinto, S. M., Getnet, D., Nirujogi, R. S., Manda, S. S., Chaerkady, R., … Pandey, A. (2014). A draft map of the human proteome. *Nature, 509*, 575–581.

Konopka, G., Friedrich, T., Davis-Turak, J., Winden, K., Oldham, M. C., Gao, F., … Geschwind, D. H. (2012). Human-specific transcriptional networks in the brain. *Neuron, 75*, 601–617.

Krumm, N., O'Roak, B. J., Shendure, J., & Eichler, E. E. (2014). A de novo convergence of autism genetics and molecular neuroscience. *Trends in Neurosciences, 37*, 95–105.

Langfelder, P., & Horvath, S. (2007). Eigengene networks for studying the relationships between co-expression modules. *BMC Systems Biology, 1*, 54.

Langfelder, P., & Horvath, S. (2008). WGCNA: an R package for weighted correlation network analysis. *BMC Bioinformatics, 9*, 559.

Langfelder, P., Luo, R., Oldham, M. C., & Horvath, S. (2011). Is my network module preserved and reproducible? *PLoS Computational Biology, 7*, e1001057.

Lara-Astiaso, D., Weiner, A., Lorenzo-Vivas, E., Zaretsky, I., Jaitin, D. A., David, E., … Amit, I. (2014). Chromatin state dynamics during blood formation. *Science, 345*, 943–949.

Lee, I., & Marcotte, E. M. (2009). Effects of functional bias on supervised learning of a gene network model. In *Computational systems biology* (pp. 463–475). Totowa, NJ: Humana Press.

Leek, J. T., Scharpf, R. B., Bravo, H. C., Simcha, D., Langmead, B., Johnson, W. E., … Irizarry, R. A. (2010). Tackling the widespread and critical impact of batch effects in high-throughput data. *Nature Reviews Genetics, 11*, 733–739.

Lein, E. S., Hawrylycz, M. J., Ao, N., Ayres, M., Bensinger, A., Bernard, A., … Jones, A. R. (2006). Genome-wide atlas of gene expression in the adult mouse brain. *Nature, 445*, 168–176.

Levy, D., Ronemus, M., Yamrom, B., Lee, Y.-H., Leotta, A., Kendall, J., … Wigler, M. (2011). Rare de novo and transmitted copy-number variation in autistic spectrum disorders. *Neuron, 70*, 886–897.

Li, J., Shi, M., Ma, Z., Zhao, S., Euskirchen, G., Ziskin, J., … Snyder, M. (2014). Integrated systems analysis reveals a molecular network underlying autism spectrum disorders. *Molecular Systems Biology, 10*, 774.

Lieberman-Aiden, E., van Berkum, N. L., Williams, L., Imakaev, M., Ragoczy, T., Telling, A., … Dekker, J. (2009). Comprehensive mapping of long-range interactions reveals folding principles of the human genome. *Science, 326*, 289–293.

Lui, J. H., Nowakowski, T. J., Pollen, A. A., Javaherian, A., Kriegstein, A. R., & Oldham, M. C. (2014). Radial glia require PDGFD–PDGFRβ signalling in human but not mouse neocortex. *Nature, 515*, 264–268.

Masel, J., & Siegal, M. L. (2009). Robustness: mechanisms and consequences. *Trends in Genetics, 25*, 395–403.

Maze, I., Shen, L., Zhang, B., Garcia, B. A., Shao, N., Mitchell, A., … Nestler, E. J. (2014). Analytical tools and current challenges in the modern era of neuroepigenomics. *Nature Neuroscience, 17*, 1476–1490.

Miller, J. A., Ding, S.-L., Sunkin, S. M., Smith, K. A., Ng, L., Szafer, A., … Lein, E. S. (2014). Transcriptional landscape of the prenatal human brain. *Nature, 508*, 199–206.

Miller, J. A., Woltjer, R. L., Goodenbour, J. M., Horvath, S., & Geschwind, D. H. (2013). Genes and pathways underlying regional and cell type changes in Alzheimer's disease. *Genome Medicine, 5*, 48.

Mirnics, K., Middleton, F. A., Marquez, A., Lewis, D. A., & Levitt, P. (2000). Molecular characterization of schizophrenia viewed by microarray analysis of gene expression in prefrontal cortex. *Neuron, 28*, 53–67.

Molyneaux, B. J., Arlotta, P., Menezes, J. R. L., & Macklis, J. D. (2007). Neuronal subtype specification in the cerebral cortex. *Nature Reviews Neuroscience, 8*, 427–437.

Molyneaux, B. J., Goff, L. A., Brettler, A. C., Chen, H.-H., Brown, J. R., Hrvatin, S., ... Arlotta, P. (2015). DeCoN: genome-wide analysis of in vivo transcriptional dynamics during pyramidal neuron fate selection in neocortex. *Neuron, 85*, 275–288.

Morgan, J. T., Chana, G., Abramson, I., Semendeferi, K., Courchesne, E., & Everall, I. P. (2012). Abnormal microglial–neuronal spatial organization in the dorsolateral prefrontal cortex in autism. *Brain Research, 1456*, 72–81.

Mostafavi, S., Ray, D., Warde-Farley, D., Grouios, C., & Morris, Q. (2008). GeneMANIA: a real-time multiple association network integration algorithm for predicting gene function. *Genome Biology, 9*, S4.

Neale, B. M., Kou, Y., Liu, L., Ma'ayan, A., Samocha, K. E., Sabo, A., ... Daly, M. J. (2012). Patterns and rates of exonic de novo mutations in autism spectrum disorders. *Nature, 485*, 242–245.

Network and Pathways Analysis Subgroup of the Psychiatric Genomics Consortium. (2015). Psychiatric genome-wide association study analyses implicate neuronal, immune and histone pathways. *Nature Neuroscience, 18*, 199–209.

Noh, H. J., Ponting, C. P., Boulding, H. C., Meader, S., Betancur, C., Buxbaum, J. D., ... Webber, C. (2013). Network topologies and convergent aetiologies arising from deletions and duplications observed in individuals with autism. *PLoS Genetics, 9*, e1003523.

Nord, A. S., Pattabiraman, K., Visel, A., & Rubenstein, J. L. R. (2015). Genomic perspectives of transcriptional regulation in forebrain development. *Neuron, 85*, 27–47.

Ogata, H., Goto, S., Sato, K., Fujibuchi, W., Bono, H., & Kanehisa, M. (1999). KEGG: kyoto encyclopedia of genes and genomes. *Nucleic Acids Research, 27*, 29–34.

Oldham, M. C., & Geschwind, D. H. (2006). Deconstructing language by comparative gene expression: from neurobiology to microarray. *Genes, Brain and Behavior, 5*, 54–63.

Oldham, M. C., Horvath, S., & Geschwind, D. H. (2006). Conservation and evolution of gene coexpression networks in human and chimpanzee brains. *Proceedings of the National Academy Sciences of the United States of America, 103*, 17973–17978.

Oldham, M. C., Konopka, G., Iwamoto, K., Langfelder, P., Kato, T., Horvath, S., ... Geschwind, D. H. (2008). Functional organization of the transcriptome in human brain. *Nature Neuroscience, 11*, 1271–1282.

O'Roak, B. J., Vives, L., Girirajan, S., Karakoc, E., Krumm, N., Coe, B. P., ... Eichler, E. E. (2012). Sporadic autism exomes reveal a highly interconnected protein network of de novo mutations. *Nature, 485*, 246–250.

Parikshak, N. N., Gandal, M. J., & Geschwind, D. H. (2015). Systems biology and gene networks in neurodevelopmental and neurodegenerative disorders. *Nature Reviews Genetics, 16*(8), 441–458.

Parikshak, N. N., Luo, R., Zhang, A., Won, H., Lowe, J. K., Chandran, V., ... Geschwind, D. H. (2013). Integrative functional genomic analyses implicate specific molecular pathways and circuits in autism. *Cell, 155*, 1008–1021.

Purcell, A. E., Jeon, O. H., Zimmerman, A. W., Blue, M. E., & Pevsner, J. (2001). Postmortem brain abnormalities of the glutamate neurotransmitter system in autism. *Neurology, 57*, 1618–1628.

Purcell, S. M., Moran, J. L., Fromer, M., Ruderfer, D., Solovieff, N., Roussos, P., ... Sklar, P. (2014). A polygenic burden of rare disruptive mutations in schizophrenia. *Nature, 506*, 185–190.

Roadmap Epigenomics Consortium, Kundaje, A., Meuleman, W., Ernst, J., Bilenky, M., Yen, A., ... Kellis, M. (2015). Integrative analysis of 111 reference human epigenomes. *Nature, 518*, 317–330.

Ronan, J. L., Wu, W., & Crabtree, G. R. (2013). From neural development to cognition: unexpected roles for chromatin. *Nature Reviews Genetics, 14*, 347–359.

Ronemus, M., Iossifov, I., Levy, D., & Wigler, M. (2014). The role of de novo mutations in the genetics of autism spectrum disorders. *Nature Reviews Genetics, 15*, 133–141.

Sakai, Y., Shaw, C. A., Dawson, B. C., Dugas, D. V., Al-Mohtaseb, Z., Hill, D. E., ... Zoghbi, H. Y. (2011). Protein interactome reveals converging molecular pathways among autism disorders. *Science Translational Medicine, 3*, 86ra49.

Salmena, L., Poliseno, L., Tay, Y., Kats, L., & Pandolfi, P. P. (2011). A ceRNA hypothesis: the Rosetta Stone of a hidden RNA language? *Cell, 146*, 353–358.

Samocha, K. E., Robinson, E. B., Sanders, S. J., Stevens, C., Sabo, A., McGrath, L. M., ... Daly, M. J. (2014). A framework for the interpretation of de novo mutation in human disease. *Nature Genetics, 46*, 944–950.

Sanders, S. J., Murtha, M. T., Gupta, A. R., Murdoch, J. D., Raubeson, M. J., Willsey, A. J., ... State, M. W. (2012). De novo mutations revealed by whole-exome sequencing are strongly associated with autism. *Nature, 485*, 237–241.

Schwanhäusser, B., Busse, D., Li, N., Dittmar, G., Schuchhardt, J., Wolf, J., ... Selbach, M. (2011). Global quantification of mammalian gene expression control. *Nature, 473*, 337–342.

Shao, L., & Vawter, M. P. (2008). Shared gene expression alterations in schizophrenia and bipolar disorder. *Biological Psychiatry, 64*, 89–97.

Stein, J. L., de la Torre-Ubieta, L., Tian, Y., Parikshak, N. N., Hernández, I. A., ... Geschwind, D. H. (2014). A quantitative framework to evaluate modeling of cortical development by neural Stem cells. *Neuron, 83*, 69–86.

Steinberg, J., & Webber, C. (2013). The roles of FMRP-regulated genes in autism spectrum disorder: single- and multiple-hit genetic etiologies. *The American Journal of Human Genetics, 93*, 825–839.

Stuart, J. M. (2003). A gene-coexpression network for Global discovery of conserved genetic modules. *Science, 302*, 249–255.

Suliman, R., Ben-David, E., & Shifman, S. (2014). Chromatin regulators, phenotypic robustness, and autism risk. *Frontiers in Genetics, 5*.

Sunkin, S. M., Ng, L., Lau, C., Dolbeare, T., Gilbert, T. L., Thompson, C. L., ... Dang, C. (2013). Allen brain atlas: an integrated spatio-temporal portal for exploring the central nervous system. *Nucleic Acids Research, 41*, D996—D1008.

Tetreault, N. A., Hakeem, A. Y., Jiang, S., Williams, B. A., Allman, E., Wold, B. J., ... Allman, J. M. (2012). Microglia in the cerebral cortex in autism. *Journal of Autism and Developmental Disorders, 42*, 2569—2584.

Tian, Y., Voineagu, I., Pasca, S. P., Won, H., Chandran, V., Horvath, S., ... Geschwind, D. H. (2014). Alteration in basal and depolarization induced transcriptional network in iPSC derived neurons from Timothy syndrome. *Genome Medicine, 6*, 75.

Torkamani, A., Dean, B., Schork, N. J., & Thomas, E. A. (2010). Coexpression network analysis of neural tissue reveals perturbations in developmental processes in schizophrenia. *Genome Research, 20*, 403—412.

Uddin, M., Tammimies, K., Pellecchia, G., Alipanahi, B., Hu, P., Wang, Z., ... Scherer, S. W. (2014). Brain-expressed exons under purifying selection are enriched for de novo mutations in autism spectrum disorder. *Nature Genetics, 46*, 742—747.

Uhlén, M., Fagerberg, L., Hallström, B. M., Lindskog, C., Oksvold, P., Mardinoglu, A., ... Pontén, F. (2015). Tissue-based map of the human proteome. *Science, 347*, 1260419.

Voineagu, I., Wang, X., Johnston, P., Lowe, J. K., Tian, Y., Horvath, S., ... Geschwind, D. H. (2011). Transcriptomic analysis of autistic brain reveals convergent molecular pathology. *Nature, 474*, 380—384.

Waddington, C. H. (1942). Canalization of development and the inheritance of acquired characters. *Nature, 150*, 563—565.

Wang, Z., Gerstein, M., & Snyder, M. (2009). RNA-Seq: a revolutionary tool for transcriptomics. *Nature Reviews Genetics, 10*, 57—63.

Wexler, E. M., Rosen, E., Lu, D., Osborn, G. E., Martin, E., Raybould, H., ... Geschwind, D. H. (2011). Genome-wide analysis of a Wnt1-regulated transcriptional network implicates neurodegenerative pathways. *Science Signaling, 4*, ra65.

Wilhelm, M., Schlegl, J., Hahne, H., Gholami, A. M., Lieberenz, M., Savitski, M. M., ... Kuster, B. (2014). Mass-spectrometry-based draft of the human proteome. *Nature, 509*, 582—587.

Willsey, A. J., Sanders, S. J., Li, M., Dong, S., Tebbenkamp, A. T., Muhle, R. A., ... State, M. W. (2013). Coexpression networks implicate human midfetal deep cortical projection neurons in the pathogenesis of autism. *Cell, 155*, 997—1007.

Winden, K. D., Oldham, M. C., Mirnics, K., Ebert, P. J., Swan, C. H., Levitt, P., ... Geschwind, D. H. (2009). The organization of the transcriptional network in specific neuronal classes. *Molecular Systems Biology, 5*, 291.

Zeng, H., Shen, E. H., Hohmann, J. G., Oh, S. W., Bernard, A., Royall, J. J., ... Jones, A. R. (2012). Large-scale cellular-resolution gene profiling in human neocortex reveals species-specific molecular signatures. *Cell, 149*, 483—496.

Zhang, C., & Darnell, R. B. (2011). Mapping in vivo protein-RNA interactions at single-nucleotide resolution from HITS-CLIP data. *Nature Biotechnology, 29*, 607—614.

Zhang, B., Gaiteri, C., Bodea, L.-G., Wang, Z., McElwee, J., Podtelezhnikov, A. A., ... Emilsson, V. (2013). Integrated systems approach identifies genetic nodes and networks in late-onset Alzheimer's disease. *Cell, 153*, 707—720.

Zhang, B., & Horvath, S. (2005). A general framework for weighted gene co-expression network analysis. *Statistical Applications in Genetics and Molecular Biology, 4*. Article17.

Chapter 12

Somatic Mosaicism and Neurological Diseases

Saumya S. Jamuar
KK Women's and Children's Hospital, Singapore; Howard Hughes Medical Institute, Boston, MA, United States; Harvard Medical School, Boston, MA, United States

Alissa M. D'Gama
Howard Hughes Medical Institute, Boston, MA, United States; Harvard Medical School, Boston, MA, United States

Christopher A. Walsh
Howard Hughes Medical Institute, Boston, MA, United States; Harvard Medical School, Boston, MA, United States; Broad Institute of MIT and Harvard, Cambridge, MA, United States

INTRODUCTION

Human genetic diseases have traditionally been thought to reflect either inherited or de novo variations in the genome. Inherited mutations are present in all cells of the affected individual and can be detected in any cell of the body, including readily available peripheral blood cells; they are referred to as "germ line." De novo mutations usually refer to mutations that occur for the first time in a parent's germ cell but are inherited by all cells at the time of fertilization, and hence are present in all cells of the offspring. On the other hand, somatic mutation is a postzygotic mutational event that leads to an individual having two or more populations of cells with distinct genotypes, despite developing from a single fertilized egg (Biesecker & Spinner, 2013; Poduri, Evrony, Cai, & Walsh, 2013), and thus represents a subset of the larger category of de novo mutations.

There is increasing recognition of the role of germ line de novo mutations in neuropsychiatric diseases, including autism spectrum disorders (ASDs), epilepsy, and schizophrenia (Michaelson et al., 2012; Neale et al., 2012; O'Roak et al., 2011; O'Roak, Vives, Girirajan, et al., 2012; Riviere et al., 2012; Sanders et al., 2012; Veltman & Brunner, 2012). These de novo mutations arise in the sperm or egg of the parent during gametogenesis and hence are absent in the blood of the parents but are present across all cell types in the affected individual. However, de novo or spontaneous mutations also can occur after fertilization, during mitotic cell divisions. These mutations lead to an individual being mosaic, with only a subset of cells harboring the mutation, which is referred to as somatic mosaicism. These mutations will also be absent in the parents, and depending on the timing of mutation with respect to embryonic development, may affect cells across all cell types or may be present in only a few specific cell types (Jamuar et al., 2014; Kennedy, Loeb, & Herr, 2012; Poduri et al., 2013). Whereas the role of somatic mutation in cancer cells is well established (Watson, Takahashi, Futreal, & Chin, 2013), appreciation of an analogous role for somatic mutations that occur randomly during normal mitotic cell divisions of embryonic development and are present in clones of cells in one or more tissues of the body has been relatively recent.

"Obligatory" somatic mutations have been described in several noncancerous disorders, including McCune Albright syndrome (caused by activating mutations in *GNAS*) (Weinstein et al., 1991), Sturge-Weber syndrome (caused by mutations in *GNAQ*) (Shirley et al., 2013), Proteus syndrome (caused by activating mutations in *AKT1*) (Lindhurst et al., 2011), and hemimegalencephaly (HMG) (which is caused by strongly activating mutations in *PIK3CA*, *PIK3R2*, and *AKT3*) (Lee et al., 2012; Poduri et al., 2012; Riviere et al., 2012). These "obligatory" somatic mutations appear to reflect

Genomics, Circuits, and Pathways in Clinical Neuropsychiatry. http://dx.doi.org/10.1016/B978-0-12-800105-9.00012-3

mutations that are lethal when present in the germ line. On the other hand, somatic mutations have also been observed in a fraction of patients with tuberous sclerosis, double cortex, periventricular nodular heterotopia, and bilateral megaloencephalic disorders (caused by mildly activating mutations in *AKT3/PIK3CA/PIK3R2*) presenting as milder versions of the germ-line condition (Lee et al., 2012; Poduri et al., 2013; Riviere et al., 2012). Whereas somatic mutations have been reported in a range of other genetic conditions (Poduri et al., 2013), the broader relevance of somatic mutations to other neurologic conditions such as autism and epilepsy in persons without structural brain malformations is not known.

With advances in genomic technologies, there has been an expansion in our knowledge of types of tissues in which somatic mutations, especially in the context of neuropsychiatric disorders, have been detected, as well as the kinds of mutations being detected (Table 12.1). Somatic mutations could reflect point mutations (also known as single-nucleotide variants [SNVs]), insertions/deletions (indels), copy number variants (CNVs) (including chromosomal aneuploidy), short tandem repeats, and transposable element variants. Here we will review the current state of data about somatic mutations in human neurological disease. We highlight the estimation of the prevalence of these somatic mutations in disorders of brain development and discuss the challenges of identifying such rare mosaic mutations. We further discuss how emerging techniques will allow more refined study of the types and rates of somatic mutation and genomic variation in the brain.

CORTICAL CLONAL ARCHITECTURE AND SOMATIC MUTATIONS

A specific progenitor cell in which a somatic mutation occurs will transmit the mutation to all of its daughter cells. The effects of this mutation will depend on the type of progenitor cell and timing during development at which the mutation occurs. Hence, understanding the clonal architecture of the developing human brain can help us understand how somatic mutations cause genetic disease. As neurons undergo layer specification through a series of fate restriction and specification processes during or after mitosis (Greig, Woodworth, Galazo, Padmanabhan, & Macklis, 2013), any disruption to this process may place an individual at risk for localized, cell type–specific or even layer-specific functional abnormalities.

The principal excitatory neurons of the cerebral cortex and hippocampus are derived from an embryonic neuroepithelium, with progenitor cells lining the ventricular surface deep in the brain (Kriegstein, Noctor, & Martinez-Cerdeno, 2006). Cerebral cortical progenitors produce postmitotic neurons that migrate radially from the deep ventricle to the superficial, outer layers of the brain. The exact details of this radial migratory path have not been established in humans, so it is not known how much dispersion of neurons with common clonal origins takes place during this radial migration (Franco & Muller, 2013). Clones of related pyramidal neurons generally maintain a funnel shape in animal models but are typically highly interspersed with neurons of diverse clonal origins (Kornack & Rakic, 1995; Magavi, Friedmann, Banks, Stolfi, & Lois, 2012; Reid, Tavazoie, & Walsh, 1997; Walsh & Cepko, 1992; Ware, Tavazoie, Reid, & Walsh, 1999). Thus, somatic mutations could produce clones of mutant cells that might concentrate in functional regions of the cortex (Rakic, 1988), but likely as a small proportion of the cells in the region harboring the mutation, so that it might be difficult to detect them using standard techniques of bulk tissue analysis. Focal cortical dysplasias appear to be funnel-shaped on neuroimaging and may represent somatic mutations of deep pyramidal-neuron progenitors, although this has not been completely worked out. Studies have shown increased activation as well as mutations of *AKT–mTOR* pathway in focal cortical dysplasias (Conti et al., 2014; Liu et al., 2014).

In contrast to pyramidal neurons, other cell types of the brain show even less evidence of clonal clustering and hence would be expected to be widely scattered throughout the human cortex, assuming human cortical development resembles that of other mammals. In animal models, inhibitory neurons that populate the cerebral cortex are formed outside the cortex in a second proliferative zone in the basal forebrain, called the ganglionic eminence, which generates the basal ganglia. These neurons migrate large distances in a nonradial direction before turning radially to enter the cortex (Pleasure et al., 2000). There is good evidence that human interneurons are formed by a similar mechanism (Hansen et al., 2013). Astrocytic glial cells arise from several sources, including progenitors that also generate principal neurons (Reid et al., 1997), whereas oligodendrocytes arise in the basal forebrain that generates cells for the entire forebrain (Kessaris et al., 2006). Hence, inhibitory cells carrying common somatic mutations would be expected to be dispersed throughout the cortex, and neighboring cells in the cortex may have diverse clonal origins. This complex architecture makes it difficult to detect somatic mutations through bulk tissue sequencing or genotyping using standard techniques, and hence it has been difficult to estimate the significance of somatic mutations in disorders of brain development. In addition, because complex neural circuits are highly interconnected throughout the cortex, localized disruptions caused by somatic mutations may affect widespread networks, leading to substantial disease (Poduri et al., 2013).

SOMATIC MUTATIONS IN NORMAL BRAIN

Mosaic aneuploidies, ie, loss or gain of chromosomes, were first reported in the cerebral cortex of normal developing mice (Rehen et al., 2001). Subsequent studies suggested that 20–30% of adult mouse and human brain cells were aneuploid (Rajendran, Wellbrock, & Zupanc, 2008; Rajendran et al., 2007; Westra et al., 2008; Westra et al., 2010). These studies used spectral karyotyping and/or fluorescent in situ hybridization (FISH), which have not been widely used as quantitative techniques to study aneuploidy, to determine the rate of aneuploidies. Other studies using single-cell genome-wide analyses have demonstrated a much lower rate of aneuploidy in the mammalian brain, with fewer than 5% of neurons reported to be aneuploid at the level of an entire chromosome (Cai et al., 2014; Knouse, Wu, Whittaker, & Amon, 2014; McConnell et al., 2013). Although aneuploidies are not as common as previously suggested, single-cell genome-wide analyses suggest that subchromosomal CNVs are common in single neurons, such that most neurons show at least one large CNV > 1 Mb in size creating a potential genomic mosaic that displays a predominance of hypoploidy over hyperploidy (Cai et al., 2014; McConnell et al., 2013).

It is also been suggested that retrotransposition of long interspersed nuclear elements (L1) is a special subset of somatic mutation that may have a critical role in nervous system development (Coufal et al., 2009; Muotri et al., 2005). Retrotransposon insertion can cause gene mutation by inactivating genes, by inserting into them or changing patterns of gene expression. Initial estimates of these events suggested that dozens of somatic L1 insertions may be present in each neuronal genome (Coufal et al., 2009). However, quantitative analysis by a single-cell sequencing approach confirmed that L1 retrotransposition occurs during neurogenesis in the human brain, but suggested a rate of less than one unique insertion per neuronal genome. This suggests that they are an occasional source of mutation but not an obligatory event in neurogenesis (Evrony et al., 2012). Further work is required to determine L1 retrotransposition rates across different regions of the human brain and assess the potential role of these and other types of somatic mutations in neurological disease.

SOMATIC MUTATION IN NEUROLOGICAL DISEASE

"Brain-Only" Somatic Mutations

Genetic disorders of neuronal migration in the brain have been previously estimated to be associated with somatic mutation in 5–10% of patients. However, application of a highly sensitive sequencing panel along with very high (>200 times) coverage, next-generation sequencing (NGS) identified as many as 30% of unexplained brain malformations that could be diagnosed as genetically reflected somatic mosaic mutations (Jamuar et al., 2014). Mutations in the gene *lissencephaly 1 (LIS1)* are typically associated with a "smooth brain" phenotype of lissencephaly, whereas mutations in *doublecortin (DCX)*, located on the X chromosome, are associated with lissencephaly in males but subcortical hand heterotopia pattern in females because X chromosome inactivation creates mosaic populations of cells in the female that have either normal or abnormal DCX function (Fig. 12.1) (Gleeson et al., 1998). Similarly, somatic mutations in *LIS1* and males with somatic mutations in *DCX* can exhibit a double cortex because only some neurons carry the mutation (Gleeson et al., 2000; Sicca et al., 2003). Brain malformations have been reported in patients when as few as 10% of blood cells carry somatic mutations (Fig. 12.1) (Jamuar et al., 2014), although how precisely the proportion of mutant blood cells predicts the proportion of brain cells carrying the mutation is not known. The presence of these mutations in the DNA derived from leukocytes suggests that the somatic mutation occurred relatively early postzygotically.

Other focal malformations of cerebral cortical development had been hypothesized to occur via somatic mutation in the developing brain, but testing required the availability of the affected brain tissue. We and others have reported on the role of somatic activating mutations in *PI3K–AKT–mTOR* pathway in the formation of brain malformations including HMG (Lee et al., 2012; Poduri et al., 2012; Riviere et al., 2012). Some of these mutations are detectable only in the affected tissue and are absent in the leukocytes, and hence represent "brain-only" somatic mutations (Poduri et al., 2012). However, somatic de novo mutations in *AKT3* and *PIK3CA* have been reported to be detectable at low levels in leukocytes as part of the megalencephaly–capillary malformation and megalencephaly–polydactyly–polymicrogyria–hydrocephalus syndromes (Lee et al., 2012; Riviere et al., 2012).

These studies suggest that brain overgrowth syndromes and neuronal migration disorders can be caused by somatic mutations in brain progenitor cells. In some, these mutations occur early enough in development to be present in many tissues, affecting cells outside the brain as well. In contrast, other mutations might be limited to the brain because they occur after the embryonic separation of brain from nonbrain tissue.

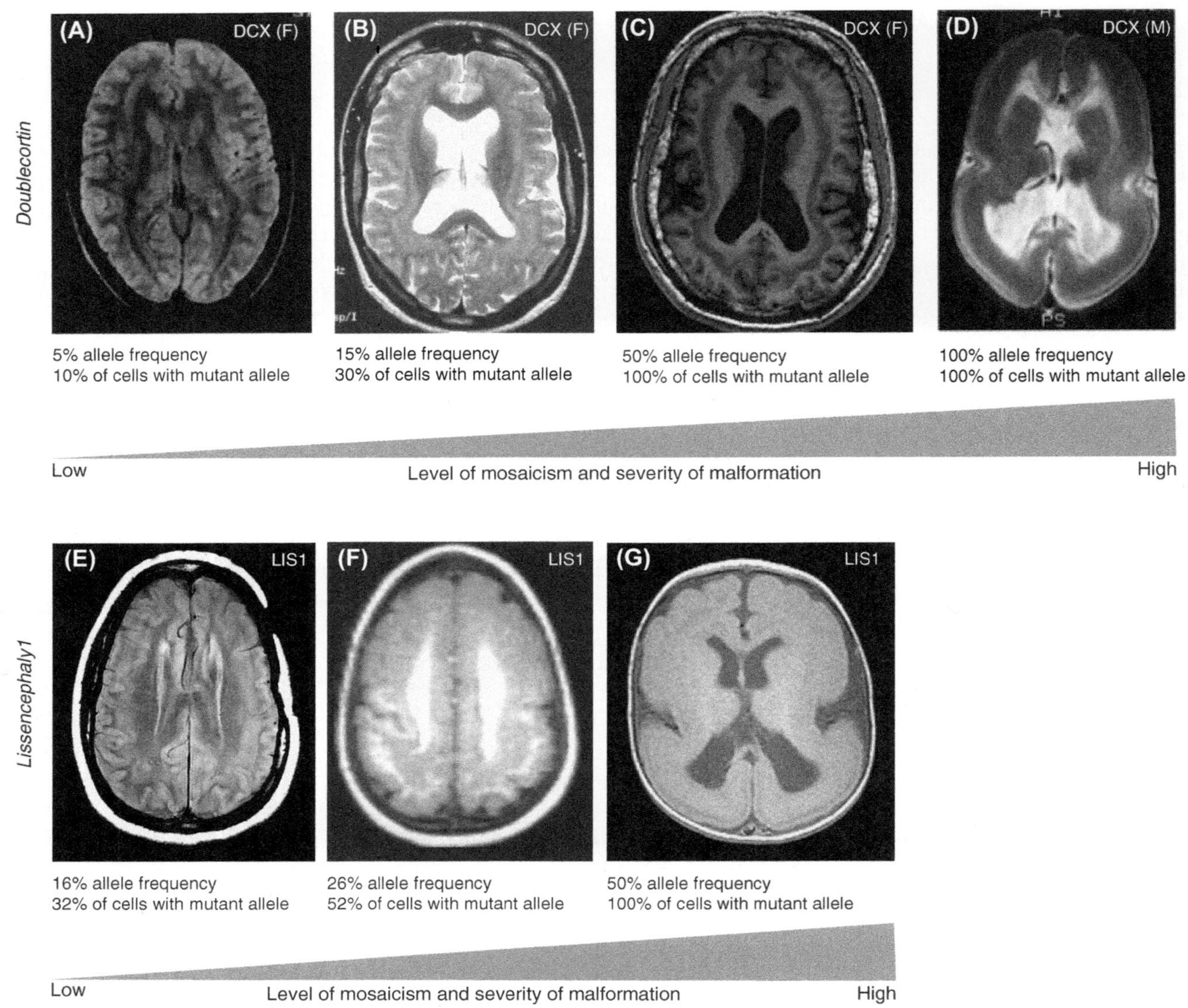

FIGURE 12.1 Level of somatic mosaicism and severity of phenotype. Axial magnetic resonance images of individuals with somatic mutations (A, B, E, and F) compared with individuals with germ-line mutations (C, D, and G) show a spectrum of the severity of the phenotype. The figures in the top row (A–D) show individuals with mutations in *DCX* and the severity of the double cortex, whereas figures in the bottom row (E–G) show individuals with mutations in *LIS1* and the severity of lissencephaly. The severity of the phenotype increases as the proportion of cells with the mutation increases. *DCX* is on the X-chromosome and for the female with a germ-line mutation (C) it appears as though the mosaic is caused by random X-inactivation in the migrating neurons.

"Second-Hit" Mutations Produce Mosaicism

In several dominantly inherited conditions, an individual exhibits a heterozygous germ-line variant, present in all cells, with a somatic second mutation leading to overgrowth of specific tissues, as per the "two-hit" model of Knudson (Knudson, 1971). For example, neurofibromatosis type 1 (NF1) is characterized by germ-line mutations in the gene *NF1* and is associated with optic gliomas and focal lesions of the skin and peripheral nervous system. A second mutation in the other *NF1* allele is known to cause neurofibromas (Garcia-Linares et al., 2011). Similarly, in a related neurocutaneous disorder, tuberous sclerosis complex (TSC), somatic second mutations have been shown in non–nervous system tumors of TSC, subependymal giant cell astrocytomas, and noncancerous cortical tubers in patients with TSC (Niida et al., 2001; Qin, Chan, et al., 2010). These "two-hit disorders" suggest how common a somatic mutation in any given gene can be, because the same gene can be independently mutated in many organs of the body. For instance, a typical patient with TSC will show dozens of cerebral hamartomas, known as tubers, in the brain, which suggests that the corresponding TSC gene underwent dozens of secondary somatic mutations on top of the inherited germ-line mutation.

Neurodevelopmental Disorders Caused by Somatic Mutations

De novo mutations have been implicated in almost all neurodevelopmental and neuropsychiatric disease, most notably intellectual disability (ID) and ASDs. A study employing whole genome sequencing (WGS) in ID identified de novo mutations as the most common causes and further noted that several of these appeared to be somatic mosaic mutations (Gilissen et al., 2014). De novo point mutations appear to be common collectively as a cause of ASD, although any given gene appears to be implicated infrequently. A large-scale study of ASD whole exome sequences identified de novo mutations in about 5% of patients, about 10% of whom appeared to be somatic mosaic (O'Roak, Vives, Fu, et al., 2012). However, because of limited coverage of typical whole exome sequence, many somatic mosaic mutations could be missed in whole exome sequence (Fig. 12.2) (Jamuar et al., 2014). De novo mutations in *MeCP2* gene cause Rett syndrome (Amir et al., 1999) and mosaic mutations have been reported in males with both classic and atypical forms of Rett syndrome (Pieras et al., 2011; Topcu et al., 2002). Our ability to detect a pathogenic somatic mutation using current clinical methods depends on its abundance in leukocytes. It is possible that some cases of autism, epilepsy, and perhaps other neuropsychiatric conditions such as schizophrenia may show roles for somatic mutations that have been overlooked by Sanger sequencing.

Neurological Diseases Caused or Modulated by Somatic Mutations

Some cases of neurodegenerative diseases have been associated with somatic mutations or can be modified by somatic mutations. Some cases of incontinentia pigmenti result from somatic mutations in the NF-κ-B essential modulator gene (Smahi et al., 2000). A case of sporadic, early-onset Alzheimer disease was attributed to a somatic mosaic *presenilin-1* mutation present in the brain (Beck et al., 2004) and another case of sporadic Creutzfeldt–Jakob disease was caused by an early embryonic somatic mutation identified by the presence of three alleles for the gene encoding major prion protein PrP (Alzualde et al., 2010). Somatic mutations in *alpha-synuclein* have been reported in association with Parkinson disease

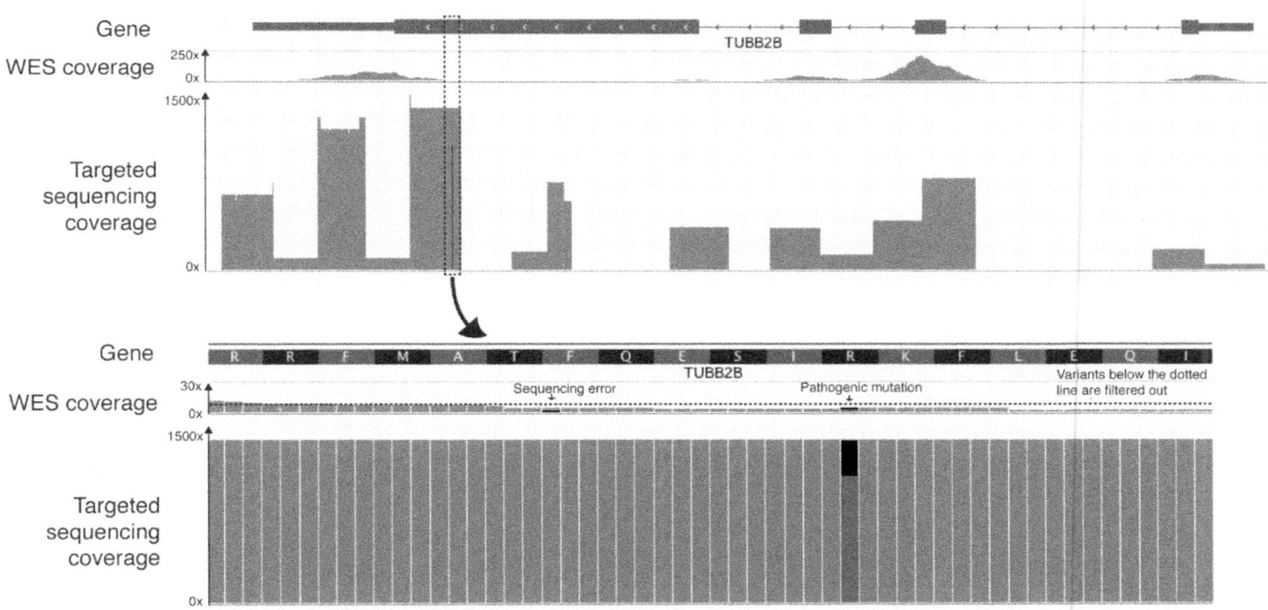

FIGURE 12.2 Comparison of coverage on WES versus deep-targeted sequencing. The figure shows coverage of the entire *TUBB2B* gene (top panel) and zoomed-in view (bottom panel) of the region near the variant (box in top panel). The top panel illustrates the variable coverage (in gray) across different exons of the same gene (*TUBB2B*), both on WES and targeted sequencing. The bottom panel demonstrates the difficulty in discerning the mutation (*black box*) from background sequencing-related error on low-coverage WES. However, the mutation (*black box*) is easily recognized on high coverage targeted sequencing. More importantly, the mutation was filtered out on WES owing to the extremely low coverage (4×). *Dotted line* represents the typical depth (10×) used as a cutoff for WES analysis.

(Proukakis, Houlden, & Schapira, 2013). Some neurodegenerative diseases, including Huntington disease, dentatorubral pallidoluysian atrophy, and Fragile X syndrome, are caused by inheritance of microsatellite repeats that are highly unstable and can exhibit marked somatic heterogeneity in repeat lengths across brain regions and tissues of affected individuals (Hashida et al., 2001; Hellenbroich, Schwinger, & Zuhlke, 2001; Ito et al., 1998).

In individuals with epilepsy, use of NGS has identified several monogenic causes of well-characterized epilepsy syndromes, predominantly showing mutations in ion channel or neurotransmitter receptor subunits. *Disheveled, Egl-10 and pleckstrin domain containing protein 5* (*DEPDC5*) has been recognized as a major cause of autosomal dominant focal epilepsy (Dibbens et al., 2013; Ishida et al., 2013). Imaging of individuals with *DEPDC5* mutations is typically normal. Scheffer et al. (2014) described a series of individuals with *DEPDC5*-related focal epilepsy, but in association with a range of brain malformations from relatively subtle bottom of sulcus focal cortical dysplasia to subcortical gray matter heterotopia. *DEPDC5* is a member of the GATOR1 complex that exerts an inhibitory effect on mTOR-mediated processes, such as cell growth and proliferation. The presence of brain malformations in individuals with *DEPDC5*-related epilepsy could be partially explained by dysregulation of the mTOR pathway at the cellular level with microscopic malformations that are undetectable by current imaging modalities. Another possibility could include a "second-hit" phenomenon, well described in TSC (Niida et al., 2001), with a somatic mutation in the other allele of *DEPDC5* or another gene in the mTOR pathway causing the malformations. Direct analysis of the brain tissues of these individuals could provide support for this hypothesis and may be achieved if these individuals require resection of the malformation to treat the epilepsy (Poduri, 2014).

TYPES OF SOMATIC VARIANTS

Large-Scale Chromosomal Abnormalities

Chromosomal alterations such as whole chromosome aneuploidy, segmental aneuploidy, and structural alterations have been historically identified by cytogenetic analyses. Chromosomal abnormalities are a fairly common cause of developmental disorders and affect one in 200 live-born individuals and up to 50% of spontaneous miscarriages (Biesecker & Spinner, 2013). Chromosomal alterations affect brain function owing to perturbation in the dosage of the genes, which negatively affects brain development (Smith et al., 2011; Toro et al., 2010).

Aneuploidy of chromosomes 13, 18, and 21 and sex chromosomes accounts for nearly all aneuploidy live births. Aneuploidy of other chromosomes is usually lethal to the developing fetus and is observed only in the mosaic state for some of the chromosomes (including 1, 8, 9, 16, 17, and 22) (Poduri et al., 2013). These individuals have a variable phenotype because the severity of the disorder is determined by the particular chromosome involved as well as by the proportion of cells in the body carrying the aneuploidy. For example, mosaic trisomy 21, which affects 1% of cases of Down syndrome, leads to a less severe phenotype compared with germ line trisomy 21 (Biesecker & Spinner, 2013; Poduri et al., 2013).

Mosaicism for structural abnormalities is less common than aneuploidy but it has been identified in the form of translocations, deletions, duplications, inversions, ring chromosomes, and isochromosomes (Conlin et al., 2011; Gijsbers et al., 2011; Hsu et al., 1996). Isochromosomes are structurally abnormal chromosomes created by the presence of two copies of one arm of a chromosome, whereas the other arm is missing. Four well-recognized syndromes include Pallister Killian syndrome (isochromosome 12p), cat-eye syndrome (isochromosome 22q), isochromosome 15q11, and isochromosome 18p; all have been associated with variable degrees of developmental delay and ID (Biesecker & Spinner, 2013).

Copy Number Variants

Advances in genomic tools such as microarray analysis have improved the resolution of cytogenomic imbalances detected at a submicroscopic level and, in some instances, to individual exons. These aberrations are referred to as CNVs. CNVs have been reported in 10−15% of patients with developmental delay and/or ASD; hence CNV analysis is recommended as the first-tier test for evaluation of patients with developmental delay and/or ASD (Miller et al., 2010).

Mosaic CNVs have been reported in 0.5−3.74% of patients with congenital and developmental anomalies (Ballif et al., 2006; Cheung et al., 2007; Conlin et al., 2010). Mosaic CNV involving chromosome 1q has been well-described in individuals with HMG (Cai et al., 2014; Poduri et al., 2012) and in FCD (Conti et al., 2014; Liu et al., 2014). The frequency of mosaic abnormalities increases with age and individuals aged more than 50 years have an estimate of 2−3% mosaic CNV per person (Jacobs et al., 2012; Laurie et al., 2012). Induced pluripotent stem cells derived from skin fibroblasts were found to contain about two CNVs per sample, half of which existed at a low frequency in the parental fibroblasts (Abyzov et al., 2012). Copy number variation has been noted across different tissues from the same individual (Piotrowski et al., 2008) and mosaic CNV has been observed in monozygotic twins with discordant phenotypes (Bruder et al., 2008).

Single-Nucleotide Variants

Mutations at the level of the nucleotide, either base substitutions or small indels, in the genomic DNA may lead to abnormal mRNA and consequent abnormal protein production. SNVs can lead to loss of function (missense or splicing variants), gain of function (missense), or absence of protein (nonsense, frameshift, or canonical splicing variants).

Mosaic SNVs have been reported in a range of disorders, including those associated with brain malformations, such as double cortex (Gleeson et al., 2000; Sicca et al., 2003), periventricular nodular heterotopia (Guerrini et al., 2004; Parrini, Mei, Wright, Dorn, & Guerrini, 2004), HMG (Lee et al., 2012; Poduri et al., 2012; Riviere et al., 2012), and Sturge Weber syndrome (Shirley et al., 2013), among others. Mosaic SNVs in the *PI3K−AKT−mTOR* pathway result in overgrowth and have been hypothesized to be lethal when constitutional because they disrupt early embryonic development. On the other hand, the mosaic variants in *DCX, LIS1, TSC1, FLNA*, etc., present as milder phenotypes compared with when they present as germ line (Poduri et al., 2013).

TISSUE TYPE CONSIDERATIONS

Because mosaicism is caused by mitotic error after fertilization, interpreting mosaicism requires an understanding of the earliest events of fetal development (Fig. 12.3). After fertilization the zygote undergoes successive mitoses to produce a

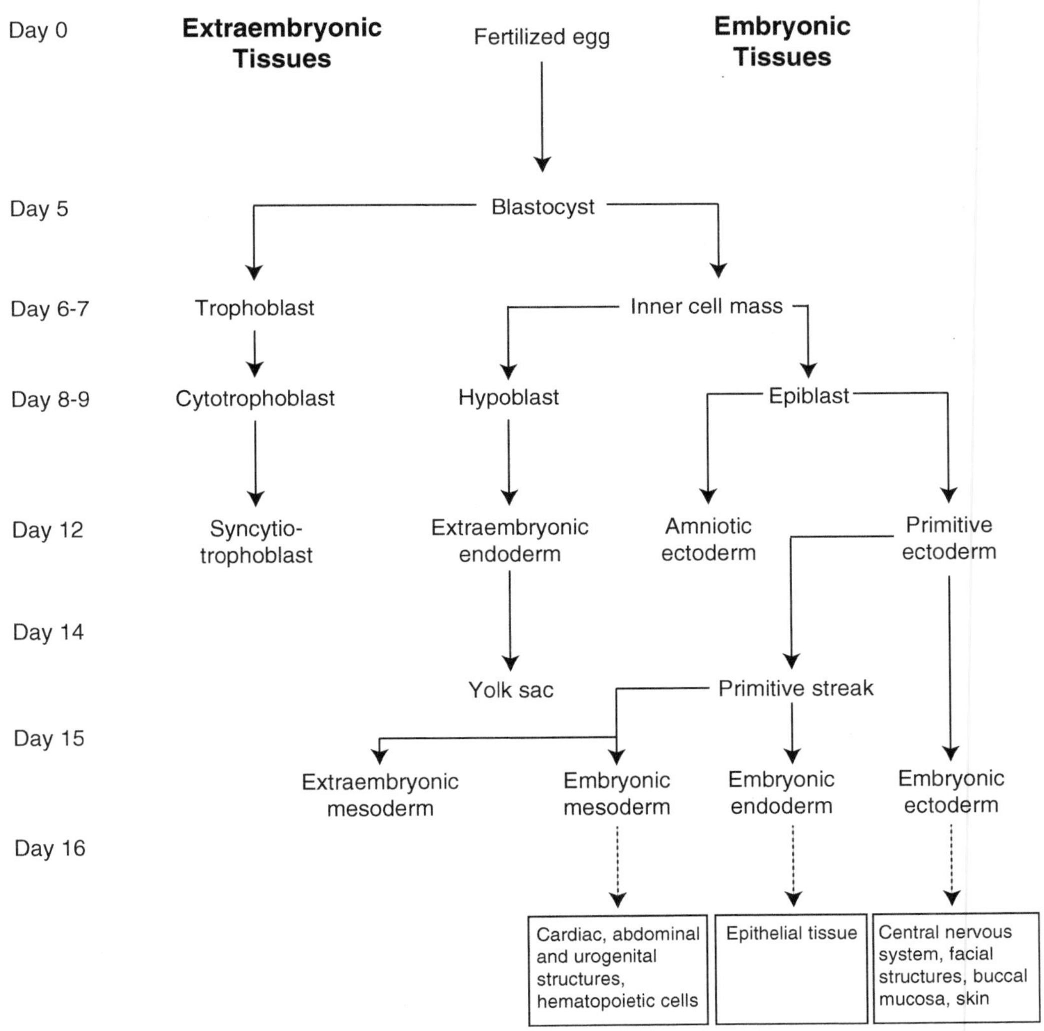

FIGURE 12.3 Development of human embryonic tissues. Schematic representation of development of human embryo postfertilization from days 0 to 15. *Adapted from Appendix A: Early Development. In Stem Cell Information <http://stemcells.nih.gov/info/scireport/pages/appendixa.aspx>. Bethesda, MD: National Institutes of Health, U.S. Department of Health and Human Services, 2009. © 2001 Terese Winslow, Lydia Kibiuk.*

ball of cells that then cavitates to produce the blastocyst. The outermost layer of the blastocyst produces the placental tissue whereas the inner cell mass results in the embryo. Initially the embryo consists of two different cell layers: epiblast and hypoblast. Upon subsequent differentiations, the epiblast gives rise to ectoderm, mesoderm, and endoderm whereas the hypoblast becomes the amniotic cavity. Structures of ectodermal origin include the central nervous system, facial structures including buccal mucosa, and the extremities including skin. Endoderm gives origin to epithelial tissues including the lining of the urinary tract and lung, whereas the mesoderm differentiates into the cardiac, abdominal, and urogenital structures and hematopoietic cells (Gardner, Sutherland, & Shaffer, 2011).

Somatic mutations that develop in the early postzygotic stage will be distributed across different tissue types and will likely be detected by analysis of the peripheral blood leukocytes. This has been well demonstrated in certain forms of brain malformations (such as double cortex and megalencephaly, among others) and has been associated with mosaicism as low as 1% in the leukocyte-derived DNA (Jamuar et al., 2014; Riviere et al., 2012).

On the other hand, some mutations develop later in fetal development, and hence analysis of peripheral blood leukocytes would be unrevealing. In such instances, direct examination of the affected tissue would provide the best chance of detecting the mutation. Mutations in the *PI3K—AKT—mTOR* pathway associated with asymmetric brain overgrowth, or HMG, have been shown to be restricted to the brain tissue and are undetectable in the peripheral blood leukocytes (Jamuar et al., 2014; Lee et al., 2012; Poduri et al., 2012). Similarly, fibroadipose hyperplasia and related syndromes (such as CLOVES and Proteus syndrome) have been associated with somatic mutation in genes in the *PI3K—AKT* pathway that are detectable only in the affected cells (Kurek et al., 2012; Lindhurst et al., 2011, 2012). Because the affected tissue (especially brain) may not be easily accessible most patients with neurological diseases, there is some evidence that analysis of embryologically related tissue such as skin fibroblasts and buccal mucosa may be more successful in detecting mutations in some of these patients (Huisman, Redeker, Maas, Mannens, & Hennekam, 2013; Jamuar et al., 2014; Riviere et al., 2012). However, some mutations are probably not detectable in skin fibroblasts or buccal swab either, and may be confined to the brain.

TOOLS TO STUDY SOMATIC VARIATION IN THE BRAIN

Copy Number Variants

Cytogenetics

Cytogenetic analysis is the study of the banded pattern of chromosomes during metaphase of the cell cycle. It is carried out on a cell-by-cell basis, and hence mosaicism is easily recognized when only a few cells carry the chromosomal abnormality. This analysis can detect large-scale chromosomal abnormalities, most commonly aneuploidy. The resolution of detection of chromosomal aberration is around 5 Mb and copy number changes below 5 Mb may not be detected on routine cytogenetic analysis. In addition, the cytogenetic detection of low-level mosaicism is challenging because adequate number of cells must be counted (Biesecker & Spinner, 2013).

FISH, on the other hand, uses tagged probes that bind to specific chromosome of interest and allows for the detection of numeric and structural abnormalities, including submicroscopic copy number changes. An advantage of FISH over chromosome analysis is that it does not require cells to be divided actively and hence it can be performed in both the interphase and metaphase of the cell cycle. However, hybridization-related artifacts are common and the high rates of mosaic aneuploidy reported with the use of karyotyping and FISH (Rehen et al., 2005) have not been replicated in single-cell studies (Cai et al., 2014; Knouse et al., 2014).

Microarrays

Over the past 10 years, microarray-based techniques have replaced conventional cytogenetic analysis because they are able to detect submicroscopic genomic imbalances or CNVs and unlike FISH, they allow for interrogation of CNVs across the entire genome. The advantages of array-based testing for detecting mosaicism are that samples do not require culturing, which itself can introduce mutations, and cells in all cell-cycle phases are analyzed. In comparison, conventional karyotyping requires culturing followed by analysis of these cells at metaphase or interphase, which may introduce artifacts (Biesecker & Spinner, 2013).

Types of microarray analyses include array-comparative genomic hybridization (aCGH), which detects copy number aberrations, and genome-wide single-nucleotide polymorphism (SNP) arrays, which can analyze both SNPs and CNVs. aCGH is able to detect somatic CNVs when variant cells constitute more than 10% of the total cell population (Ballif et al., 2006). SNP arrays are much more sensitive for mosaicism detection, and mosaicism involving less than 5% of cells has

TABLE 12.1 List of Somatic Mutations in Association With Neuropsychiatric Disorders

SNV and Microdeletions/Microduplications			
Gene	**Allele**	**% Mutant Cells**	**Comments**
FMR1	CGG repeat length in 5′ UTR	–	41% (61/148) of males with Fragile X syndrome (FXS) tested showed mosaicism for full mutation and permutation using genomic DNA from peripheral blood leukocytes (Nolin, Glicksman, Houck, Brown, & Dobkin, 1994)
		–	4 males with FXS with both premutation and full mutation alleles; 2 show significant differences in the ratio of the 2 alleles in different tissues (blood and skin) (Dobkin et al., 1996)
		28% (deletion) 72% (full mutation)	Mosaic deletion of part of the CGG repeat and 30 bp immediately 3′ (present in 28% of cells) and methylated full mutation in a male with FXS (de Graaff et al., 1996)
		20–30% (deletion) 70–80% (full mutation)	Male patient with FXS mosaic for a full mutation and a 486-bp microdeletion (present in 20–30% of leukocytes) (Schmucker, Ballhausen, & Pfeiffer, 1996)
		–	Eight male patients with FXS mosaic for full mutation and normal or reduced size allele owing to a deletion (Grasso et al., 1999)
		2% (deletion) 23% (premutation) 75% (full mutation)	Male with FXS mosaic for full mutation (75% of blood leukocytes), premutation (23% of leukocytes), and deletion (2% of leukocytes) (Petek, Kroisel, Schuster, Zierler, & Wagner, 1999)
		–	Male with FXS mosaic for full mutation and 905-bp deletion (Garcia Arocena, de Diego, Oostra, Willemsen, & Mirta Rodriguez, 2000)
		–	Male with cognitive and behavioral features of FXS mosaic for premutation and full mutation (Zeesman et al., 2004)
		–	3 males with mental retardation mosaic for full mutation and premutation (Bilgen et al., 2005)
			Female mosaic for full mutation (30%) and deletion (20%), in presence of a normal allele (Fan et al., 2005)
		–	2 males with autism mosaic for full mutation and premutation and 1 male with autism mosaic for full mutation and deletion (Reddy, 2005)
		–	Male with FXS phenotype mosaic for full mutation and premutation, with full mutation more readily detected in skin fibroblasts than peripheral blood (MacKenzie, Sumargo, & Taylor, 2006)
		15% (deletion) 25% (premutation) 60% (full mutation)	Male mosaic for full mutation (60%), premutation (25%), and deletion (15%) (Govaerts et al., 2007)
		90% (deletion) 10% (intact)	Male with cognitive and behavioral features of FXS mosaic for deletion (present in 90% of lymphocytes) and intact *FMR1* (Coffee et al., 2008)
		–	Female fetus mosaic for an intermediate allele and a full mutation, in the presence of a normal allele, confirmed in amniotic fluid, skin biopsy, and blood (Ferreira et al., 2013)

Continued

TABLE 12.1 List of Somatic Mutations in Association With Neuropsychiatric Disorders—cont'd

SNV and Microdeletions/Microduplications			
Gene	**Allele**	**% Mutant Cells**	**Comments**
		25% (premutation, blood) 75% (full mutation, blood) 5–6% (premutation, brain) 94–95% (full mutation, brain)	Male with FXS mosaic for full mutation and premutation (present in 25% of peripheral blood cells and 5–6% of brain cells) (Pretto et al., 2013)
MeCP2	p.R270X	–	Mosaic *MeCP2* mutation in a boy with classic Rett syndrome (Topcu et al., 2002)
	c.241del2	–	Mosaic *MeCP2* mutation in a boy with some features of Angelman syndrome (Hitchins et al., 2004)
	p.T158M	–	Mosaic homozygous *MeCP2* mutation in a girl with classic Rett syndrome (homozygous mutant allele in blood, mutant and wild-type allele present in cultured fibroblasts) (Karall et al., 2007)
	p.Y120X	20–30%	Mosaic *MeCP2* mutation (20–30% mosaicism) in a boy with atypical Rett syndrome (Pieras et al., 2011)
NF1	Microdeletion	–	Female with NF1 mosaic for NF1 microdeletion (Colman, Rasmussen, Ho, Abernathy, & Wallace, 1996)
	Microdeletion	–	Patient with NF1 mosaic for NF1 microdeletion (Ainsworth, Chakraborty, & Weksberg, 1997)
	Microdeletion	–	3% (2 of 67 patients with NF1 tested) mosaic for NF1 microdeletion (Rasmussen et al., 1998)
	Microdeletion	15–24% (café-au-lait spots) 0% (normal skin and peripheral blood lymphocytes)	Mosaic *NF1* microdeletion present in café-au-lait spots (deletion in 15–24% of cells) and absent from normal skin and peripheral blood lymphocytes in a boy with segmental neurofibromatosis (Tinschert et al., 2000)
	Microdeletion	70% (blood lymphocytes) 15% (fibroblast cultures)	Mother of 2 brothers with NF1 mosaic for microdeletion (Petek et al., 2003)
	Microdeletion	91–100% (peripheral leukocytes) 51–80% (buccal smear or peripheral skin fibroblasts)	40% (8 of 20) patients with sporadic NF1 microdeletions were mosaic for the deletion, present in 91–100% of peripheral leukocytes and 51–80% in buccal smears or peripheral skin fibroblasts (Kehrer-Sawatzki et al., 2004)
	p.R1968X	–	Monozygotic twins discordant with NF1, the affected twin had heterozygous mutation in all cell samples while the affected twin had mutation in lymphoblastoid and buccal cells but not fibroblasts (Kaplan et al., 2010)
	Microdeletion	94–99% (blood cells) 24–82% (urine epithelial cells)	11 patients with some features of neurofibromatosis mosaic for microdeletion present in 94–99% of blood cells and 24–82% of urine epithelial cells (Roehl et al., 2012)
	Microdeletion	–	10% (14 of 146 patients with NF1 deletions) were mosaic for microdeletion, including patients with generalized and segmental NF (Messiaen et al., 2011)

TABLE 12.1 List of Somatic Mutations in Association With Neuropsychiatric Disorders—cont'd

SNV and Microdeletions/Microduplications			
Gene	**Allele**	**% Mutant Cells**	**Comments**
	Deletion	70% (blood cells) 5% (buccal mucosal cells)	Male with cognitive disability and dysmorphic features mosaic for 2.8-Mb *NF1* deletion (70% of blood cells and 5% of buccal mucosal cells) (Taylor Tavares, Willatt, Armstrong, Simonic, & Park, 2013)
	p.Q2510X	42% (peripheral blood) 36% (oral mucosa) 12% (uroepithelium)	Patient with NF type I mosaic for mutation in peripheral blood cells (42%), oral mucosal cells (36%), and uroepithelial cells (12%) (Zhou et al., 2012)
OPRL1, NAV2, C3orf38, NTNG1, FRYL, ZNF420, GNA11, TSNARE1, FBXW9	*OPRL1* p.R157C, *NAV2* synonymous, *C3orf38* synonymous, *NTNG1* p.Y23C, *FRYL* p.E2683D, *ZNF420* p.L76P, *GNA11* synonymous, *TSNARE1* p.S466L, *FBXW9* p.G445S	–	3.6% (9 of 248) de novo mutations identified had evidence suggesting they were somatic point mutations (O'Roak, Vives, Girirajan, et al., 2012)
PIAS1, HIVEP2, KANSL2	*PIAS1*: p.D403E, *HIVEP2*:p.A1065T, *KANSL2*: Synonymous	40–44%	Whole genome sequencing of patients with intellectual disability (ID) identified 3 of 10 de novo single-nucleotide variants in candidate ID genes in a mosaic state: PIAS1 (21%), HIVEP2 (22%), and KANSL2 (20%) (Gilissen et al., 2014)
SCN1A	p.R542X, c.965 − 2A > C	–	*SCN1A* somatic and germ-line mosaicism in parents of patients with severe myoclonic epilepsy of infancy (Depienne et al., 2006)
	c.5240insAA, IVS17 + 1G > A	70–80%	*SCN1A* somatic (present in 70–80% of cells) and germ-line mosaicism in parents of patients with severe myoclonic epilepsy of infancy (Gennaro et al., 2006)
	IVS4 + 1G > A	37%	*SCN1A* somatic (present in 37% of ectodermal derivative cells) mosaicism in father, who had febrile seizures, of 2 sisters with severe myoclonic epilepsy of infancy (Marini, Mei, HelenCross, & Guerrini, 2006)
	p.V244L, p.K246X	30%	*SCN1A* somatic (present in 30% of lymphocyte cells) mosaicism in mother of 2 brothers with severe myoclonic epilepsy of infancy (Morimoto et al., 2006)
	p.Arg1329X	5%	*SCN1A* somatic mosaicism (5%) in mother of 2 half-sisters with Dravet syndrome (Selmer et al., 2009)
	p.D194N	60–70%	Mosaic *SCN1A* mutation (present in 60–70% of cells) in father (with genetic epilepsy with febrile seizures plus) of a patient with Dravet syndrome (Azmanov et al., 2010)
	c.1377 + 1G > A, c.3878delA, c.602 + 1G > A, c.965 − 2A > C, c.5493delT, p.I124N, p.N191Y, p.R542X, p.R580X, p.R712X, p.I1782M, p.R1912X, Deletion	–	*SCN1A* somatic mutations (mutation proportion in blood 0.04–85%) in parents from 13 families with children with Dravet syndrome (Depienne et al., 2010)

Continued

TABLE 12.1 List of Somatic Mutations in Association With Neuropsychiatric Disorders—cont'd

SNV and Microdeletions/Microduplications			
Gene	**Allele**	**% Mutant Cells**	**Comments**
	p.R101Q	–	Mosaic *SCN1A* mutation in mother with febrile seizures plus transmitted to daughters with Dravet syndrome (Sun et al., 2010)
	Microdeletion	20%	Mosaic *SCN1A* microdeletion (20% of cells) in father of daughters with epilepsy syndrome (Suls et al., 2010)
	p.R222X	–	*SCN1A* somatic mosaicism leading to monozygotic twins discordant for Dravet syndrome (Vadlamudi et al., 2010)
	p.Q1923R, p.I1616T	25%, 50%	*SCN2A* somatic mutations (allele frequency 12.5% and 25%, respectively) in mildly affected or unaffected parents of patients with partial epilepsy with antecedent febrile seizures (Michaelson et al., 2012)
SCN2A	p.V1326L	36%	Mosaic *SCN2A* mutation (allele frequency 18%) in a girl with Ohtahara/West syndrome (Nakamura et al., 2013)
	p.S1336Y	–	Gonadal mosaic *SCN2A* mutation in father passed to half-brothers with Ohtahara syndrome (Zerem et al., 2014)
TSC1	p.R786X	–	Mosaic *TSC1* mutation in a patient (Roberts, Jozwiak, Kwiatkowski, & Dabora, 2001)
TSC1/TSC2	*TSC1*: 942insA, 1473delC *TSC2*: 2374-2A > C, 156 + 1G > A, deletions	–	*TSC1* or *TSC2* somatic mosaicism in mildly affected parents from 5 families and *TSC1* gonadal mosaicism in an unaffected parent in 1 family, of 62 families studied (about 10%) (Verhoef et al., 1999)
TSC2	p.N1643K	30%	Mosaic *TSC2* mutation (30% of cells) in a patient with tuberous sclerosis complex (TSC) (Antonarakis, Sampson, & Cheadle, 2002)
	p.R622W	1.4–21.6%	Second-hit *TSC2* mosaic mutation (0.7–10.8% frequency) in the cortex of a patient with a *TSC2* germ-line mutation (Qin, Chan, et al., 2010)
	c.1444-1G > A, p.R1743Q	10.68%	Two TSC patients with *TSC2* mosaic mutations (Qin, Kozlowski, et al., 2010)
Copy Number/Structural Variation			
Chromosome	**Type**	**% Mutant cells**	**Comments**
2q23.3q24.3	Duplication	40–51%	Mosaic 2q duplication (present in 40–51% of blood cells) including *SCN2A* and *SCN3A* in a boy with atypical epilepsy (Vecchi et al., 2011)
4p12p16	Duplication	–	Mosaic 4p duplication in a Japanese girl diagnosed with autism at age 3 years (Kakinuma, Ozaki, Sato, & Takahashi, 2008)
16p11.2	Duplication, deletion	–	Mosaic 16p11.2 duplication in patient (61% of cells), mosaic 16p11.2 deletion in another patient (23% of cells) (Shinawi et al., 2010)
16p11.2	Deletion	–	Mosaic 16p11.2 deletion in patient's mother (Dittwald et al., 2013)
1, 9, 15, 16, 17, 18, X, Y	Loss and gain	–	16% (19 of 116) patients with idiopathic autism tested had low-level mosaic aneuploidy (Yurov et al., 2007)

been detected using these arrays (Conlin et al., 2010; Dumanski & Piotrowski, 2012; Rodriguez-Santiago et al., 2010). Because SNP arrays are able to analyze the zygosity of the SNPs, they aid in understanding the genetic mechanism by which mosaicism has occurred. An alternative approach to using genome-wide arrays is to employ targeted high-density arrays focusing on a few known and candidate regions. This allows for greater precision in detection and quantification of mosaicism (Pham et al., 2014).

There are many algorithms for analyzing array data, some of which are able to detect variable levels of mosaicism. These algorithms apply segmentation-based approaches to the alternate or B allele frequency (BAF) and log R ratio, including BAF segmentation and mosaic alteration detection (MAD) (González et al., 2011). Some of these algorithms use hidden Markov model (HMM) and include PennCNV, SNPtrio, PSCN, genoCN, MixHMM, and GPHMM (Chen, Xing, & Zhang, 2011; Li et al., 2011; Liu, Li, Schulz, Chen, & Tuck, 2010; Sun et al., 2009; Ting et al., 2007; Wang et al., 2008). Alternative approaches include Bayesian-based algorithm (Rancoita, Hutter, Bertoni, & Kwee, 2010) and a combination of parental genotypes and progeny BAF outliers (triPOD) (Baugher, Baugher, Shirley, & Pevsner, 2013). Analysis of 12 representative trios selected from the Autism Genetic Resource Exchange consortium detected mosaic CNVs as small as about 5 kb at levels as low as 2.1% with variable performance among these algorithms (Baugher et al., 2013).

Single-Cell Copy Number Analyses

Single-cell analysis allows one to isolate single nuclei (Evrony et al., 2012) that can then be subjected to amplification followed either by microarray or low-coverage WGS for CNV analysis. The most common method of amplifying DNA from single cells is multiple displacement amplification (MDA) (Dean et al., 2002; Rodrigue et al., 2009). A major technical challenge is uneven amplification across the genome, which leads to inaccurate identification of CNVs. Bias-tolerant algorithms mitigate the effects of uneven read depth (Bankevich et al., 2012; Chitsaz et al., 2011) but require high sequencing depth and may not be practical for analyzing human single cells. Other strategies to ameliorate the amplification bias include reducing the reaction volume (Marcy et al., 2007), supplementing amplification reactions with single-strand binding proteins or trehalose (Inoue, Shigemori, & Mikawa, 2006; Li et al., 2008; Pan et al., 2008), and postamplification normalization by digesting highly abundant sequences with a duplex specific nuclease (Rodrigue et al., 2009).

Other methods of amplification include multiple annealing and looping-based amplification, which uses quasilinear amplification to reduce exponential amplification bias (Zong, Lu, Chapman, & Xie, 2012), and a polymerase chain reaction (PCR)-based method such as GenomePlex (Sigma–Aldrich) and PicoPlex (Rubicon Genomics), which use random fragmentation of genomic DNA followed by amplification using universal oligonucleotide primers to produce more uniform amplification from the single cell (Cai et al., 2014; Voet et al., 2013; Yin et al., 2013). An alternative approach to amplification is the Microwell Displacement Amplification System (MIDAS) (Gole et al., 2013). In MIDAS, single cells are randomly distributed into hundreds to thousands of nanoliter wells and amplified simultaneously in physically separated nanoliter-scale reactors. Application of MIDAS to copy number analysis in five single human neuronal nuclei successfully detected an average of one CNV greater than 1 Mb in size, and possibly several smaller CNVs, although these were below the detection limit of MIDAS (Gole et al., 2013).

A study performed whole genome amplification (WGA) using both MDA and GenomePlex on 215 single cells, including 97 single neurons from three normal adults, 18 single cells from a fetus with trisomy 18, 24 cultured single lymphoblast cells from a normal adult, and 46 neurons and 30 nonneuronal cells from a patient with HMG and a somatic chromosome 1q CNV (Cai et al., 2014). Comparison of WGA of single cells showed lower amplification noise with GenomePlex compared with MDA. By using median absolute pairwise difference algorithm with a bin size of about 500 kb, multiple large (>1 Mb) clonal CNVs in lymphoblasts and in single neurons from normal human brain tissue were detected. These CNVs included one or more candidate CNVs (including 15q13.2–13.3) in neuropsychiatric conditions. These data suggest that large private and clonal somatic CNVs occur in normal and diseased human brains (Cai et al., 2014).

In another study, single neuronal cells from two sources, human induced pluripotent stem cells (hiPSC) and human postmortem frontal cortex, were examined using MDA followed by SNP array analysis or modified GenomePlex WGA (McConnell et al., 2013). Using circular binary segmentation followed by strict filtering based on the number of consecutive bins identified by segmentation, a mean CNV size detection limit of 6.7 Mb for SNP array data and 3.4 Mb for sequencing data was determined. Upon analysis of 40 hiPSC-derived neurons, 27 had copy number profiles consistent with bulk DNA analysis but 13 had unique genomes. Aberrations in the 13 single neurons included seven whole chromosome gains, four whole chromosome losses, and 12 subchromosomal CNVs (range, 7–156 Mb). Single-cell sequencing of postmortem human frontal cortex neurons identified one or more somatic CNVs in 41% of neurons, ranging from 2.9 to

75 Mb, with deletions being twice as common as duplications. Seventeen of 110 cells accounted for 73% of CNV calls and seven cells accounted for nearly half of all CNV calls. Single-cell genomic analyses of control fibroblast or neural progenitor cells did not reveal any aberration in the copy number profile. These results suggest that mosaic CNVs are abundant in human neurons and a subset of neurons are especially prone to multiple genomic alterations (McConnell et al., 2013).

Knouse et al. (2014) performed single-cell sequencing on mammalian brain and inferred copy number from sequencing read depths. Single-cell analyses in neural progenitor cells from mouse embryos did not reveal an aneuploidy. Similarly, analysis of 19 single neurons from adult mouse brain did not reveal an aneuploidy (95% confidence interval [CI], 0–17.6%), although analyses of 45 single cells from mixed adult mouse brain population showed aneuploidy in only one cell. Hence, prevalence of aneuploidy in the mouse brain is estimated to be 1% (95% CI, 0–5.6%). Analyses of single cells from postmortem neurotypical adult brain estimated the prevalence of aneuploidy at 2.2% (95% CI, 0.3–7.9%) (Knouse et al., 2014).

Single-cell genomic analyses have demonstrated that somatic aneuploidy in adult neurons is less common than previously estimated, whereas somatic CNVs are not as rare. Somatic CNVs, especially those known to be associated with neuropsychiatric disorders, could involve enough neuronal cells to cause disease, but because these CNVs affect only some of the neuronal cells, they may not be detectable in blood.

Digital Droplet Polymerase Chain Reaction

Digital droplet PCR (ddPCR) provides the ability to quantify nucleic acids with high precision and sensitivity (Pinheiro et al., 2012). Using microfluidic circuits and surfactant chemistry, sample DNA is randomly partitioned across 20,000 discrete droplets, such that each droplet contains one or no copy of the template DNA. Because copy number alterations often result from tandem gene duplications, restriction enzymes are used to separate linked copies of the template DNA predictably and efficiently such that each sequence is encapsulated in its own droplet and counted separately. PCR amplification is then performed using fluorescent Taqman probes to allow detection of "positive" droplets. Poisson statistics are then applied to the fraction of "positive" droplets to calculate the absolute concentration of the template DNA. This method allows the determination of copy number to a degree of precision far exceeding that achievable with traditional quantitative PCR and has been used successfully to detect the presence of trisomy 21 reproducibly and reliably to 2% mosaicism (Heyries et al., 2011). ddPCR was used to confirm the presence of low-frequency mosaic CNVs in parental fibroblasts. These CNVs were present at a much higher frequency in the induced pluripotent stem cells derived from the fibroblasts, but standard techniques were unable to detect the CNVs in parental fibroblasts (Abyzov et al., 2012).

Single-Nucleotide Variants, Including Insertions and Deletions

Sanger Sequencing

Sanger sequencing is the process of selective incorporation of chain-terminating dideoxynucleotides by DNA polymerase during in vitro DNA replication; it is the most widely used method for the detection of SNVs. Because both alleles of an autosomal locus are sequenced concurrently and are displayed as an analogue electropherograms, Sanger sequencing is unable to detect mosaic alleles below a threshold of 15–20% (Rohlin et al., 2009) and can miss a significant proportion of low-level mosaic mutations (Jamuar et al., 2014). In addition, mosaic mutations at higher allele fractions are miscalled "germ line," which highlights the limitations of Sanger sequencing in detecting mosaicism on both ends of the spectrum (Jamuar et al., 2014).

Subcloning Followed by Sanger Sequencing

Subcloning of the amplified PCR products into a vector followed by transformation into a bacterium such as *Escherichia coli* provides an alternative to Sanger sequencing of bulk samples. In this method, multiple colonies of the bacteria are formed, each containing either the wild type or mutant allele. Sanger sequencing of multiple individual colonies allows for confirmation of the presence of the mosaic variant and quantification of the level of mosaicism (Jamuar et al., 2014; Poduri et al., 2012).

Next-Generation Sequencing

NGS is a high throughput technique that allows parallel analysis of the multiple regions of the genome (Ng et al., 2010, 2009). NGS is a digital assay that reports read counts for each allele as integer counts, and hence is amenable to bioinformatic approaches to distinguish mosaicism from sequencing errors. Whole exome sequencing (WES) and WGS technologies have allowed an exponential increase in our understanding of human genetic disorders and have been reported to detect somatic mutations in rare instances (Gilissen et al., 2014; Pagnamenta et al., 2012; Pritchard et al., 2013). However, owing to inhomogeneities in library preparation and an average depth of 40−80×, WES/WGS may miss somatic mutations, especially if the read depth is low (Fig. 12.2). On the other hand, deep-targeted sequencing allows one to interrogate a smaller portion of the genome to much greater depths and has been used successfully to detect low-level mosaicism as low as 5% (Jamuar et al., 2014; Riviere et al., 2012). In addition, as many as 30% of mutations associated with unexplained brain malformations are somatic, most which would be missed on Sanger sequencing and even on low-coverage WES (Jamuar et al., 2014; Tapper et al., 2014). We postulate that a mean coverage of at least 500× would be required to exclude low-level mosaicism confidently (Jamuar et al., 2014). An alterative to performing deep sequencing is bioinformatic alteration of exome sequencing pipelines, including adjusting the allele frequency spectrum in SAMtools, reducing the variant calling threshold using bcftools (Tapper et al., 2014), and reducing the allele balance parameter in Genome Analysis ToolKit (GATK) (Jamuar et al., 2014; Pagnamenta et al., 2012).

Use of NGS on paired samples has been used successfully to detect somatic mutations in cancerous (Watson et al., 2013) as well as noncancerous disorders (Lee et al., 2012; Lindhurst et al., 2011). In this strategy, NGS is performed on the affected and normal tissue (in most instances, blood) and the data from the two paired samples are compared bioinformatically to detect somatic mutations. Some commonly used bioinformatic tools include MuTect (Cibulskis et al., 2013), JointSNVmix (Roth et al., 2012), SomaticSniper (Larson et al., 2012), VarScan 2 (Koboldt et al., 2012), and Virmid (Kim et al., 2013). The paired tissue approach was used successfully to detect a recurrent *PIK3CA* c.1633G > A mutation in 4 out of 20 cases of HMG at an allele frequency ranging from 8% to 40% (Lee et al., 2012).

Single-Cell Sequencing

Isolation, genome amplification, and sequencing of single cells allows for quantification of the level of mosaicism. It also allows for lineage tracing that will add to our understanding of the diverse cell types and developmental processes that build the human brain. In an individual with HMG, we identified a somatic point mutation in *AKT3* present at about 35% mosaicism based on bulk tissue analysis. After sorting, single-cell sequencing revealed the mutation in 39% of neuronal nuclei and 27% of nonneuronal nuclei, which suggests that this mutation was present in an early neocortical progenitor cell which developed into both neuronal and nonneuronal cells (Evrony et al., 2012). This was consistent with the radiological phenotype of the individual that showed involvement of both gray and white matter (Poduri et al., 2012).

Mass Spectrometry

The Sequenom MassARRAY iPLEX platform consists of an initial locus-specific PCR reaction followed by single-base extension using mass-modified dideoxynucleotide terminators of an oligonucleotide primer which anneals immediately upstream of the polymorphic site of interest. Using matrix-assisted laser desorption/ionization−time of flight mass spectrometry, the distinct mass of the extended primer identifies the SNP allele (Gabriel, Ziaugra, & Tabbaa, 2009). This allows a high-throughput method of validating and quantifying somatic point mutations in a given sample (Lee et al., 2012).

CONCLUSION

The role of somatic mutations in neurological disease is still underappreciated; however, single-cell and deep genome sequencing will allow systematic measurement of somatic mutation rates in different cell types and lineages during development of the normal as well as diseased human brain. This will allow us to estimate the prevalence of somatic mutations as a cause of neurological disorders and understand to what extent somatic mutations modify the pathogenesis of neurological diseases and how somatic mutations affect normal neurodevelopmental processes.

ACKNOWLEDGMENTS

A.M.D. is supported by the National Institute of General Medical Sciences (T32GM07753) and the National Institutes of Health Ruth L. Kirschstein National Research Service Award (5T32 GM007226-39). C.A.W. is supported by grants from the National Institute of Mental Health (R01MH083565 and 1RC2MH089952), the National Institute of Neurological Disorders and Stroke (R01NS032457, R01NS079277 and R01NS035129), the Simons Foundation, the Paul G. Allen Family Foundation, and the Manton Center for Orphan Disease Research. C.A.W. is an Investigator of the Howard Hughes Medical Institute.

REFERENCES

Abyzov, A., Mariani, J., Palejev, D., et al. (2012). Somatic copy number mosaicism in human skin revealed by induced pluripotent stem cells. *Nature, 492*, 438–442.

Ainsworth, P. J., Chakraborty, P. K., & Weksberg, R. (1997). Example of somatic mosaicism in a series of de novo neurofibromatosis type 1 cases due to a maternally derived deletion. *Human Mutation, 9*, 452–457.

Alzualde, A., Moreno, F., Martinez-Lage, P., et al. (2010). Somatic mosaicism in a case of apparently sporadic Creutzfeldt-Jakob disease carrying a de novo D178N mutation in the PRNP gene. *American Journal of Medical Genetics, Part B: Neuropsychiatric Genetics, 153B*, 1283–1291.

Amir, R. E., Van den Veyver, I. B., Wan, M., Tran, C. Q., Francke, U., & Zoghbi, H. Y. (1999). Rett syndrome is caused by mutations in X-linked MECP2, encoding methyl-CpG-binding protein 2. *Nature Genetics, 23*, 185–188.

Antonarakis, E. S., Sampson, J. R., & Cheadle, J. P. (2002). Temperature modulation of DHPLC analysis for detection of coexisting constitutional and mosaic sequence variants in TSC2. *Journal of Biochemical and Biophysical Methods, 51*, 161–164.

Azmanov, D. N., Zhelyazkova, S., Dimova, P. S., et al. (2010). Mosaicism of a missense SCN1A mutation and Dravet syndrome in a Roma/Gypsy family. *Epileptic Disorders: International Epilepsy Journal With Videotape, 12*, 117–124.

Ballif, B. C., Rorem, E. A., Sundin, K., et al. (2006). Detection of low-level mosaicism by array CGH in routine diagnostic specimens. *American Journal of Medical Genetics, Part A, 140*, 2757–2767.

Bankevich, A., Nurk, S., Antipov, D., et al. (2012). SPAdes: a new genome assembly algorithm and its applications to single-cell sequencing. *Journal of Computational Biology, 19*, 455–477.

Baugher, J. D., Baugher, B. D., Shirley, M. D., & Pevsner, J. (2013). Sensitive and specific detection of mosaic chromosomal abnormalities using the Parent-of-Origin-based Detection (POD) method. *BMC Genomics, 14*, 367.

Beck, J. A., Poulter, M., Campbell, T. A., et al. (2004). Somatic and germline mosaicism in sporadic early-onset Alzheimer's disease. *Human Molecular Genetics, 13*, 1219–1224.

Biesecker, L. G., & Spinner, N. B. (2013). A genomic view of mosaicism and human disease. *Nature Reviews Genetics, 14*, 307–320.

Bilgen, T., Keser, I., Mihci, E., Haspolat, S., Tacoy, S., & Luleci, G. (2005). Molecular analysis of fragile X syndrome in Antalya province. *Indian Journal of Medical Sciences, 59*, 150–155.

Bruder, C. E., Piotrowski, A., Gijsbers, A. A., et al. (2008). Phenotypically concordant and discordant monozygotic twins display different DNA copy-number-variation profiles. *American Journal of Human Genetics, 82*, 763–771.

Cai, X., Evrony, G. D., Lehmann, H. S., et al. (2014). Single-cell, genome-wide sequencing identifies clonal somatic copy-number variation in the human brain. *Cell Reports, 88*.

Chen, H., Xing, H., & Zhang, N. R. (2011). Estimation of parent specific DNA copy number in tumors using high-density genotyping arrays. *PLoS Computational Biology, 7*, e1001060.

Cheung, S. W., Shaw, C. A., Scott, D. A., et al. (2007). Microarray-based CGH detects chromosomal mosaicism not revealed by conventional cytogenetics. *American Journal of Medical Genetics, Part A, 143A*, 1679–1686.

Chitsaz, H., Yee-Greenbaum, J. L., Tesler, G., et al. (2011). Efficient de novo assembly of single-cell bacterial genomes from short-read data sets. *Nature Biotechnology, 29*, 915–921.

Cibulskis, K., Lawrence, M. S., Carter, S. L., et al. (2013). Sensitive detection of somatic point mutations in impure and heterogeneous cancer samples. *Nature Biotechnology, 31*, 213–219.

Coffee, B., Ikeda, M., Budimirovic, D. B., Hjelm, L. N., Kaufmann, W. E., & Warren, S. T. (2008). Mosaic FMR1 deletion causes fragile X syndrome and can lead to molecular misdiagnosis: a case report and review of the literature. *American Journal of Medical Genetics, Part A, 146A*, 1358–1367.

Colman, S. D., Rasmussen, S. A., Ho, V. T., Abernathy, C. R., & Wallace, M. R. (1996). Somatic mosaicism in a patient with neurofibromatosis type 1. *American Journal of Human Genetics, 58*, 484–490.

Conlin, L. K., Kramer, W., Hutchinson, A. L., et al. (2011). Molecular analysis of ring chromosome 20 syndrome reveals two distinct groups of patients. *Journal of Medical Genetics, 48*, 1–9.

Conlin, L. K., Thiel, B. D., Bonnemann, C. G., et al. (2010). Mechanisms of mosaicism, chimerism and uniparental disomy identified by single nucleotide polymorphism array analysis. *Human Molecular Genetics, 19*, 1263–1275.

Conti, V., Pantaleo, M., Barba, C., et al. (2014). Focal dysplasia of the cerebral cortex and infantile spasms associated with somatic 1q21.1-q44 duplication including the AKT3 gene. *Clinical Genetics, 88*(3), 241–247.

Coufal, N. G., Garcia-Perez, J. L., Peng, G. E., et al. (2009). L1 retrotransposition in human neural progenitor cells. *Nature, 460*, 1127–1131.

Dean, F. B., Hosono, S., Fang, L., et al. (2002). Comprehensive human genome amplification using multiple displacement amplification. *Proceedings of the National Academy of Sciences of the United States of America, 99*, 5261–5266.

Depienne, C., Arzimanoglou, A., Trouillard, O., et al. (2006). Parental mosaicism can cause recurrent transmission of SCN1A mutations associated with severe myoclonic epilepsy of infancy. *Human Mutation, 27*, 389.

Depienne, C., Trouillard, O., Gourfinkel-An, I., et al. (2010). Mechanisms for variable expressivity of inherited SCN1A mutations causing Dravet syndrome. *Journal of Medical Genetics, 47*, 404–410.

Dibbens, L. M., de Vries, B., Donatello, S., et al. (2013). Mutations in DEPDC5 cause familial focal epilepsy with variable foci. *Nature Genetics, 45*, 546–551.

Dittwald, P., Gambin, T., Szafranski, P., et al. (2013). NAHR-mediated copy-number variants in a clinical population: mechanistic insights into both genomic disorders and Mendelizing traits. *Genome Research, 23*, 1395–1409.

Dobkin, C. S., Nolin, S. L., Cohen, I., et al. (1996). Tissue differences in fragile X mosaics: mosaicism in blood cells may differ greatly from skin. *American Journal of Medical Genetics, 64*, 296–301.

Dumanski, J. P., & Piotrowski, A. (2012). Structural genetic variation in the context of somatic mosaicism. *Methods in Molecular Biology, 838*, 249–272.

Evrony, G. D., Cai, X., Lee, E., et al. (2012). Single-neuron sequencing analysis of L1 retrotransposition and somatic mutation in the human brain. *Cell, 151*, 483–496.

Fan, H., Booker, J. K., McCandless, S. E., Shashi, V., Fleming, A., & Farber, R. A. (2005). Mosaicism for an FMR1 gene deletion in a fragile X female. *American Journal of Medical Genetics, Part A, 136*, 214–217.

Ferreira, S. I., Pires, L. M., Ferrao, J., Sa, J., Serra, A., & Carreira, I. M. (2013). Mosaicism for FMR1 gene full mutation and intermediate allele in a female foetus: a postzygotic retraction event. *Gene, 527*, 421–425.

Franco, S. J., & Muller, U. (2013). Shaping our minds: stem and progenitor cell diversity in the mammalian neocortex. *Neuron, 77*, 19–34.

Gabriel, S., Ziaugra, L., & Tabbaa, D. (2009). SNP genotyping using the Sequenom MassARRAY iPLEX platform. *Current Protocols in Human Genetics* (Chapter 2):Unit 2 12.

Garcia-Linares, C., Fernandez-Rodriguez, J., Terribas, E., et al. (2011). Dissecting loss of heterozygosity (LOH) in neurofibromatosis type 1-associated neurofibromas: importance of copy neutral LOH. *Human Mutation, 32*, 78–90.

Garcia Arocena, D., de Diego, Y., Oostra, B. A., Willemsen, R., & Mirta Rodriguez, M. (2000). A fragile X case with an amplification/deletion mosaic pattern. *Human Genetics, 106*, 366–369.

Gardner, R. J. M., Sutherland, G. R., & Shaffer, L. G. (2011). *Chromosome abnormalities and genetic counseling*. Oxford University Press.

Gennaro, E., Santorelli, F. M., Bertini, E., et al. (2006). Somatic and germline mosaicisms in severe myoclonic epilepsy of infancy. *Biochemical and Biophysical Research Communications, 341*, 489–493.

Gijsbers, A. C., Dauwerse, J. G., Bosch, C. A., et al. (2011). Three new cases with a mosaicism involving a normal cell line and a cryptic unbalanced autosomal reciprocal translocation. *European Journal of Medical Genetics, 54*, e409–e412.

Gilissen, C., Hehir-Kwa, J. Y., Thung, D. T., et al. (2014). Genome sequencing identifies major causes of severe intellectual disability. *Nature, 511*, 344–347.

Gleeson, J. G., Allen, K. M., Fox, J. W., et al. (1998). Doublecortin, a brain-specific gene mutated in human X-linked lissencephaly and double cortex syndrome, encodes a putative signaling protein. *Cell, 92*, 63–72.

Gleeson, J. G., Minnerath, S., Kuzniecky, R. I., et al. (2000). Somatic and germline mosaic mutations in the doublecortin gene are associated with variable phenotypes. *American Journal of Human Genetics, 67*, 574–581.

Gole, J., Gore, A., Richards, A., et al. (2013). Massively parallel polymerase cloning and genome sequencing of single cells using nanoliter microwells. *Nature Biotechnology, 31*, 1126–1132.

González, J. R., Rodríguez-Santiago, B., Cáceres, A., et al. (2011). A fast and accurate method to detect allelic genomic imbalances underlying mosaic rearrangements using SNP array data. *BMC Bioinformatics, 12*, 166. http://dx.doi.org/10.1186/1471-2105-12-166.

Govaerts, L. C., Smit, A. E., Saris, J. J., et al. (2007). Exceptional good cognitive and phenotypic profile in a male carrying a mosaic mutation in the FMR1 gene. *Clinical Genetics, 72*, 138–144.

de Graaff, E., de Vries, B. B., Willemsen, R., et al. (1996). The fragile X phenotype in a mosaic male with a deletion showing expression of the FMR1 protein in 28% of the cells. *American Journal of Medical Genetics, 64*, 302–308.

Grasso, M., Faravelli, F., Lo Nigro, C., et al. (1999). Mosaicism for the full mutation and a microdeletion involving the CGG repeat and flanking sequences in the FMR1 gene in eight fragile X patients. *American Journal of Medical Genetics, 85*, 311–316.

Greig, L. C., Woodworth, M. B., Galazo, M. J., Padmanabhan, H., & Macklis, J. D. (2013). Molecular logic of neocortical projection neuron specification, development and diversity. *Nature Reviews Neuroscience, 14*, 755–769.

Guerrini, R., Mei, D., Sisodiya, S., et al. (2004). Germline and mosaic mutations of FLN1 in men with periventricular heterotopia. *Neurology, 63*, 51–56.

Hansen, D. V., Lui, J. H., Flandin, P., et al. (2013). Non-epithelial stem cells and cortical interneuron production in the human ganglionic eminences. *Nature Neuroscience, 16*, 1576–1587.

Hashida, H., Goto, J., Suzuki, T., et al. (2001). Single cell analysis of CAG repeat in brains of dentatorubral-pallidoluysian atrophy (DRPLA). *Journal of Neurological Sciences, 190*, 87–93.

Hellenbroich, Y., Schwinger, E., & Zuhlke, C. (2001). Limited somatic mosaicism for Friedreich's ataxia GAA triplet repeat expansions identified by small pool PCR in blood leukocytes. *Acta Neurologica Scandinavica, 103*, 188–192.

Heyries, K. A., Tropini, C., Vaninsberghe, M., et al. (2011). Megapixel digital PCR. *Nature Methods, 8*, 649–651.

Hitchins, M. P., Rickard, S., Dhalla, F., et al. (2004). Investigation of UBE3A and MECP2 in Angelman syndrome (AS) and patients with features of AS. *American Journal of Medical Genetics, Part A, 125A*, 167–172.

Hsu, L. Y., Yu, M. T., Richkind, K. E., et al. (1996). Incidence and significance of chromosome mosaicism involving an autosomal structural abnormality diagnosed prenatally through amniocentesis: a collaborative study. *Prenatal Diagnosis, 16*, 1–28.

Huisman, S. A., Redeker, E. J., Maas, S. M., Mannens, M. M., & Hennekam, R. C. (2013). High rate of mosaicism in individuals with Cornelia de Lange syndrome. *Journal of Medical Genetics, 50*, 339–344.

Inoue, J., Shigemori, Y., & Mikawa, T. (2006). Improvements of rolling circle amplification (RCA) efficiency and accuracy using Thermus thermophilus SSB mutant protein. *Nucleic Acids Research, 34*, e69.

Ishida, S., Picard, F., Rudolf, G., et al. (2013). Mutations of DEPDC5 cause autosomal dominant focal epilepsies. *Nature Genetics, 45*, 552–555.

Ito, Y., Tanaka, F., Yamamoto, M., et al. (1998). Somatic mosaicism of the expanded CAG trinucleotide repeat in mRNAs for the responsible gene of Machado-Joseph disease (MJD), dentatorubral-pallidoluysian atrophy (DRPLA), and spinal and bulbar muscular atrophy (SBMA). *Neurochemical Research, 23*, 25–32.

Jacobs, K. B., Yeager, M., Zhou, W., et al. (2012). Detectable clonal mosaicism and its relationship to aging and cancer. *Nature Genetics, 44*, 651–658.

Jamuar, S. S., Lam, A. T., Kircher, M., et al. (2014). Somatic mutations in cerebral cortical malformations. *New England Journal of Medicine, 371*, 733–743.

Kakinuma, H., Ozaki, M., Sato, H., & Takahashi, H. (2008). Variation in GABA-A subunit gene copy number in an autistic patient with mosaic 4 p duplication (p12p16). *American Journal of Medical Genetics Part B: Neuropsychiatric Genetics, 147B*, 973–975.

Kaplan, L., Foster, R., Shen, Y., et al. (2010). Monozygotic twins discordant for neurofibromatosis 1. *American Journal of Medical Genetics, Part A, 152A*, 601–606.

Karall, D., Haberlandt, E., Scholl-Burgi, S., Baumgartner, S., Naudo, M., & Martorell, L. (2007). Homozygosity for MECP2 gene in a girl with classical Rett syndrome. *European Journal of Medical Genetics, 50*, 465–468.

Kehrer-Sawatzki, H., Kluwe, L., Sandig, C., et al. (2004). High frequency of mosaicism among patients with neurofibromatosis type 1 (NF1) with microdeletions caused by somatic recombination of the JJAZ1 gene. *American Journal of Human Genetics, 75*, 410–423.

Kennedy, S. R., Loeb, L. A., & Herr, A. J. (2012). Somatic mutations in aging, cancer and neurodegeneration. *Mechanisms of Ageing and Development, 133*, 118–126.

Kessaris, N., Fogarty, M., Iannarelli, P., Grist, M., Wegner, M., & Richardson, W. D. (2006). Competing waves of oligodendrocytes in the forebrain and postnatal elimination of an embryonic lineage. *Nature Neuroscience, 9*, 173–179.

Kim, S., Jeong, K., Bhutani, K., et al. (2013). Virmid: accurate detection of somatic mutations with sample impurity inference. *Genome Biology, 14*, R90.

Knouse, K. A., Wu, J., Whittaker, C. A., & Amon, A. (2014). Single cell sequencing reveals low levels of aneuploidy across mammalian tissues. *Proceedings of the National Academy of Sciences of the United States of America, 111*, 13409–13414.

Knudson, A. G., Jr. (1971). Mutation and cancer: statistical study of retinoblastoma. *Proceedings of the National Academy of Sciences of the United States of America, 68*, 820–823.

Koboldt, D. C., Zhang, Q., Larson, D. E., et al. (2012). VarScan 2: somatic mutation and copy number alteration discovery in cancer by exome sequencing. *Genome Research, 22*, 568–576.

Kornack, D. R., & Rakic, P. (1995). Radial and horizontal deployment of clonally related cells in the primate neocortex: relationship to distinct mitotic lineages. *Neuron, 15*, 311–321.

Kriegstein, A., Noctor, S., & Martinez-Cerdeno, V. (2006). Patterns of neural stem and progenitor cell division may underlie evolutionary cortical expansion. *Nature Reviews Neuroscience, 7*, 883–890.

Kurek, K. C., Luks, V. L., Ayturk, U. M., et al. (2012). Somatic mosaic activating mutations in PIK3CA cause CLOVES syndrome. *American Journal of Human Genetics, 90*, 1108–1115.

Larson, D. E., Harris, C. C., Chen, K., et al. (2012). SomaticSniper: identification of somatic point mutations in whole genome sequencing data. *Bioinformatics, 28*, 311–317.

Laurie, C. C., Laurie, C. A., Rice, K., et al. (2012). Detectable clonal mosaicism from birth to old age and its relationship to cancer. *Nature Genetics, 44*, 642–650.

Lee, J. H., Huynh, M., Silhavy, J. L., et al. (2012). De novo somatic mutations in components of the PI3K-AKT3-mTOR pathway cause hemimegalencephaly. *Nature Genetics, 44*, 941–945.

Li, A., Liu, Z., Lezon-Geyda, K., et al. (2011). GPHMM: an integrated hidden Markov model for identification of copy number alteration and loss of heterozygosity in complex tumor samples using whole genome SNP arrays. *Nucleic Acids Research, 39*, 4928–4941.

Li, Y., Kim, H. J., Zheng, C., et al. (2008). Primase-based whole genome amplification. *Nucleic Acids Research, 36*, e79.

Lindhurst, M. J., Parker, V. E., Payne, F., et al. (2012). Mosaic overgrowth with fibroadipose hyperplasia is caused by somatic activating mutations in PIK3CA. *Nature Genetics, 44*, 928–933.

Lindhurst, M. J., Sapp, J. C., Teer, J. K., et al. (2011). A mosaic activating mutation in AKT1 associated with the Proteus syndrome. *New England Journal of Medicine, 365*, 611–619.

Liu, J., Reeves, C., Michalak, Z., et al. (2014). Evidence for mTOR pathway activation in a spectrum of epilepsy-associated pathologies. *Acta Neuropathologica Communications, 2*, 71.

Liu, Z., Li, A., Schulz, V., Chen, M., & Tuck, D. (2010). MixHMM: inferring copy number variation and allelic imbalance using SNP arrays and tumor samples mixed with stromal cells. *PLoS One, 5*, e10909.

MacKenzie, J. J., Sumargo, I., & Taylor, S. A. (2006). A cryptic full mutation in a male with a classical fragile X phenotype. *Clinical Genetics, 70*, 39–42.

Magavi, S., Friedmann, D., Banks, G., Stolfi, A., & Lois, C. (2012). Coincident generation of pyramidal neurons and protoplasmic astrocytes in neocortical columns. *Journal of Neuroscience, 32*, 4762–4772.

Marcy, Y., Ishoey, T., Lasken, R. S., et al. (2007). Nanoliter reactors improve multiple displacement amplification of genomes from single cells. *PLoS Genetics, 3*, 1702–1708.

Marini, C., Mei, D., Helen Cross, J., & Guerrini, R. (2006). Mosaic SCN1A mutation in familial severe myoclonic epilepsy of infancy. *Epilepsia, 47*, 1737–1740.

McConnell, M. J., Lindberg, M. R., Brennand, K. J., et al. (2013). Mosaic copy number variation in human neurons. *Science, 342*, 632–637.

Messiaen, L., Vogt, J., Bengesser, K., et al. (2011). Mosaic type-1 NF1 microdeletions as a cause of both generalized and segmental neurofibromatosis type-1 (NF1). *Human Mutation, 32*, 213–219.

Michaelson, J. J., Shi, Y., Gujral, M., et al. (2012). Whole-genome sequencing in autism identifies hot spots for de novo germline mutation. *Cell, 151*, 1431–1442.

Miller, D. T., Adam, M. P., Aradhya, S., et al. (2010). Consensus statement: chromosomal microarray is a first-tier clinical diagnostic test for individuals with developmental disabilities or congenital anomalies. *American Journal of Human Genetics, 86*, 749–764.

Morimoto, M., Mazaki, E., Nishimura, A., et al. (2006). SCN1A mutation mosaicism in a family with severe myoclonic epilepsy in infancy. *Epilepsia, 47*, 1732–1736.

Muotri, A. R., Chu, V. T., Marchetto, M. C., Deng, W., Moran, J. V., & Gage, F. H. (2005). Somatic mosaicism in neuronal precursor cells mediated by L1 retrotransposition. *Nature, 435*, 903–910.

Nakamura, K., Kato, M., Osaka, H., et al. (2013). Clinical spectrum of SCN2A mutations expanding to Ohtahara syndrome. *Neurology, 81*, 992–998.

Neale, B. M., Kou, Y., Liu, L., et al. (2012). Patterns and rates of exonic de novo mutations in autism spectrum disorders. *Nature, 485*, 242–245.

Ng, S. B., Buckingham, K. J., Lee, C., et al. (2010). Exome sequencing identifies the cause of a mendelian disorder. *Nature Genetics, 42*, 30–35.

Ng, S. B., Turner, E. H., Robertson, P. D., et al. (2009). Targeted capture and massively parallel sequencing of 12 human exomes. *Nature, 461*, 272–276.

Niida, Y., Stemmer-Rachamimov, A. O., Logrip, M., et al. (2001). Survey of somatic mutations in tuberous sclerosis complex (TSC) hamartomas suggests different genetic mechanisms for pathogenesis of TSC lesions. *American Journal of Human Genetics, 69*, 493–503.

Nolin, S. L., Glicksman, A., Houck, G. E., Jr., Brown, W. T., & Dobkin, C. S. (1994). Mosaicism in fragile X affected males. *American Journal of Medical Genetics, 51*, 509–512.

O'Roak, B. J., Deriziotis, P., Lee, C., et al. (2011). Exome sequencing in sporadic autism spectrum disorders identifies severe de novo mutations. *Nature Genetics, 43*, 585–589.

O'Roak, B. J., Vives, L., Fu, W., et al. (2012). Multiplex targeted sequencing identifies recurrently mutated genes in autism spectrum disorders. *Science, 338*, 1619–1622.

O'Roak, B. J., Vives, L., Girirajan, S., et al. (2012). Sporadic autism exomes reveal a highly interconnected protein network of de novo mutations. *Nature, 485*, 246–250.

Pagnamenta, A. T., Lise, S., Harrison, V., et al. (2012). Exome sequencing can detect pathogenic mosaic mutations present at low allele frequencies. *Journal of Human Genetics, 57*, 70–72.

Pan, X., Urban, A. E., Palejev, D., et al. (2008). A procedure for highly specific, sensitive, and unbiased whole-genome amplification. *Proceedings of the National Academy of Sciences of the United States of America, 105*, 15499–15504.

Parrini, E., Mei, D., Wright, M., Dorn, T., & Guerrini, R. (2004). Mosaic mutations of the FLN1 gene cause a mild phenotype in patients with periventricular heterotopia. *Neurogenetics, 5*, 191–196.

Petek, E., Jenne, D. E., Smolle, J., et al. (2003). Mitotic recombination mediated by the JJAZF1 (KIAA0160) gene causing somatic mosaicism and a new type of constitutional NF1 microdeletion in two children of a mosaic female with only few manifestations. *Journal of Medical Genetics, 40*, 520–525.

Petek, E., Kroisel, P. M., Schuster, M., Zierler, H., & Wagner, K. (1999). Mosaicism in a fragile X male including a de novo deletion in the FMR1 gene. *American Journal of Medical Genetics, 84*, 229–232.

Pham, J., Shaw, C., Pursley, A., et al. (2014). Somatic mosaicism detected by exon-targeted, high-resolution aCGH in 10,362 consecutive cases. *European Journal of Human Genetics, 22*, 969–978.

Pieras, J. I., Munoz-Cabello, B., Borrego, S., et al. (2011). Somatic mosaicism for Y120X mutation in the MECP2 gene causes atypical Rett syndrome in a male. *Brain & Development, 33*, 608–611.

Pinheiro, L. B., Coleman, V. A., Hindson, C. M., et al. (2012). Evaluation of a droplet digital polymerase chain reaction format for DNA copy number quantification. *Analytical Chemistry, 84*, 1003–1011.

Piotrowski, A., Bruder, C. E., Andersson, R., et al. (2008). Somatic mosaicism for copy number variation in differentiated human tissues. *Human Mutation, 29*, 1118–1124.

Pleasure, S. J., Anderson, S., Hevner, R., et al. (2000). Cell migration from the ganglionic eminences is required for the development of hippocampal GABAergic interneurons. *Neuron, 28*, 727–740.

Poduri, A. (2014). DEPDC5 does it all: shared genetics for diverse epilepsy syndromes. *Annals of Neurology, 75*, 631–633.

Poduri, A., Evrony, G. D., Cai, X., et al. (2012). Somatic activation of AKT3 causes hemispheric developmental brain malformations. *Neuron, 74*, 41–48.

Poduri, A., Evrony, G. D., Cai, X., & Walsh, C. A. (2013). Somatic mutation, genomic variation, and neurological disease. *Science, 341*, 1237758.

Pretto, D. I., Hunsaker, M. R., Cunningham, C. L., et al. (2013). Intranuclear inclusions in a fragile X mosaic male. *Translational Neurodegeneration, 2*, 10.

Pritchard, C. C., Smith, C., Marushchak, T., et al. (2013). A mosaic PTEN mutation causing Cowden syndrome identified by deep sequencing. *Genetics in Medicine, 15*, 1004–1007.

Proukakis, C., Houlden, H., & Schapira, A. H. (2013). Somatic alpha-synuclein mutations in Parkinson's disease: hypothesis and preliminary data. *Movement Disorders, 28*, 705–712.

Qin, W., Chan, J. A., Vinters, H. V., et al. (2010). Analysis of TSC cortical tubers by deep sequencing of TSC1, TSC2 and KRAS demonstrates that small second-hit mutations in these genes are rare events. *Brain Pathology, 20*, 1096–1105.

Qin, W., Kozlowski, P., Taillon, B. E., et al. (2010). Ultra deep sequencing detects a low rate of mosaic mutations in tuberous sclerosis complex. *Human Genetics, 127*, 573–582.

Rajendran, R. S., Wellbrock, U. M., & Zupanc, G. K. (2008). Apoptotic cell death, long-term persistence, and neuronal differentiation of aneuploid cells generated in the adult brain of teleost fish. *Developmental Neurobiology, 68*, 1257–1268.

Rajendran, R. S., Zupanc, M. M., Losche, A., Westra, J., Chun, J., & Zupanc, G. K. (2007). Numerical chromosome variation and mitotic segregation defects in the adult brain of teleost fish. *Developmental Neurobiology, 67*, 1334–1347.

Rakic, P. (1988). Specification of cerebral cortical areas. *Science, 241*, 170–176.

Rancoita, P. M., Hutter, M., Bertoni, F., & Kwee, I. (2010). An integrated Bayesian analysis of LOH and copy number data. *BMC Bioinformatics, 11*, 321.

Rasmussen, S. A., Colman, S. D., Ho, V. T., et al. (1998). Constitutional and mosaic large NF1 gene deletions in neurofibromatosis type 1. *Journal of Medical Genetics, 35*, 468–471.

Reddy, K. S. (2005). Cytogenetic abnormalities and fragile-X syndrome in autism spectrum Disorder. *BMC Medical Genetics, 6*, 3.

Rehen, S. K., McConnell, M. J., Kaushal, D., Kingsbury, M. A., Yang, A. H., & Chun, J. (2001). Chromosomal variation in neurons of the developing and adult mammalian nervous system. *Proceedings of the National Academy of Sciences of the United States of America, 98*, 13361–13366.

Rehen, S. K., Yung, Y. C., McCreight, M. P., et al. (2005). Constitutional aneuploidy in the normal human brain. *Journal of Neuroscience, 25*, 2176–2180.

Reid, C. B., Tavazoie, S. F., & Walsh, C. A. (1997). Clonal dispersion and evidence for asymmetric cell division in ferret cortex. *Development, 124*, 2441–2450.

Riviere, J. B., Mirzaa, G. M., O'Roak, B. J., et al. (2012). De novo germline and postzygotic mutations in AKT3, PIK3R2 and PIK3CA cause a spectrum of related megalencephaly syndromes. *Nature Genetics, 44*, 934–940.

Roberts, P. S., Jozwiak, S., Kwiatkowski, D. J., & Dabora, S. L. (2001). Denaturing high-performance liquid chromatography (DHPLC) is a highly sensitive, semi-automated method for identifying mutations in the TSC1 gene. *Journal of Biochemical and Biophysical Methods, 47*, 33–37.

Rodrigue, S., Malmstrom, R. R., Berlin, A. M., Birren, B. W., Henn, M. R., & Chisholm, S. W. (2009). Whole genome amplification and de novo assembly of single bacterial cells. *PLoS One, 4*, e6864.

Rodriguez-Santiago, B., Malats, N., Rothman, N., et al. (2010). Mosaic uniparental disomies and aneuploidies as large structural variants of the human genome. *American Journal of Human Genetics, 87*, 129–138.

Roehl, A. C., Mussotter, T., Cooper, D. N., et al. (2012). Tissue-specific differences in the proportion of mosaic large NF1 deletions are suggestive of a selective growth advantage of hematopoietic del(+/-) stem cells. *Human Mutation, 33*, 541–550.

Rohlin, A., Wernersson, J., Engwall, Y., Wiklund, L., Bjork, J., & Nordling, M. (2009). Parallel sequencing used in detection of mosaic mutations: comparison with four diagnostic DNA screening techniques. *Human Mutation, 30*, 1012–1020.

Roth, A., Ding, J., Morin, R., et al. (2012). JointSNVMix: a probabilistic model for accurate detection of somatic mutations in normal/tumour paired next-generation sequencing data. *Bioinformatics, 28*, 907–913.

Sanders, S. J., Murtha, M. T., Gupta, A. R., et al. (2012). De novo mutations revealed by whole-exome sequencing are strongly associated with autism. *Nature, 485*, 237–241.

Scheffer, I. E., Heron, S. E., Regan, B. M., et al. (2014). Mutations in mammalian target of rapamycin regulator DEPDC5 cause focal epilepsy with brain malformations. *Annals of Neurology, 75*, 782–787.

Schmucker, B., Ballhausen, W. G., & Pfeiffer, R. A. (1996). Mosaicism of a microdeletion of 486 bp involving the CGG repeat of the FMR1 gene due to misalignment of GTT tandem repeats at chi-like elements flanking both breakpoints and a full mutation. *Human Genetics, 98*, 409–414.

Selmer, K. K., Eriksson, A. S., Brandal, K., Egeland, T., Tallaksen, C., & Undlien, D. E. (2009). Parental SCN1A mutation mosaicism in familial Dravet syndrome. *Clinical Genetics, 76*, 398–403.

Shinawi, M., Liu, P., Kang, S. H., et al. (2010). Recurrent reciprocal 16p11.2 rearrangements associated with global developmental delay, behavioural problems, dysmorphism, epilepsy, and abnormal head size. *Journal of Medical Genetics, 47*, 332–341.

Shirley, M. D., Tang, H., Gallione, C. J., et al. (2013). Sturge-Weber syndrome and port-wine stains caused by somatic mutation in GNAQ. *New England Journal of Medicine, 368*, 1971–1979.

Sicca, F., Kelemen, A., Genton, P., et al. (2003). Mosaic mutations of the LIS1 gene cause subcortical band heterotopia. *Neurology, 61*, 1042–1046.

Smahi, A., Courtois, G., Vabres, P., et al. (2000). Genomic rearrangement in NEMO impairs NF-kappaB activation and is a cause of incontinentia pigmenti. The international incontinentia pigmenti (IP) consortium. *Nature, 405*, 466–472.

Smith, S. E., Zhou, Y. D., Zhang, G., Jin, Z., Stoppel, D. C., & Anderson, M. P. (2011). Increased gene dosage of Ube3a results in autism traits and decreased glutamate synaptic transmission in mice. *Science Translational Medicine, 3*, 103ra97.

Suls, A., Velizarova, R., Yordanova, I., et al. (2010). Four generations of epilepsy caused by an inherited microdeletion of the SCN1A gene. *Neurology, 75*, 72–76.

Sun, H., Zhang, Y., Liu, X., et al. (2010). Analysis of SCN1A mutation and parental origin in patients with Dravet syndrome. *Journal of Human Genetics, 55*, 421–427.

Sun, W., Wright, F. A., Tang, Z., et al. (2009). Integrated study of copy number states and genotype calls using high-density SNP arrays. *Nucleic Acids Research, 37*, 5365–5377.

Tapper, W. J., Foulds, N., Cross, N. C., et al. (2014). Megalencephaly syndromes: exome pipeline strategies for detecting low-level mosaic mutations. *PLoS One, 9*, e86940.

Taylor Tavares, A. L., Willatt, L., Armstrong, R., Simonic, I., & Park, S. M. (2013). Mosaic deletion of the NF1 gene in a patient with cognitive disability and dysmorphic features but without diagnostic features of NF1. *American Journal of Medical Genetics, Part A, 161A*, 1185–1188.

Ting, J. C., Roberson, E. D., Miller, N. D., et al. (2007). Visualization of uniparental inheritance, Mendelian inconsistencies, deletions, and parent of origin effects in single nucleotide polymorphism trio data with SNPtrio. *Human Mutation, 28*, 1225–1235.

Tinschert, S., Naumann, I., Stegmann, E., et al. (2000). Segmental neurofibromatosis is caused by somatic mutation of the neurofibromatosis type 1 (NF1) gene. *European Journal of Human Genetics, 8*, 455–459.

Topcu, M., Akyerli, C., Sayi, A., et al. (2002). Somatic mosaicism for a MECP2 mutation associated with classic Rett syndrome in a boy. *European Journal of Human Genetics: EJHG, 10*, 77–81.

Toro, R., Konyukh, M., Delorme, R., et al. (2010). Key role for gene dosage and synaptic homeostasis in autism spectrum disorders. *Trends in Genetics, 26*, 363–372.

Vadlamudi, L., Dibbens, L. M., Lawrence, K. M., et al. (2010). Timing of de novo mutagenesis—a twin study of sodium-channel mutations. *New England Journal of Medicine, 363*, 1335–1340.

Vecchi, M., Cassina, M., Casarin, A., et al. (2011). Infantile epilepsy associated with mosaic 2q24 duplication including SCN2A and SCN3A. *Seizure, 20*, 813–816.

Veltman, J. A., & Brunner, H. G. (2012). De novo mutations in human genetic disease. *Nature Reviews Genetics, 13*, 565–575.

Verhoef, S., Bakker, L., Tempelaars, A. M., et al. (1999). High rate of mosaicism in tuberous sclerosis complex. *American Journal of Human Genetics, 64*, 1632–1637.

Voet, T., Kumar, P., Van Loo, P., et al. (2013). Single-cell paired-end genome sequencing reveals structural variation per cell cycle. *Nucleic Acids Research, 41*, 6119–6138.

Walsh, C., & Cepko, C. L. (1992). Widespread dispersion of neuronal clones across functional regions of the cerebral cortex. *Science, 255*, 434–440.

Wang, K., Chen, Z., Tadesse, M. G., et al. (2008). Modeling genetic inheritance of copy number variations. *Nucleic Acids Research, 36*, e138.

Ware, M. L., Tavazoie, S. F., Reid, C. B., & Walsh, C. A. (1999). Coexistence of widespread clones and large radial clones in early embryonic ferret cortex. *Cerebral Cortex, 9*, 636–645.

Watson, I. R., Takahashi, K., Futreal, P. A., & Chin, L. (2013). Emerging patterns of somatic mutations in cancer. *Nature Reviews Genetics, 14*, 703–718.

Weinstein, L. S., Shenker, A., Gejman, P. V., Merino, M. J., Friedman, E., & Spiegel, A. M. (1991). Activating mutations of the stimulatory G protein in the McCune-Albright syndrome. *New England Journal of Medicine, 325*, 1688–1695.

Westra, J. W., Peterson, S. E., Yung, Y. C., Mutoh, T., Barral, S., & Chun, J. (2008). Aneuploid mosaicism in the developing and adult cerebellar cortex. *Journal of Comparative Neurology, 507*, 1944–1951.

Westra, J. W., Rivera, R. R., Bushman, D. M., et al. (2010). Neuronal DNA content variation (DCV) with regional and individual differences in the human brain. *Journal of Comparative Neurology, 518*, 3981–4000.

Yin, X., Tan, K., Vajta, G., et al. (2013). Massively parallel sequencing for chromosomal abnormality testing in trophectoderm cells of human blastocysts. *Biology of Reproduction, 88*, 69.

Yurov, Y. B., Vorsanova, S. G., Iourov, I. Y., et al. (2007). Unexplained autism is frequently associated with low-level mosaic aneuploidy. *Journal of Medical Genetics, 44*, 521–525.

Zeesman, S., Zwaigenbaum, L., Whelan, D. T., Hagerman, R. J., Tassone, F., & Taylor, S. A. (2004). Paternal transmission of fragile X syndrome. *American Journal of Medical Genetics, Part A, 129A*, 184–189.

Zerem, A., Lev, D., Blumkin, L., et al. (2014). Paternal germline mosaicism of a SCN2A mutation results in Ohtahara syndrome in half siblings. *European Journal of Paediatric Neurology, 18*, 567–571.

Zhou, C. Y., Li, J., Guo, X. Y., Liao, J., Wang, Z. H., & Lan, F. H. (2012). Identification of a mosaic mutation of NF1 gene in a sporadic case of neurofibromatosis type 1. *Zhonghua yi xue yi chuan xue za zhi = Zhonghua yixue yichuanxue zazhi = Chinese Journal of Medical Genetics, 29*, 529–532.

Zong, C., Lu, S., Chapman, A. R., & Xie, X. S. (2012). Genome-wide detection of single-nucleotide and copy-number variations of a single human cell. *Science, 338*, 1622–1626.

Section II

A New Neuroanatomy

Chapter 13

The Molecular Landscape of the Developing Human Central Nervous System

Ying Zhu[a], John C. Silbereis[a], Sirisha Pochareddy, Mingfeng Li and Nenad Sestan
Yale School of Medicine, New Haven, CT, United States

INTRODUCTION

The development of the central nervous system (CNS) is an immensely complex and strictly regulated process that proceeds over a remarkably protracted period of time in humans (Fig. 13.1) (Bayer & Altman, 2007; Bystron, Blakemore, & Rakic, 2008; Conel, 1939; Flechsig Of Leipsic, 1901; Giedd & Rapoport, 2010; His, 1889; Johnson, 2001; Kostovic & Judas, 2006; Meyer, 2007; Retzius, 1896; Silbereis, Pochareddy, Zhu, Li, & Sestan, 2016; Sidman & Rakic, 1973; Sousa, Meyer, & Sestan, 2014; Sun & Walsh, 2006; Tau & Peterson, 2010; Toga, Thompson, & Sowell, 2006; Yakovlev & Lecours, 1967). Human neurodevelopment is highly vulnerable to genetic mutations and environmental factors, which cause a wide array of neurological and psychiatric conditions ranging from intellectual disability, autism, and epilepsy to bipolar disorder and schizophrenia (Dixon-Salazar & Gleeson, 2010; Hu, Chahrour, & Walsh, 2014; Jeste & Geschwind, 2015; Orr & Zoghbi, 2007; Tebbenkamp, Willsey, State, & Sestan, 2014; Thompson, Levitt, & Stanwood, 2009). The genomic revolution has generated a remarkable level of knowledge about the genetic variations that confer risk for various neurological and psychiatric disorders. However, advancing our understanding of neurodevelopment and the pathogenesis of neurological and psychological conditions requires a thorough knowledge of the molecular and cellular processes in the developing and adult human brain (Fig. 13.1), as well as insight into how "normal" and disease-associated genetic variations affect these processes. Through further innovations in "-omic" technologies and systems biology, comprehensive characterization and detailed spatiotemporal mapping of the transcriptional, epigenetic, regulatory, and proteomic landscapes of human tissues, including the CNS, is within reach. As a result, public sequence-read archives contain a huge amount of genomic- and proteomic-level data from a variety of human tissues (NCBI Resource Coordinators, 2014) that provide key insights into the molecular and cellular pathology of complex disorders.

In this chapter, we will first briefly review the basic principles of the organization and development of the human brain with an emphasis on the cerebral neocortex, and then detail advances in our understanding of the transcriptional, epigenomic and regulatory landscapes of the human neocortex and other regions of the CNS. We will also update and elaborate on earlier reviews focusing on the molecular pathogenesis of certain neurodevelopmental disorders provided by these studies.

THE CELLULAR AND STRUCTURAL COMPLEXITY OF THE HUMAN BRAIN

The human brain is almost certainly the most complex biological tissue known, composed of approximately 86 billion neurons along with an equal number of glial cells (Azevedo et al., 2009; Walloe, Pakkenberg, & Fabricius, 2014). It has

a. Equal contributors.

Genomics, Circuits, and Pathways in Clinical Neuropsychiatry. http://dx.doi.org/10.1016/B978-0-12-800105-9.00013-5

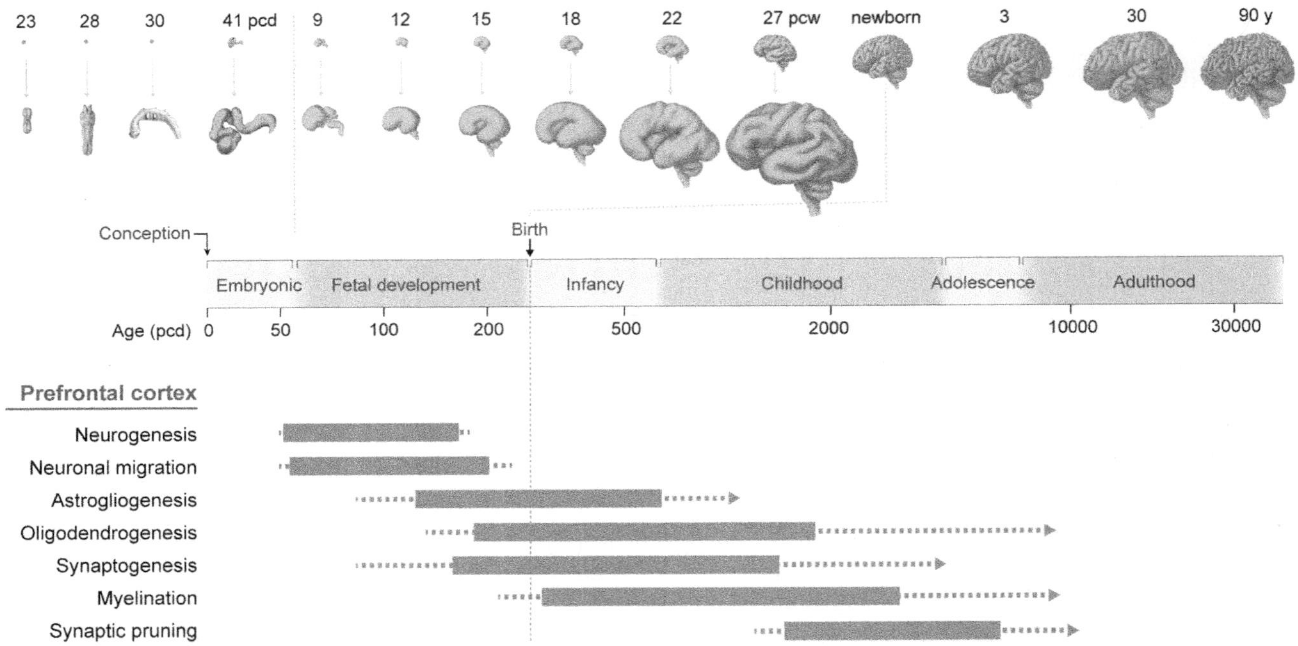

FIGURE 13.1 Timeline of key processes in human CNS development (data shown for the prefrontal neocortex). Illustrations in top panel demonstrate dramatic changes in the size and shape of the human CNS over the course of prenatal development in the human. A schematic shows the age in postconception days when essential neurodevelopmental processes take place over the prolonged course of human development. *Bars* in the bottom panel indicate the peak period when each major neurodevelopmental process (left column) occurs; *dotted lines* indicate additional ages when these processes are observed. *Based on a figure from Sousa et al. (2014).* Developmental and Evolutionary Cognitive Neuroscience *and Silbereis, J. C., Pochareddy, S., Zhu, Y., Li, M., & Sestan, N. (2016). The cellular and molecular landscapes of the developing human central nervous system.* Neuron, 89, *248–268; Brain images are adapted from Kolb and Whishaw (2009),* Fundamentals of Human Neuropsychology.

been estimated that a typical neuron receives thousands of synaptic inputs and that in total there are at least several hundred trillion local and long-distance synaptic connections in the adult human brain (Silbereis et al., 2016; Tang, Nyengaard, De Groot, & Gundersen, 2001). How this complex neural circuitry is assembled during development remains one of the great scientific mysteries.

Although the organization of the brain is generally conserved across mammals, humans most notably possess one of the largest and most highly gyrified brains and a vastly expanded cerebral cortex relative to body size (Hofman, 2012; Krubitzer & Kaas, 2005). The human brain is further characterized by reorganization of neural circuitry, particularly within intracortical projection systems (Rilling, Glasser, Jbabdi, Andersson, & Preuss, 2011; Sousa, Meyer, & Sestan, 2014) compared with even our most closely related nonhuman primate relatives. Therefore, understanding the developmental neurobiology of the human neocortex is especially dependent on studies of human cells and tissue, in addition to model organisms.

The basic features of brain development are conserved across mammals, but there are several important and distinctive characteristics of human neurodevelopment. First, it is a highly protracted process that encompasses approximately 3 decades, characterized by a remarkably long childhood and adolescence with an exceptionally protracted period of juvenile dependence (Kuzawa et al., 2014) (Fig. 13.1). Second, novel aspects of the developmental neuroanatomy and proliferative capacity of neural progenitors have emerged in evolution to enable brain expansion among mammals, especially in neocortex (Dehay, Kennedy, & Kosik, 2015; Gulden & Sestan, 2014; Lui, Hansen, & Kriegstein, 2011; Taverna, Gotz, & Huttner, 2014). Third, the appearance of major neurological and psychiatric diseases at distinct age ranges (Lee et al., 2014) suggests a dysregulation in temporally regulated molecular processes in the human brain (Fig. 13.2) (Licatalosi & Darnell, 2006; Mehler & Mattick, 2007; Mirnics & Pevsner, 2004; Tebbenkamp et al., 2014), which highlights the complexity and importance of well-regulated developmental pathways over the first 20 years of human life (Kessler et al., 2007).

This complexity is in large part the result of coordinated transcriptional and posttranscriptional processes occurring throughout development (Colantuoni et al., 2011; Johnson et al., 2009; Kang et al., 2011; Miller et al., 2014; Pletikos et al., 2014; Sousa, Meyer, & Sestan, 2014; Walsh & Engle, 2010). However, the mechanisms underlying the selective vulnerability of specific cell types, neural circuits, and brain regions to mutations which are associated with specific

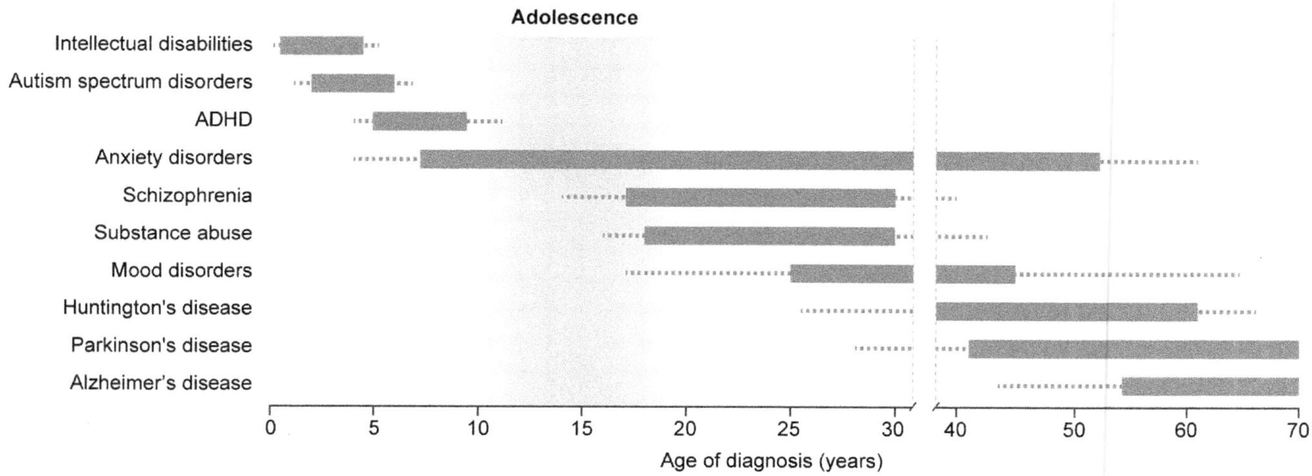

FIGURE 13.2 Many brain disorders appear at different times during development and adulthood. The age of onset of brain disorders varies substantially. This indicates the necessity for tightly controlled gene expression throughout development and adulthood and highlights the importance of defining the spatiotemporal molecular and epigenetic landscape in the healthy and diseased human brain. The gray box illustrates the peak of the age at diagnosis. *Based in part on data from Kessler, R. C., Amminger, G. P., Aguilar-Gaxiola, S., Alonso, J., Lee S., & Ustun T. B. (2007). Age of onset of mental disorders: a review of recent literature.* Current Opinion in Psychiatry, 20, *359–364 and Lee, F. S., Heimer, H., Giedd, J. N., Lein, E. S., Sestan, N., Weinberger, D. R., & Casey, B. J. (2014). Mental health. Adolescent mental health—opportunity and obligation.* Science, 346, *547–549.*

neuropsychiatric diseases remain largely unknown (Morrison & Hof, 1997; Orr & Zoghbi, 2007; Saxena & Caroni, 2011). Thus, detailed spatiotemporal maps of the molecular properties in the human brain across development and through adulthood are essential for improving our understanding of neural development, function, and dysfunction.

GENERAL PRINCIPLES OF HUMAN NEOCORTICAL DEVELOPMENT

The development of the human neocortex is a highly dynamic process during which myriads of distinct cell types, neural circuits, and regions are formed and undergo maturational changes. Moreover, transient cell types, synaptic circuits, and structures arise and disappear during specific time points in development, and are associated with distinct cell types and cellular mechanisms necessary for the formation of the mature brain (see Fig. 13.3 for an example in neocortical development). Here we briefly discuss the cellular processes by which the neocortex is built and organized prenatally and how neural circuitry is established postnatally, as summarized in Fig. 13.1.

Neurogenesis and Neuronal Migration

Early development of the neocortical anlage is characterized by the robust proliferation of neuroepithelial neural stem/progenitor cells which are arranged in a sheet forming the ventricular zone (VZ). Initially, each cell division results in two progenitor daughter cells. This process of symmetric division exponentially expands the number of progenitor cells capable of giving rise to neurons and glia. Beginning around the fifth postconception week (pcw), these cells begin to generate the earliest neurons of the emerging cortical plate (Meyer, Schaaps, Moreau, & Goffinet, 2000). Early on in embryonic neurogenesis, neuroepithelial progenitor cells transition into another form of neural stem/progenitor cell called radial glia (RGs), which extend apico-basal processes from the ventricle to the pial surface. RG cell bodies largely reside in the VZ, where they asymmetrically proliferate, giving rise to a daughter radial glial cell and a transit amplifying cell or postmitotic neuroblast.

Two major mechanisms appear to have emerged to enable expansion of the neuron population in the neocortex, in addition to the protracted period of neurogenesis. First, human transit amplifying cells appear to undergo more rounds of cell division than in other organisms. Second, a novel type of RG, called "outer RG" cells, whose cell bodies are located dorsal to the subventricular zone rather than in the VZ, has been characterized (Hansen, Lui, Parker, & Kriegstein, 2010). These cells, which are abundantly present in the human fetal neocortical wall, have the ability to self-renew and are therefore thought to contribute greatly to the increase in neuron number by increasing the absolute number of neuronal progenitor cells.

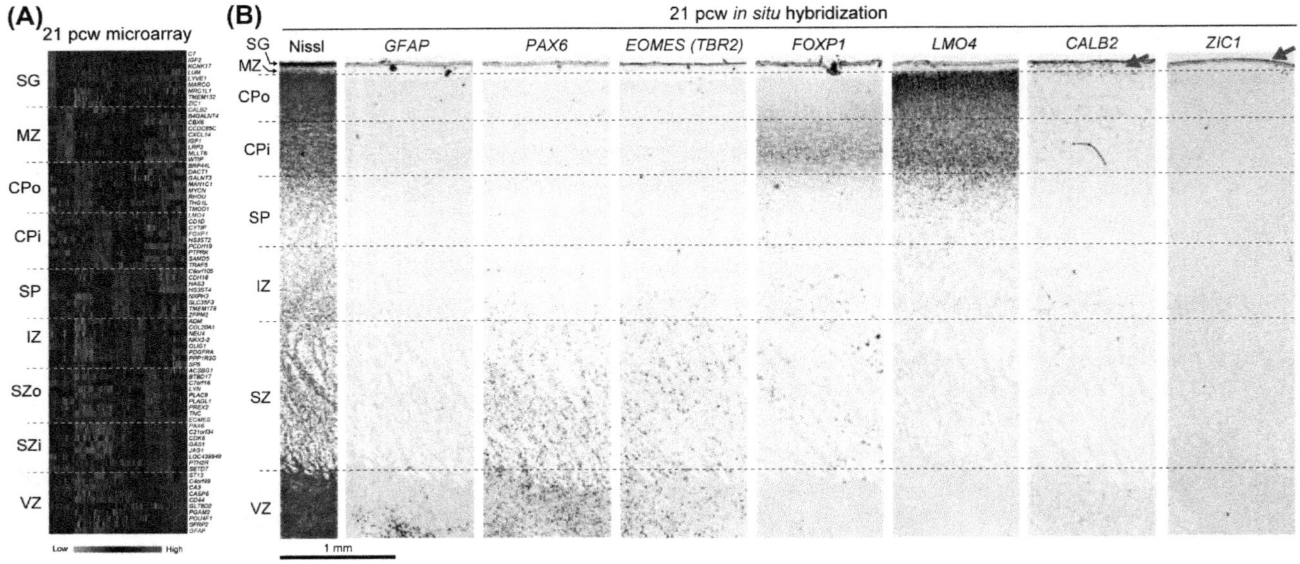

FIGURE 13.3 Transcriptional differences between zones of the midfetal human neocortical wall. Neocortical fetal zone-specific gene expression patterns have been identified in the human brain, revealing genes essential for the distinct cell types and neurodevelopmental processes that predominate in each zone: eg, neurogenesis in the VZ versus deep-layer neuronal maturation in the inner cortical plate (CPi). (A) Heat map showing fetal neocortical zone-enriched gene expression determined by correlation to binary templates at 21 pcw. Differential gene expression corresponds to the distinct cell types and stages of maturation in each zone. (B) Nissl stain on the left delineates each fetal neocortical zone; in subsequent panels are examples of the expression of essential zone-enriched genes by in situ hybridization. *i*, inner; *IZ*, intermediate zone; *MZ*, marginal zone; *o*, outer; *SG*, subpial granular layer; *SP*, subplate zone; *SZ*, subventricular zone. *Adapted from Miller, J. A., Ding, S. L., Sunkin, S. M., Smith, K. A., Ng, L., Szafer, A., ... Lein, E. S. (2014). Transcriptional landscape of the prenatal human brain.* Nature, 508, *199–206.*

As is the case throughout the brain, the precise origin and migration modes of different neocortical neurons vary by subtype. Neocortical neurons can be broadly categorized by two different cell types: excitatory, glutamatergic projection neurons (aka pyramidal neurons) that project either outside the neocortex or to different neocortical regions and layers, and locally projecting gamma-aminobutyric acid-ergic, inhibitory neurons. All cortical projection neurons are derived from progenitor cells of the dorsal pallium and migrate radially into the cortical plate. In contrast, interneurons are largely derived ventrally from the ganglionic eminences in the striatal primordium and migrate tangentially into the cortical plate (Meyer, 2007).

Glial Cell Genesis and Differentiation

The generation of glial cells follows neurogenesis, peaks around birth, and is also a protracted process in human beings (Miller et al., 2012; Roessmann & Gambetti, 1986). Astrocytes and oligodendrocyte precursor cells are thought to also derive from RGs beginning in midgestation (Amunts, Schleicher, Ditterich, & Zilles, 2003; Jakovcevski, Filipovic, Mo, Rakic, & Zecevic, 2009; Miller et al., 2012). Oligodendrocytes develop from proliferating oligodendrocyte precursor cells that arise in proliferative zones in multiple regions of the ventral and dorsal forebrain, migrate throughout the developing white matter, and divide a limited number of times before they terminally differentiate (Jakovcevski et al., 2009; Miller et al., 2012). Oligodendrocytes continue to be robustly generated and migrate extensively through the first 2 years of human life, whereas myelination continues into early adulthood (Jakovcevski et al., 2009; Miller et al., 2012). The earliest astrocytes are generated from the direct transformation of human RG cells followed by subsequent rounds of proliferation. Astrocytes that appear mature morphologically are observed as early as 15 pcw (Howard et al., 2008) and the astroglial population is likely fully differentiated by the end of the first year of life in humans (Sanai et al., 2011).

As a result of these patterns of cell proliferation, migration, and differentiation of neurons and glial cells, the embryonic and fetal neocortical wall is organized into distinct and often transient compartments (also known as zones or layers), each carrying out unique cellular processes and each composed of a different makeup of progenitor cells (Bystron et al., 2008; His, 1889). Some of these domains, such as the subplate, which is important in establishing transient early cortical neural connections, are especially enlarged in the human brain (Allendoerfer & Shatz, 1994; Hoerder-Suabedissen & Molnar, 2015; Judas, Sedmak, & Kostovic, 2013).

Neural Circuit Formation and Developmental Plasticity

At midgestation, immature neurons have extended axons and begun to elaborate dendrites, initiating a protracted period of axon growth, dendritic arborization, and synaptogenesis that extends into early childhood (Huttenlocher & Dabholkar, 1997). The refinement of synaptic connections, mediated by the convergence of influences from both intrinsic and extrinsic factors, seems a major task of the developing brain from toddlerhood through early adulthood (Petanjek et al., 2011). Remarkably, robust synaptic elimination and dendritic pruning are observed from infancy through adolescence and are likely regulated by multiple mechanisms (Katz & Shatz, 1996). Developmental changes in synaptic density, myelin, and axonal patterning have been noted in many neuropsychiatric disorders, which highlights the importance of the appropriate regulation of these prolonged human neurodevelopmental processes.

Regional Patterning and Interhemispheric Lateralization

In all mammals including humans, at least two types of spatial information must be encoded in nascent projection neurons in the emerging neocortex: (1) their position in the radial direction, corresponding to their laminar position, with distinct cell types, synaptic connectivities, and functions within each lamina or layer (Kwan, Sestan, & Anton, 2012; Leone, Srinivasan, Chen, Alcamo, & McConnell, 2008; Molyneaux, Arlotta, Menezes, & Macklis, 2007); and (2) their position in the tangential plane, corresponding to their particular cortical region or area identity (O'Leary, Chou, & Sahara, 2007; Rash & Grove, 2006; Sur & Rubenstein, 2005). The physical separation of layers and areas is functionally determined and maintained through their distinct composition of neuronal cell types and a unique set of afferent and efferent synaptic connections. The laminar identity of projection neurons reflects their birth order, with first-born neurons occupying the deepest layers and later-born neurons present in more superficial layers. The upper cortical layers, which principally make intracortical projections, are overrepresented in primates and have been proposed to contribute to some of the cognitive and motor abilities that are unique to humans (Marin-Padilla, 1978). Work in mice has identified a number of transcription factors that specify the identities of distinct subtypes of cortical projection neurons (Kwan et al., 2012; Leone et al., 2008; Molyneaux et al., 2007) and interneurons (Southwell et al., 2014), and many are conserved in the developing human neocortex (Bayatti et al., 2008; Hansen et al., 2013; Hevner, 2007; Kwan et al., 2008; Radonjic et al., 2014).

Regional patterning of the neocortex is defined in part by the emergence of patterning centers defined by gradients in the expression of morphogens and signaling molecule targets (O'Leary & Sahara, 2008) and inputs from subcortical structures (O'Leary et al., 2007; Rash & Grove, 2006; Sur & Rubenstein, 2005). A hallmark of early neocortical patterning across species is the formation of global rostrocaudal and mediolateral neurogenetic gradients (O'Leary et al., 2007; Rash & Grove, 2006; Sur & Rubenstein, 2005). Intriguingly, studies suggest that embryonic and fetal human brain patterning employs many of the same principles observed across mammalian species, which suggests a degree of homology (Bayatti et al., 2008; Johnson et al., 2009; Kang et al., 2011; Miller et al., 2014). However, the same studies have identified differences among humans, nonhuman primates, and rodents in the expression of genes involved in regional patterning of mouse embryonic neocortex, which highlights the role of species differences in the early patterning of the neocortex.

Another key feature of the human developing neocortex is asymmetry between left and right hemispheres, each comprising a topographically matched, although slightly structurally and functionally asymmetric, areal map (Amunts et al., 2003; Gazzaniga, Bogen, & Sperry, 1962; Geschwind & Levitsky, 1968; Sun & Walsh, 2006). This asymmetric organization has a crucial role in functional lateralization of many cognitive and motor functions, such as language and handedness. Several lines of evidence indicate that these asymmetries are reflected at the molecular (Sun & Walsh, 2006) and cellular (Amunts et al., 2003; Hayes & Lewis, 1993) levels. Structural asymmetry first appears during the late midfetal period (Chi, Dooling, & Gilles, 1977) and becomes more prominent during early postnatal development when functional asymmetries become noticeable (Amunts et al., 2003; Hayes & Lewis, 1993; Hill et al., 2010).

Multiple lines of evidence also indicate that distinct human neocortical areas, nuclei, and the hemispheres as a whole, mature at different rates (Flechsig Of Leipsic, 1901; Giedd & Rapoport, 2010; Huttenlocher & Dabholkar, 1997; Toga et al., 2006; Yakovlev & Lecours, 1967). For example, axons in primary sensorimotor areas start to myelinate before those in the association areas (Flechsig Of Leipsic, 1901; Yakovlev & Lecours, 1967). Other processes such as synaptogenesis also exhibit prominent interareal differences in their maturational trajectories (Huttenlocher & Dabholkar, 1997). Furthermore, the right hemisphere appears to mature faster than the left during late fetal and early postnatal development (Taylor, 1969; Thatcher, Walker, & Giudice, 1987).

The recruitment of neurons from different layers, areas, and hemispheres into complex synaptic circuits underlies cognition and other forms of complex behavior. Advancements in functional genomic technologies offer unbiased insights into normal and abnormal molecular process in postmortem human CNS tissues and neural cell culture systems, such as neural cells derived from induced pluripotent stem cells (van den Ameele, Tiberi, Vanderhaeghen, & Espuny-Camacho, 2014; Gage & Temple, 2013; Mariani et al., 2012). The remainder of this chapter will focus on our understanding of the spatiotemporal landscape of the RNA species (ie, transcriptome); epigenetic features such as DNA methylation, histone modifications, chromatin accessibility, and noncoding regulatory RNAs (ie, "the epigenome") (see also an alternative view by Mark Ptashne (2013)); and DNA regulatory elements active (ie, regulome) in the developing human neocortex and other brain structures.

TRANSCRIPTIONAL LANDSCAPE OF THE DEVELOPING HUMAN BRAIN

Transcription is the first step in transferring genetic information into specific phenotypes and establishing unique molecular and subsequently cellular properties. The development of high-throughput microarray and sequencing technologies has greatly advanced our ability to explore transcriptomic dynamics that establish regional anatomical properties and unique cell types in the prenatal and postnatal developing human brain (Abrahams et al., 2007; He, Bammann, Han, Xie, & Khaitovich, 2014; Ip et al., 2010; Li et al., 2013; Somel et al., 2009; Sun et al., 2005). By including multiple brain regions (Hawrylycz et al., 2012; Kang et al., 2011; Miller et al., 2014; Pletikos et al., 2014) and developmental periods (Jaffe et al., 2015; Johnson et al., 2009; Kang et al., 2011; Pletikos et al., 2014), studies have begun to build a comprehensive spatiotemporal profiles of the human brain transcriptome using postmortem tissue and provided publicly accessible resources: www.hbatlas.org (Johnson et al., 2009), www.humanbraintranscriptome.org (Kang et al., 2011), www.braincloud.jhmi.edu (Colantuoni et al., 2011), and www.brainspan.org (Kang et al., 2011; Miller et al., 2014). Importantly, novel and valuable insights into normal and abnormal brain development have been obtained by the analysis of these data.

Spatiotemporal Dynamics of Human Brain Transcriptome

Large-scale characterizations of gene expression in the developing human brain by several groups have revealed intricate features of the organization and complexity of the transcriptional architecture that reflect the underlying neurobiological processes (Colantuoni et al., 2011; Hawrylycz et al., 2012; Ip et al., 2010; Johnson et al., 2009; Kang et al., 2011; Lambert et al., 2011; Mazin et al., 2013; Miller et al., 2014; Pletikos et al., 2014; Somel et al., 2009). A common finding of these studies is that gene expression is highly spatiotemporally dynamic in the human brain (Fig. 13.4). Global analyses have revealed that the transcriptomes differ more prominently across time and space than they do between sexes, ethnicities, or individuals despite their underlying genetic differences. Consistent with this global pattern, transcriptome differences between males and females are more prominent during prenatal development than during postnatal life, with the adult brain having the greatest similarity (Fig. 13.5). Genes with sex-biased expression include those previously implicated in brain development and function as well as several disease-related genes, which offers one of several possible mechanisms underlying sex differences in the incidence, prevalence, and severity of some brain disorders.

Another key finding of these studies is that a high percentage of protein-coding genes analyzed (over 86%, according to Kang et al. (2011)) were expressed in at least one region of the developing or adult brain. Of these, nine of 10 genes were differentially regulated at the whole-transcript or exon level across brain regions and/or time, indicating widespread and complex splicing changes during development (Kang et al., 2011; Mazin et al., 2013). The bulk of these transcriptional differences occurred in prenatal development. Compared with dorsal pallial structures (amygdala, hippocampus, and neocortex), the cerebellum is the most transcriptionally distinctive region in the developing brain, followed by thalamus and striatum (Kang et al., 2011; Numata et al., 2012). Transcriptional differences between putative areas of the developing neocortex were less robust than those between neocortex and other brain regions that were analyzed. Interestingly, among neocortical areas, strong transcriptional differences were particularly prominent during fetal development and included specific transcriptional signatures associated with prefrontal and perisylvian areas, which are involved in some of the most distinctly human aspects of cognition and behavior (Johnson et al., 2009; Kang et al., 2011; Pletikos et al., 2014). These strong prenatal neocortical transcriptional differences diminished during infancy and childhood, and increased again in adolescence (Fig. 13.6) (Pletikos et al., 2014).

Gene coexpression analyses also revealed that the developing human brain transcriptome is organized in distinct coexpression networks enriched for specific biological functions (Johnson et al., 2009; Kang et al., 2011; Miller et al., 2014). Interestingly, genetic variation in some of the most well-connected genes in these modules has been linked to psychiatric or neurological disorders including schizophrenia and autism spectrum disorder (ASD), which suggests that they may have converging functions in specific brain regions and developmental periods (Tebbenkamp, 2014; Jeste & Geschwind, 2015).

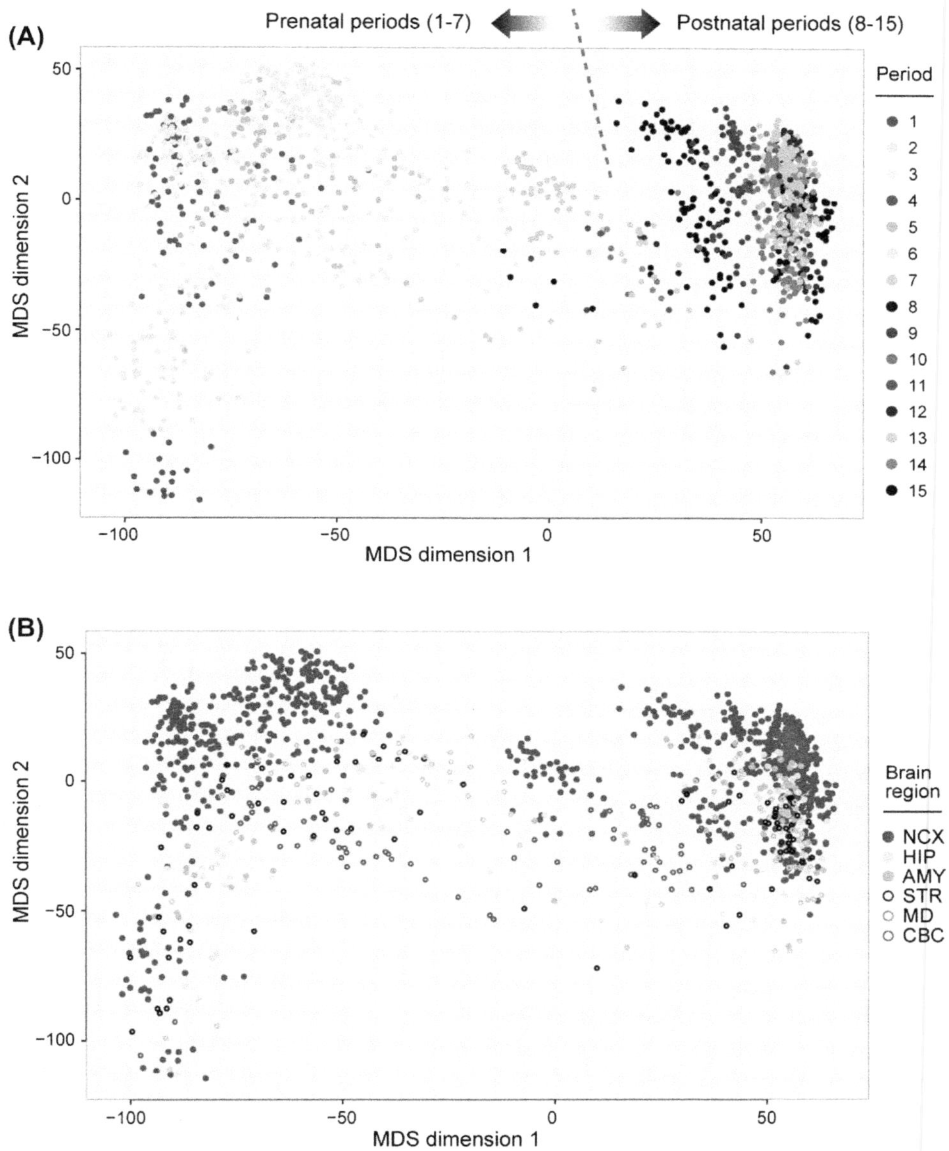

FIGURE 13.4 Global temporal and spatial dynamics of the human brain transcriptome. Global spatiotemporal dynamics of the human brain transcriptome revealed by multidimensional scaling (MDS) plots. Each analyzed tissue sample is colored according to period (A) or region (B). The bulk of transcriptional differences occurred during prenatal development. *NCX*, neocortex; *HIP*, hippocampus; *AMY*, amygdala; *STR*, striatum; *MD*, mediodorsal nucleus of thalamus; *CBC*, cerebellar cortex. *Adapted from Kang, H. J., Kawasawa, Y. I., Cheng, F., Zhu, Y., Xu, X., Li, M., … Sestan, N. (2011). Spatio-temporal transcriptome of the human brain.* Nature, *478, 483–489.*

Transcriptional Architecture Underlying Cellular and Regional Specification of the Human Brain

Differences in the transcriptional architecture of brain regions and neocortical areas reflect differences in the cellular makeup, biological process, and developmental timing (Colantuoni et al., 2011; Hawrylycz et al., 2012; Ip et al., 2010; Johnson et al., 2009; Lambert et al., 2011; Mazin et al., 2013; Miller et al., 2014; Pletikos et al., 2014; Somel et al., 2009). Moreover, in-depth transcriptome analysis of the fetal neocortical wall has also identified transcriptional signatures of different transient zones and cell types within them and shown conserved patterns and species differences (Fietz et al., 2012; Miller et al., 2014; Pollen et al., 2014). Previous work in mice has shown that early patterning of the CNS, including neocortical regions and areas, is governed by graded expression of transcription factors during early embryonic development followed by extrinsic signaling from thalamic axonal inputs (O'Leary et al., 2007; Rakic, Ayoub, Breunig, &

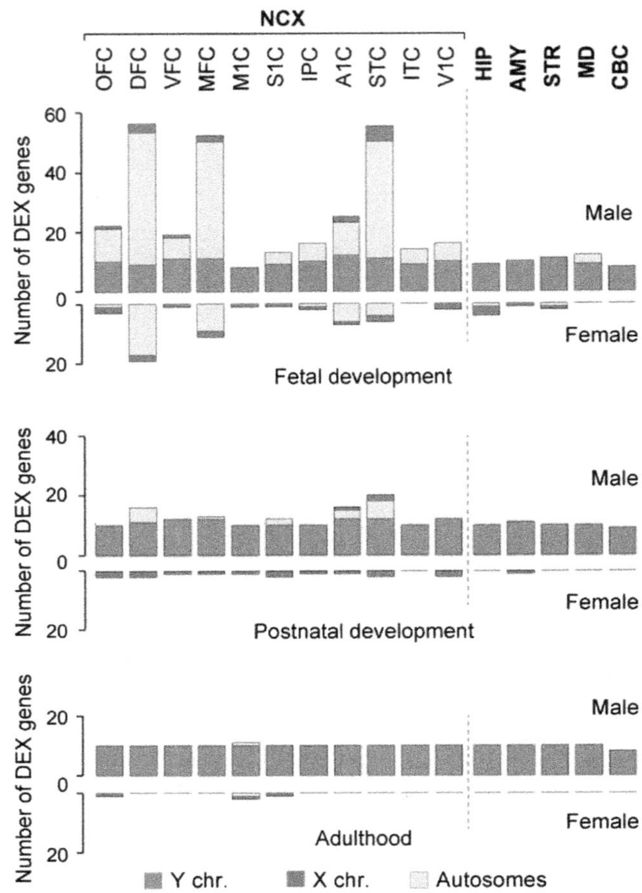

FIGURE 13.5 **Sex differences in developmental and adult human brain gene expression.** Sex differences in gene expression are highest in fetal development and decline over the course of postnatal development. Number of sex-biased DEX genes in brain regions/neocortical areas during fetal development (top), postnatal development (middle), and adulthood (bottom). *OFC*, orbital; *MFC*, medial prefrontal cortex; *M1C*, primary motor (M1) cortex; *S1C*, somatosensoprefrontal cortex; *DFC*, dorsolateral prefrontal cortexry (S1) cortex; *IPC*, posterior inferior parietal cortex; *A1C*, primary auditory (A1C) cortex; *STC*, superior temporal cortex; *ITC*, inferior temporal cortex; *V1C*, primary visual (V1) cortex; *HIP*, hippocampus; *AMY*, amygdala; *STR*, striatum; *MD*, mediodorsal nucleus of thalamus; *CBC*, cerebellar cortex. *chr.*, chromosome. *Adapted from Kang, H. J., Kawasawa, Y. I., Cheng, F., Zhu, Y., Xu, X., Li, M., … Sestan, N. (2011). Spatio-temporal transcriptome of the human brain.* Nature, 478, *483–489.*

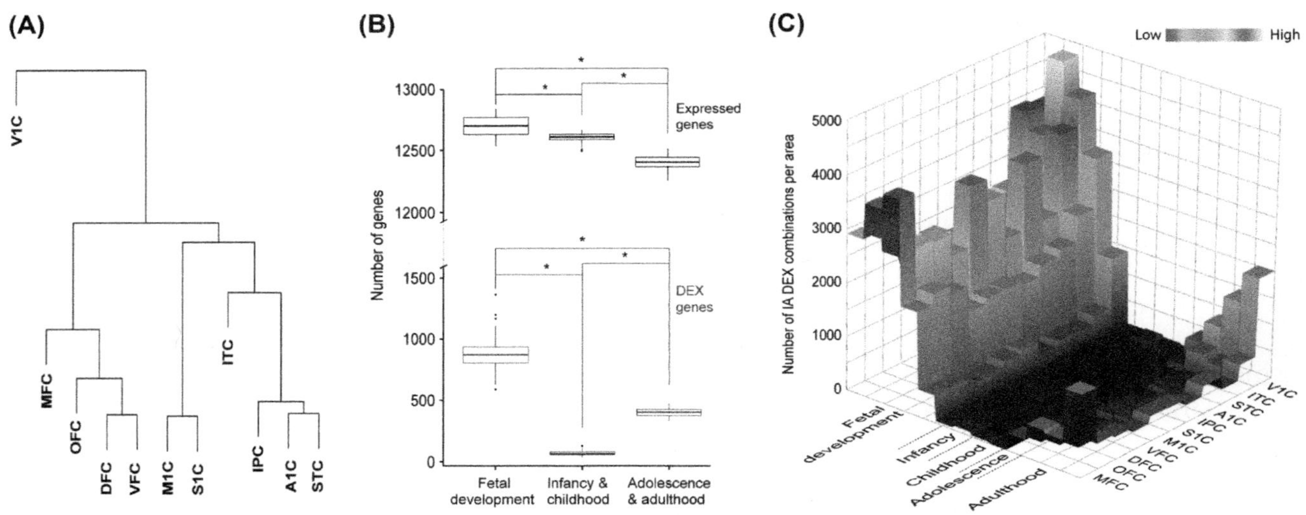

FIGURE 13.6 **Regional/areal variations in the human neocortical transcriptome exhibit a temporal hourglass pattern.** (A) Unsupervised hierarchical clustering of the 11 putative neocortical areas and regions profiled by Pletikos et al. (2014), based on the transcriptome of each area from the period of fetal development throughout life, showing relative transcriptional differences. (B) Box plots of subsampling permutations show the number of expressed (up) and differentially expressed (down) genes among neocortical areas across fetal development (periods 3–7), infancy (periods 8 and 9), childhood (periods 10 and 11), adolescence (period 12), and adulthood (periods 13–15). (C) Manhattan plot showing the sum of the number of interareal DEX genes between each neocortical area and any of the other areas over time (post hoc Tukey tests). *OFC*, orbital prefrontal cortex; *DFC*, dorsolateral prefrontal cortex; *MFC*, medial prefrontal cortex; *M1C*, primary motor (M1) cortex; *S1C*, somatosensory (S1) cortex; *IPC*, posterior inferior parietal cortex; *A1C*, primary auditory (A1C) cortex; *STC*, superior temporal cortex; *ITC*, inferior temporal cortex; *V1C*, primary visual (V1) cortex; *HIP*, hippocampus; *AMY*, amygdala; *STR*, striatum; *MD*, mediodorsal nucleus of thalamus; *CBC*, cerebellar cortex. *Adapted from Pletikos, M., Sousa, A. M., Sedmak, G., Meyer, K. A., Zhu, Y., Cheng F., … Sestan N. (2014). Temporal specification and bilaterality of human neocortical topographic gene expression.* Neuron, 81, *321–332.*

Dominguez, 2009; Rash & Grove, 2006; Sur & Rubenstein, 2005). Human brain transcriptome data have opened a door for characterizing those early and later regional patterns in the human neocortex.

Genes that are expressed in gradients and/or well-defined compartments in the embryonic and fetal neocortical wall are present in proliferative zones or postmigratory neurons of the cortical plate and subplate (Johnson et al., 2009; Kang, 2011; Miller et al., 2014; Pletikos et al., 2014). Transcriptome studies have also identified prefrontal/frontal-enriched graded expression along the anteroposterior axis and gradients with enrichment in temporal, occipital, occipitotemporal, perisylvian, and ventromedial areas (Kang et al., 2011). Expression of most gradient genes peaked in the frontal or temporal lobes, indicating that a fronto–posterior–temporal rather than fronto-occipital axis exists in the human brain (Johnson et al., 2009; Kang, 2011; Miller et al., 2014; Pletikos et al., 2014). These gradients include both conserved and divergent genes across species (Fig. 13.7). For example, *CBLN2* shows rostral enrichment, which is conserved in humans and mice, whereas *NPY* is expressed posteriorly in V1C of human neocortex in addition to expression in anterior regions as in mouse cortex.

Taken together, these data indicate that spatiotemporal expression patterns in the human neocortex (NCX) are in part shaped by regulatory programs that differ between humans and model species, and indicate the importance of delineating gene expression gradients in the human brain.

Transcriptional Insights Into Maturational Trajectories of the Human Neocortex

Transcriptome studies have also provided new insights into the progression of maturation of distinct brain regions and neocortical areas (Kang et al., 2011; Pletikos et al., 2014). For example, the trajectories (Fig. 13.8) of regional and areal gene expression during brain maturation were shown to have similar shapes across the neocortex, with steep increases

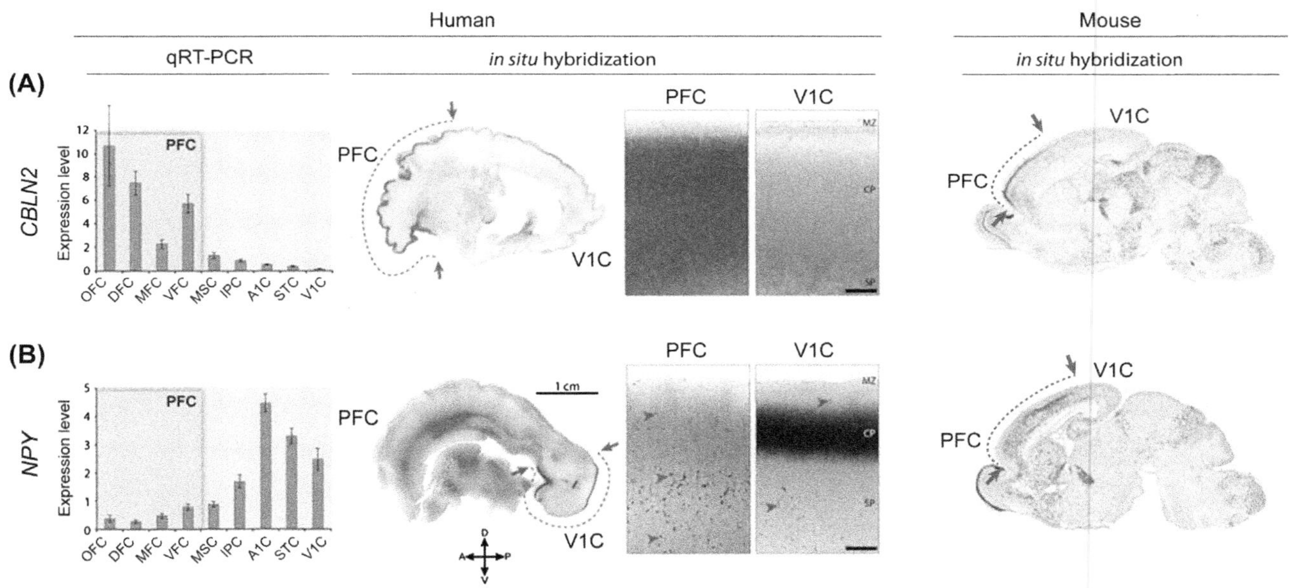

FIGURE 13.7 Commonalities and differences in developmental neocortical regional/areal expression between humans and mice. Differential patterns of gene expression may define functional differences between different areas and layers of fetal neocortex. Highlighting the importance of studying region-specific gene expression in human tissue, differences in gene expression patterns can be highly divergent between humans and model organisms, such as the mouse. (A) *CBLN2* is an example of a gene enriched in prefrontal cortex (orbital prefrontal cortex, *OFC*; dorsolateral prefrontal cortex, *DFC*; medial prefrontal cortex, *MFC*; and ventrolateral prefrontal cortex, *VFC*), as shown by reverse-transcriptase (RT)-PCR (left) and in situ hybridization (ISH) (middle) on a whole sagittal late-midfetal human brain. Analysis at higher magnification reveals that *CBLN2* mRNA is enriched throughout the prefrontal cortical plate (PFC) and subplate (SP) zone but absent from the marginal zone. (Right) ISH on a whole brain sagittal section of postnatal day (P) 4 mouse, an age equivalent to human midfetal development, which suggests that the regional pattern of *CBLN2* is relatively conserved between humans and mice. Interestingly, unlike its expression throughout the human fetal PFC, the mouse *Cbln2* ortholog is enriched in the upper parts of the frontal CP/cortex. (B) The same analysis of *NPY* expression indicated enrichment in the human midfetal occipitotemporal cortical plate (CP). (Middle) Analysis at higher magnification reveals specific enrichment in the occipital CP and high expression in scattered cells (likely migrating interneurons) throughout SP (*arrowheads*) in the occipital cortex (a posterior region). In contrast, *NPY* is sparsely expressed throughout the subplate and CP in frontal regions. ISH on a whole brain sagittal section of P4 mouse demonstrates opposing regional neocortical expression patterns for *NPY* in mice versus humans. *OFC*, orbital prefrontal cortex; *DFC*, dorsolateral prefrontal cortex; *VFC*, ventrolateral prefrontal cortex; *MFC*, medial prefrontal cortex; *MSC*, primary motor (M1)/somatosensory (S1) cortex; *IPC*, posterior inferior parietal cortex; *A1C*, primary auditory (A1C) cortex; *STC*, superior temporal cortex; *V1C*, primary visual (V1) cortex. *Adapted from Johnson, M. B., Kawasawa, Y. I., Mason, C. E., Krsnik, Z., Coppola, G., Bogdanovic, D., … Sestan, N. (2009). Functional and evolutionary insights into human brain development through global transcriptome analysis.* Neuron, 62, *494–509 and* Allen Brain Atlas (mouse data).

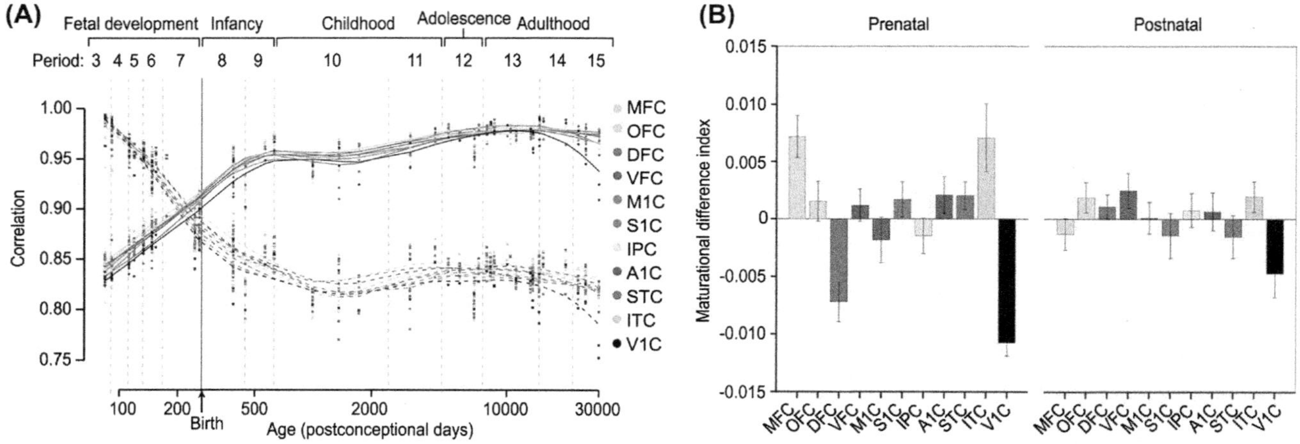

FIGURE 13.8 Neocortical regional/areal transcriptional trajectories become more synchronized during postnatal development. Prenatal development exhibits more deviation than postnatal development. (A) A maturational trajectory plot showing the Pearson correlation of gene expression in each sample to the corresponding averaged gene expression in young adulthood (*solid line*) or mid-fetal development (*dashed line*). (B) Bar plots showing the average deviation of the areal trajectory from the average overall maturational trajectory (maturational difference index). Error bars represent standard deviation. *MFC*, medial prefrontal cortex; *OFC*, orbital prefrontal cortex; *DFC*, dorsolateral prefrontal cortex, *M1C*, primary motor (M1) cortex; *S1C*, somatosensory (S1) cortex; *IPC*, posterior inferior parietal cortex; *A1C*, primary auditory (A1C) cortex; *STC*, superior temporal cortex; *ITC*, inferior temporal cortex; *V1C*, primary visual (V1) cortex. *Adapted from Pletikos, M., Sousa, A. M., Sedmak, G., Meyer, K. A., Zhu, Y., Cheng F., Sestan N. (2014). Temporal specification and bilaterality of human neocortical topographic gene expression.* Neuron, 81, *321–332.*

during mid- and late fetal development. The major inflection point on the curve detailing these trajectories is observed during late infancy. These trajectories are generally synchronized, with varying degrees of deviations in different development periods. The average deviation of the areal trajectory varied more among neocortical areas during fetal than postnatal periods. Medial prefrontal cortex and inferior temporal cortex appear to mature faster than other areas prenatally, whereas the dorsolateral prefrontal cortex transcriptome reached mature levels more slowly. During postnatal development, global levels of transcriptome maturation become more synchronized among neocortical areas.

Whereas global transcriptome trajectories are generally synchronized across the brain, the trajectories of certain neurodevelopmental processes were observed to vary dramatically from each other and among brain regions. For example, major neurodevelopmental processes, including neural cell proliferation and migration, dendrite and synapse development, and myelination exhibit differences in their onset times, rates of increase and decrease, and shapes between each other and among brain regions, which provides a molecular basis for observations of regional differences in these developmental processes (Kang et al., 2011). Similarly, the expression trajectories of genes encoding neuronal markers and neurotransmitter receptors show differences across brain regions (Kang et al., 2011). Notably, some of the processes develop in a species-dependent manner. For example, the delayed expression of genes associated with synaptic functions has been observed in the postnatal human prefrontal cortex, compared with chimpanzee and rhesus macaque (Somel et al., 2009).

Interhemispheric Neocortical Transcriptomes

Surprisingly, despite the clear functional specializations of the brain hemispheres, few statistically significant interhemispheric transcriptome differences were identified across fetal and postnatal development and adulthood, despite extensive analysis using microarray, mRNA sequencing, and quantitative polymerase chain reaction (PCR) of bulk tissue (Hawrylycz et al., 2012; Johnson et al., 2009; Kang et al., 2011; Lambert et al., 2011). Similarly, although previous studies have proposed that the left hemisphere reaches developmental stages sooner than does the left hemisphere (Taylor, 1969; Thatcher et al., 1987), no significant lateral differences in maturation rate were identified when comparing areal transcriptional trajectories of transcriptomes. All of these findings indicate that at the population level, areal transcriptomes are globally symmetric across the full course of human neocortical fetal development and adulthood. Intriguingly, observations at gene expression levels seem to complement structural imaging studies showing strong bilateral symmetry of the neocortex (Chen et al., 2011), as well as reports of an unexpectedly high percentage of monozygotic twin pairs that are discordant for handedness (McManus & Bryden, 1991). However, these findings do not rule out the possibility that hemispheric asymmetry is diluted at the population level by individual differences or driven by more subtle changes in specific cellular components that cannot be detected at the gene expression level, at least with current techniques.

EPIGENOMIC AND REGULATORY LANDSCAPES OF THE DEVELOPING HUMAN BRAIN

The spatiotemporal patterns of gene expression discussed previously are achieved through multiplayer but combinatorial regulatory mechanisms. Key components of this regulatory circuitry include *trans*-acting regulatory proteins, *cis*-regulatory elements (CREs), and epigenetic modifications such as DNA methylation, histone modifications, and changes in DNA accessibility (Shibata, Gulden, & Sisien, 2015). Regulatory proteins include transcription factors that bind to their sequence-specific cognate CREs (Vaquerizas, Kummerfeld, Teichmann, & Luscombe, 2009). CREs are short sequences in the genome that are bound by regulatory proteins. Based on their function, CREs can be divided into different classes; proximal promoters, enhancers, insulators, and silencers are the most widely studied (Barski et al., 2007; Hardison & Taylor, 2012; Nord et al., 2013; Pennacchio, Bickmore, Dean, Nobrega, & Bejerano, 2013; Wittkopp & Kalay, 2012). CREs are scattered across the genome and can function across considerable genomic distances, from locations as remote as another chromosome (Bickmore, 2013; Sexton, Schober, Fraser, & Gasser, 2007). In addition to these two components, epigenetic mechanisms also have a critical role in regulating transcription. DNA methylation, posttranslational modifications of histone tails, and noncoding RNA- including micro-RNA-mediated processes are the most common epigenetic mechanisms. Here we will give an overview of the current understanding of transcriptional regulation in the developing human brain.

DNA Methylation in Human Brain Development

DNA methylation refers to covalent modification of cytosine with a methyl group, mostly in CpG dinucleotides of the genomic DNA. DNA methylation has been implicated in neural development, plasticity, and learning and memory (Borrelli, Nestler, Allis, & Sassone-Corsi, 2008; Dulac, 2010; Lubin, Gupta, Parrish, Grissom, & Davis, 2011; Ma et al., 2010; MacDonald & Roskams, 2009; Mikaelsson & Miller, 2011; Miller, Campbell, & Sweatt, 2008; Miller & Sweatt, 2007; Sultan & Day, 2011). Several studies have addressed global DNA methylation in postmortem human brain using microarrays and identified regional and sex differences in the brain (Iwamoto et al., 2011; Ladd-Acosta et al., 2007; Numata et al., 2012; Siegmund et al., 2007; Xu et al., 2014). In the human prefrontal cortex, global methylation has also been shown to be age dependent, with fetal samples being the most distinct from postnatal and adult samples (Numata et al., 2012). Changes were rapid in fetal samples; they then slowed down after birth and then slowed further with aging. However, these studies were limited by the number of CpGs that could be assayed and by the design of the microarray.

The application of sequencing technologies that allow profiling of DNA methylation at millions of CpGs in the human genome at single-base resolution provides greater insights into DNA methylation patterns in the developing human brain. Lister et al. (2013) studied methylation patterns in the human prefrontal cortex during development by genome sequencing and reported widespread methylome reconfiguration during fetal to young adult development. They also observed accumulation of methylation in a non-CpG context (mCH) during early postnatal development (first 2 years after birth), coincident with synaptogenesis, then through adolescence, followed by a slight decrease thereafter. Considerable mCH was observed in adult human prefrontal cortex, with near negligible levels in the fetal cortex (Guo et al., 2014; Kato & Iwamoto, 2014; Lister et al., 2013). Accumulation of mCH was negatively correlated with gene expression. Interestingly, mCH was more abundant than CpG methylation in adult neurons but was at insignificant levels in glial cells. Furthermore, distinct patterns of methylation in cell type—specific genes were observed; genes highly expressed in adult neurons lost both CpG and non-CpG methylation progressively during development. Taken together, these data indicate that dynamics in methylation of genomic DNA is an essential feature of gene regulation in human neurodevelopment. Moreover, it is intriguing to speculate that the acceleration of DNA methylation in infancy and childhood may be a mechanism underlying the reduction in interareal differentially expressed gene levels and predominance in the expression of genes associated with synapse formation and regulation observed during these periods.

Histone Modification and Functional Regulatory Elements in Human Brain Development

Histone tails in the nucleosome can undergo posttranslational modifications at more than 100 amino acids. Methylation (mono-, di-, and tri-) and acetylation are the most extensively studied modifications, although many more including phosphorylation, hydroxylation, and citrullination have been observed (Kouzarides, 2007; Tan et al., 2011; Tessarz & Kouzarides, 2014). Histone tail modifications in combinations of type and position, referred to as chromatin state, affect the chromatin conformation, which in turn affects the accessibility of regulatory proteins and RNAs to CREs. Many studies have profiled chromatin states in human cancer lines or normal cells from nonbrain tissues using chromatin immunoprecipitation coupled with deep sequencing (ChIP-seq) (Encode Project Consortium et al., 2012). ChIP-seq studies in postmortem human brain have been hampered owing to difficulty in obtaining the large amounts of tissue required. However, a few studies have been carried out in adult human brain tissue (Roadmap Epigenomics et al., 2015; Zhu et al., 2013) and in fluorescent activated cell sorting−based neuronal and nonneuronal nuclei (Cheung et al., 2010; Shulha et al., 2012). Cheung et al. (2010) profiled histone-3-lysine-4 trimethylation (H3K4me3) across development in neuronal and nonneuronal cells of the human prefrontal cortex of 11 individuals aged 0.5 to 69 years. Significant remodeling of H3K4me3 was observed during postnatal development and aging of prefrontal neurons. Enrichment of H3K4me3 in the promoters of key genes in infants (aged less than 1 year) compared with the oldest samples (aged greater than 60 years) was observed. Some of these key genes include NEUROD1 and several members of cadherin and semaphorin families.

Certain chromatin states have also been shown to be signature marks for identifying CREs. For example, enrichment of H3K4me1 and histone-3-lysine-27 acetylation is a characteristic mark of active enhancers (Encode Project Consortium et al., 2012). With this approach, Vermunt et al. (2014) identified enhancers on a genome-wide scale in 136 brain regions in the adult human brain. Region-specific enhancers along with coregulated enhancers that form cell type− and context-specific networks were also identified. Visel et al. (2013) determined the genome-wide occupancy of the enhancer-associated protein p300/CBP in human fetal cortex to identify active enhancers. Enrichment of p300/CBP binding sites was observed near genes highly expressed in the fetal cortex. However, studies across key developmental periods and multiple brain regions are necessary to understand transcriptional regulation in human brain development.

Noncoding RNAs in Human Brain Development

In addition to DNA methylation and chromatin accessibility, noncoding RNAs, especially microRNAs (miRNAs), likely have important regulatory roles during human brain development. Canonically, miRNAs regulate gene expression by binding substrate mRNAs and inhibiting translation or targeting them for degradation. It has been shown in humans that one miRNA can target hundreds of mRNAs, and single mRNA transcripts can be targeted by many miRNAs (O'Carroll & Schaefer, 2013). Thus miRNAs represent a remarkably complex and powerful mechanism to control dynamic gene expression levels.

Fifty percent of all identified miRNAs are expressed in mammalian brain. Region- and cell type−specific differences in the expression of miRNAs in the brain have been shown (O'Carroll & Schaefer, 2013). Interestingly, enrichment of miRNAs in different cellular compartments—perinuclear soma, axons, and dendrites—has also been observed and implicated in regulation of local protein expression, synapse maturation, and function (Landgraf et al., 2007; Olsen, Klausen, Helboe, Nielsen, & Werge, 2009). In human dorsofrontal cortex, several hundred miRNAs were identified, including 197 putative novel miRNAs (Shao et al., 2010). miRNA expression was observed to be influenced in aging; 31% were affected by age (Somel et al., 2010). These age-related miRNA expression profiles were negatively correlated with the expression of their predicted targets, which suggests a direct role of miRNA in regulating gene expression. Moreover, in some cases the same miRNAs that affect expression changes during postnatal development drive the changes during aging. For example, miR-34a, miR-222, and miR-433 were correlated with their targets in both development and aging, and thus may regulate gene expression changes in both periods.

A significant challenge in the field is to identify true miRNA-target complexes directly rather than through computational predictions. Boudreau et al. (2014) used high-throughput sequencing of RNA isolated by cross-linking immunoprecipitation against argonaute proteins that are the core components of miRNA-silencing complexes (Meister, 2013), to identify miRNA-target complexes in the adult motor cortex and cingulate gyrus of human postmortem brains. A total of 1900 unique miRNAs, of which Let-7, miR-125, miR-124, miR-9, and miR-29 were the most abundant, were identified. They also identified six new targets for miR-137, an miRNA that has been associated with schizophrenia (Schizophrenia Psychiatric Genome-Wide Association Study (GWAS) Consortium, 2011). This resource will be valuable in the future for querying the basic biological functions of miRNAs in the brain and their role in differential spatiotemporal expression patterns.

miRNA-mediated regulation has also been implicated in the evolution of the human brain (Li et al., 2013; Somel et al., 2011). Somel et al. (2011) analyzed mRNA and miRNA expression in prefrontal cortex and cerebellum within humans, macaques, and chimpanzees across development. They observed that genes showing constant expression or no developmental pattern changes in the three species reflected the three species' known phylogenetic relationship. However, among genes that exhibit developmental pattern differences, humans display an evolutionary rate that is three to five times faster. A greater number of miRNAs were present in this gene set and showed more divergence, particularly in the cortex than cerebellum, in humans than chimpanzees. Twelve miRNAs and their 140 target genes constituted approximately 10% of the developmental remodeling events identified in the cortex. Taken together, these data indicate that miRNAs are an essential regulatory mechanism in establishing the regional and developmental landscape of gene expression in the human brain.

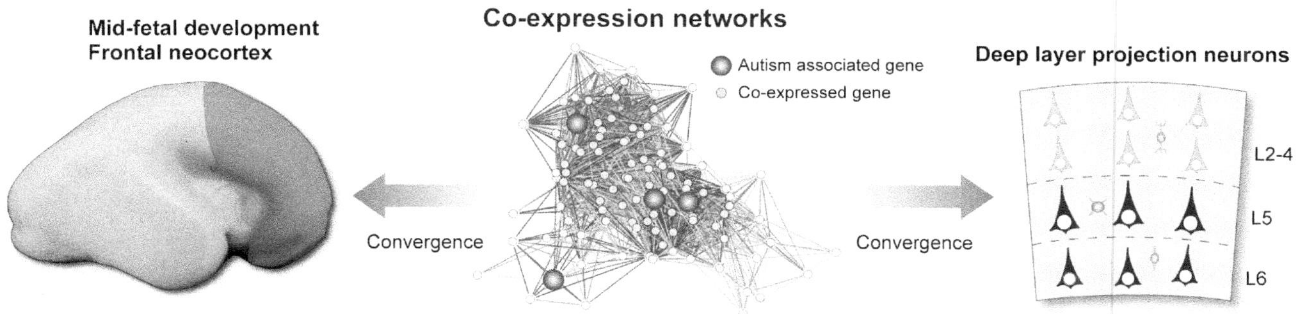

FIGURE 13.9 Integrative analysis can infer coexpression networks of disease-associated genes and implicate specific developmental periods, brain regions, and cell types. Advances in transcriptome analysis have enabled unbiased and sensitive measurements of gene expression that have produced comprehensive spatiotemporal transcriptome profiles characterizing healthy brain development. In addition, insights into disease pathology can be obtained using these data through either direct comparison of gene expression in control versus diseased tissue or statistical modeling and coexpression analysis with disease-associated genes. For example, this illustration shows how a coexpression network analysis of autism-associated genes with a spatiotemporal human transcriptome database inferred a convergence of autism-associated genes in frontal neocortex and deep-layer projection neurons, which suggests that abnormalities in this cell type may have a role in the pathogenesis of autism. *Adapted from Tebbenkamp, A. T., Willsey, A. J., State, M. W., & Sestan, N. (2014). The developmental transcriptome of the human brain: implications for neurodevelopmental disorders.* Current Opinion in Neurology, *27, 149–156; Willsey, A. J., Sanders, S. J., Li, M., Dong, S., Tebbenkamp, A. T., Muhl, R. A., … State, M. W. (2013). Coexpression networks implicate human midfetal deep cortical projection neurons in the pathogenesis of autism.* Cell, 155*, 997–1007.*

INSIGHTS INTO PSYCHIATRIC AND NEUROLOGICAL DISORDERS

Psychiatric and neurological disorders are devastating illnesses with considerable morbidity and mortality. Many of these are complex diseases with multiple genes contributing to the risk of the disease. This polygenicity poses a substantial challenge to elucidating and targeting the underlying molecular mechanisms of pathogenicity. Functional genomics studies of the developing human brain can provide great insights into molecular mechanisms underlying psychiatric diseases. Transcriptome comparisons between the brains of patients and control subjects can identify genes that show altered expression in diseased states. Moreover, the expression trajectories of genes previously associated with disease can be explored. For example, the expression trajectories of genes previously associated with ASDs and schizophrenia have been shown to exhibit distinct and dynamic expression patterns, especially among neocortex areas (Kang et al., 2011). In addition, by building coexpression networks of these genes, one can determine other coexpressed genes and how these genes converge on specific biological processes. Taking this approach, studies have determined that genes with risk alleles for ASDs are highly enriched in cortical projection neurons during midfetal development (Willsey et al., 2013) (Fig. 13.9).

Spatiotemporal profiles of transcriptome, epigenome, regulome, and proteome can enable an understanding of the functional role of genetic variations associated with psychiatric disorders by genome-wide association studies. Loci associated with ASDs and schizophrenia show enrichment of brain expression quantitative trait loci (eQTLs) (Davis et al., 2012; Richards et al., 2012). Identification of epigenetic QTLs and protein QTLs will further advance our understanding of the molecular mechanisms involved in psychiatric diseases. Functional genomics data will enable integrative analysis to reveal the convergence of genes implicated in psychiatric disorders on specific biological functions and pathways, and allow researchers to prioritize these studies for developing targeted therapy.

CONCLUSIONS AND FUTURE DIRECTIONS

Over a century of neuroscience has led to an increasing understanding of the structural and functional anatomy of the human brain. We are growing ever closer to realizing the goal of a comprehensive characterization of gene expression, epigenetic chromosomal modifications, and noncoding RNAs in distinct brain regions over the full course of human development and in aging. As our understanding of the molecular landscape of the human brain grows, neuroscientists will be better able to use these resources to research the molecular basis of specific aspects of human neurobiology. Expanding on this work will require harnessing advancements in genomic technologies, expanding the research community devoted to these efforts, acquiring high-quality tissue specimens, and increasing brain-banking efforts. A particularly important objective moving forward is deciphering the molecular basis of the remarkable cellular heterogeneity of the human brain using emerging technologies such as single-cell sequencing and improved methods for isolating distinct types of human cells. We must also decipher how epigenetic and environmental influences affect gene expression and ultimately the proteome of the human brain.

ACKNOWLEDGMENTS

Work in the authors' laboratory on the topic of this chapter is supported by grants from the National Institutes of Health, the Kavli Foundation, The Simons Foundation, and the James S. McDonnell Foundation. We apologize to all colleagues whose important work was not cited because of space limitations. Much of this chapter was adapted for a genomics and psychiatric audience from a review article by Silbereis et al. (2016).

REFERENCES

Abrahams, B. S., Tentler, D., Perederiy, J. V., Oldham, M. C., Coppola, G., & Geschwind, D. H. (2007). Genome-wide analyses of human perisylvian cerebral cortical patterning. *Proceedings of the National Academy of Sciences of the United States of America, 104*, 17849–17854.

Allendoerfer, K. L., & Shatz, C. J. (1994). The subplate, a transient neocortical structure: its role in the development of connections between thalamus and cortex. *Annual Review of Neuroscience, 17*, 185–218.

van den Ameele, J., Tiberi, L., Vanderhaeghen, P., & Espuny-Camacho, I. (2014). Thinking out of the dish: what to learn about cortical development using pluripotent stem cells. *Trends in Neurosciences, 37*, 334–342.

Amunts, K., Schleicher, A., Ditterich, A., & Zilles, K. (2003). Broca's region: cytoarchitectonic asymmetry and developmental changes. *The Journal of Comparative Neurology, 465*, 72–89.

Azevedo, F. A., Carvalho, L. R., Grinberg, L. T., Farfel, J. M., Ferretti, R. E., Leite, R. E., … Herculano-Houzel, S. (2009). Equal numbers of neuronal and nonneuronal cells make the human brain an isometrically scaled-up primate brain. *The Journal of Comparative Neurology, 513*, 532–541.

Barski, A., Cuddapah, S., Cui, K., Roh, T. Y., Schones, D. E., Wang, Z., … Zhao, K. (2007). High-resolution profiling of histone methylations in the human genome. *Cell, 129*, 823–837.

Bayatti, N., Sarma, S., Shaw, C., Eyre, J. A., Vouyiouklis, D. A., Lindsay, S., & Clowry, G. J. (2008). Progressive loss of PAX6, TBR2, NEUROD and TBR1 mRNA gradients correlates with translocation of EMX2 to the cortical plate during human cortical development. *The European Journal of Neuroscience, 28*, 1449–1456.

Bayer, S. A., & Altman, J. (2007). *Atlas of human central nervous system development*. CRC Pr I Llc.

Bickmore, W. A. (2013). The spatial organization of the human genome. *Annual Review of Genomics and Human Genetics, 14*, 67–84.

Borrelli, E., Nestler, E. J., Allis, C. D., & Sassone-Corsi, P. (2008). Decoding the epigenetic language of neuronal plasticity. *Neuron, 60*, 961–974.

Boudreau, R. L., Jiang, P., Gilmore, B. L., Spengler, R. M., Tirabassi, R., Nelson, J. A., … Davidson, B. L. (2014). Transcriptome-wide discovery of microRNA binding sites in human brain. *Neuron, 81*, 294–305.

Bystron, I., Blakemore, C., & Rakic, P. (2008). Development of the human cerebral cortex: Boulder committee revisited. *Nature Reviews Neuroscience, 9*, 110–122.

Chen, C. H., Panizzon, M. S., Eyler, L. T., Jernigan, T. L., Thompson, W., Fennema-Notestine, C., … Dale, A. M. (2011). Genetic influences on cortical regionalization in the human brain. *Neuron, 72*, 537–544.

Cheung, I., Shulha, H. P., Jiang, Y., Matevossian, A., Wang, J., Weng, Z., & Akbarian, S. (2010). Developmental regulation and individual differences of neuronal H3K4me3 epigenomes in the prefrontal cortex. *Proceedings of the National Academy of Sciences of the United States of America, 107*, 8824–8829.

Chi, J. G., Dooling, E. C., & Gilles, F. H. (1977). Left-right asymmetries of the temporal speech areas of the human fetus. *Archives of Neurology, 34*, 346–348.

Colantuoni, C., Lipska, B. K., Ye, T., Hyde, T. M., Tao, R., Leek, J. T., … Kleinman, J. E. (2011). Temporal dynamics and genetic control of transcription in the human prefrontal cortex. *Nature, 478*, 519–523.

Conel, J. L. (1939). *The postnatal development of the human cerebral cortex*. Cambridge, MA: Harvard University Press.

Davis, L. K., Gamazon, E. R., Kistner-Griffin, E., Badner, J. A., Liu, C., Cook, E. H., … Cox, N. J. (2012). Loci nominally associated with autism from genome-wide analysis show enrichment of brain expression quantitative trait loci but not lymphoblastoid cell line expression quantitative trait loci. *Molecular Autism, 3*, 3.

Dehay, C., Kennedy, H., & Kosik, K. S. (2015). The outer subventricular zone and primate-specific cortical complexification. *Neuron, 85*, 683–694.

Dixon-Salazar, T. J., & Gleeson, J. G. (2010). Genetic regulation of human brain development: lessons from Mendelian diseases. *Annals of the New York Academy of Sciences, 1214*, 156–167.

Dulac, C. (2010). Brain function and chromatin plasticity. *Nature, 465*, 728–735.

Encode Project Consortium, Bernstein, B. E., Birney, E., Dunham, I., Green, E. D., Gunter, C., & Snyder, M. (2012). An integrated encyclopedia of DNA elements in the human genome. *Nature, 489*, 57–74.

Fietz, S. A., Lachmann, R., Brandl, H., Kircher, M., Samusik, N., Schroder, R., … Huttner, W. B. (2012). Transcriptomes of germinal zones of human and mouse fetal neocortex suggest a role of extracellular matrix in progenitor self-renewal. *Proceedings of the National Academy of Sciences of the United States of America, 109*, 11836–11841.

Flechsig Of Leipsic, P. (1901). Developmental (myelogenetic) localisation of the cerebral cortex in the human subject. *The Lancet (British Edition), 158*, 1027–1030.

Gage, F. H., & Temple, S. (2013). Neural stem cells: generating and regenerating the brain. *Neuron, 80*, 588–601.

Gazzaniga, M. S., Bogen, J. E., & Sperry, R. W. (1962). Some functional effects of sectioning the cerebral commissures in man. *Proceedings of the National Academy of Sciences of the United States of America, 48*, 1765–1769.

Geschwind, N., & Levitsky, W. (1968). Human brain: left-right asymmetries in temporal speech region. *Science, 161*, 186–187.

Giedd, J. N., & Rapoport, J. L. (2010). Structural MRI of pediatric brain development: what have we learned and where are we going? *Neuron, 67*, 728–734.

Gulden, F. O., & Sestan, N. (2014). Neurobiology: building a bigger brain. *Nature, 515*, 206–207.

Guo, J. U., Su, Y., Shin, J. H., Shin, J., Li, H., Xie, B., … Song, H. (2014). Distribution, recognition and regulation of non-CpG methylation in the adult mammalian brain. *Nature Neuroscience, 17*, 215–222.

Hansen, D. V., Lui, J. H., Flandin, P., Yoshikawa, K., Rubenstein, J. L., Alvarez-Buylla, A., & Kriegstein, A. R. (2013). Non-epithelial stem cells and cortical interneuron production in the human ganglionic eminences. *Nature Neuroscience, 16*, 1576–1587.

Hansen, D. V., Lui, J. H., Parker, P. R., & Kriegstein, A. R. (2010). Neurogenic radial glia in the outer subventricular zone of human neocortex. *Nature, 464*, 554–561.

Hardison, R. C., & Taylor, J. (2012). Genomic approaches towards finding cis-regulatory modules in animals. *Nature Reviews Genetics, 13*, 469–483.

Hawrylycz, M. J., Lein, E. S., Guillozet-Bongaarts, A. L., Shen, E. H., Ng, L., Miller, J. A., … Jones, A. R. (2012). An anatomically comprehensive atlas of the adult human brain transcriptome. *Nature, 489*, 391–399.

Hayes, T. L., & Lewis, D. A. (1993). Hemispheric differences in layer III pyramidal neurons of the anterior language area. *Archives of Neurology, 50*, 501–505.

He, Z., Bammann, H., Han, D., Xie, G., & Khaitovich, P. (2014). Conserved expression of lincRNA during human and macaque prefrontal cortex development and maturation. *RNA (New York, NY), 20*, 1103–1111.

Hevner, R. F. (2007). Layer-specific markers as probes for neuron type identity in human neocortex and malformations of cortical development. *Journal of Neuropathology and Experimental Neurology, 66*, 101–109.

Hill, J., Inder, T., Neil, J., Dierker, D., Harwell, J., & Van Essen, D. (2010). Similar patterns of cortical expansion during human development and evolution. *Proceedings of the National Academy of Sciences of the United States of America, 107*, 13135–13140.

His, W. (1889). Die Neuroblasten und deren Entstehung im embryonalen Mark. *Abhandl, Math Phys Kl Konigl Sachs Ges Wiss, 26*, 313–372.

Hoerder-Suabedissen, A., & Molnar, Z. (2015). Development, evolution and pathology of neocortical subplate neurons. *Nature Reviews Neuroscience, 16*, 133–146.

Hofman, M. A. (2012). Design principles of the human brain: an evolutionary perspective. *Progress in Brain Research, 195*, 373–390.

Howard, B. M., Zhicheng, M., Filipovic, R., Moore, A. R., Antic, S. D., & Zecevic, N. (2008). Radial glia cells in the developing human brain. *The Neuroscientist: A Review Journal Bringing Neurobiology, Neurology and Psychiatry, 14*, 459–473.

Hu, W. F., Chahrour, M. H., & Walsh, C. A. (2014). The diverse genetic landscape of neurodevelopmental disorders. *Annual Review of Genomics and Human Genetics, 15*, 195–213.

Huttenlocher, P. R., & Dabholkar, A. S. (1997). Regional differences in synaptogenesis in human cerebral cortex. *The Journal of Comparative Neurology, 387*, 167–178.

Ip, B. K., Wappler, I., Peters, H., Lindsay, S., Clowry, G. J., & Bayatti, N. (2010). Investigating gradients of gene expression involved in early human cortical development. *Journal of Anatomy, 217*, 300–311.

Iwamoto, K., Bundo, M., Ueda, J., Oldham, M. C., Ukai, W., Hashimoto, E., … Kato, T. (2011). Neurons show distinctive DNA methylation profile and higher interindividual variations compared with non-neurons. *Genome Research, 5*.

Jaffe, A. E., Shin, J., Collado-Torres, L., Leek, J. T., Tao, R., Li, C., … Weinberger, D. R. (2015). Developmental regulation of human cortex transcription and its clinical relevance at single base resolution. *Nature Neuroscience, 18*, 154–161.

Jakovcevski, I., Filipovic, R., Mo, Z., Rakic, S., & Zecevic, N. (2009). Oligodendrocyte development and the onset of myelination in the human fetal brain. *Frontiers in Neuroanatomy, 3*, 5.

Jeste, S. S., & Geschwind, D. H. (2015). Developmental disorders. *Current Opinion in Neurology, 28*(2).

Johnson, M. B., Kawasawa, Y. I., Mason, C. E., Krsnik, Z., Coppola, G., Bogdanovic, D., … Sestan, N. (2009). Functional and evolutionary insights into human brain development through global transcriptome analysis. *Neuron, 62*, 494–509.

Johnson, M. H. (2001). Functional brain development in humans. *Nature Reviews Neuroscience, 2*, 475–483.

Judas, M., Sedmak, G., & Kostovic, I. (2013). The significance of the subplate for evolution and developmental plasticity of the human brain. *Frontiers in Human Neuroscience, 7*, 423.

Kang, H. J., Kawasawa, Y. I., Cheng, F., Zhu, Y., Xu, X., Li, M., … Sestan, N. (2011). Spatio-temporal transcriptome of the human brain. *Nature, 478*, 483–489.

Kato, T., & Iwamoto, K. (2014). Comprehensive DNA methylation and hydroxymethylation analysis in the human brain and its implication in mental disorders. *Neuropharmacology, 80*.

Katz, L. C., & Shatz, C. J. (1996). Synaptic activity and the construction of cortical circuits. *Science, 274*, 1133–1138.

Kessler, R. C., Amminger, G. P., Aguilar-Gaxiola, S., Alonso, J., Lee, S., & Ustun, T. B. (2007). Age of onset of mental disorders: a review of recent literature. *Current Opinion in Psychiatry, 20*, 359–364.

Kostovic, I., & Judas, M. (2006). Prolonged coexistence of transient and permanent circuitry elements in the developing cerebral cortex of fetuses and preterm infants. *Developmental Medicine and Child Neurology, 48*, 388–393.

Kouzarides, T. (2007). Chromatin modifications and their function. *Cell, 128*, 693–705.

Krubitzer, L., & Kaas, J. (2005). The evolution of the neocortex in mammals: how is phenotypic diversity generated? *Current Opinion in Neurobiology, 15*, 444–453.

Kuzawa, C. W., Chugani, H. T., Grossman, L. I., Lipovich, L., Muzik, O., Hof, P. R., … Lange, N. (2014). Metabolic costs and evolutionary implications of human brain development. *Proceedings of the National Academy of Sciences of the United States of America, 111*, 13010–13015.

Kwan, K. Y., Lam, M. M., Krsnik, Z., Kawasawa, Y. I., Lefebvre, V., & Sestan, N. (2008). SOX5 postmitotically regulates migration, postmigratory differentiation, and projections of subplate and deep-layer neocortical neurons. *Proceedings of the National Academy of Sciences of the United States of America, 105*, 16021–16026.

Kwan, K. Y., Sestan, N., & Anton, E. S. (2012). Transcriptional co-regulation of neuronal migration and laminar identity in the neocortex. *Development, 139*, 1535–1546.

Ladd-Acosta, C., Pevsner, J., Sabunciyan, S., Yolken, R. H., Webster, M. J., Dinkins, T., … Feinberg, A. P. (2007). DNA methylation signatures within the human brain. *American Journal of Human Genetics, 81*, 1304–1315.

Lambert, N., Lambot, M. A., Bilheu, A., Albert, V., Englert, Y., Libert, F., … Vanderhaeghen, P. (2011). Genes expressed in specific areas of the human fetal cerebral cortex display distinct patterns of evolution. *PLoS One, 6*, e17753.

Landgraf, P., Rusu, M., Sheridan, R., Sewer, A., Iovino, N., Aravin, A., … Tuschl, T. (2007). A mammalian microRNA expression atlas based on small RNA library sequencing. *Cell, 129*, 1401–1414.

Lee, F. S., Heimer, H., Giedd, J. N., Lein, E. S., Sestan, N., Weinberger, D. R., & Casey, B. J. (2014). Mental health. Adolescent mental health—opportunity and obligation. *Science, 346*, 547–549.

Leone, D. P., Srinivasan, K., Chen, B., Alcamo, E., & McConnell, S. K. (2008). The determination of projection neuron identity in the developing cerebral cortex. *Current Opinion in Neurobiology, 18*, 28–35.

Li, G., & Pleasure, S. J. (2014). The development of hippocampal cellular assemblies. *Wiley Interdisciplinary Reviews Developmental Biology, 3*, 165–177.

Li, Z., Bammann, H., Li, M., Liang, H., Yan, Z., Phoebe Chen, Y. P., … Khaitovich, P. (2013). Evolutionary and ontogenetic changes in RNA editing in human, chimpanzee, and macaque brains. *RNA (New York, NY), 19*, 1693–1702.

Licatalosi, D. D., & Darnell, R. B. (2006). Splicing regulation in neurologic disease. *Neuron, 52*, 93–101.

Lister, R., Mukamel, E. A., Nery, J. R., Urich, M., Puddifoot, C. A., Johnson, N. D., … Ecker, J. R. (2013). Global epigenomic reconfiguration during mammalian brain development. *Science, 341*, 1237905.

Lubin, F. D., Gupta, S., Parrish, R. R., Grissom, N. M., & Davis, R. L. (2011). Epigenetic mechanisms: critical contributors to long-term memory formation. *The Neuroscientist: A Review Journal Bringing Neurobiology, Neurology and Psychiatry, 17*, 616–632.

Lui, J. H., Hansen, D. V., & Kriegstein, A. R. (2011). Development and evolution of the human neocortex. *Cell, 146*, 18–36.

Ma, D. K., Marchetto, M. C., Guo, J. U., Ming, G. L., Gage, F. H., & Song, H. (2010). Epigenetic choreographers of neurogenesis in the adult mammalian brain. *Nature Neuroscience, 13*, 1338–1344.

MacDonald, J. L., & Roskams, A. J. (2009). Epigenetic regulation of nervous system development by DNA methylation and histone deacetylation. *Progress in Neurobiology, 88*, 170–183.

Mariani, J., Simonini, M. V., Palejev, D., Tomasini, L., Coppola, G., Szekely, A. M., … Vaccarino, F. M. (2012). Modeling human cortical development in vitro using induced pluripotent stem cells. *Proceedings of the National Academy of Sciences of the United States of America, 109*, 12770–12775.

Marin-Padilla, M. (1978). Dual origin of the mammalian neocortex and evolution of the cortical plate. *Anatomy and Embryology, 152*, 109–126.

Mazin, P., Xiong, J., Liu, X., Yan, Z., Zhang, X., Li, M., … Khaitovich, P. (2013). Widespread splicing changes in human brain development and aging. *Molecular Systems Biology, 9*, 633.

McManus, I. C., & Bryden, M. P. (1991). Geschwind's theory of cerebral lateralization: developing a formal, causal model. *Psychological Bulletin, 110*, 237–253.

Mehler, M. F., & Mattick, J. S. (2007). Noncoding RNAs and RNA editing in brain development, functional diversification, and neurological disease. *Physiological Reviews, 87*, 799–823.

Meister, G. (2013). Argonaute proteins: functional insights and emerging roles. *Nature Reviews Genetics, 14*, 447–459.

Meyer, G. (2007). Genetic control of neuronal migrations in human cortical development. *Advances in Anatomy, Embryology, and Cell Biology, 189*, 1 p preceding 1, 1–111.

Meyer, G., Schaaps, J. P., Moreau, L., & Goffinet, A. M. (2000). Embryonic and early fetal development of the human neocortex. *The Journal of Neuroscience: The Official Journal of the Society for Neuroscience, 20*, 1858–1868.

Mikaelsson, M. A., & Miller, C. A. (2011). DNA methylation: a transcriptional mechanism co-opted by the developed mammalian brain? *Epigenetics, 6*, 548–551.

Miller, C. A., Campbell, S. L., & Sweatt, J. D. (2008). DNA methylation and histone acetylation work in concert to regulate memory formation and synaptic plasticity. *Neurobiology of Learning and Memory, 89*, 599–603.

Miller, C. A., & Sweatt, J. D. (2007). Covalent modification of DNA regulates memory formation. *Neuron, 53*, 857–869.

Miller, D. J., Duka, T., Stimpson, C. D., Schapiro, S. J., Baze, W. B., McArthur, M. J., … Sherwood, C. C. (2012). Prolonged myelination in human neocortical evolution. *Proceedings of the National Academy of Sciences of the United States of America, 109*, 16480–16485.

Miller, J. A., Ding, S. L., Sunkin, S. M., Smith, K. A., Ng, L., Szafer, A., … Lein, E. S. (2014). Transcriptional landscape of the prenatal human brain. *Nature, 508*, 199–206.

Mirnics, K., & Pevsner, J. (2004). Progress in the use of microarray technology to study the neurobiology of disease. *Nature Neuroscience, 7*, 434–439.

Molyneaux, B. J., Arlotta, P., Menezes, J. R., & Macklis, J. D. (2007). Neuronal subtype specification in the cerebral cortex. *Nature Reviews Neuroscience, 8*, 427–437.

Morrison, J. H., & Hof, P. R. (1997). Life and death of neurons in the aging brain. *Science, 278*, 412–419.

NCBI Resource Coordinators. (2014). Database resources of the national center for biotechnology information. *Nucleic Acids Research, 42*, D7–D17.

Nord, A. S., Blow, M. J., Attanasio, C., Akiyama, J. A., Holt, A., Hosseini, R., … Visel, A. (2013). Rapid and pervasive changes in genome-wide enhancer usage during mammalian development. *Cell, 155*, 1521–1531.

Numata, S., Ye, T., Hyde, T. M., Guitart-Navarro, X., Tao, R., Wininger, M., … Lipska, B. K. (2012). DNA methylation signatures in development and aging of the human prefrontal cortex. *American Journal of Human Genetics, 90*, 260–272.

O'Carroll, D., & Schaefer, A. (2013). General principals of miRNA biogenesis and regulation in the brain. *Neuropsychopharmacology, 38*, 39–54.

O'Leary, D. D., Chou, S. J., & Sahara, S. (2007). Area patterning of the mammalian cortex. *Neuron, 56*, 252–269.

O'Leary, D. D., & Sahara, S. (2008). Genetic regulation of arealization of the neocortex. *Current Opinion in Neurobiology, 18*, 90–100.

Olsen, L., Klausen, M., Helboe, L., Nielsen, F. C., & Werge, T. (2009). MicroRNAs show mutually exclusive expression patterns in the brain of adult male rats. *PloS one, 4*, e7225.

Orr, H. T., & Zoghbi, H. Y. (2007). Trinucleotide repeat disorders. *Annual Review of Neuroscience, 30*, 575–621.

Pennacchio, L. A., Bickmore, W., Dean, A., Nobrega, M. A., & Bejerano, G. (2013). Enhancers: five essential questions. *Nature Reviews Genetics, 14*, 288–295.

Petanjek, Z., Judas, M., Simic, G., Rasin, M. R., Uylings, H. B., Rakic, P., & Kostovic, I. (2011). Extraordinary neoteny of synaptic spines in the human prefrontal cortex. *Proceedings of the National Academy of Sciences of the United States of America, 108*, 13281–13286.

Pletikos, M., Sousa, A. M., Sedmak, G., Meyer, K. A., Zhu, Y., Cheng, F., … Sestan, N. (2014). Temporal specification and bilaterality of human neocortical topographic gene expression. *Neuron, 81*, 321–332.

Pollen, A. A., Nowakowski, T. J., Shuga, J., Wang, X., Leyrat, A. A., Lui, J. H., … West, J. A. (2014). Low-coverage single-cell mRNA sequencing reveals cellular heterogeneity and activated signaling pathways in developing cerebral cortex. *Nature Biotechnology, 32*, 1053–1058.

Ptashne, M. (2013). Epigenetics: core misconcept. *Proceedings of the National Academy of Sciences of the United States of America, 110*, 7101–7103.

Radonjic, N. V., Ayoub, A. E., Memi, F., Yu, X., Maroof, A., Jakovcevski, I., … Zecevic, N. (2014). Diversity of cortical interneurons in primates: the role of the dorsal proliferative niche. *Cell Reports, 9*, 2139–2151.

Rakic, P., Ayoub, A. E., Breunig, J. J., & Dominguez, M. H. (2009). Decision by division: making cortical maps. *Trends in Neurosciences, 32*, 291–301.

Rash, B. G., & Grove, E. A. (2006). Area and layer patterning in the developing cerebral cortex. *Current Opinion in Neurobiology, 16*, 25–34.

Retzius, M. G. (1896). *Das Menschenhirn: Studien in der Makroskopischen Morphologie*. Text und Atlas. Stockholm: Königliche Buchdruckerei P. A. Norstedt & Söner.

Richards, A. L., Jones, L., Moskvina, V., Kirov, G., Gejman, P. V., Levinson, D. F., … O'Donovan, M. C. (2012). Schizophrenia susceptibility alleles are enriched for alleles that affect gene expression in adult human brain. *Molecular Psychiatry, 17*, 193–201.

Rilling, J. K., Glasser, M. F., Jbabdi, S., Andersson, J., & Preuss, T. M. (2011). Continuity, divergence, and the evolution of brain language pathways. *Frontiers in Evolutionary Neuroscience, 3*, 11.

Roadmap Epigenomics, C., Kundaje, A., Meuleman, W., Ernst, J., Bilenky, M., Yen, A., … Kellis, M. (2015). Integrative analysis of 111 reference human epigenomes. *Nature, 518*, 317–330.

Roessmann, U., & Gambetti, P. (1986). Astrocytes in the developing human brain. An immunohistochemical study. *Acta Neuropathologica, 70*, 308–313.

Sanai, N., Nguyen, T., Ihrie, R. A., Mirzadeh, Z., Tsai, H. H., Wong, M., … Alvarez-Buylla, A. (2011). Corridors of migrating neurons in the human brain and their decline during infancy. *Nature, 478*, 382–386.

Saxena, S., & Caroni, P. (2011). Selective neuronal vulnerability in neurodegenerative diseases: from stressor thresholds to degeneration. *Neuron, 71*, 35–48.

Schizophrenia Psychiatric Genome-Wide Association Study (GWAS) Consortium. Genome-wide association study identifies five new schizophrenia loci. *Nature Genetics, 43*, 969–976.

Sexton, T., Schober, H., Fraser, P., & Gasser, S. M. (2007). Gene regulation through nuclear organization. *Nature Structural & Molecular Biology, 14*, 1049–1055.

Shao, N. Y., Hu, H. Y., Yan, Z., Xu, Y., Hu, H., Menzel, C., … Khaitovich, P. (2010). Comprehensive survey of human brain microRNA by deep sequencing. *BMC Genomics, 11*, 409.

Shibata, M., Gulden, F. O., & Sestan, N. (2015). From trans to cis: transcriptional regulatory networks in neocortical development. *Trends in Genetics: TIG, 31*, 77–87.

Shulha, H. P., Crisci, J. L., Reshetov, D., Tushir, J. S., Cheung, I., Bharadwaj, R., … Akbarian, S. (2012). Human-specific histone methylation signatures at transcription start sites in prefrontal neurons. *PLoS Biology, 10*, e1001427.

Sidman, R. L., & Rakic, P. (1973). Neuronal migration, with special reference to developing human brain: a review. *Brain Research, 62*, 1–35.

Siegmund, K. D., Connor, C. M., Campan, M., Long, T. I., Weisenberger, D. J., Biniszkiewicz, D., … Akbarian, S. (2007). DNA methylation in the human cerebral cortex is dynamically regulated throughout the life span and involves differentiated neurons. *PLoS One, 2*, e895.

Silbereis, J. C., Pochareddy, S., Zhu, Y., Li, M., & Sestan, N. (2016). The cellular and molecular landscapes of the developing human central nervous system. *Neuron, 89*, 248–268.

Somel, M., Franz, H., Yan, Z., Lorenc, A., Guo, S., Giger, T., … Khaitovich, P. (2009). Transcriptional neoteny in the human brain. *Proceedings of the National Academy of Sciences of the United States of America, 106*, 5743–5748.

Somel, M., Guo, S., Fu, N., Yan, Z., Hu, H. Y., Xu, Y., … Khaitovich, P. (2010). MicroRNA, mRNA, and protein expression link development and aging in human and macaque brain. *Genome Research, 20*, 1207–1218.

Somel, M., Liu, X., Tang, L., Yan, Z., Hu, H., Guo, S., … Khaitovich, P. (2011). MicroRNA-driven developmental remodeling in the brain distinguishes humans from other primates. *PLoS Biology, 9*, e1001214.

Sousa, A. M. M., Meyer, K. A., & Sestan, N. (2014). Molecular and cellular mechanisms of human brain development and evolution. In M. S. Gazzaniga, & G. R. Mangun (Eds.), *The cognitive neurosciences* (pp. 67–74). The MIT Press.

Southwell, D. G., Nicholas, C. R., Basbaum, A. I., Stryker, M. P., Kriegstein, A. R., Rubenstein, J. L., & Alvarez-Buylla, A. (2014). Interneurons from embryonic development to cell-based therapy. *Science, 344*, 1240622.

Sultan, F. A., & Day, J. J. (2011). Epigenetic mechanisms in memory and synaptic function. *Epigenomics, 3*, 157–181.

Sun, T., Patoine, C., Abu-Khalil, A., Visvader, J., Sum, E., Cherry, T. J., … Walsh, C. A. (2005). Early asymmetry of gene transcription in embryonic human left and right cerebral cortex. *Science, 308*, 1794–1798.

Sun, T., & Walsh, C. A. (2006). Molecular approaches to brain asymmetry and handedness. *Nature Reviews Neuroscience, 7*, 655–662.

Sur, M., & Rubenstein, J. L. (2005). Patterning and plasticity of the cerebral cortex. *Science, 310*, 805–810.

Tan, M., Luo, H., Lee, S., Jin, F., Yang, J. S., Montellier, E., … Zhao, Y. (2011). Identification of 67 histone marks and histone lysine crotonylation as a new type of histone modification. *Cell, 146*, 1016–1028.

Tang, Y., Nyengaard, J. R., De Groot, D. M., & Gundersen, H. J. (2001). Total regional and global number of synapses in the human brain neocortex. *Synapse, 41*, 258–273.

Tau, G. Z., & Peterson, B. S. (2010). Normal development of brain circuits. *Neuropsychopharmacology: Official Publication of the American College of Neuropsychopharmacology, 35*, 147–168.

Taverna, E., Gotz, M., & Huttner, W. B. (2014). The cell biology of neurogenesis: toward an understanding of the development and evolution of the neocortex. *Annual Review of Cell and Developmental Biology, 30*, 465–502.

Taylor, D. C. (1969). Differential rates of cerebral maturation between sexes and between hemispheres. Evidence from epilepsy. *Lancet, 2*, 140–142.

Tebbenkamp, A. T., Willsey, A. J., State, M. W., & Sestan, N. (2014). The developmental transcriptome of the human brain: implications for neurodevelopmental disorders. *Current Opinion in Neurology, 27*, 149–156.

Tessarz, P., & Kouzarides, T. (2014). Histone core modifications regulating nucleosome structure and dynamics. *Nature Reviews Molecular Cell Biology, 15*, 703–708.

Thatcher, R. W., Walker, R. A., & Giudice, S. (1987). Human cerebral hemispheres develop at different rates and ages. *Science, 236*, 1110–1113.

Thompson, B. L., Levitt, P., & Stanwood, G. D. (2009). Prenatal exposure to drugs: effects on brain development and implications for policy and education. *Nature Reviews Neuroscience, 10*, 303–312.

Toga, A. W., Thompson, P. M., & Sowell, E. R. (2006). Mapping brain maturation. *Trends in Neurosciences, 29*, 148–159.

Vaquerizas, J. M., Kummerfeld, S. K., Teichmann, S. A., & Luscombe, N. M. (2009). A census of human transcription factors: function, expression and evolution. *Nature Reviews Genetics, 10*, 252–263.

Vermunt, M. W., Reinink, P., Korving, J., de Bruijn, E., Creyghton, P. M., Basak, O., … Creyghton, M. P. (2014). Large-scale identification of coregulated enhancer networks in the adult human brain. *Cell Reports, 9*, 767–779.

Visel, A., Taher, L., Girgis, H., May, D., Golonzhka, O., Hoch, R. V., … Rubenstein, J. L. (2013). A high-resolution enhancer atlas of the developing telencephalon. *Cell, 152*, 895–908.

Walloe, S., Pakkenberg, B., & Fabricius, K. (2014). Stereological estimation of total cell numbers in the human cerebral and cerebellar cortex. *Frontiers in Human Neuroscience, 8*, 508.

Walsh, C. A., & Engle, E. C. (2010). Allelic diversity in human developmental neurogenetics: insights into biology and disease. *Neuron, 68*, 245–253.

Willsey, A. J., Sanders, S. J., Li, M., Dong, S., Tebbenkamp, A. T., Muhl, R. A., … State, M. W. (2013). Coexpression networks implicate human midfetal deep cortical projection neurons in the pathogenesis of autism. *Cell, 155*, 997–1007.

Wittkopp, P. J., & Kalay, G. (2012). Cis-regulatory elements: molecular mechanisms and evolutionary processes underlying divergence. *Nature Reviews Genetics, 13*, 59–69.

Xu, H., Wang, F., Liu, Y., Yu, Y., Gelernter, J., & Zhang, H. (2014). Sex-biased methylome and transcriptome in human prefrontal cortex. *Human Molecular Genetics, 23*, 1260–1270.

Yakovlev, P. I., & Lecours, A. R. (1967). *The myelogenetic cycles of regional maturation of the brain. Regional development of the brain in early life.* Philadelphia: F.A. Davis Co.

Zhu, J., Adli, M., Zou, J. Y., Verstappen, G., Coyne, M., Zhang, X., … Bernstein, B. E. (2013). Genome-wide chromatin state transitions associated with developmental and environmental cues. *Cell, 152*, 642–654.

Chapter 14

Optogenetic Approaches to Neural Circuit Analysis in the Mammalian Brain

Stephan Lammel
University of California Berkeley, Berkeley, CA, United States

Gül Dölen
Johns Hopkins University, Baltimore, MD, United States

Robert C. Malenka
Stanford University School of Medicine, Stanford, CA, United States

INTRODUCTION

The ability to control genetically specified cell populations optically in the nervous system has transformed neuroscience (Boyden, Zhang, Bamberg, Nagel, & Deisseroth, 2005; Lima & Miesenböck, 2005). This approach, termed "optogenetics," uses photostimulation to control the activity of specific cell populations experimentally, which have been genetically transduced to express proteins that are activated by light. Importantly, this approach can be used to manipulate neurons on a millisecond time scale and with high spatial precision. Optogenetics has become a viable alternative to electrical microstimulation and pharmacological methods. Several optogenetic strategies for controlling neuronal function have been widely implemented for dissecting neural connectivity, function, and dysfunction in the rodent and primate brain (Deisseroth, 2014; Deisseroth & Schnitzer, 2013; Häusser, 2014; Tye & Deisseroth, 2012; Yizhar, Fenno, Davidson, Mogri, & Deisseroth, 2011).

Optogenetic experiments begin with the careful selection of genetic targeting strategies, which facilitate the delivery of light-sensitive optogenetic constructs to molecularly defined cell populations in the intact brain. These approaches include the use of transgenic rodent lines that express Cre recombinase in specific cell populations, cell type–specific promoters, or retrograde viruses. Once the specified neurons have been sensitized to light by expression of light-activated proteins, integrated fiber-optic cables and LEDs or lasers are used to deliver the photostimulus to neural pathways in a temporally precise manner in either in vivo or ex vivo experimental settings. In the following sections we will summarize and discuss some prominent optogenetic strategies that are used for the functional dissection of neural circuits. Moreover, we will review the impact of optogenetic-driven research on understanding the pathogenesis of diseases such as depression, drug addiction, anxiety, and Parkinson disease, as well as the implications for future therapeutic development.

TOOLS FOR FUNCTIONAL NEURAL CIRCUIT DISSECTION

Definition of Opsins

Opsins are photoreceptive proteins that, when expressed by a neuron, render these cells' plasma membrane sensitive to light. They are either naturally occurring or are engineered by fusing a selected effector domain to a photoreceptive domain that enables depolarization or hyperpolarization (by the flow of charged molecules across the plasma membrane) of excitable cells expressing the opsin. For example, channelrhodopsin-2 (ChR2), which was isolated from the single-cell green alga, *Chlamydomonas reinhardtii*, is a light-sensitive cation channel that is maximally activated by blue light

Genomics, Circuits, and Pathways in Clinical Neuropsychiatry. http://dx.doi.org/10.1016/B978-0-12-800105-9.00014-7

power densities of 1–5 mW/mm^2 at 473 nm. It can be used to depolarize neurons with millisecond precision and thereby drive precisely timed action potentials up to 30–50 Hz (Mattis et al., 2012; Tye & Deisseroth, 2012; Yizhar et al., 2011; Zhang et al., 2010). Conversely, halorhodopsin (NpHR) was isolated from *Natronomonas pharaonis*, an aerobic archaeon, and is a light-sensitive inward chloride pump that is maximally activated by amber light power densities of about 5 mW/mm^2 at 593 nm and leads to hyperpolarization of neurons. Molecular engineering including addition of a neurite trafficking sequence from the Kir2.1 potassium channel have resulted into enhanced NpHR, which can be used effectively for in vivo experiments (Gradinaru et al., 2010; Mattis et al., 2012). Nevertheless, in contrast to the excitatory ChR2, these pumps require constant photostimulation to maintain hyperpolarization. Because constant light produces a significant amount of heat, which may be enough to alter physiological processes (Yizhar et al., 2011), experiments using this approach must control for the possible nonspecific effects of heat on the physiological processes under examination.

The high-resolution crystal structure of a channelrhodopsin has been solved and has guided the generation of viable alternatives for achieving temporally precise inhibition in loss-of-function experiments (Kato et al., 2012). Specifically, molecular reengineering of the channel pore of channelrhodopsin from cation to chloride conducting has enabled conversion of the opsin from an excitatory to an inhibitory channel which no longer requires constant photoillumination (Berndt, Lee, Ramakrishnan, & Deisseroth, 2014; Wietek et al., 2014). Other engineered opsins include channelrhodopsins with extended open-state lifetimes (step-function opsins), which provide prolonged depolarization and drive action potential trains at high frequencies up to 200 Hz (Berndt, Yizhar, Gunaydin, Hegemann, & Deisseroth, 2009; Yizhar et al., 2011). Other molecular configurations allow termination of the stable blue light–triggered depolarized state by application of a pulse of amber light at 560–590 nm (Berndt et al., 2009).

Redshifted activation wavelength opsins, known as C1V1 (Gradinaru et al., 2010; Mattis et al., 2012; Yizhar et al., 2011), represent another class of engineered photosensitive molecules, which are activated by amber light at 590 nm. Because this wavelength does not overlap with the wavelengths activating ChR2 (473 nm), it is theoretically possible to use these two opsins in the same brain tissue specifically to activate distinct neuronal populations without cross-excitation. Another redshifted variant is ReaChR (red-activatable channelrhodopsin), which is activated by amber light at 590–630 nm and offers improved membrane trafficking, higher photocurrents, and faster kinetics compared with existing redshifted opsins. Importantly, ReaChR can be activated in deep brain layers or even through the intact skull (Lin, Knutsen, Muller, Kleinfeld, & Tsien, 2013), which renders this variant particularly suitable to in vivo studies.

Expression of Opsins in the Mammalian Brain

The power of optogenetic strategies is derived in large part by the ability to drive the expression of optogenetic proteins using cells' intrinsic molecular genetic machinery. Harnessing this power for neuroscience requires delivery of genetic material to the brain. Access to the brain is experimentally restricted by the skull and the blood–brain barrier, which compels researchers to develop a number of different strategies to deliver optogenetic constructs to cells within the central nervous system. By far the most common strategy is to use viral-mediated gene transfer, which is routinely accomplished by direct injection of viruses into the brain region of interest. A number of genetically engineered optogenetic mouse lines also exist. These drive opsin expression under the control of various promoters or in a conditional manner such that the opsin is expressed only in cells that also express Cre recombinase.

Adeno-associated virus (AAV) and lentivirus (LV) are the most common means by which to deliver and target optogenetic constructs to specific brain regions (Tye & Deisseroth, 2012; Yizhar et al., 2011; Zalocusky & Deisseroth, 2013). AAVs or LVs can be injected into specific brain regions using a micropump at a slow rate (100–150 nL/min) while the animal is under general anesthesia and fixed in a stereotaxic instrument. LVs can be produced using standard tissue culture techniques, but they require biosafety level 2 surgical and husbandry facilities. After in vivo injection of a small volume of LV into the brain (500–1000 nL is considered standard by many laboratories that use virus titers of about 10^{12} infection units per milliliter), the diffusion of the LV is usually restricted, and depending on the titer and other factors, only a proportion (10–70%) of neurons within this volume is infected. Injection of AAVs, in contrast, commonly results in larger infected tissue volumes (typically on a millimeter scale), and a much greater percentage (about 90–99%) of neurons within that volume is infected. Importantly, the speed and efficiency of this delivery method depend largely on the serotype of the AAV used (Tye & Deisseroth, 2012; Xu & Sudhof, 2013). AAVs are often produced and ordered through virus vector core facilities such as those at the University of Pennsylvania (http://www.med.upenn.edu/gtp/vectorcore), Stanford University (http://med.stanford.edu/gvvc), and the University of North Carolina (http://www.med.unc.edu/genetherapy/vectorcore), and usually can be employed in biosafety level 1 laboratories. An important consideration is that opsins will be expressed in all neurons and even in glial cells if AAVs or LVs are used that lack cell-specific promoters or are not expressed under the control of Cre recombinase (Fig. 14.1A).

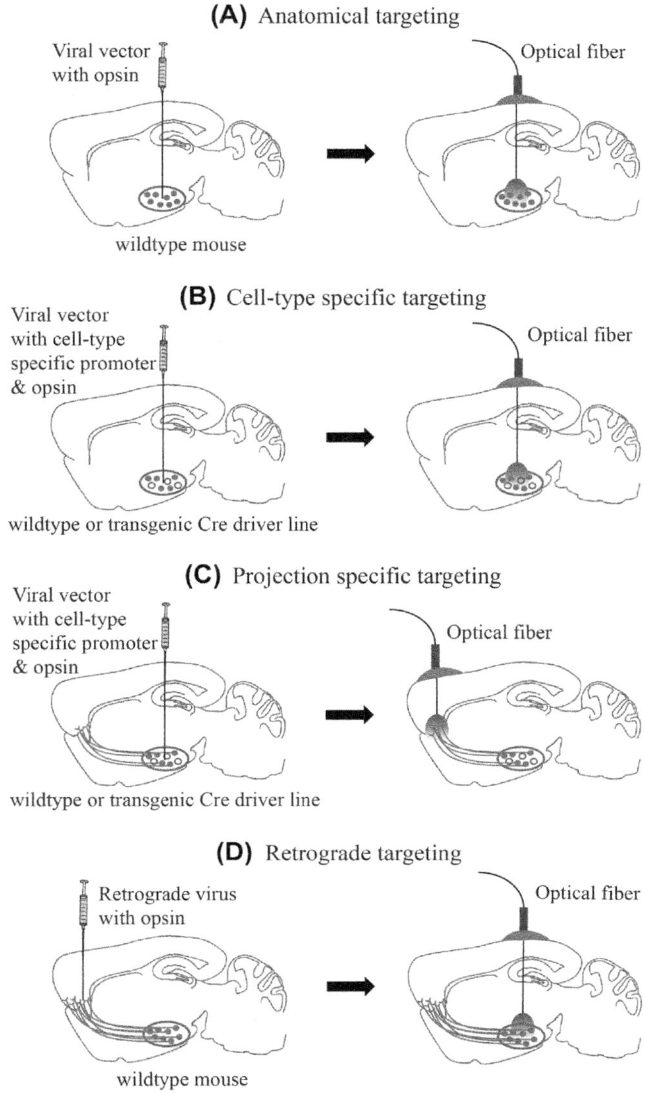

FIGURE 14.1 Targeting strategies using optogenetic tools. Schematic drawings of different targeting strategies for optogenetic manipulations in rodents. (A) Anatomical targeting. The optogenetic construct is delivered directly into the target brain region by stereotaxic injection of a viral vector (eg, adeno-associated virus or lentivirus) and the opsin is subsequently expressed in all cells (ie, neurons and glial cells) (left). The optical fiber is implanted above the target brain region (right). (B) Cell type–specific targeting. A viral vector carrying an optogenetic construct with a cell type–specific promoter will only infect genetically defined neurons (*filled circles*) but not other surrounding neurons (*open circles*) (left). The optical fiber is implanted directly above the viral injection site (right). An alternative approach is to use transgenic animals that express Cre recombinase in genetically defined neurons. (C) Projection targeting. A viral vector carrying a cell type–specific promoter and optogenetic construct is infused into the target brain region of a wild-type or transgenic mouse (left). The optical fiber is implanted above the brain region that contains axon terminals originating from the neurons in the viral injection site (right). (D) Retrograde targeting. A retrograde virus (eg, rabies virus) expressing an opsin is injected into a brain region which is innervated by neurons from a certain brain region (left). The optical fiber is implanted directly above this brain region and light stimulation will activate all neurons that project to the viral injection site (right).

Cell Type–Specific Targeting of Opsins Using Promoters

For viruses that deliver genetic material to the nucleus (ie, DNA viruses such as AAV and retroviruses such as LV), it is possible to use the cells' promoter-specific transcriptional machinery to drive the expression of genes. Using this strategy, the genetic construct carried by the virus contains a promoter (eg, synapsin, Thy1, CaMKII) followed by the opsin transgene (eg, ChR2 or enhanced yellow fluorescent protein [EYFP]). Whereas cellular uptake of the virus is relatively nonspecific (although specific viral tropisms exist), expression of the transgene only or predominantly occurs in cells that recognize the promoter (Fig. 14.1B). For instance, human synapsin I or human Thy1 selectively expresses transgenes in neurons but not in glial cells (Diester et al., 2011; Yizhar et al., 2011). Further specificity can be obtained by using promoters that target cells based on their neurochemical identity. For example, glutamatergic neurons in the cortex and hippocampus can be targeted using the Ca^{2+}/calmodulin-dependent kinase IIα (CaMKIIα) promoter (Dittgen et al., 2004; Lee et al., 2010), serotonergic neurons using the TPH-2 promoter (Benzekhroufa, Liu, Tang, Teschemacher, & Kasparov, 2009) somatostatin (SOM)-expressing neurons using the SST promoter (Tan et al., 2008), and oxytocin-expressing neurons using the OT promoter (Fields, Ponzio, Kawasaki, & Gainer, 2012; Knobloch et al., 2012).

In addition to viral-mediated gene transfer, transgenic mouse lines that constitutively express opsin genes under local promoter-enhancer regions have been used. For example, the Thy1:ChR2-EYFP mouse lines directly express ChR2 under control of the Thy1 promoter (Arenkiel et al., 2007). Furthermore, other transgenic mouse lines have been generated to enable blue light–induced activation of ChR2 in gamma-aminobutyric acid-ergic (GABAergic) (VGAT-ChR2(H134R)-EYFP), cholinergic (ChAT-ChR2(H134R)-EYFP), serotonergic (Tph2-ChR2(H134R)-EYFP), and parvalbumin (PV)-expressing (Pvalb(H134R)-ChR2-EYFP) neurons (Zhao et al., 2011).

Cell Type–Specific Targeting of Opsins Using Cre/loxP Recombinase

The most common approach to achieving cell type–specific genetic manipulations is through the Cre/loxP recombinase system (Gelman et al., 2003; Madisen et al., 2012; Saunders, Johnson, & Sabatini, 2012; Taniguchi et al., 2011; Tye & Deisseroth, 2012; Yizhar et al., 2011) (Fig. 14.1B). Here, the viruses are designed so that they express only functional transgenes in the presence of Cre recombinase. Then the virus is infused into the brain of transgenic mouse lines that have been generated to express Cre recombinase in a wide range of cell types. These cell types are defined molecularly because the presence of Cre recombinase is restricted to cells expressing a particular gene. To date, there are over 250 Cre-recombinase mouse driver lines available through the Gene Expression of the Nervous System Atlas project, in collaboration with the Intramural Program of the National Institute of Mental Health (Gong et al., 2007) (http://www.gensat.org) as well as commercial repositories such as the Jackson Laboratory (http://jaxmice.jax.org).

Although transgenic mouse lines have proven to be a powerful tool for functional studies of cell populations and circuits in the mammalian brain, an important caveat of all transgenic Cre driver lines is the actual cell type specificity, which must be confirmed individually for each line. Bacterial artificial chromosome transgenics in particular are not generated by targeted mutation, and the cell type specificity in the adult brain may not be reflective of expression patterns during development (Gong et al., 2007). Therefore, accurate interpretation of experiments involving transgenic mouse lines crucially depends on the degree to which transgene expression faithfully reproduces native gene expression patterns. Insofar as Cre expression extends beyond the cellular population of interest, the effects of manipulating these unintentionally targeted cells may be erroneously attributed.

This issue is especially significant in complex brain regions where neighboring cells are anatomically and functionally diverse, such as the ventral tegmental area (VTA), which contains a heterogeneous population of dopaminergic neurons (Fields, Hjelmstad, Margolis, & Nicola, 2007; Lammel, Lim, & Malenka, 2014; Roeper, 2013) and subpopulations of heterogeneous GABAergic and glutamatergic neurons (Brown et al., 2012; Hnasko, Hjelmstad, Fields, & Edwards, 2012; Li, Qi, Yamaguchi, Wang, & Morales, 2012; Margolis, Toy, Himmels, Morales, & Fields, 2012; Nair-Roberts et al., 2008; Olson & Nestler, 2007). The most widely implemented approach for targeting midbrain dopaminergic neurons involves the use of transgenic Cre driver mouse lines in which Cre recombinase is expressed under the control of tyrosine hydroxylase (TH) or dopamine transporter (DAT) promoters (Lindeberg et al., 2004; Savitt, 2005; Zhuang, Masson, Gingrich, Rayport, & Hen, 2005). Significantly, Cre driver lines employing the TH but not DAT promoter exhibit substantial ectopic transgene expression patterns in nondopaminergic neurons, possibly as a result of inappropriate expression driven by an exogenous promoter, or because of TH promoter activity in precursor cell populations that either never produced TH protein or subsequently lost this ability (Lindeberg et al., 2004; Min, Joh, Kim, Peng, & Son, 1994; Savitt, 2005). Moreover, inherent limitations in the ability to restrict the spread of viral infection and the light path to a precisely circumscribed area underscore the importance of cell type specification for the interpretation of experimental manipulations (Lammel et al., 2015). In addition, validating key assumptions and conclusions with complementary methodologies, such as pharmacological manipulations, careful immunohistochemical studies, or mRNA readouts will considerably strengthen conclusions drawn from the use of transgenic mouse lines, which certainly will continue to be a critically important tool for cell type–specific analyses and manipulations in the mammalian brain.

Intersectional Approaches

Neuronal cell types are specified not only by their genetic expression patterns but also by their target projections. Moreover, a single cell type may be defined by more than one molecular fingerprint and often sends axonal projections to more than one brain region. Thus an iteration of promoter-driven, Cre recombinase–dependent, and viral-mediated gene transfer strategies has been to use these strategies in combination to achieve intersectional cell type specificity for functional circuit analysis.

The simplest version of the intersectional approach uses a combination of cell type and anatomical targeting specificity, called projection targeting (Tye & Deisseroth, 2012; Yizhar et al., 2011) (Fig. 14.1C). It is based on the property of the opsins to traffic efficiently along neural processes and to reach even distant axon terminals. Consequently, axon terminals just like cell bodies become photosensitive and can be manipulated (ie, activated or inhibited) by illumination. Several studies have taken advantage of projection targeting to elucidate the role of defined axonal projections in reward, reinforcement, motivation, aversion, depression, social interaction, anxiety, and other aspects of normal and maladaptive brain function (Chaudhury et al., 2013; Ilango et al., 2014; Lammel et al., 2012; Rothermel, Brunert, Zabawa, Diaz-Quesada, & Wachowiak, 2013; Steinberg et al., 2013; Tye et al., 2011, 2012).

A second intersectional approach, called retrograde targeting, takes advantage of retrogradely propagating viruses that transduce presynaptic terminals, which can be infused into the brain region that receives projections from a particular cell population (Fig. 14.1D). Such viruses include herpes simplex virus 1 (HSV1) (Zou, De Koninck, Neve, & Friedrich, 2014), vesicular stomatitis virus (Beier et al., 2011), and rabies virus (Lammel et al., 2012), which after infection of axon terminals are retrogradely transported and express photosensitive opsins selectively in cells that target the injection site of the virus. These viruses often exhibit higher levels of toxicity than LV or AAV and therefore can require careful timing of the analysis of the infected neurons.

The repertoire of conditional expression vectors for optogenetics has been expanded and includes Cre DIO (cDIO), Dre DIO (dDIO), and Flp DIO (fDIO), which recognize lox, FRT, and rox sites, respectively (Fenno et al., 2014). Combining these new constructs with retrograde targeting strategies has allowed for further cell type specification. For example, using the retrograde HSV to drive Cre-dependent Frt expression in Th-Cre mice allows for Flp-dependent expression of ChR2-eYFP in the VTA of the subset of Th-positive neurons that project to the nucleus accumbens (Fenno et al., 2014). This approach is conceptually similar to the rabies virus approach described previously (Lammel et al., 2012), except that here the retrograde HSV is a DNA virus which can be driven in a promoter-specific manner.

Combining cDIO and fDIO constructs for intersectional strategies has been enabled by DIO lox insertion within intronic regions (added to the opsin or EYFP gene) and DIO Frt cassettes before and after the ChR2-EYFP coding sequences, allowing Flp to control the direction of the entire target, and Cre to control the expression of the second exon. Thus for this C_{on}/F_{on}–ChR2–eYFP construct, only neurons expressing both Cre and Flp are competent for conditional expression of ChR2–eYFP (Fenno et al., 2014). The power of this "inclusion criteria" approach is exemplified by virally (DJ-AAV) delivering this construct to the hippocampus of double-transgenic mice (PV-p2A-Cre; SOM-IRES-Flp) expressing Cre in PV cells and Flp in SOM cells, to optogenetically interrogate the function of the minority subset of inhibitory interneurons that are both PV and SOM expressing (Fenno et al., 2014).

Finally, additional "exclusion criteria" constructs, C_{on}/F_{off}-ChR2-eYFP and C_{off}/F_{on}-ChR2-eYFP, have been generated and can be used to specify cells molecularly that express one and not the other recombinase (Fenno et al., 2014). This approach has the potential to be useful, for example, in the nucleus accumbens, where D1 receptor- and D2 receptor-expressing medium spiny neurons (D1-MSN and D2-MSN, respectively) are not mutually exclusive. Thus, in a mouse in which D1-MSNs express Cre and D2-MSNs express Flp, MSNs that are D1 (and not D2) expressing could be targeted with C_{on}/F_{off}-ChR2-eYFP constructs whereas, MSNs that are D2 (and not D1) expressing could be targeted with C_{off}/F_{on}-ChR2-eYFP. Future iterations of this approach will undoubtedly increase the number of intersections (eg, marker 1 AND marker 2 and not marker 3) and allow for ever-increasing cellular specification for functional circuit analysis. Development of additional Dre and Flp mouse lines will be of obvious importance for the successful implementation of this approach.

FUNCTIONAL CIRCUIT ANALYSIS

The implementation of optogenetics and other modern molecular tools that allow the manipulation of neural activity such as DREADDs (designer receptors exclusively activated by designer drugs) (Sternson & Roth, 2014) have generated an explosion of interest in functional circuit mapping. Early attempts to dissect the network complexity of the brain had been limited to sensory systems in which the receptive fields of the circuit under interrogation were accessible to manipulation of the external environment (eg, visual and somatosensory stimulation) (Bishop, 1933; Gasser & Erlanger, 1929). In contrast, for brain circuits in which the "receptive field" is an "internal state" such as mood or reward, stimulating electrodes placed in the brain, lesion experiments, and pharmacological manipulation were until recently the mainstay of this type of functional circuit mapping. The advent of techniques that used, for example, antibody labeling, in situ hybridization, or knockout mice brought a level of molecular specificity to this type of analysis, but studies using these approaches were limited to conclusions that were largely correlational. Although optogenetic approaches do not yet allow the experimenter to prove causality, this strategy offers an opportunity to interrogate necessity and sufficiency, moving the field closer than ever before toward this goal. In the next section we will provide a small sampling of biological questions at the level of synaptic and whole animal circuit analysis, which have been addressed using optogenetic approaches.

Ex Vivo Synaptic Input Specificity

Whole-cell patch clamp recording is a powerful technique for interrogating the cellular response to stimulation of inputs. This approach, combined with pharmacological manipulations, can be used to infer the relative contribution of inhibitory, excitatory, modulatory, and peptidergic inputs. However, individual neurons commonly receive inputs from a wide variety of brain regions. Stimulation of these inputs with metal or glass electrodes can resolve individual input pathways only under the exceptional circumstance in which these inputs are physically segregated (eg, Schafer collateral versus perforant path inputs to the CA1 region of the hippocampus). In the vast majority of brain regions, inputs are intermingled. Furthermore, recognition of transmitter co-release and novel, molecularly defined parallel pathways (Graves et al., 2012; Tritsch, Ding, & Sabatini, 2012; Varga et al., 2009) highlight the importance of molecular isolation of inputs for stimulation. The combination of optogenetic manipulations with patch clamp recordings from brain slice preparations has greatly advanced the possibilities of analyzing pathway-specific synaptic properties and connectivity (Hjelmstad, Xia, Margolis, & Fields, 2013; Lammel et al., 2012; Matsui, Jarvie, Robinson, Hentges, & Williams, 2014; Stuber et al., 2011). With further advances it should soon be possible to express opsins with sufficiently different activation parameters in different brain regions so that multiple different inputs onto an individual cell can be independently activated with great molecular specificity. This type of approach can also be used in vivo while making extracellular or even whole cell recordings from awake behaving animals so the detailed electrophysiological consequences of activating specific inputs in vivo can be elucidated.

In Vivo Neural Circuit Analysis in Freely Moving Animals

Cell type—specific and projection-specific targeting strategies allow unprecedented precision in the manipulation of neural circuits underlying behavior in both health and disease. To perform optogenetic experiments in freely moving animals, the experimental setup typically includes a pulse generator to control a light source (diode-pumped solid-state laser or LED), which then connects to an optical commutator via a fiber-optic cable (Yizhar et al., 2011). The commutator allows the animals to move relatively freely in a behavioral assay without tangling the fiber-optic cable. Another fiber-optic cable connects the optical commutator with an optical fiber that is implanted in the animal's brain. Intracerebral placement of optical fiber can be performed either short- or long-term. For short-term placement, the optical fiber is normally inserted through a guide cannula. The advantage is that pharmacological agents can be infused through the guide cannula to the same area as the light, which allows a combination of optogenetic manipulations and pharmacological experiments. However, the risk of damaging the optical fiber when inserting it into the guide cannula is a major disadvantage. An alternative approach is to implant the optical fiber in the animal's brain long term, which is connected to a ferrule affixed to the animal's skull. However, for pharmacological interventions this approach would also require the implantation of an infusion cannula.

The intracerebral location of the optical fiber depends on the targeting strategy (Sparta et al., 2012; Yizhar et al., 2011; Zalocusky & Deisseroth, 2013; Zhang et al., 2010) (Fig. 14.1). For both non—cell type—specific and cell type—specific targeting experiments, the optical fiber is implanted directly above the somata of opsin-expressing cells, ie, the brain region in which the optogenetic construct was delivered (Fig. 14.1A and B). For projection-specific targeting experiments, the optical fiber is implanted directly above the brain region that contains opsin-expressing axon terminals (Fig. 14.1C). It is important to take into consideration that ChR2-induced action potentials can be propagated in both anterograde and retrograde directions. Axon collaterals of ChR2-expressing cells can also be activated by light stimulation of terminal regions. Thus additional experiments may be necessary; for example, chemical inactivation of the cell bodies where the virus was injected and pharmacological or genetic inhibition of the effects of light stimulation of receptors at the projection site (Yizhar et al., 2011). For projection-specific targeting using retrograde viruses, the optical fiber is implanted directly above the transgene-expressing cells that project to the presynaptic virus injection site (Fig. 14.1D). It is also possible to combine optogenetic stimulation with electrophysiological recordings in freely moving animals using "optrodes." An elegant application of optrodes involves optogenetic tagging of cells. This involves expressing ChR2 in the cells of interest so that their activation by light can be used to identify the phenotype of a recorded cell unequivocally (Cohen, Haesler, Vong, Lowell, & Uchida, 2012).

Although the use of fiber-optic cables to activate opsins in awake behaving animals has enabled the generation of a plethora of important results (Steinberg, Christoffel, Deisseroth, & Malenka, 2014; Tye & Deisseroth, 2012), for many behavioral experiments, the fiber-optic cable can be a hindrance. An exciting technical advance involves the stereotaxic implantation of flexible light-emitting devices, which can be activated wirelessly, into the parenchyma of the brain (Kim et al., 2013; McCall et al., 2013). It seems likely that such devices will eventually replace the use of fiber-optic cables so that optogenetic manipulation of neural activity can be achieved wirelessly in awake behaving animals. This should increase the range of behaviors that can be studied and allow activation and inhibition of multiple circuit elements simultaneously in individual subjects.

TRANSLATIONAL MEDICINE

An important application of optogenetics has been to identify clinically relevant pathways in animal models (Steinberg et al., 2014; Tye & Deisseroth, 2012) and to understand in more detail which brain elements need to be electrically stimulated to achieve a desired therapeutic effect (Chow & Boyden, 2013; Williams & Denison, 2013). Indeed, it is possible that optogenetics will be used clinically in humans to treat a wide range of neuropsychiatric disorders because of its many advantages over currently available neuromodulatory methods (eg, deep brain stimulation [DBS] or transcranial magnetic stimulation [TMS]). Using electrical stimulation or TMS, it is impossible to be certain that only a local brain area or a particular pathway is affected. As described earlier, optogenetic manipulations provide great precision in the circuit elements that are activated or inhibited. This in turn facilitates the identification of disease-related pathways in the brain, information that is critical for improving the targeting and efficacy of neuromodulation using DBS or TMS. Furthermore, optogenetic approaches can be used to advance understanding of the consequences of currently available neuromodulation approaches. For example, using optogenetics, the therapeutic efficacy of DBS of the subthalamic nucleus (STN) for Parkinson disease was suggested to result from activation of afferent inputs to the STN (Gradinaru, Mogri, Thompson, Henderson, & Deisseroth, 2009). Another contribution of optogenetics to translational medicine will be to facilitate the discovery of druggable molecular targets that are preferentially associated with an identified cell population, which is embedded in a neural circuit pathway, the activation or inhibition of which is therapeutically beneficial. It is routinely possible to identify the gene expression profile of single cells (Citri, Pang, Südhof, Wernig, & Malenka, 2012) and improvements in these methods are occurring rapidly. We anticipate that modern therapeutic psychopharmacology will involve the development of drugs that activate or inhibit specific circuits, the detailed functions of which have been elucidated using optogenetics combined with other tools of modern neuroscience.

The development of optogenetic based therapies in humans will require using viruses to deliver light-sensitive opsins into the human brain as well as some type of implantable light delivery system. Although gene therapy is not yet routine in the United States, many viral-based gene therapy trials are currently being conducted. Indeed, the European Medicines Agency has approved the AAV-mediated gene therapy Alipogene tiparvovec (marketed under the trade name Glybera) for lipoprotein lipase deficiency (Richards, 2012). There are, of course, several major obstacles to the development of optogenetic-based therapies. The human immune system may detect opsins as foreign. This may be less of a problem when targeting elements within the brain parenchyma, but it may be a major obstacle when targeting the peripheral nervous system. It is also unclear whether long-term stable expression of opsins can be achieved with minimal cellular toxicity. For example, viral-driven, long-term expression of ChR2 may cause abnormal axonal morphology (Miyashita, Shao, Chung, Pourzia, & Feldman, 2013). Clearly rational virus engineering and careful assessment of the long-term safety and efficacy of viral delivery systems must be performed before optogenetic therapies can be tested for the treatment of human neuropsychiatric diseases. Additional important challenges include achieving cell type—specific expression of opsins and engineering stable light delivery systems.

As other chapters describe, discoveries from human genetics have significantly advanced understanding of the pathogenesis of neuropsychiatric diseases. Identification of susceptibility genes provide important molecular entry points for the modeling of the respective diseases in genetically modified animals, most commonly mice. However, the translation of insights from pathogenic mechanisms obtained in such model organisms back to the clinic will be challenging. One major limitation is that the broad spectrum of methods and readouts that can be employed in animal studies usually can not be recapitulated or emulated in patients, for whom noninvasive methods are required. A second challenge is that some dysfunctional brain areas and circuits that are likely critical substrates for major mental illness symptoms, such as certain areas of prefrontal cortex, may be uniquely human. Thus the use of additional model organisms may be advantageous. For example, the common marmoset, *Callithrix jacchus*, a New World monkey, has significant advantages from a genetic and neuroanatomical perspective. It is small in size (about 400 g), reaches sexual maturity at 12 months, and breeds rapidly in captivity, typically producing two pairs of fraternal twins per year. The neuroanatomy of the common marmoset is well described: Like macaques, but unlike rodents, marmosets have a well-developed prefrontal cortex, a region that is critical for many cognitive functions that are impaired in human psychiatric disorders. Furthermore, marmosets are social and communicative and can perform some higher cognitive tasks developed for macaque monkeys and humans. The marmoset genome has been sequenced, which lays the necessary groundwork for genetic manipulations. Moreover, the optogenetic tools described here are also being developed for use in primates (Diester et al., 2011; Galvan, Hu, Smith, & Wichmann, 2012).

CONCLUSIONS

A major appeal of optogenetics is that it has provided the ability to perform previously unimaginable experiments that bring critical insights into fundamental questions of brain function and the pathophysiology of nervous system diseases. Combined with complementary approaches including viral-based tracing strategies to define novel circuit elements (Wall, Wickersham, Cetin, De La Parra, & Callaway, 2010; Wickersham & Feinberg, 2012) and the genetically based cell type—specific expression of other molecular entities to manipulate neural activity such as DREADDs (Sternson & Roth, 2014), it is certain that our understanding of the neural circuit basis of adaptive and pathological behaviors in model organisms will advance greatly over the ensuing decade. It is also certain that scientific and clinical interest in the use of these approaches along with noninvasive brain stimulation and imaging will expand because the spectrum of their applications is nearly inexhaustible. The most transformative approach in this context would be to truly integrate technological, basic, and clinical neuroscience with therapeutic efforts. This may lead, for example, to an exploration of combined behavioral and brain stimulation interventions that are individually targeted, perhaps based on a more sophisticated understanding of individual circuit dysfunctions as well as individual genetic and epigenetic factors.

REFERENCES

Arenkiel, B. R., Peca, J., Davison, I. G., Feliciano, C., Deisseroth, K., Augustine, G. J., ... Feng, G. (2007). In vivo light-induced activation of neural circuitry in transgenic mice expressing channelrhodopsin-2. *Neuron, 54*(2), 205—218. http://dx.doi.org/10.1016/j.neuron.2007.03.005.

Beier, K. T., Saunders, A., Oldenburg, I. A., Miyamichi, K., Akhtar, N., Luo, L., ... Cepko, C. L. (2011). Anterograde or retrograde transsynaptic labeling of CNS neurons with vesicular stomatitis virus vectors. *Proceedings of the National Academy of Sciences of the United States of America, 108*(37), 15414—15419. http://dx.doi.org/10.1073/pnas.1110854108.

Benzekhroufa, K., Liu, B., Tang, F., Teschemacher, A. G., & Kasparov, S. (2009). Adenoviral vectors for highly selective gene expression in central serotonergic neurons reveal quantal characteristics of serotonin release in the rat brain. *BMC Biotechnology, 9*, 23. http://dx.doi.org/10.1186/1472-6750-9-23.

Berndt, A., Lee, S. Y., Ramakrishnan, C., & Deisseroth, K. (2014). Structure-guided transformation of channelrhodopsin into a light-activated chloride channel. *Science (New York, N.Y.), 344*(6182), 420—424. http://dx.doi.org/10.1126/science.1252367.

Berndt, A., Yizhar, O., Gunaydin, L. A., Hegemann, P., & Deisseroth, K. (2009). Bi-stable neural state switches. *Nature Neuroscience, 12*(2), 229—234. http://dx.doi.org/10.1038/nn.2247.

Bishop, G. H. G. (1933). Fiber groups in the optic nerve. *American Journal of Physiology, 106*(2), 460—470.

Boyden, E. S., Zhang, F., Bamberg, E., Nagel, G., & Deisseroth, K. (2005). Millisecond-time scale, genetically targeted optical control of neural activity. *Nature Neuroscience, 8*(9), 1263—1268. http://dx.doi.org/10.1038/nn1525.

Brown, M. T. C., Tan, K. R., O'Connor, E. C., Nikonenko, I., Muller, D., & Lüscher, C. (2012). Ventral tegmental area GABA projections pause accumbal cholinergic interneurons to enhance associative learning. *Nature, 492*(7429), 452—456. http://dx.doi.org/10.1038/nature11657.

Chaudhury, D., Walsh, J. J., Friedman, A. K., Juarez, B., Ku, S. M., Koo, J. W., ... Han, M.-H. (2013). Rapid regulation of depression-related behaviours by control of midbrain dopamine neurons. *Nature, 493*(7433), 532—536. http://dx.doi.org/10.1038/nature11713.

Chow, B. Y., & Boyden, E. S. (2013). Optogenetics and translational medicine. *Science Translational Medicine, 5*(177), 177ps5. http://dx.doi.org/10.1126/scitranslmed.3003101.

Citri, A., Pang, Z. P., Südhof, T. C., Wernig, M., & Malenka, R. C. (2012). Comprehensive qPCR profiling of gene expression in single neuronal cells. *Nature Protocols, 7*(1), 118—127. http://dx.doi.org/10.1038/nprot.2011.430.

Cohen, J. Y., Haesler, S., Vong, L., Lowell, B. B., & Uchida, N. (2012). Neuron-type-specific signals for reward and punishment in the ventral tegmental area. *Nature, 482*(7383), 85—88. http://dx.doi.org/10.1038/nature10754.

Deisseroth, K. (2014). Circuit dynamics of adaptive and maladaptive behaviour. *Nature, 505*(7483), 309—317. http://dx.doi.org/10.1038/nature12982.

Deisseroth, K., & Schnitzer, M. J. (2013). Engineering approaches to illuminating brain structure and dynamics. *Neuron, 80*(3), 568–577. http://dx.doi.org/10.1016/j.neuron.2013.10.032.

Diester, I., Kaufman, M. T., Mogri, M., Pashaie, R., Goo, W., Yizhar, O., … Shenoy, K. V. (2011). An optogenetic toolbox designed for primates. *Nature Neuroscience, 14*(3), 387–397. http://dx.doi.org/10.1038/nn.2749.

Dittgen, T., Nimmerjahn, A., Komai, S., Licznerski, P., Waters, J., Margrie, T. W., … Osten, P. (2004). Lentivirus-based genetic manipulations of cortical neurons and their optical and electrophysiological monitoring in vivo. *Proceedings of the National Academy of Sciences of the United States of America, 101*(52), 18206–18211. http://dx.doi.org/10.1073/pnas.0407976101.

Fenno, L. E., Mattis, J., Ramakrishnan, C., Hyun, M., Lee, S. Y., He, M., … Deisseroth, K. (2014). Targeting cells with single vectors using multiple-feature Boolean logic. *Nature Methods, 11*(7), 763–772. http://dx.doi.org/10.1038/nmeth.2996.

Fields, H. L., Hjelmstad, G. O., Margolis, E. B., & Nicola, S. M. (2007). Ventral tegmental area neurons in learned appetitive behavior and positive reinforcement. *Annual Review of Neuroscience, 30*(1), 289–316. http://dx.doi.org/10.1146/annurev.neuro.30.051606.094341.

Fields, R. L., Ponzio, T. A., Kawasaki, M., & Gainer, H. (2012). Cell-type specific oxytocin gene expression from AAV delivered promoter deletion constructs into the rat supraoptic nucleus in vivo. *PLoS One, 7*(2), e32085. http://dx.doi.org/10.1371/journal.pone.0032085.

Galvan, A., Hu, X., Smith, Y., & Wichmann, T. (2012). In vivo optogenetic control of striatal and thalamic neurons in non-human primates. *PLoS One, 7*(11), e50808. http://dx.doi.org/10.1371/journal.pone.0050808.

Gasser, H. S., & Erlanger, J. (1929). The role of fiber size in the establishment of a nerve block by pressure or cocaine. *Nature Reviews Neuroscience, 88*, 581–591.

Gelman, D. M., Noain, D., Avale, M. E., Otero, V., Low, M. J., & Rubinstein, M. (2003). Transgenic mice engineered to target Cre/loxP-mediated DNA recombination into catecholaminergic neurons. *Genesis, 36*(4), 196–202. http://dx.doi.org/10.1002/gene.10217.

Gong, S., Doughty, M., Harbaugh, C. R., Cummins, A., Hatten, M. E., Heintz, N., & Gerfen, C. R. (2007). Targeting cre recombinase to specific neuron populations with bacterial artificial chromosome constructs. *Journal of Neuroscience, 27*(37), 9817–9823. http://dx.doi.org/10.1523/JNEUROSCI.2707-07.2007.

Gradinaru, V., Mogri, M., Thompson, K. R., Henderson, J. M., & Deisseroth, K. (2009). Optical deconstruction of parkinsonian neural circuitry. *Science (New York, N.Y.), 324*(5925), 354–359. http://dx.doi.org/10.1126/science.1167093.

Gradinaru, V., Zhang, F., Ramakrishnan, C., Mattis, J., Prakash, R., Diester, I., … Deisseroth, K. (2010). Molecular and cellular approaches for diversifying and extending optogenetics. *Cell, 141*, 154–165. http://dx.doi.org/10.1016/j.cell.2010.02.037.

Graves, A. R., Moore, S. J., Bloss, E. B., Mensh, B. D., Kath, W. L., & Spruston, N. (2012). Hippocampal pyramidal neurons comprise two distinct cell types that are countermodulated by metabotropic receptors. *Neuron, 76*(4), 776–789. http://dx.doi.org/10.1016/j.neuron.2012.09.036.

Häusser, M. (2014). Optogenetics: the age of light. *Nature Methods, 11*(10), 1012–1014. http://dx.doi.org/10.1038/nmeth.3111.

Hjelmstad, G. O., Xia, Y., Margolis, E. B., & Fields, H. L. (2013). Opioid modulation of ventral pallidal afferents to ventral tegmental area neurons. *Journal of Neuroscience, 33*(15), 6454–6459. http://dx.doi.org/10.1523/JNEUROSCI.0178-13.2013.

Hnasko, T. S., Hjelmstad, G. O., Fields, H. L., & Edwards, R. H. (2012). Ventral tegmental area glutamate neurons: electrophysiological properties and projections. *Journal of Neuroscience, 32*(43), 15076–15085. http://dx.doi.org/10.1523/JNEUROSCI.3128-12.2012.

Ilango, A., Kesner, A. J., Keller, K. L., Stuber, G. D., Bonci, A., & Ikemoto, S. (2014). Similar roles of substantia nigra and ventral tegmental dopamine neurons in reward and aversion. *Journal of Neuroscience, 34*(3), 817–822. http://dx.doi.org/10.1523/JNEUROSCI.1703-13.2014.

Kato, H. E., Zhang, F., Yizhar, O., Ramakrishnan, C., Nishizawa, T., Hirata, K., … Nureki, O. (2012). Crystal structure of the channelrhodopsin light-gated cation channel. *Nature, 482*(7385), 369–374. http://dx.doi.org/10.1038/nature10870.

Kim, T.-i., McCall, J. G., Jung, Y. H., Huang, X., Siuda, E. R., Li, Y., … Bruchas, M. R. (2013). Injectable, cellular-scale optoelectronics with applications for wireless optogenetics. *Science, 340*(6129), 211–216. http://dx.doi.org/10.1126/science.1232437.

Knobloch, H. S., Charlet, A., Hoffmann, L. C., Eliava, M., Khrulev, S., Cetin, A. H., … Grinevich, V. (2012). Evoked axonal oxytocin release in the central amygdala attenuates fear response. *Neuron, 73*(3), 553–566. http://dx.doi.org/10.1016/j.neuron.2011.11.030.

Lammel, S., Lim, B. K., & Malenka, R. C. (2014). Reward and aversion in a heterogeneous midbrain dopamine system. *Neuropharmacology, 76*(Pt B), 351–359. http://dx.doi.org/10.1016/j.neuropharm.2013.03.019.

Lammel, S., Lim, B. K., Ran, C., Huang, K. W., Betley, M. J., Tye, K. M., … Malenka, R. C. (2012). Input-specific control of reward and aversion in the ventral tegmental area. *Nature, 491*(7423), 212–217. http://dx.doi.org/10.1038/nature11527.

Lammel, S., Steinberg, E., Földy, C., Walls, N., Beier, K., Luo, L., & Malenka, R. C. (2015). Diversity of transgenic mouse models for selective targeting of midbrain dopamine neurons. *Neuron, 85*(2), 429–438. in press.

Lee, J. H., Durand, R., Gradinaru, V., Zhang, F., Goshen, I., Kim, D.-S., … Deisseroth, K. (2010). Global and local fMRI signals driven by neurons defined optogenetically by type and wiring. *Nature, 465*(7299), 788–792. http://dx.doi.org/10.1038/nature09108.

Li, X., Qi, J., Yamaguchi, T., Wang, H.-L., & Morales, M. (2012). Heterogeneous composition of dopamine neurons of the rat A10 region: molecular evidence for diverse signaling properties. *Brain Structure and Function, 218*(5), 1159–1176. http://dx.doi.org/10.1007/s00429-012-0452-z.

Lima, S. Q., & Miesenböck, G. (2005). Remote control of behavior through genetically targeted photostimulation of neurons. *Cell, 121*(1), 141–152. http://dx.doi.org/10.1016/j.cell.2005.02.004.

Lin, J. Y., Knutsen, P. M., Muller, A., Kleinfeld, D., & Tsien, R. Y. (2013). ReaChR: a red-shifted variant of channelrhodopsin enables deep transcranial optogenetic excitation. *Nature Neuroscience, 16*(10), 1499–1508. http://dx.doi.org/10.1038/nn.3502.

Lindeberg, J., Usoskin, D., Bengtsson, H., Gustafsson, A., Kylberg, A., Söderström, S., & Ebendal, T. (2004). Transgenic expression of Cre recombinase from the tyrosine hydroxylase locus. *Genesis, 40*(2), 67–73. http://dx.doi.org/10.1002/gene.20065.

Madisen, L., Mao, T., Koch, H., Zhuo, J. M., Berenyi, A., Fujisawa, S., ... Zeng, H. (2012). A toolbox of Cre-dependent optogenetic transgenic mice for light-induced activation and silencing. *Nature Neuroscience, 15*, 793–802. http://dx.doi.org/10.1038/nn.3078.

Margolis, E. B., Toy, B., Himmels, P., Morales, M., & Fields, H. L. (2012). Identification of rat ventral tegmental area GABAergic neurons. *PLoS One, 7*(7), e42365. http://dx.doi.org/10.1371/journal.pone.0042365.

Matsui, A., Jarvie, B. C., Robinson, B. G., Hentges, S. T., & Williams, J. T. (2014). Separate GABA afferents to dopamine neurons mediate acute action of opioids, development of tolerance, and expression of withdrawal. *Neuron, 82*(6), 1346–1356. http://dx.doi.org/10.1016/j.neuron.2014.04.030.

Mattis, J., Tye, K. M., Ferenczi, E. A., Ramakrishnan, C., O'Shea, D. J., Prakash, R., ... Deisseroth, K. (2012). Principles for applying optogenetic tools derived from direct comparative analysis of microbial opsins. *Nature Methods, 9*, 159–172. http://dx.doi.org/10.1038/nmeth.1808.

McCall, J. G., Kim, T., Shin, G., Huang, X., Jung, Y. H., Al-Hasani, R., ... Rogers, J. A. (2013). Fabrication and application of flexible, multimodal light-emitting devices for wireless optogenetics. *Nature Protocols, 8*(12), 2413–2428. http://dx.doi.org/10.1038/nprot.2013.158.

Min, N., Joh, T. H., Kim, K. S., Peng, C., & Son, J. H. (1994). 5′ upstream DNA sequence of the rat tyrosine hydroxylase gene directs high-level and tissue-specific expression to catecholaminergic neurons in the central nervous system of transgenic mice. *Molecular Brain Research, 27*(2), 281–289.

Miyashita, T., Shao, Y. R., Chung, J., Pourzia, O., & Feldman, D. E. (2013). Long-term channelrhodopsin-2 (ChR2) expression can induce abnormal axonal morphology and targeting in cerebral cortex. *Frontiers in Neural Circuits, 7*, 8. http://dx.doi.org/10.3389/fncir.2013.00008.

Nair-Roberts, R. G., Chatelain-Badie, S. D., Benson, E., White-Cooper, H., Bolam, J. P., & Ungless, M. A. (2008). Stereological estimates of dopaminergic, GABAergic and glutamatergic neurons in the ventral tegmental area, substantia nigra and retrorubral field in the rat. *Neuroscience, 152*, 1024–1031. http://dx.doi.org/10.1016/j.neuroscience.2008.01.046.

Olson, V. G., & Nestler, E. J. (2007). Topographical organization of GABAergic neurons within the ventral tegmental area of the rat. *Synapse, 61*(2), 87–95. http://dx.doi.org/10.1002/syn.20345.

Richards, S. (November 06, 2012). *Gene therapy arrives in Europe*. The Scientist.

Roeper, J. (2013). Dissecting the diversity of midbrain dopamine neurons. *Trends in Neurosciences, 36*(6), 336–342. http://dx.doi.org/10.1016/j.tins.2013.03.003.

Rothermel, M., Brunert, D., Zabawa, C., Diaz-Quesada, M., & Wachowiak, M. (2013). Transgene expression in target-defined neuron populations mediated by retrograde infection with adeno-associated viral vectors. *Journal of Neuroscience, 33*(38), 15195–15206. http://dx.doi.org/10.1523/JNEUROSCI.1618-13.2013.

Saunders, A., Johnson, C. A., & Sabatini, B. L. (2012). Novel recombinant adeno-associated viruses for Cre activated and inactivated transgene expression in neurons. *Frontiers in Neural Circuits, 6*, 47. http://dx.doi.org/10.3389/fncir.2012.00047.

Savitt, J. M. (2005). Bcl-x is required for proper development of the mouse substantia nigra. *Journal of Neuroscience, 25*(29), 6721–6728. http://dx.doi.org/10.1523/JNEUROSCI.0760-05.2005.

Sparta, D. R., Stamatakis, A. M., Phillips, J. L., Hovelso, N., van Zessen, R., & Stuber, G. D. (2012). Construction of implantable optical fibers for long-term optogenetic manipulation of neural circuits. *Nature Protocols, 7*, 12–23. http://dx.doi.org/10.1038/nprot.2011.413.

Steinberg, E. E., Christoffel, D. J., Deisseroth, K., & Malenka, R. C. (2014). Illuminating circuitry relevant to psychiatric disorders with optogenetics. *Current Opinion in Neurobiology, 30C*, 9–16. http://dx.doi.org/10.1016/j.conb.2014.08.004.

Steinberg, E. E., Keiflin, R., Boivin, J. R., Witten, I. B., Deisseroth, K., & Janak, P. H. (2013). A causal link between prediction errors, dopamine neurons and learning. *Nature Neuroscience, 16*(7), 966–973. http://dx.doi.org/10.1038/nn.3413.

Sternson, S. M., & Roth, B. L. (2014). Chemogenetic tools to interrogate brain functions. *Annual Review of Neuroscience, 37*, 387–407. http://dx.doi.org/10.1146/annurev-neuro-071013-014048.

Stuber, G. D., Sparta, D. R., Stamatakis, A. M., van Leeuwen, W. A., Hardjoprajitno, J. E., Cho, S., ... Bonci, A. (2011). Excitatory transmission from the amygdala to nucleus accumbens facilitates reward seeking. *Nature, 475*, 377–380. http://dx.doi.org/10.1038/nature10194.

Tan, W., Janczewski, W. A., Yang, P., Shao, X. M., Callaway, E. M., & Feldman, J. L. (2008). Silencing preBötzinger complex somatostatin-expressing neurons induces persistent apnea in awake rat. *Nature Neuroscience, 11*(5), 538–540. http://dx.doi.org/10.1038/nn.2104.

Taniguchi, H., He, M., Wu, P., Kim, S., Paik, R., Sugino, K., ... Huang, Z. J. (2011). A resource of Cre driver lines for genetic targeting of GABAergic neurons in cerebral cortex. *Neuron, 71*, 995–1013. http://dx.doi.org/10.1016/j.neuron.2011.07.026.

Tritsch, N. X., Ding, J. B., & Sabatini, B. L. (2012). Dopaminergic neurons inhibit striatal output through non-canonical release of GABA. *Nature, 490*(7419), 262–266. http://dx.doi.org/10.1038/nature11466.

Tye, K. M., & Deisseroth, K. (2012). Optogenetic investigation of neural circuits underlying brain disease in animal models. *Nature Reviews Neuroscience, 13*(4), 251–266. http://dx.doi.org/10.1038/nrn3171.

Tye, K. M., Mirzabekov, J. J., Warden, M. R., Ferenczi, E. A., Tsai, H.-C., Finkelstein, J., ... Deisseroth, K. (2012). Dopamine neurons modulate neural encoding and expression of depression-related behaviour. *Nature, 493*(7433), 537–541. http://dx.doi.org/10.1038/nature11740.

Tye, K. M., Prakash, R., Kim, S. Y., Fenno, L. E., Grosenick, L., Zarabi, H., ... Deisseroth, K. (2011). Amygdala circuitry mediating reversible and bidirectional control of anxiety. *Nature, 471*, 358–362. http://dx.doi.org/10.1038/nature09820.

Varga, V., Losonczy, A., Zemelman, B. V., Borhegyi, Z., Nyiri, G., Domonkos, A., ... Freund, T. F. (October 2009). Fast synaptic subcortical control of hippocampal circuits. *Science, 326*, 449–453.

Wall, N. R., Wickersham, I. R., Cetin, A., De La Parra, M., & Callaway, E. M. (2010). Monosynpatic circuit tracing in vivo through Cre-dependent targeting and complementation of modified rabies virus. *Proceedings of the National Academy of Sciences of the United States of America, 107*, 21848–21853.

Wickersham, I. R., & Feinberg, E. H. (2012). New technologies for imaging synaptic partners. *Current Opinion in Neurobiology, 22*, 121–127.

Wietek, J., Wiegert, J. S., Adeishvili, N., Schneider, F., Watanabe, H., Tsunoda, S. P., ... Hegemann, P. (2014). Conversion of channelrhodopsin into a light-gated chloride channel. *Science, 344*(6182), 409–412.

Williams, J. C., & Denison, T. (2013). From optogenetic technologies to neuromodulation therapies. *Science Translational Medicine, 5*(177), 177ps6. http://dx.doi.org/10.1126/scitranslmed.3003100.

Xu, W., & Sudhof, T. C. (2013). A neural circuit for memory specificity and generalization. *Science, 339*(6125), 1290–1295. http://dx.doi.org/10.1126/science.1229534.

Yizhar, O., Fenno, L. E., Davidson, T. J., Mogri, M., & Deisseroth, K. (2011). Optogenetics in neural systems. *Neuron, 71*(1), 9–34. http://dx.doi.org/10.1016/j.neuron.2011.06.004.

Zalocusky, K., & Deisseroth, K. (2013). Optogenetics in the behaving rat: integration of diverse new technologies in a vital animal model. *Optogenetics, 1*, 1–17. http://dx.doi.org/10.2478/optog-2013-0001.

Zhang, F., Gradinaru, V., Adamantidis, A. R., Durand, R., Airan, R. D., de Lecea, L., & Deisseroth, K. (2010). Optogenetic interrogation of neural circuits: technology for probing mammalian brain structures. *Nature Protocols, 5*(3), 439–456. http://dx.doi.org/10.1038/nprot.2009.226.

Zhao, S., Ting, J. T., Atallah, H. E., Qiu, L., Tan, J., Gloss, B., ... Feng, G. (2011). Cell type–specific channelrhodopsin-2 transgenic mice for optogenetic dissection of neural circuitry function. *Nature Methods, 8*(9), 745–752. http://dx.doi.org/10.1038/nmeth.1668.

Zhuang, X., Masson, J., Gingrich, J. A., Rayport, S., & Hen, R. (2005). Targeted gene expression in dopamine and serotonin neurons of the mouse brain. *Journal of Neuroscience Methods, 143*(1), 27–32. http://dx.doi.org/10.1016/j.jneumeth.2004.09.020.

Zou, M., De Koninck, P., Neve, R. L., & Friedrich, R. W. (2014). Fast gene transfer into the adult zebrafish brain by herpes simplex virus 1 (HSV-1) and electroporation: methods and optogenetic applications. *Frontiers in Neural Circuits, 8*, 41. http://dx.doi.org/10.3389/fncir.2014.00041.

Chapter 15

Brain Imaging With Magnetoencephalography During Rest and During Speech and Language Processing

Srikantan S. Nagarajan, Kamalini G. Ranasinghe and Keith A. Vossel
University of California San Francisco, San Francisco, CA, United States

BRAIN IMAGING WITH MAGNETOENCEPHALOGRAPHY (MEG)

Multiple modalities of noninvasive functional brain imaging have made a tremendous impact in improving our understanding of the human brain. Ever since its advent in 1991, functional magnetic resonance imaging (fMRI) has emerged as the predominant modality for imaging of the functioning brain, for several reasons. fMRI uses magnetic resonance imaging (MRI) to measure changes in blood oxygenation level dependent (BOLD) signals owing to neuronal activation. It is a safe, noninvasive method that allows for whole-brain coverage, including the ability to examine activity in deep brain structures. Importantly, the widespread availability of commercial and open-source tools for analysis of fMRI data has enabled many researchers to embrace this technology easily. However, because the BOLD signal is only an indirect measure of neural activity and is fundamentally limited by the rate of oxygen consumption and subsequent blood flow mechanism, fMRI lacks the temporal resolution required to image the dynamic and oscillatory spatiotemporal patterns that are associated with cognitive processes. The limited temporal resolution particularly constrains studies of speech and language neural systems because the auditory stimuli and the corresponding neural responses have inherently fast dynamics that cannot be readily assessed with fMRI. Furthermore, because the BOLD signal is only an approximate, indirect measure of neural activity, it might not reflect true neuronal processes accurately, especially in regions of altered vasculature. In fact, the exact frequency band of neuronal processes that corresponds to the BOLD signal is still being actively debated (Logothetis, Pauls, Augath, Trinath, & Oeltermann, 2001; Niessing et al., 2005). Furthermore, because fMRI measurements involve loud scans, caused by fast forces on magnetic resonance gradient coils, the scans themselves will invoke auditory responses that have to be deconvolved from other signals to examine external speech and language stimulus-related activity. Hence, to image brain activity noninvasively on a neurophysiologically relevant time scale and observe neurophysiological processes more directly, silent imaging techniques are needed that have both high temporal and adequate spatial resolution.

Temporal changes of the brain activity can be measured noninvasively using methods with high (eg, millisecond) temporal resolution, namely magnetoencephalography (MEG) and electroencephalography (EEG). MEG measures tiny magnetic fields outside the head that are generated by neural activity. EEG is the measurement of electric potentials generated by neural activity using an electrode array placed directly on the scalp. In contrast to fMRI, both MEG and EEG directly measure electromagnetic fields emanating from the brain with excellent temporal resolution (<1 ms) and allow the study of neural oscillatory processes over a wide frequency range (at least 1–600 Hz). MEG and EEG also provide complementary information about brain activity because of their differing sensitivity to current sources

Genomics, Circuits, and Pathways in Clinical Neuropsychiatry. http://dx.doi.org/10.1016/B978-0-12-800105-9.00015-9

within the brain. Whereas MEG is primarily sensitive to tangential currents in the brain closer to the surface and relatively unaffected by poor conductive properties of the skull, EEG is primarily sensitive to radial sources and is highly affected by the conductive properties of the brain, skull, and scalp. Because bioelectric currents produced by neurons also generate magnetic fields which are not distorted by the heterogeneous environment, measurements of these magnetic fields using MEG can be considered to result in an undistorted signature of underlying cortical activity. Therefore, MEG and EEG can be viewed as being complementary in terms of the sensitivity to underlying neural activity.

This chapter initially reviews how brain activity can be reconstructed from MEG measurements with implications for spatial and temporal resolution of such reconstructions. Subsequently, a review of MEG studies in dementia is presented.

SENSING THE BRAIN'S MAGNETIC FIELDS

Biomagnetic fields detected by MEG are extremely small, in the tens to hundreds of femto-Tesla range, seven orders of magnitude smaller than the earth's magnetic field; as a result, appropriate data collection necessitates a magnetically shielded room and highly sensitive detectors called superconducting quantum interference devices (SQUIDs). The fortuitous anatomical arrangement of cortical pyramidal cells allows the noninvasive detection of their activity by MEG. The long apical dendrites of these cells are arranged perpendicularly to the cortical surface and parallel to each other, allowing their electromagnetic fields often to sum up to magnitudes large enough to detect at the scalp. Synchronously fluctuating dendritic currents result in electric and magnetic dipoles that produce these electromagnetic fields (Nunez & Srinivasan, 2006). These dendritic currents from the brain are typically sensed using detection coils called flux transformers or magnetometers, which are positioned closely to the scalp and connected to SQUIDs. SQUIDs act as a magnetic field—to-voltage converter, and their typically nonlinear responses are linearized by flux-locked loop electronic circuits with a sensitivity of about 10 femto-Tesla per square root of Hertz, which is adequate for the detection of the brain's magnetic fields (Vrba & Robinson, 2002).

MEG sensors are often configured for differential magnetic field measurements to reduce ambient noise in measurements. MEG sensors are also referred to as gradiometers, although some MEG systems are also built of magnetometers and rely on magnetic shielding and clever electronics for noise cancellation. The two commonly used gradiometer configurations are axial and planar gradiometers. Axial gradiometers consist of two coils that share an axis, whereas planar gradiometers measure gradients (or differences) of magnetic fields in a given plane. The sensitivity profile of planar gradiometer sensors is similar to EEG in the sense that a sensor is maximally sensitive to a source closest on the cortical surface to it. However, the sensitivity profile of an axial gradiometer can be counterintuitive because it is not maximally sensitive to sources closest to the sensors. Furthermore, both planar and axial gradiometers are sensitive to the orientation of the sources in a counterintuitive manner.

Modern MEG systems often consist of simultaneous recordings from many differential sensors that cover the whole head, and total number of sensors varies from 100 to 300. The advent of such array systems has significantly advanced MEG studies. Typical MEG systems have sensors that are spaced approximately 2.2—3.6 cm apart. Although the maximum sampling rate for many MEG systems is approximately 12 kHz, most MEG data are usually recorded at about 1000 Hz, thereby still providing an excellent temporal resolution to measure the dynamics of cortical neuronal activity with millisecond precision.

Most studies published to date using MEG have mainly used it as an electrophysiological assay of sensitive brain regions. These studies focus on the response properties of specific sensors within an array of sensors (or sometimes spatial averages of specific groups of sensors) and examine component peaks in sensor waveforms. Fig. 15.1 shows a typical sensor configuration and magnetic field sensor responses to simple auditory stimuli. Fig. 15.1B shows typical position of magnetic field sensors relative to the head and brain surfaces. Fig. 15.1C shows the topographic layout of the magnetic field response recorded at 100 and 200 ms after the onset of an acoustic stimulus.

There are many reasons why neuroscientists have embraced MEG. First, MEG setup time is very short and convenient for both experimenters and subjects. A participant or patient can be in the scanner within 10—15 min from entering the laboratory because, unlike EEG, the lengthy time necessary to apply and check electrodes is obviated. Second, the anatomical location of large parts of the neocortex in the human brain makes MEG ideally suited for electrophysiological studies, especially for sensory, speech, and language processing. Furthermore, with whole-head sensor arrays, MEG is also well-suited to investigate hemispheric lateralization effects based on sensor waveforms. In contrast to evoked responses measured with EEG, which are maximal at midline electrodes, making hemispheric effects difficult to characterize, MEG responses are well-lateralized. Distinct groups of MEG sensors are sensitive to lateralized temporal lobe activity that allows for hemisphere-specific assessments.

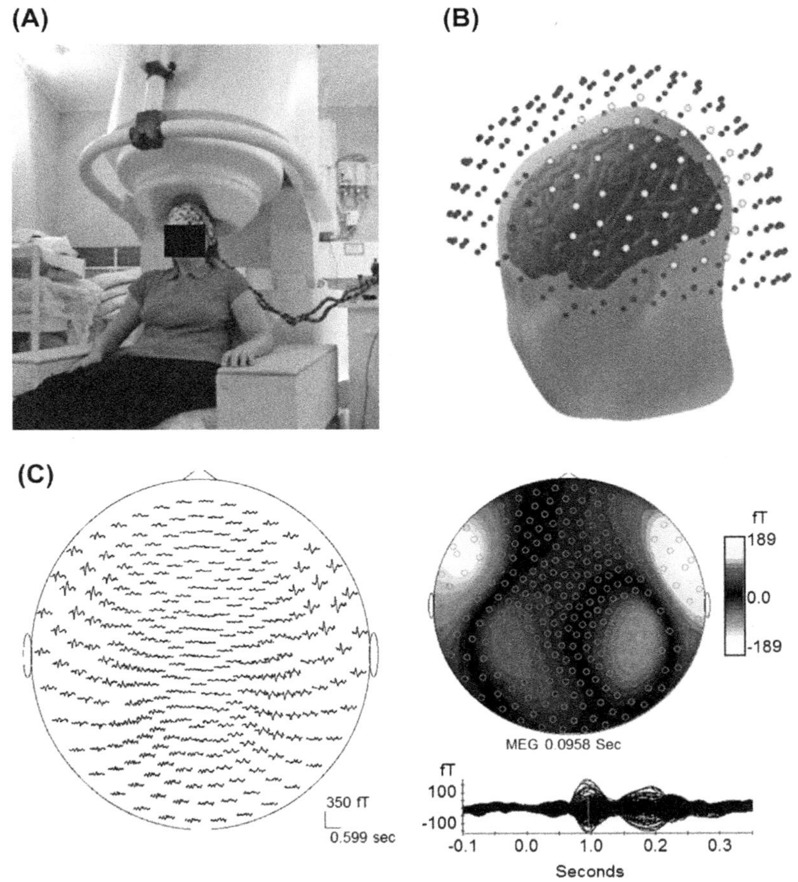

FIGURE 15.1 (A) A subject seated just below a whole-head MEG sensor array. The helmet above the volunteer contains the SQUID-based magnetic field sensors, whereas the large cylindrical Dewar flask contains liquid helium to cool them to their superconducting operation temperature. The subject is shown wearing a high-density electrode cap for optional simultaneous recording of EEG. (B) Positioning of the head within the MEG sensor array. Left hemisphere sensor locations are highlighted in yellow to provide an example of head positioning within the sensor array to provide broad coverage over the cortical regions underlying speech perception. (C) Left: Auditory-evoked magnetic field responses of each MEG sensor are shown. Right top: Topographic plot of the magnetic field across all sensors at the time latency corresponding to the dominant response peak, which occurs around 100 ms, called the M100 or N100 peak. This "butterfly-shaped" magnetic field pattern suggests that two sources located in each auditory cortex can account for these response peaks. Right bottom: An overlay of the magnetic field waveforms of all of the sensors. For this subject, a prominent peak can also be observed around 200 ms.

FROM SENSING TO IMAGING

MEG sensor data analysis provides only qualitative information about underlying brain regions whose activity is observed on the sensor array based on experienced users' intuitions about the sensitivity profile of the sensors. To interpret observed sensor data more precisely in terms of the underlying brain activity, it is possible to reconstruct brain activity from MEG data. Reconstruction of brain activity from MEG data typically involves two major components: a forward model and an inverse model.

Forward Models Describing Brain Activity and Measurements

The forward model consists of three sub-components: a source model, a volume conductor, and a measurement model. Typical source models assume that the MEG measurements outside the head are generated primarily by electric current dipoles located in the brain. This model is consistent with available measurements of coherent synaptic and intracellular currents in cortical columns that are thought to be major contributors to MEG and EEG signals. Although several more complex source models have been proposed, the equivalent current dipole is still the dominant source model in the literature. Because of the distance between the sources in the brain and the sensors outside the head, the dipole is still a reasonable approximation of the sources.

Volume conductor models refer to the equations that govern the relation between the source model and the sensor measurements, ie, the electric potentials or the magnetic fields. These surface integral equations, obtained by solving Maxwell's equations under quasistatic conditions, can be solved analytically for special geometries of the volume conductor, such as a sphere and ellipsoids. For realistic volume conductors, various numerical techniques such as finite-element and boundary-element methods are employed. These methods are time-consuming and their use may appear impractical in many settings because of the lack of knowledge about specific parameters used in these models (Mosher, Leahy, & Lewis, 1999).

Measurement models refer to the specific measurement systems used in EEG and MEG, including the position of the sensors relative to the head. For instance, different MEG systems measure axial versus planar gradients of the magnetic fields with respect to different locations of reference sensors. The measurement model incorporates such information about the type of measurement and the geometry of the reference sensors. Because MEG sensor arrays are fixed relative to the head of a subject, it is necessary to measure the position of head relative to the sensor array. Typically this is accomplished by attaching head localization coils to fiducial landmarks on the scalp, passing current through these coils, measuring the magnetic field created by the currents passed, and triangulating to locate the head position relative to the sensor array. In many MEG systems, head localization is accomplished every 5–10 min because it disrupts normal data collection. Within a block of 10 min, with the subject lying supine and the head securely positioned in the array, head movements are typically found to be less than 5 mm. However, more modern systems are sometimes equipped with continuous head localization procedures that enable constant updating of the sensor locations relative to the head, and correcting for subjects' head movements.

The source, volume conductor, and measurement models are typically combined and embodied in the idea called the "forward field" that describes a linear relationship between sources and the measurements. Usually, we assume that the forward field matrix is known. We can easily calculate the forward field for equivalent electric current dipoles in a spherical volume conductor model for a whole-head axial gradiometer MEG system. In this model, MEG is sensitive only to the tangential component of the primary current dipoles, whereas EEG is sensitive to all components but sensitive to uncertainties in the head model. Simultaneous MEG and EEG can be acquired in most modern MEG systems and require some modification to the forward field matrix for combined MEG/EEG measurements especially for more realistic source, volume conductor, and measurement models.

Coregistration is an integral part of forward model construction. Coregistration involves defining three fiducial points on an individual subject's head surface, which creates the x,y,z coordinate system that includes the brain and the position of the MEG sensors relative to it. Based on these fiducial landmarks, a transformation matrix is obtained that enables coregistration with the subject's MRI. This allows the source locations and sensors to be defined in MRI coordinates and enables interpretation of inverse model reconstructions in terms of the underlying brain anatomy provided by MRI.

Inverse Models for Reconstructing Brain Activity From Measurements

Inverse algorithms are used to solve the bioelectromagnetic inverse problem, ie, estimating neural source model parameters from MEG and EEG measurements obtained outside the human head. Because the source distributions are inherently four-dimensional (three in space and one in time) and only a few measurements are made outside the head, estimation is ill-posed; in other words, there are no unique solutions for a given set of measurements. To circumvent this problem of nonuniqueness, various estimation procedures incorporate prior knowledge and constraints about source characteristics, such as possible source locations, the source spatial extent, the total number of sources, or the source–frequency/time–frequency characteristics.

Inverse algorithms can be broadly classified into two categories: parametric dipole fitting and tomographic imaging methods. Parametric dipole fitting methods assume that a small set of current dipoles (usually two to five) can adequately represent some unknown source distribution. In this case, the dipole locations and moments form a set of unknown parameters which are typically found using either a nonlinear least-square fit or multiple signal classification algorithms or maximum likelihood estimation methods (Mosher, Baillet, & Leahy, 1999). Parametric dipole fitting has been successfully used clinically for localization of early sensory responses in somatosensory and auditory cortices. Fig. 15.2 shows an example of parametric dipole localization in the context of auditory-evoked responses and shows that responses to early auditory peaks can often be localized to activity arising from a source located in the superior temporal plane, from the auditory cortex and its immediate environs. However, the localization of higher-order auditory cortical functions is not always consistent and reliable with these methods across paradigms or subjects.

Two major problems exist in dipole fitting procedures. First, because of nonlinear optimization there are problems of local minima when more than two dipole parameters are estimated. This is usually manifested by sensitivity to

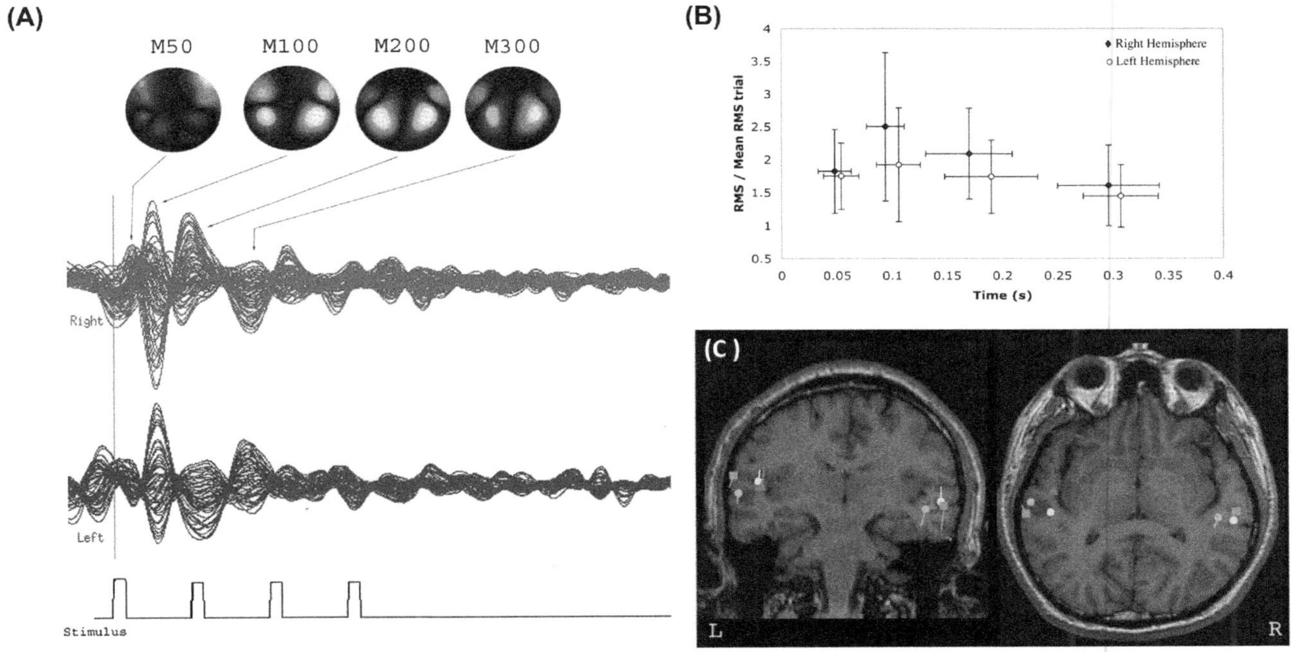

FIGURE 15.2 (A) Auditory-evoked responses to a train of tone pips occurring 200 ms apart. Blue waveforms correspond to the right hemisphere and the purple waveforms correspond to the left hemisphere. The magnetic field topography on the sensor array is shown as colored circles above for the first four peak responses. (B) Amplitude and latencies of the first four response peaks showing hemispheric similarities in latency and amplitudes. (C) Dipole localization of each of the four peaks shows activity arising from auditory cortex and its immediate environs. *From Hairston, I. S., & Nagarajan, S. S. (2007). Neural mechanisms of the time-order error: an MEG study.* Journal of Cognitive Neuroscience, 19, *1163–1174.*

initialization, and some subjectivity is involved in evaluating the validity of solutions. Brute force search methods have a huge computational burden—exponential in the number of parameters. A second, more difficult, problem in parametric methods is that these methods often require a priori knowledge of the number of dipoles. However, such information about model order is not always known a priori, especially for complex brain-mapping conditions. Although information and decision theoretic criteria have been proposed to address this problem, the success of these approaches is currently unclear, especially in real data sets. Whereas parametric dipole methods are ideal for point or focal sources, they perform poorly for distributed clusters of sources. Nevertheless, many studies to date using MEG have used dipole fitting procedures to make inferences about cortical activity.

Tomographic imaging is an alternative approach to the inverse problem. These methods impose constraints on source locations based on anatomical and physiological information that can be derived from information obtained with other imaging modalities. Anatomical MRI provides excellent spatial resolution of head and brain anatomy, whereas fMRI techniques provide an alternative measure of neural activation based on associated hemodynamic changes. Because of the high degree of overlap in activity measured using multiple modalities, such information can be used to improve solutions to the inverse problem. If we assume that the dominant sources are the transmembrane and intracellular currents in the apical dendrites of the cortical pyramidal cells, the source image can be constrained to the cortex, which can be extracted from a registered volume MRI of the subjects' head. Furthermore, the orientation of the cells normal to the cortical surface can be used to constrain the orientation of the cortical current sources. By tessellating the cortex into disjoint regions and representing sources in each region by an equivalent current dipole oriented normal to the surface, the forward model relating the sources and the measurements can be written as a linear model with additive noise. Such a formulation transforms the inverse problem into a linear imaging method because it involves estimating electrical activity at discrete locations over a finely sampled reconstruction grid based on discrete measurements. This imaging problem, although linear, is also highly ill-posed because of the limited number of sensor measurements available compared with the number of elements used in the tessellation grid.

Various solutions have been proposed for solving the tomographic imaging problem, and because there are many more unknowns to estimate simultaneously (source amplitude and time courses) than there are sensor data, the problem is undetermined.

Instead of simultaneously estimating all sources, a popular alternative is to scan the brain and estimate source amplitude at each source location independently. It can be shown that such scanning methods are closely related to whole-brain tomographic methods, and the most popular scanning algorithms are adaptive spatial filtering techniques, more commonly referred to as "adaptive beamformers", or just "beamformers" (Sekihara & Nagarajan, 2008).

Adaptive beamformers have been shown to be simple to implement and are powerful techniques for characterizing cortical oscillations and are closely related to other tomographic imaging methods. However, one major problem with adaptive beamformers is that they are extremely sensitive to the presence of strongly correlated sources. Although they are robust to moderate correlations, in the case of auditory studies, because auditory cortices are largely synchronous in their activity across the two hemispheres, these algorithms tend to perform poorly for auditory evoked data sets without work-arounds (Fig. 15.5), and many modifications have been proposed for reducing the influence of correlated sources (Dalal, Sekihara, & Nagarajan, 2006). The simplest such work-around is to use half the sensors corresponding to each hemisphere separately, and this approach works surprisingly well for cross-hemispheric interactions. Other modifications to the original algorithms have been proposed in the literature, which require some knowledge about the location of the correlated source region (Dalal et al., 2006; Quraan & Cheyne, 2010).

Many algorithms have also been proposed for simultaneous estimation of all source amplitudes, and such solutions require specification of prior knowledge about the sources, either implicitly or explicitly specified in the form of probability distributions. In these cases, the solutions often require a Bayesian inference procedure of estimating some aspect of the posterior distribution given the data and the priors. We showed that the many seemingly disparate algorithms for tomographic source imaging can be unified and shown, in some cases to be equivalent, using a hierarchical Bayesian modeling framework with a general form of prior distribution (called Gaussian scale mixture) and two different types of inferential procedures (Wipf & Nagarajan, 2008). These insights allow for continued development of novel algorithms for tomographic imaging in relation to prior efforts in this enterprise. Algorithms have shown that significant improvements in performance can be achieved by modern Bayesian inference methods that allow for accurate reconstructions of a large number of sources from typical configurations of MEG sensors (Wipf, Owen, Attias, Sekihara, & Nagarajan, 2010; Zumer, Attias, Sekihara, & Nagarajan, 2007, 2008). Fig. 15.3 shows source reconstructions of auditory-evoked responses using one such novel algorithm, called Champagne, as well as reconstructions from popular benchmark algorithms for comparisons that highlight inconsistent localizations, poorer spatial resolution, and sensitivity to correlated sources and noise.

Sources of Noise in Magnetoencephalography

Although significant breakthroughs have occurred in the source reconstruction algorithm development effort, an enduring problem in MEG- and EEG-based imaging is that brain responses to sensory or cognitive events are small compared with the large number of sources of noise, artifacts (biological and nonbiological), and interference from spontaneous brain activity unrelated to the sensory or cognitive task of interest. All existing methods for brain source localization are hampered by these many sources of noise present in MEG/EEG data. The magnitude of the stimulus-evoked cortical sources are on the order of noise on a single trial, and so typically 75–200 averaged trials are at least needed to distinguish the sources clearly above noise. This limits the type of questions that can be asked, and is prohibitive for examining processes such as learning that can occur over just one or several trials. Needing to average trials is time-consuming and therefore it is difficult for a subject to hold still or pay attention through the duration of the experiment. Gaussian thermal noise or Gaussian electrical noise is present at the MEG or EEG sensors themselves. Background room interference such as from power lines and electronic equipment can be problematic. Biological noise such as heartbeats, eye blinks, or other muscle artifacts can also be present. Ongoing brain activity itself, including the drowsy-state alpha (about 10 Hz) rhythm can drown out evoked brain sources.

Noise in MEG and EEG data is typically reduced by a variety of preprocessing algorithms before being used by source localization algorithms. Simple forms of preprocessing include filtering out frequency bands not containing a brain signal of interest. In addition, independent component analysis has been used to remove artifactual components such as eye blinks (Delorme & Makeig, 2004; Makeig, Jung, Bell, Ghahremani, & Sejnowski, 1997). More sophisticated techniques have also been developed using graphical models for preprocessing before source localization (Nagarajan, Attias, Hild, & Sekihara, 2006, 2007). Therefore, algorithms for source localization from MEG and EEG data typically use a two-stage procedure: the first for noise/interference removal and the second for source localization. However, more recent algorithms that integrate interference suppression with source localization have been proposed and provide robust source reconstructions (Wipf et al., 2010; Zumer et al., 2007).

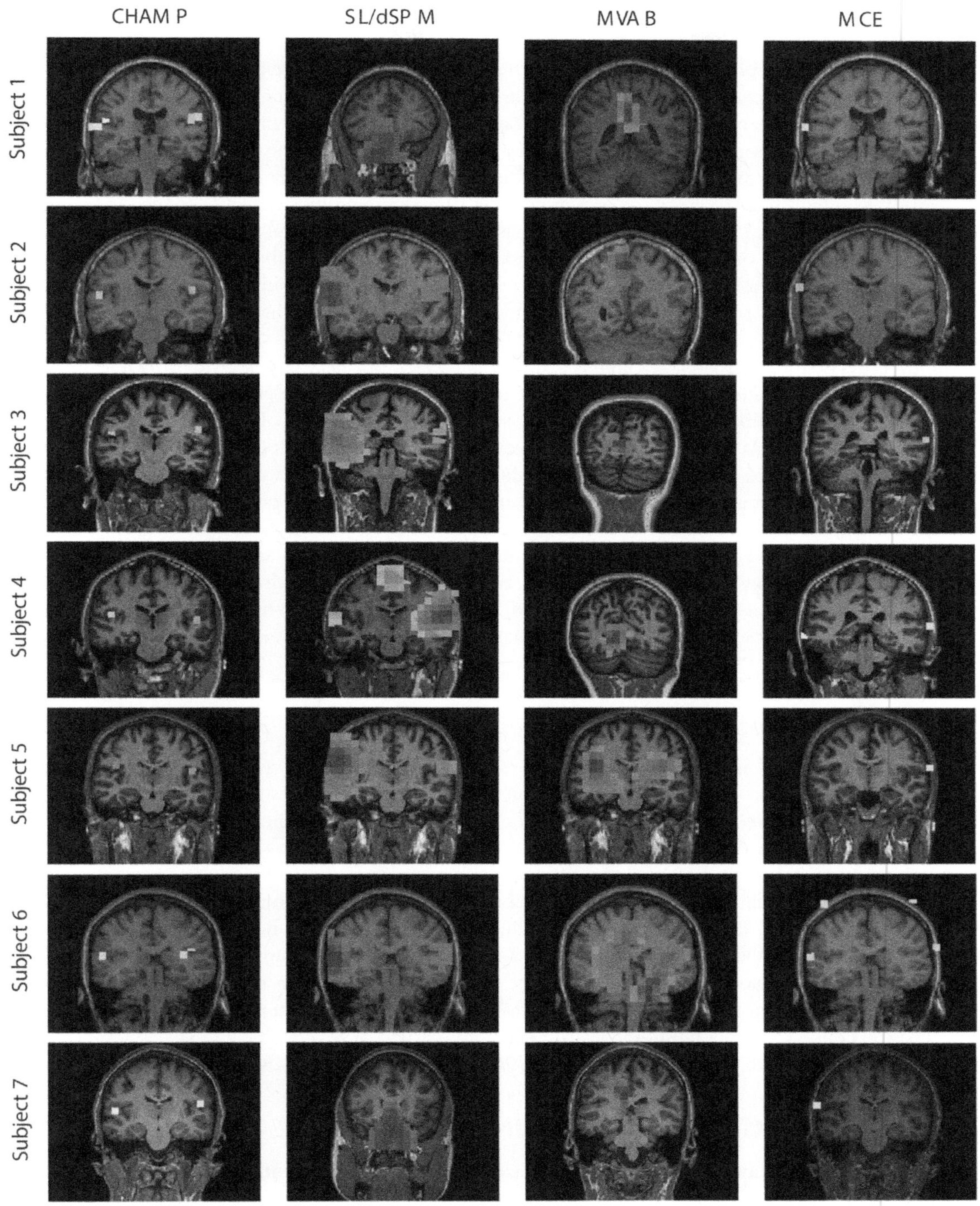

FIGURE 15.3 Auditory-evoked field results for seven different subjects. Results from Champagne (CHAM P) are shown in the left-most column and results from the benchmark algorithms (minimum variance adaptive beamformers (MVA B), sLORETA (SL/dSP M) and MCE) are shown in the other three columns. Only Champagne shows consistent localization of auditory cortices in both hemispheres for all subjects.

Temporal and Spatial Resolution of Magnetoencephalography Imaging

Because MEG data can be acquired on a submillisecond time scale, temporal resolution of MEG imaging is limited only by the sampling rate, typically about 1 kHz; and in principle, cortical oscillations can be observed up to 500 Hz. In contrast to its temporal resolution, determining the spatial resolution of MEG imaging is challenging since it depends highly on the reconstruction algorithm chosen as well as variety of other factors such as the signal-to-noise and interference ratio, model formulation, forward-model accuracy, coregistration errors, accuracy of priors, etc. In general, it can be easily shown that the spatial resolution of MEG reconstruction is not limited by sensor spacing, because many adaptive methods can perform better than estimates based on spatial sampling criteria. For instance, whereas sensor spacing in many axial gradiometer systems is 2.2 cm, reconstruction accuracy can in some cases be as small as 3 mm. In general, coregistration errors alone can account for about 3 mm accuracy in localization information for dipole fitting procedures. Tomographic imaging algorithms such as minimum-norm methods have poor spatial resolution on the order of a few centimeters, but the spatial resolution of adaptive spatial filtering methods and more recent tomographic reconstruction methods based on machine learning techniques are difficult to compute generally because these estimates depend on the data and factors contributing to data quality and so forth. As a rule of thumb, for typical data sets, these newer methods can reconstruct tens to hundreds of sources about 0.5 cm apart (assuming time—frequency separation and detectability), and this can be considered an approximate spatial resolution for MEG, keeping in mind that under certain circumstances the spatial resolution can be even greater.

A common myth related to the spatial resolution of MEG is its lack of sensitivity to gyral crown activity and relative insensitivity to deep sources. Although it is true that for single spherical volume conductor models MEG sensors are insensitive to radially pointing dipoles, this does not necessarily translate to gyral sources. It has been shown that, using realistic volume conductor models (such as boundary element methods or multiple local-sphere models), some sensitivity to radial sources can be recovered and there is no predominant loss of sensitivity to gyral sources (Hillebrand & Barnes, 2002). Furthermore, whereas there is a significant drop in sensitivity to deeper sources because their contributions will fall by approximately the square of the distance to the sensors, recovery of deep sources is an issue of the signal-to-noise ratio. In general, if high signal-to-noise ratio data are recorded, there is no inherent problem in recovery of deep sources with some of the newer Bayesian reconstruction methods. However, midbrain sources have two additional problems. First, they may not have dipolar organization owing to the architecture, although dipole approximation may not be inaccurate because of the distance to the sensors. Second, the uncertainties in the lead-field increase for deep brain sources, which makes them more difficult to reconstruct.

From Single-Subject Reconstructions to Group-Level Inference

Whereas the advantage of MEG imaging is its ability to reconstruct the timing of activation accurately across different frequency bands in single subjects, inferences across subjects require group-level statistical analyses (Fig. 15.4) (Dalal et al., 2008). The most ubiquitous form of group analysis of MEG studies is based on parameters, obtained from dipole fitting of typical component peaks in the response, such as timing, amplitude, location, and sometimes orientation. For the less common tomographic and scanning-based algorithms, group analysis of data across subjects has typically paralleled similar procedures for whole-brain analysis based on fMRI and positron emission tomography studies (Singh, Barnes, & Hillebrand, 2003; Singh, Barnes, Hillebrand, Forde, & Williams, 2002). These procedures include spatial normalization to template brains, general linear modeling of experimental effects, parametric and nonparametric inference procedures, and corrections for multiple comparisons. Notably, group-level statistical corrections for multiple comparisons are not yet as well developed for MEG imaging studies as they are for fMRI, and fMRI correction procedures such as familywise error can sometimes be too conservative for MEG reconstructions for a variety of reasons, including that spatial correlations in reconstructed images are higher than in fMRI (Dalal et al., 2008; Darvas, Pantazis, Kucukaltun-Yildirim, & Leahy, 2004).

MAGNETOENCEPHALOGRAPHY STUDIES IN AGING AND DEMENTIA

Neuronal oscillations are a fundamental property of cortical networks and represent the coordinated activity of precise temporal correlations (Buzsaki, 2011; Singer, 1999). Synchronized oscillations signify an effective means of communication between distributed network components (Fries, 2005). At the level of large-scale networks, synchronized oscillations have been directly linked to cognitive abilities ranging from sensory perception to higher-order memory and attention (Gray, Konig, Engel, & Singer, 1989; Hipp, Engel, & Siegel, 2011). Such functionally connected large-scale brain networks denote a wider paradigm shift in the neurosciences in which the brain is viewed as a distributed system of dynamic interactions (Siegel, Donner, & Engel, 2012). Brain network connectivity, the coordinated firing of connected neural networks, is emerging as an important tool for our basic understanding of human cognition in health as well as in neurodegenerative diseases that cause cognitive dysfunction (Jagust, 2013; Wang et al., 2013).

Large-scale brain networks that are developmentally formed to regulate specific cognitive behaviors are selectively targeted by different neurodegenerative diseases (Seeley, Crawford, Zhou, Miller, & Greicius, 2009). For example,

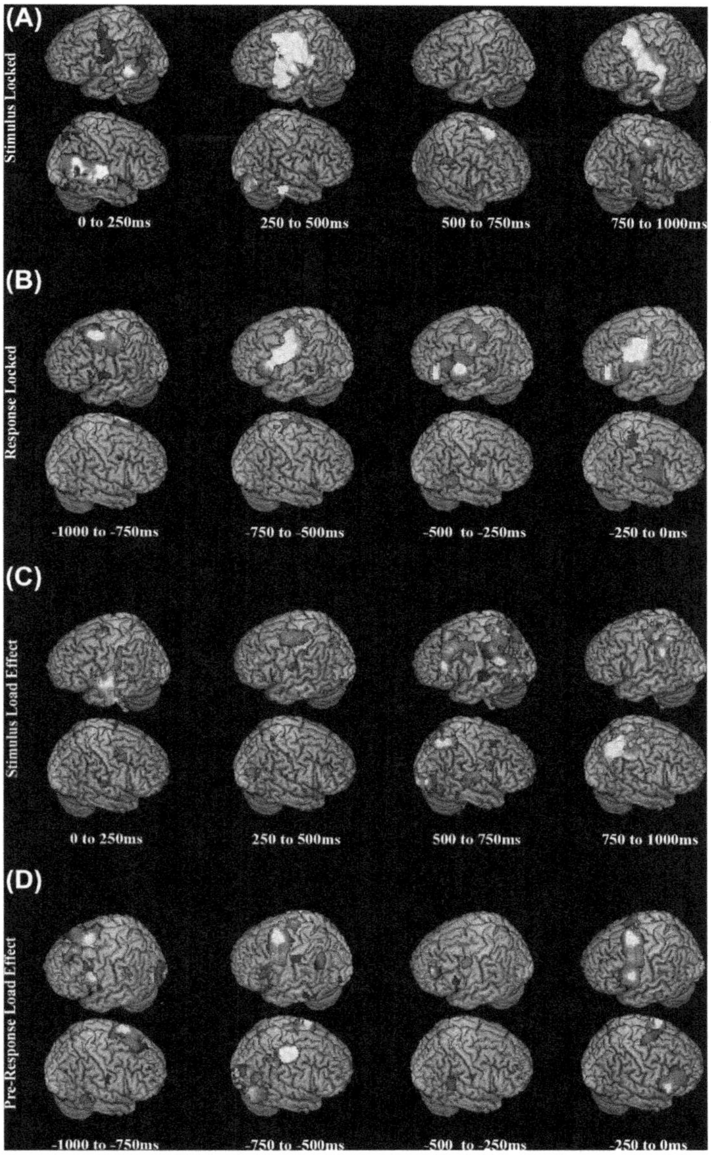

FIGURE 15.4 Oscillatory modulations during a speech syllable reproduction task. Induced oscillatory power changes from MEG are imaged and shown for 250-ms segments that are analyzed with respect to: (A) stimulus locked (ie, aligned to onset of speech sound); (B) response locked (ie, aligned to speech production) syllable encoding, speech preparation; (C) stimulus-aligned working memory load (ie, contrasting induced power for high versus low syllable load); and (D) response-aligned working memory load (ie, contrasting induced power for high versus low syllable load but aligned to speech production). Results in each figure depict activation of phonological network across all conditions. Brain renderings depict all θ, α, and β power decreases and high-gamma power increases overlaid as activations, with absolute T-values rendered additively. Activations and load-effects oscillated between frontal and posterior regions of the dorsal speech stream.

Alzheimer disease (AD), the most common age-related dementia, is associated with early disruption of functional neural networks. Structural, molecular, and metabolic changes in the AD brain mirror to a large extent the distributed intrinsic network collectively called the default mode network—regions characterized by task-related deactivation on resting-state fMRI (Buckner et al., 2005; Seeley et al., 2009). Regions that show the strongest functional connectivity are often the most vulnerable to neurodegenerative disease (Seeley et al., 2009; Zhou, Gennatas, Kramer, Miller, & Seeley, 2012). MEG studies aimed at characterizing spatiotemporal dynamics of distributed networks have demonstrated disrupted functional connectivity in AD and other dementia syndromes. Numerous MEG studies in aging and dementia have reported brain activity during rest as well as during task-engaged paradigms, specifically focusing on spectral changes and functional dysconnectivity.

Resting Magnetoencephalography Studies in Aging and Dementia

It is well documented in electroencephalographic recordings that AD-spectrum patients have diffuse slowing in their brain rhythms. A wealth of MEG studies has revealed distinct spectral changes in AD patients demonstrating consistent slowing of brain activity during rest. In a pilot study of MEG in AD, Berendse, Verbunt, Scheltens, van Dijk, and Jonkman (2000) first demonstrated the general slowing of MEG background activity, consisting of increased delta and theta power, with a maximum over frontal and central areas, and a decreased alpha and theta power, mainly over posterior temporal and occipital areas. Several studies replicated these results and showed a consistent increased delta–theta and decreased alpha–beta power spectrum in AD (de Haan et al., 2008; Osipova, Ahveninen, Jensen, Ylikoski, & Pekkonen, 2005; Ranasinghe et al., 2014; Stam et al., 2006). Poza et al. assessed the diagnostic properties of MEG spectral changes in AD and reported that both spectral mean frequency and the ratio of relative spectral power (delta–theta versus alpha–beta–gamma) are sensitive indices of AD (Poza, Hornero, Abasolo, Fernandez, & Garcia, 2007; Poza, Hornero, Abasolo, Fernandez, & Mayo, 2008). The spectral mean frequency reported a sensitivity of 85% and a specificity of 86% for an AD diagnosis, whereas the ratio of spectral power reported a sensitivity of 75% and a specificity of 95%.

With advanced sensor technology and source reconstructions, MEG has proven to be a useful tool for quantifying the strength of resting state functional connectivity in neurodegenerative diseases. In these experiments, statistical interdependencies of MEG signals from distinct brain regions were analyzed using different measures including coherence, imaginary coherence, neural complexity, synchronization likelihood, and phase lag index. The functional connectivity analyses in AD spectrum have demonstrated unique spatiotemporal patterns of altered global neural synchrony indicating failing connectivity within functional networks. For example, based on synchronization likelihood, AD patients showed increased theta band central-parietal connectivity and decreased alpha band frontotemporal and frontoparietal connectivity (Stam et al., 2006). Evidence from different connectivity indices demonstrated that impaired neural networks broadly associate with global cognitive deficits such as mini mental state exam and clinical dementia rating in AD (de Haan et al., 2008; Ranasinghe et al., 2014; Stam et al., 2006). In a study of AD patients using an unbiased, whole-brain MEG imaging approach that measured the connectivity of individual voxels to the rest of the brain, it was further shown that reductions

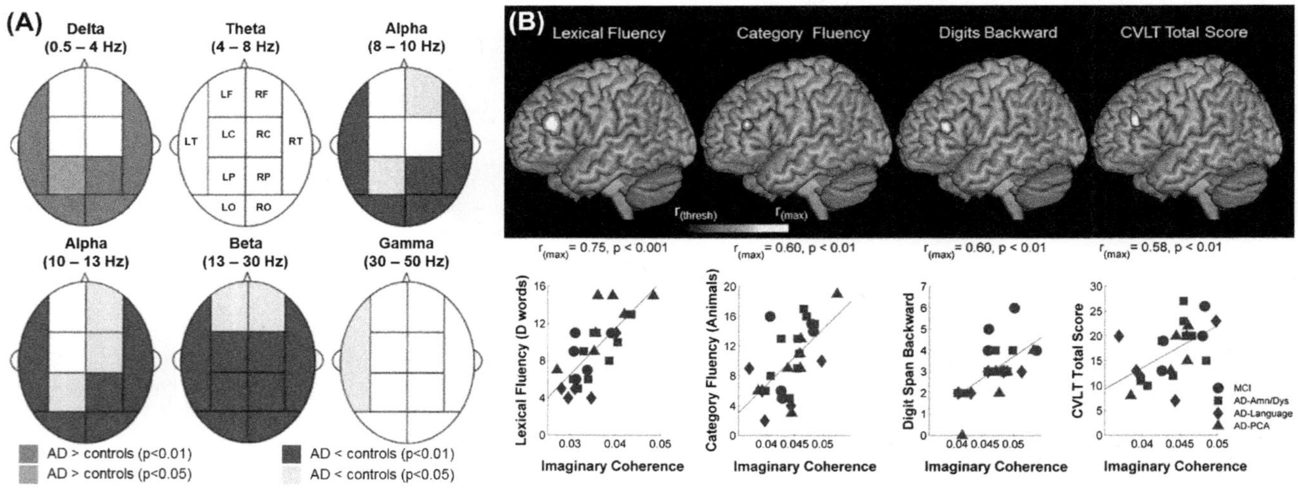

FIGURE 15.5 Spectral power changes and correlations between alpha band functional connectivity deficits and cognitive defects in AD. (A) AD patients show distinct spatiotemporal patterns of power compared with control subjects. In general, AD shows a left shift of overall spectral power distribution with an increased power in low-frequency bands and decreased power in higher-frequency bands. AD (n = 18), age-matched control subjects (n = 18). *LF*, left frontal; *RF*, right frontal; *LC*, left central; *RC*, right central; *LT*, left temporal; *RT*, right temporal; *LP*, left parietal; *RP*, right parietal; *LO*, left occipital; *RO*, right occipital. *(From de Haan, W., Stam, C. J., Jones, B. F., Zuiderwijk, I. M., van Dijk, B. W., Scheltens, P. (2008). Resting-state oscillatory brain dynamics in Alzheimer disease.* Journal of Clinical Neurophysiology: Official Publication of the American Electroencephalographic Society, 25, *187–193.)* (B) Resting state functional connectivity, as measured by imaginary coherence, of the left dorsolateral prefrontal cortex correlated with performance of lexical fluency (D words), category fluency (animals), digit span backward, and CVLT total score. Scatterplots illustrate the peak voxel correlations corresponding to the cognitive tasks shown above them. *MCI*, mild cognitive impairment; *AD*, Alzheimer disease; *PCA*, posterior cortical atrophy; *Amn/Dys*, amnestic/dysexecutive; *AD-Language*, logopenic variant primary progressive aphasia, *CVLT*, California Verbal Learning Test, *r(thresh)*, correlation coefficient at the 5% false discovery rate threshold; *r(max)*, correlation coefficient of the peak voxel. *(From Ranasinghe, K. G., Hinkley, L. B., Beagle, A. J., Mizuiri, D., Dowling, A. F., Honma, S. M., … Vossel, K. A. (2014). Regional functional connectivity predicts distinct cognitive impairments in Alzheimer's disease spectrum.* NeuroImage: Clinical 5, *385–395.)*

in resting state alpha-band connectivity in localized regions significantly correlate with specific cognitive deficits (Ranasinghe et al., 2014). For example, the resting state alpha-band connectivity deficits of dorsolateral prefrontal cortex were significantly correlated with executive, working memory and verbal learning impairments in AD-spectrum patients. In these cohorts, MEG imaging-derived functional connectivity was a more sensitive predictor of cognitive dysfunction than MRI-derived atrophy patterns.

As opposed to AD, resting state MEG neurophysiological experiments in other dementia syndromes such as frontotemporal dementia is almost nonexistent. Franciotti et al. (2006) examined a small cohort of patients with dementia with Lewy bodies (DLB) and showed impaired alpha band long-range coherence in DLB patients compared with age-matched normal control subjects.

Activation Magnetoencephalography Studies in Aging and Dementia

MEG has demonstrated brain activity patterns with high spatiotemporal resolution during a variety of cognitive functions in task-related paradigms under physiological conditions. However, only a handful of studies attempted to record from clinical populations diagnosed with neurodegenerative diseases. Kurimoto et al. (2012) examined MEG-derived induced oscillatory patterns during the retention period of a modified version of Sternberg's memory recognition task performed by patients with AD, mild cognitive impairment (MCI), and age-matched healthy control subjects. The authors reported reduced event-related desynchronizations in AD patients compared with MCI and healthy control subjects in high-frequency bands (ie, beta and gamma). It has long been known that specific cognitive tasks engage unique oscillatory frequencies. These findings indicate that MEG with its high temporal resolution provides a potential tool to map abnormalities of specific frequency bands and relate them to distinct focal cognitive deficits. A case study demonstrated MEG-derived activity patterns during auditory encoding and a verb-generation task in a patient with semantic variant primary progressive aphasia (svPPA) (Miller et al., 2015). The right hemispheric—dominant beta band activity during the auditory-encoding phase as well as during the motor articulatory phase substantiated the anomalous right hemisphere language dominance in a patient with atypical right temporal svPPA. MEG in these experiments proved a promising tool to demonstrate the functional imaging of highly time-sensitive tasks involving speech and language networks.

SUMMARY AND CONCLUSIONS

In this chapter, the technological capabilities and limits of neuroimaging with MEG are discussed. During the brief history of MEG as a promising noninvasive imaging tool in neurodegenerative diseases, we have successfully used it to gauge dysfunctional neural circuits. With the advent of more advanced and sophisticated techniques for reconstructing cortical responses with greater fidelity and robustness, it is expected that the next wave of studies using MEG will exploit the full power of reconstruction algorithms for MEG imaging and pave the way for greater understanding of neural substrates of cognition.

REFERENCES

Berendse, H. W., Verbunt, J. P., Scheltens, P., van Dijk, B. W., & Jonkman, E. J. (2000). Magnetoencephalographic analysis of cortical activity in Alzheimer's disease: a pilot study. *Clinical Neurophysiology: Official Journal of the International Federation of Clinical Neurophysiology, 111*, 604—612.

Buckner, R. L., Snyder, A. Z., Shannon, B. J., LaRossa, G., Sachs, R., Fotenos, A. F., ... Mintun, M. A. (2005). Molecular, structural, and functional characterization of Alzheimer's disease: evidence for a relationship between default activity, amyloid, and memory. *The Journal of Neuroscience: The Official Journal of the Society for Neuroscience, 25*, 7709—7717.

Buzsaki, G. (2011). *Rhythms of the brain*. Oxford University Press.

Dalal, S. S., Guggisberg, A. G., Edwards, E., Sekihara, K., Findlay, A. M., Canolty, R. T., ... Nagarajan, S. S. (2008). Five-dimensional neuroimaging: localization of the time-frequency dynamics of cortical activity. *NeuroImage, 40*, 1686—1700.

Dalal, S. S., Sekihara, K., & Nagarajan, S. S. (2006). Modified beamformers for coherent source region suppression. *IEEE Transactions on Biomedical Engineering, 53*, 1357—1363.

Darvas, F., Pantazis, D., Kucukaltun-Yildirim, E., & Leahy, R. M. (2004). Mapping human brain function with MEG and EEG: methods and validation. *NeuroImage, 23*(Suppl. 1), S289—S299.

Delorme, A., & Makeig, S. (2004). EEGLAB: an open source toolbox for analysis of single-trial EEG dynamics including independent component analysis. *Journal of Neuroscience Methods, 134*, 9—21.

Franciotti, R., Iacono, D., Della Penna, S., Pizzella, V., Torquati, K., Onofrj, M., & Romani, G. L. (2006). Cortical rhythms reactivity in AD, LBD and normal subjects: a quantitative MEG study. *Neurobiology of Aging, 27*, 1100—1109.

Fries, P. (2005). A mechanism for cognitive dynamics: neuronal communication through neuronal coherence. *Trends in Cognitive Sciences, 9*, 474–480.

Gray, C. M., Konig, P., Engel, A. K., & Singer, W. (1989). Oscillatory responses in cat visual cortex exhibit inter-columnar synchronization which reflects global stimulus properties. *Nature, 338*, 334–337.

de Haan, W., Stam, C. J., Jones, B. F., Zuiderwijk, I. M., van Dijk, B. W., & Scheltens, P. (2008). Resting-state oscillatory brain dynamics in Alzheimer disease. *Journal of Clinical Neurophysiology: Official Publication of the American Electroencephalographic Society, 25*, 187–193.

Hairston, I. S., & Nagarajan, S. S. (2007). Neural mechanisms of the time-order error: an MEG study. *Journal of Cognitive Neuroscience, 19*, 1163–1174.

Hillebrand, A., & Barnes, G. R. (2002). A quantitative assessment of the sensitivity of whole-head MEG to activity in the adult human cortex. *NeuroImage, 16*, 638–650.

Hipp, J. F., Engel, A. K., & Siegel, M. (2011). Oscillatory synchronization in large-scale cortical networks predicts perception. *Neuron, 69*, 387–396.

Jagust, W. (2013). Biomarkers and brain connectivity. *JAMA Neurology, 70*, 1233–1234.

Kurimoto, R., Ishii, R., Canuet, L., Ikezawa, K., Iwase, M., Azechi, M., … Takeda, M. (2012). Induced oscillatory responses during the Sternberg's visual memory task in patients with Alzheimer's disease and mild cognitive impairment. *NeuroImage, 59*, 4132–4140.

Logothetis, N. K., Pauls, J., Augath, M., Trinath, T., & Oeltermann, A. (2001). Neurophysiological investigation of the basis of the fMRI signal. *Nature, 412*, 150–157.

Makeig, S., Jung, T. P., Bell, A. J., Ghahremani, D., & Sejnowski, T. J. (1997). Blind separation of auditory event-related brain responses into independent components. *Proceedings of the National Academy of Sciences of the United States of America, 94*, 10979–10984.

Miller, Z. A., Hinkley, L. B., Herman, A., Honma, S., Findlay, A., Block, N., … Gorno-Tempini, M. L. (2015). Anomalous functional language lateralization in semantic variant PPA. *Neurology, 84*, 204–206.

Mosher, J. C., Baillet, S., & Leahy, R. M. (1999). EEG source localization and imaging using multiple signal classification approaches. *Journal of Clinical Neurophysiology, 16*, 225–238.

Mosher, J. C., Leahy, R. M., & Lewis, P. S. (1999). EEG and MEG: forward solutions for inverse methods. *IEEE Transactions on Biomedical Engineering, 46*, 245–259.

Nagarajan, S. S., Attias, H. T., Hild, K. E., 2nd, & Sekihara, K. (2006). A graphical model for estimating stimulus-evoked brain responses from magnetoencephalography data with large background brain activity. *NeuroImage, 30*, 400–416.

Nagarajan, S. S., Attias, H. T., Hild, K. E., 2nd, & Sekihara, K. (2007). A probabilistic algorithm for robust interference suppression in bioelectromagnetic sensor data. *Statistics in Medicine, 26*(21), 3886–3910.

Niessing, J., Ebisch, B., Schmidt, K. E., Niessing, M., Singer, W., & Galuske, R. A. (2005). Hemodynamic signals correlate tightly with synchronized gamma oscillations. *Science, 309*, 948–951.

Nunez, P. L., & Srinivasan, R. (2006). A theoretical basis for standing and traveling brain waves measured with human EEG with implications for an integrated consciousness. *Clinical Neurophysiology, 117*, 2424–2435.

Osipova, D., Ahveninen, J., Jensen, O., Ylikoski, A., & Pekkonen, E. (2005). Altered generation of spontaneous oscillations in Alzheimer's disease. *NeuroImage, 27*, 835–841.

Poza, J., Hornero, R., Abasolo, D., Fernandez, A., & Garcia, M. (2007). Extraction of spectral based measures from MEG background oscillations in Alzheimer's disease. *Medical Engineering & Physics, 29*, 1073–1083.

Poza, J., Hornero, R., Abasolo, D., Fernandez, A., & Mayo, A. (2008). Evaluation of spectral ratio measures from spontaneous MEG recordings in patients with Alzheimer's disease. *Computer Methods and Programs in Biomedicine, 90*, 137–147.

Quraan, M. A., & Cheyne, D. (2010). Reconstruction of correlated brain activity with adaptive spatial filters in MEG. *NeuroImage, 49*, 2387–2400.

Ranasinghe, K. G., Hinkley, L. B., Beagle, A. J., Mizuiri, D., Dowling, A. F., Honma, S. M., … Vossel, K. A. (2014). Regional functional connectivity predicts distinct cognitive impairments in Alzheimer's disease spectrum. *NeuroImage: Clinical, 5*, 385–395.

Seeley, W. W., Crawford, R. K., Zhou, J., Miller, B. L., & Greicius, M. D. (2009). Neurodegenerative diseases target large-scale human brain networks. *Neuron, 62*, 42–52.

Sekihara, K., & Nagarajan, S. S. (2008). *Adaptive spatial filters for electromagnetic brain imaging*. Springer Verlag.

Siegel, M., Donner, T. H., & Engel, A. K. (2012). Spectral fingerprints of large-scale neuronal interactions. *Nature Reviews Neuroscience, 13*, 121–134.

Singer, W. (1999). Neuronal synchrony: a versatile code for the definition of relations? *Neuron, 24*, 49–65, 111–25.

Singh, K. D., Barnes, G. R., & Hillebrand, A. (2003). Group imaging of task-related changes in cortical synchronisation using nonparametric permutation testing. *NeuroImage, 19*, 1589–1601.

Singh, K. D., Barnes, G. R., Hillebrand, A., Forde, E. M., & Williams, A. L. (2002). Task-related changes in cortical synchronization are spatially coincident with the hemodynamic response. *NeuroImage, 16*, 103–114.

Stam, C. J., Jones, B. F., Manshanden, I., van Cappellen van Walsum, A. M., Montez, T., Verbunt, J. P., … Scheltens, P. (2006). Magnetoencephalographic evaluation of resting-state functional connectivity in Alzheimer's disease. *NeuroImage, 32*, 1335–1344.

Vrba, J., & Robinson, S. E. (2002). SQUID sensor array configurations for magnetoencephalography applications. *Superconducting Science and Technology, 15*, 51–89.

Wang, L., Brier, M. R., Snyder, A. Z., Thomas, J. B., Fagan, A. M., Xiong, C., … Ances, B. M. (2013). Cerebrospinal fluid Abeta42, phosphorylated Tau181, and resting-state functional connectivity. *JAMA Neurology, 70*, 1242–1248.

Wipf, D., & Nagarajan, S. (2008). A unified Bayesian framework for MEG/EEG source imaging. *NeuroImage, 44*(3), 947–966.

Wipf, D. P., Owen, J. P., Attias, H. T., Sekihara, K., & Nagarajan, S. S. (2010). Robust Bayesian estimation of the location, orientation, and time course of multiple correlated neural sources using MEG. *NeuroImage, 49*, 641–655.

Zhou, J., Gennatas, E. D., Kramer, J. H., Miller, B. L., & Seeley, W. W. (2012). Predicting regional neurodegeneration from the healthy brain functional connectome. *Neuron, 73*, 1216–1227.

Zumer, J. M., Attias, H. T., Sekihara, K., & Nagarajan, S. S. (August 1, 2007). A probabilistic algorithm integrating source localization and noise suppression for MEG and EEG data. *NeuroImage, 37*(1), 102–115.

Zumer, J. M., Attias, H. T., Sekihara, K., & Nagarajan, S. S. (2008). Probabilistic algorithms for MEG/EEG source reconstruction using temporal basis functions learned from data. *NeuroImage, 41*, 924–940.

Chapter 16

Resting-State Functional MRI: A Novel Tool for Understanding Brain Networks in Neuropsychiatric Disorders

Michael Greicius

Stanford University School of Medicine, Stanford, CA, United States

This chapter considers the advent of a remarkable imaging approach, known as resting-state functional MRI (rs-fMRI), and how it has helped usher in the era of network-based approaches to neuropsychiatric disorders. The chapter begins with some early challenges and obstacles facing this novel methodology, how its proponents overcame initial skepticism through a series of multimodal imaging studies, animal studies, and patient lesion studies, and how these studies have helped us glean some preliminary insights into the molecular underpinnings of rs-fMRI. We then look into the stunning complexity of the brain's functional network architecture as elucidated by rs-fMRI and related imaging techniques. The focus will be on the brain's segregation into a dozen or more large-scale distributed networks and how that is balanced with the brain's need to integrate information across these networks. Next, we will take a still deeper look at two specific resting-state networks (RSNs): the default-mode network (DMN) and the emotional salience network, which arguably have the most consistently demonstrated relevance to neuropsychiatric disease. In the penultimate section we will follow this clinical application thread further by examining the explosion of interest in leveraging RSNs in the study of a wide variety of brain diseases such Alzheimer disease (AD), frontotemporal dementia (FTD), depression, schizophrenia, autism, and chronic pain (Fox & Greicius, 2010). The chapter closes with a look at ongoing research into the how RSNs work (molecular underpinnings) and why (evolutionary advantage).

A BRIEF HISTORY OF RESTING-STATE FUNCTIONAL MRI

In 1995, Bharat Biswal, working with James Hyde at the Medical College of Wisconsin (MCW), published a seminal set of findings that constitute the first demonstration of resting-state functional connectivity (Biswal et al., 1995). Starting with a standard on—off task fMRI approach, the authors identified the bilateral motor cortices with a finger-tapping task. Then, using a seed region in the left motor cortex, they showed that resting fluctuations in the blood oxygenation level—dependent (BOLD) signal of the left motor cortex were strongly temporally correlated with resting BOLD signal fluctuations in the right motor cortex, the bilateral sensory cortices, and a number of other cortical regions known to be crucial to sensorimotor function, such as the dorsal motor cingulate cortex. Several other critical features of rs-fMRI were identified in this first study. The spontaneous fluctuations that were correlated across these regions were remarkably slow, on the order of once every 30 s or so for a full cycle. The map generated by the rs-fMRI analysis bore a striking resemblance to the map generated by the finger-tapping task. As of this writing, the article by Biswal et al. has been cited more than 3000 times, which testifies to the foundational nature of their discovery. The standard region-of-interest approach to deriving an rs-fMRI map is shown in Fig. 16.1.

Several other early rs-fMRI articles followed with important additional contributions from the group at MCW as well as from groups at the University of Wisconsin, Madison and Oulu University in Finland (Hampson et al., 2002; Kiviniemi et al., 2000; Raichle et al., 2001; Stein et al., 2000). The motor network was replicated and groups began tentatively exploring other basic networks, including visual and auditory. An early, intriguing finding from the group at Oulu was that

Genomics, Circuits, and Pathways in Clinical Neuropsychiatry. http://dx.doi.org/10.1016/B978-0-12-800105-9.00016-0

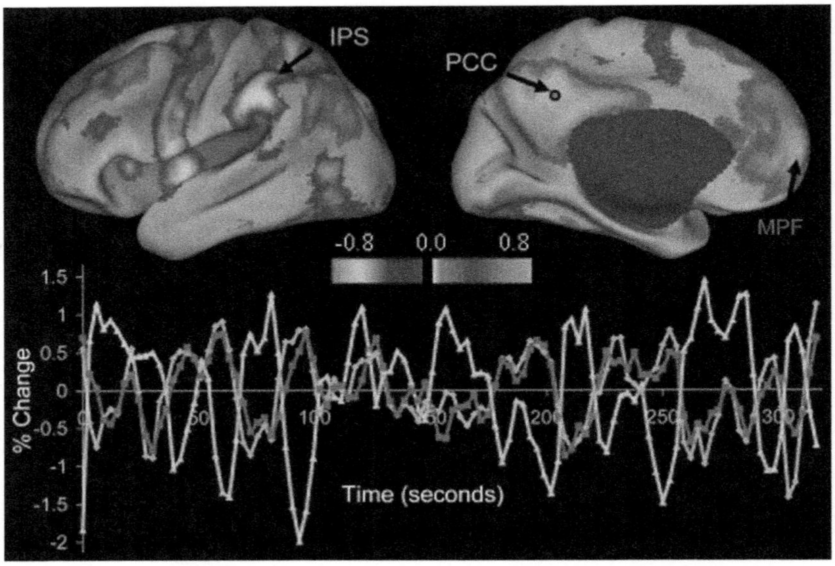

FIGURE 16.1 Deriving an rs-fMRI connectivity map. The time series from a seed region in the PCC (*arrow* points to green (gray in print versions) dot in top right panel) is shown in yellow (light gray in print versions). A computer program then extracts the time series from all other brain voxels and highlights those that are strongly positively or negatively correlated with the seed region. The medial prefrontal (MPF) time series is shown in orange and is strongly correlated with the PCC time series. The intraparietal sulcus (IPS) time series is shown in blue (dark gray in print versions) and is negatively correlated with the PCC. *Adapted from Fox, M. D., & Greicius, M. (2010). Clinical applications of resting state functional connectivity.* Frontiers in Systems Neuroscience, 4, *19.*

the primary visual and sensorimotor networks were detectable in children being scanned for clinical reasons, under light sedation with thiopental (Kiviniemi et al., 2000). Thus from the earliest days, the relationship of RSN activity and conscious processing was not straightforward (an issue addressed later under Attributes). Subsequently investigators began stepping gingerly into higher cognitive realms. Resting-state functional connectivity was shown between the left and right hippocampus (Stein et al., 2000). Then, in 2002, Hampson and colleagues published an article that was bound to draw cognitive neuroscientists into the fray. Using a language task-based fMRI paradigm to identify Broca's and Wernicke's areas in the task setting, they showed that these two canonical language centers are functionally connected at rest in the absence of a task, and are also functionally coupled to homologous language-related regions in the right hemisphere (Hampson et al., 2002). This study moved rs-fMRI out of the realm of MRI physics and firmly into the realm of cognitive neuroscience. It also helped to allay some of the early doubters (see Healthy Skepticism section, discussed later) because everyone understands that Wernicke's areas should be coupled with Broca's.

My interest in this approach had been burgeoning since my initial exposure to it at the Organization for Human Brain Mapping meeting in Brighton, England in 2001. As a behavioral neurologist, it struck me as an incredibly potent way to explore cognitive brain networks in vivo. At Stanford, Vinod Menon and I were experimenting with different approaches to extracting these networks from rs-fMRI data and casting about for a way to demonstrate the importance of this approach to cognitive neuroscience. Marc Raichle at Washington University in Saint Louis had published a compelling initial study of the default mode of brain function (Raichle et al., 2001). Working with task-activation data from a series of positron emission tomography (PET) studies, Raichle and colleagues demonstrated that the same set of regions (including the posterior cingulate cortex [PCC], lateral parietal cortex, and medial prefrontal cortex [MPFC]) tended to "deactivate" during the performance of cognitive tasks. That is, their activity was higher during the resting "control" blocks of the experiment and came down during the "experimental" blocks. This suggested to them that these regions constituted a network that was tonically active as subjects rested quietly and whose activity needed to be attenuated to perform a challenging cognitive task correctly. This finding generated a great deal of interest, in part because so little was known about the function of the PCC, and it struck us as a perfect point of departure for our foray into rs-fMRI. In our first experiment, we used a standard on—off working memory task activation paradigm to generate a typical deactivation map of regions implicated by Raichle as supporting the brain's default mode (Greicius et al., 2003). Then, working on rs-fMRI data from the same set of subjects, we used the PCC, identified in the task deactivation map as a seed region, and asked what other regions were functionally coupled with it as subjects rested quietly in the scanner for 5 min. The resting-state map overlapped substantially with the deactivation map, which lent support to Raichle's hypothesis that these regions constituted a functional network, the DMN. We made an additional observation in that study that merits consideration. In addition to examining resting-state connectivity of the PCC and typically deactivated regions, we explored the connectivity of three regions (left and right ventrolateral prefrontal cortex and right dorsolateral prefrontal cortex) that are typically activated by tasks and that were activated by our working memory task. What we found was that each of these three regions was inversely correlated with the PCC during rest. We concluded that at rest, as during a task, the DMN appears to be "maintained in a dynamic equilibrium with lateral prefrontal regions that commonly show task-related increases in activity." This finding paved the way for the lingering controversy over anticorrelations between networks and whether they are true neural phenomena or simply reflect a data-processing artifact (more subsequently under Attributes).

HEALTHY SKEPTICISM DRIVES DEEPER EXPLORATIONS

As with any new methodology, rs-fMRI was met with a mixture of heady optimism and healthy skepticism. Both perspectives were nicely summarized in a 2001 editorial that raised a number of legitimate concerns while pointing to the latent promise in this technique (Maldjian, 2001). Among the concerns was the question of whether this approach was picking up on correlated signals in blood flow responses to spontaneous neural activity or whether there were other, nonneural sources of fluctuation in the BOLD signal, such as cardiac and spinal fluid pulsations. If it were truly neural, the author suggested that it should be modulated by consciousness and depend at some level on underlying white matter structures. Over the next several years, each of these concerns and questions were addressed by a wide variety of groups beginning to adapt rs-fMRI. The careful recording and analysis of respiratory and cardiac cycle signals acquired during the acquisition of rs-fMRI scans revealed that these nonneural sources contributed somewhat to the connectivity maps (Birn et al., 2006; Shmueli et al., 2007). However, subsequent studies have demonstrated that a variety of processing techniques can be used to remove these sources of noise and that the resulting array of detectable networks are cleaner but essentially the same in terms of their spatial patterns (Chang, Cunningham, & Glover, 2009; Chang and Glover, 2009a; Glover, Li, & Ress, 2000).

The effect of different levels of consciousness on rs-fMRI networks is parametric rather than all-or-none. Several studies have shown that RSNs are readily detected in a variety of states notable for reduced consciousness. They can be detected in light sedation states, such as conscious sedation with propofol, midazolam, or thiopental (Boveroux et al., 2010; Greicius et al., 2008; Kiviniemi et al., 2000). RSNs can also be detected in the early stages of sleep using simultaneous fMRI and EEG (Horovitz et al., 2008). In monkeys, putative cognitive networks such as the DMN and dorsal attention (also called visuospatial), are detectable under light sedation (Vincent et al., 2007). Thus, there is no clear one-to-one correlation between the presence of RSNs and wakefulness. Seeing an intact DMN map, in other words, does not tell you that the subject is awake. This may not be surprising if we consider that the frequency of rs-fMRI fluctuations (<0.1 Hz) would appear to be too slow to reflect ongoing conscious processing. Although there is no binary relationship between RSNs and consciousness, several studies have shown a parametric reduction in RSN connectivity with decreasing levels of consciousness. Comparing the same subjects during wakefulness and light sedation, RSN connectivity is typically reduced (although still present) in light sedation (Boveroux et al., 2010; Greicius et al., 2009). Further deepening of sedation results in a further reduction of RSN connectivity (Boveroux et al., 2010; Vincent et al., 2007). Although detectable in light sleep, in deeper stages of sleep RSN connectivity begins to deteriorate (Horovitz et al., 2009). In the clinical realm it has been demonstrated that disorders of consciousness affect RSN connectivity in a similar parametric fashion. DMN connectivity in a locked-in patient presumed to have normal consciousness was no different from DMN connectivity in healthy control subjects. DMN connectivity in minimally conscious patients was less than normal but greater than DMN connectivity in vegetative patients (Vanhaudenhuyse et al., 2010).

Work on the relation between structural connectivity and rs-fMRI has made it clear that the latter depends in large part, but not entirely, on the former. One of the most compelling studies in this series of method-vetting articles examined rs-fMRI in a patient before and shortly after a callosotomy (Johnston et al., 2008). Several functional networks with interhemispheric connections were identified before the procedure, all of which were restricted to intrahemispheric connections after the callosotomy (Fig. 16.2). This strongly supported the notion that functional connectivity depends heavily on the underlying "wiring" or structural connectivity. Our group made a similar argument using a group of healthy control subjects and comparing their functional connections in the DMN with their structural connections (as measured with diffusion tensor imaging [DTI]) (Greicius et al., 2009). In most instances, functional connectivity between a pair of brain regions, say medial temporal lobe (MTL) and PCC, was supported by a large white matter tract, the descending cingulum bundle in this example. Certain functional connections, however, did not appear to have a clear structural connection.

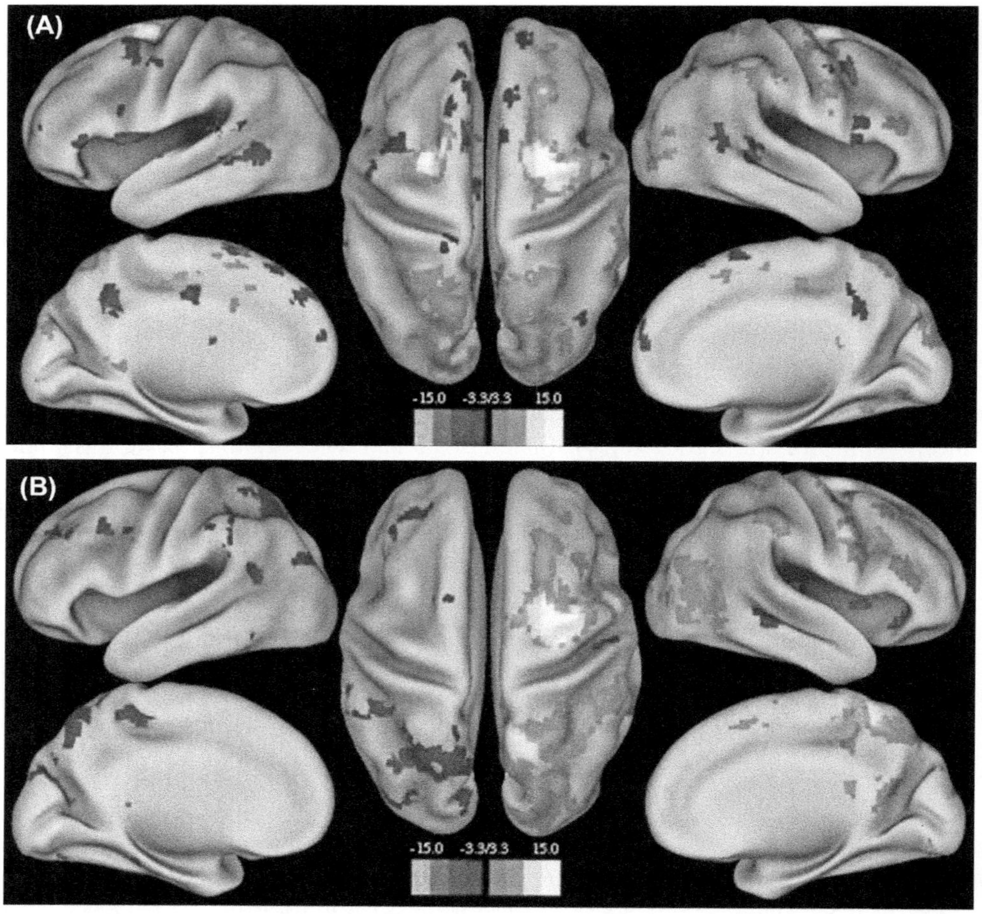

FIGURE 16.2 Functional connectivity before and after callosotomy. A seed in the right frontal eye field is used to demonstrate resting-state functional connectivity to the contralateral frontal eye field, the parietal eye fields, and area medial temporal in a single subject before corpus callosotomy for refractory seizures (A). After the procedure, the same right frontal eye field seed region yields only intrahemispheric connections (B). *Adapted from Johnston, J. M., Vaishnavi, S. N., Smyth, M. D., et al. (2008). Loss of resting interhemispheric functional connectivity after complete section of the corpus callosum.* Journal of Neuroscience, 28, *6453–6458.*

The MTL and the MPFC were functionally connected in the rs-fMRI map of the DMN (Fig. 16.3), but we could not detect a white matter tract connecting these two regions in the diffusion imaging data. This lack of correspondence between structural connectivity and functional connectivity highlights the fact that functional connectivity can reflect functional coupling across more than one synapse. In this example, both the MTL and MPFC show functional and structural connectivity to the PCC, which suggests that functional connectivity between the MTL and MPFC is mediated by the PCC. Subsequently, a number of studies examining a much larger set of regions have replicated and expanded upon this tight but imperfect association between functional connectivity and structural connectivity (Damoiseaux & Greicius, 2009; Honey et al., 2009).

ATTRIBUTES OF RESTING-STATE NETWORKS

Having weathered the storm of healthy skepticism, rs-fMRI has flourished as an imaging technique. As consensus has developed around issues such as optimal acquisition parameters, standard processing approaches, and artifact reduction techniques (particularly related to subject movement), the field has seen more robust and replicable findings. Several remarkable features of RSNs have been demonstrated, and these deserve deeper examination.

A disproportionate number of studies in the early rs-fMRI days focused exclusively on the DMN to the point that many people equated rs-fMRI with the DMN. This tendency overlooked the fact that rs-fMRI can be used to identify the complex segregation of the resting brain into not one, but closer to 15 large-scale distributed networks (Beckmann et al., 2005;

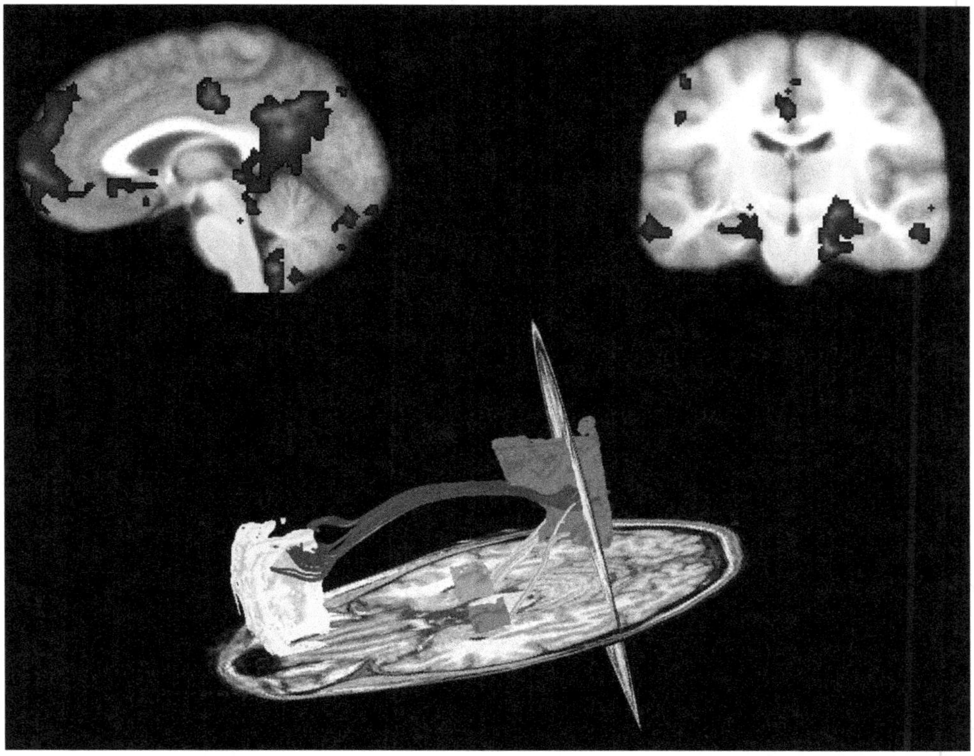

FIGURE 16.3 Functional connectivity reflects structural connectivity, but not always. The top row shows functional connectivity within the DMN. The bottom figure shows white matter tracts including the cingulum bundle in blue (dark gray in print versions), connecting PCC to MPFC, and the descending cingulum bundles in gold (light gray in print versions) connecting PCC to MTL. Although MPFC and MTL are functionally connected in the top row, there are no tracts connecting these two regions in the bottom figure. *Adapted from Greicius, M. D., Supekar, K., Menon, V., et al. (2009). Resting-state functional connectivity reflects structural connectivity in the default mode network.* Cerebral Cortex, 19, *72–78.*

Shirer et al., 2012). For the most part, if a cognitive or sensorimotor domain has a name, there appears to be an RSN that is presumed to serve that domain. Thus, distinct RSNs have been identified (reproducibly) and linked to the following functions: vision, hearing, sensation, movement, episodic memory, executive function, visuospatial attention, and emotional salience detection (Biswal et al., 1995; Damoiseaux et al., 2006; Fox et al., 2005; Greicius et al., 2003; Kiviniemi et al., 2003; Seeley, Menon, et al., 2007; Vincent et al., 2007). Assigning functional roles to these RSNs is inherently challenging because they are detected when subjects are not performing a specific functional task. For the most part, investigators have assumed a given function for an RSN based on what is known about the functional role of one or more of the regions in the network. The assignment of function is more concrete when an RSN is generated from a seed region that was first defined in a functional task. The language network described earlier is a nice example of this approach. We assume this RSN is related to language because it can be generated by seeding a region that was activated by a language task (Hampson et al., 2002). One study from the group at Oxford pursued a spatial correlation approach to match RSNs with task activation maps taken from a large repository of functional studies (Smith et al., 2009). Unfortunately, the field has not decided on a uniform nomenclature, and so the same network may be referred to differently by different laboratories. As an example, our laboratory uses the term "visuospatial network" to describe an RSN to which another group refers as the "dorsal attention network."

In addition to the naming issue, there is no consensus in the field regarding how many networks there are. Statistical approaches such as independent component analysis have shown that a given rs-fMRI dataset can generate 10 networks but can be further subdivided to generate 15–20 (or more) reasonable-looking RSNs (Abou-Elseoud et al., 2010). That is, if the dataset is decomposed into 20 components, 10 may be noise components related to scanner issues, subject motion, etc., and 10 may correspond to RSNs including the DMN. If the same dataset is then decomposed into 40 components, 25 may be noise-related and 15 may correspond to RSNs. A careful comparison of the thinly sliced data (40 components) and the thickly sliced data (20 components) might show that the DMN generated in the 20-component decomposition has been divided into two components in the 40-component decomposition. Is there one DMN or are there two (a ventral and dorsal DMN)? Is there a single executive control network or are there two (left and right hemisphere executive control networks)?

We do not have a reference standard to help determine the "optimal" number of networks, and so we are left with some uncertainty here. If one overlooks the frustrating aspect of this uncertainty, it has the potential to reveal which are the strongest subnetworks within networks. Noting, for example, that a given network splits anterior—posterior rather than left to right across the hemispheres suggests that interhemispheric connectivity is stronger than intrahemispheric connectivity in that network. Formal, quantitative approaches to subnetwork definition will ultimately provide a more detailed and nuanced understanding of functional brain networks (Cole, Smith, & Beckmann, 2010).

The ability of rs-fMRI to establish functional distinctions between neighboring regions has been used to good effect in allowing researchers to subdivide complex gray matter structures into their distinct functional components. This was demonstrated initially in studies of the thalamus in which rs-fMRI, in conjunction at times with DTI, was used to identify several distinct functional parcels of the thalamus based on their distinct connectivity to cortical regions (Zhang et al., 2008, 2010). Subsequent work has also shown that rs-fMRI has the capacity to delineate functional clusters within the amygdala that coarsely correspond to known amygdalar nuclei (Etkin et al., 2009; Roy et al., 2009). Perhaps the most compelling such application of rs-fMRI has been in the functional segregation of the cerebellum. This region is frequently oversimplified and represented as having a role mainly in refining motor movements. Increasingly, however, there is a growing consensus, based on painstaking work in patients, that the cerebellum is important not only for movement but also for cognition and emotion (Schmahmann, 2004). Three articles on rs-fMRI papers in close succession (and largely replicating each other's findings) lent considerable support to the cognitive role of the cerebellum by demonstrating that distinct cerebellar clusters were functionally coupled to cortical networks related to executive function, memory, and emotional salience detection (Fig. 16.4) (Habas et al., 2009; Krienen & Buckner, 2009; O'Reilly et al., 2010). As rs-fMRI acquisition and analysis techniques continue to improve, we can anticipate increasingly fine-grained functional neuroanatomy maps that begin to approach the reference standard of functional maps based on tracer studies in animals and postmortem cytoarchitectonic work in humans.

Before rs-fMRI can be understood in a clinical setting, the relationship between RSN connectivity and behavior in healthy controls should be established. The field has seen a number of studies in which functional connectivity in a given RSN is correlated with scores on cognitive tests or behavioral rating scales. Our group demonstrated that connectivity in

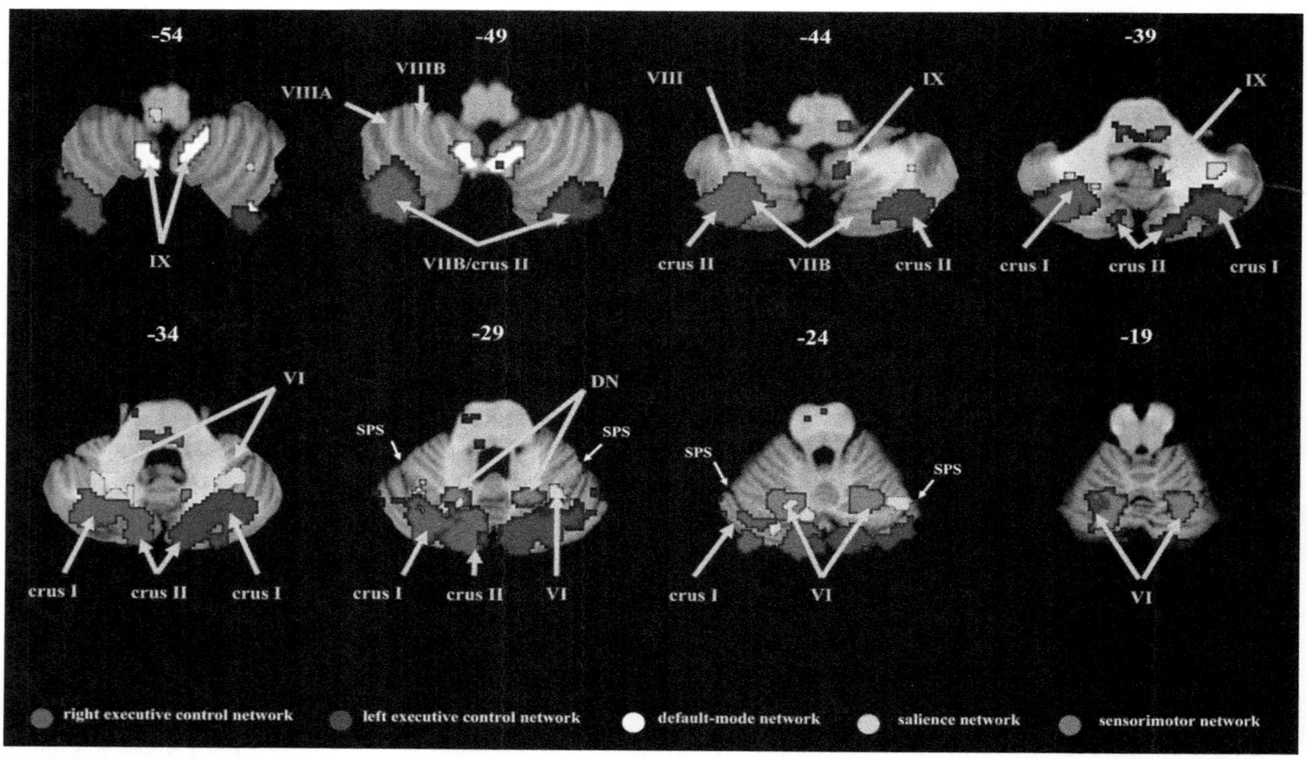

FIGURE 16.4 Functional parcellation of the cerebellum using rs-fMRI. The cerebellar clusters that participate in five different cortical resting-state networks are shown in five different colors. Cortical network labels are at the bottom. *Adapted from Habas, C., Kamdar, N., Nguyen, D., et al. (2009). Distinct cerebellar contributions to intrinsic connectivity networks.* Journal of Neuroscience, *29, 8586–8594.*

the executive control network (ECN) was correlated with performance in the Trails B set-shifting task (Seeley, Menon, et al., 2007). In the same subjects, an anxiety measure was correlated with connectivity in the salience network. These network correlations showed a compelling double dissociation in that anxiety scores were not correlated with ECN connectivity and Trails B performance was not correlated with salience network connectivity. Although not a perfect replication, a large study in China correlated full-scale IQ scores to ECN connectivity (Song et al., 2008). In children with a mean age 6 years, connectivity within the right side of the ECN is correlated with nonverbal IQ performance (Langeslag et al., 2013). Verbal IQ was not assessed in this young age group, but it is anatomically satisfying to note that left ECN connectivity did not correlate significantly with nonverbal IQ in this study. In addition, the salience network (which showed no connectivity with Trails B performance in our study) was not correlated with cognitive performance in these young children. In addition to executive function and intelligence correlating with ECN connectivity, there are convergent data linking DMN connectivity to episodic memory performance. The group at Harvard showed that DMN connectivity is correlated with episodic memory performance in older adults (Wang, LaViolette, et al., 2010). In young adults, the same group showed that interhemispheric connectivity between the MTLs is correlated with free recall of recently learned information (and in an elegant control experiment demonstrating the anatomic specificity of this effect, they showed that free recall was not correlated with interhemispheric motor cortex connectivity) (Wang, Negreira, et al., 2010). A compelling extension of this work in patients with MTL epilepsy showed that greater DMN connectivity to the epileptogenic hippocampus predicted greater memory deficits after therapeutic resection of that hippocampus (McCormick et al., 2013).

Many of the early strikes against rs-fMRI came from neurophysiologists. This field has always endured the spatial myopia of unit recordings for the high temporal resolution this approach affords. Neurophysiologists were already unimpressed with standard task activation fMRI, which requires about 1–2 s to obtain a brain volume and which relies on a blood flow response that lags about 6 s behind neuronal activity. They were even less impressed with rs-fMRI owing to the extremely low frequency (<0.1 Hz) of the spontaneous BOLD signal fluctuations corresponding to a cycle of 30 s or more. Over the past decade, however, even these lingering skeptics have begun to warm to the idea that these low-frequency BOLD signal fluctuations have neurophysiological counterparts. One of the first articles to bridge the gap came from neurophysiologists exploring the potential of rs-fMRI. Leopold, Murayama, and Logothetis (2003) demonstrated that the power of high gamma activity in local field potentials (LFPs) exhibited low-frequency (<0.1-Hz) fluctuations and that these high–gamma power fluctuations were strongly correlated across "distant" (meaning about 1 cm away) electrodes in monkey visual cortex. Subsequent work by this group using simultaneous rs-fMRI and electrophysiological recordings showed that gamma power fluctuations in LFPs were correlated with BOLD signal fluctuations (Shmuel & Leopold, 2008). Working with epilepsy patients, a team in Israel demonstrated that low-frequency gamma power fluctuations from subdural electrodes were correlated between left and right auditory cortices (Nir et al., 2008). As it turns out, these infraslow oscillations (ISOs) were not new to the field of neurophysiology but had merely been rediscovered by the broader field. In fact, ISOs have been the focus of a small, esoteric subfield of neurophysiology that dates to the 1950s (Aladjalova, 1957). Other work further shored up support for the argument that the low-frequency BOLD signal fluctuations underlying rs-fMRI connectivity reflect cortical ISOs (Hiltunen et al., 2014). The study of ISOs and rs-fMRI fluctuations converged, most intriguingly, in a number of studies suggesting that these low-frequency fluctuations in cortical excitability have a large role in event-related signals and behavioral responses (Boly et al., 2007; Fox et al., 2007; Hesselmann et al., 2008; Monto et al., 2008). In other words, a cortical regions response to a stimulus will vary depending on when that stimulus arrives in relation to the spontaneous low-frequency fluctuations in local excitability. Similarly, the likelihood that a borderline detectable stimulus will or will not be registered appears to depend on where the relevant cortical region is in its low-frequency cycle. The implications of these effects have yet to be fully realized and should make for stunning advances in our ability to measure brain signals and to understand, if not influence, perception, cognition, and emotion.

In terms of its translational potential, the capacity to help equate animal model findings with human disease findings, rs-fMRI has some obvious advantages over standard task activation fMRI (Hoyer et al., 2014). The latter can be challenging to institute in animals even in basic efforts to activate primary sensory or motor cortices; activation of higher cognitive networks is more challenging still in the scanner environment. By contrast, rs-fMRI has already been used in several species and allows for the interrogation of multiple networks, both primary sensory and higher-order cognitive, from the same 10 min of data. To date, rs-fMRI has been applied in mice, rats, pigeons, marmosets, and macaques (Belcher et al., 2013; De Groof et al., 2013; Jonckers et al., 2011; Lu et al., 2007; Vincent et al., 2007). Most of these studies, particularly the initial ones, were performed with anesthesia; but in rats, pigeons, marmosets, and macaques rs-fMRI has been acquired in awake animals as well (Belcher et al., 2013; De Groof et al., 2013; Hadj-Bouziane et al., 2014; Liang, Liu, & Zhang, 2014). Groups have begun to investigate the use of rs-fMRI as a tool in animal models of disease

(Grandjean et al., 2014; Zerbi et al., 2014). As artifact reduction methods continue to improve, the ability of rs-fMRI to translate findings from animal models of disease to the disease itself in humans has considerable potential to speed drug development.

Whereas initially the focus was on intranetwork connectivity, it has always been apparent that these RSNs, although defined by their segregation from one another, cannot operate in isolation. Rather, they must constantly interact to allow, for example in a verbal learning task, the transfer of newly heard information from primary auditory cortex to the language network, and on to the episodic memory network. Several elegant studies using graph theory approaches have highlighted regions that serve as hubs or switches between two or more networks (van den Heuvel & Sporns, 2011). In addition to working together, at times some networks most likely need to act strictly independent of one another to optimize performance. The original description of the DMN suggested that in the task setting there was a sort of antagonism between brain regions activated by the task and DMN regions deactivated by the task. In our initial rs-fMRI article on the DMN, we showed that even at rest, this "dynamic equilibrium" appears to hold true in that DMN regions tend to be negatively correlated with regions in the salience network and executive control networks (Greicius et al., 2003). Subsequently, two other articles replicated this finding and suggested that at rest, as in the task setting, when DMN regions are down, executive control, salience, and attention regions are up, and vice versa (Fox et al., 2005; Fransson, 2005). Intuitively it seems logical that regions that are anticorrelated in the setting of a task would also be anticorrelated at rest, but there is an unresolved controversy surrounding anticorrelations in resting-state data. To remove noise from the images, the global signal (across all voxels in the brain) is measured with each 2-s brain volume acquisition. This global signal is then regressed out of every individual voxel's signal. This step mathematically mandates a shift in correlation values such that some near-zero correlations between voxels get shifted to become negative correlations (Fox et al., 2009; Murphy et al., 2009). Some studies have gotten around the global signal confound (by carefully regressing out physiological artifacts owing to cardiac and respiratory cycle fluctuations) and have still found prominent anticorrelations between the DMN and the salience network (Chang and Glover, 2009b). Analysis of electrocorticography data has lent some middling support to this issue by showing rare instances of anticorrelations in low-frequency fluctuations of high gamma power (Keller et al., 2013). As described in more detail later (see Default-Mode and Salience Networks), there is also evidence from neurodegenerative diseases to support a dynamic equilibrium between the DMN and the salience network (Zhou et al., 2010). For the time being, however, the existence of anticorrelations between RSNs remains enticing but speculative.

Another advance in rs-fMRI also pertains to the temporal rather than spatial properties of RSNs. The vast majority of early rs-fMRI studies estimated the functional connectivity between two regions by taking the mean correlation of their two time series. A groundbreaking study by Catie Chang showed that the degree to which two regions are correlated varies over the typical 8-min resting-state scan (Chang & Glover, 2010). Thus, whereas the mean correlation between two regions may be $r = 0.7$, there can be shorter runs of, say, 1 or 2 min in which the correlation is considerably higher ($r = 0.9$) or lower ($r = 0.5$). The dynamic nature of these functional couplings is lost when the mean over the entire rs-fMRI run is used. The article by Chang (cited more than 300 times to date) has resulted in the subfield of dynamic functional connectivity. The hope is that the degree to which functional connectivity varies over time may have important implications related to normal cognition (Kucyi & Davis, 2014) as well as cognitive dysfunction in neuropsychiatric disorders (Damaraju et al., 2014; Jones et al., 2012). For the time being, however, these early explorations of dynamic functional connectivity should be read with caution as this promising new subfield of rs-fMRI evolves toward a consensus on important methodological issues (Hutchison et al., 2013; Leonardi & Van De Ville, 2014).

DEFAULT-MODE AND SALIENCE NETWORKS

As a prelude to the section on clinical applications presented subsequently, it will be helpful to take a closer look at two specific RSNs: the DMN and the salience network. As noted earlier, the DMN was initially identified as a set of brain regions that consistently deactivate during the performance of a wide variety of cognitively demanding tasks (Raichle et al., 2001). However, there are a few exceptions to this rule. Tasks that require retrieval of episodic memories, for example, tend to activate the DMN (Cabeza et al., 2002; Fujii et al., 2002; Maddock, Garrett, & Buonocore, 2001; Maguire & Mummery, 1999). The MPFC and PCC portions of the DMN, in particular, tend to be activated by tasks that require self-referential reflection: for example, deciding whether a particular adjective pertains to oneself (Gusnard et al., 2001; Johnson et al., 2002). Insights into the DMN from lesion studies are few and far between. The PCC, in particular, appears to be relatively protected in that it is not in a single vascular distribution (and so is not commonly infarcted) and is deeply seated so as not to be prone to trauma. The few cases of damage to this region that have been described in the literature result in "retrosplenial amnesia," which again points to the importance of this network in episodic memory encoding and retrieval (Valenstein et al., 1987; Yasuda et al., 1997). Tract-tracing studies in macaques, as well as diffusion tensor studies in humans, have demonstrated that the PCC is highly connected to MTL structures including the entorhinal cortex and hippocampus (Greicius et al., 2009; Kobayashi & Amaral, 2003, 2007). Therefore, whereas the preponderance of evidence supports the hypothesis that the DMN is mainly involved in episodic memory function, and to a lesser degree self-referential processing, additional roles related to visuospatial processing have been suggested (Gusnard & Raichle, 2001). Several detailed reviews of the PCC and nearby medial parietal regions such as the precuneus can be recommended for a more in-depth exploration of this fascinating brain region (Cavanna & Trimble, 2006; Gusnard & Raichle, 2001; Leech & Sharp, 2014).

Key regions in the salience network include the bilateral dorsal anterior cingulate cortex (dACC), the frontoinsular cortex, and the frontopolar cortex. These cortical regions are linked to a number of critical subcortical regions in the amygdala, hypothalamus, thalamus, and brain stem (Seeley, Menon, et al., 2007; Zhou & Seeley, 2014). Although our group was the first to focus on this network in rs-fMRI data (Seeley, Menon, et al., 2007), it had been fairly well-characterized in task activation studies. To oversimplify, tasks that increase sympathetic tone tend to increase activation in salience network regions. This has been shown to best effect in the work of Critchley and colleagues. In a series of articles, this group employed a number of different measures of sympathetic tone (pupil dilatation, heart rate variability, blood pressure, etc.) in the setting of functional imaging tasks to show that activation in core salience network regions tracks with autonomic nervous system changes (Critchley et al., 2000, 2003, 2005). The salience network would appear to be agnostic to the specific stimuli or tasks at hand, provided they affect sympathetic tone. Thus, salience network activation can be seen across various challenging cognitive tasks (Duncan & Owen, 2000), in studies of pain perception (Ingvar, 1999), and in studies of sexual arousal (Arnow et al., 2002). The use of autonomic regressors by Critchley and colleagues helped point to the common denominator across these various and sundry tasks. Work by Seeley and colleagues expanded considerably on these prior studies, emphasizing the role of this network in gauging the valence of internal and external events and modulating the body's autonomic system accordingly (Seeley, Allman, et al., 2007; Seeley, Menon, et al., 2007). Seeley also made a compelling case that salience network function is linked tightly to the role of an unusual neuronal subtype, the Von Economo neuron, which is found almost exclusively in network hubs such as frontoinsular cortex and dACC (Kim et al., 2012; Seeley et al., 2006, 2012).

APPLICATIONS TO BRAIN DISEASE

The list of brain diseases that putatively affect RSNs is nearly as long as the list of brain diseases (Greicius, 2008). One of the main advantages of rs-fMRI in clinical populations, as opposed to standard task activation fMRI, is that it is much easier to acquire in patients. Task-based fMRI usually requires patients to be able to perform a particular task correctly in the anxiety-provoking setting of an MRI tunnel. It can be challenging to compare patient task activation maps with control task activation maps when performance on the task differs (as it commonly does). This problem is obviated in rs-fMRI because there no task is performed. This singular advantage has led to the explosion of rs-fMRI studies in patients with neuropsychiatric disorders. Unfortunately, as is often the case in the application of novel approaches to brain disorders, efforts have been focused more on novel findings than on the firm replication of findings. In addition, many studies of rs-fMRI were applied before the field had developed some consensus on optimized parameters for the acquisition and analysis of rs-fMRI data. For example, it has become clear that basic corrections for subject movement may not be sufficient, leaving residual movement-related noise in data that can result in artifactual differences in "connectivity" between a patient group and control subjects (Power et al., 2012). Caveat emptor, therefore, when diving into the rs-fMRI clinical literature. I will focus here on two disorders which, in my less than objective view, have generated the most replicable rs-fMRI results to date: AD and behavioral variant FTD (bvFTD).

In our initial explorations of the DMN in young control subjects, it gradually dawned on us that the posterior aspects of this network bore a strong resemblance to brain regions that typically show reduced glucose metabolism in patients with AD who were studied with fluorodeoxyglucose (FDG) PET scans. Numerous groups had shown that compared with age-matched older control subjects, patients with AD had reductions in FDG PET signals in the PCC and the lateral temporal and parietal lobes (Minoshima et al., 1995, 1997; Montez et al., 2009; Reiman et al., 1996). The next obvious step was to examine DMN connectivity in patients with AD. In so doing, we found that a number of regions in the network including the PCC, the lateral parietal cortex, and the MTL showed reduced connectivity within the DMN of patients with AD compared with control subjects (Greicius et al., 2004). This finding has been replicated many times since (Binnewijzend et al., 2012; Gili et al., 2011; Koch et al., 2012; Zhou et al., 2010) and moved forward in time along the clinical continuum. Patients with mild cognitive impairment (MCI) have reduced DMN connectivity (Gili et al., 2011; Koch et al., 2012; Petrella et al., 2011; Sorg et al., 2007; Wang et al., 2013). Moving further forward in time, DMN connectivity has been assessed in well-screened, cognitively normal older control subjects with PET scan evidence for amyloid plaque deposition. Two studies have shown that amyloid-positive older control subjects have reduced DMN connectivity compared with matched amyloid-negative older control subjects (Hedden et al., 2009; Sheline, Raichle, et al., 2010). Asymptomatic carriers of autosomal dominant mutations that cause AD show reduced DMN connectivity. The data for asymptomatic carriers of the APOE4 AD risk gene are slightly mixed on first glance; some studies show reduced DMN connectivity in carriers and others do not (Machulda et al., 2011; Sheline, Morris, et al., 2010; Trachtenberg et al., 2012; Westlye et al., 2011). Our take on these discrepant findings is that they neglected the interaction between APOE and sex, which results in more AD-like effects of APOE4 in women. We examined this APOE × sex interaction with rs-fMRI and found that healthy older women carrying an APOE4 allele show reduced DMN connectivity in a large swath of the PCC, whereas male APOE4 carriers showed little to no difference in DMN connectivity compared with male noncarriers (Damoiseaux, Seeley, et al., 2012). This same interaction (greater AD-like changes in APOE4--carrying women) has been replicated by another group (Heise et al., 2014). DMN connectivity may also have some predictive value. Patients with MCI who subsequently convert to AD have reduced DMN connectivity compared with patients with MCI who do not convert to AD (Petrella et al., 2011). In a related study, patients with MCI who were clinically stable over time had greater DMN connectivity than did patients with AD, whereas patients with MCI who subsequently converted to AD had DMN connectivity that was indistinguishable from that of patients with AD (Binnewijzend et al., 2012). Finally, longitudinal studies are few for the time being, but the first of these has shown that DMN connectivity continues to decline as patients with AD progress into more severe stages of dementia (Damoiseaux, Prater, et al., 2012). Pseudo-longitudinal (cross-sectional) studies support this in showing a stepwise decline in DMN connectivity from healthy aging to MCI to mild AD using the clinical dementia rating as the marker of severity (Brier et al., 2012; Sohn et al., 2014). DMN connectivity in AD appears to be sensitive to preclinical disease, to offer potential value in predicting conversion from MCI to AD, and to track clinical deterioration over time.

If AD targets the DMN, bvFTD targets the salience network. Seeley et al. (2009) were the first to point to the similarities between atrophy maps in FTD and the salience network. Subsequent work by the same group showed that patients with bvFTD show reduced salience network connectivity even after correcting for the loss of gray matter in these regions (Zhou et al., 2010). This study also lends some support to the hypothesis that the DMN and salience network are anticorrelated. As mentioned previously, earlier work in control subjects had shown negative correlations between the PCC and salience network regions, but these studies were confounded by the global signal regression preprocessing step (Fox et al., 2005; Fransson, 2005; Greicius et al., 2003). In examining patients with bvFTD whose salience network was deteriorating, Seeley and colleagues found that their DMN actually showed increased connectivity compared with control subjects. Conversely, in a group of patients with AD whose DMN predictably showed reduced connectivity, they found that salience network connectivity was increased compared with control subjects (Zhou et al., 2010). In other words, for each disorder, as the disease-targeted network begins to deteriorate and show reduced connectivity, the anticorrelated network appears to be disinhibited and shows increased connectivity. Importantly, this finding is not compromised by the global signal regression step, because only positive correlations are being compared between patient groups and control subjects. The main finding of reduced salience network connectivity in patients with bvFTD was replicated by the Seeley group (Lee et al., 2014) and other groups (Borroni et al., 2012; Farb et al., 2013; Filippi et al., 2013; Whitwell et al., 2011). The finding of increased DMN connectivity in patients with bvFTD was also replicated by the Seeley group (Lee et al., 2014) and other groups (Borroni et al., 2012; Farb et al., 2013; Whitwell et al., 2011). It was also shown that APOE4 carriers who are at risk for AD and show reduced DMN connectivity also have increased salience network connectivity (Machulda et al., 2011). As in the AD literature, FTD research groups have used autosomal dominant mutation carriers to determine whether salience network connectivity is detectable in asymptomatic mutation carriers. Presymptomatic carriers of the MAPT gene mutation that causes FTD show reduced salience network connectivity (Dopper et al., 2014). By contrast, presymptomatic carriers of the GRN mutation that causes FTD have increased connectivity in the salience network (Borroni et al., 2012). It should be pointed out, however, that the MAPT report had four times as many asymptomatic carriers as the GRN article, and so is probably the more statistically robust. Although there are not yet longitudinal studies of rs-fMRI in FTD, several pseudo-longitudinal studies have shown that salience network connectivity is lower in sicker patients with FTD using either the clinical dementia rating (Zhou et al., 2010) or the neuropsychiatric inventory (Lee et al., 2014) as a measure of severity.

Thus for both AD and FTD there is a reasonable and encouraging degree of replicability regarding the key findings: The DMN is affected early in AD, the salience network is affected early in FTD, and as the salience network deteriorates in FTD the DMN shows increased connectivity (and vice versa). This predilection for a specific neurodegenerative disease to target a specific network may not be restricted to AD and FTD. In fact, this may be a unifying theme in neurodegenerative disease such that clinically distinct neurodegenerative syndromes each march along distinct brain networks (Seeley et al., 2009). Even clinical variants of the same disease may target distinct networks. For example, whereas the classic memory presentation of AD targets the DMN, atypical language and visuospatial variants of AD appear, unsurprisingly perhaps, to target networks related to these functions (Lehmann, Ghosh, et al., 2013; Lehmann, Madison, et al., 2013). This unifying hypothesis at the systems level converges nicely with ongoing work at the synapse level, in which animal studies suggested that a number of different proteins (including tau and alpha-synuclein) implicated in neurodegenerative diseases have the capacity to spread trans-synaptically and so might be expected to progress across a network (Holmes & Diamond, 2014).

Despite the focus here on neurodegenerative disease, I am optimistic that rs-fMRI will help elucidate other neuropsychiatric disorders as well. Schizophrenia and autism, for example, have long been conceptualized as disorders of connectivity (Friston & Frith, 1995; Frith, 2004). To date, however, owing perhaps to heterogeneity of patient samples or to movement artifacts, differential processing across sites, or any number of other variables, it has been challenging to come up with a replicable set of network findings in either of these two disorders. Epilepsy is another disorder likely to have network-related properties, and some work suggests that rs-fMRI may prove useful in important clinical roles such as seizure focus localization (Stufflebeam et al., 2011). Other challenging disorders such as chronic pain, addiction, and obsessive-compulsive disorder may also ultimately offer up some secrets to rs-fMRI analyses. Conversely our understanding of certain disorders that tend to affect the brain in a more willy-nilly fashion, such as multiple sclerosis or brain tumors, is unlikely to advance with the application of rs-fMRI. Then there are diagnostic terms such as attention-deficit disorder and chronic fatigue syndrome, which in my view are unlikely to be brain disorders at all but which are regrettably being subjected to rs-fMRI studies. It is hoped that as the field matures and as editors and reviewers and readers of journal articles develop more savvy about rs-fMRI, the wheat will be separated from the chaff in terms of where this powerful technique can help advance knowledge and where it cannot.

FUTURE DIRECTIONS

A number of compelling questions about RSNs loom large; it is hoped that they will be answered over the next 5–10 years. What is the source of the ISOs that underlie RSNs? What function do RSNs serve? What are the cellular and molecular mechanisms that support RSNs? It will be satisfying to answer these questions from a basic neuroscience standpoint but it should also help clinical investigators to apply rs-fMRI more judiciously and effectively. The answers will likely come from a combination of animal and human studies. Animal research offers the obvious advantage of having more flexibility in terms of perturbing RSNs with genetic and pharmacologic manipulations. Optogenetic fMRI (Lee et al., 2010), for example, which allows the rapid and discrete manipulation of a specific set of neuronal or glial cells, should prove particularly useful in understanding basic mechanisms supporting RSNs. That said, human rs-fMRI studies can certainly involve pharmacologic perturbation (Cole et al., 2013; Khalili-Mahani et al., 2013), and there is a growing number of good-sized, shareable datasets with rs-fMRI and genotype data that can be downloaded and analyzed (Satterthwaite et al., 2014; Schumann et al., 2010; Van Essen et al., 2012; Weiner et al., 2010). The application of rs-fMRI to clinical questions will benefit considerably from these inevitable advances. Whether rs-fMRI will ever be clinically relevant at the single-patient level remains an open question. Certain clinical applications, gauging prognosis in coma, for example (Norton et al., 2012; Vanhaudenhuyse et al., 2010), seem more tractable than others, but it is certainly conceivable that quantitative data from rs-fMRI scans may be brought to bear one day on patients being seen for "possible schizophrenia" or to "rule out AD."

REFERENCES

Abou-Elseoud, A., Starck, T., Remes, J., et al. (2010). The effect of model order selection in group PICA. *Human Brain Mapping, 31*, 1207–1216.

Aladjalova, N. A. (1957). Infra-slow rhythmic oscillations of the steady potential of the cerebral cortex. *Nature, 179*, 957–959.

Arnow, B. A., Desmond, J. E., Banner, L. L., et al. (2002). Brain activation and sexual arousal in healthy, heterosexual males. *Brain, 125*, 1014–1023.

Beckmann, C. F., DeLuca, M., Devlin, J. T., et al. (2005). Investigations into resting-state connectivity using independent component analysis. *Philosophical Transactions of the Royal Society of London B Biological Sciences, 360*, 1001–1013.

Belcher, A. M., Yen, C. C., Stepp, H., et al. (2013). Large-scale brain networks in the awake, truly resting marmoset monkey. *Journal of Neuroscience, 33*, 16796–16804.

Binnewijzend, M. A., Schoonheim, M. M., Sanz-Arigita, E., et al. (2012). Resting-state fMRI changes in Alzheimer's disease and mild cognitive impairment. *Neurobiology of Aging, 33*, 2018–2028.

Birn, R. M., Diamond, J. B., Smith, M. A., et al. (2006). Separating respiratory-variation-related fluctuations from neuronal-activity-related fluctuations in fMRI. *NeuroImage, 31*, 1536–1548.

Biswal, B., Yetkin, F. Z., Haughton, V. M., et al. (1995). Functional connectivity in the motor cortex of resting human brain using echo-planar MRI. *Magnetic Resonance in Medicine, 34*, 537–541.

Boly, M., Balteau, E., Schnakers, C., et al. (2007). Baseline brain activity fluctuations predict somatosensory perception in humans. *Proceedings of the National Academy of Sciences of the United States of America, 104*, 12187–12192.

Borroni, B., Alberici, A., Cercignani, M., et al. (2012). Granulin mutation drives brain damage and reorganization from preclinical to symptomatic FTLD. *Neurobiology of Aging, 33*, 2506–2520.

Boveroux, P., Vanhaudenhuyse, A., Bruno, M. A., et al. (2010). Breakdown of within- and between-network resting state functional magnetic resonance imaging connectivity during propofol-induced loss of consciousness. *Anesthesiology, 113*, 1038–1053.

Brier, M. R., Thomas, J. B., Snyder, A. Z., et al. (2012). Loss of intranetwork and internetwork resting state functional connections with Alzheimer's disease progression. *Journal of Neuroscience, 32*, 8890–8899.

Cabeza, R., Dolcos, F., Graham, R., et al. (2002). Similarities and differences in the neural correlates of episodic memory retrieval and working memory. *NeuroImage, 16*, 317–330.

Cavanna, A. E., & Trimble, M. R. (2006). The precuneus: a review of its functional anatomy and behavioural correlates. *Brain, 129*, 564–583.

Chang, C., Cunningham, J. P., & Glover, G. H. (2009). Influence of heart rate on the BOLD signal: the cardiac response function. *NeuroImage, 44*, 857–869.

Chang, C., & Glover, G. H. (2009a). Relationship between respiration, end-tidal CO_2, and BOLD signals in resting-state fMRI. *NeuroImage, 47*, 1381–1393.

Chang, C., & Glover, G. H. (2009b). Effects of model-based physiological noise correction on default mode network anti-correlations and correlations. *NeuroImage, 47*, 1448–1459.

Chang, C., & Glover, G. H. (2010). Time-frequency dynamics of resting-state brain connectivity measured with fMRI. *NeuroImage, 50*, 81–98.

Cole, D. M., Beckmann, C. F., Oei, N. Y., et al. (2013). Differential and distributed effects of dopamine neuromodulations on resting-state network connectivity. *NeuroImage, 78*, 59–67.

Cole, D. M., Smith, S. M., & Beckmann, C. F. (2010). Advances and pitfalls in the analysis and interpretation of resting-state FMRI data. *Frontiers in Systems Neuroscience, 4*, 8.

Critchley, H. D., Corfield, D. R., Chandler, M. P., et al. (2000). Cerebral correlates of autonomic cardiovascular arousal: a functional neuroimaging investigation in humans. *Journal of Physiology, 523*(Pt 1), 259–270.

Critchley, H. D., Mathias, C. J., Josephs, O., et al. (2003). Human cingulate cortex and autonomic control: converging neuroimaging and clinical evidence. *Brain, 126*, 2139–2152.

Critchley, H. D., Tang, J., Glaser, D., et al. (2005). Anterior cingulate activity during error and autonomic response. *NeuroImage, 27*, 885–895.

Damaraju, E., Allen, E. A., Belger, A., et al. (2014). Dynamic functional connectivity analysis reveals transient states of dysconnectivity in schizophrenia. *NeuroImage: Clinical, 5*, 298–308.

Damoiseaux, J. S., & Greicius, M. D. (2009). Greater than the sum of its parts: a review of studies combining structural connectivity and resting-state functional connectivity. *Brain Structure & Function, 213*, 525–533.

Damoiseaux, J. S., Prater, K. E., Miller, B. L., et al. (2012). Functional connectivity tracks clinical deterioration in Alzheimer's disease. *Neurobiology of Aging, 33*(828), e19–e30.

Damoiseaux, J. S., Rombouts, S. A., Barkhof, F., et al. (2006). Consistent resting-state networks across healthy subjects. *Proceedings of the National Academy of Sciences of the United States of America, 103*, 13848–13853.

Damoiseaux, J. S., Seeley, W. W., Zhou, J., et al. (2012). Gender modulates the APOE epsilon4 effect in healthy older adults: convergent evidence from functional brain connectivity and spinal fluid tau levels. *Journal of Neuroscience, 32*, 8254–8262.

De Groof, G., Jonckers, E., Gunturkun, O., et al. (2013). Functional MRI and functional connectivity of the visual system of awake pigeons. *Behavioural Brain Research, 239*, 43–50.

Dopper, E. G., Rombouts, S. A., Jiskoot, L. C., et al. (2014). Structural and functional brain connectivity in presymptomatic familial frontotemporal dementia. *Neurology, 83*, e19–e26.

Duncan, J., & Owen, A. M. (2000). Common regions of the human frontal lobe recruited by diverse cognitive demands. *Trends in Neurosciences, 23*, 475–483.

Etkin, A., Prater, K. E., Schatzberg, A. F., et al. (2009). Disrupted amygdalar subregion functional connectivity and evidence of a compensatory network in generalized anxiety disorder. *Archives of General Psychiatry, 66*, 1361–1372.

Farb, N. A., Grady, C. L., Strother, S., et al. (2013). Abnormal network connectivity in frontotemporal dementia: evidence for prefrontal isolation. *Cortex, 49*, 1856–1873.

Filippi, M., Agosta, F., Scola, E., et al. (2013). Functional network connectivity in the behavioral variant of frontotemporal dementia. *Cortex, 49*, 2389–2401.

Fox, M. D., & Greicius, M. (2010). Clinical applications of resting state functional connectivity. *Frontiers in Systems Neuroscience, 4*, 19.

Fox, M. D., Snyder, A. Z., Vincent, J. L., et al. (2005). The human brain is intrinsically organized into dynamic, anticorrelated functional networks. *Proceedings of the National Academy of Sciences of the United States of America, 102*, 9673–9678.

Fox, M. D., Snyder, A. Z., Vincent, J. L., et al. (2007). Intrinsic fluctuations within cortical systems account for intertrial variability in human behavior. *Neuron, 56*, 171–184.

Fox, M. D., Zhang, D., Snyder, A. Z., et al. (2009). The global signal and observed anticorrelated resting state brain networks. *Journal of Neurophysiology, 101*(6), 3270–3283.

Fransson, P. (2005). Spontaneous low-frequency BOLD signal fluctuations: an fMRI investigation of the resting-state default mode of brain function hypothesis. *Human Brain Mapping, 26*, 15–29.

Friston, K. J., & Frith, C. D. (1995). Schizophrenia: a disconnection syndrome? *Clinical Neuroscience, 3*, 89–97.

Frith, C. (2004). Is autism a disconnection disorder? *Lancet Neurology, 3*, 577.

Fujii, T., Okuda, J., Tsukiura, T., et al. (2002). The role of the basal forebrain in episodic memory retrieval: a positron emission tomography study. *NeuroImage, 15*, 501–508.

Gili, T., Cercignani, M., Serra, L., et al. (2011). Regional brain atrophy and functional disconnection across Alzheimer's disease evolution. *Journal of Neurology, Neurosurgery, and Psychiatry, 82*, 58–66.

Glover, G. H., Li, T. Q., & Ress, D. (2000). Image-based method for retrospective correction of physiological motion effects in fMRI: RETROICOR. *Magnetic Resonance in Medicine, 44*, 162–167.

Grandjean, J., Schroeter, A., He, P., et al. (2014). Early alterations in functional connectivity and white matter structure in a transgenic mouse model of cerebral amyloidosis. *Journal of Neuroscience, 34*, 13780–13789.

Greicius, M. (2008). Resting-state functional connectivity in neuropsychiatric disorders. *Current Opinion in Neurology, 21*, 424–430.

Greicius, M. D., Kiviniemi, V., Tervonen, O., et al. (2008). Persistent default-mode network connectivity during light sedation. *Human Brain Mapping, 29*, 839–847.

Greicius, M. D., Krasnow, B., Reiss, A. L., et al. (2003). Functional connectivity in the resting brain: a network analysis of the default mode hypothesis. *Proceedings of the National Academy of Sciences of the United States of America, 100*, 253–258.

Greicius, M. D., Srivastava, G., Reiss, A. L., et al. (2004). Default-mode network activity distinguishes Alzheimer's disease from healthy aging: evidence from functional MRI. *Proceedings of the National Academy of Sciences of the United States of America, 101*, 4637–4642.

Greicius, M. D., Supekar, K., Menon, V., et al. (2009). Resting-state functional connectivity reflects structural connectivity in the default mode network. *Cerebral Cortex, 19*, 72–78.

Gusnard, D. A., Akbudak, E., Shulman, G. L., et al. (2001). Medial prefrontal cortex and self-referential mental activity: relation to a default mode of brain function. *Proceedings of the National Academy of Sciences of the United States of America, 98*, 4259–4264.

Gusnard, D. A., & Raichle, M. E. (2001). Searching for a baseline: functional imaging and the resting human brain. *Nature Reviews Neuroscience, 2*, 685–694.

Habas, C., Kamdar, N., Nguyen, D., et al. (2009). Distinct cerebellar contributions to intrinsic connectivity networks. *Journal of Neuroscience, 29*, 8586–8594.

Hadj-Bouziane, F., Monfardini, E., Guedj, C., et al. (2014). The helmet head restraint system: a viable solution for resting state fMRI in awake monkeys. *NeuroImage, 86*, 536–543.

Hampson, M., Peterson, B. S., Skudlarski, P., et al. (2002). Detection of functional connectivity using temporal correlations in MR images. *Human Brain Mapping, 15*, 247–262.

Hedden, T., Van Dijk, K. R., Becker, J. A., et al. (2009). Disruption of functional connectivity in clinically normal older adults harboring amyloid burden. *Journal of Neuroscience, 29*, 12686–12694.

Heise, V., Filippini, N., Trachtenberg, A. J., et al. (2014). Apolipoprotein E genotype, gender and age modulate connectivity of the hippocampus in healthy adults. *NeuroImage, 98*, 23–30.

Hesselmann, G., Kell, C. A., Eger, E., et al. (2008). Spontaneous local variations in ongoing neural activity bias perceptual decisions. *Proceedings of the National Academy of Sciences of the United States of America, 105*, 10984–10989.

van den Heuvel, M. P., & Sporns, O. (2011). Rich-club organization of the human connectome. *Journal of Neuroscience, 31*, 15775–15786.

Hiltunen, T., Kantola, J., Abou Elseoud, A., et al. (2014). Infra-slow EEG fluctuations are correlated with resting-state network dynamics in fMRI. *Journal of Neuroscience, 34*, 356–362.

Holmes, B. B., & Diamond, M. I. (2014). Prion-like properties of Tau protein: the importance of extracellular Tau as a therapeutic target. *Journal of Biological Chemistry, 289*, 19855–19861.

Honey, C. J., Sporns, O., Cammoun, L., et al. (2009). Predicting human resting-state functional connectivity from structural connectivity. *Proceedings of the National Academy of Sciences of the United States of America, 106*, 2035–2040.

Horovitz, S. G., Braun, A. R., Carr, W. S., et al. (2009). Decoupling of the brain's default mode network during deep sleep. *Proceedings of the National Academy of Sciences of the United States of America, 106*, 11376–11381.

Horovitz, S. G., Fukunaga, M., de Zwart, J. A., et al. (2008). Low frequency BOLD fluctuations during resting wakefulness and light sleep: a simultaneous EEG-fMRI study. *Human Brain Mapping, 29*, 671–682.

Hoyer, C., Gass, N., Weber-Fahr, W., et al. (2014). Advantages and challenges of small animal magnetic resonance imaging as a translational tool. *Neuropsychobiology, 69*, 187–201.

Hutchison, R. M., Womelsdorf, T., Allen, E. A., et al. (2013). Dynamic functional connectivity: promise, issues, and interpretations. *NeuroImage, 80*, 360–378.

Ingvar, M. (1999). Pain and functional imaging. *Philosophical Transactions of the Royal Society of London B Biological Sciences, 354*, 1347–1358.

Johnston, J. M., Vaishnavi, S. N., Smyth, M. D., et al. (2008). Loss of resting interhemispheric functional connectivity after complete section of the corpus callosum. *Journal of Neuroscience, 28*, 6453–6458.

Johnson, S. C., Baxter, L. C., Wilder, L. S., et al. (2002). Neural correlates of self-reflection. *Brain, 125*, 1808–1814.

Jonckers, E., Van Audekerke, J., De Visscher, G., et al. (2011). Functional connectivity fMRI of the rodent brain: comparison of functional connectivity networks in rat and mouse. *PLoS One, 6*, e18876.

Jones, D. T., Vemuri, P., Murphy, M. C., et al. (2012). Non-stationarity in the "resting brain's" modular architecture. *PLoS One, 7*, e39731.

Keller, C. J., Bickel, S., Honey, C. J., et al. (2013). Neurophysiological investigation of spontaneous correlated and anticorrelated fluctuations of the BOLD signal. *Journal of Neuroscience, 33*, 6333–6342.

Khalili-Mahani, N., Chang, C., van Osch, M. J., et al. (2013). The impact of "physiological correction" on functional connectivity analysis of pharmacological resting state fMRI. *NeuroImage, 65*, 499–510.

Kim, E. J., Sidhu, M., Gaus, S. E., et al. (2012). Selective frontoinsular von Economo neuron and fork cell loss in early behavioral variant frontotemporal dementia. *Cerebral Cortex, 22*, 251–259.

Kiviniemi, V., Jauhiainen, J., Tervonen, O., et al. (2000). Slow vasomotor fluctuation in fMRI of anesthetized child brain. *Magnetic Resonance in Medicine, 44*, 373–378.

Kiviniemi, V., Kantola, J. H., Jauhiainen, J., et al. (2003). Independent component analysis of nondeterministic fMRI signal sources. *NeuroImage, 19*, 253–260.

Kobayashi, Y., & Amaral, D. G. (2003). Macaque monkey retrosplenial cortex: II. Cortical afferents. *Journal of Comparative Neurology, 466*, 48–79.

Kobayashi, Y., & Amaral, D. G. (2007). Macaque monkey retrosplenial cortex: III. Cortical efferents. *Journal of Comparative Neurology, 502*, 810–833.

Koch, W., Teipel, S., Mueller, S., et al. (2012). Diagnostic power of default mode network resting state fMRI in the detection of Alzheimer's disease. *Neurobiology of Aging, 33*, 466–478.

Krienen, F. M., & Buckner, R. L. (2009). Segregated fronto-cerebellar circuits revealed by intrinsic functional connectivity. *Cerebral Cortex, 19*, 2485–2497.

Kucyi, A., & Davis, K. D. (2014). Dynamic functional connectivity of the default mode network tracks daydreaming. *NeuroImage, 100*, 471–480.

Langeslag, S. J., Schmidt, M., Ghassabian, A., et al. (2013). Functional connectivity between parietal and frontal brain regions and intelligence in young children: the generation R study. *Human Brain Mapping, 34*, 3299–3307.

Lee, J. H., Durand, R., Gradinaru, V., et al. (2010). Global and local fMRI signals driven by neurons defined optogenetically by type and wiring. *Nature, 465*, 788–792.

Lee, S. E., Khazenzon, A. M., Trujillo, A. J., et al. (2014). Altered network connectivity in frontotemporal dementia with C9orf72 hexanucleotide repeat expansion. *Brain, 137*, 3047–3060.

Leech, R., & Sharp, D. J. (2014). The role of the posterior cingulate cortex in cognition and disease. *Brain, 137*, 12–32.

Lehmann, M., Ghosh, P. M., Madison, C., et al. (2013). Diverging patterns of amyloid deposition and hypometabolism in clinical variants of probable Alzheimer's disease. *Brain, 136*, 844–858.

Lehmann, M., Madison, C. M., Ghosh, P. M., et al. (2013). Intrinsic connectivity networks in healthy subjects explain clinical variability in Alzheimer's disease. *Proceedings of the National Academy of Sciences of the United States of America, 110*, 11606–11611.

Leonardi, N., & Van De Ville, D. (2014). On spurious and real fluctuations of dynamic functional connectivity during rest. *NeuroImage, 104*, 430–436.

Leopold, D. A., Murayama, Y., & Logothetis, N. K. (2003). Very slow activity fluctuations in monkey visual cortex: implications for functional brain imaging. *Cerebral Cortex, 13*, 422–433.

Liang, Z., Liu, X., & Zhang, N. (2014). Dynamic resting state functional connectivity in awake and anesthetized rodents. *NeuroImage, 104*, 89–99.

Lu, H., Zuo, Y., Gu, H., et al. (2007). Synchronized delta oscillations correlate with the resting-state functional MRI signal. *Proceedings of the National Academy of Sciences of the United States of America, 104*, 18265–18269.

Machulda, M. M., Jones, D. T., Vemuri, P., et al. (2011). Effect of APOE epsilon4 status on intrinsic network connectivity in cognitively normal elderly subjects. *Archives of Neurology, 68*, 1131–1136.

Maddock, R. J., Garrett, A. S., & Buonocore, M. H. (2001). Remembering familiar people: the posterior cingulate cortex and autobiographical memory retrieval. *Neuroscience, 104*, 667–676.

Maguire, E. A., & Mummery, C. J. (1999). Differential modulation of a common memory retrieval network revealed by positron emission tomography. *Hippocampus, 9*, 54–61.

Maldjian, J. A. (2001). Functional connectivity MR imaging: fact or artifact? *AJNR. American Journal of Neuroradiology, 22*, 239–240.

McCormick, C., Quraan, M., Cohn, M., et al. (2013). Default mode network connectivity indicates episodic memory capacity in mesial temporal lobe epilepsy. *Epilepsia, 54*, 809–818.

Minoshima, S., Frey, K. A., Koeppe, R. A., et al. (1995). A diagnostic approach in Alzheimer's disease using three-dimensional stereotactic surface projections of fluorine-18-FDG PET. *Journal of Nuclear Medicine, 36*, 1238–1248.

Minoshima, S., Giordani, B., Berent, S., et al. (1997). Metabolic reduction in the posterior cingulate cortex in very early Alzheimer's disease. *Annals of Neurology, 42*, 85–94.

Montez, T., Poil, S. S., Jones, B. F., et al. (2009). Altered temporal correlations in parietal alpha and prefrontal theta oscillations in early-stage Alzheimer disease. *Proceedings of the National Academy of Sciences of the United States of America, 106*, 1614–1619.

Monto, S., Palva, S., Voipio, J., et al. (2008). Very slow EEG fluctuations predict the dynamics of stimulus detection and oscillation amplitudes in humans. *Journal of Neuroscience, 28*, 8268–8272.

Murphy, K., Birn, R. M., Handwerker, D. A., et al. (2009). The impact of global signal regression on resting state correlations: are anti-correlated networks introduced? *NeuroImage, 44*, 893–905.

Nir, Y., Mukamel, R., Dinstein, I., et al. (2008). Interhemispheric correlations of slow spontaneous neuronal fluctuations revealed in human sensory cortex. *Nature Neuroscience, 11*, 1100–1108.

Norton, L., Hutchison, R. M., Young, G. B., et al. (2012). Disruptions of functional connectivity in the default mode network of comatose patients. *Neurology, 78*, 175–181.

O'Reilly, J. X., Beckmann, C. F., Tomassini, V., et al. (2010). Distinct and overlapping functional zones in the cerebellum defined by resting state functional connectivity. *Cerebral Cortex, 20*, 953–965.

Petrella, J. R., Sheldon, F. C., Prince, S. E., et al. (2011). Default mode network connectivity in stable vs progressive mild cognitive impairment. *Neurology, 76*, 511–517.

Power, J. D., Barnes, K. A., Snyder, A. Z., et al. (2012). Spurious but systematic correlations in functional connectivity MRI networks arise from subject motion. *NeuroImage, 59*, 2142–2154.

Raichle, M. E., MacLeod, A. M., Snyder, A. Z., et al. (2001). A default mode of brain function. *Proceedings of the National Academy of Sciences of the United States of America, 98*, 676–682.

Reiman, E. M., Caselli, R. J., Yun, L. S., et al. (1996). Preclinical evidence of Alzheimer's disease in persons homozygous for the epsilon 4 allele for apolipoprotein E. *New England Journal of Medicine, 334*, 752–758.

Roy, A. K., Shehzad, Z., Margulies, D. S., et al. (2009). Functional connectivity of the human amygdala using resting state fMRI. *NeuroImage, 45*, 614–626.

Satterthwaite, T. D., Elliott, M. A., Ruparel, K., et al. (2014). Neuroimaging of the Philadelphia neurodevelopmental cohort. *NeuroImage, 86*, 544–553.

Schmahmann, J. D. (2004). Disorders of the cerebellum: ataxia, dysmetria of thought, and the cerebellar cognitive affective syndrome. *Journal of Neuropsychiatry and Clinical Neurosciences, 16*, 367–378.

Schumann, G., Loth, E., Banaschewski, T., et al. (2010). The IMAGEN study: reinforcement-related behaviour in normal brain function and psychopathology. *Molecular Psychiatry, 15*, 1128–1139.

Seeley, W. W., Allman, J. M., Carlin, D. A., et al. (2007). Divergent social functioning in behavioral variant frontotemporal dementia and Alzheimer disease: reciprocal networks and neuronal evolution. *Alzheimer Disease and Associated Disorders, 21*, S50–S57.

Seeley, W. W., Carlin, D. A., Allman, J. M., et al. (2006). Early frontotemporal dementia targets neurons unique to apes and humans. *Annals of Neurology, 60*, 660–667.

Seeley, W. W., Crawford, R. K., Zhou, J., et al. (2009). Neurodegenerative diseases target large-scale human brain networks. *Neuron, 62*, 42–52.

Seeley, W. W., Menon, V., Schatzberg, A. F., et al. (2007). Dissociable intrinsic connectivity networks for salience processing and executive control. *Journal of Neuroscience, 27*, 2349–2356.

Seeley, W. W., Merkle, F. T., Gaus, S. E., et al. (2012). Distinctive neurons of the anterior cingulate and frontoinsular cortex: a historical perspective. *Cerebral Cortex, 22*, 245–250.

Sheline, Y. I., Morris, J. C., Snyder, A. Z., et al. (2010). APOE4 allele disrupts resting state fMRI connectivity in the absence of amyloid plaques or decreased CSF Abeta42. *Journal of Neuroscience, 30*, 17035–17040.

Sheline, Y. I., Raichle, M. E., Snyder, A. Z., et al. (2010). Amyloid plaques disrupt resting state default mode network connectivity in cognitively normal elderly. *Biological Psychiatry, 67*, 584–587.

Shirer, W. R., Ryali, S., Rykhlevskaia, E., et al. (2012). Decoding subject-driven cognitive states with whole-brain connectivity patterns. *Cerebral Cortex, 22*, 158–165.

Shmuel, A., & Leopold, D. A. (2008). Neuronal correlates of spontaneous fluctuations in fMRI signals in monkey visual cortex: implications for functional connectivity at rest. *Human Brain Mapping, 29*, 751–761.

Shmueli, K., van Gelderen, P., de Zwart, J. A., et al. (2007). Low-frequency fluctuations in the cardiac rate as a source of variance in the resting-state fMRI BOLD signal. *NeuroImage, 38*, 306–320.

Smith, S. M., Fox, P. T., Miller, K. L., et al. (2009). Correspondence of the brain's functional architecture during activation and rest. *Proceedings of the National Academy of Sciences of the United States of America, 106*, 13040–13045.

Sohn, W. S., Yoo, K., Na, D. L., et al. (2014). Progressive changes in hippocampal resting-state connectivity across cognitive impairment: a cross-sectional study from normal to Alzheimer disease. *Alzheimer Disease and Associated Disorders, 28*, 239–246.

Song, M., Zhou, Y., Li, J., et al. (2008). Brain spontaneous functional connectivity and intelligence. *NeuroImage, 41*, 1168–1176.

Sorg, C., Riedl, V., Muhlau, M., et al. (2007). Selective changes of resting-state networks in individuals at risk for Alzheimer's disease. *Proceedings of the National Academy of Sciences of the United States of America, 104*, 18760–18765.

Stein, T., Moritz, C., Quigley, M., et al. (2000). Functional connectivity in the thalamus and hippocampus studied with functional MR imaging. *AJNR. American Journal of Neuroradiology, 21*, 1397–1401.

Stufflebeam, S. M., Liu, H., Sepulcre, J., et al. (2011). Localization of focal epileptic discharges using functional connectivity magnetic resonance imaging. *Journal of Neurosurgery, 114*, 1693–1697.

Trachtenberg, A. J., Filippini, N., Ebmeier, K. P., et al. (2012). The effects of APOE on the functional architecture of the resting brain. *NeuroImage, 59*, 565–572.

Valenstein, E., Bowers, D., Verfaellie, M., et al. (1987). Retrosplenial amnesia. *Brain, 110*(Pt 6), 1631–1646.

Van Essen, D. C., Ugurbil, K., Auerbach, E., et al. (2012). The human connectome project: a data acquisition perspective. *NeuroImage, 62*, 2222–2231.

Vanhaudenhuyse, A., Noirhomme, Q., Tshibanda, L. J., et al. (2010). Default network connectivity reflects the level of consciousness in non-communicative brain-damaged patients. *Brain, 133*, 161–171.

Vincent, J. L., Patel, G. H., Fox, M. D., et al. (2007). Intrinsic functional architecture in the anaesthetized monkey brain. *Nature, 447*, 83–86.

Wang, L., LaViolette, P., O'Keefe, K., et al. (2010). Intrinsic connectivity between the hippocampus and posteromedial cortex predicts memory performance in cognitively intact older individuals. *NeuroImage, 51*, 910–917.

Wang, L., Negreira, A., LaViolette, P., et al. (2010). Intrinsic interhemispheric hippocampal functional connectivity predicts individual differences in memory performance ability. *Hippocampus, 20*, 345–351.

Wang, Y., Risacher, S. L., West, J. D., et al. (2013). Altered default mode network connectivity in older adults with cognitive complaints and amnestic mild cognitive impairment. *Journal of Alzheimers Disease, 35*, 751–760.

Weiner, M. W., Aisen, P. S., Jack, C. R., Jr., et al. (2010). The Alzheimer's disease neuroimaging initiative: progress report and future plans. *Alzheimers & Dementia, 6*, 202–211, e7.

Westlye, E. T., Lundervold, A., Rootwelt, H., et al. (2011). Increased hippocampal default mode synchronization during rest in middle-aged and elderly APOE epsilon4 carriers: relationships with memory performance. *Journal of Neuroscience, 31*, 7775–7783.

Whitwell, J. L., Josephs, K. A., Avula, R., et al. (2011). Altered functional connectivity in asymptomatic MAPT subjects: a comparison to bvFTD. *Neurology, 77*, 866–874.

Yasuda, Y., Watanabe, T., Tanaka, H., et al. (1997). Amnesia following infarction in the right retrosplenial region. *Clinical Neurology and Neurosurgery, 99*, 102–105.

Zerbi, V., Wiesmann, M., Emmerzaal, T. L., et al. (2014). Resting-state functional connectivity changes in aging apoE4 and apoE-KO mice. *Journal of Neuroscience, 34*, 13963–13975.

Zhang, D., Snyder, A. Z., Fox, M. D., et al. (2008). Intrinsic functional relations between human cerebral cortex and thalamus. *Journal of Neurophysiology, 100*, 1740–1748.

Zhang, D., Snyder, A. Z., Shimony, J. S., et al. (2010). Noninvasive functional and structural connectivity mapping of the human thalamocortical system. *Cerebral Cortex, 20*, 1187–1194.

Zhou, J., Greicius, M. D., Gennatas, E. D., et al. (2010). Divergent network connectivity changes in behavioural variant frontotemporal dementia and Alzheimer's disease. *Brain, 133*, 1352–1367.

Zhou, J., & Seeley, W. W. (2014). Network dysfunction in Alzheimer's disease and frontotemporal dementia: implications for psychiatry. *Biological Psychiatry, 75*, 565–573.

Chapter 17

Neuroimaging Advances in Alzheimer's Disease

Daniel R. Schonhaut and Gil D. Rabinovici
University of California San Francisco, San Francisco, CA, United States

ABBREVIATIONS

Aβ Amyloid-β
AD Alzheimer's disease
APOE Apolipoprotein E
APP Amyloid precursor protein
bvFTD Behavioral-variant frontotemporal dementia
CBD Corticobasal degeneration
CSF Cerebrospinal fluid
CTE Chronic traumatic encephalopathy
DTI Diffusion tensor imaging
FDG [^{18}F]Fluoro-2-deoxy-D-glucose
FLAIR Fluid attenuation inversion recovery
fMRI Functional magnetic resonance imaging
MCI Mild cognitive impairment
MRI Magnetic resonance imaging
NFT Neurofibrillary tangle
PIB [^{11}C]Pittsburgh Compound-B
PET Positron emission tomography
PSEN1 Presenilin 1
PSEN2 Presenilin 2
PSP Progressive supranuclear palsy

INTRODUCTION

Advances in neuroimaging have enabled the in vivo study of many neurological and psychiatric conditions. Previously, researchers could make sense of biological changes related to human brain disorders only through indirect means: by using animal models of disease, studying basic mechanisms in vitro, or observing neuropathological effects on postmortem brain tissue. While these approaches remain essential to clinical research, the advent of neuroimaging has provided a means to study the evolution of pathological processes in ways that were never before possible. Many hypotheses from basic science approaches can now be tested and validated in humans. Compared with postmortem research, neuroimaging offers the potential to conduct larger-sample studies, adopt fully quantitative methods of analysis, and make fewer a priori assumptions about which areas of the brain should be examined. Importantly, imaging studies can also observe subjects longitudinally, allowing pathological progression to be directly observed rather than surmised from cross-sectional examination. From a practical standpoint, advances in neuroimaging have the potential to accelerate drug development and improve clinical diagnosis and disease monitoring.

Genomics, Circuits, and Pathways in Clinical Neuropsychiatry. http://dx.doi.org/10.1016/B978-0-12-800105-9.00017-2

In this chapter, we use Alzheimer's disease (AD) as a model to explore how neuroimaging has influenced clinical neuroscience research, emphasizing the most recent advances in magnetic resonance imaging (MRI) and positron emission tomography (PET). We start with a brief overview of AD as it is described clinically and pathologically, and in terms of its genetic associations. Next, we describe prototypical abnormalities of MRI and PET modalities in AD patients, giving a detailed account of recent developments in tau-PET imaging. We then show how multimodal neuroimaging is being used to test the conventional model of AD pathogenesis, the amyloid cascade hypothesis, by tracking core features of the disease through varying stages of severity. Finally, we discuss how the combination of basic science and imaging approaches has led to novel insights into the mechanisms of AD pathophysiology.

ALZHEIMER'S DISEASE DEFINED

AD is a progressive neurological condition associated with the degeneration of brain regions that are essential for cognitive function (see chapter The Genetic Basis of Alzheimer's Disease: Findings From Genome-wide Studies for additional background). The disease affects about 35 million people and is the leading cause of dementia worldwide (Brookmeyer, Johnson, Ziegler-Graham, & Arrighi, 2007). Approximately 1% of AD is related to a presenile dementia driven by autosomal dominant mutations in amyloid precursor protein (APP), presenilin 1 or presenilin 2, whereas the prevalence of sporadic AD increases exponentially with age and is strongly associated with the ε4 allele of the apolipoprotein E (APOE) gene (Fargo, 2014). Episodic memory impairment is the hallmark clinical symptom of AD, but atypical clinical variants exist with predominant deficits in other cognitive domains including language (Gorno-Tempini et al., 2011), visuospatial ability (Crutch et al., 2012), and executive function (Ossenkoppele, Pijnenburg, et al., 2015). Pathologically, AD is defined by two lesions evident on microscopic examination of postmortem tissue: extracellular plaques composed primarily of aggregated amyloid-β (Aβ) 1−42 amino acid polypeptide, and intracellular neurofibrillary tangles (NFTs) formed from hyperphosphorylated aggregated forms of the microtubule-associated protein tau (Montine et al., 2012).

These proteins aggregate into highly toxic, soluble oligomeric structures and ultimately into microscopically detectable, fibrillar macrostructures such as plaques and tangles. This is associated with synaptic dysfunction and synapse loss, disruption of neurotransmitter systems, and neuronal death in regions where one or both protein aggregates are found (Selkoe, 2002; Spires-Jones & Hyman, 2014). Ultimately, neurodegeneration gives way to cognitive decline and dementia (Hardy & Selkoe, 2002).

IMAGING CHARACTERISTICS OF ALZHEIMER'S DISEASE

MRI and PET are the two flagship neuroimaging technologies in AD research, and both have the advantage of being substantially versatile. Variations in MRI pulse sequences can alter the contrast between brain tissue types to produce images of neuroanatomical structure at fairly high spatial resolution (around 1 mm), whereas more advanced techniques can measure correlates of neuronal activity (Huettel, Song, & McCarthy, 2014). PET uses intravenously injected radiotracers that can mirror the actions of endogenous compounds associated with cell activity or can bind to specific protein targets. Images are then created from spatial patterns of radiotracer uptake. MRI and PET provide complementary measures of neurodegeneration, and developments in PET radiotracers have enabled the direct imaging of fibrillar amyloid and tau pathology. Combining these modalities creates the opportunity to study spatial and temporal relationships between the presumed molecular causes of AD and their accompanying effects on brain structure and function. In this section we highlight abnormalities detected by MRI and PET modalities in AD.

Structural MRI

Structural MRI can be used to quantify spatial patterns of brain atrophy using a T1-weighted sequence, which discriminates well between gray and white matter. This contrast has been widely used in AD research since the mid-1980s, when magnetic resonance became a viable method for noninvasive brain imaging (Besson et al., 1985; Fazekas, Chawluk, & Alavi, 1987; Reiman & Jagust, 2012). Studies point to bilateral hippocampus as a hallmark region of AD-related atrophy (Fig. 17.1). Average hippocampal volumes are about 25% lower in patients than in control subjects (Shi, Liu, Zhou, Yu, & Jiang, 2009), and clinically meaningful hippocampal atrophy can be reproducibly discerned by visual inspection using standardized scales (Scheltens, Leys, et al., 1992). Volumetric reductions are also observed in neighboring medial temporal lobe structures, notably entorhinal cortex (Dickerson et al., 2001; Juottonen et al., 1998), which acts as the main relay between the hippocampus and isocortex. These results match postmortem descriptions of medial temporal lobe being an early site of neuronal loss (Ball et al., 1985; Bobinki et al., 1999) and tau

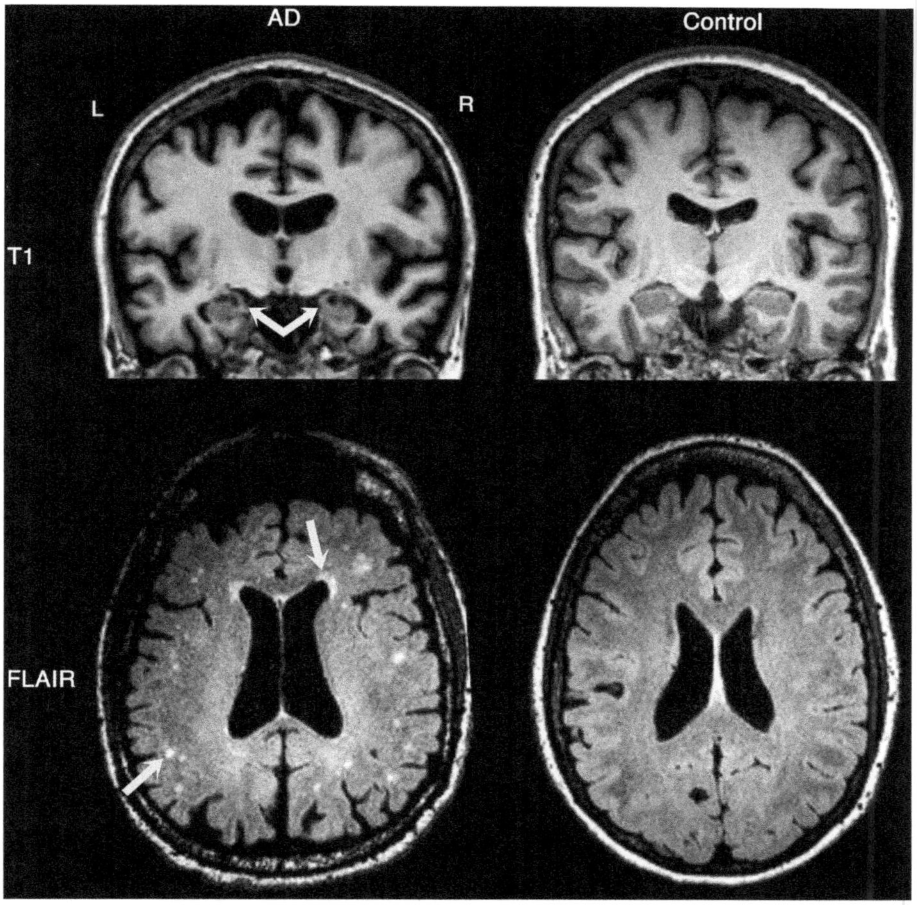

FIGURE 17.1 T1-weighted MRI of a patient with AD shows prominent, bilateral atrophy of the hippocampal formation and surrounding temporal cortex (top left). FLAIR MRI shows hyperintensities around the anterior horns of the lateral ventricles and throughout subcortical white matter (bottom left). MRIs of a cognitively normal control are shown for comparison (right).

pathology (Braak & Braak, 1991), and agree with clinical descriptions of memory impairment being the most frequent initial symptom of AD (Albert et al., 2011). Outside the medial temporal lobe, atrophy is observed in posterolateral temporal cortex and posterior cingulate/precuneus, with involvement of most isocortical regions (posterior more than frontal) in late disease stages (Frisoni et al., 2007; Lerch, 2004; Ossenkoppele, Cohn-Sheehy, et al., 2015; Thompson et al., 2003).

In addition to gray matter atrophy, AD may feature white matter lesions that are associated with demyelination and axonal loss along with small vessel ischemic disease, which often co-occurs with AD (Fig. 17.1) (Prins & Scheltens, 2015; Scheltens et al., 1995). T2-weighted structural and fluid attenuation inversion recovery (FLAIR) MRI sequences can be used to detect these lesions, called white matter hyperintensities for their bright, washed-out appearance. Hyperintensities tend to form around the periventricular zones and in subcortical white matter. Whereas these lesions are also prevalent in normal aging, they are more severe in AD (Scheltens, Barkhof, et al., 1992), with clearest separation between patients and control subjects in posterior periventricular regions and posterior corpus callosum (Yoshita et al., 2004). White matter integrity can also be assessed through diffusion tensor imaging (DTI), an MRI modality that measures the degree of directionality associated with water molecule diffusion. Because healthy white matter tracts constrain the free movement of water, white matter injury can be reflected by a decrease in anisotropic movement or an increase in overall diffusivity. In AD, DTI studies have shown that white matter changes occur along all major cortical tracts, but most prominently in connections between regions of acute gray matter atrophy (eg, hippocampal-cortical projections, temporal and parietal tracts), with relatively less severe changes to frontal and occipital white matter (Agosta, Pievani, Sala, & Geroldi, 2011; Sexton, Kalu, Filippini, Mackay, & Ebmeier, 2011).

Functional MRI

Functional MRI (fMRI) is an indirect measure of neuronal activity that takes advantage of the paramagnetic properties of deoxygenated hemoglobin, whose local concentrations are reduced by blood flow increases to active brain regions (Huettel et al., 2014). fMRI records snapshots of the brain in rapid temporal succession (about every 2 s) to capture activity changes over time. By observing regional correlations between these activation patterns, researchers discovered that the brain can be organized into networks of functionally related regions (see chapter Resting-State Functional MRI: A Novel Tool for Understanding Brain Networks in Neuropsychiatric Disorders for an in-depth review) (Power et al., 2011; Yeo et al., 2011). AD is associated with changes in several of these functional networks (Agosta, Pievani, Geroldi, et al., 2011). However, most attention has been focused on characterizing disruptions of the default mode network (Damoiseaux, Prater, Miller, & Greicius, 2012; Greicius, Srivastava, Reiss, & Menon, 2004; Liu et al., 2008), which is composed of regions that are deactivated during most cognitively engaging tasks and that may be activated during episodic memory encoding/retrieval and introspection (Andrews-Hanna, Smallwood, & Spreng, 2014; Buckner, Andrews-Hanna, & Schacter, 2008). Core default network regions include posterior cingulate, inferior parietal lobule, and medial prefrontal cortices, with weaker connections to lateral temporal and medial temporal lobe (Buckner et al., 2008; Greicius, Krasnow, Reiss, & Menon, 2003). Global activity in the default network is reduced in AD (Greicius et al., 2004), and there is disrupted synchrony between posterior and anterior parts of the network (Jones et al., 2011). Outside the default network, functional connectivity in task-specific cognitive networks may be associated with clinical variability in AD. For example, regions that are specifically atrophied in patients with primary progressive aphasia overlap with the language network, whereas specifically atrophied regions in patients with visuospatial deficits overlap with the higher-order visual network (Lehmann, Madison, et al., 2013). The default network may be a zone of convergence across these clinical variants.

In addition, functional connectivity between the hippocampus and the rest of the brain (largely including, but not limited to the default network) is dramatically reduced in AD (Allen et al., 2007; Wang et al., 2006). One study found a nearly complete absence of connections between the hippocampus and frontal and subcortical regions among patients in a moderate stage of dementia (Allen et al., 2007). This finding offers intriguing support for a theory of hippocampal isolation that was initially proposed by Hyman, Van Hoesen, Damasio, and Barnes (1984), who noted that tau NFTs in the hippocampus accumulate disproportionately in neurons that maintain projections to and from isocortex. Considering the importance of hippocampal-cortical connectivity in memory consolidation (Tambini, Ketz, & Davachi, 2010), this decoupling of connectivity could help explain the severity of episodic memory deficits in AD beyond what hippocampal atrophy alone would predict.

[^{18}F]Fluoro-2-Deoxy-D-Glucose Positron Emission Tomography

[^{18}F]Fluoro-2-deoxy-D-glucose (FDG) measures brain glucose metabolism, competing with 2-deoxy-D-glucose for phosphorylation by hexokinase at the start of the glycolytic pathway (Reivich et al., 1979). Glucose metabolism is a close correlate of neuronal activity, because about 80% of brain glucose consumption is directed toward cell signaling (Sibson et al., 1998), and lower FDG signal corresponds to regional reductions in synaptic density and activity (Rocher, Chapon, Blaizot, Baron, & Chavoix, 2003). FDG was the first PET tracer applied to dementia and aging research, and initial studies reported proof-of-point metabolic reductions in AD (Ferris et al., 1980; Friedland, Budinger, Koss, & Ober, 1985; Haxby, Duara, Grady, Cutler, & Rapoport, 1985) that mirrored earlier findings from cerebral blood flow measures (Freyhan, Woodford, & Kety, 1951; Obrist, Chivian, Cronqvist, & Ingvar, 1970). As the spatial resolution of PET scanners has improved with advances in methods and technologies, FDG studies have identified a clear set of regions that are hypometabolic in AD, including lateral temporoparietal cortex, posterior cingulate/precuneus, and, to a lesser degree, dorsolateral prefrontal cortex (Fig. 17.2) (Herholz et al., 2002; Hoffman et al., 2000). Longitudinally, these regions continue to decline throughout progression of the disease, with posterior parietal cortex remaining most affected (Alexander, Chen, Pietrini, Rapoport, & Reiman, 2002; Jagust, Friedland, Budinger, Koss, & Ober, 1988). Glucose metabolism also decreases in normal aging, but more anteriorly and far less extensively than in AD (Herholz et al., 2002).

Amyloid Positron Emission Tomography

The AD field gained a critical addition in 2004 with the development of *N*-methyl-[^{11}C]2-(4′-methylaminophenyl)-6-hydroxybenzothiazole, or Pittsburgh Compound-B (PIB), a tracer that binds to amyloid plaques with high affinity and specificity (Klunk et al., 2004) and closely matches histopathological descriptions of plaque distribution (Braak & Braak, 1991; Thal, Rüb, Orantes, & Braak, 2002). PIB was developed as an analog of Thioflavin-T, a dye used to stain amyloid

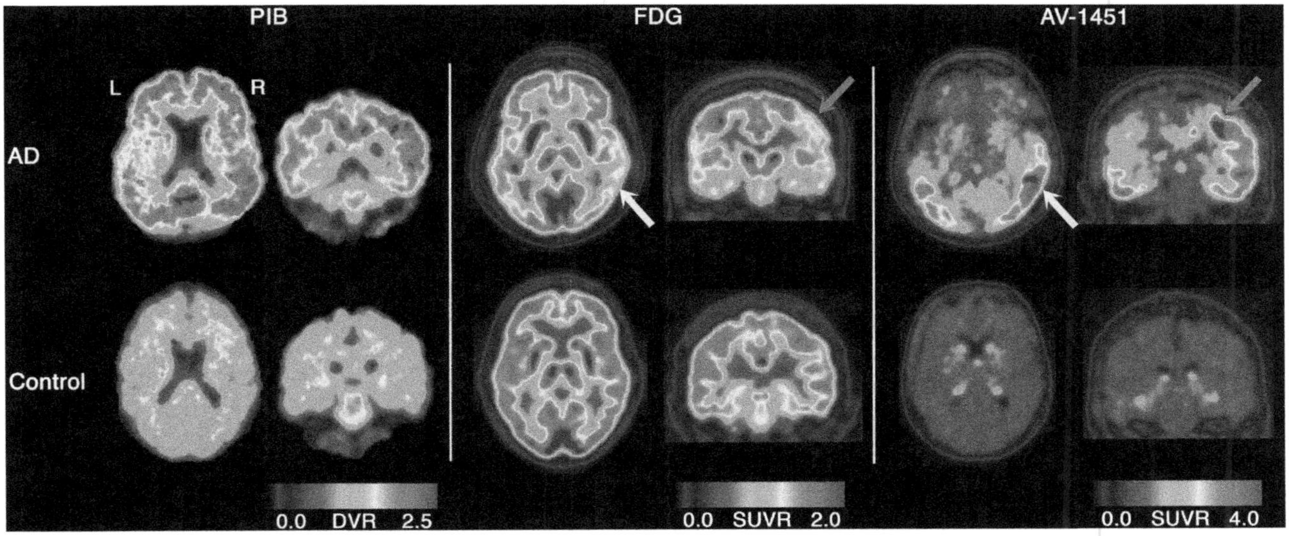

FIGURE 17.2 PET imaging (PIB, FDG, and AV-1451) of a patient with AD reveals spatial patterns of amyloid plaques (top left), glucose metabolism (top center), and pathological tau (top right) compared against a cognitively normal control (bottom). Amyloid plaques are present throughout association neocortex, whereas patterns of hypometabolism and tau are constrained to posterior temporoparietal (right > left) cortex and show striking regional overlap.

structures in histopathological specimens. In patients observed from PET to autopsy, the distribution and burden of PIB PET signal during life correlate strongly with postmortem measures of fibrillar amyloid (Ikonomovic et al., 2008; Villeneuve et al., 2015). AD patients typically have highest PIB binding in frontal and medial parietal regions, followed by lateral temporoparietal cortex and striatum, with relative sparing of medial temporal, occipital, and primary sensory cortices (Fig. 17.2) (Klunk et al., 2004; Price et al., 2005). Whereas FDG hypometabolism is commonly interpreted as a continuous measure of regional disease severity, PIB is often dichotomized as positive or negative at the whole-brain level. This is used as a proxy for global amyloid burden that indicates the presence or absence of a significant burden of amyloid pathology. The determination of PIB positivity is either made visually by a trained clinician or quantitatively by setting a threshold based on average PIB binding in cortical regions that frequently harbor amyloid plaques (Ng et al., 2007; Villeneuve et al., 2015). Although the short radioactive half-life of C-11–labeled PIB (20 min) prohibits its use outside research settings that can synthesize the tracer on-site, several newer amyloid PET tracers have been developed with the longer-lived F-18 isotope (110-min half-life) and have been validated compared with autopsy (Clark et al., 2011; Curtis et al., 2015; Sabri et al., 2015) and approved for clinical use by regulatory agencies. This opens up the possibility for using amyloid PET as a diagnostic biomarker in clinical practice (Johnson et al., 2013). In addition, the success of these tracers in autopsy confirmation studies makes amyloid PET a practical "standard of truth" for evaluating other methods of amyloid pathology detection in vivo, including cerebrospinal fluid (CSF) (Landau et al., 2013) and blood-based biomarkers (Lui et al., 2010).

Tau Positron Emission Tomography

Our understanding of amyloid plaque development and AD-related neurodegenerative changes has surged forward in recent years. However, much of what we know about tau pathology has been inferred from basic science, and the causes and effects of regional tau accumulation are incompletely understood. Heterogeneity also poses a major obstacle in our ability to study and treat dementia. Aside from AD, a number of neurodegenerative tauopathies have been characterized that are associated with distinct isoforms and pathological conformations of tau, and that target different sets of brain regions (Lee, Goedert, & Trojanowski, 2001). Pathological lesions other than tau and Aβ (eg, TAR DNA-binding protein 43, α-synuclein, ubiquitin) can also cause overlapping clinical syndromes and neurodegenerative profiles on MRI and FDG, and there is a scarcity of biomarkers that can tease apart these underlying diseases. Although CSF biomarkers that measure global tau burden can serve as a reasonable starting point for in vivo research (Buerger et al., 2006; Ishiguro et al., 1999), the clinical and research communities would benefit greatly from the development of PET tracers that can image regional tau pathology, both in AD and other neurodegenerative tauopathies. In AD, this is especially important because tau pathology does not develop in a global, diffuse manner (as may be the case with amyloid plaque development), but instead follows a characteristic staging pattern that begins in the medial temporal lobe, spreads to basolateral temporal and limbic cortices, and eventually implicates most of association isocortex, posterior more so than frontal (Braak & Braak, 1991). This description is derived from cross-sectional neuropathology and should be refined through longitudinal study in vivo.

Several families of putative tau PET tracers have come to the forefront with others still in development, and validation of these tracers is being performed through autoradiography, tau transgenic mice models, and in vivo human studies (Okamura et al., 2014; Villemagne, Fodero-Tavoletti, Masters, & Rowe, 2015). Promising candidates include several quinoline-based tracers from Tohoku University ([^{18}F]THK-5105, [^{18}F]THK-5117, and [^{18}F]THK-5351) (Harada et al., 2015; Ishiki et al., 2015; Okamura et al., 2013, 2005), a tracer structured around a phenyl/pyridinyl-butadienyl-benzothiazole/benzothiazolium modification of Thioflavin-T ([^{11}C]PBB3) (Hashimoto et al., 2014; Maruyama et al., 2013), and two tracers derived from benzimidazole pyrimidine ([^{18}F]AV-1451 [formerly T807] and [^{18}F]T808) (Chien et al., 2013; Xia et al., 2013). In vitro assays with AV-1451 have demonstrated high affinity and selectivity for paired helical filaments of tau (precursors of NFTs) (Marquié et al., 2015). In addition, Johnson et al. (2015) reported in vivo results with this tracer in a moderately sized group of subjects ($n = 75$) ranging from cognitively normal to demented. They found clear separation between AD patients and control subjects in most regions of association cortex (posterior > frontal), and uptake patterns were largely consistent with neuropathological tau staging with the exception of hippocampus, which had low tracer retention in patients and no statistical difference versus control subjects. These reports are similar to our findings in a group of clinically heterogeneous patients with AD who were recruited through the memory clinic at the University of California—San Francisco and scanned with AV-1451 in Berkeley. We observed clear separation in uptake patterns and intensities between patients with AD and control subjects in expected regions of tau pathology, and we noted a striking overlap between regions with high AV-1451 uptake and low glucose metabolism in patients with AD (Figs. 17.3 and 17.6). We have also collected AV-1451 data in several patients with suspected non-AD tauopathies, including progressive supranuclear palsy (PSP), behavioral-variant frontotemporal dementia (bvFTD) caused by a tau genetic mutation, chronic traumatic encephalopathy (CTE), and corticobasal degeneration (CBD) (Fig. 17.3). These scans show promising patterns of uptake that suggest AV-1451 may bind to variable conformations of tau. However, tracer signal in these patients is much more subtle than in patients with AD, and postmortem autoradiography of non-AD tau pathology has found little to no labeling of the straight filaments of tau that characterize these diseases (Marquié et al., 2015). PET-to-autopsy studies are needed to better characterize the in vivo binding properties of AV-1451 and other tau PET tracers.

PROGRESSION OF NEUROIMAGING ABNORMALITIES: AN AMYLOID CASCADE?

In the early 1990s, Dennis Selkoe, John Hardy, and others introduced a hypothetical model of AD pathogenesis that pointed to Aβ accumulation as the initiating event in the disease process that drives downstream NFT formation and neurodegeneration, ultimately resulting in dementia (Hardy & Higgins, 1992; Selkoe, 1991). This theory, dubbed the amyloid cascade hypothesis (Fig. 17.4), was founded on the observation that mutations in genes involved in the production of Aβ cause early-onset familial AD whereas APP, coded on chromosome 21, is overexpressed in Down syndrome, leading to a high incidence of AD. Expression of AD-associated mutations in transgenic mice reproduces many if not all of the elements of the human disease, including Aβ aggregation and plaque deposition, synaptic and network dysfunction, and anterograde memory impairment (Götz & Ittner, 2008). This model has helped uncover, if not fully explain, some of the complicated mechanisms of Aβ-induced neurotoxicity (Selkoe, 2002; Spires-Jones & Hyman, 2014; Yankner, 1996).

Genome-wide association studies have uncovered additional genetic risk factors for sporadic AD, many (but not all) of which are related to Aβ accumulation (Lambert et al., 2013). The APOE ε4 allele, which is present in about 20% of the general population, is the largest of these risk factors, with each copy of the allele conferring a three- to fourfold increase in the odds of developing AD compared with the "neutral" ε3 allele (Farrer et al., 1997). APOE is a multifunctioning gene that is largely responsible for lipid transport and regulation of cholesterol metabolism in the central nervous system. The ε4 allele increases vulnerability for amyloid and tau pathology through several independent pathways, which can be conceptualized as promoting either a gain of toxic function (eg, increased Aβ production and tau hyperphosphorylation, mitochondrial toxicity) or a loss of protective function (decreased Aβ clearance, less effective responses to a variety of neural injuries) (Liu, Kanekiyo, Xu, & Bu, 2013). Neuropathological examination of AD-affected brains shows that APOE ε4-carriers have a higher burden of cortical amyloid plaques and neurofibrillary tangles, but the association is strongest between ε4 and amyloid (Tiraboschi et al., 2004). In light of this evidence, antiamyloid therapies have been at the center of most AD drug development efforts, which emphasizes the importance of understanding amyloid's role in the disease (Mangialasche, Solomon, Winblad, Mecocci, & Kivipelto, 2010).

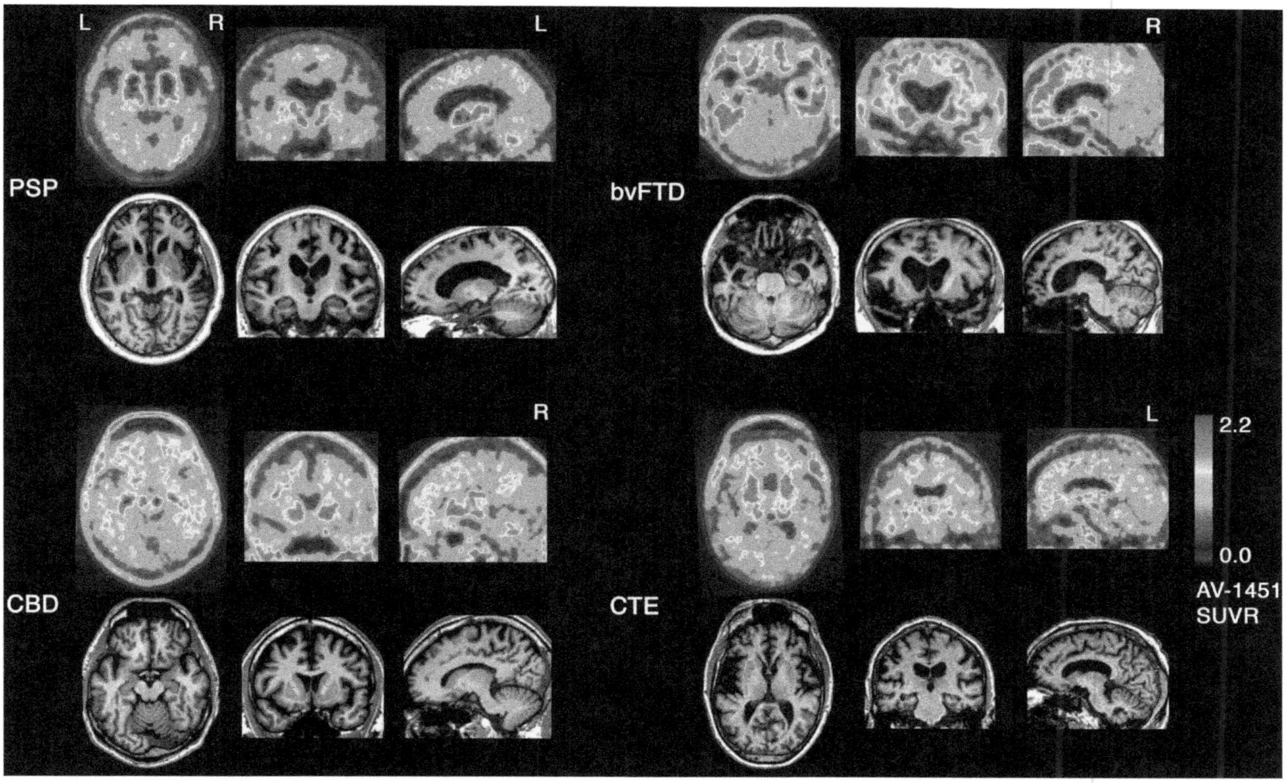

FIGURE 17.3 Tau-PET (AV-1451) and structural MRI images of four patients with suspected non-AD tauopathies show patterns of tau PET uptake and atrophy that are largely consistent with neuropathological descriptions of each disease. Clockwise from top left: progressive supranuclear palsy (PSP), behavioral-variant frontotemporal dementia (bvFTD) caused by a tau genetic mutation, chronic traumatic encephalopathy (CTE), and corticobasal degeneration (CBD).

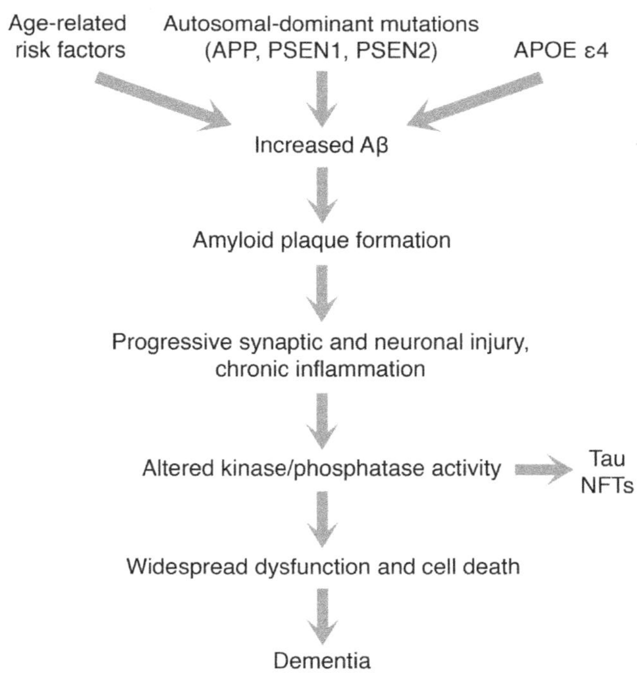

FIGURE 17.4 The amyloid cascade hypothesis as originally conceived by Hardy and Selkoe. Accumulation of amyloid pathology is modeled as the initiating event in AD and is thought to drive tau pathology, neurodegeneration, and cognitive decline.

Neuroimaging approaches present an opportunity to test the amyloid cascade hypothesis and other models of AD pathogenesis in vivo. Multiple imaging biomarkers can be assessed simultaneously, and their progression can be tracked longitudinally. This provides a major advantage of neuroimaging over postmortem assessment, but in combination these methods offer an opportunity to observe the full time-course of AD from preclinical stages through the end of life. In addition, neuroimaging can be paired with genetic information to study the biological effects of known genetic risk factors. In this section we review how multiple imaging modalities are being used to advance our understanding of the causes of AD and its progression across varying stages of disease.

Preclinical Alzheimer's Disease

The amyloid cascade hypothesis predicts an early presence of amyloid plaques that precede neurodegeneration and cognitive symptoms. Consistent with this prediction, a substantial minority (about 30%) of cognitively normal elderly people are found to have "AD-like" levels of amyloid plaque deposition at autopsy in lieu of any clinically apparent cognitive deficits (Bennett et al., 2006; Schneider, Arvanitakis, Leurgans, & Bennett, 2009). This finding has been replicated in vivo, with a similar proportion of older adults showing elevated cortical amyloid PET binding in the absence of cognitive deficits (Aizenstein et al., 2008; Mintun et al., 2006; Mormino et al., 2009; Villeneuve et al., 2015). In addition, amyloid PET studies have demonstrated that the prevalence of amyloid positivity in normal elderly people increases as a function of age (ranging from about 10% by age 60 to over 40% by age 80) and APOE status, with ε4 carriers having two to three times the odds of being amyloid positive compared with ε4 noncarriers (Jagust et al., 2015; Jansen et al., 2015). In large-sample, multimodal imaging studies, cortical amyloid-PET values appear to begin increasing a decade or more before the onset of clinical symptoms, and (with some exceptions) generally precede neurodegenerative changes measured with structural MRI or FDG PET (Jack, Knopman, et al., 2013; Villemagne et al., 2013). These findings have also been replicated across the clinical spectrum of individuals harboring APP and presenilin mutations (Bateman et al., 2012).

The high prevalence of amyloid pathology in cognitively normal elderly people raises the question of how much meaning should be placed on a positive amyloid PET scan in this population. Does amyloid positivity on its own predict decline and conversion to AD? Retrospective autopsy and cross-sectional neuroimaging studies have given equivocal answers to this question, with some studies finding associations between the presence of amyloid pathology and worse performance on episodic memory or other cognitive domains in healthy elderly individuals (Hulette et al., 1998; Kantarci et al., 2012; Oh, Madison, Haight, Markley, & Jagust, 2012; Pike et al., 2007) and other studies finding weak or no such relations (Aizenstein et al., 2008; Bennett et al., 2006; Driscoll et al., 2006; Wirth et al., 2013). However, a clearer picture emerges from longitudinal imaging studies, which have consistently reported that amyloid PET positivity in healthy elderly people predicts more rapid episodic memory loss, global cognitive decline, and gray matter atrophy (Chételat et al., 2012; Doraiswamy et al., 2012; Landau et al., 2012; Villemagne et al., 2013). Further evidence suggests that amyloid may exert its effects on cognition indirectly and through several mechanisms, including mediation by hippocampal atrophy (Mormino et al., 2009), cortical thinning (Villeneuve et al., 2014), and sleep disruption (Mander et al., 2015). In a related manner, the joint presence of amyloid PET positivity and neurodegeneration predicts more rapid cognitive decline than does either measure alone (Mormino, Betensky, Hedden, Schultz, Amariglio, et al., 2014), which emphasizes the need for multimodal imaging approaches to characterize the risk of incipient AD accurately in healthy elderly individuals.

Imaging studies also point to several effects of APOE ε4 on cognitive decline in the preclinical population. Along with their increased susceptibility for developing amyloid pathology, healthy elderly ε4 carriers have lower glucose metabolism (measured with FDG) in AD-typical regions compared with ε4 noncarriers, independent of amyloid burden (Jagust & Landau, 2012; Reiman et al., 1996). These baseline differences are associated with longitudinal metabolic and cognitive decline, indicating that metabolic differences in elderly ε4 carriers are not benign (Small et al., 2000). In addition, elderly ε4 carriers show disrupted functional connectivity patterns (Sheline et al., 2010) and have reduced cortical thickness in medial temporal lobe regions that are among the earliest affected in AD (Burggren et al., 2008). Whereas these alterations could indicate incipient disease, they could also reflect inherent variability in brain organization across APOE genotypes. Differences in gray and white matter volume between ε4 carriers and noncarriers have been found in infants (Dean et al., 2014), and functional connectivity differences have been reported in young adults (Agosta et al., 2009). It is difficult to know whether and to what extent imaging biomarkers of neural integrity represent systemic developmental differences or subtle pathological effects of APOE ε4. Nevertheless, neuroimaging studies support a profound influence of the ε4 allele on increased pathological burden in healthy elderly people that dovetails with a likely decreased resistance to this pathology, resulting in a greater risk of decline (Mormino, Betensky, Hedden, Schultz, Ward, et al., 2014).

Despite the adverse long-term outcomes associated with amyloid positivity, evidently it is possible to have a striking amount of cortical amyloid pathology and little to no short-term loss of cognitive function (Aizenstein et al., 2008; Johnson et al., 2014). This is difficult to reconcile with the notion that amyloid is the driving force behind neurodegeneration and dementia. One possible explanation for this disconnect is reserve capacity, which refers to the brain's ability to maintain function despite the loss of synapses and neural resources (brain reserve) or to reorganize functionally around degenerating regions (cognitive reserve) (Stern, 2002). Reserve capacity is thought to capture a combination of protective genetic and environmental influences. Surrogate measures of reserve, including premorbid intelligence and educational attainment, have been linked to greater resistance to AD-related neurodegeneration and slower cognitive decline among healthy elderly people (Arenaza-Urquijo, Wirth, & Chételat, 2015; Stern, 2012). It is likely that there are additional biological or genetic factors that might render an individual relatively resilient to the toxicity of Aβ. Furthermore, fMRI and FDG studies have found some cases in which functional activation increases and hypermetabolism correlate with better memory performance in cognitively normal, PIB-positive subjects, which suggests a compensatory response consistent with cognitive reserve (Elman et al., 2014; Mormino et al., 2012; Ossenkoppele, Madison, et al., 2014).

Whereas reserve capacity likely explains part of the dissociation between amyloid accumulation and AD-related decline, a second possibility is that Aβ is not in itself sufficiently neurotoxic to produce extensive cognitive impairment. Neuropathological studies have long indicated that tau pathology shows closer associations than amyloid pathology with regional neurodegeneration and antemortem disease severity (Gomez-Isla et al., 1997; Nelson et al., 2012). Rather than directly causing neurodegeneration and dementia, amyloid might exert its effects indirectly by triggering or facilitating the accumulation of pathological tau, which might then serve as the primary agent of neurodegeneration (Bloom, 2014). Future PET studies, especially longitudinal, will test this explanation by exploring regional relations between tau and amyloid burden and their effects on cognitive decline in preclinical AD. In a case example, an individual presented to our center reporting mild memory loss, but performed in the normal range on neuropsychological testing. PET imaging offered a rare glimpse into in vivo pathological burden at this early disease stage and showed focal PIB-PET binding in middle frontal cortex that coincided with slightly elevated tau PET signal and subtle FDG hypometabolism in medial and inferior temporal cortex, which was consistent with early tau staging described at autopsy (Fig. 17.5). Additional PET scans in preclinical elderly individuals will help uncover group-level associations and potential interactions between emergent amyloid and tau pathology.

Mild Cognitive Impairment

To help classify the transition from normal cognition to clinical AD, Petersen and colleagues devised an intermediary category, mild cognitive impairment (MCI), in which memory decline is the most common feature, global cognition is still relatively intact, and independent day-to-day function is preserved (Albert et al., 2011; Petersen et al., 1999). This category represents a potentially critical period for drug intervention, but there is a relatively low rate of conversion from MCI to AD (about 8−15% per year, depending on selection criteria) (Mitchell & Shiri-Feshki, 2009), which has led to much interest in predicting accurately who will convert and when.

Several neuroimaging biomarkers and cognitive performance indices have been shown to differentiate rapid MCI converters from nonconverters, with varying success. These include hippocampal atrophy (Jack et al., 1999; Koivunen et al., 2011; Landau et al., 2010), hypometabolism of posterior cingulate and lateral temporoparietal cortex (Arnaiz et al., 2001; Chételat et al., 2003; Drzezga et al., 2003; Landau et al., 2010), and several measures of cognitive performance (Amieva et al., 2004; Landau et al., 2010; Tabert et al., 2006). In addition, although the prevalence of amyloid PET positivity is higher in MCI than among healthy elderly people (40−60% compared with about 30%), the significant fraction of amyloid-negative patients with MCI highlights the fact that memory loss and cognitive decline are not exclusive to AD. Longitudinal studies show that combinations of amyloid positivity, MRI atrophy or FDG PET hypometabolism, and cognitive impairment are better predictors of conversion to AD than are any of these factors alone (Arnaiz et al., 2001; Landau et al., 2010; Mormino, Betensky, Hedden, Schultz, Amariglio, et al., 2014), which emphasizes the importance of combining multiple imaging modalities and clinical information in developing a clear understanding of individual pathophysiology.

Dementia and Limitations of the Amyloid Cascade Hypothesis

Most evidence points to Aβ having a key role in the nascent stages of AD, but at some point amyloid pathology becomes disentangled from neurodegeneration and cognitive impairment. Longitudinal increases in amyloid PET appear to plateau around the time of dementia onset and perhaps even decrease with disease progression, whereas neurodegeneration

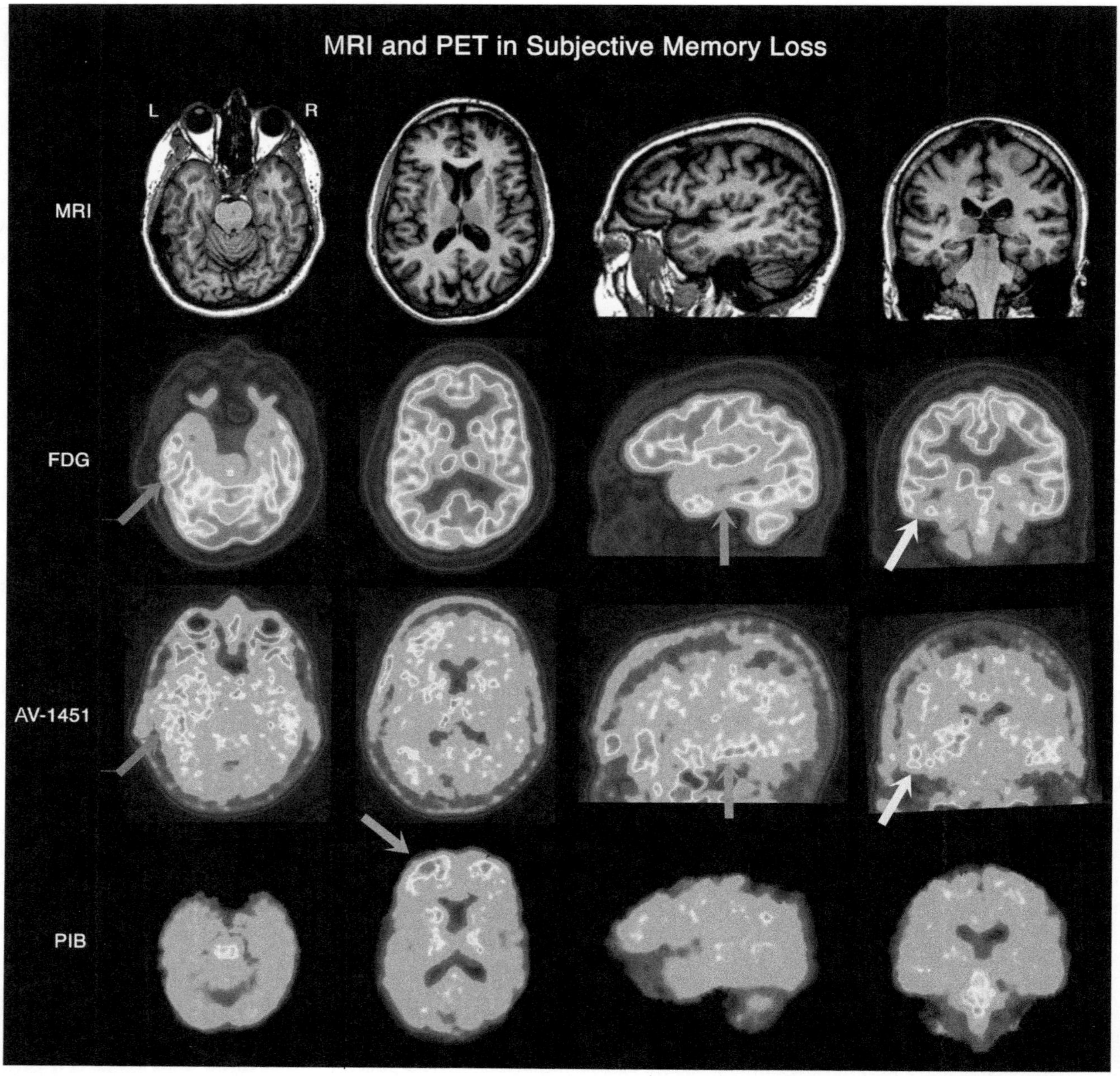

FIGURE 17.5 An individual with subjective memory loss underwent structural MRI, FDG PET, tau PET (AV-1451), and amyloid PET (PIB). Focal amyloid PET uptake is observed in middle frontal cortex (left > right), whereas elevated tau PET and lower metabolism are both observed in medial and inferior temporal cortex (left > right).

continues to compound (Engler et al., 2006; Jack, Wiste, et al., 2013; Koivunen et al., 2011). In this we see a continuation of a theme discussed earlier: that there is a disconnect between presumed amyloid neurotoxicity and the relative absence of neuroimaging and clinical evidence to support it. On the one hand, healthy elderly adults accumulate amyloid PET pathology many years before widespread neurodegeneration and global cognitive decline. On the other hand, patients with AD experience a progression of neurodegeneration and cognitive decline in the absence of increasing amyloid pathology.

A simple explanation for the second point is that at peak levels, amyloid burden might be sufficiently high to continue driving neurodegeneration. However, several lines of evidence argue against this. First, most patients with AD have substantial PIB binding in frontal cortex but frontal metabolism is relatively preserved (Edison et al., 2007), as are many aspects of cognition that are associated with frontal lobe function (Bozeat, Gregory, Ralph, & Hodges, 2000). In addition, cognitive domain performance among patients with AD correlates with the severity of hypometabolism in expected regions, but neither cognitive performance nor regional hypometabolism is associated with regional PIB (Altmann, Ng, Landau, Jagust, & Greicius, 2015; Furst et al., 2012). Finally, patients who develop dementia at an unusually early age or who present with an atypical clinical phenotype tend to have distinct patterns of atrophy and hypometabolism in regions that underlie their cognitive deficits, but their amyloid PET scans indicate that the burden and distribution of amyloid plaques is essentially the same across extreme clinical variability (Lehmann, Ghosh, et al., 2013; Ossenkoppele, Cohn-Sheehy, et al., 2015; Rabinovici et al., 2010; Rabinovici, Jagust, & Furst, 2008; Rosenbloom et al., 2011). Thus, amyloid appears to have a necessary but insufficient role in explaining the pathogenesis of AD, with little relevance at the dementia stage.

The amyloid cascade hypothesis proposed tau NFTs to be the byproduct of a process in which Aβ is the primary actor, but the weight of evidence suggests that tau may have a much more central role than was originally thought. We previously noted that there is a rough correspondence between AD hypometabolic regions and distributions of tau NFTs at autopsy (Braak & Braak, 1991) and that postmortem NFTs correlate more strongly than amyloid plaques to measures of neurodegeneration and antemortem disease severity (Gomez-Isla et al., 1997; Nelson et al., 2012). Studies in mice have reported that strains of pathological tau can induce templated misfolding of regularly functioning tau and can spread trans-synaptically through functionally connected regions, which suggests an amyloid-independent mechanism of tau propagation (see Section 5.2) (De Calignon et al., 2012; Sanders et al., 2014). Finally, several mechanisms of tau-induced neurotoxicity have been proposed that implicate abnormal tau deposits in the disruption of axonal transport and synaptic transmission, which could set off a sequence of events that results in cell death (Ballatore, Lee, & Trojanowski, 2007; Iqbal, Liu, Gong, del Alonso, & Grundke-Iqbal, 2009; Spires-Jones & Hyman, 2014). These observations are consistent with our initial tau PET results in patients with AD, which show strong regional overlap between patterns of tau PET and FDG hypometabolism that vary in accordance with clinical phenotype and appear to underlie specific cognitive deficits (Fig. 17.6) (Ossenkoppele, Schonhaut, et al., 2014). In stark contrast, regional amyloid PET uptake in the same patients does not predict hypometabolism and shows no clear associations with cognition.

UNDERLYING MECHANISMS OF ALZHEIMER'S DISEASE PATHOLOGY

PET and neuropathology studies have described where amyloid plaques and tau NFTs tend to aggregate, but the mechanisms that underlie selective regional vulnerability to these pathologies remain a mystery. Discoveries spanning work in multimodal neuroimaging, tissue cultures, and transgenic mice have begun to converge on plausible explanations for this regional variability in pathological buildup. Owing to these efforts, we are at the beginning of a major shift in our ability to describe AD in humans, moving away from surface-level descriptions of where pathology develops and what symptoms manifest, and toward models of the disease that use neuroimaging to explore basic mechanisms. These studies also highlight how neuroimaging approaches fit into the larger picture of clinical neuroscience research, and how a combination of research methods is needed to broach complicated questions of causality.

Amyloid Plaques Form in Regional Hubs

Buckner et al. (2009) reported a remarkable parallel between patterns of PIB binding in patients with AD and fMRI-derived functional connectivity in normal adults. Patients tended to accumulate highest amyloid pathology in regions that maintain a large number of functional connections throughout life. These so-called functional hubs span most of the default network, as well as lateral frontal and parietal association cortices (Seeley et al., 2007). This discovery suggests that amyloid accumulation may be influenced, in part, by a trade-off between adaptive function and vulnerability to disease.

A 2010 study shed further light on the possible biological mechanisms of endogenous vulnerability to amyloid plaque formation (Vlassenko et al., 2010). The authors used separate PET measures of glucose and oxygen metabolism to quantify regional aerobic glycolysis (glucose consumption outside of oxidative phosphorylation, despite the availability of oxygen) in healthy young adults. Disproportionately high rates of aerobic glycolysis were reported in functional hubs and in regions with high amyloid plaque burden in AD (Raichle, 2010; Vlassenko et al., 2010). Although the reasons for regional differences in aerobic glycolysis are not well understood, the process has been linked to synaptic activity and measures of neuroplasticity (Bauernfeind et al., 2014). This fits with the concept of hubs as regions of multimodal integration and synchronization (van den Heuvel & Sporns, 2013). Critically, this finding suggested that conditions of high activity and high plasticity might have a causal role in the development of amyloid pathology.

Strong experimental evidence has been found for this ostensible relationship (Bero et al., 2011). Using in vivo microdialysis in a transgenic mouse model that overexpresses a mutant form of APP, Bero and colleagues found that endogenous differences in extracellular Aβ levels of young AD transgenic mice predicted the degree of subsequent plaque

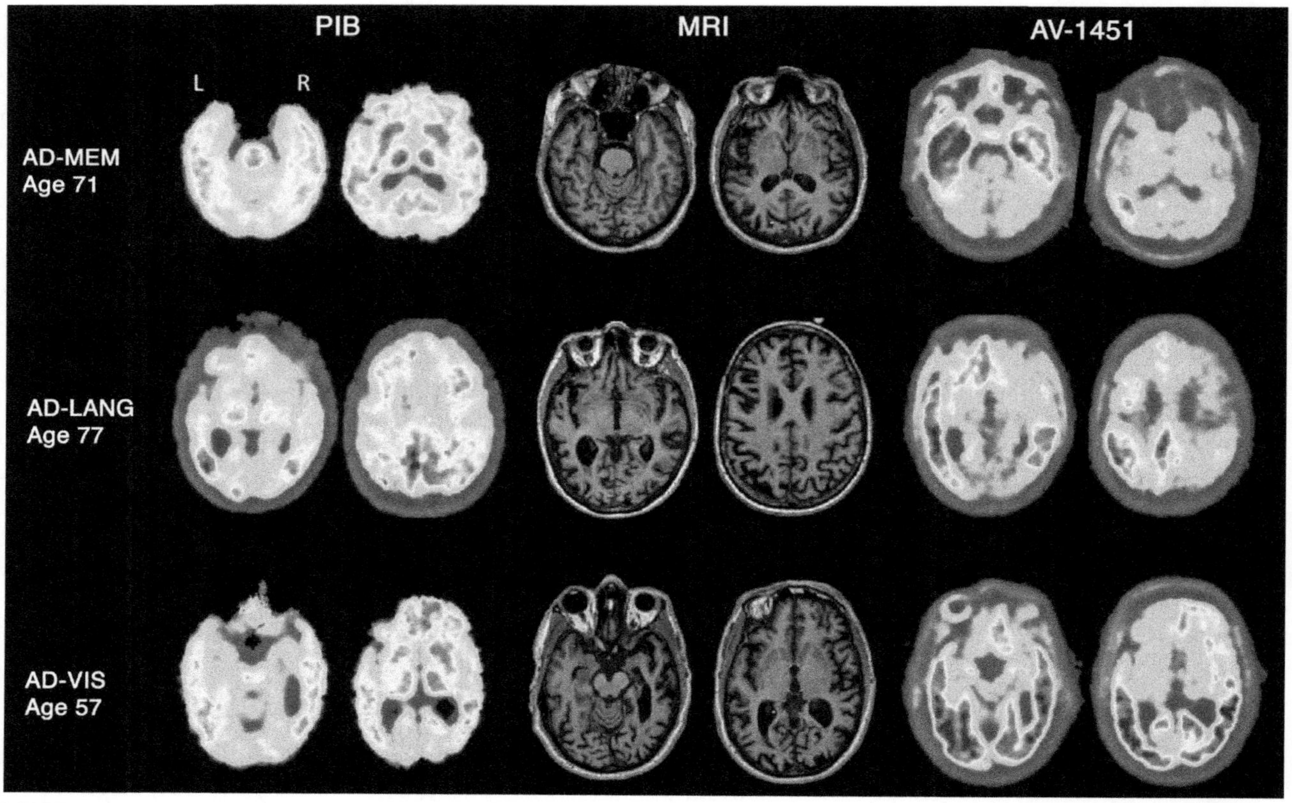

FIGURE 17.6 Amyloid PET (PIB), structural MRI, and tau PET (AV-1451) are shown for patients with AD with variable clinical presentations, including a patient with predominant memory deficits (AD-MEM), a patient with language deficits associated with the logopenic variant of primary progressive aphasia (AD-LANG), and a patient with visuospatial deficits associated with posterior cortical atrophy (AD-VIS). Amyloid scans for all three patients show similarly diffuse cortical binding, whereas atrophy and tau PET patterns substantially overlap and vary according to clinical phenotype: prominent temporal (notably medial) involvement in the memory-impaired patient, asymmetric left temporoparietal involvement in the language-impaired patient, and occipital extending into temporoparietal (right > left) involvement in the visuospatial-impaired patient.

formation in older mice. These endogenous differences were linked to neuronal activity; by increasing or decreasing activity through pharmacological manipulation, the authors induced focal, corresponding changes in extracellular Aβ levels. Taken together, this research across neuroscience approaches has established concrete links between neuronal activity, aerobic glycolysis, functional hubs, and amyloid pathology, with potential implications for drug intervention and development of early-disease biomarkers that could be sensitive to maladaptive neuronal hyperactivity.

Tau Spreads Through Functional Networks

Seeley, Crawford, Zhou, Miller, and Greicius (2009) reported that regional gray matter atrophy (a surrogate for molecular pathology) in five dementia syndromes overlapped with specific fMRI connectivity networks that were defined in healthy controls (eg, default network in AD, dorsal sensorimotor association network in corticobasal syndrome, salience network in bvFTD). Several protein pathologies (including but not limited to tau) have been associated with these clinical syndromes, which suggests that distinct proteinopathies might spread in similar ways. An explanation for the link between functional networks and syndrome-specific atrophy patterns thus could not only help explain how tau pathology develops but could reveal general principles of neurodegenerative disease progression.

In a later paper from the same group, Zhou et al. (2010) elaborated on their original results by developing a model for pathological spread based on a priori assumptions about where a neurodegenerative disease begins. The authors defined disease epicenters (essentially regions of maximal atrophy) in each of their five clinical syndromes, and then used these epicenters as seeds to test hypotheses about how atrophy patterns in patients and functional connectivity networks in healthy adults might be related. They found evidence for a "transneuronal spread" hypothesis in which atrophy was highest in regions that normally maintain close functional connections to epicenters, and progressively less atrophy was observed in regions that were more functionally remote. This helped explain the initial observation that clinical syndromes were associated with distinct networks; assuming that the diseases that underlie these syndromes begin in different brain regions, they could all be expected to spread to closely connected, within-network regions before involving other networks. This transneuronal spread hypothesis has since been further refined and validated using a diffusion model of disease progression that predicts longitudinal atrophy from baseline atrophy and connectivity patterns (Raj, Kuceyeski, & Weiner, 2012; Raj et al., 2015).

Studies in vitro and in mice have found converging evidence that distinct conformations of tau and other misfolded protein pathologies can spread transneuronally in a self-propagating manner, thus demonstrating prion-like properties (Frost & Diamond, 2010; Walker & Jucker, 2015). In one study, several strains of misfolded tau protein were injected into the hippocampal formations of tau-transgenic mice and were shown to induce NFT-like pathology in synaptically connected regions (Sanders et al., 2014). In addition, these tau strains induced stable conformational changes via templated misfolding of normal functioning tau after exposure to misfolded tau, and the same strain could be transferred through several generations of mice, with similar results. The authors speculated that the existence of stable tau stains could help explain how the same protein pathology can cause many clinical syndromes.

As with amyloid, a combination of imaging and basic science approaches has facilitated rapid progress in our understanding of the biological processes that drive tau accumulation and place its characteristic staging in context. Tau PET studies will seek to validate and extend theories of transneuronal spread and may explore concepts of individual disease vulnerability based on network connectivity measures and patterns of early molecular pathology. The self-propagating qualities of tau strains could mean that amyloid-targeting drugs will have greatest efficacy in preclinical disease, between the emergence of amyloid plaques and early aggregation of NFTs. By the time cognitive impairment becomes apparent, tau pathology may be on an irreversible course that will require therapies for tau, instead of or in addition to amyloid, to slow or halt the progression of disease.

CONCLUSIONS

Neuroimaging has been applied to many avenues of AD research, and technological advances are continually moving the needle on the kinds of questions that can be addressed through imaging studies. In this chapter we discussed some of the ways in which MRI and PET have been used to characterize disease-related changes in brain structure and function, including through measurements of atrophy, white matter damage, hypometabolism, and functional connectivity disruption. We also reported on the current status of PET tracers that can visualize regional accumulations of AD pathology. Combined, these imaging measures have had an essential role in confirming aspects of the amyloid cascade hypothesis while helping to modify others, with growing interest being placed on understanding the role of tau in AD. In the final section, we considered questions about why AD pathology preferentially targets certain brain regions and through what mechanisms pathological lesions might spread. A convergence of neuroimaging and basic science studies has pointed to the importance of functional hubs and neuronal activity in determining amyloid plaque accumulation, whereas tau pathology has been linked to a prion-like spread through functional brain networks.

Stepping back, it is worth considering what impact this work may have on our understanding and eventual treatment of AD, as well as what relevance these imaging methods have to other topics in clinical neuroscience. The research we have reviewed suggests that the integration of multimodal neuroimaging and clinical information can improve the diagnostic accuracy of AD, and even in the absence of a cure this can carry immense consequences for patients and their families, who must plan for their future and seek appropriate care. As our ability to image molecular pathology improves, we will be able to diagnose AD earlier and develop individualized predictions for future disease course. With the emergence of effective disease-modifying medications, this information will be vital to help clinicians develop optimal treatment plans that cater to the needs of each patient. At the same time, neuroimaging will facilitate clinical trial development by improving diagnostic screening and enabling direct confirmation of target engagement, pharmacodynamic effects, and ultimately disease modification in relation to cognitive change.

Although most of our discussion has focused on neuroimaging advances in AD, we conclude by noting that many if not all of the technologies and approaches discussed in this chapter are being applied to research across neurological and psychiatric disorders. Neuroimaging offers powerful methods for studying multiple aspects of a disease simultaneously, relating biomarker findings to clinical observations, and tracking disease progression longitudinally. However, neuroimaging has limitations, and it is ultimately by combining imaging and other approaches that we can develop a more complete understanding of neurological diseases and their optimal treatments.

REFERENCES

Agosta, F., Pievani, M., Geroldi, C., Copetti, M., Frisoni, G. B., & Filippi, M. (2011). Resting state fMRI in Alzheimer's disease: beyond the default mode network. *Neurobiology of Aging, 33*, 1564–1578. S0197-4580(11)00223-5 [pii]\r10.1016/j.neurobiolaging.2011.06.007.

Agosta, F., Pievani, M., Sala, S., & Geroldi, C. (2011). White matter damage in Alzheimer disease and its relationship to gray matter atrophy. *Radiology, 258*, 853–863. http://dx.doi.org/10.1148/radiol.10101284/-/DC1.

Agosta, F., Vossel, K. A., Miller, B. L., Migliaccio, R., Bonasera, S. J., Filippi, M., ... Gorno-Tempini, M. L. (2009). Apolipoprotein E epsilon4 is associated with disease-specific effects on brain atrophy in Alzheimer's disease and frontotemporal dementia. *Proceedings of the National Academy of Sciences of the United States of America, 106*, 2018–2022. http://dx.doi.org/10.1073/pnas.0812697106.

Aizenstein, H., Nebes, R. D., Saxton, J. A., Price, J. C., Mathis, C. A., Tsopelas, N. D., ... Klunk, W. E. (2008). Frequent amyloid deposition without significant cognitive impairment among the elderly. *Archives of Neurology, 65*, 1509–1517. http://dx.doi.org/10.1001/archneur.65.11.1509.Frequent.

Albert, M. S., DeKosky, S. T., Dickson, D. W., Dubois, B., Feldman, H. H., Fox, N. C., ... Phelps, C. H. (2011). The diagnosis of mild cognitive impairment due to Alzheimer's disease: recommendations from the national institute of aging-Alzheimer's asssociation workgroups on diagnostic guidelines for Alzheimer's disease. *Alzheimer's Dementia, 7*, 270–279. http://dx.doi.org/10.1016/j.jalz.2011.03.008.

Alexander, G. E., Chen, K., Pietrini, P., Rapoport, S. I., & Reiman, E. M. (2002). Longitudinal PET evaluation of cerebral metabolic decline in dementia: a potential outcome measure in Alzheimer's disease treatment studies. *American Journal of Psychiatry, 159*, 738–745. http://dx.doi.org/10.1176/appi.ajp.159.5.738.

Allen, G., Barnard, H., McColl, R., Hester, A. L., Fields, J. A., Weiner, M. F., ... Cullum, C. M. (2007). Reduced hippocampal functional connectivity in Alzheimer disease. *Archives of Neurology, 64*, 1482–1487. http://dx.doi.org/10.1001/archneur.64.10.1482.

Altmann, A., Ng, B., Landau, S. M., Jagust, W. J., & Greicius, M. D. (2015). Regional brain hypometabolism is unrelated to regional amyloid plaque burden. *Brain*, 1–13. http://dx.doi.org/10.1093/brain/awv278.

Amieva, H., Letenneur, L., Dartigues, J. F., Rouch-Leroyer, I., Sourgen, C., D'Alchee-Biree, F., ... Fabrigoule, C. (2004). Annual rate and predictors of conversion to dementia in subjects presenting mild cognitive impairment criteria defined according to a population-based study. *Dementia and Geriatric Cognitive Disorders, 18*, 87–93. http://dx.doi.org/10.1159/000077815.

Andrews-Hanna, J. R., Smallwood, J., & Spreng, R. N. (2014). The default network and self-generated thought: component processes, dynamic control, and clinical relevance. *Annals of the New York Academy of Sciences, 1316*, 29–52. http://dx.doi.org/10.1111/nyas.12360.

Arenaza-Urquijo, E. M., Wirth, M., & Chételat, G. (2015). Cognitive reserve and lifestyle: moving towards preclinical Alzheimer's disease. *Frontiers in Aging Neuroscience, 7*, 1–12. http://dx.doi.org/10.3389/fnagi.2015.00134.

Arnaiz, E., Jelic, V., Almkvist, O., Wahlund, L. O., Winblad, B., Valind, S., & Nordberg, A. (2001). Impaired cerebral glucose metabolism and cognitive functioning predict deterioration in mild cognitive impairment. *NeuroReport, 12*, 851–855. http://dx.doi.org/10.1097/00001756-200103260-00045.

Ball, M. J., Hachinski, V., Fox, A., Kirshen, A. J., Fisman, M., Blume, W., ... Fox, H. (1985). A new definition of Alzheimer's disease: a hippocampal dementia. *Lancet, 325*, 14–16. http://dx.doi.org/10.1016/S0140-6736(85)90965-1.

Ballatore, C., Lee, V. M. Y., & Trojanowski, J. Q. (2007). Tau-mediated neurodegeneration in Alzheimer's disease and related disorders. *Nature Reviews Neuroscience, 8*, 663–672. http://dx.doi.org/10.1038/nrn2194.

Bateman, R. J., Xiong, C., Benzinger, T. L. S., Fagan, A. M., Goate, A. M., Fox, N. C., ... Morris, J. C. (2012). Clinical and biomarker changes in dominantly inherited Alzheimer's disease. *New England Journal of Medicine, 367*, 795–804. http://dx.doi.org/10.1056/NEJMoa1202753.

Bauernfeind, A. L., Barks, S. K., Duka, T., Grossman, L. I., Hof, P. R., & Sherwood, C. C. (2014). Aerobic glycolysis in the primate brain: reconsidering the implications for growth and maintenance. *Brain Structure & Function, 219*, 1149–1167. http://dx.doi.org/10.1007/s00429-013-0662-z.

Bennett, D. A., Schneider, J. A., Arvanitakis, Z., Kelly, J. F., Aggarwal, N. T., Shah, R. C., & Wilson, R. S. (2006). Neuropathology of older persons without cognitive impairment from two community-based studies. *Neurology, 66*, 1837–1844. http://dx.doi.org/10.1212/01.wnl.0000219668.47116.e6.

Bero, A. W., Yan, P., Roh, J. H., Cirrito, J. R., Stewart, F. R., Raichle, M. E., ... Holtzman, D. M. (2011). Neuronal activity regulates the regional vulnerability to amyloid-β deposition. *Nature Neuroscience, 14*, 750–756. http://dx.doi.org/10.1038/nn.2801.

Besson, J. A. O., Corrigan, F. M., Foreman, E. I., Eastwood, L. M., Smith, F. W., & Ashcroft, G. W. (1985). Nuclear magnetic resonance (NMR) II. Imaging in dementia. *British Journal of Psychiatry, 146*, 31–35.

Bloom, G. S. (2014). Amyloid-β and Tau: the trigger and bullet in Alzheimer disease pathogenesis. *JAMA Neurology, 71*, 505. http://dx.doi.org/10.1001/jamaneurol.2013.5847.

Bobinki, M., de Leon, M. J., Convit, A., De Santi, S., Wegiel, J., Tarshish, C. Y., ... Wisniewski, H. M. (1999). MRI of entorhinal cortex in mild Alzheimer's disease. *Lancet, 353*, 38–40. http://dx.doi.org/10.1016/S0140-6736(05)74870-4.

Bozeat, S., Gregory, C. A., Ralph, M. A., & Hodges, J. R. (2000). Which neuropsychiatric and behavioural features distinguish frontal and temporal variants of frontotemporal dementia from Alzheimer's disease? *Journal of Neurology Neurosurgery and Psychiatry, 69*, 178–186. http://dx.doi.org/10.1136/jnnp.69.2.178.

Braak, H., & Braak, E. (1991). Neuropathological staging of Alzheimer-related changes. *Acta Neuropathologica, 82*, 239–259.

Brookmeyer, R., Johnson, E., Ziegler-Graham, K., & Arrighi, H. M. (2007). Forecasting the global burden of Alzheimer's disease. *Alzheimer's Dementia, 3*, 186–191. http://dx.doi.org/10.1016/j.jalz.2007.04.381.

Buckner, R. L., Andrews-Hanna, J. R., & Schacter, D. L. (2008). The brain's default network: anatomy, function, and relevance to disease. *Annals of the New York Academy of Sciences, 1124*, 1–38. http://dx.doi.org/10.1196/annals.1440.011.

Buckner, R. L., Sepulcre, J., Talukdar, T., Krienen, F. M., Liu, H., Hedden, T., ... Johnson, K. A. (2009). Cortical hubs revealed by intrinsic functional connectivity: mapping, assessment of stability, and relation to Alzheimer's disease. *Journal of Neuroscience, 29*, 1860–1873. http://dx.doi.org/10.1523/JNEUROSCI.5062-08.2009.

Buerger, K., Ewers, M., Pirttila, T., Zinkowski, R., Alafuzoff, I., Teipel, S. J., … Hampel, H. (2006). CSF phosphorylated tau protein correlates with neocortical neurofibrillary pathology in Alzheimer's disease. *Brain, 129*, 3035–3041. http://dx.doi.org/10.1093/brain/awl269.

Burggren, A. C., Zeineh, M. M., Ekstrom, A. D., Braskie, M. N., Thompson, P. M., Small, G. W., & Bookheimer, S. Y. (2008). Reduced cortical thickness in hippocampal subregions among cognitively normal apolipoprotein E e4 carriers. *NeuroImage, 41*, 1177–1183. http://dx.doi.org/10.1016/j.neuroimage.2008.03.039.

Chételat, G., Desgranges, B., de la Sayette, V., Viader, F., Eustache, F., & Baron, J. C. (2003). Mild cognitive impairment: can FDG-PET predict who is to rapidly convert to Alzheimer's disease? *Neurology, 60*, 1374–1377. http://dx.doi.org/10.1212/01.WNL.0000055847.17752.E6.

Chételat, G., Villemagne, V. L., Villain, N., Jones, G., Ellis, K. A., Ames, D., … Group, A. R. (2012). Accelerated cortical atrophy in cognitively normal elderly with high β-amyloid deposition. *Neurology, 78*, 477–484. http://dx.doi.org/10.1212/WNL.0b013e318246d67a.

Chien, D. T., Bahri, S., Szardenings, A. K., Walsh, J. C., Mu, F., Su, M.-Y., … Kolb, H. C. (2013). Early clinical PET imaging results with the novel PHF-tau radioligand [F-18]-T807. *Journal of Alzheimer's Disease, 34*, 457–468. http://dx.doi.org/10.3233/JAD-122059.

Clark, C. M., Schneider, J. A., Bedell, B. J., Beach, T. G., Bilker, W. B., Mintun, M. A., … Skovronsky, D. M. (2011). Use of florbetapir-PET for imaging β-amyloid pathology. *JAMA, 305*, 275–283. http://dx.doi.org/10.1001/jama.2010.2008.

Crutch, S. J., Lehmann, M., Schott, J. M., Rabinovici, G. D., Rossor, M. N., & Fox, N. C. (2012). Posterior cortical atrophy. *Lancet Neurology, 11*, 170–178. http://dx.doi.org/10.1016/S1474-4422(11)70289-7.

Curtis, C., Gamez, J. E., Singh, U., Sadowsky, C. H., Villena, T., Sabbagh, M. N., … Salloway, S. (2015). Phase 3 trial of flutemetamol labeled with radioactive fluorine 18 imaging and neuritic plaque density. *JAMA Neurology, 72*, 287–294. http://dx.doi.org/10.1001/jamaneurol.2014.4144.

Damoiseaux, J. S., Prater, K. E., Miller, B. L., & Greicius, M. D. (2012). Functional connectivity tracks clinical deterioration in Alzheimer's disease. *Neurobiology of Aging, 33*, 828. http://dx.doi.org/10.1016/j.neurobiolaging.2011.06.024. e19–30.

De Calignon, A., Polydoro, M., Suárez-Calvet, M., William, C., Adamowicz, D. H., Kopeikina, K. J., … Hyman, B. T. (2012). Propagation of tau pathology in a model of early Alzheimer's disease. *Neuron, 73*, 685–697. http://dx.doi.org/10.1016/j.neuron.2011.11.033.

Dean, D. C., Jerskey, B. A., Chen, K., Protas, H., Thiyyagura, P., Roontiva, A., … Reiman, E. M. (2014). Brain differences in infants at differential genetic risk for late-onset Alzheimer disease: a cross-sectional imaging study. *JAMA Neurology, 71*, 11–22. http://dx.doi.org/10.1001/jamaneurol.2013.4544.

Dickerson, B. C., Goncharova, I., Sullivan, M. P., Forchetti, C., Wilson, R. S., Bennett, D. A., … DeToledo-Morrell, L. (2001). MRI-derived entorhinal and hippocampal atrophy in incipient and very mild Alzheimer's disease. *Neurobiology of Aging, 22*, 747–754. http://dx.doi.org/10.1016/S0197-4580(01)00271-8.

Doraiswamy, P. M., Sperling, R. A., Coleman, R. E., Johnson, K. A., Reiman, E. M., Davis, M. D., … Pontecorvo, M. J. (2012). Amyloid-β assessed by florbetapir F 18 PET and 18-month cognitive decline: a multicenter study. *Neurology, 79*, 1636–1644. http://dx.doi.org/10.1212/WNL.0b013e3182661f74.

Driscoll, I., Resnick, S. M., Troncoso, J. C., An, Y., O'Brien, R., & Zonderman, A. B. (2006). Impact of Alzheimer's pathology on cognitive trajectories in nondemented elderly. *Annals of Neurology, 60*, 688–695. http://dx.doi.org/10.1002/ana.21031.

Drzezga, A., Lautenschlager, N., Siebner, H., Riemenschneider, M., Willoch, F., Minoshima, S., … Kurz, A. (2003). Cerebral metabolic changes accompanying conversion of mild cognitive impairment into Alzheimer's disease: a PET follow-up study. *European Journal of Nuclear Medicine and Molecular Imaging, 30*, 1104–1113. http://dx.doi.org/10.1007/s00259-003-1194-1.

Edison, P., Archer, H., Hinz, R., Hammers, A., Pavese, N., Tai, Y., … Brooks, D. J. (2007). Amyloid, hypometabolism, and cognition in Alzheimer's disease: an [11C]PIB and [18F]FDG PET study. *Neurology, 68*, 501–508.

Elman, J. A., Oh, H., Madison, C. M., Baker, S. L., Vogel, J. W., Marks, S. M., … Jagust, W. J. (2014). Neural compensation in older people with brain amyloid-β deposition. *Nature Neuroscience, 17*, 1316–1318. http://dx.doi.org/10.1038/nn.3806.

Engler, H., Forsberg, A., Almkvist, O., Blomquist, G., Larsson, E., Savitcheva, I., … Nordberg, A. (2006). Two-year follow-up of amyloid deposition in patients with Alzheimer's disease. *Brain, 129*, 2856–2866. http://dx.doi.org/10.1093/brain/awl178.

Fargo, K. (2014). Alzheimer's association report: 2014 Alzheimers disease facts and figures. *Alzheimer's Dementia, 10*, e47–e92. http://dx.doi.org/10.1016/j.jalz.2014.02.001.

Farrer, L. A., Cupples, L. A., Haines, J. L., Hyman, B. T., Kukull, W. A., Mayeux, R., … van Duijn, C. M. (1997). Effects of age, sex, and ethnicity on the association between apolipoprotein e genotype and alzheimer disease. *JAMA, 278*, 1349–1356. http://dx.doi.org/10.1001/jama.1997.03550160069041.

Fazekas, F., Chawluk, J. B., & Alavi, A. (1987). MR signal abnormalities at 1.5 T in Alzheimer's dementia and normal aging. *American Journal of Neuroradiology, 8*, 421–426. http://dx.doi.org/10.2214/ajr.149.2.351.

Ferris, S. H., de Leon, M. J., Wolf, A. P., Farkas, T., Christman, D. R., Reisberg, B., … Rampal, S. (1980). Positron emission tomography in the study of aging and senile dementia. *Neurobiology of Aging, 1*, 127–131.

Freyhan, F., Woodford, R., & Kety, S. (1951). Cerebral blood flow and metabolism in psychoses of senility. *Journal of Nervous and Mental Disease, 113*, 449–456.

Friedland, R. P., Budinger, T. F., Koss, E., & Ober, B. A. (1985). Alzheimer's disease: anterior-posterior and lateral hemispheric alterations in cortical glucose utilization. *Neuroscience Letters, 53*, 235–240.

Frisoni, G. B., Pievani, M., Testa, C., Sabattoli, F., Bresciani, L., Bonetti, M., … Thompson, P. M. (2007). The topography of grey matter involvement in early and late onset Alzheimer's disease. *Brain, 130*, 720–730. http://dx.doi.org/10.1093/brain/awl377.

Frost, B., & Diamond, M. I. (2010). Prion-like mechanisms in neurodegenerative diseases. *Nature Reviews Neuroscience, 11*, 155–159. http://dx.doi.org/10.1038/nrn2786.

Furst, A. J., Rabinovici, G. D., Rostomian, A. H., Steed, T., Alkalay, A., Racine, C., … Jagust, W. J. (2012). Cognition, glucose metabolism and amyloid burden in Alzheimer's disease. *Neurobiology of Aging, 33*, 215–225. http://dx.doi.org/10.1016/j.biotechadv.2011.08.021.Secreted.

Gomez-Isla, T., Hollister, R., West, H., Mui, S., Growdon, J. H., Petersen, R. C., ... Hyman, B. T. (1997). Neuronal loss correlates with but exceeds neurofibrillary tangles in Alzheimer's disease. *Annals of Neurology*, 17–24.

Gorno-Tempini, M. L., Hillis, A. E., Weintraub, S., Kertesz, A., Mendez, M., Cappa, S. F., ... Grossman, M. (2011). Classification of primary progressive aphasia and its variants. *Neurology, 76*, 1006–1014. http://dx.doi.org/10.1212/WNL.0b013e31821103e6.

Götz, J., & Ittner, L. M. (2008). Animal models of Alzheimer's disease and frontotemporal dementia. *Nature Reviews Neuroscience, 9*, 532–544. http://dx.doi.org/10.1038/nrn2420.

Greicius, M. D., Krasnow, B., Reiss, A. L., & Menon, V. (2003). Functional connectivity in the resting brain: a network analysis of the default mode hypothesis. *Proceedings of the National Academy of Sciences of the United States of America, 100*, 253–258. http://dx.doi.org/10.1073/pnas.0135058100.

Greicius, M. D., Srivastava, G., Reiss, A. L., & Menon, V. (2004). Default-mode network activity distinguishes Alzheimer's disease from healthy aging : evidence from functional MRI. *Proceedings of the National Academy of Sciences of the United States of America, 101*, 4637–4642.

Harada, R., Okamura, N., Furumoto, S., Furukawa, K., Ishiki, A., Tomita, N., ... Kudo, Y. (2015). [(18)F]THK-5117 PET for assessing neurofibrillary pathology in Alzheimer's disease. *European Journal of Nuclear Medicine and Molecular Imaging, 42*, 1052–1061. http://dx.doi.org/10.1007/s00259-015-3035-4.

Hardy, J., & Higgins, G. A. (1992). Alzheimer's disease: the amyloid Cascade hypothesis. *Science, 256*, 184–185.

Hardy, J., & Selkoe, D. J. (2002). The amyloid hypothesis of Alzheimer's disease: progress and problems on the road to therapeutics. *Science, 297*, 353–356. http://dx.doi.org/10.1126/science.1072994.

Hashimoto, H., Kawamura, K., Igarashi, N., Takei, M., Fujishiro, T., Aihara, Y., ... Zhang, M.-R. (2014). Radiosynthesis, photoisomerization, biodistribution, and metabolite analysis of 11C-PBB3 as a clinically useful PET probe for imaging of tau pathology. *Journal of Nuclear Medicine, 55*(9), 1532–1538. http://dx.doi.org/10.2967/jnumed.114.139550.

Haxby, J. V., Duara, R., Grady, C. L., Cutler, N. R., & Rapoport, S. I. (1985). Relations between neuropsychological and cerebral metabolic asymmetries in early Alzheimer's disease. *Journal of Cerebral Blood Flow and Metabolism, 5*, 193–200. http://dx.doi.org/10.1038/jcbfm.1985.25.

Herholz, K., Salmon, E., Perani, D., Baron, J. C., Holthoff, V., Frölich, L., ... Heiss, W. D. (2002). Discrimination between Alzheimer dementia and controls by automated analysis of multicenter FDG PET. *NeuroImage, 17*, 302–316. http://dx.doi.org/10.1006/nimg.2002.1208.

van den Heuvel, M. P., & Sporns, O. (2013). Network hubs in the human brain. *Trends in Cognitive Sciences, 17*, 683–696. http://dx.doi.org/10.1016/j.tics.2013.09.012.

Hoffman, J. M., Welsh-Bohmer, K. A., Hanson, M., Crain, B., Hulette, C. M., Earl, N., & Coleman, R. E. (2000). FDG PET imaging in patients with pathologically verified dementia. *Journal of Nuclear Medicine, 41*, 1920–1928.

Huettel, S. A., Song, A. W., & McCarthy, G. (2014). *Functional magnetic resonance imaging* (3rd ed.). Sunderland: Sinauer Associates.

Hulette, C. M., Welsh-Bohmer, K. A., Murray, M. G., Saunders, A. M., Mash, D. C., & McIntyre, L. M. (1998). Neuropathological and neuropsychological changes in "normal" aging: evidence for preclinical Alzheimer disease in cognitively normal individuals. *Journal of Neuropathology & Experimental Neurology, 57*, 1168–1174. http://dx.doi.org/10.1097/00005072-199812000-00009.

Hyman, B. T., Van Hoesen, G. W., Damasio, A. R., & Barnes, C. L. (1984). Alzheimer's disease: cell-specific pathology isolates the hippocampal formation. *Science, 225*, 1168–1170. http://dx.doi.org/10.1038/098448b0.

Ikonomovic, M. D., Klunk, W. E., Abrahamson, E. E., Mathis, C. A., Price, J. C., Tsopelas, N. D., ... DeKosky, S. T. (2008). Post-mortem correlates of in vivo PiB-PET amyloid imaging in a typical case of Alzheimer's disease. *Brain, 131*, 1630–1645. http://dx.doi.org/10.1093/brain/awn016.

Iqbal, K., Liu, F., Gong, C. X., del Alonso, A. C., & Grundke-Iqbal, I. (2009). Mechanisms of tau-induced neurodegeneration. *Acta Neuropathologica, 118*, 53–69. http://dx.doi.org/10.1007/s00401-009-0486-3.

Ishiguro, K., Ohno, H., Arai, H., Yamaguchi, H., Uwakami, K., Park, J.-M., ... Imahori, K. (1999). Phosphorylated tau in human cerebrospinal fluid is a diagnostic marker for Alzheimer's disease. *Neuroscience Letters, 270*, 91–94. http://dx.doi.org/10.1016/S0304-3940(99)00476-0.

Ishiki, A., Okamura, N., Furukawa, K., Furumoto, S., Harada, R., Tomita, N., ... Arai, H. (2015). Longitudinal assessment of tau pathology in patients with Alzheimer's disease using [18F]THK-5117 positron emission tomography. *PLoS One, 10*, e0140311. http://dx.doi.org/10.1371/journal.pone.0140311.

Jack, C. R., Knopman, D. S., Jagust, W. J., Petersen, R. C., Weiner, M. W., Aisen, P. S., ... Trojanowski, J. Q. (2013). Tracking pathophysiological processes in Alzheimer's disease: an updated hypothetical model of dynamic biomarkers. *Lancet Neurology, 12*, 207–216.

Jack, C. R., Petersen, R. C., Xu, Y. C., O'Brien, P. C., Smith, G. E., Ivnik, R. J., ... Kokmen, E. (1999). Prediction of AD with MRI-based hippocampal volume in mild cognitive impairment. *Neurology, 52*, 1397–1403. http://dx.doi.org/10.1212/WNL.52.7.1397.

Jack, C. R., Wiste, H. J., Lesnick, T. G., Weigand, S. D., Knopman, D. S., Vemuri, P., ... Petersen, R. C. (2013). Brain β-amyloid load approaches a plateau. *Neurology, 80*, 890–896. http://dx.doi.org/10.1212/WNL.0b013e3182840bbe.

Jagust, W. J., Friedland, R. P., Budinger, T. F., Koss, E., & Ober, B. (1988). Longitudinal studies of regional cerebral metabolism in Alzheimer's disease. *Neurology, 38*, 909–912.

Jagust, W. J., & Landau, S. M. (2012). Apolipoprotein E, not fibrillar β-amyloid, reduces cerebral glucose metabolism in normal aging. *Journal of Neuroscience, 32*, 18227–18233. http://dx.doi.org/10.1523/JNEUROSCI.3266-12.2012.

Jagust, W. J., Landau, S. M., Koeppe, R. A., Reiman, E. M., Chen, K., Mathis, C. A., ... Wang, A. Y. (2015). The Alzheimer's disease neuroimaging initiative 2 PET core: 2015. *Alzheimer's Dementia, 11*, 757–771. http://dx.doi.org/10.1016/j.jalz.2015.05.001.

Jansen, W. J., Ossenkoppele, R., Knol, D. L., Tijms, B. M., Scheltens, P., Verhey, F. R. J., ... Zetterberg, H. (2015). Prevalence of cerebral amyloid pathology in persons without dementia. *JAMA, 313*, 1924. http://dx.doi.org/10.1001/jama.2015.4668.

Johnson, K. A., Minoshima, S., Bohnen, N. I., Donohoe, K. J., Foster, N. L., Herscovitch, P., ... Thies, W. H. (2013). Appropriate use criteria for amyloid PET: a report of the amyloid imaging task force, the society of nuclear medicine and molecular imaging, and the Alzheimer's association. *Journal of Nuclear Medicine, 54*, 476–490. http://dx.doi.org/10.2967/jnumed.113.120618.

Johnson, K. A., Schultz, A. P., Betensky, R. A., Becker, J. A., Sepulcre, J., Rentz, D. M., ... Sperling, R. A. (2015). Tau PET imaging in aging and early Alzheimer's disease. *Annals of Neurology*. http://dx.doi.org/10.1111/peps.12055.

Johnson, S. C., Christian, B. T., Okonkwo, O. C., Oh, J. M., Harding, S., Xu, G., ... Sager, M. A. (2014). Amyloid burden and neural function in people at risk for Alzheimer's Disease. *Neurobiology of Aging, 35*, 576–584. http://dx.doi.org/10.1016/j.neurobiolaging.2013.09.028.

Jones, D. T., MacHulda, M. M., Vemuri, P., McDade, E., Zeng, G., Senjem, M. L., ... Jack, C. R. (2011). Age-related changes in the default mode network are more advanced in Alzheimer disease. *Neurology, 77*, 1524–1531. http://dx.doi.org/10.1212/WNL.0b013e318233b33d.

Juottonen, K., Laakso, M. P., Insausti, R., Lehtovirta, M., Pitkänen, A., Partanen, K., & Soininen, H. (1998). Volumes of the entorhinal and perirhinal cortices in Alzheimer's disease. *Neurobiology of Aging, 19*, 15–22. http://dx.doi.org/10.1016/S0197-4580(98)00007-4.

Kantarci, K., Lowe, V., Przybelski, S. A., Weigand, S. D., Senjem, M. L., Ivnik, R. J., ... Jack, C. R. (2012). APOE modifies the association between Aβ load and cognition in cognitively normal older adults. *Neurology, 78*, 232–240. http://dx.doi.org/10.1212/WNL.0b013e31824365ab.

Klunk, W. E., Engler, H., Nordberg, A., Wang, Y., Blomquist, G., Holt, D. P., ... Långström, B. (2004). Imaging brain amyloid in Alzheimer's disease with Pittsburgh compound-b. *Annals of Neurology, 55*, 306–319. http://dx.doi.org/10.1002/ana.20009.

Koivunen, J., Scheinin, N., Virta, J. R., Aalto, S., Vahlberg, T., Någren, K., ... Rinne, J. O. (2011). Amyloid PET imaging in patients with mild cognitive impairment: a 2-year follow-up study. *Neurology, 76*, 1085–1090. http://dx.doi.org/10.1212/WNL.0b013e318212015e.

Lambert, J.-C., Ibrahim-Verbaas, C. A., Harold, D., Naj, A. C., Sims, R., Bellenguez, C., ... Amouyel, P. (2013). Meta-analysis of 74,046 individuals identifies 11 new susceptibility loci for Alzheimer's disease. *Nature Genetics, 45*, 1452–1458. http://dx.doi.org/10.1038/ng.2802.

Landau, S. M., Harvey, D., Madison, C. M., Reiman, E. M., Foster, N. L., Aisen, P. S., ... Jagust, W. J. (2010). Comparing predictors of conversion and decline in mild cognitive impairment. *Neurology, 75*, 230–238. http://dx.doi.org/10.1212/WNL.0b013e3181e8e8b8.

Landau, S. M., Lu, M., Joshi, A. D., Pontecorvo, M. J., Mintun, M.a, Trojanowski, J. Q., ... Jagust, W. J. (2013). Comparing positron emission tomography imaging and cerebrospinal fluid measurements of β-amyloid. *Annals of Neurology, 74*, 826–836. http://dx.doi.org/10.1002/ana.23908.

Landau, S. M., Mintun, M. A., Joshi, A. D., Koeppe, R. A., Petersen, R. C., Aisen, P. S., ... Jagust, W. J. (2012). Amyloid deposition, hypometabolism, and longitudinal cognitive decline. *Annals of Neurology, 72*, 578–586. http://dx.doi.org/10.1002/ana.23650.

Lee, V. M., Goedert, M., & Trojanowski, J. Q. (2001). Neurodegenerative tauopathies. *Annual Review of Neuroscience, 24*, 1121–1159.

Lehmann, M., Ghosh, P. M., Madison, C. M., Laforce, R., Corbetta-Rastelli, C., Weiner, M. W., ... Rabinovici, G. D. (2013). Diverging patterns of amyloid deposition and hypometabolism in clinical variants of probable Alzheimer's disease. *Brain, 136*, 844–858.

Lehmann, M., Madison, C. M., Ghosh, P. M., Seeley, W. W., Mormino, E. C., Greicius, M. D., ... Rabinovici, G. D. (2013). Intrinsic connectivity networks in healthy subjects explain clinical variability in Alzheimer's disease. *Proceedings of the National Academy of Sciences of the United States of America, 110*, 11606–11611. http://dx.doi.org/10.1073/pnas.1221536110.

Lerch, J. P. (2004). Focal decline of cortical thickness in Alzheimer's disease identified by computational neuroanatomy. *Cerebral Cortex, 15*, 995–1001. http://dx.doi.org/10.1093/cercor/bhh200.

Liu, C.-C., Kanekiyo, T., Xu, H., & Bu, G. (2013). Apolipoprotein E and Alzheimer disease: risk, mechanisms, and therapy. *Nature Reviews Neurology, 9*, 106–118. http://dx.doi.org/10.1016/j.biotechadv.2011.08.021.Secreted.

Liu, Y., Wang, K., Yu, C., He, Y., Zhou, Y., Liang, M., ... Jiang, T. (2008). Regional homogeneity, functional connectivity and imaging markers of Alzheimer's disease: a review of resting-state fMRI studies. *Neuropsychologia, 46*, 1648–1656. http://dx.doi.org/10.1016/j.neuropsychologia.2008.01.027.

Lui, J. K., Laws, S. M., Li, Q.-X., Villemagne, V. L., Ames, D., Brown, B., ... AIBL Research Group. (2010). Plasma amyloid-beta as a biomarker in Alzheimer's disease: the AIBL study of aging. *Journal of Alzheimer's Disease, 20*, 1233–1242. http://dx.doi.org/10.3233/JAD-2010-090249.

Mander, B. A., Marks, S. M., Vogel, J. W., Rao, V., Lu, B., Saletin, J. M., ... Walker, M. P. (2015). β-amyloid disrupts human NREM slow waves and related hippocampus-dependent memory consolidation. *Nature Neuroscience, 18*, 1051–1057. http://dx.doi.org/10.1038/nn.4035.

Mangialasche, F., Solomon, A., Winblad, B., Mecocci, P., & Kivipelto, M. (2010). Alzheimer's disease: clinical trials and drug development. *Lancet Neurology, 9*, 702–716. http://dx.doi.org/10.1016/S1474-4422(10)70119-8.

Marquié, M., Normandin, M. D., Vanderburg, C. R., Costantino, I. M., Bien, E. A., Rycyna, L. G., ... Gómez-Isla, T. (2015). Validating novel tau positron emission tomography tracer [F-18]-AV-1451 (T807) on postmortem brain tissue. *Annals of Neurology, 1451*. http://dx.doi.org/10.1002/ana.24517. n/a–n/a.

Maruyama, M., Shimada, H., Suhara, T., Shinotoh, H., Ji, B., Maeda, J., ... Higuchi, M. (2013). Imaging of tau pathology in a Tauopathy mouse model and in alzheimer patients compared to Normal controls. *Neuron, 79*, 1094–1108. http://dx.doi.org/10.1016/j.neuron.2013.07.037.

Mintun, M. A., LaRossa, G. N., Sheline, Y. I., Dence, C., Lee, S., Mach, R., ... Morris, J. C. (2006). [11C] PIB in a nondemented population: potential antecedent marker of Alzheimer disease. *Neurology, 67*, 446–452. http://dx.doi.org/10.1212/01.wnl.0000228230.26044.a4.

Mitchell, A. J., & Shiri-Feshki, M. (2009). Rate of progression of mild cognitive impairment to dementia – meta-analysis of 41 robust inception cohort studies. *Acta Psychiatrica Scandinavica, 119*, 252–265. http://dx.doi.org/10.1111/j.1600-0447.2008.01326.x.

Montine, T. J., Phelps, C. H., Beach, T. G., Bigio, E. H., Cairns, N. J., Dickson, D. W., ... Hyman, B. T. (2012). National institute on aging-Alzheimer's association guidelines for the neuropathologic assessment of Alzheimer's disease: a practical approach. *Acta Neuropathologica, 123*, 1–11. http://dx.doi.org/10.1007/s00401-011-0910-3.

Mormino, E. C., Betensky, R. A., Hedden, T., Schultz, A. P., Amariglio, R. E., Rentz, D. M., ... Sperling, R. A. (2014). Synergistic effect of β-amyloid and neurodegeneration on cognitive decline in clinically normal individuals. *JAMA Neurology, 71*, 1379. http://dx.doi.org/10.1001/jamaneurol.2014.2031.

Mormino, E. C., Betensky, R. A., Hedden, T., Schultz, A. P., Ward, A., Huijbers, W., ... Sperling, R. A. (2014). Amyloid and APOE ε4 interact to influence short-term decline in preclinical Alzheimer disease. *Neurology, 82*(20), 1760–1767. http://dx.doi.org/10.1212/WNL.0000000000000431.

Mormino, E. C., Brandel, M. G., Madison, C. M., Marks, S. M., Baker, S. L., & Jagust, W. J. (2012). Aβ deposition in aging is associated with increases in brain activation during successful memory encoding. *Cerebral Cortex, 22*, 1813–1823. http://dx.doi.org/10.1093/cercor/bhr255.

Mormino, E. C., Kluth, J. T., Madison, C. M., Rabinovici, G. D., Baker, S. L., Miller, B. L., … Jagust, W. J. (2009). Episodic memory loss is related to hippocampal-mediated β-amyloid deposition in elderly subjects. *Brain, 132*, 1310–1323. http://dx.doi.org/10.1093/brain/awn320.

Nelson, P. T., Alafuzoff, I., Bigio, E. H., Bouras, C., Braak, H., Cairns, N. J., … Beach, T. G. (2012). Correlation of Alzheimer disease neuropathologic changes with cognitive status: a review of the literature. *Journal of Neuropathology & Experimental Neurology, 71*, 362–381. http://dx.doi.org/10.1097/NEN.0b013e31825018f7.

Ng, S., Villemagne, V. L., Berlangieri, S., Lee, S.-T., Cherk, M., Gong, S. J., … Rowe, C. C. (2007). Visual assessment versus quantitative assessment of 11C-PIB PET and 18F-FDG PET for detection of Alzheimer's disease. *Journal of Nuclear Medicine, 48*, 547–552. http://dx.doi.org/10.2967/jnumed.106.037762.

Obrist, W. D., Chivian, E., Cronqvist, S., & Ingvar, D. H. (1970). Regional cerebral blood flow in senile and presenile dementia. *Neurology, 20*, 315. http://dx.doi.org/10.1212/WNL.20.4.315.

Oh, H., Madison, C., Haight, T. J., Markley, C., & Jagust, W. J. (2012). Effects of age and β-amyloid on cognitive changes in normal elderly people. *Neurobiology of Aging, 33*, 2746–2755. http://dx.doi.org/10.1016/j.neurobiolaging.2012.02.008.

Okamura, N., Furumoto, S., Harada, R., Tago, T., Yoshikawa, T., Fodero-Tavoletti, M. T., … Kudo, Y. (2013). Novel 18F-Labeled arylquinoline derivatives for noninvasive imaging of tau pathology in Alzheimer disease. *Journal of Nuclear Medicine, 54*, 1420–1427. http://dx.doi.org/10.2967/jnumed.112.117341.

Okamura, N., Harada, R., Furumoto, S., Arai, H., Yanai, K., & Kudo, Y. (2014). Tau PET imaging in Alzheimer's disease. *Current Neurology and Neuroscience Reports, 14*, 500. http://dx.doi.org/10.1007/s11910-014-0500-6.

Okamura, N., Suemoto, T., Furumoto, S., Suzuki, M., Shimadzu, H., Akatsu, H., … Kudo, Y. (2005). Quinoline and benzimidazole derivatives: candidate probes for in vivo imaging of tau pathology in Alzheimer's disease. *Journal of Neuroscience, 25*, 10857–10862. http://dx.doi.org/10.1523/JNEUROSCI.1738-05.2005.

Ossenkoppele, R., Cohn-Sheehy, B. I., La Joie, R., Vogel, J. W., Möller, C., Lehmann, M., … Rabinovici, G. D. (2015). Atrophy patterns in early clinical stages across distinct phenotypes of Alzheimer's disease. *Human Brain Mapping, 36*(11), 4421–4437. http://dx.doi.org/10.1002/hbm.22927.

Ossenkoppele, R., Madison, C. M., Oh, H., Wirth, M., van Berckel, B. N. M., & Jagust, W. J. (2014). Is verbal episodic memory in elderly with amyloid deposits preserved through altered neuronal function? *Cerebral Cortex, 24*, 2210–2218. http://dx.doi.org/10.1093/cercor/bht076.

Ossenkoppele, R., Pijnenburg, Y., Perry, D. C., Cohn-Sheehy, B. I., Scheltens, N. M. E., Vogel, J. W., … Rabinovici, G. D. (2015). The behavioural/dysexecutive variant of Alzheimer's disease: clinical, neuroimaging and pathological features. *Brain, 138*(Pt 9), 2732–2749. http://dx.doi.org/10.1093/brain/awv191. awv191.

Ossenkoppele, R., Schonhaut, D. R., Baker, S. L., O'Neil, J. P., Janabi, M., Ghosh, P. M., … Rabinovici, G. D. (2014). Tau, amyloid and hypometabolism in a patient with posterior cortical atrophy. *Annals of Neurology*, 1–5. http://dx.doi.org/10.1002/ana.24321.

Petersen, R. C., Smith, G. E., Waring, S. C., Ivnik, R. J., Tangalos, E. G., & Kokmen, E. (1999). Mild cognitive impairment: clinical characterization and outcome. *Archives of Neurology, 56*, 303–309. http://dx.doi.org/10.1001/archneur.56.3.303.

Pike, K. E., Savage, G., Villemagne, V. L., Ng, S., Moss, S. A., Maruff, P., … Rowe, C. C. (2007). β-amyloid imaging and memory in non-demented individuals: evidence for preclinical Alzheimer's disease. *Brain, 130*, 2837–2844. http://dx.doi.org/10.1093/brain/awm238.

Power, J. D., Cohen, A. L., Nelson, S. M., Wig, G. S., Barnes, K. A., Church, J. A., … Petersen, S. E. (2011). Functional network organization of the human brain. *Neuron, 72*, 665–678. http://dx.doi.org/10.1016/j.neuron.2011.09.006.

Price, J. C., Klunk, W. E., Lopresti, B. J., Lu, X., Hoge, J. A., Ziolko, S. K., … Mathis, C. A. (2005). Kinetic modeling of amyloid binding in humans using PET imaging and Pittsburgh Compound-B. *Journal of Cerebral Blood Flow and Metabolism, 25*, 1528–1547. http://dx.doi.org/10.1038/sj.jcbfm.9600146.

Prins, N. D., & Scheltens, P. (2015). White matter hyperintensities, cognitive impairment and dementia: an update. *Nature Reviews Neurology*, 1–9. http://dx.doi.org/10.1038/nrneurol.2015.10.

Rabinovici, G. D., Furst, A. J., Alkalay, A., Racine, C., O'Neil, J. P., Janabi, M., … Jagust, W. J. (2010). Increased metabolic vulnerability in early-onset Alzheimer's disease is not related to amyloid burden. *Brain, 133*, 512–528.

Rabinovici, G. D., Jagust, W. J., & Furst, A. J. (2008). Aß amyloid and glucose metabolism in three variants of primary progressive aphasia. *Annals of Neurology, 64*, 388–401. http://dx.doi.org/10.1002/ana.21451.A.

Raichle, M. E. (2010). Two views of brain function. *Trends in Cognitive Sciences, 14*, 180–190. http://dx.doi.org/10.1016/j.tics.2010.01.008.

Raj, A., Kuceyeski, A., & Weiner, M. W. (2012). A network diffusion model of disease progression in dementia. *Neuron, 73*, 1204–1215. http://dx.doi.org/10.1016/j.neuron.2011.12.040.

Raj, A., LoCastro, E., Kuceyeski, A., Tosun, D., Relkin, N., & Weiner, M. W. (2015). Network diffusion model of progression predicts longitudinal patterns of atrophy and metabolism in Alzheimer's disease. *Cell Reports, 10*, 359–369. http://dx.doi.org/10.1016/j.celrep.2014.12.034.

Reiman, E. M., Caselli, R. J., Yun, L. S., Chen, K., Bandy, D., Minoshima, S., … Osborne, D. (1996). Preclinical evidence of Alzheimer's disease in persons homozygous for the epsilon 4 allele for apolipoprotein E. *New England Journal of Medicine, 334*, 752–758. http://dx.doi.org/10.1056/NEJM199603213341202.

Reiman, E. M., & Jagust, W. J. (2012). Brain imaging in the study of Alzheimer's disease. *NeuroImage, 61*, 505–516. http://dx.doi.org/10.1016/j.neuroimage.2011.11.075.

Reivich, M., Kuhl, D., Wolf, A., Greenberg, J., Phelps, M. E., Ido, T., … Sokoloff, L. (1979). The [18F]fluorodeoxyglucose method for the measurement of local cerebral glucose utilization in man. *Circulation Research, 44*, 127–137. http://dx.doi.org/10.1161/01.RES.44.1.127.

Rocher, A. B., Chapon, F., Blaizot, X., Baron, J.-C., & Chavoix, C. (2003). Resting-state brain glucose utilization as measured by PET is directly related to regional synaptophysin levels: a study in baboons. *NeuroImage, 20*, 1894–1898. http://dx.doi.org/10.1016/j.neuroimage.2003.07.002.

Rosenbloom, M. H., Alkalay, A., Agarwal, N., Baker, S. L., O'Neil, J. P., Janabi, M., ... Rabinovici, G. D. (2011). Distinct clinical and metabolic deficits in PCA and AD are not related to amyloid distribution. *Neurology, 76*, 1789—1796. http://dx.doi.org/10.1212/WNL.0b013e31821cccad.

Sabri, O., Sabbagh, M. N., Seibyl, J., Barthel, H., Akatsu, H., Ouchi, Y., ... Schulz-Schaeffer, W. J. (2015). Florbetaben PET imaging to detect amyloid beta plaques in Alzheimer's disease: phase 3 study. *Alzheimer's Dementia, 11*, 964—974. http://dx.doi.org/10.1016/j.jalz.2015.02.004.

Sanders, D. W., Kaufman, S. K., DeVos, S. L., Sharma, A. M., Mirbaha, H., Li, A., ... Diamond, M. I. (2014). Distinct tau prion strains propagate in cells and mice and define different tauopathies. *Neuron, 82*, 1271—1288. http://dx.doi.org/10.1016/j.neuron.2014.04.047.

Scheltens, P., Barkhof, F., Leys, D., Wolters, E. C., Ravid, R., & Kamphorst, W. (1995). Histopathologic correlates of white matter changes on MRI in Alzheimer's disease and normal aging. *Neurology, 45*, 883—888.

Scheltens, P., Barkhof, F., Valk, J., Algra, P. R., Van der Hoop, R. G., Nauta, J., & Wolters, E. C. (1992). White matter lesions on magnetic resonance imaging in clinically diagnosed Alzheimer's disease. Evidence for heterogeneity. *Brain, 115*, 735—748.

Scheltens, P., Leys, D., Barkhof, F., Huglo, D., Weinstein, H. C., Vermersch, P., ... Valk, J. (1992). Atrophy of medial temporal lobes on MRI in "probable" Alzheimer's disease and normal ageing: diagnostic value and neuropsychological correlates. *Journal of Neurology Neurosurgery and Psychiatry, 55*, 967—972. http://dx.doi.org/10.1136/jnnp.55.10.967.

Schneider, J. A., Arvanitakis, Z., Leurgans, S. E., & Bennett, D. A. (2009). The neuropathology of probable Alzheimer disease and mild cognitive impairment. *Annals of Neurology, 66*, 200—208. http://dx.doi.org/10.1002/ana.21706.

Seeley, W. W., Crawford, R. K., Zhou, J., Miller, B. L., & Greicius, M. D. (2009). Neurodegenerative diseases target large-scale human brain networks. *Neuron, 62*, 42—52.

Seeley, W. W., Menon, V., Schatzberg, A. F., Keller, J., Glover, G. H., Kenna, H., ... Greicius, M. D. (2007). Dissociable intrinsic connectivity networks for salience processing and executive control. *Journal of Neuroscience, 27*, 2349—2356. http://dx.doi.org/10.1523/JNEUROSCI.5587-06.2007.

Selkoe, D. J. (1991). The molecular pathology of Alzheimer's disease. *Neuron, 6*, 487—498. http://dx.doi.org/10.1016/0896-6273(91)90052-2.

Selkoe, D. J. (2002). Alzheimer's disease is a synaptic failure. *Science, 298*, 789—791. http://dx.doi.org/10.1126/science.1074069.

Sexton, C. E., Kalu, U. G., Filippini, N., Mackay, C. E., & Ebmeier, K. P. (2011). A meta-analysis of diffusion tensor imaging in mild cognitive impairment and Alzheimer's disease. *Neurobiology of Aging, 32*, 2322. http://dx.doi.org/10.1016/j.neurobiolaging.2010.05.019. e5—2322.e18.

Sheline, Y. I., Morris, J. C., Snyder, A. Z., Price, J. L., Yan, Z., D'Angelo, G., ... Mintun, M. A. (2010). APOE4 allele disrupts resting state fMRI connectivity in the absence of amyloid plaques or decreased CSF Aβ42. *Journal of Neuroscience, 30*, 17035—17040. http://dx.doi.org/10.1523/JNEUROSCI.3987-10.2010.

Shi, F., Liu, B., Zhou, Y., Yu, C., & Jiang, T. (2009). Hippocampal volume and asymmetry in mild cognitive impairment and Alzheimer's disease: meta-analyses of MRI studies. *Hippocampus, 19*, 1055—1064. http://dx.doi.org/10.1002/hipo.20573.

Sibson, N. R., Dhankhar, A., Mason, G. F., Rothman, D. L., Behar, K. L., & Shulman, R. G. (1998). Stoichiometric coupling of brain glucose metabolism and glutamatergic neuronal activity. *Proceedings of the National Academy of Sciences of the United States of America, 95*, 316—321. http://dx.doi.org/10.1073/pnas.95.1.316.

Small, G. W., Ercoli, L. M., Silverman, D. H., Huang, S. C., Komo, S., Bookheimer, S. Y., ... Phelps, M. E. (2000). Cerebral metabolic and cognitive decline in persons at genetic risk for Alzheimer's disease. *Proceedings of the National Academy of Sciences of the United States of America, 97*, 6037—6042. http://dx.doi.org/10.1073/pnas.090106797.

Spires-Jones, T. L., & Hyman, B. T. (2014). The Intersection of amyloid beta and tau at synapses in Alzheimer's disease. *Neuron, 82*, 756—771. http://dx.doi.org/10.1016/j.neuron.2014.05.004.

Stern, Y. (2002). What is cognitive reserve? Theory and research application of the reserve concept. *Journal of the International Neuropsychological Society, 8*, 448—460. http://dx.doi.org/10.1017/S1355617702813248.

Stern, Y. (2012). Cognitive reserve in ageing and Alzheimer's disease. *Lancet Neurology, 11*, 1006—1012. http://dx.doi.org/10.1016/S1474-4422(12)70191-6.

Tabert, M. H., Manly, J. J., Liu, X., Pelton, G. H., Rosenblum, S., Jacobs, M., ... Devanand, D. P. (2006). Neuropsychological prediction of conversion to Alzheimer disease in patients with mild cognitive impairment. *Archives of General Psychiatry, 63*, 916—924. http://dx.doi.org/10.1001/archpsyc.63.8.916.

Tambini, A., Ketz, N., & Davachi, L. (2010). Enhanced brain correlations during rest are related to memory for recent experiences. *Neuron, 65*, 280—290. http://dx.doi.org/10.1016/j.neuron.2010.01.001.

Thal, D. R., Rüb, U., Orantes, M., & Braak, H. (2002). Phases of Aβ-deposition in the human brain and its relevance for the development of AD. *Neurology, 58*, 1791—1800. http://dx.doi.org/10.1212/WNL.58.12.1791.

Thompson, P. M., Hayashi, K. M., de Zubicaray, G., Janke, A. L., Rose, S. E., Semple, J., ... Toga, A. W. (2003). Dynamics of gray matter loss in Alzheimer's disease. *Journal of Neuroscience, 23*, 994—1005. pii:23/3/994.

Tiraboschi, P., Hansen, L. A., Masliah, E., Alford, M., Thal, L. J., & Corey-Bloom, J. (2004). Impact of APOE genotype on neuropathologic and neurochemical markers of Alzheimer disease. *Neurology, 62*, 1977—1983. http://dx.doi.org/10.1212/01.WNL.0000128091.92139.0F.

Villemagne, V. L., Burnham, S., Bourgeat, P., Brown, B., Ellis, K. A., Salvado, O., ... Masters, C. L. (2013). Amyloid β deposition, neurodegeneration, and cognitive decline in sporadic Alzheimer's disease: a prospective cohort study. *Lancet Neurology, 12*, 357—367. http://dx.doi.org/10.1016/S1474-4422(13)70044-9.

Villemagne, V. L., Fodero-Tavoletti, M. T., Masters, C. L., & Rowe, C. C. (2015). Tau imaging: early progress and future directions. *Lancet Neurology, 14*, 114—124. http://dx.doi.org/10.1016/S1474-4422(14)70252-2.

Villeneuve, S., Rabinovici, G. D., Cohn-Sheehy, B. I., Madison, C. M., Ayakta, N., Ghosh, P. M., ... Jagust, W. J. (2015). Existing Pittsburgh Compound-B positron emission tomography thresholds are too high: statistical and pathological evaluation. *Brain, 138*(Pt 7), 2020—2033. http://dx.doi.org/10.1093/brain/awv112. awv112—.

Villeneuve, S., Reed, B. R., Wirth, M., Haase, C. M., Madison, C. M., Ayakta, N., ... Jagust, W. J. (2014). Cortical thickness mediates the effect of β-amyloid on episodic memory. *Neurology, 82*, 761–767. http://dx.doi.org/10.1212/WNL.0000000000000170.

Vlassenko, A. G., Vaishnavi, S. N., Couture, L., Sacco, D., Shannon, B. J., Mach, R. H., ... Mintun, M. A. (2010). Spatial correlation between brain aerobic glycolysis and amyloid-β (Aβ) deposition. *Proceedings of the National Academy of Sciences of the United States of America, 107*, 17763–17767. http://dx.doi.org/10.1073/pnas.1010461107.

Walker, L. C., & Jucker, M. (2015). Neurodegenerative diseases: expanding the prion concept. *Annual Review of Neuroscience, 38*, 87–103. http://dx.doi.org/10.1146/annurev-neuro-071714-033828.

Wang, L., Zang, Y., He, Y., Liang, M., Zhang, X., Tian, L., ... Li, K. (2006). Changes in hippocampal connectivity in the early stages of Alzheimer's disease: evidence from resting state fMRI. *NeuroImage, 31*, 496–504. http://dx.doi.org/10.1016/j.neuroimage.2005.12.033.

Wirth, M., Madison, C. M., Rabinovici, G. D., Oh, H., Landau, S. M., & Jagust, W. J. (2013). Alzheimer's disease neurodegenerative biomarkers are associated with decreased cognitive function but not β-amyloid in cognitively normal older individuals. *Journal of Neuroscience, 33*, 5553–5563. http://dx.doi.org/10.1523/JNEUROSCI.4409-12.2013. pii:r33/13/5553.

Xia, C., Arteaga, J., Chen, G., Gangadharmath, U., Gomez, L. F., Kasi, D., ... Kolb, H. C. (2013). [18F]T807, a novel tau positron emission tomography imaging agent for Alzheimer's disease. *Alzheimers Dementia, 1*, 1–11.

Yankner, B. A. (1996). Mechanisms of neuronal degeneration in Alzheimer's disease. *Neuron, 16*, 921–932.

Yeo, B. T. T., Krienen, F. M., Sepulcre, J., Sabuncu, M. R., Lashkari, D., Hollinshead, M., ... Buckner, R. L. (2011). The organization of the human cerebral cortex estimated by intrinsic functional connectivity. *Journal of Neurophysiology, 106*, 1125–1165. http://dx.doi.org/10.1152/jn.00338.2011.

Yoshita, M., Fletcher, E., Harvey, D., Ortega, M., Martinez, O., Mungas, D., ... DeCarli, C. (2004). Extent and distribution of white matter hyperintensities in normal aging, MCI, and AD. *Stroke*, 2813–2819. http://dx.doi.org/10.1161/01.STR.0000147034.25760.3d.

Zhou, J., Greicius, M. D., Gennatas, E. D., Growdon, M. E., Jang, J. Y., Rabinovici, G. D., ... Seeley, W. W. (2010). Divergent network connectivity changes in behavioural variant frontotemporal dementia and Alzheimer's disease. *Brain, 133*, 1352–1367. http://dx.doi.org/10.1093/brain/awq075.

Chapter 18

Progressive Supranuclear Palsy and Related Parkinsonian Disorders

Jesse A. Brown and William W. Seeley
University of California San Francisco, San Francisco, CA, United States

INTRODUCTION

Parkinsonism refers to a complex of neurological symptoms including tremor, rigidity, bradykinesia, and postural instability. Neurodegenerative syndromes that involve parkinsonism are united by progressive damage to the nigrostriatal dopamine system. Historically the parkinsonian disorders have been divided into (typical) Parkinson disease (PD) on the one hand and atypical parkinsonian or "Parkinson plus" disorders on the other. This distinction helped identify patients whose parkinsonism was more or less likely to respond to levodopa therapy. Increasingly, however, these terms are being abandoned as our understanding of the "atypical" disorders matures. The non-PD parkinsonian disorders have been recognized as clinically and conceptually important in their own right, no longer requiring that they simply be differentiated from PD.

Parkinsonian disorders less prevalent than PD include the tauopathies, progressive supranuclear palsy (PSP) and corticobasal degeneration (CBD), and the other synucleinopathies, diffuse Lewy body disease (LBD) and multiple system atrophy (MSA). In this chapter, we focus on the tauopathies (Table 18.1). PSP and CBD are neurodegenerative four-repeat tauopathies, which differentiates them from disorders in which predominantly three-repeat tau aggregates are seen (Pick's disease) and those in which inclusions are composed of both three-repeat and four-repeat tau (Alzheimer disease ad some inherited tauopathies). PSP and CBD begin to produce symptoms around the same age as PD but typically lead to a more rapid progression.

When discussing the parkinsonian disorders or any neurodegenerative disease, it is critical to disambiguate clinical from pathological terms. PSP and CBD refer to histopathological entities. PSP syndrome (PSP-S) (also known as Richardson syndrome) and corticobasal syndrome (CBS), on the other hand, are terms that refer to specific symptom-deficit profiles. These terms arose to separate the syndromes from their most common underlying pathological causes and allow for the observation that both PSP-S and CBS are associated with a pathological differential diagnosis. Owing to prominent substantia nigra (SN) degeneration, PSP-S and CBS both involve extrapyramidal motor impairments, as seen in PD, but there are important differences in the affected motor and extramotor domains, as described subsequently. Whereas PSP-S is a falling disease with prominent oculomotor dysfunction, CBS is an asymmetric akinetic-rigid syndrome with progressive loss of limb control (Table 18.1).

Anatomically, typical PD can be distinguished from PSP and CBD based on the specific sites of anatomical involvement. PD onset typically occurs in the lower brain stem. PSP, in contrast, most likely begins in the upper brain stem, subthalamic nucleus, and pallidum, whereas in CBD the early targets include perirolandic cortex and striatum. This distinction in the pathological epicenter for each disease will help define the each disease's primary network target.

The goal of this chapter is to provide a modern synthetic view of PSP-S and CBS from a network-based perspective. To set the stage, we will introduce intrinsic connectivity networks (ICNs) as defined by task-free functional MRI (fMRI). This information will provide background for the recurring theme of this chapter, that each of these parkinsonian neurodegenerative syndromes targets a specific large-scale network. In the second section, we will briefly summarize typical

Genomics, Circuits, and Pathways in Clinical Neuropsychiatry. http://dx.doi.org/10.1016/B978-0-12-800105-9.00018-4

TABLE 18.1 Similarities and Difference Between Parkinson Disease (PD), Progressive Supranuclear Palsy Syndrome (PSP-S), and Corticobasal Syndrome (CBS)

Syndrome	Path DDX	Major Disease Protein in Most Common Path DDX	Dominant Symptom in Typical Patient	Core Affected Anatomy	Genes
PD	**LBD** MSA **PSP** **CBD**	Alpha synuclein	Tremor/bradykinesia	**Substantia nigra**, globus pallidus, subthalamic nucleus	SNCA,* LRRK2,* PARK2,* PINK1*
PSP-S	**PSP** **CBD** **PiD**	**Four-repeat tau**	Falls	Rostral midbrain tegmentum, tectum, dentate nucleus, globus pallidus, subthalamic nucleus **substantia nigra**	**MAPT,*** STX6, EIF2AK3, **MOBP**
CBS	**CBD** AD **PSP** TDP-A **PiD** **(LBD)** (CJD)	**Four-repeat tau**	Progressive loss of limb controls	Perirolandic cortex, striatum, **substantia nigra**	**MAPT*,** **MOBP,** lnc-KIF13B-1, SOS1

Items in bold are common to more than one of the three syndromes listed. Genes with asterisks are rare monogenic causes; the remainder are risk genes. Path DDXs in parentheses are rare diagnoses. *DDX*, disease diagnosis; *TDP-A*, TDP type A; *CJD*, Creutzfeldt–Jacob disease; *LBD*, Lewy body disease; *MSA*, multiple system atrophy; *PiD*, Pick disease; *AD*, Alzheimer disease.

idiopathic PD. We will then introduce the diagnostic criteria for PSP-S and CBS and more deeply characterize these syndromes. Next, we will describe the underlying ICNs that mirror the patterns of gross atrophy in PSP-S, CBS, and related syndromes. In the third section we will expand on a model of transneuronal spread from a disease "epicenter," describing evidence that these diseases progress from a selectively vulnerable brain region in a manner that can be predicted by that brain region's structural and functional intrinsic connections. We will then examine the clinicopathological correlations, including the proteinopathies that cause PSP-S and CBS and the diversity of syndromes that result from PSP and CBD pathology. Finally, we will review genetic causes and risk factors in PSP and CBD.

EACH NEURODEGENERATIVE SYNDROME REFLECTS A NETWORK

Class-Wide Principles of Network-Based Neurodegeneration

That neurodegenerative diseases represent organized network degenerations has long been postulated (Braak & Braak, 1991; Pearson, Esiri, Hiorns, Wilcock, & Powell, 1985; Saper, Wainer, & German, 1987). Advances in human network mapping techniques have enabled researchers to clarify the network architecture of the human brain and use this information to deepen our understanding of the spatial patterning of neurodegenerative disease. Novel network imaging methods include fMRI-based functional intrinsic connectivity mapping and diffusion tractography, among others. Studies to date have shown that anatomically distinct ICNs, defined by task-free fMRI, span the same distributed set of brain regions that undergo atrophy in Alzheimer disease and the frontotemporal dementias (Buckner et al., 2009; Seeley, Crawford, Zhou, Miller, & Greicius, 2009; Zhou, Gennatas, Kramer, Miller, & Seeley, 2012). Convergent findings have been derived from diffusion tensor imaging (Raj, Kuceyeski, & Weiner, 2012), which suggests that both functional and structural connections may predict vulnerability and spread (Raj et al., 2015). The network-based model of regional vulnerability has been further refined by identifying a focal "epicenter" for each syndrome, defined as the brain region or regions whose intrinsic connectivity pattern in healthy individuals best matches the spatial pattern of disease-related atrophy. The intrinsic functional connections among all regions within a given ICN can be determined to derive an intranetwork graph. As predicted by the network spread model, in all frontotemporal dementia (FTD) syndromes investigated to date, more severe atrophy is seen in nodes with shorter network path lengths to the syndrome-specific epicenter. In addition to intranetwork disease spread from an epicenter, evidence suggests that these diseases can spread between ICNs in a process of transnetwork spread (Zhou et al., 2012). Thus a given disease is not confined to a specific set of regions within a single ICN. Instead, there appear to be "target" (or "core") and "off-target" (or "periphery") networks with predictable gradations in vulnerability based on the sites of regional onset.

To understand the regional vulnerability landscape for each syndrome, it is useful to define the extent of the major ICNs and their place within the whole-brain network. A whole-brain graph of connections within and between ICNs in the healthy brain has been described by several groups. The modular composition of one such network (Power et al., 2011) is shown in Fig. 18.1. This network was determined by using task-free fMRI data from a large set of healthy adults and calculating functional connectivity between 264 nodes in the cortex, basal ganglia, thalamus, midbrain, and cerebellum. A modularity analysis of this whole-brain network revealed approximately 10 intrinsic connectivity networks (Fig. 18.1): default mode, frontoparietal task control, dorsal attention, ventral attention (bearing some similarity to the salience network, as previously described) (Seeley et al., 2007), cingulo-opercular, sensory-somatomotor, visual, subcortical, and anterior temporal. The cognitive and behavioral processes supported by these different networks in health translate directly into the cardinal symptoms of specific neurodegenerative syndromes. Superimposed on this graph in Fig. 18.1 are boxes indicating the affected networks of regions showing most substantial atrophy and dysconnectivity in PSP-S (dotted squares—edged box), CBS (solid-edged box), and bvFTD (dotted circles—edged box), the most common FTD syndrome; we will return to the discussion of these boxes in section "Clinicopathological Correlation in Progressive Supranuclear Palsy and Corticobasal Degeneration." The sensory-somatomotor network includes primary cortices for processing sensory input and motor output, particularly along the dorsal/medial surface of the brain where the body and hands are represented. The cingulo-opercular network covering the posterior medial frontal cortex, anterior cingulate, frontal operculum, and dorsal anterior insula defines the task control system, which maintains the appropriate mental set during goal-directed behavior (Dosenbach et al., 2007). The salience network integrates interoceptive input from sensory, visceral, and autonomic streams in the ventral anterior insula to represent subjective feeling states, which can then mobilize appropriate emotional, cognitive, and behavioral responses via the dorsal anterior cingulate (Zhou & Seeley, 2014). The frontoparietal network spanning the dorsolateral prefrontal cortex and intraparietal sulcus is part of an executive control network involved in prioritizing and maintaining set during goal-directed thinking or behavior (Dosenbach et al., 2006; Seeley et al., 2007).

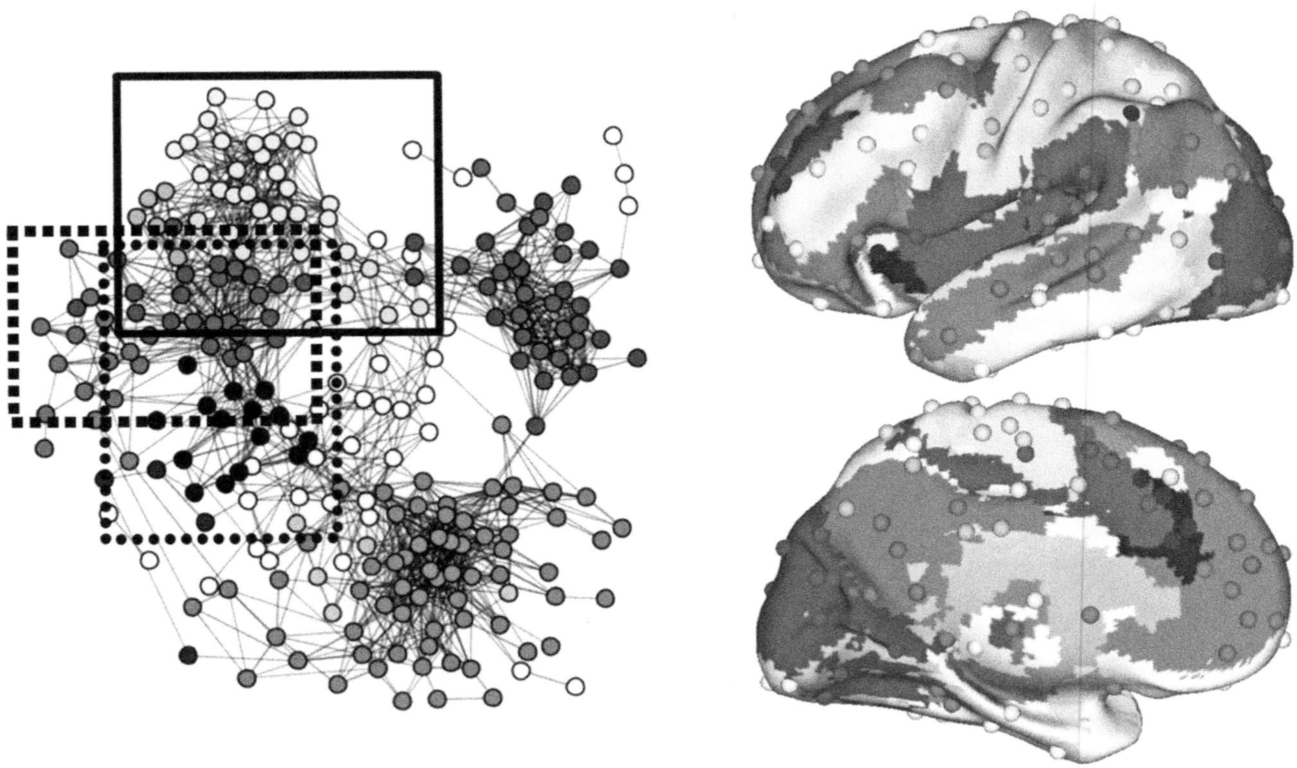

FIGURE 18.1 Functional network graph with boxes covering regions exhibiting atrophy in PSP-S (*dotted squares—edged box*), CBS (*solid-edged box*), and bvFTD (*dotted circles—edged box*). *Modified from Power, J.D., Cohen, A.L., Nelson, S.M., Wig, G.S., Barnes, K.A., Church, J.A., ... Petersen, S.E. (2011). Functional network organization of the human brain.* Neuron, *72(4), 665—678. http://doi.org/10.1016/j.neuron.2011.09.006.*

The dorsal attention network encompasses the secondary sensory and motor association cortices, which support top-down selection of stimuli and responses (Corbetta & Shulman, 2002). The basal ganglia, thalamus, brain stem, and cerebellum are sparsely represented in this cortico-centric analysis, although in many parkinsonian syndromes the disease epicenter is a subcortical nucleus and spreads to the cortex later in the disease process.

PARKINSONIAN SYNDROMES: PARKINSON DISEASE AND OTHERS

Parkinson Disease: Clinical and Anatomical Features

Clinical PD is a sporadic movement disorder syndrome with a mean age at onset of 62 years. In addition to the hallmark parkinsonian features, there are a host of nonmotor features in PD including anosmia, constipation and other forms of autonomic dysfunction, sleep disturbances (particularly excessive daytime sleepiness and rapid eye movement sleep behavior disorder), depression, anxiety, and frontal/executive dysfunction. Presenting symptoms typically involve unilateral parkinsonism, including tremor. PD results from a progressive neurodegenerative process that ascends from the peripheral enteric, autonomic, and olfactory nervous systems before gaining entry to the central nervous system via the dorsal motor nucleus of the vagus or the olfactory apparatus. The ascent continues to involve the major aminergic nuclei in the brain stem, with prominent SN involvement, the basal ganglia, and eventually the limbic system and cerebral cortex (Fig. 18.2). Most patients experience a dramatic clinical benefit from levodopa, which increases dopaminergic neurotransmission at nigrostriatal projection terminals. The underlying pathology for the PD syndrome is usually LBD, recognized by characteristic neuronal inclusions containing misfolded alpha-synuclein. Aggregates appear as neuronal cytoplasmic and neuritic inclusions, known respectively as Lewy bodies and Lewy neurites. Braak and colleagues proposed a staging scheme for LBD that identifies alpha-synuclein deposits in six sets of brain areas (Heiko Braak et al., 2003): (1) the anterior olfactory nucleus and dorsal motor nucleus of vagus; (2) pontine tegmentum; (3) SN pars compacta and pedunculopontine nucleus; (4) hypothalamus, thalamus, and anterior medial temporal cortex; (5) high-order association cortex; and (6) first-order association and primary cortices. The cardinal neuroanatomical feature of PD is loss of dopaminergic neurons in the SN. The resultant disruption of functional motor circuitry is well characterized. Loss of SN inputs from these dopaminergic neurons has two major consequences. In the direct pathway, decreased excitation of D1 dopamine receptor-expressing neurons in the putamen causes diminished inhibition of the globus pallidus interna (GPi), releasing the GPi to overinhibit the thalamus, reducing excitation of the motor cortex. In the indirect pathway, decreased inhibition of D2 neurons in the putamen causes overinhibition of the globus pallidus externa, disinhibiting the subthalamic nucleus (STN) and releasing the GPi, which in turn suppresses the thalamus, again resulting in underexcitation of the cortex (Albin, Young, & Penney, 1989; Alexander, DeLong, & Strick, 1986; Blandini, Nappi, Tassorelli, & Martignoni, 2000). Whole-brain functional MRI findings echo these circuit disruptions. Hacker and colleagues found connectivity deficits in PD among the striatum, thalamus, and midbrain (Hacker, Perlmutter, Criswell, Ances, & Snyder, 2012). This work pinpointed the lower brain stem as the epicenter of PD onset and attributed the predominant loss of posterior putamen connectivity among striatal regions to its tighter integration with lower brain stem structures in a healthy functional network. Other studies have highlighted fMRI hyperconnectivity in patients with PD, both in cortical-STN motor circuits (Baudrexel et al., 2011) and prefrontal-premotor action selection circuits (Rowe, Hughes, Barker, & Owen, 2010). Structural MRI studies in PD have revealed relatively spared subcortical areas and more substantial atrophy in prefrontal,

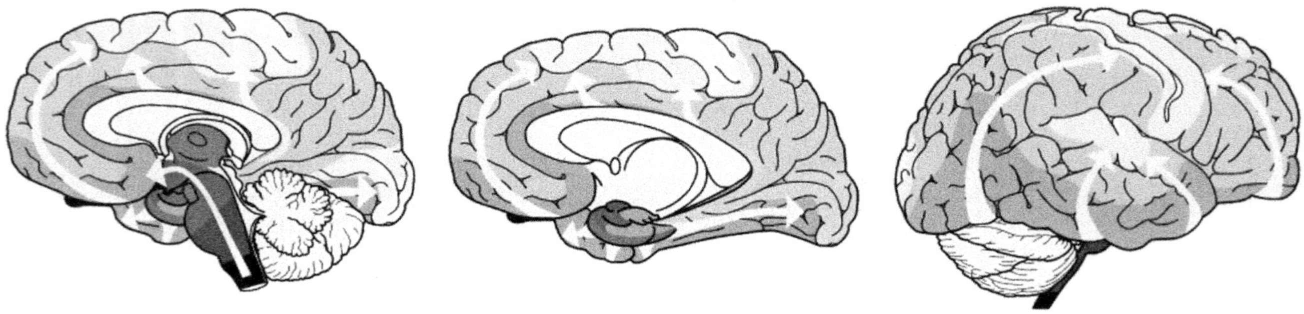

FIGURE 18.2 Braak staging of PD suggests ascent from the caudal medulla (*along white arrows*), from areas of early involvement (*darker*) to those in which inclusions are seen at later stages (in order of increasing lightness) (Heiko Braak et al., 2003). Although the scheme is based on cross-sectional data, it predicts pathways of disease spread between interconnected brain structures.

parietal, and occipital cortices (Weintraub et al., 2011), although these findings could relate to a lesser methodological sensitivity to atrophy in small deep brain structures or the occurrence of comorbid Alzheimer disease in some patients.

Although PD is most commonly sporadic, 10% of familial cases of PD have been linked to a single monogenic mutation (Trinh & Farrer, 2013). One Mendelian form of the disease relates to mutations in *SNCA* (encoding alpha-synuclein), which confirms the genotype–phenotype link between SNCA and alpha-synuclein. Other Mendelian disease-causing genes are *LRRK2*, *PARK2* (encoding parkin), *PINK1*, *PARK7* (encoding DJ-1), and *ATP13A2* (Satake et al., 2009; Simón-Sánchez et al., 2009). The genetic diversity present in PD suggests that the disease may be differentiated into specific subtypes based on the specific pathogenic pathways involved; however, all of them converge on the PD-related anatomy.

Progressive Supranuclear Palsy Syndrome: Clinical Features

PSP was originally characterized as a distinct clinicopathological entity (Steele, Richardson, & Olszewski, 1964). The authors described a progressive syndrome that often began with subtle executive functioning and personality changes followed by vertical gaze slowing, postural instability, pseudobulbar palsy, and rigidity of the neck and upper trunk. Observed damage to nuclei of the upper midbrain, rostral to the oculomotor nucleus (superior colliculus and pretectum), combined with the resultant ophthalmoparesis, warranted the description of the syndrome as a "supranuclear palsy." The age of onset in these patients was the fifth and sixth decades and the disease had a course of 5 to 7 years before death. At autopsy, nerve cell death was apparent in the brain stem, basal ganglia, and cerebellum. The most apparent histopathological changes were neurofibrillary tangles, loss of nerve cells, granulovacuolar degeneration, and gliosis.

A half-century later, we now know that patients with typical PSP-S develop a mild neuropsychiatric/executive prodrome; apathy, mental rigidity, and multitasking problems are the most common. Shortly thereafter, disabling motor symptoms emerge. Early falls occur during ambitious tasks such as changing light bulbs on a ladder or descending stairs while carrying objects. The gait is stiff owing to axial rigidity and toppling, with severe retropulsive and at times propulsive instability. Bradykinesia and especially tremor are less common at onset than in PD. Falls are exacerbated by poor judgment, impulsivity, and the signature gaze abnormalities, which follow a predictable progression from square wave jerks to slowed vertical worse than horizontal saccades, to a full-blown supranuclear gaze palsy. Vestibulo-ocular reflexes remain intact, demonstrating the supranuclear nature of the ophthalmoparesis. Spastic dysarthria and dysphagia, pseudobulbar affect, and a fixed stare are all common features. These clinical deficits are accompanied by magnetic resonance–detectable atrophy in the midbrain, superior cerebellar peduncles, and posterior medial frontal cortex (Fig. 18.3). Disease progression is rapid, with survival time from symptom onset averaging 7 years (Golbe & Ohman-Strickland, 2007). Response to levodopa is typically lacking (Tsai & Boxer, 2014). Although selected patients may

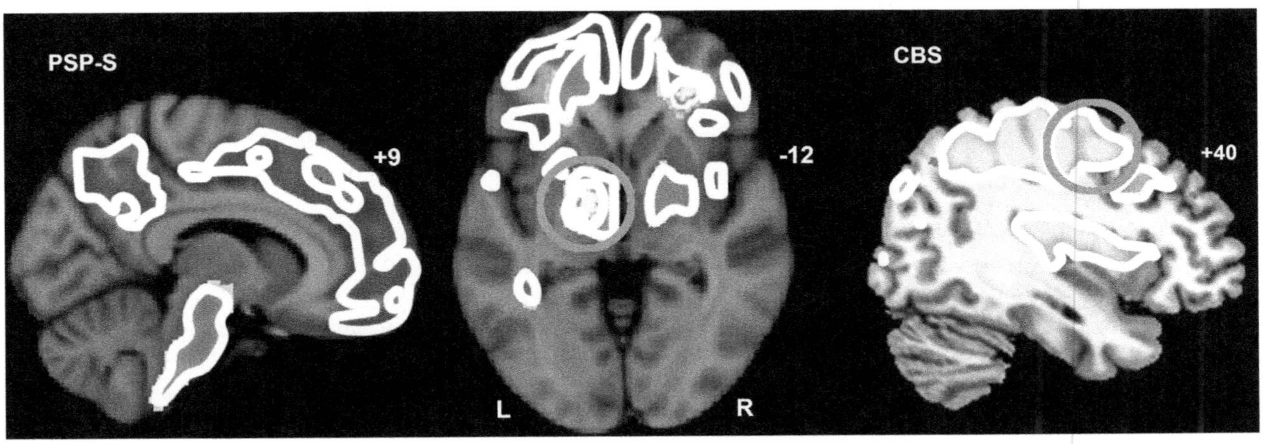

FIGURE 18.3 Patterns of atrophy in patients with PSP-S and CBS. In PSP-S, strongest atrophy is apparent in the midbrain, pallidum, SMA, preSMA, medial prefrontal cortex, and precuneus. Results are shown as white outlines around areas of high and moderate statistical significance for white matter (high: $P < 0.05$, familywise error rate corrected; moderate, $P < 0.01$ cluster corrected) and gray matter. The atrophy peak was in the mesothalamic junction and pallidum (*circled*). Results are from an unpublished analysis of 12 PSP-S patients and 20 healthy control subjects. For CBS, atrophy is apparent in the dorsal frontoparietal sensorimotor association areas, primary motor and sensory cortices, and dorsal insula. In this sample the atrophy peak was in the right premotor cortex (*circled*). *From Seeley, W.W., Crawford, R.K., Zhou, J., Miller, B.L., Greicius, M.D. (2009). Neurodegenerative diseases target large-scale human brain networks.* Neuron, *62(1), 42–52. http://doi.org/10.1016/j.neuron.2009.03.024.*

experience a partial response at higher doses, limited data are available to predict this response or about neuropathological diagnoses made in levodopa responders.

In 1996, Litvan and colleagues defined the National Institute for Neurological Disorders and Stroke-Society for Progressive Supranuclear Palsy Criteria (Litvan et al., 1996). The diagnostic criteria for "PSP probable" include age over 40 years at onset, vertical gaze palsy, slowing of vertical saccades, prominent postural and gait instability, and falls within the first year since onset (Litvan et al., 1996). The disorder must be gradually progressive and occur in the absence of other diseases or conditions that may cause the symptoms. A "PSP possible" diagnosis extends to patients who have vertical gaze palsy or a combination of slowing of vertical saccades and postural instability with falls in the first year of symptoms. Definite PSP requires a history of probable or possible PSP and histopathological confirmation of PSP at autopsy. In 2007, Golbe and Ohman-Strickland published a clinical rating scale for PSP (PSP-RS) that assesses symptom severity in six domains and provides predictive information about subsequent survival (Golbe & Ohman-Strickland, 2007). The scale is based on a 10-minute interview and examination conducted by a neurologist that measures impairment in six domains: daily activities (as measured by patient history), behavior, bulbar, oculomotor, limb motor, and gait/midline. The scale yields a score between 0 and 100. In the original publication, the observed rate of increase in PSP-RS was roughly linear at 10 points per year, which would extrapolate to a PSP-RS change from 0 to 70 between the time of symptom onset to death.

Numerous neuroimaging correlates of PSP-RS have been reported, including functional connectivity strength of the rostral midbrain (Gardner et al., 2013) (see section: Network Architecture of Neurodegenerative Parkinsonian Syndromes), longitudinal atrophy of the frontal lobe and midbrain (Josephs et al., 2013), and white matter integrity in the superior cerebellar peduncle (Whitwell, Master, et al., 2011). Specific motor and nonmotor symptoms tend to be associated with specific alterations in corresponding brain systems. Atrophy rates in the midbrain are correlated with worsening motor deficits (Josephs et al., 2013; Paviour, Price, Jahanshahi, Lees, & Fox, 2006), whereas eye movement abnormalities have been linked to lower fractional anisotropy of the superior longitudinal fasciculus (Whitwell, Master, et al, 2011). Cognitive and behavioral deficits in PSP are predominantly related to frontal processes. Apathy and impulsivity are hallmark symptoms, both of which are routinely measured using the Neuropsychiatric Inventory (Cummings et al., 1994). Patients have executive dysfunctions including set shifting difficulty, mental inflexibility, reduced processing speed (bradyphrenia), impaired verbal fluency, and difficulty with planning (Burrell, Hodges, & Rowe, 2014). Many of these deficits are likely to result from atrophy and dysfunction in the supplementary and presupplementary motor areas, dorsolateral prefrontal cortex (superior and middle frontal gyri), frontal and central operculum, and medial frontal cortex (Gardner et al., 2013; Josephs et al., 2013; Paviour et al., 2006; Whitwell, Master, et al., 2011). Social cognition is also impaired in PSP, negatively affecting emotion recognition and Theory of Mind, and has been shown to correlate with atrophy in this same distributed set of frontal regions (Ghosh et al., 2012).

PSP-S strongly predicts underlying tau pathology (Table 18.1). The large majority are 4R tauopathies, primarily PSP and less frequently CBD. In a meta-analysis of PSP-S, 13% of patients were found to have an underlying CBD pathology (Wadia & Lang, 2007).

Corticobasal Syndrome: Clinical Features

CBS is the term used for the clinical disorder originally described in 1968 by Rebeiz and colleagues, who reported three patients with progressive asymmetric rigidity and apraxia (Rebeiz, Kolodny, Richardson, 1968). They labeled the disorder "corticodentatonigral degeneration with neuronal achromasia" in reference to cortical atrophy and the loss of pigmentation in neurons from the cortex and substantia nigra. In the 1990s the disorder came to be known as "corticobasal degeneration," a term recognizing both the cortical symptoms of the disease (asymmetrical rigidity, apraxia, alien limb phenomenon, cortical sensory loss, myoclonus, and speech deficits) as well as its basal ganglia–related motor symptoms (bradykinesia, limb dystonia, action tremor, postural instability, and gait disturbance). In 2003, Boeve and colleagues clarified the distinction between this cluster of symptoms, which they called CBS, and the most common underlying pathology, for which the term CBD was reserved.

CBS typically begins as an akinetic-rigid syndrome that asymmetrically affects one hand or foot before progressing up the onset limb, into whichever of the two ipsilateral limbs was not affected at first, and ultimately into the orobuccal apparatus and contralateral limbs. Dystonia, myoclonus (which may be stimulus-sensitive), and alien limb phenomenon may accompany the core problem. Cognitive and behavioral symptoms in CBS primarily result from damage to frontal lobe-anchored networks. Patients have shown deficits in attention and concentration, processing speed, executive functioning, verbal fluency, language, and visuospatial function (Pillon et al., 1995; VanVoorst et al., 2008). A research diagnosis of probable CBS requires asymmetric presentation of at least two symptoms of limb rigidity or akinesia, dystonia, and myoclonus plus at least two symptoms of orobuccal or limb apraxia, cortical sensory deficit, and alien limb phenomenon (Armstrong et al., 2013). A possible CBS diagnosis allows for symptoms to be symmetric if patients have at least one symptom in each of the two core symptom clusters. Criteria for a diagnosis of probable CBS resulting from CBD include insidious onset and gradual progression for 1 year, age 50 years or greater at onset, no family history of disease or known mutations in tau, and a clinical phenotype of probable CBS, but expert clinicians recognize that patients with non-CBD underlying pathological diagnoses often meet all of these criteria.

In patients with CBS, as in PSP, it is critical to recognize that a number of different proteinopathies can infiltrate the CBS-vulnerable network and result in a similar set of symptoms (Table 18.1). Whereas CBD is most common, there are cases of underlying PSP with a clinical presentation of CBS (Josephs, Katsuse, et al., 2006; Tsuboi et al., 2005). A subtle distinction can be made between brain atrophy patterns in pathology-proven CBD cases versus cases of CBS in which the underlying pathology is unknown (Lee et al., 2011). Neuroimaging studies of patients with CBS with unknown underlying pathology, in whom the two candidates are 4R tau and TDP-43 (TAR DNA-binding protein 43), have consistently reported a mixture of atrophy in frontal and parietal sensorimotor association areas, primary sensory and motor areas in perirolandic cortex, and dorsal insula (Fig. 18.3, right) (Boxer et al., 2006; Seeley et al., 2009; Zhou et al., 2012). In two similar studies published in 2010 and 2011, patients with CBS resulting from CBD were found to have predominant atrophy in the frontal lobe and striatum although the parietal lobe was largely spared (Lee et al., 2011; Whitwell et al., 2010). Imaging studies of patients with CBS caused by frontotemporal lobar dementia with TAR DNA-binding protein 43 (FTLD-TDP) show preferential atrophy in the inferior frontal gyrus and insula aligned with the perisylvian component of the cingulo-opercular network (Whitwell et al., 2010), whereas CBS caused by Alzheimer disease shows great parietal atrophy (Lee et al., 2011).

NETWORK ARCHITECTURE OF NEURODEGENERATIVE PARKINSONIAN SYNDROMES

The network model of neurodegenerative disease vulnerability provides a parsimonious explanation for the progression of a diverse set of syndromes. In addition, the model offers two potential clinical-translational advances. First, a network model can aid in earlier diagnosis by helping to pinpoint an epicenter where the disease originates. Second, the model can enable more precise monitoring of the disease progression, either its natural history or during treatment, by predicting where the disease is most likely to arise next.

Network Architecture of Progressive Supranuclear Palsy Syndrome

Building on previous studies linking each major neurodegenerative dementia syndrome to a specific ICN, Gardner et al., mapped the brain regions with intrinsic connectivity to the rostral midbrain tegmentum (rMT), identified in previous studies as an epicenter of PSP-S atrophy (Boxer et al., 2006). The rMT-anchored ICN topology in healthy control subjects (Fig. 18.4) included cortical, subcortical, brain stem, and cerebellar regions with known vulnerability in PSP-S (Gardner et al., 2013) and canvassed portions of the cingulo-opercular, frontoparietal, default mode, and salience networks. Patients with probable PSP-S showed widespread functional connectivity disruptions throughout this network, both in edgewise connectivity strength (Fig. 18.4, bottom left) and in nodewise weighted degree, the sum of connection strengths to a given node (Fig. 18.4, bottom right). Among the most severely impacted nodes were the mesothalamic junction (MTJ) (the node encompassing the rMT epicenter) and the presupplementary motor area (preSMA). Importantly, greater functional connectivity reductions in this network predicted greater clinical impairment on the Clinical Dementia Rating sum of boxes scale and the PSP-RS, as well as slower downward saccade velocity. An examination of the relative effect size of rMT-ICN impairment versus structural atrophy found that many areas showed PSP-related reductions in functional connectivity despite no atrophy, which suggested that fMRI ICN analysis may be more sensitive to early-stage system-level dysfunction that precedes eventual regional atrophy. Intrinsic connectivity deficits have been reported in a thalamocortical network, suggestive of degeneration of the dentatorubrothalamic tract (Whitwell, Avula, et al., 2011). Ongoing longitudinal studies should help determine the efficacy of rMT ICN strength and related measures as biomarkers for disease monitoring.

A critical test of the transneuronal spread model is to assess how well postmortem neuropathology follows the same network trajectory seen for antemortem structural atrophy and ICN disruption. An important study (Fig. 18.5) staged 33 cases of PSP into five levels of severity based on the extensiveness of tau inclusion pathology (Williams, Holton, Strand, Pittman, et al., 2007). The five stages were: (1) pallido-luyso (subthalamic nucleus)-nigral with sparse premotor cortex involvement; (2) moderate basal ganglia, pontine nuclei, dentate nucleus, and posterior frontal lobe, without parietal lobe; (3) severe basal ganglia and dentate nucleus, moderate frontal, and parietal lobe; (4) severe basal ganglia, pontine nuclei, and frontal and parietal lobe; and (5) severe subthalamic nucleus, substantia nigra, globus pallidus interna, pontine nuclei, cerebellar structures, and neocortex, with negligible pathology in caudate, putamen, and temporal lobe. The trajectory of this pathology would roughly affect subcortical and cingulo-opercular ICNs at stage 1, default mode,

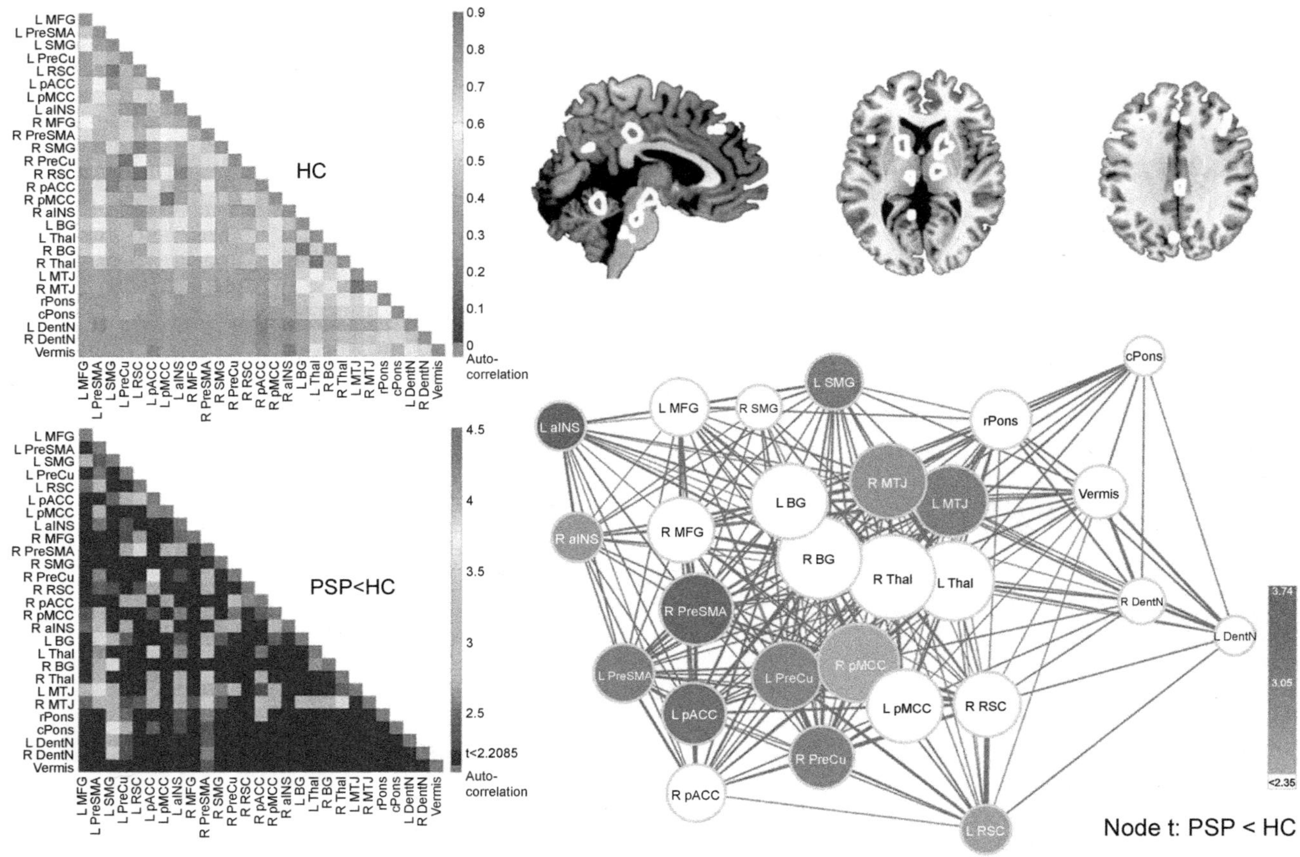

FIGURE 18.4 The rMT-anchored intrinsic connectivity network characterized in Gardner et al. (2013). The left panel shows the mean connectivity matrix representing the connections between the 27 nodes (*clusters shown at top right*) in the network in healthy controls and for connections significantly diminished in PSP-S. The bottom right shows the force-directed, spring-embedded network diagram for healthy control subjects with nodes showing significantly reduced total flow (mean connection strength) in black. *HC*, healthy controls; *aINS*, anterior insula; *BG*, basal ganglia; *DentN*, dentate nucleus; *L*, left; *MFG*, middle frontal gyrus; *MTJ*, mesothalamic junction; *pACC*, pregenual anterior cingulate cortex; *pMCC*, posterior midcingulate cortex; *PreCu*, precuneus; *PreSMA*, presupplementary motor area; *R*, right; *RSC*, retrosplenial cortex; *Thal*, thalamus.

sensory-somatomotor, and frontoparietal networks at stage 3, and worsening pathology in all nontemporal, nonoccipital ICNs at higher stages. This trajectory is extrapolate from the available data, based on a broad but incomplete subset of brain regions sampled. This observation of stages of increasingly widespread pathology, akin to the Braak staging of Alzheimer disease and PD, is consistent with a model of transsynaptic propagation of hyperphosphorylated tau fragments throughout the network. At subsequent stages the disease would radiate out further from the disease epicenter into neighboring ICNs, causing impairments in multiple domains and progressive structural atrophy (Josephs et al., 2013; Paviour et al., 2006; Sanders et al., 2014).

Network Architecture of Corticobasal Syndrome

Patients with CBS show atrophy involving epicenters of degeneration in the precentral and postcentral gyri (Fig. 18.3) (Seeley et al., 2009; Zhou et al., 2012). The intrinsic functional connectivity pattern in healthy individuals that best matches the CBS atrophy pattern spans the primary and secondary somatosensory cortex (Zhou et al., 2012, Fig. 18.6A). Nodes within this ICN were used to define the CBS vulnerable network (Fig. 18.6B) and determine the shortest path lengths for each node to the epicenters. Nodes with longer path lengths to the epicenter had less atrophy than those closer connected to the epicenter (Fig. 18.6C), which supports a "transneuronal spread" model of disease propagation. Critically, this correlation remained after controlling for the Euclidean distance between brain regions, indicating that network path length from the epicenter was the optimal model for predicting regional vulnerability.

The regions within the CBS-vulnerable network are circumscribed by the top box in Fig. 18.1 and largely cover five ICNs: sensory-somatomotor, dorsal attention, cingulo-opercular, subcortical, and frontoparietal task control. Overlap

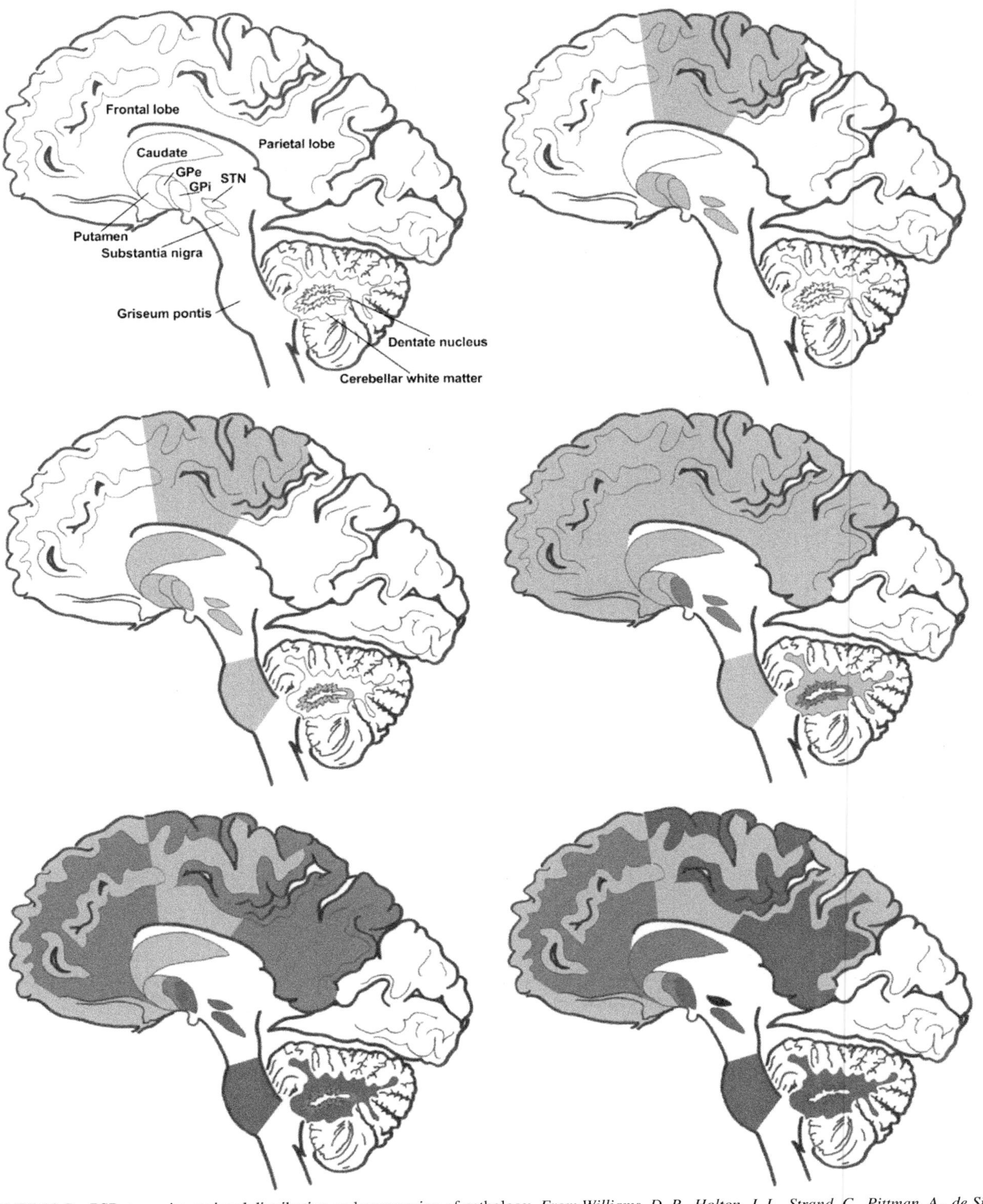

FIGURE 18.5 PSP stagewise regional distribution and progression of pathology. *From Williams, D. R., Holton, J. L., Strand, C., Pittman, A., de Silva, R., Lees, A. J., & Revesz, T. (2007). Pathological tau burden and distribution distinguishes progressive supranuclear palsy-parkinsonism from Richardson's syndrome.* Brain, 130*(6), 1566–1576. http://doi.org/10.1093/brain/awm104.*

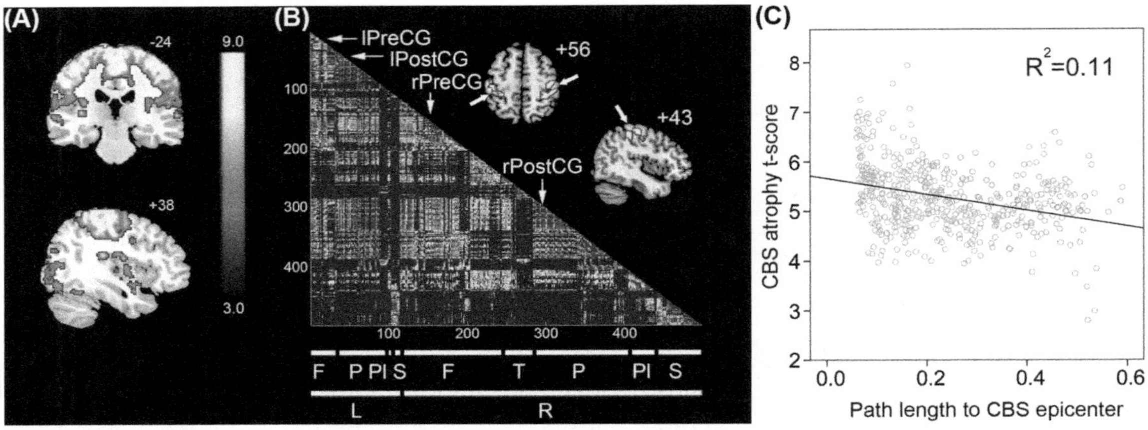

FIGURE 18.6 Intrinsic connectivity network correspondence with atrophy patterns in CBS. (A) The task-free functional MRI seed connectivity network (from $n = 16$ healthy subjects) whose spatial layout had the best goodness of fit to the gray matter atrophy pattern in CBS (from $n = 17$ patients) is a primary and secondary somatomotor network. (B) Functional connectivity matrix depicting the connectivity among all 499 regions canvassing the network shown in (A). The nodes defined as epicenters are located in the rolandic and perirolandic cortices indicated with arrows in the inset. (C) Correlation between path length from the CBS epicenter and CBS atrophy score. *F*, frontal; *T*, temporal; *P*, parietal; *PI*, paralimbic; *S*, subcortical; *L*, left hemisphere; *R*, right hemisphere; *PreCG*, precentral gyrus; *PostCG*, postcentral gyrus. *Modified from Zhou, J., Gennatas, E.D., Kramer, J.H., Miller, B.L., Seeley, W.W. (2012). Predicting regional neurodegeneration from the healthy brain functional connectome.* Neuron, *73(6), 1216–1227. http://doi.org/10.1016/j.neuron.2012.03.004.*

between these regions and areas vulnerable in PSP-S and behavioral variant FTD (bvFTD) is most evident in the cingulo-opercular network. An important goal for future studies of CBS and related syndromes will be to elucidate the temporal sequence of fMRI connectivity disruption and structural atrophy during disease progression, in a manner comparable to the insights emerging for Alzheimer disease (Jack et al., 2013; Raj et al., 2015). A complementary aim will be to study how well characteristic profiles of functional connectivity alteration track along with subject-specific symptoms or, better yet, anticipate them.

CLINICOPATHOLOGICAL CORRELATION IN PROGRESSIVE SUPRANUCLEAR PALSY AND CORTICOBASAL DEGENERATION

Multiple Proteinopathies Can Cause Progressive Supranuclear Palsy Syndrome and Corticobasal Syndrome

The most important concept for understanding neurodegenerative disease diagnosis is that each syndrome reflects where the disease is present (ie, which network), not which disease is present (ie, which proteinopathy). A patient with typical PSP-S (also known as Richardson syndrome) most likely has PSP as the underlying pathology (Dickson, Rademakers, & Hutton, 2007), but exceptions occur. In contrast, the link between CBS and CBD is more tenuous (Armstrong et al., 2013; Boeve, Lang, & Litvan, 2003; Lee et al., 2011). In this section, we focus on the pathological differential diagnoses for PSP-S and CBS. Changing viewpoints, we next discuss the spectrum of clinical syndromes beyond PSP-S and CBS associated with a pathological diagnosis of PSP or CBD.

Pathological Causes of Progressive Supranuclear Palsy Syndrome

PSP-S can be caused by a short list of related tauopathies. By far the most common cause is PSP pathology, which makes PSP-S one of the strongest predictors of a single histopathological entity across the entire neurodegenerative disease spectrum. Rare causes of PSP-S include CBD, Pick disease, and a scattering of other protean disorders. In a patient with typical PSP-S owing to PSP pathology, the burden of tau pathology and neurodegeneration is most severe in the midbrain, subthalamic nucleus, globus pallidus, and dentate nucleus of the cerebellum. Globose tau-positive tangles and other neuronal cytoplasmic inclusions in these regions are variably accompanied by neuropil threads and glial inclusions, which may include tufted astrocytes, coiled bodies (curvilinear oligodendroglial cytoplasmic inclusions), and thorny astrocytes (Fig. 18.7C) (Dickson, Ahmed, Algom, Tsuboi, & Josephs, 2010). In the cortex, the tauopathy is milder and distributed across premotor and supplementary motor, primary motor, affective-motivational, and relevant cognition-associated regions.

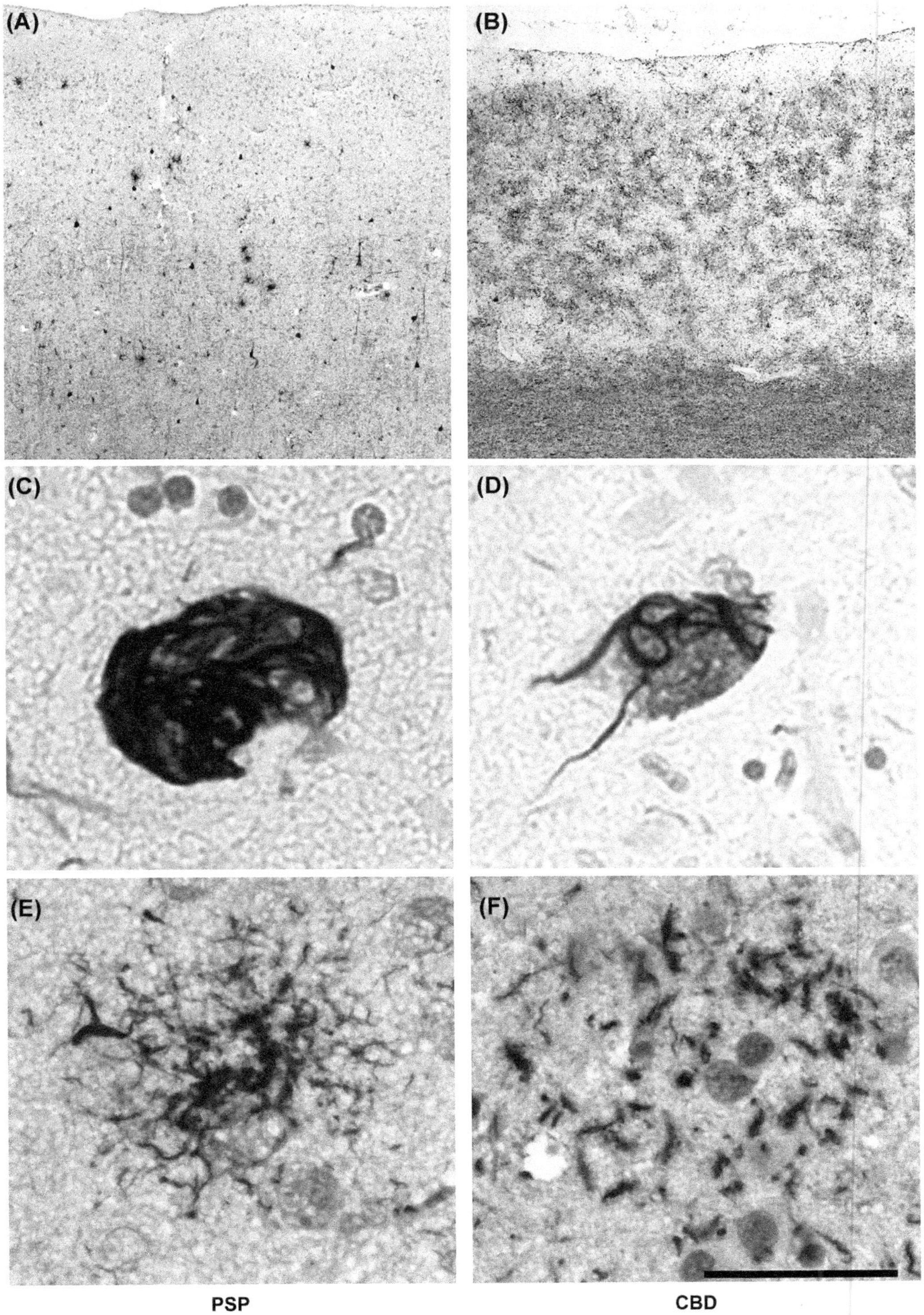

FIGURE 18.7 Histopathology in PSP and CBD. In PSP (A, C, and E), the cortical burden is mild but most prominent in deep layers and consists of neuronal cytoplasmic, tangle-like inclusions, tufted astrocytes (A), and coiled oligodendroglial inclusions, among others. The tau burden is most pronounced in subcortical, diencephalic, and brain stem nuclei, where globose tangles (C) are observed. CBD (B, D, and F), in contrast, features characteristic teeming white matter threads and coiled oligodendroglial tau inclusions (B). Neuronal inclusions may form as coiled tangles (D) or other neuronal cytoplasmic inclusions. Astrocytic plaques (F) are a signature feature and are composed of short, stubby, tau-filled astroglial processes. Sections were stained with antibodies to tau phosphorylated at Ser202 (CP-13, courtesy of Peter Davies, A, B, E, and F) or the Gallyas silver stain (C and D). Scale bar in F applies to all panels and represents 1 mm in A and B and 50 μM in C–F.

Pathological Causes of Corticobasal Syndrome

Although CBD is the most common histopathological cause of CBS, it represents barely a majority in modern series (Boeve et al., 1999; Lee et al., 2011; Litvan et al., 1997; Murray et al., 2007). Other important causes include Alzheimer disease, Pick disease, PSP, and FTLD with TDP-43 inclusions (type A), and, less commonly, inherited tauopathies, LBD, or even Creutzfeldt–Jakob disease. Most CBS cases resulting from FTLD-TDP pathology occur in the setting of a mutation in progranulin (*GRN*), although sporadic cases have been reported (Neumann et al., 2006; Tartaglia et al., 2010). This diversity of proteinopathies causing CBS suggests that the perirolandic network is vulnerable to multiple misfolded protein "seeds," although the pattern of spread from within the target CBS network into off-target networks may differ subtly across these pathological entities.

The essential features of a CBD pathological diagnosis are tau-immunoreactive lesions in neurons and glia in the cortex and striatum (Fig. 18.7). Both neuronal and glial lesions are prominent. Neuronal inclusions are diffuse and granular, filling the cytosol, and are most prominent in layer 5. Astrocytic plaques are the most characteristic glial lesion. The most essential diagnostic feature is the presence of copious tau-positive threads in white matter subjacent to affected cortices accompanied by teeming coiled oligodendroglial inclusions. The highest burden of CBD pathology is found in the perirolandic cortex, in contrast to PSP, where the concentration is highest in basal ganglia, diencephalon, and brain stem.

CBD and PSP tau aggregates are made up of the hyperphosphorylated four-repeat isoforms of tau. Tau, a microtubule-associated protein, normally functions to assemble and stabilize microtubules and is highly concentrated in axons (Grundke-Iqbal et al., 1986). Tau aggregates isolated from postmortem tissue samples in both diseases have been shown to undergo self-propagating prion-like spread between cells in culture and in living mice (Clavaguera et al., 2013; Sanders et al., 2014). These findings and increasing additional support from the field are consistent with the candidate mechanism of disease protein spread via intercellular transmission from a selectively vulnerable epicenter out through an intrinsic connectivity network (de Calignon et al., 2012; Frost & Diamond, 2010; Zhou et al., 2012).

Progressive Supranuclear Palsy and Corticobasal Degeneration Histopathology Can Cause Multiple Clinical Syndromes

General Principles of Syndromic Diversity

Syndromes other than PSP-S and CBS caused by an underlying pathology of PSP or CBD manifest when the disease originates outside the most commonly affected network. The likelihood of the pathology affecting another network can be determined by looking to the nearest neighbors of the core disease network. This is illustrated in Fig. 18.1, in which the dotted circles—edged box outlines the regions affected in bvFTD, the most common syndrome other than PSP-S and CBS caused by 4R tauopathy. Patients with bvFTD experience apathy, disinhibition, and loss of awareness of both self and others. They become detached and socially tactless, and lose empathy. The disease primarily affects the anterior insula, anterior cingulate, orbitofrontal cortex, striatum, thalamus, and amygdala, regions that are all part of the salience network vital for social and emotional processing (Seeley et al., 2007). The salience network has strong transnetwork connectivity with the cingulo-opercular network, a core affected network in PSP-S and CBS. It is likely that the vulnerability of cells in the salience network to four-repeat tau aggregation is related in some way to the vulnerability in the neighboring cingulo-opercular network. Graded vulnerability to 4R tau is likely to be driven by the concentration of vulnerable cell populations in these paralimbic areas with similar genetic expression profiles. Neurodegenerative diseases typically have a subclass of cells that show early vulnerability, such as von Economo neurons and fork cells in bvFTD (Kim et al., 2012), entorhinal cortex layer II pyramidal neurons in Alzheimer disease (Gómez-Isla et al., 1996), or upper and lower motor neurons in amyotrophic lateral sclerosis (Cleveland & Rothstein, 2001), but the precise neuronal identities most vulnerable in to PSP and CBD remain uncertain.

Other Syndromes Caused by Progressive Supranuclear Palsy or Corticobasal Degeneration

A study by Williams and colleagues in 2005 reported that of 103 cases of definite PSP under examination, 32% clustered into a subgroup that was typified by asymmetric motor onset, tremor, response to treatment with levodopa, and less dementia (Williams et al., 2005). This subgroup was classified as "PSP-parkinsonism" (PSP-P), in distinction from patients with classical Richardson syndrome. The PSP-rs has been shown not differentiate between these patients subgroups (Golbe & Ohman-Strickland, 2007). However, patients with PSP-P exhibit less widespread atrophy in gray and white matter, less involvement of infratentorial brain structures, and less tau deposition (Longoni et al., 2011; Williams, Holton, Strand,

Pittman, et al., 2007). Thus, neuroimaging biomarkers may have an important role in differentiating PSP-P and typical PSP-S (Richardson syndrome).

Pure akinesia with gait freezing (PAGF) is a syndrome closely related to PSP-S in which patients exhibit early gait disturbance that eventually leads to gait freezing, micrographia, and hypophonia (Williams, Holton, Strand, Revesz, & Lees, 2007). Patients with PAGF have focal atrophy and neuronal loss almost exclusively in subcortical and brain stem regions including the globus pallidus, substantia nigra, and subthalamic nucleus (Ahmed, Josephs, Gonzalez, DelleDonne, & Dickson, 2008).

The most comprehensive study of phenotypic diversity in PSP was an examination of 100 autopsy-confirmed patients by Respondek and colleagues (Respondek et al., 2014). One hundred subjects were classified by "predominance type," a summary of the predominant clinical features in that group during the first 2 years of the disease. Only 24% exhibited classical Richardson syndrome, as distinguished by falls and supranuclear gaze palsy. The remainder was distributed between PSP-P (19% with tremor and asymmetric onset), postural instability (18%), frontotemporal dysfunction (12% with frontal and cognitive dysfunction), corticobasal syndrome (7%), occulomotor dysfunction (7%), and a clinically heterogeneous set that was unclassifiable (13%).

The most frequent FTLD syndromes that can result from either PSP or CBD are nonfluent variant primary progressive aphasia (nfvPPA) and bvFTD. nfvPPA and apraxia of speech (AOS) are related disorders in which speech production is impaired. Patients with nfvPPA exhibit a nonfluent aphasia with effortful and agrammatic speech (Gorno-Tempini et al., 2011). Atrophy is marked in perisylvian regions including Broca's area in the frontal operculum, along with the precentral gyrus, supplemental motor area, and dorsal anterior insula (Seeley et al., 2009), implicating the cingulo-opercular network, ventral attention network, and anterior components of the sensory-somatomotor network, especially in the dominant hemisphere. AOS is characterized by slow speech rate, abnormal prosody, and distorted sound selections, all related to a deficiency in speech motor planning (Josephs, Duffy, et al., 2006). The underlying atrophy is most prominent in the superior lateral premotor cortex and supplementary motor area (Josephs et al., 2012). These areas belong to cingulo-opercular network and premotor portions of the network adjacent ICNs. Interestingly, longitudinal study of patients with AOS have shown that a substantial portion later develop a PSP-like syndrome involving parkinsonism, vertical gaze palsy, and balance problems (Josephs et al., 2014). Both nfvPPA and AOS are often caused by an underlying tauopathy, CBD more commonly than PSP (Josephs et al., 2005; Karageorgiou & Miller, 2014; Lee et al., 2011; Rohrer et al., 2010). bvFTD, the FTLD syndrome with the most diverse set of underlying proteinopathies, can be also be caused by CBD or PSP (Bigio, Brown, & White, 1999; Hassan, Parisi, & Josephs, 2012; Rankin et al., 2011).

GENETIC FACTORS IN PROGRESSIVE SUPRANUCLEAR PALSY AND CORTICOBASAL DEGENERATION

As at the molecular, cellular, and systems levels, striking similarities in genetic risk factors for PSP and CBD suggest overlapping pathogenic mechanisms.

Rare Monogenic Causes

Although most cases of PSP and CBD are sporadic, cases of familial aggregation have been reported. In one study, 12 of 172 patients with PSP fulfilled the criteria for autosomal dominant inheritance, in which at least three first- or second-degree relatives had the disease. One of these cases revealed a P301L *MAPT* mutation (Donker Kaat et al., 2009). An instance of familial CBS was discovered in a Canadian family of Chinese origin with a splice donor site mutation in *PGRN* (Masellis et al., 2006).

Risk Factors

The strongest genetic association for PSP is the H1 haplotype of the *MAPT* gene on chromosome 17 and is seen in over 90% of patients (Fogel, Clark, & Geschwind, 2014). This gene encodes the tau protein, and H1 results in alternative splicing of exon 10, which may cause a shift toward more 4R tau and less 3R tau. Possession of the H1 haplotype is also the most well-established genetic risk factor for CBD (Di Maria et al., 2000). In a study of 38 patients with CBD, inheritance of the H1/H1 haplotype was associated with the severity of motor dysfunction, which suggests a dose-dependent genotype–phenotype association (Litvan, Chism, Litvan, Cambon, & Hutton, 2010). Assessment of point mutations in *MAPT* in 109 pathologically confirmed CBD cases found an association between p.N410H that increased the ratio of 4 and 3R tau (Kouri et al., 2013). In vitro experiments on recombinant tau with the p.N410H mutation showed increased tau filament formation and a slower rate of microtubule formation. An additional risk factor related to tau, the rare tau variant p.A152T, has been found to increase the risk for PSP, Alzheimer disease, and FTD (Coppola et al., 2012). The exact mechanisms of pathogenesis of this variant remain a topic of intense investigation.

Genome-wide Association Studies

A landmark genome-wide association study (GWAS) by Höglinger and colleagues in 2011 assessed two samples, one with 1114 autopsy-confirmed PSP cases and 3287 population-based controls and a second with 1051 patients clinically diagnosed with PSP-S and 3560 controls, with no overlap in individuals (Höglinger et al., 2011). This work confirmed two independent variants of MAPT that increased risk for PSP, along with three additional genes: *STX6*, *EIF2AK3*, and *MOBP*, whose respective cellular pathways are intracellular tracking, endoplasmic reticulum—mediated clearance of misfolded proteins, and myelination. No functional links have yet been established between these genes and disease pathophysiology. An epigenetic association has been established between the level of methylation at the 17q21.31 region and the H1 haplotype, indicating that epigenetic mediators have a contribution to disease risk (Li et al., 2014).

A GWAS study was completed in 152 cases and 3311 controls of CBD in a discovery phase and 67 cases and 439 controls in a replication phase (Kouri et al., 2015). Associations were found at the 17q21 locus of *MAPT*, *lnc-KIF13B-1*, a long noncoding RNA, and *SOS1*, a potential tau phosphatase. Tests also revealed associations at known PSP SNP sites for *MOBP* and *MAPT H1c*. This study confirmed that CBD and PSP have shared genetic risk factors for both *MAPT* and *MOBP*. This study strengthened the known association between variants in *MAPT* and tau pathology evident in both diseases. Importantly, it also revealed a novel pathogenic linkage common to both diseases between *MOBP* and white matter oligodendrocyte pathology. Studies in the near future have an opportunity to understand how failure of MOBP's normal role in myelin sheath stabilization may contribute to a common disease mechanism, and what relationships may exist for transsynaptic spread of misfolded tau.

CONCLUSION

Substantial evidence suggests that PSP-S and CBS are caused by progressive loss of function within the cingulo-opercular, sensory-somatomotor, and subcortical intrinsic connectivity networks. The selective vulnerability of neurons, astrocytes, and oligodendrocytes in these networks is highest to 4R tauopathy. Conversely, 4R tauopathy tends to originate in these networks, with a decreasing gradient of likelihood farther out from the network epicenter. Earlier diagnosis and prediction of disease progression will depend on monitoring the network core for signs of abnormality, in which functional imaging may provide an advantage over structural imaging or clinical evaluation. Uncertainty about the underlying pathology should be clarified with further development and a combination of genetic screening, molecular imaging, and cerebrospinal fluid—based biomarkers. Finally, advancement in understanding biological disease mechanisms will depend on exhaustive next-generation efforts to map the systems-level cellular composition and genetic expression profiles of vulnerable brain networks.

ACKNOWLEDGMENT

We thank Dr Suzee Lee for helpful discussions while drafting this work.

REFERENCES

Ahmed, Z., Josephs, K. A., Gonzalez, J., DelleDonne, A., & Dickson, D. W. (2008). Clinical and neuropathologic features of progressive supranuclear palsy with severe pallido-nigro-luysial degeneration and axonal dystrophy. *Brain: A Journal of Neurology, 131*(pt. 2), 460—472.

Albin, R. L., Young, A. B., & Penney, J. B. (1989). The functional anatomy of basal ganglia disorders. *Trends in Neurosciences, 12*(10), 366—375.

Alexander, G. E., DeLong, M. R., & Strick, P. L. (1986). Parallel organization of functionally segregated circuits linking basal ganglia and cortex. *Annual Review of Neuroscience, 9*, 357—381. http://doi.org/10.1146/annurev.ne.09.030186.002041.

Armstrong, M. J., Litvan, I., Lang, A. E., Bak, T. H., Bhatia, K. P., Borroni, B., ... Weiner, W. J. (2013). Criteria for the diagnosis of corticobasal degeneration. *Neurology, 80*(5), 496—503. http://doi.org/10.1212/WNL.0b013e31827f0fd1.

Baudrexel, S., Witte, T., Seifried, C., von Wegner, F., Beissner, F., Klein, J. C., ... Hilker, R. (2011). Resting state fMRI reveals increased subthalamic nucleus—motor cortex connectivity in Parkinson's disease. *NeuroImage, 55*(4), 1728—1738. http://doi.org/10.1016/j.neuroimage.2011.01.017.

Bigio, E. H., Brown, D. F., & White, C. L. (1999). Progressive supranuclear palsy with dementia: cortical pathology. *Journal of Neuropathology and Experimental Neurology, 58*(4), 359—364.

Blandini, F., Nappi, G., Tassorelli, C., & Martignoni, E. (2000). Functional changes of the basal ganglia circuitry in Parkinson's disease. *Progress in Neurobiology, 62*(1), 63–88.

Boeve, B. F., Lang, A. E., & Litvan, I. (2003). Corticobasal degeneration and its relationship to progressive supranuclear palsy and frontotemporal dementia. *Annals of Neurology, 54*(S5), S15–S19. http://doi.org/10.1002/ana.10570.

Boeve, B. F., Maraganore, D. M., Parisi, J. E., Ahlskog, J. E., Graff-Radford, N., Caselli, R. J., … Petersen, R. C. (1999). Pathologic heterogeneity in clinically diagnosed corticobasal degeneration. *Neurology, 53*(4), 795–800.

Boxer, A. L., Geschwind, M. D., Belfor, N., Gorno-Tempini, M. L., Schauer, G. F., Miller, B. L., … Rosen, H. J. (2006). Patterns of brain atrophy that differentiate corticobasal degeneration syndrome from progressive supranuclear palsy. *Archives of Neurology, 63*(1), 81–86. http://doi.org/10.1001/archneur.63.1.81.

Braak, H., & Braak, E. (1991). Neuropathological stageing of Alzheimer-related changes. *Acta Neuropathologica, 82*(4), 239–259.

Braak, H., Tredici, K. D., Rüb, U., de Vos, R. A. I., Jansen Steur, E. N. H., & Braak, E. (2003). Staging of brain pathology related to sporadic Parkinson's disease. *Neurobiology of Aging, 24*(2), 197–211. http://doi.org/10.1016/S0197-4580(02)00065-9.

Buckner, R. L., Sepulcre, J., Talukdar, T., Krienen, F. M., Liu, H., Hedden, T., … Johnson, K. A. (2009). Cortical hubs revealed by intrinsic functional connectivity: mapping, assessment of stability, and relation to Alzheimer's disease. *Journal of Neuroscience, 29*(6), 1860–1873. http://doi.org/10.1523/JNEUROSCI.5062-08.2009.

Burrell, J. R., Hodges, J. R., & Rowe, J. B. (2014). Cognition in corticobasal syndrome and progressive supranuclear palsy: a review. *Movement Disorders, 29*(5), 684–693. http://doi.org/10.1002/mds.25872.

Clavaguera, F., Akatsu, H., Fraser, G., Crowther, R. A., Frank, S., Hench, J., … Tolnay, M. (2013). Brain homogenates from human tauopathies induce tau inclusions in mouse brain. *Proceedings of the National Academy of Sciences, 110*(23), 9535–9540. http://doi.org/10.1073/pnas.1301175110.

Cleveland, D. W., & Rothstein, J. D. (2001). From charcot to lou gehrig: deciphering selective motor neuron death in als. *Nature Reviews Neuroscience, 2*(11), 806–819. http://doi.org/10.1038/35097565.

Coppola, G., Chinnathambi, S., Lee, J. J., Dombroski, B. A., Baker, M. C., Soto-Ortolaza, A. I., … Geschwind, D. H. (2012). Evidence for a role of the rare p.A152T variant in MAPT in increasing the risk for FTD-spectrum and Alzheimer's diseases. *Human Molecular Genetics, 21*(15), 3500–3512. http://doi.org/10.1093/hmg/dds161.

Corbetta, M., & Shulman, G. L. (2002). Control of goal-directed and stimulus-driven attention in the brain. *Nature Reviews Neuroscience, 3*(3), 201–215. http://doi.org/10.1038/nrn755.

Cummings, J. L., Mega, M., Gray, K., Rosenberg-Thompson, S., Carusi, D. A., & Gornbein, J. (1994). The neuropsychiatric inventory: comprehensive assessment of psychopathology in dementia. *Neurology, 44*(12), 2308–2314.

de Calignon, A., Polydoro, M., Suárez-Calvet, M., William, C., Adamowicz, D. H., Kopeikina, K. J., … Hyman, B. T. (2012). Propagation of tau pathology in a model of early Alzheimer's disease. *Neuron, 73*(4), 685–697. http://doi.org/10.1016/j.neuron.2011.11.033.

Di Maria, E., Tabaton, M., Vigo, T., Abbruzzese, G., Bellone, E., Donati, C., … Mandich, P. (2000). Corticobasal degeneration shares a common genetic background with progressive supranuclear palsy. *Annals of Neurology, 47*(3), 374–377. http://dx.doi.org/10.1002/1531-8249(200003)47:3<374::AID-ANA15>3.0.CO;2-8.

Dickson, D. W., Ahmed, Z., Algom, A. A., Tsuboi, Y., & Josephs, K. A. (2010). Neuropathology of variants of progressive supranuclear palsy. *Current Opinion in Neurology, 23*(4), 394–400. http://doi.org/10.1097/WCO.0b013e32833be924.

Dickson, D. W., Rademakers, R., & Hutton, M. L. (2007). Progressive supranuclear palsy: pathology and genetics. *Brain Pathology (Zurich, Switzerland), 17*(1), 74–82. http://doi.org/10.1111/j.1750-3639.2007.00054.x.

Donker Kaat, L., Boon, A. J. W., Azmani, A., Kamphorst, W., Breteler, M. M. B., Anar, B., … van Swieten, J. C. (2009). Familial aggregation of parkinsonism in progressive supranuclear palsy. *Neurology, 73*(2), 98–105. http://doi.org/10.1212/WNL.0b013e3181a92bcc.

Dosenbach, N. U. F., Fair, D. A., Miezin, F. M., Cohen, A. L., Wenger, K. K., Dosenbach, R. A. T., … Petersen, S. E. (2007). Distinct brain networks for adaptive and stable task control in humans. *Proceedings of the National Academy of Sciences, 104*(26), 11073–11078. http://doi.org/10.1073/pnas.0704320104.

Dosenbach, N. U. F., Visscher, K. M., Palmer, E. D., Miezin, F. M., Wenger, K. K., Kang, H. C., … Petersen, S. E. (2006). A core system for the implementation of task sets. *Neuron, 50*(5), 799–812. http://doi.org/10.1016/j.neuron.2006.04.031.

Fogel, B. L., Clark, M. C., & Geschwind, D. H. (2014). The neurogenetics of atypical parkinsonian disorders. *Seminars in Neurology, 34*(2), 217–224. http://doi.org/10.1055/s-0034-1381738.

Frost, B., & Diamond, M. I. (2010). Prion-like mechanisms in neurodegenerative diseases. *Nature Reviews Neuroscience, 11*(3), 155–159. http://doi.org/10.1038/nrn2786.

Gardner, R. C., Boxer, A. L., Trujillo, A., Mirsky, J. B., Guo, C. C., Gennatas, E. D., … Seeley, W. W. (2013). Intrinsic connectivity network disruption in progressive supranuclear palsy. *Annals of Neurology, 73*(5), 603–616. http://doi.org/10.1002/ana.23844.

Ghosh, B. C. P., Calder, A. J., Peers, P. V., Lawrence, A. D., Acosta-Cabronero, J., Pereira, J. M., … Rowe, J. B. (2012). Social cognitive deficits and their neural correlates in progressive supranuclear palsy. *Brain: A Journal of Neurology, 135*(Pt 7), 2089–2102. http://doi.org/10.1093/brain/aws128.

Golbe, L. I., & Ohman-Strickland, P. A. (2007). A clinical rating scale for progressive supranuclear palsy. *Brain, 130*(6), 1552–1565. http://doi.org/10.1093/brain/awm032.

Gómez-Isla, T., Price, J. L., McKeel, D. W., Jr., Morris, J. C., Growdon, J. H., & Hyman, B. T. (1996). Profound loss of layer II entorhinal cortex neurons occurs in very mild Alzheimer's disease. *The Journal of Neuroscience: The Official Journal of the Society for Neuroscience, 16*(14), 4491–4500.

Gorno-Tempini, M. L., Hillis, A. E., Weintraub, S., Kertesz, A., Mendez, M., Cappa, S. F., … Grossman, M. (2011). Classification of primary progressive aphasia and its variants. *Neurology, 76*(11), 1006–1014. http://doi.org/10.1212/WNL.0b013e31821103e6.

Grundke-Iqbal, I., Iqbal, K., Quinlan, M., Tung, Y. C., Zaidi, M. S., & Wisniewski, H. M. (1986). Microtubule-associated protein tau. A component of Alzheimer paired helical filaments. *Journal of Biological Chemistry, 261*(13), 6084–6089.

Hacker, C. D., Perlmutter, J. S., Criswell, S. R., Ances, B. M., & Snyder, A. Z. (2012). Resting state functional connectivity of the striatum in Parkinson's disease. *Brain, 135*(12), 3699–3711. http://doi.org/10.1093/brain/aws281.

Hassan, A., Parisi, J. E., & Josephs, K. A. (2012). Autopsy-proven progressive supranuclear palsy presenting as behavioral variant frontotemporal dementia. *Neurocase, 18*(6), 478–488. http://doi.org/10.1080/13554794.2011.627345.

Höglinger, G. U., Melhem, N. M., Dickson, D. W., Sleiman, P. M. A., Wang, L.-S., Klei, L., ... Schellenberg, G. D. (2011). Identification of common variants influencing risk of the tauopathy progressive supranuclear palsy. *Nature Genetics, 43*(7), 699–705. http://doi.org/10.1038/ng.859.

Jack, C. R., Jr., Knopman, D. S., Jagust, W. J., Petersen, R. C., Weiner, M. W., Aisen, P. S., ... Trojanowski, J. Q. (2013). Tracking pathophysiological processes in Alzheimer's disease: an updated hypothetical model of dynamic biomarkers. *The Lancet Neurology, 12*(2), 207–216. http://dx.doi.org/10.1016/S1474-4422(12)70291-0.

Josephs, K. A., Boeve, B. F., Duffy, J. R., Smith, G. E., Knopman, D. S., Parisi, J. E., ... Dickson, D. W. (2005). Atypical progressive supranuclear palsy underlying progressive apraxia of speech and nonfluent aphasia. *Neurocase, 11*(4), 283–296. http://doi.org/10.1080/13554790590963004.

Josephs, K. A., Duffy, J. R., Strand, E. A., Machulda, M. M., Senjem, M. L., Gunter, J. L., ... Whitwell, J. L. (2014). The evolution of primary progressive apraxia of speech. *Brain, 137*(Pt 10), 2783–2795. awu223 http://doi.org/10.1093/brain/awu223.

Josephs, K. A., Duffy, J. R., Strand, E. A., Machulda, M. M., Senjem, M. L., Master, A. V., ... Whitwell, J. L. (2012). Characterizing a neurodegenerative syndrome: primary progressive apraxia of speech. *Brain: A Journal of Neurology, 135*(Pt 5), 1522–1536. http://doi.org/10.1093/brain/aws032.

Josephs, K. A., Duffy, J. R., Strand, E. A., Whitwell, J. L., Layton, K. F., Parisi, J. E., ... Petersen, R. C. (2006). Clinicopathological and imaging correlates of progressive aphasia and apraxia of speech. *Brain, 129*(6), 1385–1398. http://doi.org/10.1093/brain/awl078.

Josephs, K. A., Katsuse, O., Beccano-Kelly, D. A., Lin, W.-L., Uitti, R. J., Fujino, Y., ... Dickson, D. W. (2006). Atypical progressive supranuclear palsy with corticospinal tract degeneration. *Journal of Neuropathology and Experimental Neurology, 65*(4), 396–405. http://doi.org/10.1097/01.jnen.0000218446.38158.61.

Josephs, K. A., Xia, R., Mandrekar, J., Gunter, J. L., Senjem, M. L., Jack, C. R., & Whitwell, J. L. (2013). Modeling trajectories of regional volume loss in progressive supranuclear palsy. *Movement Disorders, 28*(8), 1117–1124. http://doi.org/10.1002/mds.25437.

Karageorgiou, E., & Miller, B. (2014). Frontotemporal lobar degeneration: a clinical approach. *Seminars in Neurology, 34*(02), 189–201. http://doi.org/10.1055/s-0034-1381735.

Kim, E.-J., Sidhu, M., Gaus, S. E., Huang, E. J., Hof, P. R., Miller, B. L., ... Seeley, W. W. (2012). Selective frontoinsular von Economo neuron and fork cell loss in early behavioral variant frontotemporal dementia. *Cerebral Cortex (New York, NY: 1991), 22*(2), 251–259. http://doi.org/10.1093/cercor/bhr004.

Kouri, N., Carlomagno, Y., Baker, M., Liesinger, A. M., Caselli, R. J., Wszolek, Z. K., ... Rademakers, R. (2013). Novel mutation in MAPT exon 13 (p.N410H) causes corticobasal degeneration. *Acta Neuropathologica, 127*(2), 271–282. http://doi.org/10.1007/s00401-013-1193-7.

Kouri, N., Ross, O. A., Dombroski, B., Younkin, C. S., Serie, D. J., Soto-Ortolaza, A., ... Dickson, D. W. (2015). Genome-wide association study of corticobasal degeneration identifies risk variants shared with progressive supranuclear palsy. *Nature Communications, 6*, 7247. http://doi.org/10.1038/ncomms8247.

Lee, S. E., Rabinovici, G. D., Mayo, M. C., Wilson, S. M., Seeley, W. W., DeArmond, S. J., ... Miller, B. L. (2011). Clinicopathological correlations in corticobasal degeneration. *Annals of Neurology, 70*(2), 327–340. http://doi.org/10.1002/ana.22424.

Li, Y., Chen, J. A., Sears, R. L., Gao, F., Klein, E. D., Karydas, A., ... Coppola, G. (2014). An epigenetic signature in peripheral blood associated with the haplotype on 17q21.31, a risk factor for neurodegenerative tauopathy. *PLoS Genetic, 10*(3), e1004211. http://doi.org/10.1371/journal.pgen.1004211.

Litvan, I., Agid, Y., Calne, D., Campbell, G., Dubois, B., Duvoisin, R. C., ... Zee, D. S. (1996). Clinical research criteria for the diagnosis of progressive supranuclear palsy (Steele-Richardson-Olszewski syndrome) report of the NINDS-SPSP international workshop. *Neurology, 47*(1), 1–9. http://doi.org/10.1212/WNL.47.1.1.

Litvan, I., Agid, Y., Goetz, C., Jankovic, J., Wenning, G. K., Brandel, J. P., ... Bartko, J. J. (1997). Accuracy of the clinical diagnosis of corticobasal degeneration: a clinicopathologic study. *Neurology, 48*(1), 119–125.

Litvan, I., Chism, A., Litvan, J., Cambon, A., & Hutton, M. (2010). H1/H1 genotype influences symptom severity in corticobasal syndrome. *Movement Disorders, 25*(6), 760–763. http://doi.org/10.1002/mds.22804.

Longoni, G., Agosta, F., Kostić, V. S., Stojković, T., Pagani, E., Stošić-Opinćal, T., & Filippi, M. (2011). MRI measurements of brainstem structures in patients with Richardson's syndrome, progressive supranuclear palsy-parkinsonism, and Parkinson's disease. *Movement Disorders, 26*(2), 247–255. http://doi.org/10.1002/mds.23293.

Masellis, M., Momeni, P., Meschino, W., Heffner, R., Elder, J., Sato, C., ... Rogaeva, E. (2006). Novel splicing mutation in the progranulin gene causing familial corticobasal syndrome. *Brain: A Journal of Neurology, 129*(Pt 11), 3115–3123. http://doi.org/10.1093/brain/awl276.

Murray, R., Neumann, M., Forman, M. S., Farmer, J., Massimo, L., Rice, A., ... Grossman, M. (2007). Cognitive and motor assessment in autopsy-proven corticobasal degeneration. *Neurology, 68*(16), 1274–1283. http://doi.org/10.1212/01.wnl.0000259519.78480.c3.

Neumann, M., Sampathu, D. M., Kwong, L. K., Truax, A. C., Micsenyi, M. C., Chou, T. T., ... Lee, V. M.-Y. (2006). Ubiquitinated TDP-43 in frontotemporal lobar degeneration and amyotrophic lateral sclerosis. *Science, 314*(5796), 130–133. http://doi.org/10.1126/science.1134108.

Paviour, D. C., Price, S. L., Jahanshahi, M., Lees, A. J., & Fox, N. C. (2006). Longitudinal MRI in progressive supranuclear palsy and multiple system atrophy: rates and regions of atrophy. *Brain, 129*(4), 1040–1049. http://doi.org/10.1093/brain/awl021.

Pearson, R. C., Esiri, M. M., Hiorns, R. W., Wilcock, G. K., & Powell, T. P. (1985). Anatomical correlates of the distribution of the pathological changes in the neocortex in Alzheimer disease. *Proceedings of the National Academy of Sciences United States of America, 82*(13), 4531–4534.

Pillon, B., Blin, J., Vidailhet, M., Deweer, B., Sirigu, A., Dubois, B., & Agid, Y. (1995). The neuropsychological pattern of corticobasal degeneration: comparison with progressive supranuclear palsy and Alzheimer's disease. *Neurology, 45*(8), 1477–1483.

Power, J. D., Cohen, A. L., Nelson, S. M., Wig, G. S., Barnes, K. A., Church, J. A., ... Petersen, S. E. (2011). Functional network organization of the human brain. *Neuron, 72*(4), 665–678. http://doi.org/10.1016/j.neuron.2011.09.006.

Raj, A., Kuceyeski, A., & Weiner, M. (2012). A network diffusion model of disease progression in dementia. *Neuron, 73*(6), 1204–1215. http://doi.org/10.1016/j.neuron.2011.12.040.

Raj, A., LoCastro, E., Kuceyeski, A., Tosun, D., Relkin, N., & Weiner, M. (2015). Network diffusion model of progression predicts longitudinal patterns of atrophy and metabolism in Alzheimer's disease. *Cell Reports, 10*(3), 359–369. http://doi.org/10.1016/j.celrep.2014.12.034.

Rankin, K. P., Mayo, M. C., Seeley, W. W., Lee, S., Rabinovici, G., Gorno-Tempini, M. L., ... Miller, B. L. (2011). Behavioral-variant frontotemporal dementia with corticobasal degeneration pathology: phenotypic comparison to bvFTD with Pick's disease. *Journal of Molecular Neuroscience: MN, 45*(3), 594–608. http://doi.org/10.1007/s12031-011-9615-2.

Rebeiz, J. J., Kolodny, E. H., & Richardson, E. P., Jr. (1968). Corticodentatonigral degeneration with neuronal achromasia. *Archives of Neurology, 18*(1), 20–33. http://doi.org/10.1001/archneur.1968.00470310034003.

Respondek, G., Stamelou, M., Kurz, C., Ferguson, L. W., Rajput, A., Chiu, W. Z., ... for the Movement Disorder Society—endorsed PSP Study Group. (2014). The phenotypic spectrum of progressive supranuclear palsy: a retrospective multicenter study of 100 definite cases. *Movement Disorders, 29*(14), 1758–1766. http://doi.org/10.1002/mds.26054.

Rohrer, J. D., Paviour, D., Bronstein, A. M., O'Sullivan, S. S., Lees, A., & Warren, J. D. (2010). Progressive supranuclear palsy syndrome presenting as progressive nonfluent aphasia: a neuropsychological and neuroimaging analysis. *Movement Disorders: Official Journal of the Movement Disorder Society, 25*(2), 179–188. http://doi.org/10.1002/mds.22946.

Rowe, J. B., Hughes, L. E., Barker, R. A., & Owen, A. M. (2010). Dynamic causal modelling of effective connectivity from fMRI: are results reproducible and sensitive to Parkinson's disease and its treatment? *NeuroImage, 52*(3), 1015–1026. http://doi.org/10.1016/j.neuroimage.2009.12.080.

Sanders, D. W., Kaufman, S. K., DeVos, S. L., Sharma, A. M., Mirbaha, H., Li, A., ... Diamond, M. I. (2014). Distinct tau prion strains propagate in cells and mice and define different tauopathies. *Neuron, 82*(6), 1271–1288. http://doi.org/10.1016/j.neuron.2014.04.047.

Saper, C. B., Wainer, B. H., & German, D. C. (1987). Axonal and transneuronal transport in the transmission of neurological disease: potential role in system degenerations, including Alzheimer's disease. *Neuroscience, 23*(2), 389–398. http://doi.org/10.1016/0306-4522(87)90063-7.

Satake, W., Nakabayashi, Y., Mizuta, I., Hirota, Y., Ito, C., Kubo, M., ... Toda, T. (2009). Genome-wide association study identifies common variants at four loci as genetic risk factors for Parkinson's disease. *Nature Genetics, 41*(12), 1303–1307. http://doi.org/10.1038/ng.485.

Seeley, W. W., Crawford, R. K., Zhou, J., Miller, B. L., & Greicius, M. D. (2009). Neurodegenerative diseases target large-scale human brain networks. *Neuron, 62*(1), 42–52. http://doi.org/10.1016/j.neuron.2009.03.024.

Seeley, W. W., Menon, V., Schatzberg, A. F., Keller, J., Glover, G. H., Kenna, H., ... Greicius, M. D. (2007). Dissociable intrinsic connectivity networks for salience processing and executive control. *The Journal of Neuroscience, 27*(9), 2349–2356. http://doi.org/10.1523/JNEUROSCI.5587-06.2007.

Simón-Sánchez, J., Schulte, C., Bras, J. M., Sharma, M., Gibbs, J. R., Berg, D., ... Gasser, T. (2009). Genome-wide association study reveals genetic risk underlying Parkinson's disease. *Nature Genetics, 41*(12), 1308–1312. http://doi.org/10.1038/ng.487.

Steele, J. C., Richardson, J. C., & Olszewski, J. (1964). Progressive supranuclear palsy. A heterogeneous degeneration involving the brain stem, basal ganglia and cerebellum with vertical gaze and pseudobulbar palsy, nuchal dystonia and dementia. *Archives of Neurology, 10*, 333–359.

Tartaglia, M. C., Sidhu, M., Laluz, V., Racine, C., Rabinovici, G. D., Creighton, K., ... Seeley, W. W. (2010). Sporadic corticobasal syndrome due to FTLD-TDP. *Acta Neuropathologica, 119*(3), 365–374. http://doi.org/10.1007/s00401-009-0605-1.

Trinh, J., & Farrer, M. (2013). Advances in the genetics of Parkinson disease. *Nature Reviews Neurology, 9*(8), 445–454. http://doi.org/10.1038/nrneurol.2013.132.

Tsai, R. M., & Boxer, A. L. (2014). Treatment of frontotemporal dementia. *Current Treatment Options in Neurology, 16*(11), 1–14. http://doi.org/10.1007/s11940-014-0319-0.

Tsuboi, Y., Josephs, K. A., Boeve, B. F., Litvan, I., Caselli, R. J., Caviness, J. N., ... Dickson, D. W. (2005). Increased tau burden in the cortices of progressive supranuclear palsy presenting with corticobasal syndrome. *Movement Disorders: Official Journal of the Movement Disorder Society, 20*(8), 982–988. http://doi.org/10.1002/mds.20478.

VanVoorst, W. A., Greenaway, M. C., Boeve, B. F., Ivnik, R. J., Parisi, J. E., Ahlskog, J. E., ... Josephs, K. A. (2008). Neuropsychological findings in clinically atypical autopsy confirmed corticobasal degeneration and progressive supranuclear palsy. *Parkinsonism & Related Disorders, 14*(4), 376–378.

Wadia, P. M., & Lang, A. E. (2007). The many faces of corticobasal degeneration. *Parkinsonism & Related Disorders, 13*(Suppl. 3), S336–S340. http://doi.org/10.1016/S1353-8020(08)70027-0.

Weintraub, D., Doshi, J., Koka, D., Davatzikos, C., Siderowf, A. D., Duda, J. E., ... Clark, C. M. (2011). Neurodegeneration across stages of cognitive decline in parkinson disease. *Archives of Neurology, 68*(12), 1562–1568. http://doi.org/10.1001/archneurol.2011.725.

Whitwell, J. L., Avula, R., Master, A., Vemuri, P., Senjem, M. L., Jones, D. T., ... Josephs, K. A. (2011). Disrupted thalamocortical connectivity in PSP: a resting-state fMRI, DTI, and VBM study. *Parkinsonism & Related Disorders, 17*(8), 599–605. http://doi.org/10.1016/j.parkreldis.2011.05.013.

Whitwell, J. L., Jack, C. R., Boeve, B. F., Parisi, J. E., Ahlskog, J. E., Drubach, D. A., ... Josephs, K. A. (2010). Imaging correlates of pathology in corticobasal syndrome. *Neurology, 75*(21), 1879–1887. http://doi.org/10.1212/WNL.0b013e3181feb2e8.

Whitwell, J. L., Master, A. V., Avula, R., Kantarci, K., Eggers, S. D., Edmonson, H. A., ... Josephs, K. A. (2011). Clinical correlates of white matter tract degeneration in progressive supranuclear palsy. *Archives of Neurology, 68*(6), 753–760. http://doi.org/10.1001/archneurol.2011.107.

Williams, D. R., de Silva, R., Paviour, D. C., Pittman, A., Watt, H. C., Kilford, L., … Lees, A. J. (2005). Characteristics of two distinct clinical phenotypes in pathologically proven progressive supranuclear palsy: Richardson's syndrome and PSP-parkinsonism. *Brain, 128*(6), 1247–1258. http://doi.org/10.1093/brain/awh488.

Williams, D. R., Holton, J. L., Strand, C., Pittman, A., de Silva, R., Lees, A. J., & Revesz, T. (2007). Pathological tau burden and distribution distinguishes progressive supranuclear palsy-parkinsonism from Richardson's syndrome. *Brain, 130*(6), 1566–1576. http://doi.org/10.1093/brain/awm104.

Williams, D. R., Holton, J. L., Strand, K., Revesz, T., & Lees, A. J. (2007). Pure akinesia with gait freezing: a third clinical phenotype of progressive supranuclear palsy. *Movement Disorders: Official Journal of the Movement Disorder Society, 22*(15), 2235–2241. http://doi.org/10.1002/mds.21698.

Zhou, J., Gennatas, E. D., Kramer, J. H., Miller, B. L., & Seeley, W. W. (2012). Predicting regional neurodegeneration from the healthy brain functional connectome. *Neuron, 73*(6), 1216–1227. http://doi.org/10.1016/j.neuron.2012.03.004.

Zhou, J., & Seeley, W. W. (2014). Network dysfunction in Alzheimer's disease and frontotemporal dementia: implications for psychiatry. *Biological Psychiatry, 75*(7), 565–573. http://doi.org/10.1016/j.biopsych.2014.01.020.

Chapter 19

A New Neuroanatomy of Basal Ganglia Circuitry

Nicholas AuYong, Adrienne M. Keener, Yvette M. Bordelon, Carlos Portera-Cailliau and Nader Pouratian
University of California Los Angeles, Los Angeles, CA, United States

ABBREVIATIONS

2PCI Two-photon calcium imaging
2PE Two-photon excitation
AChE Acetylcholinesterase
BG Basal ganglia
ChR2 Channelrhodopsin-2
DBS Deep brain stimulation
DREADD Designer Receptor Exclusively Activated by Designer Drugs
GPi/Gpe Globus pallidus, pars interna/externa
LFP Local field potential
MSN Medium sized spiny neuron
PD Parkinson disease
SNc Substantia nigra pars compacta
SNr Substantia nigra pars reticulata
STN Subthalamic nucleus
tTA Tetracycline transactivator
VTA Ventral tegmental area

The basal ganglia (BG) are a set of interconnected deep gray matter nuclei. Their function in executing control of motor output has been widely studied and their role in modulating aspects of cognitive and limbic function has been increasingly recognized and evaluated. The subcortical structures that comprise the basal ganglia include the striatum (including the putamen and caudate), the globus pallidus externus (GPe) and globus pallidus externus internus (GPi), the subthalamic nucleus (STN), and the substantia nigra pars compacta (SNc) and substantia nigra pars reticulate (SNr). These structures are widely connected, both structurally and functionally, to various brain regions including diverse regions of the cerebral cortex, the thalamus, and the brain stem, forming a series of functional circuits. In this chapter, we review the functional neuroanatomy of the BG, including historical studies that formed the foundation of our understanding of BG structure and function as well as advances in research techniques that enable us to reshape our understanding of the important role of BG.

ANATOMIC OVERVIEW OF BASAL GANGLIA

The striatum and STN are the primary input nuclei of the BG. The striatum receives input from diverse cortical areas, which can be broadly divided into those subserving sensorimotor, cognitive, and limbic functions. These various cortical inputs were initially thought to maintain topographical organization within the striatum, comprising parallel basal

Genomics, Circuits, and Pathways in Clinical Neuropsychiatry. http://dx.doi.org/10.1016/B978-0-12-800105-9.00019-6

ganglia-thalamocortical loops (Alexander, DeLong, & Strick, 1986). Anatomic and structural overlap between these loops has now been recognized as a potentially key mechanism to allow for integration of information within the BG (Mailly, Aliane, Groenewegen, Haber, & Deniau, 2013). The striatum also receives excitatory projections from thalamic nuclei, including the intralaminar nuclei, the centromedian and parafascicular complex, and the mediodorsal and ventromedial nuclei. Thalamostriatal projections include distinct yet overlapping projections involved in motor, cognitive, and limbic functions (Smith, Raju, Pare, & Sidibe, 2004). The striatum also receives important dopaminergic input from the SNc and the ventral tegmental area in the midbrain (Fig. 19.1).

The gamma-aminobutyric acid-ergic (GABAergic) projection neurons of the striatum can be primarily divided into two populations. Approximately half project to the basal ganglia output nuclei, GPi and SNr, forming the direct pathway, whereas the other half project to the GPe and form the polysynaptic indirect pathway. The differential effect of dopamine on these two pathways is mediated by distinct G protein−coupled dopamine receptor types, D1 and D2 receptors in the direct and indirect pathways, respectively, which is described in greater detail subsequently. This is a key feature of the classical rate model of basal ganglia circuitry (Albin, Young, & Penney, 1989).

The STN is the other primary BG input nucleus; it receives distinct yet overlapping projections from sensorimotor, cognitive, and limbic cortices (Haynes & Haber, 2013). These "hyperdirect" direct cortical-STN projections are thought to mediate executive control over action (reviewed in Jahanshahi, Obeso, Baunez, Alegre, and Krack (2014)). The indirect pathway also projects to the STN via the GPe. The STN contains glutamatergic neurons which project to BG output nuclei. In addition, feedback connections from the STN to the GPe and the striatum may serve an intrinsic pacemaker function (Plenz & Kital, 1999).

GPi and SNr are the main output nuclei of the BG (Fig. 19.1). In the direct pathway, GPi and SNr receive inhibitory input from the striatum. In the indirect pathway, they receive excitatory input from the STN, which itself receives inhibitory input from GPe. GPi is composed of GABAergic neurons, which send inhibitory projections to the thalamus. The thalamus then projects back to the cortex, creating a closed-loop circuit. The net effect of the direct pathway is to excite the cortex, whereas the indirect pathway results in overall inhibition. The classical rate model of the BG motor circuit proposes that control of movement therefore results from a balance between the direct and indirect pathways. In addition to these pathways, BG also sends outputs to the brain stem, including the superior colliculus and the pedunculopontine nucleus, which allow a direct effect on behavior without looping back through the cortex.

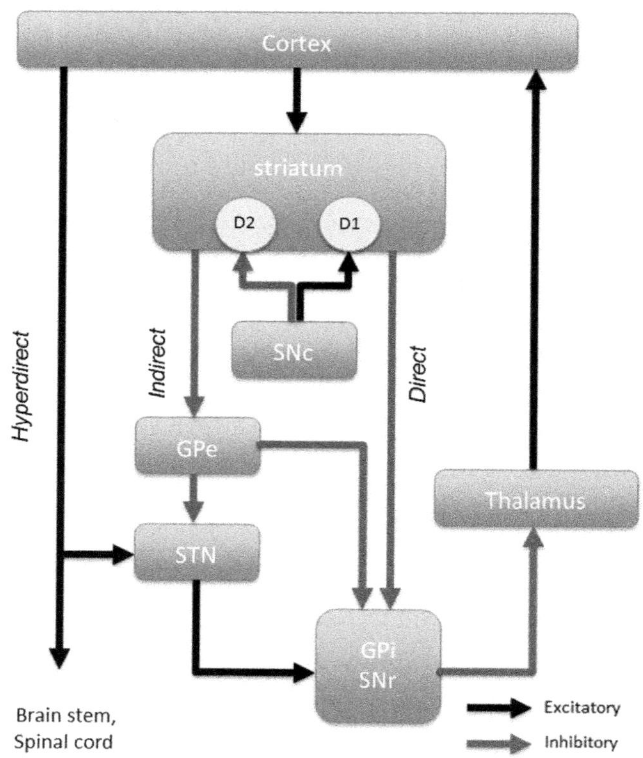

FIGURE 19.1 Basal ganglia circuit diagram schematic highlighting direct, hyperdirect, and indirect pathways.

CLINICAL BASAL GANGLIA DISORDERS

Human neurological diseases have served as the primary tool and impetus for studying the function of the BG, particularly movement disorders such as Parkinson disease (PD) and Huntington disease (HD). BG has also been implicated in neuropsychiatric disorders including obsessive compulsive disorder, Tourette syndrome, and depression, which have opened new avenues of exploration. Two diseases are briefly summarized here to provide a clinical context for subsequent discussion.

Parkinson Disease

Parkinsonism is the prototypical hypokinetic movement disorder. It manifests clinically with motor symptoms of rigidity, bradykinesia, rest tremor, and postural instability. The most well-known cause of parkinsonism is idiopathic PD, which is a neurodegenerative disease characterized pathologically by α-synuclein–containing neuronal inclusion bodies, termed Lewy bodies, and Lewy neurites. This leads to progressive cell loss initially involving brain stem and subcortical nuclei, and eventually ascending to the cortex (Braak et al., 2003). Motor symptoms result from loss of dopaminergic neurons in SNc. This is thought to alter BG motor circuits, leading to increased activity in the indirect pathway and decreased activity in the direct pathway, and to overall inhibition of thalamocortical projections and therefore paucity of movement (Albin et al., 1989). Nonmotor symptoms are now well-recognized in PD, including autonomic dysfunction, sleep disorders, depression and anxiety, and cognitive impairment, likely related to pathology of nonmotor components of BG as well as disease extension beyond BG.

Huntington Disease

HD is an autosomal dominant neurodegenerative disease caused by an expanded CAG repeat in the huntingtin gene on chromosome 4. It is clinically characterized by adult-onset, psychiatric, cognitive, and motor symptoms. Whereas longer CAG repeat expansion correlates with earlier age at onset, it accounts for only 50–60% of the predicted age at onset (Wexler et al., 2004). Research into other genetic and environmental influences is ongoing. The typical motor feature in HD is chorea, defined as a hyperkinetic movement disorder involving involuntary movements flowing from one muscle group to another. The brains of HD patients demonstrate prominent atrophy involving the caudate as well as extrastriatal structures. It has been proposed that striatal neuronal loss affects the motor circuit of the basal ganglia by causing preferential loss of medium spiny neurons (MSNs) in the indirect pathway, resulting in enhanced excitation of the cortex and therefore extraneous, uncontrolled movements (Albin et al., 1989). As the disease progresses, the direct pathway becomes involved, which leads to the relatively hypokinetic state which characterizes the late stages of HD (Deng et al., 2004). Psychiatric and cognitive symptoms in HD are thought to be primarily attributable to disruption of frontal-subcortical circuits.

CLASSICAL BASAL GANGLIA MICROANATOMY

The classical heuristic model of BG circuitry, as proposed by the work of Albin and DeLong (Albin et al., 1989; Alexander et al., 1986), has been extremely valuable in guiding both therapeutic developments and research into extrapyramidal movement disorders. According to the proposed functional connectivity, motor symptoms arise from an imbalance between the direct and indirect pathways that respectively facilitate and inhibit movement.

Anatomical studies were crucial in defining the connectivity between BG components, the microanatomy within each structure, and the input–output fiber tracts of the BG complex. The striatum was found to receive cortical afferents diffusely with a general anteroposterior organization (Kemp, 1970; Kemp & Powell, 1970). Tract-tracing studies from lesion-induced fiber degeneration at the striatum revealed distinct striatal efferent projections to the GPi, GPe, and SN (Nauta & Mehler, 1966). The GPi was found to project primarily to the thalamus and midbrain tegmentum, whereas the GPe projected to the STN. In addition to pallidal input, the STN also receives somatotopically organized ipsilateral motor cortical afferents, with a lesser degree from premotor and prefrontal areas (Monakow, Akert, & Kunzle, 1978), in what is now recognized as the hyperdirect pathway. Contralateral cortical projections to the STN have also been recognized in studies using micro-iontophoretic pharmacologic stimulation of bilateral cortical areas resulting in excitation of STN neurons which was suppressed with the application of antagonists of excitatory amino acids (Rouzaire-Dubois & Scarnati, 1987). Autoradiographic studies of STN efferents in the monkey and cat demonstrated projections to GPi and GPe as well as SNc and SNr (Nauta & Cole, 1978), which in addition to excitatory projections include GABAergic projections from the STN to pallidal segments and SN (Nauta & Cuenod, 1982). Efferent projections from BG emerge from GPi and SNr to the ventral thalamic nucleus (Kim, Nakano, Jayaraman, & Carpenter, 1976; Nauta & Mehler, 1966). Retrograde tracers were used to confirm that these thalamic nuclei (VA and VL) then project to the prefrontal, frontal, and motor cortices (Barbas, Henion, & Dermon, 1991; Goldman-Rakic & Porrino, 1985; Inase, Tokuno, Nambu, Akazawa, & Takada, 1996).

A main control point of BG operation is the differential effect of SNc dopaminergic projections to the striatum on the direct and indirect pathway. The striatum is composed of MSNs, which are GABAergic projection neurons with spiny dendritic processes (Graveland, Williams, & DiFiglia, 1985; Kemp & Powell, 1971). Cortical afferents primarily synapse on the head of the spines, whereas other inputs terminate on dendritic shafts (Smith & Bolam, 1990). Differential expression of dopamine (DA) receptors and neuropeptides also functionally separates portions of the striatum (Gerfen et al., 1990). In the direct pathway, activation of D1 receptors results in excitation of inhibitory striatal projection neurons and therefore increased inhibition of GPe whereas activation of D2 receptors in the indirect pathway results in inhibition of the inhibitory striatal projections to the GPe. These findings can account for the suppression and exacerbation of hyperkinetic movements (eg, chorea, tics, ballism) by D2 receptor antagonists and agonists, respectively. Although functionally discrete, distribution of these distinct DA receptors overlaps, with both classes of receptors present in neurons receiving dopaminergic afferents (Aizman et al., 2000). Within the striatum, a sparse collection of large cholinergic interneurons and GABAergic interneurons exists (Phelps, Houser, & Vaughn, 1985) whose function is an area of ongoing research (discussed subsequently).

The organization of the striatum into striosome and matrix, as well as the regional specificity of cortical–striatal connections, serves as the anatomical basis for parallel circuitry (Gerfen, 1984; Ragsdale & Graybiel, 1990). Areas of the striatum that densely stain for μ-opiate receptors and exhibited enkephalin and substance P immunoreactivity are termed striosomes or patches. Striosomes receive limbic-related cortical input and are surrounded by the "matrix" composed of cells that contain acetylcholinesterase and somatostatin-like immunoreactivity and receive both sensory and motor cortical input (Graybiel & Ragsdale, 1978). Anterograde fiber tracing demonstrated that corticostriatal fibers from deep cortical layer 5 preferentially terminate within the striosomal compartment, whereas projections from superficial layer 5 and supragranular layer terminate within striatal matrix (Gerfen & Sawchenko, 1984). This microanatomical feature conforms to the emphasis of the classical model on five parallel cortico–BG–thalamic loops, each segregated from the other anatomically and functionally (Alexander & Crutcher, 1990). These loops include motor, oculomotor, limbic, dorsolateral prefrontal, and lateral-orbitofrontal loops.

Whereas anatomical studies helped define the classical model's physical interconnections, neurophysiological studies defined the functional significance of these connections. The pioneering studies of DeLong, Crutcher, and Georgopoulos, (1985) established the presence of movement-related activity in intact behaving nonhuman primates. In contrast to motor cortical neurons that fire in anticipation of movement, movement-related BG activity largely follows that of the motor cortex (Mink & Thach, 1993). The association of STN lesions with hemiballism was established earliest and was highly congruent with clinical findings (Carpenter, Whittier, & Mettler, 1950; Whittier & Mettler, 1949). Reversible lesions in the putamen and GP in nonhuman primates resulted in the impairment of dynamic control of contralateral arm movements when the putamen and GPi were involved, whereas GPe lesions mainly caused the arm to assume a flexed posture (Kato & Kimura, 1992). Kinematics were affected with GPi lesions but reaction time was spared, which indicated that movement initiation is not driven by the BG circuit (Desmurget & Turner, 2008; Mink & Thach, 1991; Turner & Desmurget, 2010). GPi lesions primarily affected the velocity and magnitude in both overlearned and random movement sequences (Desmurget & Turner, 2010). Dissecting the role of the BG in both motor and nonmotor behavior is an ongoing enterprise requiring refinement to the classical model.

PROBING BASAL GANGLIA CIRCUITRY IN ANIMAL MODELS

BG anatomy is highly evolutionarily conserved across species, which reflects the critical function of this collection of subcortical nuclei. Although the ultimate goal is to understand human BG neurophysiology, investigations in animal models provide much greater flexibility and opportunity to dissect and manipulate BG circuitry to gain insight into its function. Not surprisingly, a wealth of knowledge about BG circuitry has come from studies in rodents and monkeys (comprehensive reviews include Kreitzer and Berke (2011) and Nelson and Kreitzer (2014)).

The most important animal studies that have informed our knowledge about BG circuitry have either examined the consequences of lesioning and silencing striatal neurons or have used electrophysiological recordings from striatal neurons to infer their function. In particular, transgenic mice using bacterial artificial chromosomes allow for transgene expression in direct pathway MSNs (D1 or M4 lines) or indirect pathway MSNs (D2 or A2A lines), making it possible to target recordings (and optogenetic manipulations) of these specific cell types. This has taught us that D1 and D2 MSNs differ in their gene expression profiles, their dendritic morphology and excitability, and their function within the network.

Recordings in Animals

Electrophysiological recordings in the 1-methyl-4-phenyl-1,2,3,6-tetrahydropyridine (MPTP) rodent and nonhuman primate experimental paradigms have consistently supported the classical model of BG function. For example, depletion of dopamine by MPTP produces significant increases in firing rates of STN neurons (Bergman, Wichmann, Karmon, & DeLong, 1994) and in the BG output nuclei, the GPi and the SNr (Filion, Tremblay, & Bedard, 1991), the latter of which can be reversed by administration of a dopamine agonist.

Ablation Studies

Cell type–specific inactivation has been achieved with selective expression of neurotoxins in D1 and D2 pathway MSNs, which provides an opportunity to test the validity of the classical BG model. Targeted ablation of D1 neurons with diphtheria toxin led to motor slowing in early postnatal mice (Drago et al., 1998) but not at older ages (Wong, Padungchaichot, Massalas, & Drago, 2000). Conversely, killing D2 MSNs with diphtheria toxin (Durieux et al., 2009; Saito et al., 2001) or with a different immunotoxin approach (Sano et al., 2003) resulted in enhanced motor activity, essentially as predicted by the Albin–DeLong model of BG function.

MSN neurons can also be shut down with pharmacogenetic approaches. In one study, tetanus toxin was expressed conditionally and reversibly in D1 or D2 MSNs via the substance P or encephalin promoter, respectively, using the tetracycline transactivator (tTA) method (Hikida, Kimura, Wada, Funabiki, & Nakanishi, 2010). The results were again consistent with the classical model in which the direct pathway facilitates movement whereas the indirect pathway inhibits it. Similar results were obtained using the Designer Receptor Exclusively Activated by Designer Drugs (DREADD) system (Armbruster, Li, Pausch, Herlitze, & Roth, 2007) for a more rapid manipulation of striatal MSNs (Ferguson et al., 2011).

Although largely supporting the classical BG model, interpretation of these studies must take into account methodological caveats, including concerns about the potential unwanted effects of overexpressing G protein–coupled receptors in neurons (DREADDs), the slow time course of cell inactivation (tTA), or the sequelae of large-scale cell death (diphtheria toxin).

Optogenetics

Studies in rodents using optogenetics, the method of choice for interrogating network activity both in acute brain slices and in the intact brain in awake, behaving animals, have provided arguably the best evidence in support of the classical rate model. This technique enables researchers to silence or stimulate genetically specified classes of neurons (or other electrically excitable cells) with exquisite temporal precision using light-sensitive molecules (Bernstein, Garrity, & Boyden, 2012; Fenno, Yizhar, & Deisseroth, 2011). For example, channelrhodopsin-2 (ChR2) is a light-activated cation channel that depolarizes neurons, whereas halorhodopsin and archaerhodopsin-3 are a chloride channel and an outward proton pump, respectively, that enable almost complete silencing of neurons. The opportunity to inactivate targeted neurons dynamically has helped to shed light on causal relationships between neural activity and animal behavior.

Consistent with classic BG models, investigators used ChR2 to show that activation of D2 MSNs (indirect pathway) in mice decreases their locomotion and leads to a parkinsonian state, whereas activation of D1 MSNs (direct pathway) leads to increased locomotion (Kravitz et al., 2010). In contrast to what was expected, optogenetic studies evaluating the therapeutic mechanisms of STN DBS found that symptomatic improvement was not mediated by disruption of the classical indirect pathway but rather activation of afferent from the hyperdirect pathway, which exemplifies how this elegant methodology can help shed light on BG circuitry and therapeutic mechanisms (Gradinaru, Mogri, Thompson, Henderson, & Deisseroth, 2009; Kravitz et al., 2010).

Despite its specificity and other advantages, interpretation of optogenetics experiments with ChR2 must always consider some of its limitations (Yizhar, Fenno, Davidson, Mogri, & Deisseroth, 2011), including problems with desensitization and calcium permeability. In addition, the artificial pattern of stimulation resulting from optogenetics cannot replicate the activity of the circuit under normal conditions. Thus, strong ChR2 stimulation of MSNs likely produces highly synchronous activity that could have unexpected effects on the striatal circuitry, including various degrees of unwanted lateral inhibition (Kreitzer & Berke, 2011).

Animal Studies Reveal Flaws in the Classical Model of Basal Ganglia Circuitry

For the most part, the classical model of BG circuitry has stood the test of time. However, animal studies have also found conflicting results that raise questions about the complete validity of the classical model of BG circuitry. For instance, the SNr shows no change in firing rate after MPTP lesions in monkeys (Wichmann et al., 1999). Instead, MPTP-lesioned animals show increases in the bursting of STN neurons (Bergman et al., 1994) and higher synchrony in GPe and GPi (Raz, Vaadia, & Bergman, 2000; Wichmann, Bergman, & DeLong, 1994). Thus, rather than simply changes in firing rates, it is likely that changes in the patterns of activity of individual neurons may describe more accurately how the BG affects movement, and that cooperative activity of subsets of neurons in the direct and indirect pathways are at play during normal voluntary movement. This may indeed lend credence to the study of local field potentials (LFPs) in humans, which provide a population measure of activity.

In addition, studies recording genetically identified subsets of direct and indirect pathway MSNs have shown that both pathways are simultaneously activated during the initiation of movement (Cui et al., 2013; Isomura et al., 2013). Although this seems to contradict the Albin–DeLong model, it makes intuitive sense that when direct pathways MSNs are activated at the initiation of a particular movement, an ensemble of indirect pathway MSNs must also be recruited to suppress competing motor programs. This is consistent with the action selection, classical model of BG circuitry in which for every direct pathway neuronal ensemble that is active during a particular motor execution, several indirect pathway modules would be activated to suppress opposing actions (Nelson & Kreitzer, 2014). Nevertheless, action selection might be too simplistic a model. BG circuits probably influence distinct aspects of actions, including the initiation, execution, and termination of specific motor sequences. The BG might be viewed as critical for parsing and sequencing motor commands from upstream ensembles in neocortex (Jin, Tecuapetla, & Costa, 2014).

EMERGING TECHNIQUES TO STUDY BASAL GANGLIA CIRCUITRY IN ANIMALS

Perhaps the most promising tools on the neuroscience horizon that will help solve the puzzle of BG circuitry are the ever more sophisticated approaches to record and manipulate the activity of populations of neurons noninvasively (Kreitzer & Berke, 2011). The main advantage of noninvasive imaging is that activity can be measured without potentially harmful penetrating electrodes, and chronic imaging over months or years can be an option. Whereas techniques such as functional MRI (fMRI), intrinsic optical signal imaging, new imaging, and voltage-sensitive dye imaging have been used, none simultaneously offers sufficient temporal resolution (to detect single action potentials or subthreshold changes in membrane potential) and spatial resolution to resolve individual neurons. Because of this limitation, new imaging methods have been developed to take advantage of the benefits of fluorescence microscopy. Of these, two-photon calcium imaging (2PCI) is one of the most promising for monitoring neuronal ensemble activity in vivo.

Calcium Imaging

When neurons fire action potentials, the intracellular concentration of calcium ions increases. One can record neuronal activity optically by using fluorescent dyes or proteins that respond to binding of calcium by changing their spectral properties. Calcium imaging using two-photon excitation (2PE) microscopy is an ideal tool for interrogating large ensembles of neurons in the intact brain (Grienberger & Konnerth, 2012). 2PCI makes it possible to record signals from hundreds (potentially thousands) of neurons simultaneously. Combined with mouse genetics to label individual subpopulations of neurons (eg, with a red fluorescent protein), one can also be certain of the cell types from which the calcium signals are being recorded. Second, calcium imaging is less invasive and circuits can be recorded without inserting bundles of penetrating electrodes (tetrodes) that might disrupt normal activity. Third, unlike tetrode recordings, calcium imaging is not biased toward active units and one can also visualize activity from less active or even silent neurons.

Unfortunately, 2PCI has important drawbacks such as poor temporal resolution and low signal-to-noise ratio (Gobel & Helmchen, 2007; Grewe & Helmchen, 2009). Newer-generation, genetically encoded calcium indicators with improved signal-to-noise will overcome some of these limitations. Importantly, developments in faster scanning (eg, acousto-optical deflectors) (Grewe, Langer, Kasper, Kampa, & Helmchen, 2010), parallelization of 2PE (eg, multifocal multiphoton microscopy) (Cheng, Goncalves, Golshani, Arisaka, & Portera Cailliau, 2011), improved photodetectors, and depth penetration with either microendoscopy (eg, GRIN lenses) (Wilt et al., 2009) or regenerative amplifiers (Mittmann et al., 2011; Theer, Hasan, & Denk, 2003) suggest that over the next decade, optical probing of neural activity in the BG with calcium imaging could be an excellent alternative to electrophysiology.

Voltage Sensors

Another way to probe neuronal activity optically is to image changes in membrane potential. For this purpose, several voltage sensing proteins (eg, FLaSh, VSFP2, SPARC, Flare) have been designed that work well with 2PE microscopy and could theoretically be useful to monitor the activity of thousands of individual neurons simultaneously (Chanda et al., 2005; Perron et al., 2009). These sensors are usually fluorescence resonance energy transfer–based and can report both subthreshold changes in membrane potential and spiking activity of neurons. Unfortunately, current approaches for voltage sensing with genetically encoded or synthetic indicators have a low signal-to-noise ratio and require the averaging of many stimuli to detect responses.

Optogenetics

In vivo imaging with calcium or voltage-sensitive dyes can also be combined with optogenetics to modulate and record neuronal activity with phenomenal temporal and spatial precision (Peron & Svoboda, 2011). The goal would be to mimic the normal activity of MSNs (including bursting behavior) and reveal a more accurate model of BG connectivity. As improved red-shifted calcium indicator dyes become available in the near future, it will be easier to combine optogenetics with 2PCI. Furthermore, 2PE of ChR2 is also possible in vivo (Papagiakoumou et al., 2010) and this option may be preferable when precise stimulation of individuals' MSNs (or even single synapses) is desired. Newer versions of opsins are also constantly being developed, including red-shifted opsins (eg, Jaws) that can allow for noninvasive silencing of deep brain structures (Chuong et al., 2014). In addition to optogenetics, other chemical methods for cell inactivation are being developed (beyond DREADDs) that could be applied to the study of BG circuitry.

Striatal Interneurons

As mentioned, a major advantage of calcium imaging is that it allows one to record activity from genetically identified neuronal subtypes. Thus, investigators can apply Cre-Lox approaches to record from (or optogenetically manipulate) different subtypes of striatal interneurons selectively using specific promoters. Optogenetic manipulations of cholinergic interneurons in the striatum do not produce obvious motor deficits but they can have significant effects on the excitability of MSNs (Witten et al., 2011). Even less is known about the role of GABAergic interneurons, but similar to cholinergic neurons, they, too, mainly modulate the activity of networks of MSNs. For example, pharmacological blockade of fast-spiking interneurons in mice leads to dystonia and dyskinesias (Gittis, Nelson, Thwin, Palop, & Kreitzer, 2010). Future calcium imaging studies targeting different interneuron subtypes will elucidate their role in normal BG circuit function and in movement disorders.

PROBING BASAL GANGLIA CIRCUITRY IN HUMANS: INVASIVE PHYSIOLOGY

Although animal studies are critical, the ultimate goal is to understand human BG physiology, to understand disease and develop therapeutics. Although invasive studies of human neurophysiology are limited with respect to nuclei from which one can record and the fact that they are done in patients with chronic neurodegenerative disorders in which the circuits have endured years of dysfunction, they are no doubt extremely informative and have begun to reshape our understanding of how BG modulates cortical activity.

Both microelectrode recordings and stimulations during early stereotactic pallidotomy and thalamotomy allowed for early investigation into the neurophysiology of the human BG. In contemporary neurosurgery, investigations are primarily conducted during surgical implantation of deep brain stimulation (DBS) electrodes, during which intraoperative neuronal spiking activity can be investigated with temporary microelectrodes while LFPs can be recorded from both microelectrodes as well through the DBS electrode contacts. LFPs can also be obtained during subacute recordings while DBS leads are externalized, a practice which is increasingly less common.

Whereas studies of unit activity in BG have been reported extensively, examination of the BG LFP has become increasingly popular, owing largely to its signal fidelity, ease of capture via the DBS contacts, and therapeutic potential. Here, we highlight these studies because they shed new light on approaches to understanding BG physiology, both locally and at the network level. In contrast, single/multiunit spiking activity requires high-impedance microelectrodes and associated amplifier circuits often relegated to the intraoperative environment. Moreover, consistent unit activity is difficult to maintain for extended periods of time, which makes longer-term studies challenging. Studies of the BG LFPs have largely concentrated on the STN recordings because they are a more common DBS target for PD (Deuschl et al., 2006).

Studies of GPi LFPs are growing because GPi has been shown to be an equally effective target in PD (Follett et al., 2010; Rodriguez-Oroz et al., 2005).

Abnormal synchrony (or rhythms) has been consistently demonstrated throughout the BG–thalamocortical loop in PD (Hammond, Bergman, & Brown, 2007; Stein & Bar-Gad, 2013). Distinct frequency components of LFP oscillations likely reflect different aspects of the BG operation, whether normal or pathological, with the greatest focus on beta (about 12–35 Hz) and gamma (>40 Hz) band activity. The frequency subbands of LFP oscillations can be differentially modulated behaviorally (Brittain, Sharott, & Brown, 2014) and with pharmacological interventions (Priori et al., 2004). These oscillatory signals may indicate the operational state of the BG motor control circuit (Nambu, 2004). The spatial extent of oscillatory involvement within the STN has been shown to be associated with a patient's responses to DBS (Zaidel, Spivak, Grieb, Bergman, & Israel, 2010) and the site of maximal beta activity is associated with the optimal site of therapeutic DBS (Ince et al., 2010).

Beta and gamma band activity has been tied to movement generation, but they have contrasting roles. Beta oscillations are most pronounced in a resting state and are attenuated with movement (Brittain & Brown, 2014). Gamma oscillations, on the other hand, are often considered prokinetic and are most pronounced during movement and attenuated during rest (Jenkinson, Kühn, & Brown, 2013). Movement-related modulation occurs in bilateral STN in both the beta and gamma range, but mostly with movement of the contralateral side (Alegre et al., 2005). During cued arm reaching, beta activity within the STN desynchronizes (or decreases) before movement whereas gamma power increases with reaching speed (Joundi, Jenkinson, Brittain, Aziz, & Brown, 2012). Subcomponents of beta band activity have been shown to be differentially modulated during movement by finger-tapping velocity, with periods of resynchronization during lower-velocity repeated movements (Joundi et al., 2013). Resynchronization, or increased beta band power, between even brief periods of movement supports the supposition that beta oscillations functionally indicate an "akinetic" state. Interestingly, the strength of beta activity in the STN has been shown to be associated with the degree of akinesia and rigidity in PD subjects (Chen et al., 2010). Furthermore, microelectrode recordings have demonstrated that a proportion of STN neuronal spiking activity is locked to beta band in PD subjects (Kühn et al., 2005; Levy et al., 2002; Moran, Bergman, Israel, & Bar-Gad, 2008; Shimamoto et al., 2013). This concept is a key departure from the classic BG model, in that this nested coupling of signals across nodes in the motor network suggests a dynamic interplay between activities across regions in the brain. Interestingly, Shimamoto et al. (2013) reported that STN spiking is coupled to gamma oscillations as well, which themselves are coupled to local cortical beta power, indicating an even more complex layered nesting of signals. Further exemplifying the coordination of activity across the BG network are studies that demonstrate interhemispheric coupling of beta activity across STN (Little et al., 2013; de Solages, Hill, Koop, Henderson, & Bronte-Stewart, 2010).

The akinetic nature of beta oscillations also applies to the motor cortex. Beta range stimulation of the motor cortex with transcranial alternating current stimulation leads to a reduction in initial force generation during intended movement (Joundi et al., 2012). Enhanced beta oscillations in bilateral primary sensorimotor cortices are seen in early-stage PD along with a damped attenuation during movement attenuation (Pollok et al., 2012).

In contrast to beta band activity, gamma oscillations appears to be "prokinetic" in nature (Jenkinson et al., 2013), increasing in power with the onset of movement and attenuated during rest. The correlation of gamma power with movement features including size and velocity suggests that underlying neuronal networks may participate in the scaling of ongoing movement actions, rather than just gating, as might be implied by the classical model of BG circuitry (Brücke et al., 2012). Movement-related gamma oscillations as high as 200–300 Hz have also been reported in GPi (Tsiokos, Hu, & Pouratian, 2013) and STN (Foffani et al., 2003; Lopez-Azcarate et al., 2010; Ozkurt et al., 2011). These very–high frequency oscillations are susceptible to pharmacological interventions, with changes in baseline characteristics with both rest and movement. Whereas broadband gamma activity correlates well with local neuronal population activity, finely tuned gamma correlates with, or binds, the activity across spatially separated networks (Manning, Jacobs, Fried, & Kahana, 2009; Singer, 1993). For example, movement-modulated finely tuned gamma (60- to 90-Hz) activity in the STN has been reported in some subjects with PD across both hemispheres (Cagnan, Kühn, & Brown, 2014). Magnetoencephalography and direct recordings suggest that STN gamma (60–90 Hz) drives this M1 gamma signal. In addition, levodopa alters both local and long-range synchronization at 60–90 Hz corresponding to clinical improvements (Litvak et al., 2012).

A key concept to emerge from invasive human LFP studies of BG physiology is the concept of cross-frequency coupling (CFC). Studies have revealed not only coupling of spikes to LFPs but also nonlinear coupling between low-frequency (beta) and high-frequency oscillations which occur in the off-state within STN and the motor cortex and are abolished with medication and stimulation (de Hemptinne et al., 2013; Lopez-Azcarate et al., 2010; Marceglia et al., 2006; Yang, Vanegas, Lungu, & Zaghloul, 2014). Persistence of CFC owing to the diseased BG is considered a potentially critical pathophysiological mechanism of disease.

Computational models suggest that DBS reduces the disordered activity within the BG and increases the fidelity of information transmission between the GPi and thalamus (Dorval, Kuncel, Birdno, Turner, & Grill, 2010; Guo, Rubin, McIntyre, Vitek, & Terman, 2008). Indeed, STN DBS decouples neuronal oscillations between BG nuclei (Moran, Stein, Tischler, & Bar-Gad, 2012) and decreases the entropy in the neuronal firing pattern within the pallidum and thalamus (Dorval et al., 2008). STN DBS has also been suggested to increase the sensitivity of cortical neurons to thalamic input (Kahan et al., 2014). Computational models have suggested that both the frequency and pattern of DBS can differentially alter BG network activity. High-frequency stimulation has been suggested to fix the weight of synaptic transmission whereas desynchronization patterns can undo these weights (Tass & Majtanik, 2006). Therefore, a better grasp of the oscillatory activity within the BG LFPs can help illuminate the underlying circuit and further improve the efficacy of DBS, potentially driving a closed-loop system for PD (Little & Brown, 2012; Rosin et al., 2011).

NONINVASIVE STUDIES OF BASAL GANGLIA CIRCUITRY IN HUMANS

Unlike invasive human studies, neuroimaging allows noninvasive characterization of circuit structure and function across the entire brain in both affected and healthy subjects across a large number of subjects. Although neurochemical imaging with either positron emission tomography or single-photon emission computed tomography has contributed tremendously to characterization of BG function, we focus here on using connectivity-based imaging, including magnetic resonance diffusion tensor imaging (DTI) tractography or fMRI which respectively shed light on structural and functional connectivity.

One key question explored using neuroimaging is whether distinct behavioral and cognitive functions are subserved by discrete parallel circuits (Alexander, Crutcher, & DeLong, 1990). Using DTI, Lehericy et al. (2004) reported the discrete organization of corticostriatal fibers that target striatum, consistent with a parallel architecture seen in primate studies. Prefrontal corticostriatal fibers terminated most anteriorly in striatum, supplementary motor, primary motor, primary sensory, and the mesencephalon fibers were directed toward the posterior portion, and orbitomedial frontal cortical fibers terminated mainly in the ventral striatum, with some overlap between them. Task-related functional connectivity studies examining coactivation between cortex and the striatum in normal subjects also supports a parallel architecture and a tripartite striatal organization into motor, associative, and limbic divisions (Postuma & Dagher, 2006).

Resting state functional connectivity has gained traction as a means to assess the organization of the brain (Fox et al., 2005). Comparison of the resting state functional connectivity between PD and control subjects demonstrates a reduction in the spatial segregation in parallel loops organization (Helmich et al., 2010). Furthermore, there was a reduction in functional connectivity between the posterior putamen and the cortical sensorimotor system with a possibly compensatory concomitant increase in coupling between inferior parietal cortex and anterior putamen. This suggests that corticostriatal functional connectivity is altered in the parkinsonian dopamine-depleted state, leading to dysfunctional sensorimotor integration. Conversely, levodopa administration in drug-naive subjects with PD leads to an enhancement of sensorimotor connectivity within the supplementary motor area (Esposito et al., 2013). This results in partial renormalization of the hypoactivity typically seen in untreated subjects with PD within this network. Resting state fMRI in subjects with PD also demonstrated reduced functional connectivity within thalamus, midbrain, pons, and cerebellum, which suggests that brain stem circuitry may be involved (Hacker, Perlmutter, Criswell, Ances, & Snyder, 2012) and that the structural circuitry of the classical model does not entirely account for the widespread coordination of function across the cortical—subcortical motor network.

Connectivity-based imaging also provides an opportunity to evaluate the role of BG in delineating specific symptom etiology. For example, an investigation of tremor mechanisms in PD revealed an increase in functional connectivity between the cerebellothalamic circuit and the GPi/putamen (Helmich, Janssen, Oyen, Bloem, & Toni, 2011), which suggests that tremor amplitude may be driven at least in part by the cerebellothalamic circuit. Studies in dystonia, a distinct movement disorder from HD and PD, demonstrate reduced activation throughout the cortical, striatal, and thalamic regions, including cerebellothalamic or thalamocortical pathways (Bonilha et al., 2009; Castrop, Dresel, Hennenlotter, Zimmer, & Haslinger, 2012; Delmaire et al., 2009). These studies lend credence to the notion that the interconnectivity between the cerebellum and the BG is a putative underlying mechanism of dystonia (Bostan & Strick, 2010; Neychev, Gross, Lehericy, Hess, & Jinnah, 2011).

Perhaps one of the greatest opportunities is to use noninvasive brain mapping to shed light on the nonmotor aspect of BG disease. For example, increased connectivity strength between subcortical and motor as well as sensory cortex was correlated with poorer performance on a neurocognitive task (Marchand et al., 2011). Similarly, fractional anisotropy in distinct subcortical white matter projection regions has been shown to correlate with cognitive impairment in distinct domains (Zheng et al., 2014). Investigation of the BG role in nonmotor function is not limited to PD. Abnormalities in the cortico-BG networks involving anterior cingulate, inferior frontal gyrus, and sensorimotor cortex have also been shown in subjects with major depressive disorder (Worbe et al., 2012).

CONCLUSIONS

The classical model of BG circuitry with competing direct and indirect pathways, based on careful anatomic studies and meticulous neurophysiologic studies in animal models, has provided a solid foundation to begin to understand BG function. Investigations in the past decade have begun to shed light on the increased complexity of BG circuitry, with an increased understanding of the importance of other pathways, such as the hyperdirect pathways from cortex to STN, and the complex network level of processing that appears to oscillate through the BG—thalamocortical circuit. Techniques and computational approaches that allow careful and dynamic micromanipulation of the circuit (such as optogenetics), high-resolution monitoring of populations of neurons (such as 2PCI), and network-level analyses of neural dynamics at multiple nodes in the network will surely elucidate complexities and nonlinearities in the system that previously have not been considered extensively. These emerging concepts of BG dynamic function and interaction with the rest of the brain will also open avenues to understanding the role of BG in functions beyond the motor system, including critical cognitive functions.

REFERENCES

Aizman, O., Brismar, H., Uhlén, P., Zettergren, E., Levey, A. I., Forssberg, H., ... Aperia, A. (2000). Anatomical and physiological evidence for d1 and d2 dopamine receptor colocalization in neostriatal neurons. *Nature Neuroscience, 3*(3), 226—230. http://dx.doi.org/10.1038/72929.

Albin, R. L., Young, A. B., & Penney, J. B. (1989). The functional-anatomy of basal ganglia disorders. *Trends in Neurosciences, 12*(10), 366—375. http://dx.doi.org/10.1016/0166-2236(89)90074-x.

Alegre, M., Alonso-Frech, F., Rodriguez-Oroz, M. C., Guridi, J., Zamarbide, I., Valencia, M., ... Artieda, J. (2005). Movement-related changes in oscillatory activity in the human subthalamic nucleus: ipsilateral vs. contralateral movements. *European Journal of Neuroscience, 22*(9), 2315—2324. http://dx.doi.org/10.1111/j.1460-9568.2005.04409.x.

Alexander, G. E., & Crutcher, M. D. (1990). Functional architecture of basal ganglia circuits: neural substrates of parallel processing. *Trends in Neurosciences, 13*(7), 266—271. http://dx.doi.org/10.1016/0166-2236(90)90107-L.

Alexander, G. E., Crutcher, M. D., & DeLong, M. R. (1990). Basal ganglia-thalamocortical circuits - parallel substrates for motor, oculomotor, prefrontal and limbic functions. *Progress in Brain Research, 85*, 119—146.

Alexander, G. E., DeLong, M. R., & Strick, P. L. (1986). Parallel organization of functionally segregated circuits linking basal ganglia and cortex. *Annual Review of Neuroscience, 9*, 357—381. http://dx.doi.org/10.1146/annurev.ne.09.030186.002041.

Armbruster, B. N., Li, X., Pausch, M. H., Herlitze, S., & Roth, B. L. (2007). Evolving the lock to fit the key to create a family of g protein-coupled receptors potently activated by an inert ligand. *Proceedings of the National Academy of Sciences of the United States of America, 104*(12), 5163—5168. http://dx.doi.org/10.1073/pnas.0700293104.

Barbas, H., Henion, T. H., & Dermon, C. R. (1991). Diverse thalamic projections to the prefrontal cortex in the rhesus monkey. *Journal of Comparative Neurology, 313*(1), 65—94. http://dx.doi.org/10.1002/cne.903130106.

Bergman, H., Wichmann, T., Karmon, B., & DeLong, M. R. (1994). The primate subthalamic nucleus. Ii. Neuronal activity in the mptp model of parkinsonism. *Journal of Neurophysiology, 72*(2), 507—520.

Bernstein, J. G., Garrity, P. A., & Boyden, E. S. (2012). Optogenetics and thermogenetics: technologies for controlling the activity of targeted cells within intact neural circuits. *Current Opinion in Neurobiology, 22*(1), 61—71. http://dx.doi.org/10.1016/j.conb.2011.10.023.

Bonilha, L., de Vries, P. M., Hurd, M. W., Rorden, C., Morgan, P. S., Besenski, N., ... Hinson, V. K. (2009). Disrupted thalamic prefrontal pathways in patients with idiopathic dystonia. *Parkinsonism & Related Disorders, 15*(1), 64—67. http://dx.doi.org/10.1016/j.parkreldis.2008.01.018.

Bostan, A. C., & Strick, P. L. (2010). The cerebellum and basal ganglia are interconnected. *Neuropsychology Review, 20*(3), 261—270. http://dx.doi.org/10.1007/s11065-010-9143-9.

Braak, H., Del Tredici, K., Rub, U., de Vos, R. A., Jansen Steur, E. N., & Braak, E. (2003). Staging of brain pathology related to sporadic Parkinson's disease. *Neurobiology of Aging, 24*(2), 197—211.

Brittain, J.-S., & Brown, P. (2014). Oscillations and the basal ganglia: motor control and beyond. *NeuroImage, 85*(Pt 2), 637—647. http://dx.doi.org/10.1016/j.neuroimage.2013.05.084.

Brittain, J.-S., Sharott, A., & Brown, P. (2014). The highs and lows of beta activity in cortico-basal ganglia loops. *European Journal of Neuroscience, 39*(11), 1951—1959. http://dx.doi.org/10.1111/ejn.12574.

Brücke, C., Huebl, J., Schönecker, T., Neumann, W.-J., Yarrow, K., Kupsch, A., ... Kühn, A. A. (2012). Scaling of movement is related to pallidal gamma oscillations in patients with dystonia. *Journal of Neuroscience, 32*(3), 1008—1019. http://dx.doi.org/10.1523/JNEUROSCI.3860-11.2012.

Cagnan, H., Kühn, A. A., & Brown, P. (2014). Co-modulation of finely tuned high-gamma band activity across hemispheres in Parkinson's disease. *Clinical Neurophysiology: Official Journal of the International Federation of Clinical Neurophysiology, 125*(4), 777—785. http://dx.doi.org/10.1016/j.clinph.2013.10.001.

Carpenter, M. B., Whittier, J. R., & Mettler, F. A. (1950). Analysis of choreoid hyperkinesia in the rhesus monkey; surgical and pharmacological analysis of hyperkinesia resulting from lesions in the subthalamic nucleus of luys. *Journal of Comparative Neurology, 92*(3), 293–331.

Castrop, F., Dresel, C., Hennenlotter, A., Zimmer, C., & Haslinger, B. (2012). Basal ganglia-premotor dysfunction during movement imagination in writer's cramp. *Movement Disorders: Official Journal of the Movement Disorder Society, 27*(11), 1432–1439. http://dx.doi.org/10.1002/mds.24944.

Chanda, B., Blunck, R., Faria, L. C., Schweizer, F. E., Mody, I., & Bezanilla, F. (2005). A hybrid approach to measuring electrical activity in genetically specified neurons. *Nature Neuroscience, 8*(11), 1619–1626. http://dx.doi.org/10.1038/nn1558.

Chen, C. C., Hsu, Y. T., Chan, H. L., Chiou, S. M., Tu, P. H., Lee, S. T., … Brown, P. (2010). Complexity of subthalamic 13-35 hz oscillatory activity directly correlates with clinical impairment in patients with Parkinson's disease. *Experimental Neurology, 224*(1), 234–240. http://dx.doi.org/10.1016/j.expneurol.2010.03.015.

Cheng, A., Goncalves, J. T., Golshani, P., Arisaka, A., & Portera Cailliau, C. (January 9, 2011). Simultaneous two-photon calcium imaging at different depths with spatiotemporal multiplexing. *Nature Methods, 8*, 139–142. http://dx.doi.org/10.1038/nmeth.1552.

Chuong, A. S., Miri, M. L., Busskamp, V., Matthews, G. A., Acker, L. C., Sorensen, A. T., … Boyden, E. S. (2014). Noninvasive optical inhibition with a red-shifted microbial rhodopsin. *Nature Neuroscience, 17*(8), 1123–1129. http://dx.doi.org/10.1038/nn.3752.

Cui, G. H., Jun, S. B., Jin, X., Pham, M. D., Vogel, S. S., Lovinger, D. M., & Costa, R. M. (2013). Concurrent activation of striatal direct and indirect pathways during action initiation. *Nature, 494*(7436), 238–242. http://dx.doi.org/10.1038/nature11846.

Delmaire, C., Vidailhet, M., Wassermann, D., Descoteaux, M., Valabregue, R., Bourdain, F., … Lehericy, S. (2009). Diffusion abnormalities in the primary sensorimotor pathways in writer's cramp. *Archives of Neurology, 66*(4), 502–508. http://dx.doi.org/10.1001/archneurol.2009.8.

DeLong, M. R., Crutcher, M. D., & Georgopoulos, A. P. (1985). Primate globus pallidus and subthalamic nucleus: functional organization. *Journal of Neurophysiology, 53*(2), 530–543.

Deng, Y. P., Albin, R. L., Penney, J. B., Young, A. B., Anderson, K. D., & Reiner, A. (2004). Differential loss of striatal projection systems in Huntington's disease: a quantitative immunohistochemical study. *Journal of Chemical Neuroanatomy, 27*(3), 143–164. http://dx.doi.org/10.1016/j.jchemneu.2004.02.005.

Desmurget, M., & Turner, R. S. (2008). Testing basal ganglia motor functions through reversible inactivations in the posterior internal globus pallidus. *Journal of Neurophysiology, 99*(3), 1057–1076. http://dx.doi.org/10.1152/jn.01010.2007.

Desmurget, M., & Turner, R. S. (2010). Motor sequences and the basal ganglia: kinematics, not habits. *Journal of Neuroscience, 30*(22), 7685–7690. http://dx.doi.org/10.1523/JNEUROSCI.0163-10.2010.

Deuschl, G., Schade-Brittinger, C., Krack, P., Volkmann, J., Schäfer, H., Bötzel, K., … German Parkinson Study Group, N. S. (2006). A randomized trial of deep-brain stimulation for Parkinson's disease. *New England Journal of Medicine, 355*(9), 896–908. http://dx.doi.org/10.1056/NEJMoa060281.

Dorval, A. D., Kuncel, A. M., Birdno, M. J., Turner, D. A., & Grill, W. M. (2010). Deep brain stimulation alleviates parkinsonian bradykinesia by regularizing pallidal activity. *Journal of Neurophysiology, 104*(2), 911–921. http://dx.doi.org/10.1152/jn.00103.2010.

Dorval, A. D., Russo, G. S., Hashimoto, T., Xu, W., Grill, W. M., & Vitek, J. L. (2008). Deep brain stimulation reduces neuronal entropy in the mptp-primate model of Parkinson's disease. *Journal of Neurophysiology, 100*(5), 2807–2818. http://dx.doi.org/10.1152/jn.90763.2008.

Drago, J., Padungchaichot, P., Wong, J. Y., Lawrence, A. J., McManus, J. F., Sumarsono, S. H., … Finkelstein, D. I. (1998). Targeted expression of a toxin gene to d1 dopamine receptor neurons by cre-mediated site-specific recombination. *Journal of Neuroscience, 18*(23), 9845–9857.

Durieux, P. F., Bearzatto, B., Guiducci, S., Buch, T., Waisman, A., Zoli, M., … de Kerchove d'Exaerde, A. (2009). D2r striatopallidal neurons inhibit both locomotor and drug reward processes. *Nature Neuroscience, 12*(4), 393–395. http://dx.doi.org/10.1038/nn.2286.

Esposito, F., Tessitore, A., Giordano, A., De Micco, R., Paccone, A., Conforti, R., … Tedeschi, G. (2013). Rhythm-specific modulation of the sensorimotor network in drug-naive patients with Parkinson's disease by levodopa. *Brain, 136*(Pt 3), 710–725. http://dx.doi.org/10.1093/brain/awt007.

Fenno, L., Yizhar, O., & Deisseroth, K. (2011). The development and application of optogenetics. *Annual Review of Neuroscience, 34*, 389–412. http://dx.doi.org/10.1146/annurev-neuro-061010-113817.

Ferguson, S. M., Eskenazi, D., Ishikawa, M., Wanat, M. J., Phillips, P. E., Dong, Y., … Neumaier, J. F. (2011). Transient neuronal inhibition reveals opposing roles of indirect and direct pathways in sensitization. *Nature Neuroscience, 14*(1), 22–24. http://dx.doi.org/10.1038/nn.2703.

Filion, M., Tremblay, L., & Bedard, P. J. (1991). Effects of dopamine agonists on the spontaneous activity of globus pallidus neurons in monkeys with mptp-induced parkinsonism. *Brain Research, 547*(1), 152–161.

Foffani, G., Priori, A., Egidi, M., Rampini, P., Tamma, F., Caputo, E., … Barbieri, S. (2003). 300-hz subthalamic oscillations in Parkinson's disease. *Brain, 126*(Pt 10), 2153–2163. http://dx.doi.org/10.1093/brain/awg229.

Follett, K. A., Weaver, F. M., Stern, M., Hur, K., Harris, C. L., Luo, P., … Group, C. S. (2010). Pallidal versus subthalamic deep-brain stimulation for Parkinson's disease. *New England Journal of Medicine, 362*(22), 2077–2091. http://dx.doi.org/10.1056/NEJMoa0907083.

Fox, M. D., Snyder, A. Z., Vincent, J. L., Corbetta, M., Van Essen, D. C., & Raichle, M. E. (2005). The human brain is intrinsically organized into dynamic, anticorrelated functional networks. *Proceedings of the National Academy of Sciences of the United States of America, 102*(27), 9673–9678. http://dx.doi.org/10.1073/pnas.0504136102.

Gerfen, C. R. (1984). The neostriatal mosaic: compartmentalization of corticostriatal input and striatonigral output systems. *Nature, 311*(5985), 461–464. http://dx.doi.org/10.1038/311461a0.

Gerfen, C. R., Engber, T. M., Mahan, L. C., Susel, Z., Chase, T. N., Monsma, F. J., & Sibley, D. R. (1990). D1 and d2 dopamine receptor-regulated gene expression of striatonigral and striatopallidal neurons. *Science, 250*(4986), 1429–1432.

Gerfen, C. R., & Sawchenko, P. E. (1984). An anterograde neuroanatomical tracing method that shows the detailed morphology of neurons, their axons and terminals: immunohistochemical localization of an axonally transported plant lectin, phaseolus vulgaris leucoagglutinin (pha-l). *Brain Research, 290*(2), 219–238.

Gittis, A. H., Nelson, A. B., Thwin, M. T., Palop, J. J., & Kreitzer, A. C. (2010). Distinct roles of gabaergic interneurons in the regulation of striatal output pathways. *Journal of Neuroscience, 30*(6), 2223–2234. http://dx.doi.org/10.1523/JNEUROSCI.4870-09.2010.

Gobel, W., & Helmchen, F. (2007). In vivo calcium imaging of neural network function. *Physiology (Bethesda), 22*, 358–365.

Goldman-Rakic, P. S., & Porrino, L. J. (1985). The primate mediodorsal (md) nucleus and its projection to the frontal lobe. *Journal of Comparative Neurology, 242*(4), 535–560. http://dx.doi.org/10.1002/cne.902420406.

Gradinaru, V., Mogri, M., Thompson, K. R., Henderson, J. M., & Deisseroth, K. (2009). Optical deconstruction of parkinsonian neural circuitry. *Science, 324*(5925), 354–359. http://dx.doi.org/10.1126/science.1167093.

Graveland, G. A., Williams, R. S., & DiFiglia, M. (1985). A golgi study of the human neostriatum: neurons and afferent fibers. *Journal of Comparative Neurology, 234*(3), 317–333. http://dx.doi.org/10.1002/cne.902340304.

Graybiel, A. M., & Ragsdale, C. W. (1978). Histochemically distinct compartments in the striatum of human, monkeys, and cat demonstrated by acetylthiocholinesterase staining. *Proceedings of the National Academy of Sciences of the United States of America, 75*(11), 5723–5726.

Grewe, B. F., & Helmchen, F. (2009). Optical probing of neuronal ensemble activity. *Current Opinion in Neurobiology, 19*(5), 520–529. http://dx.doi.org/10.1016/j.conb.2009.09.003. pii:S0959-4388(09)00125-1.

Grewe, B. F., Langer, D., Kasper, H., Kampa, B. M., & Helmchen, F. (2010). High-speed in vivo calcium imaging reveals neuronal network activity with near-millisecond precision. *Nature Methods, 7*(5), 399–405. http://dx.doi.org/10.1038/nmeth.1453. pii:nmeth.1453.

Grienberger, C., & Konnerth, A. (2012). Imaging calcium in neurons. *Neuron, 73*(5), 862–885. http://dx.doi.org/10.1016/j.neuron.2012.02.011.

Guo, Y., Rubin, J. E., McIntyre, C. C., Vitek, J. L., & Terman, D. (2008). Thalamocortical relay fidelity varies across subthalamic nucleus deep brain stimulation protocols in a data-driven computational model. *Journal of Neurophysiology, 99*(3), 1477–1492. http://dx.doi.org/10.1152/jn.01080.2007.

Hacker, C. D., Perlmutter, J. S., Criswell, S. R., Ances, B. M., & Snyder, A. Z. (2012). Resting state functional connectivity of the striatum in Parkinson's disease. *Brain, 135*(Pt 12), 3699–3711. http://dx.doi.org/10.1093/brain/aws281.

Hammond, C., Bergman, H., & Brown, P. (2007). Pathological synchronization in Parkinson's disease: networks, models and treatments. *Trends in Neurosciences, 30*(7), 357–364. http://dx.doi.org/10.1016/j.tins.2007.05.004.

Haynes, W. I. A., & Haber, S. N. (2013). The organization of prefrontal-subthalamic inputs in primates provides an anatomical substrate for both functional specificity and integration: implications for basal ganglia models and deep brain stimulation. *Journal of Neuroscience, 33*(11), 4804–4814. http://dx.doi.org/10.1523/JNEUROSCI.4674-12.2013.

Helmich, R. C., Derikx, L. C., Bakker, M., Scheeringa, R., Bloem, B. R., & Toni, I. (2010). Spatial remapping of cortico-striatal connectivity in Parkinson's disease. *Cerebral Cortex, 20*(5), 1175–1186. http://dx.doi.org/10.1093/cercor/bhp178.

Helmich, R. C., Janssen, M. J. R., Oyen, W. J. G., Bloem, B. R., & Toni, I. (2011). Pallidal dysfunction drives a cerebellothalamic circuit into parkinson tremor. *Annals of Neurology, 69*(2), 269–281. http://dx.doi.org/10.1002/ana.22361.

de Hemptinne, C., Ryapolova-Webb, E. S., Air, E. L., Garcia, P. A., Miller, K. J., Ojemann, J. G., … Starr, P. A. (2013). Exaggerated phase-amplitude coupling in the primary motor cortex in Parkinson disease. *Proceedings of the National Academy of Sciences of the United States of America, 110*(12), 4780–4785. http://dx.doi.org/10.1073/pnas.1214546110.

Hikida, T., Kimura, K., Wada, N., Funabiki, K., & Nakanishi, S. (2010). Distinct roles of synaptic transmission in direct and indirect striatal pathways to reward and aversive behavior. *Neuron, 66*(6), 896–907. http://dx.doi.org/10.1016/j.neuron.2010.05.011.

Inase, M., Tokuno, H., Nambu, A., Akazawa, T., & Takada, M. (1996). Origin of thalamocortical projections to the presupplementary motor area (pre-sma) in the macaque monkey. *Neuroscience Research, 25*(3), 217–227.

Ince, N. F., Gupte, A., Wichmann, T., Ashe, J., Henry, T., Bebler, M., … Abosch, A. (2010). Selection of optimal programming contacts based on local field potential recordings from subthalamic nucleus in patients with Parkinson's disease. *Neurosurgery, 67*(2), 390–397. http://dx.doi.org/10.1227/01.NEU.0000372091.64824.63.

Isomura, Y., Takekawa, T., Harukuni, R., Handa, T., Aizawa, H., Takada, M., & Fukai, T. (2013). Reward-modulated motor information in identified striatum neurons. *Journal of Neuroscience, 33*(25), 10209–10220. http://dx.doi.org/10.1523/JNEUROSCI.0381-13.2013.

Jahanshahi, M., Obeso, I., Baunez, C., Alegre, M., & Krack, P. (2014). Parkinson's disease, the subthalamic nucleus, inhibition, and impulsivity. *Movement Disorders, 30*, 128–140.

Jenkinson, N., Kühn, A. A., & Brown, P. (2013). U oscillations in the human basal ganglia. *Experimental Neurology, 245*, 72–76. http://dx.doi.org/10.1016/j.expneurol.2012.07.005.

Jin, X., Tecuapetla, F., & Costa, R. M. (2014). Basal ganglia subcircuits distinctively encode the parsing and concatenation of action sequences. *Nature Neuroscience, 17*(3), 423–430. http://dx.doi.org/10.1038/nn.3632.

Joundi, R. A., Brittain, J.-S., Green, A. L., Aziz, T. Z., Brown, P., & Jenkinson, N. (2013). Persistent suppression of subthalamic beta-band activity during rhythmic finger tapping in Parkinson's disease. *Clinical Neurophysiology, 124*(3), 565–573. http://dx.doi.org/10.1016/j.clinph.2012.07.029.

Joundi, R. A., Jenkinson, N., Brittain, J.-S., Aziz, T. Z., & Brown, P. (2012). Driving oscillatory activity in the human cortex enhances motor performance. *Current Biology, 22*(5), 403–407. http://dx.doi.org/10.1016/j.cub.2012.01.024.

Kahan, J., Urner, M., Moran, R., Flandin, G., Marreiros, A., Mancini, L., … Foltynie, T. (2014). Resting state functional mri in Parkinson's disease: the impact of deep brain stimulation on 'effective' connectivity. *Brain, 137*(Pt 4), 1130–1144. http://dx.doi.org/10.1093/brain/awu027.

Kato, M., & Kimura, M. (1992). Effects of reversible blockade of basal ganglia on a voluntary arm movement. *Journal of Neurophysiology, 68*(5), 1516–1534.

Kemp, J. M. (1970). The site of termination of afferent fibres on the neurones of the caudate nucleus. *Journal of Physiology-London, 210*(1), 17P–18P.

Kemp, J. M., & Powell, T. P. (1970). The cortico-striate projection in the monkey. *Brain, 93*(3), 525–546.

Kemp, J. M., & Powell, T. P. (1971). The structure of the caudate nucleus of the cat: light and electron microscopy. *Philosophical Transactions of the Royal Society, B: Biological Sciences, 262*(845), 383–401.

Kim, R., Nakano, K., Jayaraman, A., & Carpenter, M. B. (1976). Projections of the globus pallidus and adjacent structures: an autoradiographic study in the monkey. *Journal of Comparative Neurology, 169*(3), 263–290. http://dx.doi.org/10.1002/cne.901690302.

Kravitz, A. V., Freeze, B. S., Parker, P. R. L., Kay, K., Thwin, M. T., Deisseroth, K., & Kreitzer, A. C. (2010). Regulation of parkinsonian motor behaviours by optogenetic control of basal ganglia circuitry. *Nature, 466*(7306), 622–627. http://dx.doi.org/10.1038/nature09159.

Kreitzer, A. C., & Berke, J. D. (2011). Investigating striatal function through cell-type-specific manipulations. *Neuroscience, 198*, 19–26. http://dx.doi.org/10.1016/j.neuroscience.2011.08.018.

Kühn, A. A., Trottenberg, T., Kivi, A., Kupsch, A., Schneider, G.-H., & Brown, P. (2005). The relationship between local field potential and neuronal discharge in the subthalamic nucleus of patients with Parkinson's disease. *Special Issue: Neuronal Oscillations in Movement Disorders, 194*(1), 212–220. http://dx.doi.org/10.1016/j.expneurol.2005.02.010.

Lehericy, S., Ducros, M., Van de Moortele, P.-F., François, C., Thivard, L., Poupon, C., ... Kim, D.-S. (2004). Diffusion tensor fiber tracking shows distinct corticostriatal circuits in humans. *Annals of Neurology, 55*(4), 522–529. http://dx.doi.org/10.1002/ana.20030.

Levy, R., Ashby, P., Hutchison, W. D., Lang, A. E., Lozano, A. M., & Dostrovsky, J. O. (2002). Dependence of subthalamic nucleus oscillations on movement and dopamine in Parkinson's disease. *Brain, 125*, 1196–1209. http://dx.doi.org/10.1093/brain/awf128.

Little, S., & Brown, P. (2012). What brain signals are suitable for feedback control of deep brain stimulation in Parkinson's disease? In T. A. Pedley, S. H. Lisanby, H. S. Mayberg, & N. D. Schiff (Eds.), *Annals of the New York Academy of Sciences, Vol. 1265* (pp. 9–24).

Little, S., Tan, H., Anzak, A., Pogosyan, A., Kühn, A., & Brown, P. (2013). Bilateral functional connectivity of the basal ganglia in patients with Parkinson's disease and its modulation by dopaminergic treatment. *PLoS One, 8*(12), e82762. http://dx.doi.org/10.1371/journal.pone.0082762.

Litvak, V., Eusebio, A., Jha, A., Oostenveld, R., Barnes, G., Foltynie, T., ... Brown, P. (2012). Movement-related changes in local and long-range synchronization in Parkinson's disease revealed by simultaneous magnetoencephalography and intracranial recordings. *Journal of Neuroscience, 32*(31), 10541–10553. http://dx.doi.org/10.1523/jneurosci.0767-12.2012.

Lopez-Azcarate, J., Tainta, M., Rodriguez-Oroz, M. C., Valencia, M., Gonzalez, R., Guridi, J., ... Alegre, M. (2010). Coupling between beta and high-frequency activity in the human subthalamic nucleus may be a pathophysiological mechanism in Parkinson's disease. *Journal of Neuroscience, 30*(19), 6667–6677. http://dx.doi.org/10.1523/jneurosci.5459-09.2010.

Mailly, P., Aliane, V., Groenewegen, H. J., Haber, S. N., & Deniau, J. M. (2013). The rat prefrontostriatal system analyzed in 3d: evidence for multiple interacting functional units. *Journal of Neuroscience, 33*(13), 5718–5727. http://dx.doi.org/10.1523/JNEUROSCI.5248-12.2013.

Manning, J. R., Jacobs, J., Fried, I., & Kahana, M. J. (2009). Broadband shifts in local field potential power spectra are correlated with single-neuron spiking in humans. *Journal of Neuroscience, 29*(43), 13613–13620. http://dx.doi.org/10.1523/JNEUROSCI.2041-09.2009.

Marceglia, S., Foffani, G., Bianchi, A. M., Baselli, G., Tamma, F., Egidi, M., & Priori, A. (2006). Dopamine-dependent non-linear correlation between subthalamic rhythms in Parkinson's disease. *Journal of Physiology-London, 571*(Pt 3), 579–591. http://dx.doi.org/10.1113/jphysiol.2005.100271.

Marchand, W. R., Lee, J. N., Suchy, Y., Garn, C., Johnson, S., Wood, N., & Chelune, G. (2011). Age-related changes of the functional architecture of the cortico-basal ganglia circuitry during motor task execution. *NeuroImage, 55*(1), 194–203. http://dx.doi.org/10.1016/j.neuroimage.2010.12.030.

Mink, J. W., & Thach, W. T. (1991). Basal ganglia motor control. 3. Pallidal ablation - normal reaction-time, muscle cocontraction, and slow movement. *Journal of Neurophysiology, 65*(2), 330–351.

Mink, J. W., & Thach, W. T. (1993). Basal ganglia intrinsic circuits and their role in behavior. *Current Opinion in Neurobiology, 3*(6), 950–957. http://dx.doi.org/10.1016/0959-4388(93)90167-W.

Mittmann, W., Wallace, D. J., Czubayko, U., Herb, J. T., Schaefer, A. T., Looger, L. L., ... Kerr, J. N. (2011). Two-photon calcium imaging of evoked activity from l5 somatosensory neurons in vivo. *Nature Neuroscience, 14*(8), 1089–1093. http://dx.doi.org/10.1038/nn.2879.

Monakow, K. H., Akert, K., & Kunzle, H. (1978). Projections of the precentral motor cortex and other cortical areas of the frontal lobe to the subthalamic nucleus in the monkey. *Experimental Brain Research, 33*(3–4), 395–403.

Moran, A., Stein, E., Tischler, H., & Bar-Gad, I. (2012). Decoupling neuronal oscillations during subthalamic nucleus stimulation in the parkinsonian primate. *Neurobiology of Disease, 45*(1), 583–590. http://dx.doi.org/10.1016/j.nbd.2011.09.016.

Moran, A. A., Bergman, H. H., Israel, Z. Z., & Bar-Gad, I. I. (2008). Subthalamic nucleus functional organization revealed by parkinsonian neuronal oscillations and synchrony. *Audio, Transactions of the IRE Professional Group on, 131*(Pt 12), 3395–3409. http://dx.doi.org/10.1093/brain/awn270.

Nambu. (2004). A new dynamic model of the cortico-basal ganglia loop. *Progress in Brain Research, 143*, 461–466. http://dx.doi.org/10.1016/S0079-6123(03)43043-4.

Nauta, H. J., & Cole, M. (1978). Efferent projections of the subthalamic nucleus: an autoradiographic study in monkey and cat. *Journal of Comparative Neurology, 180*(1), 1–16. http://dx.doi.org/10.1002/cne.901800102.

Nauta, H. J., & Cuenod, M. (1982). Perikaryal cell labeling in the subthalamic nucleus following the injection of 3h-gamma-aminobutyric acid into the pallidal complex: an autoradiographic study in cat. *Neuroscience, 7*(11), 2725–2734.

Nauta, W. J., & Mehler, W. R. (1966). Projections of the lentiform nucleus in the monkey. *Brain Research, 1*(1), 3–42.

Nelson, A. B., & Kreitzer, A. C. (2014). Reassessing models of basal ganglia function and dysfunction. *Annual Review of Neuroscience, 37*, 117–135. http://dx.doi.org/10.1146/annurev-neuro-071013-013916.

Neychev, V. K., Gross, R. E., Lehericy, S., Hess, E. J., & Jinnah, H. A. (2011). The functional neuroanatomy of dystonia. *Neurobiology of Disease, 42*(2), 185–201. http://dx.doi.org/10.1016/j.nbd.2011.01.026.

Ozkurt, T. E., Butz, M., Homburger, M., Elben, S., Vesper, J., Wojtecki, L., & Schnitzler, A. (2011). High frequency oscillations in the subthalamic nucleus: a neurophysiological marker of the motor state in Parkinson's disease. *Experimental Neurology, 229*(2), 324–331. http://dx.doi.org/10.1016/j.expneurol.2011.02.015.

Papagiakoumou, E., Anselmi, F., Begue, A., de Sars, V., Gluckstad, J., Isacoff, E. Y., & Emiliani, V. (2010). Scanless two-photon excitation of channelrhodopsin-2. *Nature Methods, 7*(10), 848–854. http://dx.doi.org/10.1038/nmeth.1505.

Peron, S., & Svoboda, K. (2011). From cudgel to scalpel: toward precise neural control with optogenetics. *Nature Methods, 8*(1), 30–34. http://dx.doi.org/10.1038/nmeth.f.325.

Perron, A., Mutoh, H., Akemann, W., Gautam, S. G., Dimitrov, D., Iwamoto, Y., & Knopfel, T. (2009). Second and third generation voltage-sensitive fluorescent proteins for monitoring membrane potential. *Frontiers in Molecular Neuroscience, 2*, 5. http://dx.doi.org/10.3389/neuro.02.005.2009.

Phelps, P. E., Houser, C. R., & Vaughn, J. E. (1985). Immunocytochemical localization of choline acetyltransferase within the rat neostriatum: a correlated light and electron microscopic study of cholinergic neurons and synapses. *Journal of Comparative Neurology, 238*(3), 286–307. http://dx.doi.org/10.1002/cne.902380305.

Plenz, D., & Kital, S. T. (1999). A basal ganglia pacemaker formed by the subthalamic nucleus and external globus pallidus. *Nature, 400*(6745), 677–682. http://dx.doi.org/10.1038/23281.

Pollok, B., Krause, V., Martsch, W., Wach, C., Schnitzler, A., & Sudmeyer, M. (2012). Motor-cortical oscillations in early stages of Parkinson's disease. *Journal of Physiology-London, 590*(13), 3203–3212. http://dx.doi.org/10.1113/jphysiol.2012.231316.

Postuma, R. B., & Dagher, A. (2006). Basal ganglia functional connectivity based on a meta-analysis of 126 positron emission tomography and functional magnetic resonance imaging publications. *Cerebral Cortex, 16*(10), 1508–1521. http://dx.doi.org/10.1093/cercor/bhj088.

Priori, A., Foffani, G., Pesenti, A., Tamma, F., Bianchi, A. M., Pellegrini, M., ... Villani, R. M. (2004). Rhythm-specific pharmacological modulation of subthalamic activity in Parkinson's disease. *Special Issue: Neuronal Oscillations in Movement Disorders, 189*(2), 369–379. http://dx.doi.org/10.1016/j.expneurol.2004.06.001.

Ragsdale, C. W., & Graybiel, A. M. (1990). A simple ordering of neocortical areas established by the compartmental organization of their striatal projections. *Proceedings of the National Academy of Sciences of the United States of America, 87*(16), 6196–6199.

Raz, A., Vaadia, E., & Bergman, H. (2000). Firing patterns and correlations of spontaneous discharge of pallidal neurons in the normal and the tremulous 1-methyl-4-phenyl-1,2,3,6-tetrahydropyridine vervet model of parkinsonism. *Journal of Neuroscience, 20*(22), 8559–8571.

Rodriguez-Oroz, M. C., Obeso, J. A., Lang, A. E., Houeto, J. L., Pollak, P., Rehncrona, S., ... Van Blercom, N. (2005). Bilateral deep brain stimulation in Parkinson's disease: a multicentre study with 4 years follow-up. *Brain, 128*, 2240–2249. http://dx.doi.org/10.1093/brain/awh571.

Rosin, B., Slovik, M., Mitelman, R., Rivlin-Etzion, M., Haber, S. N., Israel, Z., ... Bergman, H. (2011). Closed-loop deep brain stimulation is superior in ameliorating parkinsonism. *Neuron, 72*(2), 370–384. http://dx.doi.org/10.1016/j.neuron.2011.08.023.

Rouzaire-Dubois, B., & Scarnati, E. (1987). Pharmacological study of the cortical-induced excitation of subthalamic nucleus neurons in the rat: evidence for amino acids as putative neurotransmitters. *Neuroscience, 21*(2), 429–440.

Saito, M., Iwawaki, T., Taya, C., Yonekawa, H., Noda, M., Inui, Y., ... Kohno, K. (2001). Diphtheria toxin receptor-mediated conditional and targeted cell ablation in transgenic mice. *Nature Biotechnology, 19*(8), 746–750. http://dx.doi.org/10.1038/90795.

Sano, H., Yasoshima, Y., Matsushita, N., Kaneko, T., Kohno, K., Pastan, I., & Kobayashi, K. (2003). Conditional ablation of striatal neuronal types containing dopamine d2 receptor disturbs coordination of basal ganglia function. *Journal of Neuroscience, 23*(27), 9078–9088.

Shimamoto, S. A., Ryapolova-Webb, E. S., Ostrem, J. L., Galifianakis, N. B., Miller, K. J., & Starr, P. A. (2013). Subthalamic nucleus neurons are synchronized to primary motor cortex local field potentials in Parkinson's disease. *Journal of Neuroscience, 33*(17), 7220–7233. http://dx.doi.org/10.1523/JNEUROSCI.4676-12.2013.

Singer, W. (1993). Synchronization of cortical activity and its putative role in information processing and learning. *Annual Review of Physiology, 55*(1), 349–374. http://dx.doi.org/10.1146/annurev.ph.55.030193.002025.

Smith, A. D., & Bolam, J. P. (1990). The neural network of the basal ganglia as revealed by the study of synaptic connections of identified neurones. *Trends in Neurosciences, 13*(7), 259–265.

Smith, Y., Raju, D. V., Pare, J. F., & Sidibe, M. (2004). The thalamostriatal system: a highly specific network of the basal ganglia circuitry. *Trends in Neurosciences, 27*(9), 520–527. http://dx.doi.org/10.1016/j.tins.2004.07.004.

de Solages, C., Hill, B. C., Koop, M. M., Henderson, J. M., & Bronte-Stewart, H. (2010). Bilateral symmetry and coherence of subthalamic nuclei beta band activity in Parkinson's disease. *Experimental Neurology, 221*(1), 260–266. http://dx.doi.org/10.1016/j.expneurol.2009.11.012.

Stein, E., & Bar-Gad, I. (2013). Beta oscillations in the cortico-basal ganglia loop during parkinsonism. *Special Issue: Neuronal Oscillations in Movement Disorders, 245*(0), 52–59.

Tass, P. A., & Majtanik, M. (2006). Long-term anti-kindling effects of desynchronizing brain stimulation: a theoretical study. *Biological Cybernetics, 94*(1), 58–66. http://dx.doi.org/10.1007/s00422-005-0028-6.

Theer, P., Hasan, M. T., & Denk, W. (2003). Two-photon imaging to a depth of 1000 microm in living brains by use of a ti:Al2o3 regenerative amplifier. *Optics Letters, 28*(12), 1022–1024.

Tsiokos, C., Hu, X., & Pouratian, N. (2013). 200-300hz movement modulated oscillations in the internal globus pallidus of patients with Parkinson's disease. *Neurobiology of Disease, 54*, 464–474. http://dx.doi.org/10.1016/j.nbd.2013.01.020.

Turner, R. S., & Desmurget, M. (2010). Basal ganglia contributions to motor control: a vigorous tutor. *Current Opinion in Neurobiology, 20*(6), 704–716. http://dx.doi.org/10.1016/j.conb.2010.08.022.

Wexler, N. S., Lorimer, J., Porter, J., Gomez, F., Moskowitz, C., Shackell, E., ... Project, U. S.-V. C. R. (2004). Venezuelan kindreds reveal that genetic and environmental factors modulate Huntington's disease age of onset. *Proceedings of the National Academy of Sciences of the United States of America, 101*(10), 3498–3503. http://dx.doi.org/10.1073/pnas.0308679101.

Whittier, J. R., & Mettler, F. A. (1949). Studies on the subthalamus of the rhesus monkey; hyperkinesia and other physiologic effects of subthalamic lesions; with special reference to the subthalamic nucleus of luys. *Journal of Comparative Neurology, 90*(3), 319–372.

Wichmann, T., Bergman, H., & DeLong, M. R. (1994). The primate subthalamic nucleus. Iii. Changes in motor behavior and neuronal activity in the internal pallidum induced by subthalamic inactivation in the mptp model of parkinsonism. *Journal of Neurophysiology, 72*(2), 521–530.

Wichmann, T., Bergman, H., Starr, P. A., Subramanian, T., Watts, R. L., & DeLong, M. R. (1999). Comparison of mptp-induced changes in spontaneous neuronal discharge in the internal pallidal segment and in the substantia nigra pars reticulata in primates. *Experimental Brain Research, 125*(4), 397–409.

Wilt, B. A., Burns, L. D., Wei Ho, E. T., Ghosh, K. K., Mukamel, E. A., & Schnitzer, M. J. (2009). Advances in light microscopy for neuroscience. *Annual Review of Neuroscience, 32*, 435–506. http://dx.doi.org/10.1146/annurev.neuro.051508.135540.

Witten, I. B., Steinberg, E. E., Lee, S. Y., Davidson, T. J., Zalocusky, K. A., Brodsky, M., ... Deisseroth, K. (2011). Recombinase-driver rat lines: tools, techniques, and optogenetic application to dopamine-mediated reinforcement. *Neuron, 72*(5), 721–733. http://dx.doi.org/10.1016/j.neuron.2011.10.028.

Wong, J. Y., Padungchaichot, P., Massalas, J. S., & Drago, J. (2000). Late direct and transneuronal effects in mice with targeted expression of a toxin gene to d1 dopamine receptor neurons. *Neuroscience, 95*(4), 1035–1041.

Worbe, Y., Malherbe, C., Hartmann, A., Pélégrini-Issac, M., Messé, A., Vidailhet, M., ... Benali, H. (2012). Functional immaturity of cortico-basal ganglia networks in gilles de la tourette syndrome. *Brain, 135*(Pt 6), 1937–1946. http://dx.doi.org/10.1093/brain/aws056.

Yang, A. I., Vanegas, N., Lungu, C., & Zaghloul, K. A. (2014). Beta-coupled high-frequency activity and beta-locked neuronal spiking in the subthalamic nucleus of Parkinson's disease. *Journal of Neuroscience, 34*(38), 12816–12827. http://dx.doi.org/10.1523/JNEUROSCI.1895-14.2014.

Yizhar, O., Fenno, L. E., Davidson, T. J., Mogri, M., & Deisseroth, K. (2011). Optogenetics in neural systems. *Neuron, 71*(1), 9–34. http://dx.doi.org/10.1016/j.neuron.2011.06.004.

Zaidel, A., Spivak, A., Grieb, B., Bergman, H., & Israel, Z. (2010). Subthalamic span of beta oscillations predicts deep brain stimulation efficacy for patients with Parkinson's disease. *Brain, 133*(Pt 7), 2007–2021. http://dx.doi.org/10.1093/brain/awq144.

Zheng, Z., Shemmassian, S., Wijekoon, C., Kim, W., Bookheimer, S. Y., & Pouratian, N. (2014). Dti correlates of distinct cognitive impairments in Parkinson's disease. *Human Brain Mapping, 35*(4), 1325–1333. http://dx.doi.org/10.1002/hbm.22256.

Chapter 20

Brainstem Circuitry and Emotions

Panos Theofilas
University of California San Francisco, San Francisco, CA, United States

Helmut Heinsen
University of São Paulo Medical School, São Paulo, Brazil; University Hospital Würzburg, Würzburg, Germany

Lea T. Grinberg
University of California San Francisco, San Francisco, CA, United States; University of São Paulo Medical School, São Paulo, Brazil

THE STRUCTURAL ORGANIZATION OF THE BRAINSTEM

Embodying one of the earliest regions in the evolving human brain, the brainstem resembles the entire brain of present-day reptiles. Connected rostrally to the diencephalon and caudally to the spinal cord, it is divided into three regions including the mesencephalon or midbrain, pons, and medulla oblongata (Nieuwenhuys, Voogd, & van Huijzen, 2008; Olszewski & Baxter, 1982) (Fig. 20.1). Despite its relatively small size, the brainstem regulates vital functions for survival such as breathing, heartbeat, blood pressure, and control of the sleep—wake cycle.

The brainstem is subdivided into three regions: the roof in the posterior, the basis in the ventral, and the tegmentum in the intermediate side. All regions display a systematic distribution of gray and white matter extending throughout its length (Nieuwenhuys et al., 2008; Noback, Strominger, Demarest, & Laemle, 2012). The roof represents the dorsal border of the brainstem, called tectum at the level of the midbrain. It includes two major centers for visual reflexes and audition: the superior (Cs; Fig. 20.2A) and inferior colliculi, respectively. The tectum lacks cranial nerve nuclei, reticular nuclei, or longitudinal tracts. Caudally, the posterior borders are formed by non—brainstem structures including the cerebellum at the level of the pons and the choroid plexus at the medulla oblongata.

The volume of the basis increases progressively during brain evolution of primates, reflecting the increasing telencephalic control of brainstem and spinal cord centers. The basis (B; Fig. 20.2) is composed of descending white matter tracts that link the cortex with the tegmentum (corticobulbar tracts) or the spinal cord (corticospinal tracts), both part of the pyramidal tract, as well as fiber connecting the cortex to the pons (corticopontine fibers) and the cerebellum (pontocerebellar fibers Pyr; Fig. 20.3). At the pons, the basis occupies the largest area in the brainstem owing to numerous nuclei receiving corticopontine fibers and subsequently projecting to the cerebellum via the corticopontocerebellar pathway. The basis occupies less space in the medulla as fiber tracts terminate in the rostral regions of the brainstem (Nieuwenhuys et al., 2008).

The tegmentum or middle zone extends between the ventricular systems of the brainstem's core where it occupies the central region in all of its subdivisions. The tegmentum incorporates a highly complex nuclear organization, including the motor and sensory cranial nerve nuclei; and a zone for higher association centers, including the inferior olivary nuclei (in the medulla), the red nucleus, and the substantia nigra (SN) (in the midbrain). Finally, sensory tracts ascend to the thalamus (LM, LL, and V; Figs. 20.2 and 20.3). Signals transmitted through fiber tracts crossing the brainstem can be modified by relay nuclei within the brainstem (Nieuwenhuys, 1974; Nieuwenhuys et al., 2008; Ramon-Moliner & Nauta, 1966).

With the exception of the olfactory (I) and optic (II) nerves that emerge from the forebrain, the tegmentum originates 10 cranial nerves (III—XII) responsible for sensory and somatic-autonomic motor functions of the head, neck, and trunk. Each nerve is associated with a cranial nucleus (i), classified based on its anatomical connection to the cranial nerve, by receiving primary afferents from or sending out primary efferents to a common projection area (Olszewski & Baxter, 1982; Paxinos & Mai, 2004).

Cranial nerves III (oculomotor), IV (trochlear), and VI (abducens) in the myelencephalon coordinate eye movements; V, VII, and VIII (trigeminal, facial, and vestibulocochlear) emerge from the pontine region and control facial sensation and

Genomics, Circuits, and Pathways in Clinical Neuropsychiatry. http://dx.doi.org/10.1016/B978-0-12-800105-9.00020-2

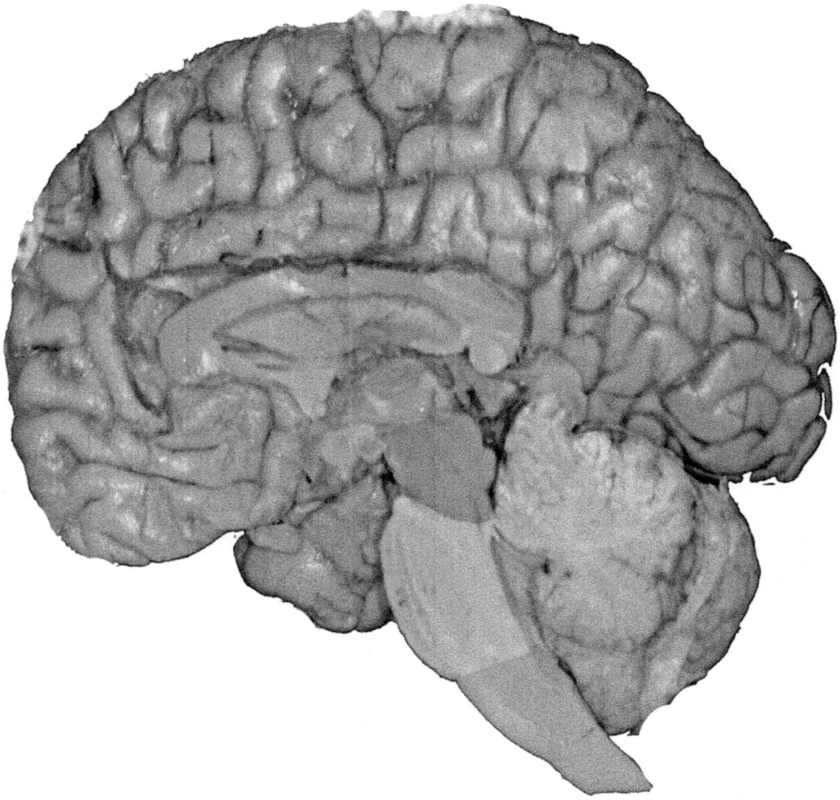

FIGURE 20.1 Medial view of the human right hemisphere, right cerebellum, and hemi-brainstem. The brainstem is divided into three regions according to their embryonic origin. The mesencephalon originates in the midbrain (orange—dark gray in print versions), the metencephalon originates in the pons and cerebellum (green—gray in print versions), and the myelencephalon originates in the medulla oblongata (blue—light gray in print versions).

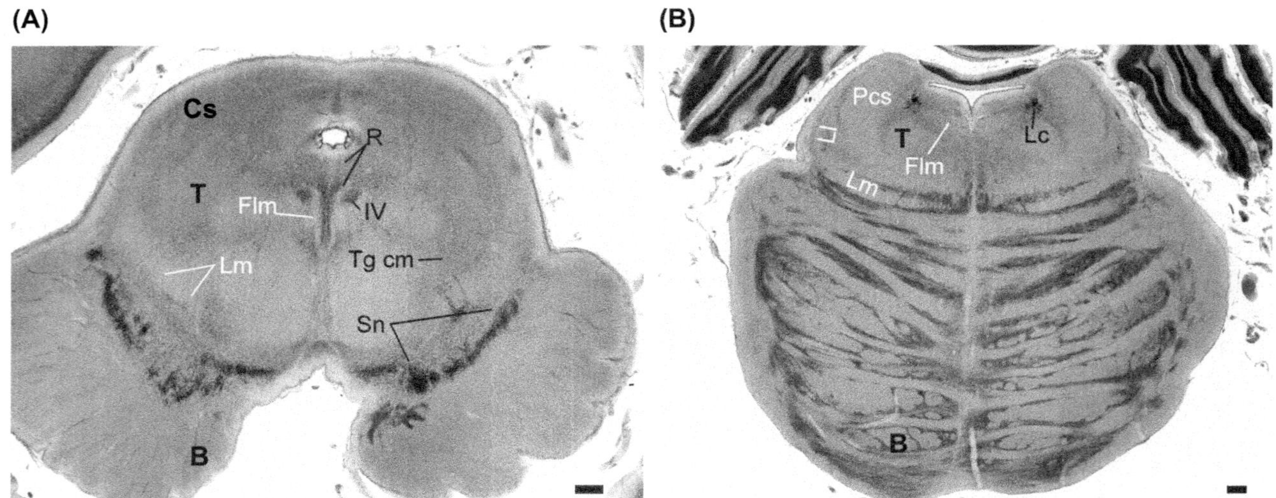

FIGURE 20.2 Horizontal section through the brainstem at the level of ventral tegmental decussation (A) and the entrance of the superior cerebellar peduncle into the tegmenum (B); bars correspond to 1 mm. *IV*, trochlear nucleus; *B*, basis; *Cs*, superior colliculus; *Flm*, medial longitudinal fascicle; *Lc*, locus caeruleus; *Ll*, lateral lemniscus; *Lm*, medial lemniscus; *Pcs*, superior cerebellar peduncle; *R*, raphe nuclei; *Sn*, substantia nigra; *T*, tegmentum; *Tg cm*, compact part of pedunculopontine tegmental nucleus.

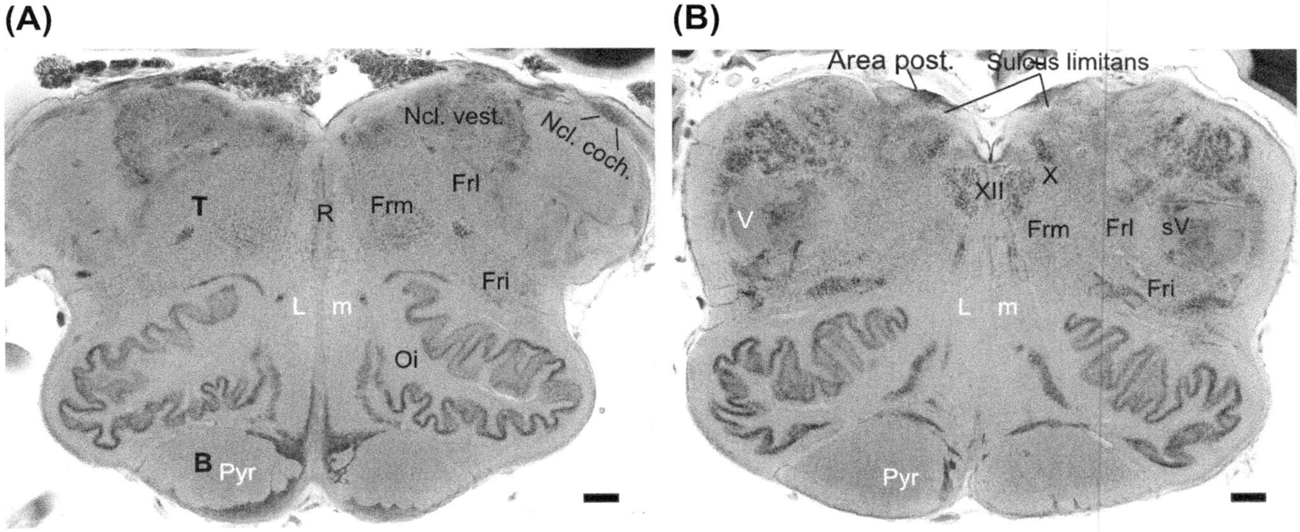

FIGURE 20.3 Horizontal section through the brainstem at the level the entrance of the vestibulocochlear nerve (A) and the caudal limit of the IVth ventricle (B); bars correspond to 1 mm. *V*, spinal tract of trigeminal nerve; *X*, dorsal nucleus of vagus; *XII*, hypoglossal nucleus; *Area post*, postreme area; *B*, basis; *Fri*, intermediate reticular formation (rf); *Frl*, lateral (parvocellular) rf; *Frm*, medial (magnocellular) rf; *Oi*, principal inferior olive nucleus; *Lm*, medial lemniscus; *Ncl. coch.*, cochlear nucleus; *Ncl. vest.*, vestibular nuclei; *Pyr*, prymidal tract; *R*, raphe pallidus; *sV*, spinal nucleus of trigeminal nerve; *T*, tegmentum.

expression, hearing, and balance. Cranial nerves IX–XII (glossopharyngeal, vagus, accessory, and hypoglossal) originate from medullary regions (Fig. 20.3). Accessory (XI) and hypoglossal (XII) represent motor cranial nerves, whereas VII, IX, and X are mixed nerves with motor, sensory, and visceral para sympathetic functions. They innervate the nasal and oral cavities, the throat and the intestines of chest and abdomen, up to the left colonic flexure. Finally, ascending pathways in the tegmentum include the sensory posterior column/medial lemniscus, spinothalamic tract, trigerminal pathway, medial longitudinal fasciculus, and auditory pathways (Nieuwenhuys et al., 2008; Noback et al., 2012).

RETICULAR FORMATION

The reticular formation (RF) represents a phylogenetically older part of the brainstem's tegmentum that extends from the midbrain to the medulla oblongata. The reticular neurons, which are predominantly multipolar, are distributed among ascending and descending fibers throughout the brainstem's tegmental core (Hobson & Brazier, 1980; Ramon-Moliner & Nauta, 1966). Functionally, it represents a critical component of the human central nervous system (CNS) and lesions of its components can lead to death.

A common morphological feature of the reticular nuclei is their netlike dendritic pattern, combined with an absence of clear cytoarchitectonic boundaries, that justifies the name "reticular" (the Latin word *rete* means "net") (Paxinos & Mai, 2004). Extending widely between passing tracts and the cranial nuclei, they receive heterogeneous input from multiple sources (Nieuwenhuys et al., 2008; Ramon-Moliner & Nauta, 1966; Theofilas, Dunlop, Heinsen, & Grinberg, 2015). These include the peripheral sensory systems that enter the spinal cord and brainstem as well as output motor and sensory signals from the cortex, thalamus, and hypothalamus. Such characteristics make the RF a converging hub for afferent and central efferent communication. On the other hand, they make RF impractical to study via traditional degeneration methods because complete severance of each connection can rarely be achieved. Therefore, earlier studies based solely on morphology were restricted to a superficial understanding of the RF function. Only recently, after the development of advanced imaging and neuropathological and tracing techniques, have key functional properties of the RF network been revealed (Hobson & Brazier, 1980).

At the functional level, the RF induces alertness and responsiveness to incoming sensory input via the ascending reticular activating system (ARAS) and maintains a waking state in both humans and animals (Hobson & Brazier, 1980). This is accomplished by the RF's ascending projections to the cortex via a dorsal pathway that projects to the thalamus or a ventral pathway terminating to the hypothalamus and basal forebrain (Moruzzi & Magoun, 1949; Starzl, Taylor, & Magoun, 1951). Furthermore, via the RF funicular trajectories descending to the spinal cord, the reticular neurons mediate motor reflexes as well as voluntary control of basic behavior such as eating, breathing, and drinking (Carlton, Chung, Leonard, & Willis, 1985; Paxinos & Mai, 2004).

Microscopically, the RF is divided into several longitudinal zones that are well-preserved across species (Hobson & Brazier, 1980). In mediolateral orientation, these include: (1) a discrete column of multipolar neurons in which the raphe nuclei are located with functions including wakefulness and consciousness; (2) the medial column with magnocellular (large) nuclei that become reticulospinal tracts involved in motor coordination; (3) an intermediate zone that shows early tau pathology in Alzheimer disease (Rüb et al., 2001); and (4) a lateral or sensory column with parvocellular (small) nuclei regulating a wide range of activities including breathing, vomiting, and swallowing (Holstege, 1991).

Studies using immunohistochemistry and tracing techniques demonstrate distinct functions and complex interconnectivity among the reticular nuclei. Moreover, thick histological sections of the brainstem, where the three-dimensional distribution of the reticular neurons is preserved, reveal the cytoarchitectonic borders of the RF network, a valuable feature lost in thinner sections (Theofilas et al., 2014) (Fig. 20.3). In this context, the reticular nuclei project to extensive regions within the brain including the cortex, cerebellum, and spinal cord and release neuromodulatory neurotransmitters via volume transmission, a nonjunctional form of neuronal communication in which a single neuron projects to vast cortical regions (Paxinos & Mai, 2004; Saper, Chou, & Scammell, 2001). The neuromodulatory neurotransmitters originating in the RF are organized in five groups: the noradrenergic locus coeruleus, the serotonergic raphe nuclei, the cholinergic pedunculopontine tegmental nucleus and the dorsolateral tegmental nucleus, the dopaminergic substantia nigra and ventral tegmental area (VTA) cell groups in the midbrain, as well as the cholinergic nucleus basalis of Meynert (NbM) in the basal forebrain (Courville, Walsh, & Cordeau, 1962; Olszewski & Baxter, 1982; Paxinos & Mai, 2004).

Chemical Properties of the Reticular Nuclei

Noradrenergic Cell Groups

The pontine locus coeruleus (LC) is the major source of noradrenaline in the brain, with minor contributions from nuclei in the lateral tegmentum of the pons and medulla. The LC ("blue spot" in Latin) is composed of a relatively homogeneous group of neurons (Baker, Tork, Hornung, & Halasz, 1989; Paxinos & Mai, 2004) clustered bilaterally in the periventricular gray at the mid pons, with its dendrites extending to the adjacent tegmentum. Extending rostocaudally into a continuous 14- to 17-mm column from the upper pons to caudal midbrain, it innervates the entire brain including the cortex and the spinal cord. Ascending projections can also innervate subcortical relay stations in the thalamus, hypothalamus, and the basal forebrain (Samuels & Szabadi, 2008).

The LC consists of two main neuronal types including medium multipolar neurons with round or oval somata and long dendrites that often extend to adjoining structures, and small, spindle-shaped neurons, both with highly ubiquitous synaptic connections throughout the CNS (Mai, Assheuer, & Paxinos, 1997; Olszewski & Baxter, 1982). By releasing norepinephrine (NE), which upon its synthesis from dopamine (DA) by dopamine β-hydroxylase binds to $\alpha 1$ and $\alpha 2$ adrenergic receptors that are highly expressed in CNS, the LC neurons maintain brain homeostasis by modulating multiple biological functions including wakefulness, attention, stress response, and the reward system in both humans and animals (Berridge & Waterhouse, 2003; Paxinos & Mai, 2004; Sara & Bouret, 2012). Neurotransmitter release occurs via volume transmission. In turn, most structures innervated by the LC also send projections back to the nucleus (Samuels & Szabadi, 2008).

Studies in rats and monkeys found the LC to be a critical modulator of stress responses, including the states of attention, arousal, and vigilance (Aston-Jones, Foote, & Segal, 1985; Sara & Bouret, 2012). Activity of the LC during stress occurs via its excitatory projections to sympathetic neurons, leading to upregulation of NE levels in the cortex that increase arousal, which also depends on dopamine release (see section on Dopaminergic Cell Groups), whereas its inhibiting projection minimizes the response of the parasympathetic neurons. In addition to its role during stress, the LC regulates cardiovascular functions, nociception, and circadian rhythms. For example, there is a significant reduction in LC discharge during sleep. This discharge becomes completely absent during rapid eye movement (REM) (see section on Sleep Cycle) (Aston-Jones et al., 1985; Samuels & Szabadi, 2008).

Serotonergic Cell Groups

The serotonergic (5-hydroxytryptamine [5-HT]) raphe nuclei derive their name from the Latin word for "seam" because their cell bodies are distributed through the midline of the brainstem. Situated along the medial tegmental field ventral to the aqueduct (rostral) and the fourth ventricle (caudal), the raphe nuclei are divided rostrocaudally to dorsal raphe nucleus (DRN), linear nucleus, median raphe nucleus, pontine nucleus, magnus raphe nucleus, pallidal raphe nucleus, and obscurus raphe nucleus (Olszewski & Baxter, 1982). The rostral raphe nuclei in the midbrain and rostral pons form ascending serotonergic projections to the forebrain, including the hippocampus, striatum, hypothalamus, and amygdala, as well as local and cerebellar projections. The caudal group, located in the caudal pons and medulla, connects with the spinal cord and cerebellum.

Studies in rats using fluorescence labeling techniques, later confirmed in humans (Dahlstroem & Fuxe, 1964; Hornung, 2003), showed that the 5-HT—positive neuronal groups outnumber the other monoaminergic systems of the brainstem and that all groups receive afferent projections primarily from limbic structures (Baker, Halliday, & Tork, 1990; Hornung, 2003).

Through activating serotonergic receptors (5-HT_1 through 5-HT_5) expressed throughout the CNS, the raphe nuclei allow for modulation of major homeostatic and behavioral functions including appetitive, emotional, motor, cognitive, and autonomic (Hornung, 2003). For example, stimulation of the dorsal raphe nucleus in cats induced behavioral inhibition and eating in waking stage, whereas pharmacological reduction of serotonin led to insomnia, arousal, and sexual behavior (Trulson & Jacobs, 1979; Ursin, 1976).

Dopaminergic Cell Groups

Neurons using dopamine as their primary neurotransmitter are located in the substantia nigra, the ventral tegmental area, and the rostral linear nucleus of the ventral midbrain (Halliday & Tork, 1986; Paxinos & Mai, 2004). They form ascending projections to basal forebrain and the striatum/basal ganglia, and are crucially involved in motor behavior, motivational processes underlying the learning and execution of goal-directed behavior, possibly by adjusting synaptic plasticity (Bjorklund & Dunnett, 2007; Schultz, Dayan, & Montague, 1997).

Electrical stimulation of this network in rats induces arousal as well as a positive reward state, with similar effects observed after administration of DA agonists including amphetamine and cocaine acting through the DA metabotropic receptors, D_1 and D_2 (Miller, Farber, Gatz, Roffwarg, & German, 1983). In contrast, DA antagonists reduce the reward states during food or drug intake (Ikemoto & Wise, 2002). In humans, through the mesolimbic dopaminergic pathway, DA is similarly associated with the development of addiction and the expression of compulsive behaviors, including food craving and binge eating. In rats and humans, VTA, through mesolimbic projections to the substantia innominata in the basal forebrain, as well as the striatum and to a lesser extent the amygdala, was found to mediate a broad range of functions including movement, reward and motivation, novelty detection, and memory and learning (Beckstead, Domesick, & Nauta, 1979; Grove, 1988). Finally, motor dysfunction characterizing Parkinson disease is caused by loss of inhibitory projections to the striatum after neuropathological lesions to the SN (Seidel et al., 2015).

Cholinergic Cell Groups

The cholinergic neurons of the basal forebrain, including the NbM, the medial habenular nucleus, and the medial septal nucleus, represent major outputs of acetylcholine (ACh) in the CNS and expression sites of its synthesizing enzyme choline acetyltransferase often used as a marker of the cholinergic cells (Mesulam, Mufson, Wainer, & Levey, 1983). These nuclei receive reciprocal connections from limbic structures and the midbrain and their morphological features include radiating dendrites that reach the medial forebrain bundle (Nieuwenhuys et al., 2008). In turn, they form outgoing fibers that terminate at the thalamus and promote alertness in the cortex via the widespread release of ACh. Although not part of the brainstem's reticular network, the cholinergic neurons of the basal forebrain are considered components of the ARAS because they exhibit abundant interconnections and functional overlapping with the brainstem's reticular system (Paxinos & Mai, 2004; Vincent et al., 1986).

In parallel with the other RF groups, experiments in cats implicate the basal forebrain nuclei in generating wakefulness and REM sleep (Marrosu et al., 1995). Together with the dopaminergic system, ACh neurons also have a role in positively rewarding activities and reinforcement during waking stage, including food intake (Inglis, Day, & Fibiger, 1994; Mufson, Ginsberg, Ikonomovic, & DeKosky, 2003). ACh-based neuromodulation is also implicated in synaptic plasticity, memory, and learning, whereas damage to basal forebrain structures can induce learning impairments and amnesia (McLin, Miasnikov, & Weinberger, 2002). Both Alzheimer disease and Parkinson disease lesions target these structures in the early stages of each disease, as shown by neuropathological studies and imaging biomarkers (Mesulam, Shaw, Mash, & Weintraub, 2004; Muller et al., 2013; Teipel et al., 2014).

RETICULAR FORMATION MODULATES BEHAVIOR

The RF is crucial for multiple neuromodulatory functions, orchestrated via diffuse projections throughout the forebrain, brainstem, and spinal cord (Hobson, McCarley, & Wyzinski, 1975). RF activity includes primary homeostatic regulation, such as wakefulness and sleep, as well as behavioral modulation including mood, attention, and motivation. The contribution of the RF to the sleep cycle and mood regulation is described next.

Sleep Cycle

Sleep and wake cycles in mammals are active states of consciousness mediated via neuromodulatory nuclei of the brainstem and hypothalamus, by promoting synchronization of the electroencephalographic activity in the brain during arousal (Moruzzi & Magoun, 1995; Steriade, McCormick, & Sejnowski, 1993). Several brainstem nuclei of the RF are critically involved in regulating the sleep cycle by forming excitatory projections to the thalamus, cortex, and basal forebrain. These include the LC and DRN and their respective neurotransmitters, NE and 5-HT (Aston-Jones & Bloom, 1981; Boeve et al., 2007; Monti, 2010; Saper, Fuller, Pedersen, Lu, & Scammell, 2010).

Brainstem studies in cats and rodents demonstrate a reduction of firing activity in the DRN as the animal falls asleep. This is followed by minimal activity during REM sleep (Trulson & Jacobs, 1979), which suggests a significant role for the DRN in cortical activation as well as sensorimotor activity during the awake state (Monti, Jantos, & Lagos, 2010). However, unlike ACh release by the NbM that initiates and maintains REM sleep (Vazquez & Baghdoyan, 2001), 5-HT (and similarly NE) do not induce cortical arousal that accompanies REM and their levels during sleep remain low or absent. Furthermore, 5-HT receptors are expressed in structures relevant to sleep and wake cycle, including the LC, DRN, cerebral cortex, hippocampus, and thalamus. In this context, mutant mice lacking 5-HT_{1A} or 5-HT_{1B} receptors show extended stages of REM compared to wild-type controls (Boutrel, Franc, Hen, Hamon, & Adrien, 1999). Conversely, activation of 5-HT_{1B} via receptor agonist administration induced waking and negatively influenced REM sleep (Monti et al., 2010).

In humans, reduced levels of 5-HT were detected in ventricular CSF during non-REM, and during REM stages compared with high 5-HT levels during waking (Zeitzer et al., 2002). Furthermore, individuals with neurodegenerative and mental disorders affecting the DRN and 5-HT transmission, such as Alzheimer disease, Parkinson disease (Grinberg et al., 2009; Grinberg, Rueb, Alho, & Heinsen, 2010; Theofilas et al., 2015), and major depression (Sutton, 2014), exhibit sleep disruptions at early stages of the disease, including insomnia. Anxiolytic medications in individuals with depression increased 5-HT firing/tone, improved sleep, and return sleep—wake stages to their restorative functions (Sharpley & Cowen, 1995).

NE, released from the pontine nucleus LC to the thalamus, cortex, and spinal cord, has a crucial role in waking, vigilance, and attention via activation of the adrenoreceptors α_1 and α_2. Similar to the DRN, LC neurons are active during waking, decrease discharge during non-REM sleep, and cease firing during the REM stage. Excitation or inhibition of the targeted neurons depends on NE receptor type. In canine studies, for example, blocking postsynaptic NE activity via administration of a_1 receptor antagonists induced sleep and narcoleptic symptoms (Nishino, 2007).

Studies in primates and rats demonstrate a shift from low LC tone related to inattentiveness, followed by increased firing when the animal focused on a task (Aston-Jones & Bloom, 1981; Aston-Jones, Rajkowski, Kubiak, Valentino, & Shipley, 1996; Hobson et al., 1975). In this context, NE levels increase most when a novel stimulus is introduced to a situation, especially when a focused cognitive effort is required (Aston-Jones et al., 1996). On the other hand, 5-HT levels show an increase during sustained levels of cortical activation and a decrease in response to novel stimuli, thus maintaining cortical arousal and concentration while suppressing distracting stimuli (Schwartz & Roth, 2008). In this context, individuals with a high degree of inattentiveness owing to sleep disruption, who are often regarded as having attention-deficit hyperactivity disorder, benefit from noradrenergic-enhancing treatments that help them regain normal sleep pattern and attention levels during waking (Siever & Davis, 1985).

Ingrained within the brainstem's ARAS, the cholinergic pedunculopontine nucleus (PPN) shows sleep—wake state—related changes in firing activity while it promotes and maintains REM sleep via release of ACh and glutamate to the cortex (Garcia-Rill, 1991; Hobson & Brazier, 1980; Saper et al., 2010). Lesions in the PPN in cats and rodents reduced REM sleep and impaired the acquisition of several learning tasks, including spatial navigation (Dellu, Mayo, Cherkaoui, Le Moal, & Simon, 1991; Shouse & Siegel, 1992). Furthermore, the gamma-aminobutyric acid-ergic and glycinergic gigantocellular nucleus (Gi) of central pons and medulla fires at very slow rates during waking and slow-wave sleep, whereas it increases activity in the REM stage. In cats, lesions of the Gi resulted in the complete elimination of REM and loss of tonic muscular atonia which normally accompanies this stage (Hobson & Brazier, 1980; Jones, 1979).

Mood

Mood spectrum disorders, including major depression and bipolar disorders, are heterogeneous in nature and result from a complex interplay of genetic, environmental, and social variables (American Psychiatric Association, 2013).

Animals studies and brain imaging in humans demonstrate the significance of subcortical monoaminergic circuitry in mood regulation via ascending projections to brain regions including prefrontal cortex, hippocampus, and amygdala, areas classically related to emotion regulation, reward, and executive function, all deregulated in individuals with depression and bipolar disorders (Chaudhury et al., 2013; Krishnan & Nestler, 2008).

It is well accepted that mood disorders result from impairments in neurotransmission at the level of the neuron, primary involving neuromodulatory systems of subcortical origin (Manji, Drevets, & Charney, 2001; Nugent, Davis, Zarate, & Drevets, 2013). Although not associated with gross brain pathology, mood disorders are often accompanied by region-specific neuronal loss, as demonstrated by imaging studies showing a reduction in gray matter in forebrain limbic regions (Drevets, Price, & Furey, 2008). Mood disorders are associated with abnormalities in monoaminergic neurotransmitter systems, which is highlighted by the fact that mood stabilizers and antidepressant medications act via these systems and are crucial in maintaining disease remission.

The monoamine hypothesis for major depression resulted from findings in which 5-HT and NE upregulation showed antidepressant effects in patients. Currently most antidepressant medications enhance monoaminergic neurotransmission by inhibiting neurotransmitter reuptake or degradation, as exemplified by 5-HT reuptake inhibitors and oxidase inhibitors, respectively (Nugent, Carlson, et al., 2013; Ruhe, Mason, & Schene, 2007). In this line, selective serotonin reuptake inhibitors and other antidepressant drugs minimize symptoms of depression by elevating 5-HT transmission and receptor levels (Haddjeri, Blier, & de Montigny, 1998). Upregulation of postsynaptic 5-HT_{1A} receptor, which is decreased in prefrontal regions in individuals with depression and bipolar disorders, is also a therapeutic outcome of antidepressant response (Sargent et al., 2000, 2010). In parallel with 5-HT receptor changes, elevated 5-HT transporter function is also considered a factor mediating pathophysiology in patients with unipolar and bipolar disorder, including adolescents, by decreasing 5-HT availability in the brain, especially during stressful events (Cannon et al., 2007; Dahlstrom et al., 2000).

The monoamines dopamine and norepinephrine, which are supplied by brainstem's SN and LC, respectively, are also involved in regulating mood; and similarly to 5-HT, their levels are altered by antidepressant drugs, including receptor agonists and reuptake inhibitors (van Enkhuizen et al., 2015; Zarate et al., 2004). Accumulating data from human imaging studies and animal models of the disease suggest that increased activation of DA and NE underlie the positive/manic states of the bipolar spectrum, followed by increased cholinergic functioning that underlies depression. Because these nuclei are also targeted during the initial stages of neurodegenerative diseases, including Parkinson and Alzheimer diseases, depression is an early symptom of these conditions, years before the onset of disease-defining symptoms (Chan-Palay & Asan, 1989; Grinberg et al., 2009; Schrag, 2004; Theofilas et al., 2015).

In contrast to the reduced monoaminergic activity in depression and bipolar disorder, the cholinergic system appears overresponsive in mood disorders (Picciotto, Higley, & Mineur, 2012). For example, individuals who are bipolar at the manic stage experience depressive symptoms by increasing the levels of ACh by inhibiting ACh-downregulating enzyme acetylcholine esterase. Furthermore, depressed individuals became worse after administration of ACh agonists or muscarinic receptor agonists, whereas administration of scopolamine, a cholinergic receptor antagonist, had antidepressant effects in patients with both unipolar and bipolar conditions (Drevets & Furey, 2010; Janowsky, Overstreet, & Nurnberger, 1994). These also improved cognitive performance, including increased concentration and higher scores in memory tasks, both of which were significantly influenced by cholinergic neuromodulation.

CONCLUSION

Forming interconnections widely within the CNS, the reticular formation of the brainstem orchestrates neural circuits of major clinical significance. These extend to basic homeostatic functions including cranial nerve reflexes, autonomic control of heart rate, breathing, and control of the sleep—wake cycle. Convergent results from clinical and animal studies support the crucial role of the reticular system in the modulation of behavioral and emotional processing and experience via release of a wide range of neuromodulatory neurotransmitters, whereas dysfunction in this network is implicated in the pathophysiology of mood disorders and major neurodegenerative diseases. Overall, this evidence emphasizes the importance of subcortical systems in health and disease and highlights the brainstem as a valuable target for therapeutic investigation.

REFERENCES

American Psychiatric Association. (2013). *Diagnostic and statistical manual of mental disorders* (5th ed.). Arlington, VA: American Psychiatric Association.

Aston-Jones, G., & Bloom, F. E. (1981). Activity of norepinephrine-containing locus coeruleus neurons in behaving rats anticipates fluctuations in the sleep-waking cycle. *The Journal of Neuroscience: The Official Journal of the Society for Neuroscience, 1*(8), 876—886.

Aston-Jones, G., Foote, S. L., & Segal, M. (1985). Impulse conduction properties of noradrenergic locus coeruleus axons projecting to monkey cerebrocortex. *Neuroscience, 15*(3), 765—777.

Aston-Jones, G., Rajkowski, J., Kubiak, P., Valentino, R. J., & Shipley, M. T. (1996). Role of the locus coeruleus in emotional activation. *Progress in Brain Research, 107*, 379—402.

Baker, K. G., Halliday, G. M., & Tork, I. (1990). Cytoarchitecture of the human dorsal raphe nucleus. *The Journal of Comparative Neurology, 301*(2), 147–161. http://dx.doi.org/10.1002/cne.903010202.

Baker, K. G., Tork, I., Hornung, J. P., & Halasz, P. (1989). The human locus coeruleus complex: an immunohistochemical and three dimensional reconstruction study. *Experimental Brain Research, 77*(2), 257–270.

Beckstead, R. M., Domesick, V. B., & Nauta, W. J. (1979). Efferent connections of the substantia nigra and ventral tegmental area in the rat. *Brain Research, 175*(2), 191–217. pii:0006-8993(79)91001-1.

Berridge, C. W., & Waterhouse, B. D. (2003). The locus coeruleus-noradrenergic system: modulation of behavioral state and state-dependent cognitive processes. *Brain Research Brain Research Reviews, 42*(1), 33–84.

Bjorklund, A., & Dunnett, S. B. (2007). Dopamine neuron systems in the brain: an update. *Trends in Neurosciences, 30*(5), 194–202. pii:S0166-2236(07)00067-7.

Boeve, B. F., Silber, M. H., Saper, C. B., Ferman, T. J., Dickson, D. W., Parisi, J. E., ... Braak, H. (2007). Pathophysiology of REM sleep behaviour disorder and relevance to neurodegenerative disease. *Brain: A Journal of Neurology, 130*(Pt 11), 2770–2788. pii:wm056.

Boutrel, B., Franc, B., Hen, R., Hamon, M., & Adrien, J. (1999). Key role of 5-HT1B receptors in the regulation of paradoxical sleep as evidenced in 5-HT1B knock-out mice. *The Journal of Neuroscience: The Official Journal of the Society for Neuroscience, 19*(8), 3204–3212.

Cannon, D. M., Ichise, M., Rollis, D., Klaver, J. M., Gandhi, S. K., Charney, D. S., ... Drevets, W. C. (2007). Elevated serotonin transporter binding in major depressive disorder assessed using positron emission tomography and [11C]DASB; comparison with bipolar disorder. *Biological Psychiatry, 62*(8), 870–877. pii:S0006-3223(07)00236-3.

Carlton, S. M., Chung, J. M., Leonard, R. B., & Willis, W. D. (1985). Funicular trajectories of brainstem neurons projecting to the lumbar spinal cord in the monkey (*Macaca fascicularis*): a retrograde labeling study. *The Journal of Comparative Neurology, 241*(3), 382–404. http://dx.doi.org/10.1002/cne.902410310.

Chan-Palay, V., & Asan, E. (1989). Alterations in catecholamine neurons of the locus coeruleus in senile dementia of the Alzheimer type and in Parkinson's disease with and without dementia and depression. *The Journal of Comparative Neurology, 287*(3), 373–392. http://dx.doi.org/10.1002/cne.902870308.

Chaudhury, D., Walsh, J. J., Friedman, A. K., Juarez, B., Ku, S. M., Koo, J. W., ... Han, M. H. (2013). Rapid regulation of depression-related behaviours by control of midbrain dopamine neurons. *Nature, 493*(7433), 532–536. http://dx.doi.org/10.1038/nature11713.

Courville, J., Walsh, J., & Cordeau, J. P. (1962). Functional organization of the brain stem reticular formation and sensory input. *Science (New York, N.Y.), 138*(3544), 973–975.

Dahlstroem, A., & Fuxe, K. (1964). Evidence for the existence of monoamine-containing neurons in the central nervous system. I. demonstration of monoamines in the cell bodies of brain stem neurons. *Acta Physiologica Scandinavica Supplementum*, (Suppl. 232), 1–55.

Dahlstrom, M., Ahonen, A., Ebeling, H., Torniainen, P., Heikkila, J., & Moilanen, I. (2000). Elevated hypothalamic/midbrain serotonin (monoamine) transporter availability in depressive drug-naive children and adolescents. *Molecular Psychiatry, 5*(5), 514–522.

Dellu, F., Mayo, W., Cherkaoui, J., Le Moal, M., & Simon, H. (1991). Learning disturbances following excitotoxic lesion of cholinergic pedunculo-pontine nucleus in the rat. *Brain Research, 544*(1), 126–132. pii:0006-8993(91)90893-Z.

Drevets, W. C., & Furey, M. L. (2010). Replication of scopolamine's antidepressant efficacy in major depressive disorder: a randomized, placebo-controlled clinical trial. *Biological Psychiatry, 67*(5), 432–438. http://dx.doi.org/10.1016/j.biopsych.2009.11.021.

Drevets, W. C., Price, J. L., & Furey, M. L. (2008). Brain structural and functional abnormalities in mood disorders: implications for neurocircuitry models of depression. *Brain Structure & Function, 213*(1–2), 93–118. http://dx.doi.org/10.1007/s00429-008-0189-x.

van Enkhuizen, J., Janowsky, D. S., Olivier, B., Minassian, A., Perry, W., Young, J. W., & Geyer, M. A. (2015). The catecholaminergic-cholinergic balance hypothesis of bipolar disorder revisited. *European Journal of Pharmacology, 753*, 114–126. http://dx.doi.org/10.1016/j.ejphar.2014.05.063.

Garcia-Rill, E. (1991). The pedunculopontine nucleus. *Progress in Neurobiology, 36*(5), 363–389. pii:0301-0082(91)90016-T.

Grinberg, L. T., Rub, U., Ferretti, R. E., Nitrini, R., Farfel, J. M., Polichiso, L., ... Brazilian Brain Bank Study Group. (2009). The dorsal raphe nucleus shows phospho-tau neurofibrillary changes before the transentorhinal region in Alzheimer's disease. A precocious onset? *Neuropathology and Applied Neurobiology, 35*(4), 406–416. http://dx.doi.org/10.1111/j.1365-2990.2009.00997.x.

Grinberg, L. T., Rueb, U., Alho, A. T., & Heinsen, H. (2010). Brainstem pathology and non-motor symptoms in PD. *Journal of the Neurological Sciences, 289*(1–2), 81–88. http://dx.doi.org/10.1016/j.jns.2009.08.021.

Grove, E. A. (1988). Efferent connections of the substantia innominata in the rat. *The Journal of Comparative Neurology, 277*(3), 347–364. http://dx.doi.org/10.1002/cne.902770303.

Haddjeri, N., Blier, P., & de Montigny, C. (1998). Long-term antidepressant treatments result in a tonic activation of forebrain 5-HT1A receptors. *The Journal of Neuroscience: The Official Journal of the Society for Neuroscience, 18*(23), 10150–10156.

Halliday, G. M., & Tork, I. (1986). Comparative anatomy of the ventromedial mesencephalic tegmentum in the rat, cat, monkey and human. *The Journal of Comparative Neurology, 252*(4), 423–445. http://dx.doi.org/10.1002/cne.902520402.

Hobson, J. A., & Brazier, M. A. B. (Eds.). (1980). *The reticular formation revisited: Specifying function for a nonspecific system.* New York: Raven Press.

Hobson, J. A., McCarley, R. W., & Wyzinski, P. W. (1975). Sleep cycle oscillation: reciprocal discharge by two brainstem neuronal groups. *Science (New York, N.Y.), 189*(4196), 55–58.

Holstege, G. (1991). Descending motor pathways and the spinal motor system: limbic and non-limbic components. *Progress in Brain Research, 87*, 307–421.

Hornung, J. P. (2003). The human raphe nuclei and the serotonergic system. *Journal of Chemical Neuroanatomy, 26*(4), 331–343.

Ikemoto, S., & Wise, R. A. (2002). Rewarding effects of the cholinergic agents carbachol and neostigmine in the posterior ventral tegmental area. *The Journal of Neuroscience: The Official Journal of the Society for Neuroscience, 22*(22), 9895–9904. pii:22/22/9895.

Inglis, F. M., Day, J. C., & Fibiger, H. C. (1994). Enhanced acetylcholine release in hippocampus and cortex during the anticipation and consumption of a palatable meal. *Neuroscience, 62*(4), 1049–1056. pii:0306-4522(94)90342-5.

Janowsky, D. S., Overstreet, D. H., & Nurnberger, J. I., Jr. (1994). Is cholinergic sensitivity a genetic marker for the affective disorders? *American Journal of Medical Genetics, 54*(4), 335–344. http://dx.doi.org/10.1002/ajmg.1320540412.

Jones, B. E. (1979). Elimination of paradoxical sleep by lesions of the pontine gigantocellular tegmental field in the cat. *Neuroscience Letters, 13*(3), 285–293.

Krishnan, V., & Nestler, E. J. (2008). The molecular neurobiology of depression. *Nature, 455*(7215), 894–902. http://dx.doi.org/10.1038/nature07455.

Mai, J. K., Assheuer, J., & Paxinos, G. (1997). *Atlas of the human brain.* San Diego, CA: Academic Press.

Manji, H. K., Drevets, W. C., & Charney, D. S. (2001). The cellular neurobiology of depression. *Nature Medicine, 7*(5), 541–547. http://dx.doi.org/10.1038/87865.

Marrosu, F., Portas, C., Mascia, M. S., Casu, M. A., Fa, M., Giagheddu, M., … Gessa, G. L. (1995). Microdialysis measurement of cortical and hippocampal acetylcholine release during sleep-wake cycle in freely moving cats. *Brain Research, 671*(2), 329–332. pii:0006-8993(94)01399-3.

McLin, D. E., 3rd, Miasnikov, A. A., & Weinberger, N. M. (2002). Induction of behavioral associative memory by stimulation of the nucleus basalis. *Proceedings of the National Academy of Sciences of the United States of America, 99*(6), 4002–4007. http://dx.doi.org/10.1073/pnas.062057099.

Mesulam, M., Shaw, P., Mash, D., & Weintraub, S. (2004). Cholinergic nucleus basalis tauopathy emerges early in the aging-MCI-AD continuum. *Annals of Neurology, 55*(6), 815–828. http://dx.doi.org/10.1002/ana.20100.

Mesulam, M. M., Mufson, E. J., Wainer, B. H., & Levey, A. I. (1983). Central cholinergic pathways in the rat: an overview based on an alternative nomenclature (Ch1-Ch6). *Neuroscience, 10*(4), 1185–1201. pii:0306-4522(83)90108-2.

Miller, J. D., Farber, J., Gatz, P., Roffwarg, H., & German, D. C. (1983). Activity of mesencephalic dopamine and non-dopamine neurons across stages of sleep and walking in the rat. *Brain Research, 273*(1), 133–141.

Monti, J. M. (2010). The role of dorsal raphe nucleus serotonergic and non-serotonergic neurons, and of their receptors, in regulating waking and rapid eye movement (REM) sleep. *Sleep Medicine Reviews, 14*(5), 319–327. http://dx.doi.org/10.1016/j.smrv.2009.10.003.

Monti, J. M., Jantos, H., & Lagos, P. (2010). Activation of serotonin 5-HT(1B) receptor in the dorsal raphe nucleus affects REM sleep in the rat. *Behavioural Brain Research, 206*(1), 8–16. http://dx.doi.org/10.1016/j.bbr.2009.08.037.

Moruzzi, G., & Magoun, H. W. (1949). Brain stem reticular formation and activation of the EEG. *Electroencephalography and Clinical Neurophysiology, 1*(4), 455–473.

Moruzzi, G., & Magoun, H. W. (1995). Brain stem reticular formation and activation of the EEG. 1949. *The Journal of Neuropsychiatry and Clinical Neurosciences, 7*(2), 251–267.

Mufson, E. J., Ginsberg, S. D., Ikonomovic, M. D., & DeKosky, S. T. (2003). Human cholinergic basal forebrain: chemoanatomy and neurologic dysfunction. *Journal of Chemical Neuroanatomy, 26*(4), 233–242. pii:S0891061803000681.

Muller, M. L., Albin, R. L., Kotagal, V., Koeppe, R. A., Scott, P. J., Frey, K. A., & Bohnen, N. I. (2013). Thalamic cholinergic innervation and postural sensory integration function in Parkinson's disease. *Brain: A Journal of Neurology, 136*(Pt 11), 3282–3289. http://dx.doi.org/10.1093/brain/awt247.

Nieuwenhuys, R. (1974). Topological analysis of the brain stem: a general introduction. *The Journal of Comparative Neurology, 156*(3), 255–276. http://dx.doi.org/10.1002/cne.901560302.

Nieuwenhuys, R., Voogd, J., & van Huijzen, C. (2008). *The human central nervous system* (4th ed.). Berlin; New York: Springer.

Nishino, S. (2007). Clinical and neurobiological aspects of narcolepsy. *Sleep Medicine, 8*(4), 373–399. pii:S1389-9457(07)00078-0.

Noback, C. R., Strominger, N. L., Demarest, R. J., & Laemle, L. (2012). *Noback's human nervous system structure and function.*

Nugent, A. C., Carlson, P. J., Bain, E. E., Eckelman, W., Herscovitch, P., Manji, H., … Drevets, W. C. (2013). Mood stabilizer treatment increases serotonin type 1A receptor binding in bipolar depression. *Journal of Psychopharmacology (Oxford, England), 27*(10), 894–902. http://dx.doi.org/10.1177/0269881113499204.

Nugent, A. C., Davis, R. M., Zarate, C. A., Jr., & Drevets, W. C. (2013). Reduced thalamic volumes in major depressive disorder. *Psychiatry Research, 213*(3), 179–185. http://dx.doi.org/10.1016/j.pscychresns.2013.05.004.

Olszewski, J., & Baxter, D. (1982). *Cytoarchitecture of the human brain stem* (2nd ed.). Basel; New York: Karger.

Paxinos, G., & Mai, J. K. (2004). *The human nervous system* (2nd ed.). Amsterdam; Boston: Elsevier Academic Press.

Picciotto, M. R., Higley, M. J., & Mineur, Y. S. (2012). Acetylcholine as a neuromodulator: cholinergic signaling shapes nervous system function and behavior. *Neuron, 76*(1), 116–129. http://dx.doi.org/10.1016/j.neuron.2012.08.036.

Ramon-Moliner, E., & Nauta, W. J. (1966). The isodendritic core of the brain stem. *The Journal of Comparative Neurology, 126*(3), 311–335. http://dx.doi.org/10.1002/cne.901260301.

Rüb, U., Del Tredici, K., Schultz, C., Thal, D. R., Braak, E., & Braak, H. (2001). The autonomic higher order processing nuclei of the lower brain stem are among the early targets of the Alzheimer's disease-related cytoskeletal pathology. *Acta Neuropathologica, 101*, 555–564.

Ruhe, H. G., Mason, N. S., & Schene, A. H. (2007). Mood is indirectly related to serotonin, norepinephrine and dopamine levels in humans: a meta-analysis of monoamine depletion studies. *Molecular Psychiatry, 12*(4), 331–359. pii:4001949.

Samuels, E. R., & Szabadi, E. (2008). Functional neuroanatomy of the noradrenergic locus coeruleus: its roles in the regulation of arousal and autonomic function part I: principles of functional organisation. *Current Neuropharmacology, 6*(3), 235–253. http://dx.doi.org/10.2174/157015908785777229.

Saper, C. B., Chou, T. C., & Scammell, T. E. (2001). The sleep switch: hypothalamic control of sleep and wakefulness. *Trends in Neurosciences, 24*(12), 726–731. pii:S0166-2236(00)02002-6.

Saper, C. B., Fuller, P. M., Pedersen, N. P., Lu, J., & Scammell, T. E. (2010). Sleep state switching. *Neuron, 68*(6), 1023–1042. http://dx.doi.org/10.1016/j.neuron.2010.11.032.

Sara, S. J., & Bouret, S. (2012). Orienting and reorienting: the locus coeruleus mediates cognition through arousal. *Neuron, 76*(1), 130–141. http://dx.doi.org/10.1016/j.neuron.2012.09.011.

Sargent, P. A., Kjaer, K. H., Bench, C. J., Rabiner, E. A., Messa, C., Meyer, J., … Cowen, P. J. (2000). Brain serotonin1A receptor binding measured by positron emission tomography with [11C]WAY-100635: effects of depression and antidepressant treatment. *Archives of General Psychiatry, 57*(2), 174–180.

Sargent, P. A., Rabiner, E. A., Bhagwagar, Z., Clark, L., Cowen, P., Goodwin, G. M., & Grasby, P. M. (2010). 5-HT(1A) receptor binding in euthymic bipolar patients using positron emission tomography with [carbonyl-(11)C]WAY-100635. *Journal of Affective Disorders, 123*(1–3), 77–80. http://dx.doi.org/10.1016/j.jad.2009.07.015.

Schrag, A. (2004). Psychiatric aspects of Parkinson's disease—an update. *Journal of Neurology, 251*(7), 795–804. http://dx.doi.org/10.1007/s00415-004-0483-3.

Schultz, W., Dayan, P., & Montague, P. R. (1997). A neural substrate of prediction and reward. *Science (New York, N.Y.), 275*(5306), 1593–1599.

Schwartz, J. R., & Roth, T. (2008). Neurophysiology of sleep and wakefulness: basic science and clinical implications. *Current Neuropharmacology, 6*(4), 367–378. http://dx.doi.org/10.2174/157015908787386050.

Seidel, K., Mahlke, J., Siswanto, S., Kruger, R., Heinsen, H., Auburger, G., … Rub, U. (2015). The brainstem pathologies of Parkinson's disease and dementia with lewy bodies. *Brain Pathology (Zurich, Switzerland), 25*(2), 121–135. http://dx.doi.org/10.1111/bpa.12168.

Sharpley, A. L., & Cowen, P. J. (1995). Effect of pharmacologic treatments on the sleep of depressed patients. *Biological Psychiatry, 37*(2), 85–98. pii:0006-3223(94)00135-P.

Shouse, M. N., & Siegel, J. M. (1992). Pontine regulation of REM sleep components in cats: integrity of the pedunculopontine tegmentum (PPT) is important for phasic events but unnecessary for atonia during REM sleep. *Brain Research, 571*(1), 50–63. pii:0006-8993(92)90508-7.

Siever, L. J., & Davis, K. L. (1985). Overview: toward a dysregulation hypothesis of depression. *The American Journal of Psychiatry, 142*(9), 1017–1031.

Starzl, T. E., Taylor, C. W., & Magoun, H. W. (1951). Ascending conduction in reticular activating system, with special reference to the diencephalon. *Journal of Neurophysiology, 14*(6), 461–477.

Steriade, M., McCormick, D. A., & Sejnowski, T. J. (1993). Thalamocortical oscillations in the sleeping and aroused brain. *Science (New York, N.Y.), 262*(5134), 679–685.

Sutton, E. L. (2014). Psychiatric disorders and sleep issues. *The Medical Clinics of North America, 98*(5), 1123–1143. http://dx.doi.org/10.1016/j.mcna.2014.06.009.

Teipel, S., Heinsen, H., Amaro, E., Jr., Grinberg, L. T., Krause, B., Grothe, M., & Alzheimer's Disease Neuroimaging Initiative. (2014). Cholinergic basal forebrain atrophy predicts amyloid burden in Alzheimer's disease. *Neurobiology of Aging, 35*(3), 482–491. http://dx.doi.org/10.1016/j.neurobiolaging.2013.09.029.

Theofilas, P., Dunlop, S., Heinsen, H., & Grinberg, L. T. (2015). Turning on the light within: subcortical nuclei of the isodentritic core and their role in Alzheimer's disease pathogenesis. *Journal of Alzheimer's Disease: JAD, 46*(1), 17–34. pii:014121K244054223.

Theofilas, P., Polichiso, L., Wang, X., Lima, L. C., Alho, A. T., Leite, R. E., … Grinberg, L. T. (2014). A novel approach for integrative studies on neurodegenerative diseases in human brains. *Journal of Neuroscience Methods, 226*, 171–183. http://dx.doi.org/10.1016/j.jneumeth.2014.01.030.

Trulson, M. E., & Jacobs, B. L. (1979). Raphe unit activity in freely moving cats: correlation with level of behavioral arousal. *Brain Research, 163*(1), 135–150. pii:0006-8993(79)90157-4.

Ursin, R. (1976). The effects of 5-hydroxytryptophan and L-tryptophan on wakefulness and sleep patterns in the cat. *Brain Research, 106*(1), 105–115. pii:0006-8993(76)90076-7.

Vazquez, J., & Baghdoyan, H. A. (2001). Basal forebrain acetylcholine release during REM sleep is significantly greater than during waking. *American Journal of Physiology.Regulatory, Integrative and Comparative Physiology, 280*(2), R598–R601.

Vincent, S. R., Satoh, K., Armstrong, D. M., Panula, P., Vale, W., & Fibiger, H. C. (1986). Neuropeptides and NADPH-diaphorase activity in the ascending cholinergic reticular system of the rat. *Neuroscience, 17*(1), 167–182.

Zarate, C. A., Jr., Payne, J. L., Singh, J., Quiroz, J. A., Luckenbaugh, D. A., Denicoff, K. D., … Manji, H. K. (2004). Pramipexole for bipolar II depression: a placebo-controlled proof of concept study. *Biological Psychiatry, 56*(1), 54–60. http://dx.doi.org/10.1016/j.biopsych.2004.03.013.

Zeitzer, J. M., Maidment, N. T., Behnke, E. J., Ackerson, L. C., Fried, I., Engel, J., Jr., & Wilson, C. L. (2002). Ultradian sleep-cycle variation of serotonin in the human lateral ventricle. *Neurology, 59*(8), 1272–1274.

Chapter 21

Apathy: Frontal and Basal Ganglia Circuits

Pongsatorn Paholpak
Khon Kaen University, Khon Kaen, Thailand; University of California Los Angeles, Los Angeles, CA, United States

Mario F. Mendez
University of California Los Angeles, Los Angeles, CA, United States; Greater Los Angeles VA Healthcare System, Los Angeles, CA, United States

INTRODUCTION

Apathy is a prominent and important behavioral manifestation of many neurological and psychiatric diseases. Apathetic patients lack feeling or emotion and show a lack of interest or concern. Apathetic behavior can be a difficult problem among patients with neuropsychiatric illnesses such as depression, schizophrenia, traumatic brain injury (TBI), stroke, Parkinson disease (PD), and the dementing illnesses. Apathy can also result from neuropsychiatric interventions such as antipsychotic medications and psychosurgery, including cingulotomies and historically, prefrontal leukotomies. We can now begin to unravel the underlying molecular and neuroanatomical basis of this behavioral disorder with the rapid developments in neuroscience. This chapter discusses the definition and diagnosis of apathy, its molecular and neuroanatomical basis, and the manifestations in neuropsychiatric diseases.

DEFINITION AND DIAGNOSIS OF APATHY

Etymologically, "apathy" is derived from Greek *apathēs*, from *a* ("without") + "pathos" (suffering or emotion); it literally refers to a state of lack of emotion. Rather than a general loss of emotion, this chapter focuses on apathy as a disorder of motivation and initiation not attributable to diminished level of consciousness, and based on distinct clinical phenotype and neural substrates. Marin (1991), who originally proposed apathy as an independent neuropsychiatric syndrome, defined it as a disorder of motivation which leads to a loss of interest and a loss of emotional responses toward goal-directed behavior. Levy and Dubois (2006) further defined this impairment in motivation as involving three different domains: emotional-affective processing, cognitive processing, and autoactivation. Besides an emphasis on reduced motivation and goal-directed behavior, other authors (Robert et al., 2009; Starkstein, Ingram, Garau, & Mizrahi, 2005) have stressed deficits in emotional reactivity (Oyebode FSACP, 2008) and include a broader range of behaviors, for example, restricted response to important life events, emotional indifference, and a lack of concern about one's own health or functional status.

Although apathy is not included as a separate syndrome in the *Diagnostic and Statistical Manual of Mental Disorders*, Fifth Edition (American Psychiatric Association APADSMTF, 2013) and the *International Classification of Diseases*, 10th Revision (World Health Organization, 1992), apathetic behaviors can occur in the context of multiple neurological and psychiatric conditions (Levy & Dubois, 2006). Apathy is arguably a central feature of other behavioral symptoms such as depression and anhedonia (a lack of pleasure). Associated apathetic behaviors include anosodiaphoria (lack of concern), anosognosia (lack of knowledge, usually for disease), alexithymia (inability to recognize one's emotions), and abulia (a lack of behavioral initiation) or akinetic mutism (Feil, Razani, Boone, & Lesser, 2003; Kirsch-Darrow, Marsiske, Okun, Bauer, & Bowers, 2011; Marin, Butters, Mulsant, Pollock, & Reynolds, 2003; Nakaaki et al., 2008; Starkstein et al., 2005, 2009). In addition, apathy has strong association with cognitive decline, particularly with frontal lobe pathology and among patients with dementia (Eslinger, Moore, Antani, Anderson, & Grossman, 2012; Landes, Sperry, & Strauss, 2005;

Genomics, Circuits, and Pathways in Clinical Neuropsychiatry. http://dx.doi.org/10.1016/B978-0-12-800105-9.00021-4

Mortby, Maercker, & Forstmeier, 2012; Rapoport et al., 2001). There is evidence of some degree of apathy among 38–58% of community-dwelling elderly populations with no cognitive impairment, and apathy can occur among healthy elderly people who have subtle changes in frontal white matter integrity and volume (Grool et al., 2014). In sum, apathy can be a specific manifestation of a neuropsychiatric disorder or a secondary symptom associated with other related neuropsychiatric symptoms or cognitive impairment.

CLINICAL VARIANTS, DIAGNOSIS, AND MEASUREMENT OF APATHY

In general, patients with apathy will have a significant reduction in initiating goal-directed behaviors along with decreased psychomotor speed, a lack of emotional expression, and a lack of interest. The associated apathetic behaviors need further definition.

1. Anosodiaphoria is a lack of emotional insight for the significance of a neurological deficit or a brain lesion. Patients might be aware that they have a certain disability from neurological deficit, but they fail to show concern about it. Anosodiaphoria is associated with frontal or parietal dysfunction, particularly from disturbances in the right hemisphere (Mendez & Shapira, 2011; Prigatano, 2013).
2. Anosognosia is a deficit of self-awareness regarding personal illness. Patients with this clinical variant do not consciously know that they have an illness, disability, or neurological deficit. This symptom usually occurs after the lesion at right parietal lobe, and in combination with neglect (Jenkinson, Preston, & Ellis, 2011).
3. Alexithymia is an inability to identify and appreciate one's own emotions. The patients may have a severe impairment in emotional awareness that finally affects interpersonal relationships and social interaction. There may be difficulty in appreciating the emotions of others, as well as the self, with a lack of empathy and decreased emotional responsiveness. Individuals with alexithymia may have an impairment of semantic associations between feelings and the bodily sensations of emotional arousal (Ramirez et al., 2001).
4. Abulia is a lack of motivation specifically for action or motor initiation that leads to global underactivity. Abulia may lie on a spectrum of motivational impairment; it is more severe than apathy but less extreme than akinetic mutism (Ghoshal, Gokhale, Rebovich, & Caplan, 2011; Vijayaraghavan, Krishnamoorthy, Brown, & Trimble, 2002).
5. Akinetic mutism (Athymhormia) (Habib, 2004) is the most severe form of apathetic behavior (Levy & Dubois, 2006). These patients have an almost depleted will and motivation. They tend not to move (akinesia) or speak (mutism). Although the patients demonstrate an extreme reduction in most motor functions, including movement, facial expressions, and verbal output, they may occasionally track with their eyes and utter a word or short phrase. Some of these patients are even unresponsive to painful stimuli (Ackermann & Ziegler, 1995; Demirtas-Tatlidede, Bahar, & Gurvit, 2013). This behavioral syndrome has multiple overlapping features with severe catatonia, which occurs in many psychiatric conditions.

MEASUREMENT AND RATING SCALE

The clinical diagnosis of apathy usually relies on thorough neuropsychiatric interviews and mental status examinations. Actigraph, a device for motor activity monitoring worn by the patients, may be used as a proxy for quantification of psychomotor activity, but it lacks the sensitivity to detect differences among healthy control subjects and mild and moderate apathetic patients (Muller, Czymmek, Thone-Otto, & Von Cramon, 2006). The most widely used measurement scales for apathy are the Apathy Evaluation Scale (AES) and its variations, Apathy Scale 14 items, 10 items, and 7 items. These scales have good validity and reliability (Clarke et al., 2011). Other apathy measures used in cognitively impaired population include the Dementia Apathy Interview and Rating and the Apathy Inventory (Clarke et al., 2011; Leontjevas et al., 2012; Robert et al., 2002). The Lille Apathy Rating Scale assesses apathy, particularly among patients with PD (Dujardin, Sockeel, Carette, Delliaux, & Defebvre, 2013; Dujardin, Sockeel, Delliaux, Destee, & Defebvre, 2008; Sockeel et al., 2006), and the Irritability–Apathy Scale is useful for patients with Huntington disease (HD) and Alzheimer disease (AD) (Clarke et al., 2011). Other psychometric scales that can measure apathy from its subscale include the Brief Psychiatric Rating Scale, Positive and Negative Syndrome Scale, Scale for the Assessment of Negative Symptoms, Frontal System Behavior Scale (FrsBe), Key Behavior Change Inventory, and Neuropsychiatric Inventory (NPI) (Oyebode FSACP, 2008).

MOLECULAR AND NEUROANATOMICAL BASIS OF APATHY

Apathy can result from damage to prefrontal areas of the brain. Regardless of the pathology, lesions in the anterior cingulate cortex (ACC) or adjacent ventromedial prefrontal cortex (PFC) (vmPFC) are highly associated with apathy, and injury to the region of the orbitofrontal cortex (OFC) can change a patient's personality from high-achieving, exuberant, and outgoing to nonproductive, reticent to action, and lacking spontaneity and interest in socialization (Peters et al., 2006). Patients with frontotemporal dementia (FTD), which results in early neurodegenerative in the vmPFC and other dementia syndromes that involve these prefrontal areas, usually seek medical attention because of inertia and executive dysfunction (Eslinger & Damasio, 1985) and decreased emotional reactivity (Boone, Miller, Swartz, Lu, & Lee, 2003; Mendez et al., 2006; Rosen et al., 2004). Apathetic behaviors in patients with AD, mild cognitive impairment, schizophrenia, and bipolar disorder may result from pathology especially affecting the dorsal part of the ACC (Bonelli & Cummings, 2007).

In addition to damage to prefrontal areas of the brain, apathy can result from disorders of the basal ganglia (BG), such as PD, HD, and progressive supranuclear palsy. Patients with these neurological conditions often exhibit cognitive inertia, executive dysfunction, and psychomotor slowness and may exhibit blunt facial expressions with reduction of emotional reactivity (Levy & Dubois, 2006). Although some meta-analysis studies of poststroke apathy failed to document an association with a specific location (van Dalen, Moll van Charante, Nederkoorn, van Gool, & Richard, 2013; Tang et al., 2013), specific reports of patients with stroke or other lesions describe apathy from lesions in the bilateral caudate nuclei, globus pallidus interna, and medial-dorsal thalamus (Adam et al., 2013; Quattrocchi & Bestmann, 2014; Rochat et al., 2013; Singh, Mahgoub, & Klimstra, 2011; Vijayaraghavan, Vaidya, Humphreys, Beglinger, & Paradiso, 2008).

PREFRONTAL CORTEX IN PRIMATES

Much of our understanding of the brain's motivation and reward systems is based on results from studies in nonhuman primates. Those areas include ACC (Brodmann's areas 24, 25, and 32), vmPFC (11, 10, and 32), and OFC (areas 11, 12, 13, and 14). Some areas overlap these prefrontal structures, especially 11 and 32 (Haber & Knutson, 2010).

In primates, the amygdala and hypothalamus, limbic areas representing basic instinct and reward processing, are closely connected to these prefrontal structures (Levy & Dubois, 2006). These limbic areas and the brain stem send dopaminergic inputs to ACC and vmPFC (Haber & Knutson, 2010; Kurniawan, Guitart-Masip, & Dolan, 2011). In the "emotional-affective" model, the vmPFC and OFC are responsible for adding affective meaning to the motivational process (Levy & Dubois, 2006) and their neuronal firing rate encodes value to stimuli. The OFC also encodes a wide range of variables including positive and negative expected outcomes (Schultz, 2000), the cost of achieving an outcome, and the chance of being successful with the choice of an outcome (Wallis & Kennerley, 2011). In addition, the ACC heavily connects with the supplement motor area (SMA) and primary motor area; therefore, it is an area responsible for encoding and selecting action that leads to reward-seeking behavior (Haber & Knutson, 2010; Kurniawan, Guitart-Masip, Dayan, & Dolan, 2013). In monkeys, direct damage to the ACC and adjacent areas can lead to a significant reduction in internally driven motor initiation compared with intact motor behavior triggered by external stimuli (Hadland, Rushworth, Gaffan, & Passingham, 2003; Kennerley & Wallis, 2009; Thaler, Chen, Nixon, Stern, & Passingham, 1995).

PREFRONTAL CORTEX IN HUMANS

Functional neuroimaging techniques such as positron emission tomography (PET) and functional MRI (fMRI) demonstrate the functional state of certain brain regions while patients undergo cognitive paradigms that elucidate motivation and reward. Using these techniques, the vmPFC and OFC demonstrate increased activity from both sensory rewards (eg, pleasant tastes, music, beautiful faces, visual stimuli) and abstract rewards (eg, money from a game) (Aharon et al., 2001; Gottfried, O'Doherty, & Dolan, 2003; Haber & Knutson, 2010; Knutson, Wimmer, Kuhnen, & Winkielman, 2008; O'Doherty, Hampton, & Kim, 2007). Whereas OFC is likely to respond to sensory stimuli and punishment (O'Doherty, Dayan, Friston, Critchley, & Dolan, 2003), a medial subregion in vmPFC is more sensitive to abstract rewards (Haber & Knutson, 2010; Knutson & Cooper, 2005). Taken altogether, in humans, OFC and vmPFC add an emotional component to the context, a necessary part of the drive to initiate behavior. Thus, focal lesion in these brain regions can decrease the motivational and reward value of stimuli and result in decreased emotional reactivity and "emotional-affective" apathy (Levy & Dubois, 2006).

The ACC may have diverse roles in higher cognitive function and motor control. The ACC is an area involved in choice prediction (Wallis & Kennerley, 2011), conflicts monitoring, and encoding action-reward associations (Croxson, Walton, O'Reilly, Behrens, & Rushworth, 2009; Haber & Knutson, 2010; Kurniawan et al., 2013). Moreover, the ACC projects to striatum, particularly caudate and ventral striatum, and this ACC–BG connection heavily implies that the ACC has a support role in initiating and selecting motor actions. Patients with lesions in the ACC can present with the "auto-activation" apathy subtype with decreased psychomotor activity, delayed initiation, and hesitation (Levy & Dubois, 2006), and event akinetic mutism when the condition is severe (Eslinger & Damasio, 1985; Grunsfeld & Login, 2006; Njomboro, Deb, & Humphreys, 2012).

The dorsolateral PFC (DLPFC) (Brodmann's areas 9 and 46) is an area representing executive function. Decreases in quantity and quality of executive function such as poor planning, hesitation, and slowness in rule-finding, set-shifting, and memory retrieval occur in many patients with DLPFC lesions (Levy & Dubois, 2006). Lesions that cause disrupted interconnections between the DLPFC and the caudate nucleus may cause deficits in motivation and goal-directed behavior from executive dysfunction (Haber & Knutson, 2010). vmPFC and OFC are responsible for adding context-dependent value to stimuli. The ACC supports initiating and selecting action—outcome association, and the DLPFC helps manage executive-choice of action. All are crucial steps for motivation.

MOTIVATION AND REWARD SYSTEM: BASAL GANGLIA STRUCTURES AND CORTICOSTRIATAL LOOP

The BG are complex subcortical nuclei that can be segmented into dorsal and ventral aspects. The dorsal aspect of the BG include the caudate nucleus and putamen; the ventral aspect includes the nucleus accumbens (NAcc), globus pallidus interna (GPi) and globus pallidus externa (GPe), the substantia nigra pars compacta and pars reticulate, and the subthalamic nucleus (STN). Anatomically, they receive signals from different frontal cortices and project to the thalamic nuclei, which in turn feedback to the prefrontal cortex. This "corticostriatal loop" has a critical role in effecting motivation and reward.

Afferent signals from PFC, amygdala, and thalamus to BG structures are major contributing factors in the reward and motivation system (Humphries & Prescott, 2010). The act of choosing one action over another and appreciating outcomes results in a reinforcement-activated blood oxygen level—dependent (BOLD) signal in both the dorsal and ventral striatum (O'Doherty, 2004). Although the dorsal and ventral striatum resemble each other in many regards (Kurniawan et al., 2013, 2011), there are distinguishable neural circuits between the two divisions. Whereas both receive input from the PFC, thalamus, and dopaminergic neurons in the ventral tegmentum (VTA), the ventral striatum uniquely receives input from the hippocampus and amygdala complex. The neurons responsible for effortful processing, particularly for choosing low compared with high physical effort options (Kurniawan et al., 2013, 2011), lie in the dorsal striatum. In contrast, mental effort tasks tend to be associated with the ventral striatum (Haber, 2011). The ventral striatum is activated during reward anticipation and the prediction of benefits from action (Croxson et al., 2009), even before the action is performed (Kurniawan et al., 2013), and either unilateral or bilateral lesion at globus pallidus, especially GPi, can cause apathetic behaviors including marked reduction in speech, movement, emotional expression, and motivation (Adam et al., 2013; Singh et al., 2011; Vijayaraghavan et al., 2008).

STRUCTURAL AND FUNCTIONAL CONNECTIVITY IN APATHY

Advances in neuroimaging, particularly diffusion tensor imaging (DTI) from MRI data and fMRI, have disclosed disconnections of networks between brain regions in apathy. The results of DTI studies show a correlation between apathetic behaviors and white matter disintegration in either the corticolimbic or corticobasal loops. One study found that the severity of apathy among 72 healthy adults, significantly among women, was correlated with decreased fractional anisotropy (FA), a measure of white matter integrity, in anterior thalamic radiations, superior longitudinal fasciculi, internal capsule, and corpus callosum (Spalletta, Fagioli, Caltagirone, & Piras, 2013). Another study in patients with AD showed a negative correlation with apathy score from NPI and DTI measures of white matter integrity in the left anterior cingulum, right superior longitudinal fasciculus, splenium, body and genu of the corpus callosum, and bilateral uncinate fasciculi (Hahn et al., 2013). A study in a patient with mild cognitive impairment from AD also showed a strong relationship between severity of apathy and white matter abnormalities in the uncinate fasciculus and superior longitudinal fasciculus (Cacciari et al., 2010), as well as white matter in the rostral ACC (Ota, Sato, Nakata, Arima, & Uno, 2012; Tighe et al., 2012) and the right thalamus (Ota et al., 2012). Furthermore, aberrant connections between medial OFC and rostral ACC were associated with the severity of apathy in schizophrenia (Ohtani et al., 2014), and abnormal projections from the vmPFC probably accounted for some of the apathy in HD (Delmaire et al., 2013). Taken altogether, disconnection in the structural connectivity of the corticolimbic system, especially the uncinate fasciculus, and the corticobasal system, especially the cingulum, may impair motivation and lead to apathy.

Whereas DTI analysis demonstrates structural connection via white matter, fMRI shows connectivity based on temporal statistical coincidence of the MRI BOLD signal. Intrinsic connectivity networks (ICN) emerge from the patterns of fluctuated signals of BOLD activity during a task-free resting state, or task-related paradigm and coactivation between brain regions (van den Heuvel & Hulshoff Pol, 2010). Many stable ICNs have been identified, but the three core ICNs are the central executive network (CEN), the default mode network (DMN), and the salience network (SN). These ICNs demonstrate strong relationships in both resting-state and task-related connectivity patterns. The DMN has posterior cingulate cortex (PCC) and medial PFC (mPFC) as key nodes for working memory and attention, especially during task-free and internally focused tasks. The CEN, which has an important role in cognitive demand tasks, has DLPFC and the posterior parietal cortex (PPC) as key nodes (Menon, 2011).

Apathy may be particularly related to dysfunction of the SN. The SN, which is anchored in the dorsal ACC and frontoinsular cortex, contributes to emotional salience by integrating interoceptive autonomic and emotional information (Menon, 2011; Seeley et al., 2007). One study showed that apathetic patients with depressive disorder had lower ICNs involving NAcc with amygdala, caudate, putamen, and thalamus compared with nonapathetic depressed patients (Alexopoulos et al., 2013). Another study found that compared with nonapathetic depressed patients and control subjects, apathetic depressed patients were distinguished by decreased intrinsic SN connectivity, especially on the right hemisphere, and increased connectivity between right anterior insula (AI) node in SN and right DLPFC in CEN (Yuen et al., 2014). The same pattern of decreased intrinsic SN connectivity and increased connectivity between SN and CEN occurred among patients with schizophrenia and negative symptoms (Manoliu et al., 2013). Other studies using fMRI with an economic game paradigm emphasized that the SN, particularly the AI node, may help in reward processing (Gradin et al., 2013), especially in anticipation of reward magnitude (Croxson et al., 2009; Stoppel et al., 2011; Tanaka, Yamada, Yoneda, & Ohtake, 2014), and in switching and linking between CEN and DMN (Palaniyappan, White, & Liddle, 2012). fMRI studies suggest that the SN has a crucial role in reward processing, and therefore a dysfunction within SN or disconnection between SN and other brain regions may cause an impairment in emotion processing, decreased perception of reward, and apathy.

The diagram in Fig. 21.1 integrates all data from both structural and functional studies to summarize areas representing different aspects of apathy.

DOPAMINE REGULATION IN MOTIVATION AND REWARD SYSTEM

Dopamine strongly modulates the interconnections of prefrontal cortex and the corticobasal loop, critical neural correlates in the motivation and reward system. There are projections from dopamine neurons in the VTA to ventral striatum (NAcc), dorsal striatum, PFC, amygdala, and hippocampus, which comprise the mesolimbic system. Dopamine transmission to the PFC and striatum has a major role in reward-seeking behaviors. Most projections from prefrontal and prelimbic cortices, which project to ventral striatum, are involved in adding an affective component to reward, emotional reaction to

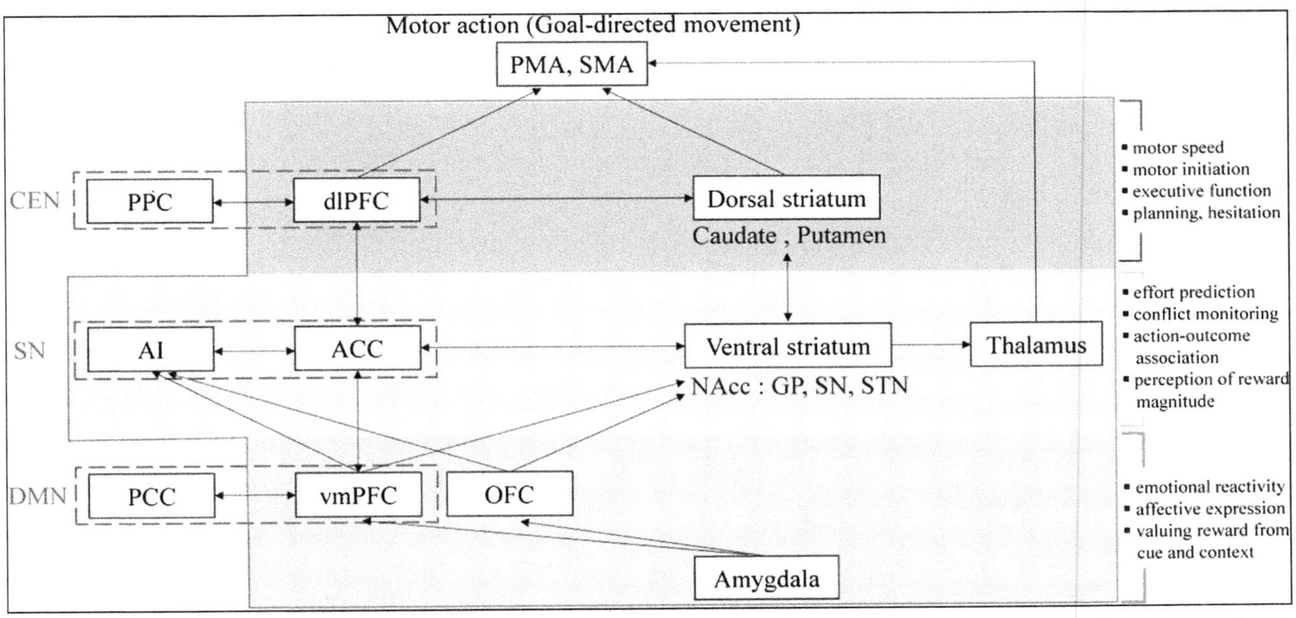

FIGURE 21.1 Brain areas associated with different aspects of apathy. Different brain regions have been associated with different aspects of apathy on the right. *PMA*, premotor area; *SMA*, supplementary motor area; *PPC*, posterior parietal cortex; *dlPFC*, dorsolateral prefrontal cortex; *AI*, anterior insular; *ACC*, anterior cingulate cortex; *NAcc*, nucleus accumbens; *GP*, globus pallidus; *SN*, substantia nigra; *STN*, subthalamic nucleus; *PCC*, posterior cingulate cortex; *vmPFC*, ventromedial prefrontal cortex; *OFC*, orbitofrontal cortex; *CEN*, central executive network; *SN*, salient network; *DMN*, default mode network.

environment, and a sense of reward magnitude; whereas the corticostriatal projections from dorsolateral and cingulate cortices to dorsal and ventral striatum are predominantly involved in more cognitive oriented processes, for example, executive and motor control. These phenomena are largely driven by the dopaminergic pathway with D1 and D2 receptors. Phasic dopamine bursts from cortical regions to striatum, by activating long-term potentiation, promote "acting and performing" via D1 receptor signaling in the "direct" nigrostriatal pathway. Tonic dopaminergic activity activates D2 receptor signaling in the "indirect" striatum–pallidum pathway and halts "moving and executive functioning" by long-term depression of neurons, mainly in mPFC. Therefore, coordination of D1 and D2 activations will modulate striatal plasticity and influence motor movements and cognitive functioning to achieve goal-directed behaviors (Wickham et al., 2013). A deficit in dopaminergic activity in these brain regions usually causes impairments of motivation in one aspect or more. Examples of this model are PD, HD, and schizophrenia, diseases which directly affect dopaminergic activity in SN, STN, and PFC, respectively. These disorders cause not only executive dysfunction but also a reduction in emotion expression, which are symptoms of apathy. These findings suggested that dopamine is a major neurotransmitter that is responsible for motivation-driven behaviors, and a dysregulation in dopaminergic activity may lead to apathetic behaviors.

APATHY IN NEUROPSYCHIATRIC DISEASES

Traumatic Brain Injury

The prevalence of apathy caused by TBI with one or more episode of loss of consciousness ranges from 20% to 70% of patients (Al-Adawi et al., 2004; Landes et al., 2005). This condition is distinct from the 3–9% of apathy reported with chronic traumatic encephalopathy, a neurodegenerative condition that is more specific to professional athletes with repeated exposure to brain trauma regardless of loss of consciousness (Gavett et al., 2011; McKee et al., 2013; Stern et al., 2013; Victoroff, 2013). The discrepancy of prevalence of apathy among TBI studies probably results from selection bias and different measurements of apathy. Some studies indicate that TBI-induced apathy is associated with depression (Andersson, Krogstad, & Finset, 1999; Kant, Duffy, & Pivovarnik, 1998), whereas others do not show such a relationship (Knutson et al., 2014). According to the level of cognitive function, some studies demonstrate an increasing frequency of apathy with greater cognitive impairment (Starkstein & Leentjens, 2008). Apathy also tends to develop in patients with TBI who have severe disability on the Glasgow Outcome Scale (Ciurli, Formisano, Bivona, Cantagallo, & Angelelli, 2011). Other factors associated with apathy after TBI are sleep disturbances, memory deficits, executive dysfunction, and psychomotor slowness (Andersson & Bergedalen, 2002; Rao et al., 2013).

Results from studies on the neuroanatomical correlation of TBI-induced apathy show involvement of the ACC, insulae, and corticobasal loop. One study shows associations between TBI-induced apathy and damage to subcortical areas and the DLPFC (Finset & Andersson, 2000). Both mPFC and DLPFC are associated with a reduction in motivation (Paradiso, Chemerinski, Yazici, Tartaro, & Robinson, 1999). In a study using voxel-based lesion-symptom mapping in TBI, the ACC, insulae, and inferior and middle PFC are more damaged in apathetic veterans compared with nonapathetic control subjects, whereas loss of white matter integrity in the corona radiata and corpus callosum correlated with the severity of apathy (Knutson et al., 2014).

Schizophrenia

Schizophrenia is a major psychiatric disorder which affects patient's perception, thoughts, and behavior. Symptoms in schizophrenia can be positive or negative. Positive symptoms include perceptual disturbances, thought disturbances, and disorganized behaviors. Negative or deficit symptoms include apathy as well as blunted affect, emotional withdrawal, social withdrawal, and lack of spontaneity. Negative symptoms and apathy reduce quality of life, increase functional disability and burden of illness, and worsen long-term outcomes (Chue & Lalonde, 2014). Apathy can occur right after the first episode of psychosis or can manifest later as part of a deficit syndrome. Executive dysfunction can also occur in either first-episode psychosis or as part of chronic psychosis (Faerden et al., 2009). Data from one long-term follow-up study demonstrate that nearly 30% of patients may experience apathy 10 years after the onset of schizophrenia (Evensen et al., 2012).

The negative symptoms of schizophrenia usually correlate with reduced activation or hypometabolism of the PFC (Gruber, Chadha Santuccione, & Aach, 2014). Decreased activation of the ventral striatum is consistently correlated with the severity of apathy and negative symptoms when anticipating a reward from an economical game (Gradin et al., 2013; Juckel et al., 2006; Simon et al., 2010). In apathetic schizophrenics, there is also a higher volume of loss in the ACC compared with nonapathetic schizophrenic people and healthy control subjects (Takayanagi et al., 2013). These studies suggest that apathy among patients with schizophrenia may be associated more with decreased motivation and reward processing (ACC and ventral striatum) versus a less specific executive dysfunction. Among individuals with schizophrenia, white matter disintegration between medial OFC and rostral ACC is associated with anhedonia-asociality and avolition-apathy in one DTI study (Ohtani et al., 2014); whereas another DTI study demonstrates correlations between negative symptoms and abnormalities in white matter tracts between ventral striatum and medial OFC (Bracht et al., 2014). On fMRI studies, negative symptoms in schizophrenia correlate with decreased intrinsic connectivity in SN and increased connectivity between SN and CEN (Manoliu et al., 2013), whereas in another study the negative symptoms correlate with reduced functional connectivity from anterior insula to other brain regions (Gradin et al., 2013). These studies suggest that dysfunction within the SN, and the disconnection between the SN and other networks, may have a crucial role in apathy among people with schizophrenia.

Stroke

Apathy is one of the most important behavioral sequelae of stroke. The prevalence of stroke-induced apathy ranges from 15.2% to 71.1% (Caeiro, Ferro, & Costa, 2013). Apathy can occur as early as the short-term phase (within a day) or as late as 15 months after the stroke event. As many as 41% of patients with apathy after an acute stroke remained apathetic after 1 year of follow-up (Caeiro, Ferro, Pinho, Canhao, & Figueira, 2013). In a study that observed patients up to 5 years after stroke, investigators report that numbers of patients with apathy increased from 26.7% to 38.6% (Brodaty, Liu, Withall, & Sachdev, 2013). The type of stroke, whether ischemic or hemorrhagic, did not influence the incidence of poststroke apathy. Apathy can also result from gradually accumulated subcortical microvascular pathology, for example, cerebral arteriopathy autosomal dominant with subcortical infarcts and leukoencephalopathy (Reyes et al., 2009). Poststroke apathy results in decreased quality of life for patients and increased burden and distress in caregivers (Tang, Lau, Mok, Ungvari, & Wong, 2014). Clinically significant depression can coexist in 10–20% of patients with poststroke apathy, but the severity of depression in apathetic patients is not different from that for nonapathetic patients. Cognitive impairment is more common in patients with apathy than for patients without apathy, and the severity of cognitive impairment in apathetic patients tends to be more severe than in nonapathetic patients (van Almenkerk, Smalbrugge, Depla, Eefsting, & Hertogh, 2014). Many case reports (Ioannidis et al., 2013; Spalletta, Cravello, et al., 2013) and case–control studies emphasize a strong association between poor reward sensitivity and apathetic behaviors with stroke locations involving the left hemisphere or subcortical areas of the corticobasal loop (Hama et al., 2007; Onoda et al., 2011; Rochat et al., 2013). Results from two *meta*-analyses, however, do not show a stronger association between those areas and poststroke apathy compared with other locations (Caeiro, Ferro, & Costa, 2013; van Dalen et al., 2013)

To date, there is only one positive randomized, controlled drug trial for the specific treatment of poststroke apathy. In that study, nefiracetam shows superior efficacy in reducing apathy compared with placebo among patients with poststroke depression (Robinson, Jorge, Clarence-Smith, & Starkstein, 2009). Donepezil has modest efficacy in reducing apathetic behaviors from open-labeled studies, and there are case reports of favorable effects on apathy from bromocriptine, methylphenidate, ropinirole, zolpidem, and selegiline (van Dalen et al., 2013). Escitalopram and problem-solving therapy may be useful in preventing the emergence of poststroke apathy within 3 months after the acute stroke. Data from that study show that participants who received placebo were 3.47 and 1.84 times more likely to develop poststroke apathy compared with patients who received escitalopram and problem-solving therapy, respectively (Mikami et al., 2013).

Parkinson Disease

PD is a neurodegenerative disease that affects the substantia nigra, especially par compacta, and it results in motor and nonmotor symptoms. Patients usually acquire nonmotor symptoms, for example, executive dysfunction, dementia, depression, and apathy in later stages of the disease. Apathy can occur as an independent nonmotor symptom or may be associated with depression, severity of motor symptoms (Santangelo et al., 2013), and dementia (Emre et al., 2007; Dujardin, Sockeel, Delliaux, Destee, & Defebvre, 2009). Up to 47.9% of patients with PD may have apathy without other nonmotor symptoms; however, the prevalence of apathy coexisting with other nonmotor symptoms may be as high as 70% (Santangelo et al., 2014). The neuroanatomy of apathy in PD is linked to cortical atrophy in the inferior frontal gyrus, insular, right ACC, and right precuneus in one study using voxel-based morphometry (VBM) (Reijnders et al., 2010). Another study, using shape analysis from spherical parameterization, reports that the degree of apathy correlates with volume loss in the ventral striatum and head of caudate nucleus (Carriere et al., 2014). Data from a study using ^{11}C-raclopride PET concludes that patients with PD who have apathy have lower dopamine transmission in DLPFC, OFC, ACC, and PCC (Thobois et al., 2010). In conclusion, studies show that apathy among patients with PD is a complex behavior involving many brain regions. This circuit includes the inferior frontal (vmPFC and OFC), DLPFC, insular, ACC, ventral striatum, and caudate nuclei, all of which are related to the reward and motivation pathway. However, to date there is a lack of data about neural correlates of apathy from connectivity studies. Further evidence from fMRI and DTI analysis of apathy in PD is needed.

Treatment of PD with deep brain stimulation (DBS), particularly of the STN but also the globus pallidus (GP), can cause apathy. Nonmotor fluctuations in everyday life (Thobois et al., 2010) and preoperative dyskinesia (Higuchi et al., 2014) are independent predictors for the development of postoperative apathy. Among patients with PD who underwent DBS, it is possible that apathy results from dopamine dysregulation in the mesolimbic pathway but not as a direct effect of stimulation of STN. Because apathy usually occurs after 3 months of DBS and after the cessation of dopaminergic medication (Castrioto, Lhommee, Moro, & Krack, 2014), it may be difficult to distinguish between apathy induced by the cessation of dopaminergic medication after DBS, or the direct effect of stimulation of limbic parts of the STN (Drapier et al., 2006). One study directly compared the prevalence of apathy between the best medical treatment and DBS and the best medical treatment without DBS; the results showed no difference in the prevalence of apathy between the groups (Schuepbach et al., 2013). Moreover, data from drug trials, both controlled and open, demonstrate that apathy is reversible after the introduction of dopaminergic medication such as piribedil (Thobois et al., 2013) and ropinirole (Czernecki et al., 2008).

Medications may decrease apathy in PD. Data from a controlled trial with a small number of patients showed that L-dopa may increase motivation in some apathetic patients with PD (Dujardin et al., 2009). A meta-analysis showed that pramipexole may be useful in helping motivational symptoms according to the Unified Parkinson Disease Rating Scale (Leentjens et al., 2009), and rotigotine may improve certain apathetic behaviors measured by the Non-Motor Symptoms Scale (Ray Chaudhuri et al., 2013). In randomized controlled trials with small numbers of patients with PD who underwent STN-DBS, piribedil (Thobois et al., 2013) and methylphenidate (Moreau et al., 2012) may reduce apathy and depression. Results from a multicenter, randomized, placebo-controlled trial also suggest that rivastigmine may be helpful in alleviating moderate to severe symptoms of apathy in patients with PD who do not have dementia or depression and who are already optimized with a dopaminergic agent (Devos et al., 2014).

Huntington Disease

HD is an inheritable trinucleotide-repeat neurodegenerative disease that predominantly affects striatum in BG. Patients carrying the mutation display chorea in early stages followed by akinesia and sometimes dystonia in late stages. Because both caudate and putamen usually become atrophic over the course of the disease, apathy can occur as a neuropsychiatric symptom in HD along with irritability/aggression, depression, and psychosis. Apathy and neuropsychiatric symptoms tend to be more common as the disease progresses (Tabrizi et al., 2013). One large study with 1993 HD mutation participants reported moderate to severe apathy, measured by the Unified HD Rating Scale, among 11.8% patients in stage 1 and 54.6% patients in stages 4 to 5 (van Duijn et al., 2014). The prevalence can be as high as 76% in symptomatic patients (van Duijn, Kingma, & van der Mast, 2007). Even in asymptomatic stages, many pre-HD participants (asymptomatic patients with mutation) exhibit higher apathy scores compared with normal control subjects (Duff et al., 2010). Many studies consistently demonstrate that apathetic behaviors in HD are not correlated with depression (Naarding, Janzing, Eling, van der Werf, & Kremer, 2009). Apathy can occur together with executive dysfunction and other neuropsychiatric symptoms (Rosenblatt, 2007). Although the prevalence of apathy correlates with the duration of the illness, the stability of apathy over time is unclear. One study shows that 41% of patients with HD who are diagnosed with apathy at baseline may be free of apathy after 2 years' follow-up (Reedeker et al., 2011), whereas another study from the same group of authors finds no significant change in the prevalence and severity of apathy after 2 years of observation (van Duijn et al., 2014).

Neuroimaging studies in HD provide evidence of reduced dopamine in patients who are either pre-HD or symptomatic. Both striatal D1 and D2 receptor levels are reduced by 45–50% and the loss progresses by 3–5% per year. Because striatal networking for motivation and reward processing are mainly driven by dopaminergic activity, apathy can result from lower dopamine levels in this circuitry (Ginovart et al., 1997; van Oostrom et al., 2009). Regarding neural correlates with apathy in HD, one study explores structural connectivity and apathetic symptoms. Using problem behaviors assessment for HD, this study finds that the severity of apathy correlates with the degree of loss of white matter integrity in the gyrus rectus and vmPFC (Delmaire et al., 2013). The authors hypothesize that this location is responsible for reward and motivation, or reward insensitivity.

Alzheimer Dementia

AD is the most common neurodegenerative disease. The typical process of AD usually occurs in elderly individuals, starting with memory impairment (amnesia) and progressing to other neurocognitive domains. This pattern of deterioration reflects the neuropathology, amyloid plaques and tau-positive neurofibrillary tangles, usually starting in the medial entorhinal cortex and hippocampus, and later spreading to other brain regions. Apathy is one of the most frequent behavioral and psychological symptoms in dementia present in AD. Apathy is usually associated with a higher severity of the disease, with reports of prevalence of apathy in as many as 72% of patients with AD (Mega, Cummings, Fiorello, & Gornbein, 1996). Using single-photon emission computed tomography and PET, many studies show a correlation of atrophy with hypoperfusion and hypometabolism in the ACC and cingulate gyrus. Some studies also demonstrate correlations of atrophy with changes in the OFC or DLPFC. On structural MRI data, there is a significant correlation between apathy and atrophy in the ACC and OFC. In addition, the presence of white matter hyperintensities in frontal areas correlates with the presence of apathetic behaviors (Theleritis, Politis, Siarkos, & Lyketsos, 2014).

The first DTI study of apathy in mild AD showed a correlation of apathy with decreased measures of white matter integrity in the left anterior cingulum (Kim et al., 2011). Another report in a patient with amnestic mild cognitive impairment found strong relationships between the severity of apathy and changes in the uncinate fasciculus, superior longitudinal fasciculus (Cacciari et al., 2010), and white matter in rostral ACC (Ota et al., 2012; Tighe et al., 2012) and right thalamus (Ota et al., 2012). A further study shows a negative correlation between apathy scores, using NPI, and the left anterior cingulum, right superior longitudinal fasciculus, splenium, body and genu of the corpus callosum, and bilateral uncinate fasciculus (Hahn et al., 2013).

In studies of functional connectivity, alterations in DMN and SN are major findings in AD and other neurodegenerative diseases (Seeley, Crawford, Zhou, Miller, & Greicius, 2009). One study correlates increased activity of the SN with hyperactivity syndrome (agitation, irritability, aberrant motor behavior, euphoria, and disinhibition) (Balthazar et al., 2014). However, investigators have yet to link apathy to a specific network in AD.

There have been a number of studies of treatments for apathy in AD. The Apathy in Dementia Methylphenidate Trial, a 6-week, randomized, placebo-controlled trial with methylphenidate, reports improvement in apathy on the NPI but not the AES, in 21% of the treated patients, compared with 3% in the placebo group (Rosenberg et al., 2013). Modafinil, another dopamine modulating agent, failed to improve symptoms of apathy significantly in a randomized controlled trial (Frakey, Salloway, Buelow, & Malloy, 2012). Donepezil may be helpful in reducing apathy as a secondary outcome in one randomized controlled trial (Gauthier et al., 2002). In another study, donepezil does not reduce apathetic behaviors but is better than placebo in preventing the emergence of apathy (Waldemar et al., 2011). Finally, data from a large open-labeled study show that rivastigmine can reduce apathetic behaviors over 12 months of treatment (Gauthier, Juby, Dalziel, Rehel, & Schecter, 2010).

Frontotemporal Dementia

Frontotemporal lobar degeneration (FTLD) is a group of neurodegenerative disorders that have a distinct degenerative pattern. Behavioral variant frontotemporal dementia (bvFTD), with early involvement in medial frontal lobes, can result in a significant change in behavior and personality, especially apathy, which occurs as an early manifestation in up to 84% of patients (Rascovsky et al., 2011). At the same stage of disease, patients with bvFTD have a higher rate of neuropsychiatric symptoms, including apathy, compared to patients with AD (Joshi et al., 2014). Apathy in bvFTD is associated with less daytime activity measured by ActiGraph, and a higher caregiver burden (Merrilees et al., 2013). Regarding its neuroanatomical correlation, VBM data in FTLD patients from one study showed an association between apathy, measured by the FrsBe, and reduced gray matter density in bilateral DLPFC (especially on the right), right OFC, right temporoparietal junction, anterior cingulate cortex, and right putamen (Zamboni, Huey, Krueger, Nichelli, & Grafman, 2008). Results from another study also indicated dorsal ACC and DLPFC as areas associated with apathetic behaviors (Massimo et al., 2009). Using the AES, the right caudate, posterior middle and posterior inferior temporal gyri, and right temporoparietal junction were associated with severity of apathy (Eslinger et al., 2012). In an MRI study using tensor-based morphometry, the emotional apathy score from the Scale for Emotional Blunting (Joshi et al., 2014), correlated with volume loss in the right anterior temporal region (Lee et al., 2014). Results from two FDG-PET studies, which used the apathy subscale from the NPI, showed conflicting results. One study failed to show a significant correlation of severity of apathy with a hypometabolic pattern (Peters et al., 2006), whereas the other found decreased glucose uptake in bilateral frontomedial and DLPFC (Franceschi et al., 2005). DTI analysis of data from our own cohort showed that the emotional apathy subscale from FrsBe correlated with decreased FA value of uncinated fasciculus (unpublished data). These results demonstrate a trend of association of dysfunctions in amygdala, DLPFC, OFC, ACC, vmPFC, striatum, and temporoparietal junction with apathy in FTD.

GENETICS STUDIES OF APATHY

There are limited numbers of studies on the genetics of apathy. Although dopaminergic neurons have been the center of attention in studies on the motivation system for many years, a correlation between dopamine-related genes and severity of apathy is not established. The only positive genetic association came from a study of 963 healthy participants, 213 of whom had apathy, which showed an association between the single nucleotide polymorphism (SNP) in the catechol-*O*-methyltransferase (COMT) gene (rs4680) and a lower risk of apathy (Mitaki et al., 2013). The authors concluded that the SNP in the COMT gene leads to a reduction in COMT activity and increased dopamine in the PFC. Those with apathy also had more severe depression, so it was possible that this gene affected not only motivation but also the mood state (Mitaki et al., 2013) (Table 21.1).

Another way to understand the genetic aspects of apathy is to study diseases associated with apathy. The next section summarizes the genetic association with apathy in each disease. Details about its prevalence, clinical association, and neuroanatomical correlation are described in the previous section on apathy in neuropsychiatric diseases.

Huntington Disease

The Huntingtin gene is responsible for HD. It has a segment containing a cytosine—adenine—guanine (CAG) trinucleotide repeat. Usually an individual who has fewer than 36 repeats will not develop symptoms whereas a person with 36 to 38 repeats may or may not develop symptoms (Walker, 2007). In asymptomatic carriers who are CAG expansion positive (repeat $\geq$ 39), there is a higher level of apathetic behaviors on the FrsBe apathy subscale compared with normal control subjects. In that study, although the degree of apathy in carriers did not meet a pathological threshold, the findings reflected a major genetic influence for apathetic behaviors in HD even in the asymptomatic phase (Duff et al., 2010).

Parkinson Disease and Parkinsonism

Most PD is idiopathic; only a minority has an inheritable form of PD. To date, no specific gene has been linked to apathy in PD; however, genetic loci have been associated with specific phenotypes of PD. A dominantly inherited late-onset PD with Lewy body pathology, which is clinically indistinguishable from idiopathic PD, can occur with mutation of SNCA and LRK2. Mutations in PARK2 and PINK1 may cause recessively inherited, early-onset or X-linked atypical parkinsonism, whereas mutations in microtubule-associated protein tau (MAPT) gene can cause familial parkinsonism and FTD (Lin & Farrer, 2014).

TABLE 21.1 Strong Candidate Genes Associated With Apathy in Pertinent Diseases

Disease	Gene	Location	Full Name
AD	PSEN1	14q24.3	Presenilin 1
	PSEN2	1q42.13	Presenilin 2
	APP	21q21.3	Amyloid precursor protein
	APOε4	19q13.32	Apolipoprotein ε4
FTD	C9ORF72	9p21.2	Chromosome 9 open reading frame 72
	GRN	17q21.31	Progranulin
	MAPT	17q21.31	Microtubule-associated protein tau gene
HD	HTT	4p16.3	Huntingtin
PD	GBA	1q22	Glucosidase beta acid
	SYT11/RAB25	1q21	Synaptotagmin
	RAB25	1q22	RAS-associated protein
	LRRK2	12q12	Leucine-rich repeat kinase 2
	SNCA	4q22.1	Synuclein alpha
	PARK2	6q26	Parkin
	PINK1	1p36.12	PTEN-induced putative kinase
	MAPT	17q21.31	Microtubule-associated protein tau
Schizophrenia	DRD2	11q23.2	Dopamine receptor D2
	DGS/COMT	22q11.21	DiGeorge syndrome/Catechol-*O*-methyltransferase

Alzheimer Dementia

Most AD is sporadic. However, 1−15% of early-onset AD has an autosomal dominant inheritance caused by one of three genes; amyloid precursor protein, presenilin 1, or presenilin 2 (Jarmolowicz, Chen, & Panegyres, 2014). In AD, there is no clear association of any of these genes with apathy; however, apolipoprotein ε4 carriers may show a higher frequency of apathy than do noncarriers (Monastero et al., 2006). One study demonstrated a negative association of apathy with variable number of tandem repeats of the serotonin transporter VNTR 5HTTLPR (Proitsi et al., 2012). The results from another study found that T-allele carriers of the 3′ untranslated region prion-like protein had more apathy along with many other neuropsychiatric symptoms (Flirski et al., 2012).

Frontotemporal Dementia

As noted, apathy is one of the most common manifestations in an early stage of bvFTD (Rascovsky et al., 2011). Most FTD cases are sporadic; currently only 10−15% of patients with bvFTD have a known mutation, which has autosomal dominant transmission (Snowden et al., 2013). These familial forms usually result from mutations in the progranulin (GRN) gene, in noncoding expanded hexanucleotide repeats in the chromosome 9 open reading frame 72 (C9ORF72), or in the MAPT gene. Rarer genetic forms of bvFTD involve valosin-containing protein, charged multivesicular body protein 2B, TAR DNA-binding protein, or Ubiquilin 2 (Jarmolowicz et al., 2014). GRN encodes the progranulin protein, a growth factor involved in wound healing and inflammation. As mentioned in PD, MAPT mutations can occur in families with FTD and parkinsonism. C9ORF72, or the presence of expanded GGGGCC hexanucleotide repeats, is also highly associated with motor neuron disease, particularly amyotrophic lateral sclerosis. Carriers of C9ORF72 may display clinical syndromes of bvFTD or FTD-motor neuron disease and may have more apathetic behaviors than do noncarriers (Snowden et al., 2013; Sha et al., 2012). One review indicated that GRN mutation was highly specific to apathy and social withdrawal (Seelaar, Rohrer, Pijnenburg, Fox, & van Swieten, 2011); another review concluded that phenotypes from these mutations were heterogeneous (Petkau & Leavitt, 2014).

Schizophrenia

Investigators have linked a number of genetic mutations to schizophrenia; however, most of these loci studies yield conflicting results. These studies did not investigate apathy as an isolated syndrome; rather, they integrated apathy with other negative symptoms. One genetic condition that is probably highly associated with negative symptoms is the 22q11.2 deletion syndrome (DiGeorge syndrome/velocardiofacial syndrome) (Schneider et al., 2014). This condition is also associated with the Val/Met polymorphism in the COMT gene (Schneider et al., 2012). An investigation in a Japanese population suggested that there was an association between the negative symptoms with the Cys311 allele in the DRD2 gene, and this result has since been replicated in a Malaysian cohort (Zahari, Teh, Ismail, & Razali, 2011). In addition, the Val-allele of rs4680 and rs4818 polymorphisms in the COMT gene may correlate with negative symptoms in Chinese and Korean populations (Kang et al., 2012; Li et al., 2012). In contrast, results from a meta-analysis from two genome-wide association studies did not show an association between negative symptoms and these genes (Xu et al., 2013).

At this point, linkages between apathy and human genes are not completely established. There are still many gaps in knowledge that require future research. One question is how dopamine genes are associated with apathy in the general population. Another mystery is why specific mutations in dopamine gene do not always affect motivation and apathy. Defining the interactions between apathy-related genes is important and can clarify the genetic relationship to apathy.

SUMMARY

Executing an intact goal-directed behavior requires multiple-step cognitive processing. Therefore, different apathetic behaviors result from different impaired processes along motivation−action−reward pathways. Critical regions in the PFC and BG have to work in concert as a "corticobasal loop" to evaluate motivation and reward, value the stimuli, recruit a higher cognitive function to plan for an action, choose the appropriate amount of effort to achieve a certain goal, and give feedback on the action. Future research using an integrated model between genetic findings, neural correlates, and neurotransmitters may be helpful in understanding this heterogeneous syndrome in a more comprehensive way.

REFERENCES

Ackermann, H., & Ziegler, W. (1995). Akinetic mutism—a review of the literature. *Fortschritte der Neurologie-Psychiatrie, 63*(2), 59—67.

Adam, R., Leff, A., Sinha, N., Turner, C., Bays, P., Draganski, B., ... Husain, M. (2013). Dopamine reverses reward insensitivity in apathy following globus pallidus lesions. *Cortex; A Journal Devoted to the Study of the Nervous System and Behavior, 49*(5), 1292—1303.

Aharon, I., Etcoff, N., Ariely, D., Chabris, C. F., O'Connor, E., & Breiter, H. C. (2001). Beautiful faces have variable reward value: fMRI and behavioral evidence. *Neuron, 32*(3), 537—551.

Al-Adawi, S., Dorvlo, A. S., Burke, D. T., Huynh, C. C., Jacob, L., Knight, R., ... Al-Hussaini, A. (2004). Apathy and depression in cross-cultural survivors of traumatic brain injury. *The Journal of Neuropsychiatry and Clinical Neurosciences, 16*(4), 435—442.

Alexopoulos, G. S., Hoptman, M. J., Yuen, G., Kanellopoulos, D., Seirup, J. K., Lim, K. O., & Gunning, F. M. (2013). Functional connectivity in apathy of late-life depression: a preliminary study. *Journal of Affective Disorders, 149*(1—3), 398—405.

van Almenkerk, S., Smalbrugge, M., Depla, M. F., Eefsting, J. A., & Hertogh, C. M. (2014). Apathy among Institutionalized stroke patients: prevalence and clinical correlates. *The American Journal of Geriatric Psychiatry: Official Journal of the American Association for Geriatric Psychiatry, 23*(2), 180—188.

American Psychiatric Association APADSMTF. (2013). *Diagnostic and statistical manual of mental disorders: DSM-5*. Washington, DC: American Psychiatric Association.

Andersson, S., & Bergedalen, A. M. (2002). Cognitive correlates of apathy in traumatic brain injury. *Neuropsychiatry, Neuropsychology, and Behavioral Neurology, 15*(3), 184—191.

Andersson, S., Krogstad, J. M., & Finset, A. (1999). Apathy and depressed mood in acquired brain damage: relationship to lesion localization and psychophysiological reactivity. *Psychological Medicine, 29*(2), 447—456.

Balthazar, M. L., Pereira, F. R., Lopes, T. M., da Silva, E. L., Coan, A. C., Campos, B. M., ... Cendes, F. (2014). Neuropsychiatric symptoms in Alzheimer's disease are related to functional connectivity alterations in the salience network. *Human Brain Mapping, 35*(4), 1237—1246.

Bonelli, R. M., & Cummings, J. L. (2007). Frontal-subcortical circuitry and behavior. *Dialogues in Clinical Neuroscience, 9*(2), 141—151.

Boone, K. B., Miller, B. L., Swartz, R., Lu, P., & Lee, A. (2003). Relationship between positive and negative symptoms and neuropsychological scores in frontotemporal dementia and Alzheimer's disease. *Journal of the International Neuropsychological Society: JINS, 9*(5), 698—709.

Bracht, T., Horn, H., Strik, W., Federspiel, A., Razavi, N., Stegmayer, K., ... Walther, S. (2014). White matter pathway organization of the reward system is related to positive and negative symptoms in schizophrenia. *Schizophrenia Research, 153*(1—3), 136—142.

Brodaty, H., Liu, Z., Withall, A., & Sachdev, P. S. (2013). The longitudinal course of post-stroke apathy over five years. *The Journal of Neuropsychiatry and Clinical Neurosciences, 25*(4), 283—291.

Cacciari, C., Moraschi, M., Di Paola, M., Cherubini, A., Orfei, M. D., Giove, F., ... Spalletta, G. (2010). White matter microstructure and apathy level in amnestic mild cognitive impairment. *Journal of Alzheimer's Disease: JAD, 20*(2), 501—507.

Caeiro, L., Ferro, J. M., & Costa, J. (2013). Apathy secondary to stroke: a systematic review and meta-analysis. *Cerebrovascular Diseases (Basel, Switzerland), 35*(1), 23—39.

Caeiro, L., Ferro, J. M., Pinho, E. M. T., Canhao, P., & Figueira, M. L. (2013). Post-stroke apathy: an exploratory longitudinal study. *Cerebrovascular Diseases (Basel, Switzerland), 35*(6), 507—513.

Carriere, N., Besson, P., Dujardin, K., Duhamel, A., Defebvre, L., Delmaire, C., & Devos, D. (2014). Apathy in Parkinson's disease is associated with nucleus accumbens atrophy: a magnetic resonance imaging shape analysis. *Movement Disorders: Official Journal of the Movement Disorder Society, 29*(7), 897—903.

Castrioto, A., Lhommee, E., Moro, E., & Krack, P. (2014). Mood and behavioural effects of subthalamic stimulation in Parkinson's disease. *Lancet Neurology, 13*(3), 287—305.

Chue, P., & Lalonde, J. K. (2014). Addressing the unmet needs of patients with persistent negative symptoms of schizophrenia: emerging pharmacological treatment options. *Neuropsychiatric Disease and Treatment, 10*, 777—789.

Ciurli, P., Formisano, R., Bivona, U., Cantagallo, A., & Angelelli, P. (2011). Neuropsychiatric disorders in persons with severe traumatic brain injury: prevalence, phenomenology, and relationship with demographic, clinical, and functional features. *The Journal of Head Trauma Rehabilitation, 26*(2), 116—126.

Clarke, D. E., Ko, J. Y., Kuhl, E. A., van Reekum, R., Salvador, R., & Marin, R. S. (2011). Are the available apathy measures reliable and valid? A review of the psychometric evidence. *Journal of Psychosomatic Research, 70*(1), 73—97.

Croxson, P. L., Walton, M. E., O'Reilly, J. X., Behrens, T. E., & Rushworth, M. F. (2009). Effort-based cost-benefit valuation and the human brain. *The Journal of Neuroscience: The Official Journal of the Society for Neuroscience, 29*(14), 4531—4541.

Czernecki, V., Schupbach, M., Yaici, S., Levy, R., Bardinet, E., Yelnik, J., ... Agid, Y. (2008). Apathy following subthalamic stimulation in Parkinson disease: a dopamine responsive symptom. *Movement Disorders: Official Journal of the Movement Disorder Society, 23*(7), 964—969.

van Dalen, J. W., Moll van Charante, E. P., Nederkoorn, P. J., van Gool, W. A., & Richard, E. (2013). Poststroke apathy. *Stroke; A Journal of Cerebral Circulation, 44*(3), 851—860.

Delmaire, C., Dumas, E. M., Sharman, M. A., van den Bogaard, S. J., Valabregue, R., Jauffret, C., ... Lehericy, S. (2013). The structural correlates of functional deficits in early Huntington's disease. *Human Brain Mapping, 34*(9), 2141—2153.

Demirtas-Tatlidede, A., Bahar, S. Z., & Gurvit, H. (2013). Akinetic mutism without a structural prefrontal lesion. *Cognitive and Behavioral Neurology: Official Journal of the Society for Behavioral and Cognitive Neurology, 26*(2), 59—62.

Devos, D., Moreau, C., Maltete, D., Lefaucheur, R., Kreisler, A., Eusebio, A., ... Kathy, D. (2014). Rivastigmine in apathetic but dementia and depression-free patients with Parkinson's disease: a double-blind, placebo-controlled, randomised clinical trial. *Journal of Neurology, Neurosurgery, and Psychiatry, 85*(6), 668—674.

Drapier, D., Drapier, S., Sauleau, P., Haegelen, C., Raoul, S., Biseul, I., … Rivier, I. (2006). Does subthalamic nucleus stimulation induce apathy in Parkinson's disease? *Journal of Neurology, 253*(8), 1083–1091.

Duff, K., Paulsen, J. S., Beglinger, L. J., Langbehn, D. R., Wang, C., Stout, J. C., … Quellar, S. (2010). "Frontal" behaviors before the diagnosis of Huntington's disease and their relationship to markers of disease progression: evidence of early lack of awareness. *The Journal of Neuropsychiatry and Clinical Neurosciences, 22*(2), 196–207.

Dujardin, K., Sockeel, P., Carette, A. S., Delliaux, M., & Defebvre, L. (2013). Assessing apathy in everyday clinical practice with the short-form Lille Apathy Rating Scale. *Movement Disorders: Official Journal of the Movement Disorder Society, 28*(14), 2014–2019.

Dujardin, K., Sockeel, P., Delliaux, M., Destee, A., & Defebvre, L. (2008). The Lille Apathy Rating Scale: validation of a caregiver-based version. *Movement Disorders: Official Journal of the Movement Disorder Society, 23*(6), 845–849.

Dujardin, K., Sockeel, P., Delliaux, M., Destee, A., & Defebvre, L. (2009). Apathy may herald cognitive decline and dementia in Parkinson's disease. *Movement Disorders: Official Journal of the Movement Disorder Society, 24*(16), 2391–2397.

van Duijn, E., Craufurd, D., Hubers, A. A., Giltay, E. J., Bonelli, R., Rickards, H., … European Huntington's Disease Network Behavioural Phenotype Working Group. (2014). Neuropsychiatric symptoms in a European Huntington's disease cohort (REGISTRY). *Journal of Neurology, Neurosurgery, and Psychiatry, 85*(12), 1411–1418.

van Duijn, E., Kingma, E. M., & van der Mast, R. C. (2007). Psychopathology in verified Huntington's disease gene carriers. *The Journal of Neuropsychiatry and Clinical Neurosciences, 19*(4), 441–448.

van Duijn, E., Reedeker, N., Giltay, E. J., Eindhoven, D., Roos, R. A., & van der Mast, R. C. (2014). Course of irritability, depression and apathy in Huntington's disease in relation to motor symptoms during a two-year follow-up period. *Neuro-Degenerative Diseases, 13*(1), 9–16.

Emre, M., Aarsland, D., Brown, R., Burn, D. J., Duyckaerts, C., Mizuno, Y., … Dubois, B. (2007). Clinical diagnostic criteria for dementia associated with Parkinson's disease. *Movement Disorders: Official Journal of the Movement Disorder Society, 22*(12), 1689–1707. quiz 837.

Eslinger, P. J., & Damasio, A. R. (1985). Severe disturbance of higher cognition after bilateral frontal lobe ablation: patient EVR. *Neurology, 35*(12), 1731–1741.

Eslinger, P. J., Moore, P., Antani, S., Anderson, C., & Grossman, M. (2012). Apathy in frontotemporal dementia: behavioral and neuroimaging correlates. *Behavioural Neurology, 25*(2), 127–136.

Evensen, J., Rossberg, J. I., Barder, H., Haahr, U., Hegelstad, W., Joa, I., … McGlashan, T. (2012). Apathy in first episode psychosis patients: a ten year longitudinal follow-up study. *Schizophrenia Research, 136*(1–3), 19–24.

Faerden, A., Vaskinn, A., Finset, A., Agartz, I., Ann Barrett, E., Friis, S., … Melle, I. (2009). Apathy is associated with executive functioning in first episode psychosis. *BMC Psychiatry, 9*, 1.

Feil, D., Razani, J., Boone, K., & Lesser, I. (2003). Apathy and cognitive performance in older adults with depression. *International Journal of Geriatric Psychiatry, 18*(6), 479–485.

Finset, A., & Andersson, S. (2000). Coping strategies in patients with acquired brain injury: relationships between coping, apathy, depression and lesion location. *Brain Injury: BI, 14*(10), 887–905.

Flirski, M., Sieruta, M., Golanska, E., Kloszewska, I., Liberski, P. P., & Sobow, T. (2012). PRND 3′UTR polymorphism may be associated with behavioral disturbances in Alzheimer disease. *Prion, 6*(1), 73–80.

Frakey, L. L., Salloway, S., Buelow, M., & Malloy, P. (2012). A randomized, double-blind, placebo-controlled trial of modafinil for the treatment of apathy in individuals with mild-to-moderate Alzheimer's disease. *The Journal of Clinical Psychiatry, 73*(6), 796–801.

Franceschi, M., Anchisi, D., Pelati, O., Zuffi, M., Matarrese, M., Moresco, R. M., … Perani, D. (2005). Glucose metabolism and serotonin receptors in the frontotemporal lobe degeneration. *Annals of Neurology, 57*(2), 216–225.

Gauthier, S., Feldman, H., Hecker, J., Vellas, B., Ames, D., Subbiah, P., … Emir, B. (2002). Efficacy of donepezil on behavioral symptoms in patients with moderate to severe Alzheimer's disease. *International Psychogeriatrics/IPA, 14*(4), 389–404.

Gauthier, S., Juby, A., Dalziel, W., Rehel, B., & Schecter, R. (2010). Effects of rivastigmine on common symptomatology of Alzheimer's disease (EXPLORE). *Current Medical Research and Opinion, 26*(5), 1149–1160.

Gavett, B. E., Cantu, R. C., Shenton, M., Lin, A. P., Nowinski, C. J., McKee, A. C., & Stern, R. A. (2011). Clinical appraisal of chronic traumatic encephalopathy: current perspectives and future directions. *Current Opinion in Neurology, 24*(6), 525–531.

Ghoshal, S., Gokhale, S., Rebovich, G., & Caplan, L. R. (2011). The neurology of decreased activity: abulia. *Reviews in Neurological Diseases, 8*(3–4), e55–e67.

Ginovart, N., Lundin, A., Farde, L., Halldin, C., Backman, L., Swahn, C. G., … Sedvall, G. (1997). PET study of the pre- and post-synaptic dopaminergic markers for the neurodegenerative process in Huntington's disease. *Brain: A Journal of Neurology, 120*(Pt 3), 503–514.

Gottfried, J. A., O'Doherty, J., & Dolan, R. J. (2003). Encoding predictive reward value in human amygdala and orbitofrontal cortex. *Science, 301*(5636), 1104–1107.

Gradin, V. B., Waiter, G., O'Connor, A., Romaniuk, L., Stickle, C., Matthews, K., … Douglas Steele, J. (2013). Salience network-midbrain dysconnectivity and blunted reward signals in schizophrenia. *Psychiatry Research, 211*(2), 104–111.

Grool, A. M., Geerlings, M. I., Sigurdsson, S., Eiriksdottir, G., Jonsson, P. V., Garcia, M. E., … Launer, L. J. (2014). Structural MRI correlates of apathy symptoms in older persons without dementia: AGES-Reykjavik Study. *Neurology, 82*(18), 1628–1635.

Gruber, O., Chadha Santuccione, A., & Aach, H. (2014). Magnetic resonance imaging in studying schizophrenia, negative symptoms, and the glutamate system. *Frontiers in Psychiatry, 5*, 32.

Grunsfeld, A. A., & Login, I. S. (2006). Abulia following penetrating brain injury during endoscopic sinus surgery with disruption of the anterior cingulate circuit: case report. *BMC Neurology, 6*, 4.

Haber, S. N. (2011). Neuroanatomy of reward: a view from the ventral striatum. In J. A. Gottfried (Ed.), *Neurobiology of sensation and reward. Frontiers in neuroscience.* Boca Raton (FL).

Haber, S. N., & Knutson, B. (2010). The reward circuit: linking primate anatomy and human imaging. *Neuropsychopharmacology: Official Publication of the American College of Neuropsychopharmacology, 35*(1), 4–26.

Habib, M. (2004). Athymhormia and disorders of motivation in basal ganglia disease. *The Journal of Neuropsychiatry and Clinical Neurosciences, 16*(4), 509–524.

Hadland, K. A., Rushworth, M. F., Gaffan, D., & Passingham, R. E. (2003). The anterior cingulate and reward-guided selection of actions. *Journal of Neurophysiology, 89*(2), 1161–1164.

Hahn, C., Lim, H. K., Won, W. Y., Ahn, K. J., Jung, W. S., & Lee, C. U. (2013). Apathy and white matter integrity in Alzheimer's disease: a whole brain analysis with tract-based spatial statistics. *PLoS One, 8*(1), e53493.

Hama, S., Yamashita, H., Shigenobu, M., Watanabe, A., Kurisu, K., Yamawaki, S., & Kitaoka, T. (2007). Post-stroke affective or apathetic depression and lesion location: left frontal lobe and bilateral basal ganglia. *European Archives of Psychiatry and Clinical Neuroscience, 257*(3), 149–152.

van den Heuvel, M. P., & Hulshoff Pol, H. E. (2010). Exploring the brain network: a review on resting-state fMRI functional connectivity. *European Neuropsychopharmacology: The Journal of the European College of Neuropsychopharmacology, 20*(8), 519–534.

Higuchi, M. A., Tsuboi, Y., Inoue, T., Fukuyama, K., Abe, H., Baba, Y., & Yamada, T. (2014). Predictors of the emergence of apathy after bilateral stimulation of the subthalamic nucleus in patients with Parkinson's disease. *Neuromodulation: Journal of the International Neuromodulation Society, 18*(2), 113–117.

Humphries, M. D., & Prescott, T. J. (2010). The ventral basal ganglia, a selection mechanism at the crossroads of space, strategy, and reward. *Progress in Neurobiology, 90*(4), 385–417.

Ioannidis, A. E., Kimiskidis, V. K., Loukopoulou, E., Geroukis, T., Vlaikidis, N., & Kosmidis, M. H. (2013). Apathy, cognitive dysfunction and impaired social cognition in a patient with bilateral thalamic infarction. *Neurocase, 19*(5), 513–520.

Jarmolowicz, A. I., Chen, H. Y., & Panegyres, P. K. (2014). The patterns of inheritance in early-onset dementia: Alzheimer's disease and frontotemporal dementia. *American Journal of Alzheimer's Disease and Other Dementias, 30*(3), 299–306.

Jenkinson, P. M., Preston, C., & Ellis, S. J. (2011). Unawareness after stroke: a review and practical guide to understanding, assessing, and managing anosognosia for hemiplegia. *Journal of Clinical and Experimental Neuropsychology, 33*(10), 1079–1093.

Joshi, A., Barsuglia, J. P., Mather, M. J., Jimenez, E. E., Shapira, J., & Mendez, M. F. (2014). Evaluation of emotional blunting in behavioral variant frontotemporal dementia compared to Alzheimer's disease. *Dementia and Geriatric Cognitive Disorders, 38*(1–2), 79–88.

Juckel, G., Schlagenhauf, F., Koslowski, M., Wustenberg, T., Villringer, A., Knutson, B., … Heinz, A. (2006). Dysfunction of ventral striatal reward prediction in schizophrenia. *NeuroImage, 29*(2), 409–416.

Kang, W. S., Park, J. K., Park, H. J., Kim, S. K., Cho, A. R., Lee, S. M., … Song, J. Y. (2012). Association between catechol-O-methyltransferase Val158Met polymorphism and negative symptoms of schizophrenia in the Korean population. *Psychiatry Research, 198*(2), 338–339.

Kant, R., Duffy, J. D., & Pivovarnik, A. (1998). Prevalence of apathy following head injury. *Brain Injury: BI, 12*(1), 87–92.

Kennerley, S. W., & Wallis, J. D. (2009). Encoding of reward and space during a working memory task in the orbitofrontal cortex and anterior cingulate sulcus. *Journal of Neurophysiology, 102*(6), 3352–3364.

Kim, J. W., Lee, D. Y., Choo, I. H., Seo, E. H., Kim, S. G., Park, S. Y., & Woo, J. I. (2011). Microstructural alteration of the anterior cingulum is associated with apathy in Alzheimer disease. *The American Journal of Geriatric Psychiatry: Official Journal of the American Association for Geriatric Psychiatry, 19*(7), 644–653.

Kirsch-Darrow, L., Marsiske, M., Okun, M. S., Bauer, R., & Bowers, D. (2011). Apathy and depression: separate factors in Parkinson's disease. *Journal of the International Neuropsychological Society: JINS, 17*(6), 1058–1066.

Knutson, B., & Cooper, J. C. (2005). Functional magnetic resonance imaging of reward prediction. *Current Opinion in Neurology, 18*(4), 411–417.

Knutson, B., Wimmer, G. E., Kuhnen, C. M., & Winkielman, P. (2008). Nucleus accumbens activation mediates the influence of reward cues on financial risk taking. *NeuroReport, 19*(5), 509–513.

Knutson, K. M., Monte, O. D., Raymont, V., Wassermann, E. M., Krueger, F., & Grafman, J. (2014). Neural correlates of apathy revealed by lesion mapping in participants with traumatic brain injuries. *Human Brain Mapping, 35*(3), 943–953.

Kurniawan, I. T., Guitart-Masip, M., Dayan, P., & Dolan, R. J. (2013). Effort and valuation in the brain: the effects of anticipation and execution. *The Journal of Neuroscience: The Official Journal of the Society for Neuroscience, 33*(14), 6160–6169.

Kurniawan, I. T., Guitart-Masip, M., & Dolan, R. J. (2011). Dopamine and effort-based decision making. *Frontiers in Neuroscience, 5*, 81.

Landes, A. M., Sperry, S. D., & Strauss, M. E. (2005). Prevalence of apathy, dysphoria, and depression in relation to dementia severity in Alzheimer's disease. *The Journal of Neuropsychiatry and Clinical Neurosciences, 17*(3), 342–349.

Lee, G. J., Lu, P. H., Mather, M. J., Shapira, J., Jimenez, E., Leow, A. D., … Mendez, M. F. (2014). Neuroanatomical correlates of emotional blunting in behavioral variant frontotemporal dementia and early-onset Alzheimer's disease. *Journal of Alzheimer's Disease: JAD, 41*(3), 793–800.

Leentjens, A. F., Koester, J., Fruh, B., Shephard, D. T., Barone, P., & Houben, J. J. (2009). The effect of pramipexole on mood and motivational symptoms in Parkinson's disease: a meta-analysis of placebo-controlled studies. *Clinical Therapeutics, 31*(1), 89–98.

Leontjevas, R., Evers-Stephan, A., Smalbrugge, M., Pot, A. M., Thewissen, V., Gerritsen, D. L., … Koopmans, R. T. (2012). A comparative validation of the abbreviated apathy evaluation scale (AES-10) with the neuropsychiatric inventory apathy subscale against diagnostic criteria of apathy. *Journal of the American Medical Directors Association, 13*(3), 308, e1–e6.

Levy, R., & Dubois, B. (2006). Apathy and the functional anatomy of the prefrontal cortex-basal ganglia circuits. *Cerebral Cortex, 16*(7), 916–928.

Li, W. J., Kou, C. G., Yu, Y., Sun, S., Zhang, X., Kosten, T. R., … Zhang, X. Y. (2012). Association of catechol-O-methyltransferase gene polymorphisms with schizophrenia and negative symptoms in a Chinese population. *American Journal of Medical Genetics Part B, Neuropsychiatric Genetics: The Official Publication of the International Society of Psychiatric Genetics, 159b*(4), 370–375.

Lin, M. K., & Farrer, M. J. (2014). Genetics and genomics of Parkinson's disease. *Genome Medicine, 6*(6), 48.

Manoliu, A., Riedl, V., Doll, A., Bauml, J. G., Muhlau, M., Schwerthoffer, D., … Sorg, C. (2013). Insular dysfunction reflects altered between-network connectivity and severity of negative symptoms in schizophrenia during psychotic remission. *Frontiers in Human Neuroscience, 7*, 216.

Marin, R. S. (1991). Apathy: a neuropsychiatric syndrome. *The Journal of Neuropsychiatry and Clinical Neurosciences, 3*(3), 243–254.

Marin, R. S., Butters, M. A., Mulsant, B. H., Pollock, B. G., & Reynolds, C. F., 3rd. (2003). Apathy and executive function in depressed elderly. *Journal of Geriatric Psychiatry and Neurology, 16*(2), 112–116.

Massimo, L., Powers, C., Moore, P., Vesely, L., Avants, B., Gee, J., … Grossman, M. (2009). Neuroanatomy of apathy and disinhibition in frontotemporal lobar degeneration. *Dementia and Geriatric Cognitive Disorders, 27*(1), 96–104.

McKee, A. C., Stern, R. A., Nowinski, C. J., Stein, T. D., Alvarez, V. E., Daneshvar, D. H., … Cantu, R. C. (2013). The spectrum of disease in chronic traumatic encephalopathy. *Brain: A Journal of Neurology, 136*(Pt 1), 43–64.

Mega, M. S., Cummings, J. L., Fiorello, T., & Gornbein, J. (1996). The spectrum of behavioral changes in Alzheimer's disease. *Neurology, 46*(1), 130–135.

Mendez, M. F., McMurtray, A., Licht, E., Shapira, J. S., Saul, R. E., & Miller, B. L. (2006). The scale for emotional blunting in patients with frontotemporal dementia. *Neurocase, 12*(4), 242–246.

Mendez, M. F., & Shapira, J. S. (2011). Loss of emotional insight in behavioral variant frontotemporal dementia or "frontal anosodiaphoria". *Consciousness and Cognition, 20*(4), 1690–1696.

Menon, V. (2011). Large-scale brain networks and psychopathology: a unifying triple network model. *Trends in Cognitive Sciences, 15*(10), 483–506.

Merrilees, J., Dowling, G. A., Hubbard, E., Mastick, J., Ketelle, R., & Miller, B. L. (2013). Characterization of apathy in persons with frontotemporal dementia and the impact on family caregivers. *Alzheimer Disease and Associated Disorders, 27*(1), 62–67.

Mikami, K., Jorge, R. E., Moser, D. J., Arndt, S., Jang, M., Solodkin, A., … Robinson, R. G. (2013). Prevention of poststroke apathy using escitalopram or problem-solving therapy. *The American Journal of Geriatric Psychiatry: Official Journal of the American Association for Geriatric Psychiatry, 21*(9), 855–862.

Mitaki, S., Isomura, M., Maniwa, K., Yamasaki, M., Nagai, A., Nabika, T., & Yamaguchi, S. (2013). Apathy is associated with a single-nucleotide polymorphism in a dopamine-related gene. *Neuroscience Letters, 549*, 87–91.

Monastero, R., Mariani, E., Camarda, C., Ingegni, T., Averna, M. R., Senin, U., … Mecocci, P. (2006). Association between apolipoprotein E epsilon4 allele and apathy in probable Alzheimer's disease. *Acta Psychiatrica Scandinavica, 113*(1), 59–63.

Moreau, C., Delval, A., Defebvre, L., Dujardin, K., Duhamel, A., Petyt, G., … Parkqait-II, Study Group. (2012). Methylphenidate for gait hypokinesia and freezing in patients with Parkinson's disease undergoing subthalamic stimulation: a multicentre, parallel, randomised, placebo-controlled trial. *Lancet Neurology, 11*(7), 589–596.

Mortby, M. E., Maercker, A., & Forstmeier, S. (2012). Apathy: a separate syndrome from depression in dementia? A critical review. *Aging Clinical and Experimental Research, 24*(4), 305–316.

Muller, U., Czymmek, J., Thone-Otto, A., & Von Cramon, D. Y. (2006). Reduced daytime activity in patients with acquired brain damage and apathy: a study with ambulatory actigraphy. *Brain Injury: BI, 20*(2), 157–160.

Naarding, P., Janzing, J. G., Eling, P., van der Werf, S., & Kremer, B. (2009). Apathy is not depression in Huntington's disease. *The Journal of Neuropsychiatry and Clinical Neurosciences, 21*(3), 266–270.

Nakaaki, S., Murata, Y., Sato, J., Shinagawa, Y., Hongo, J., Tatsumi, H., … Furukawa, T. A. (2008). Association between apathy/depression and executive function in patients with Alzheimer's disease. *International Psychogeriatrics/IPA, 20*(5), 964–975.

Njomboro, P., Deb, S., & Humphreys, G. W. (2012). Apathy and executive functions: insights from brain damage involving the anterior cingulate cortex. *BMJ Case Reports, 2012*.

O'Doherty, J. P. (2004). Reward representations and reward-related learning in the human brain: insights from neuroimaging. *Current Opinion in Neurobiology, 14*(6), 769–776.

O'Doherty, J. P., Dayan, P., Friston, K., Critchley, H., & Dolan, R. J. (2003). Temporal difference models and reward-related learning in the human brain. *Neuron, 38*(2), 329–337.

O'Doherty, J. P., Hampton, A., & Kim, H. (2007). Model-based fMRI and its application to reward learning and decision making. *Annals of the New York Academy of Sciences, 1104*, 35–53.

Ohtani, T., Bouix, S., Hosokawa, T., Saito, Y., Eckbo, R., Ballinger, T., … Kubicki, M. (2014). Abnormalities in white matter connections between orbitofrontal cortex and anterior cingulate cortex and their associations with negative symptoms in schizophrenia: a DTI study. *Schizophrenia Research, 157*(1–3), 190–197.

Onoda, K., Kuroda, Y., Yamamoto, Y., Abe, S., Oguro, H., Nagai, A., … Yamaguchi, S. (2011). Post-stroke apathy and hypoperfusion in basal ganglia: SPECT study. *Cerebrovascular Diseases (Basel, Switzerland), 31*(1), 6–11.

van Oostrom, J. C., Dekker, M., Willemsen, A. T., de Jong, B. M., Roos, R. A., & Leenders, K. L. (2009). Changes in striatal dopamine D2 receptor binding in pre-clinical Huntington's disease. *European Journal of Neurology: The Official Journal of the European Federation of Neurological Societies, 16*(2), 226–231.

Ota, M., Sato, N., Nakata, Y., Arima, K., & Uno, M. (2012). Relationship between apathy and diffusion tensor imaging metrics of the brain in Alzheimer's disease. *International Journal of Geriatric Psychiatry, 27*(7), 722–726.

Oyebode FSACP. (2008). *Sims' symptoms in the mind an introduction to descriptive psychopathology Edinburgh*. New York: Saunders/Elsevier. Available from: http://site.ebrary.com/id/10511903.

Palaniyappan, L., White, T. P., & Liddle, P. F. (2012). The concept of salience network dysfunction in schizophrenia: from neuroimaging observations to therapeutic opportunities. *Current Topics in Medicinal Chemistry, 12*(21), 2324–2338.

Paradiso, S., Chemerinski, E., Yazici, K. M., Tartaro, A., & Robinson, R. G. (1999). Frontal lobe syndrome reassessed: comparison of patients with lateral or medial frontal brain damage. *Journal of Neurology, Neurosurgery, and Psychiatry, 67*(5), 664–667.

Peters, F., Perani, D., Herholz, K., Holthoff, V., Beuthien-Baumann, B., Sorbi, S., ... Salmon, E. (2006). Orbitofrontal dysfunction related to both apathy and disinhibition in frontotemporal dementia. *Dementia and Geriatric Cognitive Disorders, 21*(5–6), 373–379.

Petkau, T. L., & Leavitt, B. R. (2014). Progranulin in neurodegenerative disease. *Trends in Neurosciences, 37*(7), 388–398.

Prigatano, G. P. (2013). Denial, anosodiaphoria, and emotional reactivity in anosognosia. *Cognitive Neuroscience, 4*(3–4), 201–202.

Proitsi, P., Lupton, M. K., Reeves, S. J., Hamilton, G., Archer, N., Martin, B. M., ... Powell, J. F. (2012). Association of serotonin and dopamine gene pathways with behavioral subphenotypes in dementia. *Neurobiology of Aging, 33*(4), 791–803.

Quattrocchi, G., & Bestmann, S. (2014). Journal club: possible role of the basal ganglia in poor reward sensitivity and apathy after stroke. *Neurology, 82*(20), e171–e173.

Ramirez, S. M., Glover, H., Ohlde, C., Mercer, R., Goodnick, P., Hamlin, C., & Perez-Rivera, M. I. (2001). Relationship of numbing to alexithymia, apathy, and depression. *Psychological Reports, 88*(1), 189–200.

Rao, V., McCann, U., Bergey, A., Han, D., Brandt, J., & Schretlen, D. J. (2013). Correlates of apathy during the first year after traumatic brain injury. *Psychosomatics, 54*(4), 403–404.

Rapoport, M. J., van Reekum, R., Freedman, M., Streiner, D., Simard, M., Clarke, D., ... Conn, D. (2001). Relationship of psychosis to aggression, apathy and function in dementia. *International Journal of Geriatric Psychiatry, 16*(2), 123–130.

Rascovsky, K., Hodges, J. R., Knopman, D., Mendez, M. F., Kramer, J. H., Neuhaus, J., ... Miller, B. L. (2011). Sensitivity of revised diagnostic criteria for the behavioural variant of frontotemporal dementia. *Brain: A Journal of Neurology, 134*(Pt 9), 2456–2477.

Ray Chaudhuri, K., Martinez-Martin, P., Antonini, A., Brown, R. G., Friedman, J. H., Onofrj, M., ... Trenkwalder, C. (2013). Rotigotine and specific non-motor symptoms of Parkinson's disease: post hoc analysis of RECOVER. *Parkinsonism & Related Disorders, 19*(7), 660–665.

Reedeker, N., Bouwens, J. A., van Duijn, E., Giltay, E. J., Roos, R. A., & van der Mast, R. C. (2011). Incidence, course, and predictors of apathy in Huntington's disease: a two-year prospective study. *The Journal of Neuropsychiatry and Clinical Neurosciences, 23*(4), 434–441.

Reijnders, J. S., Scholtissen, B., Weber, W. E., Aalten, P., Verhey, F. R., & Leentjens, A. F. (2010). Neuroanatomical correlates of apathy in Parkinson's disease: a magnetic resonance imaging study using voxel-based morphometry. *Movement Disorders: Official Journal of the Movement Disorder Society, 25*(14), 2318–2325.

Reyes, S., Viswanathan, A., Godin, O., Dufouil, C., Benisty, S., Hernandez, K., ... Chabriat, H. (2009). Apathy: a major symptom in CADASIL. *Neurology, 72*(10), 905–910.

Robert, P., Onyike, C. U., Leentjens, A. F., Dujardin, K., Aalten, P., Starkstein, S., ... Byrne, J. (2009). Proposed diagnostic criteria for apathy in Alzheimer's disease and other neuropsychiatric disorders. *European Psychiatry: The Journal of the Association of European Psychiatrists, 24*(2), 98–104.

Robert, P. H., Clairet, S., Benoit, M., Koutaich, J., Bertogliati, C., Tible, O., ... Bedoucha, P. (2002). The apathy inventory: assessment of apathy and awareness in Alzheimer's disease, Parkinson's disease and mild cognitive impairment. *International Journal of Geriatric Psychiatry, 17*(12), 1099–1105.

Robinson, R. G., Jorge, R. E., Clarence-Smith, K., & Starkstein, S. (2009). Double-blind treatment of apathy in patients with poststroke depression using nefiracetam. *The Journal of Neuropsychiatry and Clinical Neurosciences, 21*(2), 144–151.

Rochat, L., Van der Linden, M., Renaud, O., Epiney, J. B., Michel, P., Sztajzel, R., ... Annoni, J. M. (2013). Poor reward sensitivity and apathy after stroke: implication of basal ganglia. *Neurology, 81*(19), 1674–1680.

Rosen, H. J., Pace-Savitsky, K., Perry, R. J., Kramer, J. H., Miller, B. L., & Levenson, R. W. (2004). Recognition of emotion in the frontal and temporal variants of frontotemporal dementia. *Dementia and Geriatric Cognitive Disorders, 17*(4), 277–281.

Rosenberg, P. B., Lanctot, K. L., Drye, L. T., Herrmann, N., Scherer, R. W., Bachman, D. L., ... The ADMET Investigators. (2013). Safety and efficacy of methylphenidate for apathy in Alzheimer's disease: a randomized, placebo-controlled trial. *The Journal of Clinical Psychiatry, 74*(8), 810–816.

Rosenblatt, A. (2007). Neuropsychiatry of Huntington's disease. *Dialogues in Clinical Neuroscience, 9*(2), 191–197.

Santangelo, G., Trojano, L., Barone, P., Errico, D., Grossi, D., & Vitale, C. (2013). Apathy in Parkinson's disease: diagnosis, neuropsychological correlates, pathophysiology and treatment. *Behavioural Neurology, 27*(4), 501–513.

Santangelo, G., Vitale, C., Trojano, L., Picillo, M., Moccia, M., Pisano, G., ... Barone, P. (2014). Relationship between apathy and cognitive dysfunctions in de novo untreated Parkinson's disease: a prospective longitudinal study. *European Journal of Neurology: The Official Journal of the European Federation of Neurological Societies, 22*(2), 253–260.

Schneider, M., Van der Linden, M., Glaser, B., Rizzi, E., Dahoun, S. P., Hinard, C., ... Eliez, S. (2012). Preliminary structure and predictive value of attenuated negative symptoms in 22q11.2 deletion syndrome. *Psychiatry Research, 196*(2–3), 277–284.

Schneider, M., Van der Linden, M., Menghetti, S., Glaser, B., Debbane, M., & Eliez, S. (2014). Predominant negative symptoms in 22q11.2 deletion syndrome and their associations with cognitive functioning and functional outcome. *Journal of Psychiatric Research, 48*(1), 86–93.

Schuepbach, W. M., Rau, J., Knudsen, K., Volkmann, J., Krack, P., Timmermann, L., ... Deuschl, G. (2013). Neurostimulation for Parkinson's disease with early motor complications. *The New England Journal of Medicine, 368*(7), 610–622.

Schultz, W. (2000). Multiple reward signals in the brain. *Nature Reviews Neuroscience, 1*(3), 199–207.

Seelaar, H., Rohrer, J. D., Pijnenburg, Y. A., Fox, N. C., & van Swieten, J. C. (2011). Clinical, genetic and pathological heterogeneity of frontotemporal dementia: a review. *Journal of Neurology, Neurosurgery, and Psychiatry, 82*(5), 476–486.

Seeley, W. W., Crawford, R. K., Zhou, J., Miller, B. L., & Greicius, M. D. (2009). Neurodegenerative diseases target large-scale human brain networks. *Neuron, 62*(1), 42–52.

Seeley, W. W., Menon, V., Schatzberg, A. F., Keller, J., Glover, G. H., Kenna, H., … Greicius, M. D. (2007). Dissociable intrinsic connectivity networks for salience processing and executive control. *The Journal of Neuroscience: The Official Journal of the Society for Neuroscience, 27*(9), 2349–2356.

Sha, S. J., Takada, L. T., Rankin, K. P., Yokoyama, J. S., Rutherford, N. J., Fong, J. C., … Boxer, A. L. (2012). Frontotemporal dementia due to C9ORF72 mutations: clinical and imaging features. *Neurology, 79*(10), 1002–1011.

Simon, J. J., Biller, A., Walther, S., Roesch-Ely, D., Stippich, C., Weisbrod, M., … Kaiser, S. (2010). Neural correlates of reward processing in schizophrenia—relationship to apathy and depression. *Schizophrenia Research, 118*(1–3), 154–161.

Singh, A., Mahgoub, N., & Klimstra, S. (2011). Apathy associated with a unilateral globus pallidus lesion: a case report. *International Journal of Geriatric Psychiatry, 26*(7), 765–766.

Snowden, J. S., Harris, J., Richardson, A., Rollinson, S., Thompson, J. C., Neary, D., … Pickering-Brown, S. (2013). Frontotemporal dementia with amyotrophic lateral sclerosis: a clinical comparison of patients with and without repeat expansions in C9orf72. *Amyotrophic Lateral Sclerosis & Frontotemporal Degeneration, 14*(3), 172–176.

Sockeel, P., Dujardin, K., Devos, D., Deneve, C., Destee, A., & Defebvre, L. (2006). The Lille apathy rating scale (LARS), a new instrument for detecting and quantifying apathy: validation in Parkinson's disease. *Journal of Neurology, Neurosurgery, and Psychiatry, 77*(5), 579–584.

Spalletta, G., Cravello, L., Piras, F., Iorio, M., Sancesario, G., Marchi, A., … Cacciari, C. (2013). Rapid-onset apathy may be the only clinical manifestation after dorsal striatum hemorrhagic lesion: a case report. *Alzheimer Disease and Associated Disorders, 27*(2), 192–194.

Spalletta, G., Fagioli, S., Caltagirone, C., & Piras, F. (2013). Brain microstructure of subclinical apathy phenomenology in healthy individuals. *Human Brain Mapping, 34*(12), 3193–3203.

Starkstein, S. E., Ingram, L., Garau, M. L., & Mizrahi, R. (2005). On the overlap between apathy and depression in dementia. *Journal of Neurology, Neurosurgery, and Psychiatry, 76*(8), 1070–1074.

Starkstein, S. E., & Leentjens, A. F. (2008). The nosological position of apathy in clinical practice. *Journal of Neurology, Neurosurgery, and Psychiatry, 79*(10), 1088–1092.

Starkstein, S. E., Mizrahi, R., Capizzano, A. A., Acion, L., Brockman, S., & Power, B. D. (2009). Neuroimaging correlates of apathy and depression in Alzheimer's disease. *The Journal of Neuropsychiatry and Clinical Neurosciences, 21*(3), 259–265.

Stern, R. A., Daneshvar, D. H., Baugh, C. M., Seichepine, D. R., Montenigro, P. H., Riley, D. O., … McKee, A. C. (2013). Clinical presentation of chronic traumatic encephalopathy. *Neurology, 81*(13), 1122–1129.

Stoppel, C. M., Boehler, C. N., Strumpf, H., Heinze, H. J., Hopf, J. M., & Schoenfeld, M. A. (2011). Neural processing of reward magnitude under varying attentional demands. *Brain Research, 1383*, 218–229.

Tabrizi, S. J., Scahill, R. I., Owen, G., Durr, A., Leavitt, B. R., Roos, R. A., … The Track-HD Investigators. (2013). Predictors of phenotypic progression and disease onset in premanifest and early-stage Huntington's disease in the TRACK-HD study: analysis of 36-month observational data. *Lancet Neurology, 12*(7), 637–649.

Takayanagi, M., Wentz, J., Takayanagi, Y., Schretlen, D. J., Ceyhan, E., Wang, L., … Cascella, N. G. (2013). Reduced anterior cingulate gray matter volume and thickness in subjects with deficit schizophrenia. *Schizophrenia Research, 150*(2–3), 484–490.

Tanaka, S. C., Yamada, K., Yoneda, H., & Ohtake, F. (2014). Neural mechanisms of gain-loss asymmetry in temporal discounting. *The Journal of Neuroscience: The Official Journal of the Society for Neuroscience, 34*(16), 5595–5602.

Tang, W. K., Chen, Y. K., Liang, H. J., Chu, W. C., Mok, V. C., Ungvari, G. S., & Wong, K. S. (2013). Location of infarcts and apathy in ischemic stroke. *Cerebrovascular Diseases (Basel, Switzerland), 35*(6), 566–571.

Tang, W. K., Lau, C. G., Mok, V., Ungvari, G. S., & Wong, K. S. (2014). Apathy and health-related quality of life in stroke. *Archives of Physical Medicine and Rehabilitation, 95*(5), 857–861.

Thaler, D., Chen, Y. C., Nixon, P. D., Stern, C. E., & Passingham, R. E. (1995). The functions of the medial premotor cortex. I. Simple learned movements. *Experimental Brain Research, 102*(3), 445–460.

Theleritis, C., Politis, A., Siarkos, K., & Lyketsos, C. G. (2014). A review of neuroimaging findings of apathy in Alzheimer's disease. *International Psychogeriatrics/IPA, 26*(2), 195–207.

Thobois, S., Ardouin, C., Lhommee, E., Klinger, H., Lagrange, C., Xie, J., … Krack, P. (2010). Non-motor dopamine withdrawal syndrome after surgery for Parkinson's disease: predictors and underlying mesolimbic denervation. *Brain: A Journal of Neurology, 133*(Pt 4), 1111–1127.

Thobois, S., Lhommee, E., Klinger, H., Ardouin, C., Schmitt, E., Bichon, A., … Krack, P. (2013). Parkinsonian apathy responds to dopaminergic stimulation of D2/D3 receptors with piribedil. *Brain: A Journal of Neurology, 136*(Pt 5), 1568–1577.

Tighe, S. K., Oishi, K., Mori, S., Smith, G. S., Albert, M., Lyketsos, C. G., & Mielke, M. M. (2012). Diffusion tensor imaging of neuropsychiatric symptoms in mild cognitive impairment and Alzheimer's dementia. *The Journal of Neuropsychiatry and Clinical Neurosciences, 24*(4), 484–488.

Victoroff, J. (2013). Traumatic encephalopathy: review and provisional research diagnostic criteria. *NeuroRehabilitation, 32*(2), 211–224.

Vijayaraghavan, L., Krishnamoorthy, E. S., Brown, R. G., & Trimble, M. R. (2002). Abulia: a delphi survey of British neurologists and psychiatrists. *Movement Disorders: Official Journal of the Movement Disorder Society, 17*(5), 1052–1057.

Vijayaraghavan, L., Vaidya, J. G., Humphreys, C. T., Beglinger, L. J., & Paradiso, S. (2008). Emotional and motivational changes after bilateral lesions of the globus pallidus. *Neuropsychology, 22*(3), 412–418.

Waldemar, G., Gauthier, S., Jones, R., Wilkinson, D., Cummings, J., Lopez, O., ... Mackell, J. (2011). Effect of donepezil on emergence of apathy in mild to moderate Alzheimer's disease. *International Journal of Geriatric Psychiatry, 26*(2), 150—157.

Walker, F. O. (2007). Huntington's disease. *Lancet, 369*(9557), 218—228.

Wallis, J. D., & Kennerley, S. W. (2011). Contrasting reward signals in the orbitofrontal cortex and anterior cingulate cortex. *Annals of the New York Academy of Sciences, 1239*, 33—42.

Wickham, R. J., Solecki, W., Rathbun, L. R., Neugebauer, N. M., Wightman, R. M., & Addy, N. A. (2013). Advances in studying phasic dopamine signaling in brain reward mechanisms. *Frontiers in Bioscience, 5*, 982—999.

World Health Organization. (1992). *ICD-10: International statistical classification of diseases and related health problems*. Geneva: World Health Organization.

Xu, C., Aragam, N., Li, X., Villla, E. C., Wang, L., Briones, D., ... Wang, K. (2013). BCL9 and C9orf5 are associated with negative symptoms in schizophrenia: meta-analysis of two genome-wide association studies. *PLoS One, 8*(1), e51674.

Yuen, G. S., Gunning-Dixon, F. M., Hoptman, M. J., AbdelMalak, B., McGovern, A. R., Seirup, J. K., ... Alexopoulos, G. S. (2014). The salience network in the apathy of late-life depression. *International Journal of Geriatric Psychiatry, 29*(11), 1116—1124.

Zahari, Z., Teh, L. K., Ismail, R., & Razali, S. M. (2011). Influence of DRD2 polymorphisms on the clinical outcomes of patients with schizophrenia. *Psychiatric Genetics, 21*(4), 183—189.

Zamboni, G., Huey, E. D., Krueger, F., Nichelli, P. F., & Grafman, J. (2008). Apathy and disinhibition in frontotemporal dementia: insights into their neural correlates. *Neurology, 71*(10), 736—742.

Chapter 22

Emotional Dysfunction in Psychopathology and Neuropathology: Neural and Genetic Pathways

Virginia E. Sturm
University of California San Francisco, San Francisco, CA, United States

Claudia M. Haase
Northwestern University, Evanston, IL, United States

Robert W. Levenson
University of California Berkeley, Berkeley, CA, United States

EMOTIONAL DYSFUNCTION IN PSYCHOPATHOLOGY AND NEUROPATHOLOGY: NEURAL AND GENETIC PATHWAYS

In our view, emotions are "short-lived psychological-physiological phenomena that represent efficient modes of adaptation to changing environmental demands" (Levenson, 1994). As depicted in Fig. 22.1, emotions are initiated by appraisal processes, often rapid and unconscious but sometimes more protracted and conscious, and produce coordinated changes in disparate physiological systems including facial expression, voice, somatic muscles, and autonomic nervous system (Levenson, 2003). Subjective emotional experience (the "feelings" we have when in the throes of an emotion) arises from interoceptive and proprioceptive processing of visceral and somatic information and sensations that are produced when these physiological systems are activated (Levenson, 2003).

Emotional functioning in humans is incredibly durable. Emotions are generated by phylogenetically ancient, well-conserved neural circuits (Rosen & Levenson, 2009) and can persist even in decorticate brains (Shewmon, Holmes, & Byrne, 1999). Emotions appear early in ontogeny; infant distress cries and smiles provide critical avenues for bidirectional communication and influence between child and caregiver. Emotions are also built for the long haul. Emotional functioning is sustained at high levels at the outreaches of normal aging, long after cognitive and physical abilities have shown

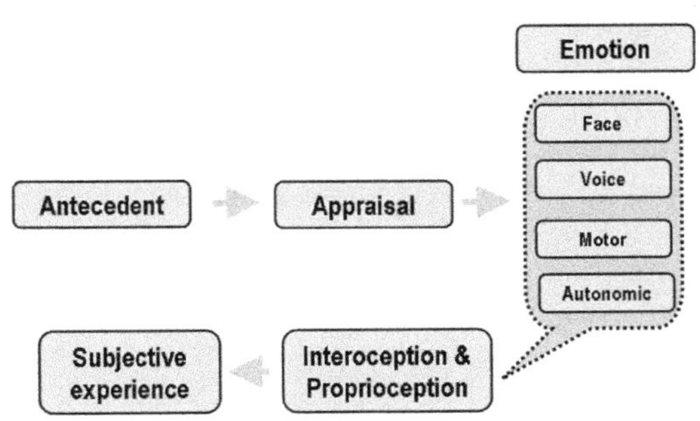

FIGURE 22.1 The emotional system.

Genomics, Circuits, and Pathways in Clinical Neuropsychiatry. http://dx.doi.org/10.1016/B978-0-12-800105-9.00022-6

dramatic declines (Levenson, 2000; Salthouse, 2004). Despite these indicators of durability, human emotion is also extremely vulnerable to dysfunction. In a large number of psychiatric and neurological diseases, emotions change in ways that interfere with daily living, work, and relationships and create enormous problems for patients, loved ones, and society.

In this chapter, we first review the current state of knowledge about neural and genetic influences on three aspects of emotional functioning (emotional reactivity, emotion regulation, and emotional affiliation). Emotional reactivity, regulation, and affiliation are all critical for daily living and maintaining mental and physical health. Importantly, each is highly vulnerable to disruption in psychopathology and neuropathology. In the final part of the chapter, we consider ways in which tools and concepts derived from modern affective science can help elucidate the relationships among neural circuits, genes, emotional functioning, and pathology and provide insights about etiology, diagnosis, and treatment that can be useful in clinical contexts.

EMOTIONS AND EMOTIONAL PROCESSES EMOTIONS: SHORT AND DISCRETE

Before initiating our discussion of neural and genetic pathways related to emotional functioning, it is important to clarify the kinds of emotional phenomena upon which we will be focusing. We have presented a model of emotion and emotional elicitation that would be characterized in the current marketplace of emotion theories as "evolutionary," "functional," and "peripheralist." In addition to theoretical diversity, affective scientists also differ in the temporal aspects of the affective phenomenon of interest. Whereas we might focus on a 5-s burst of sadness, others would focus on longer-lasting sad moods that could last for hours or days, and yet others would focus on even longer-lasting sad traits that could last for years or a lifetime (Ekman, 1984). Thus, the affective phenomena that we will be primarily considering in this chapter are short-lived, typically lasting for a matter of seconds, although they can be reactivated repeatedly over longer periods. In our view, these kinds of elemental, brief emotions are: (1) well-suited for precise study in the laboratory, (2) well-matched in grain of measurement for making functional links with neural circuits and genes, and (3) clinically relevant by virtue of being vulnerable to disruptions in psychopathology and neuropathology.

In addition to temporal considerations, affective scientists differ in their views about how best to parse affective space. In our work, we have followed a "discrete emotions" approach (Ekman, 1992), which proposes that there are particular emotions (eg, fear, anger, sadness, happiness, disgust, contempt) that differ from each other in terms of their associated physiological, expressive, and subjective qualities and that represent generalized evolved solutions (Tooby & Cosmides, 1990) to timeless problems, challenges, and opportunities (eg, sadness for dealing with loss, happiness for dealing with gains). For us, discrete emotions provide a meaningful and highly useful bridge to different forms of psychopathology and neuropathology. For example, a patient with a specific phobia is more likely to have problems with fear than with sadness, whereas the opposite would be true for a patient with depression. Likewise, a patient with amyotrophic lateral sclerosis (ALS) who is experiencing pseudobulbar affect may show expressive displays that resemble sadness and/or happiness but not those that resemble fear or disgust (Olney et al., 2011). There is a different view that eschews discrete, evolved emotions, and instead envisions emotional states as existing in a multidimensional space (eg, a two-dimensioned circumplex based on valence and arousal) and as being largely socially constructed (Russell & Barrett, 1999). In work with clinical syndromes using categorical systems for diagnosing psychopathology such as the *Diagnostic and Statistical Manual* (DSM) (American Psychiatric Association, 1994), historically the discrete emotions view has been more prominent. However, the dimensional view may make greater inroads with the introduction of the Research Domain Criteria (RDoC) (Insel et al., 2010), which views psychopathology in a dimensional rather than categorical manner.

EMOTIONAL PROCESSES

Emotion is often treated as if it were a single entity, a practice that can lead to difficulties when exploring links with particular neural circuits and genes, which arguably are better matched with finer-grained behavioral phenomena. Just as cognition is best considered as consisting of multiple dissociable processes (eg, short-term memory, executive control, computation), emotion needs to be broken down into smaller functional units. In this chapter we focus on three emotional processes that are particularly important when considering normal and pathological emotional functioning: (1) emotional reactivity, (2) emotional regulation, and (3) emotional affiliation.

Emotional Reactivity

Emotional reactivity refers to the capacity to generate emotions in response to salient challenges, threats, and opportunities. In the laboratory, emotional reactivity is typically assessed by exposing individuals to carefully selected sensory stimuli (eg, films, still images, sounds) that are known to elicit strong emotional reactions in most people. Emotional reactivity is then quantified by measuring the magnitude and duration of the emotional response in behavior, physiology, and subjective experience.

Emotional Regulation

Emotional regulation refers to the capacity to adjust aspects of the emotion response in accordance with personal, interpersonal, and social goals and standards. Most typically this involves downregulating emotion (eg, reducing anger in a dispute with a loved one) but it can also involve amplifying emotion (eg, increasing sadness to communicate distress clearly to another person). In the laboratory, emotional regulation is typically assessed by exposing individuals to emotion-eliciting situations and either giving them explicit regulatory instructions (eg, do not allow your emotions to show) or leaving them free to regulate or not using their own strategies. Emotional regulation is then quantified by measuring the magnitude and duration of emotional response (usually compared with a condition in which regulation is minimized).

Emotional Affiliation

Emotional affiliation refers to the capacity to use emotion to create and maintain social connections. It is a process that reflects the strongly social and interpersonal nature of human emotion (Keltner & Kring, 1998). In the laboratory, emotional affiliation can be assessed by exposing individuals to a social situation in which they have to process emotional information from other people and/or respond to the emotions of others. Emotional affiliation can be quantified by measuring (1) the ability to recognize (Bowers, Blonder, & Heilman, 1991) or track (Goodkind et al., 2012) emotions in others; (2) the emotional reactions that occur in response to the emotions of others (Sze, Gyurak, Goodkind, & Levenson, 2012); and (3) the patterns of emotional reactivity and regulation that occur in interactions with intimate others (Gottman & Levenson, 1986; Levenson, Haase, Bloch, Holley, & Seider, 2013).

EMOTIONAL FUNCTIONING: NEURAL PATHWAYS

In this section, we will discuss the neural circuitry that subserves emotional reactivity, regulation, and affiliation. Studying this circuitry in both normal and clinical populations provides a basis for understanding how the nervous system not only supports these emotional functions but also provides important clues for beginning to understand the basis of individual differences in emotional functioning. These individual differences exist both in the "normal" range (eg, some people are more emotionally reactive than others) and "abnormal" range. The latter are particularly important for understanding the neural pathways that are associated with emotional dysfunctions associated with neuropathology and psychopathology.

EMOTIONAL REACTIVITY

Distributed neural systems initiate the coordinated changes in autonomic nervous system reactivity and facial expression that characterize emotion. Consistent with the schematic of emotional response presented earlier, after appraisal of a salient stimulus, a network of brain regions that includes the anterior cingulate cortex, central nucleus of the amygdala, hypothalamus, and brain stem work with the anterior insula to initiate the emotional response and monitor the cascade of visceral and motor changes that accompany that response. For example, a state of fear might activate a pattern of bodily changes in which heart rate speeds, hands become sweaty, and blood is redirected from the periphery to large-muscle groups (eg, the legs) that can help us escape a predator or intruder. These rapid physiological changes are coupled with action tendencies that make certain behaviors (eg, fleeing) more likely than others. In addition to these intrapersonal changes, emotions also serve important interpersonal functions. The expressive changes that occur (in facial expression, appearance, and vocalizations) alert conspecifics to the dangerous situation and help prepare them to deal effectively with the situation (Levenson, 1999). We turn next to a review of the brain structures that support the initiation of this complex multisystem emotional response.

Anterior Cingulate Cortex

The cingulate gyrus is a band of cortex that surrounds the corpus callosum. The cingulate can be further divided into distinct subregions that are based on cytoarchitectonic, connectivity, and functional divisions (Bush, Luu, & Posner,

2000). The posterior cingulate cortex is a key node in the default mode network that is typically active at rest and during episodic and prospective memory tasks. The midcingulate (or dorsal anterior cingulate cortex) is often engaged during attention and executive control, and the anterior cingulate cortex (which includes pregenual and subgenual subregions) is critical for emotion generation (Buckner, Andrews-Hanna, & Schacter, 2008; Ochsner, Silvers, & Buhle, 2012; Seeley, Zhou, & Kim, 2012; Sturm et al., 2013; Vogt, 2005). The anterior cingulate cortex is reciprocally connected with anterior insula, amygdala, periaqueductal gray, hypothalamus, nucleus accumbens, and orbitofrontal cortex, regions important for salience detection, emotional reactivity, and social regulation (Seeley et al., 2007). Through connections with orbitofrontal cortex, the anterior cingulate cortex is a centrally positioned hub that receives salient information about the environment and triggers visceromotor activity via subcortical structures and the brain stem (Ongur, An, & Price, 1998).

Amygdala

The amygdala is a structure in the medial temporal lobe that has long been implicated in emotion (Davis & Whalen, 2001). Although the amygdala gained notoriety for its role in fear (LeDoux, 1992), it is now known that the amygdala activates in response to a wide range of salient stimuli, including both negative and positive antecedent events (Murray, 2007), and is also essential for the formation of emotional memories (Cahill & McGaugh, 1998). The amygdala is composed of multiple subnuclei that have been largely delineated in nonhuman animals via tract-tracing studies and have been cross-validated in humans using structural and functional imaging parcellation techniques (Etkin, Prater, Schatzberg, Menon, & Greicius, 2009; McDonald, 1998). The basolateral amygdala receives sensory information from the thalamus, hippocampus, and cortex, relaying this information to the central nucleus as well as other regions involved in appraisal, memory, and approach or avoidance behavior (Davis & Whalen, 2001). The central nucleus, through direct projections to the hypothalamus and via brain stem nuclei that support autonomic functions such as respiration, heart rate, and sweating, initiates the autonomic cascade that accompanies emotion and alters attention and vigilance through reciprocal connections with orbitofrontal cortex (Ongur & Price, 2000; Price & Amaral, 1981). With direct projections to the trigeminal and facial motor nuclei, the central nucleus also has an integral role in emotional facial expression (Davis & Whalen, 2001).

Hypothalamus

The hypothalamus has a key role in thermoregulation, appetite, sleep, sexual behavior, and emotion. Composed of multiple subnuclei (eg, paraventricular nucleus, dorsomedial nucleus, lateral hypothalamic area), the hypothalamus receives afferent information from multiple regions including medial orbitofrontal cortex, amygdala, and anterior cingulate cortex (Ongur et al., 1998). The hypothalamus has efferent projections to autonomic nuclei in the brain stem including rostral ventrolateral medulla, dorsal motor nucleus of the vagus, and the nucleus of the solitary tract as well as the intermediolateral column of the spinal cord (Guyenet, 2006; Price & Amaral, 1981; Saper, Loewy, Swanson, & Cowan, 1976; Tucker & Saper, 1985). There is evidence that each subregion of the hypothalamus has a distinct role in autonomic reactivity, with some areas increasing and others decreasing cardiovascular arousal (Fontes, Xavier, de Menezes, & Dimicco, 2011). Although stimulation of the hypothalamus can lead to both autonomic and behavioral defense reactions in decerebrate nonhuman animals (Bard, 1934), other studies give greater emphasis to the role the hypothalamus has in generating the autonomic response in emotion rather than a behavioral response (LeDoux, Iwata, Cicchetti, & Reis, 1988).

Brain Stem

The brain stem contains multiple nuclei that have important roles in homeostatic regulation and emotion. The periaqueductal gray is a structure that surrounds the cerebral aqueduct and has a columnar organization (Bandler & Keay, 1996). The lateral column, which receives projections from the central nucleus of the amygdala, has a role in sympathetic nervous system activity, whereas the ventrolateral column has the opposite effect and slows the heart and respiration through inhibitory vagally mediated parasympathetic pathways (Behbehani, 1995). Stimulation of the periaqueductal gray can elicit various behavioral responses including vocalizations, fight, flight, or freezing (Bandler, Keay, Floyd, & Price, 2000) and trigger patterned, coordinated autonomic responses involving multiple organ systems (Carrive & Bandler, 1991; Dampney, Furlong, Horiuchi, & Iigaya, 2013).

Other brain stem regions also have important roles in the coordination of somatic, respiratory, electrodermal, and cardiovascular events. Rostral ventrolateral medulla, which receives afferent projections from the periaqueductal gray, contributes most notably to sympathetic nervous system activity that increases heart rate, blood pressure, and respiration (Fontes et al., 2011; Menezes & Fontes, 2007). The nucleus of the solitary tract has a role in efferent visceromotor pattern generation in addition to being a central afferent way station for incoming sensory information from the body (Andresen & Kunze, 1994). The nucleus of the solitary tract and the parabrachial nucleus, a region that also receives substantial afferent viscerosensory inputs, integrates incoming interoceptive information regarding the physiological state of the viscera. This continuous stream of sensory information arrives at the brain stem via the lamina I spinothalamocortical pathway and vagal afferents and is then relayed on to the thalamus and insula, where it undergoes more comprehensive processing (Benarroch, 2006; Craig, 2002).

Whereas sympathetic nerves emerging from the intermediolateral column of the spinal cord support sympathetic functioning, parasympathetic nervous system activity is primarily governed by the activity of the vagus (cranial nerve X). The vagus originates in two nuclei in the medulla: the dorsal motor nucleus and the nucleus ambiguus. Whereas the branch of the vagus that originates in the dorsal motor nucleus predominantly supports autonomic functions below the diaphragm (eg, intestines), there is a phylogenetically newer branch of the vagus that arises from the nucleus ambiguus, which has the central role of slowing the heart from the pace that is set by the sinoatrial node (Dergacheva, Griffioen, Neff, & Mendelowitz, 2010; Porges, 2001).

This newer branch of the vagus is myelinated and is important for respiratory sinus arrhythmia, a modulation of heart rate that is linked to the respiratory cycle. Levels of respiratory sinus arrhythmia have been linked to individual differences in empathy and prosocial behavior (Kogan et al., 2014). Through its close connections with other cranial nerves that support facial expression, the branch of the vagus that emerges from the nucleus ambiguus is considered to be essential for fostering a calm physiological state in mammals that is necessary for socioemotional attunement and communication (Porges, 2001).

Insula

The insula, a structure located deep in the Sylvian fissure, has a key role in numerous cognitive and affective processes. The insula can be divided into several subregions (Kurth, Zilles, Fox, Laird, & Eickhoff, 2010). Through its projections to somatosensory and motor cortex, the midposterior insula is a key hub for sensorimotor processing and has an important role in sensing physical cues from the body and motor acts. The dorsal anterior insula, which has connections to ventrolateral prefrontal cortex, supplementary motor area, striatum, and subthalamic nucleus, has a predominant role in executive control, an ability that requires attention, working memory, response inhibition, and task-set maintenance (Aron, 2007). The ventral midinsula is integral for processing olfactory and gustatory stimuli, and the ventral anterior insula (or "frontoinsula") has strong connections with the anterior cingulate cortex, amygdala, hypothalamus, and brain stem and is a key node in social behavior, empathy, and emotion (Seeley et al., 2008). Although the insula and anterior cingulate cortex are often coactive during functional neuroimaging tasks of emotion and empathy, there is accumulating evidence that the primary role of the anterior cingulate cortex is initiation of motor behavior whereas that of the insula is sensing the body's internal milieu (Craig, 2009; Seeley et al., 2012).

The anterior insula is theorized to be one of the final sites for integrating visceral information from the body with homeostatic, hedonic, and motivational factors (Craig, 2009). The anterior insula is thus considered to be a primary center for assembling the subjective feeling states that accompany emotion. The insula maintains an online representation of the body's internal states through a pathway dedicated to interoception (Craig, 2002; Critchley & Harrison, 2013). The lamina I spinothalamocortical tract is a pathway through which information about the internal organs is relayed via small-fiber sympathetic afferents to brain stem centers including the nucleus of the solitary tract and parabrachial nucleus. From there, this information is shuttled to the ventroposterior medial nucleus of the thalamus and then to the posterior dorsal insula, midinsula, and finally, to the anterior insula. Parasympathetic afferent information is relayed via a similar pathway, but it is sent directly from the periphery to the brain stem via vagal afferents. Whereas sympathetic afferents may be more heavily represented in the right anterior insula, parasympathetic afferents may be more heavily represented in the left anterior insula (Craig, 2005). Patients who have lesions of the anterior insula (such as those with behavioral variant frontotemporal dementia) exhibit profound deficits in empathy and emotion (Seeley et al., 2008), which is likely because these patients are unable to access internal cues that typically promote understanding of and responsiveness to the emotions of others.

EMOTION REGULATION

Humans regulate emotions by choosing the situations they encounter, altering these situations, attending to certain aspects of the environment and ignoring others, changing the ways they appraise a situation, and modifying their overexpression of emotion (Gross, 1998). Regulation strategies that occur early in the cascade of emotion (eg, attentional deployment and reappraisal) can effectively attenuate emotion generation and decrease emotional experience, whereas regulatory strategies that occur later in the emotional cascade (eg, suppression) can actually accentuate emotional experience and autonomic reactivity (Gross & Levenson, 1993).

Most research on the neural systems that support emotion regulation has focused on conscious (ie, explicit) emotion regulation strategies, and these studies most often have examined the neural correlates of reappraisal (Ochsner et al., 2012). By contrast, an emerging literature on more automatic (ie, implicit) emotion regulation suggests that both kinds of emotion regulation recruit overlapping distributed neural systems. Functional neuroimaging studies of reappraisal typically find that prefrontal cortex is important for both cognitive control and emotion regulation. When participants engage prefrontal regions during reappraisal tasks that are aimed at lowering negative emotion, they exhibit lower activity in subcortical brain regions such as the amygdala when they are presented with emotional stimuli (McRae et al., 2010; Wager, Davidson, Hughes, Lindquist, & Ochsner, 2008). Here we will provide more details about brain regions that are pertinent for emotion regulation.

Ventrolateral Prefrontal Cortex

Ventrolateral prefrontal cortex is a region of the frontal lobes that is typically associated with response inhibition and goal-appropriate response selection (Aron, Robbins, & Poldrack, 2004). This region is often recruited during reappraisal tasks in which participants alter their interpretation of a negative stimulus to minimize its emotional impact. Reappraisal studies that invoke reinterpretation of a stimulus typically engage ventrolateral prefrontal cortex in conjunction with other regions (Goldin, McRae, Ramel, & Gross, 2008; McRae et al., 2010). Suppression, a form of response modulation in which participants are instructed to clamp down on their emotions, has also been linked to ventrolateral prefrontal cortex activity as well as activity in dorsal anterior insula (Hayes et al., 2010). As described previously (see the "Insula" section), the dorsal anterior insula is a key hub in the "stop signal" network that enforces behavioral control and response inhibition. Implicit emotion regulation, which can be measured simply by having participants label their affective states, also engages the ventrolateral prefrontal cortex. Whereas suppression may not attenuate activity effectively in emotional reactivity networks, implicit emotion regulation strategies can diminish activity in emotion generators such as the amygdala (Lieberman, 2007; Payer, Baicy, Lieberman, & London, 2012). Consistent with its role in emotional control, atrophy in ventrolateral prefrontal cortex has been linked to emotion dysregulation in patients with frontotemporal dementia, which suggests that volume loss in this region may unleash affective responsiveness (Sturm et al., 2014).

Dorsolateral Prefrontal Cortex

Dorsolateral prefrontal cortex, together with ventrolateral prefrontal cortex and dorsomedial prefrontal cortex (see the subsequent Affiliation section for more details about this region), can also be recruited during emotion regulation (Silvers, Wager, Weber, & Ochsner, 2015; Staudinger, Erk, & Walter, 2011). The dorsolateral prefrontal cortex is a region of the frontal lobes that is most typically associated with executive functions including working memory and selective attention (Curtis & D'Esposito, 2003). Through connections with parietal cortex, dorsolateral prefrontal cortex is a key node in dorsal attention networks that support basic cognitive selection of sensory information and response (Corbetta & Shulman, 2002). Although this region does not project directly to emotion generators, it may influence emotional reactivity by altering higher-order perceptual attention systems (Ochsner et al., 2012).

Anterior Midcingulate Cortex

The anterior midcingulate cortex has strong connections with lateral prefrontal cortex, parietal cortex, premotor cortex, and supplementary motor area (Bush et al., 2000) and has an important role in conflict detection, attention, and cognitive control (Seeley et al., 2007). This region is typically activated during demanding cognitive tasks that engage executive functions including working memory, response selection and inhibition, competition monitoring, and error detection (Botvinick, Cohen, & Carter, 2004; Carter et al., 1998). Often active during reappraisal, the dorsal anterior cingulate may promote performance monitoring by detecting mismatches between one's intended and actual emotional states, thus fueling further emotional downregulation.

Inferior Parietal Lobe

The inferior parietal lobule is a posterior region of the brain that is most well-known for its role in visuospatial processing. Parietal cortex is also involved in other, nonspatial processes including perspective-taking and judgment of social closeness between people (Parkinson, Liu, & Wheatley, 2014; Yamazaki, Hashimoto, & Iriki, 2009).

Distancing, a form of reappraisal in which individuals create more psychological space between themselves and an emotional stimulus, seems to rely relatively more upon parietal cortex than other forms of emotion regulation (Koenigsberg et al., 2010). It is possible that distancing requires participants to alter their spatiotemporal perspective to decrease their feelings of connection with the stimulus (Ochsner et al., 2012). This region has also been found to be important for overcoming our own emotions to take the perspective of another, an ability that may also depend on a separation of self from other through distancing (Silani, Lamm, Ruff, & Singer, 2013).

EMOTIONAL AFFILIATION

Humans possess the neural circuitry necessary to support the formation and maintenance of meaningful, enduring social relationships (Baumeister & Leary, 1995; Beckes & Coan, 2011). Social contact, which promotes a sense of calm in the brain and autonomic nervous system, is essential to health and longevity (Beckes & Coan, 2011; Porges, 2001). Social support attenuates neural activity in regions that detect and respond to threat and danger (Coan, Schaefer, & Davidson, 2006; Zaki, Schirmer, & Mitchell, 2011) and increases activity in autonomic and neuroendocrine systems that support positive emotions, bonding, and empathy (Carter, Williams, Witt, & Insel, 1992; Porges, 2001; Stellar, Cohen, Oveis, & Keltner, 2015).

As noted earlier, emotional affiliation is a broad construct that encompasses a number of different subprocesses. In this section we will focus on two aspects of emotional affiliation: (1) emotion recognition (the ability to recognize the emotions of others), and (2) emotions that occur in intimate relationships (focusing on nurturant love and compassion, two emotions that promote social relationships and prosociality). Responding emotionally to the emotions of others (sometimes called "emotional empathy"), another important aspect of emotional affiliation, will not be included in this section. Not surprisingly, research has shown that the neural systems that support emotional empathy largely overlap with those involved in emotion generation (Engen & Singer, 2013). These circuits were reviewed in the section on emotional reactivity presented earlier.

Superior Temporal Sulcus

The superior temporal sulcus, and often the temporoparietal junction, is a brain region that is important for numerous aspects of social cognition. This region is typically active during tasks of cognitive empathy and perspective-taking (Frith & Frith, 2006; Saxe & Kanwisher, 2003). The superior temporal sulcus is also important for the detection of social cues including prosody, faces, trustworthiness, and intention (Ethofer et al., 2006; Sabatinelli et al., 2011; Winston, Strange, O'Doherty, & Dolan, 2002). Activity in this region has also been linked to altruism and higher levels of prosocial behaviors including fairness and generosity, which suggests that being able to understand the minds of others may promote an other-oriented focus and feelings of selflessness (Morishima, Schunk, Bruhin, Ruff, & Fehr, 2012; Takagishi, Kameshima, Schug, Koizumi, & Yamagishi, 2010).

Dorsomedial Prefrontal Cortex

The medial prefrontal cortex is a midline frontal region typically associated with self-processing and other-processing. The medial prefrontal cortex can be parcellated into dorsal and ventral parts, with the dorsomedial prefrontal cortex responding more to stimuli about other people and the ventromedial prefrontal cortex responding more to stimuli that are self-related (Denny, Kober, Wager, & Ochsner, 2012). Dorsomedial prefrontal cortex, working with other structures (eg, posterior cingulate cortex, posterior superior temporal sulcus, and medial temporal lobes), is considered to be a necessary node in the default mode network that enables humans to project themselves outside the present moment and focus on things other than the self in the here-and-now (Buckner & Carroll, 2007; Mitchell, 2009). Consistent with this idea, dorsomedial prefrontal cortex is often active during tasks in which participants must focus on the perspectives and feelings of other people (Abu-Akel & Shamay-Tsoory, 2011; Iacoboni et al., 2004; Mitchell, 2009). The dorsomedial prefrontal cortex is also active during certain types of emotion regulation (eg, reappraisal), which may reflect the fact that reappraisal requires people to reconsider the perspectives of others and the significance of emotional situations (Silvers et al., 2015).

Ventral Striatum

The ventral striatum has a prominent role in reward processing. With strong connections with other emotion generators, this region (and especially the nucleus accumbens) is active during the anticipation and receipt of monetary and social rewards (Izuma, Saito, & Sadato, 2008; Knutson, Adams, Fong, & Hommer, 2001). Positive emotional stimuli, such as positive faces and pleasant scenes, also activate the ventral striatum (Wager et al., 2008), as do feelings of social connection, interpersonal warmth, and feeling understood (Inagaki & Eisenberger, 2012, 2013; Morelli, Torre, & Eisenberger, 2014). Studies of attachment and nurturant love find higher activity in ventral striatum (as well as other emotion-relevant regions including the anterior insula, anterior cingulate cortex, and periaqueductal gray) when participants view photographs of people for whom they have strong feelings of love and connection, compared with when they view people with whom they are less close (Acevedo, Aron, Fisher, & Brown, 2012; Bartels & Zeki, 2004).

Interestingly, individuals who have undergone compassion training exhibit increased activity in the ventral striatum in response to distressing photographs, which suggests that an other-oriented compassionate focus promotes prosocial helping feelings rather than self-oriented feelings of personal distress (Klimecki, Leiberg, Ricard, & Singer, 2014).

Medial Orbitofrontal Cortex

The orbitofrontal cortex is a ventral region of the frontal lobes that is important for the detection and tracking the value of a stimulus (Rolls, 2000). Whereas the lateral orbitofrontal cortex is essential for monitoring information that is relevant for tracking potential punishments, the medial orbitofrontal cortex is integral for processing reward-related cues (Kringelbach & Rolls, 2004). Via projections to ventral striatum and emotion-generating systems, the medial orbitofrontal cortex facilitates online decoding and monitoring a stimulus's reward value.

This region is integral in evaluating the meaning of primary rewards, such as odors, and social rewards, such as observing others acting prosocially (Fehr & Camerer, 2007; Rolls, 2008). The medial orbitofrontal cortex is active not only when people themselves receive a reward but also when they view others receiving rewards, which suggests that this region enables us to experience reward for ourselves and it also facilitates our ability to share the positive experiences of others vicariously (Hare, Camerer, Knoepfle, & Rangel, 2010; Morelli, Sacchet, & Zaki, 2014).

Septal Area

The septal area is a subcortical region that has strong projections to emotion-generating areas and has a key role in feelings of social connectedness and bonding. In rats, oxytocin binding in the septal area has been associated with maternal behaviors that promote kinship bonds (Francis, Champagne, & Meaney, 2000). In humans, the septal area is active when an individual has positive feelings toward others, including experiences of unconditional trust, empathy, and social connection. Activity in this region during a scanner-based empathy task predicts real-world prosocial helping behavior (Krueger et al., 2007; Morelli, Rameson, & Lieberman, 2014). Consistent with its strong role in affiliative behavior, atrophy in the septal area has been linked to a decline in prosocial feelings in patients with frontotemporal dementia (Moll et al., 2011).

EMOTION FUNCTIONING: GENETIC PATHWAYS

Researchers have used a number of different strategies to study the genetic bases of emotional functioning in humans, including (1) adoption and twin studies, (2) genome-wide association studies (GWAS), and (3) studies of common genetic polymorphisms. Adoption and twin studies seek to apportion variance in emotion-relevant factors between genetic and environmental sources. GWAS examine large parts of the human genome (often assaying up to a million single-nucleotide polymorphisms [SNPs] in thousands of individuals)[1] and have studied the genetic architecture of emotion-relevant personality traits and psychopathologies, such as neuroticism (de Moor et al., 2012) and affective disorders (Liu et al., 2011). Studies of common genetic variants or polymorphisms (ie, those present in greater than 1% of the population) often follow a hypothesis-driven approach and study links between variations in candidate genes and distal emotion-related traits or psychopathologies such as neuroticism (Lesch et al., 1996) or depression (Caspi et al., 2003), or more proximal emotion processes such as emotional reactivity (Gyurak et al., 2013).

Research linking genes with emotional processes of reactivity, regulation, and affiliation, the focus of the current chapter, have largely come from the candidate gene tradition[2] and thus will be the focus of our review. However, there is ongoing debate over the validity and usefulness of the different strategies for studying genetic pathways, and the candidate gene approach has received significant criticism (Flint & Munafò, 2013; Manolio et al., 2009).

MODEL OF THE GENETIC PATHWAY

In our view, the pathway that links candidate genes, emotion processes, and health outcomes in humans starts with genes that regulate neurotransmitter or hormone systems (eg, serotonin, dopamine, oxytocin) that influence the neural systems described earlier in this chapter (eg, amygdala, prefrontal cortex) that are critical for emotional processes such as reactivity, regulation, and affiliation. Polymorphisms in these genes create slight "biases" in these emotional functions (eg, greater capacity for emotion downregulation, tendency to respond to particular kinds of environmental events with larger-magnitude emotional responses). Because these biases are slight, they require well-designed laboratory studies with tightly controlled stimuli, precise measurement of emotional responding, thoughtful participant sampling, control for other contributing genetic influences, and careful replication to afford confidence in findings.

Such studies can reveal the "proximal" effects of genetic polymorphisms on emotional functioning. It is likely that these biases contribute to the individual differences in emotional functioning that appear in trait measures (eg, neuroticism) and in more casual observations about individual differences (eg, one person being seen as more emotionally labile than another). These biases also have effects as they interact with different kinds of environments (eg, abusive versus supportive parenting, high-stress versus low-stress lives) over the life course, contributing to more "distal" outcomes related to health and wellness (eg, depression, attention-related disorders, cardiovascular disease).

In the realm of candidate gene studies, many studies have demonstrated pathways from genetic polymorphisms to distal outcomes with moderation by environmental factors (eg, polymorphisms of the serotonin transporter (5-HTT) gene linked to depression, moderated by life stress) (Caspi et al., 2003). A number of studies have also demonstrated pathways from genetic polymorphisms to more proximal effects on emotional functioning (Gyurak et al., 2013). Studies that incorporate both the proximal and distal parts of this genetic pathway in humans would have significant advantages but are extremely challenging to mount. Such studies would require careful longitudinal assessments of genes, emotional functioning, environments, and disease processes and would need to deal with other complexities as well. At every point along any putative pathway linking a particular genetic polymorphism with a particular outcome as moderated by a particular environment, other genes and environmental factors exert influences (Manuck & McCaffery, 2014). For example, epigenetic work has shown that environmental factors (eg, maternal care) can modulate genetic effects by altering DNA methylation and gene expression (Weaver et al., 2004).

Consistent with the focus of this chapter on emotional reactivity, regulation, and affiliation, we will review the literature that has explored the proximal effects of common genetic polymorphisms involved in regulating serotonergic, dopaminergic, and oxytocinergic systems on these aspects of emotional functioning.

1. GWAS approaches typically examine SNPs but not other kinds of genetic structural variations (eg, repeat polymorphisms such as 5-HTTLPR). Some have proposed that studying these structural variations could help explain some of the "missing heritability" produced by SNP-based GWAS (Manolio et al., 2009).

2. An important exception includes a series of SNP-based GWAS (Iacono, Malone, Vaidyanathan, & Vrieze, 2014) that examined the genetic architecture of a number of psychophysiological phenomena (ie, antisaccade eye-tracking performance, resting-state EEG, P3 event-related potential amplitude, electrodermal habituation) in a large sample of twins who were assessed at age 17 or 20 years. Notably, the authors also examined the genetic architecture of startle eye blink modulation, an aspect of spontaneous emotion regulation, using a GWAS approach and found "little evidence of heritability in either biometric or molecular genetic analyses" (Vaidyanathan, Malone, Miller, McGue, & Iacono, 2014).

EMOTIONAL REACTIVITY

Serotonin-Related Genes

Serotonin is a neurotransmitter that is centrally involved in emotional reactivity (Carver, Johnson, & Joormann, 2008). A key regulator of serotonergic functioning is the 5-HTT protein, which removes serotonin released into the synaptic cleft. The serotonin transporter protein is encoded by a single gene (*SLC6A4*) on chromosome 17. Transcriptional activity of the *SLC6A4* gene is modulated by several common variants, including variations in the serotonin transporter–linked polymorphic region (5-HTTLPR). 5-HTTLPR is a repeat-length polymorphism with two primary variants. The short allele variant is associated with lower 5-HTT expression and thus with less uptake of serotonin from the synaptic cleft. The long allele variant is associated with higher transporter expression and thus with greater uptake of serotonin from the synaptic cleft (for a review, see Canli and Lesch (2007) and Lesch et al. (1996)).

There is growing evidence that the short allele of 5-HTTLPR amplifies emotional reactivity in positive (eg, amusement), negative (eg, fear), and self-conscious (eg, embarrassment) emotions. Numerous studies document a link between the short allele of 5-HTTLPR and heightened amygdala reactivity (Hariri et al., 2002 but see Bastiaansen et al., 2014), cortisol reactivity (Miller, Wankerl, Stalder, Kirschbaum, & Alexander, 2013), startle reactivity (Brocke et al., 2006), and subjective, behavioral, and physiological reactivity (eg, Gyurak et al. (2013), Study 1) to different kinds of negative emotional stimuli. Evidence for heightened stress reactivity associated with the short allele is further corroborated by a number of nonhuman animal studies with rhesus monkeys, mice, and rats (Caspi, Hariri, Holmes, Uher, & Moffitt, 2010). Several studies (Way & Taylor, 2010) have shown a link between the short allele of 5-HTTLPR and heightened reactivity in social-evaluative situations that often elicit self-conscious emotions powerfully (Tracy & Robins, 2007). We (Gyurak et al. (2013), Study 2) also have found evidence that the short allele is linked to heightened self-conscious emotional reactivity (ie, heightened embarrassment-related emotional reactions when watching oneself singing in a karaoke task). Research has also shown that the short allele is linked to positive emotional reactivity (Haase et al., 2013), heightened attention to positive images (Beevers et al., 2011), heightened self-reported positive affect in response to positive spousal affect (Schoebi, Way, Karney, & Bradbury, 2012), and heightened positive emotional expressions in response to positive and ambiguous stimuli (Haase et al., 2015). Enhanced reactivity to positive environmental conditions associated with reduced 5-HTT expression has also been demonstrated in 5-HTT knockout mice (Kästner et al., 2015). Besides 5-HTTLPR, other genes that involved in serotonergic functioning have emerged as important sources of individual differences in emotional reactivity (eg, monoamine oxidase-A gene and reactivity to negative socioemotional experiences) (Eisenberger, Way, Taylor, Welch, & Lieberman, 2007).

Other Genes

Genes involved in oxytocinergic functioning have also been implicated in emotion reactivity. Variations in the oxytocin-receptor gene have been linked to structural differences in brain regions implicated in emotion reactivity (eg, amygdala, hypothalamus) (Meyer-Lindenberg, Domes, Kirsch, & Heinrichs, 2011) and to individual differences in stress reactivity (Rodrigues, Saslow, Garcia, John, & Keltner, 2009).

EMOTION REGULATION

Dopamine-Related Genes

Dopamine is a neurotransmitter that plays an important role in emotion regulation. The protein catechol-*O*-methyltransferase (COMT) is a key regulator of dopamine in the prefrontal cortex (Meyer-Lindenberg et al., 2005). This polymorphism has two allele variants, the valine (val) allele and the methionine (met) allele, which are linked to sizeable variations in COMT enzyme activity (Lachman et al., 1996). Compared to the val allele, the met allele is associated with decreased COMT enzymatic activity and thus greater extracellular prefrontal dopamine (Tunbridge, Bannerman, Sharp, & Harrison, 2004).

Most studies that have examined COMT's role in prefrontal functioning have done so in the context of cognitive tasks, including executive functioning, working memory, fluid intelligence, and attentional control (Dickinson & Elvevag, 2009). These studies show that, overall, the met allele tends to confer a cognitive advantage in healthy populations.

The role of the COMT gene in emotional functioning in general and emotion regulation in particular is less clear. The COMT gene has been implicated in distal outcomes of psychopathology related to emotional dysregulation, but findings have not been consistent (Caspi et al., 2008) (for a review, see, for example, Witte and Floel (2012)). In terms of proximal outcomes, some studies have found that the met allele leads to reduced emotion regulation and lower top-down neural regulatory control (Bishop, Cohen, Fossella, Casey, & Farah, 2006) (for a review, see Canli, Ferri, and Duman (2009)). In a related vein, meta-analytic results from functional MRI studies have been interpreted as indicating that the met allele confers a cognitive advantage but an emotion-processing disadvantage (Mier, Kirsch, & Meyer-Lindenberg, 2010). In contrast, another set of studies reached the opposite conclusion: namely, that the met allele enhances emotion regulation (Canli et al., 2009). For example the met allele has also been found to predict higher self-reported emotion regulation (eg, "It is easy for me to get over a disappointing experience") (Weiss et al., 2014) and lower self-reported personal distress in emotional situations (eg, "Being in a tense emotional situation scares me") (Poletti et al., 2013).

To date, few studies have examined the effect of COMT on emotion regulation using laboratory paradigms derived from modern affective science. In two studies, we (Sapozhnikova et al., in preparation) have found that individuals with the met allele show greater emotion regulation abilities (ie, indexed by greater downregulation of emotional experience and emotional behavior) in several well-established emotion regulation tasks (ie, anticipated acoustic startle; instructed suppression; instructed cognitive reappraisal). Moreover, other studies by our group (Gyurak, Goodkind, Kramer, Miller, & Levenson, 2011; Gyurak et al., 2009) have found close links between emotion-regulating abilities and aspects of cognitive functioning (eg, executive functioning) that have been found to be enhanced in met allele carriers.

Other Genes

Emotion regulation is thought to be a function of bidirectional activity of both emotion-generating and emotion-regulating neural circuits. Thus, key candidate genes involved in emotion reactivity have also been implicated in emotion regulation (for reviews, see Canli et al. (2009) and Pezawas et al. (2005)).

EMOTIONAL AFFILIATION

Oxytocin-Related Genes

Oxytocin functions as a neurotransmitter and is thought to be centrally involved in affiliative behaviors associated with the caregiving-attachment system (Taylor et al., 2000). Oxytocinergic functioning depends on the availability of oxytocin receptors (OXTR). The OXTR is encoded by a single gene which is located on chromosome 3p25. It contains several dozen SNPs whose functionality is not yet fully understood. Among those, a common SNP (rs53576) in the OXTR gene has received particularly widespread attention.

Studies have shown that genetic variations in OXTR are implicated in a wide variety of affiliative behaviors, including maternal sensitivity (Bakermans-Kranenburg & van Ijzendoorn, 2008), prosocial temperament (Tost et al., 2010), prosocial behavioral cues (Kogan et al., 2011), seeking support under distress (Kim et al., 2010), loneliness (Lucht et al., 2009), altruistic behavior in economic tasks (Israel et al., 2009), and real-world prosocial behavior under threat (eg, volunteer work, charitable activities) (Poulin, Holman, & Buffone, 2012). However, a meta-analysis (which did not differentiate among different aspects of social behavior) failed to find support for an association between two commonly studied OXTR SNPs (including rs53576) and social behavior (Bakermans-Kranenburg & van Ijzendoorn, 2014).

It is possible that the link between OXTR and affiliation is more specific. Early nonhuman animal studies (examining knockout mice) suggested that OXTR has a crucial role in social recognition (Ferguson et al., 2000). Experimental studies in which oxytocin was administered intranasally likewise pointed to a role in boosting social cognition and cognitive empathy (but see Bartz et al. (2010), and Domes, Heinrichs, Michel, Berger, and Herpertz (2007)). Consistent with this, Rodrigues et al. (2009) showed that individuals with the GG variant of OXTR rs53576 had higher cognitive empathy, as measured by the "Reading the Mind in the Eyes" test and an other-oriented empathy scale. In a similar vein, individuals with the GG variant of rs53576 were found to have higher emotional empathy, indicated by increased levels of sympathetic and subjective arousal when perceiving harm to others (Smith, Porges, Norman, Connelly, & Decety, 2014).

A variety of other OXTR SNPs have been implicated in self-reported emotional and cognitive empathy (Wu, Li, & Su, 2012), but results have not always been consistent (Skuse et al., 2014). Studies linking OXTR genetic polymorphisms to objective, performance-based measures of cognitive empathy (Goodkind et al., 2012) or emotional empathy (Sze et al., 2012) have been rare. Yet, nonhuman animal studies point to a pivotal role of OXTR in social behavior and social recognition (Ferguson et al., 2000; Sala et al., 2013). It is hoped that future research will clarify OXTR's role in cognitive and emotional empathy in humans using more objective measures.

Other Genes and Affiliation

There is also evidence that polymorphisms associated with the vasopressin system are implicated in affiliative behaviors (Walum et al., 2008). This is consistent with the well-documented role that oxytocin and vasopressin have in social behaviors (for a review, see Meyer-Lindenberg et al. (2011)).

NEURAL AND GENETIC PATHWAYS: CLINICAL IMPLICATIONS

Thus far in this chapter, we have focused on the neural and genetic pathways that influence three key emotional processes: emotional reactivity, emotion regulation, and emotional affiliation. We now turn to a consideration of how these processes are disrupted in psychopathology and neuropathology.

EMOTIONAL DYSFUNCTION IN PSYCHOPATHOLOGY AND NEUROPATHOLOGY

Emotional Reactivity

Abnormalities in emotional reactivity are found in emotion magnitude (emotions that are too large or too small), duration (emotions that are too long-lasting or too short-lasting), and onset (emotions that are too slow or too quick to onset). Problems in emotional reactivity can manifest in a particular emotion (eg, overly large fear response in phobias) or in multiple emotions (eg, diminished reactivity in numerous emotions in frontotemporal dementia) (Eckart, Sturm, Miller, & Levenson, 2012; Sturm, Ascher, Miller, & Levenson, 2008). Dysfunction can also manifest in a lack of coherence among different aspects of the emotional response. Thus, for example, in schizophrenia expressive behavior can be profoundly blunted whereas subjective emotional experience remains strong (Kring & Neale, 1996). The inverse is seen in patients with ALS who have pseudobulbar affect, in which expressive behavior and physiological responses can be large whereas subjective experience is much smaller (Olney et al., 2011). Low coherence among aspects of the emotional response can be confusing and uncomfortable for patients and can interfere greatly with others' ability to understand the patient's emotional state and respond appropriately.

Emotion Regulation

Abnormalities in emotion regulation occur when emotions are either underregulated or overregulated. These problems may occur with particular emotions or with multiple emotions. The intimate connection between emotional reactivity and emotion regulation (Ochsner et al., 2004) can make it difficult to apportion responsibility for a particular symptom to different processes. Thus, for example, a patient with obsessive-compulsive disorder who reacts with inordinately high levels of disgust in response to lack of cleanliness likely reflects some combination of high levels of emotional reactivity and low levels of emotion regulation. In neurological patients, dissociations are sometimes found between instructed and spontaneous regulation. For example, patients with frontotemporal dementia do well at downregulating emotional responses when they are told exactly what to do but show markedly less downregulation when placed in a situation in which downregulation is the norm and they are not told what to do (Goodkind, Gyurak, McCarthy, Miller, & Levenson, 2010).

Emotional Affiliation

Abnormalities in emotional affiliation take a number of different forms, including poor recognition of emotions in others, lack of emotional response to the emotions of others, difficulty sustaining emotional interactions with intimate others, and overenmeshment with the emotions of others. These problems are major contributors to disruptions in social relationships that are found in many kinds of psychopathology (Ware, Hopper, Tugenberg, Dickey, & Fisher, 2007). Dissociations among different aspects of emotional affiliation are often hallmark features of pathology. For example, individuals with antisocial personality disorder may be good at recognizing the emotion that another person is experiencing but may have abnormal emotional responses to that person's emotion (eg, recognizing that another person is sad, not feeling sympathy, and then using the emotional information in exploitative ways).

EMOTION IN DIAGNOSIS

Psychopathology: *Diagnostic and Statistical Manual*

In the DSM-IV (American Psychiatric Association, 1994), alternations in emotional functioning are part of the descriptions of most Axis I and many Axis II disorders. Unfortunately, these descriptions are general and do not take advantage of the more differentiated constructs used in modern affective science. For example, among Axis I disorders in the DSM, emotional dysfunctions are described as follows (with italics added):

- Schizophrenia: "… may display *inappropriate affect* (eg, smiling, laughing, or a silly facial expression in the absence of an appropriate stimulus)"
- Autistic disorder: "*abnormalities of mood or affect* (eg, giggling or weeping for no apparent reason …)"
- Major depressive episode: "*depressed mood* most of the day, nearly every day"
- Manic episode: "a distinct period of abnormally and persistently elevated, expansive, or irritable mood …"
- Posttraumatic stress disorder: "*restricted range of affect* (eg, unable to have loving feelings)"
- Bulimia nervosa: "Binge eating is typically triggered by *dysphoric mood* states …"
- Dissociative identity disorder: "Particular identities may emerge in specific circumstances and may differ in reported age and gender, vocabulary, general knowledge, or *predominant affect*."

Psychopathology: Research Domain Criteria

Longstanding problems with the DSM (eg, intradiagnostic heterogeneity, interdiagnostic comorbidity, arbitrary cutoffs, too many syndromes) stimulated the development of the RDoC (Insel et al., 2010), a more dimensional, construct-based system of diagnosis designed for research use that is grounded in neurobiology and genetics rather than clinical observation. RDoC replaces syndromes with a set of five domains (negative valence systems, positive valence systems, cognitive systems, systems for social processes, and arousal/regulatory systems). Within each domain are a set of constructs (eg, acute threat in the negative valence domain) which must have a plausible associated neural circuitry and which span a range of normal to abnormal functioning. RDoC constructs were generated from a set of meetings of scientists and refined by the oversight committee with the door left open for future changes, additions, and deletions. Because of these origins, it is unsurprising that the initial set of constructs is more a snapshot of the current state of neuroscience than of the current state of clinical symptomatology. We believe that the RDoC approach holds great promise for producing meaningful, clinically relevant research in the coming years and for ultimately leading to a different kind of nosology for diagnosis and treatment.

RDoC has not yet fully elaborated its constructs in terms of its "units of analysis" (which include behavior, self-reports, and physiology, the critical trio for emotion) and measurement paradigms, but it certainly seems amenable to the ideas and organizational scheme presented in the current chapter. Emotional reactivity fits into the Negative and Positive valence systems domains (although RDoC seems to be more focused on "fear" than on other emotions). Emotion regulation fits into the arousal/regulatory systems domain (which currently has fewer constructs than the other domains and thus may expand in later versions). Emotional affiliation fits into the systems for social processes domain (which has "affiliation/separation" as one of its constructs).

Neuropathology

In neurology and neuropsychology, historically there has been much greater focus on the measurement and precise description of cognitive, language, and motor deficits than on emotional deficits. This is understandable given that many lesions and neurodegenerative disorders produce dramatic impairments in cognitive, language, and motor domains. However, these disorders often produce significant alternations in emotional functioning as well. When emotional functioning has been considered in neuropathology, it has often been conceptualized in general terms akin to those found in the DSM. For example, in the original consensus diagnostic criteria for frontotemporal dementia (Neary et al., 1998), a disorder that has profound effects on emotional functioning (Levenson & Miller, 2007), "early emotional blunting" is one of five core diagnostic features, the dominant features of the disorder are said to be "character change and disordered social conduct," and emotion is not mentioned in any of 15 supportive diagnostic features or any of the supportive neuropsychological test results.

EMOTION IN TREATMENT

Most of the major treatments for clinical disorders, whether biological, pharmacological, or psychotherapeutic, implicitly or explicitly target emotional functioning. Biological treatments such as deep brain stimulation and transcranial magnetic stimulation (George et al., 1996; Holtzheimer et al., 2012) are directed toward influencing the activity of the key neural circuits described earlier in the context of neural pathways. Pharmacological treatments target key neurotransmitter systems, including those discussed earlier in the context of genetic pathways. Psychotherapy has long focused on influencing emotional processes. This began with catharsis in psychoanalysis (Freud, 1910) and has continued with modern cognitive therapies (Beck, 1976), which target appraisal processes involved in both emotional reactivity and emotional regulation. A focus on emotional affiliation is seen in many therapies for couples (Johnson, Hunsley, Greenberg, & Schindler, 1999).

Arguably, one of the greatest challenges for the effective treatment of emotional dysfunction is lack of precision and specificity. As noted earlier, DSM clinical disorders are extremely complex and replete with heterogeneity and comorbidity, and define emotional dysfunctions in general ways. Most existing treatments are similarly imprecise. Thus, it is extremely difficult to isolate and reach certain circuits with biological treatments and to target particular neurotransmitters in particular brain regions with pharmacological treatments. It has similarly been difficult to isolate specific active ingredients in different psychotherapies that are effective with particular disease mechanisms.

There appears to be a clear movement in the treatment arena toward targeting more specific mechanisms (eg, RDoC) and toward developing treatments that ameliorate particular symptoms across a wide range of disorders (Harvey, Watkins, Mansell, & Shafran, 2004). This points to the need for better and more precise characterization of treatment targets. In the realm of emotional functioning, such precision has long been the concern of affective scientists studying emotional processes such as reactivity, regulation, and affiliation. Ironically, the insights and tools derived from laboratory studies of emotion have been slow to find their way into clinical practice. Nonetheless, we believe that building bridges that connect the laboratory to the bedside (and the bedside to the laboratory) will provide valuable clues that will be useful in understanding the etiology of emotional dysfunctions that are found in neurological, psychological, and psychiatric disorders and in identifying promising neural, genetic, and behavioral targets for future treatments.

REFERENCES

Abu-Akel, A., & Shamay-Tsoory, S. (2011). Neuroanatomical and neurochemical bases of theory of mind. *Neuropsychologia, 49*, 2971–2984.

Acevedo, B. P., Aron, A., Fisher, H. E., & Brown, L. L. (2012). Neural correlates of long- term intense romantic love. *Social Cognitive and Affective Neuroscience, 7*, 145–159.

American Psychiatric Association. (1994). *Diagnostic and statistical manual of mental disorders (4th ed.) (DSM-IV)*. Washington, DC: American Psychiatric Association.

Andresen, M. C., & Kunze, D. L. (1994). Nucleus tractus solitarius—gateway to neural circulatory control. *Annual Review of Physiology, 56*, 93–116.

Aron, A. R. (2007). The neural basis of inhibition in cognitive control. *The Neuroscientist, 13*, 214–228.

Aron, A. R., Robbins, T. W., & Poldrack, R. A. (2004). Inhibition and the right inferior frontal cortex. *Trends in Cognitive Sciences, 8*, 170–177.

Bakermans-Kranenburg, M. J., & van Ijzendoorn, M. H. (2008). Oxytocin receptor (OXTR) and serotonin transporter (5-HTT) genes associated with observed parenting. *Social Cognitive and Affective Neuroscience, 3*, 128–134.

Bakermans-Kranenburg, M. J., & van Ijzendoorn, M. H. (2014). A sociability gene? Meta- analysis of oxytocin receptor genotype effects in humans. *Psychiatric Genetics, 24*, 45–51.

Bandler, R., & Keay, K. A. (1996). Columnar organization in the midbrain periaqueductal gray and the integration of emotional expression. *Progress in Brain Research, 107*, 285–300.

Bandler, R., Keay, K. A., Floyd, N., & Price, J. (2000). Central circuits mediating patterned autonomic activity during active vs. passive emotional coping. *Brain Research Bulletin, 53*, 95–104.

Bard, P. (1934). On emotional expression after decortication with some remarks on certain theoretical views: part II. *Psychological Review, 41*, 424–449.

Bartels, A., & Zeki, S. (2004). The neural correlates of maternal and romantic love. *NeuroImage, 21*, 1155–1166.

Bartz, J. A., Zaki, J., Bolger, N., Hollander, E., Ludwig, N. N., Kolevzon, A., & Ochsner, K. N. (2010). Oxytocin selectively improves empathic accuracy. *Psychological Science, 21*, 1426–1428.

Bastiaansen, J. A., Servaas, M. N., Marsman, J. B. C., Ormel, J., Nolte, I. M., Riese, H., & Aleman, A. (2014). Filling the gap: relationship between the serotonin- transporter-linked polymorphic region and amygdala activation. *Psychological Science, 25*(11), 2058–2066.

Baumeister, R. F., & Leary, M. R. (1995). The need to belong: desire for interpersonal attachments as a fundamental human motivation. *Psychological Bulletin, 117*, 497–529.

Beck, A. T. (1976). *Cognitive therapy and the emotional disorders*. England: International Universities Press.

Beckes, L., & Coan, J. A. (2011). Social baseline theory: the role of social proximity in emotion and economy of action. *Social and Personality Psychology Compass, 5*, 976–988.

Beevers, C. G., Marti, C. N., Lee, H.-J., Stote, D. L., Ferrell, R. E., Hariri, A. R., & Telch, M. J. (2011). Associations between serotonin transporter gene promoter region (5-HTTLPR) polymorphism and gaze bias for emotional information. *Journal of Abnormal Psychology, 120*, 187–197.

Behbehani, M. M. (1995). Functional characteristics of the midbrain periaqueductal gray. *Progress in Neurobiology, 46*, 575–605.

Benarroch, E. E. (2006). Pain-autonomic interactions. *Neurological Sciences, 27*(Suppl. 2), S130–S133.

Bishop, S. J., Cohen, J. D., Fossella, J., Casey, B. J., & Farah, M. J. (2006). COMT genotype influences prefrontal response to emotional distraction. *Cognitive, Affective & Behavioral Neuroscience, 6*, 62–70.

Botvinick, M. M., Cohen, J. D., & Carter, C. S. (2004). Conflict monitoring and anterior cingulate cortex: an update. *Trends in Cognitive Sciences, 8*, 539–546.

Bowers, D., Blonder, L. X., & Heilman, K. M. (1991). *Florida affect battery*. Tallahassee: Center for Neuropsychological Studies, Cognitive Neuroscience Laboratory, University of Florida.

Brocke, B., Armbruster, D., Müller, J., Hensch, T., Jacob, C. P., Lesch, K. P., … Strobel, A. (2006). Serotonin transporter gene variation impacts innate fear processing: acoustic startle response and emotional startle. *Molecular Psychiatry, 11*, 1106–1112.

Buckner, R. L., Andrews-Hanna, J. R., & Schacter, D. L. (2008). The brain's default network: anatomy, function, and relevance to disease. *Annals of the New York Academy of Sciences, 1124*, 1–38.

Buckner, R. L., & Carroll, D. C. (2007). Self-projection and the brain. *Trends in Cognitive Sciences, 11*, 49–57.

Bush, G., Luu, P., & Posner, M. I. (2000). Cognitive and emotional influences in anterior cingulate cortex. *Trends in Cognitive Sciences, 4*, 215–222.

Cahill, L., & McGaugh, J. L. (1998). Mechanisms of emotional arousal and lasting declarative memory. *Trends in Neurosciences, 21*, 294–299.

Canli, T., Ferri, J., & Duman, E. A. (2009). Genetics of emotion regulation. *Neuroscience, 164*, 43–54.

Canli, T., & Lesch, K.-P. (2007). Long story short: the serotonin transporter in emotion regulation and social cognition. *Nature Neuroscience, 10*, 1103–1109.

Carrive, P., & Bandler, R. (1991). Control of extracranial and hindlimb blood flow by the midbrain periaqueductal grey of the cat. *Experimental Brain Research, 84*, 599–606.

Carter, C. S., Braver, T. S., Barch, D. M., Botvinick, M. M., Noll, D., & Cohen, J. D. (1998). Anterior cingulate cortex, error detection, and the online monitoring of performance. *Science, 280*, 747–749.

Carter, C. S., Williams, J. R., Witt, D. M., & Insel, T. R. (1992). Oxytocin and social bonding. *Annals of the New York Academy of Sciences, 652*, 204–211.

Carver, C. S., Johnson, S. L., & Joormann, J. (2008). Serotonergic function, two-mode models of self-regulation, and vulnerability to depression: what depression has in common with impulsive aggression. *Psychological Bulletin, 134*, 912–943.

Caspi, A., Hariri, A. R., Holmes, A., Uher, R., & Moffitt, T. E. (2010). Genetic sensitivity to the environment: the case of the serotonin transporter gene and its implications for studying complex diseases and traits. *The American Journal of Psychiatry, 167*, 509–527.

Caspi, A., Langley, K., Milne, B., Moffitt, T. E., O'Donovan, M., Owen, M. J., … Thapar, A. (2008). A replicated molecular genetic basis for subtyping antisocial behavior in children with attention-deficit/hyperactivity disorder. *Archives of General Psychiatry, 65*, 203–210.

Caspi, A., Sugden, K., Moffitt, T. E., Taylor, A., Craig, I. W., Harrington, H., … Poulton, R. (2003). Influence of life stress on depression: moderation by a polymorphism in the 5-HTT gene. *Science, 301*, 386–389.

Coan, J. A., Schaefer, H. S., & Davidson, R. J. (2006). Lending a hand: social regulation of the neural response to threat. *Psychological Science, 17*, 1032–1039.

Corbetta, M., & Shulman, G. L. (2002). Control of goal-directed and stimulus-driven attention in the brain. *Nature Reviews Neuroscience, 3*, 201–215.

Craig, A. D. (2002). How do you feel? Interoception: the sense of the physiological condition of the body. *Nature Reviews Neuroscience, 3*, 655–666.

Craig, A. D. (2005). Forebrain emotional asymmetry: a neuroanatomical basis? *Trends in Cognitive Sciences, 9*, 566–571.

Craig, A. D. (2009). How do you feel—now? The anterior insula and human awareness. *Nature Reviews Neuroscience, 10*, 59–70.

Critchley, H. D., & Harrison, N. A. (2013). Visceral influences on brain and behavior. *Neuron, 77*, 624–638.

Curtis, C. E., & D'Esposito, M. (2003). Persistent activity in the prefrontal cortex during working memory. *Trends in Cognitive Sciences, 7*, 415–423.

Dampney, R. A., Furlong, T. M., Horiuchi, J., & Iigaya, K. (2013). Role of dorsolateral periaqueductal grey in the coordinated regulation of cardiovascular and respiratory function. *Autonomic Neuroscience, 175*, 17–25.

Davis, M., & Whalen, P. J. (2001). The amygdala: vigilance and emotion. *Molecular Psychiatry, 6*, 13–34.

Denny, B. T., Kober, H., Wager, T. D., & Ochsner, K. N. (2012). A meta-analysis of functional neuroimaging studies of self- and other judgments reveals a spatial gradient for mentalizing in medial prefrontal cortex. *Journal of Cognitive Neuroscience, 24*, 1742–1752.

Dergacheva, O., Griffioen, K. J., Neff, R. A., & Mendelowitz, D. (2010). Respiratory modulation of premotor cardiac vagal neurons in the brainstem. *Respiratory Physiology & Neurobiology, 174*, 102–110.

Dickinson, D., & Elvevag, B. (2009). Genes, cognition and brain through a COMT lens. *Neuroscience, 164*, 72–87.

Domes, G., Heinrichs, M., Michel, A., Berger, C., & Herpertz, S. C. (2007). Oxytocin improves "mind-reading" in humans. *Biological Psychiatry, 61*, 731–733.

Eckart, J. A., Sturm, V. E., Miller, B. L., & Levenson, R. W. (2012). Diminished disgust reactivity in behavioral variant frontotemporal dementia. *Neuropsychologia, 50*, 786–790.

Eisenberger, N. I., Way, B. M., Taylor, S. E., Welch, W. T., & Lieberman, M. D. (2007). Understanding genetic risk for aggression: clues from the brain's response to social exclusion. *Biological Psychiatry, 61*, 1100–1108.

Ekman, P. (1984). Expression and the nature of emotion. In K. Scherer, & P. Ekman (Eds.), *Approaches to emotion* (pp. 319–344). Hillsdale, NJ: Lawrence Erlbaum.

Ekman, P. (1992). An argument for basic emotions. *Cognition and Emotion, 6*, 169–200.

Engen, H. G., & Singer, T. (2013). Empathy circuits. *Current Opinion in Neurobiology, 23*, 275–282.

Ethofer, T., Anders, S., Erb, M., Herbert, C., Wiethoff, S., Kissler, J., ... Wildgruber, D. (2006). Cerebral pathways in processing of affective prosody: a dynamic causal modeling study. *NeuroImage, 30*, 580–587.

Etkin, A., Prater, K. E., Schatzberg, A. F., Menon, V., & Greicius, M. D. (2009). Disrupted amygdalar subregion functional connectivity and evidence of a compensatory network in generalized anxiety disorder. *Archives of General Psychiatry, 66*, 1361–1372.

Fehr, E., & Camerer, C. F. (2007). Social neuroeconomics: the neural circuitry of social preferences. *Trends in Cognitive Sciences, 11*, 419–427.

Ferguson, J. N., Young, L. J., Hearn, E. F., Matzuk, M. M., Insel, T. R., & Winslow, J. T. (2000). Social amnesia in mice lacking the oxytocin gene. *Nature Genetics, 25*, 284–288.

Flint, J., & Munafò, M. R. (2013). Candidate and non-candidate genes in behavior genetics. *Current Opinion in Neurobiology, 23*, 57–61.

Fontes, M. A., Xavier, C. H., de Menezes, R. C., & Dimicco, J. A. (2011). The dorsomedial hypothalamus and the central pathways involved in the cardiovascular response to emotional stress. *Neuroscience, 184*, 64–74.

Francis, D. D., Champagne, F. C., & Meaney, M. J. (2000). Variations in maternal behaviour are associated with differences in oxytocin receptor levels in the rat. *Journal of Neuroendocrinology, 12*, 1145–1148.

Freud, S. (1910). The origin and development of psychoanalysis. *The American Journal of Psychology, 21*, 181–218.

Frith, C., & Frith, U. (2006). The neural basis of mentalizing. *Neuron, 50*, 531–534.

George, M. S., Wassermann, E. M., Williams, W. A., Steppel, J., Pascual-Leone, A., Basser, P., ... Post, R. M. (1996). Changes in mood and hormone levels after rapid-rate transcranial magnetic stimulation (rTMS) of the prefrontal cortex. *Journal of Neuropsychiatry and Clinical Neurosciences, 8*, 172–180.

Goldin, P. R., McRae, K., Ramel, W., & Gross, J. J. (2008). The neural bases of emotion regulation: reappraisal and suppression of negative emotion. *Biological Psychiatry, 63*, 577–586.

Goodkind, M. S., Gyurak, A., McCarthy, M., Miller, B. L., & Levenson, R. W. (2010). Emotion regulation deficits in frontotemporal lobar degeneration and Alzheimer's disease. *Psychology and Aging, 25*, 30–37.

Goodkind, M. S., Sollberger, M., Gyurak, A., Rosen, H. J., Rankin, K. P., Miller, B., & Levenson, R. (2012). Tracking emotional valence: the role of the orbitofrontal cortex. *Human Brain Mapping, 33*, 753–762.

Gottman, J. M., & Levenson, R. W. (1986). Assessing the role of emotion in marriage. *Behavioral Assessment, 8*, 31–48.

Gross, J. J. (1998). The emerging field of emotion regulation: an integrative review. *Review of General Psychology, 2*, 271–299.

Gross, J. J., & Levenson, R. W. (1993). Emotional suppression: physiology, self-report, and expressive behavior. *Journal of Personality and Social Psychology, 64*, 970–986.

Guyenet, P. G. (2006). The sympathetic control of blood pressure. *Nature Reviews Neuroscience, 7*, 335–346.

Gyurak, A., Goodkind, M. S., Kramer, J. H., Miller, B. L., & Levenson, R. W. (2011). Executive functions and the down-regulation and up-regulation of emotion. *Cognition and Emotion, 26*, 103–118.

Gyurak, A., Goodkind, M. S., Madan, A., Kramer, J. H., Miller, B. L., & Levenson, R. W. (2009). Do tests of executive functioning predict ability to downregulate emotions spontaneously and when instructed to suppress? *Cognitive, Affective, & Behavioral Neuroscience, 9*, 144–152.

Gyurak, A., Haase, C. M., Sze, J., Goodkind, M. S., Coppola, G., Lane, J., ... Levenson, R. W. (2013). The effect of the serotonin transporter polymorphism (5-HTTLPR) on empathic and self-conscious emotional reactivity. *Emotion, 13*, 25–35.

Haase, C. M., Beermann, U., Saslow, L. R., Shiota, M. N., Saturn, S. R., Lwi, S., ... Levenson, R. W. (2015). Short alleles, bigger smiles? Effects of 5-HTTLPR on positive emotional expressions. *Emotion, 15*(4), 438–448. in press.

Haase, C. M., Saslow, L. R., Bloch, L., Saturn, S. R., Casey, J. J., Seider, B. H., ... Levenson, R. W. (2013). The 5-HTTLPR polymorphism in the serotonin transporter gene moderates the association between emotional behavior and changes in marital satisfaction over time. *Emotion, 13*, 1068–1079.

Hare, T. A., Camerer, C. F., Knoepfle, D. T., & Rangel, A. (2010). Value computations in ventral medial prefrontal cortex during charitable decision making incorporate input from regions involved in social cognition. *Journal of Neuroscience, 30*, 583–590.

Hariri, A. R., Mattay, V. S., Tessitore, A., Kolachana, B., Fera, F., Goldman, D., ... Weinberger, D. R. (2002). Serotonin transporter genetic variation and the response of the human amygdala. *Science, 297*, 400–403.

Harvey, A. G., Watkins, E., Mansell, W., & Shafran, R. (2004). *Cognitive behavioural processes across psychological disorders: A transdiagnostic approach to research and treatment*. New York: Oxford University Press.

Hayes, J. P., Morey, R. A., Petty, C. M., Seth, S., Smoski, M. J., McCarthy, G., & Labar, K. S. (2010). Staying cool when things get hot: emotion regulation modulates neural mechanisms of memory encoding. *Frontiers in Human Neuroscience, 4*, 230.

Holtzheimer, P. E., Kelley, M. E., Gross, R. E., Filkowski, M. M., Garlow, S. J., Barrocas, A., ... Mayberg, H. S. (2012). Subcallosal cingulate deep brain stimulation for treatment-resistant unipolar and bipolar depression. *Archives of General Psychiatry, 69*, 150–158.

Iacoboni, M., Lieberman, M. D., Knowlton, B. J., Molnar-Szakacs, I., Moritz, M., Throop, C. J., & Fiske, A. P. (2004). Watching social interactions produces dorsomedial prefrontal and medial parietal BOLD fMRI signal increases compared to a resting baseline. *NeuroImage, 21*, 1167–1173.

Iacono, W. G., Malone, S. M., Vaidyanathan, U., & Vrieze, S. I. (2014). Genome-wide scans of genetic variants for psychophysiological endophenotypes: a methodological overview. *Psychophysiology, 51*, 1207–1224.

Inagaki, T. K., & Eisenberger, N. I. (2012). Neural correlates of giving support to a loved one. *Psychosomatic Medicine, 74*, 3–7.

Inagaki, T. K., & Eisenberger, N. I. (2013). Shared neural mechanisms underlying social warmth and physical warmth. *Psychological Sciences, 24*, 2272–2280.

Insel, T., Cuthbert, B., Garvey, M., Heinssen, R., Pine, D. S., Quinn, K., ... Wang, P. (2010). Research domain criteria (RDoC): toward a new classification framework for research on mental disorders. *The American Journal of Psychiatry, 167*, 748–751.

Israel, S., Lerer, E., Shalev, I., Uzefovsky, F., Riebold, M., Laiba, E., ... Ebstein, R. P. (2009). The oxytocin receptor (OXTR) contributes to prosocial fund allocations in the dictator game and the social value orientations task. *PLoS One, 4*, e5535.

Izuma, K., Saito, D. N., & Sadato, N. (2008). Processing of social and monetary rewards in the human striatum. *Neuron, 58*, 284–294.

Johnson, S. M., Hunsley, J., Greenberg, L., & Schindler, D. (1999). Emotionally focused couples therapy: status and challenges. *Clinical Psychology: Science and Practice, 6*, 67–79.

Kästner, N., Richter, S. H., Lesch, K. P., Schreiber, R. S., Kaiser, S., & Sachser, N. (2015). Benefits of a "vulnerability gene"? A study in serotonin transporter knockout mice. *Behavioural Brain Research, 283*, 116–120.

Keltner, D., & Kring, A. M. (1998). Emotion, social function, and psychopathology. *Review of General Psychology, 2*, 320–342.

Kim, H. S., Sherman, D. K., Sasaki, J. Y., Xu, J., Chu, T. Q., Ryu, C., ... Taylor, S. E. (2010). Culture, distress, and oxytocin receptor polymorphism(OXTR) interact to influence emotional support seeking. *Proceedings of the National Academy of Sciences of the United States of America, 107*, 15717–15721.

Klimecki, O. M., Leiberg, S., Ricard, M., & Singer, T. (2014). Differential pattern of functional brain plasticity after compassion and empathy training. *Social Cognitive and Affective Neuroscience, 9*, 873–879.

Knutson, B., Adams, C. M., Fong, G. W., & Hommer, D. (2001). Anticipation of increasing monetary reward selectively recruits nucleus accumbens. *Journal of Neuroscience, 21*, RC159.

Koenigsberg, H. W., Fan, J., Ochsner, K. N., Liu, X., Guise, K., Pizzarello, S., ... Siever, L. J. (2010). Neural correlates of using distancing to regulate emotional responses to social situations. *Neuropsychologia, 48*, 1813–1822.

Kogan, A., Oveis, C., Carr, E. W., Gruber, J., Mauss, I. B., Shallcross, A., ... Keltner, D. (2014). Vagal activity is quadratically related to prosocial traits, prosocial emotions, and observer perceptions of prosociality. *Journal of Personality and Social Psychology, 107*, 1051–1063.

Kogan, A., Saslow, L. R., Impett, E. A., Oveis, C., Keltner, D., & Rodrigues Saturn, S. (2011). Thin-slicing study of the oxytocin receptor (OXTR) gene and the evaluation and expression of the prosocial disposition. *Proceedings of the National Academy of Sciences of the United States of America, 108*, 19189–19192.

Kring, A. M., & Neale, J. M. (1996). Do schizophrenic patients show a disjunctive relationship among expressive, experiential, and psychophysiological components of emotion? *Journal of Abnormal Psychology, 105*, 249–257.

Kringelbach, M. L., & Rolls, E. T. (2004). The functional neuroanatomy of the human orbitofrontal cortex: evidence from neuroimaging and neuropsychology. *Progress in Neurobiology, 72*, 341–372.

Krueger, F., McCabe, K., Moll, J., Kriegeskorte, N., Zahn, R., Strenziok, M., ... Grafman, J. (2007). Neural correlates of trust. *Proceedings of the National Academy of Sciences of the United States of America, 104*, 20084–20089.

Kurth, F., Zilles, K., Fox, P. T., Laird, A. R., & Eickhoff, S. B. (2010). A link between the systems: functional differentiation and integration within the human insula revealed by meta-analysis. *Brain Structure and Function, 214*, 519–534.

Lachman, H. M., Papolos, D. F., Saito, T., Yu, Y. M., Szumlanski, C. L., & Weinshilboum, R. M. (1996). Human catechol-O-methyltransferase pharmacogenetics: description of a functional polymorphism and its potential application to neuropsychiatric disorders. *Pharmacogenetics, 6*, 243–250.

LeDoux, J. E. (1992). Emotion and the amygdala. In J. P. Aggleton (Ed.), *The amygdala: Neurobiological aspects of emotion, memory, and mental dysfunction* (pp. 339–351). New York, NY, USA: Wiley-Liss.

LeDoux, J. E., Iwata, J., Cicchetti, P., & Reis, D. J. (1988). Different projections of the central amygdaloid nucleus mediate autonomic and behavioral correlates of conditioned fear. *Journal of Neuroscience, 8*, 2517–2529.

Lesch, K.-P., Bengel, D., Heils, A., Sabol, S. Z., Greenberg, B. D., Petri, S., ... Murphy, D. L. (1996). Association of anxiety- related traits with a polymorphism in the serotonin transporter gene regulatory region. *Science, 274*, 1527–1531.

Levenson, R. W. (1994). Human emotion: a functional view. In P. Ekman, & R. J. Davidson (Eds.), *The nature of emotion: Fundamental questions* (pp. 123–126). New York: Oxford.

Levenson, R. W. (1999). The intrapersonal functions of emotion. *Cognition and Emotion, 13*, 481–504.

Levenson, R. W. (2000). Expressive, physiological, and subjective changes in emotion across adulthood. In S. H. Qualls, & N. Abeles (Eds.), *Psychology and the aging revolution: How we adapt to longer life* (pp. 123–140). Washington, DC, USA: American Psychological Association.

Levenson, R. W. (2003). Blood, sweat, and fears: the autonomic architecture of emotion. *Annals of the New York Academy of Sciences, 1000*, 348–366.

Levenson, R. W., Haase, C. M., Bloch, L., Holley, S. R., & Seider, B. H. (2013). Emotion regulation in couples. In J. J. Gross (Ed.), *Handbook of emotion regulation* (2nd ed., pp. 267–283). New York: Guilford Press.

Levenson, R. W., & Miller, B. M. (2007). Loss of cells—loss of self. Frontotemporal lobar degeneration and human emotion. *Current Directions in Psychological Science, 15*, 289–294.

Lieberman, M. D. (2007). Social cognitive neuroscience: a review of core processes. *Annual Review of Psychology, 58*, 259–289.

Liu, Y., Blackwood, D. H., Caesar, S., de Geus, E. J., Farmer, A., Ferreira, M. A., ... Wellcome Trust Case-Control, C.. (2011). Meta-analysis of genome-wide association data of bipolar disorder and major depressive disorder. *Molecular Psychiatry, 16*, 2–4.

Lucht, M. J., Barnow, S., Sonnenfeld, C., Rosenberger, A., Grabe, H. J., Schroeder, W., ... Rosskopf, D. (2009). Associations between the oxytocin receptor gene (OXTR) and affect, loneliness and intelligence in normal subjects. *Progress in Neuro-Psychopharmacology & Biological Psychiatry, 33*, 860–866.

Manolio, T. A., Collins, F. S., Cox, N. J., Goldstein, D. B., Hindorff, L. A., Hunter, D. J., ... Visscher, P. M. (2009). Finding the missing heritability of complex diseases. *Nature, 461*, 747–753.

Manuck, S. B., & McCaffery, J. M. (2014). Gene-environment interaction. *Annual Review of Psychology, 65*, 41–70.

McDonald, A. J. (1998). Cortical pathways to the mammalian amygdala. *Progress in Neurobiology, 55*, 257–332.

McRae, K., Hughes, B., Chopra, S., Gabrieli, J. D., Gross, J. J., & Ochsner, K. N. (2010). The neural bases of distraction and reappraisal. *Journal of Cognitive Neuroscience, 22*, 248–262.

Menezes, R. C., & Fontes, M. A. (2007). Cardiovascular effects produced by activation of GABA receptors in the rostral ventrolateral medulla of conscious rats. *Neuroscience, 144*, 336–343.

Meyer-Lindenberg, A., Domes, G., Kirsch, P., & Heinrichs, M. (2011). Oxytocin and vasopressin in the human brain: social neuropeptides for translational medicine. *Nature Reviews Neuroscience, 12*, 524–538.

Meyer-Lindenberg, A., Kohn, P. D., Kolachana, B., Kippenhan, S., McInerney-Leo, A., Nussbaum, R., ... Berman, K. F. (2005). Midbrain dopamine and prefrontal function in humans: interaction and modulation by COMT genotype. *Nature Neuroscience, 8*, 594–596.

Mier, D., Kirsch, P., & Meyer-Lindenberg, A. (2010). Neural substrates of pleiotropic action of genetic variation in COMT: a meta-analysis. *Molecular Psychiatry, 15*, 918–927.

Miller, R., Wankerl, M., Stalder, T., Kirschbaum, C., & Alexander, N. (2013). The serotonin transporter gene-linked polymorphic region (5-HTTLPR) and cortisol stress reactivity: a meta-analysis. *Molecular Psychiatry, 18*, 1018–1024.

Mitchell, J. P. (2009). Inferences about mental states. *Philosophical Transactions of the Royal Society of London B Biological Sciences, 364*, 1309–1316.

Moll, J., Zahn, R., de Oliveira-Souza, R., Bramati, I. E., Krueger, F., Tura, B., ... Grafman, J. (2011). Impairment of prosocial sentiments is associated with frontopolar and septal damage in frontotemporal dementia. *NeuroImage, 54*, 1735–1742.

de Moor, M. H., Costa, P. T., Terracciano, A., Krueger, R. F., de Geus, E. J., Toshiko, T., ... Boomsma, D. I. (2012). Meta-analysis of genome-wide association studies for personality. *Molecular Psychiatry, 17*, 337–349.

Morelli, S. A., Rameson, L. T., & Lieberman, M. D. (2014). The neural components of empathy: predicting daily prosocial behavior. *Social Cognitive and Affective Neuroscience, 9*, 39–47.

Morelli, S. A., Sacchet, M. D., & Zaki, J. (2014). Common and distinct neural correlates of personal and vicarious reward: a quantitative meta-analysis. *NeuroImage, 112*, 244–253.

Morelli, S. A., Torre, J. B., & Eisenberger, N. I. (2014). The neural bases of feeling understood and not understood. *Social Cognitive and Affective Neuroscience, 9*, 1890–1896.

Morishima, Y., Schunk, D., Bruhin, A., Ruff, C. C., & Fehr, E. (2012). Linking brain structure and activation in temporoparietal junction to explain the neurobiology of human altruism. *Neuron, 75*, 73–79.

Murray, E. A. (2007). The amygdala, reward and emotion. *Trends in Cognitive Sciences, 11*, 489–497.

Neary, D., Snowden, J. S., Gustafson, L., Passant, U., Stuss, D., Black, S., ... Benson, D. F. (1998). Frontotemporal lobar degeneration: a consensus on clinical diagnostic criteria. *Neurology, 51*, 1546–1554.

Ochsner, K. N., Ray, R. D., Cooper, J. C., Robertson, E. R., Chopra, S., Gabrieli, J. D., & Gross, J. J. (2004). For better or for worse: neural systems supporting the cognitive down- and up-regulation of negative emotion. *NeuroImage, 23*, 483–499.

Ochsner, K. N., Silvers, J. A., & Buhle, J. T. (2012). Functional imaging studies of emotion regulation: a synthetic review and evolving model of the cognitive control of emotion. *Annals of the New York Academy of Sciences, 1251*, E1–E24.

Olney, N. T., Goodkind, M. S., Lomen-Hoerth, C., Whalen, P. K., Williamson, C. A., Holley, D. E., ... Rosen, H. J. (2011). Behaviour, physiology and experience of pathological laughing and crying in amyotrophic lateral sclerosis. *Brain, 134*, 3458–3469.

Ongur, D., An, X., & Price, J. L. (1998). Prefrontal cortical projections to the hypothalamus in macaque monkeys. *Journal of Comparative Neurology, 401*, 480–505.

Ongur, D., & Price, J. L. (2000). The organization of networks within the orbital and medial prefrontal cortex of rats, monkeys, and humans. *Cerebral Cortex, 10*, 206–219.

Parkinson, C., Liu, S., & Wheatley, T. (2014). A common cortical metric for spatial, temporal, and social distance. *Journal of Neuroscience, 34*, 1979–1987.

Payer, D. E., Baicy, K., Lieberman, M. D., & London, E. D. (2012). Overlapping neural substrates between intentional and incidental down-regulation of negative emotions. *Emotion, 12*, 229–235.

Pezawas, L., Meyer-Lindenberg, A., Drabant, E. M., Verchinski, B. A., Munoz, K. E., Kolachana, B. S., ... Weinberger, D. R. (2005). 5-HTTLPR polymorphism impacts human cingulate-amygdala interactions: a genetic susceptibility mechanism for depression. *Nature Neuroscience, 8*, 828–834.

Poletti, S., Radaelli, D., Cavallaro, R., Bosia, M., Lorenzi, C., Pirovano, A., ... Benedetti, F. (2013). Catechol-O-methyltransferase (COMT) genotype biases neural correlates of empathy and perceived personal distress in schizophrenia. *Comprehensive Psychiatry, 54*, 181–186.

Porges, S. W. (2001). The polyvagal theory: phylogenetic substrates of a social nervous system. *International Journal of Psychophysiology, 42*, 123–146.

Poulin, M. J., Holman, E. A., & Buffone, A. (2012). The neurogenetics of nice: receptor genes for oxytocin and vasopressin interact with threat to predict prosocial behavior. *Psychological Science, 23*, 446–452.

Price, J. L., & Amaral, D. G. (1981). An autoradiographic study of the projections of the central nucleus of the monkey amygdala. *Journal of Neuroscience, 1*, 1242–1259.

Rodrigues, S. M., Saslow, L. R., Garcia, N., John, O. P., & Keltner, D. (2009). Oxytocin receptor genetic variation relates to empathy and stress reactivity in humans. *Proceedings of the National Academy of Sciences of the United States of America, 106*, 21437–21441.

Rolls, E. T. (2000). The orbitofrontal cortex and reward. *Cerebral Cortex, 10*, 284–294.

Rolls, E. T. (2008). Functions of the orbitofrontal and pregenual cingulate cortex in taste, olfaction, appetite and emotion. *Acta Physiologica Hungarica, 95*, 131–164.

Rosen, H. J., & Levenson, R. W. (2009). The emotional brain: combining insights from patients and basic science. *Neurocase, 15*, 173–181.

Russell, J. A., & Barrett, L. F. (1999). Core affect, prototypical emotional episodes, and other things called emotion: dissecting the elephant. *Journal of Personality and Social Psychology, 76*, 805–819.

Sabatinelli, D., Fortune, E. E., Li, Q., Siddiqui, A., Krafft, C., Oliver, W. T., … Jeffries, J. (2011). Emotional perception: meta-analyses of face and natural scene processing. *NeuroImage, 54*, 2524–2533.

Sala, M., Braida, D., Donzelli, A., Martucci, R., Busnelli, M., Bulgheroni, E., … Chini, B. (2013). Mice heterozygous for the oxytocin receptor gene (Oxtr(+/-)) show impaired social behaviour but not increased aggression or cognitive inflexibility: evidence of a selective haploinsufficiency gene effect. *Journal of Neuroendocrinology, 25*, 107–118.

Salthouse, T. A. (2004). What and when of cognitive aging. *Current Directions in Psychological Science, 13*, 140–144.

Saper, C. B., Loewy, A. D., Swanson, L. W., & Cowan, W. M. (1976). Direct hypothalamo-autonomic connections. *Brain Research, 117*, 305–312.

Sapozhnikova, A., Haase, C. M., Shiota, M. N., Coppola, G., Miller, B. L., & Levenson, R. W. (in preparation). The role of catechol O-methyltransferase (COMT) polymorphism in emotional regulation.

Saxe, R., & Kanwisher, N. (2003). People thinking about thinking people. The role of the temporo-parietal junction in "theory of mind". *NeuroImage, 19*, 1835–1842.

Schoebi, D., Way, B. M., Karney, B. R., & Bradbury, T. N. (2012). Genetic moderation of sensitivity to positive and negative affect in marriage. *Emotion, 12*, 208–212.

Seeley, W. W., Crawford, R., Rascovsky, K., Kramer, J. H., Weiner, M., Miller, B. L., & Gorno-Tempini, M. L. (2008). Frontal paralimbic network atrophy in very mild behavioral variant frontotemporal dementia. *Archives of Neurology, 65*, 249–255.

Seeley, W. W., Menon, V., Schatzberg, A. F., Keller, J., Glover, G. H., Kenna, H., … Greicius, M. D. (2007). Dissociable intrinsic connectivity networks for salience processing and executive control. *Journal of Neuroscience, 27*, 2349–2356.

Seeley, W. W., Zhou, J., & Kim, E. J. (2012). Frontotemporal dementia: what can the behavioral variant teach us about human brain organization? *Neuroscientist, 18*, 373–385.

Shewmon, D. A., Holmes, G. L., & Byrne, P. A. (1999). Consciousness in congenitally decorticate children: developmental vegetative state as self-fulfilling prophecy. *Developmental Medicine & Child Neurology, 41*, 364–374.

Silani, G., Lamm, C., Ruff, C. C., & Singer, T. (2013). Right supramarginal gyrus is crucial to overcome emotional egocentricity bias in social judgments. *Journal of Neuroscience, 33*, 15466–15476.

Silvers, J. A., Wager, T. D., Weber, J., & Ochsner, K. N. (2015). The neural bases of uninstructed negative emotion modulation. *Social Cognitive and Affective Neuroscience, 10*, 10–18.

Skuse, D. H., Lori, A., Cubells, J. F., Lee, I., Conneely, K. N., Puura, K., … Young, L. J. (2014). Common polymorphism in the oxytocin receptor gene (OXTR) is associated with human social recognition skills. *Proceedings of the National Academy of Sciences of the United States of America, 111*, 1987–1992.

Smith, K. E., Porges, E. C., Norman, G. J., Connelly, J. J., & Decety, J. (2014). Oxytocin receptor gene variation predicts empathic concern and autonomic arousal while perceiving harm to others. *Society for Neuroscience, 9*, 1–9.

Staudinger, M. R., Erk, S., & Walter, H. (2011). Dorsolateral prefrontal cortex modulates striatal reward encoding during reappraisal of reward anticipation. *Cerebral Cortex, 21*, 2578–2588.

Stellar, J. E., Cohen, A., Oveis, C., & Keltner, D. (2015). Affective and physiological responses to the suffering of others: compassion and vagal activity. *Journal of Personality and Social Psychology, 108*(4), 572–585.

Sturm, V. E., Ascher, E. A., Miller, B. L., & Levenson, R. W. (2008). Diminished self- conscious emotional responding in frontotemporal lobar degeneration patients. *Emotion, 8*, 861–869.

Sturm, V. E., Sollberger, M., Seeley, W. W., Rankin, K. P., Ascher, E. A., Rosen, H. J., … Levenson, R. W. (2013). Role of right pregenual anterior cingulate cortex in self-conscious emotional reactivity. *Social Cognitive and Affective Neuroscience, 8*, 468–474.

Sturm, V. E., Yokoyama, J. S., Eckart, J. A., Zakrzewski, J., Rosen, H. J., Miller, B. L., … Levenson, R. W. (2014). Damage to left frontal regulatory circuits produces greater positive emotional reactivity in frontotemporal dementia. *Cortex, 64C*, 55–67.

Sze, J. A., Gyurak, A., Goodkind, M. S., & Levenson, R. W. (2012). Greater emotional empathy and prosocial behavior in late life. *Emotion, 12*, 1129–1140.

Takagishi, H., Kameshima, S., Schug, J., Koizumi, M., & Yamagishi, T. (2010). Theory of mind enhances preference for fairness. *Journal of Experimental Child Psychology, 105*, 130–137.

Taylor, S. E., Klein, L. C., Lewis, B. P., Gruenewald, T. L., Gurung, R. A., & Updegraff, J. A. (2000). Biobehavioral responses to stress in females: tend-and-befriend, not fight-or-flight. *Psychological Review, 107*, 411–429.

Tooby, J., & Cosmides, L. (1990). The past explains the present: emotional adaptations and the structure of ancestral environments. *Ethology & Sociobiology, 11*, 375–424.

Tost, H., Kolachana, B., Hakimi, S., Lemaitre, H., Verchinski, B. A., Mattay, V. S., … Meyer-Lindenberg, A. (2010). A common allele in the oxytocin receptor gene (OXTR) impacts prosocial temperament and human hypothalamic-limbic structure and function. *Proceedings of the National Academy of Sciences of the United States of America, 107*, 13936–13941.

Tracy, J. L., & Robins, R. W. (2007). Self-conscious emotions: where self and emotion meet. In C. Sedikides, & S. Spence (Eds.), *The self in social psychology. Frontiers of social psychology series* (pp. 187–209). New York: Psychology Press.

Tucker, D. C., & Saper, C. B. (1985). Specificity of spinal projections from hypothalamic and brainstem areas which innervate sympathetic preganglionic neurons. *Brain Research, 360*, 159–164.

Tunbridge, E. M., Bannerman, D. M., Sharp, T., & Harrison, P. J. (2004). Catechol-o-methyltransferase inhibition improves set-shifting performance and elevates stimulated dopamine release in the rat prefrontal cortex. *Journal of Neuroscience, 24*, 5331–5335.

Vaidyanathan, U., Malone, S. M., Miller, M. B., McGue, M., & Iacono, W. G. (2014). Heritability and molecular genetic basis of acoustic startle eye blink and affectively modulated startle response: a genome-wide association study. *Psychophysiology, 51*, 1285–1299.

Vogt, B. A. (2005). Pain and emotion interactions in subregions of the cingulate gyrus. *Nature Reviews Neuroscience, 6*, 533–544.

Wager, T. D., Davidson, M. L., Hughes, B. L., Lindquist, M. A., & Ochsner, K. N. (2008). Prefrontal-subcortical pathways mediating successful emotion regulation. *Neuron, 59*, 1037–1050.

Walum, H., Westberg, L., Henningsson, S., Neiderhiser, J. M., Reiss, D., Igl, W., ... Lichtenstein, P. (2008). Genetic variation in the vasopressin receptor 1a gene (AVPR1A) associates with pair-bonding behavior in humans. *Proceedings of the National Academy of Sciences of the United States of America, 105*, 14153–14156.

Ware, N. C., Hopper, K., Tugenberg, T., Dickey, B., & Fisher, D. (2007). Connectedness and citizenship: redefining social integration. *Psychiatric Services, 58*, 469–474.

Way, B. M., & Taylor, S. E. (2010). The serotonin transporter promoter polymorphism is associated with cortisol response to psychosocial stress. *Biological Psychiatry, 67*, 487–492.

Weaver, I. C. G., Cervoni, N., Champagne, F. A., D'Alessio, A. C., Sharma, S., Seckl, J. R., ... Meaney, M. J. (2004). Epigenetic programming by maternal behavior. *Nature Neuroscience, 7*, 847–854.

Weiss, E. M., Freudenthaler, H. H., Fink, A., Reiser, E. M., Niederstatter, H., Nagl, S., ... Papousek, I. (2014). Differential influence of 5-HTTLPR - polymorphism and COMT Val158Met – polymorphism on emotion perception and regulation in healthy women. *Journal of the International Neuropsychological Society, 20*, 516–524.

Winston, J. S., Strange, B. A., O'Doherty, J., & Dolan, R. J. (2002). Automatic and intentional brain responses during evaluation of trustworthiness of faces. *Nature Neuroscience, 5*, 277–283.

Witte, A. V., & Floel, A. (2012). Effects of COMT polymorphisms on brain function and behavior in health and disease. *Brain Research Bulletin, 88*, 418–428.

Wu, N., Li, Z., & Su, Y. (2012). The association between oxytocin receptor gene polymorphism (OXTR) and trait empathy. *Journal of Affective Disorders, 138*, 468–472.

Yamazaki, Y., Hashimoto, T., & Iriki, A. (2009). The posterior parietal cortex and non- spatial cognition. *F1000Biology Reports, 1*, 74.

Zaki, J., Schirmer, J., & Mitchell, J. P. (2011). Social influence modulates the neural computation of value. *Psychological Science, 22*, 894–900.

Chapter 23

The Anatomy of Delusion

Georges Naasan
University of California San Francisco, San Francisco, CA, United States

INTRODUCTION AND DEFINITION

A 75-year-old woman accuses her husband of conspiring with a group of drug dealers and manufacturing illegal products in the bathroom of the house where they have lived together for more than 20 years. As proof, she shows him photographs of the bathroom walls, floor, and ceiling, pointing out cracks and dust as irrevocable evidence that someone has been manipulating dangerous chemicals that have altered the structure of the room. Her husband reminds her that their bathroom always looked like that and that he is with her at all times; however, she does not budge. A year earlier, she was hospitalized for a hemorrhagic stroke.

A 63-year-old man is convinced that his wife was kidnapped and the woman who currently lives with him is an alien imposter who looks like her but really is not her. She shows him her driving license and their marriage certificate but he is not convinced. He does not accept food or medications from her, is not comfortable sleeping with her in the same bed, and chases her around the house, asking her to leave. He carries a diagnosis of early-onset Alzheimer disease (AD).

Both of these are examples of delusions. According to the *Diagnostic and Statistical Manual of Mental Disorders*, Fifth Edition (American Psychiatric Association & American Psychiatric Association. DSM-5 Task Force, 2013), delusions are "fixed beliefs that are not amenable to change in light of conflicting evidence." Furthermore, delusions are termed bizarre when "they are clearly implausible and not understandable to same-culture peers and do not derive from ordinary life experiences." Delusions should be distinguished from hallucinations, which require a false perception from a sensory modality (visual, olfactory, etc.) and from confabulations, which are spoken spontaneous unreal stories but without a fixed belief. Often confabulatory stories change with multiple interviews, whereas delusions are consistent and reproducible.

Early on, delusions were deemed mysterious and incomprehensible phenomena, probably because of their bizarreness and complexity. However, evidence from research has shed much light into the science, psychology, biology, and anatomy of delusions, especially with increasing recognition of the neurological origin of delusions and their presence in many neurological conditions such as strokes and neurodegenerative diseases.

Delusions are prevalent in psychiatric diagnoses, particularly within the psychosis spectrum disorders. Up to 70% of patients with schizophrenia can present with persecutory delusions or ideas of reference, and up to 30% with religious or grandiose delusions (Gilleen & David, 2005). In neurological conditions, up to 40% of patients with AD have psychosis (Lyketsos et al., 2001; Ropacki & Jeste, 2005). Previously thought to be less prevalent in studies looking at clinically defined frontotemporal dementia (FTD) (Mendez, Shapira, Woods, Licht, & Saul, 2008), psychosis was found in 21–38% of patients with behavioral variant FTD (bvFTD) or FTD with motor neuron disease, especially when an expansion in the C9orf72 gene was involved (Sha et al., 2012; Snowden et al., 2012). Delusions are also seen in around 4% of patients with acute stroke within 24 h of admission (Kumral & Ozturk, 2004).

Some people do not necessarily act on their delusions, which has led some scientists to suggest that delusions may be "empty speech" (Berrios, 1991). However, studies have shown that many people with delusions may act on their false beliefs in a manner that is destructive, if not catastrophic, to themselves and to others in their surroundings (Gilleen & David, 2005). For example, patients with Capgras delusion, the belief that a person has been replaced by an identical-looking imposter, usually misidentify people to whom they are emotionally attached, often spouses. The consequences of that can be tragic and include violence and homicide (Bourget & Whitehurst, 2004; Silva, Leong, Weinstock, & Wine, 1993). Furthermore, psychosis is a major source of stress, anxiety, and depression, and places a significant burden

Genomics, Circuits, and Pathways in Clinical Neuropsychiatry. http://dx.doi.org/10.1016/B978-0-12-800105-9.00023-8

on people experiencing the disorder and on their caregivers (Lyketsos, Miller, & Neuropsychiatric Syndromes Professional Interest Area of the International Society to Advance Alzheimer's Research and Treatment, 2012). In addition, the disease is associated with earlier institutionalization and a higher cost of care (Herrmann et al., 2006).

In this chapter, we will focus primarily on understanding the phenomenology of delusions, exploring what is known about their biology and associated neural circuitries, in an attempt to better delineate and define the brain anatomy involved in their manifestations.

PHENOMENOLOGY AND ANATOMY

The Two-Factor Hypothesis and the Role of the Right Prefrontal Cortex

Modern psychological theories of understanding delusions focus on a **two-factor hypothesis** whereby a delusion requires (1) a neuropsychological impairment that leads to the formation of the false belief, in addition to (2) an impairment in processes that would normally lead to the rejection of the belief in question (Coltheart, 2010). It is further postulated that the first impairment is what dictates the nature or content of the delusion, whereas the second impairment is common to all delusions, leading to the fixation and persistence of the false belief.

As an example, the origins of the Capgras delusion are thought to be neurological in nature (see subsequent discussion for details). Many studies have demonstrated that people with the Capgras delusion do not have an autonomic response to familiar faces compared with unfamiliar faces, a phenomenon normally seen in healthy individuals (Ellis, Young, Quayle, & De Pauw, 1997; Hirstein & Ramachandran, 1997). However, not all patients with this impairment develop the Capgras delusion (Tranel, Damasio, & Damasio, 1995). In fact, patients with occipitotemporal damage are able to recognize the identity of familiar faces correctly, yet fail to mount the normal skin conductance response compared with control subjects (Tranel et al., 1995). In addition, contrary evidence to the Capgras belief, such as information that only the spouse would know, or the proffering of a wedding band engraved with the spouse's initials, fails to stir the patient with the delusion away from his or her belief. Therefore, some neurological impairment must also exist that leads to the inability to reject the delusion.

The anatomy of the first impairment leading to the content of the delusion varies according to the specific content in question. By contrast, the second impairment must be anatomically constant across all delusions and should explain the failure of the brain in refuting the implausible hypothesis that is the content of the delusion. Let us try to understand that common denominator from a phenomenological–anatomical perspective.

The task of hypothesis formation and hypothesis evaluation was studied using functional MRI (fMRI) techniques (Fletcher et al., 2001). Normal control research subjects were asked to formulate a hypothesis regarding what a medical condition might be in a patient taking a certain medication. During the task, subjects received consistent feedback that allowed them to formulate hypotheses accurately and predict medical conditions, except occasionally when they were given "surprising" feedback that contradicted prior information. Results revealed that there was high activation in the **right dorsolateral prefrontal cortex** (PFC) during the task of hypothesis evaluation, specifically when feedback contradicting prior hypotheses was given to subjects. Other algorithms for hypothesis generation and hypothesis evaluation, using a paradigm of predicting whether a person will have an allergic reaction to a meal based on hypotheses of what the allergen was, were studied by means of fMRI and pointed to involvement of the right dorsolateral PFC in hypothesis formation and error monitoring (Corlett et al., 2004; Turner et al., 2004). Moreover, the same paradigm was given to patients with delusions, and showed that the right dorsolateral PFC was activated both when they were given conflicting feedback and needed to reevaluate their hypothesis and when they were given consistent feedback that did not necessarily require them to question their prior hypothesis (Corlett et al., 2007).

In support of this theory, the PFC was implicated in prediction-error processing (Corlett et al., 2004; Turner et al., 2004) and reality-monitoring tasks (Lagioia et al., 2011). Structural volume loss in the PFC also correlated with symptoms of delusion, regardless of their content (Buchy et al., 2012; Tost et al., 2010). A dysfunction in PFC signaling by fMRI techniques also correlated with delusion formation (Blackwood et al., 2000; Corlett et al., 2007). Finally, individuals with surgical excisions of the PFC were reported to have a greater "jumping to conclusions" reasoning bias than were control groups (Lunt et al., 2012).

Lesion-based studies suggest that the emergence of new delusional ideation is associated with a new right hemispheric injury, potentially superimposed on preexisting parenchymal volume loss (Levine & Grek, 1984). Delusions were seen in 15 of 360 patients (4%) with acute stroke within 24 h of admissions, all of whom had right hemispheric lesions, most of whom had persecutory-type delusions (Kumral & Ozturk, 2004). A review of the literature (Devinsky, 2009) on the anatomy of delusions when it occurs in neurological diseases such as stroke and degenerative disease similarly found overwhelming evidence for a dysfunction of the right hemisphere coupled with the release of an intact and creatively narrative left hemisphere, and concluded that "delusions result from right hemisphere lesions; but it is the left hemisphere that is deluded."

Studies conducted to hone in on a common anatomy for patients developing delusions after a stroke found an overlapping region of the right inferior frontal gyrus in three patients with delusions after stroke (Devine et al., 2014). An association between right caudate stroke and delusions was observed in eight patients in whom there was also hypometabolism in the right inferior frontal gyrus structure (McMurtray, Sultzer, Monserratt, Yeo, & Mendez, 2008). Finally, the PFC demonstrated volume loss in patients with bvFTD and psychosis (Mendez, Fras, Kremen, & Tsai, 2011), it was affected both on imaging and pathological analysis in patients with AD and psychosis (Ismail et al., 2011; Lopez, Smith, Becker, Meltzer, & DeKosky, 2001; Nakano, Yamashita, Matsuda, Kodama, & Yamada, 2006; Sultzer et al., 2003), and its hypometabolism predicted delusions and was associated with the formation of hallucinations in patients with dementia with Lewy body (Perneczky et al., 2008, 2009).

According to the two-factor model, delusions arise from (1) a neurological dysfunction which explains the content of the delusion, with the anatomy varying across different types of delusions. In addition, there is (2) a dysfunction in hypothesis-evaluation and error-monitoring systems which is common to all delusions and localizes to the right PFC.

Attention Bias and Faulty Assignment of Salience to External Stimuli: The "Aberrant Salience" Model

There is evidence in the literature to suggest that patients with delusions, especially persecutory delusions, may have a preferential bias to attend to environmental cues and stimuli that appear threatening, especially when they threaten the image or the concept of the self. For example, patients with schizophrenia who were presented with photographs that were threatening, neutral, or ambiguous quickly found threat in ambiguous photographs and spent more time appraising the threatening elements of the picture than did control subjects and other schizophrenic individuals without persecutory delusions (Blackwood, Howard, Bentall, & Murray, 2001).

The "aberrant salience" model proposes that delusions arise in the setting of the inappropriate processing of stimuli that would otherwise be considered irrelevant. This promotes a faulty representation of the environment leading to erroneous inferences. Animal models suggest that this is mediated by abnormal dopaminergic signaling in the ventral striatum and involves abnormal regulation of dopamine transmission by the PFC and hippocampus (see subsequent discussion on paranoid delusion). Many studies demonstrate increased striatal dopamine synthesis and release in patients with psychosis or prodromal signs of psychosis (O. Howes et al., 2011; O.D. Howes et al., 2011; O.D. Howes, Egerton, et al., 2009; O.D. Howes, Montgomery et al., 2009). Studies have shown that unmedicated patients at risk for delusions attach motivational salience to otherwise irrelevant stimuli while increasing striatal dopamine synthesis (Roiser, Howes, Chaddock, Joyce, & McGuire, 2013).

The role of dopamine in delusion stems from the association between dopamine and the manifestation of psychosis. In fact, antipsychotic medications work by blocking dopamine receptors. Also, dopamine mediates reward processing and hence may have a role in the representation of stimuli as rewarding or aversive. Consequently, impairment in dopamine regulation may result in a faulty assignment of salience to external stimuli.

Additional evidence for the aberrant salience hypothesis comes from a study that investigated gray matter volume differences between two groups of patients with bipolar disease, one with and one without delusions (Radaelli et al., 2014). Lower gray matter volume was found in the dorsolateral PFC and insula of patients with delusions compared with those without. Both of these anatomical structures are involved in the salience and executive-control network (Seeley et al., 2007).

Certain types of delusions, especially paranoid and persecutory delusions, may arise from an aberrant network of salience to environmental stimuli, with the anatomy localizing to the ventral striatum, PFC, and insula.

The Theory of Mind in People With Delusions

One proposed theory of delusion formation is that of a malfunction of theory of mind, in which people with delusions, especially of the persecutory and paranoid type, are unable to draw accurate inferences of other people's intentions, thoughts, and beliefs. This leads to false conclusions regarding the motives of people and the truth of a particular situation, and this in turn predisposes to a state of paranoia. In fact, in a study assessing the ability of subjects to infer the intention behind various samples of indirect speech (for example, "It's hot in here" would mean "Please open a window"), difficulty performing the tasks was observed in patients with schizophrenia and persecutory delusions, compared with those in remission (Corcoran, Mercer, & Frith, 1995). Other studies show that patients with schizophrenia indeed perform poorly on tasks of theory of mind (Frith & Corcoran, 1996; Lee, Quintana, Nori, & Green, 2011), which implicates a role for the temporoparietal junction and the medial PFC (Lee et al., 2011).

Although dysfunction in the theory of mind potentially explains why patients may become suspicious of other people's actions or speech, it does not necessarily explain why this has to progress into a full-blown fixed belief irrefutable by contrary evidence. Therefore, and following the two-factor hypothesis explored previously, impairment in the theory of mind needs to be coupled with another injury, such as to the right PFC, to lead to paranoid and persecutory delusions.

These proposed theories are not necessarily comprehensive, nor is there a consensus as to what constitutes the neurological circuitry of delusions. It is probable that delusions do not necessarily represent a homogeneous group of symptoms and that different delusions have different anatomies. Therefore, let us examine our knowledge of some specific delusions as examples.

CONTENT-SPECIFIC DELUSIONS

In this next section, we will explore two examples of specific delusions: paranoid delusions and the Capgras delusions, summarizing what we know about them in terms of their neurological basis and anatomical correlation. This is meant to provide examples illustrating the association of delusions to neurological malfunction and underlining the complexities behind the phenomenology of delusions. Table 23.1 provides a list of commonly encountered content-specific delusions.

Paranoid Delusions

Theories of the anatomical localization of paranoid delusions seem to implicate the limbic system, possibly including the entorhinal cortex and the hippocampus. In animal models, these structures were found to be involved in the processing of sensory information that converge there from other primary sensory cortices and modality-specific association areas (Jones & Powell, 1970; Mesulam, Van Hoesen, Pandya, & Geschwind, 1977). In 1997, Bogerts described the pathological findings on autopsy of a man who developed an isolated paranoid delusion that people knew a secret of his and were mocking him. This led him to engage in the serial killing of his wife, children, and eight other people. On autopsy, he was found to have a developmental abnormality within the posterior parahippocampal gyrus (Bogerts, 1997). The author proposed that lesions in the limbic system, particularly the hippocampus, altered the integration of sensory information with personal, emotional, and social experiences.

These theories were extrapolated and investigated in patients with AD, a disease that preferentially involves the hippocampi in elderly individuals. It was found that people with clinical AD who have paranoid delusions had a more pronounced volume loss in the right hippocampus than do those without paranoid delusions (Geroldi et al., 2000). Despite these early studies, the role of the hippocampus in the formation of paranoid delusions in patients with AD has not been

TABLE 23.1 Examples of Content-Specific Delusions

Delusions Name	Delusion Content
Paranoid delusion	One is being persecuted or other people are plotting against one
De Clerambault syndrome	One is loved by someone of a higher socioeconomic status
Othello syndrome	One is being cheated on by one's spouse
Grandiose delusion	One has an inflated worth, or knowledge/power
Parasitosis syndrome	One is infested by parasites (insects, for example)
Capgras delusion	One's spouse has been replaced by an identical-looking imposter
Reduplicative paramnesia	A place exists in two different locations simultaneously
Fregoli delusion	Different people are in fact a single person in disguise
Intermetamorphosis	People can change into someone else in appearance/personality
Dorian Gray syndrome	One is not aging
Cotard delusion	One is deceased

firmly defined. It may be that hippocampal dysfunction leads to the first hit needed in the two-factor model proposed earlier, which explains the content of paranoia. A second factor of the right frontal structure would still be needed for the delusion to form. Indeed, in support of this, a review of 25 studies on neuroimaging in AD and delusions found a common implication of the right frontal structures in the formation of paranoid delusions in this group of patients (Ismail, Nguyen, Fischer, Schweizer, & Mulsant, 2012).

Finally, patients with schizophrenia who have paranoid delusions were found to have higher indices of anxiety and greater anticipation of threat, which highlights the role and influence of affect in delusions (Freeman et al., 2013).

The Capgras Delusion

In March 1895, a cooper lived in a small Irish town with his wife and her father. Their house was built on the site of a fairy ring fort. After growing suspicious that there were witches roaming around, the cooper became convinced, along with his father-in-law, that his then-ill wife was possessed by a changeling and that she looked like, but was not, his wife. As superstitious beliefs had it at the time, the only way to bring his wife back was to burn the changeling out of her. He gathered at home with eight neighbors and they collectively set the woman on fire. Some of the details of that story were invariably altered by time and rumor, but the homicide of Bridget Cleary by her husband in 19th-century Ireland is undeniably true.

A few decades later, the Capgras delusion was reported in 1923, when Jean-Marie Joseph Capgras and his intern Jean Reboul-Lachaux described the case of a 53-year-old woman who believed that her husband, her children, the police, and her neighbors all had been replaced by identical-looking imposters (Capgras & Reboul-Lachaux, 1923). Long thought to be a psychiatric phenomenon with no neurological basis, the Capgras delusion has increasingly come to be seen as originating from a brain disorder with increasing evidence for its occurrence after head trauma or stroke, and in neurodegenerative illnesses (Alexander, Stuss, & Benson, 1979; Collins, Hawthorne, Gribbin, & Jacobson, 1990; Gluckman, 1968; Goldfarb & Weiner, 1977; Horikawa et al., 2006; Merrin & Silberfarb, 1976; Weston & Whitlock, 1971). Misidentification syndromes were found to occur in 15.8% of AD and 16.6% of clinically defined patients with dementia with Lewy body, of which the Capgras delusion is the most common (Harciarek & Kertesz, 2008).

Many neuropsychological theories are proposed to explain the phenomenon of Capgras delusions. Staton, Brumback, and Wilson (1982) described the case of a patient with a right posterior hippocampus and right temporoparietal—occipital junction lesion resulting in the Capgras delusion as well as reduplicative paramnesia, the delusion that a place has been doubled and exists in two places. They postulated that the syndrome occurred as the result of a disconnection between remote memory stores and new memory formation. Unable to lay down new memories of subtle changes in physical appearance, the subject relies on stored images from the past to process present visual cues; the illusion of doubles is created. However, this theory accounts for neither the laterality of the lesion nor the specificity of Capgras delusion to attached faces. In addition, an unpublished series of three patients with Capgras delusion failed to demonstrate a clear temporal cutoff when patients were presented with backdated pictures of their misidentified relative (Edelstyn & Oyebode, 1999).

Other proposed neurological theories include a disconnect between two brain hemispheres, each generating its own image of a known face (David, 1994; Edelstyn et al., 1997; Joseph, 1986); a malfunction of the right hemisphere, thought to be specialized in the "recognition of uniqueness" (Cutting, 1990; Kosslyn, 1987); and a disconnection between "overt" and "covert" face visual-processing pathways (Bauer, 1984; Ellis & Young, 1990; Ellis et al., 1997). According to the latter theory, the Capgras delusion may behave as the "mirror image of prosopagnosia," defined as a general inability to recognize faces. In fact, patients with prosopagnosia retain the appropriate skin conductance response expected when looking at familiar faces (Bauer, 1984); the opposite is true of patients with the Capgras delusion (Ellis et al., 1997; Hirstein & Ramachandran, 1997). Bauer (1986, 1984) described a dual pathway of face processing: overt and a covert, or affective, recognition. While he entertained the idea that prosopagnosia results from damage to the overt route with preservation of the covert route, Ellis and Young (1990) proposed that the opposite results in the Capgras delusion.

However, as Hirstein and Ramachandran (1997) point out, this theory has a few fallacies, notably that lesions resulting in the Capgras delusion are more commonly ventral (occipitotemporal) and that the ventral pathway is strongly connected with the amygdala. Ramachandran proposes that the delusion results from a disconnection between face-processing areas and the limbic system. This leads to the creation of separate memory "files" of the same person, because patients fail to link together successive episodic memories. Most lesions studies point to a dysfunction in the right hemisphere with a possible predilection for lesions in the right temporo-parieto-occipital region, which suggests that a dysfunction of face processing with possible limbic dysfunction could underlie the Capgras phenomenon. Infarction in the right hemisphere was also found to be implicated in patients with Cotard delusion, the belief that one is not alive (Nishio & Mori, 2012).

Regardless of their validity, these theories explain neither the specificity of the Capgras delusion to closely loved ones nor the formation of a fixed belief that is not amenable to change when provided with evidence to the contrary. Other caveats to theories of visual face processing in the misidentification of loved ones are reports of the Capgras delusion in blind patients (Dalgalarrondo, Fujisawa, & Banzato, 2002; Hermanowicz, 2002; Rojo, Caballero, Iruela, & Baca, 1991) and cases in which it manifests over the phone in the absence of the visual stimulus of the loved one's face (Dietl, Herr, Brunner, & Friess, 2003). It might be that Capgras delusion finds its neurophysiology at a larger scale than a simplistic visual face-processing dysfunction or that these cases are physiologically distinct with a seemingly common phenotype.

GENETICS

Little is known about the genetic implications of delusions. The discovery of expansions in the C9orf72 gene as culprits in some familial forms of FTD has resulted in a slurry of articles determining the clinical phenotype of patients who carry those expansions. Consistently, it was shown that patients with FTD and who carry the C9orf72 expansion were more likely to have symptoms of psychosis including delusions than were those without the expansion (Sha et al., 2012; Snowden et al., 2012). The delusions associated with the C9orf72 expansion were often described as bizarre, and at times related somehow to the patient's own body, such as the delusion of pregnancy (Larner, 2013) or of foreign objects present in the body (Snowden et al., 2012).

There has also been some evidence for the presence of delusions in patients with mutations in the progranulin gene (Le Ber et al., 2008; Rohrer et al., 2011). Some of the specific content of delusions reported in patients with FTD and a progranulin gene mutation are delusions of persecution and jealousy, which is interestingly different from some of the delusion content that is often reported in C9orf72 carriers (see previous discussion). This may speak for the difference in anatomy regarding the first factor of the two-factor hypothesis for delusions, with a common anatomy of involvement of the right frontal cortex in both genetic phenotypes.

In AD, there are some conflicting results regarding the role of apolipoprotein E4 (APOE4) in the development of delusions. In some studies, the possession of an APOE4 allele was associated with the early development of delusion in the late-onset AD (Chen et al., 2012; Spalletta et al., 2006), with a more important effect seen in homozygosity (van der Flier et al., 2007). However, in other studies, hallucinations were found to be significantly more frequent in patients with no APOE4 alleles compared with those with two APOE4 alleles. However, no statistical difference was found for the occurrence of delusions (Christie et al., 2012). The same conflicting evidence is encountered in studies examining the effect of 5-hydroxytryptamine-2A T102C receptor polymorphism on the occurrence of delusions in patients with AD (Angelucci et al., 2009; Craig, Donnelly, Hart, Carson, & Passmore, 2007; Pritchard et al., 2008). Finally, the serotonin transporter–linked polymorphic region polymorphism was found to be associated with delusions in dementia with Lewy body disease (Creese et al., 2014).

REFERENCES

Alexander, M. P., Stuss, D. T., & Benson, D. F. (1979). Capgras syndrome: a reduplicative phenomenon. *Neurology, 29*(3), 334–339.

American Psychiatric Association, & American Psychiatric Association. DSM-5 Task Force. (2013). *Diagnostic and statistical manual of mental disorders: DSM-5* (5th ed.). Arlington, VA: American Psychiatric Association.

Angelucci, F., Bernardini, S., Gravina, P., Bellincampi, L., Trequattrini, A., Di Iulio, F., ... Spaletta, G. (2009). Delusion symptoms and response to antipsychotic treatment are associated with the 5-HT2A receptor polymorphism (102T/C) in Alzheimer's disease: a 3-year follow-up longitudinal study. *Journal of Alzheimer's Disease: JAD, 17*(1), 203–211.

Bauer, R. M. (1984). Autonomic recognition of names and faces in prosopagnosia: a neuropsychological application of the guilty knowledge test. *Neuropsychologia, 22*(4), 457–469.

Bauer, R. M. (1986). In H. D. Ellis, M. A. Jeeves, F. Newcombe, & A. Young (Eds.), *The cognitive psychophysiology of prosopagnosia* (pp. 253–267). Boston: Martinus Nijhoff Ed.

Berrios, G. E. (1991). Delusions as "wrong beliefs": a conceptual history. *The British Journal of Psychiatry. Supplement, 14*(14), 6–13.

Blackwood, N. J., Howard, R. J., Bentall, R. P., & Murray, R. M. (2001). Cognitive neuropsychiatric models of persecutory delusions. *The American Journal of Psychiatry, 158*(4), 527–539.

Blackwood, N. J., Howard, R. J., ffytche, D. H., Simmons, A., Bentall, R. P., & Murray, R. M. (2000). Imaging attentional and attributional bias: an fMRI approach to the paranoid delusion. *Psychological Medicine, 30*(4), 873–883.

Bogerts, B. (1997). The temporolimbic system theory of positive schizophrenic symptoms. *Schizophrenia Bulletin, 23*(3), 423–435.

Bourget, D., & Whitehurst, L. (2004). Capgras syndrome: a review of the neurophysiological correlates and presenting clinical features in cases involving physical violence. *Canadian Journal of Psychiatry. Revue Canadienne De Psychiatrie, 49*(11), 719–725.

Buchy, L., Ad-Dab'bagh, Y., Lepage, C., Malla, A., Joober, R., Evans, A., & Lepage, M. (2012). Symptom attribution in first episode psychosis: a cortical thickness study. *Psychiatry Research, 203*(1), 6–13.

Capgras, J. M., & Reboul-Lachaux, J. (1923). L'illusion des "sosies" dans un delire systematise chronique. *Bulletin De La Societe Clinique De Medecine Mentale, 2*, 6–16.

Chen, C. S., Ouyang, P., Yeh, Y. C., Lai, C. L., Liu, C. K., Yen, C. F., … Juo, S. H. (2012). Apolipoprotein E polymorphism and behavioral and psychological symptoms of dementia in patients with Alzheimer disease. *Alzheimer Disease and Associated Disorders, 26*(2), 135–139.

Christie, D., Shofer, J., Millard, S. P., Li, E., Demichele-Sweet, M. A., Weamer, E. A., … Tsuang, D. (2012). Genetic association between APOE*4 and neuropsychiatric symptoms in patients with probable Alzheimer's disease is dependent on the psychosis phenotype. *Behavioral and Brain Functions: BBF, 8*, 62-9081-8-62.

Collins, M. N., Hawthorne, M. E., Gribbin, N., & Jacobson, R. (1990). Capgras' syndrome with organic disorders. *Postgraduate Medical Journal, 66*(782), 1064–1067.

Coltheart, M. (2010). The neuropsychology of delusions. *Annals of the New York Academy of Sciences, 1191*, 16–26.

Corcoran, R., Mercer, G., & Frith, C. D. (1995). Schizophrenia, symptomatology and social inference: investigating "theory of mind" in people with schizophrenia. *Schizophrenia Research, 17*(1), 5–13.

Corlett, P. R., Aitken, M. R., Dickinson, A., Shanks, D. R., Honey, G. D., Honey, R. A., … Fletcher, P. C. (2004). Prediction error during retrospective revaluation of causal associations in humans: fMRI evidence in favor of an associative model of learning. *Neuron, 44*(5), 877–888.

Corlett, P. R., Murray, G. K., Honey, G. D., Aitken, M. R., Shanks, D. R., Robbins, T. W., … Fletcher, P. C. (2007). Disrupted prediction-error signal in psychosis: evidence for an associative account of delusions. *Brain: A Journal of Neurology, 130*(Pt 9), 2387–2400.

Craig, D., Donnelly, C., Hart, D., Carson, R., & Passmore, P. (2007). Analysis of the 5HT-2A T102C receptor polymorphism and psychotic symptoms in Alzheimer's disease. *American Journal of Medical Genetics. Part B, Neuropsychiatric Genetics: The Official Publication of the International Society of Psychiatric Genetics, 144B*(1), 126–128.

Creese, B., Ballard, C., Aarsland, D., Londos, E., Sharp, S., & Jones, E. (2014). Determining the association of the 5HTTLPR polymorphism with delusions and hallucinations in Lewy body dementias. *The American Journal of Geriatric Psychiatry: Official Journal of the American Association for Geriatric Psychiatry, 22*(6), 580–586.

Cutting, J. (1990). *The right cerebral hemisphere and psychiatric disorders*. Oxford: Oxford University Press.

Dalgalarrondo, P., Fujisawa, G., & Banzato, C. E. (2002). Capgras syndrome and blindness: against the prosopagnosia hypothesis. *Canadian Journal of Psychiatry. Revue Canadienne De Psychiatrie, 47*(4), 387–388.

David, A. S. (1994). Schizophrenia and the corpus callosum: developmental, structural and functional relationships. *Behavioural Brain Research, 64*(1–2), 203–211.

Devine, M. J., Bentley, P., Jones, B., Hotton, G., Greenwood, R. J., Jenkins, I. H., … Malhotra, P. A. (2014). The role of the right inferior frontal gyrus in the pathogenesis of post-stroke psychosis. *Journal of Neurology, 261*(3), 600–603.

Devinsky, O. (2009). Delusional misidentifications and duplications: right brain lesions, left brain delusions. *Neurology, 72*(1), 80–87.

Dietl, T., Herr, A., Brunner, H., & Friess, E. (2003). Capgras syndrome—out of sight, out of mind? *Acta Psychiatrica Scandinavica, 108*(6), 460–462. discussion 462–463.

Edelstyn, N. M., & Oyebode, F. (1999). A review of the phenomenology and cognitive neuropsychological origins of the capgras syndrome. *International Journal of Geriatric Psychiatry, 14*(1), 48–59.

Edelstyn, N. M., Oyebode, F., Riddoch, M. J., Soppitt, R., Moselhy, H., & George, M. (1997). A neuropsychological perspective on three schizophrenic patients with midline structural defects. *The British Journal of Psychiatry: The Journal of Mental Science, 170*, 416–421.

Ellis, H. D., & Young, A. W. (1990). Accounting for delusional misidentifications. *The British Journal of Psychiatry: The Journal of Mental Science, 157*, 239–248.

Ellis, H. D., Young, A. W., Quayle, A. H., & De Pauw, K. W. (1997). Reduced autonomic responses to faces in capgras delusion. *Proceedings. Biological Sciences/the Royal Society, 264*(1384), 1085–1092.

Fletcher, P. C., Anderson, J. M., Shanks, D. R., Honey, R., Carpenter, T. A., Donovan, T., … Bullmore, E. T. (2001). Responses of human frontal cortex to surprising events are predicted by formal associative learning theory. *Nature Neuroscience, 4*(10), 1043–1048.

van der Flier, W. M., Staekenborg, S., Pijnenburg, Y. A., Gillissen, F., Romkes, R., Kok, A., … Scheltens, P. (2007). Apolipoprotein E genotype influences presence and severity of delusions and aggressive behavior in Alzheimer disease. *Dementia and Geriatric Cognitive Disorders, 23*(1), 42–46.

Freeman, D., Dunn, G., Fowler, D., Bebbington, P., Kuipers, E., Emsley, R., … Garety, P. (2013). Current paranoid thinking in patients with delusions: the presence of cognitive-affective biases. *Schizophrenia Bulletin, 39*(6), 1281–1287.

Frith, C. D., & Corcoran, R. (1996). Exploring 'theory of mind' in people with schizophrenia. *Psychological Medicine, 26*(3), 521–530.

Geroldi, C., Akkawi, N. M., Galluzzi, S., Ubezio, M., Binetti, G., Zanetti, O., … Frisoni, G. B. (2000). Temporal lobe asymmetry in patients with Alzheimer's disease with delusions. *Journal of Neurology, Neurosurgery, and Psychiatry, 69*(2), 187–191.

Gilleen, J., & David, A. S. (2005). The cognitive neuropsychiatry of delusions: from psychopathology to neuropsychology and back again. *Psychological Medicine, 35*(1), 5–12.

Gluckman, L. K. (1968). A case of capgras syndrome. *Australian and New Zealand Journal of Psychiatry, 2*, 39–43.

Goldfarb, A. I., & Weiner, M. B. (1977). The capgras syndrome as an adaptational maneuver in old age. *The American Journal of Psychiatry, 134*(12), 1434–1436.

Harciarek, M., & Kertesz, A. (2008). The prevalence of misidentification syndromes in neurodegenerative diseases. *Alzheimer Disease and Associated Disorders, 22*(2), 163–169.

Hermanowicz, N. (2002). A blind man with Parkinson's disease, visual hallucinations, and capgras syndrome. *The Journal of Neuropsychiatry and Clinical Neurosciences, 14*(4), 462–463.

Herrmann, N., Lanctot, K. L., Sambrook, R., Lesnikova, N., Hebert, R., McCracken, P., ... Nguyen, E. (2006). The contribution of neuropsychiatric symptoms to the cost of dementia care. *International Journal of Geriatric Psychiatry, 21*(10), 972–976.

Hirstein, W., & Ramachandran, V. S. (1997). Capgras syndrome: a novel probe for understanding the neural representation of the identity and familiarity of persons. *Proceedings. Biological Sciences/the Royal Society, 264*(1380), 437–444.

Horikawa, H., Monji, A., Sasaki, M., Maekawa, T., Onitsuka, T., Nitazaka, Y., ... Kanba, S. (2006). Different SPECT findings before and after capgras' syndrome in interictal psychosis. *Epilepsy & Behavior: E&B, 9*(1), 189–192.

Howes, O., Bose, S., Turkheimer, F., Valli, I., Egerton, A., & Stahl, D. (2011). Progressive increase in striatal dopamine synthesis capacity as patients develop psychosis: a PET study. *Molecular Psychiatry, 16*(9), 885–886.

Howes, O. D., Bose, S. K., Turkheimer, F., Valli, I., Egerton, A., Valmaggia, L. R., ... McGuire, P. (2011). Dopamine synthesis capacity before onset of psychosis: a prospective [18F]-DOPA PET imaging study. *The American Journal of Psychiatry, 168*(12), 1311–1317.

Howes, O. D., Egerton, A., Allan, V., McGuire, P., Stokes, P., & Kapur, S. (2009). Mechanisms underlying psychosis and antipsychotic treatment response in schizophrenia: insights from PET and SPECT imaging. *Current Pharmaceutical Design, 15*(22), 2550–2559.

Howes, O. D., Montgomery, A. J., Asselin, M. C., Murray, R. M., Valli, I., Tabraham, P., ... Grasby, P. M. (2009). Elevated striatal dopamine function linked to prodromal signs of schizophrenia. *Archives of General Psychiatry, 66*(1), 13–20.

Ismail, Z., Nguyen, M. Q., Fischer, C. E., Schweizer, T. A., & Mulsant, B. H. (2012). Neuroimaging of delusions in Alzheimer's disease. *Psychiatry Research, 202*(2), 89–95.

Ismail, Z., Nguyen, M. Q., Fischer, C. E., Schweizer, T. A., Mulsant, B. H., & Mamo, D. (2011). Neurobiology of delusions in Alzheimer's disease. *Current Psychiatry Reports, 13*(3), 211–218.

Jones, E. G., & Powell, T. P. (1970). An anatomical study of converging sensory pathways within the cerebral cortex of the monkey. *Brain: A Journal of Neurology, 93*(4), 793–820.

Joseph, A. B. (1986). Focal central nervous system abnormalities in patients with misidentification syndromes. *Bibliotheca Psychiatrica, 164*, 68–79.

Kosslyn, S. M. (1987). Seeing and imagining in the cerebral hemispheres: a computational approach. *Psychological Review, 94*(2), 148–175.

Kumral, E., & Ozturk, O. (2004). Delusional state following acute stroke. *Neurology, 62*(1), 110–113.

Lagioia, A., Eliez, S., Schneider, M., Simons, J. S., Van der Linden, M., & Debbane, M. (2011). Neural correlates of reality monitoring during adolescence. *NeuroImage, 55*(3), 1393–1400.

Larner, A. J. (2013). Delusion of pregnancy: a case revisited. *Behavioural Neurology, 27*(3), 293–294.

Le Ber, I., Camuzat, A., Hannequin, D., Pasquier, F., Guedj, E., Rovelet-Lecrux, A., ... French Research Network on FTD/FTD-MND. (2008). Phenotype variability in progranulin mutation carriers: a clinical, neuropsychological, imaging and genetic study. *Brain: A Journal of Neurology, 131*(Pt 3), 732–746.

Lee, J., Quintana, J., Nori, P., & Green, M. F. (2011). Theory of mind in schizophrenia: exploring neural mechanisms of belief attribution. *Social Neuroscience, 6*(5–6), 569–581.

Levine, D. N., & Grek, A. (1984). The anatomic basis of delusions after right cerebral infarction. *Neurology, 34*(5), 577–582.

Lopez, O. L., Smith, G., Becker, J. T., Meltzer, C. C., & DeKosky, S. T. (2001). The psychotic phenomenon in probable Alzheimer's disease: a positron emission tomography study. *The Journal of Neuropsychiatry and Clinical Neurosciences, 13*(1), 50–55.

Lunt, L., Bramham, J., Morris, R. G., Bullock, P. R., Selway, R. P., Xenitidis, K., & David, A. S. (2012). Prefrontal cortex dysfunction and 'jumping to conclusions': bias or deficit? *Journal of Neuropsychology, 6*(1), 65–78.

Lyketsos, C. G., Miller, D. S., & Neuropsychiatric Syndromes Professional Interest Area of the International Society to Advance Alzheimer's Research and Treatment. (2012). Addressing the Alzheimer's disease crisis through better understanding, treatment, and eventual prevention of associated neuropsychiatric syndromes. *Alzheimer's & Dementia: The Journal of the Alzheimer's Association, 8*(1), 60–64.

Lyketsos, C. G., Sheppard, J. M., Steinberg, M., Tschanz, J. A., Norton, M. C., Steffens, D. C., & Breitner, J. C. (2001). Neuropsychiatric disturbance in Alzheimer's disease clusters into three groups: the cache county study. *International Journal of Geriatric Psychiatry, 16*(11), 1043–1053.

McMurtray, A. M., Sultzer, D. L., Monserratt, L., Yeo, T., & Mendez, M. F. (2008). Content-specific delusions from right caudate lacunar stroke: association with prefrontal hypometabolism. *The Journal of Neuropsychiatry and Clinical Neurosciences, 20*(1), 62–67.

Mendez, M. F., Fras, I. A., Kremen, S. A., & Tsai, P. H. (2011). False reports from patients with frontotemporal dementia: delusions or confabulations? *Behavioural Neurology, 24*(3), 237–244.

Mendez, M. F., Shapira, J. S., Woods, R. J., Licht, E. A., & Saul, R. E. (2008). Psychotic symptoms in frontotemporal dementia: prevalence and review. *Dementia and Geriatric Cognitive Disorders, 25*(3), 206–211.

Merrin, E. L., & Silberfarb, P. M. (1976). The capgras phenomenon. *Archives of General Psychiatry, 33*(8), 965–968.

Mesulam, M. M., Van Hoesen, G. W., Pandya, D. N., & Geschwind, N. (1977). Limbic and sensory connections of the inferior parietal lobule (area PG) in the rhesus monkey: a study with a new method for horseradish peroxidase histochemistry. *Brain Research, 136*(3), 393–414.

Nakano, S., Yamashita, F., Matsuda, H., Kodama, C., & Yamada, T. (2006). Relationship between delusions and regional cerebral blood flow in Alzheimer's disease. *Dementia and Geriatric Cognitive Disorders, 21*(1), 16–21.

Nishio, Y., & Mori, E. (2012). Delusions of death in a patient with right hemisphere infarction. *Cognitive and Behavioral Neurology: Official Journal of the Society for Behavioral and Cognitive Neurology, 25*(4), 216–223.

Perneczky, R., Drzezga, A., Boecker, H., Forstl, H., Kurz, A., & Haussermann, P. (2008). Cerebral metabolic dysfunction in patients with dementia with Lewy bodies and visual hallucinations. *Dementia and Geriatric Cognitive Disorders, 25*(6), 531–538.

Perneczky, R., Drzezga, A., Boecker, H., Wagenpfeil, S., Forstl, H., Kurz, A., & Haussermann, P. (2009). Right prefrontal hypometabolism predicts delusions in dementia with Lewy bodies. *Neurobiology of Aging, 30*(9), 1420–1429.

Pritchard, A. L., Harris, J., Pritchard, C. W., Coates, J., Haque, S., Holder, R., … Lendon, C. L. (2008). Role of 5HT 2A and 5HT 2C polymorphisms in behavioural and psychological symptoms of Alzheimer's disease. *Neurobiology of Aging, 29*(3), 341–347.

Radaelli, D., Poletti, S., Gorni, I., Locatelli, C., Smeraldi, E., Colombo, C., & Benedetti, F. (2014). Neural correlates of delusion in bipolar depression. *Psychiatry Research, 221*(1), 1–5.

Rohrer, J. D., Lashley, T., Schott, J. M., Warren, J. E., Mead, S., Isaacs, A. M., … Warren, J. D. (2011). Clinical and neuroanatomical signatures of tissue pathology in frontotemporal lobar degeneration. *Brain: A Journal of Neurology, 134*(Pt 9), 2565–2581.

Roiser, J. P., Howes, O. D., Chaddock, C. A., Joyce, E. M., & McGuire, P. (2013). Neural and behavioral correlates of aberrant salience in individuals at risk for psychosis. *Schizophrenia Bulletin, 39*(6), 1328–1336.

Rojo, V. I., Caballero, L., Iruela, L. M., & Baca, E. (1991). Capgras' syndrome in a blind patient. *The American Journal of Psychiatry, 148*(9), 1271–1272.

Ropacki, S. A., & Jeste, D. V. (2005). Epidemiology of and risk factors for psychosis of Alzheimer's disease: a review of 55 studies published from 1990 to 2003. *The American Journal of Psychiatry, 162*(11), 2022–2030.

Seeley, W. W., Menon, V., Schatzberg, A. F., Keller, J., Glover, G. H., Kenna, H., … Greicius, M. D. (2007). Dissociable intrinsic connectivity networks for salience processing and executive control. *The Journal of Neuroscience: The Official Journal of the Society for Neuroscience, 27*(9), 2349–2356.

Sha, S. J., Takada, L. T., Rankin, K. P., Yokoyama, J. S., Rutherford, N. J., Fong, J. C., … Boxer, A. L. (2012). Frontotemporal dementia due to C9ORF72 mutations: clinical and imaging features. *Neurology, 79*(10), 1002–1011.

Silva, J. A., Leong, G. B., Weinstock, R., & Wine, D. B. (1993). Delusional misidentification and dangerousness: a neurobiologic hypothesis. *Journal of Forensic Sciences, 38*(4), 904–913.

Snowden, J. S., Rollinson, S., Thompson, J. C., Harris, J. M., Stopford, C. L., Richardson, A. M., … Pickering-Brown, S. M. (2012). Distinct clinical and pathological characteristics of frontotemporal dementia associated with C9ORF72 mutations. *Brain: A Journal of Neurology, 135*(Pt 3), 693–708.

Spalletta, G., Bernardini, S., Bellincampi, L., Federici, G., Trequattrini, A., & Caltagirone, C. (2006). Delusion symptoms are associated with ApoE epsilon4 allelic variant at the early stage of Alzheimer's disease with late onset. *European Journal of Neurology: The Official Journal of the European Federation of Neurological Societies, 13*(2), 176–182.

Staton, R. D., Brumback, R. A., & Wilson, H. (1982). Reduplicative paramnesia: a disconnection syndrome of memory. *Cortex; a Journal Devoted to the Study of the Nervous System and Behavior, 18*(1), 23–35.

Sultzer, D. L., Brown, C. V., Mandelkern, M. A., Mahler, M. E., Mendez, M. F., Chen, S. T., & Cummings, J. L. (2003). Delusional thoughts and regional frontal/temporal cortex metabolism in Alzheimer's disease. *The American Journal of Psychiatry, 160*(2), 341–349.

Tost, H., Ruf, M., Schmal, C., Schulze, T. G., Knorr, C., Vollmert, C., … Reitschel, M. (2010). Prefrontal-temporal gray matter deficits in bipolar disorder patients with persecutory delusions. *Journal of Affective Disorders, 120*(1–3), 54–61.

Tranel, D., Damasio, H., & Damasio, A. R. (1995). Double dissociation between overt and covert face recognition. *Journal of Cognitive Neuroscience, 7*(4), 425–432.

Turner, D. C., Aitken, M. R., Shanks, D. R., Sahakian, B. J., Robbins, T. W., Schwarzbauer, C., & Fletcher, P. C. (2004). The role of the lateral frontal cortex in causal associative learning: exploring preventative and super-learning. *Cerebral Cortex (New York, N.Y.: 1991), 14*(8), 872–880.

Weston, M. J., & Whitlock, F. A. (1971). The capgras syndrome following head injury. *The British Journal of Psychiatry: The Journal of Mental Science, 119*(548), 25–31.

Chapter 24

Beyond Dopamine: The Role of the Serotonergic System and Treatments in Understanding and Treating Visual Hallucinations in Parkinson Disease

Dennis Velakoulis
Royal Melbourne Hospital, Melbourne, VIC, Australia; University of Melbourne, Melbourne, VIC, Australia

Ramon Mocellin
Royal Melbourne Hospital, Melbourne, VIC, Australia; University of Melbourne and Melbourne Health, Melbourne, VIC, Australia

Andrew Evans
Royal Melbourne Hospital, Melbourne, VIC, Australia

Mark Walterfang
Royal Melbourne Hospital, Melbourne, VIC, Australia; University of Melbourne, Melbourne, VIC, Australia

Visual hallucinations (VH) have long been associated with organic brain disease and are a common presentation in a range of reversible and degenerative brain disorders. One of the more common diseases associated with visual hallucinations is Parkinson disease (PD), and visual hallucinations in PD are often complex and detailed. Because patients are frequently distressed by these experiences, treatment of VH is often required. The pharmacological mainstay of the treatment of VH in PD has been antipsychotic medications such as quetiapine, olanzapine, and clozapine. The evidence base for using these medications is limited and their use is based on an underlying assumption that VH result from dopaminergic overactivity. This chapter reviews the literature regarding the neurobiological basis of VH in PD and uses our current understanding of anatomy and neurotransmitter systems to guide the pharmacological treatment of VH in this disorder.

We have previously summarized three mechanisms (deafferentation, disinhibition, and central) for the generation of complex VH (Mocellin, Walterfang, & Velakoulis, 2006); a brief summary of this work is outlined subsequently. This is followed by a more detailed elaboration of the disinhibition mechanism of VH arising from brain stem structures and the contributions of serotonergic ascending systems. This understanding of the neuroanatomical and neurotransmitter underpinnings of VH then forms the foundation for a discussion regarding available and potential pharmacological treatments of complex VH in PD.

SIMPLE VERSUS COMPLEX VISUAL HALLUCINATIONS

In 1838, Esquirol provided the earliest and simplest definition of a hallucination as a percept without an object, perceived as real and in the external environment (Esquirol, 1838). Slade refined this definition to refer to a percept-like experience in the absence of an external stimulus that is experienced as a true percept, is spontaneous and unwilled, and cannot readily be controlled (Slade, 1976).

The initial understanding of the pathophysiology of VH postulated that their origin lay in irritative foci, such as those causing epileptiform activity after infection or trauma in cortical regions involved in vision (Walsh & Hoyt, 1969). These VH are usually brief, intermittent, and stereotyped, meeting criteria for "simple" VH (Cummings & Miller, 1987), and were informed by the cortical stimulation studies by Penfield and Perot in which occipital stimulation induced simple VH, with temporo-occipital and parieto-occipital stimulation resulting in more complex scenes (Penfield & Perot, 1963).

Genomics, Circuits, and Pathways in Clinical Neuropsychiatry. http://dx.doi.org/10.1016/B978-0-12-800105-9.00024-X

Simple VH are perhaps the least common form of VH, and this model did not account for more complex, non-stereotyped, and continuous (or more "complex") VH. These types of hallucinations were most particularly seen in blind people, which led to the theory first postulated by West (1962) that the loss of afferent input "releases" the visual and nearby cortex. The theory of VH as release phenomena, elaborated by Cogan (1973), provides a model for understanding how VH occur after lesions to more "proximal" structures, particularly in circumstances of visual loss such as Charles Bonnet syndrome (CBS) (Ffytche & Howard, 1999; Manford & Andermann, 1998; Mocellin et al., 2006; Santhouse, Howard, & Ffytche, 2000). This has been supported by functional magnetic resonance imaging (fMRI) studies that show selective tonic activity in visual cortical regions even when an individual is not hallucinating (Ffytche et al., 1998; Howard et al., 1997).

Complex visual hallucinations (CVH) are differentiated from more simple, unformed hallucinations such as crudely formed flashes of light and color, or other indistinct forms. Although the form of CVH in organic brain disease can be heterogeneous, common features can be identified. The hallucinations may be vivid and colorful; include small people or children [Lilliputian hallucinations (LHs)]; involve regular patterns or disembodied faces or limbs; or include branching or tessellated patterns, vivid and colorful formed animals (real or bizarre), soldiers or others in uniform, or landscapes and complex scenes. CVH are frequently associated with a variety of visual distortions such as palinopsia, in which repeated images of a perceived image are seen or distortions of a face or head (prosometamorphopsia) (Manford & Andermann, 1998).

CVH have been reported in a wide variety of clinical and nonclinical circumstances but are best described and characterized in peduncular hallucinosis, CBS, and the synucleinopathies such as PD and Lewy body dementia (LBD). VH in the presence of frontal dementia or atypical parkinsonian syndromes may indicate testing for genetic syndromes such as PRGN, C9ORF72, FUS, CADASIL, and Niemann Pick type C under the appropriate clinical circumstances. They may also occur as hypnopompic or hypnagogic phenomena in healthy individuals or in those in sensory deprivation states (Merabet et al., 2004) or with narcolepsy-cataplexy syndrome, delirium tremens, intoxication with psychoactive substances, or temporal lobe epilepsy (Manford & Andermann, 1998).

THE VISUAL PATHWAY AND MECHANISMS OF DISRUPTION LEADING TO VISUAL HALLUCINATIONS

The anatomy of the primary visual pathway has been well described. Information from the retina passes along the optic nerve, chiasm, and tract to the lateral geniculate nucleus (LGN) in the thalamus, and thence to the optic radiation through the temporal lobe to the primary and secondary visual cortex (Fig. 24.1). The flow of visual information is modulated by ascending input from pedunculopontine and parabrachial nuclei, and raphe nuclei via the superior colliculi (Fig. 24.2), and involves the serotonergic, cholinergic, gamma-aminobutyric acid-ergic (GABAergic), and glutamatergic systems (Fig. 24.2). Interruptions to this system at any point, either in the primary direct pathway or in its ascending modulatory

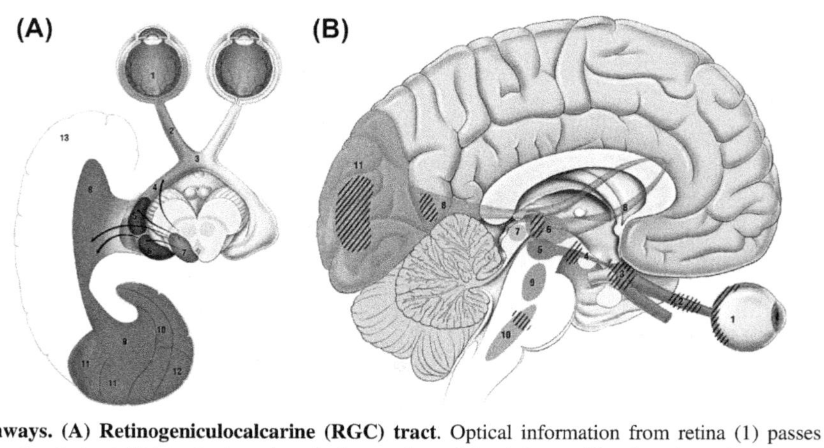

FIGURE 24.1 Visual pathways. (A) Retinogeniculocalcarine (RGC) tract. Optical information from retina (1) passes along the optic nerve (2) through the optic chiasm (3) and optic tract (4) into the lateral geniculate nucleus of the thalamus (5), where it receives input from the superior colliculus (7) via the pulvinar (6) and then traverses the optic radiation (8 and 9) through temporal lobe (13) into visual cortex (10–12). **(B) Intersection of ascending pathways.** Optical information in RGC (1–8 and 11, as in Fig. 24.3) is modulated by ascending input from pedunculopontine and parabrachial nuclei (9) and raphe nuclei (10) via the superior colliculus (7). Hash-marked areas show regions where interruptions are known to produce VH: in the RGC tract via deafferentation; in the thalamus through a reduced signal-to-noise ratio, and in the ascending pathways via the removal of inhibitory control.

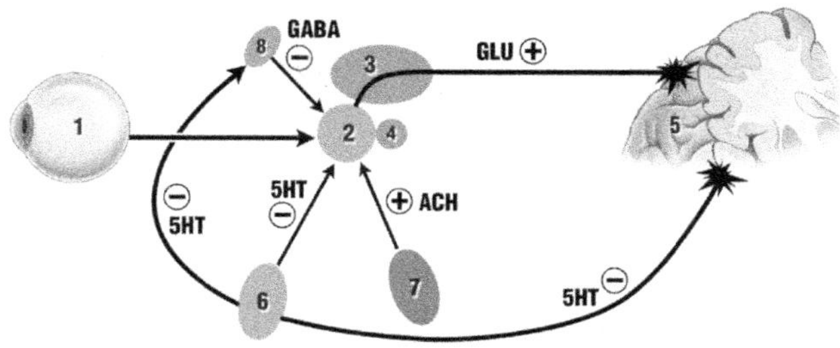

FIGURE 24.2 Neurochemistry of vision. Input from the retina (1) reaches the lateral geniculate nucleus of the thalamus (2). This structure and the adjacent pulvinar of the thalamus (3), an accessory visual structure that may act to filter out eye movement "noise," act as a junction between reticulo-geniculocalcarine and ascending brain stem circuits, receiving inhibitory serotonergic inputs from the raphe nuclei (6) and excitatory cholinergic inputs from the pedunculopontine and parabrachial nuclei (7). The reticular nucleus of the thalamus (8) also provides inhibitory GABAergic innervation to the geniculate, which is itself modulated by the same ascending cholinergic and serotonergic input. The glutamatergic excitatory circuits from the geniculate to the occipital cortex (5) are also modulated by the superior colliculus (4).

projections, may lead to VH. A series by Braun, Dumont, Duval, Hamel-Hebert, and Godbout (2003) suggested that occipital and occipitotemporal regions were the most commonly implicated cortical regions, and midbrain, cerebral peduncles, pons, and thalamus were the typically affected subcortical regions (Braun et al., 2003).

The exact mechanisms underlying VH remain unclear but may involve "cortical release" or "deafferentation" phenomena (West, 1962) (Fig. 24.3A) and/or the disinhibition of projections from ascending pathways or intact nearby

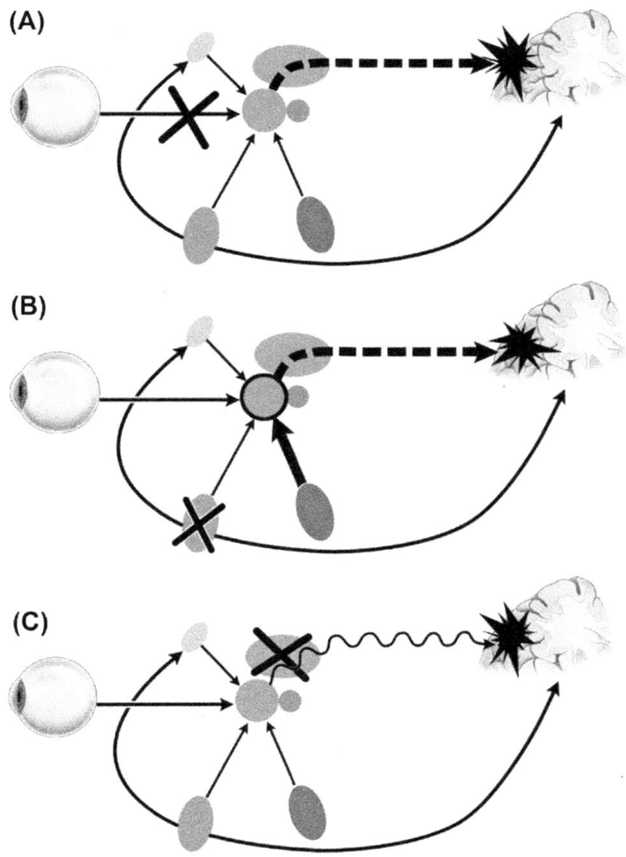

FIGURE 24.3 (A) Deafferentation: Lesions responsible for pathway CVH in which deafferentation from ocular input results in "release" activity in the cortex. (B) Disinhibition: Lesions responsible for ascending CVH in which a loss of ascending inhibition to the geniculate results in a hyperexcited geniculate and excess glutamatergic activity in the optic radiation, with resultant poor-quality signal to the cortex. (C) Central: Lesions producing central CVH in which damage to the geniculate may again "deafferent" the striate cortex and lesions to the pulvinar of the thalamus may reduce the signal-to-noise ratio of cortical input owing to a loss of the pulvinar's "visual filter" function.

visual cortex. Disruption of ascending inputs, for example at the level of the LGN, may lead to aberrant projections forward to the visual cortex (Fig. 24.3B) or a loss of central sensory filtering function and degradation of signal-to-noise (Fig. 24.3C).

Lesions anywhere in the visual system, from ocular structures through optic nerve, chiasm, and tract structures, including ascending modulatory midbrain structures, can produce VH, which are usually complex in form (Galasko, Kwo-On-Yuen, & Thal, 1988; Lepore, 1990; Mocellin et al., 2006). To correlate the location of the lesion to the type of VH, Santhouse et al. used factor analysis in patients with CVH to establish anatomical–phenomenological correlates. Their study found that landscapes and groups of figures were associated with pathologically increased activity in the ventral temporal lobe, distorted faces with activity around the superior temporal sulcus, and visual perseveration and palinopsia with activity of the visual parietal lobe (Santhouse et al., 2000). Thus, at an anatomical level, whereas simple VH are most likely related to focal lesions of the ocular apparatus or occipital cortex, CVH occur when the quality or flow of information moving through the visual system is disrupted.

Although lesion-based models afford some understanding of the anatomy of normal and abnormal visual processing, they do not shed light on functional or neurochemical changes such as are seen in substance intoxication or withdrawal, medication-related VH, or VH seen in global neurometabolic or neurodegenerative disorders such as delirium and dementia. Medications with anticholinergic (particularly antimuscarinic) properties are the most visually hallucinogenic (Perry & Perry, 1995), particularly in elderly people who generally have lower cholinergic tone than younger adults (Perry et al., 1992). Altered dopaminergic transmission in stimulant misuse and dopaminergic treatment of PD and other synucleinopathies suggest a role for dopamine transmission in VH (Angrist, Sathananthan, Wilk, & Gershon, 1974; Goetz, Vogel, Tanner, & Stebbins, 1998). Alterations in the GABAergic system that occur in benzodiazepine and alcohol withdrawal, which are often associated with CVH, implicate a loss of GABAergic cortical inhibition in withdrawal-associated VH (Nevo & Hamon, 1995), although this is likely to be mediated through other monoaminergic systems (Manford & Andermann, 1998). Finally, but less often considered, VH may also occur with perturbations to serotonergic transmission, an area which forms the focus of subsequent discussion.

These otherwise disparate anatomical and neurochemical models of VH have been united in the perception and attentional-deficit model of Collerton et al., which focuses on deficits in object-based attention resulting from dysfunction in lateral frontal cortical systems combined with object-based perceptual deficits due to dysfunction in the ventral ("what") as opposed to dorsal ("where") visual stream (Collerton, Perry, & Mckeith, 2005). This can also be understood as a failure to integrate current sensory input with prior expectations and is consistent with intrusion of previously generated proto-objects into a currently perceived scene. Generation of multiple copies of previously perceived objects in this manner can explain repetitive VH or palinopsias. Unlike other models, this can account for VH whose origin is either predominantly lesion based or neurochemically driven, in addition to VH that occur in states of sensory deprivation or in hypnagogic or hypnopompic states.

COMPLEX VISUAL HALLUCINATIONS: THE ROLE OF LESION LOCATION

The anatomy of the retinogeniculocalcarine tract is well understood, including how lesions of the tract at different levels produce classically described visual field defects (Fig. 24.3). The thalamic structures, the dorsal lateral geniculate nucleus, and the lateral pulvinar, lie at the center of this system and receive a number of inputs, predominately cholinergic and serotonergic, from the brain stem (Manford & Andermann, 1998). Although these inputs are not completely characterized in mammals, it appears that cholinergic input is predominately excitatory, originating from the parabrachial and parabigeminal nuclei, and serotonergic inputs are inhibitory. These mainly arise from the serotonergic nuclei of the dorsal raphe nuclei and also inhibit retinal input (Fig. 24.3B).

Brain stem lesions may disrupt these serotonergic inhibitory raphe nucleus inputs, resulting in excitation of the dorsal LGN and dysregulation of retinal inputs resulting in CVH (Fig. 24.3B). Many hallucinogenic compounds, such as lysergic acid, are serotonergic agonists and may produce CVH (Siegel & Jarvik, 1975). The dorsal raphe nuclei have also been implicated in both the sleep–wake cycle (Abrams, Johnson, Hollis, & Lowry, 2004) and regulation of REM and non-REM sleep (Wu et al., 2004) in mammals. Sleep-associated CVH may be seen in normal sleep (hypnopompic and hypnagogic hallucinations), sleep disorders (narcolepsy-cataplexy syndromes), delirium, and LBD. CVH in CBS are often accentuated in states of reduced consciousness or fading light. These clinical findings further suggest an important role of the dorsal raphe system in generating CVH.

Thalamic lesions may directly affect critical structures such as the pulvinar or the associated brain stem inputs described previously. The primate pulvinar has an important role in simple visual processing but also visual salience (generating signals related to the salience of visual objects) and linking eye movement to this function (Grieve, Acuna, & Cudeiro,

2000). Lesions in this area may generate CVH by disrupting these functions (Fig. 24.3C). Damage to retinal inputs in this area may also operate in the same fashion as a more proximal differentiating lesion.

Whereas differing pathologies may lead to CVH, the form these hallucinations take may relate to particular cortical locations. Using factor analysis of clinical information collected from patients with CBS linked with predominately primate fMRI and neurophysiological information, Santhouse et al. (2000) postulated that CVH in CBS can be linked to hierarchical visual pathway streams. These workers suggested that hallucinations of extended scenes and people and objects (including LHs) are associated with the ventral occipitotemporal cortex, hallucinations of faces and facial distortions (prosopometamorphosia) with the superior temporal sulcus, and visual perseveration and delayed palinopsia (reappearance of a percept with shift in gaze after time delay) with visual parietal regions. These findings await further study and clarification.

SYNUCLEINOPATHY AND VISUAL HALLUCINATIONS

Synucleinopathies are a diverse group of related neurodegenerative diseases characterized by abnormal α-synuclein metabolism, which in some instances result in the formation of intracellular inclusions known as Lewy bodies. PD-related dementia (PDD) and LBD are associated with cortical Lewy bodies and marked cholinergic deficit in areas involved in visual perception (Bohnen et al., 2003; Harding, Broe, & Halliday, 2002), loss of serotonergic and cholinergic neurons in brain stem nuclei that modulate transmission of visual information (Halliday, Blumbergs, Cotton, Blessing, & Geffen, 1990; Jellinger, 1990), and impaired contrast vision resulting from disrupted retinal dopaminergic function (Bodis-Wollner & Tagliati, 1993), each of which is a factor in the development of VH.

Lewy bodies in mesial temporal structures (parahippocampal gyrus and amygdala) are particularly associated with an increased incidence of VH (Harding et al., 2002). VH may also relate to synuclein deposition in visual areas and altered ascending input from loss of serotonergic and cholinergic brain stem nuclei, whereas the use of dopaminergic medications in PD may worsen or bring forward the development of hallucinations.

The Origins of Visual Hallucinations in Parkinson Disease

In a seminal study of VH in PD, Fenelon et al. found that VH are present in up to 40% of patients with PD and may consist of passage hallucinations, presence hallucinations, and formed or CVH (Fenelon, Mahieux, Huon, & Ziegler, 2000). These authors noted that dopaminergic treatments are not enough to explain the VH of PD and that the main predictors of CVH were cognitive impairment, daytime somnolence, and longer duration of PD (Fenelon et al., 2000). Based on their findings and the literature, they argued that the role of dopamine in VH in PD is complex.

1. Hallucinations were described in PD before the introduction of L-dopa.
2. Early studies varied in terms of inclusion criteria, doses of dopa, and the description of psychiatric symptoms.
3. There is no clear relationship between dose and hallucinations in PD.
4. Anticholinergics can produce hallucinations in PD.
5. Patients with non-LBD disorders treated with L-dopa rarely develop VH.
6. VH occur early in the course of LBD (Fenelon et al., 2000).

In patients with unclassifiable parkinsonism, the presence of VH has been regarded as a "red flag" for underlying Lewy body pathology (Williams & Lees, 2005; Williams, Warren, & Lees, 2008). The same group identified that VH correlate highly with Lewy body pathology but not other parkinsonian syndromes such as progressive supranuclear palsy or multisystem atrophy (Williams & Lees, 2005). This latter study examined 788 cases of parkinsonism (445 patients with PD, 44 with LBD, and 255 with non−Lewy body parkinsonism. The latter group included 127 with progressive supranuclear palsy, 91 with multisystem atrophy, 27 with vascular parkinsonism, nine with Alzheimer disease, nine with cortical-basal ganglionic degeneration, and eight with postencephalitic parkinsonism) (Williams & Lees, 2005). VH were identified in 50%, 73%, and 7% of patients with PD, LBD, and non−Lewy body syndromes, respectively, and were best associated with cognitive dysfunction, autonomic dysfunction, early axial rigidity, and age at onset. VH were not correlated with the use of L-dopa or anticholinergics. The authors hypothesized that VH are not directly related to dopaminergic medication but could arise from an interaction between dopaminergic treatment and progressive Lewy body involvement of the visuoperceptual systems.

In a postmortem study of patients with PD and VH, a strong relationship was identified between the presence of Lewy bodies in the amygdala and parahippocampal regions and CVH (Harding et al., 2002).

A patient with severe, recurrent complex VH was scanned using fMRI and found to have increased activation in anterior regions (cingulate, insula, and frontal lobe), thalamus, and brain stem and decreased activation in visual processing posterior cortical regions (lingual, fusiform gyri, inferior occipital gyrus, and superior temporal lobes). The authors proposed that there is a desynchronization between posterior and anterior cortical areas during VH (Goetz, Vaughan, Goldman, & Stebbins, 2014).

VH in the synucleinopathies are associated with cognitive impairment and disease severity but generally not medication (Barnes & David, 2001). In a study matching patients with DLB, PD, and Alzheimer disease on degree of cognitive impairment and visual acuity, individuals who had PDD and DLB had significantly impaired visual processing compared with those who had Alzheimer disease, and commensurately higher rates of VH (PDD, 75%; DLB, 90%; and AD, 8%) (Mosimann et al., 2004).

THE SEROTONERGIC SYSTEM AND VISUAL HALLUCINATIONS

Anatomy of the Serotonergic System

Serotonin (5-hydroxy tryptamine [5-HT]) was first linked to psychiatric symptoms, particularly depression and anxiety, in the 1950s. 5-HT cell bodies are found in the dorsal and basal raphe nuclei of the brain stem and provide serotonergic innervation to entire brain. To date, seven 5-HT receptors have been identified, most of which are postsynaptic receptors (Alex & Pehek, 2007). 5-HT neurons of the raphe nucleus innervate dopaminergic cell bodies and terminal regions within the midbrain and basal ganglia (Alex & Pehek, 2007).

Of the seven receptor subtypes, the 5-HT_{2a} and the 5-HT_{2c} receptors are most relevant to models of VH in PD and are the best characterized with regard to medications used in the treatment of VH. 5-HT_{2a} receptors are normally greatest in the middle layers of prefrontal and cingulate cortex, with lower levels identified in the striatum (Hall, Farde, Halldin, Lundkvist, & Sedvall, 2000). 5-HT_{2a} receptor activation stimulates dopamine release from all three major dopaminergic pathways (nigrostriatal, mesocortical, and mesolimbic) (Alex & Pehek, 2007). In patients with PD, reduced levels of 5-HT_{2a} receptor are found in temporal cortex, although increased levels of 5-HT_{2a} receptors have been identified in the temporal cortex (Huot et al., 2010) and the ventral visual pathway (Ballanger et al., 2010) of patients with PD and VH. These latter findings have potential treatment implications for patients with PD and VH.

5-HT_{2c} receptors are found in the ventral tegmentum, substantia nigra, striatum, and nucleus accumbens and the anterior cingulate cortex. These receptors are anatomically placed to modulate dopaminergic function, and in contrast to 5-HT_{2a} receptors, they predominantly act to inhibit dopamine release (Alex & Pehek, 2007).

It has been proposed that actions of antipsychotic drugs in schizophrenia may rely on a combination of dopaminergic blockade and 5-HT—driven dopaminergic modulation in prefrontal cortex, whereby the effect of differential antagonism of 5-HT receptors is to produce an elevation of tonic dopamine release and block phasic dopaminergic within the prefrontal cortex. Antagonism of 5-HT_{2c} receptors may also be the reason why atypical antipsychotic drugs have fewer extrapyramidal side effects compared with typical antipsychotic drugs (Alex & Pehek, 2007; Meltzer & Massey, 2011).

Psilocybin and the Serotonergic System

The best characterized substance relevant to the relationship between the serotonergic system and VH is psilocybin, a psychoactive alkaloid, which together with its active metabolite, psilocin, is a main ingredient of hallucinogenic mushrooms (Tyls, Palenicek, & Horacek, 2014). Psilocybin is structurally similar to serotonin and its hallucinogenic effects are associated with 5-HT_{2a} agonism (Nichols, 2004), with subsequent increased striatal dopamine release (Vollenweider, Vollenweider-Scherpenhuyzen, Babler, Vogel, & Hell, 1998).

Psilocybin was widely investigated in the 1960s as a research drug and is considered the archetypal drug for a serotonergic model of psychosis. Psilocybin induces a plethora of psychotomimetic effects including VH (but not auditory hallucinations), changes in body image, thought disorder, and religious and mystical experiences (Geyer & Vollenweider, 2008; Hasler, Grimberg, Benz, Huber, & Vollenweider, 2004). The nature of VH induced by psilocybin (tessellated and branching patterns, palinopsias, and color distortions) is similar to those experienced in a number of organic brain disorders including PD. Although psilocybin psychosis has been used as a model for schizophrenia, its clinical effects are equally relevant, if not more so, to an understanding of the neurobiology of VH in PD. Psilocybin-induced psychotic experiences are reversed by risperidone (5-HT_{2c} and D2 antagonist) and ketanserin (5-HT_{2a} antagonist) but not by haloperidol (D2 antagonist) (Kometer, Schmidt, Jancke, & Vollenweider, 2013; Vollenweider et al., 1998). The failure of D2 antagonists to treat psilocybin-induced VH has implications for the treatment of VH in PD, especially because D2 antagonists are undesirable in PD owing to their motor effects.

Parkinson Disease and the Serotonergic System

The nigrostriatal dopaminergic pathway begins with cell bodies in the substantia nigra projecting to the caudate and putamen. 5-HT neurons of the raphe nucleus innervate dopaminergic cell bodies and terminal regions within the midbrain and basal ganglia (Alex & Pehek, 2007). 5-HT neurones are able to metabolize exogenous L-dopa into dopamine and may release dopamine into the striatum. This is thought to be a possible mechanism for the induction of dyskinesias in patients with PD who are treated with L-dopa (Huot, Fox, & Brotchie, 2011).

Postmortem studies of patients with PD have identified neuronal loss and Lewy bodies within raphe 5-HT neurons (Halliday et al., 1990), up to 50% reduction in 5-HT transporter levels within cortical areas to which the neurones project (Damato et al., 1987), and reduced 5-HT levels in basal ganglia, hypothalamus, and thalamus (Huot et al., 2011). However, clinicopathological studies have failed to identify a significant relationship between the raphe pathology and clinical features of PD.

A study using a 5-HT_{2a} positron emission tomography ligand in patients who had PD with and without VH found that the patients with VH had increased 5-HT_{2a} binding in the ventral visual pathway and frontal cortex (Ballanger et al., 2010). These findings add weight to evidence for the potential use of selective 5-HT_{2a} antagonists in the treatment of VH.

PHARMACOLOGICAL INTERVENTIONS FOR VISUAL HALLUCINATIONS IN PARKINSON DISEASE

D2 Antagonism

Traditionally, dopamine D2 receptor blocking antipsychotic agents have been used to treat psychotic symptoms in PD, with a preference for atypical agents that have less extrapyramidal symptoms such as clozapine and quetiapine (Chou, Borek, & Friedman, 2007; Weintraub & Stern, 2005). Clozapine remains the only antipsychotic considered to be effective and tolerable in PD with psychosis, but its side-effect profile (agranulocytosis) and frequent local restrictions on prescribing prevents its widespread use for this indication. It is effective in PD with psychosis at doses much lower than those used in schizophrenia (25–50 mg compared with 300–800 mg). At these low doses, clozapine has a far greater 5-HT_{2a} antagonist effect than dopamine D2 effect (Meltzer, Kennedy, Dai, Parsa, & Riley, 1995).

Quetiapine is a widely used alternative to clozapine in PD with a receptor profile similar to that of clozapine, with high 5-HT_{2a} receptor antagonism and lower D2 antagonism (Gefvert et al., 2001). Despite this favorable profile, several placebo-controlled trials have not shown quetiapine to be better than placebo in the treatment of psychosis in PD (Ondo, Tintner, Voung, Lai, & Ringholz, 2005; Rabey, Prokhorov, Miniovitz, Dobronevsky, & Klein, 2007; Shotbolt, Samuel, & David, 2006). In their study, Rabey et al. attributed this finding to a high incidence of patients with delusions and hallucinations compared with an earlier study which found quetiapine to be better than placebo (Juncos et al., 2004) and which included a high proportion of patients with VH. An alternative explanation, however, may relate to the dose of L-dopa drugs used in each study. The average L-dopa dose used in the study by Rabey et al. was about 600 mg daily, compared with 300 mg in the study by Juncos et al. It could be that the psychotic symptoms of the patients receiving higher doses of L-dopa (Rabey et al., 2007) were related to L-dopa rather than to the neurobiological effect of PD, and that quetiapine is effective in those patients with PD whose VH were induced by the PD itself.

Serotonin Antagonism

Because of the problematic clinical problem of treating VH in PD with drugs which block dopaminergic receptors, attention has been drawn to medications which act at 5-HT receptors but not dopaminergic receptors (Meltzer et al., 2010; Meltzer & Roth, 2013). Pimavanserin, a 5-HT_{2a} inverse agonist, has 40 times lower affinity for the 5-HT_{2c} receptor and no activity at dopaminergic, muscarinic, adrenergic, or histaminergic receptors (Vanover et al., 2006). A number of studies have investigated its role in the treatment of psychosis in PD.

In a placebo-controlled, double-blind trial of 30 patients with PD and psychosis, pimavanserin was shown to have minimal impact on motor symptoms in PD and to improve hallucinations (visual and auditory) in patients with PD and psychosis (Meltzer et al., 2010; Meltzer & Roth, 2013). Pimavanserin was significantly better than placebo in treating delusions and thought disorder. In contrast to antipsychotic medications, the side-effect profile of pimavanserin was similar to that of placebo. The authors of that study did not report L-dopa doses of the patients, which preclude any comment on whether the patient group may have been vulnerable to the psychotogenetic effects of higher-dose L-dopa induced.

In a larger, more recent phase III study, 199 patients with PD and psychosis were randomly allocated to placebo or pimavanserin. The treated group had significantly reduced psychotic symptoms with a 37% improvement at 6 weeks compared with a 14% improvement in the placebo group. Pimavanserin treatment was associated with less carer burden, better sleep, and no motor dysfunction compared with the placebo treatment group. The authors did not comment on any differential effect of pimavanserin on hallucination subtypes.

Because of its potential antipsychotic effect, pimavanserin has also been investigated in the treatment of schizophrenia. In a double-blind, randomized control study of 423 patients with schizophrenia, the addition of pimavanserin to 2 mg risperidone produced better outcomes (efficacy and safety) than 6 mg risperidone or haloperidol (Meltzer et al., 2012). The authors advocated for its use as an adjunctive treatment in schizophrenia to reduce extrapyramidal motor side effects of traditional antipsychotic medications.

SUMMARY

The treatment of VH in PD poses a significant clinical issue for neurologists, psychiatrists, and geriatricians. To date, antipsychotic medications have been the mainstay of treatment based on the assumption that dopaminergic blockade is required for the treatment of psychotic symptoms. However, these drugs are associated with motor and other side effects which limit their dosing and compliance. Current knowledge about the visual pathways and the role of the serotonergic system suggests that serotonergic systems are affected by the PD process and may have a significant role in the genesis of hallucinations. Treatments aimed at 5-HT antagonism may offer more effective, safer, and better tolerated pharmacologic treatment for patients than may dopaminergic-blocking drugs.

REFERENCES

Abrams, J. K., Johnson, P. L., Hollis, J. H., & Lowry, C. A. (2004). Anatomic and functional topography of the dorsal raphe nucleus. *Annals of the New York Academy of Sciences, 1018*, 46–57.

Alex, K. D., & Pehek, E. A. (2007). Pharmacologic mechanisms of serotonergic regulation of dopamine neurotransmission. *Pharmacology & Therapeutics, 113*, 296–320.

Angrist, B., Sathananthan, G., Wilk, S., & Gershon, S. (1974). Amphetamine psychosis: behavioural and biochemical aspects. *Journal of Psychiatric Research, 11*, 13–23.

Ballanger, B., Strafella, A. P., Van Eimeren, T., Zurowski, M., Rusjan, P. M., Houle, S., & Fox, S. H. (2010). Serotonin 2A receptors and visual hallucinations in Parkinson disease. *Archives of Neurology, 67*, 416–421.

Barnes, J., & David, A. (2001). Visual hallucinations in Parkinson's disease: a review and phenomenological survey. *Journal of Neurology, Neurosurgery, and Psychiatry, 70*, 727–733.

Braun, C., Dumont, M., Duval, J., Hamel-Hebert, I., & Godbout, L. (2003). Brain modules of hallucination: an analysis of multiple patients with brain lesions. *Journal of Psychiatry & Neuroscience, 28*, 432–439.

Bodis-Wollner, I., & Tagliati, M. (1993). The visual system in Parkinson's disease. *Advances in Neurology, 60*, 390–394.

Bohnen, N., Kaufer, D., Ivanco, L., Lopresti, B., Koeppe, R., Davis, J., ... Dekosky, S. (2003). Cortical cholinergic function is more severely affected in parkinsonian dementia than in Alzheimer disease: an in vivo positron emission tomographic study. *Archives of Neurology & Psychiatry, 60*, 1745–1748.

Chou, K. L., Borek, L. L., & Friedman, J. H. (2007). The management of psychosis in movement disorder patients. *Expert Opinion on Pharmacotherapy, 8*, 935–943.

Cogan, D. (1973). Visual hallucinations as release phenomena. *Graefe's Archives for Clinical and Experimental Ophthalmology, 188*, 139–150.

Collerton, D., Perry, E., & Mckeith, I. (2005). Why people see things that are not there: a novel perception and attention deficit model for recurrent complex visual hallucinations. *The Behavioral and Brain Sciences, 28*, 737–794.

Cummings, J., & Miller, B. (1987). Visual hallucinations: clinical occurrence and use in differential diagnosis. *Western Journal of Medicine, 146*, 46–51.

Damato, R. J., Zweig, R. M., Whitehouse, P. J., Wenk, G. L., Singer, H. S., Mayeux, R., ... Snyder, S. H. (1987). Aminergic systems in Alzheimers-disease and Parkinsons-disease. *Annals of Neurology, 22*, 229–236.

Esquirol, E. (1838). *Des Maladies Mentales Considerees Sous les Rapports Medicaux, Hygieniques et Medico-Legaux*. Paris: Bailliere.

Fenelon, G., Mahieux, F., Huon, R., & Ziegler, M. (2000). Hallucinations in Parkinson's disease – prevalence, phenomenology and risk factors. *Brain: A Journal of Neurology, 123*, 733–745.

Ffytche, D., & Howard, R. (1999). The perceptual consequences of visual loss: 'positive' pathologies of vision. *Brain: A Journal of Neurology, 122*, 1247–1260.

Ffytche, D., Howard, R., Brammer, M., David, A., Woodruff, P., & Williams, S. (1998). The anatomy of conscious vision: an fMRI study of visual hallucinations. *Nature Neuroscience, 1*, 738–742.

Galasko, D., Kwo-On-Yuen, P., & Thal, L. (1988). Intracranial mass lesions associated with late-onset psychosis and depression. *The Psychiatric Clinics of North America, 11*, 151–166.

Gefvert, O., Lundberg, T., Wieselgren, I. M., Bergstrom, M., Langstrom, B., Wiesel, F., & Lindstrom, L. (2001). D(2) and 5HT(2A) receptor occupancy of different doses of quetiapine in schizophrenia: a PET study. *European Neuropsychopharmacology: The Journal of the European College of Neuropsychopharmacology, 11*, 105–110.

Geyer, M. A., & Vollenweider, F. X. (2008). Serotonin research: contributions to understanding psychoses. *Trends in Pharmacological Sciences, 29*, 445–453.

Goetz, C. G., Vaughan, C. L., Goldman, J. G., & Stebbins, G. T. (2014). I finally see what you see: Parkinson's disease visual hallucinations captured with functional neuroimaging. *Movement Disorders, 29*, 115–117.

Goetz, C., Vogel, C., Tanner, C., & Stebbins, G. (1998). Early dopaminergic drug-induced hallucinations in parkinsonian patients. *Neurology, 51*, 811–814.

Grieve, K. L., Acuna, C., & Cudeiro, J. (2000). The primate pulvinar nuclei: vision and action. *Trends in Neurosciences, 23*, 35–39.

Hall, H., Farde, L., Halldin, C., Lundkvist, C., & Sedvall, G. (2000). Autoradiographic localization of 5-HT_{2A} receptors in the human brain using (LH)-H-3 M100907 and C-11 M100907. *Synapse, 38*, 421–431.

Halliday, G., Blumbergs, P., Cotton, R., Blessing, W., & Geffen, L. (1990). Loss of brainstem and serotonin- and substance P-containing neurons in Parkinson's disease. *Brain Research, 510*, 104–107.

Harding, A. J., Broe, G. A., & Halliday, G. M. (2002). Visual hallucinations in Lewy body disease relate to Lewy bodies in the temporal lobe. *Brain: A Journal of Neurology, 125*, 391–403.

Hasler, F., Grimberg, U., Benz, M. A., Huber, T., & Vollenweider, F. X. (2004). Acute psychological and physiological effects of psilocybin in healthy humans: a double-blind, placebo-controlled dose-effect study. *Psychopharmacology, 172*, 145–156.

Howard, R., David, A., Woodruff, P., Mellers, I., Wright, J., & Brammer, M. (1997). Seeing visual hallucinations with functional magnetic resonance imaging. *Dementia and Geriatric Cognitive Disorders, 8*, 73–77.

Huot, P., Fox, S. H., & Brotchie, J. M. (2011). The serotonergic system in Parkinson's disease. *Progress in Neurobiology, 95*, 163–212.

Huot, P., Johnston, T. H., Darr, T., Hazrati, L. N., Visanji, N. P., Pires, D., … Fox, S. H. (2010). Increased 5-HT_{2A} receptors in the temporal cortex of parkinsonian patients with visual hallucinations. *Movement Disorders, 25*, 1399–1408.

Jellinger, K. (1990). New developments in the pathology of Parkinson's disease. *Advances in Neurology, 53*, 1–16.

Juncos, J. L., Roberts, V. J., Evatt, M. L., Jewart, R. D., Wood, C. D., Potter, L. S., … Yeung, P. P. (2004). Quetiapine improves psychotic symptoms and cognition in Parkinson's disease. *Movement Disorders, 19*, 29–35.

Kometer, M., Schmidt, A., Jancke, L., & Vollenweider, F. X. (2013). Activation of serotonin 2A receptors underlies the psilocybin-induced effects on alpha oscillations, N170 visual-evoked potentials, and visual hallucinations. *Journal of Neuroscience, 33*, 10544–10551.

Lepore, F. (1990). Spontaneous visual phenomena with visual loss: 104 patients with lesions of retinal and neural afferent pathways. *Neurology, 40*, 444–447.

Manford, M., & Andermann, F. (1998). Complex visual hallucinations: clinical and neurobiological insights. *Brain: A Journal of Neurology, 121*, 1819–1840.

Meltzer, H. Y., Elkis, H., Vanover, K., Weiner, D. M., Van Kammen, D. P., Peters, P., & Hacksell, U. (2012). Pimavanserin, a selective serotonin (5-HT) 2A-inverse agonist, enhances the efficacy and safety of risperidone, 2mg/day, but does not enhance efficacy of haloperidol, 2mg/day: comparison with reference dose risperidone, 6mg/day. *Schizophrenia Research, 141*, 144–152.

Meltzer, H. Y., Kennedy, J., Dai, J., Parsa, M., & Riley, D. (1995). Plasma clozapine levels and the treatment of L-dopa-induced psychosis in Parkinson's-disease – a high potency effect of clozapine. *Neuropsychopharmacology: Official Publication of the American College of Neuropsychopharmacology, 12*, 39–45.

Meltzer, H. Y., & Massey, B. W. (2011). The role of serotonin receptors in the action of atypical antipsychotic drugs. *Current Opinion in Pharmacology, 11*, 59–67.

Meltzer, H. Y., Mills, R., Revell, S., Williams, H., Johnson, A., Bahr, D., & Friedman, J. H. (2010). Pimavanserin, a serotonin(2A) receptor inverse agonist, for the treatment of Parkinson's disease psychosis. *Neuropsychopharmacology: Official Publication of the American College of Neuropsychopharmacology, 35*, 881–892.

Meltzer, H. Y., & Roth, B. L. (2013). Lorcaserin and pimavanserin: emerging selectivity of serotonin receptor subtype-targeted drugs. *The Journal of Clinical Investigation, 123*, 4986–4991.

Merabet, L., Maguire, D., Warde, A., Alterescu, K., Stickgold, R., & Pascual-Leone, A. (2004). Visual hallucinations during prolonged blindfolding in sighted subjects. *Journal of Neuro-Opthalmology, 24*, 109–113.

Mocellin, R., Walterfang, M., & Velakoulis, D. (2006). The neuropsychiatry of complex visual hallucinations. *The Australian and New Zealand Journal of Psychiatry, 40*, 742–751.

Mosimann, M., Mather, G., Wesnes, K., O'brien, J., Burn, D., & Mckeith, I. (2004). Visual perception in Parkinson disease dementia and dementia with Lewy bodies. *Neurology, 63*, 2091–2096.

Nevo, I., & Hamon, M. (1995). Neurotransmitter and neuromodulatory mechanisms involved in alcohol abuse and alcoholism. *Neurochemistry International, 26*, 305–336.

Nichols, D. E. (2004). Hallucinogens. *Pharmacology & Therapeutics, 101*, 131–181.

Ondo, W. G., Tintner, R., Voung, K. D., Lai, D., & Ringholz, G. (2005). Double-blind, placebo-controlled, unforced titration parallel trial of quetiapine for dopaminergic-induced hallucinations in Parkinson's disease. *Movement Disorders, 20*, 958–963.

Penfield, W., & Perot, P. (1963). The brain's record of auditory and visual experience: a final summary and discussion. *Brain: A Journal of Neurology, 86*, 595–696.

Perry, E., Johnson, M., Kerwin, J., Piggott, M., Court, J., Shaw, P., ... Perry, R. (1992). Convergent cholinergic activities in aging and Alzheimer's disease. *Neurobiology of Aging, 13*, 393–400.

Perry, E., & Perry, R. (1995). Acetylcholine and hallucinations: disease related compared to drug-induced alterations in human consciousness. *Brain and Cognition, 28*, 240–258.

Rabey, J. M., Prokhorov, T., Miniovitz, A., Dobronevsky, E., & Klein, C. (2007). Effect of quetiapine a double-blind in psychotic Parkinson's disease patients: labeled study of 3 months duration. *Movement Disorders, 22*, 313–318.

Santhouse, A. M., Howard, R. J., & Ffytche, D. H. (2000). Visual hallucinatory syndromes and the anatomy of the visual brain. *Brain: A Journal of Neurology, 123*, 2055–2064.

Shotbolt, P., Samuel, M., & David, A. (2006). Psychosis in Parkinson's disease: how should we treat it? *Expert Review of Neurotherapeutics, 6*, 1243–1246.

Siegel, R., & Jarvik, M. (1975). Drug-induced hallucinations in animals and man. In R. Seigel, & L. West (Eds.), *Hallucinations: Behavior, experience and theory*. New York: Wiley.

Slade, P. (1976). Hallucinations. *Psychological Medicine, 6*, 7–13.

Tyls, F., Palenicek, T., & Horacek, J. (2014). Psilocybin—summary of knowledge and new perspectives. *European Neuropsychopharmacology: The journal of the European College of Neuropsychopharmacology, 24*, 342–356.

Vanover, K. E., Weiner, D. M., Makhay, M., Veinbergs, I., Gardell, L. R., Lameh, J., ... Davis, R. E. (2006). Pharmacological and behavioral profile of N-(4-fluorophenylmethyl)-N-(1-methylpiperidin-4-yl)-N′-(4-(2-methylpropyloxy)phenylmethyl) carbamide (2R,3R)-dihydroxybutanedioate (2:1) (ACP-103), a novel 5-hydroxytryptamine(2A) receptor inverse agonist. *Journal of Pharmacology and Experimental Therapeutics, 317*, 910–918.

Vollenweider, F. X., Vollenweider-Scherpenhuyzen, M. F. I., Babler, A., Vogel, H., & Hell, D. (1998). Psilocybin induces schizophrenia-like psychosis in humans via a serotonin-2 agonist action. *Neuroreport, 9*, 3897–3902.

Walsh, F., & Hoyt, W. (1969). *Clinical neuro-ophthalmology*. Baltimore: Williams & Wilkins.

Weintraub, D., & Stern, M. B. (2005). Psychiatric complications in Parkinson disease. *American Journal of Geriatric Psychiatry, 13*, 844–851.

West, C. (1962). *Hallucinations*. New York: Grune & Stratton.

Williams, D. R., & Lees, A. J. (2005). Visual hallucinations in the diagnosis of idiopathic Parkinson's disease: a retrospective autopsy study. *Lancet Neurology, 4*, 605–610.

Williams, D. R., Warren, J. D., & Lees, A. J. (2008). Using the presence of visual hallucinations to differentiate Parkinson's disease from atypical parkinsonism. *Journal of Neurology, Neurosurgery, and Psychiatry, 79*, 652–655.

Wu, M. F., John, J., Boehmer, L. N., Yau, D., Nguyen, G. B., & Siegel, J. M. (2004). Activity of dorsal raphe cells across the sleep-waking cycle and during cataplexy in narcoleptic dogs. *Journal of Physiology, Paris, 554*, 202–215.

Section III

Clinical Phenomenologie

Chapter 25

Risk Overlap Between Clinical Disorders

Patrick F. Sullivan
University of North Carolina at Chapel Hill, Chapel Hill, NC, United States; Karolinska Institutet, Stockholm, Sweden

INTRODUCTION

The classification and interrelation of diseases are million-dollar questions in medicine. Are two clinical presentations the result of a common pathophysiological process? Are they etiologically distinctive despite common signs and symptoms? For example, one reason why the germ theory of disease had such a profound impact on medicine was its extraordinary explanatory capacity. Moreover, a complete understanding of classification has direct implications for treatment. Instead of elaborate taxonomies based on the clinical features of fever, fundamental causes could be identified. Therapeutic interventions could then target the root cause and clinical signs and symptoms became epiphenomena.

Classification is perhaps less of a solved issue in clinical neuroscience than in any other branch of medicine. For example, schizophrenia and epilepsy are comorbid in that they co-occur beyond chance. A sizable minority of people with schizophrenia has epilepsy or epileptiform activity and a fraction of people with epilepsy has a psychotic syndrome strongly reminiscent of schizophrenia. How do we understand clinical comorbidity? There is a range of possible interrelationships (Neale & Kendler, 1995), and a central issue to Section III is whether multiple neuropsychiatric disorders or diseases share a common pathophysiological process.

Sections I and II of *Genomics, Circuits, and Pathways for the Clinical Neuroscientist* reviewed the exceptional technological advances fueling the current epoch of discovery in genomics and neuroscience. Section IV focuses on clinical translation, the relevance of the new knowledge being generated to drug development, individualized therapeutics, and predictive genetic testing. Section III, this portion of the book, is arguably the heart of *Genomics, Circuits and Pathways for the Clinical Neuroscientist* and covers what we have learned about the pathophysiological processes of the diseases and disorders that feature prominently in clinical neuroscience.

This chapter is the overture to Section III and introduces the key themes that will be amplified in later chapters. For many of the disorders covered in Section III, until recently our knowledge of the biology has been highly uncertain and genomic studies have had and still have a key role in pointing to fundamental disease processes that serve to guide subsequent studies.

CLINICAL NEUROSCIENCE, NEUROLOGY, AND PSYCHIATRY

A longstanding tussle in medicine has been the placement of the divide between neurology and psychiatry. A salient example is Alzheimer disease, which is widely considered a neurological condition owing to the identification of a characteristic neuropathology and emerging pathophysiology. However, this condition was named after a psychiatrist, is often managed by psychiatrists, was included in the *International Statistical Classification of Diseases and Related Health Problems*, 10th Edition (ICD-10) as a psychiatric disorder, and often has prominent psychiatric comorbidities (eg, anxiety, depressive and psychotic features). The term "clinical neuroscience" is in many ways an explicit agreement to set aside traditional dichotomies (neurological versus psychiatric, organic vs functional, and hardware versus software) and is more current than the related terms of behavioral neurology and neuropsychiatry. These dichotomies have not served to enhance insight, but bluntly and inappropriately attempt to impose a simple classification scheme onto an inherently blurred continuum.

Genomics, Circuits, and Pathways in Clinical Neuropsychiatry. http://dx.doi.org/10.1016/B978-0-12-800105-9.00025-1

The core pathology of the disorders and diseases covered in this section is based on the central nervous system (CNS) and has what has been traditionally considered neurological and psychiatric components. This distinction tends to rest on an often loosely defined dichotomy between "identifiable" versus unknown pathology: for example, diseases with frank lesions (eg, primary brain tumors, white matter disorders, dementias) are typically categorized as "neurological" irrespective of the presence of associated alterations in behavior, emotion, and cognition. When there has been an unknown or markedly less clear relationship between behavioral and emotional symptomatology and pathophysiology, CNS disorders tend to be categorized as "psychiatric."

As the molecular mechanisms of "gross" lesions are illuminated and the molecular, cellular, and circuit abnormalities driving "idiopathic" conditions are revealed, these dichotomies will necessarily become less satisfactory. Indeed a central theme of this volume is that this is now occurring and that there is far more to be gained by rejecting these categorical boundaries than by attempting to place disorders in one versus another discipline.

It is also worthwhile to recall that psychiatry and neurology split into distinct disciplines in an earlier epoch in which the state of knowledge tended to support the distinction between mind and brain. There certainly is a general focus in psychiatric and psychological training and practice on behavioral and emotional dysfunction, and in neurology, a greater emphasis on motor and sensory disturbances. Importantly, the observation that disorders of the nervous system tend to defy such easy distinctions routinely does not imply that clinical neuroscience should focus exclusively on the same overt manifestations of CNS pathology. It argues that there is a core body of knowledge that all clinical neuroscientists must possess regardless of the dominant themes in their training and practice.

THE DISEASES AND DISORDERS IN SECTION III

In this chapter, the introduction to Section III, I use the traditional definitions (OED Online, 2014) for "disorder" (an illness that disrupts normal function) and "disease" (a disorder with known pathophysiology or structural pathology): hence, bipolar disorder and Alzheimer disease. One obvious goal for clinical neuroscience is to convert idiopathic disorders into pathophysiologically defined diseases.

Section III contains 18 chapters devoted to diseases and disorders of prominent interest in clinical neuroscience. Table 25.1 provides an overview and rough classification of the chapters in Section III. Section III contains classical psychiatric disorders (drug use disorders, eating disorders, mood disorders, posttraumatic stress syndrome (PTSD), schizophrenia, and psychosis) as well as classical neurological diseases (amyotrophic lateral sclerosis [ALS], epilepsy, Parkinson disease, primary brain tumors, and white matter disorders). Several entities are in the borderland (speech and language disorders, Alzheimer disease, frontotemporal dementia, neurodevelopmental syndromes, and autism). In addition, one chapter focuses on classification and another on neuroimmunology. Many of these entities have high global morbidity, particularly major depressive disorder (MDD), drug use disorders, epilepsy, dementias, and schizophrenia, as expected for relatively early-onset and often highly chronic conditions.

Fig. 25.1 shows a scatterplot of some of the data in Table 25.1. These entities have two rough clusters defined by common versus uncommon/rare prevalence and higher or lower heritability. These clusters are now known to be a poor guide to the rapidity with which genetic studies might be predicted to identify loci; the current success stories are distributed throughout Fig. 25.1. The determining factor is not the population-level characteristics of prevalence and heritability but rather the fundamental genetic architecture of the disorder or disease (Sullivan, Daly, & O'Donovan, 2012): that is, the degree of etiological heterogeneity and the number of contributing loci, and for each, the mode of action, allele frequency, and genotypic relative risk.

OVERLAP

Even a cursory review of the entities covered in Section III suggests substantial overlap. For instance, bipolar disorder is typified by the presence of depressive and manic episodes, and MDD by depressive episodes. Schizophrenia is a psychotic disorder. Epilepsy is common in autism. People with dementia can develop psychosis.

With the expanding reliance on electronic medical records, very large-scale data analyses of symptoms and diagnoses are becoming increasingly feasible, enabling systematic assessments of comorbidity in clinical populations (Blair et al., 2013 #7947; Kohane, 2015 #7967). In a complementary manner, one can begin to gain some insight into such questions by assessing the degree of overlap in the extant literature. To do so, I recorded the number of PubMed results for Medical Subject Headings (MeSH) translations for as many of the entities in Section III as possible, to obtain a rough estimate of the importance or amount of research for each. The number of citations ranged from 1229 (Frontotemporal Dementia) to 221,959 (Substance-Related Disorders). There were three groupings: entities with high (Brain Neoplasm, Epilepsy, and

TABLE 25.1 Prevalence, Heritability, and Impact of the Entities in Section III

Disease/Disorder	Lifetime Prevalence	Heritability	Impact
Drug use disorders (Breslau, Johnson, Hiripi, & Kessler, 2001; Kessler et al., 2005; Polderman et al., 2016) • Alcohol abuse/dependence • Nicotine dependence	0.132 0.24	0.41 0.44	1.17% ND
Eating disorders (Bulik et al., 2010; Hudson, Hiripi, Pope, & Kessler, 2007) • Anorexia nervosa • Bulimia nervosa	0.006 0.01	0.57 0.62	0.08% ND ND
Mood disorders (Kessler et al., 2005; Merikangas et al., 2007; Polderman et al., 2016; Sullivan, Neale, & Kendler, 2000; Weissman et al., 1996) • Bipolar disorder • Major depressive disorder	0.007 0.13	0.68 0.37	3.27% 0.48% 2.79%
Posttraumatic stress disorder (Kessler et al., 2005; Stein, Jang, Taylor, Vernon, & Livesley, 2002)	0.068	0.30	ND
Schizophrenia (Saha, Chant, Welham, & McGrath, 2005; Sullivan, Kendler, & Neale, 2003)	0.004	0.81	0.51%
Dementias (Feldman et al., 2003; Plassman et al., 2007; Polderman et al., 2016) • Alzheimer disease • Frontotemporal dementia	0.097 0.054	0.63 ND	0.66% ND ND
Pervasive developmental disorders (Fombonne, 2009; Polderman et al., 2016) • Autism spectrum disorder	0.0064 0.0026	0.60 0.80	0.29% ND
Amyotrophic lateral sclerosis (Al-Chalabi et al., 2010; Pasinelli & Brown, 2006)	0.0017	0.61	ND
Epilepsy (Kjeldsen, Kyvik, Christensen, & Friis, 2001; Ngugi, Bottomley, Kleinschmidt, Sander, & Newton, 2010)	0.0058	0.70	0.75%
Parkinson disease (Bulik et al., 2010; Hudson et al., 2007)	0.067	0.34	0.09%

The first five rows are traditional psychiatric disorders and the last three rows are traditional neurological disorders. Dementias and pervasive developmental disorders are psychiatric disorders in *The Diagnostic and Statistical Manual of Mental Disorders*, Fourth Edition and *International Statistical Classification of Diseases and Related Health Problems*, 10th Edition, but are also neurological disorders. Additional chapters in Section III not included here are on classification, neurodevelopmental disorders, neuroimmunology, speech and language disorders, psychosis, brain malformation syndromes, primary brain tumors, and white matter disorders, because these are general cross-cutting processes or highly heterogeneous topics without specific prevalence, heritability, and impact data. Impact is the percentage of disability-adjusted life-years in the world in 2012 (see URLs). Epilepsy estimates are for nonsyndromic forms. *ND*, no data.

Substance Related Disorders; 12.2–22.9%), low (Frontotemporal Dementia, Language Development Disorders, Brain Abnormalities, Amyotrophic Lateral Sclerosis, Autistic Disorder, Stress Disorders Post Traumatic, Eating Disorders, and Leukoencephalopathies; 0.1–2.5%), and intermediate numbers of citations (Bipolar Disorder, Psychotic Disorders, Parkinson Disease, Alzheimer Disease, Depressive Disorder, and Schizophrenia; 3.2–8.7%).

I next conducted PubMed queries for 153 pairwise combinations of these entities. The median number of citations for these pairwise combinations was 114 (interquartile range, 15–469). The number of pairwise citations ranged from a maximum of 9078 (Schizophrenia and Psychotic Disorders) to a minimum of zero for four pairs of terms (eg, Frontotemporal Dementia and Stress Disorders Post Traumatic).

The PubMed data are depicted as a network graph in Fig. 25.2. This is a relatively interconnected network with 97.4% of all pairwise combinations of these entities with more than one PubMed citation, which underscores the importance of understanding and evaluating the overlap of what have traditionally been considered distinct conditions (both within and between medical specialties).

Not surprisingly, these findings reflect the current neurology versus psychiatry dichotomy, because a modularity detection algorithm yielded two communities in the graph consisting of traditional neurological disorders (eg, ALS) and psychiatric disorders (schizophrenia and MDD). Almost certainly this reflects a number of confounds, including a continued categorical emphasis in extramural funding and in journal editorial policies. What is most notable, perhaps, is that despite these forces that would tend to drive the separation of reports of interest to one discipline or disorder versus another, considerable overlap is readily apparent.

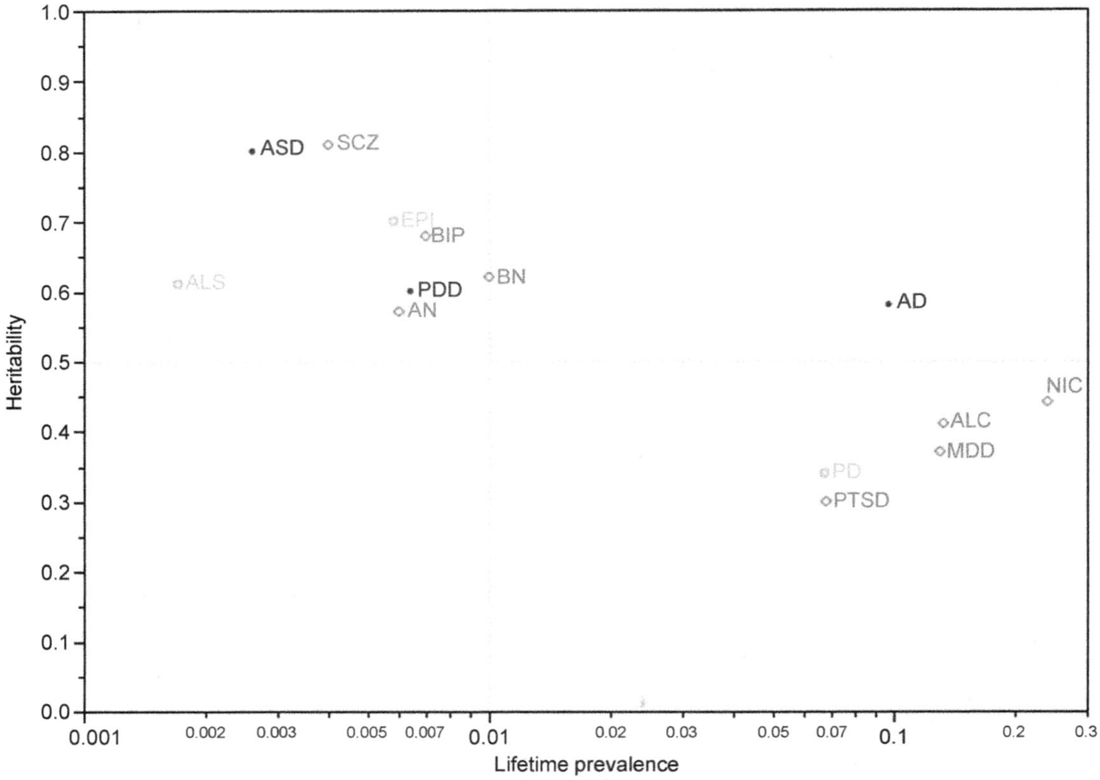

FIGURE 25.1 Scatterplot of heritability by lifetime prevalence. *ALC*, alcohol abuse/dependence; *NIC*, nicotine dependence; *AN*, anorexia nervosa; *BN*, bulimia nervosa; *BIP*, bipolar disorder; *MDD*, major depressive disorder; *PTSD*, posttraumatic stress disorder; *SCZ*, schizophrenia; *AD*, Alzheimer disease; *PDD*, pervasive developmental disorder; *ASD*, autism spectrum disorder; *ALS*, amyotrophic lateral sclerosis; *EPI*, epilepsy; *PD*, Parkinson disease. Classical psychiatric disorders are in magenta, classical neurological disorders are in cyan, and ASD, PDD, and AD are in blue.

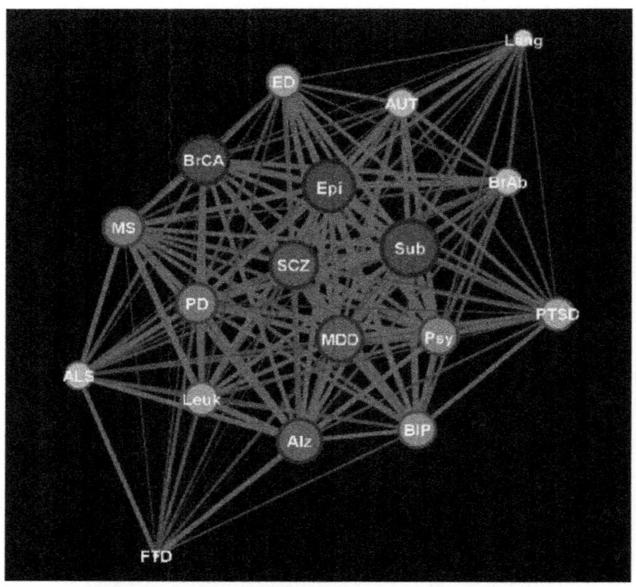

FIGURE 25.2 Network graph for PubMed citations for entities in Section III. The PubMed network graph was made using Gephi (see URLs). Each circle is a node whose diameter is proportional to the number of PubMed citations. The coloring of nodes reflects approximate sizes. The width of the lines connecting the nodes is proportional to the number of PubMed citations for the co-occurrence of the MeSH terms for the nodes. Application of a community detection/modularity algorithm yielded two classes corresponding to neurological (PD, BrCA, Leuk, and ALS) and psychiatric disorders (SCZ, Psy, AUT, BIP, MDD, Lang, PTSD, Sub, and ED). Notably, FTD and Alz clustered with neurological disorders and BrAb and Epi clustered with psychiatric disorders. *ALS*, amyotrophic lateral sclerosis; *Alz*, Alzheimer disease; *AUT*, autistic disorder; *BIP*, bipolar disorder; *BrAb*, brain abnormalities; *BrCA*, brain neoplasms (ie, cancer); *ED*, eating disorders; *Epi*, epilepsy; *FTD*, frontotemporal dementia; *Lang*, language development disorders; *Leuk*, leukoencephalopathies; *MDD*, depressive disorder; *MS*, multiple sclerosis; *PD*, Parkinson disease; *Psy*, psychotic disorders; *PTSD*, posttraumatic stress disorders; *SCZ*, schizophrenia; *Sub*, substance-related disorders.

GENOMICS AS A WAY TO UNDERSTAND RISK OVERLAP

The data in Fig. 25.2 are useful as a first approximation of evaluating the general overlap between these disorders in that these data capture aspects of the published literature. However, this does not necessarily correspond to the core purpose of this introduction to Section III, which is consideration of risk overlap.

The core issue here is that the precise mechanisms of risk are poorly defined for most of the entities in Section III. At one extreme, we have the clinical portrait, a coherent collection of signs and symptoms often refined by decades of careful observation (impeccable clinical characterization has historically been an important attribute of the clinical neurosciences). On occasion, there may even be a diagnostic test (eg, brain imaging in white matter disorders or primary brain tumors). On the other extreme there is a rapidly expanding set of genomic knowledge fueled by a converging set of remarkable technical and informatics advances (Lander, 2011; Visscher, Brown, McCarthy, & Yang, 2012). Between these extremes, DNA and clinical portrait, is often a black box. For many of the entities in Section III, genetic variation conferring risk has been identified but the impacts on molecules, cells, circuits, and organ systems have yet to be defined.

Thus, for many of the disorders and diseases in Section III, to date, confident comparison of risk overlap can occur only at the level of the genetic loci identified using genomic studies. Genomic studies may be the best way to gain entry into the risk process. This, in turn, raises two questions: What is the best way to wield genomic tools? What do we now know about the genetic architectures of these entities?

WIELDING GENOMIC TOOLS

Full descriptions of the technologies that can be brought to bear on trying to understand the genetic basis of a biomedical disease or disorder can be found in Section I of this book. However, there is now considerable experience with all of these technologies. Accumulated experience suggests a critically important caveat: that the statistical evidence should be extremely strong and in line with current best practices in human genomics.

A mnemonic for the standards of evidence in human genomics is the three *R*'s. A finding should be *robust* in that the statistical evidence exceeds chance. To account for multiple comparisons, the accepted significance threshold for a genome-wide association study is $P < 5 \times 10^{-8}$ (akin to a Bonferroni correction of 0.05 for 10^6 tests). For transmitted variation in copy number variant (CNV) and exome sequencing studies, the threshold is typically $P < 5 \times 10^{-6}$. Increasingly, researchers have begun to use false discovery rate and Bayesian approaches. A crucial point is that a finding should *replicate* in multiple independent samples, and this is generally viewed as essential. There should be unmistakable evidence that the authors were *reluctant* to accept their findings. False positives occur in genomic studies owing to batch effects (eg, cases and controls genotyped at different places and times), differences in ancestry between cases and controls (ie, population stratification), incautious quality control and analysis, and the pernicious assumption that a technology is a perfect black box. The best articles will have expert and vigorous attempts to disprove their findings.

A second caveat is that best practices in human genomics now stress the primacy of the statistical evidence. In the past, it was typical for investigators to combine soft empirical findings not meeting the "three *R*'s" with some wet bench work and to conclude that the combination equated with salience. This argument is now often viewed as weak, subjective, and prone to incorrect decisions. In general, in the absence of unequivocal knowledge of a fundamental disease process, this argument is not falsifiable. For most psychiatric disorders such as schizophrenia, this caveat applies because we do not have a clear fix on the fundamental biology. This caveat may not apply to other diseases in Section III in which the basic biology may be far clearer (eg, the genes strongly implicated in white matter disorders). This argument is counter to best practices in human genetics (Attia, 2009a, 2008b; Ioannidis, Thomas, & Daly, 2009; Kraft, Zeggini, & Ioannidis, 2009; McCarthy et al., 2008) (including rare variants of strong effect) (MacArthur et al., 2014) and is not accepted by many journals (eg, *Nature Genetics*: "the genetic and statistical evidence for association should be sound. Molecular biological evidence for a functional variant is desirable in addition to, but will not substitute for, sound genetic evidence") (Editorial, 2005).

THE GENETIC ARCHITECTURES OF SECTION III ENTITIES

The first part of defining the genetic architecture of a disease or disorder is etiological heterogeneity: How many different disease processes can lead to the same clinical outcome? There may be a few known processes (eg, one of the white matter disorders, leukoencephalopathy with vanishing white matter disease, is caused by mutations in any of five subunits of the translation initiation factor EIF2B) or a large number of hypothesized disease processes (eg, MDD being caused by different mixtures of genes and environments). For most of the entities in Section III, the number of disease processes is partly or entirely unknown and an active research topic. Even when a disease process has been defined, it is often at the level of a confident genetic association without a detailed and precise mechanism. The second part of defining the genetic architecture is to determine the number of genetic loci involved and its role. These loci can be very common or very rare, they can have genotypic relative risks ranging from probabilistic to deterministic, and they can act in a straightforward (additive, dominant, or recessive mode of action) or complex manner (eg, involving interactions with other genetic loci or environmental risk factors). These two aspects of genetic architecture are related. The latter is usually a way to elucidate the former.

The entities in Section III can be roughly classified into four different genetic risk patterns. These are disparate conditions, and this classification is required to set the stage for a discussion of risk overlap.

- Fig. 25.3A: Sets of Mendelian or near-Mendelian disorders (eg, brain malformation syndromes, primary brain tumors, and white matter disorders). These are caused by different rare but deterministic genetic mutations. Common variants of subtle effects have a small role or none at all. Environmental effects are not required but can serve to hasten or prolong disease onset.

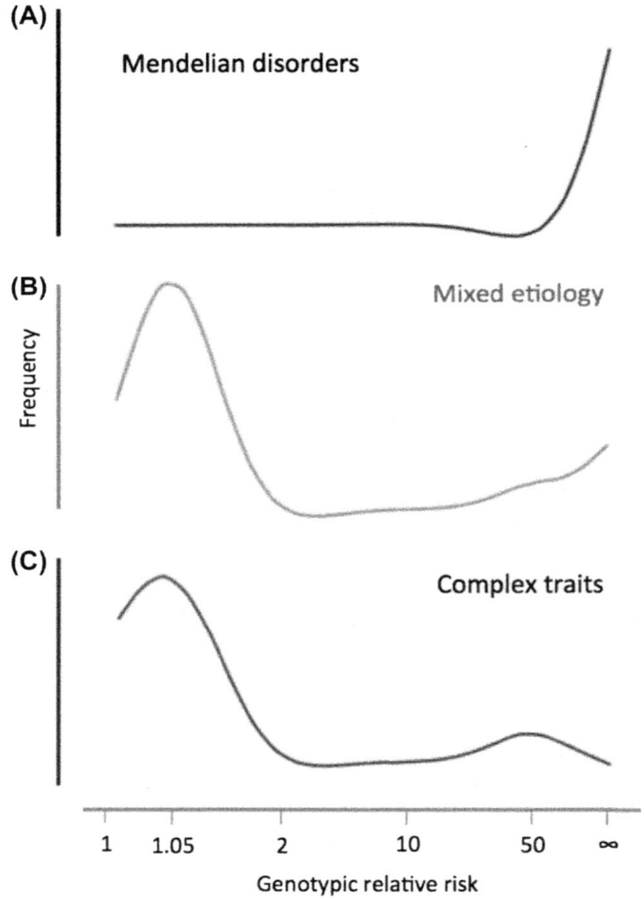

FIGURE 25.3 Genetic architectures graph for entities in Section III. Three general possibilities for genetic architectures of entities in Section III. (A) *Top area* shows a genetic architecture entirely composed of rare, deterministic Mendelian variation (ie, genotypic relative risks of ∞) (eg, white matter disorders). (C) *Bottom area* shows a genetic architecture typical of human complex diseases, disorders, and anthropometric traits: no Mendelian variation, a variable but minor contribution of rare variation of strong effect (eg, structural variation), and a dominating influence of large numbers of common variants of relatively subtle individual effects (eg, schizophrenia). (B) *Middle area* shows mixed portrait combining the top and bottom graphs (eg, autism or Alzheimer disease). Note that these graphs do not incorporate the likely role of the environment for many of the entities in Section III.

- Fig. 25.3B: Mixed etiology with the clinical entity caused by heterogeneous genetic variants (eg, Alzheimer disease, frontotemporal dementia, autism spectrum disorder, ALS, epilepsy, and Parkinson disease). The etiology of these entities is heterogeneous: subsets of cases can be attributed to a deterministic Mendelian variant (eg, *APP*, *PSEN1*, and *PSEN2* in Alzheimer disease, or *PTEN*, *FMRP*, and *TSC* in autism). Other cases have contributions of rare variants of strong effect with genotypic relative risks of 5–50 (eg, for autism, CNVs in 16p11.2 or 1q21 or rare loss of function mutations in *CHD8*, *SCN2A*, or *GRIN2B*). Still other cases have a probabilistic risk pattern from an accumulation of common variants of subtle effect.
- Fig. 25.3C: Complex traits with no known Mendelian or near-Mendelian causes, some rare variants of strong but not deterministic effect, and a predominant effect of accumulations of common variants of subtle effect (eg, schizophrenia and bipolar disorder). Environmental risks contribute to risk directly or indirectly via interactive effects.
- Although not shown in Fig. 25.3, the classification of the remaining psychiatric disorders is unclear (eg, drug use disorders, eating disorders, MDD, and PTSD). Each is heritable but genetic discovery efforts are not yet extensive and the genetic architecture remains uncertain. There may be a prominent role for environmental exposures. For instance, exposure to a severe traumatic event is required in PTSD and development of a drug use disorder requires sufficient access to the drug.

RISK OVERLAP

Fig. 25.3 considers the genetic architectures of single brain diseases and disorders. In light of the evident interconnections of many of these disorders at a phenotype level (eg, the relatively densely interconnected network shown in Fig. 25.2), a logical next question is the degree of overlap of the genetic architectures of any two of these disorders. In other words, because many of the brain diseases and disorders in Section III are wholly or partly caused by genetic variation, to what extent is rare or common genetic variation shared between any two entities? This is not an academic question but rather cuts to the core of clinical neuroscience with relevance to nosology, treatment, and etiology. Indeed, this topic has been the focus of a large number of academic studies, and is a foundational topic for neurology and psychiatry. In the next decade, I believe it likely that many of these questions will be addressed and perhaps even definitely answered with reference to empirical data. I caution that, as with many topics in clinical neuroscience, it will take a corpus of large and impeccably conducted studies until clear consensus emerges.

For both rare and common variations, genetic overlap can occur at the level of a specific gene or in multiple members of a set of genes that comprise a biological pathway. For example, common genetic variation in *CACNA1C* has been implicated in bipolar disorder and schizophrenia (Ferreira et al., 2008; Schizophrenia Psychiatric Genome-Wide Association Study Consortium, 2011) and rare genetic variation in schizophrenia and Timothy syndrome (a multi-system disorder with autistic features) (Purcell, 2014 #6192; Splawski et al., 2004 #6051). Loss of expression of Fragile X mental retardation protein, the product of *FMR1* (Fragile X mental retardation protein) is implicated in autism, and the set of genes whose mRNAs are bound by FMRP (Darnell et al., 2011 #6104) is strongly implicated in both autism and schizophrenia (De Rubeis et al., 2014 #7643; Purcell, 2014 #6192). As noted earlier, it is likely that environmental risks will have a role. For instance, head trauma is a risk factor for schizophrenia and may trigger episodes of white matter degeneration in leukodystrophies (Murray, Jones, Susser, van Os, & Cannon, 2003).

One issue noted earlier is that the genetic architectures of most of the entities considered in Section III are incompletely understood. Thus, a thorough comparison of genetic risk overlap is not currently possible. However, there are tantalizing (if incomplete) data for some disorders.

For rare genetic variation, large CNVs that alter gene dosage have been implicated in autism and schizophrenia (Sullivan et al., 2012 #5837). A notable example is a multimegabase deletion on chromosome 22 (22q11.21). If this approximately 50-gene region is present at copy number 1 (instead of the normal 2), risk for schizophrenia is markedly elevated to an odds ratio of over 20. However, the presence of this deletion is not specific to schizophrenia but rather confers increased risk for a broader range of neurodevelopmental conditions including developmental delay, intellectual disability, epilepsy, and autism. This general pattern appears to hold for other CNVs in which recurrent changes in gene dosage in 1q21.1, *NRXN1* (2p16.3), 3q29, 15q11.2, 15q13.3, 16p11.2, and 17q12 have been shown to increase risk for intellectual disability, developmental delay, epilepsy, autism, and/or schizophrenia (Sullivan et al., 2012 #5837). In many instances, these CNVs also affect other organ systems (eg, cardiac development).

Genetic variation with strong impact on protein function (eg, a single base change that creates a "stop" signal and abolishes the function of one copy of a protein) is another type of rare variation. These can be inherited from a parent or arise "de novo" in the genesis of a person. For some of the entities covered in Section III (eg, leukoencephalopathies), rare protein altering in subunits of the translation initiation factor EIF2B are demonstrably causal. However, for most of the Section III entities, rare variation in a relatively large number of genes is involved.

To date, autism has (arguably) produced the largest and most comprehensive study of rare variation that affects protein function (De Rubeis et al., 2014 #7643). As shown in Fig. 25.4A, rare variants predicted to impact proteins strongly in autism overlap with genes also implicated in psychiatric (schizophrenia and intellectual disability), neurological (epilepsy), and somatic disorders (congenital heart disease and metabolic disorders). Larger and more comprehensive studies are in progress.

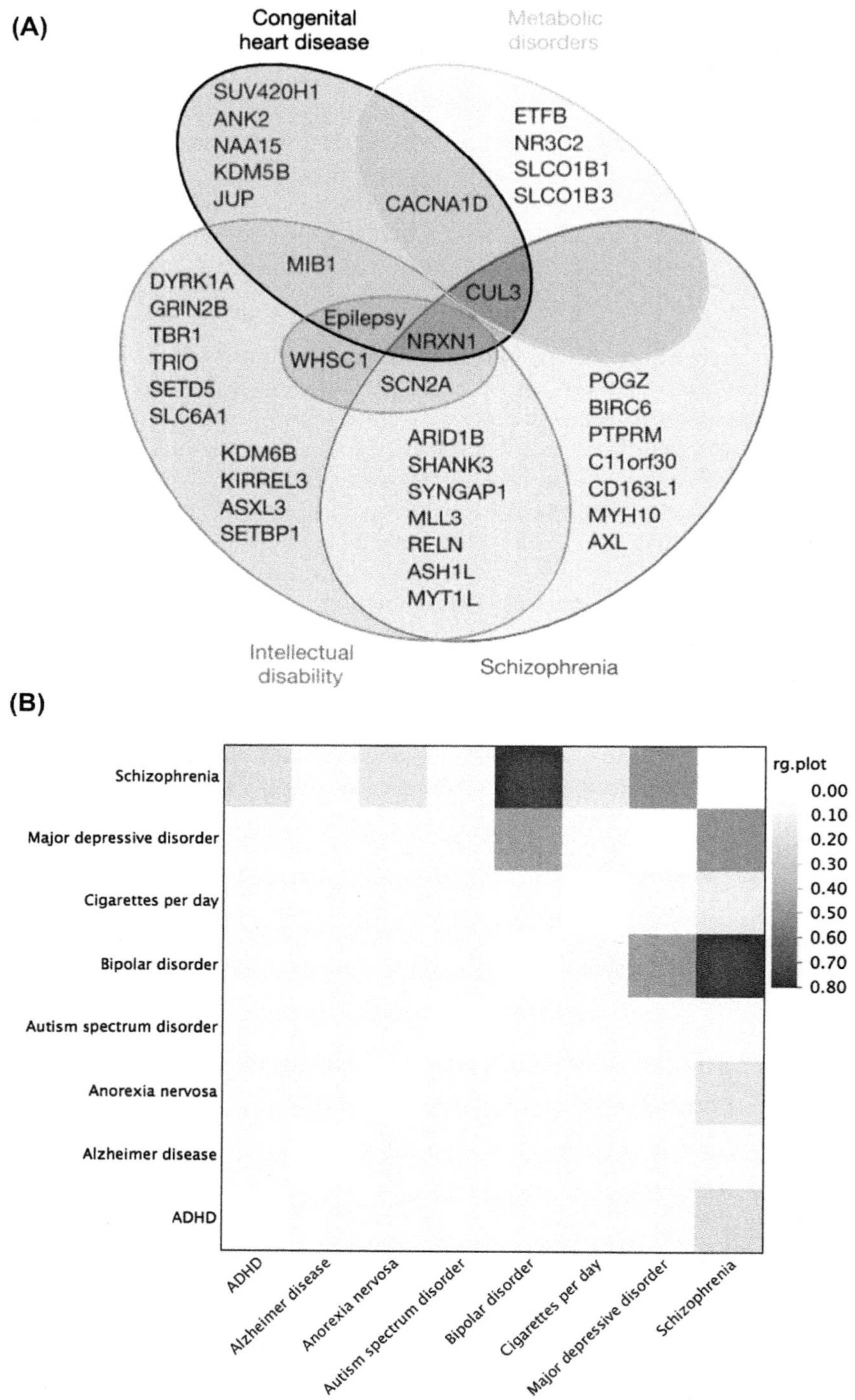

FIGURE 25.4 (A) Rare genetic variants implicated in autism predispose to multiple psychiatric, neurological, and somatic conditions. *(Reprinted with permission from Fig. 5 in De Rubeis, S., et al. (2014). Synaptic, transcriptional and chromatin genes disrupted in autism.* Nature, 515*(7526), 209–215.)* (B) Correlations of common variant genetic architectures. *(Data replotted from Bulik-Sullivan, B. K., et al. (2015). LD Score regression distinguishes confounding from polygenicity in genome-wide association studies.* Nature Genetics, 47, *291–295.)*

For common variation, methods have been developed to quantify the genetic correlation between common variants empirically shown to predispose to two disorders (Cross-Disorder Group of the Psychiatric Genomics Consortium, 2013 #6213; Bulik-Sullivan et al., 2015 #7676). A subset of results from the most comprehensive study is shown in Fig. 25.4A. There are strong and significant positive genetic correlations between "adult" psychiatric disorders (schizophrenia—bipolar disorder—major depressive disorder). Surprisingly, there was a significant positive genetic correlation between anorexia nervosa and schizophrenia. This is intriguing and raises the question of whether the markedly distorted cognitions that typify anorexia nervosa (ie, the subjective experience of being fat despite objective emaciation) are less of an "overvalued idea" and more akin to a delusion. I note that there appears to be a need for high-quality antipsychotic medication trials in anorexia nervosa.

In the "Brainstorm" project, Neale et al. (see URLs) are preparing a manuscript describing a comprehensive analysis of common variant genetic correlations from over 200,000 cases from eight psychiatric disorders and 12 neurological conditions. This widely anticipated analysis is likely to provide considerable insight into the overlap of Section III entities.

CONCLUSION

In many ways, Section III is central to *Genomics, Circuits, and Pathways for the Clinical Neuroscientist*. In this introduction to Section III, I have provided an overview of the diseases and disorders covered in this section. Using the methods and approaches described in Sections I and II, our knowledge base has advanced remarkably in the past few years. Our knowledge of the biology was often uncertain, and genomic studies have pointed to fundamental disease processes. However, new knowledge has led to a wealth of new questions, many of which are the topic of intensive current study. We now have tantalizing glimpses that predisposing genetic variations in interconnected biological processes are common to multiple disorders and diseases. This work is not complete but it is possible that there will be major advances in understanding genetic overlap in the next decade.

FINANCIAL CONFLICTS OF INTEREST

None.

LIST OF URLs

World Health Organization, global burden of disease estimates, http://www.who.int/healthinfo/global_burden_disease/estimates/en/index2.html.

Gephi, https://gephi.github.io.

"Brainstorm" project presentation, http://www.med.unc.edu/pgc/worldwide.

ACKNOWLEDGMENT

This work was supported by National Institutes of Mental Health grants R01 MH077139 and U01 MH085520.

REFERENCES

Al-Chalabi, A., et al. (2010). An estimate of amyotrophic lateral sclerosis heritability using twin data. *Journal of Neurology, Neurosurgery, and Psychiatry, 81*, 1324—1326.

Attia, J., et al. (2009a). How to use an article about genetic association: C: what are the results and will they help me in caring for my patients? *JAMA, 301*, 304—308.

Attia, J., et al. (2009b). How to use an article about genetic association: B: are the results of the study valid? *JAMA, 301*, 191—197.

Blair, D. R., et al. (2013). A nondegenerate code of deleterious variants in Mendelian loci contributes to complex disease risk. *Cell, 155*(1), 70—80.

Breslau, N., Johnson, E. O., Hiripi, E., & Kessler, R. (2001). Nicotine dependence in the United States: prevalence, trends, and smoking persistence. *Archives of General Psychiatry, 58*, 810—816.

Bulik, C. M., et al. (2010). Understanding the relation between anorexia nervosa and bulimia nervosa in a Swedish national twin sample. *Biological Psychiatry, 67*, 71—77.

Bulik-Sullivan, B. K., et al. (2015). LD Score regression distinguishes confounding from polygenicity in genome-wide association studies. *Nature Genetics, 47*, 291—295.

Cross-Disorder Group of the Psychiatric Genomics Consortium. (2013). Genetic relationship between five psychiatric disorders estimated from genome-wide SNPs. *Nature Genetics, 45*, 984—994.

Darnell, J. C., et al. (2011). FMRP stalls ribosomal translocation on mRNAs linked to synaptic function and autism. *Cell, 146*(2), 247–261.

De Rubeis, S., et al. (2014). Synaptic, transcriptional and chromatin genes disrupted in autism. *Nature, 515*(7526), 209–215.

Editorial. (2005). Framework for a fully powered risk engine. *Nature Genetics, 37*, 1153.

Feldman, H., et al. (2003). A Canadian cohort study of cognitive impairment and related dementias (ACCORD): study methods and baseline results. *Neuroepidemiology, 22*, 265–274.

Ferreira, M., et al. (2008). Collaborative genome-wide association analysis of 10,596 individuals supports a role for Ankyrin-G (*ANK*3) and the alpha-1C subunit of the L-type voltage-gated calcium channel (*CACNA1C*) in bipolar disorder. *Nature Genetics, 40*, 1056–1058.

Fombonne, E. (2009). Epidemiology of pervasive developmental disorders. *Pediatric Research, 65*, 591–598.

Hudson, J. I., Hiripi, E., Pope, H. G., Jr., & Kessler, R. C. (2007). The prevalence and correlates of eating disorders in the national comorbidity survey replication. *Biological Psychiatry, 61*, 348–358.

Ioannidis, J. P., Thomas, G., & Daly, M. J. (2009). Validating, augmenting and refining genome-wide association signals. *Nature Reviews Genetics, 10*, 318–329.

Kessler, R. C., et al. (2005). Lifetime prevalence and age-of-onset distributions of DSM-IV disorders in the national comorbidity survey replication. *Archives of General Psychiatry, 62*, 593–602.

Kjeldsen, M. J., Kyvik, K. O., Christensen, K., & Friis, M. L. (2001). Genetic and environmental factors in epilepsy: a population-based study of 11900 Danish twin pairs. *Epilepsy Research, 44*, 167–178.

Kohane, I. S. (2015). An autism case history to review the systematic analysis of large-scale data to refine the diagnosis and treatment of neuropsychiatric disorders. *Biological Psychiatry, 77*(1), 59–65.

Kraft, P., Zeggini, E., & Ioannidis, J. P. (2009). Replication in genome-wide association studies. *Statistical Science, 24*, 561–573.

Lander, E. S. (2011). Initial impact of the sequencing of the human genome. *Nature, 470*, 187–197.

MacArthur, D. G., et al. (2014). Guidelines for investigating causality of sequence variants in human disease. *Nature, 508*, 469–476.

McCarthy, M. I., et al. (2008). Genome-wide association studies for complex traits: consensus, uncertainty and challenges. *Nature Reviews Genetics, 9*, 356–369.

Merikangas, K. R., et al. (2007). Lifetime and 12-month prevalence of bipolar spectrum disorder in the national comorbidity survey replication. *Archives of General Psychiatry, 64*, 543–552.

Murray, R. M., Jones, P. B., Susser, E., van Os, J., & Cannon, M. (2003). *The Epidemiology of schizophrenia*. Cambridge, UK: Cambridge University Press.

Neale, M. C., & Kendler, K. S. (1995). Models of comorbidity for multifactorial disorders. *American Journal of Human Genetics, 57*, 935–953.

Ngugi, A. K., Bottomley, C., Kleinschmidt, I., Sander, J. W., & Newton, C. R. (2010). Estimation of the burden of active and life-time epilepsy: a meta-analytic approach. *Epilepsia, 51*, 883–890.

OED Online. (2014). *The oxford english dictionary*. Oxford, UK: Oxford University Press.

Pasinelli, P., & Brown, R. H. (2006). Molecular biology of amyotrophic lateral sclerosis: insights from genetics. *Nature Reviews Neuroscience, 7*, 710–723.

Plassman, B. L., et al. (2007). Prevalence of dementia in the United States: the aging, demographics, and memory study. *Neuroepidemiology, 29*, 125–132.

Polderman, T. J., et al. (2015). Fifty years of twin studies: a meta-analysis of the heritability of human traits. *Nature Genetics, 47*, 702–709.

Purcell, S. M., et al. (2014). A polygenic burden of rare disruptive mutations in schizophrenia. *Nature, 506*, 185–190.

Saha, S., Chant, D., Welham, J., & McGrath, J. (2005). A systematic review of the prevalence of schizophrenia. *PLoS Medicine, 2*, e141.

Schizophrenia Psychiatric Genome-Wide Association Study Consortium. (2011). Genome-wide association study identifies five new schizophrenia loci. *Nature Genetics, 43*, 969–976.

Splawski, I., et al. (2004). Ca(V)1.2 calcium channel dysfunction causes a multisystem disorder including arrhythmia and autism. *Cell, 119*(1), 19–31.

Stein, M. B., Jang, K. L., Taylor, S., Vernon, P. A., & Livesley, W. J. (2002). Genetic and environmental influences on trauma exposure and posttraumatic stress disorder symptoms: a twin study. *American Journal of Psychiatry, 159*, 1675–1681.

Sullivan, P. F., Daly, M. J., & O'Donovan, M. (2012). Genetic architectures of psychiatric disorders: the emerging picture and its implications. *Nature Reviews Genetics, 13*, 537–551.

Sullivan, P. F., Kendler, K. S., & Neale, M. C. (2003). Schizophrenia as a complex trait: evidence from a meta-analysis of twin studies. *Archives of General Psychiatry, 60*, 1187–1192.

Sullivan, P. F., Neale, M. C., & Kendler, K. S. (2000). Genetic epidemiology of major depression: review and meta-analysis. *American Journal of Psychiatry, 157*, 1552–1562.

Visscher, P. M., Brown, M. A., McCarthy, M. I., & Yang, J. (2012). Five years of GWAS discovery. *American Journal of Human Genetics, 90*, 7–24.

Weissman, M. M., et al. (1996). Cross-national epidemiology of major depression and bipolar disorder. *JAMA, 276*, 293–299.

Chapter 26

The NIMH Research Domain Criteria Project: Toward an Integrated Neuroscience of Mental Disorders

Bruce N. Cuthbert
National Institute of Mental Health, National Institutes of Health, Bethesda, MD, United States

"We must change [our ideas] when they have served their purpose, as we change a blunt lancet that we have used long enough."

Claude Bernard

INTRODUCTION

The National Institute of Mental Health (NIMH) in the United States issued a comprehensive revision of its strategic plan in September 2008 to guide its research efforts for the following 5-year period (NIMH, 2008). The plan elaborated four major strategic goals, including discoveries in basic brain and behavioral research, charting trajectories of disorders, developing new interventions, and strengthening the public health impact of NIMH-supported research. One goal in the plan (specifically, Goal 1.4) was stated as follows: "Develop, for research purposes, new ways of classifying mental disorders based on dimensions of observable behavior and neurobiological measures" (NIMH, 2008). The implementation of this plan became known as the Research Domain Criteria (RDoC) project. This chapter provides an introduction to the major framework of the RDoC project, along with an update on its current status and the implications for conducting research on mental disorders, with particular reference to genetic studies.

WHY RESEARCH DOMAIN CRITERIA?

Before describing the project itself, some background is in order to provide a perspective on why the NIMH included this element in its strategic plan. The modern system of psychiatric diagnosis began officially in 1980 with the release of the *Diagnostic and Statistical Manual of Mental Disorders*, Third Edition (DSM-III) (American Psychiatric Association, 1980). This date may seem reasonably contemporary, but in fact the field of psychiatry was only slowly emerging from decades of being in thrall to psychodynamic theories of psychopathology. Views about the nature of disorders, including major illnesses such as schizophrenia, depression, and manic-depressive illness, were riven by disagreements among various camps based on theoretical differences regarding etiology and psychopathology. Added to these problems were divergent views about the nature of such disorders as schizophrenia between US and European psychiatrists (Cooper & Sartorius, 2013). As a result, the reliability with which individual patients could be diagnosed was low, threatening the integrity and viability of the entire field.

Genomics, Circuits, and Pathways in Clinical Neuropsychiatry. http://dx.doi.org/10.1016/B978-0-12-800105-9.00026-3

These issues were addressed in the United States beginning in the 1970s, primarily through the efforts of a group at Washington University whose members perceived the need for explicit and theory-free criteria for diagnosis. The first contribution from this group came in the form of a seminal paper by Robins and Guze (1970) outlining five diagnostic criteria, some immediately useful and others aspirational. These included clinical descriptions, delimitations from other disorders, course and outcome, family studies, and laboratory tests ("when consistent with a defined clinical picture" (Robins & Guze, 1970, p. 984)). The eponymous Feighner criteria were published shortly after as the first instantiation of these desiderata (Feighner et al., 1972), laying out criteria for 14 different diagnoses that reflected a mix of well-established disorders (schizophrenia, depression, and alcoholism) and psychodynamic perspectives (anxiety neurosis and hysteria). The Feighner criteria were followed in turn by the more elaborated Research Diagnostic Criteria (Spitzer, Endicott, & Robins, 1978), which provided the immediate foundation for the structure of the DSM-III in 1980.

The new system was quickly successful in propelling the field forward in terms of both clinical utility and research. The assurance that patients could be compared meaningfully across studies led to burgeoning literatures of psychiatric research, amplified by the large number of additional new disorders available for study. It is fair to say that this system ushered in the modern era of psychiatric research. Some problems quickly emerged, largely owing to unanticipated consequences of the hierarchical way in which relationships among some disorders were expressed, that led to the relatively rapid issuance of the DSM-IIIR (revision) in 1987 and a major new edition (the DSM-IV) in 1994. However, the basic structure was unchanged, and the DSM-IV (with a minor "text revision" in 2000) remained the standard for nearly the next 2 decades.

With the wisdom of hindsight, substantive flaws in the DSM architecture can be discerned. First, notwithstanding the goals that Robins and Guze had established, the diagnostic categories were based almost entirely upon symptomatic presentations. These were subject to extensive comorbidity of disorders based on overlapping symptom criteria, leading to problems in defining "pure" disorders for study (Mineka, Watson, & Clark, 1998). A more serious structural flaw, as pointed out by former NIMH director Steven Hyman, is that the categories were at once too broad (leading to excessive heterogeneity) and too narrow (resulting in comorbidity and Not Otherwise Specified diagnoses) (Casey et al., 2013).

In addition, both clinical and early genetics data (as inferred from family history and behavioral genetics analyses) pointed to cracks in the fundamental structure of the supposedly discrete disease entities. For example, problems with the classical nosology for psychotic disorders have been debated extensively for decades. The discussions have included both the Kraepelinian distinction between schizophrenia and bipolar disorder (Greene, 2007; Lichtenstein et al., 2009) and the concept of a bipolar disorder entity (with specific subtypes) that differs from unipolar disorder (Phelps, 2012). As just one example regarding common mental disorders, it has been generally accepted since the early 1990s that major depressive episode and generalized anxiety disorder share a common genetic diathesis (Kendler, Gardner, Gatz, & Pedersen, 2007).

Given such difficulties, the need for new research paradigms has been acknowledged for some time. Consider, for instance, the following extended statement from well over a decade ago:

> *"It is our goal to translate basic and clinical neuroscience research relating brain structure, brain function, and behavior into a classification of psychiatric disorders based on etiology and pathophysiology. It is possible, even likely that such a classification will be radically different from the current DSM-IV approach. … We speculate that single genes will be discovered that map onto specific cognitive, emotional, and behavioral disturbances but will not correspond neatly to currently defined diagnostic entities. Rather, it will be discovered that specific combinations of genes will relate to constellations of abnormalities in many brain-based functions—including but not limited to the regulation of mood, anxiety, perception, learning, memory, aggression, eating, sleeping, and sexual function—that will coalesce to form disease states heretofore unrecognized. … The impact of environmental factors on gene expression will be defined. The ability to discover intermediate phenotypes will be improved with advances in techniques such as neuroimaging. This will all lead to novel therapeutic targets of greater efficacy and specificity to disease states. … Disease prevention will become a realistic goal."*
>
> Charney et al. (2002, pp. 70–71).

A focus on single genes has largely dissipated in favor of genome-wide association studies (GWAS) and gene sets. With that exception, however, this extended quotation from the DSM-V (sic) volume that introduced a research agenda for the DSM process is a fair description of the goals of the RDoC project. The fact that this ambitious goal was not realized in the eventual DSM-5 release in 2013 is a testament to the difficulty in changing theoretical and empirical approaches to mental disorders, particularly when the manual must also serve an essential clinical role.

The same considerations regarding the need to change research directions were apparent to funding agencies. For instance, in 2001 NIMH issued a set-aside funding announcement entitled "Modular Phenotypes for Mental Disorders" (NIMH, 2001). The text of this document noted that "within-group heterogeneity has been cited as a source of 'noise' in studies that search for subtle genetic or neurobiological influences on mental illness vulnerability, onset, and progression. … Progress in research areas like the genetics of mental disorders and experimental therapeutics would accelerate if methods were available to increase the phenotypic homogeneity of diagnostic samples and to define clinical treatment targets more precisely." Accordingly, the announcement called for "dissecting currently defined syndromes into simpler behavioral or biological components, thereby creating enriched samples for studying mechanisms that account for discrete symptoms or behavioral difficulties" (NIMH, 2001). It is illustrative of the problems involved with making drastic changes in research practices that although a number of grant applications were received in response to this call, not a single one was funded; this was largely due to difficulties that reviewers experienced in finding common ground to evaluate the variations in research designs and experimental approaches across the various grant applications.

This failure illustrates some of the barriers to forging fundamental shifts in research approaches to mental disorders. Well-intentioned calls for changing research practices and for picturing the future research environment cannot change the fact that the large majority of clinical research studies are funded through the traditional peer review channels, a process that is inherently conservative. Therefore, the most viable way to break this bottleneck is to create review standards that promote alternative methods of designing and conducting translational studies that can accommodate new research technologies and findings, not only at the present time but into the future as well. This is the working premise that resulted in the RDoC project.

RESEARCH DOMAIN CRITERIA PROCESS AND STRUCTURE

The NIMH organized an internal work group[1] in early 2009 to initiate implementation of the relevant goal in the Strategic Plan. The group began with a further stipulation "to initiate a process for bringing together experts in clinical and basic sciences to jointly identify the fundamental behavioral components that may span multiple disorders … and that are more amenable to neuroscience approaches" (NIMH, 2008). This statement, as well as the overall wording of Goal 1.4, made it explicit that the aim is to consider equally an understanding of behavioral functions relevant for mental health (working memory, fear, etc.) and of the neural systems that implement them. (The term "behavioral" is taken in this context to denote both overt real-world behaviors and measurements indexing cognitive or emotional processes.) In other words, the challenge is to understand complex brain–behavior relationships and how they apply to various aspects of dysfunction in mental disorders. Complaints have occasionally been lodged against RDoC for being purely reductionist, but these concerns are misplaced; the goal of RDoC is to reach an integrative understanding of how measurements in various systems relate to each other (and to symptoms) rather than to pursue an eliminative reductionism that seeks to explain behavior purely in molecular or cellular terms (Cuthbert & Insel, 2013; Cuthbert & Kozak, 2013).

The work group deliberated at length in deciding on an optimal organization for the project, which could also provide direction for the specified workshop process. A consideration of many areas of basic and clinical science led to the decision to organize the overall framework into five major domains of function (thus, the RDoC). Basic motivational systems involve two primary components, ie, aversive and appetitive, denoted as Negative and Positive Valence; cognitive and social cognitive systems are clearly major components of behavior; and finally, a number of brain systems accomplish important functions in regulating and modulating overall brain activation, such as arousal, sleep/wake, circadian, and modulatory systems. This domain structure provided a natural framework for the process of convening external experts to decide on the specific dimensions to be included, and one workshop was held for each domain.

For each workshop, the NIMH work group crafted a list of candidate dimensions as starting points for discussion, and the external experts deliberated to come up with their list of dimensions, a definition for each, and a list of elements at various Units of Analysis that had been used to measure the dimension in prior studies. There were three main criteria for including a dimension: evidence for a valid behavioral/psychological function, evidence for an implementing neural circuit or system, and a putative relationship to significant symptoms or impairments in mental disorders. This process proved successful in that at each workshop, the experts revised considerably the draft list of candidates to produce a set of dimensions that were distinct yet related, and fit within the nature of the domain.

The psychological term "constructs" was adopted for the dimensions to represent their status as theoretical concepts whose denotation changes with ongoing studies, consistent with the usual use of constructs in psychological science. The basic RDoC matrix can thus be depicted as a series of rows and columns in which the rows represent the constructs (nested within domains) and the columns represent various Units of Analysis used to measure the constructs. Circuits comprise the

1. The members of the NIMH RDoC work group are: Bruce Cuthbert (chair), Rebecca Garcia, Marjorie Garvey, Marlene Guzman, Robert Heinssen, Arina Kadam, Michael Kozak, Sarah Morris, Daniel Pine, Kevin Quinn, Matthew Rudorfer, Charles Sanislow (now at Wesleyan University), Janine Simmons, Uma Vaidyanathan, and Philip Wang. External consultants are: Deanna Barch, Will Carpenter, and Michael First.

central position of the seven columns in the Units of Analysis; the columns to the left include genes (considered to include all aspects of gene-related measures including genomics and epigenomics), molecules, and cells (ie, the constituents of circuits). The columns to the right are labeled Physiology (heart rate, cortisol, etc.), Behavior (as observed in behavioral settings or measured by laboratory tasks), and Self-reports (defined to include various interview and questionnaire instruments). A separate column was added to represent the various paradigms that are frequently used to obtain particular kinds of measurements in clinical studies. (See the RDoC Web page, http://www.nimh.nih.gov/research-priorities/rdoc/index.shtml, for an illustration of the current version.)

Although the basic RDoC matrix can be depicted in terms of the two-dimensional matrix of rows (constructs) and columns (Units of Analysis), this is only half of the full RDoC framework. Almost all mental illnesses are now regarded as neurodevelopmental disorders. In addition to traditional childhood-onset disorders such as autism and attention-deficit hyperactivity disorder (ADHD), most "adult" disorders have prodromal phases or actual onset by late adolescence, and data suggest that early signs of high risk for the psychotic spectrum may be seen in cognitive changes occurring around age 13 years (Gur, 2014). Thus, tracing trajectories of neurodevelopment represents an important third dimension in the framework (Franklin, Jamieson, Glenn, & Nock, 2014). Finally, environmental factors are important at all stages of the life span, but may be particularly critical during early development when the brain is changing rapidly.

INTERMEDIATE PHENOTYPES AND ENDOPHENOTYPES

It has sometimes been stated that RDoC simply represents a collection of endophenotypes, ie, heritable and quantitatively measured attributes typically based on laboratory tests that may be closer to fundamental disease processes and therefore represent better targets for linking with genomics as well as molecular and cellular processes (Glahn et al., 2014; Gottesman & Gould, 2003; Miller & Rockstroh, 2013). Because of the nature of the RDoC structure, this is only partially correct and is potentially misleading. RDoC constructs are the building blocks of the system, and because they are at a lower level than DSM diagnoses, it is tempting to consider them as endophenotypes in the somewhat broad sense of the term. However, RDoC calls for measurement and integration of variables obtained across multiple units of analysis, and these elements, which are measures such as P300, prepulse inhibition, functional MRI (fMRI) response patterns, or working memory performance, fall closer to the usual sense of endophenotypes. To avoid confusion, it seems heuristic to draw a distinction in RDoC between intermediate phenotypes and endophenotypes (which are used synonymously by many authors). Thus, as we noted at the outset of the project, "[constructs] might be considered to represent functional intermediate phenotypes characterized by a set of endophenotypes at different levels of analysis that, in the aggregate, define the [construct]; one advance that we hope to achieve with this approach is to forge more explicit links among these endophenotypes and their functional relevance in the [construct] of interest" (Insel & Cuthbert, 2009, p. 989).

One putative advantage of endophenotypes has been that they are posited to be closer to a fundamental biological level, and thus can assist in finding relevant genes for psychopathology. In this regard, a special issue of the journal *Psychophysiology* was devoted to an extensive genomic analysis of 17 classic psychophysiological endophenotypes (eg, the P300 event-related potential, the startle reflex, and skin conductance) as recorded in a community twin sample of nearly 5000 subjects from the Minnesota Twin-Family Study (see summary article by Iacono, Vaidyanathan, Vrieze, and Malone (2014)). Overall, a modest number of "hits" were found for the various measures, although biometric analyses confirmed prior heritability estimates (Iacono et al., 2014). Among other aspects, these findings sound a clear cautionary note about interpreting small-N molecular genetic studies assumed to be more robust owing to studying endophenotypes (Patrick, 2014). The extent of polygenicity for endophenotypes, and almost surely circuit-based RDoC constructs as well, seems comparable to that for complex diseases and thus will require the same large samples as for GWAS studies of disorders if significant numbers of associations are to be found (de Geus, 2014).

However, these considerations should not cast doubt on the use of endophenotypes for understanding mechanisms of disorders. One supposition for RDoC is that endophenotypes will prove to be more useful when they are located within a measurement system linked to their presumed functionality rather than to an empirical association with a heterogeneous disorder entity. Thus, skin conductance should exhibit particular usefulness for indexing activation, prepulse inhibition for assessing concepts of perceptual filtering, and fear-potentiated startle for evaluating variations in the nature of the fear response, all of which are considered transdiagnostically and orthogonal to traditional disorder categories.

LOOSENING THE CONSTRAINTS OF RESEARCH

The intent of the RDoC initiative is to liberate investigators to pursue research questions of interest that cut across traditional diagnostic boundaries, leading to new ideas regarding transdiagnostic disorder mechanisms that in turn prompt novel treatments that are more homogeneous with respect to the treatment targets of interest (and thus offer greater possibilities for successful trials). Some anecdotal complaints have been heard that the RDoC project is a top-down NIMH initiative that actually hinders investigators in their ability to conduct research. This concern represents a misconception, which may be understandable in the context of previous constraints on psychiatric research. Both the DSM and its close cousin, the *International Classification of Diseases* (ICD), were released as manuals for use in clinical diagnosis (notwithstanding the stated role of the DSM in particular as a research instrument). Because of this, and because the diagnoses became reified and quickly achieved the status of specific disease entities, contrary to their depiction in the introduction to the DSM-III (Hyman, 2010), investigators have become accustomed to the idea that the diagnoses listed in the manual were the only categories that can be included for study in research grant applications.

It is natural that investigators should assume that the same constraints hold true for the RDoC framework, but this is not the case. This misunderstanding may stem in part from the fact that the initial RDoC funding calls required extant constructs, a step intended to create an initial bolus of applications built around the current matrix and to evaluate the feasibility of the constructs in peer review. Overall, however, the growth of the RDoC framework depends on novel ideas (or revisions to the current set), and so investigator-initiated applications framed around new constructs are strongly encouraged.

What, then, is the distinction between extant constructs and proposed new entries? First, the constructs currently listed in the matrix are considered to be exemplars of particularly promising constructs as vetted by the workshop process, for which solid evidence is available in terms of their function and their neurobiology; as such, one of their functions is to serve as models for other constructs that might be proposed in grant applications. Second, the extant constructs are considered to be "prevetted" such that their inclusion in grant applications does not need to be justified. Obviously, because evidence for both functional and biological aspects is required for a construct to be added to the matrix, it would be impossible for any new constructs to be added if relevant research were not conducted. Thus, it is a high priority for both basic and translational research to be supported to promote the growth of the entire framework.

Another aspect of opening up the research enterprise concerns the composition of independent and dependent variables in study designs. In traditional clinical research, the independent variable typically consists of a DSM/ICD diagnosis factor, with other measures as dependent variables. In RDoC, a measure from any unit of analysis can be used as an independent variable, eg, task performance, circuit activation in a particular task, or factor scores resulting from a principal component or similar analysis. (For more information regarding research designs in RDoC, see Cuthbert and Insel (2013), Cuthbert and Kozak (2013), and Cuthbert (2014).)

One final aspect of the increased flexibility of study designs concerns the inclusion of factors regarding neurodevelopment and/or effects of the environment. Some scientists have expressed concern that these factors are not explicitly elaborated in the RDoC framework, and thus are perforce relegated to secondary status or even ignored. Again, this is a misperception. The intent is to provide maximum flexibility for explicating the role of these factors in the context of studying one or more functional constructs that can be reasonably applied to a designated clinical problem (Casey, Oliveri, & Insel, 2014). Specifying particular stages of development, or particular types of environmental events, could be interpreted to mean that these are the only such stages or types in which the Institute has interest, a misleading inference that could hamper investigators in designing studies that are most appropriate for addressing their specific research question. An equal percentage of RDoC-themed applications has been funded for adult- versus child-oriented proposals to date, which suggests that development researchers are able to use this freedom to make a successful case for the age ranges that they select.

Two hallmark features of prototypical RDoC designs are the emphasis on samples drawn from multiple DSM groups, selected to achieve the distribution needed to explore the particular construct(s) of interest, and the inclusion of control groups that represent a broader range of functioning than the "supernormal" control groups typical of "disorder group versus control" DSM designs (Cuthbert, 2014). Whereas transdiagnostic samples are preferred as a general rule, studies employing a single DSM/ICD disorder may be useful for investigators to establish the basis for a new dimension or subgroup within one current disorder before extending the concept to other diagnoses; this is also an option for clinical investigators who have access to only a single disorder category owing to the nature of their clinical setting, and who could use the RDoC approach to deconstruct the diagnostic syndrome. (With respect to the latter circumstance, however, it should be noted that in RDoC studies the inclusion of subjects who do not meet criteria for a particular disorder, or meet criteria for multiple disorders, is encouraged.) Whatever the particulars of a given experimental design, it is important to bear in mind that "The price for freedom from the DSM … is that scientists will have to describe the nature and logic of their sampling criteria with great precision and clarity so that their work is replicable" (Hyman, in Casey et al., 2013, p. 811).

The essence of all these considerations boils down to a question of peer review. The problem that RDoC was created to solve was that, owing to the hegemony of the DSM/ICD system in study sections, applications intended to pursue promising findings of transdiagnostic mechanisms rarely received fundable scores, and so were rarely even attempted. On the other hand, some standards for review are essential if study sections are to have reasonable criteria for subject inclusion and other aspects of research designs. Merely creating an alternative set of fixed categories would fail to solve this problem because of the rapid pace of advances in the science. What RDoC is intended to provide, then, is a set of criteria not tied to a particular set of categories, but rather to flexible guidelines for evaluating the potential usefulness of emerging new ideas about behavioral/cognitive functions, as related to their basis in neural systems, proposed by the applicant to advance the understanding of particular features of psychopathology.

A preliminary evaluation of the progress of the RDoC project thus relates to its reception in NIMH and National Institutes of Health (NIH) peer review committees. To date, 25 grants have been funded through the three NIMH RDoC funding set-aside announcements ("Request for Applications," [RFA]) from 2012 through 2014, with another set of applications currently under consideration. The reviewers evaluating these applications (with considerable overlap across the three meetings) have demonstrated increasing fluency and comfort with the nature and goals of the project over the three years of review. Regarding the overall picture, a search of the NIH RePORTER public database in October 2014 revealed a total of 130 grants that included "Research Domain Criteria" in the title, abstract, or keywords. Subtracting the RFA grants, this means that approximately 100 RDoC−themed grants have been funded through Center for Scientific Review committees. This tally indicates that RDoC grant applications are viable for successfully passing NIH peer review, and summary statements increasingly list the inclusion of RDoC constructs as "strength" in the Significance and Approach sections.

RESEARCH DOMAIN CRITERIA AND GENETICS

Two major shifts in understanding and conceptualizing mental disorders have taken place, the first occurring gradually over several decades and the second taking place more recently. Both shifts are relevant to the RDoC project and to genetics.

The first is the accelerating interest in viewing mental disorders in terms of dimensional aspects of functioning (Cuthbert, 2005). As recently noted, "Research that focuses on quantitative traits—including the low and high ends of normal distributions—could have far-reaching implications for the diagnosis, treatment, and prevention of the problematic extremes of these traits" (Plomin, Haworth, & Davis, 2009, p. 782). One set of approaches has stemmed from psychological studies of normal dimensions of temperament and personality, and has been applied notably to mood/anxiety disorders (Clark, Watson, & Mineka, 1994) and to personality disorders (Clark, 2005). A dimensional, personality-trait approach was proposed for DSM-5 but was voted down shortly before the final version was released (Krueger & Markon, 2014). Another major emphasis in dimensional research concerned the factor structure of the common mental disorders (Krueger & Piasecki, 2002). Furthermore, accumulating evidence indicates that perceptual and cognitive abnormalities related to psychosis also exist as a dimensional trait in the population (Johns & van Os, 2001).

RDoC implements such a dimensional orientation to psychopathology. As stated in one of the major subheadings in Goal 1.4, one aim is to "Determine the full range of variation, from normal to abnormal, among the fundamental components to improve understanding of what is typical versus pathological" (NIMH, 2008). This postulate is reflected directly in the fact that evaluation of candidate constructs was based on evidence for a basic functional aspect of behavior, as opposed to a clinical symptom. One marked advantage of this perspective is that the considerable amount of basic research on motivational, cognitive, social, and regulatory systems that NIMH (among many other funders) has supported over the years can be applied directly toward an understanding of psychopathology; in addition to behavioral neuroscience studies, this work notably includes model animal studies of genetic factors (see Simmons and Quinn (2014) for a thorough discussion of this aspect). Symptoms in this scheme (as in the earlier quotation by Plomin) are considered in terms of the tails of distributions that are associated with dysfunction in behavior. This reflects a considered stance in studying impairments and symptoms that is relatively straightforward in some instances (fear and working memory) but more complex in others (hallucinations and social cognition); in either case, however, the goal is to work toward a mechanistic understanding in terms of whatever combination of circuit-based functions is required to explain a specific aspect of psychopathology.

A related advantage is that paradigms developed for studies with animals can be applied directly to their corresponding constructs in humans (as opposed to the tenuous appeals to modeling DSM disorders, eg, the forced-swim test), speeding the translation to an understanding of abnormal behavior (Glatt & Lee, 2016). An apt example of comprehensive translation comes from the Cambridge Neuropsychological Test Automated Battery created by Robbins and Sahakian, whose developers have created touch-screen tasks for rodents that are as nearly identical as possible to the corresponding tasks with humans (Barnett et al., 2010).

Considering the RDoC matrix as a whole, many measures are available, particularly in the cognitive domain. However, there is a paucity of instruments and paradigms for several of the constructs; even where some scales exist, they have often been created either for basic research or for clinical studies, and thus lack sensitivity across "the full range of variation." Accordingly, measurement development is a high priority for RDoC.

As dimensional approaches expand in popularity, an important task in coming years will be to develop a better understanding of the relationship of measurements along various dimensions of functioning to overt clinical symptoms. In addition, it will be important to examine how developmental trajectories and environmental factors, both adverse and protective, interact to move individuals in the direction of psychopathology or resilience. Explicating the transition zones between high-risk states or mild psychopathology and marked dysfunction will be critical for prevention and early treatment, and also contribute to the understanding of behavioral and biological differences between those with overt illness and others (eg, first-degree relatives) who share many of the same measurable deficits on laboratory tasks but are clinically intact. Genomics data will be a valuable component of these projects; for instance, polygene risk scores have already begun to parse the clinical spectrum in serious mental disorders (Tesli et al., 2014).

The second major shift toward dimensionality has been the realization that many disorders fall along neurodevelopmental and affective spectra, and thus may be considered more appropriately as broad syndromes, each of which represents part of the range of a larger spectrum rather than as discrete, punctate disease entities (Craddock & Owen, 2010; Phelps, 2012). Genomics research, for which the past decade has been a watershed in proving its usefulness across all of medicine, has been a strong driver of this second dimensional trend. GWAS studies have reported ever-greater numbers of hits for common variants as sample sizes have increased (Schizophrenia Working Group of the Psychiatric Genetics Consortium, 2014) and rare variants have provided insight both into the potential penetrance of large structural variations and to some potential molecular mechanisms of disorder (Sullivan, Daly, & O'Donovan, 2012). Finally, and most relevant for RDoC, data regarding genetic overlaps between and among disorders have become almost commonplace. These include findings of common risk factors between autism and schizophrenia (Crespi, Stead, & Elliot, 2010), autism and ADHD (Rommelse, Franke, Geurts, Harman, & Buitelaar, 2010), schizophrenia and bipolar disorder (International Schizophrenia Consortium, 2009), and most recently, common overlaps among five major mental disorders (Cross-Disorder Group of the Psychiatric Genomics Consortium, 2013; Network and Pathway Analysis Subgroup of the Psychiatric Genomics Consortium, 2015).

This view was elaborated in a review of the overlaps in both genetic risk and clinical comorbidities: "... it is hard to avoid the conclusion that these disorders represent a continuum of genetic and environmentally induced neurodevelopmental impairment, rather than a set of aetiologically discrete entities, with the major clinical syndromes reflecting in part the severity and predominant pattern of abnormal brain development and resulting functional abnormalities as well as the modifying effects of other genetic and environmental factors. ... A simple conception of these findings is that severe mental illnesses occupy a gradient with the syndromes ordered by decreasing severity of neurodevelopmental impairment as follows: intellectual disability, autism, ADHD, schizophrenia, bipolar disorder" (Owen, O'Donovan, Thapar, & Craddock, 2011, p. 174). This radically new view of disorders poses a challenge to all disciplines related to mental disorders in testing hypotheses and pursuing the implications for new ideas about impairment and treatment development.

MEETING IN THE MIDDLE

Both single- and cross-disorder efforts are clearly making steady progress in elaborating the genetic architectures of mental disorders. At the same time, however, these are still salad days for genomics research, and both the nature of the genomic findings and their corresponding phenotypes fall at the outer bounds of resolution and specificity. On the genomic side, only initial hints at the nature of various small- and large-scale neural systems are provided by either GWAS hits or findings of structural alterations. There is a general consensus that gene pathways and functional gene groups are likely to be a more promising route for understanding the disruptions in synapses and neural circuits associated with mental disorders (Hyman, 2007) and thus for providing new intervention targets. Specification of these gene sets is slowly proceeding as the functions of individual genes and gene–gene interactions gradually emerge from the tedious processes of curation and are applied to psychopathology (Lips et al., 2012).

On the phenotype side, a similarly slow and iterative pace must be anticipated. It is not simply that identifying genes of very small effect requires large samples to find significance for a given disorder. In addition, the heterogeneity of syndromes as currently defined means that yet larger samples must be gathered to overcome the problem that particular genes are likely to exert even smaller (or null) effects in some subgroups compared with others. This reduces the power to find relevant genes and will also complicate the coming efforts to untangle relevant differences in synaptic structure and functioning across different subgroups. This is where a transdiagnostic framework such as RDoC comes in. The hope is that concentrating on intermediate phenotypes will provide more tractable opportunities for relating focused types of impairments to relevant synaptic perturbations, disruptions in local and distal brain networks, and other systems with putatively strong relationships to genetic variation. The need for such refined phenotypes is evident across all areas of mental disorders: the psychotic disorder spectrum (Pearlson, 2015), autism (London, 2014), the mood/anxiety spectrum (Vaidyanathan, Nelson, & Patrick, 2012), and ADHD (Fair, Bathula, Nikolas, & Nigg, 2012).

Of course, the challenge is complicated yet further by the fact that disorder heterogeneity is not determined simply by varying genomic architectures. Transactions between the organism and its environment can be considered in terms of both psychological phenomena such as learning, memory, and attachment and plasticity as observed at all levels of brain systems. There is a pressing need to delineate the relationships between neuroplasticity at the synaptic level and that observed in major brain systems, which can be observed not only as tonic changes consequent to extended environmental challenge (Di Chiara, Loddo, & Tanda, 1999) but in real time during a stressful event (Reynolds & Berridge, 2008). As aptly summarized in a review, "Our challenge is to integrate information coming from bottom-up sources such as genes, with information coming top-down from the environment, and, at the point of integration, to understand how the convergence of such information affects behavior" (Hyman, 2000, p. 272).

RESEARCH DOMAIN CRITERIA AND BIG DATA

It is apparent that exegesis of the four elements of the RDoC framework (constructs, units of analysis, neurodevelopmental trajectories, and interactions with the environment) will require massively large data sets to generate sufficient statistical power for analysis. The current set of genomics studies offers strong proof that such amalgamation is possible, but extending the effort to other units of analysis will be much more difficult. The current variety of imaging methods, behavioral tasks, and self-report instruments adds up to a welter of variables with an extensive covariance matrix to parse and varying degrees of uncertainty as to whether putative measures of the same construct converge. The results of individual studies with relatively small samples clearly cannot resolve these issues, and the only way forward will involve data sets that eventually approach GWAS samples in size.

To this end, NIMH has established the NIMH Data Archive (NDA) to house the data sets from virtually all clinical research that the Institute supports. Based on the National Database for Autism Research, which has accumulated data from 77,000 participants since its inception in 2008, the NDA includes a clinical trials database and the RDoC database (RDoCdb). (The NIMH Genetics Repository will continue to be the resource for storing genetics data.) Investigators will be strongly encouraged to deposit data into RDoCdb from clinical studies that do not involve autism or clinical trials, whether designed around RDoC constructs or DSM diagnoses, because the intent is to mine data in a transdiagnostic manner across multiple data sets. Importantly, such research will provide (among other aspects) greatly enriched phenotypic data to analyze in conjunction with the next generation of genomics data (eg, more widely available whole-genome sequencing and well-curated functional gene groups).

Exploratory analyses with these large data sets will necessitate extensive use of data-mining approaches devised to identify dimensions and subgroups not yet known. Analyses are likely to require an iterative approach given the formidable number of patterns that can be discovered in high-dimensional data sets, depending on the variables included in the analysis and the particular statistical approaches that are used. Therefore, validation against other variables not included in the original analysis, as well as cross-validation with other samples, will be essential. The eventual validity, however, will of course constitute the extent to which the results prove to be of high clinical use in identifying cohesive subgroups or dimensions that lead to enhanced clinical outcomes.

Although the first formal RDoC projects are in the initial stages, studies are appearing that exemplify the promise of these new research approaches and provide exemplars of analytic strategies. Perhaps the largest project to date is the Bipolar-Schizophrenia Network on Intermediate Phenotypes, which has explored the use of endophenotypic measures in nearly 1000 patients across the psychotic spectrum (Pearlson et al., 2014). The researchers' analysis began with a principal components analysis to derive two major factors from their assessment battery, termed Cognitive Control and Sensorimotor Reactivity. They then derived the factor scores for each subject on each factor and entered these data into a standard cluster analysis. The result revealed three "biotypes" (clusters) that cut across the schizophrenia—bipolar spectrum and have shown strong relationships with other measures, such as resting-state fMRI, that are not present in a DSM-wise analysis (Pearlson et al., 2014).

Using a smaller but comparable sample, highly compatible results have been reported with a similar set of measures (Hall et al., 2012). Using an unsupervised cluster analysis, the authors reported finding three groups, again cutting across the schizophrenia—bipolar spectrum, that they termed "globally impaired," "[enhanced] sensory processing," and "high cognitive"; as the authors concluded, "We hypothesize that each neurophysiology subgroup may share similar genetic profiles, which may increase statistical power to detect genetic risk factors" (Hall et al., 2012, p. 272).

Another study used an innovative statistical approach to parse heterogeneity in a single DSM disorder, ADHD (Fair et al., 2012). The authors first factor-analyzed their extensive neuropsychological battery to derive seven main factors, and then applied these scores to a "community [network] detection analysis" based on graph theory followed by a multivariate pattern analysis using a supervised classification algorithm (support vector machine). The results returned complex patterns, with four distinct groups in typically developing children and six related groups in the ADHD cohort. Whereas the data are obviously relevant for efforts to link phenotypes to neurophysiological systems and genetics in ADHD, two other aspects have broader relevance. First, the investigators point out that in contrast to usual assumptions that "normal" control groups are homogeneous, typically developing children may be divided into reliable subgroups, with the important corollary that some of the heterogeneity in a clinical group (here, ADHD) may be nested within the normal variation. As a second (and related) point in terms of data mining, the authors conclude that "… it should be noted that at times heritable diversity might form along a continuous dimension (i.e., unimodal dimension), whereas at other times it may form as multiple discrete strategies (i.e., multimodal distributions)" (Fair et al., 2012, p. 6772). Clearly, data-mining explorations for heterogeneity in clinical (or normal) samples need to provide appropriate analyses in this regard, which have clear implications for genetics analyses as well as other units of analysis (Wiecki, Poland, & Frank, 2015).

Another study in the same clinical area extended the analysis of heterogeneity to a transdiagnostic sample of children with ADHD, bipolar disorder, or both disorders (and also including healthy control subjects) as examined with the Conners Continuous Performance Test (Kleinman et al., 2014). A K-means clustering analysis was used, and returned two groups with high classification accuracy that differed primarily on response variability and deficits in sustained attention. Subjects with ADHD or comorbid ADHD plus bipolar disorder were split about equally between the two groups, whereas normal control subjects and bipolar-only subjects aggregated in the cluster with less severe deficits, an outcome that the investigators discussed in terms of the proximal implications for tailoring clinical treatment.

These studies provide examples of the different kinds of statistical analyses that will become common practice in uncovering new dimensions and subgroups (whether within or across traditional disorder categories). An additional point is underscored by the fact that the measures used in the studies by Fair et al. (2012) and Kleinman et al. (2014) differ substantially, which greatly complicates attempts to include the combined data in mega-analyses seeking to discover or validate subtypes or dimensions. Such divergence is, of course, common across the entire range of mental disorders research, and exemplifies the need for carefully thought-out common data elements to facilitate combined analyses. However, as the contemporary transdiagnostic research scene shows, these measures must be carefully chosen to represent wide swaths of disorder spectra (possibly the entire gamut of psychopathology) rather than being focused upon a single, presumptive DSM/ICD disorder. Obviously, balancing the extent of common data elements versus measures planned for any specific study poses a critical issue because of concerns of overall subject burden and feasibility, and will be a key concern for RDoC measurement systems moving forward.

In tandem with these important efforts to parse the current disorder spectrum, refining the RDoC constructs and relating them to traditional clinical symptoms are high-priority aspects of data mining. RDoC involves a set of intermediate phenotypes that are denoted as constructs to represent the fact that they are empirically based concepts subject to continual revision and refinement on the basis of new data. Much of the effort in revising a construct (or adding a new one) consists of a search for the appropriate "grain size," ie, the breadth of its functional aspects and neural circuitry. Grain size is important to reflect the best current understanding of complex behavioral functions and their implementing neural systems but also for the extension to symptoms in a way that proves optimal for understanding pertinent clinical problems. For instance, "anhedonia" is a traditional clinical concept characterized as an inability to experience pleasure, classically related to depression and often invoked as a transdiagnostic construct for RDoC studies of deficits in reward circuit activity. However, it is well known that the anhedonia concept is composed of several different components such as experiencing reward, exerting energy to seek rewards, and learning to associate particular situations with reward (Treadway & Zald, 2011). Thus, it is not yet clear how relatively broad or narrow the "anhedonia" concept ought to be for optimal diagnosis and treatment (and whether this may vary for different clinical purposes).

As another example, the term "anxiety" is still frequently invoked as though it were a unitary and cohesive clinical construct that can be reliably differentiated from other aversive states. However, anxiety is often not clearly distinguished from acute threat, and even granted this distinction, the cohesiveness of phenomenological anxiety per se is not clear. For instance, in one study, undergraduates carefully selected for two distinct types of self-reported anxiety, "anxious apprehension" as opposed to "anxious arousal," participated in an emotional Stroop task (naming the color of a printed word with distracting emotional content) while fMRI signals were acquired. Responses of the two groups were distinguished by activation in a left inferior frontal area (associated with speech production) as opposed to a right inferior temporal area, respectively (Engels et al., 2007). Whereas high-anxious subjects typically are elevated to some extent on both subtypes, the point is that studies validating relevant subdimensions within an RDoC construct (in this case, "potential threat") and within a traditional phenomenologically defined clinical concept such as diffuse anxiety may simultaneously advance basic knowledge and prove valuable for optimizing assessment and potential treatment.

Finally, the number and type of the units of analysis may well require several rounds of iteration to reach the optimal set for working out a given clinical problem. As one example, an unusually comprehensive series of studies conducted with patients across virtually all of the anxiety disorders has indicated that the fear-potentiated startle reflex is inversely related, in a transdiagnostic manner, to the extent of self-reported severity on mood and anxiety instruments; in other words, higher distress is related to blunted fear-potentiated startle response across all anxiety disorders (McTeague & Lang, 2012). By inference, a proximal clinical application would involve establishing the psychometric adequacy of the startle measure followed by clinical trials to test the predictive validity of the new biomarker (eg, that patients with a robust startle reflex indicative of strong fear circuit activity would respond optimally to exposure therapies whereas those with blunted startle might require an alternative treatment such as pharmacotherapy or mindfulness meditation). However, it might turn out that such an advance necessitates several additional steps before clinical utility can be established. These might include such aspects as additional measures (eg, fMRI of other motivational circuits), additional stimulus conditions (eg, positive-valence stimuli), or yet more refined assessment (eg, the anxious apprehension—anxious arousal distinction noted previously).

It is apparent that the RDoC project is rapidly expanding from a conceptual exercise to a burgeoning "Big Data" activity. There is certainly cause for circumspection moving forward, given the likelihood of unforeseen complications that will need to be overcome. However, the research results that have already emerged, along with the rapidly growing corpus of new studies and commentaries, provides ample grounds for optimism that an approach based on empirical foundations of behavioral and brain science can lead to more effective clinical applications at an accelerating rate (Doherty & Owen, 2014).

SUMMARY AND CONCLUSIONS

The purpose of this chapter has been to provide a summary of the RDoC project and the way in which it is structured to contribute to the next generation of research on mental disorders. The RDoC concept provides a framework for translational research that fosters integrative studies of behavioral/cognitive dimensions, associated brain systems, and related symptoms or impairments of mental disorders. The focus is on mechanisms that cut across traditional disorder categories or comprise subgroups within categories, with an intention to form more homogeneous clinical groups that represent promising targets for treatment development and clinical trials. The specific tactical aim of the project is to introduce a research classification system into the grant review process, providing the opportunity for investigators to exploit immediately and directly such emerging new concepts as default and salience networks, regulatory coding regions in the genome, and the nature of complex behaviors. To avoid the problem of having a system that becomes "stuck in time," RDoC eschews fixed disorder entities and instead provides criteria for evaluating the extent to which translational research designs are based on a promising foundation of current translational neuroscience and behavioral research.

A salient feature of the project is the focus on etiological processes, as reflected in the emphasis in the framework on neurodevelopment and environmental influences. To this end, emphasis is placed on the relationship between functional constructs (with their implementing neural systems) and the specific clinical impairments to which they are most likely to contribute, rather than to broad syndromes where such links are blurred. Because of the rapid pace of advances in all areas of the science, the constructs serve as a compendium of current knowledge and as exemplars for research, but are always subject to refinement as research progresses.

A perhaps underappreciated facet of the RDoC project is the need for quantitative measurement across all areas of the research matrix (including symptomatic assessments), as opposed to the relatively subjective and qualitative determination of current diagnostic categories; this is a direct consequence of the normal-to-pathological dimensional perspective. Analyses across a large number of high-dimension measurement systems pose many challenges, including the development of new tasks and assessments tailored for the constructs, exploring the covariance among measurements, and working out the best combinations of measures for effective application to actual clinical practice.

One need not belabor the significance of these various considerations as they relate to the contributions of genetics. The current generation of studies has established the salience of genomics for mental disorders. However, the success of next-generation efforts will almost certainly be related to the quality of the phenotypes with which they are compared, which in turn depends on the adequacy of measurement across various units of analysis. As the field transitions toward a nosology that is built on research into fundamental behavioral and biological systems, the extent to which advances in each area are mutually informative with the others will have a key role in determining how quickly the field can generate palpable progress toward precision medicine for mental disorders.

COMPETING INTERESTS

The author reports no biomedical financial interests or potential conflicts of interest.

REFERENCES

American Psychiatric Association. (1980). *Diagnostic and statistical manual of mental disorders* (3rd ed.). Washington, DC: American Psychiatric Association.

Barnett, J. H., Robbins, T. W., Leeson, V. C., Sahakian, B. J., Joyce, E. M., & Blackwell, A. D. (2010). Assessing cognitive function in clinical trials of schizophrenia. *Neuroscience & Biobehavioral Reviews, 34*, 1161–1177.

Casey, B. J., Craddock, N., Cuthbert, B. N., Hyman, S. E., Lee, F. S., & Ressler, K. J. (2013). DSM-5 and RDoC: progress in psychiatry research? *Nature Reviews Neuroscience, 14*, 810–814.

Casey, B. J., Oliveri, M. E., & Insel, T. (2014). A neurodevelopmental perspective on the Research Domain Criteria (RDoC) framework. *Biological Psychiatry, 76*, 350–353.

Charney, D. S., Barlow, D. H., Botteron, K., Cohen, J. D., Goldman, D., Gur, R. E., … Zalcman, S. Z. (2002). Neuroscience research agenda to guide development of a pathophysiologically based classification system. In D. J. Kupfer, M. B. First, & D. A. Regier (Eds.), *A research agenda for DSM-V* (pp. 31–84). Washington, DC: American Psychiatric Association.

Clark, L. A. (2005). Temperament as a unifying basis for personality and psychopathology. *Journal of Abnormal Psychology, 114*, 505–521.

Clark, L. A., Watson, D., & Mineka, S. (1994). Temperament, personality, and the mood and anxiety disorders. *Journal of Abnormal Psychology, 103*, 103–116.

Cooper, J. E., & Sartorius, N. (2013). *A companion to the classification of mental disorders*. Oxford: Oxford University Press.

Craddock, N., & Owen, M. J. (2010). The Kraepelinian dichotomy – going, going … but still not gone. *British Journal of Psychiatry, 196*, 92–95.

Crespi, B., Stead, P., & Elliot, M. (2010). Comparative genomics of autism and schizophrenia. *Proceedings of the National Academy of Sciences of the United States of America, 107*(Suppl. 1), 1736–1741.

Cross-Disorder Group of the Psychiatric Genetics Consortium. (2013). Identification of risk loci for five major psychiatric disorders: a genome-wide analysis. *Lancet, 381*, 1371–1379.

Cuthbert, B. N. (2005). Dimensional models of psychopathology: research agenda and clnical utility. *Journal of Abnormal Psychology, 114*, 565–569.

Cuthbert, B. N. (2014). The RDoC framework: facilitating transition from ICD/DSM to dimensional approaches that integrate neuroscience and psychopathology. *World Psychiatry, 13*, 28–35.

Cuthbert, B. N., & Insel, T. R. (2013). Toward the future of psychiatric diagnosis: the seven pillars of RDoC. *BMC Medicine, 11*, 127.

Cuthbert, B. N., & Kozak, M. J. (2013). Constructing constructs for psychopathology: the NIMH Research Domain Criteria. *Journal of Abnormal Psychology, 122*, 928–937.

Di Chiara, G., Loddo, P., & Tanda, G. (1999). Reciprocal changes in prefrontal and limbic dopamine responsiveness to aversive and rewarding stimuli after chronic mild stress: implications for the psychobiology of depression. *Biological Psychiatry, 46*, 1624–1633.

Doherty, J. L., & Owen, M. J. (2014). Genomic insights into the overlap between psychiatric disorders: implications for research and clinical practice. *Genomic Medicine, 6*, 29.

Engels, A. S., Heller, W., Mohanty, A., Herrington, J. D., Banich, M. T., Webb, G., & Miller, G. A. (2007). Specificity of regional brain activity in anxiety types during emotional processing. *Psychophysiology, 44*, 352–363.

Fair, D. A., Bathula, D., Nikolas, M. A., & Nigg, J. T. (2012). Distinct neuropsychological subgroups in typically developing youth inform heterogeneity in children with ADHD. *Proceedings of the National Academy of Sciences of the United States of America, 109*, 6769–6774.

Feighner, J. P., Robins, E., Guze, S. B., Woodruff, R. A., Winokur, G., & Munoz, R. (1972). Diagnostic criteria for use in psychiatric research. *Archives of General Psychiatry, 26*, 57–63.

Franklin, J. C., Jamieson, J. P., Glenn, C. R., & Nock, M. K. (2014). How developmental psychopathology theory and research can inform the research domain criteria (RDoC) project. *Journal of Clinical Child & Adolescent Psychology, 44*(2), 280–290. http://dx.doi.org/10.1080/15374416.2013.873981. Advance online publication.

de Geus, E. J. C. (2014). Molecular genetic psychophysiology: a perspective on the Minnesota contribution. *Psychophysiology, 51*, 1203–1204.

Glahn, D. C., Knowles, E. E., McKay, D. R., Sprooten, E., Raventós, H., Blangero, J., … Almasy, L. (2014). Arguments for the sake of endophenotpes: examining common misconceptions about the use of endophenotypes in psychiatric genetics. *American Journal of Medical Genetics. Part B, Neuropsychiatric Genetics, 165B*(2), 122–130.

Glatt, C. E., & Lee, F. S. (2016). Common polymorphisms in the age of RDoC: integration and translation. *Biological Psychiatry, 79*, 25–31. http://dx.doi.org/10.1016/j.biopsych.2014.12.020.

Gottesman, I. I., & Gould, T. D. (2003). The endophenotype concept in psychiatry: etymology and strategic intentions. *American Journal of Psychiatry, 160*, 636–645.

Greene, T. (2007). The Kraepelinian dichotomy: the twin pillars crumbling? *History of Psychiatry, 18*, 361–379.

Gur, R. E. (2014). Early prospective assessment of cognition and brain function in psychosis spectrum. In *Symposium presentation at the conference of the American College of Neuropsychopharmacology, Phoenix, December*.

Hall, M.-H., Smoller, J. W., Cook, N. R., Schulze, K., Lee, P. H., Taylor, G., … Levy, D. L. (2012). Patterns of deficits in brain function in bipolar disorder and schizophrenia: a cluster analytic study. *Psychiatry Research, 200*, 272–280.

Hyman, S. E. (2000). National Institute of Mental Health goals for behavioral science. *Experimental and Clinical Psychopharmacology, 8*, 271–272.

Hyman, S. E. (2007). Can neuroscience be integrated into the DSM-V? *Nature Reviews Neuroscience, 8*, 725–732.

Hyman, S. E. (2010). The diagnosis of mental disorders: the problem of reification. *Annual Review of Clinical Psychology, 6*, 12.1–12.25.

Iacono, W. G., Vaidyanathan, U., Vrieze, S. I., & Malone, S. M. (2014). Knowns and unknowns for psychophysiological endophenotypes: integration and response to commentaries. *Psychophysiology, 51*, 1339–1347.

Insel, T. R., & Cuthbert, B. N. (2009). Endophenotypes: bridging genomic complexity and disorder heterogeneity. *Biological Psychiatry, 66*, 988–989.

International Schizophrenia Consortium. (2009). Common polygenic variation contributes to risk of schizophrenia and bipolar disorder. *Nature, 460*, 748–752.

Johns, L., & van Os, J. (2001). The continuity of psychotic experiences in the general population. *Clinical Psychology Review, 21*, 1125–1141.

Kendler, K. S., Gardner, C. O., Gatz, M., & Pedersen, N. L. (2007). The sources of co-morbidity between major depression and generalized anxiety disorder in a Swedish national twin sample. *Psychological Medicine, 37*, 453–462.

Kleinman, A., Caetano, S. C., Brentani, H., Rocca, C., dos Santos, B., … Lafer, B. (2014). Attention-based classification pattern, a research domain criteria framework, in youths with bipolar disorder and attention-deficit/hyperactivity disorder. *Australian & New Zealand Journal of Psychiatry, 49*(3), 255–265. http://dx.doi.org/10.1177/0004867414557957. Epub ahead of print.

Krueger, R. F., & Markon, K. E. (2014). The role of the DSM-5 personality trait model in moving toward a quantitative and empirically based approach to classifying personality and pychopathology. *Annual Review of Clinical Psychology, 10*, 477–501.

Krueger, R. F., & Piasecki, T. M. (2002). Toward a dimensional and psychometrically-informed approach to conceptualizing psychopathology. *Behaviour Research & Therapy, 40*, 485–499.

Lichtenstein, P., Yip, B. H., Björk, C., Pawitan, Y., Cannon, T. D., Sullivan, P. F., & Hultman, C. M. (2009). Common genetic determinants of schizophrenia and bipolar disorder in Swedish families: a population-based study. *Lancet, 373*, 234–239.

Lips, E. S., Cornelisse, L. N., Toonen, R. F., Min, J. L., Hultman, C. M., the International Schizophrenia Consortium, … Posthuma, D. (2012). Functional gene group analysis identifies synaptic gene groups as risk factor for schizophrenia. *Molecular Psychiatry, 17*, 996–1006.

London, E. B. (2014). Categorical diagnosis: a flaw for autism research? *Trends in Neurosciences, 37*, 683–686.

McTeague, L. M., & Lang, P. J. (2012). The anxiety spectrum and the reflex physiology of defense: from circumscribed fear to broad distress. *Depression and Anxiety, 29*, 264–281.

Miller, G. A., & Rockstroh, B. (2013). Endophenotypes in psychopathology research: where do we stand? *Annual Review of Clinical Psychology, 9*, 177–213.

Mineka, S., Watson, D., & Clark, L. A. (1998). Comorbidity of anxiety and unipolar mood disorders. *Annual Review of Psychology, 49*, 377–412.

National Institute of Mental Health. (2001). *Modular phenotyping for mental disorders*. Bethesda, MD: National Institute of Mental Health. http://grants.nih.gov/grants/guide/rfa-files/RFA-MH-02-009.html Accessed 12.12.14.

National Institute of Mental Health. (2008). *The National Institute of Mental Health strategic plan*. Bethesda, MD: National Institute of Mental Health. NIH Publication 08–6368. http://www.nimh.nih.gov/about/strategic-planning-reports/index.shtml Accessed 12.12.14.

Network and Pathway Analysis Subgroup of the Psychiatric Genomics Consortium. (2015). Psychiatric genome-wide association study analyses implicate neuronal, immune and histone pathways. *Nature Neuroscience, 18*, 199–209.

Owen, M. J., O'Donovan, M. C., Thapar, A., & Craddock, N. (2011). Neurodevelopmental hypothesis of schizophrenia. *British Journal of Psychiatry, 198*, 173–175.

Patrick, C. J. (2014). Genetics, neuroscience, and psychopathology: clothing the emperor. *Psychophysiology, 51*, 1333–1334.

Pearlson, G., Meda, S., Narayanan, B., Clementz, B., Keshavan, M., Tamminga, C., & Sweeney, J. (2014). *Measures derived from resting state functional MRI and resting state EEG aggregate with psychosis biotypes more definitively than with DSM diagnoses: Findings from the Bipolar-Schizophrenia Network on Intermediate Phenotypes (BSNIP)*. Phoenix: Poster Presented at the American College of Neuropsychopharmacology. http://www.nature.com/npp/journal/v39/n1s/full/npp2014280a.html Accessed 08.01.15.

Pearlson, G. D. (2015). Etiologic, phenomenologic, and endophenotypic overlap of schizophrenia and bipolar disorder. *Annual Review of Clinical Psychology, 11*, 251–281. http://dx.doi.org/10.1146/annurev-clinpsy-032814-112915.

Phelps, J. (2012). The bipolar spectrum. In G. Parker (Ed.), *Bipolar II disorder: Modelling, measuring, and managing* (pp. 10–34). New York: Cambridge University Press.

Plomin, R., Haworth, C. M., & Davis, O. S. (2009). Common disorders are quantitative traits. *Nature Reviews Genetics, 10*, 872–878.

Reynolds, S. M., & Berridge, K. C. (2008). Emotional environments retune the valence of appetitive versus fearful functions in nucleus accumbens. *Nature Neuroscience, 11*, 423–425.

Robins, E., & Guze, S. B. (1970). Establishment of diagnostic validity in psychiatric illness: its application to schizophrenia. *American Journal of Psychiatry, 126*, 983–987.

Rommelse, N. N., Franke, B., Geurts, H. M., Harman, C. A., & Buitelaar, J. K. (2010). Shared heritability of attention-deficit/hyperactivity disorder and autism spectrum disorder. *European Child & Adolescent Psychiatry, 19*, 281–295.

Schizophrenia Working Group of the Psychiatric Genetics Consortium. (2014). Biological insights from 108 schizophrenia-associated genetic loci. *Nature, 51*, 421–427.

Simmons, J. M., & Quinn, K. J. (2014). The NIMH Research Domain Criteria (RDoC) project: implications for genetics research. *Mammalian Genome, 25*, 23–31.

Spitzer, R. L., Endicott, J., & Robins, E. (1978). Research diagnostic criteria: rationale and reliability. *Archives of General Psychiatry, 35*, 773–782.

Sullivan, P. F., Daly, M. J., & O'Donovan, M. (2012). Genetic architectures of psychiatric disorders: the emerging picture and its implications. *Nature Reviews Genetics, 13*, 537–551.

Tesli, M., Espeseth, T., Bettella, F., Mattingsdal, M., Aas, M., Melle, I., ... Andreassen, O. A. (2014). Polygenic risk score and the psychosis continuum model. *Acta Psychiatrica Scandinavica, 130*, 311–317.

Treadway, M. T., & Zald, D. H. (2011). Reconsidering anhedonia in depression: lessons from translational neuroscience. *Neuroscience and Biobehavioral Reviews, 35*, 537–555.

Vaidyanathan, U., Nelson, L. D., & Patrick, C. J. (2012). Clarifying domains of internalizing psychopathology using neurophysiology. *Psychological Medicine, 42*, 447–459.

Wiecki, T. V., Poland, J. S., & Frank, M. J. (2015). Model-based cognitive neuroscience approaches to computational psychiatry: clustering and classification. *Clinical Psychological Science, 3*, 378–399.

Chapter 27

Schizophrenia

Aiden Corvin
Trinity College Dublin, Dublin, Ireland

INTRODUCTION

Schizophrenia (SCZ) is an enigmatic but heritable brain disorder that poses remarkable challenges because it forces patients to confront their perception and understanding of the world around them. The experience for the individual is highly variable: Patients can have different clinical symptom profiles and functional outcomes ranging from recovery to persistent symptoms and cognitive deficits. About 80% of patients experience multiple relapses within the first 5 years of treatment, and the long-term course is typically characterized by an episodic or continuous illness with significant functional impairment (Andreasen et al., 2005; Bleuler, 1980; Hegarty et al., 1994). Despite interindividual variation, for the almost 1% of the affected adults, SCZ is associated with significant morbidity and mortality: studies indicate that average life expectancy is reduced by more than 15 years (Tiihonen et al., 2009; World Health Organization, 1992).

Since the 19th century, the entity that we call "schizophrenia" has been conceptually defined by the work of key researchers. Their work (reviewed in Kendler (2014)) has molded the two operationalized classifications systems commonly used today, the American Psychiatric Association's *Diagnostic and Statistical Manual of Mental Disorders*, Fifth Edition and the World Health Organization's *International Statistical Classification of Diseases and Related Health Problems*, 10th Revision. Both systems produce highly similar, reliable diagnostic results based on the presence and endurance (for at least 6 months) of characteristic clinical symptoms and behaviors that generally emerge in early adulthood. Conceptually the symptoms are grouped to represent core domains involving "reality distortion" (delusions and hallucinations, also termed psychosis), "disorganization" affecting speech and self-monitoring of behavior (eg, catatonic behavior), and "negative" symptoms representing restrictions in emotional expression, in the fluency and productivity of thought and speech (alogia), and in the initiation of purposeful activity (avolition). Many people who later develop SCZ have subtle neurodevelopmental anomalies in childhood affecting motor, cognitive, and social functioning. The emergence of the syndrome is often preceded by a "prodromal" period of attenuated psychotic symptoms, anxiety, mood disturbance, and social withdrawal.

The modern conceptualization of SCZ is rooted in the work of Emil Kraepelin, who postulated a unifying pathophysiological process to explain the psychopathology that he observed in patients. He identified this process as a disease, dementia praecox, and made a distinction between this and manic-depressive disease. This dichotomy has since been central to the classification of the major psychotic disorders but is increasingly challenged. His concept of this disease entity was shared and shaped by Eugene Bleuler, who coined the term "schizophrenia," but also described SCZ as a syndroma diagnosis capturing more than one disease by referring to "the group of schizophrenias." An alternative view, based on observation of psychotic experience in the general population, suggests that there is a continuum of symptomatology requiring dimensional rather than categorical classification of the core symptoms or behaviors across the population rather than conceptualizing SCZ as one or more discrete diseases (van Os et al., 2009; Strauss, 1969).

A century of research has taught us that SCZ is likely to be of neurodevelopmental etiology in many (if not most) cases; that the syndrome is largely heritable but that social factors (eg, childhood adversity and migration) also contribute to risk; and that the condition is associated with structural, functional, and neurochemical brain changes particularly involving the dopaminergic system (Tandon, Keshavan, & Nasrallah, 2008). Current pharmacotherapies, first introduced in the 1950s, are thought to act primarily through blockage of the type 2 dopamine (DA) receptor. The original DA hypothesis has evolved to identify overactivity of mesolimbic DA circuitry as the basis of positive symptoms, and diminished

Genomics, Circuits, and Pathways in Clinical Neuropsychiatry. http://dx.doi.org/10.1016/B978-0-12-800105-9.00027-5

mesocortical DA function as the cause of negative and cognitive symptoms. More recently, it has been argued that altered DA is a consequence of *N*-methyl-D-aspartate glutamate (NMDA) receptor hypofunction involving NMDA receptors located on gamma aminobutyric acid (GABA) interneurons (Schwartz, Sachdeva, & Stahl, 2012). With greater appreciation of the complex, dynamic nature of neural network function and better investigative tools, SCZ has also been conceptualized as a disorder of connectivity in which DA or glutamate dysfunction may be an end result of neural circuit dysfunction (Bernardinelli, Nikonenko, & Muller, 2014; Friston & Frith, 1995).

Despite this progress, a cohesive understanding of pathophysiology sufficient to improve patient care is lacking. Antipsychotic medications have proven effective for treating positive symptoms but they have little impact on debilitating negative symptoms or cognitive deficits, which may explain why illness outcomes have not improved substantially since their introduction (Leucht et al., 2013). Reaching beyond a clinically defined syndrome to grasp disease etiology may require dissecting a number of different disease mechanisms with an appreciation of how these mechanisms overlap across current clinical diagnostic entities and, on a deeper level, an understanding of how these mechanisms contribute to normal development and function of the complex human brain.

HERITABILITY AND GENETIC ARCHITECTURE

Kraepelin recognized that SCZ clustered within the families that he studied and identified a general hereditary predisposition to mental disorders in about 70% of the cases he described. This was first formally investigated by Heron in 1907 in his study of 331 pedigrees, in which he concluded that: "The insane diathesis is inherited with at least as great an intensity as any physical or mental character in man." Such early studies identified familial clustering of SCZ but failed to address to what extent this reflected a contribution from shared genetic or environmental effects, or some combination of both. The significance of genetics was confirmed by a series of pivotal twin and adoption studies beginning in the 1960s (Cardno & Gottesman, 2000; Ingraham & Kety, 2000) and discussed in chapter "Contribution of Genetic Epidemiology to Our Understanding of Psychiatric Disorders." Beyond confirming that SCZ was heritable, from twin studies, the heritability could be estimated. The data have been remarkably consistent in the estimates generated by individual twin studies and meta-analyses (80−85%) (Sullivan, Kendler, & Neale, 2003). With heritability confirmed, speculation moved to the genetic architecture involved: how many genomic loci contributing to risk, their population frequencies, their effect sizes, and their interactions with each other and with environmental risk.

In the pregenomic research era, two rival theories emerged as the genetic etiology of SCZ. The first, based on model fitting to risk data from relatives of affected probands, proposed that a large proportion of cases were polygenic involving many common genetic variants of small effect. The second, stemming from work by Rubin and others to identify Mendelian segregation ratios in SCZ, proposed a single major locus, or a number of rare high-penetrance risk loci. In the absence of tools to identify the genes involved, the pregenomics era of molecular genetics was ill-equipped to resolve this debate. As molecular genetics evolved, linkage analysis methods were powered to identify large genetic effects and localize them to relatively large genomic regions (Ng et al., 2009). Such studies resolved that SCZ was not a single gene disorder and provided a small number of signals from meta-analytic efforts. Association studies, limited by the number of markers that could be tested, investigated hundreds of potential functional or positional candidate genes. As described in the SZGene resource (http://www.szgene.org/), these studies established definitively that SCZ was a polygenic rather than Mendelian disorder but had limited power to identify definitively the common, small genetic effects that were their focus (Sullivan, Daly, & O'Donovan, 2012).

A secondary question, recognized early in the history of the field, related to the boundaries of the SCZ syndrome. In his work, Kraepelin acknowledged cases of "latent schizophrenia" within family members of affected probands and described work by Rudin identifying an increased prevalence of what we now call bipolar disorder, and what he termed "eccentric personalities" among the relatives of affected probands. Investigating this concept of "latent" SCZ, Kendler and colleagues identified a familial predisposition to a spectrum of clinical syndromes in relatives of SCZ probands including schizoaffective disorder, other nonaffective psychosis, schizotypal personality disorder, and psychotic affective illness (Kendler et al., 1993a, 1993b). This work did not support overlap with affective disorders (Kendler et al., 1993c) but others studies did (Maier et al., 1993). The debate remained open until a large, register-based investigation from Sweden was reported in 2009. This study of more than two million nuclear families found that first-degree relatives of probands with either SCZ (n = 35,985) or bipolar disorder (n = 40,487) were at substantially increased risk of both these disorders and the comorbidity between disorders mostly (63%) resulted from the additive genetic effects common to both disorders (Lichtenstein et al., 2009).

Prompted by this study but also by the findings of studies of structural genomic variants (discussed subsequently), there has been renewed interest in examining genetic overlap between SCZ and other brain disorders. Inevitably, Kraepelin was the first to note an overlap between SCZ and intellectual disability (ID). He presciently estimated that ID formed the basis for at least 3.5% of cases of SCZ (Kraepelin, 1919): studies indicate a three- to fivefold increased risk of ID in SCZ populations (Hemmings, 2006; Morgan et al., 2008). An almost threefold increased risk of autism spectrum disorder (ASD) in offspring who have one parent affected with SCZ has been reported in data from four different cohorts (Larsson et al., 2005; Sullivan et al., 2012). Finally, individuals with a parental history of epilepsy also have a twofold increase in the risk of developing a psychotic disorder, and individuals with a parental history of psychosis have a 2.7-fold increased risk of epilepsy (Clarke et al., 2012). Genetic liability to SCZ is not straightforward but overlaps significantly with risk for other psychiatric and developmental disorders including, but possibly not limited to, ID, ASD, and epilepsy.

THE GENOMICS ERA

The development of high-throughput array technologies, genome sequencing, and unprecedented collaborative research in the past decade has facilitated empirical studies to map the genetic architecture of SCZ. In the vanguard, genome-wide association studies (GWAS) (discussed in chapter: Natural Selection and Neuropsychiatric Disease: Theory, Observation, and Emerging Genetic Findings) test common genetic risk variants of small effect and have defined a substantial contribution (30–50%) to SCZ risk. This may involve thousands of loci of which more than 100 have been confirmed. Studies on a similar scale have also confirmed structural risk variants [copy number variants (CNVs)] of large effect, which are rare and so make a smaller cumulative contribution to population risk (< 5%) (see chapter: Genome Tools and Methods: Rare Genetic Variation). Through exome sequencing studies, a wider spectrum of variants is being assayed, but studies reported to date are an order of magnitude smaller than the largest GWAS analysis. The contribution from rare sequence mutations is unclear because we have yet to reach a point at which the full spectrum of exomic variation is being systematically investigated in well-powered studies. In addition, the protein-coding genome represents a very small fraction (1.5%) of all human genetic variation and few whole-genome data have been published (Hoischen, Krumm, & Eichler, 2014). Achieving a complete picture of architecture will also require careful analysis of gene–gene (epistasis) and gene–environment interaction. Therefore, being mindful of these limitations, what have we learned thus far?

GENOME-WIDE ASSOCIATION STUDIES

The advent of arrays capturing most common single-nucleotide polymorphism (SNP) variations [minor allele frequency (MAF) > 5%] in the genome provided a framework for systematic GWAS. Notable successes for other common disorders (eg, diabetes, inflammatory bowel disease) spurred international collaborative efforts to generate the large sample sizes required to perform studies in SCZ (Corvin, Craddock, & Sullivan, 2010). In 2009, three GWAS reported robust, replicable association signals in a 5.5-megabase region around the major histocompatibility complex (MHC) on chromosome 6p (Purcell et al., 2009; Shi et al., 2009; Stefansson et al., 2009). This locus, one of the most of the most complex, genetically diverse genomic regions, encodes about 250 genes including the classical and transplantation human leukocyte antigen alleles, but also many immune and nonimmune genes. Subsequent studies have provided further association evidence while highlighting the challenge this complex locus poses (Corvin & Morris, 2014). Most profoundly, these studies proved that collaborative efforts successfully applied elsewhere in human genetics could fruitfully translate to psychiatric disorders. This gave impetus to the Psychiatric GWAS Consortium (PGC), an even larger collaborative effort to combine all available GWAS data for five psychiatric disorders, with the largest sample size being for SCZ (Schizophrenia Working Group of the Psychiatric Genomics Consortium, 2014; World Health Organization, 1992).

The Schizophrenia Working Group of the PGC has seen a trajectory of discovery similar to that of other common disorders, with seven significant loci being identified by PGC1 in a sample of 9394 cases published in 2011 (Ripke et al., 2011) and increasing to 108 loci in an analysis of 36,989 cases and 113,075 controls. This was published in 2014 by the then renamed Psychiatric Genomics Consortium (PGC2) (Schizophrenia Working Group of the Psychiatric Genomics Consortium, 2014) (Fig. 27.1). Taken together, the GWAS analyses of SNP data yielded a number of major insights into the genetic basis of SCZ.

Genetic Architecture of Schizophrenia

GWAS arrays made hypothesis-free SNP association analysis feasible and allowed empirical testing of the hypothetical role of polygenic inheritance in SCZ risk. This was achieved using a "polygene score" method that summed variation across a large number of nominally associated loci into quantitative scores to ask whether these scores could predict disease state in independent samples (Wray, Goddard, & Visscher, 2007). From such analysis, up to a third of total variation in

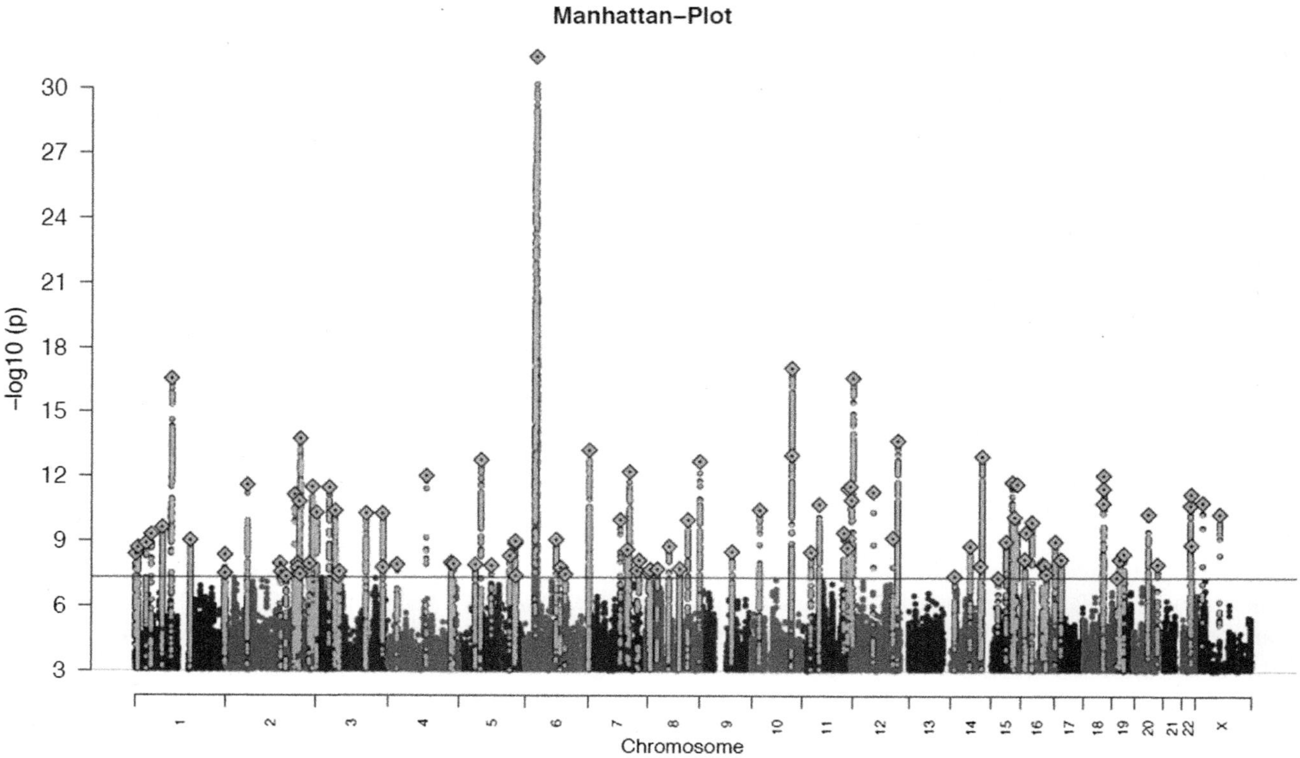

FIGURE 27.1 Manhattan plot of the discovery GWAS meta-analysis of 34,241 cases and 45,604 controls, where the *x*-axis is chromosome position and the *y*-axis is the significance of association. The *red line* (dark gray in print versions) represents the threshold for genome-wide significant levels of association (5×10^{-8}).

genetic liability could be explained by common risk variants. A subsequent analysis using a different approach (Genome-wide Complex Trait Analysis) provided further evidence that polygenic inheritance contributes to SCZ (Lee et al., 2012). By examining GWAS data from 9087 affected individuals and 12,171 control subjects from the PGC1 data set, the authors quantified the lower limit of the genetic contribution to SCZ from common variants as being 23% based on the GWAS platforms of the time. They showed that the variance explained per chromosome was linearly related to the length of the chromosome, another expected feature of polygenic disorders. More recently, a larger study, including additional Swedish samples (16,245 cases and 31,829 controls) applied approximate Bayesian polygenic analysis and estimated that 8300 independent common loci explain up to 50% of the variance in genetic risk of SCZ (Ripke et al., 2013).

Overlap Across Disorders

A second important observation from the International Schizophrenia Consortium was that the polygenic common risk variant contribution is shared with bipolar disorder but not with six common nonpsychiatric diseases, which provides molecular validation for the epidemiological evidence of overlap reported from the Swedish cohort (Lichtenstein et al., 2009; Purcell et al., 2009). Furthermore, the availability of the PGC data set made it possible to investigate to what extent genetic variation is unique to individual disorders or is shared across SCZ, bipolar disorder, major depressive disorder (MDD), ASD, and attention-deficit hyperactivity disorder. In addition to the expected substantial covariance in liability to SCZ and bipolar disorder, this identified moderate levels of covariance between SCZ and MDD, with modest overlap between SCZ and childhood-onset disorders such as ASD (Fig. 27.2) (Lee et al., 2013).

Distinctions between these disorders are based on clinical classification mask overlapping molecular etiology. With the substantially expanded number of associated GWAS loci identified in PGC2, it has been possible to test for relationships between SCZ and more severe neurodevelopmental phenotypes at individual loci. Here, the key finding is that SCZ GWAS loci significantly overlap with genes having nonsynonymous mutations in ID ($P = 0.00024$) and ASD ($P = 0.035$); specific examples are at *CACNA1C* (Timothy syndrome), *TCF4* (Pitt–Hopkins syndrome), *EP300* (Rubenstein–Taybi syndrome), *AKT3* (a megaloencephalic syndrome), *SATB2* (cleft palate and mental retardation), and the *CNTN4* and *NLGN4X* ASD loci. One potential explanation is that certain genes have critical "hub" functions in the development and

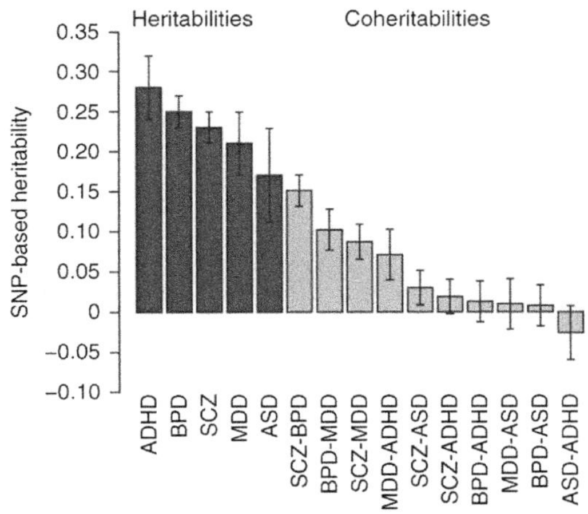

FIGURE 27.2 Proportion of variance in liability (SNP-based heritability) and proportion of covariance in liability between disorders (SNP-based coheritability) for five major psychiatric disorders. The 95% error bars represent the estimates ±1.96 standard error. *SCZ*, schizophrenia; *MDD*, major depressive disorder; *BPD*, bipolar disorder.

function of brain networks, and thus their disruption is more likely to have an impact at the level of phenotype. The clinical consequences of perturbation of function of these hub genes may depend on the nature of the genetic variant [from subtle common SNP effects to loss-of-function (LOF) mutation], its developmental timing, and the impact of environmental exposures.

Localizing Risk Genes

Examining the loci implicated by the PGC2 analysis, many findings support existing or emerging hypotheses relevant to SCZ etiology. The best example is association at the *DRD2* receptor gene (the target for current antipsychotic drug treatments). A robust literature implicates glutamatergic neurotransmission in SCZ, and the GWAS data implicate genes involved in the structure (*GRIA1*, *GRM3*, and *GRIN2A*) and function (*SRR*, *CLCN3*, and *SLC38A7*) of glutamatergic synapses (Kantrowitz & Javitt, 2012). Associations at the genes *CACNA1C*, *CACNA1I*, and *CACNB2* extend findings from earlier GWAS analyses, implicating voltage-gated calcium channel subunits in SCZ and other psychiatric disorders (Cross-Disorder Group of the Psychiatric Genomics Consortium, 2013). Other candidate genes have more general roles in immune function (eg, MHC locus), synaptic plasticity (eg, *NLGN4X*, *IGSF9B*, *CNTN4*, and *PTN*), or neurodevelopmental mechanisms (eg, *FXR1*, *SATB2*, and *TLE1*). However, the role of most loci is far from clear, and even for the examples provided, additional work is required to understand the functional mechanism involved at individual loci or how risk genes might relate at the level of gene networks.

Functional Effects of Common Variants

Of the 108 conservatively defined association loci confirmed by the PGC2 study (Schizophrenia Working Group of the Psychiatric Genomics Consortium, 2014), 25 had been previously reported but 83 were novel or had not previously received robust support. Scattered through the genome, most overlapped with genic regions: 75% included protein-coding genes and a further 8% were within 20 kb of a gene. Examining the loci in more detail, 40% intersected a single gene but almost 10% of loci included five or more genes (Fig. 27.3).

It is a formidable task to refine which genes are involved. Of the 108 SCZ loci identified, in only 10 cases could an association signal be attributed to probable nonsynonymous exonic polymorphisms at specific genes (*PRRG2*, *FAM5B*, *WBP2NL*, *EP300*, *LRRC48*, *MYO15A*, *ITIH3*, *ADAMTSL3*, *SLC39A8*, and *ATXN7*). This accords with the results of GWAS for other complex traits and it is likely that most common risk variants have subtle, regulatory effects on the function rather than the structure of proteins (Maurano et al., 2012; Schork et al., 2013). In SCZ, risk variants identified by polygene score methods are significantly enriched among genic elements, particularly the 5′ untranslated region (UTR), exon, and 3′ UTR compared with intergenic SNPs, which suggests that associated loci are likely to affect gene function. This pool of variants is enriched for expression quantitative trait loci, specifically loci that affect gene expression in the human brain (Richards et al., 2012). Disambiguating how individual common risk variants contribute to molecular risk mechanisms as part of biological systems will be a major challenge for the field. Even where association peaks

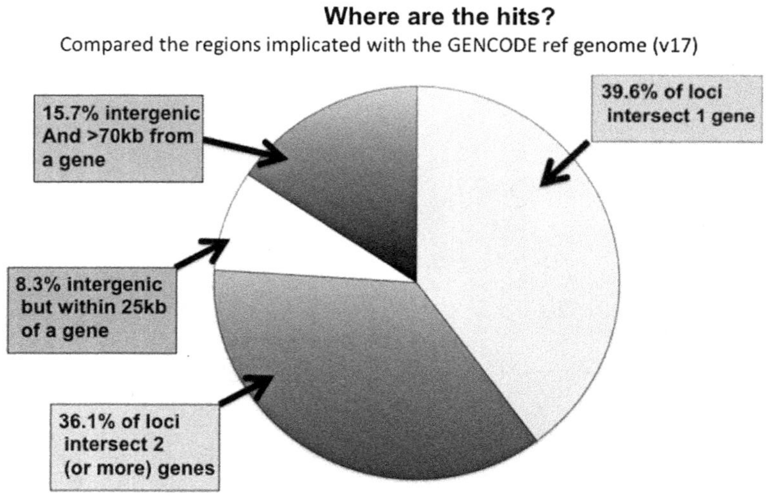

FIGURE 27.3 Breakdown of significant loci from the PGC2 GWAS based on their location to functional genomic regions (from GENCODE ref genome (v17)).

unambiguously suggest localization to a particular gene, in some cases the functional effects will not be local to the gene (Smemo et al., 2014).

Risk Prediction

Risk profile scores (RPS) based on alleles that are statistically confirmed as associated with a standard genome-wide correction explain 3.4% of the variance in SCZ; this doubles (to 7%) when less rigorous evidence for association is included. By grouping individuals into deciles and estimating the odds ratio (OR) for affected status at each decile compared with the lowest decile, the PGC2 group evaluated the capacity of RPS to predict case–control status in a number of different populations. The ORs increased to maximum for the 10th decile in all samples, which indicated that carrying more risk loci was associated with a demonstrable and progressive increase in risk. This was largest for a Swedish sample based only on individuals who had been hospitalized (OR, 15; 95% confidence interval (CI), 12.1–18.7) and least for a population-based Danish sample (from inpatient and outpatient facilities) (OR, 7.8; 95% CI, 4.4–13.9) (Fig. 27.4). Even in the Swedish sample, the sensitivity and specificity of RPS does not support its use as a predictive test. Taking the standard method to evaluate clinical efficacy of risk scores, the area under the receiver operator characteristic curve (AUC) is greater than 0.7. To put this in context, random prediction corresponds to an AUC value of 0.5, values greater than 0.75 can usefully identify high-risk groups for screening, but values of 0.99 can reliability diagnose a disease in the general population (Janssens et al., 2007). The AUC value can be (modestly) boosted by including family history information and the RPS may be useful in stratifying individuals in epidemiological research: for example, in prospective studies of "at risk" patient groups (Chatterjee et al., 2013; Iyegbe et al., 2014).

STRUCTURAL VARIATION STUDIES

CNVs are chromosomal rearrangements involving the deletion, duplication, inversion, or translocation of segments of DNA from 1000 to several million base pairs in length. By the 1990s, using cytogenetic methods, deletions involving the chr22q11.2 locus (velocardiofacial syndrome) and a reciprocal chromosomal t(1:11) translocation disrupting the *DISC1* locus had been implicated in the etiology of SCZ. Both were rare events; the former occurred at a rate of 1 in 4000 live births and the latter was identified in a single pedigree. The working assumption in the field was that CNVs were unlikely to contribute significantly to SCZ etiology. This view changed after it was demonstrated that not only could submicroscopic variations (<500 kb in size) be successfully assayed using array based methods but these were widespread in normal human genomes (Iafrate et al., 2004; Sebat et al., 2004). In the first of a series of SCZ studies, by examining CNVs greater than 100 kb in size, Walsh and colleagues (Walsh et al., 2008) identified an excess of CNV events in patients, particularly those with childhood onset, compared with control populations. Two much larger studies confirmed an excess burden of CNVs in SCZ, but were also powered to identify specific loci that were significantly associated with disease. In doing so, they confirmed the known association with chr22q11.2 deletions and identified novel associations with deletions

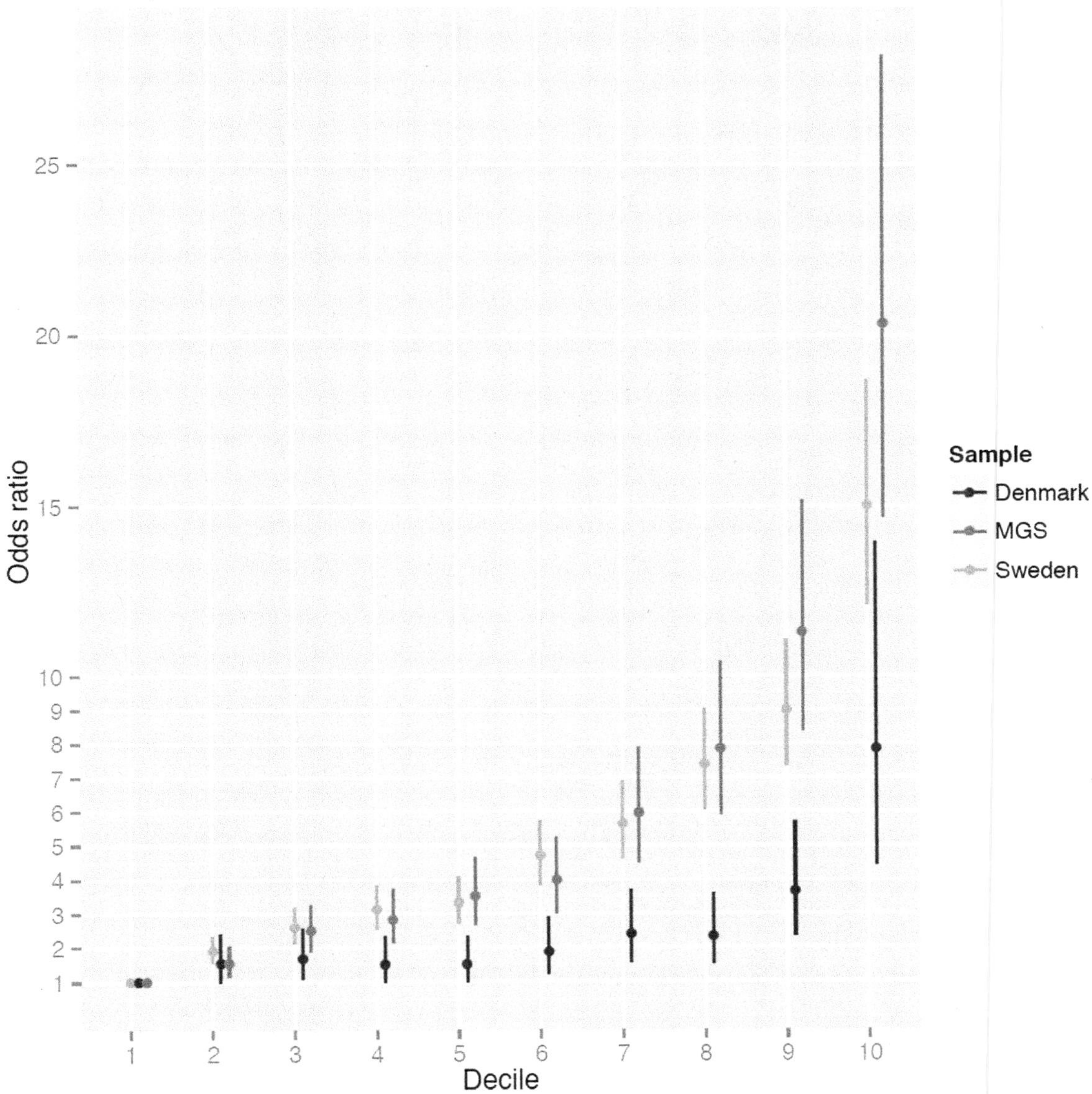

FIGURE 27.4 Odds ratio for schizophrenia by risk profile score (RPS) decile in the Sweden (SW1–6), Denmark (Aarhus), and Molecular Genetics of Schizophrenia (MGS) studies. Odds ratios and 95% confidence intervals (bars) were estimated using logistic regression with principal components to control for population stratification.

at 1q21.1, 15q11.2, and 15q13.3 (International Schizophrenia Consortium, 2008; Stefansson et al., 2008). Since then further studies have extended the list to include more than a dozen loci including duplications (Williams–Buren syndrome region, Angelman/Praeder–Willi syndrome region, at *VIPR2*, at *PAK7*, and at 16p13.11) and deletions at 3q29, 16p11.2, 17q2, and 17p12 but also evidence that the reciprocal duplication at the 22q11.2 region may be a protective mutation for SCZ (see table in Rees et al., 2014a).

Structurally, most implicated CNVs are events that recur in the population at the same genomic locations mediated through nonallelic homologous recombination (NAHR). This is because of the local genomic features of these loci, specifically that the rearrangement region is flanked by repeat sequences, which makes the region prone to repeated CNVs in unrelated individuals. Although rare, such events can be identified by association analysis of large cohorts. However, this is only one of a number of proposed mechanisms for genome rearrangements (reviewed in Liu et al. (2012) and discussed further in chapter "Genome Tools and Methods: Rare Genetic Variation"). The genes *NRXN1* and *VIPR2* have demonstrated evidence of association based on disruption by different CNV events with different breakpoints in individual carriers rather than a more universal mechanism. CNVs at chr20p12.2 (*PAK7*) and chr22q11.22 (*TOP3B*) are examples in which there is a likely common founder mutation shared by all carrier individuals (Morris et al., 2014; Stoll et al., 2013). These data suggest that a number of different mechanisms, including complex rearrangements, may contribute to SCZ etiology, with the NAHR mechanism being the most accessible for discovery using association analysis across large populations. A large-scale CNV association analysis of the PGC2 data set is currently under way.

Individually the CNVs identified are of moderate penetrance for SCZ (OR, 2–30) (Malhotra & Sebat, 2012; Morris et al., 2014; Rees et al., 2014b; Rujescu et al., 2009; Vacic et al., 2011). However, this does not directly address the question of their pathogenicity. In almost all cases these mutations confer risk across diagnostic boundaries for developmental phenotypes including ASD, ID, epilepsy, and other congenital malformations. In most cases the risk for developing any significant developmental phenotype is high (ranging from 10% to 100%) with most of the risk being for the development of an early-onset disorder (eg, ID, ASD, developmental delay) rather than SCZ (Kirov et al., 2014), and with lower risk again in bipolar disorder and the position less clear for other psychiatric disorders (Georgieva et al., 2014). Investigation of risk CNVs in individuals defined as control subjects indicate that carriers perform at a level between patients with SCZ and population controls, but the CNV events do not all affect the same cognitive domains. For example, control subjects with the chr15q11.2 deletion have increased the rates of dyslexia and dyscalculia, even after adjusting for IQ, with relative preservation of other cognitive functions. By contrast, the same study found that 10 carriers of the 1q21.1 duplication in 10 control subjects did not have significant neurocognitive impairment or learning difficulties (Stefansson et al., 2014). This has important implications. Understanding the phenotypic impact and penetrance of identified CNVs is likely to require detailed clinical assessment of all carriers, not just clinically defined patients. As such, the CNV findings represent the first high-penetrance genetic findings for SCZ and represent useful genetic models for understanding the disorder. Because most CNVs have pleiotropic neurodevelopmental effects, this may be important in investigating etiological overlap between SCZ and other clinical syndromes and how phenotypic outcomes are determined.

SEQUENCING STUDIES

Exome and genome sequencing makes it possible to assay rare coding point mutations and small insertions and deletions: a huge reservoir of rare or even private sequence mutations in the human genome (MacArthur et al., 2012). Such mutations are strongly selected against and represent de novo events or have recent founders. As a class, rare mutations may have a more direct impact on function and phenotype (higher penetrance) than the subtle regulatory effects being elucidated for common risk variants. This makes them potentially important in informing models for functional follow-up in animal and cellular systems. Defining convergent pathophysiological mechanisms across a small (Karayiorgou et al., 2012) or large (McCarthy, McCombie, & Corvin, 2014) number of mutant models based on these rare mutations could be a critical step in dissecting the molecular etiology of SCZ. Sequencing is accelerating the discovery of contributory genes for other neurodevelopmental disorders such as ID (de Ligt et al, 2012; Rauch et al., 2012), autism (De Rubeis et al., 2014; O'Roak et al., 2012), and epilepsy (Allen et al., 2013). Progress has been most evident where the underlying genetic architecture is less complex, with mutations being spread across a small number of risk genes (Hoischen et al., 2014; Samocha et al., 2014) or where disorders can be subtyped into component etiologies based on syndromal features, such as macrocephaly for *CHD8* mutations (Zaidi et al., 2013) and facial dysmorphology for *PACS1* mutations (Helsmoortel et al., 2014). Such assumptions may not apply in most cases of SCZ. However, results from other neurodevelopmental disorders that help us to understand critical molecular processes underlying neuronal migration (*KIF2A* and *TUBA1A*) or brain growth (eg, microcephaly with *WDR62* mutations) may offer a framework to interpret emergent genetic findings in SCZ (Hu, Chahrour, & Walsh, 2014).

In SCZ, risk mutations may be accumulated across hundreds of genes, which makes it difficult to draw statistical inference from a small number of rare observed events. Only very large sample sizes or the use of familial relationship data in homogeneous populations (to identify founder effects) are likely to yield sufficient observations to implicate individual genes reliably in disease (Bahlo et al., 2014; Zuk et al., 2014). Both approaches are being applied by large studies within the field.

Early reported studies were generally small; they took as their focus the total mutational burden in SCZ or performed analyses at a pathway level to identify mutation enrichment, rather than to test specific candidate genes. After smaller, inconclusive studies, increased rates of de novo mutation (DNMs) were reported in the largest SCZ study (Fromer et al., 2014), as has been identified for other neurodevelopmental disorders (de Ligt et al, 2012). This accords with the epidemiological evidence that increased paternal age is a risk factor for ASD, ID, and SCZ, presumably owing to greater number of cell divisions in the male germ line lineage, because 75–80% of DNMs arise paternally and higher mutation rates are evident in older males (Kong et al., 2012). In this study, LOF DNMs were significantly more common in patients with relatively poor school performance, but across the whole population, and in a meta-analysis including previous studies, the overall rate was the same as that reported in control populations. An interesting aspect was that the sample population included only patients who had successfully graduated from mainstream schools, so the effect cannot be exclusively attributed to misclassification of individuals with severe ID in the study. A similar association between LOF DNMs and lower cognitive function has also been reported in patients with ASD (Samocha et al., 2014). This suggests that IQ or general cognitive function may mediate differences in genetic architecture within patients with these disorders.

More relevant than the total burden of mutations are the location and genes affected by mutation. In their analysis, Purcell and colleagues evaluated rare coding variants from 2546 genes selected for prior involvement in SCZ etiology based on unbiased GWAS, CNV, and earlier exome sequencing studies (Purcell et al., 2014). In more than 2500 patients and a matching number of control subjects, they observed a significantly higher number of rare (MAF $< 0.1\%$) disruptive mutations in patients compared to control subjects ($P = 10^{-4}$). Breaking this down, they observed similar results for the most rare mutations (singletons), and the total count for mutations was 1527 in patients versus 1383 in control subjects. Observing recurrent events in patients but not in control subjects represent better evidence, and it has been suggested that observing at least three recurrent de novo LOF events at a given gene only in patients is unlikely to occur by chance (Neale et al., 2012). However, this will depend not only on the number of observations (from patients and control subjects) but also on the length and local sequence content of the gene, which will determine the mutation rate at that gene. In the Purcell study, there was evidence for association at the *KYNU* (kynureninase) locus, in which the authors reported 10 disruptive variants in patients and none in control subjects (Purcell et al., 2014). A novel nonsense mutation at chr2: 143,713,804 was also observed in seven patients but was not seen in control data from the Exome Variant Server (http://evs.gs.washington.edu/EVS/) and 1000 Genomes Project (http://www.1000genomes.org/). Across the published studies to date, there are other examples of genes in which at least two de novo LOF mutations have been observed, but more data will be required to confirm pathogenicity (eg, at *TAF13*, *INADL*, *DST*, *SETD1A*, *SLC1A2*, *KL*, *EP*, *PITPNM*, and *CACNA1C*) (Fromer et al., 2014; Gulsuner et al., 2013; McCarthy et al., 2014; Need et al., 2012; Purcell et al., 2014; Xu et al., 2012). Prioritizing loci with higher prior probability of involvement in SCZ may facilitate mutation discovery using cheaper targeted sequencing strategies perhaps targeting known SCZ loci from GWAS or CNV studies (eg, LOF mutations have been reported at *DPYD* [2], *ESAM*, *ZDHHC5*, and *LRP1*). By extension, one could also prioritize mutations in genes already implicated in other neurodevelopmental disorders, such as ID or ASD (eg, *SCN2A*, *POGZ*, *DLG2*, *SHANK1*, *CHD8*, *AUTS2*, and *MLL2*). An alternative approach is to focus analytical effort on genes that are under processes of selective constraint, meaning that they have reduced rates of functional variation and consequently, mutations detected are more likely to be deleterious (Petrovski et al., 2013; Samocha et al., 2014).

NEXT STEPS IN GENE DISCOVERY

The coming decade will see well-powered studies to test the full spectrum of genetic variation contributing to SCZ. Based on the experience of other disorders, further GWAS analysis, particularly capturing wider ethnic diversity, will yield many more common risk loci. Low-cost custom array analysis in large sample numbers, currently being conducted as part of the Psych Chip initiative (http://www.med.unc.edu/pgc/psychchip), may be informative on less frequent risk variants not well captured by GWAS platforms. This will be the prequel to exome and whole-genome studies of many thousands of patients. Gaps will still remain. Many, possibly even most, common small effects will elude detection even as case numbers approach 100,000, and very rare or private mutations, even of large effect, may be difficult to confirm statistically because of similar study power issues. At a greater level of complexity, the genetic architecture of SCZ is also likely to involve the concerted action of many risk variants, interaction with environmental risk factors, and a potential contribution from other poorly captured genomic mechanisms.

It is unclear to what extent risk variants act independently or have effects dependent on other variants. Epistasis can be functionally defined by the observation that the effect of one risk variant depends on the genotype at another variant, or may statistically refer to the interaction variance that can be explained by a combination of risk variants that does not result from their additive, individual effects. Evidence from other complex disorders, as well as in SCZ, suggests that epistasis contributes to risk. However, testing for statistical interaction between all pairs of genome-wide significant autosomal SNPs in the PGC2 failed to identify interactions that were significant after correction for multiple comparisons. This finding could not exclude such effects between other loci or the presence of higher-order interactions and may reflect limited study power. Numerous parametric and nonparametric approaches have been proposed to identify gene–gene interactions (Wei, Hemani, & Haley, 2014) and are currently being applied to investigate large SCZ data sets in more detail.

The effect of risk genotypes may depend on exposure to environmental risk factors, or such exposures may contribute to disease only in individuals with a particular genetic architecture. Studies taking family history as a proxy measure support gene (G) × environment (E) effects for prenatal infection, trauma exposure, urbanicity, and cannabis use in SCZ. A number of G × E studies have been reported with a particular focus on individual small genetic effects mostly emergent from the candidate gene association study era with inconsistent evidence of replication (Modinos et al., 2013; Duncan & Keller, 2011). Success in SCZ genomics has prompted international collaborative efforts to achieve progress in this field (van Os et al., 2014). Large sample sizes with consistent validated measurement of environmental risk factors will be required. The analytical challenges are considerable. If thousands of common risk variants contribute to disease, how important are the interactions with individual SNPs likely to be to risk and how to you select which SNPs to investigate? Genome-wide G × E methods have been developed and may be useful in identifying the most significant interactive effects. For example, a study identified a novel association between a gene at the *CTNNA3* locus and maternal cytomegalovirus infection in SCZ (Borglum et al., 2014). For rare, highly penetrant mutations (eg, CNVs), identifying sufficient carrier numbers for G × E analysis is similarly problematic. An area of great potential is the ability to derive continuous reliable measures for genetic (eg, polygenic RPS) and environmental (eg, cannabis exposure and dose) risk at an individual level to investigate G × E relationships. This could be tailored to test specific hypotheses: for example, to model RPS for immune dysfunction pathways (including common and rare gene effects) in patients with specific environmental exposures (eg, infection exposures) (McGrath et al., 2013).

New discoveries continue to challenge our understanding of the genome, even how we define "the genome." Implicit in the studies discussed is the assumption that the genetic risk variants, and in particular mutations, are present from fertilization. Some de novo mutations are known to occur after fertilization, during the mitotic cell divisions of early embryonic development. As such, these will be present in clones of cells in only selected tissues, and this can include the brain. Examples from the neurogenetics literature include somatic mutations in which equivalent mutations in the germ line would be lethal (eg, hemimegalencephaly) or in which they have less severe phenotypic manifestations of diseases than equivalent germ line mutations (eg, tuberous sclerosis). How frequent these somatic mutations are and their relevance to the etiology of neurodevelopmental disorders without gross structural malformations, including SCZ, is unknown (Jamuar & Walsh, 2014) but is further discussed in chapter "The Molecular Landscape of the Developing Human Central Nervous System."

CIRCUITS AND PATHWAYS

Ultimately, the goal of genomics research is to use gene discovery as a tool to understand pathophysiological processes. With an understanding that most of the pathway exploration is based on a relatively small number of confirmed risk genes, two interesting themes are emerging. First is that some risk loci may be of substantial etiological importance through their role in influencing function of other genes. An example from another complex brain disorder is the Fragile X mental retardation protein (FMRP), in which LOF mutations represented the first link between RNA regulation and human cognitive dysfunction (Verkerk et al., 1991). This RNA binding protein is involved in translational regulation, which underpins the protein synthesis required for the formation and persistence of memory (Malenka & Bear, 2004). By identifying FMRP—messenger RNA (mRNA) interactions initially in mouse brain, it has been possible to annotate a list of FMRP targets (Darnell et al., 2011). Sequencing studies have shown that this list captures a network of genes enriched for mutations in ASD but also SCZ (Fromer et al., 2014; Iossifov et al., 2014; Purcell et al., 2014). Expression data from human brains reveals different subgroups of FMRP targets with distinct spatiotemporal patterns of expression and biological functions, a possible explanation for differences in phenotypic outcomes across these overlapping neurodevelopmental disorders (Steinberg & Webber, 2013). Several other examples have emerged from the SCZ GWAS data. One of the strongest findings reported by the PGC2 group was evidence of association at a locus for the gene encoding the microRNA miR-137. By overexpressing miR-137 in human neural stem cell lines, Collins et al. (2014) reported that the 157 genes downregulated after analysis at 48 hours were significantly enriched for genes implicated in SCZ GWAS data sets. Another of the earliest confirmed GWAS loci is a risk allele at the zinc finger protein 804A gene (O'Donovan et al., 2008). This brain-expressed gene encodes a protein with a C2H2 domain, which suggests that it, too, may have a role in regulating gene expression through DNA and/or RNA binding.

A second, related hypothesis is that the many risk genes implicated will converge on some number of defined molecular pathways rather than individual regulatory "hub" genes. This acknowledges the complexity of the regulation of brain processes and suggests that disruption at many parts of a complex network may lead to illness. This is supported, in the broadest sense, by analysis of large numbers of unconfirmed genetic markers contributing to SCZ genetic variance and captured by the "polygene" score method. Although not determining specific pathways, it has been shown that the genes contributing to RPS are more likely to affect gene expression in adult brain than would be expected by chance (Richards et al., 2012) and are disproportionately attributable to 2725 genes expressed in the central nervous system (CNS)

($P = 7.6 \times 10^{-8}$) (Lee et al., 2012). More direct approaches to hypothesis-free pathway testing in GWAS data sets have also been applied (Weng et al., 2011). The results have not been particularly consistent, with the possible exception of evidence for enrichment of association in cell adhesion molecular pathway genes based on data from experimentally validated pathway databases (Lips et al., 2012; O'Dushlaine et al., 2011) and protein interaction networks (Yu et al., 2014) and in pathways defined by transcriptome sequencing (Zhao et al., 2014). Systematic hypothesis-free pathway-based methods are being applied to the PGC2 data set, but the absence of consensus methodology or established significance thresholds for including loci and the variable quality of pathway annotation have proved challenging.

With PGC2 analysis, a subset of confirmed genetic findings is supportive of existing hypotheses: for example, implicating dopaminergic and glutamatergic circuitry in SCZ risk. Although most identified risk variants have emerged from GWAS, evidence from CNV and sequencing studies also supports convergence of single-gene hits on a number of molecular mechanisms. Broadly these can be grouped as genes involved in:

1. Plasticity and synaptic function including neurotransmitter receptors and ion channels, secondary messenger systems, and scaffolding proteins
2. Immune function including, but not exclusive to, the MHC region
3. Chromatin remodeling and gene transcription functions.

The evidence and potential mechanisms are considered more fully subsequently. The current data capture a small fraction of total variance and are far from a complete representation of molecular etiology. In addition, these three groups are not independent and may represent many overlapping or independent molecular processes that impair the development and function of neural circuitry dynamically through development and the lifespan.

Dopaminergic Signaling

PGC2 analysis provided strong evidence that common variation of small effect at the *DRD2* receptor gene is associated with SCZ. Neuronal activity, including that regulated by the DRD2 receptor, is developmentally important in regulating axonal projections and establishing connectivity between different brain regions (Cazorla et al., 2014). Once established, dopaminergic neurotransmission is critical to normal motivation, cognition, and learning. Antagonism of the D2 receptor is a common feature of all current antipsychotic medications, and this is thought to act by lowering the firing and activity level of the mesolimbic dopamine pathway. This circuit is vulnerable to developmental insults with persistently increased striatal DA release, increased behavioral response to amphetamine, and a greater DA response to stress, all of which are reported in developmental models relevant to SCZ (eg, of perinatal hypoxia) (Howes & Murray, 2014). Increased presynaptic DA synthesis and release, possibly coupled to postsynaptic hypersensitivity to DA, remain the most likely explanations for this dysfunction.

Mesolimbic DA firing and circuit development does not occur in isolation but is a dynamic process with opposing or coordinated function across neurotransmitter systems. In the simplest model of this system, increasing the tone of glutamatergic neurons from the frontal cortex increases mesolimbic DA firing and this effect is dampened by the action of GABAergic interneuron tone. That this may be an important molecular risk mechanism is supported by the fact that genes involved in the formation of GABAergic interneurons (*IGSF9B*), the formation of glutamatergic and GABAergic presynapses (*NLGN4X*), and the function of GABAergic interneurons (*CLCN3*) represent the most likely candidate genes at three of the loci identified by the PGC2 GWAS. Questions remain as to the role and extent of genetic vulnerability present within this circuitry in patient populations and the molecular mechanisms underpinning circuit dysfunction.

Calcium Channel Genes

Voltage-gated calcium channels are formed as a complex of subunits (α_1, $\alpha_2\delta$, $\beta_{1\text{-}4}$, and γ) in which the α_1 subunit forms the ion-conducting pore and the other, associated subunits, have modulatory functions. Calcium channels can be divided into high voltage-activated (HVA), intermediate, and low voltage-activated (LVA) subgroups in which the most common form of the HVA channel is the long-lasting (L-type), and the LVA group represents transient opening (T-type) channels. With the NMDA receptor (NMDAR), they have an important role in synaptic function by regulating neuronal firing and neurotransmitter release and are also involved in gene transcription, neurite outgrowth, and activation of calcium-dependent enzymes (eg, calmodulin-dependent protein kinase II and protein kinase C) (Clapham, 2007). The diversity of channel subtypes allows for highly specialized functions in particular neuronal cell types. For example, T-type calcium channels are particularly important for repetitive firing of action potentials in thalamic neurons involved in regulating arousal and sleep/wake patterns.

The genes mostly strongly implicated in SCZ by GWAS data encode α_1 subunits of an L-type (*CACNA1C*, Cav1.2) and a T-type (*CACNA1I*, Cav3.3 subunit) calcium channel and an isoform of the β subunit (*CACNB2*) involved in trafficking α_1 subunits to the cell membrane and modulating their electrophysiological properties. Other implicated genes tether calcium channels at synapses (*RIMS1*), maintain low cytosolic calcium (*ATP2A2*), and regulate calmodulin availability and function in neurons (*NRGN* and *CAMKK2*) (Luo et al., 2014). The largest reported sequencing study identified significant enrichment for potentially damaging mutations among the 26 voltage-gated calcium ion channel genes, including two mutations in *CACNA1C* (Purcell et al., 2014). These data suggest a general role for calcium dysfunction in neurodevelopmental disorder pathogenesis rather than a highly specialized single etiological mechanism in SCZ. The role of calcium channel function in neurodevelopment has been poorly studied, but this is changing. Induced pluripotent stem cells (iPS) from patients with the severe neurodevelopmental disorder, Timothy syndrome, have facilitated studies of the developmental effects of *CACNA1C* mutations. Here, it has been reported that mutant Cav1.2 channels produce activity-dependent retraction of neuronal processes, which suggests that Cav1.2 channels are critical for the development of normal neuronal circuitry (Krey et al., 2013; Pasca et al., 2011). Furthermore, investigation of transcriptional networks in a similar *CACNA1C* iPS model is beginning to identify the molecular mechanisms involved (Tian et al., 2014). This approach represents a model for investigation of other calcium channel mutations in patient carriers to refine the etiology involved.

Glutamatergic Signaling

PGC2 analysis reported the association of genes involved in both the structure (*GRIA1*, *GRM3*, and *GRIN2A*) and function (*SRR*, *CLCN3*, and *SLC38A7*) of glutamatergic synapses (Harrison, 2014; Kantrowitz & Javitt, 2012). The role of the NMDAR is more than just its functional interactions with G protein–coupled receptors such as the DA receptors. Through its interactions with intracellular and extracellular proteins and other membrane receptors, it is critical to many CNS functions involved in diverse processes from neurodevelopment to synaptic plasticity and learning, and to cell survival (Glasgow, Siegler Retchless, & Johnson, 2015). In a gene-set analysis of CNV data, Kirov and colleagues examined whether 34 confirmed de novo risk CNV events were enriched for genes identified from proteomic data as corresponding to particular neuronal cellular components (Kirov et al., 2012). They found that compared with controls, de novo CNVs from patients were significantly enriched for the postsynaptic density (PSD) proteome. Trying to localize this enrichment, they found that the signal was largely explained by the NMDAR signaling complex and a smaller set of synaptic protein interactors of the activity-regulated, cytoskeleton-associated (ARC) protein. Looking further, these gene sets overlap but are partially independent of each other. Interestingly, the NMDAR and ARC gene sets showed significant enrichment for nonsynonymous or predicted damaging mutations in the two largest SCZ exome studies reported to date, albeit not in a pathway analysis of the common variant GWAS data. This raises the possibility that Arc represents one, but not the only, mechanism by which the integration of signaling cascades could be disrupted to affect brain function deleteriously in SCZ.

Activity-Regulated, Cytoskeleton-Associated Protein Regulation

ARC is found in the postsynaptic density of glutamatergic neurons in the hippocampus and neocortex. The protein is rapidly expressed as learning and information are encoded in neuronal circuitry and acts as an effector protein for multiple neuronal signaling pathways (Shepherd & Bear, 2011). Transcription of the ARC gene may be important in reshaping synapse function to maintain changes in long-term potentiation, long-term depression, and associative memory processes (Guzowski et al., 1999; Plath et al., 2006). This transcription is regulated by calcium influx through voltage-sensitive calcium channels (discussed subsequently) and the group I metabotropic glutamate receptors (mGluRs), but the mechanisms involved are poorly understood and represent a compelling target for future study in model systems informed by emerging genetic findings. A promising point of intersection emerging from the genetic findings is the gene *CYFIP1* at the 15q11.2 CNV risk locus. This is because CYFIP1 is a key regulator of F-actin assembly, which is an important downstream function in regulating dendrite morphology. CYFIP1 is also a protein-binding partner of FMRP and is a negative regulator of ARC translation (Park et al., 2008).

Fragile X Mental Retardation Protein Interactors

An expansion of CGG repeats at the gene encoding FMRP (*FMR1*) is causative in most cases of Fragile X syndrome and the functions of this RNA-binding protein have been subject to much scrutiny. It has been estimated that FMRP binds to about 4% of all mRNAs in the brain, representing a list of more than 800 targets (derived from mouse brain data), inevitably overlapping with PSD genes including members of the NMDAR and ARC complexes. FMRP generally suppresses the translation of target mRNAs and is implicated in neurodevelopment and adult neurogenesis. It is also required for normal synaptic function and plasticity through mGluR1/5 signaling. In addition, it is emerging that FMRP is involved in upregulating protein synthesis and regulating of protein–protein interactions and non-mGluR1/5–related mechanisms of synaptic plasticity. To disambiguate the functions of FMRP, specific subpopulations of targets have been identified with distinct spatiotemporal expression and functional biases (Steinberg & Webber, 2013). These data suggest that the two largest subpopulations cluster separately into modules: (1) upregulated during fetal development and (2) in adolescence/adulthood, respectively. Targets of FMRP are enriched for de novo mutations in autism, and a model has been proposed for a "single-hit" etiology for mutations of module 1 genes, whereas a "multiple-hit" etiology is implicated for module 2 genes. Both of the largest SCZ sequencing studies published to date found evidence for enrichment of non-synonymous mutations (de novo mutations and disruptive singletons) in FMRP targets (Fromer et al., 2014; Purcell et al., 2014). In both studies, this enrichment was still evident when overlapping PSD genes were removed, which indicates that the role of FMRP targets is partially independent of PSD function. Pathway analysis of the PGC2 data set identified significant enrichment for common variant association in the FMRP set using two different methodological approaches: ALIGATOR ($P = 0.0066$) and INRICH ($P = 5 \times 10^{-5}$).

Immune Function

It has long been suspected that the involvement of immune dysfunction in SCZ is based on evidence of increased markers of infection (eg, cytokines) and association with specific pathogens in SCZ and the more identification of phenocopies caused by autoimmunity (eg, NMDAR autoantibody encephalitis) (Carter, Bullmore, & Harrison, 2014). The most significant common variation signal for SCZ is at the MHC locus, but there is also evidence of involvement of immune genes outside this locus. One analysis performed by the PGC2 group was to determine whether SCZ-associated common variants are concentrated in regulatory elements marked as activating gene expression in particular tissues or cell lines. Importantly, though largely as predicted, associations were enriched in these regulatory elements in various brain tissues and in genes showing high expression in neurons/interneurons. A more novel and potentially important finding was that they were also enriched in these regulatory elements in the immune system, particularly B-lymphocyte cell lineages. This provides strong support for a role of immune dysfunction at a time when there is a growing appreciation that neurons and microglia express MHC molecules and immune-related molecules have an important role in neurodevelopment and activity-dependent synaptic plasticity. At a mechanistic level, the pathways involved have not been elucidated, but this may be mediated at least in part by NMDAR signaling, which could be experimentally modeled (Fourgeaud et al., 2010; Nelson et al., 2013; Shatz, 2009).

Chromatin Regulators

Chromatin regulators provide both the stability and flexibility required for normal brain development. They provide the heritable states of gene expression necessary to allow the programmed development of neural circuitry and contribute to dynamic changes in gene expression required for plasticity and learning. This is achieved through chromatin remodeling or modification. The former is an adenosine triphosphate–dependent process that physically alters chromatin by moving nucleosomes or exchanging them into or out of DNA. The latter are enzymes that alter the tails of histones projecting from nucleosomes, making DNA more or less accessible to regulatory mechanisms including chromatin remodelers or other proteins. In SCZ, predicted LOF mutations have been reported for both chromatin remodelers (eg, *CHD8*) and modifiers of histone function (eg, *MLL2*, *HUWE*)/DNA methylation (eg, *MECP2*) in genes previously implicated in severe Mendelian neurodevelopmental disorders (McCarthy et al., 2014). Consequently some of these mechanisms have been reasonably well investigated. Two other studies have identified enrichment of mutations implicating other chromatin remodelers (Gilman et al., 2012; Takata et al., 2014). Much is still to be learned about the role of chromatin regulators in facilitating the development of stable neural circuitry and the flexibility implicit in synaptic plasticity. For example, it would be interesting to know whether modifiers converge on regulating expression of the same signaling pathways or are themselves regulated by the function of other proteins (eg, FMRP).

FUTURE DIRECTIONS

Advances in genomics promise models of SCZ rooted in molecular etiology. However, SCZ is a clinical syndrome. Bridging this gap between descriptive phenomenology and molecular etiology is a major, but not insurmountable, challenge. Conceptual mechanisms such as "neural circuit dysfunction" have meaning in the real world and can be usefully assayed. Reduced beta and gamma oscillations and synchronization have been reported in patients with SCZ from electroencephalography or magnetoencephalography studies (Uhlhaas & Singer, 2010). These assays translate into animal model research more directly than do human symptoms or behaviors and can be used to refine different etiological mechanisms; they are also informative for drug trials. Similarly, genes or mechanisms relevant to synaptic plasticity can be assayed at the level of cognitive function in patients with risk mutations and in mouse models (Nithianantharajah et al., 2013). Advancing methods for electrophysiological measurement and imaging of circuits are discussed more fully in chapters "Resting-State Functional MRI: a Novel Tool for Understanding Brain Networks in Neuropsychiatric Disorders" and "Neuroimaging Advances in Alzheimer Disease," respectively.

Some elements of the syndrome may require other approaches. The Research Domain Criteria framework may prove useful in integrating different levels of data to identify specific pathophysiological mechanisms in human studies (Insel, 2014); this is discussed further in chapter "The Emergence and Underlying Neurobiology of Psychosis." For example, the construct of "agency" may be important in investigating patients with delusions of control, because this is known to be correlated with function in the right parietal, right insula, and right inferior frontal areas. Coupling clinical symptoms, structural/functional neuroimaging, and molecular pathway information may be powerful in dissecting such etiology.

As more is understood about the genetic architecture of SCZ, it will be important to incorporate information from other sources (eg, methlome or transcriptome) to define and refine the mechanisms involved (Kumar et al., 2014). Cataloging risk variation is a beginning. We need to understand how genes are expressed and how this is functionally orchestrated to cause disease. Having an understanding of dynamic epigenetic regulation of gene expression would be incredibly informative, and large-scale epigenomics studies are beginning to happen (Dempster et al., 2013). Transcriptome databases of the human brain have become more sophisticated, making it possible to examine how potential candidate genes cluster and are coexpressed spatially and temporally during neurodevelopment. This could provide a deeper understanding of the currently known genes in addressing whether SCZ patients can be differentiated or grouped based on early developmental or later synaptic etiological mechanisms. Similar questions could be asked at the level of the molecular risk pathways, as described previously in the FMRP modular analysis of ASD (Steinberg & Webber, 2013). In one of the first such reports in SCZ, identifying a modest enrichment of DNMs in patients with SCZ compared with control subjects, Gulsuner and colleagues reported that the DNMs in patients were significantly more likely to cluster in protein–protein interaction networks. The DNMs in patients were more likely to be coexpressed in fetal frontal cortex (Gulsuner et al., 2013). The overlapping network of interacting and coexpressed genes had known functions including neuronal migration, synaptic transmission, signaling, transcriptional regulation, and transport.

In future analyses it will be important to understand overlapping etiology with other developmental disorders. Pathway analysis across the PGC SCZ, bipolar disorder, and major depression phenotypes found strong evidence of a common risk pathway involving histone methylation, but also that synapse and downstream signaling mechanisms had a more prominent role in SCZ than in the other two disorders. Analysis of coexpression relationships between the genes in the histone methylation pathway found that these genes had on average threefold higher expression during early prenatal development, which suggests a particular role in neuronal differentiation and cell-fate commitment. There may be common etiological mechanisms that put the brain at risk of neurodevelopmental disorder, but later molecular events may determine the development of SCZ.

Through these approaches we can begin to learn when, where, and how aberrant network function happens and model how this contributes to disease pathogenesis. This work is happening against the backdrop of efforts to understand the genetic mechanisms that shape the interaction between cortical and subcortical brain regions. Improved methods of defining molecular circuitry (eg, with CLARITY) (Chung & Deisseroth, 2013) coupled with new methods of recording and manipulating neural activity, are revolutionizing understanding of the molecular mechanisms of normal circuit development and function (Akil et al., 2010). Optogenetics methods have been applied to study mesolimbic dopamine circuitry during behavioral tasks in rodents and would add substantial value to studies of mutant models, but they could also be applied to study signal transduction in iPS models (Saddoris et al., 2014; Steinberg et al., 2015; Yan et al., 2014). The intersection of genomics and new experimental methods offers the promise that this emerging biology of SCZ can lead us to more informed diagnostics, treatment, or prevention for future generations.

REFERENCES

Akil, H., et al. (2010). Medicine. The future of psychiatric research: genomes and neural circuits. *Science, 327*(5973), 1580–1581.

Allen, A. S., et al. (2013). De novo mutations in epileptic encephalopathies. *Nature, 501*(7466), 217–221.

Andreasen, N. C., et al. (2005). Remission in schizophrenia: proposed criteria and rationale for consensus. *American Journal of Psychiatry, 162*(3), 441–449.

Bahlo, M., et al. (2014). Using familial information for variant filtering in high-throughput sequencing studies. *Human Genetics, 133*(11), 1331–1341.

Bernardinelli, Y., Nikonenko, I., & Muller, D. (2014). Structural plasticity: mechanisms and contribution to developmental psychiatric disorders. *Frontiers in Neuroanatomy, 8*, 123.

Bleuler, M. (1980). Chronicity or chronification among our schizophrenics. *Schweizer Archiv fuer Neurologie Neurochirurgie und Psychiatrie, 126*(2), 245–253.

Borglum, A. D., et al. (2014). Genome-wide study of association and interaction with maternal cytomegalovirus infection suggests new schizophrenia loci. *Molecular Psychiatry, 19*(3), 325–333.

Cardno, A. G., & Gottesman, I. I. (2000). Twin studies of schizophrenia: from bow-and-arrow concordances to star wars Mx and functional genomics. *American Journal of Medical Genetics, 97*(1), 12–17.

Carter, C. S., Bullmore, E. T., & Harrison, P. (2014). Is there a flame in the brain in psychosis? *Biological Psychiatry, 75*(4), 258–259.

Cazorla, M., et al. (2014). Dopamine D2 receptors regulate the anatomical and functional balance of basal ganglia circuitry. *Neuron, 81*(1), 153–164.

Chatterjee, N., et al. (2013). Projecting the performance of risk prediction based on polygenic analyses of genome-wide association studies. *Nature Genetics, 45*(4), 400–405, 405e1–405e3.

Chung, K., & Deisseroth, K. (2013). CLARITY for mapping the nervous system. *Nature Methods, 10*(6), 508–513.

Clapham, D. E. (2007). Calcium signaling. *Cell, 131*(6), 1047–1058.

Clarke, M. C., et al. (2012). Evidence for shared susceptibility to epilepsy and psychosis: a population-based family study. *Biological Psychiatry, 71*(9), 836–839.

Collins, A. L., et al. (2014). Transcriptional targets of the schizophrenia risk gene MIR137. *Translational Psychiatry, 4*, e404.

Corvin, A., Craddock, N., & Sullivan, P. F. (2010). Genome-wide association studies: a primer. *Psychologie Medicale, 40*(7), 1063–1077.

Corvin, A., & Morris, D. W. (2014). Genome-wide association studies: findings at the major histocompatibility complex locus in psychosis. *Biological Psychiatry, 75*(4), 276–283.

Cross-Disorder Group of the Psychiatric Genomics Consortium. (2013). Identification of risk loci with shared effects on five major psychiatric disorders: a genome-wide analysis. *Lancet, 381*(9875), 1371–1379.

Darnell, J. C., et al. (2011). FMRP stalls ribosomal translocation on mRNAs linked to synaptic function and autism. *Cell, 146*(2), 247–261.

De Rubeis, S., et al. (2014). Synaptic, transcriptional and chromatin genes disrupted in autism. *Nature, 515*(7526), 209–215.

Dempster, E., et al. (2013). Epigenetic studies of schizophrenia: progress, predicaments, and promises for the future. *Schizophrenia Bulletin, 39*(1), 11–16.

Duncan, L. E., & Keller, M. C. (2011). A critical review of the first 10 years of candidate gene-by-environment interaction research in psychiatry. *American Journal of Psychiatry, 168*(10), 1041–1049.

Fourgeaud, L., et al. (2010). MHC class I modulates NMDA receptor function and AMPA receptor trafficking. *Proceedings of the National Academy of Sciences of the United States of America, 107*(51), 22278–22283.

Friston, K. J., & Frith, C. D. (1995). Schizophrenia: a disconnection syndrome? *Journal of Clinical Neuroscience, 3*(2), 89–97.

Fromer, M., et al. (2014). De novo mutations in schizophrenia implicate synaptic networks. *Nature, 506*(7487), 179–184.

Georgieva, L., et al. (2014). De novo CNVs in bipolar affective disorder and schizophrenia. *Human Molecular Genetics, 23*(24), 6677–6683.

Gilman, S. R., et al. (2012). Diverse types of genetic variation converge on functional gene networks involved in schizophrenia. *Nature Neuroscience, 15*(12), 1723–1728.

Glasgow, N. G., Siegler Retchless, B., & Johnson, J. W. (2015). Molecular bases of NMDA receptor subtype-dependent properties. *Journal of Physiology, 593*(1), 83–95.

Gulsuner, S., et al. (2013). Spatial and temporal mapping of de novo mutations in schizophrenia to a fetal prefrontal cortical network. *Cell, 154*(3), 518–529.

Guzowski, J. F., et al. (1999). Environment-specific expression of the immediate-early gene Arc in hippocampal neuronal ensembles. *Nature Neuroscience, 2*(12), 1120–1124.

Harrison, P. J. (2014). Recent genetic findings in schizophrenia and their therapeutic relevance. *Journal of Psychopharmacology, 29*(2), 85–96.

Hegarty, J. D., et al. (1994). One hundred years of schizophrenia: a meta-analysis of the outcome literature. *American Journal of Psychiatry, 151*(10), 1409–1416.

Helsmoortel, C., et al. (2014). A SWI/SNF-related autism syndrome caused by de novo mutations in ADNP. *Nature Genetics, 46*(4), 380–384.

Hemmings, C. P. (2006). Schizophrenia spectrum disorders in people with intellectual disabilities. *Current Opinion in Psychiatry, 19*(5), 470–474.

Hoischen, A., Krumm, N., & Eichler, E. E. (2014). Prioritization of neurodevelopmental disease genes by discovery of new mutations. *Nature Neuroscience, 17*(6), 764–772.

Howes, O. D., & Murray, R. M. (2014). Schizophrenia: an integrated sociodevelopmental-cognitive model. *Lancet, 383*(9929), 1677–1687.

Hu, W. F., Chahrour, M. H., & Walsh, C. A. (2014). The diverse genetic landscape of neurodevelopmental disorders. *Annu Rev Genomics Hum Genet, 15*, 195–213.

Iafrate, A. J., et al. (2004). Detection of large-scale variation in the human genome. *Nature Genetics, 36*(9), 949–951.

Ingraham, L. J., & Kety, S. S. (2000). Adoption studies of schizophrenia. *American Journal of Medical Genetics, 97*(1), 18–22.

Insel, T. R. (2014). The NIMH research domain criteria (RDoC) project: precision medicine for psychiatry. *American Journal of Psychiatry, 171*(4), 395–397.

International Schizophrenia Consortium. (2008). Rare chromosomal deletions and duplications increase risk of schizophrenia. *Nature, 455*(7210), 237–241.

Iossifov, I., et al. (2014). The contribution of de novo coding mutations to autism spectrum disorder. *Nature, 515*(7526), 216–221.

Iyegbe, C., et al. (2014). The emerging molecular architecture of schizophrenia, polygenic risk scores and the clinical implications for GxE research. *Social Psychiatry and Psychiatric Epidemiology, 49*(2), 169–182.

Jamuar, S. S., & Walsh, C. A. (2014). Somatic mutations in cerebral cortical malformations. *New England Journal of Medicine, 371*(21), 2038.

Janssens, A. C., et al. (2007). The impact of genotype frequencies on the clinical validity of genomic profiling for predicting common chronic diseases. *Genetics in Medicine, 9*(8), 528–535.

Kantrowitz, J., & Javitt, D. C. (2012). Glutamatergic transmission in schizophrenia: from basic research to clinical practice. *Current Opinion in Psychiatry, 25*(2), 96–102.

Karayiorgou, M., et al. (2012). The best of times, the worst of times for psychiatric disease. *Nature Neuroscience, 15*(6), 811–812.

Kendler, K. S. (2014). A joint history of the nature of genetic variation and the nature of schizophrenia. *Molecular Psychiatry, 20*, 77–83.

Kendler, K. S., et al. (1993a). The Roscommon Family Study. III. Schizophrenia-related personality disorders in relatives. *Archives of General Psychiatry, 50*(10), 781–788.

Kendler, K. S., et al. (1993b). The Roscommon Family Study. I. Methods, diagnosis of probands, and risk of schizophrenia in relatives. *Archives of General Psychiatry, 50*(7), 527–540.

Kendler, K. S., et al. (1993c). The Roscommon Family Study. IV. Affective illness, anxiety disorders, and alcoholism in relatives. *Archives of General Psychiatry, 50*(12), 952–960.

Kirov, G., et al. (2012). De novo CNV analysis implicates specific abnormalities of postsynaptic signalling complexes in the pathogenesis of schizophrenia. *Molecular Psychiatry, 17*(2), 142–153.

Kirov, G., et al. (2014). The penetrance of copy number variations for schizophrenia and developmental delay. *Biological Psychiatry, 75*(5), 378–385.

Kong, A., et al. (2012). Rate of de novo mutations and the importance of father's age to disease risk. *Nature, 488*(7412), 471–475.

Krey, J. F., et al. (2013). Timothy syndrome is associated with activity-dependent dendritic retraction in rodent and human neurons. *Nature Neuroscience, 16*(2), 201–209.

Kumar, G., et al. (2014). Refinement of schizophrenia GWAS loci using methylome-wide association data. *Human Genetics, 134*(1), 77–87.

Larsson, H. J., et al. (2005). Risk factors for autism: perinatal factors, parental psychiatric history, and socioeconomic status. *American Journal of Epidemiology, 161*(10), 916–925. discussion 926–928.

Lee, S. H., et al. (2012). Estimating the proportion of variation in susceptibility to schizophrenia captured by common SNPs. *Nature Genetics, 44*(3), 247–250.

Lee, S. H., et al. (2013). Genetic relationship between five psychiatric disorders estimated from genome-wide SNPs. *Nature Genetics, 45*(9), 984–994.

Leucht, S., et al. (2013). Comparative efficacy and tolerability of 15 antipsychotic drugs in schizophrenia: a multiple-treatments meta-analysis. *Lancet, 382*(9896), 951–962.

Lichtenstein, P., et al. (2009). Common genetic determinants of schizophrenia and bipolar disorder in Swedish families: a population-based study. *Lancet, 373*(9659), 234–239.

de Ligt, J., et al. (2012). Diagnostic exome sequencing in persons with severe intellectual disability. *New England Journal of Medicine, 367*(20), 1921–1929.

Lips, E. S., et al. (2012). Functional gene group analysis identifies synaptic gene groups as risk factor for schizophrenia. *Molecular Psychiatry, 17*(10), 996–1006.

Liu, P., et al. (2012). Mechanisms for recurrent and complex human genomic rearrangements. *Current Opinion in Genetics and Development, 22*(3), 211–220.

Luo, X. J., et al. (2014). Convergent lines of evidence support CAMKK2 as a schizophrenia susceptibility gene. *Molecular Psychiatry, 19*(7), 774–783.

MacArthur, D. G., et al. (2012). A systematic survey of loss-of-function variants in human protein-coding genes. *Science, 335*(6070), 823–828.

Maier, W., et al. (1993). Continuity and discontinuity of affective disorders and schizophrenia. Results of a controlled family study. *Archives of General Psychiatry, 50*(11), 871–883.

Malenka, R. C., & Bear, M. F. (2004). LTP and LTD: an embarrassment of riches. *Neuron, 44*(1), 5–21.

Malhotra, D., & Sebat, J. (2012). CNVs: harbingers of a rare variant revolution in psychiatric genetics. *Cell, 148*(6), 1223–1241.

Maurano, M. T., et al. (2012). Systematic localization of common disease-associated variation in regulatory DNA. *Science, 337*(6099), 1190–1195.

McCarthy, S. E., et al. (2014). De novo mutations in schizophrenia implicate chromatin remodeling and support a genetic overlap with autism and intellectual disability. *Molecular Psychiatry, 19*(6), 652–658.

McCarthy, S. E., McCombie, W. R., & Corvin, A. (2014). Unlocking the treasure trove: from genes to schizophrenia biology. *Schizophrenia Bulletin, 40*(3), 492–496.

McGrath, J. J., et al. (2013). Where GWAS and epidemiology meet: opportunities for the simultaneous study of genetic and environmental risk factors in schizophrenia. *Schizophrenia Bulletin, 39*(5), 955–959.

Modinos, G., et al. (2013). Molecular genetic gene-environment studies using candidate genes in schizophrenia: a systematic review. *Schizophrenia Research, 150*(2–3), 356–365.

Morgan, V. A., et al. (2008). Intellectual disability co-occurring with schizophrenia and other psychiatric illness: population-based study. *British Journal of Psychiatry, 193*(5), 364–372.

Morris, D. W., et al. (2014). An inherited duplication at the gene p21 Protein-Activated Kinase 7 (PAK7) is a risk factor for psychosis. *Human Molecular Genetics, 23*(12), 3316–3326.

Neale, B. M., et al. (2012). Patterns and rates of exonic de novo mutations in autism spectrum disorders. *Nature, 485*(7397), 242–245.

Need, A. C., et al. (2012). Exome sequencing followed by large-scale genotyping suggests a limited role for moderately rare risk factors of strong effect in schizophrenia. *American Journal of Human Genetics, 91*(2), 303–312.

Nelson, P. A., et al. (2013). MHC class I immune proteins are critical for hippocampus-dependent memory and gate NMDAR-dependent hippocampal long-term depression. *Learning and Memory, 20*(9), 505–517.

Ng, M. Y., et al. (2009). Meta-analysis of 32 genome-wide linkage studies of schizophrenia. *Molecular Psychiatry, 14*(8), 774–785.

Nithianantharajah, J., et al. (2013). Synaptic scaffold evolution generated components of vertebrate cognitive complexity. *Nature Neuroscience, 16*(1), 16–24.

O'Donovan, M. C., et al. (2008). Identification of loci associated with schizophrenia by genome-wide association and follow-up. *Nature Genetics, 40*(9), 1053–1055.

O'Dushlaine, C., et al. (2011). Molecular pathways involved in neuronal cell adhesion and membrane scaffolding contribute to schizophrenia and bipolar disorder susceptibility. *Molecular Psychiatry, 16*(3), 286–292.

O'Roak, B. J., et al. (2012). Multiplex targeted sequencing identifies recurrently mutated genes in autism spectrum disorders. *Science, 338*(6114), 1619–1622.

van Os, J., et al. (2009). A systematic review and meta-analysis of the psychosis continuum: evidence for a psychosis proneness-persistence-impairment model of psychotic disorder. *Psychologie Medicale, 39*(2), 179–195.

van Os, J., et al. (2014). Identifying gene-environment interactions in schizophrenia: contemporary challenges for integrated, large-scale investigations. *Schizophrenia Bulletin, 40*(4), 729–736.

Park, S., et al. (2008). Elongation factor 2 and fragile X mental retardation protein control the dynamic translation of Arc/Arg3.1 essential for mGluR-LTD. *Neuron, 59*(1), 70–83.

Pasca, S. P., et al. (2011). Using iPSC-derived neurons to uncover cellular phenotypes associated with Timothy syndrome. *Nature Medicine, 17*(12), 1657–1662.

Petrovski, S., et al. (2013). Genic intolerance to functional variation and the interpretation of personal genomes. *PLoS Genetics, 9*(8), e1003709.

Plath, N., et al. (2006). Arc/Arg3.1 is essential for the consolidation of synaptic plasticity and memories. *Neuron, 52*(3), 437–444.

Purcell, S. M., et al. (2009). Common polygenic variation contributes to risk of schizophrenia and bipolar disorder. *Nature, 460*(7256), 748–752.

Purcell, S. M., et al. (2014). A polygenic burden of rare disruptive mutations in schizophrenia. *Nature, 506*(7487), 185–190.

Rauch, A., et al. (2012). Range of genetic mutations associated with severe non-syndromic sporadic intellectual disability: an exome sequencing study. *Lancet, 380*(9854), 1674–1682.

Rees, E., et al. (2014a). Evidence that duplications of 22q11.2 protect against schizophrenia. *Molecular Psychiatry, 19*(1), 37–40.

Rees, E., et al. (2014b). CNV analysis in a large schizophrenia sample implicates deletions at 16p12.1 and SLC1A1 and duplications at 1p36.33 and CGNL1. *Human Molecular Genetics, 23*(6), 1669–1676.

Richards, A. L., et al. (2012). Schizophrenia susceptibility alleles are enriched for alleles that affect gene expression in adult human brain. *Molecular Psychiatry, 17*(2), 193–201.

Ripke, S., et al. (2011). Genome-wide association study identifies five new schizophrenia loci. *Nature Genetics, 43*(10), 969–976.

Ripke, S., et al. (2013). Genome-wide association analysis identifies 13 new risk loci for schizophrenia. *Nature Genetics, 45*(10), 1150–1159.

Rujescu, D., et al. (2009). Disruption of the neurexin 1 gene is associated with schizophrenia. *Human Molecular Genetics, 18*(5), 988–996.

Saddoris, M. P., et al. (2014). Mesolimbic dopamine dynamically tracks, and is causally linked to, discrete aspects of value-based decision making. *Biological Psychiatry, 77*(10), 903–911.

Samocha, K. E., et al. (2014). A framework for the interpretation of de novo mutation in human disease. *Nature Genetics, 46*(9), 944–950.

Schizophrenia Working Group of the Psychiatric Genomics Consortium. (2014). Biological insights from 108 schizophrenia-associated genetic loci. *Nature, 511*(7510), 421–427.

Schork, A. J., et al. (2013). All SNPs are not created equal: genome-wide association studies reveal a consistent pattern of enrichment among functionally annotated SNPs. *PLoS Genetics, 9*(4), e1003449.

Schwartz, T. L., Sachdeva, S., & Stahl, S. M. (2012). Glutamate neurocircuitry: theoretical underpinnings in schizophrenia. *Frontiers in Pharmacology, 3*, 195.

Sebat, J., et al. (2004). Large-scale copy number polymorphism in the human genome. *Science, 305*(5683), 525–528.

Shatz, C. J. (2009). MHC class I: an unexpected role in neuronal plasticity. *Neuron, 64*(1), 40–45.

Shepherd, J. D., & Bear, M. F. (2011). New views of Arc, a master regulator of synaptic plasticity. *Nature Neuroscience, 14*(3), 279–284.

Shi, J., et al. (2009). Common variants on chromosome 6p22.1 are associated with schizophrenia. *Nature, 460*(7256), 753–757.

Smemo, S., et al. (2014). Obesity-associated variants within FTO form long-range functional connections with IRX3. *Nature, 507*(7492), 371–375.

Stefansson, H., et al. (2008). Large recurrent microdeletions associated with schizophrenia. *Nature, 455*(7210), 232–236.

Stefansson, H., et al. (2009). Common variants conferring risk of schizophrenia. *Nature, 460*(7256), 744–747.

Stefansson, H., et al. (2014). CNVs conferring risk of autism or schizophrenia affect cognition in controls. *Nature, 505*(7483), 361–366.

Steinberg, E. E., et al. (2015). Illuminating circuitry relevant to psychiatric disorders with optogenetics. *Current Opinion in Neurobiology, 30c*, 9–16.

Steinberg, J., & Webber, C. (2013). The roles of FMRP-regulated genes in autism spectrum disorder: single- and multiple-hit genetic etiologies. *American Journal of Human Genetics, 93*(5), 825–839.

Stoll, G., et al. (2013). Deletion of TOP3beta, a component of FMRP-containing mRNPs, contributes to neurodevelopmental disorders. *Nature Neuroscience, 16*(9), 1228–1237.

Strauss, J. S. (1969). Hallucinations and delusions as points on continua function. Rating scale evidence. *Archives of General Psychiatry, 21*(5), 581–586.

Sullivan, P. F., Daly, M. J., & O'Donovan, M. (2012). Genetic architectures of psychiatric disorders: the emerging picture and its implications. *Nature Reviews Genetics, 13*(8), 537–551.

Sullivan, P. F., et al. (2012). Family history of schizophrenia and bipolar disorder as risk factors for autism. *Archives of General Psychiatry, 69*(11), 1099–1103.

Sullivan, P. F., Kendler, K. S., & Neale, M. C. (2003). Schizophrenia as a complex trait: evidence from a meta-analysis of twin studies. *Archives of General Psychiatry, 60*(12), 1187–1192.

Takata, A., et al. (2014). Loss-of-function variants in schizophrenia risk and SETD1A as a candidate susceptibility gene. *Neuron, 82*(4), 773–780.

Tandon, R., Keshavan, M. S., & Nasrallah, H. A. (2008). Schizophrenia, "Just the Facts": what we know in 2008 part 1: overview. *Schizophrenia Research, 100*(1–3), 4–19.

Tian, Y., et al. (2014). Alteration in basal and depolarization induced transcriptional network in iPSC derived neurons from Timothy syndrome. *Genome Medicine, 6*(10), 75.

Tiihonen, J., et al. (2009). 11-year follow-up of mortality in patients with schizophrenia: a population-based cohort study (FIN11 study). *Lancet, 374*(9690), 620–627.

Uhlhaas, P. J., & Singer, W. (2010). Abnormal neural oscillations and synchrony in schizophrenia. *Nature Reviews Neuroscience, 11*(2), 100–113.

Vacic, V., et al. (2011). Duplications of the neuropeptide receptor gene VIPR2 confer significant risk for schizophrenia. *Nature, 471*(7339), 499–503.

Verkerk, A. J., et al. (1991). Identification of a gene (FMR-1) containing a CGG repeat coincident with a breakpoint cluster region exhibiting length variation in fragile X syndrome. *Cell, 65*(5), 905–914.

Walsh, T., et al. (2008). Rare structural variants disrupt multiple genes in neurodevelopmental pathways in schizophrenia. *Science, 320*(5875), 539–543.

Wei, W. H., Hemani, G., & Haley, C. S. (2014). Detecting epistasis in human complex traits. *Nature Reviews Genetics, 15*(11), 722–733.

Weng, L., et al. (2011). SNP-based pathway enrichment analysis for genome-wide association studies. *BMC Bioinformatics, 12*, 99.

World Health Organization. (1992). *The ICD-10 Classification of Mental and Behavioural Disorders*. Geneva: World Health Organization.

Wray, N. R., Goddard, M. E., & Visscher, P. M. (2007). Prediction of individual genetic risk to disease from genome-wide association studies. *Genome Research, 17*(10), 1520–1528.

Xu, B., et al. (2012). De novo gene mutations highlight patterns of genetic and neural complexity in schizophrenia. *Nature Genetics, 44*(12), 1365–1369.

Yan, Z., et al. (2014). Human pluripotent stem cell tools for cardiac optogenetics. *Conference Proceedings: IEEE Engineering in Medicine and Biology Society, 2014*, 6171–6174.

Yu, H., et al. (2014). Protein-interaction-network-based analysis for genome-wide association analysis of schizophrenia in Han Chinese population. *Journal of Psychiatric Research, 50*, 73–78.

Zaidi, S., et al. (2013). De novo mutations in histone-modifying genes in congenital heart disease. *Nature, 498*(7453), 220–223.

Zhao, Z., et al. (2014). Transcriptome sequencing and genome-wide association analyses reveal lysosomal function and actin cytoskeleton remodeling in schizophrenia and bipolar disorder. *Molecular Psychiatry, 20*(5), 563–572.

Zuk, O., et al. (2014). Searching for missing heritability: designing rare variant association studies. *Proceedings of the National Academy of Sciences of the United States of America, 111*(4), E455–E464.

Chapter 28

The Emergence and Underlying Neurobiology of Psychosis

Raquel E. Gur and Ruben C. Gur
University of Pennsylvania, Philadelphia, PA, United States

INTRODUCTION

A successful effort in elucidating the pathophysiology of a disorder culminates in identifying genes that affect specific pathways producing manifested symptomatology. Molecular biology and genomics are transforming medicine from symptom-based disease models to "precision medicine" in which mechanistic elucidation of molecular pathways permits early identification, prevention, and intervention. This transformation is fueled by large genomic and medical databases with information on "biomarkers" of disease vulnerability.

Psychiatry and behavioral neuroscience disciplines, whose organ is the nervous system and whose outcome measures are "behavior," are hampered by limited information on behavioral parameters including symptoms and neurocognitive performance. The paucity of such data in electronic medical databases is especially apparent for age ranges that span adolescence, a critical period for brain maturation and the emergence of psychopathology including psychosis. As highlighted in other chapters in the book, progress in genomic research in schizophrenia has been achieved with integration of large samples. Thus far, schizophrenia genomic research has capitalized predominantly on established phenotypic diagnosis that is dichotomous: namely, participants met clinical categorical diagnostic criteria requiring a constellation of significant symptoms present for a specified duration and affecting functioning. The identification of common variants across several disorders such as bipolar, autism spectrum disorders, and intellectual disability exposes the limitation of the categorical approach. If different disorders share genomic variance, the mechanistic explanation requires genomic parameters to be linked to behavioral measures related to dysfunction in brain circuitries that span disorders. Thus, the field is turning its attention to examining brain circuitry and quantitative dimensional measures of liability rather than arbitrarily thresholded disease diagnostic classifications. A dimensional approach is more likely to reveal links between brain systems and adaptive behavioral domains that are genetically modulated ones. It also makes better methodological sense because of the improved power of continuous measures (Royston, Altman, & Sauerbrei, 2006). Such measures are proposed in the Research Domain Criteria approach (Insel et al., 2010).

The investigation of schizophrenia as a brain disorder began with studies that evaluated patients who had the disease for a long time, establishing neurocognitive deficits associated with abnormalities in parameters of brain structure and function. With the increased availability of noninvasive neuroimaging methodology and neurocognitive testing, studies of younger patients became feasible and this work indicated that the neurocognitive deficits and brain abnormalities are already evident in first-episode individuals. This work has produced two shifts of emphasis in the study of psychosis that can advance understanding of the underlying neurobiology. The first is a shift to study earlier phases of psychosis, the clinical risk stage; the second is a shift from dichotomous symptom categorization to quantitative dimensional endophenotypes. In some ways the first shift necessitated the second, because at initial symptom presentation the intensity may hover below the clinical threshold, and boundaries between developmentally appropriate upheavals and worrisome signs of psychosis are more challenging to discern clinically. This development will benefit genomic investigations because such vulnerability indices are heritable brain and neurobehavioral parameters that are trait markers related to illness (Braff, Freedman, Schork, & Gottesman, 2007; Glahn et al., 2014).

Genomics, Circuits, and Pathways in Clinical Neuropsychiatry. http://dx.doi.org/10.1016/B978-0-12-800105-9.00028-7

In this chapter, we will begin by highlighting the study of the emergence of psychosis. We will then summarize progress in research efforts that examine neurocognitive and neuroimaging measures in individuals at clinical risk for psychosis and the underlying neurobiological pathways. We will conclude by considering how the study of clinical risk may be integrated with progress in genomics.

THE CLINICAL EMERGENCE OF PSYCHOSIS

Similar to other fields in medicine that emphasize the identification of disease liability, there has been an increased effort to identify individuals at risk for psychosis (Cannon et al., 2008; Fusar-Poli, Bonoldi, et al., 2012; Goulding et al., 2013; Gur, 2013; Kelleher et al., 2012; Miller et al., 2003; Ruhrmann et al., 2010). The literature on early antecedents of schizophrenia documented premorbid features in retrospective studies and prospective longitudinal studies (Addington & Heinssen, 2012; Walker et al., 2013). It is hoped that the increased emphasis on identifying people at clinical high risk and examining endophenotypic measures will contribute to improved detection, inform on the underlying neurobiology, and facilitate early intervention (Heinssen & Insel, 2014; McGorry, 2011).

Psychosis is a process that commonly emerges in adolescence and early adulthood, a pivotal period in brain maturation characterized predominantly by axonal myelination and neuronal pruning (Giedd et al., 1996; Huttenlocher, de Courten, Garey, & Van Der Loos, 1982; Jernigan & Tallal, 1990; Yakovlev & Lecours, 1967). The traditional clinical diagnostic approach relies solely on a constellation of symptoms and evaluates the duration, severity, and impact of aberrations in behavior that disrupt functioning (American Psychiatric Association, 2013). This system of classification is not "brain-based" and therefore is unable to contribute to elucidating how the effects of myelination and pruning on behavior relate to the emergence of symptoms. By the time diagnostic criteria are met, the underlying process has likely been in progress and irreversible loss of function has already occurred.

Early detection of psychotic symptoms is commonly challenging because early presentation includes subtle changes in several domains (Miller et al., 2003) and frequently such changes are attributed to transitions of the developmental stage: adolescence and young adulthood. A person may seem perplexed, with decreased concentration or motivation and difficulties performing in school or at work. A mood component typically is evident at the early stage, manifested as decreased social engagement and diminished interest in previous activities and hobbies. Anxiety, misperception, and suspiciousness are associated with increased guardedness. Thus, the core features of psychosis—delusions, hallucinations, and disorganized thinking—are present but in a mild subthreshold form. They may increase in frequency and severity, causing distress and impairment, or in some individuals they may stay at the subthreshold level or diminish and even abate (Fusar-Poli, Bonoldi, et al., 2012).

In the psychosis continuum, this clinical risk stage or prodromal phase has become incorporated into the *Diagnostic and Statistical Manual of Mental Disorders*, Fifth Edition (Section III: Emerging Measures and Models) as Attenuated Psychosis Syndrome, which indicates that further study is required to determine whether it should be included as a diagnostic category in future revisions (Fusar-Poli, Carpenter, Woods, & McGlashan, 2014; Tsuang et al., 2013). Multiple considerations guided the decision, such as the lack of certainty of progression to schizophrenia and the stigma associated with the diagnosis. Fig. 28.1 provides a schematic illustration of the evolution of psychosis.

Paralleling the shift in interest in phenomenological research on early stages of psychosis, the study of brain and behavior in schizophrenia has moved from investigating chronically ill individuals to those with shorter illness duration, first episode (Andreasen et al., 2011; Gur, Cowell, et al., 2000; Gur, Turetsky, et al., 2000), and now prodromal (Fusar-Poli, Bonoldi, et al., 2012; Giuliano et al., 2012). This has been a lengthy scientific journey that required grappling with potential confounding effects of multiple factors such as psychoactive medications, limited functioning, and social isolation of patients. Furthermore, as noted earlier, symptoms emerge during a dynamic period of brain maturation, resulting in a fluid clinical presentation. Advances and availability of tools to examine brain and behavior have stimulated the integration of such measures into the study of clinical risk.

CLINICAL PHENOTYPE TO ENDOPHENOTYPE

An extensive literature has documented the nature and extent of neurobehavioral deficits in schizophrenia (Gur et al., 2015; Heinrichs & Zakzanis, 1998; Kahn & Keefe, 2013; Saykin et al., 1991) and abnormal brain parameters (Andreasen et al., 2011; Gur, Cowell, et al., 2000; Gur, Turetsky, et al., 2000). Against a background of diffuse impairment, some neurocognitive domains and brain regions and systems have shown greater vulnerability (Gur & Gur, 2013). Notably, as studies shifted to patients who have had a first episode of schizophrenia, including neuroleptic-naïve participants, it became evident that the pattern of cognitive deficits (Saykin et al., 1994) and brain dysfunction that was observed in patients who

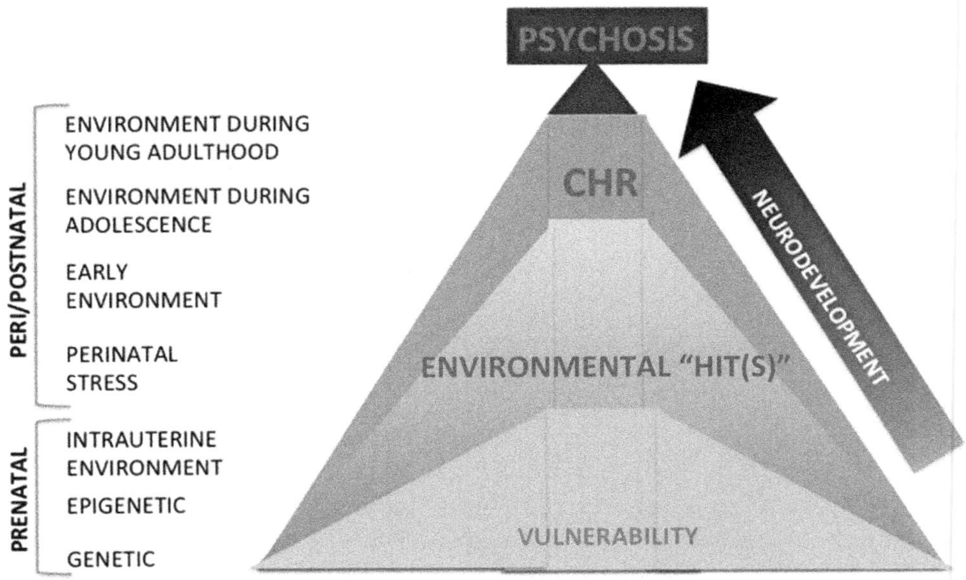

FIGURE 28.1 Schematic of psychosis vulnerability and progression. *CHR*, Clinical high risk.

have had the disease for a long time was present early in the disease (Gur, Cowell, et al., 2000; Gur, Turetsky, et al., 2000; Ho, Mola, & Andreasen, 2004). This consistency supports the application of quantitative measures in clinical risk samples as potential vulnerability markers. Furthermore, when such endophenotypic measures (Gottesman & Gould, 2003) are administered to family members, they demonstrate heritability and intermediate impairment compared with healthy participants with no family history of psychosis (Calkins et al., 2010; Greenwood et al., 2007, 2013; Gur et al., 2007). Thus, with established paradigms that documented the nature and extent of brain abnormalities in schizophrenia, a growing literature examines individuals at clinical high risk during the prodromal phase of illness. The goal of such efforts is to evaluate whether the predictability of the future course of psychosis can be enhanced with multimodal brain behavior measures. Here we will highlight lines of research and summarize emerging findings pertinent to circuitry.

NEUROCOGNITION

Neurocognitive deficits are a hallmark of schizophrenia (Barch & Ceaser, 2012; Kahn & Keefe, 2013) and various tests that have been applied in schizophrenia research gauge the presence, pattern, and extent of deficits in clinical risk studies as psychosis emerges. The rapidly growing literature (Dickson, Laurens, Cullen, & Hodgins, 2012), although different in sample sizes, rigor of reporting inclusion and exclusion criteria, and tests administered, affords quantitative meta-analyses that examine neurocognitive domains.

In a meta-analysis of 14 studies, 1214 individuals at risk for psychosis were compared with 851 healthy control subjects (Giuliano et al., 2012). Small to medium effect sizes of neurocognitive impairment in the psychosis risk group were observed. Significant deficits were noted in general cognitive abilities, attention, working memory, episodic memory, language functions and visuospatial abilities. The only domain that did not differ between groups was motor skills. Seven of these studies conducted longitudinal follow-up, demonstrating that participants in the psychosis risk group, who transitioned to psychosis at follow-up, had medium to large effect sizes of neurocognitive deficits at baseline compared with healthy participants, which supported the utility of neurocognitive assessment.

Another meta-analysis (Fusar-Poli, Deste, et al., 2012) included 19 studies with a sample of 1188 participants at clinical risk and 1029 healthy comparison participants. The clinical risk group manifested lower general intelligence and deficits in several domains were observed: executive functions, attention, working memory, verbal fluency, verbal and spatial memory and social cognition. Processing speed was not different between the groups. Transition to psychosis was examined in a subset of seven longitudinal studies with 19 months' mean follow-up (Becker et al., 2010; Brewer et al., 2005; Koutsouleris et al., 2012; Pukrop et al., 2007; Riecher-Rossler et al., 2009; Seidman et al., 2010; Woodberry et al., 2010). Findings indicated that individuals who transitioned to schizophrenia, compared with those who did not develop psychosis at follow-up, were more impaired at baseline. They had lower general intelligence and poorer performance in verbal fluency, verbal and visual memory, and working memory.

Most studies on clinical risk for psychosis have examined "cold" cognition and relatively few have focused on social cognition. Impaired social functioning has long been evident in people with schizophrenia, including premorbidly. Systematic studies evaluating affective processes have been more limited. The development of measures that relate to the perception, interpretation, and response to display of emotions is a relatively recent addition to the range of neurobehavioral probes available to evaluate this capacity. The first meta-analysis summarized earlier (Giuliano et al., 2012) included three studies that examined social cognition. Deficits in emotion processing and "theory of mind" tasks were noted in the group at clinical risk (Addington, Penn, Woods, Addington, & Perkins, 2008; Chung, Kang, Shin, Yoo, & Kwon, 2008; Pinkham, Perkins, Graham, & Siegel, 2007). In the second meta-analysis (Fusar-Poli, Deste, et al., 2012), data from six studies, some overlapping, with measures of the social cognition were included (Addington et al., 2008; An et al., 2010; Chung et al., 2008; Green et al., 2012; van Rijn et al., 2011; Szily & Keri, 2009). Significant impairment in clinical risk participants compared with healthy control subjects was noted. This literature is growing (Kohler et al., 2014), which indicates that the domain of social cognition is important in transitioning to schizophrenia and is related to the level of functioning.

NEUROIMAGING

Extensive research using MRI has documented aberrations in brain structure and function in schizophrenia, already evident in patients who have experienced a first episode (Fusar-Poli, McGuire, & Borgwardt, 2012). With the shift to studying earlier stages in the psychosis process, this technology has been applied to people at risk for psychosis, enabling examination of brain integrity as psychosis unfolds. Measures obtained include structural parameters such as gray matter and white matter volumes, cortical thickness and diffusion tensor imaging (DTI) measures of structural connectivity, as well as functional parameters including functional connectivity and activation in response to neurobehavioral tasks designed to probe a specific circuitry. The neuroimaging literature on clinical risk for psychosis is growing, although it is still relatively limited in size of samples examined and follow-up (Fusar-Poli, Bonoldi, et al., 2012). The largest body of studies has examined structural MRI focusing on gray matter (Brent et al., 2013).

A meta-analysis of 14 voxel-based morphometry studies, most using a 1.5-T scanner, compared patients who were at risk for psychosis and had experienced their first episode of schizophrenia with healthy control subjects (Fusar-Poli, McGuire, et al., 2012). The clinical risk group had lower gray matter volume in several regions including the right temporal, limbic, and prefrontal cortex whereas the group who had had a first episode had lower volumes in the temporal insular cortex and cerebellum. Notably, the onset of psychosis was associated with decreased gray matter volume in temporal, anterior cingulate, cerebellar, and insular regions. These regions are implicated in cognitive and emotion processing functions that are aberrant in schizophrenia, and volume reduction in these regions has likewise been reported in multiple studies of schizophrenia.

There are several points to consider when evaluating the highlighted finding, such as methodological limitations involved in MRI meta-analytic approaches and the cross-sectional nature of most studies. Indeed, most participants at clinical risk did not yet transition to psychosis. Nonetheless, it is informative that brain regions that show a reduction in volume in schizophrenia also show abnormalities in those at risk for psychosis (Fusar-Poli, McGuire, et al., 2012). Larger samples in a longitudinal design will be important to advance the understanding of underlying neuroanatomical differences between groups.

Other brain parameters have been evaluated in fewer studies. Thus, white matter abnormalities have been reported in schizophrenia early in the course of illness, as well as in individuals at risk for psychosis (Carletti et al., 2012; Fusar-Poli, Borgwardt, et al., 2011).

The resting blood oxygenation level—dependent (BOLD) signal in functional MRI (fMRI) paradigms provides a measure of connectivity, reflecting "cross-talk" integration among brain regions. It examines the time—series correlations among brain regions, indicating which regions show synchronized activation. Aberrations in schizophrenia in frontotemporal connectivity have been reported and have also been seen in those at clinical risk (Crossley et al., 2009). This literature is preliminary and limited.

DTI quantifies restricted water diffusivity in white matter, enabling noninvasive detection of subtle white matter abnormalities and facilitating the understanding of complex large-scale brain networks. Abnormalities in DTI have been reported in schizophrenia, both in patients who have had the disease for a long time and in first-episode presentation (Peters & Karlsgodt, 2014; Roalf et al., 2013), with reduced white matter integrity in frontotemporal tracts. The literature on psychosis risk is limited to several cross-sectional studies, with differing findings such as reduced fractional anisotropy in frontal lobe (Bloemen et al., 2010) and in the superior longitudinal fasciculus (Borgwardt, McGuire, & Fusar-Poli, 2011). In a longitudinal study (Carletti et al., 2012), individuals at risk for psychosis ($n = 32$) were compared with healthy control subjects ($n = 32$) and patients who had had a first episode of schizophrenia ($n = 15$), on a 1.5-T scanner. The psychosis risk and control participants were rescanned after 28 months. At baseline, the first-episode group had decreased fractional anisotropy and increased diffusivity relative to control subjects, and the psychosis risk group was intermediate between the other two groups. At follow-up, further reduction in fractional anisotropy was evident in left frontal region only in those individuals at risk for psychosis ($n = 8$) who transitioned to psychosis. This suggests that progressive changes occur at disease onset, which has been reported before for gray matter (Andreasen et al., 2011; Borgwardt et al., 2007; Gur, Cowell, et al., 2000; Gur, Turetsky, et al., 2000; Smeiskova et al., 2010). Again, however, the available data are meager and preliminary.

fMRI has been applied to individuals at risk for psychosis, commonly in small samples with neurobehavioral probes that have shown differences between patients with schizophrenia and control subjects. Neurobehavioral domains examined include working memory, using the n-back paradigm. Overall, psychosis risk groups show decreased activation in the BOLD response in dorsolateral and medial prefrontal regions (Fusar-Poli, McGuire, et al., 2012). The pattern of activity is similar to that seen early in the course of schizophrenia, but less pronounced abnormalities are observed. To evaluate activation changes with disease progression, longitudinal designs are necessary. Such designs have been applied in several fMRI studies (Smieskova et al., 2010). This small literature suggests that individuals who transition to psychosis differ from those who do not, with the latter group showing normalization. Thus, the application of fMRI holds promise as a tool that may facilitate elucidating the brain circuitry dysfunction underlying the psychotic process.

Several neurotransmitters that have been related to the pathophysiology of schizophrenia have been examined in those at risk for psychosis. Dopamine dysregulation has been linked to psychosis (Bonoldi & Howes, 2013), and positron emission tomography studies have shown increased dopamine striatal activity in schizophrenia (Fusar-Poli & Meyer-Lindenberg, 2013; Howes et al., 2009). Striatal 6-fluoro-L-dopa F18-dopa was also elevated in individuals who were at risk for psychosis and was related to symptom severity (Tibbo, Valiakalayil, & Allen, 2004).

Glutamatergic abnormalities have been implicated in the pathophysiology of schizophrenia, and some studies have examined the parameters of glutamate function in individuals at genetic risk for psychosis using magnetic resonance spectroscopy (MRS). An increased glutamine/glutamate ratio in medial frontal cortex was reported in adolescents at genetic risk (Tibbo et al., 2004). In a study integrating fMRI and MRS, 24 individuals at risk for psychosis were compared with 17 healthy control subjects (Fusar-Poli, Howes, et al., 2011). BOLD response to a verbal fluency task showed that the psychosis risk group had greater bilateral activation than did control subjects in midfrontal gyrus. Glutamate levels in the thalamus were lower in individuals at risk for psychosis. Furthermore, the pattern of correlations with activation suggests that prefrontal, hippocampal, and temporal lobe functioning is related to thalamic glutamate levels and differentiates those at risk from control subjects.

NEURODEVELOPMENTAL PERSPECTIVE

Converging lines of evidence indicate that schizophrenia spectrum disorders are neurodevelopmental, most likely a result of combined genetic vulnerability and environmental insult that may occur prenatally, perinatally, or during early development. Regardless of the timing of the insult, the symptoms commonly evolve during adolescence and therefore brain behavior measures need to be acquired at that time. A difficulty that arises in interpreting such endophenotypes is that normative values are highly age-dependent in youth. When studying an adult population, neurocognitive performance and neuroimaging parameters are stable until the onset of senescence. However, neurocognitive performance shows substantial improvement in both accuracy and speed between childhood and early adulthood. During the same developmental epoch, some brain parameters show a linear decline with increased age and other show an increase. For example, the volume of gray matter decreases whereas white matter volume and fractional anisotropy show age-related increases, presumably reflecting pruning and myelination. Furthermore, there are sex differences in neurocognitive performance and regional brain parameters, as well as sex differences in the predicted trajectories of age-related changes during this period. To link brain behavior parameters to genomic data, it is necessary to take these effects into account both conceptually and statistically.

These effects can be illustrated in our efforts to examine neurocognitive and neuroimaging parameters related to psychosis risk in a large-scale prospective community-based sample of youths. The Philadelphia Neurodevelopmental Cohort (PNC) was designed to address the gap in brain behavior data available on genotyped youths aged 8–21 years. Clinical phenotyping and a computerized neurocognitive battery (CNB) (Gur et al., 2010) were administered to about 9500 participants and multimodal neuroimaging was performed on a subsample of about 1600 individuals (Gur et al., 2012; Satterthwaite et al., 2014).

It became evident from examining the data that nearly all measures showed sex and age-related differences. Among about 5000 medically healthy youths, 3.7% reported threshold psychotic symptoms (delusions or hallucinations). An additional 12.3% reported significant subpsychotic positive symptoms. Odd or unusual thoughts and auditory misperceptions, followed by reality confusion, were the most discriminating and widely endorsed attenuated symptoms. A minority of youths (2.3%) endorsed subclinical negative or disorganized symptoms in the absence of positive symptoms. Male gender, younger age, and non-Caucasian ethnicity were significant predictors of psychosis spectrum status. Youths with spectrum symptoms had reduced performance accuracy on the CNB across neurocognitive domains, reduced global functioning, and increased odds of depression, anxiety, behavioral disorders, substance use, and suicidal ideation (Calkins et al., 2014). Thus, a dimensional clinical phenotype will need to be interpreted relative to age and sex in such developmental cohorts.

The neurocognitive measures likewise showed marked sex differences and age effects in this sample (Gur et al., 2012). Males and females differed in the accuracy and speed of performance of different neurocognitive domains (Fig. 28.2). For example, males outperformed females on both accuracy and speed of spatial processing and in motor speed, whereas females were more accurate and faster in word and face memory and all social cognition tests.

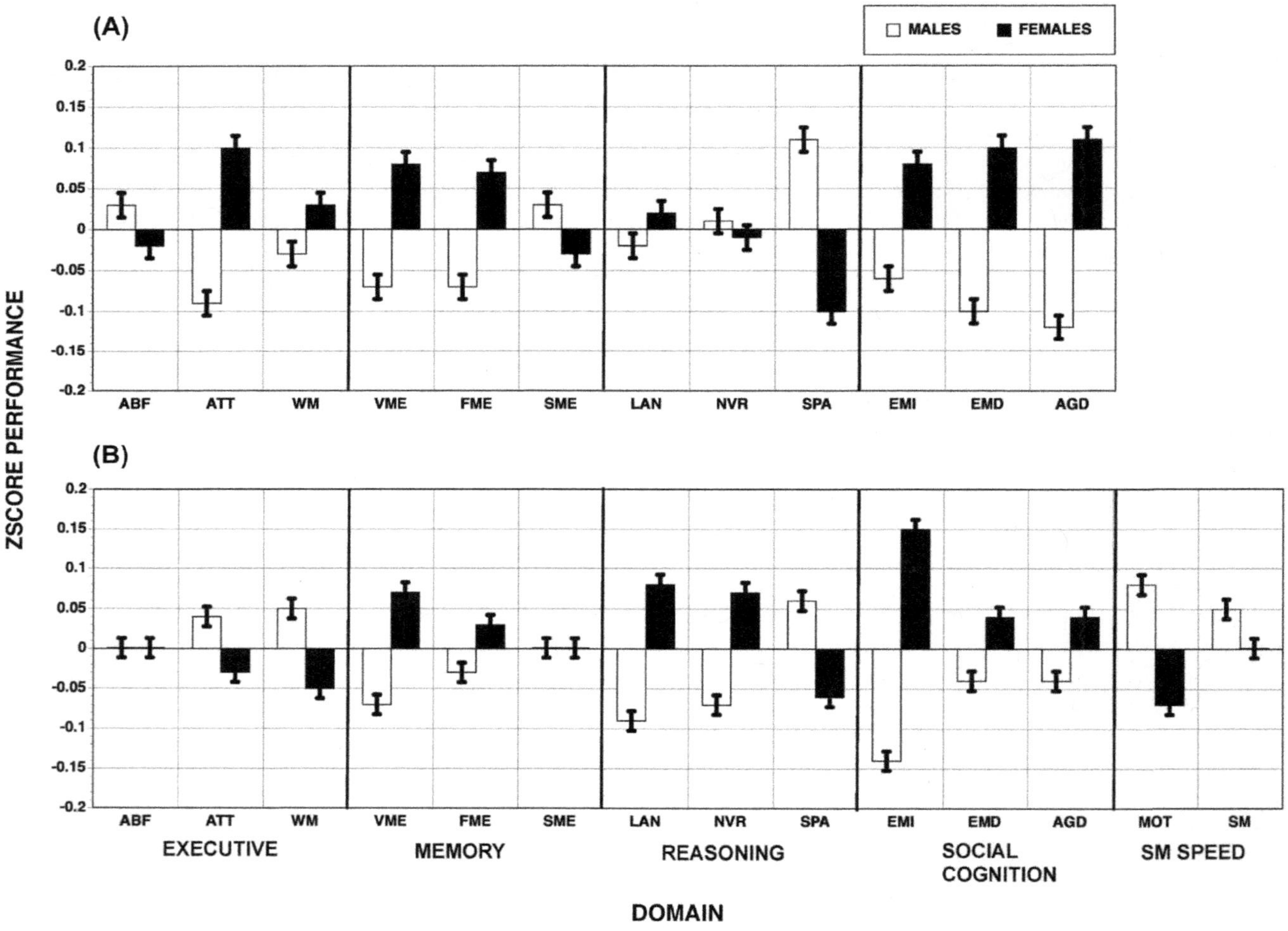

FIGURE 28.2 Means (± standard error of the mean (SEM)) of age-adjusted z-scores for accuracy (A) and speed (B) in females (*filled bars*) and males (*open bars*) across the entire sample ($n = 9010$) on executive control, episodic memory, complex cognition, social cognition, and sensorimotor speed. Sample sizes vary by task: females (4437–4679); males (4146–4322). *ABF*, abstraction and mental flexibility; *ATT*, attention; *WM*, working memory; *VMEM*, verbal; *FMEM*, facial; *SMEM*, spatial; *LAN*, language reasoning; *NVR*, nonverbal (matrix) reasoning; *SPA*, spatial ability; *EMI*, emotion identification; *EMD*, emotion differentiation; *AGD*, age differentiation; *SM*, sensorimotor speed; *MOT*, simple motor speed. *From Roalf, D. R., Gur, R. E., Ruparel, K., Calkins, M. E., Satterthwaite, T. D., Bilker, W. B., ... Gur, R. C. (2014). Within-individual variability in neurocognitive performance: age- and sex-related differences in children and youths from ages 8 to 21.* Neuropsychology, *28(4), 506–518.*

The sex differences were prominent not only in the profile of scores but also in the trajectory of age-related differences, with females reaching plateau earlier than males (Fig. 28.3). Another measure that showed robust sex differences and age-related effects was the variability of neurocognitive performance (Roalf et al., 2014). Developmental investigators have long observed the uneven characteristics of maturation: some abilities showed precocious evolution in specific individuals although eventually abilities even out. This feature of cognitive development can be operationalized in the CNB data by examining the within-individual variability (WIV) in performance, as reflected in the standard deviation of an individual's z-scores on the neurocognitive domains. As expected, WIV showed a steep age-associated decline in both accuracy and speed between ages 8 and 14, which indicates that older cohorts had more even within-individual performance across domains. Unexpectedly, subsequent to that age, older cohorts showed an increase within individual variability, which is reflected more strongly in speed than in accuracy (Fig. 28.4). Also of note, across all age cohorts and for both accuracy and speed, males had higher WIV than did females. This finding indicates that males are cognitive specialists whereas females, on average, are cognitive generalists.

These age effects and sex differences need to be kept in mind when examining neurocognitive deficits in clinical risk samples. When age effects and sex differences were statistically controlled, the sample of participants who reported psychotic symptoms showed neurocognitive deficits in performance accuracy across domains, which implicate frontotemporal dysfunction (Fig. 28.5). Whether the at-risk group also has significantly higher WIV, needs to be determined. Increased WIV in cognitive performance has been reported in schizophrenia (Roalf et al., 2013) and in developmental transitions as well as in association with brain insult.

The importance of considering age in this developmental epoch has been recognized since the inception of psychometric testing, and IQ tests have long permitted conversion of normative scores to "mental age." Such conversion permits analysis across measurement domains and would offer a convenient parameter for gauging developmental precocity or delay. The PNC sample size permitted robust calculation of "neurocognitive age" based on regressing performance on chronological age (Gur et al., 2014). This procedure generated growth charts showing normative development of general performance and for specific domains such as executive functions, memory, complex cognition, social cognition, and sensorimotor speed. Notably, these growth charts demonstrated the normative sex difference in development of general and specific neurocognitive domains (Fig. 28.6).

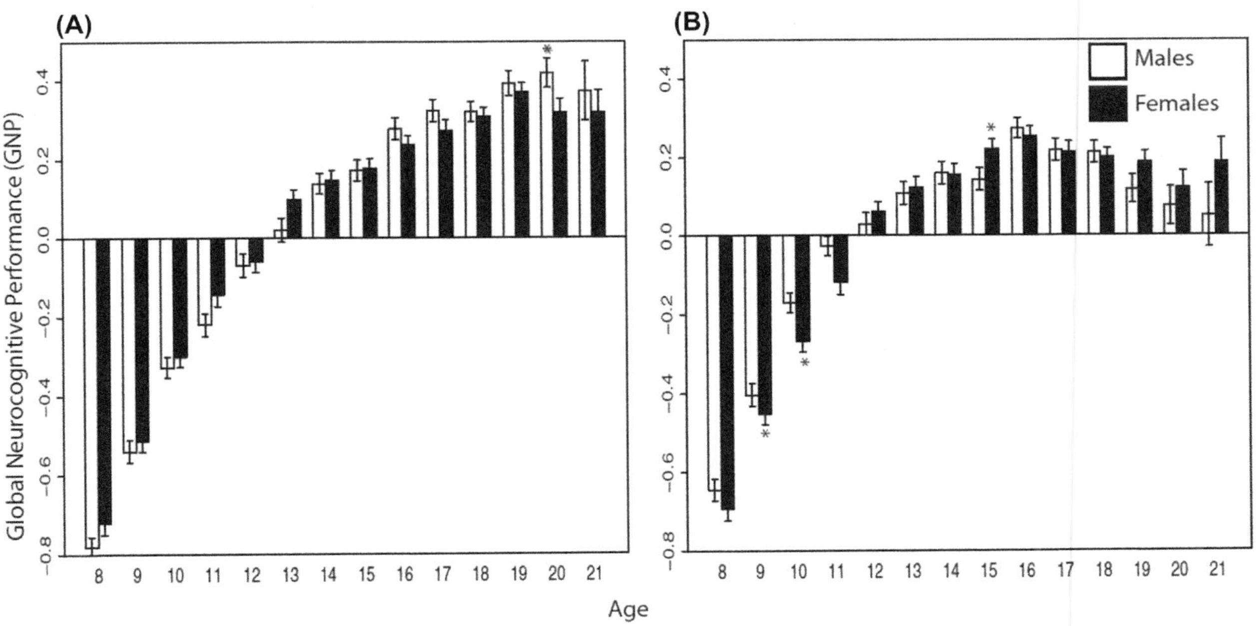

FIGURE 28.3 Means (± SEM) global neurocognitive score (GNP) for accuracy (A) and speed (B) in females (*filled bars*) and males (*open bars*) across the entire sample ($n = 9010$). As expected, GNP accuracy and speed improved with age. Overall, females had higher GNP for accuracy and speed scores than did males. Females reached mature performance earlier; however, young adult males outperformed females in accuracy but not speed. Asterisks denote age-specific sex differences. *From Roalf, D. R., Gur, R. E., Ruparel, K., Calkins, M. E., Satterthwaite, T. D., Bilker, W. B., … Gur, R. C. (2014). Within-individual variability in neurocognitive performance: age- and sex-related differences in children and youths from ages 8 to 21.* Neuropsychology, *28(4), 506–518.*

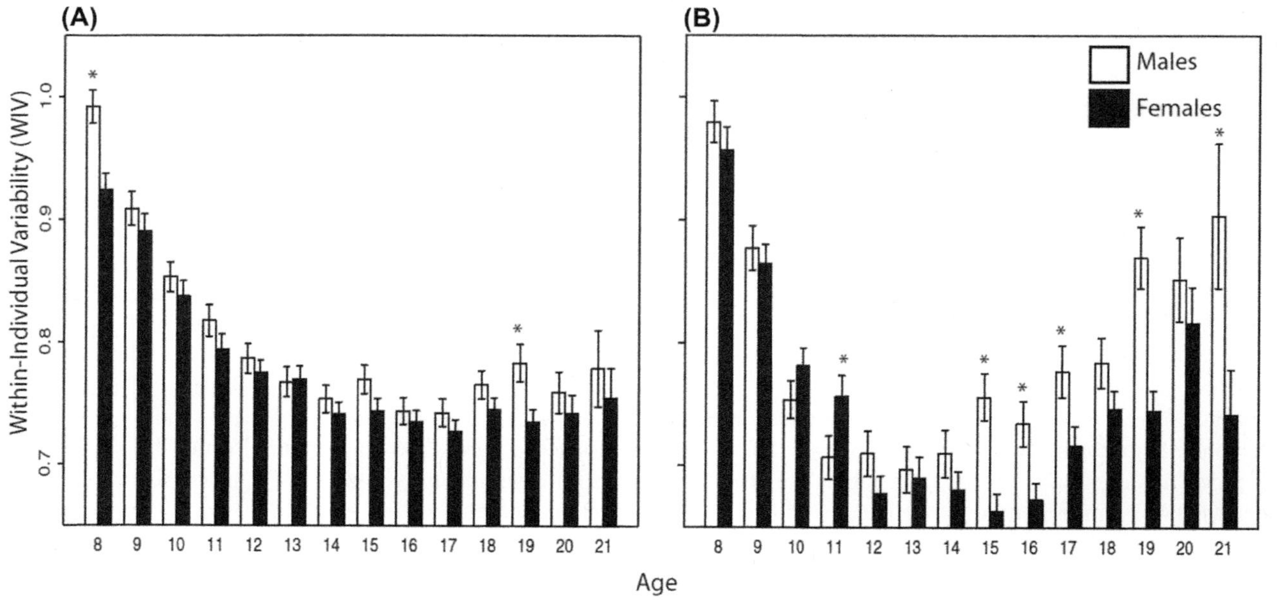

FIGURE 28.4 Means (± SEM) for across-test WIV for accuracy (A) and speed (B) in females (*filled bars*) and males (*open bars*) across the entire sample ($n = 9010$). As expected, males had higher accuracy and speed WIV compared with females. In general, accuracy WIV decreases with age and younger males (aged 8 years) show the highest variability. For speed WIV, early-adolescent females (aged 11–12 years) are more variable (speed) than males; however, this pattern reverses by age 15 years. WIV is dramatically higher in late-adolescent/early adult males compared with females. Higher values represent higher variability. Asterisks denote age-specific sex differences. Scale begins at 0.7. *From Roalf, D. R., Gur, R. E., Ruparel, K., Calkins, M. E., Satterthwaite, T. D., Bilker, W. B., … Gur, R. C. (2014). Within-individual variability in neurocognitive performance: age- and sex-related differences in children and youths from ages 8 to 21.* Neuropsychology, *28(4), 506–518.*

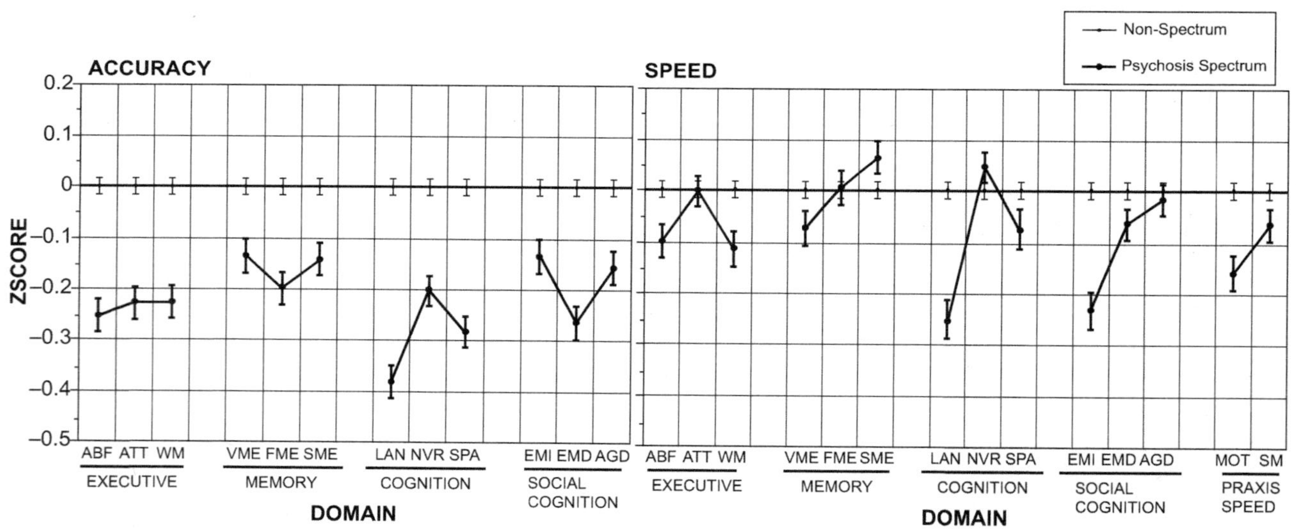

FIGURE 28.5 CNB profiles of youth diagnosed as being on the psychosis spectrum and those who are not on the spectrum and are physically healthy. *ABF*, abstraction/mental flexibility; *ATT*, attention; *WM*, working memory; *VME*, verbal memory; *FME*, face memory; *SME*, spatial memory; *LAN*, language; *NVR*, nonverbal reasoning; *SPA*, spatial processing; *EMI*, emotion identification; *EMD*, emotion differentiation; *AGD*, age differentiation; *MOT*, motor; *SM*, sensorimotor. *From Calkins, M. E., Moore, T. M., Merikangas, K. R., Burstein, M., Satterthwaite, T. D., Bilker, W. B., … Gur, R.E. (2014). The psychosis spectrum in a young U.S. community sample: findings from the Philadelphia Neurodevelopmental Cohort.* World Psychiatry, 13*(3), 296–305.*

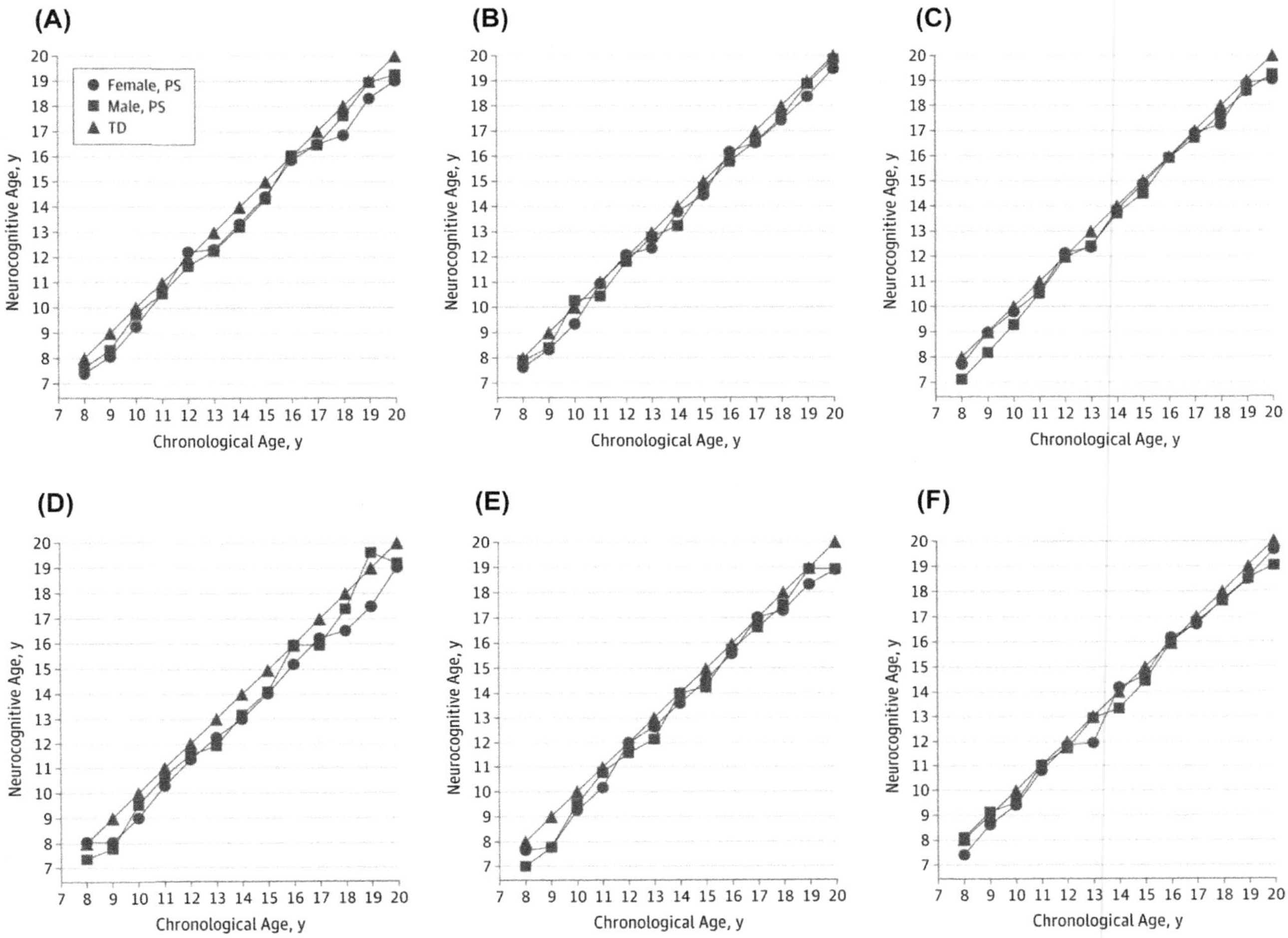

FIGURE 28.6 Chronological age compared with predicted neurocognitive age for female and male participants on the psychosis spectrum and typically developing participants. (A) All domains, (B) executive, (C) memory, (D) complex cognition, (E) social cognition, and (F) sensorimotor. *From Gur, R. C., Calkins, M. E., Satterthwaite, T. D., Ruparel, K., Bilker, W. B., Moore, T. M., ... Gur, R. E. (2014). Neurocognitive growth charting in psychosis spectrum youths.* JAMA Psychiatry, 71*(4), 366–374.*

Considering these growth charts can help identify at-risk individuals potentially at an early age. Thus, when comparing predicted age relative to chronological age in PNC participants who endorsed psychotic symptoms and typically developing participants, the former showed a developmental lag already in the age 8 to 9 cohort. This group lagged behind across the age cohorts, with the differences widening in the oldest age groups. Furthermore, the delay was more pronounced for executive and social cognition domains and least pronounced for sensorimotor speed, a pattern similar to that observed in help-seeking clinical risk samples. Importantly, the lag was wider across the age cohorts for the group of participants reporting more severe and distressing psychotic symptoms compared with that reporting milder subthreshold symptomatology (Fig. 28.7).

LINKS TO GENOMICS: CHALLENGES AND FUTURE DIRECTIONS

The genomic architecture of schizophrenia has been investigated in large samples that have relied on the clinical phenotype, with more recent efforts including endophenotypes. As detailed in other chapters, it has become increasingly clear that this heritable, heterogeneous disorder does not have shared genetic contributions of relatively few genes. Thus, in samples of over 20,000 cases and controls per group, Ripke et al. (2013) reported on 13 risk alleles estimating that about 6000 to 10,000 independent and largely common single-nucleotide polymorphisms (SNPs) contribute to the heritability and etiology of schizophrenia. Further efforts by the Schizophrenia Working Group of the Psychiatric Genomics Consortium identified 108 loci with small effects associated with the disorder (Fromer et al., 2014) and 128 established and

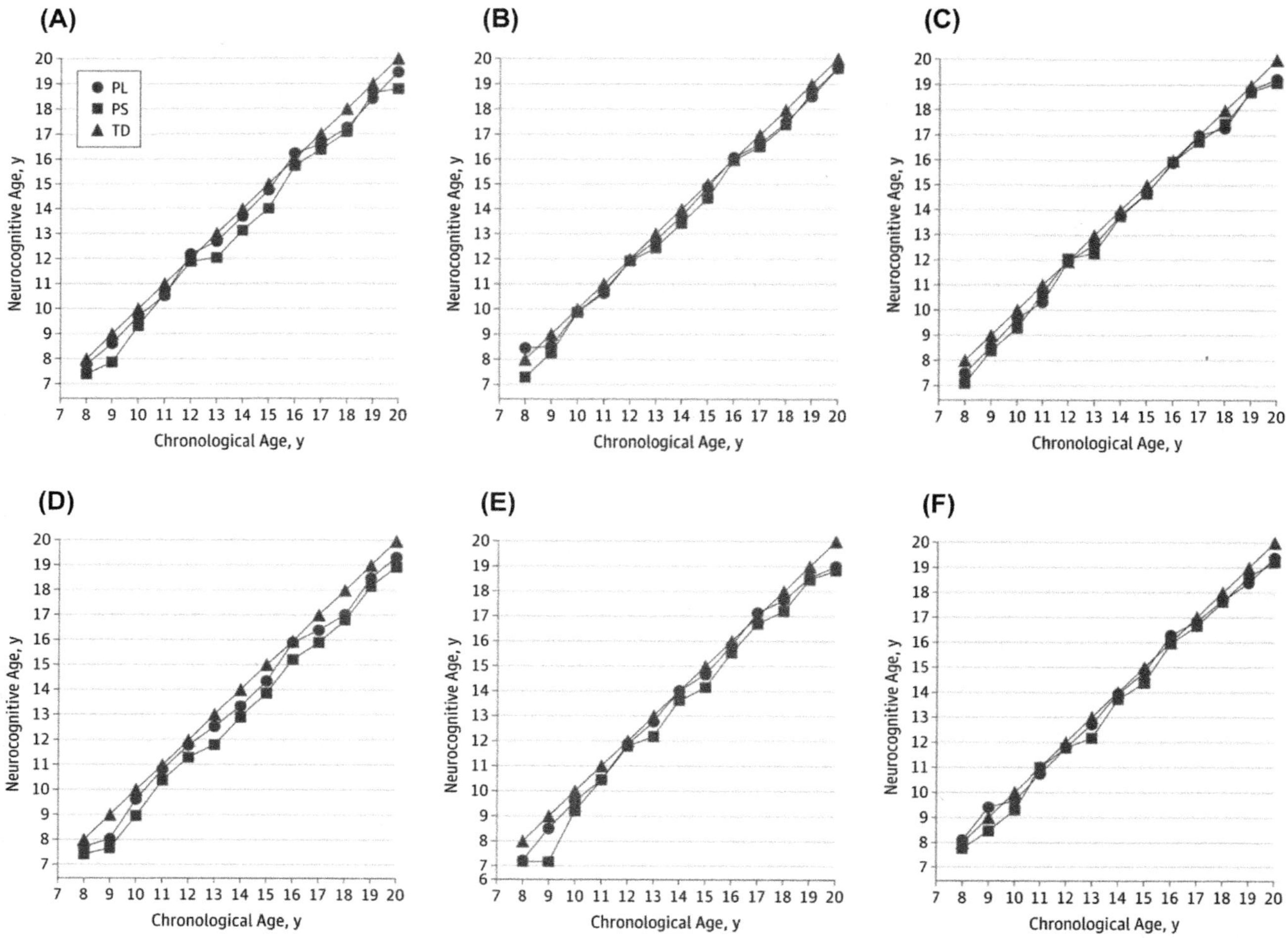

FIGURE 28.7 Chronological age compared with predicted neurocognitive age in years for typically developing (TD), psychosis spectrum (PS), and psychosis limited (PL) groups. Growth charts are provided for predicted age based on all scores (all domains) (A) and based on tests grouped by each of the five domains, including executive (B), memory (C), complex cognition (D), social cognition (E), and sensorimotor (F). *From Gur, R. C., Calkins, M. E., Satterthwaite, T. D., Ruparel, K., Bilker, W. B., Moore, T. M., … Gur, R. E. (2014). Neurocognitive growth charting in psychosis spectrum youths.* JAMA Psychiatry, *71(4), 366–374.*

novel loci have been identified (Schizophrenia Working Group of the Psychiatric Genomics Consortium, 2014). Increased sample size has added power to detect genes with small effect sizes. Notably, common variants between schizophrenia, bipolar disorder, autism spectrum disorders, and intellectual disability were observed. Especially early in the course of the emergence of psychosis, as highlighted earlier, the presenting clinical features may be less distinct and suggest several possible pathways. Longitudinal efforts are necessary to obtain data on developmental trajectories of dimensional endophenotypic parameters.

Efforts to link genomics to neurocognitive parameters will require a better understanding of the genetic architecture of cognitive abilities. We need to know the magnitude of common genetic effects across and within these abilities as well as the patterns of shared and unique genetic influences. Robinson et al. (2014) examined the PNC using genome-wide complex trait analysis to estimate the SNP-based heritability of each neurocognitive domain as well as the genetic correlation between all domains showing a significant genetic influence. Several of the individual domains indicated a strong influence of common genetic variance. The genetic correlations highlighted neurocognitive domains that are candidates for joint interrogation in future genetic studies. For example, complex reasoning, language and spatial, showed $r(g) > 0.7$. These results can be used to structure future genomic investigation of complex traits.

As efforts at early identification with convergence of endophenotypic measures are under way, larger samples of individuals at clinical risk become available for genomic studies. Applying tools established in the large-scale schizophrenia consortium, such as the polygenic risk score (Purcell et al., 2014), to these samples will extend the approach to the full spectrum of psychosis. As clinical risk samples are collecting endophenotypic measures, the use of neurocognitive, neuroimaging, and neurophysiologic parameters can be examined in efforts to create gene networks explicating the underlying neurobiology of schizophrenia. Many genes implicated (for example, *GRM3*, *GRIN2A*, *SRR*, and *GRIA1*) are involved in glutamatergic neurotransmission and synaptic plasticity, which corroborates a growing literature on underlying aberrations in schizophrenia. Both genome-wide association investigations of common variants and rare genetic variation studies converge in providing a mechanistic understanding of the etiology of schizophrenia (Fromer et al., 2014; Gulsuner et al., 2013; Owen, Craddock, & O'Donovan, 2010).

The extension of genomic research to earlier phases of the psychotic process also can contribute to investigations of gene–environment interactions. Multiple environmental risk factors contribute to schizophrenia (Iyegbe, Desmond Campbell, Butler, Ajnakina, & Sham, 2014; van Os, Linscott, Myin-Germeys, Delespaul, & Krabbendam, 2009; Walker et al., 2013). The study of large samples of youths, in informative and integrated epidemiological, genomic, and endophenotypic paradigms, can advance the field and elucidate the pathophysiology of the disorder. Such advances will facilitate the development of interventions that can affect the developmental trajectory of individuals as psychosis emerges.

REFERENCES

Addington, J., & Heinssen, R. (2012). Prediction and prevention of psychosis in youth at clinical high risk. *Annual Review of Clinical Psychology, 8*, 269–289.

Addington, J., Penn, D., Woods, S. W., Addington, D., & Perkins, D. (2008). Facial affect recognition in individuals at clinical high risk for psychosis. *British Journal of Psychiatry, 192*, 67–68.

American Psychiatric Association. (2013). *Diagnostic and statistical manual of mental disorders* (5th ed.). Washington, DC: American Psychiatric Publishing.

An, S. K., Kang, J. I., Park, J. Y., Kim, K. R., Lee, S. Y., & Lee, E. (2010). Attribution bias in ultra-high risk for psychosis and first-episode schizophrenia. *Schizophrenia Research, 118*, 54–61.

Andreasen, N. C., Nopoulos, P., Magnotta, V., Pierson, R., Ziebell, S., & Ho, B. C. (2011). Progressive brain change in schizophrenia: a prospective longitudinal study of first-episode schizophrenia. *Biological Psychiatry, 70*(7), 672–679.

Barch, D. M., & Ceaser, A. (2012). Cognition in schizophrenia: core psychological and neural mechanisms. *Trends in Cognitive Science, 16*, 27–33.

Becker, H. E., Nieman, D. H., Wiltink, S., Dingemans, P. M., van de Fliert, J. R., Velthorst, E., ... Linszen, D. H. (2010). Neurocognitive functioning before and after the first psychotic episode: does psychosis result in cognitive deterioration? *Psychological Medicine, 40*, 1599–1606.

Bloemen, O. J., de Koning, M. B., Schmitz, N., Nieman, D. H., Becker, H. E., de Haan, L., ... van Amelsvoort, T. A. (2010). White-matter markers for psychosis in a prospective ultra-high-risk cohort. *Psychological Medicine, 40*, 1297–1304.

Bonoldi, I., & Howes, O. D. (2013). The enduring centrality of dopamine in the pathophysiology of schizophrenia: in vivo evidence from the prodrome to the first psychotic episode. *Advances in Pharmacology, 68*, 199–220.

Borgwardt, S., McGuire, P. K., & Fusar-Poli, P. (2011). Gray matters! Mapping the transition to psychosis. *Schizophrenia Research, 133*, 63–67.

Borgwardt, S. J., Riecher-Rossler, A., Dazzan, P., Chitnis, X., Aston, J., Drewe, M., ... McGuire, P. K. (2007). Regional gray matter volume abnormalities in the at-risk mental state. *Biological Psychiatry, 61*, 1148–1156.

Braff, D. L., Freedman, R., Schork, N. J., & Gottesman, I. I. (2007). Deconstructing schizophrenia: an overview of the use of endophenotypes in order to understand a complex disorder. *Schizophrenia Bulletin, 33*(1), 21–32.

Brent, B. K., Thermenos, H. W., Keshavan, M. S., & Seidman, L. J. (2013). Gray matter alterations in schizophrenia high-risk youth and early-onset schizophrenia: a review of structural MRI findings. *Child and Adolescence Psychiatric Clinics North America, 22*(4), 689–714.

Brewer, W. J., Francey, S. M., Wood, S. J., Jackson, H. J., Pantelis, C., Phillips, L. J., ... McGorry, P. D. (2005). Memory impairments identified in people at ultra-high risk for psychosis who later develop first-episode psychosis. *American Journal of Psychiatry, 162*, 71–78.

Calkins, M. E., Moore, T. M., Merikangas, K. R., Burstein, M., Satterthwaite, T. D., Bilker, W. B., ... Gur, R. E. (2014). The psychosis spectrum in a young U.S. community sample: findings from the Philadelphia Neurodevelopmental Cohort. *World Psychiatry, 13*(3), 296–305.

Calkins, M. E., Tepper, P., Gur, R. C., Ragland, J. D., Klei, L., Wiener, H. W., ... Gur, R. E. (2010). Project among African-Americans to explore risks for schizophrenia (PAARTNERS): evidence for impairment and heritability of neurocognitive functioning in families of schizophrenia patients. *American Journal of Psychiatry, 167*(4), 459–472.

Cannon, T. D., Cadenhead, K., Cornblatt, B., Woods, S. W., Addington, J., Walker, E., ... Heinssen, R. (2008). Prediction of psychosis in youth at high clinical risk. *Archives of General Psychiatry, 65*, 28–37.

Carletti, F., Woolley, J. B., Bhattacharyya, S., Perez-Iglesias, R., Fusar-Poli, P., Valmaggia, L., ... McGuire, P. K. (2012). Alterations in white matter evident before the onset of psychosis. *Schizophrenia Bulletin, 38*(6), 1170–1179.

Chung, Y. S., Kang, D. H., Shin, N. Y., Yoo, S. Y., & Kwon, J. S. (2008). Deficit of theory of mind in individuals at ultra-high-risk for schizophrenia. *Schizophenia Research, 99*, 111–118.

Crossley, N. A., Mechelli, A., Fusar-Poli, P., Broome, M. R., Matthiasson, P., Johns, L. C., ... McGuire, P. K. (2009). Superior temporal lobe dysfunctionand fronto-temporal dysconnectivity in subjects at risk of psychosis and in first-episode psychosis. *Human Brain Mapping, 30*, 4129–4137.

Dickson, H., Laurens, K. R., Cullen, A. E., & Hodgins, S. (2012). Meta-analyses of cognitive and motor function in youth aged 16 years and younger who subsequently develop schizophrenia. *Psychological Medicine, 42*(4), 743–755.

Fromer, M., Pocklington, A. J., Kavanagh, D. H., Williams, H. J., Dwyer, S., Gormley, P., … O'Donovan, M. C. (2014). De novo mutations in schizophrenia implicate synaptic networks. *Nature, 506*(7487), 179–184.

Fusar-Poli, P., Bonoldi, I., Yung, A. R., Borgwardt, S., Kempton, M. J., Valmaggia, L., … McGuire, P. (2012). Predicting psychosis: meta-analysis of transition outcomes in individuals at high clinical risk. *Archives of General Psychiatry, 69*, 220–229.

Fusar-Poli, P., Borgwardt, S., Crescini, A., Deste, G., Kempton, M. J., Lawrie, S., … Sacchetti, E. (2011). Neuroanatomical correlates of vulnerability to psychosis: a voxel-based meta-analysis. *Neuroscience Biobehavior Review, 35*, 1175–1185.

Fusar-Poli, P., Carpenter, W. T., Woods, S. W., & McGlashan, T. H. (2014). Attenuated psychosis syndrome: ready for DSM-5.1? *Annual Review of Clinical Psychology, 10*, 155–192.

Fusar-Poli, P., Deste, G., Smieskova, R., Barlati, S., Yung, A. R., Howes, O., … Borgwardt, S. (2012). Cognitive functioning in prodromal psychosis: a meta-analysis. *Archives of General Psychiatry, 69*, 562–571.

Fusar-Poli, P., Howes, O. D., Allen, P., Broome, M., Valli, I., Asselin, M. C., … McGuire, P. K. (2011). Abnormal prefrontal activation directly related to pre-synaptic striatal dopamine dysfunction in people at clinical high risk for psychosis. *Molecular Psychiatry, 16*, 67–75.

Fusar-Poli, P., McGuire, P., & Borgwardt, S. (2012). Mapping prodromal psychosis: a critical review of neuroimaging studies. *European Journal of Psychiatry, 27*, 181–191.

Fusar-Poli, P., & Meyer-Lindenberg, A. (2013). Striatal presynaptic dopamine in schizophrenia, part II: meta-analysis of [18F]/[11C] DOPA PET studies. *Schizophrenai Bulletin, 39*(1), 33–42.

Giedd, J. N., Snell, J. W., Lange, N., Rajapakse, J. C., Casey, B. J., Kozuch, P. L., … Rapoport, J. L. (1996). Quantitative magnetic resonance imaging of human brain development: ages 4–18. *Cereb Cortex, 6*, 551–560.

Giuliano, A. J., Li, H., Mesholam-Gately, R. I., Sorenson, S. M., Woodberry, K. A., & Seidman, L. J. (2012). Neurocognition in the psychosis risk syndrome: a quantitative and qualitative review. *Current Pharmaceutical Design, 18*, 399–415.

Glahn, D. C., Knowles, E. E. M., McKay, D. R., Sprooten, E., Raventos, H., Blangero, J., … Almasy, L. (2014). Arguments for the sake of endophenotypes: examining common misconceptions about the use of endophenotypes in psychiatric genetics. *American Journal of Medical Genetics Part B: Neuropsychiatric Genetics, 165*(2), 122–130.

Gottesman, I. I., & Gould, T. D. (2003). The endophenotype concept in psychiatry: etymology and strategic intentions. *The American Journal of Psychiatry, 160*, 636–645.

Goulding, S. M., Holtzman, C. W., Trotman, H. D., Ryan, A. T., Macdonald, A. N., Shapiro, D. I., … Walker, E. F. (2013). The prodrome and clinical risk for psychotic disorders. *Child and Adolescent Psychiatric Clinics of North America, 22*(4), 557–567.

Green, M. F., Bearden, C. E., Cannon, T. D., Fiske, A. P., Hellemann, G. S., Horan, W. P., … Nuechterlein, K. H. (2012). Social cognition in schizophrenia, part 1: performance across phase of illness. *Schizophrenia Bulletin, 38*(4), 854–864.

Greenwood, T. A., Braff, D. L., Light, G. A., Cadenhead, K. S., Calkins, M. E., Dobie, D. J., … Schork, N. J. (2007). Initial heritability analyses of endophenotypic measures for schizophrenia: the consortium on the genetics of schizophrenia. *Archives of General Psychiatry, 64*(11), 1242–1250.

Greenwood, T. A., Swerdlow, N. R., Gur, R. E., Cadenhead, K. S., Calkins, M. E., Dobie, D. J., … Braff, D. L. (2013). Genome-wide linkage analyses of 12 endophenotypes for schizophrenia from the consortium on the genetics of schizophrenia. *The American Journal of Psychiatry, 170*(5), 521–532.

Gulsuner, S., Wash, T., Watts, A. C., Lee, M. K., Thornton, A. M., Casadei, S., … McClellan, J. M. (2013). Spatial and temporal mapping of de novo mutations in schizophrenia to a fetal prefrontal cortical network. *Cell, 154*(3), 518–529.

Gur, R. E. (2013). Psychosis proneness. In D. Charney, P. Sklar, & E. Nestler (Eds.), *Neurobiology of mental illness*. Oxford University Press.

Gur, R. C., Braff, D. L., Calkins, M. E., Dobiec, D. J., Freedman, R., Greene, M. F., … Gur, R. E. (2015). Neurocognitive performance in family-based and case-control studies of schizophrenia. *Schizophrenia Research, 163*(1–3), 17–23.

Gur, R. C., Calkins, M. E., Satterthwaite, T. D., Ruparel, K., Bilker, W. B., Moore, T. M., … Gur, R. E. (2014). Neurocognitive growth charting in psychosis spectrum youths. *JAMA Psychiatry, 71*(4), 366–374.

Gur, R. E., Cowell, P. E., Latshaw, A., Turetsky, B. I., Grossman, R. I., Arnold, S. E., … Gur, R. C. (2000). Reduced dorsal and orbital prefrontal gray matter volumes in schizophrenia. *Archives of General Psychiatry, 57*(8), 761–768.

Gur, R. C., & Gur, R. E. (2013). Memory in health and in schizophrenia. *Dialogues in Clinical Neuroscience, 15*, 399–410.

Gur, R. E., Nimgaonkar, V. L., Almasy, L., Calkins, M. E., Ragland, J. D., Pogue-Geile, M. F., … Gur, R. C. (2007). Neurocognitive endophenotypes in a multiplex multigenerational family study of schizophrenia. *The American Journal of Psychiatry, 164*(5), 813–819.

Gur, R. C., Richard, J., Calkins, M. E., Chiavacci, R., Hansen, J. A., Bilker, W. B., … Gur, R. E. (2012). Age group and sex differences in performance on a computerized neurocognitive battery in children age 8–21. *Neuropsychology, 26*, 251–265.

Gur, R. C., Richard, J., Hughett, P., Calkins, M. E., Macy, L., Bilker, W. B., … Gur, R. E. (2010). A cognitive neuroscience-based computerized battery for efficient measurement of individual differences: standardization and initial construct validation. *Journal of Neuroscience Methods, 187*, 254–262.

Gur, R. E., Turetsky, B. I., Cowell, P. E., Finkelman, C., Maany, V., Grossman, R. I., … Gur, R. C. (2000). Temporolimbic volume reductions in schizophrenia. *Archives of General Psychiatry, 57*(8), 769–775.

Heinrichs, R. W., & Zakzanis, K. K. (1998). Neurocognitive deficit in schizophrenia: a quantitative review of the evidence. *Neuropsychology, 12*, 426–445.

Heinssen, R. K., & Insel, T. R. (November 11, 2014). Preventing the onset of psychosis: not quite there yet. *Schizophrenia Bulletin, 161*(1), 61–69. pii: sbu161. [Epub ahead of print].

Ho, B. C., Mola, C., & Andreasen, N. C. (2004). Cerebellar dysfunction in neuroleptic naive schizophrenia patients: clinical, cognitive, and neuroanatomic correlates of cerebellar neurologic signs. *Biological Psychiatry, 55*(12), 1146–1153.

Howes, O., Montgomery, A., Asselin, M., Valli, I., Tabraham, P., Johns, L., ... Grasby, P. (2009). Elevated striatal dopamine function linked to prodromal signs of schizophrenia. *Archives of General Psychiatry, 66*, 13–20.

Huttenlocher, P. R., de Courten, C., Garey, L. J., & Van Der Loos, H. (1982). Synaptogenesis in human visual cortex: evidence for synapse elimination during normal development. *Neuroscience Letters, 33*, 247–252.

Insel, T., Cuthbert, B., Garvey, M., Heinssen, R., Pine, D. S., Quinn, K., ... Wang, P. (2010). Research domain criteria (RDoC): toward a new classification framework for research on mental disorders. *The American Journal of Psychiatry, 167*(7), 748–751.

Iyegbe, C., Desmond Campbell, D., Butler, A., Ajnakina, O., & Sham, P. (2014). The emerging molecular architecture of schizophrenia, polygenic risk scores and the clinical implications for GxE research. *Social Psychiatry and Psychiatric Epidemiology, 49*, 169–182.

Jernigan, T. L., & Tallal, P. (1990). Late childhood changes in brain morphology observable with MRI. *Developmental Medicine & Child Neurology, 32*, 379–385.

Kahn, R. S., & Keefe, R. S. (2013). Schizophrenia is a cognitive illness: time for a change in focus. *JAMA Psychiatry, 70*, 1107–1112.

Kelleher, I., Connor, D., Clarke, M. C., Devlin, N., Harley, M., & Cannon, M. (2012). Prevalence of psychotic symptoms in childhood and adolescence: a systematic review and meta-analysis of population-based studies. *Psychological Medicine, 42*(9), 1857–1863.

Kohler, C. G., Richard, J. A., Brensinger, C. M., Borgmann-Winter, K. E., Conroy, C. G., Moberg, P. J., ... Calkins, M. E. (2014). Facial emotion perception differs in young persons at genetic and clinical high-risk for psychosis. *Psychiatry Research, 216*(2), 206–212.

Koutsouleris, N., Davatzikos, C., Bottlender, R., Patschurek-Kliche, K., Scheuerecker, J., Decker, P., ... Meisenzahl, E. M. (2012). Early recognition and disease prediction in the at-risk mental states for psychosis using neurocognitive pattern classification. *Schizophrenia Bulletin, 38*, 1200–1215.

McGorry, P. (2011). Transition to adulthood: the critical period for pre-emptive, disease-modifying care for schizophrenia and related disorders. *Schizophrenia Bulletin, 37*(3), 524–530.

Miller, T. J., McGlashan, T. H., Rosen, J. L., Cadenhead, K., Cannon, T., Ventura, J., ... Woods, S. W. (2003). Prodromal assessment with the structured interview for prodromal syndromes and the scale of prodromal symptoms: predictive validity, interrater reliability, and training to reliability. *Schizophrenia Bulletin, 29*, 703–715.

van Os, J., Linscott, R. J., Myin-Germeys, I., Delespaul, P., & Krabbendam, L. (2009). A systematic review and meta-analysis of the psychosis continuum: evidence for a psychosis proneness-persistence impairment model of psychotic disorder. *Psychological Medicine, 39*(2), 179–195.

Owen, M. J., Craddock, N., & O'Donovan, M. C. (2010). Suggestion of roles for both common and rare risk variants in genome-wide studies of schizophrenia. *Archives of Geneneral Psychiatry, 67*(7), 667–673.

Peters, B. D., & Karlsgodt, K. H. (May 31, 2014). White matter development in the early stages of psychosis. *Schizophrenia Research, 161*(1), 61–69 [Epub ahead of print].

Pinkham, A. E., Penn, D. L., Perkins, D. O., Graham, K. A., & Siegel, M. (2007). Emotion perception and social skill over the course of psychosis: a comparison of individuals "at-risk" for psychosis and individuals with early and chronic schizophrenia spectrum illness. *Cognitive Neuropsychiatry, 12*, 198–212.

Pukrop, R., Ruhrmann, S., Schultze-Lutter, F., Bechdolf, A., Brockhaus-Dumke, A., & Klosterkötter, J. (2007). Neurocognitive indicators for a conversion to psychosis: comparison of patients in a potentially initial prodromal state who did or did not convert to a psychosis. *Schizophrenia Research, 92*, 116–125.

Purcell, S. M., Moran, J. L., Fromer, M., Ruderfer, D., Solovieff, N., Roussos, P., ... Sklar, P. (2014). A polygenic burden of rare disruptive mutations in schizophrenia. *Nature, 506*(7487), 185–190.

van Rijn, S., Aleman, A., de Sonneville, L., Sprong, M., Ziermans, T., Schothorst, P., ... Swaab, H. (2011). Misattribution of facial expressions of emotion in adolescents at increased risk of psychosis: the role of inhibitory control. *Psychological Medicine, 41*, 499–508.

Riecher-Rossler, A., Pflueger, M. O., Aston, J., Borgwardt, S. J., Brewer, W. J., Gschwandtner, U., & Stieglitz, R. D. (2009). Efficacy of using cognitive status in predicting psychosis: a 7-year follow-up. *Biological Psychiatry, 66*, 1023–1030.

Ripke, S., O'Dushlaine, C., Chambert, K., Moran, J. L., Kähler, A. K., Akterin, S., ... Sullivan, P. F. (2013). Genome-wide association analysis identifies 13 new risk loci for schizophrenia. *Nature Genetics, 45*(10), 1150–1159.

Roalf, D. R., Gur, R. E., Ruparel, K., Calkins, M. E., Satterthwaite, T. D., Bilker, W. B., ... Gur, R. C. (2014). Within-individual variability in neurocognitive performance: age- and sex-related differences in children and youths from ages 8 to 21. *Neuropsychology, 28*(4), 506–518.

Roalf, D. R., Ruparel, K., Verma, R., Elliott, M. A., Gur, R. E., & Gur, R. C. (2013). White matter organization and neurocognitive performance variability in schizophrenia. *Schizophrenia Research, 143*(1), 172–178.

Robinson, E. B., Kirby, A., Ruparel, K., Yang, J., McGrath, L., Anttila, V., ... Hakonarson, H. (July 15, 2014). The genetic architecture of pediatric cognitive abilities in the Philadelphia Neurodevelopmental Cohort. *Molecular Psychiatry, 20*, 454–458. http://dx.doi.org/10.1038/mp.2014.65 [Epub ahead of print].

Royston, P., Altman, D. G., & Sauerbrei, W. (2006). Dichotomizing continuous predictors in multiple regression: a bad idea. *Statistics in Medicine, 25*, 127–141.

Ruhrmann, S., Schultze-Lutter, F., Salokangas, R. K., Heinimaa, M., Linszen, D., Dingemans, P., ... Klosterkötter, J. (2010). Prediction of psychosis in adolescents and young adults at high risk: results from the prospective European prediction of psychosis study. *Archives of General Psychiatry, 67*(3), 241–251.

Satterthwaite, T. D., Elliott, M. A., Ruparel, K., Loughead, J., Prabhakaran, K., Calkins, M. E., ... Gur, R. E. (2014). Neuroimaging of the Philadelphia neurodevelopmental cohort. *Neuroimage, 86*, 544–553.

Saykin, A. J., Gur, R. C., Gur, R. E., Mozley, P. D., Mozley, L. H., Resnick, S. M., ... Stafiniak, P. (1991). Neuropsychological function in schizophrenia. Selective impairment in memory and learning. *Archives of General Psychiatry, 48*(7), 618–624.

Saykin, A. J., Shtasel, D. L., Kester, D. B., Mozley, L. H., Stafiniak, P., & Gur, R. C. (1994). Neuropsychological deficits in neuroleptic naive patients with first-episode schizophrenia. *Archives of General Psychiatry, 51*(2), 124–131.

Schizophrenia Working Group of the Psychiatric Genomics Consortium. (2014). Biological insights from 108 schizophrenia-associated genetic loci. *Nature, 511*, 421–427.

Seidman, L. J., Giuliano, A. J., Meyer, E. C., Addington, J., Cadenhead, K. S., Cannon, T. D., … Cornblatt, B. A. (2010). Neuropsychology of the prodrome to psychosis in the NAPLS Consortium: relationship to family history and conversion to psychosis. *Archives of General Psychiatry, 67*, 578–588.

Smieskova, R., Fusar-Poli, P., Allen, P., Bendfeldt, K., Stieglitz, R. D., Drewe, J., … Borgwardt, S. J. (2010). Neuroimaging predictors of transition to psychosis: a systematic review and meta-analysis. *Neuroscience and Biobehavioral Review, 38*, 1207–1222.

Szily, E., & Keri, S. (2009). Anomalous subjective experience and psychosis risk in young depressed patients. *Psychopathology, 42*, 229–235.

Tibbo, P., Hanstock, C., Valiakalayil, A., & Allen, P. (2004). 3T proton MRS investigation of glutamate and glutamine in adolescents at high genetic risk for schizophrenia. *American Journal of Psychiatry, 161*, 1116–1118.

Tsuang, M. T., Van OS, J., Tandon, R., Barch, D. M., Bustillo, J., Gaebel, W., … Carpenter, W. (2013). Attenuated psychosis syndrome in DSM-5. *Schizophrenia Research, 150*(1), 31–35.

Walker, E. F., Trotman, H. D., Goulding, S. M., Holtzman, C. W., Ryan, A. T., McDonald, A., … Brasfield, J. L. (2013). Developmental mechanisms in the prodrome to psychosis. *Developmental Psychopathololgy, 25*(4 Pt 2), 1585–1600.

Woodberry, K. A., Seidman, L. J., Giuliano, A. J., Verdi, M. B., Cook, W. L., & McFarlane, W. R. (2010). Neuropsychological profiles in individuals at clinical high risk for psychosis: relationship to psychosis and intelligence. *Schizophrenia Research, 123*, 188–198.

Yakovlev, P. L., & Lecours, A. R. (1967). The myelogenetic cycles of regional maturation of the brain. In A. Minkowski (Ed.), *Regional development of the brain in early life* (pp. 3–70). Oxford: Blackwell.

Chapter 29

Autism Spectrum Disorder: Genes to Pathways to Circuits

Rebecca A. Muhle
Yale University School of Medicine, New Haven, CT, United States

Stephan J. Sanders
University of California San Francisco, San Francisco, CA, United States

Hannah E. Reed
Yale University School of Medicine, New Haven, CT, United States

Matthew W. State
University of California San Francisco, San Francisco, CA, United States

INTRODUCTION

Similar to other psychiatric conditions, the diagnosis of autism spectrum disorder (ASD) is behavioral and carries all of the limitations associated with the absence of highly reliable biomarkers. Despite this, there has been a sea change in the genetics and genomics of ASD over the past decade. The field has transitioned from a preoccupation with identifying even a single definitive risk gene to accumulating dozens of reliable loci and genes, thus elaborating the genomic architecture of ASD, and moving forward in the process of characterizing the underlying biology and pathophysiology. Although there are considerable obstacles ahead (including tremendous genetic heterogeneity, the biological pleiotropy of risk genes, a wide range of diagnostic outcomes emerging from seemingly identical mutations, and the obvious challenges of understanding the developing human brain) the future of ASD genomic and neurobiological research has never been brighter. Gene discovery has provided and continues to provide a solid foundation for functional studies that reveal a surprising degree of convergence with regard to vulnerable molecular mechanisms, anatomical regions, and developmental epochs, and making plausible the development of novel and more effective treatments.

Phenomenology

Children displaying symptoms consistent with a modern-day conception of autism were initially described independently by Leo Kanner and Hans Asperger in 1943 and 1944, respectively (Kanner, 1943; Wing, 1981). These descriptions of largely male cohorts who manifested social difficulties, odd and restricted interests with inflexibility of thought, and poorly developed executive function and motor skills are remarkably consistent with the current conceptualizations of ASD. Moreover, Kanner noted the congenital absence of social interest during infancy, suggesting a disorder with origins early in development; Asperger and later Wing presaged the contemporary notions of familial transmission as well as the concept of the broader autism phenotype (Constantino, 2011; Pickles et al., 2000; Piven, Palmer, Jacobi, Childress, & Arndt, 1997) by describing similarly odd but less severe behavioral traits in the parents of affected individuals.

ASD is a syndrome encompassing an unknown number of etiologically diverse developmental conditions and defined by distinctive abnormalities of the central nervous system. Cardinal signs and symptoms include reduced social

Genomics, Circuits, and Pathways in Clinical Neuropsychiatry. http://dx.doi.org/10.1016/B978-0-12-800105-9.00029-9

communication, recognizable in early childhood, characterized by atypical, delayed, or absent social communication with unusually restricted interests and repetitive behaviors (Diagnostic and Statistical Manual of Mental Disorders, 2013). Early signs of atypical development suggestive of ASD include, but are not limited to, decreased eye contact, absent social smiling (Zwaigenbaum et al., 2005), reduced interest in faces, and reduced response to one's name (Dawson et al., 2002; Johnson et al., 2005; Osterling & Dawson, 1994; Osterling, Dawson, & Munson, 2002). As children with ASD progress through infancy and toddlerhood, they may have impaired imitative behaviors and delayed verbal communication (with no compensatory increase in nonverbal communication such as gestures), and engage in less pretend play (Ozonoff et al., 2010; Wan et al., 2013). More mildly affected children may not be diagnosed until school age, at which time impairments in pragmatic speech, decreased number and quality of peer relationships, decreased social and emotional reciprocity, and deficits in adaptive functioning, compared with peers, are the most typical presenting problems (Volkmar et al., 2014).

The earliest versions of the *Diagnostic and Statistic Manual of Mental Disorders* (DSM) did not distinguish autism from childhood-onset psychoses (Rapoport, Chavez, Greenstein, Addington, & Gogtay, 2009). It was not until DSM-III in 1980 that the contemporary diagnostic framework began to differentiate childhood-onset schizophrenia from ASD, almost a half century after the initial clinical descriptions were first published. At that time, diagnostic criteria specified that there must be impairment of social development, insistence on sameness, and onset before 30 months (Spitzer & Statistics, 1987). Subsequent revisions grouped classical or infantile autism, Asperger disorder, childhood disintegrative disorder, Rett syndrome, and pervasive developmental disorder not otherwise specified (PDD-NOS) under the heading of pervasive developmental disorders (PDD) (Association, 2000). In 2013, the American Psychiatric Association published DSM-5 (Diagnostic and Statistical Manual of Mental Disorders, 2013), and the diagnoses of classical autism, Asperger disorder, childhood disintegrative disorder, and PDD-NOS were subsumed under a single diagnosis of ASD. Rett syndrome, a monogenic syndrome caused by mutations in the gene *Methyl CpG Binding Protein 2*, which can present with symptoms of autism, was not maintained as a separate category but may be added as a clinical specifier based on known genetic etiology. Support for the change in diagnostic criteria increased after evidence that the individual PDD diagnoses shared core symptoms at presentation, were subject to interrater variability, and were unreliable predictors of future diagnosis (Lord & Jones, 2012).

In the DSM-5, ASD is defined by the current or prior history of disturbances in two domains: (1) social-communication ability and (2) repetitive behaviors and restricted interests. Under the current diagnostic scheme, individuals must meet all three criteria of the Social Communication domain, such that the individual has difficulty with social-emotional reciprocity, nonverbal communication, and starting, maintaining, or understanding relationships (Diagnostic and Statistical Manual of Mental Disorders, 2013). Whereas the DSM-IV-TR included language delay as a criterion of classical autism, this was reconceptualized in the DSM-5 to reflect the view that social communication deficits did not require a delay of language.

The second diagnostic domain reflects a combination of symptoms, including motor (repetitive behaviors), behavioral inflexibility, insistence on sameness, and sensory sensitivity (either increased or decreased). A decrease in mental flexibility is evident in resistance to change and difficulties in disengaging focused attention, and has been conceptualized as a reduced ability to attend to the environment in flexible ways, such that focus may be restricted to one object of interest to the exclusion of much or all else (Mottron, Dawson, Soulières, Hubert, & Burack, 2006). A significant proportion of individuals with ASD exhibit hyper- or hyposensitivities to visual, tactile, auditory, or gustatory stimuli (Marco, Hinkley, Hill, & Nagarajan, 2011), which can lead to abnormalities of multisensory integration (Lane, Molloy, & Bishop, 2014). Hypersensitivity to visual or auditory stimuli may lead to avoidance of harsh light or sounds, whereas understimulation of vestibular systems may manifest in self-stimulatory stereotyped behaviors such as spinning and rocking (Martínez-Sanchis, 2014).

In DSM-5, clinical specifiers are intended to add dimensionality to the ASD diagnosis. The severity of functional impairment may be noted in each domain, whereas associated features (such as intellectual disability [ID], language impairment, or known genetic etiology) may be added to permit the tailoring of specific criteria to the individual (Grzadzinski, Huerta, & Lord, 2013). Although challenges remain in routinely distinguishing frank regression from developmental plateaus, current evidence suggests that approximately one-third or fewer cases of "typical autism" have a loss of previously acquired verbal or adaptive skills (Goldberg et al., 2003). In addition, a separate category of social communication disorder (SCD) was included in DSM-5 to capture those individuals with functionally impairing pragmatic social communication skills but without a significant history of repetitive behaviors, restricted interests, or sensory sensitivities.

As with all common psychiatric disorders, diagnosis relies on a combination of clinical observation and patient and/or caregiver reports. Reference standard diagnostic tests for ASD at present include the Autism Diagnostic Observation Schedule (ADOS) (Lord et al., 1989), in which the patient is assessed for the ability to perform various standardized social and communication tasks, and the Autism Diagnostic Interview—Revised (Lord, Rutter, & Le Couteur, 1994), a structured interview designed for caregiver(s). The Social Responsiveness Scale (SRS) (Constantino & Todd, 2005)) and the Social Communication Questionnaire (Berument, Rutter, Lord, Pickles, & Bailey, 1999) are increasingly part of a standard diagnostic assessment. The SRS is a questionnaire that provides a quantitative measure of aspects of social behavior and may be useful, in particular, to assess social deficits in affected individuals as well as more subtle impairments in categorically unaffected relatives. The Vineland Adaptive Behavior Scale (Sparrow, Balla, & Cicchetti, 2005; Sparrow & Cicchetti, 1985) is a standard measure of the individual's daily life skills, which has importance both for diagnosis as well as treatment planning. Cognitive (IQ) testing, along with adaptive functioning are essential components of a thorough diagnostic assessment. Testing to assess speech and language, executive functioning, and impulsivity are similarly important aspects of a comprehensive evaluation.

Associated Psychiatric and Medical Features

In addition to these core diagnostic features, a range of additional psychiatric and medical findings often accompany an ASD diagnosis. Physical stigmata may include syndactyly, skin findings such as hypomelanic patches, subcutaneous nodules, and neurofibromas and other characteristic features suggesting a specific genetic syndrome. An abnormal increase in head circumference (HC) has been postulated by some to be a general characteristic of autism (Amaral, Schumann, & Nordahl, 2008). However, analyses that control well for parental head circumference have not supported this generalization (Chaste et al., 2013; Raznahan et al., 2013). At the same time, there are many examples of both micro- and macrocephaly in otherwise "idiopathic" patients with ASD, as well as autism-associated genetic syndromes with pathognomonic HC findings. Therefore a careful assessment of HC should be part of the assessment of affected children, with outliers potentially indicating specific genetic causes (van Bon et al., 2015; Sanders et al., 2015; Varga, Pastore, Prior, Herman, & McBride, 2009).

Studies have shown that individuals with autism may have as much as a 30-fold increase in risk for seizures (Robinson, 2012; Tuchman & Rapin, 2002) above the general population prevalence of approximately 1% (Jokiranta et al., 2014; Kurtz, Tookey, & Ross, 1998). Notably these estimates likely include monogenic ASD syndromes with a higher risk of seizures (such as Rett syndrome) (Bolton et al., 2011). Girls and individuals with lower IQ are more likely to develop epilepsy (Amiet et al., 2008). Researchers have reported a bimodal age of onset, such that a significant subset of children are affected before 5 years of age, whereas others have the first seizure in their teens (Nomura, Nagao, Kimura, Hachimori, & Segawa, 2010). Seizures may be generalized tonic-clonic seizures or complex partial seizures that can be difficult to distinguish from nonepileptiform repetitive behaviors of ASD.

"Neurological" symptoms apart from epilepsy include motor abnormalities, such as lack of coordination, hypotonia, and motor delay. These are common, with estimates of up to about 80% of patients (Lai, Lombardo, & Baron-Cohen, 2014; Shetreat-Klein, Shinnar, & Rapin, 2014). Sleep disruption/abnormal sleep architecture are also extremely common. Initial insomnia is reported by parents of 53–78% of children with autism, and may cause significant family distress owing to the need to monitor continually for the individual's safety (Malow et al., 2012). Poor sleep correlates with worse behavioral impairments during the day and may therefore aggravate preexisting behavioral and psychiatric comorbidities (Goldman et al., 2009; Kotagal & Broomall, 2012). Gastrointestinal difficulties are also extremely common problems ranging from constipation to esophageal reflux and persistent diarrhea (McElhanon, McCracken, Karpen, & Sharp, 2014).

Psychiatric Comorbidities

Psychiatric and behavioral difficulties are highly prevalent in individuals with autism. Elevated rates of lifetime psychiatric comorbidities have been reported in adults and children with ASD, with rates greater than 70% in population-derived samples (Salazar et al., 2015; Simonoff et al., 2008) and even higher rates in clinically referred samples (Gillberg, Helles, Billstedt, & Gillberg, 2015; Hofvander et al., 2009; Joshi et al., 2013; Lugnegård, Hallerbäck, & Gillberg, 2011). Although the previous DSM-IV-TR criteria did not allow for comorbid diagnoses such as anxiety or attention-deficit hyperactivity disorder (ADHD), current DSM-5 criteria no longer specify that a diagnosis of ASD exclude other psychiatric comorbidities (Association, 2013).

Common comorbid conditions apparent early in development include tic disorders (present in 11–50% of individuals with ASD) and ADHD, present in approximately one-quarter to three-quarters of individuals with ASD (Developmental Disabilities Monitoring Network Surveillance Year 2010 Principal InvestigatorsCenters for Disease Control and Prevention (CDC), 2014; Lai et al., 2014). Symptoms and presentation of ADHD were found to be similar in those with or without ASD, but fewer individuals with ASD were receiving pharmacologic management for ADHD than were children with ADHD alone (Joshi, Faraone, Wozniak, Tarko, et al., 2014). Symptoms of poor attention and impulsivity in individuals with autism respond to stimulant medications, but these medications may be less effective in individuals with ASD and cause more intolerable side effects such as irritability and difficulty with sleep (Mahajan et al., 2012).

Obsessive compulsive disorder (OCD) is diagnosed in up to one-quarter of clinically referred patients (Gillberg et al., 2015; Hofvander et al., 2009; Joshi, Faraone, Wozniak, Petty, et al., 2014; Joshi et al., 2013) but requires that clinicians differentiate autism-related motor stereotypies from the repetitive behaviors and compulsions characteristic of OCD.

Rates of schizophrenia are not markedly elevated in the ASD population, but psychotic spectrum disorders have been identified in up to 20% of affected individuals (Billstedt, Gillberg, Gillberg, & Gillberg, 2005; Bakken et al., 2010; Gillberg et al., 2015; Hofvander et al., 2009; Joshi, Faraone, Wozniak, Tarko, et al., 2014; Joshi et al., 2010; Joshi et al., 2013; Lugnegård et al., 2011; Stahlberg, Soderstrom, Rastam, & Gillberg, 2004). Fewer studies have examined individuals with psychotic spectrum disorders for comorbid ASD, and a wide range of rates has been reported, from at least 3.6% of a large sample drawn from an early intervention psychosis clinic (based on case review and subject interview) (Davidson, Greenwood, Stansfield, & Wright, 2014), to 40.6% of a sample of individuals with diagnosed schizophrenia spectrum and bipolar disorders (based on parent interview and case records) (Unenge Hallerbäck, Lugnegård, & Gillberg, 2012).

A case−control study of over 9000 individuals diagnosed with ASD without ID before age 16 age-matched with over 90,000 control individuals demonstrated an increased odds ratio (OR) for the development of nonaffective psychosis (OR, 5.6; 95% confidence interval [CI], 3.3−8.5) and bipolar disorder (OR, 5.8; 95% CI, 3.9−8.7) compared with their unaffected full siblings (OR, 1.8; 95% CI, 1.1−2.7, and OR, 1.7; 95% CI, 1.1−2.6, respectively) (Selten, Lundberg, Rai, & Magnusson, 2015). In addition, parents of individuals with autism showed rates of bipolar and psychotic disorders higher than those of parents of the control population, which suggest a possible heritable factor contributing to the risk for bipolar and psychotic spectrum disorders.

Electronic medical record data mining is being used in much larger patient samples than are commonly enrolled in clinical trials (Kohane et al., 2012) to evaluate comorbidities and clinical outcomes. Research has confirmed the previously mentioned association of ASD with seizures and sleep disorders, and has highlighted comorbidities such as type I diabetes mellitus (2.08% after age 18 years) and inflammatory bowel disease (1.99% after age 18 years). The ability to track individuals longitudinally through the lifespan has enabled the subclassification of individuals into distinct groups: those with elevated rates of seizures, those with a tendency for immunologic comorbidities, and those with increased rates of neurobehavioral comorbidities such as ADHD or anxiety (Kohane, 2015). These results point to the possibility that either genetic or phenotypic variables in very large patient cohorts may eventually be used as classifiers to predict prognosis or response to specific treatments.

Prevalence

Prevalence estimates of 8-year-olds with ASD in the United States have reached 1 in 68 in Centers for Disease Control and Prevention surveys (MMWR Surveill Summ, 2002), with rates approaching 1% noted in adult population screens (Brugha et al., 2011). This constitutes a marked increase over the past several decades, almost certainly owing to multiple factors including improved ascertainment, changes in diagnostic criteria, diagnostic substitution, greater public awareness, the increasing identification of autism as a special education classification, and the interplay of genetic and environmental factors that may increase true incidence (as discussed subsequently) (Fombonne, 2009; Iossifov et al., 2012; King & Bearman, 2009; Neale et al., 2012; O'Roak, Vives, Girirajan, et al., 2012; Sanders et al., 2012).

Importantly, these prevalence figures clearly overestimate the rate of increase in the population over the past 40 years (owing, as noted, to some combination of markedly different diagnostic criteria, diagnostic substitution, increased awareness, and variable access to services that favor ASD diagnoses). At the same time, they likely underestimate the true prevalence of individuals meeting contemporary diagnostic criteria across the entire autism spectrum, at least in part because of incomplete ascertainment. For example, Kim et al. conducted a rigorous epidemiological study in South Korea that evaluated more than 50,000 children aged 7−12 years, first with an initial screening questionnaire and then using direct patient assessment and reference standard diagnostic tools. The initial survey using DSM-IV-TR criteria found prevalence rates of 2.64% in the total population sample. Interestingly, the rates of children previously ascertained as having social difficulties were similar to previously documented rates. However, most individuals ascertained in the study were in fact undiagnosed at the time of screening (Kim et al., 2011), which suggests that careful population-based ascertainment will lead to uniformly higher rates.

Moreover, with regard to the changing diagnostic criteria in DSM-5, the authors reevaluated the same population based on the revised diagnostic criteria for ASD and SCD, and although most (2.2% of 2.6%) retained the diagnosis of ASD, individuals who would have previously qualified for a diagnosis of PDD-NOS under DSM IV criteria were less likely to retain the ASD diagnosis (71% of those with a PDD-NOS diagnosis compared with 98% of individuals with autistic disorder and 92% with Asperger disorder). Of those who did not meet criteria for ASD under DSM-5 criteria, 76% met criteria for SCD and the remainder showed evidence of other developmental psychopathologies such as attention-deficit disorder or anxiety (Kim et al., 2014).

Importantly, ASD is found worldwide, with only negligible differences in documented prevalence when similar ascertainment approaches and diagnostic criteria are used (Baxter et al., 2014). Families with higher socioeconomic status have been found to be more likely to seek an evaluation for children exhibiting delayed development, and therefore may show higher prevalence rates (Rai et al., 2012; Sun, Allison, Auyeung, Baron-Cohen, & Brayne, 2014). Similarly, in the United States there are marked differences in the age of ascertainment of ASD that correspond to ethnicity, with under-represented minorities typically coming to attention years later. This is likely the result of a combination of cultural and socioeconomic factors and limited access to care (Developmental Disabilities Monitoring Network Surveillance Year 2010 Principal InvestigatorsCenters for Disease Control and Prevention (CDC), 2014).

Irrespective of the somewhat variable estimates of prevalence rates among studies, ASD is consistently found more often in males than in females. This overrepresentation is generally apparent among individuals with higher IQ and remains, but is less pronounced, in lower-IQ cohorts (Lai et al., 2014). There are important factors to consider when interpreting these data: for example, higher-functioning females may be diagnosed at later ages (Giarelli et al., 2010), they may have fewer externalizing symptoms (such as verbal or physical aggression), and they may present with relatively intact social communication skills compared with community-matched boys (Hiller, Young, & Weber, 2014). A bias toward male ascertainment is likely exacerbated by the fact that screening tools have been validated predominantly in male cohorts. Nonetheless, the totality of population-based as well as clinical studies continues to support the notion of a bona fide male predominance, a finding supported in part by molecular discoveries, as described in the ensuing section.

GENETICS AND GENOMICS

The past several years have witnessed dramatic progress in elucidating the allelic architecture of ASD and clarifying specific genes and genomic loci carrying substantial risks. This progress has been the result of a combination of rapidly advancing genomic technologies; a particularly productive effort at gene discovery leveraging rare and de novo mutations; the consolidation of large patient and family cohorts of sufficient size to power systematic discovery efforts; highly productive partnerships among governmental agencies, advocacy groups, and philanthropy to support research; and an early and sustained commitment to data sharing within the ASD genomics community.

These factors have yielded, and are continuing to yield, a harvest of specific risk genes that offer important clues to the molecular, cellular, and circuit-level pathophysiology of ASD. At the same time, studies have highlighted an extraordinary degree of genetic heterogeneity and biological complexity underlying ASD and a range of related neurodevelopmental syndromes, prompting at least a partial reconceptualization of the challenges for translational neuroscience. The ensuing sections address what is currently known regarding the genetics and genomics of ASD, focusing particularly on more recent findings, and present an assessment of evolving approaches that aim to move the field from gene discovery to an actionable understanding of pathophysiological mechanisms, including pathways and circuits.

Genes and Environment in Autism Spectrum Disorder

Twin studies have uniformly supported the contention that autism is heritable (Folstein & Rutter, 1977; Lichtenstein, Carlström, Råstam, Gillberg, & Anckarsäter, 2010; Ritvo, Freeman, Mason-Brothers, Mo, & Ritvo, 1985; Rosenberg et al., 2009). Not surprisingly, there has been variability in estimates of heritability reflecting differences in ascertainment strategies, diagnostic formulations, and sample sizes. Monozygotic concordance rates have ranged from a low of slightly less than 40% to greater than 90% and dizygotic rates have ranged from 0% to 30% (Bailey et al., 1995; Bernier, Gerdts, Munson, Dawson, & Estes, 2012; Colvert et al., 2015; Hallmayer et al., 2011; Ritvo et al., 1985; Rosenberg et al., 2009; Sandin et al., 2014). Importantly, most of these studies indicate that genes contribute a greater proportion of the overall population risk for ASD than environmental factors and all agree that both genes and environment have a contributory role in etiology.

Family studies also support a strong genetic contribution, with recurrence risks of up to 26% in contemporary studies of siblings of individuals affected with autism (Bolton et al., 1994; Constantino et al., 2013; Jorde et al., 1991; Ozonoff et al., 2010; Ritvo et al., 1989; Rogers, 2009; Sandin et al., 2014; Zhao et al., 2007). These studies have also generally found that female siblings of individuals with autism are less likely to carry an ASD diagnosis than male siblings or twins (Ozonoff et al., 2011; Rosenberg et al., 2009; Werling & Geschwind, 2015). The findings may be partly explained by the diagnostic confounds noted earlier but also suggest that females require greater overall genetic risks to manifest ASD symptoms, a result that has garnered additional empirical support from molecular genetic studies (De Rubeis et al., 2014; Dong et al., 2014; Iossifov et al., 2014, 2012; Levy et al., 2011; Neale et al., 2012; O'Roak, Vives, Girirajan, et al., 2012; Sanders et al., 2011, 2015, 2012).

Although this chapter focuses on genetic risks and their implications for understanding molecular, cellular, and circuit mechanisms, in the long run, studies of environmental factors have the potential not only to illuminate risks but also to highlight particularly tractable opportunities for prevention. For instance, whereas exposure to thalidomide, valproic acid, or rubella infection accounted for only a tiny fraction of ASD risk, elimination of these agents in pregnant women and vaccination have proven to be effective prevention strategies (Ornoy, Weinstein-Fudim, & Ergaz, 2015). Similarly, a study demonstrated that well-accepted interventions such as supplementation with folate during pregnancy may significantly reduce the risk for ASD (Surén et al., 2013).

A case–control study of several hundred affected individuals and their parents (Croen, Grether, Yoshida, Odouli, & Hendrick, 2011) pointed to a twofold increase in risk for ASD for mothers ($N = 15$) who used selective serotonin reuptake inhibitors (SSRIs) during pregnancy, with a reported threefold increase in risk for children exposed during the first trimester. Because of the widespread use of SSRIs, the finding received considerable public attention; however, a subsequent study of a large Danish epidemiological cohort involving more than twice the number of children exposed during pregnancy found no association (Hviid, Melbye, & Pasternak, 2013). Based on the raw data, the Danish study suggested such a risk, yet when the authors controlled for relevant confounds, the association was no longer present. Specifically, a significant risk for ASD was found for a range of maternal psychiatric diagnoses including depression, as well as for medication use, before but not during pregnancy, which pointed to the likelihood of "confound by indication," an important consideration in these types of analyses.

Finally, because of the societal impact of fears regarding vaccine exposure and ASD, the issue deserves mention here. Although there clearly remains concern among segments of the general public, the weight of the now considerable scientific evidence does not support this. As has been widely noted, much of the initial consternation was generated by a fatally flawed publication that has subsequently been retracted (The Editors of The Lancet, 2010; Murch et al., 2004). Regardless, multiple subsequent rigorous investigations have pursued the question with no credible suggestion of an association. The multiple negative studies in humans and in primates (Gadad et al., 2015) are beyond the scope of this chapter. However, it is worth highlighting that a meta-analysis of studies totaling more than a million children found no increase in ASD risk associated either with measles, mumps, and rubella vaccine or vaccines containing the mercury derivative thimerosal (Taylor, Swerdfeger, & Eslick, 2014).

Early Efforts at Gene Discovery in Autism Spectrum Disorder

The evidence that genetics has a role in the etiology of ASD is incontrovertible, but until recently the process of identifying specific genes has been challenging. In the 1990s, attempts to leverage linkage analyses using multiple affected pedigrees led to the hypothesis that 5 to 15 genes would be found to contribute most liability (Risch et al., 1999). These estimates were a product of the best methods available at the time; however, starting around the turn of the millennium, with the maturation of approaches to study complex human disease genomics, it became clear that these estimates of locus heterogeneity were off by as much as two orders of magnitude. Moreover, throughout the 1990s and early 2000s, a host of studies sought to test ASD association for common variations in selected candidate genes. These are now appreciated to have vastly overestimated the plausible effect sizes of such variants and were consequently profoundly limited by a lack of power. This miscalculation, along with a host of now well-appreciated technical and methodological challenges, led candidate gene studies of ASD to be generally unreliable (Altshuler, Daly, & Lander, 2008) (refer to chapter: Association Strategies), as has been demonstrated for nearly all of human disease genetics.

In retrospect, it is now clear that the earliest successes in ASD gene discovery were a consequence of efforts that were not explicitly aimed at understanding autism or social disability. The study of monogenic ID syndromes yielded the very first ASD loci (see chapter: Neurodevelopmental Disorders, Causes and Consequences) including, for example, the genes responsible for fragile X, tuberous sclerosis complex, neurofibromatosis, and Rett syndromes. These monogenic disorders all have been show to carry increased risk for social disability in addition to ID. Moreover, collectively they have been an important harbinger of subsequent findings in nonsyndromic (also called idiopathic or typical) ASD. For example, from as early as the 1990s, gene discovery in monogenic syndromes pointed to the potential contribution of RNA binding proteins, synaptic function, the mammalian target of the rapamycin pathway, abnormal protein synthesis, and chromatin modification to the pathophysiology of ASD. As discussed subsequently, these hypotheses are now strongly supported by findings from microarray and next-generation sequencing studies of typical ASD cohorts.

Importantly, work in model systems focused on these monogenic ID/ASD syndromes have offered up promising results with regard to the plausibility of moving from an understanding of mechanism to developing somatic treatments. It is now widely accepted that at least partial rescue of the phenotypes resulting from these mutations is possible even in adult animals (Ghosh, Michalon, Lindemann, Fontoura, & Santarelli, 2013; Krueger & Bear, 2011; Silverman & Crawley, 2014). These results have prompted an important debate regarding the fundamental nature of ASD: The phenotypic rescue observed in adult knockouts via genetic and pharmacological manipulations (Dansie et al., 2013; Ehninger et al., 2008; Garg et al., 2013; Michalon et al., 2012) raises the question of whether autism is at its core an ongoing "cell biology" problem in which the molecular and cellular mechanisms of pathology are largely constant over time, or conversely, whether ASD is a bonafide disorder of development, necessitating an understanding of the temporal and anatomical sequencing of etiological events and compensatory responses to truly capture and ultimately address the underlying pathophysiology. Although this remains a subject of active debate, several lines of evidence, including data from gene discovery efforts, suggest that ASD reflects both ongoing functional deficits and altered neurodevelopmental processes, as discussed in more detail in the ensuing sections.

Genome-Wide Association Strategies

Tremendous advances have been made in understanding the contribution of common variation to common human disease using genome-wide association strategies (GWAS) (see chapter: Association Strategies). The emergence of these approaches ushered in an era of reliable and systematic identification of common risk alleles for disorders that previously appeared to be largely intractable. Studies of schizophrenia and bipolar disorders (Mühleisen et al., 2014; Schizophrenia Working Group of the Psychiatric Genomics Consortium, 2014) provided striking examples with regard to neurodevelopmental and psychiatric disorders. Successes for these syndromes also underscore the individually small allelic effects carried by transmitted common variants and the large magnitude of the cohorts necessary to identify and replicate risk loci reliably.

To date, GWAS of populations with ASD have not been as successful as in these examples. A handful of common polymorphisms (and nearby genes) have been reported to reach widely accepted thresholds for significance (Wang et al., 2009; Weiss, Arking, Gene Discovery Project of Johns Hopkins & the Autism Consortium, Daly, & Chakravarti, 2009). However, as larger cohorts have been consolidated and studied, these loci have not yet been replicated (Anney et al., 2012).

At the same time, multiple studies have confirmed the overall contribution of common variations to population risk for ASD. Indeed these investigations have demonstrated that the lion's share of population risk is attributable to common polymorphisms (Gaugler et al., 2014; Klei et al., 2012). This apparent disconnect between evidence for the contribution of common variation (writ large) and the difficulty in establishing replicable specific loci is almost certainly the result of still underpowered study samples. Although cohorts have now reached into the thousands, these studies of schizophrenia and bipolar disorder required tens of thousands of individuals to make substantial progress. Given the early onset of ASD and the well-appreciated reduction in fecundity (Power et al., 2013) (see chapter: Natural Selection and Neuropsychiatric Disease: Theory, Observation, and Emerging Genetic Findings) it is unsurprising that the anticipated effect sizes for individual common alleles carrying ASD risk are now as small or are smaller than those seen for these other psychiatric conditions, which reinforces the need for larger samples to power common variant discovery cohorts.

De Novo and Rare Variation in Autism Spectrum Disorder

Whereas studies of common variation have not yet led to a harvest of risk alleles, studies of rare and de novo mutations have been extremely productive not only in the study of monogenic forms of ASD but in investigations of typical or nonsyndromic ASD. Some of the first hints that this might be the case emerged as early as the 1990s via cytogenetic studies. Several case series found that large-scale rearrangements, duplications, and deletions (observable by light microscope) could be detected in approximately 1–5% of cases (Lauritsen, Mors, Mortensen, & Ewald, 1999; Li, Chen, Lai, Hsu, & Wang, 1993; Wassink, Piven, & Patil, 2001; Weidmer-Mikhail, Sheldon, & Ghaziuddin, 1998).

Although these studies collectively pointed to the importance of structural variation in ASD, through the early part of the current millennium they did not lead to the identification of a specific risk ASD gene, largely owing to the limited resolution allowed by traditional cytogenetics and the rarity of the observed events. However in 2003, Jamain et al. (Jamain et al., 2003) investigated a region on the X chromosome (Xp23) that had previously been observed to have recurrent de novo deletions in three females with ASD (Thomas et al., 1999). The authors undertook Sanger sequencing in 158 families with ASD and identified a nonsense mutation in the gene *Neuroligin 4 (NLGN4)* originating de novo in the unaffected mother and segregating to two affected children with ASD, which was not present in 350 control subjects. Similarly, the article also reported a missense mutation at a highly conserved amino acid in the gene *Neuroligin 3* in a second family with two affected siblings that was not present in control subjects.

In hindsight, this brief report was remarkably prescient in that it noted the role of a likely gene disrupting de novo rare structural and sequence variation in typical ASD, suggested female protective factors, and focused attention on synaptic proteins. Moreover, a subsequent article published less than a year later reporting a nearly identical protein-disrupting mutation in *NLGN4* in a family segregating ASD and ID, highlighted the phenotypic variability now commonly attributed to ASD mutations (Laumonnier et al., 2004). Although not widely appreciated at the time, these findings ushered in the modern era of gene discovery in ASD.

The simultaneous development of higher-resolution cytogenetic tools, in particular microarrays, and methods to query complex genomic DNA led to a second critical leap in ASD gene discovery. In 2006–2007, these approaches revealed that submicroscopic deletions and duplications (copy number variations [CNVs]) were a part of the spectrum of normal human genetic variation and that de novo events were overrepresented in individuals with ASD. Several groups in 2006 and 2007 seized on the opportunity afforded by CNV detection to study ASD and ID and provided the first evidence that this would be a highly productive approach to establishing ASD loci (Autism Genome Project Consortium et al., 2007; Jacquemont et al., 2006; Sebat et al., 2007).

Sebat and colleagues published a particularly influential article that focused specifically on de novo CNVs in typical ASD. They compared the rates of de novo CNVs in 118 families with only a single affected offspring (sporadic or simplex patients) with 77 families with multiple affected individuals and 198 control families (Sebat et al., 2007). They detected de novo CNVs in 10% of probands from simplex families (significantly different from controls; $p = 0.0005$), and in 2.6% of probands from multiplex families (not statistically different from control subjects). Two of 196 control subjects (1.0%) also showed de novo CNVs. A higher proportion of affected females harbored de novo mutations than did males with autism.

These findings were quickly followed by the important observation that just as with de novo gross cytogenetic abnormalities, de novo CNVs in some cases could be found to cluster in specific regions of the genome, a phenomenon that could be used to assess and confirm the association of specific regions to risk (Kumar et al., 2008; Marshall et al., 2008; Weiss et al., 2008). Subsequently, multiple studies confirmed that de novo CNVs are present in individuals with autism at rates of 5–10%, consistently higher than those detected in unaffected siblings or control subjects (<1–2%), and that these variants point to specific risk regions (Autism Genome Project Consortium et al., 2007; Itsara et al., 2010; Levy et al., 2011; Marshall et al., 2008; Pinto et al., 2014, 2010; Sanders et al., 2011, 2015).

Beginning in 2008, a dedicated effort to leverage de novo mutation in the hunt for autism risk loci was undertaken by the Simons Foundation, through the creation of the Simons Simplex Collection (SSC). This cohort is composed of more than 2500 carefully phenotyped simplex families consisting predominantly of quartets (two unaffected parents, a proband, and at least one unaffected sibling) and a minority of trios (two unaffected parents and a proband) (Fischbach & Lord, 2010). In 2011, two studies reported on the first approximately 1000 SSC families for which microarray data were available (Levy et al., 2011; Sanders et al., 2011). Consistent with prior findings, de novo CNVs were found in 5.9–8% of probands and in 1.7–2% of siblings, and large, multigenic CNVs were found to be more likely associated with ASD risk.

Sanders et al. (2011) focused specifically on developing a framework for the assessment of genome-wide significant association using the probability of recurrence of a de novo event within a given locus and reported two CNV loci that reached this threshold (16p11.2 and 7q11.23). Moreover, as in prior, smaller studies, girls were found to carry a greater number of large de novo CNVs than were boys and to have more genes within these CNVs by a wide margin, a result that was interpreted to support the notion of an ASD protective effect associated with female sex.

In addition, both studies found a downward (or leftward) shift in the distribution of IQ associated with having a large de novo CNV, but also observed that a substantial proportion of affected individuals with a putative risk mutation had IQ above 70. Moreover, there was no association between ADOS severity scores and the presence or absence of a de novo CNV in either males or females. Thus neither ID nor increased severity in typical autism was a reliable indicator of the presence or absence of a de novo CNV (Sanders et al., 2011). These findings provided an important counterpoint to the widespread practice at the time of restricting CNV studies in the clinical population to affected individuals with ID and/or dysmorphology. Finally, both studies calculated that several hundred distinct genomic regions would ultimately be found to contribute to autism risk. An analysis of CNVs in the entire SSC cohort replicated all of these findings, including the association of ASD with de novo CNVs, the relationship to IQ, and the excess mutation burden in female versus male probands, in both the newly analyzed group ($N = 1226$) as well as in a combined analysis ($N = 2100$) (Sanders et al., 2015).

De Novo Single-Nucleotide Variants and Insertion–Deletions in ASD

Although these efforts to characterize structural variation offered the first reliably systematic approach to identifying ASD loci, they remained suboptimal for gene discovery owing to their size. Only a small proportion of recurrent de novo CNV pointed to a single specific risk gene: eg, genic deletions in the region corresponding to the gene *Neurexin 1 (NRXN1)*. Most recurrent CNVs associated with risk contained many genes in addition to noncoding sequences, which made the process of identifying and pursuing a single ASD gene challenging.

The development of next-generation sequencing offered an avenue to overcome this obstacle by allowing for the analysis of most coding bases in the human genome at high throughput and reasonable cost. The first published report of exome sequencing in 20 individuals with ASD (O'Roak et al., 2011) was quickly followed by a series of four studies published contemporaneously (Iossifov et al., 2012; Neale et al., 2012; O'Roak, Vives, Girirajan, et al., 2012; Sanders et al., 2012) that completed exome sequencing on a combined total of 752 trio and quartet families from the SSC and 175 trio and quartet families from the Boston Autism Consortium. Both individually and collectively, these studies found that the rate of rare de novo single-nucleotide variants (SNVs) predicted to disrupt protein function (stop codons, canonical splice sites, and frame-shift mutations) was higher in affected individuals than in siblings (or compared with expectation). Moreover, through the observation of multiple rare de novo point mutations clustering in the same gene, they identified five genes reaching genome-wide significance: *CHD8, SCN2A, KATNAL2, POGZ,* and *DYRK1A*. In addition, the studies showed that older parents, fathers in particular, contributed a larger number of de novo variants than did younger parents (Iossifov et al., 2012; Neale et al., 2012; O'Roak, Vives, Girirajan, et al., 2012; Sanders et al., 2012), a finding later replicated in whole-genome sequencing data (Kong et al., 2012). O'Roak, Vives, Girirajan, et al. (2012) pursued this finding in greatest detail, showing that the vast majority of de novo mutations in patients with ASD were present on the paternal chromosome.

Expanded exome sequencing studies have now dramatically expanded the number of genes confidently associated with ASD. Iossifov and colleagues (Iossifov et al., 2014) reported completion of the sequencing of the SSC ($N = 2508$ families) and replicated the significant excess of gene-disrupting SNVs in probands compared with siblings ($p = 2 \times 10^{-5}$). A total of 27 genes were found to have de novo likely gene-disrupting (LGD) mutations in unrelated affected individuals, and six of these were in the highest confidence group because they were found more than twice: *CHD8*, *DYRK1A*, *ANK2*, *GRIN2B*, *DSCAM*, and *CHD2*. Simultaneously, the Autism Sequencing Consortium (Buxbaum et al., 2012) reported on a total of 4083 subjects with ASD, including 2410 trios (De Rubeis et al., 2014) (825 were from the SSC and consequently overlapped with the subject from Iossifov et al.). These authors used a recently developed statistical test that allowed for the integration of evidence from both rare transmitted as well as de novo variation. The approach (transmitted and de novo association [TADA]) calculates a gene-specific false discovery rate (FDR) that assesses the overall likelihood of association (He et al., 2013). The authors reported a total of 33 genes with an FDR of <0.1. Thirteen of these were very high confidence genes, with an FDR of <0.01; *ADNP*, *ANK2*, *ARID1B*, *CHD8*, *CUL3*, *DYRK1A*, *GRIN2B*, *KATNAL2*, *POGZ*, *SCN2A*, *SUV420H1*, *SYNGAP1*, and *TBR1*.

Sanders et al. (2015) integrated new CNV data from a further 1500 SSC families with the published data on de novo CNVs (Levy et al., 2011; Pinto et al., 2014) and de novo point mutations from both of these investigations. Using an extended version of TADA (He et al., 2013) that allowed for the integration of data from small de novo CNVs as well as SNVs, they confirmed a total of 71 risk loci, including 65 genes (Table 29.1) and six CNVs (Table 29.2). A total of 28 genes had an FDR of <0.01. This added the following genes to those listed previously: *ASHL1*, *KDM5B*, *KMT2C*, *NCKAP1*, *NRXN1*, *PTEN*, *SETD5*, *SHANK2*, *SHANK3*, *TCF7L2*, *TNRC6B*, *TRIP12*, and *WAC*. Protein–protein interaction (PPI) analysis of these 65 genes generated a network composed of two subnetworks: one enriched for gene ontology terms of chromatin regulation and transcription and the other for synaptic and neuronal development terms (Fig. 29.1). Moreover, the study demonstrated an inverse correlation between the size of a given CNV and the likelihood of overlap with de novo point mutations. This suggested that small de novo CNVs are likely to carry a single large effect risk gene, whereas large CNVs appear to carry multiple risk genes of modest effect. The data help in retrospect to explain the historic difficulties in fine-mapping multigenic ASD risk CNVs and support the overall finding that the risk associated with a de novo highly disruptive sequence mutation carries the same scale of risk as a CNV disrupting many genes.

Overall, the contribution of de novo mutations to all types of ASD is considerable. De Rubeis et al. (De Rubeis et al., 2014) calculated that the genes identified in their highest confidence group increase risk by 20-fold, which points to the very large biological effects accompanying some, but certainly not all, of these rare de novo mutations. Sanders et al. (2015) calculated that 10% of all clinically ascertained ASD cases carry a de novo sequence or structural mutation contributing to ASD, with an even higher rate (16.6%) in girls. Other groups have predicted that as many as 30% or more of individuals with ASD will ultimately be found to carry a contributing de novo LGD or missense mutation (Ronemus, Iossifov, Levy, & Wigler, 2014). Finally, predictive modeling of all of the large-scale sequencing studies concluded that several hundred to over 1000 genes will ultimately be found to contribute to autism risk through de novo point mutations (De Rubeis et al., 2014; Iossifov et al., 2014, 2012; Neale et al., 2012; O'Roak, Vives, Girirajan, et al., 2012).

TABLE 29.1 Statistical Evidence for Association of Autism Spectrum Disorder Genes Based on Rare De Novo and Transmitted Sequence Variation and De Novo Small Copy Number Variations

DN Loss-of-Function	FDR ≤ 0.01	0.01 < FDR ≤ 0.05	0.05 < FDR ≤ 0.1
2	*ADNP, ANK2, ARID1B, ASH1L, CHD2, CHD8, CUL3, DSCAM, DYRK1A, GRIN2B, KATNAL2, KDM5B, KMT2C, NCKAP1, POGZ, SCN2A, SUV420H1, SYNGAP1, TBR1, TCF7L2, TNRC6B, WAC*	*BCL11A, FOXP1, GIGYF1, ILF2, KDM6B, PHF2, RANBP17, SPAST, WDFY3*	*DIP2A, KMT2E*
1	*NRXN1, PTEN, SETD5, SHANK2, SHANK3, TRIP12*	*DNMT3A, GABRB3, KAT2B, MFRP, MYT1L, P2RX5*	*AKAP9, APH1A, CTTNBP2, ERB-B2IP, ETFB, INTS6, IRF2BPL, MBD5, NAA15, NINL, OR52M1, PTK7, TRIO, USP45*
0		*MIB1, SLC6A1, ZNF559*	*ACHE, CAPN12, NLGN3*

False discovery rates (FDRs) are calculated using transmitted and de novo association (TADA) from He, X., Sanders, S. J., Liu, L., De Rubeis, S., Lim, E. T., Sutcliffe, J. S., ... Roeder, K. (2013). Integrated model of de novo and inherited genetic variants yields greater power to identify risk genes. *PLoS Genetics, 9*(8), e1003671. http://dx.doi.org/10.1371/journal.pgen.1003671; Modified to incorporate structural variation from Sanders, S. J., He, X., Willsey, A. J., Ercan-Sencicek, A. G., Samocha, K. E., Cicek, A. E., ... State, M. W. (2015). Insights into autism spectrum disorder genomic architecture and biology from 71 risk loci. *Neuron, 87*(6), 1215–1233. http://dx.doi.org/10.1016/j.neuron.2015.09.016.

TABLE 29.2 Recurrent De Novo Copy Number Variations Found in Simons Simplex Collection and Autism Genome Project Cohorts

Band	Location (hg19)	De novo Single-Nucleotide Variants	Del/Dup	q Value (false discovery rate)	Schizophrenia Risk[a]
1q21.1	chr1:146,467,203–147,801,691	9	1/8	2×10^{-9}	Y (Del/Dup)
2p16.3 (NRXN1)	chr2:50,145,643–51,259,674	8	7/1	4×10^{-8}	Y (Del)
3q29	chr3:195,747,398–196,191,434	4	4/0	0.02	Y (Del)
7q11.23	chr7:72,773,570–74,144,177	5	1/4	0.0008	Y (Dup)
15q11.2-13.1	chr15:23,683,783–28,446,765	10	0/10	$<1 \times 10^{-10}$	Y (Del/Dup)
15q13.2-13.3	chr15:30,943,512–32,515,849	5	3/2	0.0008	Y (Del)
16p11.2	chr16:29,655,864–30,195,048	19	12/7	$<1 \times 10^{-10}$	Y (Dup)
22q11.21	chr22:18,889,490–21,463,730	8	4/4	1×10^{-7}	Y (Del)
22q13.33 (SHANK3)	chr22:51,123,505–51,174,548	4	4/0	0.02	

Del/Dup, deletion/duplication.
[a]*Overlap with schizophrenia risk is based on analyses by Rees et al. (2014). Variations listed have a q value of <0.1.*
Adapted from Sanders, S. J., He, X., Willsey, A. J., Ercan-Sencicek, A. G., Samocha, K. E., Cicek, A. E., ... State, M. W. (2015). Insights into autism spectrum disorder genomic architecture and biology from 71 risk loci. *Neuron, 87*(6), 1215–1233. http://dx.doi.org/10.1016/j.neuron.2015.09.016.

Recessive Alleles and Autism Spectrum Disorder

The productivity of studies of rare de novo mutations, the continued challenges associated with identifying common risk alleles, and the reduction in fecundity of individuals with a diagnosis of ASD have been interpreted to highlight the role of natural selection in the genetics of ASD (Power et al., 2013). New mutation largely, though not entirely, escapes the selective pressures that would be expected to eliminate large-effect common ASD alleles from the general population (see chapter: Natural Selection and Neuropsychiatric Disease: Theory, Observation, and Emerging Genetic Findings). Based on

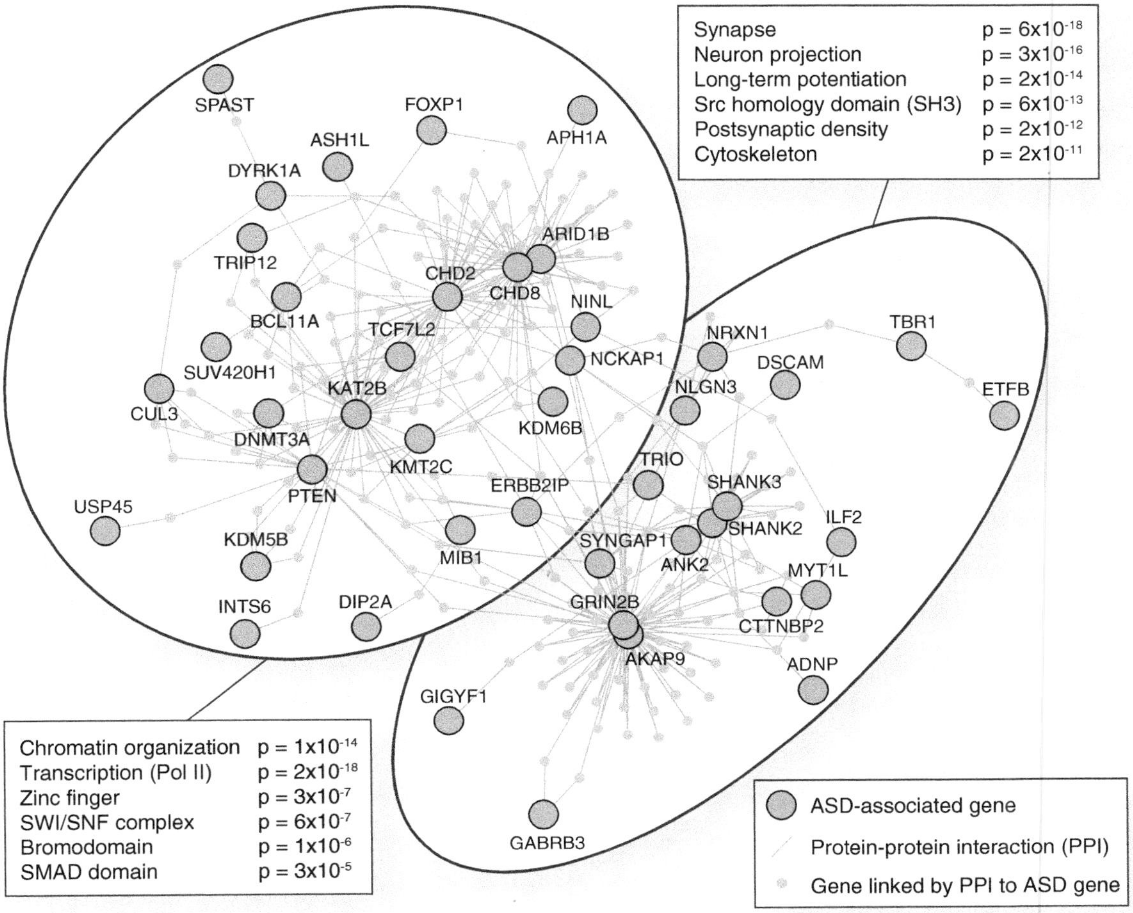

FIGURE 29.1 PPI network of ASD risk genes. The 65 ASD risk genes (Table 29.1) (*large gray circles*) form a single PPI network. The network has a clear division into two subnetworks (*shown by large ovals*). The genes in the left subnetwork are enriched for gene ontology terms associated with chromatin and transcriptional regulation. The genes in the right subnetwork are enriched for synaptic and neuronal development terms. *Adapted from Sanders, S. J., He, X., Willsey, A. J., Ercan-Sencicek, A. G., Samocha, K. E., Cicek, A. E., … State, M. W. (2015). Insights into autism spectrum disorder genomic architecture and biology from 71 risk loci.* Neuron, *87(6), 1215–1233. http://dx.doi.org/10.1016/j.neuron.2015.09.016.*

this logic, one would also expect that rare recessive mutations in extended pedigrees, population isolates, and even outbred populations could be found carrying relatively large effects. Indeed several studies have found highly deleterious recessive mutations in rare families and point to a role for these types of mutation in idiopathic ASD more broadly.

In 2006, a study leveraging traditional linkage analysis followed by Sanger sequencing identified rare recessive apparent loss-of-function homozygous mutations in the gene *Contactin Associated Protein 2* in members of the Old Amish population who had cortical dysplasia-focal epilepsy syndrome (Strauss et al., 2006). Subsequent studies of this gene have confirmed its role in central nervous system function and development. The question of whether other rare heterozygous mutations also carry measurable risk remains uncertain (Bakkaloglu et al., 2008; Murdoch et al., 2015).

An extensive molecular study of consanguineous ASD families was published in 2008 (Morrow et al., 2008). The researchers reported on more than 200 families ascertained and carefully phenotyped from across the Middle East and reported on the association of the genes *PCDH10* and *NHE9A* with ASD risk. Subsequently, the search for homozygous recessive mutations became increasingly tractable with the advent of next-generation sequencing: For example, using this approach, homozygous LGD mutations were identified in the gene *Branched Chain Ketoacid Dehydrogenase Kinase* (*BCKDK*) in multiple affected members of unrelated Middle Eastern pedigrees presenting with epilepsy, ID, and autism (Novarino et al., 2012). The *BCKDK* protein product catalyzes the rate-limiting step in the synthesis of branched chain amino acids. Homozygous LGD mutations in *BCKDK* lead to reduced serum levels of branched chain amino acids (BCAAs), and supplementation of BCAAs ameliorates the neurologic phenotype in the mouse model (Novarino et al., 2012). Studying consanguineous pedigrees, Yu et al. (2013) found recessive risk alleles in the genes *AMT*, *PEX7*, *SYNE1*, *VPS13B*, *PAH*, and *POMGNT1* with additional mutations in *VPS13B* observed among about 600 outbred simplex families from the SSC. Consistent with these findings, a study of exome sequencing data from more than 900 nonconsanguineous ASD families found a twofold enrichment of rare complete knockouts (either homozygous or compound heterozygous loss of function mutations) compared with control subjects. Combining these types of mutations on autosomes and the X chromosome, the authors estimated that approximately 5% of the risk for ASD will be explained via rare "complete knockouts" (Lim et al., 2013).

FROM GENES TO BIOLOGY IN AUTISM SPECTRUM DISORDER

The rapidly increasing number of reproducible risk genes forms a critical foundation for "bottom-up" approaches to illuminating the pathophysiology of ASD. High-confidence ASD mutations, particularly those mapping to coding regions of the genome, provide considerable traction for examining risk at the molecular and cellular levels. Moreover, germ-line variation is highly valuable in establishing cause versus effect, given the presence of the perturbation from the earliest cell divisions in the embryo.

However, it is clear that navigating the path from genes to an actionable understanding of biology is shaping up to be a formidable task. Certainly, the high degree of locus heterogeneity presents logistical challenges, even in the CRISPR/Cas genome-editing era. The combination of extraordinary variability in clinical outcome that has been widely observed for ASD risk loci (De Rubeis & Buxbaum, 2015; Malhotra & Sebat, 2012), coupled with the biological pleiotropy of many of the genes so far implicated, adds to the difficulty of linking any particular observation in an ASD model with the specific pathophysiological mechanisms relevant to human social disability. Moreover, the likelihood that neurodevelopmental disorders involve circuit level dysfunction suggests that in some if not many cases, direct translation from a mutation to a cell-autonomous phenotype, and to a drug target, as has been so central to progress in fields such as oncology, is likely to be the exception and not the rule for ASD.

Of course, the most difficult aspect of the undertaking arises as a consequence simply of studying human brain. It is a tremendously complex, relatively poorly understood, and largely inaccessible organ that shows extensive cellular diversity and developmental dynamism. Placing the functional consequences of any mutation into a meaningful context is a challenge made more formidable by the differences between the human brain and the brains of the most widely used model systems.

Functional Convergence

Despite these challenges, there has been significant progress in moving from gene discovery to at least a preliminary understanding of relevant biological mechanisms. Given the totality of findings from ASD genomic studies, the field has increasingly been focused on the notion of convergence, specifically what function(s) or mechanism(s) tie disparate ASD genes together. Of course, considerable focus remains on studying single gene mutations in traditional model systems. In fact, a cursory PubMed search of "mouse" and "autism" yielded more than 2000 papers. However, all of these challenges suggest that in addition to the deep investigation of one ASD mutation at a time, the question of which characteristics overlap among multiple ASD genes is of increasing interest.

Even with the first reports of de novo mutation in idiopathic ASD (Jamain et al., 2003), the idea of mechanistic or functional convergence began to emerge in the literature. For example, an influential viewpoint published in *Science* in 2003 (Zoghbi, 2003), entitled "Postnatal Neurodevelopmental Disorders: Meeting at the Synapse," considered the intersection of pathophysiological mechanisms between Rett syndrome and ASD, focusing on experience-dependent synaptic plasticity.

As the number of genes identified with high confidence has increased nearly exponentially over the past several years, analyses of the convergence among diverse risk loci has proliferated and begun to incorporate a number of varied analytic methods, including assessing PPI and gene ontologies (De Rubeis et al., 2014; Li et al., 2014; Liu et al., 2014; Sanders et al., 2015). Overall, despite some variation in approach, these studies have largely agreed on two groups of genes across studies: those involved in chromatin/transcriptional regulation and those related to synaptic transmission (Fig. 29.1). In addition, individual studies have highlighted genes involved in cell junction, the transforming growth factor-β pathway, neurodegeneration (De Rubeis et al., 2014), cell-signaling including mitogen-activated protein kinases (Pinto et al., 2014), cell adhesion, and ligase activity (Liu et al., 2014).

To identify gene relationships beyond PPI and gene ontology, the NETBAG+ tool also integrates data from gene expression, gene phylogeny, and genomic location (Feldman, Rzhetsky, & Vitkup, 2008; Gilman et al., 2011). This tool has been trained to identify clusters of disease-related genes within the network using known genetic variants associated with multiple nonneurological disorders. Analysis of 991 genes identified from de novo CNVs, de novo loss-of-function mutations, or de novo missense mutations (Iossifov et al., 2012; Levy et al., 2011; O'Roak, Vives, Girirajan, et al., 2012; Sanders et al., 2012) found a single ASD-associated network within which four subsets of genes were identified: (1) neuronal signaling/cytoskeleton; (2) channel activity; (3) chromatin modification/regulation; and (4) postsynaptic density (Chang, Gilman, Chiang, Sanders, & Vitkup, 2015).

Exome sequencing studies have also highlighted at least three additional nexuses of functional convergence. Targets of the fragile X mental retardation protein, identified by cross-linking immunoprecipitation in the mouse brain (Darnell et al., 2011), are highly enriched for ASD risk genes (Bernier et al., 2014; De Rubeis et al., 2014; Dong et al., 2014; Iossifov et al., 2014, 2012). Similarly, a large-scale exome study (De Rubeis et al., 2014) found enrichment of targets of RBFOX splicing factors. Also, findings from multiple exome studies have focused attention on the gene *CHD8*, a chromodomain protein that shows the greatest number of de novo putative loss of function mutations so far identified in whole-exome sequencing (WES) of ASD (Bernier et al., 2014; De Rubeis et al., 2014; O'Roak, Vives, Fu, et al., 2012; O'Roak, Vives, Girirajan, et al., 2012).

Heterozygous loss of *CHD8* has been postulated to result in decreased levels of CHD8 that then lead to widespread changes in gene expression directly, by binding to the gene promoter, or indirectly, for example, through other chromatin regulators. Using chromatin immunoprecipitation followed by high-throughput sequencing (ChIP-seq), several groups localized the protein binding sites of the CHD8 protein in human neuroprogenitor stem cells (hNSC) (Cotney et al., 2015; Sugathan et al., 2014). CHD8 was observed at active promoters, as indicated by overlapping H3K27ac and H3K4me3 histone marks. Using shRNA knockdown in hNSC, these studies independently identified sets of genes that are differentially expressed when CHD8 protein is reduced. As expected, many but not all of these differentially expressed genes are bound by CHD8, which suggest that CHD8 has direct and indirect effects on gene targets. By also mapping CHD8 sites in human fetal brain tissue and mouse embryonic cortex, Cotney et al. (Cotney et al., 2015) identified a set of conserved CHD8 targets that are bound in human NSCs, brain tissue, and mouse cortex. The genes bound across species were more significantly dysregulated by the loss of CHD8 protein. Furthermore, the conserved set of CHD8 target genes was enriched for ASD genes found to harbor de novo LGD mutations in the WES studies mentioned earlier (Iossifov et al., 2012; Neale et al., 2012; O'Roak, Vives, Girirajan, et al., 2012; Sanders et al., 2012; Willsey et al., 2013), a finding also noted in the study by Sugathan et al. (Sugathan et al., 2014). Both studies (Sugathan et al., 2014; Cotney et al., 2015) also mentioned that CHD8 ChIP-seq targets were more likely to be dysregulated, with similar numbers of genes being upregulated or downregulated, but most CHD8 ChIP-seq targets did not show robust changes in gene expression. Interestingly, gene ontology analysis of CHD8 targets shows a signal for other gene regulators, including chromatin modifiers such as methyl transferases and demethylases, which suggests that CHD8 regulation cascades through additional effector molecules, many of which (such as *ARID1B*, *POGZ*, and *DYRK1A*) have themselves been implicated in ASD (Fig. 29.2).

Mechanistic convergence can also be assessed via analysis of differential expression in postmortem brains from individuals with ASD. Owing to the scarcity of these brains, the power of such studies is limited; however, two studies have identified modules of differentially expressed genes between cases and controls (Gupta et al., 2014; Voineagu et al., 2011). Microarray analysis of 19 ASD cases (five females; median age 22 years; median postmortem interval [PMI], 20 hours) and 17 controls (one female; median age, 32 years; median PMI, 23 hours) analyzed with weighted-gene coexpression network analysis (WGCNA) identified two coexpressed modules that were differentially expressed in ASD (Voineagu et al., 2011). The first module (M12) was enriched for synaptic genes and downregulated in ASD, whereas the second module (M16) was enriched for microglial and astrocyte genes and upregulated in ASD. Of note, the M12 synaptic module was also enriched for genes with low *P*-value single-nucleotide polymorphisms in an ASD GWAS analysis ($N = 1984$ independent cases) (Voineagu et al., 2011), although no such enrichment was observed in the M16 microglia module. Similar results were observed by Gupta et al. using RNA-Seq in 32 ASD cases (eight females; median age, 20 years; median PMI, 20 hours) and 40 controls (nine females; median age, 17 years; median PMI, 18 hours) (Gupta et al., 2014). Three synaptic modules were observed (M1, M2, and M6); the M2 module was upregulated in the ASD brain and showed enrichment for rare and common variants associated with ASD. Two glial modules were observed (M5 and M7), neither of which showed enrichment for ASD-associated genetic variants.

Spatiotemporal Convergence

At the heart of any search for convergence in ASD genes is a hypothesis about the nature of the underlying pathology. For example, most analyses that focus on mechanistic convergence have searched for persistent functional deficits. Indeed, many of the studies of gene ontology or PPIs have relied on databases that do not allow for the parsing of anatomical or

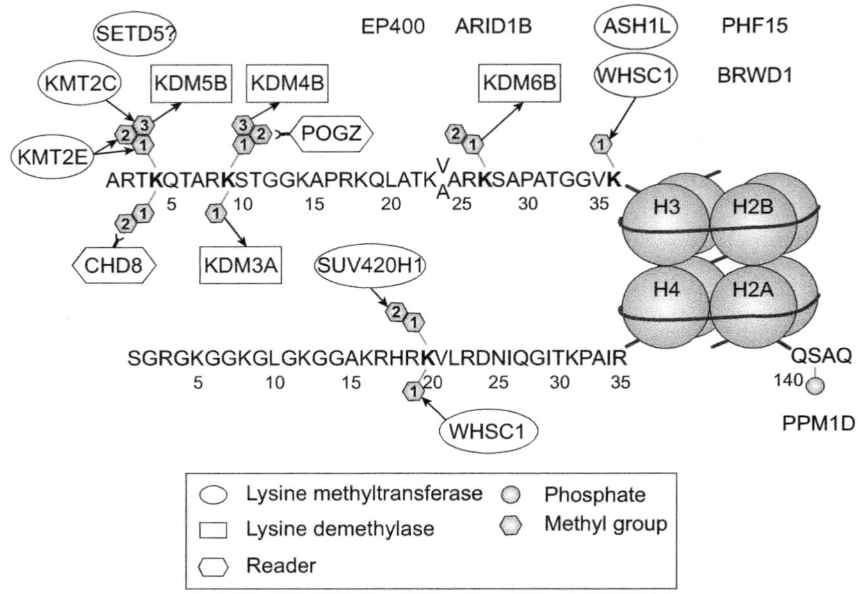

FIGURE 29.2 Histone marker targets of chromatin genes associated with ASD. Each of the four histone proteins (H2A, H2B, H3, and H4) has an amino acid tail that can be modified as part of the epigenome. Many of the genes associated with ASD (Table 29.1; Fig. 29.1) are components of chromatin remodeling complexes which modify or read specific histone marks. For example, KMT2C is a lysine methyltransferase which adds a third methyl group to the amino acid lysine (K) in position 4 on the tail of H3, ie, the H3K4me3 mark. *Adapted from De Rubeis, S., He, X., Goldberg, A. P., Poultney, C. S., Samocha, K., Ercument Cicek, A., … Buxbaum J. D. (2014). Synaptic, transcriptional and chromatin genes disrupted in autism.* Nature, *515(7526), 209–215. http://dx.doi.org/10.1038/nature13772.*

developmental variables or, in some cases, data specifically relevant to the human brain. In contrast, several studies evaluating convergence have begun to emphasize ASD as a bona fide developmental syndrome for which capturing spatial/anatomical and temporal/developmental variables is paramount (State & Sestan, 2012). Such studies have been made possible by foundational efforts developing multidimensional "-omics" data sets, reflecting the course of brain development. For example, the BrainSpan developmental transcriptome project summarizes gene expression from early fetal to late adult stages across multiple distinct anatomical regions in 57 typically developing human brains (see white paper at www.brainspan.org; Kang et al., 2011). Similar data sets are now emerging for a range of model systems from mouse (http://developingmouse.brain-map.org) to primate (http://www.blueprintnhpatlas.org).

Multiple groups have leveraged BrainSpan in the search for spatiotemporal convergence of ASD risk genes. For example, Ben-David et al. investigated 121 genes identified based on having at least one de novo LGD mutation in patients with ASD (Ben-David & Shifman, 2013) and used WGCNA (Lim et al., 2013; Zhang & Horvath, 2005). They observed two coexpressed modules: One was enriched for chromatin regulators expressed at a high level during fetal development and the other showed no enrichment for gene ontology terms and was expressed at a high level after birth (Ben-David & Shifman, 2013).

Multiple additional analyses have sought to define specific brain regions, time points, and cell types relevant for ASD pathology. For instance, Willsey et al. (2013) examined BrainSpan expression data in multiple spatiotemporal windows, constructing gene coexpression networks around the set of highest confidence ASD genes available at that time derived from WES. These seeded networks were then assessed for the presence of additional ASD risk genes, based on the reasoning that if a particular developmental expression network were related to ASD risk, it should contain a statistically significant excess of ASD risk genes beyond those used to establish the coexpression networks. Using this approach, they found that deep-layer cortical projection neurons in midfetal prefrontal cortex are an important nexus for a subset of ASD risk genes (Willsey et al., 2013). Similarly, Parikshak et al. (2013) used BrainSpan RNA-Seq data, a larger set of genes, and WGCNA. They also identified midfetal development and projection neurons associated with ASD risk, but found the greatest signal for upper rather than lower layer projection neurons in the cortex.

Xu et al. developed an approach known as cell-specific expression analysis (CSEA) (Xu, Wells, O'Brien, Nehorai, & Dougherty, 2014) for associating candidate gene lists with a particular cell type based on enrichment of cell-specific genes from mouse brains. They applied CSEA to interrogate the genes identified in exome studies (Iossifov et al., 2012; Neale et al., 2012; O'Roak, Vives, Girirajan, et al., 2012; Sanders et al., 2012) and, similar to the findings noted earlier, found enrichment in cortical projection neurons and striatal medium spiny neurons. By considering the BrainSpan data, they showed that both signals were strongest in the midfetal period of the developing brain. Moreover, several of the analyses described earlier included a spatiotemporal component. In the NETBAG+ network, both LGD mutations and genes involved in chromatin regulation had higher expression during fetal development (Chang et al., 2015). Finally, an analysis integrating PPI networks with spatiotemporal gene expression data (Kang et al., 2011) for the genes in the 16p11.2 CNV region also identified convergence in the late midfetal period (Lin et al., 2015).

A model system study focused on convergence focused on the gene *NLGN3* (Rothwell et al., 2014). In this case, the authors were interested in identifying functional and behavioral overlap among loss of function and missense mutations previously identified in individuals with ASD. They found that both the knockout and the R451C mutation identified by Jamain et al. (2003) enhanced formation of repetitive motor routines, based on rotarod performance, and that both types of mutations led to this phenotype via selective synaptic impairment in the nucleus accumbens/ventral striatum. This approach is a notable example of using multiple mutation models simultaneously to examine convergent circuit-level dysfunction related to ASD, an approach that is becoming strikingly more feasible with the development of increasingly powerful genome-editing technologies.

CONCLUSIONS

There has been a sea change in the genetics and genomics of autism over the past decade with dramatic acceleration in the past several years. The field has transitioned from a preoccupation with identifying even a single definitive risk gene to accumulating dozens of loci and genes, elaborating the genomic architecture of ASD, and beginning to move forward to the arduous process of illuminating pathophysiology and developing novel treatments.

There is little question that the difficulties with replicating the preceding era of gene discovery, coupled with a healthy appreciation of the complexity of the human brain and behavior, have continued to generate skepticism outside the field about the value of genetics in understanding psychiatric disorders in general and ASD in particular. In this regard, it is critical to convey to neuroscientists and clinical colleagues the message about the widespread convergence of results in studies of de novo variation and the general consensus regarding statistical approaches to assessing genome-wide significance.

The opportunities afforded by successful, systematic gene discovery are remarkable. Nonetheless it is worth pausing to consider the limits of current findings. For example, the failure to identify and replicate specific common risk alleles is undoubtedly a consequence of still underpowered cohorts. There is little question that common variation has a major role in population risk. Similarly, there is a clear "discovery bias" in early studies of de novo variation, with lower IQ being one of several factors that increase the likelihood of identifying damaging de novo mutations. For this reason, the first glimpses of the ASD allelic architecture have not surprisingly highlighted large effect mutations in individuals with relatively lower IQ. This observation led to some initial hypotheses that such mutations may be relevant only in comorbid ID and ASD. However, as cohort sizes have expanded, the contribution of de novo mutation to risk across the IQ spectrum has become increasingly clear (Sanders et al., 2015). In short, as samples continue to grow, a more nuanced and complete picture of the relative contribution of common and rare, de novo, and transmitted variation will emerge, as will the relationship between risk genes and the full range of phenotypic outcomes. Finally, although current studies have been highly successful in identifying rare de novo coding mutations, little is known about the contribution of noncoding mutation in ASD.

On this last point, there is ongoing debate regarding the relative value of pursuing whole-genome sequencing (WGS) because of the success and lower cost of exome studies and the difficulty in interpreting the noncoding genome. Given the specific history of ASD genomics, it is certainly likely that regulatory variants will be found to have a role. Moreover, the relatively low number of new mutations per generation in typically developing individuals promises a favorable signal-to-noise ratio for de novo variation compared with studies of transmitted mutation, as has been the case in studies of ASD exomes. However, it is safe to assume that the cumulative scale of the effect of noncoding variation will not exceed that for damaging de novo coding variants, and in fact is likely to be more modest. Consequently, sample sizes needed to begin to identify associated mutations will be comparatively large.

Apart from expanding the list of loci contributing to ASD, the most important contribution of WGS may well be as an orthogonal approach to determining the spatial and temporal distribution of ASD risk. Regulatory elements are often cell type, region, and/or developmentally restricted. Consequently, the identification of ASD-related mutations in these segments may advance efforts, similar to those discussed with regard to gene expression, to identify when and where ASD risk evolves.

Finally, although the degree of locus heterogeneity is daunting and the overt functions of ASD risk genes are tremendously diverse, the early and reproducible identification of biological clustering is reassuring. This strongly suggests that although there are as many as 1000 or more genetic risk factors, ASD pathology will resolve to a much smaller number of overlapping mechanisms. As noted, when one asks the question, "What do ASD genes do?" chromatin modification and synaptic function stand out as points of convergence. A skeptic might note that this simply suggests that ASD is a disorder of neurodevelopment. However, the degree of specificity afforded by well-validated, high-effect mutations with regard to dissecting particular aspects of chromatin modification or synaptic form or function remains to be seen. Moreover, increasingly genomic data are providing an opportunity to ask not only what, but also when and where ASD pathology exists. Here again, the early returns are surprisingly consistent in highlighting the role of midfetal cortex, glutamatergic neurons, and striatal interneurons. Almost certainly, as approaches to defining spatiotemporal convergence mature, additional brain regions and time points are likely to emerge. Nonetheless, the ability to begin to constrain experiments of specific mutations with regard to cell type, development epoch, and brain region will make critical contributions to a wide range of neurobiological studies. Indeed, the coming together of gene discovery, the elaboration of the noncoding genome, new functional genomic approaches, and rapidly advancing neurobiological tools are setting the stage for transformational advances in an understanding of ASD.

REFERENCES

Altshuler, D., Daly, M. J., & Lander, E. S. (2008). Genetic mapping in human disease. *Science (New York, NY), 322*(5903), 881–888. http://dx.doi.org/10.1126/science.1156409.

Amaral, D. G., Schumann, C. M., & Nordahl, C. W. (2008). Neuroanatomy of autism. *Trends in Neurosciences, 31*(3), 137–145. http://dx.doi.org/10.1016/j.tins.2007.12.005.

Amiet, C., Gourfinkel-An, I., Bouzamondo, A., Tordjman, S., Baulac, M., Lechat, P., ... Cohen, D. (2008). Epilepsy in autism is associated with intellectual disability and gender: evidence from a meta-analysis. *Biological Psychiatry, 64*(7), 577–582. http://dx.doi.org/10.1016/j.biopsych.2008.04.030.

Anney, R., Klei, L., Pinto, D., Almeida, J., Bacchelli, E., Baird, G., ... Devlin, B. (2012). Individual common variants exert weak effects on the risk for autism spectrum disorderspi. *Human Molecular Genetics, 21*(21), 4781–4792. http://dx.doi.org/10.1093/hmg/dds301.

Association, A. P. (2000). *Diagnostic and statistical manual of mental disorders* (4th ed.). American Psychiatric Association.

Association, A. P. (2013). *DSM 5*. American Psychiatric Association.

Autism Genome Project Consortium, Szatmari, P., Paterson, A. D., Zwaigenbaum, L., Roberts, W., Brian, J., ... Meyer, K. J. (2007). Mapping autism risk loci using genetic linkage and chromosomal rearrangements. *Nature Genetics, 39*(3), 319–328. http://dx.doi.org/10.1038/ng1985.

Bailey, A., Le Couteur, A., Gottesman, I., Bolton, P., Simonoff, E., Yuzda, E., & Rutter, M. (1995). Autism as a strongly genetic disorder: evidence from a British twin study. *Psychological Medicine, 25*(1), 63–77.

Bakkaloglu, B., O'Roak, B. J., Louvi, A., Gupta, A. R., Abelson, J. F., Morgan, T. M., ... State, M. W. (2008). Molecular cytogenetic analysis and resequencing of contactin associated protein-like 2 in autism spectrum disorders. *American Journal of Human Genetics, 82*(1), 165–173. http://dx.doi.org/10.1016/j.ajhg.2007.09.017.

Bakken, T. L., Helverschou, S. B., Eilertsen, D. E., Heggelund, T., Myrbakk, E., & Martinsen, H. (2010). Psychiatric disorders in adolescents and adults with autism and intellectual disability: a representative study in one county in Norway. *Research in Developmental Disabilities, 31*(6), 1669–1677. http://dx.doi.org/10.1016/j.ridd.2010.04.009.

Baxter, A. J., Brugha, T. S., Erskine, H. E., Scheurer, R. W., Vos, T., & Scott, J. G. (2014). The epidemiology and global burden of autism spectrum disorders. *Psychological Medicine, 45*(03), 1–13. http://dx.doi.org/10.1017/S003329171400172X.

Ben-David, E., & Shifman, S. (2013). Combined analysis of exome sequencing points toward a major role for transcription regulation during brain development in autism. *Molecular Psychiatry, 18*(10), 1054–1056. http://dx.doi.org/10.1038/mp.2012.148.

Bernier, R., Gerdts, J., Munson, J., Dawson, G., & Estes, A. (2012). Evidence for broader autism phenotype characteristics in parents from multiple-incidence autism families. *Autism Research: Official Journal of the International Society for Autism Research, 5*(1), 13–20. http://dx.doi.org/10.1002/aur.226.

Bernier, R., Golzio, C., Xiong, B., Stessman, H. A., Coe, B. P., Penn, O., ... Moon, R. (2014). Disruptive CHD8 mutations define a subtype of autism early in development. *Cell, 158*(2), 263–276. http://dx.doi.org/10.1016/j.cell.2014.06.017.

Berument, S. K., Rutter, M., Lord, C., Pickles, A., & Bailey, A. (1999). Autism screening questionnaire: diagnostic validity. *The British Journal of Psychiatry: The Journal of Mental Science, 175*, 444–451.

Billstedt, E., Gillberg, I. C., Gillberg, C., & Gillberg, C. (2005). Autism after adolescence: population-based 13- to 22-year follow-up study of 120 individuals with autism diagnosed in childhood. *Journal of Autism and Developmental Disorders, 35*(3), 351–360.

Bolton, P. F., Carcani-Rathwell, I., Hutton, J., Goode, S., Howlin, P., & Rutter, M. (2011). Epilepsy in autism: features and correlates. *The British Journal of Psychiatry: The Journal of Mental Science, 198*(4), 289–294. http://dx.doi.org/10.1192/bjp.bp.109.076877.

van Bon, B. W. M., Coe, B. P., Bernier, R., Green, C., Gerdts, J., Witherspoon, K., ... Eichler, E. E. (2015). Disruptive de novo mutations of DYRK1A lead to a syndromic form of autism and ID. *Molecular Psychiatry, 21*(1), 126–132. http://dx.doi.org/10.1038/mp.2015.5.

Bolton, P., Macdonald, H., Pickles, A., Rios, P., Goode, S., Crowson, M., ... Rutter, M. (1994). A case-control family history study of autism. *Journal of Child Psychology and Psychiatry, and Allied Disciplines, 35*(5), 877–900.

Brugha, T. S., McManus, S., Bankart, J., Scott, F., Purdon, S., Smith, J., … Meltzer, H. (2011). Epidemiology of autism spectrum disorders in adults in the community in England. *Archives of General Psychiatry, 68*(5), 459−465. http://dx.doi.org/10.1001/archgenpsychiatry.2011.38.

Buxbaum, J. D., Daly, M. J., Devlin, B., Lehner, T., Roeder, K., State, M. W., & Autism Sequencing Consortium. (2012). The autism sequencing consortium: large-scale, high-throughput sequencing in autism spectrum disorders. *Neuron, 76*, 1052−1056. http://dx.doi.org/10.1016/j.neuron.2012.12.008.

Chang, J., Gilman, S. R., Chiang, A. H., Sanders, S. J., & Vitkup, D. (2015). Genotype to phenotype relationships in autism spectrum disorders. *Nature Neuroscience, 18*(2), 191−198. http://dx.doi.org/10.1038/nn.3907.

Chaste, P., Klei, L., Sanders, S. J., Murtha, M. T., Hus, V., Lowe, J. K., … Soo-Jeong, K. (2013). Adjusting head circumference for covariates in autism: clinical correlates of a highly heritable continuous trait. *Biological Psychiatry, 74*(8), 576−584. http://dx.doi.org/10.1016/j.biopsych.2013.04.018.

Colvert, E., Tick, B., McEwen, F., Stewart, C., Curran, S. R., Woodhouse, E., … Bolton, P. (2015). Heritability of autism spectrum disorder in a UK population-based twin sample. *JAMA Psychiatry, 72*(5), 415−423. http://dx.doi.org/10.1001/jamapsychiatry.2014.3028.

Constantino, J. N. (2011). The quantitative nature of autistic social impairment. *Pediatric Research, 69*(5 Pt 2), 55R−62R. http://dx.doi.org/10.1203/PDR.0b013e318212ec6e.

Constantino, J. N., & Todd, R. D. (2005). Intergenerational transmission of subthreshold autistic traits in the general population. *Biological Psychiatry, 57*(6), 655−660. http://dx.doi.org/10.1016/j.biopsych.2004.12.014.

Constantino, J. N., Todorov, A., Hilton, C., Law, P., Zhang, Y., Molloy, E., … Geschwind, D. (2013). Autism recurrence in half siblings: strong support for genetic mechanisms of transmission in ASD. *Molecular Psychiatry, 18*(2), 137−138. http://dx.doi.org/10.1038/mp.2012.9.

Cotney, J., Muhle, R. A., Sanders, S. J., Liu, L., Willsey, A. J., Niu, W., … Noonan, J. P. (2015). The autism-associated chromatin modifier CHD8 regulates other autism risk genes during human neurodevelopment. *Nature Communications, 6*, 6404. http://dx.doi.org/10.1038/ncomms7404.

Croen, L. A., Grether, J. K., Yoshida, C. K., Odouli, R., & Hendrick, V. (2011). Antidepressant use during pregnancy and childhood autism spectrum disorders. *Archives of General Psychiatry, 68*(11), 1104−1112. http://dx.doi.org/10.1001/archgenpsychiatry.2011.73.

Dansie, L. E., Phommahaxay, K., Okusanya, A. G., Uwadia, J., Huang, M., Rotschafer, S. E., … Ethell, I. M. (2013). Long-lasting effects of minocycline on behavior in young but not adult Fragile X mice. *Neuroscience, 246*, 186−198. http://dx.doi.org/10.1016/j.neuroscience.2013.04.058.

Darnell, J. C., Van Driesche, S. J., Zhang, C., Hung, K. Y. S., Mele, A., Fraser, C. E., … Darnell, R. B. (2011). FMRP stalls ribosomal translocation on mRNAs linked to synaptic function and autism. *Cell, 146*(2), 247−261. http://dx.doi.org/10.1016/j.cell.2011.06.013.

Davidson, C., Greenwood, N., Stansfield, A., & Wright, S. (2014). Prevalence of Asperger syndrome among patients of an Early Intervention in Psychosis team. *Early Intervention in Psychiatry, 8*(2), 138−146. http://dx.doi.org/10.1111/eip.12039.

Dawson, G., Carver, L., Meltzoff, A. N., Panagiotides, H., McPartland, J., & Webb, S. J. (2002). Neural correlates of face and object recognition in young children with autism spectrum disorder, developmental delay, and typical development. *Child Development, 73*(3), 700−717.

De Rubeis, S., & Buxbaum, J. D. (2015). Genetics and genomics of autism spectrum disorder: embracing complexity. *Human Molecular Genetics, 24*(R1), R24−R31. http://dx.doi.org/10.1093/hmg/ddv273.

De Rubeis, S., He, X., Goldberg, A. P., Poultney, C. S., Samocha, K., Ercument Cicek, A., … Buxbaum, J. D. (2014). Synaptic, transcriptional and chromatin genes disrupted in autism. *Nature, 515*(7526), 209−215. http://dx.doi.org/10.1038/nature13772.

Developmental Disabilities Monitoring Network Surveillance Year 2010 Principal Investigators, Centers for Disease Control and Prevention (CDC). (2014). *Prevalence of autism spectrum disorder among children aged 8 years − autism and developmental disabilities monitoring network, 11 sites, United States, 2010*. Morbidity and Mortality Weekly Report. Surveillance Summaries (Washington, D.C.: 2002), *63*(2), 1−21.

Diagnostic and Statistical Manual of Mental Disorders. (2013). *Diagnostic and statistical manual of mental disorders*. American Psychiatric Publishing. Incorporated.

Dong, S., Walker, M. F., Carriero, N. J., DiCola, M., Willsey, A. J., Ye, A. Y., … Sanders, S. J. (2014). De novo insertions and deletions of predominantly paternal origin are associated with autism spectrum disorder. *Cell Reports, 9*(1), 16−23. http://dx.doi.org/10.1016/j.celrep.2014.08.068.

The Editors of The Lancet. (2010). Retraction−Ileal-lymphoid-nodular hyperplasia, non-specific colitis, and pervasive developmental disorder in children. *Lancet, 375*(9713), 445. http://dx.doi.org/10.1016/S0140-6736(10)60175-4.

Ehninger, D., Han, S., Shilyansky, C., Zhou, Y., Li, W., Kwiatkowski, D. J., … Silva, A. J. (2008). Reversal of learning deficits in a Tsc2+/− mouse model of tuberous sclerosis. *Nature Medicine, 14*(8), 843−848. http://dx.doi.org/10.1038/nm1788.

Feldman, I., Rzhetsky, A., & Vitkup, D. (2008). Network properties of genes harboring inherited disease mutations. *Proceedings of the National Academy of Sciences of the United States of America, 105*(11), 4323−4328. http://dx.doi.org/10.1073/pnas.0701722105.

Fischbach, G. D., & Lord, C. (2010). The Simons Simplex Collection: a resource for identification of autism genetic risk factors. *Neuron, 68*(2), 192−195. http://dx.doi.org/10.1016/j.neuron.2010.10.006.

Folstein, S., & Rutter, M. (1977). Genetic influences and infantile autism. *Nature, 265*(5596), 726−728.

Fombonne, E. (2009). Epidemiology of pervasive developmental disorders. *Pediatric Research, 65*(6), 591−598. http://dx.doi.org/10.1203/PDR.0b013e31819e7203.

Gadad, B. S., Li, W., Yazdani, U., Grady, S., Johnson, T., Hammond, J., … German, D. C. (2015). Administration of thimerosal-containing vaccines to infant rhesus macaques does not result in autism-like behavior or neuropathology. *Proceedings of the National Academy of Sciences of the United States of America, 112*(40), 12498−12503. http://dx.doi.org/10.1073/pnas.1500968112.

Garg, S. K., Lioy, D. T., Cheval, H., McGann, J. C., Bissonnette, J. M., Murtha, M. J., … Mandel, G. (2013). Systemic delivery of MeCP2 rescues behavioral and cellular deficits in female mouse models of Rett syndrome. *The Journal of Neuroscience: the Official Journal of the Society for Neuroscience, 33*(34), 13612−13620. http://dx.doi.org/10.1523/JNEUROSCI.1854-13.2013.

Gaugler, T., Klei, L., Sanders, S. J., Bodea, C. A., Goldberg, A. P., Lee, A. B., ... Buxbaum, J. D. (2014). Most genetic risk for autism resides with common variation. *Nature Genetics, 46*(8), 881−885. http://dx.doi.org/10.1038/ng.3039.

Ghosh, A., Michalon, A., Lindemann, L., Fontoura, P., & Santarelli, L. (2013). Drug discovery for autism spectrum disorder: challenges and opportunities. *Nature Reviews. Drug Discovery, 12*(10), 777−790. http://dx.doi.org/10.1038/nrd4102.

Giarelli, E., Wiggins, L. D., Rice, C. E., Levy, S. E., Kirby, R. S., Pinto-Martin, J., & Mandell, D. (2010). Sex differences in the evaluation and diagnosis of autism spectrum disorders among children. *Disability and Health Journal, 3*(2), 107−116. http://dx.doi.org/10.1016/j.dhjo.2009.07.001.

Gillberg, I. C., Helles, A., Billstedt, E., & Gillberg, C. (2015). Boys with asperger syndrome grow up: psychiatric and neurodevelopmental disorders 20 years after initial diagnosis. *Journal of Autism and Developmental Disorders*, 1−9. http://dx.doi.org/10.1007/s10803-015-2544-0.

Gilman, S. R., Iossifov, I., Levy, D., Ronemus, M., Wigler, M., & Vitkup, D. (2011). Rare de novo variants associated with autism implicate a large functional network of genes involved in formation and function of synapses. *Neuron, 70*(5), 898−907. http://dx.doi.org/10.1016/j.neuron.2011.05.021.

Goldberg, W. A., Osann, K., Filipek, P. A., Laulhere, T., Jarvis, K., Modahl, C., ... Spence, M. A. (2003). Language and other regression: assessment and timing. *Journal of Autism and Developmental Disorders, 33*(6), 607−616.

Goldman, S. E., Surdyka, K., Cuevas, R., Adkins, K., Wang, L., & Malow, B. A. (2009). Defining the sleep phenotype in children with autism. *Developmental Neuropsychology, 34*(5), 560−573. http://dx.doi.org/10.1080/87565640903133509.

Grzadzinski, R., Huerta, M., & Lord, C. (2013). DSM-5 and autism spectrum disorders (ASDs): an opportunity for identifying ASD subtypes. *Molecular Autism, 4*(1), 12. http://dx.doi.org/10.1186/2040-2392-4-12.

Gupta, S., Ellis, S. E., Ashar, F. N., Moes, A., Bader, J. S., Zhan, J., ... Arking, D. E. (2014). Transcriptome analysis reveals dysregulation of innate immune response genes and neuronal activity-dependent genes in autism. *Nature Communications, 5*, 5748. http://dx.doi.org/10.1038/ncomms6748.

Hallmayer, J., Cleveland, S., Torres, A., Phillips, J., Cohen, B., Torigoe, T., ... Risch, N. (2011). Genetic heritability and shared environmental factors among twin pairs with autism. *Archives of General Psychiatry, 68*(11), 1095−1102. http://dx.doi.org/10.1001/archgenpsychiatry.2011.76.

He, X., Sanders, S. J., Liu, L., De Rubeis, S., Lim, E. T., Sutcliffe, J. S., ... Roeder, K. (2013). Integrated model of de novo and inherited genetic variants yields greater power to identify risk genes. *PLoS Genetics, 9*(8), e1003671. http://dx.doi.org/10.1371/journal.pgen.1003671.

Hiller, R. M., Young, R. L., & Weber, N. (2014). Sex differences in autism spectrum disorder based on DSM-5 criteria: evidence from clinician and teacher reporting. *Journal of Abnormal Child Psychology, 42*(8), 1381−1393. http://dx.doi.org/10.1007/s10802-014-9881-x.

Hofvander, B., Delorme, R., Chaste, P., Nydén, A., Wentz, E., Ståhlberg, O., ... Leboyer, M. (2009). Psychiatric and psychosocial problems in adults with normal-intelligence autism spectrum disorders. *BMC Psychiatry, 9*(1), 35. http://dx.doi.org/10.1186/1471-244X-9-35.

Hviid, A., Melbye, M., & Pasternak, B. (2013). Use of selective serotonin reuptake inhibitors during pregnancy and risk of autism. *The New England Journal of Medicine, 369*(25), 2406−2415. http://dx.doi.org/10.1056/NEJMoa1301449.

Iossifov, I., O'Roak, B. J., Sanders, S. J., Ronemus, M., Krumm, N., Levy, D., ... Wigler, M. (2014). The contribution of de novo coding mutations to autism spectrum disorder. *Nature, 515*(7526), 216−221. http://dx.doi.org/10.1038/nature13908.

Iossifov, I., Ronemus, M., Levy, D., Wang, Z., Hakker, I., Rosenbaum, J., ... Wigler, M. (2012). De novo gene disruptions in children on the autistic spectrum. *Neuron, 74*(2), 285−299. http://dx.doi.org/10.1016/j.neuron.2012.04.009.

Itsara, A., Wu, H., Smith, J. D., Nickerson, D. A., Romieu, I., London, S. J., & Eichler, E. E. (2010). De novo rates and selection of large copy number variation. *Genome Research, 20*(11), 1469−1481. http://dx.doi.org/10.1101/gr.107680.110.

Jacquemont, M.-L., Sanlaville, D., Redon, R., Raoul, O., Cormier-Daire, V., Lyonnet, S., ... Philippe, A. (2006). Array-based comparative genomic hybridisation identifies high frequency of cryptic chromosomal rearrangements in patients with syndromic autism spectrum disorders. *Journal of Medical Genetics, 43*(11), 843−849. http://dx.doi.org/10.1136/jmg.2006.043166.

Jamain, S., Quach, H., Betancur, C., Råstam, M., Colineaux, C., Gillberg, I. C., ... Paris Autism Research International Sibpair Study. (2003). Mutations of the X-linked genes encoding neuroligins NLGN3 and NLGN4 are associated with autism. *Nature Genetics, 34*(1), 27−29. http://dx.doi.org/10.1038/ng1136.

Johnson, M. H., Griffin, R., Csibra, G., Halit, H., Farroni, T., de Haan, M., ... Richards, J. (2005). The emergence of the social brain network: evidence from typical and atypical development. *Development and Psychopathology, 17*(3), 599−619. http://dx.doi.org/10.1017/S0954579405050297.

Jokiranta, E., Sourander, A., Suominen, A., Timonen-Soivio, L., Brown, A. S., & Sillanpää, M. (2014). Epilepsy among children and adolescents with autism spectrum disorders: a population-based study. *Journal of Autism and Developmental Disorders, 44*(10), 2547−2557. http://dx.doi.org/10.1007/s10803-014-2126-6.

Jorde, L. B., Hasstedt, S. J., Ritvo, E. R., Mason-Brothers, A., Freeman, B. J., Pingree, C., ... Mo, A. (1991). Complex segregation analysis of autism. *American Journal of Human Genetics, 49*(5), 932−938.

Joshi, G., Faraone, S. V., Wozniak, J., Petty, C., Fried, R., Galdo, M., ... Biederman, J. (2014). Examining the clinical correlates of autism spectrum disorder in youth by ascertainment source. *Journal of Autism and Developmental Disorders, 44*(9), 2117−2126. http://dx.doi.org/10.1007/s10803-014-2063-4.

Joshi, G., Faraone, S. V., Wozniak, J., Tarko, L., Fried, R., Galdo, M., ... Biederman, J. (2014). Symptom profile of ADHD in youth with high-functioning autism spectrum disorder: a comparative study in psychiatrically referred populations. *Journal of Attention Disorders*. http://dx.doi.org/10.1177/1087054714543368.

Joshi, G., Petty, C., Wozniak, J., Henin, A., Fried, R., Galdo, M., ... Biederman, J. (2010). The heavy burden of psychiatric comorbidity in youth with autism spectrum disorders: a large comparative study of a psychiatrically referred population. *Journal of Autism and Developmental Disorders, 40*(11), 1361−1370. http://dx.doi.org/10.1007/s10803-010-0996-9.

Joshi, G., Wozniak, J., Petty, C., Martelon, M. K., Fried, R., Bolfek, A., ... Biederman, J. (2013). Psychiatric comorbidity and functioning in a clinically referred population of adults with autism spectrum disorders: a comparative study. *Journal of Autism and Developmental Disorders, 43*(6), 1314−1325. http://dx.doi.org/10.1007/s10803-012-1679-5.

Kang, H. J., Kawasawa, Y. I., Cheng, F., Zhu, Y., Xu, X., Li, M., ... Sestan, N. (2011). Spatio-temporal transcriptome of the human brain. *Nature, 478*(7370), 483–489. http://dx.doi.org/10.1038/nature10523.

Kanner, L. (1943). Autistic disturbances of affective contact. *Nervous Child, 2*, 217–250.

Kim, Y. S., Fombonne, E., Koh, Y.-J., Kim, S.-J., Cheon, K.-A., & Leventhal, B. L. (2014). A comparison of DSM-IV pervasive developmental disorder and DSM-5 autism spectrum disorder prevalence in an epidemiologic sample. *Journal of the American Academy of Child and Adolescent Psychiatry, 53*(5), 500–508. http://dx.doi.org/10.1016/j.jaac.2013.12.021.

Kim, Y. S., Leventhal, B. L., Koh, Y.-J., Fombonne, E., Laska, E., Lim, E.-C., ... Grinker, R. R. (2011). Prevalence of autism spectrum disorders in a total population sample. *The American Journal of Psychiatry, 168*(9), 904–912. http://dx.doi.org/10.1176/appi.ajp.2011.10101532.

King, M., & Bearman, P. (2009). Diagnostic change and the increased prevalence of autism. *International Journal of Epidemiology, 38*(5), 1224–1234. http://dx.doi.org/10.1093/ije/dyp261.

Klei, L., Sanders, S. J., Murtha, M. T., Hus, V., Lowe, J. K., Willsey, A. J., ... Devlin, B. (2012). Common genetic variants, acting additively, are a major source of risk for autism. *Molecular Autism, 3*(1), 9. http://dx.doi.org/10.1186/2040-2392-3-9.

Kohane, I. S. (2015). An autism case history to review the systematic analysis of large-scale data to refine the diagnosis and treatment of neuropsychiatric disorders. *Biological Psychiatry, 77*(1), 59–65. http://dx.doi.org/10.1016/j.biopsych.2014.05.024.

Kohane, I. S., McMurry, A., Weber, G., MacFadden, D., Rappaport, L., Kunkel, L., ... Churchill, S. (2012). The co-morbidity burden of children and young adults with autism spectrum disorders. *PLoS One, 7*(4), e33224. http://dx.doi.org/10.1371/journal.pone.0033224.

Kong, A., Frigge, M. L., Masson, G., Besenbacher, S., Sulem, P., Magnusson, G., ... Stefansson, K. (2012). Rate of de novo mutations and the importance of father's age to disease risk. *Nature, 488*(7412), 471–475. http://dx.doi.org/10.1038/nature11396.

Kotagal, S., & Broomall, E. (2012). Sleep in children with autism spectrum disorder. *Pediatric Neurology, 47*(4), 242–251. http://dx.doi.org/10.1016/j.pediatrneurol.2012.05.007.

Krueger, D. D., & Bear, M. F. (2011). Toward fulfilling the promise of molecular medicine in fragile X syndrome. *Annual Review of Medicine, 62*(1), 411–429. http://dx.doi.org/10.1146/annurev-med-061109-134644.

Kumar, R. A., KaraMohamed, S., Sudi, J., Conrad, D. F., Brune, C., Badner, J. A., ... Christian, S. L. (2008). Recurrent 16p11.2 microdeletions in autism. *Human Molecular Genetics, 17*(4), 628–638. http://dx.doi.org/10.1093/hmg/ddm376.

Kurtz, Z., Tookey, P., & Ross, E. (1998). Epilepsy in young people: 23 year follow up of the British national child development study. *BMJ (Clinical Research Edition), 316*(7128), 339–342.

Lai, M.-C., Lombardo, M. V., & Baron-Cohen, S. (2014). Autism. *Lancet, 383*(9920), 896–910. http://dx.doi.org/10.1016/S0140-6736(13)61539-1.

Lane, A. E., Molloy, C. A., & Bishop, S. L. (2014). Classification of children with autism spectrum disorder by sensory subtype: a case for sensory-based phenotypes. *Autism Research: Official Journal of the International Society for Autism Research, 7*(3), 322–333. http://dx.doi.org/10.1002/aur.1368.

Laumonnier, F., Bonnet-Brilhault, F., Gomot, M., Blanc, R., David, A., Moizard, M.-P., ... Briault, S. (2004). X-linked mental retardation and autism are associated with a mutation in the NLGN4 gene, a member of the neuroligin family. *American Journal of Human Genetics, 74*(3), 552–557. http://dx.doi.org/10.1086/382137.

Lauritsen, M., Mors, O., Mortensen, P. B., & Ewald, H. (1999). Infantile autism and associated autosomal chromosome abnormalities: a register-based study and a literature survey. *Journal of Child Psychology and Psychiatry, and Allied Disciplines, 40*(3), 335–345.

Levy, D., Ronemus, M., Yamrom, B., Lee, Y.-H., Leotta, A., Kendall, J., ... Wigler, M. (2011). Rare de novo and transmitted copy-number variation in autistic spectrum disorders. *Neuron, 70*(5), 886–897. http://dx.doi.org/10.1016/j.neuron.2011.05.015.

Li, S. Y., Chen, Y. C., Lai, T. J., Hsu, C. Y., & Wang, Y. C. (1993). Molecular and cytogenetic analyses of autism in Taiwan. *Human Genetics, 92*(5), 441–445.

Li, J., Shi, M., Ma, Z., Zhao, S., Euskirchen, G., Ziskin, J., ... Snyder, M. (2014). Integrated systems analysis reveals a molecular network underlying autism spectrum disorders. *Molecular Systems Biology, 10*(12), 774. http://dx.doi.org/10.15252/msb.20145487.

Lichtenstein, P., Carlström, E., Råstam, M., Gillberg, C., & Anckarsäter, H. (2010). The genetics of autism spectrum disorders and related neuropsychiatric disorders in childhood. *The American Journal of Psychiatry, 167*(11), 1357–1363. http://dx.doi.org/10.1176/appi.ajp.2010.10020223.

Lim, E. T., Raychaudhuri, S., Sanders, S. J., Stevens, C., Sabo, A., MacArthur, D. G., ... Daly, M. J. (2013). Rare complete knockouts in humans: population distribution and significant role in autism spectrum disorders. *Neuron, 77*(2), 235–242. http://dx.doi.org/10.1016/j.neuron.2012.12.029.

Lin, G. N., Corominas, R., Lemmens, I., Yang, X., Tavernier, J., Hill, D. E., ... Iakoucheva, L. M. (2015). Spatiotemporal 16p11.2 protein network implicates cortical late mid-fetal brain development and KCTD13-Cul3-RhoA pathway in psychiatric diseases. *Neuron, 85*(4), 742–754. http://dx.doi.org/10.1016/j.neuron.2015.01.010.

Liu, L., Lei, J., Sanders, S. J., Willsey, A. J., Kou, Y., Cicek, A. E., ... Roeder, K. (2014). DAWN: a framework to identify autism genes and subnetworks using gene expression and genetics. *Molecular Autism, 5*(1), 22. http://dx.doi.org/10.1186/2040-2392-5-22.

Lord, C., & Jones, R. M. (2012). Annual research review: re-thinking the classification of autism spectrum disorders. *Journal of Child Psychology and Psychiatry, and Allied Disciplines, 53*(5), 490–509. http://dx.doi.org/10.1111/j.1469-7610.2012.02547.x.

Lord, C., Rutter, M., Goode, S., Heemsbergen, J., Jordan, H., Mawhood, L., & Schopler, E. (1989). Autism diagnostic observation schedule: a standardized observation of communicative and social behavior. *Journal of Autism and Developmental Disorders, 19*(2), 185–212.

Lord, C., Rutter, M., & Le Couteur, A. (1994). Autism diagnostic interview-revised: a revised version of a diagnostic interview for caregivers of individuals with possible pervasive developmental disorders. *Journal of Autism and Developmental Disorders, 24*(5), 659–685.

Lugnegård, T., Hallerbäck, M. U., & Gillberg, C. (2011). Psychiatric comorbidity in young adults with a clinical diagnosis of Asperger syndrome. *Research in Developmental Disabilities, 32*(5), 1910–1917. http://dx.doi.org/10.1016/j.ridd.2011.03.025.

Mahajan, R., Bernal, M. P., Panzer, R., Whitaker, A., Roberts, W., Handen, B., ... Autism Speaks Autism Treatment Network Psychopharmacology Committee. (November 2012). Clinical practice pathways for evaluation and medication choice for attention-deficit/hyperactivity disorder symptoms in autism spectrum disorders. *Pediatrics, 130*(Suppl. 2), S125–S138. http://dx.doi.org/10.1542/peds.2012-0900J. American Academy of Pediatrics.

Malhotra, D., & Sebat, J. (2012). CNVs: harbingers of a rare variant revolution in psychiatric genetics. *Cell, 148*(6), 1223–1241. http://dx.doi.org/10.1016/j.cell.2012.02.039.

Malow, B., Adkins, K. W., McGrew, S. G., Wang, L., Goldman, S. E., Fawkes, D., & Burnette, C. (2012). Melatonin for sleep in children with autism: a controlled trial examining dose, tolerability, and outcomes. *Journal of Autism and Developmental Disorders, 42*(8), 1729–1737. http://dx.doi.org/10.1007/s10803-011-1418-3. author reply 1738.

Marco, E. J., Hinkley, L. B. N., Hill, S. S., & Nagarajan, S. S. (2011). Sensory processing in autism: a review of neurophysiologic findings. *Pediatric Research, 69*(5 Pt 2), 48R–54R. http://dx.doi.org/10.1203/PDR.0b013e3182130c54.

Marshall, C. R., Noor, A., Vincent, J. B., Lionel, A. C., Feuk, L., Skaug, J., ... Scherer, S. W. (2008). Structural variation of chromosomes in autism spectrum disorder. *American Journal of Human Genetics, 82*(2), 477–488. http://dx.doi.org/10.1016/j.ajhg.2007.12.009.

Martínez-Sanchis, S. (2014). Neurobiological foundations of multisensory integration in people with autism spectrum disorders: the role of the medial prefrontal cortex. *Frontiers in Human Neuroscience, 8*, 970. http://dx.doi.org/10.3389/fnhum.2014.00970.

McElhanon, B. O., McCracken, C., Karpen, S., & Sharp, W. G. (2014). Gastrointestinal symptoms in autism spectrum disorder: a meta-analysis. *Pediatrics, 133*(5), 872–883. http://dx.doi.org/10.1542/peds.2013-3995.

Michalon, A., Sidorov, M., Ballard, T. M., Ozmen, L., Spooren, W., Wettstein, J. G., ... Lindemann, L. (2012). Chronic pharmacological mGlu5 inhibition corrects fragile X in adult mice. *Neuron, 74*(1), 49–56. http://dx.doi.org/10.1016/j.neuron.2012.03.009.

Morbidity and Mortality Weekly Report. Surveillance summaries (Washington, D.C.: 2002).

Morrow, E. M., Yoo, S.-Y., Flavell, S. W., Kim, T.-K., Lin, Y., Hill, R. S., ... Walsh, C. A. (2008). Identifying autism loci and genes by tracing recent shared ancestry. *Science (New York, NY), 321*(5886), 218–223. http://dx.doi.org/10.1126/science.1157657.

Mottron, L., Dawson, M., Soulières, I., Hubert, B., & Burack, J. (2006). Enhanced perceptual functioning in autism: an update, and eight principles of autistic perception. *Journal of Autism and Developmental Disorders, 36*(1), 27–43. http://dx.doi.org/10.1007/s10803-005-0040-7.

Mühleisen, T. W., Leber, M., Schulze, T. G., Strohmaier, J., Degenhardt, F., Treutlein, J., ... Cichon, S. (2014). Genome-wide association study reveals two new risk loci for bipolar disorder. *Nature Communications, 5*, 3339. http://dx.doi.org/10.1038/ncomms4339.

Murch, S. H., Anthony, A., Casson, D. H., Malik, M., Berelowitz, M., Dhillon, A. P., ... Walker-Smith, J. A. (2004). Retraction of an interpretation. *Lancet, 363*(9411), 750. http://dx.doi.org/10.1016/S0140-6736(04)15715-2.

Murdoch, J. D., Gupta, A. R., Sanders, S. J., Walker, M. F., Keaney, J., Fernandez, T. V., ... State, M. W. (2015). No evidence for association of autism with rare heterozygous point mutations in Contactin-Associated Protein-Like 2 (CNTNAP2), or in other Contactin-Associated Proteins or Contactins. *PLoS Genetics, 11*(1), e1004852. http://dx.doi.org/10.1371/journal.pgen.1004852.

Neale, B. M., Kou, Y., Liu, L., Ma'ayan, A., Samocha, K. E., Sabo, A., ... Daly, M. J. (2012). Patterns and rates of exonic de novo mutations in autism spectrum disorders. *Nature, 485*(7397), 242–245. http://dx.doi.org/10.1038/nature11011.

Nomura, Y., Nagao, Y., Kimura, K., Hachimori, K., & Segawa, M. (2010). Epilepsy in autism: a pathophysiological consideration. *Brain & Development, 32*(10), 799–804. http://dx.doi.org/10.1016/j.braindev.2010.08.001.

Novarino, G., El-Fishawy, P., Kayserili, H., Meguid, N. A., Scott, E. M., Schroth, J., ... Gleeson, J. G. (2012). Mutations in BCKD-kinase lead to a potentially treatable form of autism with epilepsy. *Science (New York, NY), 338*(6105), 394–397. http://dx.doi.org/10.1126/science.1224631.

O'Roak, B. J., Deriziotis, P., Lee, C., Vives, L., Schwartz, J. J., Girirajan, S., ... Eichler, E. E. (2011). Exome sequencing in sporadic autism spectrum disorders identifies severe de novo mutations. *Nature Genetics, 43*(6), 585–589. http://dx.doi.org/10.1038/ng.835.

O'Roak, B. J., Vives, L., Fu, W., Egertson, J. D., Stanaway, I. B., Phelps, I. G., ... Shendure, J. (2012). Multiplex targeted sequencing identifies recurrently mutated genes in autism spectrum disorders. *Science (New York, NY), 338*(6114), 1619–1622. http://dx.doi.org/10.1126/science.1227764.

O'Roak, B. J., Vives, L., Girirajan, S., Karakoc, E., Krumm, N., Coe, B. P., ... Eichler, E. E. (2012). Sporadic autism exomes reveal a highly interconnected protein network of de novo mutations. *Nature, 485*(7397), 246–250. http://dx.doi.org/10.1038/nature10989.

Ornoy, A., Weinstein-Fudim, L., & Ergaz, Z. (2015). Prenatal factors associated with autism spectrum disorder (ASD). *Reproductive Toxicology (Elmsford, NY), 56*, 155–169. http://dx.doi.org/10.1016/j.reprotox.2015.05.007.

Osterling, J., & Dawson, G. (1994). Early recognition of children with autism: a study of first birthday home videotapes. *Journal of Autism and Developmental Disorders, 24*(3), 247–257.

Osterling, J. A., Dawson, G., & Munson, J. A. (2002). Early recognition of 1-year-old infants with autism spectrum disorder versus mental retardation. *Development and Psychopathology, 14*(2), 239–251.

Ozonoff, S., Iosif, A.-M., Baguio, F., Cook, I. C., Hill, M. M., Hutman, T., ... Young, G. S. (2010). A prospective study of the emergence of early behavioral signs of autism. *Journal of the American Academy of Child and Adolescent Psychiatry, 49*(3), 256–266. e1–e2.

Ozonoff, S., Young, G. S., Carter, A., Messinger, D., Yirmiya, N., Zwaigenbaum, L., & Stone, W. L. (2011). Recurrence risk for autism spectrum disorders: a Baby Siblings Research Consortium study. *Pediatrics, 128*(3), e488–e495. http://dx.doi.org/10.1542/peds.2010-2825.

Parikshak, N. N., Luo, R., Zhang, A., Won, H., Lowe, J. K., Chandran, V., ... Geschwind, D. H. (2013). Integrative functional genomic analyses implicate specific molecular pathways and circuits in autism. *Cell, 155*(5), 1008–1021. http://dx.doi.org/10.1016/j.cell.2013.10.031.

Pickles, A., Starr, E., Kazak, S., Bolton, P., Papanikolaou, K., Bailey, A., ... Rutter, M. (2000). Variable expression of the autism broader phenotype: findings from extended pedigrees. *Journal of Child Psychology and Psychiatry, and Allied Disciplines, 41*(4), 491–502.

Pinto, D., Delaby, E., Merico, D., Barbosa, M., Merikangas, A., Klei, L., ... Scherer, S. W. (2014). Convergence of genes and cellular pathways dysregulated in autism spectrum disorders. *American Journal of Human Genetics, 94*(5), 677–694. http://dx.doi.org/10.1016/j.ajhg.2014.03.018.

Pinto, D., Pagnamenta, A. T., Klei, L., Anney, R., Merico, D., Regan, R., ... Betancur, C. (2010). Functional impact of global rare copy number variation in autism spectrum disorders. *Nature, 466*(7304), 368–372. http://dx.doi.org/10.1038/nature09146.

Piven, J., Palmer, P., Jacobi, D., Childress, D., & Arndt, S. (1997). Broader autism phenotype: evidence from a family history study of multiple-incidence autism families. *The American Journal of Psychiatry, 154*(2), 185–190. http://dx.doi.org/10.1176/ajp.154.2.185.

Power, R. A., Kyaga, S., Uher, R., MacCabe, J. H., Långström, N., Landen, M., ... Svensson, A. C. (2013). Fecundity of patients with schizophrenia, autism, bipolar disorder, depression, anorexia nervosa, or substance abuse vs their unaffected siblings. *JAMA Psychiatry, 70*(1), 22–30. http://dx.doi.org/10.1001/jamapsychiatry.2013.268.

Rai, D., Lewis, G., Lundberg, M., Araya, R., Svensson, A., Dalman, C., ... Magnusson, C. (2012). Parental socioeconomic status and risk of offspring autism spectrum disorders in a Swedish population-based study. *Journal of the American Academy of Child and Adolescent Psychiatry, 51*(5), 467–476. http://dx.doi.org/10.1016/j.jaac.2012.02.012. e6.

Rapoport, J., Chavez, A., Greenstein, D., Addington, A., & Gogtay, N. (2009). Autism spectrum disorders and childhood-onset schizophrenia: clinical and biological contributions to a relation revisited. *Journal of the American Academy of Child and Adolescent Psychiatry, 48*(1), 10–18. http://dx.doi.org/10.1097/CHI.0b013e31818b1c63.

Raznahan, A., Wallace, G. L., Antezana, L., Greenstein, D., Lenroot, R., Thurm, A., ... Giedd, J. N. (2013). Compared to what? Early brain overgrowth in autism and the perils of population norms. *Biological Psychiatry, 74*(8), 563–575. http://dx.doi.org/10.1016/j.biopsych.2013.03.022.

Rees, E., Walters, J. T. R., Georgieva, L., Isles, A. R., Chambert, K. D., Richards, A. L., ... Kirov, G. (2014). Analysis of copy number variations at 15 schizophrenia-associated loci. *The British Journal of Psychiatry, 204*(2), 108–114. http://dx.doi.org/10.1192/bjp.bp.113.131052.

Risch, N., Spiker, D., Lotspeich, L., Nouri, N., Hinds, D., Hallmayer, J., ... Myers, R. M. (1999). A genomic screen of autism: evidence for a multilocus etiology. *American Journal of Human Genetics, 65*(2), 493–507. http://dx.doi.org/10.1086/302497.

Ritvo, E. R., Freeman, B. J., Mason-Brothers, A., Mo, A., & Ritvo, A. M. (1985). Concordance for the syndrome of autism in 40 pairs of afflicted twins. *The American Journal of Psychiatry, 142*(1), 74–77.

Ritvo, E. R., Jorde, L. B., Mason-Brothers, A., Freeman, B. J., Pingree, C., Jones, M. B., ... Mo, A. (1989). The UCLA-University of Utah epidemiologic survey of autism: recurrence risk estimates and genetic counseling. *The American Journal of Psychiatry, 146*(8), 1032–1036.

Robinson, S. J. (2012). Childhood epilepsy and autism spectrum disorders: psychiatric problems, phenotypic expression, and anticonvulsants. *Neuropsychology Review, 22*(3), 271–279. http://dx.doi.org/10.1007/s11065-012-9212-3.

Rogers, S. J. (2009). What are infant siblings teaching us about autism in infancy? *Autism Research: Official Journal of the International Society for Autism Research, 2*(3), 125–137. http://dx.doi.org/10.1002/aur.81.

Ronemus, M., Iossifov, I., Levy, D., & Wigler, M. (2014). The role of de novo mutations in the genetics of autism spectrum disorders. *Nature Reviews Genetics, 15*(2), 133–141. http://dx.doi.org/10.1038/nrg3585.

Rosenberg, R. E., Law, J. K., Yenokyan, G., McGready, J., Kaufmann, W. E., & Law, P. A. (2009). Characteristics and concordance of autism spectrum disorders among 277 twin pairs. *Archives of Pediatrics & Adolescent Medicine, 163*(10), 907–914. http://dx.doi.org/10.1001/archpediatrics.2009.98.

Rothwell, P. E., Fuccillo, M. V., Maxeiner, S., Hayton, S. J., Gokce, O., Lim, B. K., ... Südhof, T. C. (2014). Autism-associated neuroligin-3 mutations commonly impair striatal circuits to boost repetitive behaviors. *Cell, 158*(1), 198–212. http://dx.doi.org/10.1016/j.cell.2014.04.045.

Salazar, F., Baird, G., Chandler, S., Tseng, E., O'sullivan, T., Howlin, P., ... Simonoff, E. (2015). Co-occurring psychiatric disorders in preschool and elementary school-aged children with autism spectrum disorder. *Journal of Autism and Developmental Disorders, 45*(8), 2283–2294. http://dx.doi.org/10.1007/s10803-015-2361-5.

Sanders, S. J., Ercan-Sencicek, A. G., Hus, V., Luo, R., Murtha, M. T., Moreno-De-Luca, D., ... State, M. W. (2011). Multiple recurrent de novo CNVs, including duplications of the 7q11.23 Williams syndrome region, are strongly associated with autism. *Neuron, 70*(5), 863–885. http://dx.doi.org/10.1016/j.neuron.2011.05.002.

Sanders, S. J., He, X., Willsey, A. J., Ercan-Sencicek, A. G., Samocha, K. E., Cicek, A. E., ... State, M. W. (2015). Insights into autism spectrum disorder genomic architecture and biology from 71 risk loci. *Neuron, 87*(6), 1215–1233. http://dx.doi.org/10.1016/j.neuron.2015.09.016.

Sanders, S. J., Murtha, M. T., Gupta, A. R., Murdoch, J. D., Raubeson, M. J., Willsey, A. J., ... State, M. W. (2012). De novo mutations revealed by whole-exome sequencing are strongly associated with autism. *Nature, 485*(7397), 237–241. http://dx.doi.org/10.1038/nature10945.

Sandin, S., Lichtenstein, P., Kuja-Halkola, R., Larsson, H., Hultman, C. M., & Reichenberg, A. (2014). The familial risk of autism. *Jama, 311*(17), 1770–1777. http://dx.doi.org/10.1001/jama.2014.4144.

Schizophrenia Working Group of the Psychiatric Genomics Consortium. (2014). Biological insights from 108 schizophrenia-associated genetic loci. *Nature, 511*(7510), 421–427. http://dx.doi.org/10.1038/nature13595.

Sebat, J., Lakshmi, B., Malhotra, D., Troge, J., Lese Martin, C., Walsh, T., ... Wigler, M. (2007). Strong association of de novo copy number mutations with autism. *Science (New York, NY), 316*(5823), 445–449. http://dx.doi.org/10.1126/science.1138659.

Selten, J.-P., Lundberg, M., Rai, D., & Magnusson, C. (2015). Risks for nonaffective psychotic disorder and bipolar disorder in young people with autism spectrum disorder: a population-based study. *JAMA Psychiatry, 72*(5), 483–489. http://dx.doi.org/10.1001/jamapsychiatry.2014.3059.

Shetreat-Klein, M., Shinnar, S., & Rapin, I. (2014). Abnormalities of joint mobility and gait in children with autism spectrum disorders. *Brain & Development, 36*(2), 91–96. http://dx.doi.org/10.1016/j.braindev.2012.02.005.

Silverman, J. L., & Crawley, J. N. (2014). The promising trajectory of autism therapeutics discovery. *Drug Discovery Today, 19*(7), 838–844. http://dx.doi.org/10.1016/j.drudis.2013.12.007.

Simonoff, E., Pickles, A., Charman, T., Chandler, S., Loucas, T., & Baird, G. (2008). Psychiatric disorders in children with autism spectrum disorders: prevalence, comorbidity, and associated factors in a population-derived sample. *Journal of the American Academy of Child and Adolescent Psychiatry, 47*(8), 921–929. http://dx.doi.org/10.1097/CHI.0b013e318179964f.

Sparrow, S. S., Balla, D. A., & Cicchetti, D. V. (2005). *Vineland adaptive behavior scales* (2nd ed.). Circles Pines, MN: American Guidance Service.

Sparrow, S. S., & Cicchetti, D. V. (1985). Diagnostic uses of the Vineland adaptive behavior scales. *Journal of Pediatric Psychology, 10*(2), 215–225.

Spitzer, R. L., & Statistics, A.P.A.T.F.O.N.A. (1987). *Diagnostic and statistical manual of mental disorders.*

Stahlberg, O., Soderstrom, H., Rastam, M., & Gillberg, C. (2004). Bipolar disorder, schizophrenia, and other psychotic disorders in adults with childhood onset AD/HD and/or autism spectrum disorders. *Journal of Neural Transmission (Vienna, Austria: 1996), 111*(7), 891–902. http://dx.doi.org/10.1007/s00702-004-0115-1.

State, M. W., & Sestan, N. (2012). Neuroscience. The emerging biology of autism spectrum disorders. *Science (New York, NY), 337*(6100), 1301–1303. http://dx.doi.org/10.1126/science.1224989.

Strauss, K. A., Puffenberger, E. G., Huentelman, M. J., Gottlieb, S., Dobrin, S. E., Parod, J. M., ... Morton, D. H. (2006). Recessive symptomatic focal epilepsy and mutant contactin-associated protein-like 2. *The New England Journal of Medicine, 354*(13), 1370–1377. http://dx.doi.org/10.1056/NEJMoa052773.

Sugathan, A., Biagioli, M., Golzio, C., Erdin, S., Blumenthal, I., Manavalan, P., ... Talkowski, M. E. (2014). CHD8 regulates neurodevelopmental pathways associated with autism spectrum disorder in neural progenitors. *Proceedings of the National Academy of Sciences of the United States of America, 111*(42), E4468–E4477. http://dx.doi.org/10.1073/pnas.1405266111.

Sun, X., Allison, C., Auyeung, B., Baron-Cohen, S., & Brayne, C. (2014). Parental concerns, socioeconomic status, and the risk of autism spectrum conditions in a population-based study. *Research in Developmental Disabilities, 35*(12), 3678–3688. http://dx.doi.org/10.1016/j.ridd.2014.07.037.

Surén, P., Roth, C., Bresnahan, M., Haugen, M., Hornig, M., Hirtz, D., ... Stoltenberg, C. (2013). Association between maternal use of folic acid supplements and risk of autism spectrum disorders in children. *JAMA, 309*(6), 570–577. http://dx.doi.org/10.1001/jama.2012.155925.

Taylor, L. E., Swerdfeger, A. L., & Eslick, G. D. (2014). Vaccines are not associated with autism: an evidence-based meta-analysis of case-control and cohort studies. *Vaccine, 32*(29), 3623–3629. http://dx.doi.org/10.1016/j.vaccine.2014.04.085.

Thomas, N. S., Sharp, A. J., Browne, C. E., Skuse, D., Hardie, C., & Dennis, N. R. (1999). Xp deletions associated with autism in three females. *Human Genetics, 104*(1), 43–48.

Tuchman, R., & Rapin, I. (2002). Epilepsy in autism. *The Lancet Neurology, 1*(6), 352–358.

Unenge Hallerbäck, M., Lugnegård, T., & Gillberg, C. (2012). Is autism spectrum disorder common in schizophrenia? *Psychiatry Research, 198*(1), 12–17. http://dx.doi.org/10.1016/j.psychres.2012.01.016.

Varga, E. A., Pastore, M., Prior, T., Herman, G. E., & McBride, K. L. (2009). The prevalence of PTEN mutations in a clinical pediatric cohort with autism spectrum disorders, developmental delay, and macrocephaly. *Genetics in Medicine: Official Journal of the American College of Medical Genetics, 11*(2), 111–117. http://dx.doi.org/10.1097/GIM.0b013e31818fd762.

Voineagu, I., Wang, X., Johnston, P., Lowe, J. K., Tian, Y., Horvath, S., ... Geschwind, D. H. (2011). Transcriptomic analysis of autistic brain reveals convergent molecular pathology. *Nature, 474*(7351), 380–384. http://dx.doi.org/10.1038/nature10110.

Volkmar, F., Siegel, M., Woodbury-Smith, M., King, B., McCracken, J., State, M., & American Academy of Child and Adolescent Psychiatry (AACAP) Committee on Quality Issues (CQI). (2014). Practice parameter for the assessment and treatment of children and adolescents with autism spectrum disorder. *Journal of the American Academy of Child and Adolescent Psychiatry, 53*(2), 237–257. http://dx.doi.org/10.1016/j.jaac.2013.10.013.

Wan, M. W., Green, J., Elsabbagh, M., Johnson, M., Charman, T., Plummer, F., & BASIS Team. (2013). Quality of interaction between at-risk infants and caregiver at 12-15 months is associated with 3-year autism outcome. *Journal of Child Psychology and Psychiatry, and Allied Disciplines, 54*(7), 763–771. http://dx.doi.org/10.1111/jcpp.12032.

Wang, K., Zhang, H., Ma, D., Bucan, M., Glessner, J. T., Abrahams, B. S., ... Hakonarson, H. (2009). Common genetic variants on 5p14.1 associate with autism spectrum disorders. *Nature, 459*(7246), 528–533. http://dx.doi.org/10.1038/nature07999.

Wassink, T. H., Piven, J., & Patil, S. R. (2001). Chromosomal abnormalities in a clinic sample of individuals with autistic disorder. *Psychiatric Genetics, 11*(2), 57–63.

Weidmer-Mikhail, E., Sheldon, S., & Ghaziuddin, M. (1998). Chromosomes in autism and related pervasive developmental disorders: a cytogenetic study. *Journal of Intellectual Disability Research: JIDR, 42*(Pt 1), 8–12.

Weiss, L. A., Arking, D. E., Gene Discovery Project of Johns Hopkins & the Autism Consortium, Daly, M. J., & Chakravarti, A. (2009). A genome-wide linkage and association scan reveals novel loci for autism. *Nature, 461*(7265), 802–808. http://dx.doi.org/10.1038/nature08490.

Weiss, L. A., Shen, Y., Korn, J. M., Arking, D. E., Miller, D. T., Fossdal, R., ... Daly, M. J. (2008). Association between microdeletion and microduplication at 16p11.2 and autism. *The New England Journal of Medicine, 358*(7), 667–675. http://dx.doi.org/10.1056/NEJMoa075974.

Werling, D. M., & Geschwind, D. H. (2015). Recurrence rates provide evidence for sex-differential, familial genetic liability for autism spectrum disorders in multiplex families and twins. *Molecular Autism, 6*(1), 27. http://dx.doi.org/10.1186/s13229-015-0004-5.

Willsey, A. J., Sanders, S. J., Li, M., Dong, S., Tebbenkamp, A. T., Muhle, R. A., ... State, M. W. (2013). Coexpression networks implicate human midfetal deep cortical projection neurons in the pathogenesis of autism. *Cell, 155*(5), 997–1007. http://dx.doi.org/10.1016/j.cell.2013.10.020.

Wing, L. (1981). Asperger's syndrome: a clinical account. *Psychological Medicine, 11*(1), 115–129.

Xu, X., Wells, A. B., O'Brien, D. R., Nehorai, A., & Dougherty, J. D. (2014). Cell type-specific expression analysis to identify putative cellular mechanisms for neurogenetic disorders. *The Journal of Neuroscience: the Official Journal of the Society for Neuroscience, 34*(4), 1420–1431. http://dx.doi.org/10.1523/JNEUROSCI.4488-13.2014.

Yu, T. W., Chahrour, M. H., Coulter, M. E., Jiralerspong, S., Okamura-Ikeda, K., Ataman, B., ... Walsh, C. A. (2013). Using whole-exome sequencing to identify inherited causes of autism. *Neuron, 77*(2), 259–273. http://dx.doi.org/10.1016/j.neuron.2012.11.002.

Zhang, B., & Horvath, S. (2005). A general framework for weighted gene co-expression network analysis. *Statistical Applications in Genetics and Molecular Biology, 4*(1). http://dx.doi.org/10.2202/1544-6115.1128. Article17.

Zhao, X., Leotta, A., Kustanovich, V., Lajonchere, C., Geschwind, D. H., Law, K., … Wigler, M. (2007). A unified genetic theory for sporadic and inherited autism. *Proceedings of the National Academy of Sciences of the United States of America, 104*(31), 12831–12836. http://dx.doi.org/10.1073/pnas.0705803104.

Zoghbi, H. Y. (2003). Postnatal neurodevelopmental disorders: meeting at the synapse? *Science (New York, NY), 302*(5646), 826–830. http://dx.doi.org/10.1126/science.1089071.

Zwaigenbaum, L., Bryson, S., Rogers, T., Roberts, W., Brian, J., & Szatmari, P. (2005). Behavioral manifestations of autism in the first year of life. *International Journal of Developmental Neuroscience: the Official Journal of the International Society for Developmental Neuroscience, 23*(2–3), 143–152. http://dx.doi.org/10.1016/j.ijdevneu.2004.05.001.

Chapter 30

Molecular Architecture and Neurobiology of Bipolar Disorder

Carrie E. Bearden
University of California Los Angeles, Los Angeles, CA, United States

Peter Zandi
Johns Hopkins School of Medicine, Baltimore, MD, United States

Nelson B. Freimer
University of California Los Angeles, Los Angeles, CA, United States

Bipolar disorder (BP) is defined by episodic and extreme fluctuations in mood, encompassing both mania and depression. Manic and depressive episodes typically include disturbance in similar neurobehavioral domains, usually but not always in opposite directions. For example, mania is characterized by decreased need for sleep, rapid thought and speech, grandiosity, and excessive pursuit of pleasurable activities that have a high probability of negative consequences; bipolar depression, in contrast, is characterized by increased sleep, slowed thinking and speech, feelings of low self-worth, and loss of interest in pleasurable activities (anhedonia). Severe manic and depressive episodes commonly include psychotic symptoms, including both delusions and hallucinations. BP is one of the leading contributors to disability globally and is strongly associated with premature mortality; notably, up to 15% of BP-affected individuals die from suicide, for which both mania and depression display increased risk (Guze & Robins, 1970).

In this chapter we review evidence regarding the biological underpinnings of BP, framing our discussion in relation to key clinical features of the syndrome. These features, in particular the cyclicity of bipolar episodes and the dramatic shifts between mania and depression, present treatment challenges that are unique to BP; for almost 2000 years their distinctiveness has had a central role in shaping conceptions of severe mental illness (Goodwin & Jamison, 2007) and particularly influenced the initial codification of psychiatric disorders in the 19th century.

In 1854 the French psychiatrist Jean-Pierre Falret (Sedler, 1983), in conceptualizing what he termed *folie circulaire*, described the characteristic alternation of mania and depression (interspersed with periods of normal function) as a single distinct syndrome. He also noted what we now term the "switch process," the sudden shift that often occurs from one extreme mood state to the other. As discussed subsequently, this feature, which distinguishes BP from all other disorders, can have a devastating impact on affected individuals; preventing switches from occurring is therefore a major component of clinical decision making.

Following Falret, Emil Kraepelin, created a categorization system for psychiatric disorders that remains the foundation of both clinical practice and research throughout the world. He divided psychotic disorders into two categories, primarily based on course of illness. Specifically, he characterized "manic depressive insanity" as a mood disorder with a periodic course with good prognosis, differentiating it from "dementia praecox," the entity that we now term schizophrenia (SCZ), which he characterized as demonstrating a progressive downhill course. Whereas Kraepelin's conception of "manic depressive insanity" incorporated all severe, recurrent mood disorders (Tohen & Goodwin, 1995), subsequent nosologists proposed subdivisions within this category, which have gained general acceptance.

Genomics, Circuits, and Pathways in Clinical Neuropsychiatry. http://dx.doi.org/10.1016/B978-0-12-800105-9.00030-5

In the late 1950s Leonhard introduced the concept of a distinction between bipolar and unipolar depression that is commonly accepted today (Bebbington & Ramana, 1995). In modern diagnostic classifications, the occurrence of one or more episodes of mania with or without episodes of depression distinguishes BP from the more commonly occurring unipolar depression. Individuals with unipolar depression experience one or more episodes of depression without ever experiencing episodes of pathologically raised mood. Based on the work of Dunner, Gershon, and Goodwin (1976) in the 1970s, a further distinction has been made between bipolar I disorder (BPI) and bipolar II disorder (BPII). Individuals with BPI have episodes of clear-cut mania, whereas individuals with BPII experience only milder forms of mania ("hypomanias"). This distinction was codified in the *Diagnostic and Statistical Manual of Mental Disorders*, Fourth Edition (DSM-IV) (American Psychiatric Association, 1994).

The major steps in the development of BP nosology, as outlined earlier, occurred before the extensive availability of pharmacotherapies directed against symptoms of the disorder. Now that several decades have passed since the widespread implementation of drug treatment for BP, we can observe several ways in which this history has influenced our understanding of the disorder: (1) Treatment has changed the natural history of the disease and may have blurred distinctions between syndromes; observations (such as Kraepelin's) made before the availability of modern treatment must be differentiated from those made in posttreatment settings; (2) the drugs used as front-line treatment for BP, such as lithium salts or various anticonvulsant agents, appear to exert mood-stabilizing effects that are specific to this disorder, and therefore have not been typically used to treat other psychiatric disorders; (3) treatment of BP depression is itself a potent instigator of the switch process to mania, the so-called treatment-emergent affective switch (TEAS) (Salvadore et al., 2010).

The TEAS has now been observed for every class of drug used to treat depression (Salvadore et al., 2010), and in many individuals provides the first evidence that they have BP rather than unipolar depression. For most people affected by BP, depression occurs more frequently than mania, lasts longer, and may have a greater overall impact (Judd et al., 2002; Perlis et al., 2006; Post et al., 2003). In part, this situation may reflect undertreatment of BP depression based on fear of inducing TEAS, and based on a surprising paucity of data demonstrating the efficacy of treatments for this mood state (Pacchiarotti et al., 2013). The high frequency of TEAS therefore represents a clinical challenge that is unique to BP. At the same time it provides an impetus for focusing on the distinctness of BP in research aimed at understanding its biology, such as the genetic and neural systems investigations that form the main focus of this chapter.

Much as past research and clinical practice emphasized the distinctness of BP in relation to other psychiatric syndromes, perhaps the predominant view at present is that these apparent distinctions may be superficial or even artifactual. Not only are there substantial phenotypic features that cut across syndromes such as BP, unipolar depression, and SCZ, as discussed in this chapter, mounting evidence points to common underlying genetic risk factors. In our view there is much to be gained by continuing to attempt to elucidate the biological underpinnings of phenotypic features that may be considered distinct characteristics of BP, as well as those that are apparently common across psychiatric syndromes (as illustrated in Fig. 30.1). Uncertainties about the validity of categorical phenotypic constructs in psychiatry have fueled efforts to explain psychopathology from a dimensional perspective. In this chapter we therefore attempt to convey the state of progress in genetic investigations focused on BP and other syndromes, as well as progress in efforts to identify and assay quantitative measures hypothesized to represent key dimensions of psychopathology.

DESCRIPTIVE EPIDEMIOLOGY

Estimates of the lifetime prevalence of BP using modern diagnostic criteria generally range from 0.5% to 1.5%. The landmark Epidemiologic Catchment Area (ECA) study (Regier et al., 1984) surveyed five different sites across the United States using the Diagnostic Interview Schedule (Robins, Helzer, Croughan, & Ratcliff, 1981) and DSM-III (American Psychiatric Association, 1980) criteria. Although there was some variation across sites, the overall lifetime prevalence was 0.8% for BPI and 0.5% for BPII. The National Comorbidity Survey (NCS) (Kessler et al., 1994; Merikangas et al., 2007), which built on the work of the ECA, obtained higher rates. In a report from the NCS-R, a replication of the original survey, a total of 9282 individuals aged 18 years and older from noninstitutional populations in 48 states were assessed using the Composite International Diagnostic Instrument. The lifetime prevalence was found to be 1.0% for narrowly defined BPI and up to 4.4% for the more broadly defined BP spectrum that included BPI, BPII, and subthreshold BP (Merikangas et al., 2007). Studies conducted outside the United States using the same methods as part of the World Health Organization World Mental Health Survey obtained a similar range of rates, which suggests that in general, the prevalence of BP is relatively consistent across different populations (Merikangas et al., 2011).

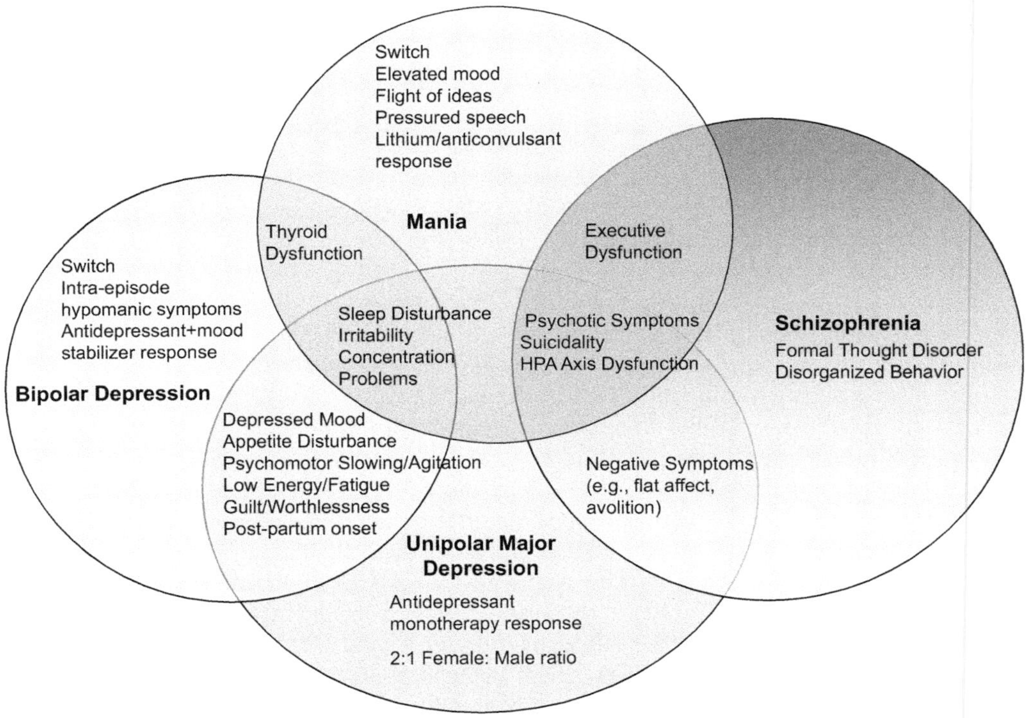

FIGURE 30.1 Illustration of distinct and overlapping phenotypic features of major psychiatric syndromes: unipolar major depression, schizophrenia, and bipolar disorder (manic and depressive phases).

The prevalence of BP does not appear to vary significantly by sex, race, or social class. The ECA study found equal rates of BP across gender and race (Weismann et al., 1991). Earlier studies, however, did find higher rates of BP in upper social classes. For example, the New Haven Community Sample (Weissman & Myers, 1978), one of the first epidemiologic surveys to use research diagnostic criteria, reported rates ranging from 4.6% in the highest social classes to 0% in the lowest social classes. Most studies have not found significant differences in BP across social classes. A notable exception is the ECA study, which found higher rates of mania in individuals with fewer years of education. The prevalence of BP was 1.1% for those with less than 12 years of education compared with 0.9% for those with more than 12 years of education (Weissman, Bruce, Leaf, Florio, & Holzer, 1991).

The ECA study also found differences in the rates of BP in urban compared with rural populations. The St. Louis site reported that the rate of BP was 1.5% in urban areas compared with 0.5% in rural areas. Similarly, the Durham site reported that the rate of BP was 0.8% in urban areas compared with 0.2% in rural areas (Weissman et al., 1991). It is uncertain whether urban living is a cause or consequence of BP.

COMORBIDITY

Both the ECA and NCS found that BP was accompanied by a comorbid disorder in considerably more than 50% of patients (Tohen & Goodwin, 1995). The ECA study reported that approximately 46% of individuals with BPI had an alcohol abuse disorder whereas 41% had a drug abuse disorder. This resulted in over 60% of individuals with BPI having any substance abuse disorder (Regier et al., 1990) and represents a sixfold increase over unaffected individuals from the community. The NCS reported strikingly similar results. In that study, individuals with BPI had a greater than eightfold increased risk of having a substance abuse disorder compared with the general population (Kessler et al., 1997). Anxiety disorders, including panic disorder, agoraphobia, social anxiety disorder, obsessive–compulsive disorder, posttraumatic stress disorder, and generalized anxiety disorder, are also frequently comorbid with BP. For example, in a cross-sectional examination of the first 500 participants in the Systematic Treatment Enhancement Program for Bipolar Disorder, a multicenter study to evaluate the longitudinal outcomes of patients with BP, over 50% were diagnosed with any anxiety disorder. Other studies have noted a significant co-occurrence of BP with DSM-III-R personality disorders (Flick, Roy-Byrne, Cowley, Shores, & Dunner, 1993) and eating disorders (Savino et al., 1993; Shisslak, Perse, & Crago, 1991; Simpson, al-Mufti, Andersen, & DePaulo, 1992).

GENETIC EPIDEMIOLOGY

Family Studies of Bipolar Disorder

Studies have consistently found that BP aggregates in families. Craddock and Jones (1999) reviewed all published family studies that used a modern concept of BP, measured lifetime risk of BP in first-degree relatives of a BP proband, and interviewed at least some of the relatives directly (Craddock & Jones, 1999). They identified 21 studies meeting these criteria, eight of which included a sample of controls. All of the studies reported an increased risk of BP in the relatives of BP probands compared with control subjects or with the baseline risk in the general population. Craddock and Jones performed a meta-analysis of the eight studies that included their own control groups. Focusing on narrowly defined BPI, they estimated that the recurrence risk ratio (λ_R, or the ratio of recurrence risk in relatives of a proband divided by the prevalence in the general population in first-degree relatives, was 7 (95% confidence interval [CI], 5−10). Adequate data were not available to estimate the recurrence risk ratios of more distant relatives accurately; however, the ratios are certainly greater than unity but less than the ratio for first-degree relatives.

A more recent review of BP family studies was carried out by Smoller and Finn (2003). It included many of the same studies as the review by Craddock and Jones (1999), and not surprisingly arrived at similar conclusions. It went further, however, and reviewed the existing evidence for differences in familial aggregation by important factors. It noted that the recurrence risk in relatives of BP probands does not appear to differ by sex of the proband or the relative (Gershon et al., 1982; Faraone, Lyons, & Tsuang, 1987; Pauls, Morton, & Egeland, 1992; Rice et al., 1987). By contrast, there is some evidence that the recurrence risk is greater in families with an early-onset BP proband (Smoller & Finn, 2003).

Several of the rigorously designed family studies separately examined the aggregation of BPI and BPII. These studies found that the recurrence risk of BPII tends to be higher in families with a BPII versus BPI proband (Andreasen et al., 1987; Coryell, Endicott, Reich, Andreasen, & Keller, 1984; Endicott et al., 1985; Gershon et al., 1982; Heun & Maier, 1993), which provides tentative evidence that BPII is a distinct entity. At the same time, however, the rates of BPI also tended to be elevated in relatives of BPII probands, which suggests a causal connection between the two spectrum disorders.

Family Studies of Bipolar Disorder and Major Depression

As noted previously, major depression (MD) and BP are closely related mood disorders that only relatively recently have been distinguished in our diagnostic nosology. In fact, many of the family studies before the 1960s lumped the two disorders together when investigating familial aggregation. In the 1980s, however, family studies began to examine these disorders separately and evaluate their cross-aggregation. The general picture to emerge from these studies is that the risk of MD is increased in relatives of BP probands, whereas the risk of BP does not appear to be increased in relatives of MD probands (Gershon et al., 1982; Maier et al., 1993; Tsuang, Winokur, & Crowe, 1980). Smoller and Finn (2003) reviewed these studies and estimated that relatives of BP probands have an almost threefold increased risk of MD compared with relatives of control probands (Craddock & Jones, 1999).

One of the largest family studies of BP in the United States to date has called into question the conclusions of the earlier family studies of BP and MD (Merikangas et al., 2014). This study recruited 447 probands with BPI, BPII, or MD and 2082 of their living first-degree relatives from a clinically enriched community screening. It then assessed the participants for mania and depression as separate nonhierarchical conditions and examined their familial aggregation and cross-aggregation. As expected from earlier studies, both mania (odds ratio [OR], 8.40; 95% CI, 3.27−20.97) and depression (OR, 2.26; 95% CI, 1.58−3.22) significantly aggregated in families. However, the investigators found no evidence of cross-aggregation of MD with BP or of BP with MD. They concluded "… that mania and major depression are largely transmitted independently in families, suggesting that these major components of BP may represent distinct underlying pathways rather than increasingly severe manifestations of a common underlying diathesis." They further concluded that the findings support the distinction between mania and depression as separate disease processes and that BP is really composed of two distinct, albeit highly comorbid syndromes (Cuellar, Johnson, & Winters, 2005). The investigators speculated that their study obtained surprising results because it was based on a community rather than clinical sample, as has been used in previous family studies, and was therefore able to draw more valid inferences about the relationship between BP and MD. However, it is unclear how the model proposed by these authors could account for the clinical observations that are the foundation of the BP syndrome: specifically, the rarity of unipolar mania and the BP switch process, including TEAS. Further study is needed to evaluate the merits of the hypothesis raised by this interesting study.

Family Studies of Bipolar Disorder and SCZ

Since Kraepelin distinguished "manic depressive insanity" and dementia praecox, there has been considerable interest in exploring the boundaries of mood disorders and psychotic disorders and evaluating the extent of their clinical and etiologic overlap. Hence, a number of family studies have examined the cross-aggregation between BP and SCZ. Some of these studies found an elevated risk of SCZ among relatives of BP probands (Taylor, Berenbaum, Jampala, & Cloninger, 1993; Tsuang et al., 1980; Valles et al., 2000) whereas several other controlled, direct interview family studies did not (Gershon et al., 1982; Maier et al., 1993; Weissman et al., 1984). Similarly whereas one controlled study found evidence of increased family risk of any mood disorder (BP and/or MD) in relatives of SCZ probands (Taylor et al., 1993), most studies have not corroborated this finding (Gershon et al., 1988; Kendler & Gardner, 1997; Maier et al., 1993; Tsuang et al., 1980). Kendler and Gardner (1997) performed a meta-analysis of three controlled direct interview studies of SCZ probands and found that the estimates of risk of BP in relatives was similar across the three studies, with a nonsignificant increase of 1.9 (0.7−5.2) (Kendler & Gardner, 1997).

Two family studies of BP and SCZ examined considerably larger samples ascertained from nationwide registries from Denmark and Sweden and reported more consistent findings (Lichtenstein et al., 2009; Mortensen, Pedersen, Melbye, Mors, & Ewald, 2003). Lichtentstein et al. (2009) examined data from the Sweden hospital discharge register which includes all public psychiatric inpatient admissions on over nine million individuals in two million nuclear families between 1973 and 2004 (Lichtenstein et al., 2009), whereas Mortensen et al. (2003) used a population based cohort of 2.1 million individuals from the Danish Civil Registration System linked with the Danish Psychiatric Central Register which included 2299 individuals diagnosed with BP. Both found remarkably similar results (Lichtenstein et al., 2009). The Denmark study reported that, compared with an individual in the general population, the risk of BP in siblings of a proband with SCZ was 4.66 (95% CI, 1.73−12.6) times greater, whereas the risk of BP in offspring of parents with SCZ ranged from 3.68 (95% CI, 1.83−7.43) to 5.79 (95% CI, 3.9−8.59) times greater for fathers and mothers, respectively. The comparable BP recurrence risk ratio estimates from the Sweden study were 3.7 (95% CI, 3.2−4.2) for siblings of a proband with SCZ and 5.2 (95% CI, 4.4−6.2) for offspring of any parent with SCZ. The Sweden study went further and estimated that the recurrence risk ratio of SCZ was 3.9 (95% CI, 3.4−4.4) for siblings of a proband with BP and 2.4 (95% CI, 2.1−2.6) for offspring of a parent with BP. These results provide compelling evidence for significant cross-aggregation of BP and SCZ and suggest substantial but not complete overlap in their genetic etiologies. These studies may provide a different picture of the familial overlap between BP and SCZ compared with earlier studies because they are truly population-based and therefore are less susceptible to bias owing to ascertainment than more clinically based samples.

Family Studies of Bipolar Disorder and Attention-Deficit Hyperactivity Disorder

Finally, a number of family studies have been carried out to examine the relationship between BP and attention-deficit hyperactivity disorder (ADHD). There is considerable comorbidity between BP and ADHD, and the differential diagnosis between the two is often complicated, especially in youth. Systematic studies have found ADHD in 57−98% of youths with BP (Geller et al., 1995; Wozniak et al., 1995; West, McElroy, Strakowski, Keck, & McConville, 199). A meta-analysis in 2005 estimated the prevalence of ADHD in youth with BP to be 62% (Kowatch, Youngstrom, Danielyan, & Findling, 2005) and significantly elevated rates of BP in youth with ADHD (Donfrancesco et al., 2011; Faraone et al., 1997; Hensch, Himmerich, & Hegerl, 2011; Lus & Mukaddes, 2009). One study reported BP in 22% of inpatients with ADHD (Butler, Arredondo, & McCloskey, 1995), with similar findings in adults with ADHD (Bernardi et al., 2012; Klassen, Katzman, & Chokka, 2010). These findings have raised questions about the common etiologic underpinnings of the two disorders and motivated family studies to address these questions.

Faraone, Biederman, and Wozniak (2012) carried out a systematic review of family studies of BP and ADHD. They identified a total of 20 BP studies which provided 37 estimates of ADHD prevalence in 4301 relatives of BP probands and 1937 control probands, and seven ADHD studies which provided 12 estimates of BP in 1877 relatives of ADHD probands and 1601 relatives of control probands. For BP probands, the weighted prevalences of ADHD were 27% in offspring, 30.1% in siblings, and 16.5% in parents, compared with control probands in which the prevalence was 9.6% in offspring, 11.6% in siblings, and 4.5% in parents. The overall relative risk was 2.6 (2.1−3.2), regardless of whether the relative also had BP. Interestingly, the relative risk of ADHD was nearly the same for pediatric BP probands (2.8; 2.1−37) and adult BDI probands (2.6; 1.9−3.4). For ADHD probands, the weighted prevalence of BP was 6.8% in offspring, 5.9% in siblings, and 5.1% in parents, compared with control probands, in which the prevalence was 3.5% in offspring, 2.8% in siblings, 3.1% in parents. The overall relative risk was 1.8 (1.3−2.6), regardless of whether the relative also had ADHD. Among studies of ADHD probands, the relative risk was 2.1 (0.9−4.6) for pediatric BP and 2.2 (1.4−3.6) for adult BP.

Twin Studies

Twin and adoption studies are natural experiments that can help determine whether the familial aggregation of BP and its related conditions is the result of genetic or environmental factors. Twin studies dating back to the 1920s have observed that monozygotic (MZ) twins are more concordant for mood disorders than are dizygotic (DZ) twins. However, as noted previously, these earlier studied did not distinguish between BP and MD. In a review of these earlier studies, Tsuang and Faraone (1990) combined data from 11 twin studies published between 1928 and 1986, composed of 195 MZ and 255 same-sex DZ pairs. They reported a proband-wise concordance of 78% for MZ and 29% for DZ twins, resulting in an estimate of heritability of 63%.

Since the 1990s, at least four twin studies using a modern concept of BP have been reported. These include studies of 486 twin pairs from the Swedish Psychiatric Twin Register (Kendler, 1993), 224 twin pairs from the Maudsley Twin Register (Cardno et al., 1999), 38 twin pairs from the Finnish Twin Register (Kieseppä, Partonen, Haukka, Kaprio, & Lonnqvist, 2004), and 303 twin pairs from the Norwegian Twin Register (Edvardsen et al., 2008). Estimates from these studies of the heritability of BP ranged from a low of 73% (Edvardsen et al., 2008) to a high of 93% (Kieseppä et al., 2004). Interestingly, all of the studies found that the remaining proportion of variation in risk for BP was explained by individual residual factors and that there was no notable contribution of shared environmental factors. The studies by Cardno et al. (1999) and Edvardsen et al. (2008) both found that the estimates of heritability increased when considering a broader spectrum of BP that included both BPI and BPII. In addition, in their study, Cardno et al. (1999) examined the overlap between BP and SCZ and found evidence for significant genetic correlation between the syndromes, 0.68 for BP and SCZ and 0.88 for BP and schizoaffective disorder (Fig. 30.2).

Adoption Studies

Two adoption studies were conducted using a modern concept of BP. Mendlewicz and Rainer (1977) examined the biological and adoptive parents of BP and normal adoptees and the biological parents of BP nonadoptees. They found a significantly greater risk of BP, MD, and schizoaffective disorders in the biological parents of the bipolar adoptees compared with the adoptive parents. The risk of affective disorders in the biological relatives of bipolar adoptees was similar to that in the biological relatives of the bipolar nonadoptees. Wender et al. (1986) conducted a smaller study, and found a similar although nonsignificant trend. Because adoptees inherit their genes from one family but are exposed to an environment from another, these results provide convincing evidence that genetic factors have an important role in the familial transmission of BP and related spectrum disorders.

MOLECULAR GENETICS

Family, twin, and adoption studies of BP provide compelling evidence that genetic factors have a leading role in the etiology of BP. In fact, estimates of the familial recurrence risk and heritability from these studies are among the highest of all psychiatric disorders. This has motivated considerable efforts to find susceptibility genes for BP using linkage, association, and more recently, sequencing approaches.

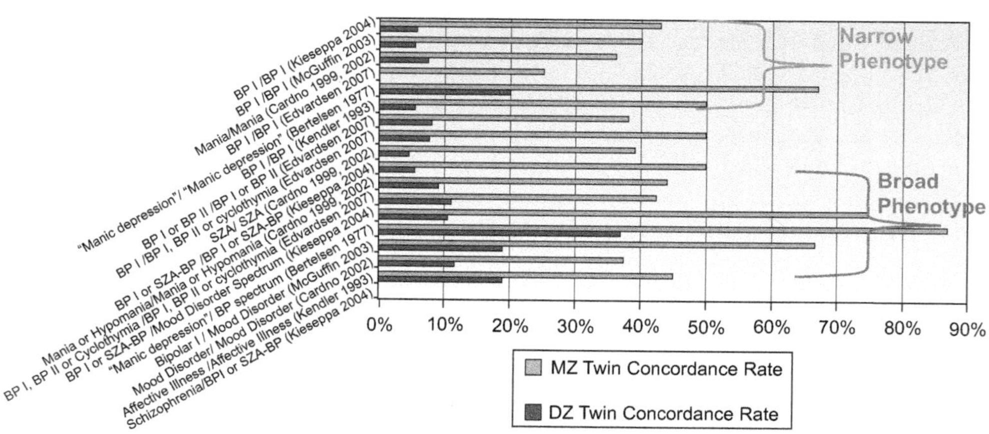

FIGURE 30.2 Twin study concordance rates for bipolar phenotypes.

Linkage Studies

Over 20 genome-wide linkage scans have been carried out with BP. Several regions have reached genome-wide significance in multifamily samples, including on 8q24 (Cichon et al., 2001), 15q14 (Turecki et al., 2001), 18q12 (Maziade et al., 2001), 21q22 (Liu et al., 2001), and 22q12 (Kelsoe et al., 2001), but none of these have been consistently replicated across studies. Other regions on 1q41, 4p16, 4q32-35, 10q21-26, 12q23-24, 13q31-33, 16p12-13, 18p11, 18q22- 23, and Xq24-28 have been implicated with at least suggestive evidence in more than one study, but again the findings are largely inconclusive. Three separate meta-analyses of prior linkage studies have been conducted (Badner & Gershon, 2002; McQueen et al., 2005; Segurado et al., 2003). Each used different approaches and each arrived at different conclusions, which highlights the inconsistency of results emerging from linkage studies of BP. More recently, a mega analysis of nearly 1000 families with SNP linkage data provided only suggestive evidence for linkage with regions on 6q21 and 9q21 (Badner et al., 2012). Thus, despite over 2 decades of effort, linkage studies have failed to identify any susceptibility genes for BP conclusively.

Association Studies

Numerous association studies have also been carried out to search for BP susceptibility genes. Over 400 candidate gene association studies of BP have been reported. These studies were motivated by various hypotheses about genes in molecular pathways, particularly related to the monamine neurotransmitter systems, thought to be involved in the etiopathogenesis of mood disorders. A meta-analysis of these studies found that none of the candidate genes we originally thought might be important were significantly associated with BP when evaluated by rigorous criteria (Seifuddin et al., 2012).

The emergence of genome-wide association studies (GWAS) (also termed common variant association studies) transformed the outlook on studying the genetics of common, complex disorders such as BP. By genotyping large numbers of single-nucleotide polymorphisms (SNPs) and leveraging linkage disequilibrium, the entire genome can be interrogated in a hypothesis-free manner for disease associations with common genetic variants with minor allele frequencies greater than 5%. Although promising, the initial GWAS of complex psychiatric disorders, which typically used samples sizes less than several thousand cases and controls, were limited in their success in identifying susceptibility variants, such as the linkage and candidate gene studies before them. The few variants that were implicated had modest effect sizes with ORs less than 1.3 (Cichon et al., 2009). As a result, it became clear that the initial GWAS were underpowered and larger sample sizes would be needed to identify susceptibility variants conclusively with effect sizes of the magnitude being observed.

It was in this context that the Psychiatric Genomics Consortium (PGC) was formed. The PGC brought together investigators from around the world with the ambitious goal of assembling existing samples in five major psychiatric disorders (BP, SCZ, depression, ADHD, and autism) to carry out the largest GWAS ever conducted with these disorders (Cichon et al., 2009).

In the initial report by the PGC on BP, data were combined from 7481 patients and 9250 control subjects and imputed to a common panel of about 2.4M SNPs based on HapMap Phase 2 CEU reference samples (Psychiatric GWAS Consortium Bipolar Disorder Working Group, 2011). The top 34 independently associated SNPs with a genomic control corrected p value $<5 \times 10^{-5}$ were then examined in a replication sample of 4496 independent cases and 42,422 independent controls. An analysis of the initial and replication samples combined together confirmed genome-wide significant evidence of association with SNPs in and around CACNA1C and ODZ4. Further analysis identified an enrichment of associations with SNPs in a pathway composed of calcium channel subunits. Intriguingly, CACNA1C and genes in the calcium channel signaling pathway were also among the strongest findings in GWAS of SCZ (Cross-Disorder Group of the Psychiatric Genomics Consortium, 2013a), suggesting a common role for disruptions in calcium channel signaling in the etiopathogenesis of both BP and SCZ.

Further analysis of BP GWAS findings confirmed a significant polygenic contribution to the disorder, ie, many alleles of small effect sizes additively increase risk. Specifically, polygenic risk scores obtained by taking the sum of alleles across multiple top associated SNPs in discovery GWAS samples weighted by their ORs were significantly associated with BP status in target GWAS samples. Interestingly, an analysis of GWAS data from all five disorders studied in the PGC revealed substantial overlap in polygenic risk among several of the disorders (Cross-Disorder Group of the Psychiatric Genomics Consortium, 2013a). In particular, highly significant overlap of polygenic risk scores was noted among BP, SC, and MD, with the strongest effects observed for BP and SCZ. By contrast, no notable overlap was observed between BP and ADHD or autism.

Similar conclusions were reported in another analysis of the PGC GWAS data using a different analytic approach (Cross-Disorder Group of the Psychiatric Genomics Consortium, 2013b). In this approach, GWAS data were used to calculate the total variance in liability explained by SNPs (ie, SNP-based heritability) by estimating genetic similarities between patients and control subjects based on the SNP genotypes, as well as the genetic correlation explained by the SNPs between case–control samples collected for two different disorders (ie, SNP based co-heritability). This analysis found that the SNP-based heritability was 0.25 for BP, whereas the SNP based co-heritability was 0.68 between BP and SCZ and 0.47 between BP and MD. Again, there was no significant evidence of co-heritability between BP and ADHD. The estimates of co-heritability among BP, SCZ, and MD were broadly consistent with findings from the earlier family and twin studies, and again suggest considerable overlap in the etiology of these disorders. The SNP-based heritability estimate for BP was notably less than the previous heritability estimates from family and twin studies discussed earlier. This phenomenon has been observed with other complex disorders and has led some investigators to question whether variation that cannot be adequately assayed by genome-wide genotyping of common variants (SNPs) could account for such "missing heritability" (Manolio et al., 2009) in GWAS data; this possibility has provided a major impetus for the initiation of exome and whole-genome sequencing studies, as discussed subsequently.

It is anticipated that as larger sample sizes are collected and studied with GWAS, our estimates of the polygenic contribution to BP will become more refined, and we will have greater power to identify additional genetic pathways that contribute to the risk of developing BP and also how these pathways do or do not contribute to risk of both SCZ and MD.

Sequencing Studies

The SNPs used in GWAS are able to tag only common genetic variants that typically have minor allele frequencies greater than 5%. Therefore, GWAS are not able to detect the contribution of rare genetic variation to the risk of BP. This will require sequencing studies, which have only recently begun to get under way. It is still early days for sequencing studies of BP, so it is not yet possible to draw from them meaningful conclusions about the genetic architecture of the disorder. However, because a number of case–control and family samples have already been sequenced, findings from these efforts should soon be forthcoming. In addition, to leverage their collective resources and maximize power, investigators from around the world have joined together to form a Bipolar Sequencing Consortium that is carrying out combined analyses of samples drawn from several independent projects. The consortium currently consists of four case–control studies with data on 5000 patients and 9000 control subjects and 10 family-based studies with data on 200 families with up to 1000 affected relatives. The consortium will continue to grow as more BP samples are sequenced around the world and as additional investigators join in the collective effort. As the results of the ongoing sequencing studies and the combined analyses from the consortium emerge, we will gain a clearer picture of how rarer genetic variation contributes alongside more common variation to risk for BP.

INTERMEDIATE TRAITS AND THE BIOLOGY OF BIPOLAR DISORDER

The handful of replicated genetic associations described earlier have begun to reveal the genetic architecture of BP, but the biology underlying the disorder remains almost largely unknown. As noted previously, the heterogeneity of BP clinical phenotypes and the overlap between BP and other disorders in both symptomatology and genetic risk profiles support the usefulness of considering psychopathology dimensionally (in relation to a set of key behavioral and neurobiological domains) rather than categorically. The National Institutes of Mental Health have established the Research Domain Criteria project to stimulate the adoption of a quantitative dimensional approach throughout all levels of psychiatric research (Insel et al., 2010). This development has coincided with increased interest in the field in tackling the heterogeneity of psychiatric syndromes by delineating and genetically investigating endophenotypes, quantitative traits (ie, dimensional measures) hypothesized to be components of the syndromes.

As initially conceived, the term "endophenotype" referred to an explanation for the geographical distributions of specific insect populations (John & Lewis, 1966): that internal, not directly observable phenotypes represent an intermediate link in a causal chain between genes and the traits that can be observed directly in the field. In the subsequent application of this concept to psychiatric genetics, categorical disease phenotypes can be considered analogous to the directly observable insect traits (Bearden & Freimer, 2006; Casey et al., 2013; Gottesman & Shields, 1973). The rationale for applying the endophenotype concept to psychiatric genetic investigations is as follows: Phenotypes that are intermediate in a causal chain between gene and disease phenotypes should have a simpler genetic basis and therefore be more amenable to genetic mapping than the disease phenotypes (Gottesman & Gould, 2003).

On the whole, the field of psychiatric genetics has not embraced endophenotype approaches. In contrast, these approaches have a key role in other fields within human genetics, where they complement investigations of disease phenotypes. The field of cardiometabolic genetics provides a good illustration. Quantitative trait analyses of blood levels of the most commonly measured lipids have led to the identification of hundreds of replicated associations to variants across the allele frequency spectrum (Service et al., 2014; Willer et al., 2013). For many of these loci the effect size of the association to the quantitative traits is much larger than for the disease traits (such as coronary artery disease) for which they are endophenotypes; this observation is not surprising given that lipid levels can be assayed precisely. Furthermore, studies that have used metabolomics to more precisely assay the components of blood lipids provide an even stronger argument for endophenotype approaches; the lipid phenotypes identified through these studies demonstrate stronger heritability and a larger number of significant SNP associations than the standard lipid measures (Kettunen et al., 2012).

The experience in cardiometabolic genetics suggests that the paucity of evidence for the use for genetic studies of psychiatric endophenotypes (Flint & Munafó, 2007) may reflect the need for better endophenotypes. In this chapter we discuss domains from which we propose that useful BP endophenotypes may emerge. We discuss traits that are quantitative, objectively measurable, and possibly translatable into animal models; these traits may be valuable as phenotypes for genetic mapping studies and eventually for investigating the "downstream" biological effects of genes.

Cognitive Endophenotypes

Traditionally, neurocognitive deficits associated with BP were regarded as state-specific, although there is now consistent evidence that deficits remain, typically in an attenuated form, during periods of euthymia (Bourne et al., 2013; Clark, Kempton, Scarna, Grasby, & Goodwin, 2005; Deckersbach et al., 2004; van Gorp, Altshuler, Theberge, Wilkins, & Dixon, 1998; Martinez-Aran et al., 2004; Rubinsztein, Michael, Paykel, & Sahakian, 2000; Thompson et al., 2005). The most prominent and consistently documented areas of cognitive difficulties in BP are in the domains of sustained attention (Arts, Jabben, Krabbendam, & van Os, 2008; Mann-Wrobel et al., 2011), working memory (Arts et al., 2008), executive function (ie, cognitive flexibility/set-shifting), and verbal declarative memory (Arts et al., 2008; Bora, Yucel, & Pantelis, 2009). Although residual mood symptoms and medication treatment may partially account for these deficits, they cannot entirely account for the observed effect sizes (Bourne et al., 2013).

Of these, executive functions (ie, working memory, response inhibition, verbal fluency), processing speed and declarative memory currently show the most promise as neurocognitive intermediate phenotypes for BP, based on evidence for stable, trait-related impairment (ie, in a euthymic mood state) (Clark, Iversen, & Goodwin, 2002; Clark, Kempton, Scarna, Grasby, & Goodwin, 2005; Swann, Pazzaglia, Nicholls, Dougherty, & Moeller, 2003; Wilder-Willis et al., 2001; Zalla et al., 2004; Zubieta, Huguelet, O'Neil, & Giordani, 2001), as well as impairment in clinically unaffected relatives (Arts et al., 2008; Bora et al., 2009) and in extended pedigrees (Glahn et al., 2010). As reviewed in Arts et al. (2008), there is some evidence for cosegregation of neurocognitive deficits in biological relatives of probands with BP, although it is limited because of relatively small studies and inconsistent findings. In particular, whereas large effect sizes ($d > 0.8$) were noted for the domains of executive functions and verbal memory in patients with BP, effect sizes were small in first-degree relatives ($d < 0.5$) but significantly different from healthy control subjects. Measures of processing speed, declarative memory, and executive functions (verbal fluency and inhibitory control) have also been shown to be substantially heritable in extended pedigrees (Fears et al., 2014; Glahn et al., 2010).

However, considerable heterogeneity exists within the BP population; some patients manifesting profound deficits and a subgroup (about 30–40%) show minimal cognitive deficits (Altshuler et al., 2004). It appears that the greater the burden of illness associated with the disorder, as reflected by the number of past manic episodes, length of illness, number of past hospitalizations (Robinson & Ferrier, 2006), and history of psychosis (Glahn et al., 2007; Martinez-Aran et al., 2008), the greater the neurocognitive deficits typically found in patients with BP. Patients with BP and a history of psychosis appear to have more severe cognitive impairment, particularly in the domains of executive functioning and spatial working memory (Glahn et al., 2007). Patients with BP and greater illness severity, particularly those with psychotic symptoms, more closely resemble patients with SCZ with regard to the level and profile of neurocognitive deficits. Notably, in a large-scale investigation in multigenerational families with heavy genetic loading for BP ($N = 738$), the neurocognitive measures that were both highly heritable and disease-associated (ie, processing speed, declarative memory, verbal fluency, and inhibitory control functions), mirrored findings from previous SCZ case–control, family, and pedigree studies (Greenwood et al., 2007; Gur et al., 2007; Hill, Harris, Herbener, Pavuluri, & Sweeney, 2008), which suggests that such phenotypes could contribute to the shared risk between these disorders suggested by GWAS (Cross-Disorder Group of the Psychiatric Genomics Consortium, 2013a).

Premorbid Cognitive Impairment in Bipolar Disorder

Epidemiologic studies consistently document associations between early cognitive deficits and increased risk of developing SCZ, but premorbid deficits do not seem to characterize BP. Two longitudinal studies have now shown that in fact, better cognitive functioning in childhood or adolescence is associated with increased risk of developing BP. One study in a Dunedin, New Zealand birth cohort (Koenen et al., 2009) and another in Sweden found that males with excellent school performance had a nearly fourfold increased risk of later BP compared with those with average grades (MacCabe et al., 2010). However, children with the poorest grades were also at moderately increased risk of BP (OR, 1.86), which suggests a possible bimodal distribution, with one at-risk group associated with high and the other with low scholastic achievement. The authors speculate that because SCZ was associated with poor scholastic achievement in the same cohort, the low-performing individuals with BP may have subtle neurodevelopmental abnormalities similar to those in SCZ. A meta-analysis (Trotta, Murray, & MacCabe, 2014) concluded that BP patients show small but significant deficits in premorbid intellectual function when assessed retrospectively using a measure such as the Wide Range Achievement Test, but not when assessed prospectively. However, moderate cognitive impairment (effect size of about 0.6) was detectable after onset, which suggests that patients with BP experience a decline in cognitive functioning that accompanies illness onset.

NEUROIMAGING INTERMEDIATE TRAITS

Because BP is characterized by a primary disturbance in mood regulation, brain systems involved in emotional behavior are likely to underlie symptoms of the disorder. In particular, the amygdala, a region of specialized nuclei in the medial temporal lobe, is a critical component of the neural circuitry involved in regulation of emotional valence. Amygdala lesions are known to cause abnormalities in emotional expression, as well as learning and memory deficits, in humans and in animal models (Gallagher & Chiba, 1996; Phelps & LeDoux, 2005). Anomalous development of limbic structures has been hypothesized to contribute to BP susceptibility through the dysregulation of brain regions involved in higher-order cognitive function (Pfeifer, Welge, Strakowski, Adler, & DelBello, 2008; Strakowski, Delbello, & Adler, 2005). However, despite intensive interest, BP disease pathophysiology is not yet well understood. Over the past several years a number of studies assessing structural neuroanatomy in BP have been published, but results have been highly variable; lateral ventricular enlargement is the most consistently reported finding (Hallahan et al., 2011). Although the etiology of ventriculomegaly in BP is unknown, there is some evidence that ventricular volumes are greater in patients with BP who have experienced multiple mood episodes (Strakowski et al., 2002), which suggests that ventricular expansion may be a consequence of repeated episodes. However, this finding has not been consistently replicated.

Small and clinically heterogeneous samples likely contribute to the variability in findings across studies. Several meta-analyses have been published to address this challenge and to more definitively answer the question of whether there are reproducible structural alterations that characterize the brains of individuals with BP (Kempton, Geddes, Ettinger, Williams, & Grasby, 2008; McDonald et al., 2004). These investigations have highlighted the substantial heterogeneity across studies for limbic structures of interest, particularly the amygdala and thalamus (McDonald et al., 2004). Notably, two meta-analyses of amygdala volume across child and adult patients with BP concluded that in fact amygdala volumes are significantly reduced in pediatric BP but not in adults (Pfeifer et al., 2008). Investigators speculate that this may be the result of a compensatory reaction in response to environmental stressors, or other aspects of disease progression including long-term medication exposure. This possibility is consistent with animal data indicating that lithium directly affects amygdala physiology (Youngs et al., 2006). Nevertheless, these findings are not based on prospective longitudinal studies, and thus another interpretation is that the underlying pathophysiology may differ between early-onset versus adult-onset BP.

In contrast to a meta-analysis, which involves combining summary results from individual studies, a mega-analytic approach involves a joint analysis of pooled participant data from available studies. Using this approach, Hallahan et al. (2011) analyzed raw data obtained from 321 individuals with BP and 442 demographically comparable healthy control subjects collected across 11 international research groups. This study found that, relative to controls, individuals with BP had increased right lateral ventricular, left temporal, and right putamen volumes. Medication effects contributed to these differences, because patients with BP who were taking lithium had significantly increased medial temporal (ie, hippocampal and amygdala) volume relative to patients not treated with lithium and healthy control subjects. In addition, illness duration was significantly associated with global cerebral volume in BP. These efforts have dramatically expanded, with current efforts such as the Enhancing NeuroImaging Genetics through Meta-Analysis Consortium (Thompson et al., 2014), a collaborative network of over 70 institutions worldwide, which is based on principles of data sharing, standardized analysis tools, and the goals of making credible discoveries regarding the genetic basis of neuroanatomic variation, in both healthy populations and in the context of disease, including BP.

Because of the high heritability of bipolar illness (Lichtenstein et al., 2009), several studies have investigated whether similar neuroanatomic alterations characterize clinically unaffected relatives. In this regard, abnormalities in white matter structure are of particular interest as endophenotype markers (Emsell et al., 2013) given early findings of increased rates of white matter hyperintensities in patients with BP (Bearden, Hoffman, & Cannon, 2001; Kempton et al., 2008) and in their clinically unaffected relatives in extended pedigrees (Ahearn et al., 1998), as well as evidence for high heritability of white matter organization (Fears et al., 2014; Kochunov et al., 2014). Diffusion tensor imaging (DTI) studies of white matter microstructure in BP have consistently observed disrupted white matter tracts in unaffected relatives of patients with BP (Chaddock et al., 2009; Mahon et al., 2013; Sprooten et al., 2011). Using anatomically driven DTI tractography methods in families multiply affected with BP-1, Emsell et al. (2013) found significantly reduced fractional anisotropy (a measure of global integrity of white matter tracts) as well as increased radial diffusivity (associated with myelin injury in animal models) (Song et al., 2002) in the cingulum, callosal splenium, and superior and inferior longitudinal fasciculus, in patients with BP compared to control subjects and their nonbipolar relatives. Moreover, increasing genetic liability was associated with increased white matter disruption in frontolimbic (uncinate fasciculus) and frontoparietal (superior longitudinal fasciculus) white matter tracts, which suggests that alterations in these tracts may be a potential marker of endophenotypic risk.

Regarding volumetric abnormalities, Hajek et al. (2013) found that both unaffected and affected relatives of BP probands, as well as individuals early in the course of illness showed larger right volume of the right inferior frontal gyrus (rIFG), a brain structure critically involved in inhibitory control, relative to healthy control subjects. The rIFG volume correlated negatively with illness duration and, relative to controls, was smaller among individuals with BP with long-term illness burden and minimal lifetime lithium exposure. Notably, several functional neuroimaging studies have also found attenuated activation in this region in both patients with BP and those at genetic high risk (Lim et al., 2013; Roberts et al., 2012; Strakowski et al., 2005). In a study of pedigrees with multiple affected family members Fears et al. (2014) found cortical thickness in this region, along with a more extensive frontotemporal cortical network, to be highly heritable, as well as differentially affected in those with BP-1 (Fig. 30.3); furthermore, the inferior frontal gyrus showed evidence of shared genetic influence with limbic regions.

Because of evidence that a large number of causal variants, each of which make a small individual contribution, cumulatively contribute to overall disease risk (Ripke et al., 2013), studies have begun investigating polygenic risk scores based on disease phenotypes relative to neuroimaging intermediate traits. In one of the first studies to use this approach, Whalley et al. (2013) investigated polygenic risk scores for BP and MD based on genome-wide association data from the Psychiatric GWAS Consortium in relation to DTI data in a cohort of unaffected youth at high familial risk of mood disorder, and found that higher polygenic risk load for MD was significantly associated with decreased white matter integrity. These findings suggest that there may be an aggregate effect of multiple SNPs associated with mood disorder risk on white matter integrity in individuals at familial high risk.

Cross-Cutting Endophenotypes

As noted earlier, certain neurocognitive anomalies have been observed to distinguish those with a psychotic from nonpsychotic form of bipolar illness. Patients with BP and SCZ generally manifest similar profiles of cognitive deficit (Hill et al., 2008; Schretlen et al., 2007; Seidman et al., 2002); attention, working memory, and verbal declarative memory are among the most prominent areas of difficulty, although deficits in BP patients tend to be milder (Ivleva et al., 2010). In contrast, other intermediate behavioral traits appear to be distinct to BP, most notably perceptual creativity; individuals diagnosed with BP tend to be overrepresented in creative occupations relative to both individuals with other psychiatric disorders and the general population (Kyaga et al., 2011), and perceptual creativity is shown to be both heritable (Fears et al., 2014) and associated with genetic vulnerability to BP (Simeonova, Chang, Strong, & Ketter, 2005).

Whereas medial temporal structural alterations are inconsistently observed in patients with BP overall, smaller hippocampi have been observed specifically among patients with BP with psychotic symptoms, but not in patients with nonpsychotic BP (Strasser et al., 2005). This raises the question of whether medial temporal alterations characterize both affective and nonaffective psychosis spectrum disorders. Findings from the Bipolar-Schizophrenia Network on Intermediate Phenotypes study indicate that indeed, hippocampal volume reductions similarly characterized patients with SCZ, schizoaffective disorder, and psychotic BP, with the most prominent differences being in cornu ammonis, which forms the area between the dentate gyrus and subiculum, and is composed primarily of pyramidal cells (Andersen, Morris, Amaral, Bliss, & O'Keefe, 2007). Interestingly, alterations in the entorhinal cortex and parahippocampal regions were limited to

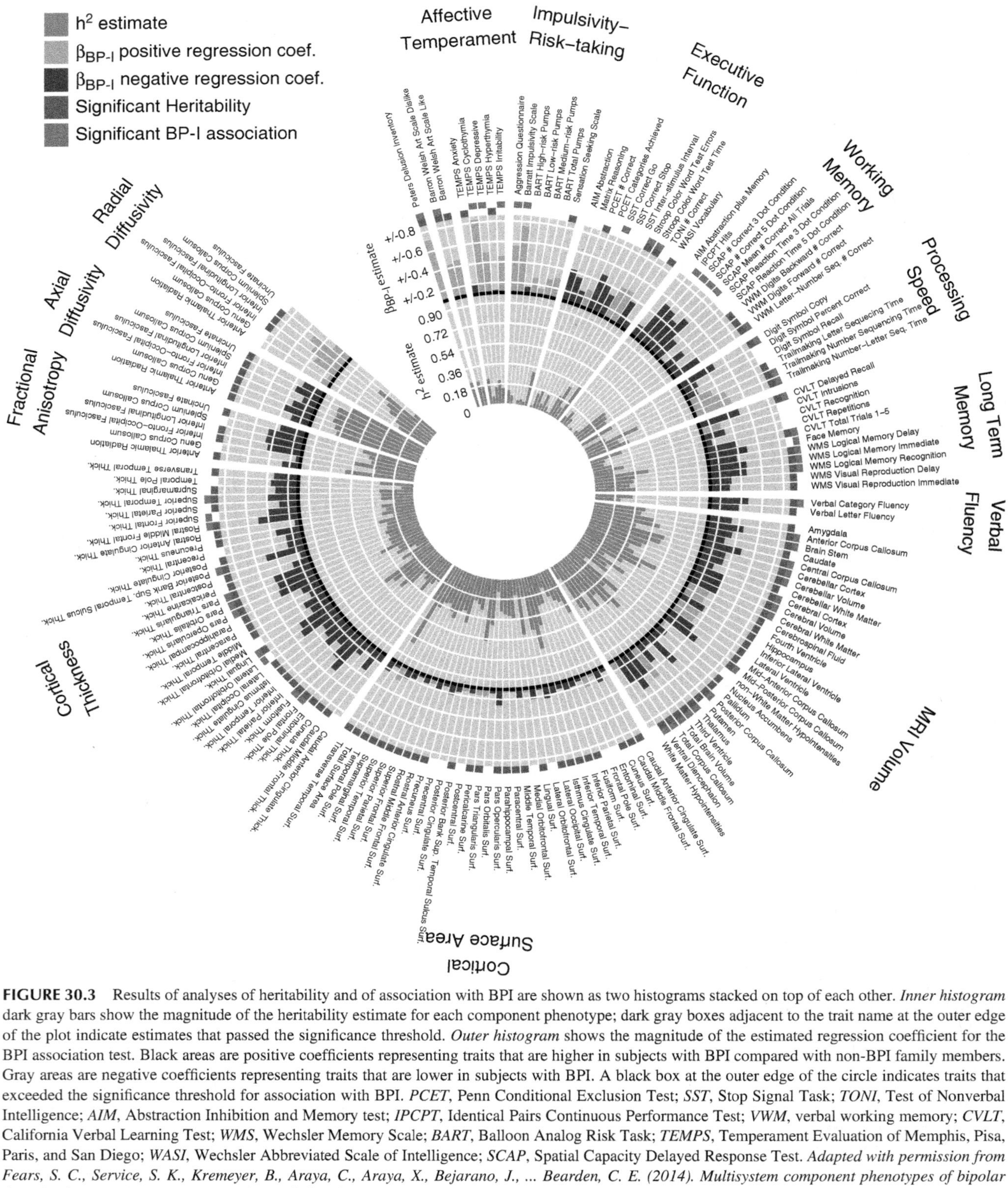

FIGURE 30.3 Results of analyses of heritability and of association with BPI are shown as two histograms stacked on top of each other. *Inner histogram* dark gray bars show the magnitude of the heritability estimate for each component phenotype; dark gray boxes adjacent to the trait name at the outer edge of the plot indicate estimates that passed the significance threshold. *Outer histogram* shows the magnitude of the estimated regression coefficient for the BPI association test. Black areas are positive coefficients representing traits that are higher in subjects with BPI compared with non-BPI family members. Gray areas are negative coefficients representing traits that are lower in subjects with BPI. A black box at the outer edge of the circle indicates traits that exceeded the significance threshold for association with BPI. *PCET*, Penn Conditional Exclusion Test; *SST*, Stop Signal Task; *TONI*, Test of Nonverbal Intelligence; *AIM*, Abstraction Inhibition and Memory test; *IPCPT*, Identical Pairs Continuous Performance Test; *VWM*, verbal working memory; *CVLT*, California Verbal Learning Test; *WMS*, Wechsler Memory Scale; *BART*, Balloon Analog Risk Task; *TEMPS*, Temperament Evaluation of Memphis, Pisa, Paris, and San Diego; *WASI*, Wechsler Abbreviated Scale of Intelligence; *SCAP*, Spatial Capacity Delayed Response Test. *Adapted with permission from Fears, S. C., Service, S. K., Kremeyer, B., Araya, C., Araya, X., Bejarano, J., ... Bearden, C. E. (2014). Multisystem component phenotypes of bipolar disorder for genetic investigations of extended pedigrees.* JAMA Psychiatry, *71(4), 375–387.*

SCZ and schizoaffective disorders. Hippocampal volumes were positively correlated with severity of psychotic symptoms and with global cognitive function, as well as with declarative memory.

Neurodevelopmental Contributions to Bipolar Disorder

Neurodevelopmental disturbances are implicated in the pathophysiology of many child or adolescent psychiatric disorders, notably SCZ (Cannon, Rosso, Bearden, Sanchez, & Hatley, 1999; Murray & Lewis, 1987; Vita, Dieci, Giobbio, Tenconi, & Invernizzi, 1997). Because of the symptomatic and genetic overlap between SCZ and BP, many investigators have speculated that similar neurodevelopmental alterations may underlie BP. However, the evidence for neurodevelopmental factors in BP is equivocal. First, as discussed earlier, there is no consistent evidence for premorbid cognitive impairment, although one population-based cohort study noted poorer premorbid social function in individuals who later developed BP (Cannon et al., 2002).

While the underlying neuropathology of structural brain anomalies in BP has not been well characterized, existing postmortem studies have observed decreases in glial density, as well as neuronal size and density, in prefrontal cortical regions in patients with mood disorders, although findings are not entirely consistent across studies (Harrison, 2002; Ongur, Drevets, & Price, 1998). Reductions in glial populations (ie, oligodendrocytes, astrocytes, and/or microglia) could underlie the observed white matter abnormalities observed in neuroimaging studies. However, there is no postmortem evidence of neuronal loss in BP. The portion of the prefrontal cortex ventral to the genu of the corpus callosum (subgenual prefrontal cortex; sg24) is a brain region of great interest in BP, because this region is implicated in the mediation of emotional responses to socially salient stimuli. An early positron emission tomography study in familial patients with unipolar disease and BP found that abnormally decreased blood flow was localized to this particular region; concomitant MRI findings indicated an almost 40% reduction in volume in this region (Drevets et al., 1997). Ongur et al. subsequently investigated the cellular correlates of this volume reduction, finding reduced glial density and number in the subgenual prefrontal cortex of familial patients with unipolar disease and BP but no differences in neuronal size or density, abnormalities which were not seen in nonfamilial patients. A subsequent, larger study from the Stanley Foundation of postmortem brain tissue from demographically well-matched and clinically well-characterized patients with unipolar disease, BP, and SCZ and control subjects confirmed these findings, again only in the subset of patients with a family history (Torrey, Webster, Knable, Johnston, & Yolken, 2000). In contrast, Benes (2000) and Benes, Vincent, and Todtenkopf (2001) found decreased interneuron density in this region but no differences in glial density or neuron size. Konopaske, Lange, Coyle, and Benes (2014) investigated postmortem brain tissue from the dorsolateral prefrontal cortex (Brodmann area 46) of patients with SCZ and BP, and healthy control subjects. Dendritic spine loss (ie, reduced spine density, number, and dendrite length) characterized both individuals with SCZ and those with BP, which suggests common pathophysiologic features across the two disorders, which the authors speculate may be related to altered *N*-methyl-D-aspartate receptor signaling. Patients with BP who were taking lithium, however, had longer basilar dendrites, which was consistent with findings in rats (Wood, Young, Reagan, Chen, & McEwen, 2004).

Collectively, these findings indicate that BP is not classically degenerative, and thus these abnormalities may occur earlier in development. Although the significance of the glial reduction observed in some studies is unknown, some investigators hypothesize that astrocytes may be the glial population primarily affected, given their predominance in gray matter (Harrison, 2002). Astrocytes are increasingly recognized to contribute to neuronal migration and synaptic plasticity, and thus observed structural and functional neuroanatomic alterations in BP may be a downstream consequence of glial pathology (Harrison, 2002; Rajkowska, 2000). This possibility remains speculative, until the underlying neuropathology of BP is better characterized.

Circadian Disturbances in Bipolar Disorder

Disruptions in biological rhythms are a hallmark of BP. Because the disorder typically displays cyclic, episodic, and often seasonal patterns, a number of investigators have speculated that circadian dysregulation may underlie its pathophysiology (Kripke, Mullaney, Atkinson, & Wolf, 1978; Lewy, Lefler, Emens, & Bauer, 2006; Mansour, Monk, & Nimgaonkar, 2005). Disruptions in the sleep–wake cycle may be the most frequent precipitants of manic and depressive episodes, regardless of whether such disruptions derive from external or endogenous influences (Frank, Swartz, & Kupfer, 2000). Furthermore, sleep disturbance is among the most commonly reported prodromal symptoms preceding the onset of an initial manic episode (Conus et al., 2008, 2010). As such, it has been hypothesized that reduced sleep may act as a "final common pathway" in triggering mania. Investigation of the molecular mechanisms underlying sleep and circadian rhythm alterations associated with BP may therefore be a key to elucidating the pathogenesis of the disorder.

A meta-analytic review compiled studies that employed actigraphy, sleep diary, polysomnography, and questionnaire measures to investigate sleep—wake patterns in individuals with interepisode BP and/or those at high risk, as defined by family history or subthreshold symptoms (Ng et al., 2014). The authors conclude that across studies, sleep onset latency, wake after sleep onset, and variability of sleep—wake measures displayed the most consistent disruption in patients with interepisode BP relative to healthy control subjects. In addition, patients with interepisode BP spend a similar amount of time trying to fall asleep as do individuals with primary insomnia. Compared with control subjects, individuals at high risk for BP demonstrated greater variability in sleep efficiency and lower relative amplitude, as evidenced by weaker and more unstable rest—activity cycles, and (according to one study) greater variability in total sleep time (Meyer & Maier, 2006). These findings are consistent with the instability model of BP originally put forth by Goodwin and Jamison (2007) and suggest that variability in sleep—wake cycle may be a possible endophenotype for BP. However, more prospective research in familial high-risk individuals is warranted to determine whether sleep—wake cycle variability is predictive of outcome.

Most tissues throughout the body incorporate endogenous molecular clocks, which have critical roles in regulating physiological processes, including those influencing mood states. Several studies in animal models have implicated specific circadian genes in regulating mood and reward responsivity. For example, mice with a Clock gene mutation have a behavioral phenotype involving hyperactivity, reduced sleep, lowered "depression-like" and anxiety behavior, and enhanced value for rewarding stimuli (cocaine, sucrose, and medial forebrain bundle stimulation) (Roybal et al., 2007), ie, one that resembles the manic phase of bipolar illness, as well as increased dopamine synthesis and dopaminergic activity (Coque et al., 2011). Administration of lithium to these mice leads to normalization in their behavior (Roybal et al., 2007) and in striatal dopamine activity (Coque et al., 2011). Investigation in mouse models generated with mutations in other circadian rhythm genes, including GSK3beta and sirtuin 1, have also described the development of "manic-like" behaviors in such mice (McClung, 2013). Knockout in mice of the circadian gene Period leads to increased mesolimbic dopamine levels and altered neuronal activity in the striatum, which may be relevant to reduction in the mutant mice in behaviors that have commonly been related to depression (reduced immobility on the forced-swim test) (Hampp et al., 2008). Chung et al. (2014) discovered that the circadian nuclear receptor REV-ERBa affects midbrain dopamine production and mood-related behavior in mice. Deletion of the REV-ERBa gene or pharmacological inhibition of REV-ERBa activity in the ventral midbrain induced mania-like behavior, concomitant with a central hyperdopaminergic state. Collectively, these findings indicate molecular links between the regulation of circadian rhythms, mesolimbic dopaminergic function, and mood.

SUMMARY AND CONCLUSIONS

More than 100 years after the first delineation of BP as a psychiatric syndrome, large-scale genetic studies are beginning to reveal its molecular underpinnings. The first sets of replicated associations include some loci overlapping with those from SCZ studies, as well as some that appear to be distinct to BP. To date, however, the progress of BP genetics has lagged behind that of SCZ genetic studies, which have now identified more than 100 associated risk loci (Schizophrenia Working Group of the Psychiatric Genomics Consortium, 2014). This may reflect the etiologic heterogeneity of the BP clinical phenotype, and has spurred efforts to refine our conception of this phenotype both to attempt to clarify the relationship between BP and other psychiatric syndromes as well as to identify the underpinnings of its distinctive features. Most notably, these include its cyclicity and episodicity, its association with disturbances in sleep and circadian rhythm, the effectiveness of mood-stabilizing agents, and the switch process between depression and mania.

Neuroimaging represents perhaps the most widely implemented approach aimed at enhancing our understanding of BP phenotypes. Because structural and functional neuroimaging measures are highly heritable and are informative regarding underlying neural dysfunction, investigation of BP-associated brain anomalies can potentially elucidate underlying biological mechanisms. However, psychotropic medications likely affect neuroanatomic measures in ways that are difficult to characterize in naturalistic studies, given nonrandom ascertainment to treatment, and highly varying lengths of treatment and dosages (Bearden et al., 2008; Kempton et al., 2008). As such, studies in clinically unaffected family members of patients with BP can be informative regarding neuroanatomic markers that index the genetic risk for the illness. Prospective longitudinal studies are also critical for answering questions regarding state versus trait anomalies. The introduction of actigraphy as a research tool has permitted the initiation of large-scale investigation of sleep and circadian rhythm traits as components of BP, but few conclusive findings have yet emerged from such research.

As genetic studies of BP extend their scope to include much larger samples and to incorporate more sophisticated phenotyping strategies, we can expect to obtain a substantially greater understanding of the genetic architecture of the syndrome. Large-scale exome and whole-genome sequencing studies of BP conducted over the next few years will provide, for the first time, the opportunity to identify low-frequency or rare variants that exert a large impact on BP and/or its endophenotypes. The results of these studies will guide a new generation of investigations into the basic biology of BP and inform the search for improved treatments and preventive interventions.

REFERENCES

Ahearn, E. P., Steffens, D. C., Cassidy, F., Van Meter, S. A., Provenzale, J. M., Seldin, M. F., … Krishnan, K. R. (1998). Familial leukoencephalopathy in bipolar disorder. *The American Journal of Psychiatry, 155*(11), 1605–1607.

Altshuler, L. L., Ventura, J., van Gorp, W. G., Green, M. F., Theberge, D. C., & Mintz, J. (2004). Neurocognitive function in clinically stable men with bipolar I disorder or schizophrenia and normal control subjects. *Biological Psychiatry, 56*(8), 560–569.

American Psychiatric Association. (1980). *Diagnostic and statistical manual of mental disorders* (3rd ed.). Washington, DC: Author.

American Psychiatric Association. (1994). *Diagnostic and statistical manual of mental disorders* (4th ed.). Washington, DC: Author.

Andersen, P., Morris, R., Amaral, D., Bliss, T., & O'Keefe, J. (2007). *The hippocampus book.* Oxford University Press.

Andreasen, N. C., Rice, J., Endicott, J., Coryell, W., Grove, W. M., & Reich, T. (1987). Familial rates of affective disorder. A report from the National Institute of Mental Health Collaborative Study. *Archives of General Psychiatry, 44*(5), 461–469.

Arts, B., Jabben, N., Krabbendam, L., & van Os, J. (2008). Meta-analyses of cognitive functioning in euthymic bipolar patients and their first-degree relatives. *Psychological Medicine, 38*(6), 771–785.

Badner, J. A., & Gershon, E. S. (2002). Meta-analysis of whole-genome linkage scans of bipolar disorder and schizophrenia. *Molecular Psychiatry, 7*(4), 405–411.

Badner, J. A., Koller, D., Foroud, T., Edenberg, H., Nurnberger, J. I., Jr., Zandi, P. P., … Byerley, W. (2012). Genome-wide linkage analysis of 972 bipolar pedigrees using single-nucleotide polymorphisms. *Molecular Psychiatry, 17*(8), 818–826.

Bearden, C. E., & Freimer, N. B. (2006). Endophenotypes for psychiatric disorders: ready for primetime? *Trends in Genetics, 22*(6), 306–313.

Bearden, C. E., Hoffman, K. M., & Cannon, T. D. (2001). The neuropsychology and neuroanatomy of bipolar affective disorder: a critical review. *Bipolar Disorder, 3*(3), 106–150.

Bearden, C. E., Thompson, P. M., Dutton, R. A., Frey, B. N., Peluso, M. A., Nicoletti, M., … Soares, J. C. (2008). Three-dimensional mapping of hippocampal anatomy in unmedicated and lithium-treated patients with bipolar disorder. *Neuropsychopharmacology, 33*(6), 1229–1238. PMID:17687266.

Bebbington, P., & Ramana, R. (1995). The epidemiology of bipolar affective disorder. *Social Psychiatry and Psychiatric Epidemiology, 30*(6), 279–292.

Benes, F. M. (2000). Emerging principles of altered neural circuitry in schizophrenia. *Brain Research Reviews, 31*(2–3), 251–269.

Benes, F. M., Vincent, S. L., & Todtenkopf, M. (2001). The density of pyramidal and nonpyramidal neurons in anterior cingulate cortex of schizophrenic and bipolar subjects. *Biological Psychiatry, 50*(6), 395–406.

Bernardi, S., Faraone, S. V., Cortese, S., Kerridge, B. T., Pallanti, S., Wang, S., & Blanco, C. (2012). The lifetime impact of attention deficit hyperactivity disorder: results from the National Epidemiologic Survey on Alcohol and Related Conditions (NESARC). *Psychologie Medicale, 42*(4), 875–887.

Bipolar Disorder Working Group of the Psychiatric Genomics Consortium. (2011). Large-scale genome-wide association analysis of bipolar disorder identifies a new susceptibility locus near ODZ4. *Nature Genetics, 43*(10), 977–983.

Bora, E., Yucel, M., & Pantelis, C. (2009). Cognitive endophenotypes of bipolar disorder: a meta-analysis of neuropsychological deficits in euthymic patients and their first-degree relatives. *Journal of Affective Disorders, 113*(1–2), 1–20.

Bourne, C., Aydemir, O., Balanza-Martinez, V., Bora, E., Brissos, S., Cavanagh, J. T., … Goodwin, G. M. (2013). Neuropsychological testing of cognitive impairment in euthymic bipolar disorder: an individual patient data meta-analysis. *Acta Psychiatrica Scandinavica, 128*(3), 149–162.

Butler, S. F., Arredondo, D. E., & McCloskey, V. (1995). Affective comorbidity in children and adolescents with attention deficit hyperactivity disorder. *Annals of Clinical Psychiatry, 7*(2), 51–55.

Cannon, M., Caspi, A., Moffitt, T. E., Harrington, H., Taylor, A., Murray, R. M., & Poulton, R. (2002). Evidence for early-childhood, pan-developmental impairment specific to schizophreniform disorder: results from a longitudinal birth cohort. *Archives of General Psychiatry, 59*(5), 449–456. PMID:11982449.

Cannon, T. D., Rosso, I. M., Bearden, C. E., Sanchez, L. E., & Hatley, T. (1999). A prospective cohort study of neurodevelopmental processes in the genesis and epigenesis of schizophrenia. *Developmental Psychopathology, 11*(3), 467–485.

Cardno, A. G., Marshall, E. J., Coid, B., Macdonald, A. M., Ribchester, T. R., Davies, N. J., … Murray, R. M. (1999). Heritability estimates for psychotic disorders: the maudsley twin psychosis series. *Archives of General Psychiatry, 56*(2), 162–168.

Casey, B. J., Craddock, N., Cuthbert, B. N., Hyman, S. E., Lee, F. S., & Ressler, K. J. (2013). DSM-5 and RDoC: progress in psychiatry research? *Nature Reviews Neuroscience, 14*(11), 810–814.

Chaddock, C. A., Barker, G. J., Marshall, N., Schulze, K., Hall, M. H., Fern, A., … McDonald, C. (2009). White matter microstructural impairments and genetic liability to familial bipolar I disorder. *British Journal of Psychiatry, 194*(6), 527–534.

Chung, S., Lee, E. J., Yun, S., Choe, H. K., Park, S. B., Son, H. J., … Kim, K. (2014). Impact of circadian nuclear receptor REV-ERBalpha on midbrain dopamine production and mood regulation. *Cell, 157*(4), 858–868.

Cichon, S., Craddock, N., Daly, M., Faraone, S. V., Gejman, P. V., Kelsoe, J., … Sullivan, P. F. (2009). Genomewide association studies: history, rationale, and prospects for psychiatric disorders. *The American Journal of Psychiatry, 166*(5), 540–556.

Cichon, S., Schumacher, J., Muller, D. J., Hurter, M., Windemuth, C., Strauch, K., … Nothen, M. M. (2001). A genome screen for genes predisposing to bipolar affective disorder detects a new susceptibility locus on 8q. *Human Molecular Genetics, 10*(25), 2933–2944.

Clark, L., Iversen, S. D., & Goodwin, G. M. (2002). Sustained attention deficit in bipolar disorder. *British Journal of Psychiatry, 180*, 313–319.

Clark, L., Kempton, M. J., Scarna, A., Grasby, P. M., & Goodwin, G. M. (2005). Sustained attention-deficit confirmed in euthymic bipolar disorder but not in first-degree relatives of bipolar patients or euthymic unipolar depression. *Biological Psychiatry, 57*(2), 183–187.

Conus, P., Ward, J., Hallam, K. T., Lucas, N., Macneil, C., McGorry, P. D., & Berk, M. (2008). The proximal prodrome to first episode mania—a new target for early intervention. *Bipolar Disorder, 10*(5), 555–565.

Conus, P., Ward, J., Lucas, N., Cotton, S., Yung, A. R., Berk, M., & McGorry, P. D. (2010). Characterisation of the prodrome to a first episode of psychotic mania: results of a retrospective study. *Journal of Affective Disorders, 124*(3), 341–345.

Coque, L., Mukherjee, S., Cao, J. L., Spencer, S., Marvin, M., Falcon, E., … McClung, C. A. (2011). Specific role of VTA dopamine neuronal firing rates and morphology in the reversal of anxiety-related, but not depression-related behavior in the ClockDelta19 mouse model of mania. *Neuropsychopharmacology, 36*(7), 1478–1488.

Coryell, W., Endicott, J., Reich, T., Andreasen, N., & Keller, M. (1984). A family study of bipolar II disorder. *British Journal of Psychiatry, 145*, 49–54.

Craddock, N., & Jones, I. (1999). Genetics of bipolar disorder. *Journal of Medical Genetics, 36*(8), 585–594.

Cross-Disorder Group of the Psychiatric Genomics Consortium. (2013a). Identification of risk loci with shared effects on five major psychiatric disorders: a genome-wide analysis. *Lancet, 381*(9875), 1371–1379.

Cross-Disorder Group of the Psychiatric Genomics,Consortium. (2013b). Genetic relationship between five psychiatric disorders estimated from genome-wide SNPs. *Nature Genetics, 45*(9), 984–994.

Cuellar, A. K., Johnson, S. L., & Winters, R. (2005). Distinctions between bipolar and unipolar depression. *Clinical Psychology Review, 25*(3), 307–339.

Deckersbach, T., Savage, C. R., Reilly-Harrington, N., Clark, L., Sachs, G., & Rauch, S. L. (2004). Episodic memory impairment in bipolar disorder and obsessive-compulsive disorder: the role of memory strategies. *Bipolar Disorder, 6*(3), 233–244.

Donfrancesco, R., Miano, S., Martines, F., Ferrante, L., Melegari, M. G., & Masi, G. (2011). Bipolar disorder co-morbidity in children with attention deficit hyperactivity disorder. *Psychiatry Research, 186*(2–3), 333–337.

Drevets, W. C., Price, J. L., Simpson, J. R., Jr., Todd, R. D., Reich, T., Vannier, M., & Raichle, M. E. (1997). *Nature, 386*(6627), 824–827.

Dunner, D. L., Gershon, E. S., & Goodwin, F. K. (1976). Heritable factors in the severity of affective illness. *Biological Psychiatry, 11*(1), 31–42.

Edvardsen, J., Torgersen, S., Roysamb, E., Lygren, S., Skre, I., Onstad, S., & Oien, P. A. (2008). Heritability of bipolar spectrum disorders. Unity or heterogeneity? *Journal of Affective Disorders, 106*(3), 229–240.

Emsell, L., Chaddock, C., Forde, N., Van Hecke, W., Barker, G. J., Leemans, A., … McDonald, C. (2013). White matter microstructural abnormalities in families multiply affected with bipolar I disorder: a diffusion tensor tractography study. *Psychologie Medicale, 44*(10), 1–12.

Endicott, J., Nee, J., Andreasen, N., Clayton, P., Keller, M., & Coryell, W. (1985). Bipolar II. Combine or keep separate? *Journal of Affective Disorders, 8*(1), 17–28.

Faraone, S. V., Biederman, J., & Wozniak, J. (2012). Examining the comorbidity between attention deficit hyperactivity disorder and bipolar I disorder: a meta-analysis of family genetic studies. *The American Journal of Psychiatry, 169*(12), 1256–1266.

Faraone, S. V., Biederman, J., Wozniak, J., Mundy, E., Mennin, D., & O'Donnell, D. (1997). Is comorbidity with ADHD a marker for juvenile-onset mania? *Journal of the American Academy of Child and Adolescent Psychiatry, 36*(8), 1046–1055.

Faraone, S. V., Lyons, M. J., & Tsuang, M. T. (1987). Sex differences in affective disorder: genetic transmission. *Genetic Epidemiology, 4*(5), 331–343.

Fears, S. C., Service, S. K., Kremeyer, B., Araya, C., Araya, X., Bejarano, J., … Bearden, C. E. (2014). Multisystem component phenotypes of bipolar disorder for genetic investigations of extended pedigrees. *JAMA Psychiatry, 71*(4), 375–387.

Flick, S. N., Roy-Byrne, P. P., Cowley, D. S., Shores, M. M., & Dunner, D. L. (1993). DSM-III-R personality disorders in a mood and anxiety disorders clinic: prevalence, comorbidity, and clinical correlates. *Journal of Affective Disorders, 27*(2), 71–79.

Flint, J., & Munafó, M. R. (2007). The endophenotype concept in psychiatric genetics. *Psychologie Medicale, 37*(2), 163–180.

Frank, E., Swartz, H. A., & Kupfer, D. J. (2000). Interpersonal and social rhythm therapy: managing the chaos of bipolar disorder. *Biological Psychiatry, 48*, 593.

Gallagher, M., & Chiba, A. A. (1996). The amygdala and emotion. *Current Opinion in Neurobiology, 6*(2), 221–227.

Geller, B., Sun, K., Zimerman, B., Luby, J., Frazier, J., & Williams, M. (1995). Complex and rapid-cycling in bipolar children and adolescents: a preliminary study. *Journal of Affective Disorders, 34*(4), 259–268.

Gershon, E. S., DeLisi, L. E., Hamovit, J., Nurnberger, J. I., Jr., Maxwell, M. E., Schreiber, J., … Guroff, J. J. (1988). A controlled family study of chronic psychoses. Schizophrenia and schizoaffective disorder. *Archives of General Psychiatry, 45*(4), 328–336.

Gershon, E. S., Hamovit, J., Guroff, J. J., Dibble, E., Leckman, J. F., Sceery, W., … Bunney, W. E., Jr. (1982). A family study of schizoaffective, bipolar I, bipolar II, unipolar, and normal control probands. *Archives of General Psychiatry, 39*(10), 1157–1167.

Glahn, D. C., Almasy, L., Barguil, M., Hare, E., Peralta, J. M., Kent, J. W., Jr., … Escamilla, M. A. (2010). Neurocognitive endophenotypes for bipolar disorder identified in multiplex multigenerational families. *Archives of General Psychiatry, 67*(2), 168–177.

Glahn, D. C., Bearden, C. E., Barguil, M., Barrett, J., Reichenberg, A., Bowden, C. L., … Velligan, D. I. (2007). The neurocognitive signature of psychotic bipolar disorder. *Biological Psychiatry, 62*(8), 910–916.

Goodwin, F. K., & Jamison, K. R. (2007). *Manic-depressive illness: Bipolar disorders and recurrent depression* (2nd ed., Vol. 1). New York: Oxford University Press, Inc.

van Gorp, W. G., Altshuler, L., Theberge, D. C., Wilkins, J., & Dixon, W. (1998). Cognitive impairment in euthymic bipolar patients with and without prior alcohol dependence. A preliminary study. *Archives of General Psychiatry, 55*(1), 41–46.

Gottesman, I., & Gould, T. D. (2003). The endophenotype concept in psychiatry: etymology and strategic intentions. *The American Journal of Psychiatry, 160*(4), 636–645.

Gottesman, I. I., & Shields, J. (1973). Genetic theorizing and schizophrenia. *British Journal of Psychiatry, 122*(566), 15–30.

Greenwood, T. A., Braff, D. L., Light, G. A., Cadenhead, K. S., Calkins, M. E., Dobie, D. J., … Schork, N. J. (2007). Initial heritability analyses of endophenotypic measures for schizophrenia: the consortium on the genetics of schizophrenia. *Archives of General Psychiatry, 64*(11), 1242–1250.

Gur, R. E., Nimgaonkar, V. L., Almasy, L., Calkins, M. E., Ragland, J. D., Pogue-Geile, M. F., … Gur, R. C. (2007). Neurocognitive endophenotypes in a multiplex multigenerational family study of schizophrenia. *The American Journal of Psychiatry, 164*(5), 813–819.

Guze, S. B., & Robins, E. (1970). Suicide and primary affective disorders. *The British Journal of Psychiatry, 117*(539), 437–438.

Hajek, T., Cullis, J., Novak, T., Kopecek, M., Blagdon, R., Propper, L., … Alda, M. (2013). Brain structural signature of familial predisposition for bipolar disorder: replicable evidence for involvement of the right inferior frontal gyrus. *Biological Psychiatry, 73*(2), 144–152.

Hallahan, B., Newell, J., Soares, J. C., Brambilla, P., Strakowski, S. M., Fleck, D. E., … McDonald, C. (2011). Structural magnetic resonance imaging in bipolar disorder: an international collaborative mega-analysis of individual adult patient data. *Biological Psychiatry, 69*(4), 326–335.

Hampp, G., Ripperger, J. A., Houben, T., Schmutz, I., Blex, C., Perreau-Lenz, S., … Albrecht, U. (2008). Regulation of monoamine oxidase A by circadian-clock components implies clock influence on mood. *Current Biology, 18*(9), 678–683.

Harrison, P. J. (2002). The neuropathology of primary mood disorder. *Brain, 125*, 1428–1449.

Hensch, T., Himmerich, H., & Hegerl, U. (2011). ADHD and bipolar disorder: common causes common cure? *Journal of Attention Disorders, 15*(2), 99–100.

Heun, R., & Maier, W. (1993). The distinction of bipolar II disorder from bipolar I and recurrent unipolar depression: results of a controlled family study. *Acta Psychiatrica Scandinavica, 87*(4), 279–284.

Hill, S. K., Harris, M. S., Herbener, E. S., Pavuluri, M., & Sweeney, J. A. (2008). Neurocognitive allied phenotypes for schizophrenia and bipolar disorder. *Schizophrenia Bulletin, 34*(4), 743–759.

Insel, T., Cuthbert, B., Garvey, M., Heinssen, R., Pine, D. S., Quinn, K., … Wang, P. (2010). Research domain criteria (RDoC): toward a new classification framework for research on mental disorders. *The American Journal of Psychiatry, 167*(7), 748–751.

Ivleva, E. I., Morris, D. W., Moates, A. F., Suppes, T., Thaker, G. K., & Tamminga, C. A. (2010). Genetics and intermediate phenotypes of the schizophrenia-bipolar disorder boundary. *Neuroscience and Biobehavioral Reviews, 34*, 897–921.

John, B., & Lewis, K. R. (1966). Chromosome variability and geographic distribution in insects. *Science, 152*(3723), 711–721.

Judd, L. L., Akiskal, H. S., Schettler, P. J., Endicott, J., Maser, J., Solomon, D. A., … Keller, M. B. (2002). The long-term natural history of the weekly symptomatic status of bipolar I disorder. *Archives of General Psychiatry, 59*(6), 530–537.

Kelsoe, J. R., Spence, M. A., Loetscher, E., Foguet, M., Sadovnick, A. D., Remick, R. A., … Luebbert, H. (2001). A genome survey indicates a possible susceptibility locus for bipolar disorder on chromosome 22. *Proceedings of the National Academy of Sciences of the United States of America, 98*(2), 585–590.

Kempton, M. J., Geddes, J. R., Ettinger, U., Williams, S. C., & Grasby, P. M. (2008). Meta-analysis, database, and meta-regression of 98 structural imaging studies in bipolar disorder. *Archives of General Psychiatry, 65*(9), 1017–1032.

Kendler, K. S. (1993). Twin studies of psychiatric illness. Current status and future directions. *Archives of General Psychiatry, 50*(11), 905–915.

Kendler, K. S., & Gardner, C. O. (1997). The risk for psychiatric disorders in relatives of schizophrenic and control probands: a comparison of three independent studies. *Psychologie Medicale, 27*(2), 411–419.

Kessler, R. C., Crum, R. M., Warner, L. A., Nelson, C. B., Schulenberg, J., & Anthony, J. C. (1997). Lifetime co-occurrence of DSM-III-R alcohol abuse and dependence with other psychiatric disorders in the National Comorbidity Survey. *Archives of General Psychiatry, 54*(4), 313–321.

Kessler, R. C., McGonagle, K. A., Zhao, S., Nelson, C. B., Hughes, M., Eshleman, S., … Kendler, K. S. (1994). Lifetime and 12-month prevalence of DSM-III-R psychiatric disorders in the United States. Results from the National Comorbidity Survey. *Archives of General Psychiatry, 51*(1), 8–19.

Kettunen, J., Tukiainen, T., Sarin, A. P., Ortega-Alonso, A., Tikkanen, E., Lyytikainen, L. P., … Ripatti, S. (2012). Genome-wide association study identifies multiple loci influencing human serum metabolite levels. *Nature Genetics, 44*(3), 269–276.

Kieseppä, T., Partonen, T., Haukka, J., Kaprio, J., & Lonnqvist, J. (2004). High concordance of bipolar I disorder in a nationwide sample of twins. *The American Journal of Psychiatry, 161*(10), 1814–1821.

Klassen, L. J., Katzman, M. A., & Chokka, P. (2010). Adult ADHD and its comorbidities, with a focus on bipolar disorder. *Journal of Affective Disorders, 124*(1–2), 1–8.

Kochunov, P., Jahanshad, N., Sprooten, E., Nichols, T. E., Mandl, R. C., Almasy, L., … Glahn, D. C. (2014). Multi-site study of additive genetic effects on fractional anisotropy of cerebral white matter: comparing meta and megaanalytical approaches for data pooling. *NeuroImage, 95*, 136–150.

Koenen, K. C., Moffitt, T. E., Roberts, A. L., Martin, L. T., Kubzansky, L., Harrington, H., … Caspi, A. (2009). Childhood IQ and adult mental disorders: a test of the cognitive reserve hypothesis. *The American Journal of Psychiatry, 166*(1), 50–57.

Konopaske, G. T., Lange, N., Coyle, J. T., & Benes, F. M. (2014). Prefrontal cortical Dendritic spine pathology in schizophrenia and bipolar disorder. *JAMA Psychiatry, 71*(12), 1323–1331.

Kowatch, R. A., Youngstrom, E. A., Danielyan, A., & Findling, R. L. (2005). Review and meta-analysis of the phenomenology and clinical characteristics of mania in children and adolescents. *Bipolar Disorder, 7*(6), 483–496.

Kripke, D. F., Mullaney, D. J., Atkinson, M., & Wolf, S. (1978). Circadian rhythm disorders in manic-depressives. *Biological Psychiatry, 13*, 335.

Kyaga, S., Lichtenstein, P., Boman, M., Hultman, C., Langstrom, N., & Landén, M. (2011). Creativity and mental disorder: family study of 300,000 people with severe mental disorder. *British Journal of Psychiatry, 199*(5), 373–379.

Lewy, A. J., Lefler, B. J., Emens, J. S., & Bauer, V. K. (2006). The circadian basis of winter depression. *Proceedings of the National Academy of Sciences, 103*, 7414.

Lichtenstein, P., Yip, B. H., Bjork, C., Pawitan, Y., Cannon, T. D., Sullivan, P. F., & Hultman, C. M. (2009). Common genetic determinants of schizophrenia and bipolar disorder in Swedish families: a population-based study. *Lancet, 373*(9659), 234–239.

Lim, C. S., Baldessarini, R. J., Vieta, E., Yucel, M., Bora, E., & Sim, K. (2013). Longitudinal neuroimaging and neuropsychological changes in bipolar disorder patients: review of the evidence. *Neuroscience & Biobehavioral Reviews, 37*(3), 418–435.

Liu, J., Juo, S. H., Terwilliger, J. D., Grunn, A., Tong, X., Brito, M., … Baron, M. (2001). A follow-up linkage study supports evidence for a bipolar affective disorder locus on chromosome 21q22. *American Journal of Medical Genetics, 105*(2), 189–194.

Lus, G., & Mukaddes, N. M. (2009). Co-morbidity of bipolar disorder in children and adolescents with attention deficit/hyperactivity disorder (ADHD) in an outpatient Turkish sample. *The World Journal of Biological Psychiatry, 10*(4 Pt 2), 488–494.

MacCabe, J. H., Lambe, M. P., Cnattingius, S., Sham, P. C., David, A. S., Reichenberg, A., … Hultman, C. M. (2010). Excellent school performance at age 16 and risk of adult bipolar disorder: national cohort study. *British Journal of Psychiatry, 196*(2), 109–115.

Mahon, K., Burdick, K. E., Ikuta, T., Braga, R. J., Gruner, P., Malhotra, A. K., & Szeszko, P. R. (2013). Abnormal temporal lobe white matter as a biomarker for genetic risk of bipolar disorder. *Biological Psychiatry, 73*(2), 177–182.

Maier, W., Lichtermann, D., Minges, J., Hallmayer, J., Heun, R., Benkert, O., & Levinson, D. F. (1993). Continuity and discontinuity of affective disorders and schizophrenia. Results of a controlled family study. *Archives of General Psychiatry, 50*(11), 871–883.

Mann-Wrobel, M. C., Carreno, J. T., & Dickinson, D. (2011). Meta-analysis of neuropsychological functioning in euthymic bipolar disorder: an update and investigation of moderator variables. *Bipolar Disorders, 13*(4), 334–342.

Manolio, T. A., Collins, F. S., Cox, N. J., Goldstein, D. B., Hindorff, L. A., Hunter, D. J., … Visscher, P. M. (2009). Finding the missing heritability of complex diseases. *Nature, 461*(7265), 747–753.

Mansour, H. A., Monk, T. H., & Nimgaonkar, V. L. (2005). Circadian genes and bipolar disorder. *Annals of Medicine, 37*, 196.

Martinez-Aran, A., Torrent, C., Tabares-Seisdedos, R., Salamero, M., Daban, C., Balanza-Martinez, V., … Vieta, E. (2008). Neurocognitive impairment in bipolar patients with and without history of psychosis. *Journal of Clinical Psychiatry, 69*(2), 233–239.

Martinez-Aran, A., Vieta, E., Colom, F., Torrent, C., Sanchez-Moreno, J., Reinares, M., … Salamero, M. (2004). Cognitive impairment in euthymic bipolar patients: implications for clinical and functional outcome. *Bipolar Disorders, 6*(3), 224–232.

Maziade, M., Roy, M. A., Rouillard, E., Bissonnette, L., Fournier, J. P., Roy, A., … Merette, C. (2001). A search for specific and common susceptibility loci for schizophrenia and bipolar disorder: a linkage study in 13 target chromosomes. *Molecular Psychiatry, 6*(6), 684–693.

McClung, C. A. (2013). How might circadian rhythms control mood? Let me count the ways. *Biological Psychiatry, 74*(4), 242–249.

McDonald, C., Zanelli, J., Rabe-Hesketh, S., Ellison-Wright, I., Sham, P., Kalidindi, S., … Kennedy, N. (2004). Meta-analysis of magnetic resonance imaging brain morphometry studies in bipolar disorder. *Biological Psychiatry, 56*(6), 411–417.

McGuffin, P., Rijsdijk, F., Andrew, M., Sham, P., Katz, R., & Cardno, A. (2003). The heritability of bipolar affective disorder and the genetic relationship to unipolar depression. *Archives of General Psychiatry, 60*(5), 497–502.

McQueen, M. B., Devlin, B., Faraone, S. V., Nimgaonkar, V. L., Sklar, P., Smoller, J. W., … Laird, N. M. (2005). Combined analysis from eleven linkage studies of bipolar disorder provides strong evidence of susceptibility loci on chromosomes 6q and 8q. *American Journal of Human Genetics, 77*(4), 582–595.

Mendlewicz, J., & Rainer, J. D. (1977). Adoption study supporting genetic transmission in manic–depressive illness. *Nature, 268*(5618), 327–329.

Merikangas, K. R., Akiskal, H., Greenberg, P. E., Hirschfeld, R. M., Petukhova, M., & Kessler, R. C. (2007). Lifetime and 12-month prevalence of bipolar spectrum disorder in the National Comorbidity Survey replication. *Archives of General Psychiatry, 64*(5), 543–552.

Merikangas, K. R., Cui, L., Heaton, L., Nakamura, E., Roca, C., Ding, J., … Angst, J. (2014). Independence of familial transmission of mania and depression: results of the NIMH family study of affective spectrum disorders. *Molecular Psychiatry, 19*(2), 214–219.

Merikangas, K. R., Jin, R., He, J. P., Kessler, R. C., Lee, S., Sampson, N. A., … Zarkov, Z. (2011). Prevalence and correlates of bipolar spectrum disorder in the world mental health survey initiative. *Archives of General Psychiatry, 68*(3), 241–251.

Meyer, T. D., & Maier, S. (2006). Is there evidence for social rhythm instability in people at risk for affective disorders? *Psychiatry Research, 141*(1), 103–114.

Mortensen, P. B., Pedersen, C. B., Melbye, M., Mors, O., & Ewald, H. (2003). Individual and familial risk factors for bipolar affective disorders in Denmark. *Archives of General Psychiatry, 60*(12), 1209–1215.

Murray, R. M., & Lewis, S. W. (1987). Is schizophrenia a neurodevelopmental disorder? *British Journal of Medicine (Clinical Research Edition), 295*(6600), 681–682.

Ng, T. H., Chung, K. F., Ho, F. Y., Yeung, W. F., Yung, K. P., & Lam, T. H. (2014). Sleep-wake disturbance in interepisode bipolar disorder and high-risk individuals: a systematic review and meta-analysis. *Sleep Medicine Reviews, 20*, 46–58.

Ongur, D., Drevets, W. C., & Price, J. L. (1998). Glial reduction in the subgenual prefrontal cortex in mood disorders. *Proceedings of the National Academy of Sciences, 95*, 13290–13295.

Pacchiarotti, I., Bond, D. J., Baldessarini, R. J., Nolen, W. A., Grunze, H., Licht, R. W., … Vieta, E. (2013). The International Society for Bipolar Disorders (ISBD) task force report on antidepressant use in bipolar disorders. *The American Journal of Psychiatry, 170*(11), 1249–1262.

Pauls, D. L., Morton, L. A., & Egeland, J. A. (1992). Risks of affective illness among first-degree relatives of bipolar I old-order amish probands. *Archives of General Psychiatry, 49*(9), 703–708.

Perlis, R. H., Ostacher, M. J., Patel, J. K., Marangell, L. B., Zhang, H., Wisniewski, S. R., … Thase, M. E. (2006). Predictors of recurrence in bipolar disorder: primary outcomes from the Systematic Treatment Enhancement Program for Bipolar Disorder (STEP-BD). *The American Journal of Psychiatry, 163*(2), 217–224.

Pfeifer, J. C., Welge, J., Strakowski, S. M., Adler, C. M., & DelBello, M. P. (2008). Meta-analysis of amygdala volumes in children and adolescents with bipolar disorder. *Journal of the American Academy of Child and Adolescent Psychiatry, 47*(11), 1289–1298.

Phelps, E. A., & LeDoux, J. E. (2005). Contributions of the amygdala to emotion processing: from animal models to human behavior. *Neuron, 48*(2), 175–187.

Post, R. M., Denicoff, K. D., Leverich, G. S., Altshuler, L. L., Frye, M. A., Suppes, T. M., … Nolen, W. A. (2003). Morbidity in 258 bipolar outpatients followed for 1 year with daily prospective ratings on the NIMH life chart method. *The Journal of Clinical Psychiatry, 64*(6), 680–690.

Rajkowska, G. (2000). Postmortem studies in mood disorders indicate altered numbers of neurons and glial cells. *Biological Psychiatry, 48*(8), 766–777.

Regier, D. A., Farmer, M. E., Rae, D. S., Locke, B. Z., Keith, S. J., Judd, L. L., & Goodwin, F. K. (1990). Comorbidity of mental disorders with alcohol and other drug abuse. Results from the Epidemiologic Catchment Area (ECA) Study. *JAMA, 264*(19), 2511–2518.

Regier, D. A., Myers, J. K., Kramer, M., Robins, L. N., Blazer, D. G., Hough, R. L., … Locke, B. Z. (1984). The NIMH epidemiologic catchment area program. Historical context, major objectives, and study population characteristics. *Archives of General Psychiatry, 41*(10), 934–941.

Rice, J., Reich, T., Andreasen, N. C., Endicott, J., Van Eerdewegh, M., Fishman, R., … Klerman, G. L. (1987). The familial transmission of bipolar illness. *Archives of General Psychiatry, 44*(5), 441–447.

Ripke, S., O'Dushlaine, C., Chambert, K., Moran, J. L., Kähler, A. K., Akterin, S., … Sullivan, P. F. (2013). Genome-wide association analysis identifies 13 new risk loci for schizophrenia. *Natural Genetics, 45*(10), 1150–1159.

Roberts, G., Green, M. J., Breakspear, M., McCormack, C., Frankland, A., Wright, A., … Mitchell, P. B. (2012). Reduced inferior frontal gyrus activation during response inhibition to emotional stimuli in youth at high risk of bipolar disorder. *Biological Psychiatry, 74*, 55–61.

Robins, L. N., Helzer, J. E., Croughan, J., & Ratcliff, K. S. (1981). National Institute of mental health diagnostic interview schedule. Its history, characteristics, and validity. *Archives of General Psychiatry, 38*(4), 381–389.

Robinson, L. J., & Ferrier, I. N. (2006). Evolution of cognitive impairment in bipolar disorder: a systematic review of cross-sectional evidence. *Bipolar Disorders, 8*(2), 103–116.

Roybal, K., Theobold, D., Graham, A., DiNieri, J. A., Russo, S. J., Krishnan, V., … McClung, C. A. (2007). Mania-like behavior induced by disruption of CLOCK. *Proceedings of the National Academy of Sciences, 104*(15), 6406–6411.

Rubinsztein, J. S., Michael, A., Paykel, E. S., & Sahakian, B. J. (2000). Cognitive impairment in remission in bipolar affective disorder. *Psychologie Medicale, 30*(5), 1025–1036.

Salvadore, G., Quiroz, J. A., Machado-Vieira, R., Henter, I. D., Manji, H. K., & Zarate, C. A. J. (2010). The neurobiology of the switch process in bipolar disorder: a review. *The Journal of Clinical Psychiatry, 71*(11), 1488–1501.

Savino, M., Perugi, G., Simonini, E., Soriani, A., Cassano, G. B., & Akiskal, H. S. (1993). Affective comorbidity in panic disorder: is there a bipolar connection? *Journal of Affective Disorders, 28*(3), 155–163.

Schizophrenia Working Group of the Psychiatric Genomics Consortium. (July 24, 2014). Biological insights from 108 schizophrenia-associated genetic loci. *Nature, 511*(7510), 421–427. PMID:25056061; PMCID:PMC4112379.

Schretlen, D. J., Cascella, N. G., Meyer, S. M., Kingery, L. R., Testa, S. M., Munro, C. A., … Pearlson, G. D. (2007). Neuropsychological functioning in bipolar disorder and schizophrenia. *Biological Psychiatry, 62*(2), 179–186.

Sedler, M. (1983). Falret's discovery: the origin of the concept of bipolar affective illness. *The American Journal of Psychiatry, 140*(9), 1127–1133.

Segurado, R., Detera-Wadleigh, S. D., Levinson, D. F., Lewis, C. M., Gill, M., Nurnberger, J. I., Jr., … Akarsu, N. (2003). Genome scan meta-analysis of schizophrenia and bipolar disorder, part III: bipolar disorder. *American Journal of Human Genetics, 73*(1), 49–62.

Seidman, L., Kremen, W. S., Koren, D., Faraone, S. V., Goldstein, J. M., & Tsuang, M. T. (2002). A comparative profile analysis of neuropsychological functioning in patients with schizophrenia and bipolar psychoses. *Schizophrenia Research, 53*(1–2), 31–44.

Seifuddin, F., Mahon, P. B., Judy, J., Pirooznia, M., Jancic, D., Taylor, J., … Zandi, P. P. (2012). Meta-analysis of genetic association studies on bipolar disorder. *American Journal of Medical Genetics Part B: Neuropsychiatric Genetics, 159b*(5), 508–518.

Service, S. K., Teslovich, T. M., Fuchsberger, C., Ramensky, V., Yajnik, P., Koboldt, D. C., … Freimer, N. B. (2014). Re-sequencing expands our understanding of the phenotypic impact of variants at GWAS loci. *PLoS Genetics, 10*(1), e1004147.

Shisslak, C. M., Perse, T., & Crago, M. (1991). Coexistence of bulimia nervosa and mania: a literature review and case report. *Comprehensive Psychiatry, 32*(2), 181–184.

Simeonova, D. I., Chang, K. D., Strong, C., & Ketter, T. A. (2005). Creativity in familial bipolar disorder. *Journal of Psychiatric Research, 39*(6), 623–631.

Simpson, S. G., al-Mufti, R., Andersen, A. E., & DePaulo, J. R., Jr. (1992). Bipolar II affective disorder in eating disorder inpatients. *The Journal of Nervous and Mental Disease, 180*(11), 719–722.

Smoller, J. W., & Finn, C. T. (2003). Family, twin, and adoption studies of bipolar disorder. *American Journal of Medical Genetics Part C, 123c*(1), 48–58. http://dx.doi.org/10.1002/ajmg.c.20013.

Song, S. K., Sun, S. W., Ramsbottom, M. J., Chang, C., Russell, J., & Cross, A. H. (2002). Dysmyelination revealed through MRI as increased radial (but unchanged axial) diffusion of water. *NeuroImage, 17*, 1429–1436.

Sprooten, E., Sussmann, J. E., Clugston, A., Peel, A., McKirdy, J., Moorhead, T. W., … McIntosh, A. M. (2011). White matter integrity in individuals at high genetic risk of bipolar disorder. *Biological Psychiatry, 70*(4), 350–356.

Strakowski, S. M., Delbello, M. P., & Adler, C. M. (2005). The functional neuroanatomy of bipolar disorder: a review of neuroimaging findings. *Molecular Psychiatry, 10*(1), 105–116.

Strakowski, S. M., DelBello, M. P., Zimmerman, M. E., Getz, G. E., Mills, N. P., Ret, J., … Adler, C. M. (2002). Ventricular and periventricular structural volumes in first- versus multiple-episode bipolar disorder. *The American Journal of Psychiatry, 159*(11), 1841–1847.

Strasser, H. C., Lilyestrom, J., Ashby, E. R., Honeycutt, N. A., Schretlen, D. J., Pulver, A. E., ... Pearlson, G. D. (2005). Hippocampal and ventricular volumes in psychotic and nonpsychotic bipolar patients compared with schizophrenia patients and community control subjects: a pilot study. *Biological Psychiatry, 57*(6), 633–639.

Swann, A. C., Pazzaglia, P., Nicholls, A., Dougherty, D. M., & Moeller, F. G. (2003). Impulsivity and phase of illness in bipolar disorder. *Journal of Affective Disorders, 73*(1–2), 105–111.

Taylor, M. A., Berenbaum, S. A., Jampala, V. C., & Cloninger, C. R. (1993). Are schizophrenia and affective disorder related? preliminary data from a family study. *The American Journal of Psychiatry, 150*(2), 278–285.

Thompson, J. M., Gallagher, P., Hughes, J. H., Watson, S., Gray, J. M., Ferrier, I. N., & Young, A. H. (2005). Neurocognitive impairment in euthymic patients with bipolar affective disorder. *British Journal of Psychiatry, 186*, 32–40.

Thompson, P. M., Stein, J. L., Medland, S. E., Hibar, D. P., Vasquez, A. A., Renteria, M. E., ... Drevets, W. (2014). The ENIGMA Consortium: large-scale collaborative analyses of neuroimaging and genetic data. *Brain Imaging and Behavior, 8*(2), 153–182.

Tohen, M., & Goodwin, F. K. (1995). Epidemiology of bipolar disorder. In M. T. Tsuang, M. Tohen, & M. Zahner (Eds.), *Textbook in psychiatric epidemiology* (1st ed., pp. 301–315). New York: John Wiley & Sons.

Torrey, E. F., Webster, M., Knable, M., Johnston, N., & Yolken, R. H. (2000). The Stanley foundation brain collection and neuropathology consortium. *Schizophrenia Research, 44*(2), 151–155.

Trotta, A., Murray, R. M., & MacCabe, J. H. (2014). Do premorbid and post-onset cognitive functioning differ between schizophrenia and bipolar disorder? A systematic review and meta-analysis. *Psychologie Medicale, 45*(2), 1–14.

Tsuang, M. T., & Faraone, S. V. (1990). *The genetics of mood disorders* (pp. 31–101). Baltimore: Johns Hopkins University Press.

Tsuang, M. T., Winokur, G., & Crowe, R. R. (1980). Morbidity risks of schizophrenia and affective disorders among first degree relatives of patients with schizophrenia, mania, depression and surgical conditions. *British Journal of Psychiatry, 137*, 497–504.

Turecki, G., Grof, P., Grof, E., D'Souza, V., Lebuis, L., Marineau, C., ... Alda, M. (2001). Mapping susceptibility genes for bipolar disorder: a pharmacogenetic approach based on excellent response to lithium. *Molecular Psychiatry, 6*(5), 570–578.

Valles, V., Van Os, J., Guillamat, R., Gutierrez, B., Campillo, M., Gento, P., & Fananas, L. (2000). Increased morbid risk for schizophrenia in families of in-patients with bipolar illness. *Schizophrenia Research, 42*(2), 83–90.

Vita, A., Dieci, M., Giobbio, G. M., Tenconi, F., & Invernizzi, G. (1997). Time course of cerebral ventricular enlargement in schizophrenia supports the hypothesis of its neurodevelopmental nature. *Schizophrenia Research, 23*(1), 25–30.

Weissman, M., Bruce, M., Leaf, P., Florio, L., & Holzer, C. (1991). *Psychiatric disorders in America*. In L. Robbins, & D. A. Regier (Eds.) (pp. 53–81). New York: Free Press.

Weissman, M. M., Gershon, E. S., Kidd, K. K., Prusoff, B. A., Leckman, J. F., Dibble, E., ... Guroff, J. J. (1984). Psychiatric disorders in the relatives of probands with affective disorders. The Yale University–National Institute of Mental Health Collaborative Study. *Archives of General Psychiatry, 41*(1), 13–21.

Weissman, M. M., & Myers, J. K. (1978). Affective disorders in a US urban community: the use of research diagnostic criteria in an epidemiological survey. *Archives of General Psychiatry, 35*(11), 1304–1311.

Wender, P. H., Kety, S. S., Rosenthal, D., Schulsinger, F., Ortmann, J., & Lunde, I. (1986). Psychiatric disorders in the biological and adoptive families of adopted individuals with affective disorders. *Archives of General Psychiatry, 43*(10), 923–929.

West, S. A., McElroy, S. L., Strakowski, S. M., Keck, P. E., Jr., & McConville, B. J. (1995). Attention deficit hyperactivity disorder in adolescent mania. *The American Journal of Psychiatry, 152*(2), 271–273.

Whalley, H. C., Sprooten, E., Hackett, S., Hall, L., Blackwood, D. H., Glahn, D. C., ... McIntosh, A. M. (2013). Polygenic risk and white matter integrity in individuals at high risk of mood disorder. *Biological Psychiatry, 74*(4), 280–286.

Wilder-Willis, K. E., Sax, K. W., Rosenberg, H. L., Fleck, D. E., Shear, P. K., & Strakowski, S. M. (2001). Persistent attentional dysfunction in remitted bipolar disorder. *Bipolar Disorders, 3*(2), 58–62.

Willer, C. J., Schmidt, E. M., Sengupta, S., Peloso, G. M., Gustafsson, S., Kanoni, S., ... Abecasis, G. R. (2013). Discovery and refinement of loci associated with lipid levels. *Nature Genetics, 45*(11), 1274–1283.

Wood, G. E., Young, L. T., Reagan, L. P., Chen, B., & McEwen, B. S. (2004). Stress-induced structural remodeling in hippocampus prevention by lithium treatment. *Proceedings of the National Academy of Sciences, 101*(110), 3973–3978.

Wozniak, J., Biederman, J., Kiely, K., Ablon, J. S., Faraone, S. V., Mundy, E., & Mennin, D. (1995). Mania-like symptoms suggestive of childhood-onset bipolar disorder in clinically referred children. *Journal of the American Academy of Child and Adolescent Psychiatry, 34*(7), 867–876.

Youngs, R. M., Chu, M. S., Meloni, E. G., Naydenov, A., Carlezon, W. A., & Konradi, C. (2006). Lithium administration to preadolescent rats causes long-lasting increases in anxiety-like behavior and has molecular consequences. *The Journal of Neuroscience, 26*, 6031–6039.

Zalla, T., Joyce, C., Szoke, A., Schurhoff, F., Pillon, B., Komano, O., ... Leboyer, M. (2004). Executive dysfunctions as potential markers of familial vulnerability to bipolar disorder and schizophrenia. *Psychiatry Research, 121*(3), 207–217.

Zubieta, J. K., Huguelet, P., O'Neil, R. L., & Giordani, B. J. (2001). Cognitive function in euthymic bipolar I disorder. *Psychiatry Research, 102*(1), 9–20.

Chapter 31

Conceptualizing Major Depression: From Genes to Neuroanatomy to Epidemiology

David C. Glahn
Hartford Hospital, Hartford, CT, United States; Yale University School of Medicine, New Haven, CT, United States

Emma E.M. Knowles and Samuel R. Mathias
Yale University School of Medicine, New Haven, CT, United States

Laura Almasy
University of Texas Rio Grand Valley, Brownsville, TX, United States

Karen Hodgson and Nailin Yao
Yale University School of Medicine, New Haven, CT, United States

Rene L. Olvera
University of Texas Health Science Center San Antonio, San Antonio, TX, United States

Joanne E. Curran and John Blangero
University of Texas Rio Grand Valley, Brownsville, TX, United States

Defining features of major depression include low mood, feelings of helplessness and hopelessness, a lack of enjoyment or interest in activities, lowered energy and concentration, and changes in appetite and sleep patterns. It is perfectly normal to experience one or several of these symptoms from time to time, perhaps in response to emotional distress or bereavement. What renders them clinically significant are their severity and duration. Are the severity and duration of the symptoms disproportionate to the instigating event, do they affect the individual's functioning, and do the symptoms persist even in the face of ameliorating external factors? If so, the depressive experience ceases to be considered a normal response to adversity and is instead considered an episode of mental illness. Major depression is common (affecting 16.2% of individuals in the United States during their lifetime) (Greenberg et al., 2003) and potentially life-threatening. It incurs great economic cost to society ($83.1 billion per annum in the United States) (Greenberg et al., 2003) and places immense burden on the affected individual as well as his or her family. The impact of major depression on well-being and functioning is in line with that seen in other major chronic conditions (eg, arthritis and diabetes mellitus) (Wells et al., 1989) and the functional impairments associated with the disorder often remain after the remission of a depressive episode (Hays, Wells, Sherbourne, Rogers, & Spritzer, 1995). Indeed, the World Health Organization predicts that major depression will be second only to ischemic heart disease as a cause of disability worldwide (Murray & Lopez, 1996; Murray et al., 2012).

Genomics, Circuits, and Pathways in Clinical Neuropsychiatry. http://dx.doi.org/10.1016/B978-0-12-800105-9.00031-7

Because its etiology and pathogenesis remain largely unexplained, major depression, like other psychiatric diseases, is understood entirely on the basis of symptomatology. Considering its high prevalence rate (see subsequent discussion), one might expect the symptomatology of major depression to be consistent from case to case. However, major depression is a highly heterogeneous disorder in which patients differ in terms of symptom presentation, response to treatment, and clinical course (Belmaker & Agam, 2008). In part, this marked heterogeneity may be exacerbated because the diagnosis of major depression, as canonized by the fifth edition of the *Diagnostic and Statistical Manual of Mental Disorders* (DSM-5) (Association AP, 2013), relies on the presence of five of nine symptoms: depressed mood, anhedonia, significant weight loss or gain, insomnia or hypersomnia, psychomotor agitation or retardation, fatigue, feelings of worthlessness, diminished ability to concentrate, and recurrent thoughts of death. Thus, it is possible that two individuals with the same diagnosis could manifest almost completely different symptoms. Indeed, our current diagnostic classification of major depression likely encompasses a group of disorders that are heterogeneous with respect to etiology and pathophysiology. It is possible that the breadth of the diagnostic criteria leads to the poor temporal reliability of this diagnosis in clinical samples (Bromet, Dunn, Connell, Dew, & Schulberg, 1986; Kendler, Neale, Kessler, Heath, & Eaves, 1993). This redundancy in the diagnostic criteria for major depression illustrates the heterogeneous nature of the disorder (Chen, Eaton, Gallo, & Nestadt, 2000; Lorant et al., 2003; Merikangas, Wicki, & Angst, 1994; Zimmermann et al., 2009), which is compounded by the fact that many individuals in a depressive episode never seek treatment (Kessler et al., 2003).

The marked heterogeneity of the illness makes it difficult to develop accurate neurobiological models of the illness. Whereas large-scale meta-analyses (Kempton et al., 2011; Schmaal et al., 2015) are beginning to dissociate patients in terms of clinical characteristics, most neuroimaging studies include all forms of major depression despite growing evidence that differences in presentation may reflect distinct etiologies and possibly unique neural systems. Similar criticisms could be raised about functional neuroimaging studies. Although the use of model organisms has fundamentally improved our understanding of the brain circuitry associated with stress, depressed mood, and psychomotor agitation/retardation (Krishnan & Nestler, 2008), because the relationship between specific symptoms and the manifestation of the illness associated with such high levels of disability in humans is unclear, the usefulness of this work for identifying the etiological bases or pathogenesis of major depression is unknown (Flint & Kendler, 2014). One argument for searching for genes that influence major depression is that such genes will provide a window into the neurobiology of the illness, potentially providing mechanistic insights (Flint & Kendler, 2014). Unfortunately, it has been difficult to identify risk genes for major depression, with relatively little progress. As discussed in more detail subsequently, an increased effort to delineate the genetic architecture of major depression is surely warranted.

It is clear that major depression is a heterogeneous disorder with a highly variable course, an inconsistent response to treatment, and no established etiology. It is unsurprising that our current treatments, which are based purely on minimizing symptoms either with pharmacological agents (Belmaker & Agam, 2008; Berman et al., 2000; Pariante & Miller, 2001; Wolkowitz et al., 1999; Zarate et al., 2006) or via psychological therapies (Butler, Chapman, Forman, & Beck, 2006; Morgan, 2003), have limited efficacy. Arguably, our current lack of understanding of the root causes of the disorder hinders improvements in prevention, diagnosis, and treatment. Given the substantial public health implications for discovering aspects of the pathophysiology of major depression that lead to prevention or therapeutic strategies, the scientific and governmental communities cannot afford to ignore this illness. Indeed, improving our understanding of the etiology and pathophysiology was identified as one of the grand challenges for mental health researchers and practitioners (Collins et al., 2011). The goal of this chapter is to review current knowledge about the epidemiology, heterogeneity, neuroanatomy, neurophysiology, and genetics of major depression to provide a more complete conceptualization of the illness.

PREVALENCE OF MAJOR DEPRESSIVE DISORDER

Using data from the National Comorbidity Survey Replication ($n = 9090$), Kessler et al. (2003) estimated that the lifetime prevalence of DSM-IV major depressive disorder is 16.2% (95% confidence interval [CI], 15.1–17.3) and that the prevalence for a depressive episode in the preceding 12 months was 6.6% (95% CI, 5.9–7.3). Mean episode duration was 16 weeks (95% CI, 15.1–17.3). Weissman et al. (1996) used similar epidemiological sampling techniques in 10 countries to estimate the rates of major depression across different nationalities ($n =$ about 38,000) and found that the lifetime prevalence for major depression varies widely across countries, ranging from 1.5% in Taiwan to 19% in Lebanon. The mean age at onset was less variable (range, 24.8–34.8 years) and in every country the rates of major depression were higher for women than men (Weissman et al., 1996). Although it is clear that rates of major depression vary between nations and between racial/ethnic groups within a nation (Riolo, Nguyen, Greden, & King, 2005; Williams et al., 2007), it is less clear whether this reflects true biological differences or is confounded with cultural and socioeconomic issues (Hirschfeld & Weissman, 2002).

In a now classic study, Bromet et al. (1986) analyzed the 18-month test-retest stability of lifetime major depression diagnosis, as determined by a standardized clinical interview, in 391 community-dwelling females. The reliability for the diagnosis was poor ($\kappa = 0.40$) and appeared to be influenced by the clinical presentation of the individual at the time of assessment rather than demographic, psychosocial, and interviewer characteristics. Even among women who reliably reported lifetime depressive episodes, the reported number of episodes, length of their longest episode, and age at first episode were found to be inconsistent. Questions about the temporal reliability of the major depression diagnoses remain a difficult issue for the field (Kendler, 1993), because current symptomatology seems to predict lifetime diagnostic status unduly. Indeed, data from test-retest reliability for major depression from the DSM-5 field trials were questionable ($\kappa = 0.25$ [0.13−0.36]), indicating that even with modern methods our ability to and reliably diagnose depression consistently is modest (Regier et al., 2013). Although there are questions about the meaning of limited diagnostic reliability and methodological issues (Kraemer, Kupfer, Clarke, Narrow, & Regier, 2012), poor temporal reliability for lifetime major depression diagnoses likely limit our understanding of the prevalence of the illness and highlight the need for more diagnostic indices that do not entirely depend on self-report.

RISK FACTORS FOR MAJOR DEPRESSIVE DISORDER

Numerous risk factors have been identified for major depression (Hirschfeld & Weissman, 2002), including sex (gender), marital status, socioeconomic status, life events (particularly early trauma), general medical history, and family history. Studies consistently show that women are almost twice as likely as men to hold a lifetime major depression diagnosis (Kendler, Gatz, Gardener, & Pedersen, 2006; Nolen-Hoeksema, 1987). However, as noted earlier, lifetime diagnoses of major depression are not terribly consistent over time and if men are less reliable when reporting depressive episodes than women (Aneshensel, Estrada, Hansell, & Clark, 1987; Wells & Horwood, 2004), differences in the rates of major depression between women and men could be exacerbated by non−depression related sex differences (eg, general accuracy for medical history) (Kriegsman, Penninx, van Eijk, Boeke, & Deeg, 1996). However the decreased reliability among men when recalling prior depressive episodes is not supported by all studies (Aneshensel et al., 1987; Wells & Horwood, 2004), and other studies do not support this hypothesis (Fennig, Schwartz, & Bromet, 1994; Kendler, Gardner, & Prescott, 2001). Interestingly, data from the National Comorbidity Survey Replication sample indicated that the probability of recurrence (eg, additional episodes) is almost identical for women as it is for men with a history of depression (Kessler, McGonagle, Swartz, Blazer, & Nelson, 1993). Similarly, women and men in the Baltimore Epidemiologic Catchment Area sample had comparable rates of recurrence and similar episode lengths (Eaton et al., 1997), which suggests that the higher prevalence among women is largely the result of higher risk of initial onset.

Divorced, separated, and widowed individuals are substantially more likely to experience depression than are married or never-married persons (about 11 times increased risk in the United States) (Weissman et al., 1996), although the direction of this association is not clear (Blazer, Kessler, McGonagle, & Swartz, 1994). In the National Comorbidity Survey sample, the odds ratios for lifetime and current major depression were significantly higher among individuals earning less than $20,000 a year and declined as income increased (Hirschfeld & Weissman, 2002). Furthermore, a prospective cohort study using the annual Belgian Household Panel Survey (1992−99) showed a clear relationship between worsening socioeconomic circumstances and increased depression (Lorant et al., 2007).

Ethnicity is risk factor for major depression (Gonzalez, Tarraf, Whitfield, & Vega, 2010; Riolo et al., 2005). Based on data from the National Health and Nutrition Examination Survey III ($n = 8449$), Riolo et al. (2005) found that prevalence of major depressive disorder was significantly higher in white people than in African-American and Mexican American individuals. However, the opposite pattern was found for dysthymic disorder, which raises questions about the diagnostic process applied (Riolo et al., 2005). Other cultural differences such as increased somatization (Kleinman, 2004) or increased stress owing to perceptions of family dysfunction (Hovey & King, 1996) have been noted in minority and immigrant populations. The finding of increased mental illness (in particular depression and substance abuse) in US-born subjects compared with immigrants, termed the "immigrant paradox," suggests that factors related to acculturation, perceived discrimination, or relative social status influence these disorders (Alegría et al., 2008). Yet, using carefully conducted face-to-face interviews, Olvera et al. (2011) reported a 35% prevalence in major depression among low-income Mexican American people from the San Antonio region, with expected sex differences, comorbid anxiety disorders, and substance misuse (Kessler et al., 2003; Weissman et al., 1996), which raises questions about the cultural competency of diagnostic measures for major depression.

Kendler et al. reported that stressful life events have a causal relationship with the onset of episodes of major depression. However, the relationship between stressful life events and depression risk is complicated and possibly bidirectional given that individuals predisposed to major depression select themselves into high-risk environments (Kendler, Karkowski, & Prescott, 1999).

In a highly influential article, Caspi et al. (2003) reported a gene-by-environment interaction influencing risk for major depression in a large epidemiological cohort. Specifically, a functional polymorphism in the promoter region of the serotonin transporter (5- HTTLPR) gene was found to moderate the influence of stressful life events on depression. Individuals with one or two copies of the short allele of the 5-HT T promoter polymorphism exhibited more depressive symptoms, diagnosable depression, and suicidality in relation to stressful life events than did individuals homozygous for the long allele (Caspi et al., 2003). However, a subsequent meta-analysis found no association between the 5-HTTLPR genotype and depression in any of the individual studies or in the weighted average (odds ratio = 1.05; 95% CI, 0.98−1.13) and no interaction effect between genotype and stressful life events on depression was observed (OR, 1.01; 95% CI, 0.94−1.10) in women alone or men alone, or in combined analyses (Risch et al., 2009). By contrast, this meta-analysis did find a significant association between the number of stressful life events and depression risk (OR, 1.41; 95% CI, 1.25−1.57). These articles demonstrate the difficulties when trying to identify a gene-by-environment interaction in candidate genes with weak effects (Glahn et al., 2014).

As with most mental illnesses, family history of depression is among the strongest risk factors. Weissman et al. (1984) reported a lifetime major depression rate (morbidity risk) of 14.7−16.4% in the first-degree relatives ($n = 2003$) of depressive probands ($n = 335$) compared with a rate of 5.1% in the relatives of control subjects (Prusoff, Weissman, Merikangas, Leckman, & Harding, 1984). A meta-analysis of five studies reported unanimous support for familial aggregation of major depression in probands compared with comparison subjects (Sullivan, Neale, & Kendler, 2000). Across the five studies, there was strong evidence for increased rates of major depression in first-degree relatives ($\chi^2 = 97.7$; degrees of freedom = 1; $P < 0.00005$), equating to an odds ratio of 2.84 (95% CI, 2.31−3.49). These data suggest a strong familial component to risk for major depression that could reflect common environment, genetic factors, or more likely a combination of the two.

Comorbidity between substance dependence and major depression is well documented in large epidemiologic studies (Grant, Hasin, et al., 2004; Grant, Stinson, et al., 2004; Kessler et al., 2005, 1994). Among individuals with alcohol or substance dependence, the likelihood of major depression and/or an anxiety disorder is significantly increased (Grant et al., 2004, 2005; Hasin, Stinson, Ogburn, & Grant, 2007), even after covarying for demographic characteristics and other psychiatric disorders (Compton, Thomas, Stinson, & Grant, 2007; Hasin et al., 2007). As the severity of substance misuse increases from use to abuse to dependence, the observed rate of depressive or anxiety disorders likewise increases (Compton et al., 2007; Merikangas et al., 1998). Psychiatric comorbidities among individuals with addictive disorders are associated with poorer outcomes, including more severe symptomatology, greater social and functional impairments, and increased suicide risk (Davis et al., 2006; Jamal, Willem Van der Does, Cuijpers, & Penninx, 2012). Although it is possible that high levels of comorbidity between major depression and substance dependence reflect an attempt to self-medicate with drugs of abuse (Khantzian, 1997; Markou, Kosten, & Koob, 1998), twin and family studies consistently report that substance use and mood disorders share overlapping genetic effects (Edwards & Kendler, 2012; Edwards, Maes, Pedersen, & Kendler, 2011; Kendler et al., 1993; Lynskey et al., 2004; Lyons et al., 2008; Maher, Marazita, Zubenko, Kaplan, & Zubenko, 2002; Olvera et al., 2011; Prescott, Aggen, & Kendler, 2000; Tsuang, Bar, Harley, & Lyons, 2001; Vink et al., 2014).

Comorbidity between major depression and other medical conditions is well documented (Anderson, Freedland, Clouse, & Lustman, 2001; Carney et al., 1988; Dowlati et al., 2010; Katon, 2003; Musselman, Evans, & Nemeroff, 1998), although a mechanistic explanation for this co-occurrence is unknown.

DIAGNOSTIC HETEROGENEITY IN MAJOR DEPRESSIVE DISORDER

As mentioned previously, heterogeneity in terms of symptom type and presentation, duration of episodes, and treatment response is substantial in major depression. One popular strategy for reducing illness heterogeneity for research and treatment purposes is through the use of clinical subtypes (Judd, 1997; Levinson, 2006). For example, by describing a distinctive phenomenology and providing evidence for specific risk factors, outcomes, and response to treatment, Prigerson et al. (2009) argued that prolonged grief constitutes a distinct mental disorder. However, prolonged grief was not included as a separate illness in DSM-5. In general, the complexity of clinical presentations makes it difficult to determine which combination of symptoms/indicators best defines subgroups of the disorder (Judd, 1997), particularly because symptoms have been shown to be independently influenced by distinct risk factors (Fried, Nesse, Zivin, Guille, & Sen, 2013).

Weissman et al. (1986) used familial aggregation to examine putative clinical subgroups for major depression, including age at onset, clinical severity, symptom patterns (eg, episode recurrence), or the presence of other disorders (comorbidity). They reported that only early age at onset, or major depression with an anxiety disorder or secondary alcoholism, were independently related to increased risk of major depression in relatives (Weissman et al., 1986), which suggests that these groupings might be particularly useful when examining the genetic underpinnings of major depression (Glahn et al., 2012; Levinson et al., 2003, 2007). Additional subtypes include melancholic depression, atypical depression, and psychotic depression, although (with the exception of the last) it is unclear whether these subtypes represent etiologically distinct disease entities or differing points on a depressive continuum (Cowen, 2013). In addition, grouping depressed individuals based entirely on their clinical presentation and history tends to have low reliability and little prognostic or research usefulness (Angst, 1985; Bromet et al., 1986; Judd & Akiskal, 2000; Merikangas et al., 1994). That being said, focusing on individuals with multiple depressive episodes tends to improve the reliability and the heritability of lifetime diagnoses (Kendler et al., 1993; Walss-Bass et al., 2005).

NEUROANATOMY OF MAJOR DEPRESSIVE DISORDER

Kempton et al. (2011) published a substantial meta-analysis comparing 9533 individuals with depression with 8846 control subjects from 225 separate published studies, and reported that major depression was associated with lateral ventricle enlargement (effect size = 0.44), increased cerebrospinal fluid volume (0.54), and smaller volumes of the basal ganglia (caudate, −0.22; putamen, −0.25; and globus pallidus, −0.31), thalamus (−0.34), hippocampus (−0.47), frontal lobe (−0.29), orbitofrontal cortex (−0.38), and gyrus rectus (−0.72). The authors argued that reduced hippocampal volume and a significant trend for increased pituitary volume provide neuroanatomic evidence for hypothalamic−pituitary−adrenal (HPA) axis involvement in major depression (Kempton et al., 2011). Alterations in HPA axis function (Pariante & Lightman, 2008; Schlesser, Winokur, & Sherman, 1980; Vreeburg et al., 2009) and increased glucocorticoid activity (Pariante & Miller, 2001) are well documented in major depression, and the authors speculated that increased pituitary volume is associated with increased adrenocorticotropic hormone production, which stimulates the adrenal cortex, increasing glucocorticoid production. Prolonged high circulating levels of glucocorticoids are known to damage hippocampal neurons (Sapolsky, 1990). In this meta-analysis, patients currently experiencing a depressive episode had significantly smaller hippocampal volume than did remitted patients (Kempton et al., 2011), which suggests potential effects of mood state on neuroanatomy. However, the vast majority of patients included in this meta-analysis were taking psychotropic medications, and disentangling pathology from medication effects was not possible.

The ENIGMA Major Depressive Disorder working group examined subcortical volumes among 1728 patients with major depression and 7199 control subjects from 15 research samples worldwide (Schmaal et al., 2015). Consistent with the Kempton meta-analysis, individuals with major depression had significantly smaller hippocampal volumes compared with control subjects (Cohen's $d = -0.14$). Interestingly, patients with multiple depressive episodes drove this effect ($d = -0.17$) and no volume differences were observed between patients who experience their first episode and control subjects. Earlier age of onset was associated with a smaller hippocampus ($d = -0.20$), a trend toward smaller amygdala ($d = -0.11$) and larger lateral ventricles ($d = 0.12$). In contrast to the Kempton report, symptom severity at study inclusion was not associated with subcortical brain volume. Subject age, proportion of antidepressant users, and proportion of remitted patients did not significantly influence these results. Samples with a higher proportion of users of antipsychotic medication showed larger caudate volumes, a finding consistent with the schizophrenia literature (Glahn et al., 2008).

Together, these two large-scale analyses clearly demonstrate neuroanatomic changes in individuals with major depression relative to healthy comparison subjects. However, as suggested by the secondary analysis conducted by the ENIGMA group, clinical characteristic (eg, recurrent versus single episode; proportion of antipsychotic medication users) could be critical for developing more precise neuroanatomic models of the illness. Yet, sample sizes are only beginning to reach levels needed to address these finer-grained questions.

FUNCTIONAL NEUROIMAGING IN MAJOR DEPRESSIVE DISORDER

Using functional neuroimaging methods such as positron emission topography (PET) and functional MRI (fMRI), investigations have provided important insights about its neural substrates of depression. For example, indexing cerebral blood flow and glucose metabolism, Drevets et al. (1997) localized an area of abnormally decreased activity in the prefrontal cortex ventral to the genu of the corpus callosum in currently depressed individuals. This region, described as sub/perigenual cingulate cortex, is linked to emotional regulation in humans (Drevets et al., 1997; Mayberg et al., 1999) and in nonhuman primates (Hamani et al., 2011; Ongur & Price, 2000; Paus, 2001). Mayberg et al. (1997) observed decreased sub/perigenual cingulate blood flow in pathological depression and normal sadness, which predicted treatment response through pharmacological and cognitive behavioral therapy (Goldapple et al., 2004). Direct stimulation of this region reduced depressive symptomatology in treatment-resistant individuals with mood disorders (Mayberg et al., 2005). Although the exact mechanism for how dysfunction in this area leads to mood disorders is still unclear, Ongur, Drevets, and Price (1998) found a reduction in glia among individuals with familial bipolar disorder in this region, which suggests that the observed neuroimaging anomalies may be sensitive to both clinical state and a predisposition to affective dysregulation.

Based on her review of cerebral blood flow and regional glucose metabolism in idiopathic depressed patients, Mayberg (1997) proposed a neural model of major depression in which limbic hyperactivation inhibits dorsal cortical activation, which reciprocally fails to limit limbic activation. Investigators examining emotion reactivity (Siegle, Thompson, Carter, Steinhauer, & Thase, 2007) or resting-state (Hamilton, Chen, Thomason, Schwartz, & Gotlib, 2011) fMRI have proposed similar models. In a meta-analysis, Hamilton et al. (2012) found that resting-state PET showed reliably greater activity in depressed participants ($n = 299$) relative to 300 comparison subjects in the pulvinar nuclei of the thalamus. The pulvinar is thought to be involved in emotional attention and awareness (Pessoa & Adolphs, 2010), given its roll in feature binding (Ward, Danziger, Owen, & Rafal, 2002), which facilitate the integration of distinct cell ensembles that code for different perceptual features (Treisman, 1999). Hamilton et al. (2012) also meta-analyzed 24 fMRI studies in which depressed individuals processed negatively valenced stimuli, showing that individuals diagnosed with major depression ($n = 351$) had reliably greater responses to negative stimuli than did comparison subjects ($n = 354$) in the amygdala, dorsal anterior cingulate cortex, insula/superior temporal gyrus, precentral gyrus, and middle temporal gyrus. In contrast, depressed individuals had relatively less activity in the dorsolateral prefrontal cortex and bilaterally caudate body (dorsal striatum) (Hamilton et al., 2012). The amygdala, dorsal anterior cingulate cortex, and insula are prominent nodes in the salience network (Seeley et al., 2007), whereas the dorsolateral prefrontal cortex is closely linked to executive processing (Koechlin & Summerfield, 2007; Miller & Cohen, 2001). Based on these meta-analyses, Hamilton et al. (2012) posited that increased baseline activity in the pulvinar nucleus, via monosynaptic projections to the amygdala, dorsal anterior cingulate, and insula, potentiates the affective subdivision of the corticostriatal—pallidal—thalamic circuit in depression. Furthermore, appraisals typically initiated by the dorsolateral prefrontal cortex to reduce the impact of negative stimuli are diminished owing to a failure at nigrostriatal relays to propagate information to the dorsal striatum and dorsolateral prefrontal cortex in major depressive disorder (Hamilton et al., 2012). Although this model must be confirmed through systematic examination, it demonstrates the utility of functional neuroimaging studies to delineate neural models of depression. Yet, this complex neurophysiological model is uninformed by the clinical heterogeneity described earlier or the complex genetics described subsequently.

HERITABILITY

Some individuals presenting with major depression have clear environmental factors that appear to be driving their symptoms (eg, prolonged bereavement, terminal illness), but it is also clear that major depression runs in families. Thus whereas environmental influences, particularly those that are specific to an individual, are likely to be etiologically significant, it is clear that genetic factors have a role in risk for the disorder. Indeed, the literature reports fairly consistently that lifetime major depression is a complex disorder that results from both genetic and environmental influences (Bienvenu, Davydow, & Kendler, 2010; Sullivan et al., 2000).

Because heritability estimates necessarily depend on the specific sample studied (Falconer & Mackay, 1996), the general consistency of heritability estimates between studies implies the robustness of phenotyping procedures. Based on a sample of 42,000 Swedish twins, Kendler, Aggen, Tambs, and Reichborn-Kjennerud (2006) estimated heritability for lifetime major depression to be $h^2 = 0.38$. That estimate was comparable to the point estimate derived from a meta-analysis ($h^2 = 0.37$) (Bienvenu et al., 2010) and from an earlier meta-analysis ($h^2 = 0.37$) (Sullivan et al., 2000). The latter study also reported a minimal contribution of environmental effects common to siblings (point estimate, 0%; 95% CI, 0—5%) and substantial individual-specific environmental effects/measurement error (point estimate, 63%; 95% CI, 58—67%). A similar heritability estimate for lifetime major depression was estimated in extended pedigrees ($h^2 = 0.39$) (Olvera et al., 2011).

As mentioned previously, defining more clinically homogeneous groups could potentially increase the genetic "signal" of the illness (Levinson et al., 2003; McGuffin, Katz, Watkins, & Rutherford, 1996; Olvera et al., 2011; Shi et al., 2010; Weissman et al., 1996). Although evidence for this supposition is limited, Kendler et al. (1996) attempted to test this hypothesis directly using a latent class analysis of 14 symptoms of depression in 1029 female twin pairs. These investigators reported that twins concordant for depression were significantly more likely than expected to share features of particular depressive syndromes, which suggests that genetic effects might operate separately on subtypes of major depression.

ENDOPHENOTYPES

Endophenotypes are measurable biomarkers that are correlated with an illness, at least in part because of shared underlying genetic influences (Glahn et al., 2014; Gottesman & Gould, 2003). According to Gottesman and Gould (2003) endophenotypes must (Greenberg et al., 2003) be associated with illness in the population; (Wells et al., 1989) be heritable; (Hays et al., 1995) be primarily state-independent; (Murray & Lopez, 1996) co-segregate with the illness in families; and (Murray et al., 2012) be found in unaffected family members at a higher rate than in the general population. Endophenotypes are thought to be critical for the functional characterization of risk genes discovered in traditional psychiatric genetic experiments and, if applied in large-scale studies, could identify novel risk genes, because of our lack of understanding of pathophysiology and gene regulation (Glahn et al., 2014). A number of putative endophenotypes have been proposed for major depression, including learning and memory impairments, reduced reward functioning, increased stress sensitivity, REM sleep abnormalities, functional and structural brain abnormalities, dysfunctions in serotonergic, catecholaminergic, HPA axis, and corticotropin-releasing hormone systems, and intracellular signal transduction measures (Hasler, Drevets, Manji, & Charney, 2004). However, few studies systematically assess the criteria needed for a measure to be considered an endophenotype set forth by Gottesman and Gould (2003). Unfortunately, this is particularly true of more biologically based endophenotypes.

Glahn et al. (2012) developed the Endophenotype Ranking Value (ERV), an empirical metric for ranking endophenotypes based on their genetic similarity to the studied illness. This approach is based on the assertion that joint genetic determination of endophenotype and disease risk is fundamental to the endophenotype concept (Glahn et al., 2012). The ERV method involves estimating the standardized genetic covariance between each putative endophenotype and a particular illness (conceptually similar to the coheritability between traits) and then ranking endophenotypes based on their genetic covariance. Formally, the ERV statistic is defined as the absolute value of the square root of the heritability of the illness multiplied by the square root of the heritability of the endophenotype multiplied by the genetic correlation between the endophenotype and the illness: $ERV_{\mathrm{ie}} = |\sqrt{h_{\mathrm{i}}^2}\sqrt{h_{\mathrm{e}}^2}\rho_{\mathrm{g}}|$. *ERV* values vary between 0 and 1, with higher values indicate stronger shared genetic influence between the endophenotype and the illness.

Applying the ERV approach to 1122 participants of the "Genetics of Brain Structure and Function" study, Glahn et al. ranked neurocognitive/behavioral, neuroanatomic and transcriptional endophenotypes for recurrent major depression. Top-ranked neurocognitive/behavioral endophenotypes included the score on the Beck Depression Inventory (BDI)II (Beck, Steer, & Brown, 1996) (ERV = 0.26; $P = 1.9 \times 10^{-5}$, genetic correlation $[\rho_g] = 0.83$) and the neuroticism questions from the Eysenck Personality Questionnaire (Eysenck & Eysenck, 1975) (ERV = 0.24; $P = 1.7 \times 10^{-4}$, $\rho_g = 0.74$). At one level, it is not surprising that measures of depressive symptomatology are strongly associated with genetic risk for major depression. However, both the BDI and EPQ are thought to index clinical state and not genetic predisposition. Nonetheless, because both questionnaires are heritable ($h^2 = 0.25$ and 0.23, respectively), they have the potential to detect trait-level variation in addition to clinical state. Highly ranked neurocognitive endophenotypes included measures of declarative (CVLT Recognition, ERV = 0.14; $P = 5.4 \times 10^{-2}$; $\rho_g = -0.34$) and working memory (Digit Span Forward, ERV = 0.14; $P = 5.6 \times 10^{-2}$; $\rho_g = -0.30$; and Letter–Number Sequencing, ERV = 0.14; $P = 6.3 \times 10^{-2}$; $\rho_g = -0.27$), but these neurocognitive endophenotypes did not reach statistical significance.

Candidate neuroimaging endophenotypes included T1-weighted neuroanatomic scans (providing subcortical volumes and cortical thickness measures), white matter hyperintensities derived from T2-weighted FLAIR sequences, and measures of white matter coherence indexed by fractional anisotropy with diffusion tensor imaging. The highest-ranked neuroanatomic measure was the volume of the ventral diencephalon (ERV = 0.24; $P = 3.9 \times 10^{-3}$; $\rho_g = -0.43$). The ventral diencephalon is primarily composed of the hypothalamus (Desikan et al., 2006). As part of the HPA axis, the hypothalamus is thought to mediate neuroendocrine and neurovegetative functions implicated in depression (Pariante & Lightman, 2008; Pariante & Miller, 2001; Nestler et al., 2002). The finding that variation in ventral diencephalon volume is genetically correlated with risk for recurrent major depression suggests that the same genetic factors influence both traits. In addition, hippocampal (ERV = 0.20; $P = 1.2 \times 10^{-2}$p; $\rho_g = -0.35$), pallidum (ERV = 0.20; $P = 1.3 \times 10^{-2}$; $\rho_g = -0.40$), and thalamic volumes (ERV = 0.17; $p = 4.8 \times 10^{-2}$; $\rho_g = -0.29$) were highly ranked (Fig. 31.1).

The cerebrovascular hypothesis of major depression posits that vascular disease or vascular risk factors predispose, precipitate, or perpetuate depressive syndromes, particularly in elderly individuals (Alexopoulos et al., 1997). Indeed, there

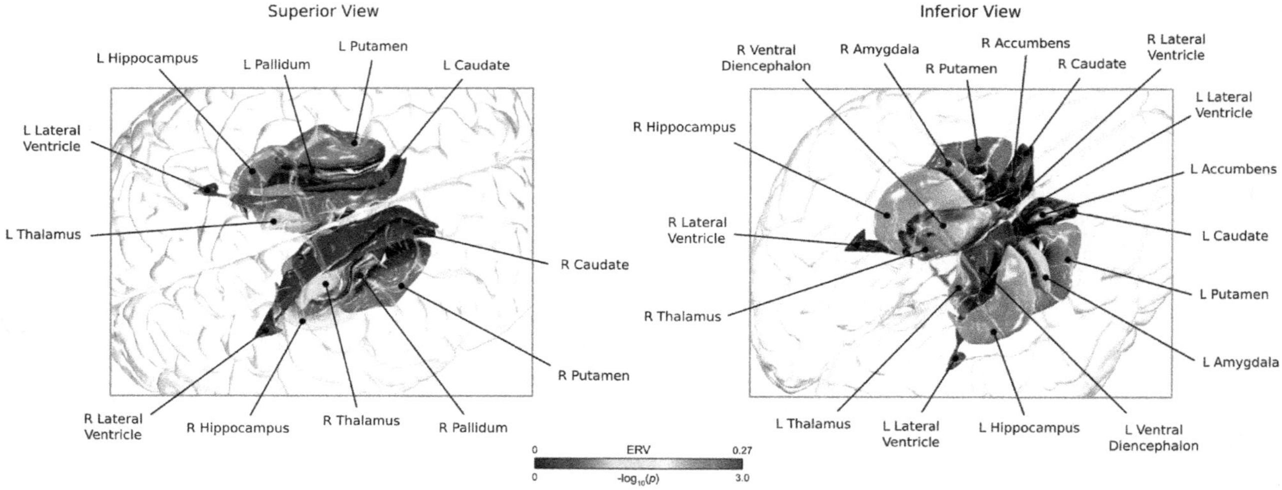

FIGURE 31.1 Endophenotype ranking value statistics for subcortical brain regions and recurrent major depression, after Glahn et al. (2012). Volume measurements of subcortical nuclei were found to share genetic variance with liability for recurrent major depression in extended pedigrees selected without regard to phenotype. *ERV*, endophenotype ranking value; *L*, left; *R*, right.

is mounting evidence in support of the vascular hypothesis (Lyness et al., 1999; Steffens, Helms, Krishnan, & Burke, 1999; Sweet et al., 2004), although no specific biological mechanisms have been identified. White matter hyperintensities are heritable (Carmelli et al., 1998; Kochunov et al., 2009) nonspecific measures of cerebrovascular functioning (Debette & Markus, 2010). The highest-ranked white matter hyperintensity measures included parietal (ERV = 0.28; $P = 7.8 \times 10^{-3}$; $\rho_g = 0.57$), frontal (ERV = 0.26; $P = 1.3 \times 10^{-2}$; $\rho_g = 0.48$), and subcortical volumes (ERV = 0.21; $P = 4.1 \times 10^{-2}$; $\rho_g = -0.46$). These results indicate that common genetic factors influence the increase in white matter hyperintensities with aging and the risk for lifetime major depression.

Variation in RNA levels measured in lymphocytes appear to be sensitive to the function of neurological disease-relevant genes (Borovecki et al., 2005), which makes lymphocyte-based transcriptional profiling a potential method for identifying illness-related genes (Goring et al., 2007). Transcript-based endophenotypes are clearly a more direct index of gene action than are the traditional behavioral phenotypes (eg, diagnoses). Many factors may influence gene expression, including the tissue sampled, age, sex, and the time of day (Borovecki et al., 2005; Radich et al., 2004; Whitney et al., 2003). However, there is also substantial genetic influence on gene expression as evidenced by numerous expression quantitative trait locus and expression genome-wide association studies (Ertekin-Taner, 2011; Zou et al., 2010). Thus, gene expression may make for excellent endophenotypes of complex disease in which variation in *cis*-regulatory polymorphisms mediates disease risk by influencing expression level (Goring et al., 2007; Rockman & Wray, 2002; Wray, 2007). At some level, this approach includes a weak assumption that lymphocyte-based RNA levels can be considered a surrogate measure for gene expression in other tissues (eg, brain). Although a number of potential confounders must be considered when interpreting lymphocyte-based transcriptional results (Cai et al., 2010), the potential utility of peripheral biomarkers/endophenotypes for psychiatry are so substantial that these measures merit study (Glahn et al., 2014).

Total RNA was isolated from lymphocytes and hybridized to Illumina Sentrix Human Whole Genome (WG-6) Series 1 BeadChips, following procedures described by Göring et al. (2007). These BeadChips simultaneously probe about 48,000 transcripts, representing more than 25,000 annotated human genes. Although they previously identified 20,413 quantitative transcripts in lymphocytes, we only examined those with heritabilities greater than or equal to 0.20 ($n = 11,337$). Numerous transcripts were significantly genetically correlated with lifetime recurrent major depression, even when controlling for multiple comparisons (Glahn et al., 2012). The top-ranking transcript, *RNF123*, is a member of the E3 ubiquitin-protein ligase family, which has diverse functions, including protein degradation and modulation of protein assembly, structure, function, and localization (Deshaies & Joazeiro, 2009; Doolittle et al., 2009). It is notable that the top-ranked lymphocyte-based transcriptional endophenotype for recurrent major depression was ranked higher than any of the behavioral/cognitive or brain imaging traits, including BDI, a quantitative index of depressive symptomatology, which suggests that transcriptional profiles may provide an important new set of markers for disease risk. However, far more transcripts were tested than any other endophenotype class (Glahn et al., 2012).

GENETICS

Unlike schizophrenia, in which there common variants associated with illness risk are now replicated (Schizophrenia Working Group of the Psychiatric Genomics C, 2014), or autism, in which rare variants and de novo mutations point to genes that increase liability (De Rubeis et al., 2014; Iossifov et al., 2014), there are no consistent genetic findings for major depression (Cohen-Woods, Craig, & McGuffin, 2013; Flint & Kendler, 2014). In an authoritative review, Flint and Kendler (2014) reported that despite one article's claim (Kohli et al., 2011), no significant common variants have been localized for major depression, and suggested that the most likely explanation for this failure is that studies have been underpowered to detect the causative loci. Ripke et al. in the Major Depressive Disorder Working Group of the Psychiatric Genetics Consortium (2013) failed to find robust evidence for loci that exceed genome-wide significance even with more than 9000 cases. Furthermore, Flint and Kendler (2014) argued that despite over 1500 published articles on the topic, candidate gene studies of major depression have likewise failed and that findings are likely false positives.

Examining genetic linkage studies of major depression (Flint & Kendler, 2014), which are influenced by both common and rare variation, five regions are reported at least twice: chromosome 11, 75–80 Mb (Breen et al., 2011; Zubenko et al., 2003); chromosome 15, 37–42 Mb (Camp et al., 2005; Zubenko et al., 2003); chromosome 15, 87–92 Mb (Breen et al., 2011; Holmans et al., 2007, 2004; Levinson et al., 2007), chromosome 3, 4–9 Mb (Breen et al., 2011; Middeldorp et al., 2008), and chromosome 2, 64–68 Mb (Middeldorp et al., 2008; Schol-Gelok et al., 2010). These putative replications are restricted to being within 5 Mb of each other, as the confidence intervals for the position of loci found by linkage studies are notoriously broad, which increases the likelihood that overlapping localizations occur by chance (Roberts, MacLean, Neale, Eaves, & Kendler, 1999). There is some thought that multiple genes or causal variants in a single region drive strong, replicable linkage signals for complex phenotypes seen across samples and studies. Yet, even in repeated linkage localizations, the exact gene(s) or casual variant(s) driving the linkage signal is unknown.

Using low-coverage whole-genome sequencing of 5303 Chinese women with recurrent major depression and 5337 control subjects, the China, Oxford, and Virginia Commonwealth University Experimental Research on Genetic Epidemiology Consortium identified two loci contributing to depression risk (Consortium C, 2015). Both variants were replicated in an independent Chinese sample and are on chromosome 10. The first variant (rs12415800; MAF 45.2%, $P = 2.53 \times 10^{-10}$) is near the *SIRT1* gene on chromosome 10 at 69.6 Mb. The second is near the *LHPP* gene on chromosome 10 at position 126.2 Mb (rs35936514; MAF 26.0%; $P = 6.45 \times 10^{-12}$). Analysis of 4509 patients with melancholia, a severe subtype of depression, yielded an increased genetic signal at the *SIRT1* locus. *SIRT1* is involved in stress tolerance, mitochondrial biogenesis, and fat metabolism, whereas the function of *LHPR* is unknown. These results are promising, but localized loci likely reflect only a small portion of the heritable risk for major depression.

TREATMENT FOR MAJOR DEPRESSION

Although a number of potential treatment options are available for individuals experiencing major depression (Association AP, 2010), a sizable proportion of patients with the illness experience only partial or no clinical response (Nelson, 2003). Typical treatment options include pharmacology (Belmaker & Agam, 2008), electroconvulsive therapy (Bouckaert et al., 2014), psychotherapy (Butler et al., 2006; Morgan, 2003), and deep brain stimulation (Morishita, Fayad, Higuchi, Nestor, & Foote, 2014), either alone or in combination. Antidepressant medication, which is thought to act on brain systems that process emotions to ameliorate depressive symptoms (Harmer, 2008), is the current standard treatment for depression (Association AP, 2010). A meta-analysis by Ma (2015) synthesized 60 fMRI studies ($n = 1569$) applying antidepressants in both healthy volunteers and patients with major depression to define the brain's response to these agents. Repeated antidepressant administration was associated with increased activation across a frontolimbic network (eg, anterior cingulate, amygdala, and thalamus) when individuals observed positively valenced emotional stimuli but those same regions show decreased activity when observing negatively valenced emotional stimuli. Antidepressants also increased activity in the dorsolateral prefrontal cortex. Ma concluded that antidepressants increase brain activity to positive stimuli and decreasing activity to negative stimuli with in the emotional network, and increase engagement of prefrontal regulatory regions to reduce depressive symptoms (168). Many of the brain regions showing antidepressant effects were part of the network discussed in Hamilton's neural model of depression (Hamilton et al., 2012), which suggest a level of convergence between brain regions disrupted by the illness and those sensitive to treatment. Currently, it is unknown whether activation within these brain regions predicts clinical response, a question that is well worth exploring.

CONCLUSIONS

There are major controversies among depression researchers as to the causes, cures, and prevalence of the illness. Whereas epidemiological results stress the heterogeneity and complex nature of the illness, neuroimaging-based models embrace this diversity only minimally. Indeed, very large-scale neuroanatomic studies currently attempt to examine differences in rudimentary aspects of clinical heterogeneity (eg, patients with a single versus multiple episodes) in a convincing manner. Ignoring the diversity of clinical factors potentially limits the utility of neuroimaging-based models of the disease. Although certainly influenced by environmental factors, there is ample evidence for a genetic component to major depression. However, no specific genomic variant or gene has been implicated in this most common mental illness. Identifying genes influencing the risk of depression would enhance understanding of the biological basis of major depression, which in turn would improve detection, enabling earlier and more effective treatment. Although it is possible that substantially increasing sample sizes could result in genome-wide significant hits for depression (Flint & Kendler, 2014), the hunt for depression genes likely has been hampered by the complex nature of the illness, particularly the genetic and phenotypic heterogeneity. Increasing the sample size could compound these problems by introducing more heterogeneity, thus rendering genetic influences more difficult to detect and requiring a combination of individuals with distinct clinical presentations, potentially obscuring the relationship between genotype and depressive symptoms. In contrast, studies designed to reduce genetic heterogeneity by focusing on population isolates or extended pedigrees, both common and rare variants, appropriately models phenotypic heterogeneity, and that using well-chosen endophenotypes could potentially identify causal genes for depression.

REFERENCES

Alegría, M., Canino, G., Shrout, P., Woo, M., Duan, N., Vila, D., … Meng, X. (2008). Prevalence of mental illness in immigrant and non-immigrant U.S. Latino groups. *American Journal of Psychiatry, 165*, 359–369.

Alexopoulos, G., Meyers, B., Young, R., Campbell, S., Silbersweig, D., & Charlson, M. (1997). 'Vascular depression' hypothesis. *Archives of General Psychiatry, 54*, 915–922.

Anderson, R. J., Freedland, K. E., Clouse, R. E., & Lustman, P. J. (2001). The prevalence of comorbid depression in adults with diabetes: a meta-analysis. *Diabetes Care, 24*, 1069–1078.

Aneshensel, C. S., Estrada, A. L., Hansell, M. J., & Clark, V. A. (1987). Social psychological aspects of reporting behavior: lifetime depressive episode reports. *Journal of Health and Social Behavior, 28*, 232–246.

Angst, J. (1985). Switch from depression to mania—a record study over decades between 1920 and 1982. *Psychopathology, 18*, 140–154.

Association AP: American psychiatric association practice guideline for the treatment of patients with major depressive disorder (2nd ed.). (2010). Arlington, VA, USA: American Psychiatric Association.

Association AP: Diagnostic and statistical manual of mental disorders (5th ed.). (2013). Washington, DC: American Psychiatric Association.

Beck, A. T., Steer, R. A., & Brown, G. K. (1996). *Manual for the beck depression inventory-II.* San Antonio, TX: Psychological Corporation.

Belmaker, R., & Agam, G. (2008). Major depressive disorder. *The New England Journal of Medicine, 358*, 55–68.

Berman, R. M., Cappiello, A., Anand, A., Oren, D. A., Heninger, G. R., Charney, D. S., & Krystal, J. H. (2000). Antidepressant effects of ketamine in depressed patients. *Biological Psychiatry, 47*, 351–354.

Bienvenu, O., Davydow, D., & Kendler, K. (2010). Psychiatric 'diseases' versus behavioral disorders and degree of genetic influence. *Psychological Medicine*, 1–8.

Blazer, D., Kessler, R., McGonagle, K., & Swartz, M. (1994). The prevalence and distribution of major depression in a national community sample: the National Comorbidity Survey. *American Journal of Psychiatry, 151*, 979–986.

Borovecki, F., Lovrecic, L., Zhou, J., Jeong, H., Then, F., Rosas, H. D., … Krainc, D. (2005). Genome-wide expression profiling of human blood reveals biomarkers for Huntington's disease. *Proceedings of the National Academy of Sciences of the United States of America, 102*, 11023–11028.

Bouckaert, F., Sienaert, P., Obbels, J., Dols, A., Vandenbulcke, M., Stek, M., & Bolwig, T. (2014). ECT: its brain enabling effects: a review of electroconvulsive therapy-induced structural brain plasticity. *The Journal of ECT, 30*, 143–151.

Breen, G., Webb, B. T., Butler, A. W., van den Oord, E. J., Tozzi, F., Craddock, N., … McGuffin, P. (2011). A genome-wide significant linkage for severe depression on chromosome 3: the depression network study. *American Journal of Psychiatry, 168*(8), 840–847.

Bromet, E. J., Dunn, L. O., Connell, M. M., Dew, M. A., & Schulberg, H. C. (1986). Long-term reliability of diagnosing lifetime major depression in a community sample. *Archives of General Psychiatry, 43*, 435–440.

Butler, A. C., Chapman, J. E., Forman, E. M., & Beck, A. T. (2006). The empirical status of cognitive-behavioral therapy: a review of meta-analyses. *Clinical Psychology Review, 26*, 17–31.

Cai, C., Langfelder, P., Fuller, T. F., Oldham, M. C., Luo, R., van den Berg, L. H., … Horvath, S. (2010). Is human blood a good surrogate for brain tissue in transcriptional studies? *BMC Genomics, 11*, 589.

Camp, N. J., Lowry, M. R., Richards, R. L., Plenk, A. M., Carter, C., Hensel, C. H., … Cannon-Albright, L. A. (2005). Genome-wide linkage analyses of extended Utah pedigrees identifies loci that influence recurrent, early-onset major depression and anxiety disorders. *American Journal of Medical Genetics Part B: Neuropsychiatric Genetics, 135B*, 85–93.

Carmelli, D., DeCarli, C., Swan, G. E., Jack, L. M., Reed, T., Wolf, P. A., & Miller, B. L. (1998). Evidence for genetic variance in white matter hyperintensity volume in normal elderly male twins. *Stroke, 29*, 1177–1181.

Carney, R. M., Rich, M. W., Freedland, K. E., Saini, J., teVelde, A., Simeone, C., & Clark, K. (1988). Major depressive disorder predicts cardiac events in patients with coronary artery disease. *Psychosomatic Medicine, 50*, 627–633.

Caspi, A., Sugden, K., Moffitt, T. E., Taylor, A., Craig, I. W., Harrington, H., … Poulton, R. (2003). Influence of life stress on depression: moderation by a polymorphism in the 5-HTT gene. *Science, 301*, 386–389.

Chen, L., Eaton, W. W., Gallo, J. J., & Nestadt, G. (2000). Understanding the heterogeneity of depression through the triad of symptoms, course and risk factors: a longitudinal, population-based study. *Journal of Affective Disorders, 59*, 1–11.

Cohen-Woods, S., Craig, I. W., & McGuffin, P. (2013). The current state of play on the molecular genetics of depression. *Psychological Medicine, 43*, 673–687.

Collins, P. Y., Patel, V., Joestl, S. S., March, D., Insel, T. R., Daar, A. S., … Stein, D. J. (2011). Grand challenges in global mental health. *Nature, 475*, 27–30.

Compton, W. M., Thomas, Y. F., Stinson, F. S., & Grant, B. F. (2007). Prevalence, correlates, disability, and comorbidity of DSM-IV drug abuse and dependence in the United States: results from the national epidemiologic survey on alcohol and related conditions. *Archives of General Psychiatry, 64*, 566–576.

Consortium C. (2015). Sparse whole-genome sequencing identifies two loci for major depressive disorder. *Nature, 523*, 588–591.

Cowen, P. J. (2013). Classification of depressive disorders. *Current Topics in Behavioral Neurosciences, 14*, 3–13.

Davis, L. L., Frazier, E., Husain, M. M., Warden, D., Trivedi, M., Fava, M., … Rush, A. J. (2006). Substance use disorder comorbidity in major depressive disorder: a confirmatory analysis of the STAR*D cohort. *American Journal on Addictions, 15*, 278–285.

De Rubeis, S., He, X., Goldberg, A. P., Poultney, C. S., Samocha, K., Cicek, A. E., … Buxbaum, J. D. (2014). Synaptic, transcriptional and chromatin genes disrupted in autism. *Nature, 515*, 209–215.

Debette, S., & Markus, H. S. (2010). The clinical importance of white matter hyperintensities on brain magnetic resonance imaging: systematic review and meta-analysis. *BMJ, 341*, c3666.

Deshaies, R., & Joazeiro, C. (2009). RING domain E3 ubiquitin ligases. *Annual Review of Biochemistry, 78*, 399–434.

Desikan, R. S., Segonne, F., Fischl, B., Quinn, B. T., Dickerson, B. C., Blacker, D., … Killiany, R. J. (2006). An automated labeling system for subdividing the human cerebral cortex on MRI scans into gyral based regions of interest. *NeuroImage, 31*, 968–980.

Doolittle, M., Ben-Zeev, O., Bassilian, S., Whitelegge, J., Péterfy, M., & Wong, H. (2009). Hepatic lipase maturation: a partial proteome of interacting factors. *Journal of Lipid Research, 50*, 1173–1184.

Dowlati, Y., Herrmann, N., Swardfager, W., Liu, H., Sham, L., Reim, E. K., & Lanctot, K. L. (2010). A meta-analysis of cytokines in major depression. *Biological Psychiatry, 67*, 446–457.

Drevets, W. C., Price, J. L., Simpson, J. R., Jr., Todd, R. D., Reich, T., Vannier, M., & Raichle, M. E. (1997). Subgenual prefrontal cortex abnormalities in mood disorders. *Nature, 386*, 824–827.

Eaton, W. W., Anthony, J. C., Gallo, J., Cai, G., Tien, A., Romanoski, A., … Chen, L. S. (1997). Natural history of diagnostic interview schedule/DSM-IV major depression. The Baltimore epidemiologic catchment area follow-up. *Archives of General Psychiatry, 54*, 993–999.

Edwards, A. C., & Kendler, K. S. (2012). A twin study of depression and nicotine dependence: shared liability or causal relationship? *Journal of Affective Disorders, 142*, 90–97.

Edwards, A. C., Maes, H. H., Pedersen, N. L., & Kendler, K. S. (2011). A population-based twin study of the genetic and environmental relationship of major depression, regular tobacco use and nicotine dependence. *Psychological Medicine, 41*, 395–405.

Ertekin-Taner, N. (2011). Gene expression endophenotypes: a novel approach for gene discovery in Alzheimer's disease. *Molecular Neurodegeneration, 6*, 31.

Eysenck, H. J., & Eysenck, S. B. G. (1975). *Manual of the eysenck personality questionnaire*. San Diego, CA: Educational and Industrial Testing Service.

Falconer, D. S., & Mackay, T. F. C. (1996). *Introduction to quantitative genetics* (4th ed.). Harlow, England; New York: Prentice Hall.

Fennig, S., Schwartz, J. E., & Bromet, E. J. (1994). Are diagnostic criteria, time of episode and occupational impairment important determinants of the female:male ratio for major depression? *Journal of Affective Disorders, 30*, 147–154.

Flint, J., & Kendler, K. S. (2014). The genetics of major depression. *Neuron, 81*, 484–503.

Fried, E. I., Nesse, R. M., Zivin, K., Guille, C., & Sen, S. (2013). Depression is more than the sum score of its parts: individual DSM symptoms have different risk factors. *Psychological Medicine*, 1–10.

Glahn, D. C., Curran, J. E., Winkler, A. M., Carless, M. A., Kent, J. W., Jr., Charlesworth, J. C., … Blangero, J. (2012). High dimensional endophenotype ranking in the search for major depression risk genes. *Biological Psychiatry, 71*, 6–14.

Glahn, D. C., Knowles, E. E., McKay, D. R., Sprooten, E., Raventos, H., Blangero, J., … Almasy, L. (2014). Arguments for the sake of endophenotypes: examining common misconceptions about the use of endophenotypes in psychiatric genetics. *American Journal of Medical Genetics Part B: Neuropsychiatric Genetics, 165*, 122–130.

Glahn, D., Laird, A., Ellison-Wright, I., Thelen, S., Robinson, J., Lancaster, J., … Fox, P. (2008). Meta-analysis of gray matter anomalies in schizophrenia: application of anatomic likelihood estimation and network analysis. *Biological Psychiatry, 64*, 774–781.

Goldapple, K., Segal, Z., Garson, C., Lau, M., Bieling, P., Kennedy, S., & Mayberg, H. (2004). Modulation of cortical-limbic pathways in major depression: treatment-specific effects of cognitive behavior therapy. *Archives of General Psychiatry, 61*, 34–41.

Gonzalez, H. M., Tarraf, W., Whitfield, K. E., & Vega, W. A. (2010). The epidemiology of major depression and ethnicity in the United States. *Journal of Psychiatric Research, 44*, 1043−1051.

Göring, H. H., Curran, J. E., Johnson, M. P., Dyer, T. D., Charlesworth, J., Cole, S. A., ... Blangero, J. (2007). Discovery of expression QTLs using large-scale transcriptional profiling in human lymphocytes. *Nature Genetics, 39*, 1208−1216.

Gottesman, I. I., & Gould, T. D. (2003). The endophenotype concept in psychiatry: etymology and strategic intentions. *American Journal of Psychiatry, 160*, 636−645.

Grant, B. F., Hasin, D. S., Chou, S. P., Stinson, F. S., & Dawson, D. A. (2004). Nicotine dependence and psychiatric disorders in the United States: results from the national epidemiologic survey on alcohol and related conditions. *Archives of General Psychiatry, 61*, 1107−1115.

Grant, B. F., Stinson, F. S., Dawson, D. A., Chou, S. P., Dufour, M. C., Compton, W., ... Kaplan, K. (2004). Prevalence and co-occurrence of substance use disorders and independent mood and anxiety disorders: results from the National Epidemiologic Survey on Alcohol and Related Conditions. *Archives of General Psychiatry, 61*, 807−816.

Grant, B. F., Stinson, F. S., Hasin, D. S., Dawson, D. A., Chou, S. P., Ruan, W. J., & Huang, B. (2005). Prevalence, correlates, and comorbidity of bipolar I disorder and axis I and II disorders: results from the National Epidemiologic Survey on Alcohol and Related Conditions. *Journal of Clinical Psychiatry, 66*, 1205−1215.

Greenberg, P. E., Kessler, R. C., Birnbaum, H. G., Leong, S. A., Lowe, S. W., Berglund, P. A., & Corey-Lisle, P. K. (2003). The economic burden of depression in the United States: how did it change between 1990 and 2000? *Journal of Clinical Psychiatry, 64*, 1465−1475.

Hamani, C., Mayberg, H., Stone, S., Laxton, A., Haber, S., & Lozano, A. M. (2011). The subcallosal cingulate gyrus in the context of major depression. *Biological Psychiatry, 69*, 301−308.

Hamilton, J. P., Chen, G., Thomason, M. E., Schwartz, M. E., & Gotlib, I. H. (2011). Investigating neural primacy in Major Depressive Disorder: multivariate Granger causality analysis of resting-state fMRI time-series data. *Molecular Psychiatry, 16*, 763−772.

Hamilton, J. P., Etkin, A., Furman, D. J., Lemus, M. G., Johnson, R. F., & Gotlib, I. H. (2012). Functional neuroimaging of major depressive disorder: a meta-analysis and new integration of base line activation and neural response data. *American Journal of Psychiatry, 169*, 693−703.

Harmer, C. J. (2008). Serotonin and emotional processing: does it help explain antidepressant drug action? *Neuropharmacology, 55*, 1023−1028.

Hasin, D., Stinson, F., Ogburn, E., & Grant, B. (2007). Prevalence, correlates, disability, and comorbidity of DSM-IV alcohol abuse and dependence in the United States: results from the National Epidemiologic Survey on Alcohol and Related Conditions. *Archives of General Psychiatry, 64*, 830−842.

Hasler, G., Drevets, W. C., Manji, H. K., & Charney, D. S. (2004). Discovering endophenotypes for major depression. *Neuropsychopharmacology, 29*, 1765−1781.

Hays, R. D., Wells, K. B., Sherbourne, C. D., Rogers, W., & Spritzer, K. (1995). Functioning and well-being outcomes of patients with depression compared with chronic general medical illnesses. *Archives of General Psychiatry, 52*, 11−19.

Hirschfeld, R. M., & Weissman, M. M. (2002). Risk factors for major depression and bipolar disorder. In K. L. Davis, D. Charney, J. T. Coyle, & C. Nemeroff (Eds.), *Neuropsychopharmacology: The fifth generation of progress*. American College of Neuropsychopharmacology.

Holmans, P., Weissman, M., Zubenko, G., Scheftner, W., Crowe, R., Depaulo, J. J., ... Levinson, D. (2007). Genetics of recurrent early-onset major depression (GenRED): final genome scan report. *American Journal of Psychiatry, 164*, 248−258.

Holmans, P., Zubenko, G. S., Crowe, R. R., DePaulo, J. R., Jr., Scheftner, W. A., Weissman, M. M., ... Levinson, D. F. (2004). Genomewide significant linkage to recurrent, early-onset major depressive disorder on chromosome 15q. *American Journal of Human Genetics, 74*, 1154−1167.

Hovey, J., & King, C. (1996). Acculturative stress, depression, and suicidal ideation among immigrant and second-generation Latino adolescents. *Journal of the American Academy of Child & Adolescent Psychiatry, 35*, 1183−1192.

Iossifov, I., O'Roak, B. J., Sanders, S. J., Ronemus, M., Krumm, N., Levy, D., ... Wigler, M. (2014). The contribution of de novo coding mutations to autism spectrum disorder. *Nature, 515*, 216−221.

Jamal, M., Willem Van der Does, A. J., Cuijpers, P., & Penninx, B. W. (2012). Association of smoking and nicotine dependence with severity and course of symptoms in patients with depressive or anxiety disorder. *Drug and Alcohol Dependence, 126*, 138−146.

Judd, L. L. (1997). The clinical course of unipolar major depressive disorders. *Archives of General Psychiatry, 54*, 989−991.

Judd, L. L., & Akiskal, H. S. (2000). Delineating the longitudinal structure of depressive illness: beyond clinical subtypes and duration thresholds. *Pharmacopsychiatry, 33*, 3−7.

Katon, W. J. (2003). Clinical and health services relationships between major depression, depressive symptoms, and general medical illness. *Biological Psychiatry, 54*, 216−226.

Kempton, M. J., Salvador, Z., Munafo, M. R., Geddes, J. R., Simmons, A., Frangou, S., & Williams, S. C. (2011). Structural neuroimaging studies in major depressive disorder. Meta-analysis and comparison with bipolar disorder. *Archives of General Psychiatry, 68*, 675−690.

Kendler, K. S. (1993). The lifetime history of major depression in women. *Archives of General Psychiatry, 50*, 863.

Kendler, K., Aggen, S., Tambs, K., & Reichborn-Kjennerud, T. (2006). Illicit psychoactive substance use, abuse and dependence in a population-based sample of Norwegian twins. *Psychological Medicine, 36*, 955−962.

Kendler, K. S., Eaves, L. J., Walters, E. E., Neale, M. C., Heath, A. C., & Kessler, R. C. (1996). The identification and validation of distinct depressive syndromes in a population-based sample of female twins. *Archives of General Psychiatry, 53*, 391−399.

Kendler, K. S., Gardner, C. O., & Prescott, C. A. (2001). Are there sex differences in the reliability of a lifetime history of major depression and its predictors? *Psychological Medicine, 31*, 617−625.

Kendler, K. S., Gatz, M., Gardner, C. O., & Pedersen, N. L. (2006). A swedish national twin study of lifetime major depression. *American Journal of Psychiatry, 163*, 109−114.

Kendler, K. S., Karkowski, L. M., & Prescott, C. A. (1999). Causal relationship between stressful life events and the onset of major depression. *American Journal of Psychiatry, 156*, 837–841.

Kendler, K. S., Neale, M. C., Kessler, R. C., Heath, A. C., & Eaves, L. J. (1993). The lifetime history of major depression in women. Reliability of diagnosis and heritability. *Archives of General Psychiatry, 50*, 863–870.

Kessler, R., Berglund, P., Demler, O., Jin, R., Koretz, D., Merikangas, K., ... Wang, P. (2003). The epidemiology of major depressive disorder: results from the National Comorbidity Survey Replication (NCS-R). *JAMA, 289*, 3095–3105.

Kessler, R., Berglund, P., Demler, O., Jin, R., Merikangas, K., & Walters, E. (2005). Lifetime prevalence and age-of-onset distributions of DSM-IV disorders in the National Comorbidity Survey Replication. *Archives of General Psychiatry, 62*, 593–602.

Kessler, R. C., McGonagle, K. A., Swartz, M., Blazer, D. G., & Nelson, C. B. (1993). Sex and depression in the National Comorbidity Survey. I: lifetime prevalence, chronicity and recurrence. *Journal of Affective Disorders, 29*, 85–96.

Kessler, R. C., McGonagle, K. A., Zhao, S., Nelson, C. B., Hughes, M., Eshleman, S., ... Kendler, K. S. (1994). Lifetime and 12-month prevalence of DSM-III-R psychiatric disorders in the United States. Results from the National Comorbidity Survey. *Archives of General Psychiatry, 51*, 8–19.

Khantzian, E. J. (1997). The self-medication hypothesis of substance use disorders: a reconsideration and recent applications. *Harvard Review of Psychiatry, 4*, 231–244.

Kleinman, A. (2004). Culture and depression. *The New England Journal of Medicine, 351*, 951–953.

Kochunov, P., Glahn, D., Winkler, A., Duggirala, R., Olvera, R., Cole, S., ... Blangero, J. (2009). Analysis of genetic variability and whole genome linkage of whole-brain, subcortical, and ependymal hyperintense white matter volume. *Stroke, 40*, 3685–3690.

Koechlin, E., & Summerfield, C. (2007). An information theoretical approach to prefrontal executive function. *Trends in Cognitive Sciences, 11*, 229–235.

Kohli, M. A., Lucae, S., Saemann, P. G., Schmidt, M. V., Demirkan, A., Hek, K., ... Binder, E. B. (2011). The neuronal transporter gene SLC6A15 confers risk to major depression. *Neuron, 70*, 252–265.

Kraemer, H. C., Kupfer, D. J., Clarke, D. E., Narrow, W. E., & Regier, D. A. (2012). DSM-5: how reliable is reliable enough? *American Journal of Psychiatry, 169*, 13–15.

Kriegsman, D. M., Penninx, B. W., van Eijk, J. T., Boeke, A. J., & Deeg, D. J. (1996). Self-reports and general practitioner information on the presence of chronic diseases in community dwelling elderly. A study on the accuracy of patients' self-reports and on determinants of inaccuracy. *Journal of Clinical Epidemiology, 49*, 1407–1417.

Krishnan, V., & Nestler, E. J. (2008). The molecular neurobiology of depression. *Nature, 455*, 894–902.

Levinson, D. (2006). The genetics of depression: a review. *Biological Psychiatry, 60*, 84–92.

Levinson, D., Evgrafov, O., Knowles, J., Potash, J., Weissman, M., Scheftner, W., ... Holmans, P. (2007). Genetics of recurrent early-onset major depression (GenRED): significant linkage on chromosome 15q25-q26 after fine mapping with single nucleotide polymorphism markers. *American Journal of Psychiatry, 164*, 259–264.

Levinson, D., Zubenko, G., Crowe, R., DePaulo, R., Scheftner, W., Weissman, M., ... Chellis, J. (2003). Genetics of recurrent early-onset depression (GenRED): design and preliminary clinical characteristics of a repository sample for genetic linkage studies. *American Journal of Medical Genetics Part B: Neuropsychiatric Genetics, 119B*, 118–130.

Lorant, V., Croux, C., Weich, S., Deliege, D., Mackenbach, J., & Ansseau, M. (2007). Depression and socio-economic risk factors: 7-year longitudinal population study. *British Journal of Psychiatry, 190*, 293–298.

Lorant, V., Deliege, D., Eaton, W., Robert, A., Philippot, P., & Ansseau, M. (2003). Socioeconomic inequalities in depression: a meta-analysis. *American Journal of Epidemiology, 157*, 98–112.

Lyness, J. M., Caine, E. D., King, D. A., Conwell, Y., Cox, C., & Duberstein, P. R. (1999). Cerebrovascular risk factors and depression in older primary care patients: testing a vascular brain disease model of depression. *The American Journal of Geriatric Psychiatry, 7*, 252–258.

Lynskey, M. T., Glowinski, A. L., Todorov, A. A., Bucholz, K. K., Madden, P. A. F., Nelson, E. C., ... Heath, A. C. (2004). Major depressive disorder, suicidal ideation, and suicide attempt in twins discordant for cannabis dependence and early-onset cannabis use. *Archives of General Psychiatry, 61*, 1026–1032.

Lyons, M., Hitsman, B., Xian, H., Panizzon, M. S., Jerskey, B. A., Santangelo, S., ... Tsuang, M. T. (2008). A twin study of smoking, nicotine dependence, and major depression in men. *Nicotine & Tobacco Research: Official Journal of the Society for Research on Nicotine and Tobacco, 10*, 97–108.

Ma, Y. (2015). Neuropsychological mechanism underlying antidepressant effect: a systematic meta-analysis. *Molecular Psychiatry, 20*, 311–319.

Maher, B. S., Marazita, M. L., Zubenko, W. N., Kaplan, B. B., & Zubenko, G. S. (2002). Genetic segregation analysis of alcohol and other substance-use disorders in families with recurrent. *Early-Onset Major Depression, 28*(4), 711–731.

Major Depressive Disorder Working Group of the Psychiatric GC, Ripke, S., Wray, N. R., Lewis, C. M., Hamilton, S. P., Weissman, M. M., ... Sullivan, P. F. (2013). A mega-analysis of genome-wide association studies for major depressive disorder. *Molecular Psychiatry, 18*, 497–511.

Markou, A., Kosten, T. R., & Koob, G. F. (1998). Neurobiological similarities in depression and drug dependence: a self-medication hypothesis. *Neuropsychopharmacology, 18*, 135–174.

Mayberg, H. S. (1997). Limbic-cortical dysregulation: a proposed model of depression. *Journal of Neuropsychiatry & Clinical Neurosciences, 9*, 471–481.

Mayberg, H. S., Brannan, S. K., Mahurin, R. K., Jerabek, P. A., Brickman, J. S., Tekell, J. L., ... Fox, P. T. (1997). Cingulate function in depression: a potential predictor of treatment response. *Neuroreport, 8*, 1057–1061.

Mayberg, H. S., Liotti, M., Brannan, S. K., McGinnis, S., Mahurin, R. K., Jerabek, P. A., … Fox, P. T. (1999). Reciprocal limbic-cortical function and negative mood: converging PET findings in depression and normal sadness. *American Journal of Psychiatry, 156*, 675–682.

Mayberg, H. S., Lozano, A. M., Voon, V., McNeely, H. E., Seminowicz, D., Hamani, C., … Kennedy, S. H. (2005). Deep brain stimulation for treatment-resistant depression. *Neuron, 45*, 651–660.

McGuffin, P., Katz, R., Watkins, S., & Rutherford, J. (1996). A hospital-based twin register of the heritability of DSM-IV unipolar depression. *Archives of General Psychiatry, 53*, 129–136.

Merikangas, K. R., Mehta, R. L., Molnar, B. E., Walters, E. E., Swendsen, J. D., Aguilar-Gaziola, S., … Kessler, R. C. (1998). Comorbidity of substance use disorders with mood and anxiety disorders: results of the International Consortium in Psychiatric Epidemiology. *Addictive Behaviors, 23*, 893–907.

Merikangas, K. R., Wicki, W., & Angst, J. (1994). Heterogeneity of depression. Classification of depressive subtypes by longitudinal course. *British Journal of Psychiatry, 164*, 342–348.

Middeldorp, C. M., Hottenga, J. J., Slagboom, P. E., Sullivan, P. F., de Geus, E. J., Posthuma, D., … Boomsma, D. I. (2008). Linkage on chromosome 14 in a genome-wide linkage study of a broad anxiety phenotype. *Molecular Psychiatry, 13*, 84–89.

Miller, E. K., & Cohen, J. D. (2001). An integrative theory of prefrontal cortex function. *Annual Review of Neuroscience, 24*, 167–202.

Morgan, D. (2003). Mindfulness-based cognitive therapy for depression: a new approach to preventing relapse. *Psychotherapy Research, 13*, 123–125.

Morishita, T., Fayad, S. M., Higuchi, M. A., Nestor, K. A., & Foote, K. D. (2014). Deep brain stimulation for treatment-resistant depression: systematic review of clinical outcomes. *Neurotherapeutics, 11*, 475–484.

Murray, C. J. L., & Lopez, A. D. (1996). *The global burden of disease: A comprehensive assessment of mortality, injuries, and risk factors in 1990 and projected to 2020.* Boston: Harvard School of Public Health and the World Health Organisation.

Murray, C. J., Vos, T., Lozano, R., Naghavi, M., Flaxman, A. D., Michaud, C., … Memish, Z. A. (2012). Disability-adjusted life years (DALYs) for 291 diseases and injuries in 21 regions, 1990–2010: a systematic analysis for the Global Burden of Disease Study 2010. *Lancet, 380*, 2197–2223.

Musselman, D. L., Evans, D. L., & Nemeroff, C. B. (1998). The relationship of depression to cardiovascular disease: epidemiology, biology, and treatment. *Archives of General Psychiatry, 55*, 580–592.

Nelson, J. C. (2003). Managing treatment-resistant major depression. *Journal of Clinical Psychiatry, 64*(Suppl. 1), 5–12.

Nestler, E., Barrot, M., DiLeone, R., Eisch, A., Gold, S., & Monteggia, L. (2002). Neurobiology of depression. *Neuron, 34*, 13–25.

Nolen-Hoeksema, S. (1987). Sex differences in unipolar depression: evidence and theory. *Psychological Bulletin, 101*, 259–282.

Olvera, R. L., Bearden, C. E., Velligan, D. I., Almasy, L., Carless, M. A., Curran, J. E., … Glahn, D. C. (2011). Common genetic influences on depression, alcohol, and substance use disorders in Mexican–American families. *American Journal of Medical Genetics Part B: Neuropsychiatric Genetics, 156B*(5), 561–568.

Ongur, D., Drevets, W. C., & Price, J. L. (1998). Glial reduction in the subgenual prefrontal cortex in mood disorders. *Proceedings of the National Academy of Sciences of the United States of America, 95*, 13290–13295.

Ongur, D., & Price, J. L. (2000). The organization of networks within the orbital and medial prefrontal cortex of rats, monkeys and humans. *Cerebral Cortex, 10*, 206–219.

Pariante, C. M., & Lightman, S. L. (2008). The HPA axis in major depression: classical theories and new developments. *Trends in Neurosciences, 31*, 464–468.

Pariante, C. M., & Miller, A. H. (2001). Glucocorticoid receptors in major depression: relevance to pathophysiology and treatment. *Biological Psychiatry, 49*, 391–404.

Paus, T. (2001). Primate anterior cingulate cortex: where motor control, drive and cognition interface. *Nature Reviews Neuroscience, 2*, 417–424.

Pessoa, L., & Adolphs, R. (2010). Emotion processing and the amygdala: from a 'low road' to 'many roads' of evaluating biological significance. *Nature Reviews Neuroscience, 11*, 773–783.

Prescott, C., Aggen, S., & Kendler, K. (2000). Sex-specific genetic influences on the comorbidity of alcoholism and major depression in a population-based sample of US twins. *Archives of General Psychiatry, 57*, 803–811.

Prigerson, H. G., Horowitz, M. J., Jacobs, S. C., Parkes, C. M., Aslan, M., Goodkin, K., … Maciejewski, P. K. (2009). Prolonged grief disorder: psychometric validation of criteria proposed for DSM-V and ICD-11. *PLoS Medicine, 6*, e1000121.

Prusoff, B. A., Weissman, M. M., Merikangas, K. R., Leckman, J. F., & Harding, P. S. (1984). Psychiatric illness in relatives of probands with delusional depression. *Psychopharmacology Bulletin, 20*, 358–361.

Radich, J. P., Mao, M., Stepaniants, S., Biery, M., Castle, J., Ward, T., … Linsley, P. S. (2004). Individual-specific variation of gene expression in peripheral blood leukocytes. *Genomics, 83*, 980–988.

Regier, D. A., Narrow, W. E., Clarke, D. E., Kraemer, H. C., Kuramoto, S. J., Kuhl, E. A., & Kupfer, D. J. (2013). DSM-5 field trials in the United States and Canada, Part II: test-retest reliability of selected categorical diagnoses. *American Journal of Psychiatry, 170*, 59–70.

Riolo, S. A., Nguyen, T. A., Greden, J. F., & King, C. A. (2005). Prevalence of depression by race/ethnicity: findings from the National Health and Nutrition Examination Survey III. *American Journal of Public Health, 95*, 998–1000.

Risch, N., Herrell, R., Lehner, T., Liang, K. Y., Eaves, L., Hoh, J., … Merikangas, K. R. (2009). Interaction between the serotonin transporter gene (5-HTTLPR), stressful life events, and risk of depression: a meta-analysis. *JAMA, 301*, 2462–2471.

Roberts, S. B., MacLean, C. J., Neale, M. C., Eaves, L. J., & Kendler, K. S. (1999). Replication of linkage studies of complex traits: an examination of variation in location estimates. *American Journal of Human Genetics, 65*, 876–884.

Rockman, M. V., & Wray, G. A. (2002). Abundant raw material for cis-regulatory evolution in humans. *Molecular Biology and Evolution, 19*, 1991–2004.

Sapolsky, R. M. (1990). Glucocorticoids, hippocampal damage and the glutamatergic synapse. *Progress in Brain Research, 86*, 13–23.

Schizophrenia Working Group of the Psychiatric Genomics C. (2014). Biological insights from 108 schizophrenia-associated genetic loci. *Nature, 511*, 421–427.

Schlesser, M. A., Winokur, G., & Sherman, B. M. (1980). Hypothalamic-pituitary-adrenal axis activity in depressive illness. Its relationship to classification. *Archives of General Psychiatry, 37*, 737–743.

Schmaal, L., Veltman, D. J., van Erp, T. G., Samann, P. G., Frodl, T., Jahanshad, N., … Hibar, D. P. (2015). Subcortical brain alterations in major depressive disorder: findings from the ENIGMA Major Depressive Disorder working group. *Molecular Psychiatry* [Epub ahead of print].

Schol-Gelok, S., Janssens, A. C., Tiemeier, H., Liu, F., Lopez-Leon, S., Zorkoltseva, I. V., … van Duijn, C. M. (2010). A genome-wide screen for depression in two independent Dutch populations. *Biological Psychiatry, 68*, 187–196.

Seeley, W., Menon, V., Schatzberg, A., Keller, J., Glover, G., Kenna, H., … Greicius, M. (2007). Dissociable intrinsic connectivity networks for salience processing and executive control. *Journal of Neuroscience, 27*, 2349–2356.

Shi, J., Potash, J., Knowles, J., Weissman, M., Coryell, W., Scheftner, W., … Levinson, D. (2010). Genome-wide association study of recurrent early-onset major depressive disorder. *Molecular Psychiatry, 16*(2), 193–201.

Siegle, G. J., Thompson, W., Carter, C. S., Steinhauer, S. R., & Thase, M. E. (2007). Increased amygdala and decreased dorsolateral prefrontal BOLD responses in unipolar depression: related and independent features. *Biological Psychiatry, 61*, 198–209.

Steffens, D. C., Helms, M. J., Krishnan, K. R., & Burke, G. L. (1999). Cerebrovascular disease and depression symptoms in the cardiovascular health study. *Stroke, 30*, 2159–2166.

Sullivan, P. F., Neale, M. C., & Kendler, K. S. (2000). Genetic epidemiology of major depression: review and meta-analysis. *American Journal of Psychiatry, 157*, 1552–1562.

Sweet, R. A., Hamilton, R. L., Butters, M. A., Mulsant, B. H., Pollock, B. G., Lewis, D. A., … Reynolds, C. F., 3rd (2004). Neuropathologic correlates of late-onset major depression. *Neuropsychopharmacology, 29*, 2242–2250.

Treisman, A. (1999). Solutions to the binding problem: progress through controversy and convergence. *Neuron, 24*, 105–110, 111–125.

Tsuang, M. T., Bar, J. L., Harley, R. M., & Lyons, M. J. (2001). The Harvard twin study of substance abuse: what we have learned. *Harvard Review of Psychiatry, 9*, 267–279.

Vink, J. M., Hottenga, J. J., de Geus, E. J., Willemsen, G., Neale, M. C., Furberg, H., & Boomsma, D. I. (2014). Polygenic risk scores for smoking: predictors for alcohol and cannabis use? *Addiction, 109*, 1141–1151.

Vreeburg, S. A., Hoogendijk, W. J., van Pelt, J., Derijk, R. H., Verhagen, J. C., van Dyck, R., … Penninx, B. W. (2009). Major depressive disorder and hypothalamic-pituitary-adrenal axis activity: results from a large cohort study. *Archives of General Psychiatry, 66*, 617–626.

Walss-Bass, C., Escamilla, M., Raventos, H., Montero, A., Armas, R., Dassori, A., … Almasy, L. (2005). Evidence of genetic overlap of schizophrenia and bipolar disorder: linkage disequilibrium analysis of chromosome 18 in the Costa Rican population. *American Journal of Medical Genetics Part B: Neuropsychiatric Genetics, 139B*, 54–60.

Ward, R., Danziger, S., Owen, V., & Rafal, R. (2002). Deficits in spatial coding and feature binding following damage to spatiotopic maps in the human pulvinar. *Nature Neuroscience, 5*, 99–100.

Weissman, M. M., Bland, R. C., Canino, G. J., Faravelli, C., Greenwald, S., Hwu, H. G., … Yeh, E. K. (1996). Cross-national epidemiology of major depression and bipolar disorder. *JAMA, 276*, 293–299.

Weissman, M. M., Gershon, E. S., Kidd, K. K., Prusoff, B. A., Leckman, J. F., Dibble, E., … Guroff, J. J. (1984). Psychiatric disorders in the relatives of probands with affective disorders. The Yale University–National Institute of Mental Health Collaborative Study. *Archives of General Psychiatry, 41*(1), 13–21. PMID:6691780.

Weissman, M. M., Merikangas, K. R., Wickramaratne, P., Kidd, K. K., Prusoff, B. A., Leckman, J. F., & Pauls, D. L. (1986). Understanding the clinical heterogeneity of major depression using family data. *Archives of General Psychiatry, 43*, 430–434.

Wells, J. E., & Horwood, L. J. (2004). How accurate is recall of key symptoms of depression? A comparison of recall and longitudinal reports. *Psychological Medicine, 34*, 1001–1011.

Wells, K. B., Stewart, A., Hays, R. D., Burnam, M. A., Rogers, W., Daniels, M., … Ware, J. (1989). The functioning and well-being of depressed patients. Results from the Medical Outcomes Study. *JAMA, 262*, 914–919.

Whitney, A. R., Diehn, M., Popper, S. J., Alizadeh, A. A., Boldrick, J. C., Relman, D. A., & Brown, P. O. (2003). Individuality and variation in gene expression patterns in human blood. *Proceedings of the National Academy of Sciences of the United States of America, 100*, 1896–1901.

Williams, D. R., Gonzalez, H. M., Neighbors, H., Nesse, R., Abelson, J. M., Sweetman, J., & Jackson, J. S. (2007). Prevalence and distribution of major depressive disorder in African Americans, Caribbean blacks, and non-Hispanic whites: results from the National Survey of American Life. *Archives of General Psychiatry, 64*, 305–315.

Wolkowitz, O. M., Reus, V. I., Keebler, A., Nelson, N., Friedland, M., Brizendine, L., & Roberts, E. (1999). Double-blind treatment of major depression with dehydroepiandrosterone. *American Journal of Psychiatry, 156*, 646–649.

Wray, G. A. (2007). The evolutionary significance of cis-regulatory mutations. *Nature Reviews Genetics, 8*, 206–216.

Zarate, C. A., Jr., Singh, J. B., Carlson, P. J., Brutsche, N. E., Ameli, R., Luckenbaugh, D. A., … Manji, H. K. (2006). A randomized trial of an *N*-methyl-D-aspartate antagonist in treatment-resistant major depression. *Archives of General Psychiatry, 63*, 856–864.

Zimmermann, P., Bruckl, T., Nocon, A., Pfister, H., Lieb, R., Wittchen, H. U., … Angst, J. (2009). Heterogeneity of DSM-IV major depressive disorder as a consequence of subthreshold bipolarity. *Archives of General Psychiatry, 66*, 1341–1352.

Zou, F., Carrasquillo, M. M., Pankratz, V. S., Belbin, O., Morgan, K., Allen, M., … Ertekin-Taner, N. (2010). Gene expression levels as endophenotypes in genome-wide association studies of Alzheimer disease. *Neurology, 74*, 480–486.

Zubenko, G. S., Maher, B., Hughes, H. B., 3rd, Zubenko, W. N., Stiffler, J. S., Kaplan, B. B., & Marazita, M. L. (2003). Genome-wide linkage survey for genetic loci that influence the development of depressive disorders in families with recurrent, early-onset, major depression. *American Journal of Medical Genetics Part B: Neuropsychiatric Genetics, 123B*, 1–18.

Chapter 32

Speech and Language Disorders

Edoardo G. Spinelli
University of California San Francisco, San Francisco, CA, United States; Vita-Salute San Raffaele University, Milan, Italy

Maria L.G. Tempini and Kevin A. Shapiro
University of California San Francisco, San Francisco, CA, United States

INTRODUCTION

Until recently, almost everything that was known about how the brain produces language was deduced from the careful study of language impairments after focal cerebral lesions, mostly as the result of stroke or traumatic injury, but also in some cases as the result of tumors or infection. The origin of the modern concept of the neuroanatomy of language is generally traced to Broca's (1861) famous report of a patient with a profound impairment in speech production but apparently spared language comprehension, attributed to a chronic lesion in the left inferior frontal lobe (Broca, 1861; Dronkers, Plaisant, Iba-Zizen, & Cabanis, 2007). (Similar, if less publicized, observations had in fact been made decades earlier by Bouillaud and Dax (Benton, 1984; Finger, 2010; Manning & Thomas-Anterion, 2011).) Over the next century and a half, various models were proposed for a "language network," centered primarily on the Sylvian fissure of the left hemisphere, with centers or areas defined by the deficits observed in patients with different patterns of acquired injury (Graves, 1997).

Only in the past few decades has our understanding of the biology of language begun to evolve beyond the focal lesion-based model of classical aphasiology. One of the most important developments in this regard has been the increasing sophistication of noninvasive tools to map functional and anatomical language networks in vivo, in both in patients and healthy control subjects. These include diffusion tensor imaging (DTI), an MRI technique which has enabled the identification and quantification of major white matter tracts in the living brain, and functional MRI (fMRI), which allows us to visualize changes in the distribution of oxygenated blood as a surrogate for neuroglial metabolism and thus presumably for neural activation.

In parallel with the emergence of new neuroimaging modalities, greater attention has been devoted to conditions that affect distributed components of the language network, not based on accidental correlations with cerebral blood supply but on functional connections between brain regions involved in particular aspects of language processing. These include neurodevelopmental differences that affect children's capacity to speak, write, and read, as well as neurodegenerative disorders, the primary progressive aphasias, that selectively degrade the ability to retrieve words, produce them, and understand their meanings.

In this chapter we will provide an overview of brain regions and pathways involved in language production and comprehension. With this as a framework, we will then discuss several neurodevelopmental and neurodegenerative conditions involving the language network, with particular attention to their clinical, neuroanatomical, pathological, and genetic correlates.

Genomics, Circuits, and Pathways in Clinical Neuropsychiatry. http://dx.doi.org/10.1016/B978-0-12-800105-9.00032-9

LANGUAGE NETWORKS AND PATHWAYS

A comprehensive overview of brain areas involved in the production and comprehension of language is beyond the scope of this chapter. The interested reader is referred to reviews and position articles on this topic (Dick, Bernal, & Tremblay, 2014; Dick & Tremblay, 2012; Friederici & Gierhan, 2013; Gow, 2012; Hickok & Poeppel, 2007; Poeppel & Hickok, 2004; Price, 2000, 2012). Here we will present only a brief sketch of cortical regions and white matter tracts important for speech and language, to serve as a point of reference for discussion of neuroanatomical abnormalities in neurodevelopmental and neurodegenerative disorders. We provide specific references only for observations that are relatively novel or controversial. We also assume here that the language network is left-lateralized, which is accurate for almost all right-handed individuals and most left-handed individuals.

It is conceptually useful to divide the language network into a dorsal component composed of structures and pathways arranged in frontal, parietal, and superior temporal regions along the superior margin of the left Sylvian fissure, and a ventral component encompassing much of the left lateral and anterior temporal lobe. As a rough heuristic, the ventral component is involved in the representation of word form and meaning, whereas the dorsal component is engaged in the combinatorial aspects of language, including computations that range from the conceptual (combining words and morphemes into meaningful utterances; ie, syntax and morphology) to the more concrete (planning of articulatory sequences for speech). The dorsal and ventral pathways converge in the region of the left temporoparietal junction, which is thought to be important specifically in the process of word retrieval and phonological assembly.

The Ventral Pathway

The manner in which word meaning is represented in the ventral part of the language network is controversial, and there is continuing debate about whether the organization of semantic knowledge is determined more by sensory modality (eg, semantic features encoding visual, auditory, or tactile information about concepts) or by cognitively salient cross-modal categories (features of animals, fruits and vegetables, nonliving things, etc.) (Mahon & Caramazza, 2009, 2011). In general, there seems to be a representational gradient according to which kinds of knowledge that are shared across concepts (or categories of concepts) are represented in more posterior and inferior regions (for example, knowledge about motion in the posterior middle temporal gyrus and knowledge about form in the inferior temporal gyrus and ventral temporo-occipital cortex), whereas information relating to particular concepts is represented more anteriorly. One widely cited model envisions the anterior temporal lobe as an amodal hub or multimodal convergence zone that binds together modality-specific information represented in the posterior and lateral temporal lobes to create a holistic concept (Patterson, Nestor, & Rogers, 2007).

Word form, both spoken (phonological) and written (orthographic), is also presumed to be stored and processed in the ventral pathway, and in the context of brain lesions, access to word form is often impaired in conjunction with retrieval of semantic knowledge. Like semantic knowledge, lexical form may be encoded in a hierarchical fashion, with distinct representations at various levels including syllable structure, consonant/vowel status, and segmental structure. The anatomical distribution of lower-level, modality-specific information appears to be closest topographically to primary sensory cortices: the regions of the superior temporal gyrus and sulcus are important for processing auditory-phonological information, whereas a visual word form area has been identified in the left fusiform gyrus. More abstract representations of word form seem to depend on regions within the middle temporal gyrus.

The brain regions involved in the ventral language pathway are interconnected within the temporal lobe by the inferior longitudinal fasciculus (ILF), which runs from the occipital lobe to the anterior part of the temporal lobe. There are also shorter-range connections between the anterior temporal lobe and more posterior regions, including the fusiform and middle and inferior temporal gyri (Binney, Parker, & Lambon Ralph, 2012). In addition, the inferior frontal occipital fasciculus traverses the temporal lobe to connect the occipital lobe to inferior frontal regions, and the anterior temporal lobe is connected directly to inferior and orbital frontal regions via the uncinate fasciculus (Fig. 32.1).

The Dorsal Pathway

Whereas the ventral part of the language network is critical for the representation and retrieval of word form and meaning, the dorsal component is heavily engaged in the production and comprehension of propositional language: that is, in the composition and decomposition of sequences of sounds and words in the context of a spoken or written message. Just as in the ventral pathway it is convenient to distinguish more conceptually laden levels of word representation from the representation of phoneme and grapheme structure, so in the dorsal pathway we typically consider the grammatical aspects

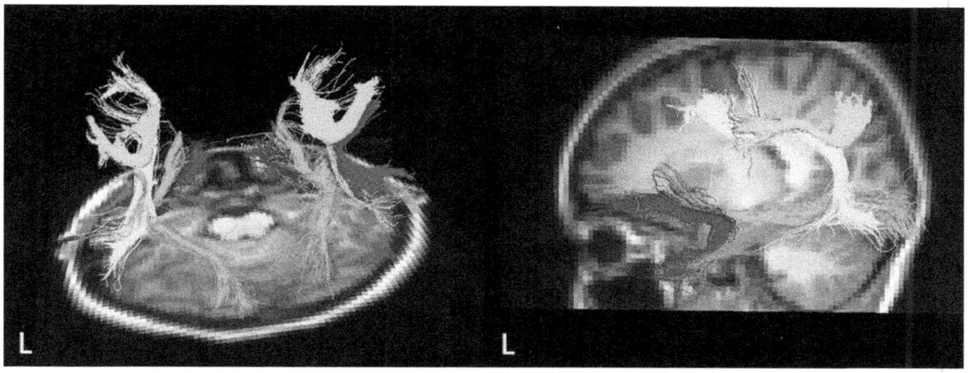

FIGURE 32.1 Major white matter tracts involved in language, reconstructed for a single participant and superimposed on an axial cut through the temporal lobes (left) and a sagittal cut at midline, showing tracts in left hemisphere (right). Yellow: arcuate fasciculus; pink: superior longitudinal fasciculus; green: superior longitudinal fasciculus (temporoparietal component); orange: inferior longitudinal fasciculus; purple: uncinate fasciculus; cyan: inferior frontal occipital fasciculus.

of language production (morphology and syntax) separately from motor speech (phonological and articulatory planning and execution) and its analogue in writing. Again, the anatomical proximity of these processes means that they are often spared or impaired in parallel, but their dissociation in some patients can be extremely informative.

The best-known and most intensively studied region within the dorsal pathway is the posterior inferior frontal cortex, roughly corresponding to Broca's area, which has a role in both motor speech and grammatical processing. Initiation of speech seems to depend on links between this region and the supplementary motor area in superior frontal cortex; dynamic speech production also engages the insula and the precentral regions representing the articulators, and the fluency and rhythm of output are regulated by frontostriatal and frontocerebellar connections. Processing of syntactic and morphological information preferentially engages Broca's area but also involves prefrontal areas anterior and superior to Broca's area, inferior parietal cortex, and frontostriatal networks (Fig. 32.2).

The major white matter tract connecting cortical regions within the dorsal pathway is the superior longitudinal fasciculus (SLF), which courses from the inferior parietal lobe to the inferior frontal lobe. The SLF has a number of constituent tracts, the most significant of which for our purposes is the arcuate, so named because it begins in the inferior frontal cortex, curves around the posterior limit of the Sylvian fissure in the region of the supramarginal and angular gyri, and terminates in the superior temporal lobe. A more recently described fiber bundle, the frontal aslant tract, connects the inferior frontal cortex to the supplementary motor area (Catani et al., 2012; Martino & De Lucas, 2014).

The area of the temporoparietal junction comprising the posterior superior temporal gyrus and angular and supramarginal gyri and the underlying arcuate fibers is the main anatomical nexus between the dorsal and ventral language

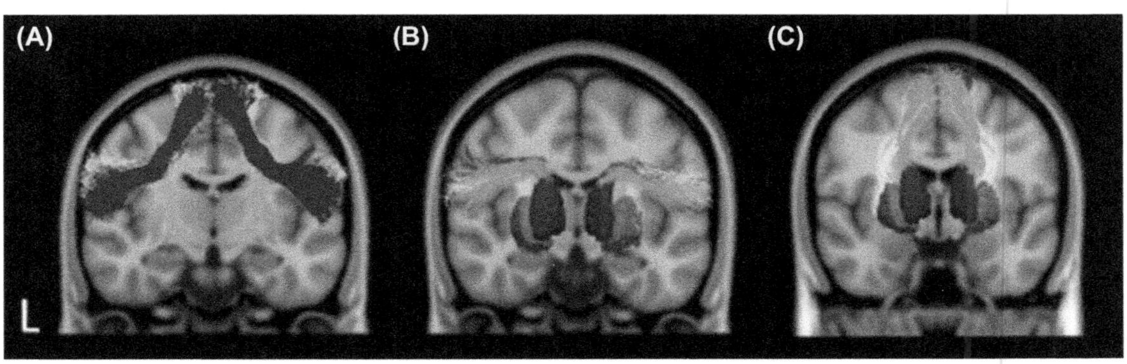

FIGURE 32.2 Three-dimensional representation of the frontocortical and frontostriatal speech production tracts reconstructed from a group of healthy control subjects ($n = 21$). (A) Tracts connecting the pre-SMA and SMA to Broca's area (blue) and the ventral premotor cortex (BA6) (green). (B) Tracts connecting the caudate to the ventral premotor cortex (BA6) (purple), the putamen to the ventral premotor cortex (BA6) (green), and the caudate to Broca's area (orange). (C) Tracts connecting the pre-SMA and SMA to the caudate (orange) and the putamen (light blue). *BA*, Brodmann area; *SMA*, supplementary motor area. *Modified from Mandelli, M. L., Caverzasi, E., Binney, R. J., Henry, M. L., Lobach, I., Block, N., ... Gorno-Tempini, M. L. (2014). Frontal white matter tracts sustaining speech production in primary progressive aphasia.* Journal of Neuroscience, *34(29), 9754–9767.*

pathways, and is generally thought to be crucial for the retrieval of words to be understood or produced (including repetition and copying). There are, however, other direct connections between the dorsal and ventral pathways, including the uncinate, an inferior frontal occipital fasciculi mentioned earlier, and a temporoparietal component of the SLF that links the posterior temporal and inferior parietal lobes (Fig. 32.1).

NEURODEVELOPMENTAL LANGUAGE DISORDERS

Approximately one in eight school-aged children is thought to have a disorder in the production or comprehension of written or spoken language (McLeod & McKinnon, 2007), and about half this number have delays in speech and language acquisition (Law, Boyle, Harris, Harkness, & Nye, 2000). Despite the frequency of these problems, and the demonstrable long-term impact of childhood language impairment on educational and psychosocial outcomes (Elbro, Dalby, & Maarbjerg, 2011), our understanding of the neurobiological and genetic bases of language disorders remains meager. Part of the reason for this is undoubtedly nosological confusion. Developmental pediatricians, child neurologists, child psychiatrist, speech-language pathologists, audiologists, and educators use a variety of overlapping and often poorly defined diagnostic terms, including "language delay," "speech delay," "specific language impairment," "speech sound disorder," "auditory processing disorder," "communication disorder," "reading disorder," and so on.

Here, we will focus on two neurodevelopmental language disorders, or more accurately, two neurodevelopmental phenotypes, which are comparatively well-characterized and are now actively being studied from the perspective of neuroscience and genetics. The first, childhood apraxia of speech, is relatively rare, accounting for no more than about 4% of children referred for evaluation of idiopathic speech delay (ASHA, 2007). However, it has the distinct scientific advantage of being linked to a known genetic pathway: the transcription factor forkhead box protein P2 (FOXP2) and its downstream target, SRPX2 (Lai, Fisher, Hurst, Vargha-Khadem, & Monaco, 2001; Roll et al., 2006, 2010). Studies of patients with genetic changes affecting *FOXP2* and *SRPX2* have thus provided a unique window into the neurobiology of the language network.

By contrast, developmental dyslexia is fairly common, occurring in around 5% of school-aged children (Lindgren, De Renzi, & Richman, 1985; Peterson & Pennington, 2012) and, by extension, a similar proportion of adults. Although the term "dyslexia" suggests a specific problem with reading, in most individuals with dyslexia the locus of impairment is at a stage of language or cognitive processing that precedes the recognition of written words. Although dyslexia is clearly hereditary, specific candidate genes remain elusive. Nevertheless, neuroimaging studies are beginning to shed light on functional and anatomical differences in the language network that distinguish dyslexic from typically developing individuals.

Childhood Apraxia of Speech

Clinical Features

Childhood apraxia of speech (CAS), also known as developmental verbal dyspraxia, is a neurological disorder of speech production in which the precision and consistency of the movements of the articulators (tongue, lips, uvula, etc.) are impaired in the absence of frank neuromuscular abnormalities, resulting in errors in the production of individual speech sounds and in the prosody of connected speech (ASHA, 2007). The modifier "childhood" is used to indicate that the speech abnormalities are usually apparent at an early age, and to distinguish this disorder from acquired apraxia of speech (AOS) resulting from stroke or neurodegenerative disease in adulthood. However, CAS is best understood as a lifelong and often familial disorder.

Although there are no universally accepted criteria for the diagnosis of CAS, several key features have been identified. These include inconsistent phonetic errors across attempts to repeat a syllable or a word: for example, pronouncing *fish* variably as "pish," "pit," "fit," and "shiff" on different repetitions (McCabe, Rosenthal, & McLeod, 1998; Seddoh et al., 1996); lengthened and disrupted transitions between sounds and syllables, giving the impression of staccato speech (Shriberg, Green, Campbell, McSweeny, & Scheer, 2003); and inappropriate prosody, especially with respect to stress, so that the listener perceives apraxic speech as uniform in tone and loudness (Davis, Jakielski, & Marquardt, 1998; Shriberg, Aram, & Kwiatkowski, 1997a,b).

Most children with apraxia of speech also have deficits in other domains, including nonspeech motor behaviors, speech perception, syntax and morphology, and reading and writing (Nijland, Terband, & Maassen, 2015). CAS is often found in conjunction with a history of generalized hypotonia and delayed motor development, as well as specific abnormalities in sensation and volitional movements of the lower face and tongue (Davis et al., 1998; McCabe et al., 1998; Shriberg et al.,

1997a,b). With regard to speech perception, children with apraxia of speech have more difficulty than do control children in discriminating between acoustically similar consonants (Groenen, Maassen, Crul, & Thoonen, 1996), vowels (Maassen, Groenen, & Crul, 2003), and meaningless sequences of sounds (Bridgeman & Snowling, 1988). They also make errors in grammatical morphology, word order, and the selection of pronouns and verbs (Ekelman & Aram, 1983). These errors are not clearly related to phonological or articulatory problems, and tend to persist despite gains in speech intelligibility (Lewis, Freebairn, Hansen, Iyengar, & Taylor, 2004).

All forms of speech delay, including CAS, are associated with difficulties in phonological awareness, the metalinguistic ability to recognize that individual words are composed of smaller segments such as syllables, vowels, and consonants (Marquardt, Sussman, Snow, & Jacks, 2002). Phonological awareness, in turn, is widely recognized as a building block of literacy (Justice & Schuele, 2004). Even compared with children with other speech disorders, however, children with CAS seem to have particular difficulty with reading and spelling (Lewis et al., 2004), for reasons which are not entirely clear but which may be related to an underlying problem in processing the syllabic framework of words (Maassen, 2002; Marquardt et al., 2002; Nijland et al., 2003).

Despite the lack of high-quality studies on treatment for CAS, there is a consensus that remediation involves specialized treatment that is much more intensive than the treatment used for children with phonological processing deficits that affect articulation (Strand, 1995). Typical treatment approaches for CAS rely on principles of motor learning, such as frequent drill to learn specific sounds, oral—motor and oral—sensory exercises (eg, using light touch and brushing of the face and articulators), and use of visual cues (such as gestural cueing and mirror work). As we will see, this emphasis on motor learning fits well with genetic and neurobiological insights into the pathophysiology of CAS.

Genetics

The symptom complex of apraxia of speech has been recognized as a component of a number of neurobehavioral disorders of clear genetic origin, including galactosemia (Potter, Nievergelt, & Shriberg, 2013; Shriberg, Potter, & Strand, 2011; Webb, Singh, Kennedy, & Elsas, 2003), fragile X syndrome (Spinelli, Rocha, Giacheti, & Richieri-Costa, 1995), and deletions or duplications at various chromosomal loci, namely 15q11-13 (Boyar et al., 2001), 16p11.2 (Raca et al., 2013), 12p13.33 (Fanizza et al., 2014; Thevenon et al., 2013), and the *BCL11A* gene at 2p15 (Peter, Matsushita, Oda, & Raskind, 2014). In our own clinical practice, we have observed severe apraxia of speech, as well as oral—buccal apraxia and limb dyspraxia, in a child with a 656-kb terminal deletion at 19p13.3 (unpublished observation).

By far the best-documented genetic cause of CAS, however, involves the *FOXP2* gene on chromosome 7q31, an association which was first identified in the KE family of London (Fisher, Vargha-Khadem, Watkins, Monaco, & Pembrey, 1998; Lai et al., 2001). Four generations of KE family members have participated in research; approximately half of them have the core features of CAS, as well as associated features of limb apraxia, orofacial apraxia, and language impairment (Alcock, Passingham, Watkins, & Vargha-Khadem, 2000a,b; Vargha-Khadem, Watkins, Alcock, Fletcher, & Passingham, 1995; Watkins, Dronkers, & Vargha-Khadem, 2002). From this pedigree, it was determined that CAS associated with *FOXP2* mutation is transmitted as an autosomal dominant, monogenetic trait. Since the initial description of the KE family, disruption of *FOXP2* has been identified in a number of other cases of apraxia of speech worldwide (Feuk et al., 2006; Lennon et al., 2007; MacDermot et al., 2005; Palka et al., 2012; Turner et al., 2013).

FOXP2 encodes a transcription factor that appears to be highly conserved in mammals and birds (Webb & Zhang, 2005), although human-specific amino acid sequence changes of potential functional significance have also been identified (Enard et al., 2002; Konopka et al., 2009). The gene is expressed in a number of structures in the developing brain, including the cortical plate, basal ganglia, thalamus, inferior olives, and cerebellum (Lai, Gerrelli, Monaco, Fisher, & Copp, 2003). In mice, mutation of *FOXP2* results in deficits in motor skill learning, accompanied by abnormal synaptic plasticity in cerebellar and motor circuits (Groszer et al., 2008). In songbirds, the expression of *FOXP2* in the basal ganglia correlates with vocal learning (Heston & White, 2015; Rochefort, He, Scotto-Lomassese, & Scharff, 2007; Teramitsu, Kudo, London, Geschwind, & White, 2004; Teramitsu, Poopatanapong, Torrisi, & White, 2010; Teramitsu & White, 2006), and knockdown of the gene in the song-learning region of the basal ganglia disrupts this process (Haesler et al., 2007; Murugan, Harward, Scharff, & Mooney, 2013).

These animal models demonstrate that *FOXP2* has an important role in the organization of frontostriatal and fronto-cerebellar circuits implicated in specific forms of motor learning, and suggest a possible neurobiological basis for the apraxic symptoms in individuals with *FOXP2* mutations. However, a motor learning deficit does not obviously account for the other cognitive and linguistic difficulties (for example, impairment in speech perception) associated with CAS, including in those with CAS who have *FOXP2* mutations.

The language deficits that co-occur with apraxia of speech may be explained, at least in part, by disruption of a larger genetic pathway involving *FOXP2*. In particular, there is increasing evidence that *FOXP2* interacts with the gene *SRPX2* at Xq22, encoding a sushi repeat protein which promotes synaptogenesis in the cerebral cortex (Roll et al., 2006, 2010; Sia, Clem, & Huganir, 2013). In mice, reduction of SPRX2 results in impaired development of ultrasonic vocalizations (Sia et al., 2013). Humans with *SRPX2* mutations present with oral and speech dyspraxia in the context of perisylvian polymicrogyria and rolandic epilepsy (Roll et al., 2006), which suggests that the protein product of this gene is particularly important for the cortical organization of the language network.

Neuroanatomy

Detailed anatomical and fMRI studies of the KE family have yielded most of our insights on the neuroanatomy of CAS. Using voxel-based morphometry, Watkins, Vargha-Khadem, et al. (2002) showed that gray matter within the caudate nucleus of the basal ganglia in particular was decreased in affected KE family members compared to unaffected members and normal control subjects, and that caudate gray matter density correlated with performance on tests of oral praxis, nonword repetition, and coding. Additional analyses demonstrated decreased gray matter density in the ventral cerebellum (lobules VIIB and VIIIB), inferior frontal gyri, precentral gyri, and temporal poles bilaterally. Interestingly, gray matter density in the posterior superior temporal gyrus and angular gyrus was increased (Belton, Salmond, Watkins, Vargha-Khadem, & Gadian, 2003). Positron emission tomography (PET) in two affected individuals showed increased cerebral blood flow (compared with unaffected individuals) during a word repetition task in the caudate nuclei and bilateral inferior frontal gyri (Vargha-Khadem et al., 1998). This pattern corresponds to the most prominent areas of reduction in gray matter density and may reflect a mechanism of functional compensation.

Functional brain activation in members of the KE family has also been studied using an fMRI paradigm in which participants were asked either to produce verbs upon hearing a noun (verb generation) or to repeat words (Liegeois et al., 2003). In contrast to the PET study described earlier, this showed a decreased task-related blood oxygen level–dependent signal in affected family members across tasks in bilateral inferior frontal regions and in the left precentral gyrus. The reason for the opposite findings using PET and fMRI is not entirely clear and is not straightforwardly attributable to differences in methodology. Possibly it may have to do with different compensatory strategies used by family members with the same underlying abnormalities in circuitry. In this regard, it would be interesting to study how performance on speech and language tasks varies as a function of brain activation in individuals with CAS.

Developmental Dyslexia

Clinical Features

Dyslexia is usually diagnosed in children who have difficulties or delays in learning to read, despite normal general intelligence and adequate peripheral visual and auditory perception. However, childhood reading difficulties are almost always only the presenting symptom of a clinically heterogeneous syndrome that may include deficits in attention and executive functioning, comprehension of spoken language, writing, and spelling, calculation, and motor coordination that persist over the lifespan. Moreover, because the specific difficulties with reading and writing depend both on educational opportunities and on the orthographic characteristics of the language of instruction (Lindgren et al., 1985; Paulesu et al., 2001), these symptoms often evade diagnosis. On the other hand, specific language impairment, typically diagnosed in younger children with speech and language delays, is highly comorbid with dyslexia (Newbury et al., 2011). In contrast to their cognitive weaknesses, dyslexic individuals may show relative strengths in nonlanguage visuospatial processing (Diehl et al., 2014).

Whereas a number of hypotheses have been proposed to explain the underlying neurocognitive differences in individuals with dyslexia (compared with typically developing individuals), it is far from clear that dyslexia is a unitary syndrome, and it is possible that different neurodevelopmental alterations can result in superficially similar clinical presentations. Ramus (2003) identified four main domains of cognitive function that have been proposed to be impaired in dyslexia: the sensorimotor domains of auditory, visual, and motor processing, and the linguistic domain of phonological processing.

A relatively consistent finding in dyslexic individuals is the presence of a phonological deficit, although whether this is the "core" deficit in the disorder or a symptom of lower-level processing issues remains a subject of debate. Dyslexic people usually have difficulties with phonological awareness (the detection and manipulation of sounds within words), verbal short-term memory, and rapid retrieval of names for pictures and overlearned stimuli (colors, numbers, and letters), or "rapid automatized naming" (Denckla & Rudel, 1974; Schatschneider, Carlson, Francis, Foorman, & Fletcher, 2002).

Some of these impairments are consistent enough that they can be used reliably to identify preliterate children at risk of developing dyslexia.

It has long been observed that many dyslexics have difficulties with auditory tasks, including word repetition and judgment of temporal order (Chiappe, Stringer, Siegel, & Stanovich, 2002; Share, Jorm, Maclean, & Matthews, 2002; Tallal, 1980), categorical perception of phonemes (Serniclaes, Sprenger-Charolles, Carre, & Demonet, 2001), and perception of frequency and intensity of sounds (Amitay, Ahissar, et al., 2002; Amitay, Ben-Yehudah, et al., 2002). Originally these data were taken to support the proposal that a fundamental deficit underlying dyslexia is in the rapid processing of auditory information (Tallal, 1980; Vandermosten et al., 2010, 2011). Subsequent studies have shown that auditory processing in people with dyslexia does not seem to be particularly sensitive to temporal variables, or that if anything, dyslexic individuals have more difficulty with slow-rate auditory processing tasks (Amitay, Ben-Yehudah, et al., 2002; Chiappe et al., 2002; Poelmans, Luts, et al., 2011; Share et al., 2002). In any case, the extent to which auditory processing problems can inform a unifying theory of dyslexia is questionable: Across various studies, only 30–60% of dyslexic individuals have impairments in auditory processing (Ramus, 2003; Tallal, 1980) and these do not correlate directly with phonological skill or reading ability (Amitay, Ahissar, et al., 2002; Law, Vandermosten, Ghesquiere, & Wouters, 2014; Ramus, Rosen, et al., 2003).

Subtle visual problems in dyslexia include eye movement and fixation anomalies (Quercia, Feiss, & Michel, 2013; Stein, 2014), excessive visual crowding (Cassim, Talcott, & Moores, 2014; Gori & Facoetti, 2015; Spinelli, De Luca, Judica, & Zoccolotti, 2002), and visuospatial attention deficits (Bogon, Finke, & Schulte-Korne, 2014; Bogon, Finke, & Stenneken, 2014; Gabay, Gabay, Schiff, Ashkenazi, & Henik, 2013; Hari, Renvall, & Tanskanen, 2001; Ruffino, Gori, Boccardi, Molteni, & Facoetti, 2014). About a third of dyslexic people show abnormal visual evoked potentials in response to a variety of stimuli (Kubova et al., 2015). As with auditory processing, these deficits have been thought to support the view that dyslexia is attributable to an impairment in rapid temporal processing of sensory information (Hari & Renvall, 2001), but on closer inspection they do not seem to be restricted to rapid tasks and are present in only a minority of people with dyslexia (Amitay, Ahissar, et al., 2002; Ramus, 2003). Likewise, deficits in motor control are found in about 30–50% of dyslexic individuals and do not seem to be causally linked to difficulties with phonological processing and reading (Ramus, 2003; Ramus, Pidgeon, & Frith, 2003; Ramus, Rosen, et al., 2003).

Neuroanatomy

The number of neuroimaging studies of dyslexia (or reading disorders) has exploded in recent years, yielding a wealth of information about possible functional and anatomical differences between dyslexic and typical brains. These differences involve various components of the left hemisphere language network, although not always in a consistent fashion across studies. A common and fundamental dilemma in interpreting these findings is that it is hard to determine whether they reflect differences in brain organization that lead to dyslexia, differences that result from differences in processing, or reduced experience with written and verbal stimuli.

Voxel-based morphometric analyses of gray matter in the brains of dyslexic individuals tend to show a reduction in gray matter density in superior temporal regions bilaterally, areas that are associated with phonological processing (Brambati et al., 2004; Dole, Meunier, & Hoen, 2013; Richlan, Kronbichler, & Wimmer, 2013; Steinbrink et al., 2008). Individual studies have also shown structural gray matter differences in the cerebellum (Brambati et al., 2004; Eckert et al., 2005; Stoodley, 2014), left parietal cortex (Eckert et al., 2005; Hoeft et al., 2007), and inferior temporo-occipital areas including the lingual and fusiform gyri (Eckert et al., 2005; Kronbichler et al., 2008; Linkersdorfer, Lonnemann, Lindberg, Hasselhorn, & Fiebach, 2012).

Among the most consistently replicated findings in fMRI and PET studies is that of reduced activation in left occipitotemporal regions and in the visual word form area of the left fusiform gyrus during reading tasks (Paulesu, Danelli, & Berlingeri, 2014; Richlan, 2012; Richlan, Kronbichler, & Wimmer, 2009, 2011). Interestingly, decreased activation in this area correlates (at least in some studies) with decreased gray matter density (Linkersdorfer et al., 2012), and decreased density and activation have been observed in preliterate children at risk for dyslexia based on family history (Raschle, Chang, & Gaab, 2011; Raschle, Zuk, & Gaab, 2012). Conversely, intensive remediation focused on reading skills results in increased left fusiform gray matter density (Krafnick, Flowers, Napoliello, & Eden, 2011), which suggests that the structure–function relationship in this region is at least partially mediated by experience.

Analyses of structural and functional connectivity in dyslexic brains have tended to highlight a role for tracts connecting the ventral and dorsal pathways of language processing. DTI shows microstructural abnormalities in white matter organization at the left temporoparietal junction around the posterior terminus of the arcuate fasciculus (Deutsch et al., 2005; Klingberg et al., 2000; Lebel et al., 2013; Niogi & McCandliss, 2006; Rimrodt, Peterson, Denckla, Kaufmann,

& Cutting, 2010; Steinbrink et al., 2008; Vandermosten, Boets, Wouters, & Ghesquiere, 2012) and in the SLF (Carter et al., 2009; Steinbrink et al., 2008). One neuroimaging study used multivoxel pattern analysis to show that activation in bilateral auditory cortex is intact in dyslexic individuals, but functional coherence and white matter tracts between this region and the inferior frontal cortex were reduced (Boets et al., 2013).

Neuropathology

In a series of postmortem examinations of the brains of dyslexic individuals, Galaburda and colleagues documented a high prevalence of microscopic abnormalities of neuronal migration, including ectopias and focal microgyri. These abnormalities tend to be clustered in the perisylvian cortex, predominantly in the left hemisphere, which suggests a cytoarchitectonic explanation for functional disruption within the language system (Galaburda & Kemper, 1979; Galaburda, Sherman, Rosen, Aboitiz, & Geschwind, 1985; Humphreys, Kaufmann, & Galaburda, 1990).

Nevertheless, neurobiological theories of dyslexia have tended to focus on abnormalities in the thalamus, and in particular in the magnocellular layers of the lateral geniculate nucleus, which are often disorganized and contain a disproportionate number of small neurons (Galaburda & Kemper, 1978; Galaburda & Livingstone, 1993; Livingstone, Rosen, Drislane, & Galaburda, 1991). The "magnocellular theory" of dyslexia seems to predict that dyslexic individuals should have difficulties with rapid temporal processing of sensory information, because this is the specialized function of the magnocellular pathway. Thus, the theory has derived support from observations of visual and auditory processing problems in dyslexia.

Unfortunately for this theory, the particular kinds of visual and auditory deficits found in dyslexic people are not always consistent with magnocellular dysfunction, as noted previously. Moreover, it is possible, and perhaps even likely, that the thalamic abnormalities observed in dyslexic brains are a developmental consequence of primary abnormalities in cortical architecture, and not the root cause of other anatomical and behavioral differences.

At least one small autopsy series has shown that people with dyslexia, compared with those who do not have it, have an increased proportion of large Purkinje neurons in the anterior and posterior lobes of the cerebellum (Finch, Nicolson, & Fawcett, 2002). This observation has been adduced to support the view that dyslexia is primarily attributable to a deficit in cerebellar function (the "cerebellar theory") (Nicolson, Fawcett, & Dean, 2001). As is the case with the magnocellular theory, however, it may be that cerebellar cortical abnormalities are merely the result of an underlying disruption in corticocerebellar connectivity, and thus may contribute to the disorder without explaining its origins.

Genetics

Variance in reading skill has a significant genetic component, with heritability estimates ranging from 0.4 to 0.8 (Schumacher, Hoffmann, Schmal, Schulte-Korne, & Nothen, 2007). To date, about 15 different candidate genes have been associated with susceptibility to developmental dyslexia across nine genetic loci, labeled DYX1−9 (Carrion-Castillo, Franke, & Fisher, 2013; Giraud & Ramus, 2013; Poelmans, Buitelaar, Pauls, & Franke, 2011). These are associated with a variety of neurodevelopmental processes, most of which seem to affect neural migration within the neocortex. There is general agreement that dyslexia is a polygenic and heterogeneous trait, with different combinations of genetic risk factors contributing to the phenotype in different individuals (Carrion-Castillo et al., 2013); the most promising genetic leads have emerged from studies of families in which alterations in single genes are associated with cognitive symptoms.

A paradigmatic example is the *DYX1C1* gene, encoding a 420−amino acid protein with three tetratricopeptide repeat domains, first identified in a Finnish family in which a chromosome 2;15 translocation cosegregated with difficulties in reading and writing (Nopola-Hemmi et al., 2000; Taipale et al., 2003). *DYX1C1* corresponds to the chromosome 15 breakpoint of the familial translocation and is located within the DYX1 candidate region at 15q21 that has been linked to reading and spelling disability in other studies (Bates et al., 2007; Chapman et al., 2004; Grigorenko et al., 1997; Platko et al., 2008; Schulte-Korne et al., 1998).

The *DYX1C1* gene is expressed in neural and glial cells in the developing neocortex and appears to affect neuronal migration specifically by regulating the assembly of dynein, a motor protein important for the function of microtubules (Tarkar et al., 2013; Wang et al., 2006). Using RNA interference techniques, it has been shown that knockdown of this gene in the neocortical ventricular zone results in cortical and hippocampal malformations similar to those sometimes observed in dyslexic people (Rosen et al., 2007). Cognitively, inactivation of *DYX1C1* seems to result in impairments in auditory function and working memory (Szalkowski et al., 2011; Threlkeld et al., 2007). Nevertheless, a link between specific human polymorphisms in *DYX1C1* and developmental dyslexia has proven elusive, with inconsistent results across population-based studies (Tran et al., 2013; Venkatesh, Siddaiah, Padakannaya, & Ramachandra, 2014; Zou et al., 2012), although it is possible that this may be attributable in part to methodological issues.

A second dyslexia susceptibility gene, *DCDC2*, has been identified within the dyslexia-associated locus DYX2 at 6p21–23 (Fisher et al., 1999; Grigorenko et al., 1997; Platko et al., 2008). Meng et al. (2005) showed that polymorphisms and a small (2.4 kb) deletion in this gene predicted performance in quantitative measures implicated in dyslexia, including discrepancy between observed and expected reading scores and homonym selection. Subsequent studies have confirmed an association between *DCDC2* and dyslexic phenotypes (Matsson et al., 2015; Schumacher et al., 2006). As with *DYX1C1*, however, links between dyslexia and specific markers in *DCDC2* have not consistently replicated in different populations (Venkatesh, Siddaiah, Padakannaya, & Ramachandra, 2013; Zou et al., 2012).

DCDC2 encodes a protein with two doublecortin domains, and like *DYX1C1* it appears to influence neuronal migration through its role in microtubule formation (Burbridge et al., 2008; Grati et al., 2015; Meng et al., 2005; Schumacher et al., 2006). In the human brain, it is expressed most robustly in the entorhinal cortex, inferior and medial temporal cortex, hypothalamus, amygdala, and hippocampus, and knockdown of expression in the rat ventricular zone produces neuronal heterotopias (Meng et al., 2005). Mutations of *DCDC2* in mice affect visual discrimination, visuospatial memory, and long-term memory (Gabel et al., 2011). Some studies have suggested that both the anatomical and behavioral effects of *DCDC2* mutations in animal models depend on concurrent disruption of a related gene, *DCX*, whose function may be partially redundant with *DCDC2* (Meng et al., 2011; Wang et al., 2011).

In humans with and without dyslexia, polymorphisms in *DCDC2* appear to predict cortical thickness in the left angular and supramarginal gyri, which form part of the language network, and in the left lateral occipital cortex (Darki, Peyrard-Janvid, Matsson, Kere, & Klingberg, 2014). Individuals with deletions in *DCDC2* intron 2 have decreased white matter integrity in the left arcuate fasciculus and the splenium of the corpus callosum (Marino et al., 2014). Those with both *DCDC2* deletions and dyslexia also appear to have decreased fractional anisotropy in the inferior longitudinal fasciculus and the genu of the corpus callosum.

The DYX2 locus at 6p21–23 contains yet a third gene, *KIAA0319*, with putative causal links to developmental reading disorders (Cope et al., 2005; Francks et al., 2004; Harold et al., 2006). In some populations, *KIAA0319* does not appear to contribute significantly to dyslexia (Schumacher et al., 2006), whereas in other populations, in which no risk was attributable to polymorphisms in *DYX1C1* or *DCDC2*, variants in *KIAA0319* appear to confer risk (Venkatesh et al., 2013; Zou et al., 2012).

KIAA0319 is a plasma membrane protein thought to function in cellular adhesion and attachment. In the human fetal brain, it is expressed in the developing neocortex, ganglionic eminence, cortical plate, and ventricular zone (Paracchini et al., 2006). In common with the other dyslexia candidate genes discussed thus far, disruption of *KIAA0319* in animal models results in aberrant neocortical migration and heterotopia formation, with a significant reduction in the volume of the corpus callosum (Szalkowski et al., 2012), possibly owing to defective axonal guidance (Poon et al., 2011). Behavioral effects observed in animals with abnormal *KIAA0319* expression include deficits in processing acoustic stimuli and impaired spatial learning (Szalkowski et al., 2013). A protein similar in structure to KIAA0319 (called KIAA0319-Like) is encoded by a gene at the 1p36–34 dyslexia susceptibility locus (DYX8) (Couto et al., 2008) and also appears to be involved in neuronal migration; disruption of this gene in rats produces large periventricular nodular heterotopias (Platt et al., 2013).

A final gene whose association with dyslexia is relatively well-characterized is *ROBO1*, found in the DYX5 region at chromosome 3p12–13 (Hannula-Jouppi et al., 2005; Nopola-Hemmi et al., 2001). Nopola-Hemmi et al. (2001) described profound reading difficulties in a four-generation Finnish family which appeared to be transmitted as an autosomal dominant, monogeneic trait mapped to this locus; Hannula-Jouppi et al. (2005) identified *ROBO1* as the gene in this region most likely to contribute to dyslexia in a patient with a de novo chromosomal translocation. More recent studies have supported a role for *ROBO1* in the pathogenesis of dyslexia (Mascheretti et al., 2014; Tran et al., 2014).

ROBO1 is expressed strongly in developing brain tissue and has been shown to be involved in axon growth and guidance (Lamminmaki, Massinen, Nopola-Hemmi, Kere, & Hari, 2012; Mire et al., 2012) as well as proliferation and neurogenesis of pyramidal neurons (Yeh et al., 2014). The behavioral effects of *ROBO1* disruption are less well understood, but at least one study has suggested that dyslexic individuals with a common *ROBO1* haplotype showed abnormal patterns of magnetoencephalographic activity in a paradigm examining auditory cortex responses to sounds played at different frequencies in the left and right ears, which suggests impaired interaural interaction (Lamminmaki et al., 2012).

Overall, genes associated with developmental dyslexia appear predominantly to result in altered patterns of cortical migration. In animal models, frank histologic abnormalities on such as heterotopias are often observed; in humans, variations in these genes may produce alterations in gray or white matter microstructure that are apparent in neuroimaging, although pathological confirmation for abnormalities linked to specific genes has not yet been obtained. Interestingly, most of these genes are expressed widely throughout the cortex and in subcortical structures, which is consistent with the observation that the neurological syndrome in dyslexia is rarely limited to difficulty with reading. Gene–gene interactions explain a significant proportion of the variability in heritability of dyslexia and its clinical phenotype (Mascheretti, Bureau, Trezzi, Giorda, & Marino, 2015).

PRIMARY PROGRESSIVE APHASIA

Neurodegenerative diseases such as Alzheimer disease (AD) and frontotemporal dementia (FTD) are commonly recognized to lead to progressive changes in memory or personality. Since these disorders were first described, however, it has also been known that impairments in language function can be a prominent feature (Alzheimer, 1907; Pick, 1892). Nonetheless, for most of the past century it was believed that language deficits would emerge only after deficits in other cognitive domains became apparent.

Beginning in the 1970s, researchers began to recognize neurodegenerative conditions in which language-related impairment was the most prominent early symptom (Warrington, 1975). The term "primary progressive aphasia" (PPA) was coined in the 1980s after Mesulam's landmark article describing a case series of patients presenting with slowly progressive aphasia associated with degeneration of the perisylvian region of the left hemisphere (Mesulam, 1982). In contrast to the previously prevailing view, in which aphasia was seen merely as part of a broader dementing syndrome, it became clear that the cognitive impairment in PPA affected language earlier and more significantly than other domains, and that this was a result of neurodegeneration in specific anatomical regions related to language function.

Although PPA was initially seen as a single syndrome, it was quickly recognized that specific aphasic symptoms varied depending on the underlying pattern of degeneration. PPA was thus initially divided into two main subtypes: a "fluent" form characterized by a progressive disorder of semantic memory (also called semantic dementia) (Snowden, Goulding, & Neary, 1989) and a "nonfluent" form of progressive aphasia characterized mainly by a disorder of grammar (progressive nonfluent aphasia) (Grossman et al., 1996). In the early 2000s, a third form was described in which the most prominent deficits are word-finding difficulties, which led to the appellation of logopenic PPA, and problems with sentence repetition (Gorno-Tempini, Dronkers, et al., 2004). These three variants of PPA differ not only in their clinical presentation but also in their neuroanatomical patterns of degeneration and underlying neuropathology (Gorno-Tempini et al., 2011) (Table 32.1).

Nonfluent Variant Primary Progressive Aphasia

Clinical Features

The presenting symptoms of nonfluent variant PPA (nfvPPA) can be heterogeneous but typically include slow and effortful speech output. These problems are sometimes described by patients or caregivers as "word-finding difficulties," but careful clinical examination reveals that these difficulties are attributable to two core impairments: the articulation planning deficit known as AOS and a deficit in processing grammatical elements of language, known as agrammatism.

As noted earlier in our discussion of CAS, AOS refers to the incoordination of subtle movements required for speech production, in the absence of neuromuscular dysfunction or frank dysarthria (Duffy, Peach, & Strand, 2007). Adults with AOS may report a sensation of knowing what they want to say, but are unable to produce the desired words with their lips and tongue. Observers may perceive uncoordinated "groping" mouth movements while these patients try to articulate. Typically, AOS is most evident in the pronunciation of particularly complex phrases or in the repetition of multisyllabic words (eg, "artillery," "hippopotamus") (Hodges, Martinos, Woollams, Patterson, & Adlam, 2008; Ogar, Slama, Dronkers, Amici, & Gorno-Tempini, 2005). As a consequence, patients with nfvPPA have slower speech than patients with other PPA syndromes, even when word pauses are taken into account (Ash et al., 2009). Speech is characterized by inconsistent sound errors such as distortions, deletions, insertions, and substitutions (Ogar, Dronkers, Brambati, Miller, & Gorno-Tempini, 2007). Prosody is frequently impaired as well (Wilson, Henry, et al., 2010).

Agrammatism in language production may manifest as abnormal syntactic construction, inappropriate use of pronouns, inability to produce appropriate subject–verb agreement, decreased verb use, and omission of grammatical morphemes such as articles, prepositions, and auxiliary verbs (Ash et al., 2009; Gunawardena et al., 2010; Wilson, Henry, et al., 2010). In early stages, grammatical difficulties may be apparent only in writing, but they will ultimately involve both written and spoken language (Grossman, 2012). Patients may try to overcome these deficits by using shortened sentences, resulting in an abbreviated mean length of utterance (Thompson et al., 2012). Comprehension of single words and simple sentences is generally spared, but there may be deficits in comprehending more complex syntactic constructions such as negative

TABLE 32.1 Classification Criteria for Primary Progressive Aphasia

General PPA Criteria
I. Inclusion criteria (all must be present): **1.** Most prominent clinical feature is difficulty with language **2.** Language deficit is the cause of impaired daily living activities **3.** Aphasia is the prominent deficit at symptom onset and for the initial phases of the disease
II. Exclusion criteria (none must be present): **1.** Other central nervous system or medical disorders that can cause symptoms **2.** Psychiatric disorders that can cause symptoms **3.** Prominent initial episodic memory, visual memory, and visuoperceptual impairments **4.** Prominent, initial behavioral disturbance
Nonfluent Variant PPA
I. Core criteria (at least one must be present): **1.** Agrammatism **2.** Effortful, halting speech production with speech sound errors (consistent with apraxia of speech)
II. Supportive features (two of three must be present): **1.** Impaired syntactic comprehension **2.** Spared single-word comprehension **3.** Spared object knowledge
III. Imaging-supported nfvPPA (both must be present): **1.** Clinical diagnosis of nfvPPA **2.** Left posterior frontoinsular atrophy on MRI and/or hypometabolism/hypoperfusion
Semantic Variant PPA
I. Core criteria (both must be present): **1.** Poor confrontation naming **2.** Impaired single-word comprehension
II. Supportive features (at least three must be present): **1.** Poor object/face knowledge **2.** Surface dyslexia/dysgraphia **3.** Spared repetition **4.** Spared motor speech and grammar
III. Imaging-supported svPPA (both must be present): **1.** Clinical diagnosis of svPPA **2.** Anterior temporal lobe atrophy on MRI and/or hypometabolism/hypoperfusion
Logopenic Variant PPA
I. Core criteria (both must be present): **1.** Poor single-word retrieval **2.** Impaired repetition of sentences
II. Supportive features (at least three must be present): **1.** Phonological errors **2.** Spared single word comprehension and semantics **3.** Spared motor speech **4.** Absence of frank agrammatism
III. Imaging-supported lvPPA (both must be present): **1.** Clinical diagnosis of lvPPA **2.** Posterior perisylvian atrophy on MRI and/or hypometabolism/hypoperfusion

PPA, primary progressive aphasia; *nfvPPA*, nonfluent variant PPA; *svPPA*, semantic variant PPA; *lvPPA*, logopenic variant PPA.
Modified from Gorno-Tempini, M. L., Hillis, A. E., Weintraub, S., Kertesz, A., Mendez, M., Cappa, S. F., ... Grossman, M. (2011). Classification of primary progressive aphasia and its variants. *Neurology, 76*(11), 1006–1014.

passive or relative sentences (eg, "The cake was not eaten by the girl") (Weintraub et al., 2009). Spelling of nonwords may also be impaired (Shim, Hurley, Rogalski, & Mesulam, 2012).

As the disease progresses, speech becomes increasingly effortful and patients may develop selective mutism even when other cognitive and motor abilities are relatively preserved (Gorno-Tempini, Dronkers, et al., 2004). However, other cognitive symptoms frequently develop over time (Sapolsky et al., 2010) as impairments in working memory and executive function lead to difficulties with concentration, multitasking, planning, and organization. Patients in the final stages of the disease may exhibit disinhibition, compulsive behaviors, apathy, diminished empathy, and lack of insight (Rohrer & Warren, 2010). Episodic memory and visuospatial functioning tend to be relatively spared even in advanced stages.

On neurological examination, it is common to note mild extrapyramidal signs, particularly involving the right side of the body. These can include mild limb rigidity, reduced hand dexterity, and bradykinesia. Limb apraxia, often asymmetric at presentation, may also be present (Zadikoff & Lang, 2005), as may dysarthria (as distinct from AOS) (Ogar et al., 2007). Features suggestive of corticobasal syndrome or progressive supranuclear palsy (PSP) often occur during the course of the disease (Gorno-Tempini, Murray, et al., 2004; Josephs et al., 2006; Nestor et al., 2007). In some cases, nfvPPA may also be comorbid with amyotrophic lateral sclerosis; these patients exhibit weakness, weight loss, fasciculations, and muscle atrophy (Seelaar, Rohrer, Pijnenburg, Fox, & van Swieten, 2011). Such nonlanguage clinical features may provide a clue as to the pathologic mechanism of disease in individual patients.

Neuroanatomy

Most patients with nfvPPA show a characteristic pattern of atrophy involving the left posterior-inferior frontal lobe (including Broca's area), premotor cortex, and anterior insula (Fig. 32.3). This pattern has been demonstrated consistently in MRI studies using volumetric and cortical thickness analyses (Agosta et al., 2015; Caso et al., 2014; Gorno-Tempini, Dronkers, et al., 2004; Josephs et al., 2006; Rogalski, Cobia, Harrison, Wieneke, Thompson, et al., 2011; Rogalski, Cobia, Harrison, Wieneke, Weintraub, et al., 2011; Rogalski et al., 2014; Rohrer, Warren, et al., 2009), and nuclear imaging shows hypometabolism in the same regions (Nestor et al., 2003). As the disease progresses, cortical atrophy extends rostrally into left prefrontal areas and ventrally into the left anterior temporal lobe, as well as to contralateral posterior frontal regions (Agosta et al., 2015; Rogalski et al., 2014). There is also atrophy of deep gray nuclei in the basal ganglia and thalamus (Garibotto et al., 2011; Gorno-Tempini et al., 2008; Rohrer, Warren, et al., 2009).

These patterns of atrophy have been implicated directly in deficits observed in patients with nfvPPA. For example, apraxia of speech in patients with nfvPPA correlates with volume loss in the left posterior frontal, anterior insular, and basal ganglia regions (Ogar et al., 2007); dysarthria co-occurs with atrophy extending to the face region of primary motor cortex and to the left caudate head. Another study showed a correlation between apraxic speech distortions and volume loss in the left frontal subcortical white matter, and a smaller homologous region on the right (Wilson, Henry, et al., 2010).

With respect to the agrammatic symptoms of nfvPPA, it has been shown that comprehension of syntactically complex sentences is impaired in patients with more atrophy in the dorsal portion of the left inferior and middle frontal gyri (Amici et al., 2007) as well as the left superior temporal gyrus (Gunawardena et al., 2010), ventral sensorimotor cortex, and

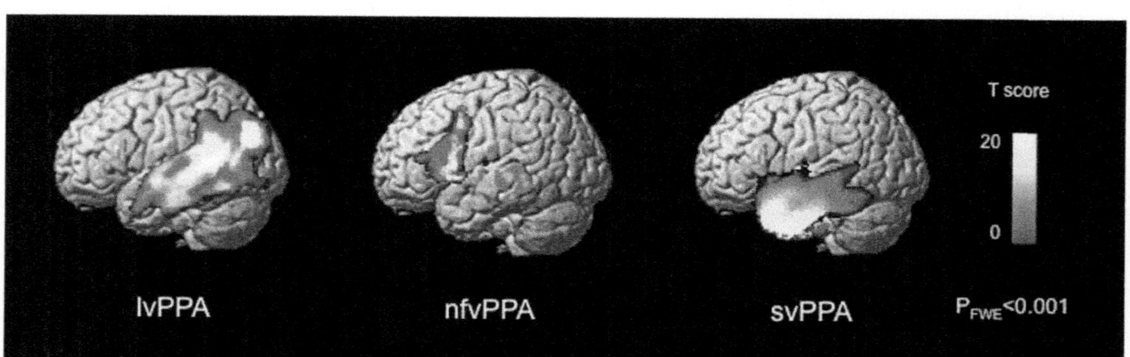

FIGURE 32.3 Patterns of gray matter atrophy in patients from the University of California San Francisco cohort with lvPPA ($n = 25$), nfvPPA ($n = 42$), and svPPA ($n = 59$) compared with healthy control groups matched for age and gender. Voxel-based morphometry results are thresholded at $P_{FWE} < 0.001$. *lvPPA*, logopenic variant primary progressive aphasia; *nfvPPA*, nonfluent variant primary progressive aphasia; *svPPA*, semantic variant primary progressive aphasia.

supramarginal gyrus (Rogalski, Cobia, Harrison, Wieneke, Thompson, et al., 2011; Rogalski, Cobia, Harrison, Wieneke, Weintraub, et al., 2011). Poor verbal fluency is associated with more focal atrophy in the posterior inferior and middle frontal gyri (Rogalski, Cobia, Harrison, Wieneke, Thompson, et al., 2011; Rogalski, Cobia, Harrison, Wieneke, Weintraub, et al., 2011).

DTI has also been used to investigate abnormalities in structural connectivity along white matter tracts involved in language processing (Agosta et al., 2013, 2011; Galantucci et al., 2011; Schwindt et al., 2013; Whitwell et al., 2010). A common finding across these studies has been the observation of significant microstructural changes in dorsal language pathways connecting the left frontal lobe with parietal and superior temporal regions, specifically, the SLF, including the arcuate component, which in turn are strongly related to deficits in syntactic production and comprehension (Wilson et al., 2011).

Connections within the frontal lobe and in frontostriatal circuits are also disrupted in nonfluent patients with motor speech disturbances (Mandelli et al., 2014). The frontal aslant tract, connecting Broca's area to the anterior supplementary motor area, appears to be particularly vulnerable in this disorder (Catani et al., 2013). Integrity of this tract correlates with performance in verbal fluency (Catani et al., 2013) as well as with the rate of speech and sentence production and the frequency of articulatory distortions (Mandelli et al., 2014).

By contrast, ventral white matter tracts connecting temporal, occipital, and orbitofrontal regions tend to be spared (Agosta et al., 2013, 2011; Galantucci et al., 2011; Schwindt et al., 2013), although some studies have shown that changes in ventral tracts (such as the inferior frontal-occipital and uncinate fasciculi) also correlate with a reduced proportion of grammatically well-formed utterances (Grossman et al., 2012).

Functional neuroimaging studies in patients with nfvPPA are more limited than structural studies. However, at least two studies have suggested that nonfluent patients show altered activation in the left inferior frontal cortex during sentence comprehension tasks compared with healthy control subjects (Cooke et al., 2003; Wilson, Dronkers, et al., 2010).

Taken together, these data outline distinct networks whose degeneration contributes to motor speech and grammatical disturbances in nfvPPA. Impairment in syntactic processing seems to correlate with atrophy and altered activation of the left inferior frontal cortex and its interhemispheric connections to parietal and temporal language areas, whereas apraxia of speech correlates with degeneration of the network linking Broca's area and precentral speech areas to brain regions involved in motor planning, including the supplementary motor area and the basal ganglia.

Neuropathology

Pathologically, the nonfluent variant of PPA is usually found to be a form of frontotemporal lobar degeneration (FTLD). About 70% of cases in clinicopathological series are associated with pathological deposits of hyperphosporylated microtubule-associated tau protein (FTLD-tau); most of the remainder have tau-negative, TDP-43–positive inclusions (FTLD-TDP) (Grossman, 2010; Josephs et al., 2011; Rohrer et al., 2011) (Fig. 32.5). Less frequently, AD pathology has been described (Caso et al., 2013), although there is controversy as to whether some of the patients purported to show this pathology should have been classified as logopenic rather than nonfluent (Chare et al., 2014).

The physiological function of the tau protein is to promote the assembly and stabilization of microtubules. Hyperphosphorylated tau has reduced affinity for microtubules, preventing their stabilization. In FTLD there is thought to be an excess of hyperphosphorylated tau (for reasons that are mostly unclear), leading to cellular degeneration. FTLD-tau can be further subdivided based on the number of repeat regions in the microtubule binding domains of the deposited tau protein. Tauopathies with three repeats (FTLD-3R) have the histopathological characteristics of Pick disease, whereas those with four repeats (FTLD-4R) have the patterns of progressive supranuclear palsy and corticobasal degeneration. Most commonly, nfvPPA is associated with FTLD-4R pathology although, as we have noted, the nonlanguage clinical symptoms of corticobasal degeneration or PSP are observed only in some patients.

FTLD-TDP pathology, which is characterized by neuronal and glial ubiquitinated inclusions of transactive-response DNA-binding protein of about 43 kD (TDP-43), is found in a minority of cases of nfvPPA (Knibb, Xuereb, Patterson, & Hodges, 2006; Snowden, Neary, & Mann, 2007). Of the four subtypes of FTLD-TDP, the one most commonly associated with nfvPPA is type A, which is characterized by relatively few inclusions, with long neuritic profiles in the superficial cortical laminae (Josephs, Stroh, Dugger, & Dickson, 2009).

The ability to make the correct premortem pathological diagnosis in patients who present clinically with nfvPPA is critical to the design of therapeutic trials that target specific molecular alterations. At least one study has suggested that the presence of extrapyramidal signs and agrammatism, combined with neuroimaging evidence of white matter volume loss, might point to the presence of an underlying 4R tauopathy (Caso et al., 2014). Likewise, FTLD-4R may be more prevalent in patients with apraxia of speech as a prominent symptom (Deramecourt et al., 2010; Josephs et al., 2006).

Genetics

Whereas other FTD syndromes are often associated with a positive family history (Goldman et al., 2011; Rohrer, Guerreiro, et al., 2009), most cases of nfvPPA are sporadic. The genetic forms of the disorder are usually more difficult to classify in the three main variants. Nevertheless, syndromes resembling nfvPPA and lvPPA have been described in patients who have mutations in various genes implicated in FTLD, including the *MAPT* gene, encoding microtubule-associated protein tau, on chromosome 17 (Boeve et al., 2005), and *PGRN* (encoding progranulin), also on chromosome 17 (Beck et al., 2008; Le Ber, Camuzat et al., 2008). The presence of *MAPT* and *PGRN* mutations does not appear to predict a clinical presentation of nfvPPA (Bird et al., 1999; Beck et al., 2008), and family members with the same mutation can have different clinical symptoms (Le Ber et al., 2008).

Some nfvPPA patients have been shown to have a hexanucleotide repeat expansion in chromosome 9 open reading frame 72, typically associated with FTLD-TDP pathology and clinical signs of motor neuron disease (Mahoney et al., 2012; Snowden et al., 2012). Other, less common genetic changes described in nfvPPA cases include mutations of *TARDBP* (encoding TDP-43) gene on chromosome 1 (Chio et al., 2010) and *VCP* on chromosome 9 (Watts et al., 2004).

Semantic Variant Primary Progressive Aphasia

Clinical Features

Patients with semantic variant primary progressive aphasia (svPPA) (also known as semantic dementia) usually present with anomia in spontaneous speech and writing, frequently described as a loss of memory for words. At the root of this impairment is a loss of conceptual knowledge of the person or object that needs to be named (Hodges, Patterson, Oxbury, & Funnell, 1992). Early in the course of the disease, only less familiar and less frequently encountered concepts are lost; high-frequency words referring to prototypical or generic objects may be preserved. Thus, patients with svPPA often use category-level terms or more frequent words of the same semantic category, sometimes incorrectly (eg, "animal" or "dog" instead of "tiger"), or very generic words ("thing").

Compared with other forms of PPA that are characterized by prominent anomic difficulties (particularly the logopenic variant), patients with svPPA are less likely to think of a target word later, and are more consistent in the words they cannot recall (Ralph & Howard, 2000). Moreover, cuing with the initial sounds of a word and offering multiple choices are less helpful (Graham, Patterson, & Hodges, 1995). Single word comprehension is also impaired, although patients may be unaware of their comprehension problems. Patients who are in the early stage can engage in simple conversations and understand simple texts with no apparent difficulties, but when confronted with an out-of-context, low-frequency word or object, comprehension deficits become apparent. These deficits are multimodal in nature, encompassing not only knowledge of words associated with objects but also visual, auditory, tactile, and olfactory features (Bozeat, Lambon Ralph, Patterson, Garrard, & Hodges, 2000; Luzzi et al., 2007). Impairment may extend to difficulties recognizing objects (agnosia) and faces (prosopagnosia), as well as difficulties in using objects correctly (ideational apraxia).

The discrepancy between the loss of naming and comprehension abilities and the preservation of motor speech, phonological, and grammatical skills can be striking, especially early on. The combination of loss of lexical-conceptual knowledge with preservation of language mechanics produces characteristic error patterns such as surface dyslexia (reading irregularly spelled words using regular phonological rules, eg, reading *sew* as/su/) and overregularization of irregularly inflected words (saying "childs" for *children*).

Behavioral alterations constitute the other major group of symptoms manifested by patients with svPPA, and they may follow or precede symptoms of semantic loss. Early stages may present with a behavioral syndrome characterized by emotional detachment, irritability, and disruption of physiologic drives such as sleep, appetite, and libido. Later, patients with svPPA often develop disinhibition, compulsions, and altered food preferences (Seeley et al., 2005).

Despite their deficits, many cognitive domains remain intact in patients with svPPA. For example, these patients may remember life events well, find their way around without difficulty, and engage in complex hobbies with retention of many practical, visuospatial, and creative skills.

Neuroanatomy

Conventional MRI scans and volumetric analyses in patients with svPPA most often show asymmetric atrophy in the anterior temporal lobes, involving the left side more than the right (Davies et al., 2009; Gorno-Tempini, Rankin, et al., 2004; Hodges & Patterson, 2007; Rohrer, Warren, et al., 2009) (Fig. 32.3). Volume loss is greatest in the polar, parahippocampal, middle, and inferior temporal regions, including the anterior fusiform gyrus. Significant hippocampal atrophy has been described; this may be as severe as the hippocampal atrophy in patients with AD matched for disease duration (Davies, Graham, Xuereb, Williams, & Hodges, 2004). The amygdala is also typically involved in svPPA, likely contributing to behavioral symptoms and impairments in reading social cues and facial emotions (Rosen et al., 2002). Over time, atrophy spreads to more posterior temporal regions, orbitofrontal cortex, insula, and anterior cingulate, as well as to homologous contralateral areas (Brambati et al., 2009; Rohrer, Warren, et al., 2009).

DTI studies have paralleled those using gray matter morphometry, showing more pronounced degeneration in ventral white matter tracts including the inferior longitudinal fasciculus, inferior frontooccipital fasciculus, and uncinate fasciculus. Dorsal degeneration is limited to the temporoparietal component of the SLF (Acosta-Cabronero et al., 2011; Agosta et al., 2013, 2010, 2011; Galantucci et al., 2011; Schwindt et al., 2013). By contrast, the frontoparietal portions of the SLF and the frontal speech network are relatively spared, which is consistent with the sparing of phonological and grammatical knowledge in these patients (Agosta et al., 2015, 2010; Galantucci et al., 2011) (Fig. 32.4).

The profound loss of conceptual semantic knowledge in patients with degeneration of the anterior temporal lobe has inspired a number of hypotheses about the role of this structure in language and cognition (Butler, Brambati, Miller, & Gorno-Tempini, 2009; Chan et al., 2001; Galton et al., 2001; Mummery et al., 2000), and in particular the proposal that this region is an amodal semantic hub (Patterson et al., 2007). Consistent with this view, resting-state fMRI studies show that anterior temporal lobe atrophy in svPPA co-occurs with reduced functional connectivity with several upstream modality-specific cortical regions, including posterior temporal structures and the fusiform gyri (Guo et al., 2013). These perturbations in functional connectivity correlate with impairments in naming (Agosta et al., 2014; Guo et al., 2013).

In about a quarter of cases of svPPA there is more pronounced atrophy of the right anterior temporal lobe than the left (Hodges et al., 2010). In these patients, behavioral alterations may be the earliest symptoms, and semantic memory impairments may initially be limited to loss of knowledge about famous people (Edwards-Lee et al., 1997; Gorno-Tempini, Rankin, et al., 2004; Henry et al., 2012). Patients who initially show semantic memory impairments in the context of left temporal atrophy eventually develop behavioral changes when the disease has progressed to involve the right temporal lobe and orbitofrontal regions (Brambati et al., 2009). The degree of atrophy in right orbitofrontal and temporal regions has been linked to the clinical development of anxiety, apathy, irritability, and eating disorders (Rohrer & Warren, 2010), as well as social-emotional symptoms such as deficits in self-awareness of empathic concern, insight, and theory of mind (Irish, Hodges, & Piguet, 2014; Shany-Ur et al., 2014; Sollberger et al., 2014).

Neuropathology

Semantic variant PPA is the most pathologically homogeneous of the PPA syndromes, because FTLD-TDP type C pathology, which is associated with many small neurites and cytoplasmic inclusions in superficial cortical layers, is found in up to 90% of affected patients. A small proportion of cases are found to have FTLD-tau or AD pathology (Hodges et al., 2010; Josephs et al., 2011; Rohrer, Geser, et al., 2010; Rohrer et al., 2011) (Fig. 32.5). In a clinical series reviewing 24 svPPA cases with autopsy data, no consistent clinical or macroscopic neuroradiological differences were found between patients with different pathological findings (Hodges et al., 2010).

Genetics

Compared with other patients found to have FTLD pathology, patients with svPPA are much less likely to have a positive family history (Goldman et al., 2005). Most of the familial cases that have been described have mutations in *GRN* (Beck et al., 2008; Pickering-Brown et al., 2008; Whitwell et al., 2007). Patients with *MAPT* mutations and semantic impairment tend to have a more pronounced behavioral disorder with evidence of frontal lobe dysfunction (Bessi et al., 2010; Pickering-Brown et al., 2008).

The puzzling absence of a familial association with svPPA has prompted researchers to search for other heritable and acquired risk factors for this disorder. Interestingly, svPPA patients have a much higher prevalence of autoimmune diseases (18%) do than age-matched controls or patients with AD (Miller, Rankin, et al., 2013). Furthermore, left-handedness is almost twice as common in patients with svPPA as in the general population (Miller, Mandelli, et al., 2013). The precise ways in which these population characteristics are related to the clinical presentation of svPPA remain unknown.

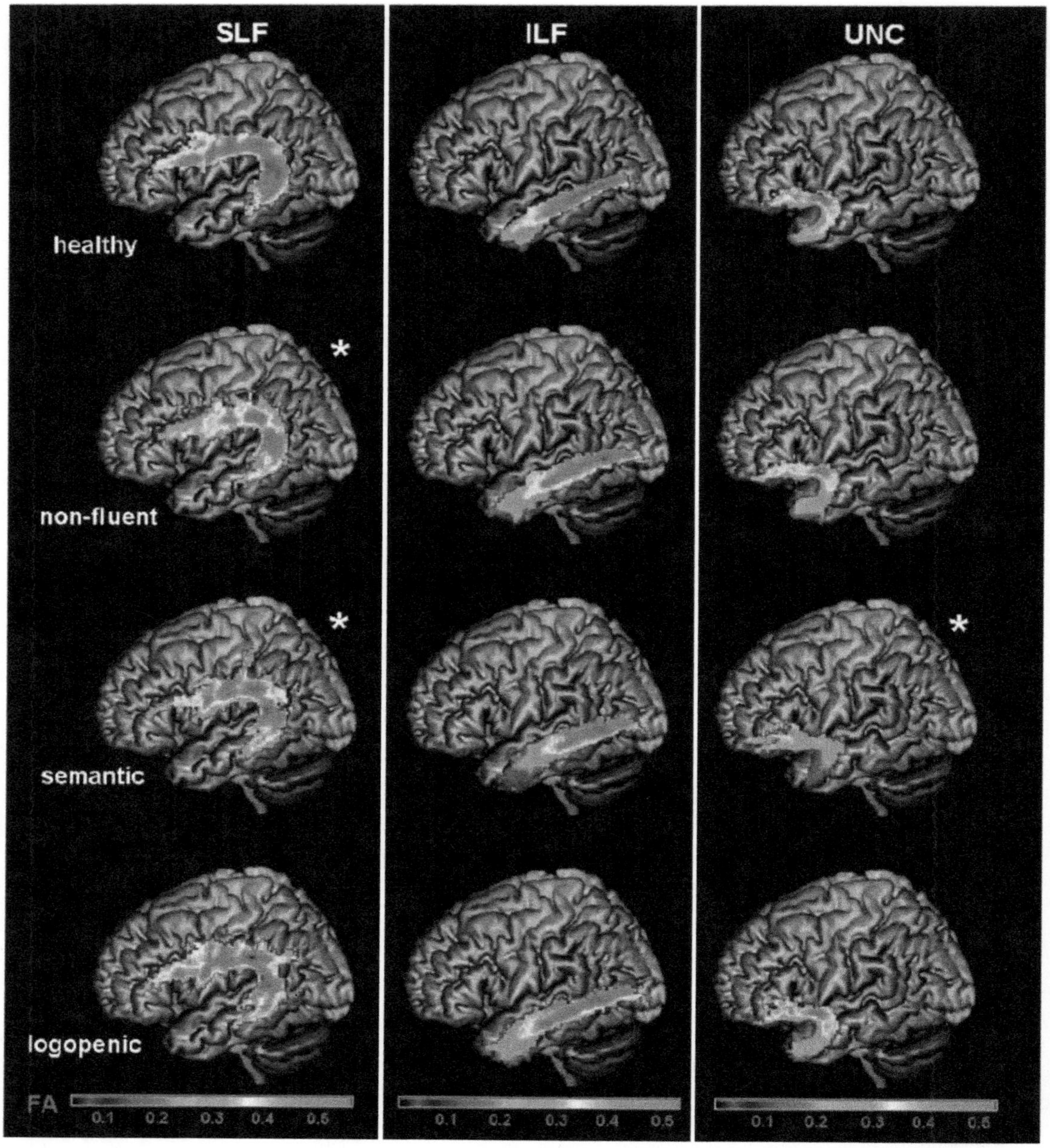

FIGURE 32.4 White matter microstructural integrity in the language-related tracts from 48 subjects: nonfluent ($n = 9$), semantic ($n = 9$), logopenic ($n = 9$), and normal control subjects ($n = 21$). Fractional anisotropy (FA) values of each group in the probability maps for left superior longitudinal fasciculus (SLF), ILF, uncinate fasciculus (UNC), are overlaid on a standard Montreal Neurological Institute (MNI) brain. Only voxels that are in common in at least 20% of the subjects in each group were included in the probability maps. Asterisk denotes significantly different relative to normal controls at $P < 0.05$. The chromatic scale represents average fractional anisotropy values ranging from lower (violet-blue) to higher values (yellow-red). *From Galantucci, S., Tartaglia, M. C., Wilson, S. M., Henry, M. L., Filippi, M., Agosta, F., … Gorno-Tempini, M. L. (October 2011). White matter damage in primary progressive aphasias: a diffusion tensor tractography study.* Brain, 134*(Pt 10), 3011–3029.*

Logopenic Variant Primary Progressive Aphasia

Clinical Features

The logopenic variant of PPA (lvPPA) is so called because it usually presents with word-finding difficulties, leading to frequent pauses in speech and a decreased rate of speech production. However, confrontation naming is generally less consistently impaired than in svPPA patients, because the locus of the cognitive deficit in logopenic patients is at the level of lexical retrieval rather than conceptual knowledge (Gorno-Tempini, Dronkers, et al., 2004).

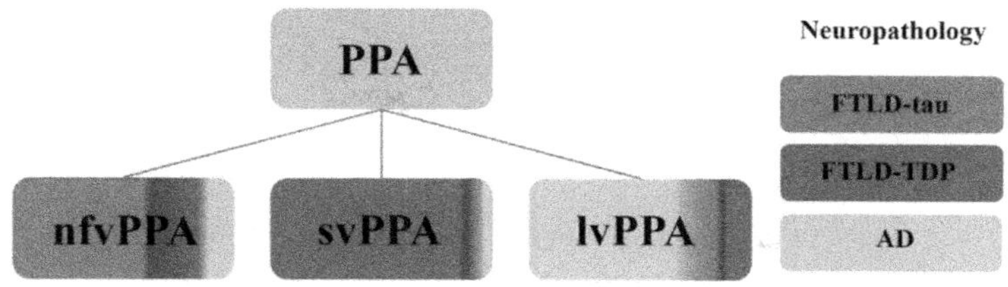

FIGURE 32.5 Clinicopathological correlates observed in patients with PPA, based on previous literature (see text for details). *AD*, Alzheimer disease; *lvPPA*, logopenic variant primary progressive aphasia; *nfvPPA*, nonfluent variant primary progressive aphasia; *svPPA*, semantic variant primary progressive aphasia.

Word retrieval deficits in patients with lvPPA may manifest as hesitation, false starts, and phonemic paraphasias, which are phonological in nature but can sometimes be difficult to distinguish from articulatory distortions owing to AOS (Ash et al., 2013; Croot, Ballard, Leyton, & Hodges, 2012). Unlike patients with nfvPPA, however, patients with logopenia do not exhibit frank grammatical errors and omissions until later stages of disease (Wilson, Henry, et al., 2010).

The other prominent symptom described by patients and caregivers as disease progresses is difficulty comprehending and repeating spoken language, especially when sentences are long or contain unfamiliar word combinations (Gorno-Tempini et al., 2008). When asked to repeat a sentence of this kind, patients with lvPPA often use circumlocutions, substituting frequent words for less frequent ones. Patients may also exhibit deficits in comprehension of grammatically complex sentences because of problems integrating long strings of words (Wilson, Dronkers, et al., 2010). These observations have led to the hypothesis that a deficit in phonological short-term memory might underlie many of the symptoms of lvPPA.

In fact, the phonological deficit in lvPPA seems to extend beyond spoken language. These patients also have problems with reading and writing, particularly with low-frequency words and nonwords, which suggests a specific impairment in matching sounds (phonemes) to letters (graphemes) (Sepelyak et al., 2011; Shim et al., 2012).

In addition to language difficulties, patients with lvPPA are more prone to general memory impairment than are those with other PPA subtypes (Mesulam et al., 2008). Moreover, they often develop problems with calculations, limb praxis, and visuospatial abilities at some point in the course of the disease (Gorno-Tempini, Dronkers, et al., 2004; Rohrer, Ridgway, et al., 2010). Mild neuropsychiatric symptoms such as apathy, anxiety, agitation, and irritability are also common findings (Rohrer & Warren, 2010).

Neuroanatomy

Neuroimaging studies using volumetric MRI and fluorodeoxyglucose PET data in patients with lvPPA have consistently shown a pattern of atrophy and hypometabolism primarily affecting the left temporoparietal junction, including posterior superior and middle temporal gyri and inferior parietal lobule (Gorno-Tempini, Dronkers, et al., 2004; Rabinovici et al., 2008) (Fig. 32.3). Cortical atrophy may also involve the medial temporal, parietal, inferior frontal, and posterior cingulate cortices, sometimes extending to the contralateral temporoparietal junction (Gorno-Tempini et al., 2008; Rohrer et al., 2013). In one longitudinal study, the rate of brain volume loss was found to be 2.3% per year in the left hemisphere and 1.6% per year in the right hemisphere (Rohrer et al., 2013). Disease progression leads to involvement of more anterior regions, including caudate, hippocampus, and medial parietal lobes. This pattern of cortical atrophy closely overlaps with some early-onset variants of AD (Migliaccio et al., 2009).

Despite this well-established pattern of atrophy, the neural basis for the core deficits in lvPPA is not completely clear. Studies have shown various associations between decreased auditory phonological working memory and atrophy in the left posterior superior temporal gyrus (Baldo, Katseff, & Dronkers, 2012; Leff et al., 2009; Leyton, Piguet, Savage, Burrell, & Hodges, 2012) and the left dorsolateral prefrontal and inferior parietal cortex (Amici et al., 2007), whereas confrontation naming impairments seem to be associated more specifically with thinning of the inferior posterior parietal cortex (Leyton et al., 2012).

White matter alterations in patients with lvPPA are generally less prominent than in the other two PPA syndromes, and like gray matter atrophy, they tend to be less more variable across studies, which have reported alterations in the superior longitudinal fasciculus (Galantucci et al., 2011) as well as in more ventral tracts, including the inferior longitudinal and uncinate fasciculi (Mahoney et al., 2013). Microstructural damage to the dorsal white matter tracts may be responsible for the phonologic working memory impairment, whereas damage to the ventral language tracts may be more related to word-finding difficulties, because lexical retrieval deficits occur along with atrophy of anterior and inferior temporal regions.

A study used resting-state fMRI to assess functional connectivity within the language, bilateral working memory, and ventral default mode networks in a large cohort of patients with lvPPA (Whitwell et al., 2015). The authors found that patients with lvPPA showed reduced connectivity in the left temporal language network as well as in the inferior parietal and prefrontal regions of the left hemisphere working memory network, compared with healthy subjects and patients with typical AD. Moreover, neuropsychological measures of aphasia severity correlated with connectivity in the working memory network. Such findings support the notion that a wide-scale disruption of the language and working memory networks is a characteristic feature of lvPPA.

Some patients who meet clinical criteria for lvPPA have patterns of cortical atrophy involving areas more commonly associated with other PPA variants. In such cases, the syndrome is sometimes diagnosed as "mixed" or "global" aphasia (Rogalski, Cobia, Harrison, Wieneke, Thompson, et al., 2011; Rogalski, Cobia, Harrison, Wieneke, Weintraub, et al., 2011).

Neuropathology

In the initial description of lvPPA, it was noted that the frequency of the ε4 allele of the gene for apolipoprotein E (*APOE*), which is highly associated with AD, was higher in patients with lvPPA than in the other two variants (Gorno-Tempini, Rankin, et al., 2004). The role of *APOE* ε4 as a risk factor in lvPPA and PPA in general is still debated (Josephs et al., 2014; Lehmann et al., 2013); however, subsequent reports have confirmed a strong association between AD pathology and lvPPA (Gefen et al., 2012; Josephs et al., 2013; Mesulam et al., 2008), and retrospective studies of patients with PPA who were later found to have AD pathology showed that these patients had significant logopenic symptoms (Rohrer, Rossor, & Warren, 2012). At least 60% of patients with lvPPA have cerebrospinal fluid biomarkers of AD (ie, elevated tau levels and reduced β-amyloid levels) (Hu et al., 2010; Teichmann, Kas, et al., 2013), and an even higher proportion (up to 90−100%) show brain amyloid deposition in PET imaging using the amyloid radioligand Pittsburgh compound B (Rabinovici et al., 2008).

The relationship between lvPPA and AD pathology is not absolute. There are documented cases of lvPPA with FTLD-tau or FTLD-TDP pathologies (Mesulam et al., 2008). In rare cases, lvPPA may be the presenting phenotype of Creutzfeldt−Jakob disease or Lewy body dementia (Martory et al., 2012; Teichmann, Migliaccio, et al., 2013).

Genetics

As noted previously, the best-documented genetic risk factors for lvPPA are those that are associated with AD. Some authors have therefore proposed that lvPPA is best understood as an asymmetric presentation of AD (Rohrer et al., 2012), which begs the question of why some patients present with this pattern of clinical and neuroanatomical deterioration as opposed to a pattern more typical for dementia of the AD type. One intriguing observation that may speak to this problem is that patients with lvPPA more often have a history of language-related learning disabilities than do patients with other PPA variants (Miller, Mandelli, et al., 2013). Thus, it may be that an underlying neurodevelopmental disorder involving the lexical-phonological system increases the vulnerability of this network when neurodegenerative disease arises later in life.

Patients with mixed clinical syndromes who meet criteria for lvPPA may be more likely to have mutations in genes associated with FTLD pathology (Fig. 32.5). For example, *PGRN* mutations have been found in patients with lvPPA who also had severe single-word comprehension deficits, anomia, and surface dyslexia, as well as anterior temporal atrophy more characteristic of svPPA (Rohrer, Ridgway, et al., 2010). Another patient with lvPPA with a *PGRN* mutation was found to have symptoms of agrammatism accompanied by atrophy in the inferior frontal region (Rohrer, Crutch, Warrington, & Warren, 2010). The logopenic-like symptoms in these patients may arise because of disruptions in the functional connections of the temporoparietal junction to dorsal networks (involved in articulation-to-sound mapping) or ventral networks (involved in processing word meaning).

REFERENCES

Acosta-Cabronero, J., Patterson, K., Fryer, T. D., Hodges, J. R., Pengas, G., Williams, G. B., & Nestor, P. J. (2011). Atrophy, hypometabolism and white matter abnormalities in semantic dementia tell a coherent story. *Brain, 134*(Pt 7), 2025−2035.

Agosta, F., Ferraro, P. M., Canu, E., Copetti, M., Galantucci, S., Magnani, G., ... Filippi, M. (2015). Differentiation between subtypes of primary progressive aphasia by using cortical thickness and diffusion-tensor MR imaging measures. *Radiology*, 141869.

Agosta, F., Galantucci, S., Canu, E., Cappa, S. F., Magnani, G., Franceschi, M., ... Filippi, M. (2013). Disruption of structural connectivity along the dorsal and ventral language pathways in patients with nonfluent and semantic variant primary progressive aphasia: a DT MRI study and a literature review. *Brain and Language, 127*(2), 157−166.

Agosta, F., Galantucci, S., Valsasina, P., Canu, E., Meani, A., Marcone, A., ... Filippi, M. (2014). Disrupted brain connectome in semantic variant of primary progressive aphasia. *Neurobiology of Aging, 35*(11), 2646–2655.

Agosta, F., Henry, R. G., Migliaccio, R., Neuhaus, J., Miller, B. L., Dronkers, N. F., ... Gorno-Tempini, M. L. (2010). Language networks in semantic dementia. *Brain, 133*(Pt 1), 286–299.

Agosta, F., Scola, E., Canu, E., Marcone, A., Magnani, G., Sarro, L., ... Filippi, M. (2011). White matter damage in frontotemporal lobar degeneration spectrum. *Cerebral Cortex, 22*(12), 2705–2714. October 10. [Epub ahead of print].

Alcock, K. J., Passingham, R. E., Watkins, K., & Vargha-Khadem, F. (2000a). Pitch and timing abilities in inherited speech and language impairment. *Brain and Language, 75*(1), 34–46.

Alcock, K. J., Passingham, R. E., Watkins, K. E., & Vargha-Khadem, F. (2000b). Oral dyspraxia in inherited speech and language impairment and acquired dysphasia. *Brain and Language, 75*(1), 17–33.

Alzheimer, A. (1907). Uber eine eigenartige Erkrankung der Hirnrinde. *Allgemeine Zeitschrift for Psychiatrie und Psychish-Gerichtlich Medicin, 64*, 146–148.

Amici, S., Brambati, S. M., Wilkins, D. P., Ogar, J., Dronkers, N. L., Miller, B. L., & Gorno-Tempini, M. L. (2007). Anatomical correlates of sentence comprehension and verbal working memory in neurodegenerative disease. *Journal of Neuroscience, 27*(23), 6282–6290.

Amitay, S., Ahissar, M., & Nelken, I. (2002). Auditory processing deficits in reading disabled adults. *Journal of the Association for Research in Otolaryngology, 3*(3), 302–320.

Amitay, S., Ben-Yehudah, G., Banai, K., & Ahissar, M. (2002). Disabled readers suffer from visual and auditory impairments but not from a specific magnocellular deficit. *Brain, 125*(Pt 10), 2272–2285.

ASHA. (2007). *Technical report: Childhood apraxia of speech*. ASHA.

Ash, S., Evans, E., O'Shea, J., Powers, J., Boller, A., Weinberg, D., ... Grossman, M. (2013). Differentiating primary progressive aphasias in a brief sample of connected speech. *Neurology, 81*(4), 329–336.

Ash, S., Moore, P., Vesely, L., Gunawardena, D., McMillan, C., Anderson, C., ... Grossman, M. (2009). Non-fluent speech in frontotemporal lobar degeneration. *Journal of Neurolinguistics, 22*(4), 370–383.

Baldo, J. V., Katseff, S., & Dronkers, N. F. (2012). Brain regions underlying repetition and auditory-verbal short-term memory deficits in aphasia: evidence from voxel-based lesion symptom mapping. *Aphasiology, 26*(3–4), 338–354.

Bates, T. C., Luciano, M., Castles, A., Coltheart, M., Wright, M. J., & Martin, N. G. (2007). Replication of reported linkages for dyslexia and spelling and suggestive evidence for novel regions on chromosomes 4 and 17. *European Journal of Human Genetics, 15*(2), 194–203.

Beck, J., Rohrer, J. D., Campbell, T., Isaacs, A., Morrison, K. E., Goodall, E. F., ... Mead, S. (2008). A distinct clinical, neuropsychological and radiological phenotype is associated with progranulin gene mutations in a large UK series. *Brain, 131*(Pt 3), 706–720.

Belton, E., Salmond, C. H., Watkins, K. E., Vargha-Khadem, F., & Gadian, D. G. (2003). Bilateral brain abnormalities associated with dominantly inherited verbal and orofacial dyspraxia. *Human Brain Mapping, 18*(3), 194–200.

Benton, A. (1984). Hemispheric dominance before Broca. *Neuropsychologia, 22*(6), 807–811.

Bessi, V., Bagnoli, S., Nacmias, B., Tedde, A., Sorbi, S., & Bracco, L. (2010). Semantic dementia associated with mutation V363I in the tau gene. *Journal of the Neurological Sciences, 296*(1–2), 112–114.

Binney, R. J., Parker, G. J., & Lambon Ralph, M. A. (2012). Convergent connectivity and graded specialization in the rostral human temporal lobe as revealed by diffusion-weighted imaging probabilistic tractography. *Journal of Cognitive Neuroscience, 24*(10), 1998–2014.

Bird, T. D., Nochlin, D., Poorkaj, P., Cherrier, M., Kaye, J., Payami, H., ... Schellenberg, G. D. (1999). A clinical pathological comparison of three families with frontotemporal dementia and identical mutations in the tau gene (P301L). *Brain, 122*(Pt 4), 741–756.

Boets, B., Op de Beeck, H. P., Vandermosten, M., Scott, S. K., Gillebert, C. R., Mantini, D., ... Ghesquiere, P. (2013). Intact but less accessible phonetic representations in adults with dyslexia. *Science, 342*(6163), 1251–1254.

Boeve, B. F., Tremont-Lukats, I. W., Waclawik, A. J., Murrell, J. R., Hermann, B., Jack, C. R., Jr., ... Ghetti, B. (2005). Longitudinal characterization of two siblings with frontotemporal dementia and parkinsonism linked to chromosome 17 associated with the S305N tau mutation. *Brain, 128*(Pt 4), 752–772.

Bogon, J., Finke, K., Schulte-Korne, G., Muller, H. J., Schneider, W. X., & Stenneken, P. (2014). Parameter-based assessment of disturbed and intact components of visual attention in children with developmental dyslexia. *Developmental Science, 17*(5), 697–713.

Bogon, J., Finke, K., & Stenneken, P. (2014). TVA-based assessment of visual attentional functions in developmental dyslexia. *Frontiers in psychology, 5*, 1172.

Boyar, F. Z., Whitney, M. M., Lossie, A. C., Gray, B. A., Keller, K. L., Stalker, H. J., ... Driscoll, D. J. (2001). A family with a grand-maternally derived interstitial duplication of proximal 15q. *Clinical Genetics, 60*(6), 421–430.

Bozeat, S., Lambon Ralph, M. A., Patterson, K., Garrard, P., & Hodges, J. R. (2000). Non-verbal semantic impairment in semantic dementia. *Neuropsychologia, 38*(9), 1207–1215.

Brambati, S. M., Rankin, K. P., Narvid, J., Seeley, W. W., Dean, D., Rosen, H. J., ... Gorno-Tempini, M. L. (2009). Atrophy progression in semantic dementia with asymmetric temporal involvement: a tensor-based morphometry study. *Neurobiology of Aging, 30*(1), 103–111.

Brambati, S. M., Termine, C., Ruffino, M., Stella, G., Fazio, F., Cappa, S. F., & Perani, D. (2004). Regional reductions of gray matter volume in familial dyslexia. *Neurology, 63*(4), 742–745.

Bridgeman, E., & Snowling, M. (1988). The perception of phoneme sequence: a comparison of dyspraxic and normal children. *British Journal of Disorders of Communication, 23*(3), 245–252.

Broca, P. (1861). Perte de la parole: ramollissement chronique et destruction partielle du lobe anterieur gauche du cerveau. *Bulletins de la Societe d'anthropologie, 2*, 235–238.

Burbridge, T. J., Wang, Y., Volz, A. J., Peschansky, V. J., Lisann, L., Galaburda, A. M., ... Rosen, G. D. (2008). Postnatal analysis of the effect of embryonic knockdown and overexpression of candidate dyslexia susceptibility gene homolog Dcdc2 in the rat. *Neuroscience, 152*(3), 723–733.

Butler, C. R., Brambati, S. M., Miller, B. L., & Gorno-Tempini, M. L. (2009). The neural correlates of verbal and nonverbal semantic processing deficits in neurodegenerative disease. *Cognitive and Behavioral Neurology, 22*(2), 73–80.

Carrion-Castillo, A., Franke, B., & Fisher, S. E. (2013). Molecular genetics of dyslexia: an overview. *Dyslexia, 19*(4), 214–240.

Carter, J. C., Lanham, D. C., Cutting, L. E., Clements-Stephens, A. M., Chen, X., Hadzipasic, M., ... Kaufmann, W. E. (2009). A dual DTI approach to analyzing white matter in children with dyslexia. *Psychiatry Research, 172*(3), 215–219.

Caso, F., Gesierich, B., Henry, M., Sidhu, M., LaMarre, A., Babiak, M., ... Gorno-Tempini, M. L. (2013). Nonfluent/agrammatic PPA with in-vivo cortical amyloidosis and pick's disease pathology. *Behavioural Neurology, 26*(1–2), 95–106.

Caso, F., Mandelli, M. L., Henry, M., Gesierich, B., Bettcher, B. M., Ogar, J., ... Gorno-Tempini, M. L. (2014). In vivo signatures of nonfluent/agrammatic primary progressive aphasia caused by FTLD pathology. *Neurology, 82*(3), 239–247.

Cassim, R., Talcott, J. B., & Moores, E. (2014). Adults with dyslexia demonstrate large effects of crowding and detrimental effects of distractors in a visual tilt discrimination task. *PLoS One, 9*(9), e106191.

Catani, M., Dell'acqua, F., Vergani, F., Malik, F., Hodge, H., Roy, P., ... Thiebaut de Schotten, M. (2012). Short frontal lobe connections of the human brain. *Cortex, 48*(2), 273–291.

Catani, M., Mesulam, M. M., Jakobsen, E., Malik, F., Martersteck, A., Wieneke, C., ... Rogalski, E. (2013). A novel frontal pathway underlies verbal fluency in primary progressive aphasia. *Brain, 136*(Pt 8), 2619–2628.

Chan, D., Fox, N. C., Scahill, R. I., Crum, W. R., Whitwell, J. L., Leschziner, G., ... Rossor, M. N. (2001). Patterns of temporal lobe atrophy in semantic dementia and Alzheimer's disease. *Annals of Neurology, 49*(4), 433–442.

Chapman, N. H., Igo, R. P., Thomson, J. B., Matsushita, M., Brkanac, Z., Holzman, T., ... Raskind, W. H. (2004). Linkage analyses of four regions previously implicated in dyslexia: confirmation of a locus on chromosome 15q. *American Journal of Medical Genetics, Part B: Neuropsychiatric Genetics, 131B*(1), 67–75.

Chare, L., Hodges, J. R., Leyton, C. E., McGinley, C., Tan, R. H., Kril, J. J., & Halliday, G. M. (2014). New criteria for frontotemporal dementia syndromes: clinical and pathological diagnostic implications. *Journal of Neurology Neurosurgery and Psychiatry, 85*(8), 865–870.

Chiappe, P., Stringer, R., Siegel, L., & Stanovich, K. (2002). Why the timing deficit hypothesis does not explain reading disability in adults. *Reading and Writing: An Interdisciplinary Journal, 15*, 73–107.

Chio, A., Calvo, A., Moglia, C., Restagno, G., Ossola, I., Brunetti, M., ... Borghero, G. (2010). Amyotrophic lateral sclerosis-frontotemporal lobar dementia in 3 families with p.Ala382Thr TARDBP mutations. *Archives of Neurology, 67*(8), 1002–1009.

Cooke, A., DeVita, C., Gee, J., Alsop, D., Detre, J., Chen, W., & Grossman, M. (2003). Neural basis for sentence comprehension deficits in frontotemporal dementia. *Brain and Language, 85*(2), 211–221.

Cope, N., Harold, D., Hill, G., Moskvina, V., Stevenson, J., Holmans, P., ... Williams, J. (2005). Strong evidence that KIAA0319 on chromosome 6p is a susceptibility gene for developmental dyslexia. *American Journal of Human Genetics, 76*(4), 581–591.

Couto, J. M., Gomez, L., Wigg, K., Cate-Carter, T., Archibald, J., Anderson, B., ... Barr, C. L. (2008). The KIAA0319-like (KIAA0319L) gene on chromosome 1p34 as a candidate for reading disabilities. *Journal of Neurogenetics, 22*(4), 295–313.

Croot, K., Ballard, K., Leyton, C. E., & Hodges, J. R. (2012). Apraxia of speech and phonological errors in the diagnosis of nonfluent/agrammatic and logopenic variants of primary progressive aphasia. *Journal of Speech and Hearing Research, 55*(5), S1562–S1572.

Darki, F., Peyrard-Janvid, M., Matsson, H., Kere, J., & Klingberg, T. (2014). DCDC2 polymorphism is associated with left temporoparietal gray and white matter structures during development. *Journal of Neuroscience, 34*(43), 14455–14462.

Davies, R. R., Graham, K. S., Xuereb, J. H., Williams, G. B., & Hodges, J. R. (2004). The human perirhinal cortex and semantic memory. *European Journal of Neuroscience, 20*(9), 2441–2446.

Davies, R. R., Scahill, V. L., Graham, A., Williams, G. B., Graham, K. S., & Hodges, J. R. (2009). Development of an MRI rating scale for multiple brain regions: comparison with volumetrics and with voxel-based morphometry. *Neuroradiology, 51*(8), 491–503.

Davis, B., Jakielski, K., & Marquardt, T. (1998). Developmental apraxia of speech: determiners of differential diagnosis. *Clinical Linguistics and Phonetics, 12*, 25–45.

Denckla, M. B., & Rudel, R. (1974). Rapid "automatized" naming of pictured objects, colors, letters and numbers by normal children. *Cortex, 10*(2), 186–202.

Deramecourt, V., Lebert, F., Debachy, B., Mackowiak-Cordoliani, M. A., Bombois, S., Kerdraon, O., ... Pasquier, F. (2010). Prediction of pathology in primary progressive language and speech disorders. *Neurology, 74*(1), 42–49.

Deutsch, G. K., Dougherty, R. F., Bammer, R., Siok, W. T., Gabrieli, J. D., & Wandell, B. (2005). Children's reading performance is correlated with white matter structure measured by diffusion tensor imaging. *Cortex, 41*(3), 354–363.

Dick, A. S., Bernal, B., & Tremblay, P. (2014). The language connectome: new pathways, new concepts. *Neuroscientist, 20*(5), 453–467.

Dick, A. S., & Tremblay, P. (2012). Beyond the arcuate fasciculus: consensus and controversy in the connectional anatomy of language. *Brain, 135*(Pt 12), 3529–3550.

Diehl, J. J., Frost, S. J., Sherman, G., Mencl, W. E., Kurian, A., Molfese, P., ... Pugh, K. R. (2014). Neural correlates of language and non-language visuospatial processing in adolescents with reading disability. *Neuroimage, 101*, 653–666.

Dole, M., Meunier, F., & Hoen, M. (2013). Gray and white matter distribution in dyslexia: a VBM study of superior temporal gyrus asymmetry. *PLoS One, 8*(10), e76823.

Dronkers, N. F., Plaisant, O., Iba-Zizen, M. T., & Cabanis, E. A. (2007). Paul Broca's historic cases: high resolution MR imaging of the brains of Leborgne and Lelong. *Brain, 130*(Pt 5), 1432–1441.

Duffy, J. R., Peach, R. K., & Strand, E. A. (2007). Progressive apraxia of speech as a sign of motor neuron disease. *American Journal of Speech-Language Pathology, 16*(3), 198–208.

Eckert, M. A., Leonard, C. M., Wilke, M., Eckert, M., Richards, T., Richards, A., & Berninger, V. (2005). Anatomical signatures of dyslexia in children: unique information from manual and voxel based morphometry brain measures. *Cortex, 41*(3), 304–315.

Edwards-Lee, T., Miller, B. L., Benson, D. F., Cummings, J. L., Russell, G. L., Boone, K., & Mena, I. (1997). The temporal variant of frontotemporal dementia. *Brain, 120*(Pt 6), 1027–1040.

Ekelman, B. L., & Aram, D. M. (1983). Syntactic findings in developmental verbal apraxia. *Journal of Communication Disorders, 16*(4), 237–250.

Elbro, C., Dalby, M., & Maarbjerg, S. (2011). Language-learning impairments: a 30-year follow-up of language-impaired children with and without psychiatric, neurological and cognitive difficulties. *International Journal of Language & Communication Disorders, 46*(4), 437–448.

Enard, W., Przeworski, M., Fisher, S. E., Lai, C. S., Wiebe, V., Kitano, T., … Paabo, S. (2002). Molecular evolution of FOXP2, a gene involved in speech and language. *Nature, 418*(6900), 869–872.

Fanizza, I., Bertuzzo, S., Beri, S., Scalera, E., Massagli, A., Sali, M. E., … Bonaglia, M. C. (2014). Genotype-phenotype relationship in a child with 2.3 Mb de novo interstitial 12p13.33-p13.32 deletion. *European Journal of Medical Genetics, 57*(7), 334–338.

Feuk, L., Kalervo, A., Lipsanen-Nyman, M., Skaug, J., Nakabayashi, K., Finucane, B., … Hannula-Jouppi, K. (2006). Absence of a paternally inherited FOXP2 gene in developmental verbal dyspraxia. *American Journal of Human Genetics, 79*(5), 965–972.

Finch, A. J., Nicolson, R. I., & Fawcett, A. J. (2002). Evidence for a neuroanatomical difference within the olivo-cerebellar pathway of adults with dyslexia. *Cortex, 38*(4), 529–539.

Finger, S. (2010). Chapter 10: the birth of localization theory. *Handbook of Clinical Neurology, 95*, 117–128.

Fisher, S. E., Marlow, A. J., Lamb, J., Maestrini, E., Williams, D. F., Richardson, A. J., … Monaco, A. P. (1999). A quantitative-trait locus on chromosome 6p influences different aspects of developmental dyslexia. *American Journal of Human Genetics, 64*(1), 146–156.

Fisher, S. E., Vargha-Khadem, F., Watkins, K. E., Monaco, A. P., & Pembrey, M. E. (1998). Localisation of a gene implicated in a severe speech and language disorder. *Nature Genetics, 18*(2), 168–170.

Francks, C., Paracchini, S., Smith, S. D., Richardson, A. J., Scerri, T. S., Cardon, L. R., … Monaco, A. P. (2004). A 77-kilobase region of chromosome 6p22.2 is associated with dyslexia in families from the United Kingdom and from the United States. *American Journal of Human Genetics, 75*(6), 1046–1058.

Friederici, A. D., & Gierhan, S. M. (2013). The language network. *Current Opinion in Neurobiology, 23*(2), 250–254.

Gabay, Y., Gabay, S., Schiff, R., Ashkenazi, S., & Henik, A. (2013). Visuospatial attention deficits in developmental dyslexia: evidence from visual and mental number line bisection tasks. *Archives of Clinical Neuropsychology, 28*(8), 829–836.

Gabel, L. A., Marin, I., LoTurco, J. J., Che, A., Murphy, C., Manglani, M., & Kass, S. (2011). Mutation of the dyslexia-associated gene Dcdc2 impairs LTM and visuo-spatial performance in mice. *Genes, Brain and Behavior, 10*(8), 868–875.

Galaburda, A. M., & Kemper, T. L. (1978). Auditory cytoarchitectonic abnormalities in a case of familial developmental dyslexia. *Transactions of the American Neurological Association, 103*, 262–265.

Galaburda, A. M., & Kemper, T. L. (1979). Cytoarchitectonic abnormalities in developmental dyslexia: a case study. *Annals of Neurology, 6*(2), 94–100.

Galaburda, A., & Livingstone, M. (1993). Evidence for a magnocellular defect in developmental dyslexia. *Annals of the New York Academy of Sciences, 682*, 70–82.

Galaburda, A. M., Sherman, G. F., Rosen, G. D., Aboitiz, F., & Geschwind, N. (1985). Developmental dyslexia: four consecutive patients with cortical anomalies. *Annals of Neurology, 18*(2), 222–233.

Galantucci, S., Tartaglia, M. C., Wilson, S. M., Henry, M. L., Filippi, M., Agosta, F., … Gorno-Tempini, M. L. (October 2011). White matter damage in primary progressive aphasias: a diffusion tensor tractography study. *Brain, 134*(Pt 10), 3011–3029.

Galton, C. J., Patterson, K., Graham, K., Lambon-Ralph, M. A., Williams, G., Antoun, N., … Hodges, J. R. (2001). Differing patterns of temporal atrophy in Alzheimer's disease and semantic dementia. *Neurology, 57*(2), 216–225.

Garibotto, V., Borroni, B., Agosti, C., Premi, E., Alberici, A., Eickhoff, S. B., … Padovani, A. (2011). Subcortical and deep cortical atrophy in Frontotemporal Lobar Degeneration. *Neurobiology of Aging, 32*(5), 875–884.

Gefen, T., Gasho, K., Rademaker, A., Lalehzari, M., Weintraub, S., Rogalski, E., … Mesulam, M. M. (2012). Clinically concordant variations of Alzheimer pathology in aphasic versus amnestic dementia. *Brain, 135*(Pt 5), 1554–1565.

Giraud, A. L., & Ramus, F. (2013). Neurogenetics and auditory processing in developmental dyslexia. *Current Opinion in Neurobiology, 23*(1), 37–42.

Goldman, J. S., Farmer, J. M., Wood, E. M., Johnson, J. K., Boxer, A., Neuhaus, J., … Miller, B. L. (2005). Comparison of family histories in FTLD subtypes and related tauopathies. *Neurology, 65*(11), 1817–1819.

Goldman, J. S., Rademakers, R., Huey, E. D., Boxer, A. L., Mayeux, R., Miller, B. L., & Boeve, B. F. (2011). An algorithm for genetic testing of frontotemporal lobar degeneration. *Neurology, 76*(5), 475–483.

Gori, S., & Facoetti, A. (2015). How the visual aspects can be crucial in reading acquisition? The intriguing case of crowding and developmental dyslexia. *Journal of Vision, 15*(1), 15.1.8.

Gorno-Tempini, M. L., Brambati, S. M., Ginex, V., Ogar, J., Dronkers, N. F., Marcone, A., … Miller, B. L. (October 14, 2008). The logopenic/phonological variant of primary progressive aphasia. *Neurology, 71*(16), 1227–1234.

Gorno-Tempini, M. L., Dronkers, N. F., Rankin, K. P., Ogar, J. M., Phengrasamy, L., Rosen, H. J., … Miller, B. L. (March 2004). Cognition and anatomy in three variants of primary progressive aphasia. *Annals of Neurology, 55*(3), 335–346.

Gorno-Tempini, M. L., Hillis, A. E., Weintraub, S., Kertesz, A., Mendez, M., Cappa, S. F., … Grossman, M. (2011). Classification of primary progressive aphasia and its variants. *Neurology, 76*(11), 1006–1014.

Gorno-Tempini, M. L., Murray, R. C., Rankin, K. P., Weiner, M. W., & Miller, B. L. (2004). Clinical, cognitive and anatomical evolution from nonfluent progressive aphasia to corticobasal syndrome: a case report. *Neurocase, 10*(6), 426–436.

Gorno-Tempini, M. L., Rankin, K. P., Woolley, J. D., Rosen, H. J., Phengrasamy, L., & Miller, B. L. (2004). Cognitive and behavioral profile in a case of right anterior temporal lobe neurodegeneration. *Cortex, 40*(4–5), 631–644.

Gow, D. W., Jr. (2012). The cortical organization of lexical knowledge: a dual lexicon model of spoken language processing. *Brain and Language, 121*(3), 273–288.

Graham, K., Patterson, K., & Hodges, J. (1995). Progressive pure anomia: insufficient activation of phonology by meaning. *Neurocase, 1*, 25–38.

Grati, M., Chakchouk, I., Ma, Q., Bensaid, M., Desmidt, A., Turki, N., … Masmoudi, S. (2015). A missense mutation in DCDC2 causes human recessive deafness DFNB66, likely by interfering with sensory hair cell and supporting cell cilia length regulation. *Human Molecular Genetics, 24*(9), 2482–2491.

Graves, R. E. (1997). The legacy of the Wernicke-Lichtheim model. *Journal of the History of the Neurosciences, 6*(1), 3–20.

Grigorenko, E. L., Wood, F. B., Meyer, M. S., Hart, L. A., Speed, W. C., Shuster, A., & Pauls, D. L. (1997). Susceptibility loci for distinct components of developmental dyslexia on chromosomes 6 and 15. *American Journal of Human Genetics, 60*(1), 27–39.

Groenen, P., Maassen, B., Crul, T., & Thoonen, G. (1996). The specific relation between perception and production errors for place of articulation in developmental apraxia of speech. *Journal of Speech Language and Hearing Research, 39*(3), 468–482.

Grossman, M. (2010). Primary progressive aphasia: clinicopathological correlations. *Nature Reviews Neurology, 6*(2), 88–97.

Grossman, M. (2012). The non-fluent/agrammatic variant of primary progressive aphasia. *Lancet Neurology, 11*(6), 545–555.

Grossman, M., Mickanin, J., Onishi, K., Hughes, E., D'Esposito, M., Ding, X. S., … Reivich, M. (1996). Progressive non-fluent aphasia: language, cognitive and PET measures contrasted with probable Alzheimer's disease. *Journal of Cognitive Neuroscience, 8*, 135–154.

Grossman, M., Powers, J., Ash, S., McMillan, C., Burkholder, L., Irwin, D., & Trojanowski, J. Q. (2012). Disruption of large-scale neural networks in non-fluent/agrammatic variant primary progressive aphasia associated with frontotemporal degeneration pathology. *Brain and Language, 127*(2), 106–120.

Groszer, M., Keays, D. A., Deacon, R. M., de Bono, J. P., Prasad-Mulcare, S., Gaub, S., … Fisher, S. E. (2008). Impaired synaptic plasticity and motor learning in mice with a point mutation implicated in human speech deficits. *Current Biology, 18*(5), 354–362.

Gunawardena, D., Ash, S., McMillan, C., Avants, B., Gee, J., & Grossman, M. (2010). Why are patients with progressive nonfluent aphasia nonfluent? *Neurology, 75*(7), 588–594.

Guo, C. C., Gorno-Tempini, M. L., Gesierich, B., Henry, M., Trujillo, A., Shany-Ur, T., … Seeley, W. W. (2013). Anterior temporal lobe degeneration produces widespread network-driven dysfunction. *Brain, 136*(Pt 10), 2979–2991.

Haesler, S., Rochefort, C., Georgi, B., Licznerski, P., Osten, P., & Scharff, C. (2007). Incomplete and inaccurate vocal imitation after knockdown of FoxP2 in songbird basal ganglia nucleus Area X. *PLoS Biology, 5*(12), e321.

Hannula-Jouppi, K., Kaminen-Ahola, N., Taipale, M., Eklund, R., Nopola-Hemmi, J., Kaariainen, H., & Kere, J. (2005). The axon guidance receptor gene ROBO1 is a candidate gene for developmental dyslexia. *PLoS Genetics, 1*(4), e50.

Hari, R., & Renvall, H. (2001). Impaired processing of rapid stimulus sequences in dyslexia. *Trends in Cognitive Sciences, 5*(12), 525–532.

Hari, R., Renvall, H., & Tanskanen, T. (2001). Left minineglect in dyslexic adults. *Brain, 124*(Pt 7), 1373–1380.

Harold, D., Paracchini, S., Scerri, T., Dennis, M., Cope, N., Hill, G., … Monaco, A. P. (2006). Further evidence that the KIAA0319 gene confers susceptibility to developmental dyslexia. *Molecular Psychiatry, 11*(12), 1085–1091, 1061.

Henry, M. L., Wilson, S. M., Ogar, J. M., Sidhu, M. S., Rankin, K. P., Cattaruzza, T., … Seeley, W. W. (2012). Neuropsychological, behavioral, and anatomical evolution in right temporal variant frontotemporal dementia: a longitudinal and post-mortem single case analysis. *Neurocase, 20*(1).

Heston, J. B., & White, S. A. (2015). Behavior-linked FoxP2 regulation enables zebra finch vocal learning. *Journal of Neuroscience, 35*(7), 2885–2894.

Hickok, G., & Poeppel, D. (2007). The cortical organization of speech processing. *Nature Reviews Neuroscience, 8*(5), 393–402.

Hodges, J. R., Martinos, M., Woollams, A. M., Patterson, K., & Adlam, A. L. (2008). Repeat and Point: differentiating semantic dementia from progressive non-fluent aphasia. *Cortex, 44*(9), 1265–1270.

Hodges, J. R., Mitchell, J., Dawson, K., Spillantini, M. G., Xuereb, J. H., McMonagle, P., … Patterson, K. (2010). Semantic dementia: demography, familial factors and survival in a consecutive series of 100 cases. *Brain, 133*(Pt 1), 300–306.

Hodges, J. R., & Patterson, K. (2007). Semantic dementia: a unique clinicopathological syndrome. *Lancet Neurology, 6*(11), 1004–1014.

Hodges, J. R., Patterson, K., Oxbury, S., & Funnell, E. (1992). Semantic dementia. Progressive fluent aphasia with temporal lobe atrophy. *Brain, 115*(Pt 6), 1783–1806.

Hoeft, F., Meyler, A., Hernandez, A., Juel, C., Taylor-Hill, H., Martindale, J. L., … Gabrieli, J. D. (2007). Functional and morphometric brain dissociation between dyslexia and reading ability. *Proceedings of the National Academy of Sciences of the United States of America, 104*(10), 4234–4239.

Hu, W. T., McMillan, C., Libon, D., Leight, S., Forman, M., Lee, V. M., … Grossman, M. (2010). Multimodal predictors for Alzheimer disease in nonfluent primary progressive aphasia. *Neurology, 75*(7), 595–602.

Humphreys, P., Kaufmann, W. E., & Galaburda, A. M. (1990). Developmental dyslexia in women: neuropathological findings in three patients. *Annals of Neurology, 28*(6), 727–738.

Irish, M., Hodges, J. R., & Piguet, O. (2014). Right anterior temporal lobe dysfunction underlies theory of mind impairments in semantic dementia. *Brain, 137*(Pt 4), 1241–1253.

Josephs, K. A., Dickson, D. W., Murray, M. E., Senjem, M. L., Parisi, J. E., Petersen, R. C., … Whitwell, J. L. (2013). Quantitative neurofibrillary tangle density and brain volumetric MRI analyses in Alzheimer's disease presenting as logopenic progressive aphasia. *Brain and Language, 127*(2), 127–134.

Josephs, K. A., Duffy, J. R., Strand, E. A., Machulda, M. M., Senjem, M. L., Lowe, V. J., … Whitwell, J. L. (2014). APOE epsilon4 influences beta-amyloid deposition in primary progressive aphasia and speech apraxia. *Alzheimer's & Dementia, 10*(6), 630–636.

Josephs, K. A., Duffy, J. R., Strand, E. A., Whitwell, J. L., Layton, K. F., Parisi, J. E., … Petersen, R. C. (2006). Clinicopathological and imaging correlates of progressive aphasia and apraxia of speech. *Brain, 129*(Pt 6), 1385–1398.

Josephs, K. A., Hodges, J. R., Snowden, J. S., Mackenzie, I. R., Neumann, M., Mann, D. M., & Dickson, D. W. (2011). Neuropathological background of phenotypical variability in frontotemporal dementia. *Acta Neuropathologica, 122*(2), 137–153.

Josephs, K. A., Stroh, A., Dugger, B., & Dickson, D. W. (2009). Evaluation of subcortical pathology and clinical correlations in FTLD-U subtypes. *Acta Neuropathologica, 118*(3), 349–358.

Justice, L. M., & Schuele, C. M. (2004). In J. E. Bernthal, & N. W. Bankson (Eds.), *Phonological awareness: Description, assessment, and intervention. Articulation and phonological disorders* (pp. 376–405). Boston: Allyn & Bacon.

Klingberg, T., Hedehus, M., Temple, E., Salz, T., Gabrieli, J. D., Moseley, M. E., & Poldrack, R. A. (2000). Microstructure of temporo-parietal white matter as a basis for reading ability: evidence from diffusion tensor magnetic resonance imaging. *Neuron, 25*(2), 493–500.

Knibb, J. A., Xuereb, J. H., Patterson, K., & Hodges, J. R. (2006). Clinical and pathological characterization of progressive aphasia. *Annals of Neurology, 59*(1), 156–165.

Konopka, G., Bomar, J. M., Winden, K., Coppola, G., Jonsson, Z. O., Gao, F., … Geschwind, D. H. (2009). Human-specific transcriptional regulation of CNS development genes by FOXP2. *Nature, 462*(7270), 213–217.

Krafnick, A. J., Flowers, D. L., Napoliello, E. M., & Eden, G. F. (2011). Gray matter volume changes following reading intervention in dyslexic children. *Neuroimage, 57*(3), 733–741.

Kronbichler, M., Wimmer, H., Staffen, W., Hutzler, F., Mair, A., & Ladurner, G. (2008). Developmental dyslexia: gray matter abnormalities in the occipitotemporal cortex. *Human Brain Mapping, 29*(5), 613–625.

Kubova, Z., Kuba, M., Kremlacek, J., Langrova, J., Szanyi, J., Vit, F., & Chutna, M. (2015). Comparison of visual information processing in school-age dyslexics and normal readers via motion-onset visual evoked potentials. *Vision Research, 111*(Pt A), 97–104.

Lai, C. S., Fisher, S. E., Hurst, J. A., Vargha-Khadem, F., & Monaco, A. P. (2001). A forkhead-domain gene is mutated in a severe speech and language disorder. *Nature, 413*(6855), 519–523.

Lai, C. S., Gerrelli, D., Monaco, A. P., Fisher, S. E., & Copp, A. J. (2003). FOXP2 expression during brain development coincides with adult sites of pathology in a severe speech and language disorder. *Brain, 126*(Pt 11), 2455–2462.

Lamminmaki, S., Massinen, S., Nopola-Hemmi, J., Kere, J., & Hari, R. (2012). Human ROBO1 regulates interaural interaction in auditory pathways. *Journal of Neuroscience, 32*(3), 966–971.

Law, J., Boyle, J., Harris, F., Harkness, A., & Nye, C. (2000). Prevalence and natural history of primary speech and language delay: findings from a systematic review of the literature. *International Journal of Language & Communication Disorders, 35*(2), 165–188.

Law, J. M., Vandermosten, M., Ghesquiere, P., & Wouters, J. (2014). The relationship of phonological ability, speech perception, and auditory perception in adults with dyslexia. *Frontiers in Human Neuroscience, 8*, 482.

Le Ber, I., Camuzat, A., Hannequin, D., Pasquier, F., Guedj, E., Rovelet-Lecrux, A., … French Research Network on FTD/FTD-MND. (2008). Phenotype variability in progranulin mutation carriers: a clinical, neuropsychological, imaging and genetic study. *Brain, 131*(Pt 3), 732–746.

Lebel, C., Shaywitz, B., Holahan, J., Shaywitz, S., Marchione, K., & Beaulieu, C. (2013). Diffusion tensor imaging correlates of reading ability in dysfluent and non-impaired readers. *Brain and Language, 125*(2), 215–222.

Leff, A. P., Schofield, T. M., Crinion, J. T., Seghier, M. L., Grogan, A., Green, D. W., & Price, C. J. (2009). The left superior temporal gyrus is a shared substrate for auditory short-term memory and speech comprehension: evidence from 210 patients with stroke. *Brain, 132*(Pt 12), 3401–3410.

Lehmann, M., Ghosh, P. M., Madison, C., Laforce, R., Jr., Corbetta-Rastelli, C., Weiner, M. W., … Rabinovici, G. D. (2013). Diverging patterns of amyloid deposition and hypometabolism in clinical variants of probable Alzheimer's disease. *Brain, 136*(Pt 3), 844–858.

Lennon, P. A., Cooper, M. L., Peiffer, D. A., Gunderson, K. L., Patel, A., Peters, S., … Bacino, C. A. (2007). Deletion of 7q31.1 supports involvement of FOXP2 in language impairment: clinical report and review. *American Journal of Medical Genetics Part A, 143A*(8), 791–798.

Lewis, B. A., Freebairn, L. A., Hansen, A. J., Iyengar, S. K., & Taylor, H. G. (2004). School-age follow-up of children with childhood apraxia of speech. *Language, Speech, and Hearing Services in Schools, 35*, 122–140.

Leyton, C. E., Piguet, O., Savage, S., Burrell, J., & Hodges, J. R. (2012). The neural basis of logopenic progressive aphasia. *Journal of Alzheimer's Disease, 32*(4), 1051–1059.

Liegeois, F., Baldeweg, T., Connelly, A., Gadian, D. G., Mishkin, M., & Vargha-Khadem, F. (2003). Language fMRI abnormalities associated with FOXP2 gene mutation. *Nature Neuroscience, 6*(11), 1230–1237.

Lindgren, S. D., De Renzi, E., & Richman, L. C. (1985). Cross-national comparisons of developmental dyslexia in Italy and the United States. *Child Development, 56*(6), 1404–1417.

Linkersdorfer, J., Lonnemann, J., Lindberg, S., Hasselhorn, M., & Fiebach, C. J. (2012). Grey matter alterations co-localize with functional abnormalities in developmental dyslexia: an ALE meta-analysis. *PLoS One, 7*(8), e43122.

Livingstone, M. S., Rosen, G. D., Drislane, F. W., & Galaburda, A. M. (1991). Physiological and anatomical evidence for a magnocellular defect in developmental dyslexia. *Proceedings of the National Academy of Sciences of the United States of America, 88*(18), 7943–7947.

Luzzi, S., Snowden, J. S., Neary, D., Coccia, M., Provinciali, L., & Lambon Ralph, M. A. (2007). Distinct patterns of olfactory impairment in Alzheimer's disease, semantic dementia, frontotemporal dementia, and corticobasal degeneration. *Neuropsychologia, 45*(8), 1823–1831.

Maassen, B. (2002). Issues contrasting adult acquired versus developmental apraxia of speech. *Seminars in Speech and Language, 23*, 257–266.

Maassen, B., Groenen, P., & Crul, T. (2003). Auditory and phonetic perception of vowels in children with apraxic speech disorders. *Clinical Linguistics & Phonetics, 17*(6), 447–467.

MacDermot, K. D., Bonora, E., Sykes, N., Coupe, A. M., Lai, C. S., Vernes, S. C., ... Fisher, S. E. (2005). Identification of FOXP2 truncation as a novel cause of developmental speech and language deficits. *American Journal of Human Genetics, 76*(6), 1074–1080.

Mahon, B. Z., & Caramazza, A. (2009). Concepts and categories: a cognitive neuropsychological perspective. *Annual Review of Psychology, 60*, 27–51.

Mahon, B. Z., & Caramazza, A. (2011). What drives the organization of object knowledge in the brain? *Trends in Cognitive Sciences, 15*(3), 97–103.

Mahoney, C. J., Beck, J., Rohrer, J. D., Lashley, T., Mok, K., Shakespeare, T., ... Warren, J. D. (2012). Frontotemporal dementia with the C9ORF72 hexanucleotide repeat expansion: clinical, neuroanatomical and neuropathological features. *Brain, 135*(Pt 3), 736–750.

Mahoney, C. J., Malone, I. B., Ridgway, G. R., Buckley, A. H., Downey, L. E., Golden, H. L., ... Warren, J. D. (2013). White matter tract signatures of the progressive aphasias. *Neurobiology of Aging, 34*(6), 1687–1699.

Mandelli, M. L., Caverzasi, E., Binney, R. J., Henry, M. L., Lobach, I., Block, N., ... Gorno-Tempini, M. L. (2014). Frontal white matter tracts sustaining speech production in primary progressive aphasia. *Journal of Neuroscience, 34*(29), 9754–9767.

Manning, L., & Thomas-Anterion, C. (2011). Marc Dax and the discovery of the lateralisation of language in the left cerebral hemisphere. *Revue Neurologique (Paris), 167*(12), 868–872.

Marino, C., Scifo, P., Della Rosa, P. A., Mascheretti, S., Facoetti, A., Lorusso, M. L., ... Perani, D. (2014). The DCDC2/intron 2 deletion and white matter disorganization: focus on developmental dyslexia. *Cortex, 57*, 227–243.

Marquardt, T. P., Sussman, H. M., Snow, T., & Jacks, A. (2002). The integrity of the syllable in developmental apraxia of speech. *Journal of Communication Disorders, 35*(1), 31–49.

Martino, J., & De Lucas, E. M. (2014). Subcortical anatomy of the lateral association fascicles of the brain: a review. *Clinical Anatomy, 27*(4), 563–569.

Martory, M. D., Roth, S., Lovblad, K. O., Neumann, M., Lobrinus, J. A., & Assal, F. (2012). Creutzfeldt-Jakob disease revealed by a logopenic variant of primary progressive aphasia. *European Neurology, 67*(6), 360–362.

Mascheretti, S., Bureau, A., Trezzi, V., Giorda, R., & Marino, C. (2015). An assessment of gene-by-gene interactions as a tool to unfold missing heritability in dyslexia. *Human Genetics, 134*(7), 749–760.

Mascheretti, S., Riva, V., Giorda, R., Beri, S., Lanzoni, L. F., Cellino, M. R., & Marino, C. (2014). KIAA0319 and ROBO1: evidence on association with reading and pleiotropic effects on language and mathematics abilities in developmental dyslexia. *Journal of Human Genetics, 59*(4), 189–197.

Matsson, H., Huss, M., Persson, H., Einarsdottir, E., Tiraboschi, E., Nopola-Hemmi, J., ... Kere, J. (2015). Polymorphisms in DCDC2 and S100B associate with developmental dyslexia. *Journal of Human Genetics, 60*(7), 399–401.

McCabe, P., Rosenthal, J. B., & McLeod, S. (1998). Features of developmental dyspraxia in the general speech-impaired population? *Clinical Linguistics & Phonetics, 12*(2), 105–126.

McLeod, S., & McKinnon, D. (2007). Prevalence of communication disorders compared with other learning needs in 14,500 primary and secondary school students. *International Journal of Language & Communication Disorders, 42*(Suppl. 1), 37–59.

Meng, H., Powers, N. R., Tang, L., Cope, N. A., Zhang, P. X., Fuleihan, R., ... Gruen, J. R. (2011). A dyslexia-associated variant in DCDC2 changes gene expression. *Behavior Genetics, 41*(1), 58–66.

Meng, H., Smith, S. D., Hager, K., Held, M., Liu, J., Olson, R. K., ... Gruen, J. R. (2005). DCDC2 is associated with reading disability and modulates neuronal development in the brain. *Proceedings of the National Academy of Sciences of the United States of America, 102*(47), 17053–17058.

Mesulam, M. M. (1982). Slowly progressive aphasia without generalized dementia. *Annals of Neurology, 11*(6), 592–598.

Mesulam, M., Wicklund, A., Johnson, N., Rogalski, E., Leger, G. C., Rademaker, A., ... Bigio, E. H. (2008). Alzheimer and frontotemporal pathology in subsets of primary progressive aphasia. *Annals of Neurology, 63*(6), 709–719.

Migliaccio, R., Agosta, F., Rascovsky, K., Karydas, A., Bonasera, S., Rabinovici, G. D., ... Gorno-Tempini, M. L. (November 10, 2009). Clinical syndromes associated with posterior atrophy: early age at onset AD spectrum. *Neurology, 73*(19), 1571–1578.

Miller, Z. A., Mandelli, M. L., Rankin, K. P., Henry, M. L., Babiak, M. C., Frazier, D. T., ... Gorno-Tempini, M. L. (2013). Handedness and language learning disability differentially distribute in progressive aphasia variants. *Brain, 136*(Pt 11), 3461–3473.

Miller, Z., Rankin, K., Graff-Radford, N., Takada, L., Sturm, V., Cleveland, C., ... Miller, B. (2013). TDP-43 frontotemporal lobar degeneration and autoimmune disease. *Journal of Neurology, Neurosurgery and Psychiatry, 84*(9), 956–962.

Mire, E., Mezzera, C., Leyva-Diaz, E., Paternain, A. V., Squarzoni, P., Bluy, L., ... Lopez-Bendito, G. (2012). Spontaneous activity regulates Robo1 transcription to mediate a switch in thalamocortical axon growth. *Nature Neuroscience, 15*(8), 1134–1143.

Mummery, C. J., Patterson, K., Price, C. J., Ashburner, J., Frackowiak, R. S., & Hodges, J. R. (2000). A voxel-based morphometry study of semantic dementia: relationship between temporal lobe atrophy and semantic memory. *Annals of Neurology, 47*(1), 36–45.

Murugan, M., Harward, S., Scharff, C., & Mooney, R. (2013). Diminished FoxP2 levels affect dopaminergic modulation of corticostriatal signaling important to song variability. *Neuron, 80*(6), 1464–1476.

Nestor, P. J., Balan, K., Cheow, H. K., Fryer, T. D., Knibb, J. A., Xuereb, J. H., & Hodges, J. R. (2007). Nuclear imaging can predict pathologic diagnosis in progressive nonfluent aphasia. *Neurology, 68*(3), 238–239.

Nestor, P. J., Graham, N. L., Fryer, T. D., Williams, G. B., Patterson, K., & Hodges, J. R. (2003). Progressive non-fluent aphasia is associated with hypometabolism centred on the left anterior insula. *Brain, 126*(Pt 11), 2406–2418.

Newbury, D. F., Paracchini, S., Scerri, T. S., Winchester, L., Addis, L., Richardson, A. J., ... Monaco, A. P. (2011). Investigation of dyslexia and SLI risk variants in reading- and language-impaired subjects. *Behavior Genetics, 41*(1), 90–104.

Nicolson, R. I., Fawcett, A. J., & Dean, P. (2001). Developmental dyslexia: the cerebellar deficit hypothesis. *Trends in Neurosciences, 24*(9), 508–511.

Nijland, L., Maassen, B., Van Der Meulen, S., Gabreels, F., Kraaimaat, F. W., & Schreuder, R. (2003). Planning of syllables in children with developmental apraxia of speech. *Clinical Linguistics & Phonetics, 17*(1), 1–24.

Nijland, L., Terband, H., & Maassen, B. (2015). Cognitive functions in childhood apraxia of speech. *Journal of Speech and Hearing Research, 58*(3), 550–565.

Niogi, S. N., & McCandliss, B. D. (2006). Left lateralized white matter microstructure accounts for individual differences in reading ability and disability. *Neuropsychologia, 44*(11), 2178–2188.

Nopola-Hemmi, J., Myllyluoma, B., Haltia, T., Taipale, M., Ollikainen, V., Ahonen, T., ... Widen, E. (2001). A dominant gene for developmental dyslexia on chromosome 3. *Journal of Medical Genetics, 38*(10), 658–664.

Nopola-Hemmi, J., Taipale, M., Haltia, T., Lehesjoki, A. E., Voutilainen, A., & Kere, J. (2000). Two translocations of chromosome 15q associated with dyslexia. *Journal of Medical Genetics, 37*(10), 771–775.

Ogar, J. M., Dronkers, N. F., Brambati, S. M., Miller, B. L., & Gorno-Tempini, M. L. (2007). Progressive nonfluent aphasia and its characteristic motor speech deficits. *Alzheimer Disease and Associated Disorders, 21*(4), S23–S30.

Ogar, J., Slama, H., Dronkers, N., Amici, S., & Gorno-Tempini, M. L. (2005). Apraxia of speech: an overview. *Neurocase, 11*(6), 427–432.

Palka, C., Alfonsi, M., Mohn, A., Cerbo, R., Guanciali Franchi, P., Fantasia, D., ... Palka, G. (2012). Mosaic 7q31 deletion involving FOXP2 gene associated with language impairment. *Pediatrics, 129*(1), e183–188.

Paracchini, S., Thomas, A., Castro, S., Lai, C., Paramasivam, M., Wang, Y., ... Monaco, A. P. (2006). The chromosome 6p22 haplotype associated with dyslexia reduces the expression of KIAA0319, a novel gene involved in neuronal migration. *Human Molecular Genetics, 15*(10), 1659–1666.

Patterson, K., Nestor, P. J., & Rogers, T. T. (2007). Where do you know what you know? The representation of semantic knowledge in the human brain. *Nature Reviews Neuroscience, 8*(12), 976–987.

Paulesu, E., Danelli, L., & Berlingeri, M. (2014). Reading the dyslexic brain: multiple dysfunctional routes revealed by a new meta-analysis of PET and fMRI activation studies. *Frontiers in Human Neuroscience, 8*, 830.

Paulesu, E., Demonet, J. F., Fazio, F., McCrory, E., Chanoine, V., Brunswick, N., ... Frith, U. (2001). Dyslexia: cultural diversity and biological unity. *Science, 291*(5511), 2165–2167.

Peter, B., Matsushita, M., Oda, K., & Raskind, W. (2014). De novo microdeletion of BCL11A is associated with severe speech sound disorder. *American Journal of Medical Genetics Part A, 164A*(8), 2091–2096.

Peterson, R. L., & Pennington, B. F. (2012). Developmental dyslexia. *Lancet, 379*(9830), 1997–2007.

Pick, A. (1892). Uber die Beziehungen der senilen Hirnatrophie zur Aphasie. *Prager Medizinische Wochenschrift, 17*, 165–167.

Pickering-Brown, S. M., Rollinson, S., Du Plessis, D., Morrison, K. E., Varma, A., Richardson, A. M., ... Mann, D. M. (2008). Frequency and clinical characteristics of progranulin mutation carriers in the Manchester frontotemporal lobar degeneration cohort: comparison with patients with MAPT and no known mutations. *Brain, 131*(Pt 3), 721–731.

Platko, J. V., Wood, F. B., Pelser, I., Meyer, M., Gericke, G. S., O'Rourke, J., ... Pauls, D. L. (2008). Association of reading disability on chromosome 6p22 in the Afrikaner population. *American Journal of Medical Genetics, Part B: Neuropsychiatric Genetics, 147B*(7), 1278–1287.

Platt, M. P., Adler, W. T., Mehlhorn, A. J., Johnson, G. C., Wright, K. A., Choi, R. T., ... Rosen, G. D. (2013). Embryonic disruption of the candidate dyslexia susceptibility gene homolog Kiaa0319-like results in neuronal migration disorders. *Neuroscience, 248*, 585–593.

Poelmans, G., Buitelaar, J. K., Pauls, D. L., & Franke, B. (2011). A theoretical molecular network for dyslexia: integrating available genetic findings. *Molecular Psychiatry, 16*(4), 365–382.

Poelmans, H., Luts, H., Vandermosten, M., Boets, B., Ghesquiere, P., & Wouters, J. (2011). Reduced sensitivity to slow-rate dynamic auditory information in children with dyslexia. *Research in Developmental Disabilities, 32*(6), 2810–2819.

Poeppel, D., & Hickok, G. (2004). Towards a new functional anatomy of language. *Cognition, 92*(1–2), 1–12.

Poon, M. W., Tsang, W. H., Chan, S. O., Li, H. M., Ng, H. K., & Waye, M. M. (2011). Dyslexia-associated kiaa0319-like protein interacts with axon guidance receptor nogo receptor 1. *Cellular and Molecular Neurobiology, 31*(1), 27–35.

Potter, N. L., Nievergelt, Y., & Shriberg, L. D. (2013). Motor and speech disorders in classic galactosemia. *JIMD Reports, 11*, 31–41.

Price, C. J. (2000). The anatomy of language: contributions from functional neuroimaging. *Journal of Anatomy, 197*(Pt 3), 335–359.

Price, C. J. (2012). A review and synthesis of the first 20 years of PET and fMRI studies of heard speech, spoken language and reading. *Neuroimage, 62*(2), 816–847.

Quercia, P., Feiss, L., & Michel, C. (2013). Developmental dyslexia and vision. *Clinical Ophthalmology, 7*, 869–881.

Rabinovici, G. D., Jagust, W. J., Furst, A. J., Ogar, J. M., Racine, C. A., Mormino, E. C., ... Gorno-Tempini, M. L. (October 2008). Abeta amyloid and glucose metabolism in three variants of primary progressive aphasia. *Annals of Neurology, 64*(4), 388–401.

Raca, G., Baas, B. S., Kirmani, S., Laffin, J. J., Jackson, C. A., Strand, E. A., ... Shriberg, L. D. (2013). Childhood Apraxia of Speech (CAS) in two patients with 16p11.2 microdeletion syndrome. *European Journal of Human Genetics, 21*(4), 455–459.

Ralph, M. A., & Howard, D. (2000). Gogi aphasia or semantic dementia? Simulating and assessing poor verbal comprehension in a case of progressive fluent aphasia. *Cognitive Neuropsychology, 17*(5), 437–465.

Ramus, F. (2003). Developmental dyslexia: specific phonological deficit or general sensorimotor dysfunction? *Current Opinion in Neurobiology, 13*(2), 212–218.

Ramus, F., Pidgeon, E., & Frith, U. (2003). The relationship between motor control and phonology in dyslexic children. *Journal of Child Psychology and Psychiatry, 44*(5), 712–722.

Ramus, F., Rosen, S., Dakin, S. C., Day, B. L., Castellote, J. M., White, S., & Frith, U. (2003). Theories of developmental dyslexia: insights from a multiple case study of dyslexic adults. *Brain, 126*(Pt 4), 841–865.

Raschle, N. M., Chang, M., & Gaab, N. (2011). Structural brain alterations associated with dyslexia predate reading onset. *Neuroimage, 57*(3), 742–749.

Raschle, N. M., Zuk, J., & Gaab, N. (2012). Functional characteristics of developmental dyslexia in left-hemispheric posterior brain regions predate reading onset. *Proceedings of the National Academy of Sciences of the United States of America, 109*(6), 2156–2161.

Richlan, F. (2012). Developmental dyslexia: dysfunction of a left hemisphere reading network. *Frontiers in Human Neuroscience, 6*, 120.

Richlan, F., Kronbichler, M., & Wimmer, H. (2009). Functional abnormalities in the dyslexic brain: a quantitative meta-analysis of neuroimaging studies. *Human Brain Mapping, 30*(10), 3299–3308.

Richlan, F., Kronbichler, M., & Wimmer, H. (2011). Meta-analyzing brain dysfunctions in dyslexic children and adults. *Neuroimage, 56*(3), 1735–1742.

Richlan, F., Kronbichler, M., & Wimmer, H. (2013). Structural abnormalities in the dyslexic brain: a meta-analysis of voxel-based morphometry studies. *Human Brain Mapping, 34*(11), 3055–3065.

Rimrodt, S. L., Peterson, D. J., Denckla, M. B., Kaufmann, W. E., & Cutting, L. E. (2010). White matter microstructural differences linked to left perisylvian language network in children with dyslexia. *Cortex, 46*(6), 739–749.

Rochefort, C., He, X., Scotto-Lomassese, S., & Scharff, C. (2007). Recruitment of FoxP2-expressing neurons to Area X varies during song development. *Developmental Neurobiology, 67*(6), 809–817.

Rogalski, E., Cobia, D., Harrison, T. M., Wieneke, C., Thompson, C. K., Weintraub, S., & Mesulam, M. M. (2011). Anatomy of language impairments in primary progressive aphasia. *Journal of Neuroscience, 31*(9), 3344–3350.

Rogalski, E., Cobia, D., Harrison, T. M., Wieneke, C., Weintraub, S., & Mesulam, M. M. (2011). Progression of language decline and cortical atrophy in subtypes of primary progressive aphasia. *Neurology, 76*(21), 1804–1810.

Rogalski, E., Cobia, D., Martersteck, A., Rademaker, A., Wieneke, C., Weintraub, S., & Mesulam, M. M. (2014). Asymmetry of cortical decline in subtypes of primary progressive aphasia. *Neurology, 83*(13), 1184–1191.

Rohrer, J. D., Caso, F., Mahoney, C., Henry, M., Rosen, H. J., Rabinovici, G., ... Gorno-Tempini, M. L. (2013). Patterns of longitudinal brain atrophy in the logopenic variant of primary progressive aphasia. *Brain and Language, 127*(2), 121–126.

Rohrer, J. D., Crutch, S. J., Warrington, E. K., & Warren, J. D. (2010). Progranulin-associated primary progressive aphasia: a distinct phenotype? *Neuropsychologia, 48*(1), 288–297.

Rohrer, J. D., Geser, F., Zhou, J., Gennatas, E. D., Sidhu, M., Trojanowski, J. Q., ... Seeley, W. W. (2010). TDP-43 subtypes are associated with distinct atrophy patterns in frontotemporal dementia. *Neurology, 75*(24), 2204–2211.

Rohrer, J. D., Guerreiro, R., Vandrovcova, J., Uphill, J., Reiman, D., Beck, J., ... Rossor, M. N. (2009). The heritability and genetics of frontotemporal lobar degeneration. *Neurology, 73*(18), 1451–1456.

Rohrer, J. D., Lashley, T., Schott, J. M., Warren, J. E., Mead, S., Isaacs, A. M., ... Warren, J. D. (2011). Clinical and neuroanatomical signatures of tissue pathology in frontotemporal lobar degeneration. *Brain, 134*(Pt 9), 2565–2581.

Rohrer, J. D., Ridgway, G. R., Crutch, S. J., Hailstone, J., Goll, J. C., Clarkson, M. J., ... Warren, J. D. (2010). Progressive logopenic/phonological aphasia: erosion of the language network. *Neuroimage, 49*(1), 984–993.

Rohrer, J. D., Rossor, M. N., & Warren, J. D. (2012). Alzheimer's pathology in primary progressive aphasia. *Neurobiology of Aging, 33*(4), 744–752.

Rohrer, J. D., & Warren, J. D. (2010). Phenomenology and anatomy of abnormal behaviours in primary progressive aphasia. *Journal of the Neurological Sciences, 293*(1–2), 35–38.

Rohrer, J. D., Warren, J. D., Modat, M., Ridgway, G. R., Douiri, A., Rossor, M. N., ... Fox, N. C. (2009). Patterns of cortical thinning in the language variants of frontotemporal lobar degeneration. *Neurology, 72*(18), 1562–1569.

Roll, P., Rudolf, G., Pereira, S., Royer, B., Scheffer, I. E., Massacrier, A., ... Szepetowski, P. (2006). SRPX2 mutations in disorders of language cortex and cognition. *Human Molecular Genetics, 15*(7), 1195–1207.

Roll, P., Vernes, S. C., Bruneau, N., Cillario, J., Ponsole-Lenfant, M., Massacrier, A., ... Szepetowski, P. (2010). Molecular networks implicated in speech-related disorders: FOXP2 regulates the SRPX2/uPAR complex. *Human Molecular Genetics, 19*(24), 4848–4860.

Rosen, G. D., Bai, J., Wang, Y., Fiondella, C. G., Threlkeld, S. W., LoTurco, J. J., & Galaburda, A. M. (2007). Disruption of neuronal migration by RNAi of Dyx1c1 results in neocortical and hippocampal malformations. *Cerebral Cortex, 17*(11), 2562–2572.

Rosen, H. J., Perry, R. J., Murphy, J., Kramer, J. H., Mychack, P., Schuff, N., ... Miller, B. L. (2002). Emotion comprehension in the temporal variant of frontotemporal dementia. *Brain, 125*(Pt 10), 2286–2295.

Ruffino, M., Gori, S., Boccardi, D., Molteni, M., & Facoetti, A. (2014). Spatial and temporal attention in developmental dyslexia. *Frontiers in Human Neuroscience, 8*, 331.

Sapolsky, D., Bakkour, A., Negreira, A., Nalipinski, P., Weintraub, S., Mesulam, M. M., ... Dickerson, B. C. (2010). Cortical neuroanatomic correlates of symptom severity in primary progressive aphasia. *Neurology, 75*(4), 358–366.

Schatschneider, C., Carlson, C. D., Francis, D. J., Foorman, B. R., & Fletcher, J. M. (2002). Relationship of rapid automatized naming and phonological awareness in early reading development: implications for the double-deficit hypothesis. *Journal of Learning Disabilities, 35*(3), 245–256.

Schulte-Korne, G., Grimm, T., Nothen, M. M., Muller-Myhsok, B., Cichon, S., Vogt, I. R., ... Remschmidt, H. (1998). Evidence for linkage of spelling disability to chromosome 15. *American Journal of Human Genetics, 63*(1), 279–282.

Schumacher, J., Anthoni, H., Dahdouh, F., Konig, I. R., Hillmer, A. M., Kluck, N., ... Kere, J. (2006). Strong genetic evidence of DCDC2 as a susceptibility gene for dyslexia. *American Journal of Human Genetics, 78*(1), 52–62.

Schumacher, J., Hoffmann, P., Schmal, C., Schulte-Korne, G., & Nothen, M. M. (2007). Genetics of dyslexia: the evolving landscape. *Journal of Medical Genetics, 44*(5), 289–297.

Schwindt, G. C., Graham, N. L., Rochon, E., Tang-Wai, D. F., Lobaugh, N. J., Chow, T. W., & Black, S. E. (2013). Whole-brain white matter disruption in semantic and nonfluent variants of primary progressive aphasia. *Human Brain Mapping, 34*(4), 973–984.

Seddoh, S. A., Robin, D. A., Sim, H. S., Hageman, C., Moon, J. B., & Folkins, J. W. (1996). Speech timing in apraxia of speech versus conduction aphasia. *Journal of Speech Language and Hearing Research, 39*(3), 590–603.

Seelaar, H., Rohrer, J. D., Pijnenburg, Y. A., Fox, N. C., & van Swieten, J. C. (2011). Clinical, genetic and pathological heterogeneity of frontotemporal dementia: a review. *Journal of Neurology, Neurosurgery and Psychiatry, 82*(5), 476–486.

Seeley, W. W., Bauer, A. M., Miller, B. L., Gorno-Tempini, M. L., Kramer, J. H., Weiner, M., & Rosen, H. J. (2005). The natural history of temporal variant frontotemporal dementia. *Neurology, 64*(8), 1384–1390.

Sepelyak, K., Crinion, J., Molitoris, J., Epstein-Peterson, Z., Bann, M., Davis, C., ... Hillis, A. E. (2011). Patterns of breakdown in spelling in primary progressive aphasia. *Cortex, 47*(3), 342–352.

Serniclaes, W., Sprenger-Charolles, L., Carre, R., & Demonet, J. F. (2001). Perceptual discrimination of speech sounds in developmental dyslexia. *Journal of Speech and Hearing Research, 44*(2), 384–399.

Shany-Ur, T., Lin, N., Rosen, H. J., Sollberger, M., Miller, B. L., & Rankin, K. P. (2014). Self-awareness in neurodegenerative disease relies on neural structures mediating reward-driven attention. *Brain. 137*(Pt. 8), 2368–2381.

Share, D., Jorm, A. F., Maclean, R., & Matthews, R. (2002). Temporal processing and reading disability. *Reading and Writing: An Interdisciplinary Journal, 15*, 151–178.

Shim, H., Hurley, R. S., Rogalski, E., & Mesulam, M. M. (2012). Anatomic, clinical, and neuropsychological correlates of spelling errors in primary progressive aphasia. *Neuropsychologia, 50*(8), 1929–1935.

Shriberg, L. D., Aram, D. M., & Kwiatkowski, J. (1997a). Developmental apraxia of speech: I. Descriptive and theoretical perspectives. *Journal of Speech and Hearing Research, 40*(2), 273–285.

Shriberg, L. D., Aram, D. M., & Kwiatkowski, J. (1997b). Developmental apraxia of speech: III. A subtype marked by inappropriate stress. *Journal of Speech and Hearing Research, 40*(2), 313–337.

Shriberg, L. D., Green, J. R., Campbell, T. F., McSweeny, J. L., & Scheer, A. R. (2003). A diagnostic marker for childhood apraxia of speech: the coefficient of variation ratio. *Clinical Linguistics & Phonetics, 17*(7), 575–595.

Shriberg, L. D., Potter, N. L., & Strand, E. A. (2011). Prevalence and phenotype of childhood apraxia of speech in youth with galactosemia. *Journal of Speech and Hearing Research, 54*(2), 487–519.

Sia, G. M., Clem, R. L., & Huganir, R. L. (2013). The human language-associated gene SRPX2 regulates synapse formation and vocalization in mice. *Science, 342*(6161), 987–991.

Snowden, J., Goulding, P., & Neary, D. (1989). Semantic dementia: a form of circumscribed cerebral atrophy. *Behavioural Neurology, 2*, 167–182.

Snowden, J., Neary, D., & Mann, D. (2007). Frontotemporal lobar degeneration: clinical and pathological relationships. *Acta Neuropathologica, 114*(1), 31–38.

Snowden, J. S., Rollinson, S., Thompson, J. C., Harris, J. M., Stopford, C. L., Richardson, A. M., ... Pickering-Brown, S. M. (2012). Distinct clinical and pathological characteristics of frontotemporal dementia associated with C9ORF72 mutations. *Brain, 135*(Pt 3), 693–708.

Sollberger, M., Rosen, H. J., Shany-Ur, T., Ullah, J., Stanley, C. M., Laluz, V., ... Rankin, K. P. (2014). Neural substrates of socioemotional self-awareness in neurodegenerative disease. *Brain and Behavior, 4*(2), 201–214.

Spinelli, D., De Luca, M., Judica, A., & Zoccolotti, P. (2002). Crowding effects on word identification in developmental dyslexia. *Cortex, 38*(2), 179–200.

Spinelli, M., Rocha, A. C., Giacheti, C. M., & Richieri-Costa, A. (1995). Word-finding difficulties, verbal paraphasias, and verbal dyspraxia in ten individuals with fragile X syndrome. *American Journal of Medical Genetics, 60*(1), 39–43.

Stein, J. (2014). Dyslexia: the role of vision and visual attention. *Current Developmental Disorders Reports, 1*(4), 267–280.

Steinbrink, C., Vogt, K., Kastrup, A., Muller, H. P., Juengling, F. D., Kassubek, J., & Riecker, A. (2008). The contribution of white and gray matter differences to developmental dyslexia: insights from DTI and VBM at 3.0 T. *Neuropsychologia, 46*(13), 3170–3178.

Stoodley, C. J. (2014). Distinct regions of the cerebellum show gray matter decreases in autism, ADHD, and developmental dyslexia. *Frontiers in Systems Neuroscience, 8*, 92.

Strand, E. A. (1995). Treatment of motor speech disorders in children. *Seminars in Speech and Language, 16*, 126–139.

Szalkowski, C. E., Fiondella, C. G., Galaburda, A. M., Rosen, G. D., Loturco, J. J., & Fitch, R. H. (2012). Neocortical disruption and behavioral impairments in rats following in utero RNAi of candidate dyslexia risk gene Kiaa0319. *International Journal of Developmental Neuroscience, 30*(4), 293–302.

Szalkowski, C. E., Fiondella, C. F., Truong, D. T., Rosen, G. D., LoTurco, J. J., & Fitch, R. H. (2013). The effects of Kiaa0319 knockdown on cortical and subcortical anatomy in male rats. *International Journal of Developmental Neuroscience, 31*(2), 116–122.

Szalkowski, C. E., Hinman, J. R., Threlkeld, S. W., Wang, Y., LePack, A., Rosen, G. D., ... Fitch, R. H. (2011). Persistent spatial working memory deficits in rats following in utero RNAi of Dyx1c1. *Genes, Brain and Behavior, 10*(2), 244–252.

Taipale, M., Kaminen, N., Nopola-Hemmi, J., Haltia, T., Myllyluoma, B., Lyytinen, H., ... Kere, J. (2003). A candidate gene for developmental dyslexia encodes a nuclear tetratricopeptide repeat domain protein dynamically regulated in brain. *Proceedings of the National Academy of Sciences of the United States of America, 100*(20), 11553–11558.

Tallal, P. (1980). Auditory temporal perception, phonics, and reading disabilities in children. *Brain and Language, 9*(2), 182–198.

Tarkar, A., Loges, N. T., Slagle, C. E., Francis, R., Dougherty, G. W., Tamayo, J. V., ... Omran, H. (2013). DYX1C1 is required for axonemal dynein assembly and ciliary motility. *Nature Genetics, 45*(9), 995–1003.

Teichmann, M., Kas, A., Boutet, C., Ferrieux, S., Nogues, M., Samri, D., ... Migliaccio, R. (2013). Deciphering logopenic primary progressive aphasia: a clinical, imaging and biomarker investigation. *Brain, 136*(Pt 11), 3474–3488.

Teichmann, M., Migliaccio, R., Kas, A., & Dubois, B. (2013). Logopenic progressive aphasia beyond Alzheimer's – an evolution towards dementia with Lewy bodies. *Journal of Neurology, Neurosurgery and Psychiatry, 84*(1), 113–114.

Teramitsu, I., Kudo, L. C., London, S. E., Geschwind, D. H., & White, S. A. (2004). Parallel FoxP1 and FoxP2 expression in songbird and human brain predicts functional interaction. *Journal of Neuroscience, 24*(13), 3152–3163.

Teramitsu, I., Poopatanapong, A., Torrisi, S., & White, S. A. (2010). Striatal FoxP2 is actively regulated during songbird sensorimotor learning. *PLoS One, 5*(1), e8548.

Teramitsu, I., & White, S. A. (2006). FoxP2 regulation during undirected singing in adult songbirds. *Journal of Neuroscience, 26*(28), 7390–7394.

Thevenon, J., Callier, P., Andrieux, J., Delobel, B., David, A., Sukno, S., ... Faivre, L. (2013). 12p13.33 microdeletion including ELKS/ERC1, a new locus associated with childhood apraxia of speech. *European Journal of Human Genetics, 21*(1), 82–88.

Thompson, C. K., Cho, S., Hsu, C. J., Wieneke, C., Rademaker, A., Weitner, B. B., ... Weintraub, S. (2012). Dissociations between fluency and agrammatism in primary progressive aphasia. *Aphasiology, 26*(1), 20–43.

Threlkeld, S. W., McClure, M. M., Bai, J., Wang, Y., LoTurco, J. J., Rosen, G. D., & Fitch, R. H. (2007). Developmental disruptions and behavioral impairments in rats following in utero RNAi of Dyx1c1. *Brain Research Bulletin, 71*(5), 508–514.

Tran, C., Gagnon, F., Wigg, K. G., Feng, Y., Gomez, L., Cate-Carter, T. D., ... Barr, C. L. (2013). A family-based association analysis and meta-analysis of the reading disabilities candidate gene DYX1C1. *American Journal of Medical Genetics, Part B: Neuropsychiatric Genetics, 162B*(2), 146–156.

Tran, C., Wigg, K. G., Zhang, K., Cate-Carter, T. D., Kerr, E., Field, L. L., ... Barr, C. L. (2014). Association of the ROBO1 gene with reading disabilities in a family-based analysis. *Genes, Brain and Behavior, 13*(4), 430–438.

Turner, S. J., Hildebrand, M. S., Block, S., Damiano, J., Fahey, M., Reilly, S., ... Morgan, A. T. (2013). Small intragenic deletion in FOXP2 associated with childhood apraxia of speech and dysarthria. *American Journal of Medical Genetics Part A, 161A*(9), 2321–2326.

Vandermosten, M., Boets, B., Luts, H., Poelmans, H., Golestani, N., Wouters, J., & Ghesquiere, P. (2010). Adults with dyslexia are impaired in categorizing speech and nonspeech sounds on the basis of temporal cues. *Proceedings of the National Academy of Sciences of the United States of America, 107*(23), 10389–10394.

Vandermosten, M., Boets, B., Luts, H., Poelmans, H., Wouters, J., & Ghesquiere, P. (2011). Impairments in speech and nonspeech sound categorization in children with dyslexia are driven by temporal processing difficulties. *Research in Developmental Disabilities, 32*(2), 593–603.

Vandermosten, M., Boets, B., Wouters, J., & Ghesquiere, P. (2012). A qualitative and quantitative review of diffusion tensor imaging studies in reading and dyslexia. *Neuroscience & Biobehavioral Reviews, 36*(6), 1532–1552.

Vargha-Khadem, F., Watkins, K., Alcock, K., Fletcher, P., & Passingham, R. (1995). Praxic and nonverbal cognitive deficits in a large family with a genetically transmitted speech and language disorder. *Proceedings of the National Academy of Sciences of the United States of America, 92*(3), 930–933.

Vargha-Khadem, F., Watkins, K. E., Price, C. J., Ashburner, J., Alcock, K. J., Connelly, A., ... Passingham, R. E. (1998). Neural basis of an inherited speech and language disorder. *Proceedings of the National Academy of Sciences of the United States of America, 95*(21), 12695–12700.

Venkatesh, S. K., Siddaiah, A., Padakannaya, P., & Ramachandra, N. B. (2013). Analysis of genetic variants of dyslexia candidate genes KIAA0319 and DCDC2 in Indian population. *Journal of Human Genetics, 58*(8), 531–538.

Venkatesh, S. K., Siddaiah, A., Padakannaya, P., & Ramachandra, N. B. (2014). Association of SNPs of DYX1C1 with developmental dyslexia in an Indian population. *Neuropsychiatric Genetics, 24*(1), 10–20.

Wang, Y., Paramasivam, M., Thomas, A., Bai, J., Kaminen-Ahola, N., Kere, J., ... Loturco, J. J. (2006). DYX1C1 functions in neuronal migration in developing neocortex. *Neuroscience, 143*(2), 515–522.

Wang, Y., Yin, X., Rosen, G., Gabel, L., Guadiana, S. M., Sarkisian, M. R., ... Loturco, J. J. (2011). Dcdc2 knockout mice display exacerbated developmental disruptions following knockdown of doublecortin. *Neuroscience, 190*, 398–408.

Warrington, E. K. (1975). The selective impairment of semantic memory. *Quarterly Journal of Experimental Psychology, 27*(4), 635–657.

Watkins, K. E., Dronkers, N. F., & Vargha-Khadem, F. (2002). Behavioural analysis of an inherited speech and language disorder: comparison with acquired aphasia. *Brain, 125*(Pt. 3), 452–464.

Watkins, K. E., Vargha-Khadem, F., Ashburner, J., Passingham, R. E., Connelly, A., Friston, ... Gadian, D. G. (2002). MRI analysis of an inherited speech and language disorder: structural brain abnormalities. *Brain, 125*(Pt. 3), 465–478.

Watts, G. D., Wymer, J., Kovach, M. J., Mehta, S. G., Mumm, S., Darvish, D., ... Kimonis, V. E. (2004). Inclusion body myopathy associated with Paget disease of bone and frontotemporal dementia is caused by mutant valosin-containing protein. *Nature Genetics, 36*(4), 377–381.

Webb, A. L., Singh, R. H., Kennedy, M. J., & Elsas, L. J. (2003). Verbal dyspraxia and galactosemia. *Pediatric Research, 53*(3), 396–402.

Webb, D. M., & Zhang, J. (2005). FoxP2 in song-learning birds and vocal-learning mammals. *Journal of Heredity, 96*(3), 212–216.

Weintraub, S., Mesulam, M. M., Wieneke, C., Rademaker, A., Rogalski, E. J., & Thompson, C. K. (2009). The northwestern anagram test: measuring sentence production in primary progressive aphasia. *American Journal of Alzheimer's Disease & Other Dementias, 24*(5), 408–416.

Whitwell, J. L., Avula, R., Senjem, M. L., Kantarci, K., Weigand, S. D., Samikoglu, A., ... Jack, C. R., Jr. (2010). Gray and white matter water diffusion in the syndromic variants of frontotemporal dementia. *Neurology, 74*(16), 1279–1287.

Whitwell, J. L., Jack, C. R., Jr., Baker, M., Rademakers, R., Adamson, J., Boeve, B. F., ... Josephs, K. A. (2007). Voxel-based morphometry in frontotemporal lobar degeneration with ubiquitin-positive inclusions with and without progranulin mutations. *Archives of Neurology, 64*(3), 371–376.

Whitwell, J. L., Jones, D. T., Duffy, J. R., Strand, E. A., Machulda, M. M., Przybelski, S. A., … Josephs, K. A. (2015). Working memory and language network dysfunctions in logopenic aphasia: a task-free fMRI comparison with Alzheimer's dementia. *Neurobiology of Aging, 36*(3), 1245–1252.

Wilson, S. M., Dronkers, N. F., Ogar, J. M., Jang, J., Growdon, M. E., Agosta, F., … Gorno-Tempini, M. L. (2010). Neural correlates of syntactic processing in the nonfluent variant of primary progressive aphasia. *Journal of Neuroscience, 30*(50), 16845–16854.

Wilson, S. M., Galantucci, S., Tartaglia, M. C., Rising, K., Patterson, D. K., Henry, M. L., … Gorno-Tempini, M. L. (2011). Syntactic processing depends on dorsal language tracts. *Neuron, 72*(2), 397–403.

Wilson, S. M., Henry, M. L., Besbris, M., Ogar, J. M., Dronkers, N. F., Jarrold, W., … Gorno-Tempini, M. L. (2010). Connected speech production in three variants of primary progressive aphasia. *Brain, 133*(Pt 7), 2069–2088.

Yeh, M. L., Gonda, Y., Mommersteeg, M. T., Barber, M., Ypsilanti, A. R., Hanashima, C., … Andrews, W. D. (2014). Robo1 modulates proliferation and neurogenesis in the developing neocortex. *Journal of Neuroscience, 34*(16), 5717–5731.

Zadikoff, C., & Lang, A. E. (2005). Apraxia in movement disorders. *Brain, 128*(Pt 7), 1480–1497.

Zou, L., Chen, W., Shao, S., Sun, Z., Zhong, R., Shi, J., … Song, R. (2012). Genetic variant in KIAA0319, but not in DYX1C1, is associated with risk of dyslexia: an integrated meta-analysis. *American Journal of Medical Genetics, Part B: Neuropsychiatric Genetics, 159B*(8), 970–976.

Chapter 33

Molecular Pathways Leading to the Clinical Phenomenology of Frontotemporal Dementia

Jennifer S. Yokoyama and Suzee E. Lee

University of California San Francisco, San Francisco, CA, United States

INTRODUCTION TO FRONTOTEMPORAL DEMENTIA AND ASSOCIATED CLINICAL SYNDROMES

Frontotemporal dementia (FTD) is a disorder featuring profound, early personality and behavioral changes caused by neurodegeneration in the frontal and temporal lobes. Arnold Pick (1982) first described a patient with progressive aphasia and frontotemporal atrophy at autopsy. Subsequently, Alois Alzheimer studied the neuropathological characteristics of this disease and named Pick bodies (distinctive round inclusions apparent on silver staining) as its defining hallmark (Alzheimer, 1911). Once thought to be an uncommon illness, FTD is a common cause of dementia before age 65 years. Over the past century, clinicians have struggled to determine which features distinguish FTD from Alzheimer disease (AD) and other dementias. Because the pattern and extent of focal degeneration within the frontal and temporal lobes vary among patients, causing different symptoms, FTD represents an umbrella term for three clinically defined syndromes, including behavioral variant FTD (bvFTD) and two primary progressive aphasias (PPA), which include the nonfluent variant (nfvPPA) and the semantic variant (svPPA). Some patients with FTD also develop parkinsonism or motor neuron disease, layering yet further complexity to the clinical presentation.

Progressive changes in personality and behavior are the fundamental feature of bvFTD. These changes include disinhibition and socially inappropriate interactions, a loss of manners or decorum, or impulsive, reckless actions (Rascovsky et al., 2007). Apathy emerges as a loss of motivation or interest in social relationships and previously rewarding activities. Caregivers of those with bvFTD note a loss of empathy with diminished responses to others' needs and emotions or waning interest in other people. Because cognitive decline emerges later in the disease course, clinicians occasionally misdiagnose patients with bvFTD with psychiatric illnesses, such as depression or bipolar disorder. The other two FTD clinical syndromes are PPA, in which insidious, progressive language impairments result in speech and language expression and comprehension problems. Patients with nfvPPA struggle with motor speech production, although language comprehension remains intact during milder disease stages. In contrast, patients with svPPA maintain fluent speech; word finding difficulties evolve in both aphasias, but svPPA also erodes knowledge about objects and their salient characteristics.

FTD arises sporadically and genetically. Interestingly, the three known autosomal dominant genetic mutations cause FTD by different pathophysiological mechanisms, leading to coinciding neuroanatomical patterns of degeneration. To complicate matters further, autopsies of FTD patients reveal different proteins (eg, tau, TAR DNA-binding protein 43 [TDP-43], fused in sarcoma) aggregated in diverse morphologies and dispersed in diverse constellations of neuroanatomical regions.

For each FTD clinical syndrome, the possible underlying neuropathological diagnoses are heterogeneous; thus, predicting underlying neuropathology, which is believed to reflect the pathophysiology of disease, remains an elusive goal.

Genomics, Circuits, and Pathways in Clinical Neuropsychiatry. http://dx.doi.org/10.1016/B978-0-12-800105-9.00033-0

For clarity, we will use the term "frontotemporal lobar degeneration" (FTLD) to denote pathological diagnoses, whereas "FTD" refers to clinical syndromes. In considering the complex and nuanced clinicoanatomical pathology seen in FTD, clinicians and scientists face many questions. For example, why do neuropathologies arising from different molecular mechanisms converge on the same brain circuitry? Conversely, what features of the cell populations underlying the brain circuits vulnerable in FTD constitute their vulnerability to multiple but selective pathological mechanisms? We will first explore the known genetic and molecular mechanisms causing genetic FTLD, review the neuroanatomical circuits associated with each gene, and finally examine how disruptions in specific brain networks provide insight into understanding clinical syndromes.

THREE UNIQUE GENETIC MECHANISMS CONVERGING ON FRONTOTEMPORAL DEMENTIA SPECTRUM DISORDERS

About 10–20% of patients with FTLD are estimated to be genetic cases caused by a single mutation inherited in an autosomal dominant fashion, and up to 40% have a family history of at least one family member with dementia or other related disorder (Rohrer et al., 2009; See, LaMarre, Lee, & Miller, 2010). Three gene mutations (in order of discovery), *MAPT*, *GRN*, and *C9ORF72*, cause most genetic cases of FTLD of known origin (rare mutations in *VCP*, *CHMP2B*, *TARDBP*, and *FUS* also cause FTD, reviewed elsewhere) (See et al., 2010; Sieben et al., 2012). In this section, we will outline the discovery and proposed pathogenic mechanisms of the three main FTLD genes.

MAPT

Mapping

The term "FTDP-17" was created at a consensus meeting in 1997 to characterize a group of 13 autosomal dominant families segregating FTD with parkinsonism that was linked to chromosome 17q21 through genetic mapping studies (Foster et al., 1997). Causative mutations in *MAPT* were identified in many of those families shortly thereafter (Dumanchin et al., 1998; Hutton et al., 1998; Poorkaj et al., 1998; Spillantini et al., 1998). Over 40 different mutations have since been reported in *MAPT*, including missense and silent mutations, single-codon deletions, and intronic mutations that alter splicing (http://www.molgen.ua.ac.be/ADMutations/). Most disease-causing mutations in *MAPT* cluster around exons 9–13; pathogenic mutations have also been reported in exons 1–2. An extended (approximately 100-kb) haplotype encompassing the entire *MAPT* region has also been identified, creating two common haplotypes termed "H1" and "H2" (Baker et al., 1999). The more common H1 haplotype is associated with risk for progressive supranuclear palsy (PSP) and corticobasal degeneration (CBD) (Baker et al., 1999), and an H1 subhaplotype is associated with risk for Parkinson disease (Skipper et al., 2004). In addition, carriers of a rare nonsynonymous variant, p.A152T, show a two- to threefold risk for AD and FTD-spectrum disorders, respectively (Coppola et al., 2012).

Molecular Pathogenicity

Microtubule-associated protein tau (tau) was first isolated in the 1970s (Cleveland, Hwo, & Kirschner, 1977a, 1977b; Weingarten, Lockwood, Hwo, & Kirschner, 1975) and is highly constitutively expressed in both the central and peripheral nervous systems (Binder, Frankfurter, & Rebhun, 1985). Goedert and colleagues identified tau as the core protein of the neurofibrillary tangle in AD pathology and then subsequently cloned and sequenced tau cDNA (Goedert, Spillantini, Potier, Ulrich, & Crowther, 1989; Goedert, Wischik, Crowther, Walker, & Klug, 1988). Tau protein is enriched in neuronal axons and promotes microtubule assembly and stabilization by binding tubulin, the subunits that comprise microtubules (Hirokawa, 1994). Tau also regulates intracellular trafficking by modulating the attachment of the motor protein, kinesin, to axonal microtubules (Ebneth et al., 1998; Sato-Harada, Okabe, Umeyama, Kanai, & Hirokawa, 1996; Trinczek, Ebneth, Mandelkow, & Mandelkow, 1999). *MAPT* contains 15 exons, and differential splicing of exons 2, 3, and 10 results in six major protein isoforms that are expressed in the brain (Rademakers, Cruts, & Broeckhoven, 2004). The microtubule-binding domain of tau occurs as imperfect repeat motifs of 31 or 32 amino acids in the C-terminal portion of the protein (coded by exons 9–13), with three or four repeats depending on whether exon 10 is included (termed 3 and 4R, respectively). The 4R isoform (originally cloned by Goedert et al. (1989)) promotes microtubule assembly more than 3R tau owing to a threefold higher affinity for tubulin (Rademakers et al., 2004).

Tau expression and posttranslational modifications change because the requirements for cytoskeletal reorganization evolve during human brain development. This change appears to transpire in two ways. First, the ratio of 3R versus 4R tau expression varies throughout the lifespan; fetal brains express only 3R tau, whereas 3 and 4R tau are expressed in approximately equal proportions in the adult brain (Goedert et al., 1989). Second, posttranslational phosphorylation of tau decreases binding affinity to microtubules. Tau in the fetal brain is more phosphorylated than in the adult brain (Kanemaru, Takio, Miura, Titani, & Ihara, 1992). Increases in 4R tau levels and reduced tau phosphorylation result in increased microtubule stability as the human brain matures. For these reasons, maintenance of specific 3R:4R ratios and modulation of tau phosphorylation are two critical mechanisms by which microtubule dynamics are regulated in different stages of life.

In general, posttranslational modifications regulate a protein's activity, localization, and interaction with other molecules, thereby expanding the functional variety and complexity of the proteome. The most common post-translational modification of tau is phosphorylation; over 25 phosphorylation sites have been identified across the protein. As mentioned, phosphorylation of tau first serves as a way of modifying microtubule assembly during development. However, hyperphosphorylated tau is a pathological hallmark of FTLD and promotes tau misfolding and aggregation (Tenreiro, Eckermann, & Outeiro, 2014). In addition to phosphorylation, posttranslational acetylation may also have a critical role in FTLD owing to tau pathology by modifying the rate of tau aggregation and clearance (Cohen et al., 2011; Cook, Stankowski, Carlomagno, Stetler, & Petrucelli, 2014; Min et al., 2010). Research suggests that acetylation may also promote the accumulation and hyperphosphorylation of TDP-43, the other frequently aggregated FTLD protein (Cohen et al., 2015).

Tau's role in binding and assembling tubulin is critical for cell homeostasis and represents a balancing act between stabilization and dynamic mobility. Indeed, *MAPT* mutations associated with disease most often alter the homeostasis of bound versus free tau within the neuron. This typically occurs through changes in the 3R:4R tau ratio or enhancement of tau self-aggregation. Most of the disease-causing *MAPT* mutations in intron 10 and exon 10 alter exon 10 splicing, thereby altering the balance of 3R:4R tau and changing microtubule assembly dynamics. Missense mutations also appear to affect microtubule assembly, likely by interfering with tau's ability to bind tubulin (Rademakers et al., 2004). Changes in the ratio of tau's binding affinity to tubulin result in altered microtubule stabilization, which results in alterations in cyto-skeletal integrity. Moreover, as a consequence of reduced binding to tubulin, there is an increase in free tau available to form pathological aggregates. As tau aggregates, the pool of free tau available to maintain microtubule stability and cytoskeletal architecture is further drained, culminating in defective axonal transport (Rademakers et al., 2004). Because of these detrimental changes, microtubule stabilizing compounds are one type of therapeutic currently being explored to treat tauopathies (Lou et al., 2014).

Disease-causing coding mutations in *MAPT* appear to increase affinity for tau aggregation through filament assembly (first described through biochemical assays using polymerization-inducing agents) (Goedert, Jakes, & Crowther, 1999; Yen, Hutton, DeTure, Ko, & Nacharaju, 1999). Misfolded tau has been shown to propagate in a prion-like manner such that misfolded tau can induce normal tau to misfold (eg, (Calignon et al., 2012; Goedert, Clavaguera, & Tolnay, 2010; Liu et al., 2012)). In addition to promoting aggregation, misfolded tau can travel from one neuron to the next, likely via macropinocytosis (Falcon et al., 2014), spreading pathological aggregation throughout connected neuronal networks (Clavaguera et al., 2009; Holmes et al., 2014). Prion-like spread has also been suggested for TDP-43, first observed in vitro (Furukawa, Kaneko, Watanabe, Yamanaka, & Nukina, 2011). Antibodies halting tau propagation are currently being investigated as one avenue for preventing spread of tau pathology throughout the brain and have been shown in mouse models to improve cognitive function by blocking propagation and reducing hyperphosphorylation, aggregation, insoluble tau levels, and microglial activation (Yanamandra et al., 2013). Identifying the earliest site of protein aggregation for targeted delivery of such antibodies will be key for this strategy to prevent widespread neuronal damage throughout the brain networks affected by disease.

Altering the 3R:4R tau ratio and tau phosphorylation represent two critical mechanisms by which tau promotes microtubule assembly during different aspects of normal brain development. Whether disease-causing mutations altering tau function also affect neuronal development remains an open question. Understanding how functional changes in tau result in neuronal dysfunction, neuropathological changes, and cell death will be critical for elucidating the underpinnings of FTD and AD pathophysiology.

GRN

Mapping

Some familial cases of FTD linked to chromosome 17q21 showed mutations in *MAPT* (Froelich et al., 1997; Rizzu et al., 1999). The remainder of cases did not have tau pathology; however; linkage for this group was narrowed to a region on chr17 that did not contain the *MAPT* locus (Rademakers et al., 2002) and this second candidate region was further refined in 2006 via a large Belgian kindred (van der Zee et al., 2006). Shortly thereafter, two groups identified autosomal dominant *GRN* mutations in tau-negative familial cases (Baker et al., 2006; Cruts et al., 2006). A total of 69 *GRN* mutations have since been reported (http://www.molgen.ua.ac.be/ADMutations/). *GRN* mutations cause FTLD through haploinsufficiency of the progranulin (PGRN) protein that it codes, primarily through nonsense or frameshift mutations (Kleinberger, Capell, Haass, & Broeckhoven, 2013). Rarely homozygous *GRN* mutation carriers develop neuronal ceroid lipofuscinosis, a lysosomal storage disorder, which in all other known cases is an autosomal recessive disease caused by a completely different mutation (Smith et al., 2012).

Molecular Pathogenicity

Mutations in *GRN* affect PGRN levels both in the nervous system and systemically. *GRN* mutation carriers show greater than 50% reduced plasma and cerebrospinal fluid levels of PGRN (Coppola et al., 2008; Finch et al., 2009; Ghidoni, Benussi, Glionna, Franzoni, & Binetti, 2008; Sleegers et al., 2009; Van Damme et al., 2008). In the body, PGRN is involved in development, wound healing, inflammation, tumor growth, and energy homeostasis (De Muynck & Van Damme, 2011; Kleinberger et al., 2013). In the nervous system, PGRN has been suggested to participate in neuroprotection. PGRN exhibits antiinflammatory effects inhibiting the tumor necrosis factor-α receptor (Tang et al., 2011). Extracellularly, PGRN is cleaved by proteases into peptides, termed granulins A–G (GRN A–G), some of which serve as neurotrophic factors (eg, E) (De Muynck et al., 2013) whereas others appear to be proinflammatory (eg, A, B) (Okura et al., 2010; Zhu et al., 2002).

Studies in multiple model systems support a role for PGRN in neuronal development and survival. In studies of zebrafish, knockdown of the *Danio rerio* paralogues of *GRN* results in truncated motor neurons and other growth abnormalities (Chitramuthu, Baranowski, Kay, Bateman, & Bennett, 2010; Laird et al., 2010). In *Caenorhabditis elegans* models of *GRN* haploinsufficiency, worms lacking PGRN undergo hastened apoptosis, which may reduce the ability of injured neurons to repair themselves and ultimately culminate in premature cell death (Kao et al., 2011). These zebrafish and worm findings are consistent with studies in rodents and human cell culture models, which suggests that PGRN promotes neurite outgrowth and neuronal survival (Kleinberger et al., 2013).

Maintaining the proper amount of PGRN and resulting GRN appears particularly important for maintaining a balance between the anti- and proinflammatory activities involved in neuronal protection. *GRN*-deficient mice demonstrate increased inflammatory responses to neuronal injury owing to activated microglia resulting in reduced neuronal survival, which suggests a mechanism by which PGRN deficiency affects responses to neuronal insult (Martens et al., 2012). The effect of *GRN* mutations across the lifespan remains unknown, but given the evidence that PGRN has important roles in inflammation, it is possible that as *GRN*-haploinsufficient patients age, the exaggerated neuroinflammatory response brought on by reduced PGRN levels begets a cascade of pathological processes after an insult such as an acute injury or chronic inflammation.

Could genetic modifiers of PGRN expression affect disease onset and clinical presentation? A variant in *TMEM106B* modifying levels of serum PGRN (Cruchaga et al., 2011; Finch et al., 2010) has been associated with risk for sporadic FTLD-TDP (Van Deerlin et al., 2010) and may modify the clinical presentation in *GRN* and *C9ORF72* mutation carriers (van Blitterswijk et al., 2014; Gallagher et al., 2014). *TMEM106B* influences lysosomal morphology and dendritic trafficking of lysosomes (Brady, Zheng, Murphy, Huang, & Hu, 2013; Schwenk et al., 2014); protective variation in *TMEM106B* may increase PGRN levels by reducing efficiency of its trafficking into degradation pathways. Molecular modifiers of PGRN expression that would boost protein levels in *GRN* mutation carriers are currently being explored as potential therapeutics. For example, PGRN's neuronal receptor, sortilin (SORT1), regulates extracellular levels of PGRN by routing it to the lysosomal pathway for degradation via endocytosis (Hu et al., 2010); studies in induced pluripotent stem cell (iPSC) neurons and lymphocytes derived from a *GRN* mutation carrier with bvFTD have demonstrated that small molecules that interfere with PGRN's interaction with SORT1 can increase extracellular PGRN levels by inhibiting PGRN endocytosis and its subsequent degradation (Lee, Almeida, et al., 2014). In addition, the histone deacetylase inhibitor, suberoylanilide hydroxamic acid, has been shown to be a potent activator of *GRN* mRNA and protein expression, normalizing PGRN levels in immortalized human lymphoblastoid cells derived from patients carrying *GRN* mutations (Cenik et al., 2011). Notably, *GRN* mutations are the only known cause of neurodegenerative disease resulting from protein haploinsufficiency rather than toxic gain of function. Thus, drugs elevating PGRN levels may abate or even prevent *GRN*-related FTLD.

C9ORF72

Mapping

A region at chromosome 9q21 was first linked to familial cases of FTD co-occurring with amyotrophic lateral sclerosis (ALS) in 2000 (Hosler et al., 2000) with subsequent fine-mapping to 9p13.2–21.3 in 2006 (Morita et al., 2006; Vance et al., 2006). The region was mapped to a minimum critical region containing 10 genes by 2011 (Boxer et al., 2011). Later that year, two groups simultaneously published the pathogenic mutation (DeJesus-Hernandez et al., 2011; Renton et al., 2011): a large GGGGCC hexanucleotide repeat expansion in the intronic region between exons 1a and 1b of *C9ORF72*, a gene that codes an uncharacterized protein. Unaffected individuals carry 2–23 repeats; pathogenic expansion occurs when 30 or more to greater than 1600 repeats are present (DeJesus-Hernandez et al., 2011; Renton et al., 2011). *C9ORF72* expansion represents the most common genetic cause of both FTD and ALS. In addition, about 4–7% of individuals with FTD and/or patients with ALS with no family history (presumed sporadic disease) also have pathologic expansions (Fong, Karydas, & Goldman, 2012). *C9ORF72* mutation appears to be autosomal dominant, with nearly all reported cases carrying one expanded copy and one normal length copy of the gene. In the literature, one patient with bvFTD has been identified as carrying two copies of the pathogenic *C9ORF72* expansion. He developed behavioral changes and cognitive impairment at 43, followed by severe apathy and declining speech output, and a disease course within the range of individuals carrying one *C9ORF72* expansion (Fratta et al., 2013).

Molecular Pathogenicity

C9ORF72 shares homology with DENN-domain containing proteins (Levine, Daniels, Gatta, Wong, & Hayes, 2013; Zhang, Iyer, He, & Aravind, 2012) involved in endolysosomal trafficking (Yokoyama, Sirkis, & Miller, 2014). There are three transcribed isoforms of *C9ORF72*, variants 1–3, which result in two predicted forms of the protein, a shorter 222–amino acid version (variants 1 and 3), and a longer 481–amino acid version (variant 2) (DeJesus-Hernandez et al., 2011). Characterization of the native activity of C9ORF72 proteins remains unknown. Hexanucleotide expansions occur either in the promoter region (variant 1) or in intron 1 (variants 2–3). Pathogenic expansion in the promoter results in silencing of that allele via methylation (Belzil et al., 2013; Xi et al., 2013). One case report of an individual carrying two *C9ORF72* repeat expansions demonstrated no obvious differences in disease course or severity compared with single expansion carriers, which argues against haploinsufficiency as a primary mode of disease. However, lack of knowledge regarding C9ORF72's regular function, particularly in light of its homology to proteins that are involved in cellular trafficking, makes it difficult to rule out that C9ORF72 depletion does not contribute to disease pathogenesis (Yokoyama et al., 2014).

In addition to haploinsufficiency, another possible mechanism of molecular pathogenicity may involve transcription of pre-mRNA of the repeat expansion in intron 1 (Polymenidou et al., 2012). After transcription of the expanded hexanucleotide repeat, RNA foci can form from the sense and antisense strands, which accumulate in the neuronal nucleus (DeJesus-Hernandez et al., 2011). These foci have the potential to sequester RNA-binding proteins, potentially leading broad dysregulation of RNA metabolism and protein production (Polymenidou et al., 2012). This form of toxic gain-of-function has also been observed in other noncoding repeat expansion disorders such as Huntington disease, fragile X–associated tremor ataxia syndrome, and myotonic dystrophy (van Blitterswijk, DeJesus-Hernandez, & Rademakers, 2012). Antisense oligonucleotides that hybridize with the *C9ORF72* transcript appear to suppress RNA foci formation and reverse pathological changes in patient-derived iPSCs and are currently being explored as a treatment for *C9ORF72*-associated ALS (Donnelly et al., 2013; Sareen et al., 2013).

Pathological studies demonstrate that *C9ORF72* repeat expansion carriers uniquely develop ubiquitin/p62+ pathology composed of the five dipeptide repeats resulting from non-ATG–mediated (RAN) translation of the hexanucleotide repeat from both DNA strands and all reading frames (Ash et al., 2013; Mori et al., 2013). RAN translation of the sense strand creates poly Gly-Arg, poly Gly-Pro, and poly Gly-Ala dipeptides; antisense RAN translation results in Pro–Ala, Pro–Gly, and Pro–Arg dipeptides. Studies suggest that dipeptide formation arises more commonly from the sense strand, whereas stabilizing three-dimensional (secondary) structures form owing to interactions between complementary G and C nucleotides in the extended strand of rGGGGCC repeats (Fratta et al., 2012; Reddy, Zamiri, Stanley, Macgregor, & Pearson, 2013). Small molecules that block the formation of these secondary structures are being developed as one avenue for preventing this pathogenicity (Su et al., 2014).

An inducible mouse model of "RNA-only" gain-of-function toxicity expressing 80 GGGGCC repeats that could not undergo RAN translation recapitulated the ubiquitin-positive but not TDP-43 pathology associated with *C9ORF72* expansion in human pathological samples, and the mice showed no clear behavioral phenotype or neuronal loss (Hukema et al., 2014). The authors pointed out that TDP-43 dysfunction could still occur without the presence of TDP-43 pathology. It is also possible that the 12-week exposure period was not long enough to induce pathological, neurodegenerative, and behavioral changes in these rodents. Early pathological studies suggest that neuronal death may not correlate with the burden of dipeptide aggregation but instead correlates with FTLD-TDP pathology (Mackenzie et al., 2013). In animal models, the arginine-rich dipeptides particularly appear to cause the most toxicity and neurodegeneration in *Drosophila* and cell culture models (Kwon et al., 2014; Mizielinska et al., 2014; Wen et al., 2014). Further studies are required to refine the effects of each *C9ORF72* expansion-associated change and their role in human disease.

NEUROANATOMICAL CIRCUITS AND CLINICAL SYNDROMES IN AUTOSOMAL DOMINANT FRONTOTEMPORAL LOBAR DEGENERATION

How do gene mutations with such diverse molecular mechanisms converge on the common structural and functional brain networks affected in FTLD? In the following section, we will examine the neuroanatomical correlations of clinically defined bvFTD, discuss how different neurodegenerative diseases target specific circuits in the healthy brain, and explore how mapping large-scale brain circuits holds promise for differentiating the unique profiles of each FTLD genetic mutation.

Behavioral Variant Frontotemporal Dementia Provides Insight Into the Anatomical Correlates of Behavior

The broad spectrum of neuropsychiatric symptoms comprising the bvFTD syndrome has greatly refined our understanding of which neuroanatomical regions are associated with impairments in behavior and social cognition. The key feature of the bvFTD syndrome is the progressive deterioration of behavior with cognitive functions spared early in the disease course (Rascovsky et al., 2011). bvFTD research criteria require that three of the six following behavioral or cognitive symptoms must be present: (1) behavioral disinhibition; (2) apathy; (3) loss of sympathy or empathy; (4) perseverative, compulsive behavior; (5) hyperorality and dietary changes; and (6) neuropsychological testing with executive function deficits with relative sparing of memory and visuospatial function.

Individual patients with neurodegenerative diseases, including bvFTD, show diverse constellations of behavioral symptoms, which allow researchers to draw correlations between a particular behavioral trait and focal regions of brain atrophy. For example, disinhibited behavior is associated with atrophy in orbitofrontal cortex (Rosen et al., 2005) whereas apathy is associated with neurodegeneration in medial frontal cortex (Rosen et al., 2006). Patients with bvFTD developing insidiously worsening empathy toward others show neurodegeneration focally in the right anterior temporal lobe (Rankin et al., 2006).

In addition to these emotional and behavioral changes, compulsions and eating behaviors are associated with focal brain regions. Compulsive, repetitive behaviors range from simple repetitive movements (finger tapping or rubbing) to ritualistic behaviors (eg, compulsive, frequent urination) or repeating stereotyped catchphrases. Simple stereotyped movements have been associated with striatal atrophy, whereas more complex repetitive behaviors involve orbitofrontal and temporal cortex and the striatum (Ames, Cummings, Wirshing, Quinn, & Mahler, 1994; Perry et al., 2012; Rosso et al., 2001). Patients with bvFTD with compulsive eating behaviors such as consuming large quantities of sweets and carbohydrates or placing inedible objects in the mouth show degeneration in the orbitofrontal cortex, right frontoinsula, striatum (Woolley et al., 2007), and hypothalamus (Piguet et al., 2011).

Younger patients in their fifties or sixties with poor performance on executive function testing and with no behavioral changes represent a common scenario for misdiagnosed bvFTD. Although patients with bvFTD eventually develop executive function impairments (Possin et al., 2013), patients with bvFTD usually show behavioral impairments before executive dysfunction (Seeley et al., 2008). Typically, patients with bvFTD have relative preservation of memory and visuospatial function in keeping with relatively spared hippocampus and parieto-occipital cortices.

Toward a Circuit-Based Understanding of Neurodegenerative Diseases

Behavioral symptomatology can be parsed into various components, each with its neuroanatomical correlate. But how is it that particular ensembles of distant brain regions are consistently targeted in different neurodegenerative diseases?

Structural and functional connectivity analyses have enabled mapping of white matter tract integrity and large-scale functional brain networks in healthy individuals and patients. In task-free functional MRI (fMRI), a participant is instructed simply to lie inside an fMRI scanner and spontaneous low-frequency fluctuations in the bold oxygen level—dependent (BOLD) signal are detectable across the brain. Brain regions whose BOLD time series are highly correlated are identified as intrinsic connectivity networks (ICNs). Neural functional connectivity may be similarly determined in electroencephalography and magnetoencephalography signals, whereas techniques such as diffusion tensor imaging measure the microstructural integrity of white matter tracts.

Each neurodegenerative disease syndrome targets a unique ICN that is detectable in the healthy brain (Seeley, Crawford, Zhou, Miller, & Greicius, 2009). For example, bvFTD causes degeneration earliest in the frontoinsula, anterior cingulate cortex, mediofrontal and orbitoprefrontal cortex, amygdala, and striatum in patients with both genetic and sporadic disease (Boccardi et al., 2005; Broe et al., 2003; Rosen et al., 2002; Seeley et al., 2008). These regions comprise the salience network, which was first delineated in healthy individuals by examining task-free fMRI connectivity to the right frontoinsula, the region most atrophied in bvFTD (Seeley et al., 2009). The salience network is believed to evaluate the emotional importance of internal and external stimuli and coordinate cognitive, behavioral, and visceroautonomic responses (Seeley et al., 2007).

In contrast, AD is associated with atrophy in the hippocampus, posterior cingulate, and temporoparietal cortices, regions comprising the default mode network (DMN). The DMN was first identified as a network of regions that deactivate in task-related fMRI during cognitive tasks (Raichle et al., 2001) and is active during memory retrieval and visual imagery (Buckner, Andrews-Hanna, & Schacter, 2008; Mason et al., 2007). In addition, other ICNs identified in the healthy brain show anatomical congruence with atrophy patterns in CBS, semantic and nonfluent PPA (Seeley et al., 2009), and PSP (Gardner et al., 2013), which suggests that spatially distant brain regions comprise networks targeted in neurodegenerative diseases.

Interestingly, ICN disruptions and enhancements emerge in bvFTD and AD compared with healthy individuals. For example, patients with bvFTD show disruption in the salience network compared with healthy control subjects (Farb et al., 2013; Whitwell et al., 2011; Zhou et al., 2010). In bvFTD, disintegration of salience network connectivity worsens with disease severity and parallels impairments in the social-emotional and visceroautonomic responses that prevail in the disease (Zhou et al., 2010). Although frontal lobe functions decline in bvFTD, abilities referable to parietal cortex such as visuospatial ability and artistic creativity show relative preservation or even enhancement in some patients (Miller et al., 1998; Viskontas et al., 2011), a duality reminiscent of the social-emotional impairments accompanied by extraordinary visuospatial abilities seen in some individuals with autism (Dakin & Frith, 2005). Such enhancements are suspected to underlie the increased DMN connectivity seen in bvFTD compared with both patients with AD and control subjects. Patients with AD, on the other hand, show reductions in DMN connectivity, which reflects the impairments in memory and loss of visuoconstruction and navigational abilities that evolve as core features of the disease. On the other hand, patients with AD show preserved social graces and emotions early in the course of the disease, with family members sometimes relating that patients may show greater emotional sensitivity, irritability, or anxiety, which correlate with salience network connectivity enhancement (Balthazar et al., 2014; Zhou et al., 2010).

Characterizing Genetic Frontotemporal Lobar Degeneration With Structural and Functional Brain Circuitry

Building on early work in sporadic FTD, studies have aimed to characterize common and distinctive atrophy patterns, white matter vulnerability, and functional network alterations for each of the three autosomal dominant FTLD genetic mutations. To complicate matters, each gene is associated with heterogeneous clinical neurodegenerative disease syndromes. Even family members with the same mutation may develop different clinical syndromes, which suggests that other genetic or environmental factors contribute to disease development.

For example, patients with *MAPT* mutations commonly present with bvFTD, which may be accompanied by CBS, or less commonly, PSP, both parkinsonian neurodegenerative diseases. Whereas neuropathological findings in *MAPT* mutation carriers consistently reveal tau protein inclusions in either neurons or in both neurons and glia, substantial variations in the morphology and distribution of tau protein inclusions by mutation type appear (van Swieten & Spillantini, 2007). In contrast to sporadic bvFTD, episodic memory deficits may be prominent in *MAPT* mutation carriers (van Swieten & Spillantini, 2007), reflective of mesial temporal lobe degeneration. Group analyses of symptomatic *MAPT* carriers show a symmetrical atrophy pattern with orbitofrontal, anterior and medial temporal, and ventral anterior insula (Rohrer et al., 2010; Whitwell, Jack, Boeve, Senjem, Baker, Rademakers, et al., 2009). Efforts to profile whether specific *MAPT* mutations target different structural or functional brain networks will be challenging owing to the rarity of the *MAPT* mutation, and different brain regions may be vulnerable for each mutation type (Whitwell, Jack, Boeve, Senjem, Baker, Ivnik, et al., 2009).

Patients with *GRN* mutations show the most heterogeneity in terms of clinical presentations. Whereas the most common presentation is bvFTD, *GRN* mutation carriers may also present with CBS, PPA, and clinical AD (Le Ber et al., 2008). *GRN* mutation carriers show FTLD pathological changes consisting of TDP-43 and ubiquitin-positive neuronal inclusions in the frontotemporal cortex and dentate gyrus (Mackenzie et al., 2011). In contrast to symptomatic *MAPT* carriers, symptomatic *GRN* carriers show atrophy not only in the anterior insula and frontotemporal cortex but also extending into parietal cortices (Rohrer et al., 2010; Whitwell, Jack, Boeve, Senjem, Baker, Rademakers, et al., 2009). Curiously, *GRN* carriers develop asymmetric brain atrophy (Rohrer et al., 2010; Whitwell, Jack, Boeve, Senjem, Baker, Rademakers, et al., 2009) and show the fastest rates of atrophy compared with *MAPT* and *C9ORF72* mutation carriers (Whitwell et al., 2015). One study of seven symptomatic *GRN* carriers suggested reductions in DMN connectivity compared with symptomatic noncarriers matched for clinical syndrome (Borroni et al., 2012); further studies with larger numbers of individuals will elucidate functional brain alterations in *GRN*.

bvFTD with or without motor neuron disease and ALS are the most common presentations of *C9ORF72* repeat expansion carriers. Less commonly, *C9ORF72* expansion carriers with clinical AD (Murray et al., 2011) and PPA (Mahoney et al., 2012; Snowden, Rollinson, Thompson, et al., 2012) have been reported. Three neuropathological changes have been yet identified in *C9ORF72* expansion carriers: (1) neuronal and oligodendroglial inclusions composed of FTLD-TDP type B pathology; (2) dipeptide repeat proteins, reflecting the translation of the expanded hexanucleotide GGGGCC sequence (Mori et al., 2013); and (3) nuclear RNA foci (DeJesus-Hernandez et al., 2011).

In contrast to sporadic and other forms of genetic FTD, patients with *C9ORF72* show a higher incidence of psychotic symptoms, including hallucinations and delusions (Boeve et al., 2012; Sha et al., 2012; Snowden, Rollinson, Lafon, et al., 2012). The atrophy pattern seen in *C9ORF72* expansion carriers mirrors the principal regions of degeneration seen in sporadic bvFTD, including the anterior insula, frontotemporal cortex, and anterior cingulate (Irwin et al., 2013; Sha et al., 2012; Whitwell et al., 2012). Interestingly, *C9ORF72* mutation carriers with bvFTD show milder atrophy in regions targeted in sporadic bvFTD, but also show atrophy extending more posteriorly into parietal cortices and posterior thalamus (Lee, Khazenzon, et al., 2014; Yokoyama & Rosen, 2012). Adding to the complexity of the clinical presentation is that some *C9ORF72* mutation carriers show a protracted disease course with slowly progressive symptoms and minimal structural brain atrophy (Boeve et al., 2012; Khan et al., 2012). Intrinsic connectivity mapping of *C9ORF72* carriers with bvFTD suggests similar alterations in the salience network, and sensorimotor network compared with sporadic bvFTD (Lee, Khazenzon, et al., 2014). Medial pulvinar atrophy may mediate salience network disruptions in *C9ORF72*, resulting in the bvFTD syndrome. On the other hand, the DMN, enhanced in sporadic bvFTD, shows no such enhancements in *C9ORF72* bvFTD. Interestingly, *C9ORF72* mutation carriers with early-stage or slowly progressive disease show DMN enhancements, which suggests that DMN connectivity may change dynamically throughout the *C9ORF72* carrier lifespan (Lee, Khazenzon, et al., 2014).

Investigating the presymptomatic phase of genetic FTLD will prove critical for early disease detection and understanding the natural course of the disease. In a study of 78 presymptomatic mutation carriers, lower neuropsychological measures and gray matter volume emerged in carriers about 5 years before symptom onset (Rohrer et al., 2015). Smaller group studies investigating task-free fMRI in genetic FTLD carriers have shown variable alterations in ICNs, with presymptomatic *GRN* carriers showing reductions in salience network connectivity and no differences in DMN connectivity in one study (Borroni et al., 2012; Dopper et al., 2014) and presymptomatic *MAPT* showing increases (lateral temporal and medial prefrontal cortex) and decreases (precuneus) in DMN connectivity, but no salience network alterations compared with control subjects (Whitwell et al., 2011). Presymptomatic *MAPT* and *GRN* carriers also demonstrate reduced white matter integrity in certain tracts (Borroni et al., 2008; Dopper et al., 2014).

Future Questions

Investigations into hereditary FTLD have broadly ranged from studies of genetic analyses to clinical syndromes, biomarkers (including neuroimaging), and neuropathology; integrating translational data from such studies presents a challenge for the future. It remains unknown why the effects of all three genetic mutations cause neurodegeneration in later life despite being present from conception, and whether FTLD mutation carriers or those with sporadic FTLD harbor neurodevelopmental differences. Advances in molecular imaging, such as tau PET, will be important for early disease characterization and diagnosis. Longitudinal studies in *MAPT*, *GRN*, and *C9ORF72* carriers have begun to quantify changes in structural brain degeneration (Mahoney et al., 2014; Whitwell et al., 2015) and collaborations to recruit and follow large numbers of subjects over many years will be important to understanding the disease course and differentiating brain changes from normal aging.

REFERENCES

Alzheimer, A. (1911). Über Eigenartige Krankheitsfälle Des Späteren alters. *Zeitschrift für die gesamte Neurologie und Psychiatrie*, 356.

Ames, D., Cummings, J. L., Wirshing, W. C., Quinn, B., & Mahler, M. (1994). Repetitive and compulsive behavior in frontal lobe degenerations. *The Journal of Neuropsychiatry and Clinical Neurosciences, 6*, 100–113.

Ash, P. E., Bieniek, K. F., Gendron, T. F., Caulfield, T., Lin, W., DeJesus-Hernandez, M., … Petrucelli, L. (2013). Unconventional translation of C9ORF72 GGGGCC expansion generates insoluble polypeptides specific to c9FTD/ALS. *Neuron, 77*, 639–646.

Baker, M., Litvan, I., Houlden, H., Adamson, J., Dickson, D., Perez-Tur, J., … Hutton, M. (1999). Association of an extended haplotype in the tau gene with progressive supranuclear palsy. *Human Molecular Genetics, 8*, 711–715.

Baker, M., Mackenzie, I. R., Pickering-Brown, S. M., Gass, J., Rademakers, R., Lindholm, C., … Hutton, M. (2006). Mutations in progranulin cause tau-negative frontotemporal dementia linked to chromosome 17. *Nature, 442*, 916–919.

Balthazar, M. L. F., Pereira, F. R. S., Lopes, T. M., da Silva, E. L., Coan, A. C., Campos, B. M., … Cendes, F. (2014). Neuropsychiatric symptoms in Alzheimer's disease are related to functional connectivity alterations in the salience network. *Human Brain Mapping, 35*, 1237–1246.

Belzil, V. V., Bauer, P. O., Prudencio, M., Gendron, T. F., Stetler, C. T., Yan, I. K., … Petrucelli, L. (2013). Reduced C9orf72 gene expression in c9FTD/ALS is caused by histone trimethylation, an epigenetic event detectable in blood. *Acta Neuropathologica, 126*, 895–905.

Binder, L. I., Frankfurter, A., & Rebhun, L. I. (1985). The distribution of tau in the mammalian central nervous system. *The Journal of Cell Biology, 101*, 1371–1378.

van Blitterswijk, M., DeJesus-Hernandez, M., & Rademakers, R. (2012). How do C9ORF72 repeat expansions cause amyotrophic lateral sclerosis and frontotemporal dementia: can we learn from other noncoding repeat expansion disorders? *Current Opinion in Neurology, 25*, 689–700.

van Blitterswijk, M., Mullen, B., Nicholson, A. M., Bieniek, K. F., Heckman, M. G., Baker, M. C., … Rademakers, R. (2014). TMEM106B protects C9ORF72 expansion carriers against frontotemporal dementia. *Acta Neuropathologica, 127*, 397–406.

Boccardi, M., Sabattoli, F., Laakso, M. P., Testa, C., Rossi, R., Beltramello, A., … Frisoni, G. B. (2005). Frontotemporal dementia as a neural system disease. *Neurobiology of Aging, 26*, 37–44.

Boeve, B. F., Boylan, K. B., Graff-Radford, N. R., DeJesus-Hernandez, M., Knopman, D. S., Pedraza, O., … Rademakers, R. (2012). Characterization of frontotemporal dementia and/or amyotrophic lateral sclerosis associated with the GGGGCC repeat expansion in C9ORF72. *Brain, 135*, 765–783.

Borroni, B., Alberici, A., Cercignani, M., Premi, E., Serra, L., Cerini, C., … Bozzali, M. (2012). Granulin mutation drives brain damage and reorganization from preclinical to symptomatic FTLD. *Neurobiology of Aging, 33*, 2506–2520.

Borroni, B., Alberici, A., Premi, E., Archetti, S., Garibotto, V., Agosti, C., … Padovani, A. (2008). Brain magnetic resonance imaging structural changes in a pedigree of asymptomatic progranulin mutation carriers. *Rejuvenation Research, 11*, 585–595.

Boxer, A. L., Mackenzie, I. R., Boeve, B. F., Baker, M., Seeley, W. W., Crook, R., … Rademakers, R. (2011). Clinical, neuroimaging and neuropathological features of a new chromosome 9p-Linked FTD-ALS family. *Journal of Neurology, Neurosurgery, and Psychiatry, 82*, 196–203.

Brady, O. A., Zheng, Y., Murphy, K., Huang, M., & Hu, F. (2013). The frontotemporal lobar degeneration risk factor, TMEM106B, regulates lysosomal morphology and function. *Human Molecular Genetics, 22*, 685–695.

Broe, M., Hodges, J. R., Schofield, E., Shepherd, C. E., Kril, J. J., & Halliday, G. M. (2003). Staging disease severity in pathologically confirmed cases of frontotemporal dementia. *Neurology, 60*, 1005–1011.

Buckner, R. L., Andrews-Hanna, J. R., & Schacter, D. L. (2008). The Brain's default network: anatomy, function, and relevance to disease. *Annals of the New York Academy of Sciences, 1124*, 1–38.

de Calignon, A., Polydoro, M., Suárez-Calvet, M., William, C., Adamowicz, D. H., Kopeikina, K. J., … Hyman, B. T. (2012). Propagation of tau pathology in a model of early Alzheimer's disease. *Neuron, 73*, 685–697.

Cenik, B., Sephton, C. F., Dewey, C. M., Xian, X., Wei, S., Yu, K., … Herz, J. (2011). Suberoylanilide hydroxamic acid (vorinostat) up-regulates progranulin transcription: rational therapeutic approach to frontotemporal dementia. *The Journal of Biological Chemistry, 286*, 16101–16108.

Chitramuthu, B. P., Baranowski, D. C., Kay, D. G., Bateman, A., & Bennett, H. P. (2010). Progranulin modulates zebrafish motoneuron development in vivo and rescues truncation defects associated with knockdown of survival motor neuron 1. *Molecular Neurodegeneration, 5*, 41.

Clavaguera, F., Bolmont, T., Crowther, R. A., Abramowski, D., Frank, S., Probst, A., … Tolnay, M. (2009). Transmission and spreading of tauopathy in transgenic mouse brain. *Nature Cell Biology, 11*, 909–913.

Cleveland, D. W., Hwo, S. Y., & Kirschner, M. W. (1977a). Purification of tau, a microtubule-associated protein that induces assembly of microtubules from purified tubulin. *Journal of Molecular Biology, 116*, 207–225.

Cleveland, D. W., Hwo, S. Y., & Kirschner, M. W. (1977b). Physical and chemical properties of purified tau factor and the role of tau in microtubule assembly. *Journal of Molecular Biology, 116*, 227–247.

Cohen, T. J., Guo, J. L., Hurtado, D. E., Kwong, L. K., Mills, I. P., Trojanowski, J. Q., & Lee, V. M. Y. (2011). The acetylation of tau inhibits its function and promotes pathological tau aggregation. *Nature Communications, 2*, 252.

Cohen, T. J., Hwang, A. W., Restrepo, C. R., Yuan, C.-X., Trojanowski, J. Q., & Lee, V. M. Y. (2015). An acetylation switch controls TDP-43 function and aggregation propensity. *Nature Communications, 6*, 5845.

Cook, C., Stankowski, J. N., Carlomagno, Y., Stetler, C., & Petrucelli, L. (2014). Acetylation: a new key to unlock tau's role in neurodegeneration. *Alzheimer's Research & Therapy, 6*, 29.

Coppola, G., Chinnathambi, S., Lee, J. J., Dombroski, B. A., Baker, M. C., Soto-Ortolaza, A. I., ... Geschwind, D. H. (2012). Evidence for a role of the rare p.A152T variant in MAPT in increasing the risk for FTD-spectrum and Alzheimer's diseases. *Human Molecular Genetics, 21*, 3500−3512.

Coppola, G., Karydas, A., Rademakers, R., Wang, Q., Baker, M., Hutton, M., ... Geschwind, D. H. (2008). Gene expression study on peripheral blood identifies progranulin mutations. *Annals of Neurology, 64*, 92−96.

Cruchaga, C., Graff, C., Chiang, H.-H., Wang, J., Hinrichs, A. L., Spiegel, N., ... Goate, A. (2011). Association of TMEM106B gene polymorphism with age at onset in granulin mutation carriers and plasma granulin protein levels. *Archives of Neurology, 68*, 581−586.

Cruts, M., Gijselinck, I., van der Zee, J., Engelborghs, S., Wils, H., Pirici, D., ... Van Broeckhoven, C. (2006). Null mutations in progranulin cause ubiquitin-positive frontotemporal dementia linked to chromosome 17q21. *Nature, 442*, 920−924.

Dakin, S., & Frith, U. (2005). Vagaries of visual perception in autism. *Neuron, 48*, 497−507.

De Muynck, L., Herdewyn, S., Beel, S., Scheveneels, W., Van Den Bosch, L., Robberecht, W., & Van Damme, P. (2013). The neurotrophic properties of progranulin depend on the granulin e domain but do not require sortilin binding. *Neurobiology of Aging, 34*, 2541−2547.

De Muynck, L., & Van Damme, P. (2011). Cellular effects of progranulin in health and disease. *Journal of Molecular Neuroscience : MN, 45*, 549−560.

DeJesus-Hernandez, M., Mackenzie, I. R., Boeve, B. F., Boxer, A. L., Baker, M., Rutherford, N. J., ... Rademakers, R. (2011). Expanded GGGGCC hexanucleotide repeat in noncoding region of C9ORF72 causes chromosome 9p-Linked FTD and ALS. *Neuron, 72*, 245−256.

Donnelly, C. J., Zhang, P.-W., Pham, J. T., Haeusler, A. R., Heusler, A. R., Mistry, N. A., ... Rothstein, J. D. (2013). RNA toxicity from the ALS/FTD C9ORF72 expansion is mitigated by antisense intervention. *Neuron, 80*, 415−428.

Dopper, E. G. P., Rombouts, S. A. R. B., Jiskoot, L. C., den Heijer, T., de Graaf, J. R. A., de Koning, I., ... van Swieten, J. C. (2014). Structural and functional brain connectivity in presymptomatic familial frontotemporal dementia. *Neurology, 83*, e19−e26.

Dumanchin, C., Camuzat, A., Campion, D., Verpillat, P., Hannequin, D., Dubois, B., ... Brice, A. (1998). Segregation of a missense mutation in the microtubule-associated protein tau gene with familial frontotemporal dementia and parkinsonism. *Human Molecular Genetics, 7*, 1825−1829.

Ebneth, A., Godemann, R., Stamer, K., Illenberger, S., Trinczek, B., & Mandelkow, E. (1998). Overexpression of tau protein inhibits kinesin-dependent trafficking of vesicles, mitochondria, and endoplasmic reticulum: implications for Alzheimer's disease. *The Journal of Cell Biology, 143*, 777−794.

Falcon, B., Cavallini, A., Angers, R., Glover, S., Murray, T. K., Barnham, L., ... Bose, S. (2014). Conformation determines the seeding potencies of native and recombinant tau aggregates. *The Journal of Biological Chemistry, 290*, 1049−1065.

Farb, N. A. S., Grady, C. L., Strother, S., Tang-Wai, D. F., Masellis, M., Black, S., ... Chow, T. W. (2013). Abnormal network connectivity in frontotemporal dementia: evidence for prefrontal isolation. *Cortex; A Journal Devoted to the Study of the Nervous System and Behavior, 49*, 1856−1873.

Finch, N., Baker, M., Crook, R., Swanson, K., Kuntz, K., Surtees, R., ... Rademakers, R. (2009). Plasma progranulin levels predict progranulin mutation status in frontotemporal dementia patients and asymptomatic family members. *Brain : A Journal of Neurology, 132*, 583−591.

Finch, N., Carrasquillo, M. M., Baker, M., Rutherford, N. J., Coppola, G., Dejesus-Hernandez, M., ... Rademakers, R. (2010). TMEM106B regulates progranulin levels and the penetrance of FTLD in GRN mutation carriers. *Neurology, 76*, 467−474.

Fong, J. C., Karydas, A. M., & Goldman, J. S. (2012). Genetic counseling for FTD/ALS caused by the C9ORF72 hexanucleotide expansion. *Alzheimer's Research & Therapy, 4*, 27.

Foster, N. L., Wilhelmsen, K., Sima, A. A., Jones, M. Z., D'Amato, C. J., & Gilman, S. (1997). Frontotemporal dementia and parkinsonism linked to chromosome 17: a consensus conference. Conference participants. *Annals of Neurology, 41*, 706−715.

Fratta, P., Mizielinska, S., Nicoll, A., Zloh, M., Fisher, E. M., Parkinson, G., & Isaacs, A. (2012). C9orf72 hexanucleotide repeat associated with amyotrophic lateral sclerosis and frontotemporal dementia forms RNA g-Quadruplexes. *Scientific Reports, 2*, 1016.

Fratta, P., Poulter, M., Lashley, T., Rohrer, J. D., Polke, J. M., Beck, J., ... Mead, S. (2013). Homozygosity for the C9orf72 GGGGCC repeat expansion in frontotemporal dementia. *Acta Neuropathologica, 126*, 401−409.

Froelich, S., Basun, H., Forsell, C., Lilius, L., Axelman, K., Andreadis, A., & Lannfelt, L. (1997). Mapping of a disease locus for familial rapidly progressive frontotemporal dementia to chromosome 17q12-21. *American Journal of Medical Genetics, 74*, 380−385.

Furukawa, Y., Kaneko, K., Watanabe, S., Yamanaka, K., & Nukina, N. (2011). A seeding reaction recapitulates intracellular formation of Sarkosyl-insoluble transactivation response element (TAR) DNA-binding protein-43 inclusions. *The Journal of Biological Chemistry, 286*, 18664−18672.

Gallagher, M. D., Suh, E., Grossman, M., Elman, L., McCluskey, L., Van Swieten, J. C., ... Chen-Plotkin, A. S. (2014). TMEM106B is a genetic modifier of frontotemporal lobar degeneration with C9orf72 hexanucleotide repeat expansions. *Acta Neuropathologica, 127*, 407−418.

Gardner, R. C., Boxer, A. L., Trujillo, A., Mirsky, J. B., Guo, C. C., Gennatas, E. D., ... Seeley, W. W. (2013). Intrinsic connectivity network disruption in progressive supranuclear palsy. *Annals of Neurology, 73*, 603−616.

Ghidoni, R., Benussi, L., Glionna, M., Franzoni, M., & Binetti, G. (2008). Low plasma progranulin levels predict progranulin mutations in frontotemporal lobar degeneration. *Neurology, 71*, 1235−1239.

Goedert, M., Clavaguera, F., & Tolnay, M. (2010). The propagation of prion-like protein inclusions in neurodegenerative diseases. *Trends in Neurosciences, 33*, 317−325.

Goedert, M., Jakes, R., & Crowther, R. A. (1999). Effects of frontotemporal dementia FTDP-17 mutations on heparin-induced assembly of tau filaments. *FEBS Letters, 450*, 306−311.

Goedert, M., Spillantini, M. G., Potier, M. C., Ulrich, J., & Crowther, R. A. (1989). Cloning and sequencing of the cDNA encoding an isoform of microtubule-associated protein tau containing four tandem repeats: differential expression of tau protein mRNAs in human brain. *The EMBO Journal, 8*, 393−399.

Goedert, M., Wischik, C. M., Crowther, R. A., Walker, J. E., & Klug, A. (1988). Cloning and sequencing of the cDNA encoding a core protein of the paired helical filament of Alzheimer disease: identification as the microtubule-associated protein tau. *Proceedings of the National Academy of Sciences of the United States of America, 85*, 4051–4055.

Hirokawa, N. (1994). Microtubule organization and dynamics dependent on microtubule-associated proteins. *Current Opinion in Cell Biology, 6*, 74–81.

Holmes, B. B., Furman, J. L., Mahan, T. E., Yamasaki, T. R., Mirbaha, H., Eades, W. C., … Diamond, M. I. (2014). Proteopathic tau seeding predicts tauopathy in vivo. *Proceedings of the National Academy of Sciences of the United States of America, 111*, E4376–E4385.

Hosler, B. A., Siddique, T., Sapp, P. C., Sailor, W., Huang, M. C., Hossain, A., … Brown, R. H. (2000). Linkage of familial amyotrophic lateral sclerosis with frontotemporal dementia to chromosome 9q21-q22. *JAMA, 284*, 1664–1669.

Hu, F., Padukkavidana, T., Vægter, C. B., Brady, O. A., Zheng, Y., Mackenzie, I. R., … Strittmatter, S. M. (2010). Sortilin-mediated endocytosis determines levels of the frontotemporal dementia protein, progranulin. *Neuron, 68*, 654–667.

Hukema, R. K., Riemslagh, F. W., Melhem, S., van der Linde, H. C., Severijnen, L.-A., Edbauer, D., … van Swieten, J. C. (2014). A new inducible transgenic mouse model for C9orf72-associated GGGGCC repeat expansion supports a gain-of-function mechanism in C9orf72 associated ALS and FTD. *Acta Neuropathologica Communications, 2*, 166.

Hutton, M., Lendon, C. L., Rizzu, P., Baker, M., Froelich, S., Houlden, H., … Heutink, P. (1998). Association of missense and 5′-splice-site mutations in tau with the inherited dementia FTDP-17. *Nature, 393*, 702–705.

Irwin, D. J., McMillan, C. T., Brettschneider, J., Libon, D. J., Powers, J., Rascovsky, K., … Grossman, M. (2013). Cognitive decline and reduced survival in C9orf72 expansion frontotemporal degeneration and amyotrophic lateral sclerosis. *Journal of Neurology, Neurosurgery, and Psychiatry, 84*, 163–169.

Kanemaru, K., Takio, K., Miura, R., Titani, K., & Ihara, Y. (1992). Fetal-type phosphorylation of the tau in paired helical filaments. *Journal of Neurochemistry, 58*, 1667–1675.

Kao, A. W., Eisenhut, R. J., Martens, L. H., Nakamura, A., Huang, A., Bagley, J. A., … Kenyon, C. (2011). A neurodegenerative disease mutation that accelerates the clearance of apoptotic cells. *Proceedings of the National Academy of Sciences of the United States of America, 108*, 4441–4446.

Khan, B., Yokoyama, J., Takada, L., Sha, S., Rutherford, N., Fong, J., … Miller, B. (2012). Atypical, slowly progressive behavioural variant frontotemporal dementia associated with C9ORF72 hexanucleotide expansion. *Journal of Neurology, Neurosurgery, and Psychiatry, 83*, 358–364.

Kleinberger, G., Capell, A., Haass, C., & Broeckhoven, C. (2013). Van mechanisms of granulin deficiency: lessons from cellular and animal models. *Molecular Neurobiology, 47*, 337–360.

Kwon, I., Xiang, S., Kato, M., Wu, L., Theodoropoulos, P., Wang, T., … McKnight, S. L. (2014). Poly-dipeptides encoded by the C9ORF72 repeats bind nucleoli, impede RNA biogenesis, and kill cells. *Science (New York, NY), 345*, 1139–1145.

Laird, A. S., Van Hoecke, A., De Muynck, L., Timmers, M., Van den Bosch, L., Van Damme, P., & Robberecht, W. (2010). Progranulin is neurotrophic in vivo and protects against a mutant TDP-43 induced axonopathy. *PLoS One, 5*, e13368.

Le Ber, I., Camuzat, A., Hannequin, D., Pasquier, F., Guedj, E., Rovelet-Lecrux, A., … French Research Network FTD/FTD-MND. (2008). Phenotype variability in progranulin mutation carriers: a clinical, neuropsychological, imaging and genetic study. *Brain : A Journal of Neurology, 131*, 732–746.

Lee, S. E., Khazenzon, A. M., Trujillo, A. J., Guo, C. C., Yokoyama, J. S., Sha, S. J., … Seeley, W. W. (2014). Altered network connectivity in frontotemporal dementia with C9orf72 hexanucleotide repeat expansion. *Brain : A Journal of Neurology, 137*, 3047–3060.

Lee, W. C., Almeida, S., Prudencio, M., Caulfield, T. R., Zhang, Y.-J., Tay, W. M., … Petrucelli, L. (2014). Targeted manipulation of the sortilin-progranulin axis rescues progranulin haploinsufficiency. *Human Molecular Genetics, 23*, 1467–1478.

Levine, T. P., Daniels, R. D., Gatta, A. T., Wong, L. H., & Hayes, M. J. (2013). The product of C9orf72, a gene strongly implicated in neurodegeneration, is structurally related to DENN Rab-GEFs. *Bioinformatics, 29*, 499–503.

Liu, L., Drouet, V., Wu, J. W., Witter, M. P., Small, S. A., Clelland, C., & Duff, K. (2012). Trans-synaptic spread of tau pathology in vivo. *PLoS One, 7*, e31302.

Lou, K., Yao, Y., Hoye, A. T., James, M. J., Cornec, A.-S., Hyde, E., … Ballatore, C. (2014). Brain-penetrant, orally bioavailable microtubule-stabilizing small molecules are potential candidate therapeutics for Alzheimer's disease and related tauopathies. *Journal of Medicinal Chemistry, 57*, 6116–6127.

Mackenzie, I. R., Arzberger, T., Kremmer, E., Troost, D., Lorenzl, S., Mori, K., … Neumann, M. (2013). Dipeptide repeat protein pathology in C9ORF72 mutation cases: clinico-pathological correlations. *Acta Neuropathologica, 126*, 859–879.

Mackenzie, I. R., Neumann, M., Baborie, A., Sampathu, D., Du Plessis, D., Jaros, E., … Lee, V. M. (2011). A harmonized classification system for FTLD-TDP pathology. *Acta Neuropathologica, 122*, 111–113.

Mahoney, C. J., Beck, J., Rohrer, J. D., Lashley, T., Mok, K., Shakespeare, T., … Warren, J. D. (2012). Frontotemporal dementia with the C9ORF72 hexanucleotide repeat expansion: clinical, neuroanatomical and neuropathological features. *Brain : A Journal of Neurology, 135*, 736–750.

Mahoney, C. J., Simpson, I. J. A., Nicholas, J. M., Fletcher, P. D., Downey, L. E., Golden, H. L., … Fox, N. C. (2014). Longitudinal diffusion tensor imaging in frontotemporal dementia. *Annals of Neurology, 77*, 33–46.

Martens, L. H., Zhang, J., Barmada, S. J., Zhou, P., Kamiya, S., Sun, B., … Farese, R. V. (2012). Progranulin deficiency promotes neuroinflammation and neuron loss following toxin-induced injury. *The Journal of Clinical Investigation, 122*, 3955–3959.

Mason, M. F., Norton, M. I., Van Horn, J. D., Wegner, D. M., Grafton, S. T., & Macrae, C. N. (2007). Wandering minds: the default network and stimulus-independent thought. *Science, 315*, 393–395.

Miller, B. L., Cummings, J., Mishkin, F., Boone, K., Prince, F., Ponton, M., & Cotman, C. (1998). Emergence of artistic talent in frontotemporal dementia. *Neurology, 51*, 978–982.

Min, S.-W., Cho, S.-H., Zhou, Y., Schroeder, S., Haroutunian, V., Seeley, W. W., … Gan, L. (2010). Acetylation of tau inhibits its degradation and contributes to tauopathy. *Neuron, 67*, 953—966.

Mizielinska, S., Grönke, S., Niccoli, T., Ridler, C. E., Clayton, E. L., Devoy, A., … Isaacs, A. M. (2014). C9orf72 repeat expansions cause neurodegeneration in *Drosophila* through arginine-rich proteins. *Science (New York, NY), 345*, 1192—1194.

Mori, K., Weng, S. M., Arzberger, T., May, S., Rentzsch, K., Kremmer, E., … Edbauer, D. (2013). The C9orf72 GGGGCC repeat is translated into aggregating dipeptide-repeat proteins in FTLD/ALS. *Science (New York, NY), 339*, 1335—1338.

Morita, M., Al-Chalabi, A., Andersen, P. M., Hosler, B., Sapp, P., Englund, E., … Brown, R. H. (2006). A locus on chromosome 9p confers susceptibility to ALS and frontotemporal dementia. *Neurology, 66*, 839—844.

Murray, M. E., DeJesus-Hernandez, M., Rutherford, N. J., Baker, M., Duara, R., Graff-Radford, N. R., … Dickson, D. W. (2011). Clinical and neuropathologic heterogeneity of c9FTD/ALS associated with hexanucleotide repeat expansion in C9ORF72. *Acta Neuropathologica, 122*, 673—690.

Okura, H., Yamashita, S., Ohama, T., Saga, A., Yamamoto-Kakuta, A., Hamada, Y., … Matsuyama, A. (2010). HDL/apolipoprotein a-i binds to macrophage-derived progranulin and suppresses its conversion into proinflammatory granulins. *Journal of Atherosclerosis and Thrombosis, 17*, 568—577.

Perry, D. C., Whitwell, J. L., Boeve, B. F., Pankratz, V. S., Knopman, D. S., Petersen, R. C., … Josephs, K. A. (2012). Voxel-based morphometry in patients with obsessive-compulsive behaviors in behavioral variant frontotemporal dementia. *European Journal of Neurology, 19*, 911—917.

Pick, A. (1892). Ueber Die Beiehungen der Senilen Hirnatrophie Zur Aphasie. *Prager Medicinische Wochenschrift, 16*, 165—167.

Piguet, O., Petersén, A., Yin Ka Lam, B., Gabery, S., Murphy, K., Hodges, J. R., & Halliday, G. M. (2011). Eating and hypothalamus changes in behavioral-variant frontotemporal dementia. *Annals of Neurology, 69*, 312—319.

Polymenidou, M., Lagier-Tourenne, C., Hutt, K. R., Bennett, C. F., Cleveland, D. W., & Yeo, G. W. (2012). Misregulated RNA processing in amyotrophic lateral sclerosis. *Brain Research, 1462*, 3—15.

Poorkaj, P., Bird, T. D., Wijsman, E., Nemens, E., Garruto, R. M., Anderson, L., … Schellenberg, G. D. (1998). Tau is a candidate gene for chromosome 17 frontotemporal dementia. *Annals of Neurology, 43*, 815—825.

Possin, K. L., Feigenbaum, D., Rankin, K. P., Smith, G. E., Boxer, A. L., Wood, K., … Kramer, J. H. (2013). Dissociable executive functions in behavioral variant frontotemporal and Alzheimer dementias. *Neurology, 80*, 2180—2185.

Rademakers, R., Cruts, M., & Broeckhoven, C. (2004). Van the role of tau (MAPT) in frontotemporal dementia and related tauopathies. *Human Mutation, 24*, 277—295.

Rademakers, R., Cruts, M., Dermaut, B., Sleegers, K., Rosso, S. M., Broeck, M., … Van Broeckhoven, C. (2002). Tau negative frontal lobe dementia at 17q21: significant finemapping of the candidate region to a 4.8 cM interval. *Molecular Psychiatry, 7*, 1064—1074.

Raichle, M. E., MacLeod, A. M., Snyder, A. Z., Powers, W. J., Gusnard, D. A., & Shulman, G. L. (2001). A default mode of brain function. *Proceedings of the National Academy of Sciences of the United States of America, 98*, 676—682.

Rankin, K., Gorno-Tempini, M., Allison, S., Stanley, C., Glenn, S., Weiner, M., & Miller, B. (2006). Structural anatomy of empathy in neurodegenerative disease. *Brain, 129*, 2945—2956.

Rascovsky, K., Hodges, J. R., Kipps, C. M., Johnson, J. K., Seeley, W. W., Mendez, M. F., … Miller, B. M. (2007). Diagnostic criteria for the behavioral variant of frontotemporal dementia (bvFTD): current limitations and future directions. *Alzheimer Disease and Associated Disorders, 21*, S14—S18.

Rascovsky, K., Hodges, J. R., Knopman, D., Mendez, M. F., Kramer, J. H., Neuhaus, J., … Miller, B. L. (2011). Sensitivity of revised diagnostic criteria for the behavioural variant of frontotemporal dementia. *Brain, 134*, 2456—2477.

Reddy, K., Zamiri, B., Stanley, S. Y., Macgregor, R. B., & Pearson, C. E. (2013). The disease-associated r(GGGGCC)n repeat from the C9ORF72 gene forms tract length-dependent uni- and multi-molecular RNA g-Quadruplex structures. *Journal of Biological Chemistry, 288*, 9860—9866.

Renton, A. E., Majounie, E., Waite, A., Simón-Sánchez, J., Rollinson, S., Gibbs, J. R., … Traynor, B. J. (2011). A hexanucleotide repeat expansion in C9ORF72 is the cause of chromosome 9p21-Linked ALS-FTD. *Neuron, 72*, 257—268.

Rizzu, P., Swieten, J. C., Van; Joosse, M., Hasegawa, M., Stevens, M., Tibben, A., … Heutink, P. (1999). High prevalence of mutations in the microtubule-associated protein tau in a population study of frontotemporal dementia in the Netherlands. *American Journal of Human Genetics, 64*, 414—421.

Rohrer, J. D., Guerreiro, R., Vandrovcova, J., Uphill, J., Reiman, D., Beck, J., … Rossor, M. N. (2009). The heritability and genetics of frontotemporal lobar degeneration. *Neurology, 73*, 1451—1456.

Rohrer, J. D., Nicholas, J. M., Cash, D. M., van Swieten, J., Dopper, E., Jiskoot, L., … Rossor, M. N. (2015). Presymptomatic cognitive and neuroanatomical changes in genetic frontotemporal dementia in the genetic frontotemporal dementia initiative (GENFI) study: a cross-sectional analysis. *Lancet Neurology, 14*, 253—262.

Rohrer, J. D., Ridgway, G. R., Modat, M., Ourselin, S., Mead, S., Fox, N., … Warren, J. D. (2010). Distinct profiles of brain atrophy in frontotemporal lobar degeneration caused by progranulin and tau mutations. *NeuroImage, 53*, 1070—1076.

Rosen, H. J., Allison, S. C., Schauer, G. F., Gorno-Tempini, M. L., Weiner, M. W., & Miller, B. L. (2005). Neuroanatomical correlates of behavioural disorders in dementia. *Brain, 128*, 2612—2625.

Rosen, H. J., Gorno-Tempini, M. L., Goldman, W. P., Perry, R. J., Schuff, N., Weiner, M., … Miller, B. L. (2002). Patterns of brain atrophy in frontotemporal dementia and semantic dementia. *Neurology, 58*, 198—208.

Rosen, H. J., Wilson, M. R., Schauer, G. F., Allison, S., Gorno-Tempini, M. L., Pace-Savitsky, C., … Miller, B. L. (2006). Neuroanatomical correlates of impaired recognition of emotion in dementia. *Neuropsychologia, 44*, 365—373.

Rosso, S. M., Roks, G., Stevens, M., de Koning, I., T, H. L. J., Kamphorst, W., … van Swieten, J. C. (2001). Complex compulsive behaviour in the temporal variant of frontotemporal dementia. *Journal of Neurology, 248*, 965—970.

Sareen, D., O'Rourke, J. G., Meera, P., Muhammad, A. K. M. G., Grant, S., Simpkinson, M., … Baloh, R. H. (2013). Targeting RNA foci in iPSC-derived motor neurons from ALS patients with a C9ORF72 repeat expansion. *Science Translational Medicine, 5*, 1422–1423.

Sato-Harada, R., Okabe, S., Umeyama, T., Kanai, Y., & Hirokawa, N. (1996). Microtubule-associated proteins regulate microtubule function as the track for intracellular membrane organelle transports. *Cell Structure and Function, 21*, 283–295.

Schwenk, B. M., Lang, C. M., Hogl, S., Tahirovic, S., Orozco, D., Rentzsch, K., … Edbauer, D. (2014). The FTLD risk factor TMEM106B and MAP6 control dendritic trafficking of lysosomes. *The EMBO Journal, 33*, 450–467.

See, T. M., LaMarre, A. K., Lee, S. E., & Miller, B. L. (2010). Genetic causes of frontotemporal degeneration. *Journal of Geriatric Psychiatry and Neurology, 23*, 260–268.

Seeley, W. W., Crawford, R., Rascovsky, K., Kramer, J. H., Weiner, M., Miller, B. L., & Gorno-Tempini, M. L. (2008). Frontal paralimbic network atrophy in very mild behavioral variant frontotemporal dementia. *Archives of Neurology, 65*, 249–255.

Seeley, W. W., Crawford, R. K., Zhou, J., Miller, B. L., & Greicius, M. D. (2009). Neurodegenerative diseases target large-scale human brain networks. *Neuron, 62*, 42–52.

Seeley, W., Menon, V., Schatzberg, A., Keller, J., Glover, G., Kenna, H., … Greicius, M. (2007). Dissociable intrinsic connectivity networks for salience processing and executive control. *Journal of Neuroscience, 27*, 2349–2356.

Sha, S. J., Takada, L. T., Rankin, K. P., Yokoyama, J. S., Rutherford, N. J., Fong, J. C., … Boxer, A. L. (2012). Frontotemporal dementia due to C9ORF72 mutations clinical and imaging features. *Neurology, 79*, 1002–1011.

Sieben, A., Van Langenhove, T., Engelborghs, S., Martin, J.-J., Boon, P., Cras, P., … Cruts, M. (2012). The genetics and neuropathology of frontotemporal lobar degeneration. *Acta Neuropathologica, 124*, 353–372.

Skipper, L., Wilkes, K., Toft, M., Baker, M., Lincoln, S., Hulihan, M., … Farrer, M. (2004). Linkage disequilibrium and association of MAPT H1 in Parkinson disease. *American Journal of Human Genetics, 75*, 669–677.

Sleegers, K., Brouwers, N., Van Damme, P., Engelborghs, S., Gijselinck, I., van der Zee, J., … Van Broeckhoven, C. (2009). Serum biomarker for progranulin-associated frontotemporal lobar degeneration. *Annals of Neurology, 65*, 603–609.

Smith, K. R., Damiano, J., Franceschetti, S., Carpenter, S., Canafoglia, L., Morbin, M., … Berkovic, S. F. (2012). Strikingly different clinicopathological phenotypes determined by progranulin-mutation dosage. *American Journal of Human Genetics, 90*, 1102–1107.

Snowden, J. S., Rollinson, S., Lafon, C., Harris, J., Thompson, J., Richardson, A. M., … Pickering-Brown, S. (2012). Psychosis, C9ORF72 and dementia with Lewy bodies. *Journal of Neurology, Neurosurgery, and Psychiatry, 83*, 1031–1032.

Snowden, J. S., Rollinson, S., Thompson, J. C., Harris, J. M., Stopford, C. L., Richardson, A. M., … Pickering-Brown, S. M. (2012). Distinct clinical and pathological characteristics of frontotemporal dementia associated with C9ORF72 mutations. *Brain : A Journal of Neurology, 135*, 693–708.

Spillantini, M. G., Murrell, J. R., Goedert, M., Farlow, M. R., Klug, A., & Ghetti, B. (1998). Mutation in the tau gene in familial multiple system tauopathy with presenile dementia. *Proceedings of the National Academy of Sciences of the United States of America, 95*, 7737–7741.

Su, Z., Zhang, Y., Gendron, T. F., Bauer, P. O., Chew, J., Yang, W.-Y., … Disney, M. D. (2014). Discovery of a biomarker and lead small molecules to target r(GGGGCC)-associated defects in c9FTD/ALS. *Neuron, 83*, 1043–1050.

van Swieten, J., & Spillantini, M. G. (2007). Hereditary frontotemporal dementia caused by tau gene mutations. *Brain Pathology (Zurich, Switzerland), 17*, 63–73.

Tang, W., Lu, Y., Tian, Q., Zhang, Y., Guo, F., Liu, G., … Liu, C. (2011). The growth factor progranulin binds to TNF receptors and is therapeutic against inflammatory arthritis in mice. *Science (New York, NY), 332*, 478–484.

Tenreiro, S., Eckermann, K., & Outeiro, T. F. (2014). Protein phosphorylation in neurodegeneration: friend or foe? *Frontiers in Molecular Neuroscience, 7*, 42.

Trinczek, B., Ebneth, A., Mandelkow, E. M., & Mandelkow, E. (1999). Tau regulates the Attachment/detachment but not the speed of motors in microtubule-dependent transport of single vesicles and organelles. *Journal of Cell Science, 112*, 2355–2367.

Van Damme, P., Van Hoecke, A., Lambrechts, D., Vanacker, P., Bogaert, E., van Swieten, J., … Robberecht, W. (2008). Progranulin functions as a neurotrophic factor to regulate neurite outgrowth and enhance neuronal survival. *The Journal of Cell Biology, 181*, 37–41.

Van Deerlin, V. M., Sleiman, P. M., Martinez-Lage, M., Chen-Plotkin, A., Wang, L. S., Graff-Radford, N. R., … Lee, V. M. (2010). Common variants at 7p21 are associated with frontotemporal lobar degeneration with TDP-43 inclusions. *Nature Genetics, 42*, 234–239.

Vance, C., Al-Chalabi, A., Ruddy, D., Smith, B. N., Hu, X., Sreedharan, J., … Shaw, C. E. (2006). Familial amyotrophic lateral sclerosis with frontotemporal dementia is linked to a locus on chromosome 9p13.2-21.3. *Brain : A Journal of Neurology, 129*, 868–876.

Viskontas, I. V., Boxer, A. L., Fesenko, J., Matlin, A., Heuer, H. W., Mirsky, J., & Miller, B. L. (2011). Visual search patterns in semantic dementia show paradoxical facilitation of binding processes. *Neuropsychologia, 49*, 468–478.

Weingarten, M. D., Lockwood, A. H., Hwo, S. Y., & Kirschner, M. W. (1975). A protein factor essential for microtubule assembly. *Proceedings of the National Academy of Sciences of the United States of America, 72*, 1858–1862.

Wen, X., Tan, W., Westergard, T., Krishnamurthy, K., Markandaiah, S. S., Shi, Y., … Trotti, D. (2014). Antisense proline-arginine RAN dipeptides linked to C9ORF72-ALS/FTD form toxic nuclear aggregates that initiate in vitro and in vivo neuronal death. *Neuron, 84*, 1213–1225.

Whitwell, J. L., Boeve, B. F., Weigand, S. D., Senjem, M. L., Gunter, J. L., Baker, M. C., … Josephs, K. A. (2015). Brain atrophy over time in genetic and sporadic frontotemporal dementia: a study of 198 serial magnetic resonance images. *European Journal of Neurology, 22*, 745–752.

Whitwell, J. L., Jack, C. R., Boeve, B. F., Senjem, M. L., Baker, M., Ivnik, R. J., … Josephs, K. A. (2009). Atrophy patterns in IVS10 + 16, IVS10 + 3, N279K, S305N, P301L, and V337M MAPT mutations. *Neurology, 73*, 1058–1065.

Whitwell, J. L., Jack, C. R., Boeve, B. F., Senjem, M. L., Baker, M., Rademakers, R., … Josephs, K. A. (2009). Voxel-based morphometry patterns of atrophy in FTLD with mutations in MAPT or PGRN. *Neurology, 72*, 813–820.

Whitwell, J. L., Josephs, K. A., Avula, R., Tosakulwong, N., Weigand, S. D., Senjem, M. L., ... Jack, C. R. (2011). Altered functional connectivity in asymptomatic MAPT subjects: a comparison to bvFTD. *Neurology, 77*, 866–874.

Whitwell, J. L., Weigand, S. D., Boeve, B. F., Senjem, M. L., Gunter, J. L., DeJesus-Hernandez, M., ... Josephs, K. A. (2012). Neuroimaging signatures of frontotemporal dementia genetics: C9ORF72, tau, progranulin and sporadics. *Brain : A Journal of Neurology, 135*, 794–806.

Woolley, J. D., Gorno-Tempini, M. L., Seeley, W. W., Rankin, K., Lee, S. S., Matthews, B. R., & Miller, B. L. (2007). Binge eating is associated with right orbitofrontal-insular-striatal atrophy in frontotemporal dementia. *Neurology, 69*, 1424–1433.

Xi, Z., Zinman, L., Moreno, D., Schymick, J., Liang, Y., Sato, C., ... Rogaeva, E. (2013). Hypermethylation of the CpG island near the G4C2 repeat in ALS with a C9orf72 Expansion. *American Journal of Human Genetics, 92*, 981–989.

Yanamandra, K., Kfoury, N., Jiang, H., Mahan, T. E., Ma, S., Maloney, S. E., ... Holtzman, D. M. (2013). Anti-tau antibodies that block tau aggregate seeding in vitro markedly decrease pathology and improve cognition in vivo. *Neuron, 80*, 402–414.

Yen, S. H., Hutton, M., DeTure, M., Ko, L. W., & Nacharaju, P. (1999). Fibrillogenesis of tau: insights from tau missense mutations in FTDP-17. *Brain Pathology (Zurich, Switzerland), 9*, 695–705.

Yokoyama, J. S., & Rosen, H. J. (2012). Neuroimaging features of C9ORF72 expansion. *Alzheimer's Research & Therapy, 4*, 45.

Yokoyama, J. S., Sirkis, D. W., & Miller, B. L. (2014). C9ORF72 hexanucleotide repeats in behavioral and motor neuron disease: clinical heterogeneity and pathological diversity. *American Journal of Neurodegenerative Disease, 3*, 1–18.

van der Zee, J., Rademakers, R., Engelborghs, S., Gijselinck, I., Bogaerts, V., Vandenberghe, R., ... Dermaut, B. (2006). A belgian ancestral haplotype harbours a highly prevalent mutation for 17q21-Linked tau-negative FTLD. *Brain : A Journal of Neurology, 129*, 841–852.

Zhang, D., Iyer, L. M., He, F., & Aravind, L. (2012). Discovery of novel DENN Proteins: implications for the evolution of eukaryotic intracellular membrane structures and human disease. *Frontiers in Genetics, 3*, 283.

Zhou, J., Greicius, M. D., Gennatas, E. D., Growdon, M. E., Jang, J. Y., Rabinovici, G. D., ... Seeley, W. W. (2010). Divergent network connectivity changes in behavioural variant frontotemporal dementia and Alzheimer's disease. *Brain, 133*, 1352–1367.

Zhu, J., Nathan, C., Jin, W., Sim, D., Ashcroft, G. S., Wahl, S. M., ... Ding, A. (2002). Conversion of proepithelin to epithelins: roles of SLPI and elastase in host defense and wound repair. *Cell, 111*, 867–878.

Chapter 34

The Genetic Basis of Alzheimer's Disease: Findings From Genome-Wide Studies

Basavaraj Hooli and Rudolph E. Tanzi
Massachusetts General Hospital, Boston, MA, United States; Harvard Medical School, Boston, MA, United States

INTRODUCTION

Alzheimer's disease (AD) is the sixth leading cause of death in the United States, with health care costs surpassing \$200 billion in 2013 (www.curealz.org; www.alz.org). With longer life expectancies in the aging population, the costs associated with this devastating illness are expected to quadruple in coming years, which poses a significant threat to the public health system. The clinical symptoms are broadly defined by a slowly progressing loss of memory and cognitive functions, dementia, and ultimately death. Neuropathologically, aggregation, and deposition of β-amyloid (Aβ) peptide aggregations in the form of extracellular or neuritic "plaques," is crucial to initiating AD pathogenesis. In the original amyloid hypothesis proposed by George Glenner (Glenner & Wong, 1984), deposition of amyloid plaques induces the accumulation of hyperphosphorlylated tau protein in the form of intracellular neurofibrillary "tangles" (NFTs) leading to cell death, neuroinflammation (beyond that initially caused by Aβ), and dementia. After 30 years of debate, Glenner's hypothesis was confirmed in a three-dimensional human neural culture system (Choi et al., 2014). Increased synaptic load of Aβ aggregates and neurofibrillary tangles in AD brains strongly correlate with neuronal dysfunction and disease progression (Hooli & Tanzi, 2009). In many cases, these cellular changes are often accompanied by various other pathological phenotypes, including TDP-43 immunoreactivity, cerebral amyloid angiopathy (CAA), and white matter lesions (Dubois et al., 2014).

Advanced age is the greatest risk factor for AD; risk doubles with every decade after 65 years of age. After advanced age, family history is the second strongest risk factor for AD. Population and twin studies estimate that up to 80% of cases of AD are attributable to genetic factors (Gatz et al., 2006). Studies conducted over the past 3 decades have shown that genetic factors leading to AD are complex and heterogeneous, which means that the risk for AD is a complex interaction of heritable (eg, genetic) and nonheritable (eg, environmental, lifestyle) factors.

AD has been traditionally classified into two dichotomous forms based on the age of onset and associated genetic factors. The less prevalent, early-onset familial form of AD (EOFAD) (<5% of patients with AD) represents patients with AD who exhibit a Mendelian mode of inheritance of highly penetrant autosomal dominant gene variants. To date, more than 200 mutations in the genes Amyloid Precursor Protein (APP), presenilin (PSEN)1, and PSEN2 are known to cause EOFAD with an onset at age <60 years. On the other hand, the more predominantly diagnosed form (>90% of patients) of AD is classified as "sporadic" late-onset AD (LOAD), with less obvious familial aggregation of genetic risk factors and a later-onset age of >60 years. The consensus in the research community posits (Lill & Bertram, 2011; Mitsui & Tsuji, 2014; Sharma, Kruger, & Gasser, 2014; Tanzi, 2012) that susceptibility for LOAD is conferred by (1) numerous genetic risk factors of relatively high allele frequency but low penetrance and small effect sizes (Table 34.1), or (2) rare and private functional gene variants with high penetrance. In LOAD, the apolipoprotein E (APOE) gene remains the most well-established risk factor, in which the ε4 allele confers between 3.7- and 14-fold risk, depending on the number of copies present in the carrier (Farrer et al., 1997). That this broad dichotomization of EOFAD and LOAD is almost certainly overly simplistic; reports of EOFAD without Mendelian transmission but with strong familial clustering are also noted for LOAD. Large-scale, whole-genome sequencing efforts should eventually reveal the extent to which LOAD is caused by highly penetrant rare functional variants.

Genomics, Circuits, and Pathways in Clinical Neuropsychiatry. http://dx.doi.org/10.1016/B978-0-12-800105-9.00034-2

TABLE 34.1 List of Genes Associated With Alzheimer's Disease Pathogenesis

Gene Category	Gene	Postulated Mechanisms in AD
EOFAD Genes		
	APP, PSEN1, and PSEN2	Aβ production/aggregation, Aβ42/Aβ40 ratio
	APOE	Aβ clearance
AD GWAS Confirmed		
	CR1	Complement system-mediated Aβ clearance, immunological response
	MS4A4A, MS4A6A	Signal transduction, IgE-mediated immune response
	PICALM	Accumulation of Aβ via APP processing, clarithin-mediated endocytosis
	SORL1	Accumulation of Aβ via APP endosome recycling pathways
	FERMT2	Integrin-mediated axonal transport, tau metabolism, and toxicity
	DSG2	Mediate Aβ and intermediate NFTs interaction
	ABCA7	Accumulation of Aβ, regulate immune response
	CD33	Aβ clearance and microglia-mediated neuroinflammation
	BIN1	Endocytosis, immune response, synaptic activity, tau-induced toxicity
	INPP5D	Immune response and inflammation
	CASS4	Cytoskeleton and axonal support transport, APP and tau toxicity
	MEF2C	Synaptic function and immunological response
	CD2AP	Cytoskeletal function, synapse formation, endocytosis
	NME8	Oxidative stress, cytoskeletal function, immune response
	ZCWPW1	Immune response and inflammation
	NYAP1	Cytoskeletal function, synapse formation
	EPHA1	Synapse formation, cytoskeletal function, immune response
	CLU	Aβ clearance, immune response and inflammation
	TREM2	Aβ clearance, immune response and inflammation
AD GWAS-Suggestive		
	DLGAP1	Postsynaptic density regulation
	ECHDC3	Synaptic activity
	TREML2	Aβ clearance, immune response, and inflammation
	CELF1	Synaptic activity, tau toxicity
	ADAMTS20	Synapse formation, Aβ accumulation
	RIN3	Endosomal trafficking
	IGHV1-67	Immune response and inflammation
	SPPL2A	Signal transduction, immune response
	TRIP4	Synapse survival, immune response
	SCIMP	Signal transduction, immune response
	ACE	Aβ clearance and inflammation
	HS3ST1	Aβ clearance and synaptic activity

TABLE 34.1 List of Genes Associated With Alzheimer's Disease Pathogenesis—cont'd

Gene Category	Gene	Postulated Mechanisms in AD
	SQSTM1	Aβ clearance and inflammation
	ATXN1	Aβ accumulation
	DPYSL2	Synaptic activity, tau toxicity
	TP53INP1	Apoptosis
	ADAM10	Aβ production

Table shows an exhaustive list of genes associated with Alzheimer's disease (AD) based on published genetic studies, including genome-wide association studies (GWAS) and meta-analysis reports. Genes are classified into three broad categories: EOFAD genes carry mutations that cause early-onset Mendelian form of AD; GWAS-confirmed gene variants have shown consistent association across multiple AD study cohort in the GWAS and meta-analyses results; GWAS-suggestive genes show the presence for variants that reach clear and consistent genome-wide significance in all replication studies.

Overall, although twin and population studies estimate heritability in AD as high as 60–80%, the genetic factors known to date explain roughly 30–50% of heritability in AD (Bertram & Tanzi, 2012). The absence of viable therapeutics reflects the fact that pathophysiological and genomic causes of the disease are not entirely understood. Moreover, the environmental and epigenetic factors that are also most likely to make significant contributions to an individual's risk remain elusive, owing mainly to difficulties in assessment techniques (Traynor & Singleton, 2010).

Genome-wide association studies (GWAS), a small number of which use next-generation sequencing (NGS) technology, have revealed more than 30 additional AD genetic loci (Bertram, 2011; Bertram et al., 2008; Hollingworth et al., 2011; Lambert et al., 2013; Naj et al., 2011; Reitz et al., 2013), which hints at a promising new era of individualized medicine in AD (Sharma et al., 2014). Future NGS studies hold an enormous potential for creating a detailed catalog of the genetic variants in AD and for consequently elucidating the missing heritability in the onset of AD. In this chapter we review implicated GWAS loci in AD, with particular emphasis on genes that show evidence for genome-wide significant association either in large studies or as a result of systematic meta-analyses.

GENETICS OF EARLY-ONSET FAMILIAL ALZHEIMER'S DISEASE

EOFAD is an autosomal-dominant Mendelian form of the disease that appears in multiple affected subjects in the carrier families and in more than one generation. The age of disease onset is consistently before age 60–65 years and often before age 55 years. Roughly 50% of reported EOFAD is caused by more than 200 mutations in three genes: *APP*, *PSEN1*, and *PSEN2* (Bird, 2008; Campion et al., 1999; Cruts, Theuns, & Van Broeckhoven, 2012b; Rocca et al., 1991). These three genes, as well as the most established late-onset gene APOE, were identified through genetic linkage studies in large multigenerational pedigrees that had AD (Tanzi & Bertram, 2005). In fact, identification of these mutations was crucial to our current understanding of the disease pathogenesis ("amyloid hypothesis") (Glenner & Wong, 1984; Hardy & Higgins, 1992; Hardy & Selkoe, 2002; Tanzi, 2013; Tanzi & Bertram, 2005).

Functionally, pathogenic variants in the three genes (*APP*, *PSEN1*, and *PSEN2*) lead to EOFAD by altering the production or rate of aggregation of Aβ. Most mutations causing AD are clustered in the vicinity of the cleavage sites in APP or in the genes (PSEN1 or PSEN2) encoding the γ-secretase complex enzyme that cleaves APP to produce Aβ (Cruts, Theuns, & Van Broeckhoven, 2012a), and increase the $A\beta_{42}$:$A\beta_{40}$ ratio. Whereas these three currently known EOFAD genes explain a large proportion of this Mendelian form of AD (13–50% of cases), incidences of EOFAD kindreds without mutations in these three genes suggest that additional unknown EOFAD-causing gene variants most likely exist.

AMYLOID PRECURSOR PROTEIN

APP (Chr 21q21) gene encodes a ubiquitously expressed type 1 transmembrane protein with three predominant splice variants: APP695, APP751, and APP770. APP695, the major isoform, is expressed in neurons, whereas the APP751 splice variant is detected mainly in astrocytes (Yoshikai, Sasaki, Doh-ura, Furuya, & Sakaki, 1991). The cell-surface APP is internalized, allowing APP to be processed by endocytic pathways, and different fragments are secreted into the extracellular space (Thinakaran & Koo, 2008). Most of APP undergoes nonamyloidogenic processing via consecutive cleavage by α- and γ-secretases within the Aβ domain, resulting in nonpathogenic fragments, sAPPα and

C-terminal fragments (CTFs) (Vetrivel & Thinakaran, 2006). Alternatively, APP undergoes sequential proteolytic cleavage by β- and γ-secretases to generate the neurotoxic Aβ peptides, as well as sAPPβ and CTF. More than 30 pathogenic APP gene variants are reported to date (causing 10−15% of EOFAD cases). In addition, APP gene duplication (three copies of APP, <1% of EOFAD) causes EOFAD with CAA in carrier families by increased Aβ production (Hooli et al., 2012).

A rare protective variant in *APP*, A673T, has been reported in Icelandic and Finnish study subjects that confers resilience against dementia (Kero et al., 2013; Maloney et al., 2014). To date, studies indicate that this rare variant is likely limited to members of the two populations in the original study (Bamne et al., 2014; Wang, Naj, et al., 2015). The association between APP and EOFAD is well established; however, there are no confirmed reports showing the presence of risk-conferring APP variants in LOAD (Hooli et al., 2012). This is in stark contrast to other neurodegenerative disorders, eg, Parkinson disease or frontotemporal lobe dementia (FTLD), in which genes carrying rare variants that cause Mendelian disease forms also show the presence of risk-conferring alleles (Lill & Bertram, 2011). Moreover, all known pathogenic variants in APP cluster in cleavage sites near exons 16 and 17, ie, the Aβ domain, which leaves the possibility of additional variants to be found elsewhere in the APP gene.

PRESENILIN GENES (*PSEN1* AND *PSEN2*)

PSEN1 (14q24.2) and its homolog, *PSEN2* (1q42.13), are structurally similar integral membrane proteins that contain nine transmembrane domains with a hydrophilic intracellular loop region. The presenilins, PSEN1 and PSEN2, together with nicastrin, anterior pharynx-defective-1, and presenilin enhancer 2, form the γ-secretase complex that catalyzes the cleavage of numerous membrane proteins, including Notch, which is crucial in the developmental stages (Donoviel et al., 1999; Wakabayashi & De Strooper, 2008). Both of these proteins localize in the endoplasmic reticulum and Golgi apparatus (De Strooper, 2003; Hansson et al., 2004; Kovacs et al., 1996) and have a key role in processing APP by the amyloidogenic pathway to release Aβ fragments. Close to 200 dominant pathogenic mutations in *PSEN1* and 14 single-nucleotide variants in *PSEN2* (40−80% and <5%, respectively) cause EOFAD (Bagyinszky, Youn, An, & Kim, 2014; Cruts et al., 2012a). The pathogenic variants in *PSEN1* and *PSEN2* are present all across the protein but show some clustering close to the transmembrane domains (Wagner, Tanzi, Mobley, & Galasko, 2012). Interestingly, there are no genomic structural variants reported in these genes, possibly owing to high cross-species conservation (Hooli, Kovacs-Vajna, et al., 2014).

OTHER EARLY-ONSET FAMILIAL ALZHEIMER'S DISEASE GENES

Efforts to identify novel EOFAD genes beyond *APP*, *PSEN1*, and *PSEN2* have led to potentially pathogenic mutations in genes implicated in FTLD, including the genes *MAPT* and *CHMP2B*. MAPT encodes Tau protein, which is a key component of the neurofibrillary tangles. *MAPT* is present in the highly variable 17q21.31 region and is associated with a broad range of disorders, including neurodevelopmental and neurodegenerative disorders (Rovelet-Lecrux & Campion, 2012). A family-based genome-wide copy number variant analysis reported a private variant in one of the early-onset National Institutes of Mental Health (NIMH) families, with all the affected members carrying an extra copy of the MAPT gene but no other adjacent genes (Hooli, Kovacs-Vajna, et al., 2014). The same study identified the pathogenic Pro301Leu variant in one family diagnosed with probable AD (not confirmed postmortem). In the same line, our laboratory has identified another postmortem-confirmed individual with dementia carrying the same P301L change but without NFTs in the brain (unpublished), which further affirms the role of *MAPT* genetic variants leading to a range of disease phenotypes. A list of additional missense mutations in the tau gene in patients exhibiting the AD/FTLD phenotype can be found on the AD&FTD mutation database (Cruts et al., 2012a). A heterozygous deletion in the *CHMP2B* gene was also associated with neurodegeneration in a family in which one copy of *CHMP2B* was absent in affected subjects (heterozygous deletion) (Hooli, Kovacs-Vajna, et al., 2014). As a precautionary note, however, before all of these mutations can be validated to cause AD, autopsy confirmation of individuals carrying these structural variants will be necessary to confirm the diagnosis of AD.

LATE-ONSET SPORADIC FORM OF ALZHEIMER'S DISEASE

In 2007, the first GWAS was reported in AD (Reiman et al., 2007). However, the family-based association analysis (Lange, DeMeo, Silverman, Weiss, & Laird, 2004) in close to 1500 subjects from 450 families from the NIMH AD initiative study cohort in 2008 was the first study to identify genetic markers showing genome-wide significance beyond APOE (Bertram et al., 2008). This 2008 GWAS (Bertram et al., 2008) reported genome-wide significance for association of three AD genes in AD, including *ATXN1* (ataxin 1), *CD33* (siglec 3), and an uncharacterized locus on chromosome 14 (GWA_14q31.2), and a highly suggestive association of AD with *DLGAP1*. Several additional GWAS and meta-analyses of large LOAD consortium data sets have since been reported, revealing close to 30 different loci showing association with AD (Table 34.1).

Some genes that either encompass confirmed AD-associated GWAS markers or are in close vicinity to them include *TREM2, BIN1, CLU, ABCA7, CR1, PICALM, MS4A6A, CD33, CD2AP, SORL1, FERMT2, DSG2, INPP5D, CASS4, MEF2C, NME8, ZCWPW1, EPHA1, HLA-DRB1/HLA-DRB5, PTK2B, CELF1*, and *RIN3* (Abraham et al., 2008; Beecham et al., 2009; Bertram et al., 2008; Bertram & Tanzi, 2009, 2012; Carrasquillo et al., 2009; Coon et al., 2007; Grupe et al., 2007; Guerreiro et al., 2013; Harold et al., 2009; Heinzen et al., 2010; Hollingworth et al., 2011; Hooli et al., 2015; Jonsson et al., 2013; Jun et al., 2015; Lambert et al.,2009, 2013; Li et al., 2008; Naj et al., 2010, 2011; Poduslo, Huang, Huang, & Smith, 2009; Potkin et al., 2009; Seshadri et al., 2010; Shaw et al., 2011). Although these gene variants have shown a significant association with AD, the functional variants that lead to AD are not fully obvious (Swerdlow & Corder, 2012). In fact, GWAS markers generally identify the genomic region associated with the disease and not necessarily the actual gene. In most cases, the genes associated markers remain elusive owing to the presence of multiple genes in close vicinity to the disease-associated markers. Therefore, most GWAS-derived AD gene candidates require more in-depth genetic and functional studies.

Nonetheless, the AD genes arising from GWAS likely affect various molecular and biochemical pathways (Fig. 34.1; Table 34.1) and aspects of AD pathologies, including Aβ production, aggregation and clearance, tangle formation, and neuroinflammation (Fig. 34.1). However, an understanding of the precise functional mechanisms leading to AD and the affected pathways remains rudimentary. Furthermore, despite strong statistical support for these genes arising from GWAS, the effects of these genes on risk for AD are exceedingly small, with allelic odds ratios (ORs) between 0.85 and 1.15: that is, increasing or decreasing risk for AD by about 15%. This risk impact in these GWAS genes is considerably low compared with the OR for *APOE ε4*, which is between fourfold and 15-fold for one or two alleles, respectively (Farrer et al., 1997).

In the following sections, we examine both general and postulated disease functions of the new AD GWAS genes and provide an overview of their potential roles in LOAD risk (Fig. 34.1; Table 34.1). We will summarize the genetic

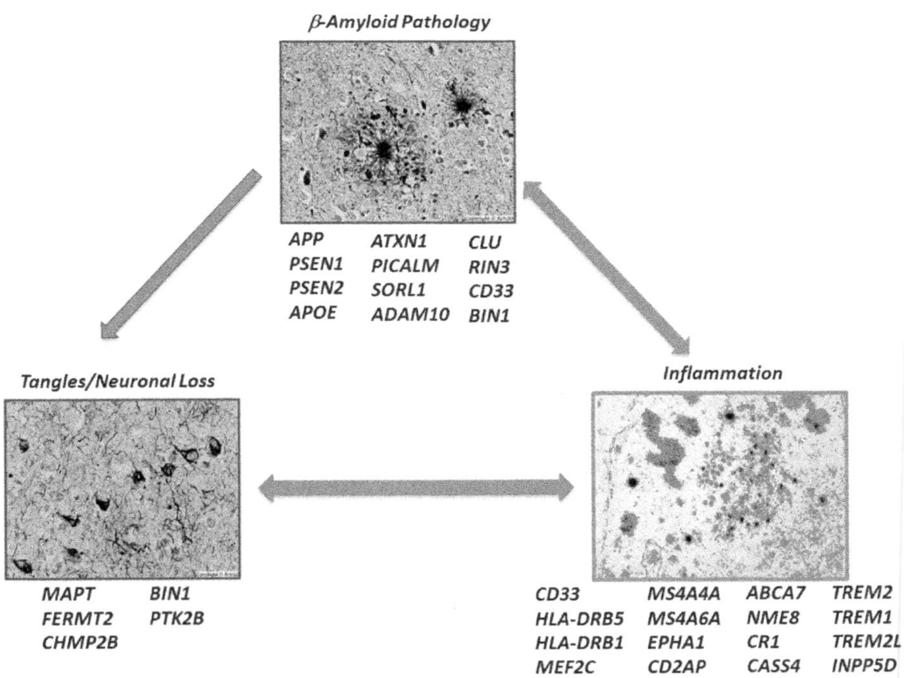

FIGURE 34.1 Overview of GWAS genes and AD neuropathological features influenced by these genes. The AD genes showing significant association in GWAS and the disease-specific neuropathological features most likely linked to their functional role are shown above. The three main pathological phenotypes shown above correlate with the relevant disease mechanisms leading to AD, ie, amyloidogenic pathways, inflammation response, and neurofibrillary tangles. β-Amyloid deposition can lead to tangle formation (Choi et al., 2014) and inflammation. Tangles and inflammation lead to neuronal cell death, and then to more inflammation and β-amyloid deposition in a vicious cycle of neurodegeneration.

epidemiological evidence, review the potential functional genomic consequences of the implicated sequence variants, and discuss these variants' possible role in the onset of this devastating disease in elderly people.

APOLIPOPROTEIN E

The *APOE* (19q13.32) gene encodes a pleiotropic glycoprotein, which is a major component of very-low density lipoproteins. There are at least three major alleles in the APOE gene (ε2, ε3, and ε4), corresponding to combinations of two amino acid changes at residues 112 and 158 (ε2: Cys112/Cys158; ε3: Cys112/Arg158; and ε4: Arg112/Arg158). The most common allele, ε3, is present in more than half of the general population (up to 70%). Family-based methods originally led to the identification of genetic linkage between AD and the *APOE* gene (Corder et al., 1994; Strittmatter et al., 1993), revealing that the APOE-ε4 allele increases risk in both familial and sporadic AD. Unlike mutations in the three known EOFAD genes discussed previously, the *APOE-ε4* allele is neither necessary nor sufficient to cause AD but instead confers risk by decreasing the age of onset in a dose-dependent manner. The risk effect of APOE-ε4 has been consistently replicated in a large number of studies across study cohorts from different ethnic groups; in addition, with ORs between fourfold and 15-fold, it remains the strongest predictor for AD risk. Studies have also revealed rare coding variants that affect risk for AD (Kamboh et al., 1999; Medway et al., 2014). APOE variants in regulatory regions that alter *APOE* expression leading to decreased plaque clearance have also been studied in mouse models (Bien-Ly, Gillespie, Walker, Yoon, & Huang, 2012; Kim et al., 2011). However, further studies are required to confirm whether these functional variants in APOE lead to AD.

The functional consequences of APOE in AD pathogenesis are complex (Holtzman, Herz, & Bu, 2012; Michaelson, 2014). APOE is expressed in liver, brain, and macrophages (Siest et al., 1995) and is involved in cholesterol mobilization and redistribution (Mahley & Rall, 2000). The APOE-ε4 allele also predisposes patients to vascular disease as a result of its association with increased plasma cholesterol levels (Huang, 2010). APOE also regulates Aβ metabolism indirectly by interacting with low-density lipoprotein receptor-related protein 1 receptors (Verghese et al., 2013). In the brains of transgenic AD mice, cholesterol has also been shown to increase Aβ production and stabilize the peptide (Kim, Basak, & Holtzman, 2009). Thus, it is possible that APOE-ε4 confers risk for AD via hypercholesterolemia, because this would also influence the clearance of soluble Aβ and Aβ aggregation (Castellano et al., 2011; Kanekiyo, Xu, & Bu, 2014). APOE-ε4 binds to Aβ more rapidly than does APOE-ε3, resulting in accelerated fibril formation (Hauser & Ryan, 2013; Stratman et al., 2005). In vivo, APOE influences the amount and structure of intraparenchymal Aβ deposits in an isoform-dependent manner (Fagan et al., 2014). Overall, data indicate that APOE most likely influences Aβ metabolism and clearance (Bertram & Tanzi, 2005; Liu, Kanekiyo, Xu, & Bu, 2013).

ADAM METALLOPEPTIDASE DOMAIN 10

ADAM Metallopeptidase Domain 10 (*ADAM10*) encodes the major α-secretase in the brain (Endres & Fahrenholz, 2012), which cleaves the APP ectodomain to preclude Aβ production (Jorissen et al., 2010; Kuhn et al., 2010; Postina et al., 2004). Two rare missense mutations conferring risk for LOAD were reported in 2009 in seven LOAD families (Kim, Suh, et al., 2009). The two *ADAM10* LOAD mutations, Q170H and R181G, are located in the prodomain region and were subsequently shown in vivo to impair the ability dramatically of ADAM10 to carry out α-secretase cleavage of APP, resulting in elevated Aβ deposition (Suh et al., 2013). Other studies have suggested that 5′-untranslated region of *ADAM10* may have an important role for posttranscriptional regulation of ADAM10 expression and consequently Aβ production (Lammich et al., 2010). These genetic and biological findings provide further support for the amyloid hypothesis implicating APP processing and Aβ generation, and also targets to identify therapeutic interventions to treat AD.

TRIGGERING RECEPTOR EXPRESSED ON MYELOID CELLS 2

Triggering Receptor Expressed on Myeloid Cells 2 (*TREM2*) harbors a missense mutation, with the second highest reported effect size after the APOE-ε4 allele (Lill et al., 2015) for LOAD. The rare TREM2-R47H codon change (OR, about 1.5), first found in an Icelandic population and later in an international cohort of a European population (Guerreiro et al., 2013; Jonsson et al., 2013), has been replicated in a meta-analysis (Benitez et al., 2013; Hooli, Parrado, et al., 2014; Pottier et al., 2013). Previously, mutations in *TREM2* were associated with Nasu-Hakola disease, a polycystic lipomembranous osteodysplasia with sclerosing leukoencephalopathy (Bianchin et al., 2004; Paloneva et al., 2003). This rare recessive disease includes progressive frontal-type dementia among its clinical features (Bock et al., 2013; Madry, Prudlo, Grgic, & Freyschmidt, 2007; Neumann & Takahashi, 2007; Paloneva, Autti, Hakola, & Haltia, 1993).

TREM2 likely influences risk for AD via innate immune-related pathogenic mechanisms. TREM2 is a type 1 transmembrane receptor protein expressed on myeloid cells including microglia, monocyte-derived dendritic cells, osteoclasts, and bone marrow–derived macrophages (Colonna, 2003; Ford & McVicar, 2009; Klesney-Tait, Turnbull, & Colonna, 2006; Lue, Schmitz, & Walker, 2014). TREM2 transduces intracellular signaling through TYRO protein tyrosine kinase binding protein. In the brain, TREM2 is primarily expressed on microglia and has been shown to control two signaling pathways: regulation of phagocytosis and suppression of inflammation reactivity, which suggests that TREM2 influences neurodegeneration through neuroinflammation and dysregulation of the immune response feature of AD (Golde, Streit, & Chakrabarty, 2013; Hickman & El Khoury, 2014; Holtzman, Morris, & Goate, 2011). Furthermore, inhibition of TREM2 activity decreases phagocytosis of these cells by approximately one-third, which suggests that the interaction between TREM2 and its ligands facilitates clearance of apoptotic neurons (Hsieh et al., 2009). However, the pathogenetic mechanism by which these variants increase risk for AD remains to be fully elucidated (Hickman & El Khoury, 2014). Interestingly, other TREM2-related genes, including *TREM1* and *TREML2*, have also been associated with AD, pending confirmation.

CD33 (MYELOID CELL SURFACE ANTIGEN CD33; SIALIC ACID-BINDING IMMUNOGLOBULIN-LIKE LECTIN 3)

The *CD33* gene, which is located on chromosome 19q13.3, is a member of a class of immune cell surface receptors called sialic acid–binding immunoglobulin (Ig)-like lectins (Siglecs) family of receptors. AD-associated variants in *CD33* were first reported in the family-based GWAS (Bertram et al., 2008) and were later confirmed by two other large GWAS (Lambert et al., 2013; Naj et al., 2011). The original family-based study led to a *CD33* single-nucleotide polymorphism (SNP) that increased risk for AD, whereas the two latter case–control GWAS meta-analyses found two protective SNPs about 1400 base pairs away. Similar to other Siglec families of receptors, CD33 is expressed on myeloid cells and microglia (Crocker, 2002; Crocker, Hartnell, Munday, & Nath, 1997). Binding of sialic acid activates CD33, leading to monocyte inhibition via immunoreceptor tyrosine-based inhibitory motif domains (Linnartz-Gerlach, Mathews, & Neumann, 2014). CD33 is also reported to have a role in clathrinin-dependent receptor-mediated endocytosis (Cao & Crocker, 2011; Tateno et al., 2007).

Functional studies in AD show that CD33 is associated with Aβ clearance and microglia-mediated neuroinflammatory pathways (Griciuc et al., 2013; Malik et al., 2013; Raj et al., 2014). CD33 mRNA expression is specifically increased in microglia, and expression in autopsy brain tissue is associated with more advanced cognitive decline (Bradshaw et al., 2013; Jiang et al., 2014). Aβ phagocytosis is inhibited in microglial cells overexpressing CD33 (Griciuc et al., 2013), and this effect is abolished in cells expressing CD33 lacking exon 2, which encodes the sialic acid–binding domain. The risk SNP rs3865444 was shown to increase expression of the CD33 splice form lacking exon 2, whereas the second protective marker rs12459419 affects exon 2 splicing efficiency. Overall, results from multiple GWAS showing significant association of CD33 variants with AD, together with evidence from functional studies, suggest that CD33 has an important role in the onset of LOAD.

CAS SCAFFOLDING PROTEIN FAMILY MEMBER 4

Cas Scaffolding Protein Family Member 4 (*CASS4*) is a member of the CAS protein family, scaffolding proteins responsible for a number of cellular activities (Nikonova, Gaponova, Kudinov, & Golemis, 2014; Tikhmyanova, Little, & Golemis, 2010). CASS4 is a docking protein in the tyrosine-kinase signaling pathway and is associated with cell adhesion and spreading (Tornillo, Defilippi, & Cabodi, 2014). First characterized in 2008, CASS4 shares up to 42% similarity in gene sequence with the other three members of the CASS family but it lacks the conserved YDYVHL motif (Singh et al., 2008). It is most highly expressed in spleen and lung tissues as well as ovarian and leukemia cells and its tissue-specific function is linked to the activity of the other CAS family members.

CASS4 is a relatively understudied gene, but it has been implicated in APP and tau metabolism (Beck, Nicolas, Kopp, & Golemis, 2014). The reported GWAS significant SNP rs7274581 shows overtransmission of minor allele in unaffected controls, indicating a protective effect. Another SNP in linkage disequilibrium (LD) with this protective SNP, rs6024870, also has been reported to show evidence of regulatory function (Wang, Lopez, et al., 2015). Dcas, the *Drosophila* homolog to CAS family proteins, interacts with integrin pathway genes (Tikhmyanova, Tulin, Roegiers, & Golemis, 2010), and CASS4 is speculated to have a role in AD through its interaction with *CD2AP*, another gene implicated in AD (Becam, Tanentzapf, Lepesant, Brown, & Huynh, 2005; Tibaldi & Reinherz, 2003).

COMPLEMENT COMPONENT (3B/4B) RECEPTOR 1

Complement receptor 1 (*CR1*), which is located on chromosome 1q32 in a cluster of complement-related proteins, is a major player in the immune system. *CR1* is one of the genes showing consistent association with AD since the early wave of large collaborative GWAS in AD (Harold et al., 2009; Hollingworth et al., 2011; Lambert et al., 2013; Naj et al., 2010, 2011; Seshadri et al., 2010). Several *CR1* SNPs, mainly in the noncoding region, show a significant association with LOAD across these studies. The *CR1* locus has a high degree of repetitive sequences, which complicates identification of the true functional risk factors. However, *CR1* encodes four isoforms that differ based on genomic duplication and deletions (Liu & Niu, 2009), and analysis of splice variants revealed a functional isoform (Brouwers et al., 2012) showing association with AD pathogenesis (Crehan et al., 2012; Krych-Goldberg, Moulds, & Atkinson, 2002). The splice form associated with AD determines the length of the CR1 protein, and with that the number of C3b or C4b cofactor activity binding sites that are important in the complement cascade (Rogers et al., 2006; Velazquez, Cribbs, Poulos, & Tenner, 1997).

Mechanistically, CR1 expression in phagocytic cells such as erythrocytes results in the ingestion and removal of complement-activated particles (Khera & Das, 2009). It serves as a B-cell receptor for fragments of complement components C3 and C4 and is involved in factor-I—mediated cleavage of C3, and thus regulates complement activation (Dunkelberger & Song, 2010; Zipfel & Skerka, 2009). CR1 expression of complement factors is reportedly upregulated in affected neurons and glia are sources of complement in the brain regions of brains with AD (Hazrati et al., 2012). Higher CR1 protein expression is associated with a higher clearance rate of immune complexes (Gibson & Waxman, 1994; Schifferli & Paccaud, 1989). On the other hand, CR1 mRNA expression in autopsy brain tissue is also associated with advanced cognitive decline (Chibnik et al., 2011). CR1 SNPs are associated with increased amyloid β1-42 concentrations in cerebrospinal fluid but require further confirmation in larger samples (Schjeide et al., 2011). Variants in the CR1 locus are also associated with neuroimaging measures in AD and neuritic plaque burden in brains with AD (Chibnik et al., 2011; Shulman et al., 2013). *CR1* SNPs are also associated with low CR1 expression in white matter and cerebellum, leading to shrinking of entorhinal cortex and further evidence for the role of CR1 in brain vasculature (Holton et al., 2013). Individuals with the *CR1*-GG and *APOE-ε4* had decreased episodic memory, an endophenotype of LOAD, which provides further evidence for CR1's role in LOAD risk (Barral et al., 2012).

BRIDGING INTEGRATOR 1

The association of Bridging Integrator 1 (*BIN1*) gene variants with AD has also emerged from GWAS and has been replicated in multiple studies in different ethnic populations worldwide (Tan, Yu, & Tan, 2013). *BIN1* yields the strongest association signal in the meta-analyses results of large GWAS data sets after APOE-ε4 (www.alzgene.org). *BIN1* has been systematically studied for its role as a tumor suppressor gene (Prokic, Cowling, & Laporte, 2014) and has been shown to be involved in a number of cancer phenotypes (Ge et al., 1999; Ghaneie et al., 2007; Pan et al., 2012). BIN1 is important in endocytosis (Pant et al., 2009; Treusch et al., 2011), and BIN1 protein levels were significantly lower in tissues from patients with LOAD than in age-matched control subjects, which suggests its role in multiple pathways involved in AD pathogenesis (Camargo et al., 2015; De Jager et al., 2014). As with the other GWAS genes, the functional gene variants that confer risk for AD and the pathogenic mechanisms leading to the disease have not yet been established.

CLUSTERIN

Clusterin (*CLU*), also referred to as apolipoprotein J, is another gene carrying variants that have a significant association with AD in most of the large GWAS in AD to date (Shuai et al., 2015). CLU has been implicated in the formation of complexes that can cross the blood—brain barrier (Calero et al., 2000) and has been proposed as one of the primary chaperones for reentry of Aβ back into the brain from the plasma (Calero, Rostagno, Frangione, & Ghiso, 2005; Rohne, Prochnow, Wolf, Renner, & Koch-Brandt, 2014).

Patients with AD have increased levels of CLU in the cortex and hippocampus (May et al., 1990; Oda et al., 1994). Thus, a link between increased levels of CLU and AD risk is both expected and observed (Nuutinen, Suuronen, Kauppinen, & Salminen, 2009; Schrijvers, Koudstaal, Hofman, & Breteler, 2011). In the same line, analysis of blood plasma revealed increased plasma CLU, which correlated with hippocampal atrophy, disease severity, and progression (Thambisetty et al., 2012). In contrast, other studies show that CLU variants are associated with increased CLU expression but decreased AD risk (Allen et al., 2012; Ling, Bhongsatiern, Simpson, Fardo, & Estus, 2012). CLU affects inflammation, immune responses, and amyloid clearance, alterations which can result in neurodegeneration (Allen et al., 2012; Li et al., 2014; Park, Mathis, & Lee, 2014; Russell, Koncarevic, & Ward, 2014).

ADENOSINE TRIPHOSPHATE-BINDING CASSETTE, SUBFAMILY A (ABC1), MEMBER 7

Adenosine triphosphate (ATP)-Binding Cassette, Subfamily A (ABC1), Member 7 (*ABCA7*) is a member of the ATP-binding cassette genes family that is responsible for lipid transport, a particularly important function in the central nervous system (Mack, Townsend, Beljanski, & Tew, 2007; Reitz, 2013). Variants in *ABCA7* have been associated with AD in numerous GWAS with significant risk for the disease and age at onset. Functional studies in mice have shown that loss of *ABCA7* is not embryonic lethal and does not produce clear irregularities in young mice, which is consistent with the late age at onset of AD (Kim et al., 2005; Kim, Guillemin, Glaros, Lim, & Garner, 2006; Kim, Weickert, & Garner, 2008). Those studies have demonstrated that knockout of ABCA7 does not affect cholesterol efflux by macrophages, nor is it sufficient to compensate when function of the homologous lipid transporter ABCA1 is lost. ABCA7 expression is highest in the hippocampus, one of the earliest affected regions in the brains of patients with AD, and microglia, cells responsible for cerebral inflammatory response (Kim et al., 2008). ABCA7 also participates in macrophage uptake of Aβ, and lack of ABCA7 results in increased levels of insoluble Aβ (Kim et al., 2013). Although ABCA7 has been shown to mediate APP processing, it remains to be established whether ABCA7 influences AD onset through Aβ-related pathways, its interaction with APOE and lipid metabolism, or its role in the immune system response (Bohm et al., 2015; Pahnke, Frohlich, Krohn, Schumacher, & Paarmann, 2013; Tanaka, Abe-Dohmae, Iwamoto, & Yokoyama, 2011). Interestingly, some of the lipids for which transport is mediated by ABCA7, eg, anionic and zwitterionic lipids, may bind and activate TREM2 on microglia to prevent neuroinflammation (Wang, Cella, et al., 2015).

PHOSPHATIDYLINOSITOL-BINDING CLATHRIN ASSEMBLY PROTEIN

Phosphatidylinositol-Binding Clathrin Assembly Protein (*PICALM*) is located at chromosome 11q14.2 and carries an SNP showing the strongest association with AD upstream of the gene. The minor allele of the SNP rs3851179 in *PICALM* shows a protective effect in AD; ie, the minor allele is undertransmitted in patients with AD patients (95% confidence interval (CI) in OR, about 0.86−0.91) (Kamboh, Minster, et al., 2012; Lambert et al., 2013). APP trafficking as well as Aβ clearance, specifically via clathrin-mediated endocytosis, is proposed to be the molecular pathway affecting risk for LOAD (Harel, Wu, Mattson, Morris, & Yao, 2008; Tebar, Bohlander, & Sorkin, 1999). In the same line, studies conducted in yeast and *Caenorhabditis elegans* also show homologs of PICALM to curb the toxic effects of Aβ (Treusch et al., 2011). Alternative splicing of PICALM yields 23 isoforms that are predominantly expressed in the brain (Xiao et al., 2012). Postmortem studies of brain samples revealed a decrease in the levels of full-length PICALM and an increase in the shorter species in cases, which indicates that abnormal proteolysis of PICALM may affect Aβ clearance (Baig et al., 2010). More recently, a link between PICALM and tau pathology in the AD brains was reported, in which PICALM was found to colocalize with tau in NFTs but not with pretangles or extracellular ghost tangles (Tebar et al., 1999); further studies are needed to confirm the role of PICALM in tau pathology (Kok et al., 2011; Schjeide et al., 2011). PICALM also appears to have a key role in export of Aβ out of the brain via low-density lipoprotein receptor-related protein (Zlokovic, Personal Communication).

MEMBRANE-SPANNING 4 DOMAIN SUBFAMILY A MEMBERS 4A AND 6A

Located at chromosomal region 11q12.2, SNPs in the Membrane-Spanning 4A (*MS4A*) gene cluster have consistently shown association with AD across multiple studies (Antunez et al., 2011; Harold et al., 2009; Hollingworth et al., 2011; Lambert et al., 2013; Naj et al., 2011; Seshadri et al., 2010). The two genes, *MS4A4A* and *MS4A6A*, are implicated in AD; however, because of the complex genomic arrangement in the region, other genes in the GWAS locus (*MS4A3*, *MS4A2*, and *MS4A6E*) may also harbor variants that influence risk for AD (Liang, Buckley, Tu, Langdon, & Tedder, 2001). Despite continued replication in multiple GWAS (OR, 95% CI, 0.88−0.93) of the intergenic markers, the exact functional effect of the gene family in AD remains to be established (Ishibashi, Suzuki, Sasaki, & Imai, 2001; Kutok et al., 2005; Liang & Tedder, 2001).

The two most implicated genes, *MS4A4A* and *MS4A6A*, but not *MS4A6E*, are expressed on primary adult microglial cells and were downregulated when microglial cells were activated to induce neuroinflammation (Kofler, Bissel, Wiley, Stauffer, & Murdoch, 2012). Thus, *MS4A4A* and *MS4A6A* variants may affect AD pathology by modulating microglial function. Furthermore, gene expression analysis of markers in these genes has been shown to correlate with tangle and plaque scores in patients with AD and with disease risk (Allen et al., 2012; Karch et al., 2012). As evident from consistent replication of the MS4A genes with AD risk, further studies are warranted to assess the role of these genes leading to AD.

CUGBP, ELAV-LIKE FAMILY MEMBER 1

CUGBP, Elav-like Family Member 1 (*CELF1*) is located at 11p11.2 in a gene-dense region with the genes *SLC39A13*, *PSMC3*, *NDUFS3*, *KBTBD4*, *PTPMT1*, *MTCH2*, *AGBL2*, *FNBP4*, and *NUP160* also located within the GWAS locus associated with AD. The original study (Lambert et al., 2013) revealed that common CELF1 variants have small effects for risk of AD (about 10%); this has been replicated by multiple studies (Hinney et al., 2014). As with the other new AD genes emerging from GWAS, the role of functional variants in this GWAS locus has not been fully elucidated. Preliminary studies indicate that the disruption of cytoskeletal transport and tau-mediated pathway owing to *CELF1* variants could lead to increased risk for AD (Shulman et al., 2014).

CELF1 has been largely studied with regard to the regulation of gene expression. CELF1 has been associated with modulating alternative polyadenylation, an evolutionarily conserved mechanism for regulating gene expression, which in turn controls cellular proliferation during biological processes such as development, oncogenesis, and T-cell activation (Beisang, Reilly, & Bohjanen, 2014). In the same line, studies have implicated CELF1 mainly in cancer and myotonic dystrophy type I (Klein, Gasnier, & Furling, 2011; Meola, Jones, Wei, & Timchenko, 2013; Talwar et al., 2013). Expression quantitative trait loci studies suggest that CELF1 variants may be acting in conjunction with or serving as a proxy for other AD-associated genes in this GWAS locus (Rosenthal & Kamboh, 2014).

SORTILIN-RELATED RECEPTOR, LDLR CLASS A REPEATS CONTAINING

Sortilin-Related Receptor, LDLR Class A Repeats Containing (*SORL1*) is located on chromosome 11q24.1 and has been studied as a candidate gene in AD since the first report of decreased expression of SORL1 in a small set of patients with AD (Scherzer et al., 2004). *SORL1* is unique in the way the gene variants show the most compelling association with AD in studies using the candidate gene approach and functional experiments, although the association signal is just below genome-wide significance in many GWAS. SORL1 belongs to a family of sorting receptors that contain a vacuolar protein sorting protein (VPS) 10 domain, and is widely expressed in the brain (Taira et al., 2001). SORL1 mediates various intracellular sorting and trafficking functions (Yamazaki et al., 1996), including APP trafficking (Andersen et al., 2006) that consequently affect APP processing and Aβ generation leading to AD. However, whether *SORL1* gene activity in the brains of patients with AD is altered is not fully clear (Sager et al., 2012).

Several *SORL1* variants were initially identified using a candidate gene approach (Rogaeva et al., 2007), and both common and rare *SORL1* variants have since been implicated for their potential role in AD risk in numerous GWAS (Bettens et al., 2008; Lambert et al., 2013; Lee et al., 2008; Miyashita et al., 2013; Reitz et al., 2011; Vardarajan et al., 2015). Although multiple different association signals throughout SORL1 are reported in these studies, many of the associated SNPs show no clear and consistent evidence of association across all data sets, possibly owing to allelic heterogeneity in the study population. In addition, rare variants in *SORL1* have been found (Pottier et al., 2012) which if replicated in other study cohorts would result in *SORL1* being the first gene in AD to carry both risk-conferring polymorphisms as well as rare disease causing functional variants. AlzGene meta-analyses demonstrate multiple SNPs showing an association with AD that confers small to modest AD risk with summary ORs ranging from 1.09 to 1.21. The SNPs that result in synonymous coding changes are in strong LD with the 3′ half of the gene, which indicates that the postulated functional effects on AD pathogenesis alter gene regulation in the 3′ region (Willnow & Andersen, 2013). This observation dovetails with functional experiments that suggest SORL1 directly influences the production of Aβ by affecting the processing or trafficking of APP by binding to a complement-type repeat domain in the SORL1 protein (Offe et al., 2006).

INOSITOL POLYPHOSPHATE-5-PHOSPHATASE

Inositol polyphosphate-5-phosphatase (*INPP5DP*) is located in the 2q37.1 region and has been extensively studied for a possible role in inflammatory response, regulation of cytokine signaling, and signal transduction in oncogenic pathways cancer (Metzner et al., 2009; Srivastava, Sudan, & Kerr, 2013). The highly frequent minor allele (>0.4 minor allele frequency (MAF)) in the marker rs35349669 located in the intronic region shows a modest association with LOAD (OR, 95% CI, 1.05−1.11) in a large GWAS meta-analyses (Lambert et al., 2013) and has since been replicated in other cohorts (Ruiz et al., 2014). *INPP5D* is yet another novel GWAS gene that highlights the role of immune-related pathways in AD. INPP5D is well-characterized in negative regulation of immune cell activation. Besides catalytic biological mechanisms,

INPP5D is associated with nonenzymatic activity in several immune pathways and in mediating protein–protein interactions (Conde, Gloire, & Piette, 2011). INPP5D interacts with CD2AP (also implicated in AD; see subsequent discussion) and in plasmacytoid dendritic cells (pDCs) and regulates degradation of the IgE receptor, FceRIc (Bao et al., 2012). In the same line, INPP5D alters B-cell longevity, especially molecules that are involved in apoptosis (the Fas/Fas-L pathway), as well as molecules that alter activation thresholds of B cells (CD19, CD21, CD22, etc.) in the development of autoimmunity (Dorner & Lipsky, 2006). Further genetic and functional studies in INPP5D might hold clues unraveling the role of these complex immune-related pathways in AD pathogenesis.

CD2-ASSOCIATED PROTEIN

CD2-Associated Protein (*CD2AP*) is located at chromosome 6p12.3 region; the original markers associated with AD were reported in a study involving collaborative efforts from several large consortia in 2011 (Hollingworth et al., 2011; Lambert et al., 2013; Naj et al., 2011; Shulman et al., 2013). Several variants in *CD2AP* show a modest 10% increased risk for AD in these studies and neuritic plaque burden in brains with AD. Although the exact pathological AD variant in *CD2AP* needs to be established, there are a number of speculations about its role in AD in published reports.

CD2AP is involved in receptor-mediated endocytosis (Kobayashi, Sawano, Nojima, Shibuya, & Maru, 2004) as well as cytoskeleton and vesicle movement (Lynch et al., 2003); hence it is associated with modulating Aβ clearance (Treusch et al., 2011). CD2AP is required for synapse formation via its interaction with Cbl, endophilin, and synaptojanin. Lysosomal function is impaired in cells from CD2AP-deficient mice, in which CD2AP has been shown to be an essential regulator of vesicular trafficking (Cormont et al., 2003). CD2AP is also involved with immune-related pathways that can have a role in AD pathogenesis owing to its interaction with two other LOAD genes, INPP5D and CASS4. CD2AP binds and clusters CD2 to facilitate junction formation between T cells and antigen-presenting cells (Hutchings, Clarkson, Chalkley, Barclay, & Brown, 2003). Although CD2AP expression levels were not altered in a *Drosophila* model (Shulman et al., 2011), loss of the fly ortholog of CD2AP and CIN85, cindr, led to increased tau neurotoxicity in transgenic flies, which suggests another mechanism by which CD2AP loss of function could influence risk for AD (Shulman et al., 2014).

MAJOR HISTOCOMPATIBILITY COMPLEX CLASS II, DRβ1 AND 5

The Major Histocompatibility Complex Class II, DRβ1 and 5 (*HLA-DRB1/HLA-DRB5*) gene locus is a member of the major histocompatibility complex, a highly polymorphic region located on chromosome 6p21.32, and has been a focal point in the studies concerning neurodegenerative diseases for decades. In fact, variants in the HLA-gene locus show the single largest genetic effect in risk for multiple sclerosis (Cree, 2014; Hauser, Chan, & Oksenberg, 2013). Since the early 1970s, the HLA locus has been investigated in AD (Henschke, Bell, & Cape, 1978), Parkinsons (Nalls et al., 2014) and FTLD (Nalls et al., 2014). (Please refer to the meta-analysis results on the disease-specific portals: MSGene Database (http://www.msgene.org/), PDGene Database (http://www.pdgene.org), and AlzGene Database (http://www.alzgene.org) for more details) (Bertram, McQueen, Mullin, Blacker, & Tanzi, 2007; Lill, Abel, Bertram, & Al-Chalabi, 2011; Lill et al., 2012).

Although the first report on HLA-locus variants showing an association with AD emerged more than 35 years ago using protein polymorphisms (Henschke et al., 1978), the GWAS from the analysis of the mega-consortium in 2013 (Lambert et al., 2013) provided a more comprehensive study of these genes. The region associated with AD includes other human leukocyte antigen genes, including *HLA-DRB6*, *HLA-DQA1*, and *HLA-DQB1* within the GWAS locus. The strongest AD-associated marker in this meta-analysis report (Lambert et al., 2013), rs9271192 (MAF ~0.3), marginally increased risk for AD by just over 10%. These results were subsequently replicated (Mansouri et al., 2015) and expanded further, leading to the observation of association of HLA-DRB1 with total brain volume ($P = 0.0006$) (Chauhan et al., 2015) and brain DNA methylation (Yu et al., 2015). The HLA genes add to the growing number of immune-related, AD-associated genes (Trowsdale & Knight, 2013).

EPH RECEPTOR A1

The EPH Receptor A1 *EPHA1* gene is located on chromosome 7q34 and contains 18 exons that span a little over 18 kb (Coulthard et al., 2001). The SNP rs11767557, located upstream of *EPHA1*, was the first marker in this gene to be associated with decreased LOAD risk in 2010 (OR, 95% CI, 0.83–0.96) (Seshadri et al., 2010) and was later confirmed by multiple reports from larger studies (Carrasquillo et al., 2011; Hollingworth et al., 2011; Lambert et al., 2013; Naj et al., 2011; Wang, Lopez, et al., 2015). EPH and EPH-related receptors comprise the largest of the receptor tyrosine kinases, which function in binding membrane-bound ephrin-A ligands on adjacent cells via bidirectional signaling (Yamazaki et al., 2009). The ephrin genes are crucial during development of the nervous system (Wilkinson, 2000; Zhou, 1998) in axonal guidance and synaptic plasticity (Lai & Ip, 2009). EPHA1 is also associated with regulation of T-cell interactions through the integrin pathway (Sharfe et al., 2008). EPHA1 may be associated with AD via the α-secretase pathway owing to its role as a substrate of ADAM10 (Deuss, Reiss, & Hartmann, 2008; Janes et al., 2005).

MYOCYTE ENHANCER FACTOR 2C

Myocyte Enhancer Factor 2C (*MEF2C*) is located on the chromosome 5q14.3 region and carries a minor allele in the SNP rs190982 (MAF about 0.4, OR, 95% CI, 0.9–0.95) that confers modest protection against the onset of AD in the mega-meta-analysis report (Lambert et al., 2013; Ruiz et al., 2014). The role of MEF2C in AD is not currently known. MEF2C is crucial during neuronal development because of its role in synapse formation during activity-dependent refinement of synapses (Janson, Chen, Li, & Leifer, 2001) and hippocampal-dependent learning and memory (Potthoff & Olson, 2007). *MEF2C* mutations are implicated in causing distinct phenotypes in del5q14 developmental syndrome, which is characterized by phenotypes similar to Rett syndrome, including seizures, severe mental retardation, and stereotypical movement (Zweier & Rauch, 2012). *MEF2C* is also associated with another brain disorder, Angelman-like syndrome (Tan, Bird, Thibert, & Williams, 2014), a clinical and behavioral aspect attributed to localized central nervous system dysfunction. A large GWAS analysis in 31 study cohorts ($N = 53{,}949$) found evidence for association of MEF2C variants previously associated with AD to influence cognitive ability (Davies et al., 2015). MEF2C also modulates transcription factors that have an antiinflammatory role in endothelial cells (Xu et al., 2015).

NME/NM23 FAMILY MEMBER 8

NME/NM23 Family Member 8 (*NME8*) is located in the 7p14.1 region and belongs to the NME gene family, a evolutionarily highly conserved gene family linked to a wide variety of key developmental and cellular processes (Desvignes, Pontarotti, & Bobe, 2010). The marker rs2718058, with a frequent minor allele (MAF about 0.37), was shown to confer modest protection against AD (OR, 95% CI, 0.90–0.95) in the large GWAS meta-analysis (Lambert et al., 2013). The role of NME8 in AD is still unknown. NME8 mutations cause primary ciliary dyskinesia type 6, a disorder characterized by abnormalities of motile cilia leading to respiratory infections, chronic inflammation, and bronchiectasis (Leigh et al., 2009). NME8 variants have also been linked to osteoarthritis (OA). Studies associate the regulatory 5′ SNPs in NME8 with increased bone mineral density and OA risk (Mahr et al., 2006; Yerges-Armstrong et al., 2014). NME8 may also be considered an AD-associated gene with a role in innate immunity and inflammation.

ZINC FINGER, CW TYPE WITH PWWP DOMAIN 1

Zinc Finger, CW Type With PWWP Domain 1 (*ZCWPW1*), located on chromosome 7q22.1, carries a variant in the intron close to the exon 12 splice site that shows a modest protective effect in AD (rs1476679; MAF about 0.287; OR, 95% CI, 0.89–0.94). Originally reported in the large GWAS meta-analysis study (Lambert et al., 2013), the same SNP was later confirmed by another group with similar effect size and direction (Ruiz et al., 2014). The AD-associated marker was also observed to show genome-wide significant association for potential regulatory functions using RegulomeDB (Rosenthal, Barmada, Wang, Demirci, & Kamboh, 2014). A study also reported a significant association of the same marker, rs1476679, in close to 5000 brain samples with postmortem neuropathologic confirmation (Beecham et al., 2014). This large brain autopsy study further confirmed the association of *ZCWPW1* variants with specific neuropathologic features in AD. The only functional clues regarding this relate to chromatin remodeling and methylation states, owing to the implication of ZCWPW1 as a histone modification reader (He et al., 2010). Thus, studies aimed at investigating a role for this gene in epigenetic-mediated risk for AD may be warranted.

FERMITIN FAMILY MEMBER 2

Fermitin Family Member 2 (*FERMT2*) is located on chromosomal region 14q22.1 and is a member of the Fermitin (also referred to as Kindlins) family of proteins, which represent a class of focal adhesion proteins implicated in integrin activation. The variant rs17125944 also emerged from the large GWAS meta-analyses study (Lambert et al., 2013) reported to increase about 15% risk for AD (OR, 95% CI, 1.09–1.19). Integrin activation mediates cell–cell and cell-matrix contact and is important in a wide range of pathways, including development, immune response, hemostasis, and wound healing (Lai-Cheong, Parsons, & McGrath, 2010). Known mutations in integrins or the major effectors of integrin signaling pathways lead to defective organ development, immunodeficiency, cancer, or autoimmune disease (Ye, Snider, & Ginsberg, 2014). Loss-of-function mutations in *FERMT1* and *FERMT3* cause Kindler syndrome and leukocyte adhesion deficiency-III syndrome, respectively (Fagerholm, Lek, & Morrison, 2014), whereas *FERMT2* association with AD remains the only disease attributed to variants in this gene (Lai-Cheong et al., 2010). Studies performed in *Drosophila melanogaster* (Shulman et al., 2011, 2014) suggest that FERMT2 influences LOAD risk by mediating Tau neurotoxicity. Nonetheless, additional genetic and functional studies are needed to elucidate the role of FERMT2 in AD.

PROTEIN TYROSINE KINASE 2β

The Protein Tyrosine Kinase 2β (*PTK2B*) gene, located at chromosome 8p21.2, encodes a cytoplasmic protein tyrosine kinase involved in calcium-induced regulation of ion channels and activation of the map kinase signaling pathway (Lev et al., 1995). *PTK2B* is located close to *CLU* and interacts with *CASS4* and *NEDD9*, two other LOAD-associated genes. *PTK2B* variants were originally observed show an association with AD (Kamboh, Demirci, et al., 2012), and eventually were confirmed in the large meta-analyses (Lambert et al., 2013). The most strongly associated SNP with AD, rs28834970 (MAF about 0.37), shows a modest (about 10%) increase in risk for AD (OR, 95% CI, 1.08−1.13). Together with the highly conserved focal adhesion kinase (about 90% sequence identity), PTK2B comprises the complete focal adhesion protein tyrosine kinase family, which is preferentially expressed in the central nervous system (Xiong, Macklem, & Parsons, 1998). Studies suggest that balanced function resulting in reorganization of actin association with focal adhesions and cell rounding of these proteins may be important for tissue organization (Mitra, Hanson, & Schlaepfer, 2005). Studies conducted in mice show that loss of protein tyrosine phosphatase-α, a regulator of PTK2B, can cause defects in *N*-methyl-D-aspartate receptor processes involving memory formation (Le, Maksumova, Wang, & Pallen, 2006). PTK2B is an important signaling intermediate between neuropeptide-activated receptors or neurotransmitters that increase calcium flux and the downstream signals that regulate neuronal activity, and associated is with depressive disorders (Beck et al., 2014). The same intracellular calcium levels stimulus for PTK2B activation is also disrupted in AD brains (Avraham, Park, Schinkmann, & Avraham, 2000; Bojarski, Herms, & Kuznicki, 2008), which suggests a possible role for PTK2B leading to risk for AD. Interestingly, PTK2B also activates GSK3 (Sayas, Ariaens, Ponsioen, & Moolenaar, 2006), which has been shown to be required for Aβ-induced neurofibrillary tangle formation (Choi et al., 2014). Thus the gene may serve to link plaque and tangle pathologies in AD.

RAS AND RAB INTERACTOR 3 (RIN3) AND SOLUTE CARRIER FAMILY 24 SODIUM/POTASSIUM/CALCIUM EXCHANGER, MEMBER 4 (SLC24A4)

Rab Interactor 3 (*RIN3*) is located in the chromosome14q32.12 region and is in the vicinity of SNP rs10498633 (MAF about 0.22; OR, 95% CI, 0.88−0.94), which exhibited a protective effect against the onset of AD in a GWAS meta-analysis report (Lambert et al., 2013). The role of RIN3 in AD is not fully elucidated, but it is widely speculated that RIN3 may influence the onset of AD via the endocytosis pathway, by altering APP trafficking and, as a result, Aβ generation. RIN3 represents a family of multifunctional proteins that have a VPS9 domain, Src homology, and Ras association domains. By acting as a guanine nucleotide exchange factor, RIN3 activates GTPase Rab5 and controls membrane budding and trafficking in the early endocytic pathway (Kajiho et al., 2011).

RIN3 is mainly found throughout the cytoplasm, and T-cell activation-dependent tyrosine phosphorylation initiates RIN3 translocation into Rab5-positive early endocytic vesicles (Kaneko, Li, & Li, 2008; Yoshikawa et al., 2008). GTP-Rab5 has an important role in transporting proteins from the plasma membrane to early endosomes, and hence could affect APP trafficking via early endosomes and consequently Aβ generation. RIN3 was shown to interact through its SH3 domain with two other LOAD genes, *CD2AP* and *BIN1*, in RIN3-expressing HeLa cells (Kajiho et al., 2003), which provides further evidence for its role in AD. In addition to AD, *RIN3* variants are associated with two other human diseases. Paget disease of bone is a common late-onset skeletal disorder which, interestingly is caused by pathogenic variants in Sequestosome-1 (another LOAD gene); additional research could provide clues to the shared disease pathway (Albagha et al., 2011). *RIN3* is also reported to carry genetic risk factors for susceptibility to chronic obstructive pulmonary disease (Cho et al., 2014), which is attributed to a combination of risk-involving genetics and long-term exposure to irritants causing an inflammatory response.

Next to RIN3 is the gene, *SLC24A4*, a solute carrier that belongs to the potassium-dependent sodium/calcium exchanging protein family and is associated with a number of traits in humans. Variants in the gene *SLC24A4* are associated with pigmentation traits in European populations (Sturm, 2009; Sulem et al., 2007), regulation of adaptation in olfactory sensory neurons (Stephan et al., 2012) and amelogenesis (Parry et al., 2013; Urzua, Ortega-Pinto, Morales-Bozo, Rojas-Alcayaga, & Cifuentes, 2011), to name a few. Additional research is necessary to unravel the role of the two genes in this locus in AD.

CONCLUSION

The genetic factors leading to AD are complex and heterogeneous. EOFAD genes, *APP*, *PSEN1*, and *PSEN2*, carry over 200 rare mutations that virtually guarantee onset of disease. In contrast, GWAS conducted in the past 8 years have revealed common genetic variants spread across multiple genomic loci showing an association with AD with modest effects on risk. In most cases, these common variants likely do not cause the disease biologically but are LD with true functional variants, common or rare.

Genome-wide agnostic studies using NGS technology will be necessary to identify the actual functional variants in most GWAS genes. Until that time, we can at least begin to see three broad categories of AD pathology into which the known and GWAS-confirmed AD genes fall: β-amyloid deposition, tangle formation, and neuroinflammation. Ultimately, to treat and prevent AD, we will need a cocktail that hits all three pathologies. Although all three pathologies can be targeted for prevention, in middle- to late-stage patients, it may be too late to treat only β-amyloid deposition, which occurs 2 decades before symptoms. Nonetheless, no matter what pathology is being therapeutically addressed, history has shown that genetics provides the best pipeline of targets. As illustrated in this chapter, we now know of many of these genetic-derived targets for addressing each of the three pillars of AD pathology.

Genetic studies of AD continue to hold enormous potential for affording a more comprehensive understanding of the genetic basis of AD, and eventually to devising treatment to this truly devastating disease in the aging population. It will be important to understand the role of any environmental or lifestyle factors that influence susceptibility to AD in tandem with the genetic risk factors. Ultimately, we will end AD by a strategy of early prediction, early detection, and early prevention. Whereas preventative strategies will almost certainly entail lifestyle changes, eg, regarding diet, exercise, sleep habits, etc., they will also require administering the right therapies aimed at the right patient at the right stage of the disease, from prodromal asymptomatic accrual of β-amyloid to late-stage disease, in which neuroinflammation is likely doing most of the damage. In each case, genetic targets such as those described here will be invaluable for devising future therapeutic strategies aimed at effective treatment and prevention of this devastating disease.

ACKNOWLEDGMENTS

We thank all patients and their families whose trust, help, and participation were crucial to making possible all of the research summarized in this review. We thank Dr L.B. Ufer for helpful discussions. We are grateful for funding from the NIMH, National Institute on Aging, Cure Alzheimer's Fund, and the JPB Foundation.

REFERENCES

Abraham, R., Moskvina, V., Sims, R., Hollingworth, P., Morgan, A., Georgieva, L., ... Kirov, G. (2008). A genome-wide association study for late-onset Alzheimer's disease using DNA pooling. *BMC Medical Genomics, 1*, 44. http://dx.doi.org/10.1186/1755-8794-1-44.

Albagha, O. M., Wani, S. E., Visconti, M. R., Alonso, N., Goodman, K., Brandi, M. L., ... Genetic Determinants of Paget's Disease (GCPD) Consortium. (2011). Genome-wide association identifies three new susceptibility loci for Paget's disease of bone. *Nature Genetics, 43*(7), 685–689. http://dx.doi.org/10.1038/ng.845.

Allen, M., Zou, F., Chai, H. S., Younkin, C. S., Crook, J., Pankratz, V. S., ... Woltjer, R. L. (2012). Novel late-onset Alzheimer disease loci variants associate with brain gene expression. *Neurology, 79*(3), 221–228. http://dx.doi.org/10.1212/WNL.0b013e3182605801.

Andersen, O. M., Schmidt, V., Spoelgen, R., Gliemann, J., Behlke, J., Galatis, D., ... Willnow, T. E. (2006). Molecular dissection of the interaction between amyloid precursor protein and its neuronal trafficking receptor SorLA/LR11. *Biochemistry, 45*(8), 2618–2628. http://dx.doi.org/10.1021/bi052120v.

Antunez, C., Boada, M., Gonzalez-Perez, A., Gayan, J., Ramirez-Lorca, R., Marin, J., ... Ruiz, A. (2011). The membrane-spanning 4-domains, subfamily A (MS4A) gene cluster contains a common variant associated with Alzheimer's disease. *Genome Medicine, 3*(5), 33. http://dx.doi.org/10.1186/gm249.

Avraham, H., Park, S. Y., Schinkmann, K., & Avraham, S. (2000). RAFTK/Pyk2-mediated cellular signalling. *Cell Signal, 12*(3), 123–133.

Bagyinszky, E., Youn, Y. C., An, S. S., & Kim, S. (2014). The genetics of Alzheimer's disease. *Clinical Interventions in Aging, 9*, 535–551. http://dx.doi.org/10.2147/CIA.S51571.

Baig, S., Joseph, S. A., Tayler, H., Abraham, R., Owen, M. J., Williams, J., ... Love, S. (2010). Distribution and expression of picalm in Alzheimer disease. *Journal of Neuropathology and Experimental Neurology, 69*(10), 1071–1077. http://dx.doi.org/10.1097/NEN.0b013e3181f52e01.

Bamne, M. N., Demirci, F. Y., Berman, S., Snitz, B. E., Rosenthal, S. L., Wang, X., ... Kamboh, M. I. (2014). Investigation of an amyloid precursor protein protective mutation (A673T) in a North American case-control sample of late-onset Alzheimer's disease. *Neurobiology of Aging, 35*(7). http://dx.doi.org/10.1016/j.neurobiolaging.2014.01.020, 1779. e1715–1776.

Bao, M., Hanabuchi, S., Facchinetti, V., Du, Q., Bover, L., Plumas, J., ... Liu, Y. J. (2012). CD2AP/SHIP1 complex positively regulates plasmacytoid dendritic cell receptor signaling by inhibiting the E3 ubiquitin ligase Cbl. *Journal of Immunology, 189*(2), 786–792. http://dx.doi.org/10.4049/jimmunol.1200887.

Barral, S., Bird, T., Goate, A., Farlow, M. R., Diaz-Arrastia, R., Bennett, D. A., ... National Institute on Aging Late-Onset Alzheimer's Disease Genetics Study. (2012). Genotype patterns at PICALM, CR1, BIN1, CLU, and APOE genes are associated with episodic memory. *Neurology, 78*(19), 1464–1471. http://dx.doi.org/10.1212/WNL.0b013e3182553c48.

Becam, I. E., Tanentzapf, G., Lepesant, J. A., Brown, N. H., & Huynh, J. R. (2005). Integrin-independent repression of cadherin transcription by talin during axis formation in *Drosophila. Nature Cell Biology, 7*(5), 510–516. http://dx.doi.org/10.1038/ncb1253.

Beck, T. N., Nicolas, E., Kopp, M. C., & Golemis, E. A. (2014). Adaptors for disorders of the brain? The cancer signaling proteins NEDD9, CASS4, and PTK2B in Alzheimer's disease. *Oncoscience, 1*(7), 486–503.

Beecham, G. W., Hamilton, K., Naj, A. C., Martin, E. R., Huentelman, M., Myers, A. J., ... Montine, T. J. (2014). Genome-wide association meta-analysis of neuropathologic features of Alzheimer's disease and related dementias. *PLoS Genetics, 10*(9), e1004606. http://dx.doi.org/10.1371/journal.pgen.1004606.

Beecham, G. W., Martin, E. R., Li, Y. J., Slifer, M. A., Gilbert, J. R., Haines, J. L., & Pericak-Vance, M. A. (2009). Genome-wide association study implicates a chromosome 12 risk locus for late-onset Alzheimer disease. *American Journal of Human Genetics, 84*(1), 35–43. http://dx.doi.org/10.1016/j.ajhg.2008.12.008.

Beisang, D., Reilly, C., & Bohjanen, P. R. (2014). Alternative polyadenylation regulates CELF1/CUGBP1 target transcripts following T cell activation. *Gene, 550*(1), 93–100. http://dx.doi.org/10.1016/j.gene.2014.08.021.

Benitez, B. A., Cooper, B., Pastor, P., Jin, S. C., Lorenzo, E., Cervantes, S., & Cruchaga, C. (2013). TREM2 is associated with the risk of Alzheimer's disease in Spanish population. *Neurobiology of Aging, 34*(6), 1711.e1715–1717. http://dx.doi.org/10.1016/j.neurobiolaging.2012.12.018.

Bertram, L. (2011). Alzheimer's genetics in the GWAS era: a continuing Story of 'replications and refutations'. *Current Neurology and Neuroscience Reports, 11*(3), 246–253. http://dx.doi.org/10.1007/s11910-011-0193-z.

Bertram, L., Lange, C., Mullin, K., Parkinson, M., Hsiao, M., Hogan, M. F., ... Tanzi, R. E. (2008). Genome-wide association analysis reveals putative Alzheimer's disease susceptibility loci in addition to APOE. *American Journal of Human Genetics, 83*(5), 623–632. http://dx.doi.org/10.1016/j.ajhg.2008.10.008.

Bertram, L., McQueen, M. B., Mullin, K., Blacker, D., & Tanzi, R. E. (2007). Systematic meta-analyses of Alzheimer disease genetic association studies: the AlzGene database. *Nature Genetics, 39*(1), 17–23. http://dx.doi.org/10.1038/ng1934.

Bertram, L., & Tanzi, R. E. (2005). Twenty years of the Alzheimer's disease amyloid hypothesis: a genetic perspective. *Cell, 120*(4), 545–555. PMID:15734686.

Bertram, L., & Tanzi, R. E. (2009). Genome-wide association studies in Alzheimer's disease. *Human Molecular Genetics, 18*(R2), R137–R145. http://dx.doi.org/10.1093/hmg/ddp406.

Bertram, L., & Tanzi, R. E. (2012). The genetics of Alzheimer's disease. *Progress in Molecular Biology and Translational Science, 107*, 79–100. http://dx.doi.org/10.1016/B978-0-12-385883-2.00008-4.

Bettens, K., Brouwers, N., Engelborghs, S., De Deyn, P. P., Van Broeckhoven, C., & Sleegers, K. (2008). SORL1 is genetically associated with increased risk for late-onset Alzheimer disease in the Belgian population. *Human Mutation, 29*(5), 769–770. http://dx.doi.org/10.1002/humu.20725.

Bianchin, M. M., Capella, H. M., Chaves, D. L., Steindel, M., Grisard, E. C., Ganev, G. G., ... Sakamoto, A. C. (2004). Nasu-Hakola disease (polycystic lipomembranous osteodysplasia with sclerosing leukoencephalopathy—PLOSL): a dementia associated with bone cystic lesions. From clinical to genetic and molecular aspects. *Cellular and Molecular Neurobiology, 24*(1), 1–24.

Bien-Ly, N., Gillespie, A. K., Walker, D., Yoon, S. Y., & Huang, Y. (2012). Reducing human apolipoprotein E levels attenuates age-dependent Aβ accumulation in mutant human amyloid precursor protein transgenic mice. *Journal of Neuroscience, 32*(14), 4803–4811. http://dx.doi.org/10.1523/JNEUROSCI.0033-12.2012.

Bird, T. D. (2008). Genetic aspects of Alzheimer disease. *Genetics in Medicine, 10*(4), 231–239. http://dx.doi.org/10.1097/GIM.0b013e31816b64dc.

Bock, V., Botturi, A., Gaviani, P., Lamperti, E., Maccagnano, C., Piccio, L., ... Salmaggi, A. (2013). Polycystic lipomembranous osteodysplasia with sclerosing leukoencephalopathy (PLOSL): a new report of an Italian woman and review of the literature. *Journal of the Neurological Sciences, 326*(1–2), 115–119. http://dx.doi.org/10.1016/j.jns.2013.01.021.

Bohm, C., Chen, F., Sevalle, J., Qamar, S., Dodd, R., Li, Y., ... St George-Hyslop, P. H. (May 2015). Current and future implications of basic and translational research on amyloid-beta peptide production and removal pathways. *Molecular and Cellular Neuroscience, 66*(Part A), 3–11. http://dx.doi.org/10.1016/j.mcn.2015.02.016.

Bojarski, L., Herms, J., & Kuznicki, J. (2008). Calcium dysregulation in Alzheimer's disease. *Neurochemistry International, 52*(4–5), 621–633. http://dx.doi.org/10.1016/j.neuint.2007.10.002.

Bradshaw, E. M., Chibnik, L. B., Keenan, B. T., Ottoboni, L., Raj, T., Tang, A., ... De Jager, P. L. (2013). CD33 Alzheimer's disease locus: altered monocyte function and amyloid biology. *Nature Neuroscience, 16*(7), 848–850. http://dx.doi.org/10.1038/nn.3435.

Brouwers, N., Van Cauwenberghe, C., Engelborghs, S., Lambert, J. C., Bettens, K., Le Bastard, N., ... Van Broeckhoven, C. (2012). Alzheimer risk associated with a copy number variation in the complement receptor 1 increasing C3b/C4b binding sites. *Molecular Psychiatry, 17*(2), 223–233. http://dx.doi.org/10.1038/mp.2011.24.

Calero, M., Rostagno, A., Frangione, B., & Ghiso, J. (2005). Clusterin and Alzheimer's disease. *Subcellular Biochemistry, 38*, 273–298.

Calero, M., Rostagno, A., Matsubara, E., Zlokovic, B., Frangione, B., & Ghiso, J. (2000). Apolipoprotein J (clusterin) and Alzheimer's disease. *Microscopy Research and Technique, 50*(4), 305–315. http://dx.doi.org/10.1002/1097-0029(20000815)50:4<305::AID-JEMT10>3.0.CO;2-L.

Camargo, L. M., Zhang, X. D., Loerch, P., Caceres, R. M., Marine, S. D., Uva, P., ... Mizuguchi, K. (2015). Pathway-based analysis of genome-wide siRNA screens reveals the regulatory landscape of APP processing. *PLoS One, 10*(2), e0115369. http://dx.doi.org/10.1371/journal.pone.0115369.

Campion, D., Dumanchin, C., Hannequin, D., Dubois, B., Belliard, S., Puel, M., ... Frebourg, T. (1999). Early-onset autosomal dominant Alzheimer disease: prevalence, genetic heterogeneity, and mutation spectrum. *American Journal of Human Genetics, 65*(3), 664–670. http://dx.doi.org/10.1086/302553.

Cao, H., & Crocker, P. R. (2011). Evolution of CD33-related siglecs: regulating host immune functions and escaping pathogen exploitation? *Immunology, 132*(1), 18–26. http://dx.doi.org/10.1111/j.1365-2567.2010.03368.x.

Carrasquillo, M. M., Belbin, O., Hunter, T. A., Ma, L., Bisceglio, G. D., Zou, F., ... Younkin, S. G. (2011). Replication of EPHA1 and CD33 associations with late-onset Alzheimer's disease: a multi-centre case-control study. *Molecular Neurodegeneration, 6*(1), 54. http://dx.doi.org/10.1186/1750-1326-6-54.

Carrasquillo, M. M., Zou, F., Pankratz, V. S., Wilcox, S. L., Ma, L., Walker, L. P., ... Younkin, S. G. (2009). Genetic variation in PCDH11X is associated with susceptibility to late-onset Alzheimer's disease. *Nature Genetics, 41*(2), 192–198. http://dx.doi.org/10.1038/ng.305.

Castellano, J. M., Kim, J., Stewart, F. R., Jiang, H., DeMattos, R. B., Patterson, B. W., ... Holtzman, D. M. (2011). Human apoE isoforms differentially regulate brain amyloid-beta peptide clearance. *Science Translational Medicine, 3*(89), 89ra57. http://dx.doi.org/10.1126/scitranslmed.3002156.

Chauhan, G., Adams, H. H., Bis, J. C., Weinstein, G., Yu, L., Toglhofer, A. M., ... Debette, S. (2015). Association of Alzheimer's disease GWAS loci with MRI markers of brain aging. *Neurobiology of Aging, 36*(4). http://dx.doi.org/10.1016/j.neurobiolaging.2014.12.028, 1765.e1767–1716.

Chibnik, L. B., Shulman, J. M., Leurgans, S. E., Schneider, J. A., Wilson, R. S., Tran, D., ... De Jager, P. L. (2011). CR1 is associated with amyloid plaque burden and age-related cognitive decline. *Annals of Neurology, 69*(3), 560–569. http://dx.doi.org/10.1002/ana.22277.

Choi, S. H., Kim, Y. H., Hebisch, M., Sliwinski, C., Lee, S., D'Avanzo, C., ... Kim, D. Y. (2014). A three-dimensional human neural cell culture model of Alzheimer's disease. *Nature, 515*(7526), 274–278. http://dx.doi.org/10.1038/nature13800.

Cho, M. H., McDonald, M. L., Zhou, X., Mattheisen, M., Castaldi, P. J., Hersh, C. P., ... Investigators, C. O. (2014). Risk loci for chronic obstructive pulmonary disease: a genome-wide association study and meta-analysis. *Lancet Respiratory Medicine, 2*(3), 214–225. http://dx.doi.org/10.1016/S2213-2600(14)70002-5.

Colonna, M. (2003). TREMs in the immune system and beyond. *Nature Reviews Immunology, 3*(6), 445–453. http://dx.doi.org/10.1038/nri1106.

Conde, C., Gloire, G., & Piette, J. (2011). Enzymatic and non-enzymatic activities of SHIP-1 in signal transduction and cancer. *Biochemical Pharmacology, 82*(10), 1320–1334. http://dx.doi.org/10.1016/j.bcp.2011.05.031.

Coon, K. D., Myers, A. J., Craig, D. W., Webster, J. A., Pearson, J. V., Lince, D. H., ... Stephan, D. A. (2007). A high-density whole-genome association study reveals that APOE is the major susceptibility gene for sporadic late-onset Alzheimer's disease. *Journal of Clinical Psychiatry, 68*(4), 613–618.

Corder, E. H., Saunders, A. M., Risch, N. J., Strittmatter, W. J., Schmechel, D. E., Gaskell, P. C., Jr., et al. (1994). Protective effect of apolipoprotein E type 2 allele for late onset Alzheimer disease. *Nature Genetics, 7*(2), 180–184. http://dx.doi.org/10.1038/ng0694-180.

Cormont, M., Meton, I., Mari, M., Monzo, P., Keslair, F., Gaskin, C., ... Le Marchand-Brustel, Y. (2003). CD2AP/CMS regulates endosome morphology and traffic to the degradative pathway through its interaction with Rab4 and c-Cbl. *Traffic, 4*(2), 97–112.

Coulthard, M. G., Lickliter, J. D., Subanesan, N., Chen, K., Webb, G. C., Lowry, A. J., ... Boyd, A. W. (2001). Characterization of the Epha1 receptor tyrosine kinase: expression in epithelial tissues. *Growth Factors, 18*(4), 303–317.

Cree, B. A. (2014). Multiple sclerosis genetics. *Handbook of Clinical Neurology, 122*, 193–209. http://dx.doi.org/10.1016/B978-0-444-52001-2.00009-1.

Crehan, H., Holton, P., Wray, S., Pocock, J., Guerreiro, R., & Hardy, J. (2012). Complement receptor 1 (CR1) and Alzheimer's disease. *Immunobiology, 217*(2), 244–250. http://dx.doi.org/10.1016/j.imbio.2011.07.017.

Crocker, P. R. (2002). Siglecs: sialic-acid-binding immunoglobulin-like lectins in cell-cell interactions and signalling. *Current Opinion in Structural Biology, 12*(5), 609–615.

Crocker, P. R., Hartnell, A., Munday, J., & Nath, D. (1997). The potential role of sialoadhesin as a macrophage recognition molecule in health and disease. *Glycoconjugate Journal, 14*(5), 601–609.

Cruts, M., Theuns, J., & Van Broeckhoven, C. (2012a). *AD&FTD mutation database*. http://www.molgen.vib-ua.be/ADMutations.

Cruts, M., Theuns, J., & Van Broeckhoven, C. (2012b). Locus-specific mutation databases for neurodegenerative brain diseases. *Human Mutation, 33*(9), 1340–1344. http://dx.doi.org/10.1002/humu.22117.

Davies, G., Armstrong, N., Bis, J. C., Bressler, J., Chouraki, V., Giddaluru, S., ... Deary, I. J. (2015). Genetic contributions to variation in general cognitive function: a meta-analysis of genome-wide association studies in the CHARGE consortium (N = 53949). *Molecular Psychiatry, 20*(2), 183–192. http://dx.doi.org/10.1038/mp.2014.188.

De Jager, P. L., Srivastava, G., Lunnon, K., Burgess, J., Schalkwyk, L. C., Yu, L., ... Bennett, D. A. (2014). Alzheimer's disease: early alterations in brain DNA methylation at ANK1, BIN1, RHBDF2 and other loci. *Nature Neuroscience, 17*(9), 1156–1163. http://dx.doi.org/10.1038/nn.3786.

De Strooper, B. (2003). Aph-1, Pen-2, and nicastrin with presenilin generate an active gamma-secretase complex. *Neuron, 38*(1), 9–12.

Desvignes, T., Pontarotti, P., & Bobe, J. (2010). Nme gene family evolutionary history reveals pre-metazoan origins and high conservation between humans and the sea anemone, *Nematostella vectensis. PLoS One, 5*(11), e15506. http://dx.doi.org/10.1371/journal.pone.0015506.

Deuss, M., Reiss, K., & Hartmann, D. (2008). Part-time alpha-secretases: the functional biology of ADAM 9, 10 and 17. *Current Alzheimer Research, 5*(2), 187–201.

Donoviel, D. B., Hadjantonakis, A. K., Ikeda, M., Zheng, H., Hyslop, P. S., & Bernstein, A. (1999). Mice lacking both presenilin genes exhibit early embryonic patterning defects. *Genes & Development, 13*(21), 2801–2810.

Dorner, T., & Lipsky, P. E. (2006). Signalling pathways in B cells: implications for autoimmunity. *Current Topics in Microbiology and Immunology, 305*, 213–240.

Dubois, B., Feldman, H. H., Jacova, C., Hampel, H., Molinuevo, J. L., Blennow, K., … Cummings, J. L. (2014). Advancing research diagnostic criteria for Alzheimer's disease: the IWG-2 criteria. *Lancet Neurology, 13*(6), 614–629. http://dx.doi.org/10.1016/S1474-4422(14)70090-0.

Dunkelberger, J. R., & Song, W. C. (2010). Complement and its role in innate and adaptive immune responses. *Cell Research, 20*(1), 34–50. http://dx.doi.org/10.1038/cr.2009.139.

Endres, K., & Fahrenholz, F. (2012). Regulation of alpha-secretase ADAM10 expression and activity. *Experimental Brain Research, 217*(3–4), 343–352. http://dx.doi.org/10.1007/s00221-011-2885-7.

Fagan, A. M., Xiong, C., Jasielec, M. S., Bateman, R. J., Goate, A. M., Benzinger, T. L., … Dominantly Inherited Alzheimer Network. (2014). Longitudinal change in CSF biomarkers in autosomal-dominant Alzheimer's disease. *Science Translational Medicine, 6*(226), 226ra230. http://dx.doi.org/10.1126/scitranslmed.3007901.

Fagerholm, S. C., Lek, H. S., & Morrison, V. L. (2014). Kindlin-3 in the immune system. *American Journal of Clinical and Experimental Immunology, 3*(1), 37–42.

Farrer, L. A., Cupples, L. A., Haines, J. L., Hyman, B., Kukull, W. A., Mayeux, R., … van Duijn, C. M. (1997). Effects of age, sex, and ethnicity on the association between apolipoprotein E genotype and Alzheimer disease. A meta-analysis. APOE and Alzheimer Disease Meta Analysis Consortium. *JAMA, 278*(16), 1349–1356.

Ford, J. W., & McVicar, D. W. (2009). TREM and TREM-like receptors in inflammation and disease. *Current Opinion in Immunology, 21*(1), 38–46. http://dx.doi.org/10.1016/j.coi.2009.01.009.

Gatz, M., Reynolds, C. A., Fratiglioni, L., Johansson, B., Mortimer, J. A., Berg, S., … Pedersen, N. L. (2006). Role of genes and environments for explaining Alzheimer disease. *Archives of General Psychiatry, 63*(2), 168–174. http://dx.doi.org/10.1001/archpsyc.63.2.168.

Ge, K., DuHadaway, J., Du, W., Herlyn, M., Rodeck, U., & Prendergast, G. C. (1999). Mechanism for elimination of a tumor suppressor: aberrant splicing of a brain-specific exon causes loss of function of Bin1 in melanoma. *Proceedings of the National Academy of Sciences of the United States of America, 96*(17), 9689–9694.

Ghaneie, A., Zemba-Palko, V., Itoh, H., Itoh, K., Sakamuro, D., Nakamura, S., … Prendergast, G. C. (2007). Bin1 attenuation in breast cancer is correlated to nodal metastasis and reduced survival. *Cancer Biology and Therapy, 6*(2), 192–194.

Gibson, N. C., & Waxman, F. J. (1994). Relationship between immune complex binding and release and the quantitative expression of the complement receptor, type 1 (CR1, CD35) on human erythrocytes. *Clinical Immunology and Immunopathology, 70*(2), 104–113.

Glenner, G. G., & Wong, C. W. (1984). Alzheimer's disease and Down's syndrome: sharing of a unique cerebrovascular amyloid fibril protein. *Biochemical and Biophysical Research, 122*(3), 1131–1135.

Golde, T. E., Streit, W. J., & Chakrabarty, P. (2013). Alzheimer's disease risk alleles in TREM2 illuminate innate immunity in Alzheimer's disease. *Alzheimer's Research and Therapy, 5*(3), 24. http://dx.doi.org/10.1186/alzrt178.

Griciuc, A., Serrano-Pozo, A., Parrado, A. R., Lesinski, A. N., Asselin, C. N., Mullin, K., … Tanzi, R. E. (2013). Alzheimer's disease risk gene CD33 inhibits microglial uptake of amyloid beta. *Neuron, 78*(4), 631–643. http://dx.doi.org/10.1016/j.neuron.2013.04.014.

Grupe, A., Abraham, R., Li, Y., Rowland, C., Hollingworth, P., Morgan, A., … Williams, J. (2007). Evidence for novel susceptibility genes for late-onset Alzheimer's disease from a genome-wide association study of putative functional variants. *Human Molecular Genetics, 16*(8), 865–873. http://dx.doi.org/10.1093/hmg/ddm031.

Guerreiro, R., Wojtas, A., Bras, J., Carrasquillo, M., Rogaeva, E., Majounie, E., … Alzheimer Genetic Analysis Group. (2013). TREM2 variants in Alzheimer's disease. *New England Journal of Medicine, 368*(2), 117–127. http://dx.doi.org/10.1056/NEJMoa1211851.

Hansson, C. A., Frykman, S., Farmery, M. R., Tjernberg, L. O., Nilsberth, C., Pursglove, S. E., … Ankarcrona, M. (2004). Nicastrin, presenilin, APH-1, and PEN-2 form active gamma-secretase complexes in mitochondria. *Journal of Biological Chemistry, 279*(49), 51654–51660. http://dx.doi.org/10.1074/jbc.M404500200.

Hardy, J. A., & Higgins, G. A. (1992). Alzheimer's disease: the amyloid cascade hypothesis. *Science, 256*(5054), 184–185.

Hardy, J., & Selkoe, D. J. (2002). The amyloid hypothesis of Alzheimer's disease: progress and problems on the road to therapeutics. *Science, 297*(5580), 353–356. http://dx.doi.org/10.1126/science.1072994.

Harel, A., Wu, F., Mattson, M. P., Morris, C. M., & Yao, P. J. (2008). Evidence for CALM in directing VAMP2 trafficking. *Traffic, 9*(3), 417–429. http://dx.doi.org/10.1111/j.1600-0854.2007.00694.x.

Harold, D., Abraham, R., Hollingworth, P., Sims, R., Gerrish, A., Hamshere, M. L., … Williams, J. (2009). Genome-wide association study identifies variants at CLU and PICALM associated with Alzheimer's disease. *Nature Genetics, 41*(10), 1088–1093. http://dx.doi.org/10.1038/ng.440.

Hauser, S. L., Chan, J. R., & Oksenberg, J. R. (2013). Multiple sclerosis: prospects and promise. *Annals of Neurology, 74*(3), 317–327. http://dx.doi.org/10.1002/ana.24009.

Hauser, P. S., & Ryan, R. O. (2013). Impact of apolipoprotein E on Alzheimer's disease. *Current Alzheimer Research, 10*(8), 809–817.

Hazrati, L. N., Van Cauwenberghe, C., Brooks, P. L., Brouwers, N., Ghani, M., Sato, C., … Rogaeva, E. (2012). Genetic association of CR1 with Alzheimer's disease: a tentative disease mechanism. *Neurobiology of Aging, 33*(12), 2949.e5–2949.e12. http://dx.doi.org/10.1016/j.neurobiolaging.2012.07.001.

Heinzen, E. L., Need, A. C., Hayden, K. M., Chiba-Falek, O., Roses, A. D., Strittmatter, W. J., … Goldstein, D. B. (2010). Genome-wide scan of copy number variation in late-onset Alzheimer's disease. *Journal of Alzheimer's Disease, 19*(1), 69–77. http://dx.doi.org/10.3233/JAD-2010-1212.

Henschke, P. J., Bell, D. A., & Cape, R. D. (1978). Alzheimer's disease and HLA. *Tissue Antigens, 12*(2), 132–135.

He, F., Umehara, T., Saito, K., Harada, T., Watanabe, S., Yabuki, T., … Muto, Y. (2010). Structural insight into the zinc finger CW domain as a histone modification reader. *Structure, 18*(9), 1127–1139. http://dx.doi.org/10.1016/j.str.2010.06.012.

Hickman, S. E., & El Khoury, J. (2014). TREM2 and the neuroimmunology of Alzheimer's disease. *Biochemical Pharmacology, 88*(4), 495–498. http://dx.doi.org/10.1016/j.bcp.2013.11.021.

Hinney, A., Albayrak, O., Antel, J., Volckmar, A. L., Sims, R., Chapman, J., … Consortium, G. (2014). Genetic variation at the CELF1 (CUGBP, elav-like family member 1 gene) locus is genome-wide associated with Alzheimer's disease and obesity. *American Journal of Medical Genetics, Part B: Neuropsychiatric Genetics, 165B*(4), 283–293. http://dx.doi.org/10.1002/ajmg.b.32234.

Hollingworth, P., Harold, D., Sims, R., Gerrish, A., Lambert, J. C., Carrasquillo, M. M., … Williams, J. (2011). Common variants at ABCA7, MS4A6A/MS4A4E, EPHA1, CD33 and CD2AP are associated with Alzheimer's disease. *Nature Genetics, 43*(5), 429–435. http://dx.doi.org/10.1038/ng.803.

Holton, P., Ryten, M., Nalls, M., Trabzuni, D., Weale, M. E., Hernandez, D., … Guerreiro, R. (2013). Initial assessment of the pathogenic mechanisms of the recently identified Alzheimer risk Loci. *Annals of Human Genetics, 77*(2), 85–105. http://dx.doi.org/10.1111/ahg.12000.

Holtzman, D. M., Herz, J., & Bu, G. (2012). Apolipoprotein E and apolipoprotein E receptors: normal biology and roles in Alzheimer disease. *Cold Spring Harbor Perspectives in Medicine, 2*(3), a006312. http://dx.doi.org/10.1101/cshperspect.a006312.

Holtzman, D. M., Morris, J. C., & Goate, A. M. (2011). Alzheimer's disease: the challenge of the second century. *Science Translational Medicine, 3*(77), 77sr71. http://dx.doi.org/10.1126/scitranslmed.3002369.

Hooli, B. V., Kovacs-Vajna, Z. M., Mullin, K., Blumenthal, M. A., Mattheisen, M., Zhang, C., … Tanzi, R. E. (2014). Rare autosomal copy number variations in early-onset familial Alzheimer's disease. *Molecular Psychiatry, 19*(6), 676–681. http://dx.doi.org/10.1038/mp.2013.77.

Hooli, B. V., Lill, C. M., Mullin, K., Qiao, D., Lange, C., Bertram, L., & Tanzi, R. E. (2015). PLD3 gene variants and Alzheimer's disease. *Nature, 520*(7545), E7–E8. http://dx.doi.org/10.1038/nature14040.

Hooli, B. V., Mohapatra, G., Mattheisen, M., Parrado, A. R., Roehr, J. T., Shen, Y., … Bertram, L. (2012). Role of common and rare APP DNA sequence variants in Alzheimer disease. *Neurology, 78*(16), 1250–1257. http://dx.doi.org/10.1212/WNL.0b013e3182515972.

Hooli, B. V., Parrado, A. R., Mullin, K., Yip, W. K., Liu, T., Roehr, J. T., … Tanzi, R. E. (2014). The rare TREM2 R47H variant exerts only a modest effect on Alzheimer disease risk. *Neurology, 83*(15), 1353–1358. http://dx.doi.org/10.1212/WNL.0000000000000855.

Hooli, B. V., & Tanzi, R. E. (2009). A current view of Alzheimer's disease. *F1000 Biology Reports, 1*(54). http://dx.doi.org/10.3410/B1–54.

Hsieh, C. L., Koike, M., Spusta, S. C., Niemi, E. C., Yenari, M., Nakamura, M. C., & Seaman, W. E. (2009). A role for TREM2 ligands in the phagocytosis of apoptotic neuronal cells by microglia. *Journal of Neurochemistry, 109*(4), 1144–1156. http://dx.doi.org/10.1111/j.1471-4159.2009.06042.x.

Huang, Y. (2010). Mechanisms linking apolipoprotein E isoforms with cardiovascular and neurological diseases. *Current Opinion in Lipidology, 21*(4), 337–345. http://dx.doi.org/10.1097/MOL.0b013e32833af368.

Hutchings, N. J., Clarkson, N., Chalkley, R., Barclay, A. N., & Brown, M. H. (2003). Linking the T cell surface protein CD2 to the actin-capping protein CAPZ via CMS and CIN85. *Journal of Biological Chemistry, 278*(25), 22396–22403. http://dx.doi.org/10.1074/jbc.M302540200.

Ishibashi, K., Suzuki, M., Sasaki, S., & Imai, M. (2001). Identification of a new multigene four-transmembrane family (MS4A) related to CD20, HTm4 and beta subunit of the high-affinity IgE receptor. *Gene, 264*(1), 87–93.

Janes, P. W., Saha, N., Barton, W. A., Kolev, M. V., Wimmer-Kleikamp, S. H., Nievergall, E., … Nikolov, D. B. (2005). Adam meets Eph: an ADAM substrate recognition module acts as a molecular switch for ephrin cleavage in trans. *Cell, 123*(2), 291–304. http://dx.doi.org/10.1016/j.cell.2005.08.014.

Janson, C. G., Chen, Y., Li, Y., & Leifer, D. (2001). Functional regulatory regions of human transcription factor MEF2C. *Brain Research. Molecular Brain Research, 97*(1), 70–82.

Jiang, T., Yu, J. T., Hu, N., Tan, M. S., Zhu, X. C., & Tan, L. (2014). CD33 in Alzheimer's disease. *Molecular Neurobiology, 49*(1), 529–535. http://dx.doi.org/10.1007/s12035-013-8536-1.

Jonsson, T., Stefansson, H., Steinberg, S., Jonsdottir, I., Jonsson, P. V., Snaedal, J., … Stefansson, K. (2013). Variant of TREM2 associated with the risk of Alzheimer's disease. *New England Journal of Medicine, 368*(2), 107–116. http://dx.doi.org/10.1056/NEJMoa1211103.

Jorissen, E., Prox, J., Bernreuther, C., Weber, S., Schwanbeck, R., Serneels, L., … Saftig, P. (2010). The disintegrin/metalloproteinase ADAM10 is essential for the establishment of the brain cortex. *Journal of Neuroscience, 30*(14), 4833–4844. http://dx.doi.org/10.1523/JNEUROSCI.5221-09.2010.

Jun, G., Ibrahim-Verbaas, C. A., Vronskaya, M., Lambert, J. C., Chung, J., Naj, A. C., … Farrer, L. A. (2015). A novel Alzheimer disease locus located near the gene encoding tau protein. *Molecular Psychiatry, 21*(1), 108–117. http://dx.doi.org/10.1038/mp.2015.23.

Kajiho, H., Saito, K., Tsujita, K., Kontani, K., Araki, Y., Kurosu, H., & Katada, T. (2003). RIN3: a novel Rab5 GEF interacting with amphiphysin II involved in the early endocytic pathway. *Journal of Cell Science, 116*(Pt 20), 4159–4168. http://dx.doi.org/10.1242/jcs.00718.

Kajiho, H., Sakurai, K., Minoda, T., Yoshikawa, M., Nakagawa, S., Fukushima, S., … Katada, T. (2011). Characterization of RIN3 as a guanine nucleotide exchange factor for the Rab5 subfamily GTPase Rab31. *Journal of Biological Chemistry, 286*(27), 24364–24373. http://dx.doi.org/10.1074/jbc.M110.172445.

Kamboh, M. I., Aston, C. E., Perez-Tur, J., Kokmen, E., Ferrell, R. E., Hardy, J., & DeKosky, S. T. (1999). A novel mutation in the apolipoprotein E gene (APOE*4 Pittsburgh) is associated with the risk of late-onset Alzheimer's disease. *Neuroscience Letters, 263*(2–3), 129–132.

Kamboh, M. I., Demirci, F. Y., Wang, X., Minster, R. L., Carrasquillo, M. M., Pankratz, V. S., … Barmada, M. M. (2012). Genome-wide association study of Alzheimer's disease. *Translational Psychiatry, 2*, e117. http://dx.doi.org/10.1038/tp.2012.45.

Kamboh, M. I., Minster, R. L., Demirci, F. Y., Ganguli, M., Dekosky, S. T., Lopez, O. L., & Barmada, M. M. (2012). Association of CLU and PICALM variants with Alzheimer's disease. *Neurobiology of Aging, 33*(3), 518–521. http://dx.doi.org/10.1016/j.neurobiolaging.2010.04.015.

Kanekiyo, T., Xu, H., & Bu, G. (2014). ApoE and Aβ in Alzheimer's disease: accidental encounters or partners? *Neuron, 81*(4), 740–754. http://dx.doi.org/10.1016/j.neuron.2014.01.045.

Kaneko, T., Li, L., & Li, S. S. (2008). The SH3 domain–a family of versatile peptide- and protein-recognition module. *Frontiers in Bioscience, 13*, 4938–4952.

Karch, C. M., Jeng, A. T., Nowotny, P., Cady, J., Cruchaga, C., & Goate, A. M. (2012). Expression of novel Alzheimer's disease risk genes in control and Alzheimer's disease brains. *PLoS One, 7*(11), e50976. http://dx.doi.org/10.1371/journal.pone.0050976.

Kero, M., Paetau, A., Polvikoski, T., Tanskanen, M., Sulkava, R., Jansson, L., … Tienari, P. J. (2013). Amyloid precursor protein (APP) A673T mutation in the elderly Finnish population. *Neurobiology of Aging, 34*(5). http://dx.doi.org/10.1016/j.neurobiolaging.2012.09.017, 1518.e1511–1513.

Khera, R., & Das, N. (2009). Complement receptor 1: disease associations and therapeutic implications. *Molecular Immunology, 46*(5), 761–772. http://dx.doi.org/10.1016/j.molimm.2008.09.026.

Kim, J., Basak, J. M., & Holtzman, D. M. (2009). The role of apolipoprotein E in Alzheimer's disease. *Neuron, 63*(3), 287–303. http://dx.doi.org/10.1016/j.neuron.2009.06.026.

Kim, W. S., Fitzgerald, M. L., Kang, K., Okuhira, K., Bell, S. A., Manning, J. J., … Freeman, M. W. (2005). Abca7 null mice retain normal macrophage phosphatidylcholine and cholesterol efflux activity despite alterations in adipose mass and serum cholesterol levels. *Journal of Biological Chemistry, 280*(5), 3989–3995. http://dx.doi.org/10.1074/jbc.M412602200.

Kim, W. S., Guillemin, G. J., Glaros, E. N., Lim, C. K., & Garner, B. (2006). Quantitation of ATP-binding cassette subfamily-A transporter gene expression in primary human brain cells. *Neuroreport, 17*(9), 891–896. http://dx.doi.org/10.1097/01.wnr.0000221833.41340.cd.

Kim, J., Jiang, H., Park, S., Eltorai, A. E., Stewart, F. R., Yoon, H., … Holtzman, D. M. (2011). Haploinsufficiency of human APOE reduces amyloid deposition in a mouse model of amyloid-beta amyloidosis. *Journal of Neuroscience, 31*(49), 18007–18012. http://dx.doi.org/10.1523/JNEUROSCI.3773-11.2011.

Kim, W. S., Li, H., Ruberu, K., Chan, S., Elliott, D. A., Low, J. K., … Garner, B. (2013). Deletion of Abca7 increases cerebral amyloid-beta accumulation in the J20 mouse model of Alzheimer's disease. *Journal of Neuroscience, 33*(10), 4387–4394. http://dx.doi.org/10.1523/JNEUROSCI.4165-12.2013.

Kim, M., Suh, J., Romano, D., Truong, M. H., Mullin, K., Hooli, B., … Tanzi, R. E. (2009). Potential late-onset Alzheimer's disease-associated mutations in the ADAM10 gene attenuate {alpha}-secretase activity. *Human Molecular Genetics, 18*(20), 3987–3996. http://dx.doi.org/10.1093/hmg/ddp323.

Kim, W. S., Weickert, C. S., & Garner, B. (2008). Role of ATP-binding cassette transporters in brain lipid transport and neurological disease. *Journal of Neurochemistry, 104*(5), 1145–1166. http://dx.doi.org/10.1111/j.1471-4159.2007.05099.x.

Klein, A. F., Gasnier, E., & Furling, D. (2011). Gain of RNA function in pathological cases: focus on myotonic dystrophy. *Biochimie, 93*(11), 2006–2012. http://dx.doi.org/10.1016/j.biochi.2011.06.028.

Klesney-Tait, J., Turnbull, I. R., & Colonna, M. (2006). The TREM receptor family and signal integration. *Nature Immunology, 7*(12), 1266–1273. http://dx.doi.org/10.1038/ni1411.

Kobayashi, S., Sawano, A., Nojima, Y., Shibuya, M., & Maru, Y. (2004). The c-Cbl/CD2AP complex regulates VEGF-induced endocytosis and degradation of Flt-1 (VEGFR-1). *FASEB Journal, 18*(7), 929–931. http://dx.doi.org/10.1096/fj.03-0767fje.

Kofler, J. K., Bissel, S., Wiley, C., Stauffer, M., & Murdoch, G. (2012). Differential microglial expression of new Alzheimer's disease associated genes MS4A4A and MS4A6A. *Alzheimer's and Dementia, 8*(Issue 4), 253. http://dx.doi.org/10.1016/j.jalz.2012.05.675.

Kok, E. H., Luoto, T., Haikonen, S., Goebeler, S., Haapasalo, H., & Karhunen, P. J. (2011). CLU, CR1 and PICALM genes associate with Alzheimer's-related senile plaques. *Alzheimer's Research and Therapy, 3*(2), 12. http://dx.doi.org/10.1186/alzrt71.

Kovacs, D. M., Fausett, H. J., Page, K. J., Kim, T. W., Moir, R. D., Merriam, D. E., … Wasco, W. (1996). Alzheimer-associated presenilins 1 and 2: neuronal expression in brain and localization to intracellular membranes in mammalian cells. *Nature Medicine, 2*(2), 224–229.

Krych-Goldberg, M., Moulds, J. M., & Atkinson, J. P. (2002). Human complement receptor type 1 (CR1) binds to a major malarial adhesin. *Trends in Molecular Medicine, 8*(11), 531–537.

Kuhn, P. H., Wang, H., Dislich, B., Colombo, A., Zeitschel, U., Ellwart, J. W., … Lichtenthaler, S. F. (2010). ADAM10 is the physiologically relevant, constitutive alpha-secretase of the amyloid precursor protein in primary neurons. *EMBO Journal, 29*(17), 3020–3032. http://dx.doi.org/10.1038/emboj.2010.167.

Kutok, J. L., Yang, X., Folkerth, R. D., Imitola, J., Raddassi, K., Yano, Y., … Adra, C. (2005). The cell cycle associated protein, HTm4, is expressed in differentiating cells of the hematopoietic and central nervous system in mice. *Journal of Molecular Histology, 36*(1–2), 77–87. http://dx.doi.org/10.1007/s10735-004-3913-8.

Lai-Cheong, J. E., Parsons, M., & McGrath, J. A. (2010). The role of kindlins in cell biology and relevance to human disease. *International Journal of Biochemistry and Cell Biology, 42*(5), 595–603. http://dx.doi.org/10.1016/j.biocel.2009.10.015.

Lai, K. O., & Ip, N. Y. (2009). Synapse development and plasticity: roles of ephrin/Eph receptor signaling. *Current Opinion in Neurobiology, 19*(3), 275–283. http://dx.doi.org/10.1016/j.conb.2009.04.009.

Lambert, J. C., Heath, S., Even, G., Campion, D., Sleegers, K., Hiltunen, M., … Amouyel, P. (2009). Genome-wide association study identifies variants at CLU and CR1 associated with Alzheimer's disease. *Nature Genetics, 41*(10), 1094–1099. http://dx.doi.org/10.1038/ng.439.

Lambert, J., Ibrahim-Verbaas, C., Harold, D., Naj, A., Sims, R., Bellenguez, C., … Amouyel, P. (2013). Meta-analysis of 74,046 individuals identifies 11 new susceptibility loci for Alzheimer's disease. *Nature Genetics, 45*(12), 1452–1458. http://dx.doi.org/10.1038/ng.2802.

Lammich, S., Buell, D., Zilow, S., Ludwig, A. K., Nuscher, B., Lichtenthaler, S. F., … Haass, C. (2010). Expression of the anti-amyloidogenic secretase ADAM10 is suppressed by its 5′-untranslated region. *Journal of Biological Chemistry, 285*(21), 15753–15760. http://dx.doi.org/10.1074/jbc.M110.110742.

Lange, C., DeMeo, D., Silverman, E. K., Weiss, S. T., & Laird, N. M. (2004). PBAT: tools for family-based association studies. *American Journal of Human Genetics, 74*(2), 367–369. http://dx.doi.org/10.1086/381563.

Lee, J. H., Cheng, R., Honig, L. S., Vonsattel, J. P., Clark, L., & Mayeux, R. (2008). Association between genetic variants in SORL1 and autopsy-confirmed Alzheimer disease. *Neurology, 70*(11), 887–889. http://dx.doi.org/10.1212/01.wnl.0000280581.39755.89.

Leigh, M. W., Pittman, J. E., Carson, J. L., Ferkol, T. W., Dell, S. D., Davis, S. D., ... Zariwala, M. A. (2009). Clinical and genetic aspects of primary ciliary dyskinesia/Kartagener syndrome. *Genetics in Medicine, 11*(7), 473–487. http://dx.doi.org/10.1097/GIM.0b013e3181a53562.

Le, H. T., Maksumova, L., Wang, J., & Pallen, C. J. (2006). Reduced NMDA receptor tyrosine phosphorylation in PTPalpha-deficient mouse synaptosomes is accompanied by inhibition of four src family kinases and Pyk2: an upstream role for PTPalpha in NMDA receptor regulation. *Journal of Neurochemistry, 98*(6), 1798–1809. http://dx.doi.org/10.1111/j.1471-4159.2006.04075.x.

Lev, S., Moreno, H., Martinez, R., Canoll, P., Peles, E., Musacchio, J. M., ... Schlessinger, J. (1995). Protein tyrosine kinase PYK2 involved in Ca(2+)-induced regulation of ion channel and MAP kinase functions. *Nature, 376*(6543), 737–745. http://dx.doi.org/10.1038/376737a0.

Liang, Y., Buckley, T. R., Tu, L., Langdon, S. D., & Tedder, T. F. (2001). Structural organization of the human MS4A gene cluster on chromosome 11q12. *Immunogenetics, 53*(5), 357–368.

Liang, Y., & Tedder, T. F. (2001). Identification of a CD20-, FcεRIβ-, and HTm4-related gene family: sixteen new MS4A family members expressed in human and mouse. *Genomics, 72*(2), 119–127. http://dx.doi.org/10.1006/geno.2000.6472.

Lill, C. M., Abel, O., Bertram, L., & Al-Chalabi, A. (2011). Keeping up with genetic discoveries in amyotrophic lateral sclerosis: the ALSoD and ALSGene databases. *Amyotrophic Lateral Sclerosis, 12*(4), 238–249. http://dx.doi.org/10.3109/17482968.2011.584629.

Lill, C. M., & Bertram, L. (2011). Towards unveiling the genetics of neurodegenerative diseases. *Seminars in Neurology, 31*(5), 531–541. http://dx.doi.org/10.1055/s-0031-1299791.

Lill, C. M., Rengmark, A., Pihlstrom, L., Fogh, I., Shatunov, A., Sleiman, P. M., ... Bertram, L. (2015). The role of TREM2 R47H as a risk factor for Alzheimer's disease, frontotemporal lobar degeneration, amyotrophic lateral sclerosis, and Parkinson's disease. *Alzheimer's and Dementia.* http://dx.doi.org/10.1016/j.jalz.2014.12.009.

Lill, C. M., Roehr, J. T., McQueen, M. B., Kavvoura, F. K., Bagade, S., Schjeide, B. M., ... Bertram, L. (2012). Comprehensive research synopsis and systematic meta-analyses in Parkinson's disease genetics: the PDGene database. *PLoS Genetics, 8*(3), e1002548. http://dx.doi.org/10.1371/journal.pgen.1002548.

Li, X., Ma, Y., Wei, X., Li, Y., Wu, H., Zhuang, J., & Zhao, Z. (2014). Clusterin in Alzheimer's disease: a player in the biological behavior of amyloid-beta. *Neuroscience Bulletin, 30*(1), 162–168. http://dx.doi.org/10.1007/s12264-013-1391-2.

Ling, I. F., Bhongsatiern, J., Simpson, J. F., Fardo, D. W., & Estus, S. (2012). Genetics of clusterin isoform expression and Alzheimer's disease risk. *PLoS One, 7*(4), e33923. http://dx.doi.org/10.1371/journal.pone.0033923.

Linnartz-Gerlach, B., Mathews, M., & Neumann, H. (2014). Sensing the neuronal glycocalyx by glial sialic acid binding immunoglobulin-like lectins. *Neuroscience, 275*, 113–124. http://dx.doi.org/10.1016/j.neuroscience.2014.05.061.

Liu, C. C., Kanekiyo, T., Xu, H., & Bu, G. (2013). Apolipoprotein E and Alzheimer disease: risk, mechanisms and therapy. *Nature Reviews Neurology, 9*(2), 106–118. http://dx.doi.org/10.1038/nrneurol.2012.263.

Liu, D., & Niu, Z. X. (2009). The structure, genetic polymorphisms, expression and biological functions of complement receptor type 1 (CR1/CD35). *Immunopharmacology and Immunotoxicology, 31*(4), 524–535. http://dx.doi.org/10.3109/08923970902845768.

Li, H., Wetten, S., Li, L., St Jean, P. L., Upmanyu, R., Surh, L., ... Roses, A. D. (2008). Candidate single-nucleotide polymorphisms from a genomewide association study of Alzheimer disease. *Archives of Neurology, 65*(1), 45–53. http://dx.doi.org/10.1001/archneurol.2007.3.

Lue, L. F., Schmitz, C., & Walker, D. G. (2014). What happens to microglial TREM2 in Alzheimer's disease: immunoregulatory turned into immunopathogenic? *Neuroscience, 302*, 138–150. http://dx.doi.org/10.1016/j.neuroscience.2014.09.050.

Lynch, D. K., Winata, S. C., Lyons, R. J., Hughes, W. E., Lehrbach, G. M., Wasinger, V., ... Daly, R. J. (2003). A cortactin-CD2-associated protein (CD2AP) complex provides a novel link between epidermal growth factor receptor endocytosis and the actin cytoskeleton. *Journal of Biological Chemistry, 278*(24), 21805–21813. http://dx.doi.org/10.1074/jbc.M211407200.

Mack, J. T., Townsend, D. M., Beljanski, V., & Tew, K. D. (2007). The ABCA2 transporter: intracellular roles in trafficking and metabolism of LDL-derived cholesterol and sterol-related compounds. *Current Drug Metabolism, 8*(1), 47–57.

Madry, H., Prudlo, J., Grgic, A., & Freyschmidt, J. (2007). Nasu-Hakola disease (PLOSL): report of five cases and review of the literature. *Clinical Orthopaedics and Related Research, 454*, 262–269. http://dx.doi.org/10.1097/01.blo.0000229364.57985.df.

Mahley, R. W., & Rall, S. C., Jr. (2000). Apolipoprotein E: far more than a lipid transport protein. *Annual Review of Genomics and Human Genetics, 1*, 507–537. http://dx.doi.org/10.1146/annurev.genom.1.1.507.

Mahr, S., Burmester, G. R., Hilke, D., Gobel, U., Grutzkau, A., Haupl, T., ... Muller, B. (2006). Cis- and trans-acting gene regulation is associated with osteoarthritis. *American Journal of Human Genetics, 78*(5), 793–803. http://dx.doi.org/10.1086/503849.

Malik, M., Simpson, J. F., Parikh, I., Wilfred, B. R., Fardo, D. W., Nelson, P. T., & Estus, S. (2013). CD33 Alzheimer's risk-altering polymorphism, CD33 expression, and exon 2 splicing. *Journal of Neuroscience, 33*(33), 13320–13325. http://dx.doi.org/10.1523/JNEUROSCI.1224-13.2013.

Maloney, J. A., Bainbridge, T., Gustafson, A., Zhang, S., Kyauk, R., Steiner, P., ... Atwal, J. K. (2014). Molecular mechanisms of Alzheimer disease protection by the A673T allele of amyloid precursor protein. *Journal of Biological Chemistry, 289*(45), 30990–31000. http://dx.doi.org/10.1074/jbc.M114.589069.

Mansouri, L., Messalmani, M., Klai, S., Bedoui, I., Derbali, H., Gritli, N., ... Fekih-Mrissa, N. (2015). Association of HLA-DR/DQ polymorphism with Alzheimer's disease. *American Journal of the Medical Sciences, 349*(4), 334–337. http://dx.doi.org/10.1097/MAJ.0000000000000416.

May, P. C., Lampert-Etchells, M., Johnson, S. A., Poirier, J., Masters, J. N., & Finch, C. E. (1990). Dynamics of gene expression for a hippocampal glycoprotein elevated in Alzheimer's disease and in response to experimental lesions in rat. *Neuron, 5*(6), 831–839.

Medway, C. W., Abdul-Hay, S., Mims, T., Ma, L., Bisceglio, G., Zou, F., … Younkin, S. G. (2014). ApoE variant p.V236E is associated with markedly reduced risk of Alzheimer's disease. *Molecular Neurodegeneration, 9*, 11. http://dx.doi.org/10.1186/1750-1326-9-11.

Meola, G., Jones, K., Wei, C., & Timchenko, L. T. (2013). Dysfunction of protein homeostasis in myotonic dystrophies. *Histology and Histopathology, 28*(9), 1089–1098.

Metzner, A., Precht, C., Fehse, B., Fiedler, W., Stocking, C., Gunther, A., … Jucker, M. (2009). Reduced proliferation of CD34(+) cells from patients with acute myeloid leukemia after gene transfer of INPP5D. *Gene Therapy, 16*(4), 570–573. http://dx.doi.org/10.1038/gt.2008.184.

Michaelson, D. M. (2014). APOE ε4: the most prevalent yet understudied risk factor for Alzheimer's disease. *Alzheimer's and Dementia, 10*(6), 861–868. http://dx.doi.org/10.1016/j.jalz.2014.06.015.

Mitra, S. K., Hanson, D. A., & Schlaepfer, D. D. (2005). Focal adhesion kinase: in command and control of cell motility. *Nature Reviews Molecular Cell Biology, 6*(1), 56–68. http://dx.doi.org/10.1038/nrm1549.

Mitsui, J., & Tsuji, S. (2014). Genomic aspects of sporadic neurodegenerative diseases. *Biochemical and Biophysical Research, 452*(2), 221–225. http://dx.doi.org/10.1016/j.bbrc.2014.07.098.

Miyashita, A., Koike, A., Jun, G., Wang, L. S., Takahashi, S., Matsubara, E., … Kuwano, R. (2013). SORL1 is genetically associated with late-onset Alzheimer's disease in Japanese, Koreans and Caucasians. *PLoS One, 8*(4), e58618. http://dx.doi.org/10.1371/journal.pone.0058618.

Naj, A. C., Beecham, G. W., Martin, E. R., Gallins, P. J., Powell, E. H., Konidari, I., … Pericak-Vance, M. A. (2010). Dementia revealed: novel chromosome 6 locus for late-onset Alzheimer disease provides genetic evidence for folate-pathway abnormalities. *PLoS Genetics, 6*(9), e1001130. http://dx.doi.org/10.1371/journal.pgen.1001130.

Naj, A. C., Jun, G., Beecham, G. W., Wang, L. S., Vardarajan, B. N., Buros, J., … Schellenberg, G. D. (2011). Common variants at MS4A4/MS4A6E, CD2AP, CD33 and EPHA1 are associated with late-onset Alzheimer's disease. *Nature Genetics, 43*(5), 436–441. http://dx.doi.org/10.1038/ng.801.

Nalls, M. A., Pankratz, N., Lill, C. M., Do, C. B., Hernandez, D. G., Saad, M., … Singleton, A. B. (2014). Large-scale meta-analysis of genome-wide association data identifies six new risk loci for Parkinson's disease. *Nature Genetics, 46*(9), 989–993. http://dx.doi.org/10.1038/ng.3043.

Neumann, H., & Takahashi, K. (2007). Essential role of the microglial triggering receptor expressed on myeloid cells-2 (TREM2) for central nervous tissue immune homeostasis. *Journal of Neuroimmunology, 184*(1–2), 92–99. http://dx.doi.org/10.1016/j.jneuroim.2006.11.032.

Nikonova, A. S., Gaponova, A. V., Kudinov, A. E., & Golemis, E. A. (2014). CAS proteins in health and disease: an update. *IUBMB Life, 66*(6), 387–395. http://dx.doi.org/10.1002/iub.1282.

Nuutinen, T., Suuronen, T., Kauppinen, A., & Salminen, A. (2009). Clusterin: a forgotten player in Alzheimer's disease. *Brain Research Reviews, 61*(2), 89–104. http://dx.doi.org/10.1016/j.brainresrev.2009.05.007.

Oda, T., Pasinetti, G. M., Osterburg, H. H., Anderson, C., Johnson, S. A., & Finch, C. E. (1994). Purification and characterization of brain clusterin. *Biochemical and Biophysical Research, 204*(3), 1131–1136. http://dx.doi.org/10.1006/bbrc.1994.2580.

Offe, K., Dodson, S. E., Shoemaker, J. T., Fritz, J. J., Gearing, M., Levey, A. I., & Lah, J. J. (2006). The lipoprotein receptor LR11 regulates amyloid beta production and amyloid precursor protein traffic in endosomal compartments. *Journal of Neuroscience, 26*(5), 1596–1603. http://dx.doi.org/10.1523/JNEUROSCI.4946-05.2006.

Pahnke, J., Frohlich, C., Krohn, M., Schumacher, T., & Paarmann, K. (2013). Impaired mitochondrial energy production and ABC transporter function-A crucial interconnection in dementing proteopathies of the brain. *Mechanisms of Ageing and Development, 134*(10), 506–515. http://dx.doi.org/10.1016/j.mad.2013.08.007.

Paloneva, J., Autti, T., Hakola, P., & Haltia, M. J. (1993). Polycystic lipomembranous osteodysplasia with sclerosing leukoencephalopathy (PLOSL). In R. A. Pagon, M. P. Adam, H. H. Ardinger, S. E. Wallace, A. Amemiya, L. J. H. Bean, et al. (Eds.), *GeneReviews®*. Seattle (WA).

Paloneva, J., Mandelin, J., Kiialainen, A., Bohling, T., Prudlo, J., Hakola, P., … Peltonen, L. (2003). DAP12/TREM2 deficiency results in impaired osteoclast differentiation and osteoporotic features. *Journal of Experimental Medicine, 198*(4), 669–675. http://dx.doi.org/10.1084/jem.20030027.

Pan, K., Liang, X. T., Zhang, H. K., Zhao, J. J., Wang, D. D., Li, J. J., … Xia, J. C. (2012). Characterization of bridging integrator 1 (BIN1) as a potential tumor suppressor and prognostic marker in hepatocellular carcinoma. *Molecular Medicine, 18*, 507–518. http://dx.doi.org/10.2119/molmed.2011.00319.

Pant, S., Sharma, M., Patel, K., Caplan, S., Carr, C. M., & Grant, B. D. (2009). AMPH-1/Amphiphysin/Bin1 functions with RME-1/Ehd1 in endocytic recycling. *Nature Cell Biology, 11*(12), 1399–1410. http://dx.doi.org/10.1038/ncb1986.

Park, S., Mathis, K. W., & Lee, I. K. (2014). The physiological roles of apolipoprotein J/clusterin in metabolic and cardiovascular diseases. *Reviews in Endocrine and Metabolic Disorders, 15*(1), 45–53. http://dx.doi.org/10.1007/s11154-013-9275-3.

Parry, D. A., Poulter, J. A., Logan, C. V., Brookes, S. J., Jafri, H., Ferguson, C. H., … Mighell, A. J. (2013). Identification of mutations in SLC24A4, encoding a potassium-dependent sodium/calcium exchanger, as a cause of amelogenesis imperfecta. *American Journal of Human Genetics, 92*(2), 307–312. http://dx.doi.org/10.1016/j.ajhg.2013.01.003.

Poduslo, S. E., Huang, R., Huang, J., & Smith, S. (2009). Genome screen of late-onset Alzheimer's extended pedigrees identifies TRPC4AP by haplotype analysis. *American Journal of Medical Genetics, Part B: Neuropsychiatric Genetics, 150B*(1), 50–55. http://dx.doi.org/10.1002/ajmg.b.30767.

Postina, R., Schroeder, A., Dewachter, I., Bohl, J., Schmitt, U., Kojro, E., … Fahrenholz, F. (2004). A disintegrin-metalloproteinase prevents amyloid plaque formation and hippocampal defects in an Alzheimer disease mouse model. *J Clin Invest, 113*(10), 1456–1464. http://dx.doi.org/10.1172/JCI20864.

Potkin, S. G., Guffanti, G., Lakatos, A., Turner, J. A., Kruggel, F., Fallon, J. H., … Alzheimer's Disease Neuroimaging Initiative. (2009). Hippocampal atrophy as a quantitative trait in a genome-wide association study identifying novel susceptibility genes for Alzheimer's disease. *PLoS One, 4*(8), e6501. http://dx.doi.org/10.1371/journal.pone.0006501.

Potthoff, M. J., & Olson, E. N. (2007). MEF2: a central regulator of diverse developmental programs. *Development, 134*(23), 4131–4140. http://dx.doi.org/10.1242/dev.008367.

Pottier, C., Hannequin, D., Coutant, S., Rovelet-Lecrux, A., Wallon, D., Rousseau, S., … Collaborators, P. G. (2012). High frequency of potentially pathogenic SORL1 mutations in autosomal dominant early-onset Alzheimer disease. *Molecular Psychiatry, 17*(9), 875–879. http://dx.doi.org/10.1038/mp.2012.15.

Pottier, C., Wallon, D., Rousseau, S., Rovelet-Lecrux, A., Richard, A. C., Rollin-Sillaire, A., … Hannequin, D. (2013). TREM2 R47H variant as a risk factor for early-onset Alzheimer's disease. *Journal of Alzheimer's Disease, 35*(1), 45–49. http://dx.doi.org/10.3233/JAD-122311.

Prokic, I., Cowling, B. S., & Laporte, J. (2014). Amphiphysin 2 (BIN1) in physiology and diseases. *Journal of Molecular Medicine (Berlin), 92*(5), 453–463. http://dx.doi.org/10.1007/s00109-014-1138-1.

Raj, T., Ryan, K. J., Replogle, J. M., Chibnik, L. B., Rosenkrantz, L., Tang, A., … Bradshaw, E. M. (2014). CD33: increased inclusion of exon 2 implicates the Ig V-set domain in Alzheimer's disease susceptibility. *Human Molecular Genetics, 23*(10), 2729–2736. http://dx.doi.org/10.1093/hmg/ddt666.

Reiman, E. M., Webster, J. A., Myers, A. J., Hardy, J., Dunckley, T., Zismann, V. L., … Stephan, D. A. (2007). GAB2 alleles modify Alzheimer's risk in APOE epsilon4 carriers. *Neuron, 54*(5), 713–720. http://dx.doi.org/10.1016/j.neuron.2007.05.022.

Reitz, C. (2013). Dyslipidemia and the risk of Alzheimer's disease. *Current Atherosclerosis Reports, 15*(3), 307. http://dx.doi.org/10.1007/s11883-012-0307-3.

Reitz, C., Cheng, R., Rogaeva, E., Lee, J. H., Tokuhiro, S., Zou, F., … Environmental Risk in Alzheimer Disease 1 Consortium. (2011). Meta-analysis of the association between variants in SORL1 and Alzheimer disease. *Archives of Neurology, 68*(1), 99–106. http://dx.doi.org/10.1001/archneurol.2010.346.

Reitz, C., Jun, G., Naj, A., Rajbhandary, R., Vardarajan, B. N., Wang, L. S., … Alzheimer Disease Genetics Consortium. (2013). Variants in the ATP-binding cassette transporter (ABCA7), apolipoprotein E 4, and the risk of late-onset Alzheimer disease in African Americans. *JAMA, 309*(14), 1483–1492. http://dx.doi.org/10.1001/jama.2013.2973.

Rocca, W. A., Hofman, A., Brayne, C., Breteler, M. M., Clarke, M., Copeland, J. R., et al. (1991). Frequency and distribution of Alzheimer's disease in Europe: a collaborative study of 1980–1990 prevalence findings. The EURODEM-Prevalence Research Group. *Annals of Neurology, 30*(3), 381–390. http://dx.doi.org/10.1002/ana.410300310.

Rogaeva, E., Meng, Y., Lee, J. H., Gu, Y., Kawarai, T., Zou, F., … St George-Hyslop, P. (2007). The neuronal sortilin-related receptor SORL1 is genetically associated with Alzheimer disease. *Nature Genetics, 39*(2), 168–177. http://dx.doi.org/10.1038/ng1943.

Rogers, J., Li, R., Mastroeni, D., Grover, A., Leonard, B., Ahern, G., … Sabbagh, M. (2006). Peripheral clearance of amyloid beta peptide by complement C3-dependent adherence to erythrocytes. *Neurobiology of Aging, 27*(12), 1733–1739. http://dx.doi.org/10.1016/j.neurobiolaging.2005.09.043.

Rohne, P., Prochnow, H., Wolf, S., Renner, B., & Koch-Brandt, C. (2014). The chaperone activity of clusterin is dependent on glycosylation and redox environment. *Cellular Physiology and Biochemistry, 34*(5), 1626–1639. http://dx.doi.org/10.1159/000366365.

Rosenthal, S. L., Barmada, M. M., Wang, X., Demirci, F. Y., & Kamboh, M. I. (2014). Connecting the dots: potential of data integration to identify regulatory SNPs in late-onset Alzheimer's disease GWAS findings. *PLoS One, 9*(4), e95152. http://dx.doi.org/10.1371/journal.pone.0095152.

Rosenthal, S. L., & Kamboh, M. I. (2014). Late-onset Alzheimer's disease genes and the potentially implicated pathways. *Current Genetic Medicine Reports, 2*, 85–101. http://dx.doi.org/10.1007/s40142-014-0034-x.

Rovelet-Lecrux, A., & Campion, D. (2012). Copy number variations involving the microtubule-associated protein tau in human diseases. *Biochemical Society Transactions, 40*(4), 672–676. http://dx.doi.org/10.1042/BST20120045.

Ruiz, A., Heilmann, S., Becker, T., Hernandez, I., Wagner, H., Thelen, M., … Ramirez, A. (2014). Follow-up of loci from the International Genomics of Alzheimer's Disease Project identifies TRIP4 as a novel susceptibility gene. *Translational Psychiatry, 4*, e358. http://dx.doi.org/10.1038/tp.2014.2.

Russell, C. L., Koncarevic, S., & Ward, M. A. (2014). Post-translational modifications in Alzheimer's disease and the potential for new biomarkers. *Journal of Alzheimer's Disease, 41*(2), 345–364. http://dx.doi.org/10.3233/JAD-132312.

Sager, K. L., Wuu, J., Herskowitz, J. H., Mufson, E. J., Levey, A. I., & Lah, J. J. (2012). Neuronal LR11 expression does not differentiate between clinically-defined Alzheimer's disease and control brains. *PLoS One, 7*(8), e40527. http://dx.doi.org/10.1371/journal.pone.0040527.

Sayas, C. L., Ariaens, A., Ponsioen, B., & Moolenaar, W. H. (2006). GSK-3 is activated by the tyrosine kinase Pyk2 during LPA1-mediated neurite retraction. *Molecular Biology of the Cell, 17*(4), 1834–1844. http://dx.doi.org/10.1091/mbc.E05-07-0688.

Scherzer, C. R., Offe, K., Gearing, M., Rees, H. D., Fang, G., Heilman, C. J., … Lah, J. J. (2004). Loss of apolipoprotein E receptor LR11 in Alzheimer disease. *Archives of Neurology, 61*(8), 1200–1205. http://dx.doi.org/10.1001/archneur.61.8.1200.

Schifferli, J. A., & Paccaud, J. P. (1989). Two isotypes of human C4, C4A and C4B have different structure and function. *Complement and Inflammation, 6*(1), 19–26.

Schjeide, B. M., Schnack, C., Lambert, J. C., Lill, C. M., Kirchheiner, J., Tumani, H., … Bertram, L. (2011). The role of clusterin, complement receptor 1, and phosphatidylinositol binding clathrin assembly protein in Alzheimer disease risk and cerebrospinal fluid biomarker levels. *Archives of General Psychiatry, 68*(2), 207–213. http://dx.doi.org/10.1001/archgenpsychiatry.2010.196.

Schrijvers, E. M., Koudstaal, P. J., Hofman, A., & Breteler, M. M. (2011). Plasma clusterin and the risk of Alzheimer disease. *JAMA, 305*(13), 1322–1326. http://dx.doi.org/10.1001/jama.2011.381.

Seshadri, S., Fitzpatrick, A. L., Ikram, M. A., DeStefano, A. L., Gudnason, V., Boada, M., ... EADI1 Consortium. (2010). Genome-wide analysis of genetic loci associated with Alzheimer disease. *JAMA, 303*(18), 1832–1840. http://dx.doi.org/10.1001/jama.2010.574.

Sharfe, N., Nikolic, M., Cimpeon, L., Van De Kratts, A., Freywald, A., & Roifman, C. M. (2008). EphA and ephrin-A proteins regulate integrin-mediated T lymphocyte interactions. *Molecular Immunology, 45*(5), 1208–1220. http://dx.doi.org/10.1016/j.molimm.2007.09.019.

Sharma, M., Kruger, R., & Gasser, T. (2014). From genome-wide association studies to next-generation sequencing: lessons from the past and planning for the future. *JAMA Neurology, 71*(1), 5–6. http://dx.doi.org/10.1001/jamaneurol.2013.3682.

Shaw, C. A., Li, Y., Wiszniewska, J., Chasse, S., Zaidi, S. N., Jin, W., ... Szigeti, K. (2011). Olfactory copy number association with age at onset of Alzheimer disease. *Neurology, 76*(15), 1302–1309. http://dx.doi.org/10.1212/WNL.0b013e3182166df5. pii:76/15/1302.

Shuai, P., Liu, Y., Lu, W., Liu, Q., Li, T., & Gong, B. (2015). Genetic associations of CLU rs9331888 polymorphism with Alzheimer's disease: a meta-analysis. *Neuroscience Letters, 591*, 160–165. http://dx.doi.org/10.1016/j.neulet.2015.02.040.

Shulman, J. M., Chen, K., Keenan, B. T., Chibnik, L. B., Fleisher, A., Thiyyagura, P., ... Bennett, D. A. (2013). Genetic susceptibility for Alzheimer disease neuritic plaque pathology. *JAMA Neurology, 70*(9), 1150–1157. http://dx.doi.org/10.1001/jamaneurol.2013.2815.

Shulman, J. M., Chipendo, P., Chibnik, L. B., Aubin, C., Tran, D., Keenan, B. T., ... De Jager, P. L. (2011). Functional screening of Alzheimer pathology genome-wide association signals in *Drosophila*. *American Journal of Human Genetics, 88*(2), 232–238. http://dx.doi.org/10.1016/j.ajhg.2011.01.006.

Shulman, J. M., Imboywa, S., Giagtzoglou, N., Powers, M. P., Hu, Y., Devenport, D., ... Feany, M. B. (2014). Functional screening in *Drosophila* identifies Alzheimer's disease susceptibility genes and implicates Tau-mediated mechanisms. *Human Molecular Genetics, 23*(4), 870–877. http://dx.doi.org/10.1093/hmg/ddt478.

Siest, G., Pillot, T., Regis-Bailly, A., Leininger-Muller, B., Steinmetz, J., Galteau, M. M., & Visvikis, S. (1995). Apolipoprotein E: an important gene and protein to follow in laboratory medicine. *Clinical Chemistry, 41*(8 Pt. 1), 1068–1086.

Singh, M. K., Dadke, D., Nicolas, E., Serebriiskii, I. G., Apostolou, S., Canutescu, A., ... Golemis, E. A. (2008). A novel Cas family member, HEPL, regulates FAK and cell spreading. *Molecular Biology of the Cell, 19*(4), 1627–1636. http://dx.doi.org/10.1091/mbc.E07-09-0953.

Srivastava, N., Sudan, R., & Kerr, W. G. (2013). Role of inositol poly-phosphatases and their targets in T cell biology. *Frontiers in Immunology, 4*, 288. http://dx.doi.org/10.3389/fimmu.2013.00288.

Stephan, A. B., Tobochnik, S., Dibattista, M., Wall, C. M., Reisert, J., & Zhao, H. (2012). The Na(+)/Ca(2+) exchanger NCKX4 governs termination and adaptation of the mammalian olfactory response. *Nature Neuroscience, 15*(1), 131–137. http://dx.doi.org/10.1038/nn.2943.

Stratman, N. C., Castle, C. K., Taylor, B. M., Epps, D. E., Melchior, G. W., & Carter, D. B. (2005). Isoform-specific interactions of human apolipoprotein E to an intermediate conformation of human Alzheimer amyloid-beta peptide. *Chemistry and Physics of Lipids, 137*(1–2), 52–61. http://dx.doi.org/10.1016/j.chemphyslip.2005.06.005.

Strittmatter, W. J., Saunders, A. M., Schmechel, D., Pericak-Vance, M., Enghild, J., Salvesen, G. S., & Roses, A. D. (1993). Apolipoprotein E: high-avidity binding to beta-amyloid and increased frequency of type 4 allele in late-onset familial Alzheimer disease. *Proceedings of the National Academy of Sciences of the United States of America, 90*(5), 1977–1981.

Sturm, R. A. (2009). Molecular genetics of human pigmentation diversity. *Human Molecular Genetics, 18*(R1), R9–R17. http://dx.doi.org/10.1093/hmg/ddp003.

Suh, J., Choi, S. H., Romano, D. M., Gannon, M. A., Lesinski, A. N., Kim, D. Y., & Tanzi, R. E. (2013). ADAM10 missense mutations potentiate beta-amyloid accumulation by impairing prodomain chaperone function. *Neuron, 80*(2), 385–401. http://dx.doi.org/10.1016/j.neuron.2013.08.035.

Sulem, P., Gudbjartsson, D. F., Stacey, S. N., Helgason, A., Rafnar, T., Magnusson, K. P., ... Stefansson, K. (2007). Genetic determinants of hair, eye and skin pigmentation in Europeans. *Nature Genetics, 39*(12), 1443–1452. http://dx.doi.org/10.1038/ng.2007.13.

Swerdlow, R. H., & Corder, E. H. (2012). For Alzheimer disease GWAS, pulling needles from the haystack is just the first step. *Neurology, 79*(3), 204–205. http://dx.doi.org/10.1212/WNL.0b013e31826058 1d.

Taira, K., Bujo, H., Hirayama, S., Yamazaki, H., Kanaki, T., Takahashi, K., ... Saito, Y. (2001). LR11, a mosaic LDL receptor family member, mediates the uptake of ApoE-rich lipoproteins in vitro. *Arteriosclerosis Thrombosis and Vascular Biology, 21*(9), 1501–1506.

Talwar, S., Balasubramanian, S., Sundaramurthy, S., House, R., Wilusz, C. J., Kuppuswamy, D., ... Palanisamy, V. (2013). Overexpression of RNA-binding protein CELF1 prevents apoptosis and destabilizes pro-apoptotic mRNAs in oral cancer cells. *RNA Biology, 10*(2), 277–286. http://dx.doi.org/10.4161/rna.23315.

Tanaka, N., Abe-Dohmae, S., Iwamoto, N., & Yokoyama, S. (2011). Roles of ATP-binding cassette transporter A7 in cholesterol homeostasis and host defense system. *Journal of Atherosclerosis and Thrombosis, 18*(4), 274–281.

Tan, W. H., Bird, L. M., Thibert, R. L., & Williams, C. A. (2014). If not Angelman, what is it? A review of Angelman-like syndromes. *American Journal of Medical Genetics, Part A, 164A*(4), 975–992.

Tan, M. S., Yu, J. T., & Tan, L. (2013). Bridging integrator 1 (BIN1): form, function, and Alzheimer's disease. *Trends in Molecular Medicine, 19*(10), 594–603. http://dx.doi.org/10.1016/j.molmed.2013.06.004.

Tanzi, R. E. (2012). The genetics of Alzheimer disease. *Cold Spring Harbor Perspectives in Medicine, 2*(10). http://dx.doi.org/10.1101/cshperspect.a006296.

Tanzi, R. E. (2013). A brief history of Alzheimer's disease gene discovery. *Journal of Alzheimer's Disease, 33*(Suppl. 1), S5–S13. http://dx.doi.org/10.3233/JAD-2012-129044.

Tanzi, R. E., & Bertram, L. (2005). Twenty years of the Alzheimer's disease amyloid hypothesis: a genetic perspective. *Cell, 120*(4), 545–555. http://dx.doi.org/10.1016/j.cell.2005.02.008.

Tateno, H., Li, H., Schur, M. J., Bovin, N., Crocker, P. R., Wakarchuk, W. W., & Paulson, J. C. (2007). Distinct endocytic mechanisms of CD22 (Siglec-2) and Siglec-F reflect roles in cell signaling and innate immunity. *Molecular and Cellular Biology, 27*(16), 5699–5710. http://dx.doi.org/10.1128/MCB.00383-07.

Tebar, F., Bohlander, S. K., & Sorkin, A. (1999). Clathrin assembly lymphoid myeloid leukemia (CALM) protein: localization in endocytic-coated pits, interactions with clathrin, and the impact of overexpression on clathrin-mediated traffic. *Molecular Biology of the Cell, 10*(8), 2687–2702.

Thambisetty, M., An, Y., Kinsey, A., Koka, D., Saleem, M., Guntert, A., … Resnick, S. M. (2012). Plasma clusterin concentration is associated with longitudinal brain atrophy in mild cognitive impairment. *Neuroimage, 59*(1), 212–217. http://dx.doi.org/10.1016/j.neuroimage.2011.07.056.

Thinakaran, G., & Koo, E. H. (2008). Amyloid precursor protein trafficking, processing, and function. *Journal of Biological Chemistry, 283*(44), 29615–29619. http://dx.doi.org/10.1074/jbc.R800019200.

Tibaldi, E. V., & Reinherz, E. L. (2003). CD2BP3, CIN85 and the structurally related adaptor protein CMS bind to the same CD2 cytoplasmic segment, but elicit divergent functional activities. *International Immunology, 15*(3), 313–329.

Tikhmyanova, N., Little, J. L., & Golemis, E. A. (2010). CAS proteins in normal and pathological cell growth control. *Cellular and Molecular Life Sciences, 67*(7), 1025–1048. http://dx.doi.org/10.1007/s00018-009-0213-1.

Tikhmyanova, N., Tulin, A. V., Roegiers, F., & Golemis, E. A. (2010). Dcas supports cell polarization and cell-cell adhesion complexes in development. *PLoS One, 5*(8), e12369. http://dx.doi.org/10.1371/journal.pone.0012369.

Tornillo, G., Defilippi, P., & Cabodi, S. (2014). Cas proteins: dodgy scaffolding in breast cancer. *Breast Cancer Research: BCR, 16*(5), 443.

Traynor, B. J., & Singleton, A. B. (2010). Nature versus nurture: death of a dogma, and the road ahead. *Neuron, 68*(2), 196–200. http://dx.doi.org/10.1016/j.neuron.2010.10.002.

Treusch, S., Hamamichi, S., Goodman, J. L., Matlack, K. E., Chung, C. Y., Baru, V., … Lindquist, S. (2011). Functional links between Aβ toxicity, endocytic trafficking, and Alzheimer's disease risk factors in yeast. *Science, 334*(6060), 1241–1245. http://dx.doi.org/10.1126/science.1213210.

Trowsdale, J., & Knight, J. C. (2013). Major histocompatibility complex genomics and human disease. *Annual Review of Genomics and Human Genetics, 14*, 301–323. http://dx.doi.org/10.1146/annurev-genom-091212-153455.

Urzua, B., Ortega-Pinto, A., Morales-Bozo, I., Rojas-Alcayaga, G., & Cifuentes, V. (2011). Defining a new candidate gene for amelogenesis imperfecta: from molecular genetics to biochemistry. *Biochemical Genetics, 49*(1–2), 104–121. http://dx.doi.org/10.1007/s10528-010-9392-6.

Vardarajan, B. N., Zhang, Y., Lee, J. H., Cheng, R., Bohm, C., Ghani, M., … Mayeux, R. (2015). Coding mutations in SORL1 and Alzheimer disease. *Annals of Neurology, 77*(2), 215–227. http://dx.doi.org/10.1002/ana.24305.

Velazquez, P., Cribbs, D. H., Poulos, T. L., & Tenner, A. J. (1997). Aspartate residue 7 in amyloid beta-protein is critical for classical complement pathway activation: implications for Alzheimer's disease pathogenesis. *Nature Medicine, 3*(1), 77–79.

Verghese, P. B., Castellano, J. M., Garai, K., Wang, Y., Jiang, H., Shah, A., … Holtzman, D. M. (2013). ApoE influences amyloid-β (Aβ) clearance despite minimal apoE/Aβ association in physiological conditions. *Proceedings of the National Academy of Sciences of the United States of America, 110*(19), E1807–E1816. http://dx.doi.org/10.1073/pnas.1220484110.

Vetrivel, K. S., & Thinakaran, G. (2006). Amyloidogenic processing of beta-amyloid precursor protein in intracellular compartments. *Neurology, 66*(2 Suppl. 1), S69–S73. http://dx.doi.org/10.1212/01.wnl.0000192107.17175.39.

Wagner, S. L., Tanzi, R. E., Mobley, W. C., & Galasko, D. (2012). Potential use of gamma-secretase modulators in the treatment of Alzheimer disease. *Archives of Neurology, 69*(10), 1255–1258. http://dx.doi.org/10.1001/archneurol.2012.540.

Wakabayashi, T., & De Strooper, B. (2008). Presenilins: members of the gamma-secretase quartets, but part-time soloists too. *Physiology (Bethesda), 23*, 194–204. http://dx.doi.org/10.1152/physiol.00009.2008.

Wang, Y., Cella, M., Mallinson, K., Ulrich, J. D., Young, K. L., Robinette, M. L., … Colonna, M. (2015). TREM2 lipid sensing sustains the microglial response in an Alzheimer's disease model. *Cell, 160*(6), 1061–1071. http://dx.doi.org/10.1016/j.cell.2015.01.049.

Wang, X., Lopez, O. L., Sweet, R. A., Becker, J. T., DeKosky, S. T., Barmada, M. M., … Kamboh, M. I. (2015). Genetic determinants of disease progression in Alzheimer's disease. *Journal of Alzheimer's Disease, 43*(2), 649–655. http://dx.doi.org/10.3233/JAD-140729.

Wang, L. S., Naj, A. C., Graham, R. R., Crane, P. K., Kunkle, B. W., Cruchaga, C., … Yu, L. (2015). Rarity of the Alzheimer disease-protective APP A673T variant in the United States. *JAMA Neurol, 72*(2), 209–216. http://dx.doi.org/10.1001/jamaneurol.2014.2157.

Wilkinson, D. G. (2000). Eph receptors and ephrins: regulators of guidance and assembly. *International Review of Cytology, 196*, 177–244.

Willnow, T. E., & Andersen, O. M. (2013). Sorting receptor SORLA–a trafficking path to avoid Alzheimer disease. *Journal of Cell Science, 126*(Pt 13), 2751–2760. http://dx.doi.org/10.1242/jcs.125393.

Xiao, Q., Gil, S. C., Yan, P., Wang, Y., Han, S., Gonzales, E., … Lee, J. M. (2012). Role of phosphatidylinositol clathrin assembly lymphoid-myeloid leukemia (PICALM) in intracellular amyloid precursor protein (APP) processing and amyloid plaque pathogenesis. *Journal of Biological Chemistry, 287*(25), 21279–21289. http://dx.doi.org/10.1074/jbc.M111.338376.

Xiong, W. C., Macklem, M., & Parsons, J. T. (1998). Expression and characterization of splice variants of PYK2, a focal adhesion kinase-related protein. *Journal of Cell Science, 111*(Pt 14), 1981–1991.

Xu, Z., Yoshida, T., Wu, L., Maiti, D., Cebotaru, L., & Duh, E. J. (2015). Transcription factor MEF2C suppresses endothelial cell inflammation via regulation of NF-kappaB and KLF2. *Journal of Cellular Physiology, 230*(6), 1310–1320. http://dx.doi.org/10.1002/jcp.24870.

Yamazaki, H., Bujo, H., Kusunoki, J., Seimiya, K., Kanaki, T., Morisaki, N., … Saito, Y. (1996). Elements of neural adhesion molecules and a yeast vacuolar protein sorting receptor are present in a novel mammalian low density lipoprotein receptor family member. *Journal of Biological Chemistry, 271*(40), 24761–24768.

Yamazaki, T., Masuda, J., Omori, T., Usui, R., Akiyama, H., & Maru, Y. (2009). EphA1 interacts with integrin-linked kinase and regulates cell morphology and motility. *Journal of Cell Science, 122*(Pt 2), 243–255. http://dx.doi.org/10.1242/jcs.036467.

Yerges-Armstrong, L. M., Yau, M. S., Liu, Y., Krishnan, S., Renner, J. B., Eaton, C. B., ... Jackson, R. D. (2014). Association analysis of BMD-associated SNPs with knee osteoarthritis. *Journal of Bone and Mineral Research, 29*(6), 1373–1379. http://dx.doi.org/10.1002/jbmr.2160.

Ye, F., Snider, A. K., & Ginsberg, M. H. (2014). Talin and kindlin: the one-two punch in integrin activation. *Frontiers of Medicine, 8*(1), 6–16. http://dx.doi.org/10.1007/s11684-014-0317-3.

Yoshikai, S., Sasaki, H., Doh-ura, K., Furuya, H., & Sakaki, Y. (1991). Genomic organization of the human-amyloid beta-protein precursor gene. *Gene, 102*(2), 291–292.

Yoshikawa, M., Kajiho, H., Sakurai, K., Minoda, T., Nakagawa, S., Kontani, K., & Katada, T. (2008). Tyr-phosphorylation signals translocate RIN3, the small GTPase Rab5-GEF, to early endocytic vesicles. *Biochemical and Biophysical Research, 372*(1), 168–172. http://dx.doi.org/10.1016/j.bbrc.2008.05.027.

Yu, L., Chibnik, L. B., Srivastava, G. P., Pochet, N., Yang, J., Xu, J., ... Bennett, D. A. (2015). Association of Brain DNA methylation in SORL1, ABCA7, HLA-DRB5, SLC24A4, and BIN1 with pathological diagnosis of Alzheimer disease. *JAMA Neurology, 72*(1), 15–24. http://dx.doi.org/10.1001/jamaneurol.2014.3049.

Zhou, R. (1998). The Eph family receptors and ligands. *Pharmacology and Therapeutics, 77*(3), 151–181.

Zipfel, P. F., & Skerka, C. (2009). Complement regulators and inhibitory proteins. *Nature Reviews Immunology, 9*(10), 729–740. http://dx.doi.org/10.1038/nri2620.

Zweier, M., & Rauch, A. (2012). The MEF2C-related and 5q14.3q15 microdeletion syndrome. *Molecular Syndromology, 2*(3–5), 164–170. http://dx.doi.org/10.1159/00337496.

Chapter 35

Posttraumatic Stress Disorder: From Circuits to Genes

Aliza P. Wingo, Lynn M. Almli and Kerry J. Ressler
Emory University, Atlanta, GA, United States; Harvard Medical School, Belmont, MA, United States

INTRODUCTION

Mental disorders are relatively common and can be seriously impairing in the United States as well as throughout the world. The World Health Organization has reported that mental illnesses are the leading causes of disability worldwide, at a cost of tens of billions in the United States alone and potentially in the trillions worldwide. In both our military and our most at-risk civilian populations, posttraumatic stress disorder (PTSD) is among the most prevalent and debilitating anxiety and fear-related disorders. About 7–8% of the US population will develop PTSD at some point in their lives. Over 5 million adults have PTSD during a given year. Yet this is only a small portion of those who have gone through a trauma both on the battlefield and at home, particularly in our inner cities and other areas of high trauma exposure. Furthermore, women are twice as likely to develop PTSD, and about 10% develop it sometime in their lives (Breslau et al., 1998).

The symptoms of PTSD most notably develop in the weeks to months after severe trauma, but they can last for years to decades (Boe, Holgersen, & Holen, 2011; Hull, Alexander, & Klein, 2002). PTSD symptoms include overwhelming fear, intrusive memories, avoidance of trauma reminders, and hyperarousal symptoms, among others (APA, 2013). These symptoms can be overwhelming to the victims, because they are talked about as an "emotional black hole" from which there is no escape. Among those with severe PTSD, the extreme fear resulting from past traumatic memories can often generalize to fear in other areas in their lives. For instance, they can become paranoid, have difficulty holding jobs and staying in relationships, or have erratic behavior. PTSD is a severe problem among war veterans, and it is at underrecognized, near endemic proportions in inner cities where violence, drugs, and danger are common. At a time when there is so much promise in our nation and world, it is shameful that a disorder of fear prevents progress for so many.

Translational Neuroscience and Psychiatry: Posttraumatic Stress Disorder as a Model

Scientifically, PTSD offers an ideal opportunity to study environmental and biological factors that result in the development of pathological fear responses. Because we know when the PTSD symptoms start, usually at the time of trauma exposure, they may be amenable to interventions initiated shortly after a trauma to prevent their development.

New, rationally designed treatments that derive from basic neuroscience and preclinical animal and human studies are currently under investigation. Remarkably, the same brain regions (the amygdala, hippocampus, and insula, to name a few) are involved in fear processing from mice to humans. Therefore, PTSD and other fear disorders may be among the most tractable targets in psychiatry. We are now beginning to dissect specific cell pathways functionally within small subregions of the amygdala that activate the "fear reflex" (the fight or flight response), as well as opposing circuits that act to "turn off" fear. Further understanding the molecular, cellular, and circuit mechanisms underlying fear will have huge implications for the millions of people experiencing PTSD and other fear and anxiety-related disorders.

Genomics, Circuits, and Pathways in Clinical Neuropsychiatry. http://dx.doi.org/10.1016/B978-0-12-800105-9.00035-4

Importance of Large Science Consortia and Public/Private Partnerships

The Framingham Heart Study, started in the 1940s, marked a watershed event in using large cross-sectional and prospective collaborative research to identify risk factors for cardiovascular disease, changing the prevalence, treatment, and prevention strategies for heart attacks, strokes, and related illness. Similarly, several ongoing initiatives between the National Institutes of Health and the military, including the Army STARRS project, STRONG STARR, and the Marine Resiliency Study, provide the beginning of similar large scientific cohorts in mental disorders.

From the perspective of large-scale genetics, the Psychiatric Genomics Consortia has made great strides in understanding the genetics of schizophrenia and autism through large-scale collaborations involving hundreds of thousands of patients, which is leading to exciting discovery and treatment approaches for these debilitating illnesses. We need to do the same for other disorders of the brain. We know that up to 40% of PTSD is genetically determined (True et al., 1993; Xian et al., 2000). Hence, consortia are under way to uncover the genomic architecture of PTSD through large-scale, collaborative studies. Identifying the molecular pathways underlying PTSD will lead to improved understanding, prevention, and treatment of this illness.

Targeted Interventions and Preventions, From Bench to Bedside

Every day, promising advances are being translated from basic science to the clinic, including methods to interfere with fear development after a trauma and prevent PTSD from forming initially, as well as techniques to augment therapy by normalizing the fearful memories (Bowers & Ressler, 2014; Gafford, Jasnow, & Ressler, 2014; Heldt et al., 2014; de Kleine, Rothbaum, & van Minnen, 2013; Pitts, Todorovic, Blank, & Takahashi, 2009; Rothbaum et al., 2012; Rothbaum, Kearns, et al., 2014). Invasive treatments that affect disrupted emotion circuits, such as transcranial magnetic stimulation and deep brain stimulation, are borrowed from neurosurgical interventions that attempt to regulate the known brain targets identified from neuroscience research.

Other approaches include drugs that are given at the time of specific learning events to enhance or disrupt emotional learning with talk therapy, derived from the neuroscience of learning, memory, and brain plasticity (Litz et al., 2012; Ressler et al., 2004; Rothbaum, Price, et al., 2014). Neuroimaging studies suggest that both biological treatments and emotional learning target common regions of the brain. By understanding the roles of specific molecules and cells, networks and circuits, underlying fear and emotion, it is hoped to target prevention of PTSD in the early hours on the battlefield and in emergency rooms as well as to use new, targeted, and powerful approaches to treatment and recovery. To do this, however, the field needs a greater understanding of the brain.

CLINICAL ASPECTS OF POSTTRAUMATIC STRESS DISORDER DIAGNOSIS

The *Diagnostic and Statistical Manual of Mental Disorders*, Fifth Edition (DSM-V) diagnostic criteria for PTSD include having trauma exposure followed by the development of a cluster of symptoms from each of the four categories of intrusion (at least one), avoidance (at least one), persistent negative alterations in cognition and mood (at least two), and dysregulated arousal and reactivity (at least two) (APA, 2013). Duration of these symptoms should be 1 month or longer, and the symptoms cause significant distress or social, occupational, or functional impairment (APA, 2013). The specific DSM-V criteria that are unique to PTSD are described in more detail as follows.

Criterion A represents exposure to a traumatic event that is required for a diagnosis of PTSD. Exposure can be direct, witnessed in person, or indirect by learning that a close relative or friend was exposed to trauma, or repeated or extreme direct exposure to aversive details of the event, usually in the course of professional duties (eg, first responders, collecting body parts, professionals repeatedly exposed to details of child abuse). A traumatic event is defined as being exposed to death, threatened death, actual or threatened serious injury, or actual or threatened sexual violence (APA, 2013). Examples include combat, violent personal assault, kidnapping, hostage-taking, terrorist attack, torture, incarceration as a POW or in a concentration camp, natural or man-made disasters, severe automobile accidents, or being diagnosed with a life-threatening illness.

Criterion B represents intrusion symptoms. The traumatic event is persistently reexperienced in at least one of the following ways: (1) having recurrent, involuntary, and intrusive memories; (2) having nightmares about the traumatic experience; (3) having flashbacks or dissociative reactions of the experience; (4) having intense or prolonged distress after exposure to reminders of the trauma; or (5) having marked physiological reactivity after exposure to trauma-related stimuli.

Criterion C represents the persistent effort to avoid having thoughts or feelings associated with the trauma, or to avoid trauma-related external reminders such as people, places, conversations, activities, objects, or situations.

Criterion D represents the primary addition in a diagnosis of PTSD in DSM-V relative to DSM-IV, recognizing the previously unaccounted for negative alterations in cognitions and mood that began or worsened after the traumatic event.

At least two of the following symptoms are required for the diagnosis: (1) inability to recall key features of the traumatic event (not due to head injury, alcohol, or drugs); (2) persistent and often distorted negative beliefs and expectations about oneself or the world (eg, "I'm bad," "The world is completely dangerous"); (3) persistent distorted blame of self or others for causing the traumatic event or for resulting consequences; (4) persistent negative trauma-related emotions (fear, horror, anger, guilt, or shame); (5) markedly diminished interest in (pretraumatic) significant activities; (6) feeling alienated from others (eg, detachment or estrangement); and (7) constricted affect, as reflected by the persistent inability to experience positive emotions.

Criterion E represents the trauma-related alterations in arousal or reactivity that began or worsened after the traumatic event. At least two of the following are required: (1) irritable or aggressive behavior; (2) self-destructive or reckless behavior; (3) hypervigilance; (4) exaggerated startle response; (5) problems with concentration; and (6) sleep disturbance.

NEURAL CIRCUITS OF FEAR AND EXTINCTION, AND THEIR DYSREGULATION IN POSTTRAUMATIC STRESS DISORDER

Fear Processing

The fear-related disorders arguably offer the best example of translational methods in psychiatry, in that animal models of fear learning and extinction have informed the treatment approaches, and the fear reflex in humans is remarkably similar behaviorally and neurobiologically to that in rodent models. Here, we briefly outline the neurobiology of fear and extinction to underscore the translational neuroscience basis of the therapeutic approaches for PTSD.

Decades of work have demonstrated that the amygdala is the hub of the fear "reflex" across mammals. Neural inputs representing the sensory conditioned stimuli (CS) (the previously neutral cues and contexts) and unconditioned stimuli (UCS) (the aversive cues) are combined within the lateral and basolateral nuclei of the amygdala (BLA) (Fig. 35.1A).

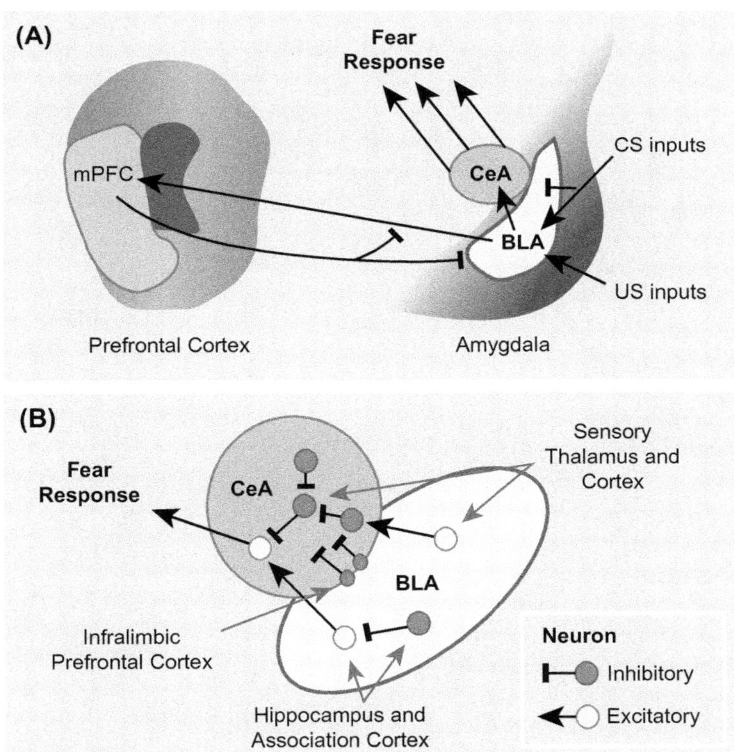

FIGURE 35.1 Schematic diagram of amygdala function. (A) The BLA and CeA nuclei of amygdala are considered the main "hub" of fear responding and activation of downstream fear reflexive behaviors and physiological reactions. The BLA receives and mediates the pairing of previously neutral CS with aversive US, so that future exposure to sensory CS alone is sufficient for the "fear reflex" behavior to occur. These processes are regulated by multiple additional brain areas, most notably the mPFC, consisting of the IL and PL cortex in rodents. (B) Within the BLA and CeA, multiple different neuronal types including excitatory and inhibitory neurons make up the microcircuitry of fear processing, that involves "fear on," "fear off," and modulatory neuronal subcircuits. *Adapted from Jovanovic, T., & Ressler, K. J. (2010). How the neurocircuitry and genetics of fear inhibition may inform our understanding of PTSD.* American Journal of Psychiatry, 67*(6), 648–662. http://dx.doi.org/10.1176/appi.ajp.2009.09071074.*

Processed information from these regions is sent to the central nucleus of the amygdala (CeA), and hard-wired outputs from the CeA are sent to multiple subcortical and brain stem areas that mediate the fear reflex reaction. In addition to receiving inputs from sensory areas, amygdala processing, expression, and extinction of fear are also modulated by contextual information from the hippocampus and regulated by medial prefrontal cortex (mPFC) projections (Fig. 35.1B). The location of amygdala, hippocampal, and prefrontal regions in the human brain and their respective roles in the fear reflex are outlined in Fig. 35.2.

The acquisition of fear involves the co-occurrence of neutral CS with aversive UCS, leading to increased conditioned fear responses such as freezing, with repetitive pairings. Extinction of fear then involves ongoing exposure to the CS stimuli in the absence of any aversive UCS signals. Drugs that enhance the extinction process or the consolidation of extinction lead to decreased memory when the CS is presented again at later times.

In contrast, reconsolidation is a process that allows prior memories to be strengthened through reminder presentations, but which also places the memory in a transient labile state by which it can again be modulated. Thus, the primary difference between extinguishing a prior aversive memory and reconsolidating that memory is whether the reminder is short

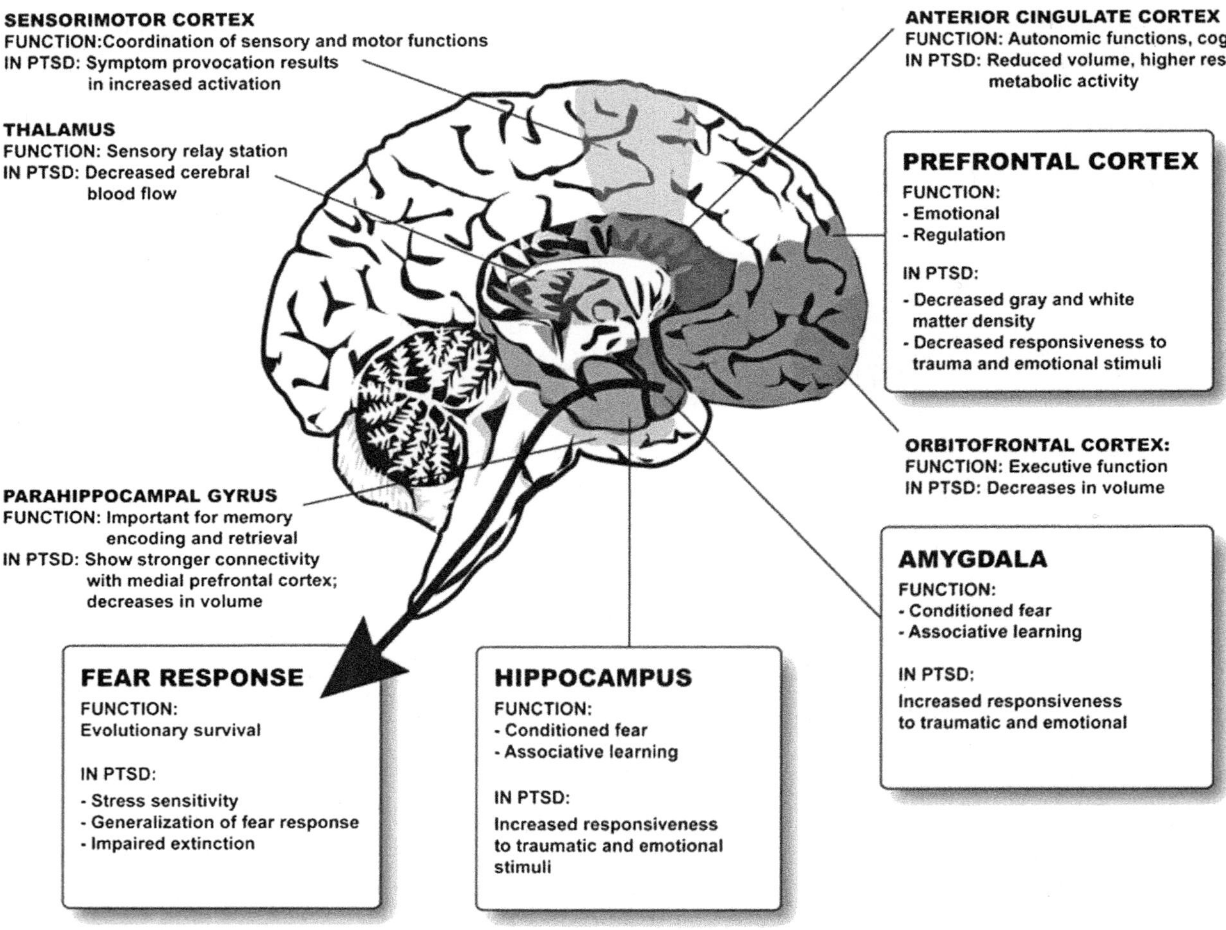

FIGURE 35.2 **Human brain schematic illustrating how the limbic system is involved in PTSD.** The prefrontal cortex (PFC) and the hippocampus both have dense connections to the amygdala, which is important for conditioned fear and associative emotional learning. The PFC is thought to be responsible for reactivating past emotional associations and is decreased in both responsiveness and density. The hippocampus is thought to have a role in explicit memories of traumatic events and in mediating learned responses to contextual cues; in PTSD, the hippocampus is decreased in volume and responsiveness to traumatic stimuli. The top-down control of the amygdala by the hippocampus and PFC might result in the increased activation of the amygdala, as is observed in subjects with PTSD. The end result of these neuroanatomical alterations is increased stress sensitivity, generalized fear responses, and impaired extinction. Other regions including the anterior cingulate cortex, orbitofrontal cortex, parahippocampal gyrus, thalamus, and sensorimotor cortex have a secondary role in the regulation of fear and PTSD. *Adapted from Mahan, A. L., & Ressler, K. J. (2012). Fear conditioning, synaptic plasticity and the amygdala: implications for posttraumatic stress disorder.* Trends in Neurosciences, *5(1), 24–35. http://dx.doi.org/10.1016/j.tins.2011.06.007.*

(reconsolidation prevails) or repetitive/long (extinction prevails) when no UCS is present. The extracellular and intracellular machinery involved in reconsolidation provide several potential targets for pharmacological manipulation. Therefore, the nascent neurobiological understanding of fear inhibition/extinction and fear enhancement/reconsolidation leads to new approaches to targeted modulation of fear memory modulation, as discussed subsequently (Singewald, Schmuckermair, Whittle, Holmes, & Ressler, 2015).

Amygdala

The amygdala (Latin for "almond") lies in the medial temporal lobe and is arguably the best understood brain region with respect to behavioral function. It has been the subject of study for several decades, from initial lesion studies to in vivo behaving electrophysiology, to inducible cell type–specific molecular genetic approaches. Together this body of work has demonstrated that the amygdala is critical for the learning and expression of many emotional behaviors, from fear to sex-aversive to -appetitive behaviors. Data suggest that amygdala function is significantly dysregulated in both fear-related disorders such as PTSD and panic disorder, as well as in substance use disorders (Fig. 35.3A,B).

Although we still generally refer to "the" amygdala in human studies, we know that at least 12 to 15 subnuclei make up the amygdaloid complex. The lateral, BLA, and central nuclei (now broken into centromedial [CeM] and centrolateral [CeL], among other subdivisions) are the areas most well understood from the perspective of fear memory formation and

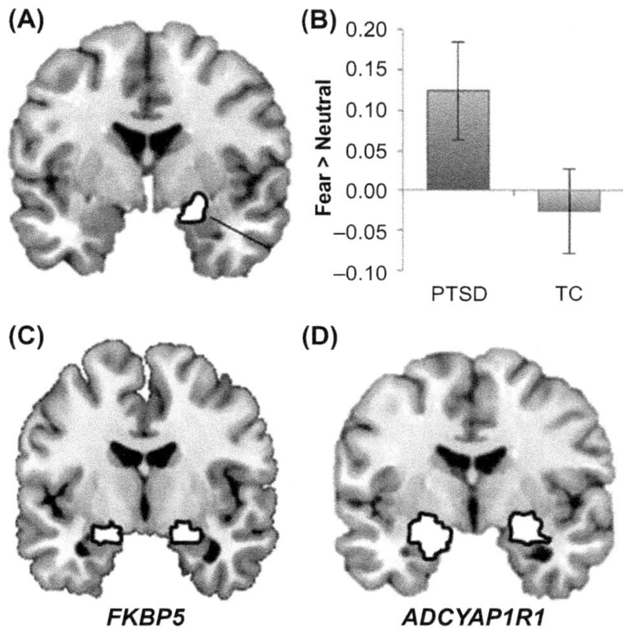

FIGURE 35.3 Amygdala hyperactivation in PTSD and as an intermediate phenotype for genetic association studies of PTSD-related polymorphisms. (A) Increased right amygdala response to fearful stimuli in the PTSD group, relative to the traumatized control (TC) group ($P_{corr} < 0.05$). Results are displayed in neurological orientation on a representative single-subject template brain in MNI space. *(Adapted from Stevens, J. S., Jovanovic, T., Fani, N., Ely, T. D., Glover, E. M., Bradley, B., & Ressler, K. J. (2013). Disrupted amygdala-prefrontal functional connectivity in civilian women with posttraumatic stress disorder.* Journal of Psychiatric Research, *47(10), 1469–1478. http://dx.doi.org/10.1016/j.jpsychires.2013.05.031.)* (B) Bar graph shows the mean contrast estimate across voxels in the right amygdala cluster, for the Fear > Neutral contrast, and error bars show standard error of the mean. *(Adapted from Stevens, J. S., Jovanovic, T., Fani, N., Ely, T. D., Glover, E. M., Bradley, B., & Ressler, K. J. (2013). Disrupted amygdala-prefrontal functional connectivity in civilian women with posttraumatic stress disorder.* Journal of Psychiatric Research, *47(10), 1469–1478. http://dx.doi.org/10.1016/j.jpsychires.2013.05.031.)* (C) In a healthy control population, *FKBP5* risk genotypes were found to be associated with increased dorsal amygdala reactivity in the context of higher emotional neglect compared with those without the risk alleles. *(Adapted from White, M. G., Bogdan, R., Fisher, P. M., Muñoz, K. E., Williamson, D. E., & Hariri, A. R. (2012). FKBP5 and emotional neglect interact to predict individual differences in amygdala reactivity.* Genes, Brain and Behavior, *11(7), 869–878. http://dx.doi.org/10.1111/j.1601-183X.2012.00837.x.)* (D) Forty-nine women who had experienced moderate to high levels of lifetime trauma participated in a functional MRI task involving passive viewing of threatening and neutral face stimuli. The risk genotype from the ADCYAP1R1, rs2267735, was associated with increased reactivity of the amygdala to threat stimuli. *(Adapted from Stevens, J. S., Almli, L. M., Fani, N., Gutman, D. A., Bradley, B., Norrholm, S. D., … Ressler, K. J. (2014). PACAP receptor gene polymorphism impacts fear responses in the amygdala and hippocampus.* Proceedings of the National Academy of Sciences of the United States of America, *111(8), 3158–3163. http://dx.doi.org/10.1073/pnas.1318954111.)*

expression. From a functional perspective, the lateral and BLA nuclei encompass a number of excitatory, pyramidal output neurons that can be divided, at least, into "fear on" versus "fear off" populations, which are thought to represent the learned fear versus learned inhibition/extinction pathways differentially. The overlay of how the appetitive-responding neurons interact/overlay with the aversive ones is not yet well understood. Information from the BLA is then routed through the CeL, where it is further processed by "fear" and "extinction" neurons and then to the CeM, where many of the hardwired outputs are processed through to the brain stem and other subcortical circuitry mediating the hardwired fear response (Fig. 35.1B).

Medial Prefrontal Cortex

Although long expected to be involved in the regulation of subcortical emotion processing, more specific roles of the mPFC have begun to be dissected through modern neuroscience approaches. With regard to fear expression and inhibition, the most robust data have come from the observation that the mPFC in rodents can be structurally and functionally divided between the prelimbic (PL) and infralimbic (IL) subregions. In humans, these regions are thought to be functionally similar to the dorsal anterior cingulate cortex (possibly analogous with PL) and the pregenual/subgenual cortex (sGC) (possibly analogous with IL) (Fig. 35.2). Initially the IL was found to be associated with inhibiting the fear response, and activation of IL was mechanistically shown to inhibit fear and enhance extinction. These rodent data are complementary with human data showing that the ventral mPFC/pregenual/sGC regions are inversely correlated with amygdala activation at the time of fear expression; furthermore, they appear to show decreased activation with regard to PTSD symptoms at a time when amygdala is hyperactivated. In rodents, the IL is also thought to activate specifically the intercalated cell masses between the BTL and CeM nuclei, which are thought to act as a "gate" between BLA processing and CeM activation. These findings provide a relatively specific and precise model for how mPFC regulation of information processing leads to amygdala-mediated inhibition of fear processing. Overall, most data in this area are consistent with a model in which PL is coactivated with amygdala for expression of learned fear and trauma-related PTSD symptoms, whereas IL is involved in amygdala-dependent fear expression, extinction of fear, and resilience in the aftermath of trauma.

Hippocampus

The hippocampus remains the most fully studied brain region relative to plasticity as well as mood and anxiety disorders. With regard to PTSD, its symptoms have been associated with decreased hippocampal volume among several studies focusing on childhood trauma and adult PTSD (Carballedo et al., 2013; Everaerd et al., 2012; Gatt et al., 2009). Whether the smaller hippocampal volume occurs as a result of PTSD or is a risk factor for PTSD remains unclear. The most direct study examining this topic included monozygotic twins, one of whom was a veteran with combat exposure; the other was a veteran without combat exposure (Gilbertson et al., 2007). The findings suggested that PTSD was associated with smaller hippocampal volume in both twins, including the twin with combat exposure and PTSD as well as his non–trauma exposed twin. These data were interpreted as evidence that genetic factors may contribute to smaller hippocampal volume, which may then be a risk factor for PTSD development after trauma. These data are consistent with animal studies showing that both hippocampal lesions and deletions of the *bdnf* neural plasticity gene are associated with decreases in fear extinction, and thus overrepresentation of fear memories, phenotypes associated with PTSD (Andero & Ressler, 2012). In addition, decreases in working memory and trauma narrative processing, both of which likely highly depend on hippocampal function, are associated with PTSD (Cisler et al., 2014; Gatt et al., 2009).

GENETIC STUDIES OF POSTTRAUMATIC STRESS DISORDER

Candidate Gene Studies

As with most areas, PTSD genetics were initially dominated by candidate gene studies, which are hypothesis-driven investigations that probe scientific questions formulated from our existing understanding of the neurobiology of PTSD. As with many other candidate gene studies, they were often underpowered, did not require replication, and were subject to file-drawer effect problems with publication. Findings so far have suggested a complex interaction between genetic and environmental factors in the manifestation of PTSD. We next briefly review genetic variants in PTSD that are categorized in the hypothalamic–pituitary–adrenal (HPA), adrenergic, serotonergic, and dopaminergic systems.

FK506 Binding Protein 5

PTSD is characterized by dysregulation of the stress response system such that activity of the HPA axis is altered, presumably through enhanced sensitivity of the glucocorticoid receptor-mediated feedback mechanism that suppresses stress-induced cortisol release (van Zuiden, Kavelaars, Geuze, Olff, & Heijnen, 2012). Consistently, four single-nucleotide polymorphisms (SNPs) in the FK506 binding protein 5 (FKBP5) gene have been observed to interact with the severity of childhood trauma in predicting PTSD symptom severity in adults (Binder et al., 2008; Klengel et al., 2013; Mehta et al., 2011; Xie et al., 2010). FKBP5 is a co-chaperone of hsp90, which binds to the glucocorticoid receptor (GR). In addition, FKBP5 is part of the mature GR heterocomplex and regulates GR sensitivity (Scammell, Denny, Valentine, & Smith, 2001). One of these four SNPs, rs1360780, was found to increase the risk of developing stress-related psychiatric disorders in adulthood, likely through changes in DNA methylation as a consequence of childhood trauma–dependent stress in a gene-specific fashion (Klengel et al., 2013). Consistently, individuals with FKBP5 risk genotypes were found to have increased amygdala reactivity to threat stimuli compared with those without the risk alleles (Fig. 35.3C), (White et al., 2012).

ADCYAP1R1

Also regulating the stress response and HPA axis is the neuropeptide pituitary adenylate cyclase-activating polypeptide (ADCYAP1), which functions in parts of the brain that mediate anxiety and fear-related behaviors. A genetic variant in the PAC1 receptor (*ADCYAP1R1*; rs2267735) that disrupts a putative estrogen response element has been found to be associated with PTSD in a primarily African American cohort of women (Ressler et al., 2011). Furthermore, PTSD associates with alterations in peripheral blood DNA methylation and messenger RNA expression of the *ADCYAP1R1* transcript. Although the association with the *ADCYAP1R1* variant was not replicated among presumably less traumatized African American and Caucasian populations (Chang, Xie, et al., 2012), an interaction between *ADCYAP1R1* genotype and childhood maltreatment predicted PTSD in females (Uddin et al., 2012).

The ADCYAP1R1 risk genotype was also associated with increased reactivity of the amygdala and hippocampus to threat stimuli and decreased functional connectivity between the amygdala and hippocampus (Fig. 35.3D; Stevens et al., 2014). Taken together, differences in *ADCYAP1R1* genotype may contribute to dysregulated fear circuitry known to have a central role in PTSD and other anxiety disorders.

Monoaminergic System Candidate Genes

Several preclinical and clinical studies suggested contributions of adrenergic and noradrenergic abnormalities to the symptomatology of patients with PTSD, particularly to physiological reactivity upon encountering reminders and hyperarousal symptoms (Southwick et al., 1999). However, only recently has direct evidence of genetic variance in noradrenergic and adrenergic function been shown to contribute to symptom severity of PTSD via the gene × environment approach (Liberzon et al., 2014). An SNP in the promoter region of the β2-adrenergic receptor (*ADRB2*) gene, rs2400707, was found to interact with childhood trauma to predict adult symptom severity of PTSD in a discovery cohort of 810 primarily male soldiers of European ancestry and was replicated in predominantly female African American, inner-city civilians in the Grady Trauma Project (Liberzon et al., 2014).

Dysregulation of brain serotonergic system has been implicated in the pathophysiology of PTSD (Anger et al., 1999; Southwick, Bremner, Krystal, & Charney, 1994). As the serotonin transporter reuptakes serotonin at the brain synapses, a tandem repeat polymorphic region in the gene that encodes this serotonin transporter protein, SLC6A4, called the serotonin transporter–linked polymorphic region (5-HTTLPR), has received much attention in PTSD studies. The *5-HTTLPR* polymorphism contains two alleles, L (long) and S (short), and the short allele is associated with reduced serotonin transporter gene expression and function leading to reduced serotonin uptake (Lesch et al., 1996). Furthermore, within the long fragment an additional common SNP exists, which makes 5-HTTLPR triallelic with S, L_A, and L_G alleles. Several studies observed an interaction between the 5-HTTLPR polymorphism and trauma load in predicting PTSD prevalence (Grabe et al., 2009; Kilpatrick et al., 2007; Koenen et al., 2009; Kolassa, Ertl, et al., 2010; Mercer et al., 2012; Wang et al., 2011; Xie, Kranzler, Farrer, & Gelernter, 2012; Xie et al., 2009).

Dysregulation of dopaminergic neurotransmission has also been implicated in the pathophysiology of PTSD (Hamner & Diamond, 1993; Yehuda, Southwick, Giller, Ma, & Mason, 1992). Solute carrier family 6, member 3 (SLC6A3, also known as DAT or DAT1) is a gene that encodes the dopamine transporter which is a member of the neurotransmitter transporter family. The 3′ untranslated region of SLC6A3 has a 40–base pair tandem repeat. Segman and colleagues observed a significant excess of nine repeat alleles of SLC6A3 3′ variable number tandem repeat among patients with PTSD compared with control subjects (Segman et al., 2002). The association of the nine-repeat allele of DAT with PTSD was replicated in a sample of preschool children (Drury, Theall, Keats, & Scheeringa, 2009), a sample of 369 Brazilian adults (Valente, Vallada, Cordeiro, Miguita, et al., 2011), and another sample of 320 adults recruited from Detroit, United States (Chang, Koenen, et al., 2012).

Polymorphism of the dopamine receptor D2 (*DRD2*), specifically its functional SNP rs1800497, has been examined in PTSD in small samples, with inconsistent results. The rs1800497 SNP was associated with PTSD in a sample of 56 Vietnam veterans (Comings, Muhleman, & Gysin, 1996) and with PTSD and heavy alcohol consumption in another 142 adults (Young et al., 2002), but not in 139 European American adults (Gelernter et al., 1999) or 200 adult survivors of an earthquake (Bailey et al., 2010).

Catechol-*O*-methyltransferase (COMT) is an enzyme that is involved in the breakdown of the catecholamine neurotransmitters, including dopamine, norepinephrine, and epinephrine. COMT contains a functional polymorphism at codon 158 (rs4680), which substitutes the amino acid valine (Val) for methionine (Met). This Val158Met polymorphism is associated with a significant reduction in COMT enzyme activity (Lachman et al., 1996) and the decreased ability to extinguish conditioned fear (Lonsdorf et al., 2009), a putative trait of PTSD. A study of 434 Brazilian adults recruited from an area of high rates of urban violence found a significant association between the Met158 allele and PTSD (Valente, Vallada, Cordeiro, Bressan, et al., 2011). In addition, in a population of 424 Rwanda refugees, the Met158 allele carriers were found to interact with trauma load in determining the risk of developing PTSD (Kolassa, Kolassa, Ertl, Papassotiropoulos, & De Quervain, 2010).

Genome-Wide Association Studies

Although most PTSD genetic research has been candidate gene studies, there is a gradual shift to genome-wide association study (GWAS). By its nature, GWAS employs a hypothesis neutral approach to identify disease-associated variants independently of what is known about PTSD pathophysiology, and thus can uncover novel genes or molecular pathways. As a caveat, for complex phenotypes such as PTSD, the sample size required to find at least one genome-wide significant finding is at least 11,000 (Sullivan, Daly, & O'Donovan, 2012). So far, there have been a handful of GWAS of PTSD, and most of these had relatively small sample sizes. Here we briefly describe the existing GWAS of PTSD and the collaborative Psychiatric Genomics Consortium—PTSD Working Group toward achieving a large sample size to obtain sufficient power and yield replicable results in GWAS of PTSD.

Retinoid-Related Orphan Receptor Alpha

The first GWAS of PTSD was performed in a sample of trauma-exposed, white, non-Hispanic veterans and their intimate partners, consisting of 295 patients and 196 control subjects (Logue et al., 2012). It found one SNP, rs8042149, located on chromosome 15 in the intronic region of the retinoid-related orphan receptor alpha gene (RORA) associated with PTSD at genome-wide significance level ($P = 2.5 \times 10^{-8}$). The risk allele for rs8042149 was associated with a two times higher risk for having a lifetime diagnosis of PTSD. Although this SNP was not significantly associated with PTSD in two independent replication samples, several SNPs in the RORA gene were (Logue et al., 2012). The protein encoded by RORA is involved in a variety of processes including brain development, neuroprotection, and regulation of circadian rhythms and steroid hormones (Logue et al., 2012). In addition, RORA had been found to be associated with depression (Garriock et al., 2010), attention-deficit hyperactivity disorder (Neale et al., 2008), bipolar disorder (Le-Niculescu et al., 2009), and autism (Nguyen, Rauch, Pfeifer, & Hu, 2010) in prior genetic studies.

COBL

The second GWAS identified an SNP on chromosome 7p12, rs406001, that is associated with PTSD at a genome-wide significance level of $P = 3.97 \times 10^{-8}$ in 1578 European Americans (Xie et al., 2013). As the first study, this association was not significant in an African American replication sample ($N = 744$) or European American sample ($N = 1658$). However, in another, larger African American cohort of 3000 samples, there was a significant gene × environment effect for rs406001 (and other SNPs in linkage disequilibrium with it), in which rs406001 interacted with childhood trauma exposure in predicting PTSD symptom severity (Almli et al., 2014).

rs406001 is an intergenic SNP with no known function. Its closest gene is *COBL*, which may be related to actin polymerization and neuronal development and function. In the replication study by Almli et al., brain imaging findings suggested that carriers of the risk allele had poorer white matter integrity in brain regions associated with emotion processing, including left inferior-fronto-occipital fasciculus, left inferior longitudinal fasciculus, white matter in the left frontal orbital cortex, and left uncinate fasciculus (Almli et al., 2014). Notably, the uncinate fasciculus has been shown to serve as a primary connection between the amygdala and ventral aspects of the prefrontal cortex (See Fig. 35.2), which is thought to have a role in the extinction of learned fear (Maren, Phan, & Liberzon, 2013; Phelps, Delgado, Nearing, & LeDoux, 2004). Thus, it is possible that carriers of the rs406001 risk allele are more vulnerable to the development of anxious psychopathology via decrements in these white matter pathways.

Long Intergenic Noncoding RNA

The third GWAS of PTSD was carried out in a sample of African American women from the Detroit Neighborhood Health Study, consisting of 94 patients with PTSD and 319 control subjects (Guffanti et al., 2013). A marker, rs10170218, located within *AC067818.1,* a novel long intergenic noncoding RNA gene located on chromosome 2, was associated with PTSD at $P = 5.1 \times 10^{-8}$. However, this marker was only marginally significant in the female replication sample of 578 patients and 1963 control subjects ($P = 0.07$) (Guffanti et al., 2013). Network-based analysis of the top GWAS loci found enrichment for pathways related to telomere maintenance and immune function (Guffanti et al., 2013). These results suggest that noncoding RNA genes may have a role in risk and resilience for PTSD.

PRTFDC1

Nievergelt and colleagues presented results from a GWAS of PTSD among 3494 combat-exposed US Marines and sailors from the Marine Resiliency Study scheduled for deployment to Iraq and/or Afghanistan (Nievergelt et al., 2015). In this prospective, longitudinal study, an SNP, rs6482463, located in the intronic region of the phosphoribosyl transferase domain containing one gene (*PRTFDC1*) was found to be significantly associated with PTSD (odds ratio, 1.47; standard error, 0.06; $P = 2.04 \times 10^{-9}$). This association was not significant, however, in a veteran replication sample of 491. PRTFDC1 is a small protein with high expression in the brain (Nievergelt et al., 2015). This GWAS was notable for including participants of various ethnic groups such as European, African, Native American/Hispanic, and other ancestries.

PGC Consortium

Overall, to date, the GWAS are consistent with the notion that the genetic architecture of PTSD is likely determined by many SNPs with small effects, and overlap with other neuropsychiatric disorders, consistent with findings from large GWAS of other psychiatric disorders. Although these candidate gene and GWAS studies of PTSD have provided valuable insights in determining genetic risk for PTSD, studies with larger sample size are needed to provide sufficient power to detect more genetic variants for the complex phenotype of PTSD. In fact, a Psychiatric Genomics Consortium—PTSD Working Group has been established to combine genetic, epigenetic, neuroimaging, and other neurobiological measures related to PTSD from investigators all over the world. A GWAS of PTSD in over 40,000 subjects is under way in civilian and veteran cohorts (Koenen, Duncan, Liberzon, & Ressler, 2013; Logue et al., 2015).

CONCLUSIONS

As with many psychiatric disorders, risks for PTSD are both genetic and environmentally based. So far, studies on genetics of PTSD have largely relied on a candidate gene approach, and reported GWAS of PTSD have relatively small sample sizes. New developments in large-scale GWAS, whole-genome sequencing, copy number variants, epigenetics, and gene expression will allow for integrative and convergent genomic approaches to uncover gene pathways further in PTSD in a hypothesis-neutral manner. Identifying and understanding the gene pathways underlying the manifestation of PTSD symptoms, and integrating them with a neural circuitry that is being rapidly dissected, will provide insights into our efforts to treat and prevent PTSD.

ACKNOWLEDGMENT

Support was provided by National Institutes of Health (NIH) (grants R01MH071537, R01MH096764, and 1R01MH094757), the Burroughs Wellcome Fund, NIH/National Center for Research Resources base grant P51RR000165 to Yerkes National Primate Research Center, and by the Department of Veterans Affairs Career Development Award IK2CX000601.

DISCLOSURES

Dr. Ressler is a founding member of Extinction Pharmaceuticals/Therapade Technologies, which exist to develop D-Cycloserine for use to augment the effectiveness of psychotherapy. He has received no equity or income from this relationship within the past 3 years.

REFERENCES

Almli, L. M., Srivastava, A., Fani, N., Kerley, K., Mercer, K. B., Feng, H., … Ressler, K. J. (2014). Follow-up and extension of a prior genome-wide association study of posttraumatic stress disorder: gene x environment associations and structural magnetic resonance imaging in a highly traumatized African-American civilian population. *Biological Psychiatry, 76*(4), e3–e4. http://dx.doi.org/10.1016/j.biopsych.2014.01.017.

Andero, R., & Ressler, K. J. (2012). Fear extinction and BDNF: translating animal models of PTSD to the clinic. *Genes Brain and Behavior, 11*(5), 503–512. http://dx.doi.org/10.1111/j.1601-183X.2012.00801.x.

Anger, W. K., Storzbach, D., Binder, L. M., Campbell, K. A., Rohlman, D. S., McCauley, L., … Davis, K. L. (1999). Neurobehavioral deficits in Persian Gulf veterans: evidence from a population-based study. Portland Environmental Hazards Research Center. *Journal of the International Neuropsychological Society, 5*(3), 203–212.

APA. (2013). *Diagnostic and statistical manual of mental disorders: DSM-5* (5th ed.). Arlington, VA: American Psychiatric Association.

Bailey, J. N., Goenjian, A. K., Noble, E. P., Walling, D. P., Ritchie, T., & Goenjian, H. A. (2010). PTSD and dopaminergic genes, DRD2 and DAT, in multigenerational families exposed to the Spitak earthquake. *Psychiatry Research, 178*(3), 507–510. http://dx.doi.org/10.1016/j.psychres.2010.04.043.

Binder, E. B., Bradley, R. G., Liu, W., Epstein, M. P., Deveau, T. C., Mercer, K. B., … Ressler, K. J. (2008). Association of FKBP5 polymorphisms and childhood abuse with risk of posttraumatic stress disorder symptoms in adults. *JAMA : The Journal of the American Medical Association, 299*(11), 1291–1305. http://dx.doi.org/10.1001/jama.299.11.1291.

Boe, H. J., Holgersen, K. H., & Holen, A. (2011). Mental health outcomes and predictors of chronic disorders after the North Sea oil rig disaster: 27-year longitudinal follow-up study. *Journal of Nervous and Mental Disease, 199*(1), 49–54. http://dx.doi.org/10.1097/NMD.0b013e31820446a8. pii:00005053-201101000-00009.

Bowers, M. E., & Ressler, K. J. (2014). Interaction between the cholecystokinin and endogenous cannabinoid systems in cued fear expression and extinction retention. *Neuropsychopharmacology, 40*(3), 688–700. http://dx.doi.org/10.1038/npp.2014.225.

Breslau, N., Kessler, R. C., Chilcoat, H. D., Schultz, L. R., Davis, G. C., & Andreski, P. (1998). Trauma and posttraumatic stress disorder in the community: the 1996 Detroit Area Survey of Trauma. *Archives of General Psychiatry, 55*(7), 626–632.

Carballedo, A., Morris, D., Zill, P., Fahey, C., Reinhold, E., Meisenzahl, E., … Frodl, T. (2013). Brain-derived neurotrophic factor Val66Met polymorphism and early life adversity affect hippocampal volume. *American Journal of Medical Genetics Part B Neuropsychiatric Genetics, 162b*(2), 183–190. http://dx.doi.org/10.1002/ajmg.b.32130.

Chang, S. C., Koenen, K. C., Galea, S., Aiello, A. E., Soliven, R., Wildman, D. E., & Uddin, M. (2012). Molecular variation at the SLC6A3 locus predicts lifetime risk of PTSD in the Detroit Neighborhood Health Study. *PLoS One, 7*(6), e39184. http://dx.doi.org/10.1371/journal.pone.0039184.

Chang, S. C., Xie, P., Anton, R. F., De Vivo, I., Farrer, L. A., Kranzler, H. R., … Koenen, K. C. (2012). No association between ADCYAP1R1 and post-traumatic stress disorder in two independent samples. *Molecular Psychiatry, 17*(3), 239–241. http://dx.doi.org/10.1038/mp.2011.118.

Cisler, J. M., Steele, J. S., Lenow, J. K., Smitherman, S., Everett, B., Messias, E., & Kilts, C. D. (2014). Functional reorganization of neural networks during repeated exposure to the traumatic memory in posttraumatic stress disorder: an exploratory fMRI study. *Journal of Psychiatric Research, 48*(1), 47–55. http://dx.doi.org/10.1016/j.jpsychires.2013.09.013.

Comings, D. E., Muhleman, D., & Gysin, R. (1996). Dopamine D2 receptor (DRD2) gene and susceptibility to posttraumatic stress disorder: a study and replication. *Biological Psychiatry, 40*(5), 368–372. http://dx.doi.org/10.1016/0006-3223(95)00519-6.

Drury, S. S., Theall, K. P., Keats, B. J., & Scheeringa, M. (2009). The role of the dopamine transporter (DAT) in the development of PTSD in preschool children. *Journal of Traumatic Stress, 22*(6), 534–539. http://dx.doi.org/10.1002/jts.20475.

Everaerd, D., Gerritsen, L., Rijpkema, M., Frodl, T., van Oostrom, I., Franke, B., … Tendolkar, I. (2012). Sex modulates the interactive effect of the serotonin transporter gene polymorphism and childhood adversity on hippocampal volume. *Neuropsychopharmacology, 37*(8), 1848–1855. http://dx.doi.org/10.1038/npp.2012.32.

Gafford, G., Jasnow, A. M., & Ressler, K. J. (2014). Grin1 receptor deletion within CRF neurons enhances fear memory. *PLoS One, 9*(10), e111009. http://dx.doi.org/10.1371/journal.pone.0111009.

Garriock, H. A., Kraft, J. B., Shyn, S. I., Peters, E. J., Yokoyama, J. S., Jenkins, G. D., … Hamilton, S. P. (2010). A genomewide association study of citalopram response in major depressive disorder. *Biological Psychiatry, 67*(2), 133–138. http://dx.doi.org/10.1016/j.biopsych.2009.08.029.

Gatt, J. M., Nemeroff, C. B., Dobson-Stone, C., Paul, R. H., Bryant, R. A., Schofield, P. R., … Williams, L. M. (2009). Interactions between BDNF Val66Met polymorphism and early life stress predict brain and arousal pathways to syndromal depression and anxiety. *Molecular Psychiatry, 14*(7), 681–695.

Gelernter, J., Southwick, S., Goodson, S., Morgan, A., Nagy, L., & Charney, D. S. (1999). No association between D2 dopamine receptor (DRD2) "A" system alleles, or DRD2 haplotypes, and posttraumatic stress disorder. *Biological Psychiatry, 45*(5), 620–625.

Gilbertson, M. W., Williston, S. K., Paulus, L. A., Lasko, N. B., Gurvits, T. V., Shenton, M. E., … Orr, S. P. (2007). Configural cue performance in identical twins discordant for posttraumatic stress disorder: theoretical implications for the role of hippocampal function. *Biological Psychiatry, 62*(5), 513–520. http://dx.doi.org/10.1016/j.biopsych.2006.12.023.

Grabe, H. J., Spitzer, C., Schwahn, C., Marcinek, A., Frahnow, A., Barnow, S., … Rosskopf, D. (2009). Serotonin transporter gene (SLC6A4) promoter polymorphisms and the susceptibility to posttraumatic stress disorder in the general population. *The American Journal of Psychiatry, 166*(8), 926–933. http://dx.doi.org/10.1176/appi.ajp.2009.08101542.

Guffanti, G., Galea, S., Yan, L., Roberts, A. L., Solovieff, N., Aiello, A. E., … Koenen, K. C. (2013). Genome-wide association study implicates a novel RNA gene, the lincRNA AC068718.1, as a risk factor for post-traumatic stress disorder in women. *Psychoneuroendocrinology, 38*(12), 3029–3038. http://dx.doi.org/10.1016/j.psyneuen.2013.08.014.

Hamner, M. B., & Diamond, B. I. (1993). Elevated plasma dopamine in posttraumatic stress disorder: a preliminary report. *Biological Psychiatry, 33*(4), 304–306.

Heldt, S. A., Zimmermann, K., Parker, K., Gaval, M., Weinshenker, D., & Ressler, K. J. (2014). BDNF deletion or TrkB impairment in amygdala inhibits both appetitive and aversive learning. *Journal of Neuroscience, 34*(7), 2444–2450. http://dx.doi.org/10.1523/jneurosci.4085-12.2014.

Hull, A. M., Alexander, D. A., & Klein, S. (2002). Survivors of the Piper Alpha oil platform disaster: long-term follow-up study. *The British Journal of Psychiatry, 181*(5), 433–438. http://dx.doi.org/10.1192/bjp.181.5.433.

Jovanovic, T., & Ressler, K. J. (2010). How the neurocircuitry and genetics of fear inhibition may inform our understanding of PTSD. *American Journal of Psychiatry, 67*(6), 648–662. http://dx.doi.org/10.1176/appi.ajp.2009.09071074.

Kilpatrick, D. G., Koenen, K. C., Ruggiero, K. J., Acierno, R., Galea, S., Resnick, H. S., … Gelernter, J. (2007). The serotonin transporter genotype and social support and moderation of posttraumatic stress disorder and depression in hurricane-exposed adults. *The American Journal of Psychiatry, 164*(11), 1693–1699.

de Kleine, R. A., Rothbaum, B. O., & van Minnen, A. (2013). Pharmacological enhancement of exposure-based treatment in PTSD: a qualitative review. *European Journal of Psychotraumatology, 4*. http://dx.doi.org/10.3402/ejpt.v4i0.21626.

Klengel, T., Mehta, D., Anacker, C., Rex-Haffner, M., Pruessner, J. C., Pariante, C. M., … Binder, E. B. (2013). Allele-specific FKBP5 DNA demethylation mediates gene-childhood trauma interactions. *Nature Neuroscience, 16*(1), 33–41. http://dx.doi.org/10.1038/nn.3275.

Koenen, K. C., Aiello, A. E., Bakshis, E., Amstadter, A. B., Ruggiero, K. J., Acierno, R., … Galea, S. (2009). Modification of the association between serotonin transporter genotype and risk of posttraumatic stress disorder in adults by County-level social environment. *American Journal of Epidemiology, 169*(6), 704–711.

Koenen, K. C., Duncan, L. E., Liberzon, I., & Ressler, K. J. (2013). From candidate genes to genome-wide association: the challenges and promise of posttraumatic stress disorder genetic studies. *Biological Psychiatry, 74*(9), 634–636. http://dx.doi.org/10.1016/j.biopsych.2013.08.022.

Kolassa, I. T., Ertl, V., Eckart, C., Glockner, F., Kolassa, S., Papassotiropoulos, A., … Elbert, T. (2010). Association study of trauma load and SLC6A4 promoter polymorphism in posttraumatic stress disorder: evidence from survivors of the Rwandan Genocide. *Journal of Clinical Psychiatry, 71*(5), 543–547.

Kolassa, I. T., Kolassa, S., Ertl, V., Papassotiropoulos, A., & De Quervain, D. J. (2010). The risk of posttraumatic stress disorder after trauma depends on traumatic load and the catechol-o-methyltransferase Val(158)Met polymorphism. *Biological Psychiatry, 67*(4), 304–308. http://dx.doi.org/10.1016/j.biopsych.2009.10.009.

Lachman, H. M., Papolos, D. F., Saito, T., Yu, Y. M., Szumlanski, C. L., & Weinshilboum, R. M. (1996). Human catechol-O-methyltransferase pharmacogenetics: description of a functional polymorphism and its potential application to neuropsychiatric disorders. *Pharmacogenetics, 6*(3), 243–250.

Le-Niculescu, H., Patel, S. D., Bhat, M., Kuczenski, R., Faraone, S. V., Tsuang, M. T., … Niculescu, A. B., 3rd (2009). Convergent functional genomics of genome-wide association data for bipolar disorder: comprehensive identification of candidate genes, pathways and mechanisms. *American Journal of Medical Genetics Part B, Neuropsychiatric Genetics: The Official Publication of the International Society of Psychiatric Genetics, 150B*(2), 155–181. http://dx.doi.org/10.1002/ajmg.b.30887.

Lesch, K. P., Bengel, D., Heils, A., Sabol, S. Z., Greenberg, B. D., Petri, S., … Murphy, D. L. (1996). Association of anxiety-related traits with a polymorphism in the serotonin transporter gene regulatory region. *Science, 274*(5292), 1527–1531.

Liberzon, I., King, A. P., Ressler, K. J., Almli, L. M., Zhang, P., Ma, S. T., … Galea, S. (2014). Interaction of the ADRB2 gene polymorphism with childhood trauma in predicting adult symptoms of posttraumatic stress disorder. *JAMA Psychiatry, 71*(10), 1174–1182. http://dx.doi.org/10.1001/jamapsychiatry.2014.999.

Litz, B. T., Salters-Pedneault, K., Steenkamp, M. M., Hermos, J. A., Bryant, R. A., Otto, M. W., & Hofmann, S. G. (2012). A randomized placebo-controlled trial of D-cycloserine and exposure therapy for posttraumatic stress disorder. *Journal of Psychiatric Research, 46*(9), 1184–1190. http://dx.doi.org/10.1016/j.jpsychires.2012.05.006.

Logue, M. W., Amstadter, A. B., Baker, D. G., Duncan, L., Koenen, K. C., Liberzon, I., … Uddin, M. (2015). The Psychiatric Genomics Consortium Posttraumatic Stress Disorder Workgroup: Posttraumatic Stress Disorder Enters the Age of Large-Scale Genomic Collaboration. *Neuropsychopharmacology, 40*(10), 2287–2297. http://dx.doi.org/10.1038/npp.2015.118.

Logue, M. W., Baldwin, C., Guffanti, G., Melista, E., Wolf, E. J., Reardon, A. F., … Miller, M. W. (2012). A genome-wide association study of post-traumatic stress disorder identifies the retinoid-related orphan receptor alpha (RORA) gene as a significant risk locus. *Molecular Psychiatry, 18*(8), 937–942. http://dx.doi.org/10.1038/mp.2012.113.

Lonsdorf, T. B., Weike, A. I., Nikamo, P., Schalling, M., Hamm, A. O., & Ohman, A. (2009). Genetic gating of human fear learning and extinction: possible implications for gene-environment interaction in anxiety disorder. *Psychological Science, 20*(2), 198–206. http://dx.doi.org/10.1111/j.1467-9280.2009.02280.x.

Mahan, A. L., & Ressler, K. J. (2012). Fear conditioning, synaptic plasticity and the amygdala: implications for posttraumatic stress disorder. *Trends in Neurosciences, 5*(1), 24–35. http://dx.doi.org/10.1016/j.tins.2011.06.007.

Maren, S., Phan, K. L., & Liberzon, I. (2013). The contextual brain: implications for fear conditioning, extinction and psychopathology. *Nature Reviews Neuroscience, 14*(6), 417–428. http://dx.doi.org/10.1038/nrn3492.

Mehta, D., Gonik, M., Klengel, T., Rex-Haffner, M., Menke, A., Rubel, J., … Binder, E. B. (2011). Using polymorphisms in FKBP5 to define biologically distinct subtypes of posttraumatic stress disorder: evidence from endocrine and gene expression studies. *Archives of General Psychiatry, 68*(9), 901–910. http://dx.doi.org/10.1001/archgenpsychiatry.2011.50.

Mercer, K. B., Orcutt, H. K., Quinn, J. F., Fitzgerald, C. A., Conneely, K. N., Barfield, R. T., … Ressler, K. J. (2012). Acute and posttraumatic stress symptoms in a prospective gene × environment study of a university campus shooting. *Archives of General Psychiatry, 69*(1), 89–97. http://dx.doi.org/10.1001/archgenpsychiatry.2011.109.

Neale, B. M., Lasky-Su, J., Anney, R., Franke, B., Zhou, K., Maller, J. B., ... Faraone, S. V. (2008). Genome-wide association scan of attention deficit hyperactivity disorder. *American Journal of Medical Genetics Part B, Neuropsychiatric Genetics: The Official Publication of the International Society of Psychiatric Genetics, 147B*(8), 1337–1344. http://dx.doi.org/10.1002/ajmg.b.30866.

Nguyen, A., Rauch, T. A., Pfeifer, G. P., & Hu, V. W. (2010). Global methylation profiling of lymphoblastoid cell lines reveals epigenetic contributions to autism spectrum disorders and a novel autism candidate gene, RORA, whose protein product is reduced in autistic brain. *FASEB Journal, 24*(8), 3036–3051. http://dx.doi.org/10.1096/fj.10-154484.

Nievergelt, C. M., Maihofer, A. X., Mustapic, M., Yurgil, K. A., Schork, N. J., Miller, M. W., ... Baker, D. G. (2015). Genomic predictors of combat stress vulnerability and resilience in U.S. Marines: a genome-wide association study across multiple ancestries implicates PRTFDC1 as a potential PTSD gene. *Psychoneuroendocrinology, 51*, 459–471. http://dx.doi.org/10.1016/j.psyneuen.2014.10.017.

Phelps, E. A., Delgado, M. R., Nearing, K. I., & LeDoux, J. E. (2004). Extinction learning in humans: role of the amygdala and vmPFC. *Neuron, 43*(6), 897–905. http://dx.doi.org/10.1016/j.neuron.2004.08.042.

Pitts, M. W., Todorovic, C., Blank, T., & Takahashi, L. K. (2009). The central nucleus of the amygdala and corticotropin-releasing factor: insights into contextual fear memory. *Journal of Neuroscience, 29*(22), 7379–7388. http://dx.doi.org/10.1523/jneurosci.0740-09.2009.

Ressler, K. J., Mercer, K. B., Bradley, B., Jovanovic, T., Mahan, A., Kerley, K., ... May, V. (2011). Post-traumatic stress disorder is associated with PACAP and the PAC1 receptor. *Nature, 470*(7335), 492–497. http://dx.doi.org/10.1038/nature09856.

Ressler, K. J., Rothbaum, B. O., Tannenbaum, L., Anderson, P., Graap, K., Zimand, E., ... Davis, M. (2004). Cognitive enhancers as adjuncts to psychotherapy: use of D-cycloserine in phobic individuals to facilitate extinction of fear. *Archives of General Psychiatry, 61*(11), 1136–1144. http://dx.doi.org/10.1001/archpsyc.61.11.1136.

Rothbaum, B. O., Kearns, M. C., Price, M., Malcoun, E., Davis, M., Ressler, K. J., ... Houry, D. (2012). Early intervention may prevent the development of posttraumatic stress disorder: a randomized pilot civilian study with modified prolonged exposure. *Biological Psychiatry, 72*(11), 957–963. http://dx.doi.org/10.1016/j.biopsych.2012.06.002.

Rothbaum, B. O., Kearns, M. C., Reiser, E., Davis, J. S., Kerley, K. A., Rothbaum, A. O., ... Ressler, K. J. (2014). Early intervention following trauma may mitigate genetic risk for PTSD in civilians: a pilot prospective emergency department study. *J Clin Psychiatry, 75*(12), 1380–1387. http://dx.doi.org/10.4088/JCP.13m08715.

Rothbaum, B. O., Price, M., Jovanovic, T., Norrholm, S. D., Gerardi, M., Dunlop, B., ... Ressler, K. J. (2014). A randomized, double-blind evaluation of D-cycloserine or alprazolam combined with virtual reality exposure therapy for posttraumatic stress disorder in Iraq and Afghanistan War veterans. *The American Journal of Psychiatry, 171*(6), 640–648. http://dx.doi.org/10.1176/appi.ajp.2014.13121625.

Scammell, J. G., Denny, W. B., Valentine, D. L., & Smith, D. F. (2001). Overexpression of the FK506-binding immunophilin FKBP51 is the common cause of glucocorticoid resistance in three New World primates. *General and Comparative Endocrinology, 124*(2), 152–165. http://dx.doi.org/10.1006/gcen.2001.7696.

Segman, R. H., Cooper-Kazaz, R., Macciardi, F., Goltser, T., Halfon, Y., Dobroborski, T., & Shalev, A. Y. (2002). Association between the dopamine transporter gene and posttraumatic stress disorder. *Molecular Psychiatry, 7*(8), 903–907. http://dx.doi.org/10.1038/sj.mp.4001085.

Singewald, N., Schmuckermair, C., Whittle, N., Holmes, A., & Ressler, K. J. (2015). Pharmacology of cognitive enhancers for exposure-based therapy of fear, anxiety and trauma-related disorders. *Pharmacology & Therapeutics, 149*, 150–190. http://dx.doi.org/10.1016/j.pharmthera.2014.12.004.

Southwick, S. M., Bremner, D., Krystal, J. H., & Charney, D. S. (1994). Psychobiologic research in post-traumatic stress disorder. *The Psychiatric Clinics of North America, 17*(2), 251–264.

Southwick, S. M., Bremner, J. D., Rasmusson, A., Morgan, C. A., 3rd, Arnsten, A., & Charney, D. S. (1999). Role of norepinephrine in the pathophysiology and treatment of posttraumatic stress disorder. *Biological Psychiatry, 46*(9), 1192–1204.

Stevens, J. S., Almli, L. M., Fani, N., Gutman, D. A., Bradley, B., Norrholm, S. D., ... Ressler, K. J. (2014). PACAP receptor gene polymorphism impacts fear responses in the amygdala and hippocampus. *Proceedings of the National Academy of Sciences of the United States of America, 111*(8), 3158–3163. http://dx.doi.org/10.1073/pnas.1318954111.

Stevens, J. S., Jovanovic, T., Fani, N., Ely, T. D., Glover, E. M., Bradley, B., & Ressler, K. J. (2013). Disrupted amygdala-prefrontal functional connectivity in civilian women with posttraumatic stress disorder. *Journal of Psychiatric Research, 47*(10), 1469–1478. http://dx.doi.org/10.1016/j.jpsychires.2013.05.031.

Sullivan, P. F., Daly, M. J., & O'Donovan, M. (2012). Genetic architectures of psychiatric disorders: the emerging picture and its implications. *Nature Reviews Genetics, 13*(8), 537–551.

True, W. R., Rice, J., Eisen, S. A., Heath, A. C., Goldberg, J., Lyons, M. J., & Nowak, J. (1993). A twin study of genetic and environmental contributions to liability for posttraumatic stress symptoms. *Archives of General Psychiatry, 50*(4), 257–264.

Uddin, M., Chang, S. C., Zhang, C., Ressler, K., Mercer, K. B., Galea, S., ... Koenen, K. C. (2012). Adcyap1r1 genotype, posttraumatic stress disorder, and depression among women exposed to childhood maltreatment. *Depression and Anxiety, 30*(3), 251–258. http://dx.doi.org/10.1002/da.22037.

Valente, N. L., Vallada, H., Cordeiro, Q., Bressan, R. A., Andreoli, S. B., Mari, J. J., & Mello, M. F. (2011). Catechol-O-methyltransferase (COMT) val158met polymorphism as a risk factor for PTSD after urban violence. *Journal of Molecular Neuroscience: MN, 43*(3), 516–523. http://dx.doi.org/10.1007/s12031-010-9474-2.

Valente, N. L., Vallada, H., Cordeiro, Q., Miguita, K., Bressan, R. A., Andreoli, S. B., ... Mello, M. F. (2011). Candidate-gene approach in posttraumatic stress disorder after urban violence: association analysis of the genes encoding serotonin transporter, dopamine transporter, and BDNF. *Journal of Molecular Neuroscience: MN, 44*(1), 59–67. http://dx.doi.org/10.1007/s12031-011-9513-7.

Wang, Z., Baker, D. G., Harrer, J., Hamner, M., Price, M., & Amstadter, A. (2011). The relationship between combat-related posttraumatic stress disorder and the 5-HTTLPR/rs25531 polymorphism. *Depression and Anxiety, 28*(12), 1067–1073. http://dx.doi.org/10.1002/da.20872.

White, M. G., Bogdan, R., Fisher, P. M., Muñoz, K. E., Williamson, D. E., & Hariri, A. R. (2012). FKBP5 and emotional neglect interact to predict individual differences in amygdala reactivity. *Genes, Brain and Behavior, 11*(7), 869–878. http://dx.doi.org/10.1111/j.1601-183X.2012.00837.x.

Xian, H., Chantarujikapong, S. I., Scherrer, J. F., Eisen, S. A., Lyons, M. J., Goldberg, J., … True, W. R. (2000). Genetic and environmental influences on posttraumatic stress disorder, alcohol and drug dependence in twin pairs. *Drug and Alcohol Dependence, 61*(1), 95–102. http://dx.doi.org/10.1016/s0376-8716(00)00127-7.

Xie, P., Kranzler, H. R., Farrer, L., & Gelernter, J. (2012). Serotonin transporter 5-HTTLPR genotype moderates the effects of childhood adversity on posttraumatic stress disorder risk: a replication study. *American Journal of Medical Genetics Part B, Neuropsychiatric Genetics: The Official Publication of the International Society of Psychiatric Genetics, 159B*(6), 644–652. http://dx.doi.org/10.1002/ajmg.b.32068.

Xie, P. X., Kranzler, H. R., Poling, J., Stein, M. B., Anton, R. F., Brady, K., … Gelernter, J. (2009). Interactive effect of stressful life events and the serotonin transporter 5-HTTLPR genotype on posttraumatic stress disorder diagnosis in 2 independent populations. *Archives of General Psychiatry, 66*(11), 1201–1209.

Xie, P., Kranzler, H. R., Poling, J., Stein, M. B., Anton, R. F., Farrer, L. A., & Gelernter, J. (2010). Interaction of FKBP5 with childhood adversity on risk for post-traumatic stress disorder. *Neuropsychopharmacology: Official Publication of the American College of Neuropsychopharmacology, 35*(8), 1684–1692. http://dx.doi.org/10.1038/npp.2010.37.

Xie, P., Kranzler, H. R., Yang, C., Zhao, H., Farrer, L. A., & Gelernter, J. (2013). Genome-wide association study identifies new susceptibility loci for posttraumatic stress disorder. *Biological Psychiatry, 74*(9), 656–663. http://dx.doi.org/10.1016/j.biopsych.2013.04.013.

Yehuda, R., Southwick, S., Giller, E. L., Ma, X., & Mason, J. W. (1992). Urinary catecholamine excretion and severity of PTSD symptoms in Vietnam combat veterans. *The Journal of Nervous and Mental Disease, 180*(5), 321–325.

Young, R. M., Lawford, B. R., Noble, E. P., Kann, B., Wilkie, A., Ritchie, T., … Shadforth, S. (2002). Harmful drinking in military veterans with post-traumatic stress disorder: association with the D2 dopamine receptor A1 allele. *Alcohol and Alcoholism, 37*(5), 451–456.

van Zuiden, M., Kavelaars, A., Geuze, E., Olff, M., & Heijnen, C. J. (2012). Predicting PTSD: pre-existing vulnerabilities in glucocorticoid-signaling and implications for preventive interventions. *Brain Behavior and Immunity, 30*, 12–21. http://dx.doi.org/10.1016/j.bbi.2012.08.015.

Chapter 36

Neurodevelopmental Disorders, Causes, and Consequences

Elliot H. Sherr
University of California San Francisco, San Francisco, CA, United States

NEURODEVELOPMENTAL DISORDERS, CAUSES, AND CONSEQUENCES

In the overlap between the fields of neurology, psychiatry, and pediatrics, it may be hard to find a term with as much generality as neurodevelopmental disorders (NDD). Hence broadly defined, "neurodevelopmental disorders are a group of disorders in which the development of the central nervous system is disturbed. This can include developmental brain dysfunction, which can manifest as neuropsychiatric problems or impaired motor function, learning, language or non-verbal communication" (http://www.nature.com/subjects/neurodevelopmental-disorders). Thus, this list of disorders includes intellectual disability (ID), autism, language disorders, sensory processing disorders, attention-deficit and related disorders, epilepsy, cerebral palsy, and the more recent recognition that psychiatric disorders such as schizophrenia, bipolar, Tourette, and obsessive-compulsive disorders, although these most often have later onsets in development, are also the result of disruption of brain development (Corvin, 2010; Innocenti, Ansermet, & Parnas, 2003; O'Dushlaine et al., 2011; Owen, O'Donovan, Thapar, & Craddock, 2011). With such a broad definition, there is likely to be a high incidence of this group of disorders (McGrath & Richards, 2009). However, this review will focus on disorders that present in childhood and include disruption in intellectual capacity. Thus, ID, global developmental delay, autism (when intellectual impairment is comorbid), and the intersection of these with cerebral palsy and epilepsy are the main themes of this chapter. The more specific disorders of autism and language disorders will be addressed in more detail in other chapters. In addition, because these disorders demonstrate considerable biological complexity, it is often difficult to discern specific mechanisms that explain how these disorders disrupt specific aspects of brain function. Thus, this chapter will focus on molecular mechanisms that are unique to certain disorders, and when applicable, to the circuit disruptions that are relevant to the behaviors and abilities in question. Moreover, because this review focuses on neurodevelopment, disorders that result in neurologic regression, such as Rett syndrome (Ben Zeev Ghidoni, 2007), infantile neuroaxonal dystrophy (El Arbi, Demant, Kohlschmidt, & Horneff, 2013), Menkes, or neurodegeneration from mutations in KIF1A (Yonekawa et al., 1998) will not be addressed.

CAUSES OF DISORDERS OF NEURODEVELOPMENT

What steps are necessary to better treat these broad-based conditions? Progress in understanding the causes of NDD is the first necessary step, and work in the epidemiology and genetics of NDD has provided considerable insight into this group of disorders. Moreover, understanding the biology of neurodevelopmental disorders has led in certain cases to the development and testing of targeted therapies, which may be the beginning of a significant breakthrough in our ability to alter the lives of these children in a meaningful way.

Genomics, Circuits, and Pathways in Clinical Neuropsychiatry. http://dx.doi.org/10.1016/B978-0-12-800105-9.00036-6

GENETICS OF NEURODEVELOPMENTAL DISORDERS

Many causes have been identified (and remain to be identified) for neurodevelopmental disorders. Discovery of the genetic causes of ID has progressed considerably since identification of the gene (*FMR1*) for Fragile X syndrome (FXS) in 1991 (Kremer et al., 1991). A search on the Online Mendelian Inheritance of Man database reveals 3186 entries for disorders of cognition, 1722 for seizures, 1604 for cerebral palsy, and 396 for autism. Many of these disorders remain clinical descriptions only, without the causative genes pinpointed. However, with tools now available for genetic discovery (principally whole-exome and whole-genome sequencing), there has been a dramatic increase in the number of novel syndromes recognized as well as more genes discovered for known, already described disorders, with a 25% increase in genes for known disorders such as ID just since 2010. There has also been an unanticipated growth in the description of novel disorders, along with quick discovery of the causative genes, enabled through the power of next-generation sequencing. However, before this new phase in gene discovery, most genes described for ID were those inherited in an X-linked manner.

Along with the identification of *FMR1* as the cause of FXS, the clearly evident familial pattern of X-linked inheritance permitted the discovery of many other genes that cause ID. Based on considerable progress in identifying and evaluating such families, there are now over 112 genes that reside on the X chromosome that causes inherited X-linked ID (Lubs, Stevenson, & Schwartz, 2012; Soden et al., 2014). Whereas the early studies for X-linked ID (XLID) identified male children with ID and syndromic features, as the power of genetic testing increased, genes were identified for additional families, many of which did not have syndromic features, that have been grouped under the category of nonsyndromic X-linked ID (NSXLID) (Shoubridge et al., 2010). For the first extended family (the disorder labeled as MRX1) systematically studied with nonsyndromic ID (Suthers, Turner, & Mulley, 1988), the cause was eventually determined to be from a mutation in the gene *IQSEC2* (Shoubridge et al., 2010). Once this gene was found in the index family, three other families, identified through unbiased X chromosome resequencing, were also found to have mutations in this gene. IQSEC2 is a guanine nucleotide exchange factor for the adenosine phosphate-ribosylation factor family of small GTPases. It is localized in the postsynaptic density of excitatory synapses and is hypothesized to regulate excitatory long-term potentiation. Another gene in the same family, *IQSEC1* (*BRAG2*) binds to the GluA2 subunit of the AMPA receptor and mediates long-term depression in excitatory synapses (Scholz et al., 2010). Additional genetic studies have found that IQSEC2 is mutated in cases of intractable epilepsy, such as Lennox–Gastaut syndrome (Allen et al., 2013) and infantile spasms (Morleo et al., 2008), or in patients who have both ID and epileptic encephalopathy (EE) (Gandomi et al., 2014; Tran Mau-Them et al., 2014). Thus, the mutations in this gene have a broad spectrum of clinical phenotypes. This notion of diverse phenotypes for genes implicated in ID may be prevalent. For example, the gene aristaless X is a transcription factor that can present with severe brain malformations, isolated agenesis of the corpus callosum with mild intellectual impairment, or severe intellectual impairment with infantile spasms and no change in brain anatomy, as well as in children with autism and NSXLID (Sherr, 2003). That single genes can present with a diversity of phenotypes is becoming increasingly clear as deeper sequencing of patient cohorts is more common. An investigation systematically examined 3334 exonic de novo mutations from 3555 trios across four neurodevelopmental disorders, autism, EE, ID and schizophrenia (Li et al., 2015). Using stringent criteria, this study demonstrated that 53 genes were present in more than one disorder, with 23 genes overlapping between autism spectrum disorder (ASD) and schizophrenia, for example. One gene, *SCN2A*, was identified as a source of de novo pathogenic mutations in all four disorders, which demonstrates the clinical complexity that can arise from a single gene.

Deeper sequencing in large cohorts continues to uncover new genes. Additional genes causing XLID remain to be discovered (Marco et al., 2008). One study tested over 700 genes in 405 families that were negative for known causes of XLID. Using deep sequencing of the X chromosome in these families, over 10 novel confirmed and candidate genes were detected to have causative mutations (Hu et al., 2015). These mutations in XLID are most often inherited from an unaffected carrier mother. In more severe cases, these conditions can arise from de novo mutations in either male or female children (Allen et al., 2013). These de novo cases are typically seen in "simplex" families, in which only one family member is affected. The inherited cases can also present in simplex families, because the mother of the male patient may have not had brother siblings, or the affected allele was never transmitted to an uncle or other male relative. Moreover, the severity of the mutation may also affect presentation of the disorder. For FXS, "carrier" females often have clinical symptoms. This ranges from common premature ovarian failure (Martin & Arici, 2008) or migraine headaches (Au et al., 2013) to autism and other cognitive impairments (Hagerman et al., 1986). This mode of inheritance is typical of the complexity seen in X-linked traits (Dobyns et al., 2004).

In addition to X-linked traits, there has been substantial progress in the understanding of recessive inherited and dominant de novo causes of ID. Advances in genetic linkage and sequencing tools have also led to an acceleration in the discovery of recessive disorders of ID, particularly when there are no physical or other unique characteristics that allow cross-comparison of families with multiple affected children (El Chehadeh et al., 2015; Law et al., 2014; Ropers et al., 2011). Indeed, one of the most significant advances in human genetics in the past few years has been the discovery of the importance that de novo mutations have in neurodevelopmental disorders. This was first shown in cohorts of children with severe ID, which demonstrated that many of these patients had truncating or other severe mutations that were likely the cause of the symptoms (de Ligt et al., 2012). This was then demonstrated in cohorts of families in which only one child in the family had ASD and at least one other unaffected child was available to participate in the study. In nearly 3000 families tested, the rate of de novo inactivating mutations (stop gained, splice mutations, or deletions) was significantly elevated in probands with ASD compared with their unaffected siblings (Dong et al., 2014; Gilman et al., 2011; Iossifov et al., 2014; O'Roak et al., 2014; Sanders et al., 2012). A similar association of rare de novo variants was seen in patients with two severe childhood epilepsies, infantile spasms, and Lennox–Gastaut syndrome, often grouped together as EEs (Allen et al., 2013; EuroEPINOMICS-RES Consortium, Epilepsy Phenome/Genome Project, & Epi4K Consortium, 2014; Esmaeeli Nieh and Sherr, 2014). More recently, other neurodevelopmental disorders such as cerebral palsy and ataxia have been shown to result from similar de novo point mutations (McMichael et al., 2015; Pyle et al., 2015). An additional contribution of exome sequencing has unraveled the causes of specific syndromes that have neurodevelopmental impairment as a key feature, but not the only one. Hundreds of new syndromes have already been identified and the National Institutes of Health–funded Centers for Mendelian Genomics (http://www.mendelian.org) added significantly to this total by grouping patients together through shared phenotypes (Bamshad et al., 2012; Pehlivan et al., 2014; Stray-Pedersen et al., 2014). In addition to demonstrating how significant de novo variants are to understanding these broad classifications of neurodevelopmental disorders and to identifying novel syndromes with specific phenotypes, exome sequencing has now shown considerable utility in clinical diagnosis of disorders of development (Lee et al., 2014; Soden et al., 2014; Yang et al., 2013, 2014). This can include disorders of other organ systems, but in two seminal reports that describe over 3000 patients who underwent clinical diagnostic exome testing, the vast majority of patients referred for testing had as their primary diagnosis a disorder of neurodevelopment; thus those patients with cognitive impairment, along with other neurologic features such as cerebral palsy, hypotonia, epilepsy, or ataxia truly predominate (Lee et al., 2014; Yang et al., 2014). Moreover, this group was more likely to find a diagnosis than the idiopathic ASD group or the idiopathic epilepsy cohort (Lee et al., 2014; Yang et al., 2014). It is clear from these large-scale exome sequencing projects that we as a scientific and clinical community have just begun this process of gene discovery. This is evident because, although some already known genes show up again in these large cohorts, many genes are still novel. Even a number of these show up in only one patient, at least initially. Follow-up cohorts often present with additional "hits" in these candidate genes, thus supporting a diagnosis of likely pathogenic (EuroEPINOMICS-RES Consortium et al., 2014). Only when the community observes a significant slowdown in the number of new disorders identified by this approach will there be a sense that gene discovery may be reaching saturation by broad-based exome sequencing.

Before the development of high-density exome sequencing, genome-wide platforms were developed that allowed for assessment of genomic copy number variants (CNVs). These tools allowed for the identification of large trisomies and monosomies that were already observed using conventional karyotyping, all the way to 50-kb deletions and duplications that were visible only with this enhanced sensitivity platform. These array-based platforms followed the use of fluorescent in situ hybridization detecting either known loci or the distal ends of the long and short arms of each of the chromosomes (subtelomeric probes). The only genomic changes that are not readily detected by this approach are truly balanced translocations and balanced inversions (Brady & Vermeesch, 2012). Because these platforms were genome-wide, it was possible to use this tool to look broadly for principally de novo genetic events that would lead to neurodevelopmental disorders (Henderson et al., 2014). This is now the first-tier test for genetic testing in children with NDD and is also being used increasingly in the prenatal setting (Hillman, McMullan, Williams, Maher, & Kilby, 2012; Miller et al., 2010; Novelli et al., 2012). These platforms can detect common variants that are present in the general population with a low frequency and more prevalent in children with neurodevelopmental disorders, but there are also many rare variants for which the pathogenicity remains uncertain (Coulter et al., 2011; Peters & Pertile, 2014; Wapner, Driscoll, & Simpson, 2012). However, it is clear that by aggregating large numbers of patients with NDD, many of these rare loci are actually recurrent and highly correlated with disease presentation (Nicholas, Baker, Eichler, & Akey, 2011; Stefansson et al., 2014). However, some loci have only incomplete penetrance, and only by having large numbers of individuals present with these genetic changes (or even population-based studies) was this insight possible (Rosenfeld, Coe, Eichler, Cuckle, & Shaffer, 2012; Stefansson et al., 2014). In addition, these loci are often large and gene-rich. For some of these loci, further analysis reveals that a single gene accounts for most of the clinical deficit observed (O'Roak et al., 2014) whereas for others, it is likely that many genes across the CNV account for the overall phenotype (Lin et al., 2015).

In addition to single-gene or genomic regions that are highly penetrant causes, there is also accumulating evidence that polygenic or oligogenic causes may mediate ASD, epilepsy, ID, and other neurodevelopmental disorders. This has been best studied in autism and epilepsy (EPICURE Consortium et al., 2012; Durner et al., 2001; Jiang et al., 2013; Schaaf et al., 2011). Because each genetic locus by definition contributes a smaller amount of burden to the underlying diagnosis, this type of genome-wide burden analysis is likely to identify significant associations only with larger numbers of patients. For example, in schizophrenia, pooling the data from 36,989 patients and 113,075 control subjects from many studies led to the identification of over 100 genetic loci that met genome-wide statistical thresholds (Maier et al., 2015). Moreover, there is evidence that combining these neuropsychiatric disorders (schizophrenia, bipolar disorder, and major depressive disorder) into a larger group for statistical analysis actually increases the ability to assess risk accurately based on allele frequencies (Maier et al., 2015). Thus, viewing all of these disorders as having some shared and some unique risk factors provides greater insight into the precision of diagnosis and then likely associated disease pathogenesis. For autism, the numbers of patients enrolled in similar large genome-wide association studies is still much smaller, by an order of magnitude or greater. Thus, whereas investigations have shown that much of risk variance in autism is explained by common variants as a whole, individual variants exert a weak effect on the overall disease burden (Anney et al., 2012; Klei et al., 2012). In ID, most of the genetic analysis has been in children with moderate or severe intellectual impairment, while children with mild ID, who are more likely to have many small genetic effects contribute to cognitive challenges, have not been studied with the same intensity (Hill et al., 2014).

With the power of genetics, it is clear that both de novo and inherited variants have an important causative role in mediating disorders of neurodevelopment. As more patients are tested, it will be possible to make progress in understanding how the common inherited variants of small effect size individually contribute to these conditions. However, we can currently investigate how these highly penetrant de novo variants lead to disease. For example, publications on disorders that involve the primary cilium of the cell (ciliopathies), such as Joubert or Meckel syndromes, as well as other disorders that involve eye or renal development, have shown that there is considerable variation in genetic etiology as well as in clinical presentation for mutations even in the same gene. While there are not yet targeted treatments for these disorders, the availability of prenatal and newborn testing does make understanding the genotype–phenotype correlation important (Akizu et al., 2014; Brancati et al., 2007, 2008; Dafinger et al., 2011; Tsurusaki et al., 2012). Thus, a publication tackled the question of genotype–phenotype correlation in Joubert and related disorders, asking how mutations in the commonly mutated gene *CEP290* can result in phenotypically variable disease using both computational and in vitro testing approaches. They demonstrated that both the site of the mutation and the degree with which exon skipping occurs can be modeled to estimate protein abundance and hence the degree of clinical severity (Drivas, Wojno, Tucker, Stone, & Bennett, 2015). This type of analysis is still in its early stage, but it must continue when larger numbers of patients are identified with mutations in the same gene, or genes in the same signaling pathway.

Nongenetic causes also have significant roles in neurodevelopmental disorders. There is substantial evidence for birth asphyxia and other perinatal infections and central nervous system injuries serving as major risk factors for neurodevelopmental disorders (Nelson & Chang, 2008). A detailed discussion of these issues is beyond the scope of this review, but the implementation of therapeutic cooling in the newborn period for children with evidence of perinatal hypoxic-ischemic injury underscores both the benefits of this intervention and the evidence linking these apparent injuries with poor neurodevelopmental outcomes (Jacobs et al., 2013). Prematurity (and associated low birth weight) is also associated with poor developmental outcome (Johnson et al., 2015; Molloy, Anderson, Anderson, & Doyle, 2015), and these deficits can persist into adulthood (Roggero, Gianni, Garbarino, & Mosca, 2013).

Exposure to toxins is also a risk factor for neurodevelopmental disorders, even if the ability to assess exposure and outcomes is challenging. Fetal alcohol syndrome (FAS) has been long appreciated as one of the more significant risk factors. However, in the absence of clear dysmorphology criteria for FAS, it has been challenging to measure the true prevalence of FAS in the general pediatric population. However, a study examined the prevalence of FASD in a large cohort of first-grade children (over 70% of the children in that age window were enrolled) in a Midwestern US community (May et al., 2014). Their analysis showed that the prevalence of FAS by strict criteria ranged from 6–9 per 1000 children, with the total rate of FASD (the broader FAS disorder) having an imputed prevalence between 24 and 48 per 1000 children. In this study both estimates were much higher than previously projected. Moreover, children with FAS were likely to perform worse on many key cognitive and behavioral tests than age-matched controls, with both cognitive and physical challenges. These findings underscore how significant FAS and the broader FASD may be to the cohort of children with neurodevelopmental impairment and enhance the need for careful assessment of FAS criteria in all children with NDD (May et al., 2014). Cocaine exposure has also been shown to correlate with poor developmental outcome (Davis et al., 1992), although for many children, exposure to drugs of abuse is linked to exposure to many drugs, confounding the linkage for any one drug. However, tobacco smoking is well linked to risk for a number of neurodevelopmental impairments. This has been shown in a number of prospective studies in which the level of tobacco exposure was biologically quantified (Cornelius & Day, 2009). Moreover, a study demonstrated effects on childhood and teenage behavior of babies exposed in utero to second-hand smoke (Gatzke-Kopp & Beauchaine, 2007). In contrast to these studies linking exposure of toxins to impairment of neurodevelopmental outcomes, no evidence has been generated linking exposure of thimerosal in vaccines to neurodevelopmental disorders. A report issued by the Institute of Medicine's Immunization Safety Review Committee found that no published epidemiological studies demonstrated an association between receiving thimerosal-containing vaccines and neurodevelopmental disorders (Stratton, Gable, & McCormick, 2001).

There has been considerable progress in the genetics of neurodevelopmental disorders over the past 10 years, and given the greater availability of whole-exome sequencing and soon likely to be whole-genome sequencing in the clinical realm, this progress is likely to continue at a rapid pace in the near term. However, we will still be confronted with a need to better understand the cellular and circuit deficits that result in NDD and will need to determine what modes of treatment, including such interventions as gene therapy, would be efficacious for these patients (Rahman, Maeder, Joung, & Cathomen, 2011).

MECHANISMS OF ACTION AND CONSEQUENCES ON THE CIRCUITS OF COGNITION AND BEHAVIOR

How do these factors (genetic, environmental, and a mixture of both) lead to intellectual and developmental impairment? What mechanisms are perturbed, in what cell types, and when during development are these processes most evident or critical for impairing normal brain maturation? Also, are there specific circuits in the brain that when perturbed lead to selective impairments? In many examples, there is a paucity of data on many of these larger-picture questions. However, there are examples of well-known disorders in which research has shed some light on the mechanisms of disease and hence possible means for intervention. In addition, there is evidence pointing to general mechanisms that are disrupted across a number of different disorders.

Perhaps one of the best-characterized disorders from a genetic/molecular perspective, as well as an understanding of the mechanisms altered, is FXS, the most common inherited single-gene neurodevelopmental disorder, with an incidence of 1:4000 in males and 1:4000–6000 in females. As mentioned earlier, the disorder is the result of expansion of a triplet repeat (CGG) in the promoter of the *FMR1* gene on the X chromosome. When the repeat is present 55 to 200 times, this is referred to as a premutation, often presenting with no or mild symptoms in childhood, whereas male carriers of greater than 200 repeats nearly always have the classic FXS clinical presentations (Reiss & Freund, 1990). This premutation is more prevalent than the classic disease, observed in as many as one in 151 women and one in 468 men. In later adulthood (aged >50 years) carriers of the premutation are also at risk for developing Fragile X–associated tremor/ataxia syndrome, which includes ataxia, tremor, autonomic and sensory neuropathy, worsening cognitive deficits (particularly executive function impairment), along with anxiety and depression (Leehey & Hagerman, 2012). The protein, FMRP, has been shown to bind to polyribosomes and downregulate the translation of messenger RNA (mRNA) located at synapses, whereas the mechanisms that specifically mediate this interaction are uncertain (Bhakar, Dolen, & Bear, 2012). As part of this mechanism, investigators led by Mark Bear and colleagues have shown that a form of synaptic plasticity (long-term depression) is unexpectedly enhanced in Fragile X mice. Moreover, this form of plasticity is in part mediated through metabotropic glutamate receptors and evidence has shown that long-term inhibition of mGluR5 enhances cognition and behavior in these mice (Michalon et al., 2012). This work in multiple animal models and in cell culture in vitro suggests that mGluR5 antagonists could ameliorate the deficits in Fragile X. This series of observations has led to testing a group of mGlur5 antagonists in Fragile X cohorts. These clinical trials were not successful, nor was a trial that tested the efficacy of arbaclofen to treat the behavioral deficits seen in FXS (Emmitte, 2013; Jacquemont et al., 2014; Mullard, 2015). However, the lack of clinical improvement in these trials may be the result of a more complex biology that underlies FXS or of less than optimal dosing, treatment regimens, or methods to assess clinical improvement. Thus, in the 24 years since the discovery of the *FMR1* gene, there has been substantial progress in assessing the molecular mechanisms that drive FXS clinical deficits, but the goal of improving the lives of these patients remains as complex and elusive as before.

FAS, as introduced earlier, unfortunately remains a common yet preventable cause of ID and behavioral concerns (American Academy of Pediatrics, 2000). The incidence ranges from 0.3 to 30 per 1000, and partial FAS (with some but not all of the features, including neurodevelopmental impairment) can range from 0.5% to 5% in school-based ascertainment cohorts. In developed countries, this remains the most significant preventable environmental cause of neurodevelopmental disorders (Pruett, Waterman, & Caughey, 2013). Alcohol exposure can have pleiotropic effects on brain development. One of the main changes in brain anatomy is microcephaly, or micrencephaly (West, Chen, & Pantazis, 1994). This smaller brain may result from *N*-methyl-D-aspartate (NMDA)-based neurotoxicity, with stimulation of the

NMDA receptors at critical periods leading to broad-based apoptosis (Ikonomidou et al., 2000; Olney, Farber, Wozniak, Jevtovic-Todorovic, & Ikonomidou, 2000; Olney, Wozniak, Jevtovic-Todorovic, & Ikonomidou, 2001; Tsai, Gastfriend, & Coyle, 1995). More advanced imaging tools such as careful volumetric studies along with diffusion tensor imaging have been applied to the study of FASD. These approaches have shown that cortical thickness is frequently decreased in FAS (Sowell et al., 2008). Moreover, in addition to microcephaly, the corpus callosum is often affected, whether measured by volumetric or diffusion tensor approaches (Lebel, Roussotte, & Sowell, 2011; Paul, 2011). This susceptibility to callosal impairment can affect executive function (Bookstein, Streissguth, Sampson, Connor, & Barr, 2002; Wozniak & Muetzel, 2011) and more tailored assays that measure interhemispheric transfer (Roebuck, Mattson, & Riley, 2002). In addition to the apoptosis seen in animal models, there are other cellular mechanisms by which exposure to alcohol in utero is likely to impair neurodevelopment. Qualitatively, the brains of FAS children resemble those of children with mutations in the gene L1, which is necessary for many aspects of axon outgrowth and fasciculation (Bearer, 2001a, 2001b). In vitro data have demonstrated that ethanol directly affects the function of L1 by altering its distribution in lipid rafts and changing the degree of tyrosine phosphorylation necessary for downstream signaling through LI and fasciculation (Dou & Charness, 2014; Littner, Tang, He, & Bearer, 2013).

A main barrier to understanding the full range of FASD and obtaining more accurate estimates of incidence has been the need to implement a biomarker to assess exposure in utero. Data suggest that measuring ethylglucuronide in meconium may be a way to assess fetal exposure to alcohol (Gauthier et al., 2015; Morini et al., 2013). This may enable a better assessment of the true incidence of fetal alcohol exposure and the dose over time necessary to result in central nervous system (CNS) impairment.

Fragile X is the most common inherited cause of ID, whereas fetal alcohol syndrome is the most common preventable neurotoxic cause of NDD. Thyroid hormone deficiency is an excellent example of how environmental and genetic causes intersect in leading to one of the most common groups of neurodevelopmental disorders. Thyroid hormone (TH) is one of the central molecules signaling growth in the developing fetus. Lack of sufficient TH impairs nearly all organ systems, leading to the syndrome cretinism (Pemberton, Franklyn, & Kilby, 2005). CNS-related deficits in cretinism include ID, ataxia, spasticity, and deafness. Morphologically in the brain, neurons are smaller and more tightly spaced and dendritic spines are fewer, and in the cerebellum both migration and differentiation of granule neurons are deficient (Thompson & Potter, 2000). In addition to neuronal effects, there is also a decrease in myelination, which may result from the need of TH for proper oligodendrocyte differentiation (Baxi et al., 2014).

What might be the molecular mechanisms that lead to these changes in brain structure and function? Many of the genes necessary for myelin production, including myelin basic protein, proteolipid protein, and myelin-associated glycoprotein mRNA levels are reduced in the absence of TH. Many growth factors necessary for brain development, including nerve growth factor, brain-derived neurotrophic factor, and neurotrophin-3, show a reduction in mRNA levels with hypothyroidism. Neuronal structure, as seen at the level of dendrites and synapses, is impaired in TH deficiency. The gene neurogranin is directly regulated by TH as is the synapse-specific protein synaptotagmin, and their deficits lead to decreased synapse production. The Srg1 protein, also expressed in the synapse, is reduced in protein abundance in TH-deficient brains. There are at least two main pathways that lead to TH deficiency in patients. In one, Allan–Herndon–Dudley syndrome, a mutation on the X chromosome gene, *MCT8*, is a transporter for thyroid hormone from the extracellular space into the nucleus. Males with mutations in *MCT8* are affected in a manner similar to TH deficiency, with significant ID. TH deficiency per se is often not genetically mediated, but is the result of iodine deficiency. Iodine deficiency that occurs during pregnancy is the most common worldwide preventable cause of neurodevelopmental disorders and ID. If iodine is supplemented in the diet in the first trimester, there is strong evidence that these neurodevelopmental impairments can be ameliorated (Pemberton et al., 2005; Skeaff, 2011). In addition, there are genetic modifiers influencing how body stores of iodine can lead to insufficient TH. The enzyme deiodinase, that removes iodine from T3 and T4, has alleles with different efficiency in this removal. Thus, this is an example of how both environment (lack of iodine in the diet) and genetics (alleles of deiodinase) lead to disease.

In these cases (FXS, FAS, and iodine or TH deficiency), there is insight into how these disorders lead to disease, but the manifestations of intellectual impairment are broad-based and the mechanisms that underlie these disorders are also pleiotropic; thus, no clear neuronal circuit or brain region is responsible for the cognitive deficits. In contrast, for Prader–Willi syndrome (PWS), regions and circuits appear to be responsible for at least some of the features of the disorder.

PWS is present in approximately one in 15,000 individuals and is noted for early-onset hypotonia and poor feeding (often requiring a nasogastric tube for supplemental feeding) and characteristic physical features that include almond-shaped eyes, a triangular mouth, short stature, small hands and feet, and small genitalia. As these children develop, behavioral problems are common, including emotional outbursts, and in opposition to the neonatal period, they develop hyperphagia and often extreme obesity (Veltman et al., 2004). PWS is caused by deletion of the paternal chromosome at 15q11 to q13, which overlaps with the region implicated in Angelman syndrome. For PWS, uniparental disomy of the maternal chromosome can also cause the disorder; two reports implicate two smaller regions within the PWS locus that might be pathogenic, including the noncoding RNA region SNORD1186 and the gene *MAGEL2* (Bieth et al., 2015; Schaaf et al., 2013). Patients with PWS have been shown to have hypothalamic abnormalities, in leptin receptor regulation and with much lower levels of oxytocin, and it has been hypothesized that a number of the abnormalities unique to PWS result from direct hypothalamic dysfunction. Thus, knockout of the *MAGEL2* gene generates mice that show deficits with the ability to feed in the neonatal period (lethal if not treated) and with impairment of social cognition and memory-based tasks. Treatment of these mice with a single dose of oxytocin was able to rescue potentially lethal feeding behavior in the newborn mice (Schaller et al., 2010) and treatment with daily doses of oxytocin (OT) for 1 week was able to restore performance of tests of social function and learning and memory (Meziane et al., 2014). These findings underscore the notion that for PWS some of the central circuits and molecules that direct the areas of behavior and cognition that are impaired are within the hypothalamus and can be treated with at least one hypothalamic peptide, OT.

FUTURE DIRECTIONS

The tools of genetics have tremendously expanded our vista of etiologies for NDD, but the more detailed analyses necessary to unpack the biology behind these disorders are progressing at a slower rate. New tools for genetically modifying cell lines and mouse strains, such as CRISPR, may accelerate these steps (Sternberg & Doudna, 2015). Even after shortening the step of generating the cell or animal model, however, considerable work is necessary. Despite this, there is still strong progress in developing animal models and devising approaches to ameliorate symptoms in these models as a first step toward rational therapeutics (Fortress et al., 2015; Gross et al., 2015; Krencik et al., 2015; Sternberg & Doudna, 2015). In parallel to this work in animal models, there has been progress in gene therapy in patients, albeit to a limited extent. There is additional evidence that enzyme replacement delivered into the CNS for lysosomal storage disorders may significantly improve the lifespan of affected engineered mice (Kan et al., 2014). There is also preliminary evidence of observable improvement in patients with aromatic L-amino acid decarboxylase (AADC) deficiency, who have responded favorably to surgically implanted, virally mediated AADC gene therapy (Hwu, Lee, Chien, Muramatsu, & Ichinose, 2013; Hwu et al., 2012), and that improvements in this approach can yield more robust treatment efficacy (San Sebastian et al., 2014). If successful, these approaches could pave the way for further gene therapy-based approaches for children with neurodevelopmental disorders.

REFERENCES

Akizu, N., Silhavy, J. L., Rosti, R. O., Scott, E., Fenstermaker, A. G., Schroth, J., ... Gleeson, J. G. (2014). Mutations in CSPP1 lead to classical Joubert syndrome. *American Journal of Human Genetics, 94*(1), 80–86. http://dx.doi.org/10.1016/j.ajhg.2013.11.015. Epub 2013/12/24. PubMed PMID:24360807; PMCID:PMC3882909.

Allen, A. S., Berkovic, S. F., Cossette, P., Delanty, N., Dlugos, D., Eichler, E. E., ... Winawer, M. R. (2013). De novo mutations in epileptic encephalopathies. *Nature, 501*(7466), 217–221. http://dx.doi.org/10.1038/nature12439. Epub 2013/08/13. PubMed PMID:23934111; PMCID:PMC3773011.

American Academy of Pediatrics. (2000). Committee on substance abuse and committee on children with disabilities. Fetal alcohol syndrome and alcohol-related neurodevelopmental disorders. *Pediatrics, 106*(2 Pt 1), 358–361. Epub 2000/08/02. PubMed PMID:10920168.

Anney, R., Klei, L., Pinto, D., Almeida, J., Bacchelli, E., Baird, G., ... Devlin, B. (2012). Individual common variants exert weak effects on the risk for autism spectrum disorderspi. *Human Molecular Genetics*. http://dx.doi.org/10.1093/hmg/dds301. Epub 2012/07/31. PubMed PMID:22843504.

Au, J., Akins, R. S., Berkowitz-Sutherland, L., Tang, H. T., Chen, Y., Boyd, A., ... Hagerman, R. (2013). Prevalence and risk of migraine headaches in adult fragile X premutation carriers. *Clinical Genetics, 84*(6), 546–551. http://dx.doi.org/10.1111/cge.12109. PubMed PMID:23373759.

Bamshad, M. J., Shendure, J. A., Valle, D., Hamosh, A., Lupski, J. R., Gibbs, R. A., ... Nickerson, D. A. (2012). The centers for Mendelian genomics: a new large-scale initiative to identify the genes underlying rare Mendelian conditions. *American Journal of Medical Genetics Part A, 158A*(7), 1523–1525. http://dx.doi.org/10.1002/ajmg.a.35470. Epub 2012/05/26. PubMed PMID:22628075.

Baxi, E. G., Schott, J. T., Fairchild, A. N., Kirby, L. A., Karani, R., Uapinyoying, P., ... Calabresi, P. A. (2014). A selective thyroid hormone beta receptor agonist enhances human and rodent oligodendrocyte differentiation. *Glia, 62*(9), 1513–1529. http://dx.doi.org/10.1002/glia.22697. Epub 2014/05/28. PubMed PMID:24863526; PMCID:PMC4107024.

Bearer, C. F. (2001a). L1 cell adhesion molecule signal cascades: targets for ethanol developmental neurotoxicity. *Neurotoxicology, 22*(5), 625–633. Epub 2002/01/05. PubMed PMID:11770884.

Bearer, C. F. (2001b). Mechanisms of brain injury: L1 cell adhesion molecule as a target for ethanol-induced prenatal brain injury. *Seminars in Pediatric Neurology, 8*(2), 100–107. Epub 2001/07/24. PubMed PMID:11464956.

Ben Zeev Ghidoni, B. (2007). Rett syndrome. *Child and Adolescent Psychiatric Clinics of North America, 16*(3), 723–743. http://dx.doi.org/10.1016/j.chc.2007.03.004. Epub 2007/06/15. PubMed PMID:17562589.

Bhakar, A. L., Dolen, G., & Bear, M. F. (2012). The pathophysiology of fragile x (and what it teaches us about synapses). *Annual Review of Neuroscience*. http://dx.doi.org/10.1146/annurev-neuro-060909-153138. Epub 2012/04/10. PubMed PMID:22483044.

Bieth, E., Eddiry, S., Gaston, V., Lorenzini, F., Buffet, A., Conte Auriol, F., … Tauber, M. (2015). Highly restricted deletion of the SNORD116 region is implicated in Prader-Willi Syndrome. *European Journal of Human Genetics: EJHG, 23*(2), 252–255. http://dx.doi.org/10.1038/ejhg.2014.103. Epub 2014/06/12. PubMed PMID:24916642; PMCID:PMC4297892.

Bookstein, F. L., Streissguth, A. P., Sampson, P. D., Connor, P. D., & Barr, H. M. (2002). Corpus callosum shape and neuropsychological deficits in adult males with heavy fetal alcohol exposure. *NeuroImage, 15*(1), 233–251. http://dx.doi.org/10.1006/nimg.2001.0977. Epub 2002/01/05. PubMed PMID:11771992.

Brady, P. D., & Vermeesch, J. R. (2012). Genomic microarrays: a technology overview. *Prenatal Diagnosis, 32*(4), 336–343. http://dx.doi.org/10.1002/pd. 2933. Epub 2012/04/03. PubMed PMID:22467164.

Brancati, F., Barrano, G., Silhavy, J. L., Marsh, S. E., Travaglini, L., Bielas, S. L., … Gleeson, J. G. (2007). CEP290 mutations are frequently identified in the oculo-renal form of Joubert syndrome-related disorders. *American Journal of Human Genetics, 81*(1), 104–113. http://dx.doi.org/10.1086/519026. Epub 2007/06/15. PubMed PMID:17564967; PMCID:1950920.

Brancati, F., Travaglini, L., Zablocka, D., Boltshauser, E., Accorsi, P., Montagna, G., … Valente, E. M. (2008). RPGRIP1L mutations are mainly associated with the cerebello-renal phenotype of Joubert syndrome-related disorders. *Clinical Genetics, 74*(2), 164–170. http://dx.doi.org/10.1111/j.1399-0004.2008.01047.x. Epub 2008/06/21. PubMed PMID:18565097; PMCID:PMC2752690.

Cornelius, M. D., & Day, N. L. (2009). Developmental consequences of prenatal tobacco exposure. *Current Opinion in Neurology, 22*(2), 121–125. http://dx.doi.org/10.1097/WCO.0b013e328326f6dc. PubMed PMID:19532034; PMCID:2745235.

Corvin, A. P. (2010). Neuronal cell adhesion genes: key players in risk for schizophrenia, bipolar disorder and other neurodevelopmental brain disorders? *Cell Adhesion & Migration, 4*(4), 511–514. Epub 2010/06/25. PubMed PMID:20574149; PMCID:3011270.

Coulter, M. E., Miller, D. T., Harris, D. J., Hawley, P., Picker, J., Roberts, A. E., … Irons, M. (2011). Chromosomal microarray testing influences medical management. *Genetics in Medicine: Official Journal of the American College of Medical Genetics, 13*(9), 770–776. http://dx.doi.org/10.1097/GIM.0b013e31821dd54a. Epub 2011/07/01. PubMed PMID:21716121.

Dafinger, C., Liebau, M. C., Elsayed, S. M., Hellenbroich, Y., Boltshauser, E., Korenke, G. C., … Bolz, H. J. (2011). Mutations in KIF7 link Joubert syndrome with sonic hedgehog signaling and microtubule dynamics. *The Journal of Clinical Investigation, 121*(7), 2662–2667. http://dx.doi.org/10.1172/JCI43639. Epub 2011/06/03. PubMed PMID:21633164; PMCID:3223820.

Davis, E., Fennoy, I., Laraque, D., Kanem, N., Brown, G., & Mitchell, J. (1992). Autism and developmental abnormalities in children with perinatal cocaine exposure. *Journal of the National Medical Association, 84*(4), 315–319. PubMed PMID:1380564; PMCID:2637680.

Dobyns, W. B., Filauro, A., Tomson, B. N., Chan, A. S., Ho, A. W., Ting, N. T., … Ober, C. (2004). Inheritance of most X-linked traits is not dominant or recessive, just X-linked. *American Journal of Medical Genetics Part A, 129A*(2), 136–143. http://dx.doi.org/10.1002/ajmg.a.30123. PubMed PMID:15316978.

Dong, S., Walker, M. F., Carriero, N. J., DiCola, M., Willsey, A. J., Ye, A. Y., … Sanders, S. J. (2014). De novo insertions and deletions of predominantly paternal origin are associated with autism spectrum disorder. *Cell Reports, 9*(1), 16–23. http://dx.doi.org/10.1016/j.celrep.2014.08.068. Epub 2014/10/07. PubMed PMID:25284784; PMCID:PMC4194132.

Dou, X., & Charness, M. E. (2014). Effect of lipid raft disruption on ethanol inhibition of l1 adhesion. *Alcoholism, Clinical and Experimental Research, 38*(11), 2707–2711. http://dx.doi.org/10.1111/acer.12556. Epub 2014/11/26. PubMed PMID:25421507; PMCID:PMC4278581.

Drivas, T. G., Wojno, A. P., Tucker, B. A., Stone, E. M., & Bennett, J. (2015). Basal exon skipping and genetic pleiotropy: a predictive model of disease pathogenesis. *Science Translational Medicine, 7*(291), 291ra97. http://dx.doi.org/10.1126/scitranslmed.aaa5370.

Durner, M., Keddache, M. A., Tomasini, L., Shinnar, S., Resor, S. R., Cohen, J., … Greenberg, D. A. (2001). Genome scan of idiopathic generalized epilepsy: evidence for major susceptibility gene and modifying genes influencing the seizure type. *Annals of Neurology, 49*(3), 328–335. PubMed PMID:11261507.

El Arbi, S., Demant, A., Kohlschmidt, N., & Horneff, G. (2013). Infantile neuroaxonal dystrophy: a rare cause of early childhood ataxia with poor prognosis. *Klinische Padiatrie, 225*(1), 41–42. http://dx.doi.org/10.1055/s-0032-1323834. PubMed PMID:23073992.

El Chehadeh, S., Bonnet, C., Callier, P., Beri, M., Dupre, T., Payet, M., … Robinet, C. (2015). Homozygous truncating intragenic duplication in TUSC3 responsible for rare autosomal recessive nonsyndromic intellectual disability with no clinical or biochemical metabolic markers. *JIMD Reports, 20*, 45–55. http://dx.doi.org/10.1007/8904_2014_390. PubMed PMID:25626710; PMCID:4375127.

Emmitte, K. A. (2013). mGlu5 negative allosteric modulators: a patent review (2010–2012). *Expert Opinion on Therapeutic Patents, 23*(4), 393–408. http://dx.doi.org/10.1517/13543776.2013.760544. Epub 2013/01/24. PubMed PMID:23339457.

EPICURE Consortium, Leu, C., de Kovel, C. G., Zara, F., Striano, P., Pezzella, M., … Sander, T. (2012). Genome-wide linkage meta-analysis identifies susceptibility loci at 2q34 and 13q31.3 for genetic generalized epilepsies. *Epilepsia, 53*(2), 308–318. http://dx.doi.org/10.1111/j.1528-1167.2011.03379.x. PubMed PMID:22242659.

Esmaeeli Nieh, S., & Sherr, E. H. (2014). Epileptic encephalopathies: new genes and new pathways. *Neurotherapeutics: The Journal of the American Society for Experimental Neurotherapeutics*. http://dx.doi.org/10.1007/s13311-014-0301-2. Epub 2014/10/01. PubMed PMID:25266964.

EuroEPINOMICS-RES Consortium, Epilepsy Phenome/Genome Project, & Epi4K Consortium. (2014). De novo mutations in synaptic transmission genes including DNM1 cause epileptic encephalopathies. *American Journal of Human Genetics, 95*(4), 360–370. http://dx.doi.org/10.1016/j.ajhg.2014.08.013. PubMed PMID:25262651; PMCID:4185114.

Fortress, A. M., Hamlett, E. D., Vazey, E. M., Aston-Jones, G., Cass, W. A., Boger, H. A., & Granholm, A. C. (2015). Designer receptors enhance memory in a mouse model of Down syndrome. *The Journal of Neuroscience: The Official Journal of the Society for Neuroscience, 35*(4), 1343–1353. http://dx.doi.org/10.1523/JNEUROSCI.2658-14.2015. PubMed PMID:25632113; PMCID:4308587.

Gandomi, S. K., Farwell Gonzalez, K. D., Parra, M., Shahmirzadi, L., Mancuso, J., Pichurin, P., … Tang, S. (2014). Diagnostic exome sequencing identifies two novel IQSEC2 mutations associated with X-linked intellectual disability with seizures: implications for genetic counseling and clinical diagnosis. *Journal of Genetic Counseling, 23*(3), 289–298. http://dx.doi.org/10.1007/s10897-013-9671-6. PubMed PMID:24306141.

Gatzke-Kopp, L. M., & Beauchaine, T. P. (2007). Direct and passive prenatal nicotine exposure and the development of externalizing psychopathology. *Child Psychiatry and Human Development, 38*(4), 255–269. http://dx.doi.org/10.1007/s10578-007-0059-4. PubMed PMID:17520361; PMCID:2711763.

Gauthier, T. W., Mohan, S. S., Gross, T. S., Harris, F. L., Guidot, D. M., & Brown, L. A. (2015). Placental fatty acid ethyl esters are elevated with maternal alcohol use in pregnancies complicated by prematurity. *PLoS One, 10*(5), e0126552. http://dx.doi.org/10.1371/journal.pone.0126552. Epub 2015/05/16. PubMed PMID:25978403; PMCID:PMC4433254.

Gilman, S. R., Iossifov, I., Levy, D., Ronemus, M., Wigler, M., & Vitkup, D. (2011). Rare de novo variants associated with autism implicate a large functional network of genes involved in formation and function of synapses. *Neuron, 70*(5), 898–907. http://dx.doi.org/10.1016/j.neuron.2011.05.021. Epub 2011/06/11. PubMed PMID:21658583.

Gross, C., Chang, C. W., Kelly, S. M., Bhattacharya, A., McBride, S. M., Danielson, S. W., … Bassell, G. J. (2015). Increased expression of the PI3K enhancer PIKE mediates deficits in synaptic plasticity and behavior in fragile x syndrome. *Cell Reports, 11*(5), 727–736. http://dx.doi.org/10.1016/j.celrep.2015.03.060. PubMed PMID:25921541; PMCID:4418204.

Hagerman, R. J., Chudley, A. E., Knoll, J. H., Jackson, A. W., 3rd, Kemper, M., & Ahmad, R. (1986). Autism in fragile X females. *American Journal of Medical Genetics, 23*(1–2), 375–380. Epub 1986/01/01. PubMed PMID:3953655.

Henderson, L. B., Applegate, C. D., Wohler, E., Sheridan, M. B., Hoover-Fong, J., & Batista, D. A. (2014). The impact of chromosomal microarray on clinical management: a retrospective analysis. *Genetics in Medicine: Official Journal of the American College of Medical Genetics, 16*(9), 657–664. http://dx.doi.org/10.1038/gim.2014.18. Epub 2014/03/15. PubMed PMID:24625444.

Hill, W. D., Davies, G., van de Lagemaat, L. N., Christoforou, A., Marioni, R. E., Fernandes, C. P. D., … Deary, I. J. (2014). Human cognitive ability is influenced by genetic variation in components of postsynaptic signalling complexes assembled by NMDA receptors and MAGUK proteins. *Translational Psychiatry, 4*, e341. http://dx.doi.org/10.1038/tp.2013.114.

Hillman, S. C., McMullan, D. J., Williams, D., Maher, E. R., & Kilby, M. D. (2012). Microarray comparative genomic hybridization in prenatal diagnosis: a review. *Ultrasound in Obstetrics & Gynecology: The Official Journal of the International Society of Ultrasound in Obstetrics and Gynecology, 40*(4), 385–391. http://dx.doi.org/10.1002/uog.11180. Epub 2012/08/14. PubMed PMID:22887694.

Hu, H., Haas, S. A., Chelly, J., Van Esch, H., Raynaud, M., de Brouwer, A. P., … Kalscheuer, V. M. (2015). X-exome sequencing of 405 unresolved families identifies seven novel intellectual disability genes. *Molecular Psychiatry*. http://dx.doi.org/10.1038/mp.2014.193. PubMed PMID:25644381.

Hwu, W. L., Lee, N. C., Chien, Y. H., Muramatsu, S., & Ichinose, H. (2013). AADC deficiency: occurring in humans, modeled in rodents. *Advances in Pharmacology (San Diego, CA), 68*, 273–284. http://dx.doi.org/10.1016/b978-0-12-411512-5.00013-0. Epub 2013/09/24. PubMed PMID:24054149.

Hwu, W. L., Muramatsu, S., Tseng, S. H., Tzen, K. Y., Lee, N. C., Chien, Y. H., … Wu, R. M. (2012). Gene therapy for aromatic L-amino acid decarboxylase deficiency. *Science Translational Medicine, 4*(134), 134ra61. http://dx.doi.org/10.1126/scitranslmed.3003640. Epub 2012/05/18. PubMed PMID:22593174.

Ikonomidou, C., Bittigau, P., Ishimaru, M. J., Wozniak, D. F., Koch, C., Genz, K., … Olney, J. W. (2000). Ethanol-induced apoptotic neurodegeneration and fetal alcohol syndrome. *Science, 287*(5455), 1056–1060. Epub 2000/02/11. PubMed PMID:10669420.

Innocenti, G. M., Ansermet, F., & Parnas, J. (2003). Schizophrenia, neurodevelopment and corpus callosum. *Molecular Psychiatry, 8*(3), 261–274. http://dx.doi.org/10.1038/sj.mp.4001205. PubMed PMID:12660799.

Iossifov, I., O'Roak, B. J., Sanders, S. J., Ronemus, M., Krumm, N., Levy, D., … Wigler, M. (2014). The contribution of de novo coding mutations to autism spectrum disorder. *Nature*. http://dx.doi.org/10.1038/nature13908. PubMed PMID:25363768.

Jacobs, S. E., Berg, M., Hunt, R., Tarnow-Mordi, W. O., Inder, T. E., & Davis, P. G. (2013). Cooling for newborns with hypoxic ischaemic encephalopathy. *The Cochrane Database of Systematic Reviews, 1*, CD003311. http://dx.doi.org/10.1002/14651858.CD003311.pub3. PubMed PMID:23440789.

Jacquemont, S., Berry-Kravis, E., Hagerman, R., von Raison, F., Gasparini, F., Apostol, G., … Mancilla, B. (2014). The challenges of clinical trials in fragile X syndrome. *Psychopharmacology, 231*(6), 1237–1250. http://dx.doi.org/10.1007/s00213-013-3289-0. Epub 2013/11/01. PubMed PMID:24173622; PMCID:PMC3932172.

Jiang, M., Ash, R. T., Baker, S. A., Suter, B., Ferguson, A., Park, J., … Smirnakis, S. M. (2013). Dendritic arborization and spine dynamics are abnormal in the mouse model of MECP2 duplication syndrome. *The Journal of Neuroscience: The Official Journal of the Society for Neuroscience, 33*(50), 19518–19533. http://dx.doi.org/10.1523/JNEUROSCI.1745-13.2013. PubMed PMID:24336718; PMCID:3858623.

Johnson, S., Evans, T. A., Draper, E. S., Field, D. J., Manktelow, B. N., Marlow, N., … Boyle, E. M. (2015). Neurodevelopmental outcomes following late and moderate prematurity: a population-based cohort study. *Archives of Disease in Childhood Fetal and Neonatal Edition*. http://dx.doi.org/10.1136/archdischild-2014-307684. PubMed PMID:25834170.

Kan, S. H., Aoyagi-Scharber, M., Le, S. Q., Vincelette, J., Ohmi, K., Bullens, S., … Neufeld, E. F. (2014). Delivery of an enzyme-IGFII fusion protein to the mouse brain is therapeutic for mucopolysaccharidosis type IIIB. *Proceedings of the National Academy of Sciences of the United States of America, 111*(41), 14870–14875. http://dx.doi.org/10.1073/pnas.1416660111. Epub 2014/10/01. PubMed PMID:25267636; PMCID:PMC4205671.

Klei, L., Sanders, S. J., Murtha, M. T., Hus, V., Lowe, J. K., Willsey, A. J., … Devlin, B. (2012). Common genetic variants, acting additively, are a major source of risk for autism. *Molecular Autism, 3*(1), 9. http://dx.doi.org/10.1186/2040-2392-3-9. Epub 2012/10/17. PubMed PMID:23067556; PMCID:3579743.

Kremer, E. J., Yu, S., Pritchard, M., Nagaraja, R., Heitz, D., Lynch, M., … Richards, R. I. (1991). Isolation of a human DNA sequence which spans the fragile X. *American Journal of Human Genetics, 49*(3), 656–661. PubMed PMID:1882843; PMCID:1683130.

Krencik, R., Hokanson, K. C., Narayan, A. R., Dvornik, J., Rooney, G. E., Rauen, K. A., … Ullian, E. M. (2015). Dysregulation of astrocyte extracellular signaling in Costello syndrome. *Science Translational Medicine, 7*(286), 286ra66. http://dx.doi.org/10.1126/scitranslmed.aaa5645. Epub 2015/05/08. PubMed PMID:25947161.

Law, R., Dixon-Salazar, T., Jerber, J., Cai, N., Abbasi, A. A., Zaki, M. S., … Gleeson, J. G. (2014). Biallelic truncating mutations in FMN2, encoding the actin-regulatory protein formin 2, cause nonsyndromic autosomal-recessive intellectual disability. *American Journal of Human Genetics, 95*(6), 721–728. http://dx.doi.org/10.1016/j.ajhg.2014.10.016. Epub 2014/12/07. PubMed PMID:25480035.

Lebel, C., Roussotte, F., & Sowell, E. R. (2011). Imaging the impact of prenatal alcohol exposure on the structure of the developing human brain. *Neuropsychology Review, 21*(2), 102–118. http://dx.doi.org/10.1007/s11065-011-9163-0. Epub 2011/03/04. PubMed PMID:21369875; PMCID:PMC3098972.

Lee, H., Deignan, J. L., Dorrani, N., Strom, S. P., Kantarci, S., Quintero-Rivera, F., … Nelson, S. F. (2014). Clinical exome sequencing for genetic identification of rare Mendelian disorders. *JAMA: The Journal of the American Medical Association, 312*(18), 1880–1887. http://dx.doi.org/10.1001/jama.2014.14604. Epub 2014/10/19. PubMed PMID:25326637; PMCID:PMC4278636.

Leehey, M. A., & Hagerman, P. J. (2012). Fragile X-associated tremor/ataxia syndrome. Edited by P.J. Vinken and G.W. Bruyn *Handbook of Clinical Neurology, 103*, 373–386. http://dx.doi.org/10.1016/b978-0-444-51892-7.00023-1. Epub 2011/08/11. PubMed PMID:21827901.

Li, J., Cai, T., Jiang, Y., Chen, H., He, X., Chen, C., … Wu, J. (2015). Genes with de novo mutations are shared by four neuropsychiatric disorders discovered from NP de novo database. *Molecular Psychiatry*. http://dx.doi.org/10.1038/mp.2015.40. Epub 2015/04/08. PubMed PMID:25849321.

de Ligt, J., Willemsen, M. H., van Bon, B. W., Kleefstra, T., Yntema, H. G., Kroes, T., … Vissers, L. E. (2012). Diagnostic exome sequencing in persons with severe intellectual disability. *The New England Journal of Medicine, 367*(20), 1921–1929. http://dx.doi.org/10.1056/NEJMoa1206524. PubMed PMID:23033978.

Lin, G. N., Corominas, R., Lemmens, I., Yang, X., Tavernier, J., Hill, D. E., … Iakoucheva, L. M. (2015). Spatiotemporal 16p11.2 protein network implicates cortical late mid-fetal brain development and KCTD13-Cul3-RhoA pathway in psychiatric diseases. *Neuron, 85*(4), 742–754. http://dx.doi.org/10.1016/j.neuron.2015.01.010. PubMed PMID:25695269; PMCID:4335356.

Littner, Y., Tang, N., He, M., & Bearer, C. F. (2013). L1 cell adhesion molecule signaling is inhibited by ethanol in vivo. *Alcoholism, Clinical and Experimental Research, 37*(3), 383–389. http://dx.doi.org/10.1111/j.1530-0277.2012.01944.x. Epub 2012/10/12. PubMed PMID:23050935; PMCID:PMC3543757.

Lubs, H. A., Stevenson, R. E., & Schwartz, C. E. (2012). Fragile X and X-linked intellectual disability: four decades of discovery. *American Journal of Human Genetics, 90*(4), 579–590. http://dx.doi.org/10.1016/j.ajhg.2012.02.018. PubMed PMID:22482801; PMCID:3322227.

Maier, R., Moser, G., Chen, G. B., Ripke, S., Coryell, W., Potash, J. B., … Lee, S. H. (2015). Joint analysis of psychiatric disorders increases accuracy of risk prediction for schizophrenia, bipolar disorder, and major depressive disorder. *American Journal of Human Genetics, 96*(2), 283–294. http://dx.doi.org/10.1016/j.ajhg.2014.12.006. Epub 2015/02/03. PubMed PMID:25640677; PMCID:PMC4320268.

Marco, E. J., Abidi, F. E., Bristow, J., Dean, W. B., Cotter, P., Jeremy, R. J., … Sherr, E. H. (2008). ARHGEF9 disruption in a female patient is associated with X linked mental retardation and sensory hyperarousal. *Journal of Medical Genetics, 45*(2), 100–105. http://dx.doi.org/10.1136/jmg.2007.052324. Epub 2007/09/26. PubMed PMID:17893116.

Martin, J. R., & Arici, A. (2008). Fragile X and reproduction. *Current Opinion in Obstetrics & Gynecology, 20*(3), 216–220. http://dx.doi.org/10.1097/GCO.0b013e3282fe7254. Epub 2008/05/08. PubMed PMID:18460934.

May, P. A., Baete, A., Russo, J., Elliott, A. J., Blankenship, J., Kalberg, W. O., … Hoyme, H. E. (2014). Prevalence and characteristics of fetal alcohol spectrum disorders. *Pediatrics, 134*(5), 855–866. http://dx.doi.org/10.1542/peds.2013-3319. Epub 2014/10/29. PubMed PMID:25349310; PMCID:PMC4210790.

McGrath, J. J., & Richards, L. J. (2009). Why schizophrenia epidemiology needs neurobiology—and vice versa. *Schizophrenia Bulletin, 35*(3), 577–581. http://dx.doi.org/10.1093/schbul/sbp004. PubMed PMID:19273583; PMCID:2669581.

McMichael, G., Bainbridge, M. N., Haan, E., Corbett, M., Gardner, A., Thompson, S., … MacLennan, A. H. (2015). Whole-exome sequencing points to considerable genetic heterogeneity of cerebral palsy. *Molecular Psychiatry*. http://dx.doi.org/10.1038/mp.2014.189. Epub 2015/02/11. PubMed PMID:25666757.

Meziane, H., Schaller, F., Bauer, S., Villard, C., Matarazzo, V., Riet, F., … Muscatelli, F. (2014). An early postnatal oxytocin treatment prevents social and learning deficits in adult mice deficient for Magel2, a gene involved in Prader-Willi syndrome and autism. *Biological Psychiatry*. http://dx.doi.org/10.1016/j.biopsych.2014.11.010. Epub 2015/01/21. PubMed PMID:25599930.

Michalon, A., Sidorov, M., Ballard, T. M., Ozmen, L., Spooren, W., Wettstein, J. G., … Lindemann, L. (2012). Chronic pharmacological mGlu5 inhibition corrects fragile X in adult mice. *Neuron, 74*(1), 49–56. http://dx.doi.org/10.1016/j.neuron.2012.03.009. Epub 2012/04/17. PubMed PMID:22500629.

Miller, D. T., Adam, M. P., Aradhya, S., Biesecker, L. G., Brothman, A. R., Carter, N. P., … Ledbetter, D. H. (2010). Consensus statement: chromosomal microarray is a first-tier clinical diagnostic test for individuals with developmental disabilities or congenital anomalies. *American Journal of Human Genetics, 86*(5), 749–764. http://dx.doi.org/10.1016/j.ajhg.2010.04.006. PubMed PMID:20466091; PMCID:2869000.

Molloy, C. S., Anderson, P. J., Anderson, V. A., & Doyle, L. W. (2015). The long-term outcome of extremely preterm (<28 weeks' gestational age) infants with and without severe retinopathy of prematurity. *Journal of Neuropsychology*. http://dx.doi.org/10.1111/jnp.12069. PubMed PMID:25809467.

Morini, L., Marchei, E., Tarani, L., Trivelli, M., Rapisardi, G., Elicio, M. R., ... Pichini, S. (2013). Testing ethylglucuronide in maternal hair and nails for the assessment of fetal exposure to alcohol: comparison with meconium testing. *Therapeutic Drug Monitoring, 35*(3), 402–407. http://dx.doi.org/10.1097/FTD.0b013e318283f719. Epub 2013/05/15. PubMed PMID:23666568.

Morleo, M., Iaconis, D., Chitayat, D., Peluso, I., Marzella, R., Renieri, A., ... Franco, B. (2008). Disruption of the IQSEC2 transcript in a female with X;autosome translocation t(X;20)(p11.2;q11.2) and a phenotype resembling X-linked infantile spasms (ISSX) syndrome. *Molecular Medicine Reports, 1*(1), 33–39. PubMed PMID:21479374.

Mullard, A. (2015). Fragile X disappointments upset autism ambitions. *Nature Reviews Drug Discovery, 14*(3), 151–153. http://dx.doi.org/10.1038/nrd4555. Epub 2015/02/28. PubMed PMID:25722228.

Nelson, K. B., & Chang, T. (2008). Is cerebral palsy preventable? *Current Opinion in Neurology, 21*(2), 129–135. http://dx.doi.org/10.1097/WCO.0-b013e3282f4958b. Epub 2008/03/05. PubMed PMID:18317269.

Nicholas, T. J., Baker, C., Eichler, E. E., & Akey, J. M. (2011). A high-resolution integrated map of copy number polymorphisms within and between breeds of the modern domesticated dog. *BMC Genomics, 12*, 414. http://dx.doi.org/10.1186/1471-2164-12-414. Epub 2011/08/19. PubMed PMID:21846351; PMCID:3166287.

Novelli, A., Grati, F. R., Ballarati, L., Bernardini, L., Bizzoco, D., Camurri, L., ... Giardino, D. (2012). Microarray application in prenatal diagnosis: a position statement from the cytogenetics working group of the Italian Society of Human Genetics (SIGU), November 2011. *Ultrasound in Obstetrics & Gynecology: The Official Journal of the International Society of Ultrasound in Obstetrics and Gynecology, 39*(4), 384–388. http://dx.doi.org/10.1002/uog.11092. Epub 2012/01/21. PubMed PMID:22262341.

O'Dushlaine, C., Kenny, E., Heron, E., Donohoe, G., Gill, M., Morris, D., & Corvin, A. (2011). Molecular pathways involved in neuronal cell adhesion and membrane scaffolding contribute to schizophrenia and bipolar disorder susceptibility. *Molecular Psychiatry, 16*(3), 286–292. http://dx.doi.org/10.1038/mp.2010.7. Epub 2010/02/17. PubMed PMID:20157312.

Olney, J. W., Farber, N. B., Wozniak, D. F., Jevtovic-Todorovic, V., & Ikonomidou, C. (2000). Environmental agents that have the potential to trigger massive apoptotic neurodegeneration in the developing brain. *Environmental Health Perspectives, 108*(Suppl. 3), 383–388. Epub 2000/06/15. PubMed PMID:10852832; PMCID:PMC1637813.

Olney, J. W., Wozniak, D. F., Jevtovic-Todorovic, V., & Ikonomidou, C. (2001). Glutamate signaling and the fetal alcohol syndrome. *Mental Retardation and Developmental Disabilities Research Reviews, 7*(4), 267–275. http://dx.doi.org/10.1002/mrdd. 1037. Epub 2002/01/05. PubMed PMID:11754521.

O'Roak, B. J., Stessman, H. A., Boyle, E. A., Witherspoon, K. T., Martin, B., Lee, C., ... Eichler, E. E. (2014). Recurrent de novo mutations implicate novel genes underlying simplex autism risk. *Nature Communications, 5*, 5595. http://dx.doi.org/10.1038/ncomms6595. Epub 2014/11/25. PubMed PMID:25418537; PMCID:PMC4249945.

Owen, M. J., O'Donovan, M. C., Thapar, A., & Craddock, N. (2011). Neurodevelopmental hypothesis of schizophrenia. *The British Journal of Psychiatry: The Journal of Mental Science, 198*(3), 173–175. http://dx.doi.org/10.1192/bjp.bp.110.084384. Epub 2011/03/02. PubMed PMID:21357874.

Paul, L. K. (2011). Developmental malformation of the corpus callosum: a review of typical callosal development and examples of developmental disorders with callosal involvement. *Journal of Neurodevelopmental Disorders, 3*(1), 3–27. http://dx.doi.org/10.1007/s11689-010-9059-y. Epub 2011/04/13. PubMed PMID:21484594; PMCID:3163989.

Pehlivan, D., Karaca, E., Aydin, H., Beck, C. R., Gambin, T., Muzny, D. M., ... Lupski, J. R. (2014). Whole-exome sequencing links TMCO1 defect syndrome with cerebro-facio-thoracic dysplasia. *European Journal of Human Genetics: EJHG, 22*(9), 1145–1148. http://dx.doi.org/10.1038/ejhg.2013.291. PubMed PMID:24424126; PMCID:4135405.

Pemberton, H. N., Franklyn, J. A., & Kilby, M. D. (2005). Thyroid hormones and fetal brain development. *Minerva Ginecologica, 57*(4), 367–378. Epub 2005/09/20. PubMed PMID:16170282.

Peters, G. B., & Pertile, M. D. (2014). Chromosome microarrays in diagnostic testing: interpreting the genomic data. *Methods in Molecular Biology (Clifton, NJ), 1168*, 117–155. http://dx.doi.org/10.1007/978-1-4939-0847-9_8. Epub 2014/05/30. PubMed PMID:24870134.

Pruett, D., Waterman, E. H., & Caughey, A. B. (2013). Fetal alcohol exposure: consequences, diagnosis, and treatment. *Obstetrical & Gynecological Survey, 68*(1), 62–69. http://dx.doi.org/10.1097/OGX.0b013e31827f238f. Epub 2013/01/17. PubMed PMID:23322082.

Pyle, A., Smertenko, T., Bargiela, D., Griffin, H., Duff, J., Appleton, M., ... Chinnery, P. F. (2015). Exome sequencing in undiagnosed inherited and sporadic ataxias. *Brain: A Journal of Neurology, 138*(Pt 2), 276–283. http://dx.doi.org/10.1093/brain/awu348. Epub 2014/12/17. PubMed PMID:25497598; PMCID:PMC4306819.

Rahman, S. H., Maeder, M. L., Joung, J. K., & Cathomen, T. (2011). Zinc-finger nucleases for somatic gene therapy: the next frontier. *Human Gene Therapy, 22*(8), 925–933. http://dx.doi.org/10.1089/hum.2011.087. Epub 2011/06/03. PubMed PMID:21631241; PMCID:3159524.

Reiss, A. L., & Freund, L. (1990). Fragile X syndrome. *Biological Psychiatry, 27*(2), 223–240. Epub 1990/01/15. PubMed PMID:2403814.

Roebuck, T. M., Mattson, S. N., & Riley, E. P. (2002). Interhemispheric transfer in children with heavy prenatal alcohol exposure. *Alcoholism, Clinical and Experimental Research, 26*(12), 1863–1871. http://dx.doi.org/10.1097/01.alc.0000042219.73648.46. Epub 2002/12/25. PubMed PMID:12500111.

Roggero, P., Gianni, M. L., Garbarino, F., & Mosca, F. (2013). Consequences of prematurity on adult morbidities. *European Journal of Internal Medicine, 24*(7), 624–626. http://dx.doi.org/10.1016/j.ejim.2013.01.011. PubMed PMID:23385012.

Ropers, F., Derivery, E., Hu, H., Garshasbi, M., Karbasiyan, M., Herold, M., ... Rajab, A. (2011). Identification of a novel candidate gene for non-syndromic autosomal recessive intellectual disability: the WASH complex member SWIP. *Human Molecular Genetics, 20*(13), 2585–2590. http://dx.doi.org/10.1093/hmg/ddr158. PubMed PMID:21498477.

Rosenfeld, J. A., Coe, B. P., Eichler, E. E., Cuckle, H., & Shaffer, L. G. (2012). Estimates of penetrance for recurrent pathogenic copy-number variations. *Genetics in Medicine: Official Journal of the American College of Medical Genetics*. http://dx.doi.org/10.1038/gim.2012.164. Epub 2012/12/22. PubMed PMID:23258348.

San Sebastian, W., Kells, A. P., Bringas, J., Samaranch, L., Hadaczek, P., Ciesielska, A., … Bankiewicz, K. S. (2014). Safety and tolerability of mri-guided infusion of AAV2-hAADC into the mid-brain of non-human primate. *Molecular Therapy Methods & Clinical Development*, 3. http://dx.doi.org/10.1038/mtm.2014.49. PubMed PMID:25541617; PMCID:4274790.

Sanders, S. J., Murtha, M. T., Gupta, A. R., Murdoch, J. D., Raubeson, M. J., Willsey, A. J., … State, M. W. (2012). De novo mutations revealed by whole-exome sequencing are strongly associated with autism. *Nature, 485*(7397), 237–241. http://dx.doi.org/10.1038/nature10945. Epub 2012/04/13. PubMed PMID:22495306; PMCID:3667984.

Schaaf, C. P., Goin-Kochel, R. P., Nowell, K. P., Hunter, J. V., Aleck, K. A., Cox, S., … Shinawi, M. (2011). Expanding the clinical spectrum of the 16p11.2 chromosomal rearrangements: three patients with syringomyelia. *European Journal of Human Genetics: EJHG, 19*(2), 152–156. http://dx.doi.org/10.1038/ejhg.2010.168. Epub 2010/10/21. PubMed PMID:20959866; PMCID:3025795.

Schaaf, C. P., Gonzalez-Garay, M. L., Xia, F., Potocki, L., Gripp, K. W., Zhang, B., … Yang, Y. (2013). Truncating mutations of MAGEL2 cause Prader-Willi phenotypes and autism. *Nature Genetics, 45*(11), 1405–1408. http://dx.doi.org/10.1038/ng. 2776. Epub 2013/10/01. PubMed PMID:24076603; PMCID:PMC3819162.

Schaller, F., Watrin, F., Sturny, R., Massacrier, A., Szepetowski, P., & Muscatelli, F. (2010). A single postnatal injection of oxytocin rescues the lethal feeding behaviour in mouse newborns deficient for the imprinted Magel2 gene. *Human Molecular Genetics, 19*(24), 4895–4905. http://dx.doi.org/10.1093/hmg/ddq424. Epub 2010/09/30. PubMed PMID:20876615.

Scholz, R., Berberich, S., Rathgeber, L., Kolleker, A., Kohr, G., & Kornau, H. C. (2010). AMPA receptor signaling through BRAG2 and Arf6 critical for long-term synaptic depression. *Neuron, 66*(5), 768–780. http://dx.doi.org/10.1016/j.neuron.2010.05.003. PubMed PMID:20547133.

Sherr, E. H. (2003). The ARX story (epilepsy, mental retardation, autism, and cerebral malformations): one gene leads to many phenotypes. *Current Opinion in Pediatrics, 15*(6), 567–571. Epub 2003/11/25. PubMed PMID:14631200.

Shoubridge, C., Tarpey, P. S., Abidi, F., Ramsden, S. L., Rujirabanjerd, S., Murphy, J. A., … Gecz, J. (2010). Mutations in the guanine nucleotide exchange factor gene IQSEC2 cause nonsyndromic intellectual disability. *Nature Genetics, 42*(6), 486–488. http://dx.doi.org/10.1038/ng.588. Epub 2010/05/18. PubMed PMID:20473311.

Skeaff, S. A. (2011). Iodine deficiency in pregnancy: the effect on neurodevelopment in the child. *Nutrients, 3*(2), 265–273. http://dx.doi.org/10.3390/nu3020265. Epub 2012/01/19. PubMed PMID:22254096; PMCID:PMC3257674.

Soden, S. E., Saunders, C. J., Willig, L. K., Farrow, E. G., Smith, L. D., Petrikin, J. E., … Kingsmore, S. F. (2014). Effectiveness of exome and genome sequencing guided by acuity of illness for diagnosis of neurodevelopmental disorders. *Science Translational Medicine, 6*(265), 265ra168. http://dx.doi.org/10.1126/scitranslmed.3010076. Epub 2014/12/05. PubMed PMID:25473036; PMCID:PMC4286868.

Sowell, E. R., Mattson, S. N., Kan, E., Thompson, P. M., Riley, E. P., & Toga, A. W. (2008). Abnormal cortical thickness and brain-behavior correlation patterns in individuals with heavy prenatal alcohol exposure. *Cerebral Cortex, 18*(1), 136–144. http://dx.doi.org/10.1093/cercor/bhm039. Epub 2007/04/20. PubMed PMID:17443018; PMCID:PMC2770438.

Stefansson, H., Meyer-Lindenberg, A., Steinberg, S., Magnusdottir, B., Morgen, K., Arnarsdottir, S., … Stefansson, K. (2014). CNVs conferring risk of autism or schizophrenia affect cognition in controls. *Nature, 505*(7483), 361–366. http://dx.doi.org/10.1038/nature12818. Epub 2013/12/20. PubMed PMID:24352232.

Sternberg, S. H., & Doudna, J. A. (2015). Expanding the Biologist's toolkit with CRISPR-Cas9. *Molecular Cell, 58*(4), 568–574. http://dx.doi.org/10.1016/j.molcel.2015.02.032. PubMed PMID:26000842.

Stratton, K., Gable, A., & McCormick, M. C. (Eds.). (2001). *Immunization safety review: Thimerosal-containing vaccines and neurodevelopmental disorders. Washington, DC.*

Stray-Pedersen, A., Backe, P. H., Sorte, H. S., Morkrid, L., Chokshi, N. Y., Erichsen, H. C., … Hanson, I. C. (2014). PGM3 mutations cause a congenital disorder of glycosylation with severe immunodeficiency and skeletal dysplasia. *American Journal of Human Genetics, 95*(1), 96–107. http://dx.doi.org/10.1016/j.ajhg.2014.05.007. PubMed PMID:24931394; PMCID:4085583.

Suthers, G. K., Turner, G., & Mulley, J. C. (1988). A non-syndromal form of X-linked mental retardation (XLMR) is linked to DXS14. *American Journal of Medical Genetics, 30*(1–2), 485–491. PubMed PMID:3177466.

Thompson, C. C., & Potter, G. B. (2000). Thyroid hormone action in neural development. *Cerebral Cortex, 10*(10), 939–945. PubMed PMID:11007544.

Tran Mau-Them, F., Willems, M., Albrecht, B., Sanchez, E., Puechberty, J., Endele, S., … Genevieve, D. (2014). Expanding the phenotype of IQSEC2 mutations: truncating mutations in severe intellectual disability. *European Journal of Human Genetics: EJHG, 22*(2), 289–292. http://dx.doi.org/10.1038/ejhg.2013.113. PubMed PMID:23674175; PMCID:3895633.

Tsai, G., Gastfriend, D. R., & Coyle, J. T. (1995). The glutamatergic basis of human alcoholism. *The American Journal of Psychiatry, 152*(3), 332–340. Epub 1995/03/01. PubMed PMID:7864257.

Tsurusaki, Y., Kobayashi, Y., Hisano, M., Ito, S., Doi, H., Nakashima, M., … Miyake, N. (2012). The diagnostic utility of exome sequencing in Joubert syndrome and related disorders. *Journal of Human Genetics*. http://dx.doi.org/10.1038/jhg.2012.117. Epub 2012/10/05. PubMed PMID:23034536.

Veltman, M. W., Thompson, R. J., Roberts, S. E., Thomas, N. S., Whittington, J., & Bolton, P. F. (2004). Prader-Willi syndrome—a study comparing deletion and uniparental disomy cases with reference to autism spectrum disorders. *European Child & Adolescent Psychiatry, 13*(1), 42–50. http://dx.doi.org/10.1007/s00787-004-0354-6. Epub 2004/03/03. PubMed PMID:14991431.

Wapner, R. J., Driscoll, D. A., & Simpson, J. L. (2012). Integration of microarray technology into prenatal diagnosis: counselling issues generated during the NICHD clinical trial. *Prenatal Diagnosis, 32*(4), 396–400. http://dx.doi.org/10.1002/pd. 3863. Epub 2012/04/03. PubMed PMID:22467170.

West, J. R., Chen, W. J., & Pantazis, N. J. (1994). Fetal alcohol syndrome: the vulnerability of the developing brain and possible mechanisms of damage. *Metabolic Brain Disease, 9*(4), 291–322. PubMed PMID:7898398.

Wozniak, J. R., & Muetzel, R. L. (2011). What does diffusion tensor imaging reveal about the brain and cognition in fetal alcohol spectrum disorders? *Neuropsychology Review, 21*(2), 133–147. http://dx.doi.org/10.1007/s11065-011-9162-1. Epub 2011/02/25. PubMed PMID:21347880.

Yang, Y., Muzny, D. M., Reid, J. G., Bainbridge, M. N., Willis, A., Ward, P. A., … Eng, C. M. (2013). Clinical whole-exome sequencing for the diagnosis of mendelian disorders. *The New England Journal of Medicine, 369*(16), 1502–1511. http://dx.doi.org/10.1056/NEJMoa1306555. PubMed PMID:24088041.

Yang, Y., Muzny, D. M., Xia, F., Niu, Z., Person, R., Ding, Y., … Eng, C. M. (2014). Molecular findings among patients referred for clinical whole-exome sequencing. *JAMA: The Journal of the American Medical Association, 312*(18), 1870–1879. http://dx.doi.org/10.1001/jama.2014.14601. PubMed PMID:25326635; PMCID:4326249.

Yonekawa, Y., Harada, A., Okada, Y., Funakoshi, T., Kanai, Y., Takei, Y., … Hirokawa, N. (1998). Defect in synaptic vesicle precursor transport and neuronal cell death in KIF1A motor protein-deficient mice. *The Journal of Cell Biology, 141*(2), 431–441. PubMed PMID:9548721; PMCID:2148442.

Chapter 37

Molecular Architecture and Neurobiology of the Epilepsies

Ryan S. Dhindsa
Columbia University Medical Center, New York, NY, United States

Daniel H. Lowenstein
University of California San Francisco, San Francisco, CA, United States

David B. Goldstein
Columbia University Medical Center, New York, NY, United States

INTRODUCTION

Epilepsy is one of the most common neurological disorders, affecting roughly one in 26 people at some point in their lifetime (England, Liverman, Schultz, & Strawbridge, 2012). The International League Against Epilepsy (ILAE) defines an epileptic seizure as "a transient occurrence of signs and/or symptoms due to abnormal excessive or synchronous neuronal activity in the brain" (Fisher et al., 2005). Conceptually, epilepsy is characterized as an enduring predisposition of the brain to generate epileptic seizures, with a broad range of potential sensorimotor, cognitive, psychological, and social consequences. For use in clinical settings, the ILAE defines epilepsy as a disease of the brain that is defined by any of the three following conditions: (1) at least two unprovoked (or reflex) seizures occurring more than 24 h apart; (2) one unprovoked (or reflex) seizure and a probability of further seizures similar to the general recurrence risk after two unprovoked seizures (at least 60%) occurring over the next 10 years; and (3) diagnosis of an epilepsy syndrome (Fisher et al., 2014).

The epilepsies are highly heterogeneous, encompassing a wide spectrum of clinical subtypes that are defined by seizure type, EEG, and brain imaging criteria (Berg et al., 2010; Speed et al., 2014). Despite this heterogeneity, the epilepsies are broadly classified into two main groups for clinical use: focal epilepsies, in which seizures originate within one cerebral hemisphere, and generalized epilepsies, in which seizures occur in bilaterally distributed networks. However, there is a smaller proportion of epilepsy-related syndromes in which the seizures are unclassifiable. Although they have diverse and heterogeneous etiologies, many forms of epilepsy are considered to be highly genetic and heritable conditions. One of the most comprehensive twin studies indicated a higher frequency of concordance in monozygotic twins (44.4%) than dizygotic twins (9.7%) in both focal and generalized epilepsies (Berkovic, Howell, Hay, & Hopper, 1998). Furthermore, the risk of epilepsy among first-degree relatives of individuals with epilepsy is 4.7%, a 3.3-fold risk increase compared with population incidence (Peljto et al., 2014).

PROGRESS IN EPILEPSY GENETICS

Linkage Analysis

Some of the first genetic epilepsy studies relied on large pedigrees, linkage analyses, and positional cloning to implicate pathogenic gene mutations in epilepsies that show Mendelian inheritance patterns. Linkage analysis is based on the identification of polymorphic genetic markers that are distributed throughout the genome, such as single-nucleotide

Genomics, Circuits, and Pathways in Clinical Neuropsychiatry. http://dx.doi.org/10.1016/B978-0-12-800105-9.00037-8

polymorphisms (SNPs) and microsatellites, in all individuals of the family. The co-inheritance of the markers and affection status is then used to localize the genomic region within which the disease-causing mutations must reside. After mapping a candidate chromosomal region using such linkage analyses, targeted sequencing is then used to identify the disease-causing gene.

By employing these methods to study a family with autosomal dominant nocturnal frontal lobe epilepsy (ADNFLE), Steinlein et al. reported the first epilepsy gene, *CHRNA4*, in 1995. Since then, a variety of linkage analysis-based studies have been performed, resulting in the discovery of over 20 epilepsy genes (Ottman et al., 2010).

Genome-Wide Association Studies

The next efforts in epilepsy gene discovery were primarily based on genome-wide association studies (GWAS), which use high-throughput microarray-based technologies to genotype hundreds of thousands of SNPs across the genome. GWAS attempt to identify polymorphisms that are significantly associated with a specific trait. If a genetic association increases susceptibility to a given disease such as epilepsy, the associated genetic variant will be seen more often than expected by chance in diseased individuals. The idea of GWAS was exciting in the field of epilepsy genetics because, unlike linkage analyses, they can be conducted in case–control populations without relying on the acquisition of multiplex families. Unfortunately, the vast majority of epilepsy GWAS that have been completed to date have revealed either no evidence or, in a few cases, modest evidence for candidate genes or genetic risk factors, likely owing to small sample sizes and insufficient power.

However, the ILAE Consortium on Complex Epilepsies combined genome-wide association data from population-based data sets and 12 cohorts of individuals with common epilepsies (2014). Meta-analyses were conducted for three phenotypic groups: genetic generalized epilepsy (GGE), focal epilepsy, and all epilepsy (consisting of all patients with a confirmed diagnosis of epilepsy, including GGE, focal epilepsy, and unclassified epilepsy). Loci at 2q24.3 and 4p15.1, harboring *SCN1A* and *PCDH7*, respectively, were implicated in the all-epilepsy analysis with genome-wide significance. *SCN1A* encodes a sodium channel that is strongly linked to other epilepsy syndromes (see section Sodium Channels). In the GGE cohort, there was a single signal at 2p16.1, implicating either *VRK2* or *FANCL*. No SNP achieved genome-wide significance for focal epilepsy.

Various hypotheses address the lack of strong signal from epilepsy GWAS. One hypothesis is that epilepsy is a highly heterogeneous disease and the successful application of GWAS requires large, well-phenotyped cohorts that are phenotypically homogeneous (Poduri & Lowenstein, 2011). Other hypotheses are covered in section Complex Epilepsies. Furthermore, epilepsy GWAS remain small compared with studies conducted for other neuropsychiatric disorders, such as schizophrenia. Obtaining larger cohorts will likely lead to the identification of more signals.

Next-Generation Sequencing

Rapid advances in next-generation sequencing (NGS), also known as massively parallel sequencing, have revolutionized the cost and speed with which human genomes can be sequenced. Unlike GWAS, NGS allows examination of nearly all of the genetic variants (rare or otherwise) in a given genome. High-throughput sequencing approaches generate millions of short sequence reads in parallel that are then aligned to the human reference genome. Computer algorithms are used to detect variants, including single-nucleotide variants (SNVs), insertions and deletions (indels), and structural variants.

Currently, one of the most popular NGS methods is whole-exome sequencing (WES), which is used to sequence only the coding portion of the genome. Before sequencing, a capture kit is used to isolate exonic and flanking intronic base pairs. Current exome-sequencing kits target only about 2% of the human genome, making it cheaper and faster than whole-genome sequencing (WGS). WES is clearly a pragmatic approach, both because most currently known disease-causing mutations occur in coding regions of the genome and because we are far better at interpreting variants in the exome than variants in the genome as a whole. The limitation of WES is that it largely focuses on protein coding exons, and therefore has limited ability to detect regulatory mutations that influence risk.

Perhaps the biggest challenge in interpreting NGS data is properly prioritizing the variants. First, one must consider broad factors such as the mode of inheritance, the frequency of the disorder, and the predicted deleterious nature of the variant. For example, if the disease is recessive, the study should focus on homozygous variants; if the disorder is rare, the causal variants should either be very rare or absent in control populations. Of course, in many cases the underlying genetic model is not known at the outset.

Furthermore, depending on the hypothesis, either a variant-based analysis or burden tests (described subsequently) can be used to interpret genetic variation. When studying rare variation, it is difficult to detect association using variant-based analyses (such as GWAS) because there are generally insufficient copies of the rare alleles to be tested as a single variant at a time. The better alternative is to aggregate statistical information across mutations within a functional unit, such as a gene or pathway. The most straightforward way to conduct this type of analysis is by using a burden test, in which the number of rare alleles in cases is tested against the number of alleles in control subjects. The choice of threshold for inclusion of these variants can be challenging. Thus, there are various burden methods that differ in the way they factor in allele frequencies of individual variants and whether they take weighted combinations of variants based on external information, such as functional class, frequency, and conservation. These tests assume that all variants act in the same direction with respect to disease risk. This assumption can be limiting in the case in which some variants increase risk whereas others decrease risk. However, methods such as the C-alpha test (Neale et al., 2011), the sequence kernel association test (Wu et al., 2011), and the estimated regression coefficient test (Lin & Tang, 2011) relax this assumption.

Various computational models help prioritize both variants and genes for these analyses. To prioritize individual variants, algorithms such as SIFT (Sim et al., 2012), PolyPhen2 (Adzhubei et al., 2010), and others are used to predict the functional effect of amino acid changes. These methods can estimate the likelihood that a given mutation will damage the structure and function of a human protein. Although these models are useful, variants must be placed in an evolutionary context to determine whether they are pathogenic for a specific condition. By using population genetics, we can generate quantitative assessments of how well genes tolerate functional genetic variation on a genome-wide scale. Two such models are the Residual Variant Intolerance Score (RVIS) (Petrovski, Wang, Heinzen, Allen, & Goldstein, 2013) and constraint scores (Samocha et al., 2014). Both of these tools illustrate that disease genes tend to harbor fewer than expected functional mutations in a population.

GENETIC ARCHITECTURE OF THE EPILEPSIES

Although the initial success in identifying epilepsy genes came from pedigree analyses in which there was evidence of segregation of a major autosomal dominant gene, these families are not representative of most human subjects with inherited epilepsies. As with most common diseases, most epilepsies show complex inheritance patterns. It is clear that the epilepsies as a whole are characterized by extreme locus and allelic heterogeneity, which means that both mutations in genes at different chromosomal loci and different mutations at the same locus can cause the disease. Furthermore, the epilepsies presumably also involve a degree of inheritance that is either oligogenic (influenced by a few genes) or polygenic (influenced by many genes). Although it is hard to identify causal oligogenic and polygenic traits, we can study epilepsies with single genes of major effect to determine critical pathways and principles regarding causation of disease. In fact, advances in these epilepsies have been accompanied by an exponential increase in our understanding of the contribution of copy number variants, susceptibility alleles, and de novo mutations to epilepsy pathogenesis.

Overall, three overlapping categories describe the genetic architecture of the epilepsies (Thomas & Berkovic, 2014). At one end of the spectrum are the epilepsies in which a single gene mutation (inherited or de novo) with a large effect size accounts for the phenotypic features. This continuum evolves into the complex epilepsies, which are putatively polygenic or oligogenic and show extreme locus and allelic heterogeneity. At the other end of the spectrum are epilepsies with a major acquired cause, such as trauma and stroke. Even acquired cases, however, such as those caused by trauma, infections, and strokes, are known to have genetic contributions (Christensen et al., 2009; Dichgans, 2007; Kariuki et al., 2013).

Copy Number Variation in the Epilepsies

Large-scale variations in the human genome are common, even among healthy individuals. By convention, a deletion or duplication larger than 1000 base pairs (1 kilobase) is called a copy number variation (CNV), but CNVs can reach up to several million base pairs (megabases [Mb]). There are around 1500 regions of variable copy number spread across 360 Mb (about 12%) of the human genome (Redon et al., 2006). CNVs can be conceptualized into two groups: recurrent and private. Recurrent CNVs occur at genomic "hot spots" that are sensitive to CNV by unequal crossing over during meiosis (Mefford & Eichler, 2009). Private (nonrecurrent) CNVs are rare and are primarily caused by de novo mutations in gametes or at a very early stage of embryonic development (Mulley & Mefford, 2011). In contrast to recurrent CNVs, private CNVs can occur anywhere across the genome and can affect many genes. When these mutations are large, they are usually directly clinically relevant.

Evidence suggests that CNVs can have a role in many different neuropsychiatric disorders. Most notably, large deletions and duplications have been associated with mental retardation, schizophrenia, and autism. Interestingly, some of these same recurrent deletions (at 15q13.3 and 16p13.11) have been associated with GGE, one of the most common forms of epilepsy (Heinzen et al., 2010; Helbig et al., 2009). Helbig et al. found that recurrent microdeletions at 15q13.3 occurred in 12 of 1223 patients with GGE (0.98%) but none of 3699 control subjects. There was a 1.5-Mb "critical region" that was contained by the deletions across all 12 patients containing seven genes, including a plausible epilepsy candidate gene, *CHRNA7*, that encodes a subunit of the nicotinic acetylcholine receptor. In addition to the two CNVs at 15q13.3 and 16p13.11, an additional recurrent CNV at 15q11.2 has been associated with common epilepsies (de Kovel et al., 2010). Collectively, these three CNVs account for an estimated 2.9% of patients with GGE (Mefford et al., 2010).

Private CNVs can also be attributed to epilepsy, but because these variations are rare, individual nonrecurrent CNVs will not account for large proportions of patients with epilepsy. Nonetheless, these CNVs may include known epilepsy genes. For example, in a study of roughly 500 patients with GGE and focal epilepsy, two patients harbored microdeletions involving *AUTS2* (also associated with autism) and one had a microdeletion involving *CNTNAP2* (previously associated with autism, cortical dysplasia-focal epilepsy syndrome, and Pitt–Hopkins-like syndrome 1) (Mefford et al., 2010). Other investigations found that roughly 4% of patients with epileptic encephalopathies, a collection of very rare and severe epileptic syndromes characterized by treatment-refractory seizures and developmental delay, harbor rare and clearly pathogenic CNVs (Mefford et al., 2011).

CNVs clearly have an important role in epilepsy susceptibility. Microarray analysis is part of routine clinical use, largely replacing karyotyping as a genetic diagnosis tool (Thomas & Berkovic, 2014). Unlike SNVs and indels, predicting mechanism of pathogenicity of CNVs is often an inexact science, which makes it harder to gain precise insight into the underlying pathophysiology. As a result, it is more challenging to devise possible targeted therapies for CNVs.

De Novo Mutations in the Epilepsies

De novo mutations are new mutations that are seen in a child but in neither of the parents, arising either during meiosis of the parents' gametes or during a very early stage of embryonic development. The human mutation rate is between 7.6×10^{-9} and 2.2×10^{-8}, equating to roughly 40 de novo mutations per individual per generation (Lynch, 2010). Overall, each individual is expected to have roughly one exonic de novo mutation. These mutations can be identified through trio-based sequencing, in which the healthy biological parents and affected children are sequenced.

De novo mutations contribute to the pathogenesis of many neurological disorders, including intellectual disability, autism spectrum disorder, and epilepsy (Neale et al., 2012; O'Roak et al., 2011; Vissers et al., 2010). Studies have clearly implicated de novo mutations in the severe epileptic encephalopathies, which are prime examples of epilepsies that are strongly influenced by single mutations. The genetic etiologies of these devastating disorders have been unraveled at an unprecedented rate, owing in large part to trio-based WES, and have opened the floodgates to examining new disease mechanisms.

In one large study conducted by the Epilepsy Phenome/Genome Project and the Epi4K Consortium, trio-based WES was performed on 264 probands and their unaffected biological parents to identify de novo mutations in two main types of epileptic encephalopathies, infantile spasms, and Lennox–Gastaut syndrome (Epi4K Consortium, 2013). Using RVIS (see section Next-Generation Sequencing), the consortium confirmed that patients with epileptic encephalopathies show a clear statistical excess of de novo mutations in genes that are intolerant to genomic changes. This project identified seven recurrently mutated genes in 10% of patients; among these genes, five were previously linked to the encephalopathies and two were identified as novel epileptic encephalopathy genes with clear statistical evidence of association. In addition, the group provided suggestive evidence for the role of several other genes.

Another study of a cohort of 356 probands, including the 264 trios analyzed in the previously-described study, representing a collaboration between the Epi4K and EuroEpinomics consortia, revealed that de novo mutations causing epileptic encephalopathy are enriched in synaptic transmission genes (Euro Epinomics- R. E. S. Consortium, Epilepsy Phenome/Genome Project, & Epi4K Consortium, 2014). In fact, 75% of the 429 de novo mutations occurred in genes involved in this pathway, which emphasizes an important role of synaptic dysfunction in the epilepsies. Furthermore, in both of these large trio-based studies, the identified mutations are also enriched in specific gene sets that are regulated by the Fragile X protein (FMRP). Interestingly, this same association has been reported for autism spectrum disorders (Iossifov et al., 2012).

The challenge of implementing WES studies in clinical practice is that mutations in a great number of genes can cause a similar phenotype. For example, in the Epi4K study, only 29 of 264 patients had mutations in the same gene as at least one other patient (nine genes total). Whereas many of the identified de novo mutations in a clinical setting are likely disease-causing, each mutation must be considered case by case. If a predicted deleterious coding mutation occurs in a previously established gene, that mutation is generally considered causal (although this will obviously not always be so). For genes that are not established as epilepsy-causing genes, a gene-based approach can be used. Essentially, if de novo mutations in phenotypically similar patients occur in the same gene more often than expected by chance (based on site-specific mutation rates), this gene is significantly associated with the phenotype (Epi4K Consortium, 2013). As an increasing number of patients are sequenced, de novo mutations will surely continue to be implicated in the epilepsies and other disorders associated with seizures. It is also hypothesized that de novo mutations may provide an alternate explanation for the absence of a family history in certain patients with GGE (Scheffer, 2011).

Complex Epilepsies

As opposed to the epileptic encephalopathies, which are often influenced by single mutations of major effect, the more common epilepsies, such as the GGEs and nonlesional focal epilepsies, do not show obvious evidence of single mutations of major effect. Although the genetic models relevant here remain largely unknown, it is likely that discovery is currently difficult because these epilepsies have a broad range of different underlying genetic architectures. Whatever the real architecture, it is clear that the epilepsies as a whole are characterized by extreme locus and allelic heterogeneity, and presumably also involve a degree of either oligogenic or polygenic inheritance (Crompton et al., 2010; Helbig, Scheffer, Mulley, & Berkovic, 2008). As a result, determining the genetic basis of complex epilepsies remains challenging.

Because epilepsy GWAS have had only limited success to date, associations with common SNPs appear to account for only a small fraction of epilepsy heritability. One hypothesis, often termed the "common disease-rare variant hypothesis," proposes that the true disease-causing variants exist at lower frequencies in the population (Petrovski & Kwan, 2013). Rare variants cannot be detected on the SNP platforms that are used in GWAS; rather, they must be identified by high-throughput sequencing of whole genomes and exomes. Thus, multiple rare causal variants of large effect may drive the diluted signals observed when assaying common variants; and even with a successful GWAS, it remains unclear whether common or rare variants are causing the given signal (Dickson, Wang, Krantz, Hakonarson, & Goldstein, 2010).

An exome-sequencing—based case—control study of 118 patients with GGE and 242 control subjects was conducted to elucidate the role of low-frequency variants with intermediate effects (sometimes referred to as "Goldilocks alleles") in the common epilepsies (Heinzen et al., 2012). This study concluded that overall, Goldilocks alleles do not have a major role in GGE. This study and the other GGE research projects conducted to date highlight the extreme genetic heterogeneity of epilepsy disorders. Thus, the variants of most interest may still be rare even in very large cohorts of cases. As a result, gene-based analyses and more phenotypically homogeneous cohorts are needed to reveal true risk factors for the common epilepsies.

Interpreting the Genetic Architecture

Further genetic studies will provide us with increased insight into this architecture. However, gene discovery is only one element (and in many ways the easier element) in understanding the overall pathophysiology of the epilepsies. The biggest challenge lies in understanding the nature and biological effects of genes and pathogenic variants involved in order to highlight the precise pathways that result in epileptogenesis. Clearly, the more genes and mutations we discover, the easier it is to delineate these key pathways. In fact, many of the genes available to date can already be organized to a degree into broad categories, which are described next. However, the list of genes described in the text that follows is not meant to be exhaustive. Rather, we have chosen a subset of strongly implicated epilepsy genes that represent common groups and pathways underlying epileptogenesis.

VOLTAGE-GATED CHANNELOPATHIES

Voltage-gated ion channels are integral membrane proteins that are essential for normal neurologic function. Epilepsy-causing mutations have been associated with a variety of channel protein coding genes; however, in this section, we have focused on only the most strongly implicated genes for which some level of functional analysis has been performed.

Sodium Channels

Voltage-gated sodium channels initiate and propagate action potentials and thus have an important role in neuronal excitability. Nine genes encode the pore-forming α subunit whereas, four genes encode the auxiliary β units involved in channel localization and interaction with cell adhesion molecules, the extracellular matrix, and the intracellular cytoskeleton (Catterall, Goldin, & Waxman, 2005). Mutations in both of these subunits have been implicated in epilepsy.

SCN1A

SCN1A encodes the α1 subunit of the sodium channel that is distributed in the axon initial segment and nodes of Ranvier and forms a fast-inactivating, voltage-dependent Na^+ channel (Duflocq, Le Bras, Bullier, Couraud, & Davenne, 2008). To date, over 650 mutations in *SCN1A* have been described in different epilepsy syndromes, including generalized epilepsy with febrile seizure plus (GEFS+), Dravet syndrome, and Lennox–Gastaut syndrome.

Missense mutations at various sites in *SCN1A* are associated with GEFS+, and most individuals have mild, self-limited phenotypes (Reid, Berkovic, & Petrou, 2009). However, only a small portion of GEFS+ cases results from mutations in *SCN1A*. Interestingly, Dravet syndrome, infantile spasms, and Lennox–Gastaut syndrome, all usually very severe epileptic syndromes, can also be caused by de novo missense and nonsense mutations (Claes et al., 2001; Epi4K Consortium, 2013). In addition, deletions of whole exons, multiple exons, and the whole gene have been linked to Dravet syndrome (Madia et al., 2006; Mulley & Mefford, 2011). In most epileptic encephalopathies no single gene accounts for most cases; however, approximately 80% of cases of Dravet syndrome can be attributed to *SCN1A* mutations.

More than half of *SCN1A* mutations in epilepsy are truncating, leading to complete loss of function of the protein. However, there is considerable controversy and multiple hypotheses about how these mutations cause seizures. One hypothesis, termed the "interneuron hypothesis," states that because *SCN1A* is predominantly found in neurons expressing gamma-aminobutyric acid (GABA), the principal inhibitory neurotransmitter in the nervous system, these mutations result in decreased inhibition and overall hyperexcitability (Oliva, Berkovic, & Petrou, 2012). Work involving human-patient–derived induced pluripotent stem cell neurons, however, illustrates that these mutations may increase intrinsic hyperexcitability of both GABAergic and glutamatergic neurons (Liu et al., 2013). Interestingly, several studies have shown that *SCN1A* haploinsufficiency results in a compensatory increase in sodium current, presumably through altering the expression of other voltage-gated sodium channels (Chopra & Isom, 2014).

SCN2A

SCN2A encodes the α2 subunit in sodium channels and is found predominantly at terminals, unmyelinated axons, and at very high levels in the axon initial segment (Gong, Rhodes, Bekele-Arcuri, & Trimmer, 1999; Hu et al., 2009; Westenbroek, Merrick, & Catterall, 1989; Whitaker et al., 2000). It is crucial for action potential initiation, propagation, and repetitive firing. Missense mutations in *SCN2A* are often implicated in benign familial neonatal infantile seizures (BFNIS), GEFS+, and epileptic encephalopathy (Berkovic et al., 2004; Hackenberg et al., 2014; Herlenius et al., 2007; Matalon, Goldberg, Medne, & Marsh, 2014). Unlike epileptic encephalopathy, the prognosis of BFNIS is excellent in terms of normal development and most patients show age-dependent seizure remission. Research suggests that this remission may correlate with developmental reorganization of the axon initial segment, in which *SCN2A* is replaced by *SCN8A*, thereby diminishing the influence of the mutated channel (Liao et al., 2010). However, further functional studies are required to explain the mechanism of pathogenicity for this broad range of phenotypes.

SCN1B

SCN1B encodes the β1 multifunctional ancillary subunit that is involved in modulating channel gating and regulating the level of channel expression (Isom, 2002). Four mutations that associate with GEFS+ have been identified in the extracellular domain of *SCN1B* (Audenaert et al., 2003; Scheffer et al., 2007; Wallace et al., 1998). In one study, the mutation *C121W* occurred in five patients who also presented with temporal lobe epilepsy (Scheffer et al., 2007). Finally, a single *SCN1B* mutation has also been reported in Dravet syndrome (Ogiwara et al., 2012).

Heterologous expression assays illustrate that reported *SCN1B* mutations are loss of function and appear to interfere with modulation of channel gating (Wallace et al., 1998). In addition, SCN1B expression decreases activity of the α subunit, which means that a mutation interfering with this mechanism will increase the activity of the subunit to which it is bound (Aman et al., 2009; Ferrera & Moran, 2006; Qu et al., 2001). If the activity of *SCN2A* were increased, this would lead to increased excitability of the system; however, if the activity of the *SCN1A* unit were increased, one would expect increased inhibition. Therefore, a clear mechanism of pathogenicity has not yet been established for *SCN1B* mutations.

Potassium Channels

Voltage-gated potassium channels are extremely diverse, with over 70 domains. They have a wide variety of functional roles in the neuron, including the modulation of neuronal firing patterns, defining resting membrane potential, and the modulation of neurotransmitter release (Reid et al., 2009).

KCNQ2 and *KCNQ3*

KCNQ2 encodes $K_V7.2$ and *KCNQ3* encodes $K_V7.3$; both of these channels contribute to the M current, which activates when an excitatory stimulus depolarizes the neuron toward spike threshold, repolarizes the membrane back toward resting potential, and suppresses firing (Rogawski, 2000; Wang et al., 1998).

Benign familial neonatal epilepsy (BFNE) is an autosomal dominant epilepsy that has a favorable outcome for seizure remission and cognitive function (Thomas & Berkovic, 2014). A vast majority of BFNE cases are associated with *KCNQ2* mutations, but a few are associated with *KCNQ3* (Biervert et al., 1998; Hirose et al., 2000; Singh et al., 1998). De novo mutations in *KCNQ2* have also been linked to an epileptic encephalopathy that resembles Ohtahara syndrome (Saitsu et al., 2012).

Because M channels stabilize the membrane potential and thereby limit action potential firing, a loss of this ability would likely increase neuronal excitability. However, the role of loss-of-function mutations in conferring network level hyperexcitability remains unclear, as this phenotype would depend on the distribution of the channels in excitatory versus inhibitory neurons. A transgenic mouse harboring a dominant-negative *KCNQ2* mutation (G279S) found in humans developed spontaneous seizures (Peters, Hu, Pongs, Storm, & Isbrandt, 2005). Electrophysiological recordings from the mouse demonstrated increased excitability, reduced spike-frequency adaptation, and attenuated medium afterhyperpolarization.

KCNA1

KCNA1 encodes the $K_V1.1$ channel that is widely expressed throughout the nervous system and localizes to axonal membrane and presynaptic nerve terminals, where it aids in repolarizing and shaping action potentials. Mutations in this gene can cause Episodic Ataxia Type 1 (EA1), which is associated with sudden attacks of ataxia and with a variety of other nervous system abnormalities including an increased incidence of epilepsy (Browne et al., 1994). Functional analyses of proteins harboring EA1 mutations suggest impaired channel assembly, trafficking, and kinetics (Reid et al., 2009).

KCNT1

KCNT1 encodes a sodium-activated potassium channel that is widely expressed in the brain, particularly the frontal cortex. It is associated with both ADNFLE and a severe epileptic encephalopathy called epilepsy in infancy with migrating focal seizures (Barcia et al., 2012; Heron et al., 2012). Patch-clamp studies in which *Xenopus laevis* oocytes were transfected with mutant *KCNT1* constructs containing two of the de novo mutations showed currents that resembled wild type in terms of voltage dependence and kinetic behavior, but had much higher amplitude, which indicates that these mutations are gain of function (Barcia et al., 2012). Evidence suggests that this phenotype can be reversed by quinidine (see section Quinidine, Phorbol-12-Myristate-13-Acetate, and Bryostatin).

Calcium Channels

Voltage-gated Ca^{2+} channels mediate calcium entry into neurons in response to membrane depolarization. Calcium influx controls a number of essential neuronal responses, including calcium-dependent enzymes, gene expression, neurotransmitter release from the presynaptic terminal, and regulation of neuronal excitability. There are six calcium channel types: L-, N-, P-, Q-, R-, and T-type, each classified by its respective electrophysiological properties.

CACNA1H

Whereas mutations have been implicated in several voltage-gated calcium channel genes, including *CACNA1A* and *CACNA1B*, the most compelling evidence for calcium channel involvement in epilepsy is for *CACNA1H*. This gene codes for the CaV3.2 subunit of the low-threshold T-type Ca^{2+} channel, which opens during depolarization. Low-voltage Ca^{2+} channels are predominantly located in neuronal dendrites, where they affect neuronal excitability (Perez-Reyes, 2003). Over 30 mutations in *CACNA1H* are associated with GGE cases and there is some evidence for association with childhood absence epilepsy (Chen et al., 2003; Heron et al., 2004). However, it is unclear whether these variants are sufficient to cause epilepsy on their own. Functional studies have indicated that several of these mutations confer gain of function through a variety of mechanisms. Biophysical analyses have also shown that many mutations affect channel properties, including rates of activation and inactivation, and current density (Vitko et al., 2005). Because most of the functional changes are small, and some mutations confer no obvious biophysical changes, it is predicted that these mutations may be only one risk factor for these presumably polygenic disorders.

LIGAND-GATED CHANNELOPATHIES

Nicotinic Acetylcholine Receptors

CHRNA4, CHRNB2, and *CHRNA2*

Nicotinic acetylcholine receptors (nAChRs) are ion channels (permeable to Na^+, K^+ and Ca^{2+}) that are modulated by acetylcholine, thus converting neurotransmitter binding into depolarization. There are a total of 17 subunits (α1–10, β1–4, γ, δ, and ε), and epilepsy mutations have been identified in the alpha and beta subunits. nAChRs localize to the synapses of pyramidal neurons and interneurons where they modulate neurotransmitter release, including glutamate and GABA.

Several mutations in *CHRNA4*, *CHRNB2*, and *CHRNA2* have been identified in ADNFLE. *CHRNA4* and *CHRNB2* code for the α4 and β2 subunits, which combine to form the most abundant nAChR in the brain (Gotti et al., 2007). *CHRNA2* encodes the α2 subunit that is expressed in GABAergic interneurons (Son & Winzer-Serhan, 2006). One common effect of the mutations identified in all three subunits is an increased sensitivity to acetylcholine (Bertrand, 2002). A potential consequence of this heightened sensitivity is an increase in presynaptic Ca^{2+} influx that would increase neurotransmitter release (Reid, Bekkers, & Clements, 1998). In addition, it has been proposed that nAChR impairment may be most pronounced at excitatory synapses, thus leading to increased excitatory transmission (Rodrigues-Pinguet et al., 2003).

Gamma-Aminobutyric Acid Receptors

GABA is the primary inhibitory transmitter in the central nervous system. There are three GABA receptor classes: $GABA_A$ and $GABA_C$ receptors are ionotropic and $GABA_B$ is metabotropic.

GABRG2 and *GABRA1*

Mutations in *GABRG2*, which encodes the γ2 subunit of the GABA receptor, have been associated in families with GEFS+, childhood absence epilepsy (CAE), Dravet syndrome, and febrile seizures (Baulac et al., 2001; Harkin et al., 2002; Kananura et al., 2002). Mutations in *GABRA1*, which encodes the α1 subunit, have been implicated in a family with juvenile myoclonic epilepsy, patients with epileptic encephalopathy, and in one case of CAE (Cossette et al., 2002; Epi4K Consortium, 2013; Maljevic et al., 2006). Many of these mutations in the $GABA_A$ receptors generally result in decreased surface expression of the receptor protein, thus resulting in a reduction of GABA-activated chloride currents and altered gating kinetics (Reid et al., 2009).

N-Methyl-D-Aspartate Receptors

The N-methyl-D-aspartate (NMDA) receptor is an ionotropic glutamate receptor that mediates excitatory neurotransmission in the mammalian brain. These receptors localize to the postsynaptic terminal and allow for the influx of sodium and calcium, which are essential for synaptic transmission of neuronal activity. In addition, NMDA receptors have a crucial role in synaptic plasticity, the molecular mechanism involved in learning and memory.

GRIN2A

Mutations in *GRIN2A*, which encodes the NMDA receptor subunit NR2A, cause idiopathic focal epilepsy with rolandic spikes and epileptic encephalopathies including Landau–Kleffner syndrome, epilepsy with continuous spike and waves during slow sleep syndrome, and nonsyndromic epilepsy associated with intellectual disability (Carvill, Regan, et al., 2013; Lemke et al., 2013; Lesca et al., 2013). All of these encephalopathies fall within the epilepsy aphasia spectrum and are characterized by a specific EEG pattern and developmental regression that particularly affects speech. As expected, many of these mutations are de novo.

Functional analysis and electrophysiological recordings of one de novo mutation (L812M) revealed an enhanced agonist potency; decreased sensitivity to negative modulators, including magnesium, protons, and zinc, prolonged synaptic response time, and increased single-channel open probability (Yuan et al., 2014).

GRIN2B

GRIN2B encodes the NMDA receptor subunit NR2B. De novo mutations in this gene are associated with West syndrome and intellectual disability with focal epilepsy, and may also be associated with infantile spasms and Lennox–Gastaut syndrome (Epi4K Consortium, 2013; Lemke et al., 2014). Interestingly, the functional consequences of mutations in *GRIN2B* seem to be far different from *GRIN2A* mutations. The mutations caused a significant loss of ion channel block by extracellular Mg^{2+} and dramatically increased Ca^{2+} permeability, consistent with gain of function. The mutation observed in the patient with focal epilepsy, a milder phenotype, caused less of a disturbance to the channel function.

VESICLE TRAFFICKING DYSREGULATION

The channelopathy paradigm has received much attention in the field of epilepsy genetics. However, data suggest that roughly 75% of de novo mutations identified in a large cohort of patients with epileptic encephalopathies disrupt proteins involved in synaptic transmission (Euro Epinomics- R. E. S. Consortium et al., 2014). A number of synaptic transmission genes specifically involved in vesicle trafficking have been identified in the epilepsies. In neurons, synaptic vesicles store various neurotransmitters that are released at the synapse when there is an influx of calcium from voltage-dependent channels. Vesicles are essential for propagating nerve impulses and are constantly recycled via exocytosis and endocytosis.

STX1B

STX1B encodes syntaxin-1b, a member of the SNARE complex that mediates synaptic vesicle fusion to the presynaptic membrane. Three mutations, one nonsense and two missense, have been identified in probands with febrile seizures and associated epilepsies (Schubert et al., 2014). When *STX1B* is knocked down in zebrafish, the animals showed seizure-like behavior and epileptiform discharges that were highly sensitive to increased temperature. However, the mechanism of pathogenicity of these mutations remains to be elucidated.

STXBP1

De novo mutations in *STXBP1*, which regulates synaptic vesicle release by binding to syntaxin-1a and the SNARE complex, have been implicated in some of the most severe epileptic encephalopathies: Ohtahara syndrome and Lennox–Gastaut syndrome (Epi4K Consortium, 2013; Saitsu et al., 2008). Functional studies show that these mutations are largely loss of function, but further research is required to understand the role of dysfunctional exocytosis in epileptic phenotypes. However, because of STXBP1's crucial role in exocytosis, these mutations likely impair synaptic vesicle release.

DNM1

DNM1 encodes dynamin-1, which is a large mechanochemical GTPase that is required for synaptic vesicle endocytosis. Upon hydrolysis of GTP, DNM1 pinches the forming clathrin-coated vesicle from the plasma membrane at the presynaptic terminal. De novo mutations in *DNM1* have been implicated in Lennox–Gastaut syndrome and infantile spasms (Euro Epinomics- R. E. S. Consortium et al., 2014). Functional studies have shown that epilepsy-causing mutations in this gene disrupt endocytosis via a dominant negative mechanism (Dhindsa et al., 2015). Further, a mouse model with a mutation in *Dnm1* shows severe seizures and electron microscopy reveals that these mice have fewer, but larger, vesicles than wild type mice (Boumil et al., 2010, Dhindsa et al., 2015).

SYN1

SYN1 encodes synapsin I, a phosphoprotein that associates with the membranes of synaptic vesicles, regulates synaptic vesicle trafficking, and affects presynaptic targeting. Mutations in *SYN1* are associated with focal epilepsies as well as autism spectrum disorder (Fassio et al., 2011). In vitro assays illustrate that the human mutations impair the release of vesicles from the readily releasable pool, thus conferring a complete loss of function.

MISCELLANEOUS GENES

Whereas some epilepsy-causing genes easily cluster into groups, other implicated genes do not group as readily. We outline genes in which functional studies have been conducted.

SLC2A1

De novo and inherited mutations in *SLC2A1*, which encodes a glucose (GLUT1) transporter that transports glucose across the blood–brain barrier and into glia, can cause GLUT1 deficiency syndromes 1 and 2 (GLUT1DS1 and GLUT1DS2)

(De Vivo et al., 1991; Suls et al., 2008; Weber et al., 2008). GLUT1DS1 is a severe infantile encephalopathy characterized by microcephaly, intellectual disability, and motor incoordination. GLUT1DS2 is associated with paroxysmal exercise-induced dyskinesia, and some patients also have early-onset childhood absence epilepsy. These patients typically show low glucose concentration in the cerebral spinal fluid. Functional studies of some GLUT1 deficiency-related mutations in *Xenopus* oocytes confirm that these mutant proteins impair glucose transport.

LGI1

LGI1 encodes a protein that is characterized by leucine-rich domains. Although its biological function is unclear, evidence suggests that it is a secreted neuronal protein that may also regulate voltage-gated potassium channels. Mutations in *LGI1* have been identified in patients with autosomal dominant partial epilepsy with auditory features (Kalachikov et al., 2002). Interestingly, autoantibodies that target LGI1 have been implicated in autoimmune limbic encephalitis. Findings in a conditional *Lgi1* knockout mouse suggest that LGI1 regulates neuronal excitability. Furthermore, loss of *LGI1* primarily dysregulates glutamatergic neurons (Boillot et al., 2014). Importantly, LGI1 is required from embryogenesis to adulthood to achieve proper circuit functioning (Boillot et al., 2014).

SYNGAP1

The SYNGAP1 protein is a synaptic Ras GTPase-activating protein that localizes to dendritic cells in neocortical pyramidal neurons, suppressing NMDA receptor–mediated synaptic plasticity and α-amino-3-hydroxy-5-methyl-4-isoxazolepropionic acid receptor membrane insertion. Mutations in *SYNGAP1* cause epileptic encephalopathy, intellectual disability, and autism via haploinsufficiency (Berryer et al., 2013; Carvill, Heavin, et al., 2013; Hamdan et al., 2011). Haploinsufficiency of *Syngap1* in a mouse model accelerated the maturation of glutamatergic synapses in the hippocampus and altered the excitatory/inhibitory balance in the hippocampus, thereby increasing excitation and seizure susceptibility (Clement et al., 2012).

DEPDC5

The protein encoded by *DEPDC5* is a member of the GATOR1 complex that inhibits mammalian targets of rapamycin–mediated processes, including cell growth and proliferation. Nonsense and missense mutations in *DEPDC5* have been identified in ADNFLE, familial temporal lobe epilepsy, and familial focal epilepsy with variable foci (Dibbens et al., 2013; Ishida et al., 2013; Scheffer et al., 2014). Evidence suggests that some, but not all, of the epilepsy-causing mutations significantly inhibit GATOR1-dependent inhibition of TORC1 signaling. These results suggest that dysfunctional neuronal signal transduction may cause epilepsy (van Kranenburg, Hoogeveen-Westerveld, & Nellist, 2014).

GNAO1

Heterotrimeric G proteins are composed of α, β, and γ subunits and transduce a variety of signals from membrane receptors to intracellular effectors. De novo mutations in *GNAO1*, which codes for a G protein α subunit, cause epileptic encephalopathy (Nakamura et al., 2013). Functional expression studies showed that three of the mutations impaired protein localization to the plasma membrane, and electrophysiology analysis illustrated that these mutations also decreased inhibition of calcium currents by norepinephrine.

INTERPRETATION

Together, the number of genes that have been implicated in the epilepsies illustrates a tremendous amount of locus heterogeneity, even in clinically well-defined epilepsy syndromes such as infantile spasms. Despite the extreme locus heterogeneity, protein–protein interaction networks show that most of these implicated epilepsy genes may be functionally connected. Thus, rare mutations across many genes may actually converge on a few key pathways, such as those mentioned previously. Oftentimes, epilepsy genetic studies identify private mutations (ie, mutated in single individuals). These mutations require functional studies and perhaps further genetic studies to determine whether they truly contribute to the phenotype.

Although many of the genetic findings thus far have been in the epilepsies that are strongly influenced by single mutations, the underlying pathways may be shared among these rare, strongly genetic epilepsies and the complex epilepsies. For example, *SCN1A* is associated with both GEFS+ and Dravet syndrome. Thus, studying rare, highly genetic epilepsies provides insights into pathophysiology that might otherwise be unattainable.

MANAGEMENT/THERAPEUTIC IMPLICATIONS

Understanding the fundamental disease mechanisms underlying epilepsy is essential for developing targeted treatments, especially because a third of all patients with epilepsy do not respond to any of the numerous antiepileptic drugs currently available, and many patients who do respond experience significant side effects from the medications. This is particularly the case in the vast majority of encephalopathies, which are essentially untreatable with current therapies. Discovering "epilepsy genes" and mutations allows us to conduct targeted functional studies to understand aberrant networks at genetic, cellular, and electrophysiological levels, thereby providing a gateway to personalized pharmacological and non-pharmacological treatment options. Although this paradigm of genetically and functionally targeted therapies remains aspirational for the vast majority of patients with epilepsy, there are already a handful of examples that emphasize the potential of this approach.

Ketogenic Diet

The ketogenic diet is a high-fat, low-carbohydrate diet that is thought to stimulate the effects of starvation by forcing the body to use fat as a fuel source. The body metabolizes fat via lipolysis and the fatty acids undergo beta-oxidation into ketone bodies that can be used as precursors to generate adenosine triphosphate (ATP). In a randomized controlled trial of 145 children with generalized or focal seizures, 38% of patients on the diet had a more than 50% reduction in seizure frequency, compared with 6% of control subjects on an unmodified diet (Neal et al., 2008). The efficacy of the ketogenic diet varies across epilepsy syndromes. However, it results in marked clinical improvement of motor and seizure symptoms in patients with GLUT1 deficiency syndromes (caused by mutations in *SLC2A1*) (section Miscellaneous Genes) (Brockmann, 2009). Given the impressive efficacy of the ketogenic diet in GLUT1-related disorders, accurate clinical diagnosis is extremely important. Furthermore, the diet should be considered early in the treatment of Dravet syndrome and is surprisingly effective in some acquired epilepsies such as lissencephaly and hypoxic-ischemic encephalopathy (Thammongkol et al., 2012). Although the mechanisms by which the ketogenic diet ameliorates seizures are not fully elucidated, work indicates that several factors may be at play, including disruption of glutamatergic synaptic transmission, inhibition of glycolysis, and activation of ATP-sensitive potassium channels (Lutas & Yellen, 2013).

Quinidine, Phorbol-12-Myristate-13-Acetate, and Bryostatin

In vitro oocyte patch-clamp assays suggest that gain-of-function mutations in *KCNT1* associated with ADNFLE and epilepsy of infancy with migrating focal seizures (EIMFS) (section Potassium Channels) can be reduced, with variable sensitivities, by the introduction of quinidine, a drug approved for the treatment of cardiac arrhythmias (Milligan et al., 2014). Shortly after the publication of this finding, a child with EIMFS was treated with quinidine in a carefully monitored clinical environment and showed reduced seizure activity and improved psychomotor development (Bearden et al., 2014). In addition, studies indicate that KCNT1 is modulated after protein kinase C (PKC) activation. In vitro patch-clamping experiments suggest that increased potassium currents caused by at least one EIMFS-causing mutation (P924L) can be inhibited by PKC activators phorbol-12-myristate-13-acetate and Bryostatin-1 (Padilla et al., 2014).

Memantine

GRIN2A mutations cause epileptic encephalopathies (section N-Methyl-D-Aspartate Receptors). Functional analysis of one particular *GRIN2A* mutation (L812M) illustrated that the mutation conferred increased intrinsic activity in response to agonists and decreased response to negative modulators, thus inducing neuronal hyperexcitability and possibly excitotoxicity. In vitro studies suggested that memantine, an NMDA receptor blocker, could reverse the effects of this mutation. The proband was then treated with memantine and experience marked seizure reduction (Pierson et al., 2014).

Retigabine

Retigabine is a commercial antiepileptic drug that opens KCNQ/K_V7 channels. Oocyte-based patch-clamp studies demonstrated that retigabine has the potential to reverse dominant negative effects of certain *KCNQ2* and *KCNQ3* mutations associated with BFNE (section Potassium Channels) by increasing currents through the potassium channel (Orhan et al., 2014). Retigabine may eventually serve as an individualized treatment option for these patients.

FUTURE STUDIES AND CONCLUSION

Genetic studies have led to great advances in understanding the neurobiology and molecular architecture of the epilepsies, particularly in epilepsies that are strongly influenced by single mutations. With decreasing costs of NGS technologies and our increasing ability to understand genetic variation, causative epilepsy genes will continue to be discovered. In vitro models allow us to discriminate among pathogenic and nonpathogenic mutations and often give us insight into the mechanism of pathogenicity associated with a mutation. It is critically important to conduct functional studies to distinguish epilepsy-causing mutations and polymorphisms that may be enriched in a specific population or ancestry.

The patch-clamping technique is often used to analyze the effect of mutations on currents through ion channels. Multielectrode arrays (MEAs) are also becoming an increasingly popular electrophysiological technique. Whereas patch-clamping examines currents at a single cell level, MEAs contain multiple electrodes and can be used to observe network-level characteristics of neurons. Zebrafish (*Danio rerio*) have also been emerging as a promising model organism for studying epilepsy (Hortopan, Dinday, & Baraban, 2010). Seizure-like neurophysiological responses can be evoked in adult and larval zebrafish via genetic and pharmacological manipulations (Baraban, Dinday, & Hortopan, 2013). Because it is inexpensive and sensitive to traditional antiepileptic drugs, this organism offers excellent opportunities to validate molecular biomarkers of epilepsy and identify novel therapeutic options. The mouse (*Mus musculus*) is an extremely good model organisms for human epilepsy. For example, *Dnm1* was discovered as an epilepsy gene in mice before it was identified as a human epilepsy gene. Clearly, mouse models can give us profound insight into epilepsy mechanisms. Importantly, drugs that ameliorate epilepsy in mice are often effective in humans. When this chapter was written, it only cost $5000 to model a single mutation using the CRISPR/Cas9 system; thus, it is realistic to envision testing an entire batch of drugs on the mouse to determine a targeted therapy for the patient.

Interestingly, the same genes associated with epilepsy are being implicated in different neurological disorders. Perhaps the strongest example of this is the overlapping genes in epileptic encephalopathies, autism spectrum disorder, and intellectual disability. Intriguingly, the FMRP regulates a significant number of these genes. Thus, targeted treatments for a gene in epilepsy may work for patients who have a different disorder but who have mutations in the same gene or pathway.

REFERENCES

Adzhubei, I. A., Schmidt, S., Peshkin, L., Ramensky, V. E., Gerasimova, A., Bork, P., … Sunyaev, S. R. (2010). A method and server for predicting damaging missense mutations. *Nature Methods, 7*(4), 248–249. http://dx.doi.org/10.1038/nmeth0410-248.

Aman, T. K., Grieco-Calub, T. M., Chen, C., Rusconi, R., Slat, E. A., Isom, L. L., & Raman, I. M. (2009). Regulation of persistent Na current by interactions between beta subunits of voltage-gated Na channels. *Journal of Neuroscience, 29*(7), 2027–2042. http://dx.doi.org/10.1523/JNEUROSCI.4531-08.2009.

Audenaert, D., Claes, L., Ceulemans, B., Lofgren, A., Van Broeckhoven, C., & De Jonghe, P. (2003). A deletion in SCN1B is associated with febrile seizures and early-onset absence epilepsy. *Neurology, 61*(6), 854–856.

Baraban, S. C., Dinday, M. T., & Hortopan, G. A. (2013). Drug screening in Scnla zebrafish mutant identifies clemizole as a potential dravet syndrome treatment. *Nature Communications, 4*, 2410.

Barcia, G., Fleming, M. R., Deligniere, A., Gazula, V. R., Brown, M. R., Langouet, M., … Nabbout, R. (2012). De novo gain-of-function KCNT1 channel mutations cause malignant migrating partial seizures of infancy. *Nature Genetics, 44*(11), 1255–1259. http://dx.doi.org/10.1038/ng.2441.

Baulac, S., Huberfeld, G., Gourfinkel-An, I., Mitropoulou, G., Beranger, A., Prud'homme, J. F., … LeGuern, E. (2001). First genetic evidence of GABA(A) receptor dysfunction in epilepsy: a mutation in the gamma2-subunit gene. *Nature Genetics, 28*(1), 46–48. http://dx.doi.org/10.1038/88254.

Bearden, D., Strong, A., Ehnot, J., DiGiovine, M., Dlugos, D., & Goldberg, E. M. (2014). Targeted treatment of migrating partial seizures of infancy with quinidine. *Annals of Neurology, 76*(3), 457–461. http://dx.doi.org/10.1002/ana.24229.

Berg, A. T., Berkovic, S. F., Brodie, M. J., Buchhalter, J., Cross, J. H., van Emde Boas, W., … Scheffer, I. E. (2010). Revised terminology and concepts for organization of seizures and epilepsies: report of the ILAE Commission on Classification and Terminology, 2005–2009. *Epilepsia, 51*(4), 676–685. http://dx.doi.org/10.1111/j.1528-1167.2010.02522.x.

Berkovic, S. F., Heron, S. E., Giordano, L., Marini, C., Guerrini, R., Kaplan, R. E., … Scheffer, I. E. (2004). Benign familial neonatal-infantile seizures: characterization of a new sodium channelopathy. *Annals of Neurology, 55*(4), 550–557. http://dx.doi.org/10.1002/ana.20029.

Berkovic, S. F., Howell, R. A., Hay, D. A., & Hopper, J. L. (1998). Epilepsies in twins: genetics of the major epilepsy syndromes. *Annals of Neurology, 43*(4), 435–445. http://dx.doi.org/10.1002/ana.410430405.

Berryer, M. H., Hamdan, F. F., Klitten, L. L., Moller, R. S., Carmant, L., Schwartzentruber, J., … Di Cristo, G. (2013). Mutations in SYNGAP1 cause intellectual disability, autism, and a specific form of epilepsy by inducing haploinsufficiency. *Human Mutation, 34*(2), 385–394. http://dx.doi.org/10.1002/humu.22248.

Bertrand, D. (2002). Neuronal nicotinic acetylcholine receptors and epilepsy. *Epilepsy Currents, 2*(6), 191–193. http://dx.doi.org/10.1046/j.1535-7597.2002.00072.x.

Biervert, C., Schroeder, B. C., Kubisch, C., Berkovic, S. F., Propping, P., Jentsch, T. J., & Steinlein, O. K. (1998). A potassium channel mutation in neonatal human epilepsy. *Science, 279*(5349), 403–406.

Boillot, M., Huneau, C., Marsan, E., Lehongre, K., Navarro, V., Ishida, S., … Baulac, S. (2014). Glutamatergic neuron-targeted loss of LGI1 epilepsy gene results in seizures. *Brain, 137*(Pt 11), 2984–2996. http://dx.doi.org/10.1093/brain/awu259.

Boumil, R. M., Letts, V. A., Roberts, M. C., Lenz, C., Mahaffey, C. L., Zhang, Z. W., … Frankel, W. N. (2010). A missense mutation in a highly conserved alternate exon of dynamin-1 causes epilepsy in fitful mice. *PLoS Genetics, 6*(8). http://dx.doi.org/10.1371/journal.pgen.1001046.

Brockmann, K. (2009). The expanding phenotype of GLUT1-deficiency syndrome. *Brain and Development, 31*(7), 545–552. http://dx.doi.org/10.1016/j.braindev.2009.02.008.

Browne, D. L., Gancher, S. T., Nutt, J. G., Brunt, E. R., Smith, E. A., Kramer, P., & Litt, M. (1994). Episodic ataxia/myokymia syndrome is associated with point mutations in the human potassium channel gene, KCNA1. *Nature Genetics, 8*(2), 136–140. http://dx.doi.org/10.1038/ng1094-136.

Carvill, G. L., Heavin, S. B., Yendle, S. C., McMahon, J. M., O'Roak, B. J., Cook, J., … Mefford, H. C. (2013). Targeted resequencing in epileptic encephalopathies identifies de novo mutations in CHD2 and SYNGAP1. *Nature Genetics, 45*(7), 825–830. http://dx.doi.org/10.1038/ng.2646.

Carvill, G. L., Regan, B. M., Yendle, S. C., O'Roak, B. J., Lozovaya, N., Bruneau, N., … Mefford, H. C. (2013). GRIN2A mutations cause epilepsy-aphasia spectrum disorders. *Nature Genetics, 45*(9), 1073–1076. http://dx.doi.org/10.1038/ng.2727.

Catterall, W. A., Goldin, A. L., & Waxman, S. G. (2005). International Union of Pharmacology. XLVII. Nomenclature and structure-function relationships of voltage-gated sodium channels. *Pharmacological Reviews, 57*(4), 397–409. http://dx.doi.org/10.1124/pr.57.4.4.

Chen, Y., Lu, J., Pan, H., Zhang, Y., Wu, H., Xu, K., … Wu, X. (2003). Association between genetic variation of CACNA1H and childhood absence epilepsy. *Annals of Neurology, 54*(2), 239–243. http://dx.doi.org/10.1002/ana.10607.

Chopra, R., & Isom, L. L. (2014). Untangling the dravet syndrome seizure network: the changing face of a rare genetic epilepsy. *Epilepsy Currents, 14*(2), 86–89. http://dx.doi.org/10.5698/1535-7597-14.2.86.

Christensen, J., Pedersen, M. G., Pedersen, C. B., Sidenius, P., Olsen, J., & Vestergaard, M. (2009). Long-term risk of epilepsy after traumatic brain injury in children and young adults: a population-based cohort study. *Lancet, 373*(9669), 1105–1110. http://dx.doi.org/10.1016/S0140-6736(09)60214-2.

Claes, L., Del-Favero, J., Ceulemans, B., Lagae, L., Van Broeckhoven, C., & De Jonghe, P. (2001). De novo mutations in the sodium-channel gene SCN1A cause severe myoclonic epilepsy of infancy. *American Journal of Human Genetics, 68*(6), 1327–1332. http://dx.doi.org/10.1086/320609.

Clement, J. P., Aceti, M., Creson, T. K., Ozkan, E. D., Shi, Y., Reish, N. J., … Rumbaugh, G. (2012). Pathogenic SYNGAP1 mutations impair cognitive development by disrupting maturation of dendritic spine synapses. *Cell, 151*(4), 709–723. http://dx.doi.org/10.1016/j.cell.2012.08.045.

Cossette, P., Liu, L., Brisebois, K., Dong, H., Lortie, A., Vanasse, M., … Rouleau, G. A. (2002). Mutation of GABRA1 in an autosomal dominant form of juvenile myoclonic epilepsy. *Nature Genetics, 31*(2), 184–189. http://dx.doi.org/10.1038/ng885.

Crompton, D. E., Scheffer, I. E., Taylor, I., Cook, M. J., McKelvie, P. A., Vears, D. F., … Berkovic, S. F. (2010). Familial mesial temporal lobe epilepsy: a benign epilepsy syndrome showing complex inheritance. *Brain, 133*(11), 3221–3231. http://dx.doi.org/10.1093/brain/awq251.

De Vivo, D. C., Trifiletti, R. R., Jacobson, R. I., Ronen, G. M., Behmand, R. A., & Harik, S. I. (1991). Defective glucose transport across the blood–brain barrier as a cause of persistent hypoglycorrhachia, seizures, and developmental delay. *New England Journal of Medicine, 325*(10), 703–709. http://dx.doi.org/10.1056/NEJM199109053251006.

Dhindsa, R. S., Bradrick, S. S., Yao, X., Heinzen, E. L., Petrovski, S., Kreuger, B. J., … Goldstein, D. B. (2015). Epileptic encephalopathy causing mutations impair neuronal synaptic endocytosis. *Neurology Genetics, 1*(1), e4. http://dx.doi.org/10.1212/01.nxg.0000464295.65736.da.

Dibbens, L. M., de Vries, B., Donatello, S., Heron, S. E., Hodgson, B. L., Chintawar, S., … Scheffer, I. E. (2013). Mutations in DEPDC5 cause familial focal epilepsy with variable foci. *Nature Genetics, 45*(5), 546–551. http://dx.doi.org/10.1038/ng.2599.

Dichgans, M. (2007). Genetics of ischaemic stroke. *Lancet Neurology, 6*(2), 149–161. http://dx.doi.org/10.1016/S1474-4422(07)70028-5.

Dickson, S. P., Wang, K., Krantz, I., Hakonarson, H., & Goldstein, D. B. (2010). Rare variants create synthetic genome-wide associations. *PLoS Biology, 8*(1), e1000294. http://dx.doi.org/10.1371/journal.pbio.1000294.

Duflocq, A., Le Bras, B., Bullier, E., Couraud, F., & Davenne, M. (2008). Nav1.1 is predominantly expressed in nodes of Ranvier and axon initial segments. *Molecular and Cellular Neuroscience, 39*(2), 180–192. http://dx.doi.org/10.1016/j.mcn.2008.06.008.

England, M. J., Liverman, C. T., Schultz, A. M., & Strawbridge, L. M. (2012). Epilepsy across the spectrum: promoting health and understanding. A summary of the Institute of Medicine report. *Epilepsy and Behavior, 25*(2), 266–276. http://dx.doi.org/10.1016/j.yebeh.2012.06.016.

Epi4K Consortium. (2013). De novo mutations in epileptic encephalopathies. *Nature, 501*(7466), 217–221. http://dx.doi.org/10.1038/nature12439.

Euro Epinomics- R. E. S. Consortium, Epilepsy Phenome/Genome Project, & Epi4K Consortium. (2014). De novo mutations in synaptic transmission genes including DNM1 cause epileptic encephalopathies. *American Journal of Human Genetics, 95*(4), 360–370. http://dx.doi.org/10.1016/j.ajhg.2014.08.013.

Fassio, A., Patry, L., Congia, S., Onofri, F., Piton, A., Gauthier, J., … Cossette, P. (2011). SYN1 loss-of-function mutations in autism and partial epilepsy cause impaired synaptic function. *Human Molecular Genetics, 20*(12), 2297–2307. http://dx.doi.org/10.1093/hmg/ddr122.

Ferrera, L., & Moran, O. (2006). Beta1-subunit modulates the Nav1.4 sodium channel by changing the surface charge. *Experimental Brain Research, 172*(2), 139–150. http://dx.doi.org/10.1007/s00221-005-0323-4.

Fisher, R. S., Acevedo, C., Arzimanoglou, A., Bogacz, A., Cross, J. H., Elger, C. E., … Wiebe, S. (2014). ILAE official report: a practical clinical definition of epilepsy. *Epilepsia, 55*(4), 475–482. http://dx.doi.org/10.1111/epi.12550.

Fisher, R. S., van Emde Boas, W., Blume, W., Elger, C., Genton, P., Lee, P., & Engel, J., Jr. (2005). Epileptic seizures and epilepsy: definitions proposed by the International League Against Epilepsy (ILAE) and the International Bureau for Epilepsy (IBE). *Epilepsia, 46*(4), 470–472. http://dx.doi.org/10.1111/j.0013-9580.2005.66104.x.

Gong, B., Rhodes, K. J., Bekele-Arcuri, Z., & Trimmer, J. S. (1999). Type I and type II Na(+) channel alpha-subunit polypeptides exhibit distinct spatial and temporal patterning, and association with auxiliary subunits in rat brain. *Journal of Comparative Neurology, 412*(2), 342–352.

Gotti, C., Moretti, M., Gaimarri, A., Zanardi, A., Clementi, F., & Zoli, M. (2007). Heterogeneity and complexity of native brain nicotinic receptors. *Biochemical Pharmacology, 74*(8), 1102–1111. http://dx.doi.org/10.1016/j.bcp.2007.05.023.

Hackenberg, A., Baumer, A., Sticht, H., Schmitt, B., Kroell-Seger, J., Wille, D., … Plecko, B. (2014). Infantile epileptic encephalopathy, transient choreoathetotic movements, and hypersomnia due to a De Novo missense mutation in the SCN2A gene. *Neuropediatrics, 45*(4), 261–264. http://dx.doi.org/10.1055/s-0034-1372302.

Hamdan, F. F., Daoud, H., Piton, A., Gauthier, J., Dobrzeniecka, S., Krebs, M. O., … Michaud, J. L. (2011). De novo SYNGAP1 mutations in nonsyndromic intellectual disability and autism. *Biological Psychiatry, 69*(9), 898–901. http://dx.doi.org/10.1016/j.biopsych.2010.11.015.

Harkin, L. A., Bowser, D. N., Dibbens, L. M., Singh, R., Phillips, F., Wallace, R. H., … Petrou, S. (2002). Truncation of the GABA(A)-receptor gamma2 subunit in a family with generalized epilepsy with febrile seizures plus. *American Journal of Human Genetics, 70*(2), 530–536. http://dx.doi.org/10.1086/338710.

Heinzen, E. L., Depondt, C., Cavalleri, G. L., Ruzzo, E. K., Walley, N. M., Need, A. C., … Goldstein, D. B. (2012). Exome sequencing followed by large-scale genotyping fails to identify single rare variants of large effect in idiopathic generalized epilepsy. *American Journal of Human Genetics, 91*(2), 293–302. http://dx.doi.org/10.1016/j.ajhg.2012.06.016.

Heinzen, E. L., Radtke, R. A., Urban, T. J., Cavalleri, G. L., Depondt, C., Need, A. C., … Goldstein, D. B. (2010). Rare deletions at 16p13.11 predispose to a diverse spectrum of sporadic epilepsy syndromes. *American Journal of Human Genetics, 86*(5), 707–718. http://dx.doi.org/10.1016/j.ajhg.2010.03.018.

Helbig, I., Mefford, H. C., Sharp, A. J., Guipponi, M., Fichera, M., Franke, A., … Sander, T. (2009). 15q13.3 microdeletions increase risk of idiopathic generalized epilepsy. *Nature Genetics, 41*(2), 160–162. http://dx.doi.org/10.1038/ng.292.

Helbig, I., Scheffer, I. E., Mulley, J. C., & Berkovic, S. F. (2008). Navigating the channels and beyond: unravelling the genetics of the epilepsies. *Lancet Neurology, 7*(3), 231–245. http://dx.doi.org/10.1016/S1474-4422(08)70039-5.

Herlenius, E., Heron, S. E., Grinton, B. E., Keay, D., Scheffer, I. E., Mulley, J. C., & Berkovic, S. F. (2007). SCN2A mutations and benign familial neonatal-infantile seizures: the phenotypic spectrum. *Epilepsia, 48*(6), 1138–1142. http://dx.doi.org/10.1111/j.1528-1167.2007.01049.x.

Heron, S. E., Phillips, H. A., Mulley, J. C., Mazarib, A., Neufeld, M. Y., Berkovic, S. F., & Scheffer, I. E. (2004). Genetic variation of CACNA1H in idiopathic generalized epilepsy. *Annals of Neurology, 55*(4), 595–596. http://dx.doi.org/10.1002/ana.20028.

Heron, S. E., Smith, K. R., Bahlo, M., Nobili, L., Kahana, E., Licchetta, L., … Dibbens, L. M. (2012). Missense mutations in the sodium-gated potassium channel gene KCNT1 cause severe autosomal dominant nocturnal frontal lobe epilepsy. *Nature Genetics, 44*(11), 1188–1190. http://dx.doi.org/10.1038/ng.2440.

Hirose, S., Zenri, F., Akiyoshi, H., Fukuma, G., Iwata, H., Inoue, T., … Mitsudome, A. (2000). A novel mutation of KCNQ3 (c.925T→C) in a Japanese family with benign familial neonatal convulsions. *Annals of Neurology, 47*(6), 822–826.

Hortopan, G. A., Dinday, M. T., & Baraban, S. C. (2010). Zebrafish as a model for studying genetic aspects of epilepsy. *Disease Models & Mechanisms, 3*(3–4), 144–148.

Hu, W., Tian, C., Li, T., Yang, M., Hou, H., & Shu, Y. (2009). Distinct contributions of Na(v)1.6 and Na(v)1.2 in action potential initiation and backpropagation. *Nature Neuroscience, 12*(8), 996–1002. http://dx.doi.org/10.1038/nn.2359.

ILAE Consortium on Complex Epilepsies. (2014). Genetic determinants of common epilepsies: a meta-analysis of genome-wide association studies. *Lancet Neurology, 13*(9), 893–903. http://dx.doi.org/10.1016/S1474-4422(14)70171-1.

Iossifov, I., Ronemus, M., Levy, D., Wang, Z., Hakker, I., Rosenbaum, J., … Wigler, M. (2012). De novo gene disruptions in children on the autistic spectrum. *Neuron, 74*(2), 285–299. http://dx.doi.org/10.1016/j.neuron.2012.04.009.

Ishida, S., Picard, F., Rudolf, G., Noe, E., Achaz, G., Thomas, P., … Baulac, S. (2013). Mutations of DEPDC5 cause autosomal dominant focal epilepsies. *Nature Genetics, 45*(5), 552–555. http://dx.doi.org/10.1038/ng.2601.

Isom, L. L. (2002). Beta subunits: players in neuronal hyperexcitability? *Novartis Foundation Symposium, 241*, 124–138. discussion 138–143, 226–132.

Kalachikov, S., Evgrafov, O., Ross, B., Winawer, M., Barker-Cummings, C., Martinelli Boneschi, F., … Gilliam, T. C. (2002). Mutations in LGI1 cause autosomal-dominant partial epilepsy with auditory features. *Nature Genetics, 30*(3), 335–341. http://dx.doi.org/10.1038/ng832.

Kananura, C., Haug, K., Sander, T., Runge, U., Gu, W., Hallmann, K., … Steinlein, O. K. (2002). A splice-site mutation in GABRG2 associated with childhood absence epilepsy and febrile convulsions. *Archives of Neurology, 59*(7), 1137–1141.

Kariuki, S. M., Rockett, K., Clark, T. G., Reyburn, H., Agbenyega, T., Taylor, T. E., … Newton, C. R. (2013). The genetic risk of acute seizures in African children with falciparum malaria. *Epilepsia, 54*(6), 990–1001. http://dx.doi.org/10.1111/epi.12173.

de Kovel, C. G., Trucks, H., Helbig, I., Mefford, H. C., Baker, C., Leu, C., … Sander, T. (2010). Recurrent microdeletions at 15q11.2 and 16p13.11 predispose to idiopathic generalized epilepsies. *Brain, 133*(Pt 1), 23–32. http://dx.doi.org/10.1093/brain/awp262.

van Kranenburg, M., Hoogeveen-Westerveld, M., & Nellist, M. (2014). Preliminary functional assessment and classification of DEPDC5 variants associated with focal epilepsy. *Human Mutation, 36*(2), 200–209. http://dx.doi.org/10.1002/humu.22723.

Lemke, J. R., Hendrickx, R., Geider, K., Laube, B., Schwake, M., Harvey, R. J., … Weckhuysen, S. (2014). GRIN2B mutations in West syndrome and intellectual disability with focal epilepsy. *Annals of Neurology, 75*(1), 147–154. http://dx.doi.org/10.1002/ana.24073.

Lemke, J. R., Lal, D., Reinthaler, E. M., Steiner, I., Nothnagel, M., Alber, M., ... von Spiczak, S. (2013). Mutations in GRIN2A cause idiopathic focal epilepsy with rolandic spikes. *Nature Genetics, 45*(9), 1067–1072. http://dx.doi.org/10.1038/ng.2728.

Lesca, G., Rudolf, G., Bruneau, N., Lozovaya, N., Labalme, A., Boutry-Kryza, N., ... Szepetowski, P. (2013). GRIN2A mutations in acquired epileptic aphasia and related childhood focal epilepsies and encephalopathies with speech and language dysfunction. *Nature Genetics, 45*(9), 1061–1066. http://dx.doi.org/10.1038/ng.2726.

Liao, Y., Anttonen, A. K., Liukkonen, E., Gaily, E., Maljevic, S., Schubert, S., ... Lehesjoki, A. E. (2010). SCN2A mutation associated with neonatal epilepsy, late-onset episodic ataxia, myoclonus, and pain. *Neurology, 75*(16), 1454–1458. http://dx.doi.org/10.1212/WNL.0b013e3181f8812e.

Lin, D. Y., & Tang, Z. Z. (2011). A general framework for detecting disease associations with rare variants in sequencing studies. *American Journal of Human Genetics, 89*(3), 354–367. http://dx.doi.org/10.1016/j.ajhg.2011.07.015.

Liu, Y., Lopez-Santiago, L. F., Yuan, Y., Jones, J. M., Zhang, H., O'Malley, H. A., ... Parent, J. M. (2013). Dravet syndrome patient-derived neurons suggest a novel epilepsy mechanism. *Annals of Neurology, 74*(1), 128–139. http://dx.doi.org/10.1002/ana.23897.

Lutas, A., & Yellen, G. (2013). The ketogenic diet: metabolic influences on brain excitability and epilepsy. *Trends in Neurosciences, 36*(1), 32–40. http://dx.doi.org/10.1016/j.tins.2012.11.005.

Lynch, M. (2010). Rate, molecular spectrum, and consequences of human mutation. *Proceedings of the National Academy of Sciences of the United States of America, 107*(3), 961–968. http://dx.doi.org/10.1073/pnas.0912629107.

Madia, F., Striano, P., Gennaro, E., Malacarne, M., Paravidino, R., Biancheri, R., ... Zara, F. (2006). Cryptic chromosome deletions involving SCN1A in severe myoclonic epilepsy of infancy. *Neurology, 67*(7), 1230–1235. http://dx.doi.org/10.1212/01.wnl.0000238513.70878.54.

Maljevic, S., Krampfl, K., Cobilanschi, J., Tilgen, N., Beyer, S., Weber, Y. G., ... Heils, A. (2006). A mutation in the GABA(A) receptor alpha(1)-subunit is associated with absence epilepsy. *Annals of Neurology, 59*(6), 983–987. http://dx.doi.org/10.1002/ana.20874.

Matalon, D., Goldberg, E., Medne, L., & Marsh, E. D. (2014). Confirming an expanded spectrum of SCN2A mutations: a case series. *Epileptic Disorders, 16*(1), 13–18. http://dx.doi.org/10.1684/epd.2014.0641.

Mefford, H. C., & Eichler, E. E. (2009). Duplication hotspots, rare genomic disorders, and common disease. *Current Opinion in Genetics and Development, 19*(3), 196–204. http://dx.doi.org/10.1016/j.gde.2009.04.003.

Mefford, H. C., Muhle, H., Ostertag, P., von Spiczak, S., Buysse, K., Baker, C., ... Eichler, E. E. (2010). Genome-wide copy number variation in epilepsy: novel susceptibility loci in idiopathic generalized and focal epilepsies. *PLoS Genetics, 6*(5), e1000962. http://dx.doi.org/10.1371/journal.pgen.1000962.

Mefford, H. C., Yendle, S. C., Hsu, C., Cook, J., Geraghty, E., McMahon, J. M., ... Scheffer, I. E. (2011). Rare copy number variants are an important cause of epileptic encephalopathies. *Annals of Neurology, 70*(6), 974–985. http://dx.doi.org/10.1002/ana.22645.

Milligan, C. J., Li, M., Gazina, E. V., Heron, S. E., Nair, U., Trager, C., ... Petrou, S. (2014). KCNT1 gain of function in 2 epilepsy phenotypes is reversed by quinidine. *Annals of Neurology, 75*(4), 581–590. http://dx.doi.org/10.1002/ana.24128.

Mulley, J. C., & Mefford, H. C. (2011). Epilepsy and the new cytogenetics. *Epilepsia, 52*(3), 423–432. http://dx.doi.org/10.1111/j.1528-1167.2010.02932.x.

Nakamura, K., Kodera, H., Akita, T., Shiina, M., Kato, M., Hoshino, H., ... Saitsu, H. (2013). De Novo mutations in GNAO1, encoding a Gαo subunit of heterotrimeric G proteins, cause epileptic encephalopathy. *American Journal of Human Genetics, 93*(3), 496–505. http://dx.doi.org/10.1016/j.ajhg.2013.07.014.

Neal, E. G., Chaffe, H., Schwartz, R. H., Lawson, M. S., Edwards, N., Fitzsimmons, G., ... Cross, J. H. (2008). The ketogenic diet for the treatment of childhood epilepsy: a randomised controlled trial. *Lancet Neurology, 7*(6), 500–506. http://dx.doi.org/10.1016/S1474-4422(08)70092-9.

Neale, B. M., Kou, Y., Liu, L., Ma'ayan, A., Samocha, K. E., Sabo, A., ... Daly, M. J. (2012). Patterns and rates of exonic de novo mutations in autism spectrum disorders. *Nature, 485*(7397), 242–245. http://dx.doi.org/10.1038/nature11011.

Neale, B. M., Rivas, M. A., Voight, B. F., Altshuler, D., Devlin, B., Orho-Melander, M., ... Daly, M. J. (2011). Testing for an unusual distribution of rare variants. *PLoS Genetics, 7*(3), e1001322. http://dx.doi.org/10.1371/journal.pgen.1001322.

O'Roak, B. J., Deriziotis, P., Lee, C., Vives, L., Schwartz, J. J., Girirajan, S., ... Eichler, E. E. (2011). Exome sequencing in sporadic autism spectrum disorders identifies severe de novo mutations. *Nature Genetics, 43*(6), 585–589. http://dx.doi.org/10.1038/ng.835.

Ogiwara, I., Nakayama, T., Yamagata, T., Ohtani, H., Mazaki, E., Tsuchiya, S., ... Yamakawa, K. (2012). A homozygous mutation of voltage-gated sodium channel beta(I) gene SCN1B in a patient with Dravet syndrome. *Epilepsia, 53*(12), e200–203. http://dx.doi.org/10.1111/epi.12040.

Oliva, M., Berkovic, S. F., & Petrou, S. (2012). Sodium channels and the neurobiology of epilepsy. *Epilepsia, 53*(11), 1849–1859. http://dx.doi.org/10.1111/j.1528-1167.2012.03631.x.

Orhan, G., Bock, M., Schepers, D., Ilina, E. I., Reichel, S. N., Loffler, H., ... Maljevic, S. (2014). Dominant-negative effects of KCNQ2 mutations are associated with epileptic encephalopathy. *Annals of Neurology, 75*(3), 382–394. http://dx.doi.org/10.1002/ana.24080.

Ottman, R., Hirose, S., Jain, S., Lerche, H., Lopes-Cendes, I., Noebels, J. L., ... Scheffer, I. E. (2010). Genetic testing in the epilepsies—report of the ILAE Genetics Commission. *Epilepsia, 51*(4), 655–670. http://dx.doi.org/10.1111/j.1528-1167.2009.02429.x.

Padilla, K. M., Antonio, B. M., Santos, S. C., Lin, Z., Theile, J. W., Chapman, M. L., ... Krafte, S. (2014). *Approaches to understanding human ion channel genetic variation and disease — An example with a KCNT1 variant and infantile epilepsy disorder.*

Peljto, A. L., Barker-Cummings, C., Vasoli, V. M., Leibson, C. L., Hauser, W. A., Buchhalter, J. R., & Ottman, R. (2014). Familial risk of epilepsy: a population-based study. *Brain, 137*(Pt 3), 795–805. http://dx.doi.org/10.1093/brain/awt368.

Perez-Reyes, E. (2003). Molecular physiology of low-voltage-activated t-type calcium channels. *Physiological Reviews, 83*(1), 117–161. http://dx.doi.org/10.1152/physrev.00018.2002.

Peters, H. C., Hu, H., Pongs, O., Storm, J. F., & Isbrandt, D. (2005). Conditional transgenic suppression of M channels in mouse brain reveals functions in neuronal excitability, resonance and behavior. *Nature Neuroscience, 8*(1), 51—60. http://dx.doi.org/10.1038/nn1375.

Petrovski, S., & Kwan, P. (2013). Unraveling the genetics of common epilepsies: approaches, platforms, and caveats. *Epilepsy and Behavior, 26*(3), 229—233. http://dx.doi.org/10.1016/j.yebeh.2012.09.007.

Petrovski, S., Wang, Q., Heinzen, E. L., Allen, A. S., & Goldstein, D. B. (2013). Genic intolerance to functional variation and the interpretation of personal genomes. *PLoS Genetics, 9*(8), e1003709. http://dx.doi.org/10.1371/journal.pgen.1003709.

Pierson, T. M., Yuan, H., Marsh, E. D., Fuentes-Fajardo, K., Adams, D. R., Markello, T., ... Gahl, W. A. (2014). GRIN2A mutation and early-onset epileptic encephalopathy: personalized therapy with memantine. *Annals of Clinical and Translational Neurology, 1*(3), 190—198. http://dx.doi.org/10.1002/acn3.39.

Poduri, A., & Lowenstein, D. (2011). Epilepsy genetics—past, present, and future. *Current Opinion in Genetics and Development, 21*(3), 325—332. http://dx.doi.org/10.1016/j.gde.2011.01.005.

Qu, Y., Curtis, R., Lawson, D., Gilbride, K., Ge, P., DiStefano, P. S., ... Scheuer, T. (2001). Differential modulation of sodium channel gating and persistent sodium currents by the beta1, beta2, and beta3 subunits. *Molecular and Cellular Neuroscience, 18*(5), 570—580. http://dx.doi.org/10.1006/mcne.2001.1039.

Redon, R., Ishikawa, S., Fitch, K. R., Feuk, L., Perry, G. H., Andrews, T. D., ... Hurles, M. E. (2006). Global variation in copy number in the human genome. *Nature, 444*(7118), 444—454. http://dx.doi.org/10.1038/nature05329.

Reid, C. A., Bekkers, J. M., & Clements, J. D. (1998). N- and P/Q-type Ca^{2+} channels mediate transmitter release with a similar cooperativity at rat hippocampal autapses. *Journal of Neuroscience, 18*(8), 2849—2855.

Reid, C. A., Berkovic, S. F., & Petrou, S. (2009). Mechanisms of human inherited epilepsies. *Progress in Neurobiology, 87*(1), 41—57. http://dx.doi.org/10.1016/j.pneurobio.2008.09.016.

Rodrigues-Pinguet, N., Jia, L., Li, M., Figl, A., Klaassen, A., Truong, A., ... Cohen, B. N. (2003). Five ADNFLE mutations reduce the Ca^{2+} dependence of the mammalian alpha4beta2 acetylcholine response. *Journal of Physiology, 550*(Pt 1), 11—26. http://dx.doi.org/10.1113/jphysiol.2002.036681.

Rogawski, M. A. (2000). KCNQ2/KCNQ3 K^{+} channels and the molecular pathogenesis of epilepsy: implications for therapy. *Trends in Neurosciences, 23*(9), 393—398.

Saitsu, H., Kato, M., Koide, A., Goto, T., Fujita, T., Nishiyama, K., ... Matsumoto, N. (2012). Whole exome sequencing identifies KCNQ2 mutations in Ohtahara syndrome. *Annals of Neurology, 72*(2), 298—300. http://dx.doi.org/10.1002/ana.23620.

Saitsu, H., Kato, M., Mizuguchi, T., Hamada, K., Osaka, H., Tohyama, J., ... Matsumoto, N. (2008). De novo mutations in the gene encoding STXBP1 (MUNC18-1) cause early infantile epileptic encephalopathy. *Nature Genetics, 40*(6), 782—788. http://dx.doi.org/10.1038/ng.150.

Samocha, K. E., Robinson, E. B., Sanders, S. J., Stevens, C., Sabo, A., McGrath, L. M., ... Daly, M. J. (2014). A framework for the interpretation of de novo mutation in human disease. *Nature Genetics, 46*(9), 944—950. http://dx.doi.org/10.1038/ng.3050.

Scheffer, I. E. (2011). Genetic testing in epilepsy: what should you be doing? *Epilepsy Currents, 11*(4), 107—111. http://dx.doi.org/10.5698/1535-7511-11.4.107.

Scheffer, I. E., Harkin, L. A., Grinton, B. E., Dibbens, L. M., Turner, S. J., Zielinski, M. A., ... Berkovic, S. F. (2007). Temporal lobe epilepsy and GEFS+ phenotypes associated with SCN1B mutations. *Brain, 130*(Pt 1), 100—109. http://dx.doi.org/10.1093/brain/awl272.

Scheffer, I. E., Heron, S. E., Regan, B. M., Mandelstam, S., Crompton, D. E., Hodgson, B. L., ... Dibbens, L. M. (2014). Mutations in mammalian target of rapamycin regulator DEPDC5 cause focal epilepsy with brain malformations. *Annals of Neurology, 75*(5), 782—787. http://dx.doi.org/10.1002/ana.24126.

Schubert, J., Siekierska, A., Langlois, M., May, P., Huneau, C., Becker, F., ... Lerche, H. (2014). Mutations in STX1B, encoding a presynaptic protein, cause fever-associated epilepsy syndromes. *Nature Genetics, 46*(12), 1327—1332. http://dx.doi.org/10.1038/ng.3130.

Sim, N. L., Kumar, P., Hu, J., Henikoff, S., Schneider, G., & Ng, P. C. (2012). SIFT web server: predicting effects of amino acid substitutions on proteins. *Nucleic Acids Research, 40*(Web Server issue), W452—W457. http://dx.doi.org/10.1093/nar/gks539.

Singh, N. A., Charlier, C., Stauffer, D., DuPont, B. R., Leach, R. J., Melis, R., ... Leppert, M. (1998). A novel potassium channel gene, KCNQ2, is mutated in an inherited epilepsy of newborns. *Nature Genetics, 18*(1), 25—29. http://dx.doi.org/10.1038/ng0198-25.

Son, J. H., & Winzer-Serhan, U. H. (2006). Postnatal expression of alpha2 nicotinic acetylcholine receptor subunit mRNA in developing cortex and hippocampus. *Journal of Chemical Neuroanatomy, 32*(2—4), 179—190. http://dx.doi.org/10.1016/j.jchemneu.2006.09.001.

Speed, D., O'Brien, T. J., Palotie, A., Shkura, K., Marson, A. G., Balding, D. J., & Johnson, M. R. (2014). Describing the genetic architecture of epilepsy through heritability analysis. *Brain, 137*(Pt 10), 2680—2689. http://dx.doi.org/10.1093/brain/awu206.

Steinlein, O. K., Mulley, J. C., Propping, P., Wallace, R. H., Phillips, H. A., Sutherland, G. R., ... Berkovic, S. F. (1995). A missense mutation in the neuronal nicotinic acetylcholine receptor alpha 4 subunit is associated with autosomal dominant nocturnal frontal lobe epilepsy. *Nature Genetics, 11*(2), 201—203. http://dx.doi.org/10.1038/ng1095-201.

Suls, A., Dedeken, P., Goffin, K., Van Esch, H., Dupont, P., Cassiman, D., ... Van Paesschen, W. (2008). Paroxysmal exercise-induced dyskinesia and epilepsy is due to mutations in SLC2A1, encoding the glucose transporter GLUT1. *Brain, 131*(Pt 7), 1831—1844. http://dx.doi.org/10.1093/brain/awn113.

Thammongkol, S., Vears, D. F., Bicknell-Royle, J., Nation, J., Draffin, K., Stewart, K. G., ... Mackay, M. T. (2012). Efficacy of the ketogenic diet: which epilepsies respond? *Epilepsia, 53*(3), e55—59. http://dx.doi.org/10.1111/j.1528-1167.2011.03394.x.

Thomas, R. H., & Berkovic, S. F. (2014). The hidden genetics of epilepsy-a clinically important new paradigm. *Nature Reviews Neurology, 10*(5), 283—292. http://dx.doi.org/10.1038/nrneurol.2014.62.

Vissers, L. E., de Ligt, J., Gilissen, C., Janssen, I., Steehouwer, M., de Vries, P., ... Veltman, J. A. (2010). A de novo paradigm for mental retardation. *Nature Genetics, 42*(12), 1109—1112. http://dx.doi.org/10.1038/ng.712.

Vitko, I., Chen, Y., Arias, J. M., Shen, Y., Wu, X. R., & Perez-Reyes, E. (2005). Functional characterization and neuronal modeling of the effects of childhood absence epilepsy variants of CACNA1H, a T-type calcium channel. *Journal of Neuroscience, 25*(19), 4844–4855. http://dx.doi.org/10.1523/JNEUROSCI.0847-05.2005.

Wallace, R. H., Wang, D. W., Singh, R., Scheffer, I. E., George, A. L., Jr., Phillips, H. A., … Mulley, J. C. (1998). Febrile seizures and generalized epilepsy associated with a mutation in the Na^+-channel beta1 subunit gene SCN1B. *Nature Genetics, 19*(4), 366–370. http://dx.doi.org/10.1038/1252.

Wang, H. S., Pan, Z., Shi, W., Brown, B. S., Wymore, R. S., Cohen, I. S., … McKinnon, D. (1998). KCNQ2 and KCNQ3 potassium channel subunits: molecular correlates of the M-channel. *Science, 282*(5395), 1890–1893.

Weber, Y. G., Storch, A., Wuttke, T. V., Brockmann, K., Kempfle, J., Maljevic, S., … Lerche, H. (2008). GLUT1 mutations are a cause of paroxysmal exertion-induced dyskinesias and induce hemolytic anemia by a cation leak. *Journal of Clinical Investigation, 118*(6), 2157–2168. http://dx.doi.org/10.1172/JCI34438.

Westenbroek, R. E., Merrick, D. K., & Catterall, W. A. (1989). Differential subcellular localization of the RI and RII Na^+ channel subtypes in central neurons. *Neuron, 3*(6), 695–704.

Whitaker, W. R., Clare, J. J., Powell, A. J., Chen, Y. H., Faull, R. L., & Emson, P. C. (2000). Distribution of voltage-gated sodium channel alpha-subunit and beta-subunit mRNAs in human hippocampal formation, cortex, and cerebellum. *Journal of Comparative Neurology, 422*(1), 123–139.

Wu, M. C., Lee, S., Cai, T., Li, Y., Boehnke, M., & Lin, X. (2011). Rare-variant association testing for sequencing data with the sequence kernel association test. *American Journal of Human Genetics, 89*(1), 82–93. http://dx.doi.org/10.1016/j.ajhg.2011.05.029.

Yuan, H., Hansen, K. B., Zhang, J., Pierson, T. M., Markello, T. C., Fajardo, K. V., … Traynelis, S. F. (2014). Functional analysis of a de novo GRIN2A missense mutation associated with early-onset epileptic encephalopathy. *Nature Communications, 5*, 3251. http://dx.doi.org/10.1038/ncomms4251.

Chapter 38

Clinical Syndromes of Substance Use Disorder

Nii A. Addy
Yale University School of Medicine, New Haven, CT, United States

Wojciech B. Solecki
Institute of Pharmacology PAS, Krakow, Poland; Jagiellonian University, Krakow, Poland

INTRODUCTION

One challenge in the diagnosis and successful treatment of substance use disorder (SUD) is the field's reliance on a diagnostic system that is based primarily on behavioral symptomatology rather than on biological pathology. It will be increasingly important for the next generation of clinical neuroscientists to understand the multiple biological processes that underlie the etiology and pathogenesis of substance abuse. Indeed, extensive preclinical and clinical research investigations have provided great insight into the cellular and molecular mechanisms that mediate addiction. However, the field has struggled with translating the knowledge gained from these studies into effective therapeutic interventions. Clearly, the manifestation of substance abuse results from a complex interaction between biological and environmental factors. With advances in genomics, new biological questions can be addressed regarding the role of genetic susceptibility in SUD. In this chapter, we will review the neurophysiological, cellular, and molecular mechanisms that mediate specific behavioral symptoms in SUD. We will also consider genetic and epigenetic factors and their ability to modulate the neurobiological processes that underlie substance abuse. In doing so, we will discuss how advances in our knowledge of genetic mechanisms can inform understanding of SUD susceptibility, and the implications of such knowledge for the successful treatment of SUD.

OVERVIEW OF SUBSTANCE USE DISORDER

SUD is a central nervous system disorder characterized by specific behaviors that lead to recurring and oftentimes compulsive drug seeking. According to the *International Classification of Diseases*, 10th revision, addiction is a mixture of physiological (sensitization, tolerance, and physical dependence), psychological (drug craving), and behavioral symptoms, among which the overwhelming desire to take the drug is proposed to be a core symptom (WHO, 1992). Another important symptom of addiction is the emergence of negative emotional states (irritability, anxiety, and dysphoria) that result from drug inaccessibility (Koob & Nestler, 1997; Kreek & Koob, 1998). The *Diagnostic and Statistical Manual of Mental Disorders*, Fifth Edition (APA, 2013), emphasizes that drug use leads to neuroadaptations that can be maintained through periods of abstinence. These neuronal adaptions are thought to have a causal role in the development of SUD symptoms, including drug craving, drug seeking, and the recurrent drug use that occurs despite adverse physical, psychological, or social consequences of that drug use.

In discussing the neurobiological processes that mediate substance abuse, it is important to consider that SUD is strongly associated with individual differences in genetic, epigenetic, and environmental factors. Many people engage in the use of legal or illicit psychoactive substances. However, only a small percentage of these individuals (7.7% for alcohol and almost 3% for illicit drugs) meet the diagnostic criteria for SUD (SAMHSA, 2009). The factors mediating this susceptibility include differences in sensitivity to the rewarding effects of a drug as well as individual differences in how

Genomics, Circuits, and Pathways in Clinical Neuropsychiatry. http://dx.doi.org/10.1016/B978-0-12-800105-9.00038-X

quickly a drug such as ethanol or nicotine is metabolized. Individual traits, such as high impulsivity, that are not directly associated with drug sensitivity can also influence susceptibility (Ernst et al., 2006).

In the early stages of drug use, subjects remain in control of their drug taking and typically do not experience negative social or legal consequences of drug taking. With repeated drug use, a strong association forms between the drug and the drug-associated cues (paraphernalia) or contexts. These pairings can eventually lead to conditioning in which drug-associated stimuli can induce craving and promote drug taking. Over time, increased episodes of drug intoxication can begin to damage an individual's health and lead to negative social consequences. In some instances, the experience of negative physical, psychological, or social consequences can lead to termination of drug taking. However, highly susceptible individuals will continue to engage in drug-taking behavior despite the negative or harmful consequences.

Repeated drug taking can also lead to the development of both tolerance and dependence. Tolerance results from neurobiological and psychological habituation to the effects of a drug that occurs with frequent exposure. After tolerance to the reinforcing effects of a drug, an individual will often take larger amounts of drug in an attempt to experience the desired euphoric effects of the drug. Chronic drug use also induces neuroadaptations that can lead to a "withdrawal syndrome" during periods of drug abstinence, manifested by somatic and psychological symptoms. Withdrawal-induced dysphoria can evoke a strong desire to reengage in drug-taking behavior to relieve these negative emotional and somatic states. Intermittent drug use (particularly psychostimulant use) can also lead to development of sensitization, defined as an enhanced biological or physiological response over time to repeated drug use. As with tolerance, sensitization results from neuroadaptations that occur with repeated drug exposure. Furthermore, tolerance and sensitization to specific effects of drugs of abuse can occur at the same time, owing to specific neuroadaptations occurring in different neuronal circuits or pathways. For instance, an individual may exhibit tolerance to the euphoric effects of cocaine while simultaneously exhibiting a sensitized response to the detrimental cardiovascular effects, which can eventually lead to drug overdose and death. The full manifestation of substance abuse is often characterized by the dominance of drug craving as well as a loss of control over drug-taking activities. In this stage, the overpowering craving and the resulting drug seeking and drug taking are prioritized over other behaviors, responsibilities, and obligations.

CIRCUITS AND PATHWAYS THAT MEDIATE SUBSTANCE ABUSE

Initial hypotheses concerning the potential brain processes that underlie substance abuse were largely based on studies from the 1950s to 1970s that investigated the neural basis of reward processing. In the 1950s, Olds and Milner demonstrated that rats would perform operant behavior to receive direct electrical stimulation (intracranial self-stimulation) of the brain (Olds & Milner, 1954). Subsequent pharmacological experimenters suggested that dopamine was a critical neurotransmitter mediating the reinforcing effects of both intracranial stimulation and drugs of abuse (Fouriezos, Hansson, & Wise, 1978; Wise, 1987). Specifically, early preclinical studies with stimulants demonstrated that lesions of the dopamine system dramatically reduced cocaine and amphetamine taking in rat drug self-administration models (Roberts, Corcoran, & Fibiger, 1977; Yokel & Wise, 1976).

Mesolimbic Dopamine System

The mesolimbic dopamine system consists of dopaminergic neurons in the ventral tegmental area (VTA) that send projections to several brain regions including the nucleus accumbens (NAc), prefrontal cortex (PFC), extended amygdala, and hippocampus (Fig. 38.1). Seminal preclinical microdialysis studies by Di Chiara and Imperato (1988) first demonstrated that administration of nicotine, cocaine, or amphetamine led to increased dopamine levels, an effect which has also been observed with several other drugs of abuse. Stimulants, including cocaine, amphetamine, and methamphetamine, act on dopamine transporter (DAT) (in the NAc and elsewhere) to prevent dopamine reuptake or reverse the action of the DAT and increase dopamine release into the synapse (reviewed in Sulzer, 2011). Nicotine, on the other hand, acts at nicotinic acetylcholine receptors (nAChRs) located on multiple cellular populations in the VTA and the NAc to increase dopamine release (Balfour, Wright, Benwell, & Birrell, 2000; Di Chiara & Imperato, 1988; Picciotto, Addy, Mineur, & Brunzell, 2008). In contrast to stimulating dopamine release directly, cannabinoids and opioids act at their respective target receptors to decrease gamma aminobutyric acid-ergic (GABAergic) inhibition in the VTA, leading to disinhibition and oftentimes allowing for greater or sustained dopamine release (Hyman, Malenka, & Nestler, 2006; Nestler, 1994). These preclinical findings are also consistent with clinical positron emission tomography (PET) findings that have shown increased dopamine activity in the striatum of human subjects after drug abuse exposure (Brody et al., 2009; Volkow et al., 1999).

Whereas dopamine clearly contributes to the initially intoxicating effects of drug use, a remaining question is whether these dopaminergic effects contribute to the pathogenesis of substance abuse. Electrochemical studies, using high spatial

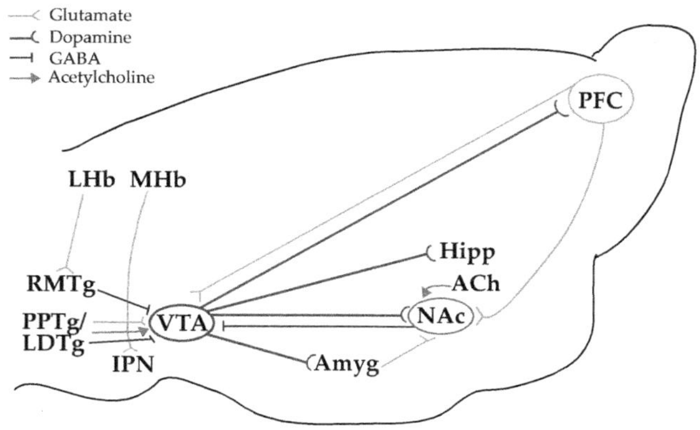

FIGURE 38.1 Circuits and pathways mediating substance abuse. Dopamine neurons in the VTA send projections to and receive input from multiple loci throughout the brain (Bromberg-Martin et al., 2010; Lammel et al., 2012). Findings have also begun to elucidate the role of the habenula in reward-related behavior and in mediating the aversive components of drug exposure (Fowler et al., 2011). *Ach*, acetylcholine; *Amyg*, amygdala; *Hipp*, hippocampus; *IPN*, interpeduncular nucleus; *LDTg*, laterodorsal tegmental nucleus; *LHb*, lateral habenula; *MHb*, medial habenula; *NAc*, nucleus accumbens; *PFC*, prefrontal cortex; *PPTg*, pedunculopontine tegmental nucleus; *RMTg*, rostromedial tegmental nucleus; *VTA*, ventral tegmental area.

and temporal resolution voltammetry techniques, have begun to highlight the dynamic and complex nature of dopamine signaling, revealing dopamine responses to drug-associated cues and highlighting the role of this neurophysiological response in drug-seeking behavior (Phillips, Stuber, Heien, Wightman, & Carelli, 2003; Stuber, Wightman, & Carelli, 2005). Clinical and preclinical studies have also highlighted a role for dopaminergic activity in the amygdala in attributing salience to drug-associated contextual and discrete cues, thus mediating the ability of these drug-associated cues to facilitate drug-seeking behavior (Childress et al., 1999; Shaham, Shalev, Lu, De Wit, & Stewart, 2003; Yahyavi-Firouz-Abadi & See, 2009). In addition, preclinical and clinical studies have shown that dopamine activity in the PFC mediates drug-induced and cue-induced drug-seeking behavior, specifically through regulation of inhibitory control and working memory (McFarland & Kalivas, 2001; See, 2009; Sun & Rebec, 2005). Several different brain structures send input into the VTA, including the PFC, NAc, laterodorsal tegmentum, and the caudal portion of the VTA (known as the rostromedial tegmental nucleus (Bromberg-Martin, Matsumoto, & Hikosaka, 2010; Volkow & Baler, 2014), (Fig. 38.1)). Thus, studies have begun to elucidate the distinct and sometimes opposing roles of specific inputs to the VTA in mediating rewarding, aversive, and drug-related behavior (Lammel et al., 2012; Solecki et al., 2013).

A major emphasis of SUD research has been on dopamine, whereas the VTA consists of multiple neuronal cell types including dopaminergic (about 65%), inhibitory GABAergic (about 30%), and excitatory glutamatergic (about 5%) neurons (Dobi, Margolis, Wang, Harvey, & Morales, 2010; Yamaguchi, Sheen, & Morales, 2007). Studies also suggest that 30–60% of VTA neurons are capable of releasing glutamate (either alone or co-released with dopamine), forming a parallel glutamatergic VTA to NAc pathway with a currently unclear functional role (Morales & Root, 2014). As will be discussed in section Synaptic Plasticity, drug-induced glutamatergic synaptic plasticity within the mesolimbic system is thought to underlie the neuroadaptations and behavioral changes associated with SUD. In contrast, activation of VTA GABA neurons decreases dopamine neuronal firing in the VTA and is sufficient to induce aversive behavior (Tan et al., 2012). Inhibition of VTA GABAergic neurons thus leads to a removal of inhibition (disinhibition) of dopamine neurons (Bocklisch et al., 2013), which is a major neuronal mechanism through which opiates (Johnson & North, 1992a), gamma-hydroxy butyric acid (Cruz et al., 2004), and cannabinoids exert their reinforcing effects (Oleson & Cheer, 2012). Acetylcholine mechanisms are also known to mediate substance abuse, an effect most clearly defined for nicotine dependence (ND). Nicotine acts at nAChRs (endogenously activated by acetylcholine) to induce neurophysiological and behavioral effects that lead to ND (Picciotto et al., 2008). Emerging evidence has also demonstrated that alcohol, cocaine, and heroin act through acetylcholine receptor mechanisms (particularly in the VTA) to regulate drug-taking and drug-seeking behavior (Hendrickson, Zhao-Shea, & Tapper, 2009; Lof et al., 2007; Solecki et al., 2013; You, Wang, Zitzman, & Wise, 2008; Zhou et al., 2007). In terms of noradrenaline, despite initial findings suggesting its role in positive reinforcement (Stein, 1962), studies have revealed that noradrenaline release from neurons in the locus coeruleus into the medial PFC is associated with salience attribution and can also modulate dopamine release in the NAc (Ventura, Morrone, & Puglisi-Allegra, 2007). Indeed, current evidence suggests that noradrenaline transmission in the PFC is necessary for attention and for attributing motivational salience to both reward-associated and aversion-associated stimuli (Feenstra, 2000; Ventura et al., 2007).

Neurophysiological Mechanisms of Reward Learning: Role in Drug Seeking

A key component of substance abuse is drug craving that is precipitated by drug-associated cues. Facilitation of drug taking by drug-associated cues is based on Pavlovian conditioning, in which an unconditioned stimulus (the drug of abuse) is repeatedly paired with a conditioned stimulus (drug paraphernalia and other discrete cues), such that subsequent exposure to those cues can elicit seeking behavior for the unconditioned stimulus. The relevance of Pavlovian conditioning to substance dependence was first proposed by Wikler in 1948 and has since been characterized extensively in the preclinical and clinical literature. In clinical research, cue reactivity serves as the metric for cue-induced craving and takes into account both peripheral (heart rate and body temperature) and central nervous system responses (as determined by craving rating scales) to drug-associated cues versus non–drug associated cues (Drummond, 2000). Several pioneering human PET studies and functional MRI studies revealed that several key dopaminergic brain regions (including the striatum, prefrontal cortex, and amygdala) show increased activity in response to cues associated with cocaine (Childress et al., 1999), alcohol (George et al., 2001), and opiates (Daglish et al., 2001). Extensive, robust preclinical studies in rat and mouse models have also demonstrated a critical role of several dopamine regions, including the PFC (McLaughlin & See, 2003), NAc (Di Ciano, Robbins, & Everitt, 2008), and amygdala (Meil & See, 1996; See, 2005) in the ability of cues to promote drug seeking and drug taking.

Much of the current understanding of the neurophysiological mechanisms that underlie cue-induced drug seeking has come from findings in the learning field. Seminal experiments by Schultz and colleagues in macaque monkeys revealed increased high-frequency (burst) firing of dopamine neurons that occurs in response to the presentation of either an unexpected reward (unconditioned stimulus) or the presentation of a cue (conditioned stimulus) that predicts reward availability (Schultz, 1998). These initial dopamine firing responses were later replicated in a rat model (Hyland, Reynolds, Hay, Perk, & Miller, 2002), and similar effects have been observed for dopamine release at the level of the NAc, both for natural rewards and for drugs of abuse (Day, Roitman, Wightman, & Carelli, 2007; Phillips et al., 2003). These findings highlighted the dynamic nature of dopamine responses to rewards and reward-associated cues and served as the foundation for the reward prediction error model of reward processing (Schultz, 1998).

In considering the neurophysiological pathology of substance abuse, it is also important to discuss the concept of incentive salience, initially proposed by Robinson and Berridge (Robinson & Berridge, 1993), in which reward-associated cues acquire incentive salience over time and become "wanted" or "desired" in and of themselves. In the incentive salience model, the desire for the cue serves as a driving force that can facilitate subsequent drug taking. An example is cigarette smokers who, over time, begin to crave the warm, burning sensation at the back of their throat that has become strongly associated with cigarette smoking. In preclinical studies, the presentation of cues that have acquired incentive salience leads to increased dopamine in the NAc core (Flagel et al., 2011), and that dopamine increase is necessary for the ability of cues to promote reward-seeking behavior (Saunders & Robinson, 2012; Saunders, Yager, & Robinson, 2013). In addition, optogenetics studies have shown that cell-specific activation of dopamine neurons that mimics dopamine firing patterns observed in response to rewards and reward-associated cues is sufficient to evoke dopamine release in the NAc and sufficient to promote reward-seeking behavior (Tsai et al., 2009). Thus a major emphasis for ongoing research is to identify the processes that regulate cue-induced dopamine activity and alter drug-seeking and drug-taking behavior (Fowler, Lu, Johnson, Marks, & Kenny, 2011; Lecca, Meye, & Mameli, 2014; Solecki et al., 2013).

In SUD, the transition from recreational drug use to harmful drug use and dependence is often accompanied by compulsive drug-seeking and drug-taking behavior. Whereas activity in the NAc mediates both drug-taking and drug-seeking behavior, current evidence suggests that activity in the dorsal striatum mediates the transition from recreational to compulsive or habitual drug use (Belin & Everitt, 2008; Willuhn, Burgeno, Groblewski, & Phillips, 2014). Compulsive drug seeking is also hypothesized to reflect the inability of executive, cortical systems to regulate behavior effectively in these individuals. Indeed, the degree of cortical hypoactivity during abstinence is correlated with severity of cocaine craving in cocaine-dependent human subjects (Goldstein & Volkow, 2011; Jasinska, Chen, Bonci, & Stein, 2015). In addition, rescuing cocaine-induced PFC hypoactivity prevents compulsive cocaine seeking in rodents (Chen et al., 2013). Such results suggest that substance abuse may be partially mediated by decreased PFC inhibitory control of drug craving. Furthermore, the strengthening of inhibitory control may underlie the efficacy of cognitive behavioral therapy for substance abuse (Beck, 2008). In addition, activity in the cortical–striatal pathways is strongly influenced by "bottom-up" signaling from brain stem nuclei, which respond to salient environmental stimuli (Chandler, Waterhouse, & Gao, 2014; Navailles, Guillem, Vouillac-Mendoza, & Ahmed, 2014). Thus, top-down and bottom-up processing, and drug-mediated alterations in this processing, can powerfully modulate behaviors implicated in SUD.

Stress can also facilitate drug-seeking and drug-taking behavior in substance-abusing individuals, especially during periods of drug abstinence or withdrawal. The neurocircuitry underlying stress-induced drug-seeking resembles the circuitry mediating the aversive characteristics of drug withdrawal. Specifically, inactivation of bed nucleus of stria terminalis (BNST), the central amygdala, the NAc shell, or the VTA is sufficient to abolish stress-induced relapse to cocaine seeking in animal models (McFarland, Davidge, Lapish, & Kalivas, 2004). Several studies have also highlighted the role of stress-mediated processes (including corticotropin-releasing factor [CRF] activity, noradrenergic signaling, and glucocorticoid activity), in drug relapse (Koob, 2015; Mantsch et al., 2014). Specifically, multiple studies have provided evidence of noradrenergic inputs in the central amygdala and BNST, as well as CRF signaling within the central amygdala, BNST, and VTA, as key mediators of stress-induced drug seeking (Erb, Salmaso, Rodaros, & Stewart, 2001; Smith & Aston-Jones, 2008; Wang, You, Rice, & Wise, 2007).

Neurocircuitry Underlying Drug Withdrawal

Chronic drug use leads to both within-system and between-system neuroadaptations that have distinct functional consequences. Within-system neuroadaptations (within the mesolimbic system) are characterized by dampened dopaminergic and serotoninergic neurotransmission (Koob, 1996; Rossetti, Melis, Carboni, & Gessa, 1992). These neurochemical changes facilitate the emergence of motivational deficits, depressed mood, psychomotor retardation, decreased motivation for natural rewards, and increased sensitivity to drugs of abuse (Barr & Phillips, 1999; Melis, Spiga, & Diana, 2005). The concept of within-system neuroadaptation is strongly supported by human imaging studies demonstrating hypoactivity in mesolimbic dopamine and cortical functioning during drug abstinence and withdrawal (Volkow, Fowler, & Wang, 2003). Furthermore, it has been proposed that within-system neuroadaptations alter reward system processing, which in turn leads to an escalation of drug taking and an enhancement of compulsive drug-seeking during periods of abstinence (Koob, 2015; Shaham & Hope, 2005). Chronic administration of all drugs of abuse also leads to hypothalamic–pituitary–adrenal axis and extra-hypothalamic dysregulation, which can subsequently lead to increased noradrenaline signaling, increased CRF activity, and increased corticosterone signaling during periods of drug abstinence (George et al., 2007; Koob, 2008; Koob & Volkow, 2010). Functionally, these neuroadaptations result in the emergence of somatic symptoms of withdrawal and negative affective states (eg, anxiety or irritability). Indeed, enhanced CRF signaling has been observed in response to withdrawal from drugs of abuse (Sarnyai, Shaham, & Heinrichs, 2001). In contrast, administration of a CRF_1 receptor antagonist decreases the somatic signs of opiate withdrawal (Navarro-Zaragoza, Nunez, Laorden, & Milanes, 2010) and attenuates anxiety-like behaviors induced by withdrawal from cocaine, nicotine, or alcohol (Basso, Spina, Rivier, Vale, & Koob, 1999; Bruijnzeel, Small, Pasek, & Yamada, 2010; George et al., 2007; Sarnyai et al., 2001). Interestingly, CRF-dependent regulation of motivational control is likely a consequence of convergence of within- and between-system neuroadaptations.

CELLULAR AND MOLECULAR PATHOLOGY OF SUBSTANCE ABUSE

Synaptic Plasticity

Many of the behavioral symptoms associated with SUD are thought to result from the synaptic plasticity that occurs after chronic drug exposure. Synaptic plasticity is broadly defined as an activity-dependent strengthening or weakening of synapses over time (Kandel, Schwartz, & Jessel, 2000). Initial studies on the ability of drugs of abuse to alter synaptic plasticity were based on previous investigations on the role of hippocampal synaptic plasticity in learning and memory. Seminal findings from Bliss and Lomo in 1973 provided the first evidence for long-term potentiation (LTP) – defined as a sustained increase in excitatory neuronal activity after a brief high-frequency stimulation (Kandel, Schwartz, & Jessel, 2000). Because of the extensive evidence for dopaminergic activity in drug-related behavior, researchers eventually began to investigate whether exposure to drugs of abuse led to synaptic plasticity in the dopamine system. In seminal work, Ungless and colleagues examined excitatory postsynaptic currents (EPSCs) in VTA dopamine neurons in brain slices from mice that had previously received injections of saline or cocaine (Ungless, Whistler, Malenka, & Bonci, 2001). In light of previous evidence for increased α-amino-3-hydroxy-5-methyl-4-isoxazolepropionic acid (AMPA) receptor-mediated contributions to EPSCs after LTP (Bonci & Malenka, 1999; Johnson & North, 1992b), the authors compared the ratio of AMPA-mediated currents with *N*-methyl-D-aspartate (NMDA)-mediated currents as a metric of synaptic plasticity. They found that a single exposure to cocaine was sufficient to induce LTP in VTA dopamine neurons (Ungless et al., 2001). Importantly, their results revealed that in addition to cocaine-mediated blockade of DATs in the NAc, cocaine could also enhance the activity of dopamine neurons at the level of the VTA. This VTA plasticity is not limited to cocaine, because subsequent studies have revealed robust LTP in VTA dopamine neurons after exposure to other drugs of abuse including nicotine, morphine, ethanol, and amphetamine (Bernier, Whitaker, & Morikawa, 2011; Mansvelder & McGehee, 2000; Saal, Dong, Bonci, & Malenka, 2003).

Initial evidence for NMDA glutamate receptor mechanisms mediating VTA plasticity facilitated subsequent investigations into the role of glutamatergic and dopaminergic interactions in drug-induced synaptic plasticity. Ungless and colleagues observed that cocaine-mediated plasticity in the VTA depended on NMDA receptors (Ungless et al., 2001). Other studies revealed that nicotine-mediated LTP in VTA dopamine neurons resulted from the activation of specific nicotinic receptors (α7-containing receptors) located on glutamate neurons in the VTA, and not the activation of nicotinic receptors on the dopamine cells themselves (Mansvelder & McGehee, 2000). These glutamatergic and dopamine interactions in the VTA also underlie drug-mediated increases in dopamine in the NAc, because blockade of VTA glutamate receptors attenuates drug-induced increases in NAc dopamine (Schilstrom, Nomikos, Nisell, Hertel, & Svensson, 1998). Behaviorally, VTA glutamate receptor mechanisms in preclinical rodent models also regulate drug taking and drug-seeking behavior for several drugs of abuse including nicotine, heroin, cocaine, and morphine (Harris & Aston-Jones, 2003; Harris, Wimmer, Byrne, & Aston-Jones, 2004; Liechti & Markou, 2007; Xi & Stein, 2002).

In addition to these acute effects, repeated cocaine exposure leads to the potentiation of synaptic strength in the VTA (Borgland, Malenka, & Bonci, 2004). This chronic drug-induced synaptic plasticity can also persist after periods of prolonged abstinence (Chen et al., 2008). Again, glutamatergic mechanisms in the VTA are thought to have a critical role because glutamate receptor neuroadaptations in the VTA are associated with prolonged drug exposure, withdrawal, and subsequent relapse to drug taking (Shaham & Hope, 2005). For instance, the cocaine-mediated neuroadaptations in the VTA that are associated with prolonged cocaine use and withdrawal also facilitate cue-induced cocaine-seeking behavior (Lu, Dempsey, Liu, Bossert, & Shaham, 2004; Lu et al., 2009; Mameli et al., 2009). Specifically, repeated cocaine exposure and withdrawal induces synaptic plasticity in the NAc (Kourrich, Rothwell, Klug, & Thomas, 2007; Mameli et al., 2009) that depends on cocaine-mediated plasticity in the VTA (Mameli et al., 2009), which again demonstrates the critical role of the VTA-to-NAc pathway. It is also important to note that drug-induced synaptic plasticity in other dopamine target regions, such as the PFC, mediates cue-induced drug seeking (Moussawi et al., 2011; Van den Oever et al., 2008; Van den Oever, Spijker, Smit, & DeVries, 2010). Specifically, glutamatergic neurons projecting from the PFC to both the VTA and the NAc can further modulate dopaminergic activity to modulate drug-taking and drug-seeking behavior.

Intracellular Signaling

Drug-induced increases in neurotransmitter release and synaptic activity are often mediated through activation or inhibition of membrane-bound receptors that subsequently alter the activity of intracellular target proteins. Several neurotransmitter receptors implicated in substance abuse (including dopamine receptors and specific types of glutamate, acetylcholine, and GABA receptors) are associated with membrane-bound G-proteins. Activation of G-protein—coupled receptors leads to conformational changes that can activate adenylyl cyclase (G_s-coupled), phospholipase C (PLC) (G_q-coupled), or inhibit adenylyl cyclase (G_i-coupled) to initiate a cascade of downstream processes within the cell (O'Connor & Adams, 2010). As described in section Circuits and Pathways That Mediate Substance Abuse, dopamine levels in the NAc increase in response to drugs of abuse (inducing euphoria) and in response to drug-associated cues and contexts (promoting craving, drug-seeking, and relapse). These increases in NAc dopamine lead to activation of the low affinity D1 dopamine receptors in the NAc (Fig. 38.2). These G_s—coupled D1 receptors activate adenylyl cyclase to modulate downstream intracellular targets and subsequently alter gene expression, membrane stabilization, and synaptic plasticity. One well-characterized drug-activated pathway in the NAc is adenylyl cyclase to the cyclic adenosine monophosphate (cAMP) pathway, whose downstream targets include protein kinase A (PKA), dopamine- and cAMP—regulated phosphoprotein 32 (DARPP-32), extracellular signal regulated kinase (ERK), and the transcription factor cAMP response element-binding protein (CREB) (Fig. 38.2). Multiple types of drugs of abuse increase DARPP-32, ERK, and CREB activity in the striatum and NAc (Greengard, Allen, & Nairn, 1999; Valjent et al., 2000). Functionally, drug-induced activation of PKA and its downstream targets promotes several drug-related behaviors, including drug-induced locomotor stimulation and sensitization, as well as drug-seeking behavior (Borgkvist, Marcellino, Fuxe, Greengard, & Fisone, 2008; Fienberg et al., 1998). In contrast inhibition of NAc PKA targets, such as ERK, attenuates the behavioral stimulating effects of drugs of abuse and attenuates drug-seeking behavior (Fienberg et al., 1998; Valjent et al., 2000).

Drug-mediated stimulation of the D1 pathway also facilitates neuroadaptations by promoting genetic alterations within the NAc. Specifically, drug-mediated increases in CREB activity leads to the increased transcription of multiple genes, including the opioid peptides dynorphin and prodynorphin, as well as the immediately early genes fos and fosB (Nestler, 2013). CREB activation is suggested to induce tolerance by decreasing sensitivity to drugs of abuse, whereas inhibition of CREB activity in the NAc is sufficient to increase drug-taking and drug-seeking behavior (Nestler, 2013). Indeed, inhibiting CREB activity specifically in D1 expressing NAc neurons also enhances the psychostimulant effects of cocaine and enhances cocaine-seeking behavior (Bilbao et al., 2014). In addition to drug-mediated activation of transcription factors, investigations have revealed drug-induced epigenetic processes that mediate behavioral responses to drugs of

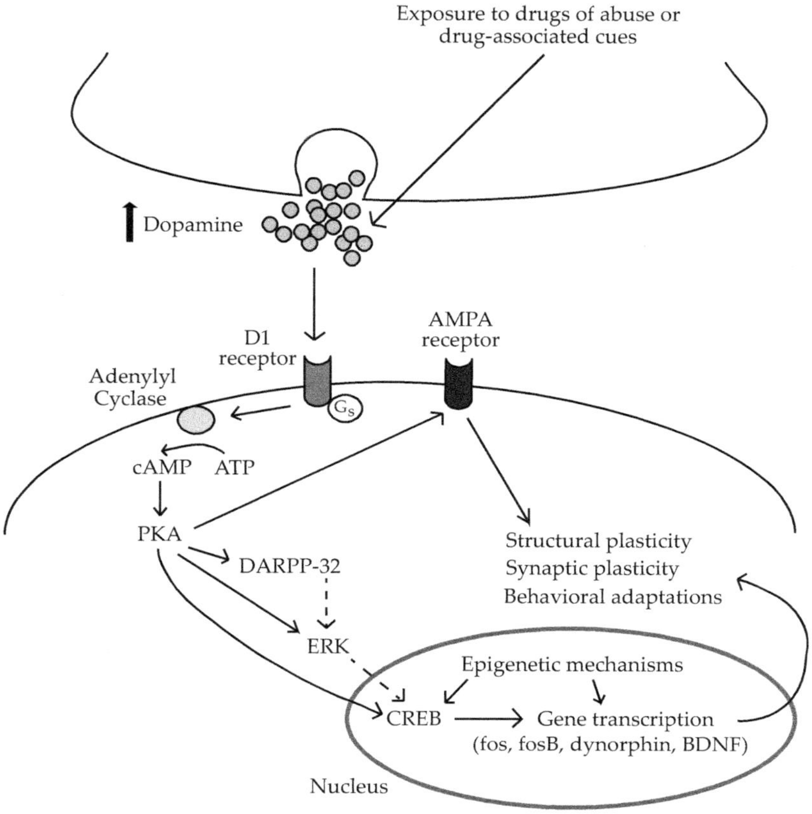

FIGURE 38.2 Nucleus accumbens D1-mediated signaling in substance abuse. Exposure to drugs of abuse or to drug-associated cues enhances dopamine levels (Di Chiara & Imperato, 1988; Phillips et al., 2003), which stimulates D1 receptors and leads to the activation of several intracellular targets, including PKA, DARPP-32, ERK, and CREB (Fienberg et al., 1998; Greengard et al., 1999; Valjent et al., 2000). Drug- and cue-induced activation of D1 targets also mediates several behavioral effects of drugs of abuse (Edenberg & Foroud, 2006; Fienberg et al., 1998; Greengard et al., 1999) and leads to gene transcript, structural, and synaptic neuroadaptations that mediate subsequent drug-taking and drug-seeking behavior (Nestler, 2013).

abuse. Epigenetics is defined as a change in gene expression that occurs in the absence of alterations to the DNA sequence. Multiple epigenetic processes including histone modification, DNA methylation, and micro-RNA mechanisms contribute to the effects of drugs of abuse in the mesolimbic system (Schmidt, McGinty, West, & Sadri-Vakili, 2013). Several studies have focused on histone modifications that either promote or prevent gene transcription. Histones exist in octamers and are essential components of the chromatin structure. DNA wraps around histones, allowing for tight DNA packaging (Schmidt et al., 2013). Histone acetylation by histone acetyl transferases (HATs) leads to conformational changes that make DNA accessible for transcription, whereas histone deacetylation by histonedeacetylases (HDACs) deactivates histones and prevents transcription (Nestler, 2013). Exposure to either cocaine or amphetamine has been shown to increase histone H4 acetylation in the NAc, thus promoting gene transcription (Kumar et al., 2005; Shen et al., 2008). In contrast, blocking the effects of CREB-binding protein (an H4 HAT which normally interacts with CREB) attenuates the psychostimulant effects of cocaine (Levine et al., 2005). In addition, direct NAc administration of an HDAC inhibitor (again promoting gene transcription) is sufficient to enhance drug-seeking behavior (Kumar et al., 2005).

Many of these observed drug-mediated epigenetic alterations are transient and likely do not mediate the key behaviors associated with prolonged drug use. Intracellular and epigenetic alterations maintained after prolonged drug abuse exposure, in contrast, are more likely to underlie the behavioral symptoms of SUD. For instance, chronic cocaine exposure is associated with increased H3 acetylation of the brain-derived neurotrophic factor (BDNF) promoter and is associated with increased BDNF transcription and expression in the NAc and PFC (Kumar et al., 2005; Sadri-Vakili et al., 2010). Furthermore, manipulating BDNF activity or histone acetylation in these brain regions alters the psychostimulant effects of cocaine and also alters cocaine-seeking behavior (Kumar et al., 2005; Sadri-Vakili et al., 2010).

In chronic drug users, the ability of these intracellular and transcriptional processes to induce neuroadaptations likely enhances behavioral responses to relapse triggers. Indeed, drug-mediated activation of NAc D1 pathway can induce downstream neuroadaptations that further promote neuronal and behavioral responses to subsequent drug exposure

(Fig. 38.2). Drug-mediated neuroadaptations in the dopamine system are also likely to enhance neuronal responses to drug-associated stimuli, as the drug-associated cues also increase dopamine levels in the NAc to activate D1 receptors. Investigations have revealed that exposure to drug-associated cues and contexts that promote drug relapse also leads to the activation of PKA and PKA target proteins in the NAc, including ERK and the AMPA receptor (Edwards, Bachtell, Guzman, Whisler, & Self, 2011; Ferrario et al., 2010; Schroeder et al., 2008). Although our discussion here has focused on D1-mediated effects, other studies have revealed roles for several other intracellular signaling pathways. Within the NAc PLC-mediated signaling through protein kinase C is suggested to underlie drug-mediated structural plasticity, whereby exposure to drugs of abuse can increase the number of dendritic contacts ("spines") and branches to further promote synaptic plasticity (Robinson & Kolb, 2004). Indeed, the structural changes associated with drug abuse exposure are strongly correlated with drug-mediated behavioral plasticity, as reflected in behavioral sensitization (Robinson & Kolb, 2004). Outside the NAc, BDNF-mediated activation in the PFC (Whitfield, Shi, Sun, & McGinty, 2011) and NMDA receptor—mediated activation of the ERK pathway in the amygdala (Lu et al., 2005) have been demonstrated to regulate cue-induced drug-seeking behavior for multiple drugs of abuse. Importantly, the interactions between these pathways can further modulate NAc activity, as the PFC and amygdala send glutamatergic input to the NAc (Fig. 38.1).

Although cellular and molecular studies have provided important insight into the mechanisms underlying responses to drugs of abuse, most are inherently limited in scope, because they typically focus on one target at a time. However, advances in genetic sampling techniques such as the use of microarrays, have allowed for broader, more high-throughput examinations of the gene products involved in substance abuse. One challenge for the field has been to determine whether the identified changes in gene expression (in animal and human studies) lead to functional changes in protein levels that ultimately mediate synaptic plasticity, intracellular signaling, circuit-level activity, and changes in behavior. To address this challenge, proteomic studies have been used to directly investigate drug-induced changes in protein levels (Bantscheff, Lemeer, Savitski, & Kuster, 2012). Together, these gene expression and proteomic studies have identified (and in many cases confirmed) specific genes and proteins involved in cellular and molecular responses to drugs of abuse.

GENETIC FACTORS CONTRIBUTING TO SUBSTANCE USE DISORDER

In an effort to assess the overall genetic contribution to susceptibility to SUD, several family studies, twin studies, and adoption studies have been performed (Agrawal & Lynskey, 2008; Uhl, Elmer, Labuda, & Pickens, 1995). These have revealed that genetic factors account for about 50—60% of the variance in SUD, whereas environmental factors account for the rest of the variance (Uhl et al., 2008). As further evidence for the genetic contributions to SUD susceptibility, traits that are known risk factors for SUD, such as hyperactivity, low self-esteem, school problems, social withdrawal, and neuroticism, have also been shown to be heritable (accounting for 34—61% of variance) (Siewert, Stallings, & Hewitt, 2004). In addition to genetic factors, several environmental factors are clearly associated with SUD. Specifically, stress is a major risk factor for multiple forms of substance abuse, including alcoholism (Aseltine & Gore, 2000; Schmidt, Dufeu, Kuhn, Smolka, & Rommelspacher, 2000). Indeed, some psychological theories of SUD postulate that drug use and abuse are a form of self-medication in an attempt to cope with, and exert control over, stressors (Baker, Piper, McCarthy, Majeskie, & Fiore, 2004; Leventhal & Cleary, 1980; Russell & Mehrabian, 1975). Susceptibility to SUD also varies across age, and adolescents are most vulnerable to initiating drug use that may later develop into substance abuse. Ultimately, however, similar to other common neuropsychiatric disorders, the etiology and pathophysiology of SUD reflect the interaction of genetic and environmental factors. Indeed, among psychiatric disorders, the impact of gene—environment interaction is most clear for SUD because substance abuse cannot occur without exposure to a drug of abuse.

Throughout the 1990s and early 2000s, efforts to identify specific risk genes for SUD relied on linkage analysis and candidate gene studies. Candidate approaches were particularly popular given a host of genes with high biological plausibility. Whereas these studies generally suffered from the considerable methodological vulnerabilities associated with the method (chapter: Association Strategies), variations in alcohol-metabolizing enzymes have remained the subject of considerable interest. *ADH1B* His48Arg and *ALDH2* Lys487 polymorphisms have long been associated with risk of alcoholism, and directly and predictably lead to alcohol-induced flushing through accumulation of acetaldehyde and release of histamine (Radel & Goldman, 2001). *ADH1B*, *ALDH2*, and *ADH4* influence alcohol consumption and have been implicated as risk factors for developing alcohol abuse or dependence (Radel & Goldman, 2001; Thomasson et al., 1994, 1991). In general however, more contemporary gene discovery approaches have turned to genome-wide association (GWA) methods, with a particular focus on alcohol dependence (AD) and ND. GWA studies (GWAS) of nicotine have been the most productive to date. These have tended to rely a composite categorization of ND as well its disaggregation into component parts, including, for example cigarettes smoked per day, smoking irritation, smoking quantity, and smoking heaviness (Wang, Yang, Ma, Payne, & Li, 2014). The most robust GWAS findings for ND have implicated a cluster of three genes in strong linkage disequilibrium on chromosome 15q25 (Bierut et al., 2008; Sherva et al., 2008; Weiss et al., 2008), encoding neuronal nAChR subunits, *CHRNA5*, *CHRNA3*, and *CHRNB4*. In these initial studies, individuals carrying the putative coding risk allele, rs16969968, show an approximately 30% increased risk for ND, reflected by increased likelihood to early onset of smoking, heaviness of smoking, and smoking-induced euphoria ("pleasurable buzz"). Those with two copies of the risk allele are more than twice as likely to develop ND (Bierut et al., 2008; Schroeder et al., 2008; Weiss et al., 2008).

Two subsequent genome-wide meta-analyses, including tens of thousands of samples, confirmed these findings and further dissected the association signals within the region. Both (Liu et al., 2010; Saccone et al., 2010) found the strongest association signal for smoking quantity (measured as cigarettes per day) mapping to the gene *CHRNA5*. Liu et al. (2010) also found evidence for a separate association signal within an intron of the gene *CHRNA3*. As strong orthogonal support for these findings, this locus has also been associated by GWAS with lung cancer, peripheral arterial disease, and chronic obstructive pulmonary disease (Amos et al., 2008; Hung et al., 2008; Pillai et al., 2009; Thorgeirsson et al., 2008). Moreover, subsequent GWAS provided further support for the CHRNA5/3/4 cluster as well as additional association signals on chromosomes 19q13 and 8p11 (Thorgeirsson et al., 2010). These loci contain genes encoding nicotine-metabolizing enzymes (*CYP2A6* and *CYP2B6*) and nicotinic acetylcholine receptor subunits (*CHRNB3* and *CHRNA6*).

GWAS findings with regard to the CHRNA5/3/4 cluster have been further illuminated by preclinical studies demonstrating that genetic manipulation of the alpha5 nAChR dramatically alters nicotine taking in rodents. Genetic deletion of the alpha5 nAChR subunit or downregulation of alpha5 receptors dramatically increases nicotine intake in mice, which is consistent with the increased association with heavy smoking in humans carrying the alpha5 risk allele (Bierut et al., 2008; Fowler & Kenny, 2014; Fowler et al., 2011). Specifically, rats and mice with genetically induced alpha5 receptor deficits show maintained nicotine self-administration behavior across multiple nicotine doses, even those that are normally aversive (Fowler et al., 2011). This increased nicotine intake is not the result of alterations in the rewarding effects of nicotine, but rather a decreased sensitivity to the inhibitory effects of the high nicotine doses (Fowler et al., 2011). Given this GWAS identified alpha5 target and the robust behavioral phenotype observed in preclinical models, investigators have used cellular, molecular, and genetic tools to delineate the brain loci, circuits, and pathways through which alpha5 mechanisms mediate susceptibility to ND. Whereas alpha5 receptor expression is relatively low in the dopamine reward system, alpha5 receptors expression is high in the medial habenula (MHb) (Fowler et al., 2011). Furthermore, experimental induction of an alpha5 deficit directly in the MHb is sufficient to increase nicotine intake, similar to the effects observed in alpha5 knockout mouse. These groundbreaking findings have thus provided novel understanding on the role of alpha5 receptors in the MHb to interpeduncular nucleus pathway in mediating aversive responses to nicotine and susceptibility to nicotine dependence (Fig. 38.1).

To date, GWAS of alcohol dependence have been less revealing: Multiple studies have been reported in the literature leveraging samples from large consortia, such as the Collaborative Studies of Genetics of Alcoholism, the Study of Addiction: Genetics and Environment, and the Australian Twin Registry. Although several genes have been found to reach genome-wide significance (Frank et al., 2012; Schumann et al., 2011; Treutlein et al., 2009), there has been no reliable overlap between studies. Moreover, several moderately sized studies (>1000 probands) (Bierut et al., 2010; Edenberg et al., 2010; Heath et al., 2011), including one of the largest GWAS of AD conducted to date (Heath et al., 2011), have failed to identify SNPs with genome-wide significance.

As noted previously, difficulties in identifying replicable loci in GWAS of psychiatric disorders does not suggest that common variants are unimportant. As noted (chapter: Association Strategies), sample size and associated power issues, small individual effects of most risk alleles, and marked locus heterogeneity commonly contribute to these challenges. Similarly, for SUD in general, additional complexities regarding the role of exposure in determining risk outcome likely contribute as well. For AD it has been hypothesized that varying commonly used phenotypic measures, including quantity and frequency of drinking, maximum drinks in 24 hours, frequency of heavy drinking, and frequency of intoxication, could have a considerable impact on the study's outcome, making replication particularly difficult (Morozova, Goldman, Mackay, & Anholt, 2012).

Because of the challenges associated with genome-wide studies of common variants and the emergence of new genomic technologies, there has been renewed interest in the study of rare variation in both alcohol and nicotine (Haller et al., 2014; Wang et al., 2014; Xie et al., 2011; Zuo et al., 2013). As noted in chapter "Association Strategies", unbiased, genome-wide studies of low-frequency mutations (whether protective or risk alleles) are likely to require sample sizes that equal or exceed those necessary for successful GWAS of common variants. Whereas effect sizes are expected to be somewhat larger for rare mutations, the low allele frequency presents its own considerable power issues, even when collapsing (burden) approaches are leveraged (chapter: Association Strategies). Thus it is likely that tens to hundreds of thousands of subjects will be required for genome-wide approaches to rare variant discovery. Certainly, the contribution of de novo variation could change this calculus, as it has for autism and other neurodevelopmental disorders, but there is scant evidence to date that new mutation has a substantial role in the overall genetic architecture of SUD.

CONCLUSION

Understanding the genetic-, molecular-, and circuit-level neuroadaptations that mediate SUD in specific individuals can provide a stronger foundation for addressing specific perturbations as an effective therapeutic strategy. Specifically, a bottom-up investigative approach can provide invaluable insight into the neurobiological factors that mediate different components of SUD for a particular substance. For example, numerous neurobiological investigations have effectively used a reductionist approach to identify different genetic, cellular and molecular, circuit, and pathway-specific mechanisms that underlie the effects of drugs of abuse. In ND, genetic studies in human populations, however, have provided an opportunity to examine the role and functional consequences of identified genetic risk factors that enhance susceptibility to specific SUDs. This ability to trace such effects from genetic susceptibility to behavioral symptomatology has provided an unprecedented opportunity to understand the critical neurobiological processes that mediate increased susceptibility in individuals carrying the $\alpha5$ nAChR risk allele. Importantly, investigators are also seeking to identify genetic factors that enhance susceptibility for other SUDs, including cocaine, alcohol, and opioid dependence. As the field moves forward, a deeper understanding of the genetic factors and the processes through which these factors mediate the etiology and pathogenesis of substance abuse will greatly enhance the development of more effective preventative and therapeutic strategies.

ACKNOWLEDGMENTS

This work was supported by the National Institutes of Health grant DA038048 (NAA), by the Foundation for Polish Science grant HOMING PLUS/2013-7/14 (WBS) and by the Polish National Science Center grant 2013/11/D/NZ4/02371 (WBS). We also thank Dr. Marina Picciotto and Dr. Robert Malison for critical reading and helpful feedback on this chapter.

REFERENCES

Agrawal, A., & Lynskey, M. T. (2008). Are there genetic influences on addiction: evidence from family, adoption and twin studies. *Addiction, 103*(7), 1069–1081.

Amos, C. I., Wu, X., Broderick, P., Gorlov, I. P., Gu, J., Eisen, T., et al. (2008). Genome-wide association scan of tag SNPs identifies a susceptibility locus for lung cancer at 15q25.1. *Nature Genetics, 40*(5), 616–622.

APA. (2013). *Diagnostic and statistical manual of mental disorders* (5th ed.). Washington, D.C.

Aseltine, R. H., Jr., & Gore, S. L. (2000). The variable effects of stress on alcohol use from adolescence to early adulthood. *Substance Use and Misuse, 35*(5), 643–668.

Baker, T. B., Piper, M. E., McCarthy, D. E., Majeskie, M. R., & Fiore, M. C. (2004). Addiction motivation reformulated: an affective processing model of negative reinforcement. *Psychological Review, 111*(1), 33–51.

Balfour, D. J., Wright, A. E., Benwell, M. E., & Birrell, C. E. (2000). The putative role of extra-synaptic mesolimbic dopamine in the neurobiology of nicotine dependence. *Behavioural Brain Research, 113*(1–2), 73–83.

Bantscheff, M., Lemeer, S., Savitski, M. M., & Kuster, B. (2012). Quantitative mass spectrometry in proteomics: critical review update from 2007 to the present. *Analytical and Bioanalytical Chemistry, 404*(4), 939–965.

Barr, A. M., & Phillips, A. G. (1999). Withdrawal following repeated exposure to d-amphetamine decreases responding for a sucrose solution as measured by a progressive ratio schedule of reinforcement. *Psychopharmacology (Berlin), 141*(1), 99–106.

Basso, A. M., Spina, M., Rivier, J., Vale, W., & Koob, G. F. (1999). Corticotropin-releasing factor antagonist attenuates the "anxiogenic-like" effect in the defensive burying paradigm but not in the elevated plus-maze following chronic cocaine in rats. *Psychopharmacology (Berlin), 145*(1), 21–30.

Beck, A. T. (2008). The evolution of the cognitive model of depression and its neurobiological correlates. *American Journal of Psychiatry, 165*(8), 969–977.

Belin, D., & Everitt, B. J. (2008). Cocaine seeking habits depend upon dopamine-dependent serial connectivity linking the ventral with the dorsal striatum. *Neuron, 57*(3), 432–441.

Bernier, B. E., Whitaker, L. R., & Morikawa, H. (2011). Previous ethanol experience enhances synaptic plasticity of NMDA receptors in the ventral tegmental area. *Journal of Neuroscience, 31*(14), 5205–5212.

Bierut, L. J., Agrawal, A., Bucholz, K. K., Doheny, K. F., Laurie, C., Pugh, E., et al. (2010). A genome-wide association study of alcohol dependence. *Proceedings of the National Academy of Sciences of the United States of America, 107*(11), 5082–5087.

Bierut, L. J., Stitzel, J. A., Wang, J. C., Hinrichs, A. L., Grucza, R. A., Xuei, X., et al. (2008). Variants in nicotinic receptors and risk for nicotine dependence. *American Journal of Psychiatry, 165*(9), 1163–1171.

Bilbao, A., Rieker, C., Cannella, N., Parlato, R., Golda, S., Piechota, M., et al. (2014). CREB activity in dopamine D1 receptor expressing neurons regulates cocaine-induced behavioral effects. *Frontiers in Behavioral Neuroscience, 8*, 212.

Bliss, T. V., & Lomo, T. (1973). Long-lasting potentiation of synaptic transmission in the dentate area of the anaesthetized rabbit following stimulation of the perforant path. *Journal of Physiology, 232*(2), 331–356.

Bocklisch, C., Pascoli, V., Wong, J. C., House, D. R., Yvon, C., de Roo, M., et al. (2013). Cocaine disinhibits dopamine neurons by potentiation of GABA transmission in the ventral tegmental area. *Science, 341*(6153), 1521–1525.

Bonci, A., & Malenka, R. C. (1999). Properties and plasticity of excitatory synapses on dopaminergic and GABAergic cells in the ventral tegmental area. *Journal of Neuroscience, 19*(10), 3723–3730.

Borgkvist, A., Marcellino, D., Fuxe, K., Greengard, P., & Fisone, G. (2008). Regulation of DARPP-32 phosphorylation by Delta9-tetrahydrocannabinol. *Neuropharmacology, 54*(1), 31–35.

Borgland, S. L., Malenka, R. C., & Bonci, A. (2004). Acute and chronic cocaine-induced potentiation of synaptic strength in the ventral tegmental area: electrophysiological and behavioral correlates in individual rats. *Journal of Neuroscience, 24*(34), 7482–7490.

Brody, A. L., Mandelkern, M. A., Olmstead, R. E., Allen-Martinez, Z., Scheibal, D., Abrams, A. L., et al. (2009). Ventral striatal dopamine release in response to smoking a regular vs a denicotinized cigarette. *Neuropsychopharmacology, 34*(2), 282–289.

Bromberg-Martin, E. S., Matsumoto, M., & Hikosaka, O. (2010). Dopamine in motivational control: rewarding, aversive, and alerting. *Neuron, 68*(5), 815–834.

Bruijnzeel, A. W., Small, E., Pasek, T. M., & Yamada, H. (2010). Corticotropin-releasing factor mediates the dysphoria-like state associated with alcohol withdrawal in rats. *Behavioural Brain Research, 210*(2), 288–291.

Chandler, D. J., Waterhouse, B. D., & Gao, W. J. (2014). New perspectives on catecholaminergic regulation of executive circuits: evidence for independent modulation of prefrontal functions by midbrain dopaminergic and noradrenergic neurons. *Frontiers in Neural Circuits, 8*, 53.

Chen, B. T., Bowers, M. S., Martin, M., Hopf, F. W., Guillory, A. M., Carelli, R. M., et al. (2008). Cocaine but not natural reward self-administration nor passive cocaine infusion produces persistent LTP in the VTA. *Neuron, 59*(2), 288–297.

Chen, B. T., Yau, H. J., Hatch, C., Kusumoto-Yoshida, I., Cho, S. L., Hopf, F. W., et al. (2013). Rescuing cocaine-induced prefrontal cortex hypoactivity prevents compulsive cocaine seeking. *Nature, 496*(7445), 359–362.

Childress, A. R., Mozley, P. D., McElgin, W., Fitzgerald, J., Reivich, M., & O'Brien, C. P. (1999). Limbic activation during cue-induced cocaine craving. *American Journal of Psychiatry, 156*(1), 11–18.

Cruz, H. G., Ivanova, T., Lunn, M. L., Stoffel, M., Slesinger, P. A., & Luscher, C. (2004). Bi-directional effects of GABA(B) receptor agonists on the mesolimbic dopamine system. *Nature Neuroscience, 7*(2), 153–159.

Daglish, M. R., Weinstein, A., Malizia, A. L., Wilson, S., Melichar, J. K., Britten, S., et al. (2001). Changes in regional cerebral blood flow elicited by craving memories in abstinent opiate-dependent subjects. *American Journal of Psychiatry, 158*(10), 1680–1686.

Day, J. J., Roitman, M. F., Wightman, R. M., & Carelli, R. M. (2007). Associative learning mediates dynamic shifts in dopamine signaling in the nucleus accumbens. *Nature Neuroscience, 10*(8), 1020–1028.

Di Chiara, G., & Imperato, A. (1988). Drugs abused by humans preferentially increase synaptic dopamine concentrations in the mesolimbic system of freely moving rats. *Proceedings of the National Academy of Sciences of the United States of America, 85*(14), 5274–5278.

Di Ciano, P., Robbins, T. W., & Everitt, B. J. (2008). Differential effects of nucleus accumbens core, shell, or dorsal striatal inactivations on the persistence, reacquisition, or reinstatement of responding for a drug-paired conditioned reinforcer. *Neuropsychopharmacology, 33*(6), 1413–1425.

Dobi, A., Margolis, E. B., Wang, H. L., Harvey, B. K., & Morales, M. (2010). Glutamatergic and nonglutamatergic neurons of the ventral tegmental area establish local synaptic contacts with dopaminergic and nondopaminergic neurons. *Journal of Neuroscience, 30*(1), 218–229.

Drummond, D. C. (2000). What does cue-reactivity have to offer clinical research? *Addiction, 95*(Suppl. 2), S129–S144.

Edenberg, H. J., & Foroud, T. (2006). The genetics of alcoholism: identifying specific genes through family studies. *Addiction Biology, 11*(3–4), 386–396.

Edenberg, H. J., Koller, D. L., Xuei, X., Wetherill, L., McClintick, J. N., Almasy, L., et al. (2010). Genome-wide association study of alcohol dependence implicates a region on chromosome 11. *Alcoholism Clinical and Experimental Research, 34*(5), 840–852.

Edwards, S., Bachtell, R. K., Guzman, D., Whisler, K. N., & Self, D. W. (2011). Emergence of context-associated GluR(1) and ERK phosphorylation in the nucleus accumbens core during withdrawal from cocaine self-administration. *Addiction Biology, 16*(3), 450–457.

Erb, S., Salmaso, N., Rodaros, D., & Stewart, J. (2001). A role for the CRF-containing pathway from central nucleus of the amygdala to bed nucleus of the stria terminalis in the stress-induced reinstatement of cocaine seeking in rats. *Psychopharmacology (Berlin), 158*(4), 360–365.

Ernst, M., Luckenbaugh, D. A., Moolchan, E. T., Leff, M. K., Allen, R., Eshel, N., et al. (2006). Behavioral predictors of substance-use initiation in adolescents with and without attention-deficit/hyperactivity disorder. *Pediatrics, 117*(6), 2030–2039.

Feenstra, M. G. (2000). Dopamine and noradrenaline release in the prefrontal cortex in relation to unconditioned and conditioned stress and reward. *Progress in Brain Research, 126*, 133–163.

Ferrario, C. R., Li, X., Wang, X., Reimers, J. M., Uejima, J. L., & Wolf, M. E. (2010). The role of glutamate receptor redistribution in locomotor sensitization to cocaine. *Neuropsychopharmacology, 35*(3), 818–833.

Fienberg, A. A., Hiroi, N., Mermelstein, P. G., Song, W., Snyder, G. L., Nishi, A., et al. (1998). DARPP-32: regulator of the efficacy of dopaminergic neurotransmission. *Science, 281*(5378), 838–842.

Flagel, S. B., Clark, J. J., Robinson, T. E., Mayo, L., Czuj, A., Willuhn, I., et al. (2011). A selective role for dopamine in stimulus-reward learning. *Nature, 469*(7328), 53–57.

Fouriezos, G., Hansson, P., & Wise, R. A. (1978). Neuroleptic-induced attenuation of brain stimulation reward in rats. *Journal of Comparative and Physiological Psychology, 92*(4), 661–671.

Fowler, C. D., & Kenny, P. J. (2014). Nicotine aversion: neurobiological mechanisms and relevance to tobacco dependence vulnerability. *Neuropharmacology, 76*(Pt B), 533–544.

Fowler, C. D., Lu, Q., Johnson, P. M., Marks, M. J., & Kenny, P. J. (2011). Habenular alpha5 nicotinic receptor subunit signalling controls nicotine intake. *Nature, 471*(7340), 597–601.

Frank, J., Cichon, S., Treutlein, J., Ridinger, M., Mattheisen, M., Hoffmann, P., et al. (2012). Genome-wide significant association between alcohol dependence and a variant in the ADH gene cluster. *Addiction Biology, 17*(1), 171–180.

George, M. S., Anton, R. F., Bloomer, C., Teneback, C., Drobes, D. J., Lorberbaum, J. P., et al. (2001). Activation of prefrontal cortex and anterior thalamus in alcoholic subjects on exposure to alcohol-specific cues. *Archives of General Psychiatry, 58*(4), 345–352.

George, O., Ghozland, S., Azar, M. R., Cottone, P., Zorrilla, E. P., Parsons, L. H., et al. (2007). CRF-CRF_1 system activation mediates withdrawal-induced increases in nicotine self-administration in nicotine-dependent rats. *Proceedings of the National Academy of Sciences of the United States of America, 104*(43), 17198–17203.

Goldstein, R. Z., & Volkow, N. D. (2011). Dysfunction of the prefrontal cortex in addiction: neuroimaging findings and clinical implications. *Nature Reviews Neuroscience, 12*(11), 652–669.

Greengard, P., Allen, P. B., & Nairn, A. C. (1999). Beyond the dopamine receptor: the DARPP-32/protein phosphatase-1 cascade. *Neuron, 23*(3), 435–447.

Haller, G., Kapoor, M., Budde, J., Xuei, X., Edenberg, H., Nurnberger, J., et al. (2014). Rare missense variants in CHRNB3 and CHRNA3 are associated with risk of alcohol and cocaine dependence. *Human Molecular Genetics, 23*(3), 810–819.

Harris, G. C., & Aston-Jones, G. (2003). Critical role for ventral tegmental glutamate in preference for a cocaine-conditioned environment. *Neuropsychopharmacology, 28*(1), 73–76.

Harris, G. C., Wimmer, M., Byrne, R., & Aston-Jones, G. (2004). Glutamate-associated plasticity in the ventral tegmental area is necessary for conditioning environmental stimuli with morphine. *Neuroscience, 129*(3), 841–847.

Heath, A. C., Whitfield, J. B., Martin, N. G., Pergadia, M. L., Goate, A. M., Lind, P. A., et al. (2011). A quantitative-trait genome-wide association study of alcoholism risk in the community: findings and implications. *Biological Psychiatry, 70*(6), 513–518.

Hendrickson, L. M., Zhao-Shea, R., & Tapper, A. R. (2009). Modulation of ethanol drinking-in-the-dark by mecamylamine and nicotinic acetylcholine receptor agonists in C57BL/6J mice. *Psychopharmacology (Berlin), 204*(4), 563–572.

Hung, R. J., McKay, J. D., Gaborieau, V., Boffetta, P., Hashibe, M., Zaridze, D., et al. (2008). A susceptibility locus for lung cancer maps to nicotinic acetylcholine receptor subunit genes on 15q25. *Nature, 452*(7187), 633–637.

Hyland, B. I., Reynolds, J. N., Hay, J., Perk, C. G., & Miller, R. (2002). Firing modes of midbrain dopamine cells in the freely moving rat. *Neuroscience, 114*(2), 475–492.

Hyman, S. E., Malenka, R. C., & Nestler, E. J. (2006). Neural mechanisms of addiction: the role of reward-related learning and memory. *Annual Review of Neuroscience, 29*, 565–598.

Jasinska, A. J., Chen, B. T., Bonci, A., & Stein, E. A. (2015). Dorsal medial prefrontal cortex (MPFC) circuitry in rodent models of cocaine use: implications for drug addiction therapies. *Addiction Biology, 20*(2), 215–226.

Johnson, S. W., & North, R. A. (1992a). Opioids excite dopamine neurons by hyperpolarization of local interneurons. *Journal of Neuroscience, 12*(2), 483–488.

Johnson, S. W., & North, R. A. (1992b). Two types of neurone in the rat ventral tegmental area and their synaptic inputs. *Journal of Physiology, 450*, 455–468.

Kandel, E. R., Schwartz, J. H., & Jessel, J. L. (Eds.). (2000). *Principles of neural science* (4th ed.). McGraw Hill.

Koob, G. F. (1996). Drug addiction: the yin and yang of hedonic homeostasis. *Neuron, 16*(5), 893–896.

Koob, G. F. (2008). A role for brain stress systems in addiction. *Neuron, 59*(1), 11–34.

Koob, G. F. (2015). The dark side of emotion: the addiction perspective. *European Journal of Pharmacology, 753*, 73–87.

Koob, G. F., & Nestler, E. J. (1997). The neurobiology of drug addiction. *Journal of Neuropsychiatry & Clinical Neurosciences, 9*(3), 482–497.

Koob, G. F., & Volkow, N. D. (2010). Neurocircuitry of addiction. *Neuropsychopharmacology, 35*(1), 217–238.

Kourrich, S., Rothwell, P. E., Klug, J. R., & Thomas, M. J. (2007). Cocaine experience controls bidirectional synaptic plasticity in the nucleus accumbens. *Journal of Neuroscience, 27*(30), 7921–7928.

Kreek, M. J., & Koob, G. F. (1998). Drug dependence: stress and dysregulation of brain reward pathways. *Drug and Alcohol Dependence, 51*(1–2), 23–47.

Kumar, A., Choi, K. H., Renthal, W., Tsankova, N. M., Theobald, D. E., Truong, H. T., et al. (2005). Chromatin remodeling is a key mechanism underlying cocaine-induced plasticity in striatum. *Neuron, 48*(2), 303–314.

Lammel, S., Lim, B. K., Ran, C., Huang, K. W., Betley, M. J., Tye, K. M., et al. (2012). Input-specific control of reward and aversion in the ventral tegmental area. *Nature, 491*(7423), 212–217.

Lecca, S., Meye, F. J., & Mameli, M. (2014). The lateral habenula in addiction and depression: an anatomical, synaptic and behavioral overview. *European Journal of Neuroscience, 39*(7), 1170–1178.

Leventhal, H., & Cleary, P. D. (1980). The smoking problem: a review of the research and theory in behavioral risk modification. *Psychological Bulletin, 88*(2), 370–405.

Levine, A. A., Guan, Z., Barco, A., Xu, S., Kandel, E. R., & Schwartz, J. H. (2005). CREB-binding protein controls response to cocaine by acetylating histones at the fosB promoter in the mouse striatum. *Proceedings of the National Academy of Sciences of the United States of America, 102*(52), 19186–19191.

Liechti, M. E., & Markou, A. (2007). Interactive effects of the mGlu5 receptor antagonist MPEP and the mGlu2/3 receptor antagonist LY341495 on nicotine self-administration and reward deficits associated with nicotine withdrawal in rats. *European Journal of Pharmacology, 554*(2–3), 164–174.

Liu, J. Z., Tozzi, F., Waterworth, D. M., Pillai, S. G., Muglia, P., Middleton, L., et al. (2010). Meta-analysis and imputation refines the association of 15q25 with smoking quantity. *Nature Genetics, 42*(5), 436–440.

Lof, E., Olausson, P., deBejczy, A., Stomberg, R., McIntosh, J. M., Taylor, J. R., et al. (2007). Nicotinic acetylcholine receptors in the ventral tegmental area mediate the dopamine activating and reinforcing properties of ethanol cues. *Psychopharmacology (Berlin), 195*(3), 333–343.

Lu, L., Dempsey, J., Liu, S. Y., Bossert, J. M., & Shaham, Y. (2004). A single infusion of brain-derived neurotrophic factor into the ventral tegmental area induces long-lasting potentiation of cocaine seeking after withdrawal. *Journal of Neuroscience, 24*(7), 1604–1611.

Lu, L., Hope, B. T., Dempsey, J., Liu, S. Y., Bossert, J. M., & Shaham, Y. (2005). Central amygdala ERK signaling pathway is critical to incubation of cocaine craving. *Nature Neuroscience, 8*(2), 212–219.

Lu, L., Wang, X., Wu, P., Xu, C., Zhao, M., Morales, M., et al. (2009). Role of ventral tegmental area glial cell line-derived neurotrophic factor in incubation of cocaine craving. *Biological Psychiatry, 66*(2), 137–145.

Mameli, M., Halbout, B., Creton, C., Engblom, D., Parkitna, J. R., Spanagel, R., et al. (2009). Cocaine-evoked synaptic plasticity: persistence in the VTA triggers adaptations in the NAc. *Nature Neuroscience, 12*(8), 1036–1041.

Mansvelder, H. D., & McGehee, D. S. (2000). Long-term potentiation of excitatory inputs to brain reward areas by nicotine. *Neuron, 27*(2), 349–357.

Mantsch, J. R., Vranjkovic, O., Twining, R. C., Gasser, P. J., McReynolds, J. R., & Blacktop, J. M. (2014). Neurobiological mechanisms that contribute to stress-related cocaine use. *Neuropharmacology, 76*(Pt B), 383–394.

McFarland, K., Davidge, S. B., Lapish, C. C., & Kalivas, P. W. (2004). Limbic and motor circuitry underlying footshock-induced reinstatement of cocaine-seeking behavior. *Journal of Neuroscience, 24*(7), 1551–1560.

McFarland, K., & Kalivas, P. W. (2001). The circuitry mediating cocaine-induced reinstatement of drug-seeking behavior. *Journal of Neuroscience, 21*(21), 8655–8663.

McLaughlin, J., & See, R. E. (2003). Selective inactivation of the dorsomedial prefrontal cortex and the basolateral amygdala attenuates conditioned-cued reinstatement of extinguished cocaine-seeking behavior in rats. *Psychopharmacology (Berlin), 168*(1–2), 57–65.

Meil, W. M., & See, R. E. (1996). Conditioned cued recovery of responding following prolonged withdrawal from self-administered cocaine in rats: an animal model of relapse. *Behavioural Pharmacology, 7*(8), 754–763.

Melis, M., Spiga, S., & Diana, M. (2005). The dopamine hypothesis of drug addiction: hypodopaminergic state. *International Review of Neurobiology, 63*, 101–154.

Morales, M., & Root, D. H. (2014). Glutamate neurons within the midbrain dopamine regions. *Neuroscience, 282C*, 60–68.

Morozova, T. V., Goldman, D., Mackay, T. F., & Anholt, R. R. (2012). The genetic basis of alcoholism: multiple phenotypes, many genes, complex networks. *Genome Biology, 13*(2), 239.

Moussawi, K., Zhou, W., Shen, H., Reichel, C. M., See, R. E., Carr, D. B., et al. (2011). Reversing cocaine-induced synaptic potentiation provides enduring protection from relapse. *Proceedings of the National Academy of Sciences of the United States of America, 108*(1), 385–390.

Navailles, S., Guillem, K., Vouillac-Mendoza, C., & Ahmed, S. H. (2014). Coordinated recruitment of cortical-subcortical circuits and ascending dopamine and serotonin neurons during inhibitory control of cocaine seeking in rats. *Cerebral Cortex, 25*(9), 3167–3181.

Navarro-Zaragoza, J., Nunez, C., Laorden, M. L., & Milanes, M. V. (2010). Effects of corticotropin-releasing factor receptor-1 antagonists on the brain stress system responses to morphine withdrawal. *Molecular Pharmacology, 77*(5), 864–873.

Nestler, E. J. (1994). Molecular neurobiology of drug addiction. *Neuropsychopharmacology, 11*(2), 77–87.

Nestler, E. J. (2013). Cellular basis of memory for addiction. *Dialogues in Clinical Neurosciences, 15*(4), 431–443.

O'Connor, C. M., & Adams, J. U. (2010). *Essentials of cell biology*. Cambridge, MA: NPG Education.

Olds, J., & Milner, P. (1954). Positive reinforcement produced by electrical stimulation of septal area and other regions of rat brain. *Journal of Comparative and Physiological Psychology, 47*(6), 419–427.

Oleson, E. B., & Cheer, J. F. (2012). A brain on cannabinoids: the role of dopamine release in reward seeking. *Cold Spring Harbor Perspectives in Medicine, 2*(8).

Phillips, P. E., Stuber, G. D., Heien, M. L., Wightman, R. M., & Carelli, R. M. (2003). Subsecond dopamine release promotes cocaine seeking. *Nature, 422*(6932), 614–618.

Picciotto, M. R., Addy, N. A., Mineur, Y. S., & Brunzell, D. H. (2008). It is not "either/or": activation and desensitization of nicotinic acetylcholine receptors both contribute to behaviors related to nicotine addiction and mood. *Progress in Neurobiology, 84*(4), 329–342.

Pillai, S. G., Ge, D., Zhu, G., Kong, X., Shianna, K. V., Need, A. C., et al. (2009). A genome-wide association study in chronic obstructive pulmonary disease (COPD): identification of two major susceptibility loci. *PLoS Genetics, 5*(3), e1000421.

Radel, M., & Goldman, D. (2001). Pharmacogenetics of alcohol response and alcoholism: the interplay of genes and environmental factors in thresholds for alcoholism. *Drug Metabolism and Disposition, 29*(4 Pt 2), 489–494.

Roberts, D. C., Corcoran, M. E., & Fibiger, H. C. (1977). On the role of ascending catecholaminergic systems in intravenous self-administration of cocaine. *Pharmacology Biochemistry and Behavior, 6*(6), 615–620.

Robinson, T. E., & Berridge, K. C. (1993). The neural basis of drug craving: an incentive-sensitization theory of addiction. *Brain Research. Brain Research Review, 18*(3), 247–291.

Robinson, T. E., & Kolb, B. (2004). Structural plasticity associated with exposure to drugs of abuse. *Neuropharmacology, 47*(Suppl. 1), 33–46.

Rossetti, Z. L., Melis, F., Carboni, S., & Gessa, G. L. (1992). Dramatic depletion of mesolimbic extracellular dopamine after withdrawal from morphine, alcohol or cocaine: a common neurochemical substrate for drug dependence. *Annals of the New York Academy of Sciences, 654*, 513–516.

Russell, J. A., & Mehrabian, A. (1975). The mediating role of emotions in alcohol use. *Journal of Studies on Alcohol, 36*(11), 1508–1536.

Saal, D., Dong, Y., Bonci, A., & Malenka, R. C. (2003). Drugs of abuse and stress trigger a common synaptic adaptation in dopamine neurons. *Neuron, 37*(4), 577–582.

Saccone, N. L., Culverhouse, R. C., Schwantes-An, T. H., Cannon, D. S., Chen, X., Cichon, S., et al. (2010). Multiple independent loci at chromosome 15q25.1 affect smoking quantity: a meta-analysis and comparison with lung cancer and COPD. *PLoS Genetics, 6*(8).

Sadri-Vakili, G., Kumaresan, V., Schmidt, H. D., Famous, K. R., Chawla, P., Vassoler, F. M., et al. (2010). Cocaine-induced chromatin remodeling increases brain-derived neurotrophic factor transcription in the rat medial prefrontal cortex, which alters the reinforcing efficacy of cocaine. *Journal of Neuroscience, 30*(35), 11735–11744.

SAMHSA. (2009). *Results from the 2008 national survey on drug use and health*. Rockville, MD: National Findings.

Sarnyai, Z., Shaham, Y., & Heinrichs, S. C. (2001). The role of corticotropin-releasing factor in drug addiction. *Pharmacological Reviews, 53*(2), 209–243.

Saunders, B. T., & Robinson, T. E. (2012). The role of dopamine in the accumbens core in the expression of Pavlovian-conditioned responses. *European Journal of Neuroscience, 36*(4), 2521–2532.

Saunders, B. T., Yager, L. M., & Robinson, T. E. (2013). Cue-evoked cocaine "craving": role of dopamine in the accumbens core. *Journal of Neuroscience, 33*(35), 13989–14000.

Schilstrom, B., Nomikos, G. G., Nisell, M., Hertel, P., & Svensson, T. H. (1998). N-methyl-D-aspartate receptor antagonism in the ventral tegmental area diminishes the systemic nicotine-induced dopamine release in the nucleus accumbens. *Neuroscience, 82*(3), 781–789.

Schmidt, H. D., McGinty, J. F., West, A. E., & Sadri-Vakili, G. (2013). Epigenetics and psychostimulant addiction. *Cold Spring Harbor Perspectives in Medicine, 3*(3), a012047.

Schmidt, L. G., Dufeu, P., Kuhn, S., Smolka, M., & Rommelspacher, H. (2000). Transition to alcohol dependence: clinical and neurobiological considerations. *Comprehensive Psychiatry, 41*(2 Suppl. 1), 90–94.

Schroeder, F. A., Penta, K. L., Matevossian, A., Jones, S. R., Konradi, C., Tapper, A. R., et al. (2008). Drug-induced activation of dopamine D(1) receptor signaling and inhibition of class I/II histone deacetylase induce chromatin remodeling in reward circuitry and modulate cocaine-related behaviors. *Neuropsychopharmacology, 33*(12), 2981–2992.

Schultz, W. (1998). Predictive reward signal of dopamine neurons. *Journal of Neurophysiology, 80*(1), 1–27.

Schumann, G., Coin, L. J., Lourdusamy, A., Charoen, P., Berger, K. H., Stacey, D., et al. (2011). Genome-wide association and genetic functional studies identify autism susceptibility candidate 2 gene (AUTS2) in the regulation of alcohol consumption. *Proceedings of the National Academy of Sciences of the United States of America, 108*(17), 7119–7124.

See, R. E. (2005). Neural substrates of cocaine-cue associations that trigger relapse. *European Journal of Pharmacology, 526*(1–3), 140–146.

See, R. E. (2009). Dopamine D1 receptor antagonism in the prelimbic cortex blocks the reinstatement of heroin-seeking in an animal model of relapse. *International Journal of Neuropsychopharmacology, 12*(3), 431–436.

Shaham, Y., & Hope, B. T. (2005). The role of neuroadaptations in relapse to drug seeking. *Nature Neuroscience, 8*(11), 1437–1439.

Shaham, Y., Shalev, U., Lu, L., De Wit, H., & Stewart, J. (2003). The reinstatement model of drug relapse: history, methodology and major findings. *Psychopharmacology (Berlin), 168*(1–2), 3–20.

Shen, H. Y., Kalda, A., Yu, L., Ferrara, J., Zhu, J., & Chen, J. F. (2008). Additive effects of histone deacetylase inhibitors and amphetamine on histone H4 acetylation, cAMP responsive element binding protein phosphorylation and DeltaFosB expression in the striatum and locomotor sensitization in mice. *Neuroscience, 157*(3), 644–655.

Sherva, R., Wilhelmsen, K., Pomerleau, C. S., Chasse, S. A., Rice, J. P., Snedecor, S. M., et al. (2008). Association of a single nucleotide polymorphism in neuronal acetylcholine receptor subunit alpha 5 (CHRNA5) with smoking status and with 'pleasurable buzz' during early experimentation with smoking. *Addiction, 103*(9), 1544–1552.

Siewert, E. A., Stallings, M. C., & Hewitt, J. K. (2004). Genetic influences on vulnerability to, and protective factors for, adolescent drinking. *Twin Research, 7*(6), 617–625.

Smith, R. J., & Aston-Jones, G. (2008). Noradrenergic transmission in the extended amygdala: role in increased drug-seeking and relapse during protracted drug abstinence. *Brain Structure and Function, 213*(1–2), 43–61.

Solecki, W., Wickham, R. J., Behrens, S., Wang, J., Zwerling, B., Mason, G. F., et al. (2013). Differential role of ventral tegmental area acetylcholine and N-methyl-D-aspartate receptors in cocaine-seeking. *Neuropharmacology, 75*, 9–18.

Stein, L. (1962). An analysis of stimulus-duration preference in self-stimulation of the brain. *Journal of Comparative and Physiological Psychology, 55*, 405–414.

Stuber, G. D., Wightman, R. M., & Carelli, R. M. (2005). Extinction of cocaine self-administration reveals functionally and temporally distinct dopaminergic signals in the nucleus accumbens. *Neuron, 46*(4), 661–669.

Sulzer, D. (2011). How addictive drugs disrupt presynaptic dopamine neurotransmission. *Neuron, 69*(4), 628–649.

Sun, W., & Rebec, G. V. (2005). The role of prefrontal cortex D1-like and D2-like receptors in cocaine-seeking behavior in rats. *Psychopharmacology (Berlin), 177*(3), 315–323.

Tan, K. R., Yvon, C., Turiault, M., Mirzabekov, J. J., Doehner, J., Labouebe, G., et al. (2012). GABA neurons of the VTA drive conditioned place aversion. *Neuron, 73*(6), 1173–1183.

Thomasson, H. R., Crabb, D. W., Edenberg, H. J., Li, T. K., Hwu, H. G., Chen, C. C., et al. (1994). Low frequency of the ADH2*2 allele among Atayal natives of Taiwan with alcohol use disorders. *Alcoholism Clinical and Experimental Research, 18*(3), 640–643.

Thomasson, H. R., Edenberg, H. J., Crabb, D. W., Mai, X. L., Jerome, R. E., Li, T. K., et al. (1991). Alcohol and aldehyde dehydrogenase genotypes and alcoholism in Chinese men. *American Journal of Human Genetics, 48*(4), 677–681.

Thorgeirsson, T. E., Geller, F., Sulem, P., Rafnar, T., Wiste, A., Magnusson, K. P., et al. (2008). A variant associated with nicotine dependence, lung cancer and peripheral arterial disease. *Nature, 452*(7187), 638–642.

Thorgeirsson, T. E., Gudbjartsson, D. F., Surakka, I., Vink, J. M., Amin, N., Geller, F., et al. (2010). Sequence variants at CHRNB3-CHRNA6 and CYP2A6 affect smoking behavior. *Nature Genetics, 42*(5), 448–453.

Treutlein, J., Cichon, S., Ridinger, M., Wodarz, N., Soyka, M., Zill, P., et al. (2009). Genome-wide association study of alcohol dependence. *Archives of General Psychiatry, 66*(7), 773–784.

Tsai, H. C., Zhang, F., Adamantidis, A., Stuber, G. D., Bonci, A., de Lecea, L., et al. (2009). Phasic firing in dopaminergic neurons is sufficient for behavioral conditioning. *Science, 324*(5930), 1080–1084.

Uhl, G. R., Drgon, T., Johnson, C., Li, C. Y., Contoreggi, C., Hess, J., et al. (2008). Molecular genetics of addiction and related heritable phenotypes: genome-wide association approaches identify "connectivity constellation" and drug target genes with pleiotropic effects. *Annals of the New York Academy of Sciences, 1141*, 318–381.

Uhl, G. R., Elmer, G. I., Labuda, M. C., & Pickens, R. W. (1995). Genetic influences in drug abuse. In F. E. Bloom, & D. J. Kupfer (Eds.), *Psychopharmacology: The fourth generation of progress* (pp. 1793–2783). New York: Raven Press.

Ungless, M. A., Whistler, J. L., Malenka, R. C., & Bonci, A. (2001). Single cocaine exposure in vivo induces long-term potentiation in dopamine neurons. *Nature, 411*(6837), 583–587.

Valjent, E., Corvol, J. C., Pages, C., Besson, M. J., Maldonado, R., & Caboche, J. (2000). Involvement of the extracellular signal-regulated kinase cascade for cocaine-rewarding properties. *Journal of Neuroscience, 20*(23), 8701–8709.

Van den Oever, M. C., Goriounova, N. A., Li, K. W., Van der Schors, R. C., Binnekade, R., Schoffelmeer, A. N., et al. (2008). Prefrontal cortex AMPA receptor plasticity is crucial for cue-induced relapse to heroin-seeking. *Nature Neuroscience, 11*(9), 1053–1058.

Van den Oever, M. C., Spijker, S., Smit, A. B., & De Vries, T. J. (2010). Prefrontal cortex plasticity mechanisms in drug seeking and relapse. *Neuroscience & Biobehavioral Reviews, 35*(2), 276–284.

Ventura, R., Morrone, C., & Puglisi-Allegra, S. (2007). Prefrontal/accumbal catecholamine system determines motivational salience attribution to both reward- and aversion-related stimuli. *Proceedings of the National Academy of Sciences of the United States of America, 104*(12), 5181–5186.

Volkow, N. D., & Baler, R. D. (2014). Addiction science: uncovering neurobiological complexity. *Neuropharmacology, 76*(Pt B), 235–249.

Volkow, N. D., Fowler, J. S., & Wang, G. J. (2003). The addicted human brain: insights from imaging studies. *Journal of Clinical Investment, 111*(10), 1444–1451.

Volkow, N. D., Wang, G. J., Fowler, J. S., Logan, J., Gatley, S. J., Wong, C., et al. (1999). Reinforcing effects of psychostimulants in humans are associated with increases in brain dopamine and occupancy of D(2) receptors. *Journal of Pharmacology and Experimental Therapeutics, 291*(1), 409–415.

Wang, S., Yang, Z., Ma, J. Z., Payne, T. J., & Li, M. D. (2014). Introduction to deep sequencing and its application to drug addiction research with a focus on rare variants. *Molecular Neurobiology, 49*(1), 601–614.

Wang, B., You, Z. B., Rice, K. C., & Wise, R. A. (2007). Stress-induced relapse to cocaine seeking: roles for the CRF(2) receptor and CRF-binding protein in the ventral tegmental area of the rat. *Psychopharmacology (Berlin), 193*(2), 283–294.

Weiss, R. B., Baker, T. B., Cannon, D. S., von Niederhausern, A., Dunn, D. M., Matsunami, N., et al. (2008). A candidate gene approach identifies the CHRNA5-A3-B4 region as a risk factor for age-dependent nicotine addiction. *PLoS Genetics, 4*(7), e1000125.

Whitfield, T. W., Jr., Shi, X., Sun, W. L., & McGinty, J. F. (2011). The suppressive effect of an intra-prefrontal cortical infusion of BDNF on cocaine-seeking is Trk receptor and extracellular signal-regulated protein kinase mitogen-activated protein kinase dependent. *Journal of Neuroscience, 31*(3), 834–842.

WHO. (1992). *The ICD-10 classification of mental and behavioural disorders*. Geneva: Clinical Descriptions and Diagnostic Guidelines.

Wikler, A. (1948). Recent progress in research on the neurophysiologic basis of morphine addiction. *American Journal of Psychiatry, 105*(5), 329–338.

Willuhn, I., Burgeno, L. M., Groblewski, P. A., & Phillips, P. E. (2014). Excessive cocaine use results from decreased phasic dopamine signaling in the striatum. *Nature Neuroscience, 17*(5), 704–709.

Wise, R. A. (1987). The role of reward pathways in the development of drug dependence. *Pharmacology & Therapeutics, 35*(1–2), 227–263.

Xi, Z. X., & Stein, E. A. (2002). Blockade of ionotropic glutamatergic transmission in the ventral tegmental area reduces heroin reinforcement in rat. *Psychopharmacology (Berlin), 164*(2), 144–150.

Xie, P., Kranzler, H. R., Krauthammer, M., Cosgrove, K. P., Oslin, D., Anton, R. F., et al. (2011). Rare nonsynonymous variants in alpha-4 nicotinic acetylcholine receptor gene protect against nicotine dependence. *Biological Psychiatry, 70*(6), 528–536.

Yahyavi-Firouz-Abadi, N., & See, R. E. (2009). Anti-relapse medications: preclinical models for drug addiction treatment. *Pharmacology & Therapeutics, 124*(2), 235–247.

Yamaguchi, T., Sheen, W., & Morales, M. (2007). Glutamatergic neurons are present in the rat ventral tegmental area. *European Journal of Neuroscience, 25*(1), 106–118.

Yokel, R. A., & Wise, R. A. (1976). Attenuation of intravenous amphetamine reinforcement by central dopamine blockade in rats. *Psychopharmacology (Berlin), 48*(3), 311–318.

You, Z. B., Wang, B., Zitzman, D., & Wise, R. A. (2008). Acetylcholine release in the mesocorticolimbic dopamine system during cocaine seeking: conditioned and unconditioned contributions to reward and motivation. *Journal of Neuroscience, 28*(36), 9021–9029.

Zhou, W., Liu, H., Zhang, F., Tang, S., Zhu, H., Lai, M., et al. (2007). Role of acetylcholine transmission in nucleus accumbens and ventral tegmental area in heroin-seeking induced by conditioned cues. *Neuroscience, 144*(4), 1209–1218.

Zuo, L., Wang, K. S., Zhang, X. Y., Li, C. S., Zhang, F., Wang, X., et al. (2013). Rare SERINC2 variants are specific for alcohol dependence in individuals of European descent. *Pharmacogenetics and Genomics, 23*(8), 395–402.

Chapter 39

Immunologic and Genetic Aspects of Type 1 Narcolepsy

Louis Jacob, Hanna M. Ollila, Joshua Mahlios and Emmanuel Mignot
Stanford University Center for Narcolepsy, Palo Alto, CA, United States

INTRODUCTION TO THE CLINICAL CONDITION AND ITS PATHOPHYSIOLOGY

Narcolepsy was first described in the late 19th century (Gélineau, 1881; Schenck, Bassetti, Arnulf, & Mignot, 2007). In 1976, during the first international meeting on narcolepsy, a panel of experts put forth the first consensus definition of narcolepsy, a disorder characterized by "excessive daytime sleepiness associated with cataplexy and other REM sleep phenomena such as sleep paralysis and hypnagogic hallucinations" (Guilleminault, 1976; Kramer, 1977). To diagnose the condition, a test, the Multiple Sleep Latency Test (MSLT) was recommended, with demonstration of at least two rapid transitions from wake to REM sleep (so-called sleep-onset REM sleep periods [SOREMPs]) during four to five daytime naps as a diagnostic for the condition. Cataplexy, a symptom characterized by episodic loss of muscle tone triggered by emotions, is now considered the hallmark of narcolepsy, because it is rarely found in other conditions (Aldrich, 1996; Anic-Labat et al., 1999; Guilleminault, Mignot, & Partinen, 1994).

A critical feature of narcolepsy is that it involves abnormal intrusions of REM sleep into wakefulness ("dissociated REM sleep events") in addition to daytime sleepiness, unlike other hypersomnias (see Table 39.1, from the *International Classification of Sleep Disorders*, Third Edition [ICSD3]). In this model, cataplexy is akin to REM sleep atonia, in which muscle weakness can be triggered by emotions, typically laughing, joking, or anger. Sleep paralysis (paralysis while awake, occurring when falling asleep or when emerging from REM sleep) and hypnagogic hallucinations (dream-like hallucinations that occur at sleep onset) are two other known manifestations of dissociated REM sleep. Although characterized as a hypersomnia, narcolepsy is better characterized as a dyssomnia, because if left alone,

TABLE 39.1 International Classification of Sleep Disorders (ICSD III): Definitions and Pathophysiology

Condition	Diagnostic Criteria	Pathophysiology
Type 1 narcolepsy	Presence of cataplexy and positive MSLT, and/or low cerebrospinal fluid hypocretin-1	Hypocretin deficiency; 97% *HLA-DQB1*06:02*
Type 2 narcolepsy	Positive MSLT most often with no or unclear cataplexy	Unknown, heterogeneous, about 16% hypocretin deficiency; about 40% *HLA-DQB1*0602*
Secondary narcolepsy	As previously, but owing to other conditions (eg, neurological)	With or without hypocretin deficiency; various disorders
Idiopathic hypersomnia	No cataplexy, No SOREMPs during MSLT	Unknown, likely heterogeneous

Abnormal Multiple Sleep Latency Test (MSLT): MSLT: sleep latency $\leq$ 8 min, $\geq$ 2 sleep-onset REM sleep periods (SOREMPs), including a nocturnal SOREMP.
For details, see ICSD3. (2014). *International classification of sleep disorders* (3rd ed.). American Academy of Sleep Medicine.

Genomics, Circuits, and Pathways in Clinical Neuropsychiatry. http://dx.doi.org/10.1016/B978-0-12-800105-9.00039-1

patients do not display large increases in sleep amounts. Rather, patients are unable to stay awake for long periods of time, napping regularly and feeling refreshed. Similarly, these patients are often unable to stay asleep all night, because sleep is disturbed by nightmares, active dreaming, insomnia, and sleep paralysis. In 2000, the cause of narcolepsy with cataplexy was found to be a lack of hypocretin (orexin), a wake-promoting neuropeptide synthesized in the hypothalamus.

Since its initial description, numerous population-based studies have been performed, seeking patients positive for cataplexy by questionnaires, and confirming diagnoses by clinical and MSLT evaluations. Remarkably, with a few exceptions, all of these studies led to an adult prevalence of approximately 0.02–0.03% in multiple countries (Mignot, 1998), and reports of onset in childhood, adolescence, or early adulthood. After the development of sleep medicine in the late 20th and early 21st centuries, however, more and more clinicians used the MSLT, and it was recognized that the test has an approximately 2–5% false-positive rate (Goldbart et al., 2014), notably in subjects who are sleep deprived or/and shift workers, but it is also positive in some subjects with unexplained sleepiness without cataplexy and with or without hypocretin deficiency.

To reflect this, the newest international classification of sleep disorders, the ICSD3 (2014) subdivides narcolepsy into two subtypes, types 1 and 2 (Table 39.1). Type 1 narcolepsy is caused by a loss of hypocretin, a neuropeptide involved in regulating wakefulness, appetite, and arousal (Mignot et al., 2002; Nishino, 2000; Peyron et al., 2000; Thannickal, 2000). In sleep clinics, almost all patients with type 1 narcolepsy display clear cataplexy. Type 2 narcolepsy, whose physiopathology is unknown and likely heterogeneous, includes patients with abnormal sleep tests (ie, MSLT with SOREMPs) but normal or unknown hypocretin levels and without or with atypical cataplexy (Pizza et al., 2013). Other cases, which have short sleep latency during the MSLT or extended periods of sleep, are referred to as “idiopathic” hypersomnia (Table 39.1).

When cataplexy or hypocretin deficiency is found, there is a tight genetic association with human leukocyte antigen (HLA) class II allele *DQB1*06:02* (Mignot et al., 2001). The association of *HLA-DQB1*06:02* with narcolepsy is so strong (97%) that it can be genotyped to support the diagnosis of type 1 narcolepsy: for example, before a lumbar puncture for cerebrospinal fluid (CSF) hypocretin-1 evaluation. The main function of HLA alleles is to present foreign peptides from pathogens to immune cells, and most diseases with a strong HLA association are autoimmune. The current pathophysiological hypothesis is that HLA molecules may be presenting peptides that resemble hypocretin or other proteins in hypocretin cells, causing the destruction of hypocretin neurons by the immune system. As described subsequently, however, although circumstantial evidence is strong, the final proof, demonstration of autoantibodies or autoreactive T cells, is still wanting.

Although there is hope that type 1 narcolepsy will eventually be prevented through immune monitoring and treated when established using hypocretin agonists, current treatment is symptomatically based (Mignot, 2012; Nishino & Mignot, 1997). Because the main symptom is excessive daytime sleepiness, patients are generally treated with stimulants such as amphetamine-like stimulants or modafinil. Although these treatments are effective and lead to a reduction in daytime sleepiness, they have no effect on cataplexy or symptoms of abnormal REM sleep such as sleep paralysis, vivid dreaming, and hypnagogic hallucinations. To treat these symptoms, antidepressants are used, typically dual-serotonin and adrenergic reuptake inhibitors known to suppress REM sleep. In addition, gammahydroxybutyric acid (sodium oxybate) is also used to treat disturbed nocturnal sleep, sleepiness, and cataplexy.

TYPE 1 NARCOLEPSY IS CAUSED BY A LOSS OF HYPOCRETIN NEURONS

Almost all patients carrying *HLA-DQB1*06:02* and displaying cataplexy have undetectable or low hypocretin-1 levels in the CSF (hypocretin-1 $\leq$ 110 pg/mL versus normal levels in healthy subjects $\geq$ 200 pg/mL) (Kanbayashi et al., 2002; Nishino, 2000; Ripley et al., 2001). Indeed, type 1 narcolepsy is caused by a loss of 70,000–90,000 hypocretin neurons, which corresponds to 90–95% of this neural population (Fig. 39.1) (Overeem, Black, & Lammers, 2008). The first link between narcolepsy and the hypocretin system came from dogs, which have been the historical model of the disorder since 1973, when canine narcolepsy was first reported (Fig. 39.2) (Knecht, Oliver, Redding, Selcer, & Johnson, 1973). Canine narcolepsy is similar to human narcolepsy. As in humans, cataplexy in dogs is triggered by positive emotions, although in this species it is the presentation of food or playing. Unlike in humans, in which the disease is generally sporadic and associated with HLA, several breeds transmit narcolepsy as an autosomal recessive trait with full penetrance. In 1999, positional cloning led to the identification of mutations in the hypocretin receptor-2 as the cause of familial canine narcolepsy (Lin et al., 1999). Interestingly, in this presentation of canine narcolepsy, the neurotransmitter hypocretin is functional and symptoms are caused by downstream mutation resulting in defective hypocretin signaling. Shortly thereafter, the link between narcolepsy and hypocretin system was

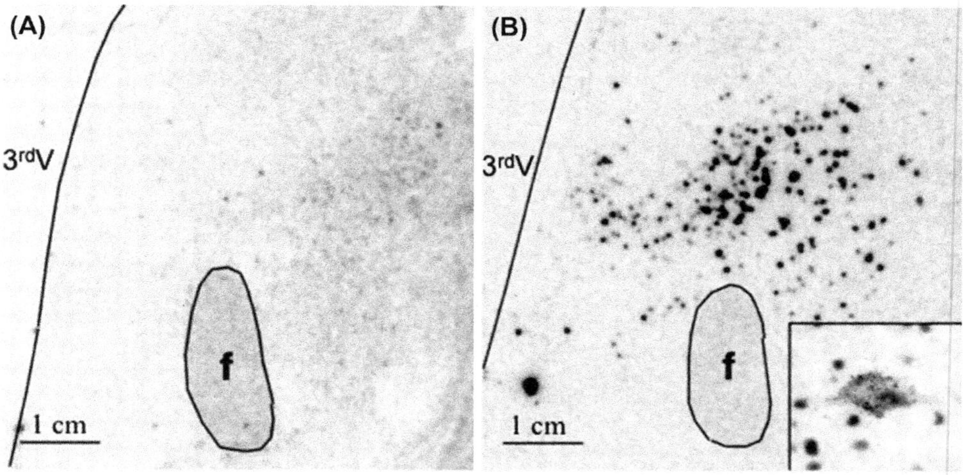

FIGURE 39.1 The cause of narcolepsy involves hypocretin neurons loss, as demonstrated using in situ hybridization. Preprohypocretin mRNA molecules are detected in the hypothalamus of a control (B) but not a narcolepsy (A) subject. Insert shows exemplar high magnification of preprohypocretin-positive neuron. *f*, fornix; $3^{rd}V$, third ventricle. *With modifications, from Peyron, C., Faraco, J., Rogers, W., Ripley, B., Overeem, S., Charnay, Y., … Mignot, E. (2000). A mutation in a case of early onset narcolepsy and a generalized absence of hypocretin peptides in human narcoleptic brains.* Nature Medicine, *6(9), 991–997. http://dx.doi.org/10.1038/79690.*

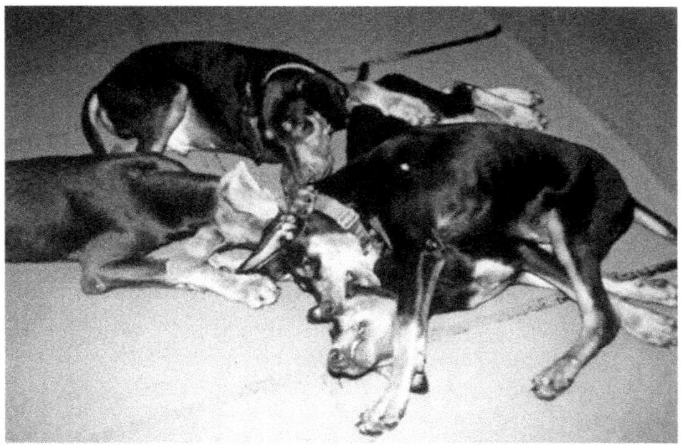

FIGURE 39.2 Narcoleptic Doberman Pinschers displaying cataplexy. Note that eyes are open. During cataplexy, subjects are awake but paralyzed.

confirmed in a murine narcolepsy model by Chemelli et al. (1999) who developed a hypocretin-deficient knockout mouse line. In contrast to dogs, narcolepsy in humans is not a fully penetrant genetic disease, and mutations in the hypocretin system leading to narcolepsy have been characterized only in a single atypical early-onset case (Peyron et al., 2000).

Hypocretin-producing neurons are found within the perifornical area of the posterior hypothalamus. Although their localization is discrete, these neurons have widespread projections that include various intrahypothalamic nuclei, cortex, brain stem, and limbic system, and of likely functional importance, monoaminergic nuclei (Beuckmann & Yanagisawa, 2002; Hungs et al., 2001; Taheri, Zeitzer, & Mignot, 2002). Hypocretin neurons synthesize hypocretin-1 and -2 (also called orexin A and B), two homologous neuropeptides involved in the regulation of wakefulness, appetite, and arousal (Peyron et al., 2000; Tsujino & Sakurai, 2009). First discovered as a hypothalamus-specific messenger RNA (mRNA) transcript, the peptides were called hypocretin-1 and hypocretin-2, and were suggested to be derived from a single preprohypocretin precursor (De Lecea et al., 1998). In parallel with this effort, Sakurai et al. (1998) used cell lines expressing various orphan G protein–coupled receptors (GPCR), screening tissue extracts for GPCR agonist activity in an attempt to "de-orphanize" these receptors. Two peptides, which were baptized orexin A and B, were isolated as agonists for the HFGAN72 GPCR cell line (hypocretin receptor 1 or orexin receptor OXR1). The name of these neuropeptides was selected using the Greek root *orexis* ("appetite") to reflect the finding that central administration increases appetite, an effect that is now considered minor. Another receptor binding these ligands and sharing homology with HFGAN72/OXR1 was also identified and called hypocretin receptor-2 or orexin receptor 2 (OXR2). Hypocretin receptor-1 selectively binds hypocretin-1 whereas hypocretin receptor-2 (OXR2) binds both hypocretin-1 and -2 with similar affinity (Sakurai et al., 1998). The binding of these neuropeptides to their receptors leads most often to an activation of a Gq protein, the opening of Ca^{2+} channels, and finally activation of targeted neurons. It is hypothesized that a loss of hypocretin neurons caused decreased activation of multiple target areas, including wake-active monoaminergic systems resulting in narcolepsy.

GENETIC ASSOCIATION OF NARCOLEPSY WITH THE HUMAN LEUKOCYTE ANTIGEN

HLA genes encompass a large family of polymorphic genes that are spread over several megabases of human chromosome 6. There are three different classes of HLA genes (I, II, and III). These genes were initially discovered by Jean Dausset in 1958 as transplantation antigens or major histocompatibility complex proteins essential to the compatibility of recipients to organ donors (Richmond, 2009). Later studies, notably performed by Hugh Mac Devitt (2000) at Stanford, showed these genes to be fundamental to explain genetic diversity of immune responses ("immune response genes"). Indeed, a large portion of interindividual variation in the response to small peptide antigens or epitopes depends on each subject's individual HLA and the corresponding ability to bind specific epitopes for presentation to the rest of the immune system, which can then mount a T-cell and B-cell (antibody) response. The great diversity of HLA subtypes allows individual immune responses to be more diverse, thus better protecting the population at large against infections. The HLA region is one of the few regions of the genome in which clear evidence of natural selection can be demonstrated. It is likely that HLA polymorphisms are involved in hundreds of known and unknown diseases, infectious and autoimmune.

The first report of an association between narcolepsy and HLA came from Japan in 1983 (Honda, 1988), after the launch of a global program focusing on potential associations between HLA and orphan disorders. In this first report, narcolepsy was found to be weakly associated with the HLA class I gene Bw35. One year later, pursuing this finding, a stronger association was discovered, with 100% of Japanese patients with narcolepsy being HLA-DR2 and DQ1 positive against 25% of control subjects (Juji, Satake, Honda, & Doi, 1984). This finding was rapidly confirmed in Europe (Billiard et al., 1985; Marcadet et al., 1985; Mueller-Eckhardt et al., 1986; Roth et al., 1988) and North America (Poirier, Montplaisir, Décary, Momège, & Lebrun, 1986), and the Stanford group found rare DR2-negative patients, which created controversy (Guilleminault & Grumet, 1986). Since then, HLA DR and DQ typing has moved from serological to molecular-based typing. Both DR and DQ molecules were found to be α/β heterodimers encoded by DRA/DRB genes and DQA1/DQB1 genes, respectively. As a consequence, additional molecular subtypes that were not detected using antisera were discovered, and the DR2 and DQ1 subtypes associated with narcolepsy were further characterized as DR15 and DQ6 subtypes, and finally as *DRA-DRB1*15:01* and *DQA1*01:02-DQB1*06:02* (whereas DRα is essentially monomorphic, both DQ genes are polymorphic and located a few kilobases apart, encoding the DQ602 heterodimer in this haplotype). In both Japanese and Caucasian populations, however, linkage disequilibrium between the DR and DQ genes is so strong (the genes are approximately 50 kb apart) that it was impossible to tell whether *DRA-DRB1*15:01* or *DQA1*01:02-DQB1*06:02*, or both, were involved in the association.

The next breakthrough came from studying African American individuals; many patients with narcolepsy in this ethnic group were found to be HLA-DR2 negative but DQ1 positive. In 1992, it was found that these patients were all positive for *HLA-DQB1*06:02*, an allele associated with *DRB1*15:01, DRB1*15:03, DRB1*11:01*, and other subtypes (Matsuki et al., 1992; Mignot et al., 1997; Neely, Rosenberg, Spire, Antel, & Arnason, 1987; Rogers, Meehan, Guilleminault, Grumet, & Mignot, 1997). All patients were also positive for *HLA-DQA1*01:02*, the alpha chain partner gene proximal to *DQB1*06:02* (Mignot et al., 1997, 1994). Further trans-ethnic fine-mapping studies showed that the association with narcolepsy rapidly decreased on both sides of the HLA-DQ locus. Therefore, it became obvious that the previous link between narcolepsy and HLA-DR2 was secondary because this gene is geographically close to the HLA-DQ locus. It is also unlikely that other HLA genes within the *DQA1* and *DQB1* segment are involved in the development of narcolepsy, because this region has been sequenced and no other genes were found (Ellis et al., 1997).

As mentioned previously, *HLA-DQB1*06:02* is found almost exclusively with *HLA-DQA1*01:02* within a single genomic segment of the HLA complex in both control subjects and patients with narcolepsy (Mignot et al., 1994). Indeed, DQA1 and DQB1 products heterodimerize to form the functional, peptide binding HLA-DQ α/β heterodimer. This nonrandom association of *DQA1*01:02* and *DQB1*06:02* in *cis* on the same haplotype has been selected by evolution, because not all *DQA1*- and *DQB1*-encoded alleles can form stable heterodimers. Interesting, subjects carrying *HLA-DQA1*01:02* but not *HLA-DQB1*06:02*: for example, *HLA-DQA1*01:02-DQB1*06:04* and

*HLA-DQA1*01:02-DQB1*06:09* (in Caucasian people) or *HLA-DQA1*01:02-DQB1*06:01* (in Southern Chinese people) are not more prone to develop narcolepsy (Mignot et al., 1997). Conversely, subjects with *HLA-DQB1*06:02* but without *HLA-DQA1*01:02* (for example, *HLA-DQA1*01:04-DQB1*06:02* or *HLA-DQA1*01:03-DQB1*06:02*) are rarely observed in control subjects, notably African American individuals but not in narcoleptic populations (Mignot et al., 1997). Therefore, it is likely that both *DQA1*01:02* and *DQB1*06:02* alleles are required to predispose individuals to narcolepsy. This hypothesis makes sense because both DQα and DQβ are needed to form the functional DQ0602 heterodimer and polymorphisms on both molecules contribute to peptide binding and thus antigen presentation.

Further genetic investigations on African American and Caucasian American groups showed that subjects who were homozygous for *HLA-DQB1*06:02* were approximately twice as susceptible to developing narcolepsy as compared to heterozygous individuals (Han, Lin, Li, Aran, et al., 2012; Han, Lin, Li, Dong, et al., 2012; Pelin, Guilleminault, Risch, Grumet, & Mignot, 1998). *HLA-DQB1*06:02* dosage therefore influences narcolepsy risk, which is in contrast to some alleles such as *DQA1*01:01*, *DQB1*01:03*, *DQB1*05:01*, *DQB1*06:01*, and *DQB1*06:03* (belonging to the DQ1 group), which confer protection against narcolepsy (Hohjoh et al., 2001; Hong et al., 2007; Hor et al., 2010), an effect we believed to be the result of "allele competition." Indeed, as mentioned earlier, not all DQα and DQβ alleles are compatible with each other and can form stable heterodimers, as reflected by the selection of specific DQA1-DQB1 haplotypes. DQ subtypes are broadly separated into DQ1 subtypes, including DQα alleles encoded by *DQA1*01* and DQβ alleles encoded by *DQB1*05* and *06* subtypes and non-DQ1 subtypes (DQ2, 3, and 4). Whereas DQ1 alleles are compatible with each other, they cannot heterodimerize with non-DQ1 subtypes (Miyadera, Ohashi, Lernmark, Kitamura, & Tokunaga, 2014). We hypothesize that in the presence of other DQ1 alleles, *DQB1*06:02* is in competition with DQα alleles other than *DQA1*01:02*, whereas the reverse is true with *DQA1*06:02*, which may be in competition with DQβ alleles other than *DQB1*06:02*. Although hypothetical, the allele competition explains well why other DQ1 alleles may be protective against narcolepsy when located in *trans* of *DQB1*01:02–DQB1*06:02*, with the various combinations with approximately the expected odds ratios (ORs) (Fig. 39.3) (Han, Lin, Li, Aran, et al., 2012; Han, Lin, Li, Dong, et al., 2012; Ollila, Fernandez-Vina, & Mignot, 2014).

More surprisingly, multiple studies have found that *DQA1*0102–DQB1*06:02/DQB1*03:01* is also associated with increased risk compared with the reference *DQA1*0102-DQB1*06:02*/other combination (Han, Lin, Li, Aran, et al., 2012; Han, Lin, Li, Dong, et al., 2012; Hong et al., 2007; Mignot et al., 2001; Ollila et al., 2014). This last result is hard to explain because *DQB1*03:01* is found in the context of multiple DQα-associated alleles (*DQA1*03:01*, *DQA1*03:02*, *DQA1*05:05*, and *DQA1*06:01*), which suggests that the effect is not mediated via DQα/β heterodimers (furthermore, *DQB1*03:01* cannot heterodimerize with *DQA1*01:02*; see previous discussion). Disequilibrium tests were used to confirm this effect (Thomsom & Dorman, 2006). This statistical approach is interesting because in patients who are DQ0602 positive, it removes the alleles that are located together with DQ0602. Although this additional effect is still poorly understood, studies have shown that *DQB1*03:01* reduces the age of onset in patients who are DQ0602 positive, which explains why the association is stronger in subjects displaying early-onset narcolepsy (Han et al., 2013).

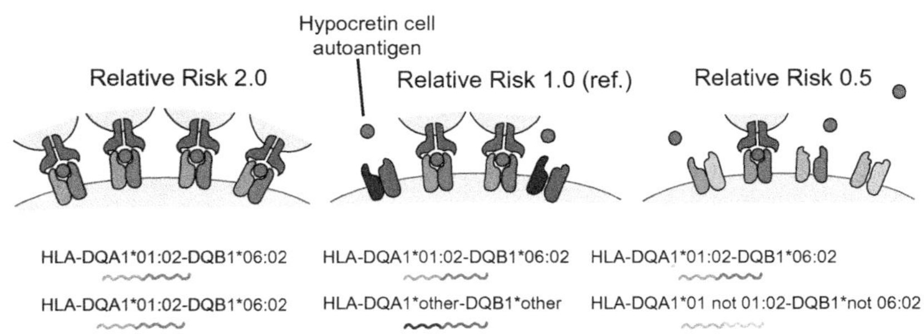

FIGURE 39.3 HLA-DQ effects in narcolepsy are explained by allele competition. Ninety-seven percent of patients with narcolepsy-cataplexy are *HLA-DQB1*06:02* positive. Therefore, this allele is almost a prerequisite for developing the disorder. *HLA-DQB1*06:02* dosage effects: *DQB1*06:02* homozygotes have about twice higher risk of developing the disorder, whereas *DQB1*06:02* heterozygotes that have other DQ1 alleles in *trans* have an approximately twice lower risk (Han, Lin, Li, Aran, et al., 2012; Han, Lin, Li, Dong, et al., 2012; Hong et al., 2007; Ollila et al., 2014; Pelin et al., 1998). We hypothesize that this is the result of allele competition (see text). *With modifications, from Mignot, E. J. M. (2014). History of narcolepsy at Stanford University.* Immunologic Research, *58(2–3), 315–339. http://dx.doi.org/10.1007/s12026-014-8513-4.*

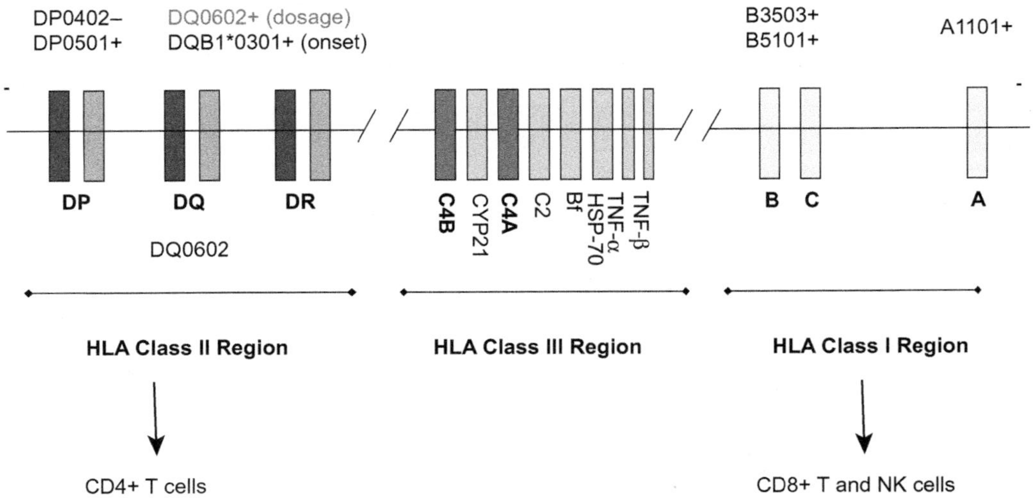

FIGURE 39.4 Effects of HLA loci other than HLA-DQ. HLA loci other than DQ also influence narcolepsy, notably HLA class II genes *DPB1*04:02* (protective) and *DPB1*05:01* (predisposing). In addition, effects of specific HLA class I gene alleles are also evident. HLA class I genes present peptides to TCRs located on $CD8^+$ cytotoxic T cells, or may interact with natural killer cells, which suggests involvement of these cells in the pathophysiology of narcolepsy. *Data derived from Ollila, H. M., Ravel, J.-M., Han, F., & Mignot, E. (2015). Novel genetic loci in HLA-DPB1 and HLA- class I region confer risk and protection for narcolepsy.* American Journal of Human Genetics, *96(1), 136–146.*

Other genes located within the extended HLA locus are also associated with narcolepsy and add further complexity (Fig. 39.4). Matching control subjects and patients with narcolepsy one to one for HLA–DQ, we found that *HLA-DPB1*04:02* protects (OR, 0.5), whereas *DPB1*05:01* (OR, 1.5) predisposes to narcolepsy (Nishida et al., 2014; Ollila et al., 2014). HLA class II–independent associations were also seen in the HLA class I region with *HLA-A*11:01* (OR, 1.3), *HLA-B*35:03* (OR, 2.0), and *HLA-B*51:01* (OR, 1.5) increasing predisposition. These complex associations are similar to those reported in other autoimmune diseases such as type I diabetes, in which there is a strong association with HLA-DR and DQ and other smaller associations with HLA-DP and class I HLA genes (Varney et al., 2010).

Although narcolepsy is strongly associated with HLA, HLA typing remains of limited interest in clinical practice. Indeed, the *DQB1*06:02* association is high (97%) only in subjects displaying narcolepsy with hypocretin deficiency, and most of these patients have clear cataplexy, a symptom highly specific to the disorder. In these patients, the diagnosis can typically be made on clinical grounds alone. Furthermore, *HLA-DQB1*06:02* is not specific to narcolepsy and large numbers of control subjects are also positive for this allele (12% in Japanese, 25% in Caucasians and Chinese, and 38% in African American people). Finally, few patients with narcolepsy who have hypocretin deficiency (2–3%) are *HLA-DQB1*06:02* negative, and half of these patients carry *DPB1*09:01* (Han et al., 2014). Finally, it is unclear whether type 1 or type 2 narcolepsy needs to be treated differentially, although it is the opinion of the authors that having a definitive unambiguous answer regarding the cause of the symptoms allows for more aggressive management. In this context, a potential indication of HLA typing may be in difficult cases, notably before a lumbar puncture assessing CSF hypocretin-1 levels is ordered.

NON–HUMAN LEUKOCYTE ANTIGEN GENES ARE ALSO ASSOCIATED WITH NARCOLEPSY

Polymorphisms in genetic regions other than HLA are involved in narcolepsy. Indeed, familial risk is increased in first-degree relatives (10-fold in Japanese people and 20- to 40-fold in Caucasian people, approximately 1% of first-degree relatives), and this increase cannot be explained by the sharing of HLA subtypes alone (two- to three-fold increased risk) (Mignot, 1998). Additional data have shown that the disease severity is also modified by a polymorphism in the catechol-*O*-methyltransferase gene, whose product is involved in the degradation of catecholamines (Dauvilliers et al., 2002; Dauvilliers, Neidhart, Lecendreux, Billiard, & Tafti, 2001). Interestingly, these families also show milder forms of narcolepsy, which do not have cataplexy but are HLA positive and a positive MSLT in 2–4% of first-degree relatives, which suggests the existence of a disease spectrum (CSF hypocretin has never been evaluated in these patients) (Mignot, 1998). The possibility that milder forms of HLA-associated hypocretin deficiency exist at higher frequency is also supported by a single epidemiological study in which repeated MSLT testing was performed in a cohort of about 1500 "healthy subjects" and one subject with cataplexy and two without were identified (Goldbart et al., 2014). In sleep clinics, however, most patients with positive MSLTs are HLA negative, which suggests that they represent a false-positive test. Repeating the MSLT and requiring two positive results after documenting sufficient sleep before testing may be helpful, because successful retesting of positive MSLTs in the general population (Goldbart et al., 2014) and in narcolepsy without cataplexy in sleep clinics is poor ($\kappa < 0.3$) (Trotti, Staab, & Rye, 2013).

Involvement of other genes and the existence of genetic heterogeneity in narcolepsy have also been suggested by the observation of rare *HLA-DQB1*06:02*—negative families. These patients rarely display low levels of CSF hypocretin-1, nor do they carry mutations in hypocretin or hypocretin receptor genes. In one family, in which subjects were *DQB1*06:02* negative and displayed low CSF hypocretin levels, a mutation in myelin oligodendrocyte glycoprotein (*MOG*) gene was found (Hor et al., 2011). Unfortunately, mutations in *MOG* were not found in any other patients with *DQB1*06:02* displaying low levels of CSF hypocretin-1, and whole-exome sequencing revealed no additional gene (Han et al., 2014). Interestingly, however, one patient carried a likely pathogenic dominant mutation in the preprohypocretin gene. This case had a very early onset, at age 6 months, with clear cataplexy and without positivity for HLA-predisposing genes (Peyron et al., 2000).

Progress in the analysis of the human genome, notably through genome-wide association studies (GWAS), has led to the discovery of other genes associated with complex diseases with genetic components such as type 1 narcolepsy (Faraco et al., 2013; Hallmayer et al., 2009; Han et al., 2013; Hor et al., 2010; Kornum et al., 2011; Miyagawa et al., 2008). Studies including a large number of Caucasians and Asians patients and control subjects found that narcolepsy was associated with polymorphisms in the T-cell receptor (TCR) α (Hallmayer et al., 2009) and β loci (Han et al., 2013). Because these two genes encode for the TCR α/β heterodimer, a receptor present on T cells that recognizes peptides presented by HLA molecules, the results support the growing body of circumstantial evidence to the autoimmune hypothesis.

GWAS studies were next extended in larger and larger samples that included more diverse ethnic groups (African Americans, Chinese, and Japanese people) (Faraco et al., 2013; Han et al., 2013; Kornum et al., 2011). Additional polymorphisms known to be involved in other autoimmune diseases were found to be significantly associated with narcolepsy, including *OX40L* (also known as the *TNFSF4*, a co-stimulatory receptor involved in the activation of $CD4^+$ T cells), cathepsin H (an enzyme known to be associated with type 1 diabetes and involved in processing of antigens before their membranous presentation), *ZNF365* (a transcription factor linked with inflammatory bowel disease [IBD]), and *IL10RB-IFNAR1*, another region associated with IBD and other autoimmune disorders (Faraco et al., 2013; Han et al., 2013; Kornum et al., 2011).

Another interesting finding was the discovery of an association between narcolepsy and the *PPAN-P2RY11-EIF3G* gene region, a 10-kb telomeric of the DNA methylase gene 1 (*DNMT1*) gene (Kornum et al., 2011). The strongest signal was seen with *P2RY11*, a gene that encodes for an adenosine triphosphate receptor that has a role in chemotaxis and immune cell survival (Bours, Swennen, Di Virgilio, Cronstein, & Dagnelie 2006; Di Virgilio, Boeynaems, & Robson 2009). Interestingly, in a parallel study, exome sequencing was conducted in a rare syndrome that includes narcolepsy as one of its core components, autosomal dominant cerebellar ataxia, deafness, and narcolepsy (ADCA-DN) leading to the identification of three mutations (A570V, G605V, and V606F) in exon 21 of the *DNMT1* gene (Winkelmann et al., 2012, p. 1). Although these signals were initially thought to be connected, later studies found that the association signal seen in type 1 narcoleptic patients was in the *P2RY11-EIF3G* region and not within the *DNMT1* gene (Kornum et al., 2014). It remains possible that regulatory elements located within the *P2RY11-EIF3G* region regulate *DNMT1*, or that these pathologies are not directly connected pathophysiologically.

In addition to the genetic associations found here, Miyagawa et al. (2008) reported that a polymorphism, rs5770917, located between the *CPT1B* and *CHKB* genes (two genes involved in the regulation of cholinergic metabolism and beta-chain fatty acid oxidation, respectively) was also associated with narcolepsy. Interestingly, the same polymorphism has been found to be associated with "essential hypersomnia syndrome," an intermediate form of narcolepsy characterized by sleepiness but no cataplexy (Miyagawa et al., 2008). Even if rs5770917 replicated weakly in Korean people (Miyagawa et al., 2008), the association has not been confirmed in Chinese or Caucasian narcoleptic subjects, which suggests that it may be a false positive or that differences in associations reflect differences in disease definition or populations (Hallmayer et al., 2009; Han, Lin, Li, Aran, et al., 2012; Han, Lin, Li, Dong, et al., 2012).

As typically observed when susceptibility genes are identified through GWAS, effect sizes of individual SNPs are small compared with the effects of HLA alleles such as *DQB1*06:02*. However, the association of *DQB1*06:02* with narcolepsy is unusually strong compared with other autoimmune diseases. In addition, none of the other GWAS performed with autoimmune disorders found associations with the TCR locus.

NARCOLEPSY IN ASSOCIATION WITH OTHER SYNDROMIC GENETIC DISORDERS

Excessive daytime sleepiness, the onset of REM sleep, or cataplexy has been reported in several genetic disorders such as ADCA-DN, MIM 604,121, and *DNMT1* mutations (Winkelmann et al., 2012, p. 1), Moebius syndrome (MIM 157,900, heterogenous disorder) (Krämer, Goldammer, & Sindern, 2014; Parkes, 1999; Tyagi & Harrington, 2003), Coffin–Lowry syndrome (MIM 303,600, RSK2 mutations) (Nelson & Hahn, 2003), Niemann–Pick type C1 (MIM 257,219, NPC1) (Oyama et al., 2006; Pedroso et al., 2012; Vankova et al., 2003), Norrie disease (MIM 310,600, Xp11.4-p11.23 deletions including NDP and MAO genes when cataplexy is present) (Parkes, 1999; Smit, Lammers, & Catsman-Berrevoets, 2006; Vossler, Wyler, Wilkus, Gardner-Walker, & Vlcek, 1996), myotonic dystrophy (MIM 160,900 and 6,022,668, MD1) (Martínez-Rodríguez et al., 2003) and Prader–Willi (MIM 176,270, caused by 15q11.2 deletions including paternal copies of imprinted small nuclear ribonucleoprotein polypeptide N [SNRPN] and nectin [NDN] genes) (Mignot et al., 2002; Parkes, 1999). Patients displaying these genetic syndromes may be either *HLA-DQB1*06:02* positive or negative and usually have normal or intermediate CSF hypocretin-1 levels (110–200 pg/mL, or >200 pg/mL) (Mignot et al., 2002). Because a better understanding of these symptoms in the context of these disorders could be informative to narcolepsy and REM sleep regulation, these disorders are briefly discussed next.

Cataplexy is atypical in Coffin–Lowry syndrome, a disorder characterized by mental retardation, tonicoclonic seizures, and cognitive impairment. In this genetic disorder, symptoms that resemble cataplexy are more likely atonic seizures, because events can include tonicoclonic manifestations and are typically triggered by surprise such as a sudden noises rather than laughter (Nelson & Hahn, 2003).

ADCA-DN, a disorder briefly mentioned earlier, was first described by Melberg et al., in 2001 (Melberg et al., 2001). ADCA-DN is a neurodegenerative disease characterized by late onset (aged 30–40 years) ataxia, deafness, and narcolepsy-cataplexy linked with intermediary or low hypocretin-1 levels in CSF. Interestingly, although narcolepsy is one of the first clinical symptoms, CSF hypocretin-1 levels remain intermediary (110–200 pg/mL) and become low (110 pg/mL) to undetectable only at later stages of the disease (Melberg et al., 2001; Moghadam et al., 2014; Winkelmann et al., 2012). Another early symptom is deafness followed by cerebellar ataxia, ocular nerve atrophy, and neurodegeneration, leading to death 5–10 years later. In 2012, it was found that the disorder was caused by mutations in exon 21 of *DNMT1*, in a regulatory region of the protein. This disorder is close to hereditary sensory and autonomic neuropathy type 2 (MIM 614,116, HSAN2), a similar disease also characterized by peripheral neuropathy and in which mutations of exon 20 are involved (Moghadam et al., 2014).

Mobius syndrome is associated with brain-stem anomalies involving the sixth and seventh cranial nerves. This leads to congenital facial palsy and impairment of ocular abduction; some patients display skeletal abnormalities. Most cases result from a 13q12.2 deletion that includes the MBS1 gene. Several cases associated with cataplexy have been reported in the literature (Parkes, 1999; Tyagi & Harrington, 2003), including one with intermediate levels of CSF hypocretin-1 (Krämer et al., 2014). Some patients may have dysregulated breathing, which complicates the clinical picture. Because hypocretin levels are not entirely absent and the disease involves brain-stem abnormalities, lesions of brain-stem REM sleep regulatory centers are more likely involved in the generation of these symptoms (Lu, Sherman, Devor, & Saper, 2006; Peever, Luppi, & Montplaisir, 2014).

Norrie disease, an X-linked recessive disorder, is characterized by early childhood blindness caused by degeneration and proliferation of the neuroretina. Norrie disease is also associated with hearing deficits and neuropsychiatric symptoms (Smith, Mullen, Graham, Sims, & Rehm, 2012). Deletion of Xp11.3-p11.4, a region including the Norrie disease gene (NRD), is the cause of most cases. Rare mutations of the NRD gene in isolation of larger deletions have been shown to produce the same syndrome, which demonstrates that NRD is causal. Since its description, it was noted that some patients with microdeletions (none with NRD point mutations) also displayed cataplexy (Parkes, 1999; Smit et al., 2006; Vossler et al., 1996). Interestingly, deletion of nearby monoamine oxidase A and B genes in isolation of the NRD gene was found to lead to developmental problems, intermittent hypotonia, and stereotypical movements of the hand (Saito et al., 2014; Whibley et al., 2010). In the presence of Dr Melberg, I saw a patient characterized by an isolated deletion of the monoamine oxidase A (MAO) genes. The patient displayed cataplexy-like symptoms and mental retardation with no associated ocular symptoms, which suggested involvement of MAO and thus abnormal monoaminergic metabolism (these subjects are predicted to have increased monoaminergic activity) in the generation of cataplexy-like symptoms. Interestingly and possibly relevant to this syndrome, intense adrenergic locus coeruleus stimulation was shown to lead to arousal followed by behavioral, cataplexy-like arrests in mice (Carter et al., 2010).

Niemann–Pick type C, a lysosomal storage disease caused by NPC1 and NPC2 mutations, is associated with abnormal cholesterol metabolism. Onset generally occurs during childhood, before age 10 years; in these patients, death occurs before age 20. Other cases can have a much later onset in adulthood for unknown reasons. Patients present with a very wide range of symptoms, with neurological deficits (cerebellar ataxia, seizures, dysphagia, dysarthria, hypotonia, and dystonia); hepatosplenomegaly is the most characteristic. Psychiatric symptoms are also observed, notably dementia and psychosis. Niemann–Pick type C often involves vertical gaze palsy with implication of the third cranial nerve, which is typical and an early manifestation of the disease. Cataplexy may occur in young children with NPC1 mutations. In this case, cataplexy is clear and is usually triggered by laughing, crying, and other emotions (Pedroso et al., 2012). Anti-cataplectic treatments can be effective in this situation. Most of these children display intermediate levels of CSF hypocretin-1 (Mignot et al., 2002; Oyama et al., 2006). How NPC causes narcolepsy is unknown and may involve effects in the brain stem, hypothalamus, or both.

Myotonic dystrophy and Prader–Willi (Martínez-Rodríguez et al., 2003; Mignot et al., 2002) are linked with both narcolepsy and sleep-disordered breathing, which makes causality inferences regarding sleepiness difficult. Primary hypersomnia may be diagnosed in these patients only if sleep-disordered breathing does not improve symptoms. DM1 mutations are involved in myotonic dystrophy, an X-linked disease affecting mostly males and showing anticipation (decreased age of onset owing to repeat expansion during successive meiosis). The disease involves a CTG trinucleotide expansion in the noncoding region of the DMPK gene; the length of this repeat is variable across tissues, typically highest in muscle, brain, and heart. Myotonic dystrophy is associated with progressive muscle weakness, cataract, cardiac conduction defects, baldness, endocrinopathies, and infertility. As for other trinucleotide expansion-related diseases, age of onset of myotonic dystrophy is highly variable. Hypoventilation and sleep-disordered breathing are caused by muscle weakness and can result in sleepiness, but do not explain it fully. Cataplexy has never been reported in this disorder, but MSLTs invariably show multiple sleep-onset REM periods. CSF hypocretin-1 levels are normal in myotonic dystrophy.

Pradder–Willi syndrome is caused by a loss of function of genes located in a critical region of 15q11.2. Although the syndrome has not been shown to involve a single gene, involvement of the SNRPN and NDN genes is likely to be most important. In most cases, the syndrome is caused by a deletion of the paternal 15q11.2 region (70%). In other cases (30%), patients inherit two maternal deletions stemming from the same maternal chromosome (maternal uniparental disomy). Pradder–Willi patients have a characteristic facial appearance, mental retardation, hypotonia, hyperphagia, and a number of other symptoms. Hypoventilation and sleep-disordered breathing are common consequences of obesity and hypotonia, and certainly contribute to sleepiness. As for myotonic dystrophy, however, ventilation often does not resolve the sleepiness and MSLT frequently shows SOREMPs, which suggests central nervous system (CNS) effects. Furthermore, emotions such as laughing can cause cataplexy in some patients with Pradder–Willi, which suggests that narcolepsy is a genuine association.

A small number of genetic syndromes have been convincingly associated with secondary narcolepsy, although in most cases hypocretin transmission is only partially impaired, which suggests indirect effects. A better understanding of these cases may lead to the discovery of novel REM sleep regulatory systems.

IS TYPE 1 NARCOLEPSY AN AUTOIMMUNE DISEASE?

That narcolepsy is associated with *HLA-DQB1*06:02* and immune response polymorphisms involved in other autoimmune diseases strongly suggests that type 1 narcolepsy is an autoimmune disease. In other autoimmune diseases associated with HLA-DQ, such as type 1 diabetes or celiac disease, antibodies toward self-proteins are typically identified: for example anti-GAD, anti-IA2 or anti-insulin antibodies (in type 1 diabetes) or anti-transglutaminases and gluten-derived peptides (in celiac disease). In type 1 narcolepsy, however, autoantibodies targeting hypocretin neurons have not been identified despite many attempts (Black, Avula, et al., 2005; Black, Silber, et al., 2005; Bergman et al., 2014; Tanaka, Honda, Inoue, & Honda, 2006), although one study identified immunoreactivity toward neighboring cell populations but not hypocretin neurons (Bergman et al., 2014). Similarly, immunostaining studies of postmortem hypothalamic tissue from narcoleptics have not revealed inflammation or T-cell infiltrates in the vicinity of hypocretin neurons (Knudsen, Mikkelsen, & Jennum, 2007; Overeem et al., 2006; Martínez-Rodríguez, Sabater, Graus, Iranzo, & Santamaria, 2007).

Functional data obtained using animal models have similarly been inconclusive. One study found changes in sleep pattern after administration of human narcolepsy sera to mice (Bergman et al., 2014). Other passive transfer experiments using human sera in wild-type mice were reported to result in cataplexy, but the possibility remains that these authors may have confused freezing seizures with cataplexy (Katzav et al., 2013; Smith, Jackson, Neufing, McEvoy, & Gordon, 2004). Indeed, conducting similar experiments, we found that two of five treated animals died and that postmortem analyses did not show hypocretin neuron destruction. One group reported that the sera of narcolepsy patients contained autoantibodies that had functional effects on spontaneous colonic migrating motor complex contractions or/and interfering with the effect of muscarinic stimulation on bladder strip contractions (Jackson, Reed, Smith, & Gordon, 2008;

Jackson, Spencer, Reed, Smith, & Gordon, 2009). Sera samples from patients and controls were sent blind to this group for the bladder strip assays, but opposite results were observed, which suggested that the results were not reproducible.

More recently, Cvetkovic-Lopes et al. (2010) isolated transcripts that were enriched in murine hypocretin cells, including a transcript encoding for the protein Tribbles homologue 2 (TRIB2). The same authors found elevated anti-TRIB2 antibodies in narcoleptic patients, raising the hope this protein could be the long sought-after narcolepsy autoantigen. Shortly after this publication, this finding was replicated by two groups using sera collected between 1990 and 2005 (Kawashima et al., 2010; Toyoda et al., 2010). Disappointingly, the finding was not confirmed in studies using more recent narcolepsy samples (Dauvilliers et al., 2010; Dougherty, Schmidt, Nakajima, & Heintz, 2010; Eriksson & Mignot, 2009; Honda et al., 2009). Interestingly, a study found that the presence of these autoantibodies correlated significantly with A/H1N1 antibody titers in younger subjects, which suggests that TRIB2 autoantibodies were an indirect marker of co-infections (Lind et al., 2014). Another report claimed that the injection of immunoglobulins extracted from narcolepsy-TRIB2–positive sera but not controls led to hypocretin cell lesions and narcolepsy when injected into the lateral hypothalamus (Katzav et al., 2013). Unfortunately, however, the results of this study are unclear because no statistics on hypocretin cells loss are provided, and the picture provided as supporting information shows a single hypothalamic section with extensive cell loss. It is also likely that the authors saw seizures rather than cataplexy, because the mean duration of reported episodes was longer than murine cataplexy (66–464 s vs 2–60 s).

The absence of autoantibodies in narcoleptic individuals suggests that the disease pathophysiology likely involves cytotoxic T cells mediating the destruction of hypocretin cells. Indeed, such mediation has been suggested for paraneoplastic ataxias, in which Purkinje cells can be destroyed with no apparent collateral damage (Graus, Saiz, & Dalmau, 2010). In addition, narcolepsy cataplexy with hypocretin deficiency is a well-established association of anti-Ma encephalitis, a paraneoplastic complication of seminomas, and this pathology was shown to be associated with a striking CD8 hypothalamic infiltration and hypocretin cell loss (Overeem et al., 2004).

Although genetic data support the hypothesis of an autoimmune process mediating the destruction of hypocretin neurons, there is currently no definitive evidence supporting this claim (Julkunen & Partinen, 2014). A likely hypothesis may be the primary involvement of CD8 T cells, which have not been studied in much detail in narcolepsy. Fig. 39.5 summarizes the current hypothesis of the autoimmune process involved in the development of narcolepsy.

INVOLVEMENT OF ENVIRONMENTAL FACTORS IN NARCOLEPSY

Only 32% of reported monozygotic twin pairs are concordant for narcolepsy (6 of 19 twin pairs), which suggests the importance of environmental triggers (Dauvilliers, Arnulf, & Mignot, 2007). Among these, an increasing number of studies suggest that upper airway infections can be involved. Indeed, as early as the 1980s, studies had shown a possible association of narcolepsy with antistreptolysin O (ASO) antibodies, which suggests that *Streptococcus pyogenes* could be a trigger (Aran et al., 2009). Studies have shown epidemiological associations of narcolepsy with a past history of diagnosed strep throat infections (Longstreth, Ton, & Koepsell, 2009) and increased ASO titers only in patients with samples collected close to disease onset (Aran et al., 2009). Because S. *pyogenes* has also been associated with other neurological disorders, more particularly with basal ganglia disorders such as Sydenham chorea and other atypical psychiatric and movement disorders (pediatric autoimmune neuropsychiatric disorders associated with streptococcal infection) (Dale & Heyman, 2002), these infections were logical candidates.

In agreement with this finding, a Chinese study in 629 patients with narcolepsy established that the incidence of narcolepsy increased between April and August, after winter months, when seasonal upper airway infections struck China (Fig. 39.6) (Han et al., 2011). More surprisingly, the same study also showed that the number of new-onset narcolepsy cases increased several-fold in 2010 after the H1N1 2009–10 pandemic swine flu season, and suggested that not only *Streptococcus* but also influenza may be involved. Even if the association between narcolepsy and H1N1 infections were strong in China and were also found in the United States (Dauvilliers et al., 2010), a study, the Narco-Flu-VF, did not confirm this association in France (Dauvilliers et al., 2013). In this study, 59 narcoleptic subjects were matched with 135 control subjects for age, sex, and medical records, and the numbers of infectious and vaccination episodes since the January 1, 2005 were compared. However, it remains possible that H1N1 infections that trigger narcolepsy are largely asymptomatic because the immune response is successful at clearing the virus.

Of notable interest is the finding that the genetic architecture of the predisposition to narcolepsy changed after the 2009 H1N1 pandemic (Han et al., 2013). Indeed, analyzing GWAS data in Chinese patients before and after 2009, a GWAS significant difference was found for an SNP located between DRB and DQA1 (Fig. 39.7). The difference was independent

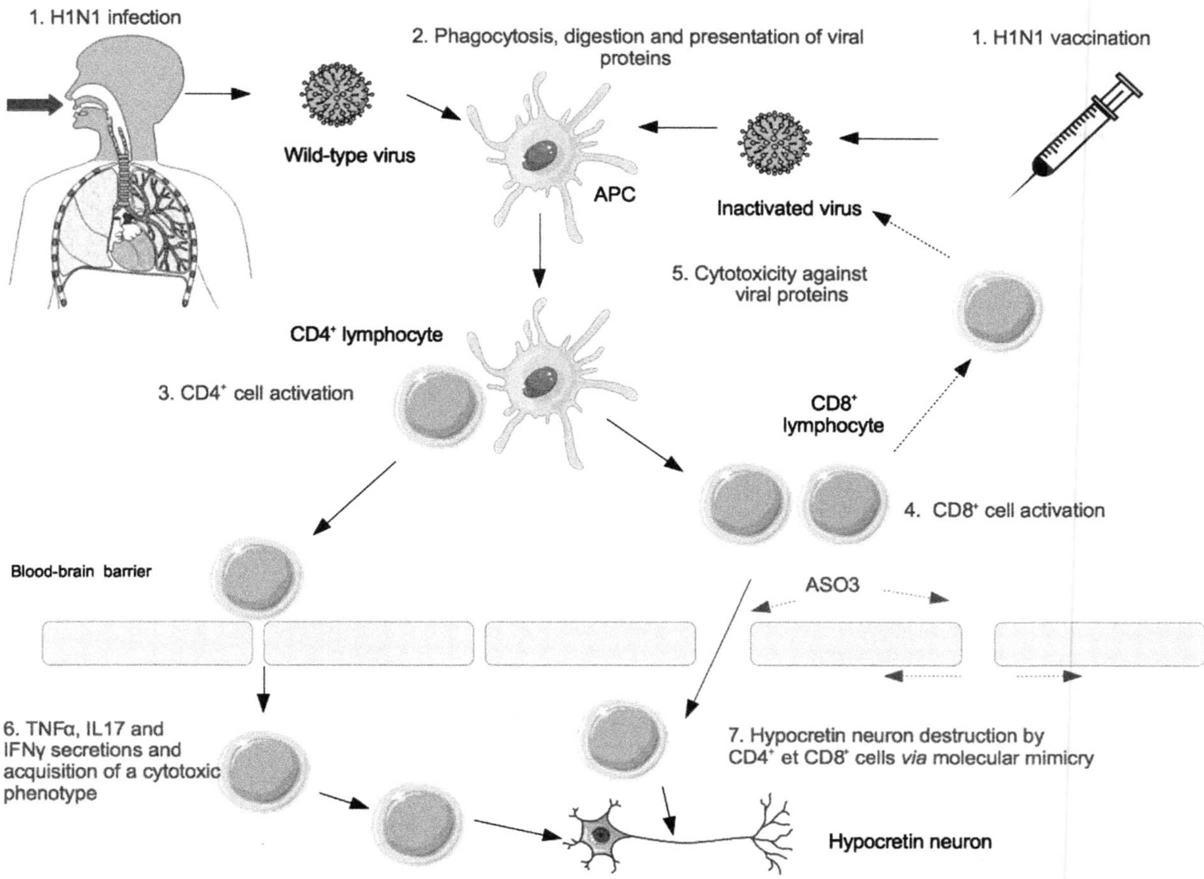

FIGURE 39.5 Destruction of hypocretin neurons by immune system after H1N1 infection or vaccination. Hypothetical model. *With modification from Jacob, L., & Dauvilliers, Y. (2014). Narcolepsy with cataplexy: an autoimmune disease?* Médecine-Sciences, 30*(12), 1136–1143.*

of other characteristics and of HLA-exon–based typing, which suggests regulatory effects. The finding suggests that GWAS data should also be analyzed in relation to chronological time, and that findings could suggest the emergence of novel environmental triggers.

H1N1 PANDEMRIX VACCINATION LEADING TO AN INCREASE IN NARCOLEPSY INCIDENCE

Discussing postvaccine complications is always delicate because passions run high and data are often difficult to interpret. However, it is reasonable to think that in rare cases, any effective immune manipulation could lead to an abnormal immune response. In 1977, vaccinations against an H1N1 strain led to an increased incidence of Guillain–Barré syndrome (Mancardi, Del Sette, Primavera, Farinelli, & Fumarola, 1989), another autoimmune disease. The H1N1 flu strain in question was similar to the 1918 Spanish strain, a strain that had been associated with "encephalitis lethargica" by Van Economo in 1930, a complication suggested by some also to be an autoimmune reaction to H1N1 (Vincent, 2004). Interestingly, although patients with encephalitis lethargica did not have cataplexy, the disease was associated with the destruction of neurons in the posterior hypothalamus, a region containing the wake-promoting hypocretin and histamine neurons. Current analyses suggest that Guillain–Barré syndrome can be a rare complication of flu vaccination, less than one case per million vaccinations, although it is still controversial (Lehmann, Hartung, Kieseier, & Hughes, 2010).

In spring 2009, the world was alarmed by the emergence of a new reassortant H1N1 strain that emerged from pigs and was associated with a high mortality rate in Mexico. Faced with the threat of heavy casualties, the World Health Organization and governments stimulated the production of new vaccines targeting this strain, to be ready for the winter of 2009. In 2010, shortly after the launch of the vaccination campaign, cases of sudden-onset narcolepsy started to be reported after vaccination with Pandemrix, an AS03-adjuvanted vaccine developed by GlaxoSmithKline (GSK) (Dresden,

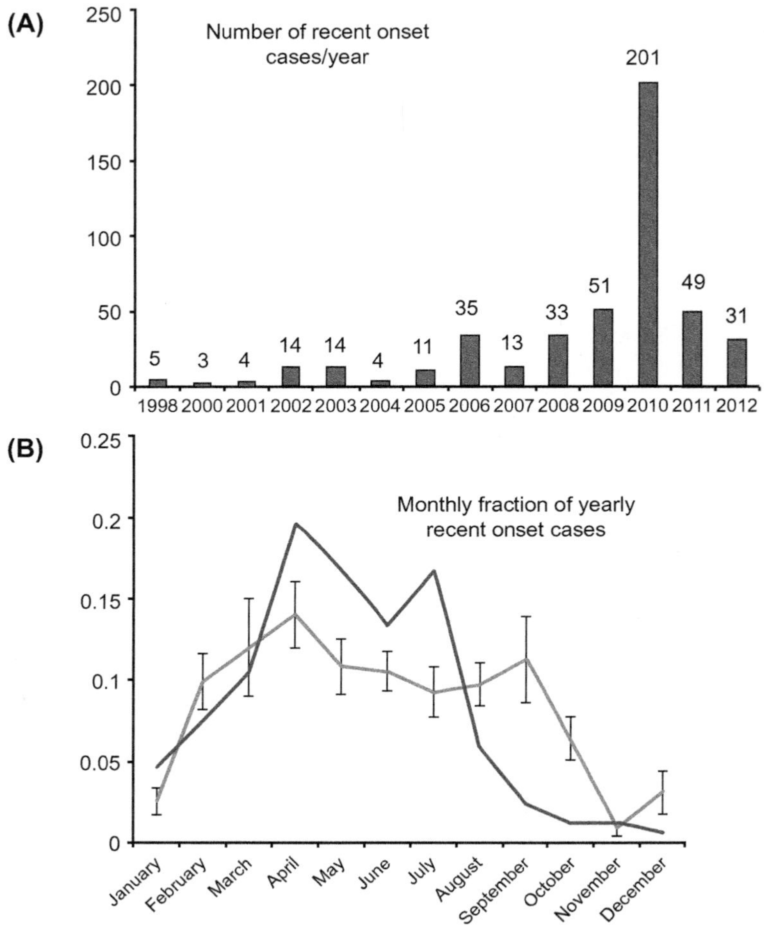

FIGURE 39.6 Evolution of narcolepsy incidence in China over the past 15 years (A) and monthly occurrence of narcolepsy onset displaying annual variation (B). (A) Yearly occurrence of onset (diagnosis within 1 year of onset) showing a dramatic increase in 2010, after the H1N1 pandemic of 2009, with return to baseline condition the following year (B) Seasonal pattern of onset of narcolepsy in Chinese patients showing highly increased risk in spring and summer versus early winter. Data are represented as a yearly fraction of 12 months; 0.083 (8.3%) would be the expected value of each month if onsets were randomly distributed across the year. Blue line represents the mean ± standard error of the mean of this yearly fraction for years 2002–2009, with very low levels of new onset in late winter. Red (gray in print versions) line represents the numbers for 2010, after the 2009 pH1N1 pandemic, and shows a more pronounced circannual pattern peaking in spring and summer. *Data updated derived from Han, F., Lin, L., Warby, S. C., Faraco, J., Li, J., Dong, S. X., … Mignot, E. (2011). Narcolepsy onset is seasonal and increased following the 2009 H1N1 pandemic in China.* Annals of Neurology, *70(3), 410–417. http://dx.doi.org/10.1002/ana.22587.*

Germany). This association has since been confirmed in many countries, with the increased risk ranging from threefold to 13-fold (Dauvilliers et al., 2013; Heier et al., 2013; O'Flanagan et al., 2014; Partinen et al., 2012; Winstone et al., 2014). Fortunately, however, the increased risk has mostly been observed in children, and not with other vaccines such as those that did not used adjuvants and have been used in the United States.

More surprisingly, Arepanrix, another AS03-adjuvanted vaccine (AS03 contains squalene and vitamin E) also developed by GSK (Ste Foy, Canada) that was used in Canada and South America, was not strongly associated with narcolepsy (Dauvilliers et al., 2010; Mendes, Neto, de Azevedo, & Caramelli, 2012). Arepanrix is virtually identical to Pandemrix, except for the antigen extraction procedure and the site of production (Jacob et al., 2014). Similarly, Focetria, another adjuvanted vaccine developed by Novartis that used a slightly different adjuvant (MF59 containing squalene alone), and a distinct antigen preparation, was not associated with large increases in cases (Leroux-Roels et al., 2010). The reason for these differences is unknown, but it may involve differences in antigen modification by detergents in some but not other vaccines (Vaarala et al., 2014). Studying Pandemrix and Arepanrix by two-dimensional difference in gel electrophoresis and mass spectrometry, we also found that in at least one batch of Arepanrix, the hemagglutinin is mutated at position 129, a position also changed in subsequent vaccines (including Focetria), because it increases the yield of

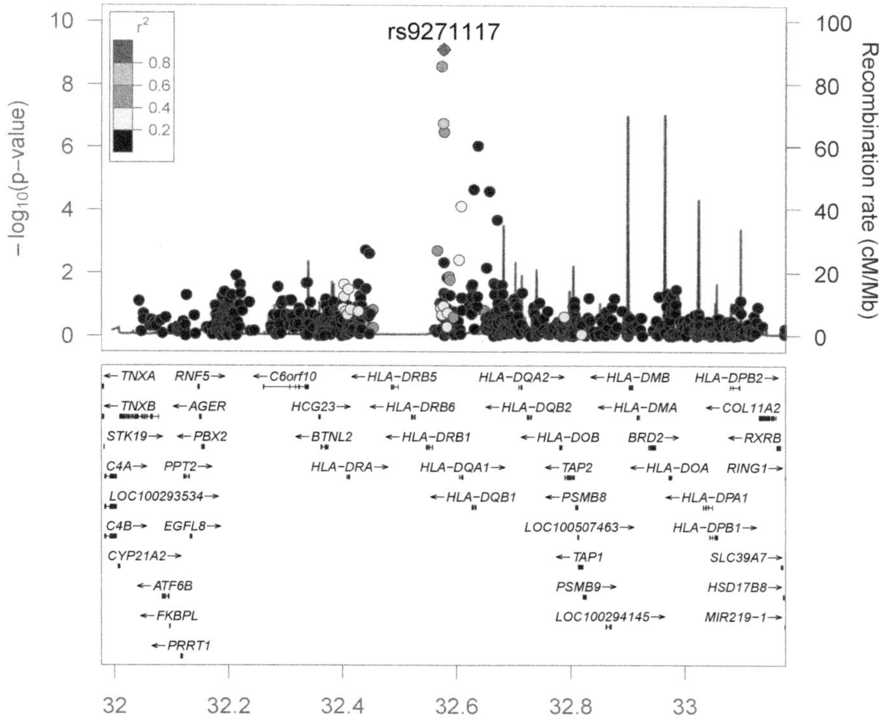

FIGURE 39.7 Comparison of HLA SNP associations between cases before and after 2009 onset. Note genome-wide significance. *With modification, from Han, F., Faraco, J., Dong, X. S., Ollila, H. M., Lin, L., Li, J., … Mignot, E. (2013). Genome wide analysis of narcolepsy in China implicates novel immune loci and reveals changes in association prior to versus after the 2009 H1N1 influenza pandemic.* PLoS Genetics, *9(10). http://dx.doi.org/10.1371/journal.pgen.1003880.*

production (Jacob et al., 2014). Whether this epitope change is important in explaining differences in risk remains to be established. Because only one in 15,000 vaccinated children developed narcolepsy, many other cofactors may be involved, including differences in immune history across regions.

Another question that has been raised is whether the AS03 adjuvant itself could have been involved in increasing risk. This possibility worries regulators a great deal, because adjuvants are critical in some cases to increase vaccine effectiveness (this may turn out to be particularly important in the use of immunotherapy for cancer, in which tolerance must be broken). Because US data shows that nonadjuvanted vaccines have not increased narcolepsy risk substantially, AS03 may indeed have had a role. As mentioned earlier, however, both Pandemrix (associated with narcolepsy) and Arepanrix (not associated with narcolepsy) have the same adjuvant but different risks; thus, the adjuvant alone must not be sufficient. It is our hypothesis that the AS03 only acted as a catalyst for an H1N1-restricted immune response. AS03 is known to induce strong immune responses compared with other adjuvants such as MF59, which was used in another pandemic H1N1 vaccine (Focetria), and this may have potentiated a reaction that occurs more rarely with the wild-type virus infection.

Although the Pandemrix-narcolepsy association has been a great tragedy for patients, it has the potential to reveal how a normal immune response can lead to autoimmunity. The leading hypothesis to explain these associations involves molecular mimicry, but the concept has never been formally proven in humans. In this model, specific viral H1N1 peptides would share homology with proteins expressed in hypocretin neurons. In some case, H1N1 infections or vaccinations can lead to the presentation of these viral epitopes by DQ0602 to T cells, which in some cases can confuse such antigens with hypocretin cell autoantigens, leading to cell destruction and narcolepsy.

PERSPECTIVES

Responses to difficult scientific problems often come from the most unexpected quarters. The discovery of the link with hypocretin came from dog genetics and is now opening novel therapeutic possibilities for insomnia and other sleep disorders. It has also increased our understanding of the neural networks that regulate sleep and wake. Similarly, based on its clinical presentation, few investigators would have thought that narcolepsy could be autoimmune and involve influenza A as a trigger. We believe that a final understanding of the immune mechanisms involved will similarly provide new insight into antineuronal autoimmunity, because little is known in this area. It is becoming increasingly clear that autoimmune diseases affecting the brain have been largely ignored, but progress in this area has been driven by work in the paraneoplastic area and the finding of autoantibodies cross-reacting with tumor antigens (Graus et al., 2010). Narcolepsy raises the possibility that other diseases that involve CNS-specific antigens (Hypocretin) may be difficult to detect because they involve more selective cellular rather than humoral immunity.

REFERENCES

Aldrich, M. S. (1996). The clinical spectrum of narcolepsy and idiopathic hypersomnia. *Neurology, 46*(2), 393–401.

Anic-Labat, S., Guilleminault, C., Kraemer, H. C., Meehan, J., Arrigoni, J., & Mignot, E. (1999). Validation of a cataplexy questionnaire in 983 sleep-disorders patients. *Sleep, 22*(1), 77–87.

Aran, A., Lin, L., Nevsimalova, S., Plazzi, G., Chul Hong, S., Weiner, K., ... Mignot, E. (2009). Elevated anti-streptococcal antibodies in patients with recent narcolepsy onset. *Sleep, 32*(8), 979–983.

Bergman, P., Adori, C., Vas, S., Kai-Larsen, Y., Sarkanen, T., Cederlund, A., ... Hokfelt, T. (2014). Narcolepsy patients have antibodies that stain distinct cell populations in rat brain and influence sleep patterns. *Proceedings of the National Academy of Sciences of the United States of America, 111*(35), E3735–E3744. http://dx.doi.org/10.1073/pnas.1412189111.

Beuckmann, C. T., & Yanagisawa, M. (2002). Orexins: from neuropeptides to energy homeostasis and sleep/wake regulation. *Journal of Molecular Medicine (Berlin, Germany), 80*(6), 329–342. http://dx.doi.org/10.1007/s00109-002-0322-x.

Billiard, M., Seignalet, J., Juji, T., Matsuki, K., Naohara, T., Satake, M., & Honda, Y. (1985). Extraordinary association between HLA-DR2 and narcolepsy. *Lancet, 1*(8422), 226–227.

Black, J. L., Avula, R. K., Walker, D. L., Silber, M. H., Krahn, L. E., Shane Pankratz, V., ... Slocumb, N. L. (2005). HLA DQB1*0602 positive narcoleptic subjects with cataplexy have CSF IgG reactive to rat hypothalamic protein extract. *Sleep, 28*(9), 1191–1192.

Black, J. L., Silber, M. H., Krahn, L. E., Avula, R. K., Walker, D. L., Shane Pankratz, V., ... Slocumb, N. L. (2005). Studies of humoral immunity to preprohypocretin in human leukocyte antigen DQB1*0602-positive narcoleptic subjects with cataplexy. *Biological Psychiatry, 58*(6), 504–509. http://dx.doi.org/10.1016/j.biopsych.2005.04.026.

Bours, M. J., Swennen, E. L., Di Virgilio, F., Cronstein, B. N., & Dagnelie, P. C. (2006). Adenosine 5'-triphosphate and adenosine as endogenous signaling molecules in immunity and inflammation. *Pharmacology and Therapeutics, 112*(2), 358–404.

Carter, M. E., Yizhar, O., Chikahisa, S., Nguyen, H., Adamantidis, A., Nishino, S., ... de Lecea, L. (2010). Tuning arousal with optogenetic modulation of locus coeruleus neurons. *Nature Neuroscience, 13*(12), 1526–1533. http://dx.doi.org/10.1038/nn.2682.

Chemelli, R. M., T Willie, J., M Sinton, C., K Elmquist, J., Scammell, T., Lee, C., ... Yanagisawa, M. (1999). Narcolepsy in orexin knockout mice: molecular genetics of sleep regulation. *Cell, 98*(4), 437–451.

Cvetkovic-Lopes, V., Bayer, L., Dorsaz, S., Maret, S., Pradervand, S., Dauvilliers, Y., ... Tafti, M. (2010). Elevated tribbles homolog 2–specific antibody levels in narcolepsy patients. *Journal of Clinical Investigation, 120*(3), 713–719. http://dx.doi.org/10.1172/JCI41366.

Dale, R. C., & Heyman, I. (2002). Post-streptococcal autoimmune psychiatric and movement disorders in children. *The British Journal of Psychiatry, 181*(3), 188–190. http://dx.doi.org/10.1192/bjp.181.3.188.

Dauvilliers, Y., Arnulf, I., Lecendreux, M., Monaca Charley, C., Franco, P., Drouot, X., ... Narcoflu-VF Study Group. (2013). Increased risk of narcolepsy in children and adults after pandemic H1N1 vaccination in France. *Brain: A Journal of Neurology, 136*(Pt 8), 2486–2496. http://dx.doi.org/10.1093/brain/awt187.

Dauvilliers, Y., Arnulf, I., & Mignot, E. (2007). Narcolepsy with cataplexy. *Lancet, 369*(9560), 499–511. http://dx.doi.org/10.1016/S0140-6736(07)60237-2.

Dauvilliers, Y., Mayer, G., Lecendreux, M., Neidhart, E., Peraita-Adrados, R., Sonka, K., ... Tafti, M. (2002). Kleine-levin syndrome: an autoimmune hypothesis based on clinical and genetic analyses. *Neurology, 59*(11), 1739–1745.

Dauvilliers, Y., Montplaisir, J., Cochen, V., Desautels, A., Einen, M., Lin, L., ... Mignot, E. (2010). Post-H1N1 narcolepsy-cataplexy. *Sleep, 33*(11), 1428–1430.

Dauvilliers, Y., Neidhart, E., Lecendreux, M., Billiard, M., & Tafti, M. (2001). MAO-a and COMT polymorphisms and gene effects in narcolepsy. *Molecular Psychiatry, 6*(4), 367–372. http://dx.doi.org/10.1038/sj.mp.4000911.

De Lecea, L., Kilduff, T. S., Peyron, C., Gao, X., Foye, P. E., Danielson, P. E., ... Sutcliffe, J. G. (1998). The hypocretins: hypothalamus-specific peptides with neuroexcitatory activity. *Proceedings of the National Academy of Sciences of the United States of America, 95*(1), 322–327.

Di Virgilio, F., Boeynaems, J. M., & Robson, S. C. (2009). Extracellular nucleotides as negative modulators of immunity. *Current Opinion in Pharmacology, 9*(4), 507–513.

Dougherty, J. D., Schmidt, E. F., Nakajima, M., & Heintz, N. (2010). Analytical approaches to RNA profiling data for the identification of genes enriched in specific cells. *Nucleic Acids Research, 38*(13), 4218–4230. http://dx.doi.org/10.1093/nar/gkq130.

Ellis, M. C., Hetisimer, A. H., Ruddy, D. A., Hansen, S. L., Kronmal, G. S., McClelland, E., ... Mignot, E. (1997). HLA class II haplotype and sequence analysis support a role for DQ in narcolepsy. *Immunogenetics, 46*(5), 410–417.

Eriksson, K. S., & Mignot, E. (December 2009). T-box 3 is expressed in the adult mouse hypothalamus and medulla. *Brain Research, 1302*, 233–239. http://dx.doi.org/10.1016/j.brainres.2009.08.101.

Faraco, J., Lin, L., Kornum, B. R., Kenny, E. E., Trynka, G., Einen, M., ... Mignot, E. (2013). ImmunoChip study implicates antigen presentation to T cells in narcolepsy. *PLoS Genetics, 9*(2). http://dx.doi.org/10.1371/journal.pgen.1003270.

Gélineau, J.-B. (1881). *De La Narcolepsie.*

Goldbart, A., Peppard, P., Finn, L., Ruoff, C. M., Barnet, J., Young, T., & Mignot, E. (2014). Narcolepsy and predictors of positive MSLTs in the Wisconsin sleep cohort. *Sleep, 37*(6), 1043–1051. http://dx.doi.org/10.5665/sleep.3758.

Graus, F., Saiz, A., & Dalmau, J. (2010). Antibodies and neuronal autoimmune disorders of the CNS. *Journal of Neurology, 257*(4), 509–517. http://dx.doi.org/10.1007/s00415-009-5431-9.

Guilleminault, C. (1976). In *Narcolepsy: Proceedings of the first international symposium on narcolepsy, July 1975*. Montpellier, France: SP Boods Division of Spectrum Publications : Distributed by Halstead Press.

Guilleminault, C., & Grumet, C. (1986). HLA-DR2 and narcolepsy: not all narcoleptic-cataplectic patients are DR2. *Human Immunology, 17*(1), 1–2.

Guilleminault, C., Mignot, E., & Partinen, M. (1994). Controversies in the diagnosis of narcolepsy. *Sleep, 17*(8 Suppl.), S1–S6.

Hallmayer, J., Faraco, J., Lin, L., Hesselson, S., Winkelmann, J., Kawashima, M., ... Mignot, E. (2009). Narcolepsy is strongly associated with the TCR alpha locus. *Nature Genetics, 41*(6), 708–711. http://dx.doi.org/10.1038/ng.372.

Han, F., Faraco, J., Dong, X. S., Ollila, H. M., Lin, L., Li, J., ... Mignot, E. (2013). Genome wide analysis of narcolepsy in China implicates novel immune loci and reveals changes in association prior to versus after the 2009 H1N1 influenza pandemic. *PLoS Genetics, 9*(10). http://dx.doi.org/10.1371/journal.pgen.1003880.

Han, F., Lin, L., Li, J., Aran, A., Dong, S. X., An, P., ... Mignot, E. (2012). TCRA, P2RY11, and CPT1B/CHKB associations in Chinese narcolepsy. *Sleep Medicine, 13*(3), 269–272. http://dx.doi.org/10.1016/j.sleep.2011.06.020.

Han, F., Lin, L., Li, J., Dong, S. X., An, P., Zhao, L., ... Mignot, E. (2012). HLA-dq association and allele competition in Chinese narcolepsy. *Tissue Antigens, 80*(4), 328–335. http://dx.doi.org/10.1111/j.1399-0039.2012.01948.x.

Han, F., Lin, L., Schormair, B., Pizza, F., Plazzi, G., Ollila, H. M., ... Mignot, E. (2014). HLA DQB1*06:02 negative narcolepsy with Hypocretin/Orexin deficiency. *Sleep, 37*(10), 1601–1608. http://dx.doi.org/10.5665/sleep.4066.

Han, F., Lin, L., Warby, S. C., Faraco, J., Li, J., Dong, S. X., ... Mignot, E. (2011). Narcolepsy onset is seasonal and increased following the 2009 H1N1 pandemic in China. *Annals of Neurology, 70*(3), 410–417. http://dx.doi.org/10.1002/ana.22587.

Heier, M. S., Gautvik, K. M., Wannag, E., Bronder, K. H., Midtlyng, E., Kamaleri, Y., & Storsaeter, J. (2013). Incidence of narcolepsy in Norwegian children and adolescents after vaccination against H1N1 Influenza A. *Sleep Medicine, 14*(9), 867–871. http://dx.doi.org/10.1016/j.sleep.2013.03.020.

Hohjoh, H., Terada, N., Nakayama, T., Kawashima, M., Miyagawa, T., Honda, Y., & Tokunaga, K. (2001). Case-control study with narcoleptic patients and healthy controls who, like the patients, possess both HLA-DRB1*1501 and -DQB1*0602. *Tissue Antigens, 57*(3), 230–235.

Honda, M., Eriksson, K. S., Zhang, S., Tanaka, S., Lin, L., Salehi, A., ... Mignot, E. (2009). IGFBP3 colocalizes with and regulates hypocretin (orexin). *PLoS One, 4*(1), e4254. http://dx.doi.org/10.1371/journal.pone.0004254.

Honda, Y. (1988). *HLA in narcolepsy.*

Hong, S.-C., Lin, L., Lo, B., Jeong, J.-H., Shin, Y.-K., Kim, Su-Y., ... Mignot, E. (2007). DQB1*0301 and DQB1*0601 modulate narcolepsy susceptibility in koreans. *Human Immunology, 68*(1), 59–68. http://dx.doi.org/10.1016/j.humimm.2006.10.006.

Hor, H., Bartesaghi, L., Kutalik, Z., Vicário, J. L., de Andrés, C., Pfister, C., ... Peraita-Adrodas, R. (2011). A missense mutation in myelin oligodendrocyte glycoprotein as a cause of familial narcolepsy with cataplexy. *American Journal of Human Genetics, 89*(3), 474–479. http://dx.doi.org/10.1016/j.ajhg.2011.08.007.

Hor, H., Kutalik, Z., Dauvilliers, Y., Valsesia, A., Lammers, G. J., Donjacour, C. E. H. M., ... Tafti, M. (2010). Genome-wide association study identifies new HLA class II haplotypes strongly protective against narcolepsy. *Nature Genetics, 42*(9), 786–789. http://dx.doi.org/10.1038/ng.647.

Hungs, M., Fan, J., Lin, L., Lin, X., Maki, R. A., & Mignot, E. (2001). Identification and functional analysis of mutations in the hypocretin (orexin) genes of narcoleptic canines. *Genome Research, 11*(4), 531–539. http://dx.doi.org/10.1101/gr.161001.

ICSD3. (2014). *International classification of sleep disorders* (3rd ed.). American Academy of Sleep Medicine.

Jackson, M. W., Reed, J. H., Smith, A. J. F., & Gordon, T. P. (2008). An autoantibody in narcolepsy disrupts colonic migrating motor complexes. *The Journal of Neuroscience: The Official Journal of the Society for Neuroscience, 28*(49), 13303–13309. http://dx.doi.org/10.1523/JNEUROSCI.4489-08.2008.

Jackson, M. W., Spencer, N. J., Reed, J. H., Smith, A. J. F., & Gordon, T. P. (2009). Potentiation of a functional autoantibody in narcolepsy by a cholinesterase inhibitor. *Laboratory Investigation; A Journal of Technical Methods and Pathology, 89*(12), 1332–1339. http://dx.doi.org/10.1038/labinvest.2009.108.

Jacob, L., & Dauvilliers, Y. (2014). Narcolepsy with cataplexy: an autoimmune disease? *Médecine-Sciences, 30*(12), 1136–1143.

Jacob, L., Leib, R., Bonvalet, M., Ollila, H. M., Adams, C., & Mignot, E. (2014). Comparison of pandemrix and arepanrix, two pH1N1 AS03-Adjuvanted vaccines differentially associated with narcolepsy development. *Brain Behavior and Immunity, 47*, 44–57.

Juji, T., Satake, M., Honda, Y., & Doi, Y. (1984). HLA antigens in Japanese patients with narcolepsy. All the patients were DR2 positive. *Tissue Antigens, 24*(5), 316–319.

Julkunen, I., & Partinen, M. (2014). Neuroimmunology: disease mechanisms in narcolepsy remain elusive. *Nature Reviews. Neurology, 10*(11), 616–617. http://dx.doi.org/10.1038/nrneurol.2014.191.

Kanbayashi, T., Inoue, Y., Chiba, S., Aizawa, R., Saito, Y., Tsukamoto, H., ... Shimizu, T. (2002). CSF Hypocretin-1 (orexin-A) concentrations in narcolepsy with and without cataplexy and idiopathic hypersomnia. *Journal of Sleep Research, 11*(1), 91–93.

Katzav, A., Arango, M. T., Kivity, S., Tanaka, S., Givaty, G., Agmon-Levin, N., ... Shoenfeld, Y. (September 2013). Passive transfer of narcolepsy: anti-TRIB2 autoantibody positive patient IgG causes hypothalamic orexin neuron loss and sleep attacks in mice. *Journal of Autoimmunity, 45*, 24–30. http://dx.doi.org/10.1016/j.jaut.2013.06.010.

Kawashima, M., Lin, L., Tanaka, S., Jennum, P., Knudsen, S., Nevsimalova, S., … Mignot, E. (2010). Anti-tribbles Homolog 2 (TRIB2) autoantibodies in narcolepsy are associated with recent onset of cataplexy. *Sleep, 33*(7), 869—874.

Knecht, C. D., Oliver, J. E., Redding, R., Selcer, R., & Johnson, G. (1973). Narcolepsy in a dog and a cat. *Journal of the American Veterinary Medical Association, 162*(12), 1052—1053.

Knudsen, S., Mikkelsen, J. D., & Jennum, P. (2007). Antibodies in narcolepsy-cataplexy patient serum bind to rat hypocretin neurons. *NeuroReport, 18*(1), 77—79. http://dx.doi.org/10.1097/WNR.0b013e328010baad.

Kornum, B. R., Holm, A., Lin, L., Faraco, J., Mostafavi, S., Battle, A., … Kornum, B. R. (2014). EIF3G is associated with narcolepsy across ethnicities. *European Journal of Human Genetics, 23*(11), 1573—1580.

Kornum, B. R., Kawashima, M., Faraco, J., Lin, L., Rico, T., Hesselson, S., … Mignot, E. (2011). Common variants in P2RY11 are associated with narcolepsy. *Nature Genetics, 43*(1), 66—71. http://dx.doi.org/10.1038/ng.734.

Kramer, M. (1977). Narcolepsy. Advances in sleep research, Vol. 3. *American Journal of Psychiatry, 134*(6), 709. http://dx.doi.org/10.1176/ajp.134.6.709.

Krämer, S., Goldammer, U., & Sindern, E. (2014). Moebius syndrome and narcolepsy. *Der Nervenarzt, 85*(12), 1569—1572. http://dx.doi.org/10.1007/s00115-014-4174-8.

Lehmann, H. C., Hartung, H.-P., Kieseier, B. C., & Hughes, R. A. C. (2010). Guillain-barré syndrome after exposure to influenza virus. *The Lancet Infectious Diseases, 10*(9), 643—651. http://dx.doi.org/10.1016/S1473-3099(10)70140-7.

Leroux-Roels, I., Roman, F., Forgus, S., Maes, C., De Boever, F., Dramé, M., … Leroux-Roels, G. (2010). Priming with AS03 A-Adjuvanted H5N1 influenza vaccine improves the kinetics, magnitude and durability of the immune response after a heterologous booster vaccination: an open non-randomised extension of a double-blind randomised primary study. *Vaccine, 28*(3), 849—857. http://dx.doi.org/10.1016/j.vaccine.2009.10.017.

Lin, L., Faraco, J., Li, R., Kadotani, H., Rogers, W., Lin, X., … Mignot, E. (1999). The sleep disorder canine narcolepsy is caused by a mutation in the hypocretin (orexin) receptor 2 gene. *Cell, 98*(3), 365—376.

Lind, A., Ramelius, A., Olsson, T., Arnheim-Dahlström, L., Lamb, F., Khademi, M., … Lernmark, A. (May 2014). A/H1N1 antibodies and TRIB2 autoantibodies in narcolepsy patients diagnosed in conjunction with the pandemrix vaccination campaign in Sweden 2009—2010. *Journal of Autoimmunity, 50*, 99—106. http://dx.doi.org/10.1016/j.jaut.2014.01.031.

Longstreth, W. T., Ton, T. G. N., & Koepsell, T. D. (2009). Narcolepsy and streptococcal infections. *Sleep, 32*(12), 1548.

Lu, J., Sherman, D., Devor, M., & Saper, C. B. (2006). A putative flip-flop switch for control of REM sleep. *Nature, 441*(7093), 589—594. http://dx.doi.org/10.1038/nature04767.

Mancardi, G. L., Del Sette, M., Primavera, A., Farinelli, M., & Fumarola, D. (1989). A prospective study of acute idiopathic neuropathy II antecedent events. *Journal of Neurology, Neurosurgery, and Psychiatry, 52*(3), 424—425.

Marcadet, A., Gebuhrer, L., Betuel, H., Seignalet, J., Freidel, A. C., Confavreux, C., … Cohen, D. (1985). DNA polymorphism related to HLA-DR2 Dw2 in patients with narcolepsy. *Immunogenetics, 22*(6), 679—683.

Martínez-Rodríguez, J. E., Lin, L., Iranzo, A., Genis, D., Martí, M. J., Santamaria, J., & Mignot, E. (2003). Decreased Hypocretin-1 (Orexin-A) levels in the cerebrospinal fluid of patients with myotonic dystrophy and excessive daytime sleepiness. *Sleep, 26*(3), 287—290.

Martínez-Rodríguez, J. E., Sabater, L., Graus, F., Iranzo, A., & Santamaria, J. (2007). Evaluation of hypothalamic-specific autoimmunity in patients with narcolepsy. *Sleep, 30*(1), 27—28.

Matsuki, K., Grumet, F. C., Lin, X., Gelb, M., Guilleminault, C., Dement, W. C., & Mignot, E. (1992). DQ (rather than DR) gene marks susceptibility to narcolepsy. *Lancet, 339*(8800), 1052.

McDevitt, H. O. (2000). Discovering the role of the major histocompatibility complex in the immune response. *Annual Review of Immunology, 18*, 1—17. http://dx.doi.org/10.1146/annurev.immunol.18.1.1.

Melberg, A., Ripley, B., Lin, L., Hetta, J., Mignot, E., & Nishino, S. (2001). Hypocretin deficiency in familial symptomatic narcolepsy. *Annals of Neurology, 49*(1), 136—137.

Mendes, M. F. S. G., Neto, V., de Azevedo, R. A., & Caramelli, P. (2012). Narcolepsy after A/H1N1 vaccination. *Clinics, 67*(1), 77—78. http://dx.doi.org/10.6061/clinics/2012(01)12.

Mignot, E. (1998). Genetic and familial aspects of narcolepsy. *Neurology, 50*(2 Suppl. 1), S16—S22.

Mignot, E., Jan Lammers, G., Ripley, B., Okun, M., Nevsimalova, S., Overeem, S., … Nishino, S. (2002). The role of cerebrospinal fluid hypocretin measurement in the diagnosis of narcolepsy and other hypersomnias. *Archives of Neurology, 59*(10), 1553—1562.

Mignot, E., Kimura, A., Lattermann, A., Lin, X., Yasunaga, S., Mueller-Eckhardt, G., … Underhill, P. (1997). Extensive HLA class II studies in 58 non-DRB1*15 (DR2) narcoleptic patients with cataplexy. *Tissue Antigens, 49*(4), 329—341.

Mignot, E., Lin, X., Arrigoni, J., Macaubas, C., Olive, F., Hallmayer, J., … Grumet, F. C. (1994). DQB1*0602 and DQA1*0102 (DQ1) are better markers than DR2 for narcolepsy in Caucasian and black Americans. *Sleep, 17*(8 Suppl.), S60—S67.

Mignot, E., Lin, L., Rogers, W., Honda, Y., Qiu, X., Lin, X., & Okun, M. (2001). Complex HLA-DR and -DQ interactions confer risk of narcolepsy-cataplexy in three ethnic groups. *American Journal of Human Genetics, 68*(3), 686—699.

Mignot, E. J. M. (2012). A practical guide to the therapy of narcolepsy and hypersomnia syndromes. *Neurotherapeutics: The Journal of the American Society for Experimental NeuroTherapeutics, 9*(4), 739—752. http://dx.doi.org/10.1007/s13311-012-0150-9.

Mignot, E. J. M. (2014). History of narcolepsy at Stanford University. *Immunologic Research, 58*(2—3), 315—339. http://dx.doi.org/10.1007/s12026-014-8513-4.

Miyadera, H., Ohashi, J., Lernmark, A., Kitamura, T., & Tokunaga, K. (December 2014). Cell-surface MHC density profiling reveals instability of autoimmunity-associated HLA. *The Journal of Clinical Investigation*. http://dx.doi.org/10.1172/JCI74961.

Miyagawa, T., Kawashima, M., Nishida, N., Ohashi, J., Kimura, R., Fujimoto, A., … Tokunaga, K. (2008). Variant between CPT1B and CHKB associated with susceptibility to narcolepsy. *Nature Genetics, 40*(11), 1324–1328. http://dx.doi.org/10.1038/ng.231.

Moghadam, K. K., Pizza, F., La Morgia, C., Franceschini, C., Tonon, C., Lodi, R., … Plazzi, G. (2014). Narcolepsy is a common phenotype in HSAN IE and ADCA-DN. *Brain: A Journal of Neurology, 137*(Pt 6), 1643–1655. http://dx.doi.org/10.1093/brain/awu069.

Mueller-Eckhardt, G., Meier-Ewert, K., Schendel, D. J., Reinecker, F. B., Multhoff, G., & Mueller-Eckhardt, C. (1986). HLA and narcolepsy in a German population. *Tissue Antigens, 28*(3), 163–169.

Neely, S., Rosenberg, R., Spire, J. P., Antel, J., & Arnason, B. G. (1987). HLA antigens in narcolepsy. *Neurology, 37*(12), 1858–1860.

Nelson, G. B., & Hahn, J. S. (2003). Stimulus-induced drop episodes in Coffin-Lowry syndrome. *Pediatrics, 111*(3), e197–202.

Nishida, N., Sawai, H., Kashiwase, K., Minami, M., Sugiyama, M., Seto, W.-K., … Mizokami, M. (2014). New susceptibility and resistance HLA-DP alleles to HBV-related diseases identified by a trans-ethnic association study in Asia. *PLoS One, 9*(2), e86449. http://dx.doi.org/10.1371/journal.pone.0086449.

Nishino, S. (2000). Hypocretin (orexin) deficiency in human narcolepsy. *Lancet, 355*, 39–40. http://dx.doi.org/10.1016/S0140-6736(99)05582-8.

Nishino, S., & Mignot, E. (1997). Pharmacological aspects of human and canine narcolepsy. *Progress in Neurobiology, 52*(1), 27–78.

O'Flanagan, D., Barret, A. S., Foley, M., Cotter, S., Bonner, C., Crowe, C., … Purcell, E. (2014). Investigation of an association between onset of narcolepsy and vaccination with pandemic influenza vaccine, Ireland April 2009–December 2010. *Euro Surveillance: Bulletin Européen Sur Les Maladies Transmissibles = European Communicable Disease Bulletin, 19*(17), 15–25.

Ollila, H. M., Fernandez-Vina, M., & Mignot, E. (October 2014). HLA-DQ Allele competition in narcolepsy: a comment on Tafti et Al. DQB1 locus alone explains most of the risk and protection in narcolepsy with cataplexy in Europe. *Sleep, 38*(1), 147–151.

Ollila, H. M., Ravel, J.-M., Han, F., & Mignot, E. (2015). Novel genetic loci in HLA-DPB1 and HLA- class I region confer risk and protection for narcolepsy. *American Journal of Human Genetics, 96*(1), 136–146.

Overeem, S., Dalmau, J., Bataller, L., Nishino, S., Mignot, E., Verschuuren, J., & Lammers, G. J. (2004). Hypocretin-1 CSF levels in anti-Ma2 associated encephalitis. *Neurology, 62*(1), 138–140.

Overeem, S., Logan Black, J., & Lammers, G. J. (2008). Narcolepsy: immunological aspects. *Sleep Medicine Reviews, 12*(2), 95–107. http://dx.doi.org/10.1016/j.smrv.2007.07.010.

Overeem, S., Verschuuren, J. J., Fronczek, R., Schreurs, L., den Hertog, H., Hegeman-Kleinn, I. M., … Lammers, G. J. (2006). Immunohistochemical screening for autoantibodies against lateral hypothalamic neurons in human narcolepsy. *Journal of Neuroimmunology, 174*(1–2), 187–191. http://dx.doi.org/10.1016/j.jneuroim.2006.02.010.

Oyama, K., Takahashi, T., Shoji, Y., Oyamada, M., Noguchi, A., Tamura, H., … Kanbayashi, T. (2006). Niemann-pick disease type C: cataplexy and hypocretin in cerebrospinal fluid. *The Tohoku Journal of Experimental Medicine, 209*(3), 263–267.

Parkes, J. D. (June 1999). Genetic factors in human sleep disorders with special reference to norrie disease, Prader–Willi syndrome and moebius syndrome. *Journal of Sleep Research, 8*(Suppl. 1), 14–22.

Partinen, M., Saarenpaa-Heikkila, O., Ilveskoski, I., Hublin, C., Linna, M., Olsen, P., … Kirjavainen, T. (2012). Increased incidence and clinical picture of childhood narcolepsy following the 2009 H1N1 pandemic vaccination campaign in Finland. *PLoS One, 7*(3). http://dx.doi.org/10.1371/journal.pone.0033723.

Pedroso, J. L., Fusão, E. F., Ladeia-Frota, C., Arita, J. H., Barsottini, O. G., Rodrigues Masruha, M., & Pereira Vilanova, L. C. (2012). Teaching Video neuroimages: gelastic cataplexy as the first neurologic manifestation of Niemann-pick disease type C. *Neurology, 79*(22), e189. http://dx.doi.org/10.1212/WNL.0b013e31827597bf.

Peever, J., Luppi, P.-H., & Montplaisir, J. (2014). Breakdown in REM sleep circuitry underlies REM sleep behavior disorder. *Trends in Neurosciences, 37*(5), 279–288. http://dx.doi.org/10.1016/j.tins.2014.02.009.

Pelin, Z., Guilleminault, C., Risch, N., Grumet, F. C., & Mignot, E. (1998). HLA-DQB1*0602 homozygosity increases relative risk for narcolepsy but not disease severity in two ethnic groups. US modafinil in narcolepsy multicenter study group. *Tissue Antigens, 51*(1), 96–100.

Peyron, C., Faraco, J., Rogers, W., Ripley, B., Overeem, S., Charnay, Y., … Mignot, E. (2000). A mutation in a case of early onset narcolepsy and a generalized absence of hypocretin peptides in human narcoleptic brains. *Nature Medicine, 6*(9), 991–997. http://dx.doi.org/10.1038/79690.

Pizza, F., Franceschini, C., Peltola, H., Vandi, S., Finotti, E., Ingravallo, F., … Plazzi, G. (2013). Clinical and polysomnographic course of childhood narcolepsy with cataplexy. *Brain: A Journal of Neurology, 136*(Pt 12), 3787–3795. http://dx.doi.org/10.1093/brain/awt277.

Poirier, G., Montplaisir, J., Décary, F., Momège, D., & Lebrun, A. (1986). HLA antigens in narcolepsy and idiopathic central nervous system hypersomnolence. *Sleep, 9*(1 Pt 2), 153–158.

Richmond, C. (2009). Jean dausset. *The Lancet, 374*(9698), 1324. http://dx.doi.org/10.1016/S0140-6736(09)61813-4.

Ripley, B., Overeem, S., Fujiki, N., Nevsimalova, S., Uchino, M., Yesavage, J., … Nishino, S. (2001). CSF Hypocretin/orexin levels in narcolepsy and other neurological conditions. *Neurology, 57*(12), 2253–2258.

Rogers, A. E., Meehan, J., Guilleminault, C., Grumet, F. C., & Mignot, E. (1997). HLA DR15 (DR2) and DQB1*0602 typing studies in 188 narcoleptic patients with cataplexy. *Neurology, 48*(6), 1550–1556.

Roth, B., Nevsímalová, S., Sonka, K., Docekal, P., Schulz, H., Geisler, P., … Scholz, S. (1988). A study of the occurrence of HLA DR2 in 124 narcoleptics: clinical aspects. *Schweizer Archiv Für Neurologie Und Psychiatrie (Zurich, Switzerland: 1985), 139*(4), 41–51.

Saito, M., Yamagata, T., Matsumoto, A., Shiba, Y., Nagashima, M., Taniguchi, S., … Momoi, M. Y. (2014). MAOA/B deletion syndrome in male siblings with severe developmental delay and sudden loss of muscle tonus. *Brain & Development, 36*(1), 64–69. http://dx.doi.org/10.1016/j.braindev.2013.01.004.

Sakurai, T., Amemiya, A., Ishii, M., Matsuzaki, I., Chemelli, R. M., Tanaka, H., … Yanagisawa, M. (1998). Orexins and orexin receptors: a family of hypothalamic neuropeptides and G protein-coupled receptors that regulate feeding behavior. *Cell, 92*(5), 1. page following 696.

Schenck, C. H., Bassetti, C. L., Arnulf, I., & Mignot, E. (2007). English translations of the first clinical reports on narcolepsy and cataplexy by Westphal and Gelineau in the late 19th century, with commentary. *Journal of Clinical Sleep Medicine : JCSM : Official Publication of the American Academy of Sleep Medicine, 3*(3), 301–311.

Smit, L. S., Jan Lammers, G., & Catsman-Berrevoets, C. E. (2006). Cataplexy leading to the diagnosis of Niemann-Pick disease type C. *Pediatric Neurology, 35*(1), 82–84. http://dx.doi.org/10.1016/j.pediatrneurol.2005.12.012.

Smith, A. J. F., Jackson, M. W., Neufing, P., McEvoy, R. D., & Gordon, T. P. (2004). A functional autoantibody in narcolepsy. *Lancet, 364*(9451), 2122–2124. http://dx.doi.org/10.1016/S0140-6736(04)17553-3.

Smith, S. E., Mullen, T. E., Graham, D., Sims, K. B., & Rehm, H. L. (2012). Norrie disease: extraocular clinical manifestations in 56 patients. *American Journal of Medical Genetics Part A, 158A*(8), 1909–1917. http://dx.doi.org/10.1002/ajmg.a.35469.

Taheri, S., Zeitzer, J. M., & Mignot, E. (2002). The role of hypocretins (orexins) in sleep regulation and narcolepsy. *Annual Review of Neuroscience, 25*, 283–313. http://dx.doi.org/10.1146/annurev.neuro.25.112701.142826.

Tanaka, S., Honda, Y., Inoue, Y., & Honda, M. (2006). Detection of autoantibodies against hypocretin, Hcrtrl, and hcrtr2 in narcolepsy: anti-Hcrt system antibody in narcolepsy. *Sleep, 29*(5), 633–638.

Thannickal, T. C. (2000). Reduced number of hypocretin neurons in human narcolepsy. *Neuron, 27*, 469–474. http://dx.doi.org/10.1016/S0896-6273(00)00058-1.

Thomsom, G., & Dorman, J. (2006). Statistical approaches for analyses of HLA-associated and other complex diseases. In J. A. Hansen (Ed.), *Immunology of the human MHC.*

Toyoda, H., Tanaka, S., Miyagawa, T., Honda, Y., Tokunaga, K., & Honda, M. (2010). Anti-tribbles Homolog 2 autoantibodies in japanese patients with narcolepsy. *Sleep, 33*(7), 875–878.

Trotti, L. M., Staab, B. A., & Rye, D. B. (2013). Test-retest reliability of the multiple sleep latency test in narcolepsy without cataplexy and idiopathic hypersomnia. *Journal of Clinical Sleep Medicine: JCSM: Official Publication of the American Academy of Sleep Medicine, 9*(8), 789–795. http://dx.doi.org/10.5664/jcsm.2922.

Tsujino, N., & Sakurai, T. (2009). Orexin/Hypocretin: a neuropeptide at the interface of sleep, energy homeostasis, and reward system. *Pharmacological Reviews, 61*(2), 162–176. http://dx.doi.org/10.1124/pr.109.001321.

Tyagi, A., & Harrington, H. (2003). Cataplexy in association with moebius syndrome. *Journal of Neurology, 250*(1), 110–111. http://dx.doi.org/10.1007/s00415-003-0924-4.

Vaarala, O., Vuorela, A., Partinen, M., Baumann, M., Freitag, T. L., Meri, S., … Kilpi, T. (2014). Antigenic differences between AS03 adjuvanted influenza a (H1N1) pandemic vaccines: implications for Pandemrix-associated narcolepsy risk. *PLoS One, 9*(12), e114361. http://dx.doi.org/10.1371/journal.pone.0114361.

Vankova, J., Stepanova, I., Jech, R., Elleder, M., Ling, L., Mignot, E., … Nevsimalova, S. (2003). Sleep disturbances and hypocretin deficiency in Niemann-Pick disease type C. *Sleep, 26*(4), 427–430.

Varney, M. D., Maria Valdes, A., Carlson, J. A., Noble, J. A., Tait, B. D., Bonella, P., … Type 1 Diabetes Genetics Consortium. (2010). HLA DPA1, DPB1 alleles and haplotypes contribute to the risk associated with type 1 diabetes: analysis of the type 1 diabetes genetics consortium families. *Diabetes, 59*(8), 2055–2062. http://dx.doi.org/10.2337/db09-0680.

Vincent, A. (2004). Encephalitis Lethargica: part of a spectrum of post-streptococcal autoimmune diseases? *Brain: A Journal of Neurology, 127*(Pt 1), 2–3. http://dx.doi.org/10.1093/brain/awh063.

Vossler, D. G., Wyler, A. R., Wilkus, R. J., Gardner-Walker, G., & Vlcek, B. W. (1996). Cataplexy and monoamine oxidase deficiency in norrie disease. *Neurology, 46*(5), 1258–1261.

Whibley, A., Urquhart, J., Dore, J., Willatt, L., Parkin, G., Gaunt, L., … Lucy Raymond, F. (2010). Deletion of MAOA and MAOB in a male patient causes severe developmental delay, intermittent hypotonia and stereotypical hand movements. *European Journal of Human Genetics: EJHG, 18*(10), 1095–1099. http://dx.doi.org/10.1038/ejhg.2010.41.

Winkelmann, J., Lin, L., Schormair, B., Kornum, B. R., Faraco, J., Plazzi, G., … Mignot, E. (2012). Mutations in DNMT1 cause autosomal dominant cerebellar ataxia, deafness and narcolepsy. *Human Molecular Genetics, 21*(10), 2205–2210. http://dx.doi.org/10.1093/hmg/dds035.

Winstone, Marie, A., Stellitano, L., Verity, C., Andrews, N., Miller, E., … Shneerson, J. (July 2014). Clinical features of narcolepsy in children vaccinated with AS03 adjuvanted pandemic A/H1N1 2009 influenza vaccine in England. *Developmental Medicine and Child Neurology*. http://dx.doi.org/10.1111/dmcn.12522.

Chapter 40

Genomic Landscape of Brain Tumors

Victoria E. Clark, Mark W. Youngblood, Hanwen Bai, Jennifer M. Günel and Murat Günel
Yale University School of Medicine, New Haven, CT, United States

INTRODUCTION

The use of next-generation sequencing technologies has transformed our understanding of the molecular pathways driving the formation of central nervous system tumors. Tumor taxonomy is increasingly based on molecular characteristics, including driver somatic gene mutations, genomic stability, epigenetic changes, and gene expression profiles. These markers define clinically relevant entities that relate to anatomic location, response to therapy, and overall survival. The major categories of both adult and pediatric brain tumors are addressed here with a focus on identifying recent genomic findings, their relationship to molecular pathways, and their implications for diagnosis, prognosis, and, ultimately, treatment.

MENINGIOMAS

Over one-third of all primary intracranial tumors are meningiomas (Ostrom et al., 2013). It has been postulated that these tumors arise from the arachnoid layer of the meninges, and thus can be found anywhere along the brain or spinal cord. According to the World Health Organization (WHO) 2007 classification, meningiomas are grouped into three grades with 15 histologic subtypes (Louis et al., 2007). Whereas the vast majority (about 80%) of meningiomas are benign (WHO grade I), the remaining tumors can exhibit more aggressive and even malignant behavior (WHO grades II and III) (Louis et al., 2007). Genomic studies have revolutionized our knowledge of meningioma pathophysiology, demonstrating that 80% of tumors can be categorized into at least three discrete, clinically relevant genetic groups with distinct histopathology, location of growth, and varying risk of malignant progression (Fig. 40.1) (Brastianos et al., 2013; Clark et al., 2013; Reuss et al., 2013).

Approximately 50% of meningiomas are characterized by somatic, biallelic loss of *neurofibromin 2* (*NF2*), a gene found on chromosome 22q12.2 (chr22) that encodes for the tumor suppressor merlin (Ruttledge et al., 1994). The association of this gene with meningioma was first identified in the inherited dominant tumor syndrome neurofibromatosis type II, in which patients harbor germline mutations of *NF2* and often present with multiple meningiomas (Rouleau et al., 1993). It was later discovered also to be an important driver in sporadic cases of all grades, with at least 75% of WHO grade II tumors harboring biallelic *NF2/chr22 loss* (Clark et al., 2013; Louis et al., 2007). Anatomically, *NF2/chr22 loss* tumors are more likely to form in the meninges encompassing the cerebral convexities, the posterior and lateral aspects of the skull base, and along the spinal cord (Clark et al., 2013). A small subset of *NF2/chr22 loss* tumors have additional biallelic loss of the tumor suppressor INI1/SNF5, encoded by *SWI/SNF-related, matrix-associated, actin-dependent regulator of chromatin, subfamily b, member 1* (*SMARCB1*), which is also located on the long arm of chromosome 22 (Schmitz et al., 2001). Mutations or loss of *SMARCB1*, which encodes a component of the SWI/SNF chromatin remodeling complex, have been reported in families with meningiomas with schwannomatosis (van den Munckhof, Christiaans, Kenter, Baas, & Hulsebos, 2012) as well as in malignant rhabdoid tumors of the brain and kidney (Biegel et al., 1999; Versteege et al., 1998).

The second group of meningiomas involves somatic mutations in *tumor necrosis factor receptor—associated factor 7* (*TRAF7*) and includes about 25% of sporadic cases. *TRAF7* encodes for a proapoptotic protein with E3 ubiquitin ligase

Genomics, Circuits, and Pathways in Clinical Neuropsychiatry. http://dx.doi.org/10.1016/B978-0-12-800105-9.00040-8

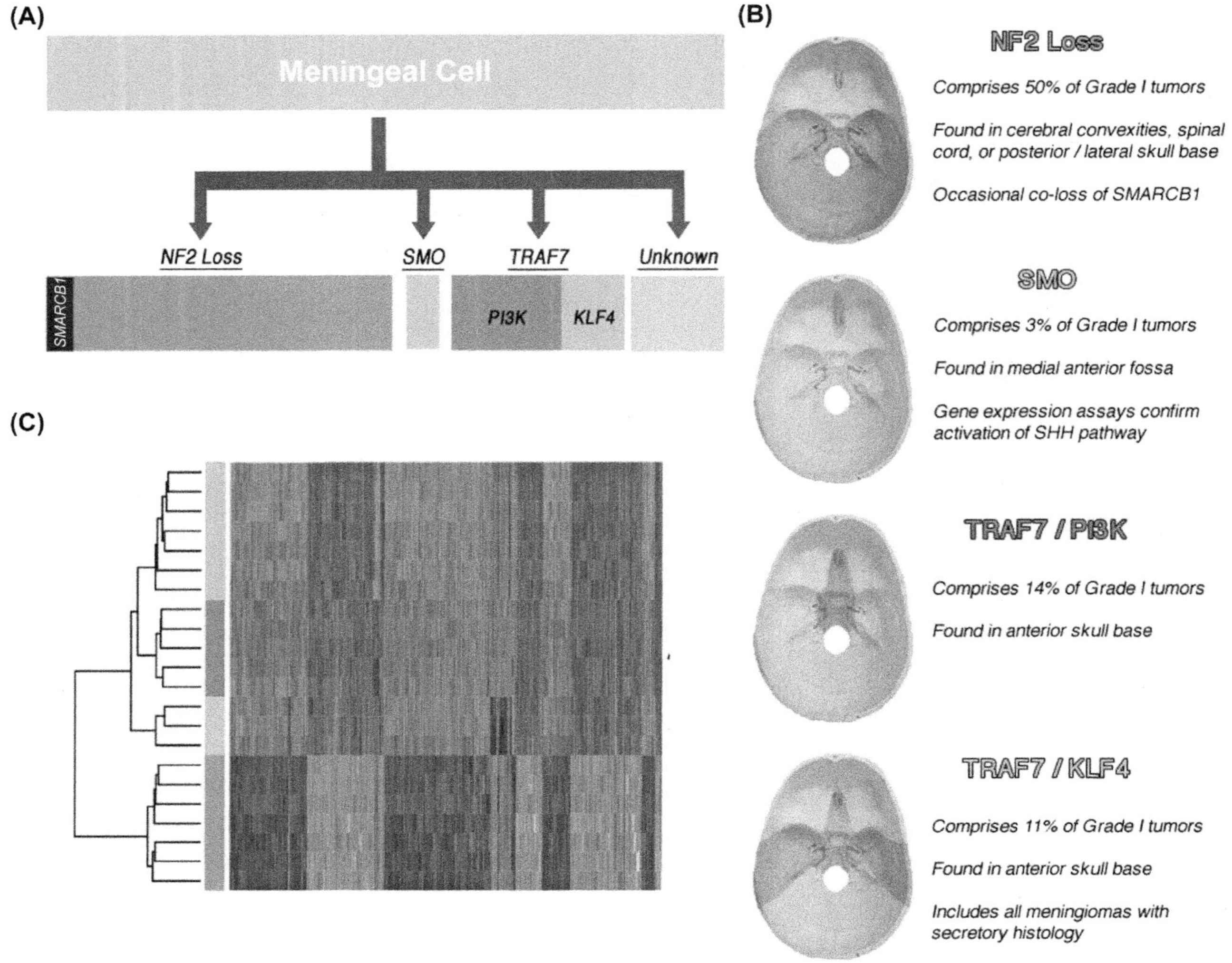

FIGURE 40.1 (A) Meningiomas are thought to arise when meningeal cells acquire alterations in one of three mutually exclusive pathways, with *NF2* loss seen in one-half of all tumors. (B) Each mutational subgroup has location, prognostic, and histological correlates. Non-*NF2* mutant tumors are restricted primarily to the skull base and are enriched among grade I meningiomas. (C) The importance of these driver mutations is illustrated by gene expression studies in which tumors from each group cluster independently. *Images modified with permission from Clark, V.E., et al. (2013). Genomic analysis of non-NF2 meningiomas reveals mutations in TRAF7, KLF4, AKT1, and SMO.* Science, *339, 1077–1080.*

activity mediated by a WD-40 repeat domain, in which the meningioma mutations cluster. Tumors with *TRAF7* mutations have a lower risk of malignant transformation (Clark et al., 2013) and commonly co-occur with activating mutations of the PI3K pathway, most frequently $AKT1^{E17K}$ (reported in 14% of grade I meningiomas) (Brastianos et al., 2013; Clark et al., 2013). A second *TRAF7* subgroup, accounting for 12% of WHO grade I meningiomas, has a co-occurring, recurrent K409Q mutation impacting the DNA-binding domain of the transcription factor *Krupple-like factor 4* (*KLF4*) (Clark et al., 2013; Reuss et al., 2013). KLF4 activity is critical during normal embryonic development and is one of four transcription factors that in combination can induce a pluripotent stem cell state from a fully differentiated cell (Takahashi & Yamanaka, 2006). Based on structural analyses, the recurrent $KLF4^{K409Q}$ mutation is predicted to impact KLF4-DNA interactions directly, and may potentially change consensus motif binding (Clark et al., 2013; Reuss et al., 2013). All meningiomas with "secretory" histology (which bear an increased risk of cerebral edema) (Regelsberger et al., 2009) harbor both $KLF4^{K409Q}$ and *TRAF7* mutations, demonstrating a close correlation between certain histological subtypes and mutational background (Clark et al., 2013; Reuss et al., 2013). In contrast to *NF2/chr22 loss* meningiomas, *TRAF7* mutant tumors grow in the midline of skull base, although they can also be found in the meninges surrounding the frontal lobes (Clark et al., 2013).

The third group of meningiomas (~3% of WHO grade I tumors) harbor somatic mutations in *smoothened, frizzled class receptor* (*SMO*) that activate Sonic Hedgehog (SHH) signaling (Brastianos et al., 2013; Clark et al., 2013). Previous studies suggest that tumors bearing the recurrent SMO^{L412F} mutation (n = 5) all localize to the midline of the anterior skull base (olfactory groove) (Clark et al., 2013), where SHH pathway has been shown to play a critical role in craniofacial patterning during development. Interestingly, inherited loss-of-function mutations in the *SHH inhibitor suppressor of fused homolog* (*Drosophila*) (*SUFU*) have been observed in a family with multiple meningiomas (Aavikko et al., 2012), thus further supporting dysregulation of SHH signaling as the driver event for a distinct subset of meningiomas.

GLIOMAS

Gliomas comprise 26.4% of all primary brain tumors and are classified by the WHO according to the cells they histologically resemble (oligodendrocytes, astrocytes, or a mix). They are grouped into four grades based on both histology and clinical aggressiveness, with malignant or high-grade gliomas (HGGs), defined as WHO grade III or IV and accounting for 19.9% of all primary brain tumors (Louis et al., 2007; Ostrom et al., 2013). HGGs are composed of two subgroups, anaplastic glioma (WHO grade III) and glioblastoma multiforme (GBM) (WHO grade IV), and have a poor prognosis. The median survival for the latter is less than 15–20 months despite multimodal treatment (Grossman et al., 2010). By contrast, the histologically benign glioma pilocytic astrocytomas (PCAs) (WHO grade I) have a favorable prognosis because they can be surgically cured. Low-grade gliomas (LGGs) (WHO grade II) are more infiltrative in nature and are heterogeneous in terms of their median survival and the probability of malignant progression to HGGs.

Adult Low-Grade Gliomas

Grade I gliomas, mainly composed of PCAs, typically occur in the pediatric population and are discussed in detail in the "Pediatric Low-Grade Gliomas" section of this chapter. WHO grade II gliomas can be further divided histologically into tumors that morphologically resemble astrocytes (diffuse astrocytomas), oligodendrocytes (oligodendrogliomas), or tumors with dual characteristics (oligoastrocytomas) (Louis et al., 2007). The triggers for a subset of adult LGGs to advance to HGGs, while others remain benign, are controversial and remain an area of promising research. Despite histologic differences, all subtypes are characterized by a recurrent driver mutation affecting the R132 residue of the *isocitrate dehydrogenase 1* (*IDH1*) gene, which encodes an enzyme responsible for the conversion of isocitrate to alpha-ketoglutarate, an important intermediate of the tricarboxylic acid cycle (Yan et al., 2009). Recurrent $IDH1^{R132}$ mutations (including *R132H/G/S/C/V/L*) have been characterized as conferring a neomorphic ability to produce 2-hydroxyglutarate, an oncometabolite (Dang et al., 2009). Several levels of evidence suggest that $IDH1^{R132}$ mutations cause global epigenetic dysregulation, including driving the DNA hypermethylation of CpG islands [CpG island methylator (G-CIMP) phenotype] observed in progressed HGGs (Noushmehr et al., 2010) and dysregulation of histone methylation (Prensner & Chinnaiyan, 2011; Rohle et al., 2013; Sturm et al., 2012). Although the $IDH1^{R132}$ driver mutation is common to all three adult LGG histological subtypes, the presence of tumor cells resembling oligodendrocytes confers an overall median survival advantage (11.6 versus 5.6 years for grade II oligodendrogliomas versus astrocytomas, respectively) (Ohgaki & Kleihues, 2007) and a markedly decreased risk of malignant transformation (45% for oligodendroglioma versus 74% for pure astrocytoma) (Ostrom et al., 2013). These differences in tumor aggressiveness observed in IDH1 mutant grade II gliomas are most likely due to additional molecular alterations, including *phosphatase and tensin homolog* (*PTEN*) loss or 1p/19q deletion (Sabha et al., 2014).

Oligodendrogliomas are frequently characterized by a loss of heterozygosity (LOH) on chromosomes 1p and 19q (Bello et al., 1994; Ransom et al., 1992; Reifenberger et al., 1994), which is commonly the result of a pericentromeric translocation (Jenkins et al., 2006). As the result of the combination of this translocation and additional somatic mutations, oligodendrogliomas commonly have biallelic loss of *capicua transcriptional repressor* (*CIC*) (located on chromosome 19q) or *far upstream element binding protein 1* (*FUBP1*), located on chromosome 1p (Bettegowda et al., 2011). In contrast, grade II astrocytomas commonly have somatic mutations in the chromatin modifier *alpha thalassemia/mental retardation syndrome X-linked* (*ATRX*) (Jiao et al., 2012) and loss of *tumor protein p53* (*TP53*) by either somatic mutation of *TP53* or LOH at chr17 (Yan et al., 2009). *IDH1-ATRX-TP53* gliomas have a worse prognosis compared with *IDH1-CIC/FUBP1-1p/19q loss* gliomas, with a median survival of 5 years (Jiao et al., 2012) versus 8 years (Jiao et al., 2012), respectively. This survival difference is partially accounted for by the increased risk of *IDH1-ATRX-TP53* to progress to malignancy via the formation of secondary GBMs, which is discussed subsequently. The third subtype of grade II gliomas, mixed oligoastrocytoma, has a combination of genomic features found in pure astrocytomas and oligoastrocytomas and is associated with an intermediate median survival (6.6 years) (Ohgaki & Kleihues, 2007).

Adult High-Grade Gliomas

Gene expression studies in combination with next-generation sequencing have been used to define four molecular subgroups of GBM, including mesenchymal, proneural, neural, and classical (Brennan et al., 2013; Verhaak et al., 2010). Importantly, previous work suggested that a tumor's subgroup may hold prognostic importance. For example, proneural tumors are often refractory to intensive therapy, which shows benefit in other subgroups (Verhaak et al., 2010). These lesions can also be assessed according to their G-CIMP phenotype, which is associated with better survival when present in *IDH1* mutant proneural GBM (Brennan et al., 2013; Parsons et al., 2008; Noushmehr et al., 2010). Furthermore, the methylation state of the *O-6-methylguanine-DNA methyltransferase* promoter can be used to predict response to therapies such as temozolomide, since lesions showing silencing at this locus are not capable of removing alkyl groups deposited by alkylating agents (Brennan et al., 2013; Erickson, Laurent, Sharkey, & Kohn, 1980; Esteller et al., 2000; Hegi et al., 2005; McLendon et al., 2008).

Compared with GBMs, the genomic architecture of anaplastic gliomas (anaplastic astrocytomas, anaplastic oligodendrogliomas, and anaplastic oligoastrocytomas; WHO grade III) is not as well elucidated. This is likely in part because of their rarity; anaplastic astrocytomas, for instance, account for only 1.7% of primary brain tumors (Ostrom et al., 2013). Analogous to other HGGs, these tumors can arise via progression from LGGs (ie, secondary anaplastic glioma) or in the absence of an LGG history (ie, de novo). Anaplastic gliomas have an intrinsic propensity to progress to GBM, although the presence of cells resembling oligodendrocytes on histopathology confers a survival advantage, such that anaplastic astrocytomas have a 5-year overall survival of 26.5% and anaplastic oligodendrogliomas have a 5-year overall survival of 50.7% (Ostrom et al., 2013).

Analogous to WHO grade II gliomas, recurrent $IDH1^{R132}$ mutations have been observed in the vast majority (75–90%) of WHO grade III gliomas (Jiao et al., 2012; Killela et al., 2014; Yan et al., 2009). Consistent with the observation that a subset of anaplastic astrocytomas have progressed from a low-grade lesion, there is significant overlap in mutations between grade II astrocytomas and grade III astrocytomas, both of which commonly have mutations in *ATRX*, *IDH1*, and loss of p53 (Jiao et al., 2012). Anaplastic astrocytomas are also characterized by alterations to the retinoblastoma (Rb) pathway (including *retinoblastoma 1 (RB1)* loss, *cyclin-dependent kinase inhibitor 2A (CDKN2A)* deletion, and *cyclin-dependent kinase 4/6 (CDK4/6)* amplification) (Dunn et al., 2012; Jiao et al., 2012; Killela et al., 2014). It is hypothesized that somatic alterations leading to activation of the receptor tyrosine kinase/Ras/phosphoinositide 3-kinase (RTK/Ras/PI3K) signaling pathway, including LOH at chr10q, drive progression to GBM (Dunn et al., 2012; Furnari et al., 2007; Ohgaki Kleihues, 2007). Anaplastic oligodendrogliomas are even less common (ie, 0.5% of primary brain tumors) (Ostrom et al., 2013) and share the defining mutations of their grade II analogs (chr1p/19q loss and *IDH1-CIC/FUBP1* mutations). The anaplastic progression from grade II oligodendroglioma is likely mediated by somatic loss of *PTEN* and *CDKN2A* (Reifenberger et al., 1994). Recurrent *telomerase reverse transcriptase (TERT)* promoter mutations (C288T or C250T) have been identified in anaplastic gliomas with varying frequency by histological subtype, ranging from 14.8% of anaplastic astrocytomas to 88.4% of anaplastic oligodendrogliomas (Killela et al., 2013). This disparity in *TERT* promoter mutations is likely due to the increased rate of *ATRX* mutations in anaplastic astrocytomas, because *ATRX* mutations and *TERT* promoter mutations are mutually exclusive in other tumors, including GBM (Killela et al., 2013).

Malignant Progression of Adult Low-Grade Gliomas

In general, patients with grade II gliomas, compared with HGGs, reveal a more favorable clinical outcome, including longer overall survival time. However, approximately 70% of grade II gliomas eventually transform into grade III and IV malignancies in 5–10 years (Furnari et al., 2007; Jaeckle et al., 2011; Maher et al., 2001). In fact, about 20% of GBM arises through the progression of a preexisting lower-grade tumor (Furnari et al., 2007; Maher et al., 2001). In addition, once transformed, secondary GBM is histologically and clinically indistinguishable from de novo GBM, with both showing similar morphological features and patient survival time (Furnari et al., 2007; Maher et al., 2001). Finally, there are currently no standard guidelines for the treatment of the progressed tumors (Raizer & Parsa, 2014). Further understanding of the genetics of glioma progression will shed light on disease pathogenesis and could also have important implications for disease treatment.

To elucidate the genetic mechanisms governing glioma malignant progression, a collection of grade II and III gliomas carrying $IDH1^{R132}$ mutations at the initial diagnosis and their progressed counterparts up to 10 years later was studied (Bai et al., 2016). By comparing the genomic landscape of progressed gliomas with that of their corresponding lower-grade counterparts, insights about how protein-altering mutations, chromosomal copy number alterations, gene expression, and DNA methylation changes work together to drive the progression of *IDH1*-mutant gliomas could be gleaned (Bai et al., 2016). Through this analysis, the oncogenic MYC and RTK/Ras/PI3K pathways were noted to become activated during progression, through, for instance, amplification of the *MYC (v-myc avian myelocytomatosis viral oncogene homolog)* gene locus on chromosome 8q and deletion of *PTEN* on 10q (Bai et al., 2016). Convergent genomic alterations also affect many cell cycle regulators, such as deletions of *CDKN2A-CDKN2B* and *RB1*, as well as transcriptional upregulation of the FOXM1-and E2F2-mediated cell cycle transitions (Bai et al., 2016). In addition, key developmental transcription factors become epigenetically silenced during glioma progression, which reprograms tumor cells into a state of perpetual self-renewal, like embryonic stem cells (Bai et al., 2016). Collectively, during progression, genomic alterations converge upon stimulating cancer cell proliferation while inhibiting its differentiation.

In addition to dissecting the underlying genomic drivers, several studies have delineated tumor cell heterogeneity and clonal dynamics during glioma progression (Bai et al., 2016; Johnson et al., 2014; Mazor et al., 2015). For example, for many patients, mutations important for the formation of the initial tumor were no longer present in the progressed tumor; instead, the progressed tumor harbored a different set of mutations (Bai et al., 2016; Johnson et al., 2014). This suggests that the progressed tumor evolved independently from clones of glioma cells different from the ones constituting the initial tumor. Interestingly, the recurrent $IDH1^{R132}$ mutations were also retained during glioma progression, which suggests that they are the founding event of gliomagenesis (Bai et al., 2016; Johnson et al., 2014) (Fig. 40.2).

Pediatric Low-Grade Gliomas

Pediatric LGGs (WHO grades I and II) are the most common brain tumors in children (Ostrom et al., 2013) and include nondiffuse PCAs and gangliogliomas (both WHO grade I), nondiffuse WHO grade II tumors [ie, pleomorphic xanthoastrocytoma (PXA)], and diffuse WHO grade II gliomas (ie, diffuse astrocytomas, oligodendrogliomas, and oligoastrocytomas). PCAs are often cystic (Louis et al., 2007), well-circumscribed, and histologically benign tumors with a low propensity for malignant progression. They present most commonly in the pediatric population; the most frequent

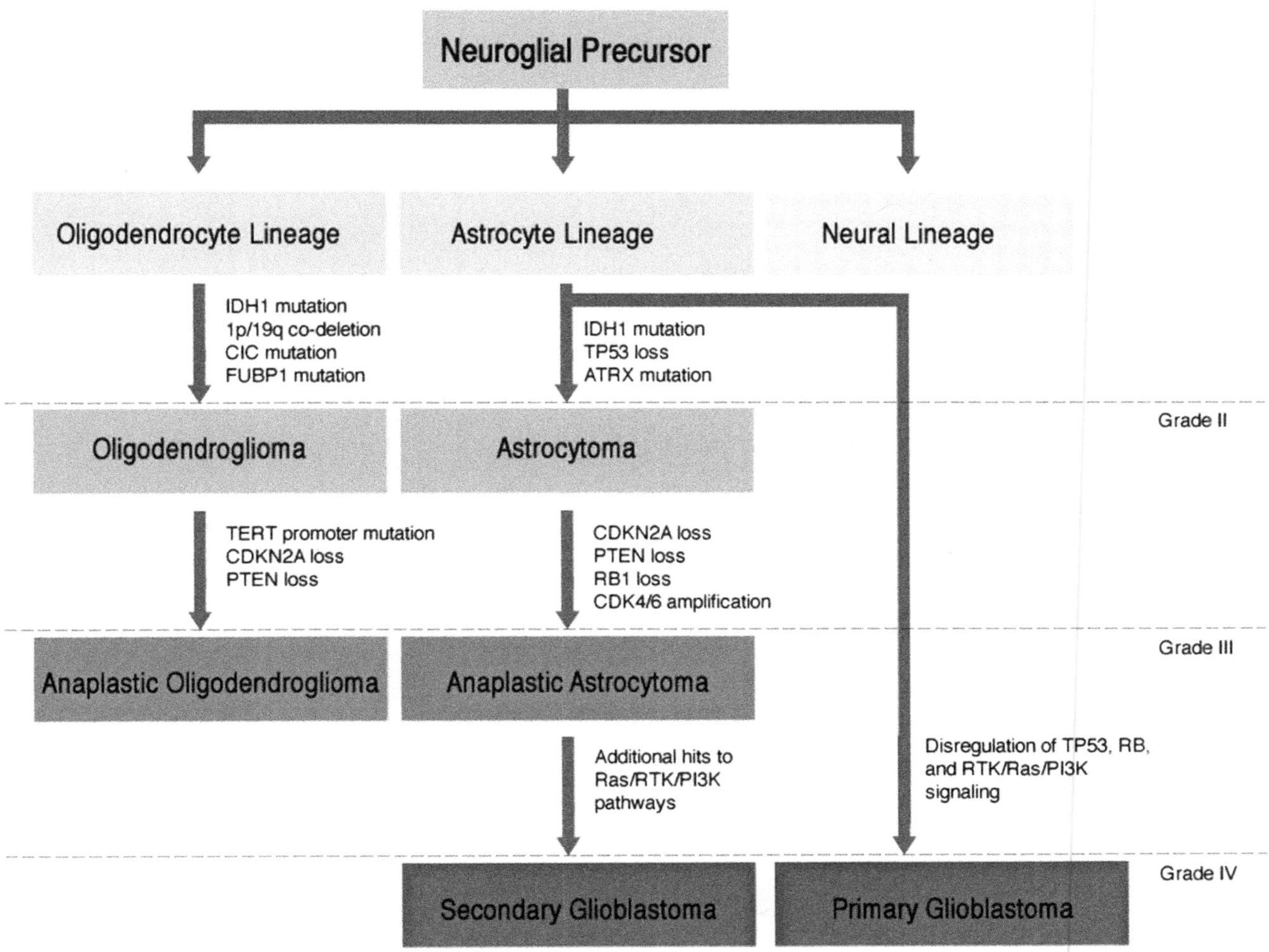

FIGURE 40.2 Adult gliomagenesis can proceed in a stepwise fashion from low- to high-grade lesions, or alternatively transform directly into glioblastoma (de novo pathway). *IDH1* mutation is observed in 75–90% of grade III gliomas and is a positive prognostic indicator in glioblastoma.

location is in the cerebellar hemispheres (67%) (Ohgaki & Kleihues, 2007). Adult PCAs tend to occur in other locations within the cerebral hemispheres. They can be cured with surgical excision.

Constitutive activation of the mitogen-activated protein kinase/extracellular signal-regulated kinase (MAPK/ERK) pathway is the hallmark of PCAs. Studies of the hereditary tumor syndrome neurofibromatosis type I first revealed the potential for alterations in MAPK/ERK signaling to cause PCA formation. In this syndrome, approximately 15% of patients have germline, loss-of-function mutations in the tumor suppressor *NF1*, a negative regulator of Ras signaling, and these patients frequently have PCAs in addition to cutaneous neurofibromas and café-au-lait spots (Listernick, Charrow, & Gutmann, 1999). In addition to inherited syndromes, activation of the MAPK/ERK pathway is crucial to sporadic PCA formation. Ninety percent of cerebellar PCAs have constitutive activation of *B-Raf proto-oncogene, serine/threonine kinase* (*BRAF*) signaling via a gene fusion event (*KIAA1549-BRAF*) that leads to truncation of the BRAF autoinhibitory domain (Forshew et al., 2009; Jones et al., 2008, 2013; Zhang et al., 2013). PCAs can also have constitutive activation of MAPK/ERK through somatic mutations [affecting $BRAF^{V600E}$, *Kirsten rat sarcoma viral oncogene homolog* (*KRAS*), or *NF1*] or additional fusions (*Raf-1 protooncogene, serine/threonine kinase* fusions) (Jones et al., 2013; Zhang et al., 2013). The approximately 20% of noncerebellar PCAs that lack the *KIAA1549-BRAF* fusion activate MAPK signaling through alternative molecular mechanisms, including alterations to *fibroblast growth factor receptor 1* (*FGFR1*) (Jones et al., 2013; Zhang et al., 2013), *protein tyrosine phosphatase, nonreceptor type 11* hotspot mutations in tumors co-mutated for *FGFR1* (Jones et al., 2013), or *neurotrophic tyrosine kinase, receptor, type 2 fusions* (Jones et al., 2013).

Other pediatric LGGs, including diffuse astrocytomas, oligodendrogliomas, oligoastrocytomas, angiocentric gliomas, and PXAs (Ramkissoon et al., 2013; Zhang et al., 2013), differ from PCAs in their more diffuse growth pattern, location of growth (generally supratentorial), and/or capacity for malignant progression (Louis et al., 2007). As demonstrated by Zhang et al. (2013), 52% (12 of 23) of diffuse astrocytomas (WHO grade II) had ERK/MAPK signaling activation (via *FGFR1/3* alterations, *BRAF* alterations, or $KRAS^{Q61H}$). This same series identified common, recurrent *FGFR1* alterations in oligodendrogliomas and oligoastrocytomas, including frequent *FGFR1* tyrosine kinase domain duplications. A minority of ERK/MAPK-activated diffuse astrocytomas (2 of 12; 16.67%) had a co-occurring *H3 histone, family 3A* (*H3F3A*) *K27M* mutation, the consequences of which are discussed in the pediatric GBM section of this chapter.

Most of other pediatric WHO grade II gliomas, including PXAs and gangliogliomas, also demonstrate MAPK/ERK activation. PXAs are uncommon supratentorial astrocytomas with a 5-year overall survival of 81% that are found in the pediatric population two-thirds of the time (Giannini & Scheithauer, 1997; Giannini et al., 1999). Owing to the rarity of these tumors, study sample sizes are small; however, a genetic analysis of PXAs and gangliogliomas showed recurrent $BRAF^{V600E}$ mutations in 7 of 10 PXA samples and five of nine gangliogliomas to have *BRAF* alterations (*V600E* in 3; *BRAF* fusion in 2) (Zhang et al., 2013).

In contrast to the well-circumscribed PCAs, approximately one-quarter of pediatric diffuse astrocytomas have disruption of *v-myb avian myeloblastosis viral oncogene homolog* (*MYB*) or *MYB-like 1* (*MYBL1*), including episome formation, gene fusion, or *MYBL1* rearrangement (Zhang et al., 2013). Analysis of structural variation reveals focal amplification causing *MYBL1* duplication/truncation in 28% (5 of 18) of pediatric diffuse astrocytomas (Ramkissoon et al., 2013). These *MYB/MYBL1* alterations were restricted to diffuse pediatric gliomas and were not observed in PCAs, although angiocentric gliomas bore *MYB* or *MYBL1* fusions (Ramkissoon et al., 2013; Zhang et al., 2013).

Pediatric High-Grade Gliomas

Whereas pediatric and adult GBMs share histopathological features, a subgroup of pediatric GBMs is genomically distinct from its adult correlates, with mutations impacting genes critical for normal epigenetic regulation. In approximately one-third of pediatric GBMs, recurrent mutations in *histone variant H3.3* ($H3F3A^{K27M;G34R/V}$), commonly co-mutated with loss-of-function mutations in chromatin remodelers (*ATRX* or *death-domain associated protein*) and *TP53*, drove gliomagenesis (Schwartzentruber et al., 2012). $H3K27^{K27M}$ is a dominant-negative mutation that alters binding with the *polycomb repressive complex 2*, ultimately resulting in a global decrease in H3K27me3 chromatin markings and genome-wide DNA hypomethylation (Bender et al., 2013). Unlike in adult GBMs, the neomorphic $IDH1^{R132H}$ mutation is rare in pediatric GBMs and is mutually exclusive with *H3.3* mutations (Schwartzentruber et al., 2012; Sturm et al., 2012). In addition to histone mutations, approximately 10% of pediatric HGGs bear the recurrent $BRAF^{V600E}$ mutation (Nicolaides et al., 2011) with common co-deletion of *CDKN2A/B* (Schiffman et al., 2010).

The proportion and distribution of pediatric HGG driver mutations vary by anatomic location. In the brain stem, approximately 80% of diffuse intrinsic pontine gliomas contained somatic *K27M* mutations in *H3F3A* or *histone cluster 1, H3b*, whereas non–brain stem pediatric GBMs contained the somatic $H3F3A^{G34R}$ mutation (Giannini & Scheithauer, 1997). In the cerebral cortex, *histone H3.3 Gly34* and $BRAF^{V600E}$ mutations frequently occurred, whereas *histone H3*

K27M mutations were found throughout the midline, with the highest frequency at the pons (Schwartzentruber et al., 2012; Sturm et al., 2012; Wu et al., 2012). On the other hand, *activin receptor type 1* mutations arise exclusively at the pons, and *FGFR1* mutations occur in the thalamus (Fontebasso et al., 2014).

In terms of genomic architecture, pediatric GBMs differ from primary adult GBMs, which typically harbor chromosome 7 amplification (74% versus 13% of adult versus pediatric GBMs) and chromosome 10 loss (80% versus 35% of adult versus pediatric GBMs) (Paugh et al., 2010). Instead, analysis of pediatric GBM structural variation reveals frequent chromosome 1q gain (30% versus 9% in pediatric versus adult GBMs), focal, homozygous loss of *CDKN2A/B* (19%), and focal amplification of *PDGFRA* (12%) (Paugh et al., 2010).

MEDULLOBLASTOMAS

Often arising in midline of the cerebellum, medulloblastomas (WHO grade IV) are the most frequent malignant brain tumor in children. These invasive, embryonic tumors commonly exhibit leptomeningeal spread via the cerebrospinal fluid, gaining access through their intimate relationship with the fourth ventricle (Louis et al., 2007). The past decade of research has identified four medulloblastoma subgroups with distinct gene expression profiles, structural variation, driver mutations, cells of origin, location of growth, prognosis, metastatic potential, and therapeutic response.

WNT signaling was first implicated in the formation of medulloblastomas by studies of Turcot syndrome (intestinal polyposis and central nervous system tumors, including medulloblastomas), which identified germline loss-of-function mutations in *adenomatous polyposis coli* (Hamilton et al., 1995). Candidate gene studies identified WNT pathway mutations in a minority of sporadic medulloblastomas (Zurawel, Chiappa, Allen, & Raffel, 1998), thus implying a role for WNT activation in a subset of nonsyndromic medulloblastomas.

With the advent of genome-wide expression profiling and next-generation sequencing, the WNT group of sporadic medulloblastomas (about 10% of tumors) was identified through unsupervised hierarchical clustering of gene expression profiles (Thompson et al., 2006). Ninety-one percent of patients in the WNT group (as defined by gene expression profile) were later found to bear somatic activating mutations in *beta-catenin*, with common co-mutations in the *RNA helicase Asp-Glu-Ala-Asp box helicase 3, X-linked* (*DDX3X*) (50% of patients), two chromatin modifiers [*SWI/SNF-related, matrix-associated, actin dependent regulator of chromatin, subfamily a, member 4* (*SMARCA4*), 26.3%; and *lysine (K)-specific methyltransferase 2D* (*MLL2*), 12.5%], and *TP53* (12.5%) (Jones et al., 2012; Northcott et al., 2012a; Pugh et al., 2012; Robinson et al., 2012). In terms of genomic architecture, this group of tumors displays loss of one copy of chromosome 6 but is otherwise genomically stable (Northcott et al., 2012b). WNT subtype medulloblastomas occur within the fourth ventricle and are prone to brain stem invasion, and animal tumor models have implicated lower rhombic lip progenitor cells as the cells of origin (Fig. 40.3) (Gibson et al., 2010). Clinically, the WNT group has an excellent prognosis, with a 95% overall 5-year survival and low rate of recurrence (Ellison et al., 2005; Grill & Dufour, 2014).

Analogous to WNT signaling, SHH signaling was originally discovered to drive a subset of medulloblastomas through the genetic study of Gorlin syndrome (basal cell carcinomas, keratocystic odontogenic tumors, and occasional medulloblastomas), a distinct hereditary tumor syndrome with SHH pathway activation due to germline loss of *patched 1* (*PTCH1*) (Hahn et al., 1996; Johnson et al., 1996). Candidate gene studies identified frequent *PTCH1* loss in somatic medulloblastomas (Raffel et al., 1997). Larger, transcriptional profiling and next-generation sequencing efforts confirmed a distinct SHH group of medulloblastomas (about 30%), defined by activated SHH signaling (Kool et al., 2012) due to somatic mutations (*PTCH1*, *SUFU*, and *SMO*) (Jones et al., 2012; Pugh et al., 2012; Robinson et al., 2012; Thompson et al., 2006) or somatic copy number alterations (SCNAs) (Northcott et al., 2012b) affecting SHH pathway members. Within the SHH group, a subset bears *TP53* mutations or deletions (13.6%) (Jones et al., 2012; Northcott et al., 2012a, 2012b; Pugh et al., 2012), which is consistent with the development of SHH medulloblastomas in Li–Fraumeni syndrome (characterized by germline *TP53* mutations) (Rausch et al., 2012). In addition to *TP53* mutations, the SHH group has additional mutational overlap with the WNT group, including mutations in *MLL2* (12.9%) and *DDX3X* (11.7%) (Jones et al., 2012; Northcott et al., 2012a; Pugh et al., 2012; Robinson et al., 2012). In contrast to the genomic stability of the WNT group, SHH tumors are characterized by focal copy number alterations impacting SHH (18%), RTK/PI3K (10%), and p53 (9.4%) signaling (Northcott et al., 2012b). SHH medulloblastomas are found in the cerebellar hemispheres, and the external granule layer progenitor cells have been proposed as the SHH cell of origin (Fig. 40.3) (Gibson et al., 2010). SHH tumors have a worse 5-year survival than do WNT tumors (75% versus 95%, respectively) (Kool et al., 2012; Northcott et al., 2012a); however, prognosis can be stratified within the SHH group by the presence of a *TP53* mutation (41% 5-year survival in *TP53* mutant versus 81% in *TP53* wild-type SHH tumors) (Jones et al., 2012; Northcott et al., 2012a, 2012b; Pugh et al., 2012; Zhukova et al., 2013). In addition, SHH tumors are more likely to have local but not

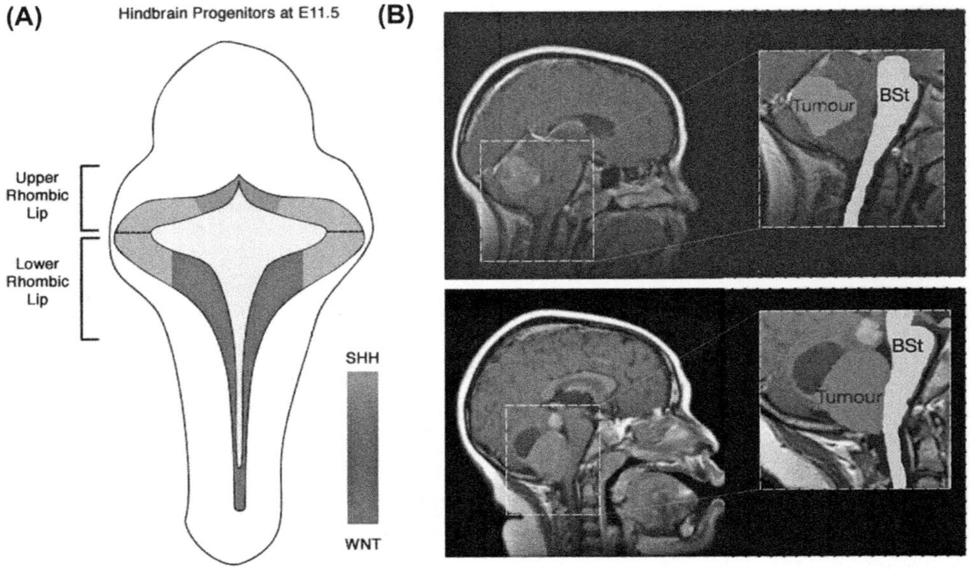

FIGURE 40.3 Gene expression data from mouse and human suggest that medulloblastoma subgroups have distinct anatomical locations and progenitor cells. (A) Schematic of hindbrain precursors during normal mouse development [mouse embryonic day (E)11.5], with cells displaying the human SHH medulloblastoma gene expression signature [blue (dark gray in print versions)] and the human WNT medulloblastoma gene expression signature [red (light gray in print versions)]. The SHH subgroup expression signature is demonstrated in the upper rhombic lip in cerebellar granular neuron precursors, whereas the WNT subgroup expression signature is observed in the lower rhombic lip progenitors. (B) The embryonic cell of origin has implications for numerous tumor characteristics, including location. SHH subtype medulloblastomas most commonly present within the cerebellum [top, shaded blue (gray in print versions)], whereas WNT medulloblastomas grow in the fourth ventricle and infiltrate the dorsal brain stem (BSt) [bottom, shaded red (light gray in print versions)]. *Images modified with permission from Gibson, P., et al. (2010). Subtypes of medulloblastoma have distinct developmental origins.* Nature, 468, *1095–1099.*

metastatic recurrence (Ramaswamy et al., 2013). Treatment of patients with SHH medulloblastomas with targeted SHH inhibitors (eg, SMO inhibitors) has not produced durable remission (Rudin et al., 2009; Yauch et al., 2009).

The pathway(s) responsible for the remaining two medulloblastoma subgroups, currently called groups 3 and 4, is an area of active investigation. These two groups, identified based on their expression profile, have a paucity of recurrent driver mutations but frequent structural abnormalities and distinct 5-year survival rates (50% for group 3; 75% for group 4) (Kool et al., 2012). Both groups are more likely to present with metastasis compared with SHH or WNT subtypes (Kool et al., 2012). "Enhancer hijacking" has been proposed as a possible oncogenic mechanism for these groups, mediated by structural variation that ties the *growth factor independent 1 family proto-oncogenes* (*GFI1* and *GFI1B*) to active enhancer elements, including superenhancers (Northcott et al., 2014). Evidence from mouse models supports the oncogenic activity of these alterations.

Group 3 tumors (about 25% of medulloblastomas) express a gene signature similar to photoreceptors/gamma-aminobutyric acid-ergic neurons, whereas group 4 tumors (about 35%) express a neuronal gene signature (Cho et al., 2011; Taylor et al., 2012). A SCNA analysis of more than 1000 medulloblastomas revealed that SCNAs found in group 3 are enriched for *MYC* amplification (including *MYC/PVT1* gene fusions) or *transforming growth factor b* signaling members, including frequent amplification of *orthodenticle homeobox 2* (Northcott et al., 2012b). In contrast, group 4 tumors have *MYCN* amplification (6.3%; also seen in about 8.2% of SHH tumors) but lack *MYC* amplification (Northcott et al., 2012a, 2012b). In terms of somatic mutations, about 12% of group 4 tumors bore damaging mutations in *KDM6A*, a histone H3K27 demethylase, and 10.4% had a tandem duplication of *synuclein, alpha interacting protein*, which has been implicated in the formation of Lewy bodies in Parkinson disease (Jones et al., 2012; Northcott et al., 2012a, 2012b; Pugh et al., 2012; Robinson et al., 2012).

EPENDYMAL TUMORS

Ependymomas are glial tumors that arise from ependymal cells covering the ventricular system, where cerebrospinal fluid is produced and circulated. The site of ependymoma growth along the ventricular system predicts a number of features, including age of presentation, survival, genomic alterations, and transcriptional profile. Similar to spinal meningiomas,

NF2/chr22 loss is associated with spinal ependymomas: Patients with neurofibromatosis type II commonly have intramedullary spinal ependymomas, and up to 95% of adult sporadic spinal cord ependymomas have LOH at chr22 (Ebert et al., 1999). In contrast, supratentorial ependymomas have intact *NF2* but are characterized by deletion of *CDKN2A* (>90% of tumors) and co-occurring amplification of *EPH receptor B2* (*Ephb2*). Posterior fossa ependymomas, which in children bear the worst prognosis, can be further classified by laterality (Witt et al., 2011). Midline posterior fossa ependymomas have widespread chromosomal instability including chr22 loss and present at an older age (Witt et al., 2011). In contrast, lateral posterior fossa tumors have a poor prognosis despite having greater chromosomal stability than do midline posterior fossa ependymomas. These tumors occur in younger patients, are more likely to metastasize, and harbor chromosome 1q gain (Witt et al., 2011). The radial glial cell has been proposed as the ependymoma cancer stem cell (Taylor et al., 2005), and transcriptional profiling of radial glial cells and ependymoma cells harvested from specific anatomic regions (supratentorial versus infratentorial versus spinal cord) shows matching, region-specific patterns of gene expression (Johnson et al., 2010).

SUMMARY

The development and proliferation of next-generation sequencing technology has rapidly increased our understanding of the molecular landscape of most intracranial tumors. Large-scale cohorts have led to the discovery of key molecular driver events, which often segregate into recurrent pathways that correlate with prognostic and other clinical features. In many cases, initial genomic studies have been validated with gene expression assays, animal models, and other functional studies that have elucidated the oncogenic mechanisms behind identified driver mutations. Further characterization of the cell-specific requirements and sequence of events underlying transformation will be an important focus of future work. However, clinicians and scientists have already begun leveraging the insights gained toward the targeted treatment of central nervous system tumors.

In a handful of large medical centers, multidisciplinary tumor boards have begun to meet regularly to review imaging, histology, and clinical findings for patients with intracranial tumors. The wealth of reported genomic findings and availability of precise pharmacological inhibitors has led to routine molecular profiling of patients who may benefit from targeted therapies. Genomic alterations observed in a patient's tumor are now routinely considered along with radiology and histology findings to direct treatment. In some cases, this information can provide important clues about the likely course of a patient's disease. Although initial results are promising, acquisition of increasingly large data sets will continue to pose challenges to scientists and clinicians attempting to draw actionable conclusions. However, as new insights and data analysis techniques are developed, patients will inevitably benefit from integration of traditional approaches with targeted therapy.

REFERENCES

Aavikko, M., et al. (2012). Loss of SUFU function in familial multiple meningioma. *The American Journal of Human Genetics, 91*, 520–526.

Bai, H., Harmancı, A. S., Erson-Omay, E. Z., Li, J., Coşkun, S., Simon, M., … Gunel, M. (2016). Integrated genomic characterization of IDH1-mutant glioma malignant progression. *Nature Genetics, 48*, 59–66.

Bello, M. J., et al. (1994). Molecular analysis of chromosome 1 abnormalities in human gliomas reveals frequent loss of 1p in oligodendroglial tumors. *International Journal of Cancer, 57*, 172–175.

Bender, S., et al. (2013). Reduced H3K27me3 and DNA hypomethylation are major drivers of gene expression in K27M mutant pediatric high-grade gliomas. *Cancer Cell, 24*, 660–672.

Bettegowda, C., et al. (2011). Mutations in CIC and FUBP1 contribute to human oligodendroglioma. *Science, 333*, 1453–1455.

Biegel, J. A., et al. (1999). Germ-line and acquired mutations of INI1 in atypical teratoid and rhabdoid tumors. *Cancer Research, 59*, 74–79.

Brastianos, P. K., et al. (2013). Genomic sequencing of meningiomas identifies oncogenic SMO and AKT1 mutations. *Nature Genetics, 45*, 285–289.

Brennan, C. W., et al. (2013). The somatic genomic landscape of glioblastoma. *Cell, 155*, 462–477. http://dx.doi.org/10.1016/j.cell.2013.09.034.

Cho, Y. J., et al. (2011). Integrative genomic analysis of medulloblastoma identifies a molecular subgroup that drives poor clinical outcome. *Journal of Clinical Oncology: Official Journal of the American Society of Clinical Oncology, 29*, 1424–1430. http://dx.doi.org/10.1200/jco.2010.28.5148.

Clark, V. E., et al. (2013). Genomic analysis of non-NF2 meningiomas reveals mutations in TRAF7, KLF4, AKT1, and SMO. *Science, 339*, 1077–1080.

Dang, L., et al. (2009). Cancer-associated IDH1 mutations produce 2-hydroxyglutarate. *Nature, 462*, 739–744.

Dunn, G. P., et al. (2012). Emerging insights into the molecular and cellular basis of glioblastoma. *Genes & Development, 26*, 756–784.

Ebert, C., et al. (1999). Molecular genetic analysis of ependymal tumors. NF2 mutations and chromosome 22q loss occur preferentially in intramedullary spinal ependymomas. *American Journal of Pathology, 155*, 627–632. http://dx.doi.org/10.1016/s0002-9440(10)65158-9.

Ellison, D. W., et al. (2005). b-Catenin status predicts a favorable outcome in childhood medulloblastoma: the United Kingdom Children's Cancer Study Group Brain Tumour Committee. *Journal of Clinical Oncology, 23*, 7951–7957.

Erickson, L. C., Laurent, G., Sharkey, N. A., & Kohn, K. W. (1980). DNA cross-linking and monoadduct repair in nitrosourea-treated human tumour cells. *Nature, 288*(5792), 727–729.

Esteller, M., et al. (2000). Inactivation of the DNA-repair gene MGMT and the clinical response of gliomas to alkylating agents. *New England Journal of Medicine, 343*, 1350–1354.

Fontebasso, A. M., et al. (2014). Recurrent somatic mutations in ACVR1 in pediatric midline high-grade astrocytoma. *Nature Genetics, 46*, 462–466. http://dx.doi.org/10.1038/ng.2950.

Forshew, T., et al. (2009). Activation of the ERK/MAPK pathway: a signature genetic defect in posterior fossa pilocytic astrocytomas. *The Journal of Pathology, 218*, 172–181.

Furnari, F. B., Fenton, T., Bachoo, R. M., Mukasa, A., Stommel, J. M., Stegh, A., … Cavenee, W. K. (2007). Malignant astrocytic glioma: genetics, biology, and paths to treatment. *Genes & Development, 21*, 2683–2710.

Giannini, C., & Scheithauer, B. W. (1997). Classification and grading of low-grade astrocytic tumors in children. *Brain Pathology, 7*, 785–798.

Giannini, C., et al. (1999). Pleomorphic xanthoastrocytoma. *Cancer, 85*, 2033–2045.

Gibson, P., et al. (2010). Subtypes of medulloblastoma have distinct developmental origins. *Nature, 468*, 1095–1099.

Grill, J., & Dufour, C. (2014). Neuro-oncology: stability of medulloblastoma subgroups at tumour recurrence. *Nature Reviews Neurology, 10*, 5–6. http://dx.doi.org/10.1038/nrneurol.2013.256.

Grossman, S. A., et al. (2010). Survival of patients with newly diagnosed glioblastoma treated with radiation and temozolomide in research studies in the United States. *Clinical Cancer Research, 16*, 2443–2449.

Hahn, H., et al. (1996). Mutations of the human homolog of *Drosophila* patched in the nevoid basal cell carcinoma syndrome. *Cell, 85*, 841–851.

Hamilton, S. R., et al. (1995). The molecular basis of Turcot's syndrome. *New England Journal of Medicine, 332*, 839–847.

Hegi, M. E., et al. (2005). MGMT gene silencing and benefit from temozolomide in glioblastoma. *New England Journal of Medicine, 352*, 997–1003.

Jaeckle, K. A., Decker, P. A., Ballman, K. V., Flynn, P. J., Giannini, C., Scheithauer, B. W., … Buckner, J. C. (2011). Transformation of low grade glioma and correlation with outcome: an NCCTG database analysis. *Journal of Neuro-Oncology, 104*, 253–259.

Jenkins, R. B., et al. (2006). A t(1;19)(q10;p10) mediates the combined deletions of 1p and 19q and predicts a better prognosis of patients with oligodendroglioma. *Cancer Research, 66*, 9852–9861.

Jiao, Y., et al. (2012). Frequent ATRX, CIC, FUBP1 and IDH1 mutations refine the classification of malignant gliomas. *Oncotarget, 3*, 709.

Johnson, B. E., Mazor, T., Hong, C., Barnes, M., Aihara, K., McLean, C. Y., … Costello, J. F. (2014). Mutational analysis reveals the origin and therapy-driven evolution of recurrent glioma. *Science, 343*, 189–193.

Johnson, R. A., et al. (2010). Cross-species genomics matches driver mutations and cell compartments to model ependymoma. *Nature, 466*, 632–636. http://dx.doi.org/10.1038/nature09173.

Johnson, R. L., et al. (1996). Human homolog of patched, a candidate gene for the basal cell nevus syndrome. *Science, 272*, 1668–1671.

Jones, D. T., et al. (2008). Tandem duplication producing a novel oncogenic BRAF fusion gene defines the majority of pilocytic astrocytomas. *Cancer Research, 68*, 8673–8677.

Jones, D. T., et al. (2012). Dissecting the genomic complexity underlying medulloblastoma. *Nature, 488*, 100–105.

Jones, D. T., et al. (2013). Recurrent somatic alterations of FGFR1 and NTRK2 in pilocytic astrocytoma. *Nature Genetics, 45*, 927–932.

Killela, P. J., et al. (2013). TERT promoter mutations occur frequently in gliomas and a subset of tumors derived from cells with low rates of self-renewal. *Proceedings of the National Academy of Sciences, 110*, 6021–6026.

Killela, P. J., et al. (2014). The genetic landscape of anaplastic astrocytoma. *Oncotarget, 5*, 1452–1457.

Kool, M., et al. (2012). Molecular subgroups of medulloblastoma: an international meta-analysis of transcriptome, genetic aberrations, and clinical data of WNT, SHH, group 3, and group 4 medulloblastomas. *Acta Neuropathologica, 123*, 473–484. http://dx.doi.org/10.1007/s00401-012-0958-8.

Listernick, R., Charrow, J., & Gutmann, D. H. (1999). Intracranial gliomas in neurofibromatosis type 1. *American Journal of Medical Genetics, 89*, 38–44.

Louis, D. N., et al. (2007). The 2007 WHO classification of tumours of the central nervous system. *Acta Neuropathologica, 114*, 97–109.

Maher, E. A., Furnari, F. B., Bachoo, R. M., Rowitch, D. H., Louis, D. N., Cavenee, W. K., & dePinho, R. A. (2001). Malignant glioma: genetics and biology of a grave matter. *Genes & Development, 15*, 1311–1333.

Mazor, T., Pankov, A., Johnson, B. E., Hong, C., Hamilton, E. G., Bell, R. J. A., … Costello, J. F. (2015). DNA methylation and somatic mutations converge on the cell cycle and define similar evolutionary histories in brain tumors. *Cancer Cell, 28*, 307–317.

McLendon, R., et al. (2008). Comprehensive genomic characterization defines human glioblastoma genes and core pathways. *Nature, 455*, 1061–1068.

van den Munckhof, P., Christiaans, I., Kenter, S. B., Baas, F., & Hulsebos, T. J. (2012). Germline SMARCB1 mutation predisposes to multiple meningiomas and schwannomas with preferential location of cranial meningiomas at the falx cerebri. *Neurogenetics, 13*, 1–7.

Nicolaides, T. P., et al. (2011). Targeted therapy for BRAFV600E malignant astrocytoma. *Clinical Cancer Research, 17*, 7595–7604.

Northcott, P. A., et al. (2012a). Medulloblastomics: the end of the beginning. *Nature Reviews Cancer, 12*, 818–834.

Northcott, P. A., et al. (2012b). Subgroup-specific structural variation across 1000 medulloblastoma genomes. *Nature, 488*, 49–56.

Northcott, P. A., et al. (2014). Enhancer hijacking activates GFI1 family oncogenes in medulloblastoma. *Nature, 511*, 428–434. http://dx.doi.org/10.1038/nature13379.

Noushmehr, H., et al. (2010). Identification of a CpG island methylator phenotype that defines a distinct subgroup of glioma. *Cancer Cell, 17*, 510–522.

Ohgaki, H., & Kleihues, P. (2007). Genetic pathways to primary and secondary glioblastoma. *The American Journal of Pathology, 170*, 1445–1453.

Ostrom, Q. T., et al. (2013). CBTRUS statistical report: primary brain and central nervous system tumors diagnosed in the United States in 2006–2010. *Neuro-Oncology, 15*, ii1–ii56.

Parsons, D. W., et al. (2008). An integrated genomic analysis of human glioblastoma multiforme. *Science, 321*, 1807–1812.

Paugh, B. S., et al. (2010). Integrated molecular genetic profiling of pediatric high-grade gliomas reveals key differences with the adult disease. *Journal of Clinical Oncology, 28*, 3061–3068.

Prensner, J. R., & Chinnaiyan, A. M. (2011). Metabolism unhinged: IDH mutations in cancer. *Nature Medicine, 17*, 291–293.

Pugh, T. J., et al. (2012). Medulloblastoma exome sequencing uncovers subtype-specific somatic mutations. *Nature, 488*, 106–110.

Raffel, C., et al. (1997). Sporadic medulloblastomas contain PTCH mutations. *Cancer Research, 57*, 842–845.

Raizer, J., & Parsa, A. (2014). *Current understanding and treatment of gliomas*. Springer.

Ramaswamy, V., et al. (2013). Recurrence patterns across medulloblastoma subgroups: an integrated clinical and molecular analysis. *The Lancet Oncology, 14*, 1200–1207. http://dx.doi.org/10.1016/s1470-2045(13)70449-2.

Ramkissoon, L. A., et al. (2013). Genomic analysis of diffuse pediatric low-grade gliomas identifies recurrent oncogenic truncating rearrangements in the transcription factor MYBL1. *Proceedings of the National Academy of Sciences of the United States of America, 110*, 8188–8193.

Ransom, D. T., et al. (1992). Cytogenetic and loss of heterozygosity studies in ependymomas, pilocytic astrocytomas, and oligodendrogliomas. *Genes, Chromosomes & Cancer, 5*, 348–356.

Rausch, T., et al. (2012). Genome sequencing of pediatric medulloblastoma links catastrophic DNA rearrangements with TP53 mutations. *Cell, 148*, 59–71. http://dx.doi.org/10.1016/j.cell.2011.12.013.

Regelsberger, J., et al. (2009). Secretory meningiomas: a benign subgroup causing life-threatening complications. *Neuro-Oncology, 11*, 819–824.

Reifenberger, J., et al. (1994). Molecular genetic analysis of oligodendroglial tumors shows preferential allelic deletions on 19q and 1p. *The American Journal of Pathology, 145*, 1175.

Reuss, D. E., et al. (2013). Secretory meningiomas are defined by combined KLF4 K409Q and TRAF7 mutations. *Acta Neuropathologica, 125*, 351–358.

Robinson, G., et al. (2012). Novel mutations target distinct subgroups of medulloblastoma. *Nature, 488*, 43–48.

Rohle, D., et al. (2013). An inhibitor of mutant IDH1 delays growth and promotes differentiation of glioma cells. *Science, 340*, 626–630.

Rouleau, G. A., et al. (1993). Alteration in a new gene encoding a putative membrane-organizing protein causes neuro-fibromatosis type 2. *Nature, 363*(6429).

Rudin, C. M., et al. (2009). Treatment of medulloblastoma with hedgehog pathway inhibitor GDC-0449. *The New England Journal of Medicine, 361*, 1173–1178. http://dx.doi.org/10.1056/NEJMoa0902903.

Ruttledge, M. H., et al. (1994). Evidence for the complete inactivation of the NF2 gene in the majority of sporadic meningiomas. *Nature Genetics, 6*, 180–184.

Sabha, N., et al. (2014). Analysis of IDH mutation, 1p/19q deletion, and PTEN loss delineates prognosis in clinical low-grade diffuse gliomas. *Neuro-Oncology, 16*, 914–923. http://dx.doi.org/10.1093/neuonc/not299.

Schiffman, J. D., et al. (2010). Oncogenic BRAF mutation with CDKN2A inactivation is characteristic of a subset of pediatric malignant astrocytomas. *Cancer Research, 70*, 512–519.

Schmitz, U., et al. (2001). INI1 mutations in meningiomas at a potential hotspot in exon 9. *British Journal of Cancer, 84*, 199.

Schwartzentruber, J., et al. (2012). Driver mutations in histone H3.3 and chromatin remodelling genes in paediatric glioblastoma. *Nature, 482*, 226–231. http://dx.doi.org/10.1038/nature10833.

Sturm, D., et al. (2012). Hotspot mutations in H3F3A and IDH1 define distinct epigenetic and biological subgroups of glioblastoma. *Cancer Cell, 22*, 425–437.

Takahashi, K., & Yamanaka, S. (2006). Induction of pluripotent stem cells from mouse embryonic and adult fibroblast cultures by defined factors. *Cell, 126*, 663–676.

Taylor, M. D., et al. (2005). Radial glia cells are candidate stem cells of ependymoma. *Cancer Cell, 8*, 323–335. http://dx.doi.org/10.1016/j.ccr.2005.09.001.

Taylor, M. D., et al. (2012). Molecular subgroups of medulloblastoma: the current consensus. *Acta Neuropathologica, 123*, 465–472. http://dx.doi.org/10.1007/s00401-011-0922-z.

Thompson, M. C., et al. (2006). Genomics identifies medulloblastoma subgroups that are enriched for specific genetic alterations. *Journal of Clinical Oncology, 24*, 1924–1931.

Verhaak, R. G., et al. (2010). Integrated genomic analysis identifies clinically relevant subtypes of glioblastoma characterized by abnormalities in PDGFRA, IDH1, EGFR, and NF1. *Cancer Cell, 17*, 98–110.

Versteege, I., et al. (1998). Truncating mutations of hSNF5/INI1 in aggressive paediatric cancer. *Nature, 394*, 203–206.

Witt, H., et al. (2011). Delineation of two clinically and molecularly distinct subgroups of posterior fossa ependymoma. *Cancer Cell, 20*, 143–157. http://dx.doi.org/10.1016/j.ccr.2011.07.007.

Wu, G., et al. (2012). Somatic histone H3 alterations in pediatric diffuse intrinsic pontine gliomas and non-brainstem glioblastomas. *Nature Genetics, 44*, 251–253. http://dx.doi.org/10.1038/ng.1102.

Yan, H., et al. (2009). IDH1 and IDH2 mutations in gliomas. *New England Journal of Medicine, 360*, 765–773.

Yauch, R. L., et al. (2009). Smoothened mutation confers resistance to a Hedgehog pathway inhibitor in medulloblastoma. *Science, 326*, 572–574. http://dx.doi.org/10.1126/science.1179386.

Zhang, J., et al. (2013). Whole-genome sequencing identifies genetic alterations in pediatric low-grade gliomas. *Nature Genetics, 45*, 602–612. http://dx.doi.org/10.1038/ng.2611.

Zhukova, N., et al. (2013). Subgroup-specific prognostic implications of TP53 mutation in medulloblastoma. *Journal of Clinical Oncology: Official Journal of the American Society of Clinical Oncology, 31*, 2927–2935. http://dx.doi.org/10.1200/jco.2012.48.5052.

Zurawel, R. H., Chiappa, S. A., Allen, C., & Raffel, C. (1998). Sporadic medulloblastomas contain oncogenic b-catenin mutations. *Cancer Research, 58*, 896–899.

Chapter 41

White Matter Disorders

Brianne M. Bettcher
University of Colorado School of Medicine, Aurora, CO, United States

Carolyn Bevan and Michael D. Geschwind
University of California San Francisco, San Francisco, CA, United States

INTRODUCTION

The rich history of neuroscience research has yielded an extensive literature on white matter clinical syndromes, ranging from vascular and demyelinating diseases to autosomally inherited white matter disorders. Although interest in white matter structure and abnormalities has spanned several hundred years, significant attention to white matter–mediated neuropsychiatric syndromes did not gain a firm foothold until the mid-20th century. Gross dissection methods gave way to more fine-grained pathological evaluations of white matter tracts, and behavioral phenotyping moved beyond focal cortical syndromes to reveal striking phenotypes associated with white matter disease (Schmahmann & Pandya, 2007). Further characterization of both white matter anatomy and clinical phenomenology has flourished with the advent of sophisticated neuroimaging methodologies, offering an in vivo window into patients' cognitive and behavioral symptoms (Filley, 2009). Although perhaps less extensive than gray matter research, investigations of white matter disorders have expanded our understanding of structural and functional anatomy while also highlighting the central role of white matter in cognitive and emotional functioning (Filley, 2005; Schmahmann, Smith, Eichler, & Filley, 2008).

The goal of this chapter is to characterize white matter clinical disorders from clinical, structural, and genetic perspectives. We will first provide a historical perspective to establish a foundation for white matter research, followed by a brief discussion of white matter structural anatomy, and will conclude with a presentation of several white matter clinical syndromes. Although we will focus primarily on autosomal dominant and recessive diseases (ie, leukodystrophies), we will also examine clinical manifestations and genetic underpinnings of non-Mendelian white matter syndromes, including cerebrovascular disease and multiple sclerosis (MS).

HISTORY OF WHITE MATTER RESEARCH

To appreciate fully the complex characterizations of white matter syndromes, it is important to garner an understanding of how white matter gained prominence in behavioral neurology research. Although Galen (c. 170) reported early descriptions of the corpus callosum in animals, the history of white matter research in humans harkens back to investigations conducted in the 16th century by Andreas Vesalius. Vesalius, who is often referred to as the father of brain anatomy, offered the first comprehensive description of the corpus callosum in humans, observing that it connected the two hemispheres of the brain together without a clear explication of its function (Filley, 2012). With the use of primitive microscopes, further delineations of basic white matter anatomy emerged, such that white matter was determined to contain "fibers" that could connect the hemispheres (ie, commissural) or connect regions within the same hemisphere (ie, association). These anatomical characterizations were later refined by Gall and Spurzheim, who in the early 19th century proposed revelatory, if not initially controversial principles of white matter organization. The accuracy of their guiding principles have been repeatedly validated and include the presence of topographically arranged commissural fibers, as well as the presence of projection fibers that link cortical brain regions to subcortical systems and the spinal cord (Schmahmann & Pandya, 2007).

In tandem with improvements in microscopes came increased precision in the characterization of white matter structure, particularly by Theodore Meynert, who provided a detailed classification system for white matter pathways

Genomics, Circuits, and Pathways in Clinical Neuropsychiatry. http://dx.doi.org/10.1016/B978-0-12-800105-9.00041-X

that is still in common parlance (Catani & ffytche, 2005). This led to newfound interest in the clinical role of white matter changes. In particular, conceptualizations of conduction aphasia (Wernicke, 1874), alexia without agraphia (Dejerine, 1892), cerebrovascular disease (Binswanger, 1894; see Hauw, 1995 for review), and MS (Charcot, 1877) also stem in part from this surge in clinical interest of white matter disorders in the late 19th century, and highlighted the critical nature of specific white matter tracts in cognitive and behavioral phenotypes. Despite these seminal behavioral neurology reports, the role of white matter in clinical syndromes remained relatively ignored in the early part of the 20th century until it was reenergized by the influential and eminent work on cerebral disconnection syndromes by Norman Geschwind (Geschwind, 1965). Geschwind's extensive influence on behavioral neurology included a characterization of syndromes through the perspective of white matter pathways, which laid the groundwork for modern conceptualizations of distributed brain networks (Catani & ffytche, 2005).

The introduction of magnetic resonance imaging (MRI) in the late 20th century ushered in the contemporary era of white matter study, because it facilitated in vivo appraisal of white matter from a macroscopic level in the setting of normal clinical practice. The use of brain MRIs has altered the practice of behavioral neurology and has helped lead to the discovery of genetic white matter diseases that were previously unidentified (eg, vanishing white matter disease) while the patient was alive. Through the use of diffusion tensor imaging (DTI), a type of MRI which capitalizes on the directional diffusion of water across brain tissue, researchers have also been able to identify both subclinical and clinical alterations in the context of a "virtual dissection of white matter pathways in the living brain" (Catani & Thiebaut de Schotten, 2008).

WHITE MATTER ANATOMY

White matter constitutes 50% of the human brain volume and is composed of interleaved groups of myelinated axons that extend within and between hemispheres and connect the neocortex to cortical, subcortical, brain stem, and spinal cord regions. Theodor Schwann offered initial descriptions of the fatty sheaths that surround axons, which were later coined "myelin," and Louis-Antoine Ranvier introduced the classic concept of "nodes" that facilitate conduction of electrical signals along white matter axons (ie, nodes of Ranvier) (Schmahmann & Pandya, 2009). In terms of composition, white matter is made in situ of 40% water and its dry mass consists of 70–80% lipids (Baumann & Pham-Dinh, 2001), which account for the striking white hue of myelinated white matter.

White matter pathways originating in the cerebral cortex typically have been grouped into five distinct categories (Schmahmann et al., 2008): cortico-cortical association fibers, corticostriatal fibers, commissural fibers, cortico-subcortical pathways to the thalamus, and cortico-subcortical pathways to the brain stem, pontocerebellar system, and/or spinal cord. More specifically, association fibers begin and terminate in cortical areas of the same hemisphere and are further subdivided into local "u-fiber" neighborhood, and long association fibers. u-Fibers connect adjacent gyri and thus are restricted to fibers that extend only to gyri that are directly proximal. Neighborhood association fibers extend farther than u-fibers but still connect nearby regions, whereas long association fibers form bundles, or "fascicles" that terminate in distant cortical regions in the same hemisphere. Because of the importance of these long association fascicles in connecting distal regions and efficiently transmitting information between lobes, they have been more intensively studied in the context of clinical syndromes. Several relevant association fibers to white matter clinical syndromes include the superior longitudinal fasciculus, extreme capsule, arcuate fasciculus, middle and inferior longitudinal fasciculi, fronto-occipital fasciculus, and cingulum bundle.

Corticostriatal fibers (ie, Muratoff bundle; extreme capsule) connect to the caudate, putamen, and claustrum, and the commissural fibers cross over the two hemispheres (ie, corpus callosum, anterior commissure, and hippocampal commissure). Finally, cortico-subcortical (projection fibers) terminate in the thalamus, or brain stem and/or spinal cord by way of the internal capsule and sagittal stratum.

A final note on structural neuroanatomy is that white matter is often erroneously assumed to be restricted to subcortical regions. Although white matter is disproportionately positioned in subcortical regions external to gray matter, it is also found in cortical gray matter and in subcortical gray matter nuclei. These small white matter fibers within the cortical mantle have been implicated in several white matter diseases, including MS, and thus may come into increasing prominence as research progresses.

CLINICAL SYNDROMES

In this section, we discuss the evaluation of a patient with suspected or known white matter disease, present a broad differential of adult-onset white matter disorders, and highlight just a few of these syndromes in detail. Because of the size limitations of this chapter, it is not intended in any way to be a comprehensive review of white matter diseases. To give an idea of the complexity and breadth of white matter disorders, the wonderful, comprehensive textbook, *Magnetic Resonance of Myelination and Myelin Disorders* (Third Edition, edited by Marjo S. van der Knapp and Jaap Valk), which covers only the subset of white matter disorders involving myelin, is a "mere" 109 chapters and 1087 pages (van der Knaap & Valk, 2005c).

Clinical Approach

Physicians typically encounter patients with brain white matter diseases either because the clinical presentation suggests involvement of the white matter and/or frontal lobes or a brain MRI is ordered and reveals white matter. When conducting a workup on a patient with white matter disease, a systematic approach is helpful. Some clinicians find a mnemonic, such as VITAMINS, useful for working-up patients with dementia. Table 41.1 presents several conditions with prominent white matter pathology using a similar mnemonic, VITAMIN D, in which the letters refer to various etiologies including Vascular, Infectious, Toxic-metabolic, Autoimmune, Mitochondrial, Iatrogenic, Neoplasms, and Demyelinating.

When working up a white matter disorder, including reviewing brain imaging, it is important to consider the notion of selective vulnerability, a recurring theme in this textbook. Just as many neurodegenerative disorders occur and progress owing to selective vulnerability of gray matter regions and their connections, white matter abnormalities can occur as a result of selective vulnerability of certain cell types, structures or regions to various insults or abnormalities (van der Knaap & Valk, 2005a, Chapter 3).

Pattern recognition is an extension of this concept of selective vulnerability; certain disorders preferentially affect certain white or gray matter brain regions. For example, symmetry is typically seen in inherited white matter disorders and toxic leukoencephalopathies, whereas asymmetry is more common in acquired disorders, such as infections or inflammatory disorders. Whether lesions are isolated, confluent, or both often can be diagnostically helpful. Most inherited, toxic (drugs, radiation, chemotherapy, etc.) causes result in confluent lesions. Isolated or multifocal lesions are more common in acquired causes, such as various infections. Some disorders such as MS, vascular disease, and cerebral autosomal dominant arteriopathy with subcortical infarcts and leukoencephalopathy (CADASIL) (early on) usually have a mix of isolated and confluent lesions. u-Fibers are often spared in many inherited disorders, vascular dementias (VaDs) (eg, Binswanger, CADASIL, etc.), and HIV encephalitis but are preferentially involved in many of the organic and amino acid—opathies. Contrast enhancement might also help differentiate disorders. Enhancement is a prominent feature of MS, Alexander disease, and X-linked adrenoleukodystrophy (the latter at the border between ongoing demyelination and complete demyelination) (van der Knaap, Breiter, Naidu, Hart, & Valk, 1999; van der Knaap et al., 2001; van der Knaap & Valk, 2005b, Chapter 109; Steenweg et al., 2010). It is important to recall that white matter is present within gray matter, and therefore demyelinating syndromes such as central and extrapontine myelinolysis, acute demyelinating encephalomyelitis, and MS can also cause abnormalities in the gray matter (Rosenbloom et al., 2015).

The history and physical examination often are critical in diagnosis. For example, certain combinations of signs and symptoms can help narrow the differential and even are virtually diagnostic in some cases. For example, in an Ashkenazi Jewish older patient who presents with slowly progressive gait disturbance, urinary incontinence, mild cognitive impairment (MCI), or dementia with or without seizures, adult polyglucosan body disease (APBD) should be considered. This diagnosis would be further supported if the MRI showed T2 white matter hyperintensities (WMH) with cavitations and a very thin medulla and cervical spinal cord. In such a case, blood lymphocyte testing for reduced glycogen branching enzyme (GBE) or genetic testing for biallelic (autosomal recessive disorder) GBE gene mutations should be done (Mochel et al., 2012). In a late middle-aged patient (about 40—60 years) with long-standing migraine headaches with auras, late-life progressive frontal-executive cognitive impairment, with strokes or transient ischemic attacks (TIAs), and with a positive family history of similar disorder, CADASIL must be high on the differential (see subsequent discussion) (Chabriat, Joutel, Dichgans, Tournier-Lasserve, & Bousser, 2009). Table 41.2 presents a few points to consider in evaluating a patient with white matter disease.

Some clinical syndromes mentioned in Table 41.1 are described next. We begin with the most common cause of white matter—related cognitive dysfunction, cerebrovascular disease, by discussing vascular cognitive impairment (VCI) and CADASIL. We then briefly discuss a few other white matter syndromes, taking examples from a few of the categories in Table 41.1, and then end with a more comprehensive look at MS, the most common demyelinating syndrome of adults.

Vascular Cognitive Impairment

VCI is a relatively new umbrella term that encapsulates a range of vascular clinical phenotypes (eg, subclinical VCI, strategic infarct dementia, lacunar state, subcortical VaD, or Binswanger disease), severity levels (ie, MCI to dementia),

TABLE 41.1 VITAMIN-D Mnemonic for Consideration of Several Etiologies of White Matter Disorders Causing Dementia[a]

- Vascular
  - Cerebrovascular disease, Binswanger's
  - Posthypoxic-ischemic leukoencephalopathy
  - **CADASIL**
  - **CARASIL**
  - Fabry's disease (FD) (not a leukodystrophy)
    - Renal insufficiency, cardiomyopathy
  - Cerebral amyloid angiopathy (CAA)
  - Other vasculopathies
- Infectious
  - HIV
  - PML
  - Lyme
  - Neurosyphilis
  - Brucellosis
  - **Subacute sclerosing panencephalitis**
- Toxic-metabolic—inborn error of metabolism
  - Vitamin B12, homocysteine, folate deficiencies (usually metabolism defects)
  - Combined methylmalonic and malonic acidemia and aciduria (MMAA)
  - Central or extrapontine myelinolysis
  - Posterior reversible encephalopathy syndrome (PRES)
  - Brain radiation therapy
  - Chemotherapy (eg, tacrolimus)
- Autoimmune
  - Bechet's
  - CNS vasculitis
  - Anti-phospholipid syndrome
  - Hashimoto's encephalopathy
- Mitochondrial disorders
  - Mitochondrial encephalopathy lactic acidosis and strokes (MELAS)
  - Myoclonic epilepsy and ragged-red fibers (MERRF)
- Inborn-error of metabolism and leukodystrophies
  - Adrenoleukodystrophy (ALD)
  - Lysosomal storage disorders, including metachromatic leukodystrophy (MLD), globoid cell leukodystrophy (GLD), and FD
  - MLD
  - **Hereditary leukodystrophy with neuronal spheroids** (**HDLS;** adult-onset leukoencephalopathy with axonal spheroids and pigmented glia (ALSP) or pigmentary orthochromatic leukodystrophy (POLD)
  - Krabbe's disease (GLD)
  - **Adult polyglucosan body disease (ABPD)**
    - Urinary incontinence, gait disorder, cognitive impairment, white matter lesions with cavitations
  - Vanishing white matter disease
  - RNA polymerase subunit mutations, including POLR3-related leukodystrophy (or 4H [hypomyelination, hypodontia, and hypogonadotropic hypogonadism) leukodystrophy] and POLR1C mutations
  - Adult-onset autosomal dominant leukodystrophy (ADLD) with autonomic dysfunction due to LMNB1 mutation (5q23)
- Neoplasms
  - Glioblastoma multiforme
  - Primary CNS lymphoma
    - Lymphomatosis cerebri
- Demyelinating disease
  - Multiple sclerosis
  - Acute demyelinating encephalomyelitis (ADEM)

[a]*Bold items are discussed in this chapter.*

and vascular mechanisms (ie, ischemia, hemorrhage, anoxia). Because of the breadth of this overarching clinical term, we will focus on mild vascular cognitive impairment (ie, MCI level of impairment) and subcortical VaD (ie, dementia level of impairment; also known as Binswanger disease), because these phenotypes are more typically seen in an outpatient neurology clinic.

TABLE 41.2 Some Key Points for Adult Onset White Matter Disorders

- Common things being common, consider and assess for cardiovascular risk factors
  - Hypertension, hyperlipidemia, diabetes, elevated homocysteine, vitamin B12, or related deficiencies, smoking, etc.
- Consider treatable causes of white matter disease, such as infiltrating tumors
  - If CNS lymphoma is in the differential, in addition to CSF cytology and flow cytometry, consider testing for send for CSF beta-2-microglobulin and anti-thrombin III levels, performing microRNA analysis. Also consider ophthalmological examination for ocular involvement.
  - Radiologically, CNS lymphoma can present as multifocal lesions or confluent white matter hyperintensity.
- Family history must be complete in order to not miss potential genetic etiology
  - Consider autosomal dominant disorders if family members in more than one generation are affected
  - Consider autosomal recessive disorders if multiple family members in a single generation are affected; always ask about co-sanguinity
  - Get ages and causes of death, age of onset of neuropsychiatric illness of all first-, second-, and some third-degree relatives (eg, first cousins)
  - Early death or family history of related neuropsychiatric or autoimmune disorders might be a clue to genetic etiology
- If considering a genetic etiology, input key features into a website or program, such as Online Mendelian Inheritance in Man (OMIM; www.omim.org) or the GeneReviews site which recently was incorporated by the NIH: http://www.ncbi.nlm.nih.gov/books/NBK1116/ or www.genereviews.org
- Always consider treatable or reversible etiologies, such as infectious (eg, HIV, PML, syphilis, Lyme disease) and autoimmune
- Have a low threshold for analyzing CSF
  - Include cell count and differential, protein, glucose (need accompanying serum glucose for ratio), IgG index, oligoclonal bands (OCBs; need accompanying serum IgG).
  - If PML is being considered, such as in known or suspected HIV or immunosuppressed case, or posterior predominant leukoencephalopathy, get JC virus PCR.
  - If cancer or tumor is suspected, if there is involvement of the corpus callosum or anterior commissure, or edema is present, get cytology.
    - if CNS lymphoma is in differential, get flow cytometry (often need three large (10 cc) CSF samples to improve likelihood of diagnosis and consider ocular exam and vitreous biopsy
- When reviewing brain MRI findings consider the following:
  - Is the T2-weighted white matter hyperintensity symmetrical (slight or very) or asymmetrical? Most leukodystrophies, toxic-metabolic disorders and inborn errors of metabolism often are relatively symmetric. Vascular, infectious, mitochondrial, and neoplastic etiologies tend to be asymmetric.
  - T2-weighted white matter hyperintensity is required for diagnosis of a leukodystrophy. T2 white matter hyperintensity is milder in hypomyelination than in demyelination and other white matter lesions.
  - T1 signal may be variable: iso- or hyper-intense T1 signal is consistent with a hypomyelinating leukodystrophy; hypointense T1 signal is consistent with a demyelinating leukodystrophy (eg, multiple sclerosis)
  - In demyelination and other lesions the T1 signal is almost always low, much lower than the cortex, whereas in hypomyelination disorders T1 signal is mildly hyperintense, isointense, or mildly hypointense relative to the cortex.
  - Cavitary lesions – consider Binswanger's, CADASIL, adult polyglucosan body disease (APBD)
  - Certain disorders involve, whereas others preferentially spare, U-fibers (prior to end-stages). In some leukodystrophies, T2 white matter hyperintensity if very homogenous, whereas in others it is somewhat heterogenous.
  - Some leukodystrophies have localization or greater prominence of white matter involvement.
    - Is the pattern more predominant in frontal, parietal, occipital, or temporal regions?
    - Is there involvement of periventricular, deep, or arcuate fibers
    - Are the T2 white matter hyperintensities only supratentorial or is the posterior fossa involved as well?

Risk Factors and Clinical Phenotypes

VCI has been defined in several ways, although it is often characterized as a cognitive disorder in the context of clinical strokes or vascular disease (Gorelick et al., 2011). To establish causality, a close temporal relationship between neuroimaging evidence of vascular burden and cognitive impairment should be observed, thus reducing the likelihood that the cognitive symptomatology stems from another etiology. VCI is strongly associated with a host of risk factors, including hypertension, hyperlipidemia, smoking, diabetes, and atrial fibrillation, which collectively increase the risk of dementia independent of stroke risk (Sahathevan, Brodtmann, & Donnan, 2012). Although many vascular risk factors have been studied in isolation, we also know that the constellation of several of these factors (metabolic syndrome) are more than the sum of the parts, and extend negative consequences not only in terms of white matter lesions, but of cognitive functioning over time (van den Berg, Biessels, de Craen, Gussekloo, & Westendorp, 2007; Crichton et al., 2012; Yaffe et al., 2007). Numerous mechanisms have been proposed for the deleterious association between vascular risk factors and cognitive impairment, ranging from chronic hypoperfusion to microinfarcts and cerebrovascular disease; it is important, however, to highlight that vascular risk factors predict poorer cognitive performance on formal neuropsychological testing even in the absence of observable cerebrovascular disease, which suggests more complex vascular- or inflammatory-mediated pathways (DeCarli, 2013; Zheng et al., 2012).

The phenotypic presentation of vascular mild cognitive impairment (VaMCI) and subcortical VaD is heterogeneous and includes cognitive, behavioral, and/or motor symptoms. Cognitively, patients often present with slowed processing speed and a dysexecutive-predominant neuropsychological profile, typified by impairments in selective attention, mental flexibility, speeded retrieval of information, and establishing and maintaining mental sets (DeCarli et al., 1995; Filley, 2005; Lamar, Price, Giovannetti, Swenson, & Libon, 2010; Libon, Price, Davis Garrett, & Giovannetti, 2004). Performance on measures of fluency often differ from that seen in patients with classic Alzheimer disease (AD); patients with AD show disproportionate impairment on category fluency relative to phonemic fluency (Rascovsky, Salmon, Hansen, Thal, & Galasko, 2007) and patients with VaMCI/VaD show impairment on both, if not slightly better performance on category fluency (Looi & Sachdev, 1999). Verbal memory test performance also tends to be impoverished, although patients typically show retrieval-based deficits that improve with semantic cueing and recognition formats rather than a frank amnestic profile.

Patients with VaMCI or VaD may show a range of personality and mood changes, including depression, abulia, psychomotor retardation, and pseudobulbar affect. Depressed mood in connection with vascular disease has received extensive research attention in the past couple of decades, leading to the term "vascular depression" (Taylor, Aizenstein, & Alexopoulos, 2013). In addition, extrapyramidal signs and gait disturbance are prevalent, depending on the extent of subcortical white matter and basal ganglia involvement.

Structural and Functional Neuroimaging

In terms of structural brain imaging, VaMCI and subcortical VaD are characterized by cerebral microbleeds, lacunar infarcts, and most commonly, WMH in the subcortical white and gray matter on fluid-attenuated inversion recovery (FLAIR) or T2-weighted MRI scans. Although WMH on MRI may appear focal or multifocal, with increasing severity they may develop into a confluence that involves larger swaths of white matter. Notably, the "rarefaction" of white matter on MRI or computed tomography scans of older people was previously termed leukoaraiosis, and remained a controversial area of research for some time owing to the "subclinical" (ie, no discrete neurological incident suggestive of a stroke) and relatively common (ie, present in up to one-third of healthy older adults) nature of the imaging findings. Periventricular and deep white matter tracts tend to be particularly vulnerable to lesions and vascular insufficiency, in part because of the smaller and more delicate vasculature in these areas as well as their location near border zones for vascular territories (Iadecola, 2013); as such, WMH are more common in these regions. Although there is some evidence that the location of WMH differentially affects cognitive functioning, considerable research also suggests that WMH predict a reduction in frontal lobe glucose metabolism and a global reduction in cortical blood flow irrespective of the location. In most VaMCI and VaD, the white matter disease tends to be more periventricular and generally spares the u-fibers (Erkinjuntti et al., 1996). Several methods or scales, such as the Fazekas scale, have been developed for grading the severity of white matter disease in VaMCI and VaD (Scheltens et al., 1998).

Functional neuroimaging in VaMCI and VaD has been limited in scope, although studies suggest that early stages of disease (ie, nondemented older adults with WMH) may be associated with increased activation in frontal brain regions, particularly the right middle frontal gyrus, as well as frontal interhemispheric connectivity during visual search tasks (Lockhart et al., 2015). This may reflect a failed compensatory response to increasing vascular burden, with the net effect being reduced neural insufficiency in early stages. Thus, whereas early stages of disease may be associated with increased frontal activation, later stages of disease and clinical dementia remain predictive of decreased frontal activation relative to healthy older adults (Li, Zheng, & Wang, 2012).

Genetics

The genetic underpinnings of sporadic VaMCI and VaD are multifactorial and likely the result of complex interactions between risk alleles and the environment. Studies suggest contributions of genetics to WMH, with overall heritability of 52–78% reported in the Framingham study (Atwood et al., 2004), and twin studies suggest 61% concordance rates in monozygotic pairs (Carmelli et al., 1998). Meta-analyses of candidate genes and genome-wide association screening (GWAS) have isolated several single-nucleotide polymorphisms associated with WMH (tripart motif containing 65 and 47, TRIM65, and TRIM 47), lacunar infarcts (protein kinase C family, PRKCH), or both (angiotensin-converting enzyme insertion/deletion), although the contribution to overall mean WMH burden is relatively small (Choi, 2015).

Pathology

Although there is no reference standard for neuropathological diagnoses of VCI, the vascular pathologies underlying white matter lesions typically include a wide range of deleterious changes to the vessel walls that ultimately lead to vacuolation, demyelination, and axonal loss. The presence of atherosclerotic plaques in small cerebral blood vessels, lipohyalinosis, arteriosclerosis, and complete loss of vascular wall integrity (ie, fibrinoid necrosis) are all possible etiologies leading to infarcts and microinfarcts (Thal, Grinberg, & Attems, 2012). Vestiges of chronic hypoxia are also common in postmortem studies, with molecular evidence of hypoxia-inducible factors colocalizing with infarcts (Fernando et al., 2006).

Cerebral Autosomal Dominant Arteriopathy With Subcortical Infarcts and Leukoencephalopathy

Clinical Phenotype

CADASIL is a genetic autosomal dominant cerebrovascular white matter disease. The acronym CADASIL describes several key features of the disorder (ie, hereditary small vessel disease leading to strokes and dementia). It is caused by mutations in the Notch3 gene on chromosome 19q12a. The worldwide prevalence of CADASIL is unknown, but studies in Europe suggest at least two to four cases per 100,000 (Moreton, Razvi, Davidson, & Muir, 2014; Narayan, Gorman, Kalaria, Ford, & Chinnery, 2012). Main clinical features of CADASIL include migraine with aura, frontal-executive cognitive impairment, ischemic strokes, behavioral changes, and T2 WMH on MRI. Around 20–50% of patients with CADASIL have migraine with aura, which is usually the first symptom, with onset on average at age 30 years (range, 6–48 years); curiously, mean age of onset for women is 10 years earlier than for men: age 26 years versus 36 years (Chabriat et al., 2009). TIAs and ischemic strokes occur in 60–85% of patients, at a mean age of 49 years (range, 20–70 years). Although TIAs and strokes often occur in the absence of other risk factors, many patients also have risk factors such as hypertension, smoking, and hyperlipidemia. Ischemic events are almost invariably subcortical; in about two-thirds of patients they present as lacunar syndromes, such as pure motor or sensory deficit, ataxic hemiparesis, and clumsy hand–dysarthria syndrome. Most of the ischemia in CADASIL is small silent strokes; years of these cause progression to significant morbidity with motor weakness, spasticity, gait difficulties, urinary urgency with or without incontinence, mood disturbances, and pseudobulbar palsy. Behavioral changes include depression in at least 20% and apathy, independent of depression, in at least 40% of patients; mania can also occur (Chabriat et al., 2009).

Cognitive impairment is the second most frequent clinical manifestation after migraine with aura, and typically includes frontal-executive dysfunction. Upon testing, patients often perform poorly on Trails switching, fluency tasks, symbol digit tests, and other tests involving processing speed. Consistent with frontal-executive dysfunction and white matter disease, patients usually show memory retrieval deficits with preservation of recognition and semantic memory (Chabriat et al., 2009; Charlton, Morris, Nitkunan, & Markus, 2006). Perhaps not surprisingly, the pattern of cognitive impairment is similar to that of small vessel ischemic vascular disease (SIVD) (Charlton et al., 2006). Executive dysfunction is usually present by the third decade. Interestingly, cognitive dysfunction appears to correlate better with lacunar strokes than with WMH burden (Liem et al., 2007; Viswanathan et al., 2007). Other less common clinical manifestations include seizures (5–10%), extrapyramidal features, and very rarely, intracerebral hemorrhages and territorial infarcts (possibly coincidental) (Chabriat et al., 2009).

A typical temporal profile of CADASIL is shown in Fig. 41.1, with migraines beginning in the teens or twenties, T2-weighted MRI hyperintensities begin in the mid-twenties to age 30 and invariably present by the mid-thirties, ischemic events and behavioral changes in the forties to sixties, and dementia and motor disabilities by the fifties to sixties. Patients often are bedridden by age 65; survival is to the mid- to late-sixties or early-seventies (earlier in men than women) (Chabriat et al., 2009).

Imaging

T2/FLAIR hyperintensities on MRI usually precede other symptoms by 10–15 years and begin as punctiform or nodular lesions in periventricular areas and in the centrum semiovale. Over years, they progress to diffuse, extensive, and somewhat symmetrical hyperintensities, often with cavitations. Involvement of the external capsule and the anterior part of the temporal lobe is characteristic, which helps to differentiate it from MS and some other white matter conditions (Chabriat et al., 2009), although these areas can be involved in other conditions as well, such as moderate-to-severe SIVD and cerebral amyloid angiopathy, respectively. Lacunar infarcts tend to appear later in the disease course and their total volume appears to correlate with cognitive decline (Viswanathan et al., 2007). Dilated perivascular spaces, which

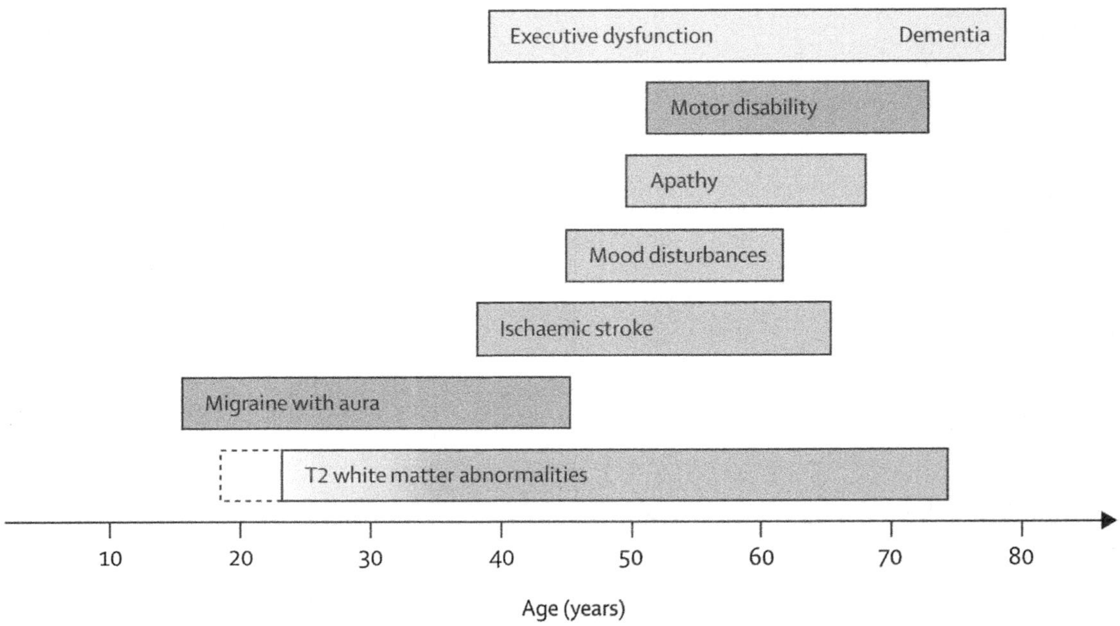

FIGURE 41.1 Natural history of the main clinical manifestations of CADASIL. The exact age at earliest onset or of first MRI abnormalities is uncertain (*dotted line*). The frequency of T2 white matter abnormalities increases progressively and becomes constant by around age 35 years in all patients. *CADASIL*, cerebral autosomal dominant arteriopathy with subcortical infarcts and leukoencephalopathy. *Reprinted with permission from Chabriat, H., Joutel, A., Dichgans, M., Tournier-Lasserve, E., & Bousser, M. G. (2009). CADASIL.* Lancet Neurology, *8(7), 643–653. http://dx.doi.org/10.1016/s1474-4422(09)70127-9.*

sometimes can be confused with lacunes, often occur; when severe, they are called "état crible," or status cribrosum, most often in the basal ganglia (Chabriat et al., 2009).

Microbleeds are seen in about 25–69% of patients, on hemosiderin-sensitive sequences, and appear to correlate with increasing age, higher blood pressure, hemoglobin A_{1C} level, and the extent of white matter disease (Chabriat et al., 2009). Global brain atrophy progresses about three times faster in CADASIL compared with normal aging, and correlates with cognitive decline and disability (Chabriat et al., 2009). Typical CADASIL MRI abnormalities are shown in Fig. 41.2.

Genetics

The Notch3 gene encodes a large transmembrane receptor with an extracellular domain containing 34 epidermal growth factor repeats (EGFR). Each EGFR region has six cysteine residues. More than 95% of the approximately150 mutations reported to date are missense mutations in exons 2–24, which lead to addition or deletion of a cysteine residue in the EGFR (Chabriat et al., 2009). There is controversy as to whether mutations outside the EGFR domains and/or without affecting cysteine residues might also cause CADASIL. In some cases, these cases have been less severe, later-onset forms, or in families with reduced penetrance (Brass, Smith, Arboleda-Velasquez, Copen, & Frosch, 2009).

Pathology

NOTCH3 is predominantly expressed in vascular smooth muscle cells of small arteries. CADASIL NOTCH3 mutations result in the gradual accumulation of its extracellular domain as microscopic aggregates around vascular smooth muscle cells and pericytes of brain arteries and capillaries, close to deposits of granular osmiophilic material (GOM). Mouse model data suggest that pericytes are the initial cells affected by Notch3 aggregation, resulting in blood–brain barrier and microvascular dysfunction (Hurth et al., 2015). Loss of Notch3, such as through double knockout mice, however, does not cause CADASIL pathology, which supports the notion that CADASIL mutations act through gain of novel function and that the change in the number of cysteine residues in NOTCH3 may be the common denominator for CADASIL mutations (Chabriat et al., 2009).

Brain pathology shows macropathologic changes consistent with small vessel ischemic disease, including diffuse myelin pallor and rarefaction of white matter in periventricular areas and centrum semiovale, lacunar infarcts in white

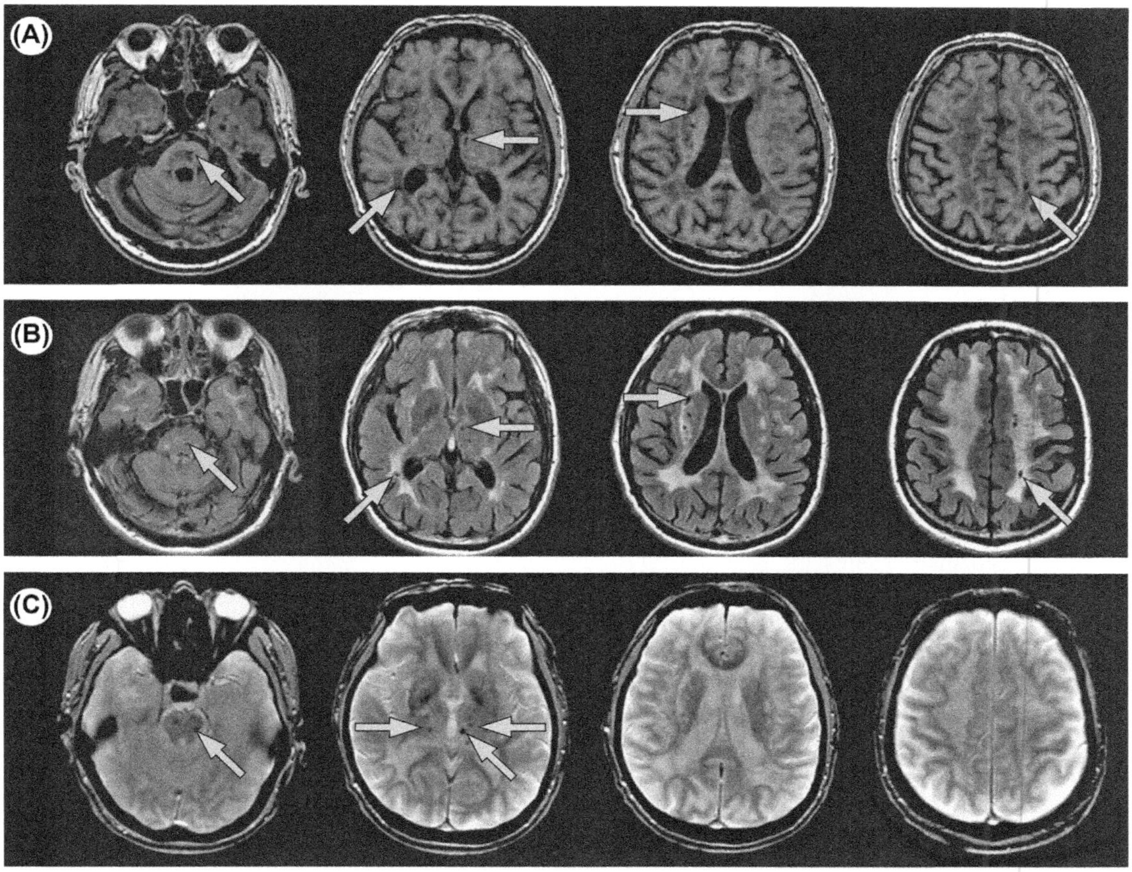

FIGURE 41.2 Main MRI changes in CADASIL. (A) Lacunar infarcts shown on T1-weighted imaging are mainly located in the brain stem (pons), thalamus, and lentiform nuclei in a 61-year-old man with a history of stroke, gait difficulties, and executive dysfunction with memory deficits. (B) Small deep infarcts are shown on fluid-attenuated inversion recovery images in association with diffuse and confluent white matter hyperintensities involving the anterior part of the temporal lobes. (C) Microbleeds are visible on T2* or gradient-echo images as small hypointense foci in the thalamus and brain stem. *CADASIL*, cerebral autosomal dominant arteriopathy with subcortical infarcts and leukoencephalopathy. *Borrowed with permission from Jouvent, E., Mangin, J. F., Herve, D., During, M., Dichgans, M., & Chabriat, H. (2012). Cortical folding influences migraine aura symptoms in CADASIL [Multicenter Study Research Support, Non-U.S. Gov't].* The Journal of Neurology, Neurosurgery, and Psychiatry, *83(2), 213–216. http://dx.doi.org/10.1136/jnnp-2011-300825.*

matter and basal ganglia, as well as dilated Virchow-Robin (perivascular) spaces. The cortex shows widespread neuronal apoptosis, especially in layers 3 and 5.

Microscopically, there is a specific arteriopathy of the small cerebral and leptomeningeal penetrating arteries, characterized by thickening of the arterial wall causing luminal stenosis, deposition of nonamyloid GOM in the media extending into the adventitia, and eventual disintegration of smooth muscle cells. This partly explains the typical pattern of white matter abnormalities found in CADASIL. Although arteriopathy is found in other organs, such as the spleen, liver, kidneys, muscle, aorta, and skin, clinical manifestations are restricted to the central nervous system.

Diagnosis

Molecular testing by NOTCH3-sequencing analysis (especially screening exons 2–24) for detection of typical cysteine-altering mutations generally is considered the reference standard for the clinical diagnosis of CADASIL (Chabriat et al., 2009). As noted previously, some patients and families with atypical Notch3 mutations have been reported (Brass et al., 2009; Wollenweber et al., 2015); these patients might not have classic medial temporal lobe involvement (Brass et al., 2009). Such genetic diagnosis is particularly true in the setting of a compatible clinical and family history and a positive brain MRI scan. If clinical or molecular results are inconclusive, immunohistochemistry and electron microscopy scan can be done on a skin biopsy to determine the presence of pathognomonic vessel wall abnormalities, including positive NOTCH3 staining, electron microscopic deposits of GOM, and vascular smooth muscle cell degeneration (Tikka et al., 2009). A skin biopsy demonstrating this GOM has a reported sensitivity of about 40–50%, although immunostaining with NOTCH3 monoclonal antibody to detect the accumulation of NOTCH3 protein in the vessel wall has higher sensitivity (85–100%) and specificity (95–100%) (Chabriat et al., 2009; Tikka et al., 2009). Some of the cysteine-sparing CADASIL mutations might also not always show GOM on skin biopsy (Wollenweber et al., 2015).

When considering genetic testing for CADASIL, as with other autosomal dominant neurodegenerative conditions without cures, the medical community uses the Huntington protocol, which includes detailed genetic counseling. It is not recommended to test children, because currently there is no disease-modifying treatment (MacLeod et al., 2013).

Treatment

Unfortunately, currently there are no cures for CADASIL and treatment of symptoms is limited. Brain volume (corrected for skull size), number of lacunes, and active smoking are predictors of subsequent stroke, whereas active smoking, disability, and brain volume are predictors of future dementia (Chabriat et al., 2015). Thus, stopping smoking and reducing cardiovascular risk factors are recommended (Chabriat et al., 2009). Migraine with aura of CADASIL is treated similarly to that of general population with the exception that ergot derivatives and triptans generally are not recommended because of their vasoconstricting property (Chabriat et al., 2009). Typical migraine prophylactic medications, such as antiepileptic drugs, tricyclic antidepressants, or antihypertensives can be used, although acetazolamide anecdotally has been found to be effective (Forteza, Brozman, Rabinstein, Romano, & Bradley, 2001).

Secondary stroke prevention similar to noncardioembolic ischemic stroke is recommended, such as the use of antiplatelet drugs (eg, aspirin or Plavix) and treatment of vascular risk factors. Use of anticoagulants or clot-busting agents generally is contraindicated because of the increased risk of intracerebral hemorrhage in the presence of cerebral microbleeds. It is not clear whether these agents are dangerous in patients who have revealed microbleeds in hemosiderin MRI sequences. A randomized, double-blinded, placebo-controlled trial of donepezil in patients with CADASIL with cognitive impairment found no benefit on the primary end point of the cognitive subscale of the vascular AD cognitive assessment scale, but improvements were found on measures of executive functions (Dichgans, 2009). Supportive measures such as physical therapy and rehabilitation, psychological support, and nursing care have an important role in the long-term management of elderly debilitated and demented individuals.

Cerebral Autosomal Recessive Arteriopathy With Subcortical Infarcts and Leukoencephalopathy

Similar to CADASIL, cerebral autosomal recessive arteriopathy with subcortical infarcts and leukoencephalopathy (CARASIL) is a single-gene disorder that affects the cerebral small blood vessels, causing ischemia. It is caused by mutations in the *HTRA1* gene, which encodes HtrA serin peptidase/protease 1. Similar to CADASIL, there is progressive cognitive impairment. MRI scans show white matter disease, lacunes (often in the deep nuclei), and severe arteriosclerosis in the small penetrating arteries, but unlike CADASIL, it is a recessive disorder found only in Japan and China (as of 2015), and is associated with alopecia and severe low back pain, and there is no GOM deposition (Fukutake, 2011).

Adult Polyglucosan Body Disease

APBD is an autosomal recessive disorder that is an adult-onset form of glycogen storage disease type IV. It is caused by homozygous or compound heterozygous mutations in the GBE gene. Mutations result in glycogen not being able to branch properly; therefore, it accumulates in the nervous system. The clinical presentation is usually a mix of key symptoms including urinary incontinence (neurogenic bladder), gait disorder owing to progressive spastic paraparesis, vibratory loss, and axonal neuropathy. Cognitive impairment occurs in about half of subjects. Seizures may also occur. Median age is 51 years for onset of neurogenic bladder symptoms, 63 years for wheelchair dependence, and 70 years for death (Mochel et al., 2012). The most common mutation has been found in Ashkenazi Jews, but other allelic mutations and ethnic groups have been identified. MRI scans usually show leukoencephalopathy with cavitations, with hyperintensities in periventricular regions, posterior limb of the internal capsule, the external capsule, the pyramidal tracts, and the medial lemniscus of the pons and medulla and a very thin medulla and cervical cord (Fig. 41.3). Diagnosis can be made by identifying reduced levels of GBE in blood leukocytes or by genetic testing for GBE mutations (Mochel et al., 2012). The disorder is currently incurable, although a serendipitous discovery of a potential mouse model might lead to additional insights into this disorder (Raben et al., 2001).

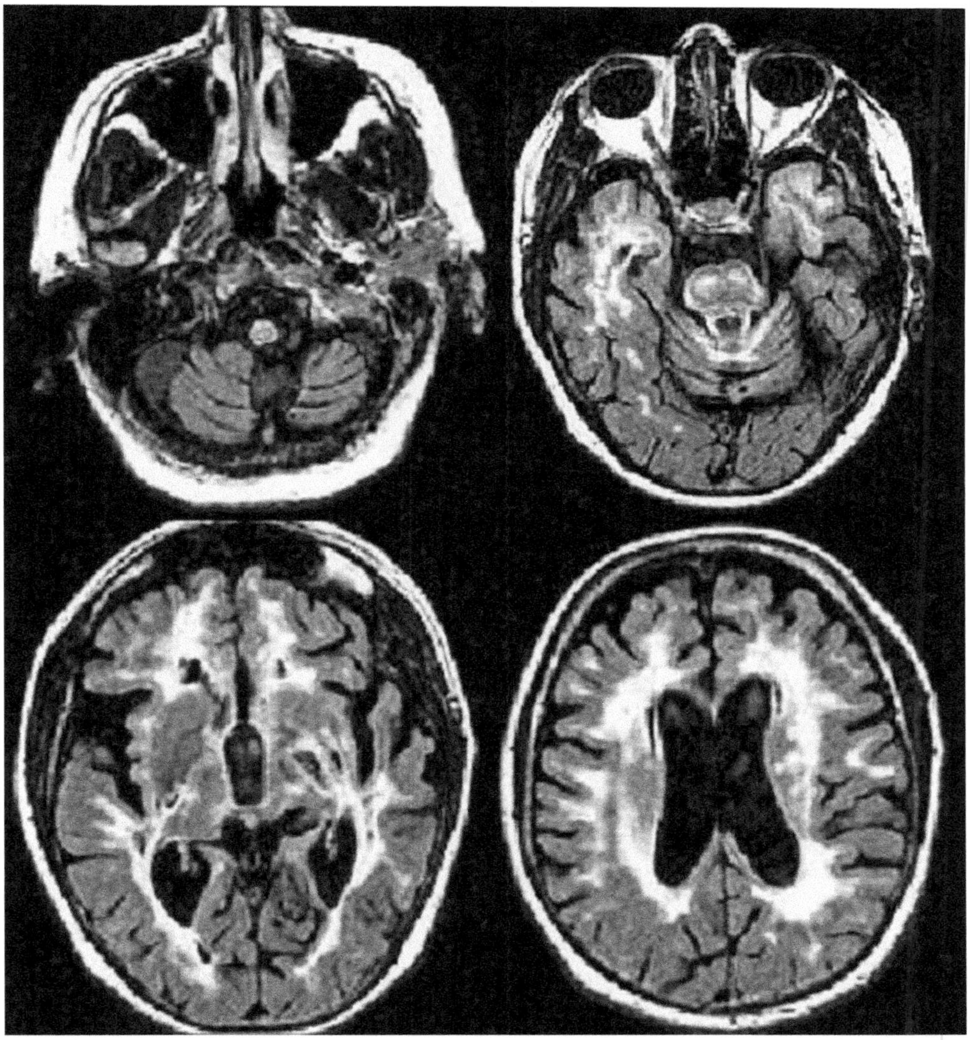

FIGURE 41.3 Brain MRI adult polyglucosan body disease (APBD). Axial FLAIR brain MRI scans in a 68-year-old woman with 4 years of cognitive decline (short-term memory loss), progressive gait disturbance, 18 years of worsening urinary incontinence, and worsening fatigue. Note some key radiological features of APBD including confluent white matter disease with cavitations and a thin cervical spinal cord (best seen in sagittal view; not shown).

Hereditary Leukodystrophy With Neuronal Spheroids

Hereditary diffuse leukoencephalopathy with spheroids (HDLS) is an autosomal dominant central nervous system white matter disease with variable clinical phenotypes, including behavioral disorders (personality changes, disinhibition, or depression), cognitive impairment (frontal executive dysfunction and memory impairment), dementia, motor impairment (parkinsonism, tremor, spastic paraparesis, or ataxia), seizures, and other presentations. Mean age of onset is in the mid-forties (range, mid-thirties to early-fifties), duration of about 6 years (range, 3−11 years), and age of death around age 48 years (range, 40−63 years) (Sundal et al., 2012). This disease is probably within the same disease spectrum as pigmentary orthochromatic leukodystrophy and adult-onset leukoencephalopathy with axonal spheroids and pigmented glia. In the past, these were considered separate entities, but now they are generally thought to be the same clinicopathologic condition because of mutations in the colony-stimulating factor 1 receptor (CSF1R) gene (Nicholson et al., 2013). CSF1R is a tyrosine kinase receptor expressed on the surface of microglia, and to a lesser extent in neurons. MRI scans in HDLS reveal focal white matter lesions often in a predominantly frontal distribution, progressing from the periventricular and deep white matter into subcortical areas. Over time, white lesions become confluent, showing bilateral frontal and white matter T2 hyperintensities, often extending from the ventricles to subcortical white matter (Fig. 41.4). Unlike typical

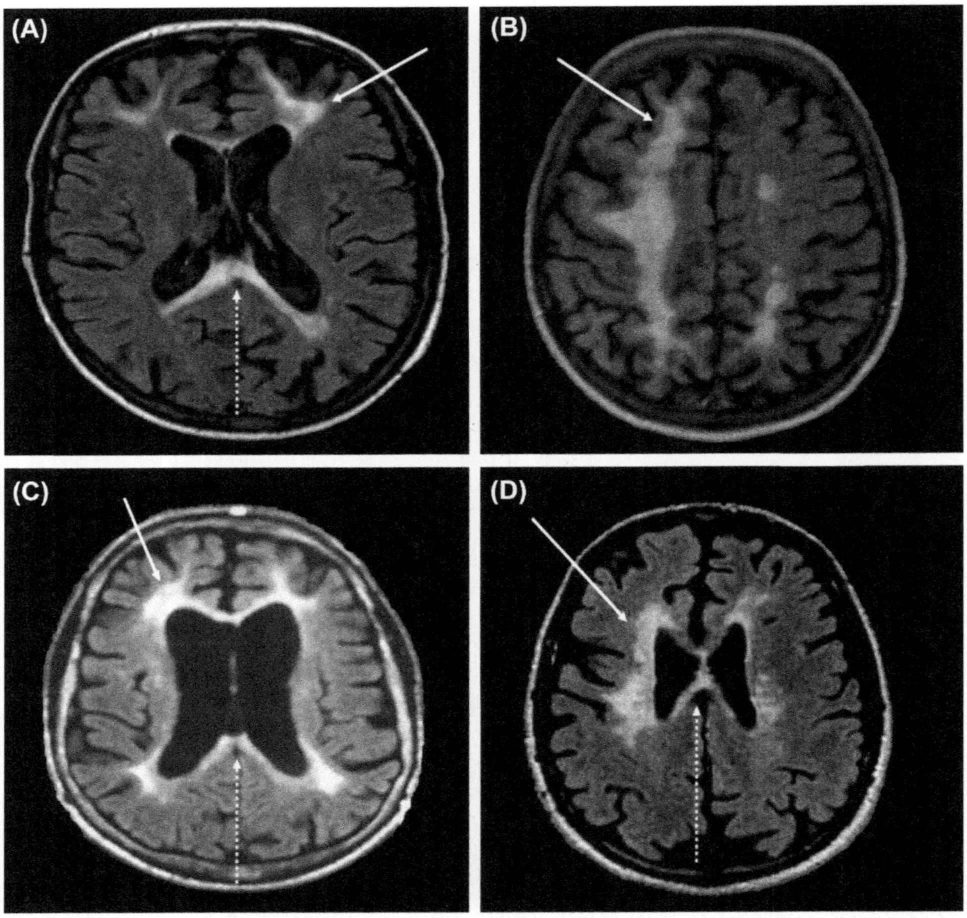

FIGURE 41.4 MRI in HDLS. Magnetic resonance images (axial sections, T2-weighted) from the four patients: (A) Patient 1 (MRI performed 1.2 years after symptoms began); localized white matter lesions (*arrow*) in both frontal and parietal hemispheres involving the corpus callosum (*dashed arrow*). (B) Patient 2 (MRI performed 1.9 years after the start of symptoms): confluent white matter lesions in both frontal and parietal hemispheres with cortical atrophy in the affected areas. (C) Patient 3 (MRI performed 3.5 years after the start of symptoms): localized periventricular lesions (*arrow*) with corresponding frontoparietal atrophy and involvement of the corpus callosum (*dashed arrow*). (D) Patient 4 (MRI performed 2.5 years after symptoms began): bilateral frontoparietal white matter changes (*arrow*) extending into the corpus callosum (*dashed arrow*). *Borrowed with permission from Sundal, C., Lash, J., Aasly, J., Oygarden, S., Roeber, S., Kretzschman, H., ... Wszolek, Z. K. (2012). Hereditary diffuse leukoencephalopathy with axonal spheroids (HDLS): a misdiagnosed disease entity.* Journal of the Neurological Sciences, 314*(1–2), 130–137. http://dx.doi.org/10.1016/j.jns.2011.10.006. pii:S0022-510X(11)00609-5.*

ischemic vascular disease, the ventricles might be enlarged and T2 hyperintensities can be seen in the corpus callosum. Because of the varied clinical presentation, patients often are misdiagnosed with AD, frontotemporal dementia, atypical parkinsonian disorders, MS, or multiple or SVID (Sundal et al., 2012).

Subacute Sclerosing Panencephalitis

Subacute sclerosing panencephalitis (SSPE) is a rare inflammatory and neurodegenerative disease caused by persistent cerebral infection with the measles virus occurring in a delayed manner after acute measles infection (Garg, 2008; Studart Neto et al., 2015). Symptoms often include behavioral changes and cognitive dysfunction, followed by myoclonus, corticospinal tract signs, and rigidity. Although typically a childhood and adolescence disease, several adult-onset cases have been identified, even at our own center, with presentations as late as age 50 years (Studart Neto et al., 2015). Death usually occurs 1–3 years after onset, but some patients have lived several years (Studart Neto et al., 2015). It is not clear why adult patients have such a long time from measles infection and onset of symptoms (Prashanth, Taly, Ravi, Sinha, & Arunodaya, 2006). The incidence estimates range from 4 to nearly 30 patients with SSPE per 100,000 cases of measles, with the higher incidences in developing countries. SSPE is now very rare in countries with effective measles vaccination programs, and perhaps because of its rarity the diagnosis is easily missed or delayed.

There are several tests used for diagnosis. Electroencephalogram (EEG) may be normal or nonspecifically slow in early disease, but by mid to later stages it usually shows synchronous, stereotyped, high-voltage periodic complexes, often temporally associated with myoclonus (Garg, 2008). As with EEG, brain MRI might be normal early on, but as the disease progresses, subcortical and periventricular white matter and basal ganglia abnormalities may occur, progressing to hemispheric, brain stem, and cerebellar atrophy (Garg, 2008; Prashanth et al., 2006). Serial brain MRI scans showing a leukoencephalopathy in a 50-year-old man with SSPE is shown in Fig. 41.5. Routine cerebrospinal fluid (CSF) studies may be entirely normal or show mild lymphocytic pleocytosis and protein elevation, with elevated CSF gamma-globulin concentration, but elevated antimeasles antibody titers of 1:256 or greater in serum and 1:4 or greater in CSF are considered diagnostic of SSPE. The characteristic ratio of CSF to serum titers ranges from 1:4 to 1:128 (<1:200) compared with the normal ratio (1:200–1:500) (Garg, 2008). Production of antibodies is intrathecal (Prashanth et al., 2006). Rarely, brain biopsy is required for diagnosis, such as in very early or advanced cases when characteristic clinical findings might not be present, or when EEG does not show periodic complexes (Garg, 2008).

Multiple Sclerosis

Background

First described by Charcot in 1868, MS is an inflammatory demyelinating disease of the central nervous system. Whereas white matter involvement is a hallmark of the disease, gray matter pathology is increasingly recognized as an important contributor to cognitive and other disability in MS.

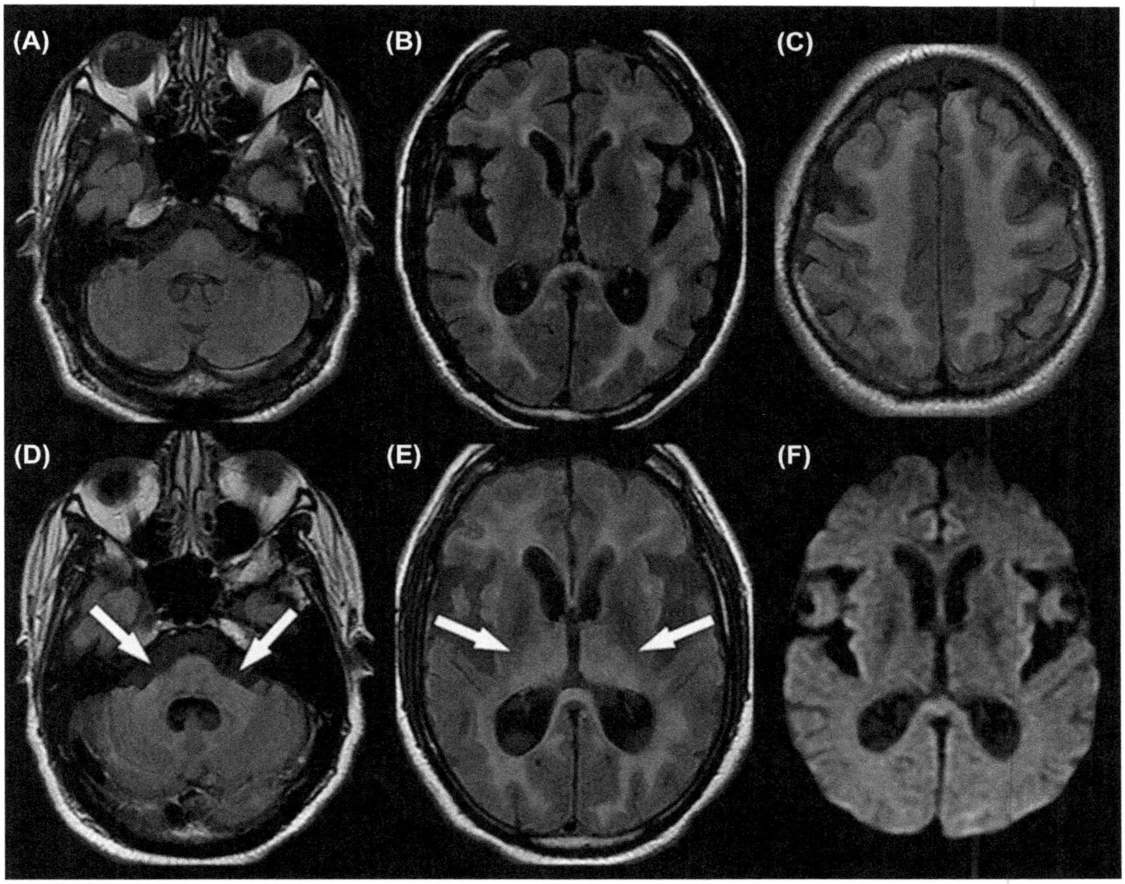

FIGURE 41.5 Brain MRI in SSPE. Serial brain MRIs in a 50-year-old man who presented with 8 years of behavioral changes and 3 years of short-term memory changes. The CSF was inflammatory with elevated protein. Diagnosis was by brain biopsy. Initial (A–C) and follow-up (17 months after the initial one) (D–F) brain MRI scan. FLAIR images from the initial exam (A–C): confluent bilateral and symmetric hyperintensities involving subcortical and deep white matter of the cerebral hemispheres and pons. Follow-up FLAIR images (D, E): more extensive signal changes involving the middle cerebellar peduncles (*arrows* in D) and thalami (*arrows* in E). Follow-up diffusion-weighted images (F) show no diffusion restriction. *Borrowed with permission from Studart Neto, A., Nobrega, P. R., Duarte, M. I., Lucato, L. T., Castro, L. H., & Nitrini, R. (2015). Adult-onset subacute sclerosing panencephalitis manifesting as slowly progressive dementia.* Journal of Neurovirology, 21*(4), 468–471. http://dx.doi.org/10.1007/s13365-015-0336-0.*

Clinical Phenotype

The diagnosis of MS requires evidence of separation in space and time. Older criteria required two discrete clinical attacks, but the 2010 International Panel Criteria for Clinically Definite Multiple Sclerosis (CDMS) allows these criteria to be met with a combination of clinical and radiologic findings (Polman et al., 2011). Elevated immunoglobulin G index and unique oligoclonal bands in the CSF are supportive of but not mandatory for the diagnosis of relapsing MS. Those who experience a clinical attack consistent with demyelinating disease but do not meet criteria for dissemination in time are diagnosed with clinically isolated syndrome (CIS). Conversion from CIS to CDMS ranges from 30% to 70% depending on clinical, imaging, and CSF markers (Miller, Barkhof, Montalban, Thompson, & Filippi, 2005). Treatment with an MS disease-modifying therapy is indicated for patients with high-risk CIS. A subgroup of patients with radiologically isolated syndrome was identified in which brain MRIs obtained for indications such as head trauma or migraine revealed lesions meeting radiologic criteria for MS but the disease was clinically silent. In available case series, some but not all of these patients go on to develop CDMS; 34% of patients in one series experienced a clinical event at 5 years (Lebrun et al., 2009; Okuda et al., 2011, 2014).

MS follows one of four disease courses: relapsing-remitting (RR), secondary progressive (SP), primary progressive (PP), or progressive-relapsing (Lublin & Reingold, 1996). In most patients (80−85%), the disease follows an RR course characterized by subacute onset of neurologic dysfunction evolving over a period of days to weeks followed by gradual, partial, or complete recovery. Some patients with RRMS go on to experience a more progressive decline without superimposed relapses. These patients are characterized as having SPMS. In a pretreatment cohort, 50% of patients with relapsing onset developed SPMS at 10 years (Weinshenker & Ebers, 1989). Newer longitudinal data will shed more light on rates of conversion to SPMS in a treated cohort. A smaller subset of patients (10−15%) presents with PPMS, defined by insidious and irreversible accumulation of disability from onset without clear superimposed attacks. PRMS is increasingly thought not to represent a truly a distinct course, but rather highlights that patients with progressive MS may experience rare attacks (Lublin, 2014).

Typical presentations of relapsing MS include optic neuritis, brain stem dysfunction (eg, internuclear ophthalmoplegia), and partial myelitis. Multifocal presentations also occur. Over time, patients may develop impairment of vision or other cranial nerve functions, spastic paresis, sensory dysfunction, ataxia, disturbance of bowel or bladder function, or cognitive dysfunction. Fatigue and neuropathic pain are common.

Between 40% and 70% of patients with MS experience cognitive dysfunction (Chiaravalloti & DeLuca, 2008). Affected domains include information processing speed, visual learning, and memory, with less frequent but still clinically important involvement of simple attention and verbal skills (DeLuca, Yates, Beale, & Morrow, 2015; Rocca et al., 2015b). MS-related cognitive impairment is considered a subcortical dementia, differing from cortical dementias such as AD in that aphasia, apraxia, and agnosia are uncommon (DeLuca et al., 2015), and memory dysfunction is usually not amnestic but more the result of a frontal lobe retrieval deficit. Symptoms are often subtle and frank dementia is rare (Chiaravalloti & DeLuca, 2008). Cognitive impairment can be detected early in MS and seems to be more prominent in progressive than relapsing forms of the disease (Achiron et al., 2013; Feuillet et al., 2007). Cognitive dysfunction increases with time. In one longitudinally followed cohort, 74% of patients were cognitively normal at baseline but only 44% of patients were considered cognitively normal at 10 years (Amato, Ponziani, Siracusa, & Sorbi, 2001). Although generally progressive, cognitive function may also worsen transiently during relapses (Benedict et al., 2014; Morrow, Jurgensen, Forrestal, Munchauer, & Benedict, 2011). Measurement of cognitive dysfunction in MS can be challenging. Patient-reported outcomes are influenced by comorbid depression and fatigue (Benedict, Carone, & Bakshi, 2004; Simioni, Ruffieux, Bruggimann, Annoni, & Schluep, 2007). Many cognitive screening assessments were not designed to detect MS-related cognitive dysfunction. The symbol digit modality test is an important screening tool for MS-related cognitive dysfunction. The Brief Repeatable Battery of Neuropsychological tests and Minimal Assessment of Cognitive Function in MS are more comprehensive, although time-intensive, metrics for quantifying cognitive impairment in MS (DeLuca et al., 2015; Rocca et al., 2015a).

Depression and anxiety are also common. Major depression occurs in patients at a rate two to five times higher than in the general population (Feinstein, Magalhaes, Richard, Audet, & Moore, 2014). It is important to rule out depression as a cause of cognitive impairment in MS. Psychosis is rare but reported (Marrie et al., 2015).

Structural and Functional Neuroimaging

MRI is a critical tool in the diagnosis and management of MS; it also has a key role in research. Magnetic Imaging in MS criteria are used to establish dissemination in space for the diagnosis of CDMS in the 2010 International Panel Criteria, whereas the presence of both enhancing and nonenhancing lesions on a single MRI scan establishes dissemination in time. These criteria require one or more lesions in at least two of four locations: periventricular, juxtacortical, infratentorial, and spinal cord, excluding symptomatic brain stem and spinal cord lesions (Polman et al., 2011). Brain MRI scan reveals characteristic ovoid T2/FLAIR hyperintense lesions predominantly involving periventricular and juxtacortical locations. This appearance is distinct from the small, round subcortical lesions of small vessel ischemic disease or migraine. The corpus callosum is a relatively specific site of MS plaques; lesions there appear to be more closely associated with cognitive dysfunction (Pelletier et al., 1993). Contrast enhancement on post–gadolinium T1 imaging occurs during periods of active disease. Enhancement may be described as nodular, punctate, or ringlike, with an open ring being more suggestive of demyelination versus other causes of enhancement (Masdeu et al., 1996). Some lesions appear hypointense on T1 imaging, also called black holes. These T1 hypointense lesions correlate histopathologically with greater tissue damage and axonal injury (van Walderveen et al., 1998). High-resolution (7T) scanning has improved detection and characterization of cortical lesion burden, but this technology is not available for routine clinical care.

In addition to white matter lesion burden, brain atrophy is an important surrogate for clinical disability and cognitive dysfunction (Chard et al., 2002). Atrophy progresses more quickly in patients with MS compared with age-matched controls. Atrophy occurs in both white and gray matter, but regional differences including cortical atrophy, thalamic, callosal, and spinal cord gray matter atrophy may correlate more accurately with measures of disability (Houtchens et al., 2007; Schlaeger et al., 2014; Yaldizli et al., 2010).

Advanced imaging provides an additional window into the underlying pathology of MS. DTI provides additional information with regard to the integrity of white matter and tractography and supports the concept that disease activity occurs even in normal-appearing white matter on standard MRI sequences such as T2 and FLAIR (Moll et al., 2011). Magnetization transfer imaging and the magnetization transfer ratio are useful surrogates for monitoring changes associated with acute inflammation and subsequent remyelination (Chen et al., 2008). Functional MRI is a useful measure of plasticity in MS. Studies of resting-state functional connectivity have revealed differences in patients with CIS, RRMS, and PPMS or SPMS, with increased connectivity in patients with CIS followed by decreases in patients with more advanced relapsing or progressive disease (Rocca et al., 2010; Roosendaal et al., 2010); this suggests a pattern of early adaptation which is ultimately overcome as the disease progresses.

Genetics

MS is a polygenic disorder with complex inheritance patterns. Risk alleles are common in the population and heritability is complex. Monozygotic twins have an MS concordance of 25–50%, which suggests that environmental or postgenomic factors have an important role in the development of MS (Nielsen et al., 2005). Human leukocyte antigen (HLA) polymorphisms were identified early as an important risk factor for MS. HLA are peripheral blood mononuclear cell surface proteins that have a key role in self-antigen recognition and immune regulation. The HLA proteins are the human form of the major histocompatibility complex (MHC). In humans, the MHC gene complex resides on chromosome 6 and consists of more than 3000 genes encoding for different HLA classes (Cree, 2014). Polymorphisms in the HLA-DRB1 gene confer most of the genetic susceptibility for MS. In European-descended populations, the HLA-DRB1*15:01 is the primary risk allele (Oksenberg et al., 2004). In African Americans, the primary risk allele is HLA-DRB1*15:03 (Cree et al., 2009). Numerous other HLA alleles have a smaller but still important role.

Although the role of HLA as a risk factor for MS is well-established, its overall contribution to MS susceptibility is small. MHC accounts for approximately 20% of MS heritability. This raises the hypothesis that loci outside the MHC must also have a role in MS genetics (International Multiple Sclerosis Genetics Consortium, 2010). The International Multiple Sclerosis Genetics Consortium (IMSGC) was founded in 2003 to collect the large sample sizes required to study these small but important genetic signals. Using GWAS, IMSGC researchers identified interleukin-2 receptor (IL2R)-α– and IL7R-α–encoding genes as important contributors, although the odds ratio for these loci were small (International Multiple Sclerosis Genetics Consortium et al., 2007). The list of non-MHC risk genes continues to expand; currently, non-MHC loci account for an additional 5% of MS heritability (Cree, 2014).

Vitamin D deficiency is a risk factor for MS susceptibility and genes related to vitamin D metabolism. Two genes, CYP27B1 and CYP24A1, which encode for enzymes involved in the synthesis and degradation of vitamin D, were identified in an IMSGC GWAS analysis (Ramagopalan et al., 2011). Vitamin D may also have a role in gene expression at MHC risk loci, with vitamin D receptor binding elements identified in the region of most MS-associated genes (Ramagopalan et al., 2010; Ramagopalan, Knight, & Ebers, 2009).

Phenotypically, HLA alleles contribute to age at onset and may also contribute to disease burden on MRI scans (Cree et al., 2009; Okuda et al., 2009). Some studies suggest a role in cognitive performance (Okuda et al., 2009). These alleles do not appear to influence overall neurologic disabilities as measured by the expanded disability status scale or disease course (relapsing versus progressive) (Romero-Pinel et al., 2011). The impact of HLA alleles on MS severity is less clear.

Pathology

Multifocal plaques, characterized by demyelination with variable gliosis and perivenular inflammation, are the pathological hallmark of MS. As with imaging patterns, the optic nerves, subpial spinal cord, brain stem, cerebellum, juxtacortical, and periventricular white matter regions are preferentially involved (Popescu & Lucchinetti, 2012). The pattern of inflammation in the MS plaque evolves, with early active plaques showing dense infiltrates of macrophages containing myelin debris, followed by a transition to chronic active plaques containing activated microglia or smaller numbers of myelin-laden macrophages (Popescu & Lucchinetti, 2012). Remyelination occurs as acute inflammation subsides, although it is often incomplete. About half of chronic plaques show evidence of remyelination, although myelin density usually remains reduced and myelin sheaths are thin (Barkhof et al., 2003). Axonal damage and loss also occur, although relative axonal sparing is important in the pathologic diagnosis of demyelinating disease and the presence of axon loss commensurate with demyelination is more suggestive of infarction (Popescu & Lucchinetti, 2012). Cortical lesions are present early in MS and appear to correlate with cognitive dysfunction. Three types of cortical lesions have been described, including subpial lesions which extend from the pial surface into the cortex, intracortical lesions which are confined within the cortex, and leukocortical lesions which involve both gray and white matter (Lucchinetti et al., 2011). Meningeal inflammation, consisting of lymphocytes and macrophages, may have a role in the formation of these lesions and may be an important driver of neurodegeneration in patients with progressive disease (Choi et al., 2012).

REFERENCES

Achiron, A., Chapman, J., Magalashvili, D., Dolev, M., Lavie, M., Bercovich, E., … Barak, Y. (2013). Modeling of cognitive impairment by disease duration in multiple sclerosis: a cross-sectional study. *Plos One, 8*(8), e71058. http://dx.doi.org/10.1371/journal.pone.0071058.

Amato, M. P., Ponziani, G., Siracusa, G., & Sorbi, S. (2001). Cognitive dysfunction in early-onset multiple sclerosis: a reappraisal after 10 years. *Archives of Neurology, 58*(10), 1602–1606.

Atwood, L. D., Wolf, P. A., Heard-Costa, N. L., Massaro, J. M., Beiser, A., D'Agostino, R. B., & DeCarli, C. (2004). Genetic variation in white matter hyperintensity volume in the Framingham Study. *Stroke: A Journal of Cerebral Circulation, 35*(7), 1609–1613. http://dx.doi.org/10.1161/01.STR.0000129643.77045.10.

Barkhof, F., Bruck, W., De Groot, C. J., Bergers, E., Hulshof, S., Geurts, J., … van der Valk, P. (2003). Remyelinated lesions in multiple sclerosis: magnetic resonance image appearance [Research Support, Non-U.S. Gov't] *Archives of Neurology, 60*(8), 1073–1081. http://dx.doi.org/10.1001/archneur.60.8.1073.

Baumann, N., & Pham-Dinh, D. (2001). Biology of oligodendrocyte and myelin in the mammalian central nervous system. *Physiological Reviews, 81*(2), 871–927.

Benedict, R. H., Carone, D. A., & Bakshi, R. (2004). Correlating brain atrophy with cognitive dysfunction, mood disturbances, and personality disorder in multiple sclerosis. *Journal of Neuroimaging: Official Journal of the American Society of Neuroimaging, 14*(Suppl. 3), 36S–45S. http://dx.doi.org/10.1177/1051228404266267.

Benedict, R. H., Morrow, S., Rodgers, J., Hojnacki, D., Bucello, M. A., Zivadinov, R., & Weinstock-Guttman, B. (2014). Characterizing cognitive function during relapse in multiple sclerosis. *Multiple Sclerosis: Clinical and Laboratory Research, 20*(13), 1745–1752. http://dx.doi.org/10.1177/1352458514533229.

Binswanger, O. (1894). Die Abgrenzung der allgemeinen progressiven Paralyse. *Berl Klin Wochenschr, 31*, 1103–1105, 1137–1139; 1180–1186.

van den Berg, E., Biessels, G. J., de Craen, A. J. M., Gussekloo, J., & Westendorp, R. G. J. (2007). The metabolic syndrome is associated with decelerated cognitive decline in the oldest old. *Neurology, 69*(10), 979–985. http://dx.doi.org/10.1212/01.wnl.0000271381.30143.75.

Brass, S. D., Smith, E. E., Arboleda-Velasquez, J. F., Copen, W. A., & Frosch, M. P. (2009). Case records of the Massachusetts General Hospital. Case 12-2009. A 46-year-old man with migraine, aphasia, and hemiparesis and similarly affected family members [Case Reports] *The New England Journal of Medicine, 360*(16), 1656–1665. http://dx.doi.org/10.1056/NEJMcpc0810839.

Carmelli, D., DeCarli, C., Swan, G. E., Jack, L. M., Reed, T., Wolf, P. A., & Miller, B. L. (1998). Evidence for genetic variance in white matter hyperintensity volume in normal elderly male twins. *Stroke: A Journal of Cerebral Circulation, 29*(6), 1177–1181.

Catani, M., & ffytche, D. H. (2005). The rises and falls of disconnection syndromes. *Brain: A Journal of Neurology, 128*(Pt 10), 2224–2239. http://dx.doi.org/10.1093/brain/awh622.

Catani, M., & Thiebaut de Schotten, M. (2008). A diffusion tensor imaging tractography atlas for virtual in vivo dissections. *Cortex: A Journal Devoted to the Study of the Nervous System and Behavior, 44*(8), 1105–1132. http://dx.doi.org/10.1016/j.cortex.2008.05.004.

Chabriat, H., Herve, D., Duering, M., Godin, O., Jouvent, E., Opherk, C., … Dichgans, M. (2015). Predictors of clinical worsening in cerebral autosomal dominant arteriopathy with subcortical infarcts and leukoencephalopathy: prospective cohort study. *Stroke: A Journal of Cerebral Circulation, 47*(1), 4–11. http://dx.doi.org/10.1161/STROKEAHA.115.010696.

Chabriat, H., Joutel, A., Dichgans, M., Tournier-Lasserve, E., & Bousser, M. G. (2009). Cadasil. *Lancet Neurology, 8*(7), 643–653. http://dx.doi.org/10.1016/s1474-4422(09)70127-9.

Charcot, J. M. (1877). *Lectures on the diseases of the nervous system delivered at La Salpêtrière*. London: New Sydenham Society.

Chard, D. T., Griffin, C. M., Parker, G. J., Kapoor, R., Thompson, A. J., & Miller, D. H. (2002). Brain atrophy in clinically early relapsing-remitting multiple sclerosis. *Brain: A Journal of Neurology, 125*(Pt 2), 327–337.

Charlton, R. A., Morris, R. G., Nitkunan, A., & Markus, H. S. (2006). The cognitive profiles of CADASIL and sporadic small vessel disease. *Neurology, 66*(10), 1523–1526.

Chen, J. T., Collins, D. L., Atkins, H. L., Freedman, M. S., Arnold, D. L., & Canadian MS/BMT Study Group. (2008). Magnetization transfer ratio evolution with demyelination and remyelination in multiple sclerosis lesions. *Annals of Neurology, 63*(2), 254–262. http://dx.doi.org/10.1002/ana.21302.

Chiaravalloti, N. D., & DeLuca, J. (2008). Cognitive impairment in multiple sclerosis. *Lancet Neurology, 7*(12), 1139–1151. http://dx.doi.org/10.1016/S1474-4422(08)70259-X.

Choi, J. C. (2015). Genetics of cerebral small vessel disease. *Journal of Stroke, 17*(1), 7–16. http://dx.doi.org/10.5853/jos.2015.17.1.7.

Choi, S. R., Howell, O. W., Carassiti, D., Magliozzi, R., Gveric, D., Muraro, P. A., … Reynolds, R. (2012). Meningeal inflammation plays a role in the pathology of primary progressive multiple sclerosis [Research Support, Non-U.S. Gov't] *Brain: A Journal of Neurology, 135*(Pt 10), 2925–2937. http://dx.doi.org/10.1093/brain/aws189.

Cree, B. A. (2014). Multiple sclerosis genetics [Review] *Handbook of Clinical Neurology, 122*, 193–209. http://dx.doi.org/10.1016/B978-0-444-52001-2.00009-1.

Cree, B. A., Reich, D. E., Khan, O., De Jager, P. L., Nakashima, I., Takahashi, T., … Oksenberg, J. R. (2009). Modification of multiple sclerosis phenotypes by African ancestry at HLA [Multicenter Study Research Support, N.I.H., Extramural Research Support, Non-U.S. Gov't] *Archives of Neurology, 66*(2), 226–233. http://dx.doi.org/10.1001/archneurol.2008.541.

Crichton, G. E., Elias, M. F., Buckley, J. D., Murphy, K. J., Bryan, J., & Frisardi, V. (2012). Metabolic syndrome, cognitive performance, and dementia. *Journal of Alzheimer's Disease: JAD, 30*(Suppl. 2), S77–S87. http://dx.doi.org/10.3233/JAD-2011-111022.

DeCarli, C. (2013). Clinically asymptomatic vascular brain injury: a potent cause of cognitive impairment among older individuals. *Journal of Alzheimer's Disease: JAD, 33*(Suppl. 1), S417–S426. http://dx.doi.org/10.3233/JAD-2012-129004.

DeCarli, C., Murphy, D. G., Tranh, M., Grady, C. L., Haxby, J. V., Gillette, J. A., … Schapiro, M. B. (1995). The effect of white matter hyperintensity volume on brain structure, cognitive performance, and cerebral metabolism of glucose in 51 healthy adults. *Neurology, 45*(11), 2077–2084.

Dejerine, J. J. (1892). Contribution à l'étude anatomopathologique et clinique des différentes variétés de cécité verbale. *Mémoires de la Société de Biologie, 4*, 61–90.

DeLuca, G. C., Yates, R. L., Beale, H., & Morrow, S. A. (2015). Cognitive impairment in multiple sclerosis: clinical, radiologic and pathologic insights. *Brain Pathology (Zurich, Switzerland), 25*(1), 79–98. http://dx.doi.org/10.1111/bpa.12220.

Dichgans, M. (2009). Cognition in CADASIL. *Stroke: A Journal of Cerebral Circulation, 40*(Suppl. 3), S45–S47. http://dx.doi.org/10.1161/strokeaha.108.534412.

Erkinjuntti, T., Benavente, O., Eliasziw, M., Munoz, D. G., Sulkava, R., Haltia, M., & Hachinski, V. (1996). Diffuse vacuolization (spongiosis) and arteriolosclerosis in the frontal white matter occurs in vascular dementia. *Archives of Neurology, 53*(4), 325–332.

Feinstein, A., Magalhaes, S., Richard, J. F., Audet, B., & Moore, C. (2014). The link between multiple sclerosis and depression. *Nature Reviews. Neurology, 10*(9), 507–517. http://dx.doi.org/10.1038/nrneurol.2014.139.

Fernando, M. S., Simpson, J. E., Matthews, F., Brayne, C., Lewis, C. E., Barber, R., … MRC Cognitive Function and Ageing Neuropathology Study Group. (2006). White matter lesions in an unselected cohort of the elderly: molecular pathology suggests origin from chronic hypoperfusion injury. *Stroke: A Journal of Cerebral Circulation, 37*(6), 1391–1398. http://dx.doi.org/10.1161/01.STR.0000221308.94473.14.

Feuillet, L., Reuter, F., Audoin, B., Malikova, I., Barrau, K., Cherif, A. A., & Pelletier, J. (2007). Early cognitive impairment in patients with clinically isolated syndrome suggestive of multiple sclerosis. *Multiple Sclerosis: Clinical and Laboratory Research, 13*(1), 124–127.

Filley, C. M. (2005). White matter and behavioral neurology. *Annals of the New York Academy of Sciences, 1064*, 162–183. http://dx.doi.org/10.1196/annals.1340.028.

Filley, C. M. (2009). Exploring white matter microstructure: new insights from diffusion tensor imaging. *Neurology, 73*(21), 1718–1719. http://dx.doi.org/10.1212/WNL.0b013e3181c2936b.

Filley, C. M. (2012). *The behavioral neurology of white matter* (2nd ed.). Oxford; New York: Oxford University Press.

Forteza, A. M., Brozman, B., Rabinstein, A. A., Romano, J. G., & Bradley, W. G. (2001). Acetazolamide for the treatment of migraine with aura in CADASIL [Case Reports] *Neurology, 57*(11), 2144–2145.

Fukutake, T. (2011). Cerebral autosomal recessive arteriopathy with subcortical infarcts and leukoencephalopathy (CARASIL): from discovery to gene identification [Review] *Journal of Stroke and Cerebrovascular Diseases: The Official Journal of National Stroke Association, 20*(2), 85–93. http://dx.doi.org/10.1016/j.jstrokecerebrovasdis.2010.11.008.

Garg, R. K. (2008). Subacute sclerosing panencephalitis [Review] *Journal of Neurology, 255*(12), 1861–1871. http://dx.doi.org/10.1007/s00415-008-0032-6.

Geschwind, N. (1965). Disconnexion syndromes in animals and man. II. *Brain: A Journal of Neurology, 88*(3), 585–644.

Gorelick, P. B., Scuteri, A., Black, S. E., Decarli, C., Greenberg, S. M., Iadecola, C., … Anesthesia. (2011). Vascular contributions to cognitive impairment and dementia: a statement for healthcare professionals from the American Heart Association/American Stroke Association [Review] *Stroke: A Journal of Cerebral Circulation, 42*(9), 2672–2713. http://dx.doi.org/10.1161/STR.0b013e3182299496.

Hauw, J. J. (1995). The history of lacunes. In G. A. Donnan, B. Norrving, J. M. Bamford, & J. Bogousslavsky (Eds.), *Lacunar and other subcortical infarctions* (pp. 3–15). New York: Oxford University Press.

Houtchens, M. K., Benedict, R. H., Killiany, R., Sharma, J., Jaisani, Z., Singh, B., ... Bakshi, R. (2007). Thalamic atrophy and cognition in multiple sclerosis. *Neurology, 69*(12), 1213–1223. http://dx.doi.org/10.1212/01.wnl.0000276992.17011.b5.

Hurth, K., Tarawneh, R., Ghoshal, N., Benzinger, T. L., Clifford, D. B., Geschwind, M., ... Cairns, N. J. (2015). Whipple's disease masquerades as dementia with Lewy bodies [Research Support, N.I.H., Extramural Research Support, Non-U.S. Gov't] *Alzheimer Disease and Associated Disorders, 29*(1), 85–89. http://dx.doi.org/10.1097/WAD.0b013e3182a715da.

Iadecola, C. (2013). The pathobiology of vascular dementia. *Neuron, 80*(4), 844–866. http://dx.doi.org/10.1016/j.neuron.2013.10.008.

International Multiple Sclerosis Genetics Consortium. (2010). Comprehensive follow-up of the first genome-wide association study of multiple sclerosis identifies KIF21B and TMEM39A as susceptibility loci. *Human Molecular Genetics, 19*(5), 953–962. http://dx.doi.org/10.1093/hmg/ddp542.

International Multiple Sclerosis Genetics Consortium, Hafler, D. A., Compston, A., Sawcer, S., Lander, E. S., Daly, M. J., ... Hauser, S. L. (2007). Risk alleles for multiple sclerosis identified by a genomewide study. *The New England Journal of Medicine, 357*(9), 851–862. http://dx.doi.org/10.1056/NEJMoa073493.

Jouvent, E., Mangin, J. F., Herve, D., During, M., Dichgans, M., & Chabriat, H. (2012). Cortical folding influences migraine aura symptoms in CADASIL [Multicenter Study Research Support, Non-U.S. Gov't] *The Journal of Neurology, Neurosurgery, and Psychiatry, 83*(2), 213–216. http://dx.doi.org/10.1136/jnnp-2011-300825.

van der Knaap, M. S., Breiter, S. N., Naidu, S., Hart, A. A., & Valk, J. (1999). Defining and categorizing leukoencephalopathies of unknown origin: MR imaging approach. *Radiology, 213*(1), 121–133.

van der Knaap, M. S., Naidu, S., Breiter, S. N., Blaser, S., Stroink, H., Springer, S., ... Powers, J. M. (2001). Alexander disease: diagnosis with MR imaging. *AJNR. American Journal of Neuroradiology, 22*(3), 541–552.

van der Knaap, M. S., & Valk, J. (2005a). Selective vulnerability. In M. S. van der Knaap, & J. Valk (Eds.), *Magnetic resonance of myelination and myelin disorders* (3rd ed., pp. 25–36). Berlin; New York: Springer.

van der Knaap, M. S., & Valk, J. (2005b). Pattern recognition in white matter disorders. In M. S. van der Knaap, & J. Valk (Eds.), *Magnetic resonance of myelination and myelin disorders* (3rd ed., pp. 881–904). Berlin; New York: Springer.

van der Knaap, M. S., & Valk, J. (2005c). *Magnetic resonance of myelination and myelin disorders* (3rd ed.). Berlin; New York: Springer.

Lamar, M., Price, C. C., Giovannetti, T., Swenson, R., & Libon, D. J. (2010). The dysexecutive syndrome associated with ischaemic vascular disease and related subcortical neuropathology: a Boston process approach. *Behavioural Neurology, 22*(1–2), 53–62. http://dx.doi.org/10.3233/BEN-2009-0237.

Lebrun, C., Bensa, C., Debouverie, M., Wiertlevski, S., Brassat, D., de Seze, J., ... Club Francophone de la Sclérose en Plaques. (2009). Association between clinical conversion to multiple sclerosis in radiologically isolated syndrome and magnetic resonance imaging, cerebrospinal fluid, and visual evoked potential: follow-up of 70 patients. *Archives of Neurology, 66*(7), 841–846. http://dx.doi.org/10.1001/archneurol.2009.119.

Li, C., Zheng, J., & Wang, J. (2012). An fMRI study of prefrontal cortical function in subcortical ischemic vascular cognitive impairment [Research Support, Non-U.S. Gov't] *American Journal of Alzheimer's Disease and Other Dementias, 27*(7), 490–495. http://dx.doi.org/10.1177/1533317512455841.

Libon, D. J., Price, C. C., Davis Garrett, K., & Giovannetti, T. (2004). From Binswanger's disease to leuokoaraiosis: what we have learned about subcortical vascular dementia. *The Clinical Neuropsychologist, 18*(1), 83–100.

Liem, M. K., van der Grond, J., Haan, J., van den Boom, R., Ferrari, M. D., Knaap, Y. M., ... Lesnik Oberstein, S. A. (2007). Lacunar infarcts are the main correlate with cognitive dysfunction in CADASIL. *Stroke: A Journal of Cerebral Circulation, 38*(3), 923–928.

Lockhart, S. N., Luck, S. J., Geng, J., Beckett, L., Disbrow, E. A., Carmichael, O., ... DeCarli, C. (2015). White matter hyperintensities among older adults are associated with futile increase in frontal activation and functional connectivity during spatial search. *Plos One, 10*(3), e0122445. http://dx.doi.org/10.1371/journal.pone.0122445.

Looi, J. C., & Sachdev, P. S. (1999). Differentiation of vascular dementia from AD on neuropsychological tests. *Neurology, 53*(4), 670–678.

Lublin, F. D. (2014). New multiple sclerosis phenotypic classification. *European Neurology, 72*(Suppl. 1), 1–5. http://dx.doi.org/10.1159/000367614.

Lublin, F. D., & Reingold, S. C. (1996). Defining the clinical course of multiple sclerosis: results of an international survey. National Multiple Sclerosis Society (USA) Advisory Committee on Clinical Trials of New Agents in Multiple Sclerosis. *Neurology, 46*(4), 907–911.

Lucchinetti, C. F., Popescu, B. F., Bunyan, R. F., Moll, N. M., Roemer, S. F., Lassmann, H., ... Ransohoff, R. M. (2011). Inflammatory cortical demyelination in early multiple sclerosis [Research Support, N.I.H., Extramural Research Support, Non-U.S. Gov't] *The New England Journal of Medicine, 365*(23), 2188–2197. http://dx.doi.org/10.1056/NEJMoa1100648.

MacLeod, R., Tibben, A., Frontali, M., Evers-Kiebooms, G., Jones, A., Martinez-Descales, A., ... Editorial Committee and Working Group 'Genetic Testing Counselling' of the European Huntington Disease Network. (2013). Recommendations for the predictive genetic test in Huntington's disease. *Clinical Genetics, 83*(3), 221–231. http://dx.doi.org/10.1111/j.1399-0004.2012.01900.x.

Marrie, R. A., Reingold, S., Cohen, J., Stuve, O., Trojano, M., Sorensen, P. S., ... Reider, N. (2015). The incidence and prevalence of psychiatric disorders in multiple sclerosis: a systematic review. *Multiple Sclerosis: Clinical and Laboratory Research, 21*(3), 305–317. http://dx.doi.org/10.1177/1352458514564487.

Masdeu, J. C., Moreira, J., Trasi, S., Visintainer, P., Cavaliere, R., & Grundman, M. (1996). The open ring. A new imaging sign in demyelinating disease. *Journal of Neuroimaging: Official Journal of the American Society of Neuroimaging, 6*(2), 104–107.

Miller, D., Barkhof, F., Montalban, X., Thompson, A., & Filippi, M. (2005). Clinically isolated syndromes suggestive of multiple sclerosis, part I: natural history, pathogenesis, diagnosis, and prognosis. *Lancet Neurology, 4*(5), 281–288. http://dx.doi.org/10.1016/S1474-4422(05)70071-5.

Mochel, F., Schiffmann, R., Steenweg, M. E., Akman, H. O., Wallace, M., Sedel, F., … Lossos, A. (2012). Adult polyglucosan body disease: natural history and key magnetic resonance imaging findings. *Annals of Neurology, 72*(3), 433–441. http://dx.doi.org/10.1002/ana.23598.

Moll, N. M., Rietsch, A. M., Thomas, S., Ransohoff, A. J., Lee, J. C., Fox, R., … Fisher, E. (2011). Multiple sclerosis normal-appearing white matter: pathology-imaging correlations. *Annals of Neurology, 70*(5), 764–773. http://dx.doi.org/10.1002/ana.22521.

Moreton, F. C., Razvi, S. S., Davidson, R., & Muir, K. W. (2014). Changing clinical patterns and increasing prevalence in CADASIL. *Acta Neurologica Scandinavica, 130*(3), 197–203. http://dx.doi.org/10.1111/ane.12266.

Morrow, S. A., Jurgensen, S., Forrestal, F., Munchauer, F. E., & Benedict, R. H. (2011). Effects of acute relapses on neuropsychological status in multiple sclerosis patients. *Journal of Neurology, 258*(9), 1603–1608. http://dx.doi.org/10.1007/s00415-011-5975-3.

Narayan, S. K., Gorman, G., Kalaria, R. N., Ford, G. A., & Chinnery, P. F. (2012). The minimum prevalence of CADASIL in northeast England [Research Support, N.I.H., Extramural Research Support, Non-U.S. Gov't] *Neurology, 78*(13), 1025–1027. http://dx.doi.org/10.1212/WNL.0b013e31824d586c.

Nicholson, A. M., Baker, M. C., Finch, N. A., Rutherford, N. J., Wider, C., Graff-Radford, N. R., … Rademakers, R. (2013). CSF1R mutations link POLD and HDLS as a single disease entity [Case Reports Research Support, N.I.H., Extramural Research Support, Non-U.S. Gov't] *Neurology, 80*(11), 1033–1040. http://dx.doi.org/10.1212/WNL.0b013e31828726a7.

Nielsen, N. M., Westergaard, T., Rostgaard, K., Frisch, M., Hjalgrim, H., Wohlfahrt, J., … Melbye, M. (2005). Familial risk of multiple sclerosis: a nationwide cohort study [Comparative Study Research Support, Non-U.S. Gov't] *American Journal of Epidemiology, 162*(8), 774–778. http://dx.doi.org/10.1093/aje/kwi280.

Oksenberg, J. R., Barcellos, L. F., Cree, B. A., Baranzini, S. E., Bugawan, T. L., Khan, O., … Hauser, S. L. (2004). Mapping multiple sclerosis susceptibility to the HLA-DR locus in African Americans [Research Support, Non-U.S. Gov't Research Support, U.S. Gov't, P.H.S.] *American Journal of Human Genetics, 74*(1), 160–167. http://dx.doi.org/10.1086/380997.

Okuda, D. T., Mowry, E. M., Cree, B. A., Crabtree, E. C., Goodin, D. S., Waubant, E., & Pelletier, D. (2011). Asymptomatic spinal cord lesions predict disease progression in radiologically isolated syndrome. *Neurology, 76*(8), 686–692. http://dx.doi.org/10.1212/WNL.0b013e31820d8b1d.

Okuda, D. T., Siva, A., Kantarci, O., Inglese, M., Katz, I., Tutuncu, M., … Club Francophone de la Sclérose en Plaques (CFSEP). (2014). Radiologically isolated syndrome: 5-year risk for an initial clinical event. *Plos One, 9*(3), e90509. http://dx.doi.org/10.1371/journal.pone.0090509.

Okuda, D. T., Srinivasan, R., Oksenberg, J. R., Goodin, D. S., Baranzini, S. E., Beheshtian, A., … Pelletier, D. (2009). Genotype-phenotype correlations in multiple sclerosis: HLA genes influence disease severity inferred by 1HMR spectroscopy and MRI measures. *Brain: A Journal of Neurology, 132*(Pt 1), 250–259. http://dx.doi.org/10.1093/brain/awn301.

Pelletier, J., Habib, M., Lyon-Caen, O., Salamon, G., Poncet, M., & Khalil, R. (1993). Functional and magnetic resonance imaging correlates of callosal involvement in multiple sclerosis. *Archives of Neurology, 50*(10), 1077–1082.

Polman, C. H., Reingold, S. C., Banwell, B., Clanet, M., Cohen, J. A., Filippi, M., … Wolinsky, J. S. (2011). Diagnostic criteria for multiple sclerosis: 2010 revisions to the McDonald criteria [Research Support, Non-U.S. Gov't] *Annals of Neurology, 69*(2), 292–302. http://dx.doi.org/10.1002/ana.22366.

Popescu, B. F., & Lucchinetti, C. F. (2012). Pathology of demyelinating diseases [Research Support, N.I.H., Extramural Research Support, Non-U.S. Gov't Review] *Annual Review of Pathology, 7*, 185–217. http://dx.doi.org/10.1146/annurev-pathol-011811-132443.

Prashanth, L. K., Taly, A. B., Ravi, V., Sinha, S., & Arunodaya, G. R. (2006). Adult onset subacute sclerosing panencephalitis: clinical profile of 39 patients from a tertiary care centre [Comparative Study] *The Journal of Neurology, Neurosurgery, and Psychiatry, 77*(5), 630–633. http://dx.doi.org/10.1136/jnnp.2005.085829.

Raben, N., Danon, M., Lu, N., Lee, E., Shliselfeld, L., Skurat, A. V., … Plotz, P. (2001). Surprises of genetic engineering: a possible model of polyglucosan body disease. *Neurology, 56*(12), 1739–1745.

Ramagopalan, S. V., Dyment, D. A., Cader, M. Z., Morrison, K. M., Disanto, G., Morahan, J. M., … Ebers, G. C. (2011). Rare variants in the CYP27B1 gene are associated with multiple sclerosis [Research Support, Non-U.S. Gov't] *Annals of Neurology, 70*(6), 881–886. http://dx.doi.org/10.1002/ana.22678.

Ramagopalan, S. V., Hanwell, H. E., Giovannoni, G., Knappskog, P. M., Nyland, H. I., Myhr, K. M., … Torkildsen, O. (2010). Vitamin D-dependent rickets, HLA-DRB1, and the risk of multiple sclerosis [Case Reports Letter Research Support, Non-U.S. Gov't] *Archives of Neurology, 67*(8), 1034–1035. http://dx.doi.org/10.1001/archneurol.2010.182.

Ramagopalan, S. V., Knight, J. C., & Ebers, G. C. (2009). Multiple sclerosis and the major histocompatibility complex [Research Support, Non-U.S. Gov't Review] *Current Opinion in Neurology, 22*(3), 219–225. http://dx.doi.org/10.1097/WCO.0b013e32832b5417.

Rascovsky, K., Salmon, D. P., Hansen, L. A., Thal, L. J., & Galasko, D. (2007). Disparate letter and semantic category fluency deficits in autopsy-confirmed frontotemporal dementia and Alzheimer's disease. *Neuropsychology, 21*(1), 20–30.

Rocca, M. A., Absinta, M., Moiola, L., Ghezzi, A., Colombo, B., Martinelli, V., … Filippi, M. (2010). Functional and structural connectivity of the motor network in pediatric and adult-onset relapsing-remitting multiple sclerosis. *Radiology, 254*(2), 541–550. http://dx.doi.org/10.1148/radiol.09090463.

Rocca, M. A., Amato, M. P., De Stefano, N., Enzinger, C., Geurts, J. J., Penner, I. K., … Group, M. S. (2015a). Clinical and imaging assessment of cognitive dysfunction in multiple sclerosis. *Lancet Neurology, 14*(3), 302–317. http://dx.doi.org/10.1016/S1474-4422(14)70250-9.

Rocca, M. A., Morelli, M. E., Amato, M. P., Moiola, L., Ghezzi, A., Veggiotti, P., … Filippi, M. (2015b). Regional hippocampal involvement and cognitive impairment in pediatric multiple sclerosis. *Multiple Sclerosis: Clinical and Laboratory Research*. http://dx.doi.org/10.1177/1352458515598569.

Romero-Pinel, L., Pujal, J. M., Martinez-Yelamos, S., Gubieras, L., Matas, E., Bau, L., … Arbizu, T. (2011). HLA-DRB1: genetic susceptibility and disability progression in a Spanish multiple sclerosis population. *European Journal of Neurology: The Official Journal of the European Federation of Neurological Societies, 18*(2), 337–342. http://dx.doi.org/10.1111/j.1468-1331.2010.03148.x.

Roosendaal, S. D., Hulst, H. E., Vrenken, H., Feenstra, H. E., Castelijns, J. A., Pouwels, P. J., ... Geurts, J. J. (2010). Structural and functional hippocampal changes in multiple sclerosis patients with intact memory function. *Radiology, 255*(2), 595–604. http://dx.doi.org/10.1148/radiol.10091433.

Rosenbloom, M. H., Tartaglia, M. C., Forner, S. A., Wong, K. K., Kuo, A., Johnson, D. Y., ... Geschwind, M. D. (2015). Metabolic disorders with clinical and radiologic features of sporadic Creutzfeldt-Jakob disease. *Neurology: Clinical Practice, 5*(2), 108–115. http://dx.doi.org/10.1212/CPJ.0000000000000114.

Sahathevan, R., Brodtmann, A., & Donnan, G. A. (2012). Dementia, stroke, and vascular risk factors; a review. *International Journal of Stroke: Official Journal of the International Stroke Society, 7*(1), 61–73. http://dx.doi.org/10.1111/j.1747-4949.2011.00731.x.

Scheltens, P., Erkinjunti, T., Leys, D., Wahlund, L. O., Inzitari, D., del Ser, T., ... Pantoni, L. (1998). White matter changes on CT and MRI: an overview of visual rating scales. European Task Force on Age-Related White Matter Changes [Review] *European Neurology, 39*(2), 80–89.

Schlaeger, R., Papinutto, N., Panara, V., Bevan, C., Lobach, I. V., Bucci, M., ... Henry, R. G. (2014). Spinal cord gray matter atrophy correlates with multiple sclerosis disability. *Annals of Neurology, 76*(4), 568–580. http://dx.doi.org/10.1002/ana.24241.

Schmahmann, J. D., & Pandya, D. N. (2007). Cerebral white matter—historical evolution of facts and notions concerning the organization of the fiber pathways of the brain. *Journal of the History of the Neurosciences, 16*(3), 237–267. http://dx.doi.org/10.1080/09647040500495896.

Schmahmann, J. D., & Pandya, D. N. (2009). *Fiber pathways of the brain* (Vol. 1). Oxford, United Kingdom: Oxford University Press.

Schmahmann, J. D., Smith, E. E., Eichler, F. S., & Filley, C. M. (2008). Cerebral white matter: neuroanatomy, clinical neurology, and neurobehavioral correlates. *Annals of the New York Academy of Sciences, 1142*, 266–309. http://dx.doi.org/10.1196/annals.1444.017.

Simioni, S., Ruffieux, C., Bruggimann, L., Annoni, J. M., & Schluep, M. (2007). Cognition, mood and fatigue in patients in the early stage of multiple sclerosis. *Swiss Medical Weekly, 137*(35–36), 496–501. doi: 2007/35/smw-11874.

Steenweg, M. E., Vanderver, A., Blaser, S., Bizzi, A., de Koning, T. J., Mancini, G. M., ... van der Knaap, M. S. (2010). Magnetic resonance imaging pattern recognition in hypomyelinating disorders [Research Support, Non-U.S. Gov't] *Brain: A Journal of Neurology, 133*(10), 2971–2982. http://dx.doi.org/10.1093/brain/awq257.

Studart Neto, A., Nobrega, P. R., Duarte, M. I., Lucato, L. T., Castro, L. H., & Nitrini, R. (2015). Adult-onset subacute sclerosing panencephalitis manifesting as slowly progressive dementia. *Journal of Neurovirology, 21*(4), 468–471. http://dx.doi.org/10.1007/s13365-015-0336-0.

Sundal, C., Lash, J., Aasly, J., Oygarden, S., Roeber, S., Kretzschman, H., ... Wszolek, Z. K. (2012). Hereditary diffuse leukoencephalopathy with axonal spheroids (HDLS): a misdiagnosed disease entity. *Journal of the Neurological Sciences, 314*(1–2), 130–137. http://dx.doi.org/10.1016/j.jns.2011.10.006. pii:S0022-510X(11)00609-5.

Taylor, W. D., Aizenstein, H. J., & Alexopoulos, G. S. (2013). The vascular depression hypothesis: mechanisms linking vascular disease with depression. *Molecular Psychiatry, 18*(9), 963–974. http://dx.doi.org/10.1038/mp.2013.20.

Thal, D. R., Grinberg, L. T., & Attems, J. (2012). Vascular dementia: different forms of vessel disorders contribute to the development of dementia in the elderly brain. *Experimental Gerontology, 47*(11), 816–824. http://dx.doi.org/10.1016/j.exger.2012.05.023.

Tikka, S., Mykkanen, K., Ruchoux, M. M., Bergholm, R., Junna, M., Poyhonen, M., ... Kalimo, H. (2009). Congruence between NOTCH3 mutations and GOM in 131 CADASIL patients. *Brain: A Journal of Neurology, 132*(Pt 4).

Viswanathan, A., Gschwendtner, A., Guichard, J. P., Buffon, F., Cumurciuc, R., O'Sullivan, M., ... Chabriat, H. (2007). Lacunar lesions are independently associated with disability and cognitive impairment in CADASIL. *Neurology, 69*(2), 172–179.

van Walderveen, M. A., Kamphorst, W., Scheltens, P., van Waesberghe, J. H., Ravid, R., Valk, J., ... Barkhof, F. (1998). Histopathologic correlate of hypointense lesions on T1-weighted spin-echo MRI in multiple sclerosis. *Neurology, 50*(5), 1282–1288.

Weinshenker, B. G., & Ebers, G. C. (1989). Systemic interferon alfa in multiple sclerosis. *Archives of Neurology, 46*(3), 251–252.

Wernicke, C. (1874). *Der aphasische Symptomencomplex. Eine psychologische Studie auf anatomischer Basis* (p. 28). Breslau: Cohn.

Wollenweber, F. A., Hanecker, P., Bayer-Karpinska, A., Malik, R., Bazner, H., Moreton, F., ... Duering, M. (2015). Cysteine-sparing CADASIL mutations in NOTCH3 show proaggregatory properties in vitro [Case Reports Research Support, Non-U.S. Gov't] *Stroke: A Journal of Cerebral Circulation, 46*(3), 786–792. http://dx.doi.org/10.1161/STROKEAHA.114.007472.

Yaffe, K., Haan, M., Blackwell, T., Cherkasova, E., Whitmer, R. A., & West, N. (2007). Metabolic syndrome and cognitive decline in elderly Latinos: findings from the Sacramento Area Latino Study of Aging study. *Journal of the American Geriatrics Society, 55*(5), 758–762. http://dx.doi.org/10.1111/j.1532-5415.2007.01139.x.

Yaldizli, O., Atefy, R., Gass, A., Sturm, D., Glassl, S., Tettenborn, B., & Putzki, N. (2010). Corpus callosum index and long-term disability in multiple sclerosis patients. *Journal of Neurology, 257*(8), 1256–1264. http://dx.doi.org/10.1007/s00415-010-5503-x.

Zheng, L., Mack, W. J., Chui, H. C., Heflin, L., Mungas, D., Reed, B., ... Kramer, J. H. (2012). Coronary artery disease is associated with cognitive decline independent of changes on magnetic resonance imaging in cognitively normal elderly adults. *Journal of the American Geriatrics Society, 60*(3), 499–504. http://dx.doi.org/10.1111/j.1532-5415.2011.03839.x.

Chapter 42

Amyotrophic Lateral Sclerosis 1 and Many Diseases

Elena Ratti and James D. Berry
Neurological Clinical Research Institute (NCRI), Massachusetts General Hospital, Boston, MA, United States

INTRODUCTION

In 1869, French neurologist Dr Jean-Martin Charcot first described a clinical syndrome initially named Charcot sclerosis, now known as amyotrophic lateral sclerosis (ALS) (Charcot & Joffory, 1869). "Amyotrophy" derives from the Greek for "lack of nourishment to the muscle"; "lateral sclerosis" refers to the "scarring" of the corticospinal tracts containing the axons of upper motor neurons.

ALS affects both upper and lower motor neurons, leading to progressive paralysis and ultimately to death, on average 3 to 5 years after the onset of weakness, although variability is striking (Turner, Parton, Shaw, Leigh, & Al-Chalabi, 2003). It belongs to the broader category of motor neuron diseases, which includes diseases presenting with isolated lower or upper motor neuron involvement (Fig. 42.1). Whereas traditionally ALS has been construed as a disease of exclusively the motor neurons, important cognitive and behavioral effects of the disease are now well appreciated, recasting the disease as a neurodegenerative disease, albeit one marked primarily by motor neuron loss, which is a paradigm shift in the disease conceptualization.

The diagnosis of ALS remains clinical, although electromyography (EMG) lends electrodiagnostic support. The revised El Escorial Criteria are consensus guidelines that define levels of certainty of diagnosis, more or less based on the number of body regions affected clinically and by EMG (Brooks, Miller, Swash, & Munsat, 2000). Although these criteria were

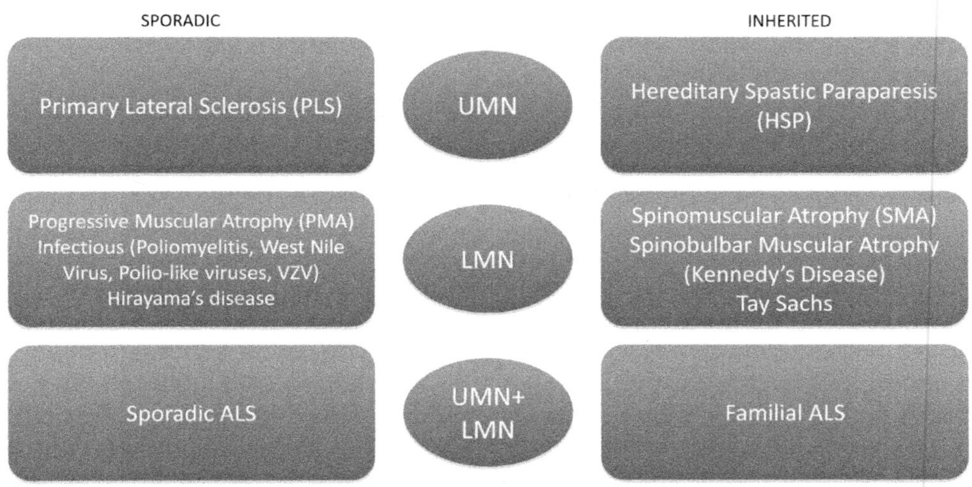

FIGURE 42.1 Inherited and sporadic motor neuron diseases (MND). *UMN*, upper motor neuron; *LMN*, lower motor neuron; *VZV*, varicella zoster virus.

Genomics, Circuits, and Pathways in Clinical Neuropsychiatry. http://dx.doi.org/10.1016/B978-0-12-800105-9.00042-1

established for clinical research, they provide a framework for clinical diagnosis. Unfortunately, they are highly specific but not sensitive (Makki & Benatar, 2007). Furthermore, the motor neuron loss and resultant weakness of ALS can strike first in any body region, and its progression is highly variable, even in people sharing an identical gene mutation (Turner, Hardiman, et al., 2013). As a result, there remains a substantial delay in diagnosis after onset of symptoms, averaging approximately 1 year (Mitchell, Callagher, et al., 2010; Paganoni et al., 2014). For now, ALS remains a disease orphan of cure, although there is one disease-modifying therapy, riluzole, which has been shown to have the modest effect of prolonging survival by 2 to 3 months (Miller, Mitchell, & Moore, 2012).

ALS genetics has been a fertile area of discovery in recent years. In the early 1990s, *superoxide dismutase 1 (SOD1)* became the first ALS-causative gene discovered; it led to the first and most robust animal models. Since then, a panoply of causative genes has been discovered and new animal and cell models are being generated at an astonishing rate. The causative genes do not appear to have a single obvious point of convergence in cell biology, but as more genes are discovered, functional groups are becoming clear. These functional groups are based on the biological pathways that are disrupted within the cell. The identification of these pathways is, in turn, informing biomarker discovery and therapeutic target identification.

In this chapter we will review the current status of knowledge of ALS genetics and connections with neuropathology, cellular pathways, and clinical manifestations of ALS.

EPIDEMIOLOGY

Average age at onset of ALS is 55 to 65 years with a male to female predominance of 1.2–1.5, an incidence of approximately 2–3/100,000 per year, a prevalence of 3–5/100,000 (Al-Chalabi & Hardiman, 2013). This translates to a lifetime risk of 1:350 for men and 1:400 for women (Al-Chalabi & Hardiman, 2013).

Approximately 90–95% of ALS is sporadic (sALS), which is defined as ALS occurring in a patient with no known family history. Five percent to 10% of people with ALS have a positive family history of ALS and are classified as familial (fALS). There is a lack of consensus about the exact definition of fALS and whether a first-, second-, or third-degree relative can be affected to designate a person as having fALS (Byrne, Elamin, Bede, & Hardiman, 2012). The distinction between fALS and sALS is becoming increasingly blurred as we learn more about the numerous causative ALS genes. Whereas nearly all of the genes seem to be autosomal dominant, they are not all completely penetrant, and at least one gene, *C9orf72*, can cause either ALS or frontotemporal dementia (FTD). That *C9orf72* expansions are being found in people with otherwise "sporadic" ALS points to an important complication in the current definition. In fact, known genes now explain the genetic underpinning of about 68% of patients with fALS and approximately 11% of patients with sALS (Renton, Chio, & Traynor, 2014). These definition challenges are resulting in new uses of the term fALS to mean either ALS in a person with an ALS-causative gene and/or a family history of ALS or FTD. This transition in the usage of the term "fALS" has been essentially imperceptible, but use of the term "fALS" will become increasingly complex as genes that confer risk and disease-modifying genes are discovered.

Thus, ALS is an uncommon disease by incidence but rare by prevalence because of its rapid progression. Clinical and genetic subsets of the disease exist and divide the population into ever smaller groupings. As a result, gene association studies require international collaborative efforts, and even biomarker studies and small trials typically require multiple centers to reach enrollment goals quickly. When subsets of patients are enrolled for trials of gene-targeted therapy or studies aimed at gene discovery in a clinical subpopulation, these collaborations become even more critical.

CLINICAL FEATURES

Clinical manifestations of ALS begin in adulthood with rare exception; however, it is unclear when subclinical changes begin. Serendipitous EMG results have shown that patients may be electrophysiologically normal even shortly before the onset of symptoms, although the EMG changes likely predate the clinical onset by a brief horizon (Aggarwal & Nicholson, 2002; Mancuso, Osta, & Navarro, 2014). Patients experience symptoms resulting from a combination of upper and lower motor neuron symptoms, and examination discloses the corresponding signs.

Upper motor signs include increased muscle tone (spasticity) and reflexes (hyperreflexia) with extensor plantar response, whereas lower motor neuron signs include decreased muscle tone, muscle atrophy (Figs. 42.2 and 42.3), and fasciculations.

Symptoms begin in the bulbar, cervical, and lumbar regions with approximately equal frequency, whereas generalized and respiratory onset is described in a few percent of patients (Swinnen & Robberecht, 2014). Symptoms generally spread within a body region and then to adjacent regions, and detailed pathological study reveals a pattern of spread from one cell

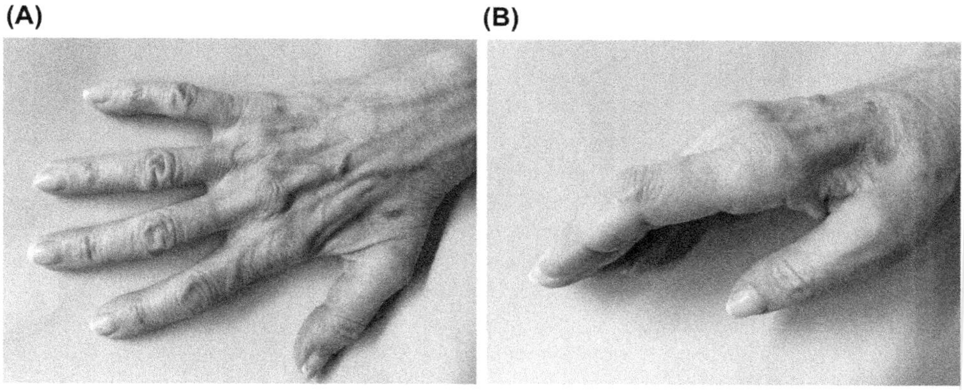

FIGURE 42.2 Hand atrophy, particularly involving the first dorsal interosseous muscle. (A) dorsal view and (B) lateral view.

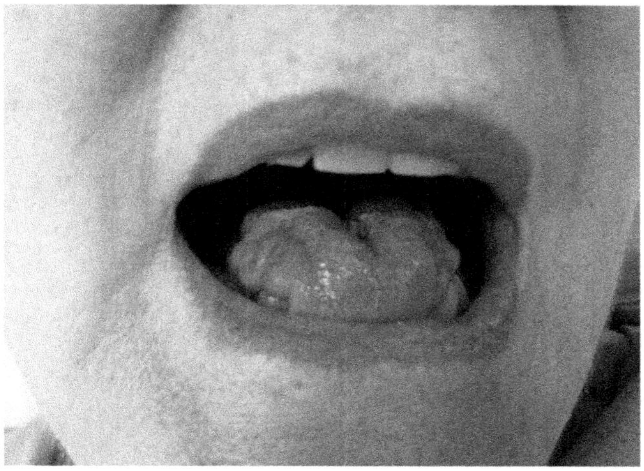

FIGURE 42.3 Tongue atrophy.

to surrounding cells between upper and lower motor neurons (UMN and LMN, respectively) (Ravits, 2014), supporting both theories about protein or RNA-mediated local spread and excitotoxicity-mediated UMN/LMN spread.

Over time, the disease causes progressive immobility, difficulty with activities of daily living, loss of independence, and dysarthria, dysphagia, protecting the airway, and (ultimately) breathing.

Weight loss appears to be nearly ubiquitous, and more rapid weight loss correlates with more rapid disease progression (Ngo, Steyn, & McCombe, 2014; Reich-Slotky et al., 2013; Shimizu et al., 2012). The cause of this weight loss is uncertain; in some cases bulbar dysfunction has a part, but in other cases the weight loss occurs despite normal or increased caloric intake, perhaps suggesting dysfunction in cellular energy use.

CLINICAL SUBTYPES OF AMYOTROPHIC LATERAL SCLEROSIS

Clinically, ALS might be divided into any number of subcategories. Common clinical distinctions include bulbar onset versus limb onset, rapid versus slow progression, UMN versus LMN predominant, and the flail arm and flail leg variants (Swinnen & Robberecht, 2014), but the relationship of these clinical variants to genetic or pathophysiologic distinctions is highly uncertain.

Primary lateral sclerosis (PLS) is a sporadically occurring motor neuron disease characterized by exclusive involvement of the upper motor neurons. It is either an ALS subtype or a unique but related motor neuron disease. Progressive muscular atrophy is the analogous subgroup or related motor neuron disease with exclusively LMN findings. Neither has a clear genetic cause nor is there evidence for unique pathophysiology or motor neuron dysfunction in these patients, relative to those with ALS.

Patients with ALS can also develop a variable degree of cognitive and/or behavioral impairment (Phukan, Pender, & Hardiman, 2007; Ringholz et al., 2005; Strong et al., 2009). On detailed cognitive testing, more than 40% of people with ALS have cognitive impairment at the initial presentation, predominantly executive dysfunction (Phukan et al., 2012). Approximately 20% of people with ALS show moderate to severe behavioral changes (apathy or disinhibition) (Lillo, Mioshi, Zoing, Kiernan, & Hodges, 2011). In fact, 5–15% of people with ALS meet the diagnostic criteria for FTD (Phukan et al., 2012). Both executive dysfunction (Elamin et al., 2011) and behavioral (Hu et al., 2013) changes are considered negative prognostic factors and confer shorter survival.

Cognitive and behavioral dysfunction is more common in patients with a hexanucleotide expansion in the *C9orf72* gene, providing one of the most striking examples of a link between ALS genetics and clinical phenotype. At the same time, not all those with *C9orf72* expansions have cognitive or behavioral dysfunction, and not all those with cognitive or behavioral dysfunction have *C9orf72* expansions.

Other examples of genotype–phenotype correlation include the *SOD1 A4V* mutation, which confers a rapid progression with LMN predominance; even so, the age and site of onset vary dramatically within a pedigree (Cudkowicz et al., 1997). More commonly, two patients with the same genetic mutation can have widely divergent clinical presentations with regard to clinical phenotype. This may highlight a role for disease-modifying factors, whether environmental, genetic, or a combination of both (Al-Chalabi & Hardiman, 2013).

AMYOTROPHIC LATERAL SCLEROSIS ANATOMY

Although degeneration of motor neurons occurs in the brain and spinal cord, the loss of motor neurons in ALS affects both the central and the peripheral nervous system. Macroscopically, motor neuron loss is seen in the motor cortex and in the motor nuclei of the brain stem and the anterior horn of the spinal cord. Concomitant degeneration of axons within the central nervous system (corticospinal tracts) and the peripheral nerves leads directly to muscle atrophy. Corticospinal tract abnormalities may be appreciated on traditional MRI (Fig. 42.4), although this is neither sensitive nor specific, and traditional imaging does not reveal other clear abnormalities. However, novel brain imaging techniques highlight patterns of cortical and subcortical atrophy, altered brain metabolism, biochemistry, receptor distribution, blood flow, and white matter integrity and connectivity even outside motor regions (Chio et al., 2014; Foerster, Welsh, & Feldman, 2013). Cortical atrophy and white matter damage can extend from motor cortex and corticospinal tracts into the frontal, temporal, and parietal regions and may also be associated with subcortical structures involvement (Agosta et al., 2012; Bede, Elamin,

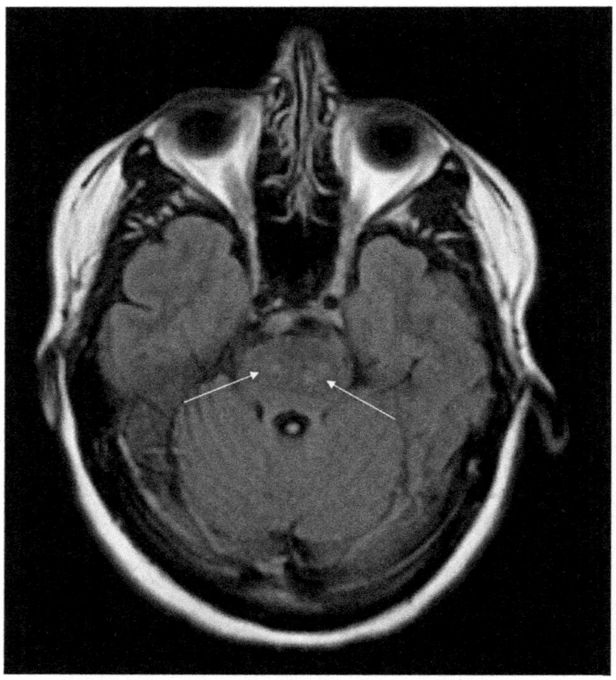

FIGURE 42.4 Bilateral T2 MRI signal hyperintensity representing the corticospinal tracts in ALS (indicated by *arrows*).

et al., 2013; Bede et al., 2014; Chio et al., 2014; Mezzapesa et al., 2013; Mioshi et al., 2013; Schuster et al., 2014; Walhout et al., 2015). Resting-state functional MRI (fMRI) evaluations of ALS patients have shown decreased functional connectivity in both sensorimotor and cognitive and behavioral associated networks and increased functional connectivity in somatosensory and extramotor areas, whereas altered patterns of brain activation were observed in task-associated fMRI studies (Chio et al., 2014). In addition, patients with *C9orf72* show characteristic atrophy of the thalamus and the cerebellum, and a range of gray and white matter abnormalities typical of both ALS and FTD (Bede, Bokde, et al., 2013; Mahoney et al., 2012; Rohrer et al., 2015; Whitwell et al., 2012). This suggests a specific neuronal vulnerability associated with this genetic background supporting an ALS-FTD continuum.

AMYOTROPHIC LATERAL SCLEROSIS PATHOLOGY

The overriding pathological signature of ALS is the loss of motor neurons in the brain, hypoglossal nucleus, and spinal cord. Microscopically, ALS is characterized by protein aggregation and deposition within motor neurons (Al-Chalabi et al., 2012; Blokhuis, Groen, Koppers, van den Berg, & Pasterkamp, 2013). Ubiquitin-positive and negative inclusions are seen in motor neurons. Ubiquitin-positive aggregates have been classified as Lewy body-like hyaline or skein-like inclusions and they are often close to neurofilamentous accumulations within the axon hillock area (Al-Chalabi et al., 2012; Blokhuis et al., 2013; Mackenzie et al., 2007). With the exception of *SOD1* and *Fused in Sarcoma/Translated in Liposarcoma Protein* (*FUS*) cases, these inclusions are TDP-43—positive, a potential common downstream process. Curiously, *SOD1*- and *FUS*-mediated ALS shows SOD1 or FUS protein accumulations, rather than TDP-43 inclusions in the motor neurons. Other pathological aggregates associated with ALS, including sporadic cases, may contain proteins or products deriving from genes that have been found mutated in ALS, such as *FUS*, *OPTN*, *UBQLN2*, and *C9orf72*. How and why these aggregates originate is a subject of active investigation.

In general, the intracellular protein aggregation of ALS appears to originate from the presence of at least one of three conditions in the diseased motor neuron, and likely an interplay of all three: (1) an increased propensity for proteins to aggregate, (2) reductions in the protein degradation pathways, and (3) increased cellular stress. Structural changes at the protein level, such as protein misfolding, particularly in low-complexity domains (referred to in some places as "prion-like domains"), appear to facilitate aggregation (Udan-Johns et al., 2014). Impairment of the normal protein degradation pathways may lead to the accumulation of aggregated proteins (Alami et al., 2014; Bosco, Lemay, et al., 2010). The aggregation of proteins can lead to cellular stress and activation of inflammatory pathways within the neuron and surrounding cells, which can hasten protein aggregation (Robberecht & Philips, 2013).

Ubiquitin-negative inclusions, Bunina bodies, found in the cytoplasm of LMNs, are also characteristic in ALS. They are small and round cystatin C—positive eosinophilic inclusions containing amorphous electron-dense material associated with tubular and vesicular structures visualized by electron microscopy (Okamoto, Mizuno, & Fujita, 2008). Whereas their exact nature and significance remain unclear, it is speculated that they may derive from cellular organelles, such as the endoplasmic reticulum, or result from abnormal protein metabolism.

Intracellular aggregates have also been observed in extramotor regions, such as frontal and temporal cortices, hippocampus, and cerebellum. They are present in both fALS and sALS and in other neurodegenerative diseases, such as frontotemporal lobar degeneration (FTLD). This suggests that protein aggregation may link fALS and sALS and neurodegenerative diseases in general. Whether the presence of protein aggregates is in itself the key deleterious event remains a subject of debate (Brotherton, Li, & Glass, 2013), although the accumulations clearly reflect abnormalities in normal cellular processes. Some evidence suggests that these protein aggregations are necessary but not sufficient to cause neurodegenerative disease, and that the resultant loss of protein function from sequestered proteins or the inflammation induced within the neuron and surrounding glia is to blame for the degenerative process (Robberecht & Philips, 2013).

AMYOTROPHIC LATERAL SCLEROSIS GENETICS BY PUTATIVE MECHANISM

Since the discovery in 1993 of the first gene associated with ALS, *SOD1* (Rosen, 1993), the field of ALS genetics has blossomed. Over the past decade, the pace of genetic discovery has hastened owing to an increasing scientific focus on ALS genetics and improvements in sequencing technologies and computational genetics methods. This led to the identification of more than 20 genes associated with ALS (Table 42.1). The discovery of this large complement of causative genes was initially unexpected, but as the number of genes discovered has grown, several potential molecular pathways involved in the pathogenesis of ALS have been highlighted, including oxidative stress, mitochondrial dysfunction, excitotoxicity, neuroinflammation, cytoskeletal disruption and axonal transport dysfunction, altered RNA

TABLE 42.1 Genes Associated With ALS by Putative Disease Mechanism (Berry & Cudkowicz, 2011; Leblond et al., 2014; Marangi & Traynor, 2014; Ratti, Cudkowicz, & Berry, 2014; Robberecht & Philips, 2013; Turner, Hardiman, et al., 2013)

Gene	Location	Gene Product	Known Normal Function	Known Inheritance Pattern	Causative Versus Putative Disease Modifying Genes (Leblond et al., 2014)	Positive Replication Studies (Leblond et al., 2014)	Known Phenotype
Oxidative Stress and Mitochondrial Dysfunction							
SOD1	21q22	Cu/Zn *Superoxide dismutase 1*	Superoxide radicals metabolism	AD, AR	Causative[a]	Yes	LMN (more often), UMN
CHCHD10	22q11	*Coiled-coil helix coiled-coil helix protein*	Mitochondrial genome stability and mitochondrial cristae junctions maintenance	AD	–	–	AD, FTD, ALS, cerebellar ataxia, myopathy
Protein Degradation and Autophagy Impairment							
UBQLN2	Xp11	*Ubiquilin 2*	Ubiquitinated proteins proteasome degradation	X-linked dominant	Causative	Yes	ALS, FTD, ALS-FTD, juvenile and adult onset
SQSTM1	5q35	*Sequestosome 1*/p62	Ubiquitinated proteins autophagy and proteasome degradation	AD	Susceptibility risk factor[b]	Yes	ALS, ALS-FTD, FTD, Paget disease of bone
VCP	9p13	Valosin-containing protein	Ubiquitin-sensitive chaperone ATPase participating in protein unfolding, complexes disassembly, autophagy, vesicle trafficking	AD	Causative	No	ALS, ALS-FTD, FTD, IBMPFD
CHMP2B	3p11	Charged Multivescicular body protein 2B or chromatin-modifying protein 2B	Component of endosomal ESCRT-III complex, involved in the sorting of membrane proteins of multivesicular bodies and their autophagy	AD	Susceptibility risk factor	Yes	ALS, LMN predominant ALS, ALS-FTD, FTD
OPTN	10p13	*Optineurin*	Regulates nuclear factor-κB pathway; involved in vesicular trafficking, immune response, and transcription regulation	AR, AD	Causative	Yes	ALS, ALS-FTD, FTD, open angle glaucoma, Paget disease of bone
FIG4	6q21	PI(3,5)P(2) 5-phosphatase	Located in vacuolar membranes, essential for lysosome function and endosomal vesicle trafficking	AD, AR	Susceptibility risk factor	No	ALS, PLS, CMT IV
TBK1	12q14	*TANK-Binding kinase 1*	Autophagy and neuroinflammation	AD	–	–	ALS, ALS-FTD, FTD

RNA Metabolism and Processing Impairment							
TARDBP	1p36	*TAR DNA binding protein 43*	DNA and RNA binding protein, containing prion-like domain and involved in RNA and DNA metabolism	AD	Causative	Yes	ALS, FTD, ALS-FTD
FUS/TLS	16p11	*Fused in Sarcoma/ Translated in Liposarcoma protein*	RNA binding protein, containing prion-like domain and involved in RNA and DNA metabolism	AD, AR	Causative	Yes	ALS, FTD, ALS-FTD possible juvenile onset and aggressive ALS phenotype
C9ORF72	9p21	Hexanucleotide repeat expansion in intronic region involved in non-ATG -dependent translation producing sense and antisense dipeptide repeat proteins (DPR)	Unknown, may act as guanine exchange factor (GEF) activating Rab GTPases regulating membrane trafficking	AD	Causative	Yes	ALS, FTD, ALS-FTD
TAF15	17q11	*TATA-Binding Protein-Associated Factor 15*	Component of transcription factor IID of RNA polymerase, participating in initiation of transcription	AD, AR	Susceptibility risk factor	No	ALS
EWSR1	22q12	*Ewing Sarcoma Breakpoint Region 1*	RNA-binding protein	AD	–	–	ALS
ANG	14q11	*Angiogenin*	RNAse; stimulates angiogenesis	AD	Susceptibility risk factor	Yes	ALS
SETX	9q34	Senataxin	DNA/RNA helicase	AD	Causative	Yes	Juvenile onset, slowly progressive, LMN predominant, ataxia and oculomotor apraxia type 2
ATXN2	12q24	Ataxin-2	Unknown; may participate in RNA metabolism/translation, interact with endoplasmic reticulum, modify TDP-43 toxicity	AD	Susceptibility risk factor	Yes	ALS and ALS-FTD if intermediate CAG repeats length (30–33), SCA2 when >34 CAG repeats
hnRNPA2B1 *hnRNPA1*	7p15, 12q13	Heterogeneous nuclear ribonucleoproteins 2B1 and A1	RNA binding proteins, regulate RNA metabolism and transport	AD	Causative	No	ALS, IBMPFD
MATR3	5q31	*Matrin 3*	RNA and DNA binding protein, interacting with TDP-43	AD	–	–	ALS

Continued

TABLE 42.1 Genes Associated With ALS by Putative Disease Mechanism (Berry & Cudkowicz, 2011; Leblond et al., 2014; Marangi & Traynor, 2014; Ratti, Cudkowicz, & Berry, 2014; Robberecht & Philips, 2013; Turner, Hardiman, et al., 2013)—cont'd

Gene	Location	Gene Product	Known Normal Function	Known Inheritance Pattern	Causative Versus Putative Disease Modifying Genes (Leblond et al., 2014)	Positive Replication Studies (Leblond et al., 2014)	Known Phenotype
Excitotoxicity							
DAO	12q22	D-Amino-acid oxidase	Controls levels of its substrate D-serine, a modifier of glutamate transmission	AD	Causative	No	ALS
Cytoskeleton Disruption and Cellular and Axonal Transport Dysfunction							
NEFH	22q12	Neurofilament-heavy subunit	Participates in axonal transport and maintenance	AD	Susceptibility risk factor	No	ALS
PRPH	12q13	Peripherin	Cytoskeletal protein Type III intermediate filament protein	AD	Susceptibility risk factor	Yes	ALS
SPG4	2q22	Spastin	Microtubule-severing protein	AD	–	–	HSP, ALS
DCTN1	2p13	Dynactin subunit 1	Axonal transport	AD	Susceptibility risk factor	Yes	LMN syndrome with vocal cord paralysis, ALS
PFN1	17p13	Profilin 1	Actin polymerization	AD	Causative	No	ALS
VAPB	20q13	*Vesicle-associated membrane protein—associated proteins B and C*	Vesicle trafficking; modulates ephrin-induced signaling, interacting with EPHA4	AD	Causative	Yes	ALS
TUBA4A	2q36	Tubulin alpha 4A protein	Microtubule component	AD	–	–	fALS
Uncertain Mechanism							
SPG11	15q14	Spatacsin	Unknown; possible role in protein trafficking, gene expression, axon maintenance, DNA damage repair	AR	Causative	Yes	ALS, including juvenile onset, HSP
ALS2	2q33	Alsin	Guanine nucleotide exchange factor for GTPases	AR	Causative	Yes	ALS, mostly UMN involvement, LMN, possible juvenile onset

Disease Modifiers							
EPHA4	2q36	Ephrin receptor tyrosine kinase EPHA4	Receptor of ephrin axonal repellent system, induces cytoskeletal rearrangements and may participate in synapse formation	–	Modifier	–	Reduced expression is associated with later onset and longer survival
KIFAP3	1q24	*Kinesin-Associated Protein 3*	Participate in formation of trimeric motor complex (KIF3) that participates in chromosomal cytokinesis and anterograde transport	–	Susceptibility risk factor	No	ALS, reduced expression in sporadic ALS is associated with longer survival
UNC13A	19p13	Protein unc-13 homolog A	Regulates neurotrasmitters release (ie, glutamate) at central and neuromuscular synapses	–	Susceptibility risk factor	Yes	ALS, associated with shorter survival
ELP3	8p21	Elongator acetyltransferase complex subunit 3	Component of RNA polymerase II, involved in RNA processing and transcript elongation	–	Susceptibility risk factor	No	ALS

LMN, lower motor neuron involvement; *UMN*, upper motor neuron involvement; *FTD*, frontotemporal dementia; *ALS*, amyotrophic lateral sclerosis; *HSP*, hereditary spastic paraplegia; *CMT IV*, Charcot–Marie–Tooth type IV; *IBMPFD*, inclusion body myopathy with early-onset Paget disease and frontotemporal dementia; *SCA-2*, spinocerebellar ataxia type 2; *polyQ*, polyglutamine; *AD*, autosomal dominant; *AR*, autosomal recessive.

[a]*Causative: genes associated with high risk of ALS.*

[b]*Susceptibility risk factor: genes associated with low risk of ALS (Leblond et al., 2014).*

metabolism and processing, autophagy dysregulation, and sequestration of normal proteins and endoplasmic reticulum stress (Ajroud-Driss & Siddique, 2014; Renton et al., 2014; Robberecht & Philips, 2013). An examination of each mutation, grouped by major putative disease mechanism, can underline shared pathology across mutations.

Oxidative Stress and Mitochondrial Dysfunction

Although oxidative damage and mitochondrial dysfunction are recognized phenomena associated with ALS, their exact role in the disease pathogenesis remains uncertain (Cozzolino, Rossi, Mirra, & Carri, 2015).

The association between ALS and mutations in *SOD1* and *CHCHD10* genes provides genetic evidence of the significance of mitochondrial involvement in motor neuron disease.

Superoxide Dismutase

SOD1, which was discovered to harbor an ALS-causing mutation in 1993 (Rosen, 1993), encodes the SOD1 protein, a cytoplasmic homodimer enzyme made by 153 amino acids and ubiquitously expressed. Its main function is to metabolize superoxide radicals, protecting cells from oxidative damage. More than 160 mutations have been identified in this gene, and with the exception of the recessive D90A mutation (Andersen et al., 1995), they are inherited in a highly penetrant autosomal dominant manner. Most pathogenic mutations are missense mutations, and nonsense and deletion mutations have only rarely been associated with disease. Mutations are not concentrated in any one location on the protein, and the normal function of the SOD1 protein may or may not be affected by mutations (Prudencio, Hart, Borchelt, & Andersen, 2009). This points to a toxic-gain-of-function, potentially mediated through its misfolding and aggregation (Bruijn et al., 1998), leading to mitochondrial dysfunction and oxidative stress. Excitotoxicity, endoplasmic reticulum stress, disruption of axonal and endosomal trafficking, and autophagy dysregulation may follow (Ilieva, Polymenidou, & Cleveland, 2009). Patients with *SOD1* mutations lack TDP-43 accumulations and instead have SOD1 accumulations (Mackenzie et al., 2007; Tan et al., 2007). Misfolded wild-type SOD1 protein aggregates have also been found in sALS and non-*SOD1* fALS (Bosco, Morfini, et al., 2010; Rotunno & Bosco, 2013), which suggests that SOD1 misfolding may have a general role in ALS.

SOD1 mutations account for approximately 12–20% of fALS cases and 1% of sALS (Renton et al., 2014). *SOD1*-related cases tend to be LMN predominant (Cudkowicz, McKenna-Yasek, Chen, Hedley-Whyte, & Brown, 1998) and are rarely associated with comorbid cognitive impairment or FTD (Wicks et al., 2009). A recessive form of ALS-associated *SOD1* D90A mutation seen in Scandinavia is characterized by a slowly progressive and ascending paralysis with onset in the lower extremities and with occasional difficulties with micturition (Andersen et al., 1996). The A4V mutation is the most common mutation in the United States and is associated with an aggressive form of ALS, leading to death within 1 year from symptom onset (Cudkowicz et al., 1997). The age of onset is highly variable and also tends to be LMN predominant.

Coiled-Coil-Helix-Coiled-Coil-Helix Domain-Containing Protein 10

Coiled-coil-helix-coiled-coil-helix domain-containing protein 10 (*CHCHD10*) encodes a coiled-coil helix coiled-coil helix protein located in the mitochondrial intermembrane space that may be involved in mitochondrial genome stability and mitochondrial cristae junctions maintenance. *CHCHD10* mutations have been identified in FTD-ALS (Chaussenot et al., 2014), sporadic ALS (Chio, Mora, et al., 2015), and patients with fALS with (Bannwarth et al., 2014) or without FTD, cerebellar ataxia, and myopathy (Johnson, Glynn, et al., 2014; Muller et al., 2014). Its clear involvement in mitochondrial function may provide a functional link between *SOD1* and *CHCHD10*.

Protein Degradation and Autophagy Impairment

In addition to the pathological evidence of intraneuronal protein aggregates, the association between ALS and genes associated with protein degradation pathways involving the proteasome and autophagy further support the role of a deficient protein clearance in ALS pathogenesis.

Ubiquilin 2

Ubiquilin 2 (*UBQNL2*) encodes a ubiquitin-like protein, ubiquilin-2, involved mainly in ubiquinated protein degradation through the proteasome system and autophagy. Isolated ALS and juvenile-onset ALS are most frequently associated with *UBQNL2* mutations; however, concomitant FTD has been reported (Deng et al., 2011). *UBQLN2* mutations explain less than 1% of both fALS and sALS (Renton et al., 2014). Because ubiquilin-2 inclusions are observed in the absence of *UBQLN2* mutations (Deng et al., 2011), this mechanism may have a role in sporadic ALS, as well.

Sequestosome 1

Sequestosome 1 (*SQSTM1*) encodes for sequestosome 1/p62, which, like ubiquilin-2, regulates ubiquinated protein degradation (Bjorkoy, Lamark, & Johansen, 2006). It also regulates nuclear factor-κB signaling and can bind ubiquitin and TDP-43 (Teyssou et al., 2013). *SQSTM1* mutations have been associated with approximately 1% of fALS and less than 1% of sALS (Fecto et al., 2011; Renton et al., 2014). Paget disease of bone (Laurin, Brown, Morissette, & Raymond, 2002; Ralston & Albagha, 2011), FTD, and FTD-ALS (Le Ber et al., 2013; Rubino et al., 2012; van der Zee et al., 2014) have also been associated with *SQSTM1* mutations. Pathologically, carriers of *SQSTM1* mutations showed large round p62 inclusions in motor neurons and increased p62 and TDP-43 protein levels in the spinal cord (Teyssou et al., 2013). Curiously, *C9orf72* hexanucleotide expansions lead to accumulations of p62 in neurons and in the cerebellum.

Valosin-Containing Protein

As a ubiquitin-sensitive chaperone adenosine triphosphatase (ATPase) that participates in protein unfolding and complex disassembly, valosin-containing protein (VCP) has been implicated in proteasome pathway, autophagy, and vesicle trafficking (Meyer, Bug, & Bremer, 2012; Weihl, Pestronk, & Kimonis, 2009). Mutated VCP has been involved in decreased ATP production through mitochondrial uncoupling (Bartolome et al., 2013). However, perhaps one of the most intriguing functions of VCP is its role in the dissolution of stress granules (Buchan, Kolaitis, Taylor, & Parker, 2013; Seguin et al., 2014). Mutated *VCP* was found in FTD associated with Paget disease of the bone and inclusion body myopathy (IBMPFD) (Watts et al., 2004) and it explains 1% of fALS and 1% of sALS (Johnson et al., 2010; Renton et al., 2014). Its place as a multisystem disease may suggest less cell specificity for *VCP* than for other mutations causing ALS, which could provide an avenue for further ALS research.

Charged Multivesicular Body Protein 2b or Chromatin-Modifying Protein 2B

Charged Multivesicular Body Protein 2b or *Chromatin-Modifying Protein 2B* (*CHMP2B*) is a component of the ESCRT-III complex involved in the sorting of membrane proteins of multivesicular bodies and their autophagy (Filimonenko et al., 2007). Mutations in *CHMP2B* have been reported in FTD, ALS, and ALS-FTD (Cox et al., 2010; Parkinson et al., 2006; Skibinski et al., 2005).

Optineurin

Optineurin (*OPTN*) is involved in multiple cellular functions; it regulates the nuclear factor-κB—signaling pathway (Zhu, Wu, Zhao, & Ashwell, 2007), and it has a role in autophagy, mitosis, and immune response, and in transcription and vesicular trafficking regulation (Chalasani, Kumari, Radha, & Swarup, 2014; Kachaner, Genin, Laplantine, & Weil, 2012). Mechanisms of diseases mediated by mutated *OPTN* remain unknown, in part because it is central to numerous pathways implicated in ALS pathology by other genes. Rare cases (<1%) of both sALS and fALS have been associated with *OPTN* mutations (Maruyama et al., 2010; Renton et al., 2014). In addition to ALS, mutations in *OPTN* have been associated with primary open angle glaucoma (Rezaie et al., 2002), Paget disease of the bone (Albagha et al., 2010), and FTD (Pottier et al., 2015), which underline its broad importance and suggest that it can cause multisystem disease such as *VCP*.

FIG4

FIG4 encodes phosphatidylinositol 3,5-biphosphate 5-phosphatase (PI(3,5)P(2)5-phosphatase), which is located in vacuolar membranes and has an important role in lysosomal function and endosomal trafficking (Gary et al., 2002). ALS has been associated with dominant *FIG4* mutations, whereas heterozygosity for a deleterious allele of *FIG4* appears to be a risk factor for ALS and PLS (Chow et al., 2009). Charcot—Marie—Tooth disease IV has been associated with mutations in *FIG4* recessively inherited.

TANK-Binding Kinase 1

A large exome-sequencing effort led to the identification of *TANK-Binding Kinase 1* (*TBK1*) as an ALS gene (Cirulli et al., 2015). *TBK1* encodes TANK-binding kinase 1, which interacts with proteins involved in innate immunity and autophagy such as *OPTN* and p62. *TBK1* mutations, alone and in association with *OPTN* mutations, have been associated with FTLD (Pottier et al., 2015), which suggests a potential additional oligogenic role in FTD pathogenesis. *TBK1* is associated with TDP-43 pathology and may mediate fALS and FTD through haploinsufficiency (Freischmidt et al., 2015). The associations of TBK1 with *OPTN*, p62, and TDP-43 provide key links between ALS-causing genes and add credence to the idea that disruption of these pathways is central to the causation of at least some types of ALS and FTD.

RNA Metabolism and Processing Impairment

Alterations in RNA processing appear to have a fundamental role in ALS pathophysiology. Both TDP-43 and FUS proteins are involved in RNA splicing, transport, and translation (Lagier-Tourenne, Polymenidou, & Cleveland, 2010). The proteins share a low-complexity prion-like domain that promotes the formation of stress granules, cytoplasmic aggregates containing proteins, and mRNA that develop in situations of cell stress to protect RNA and temporarily repress RNA translation. Either increased propensity to form granules or inefficient dissolution of the granules (eg, with mutations in *VCP*) can create permanent cytoplasmic aggregates. As such, mutations in proteins that ultimately increase the occurrence or decrease disassembly of stress granules may sequester critical proteins such as TDP-43 and FUS, and induce loss of function in these proteins.

TAR DNA-Binding Protein

The discovery of the association between TAR DNA-binding protein (TDP-43) and ubiquitin-positive neuronal inclusions in both ALS and FTD in 2006 (Neumann et al., 2006) marked a remarkable step forward in understanding ALS mechanisms and provided an incontrovertible link between ALS and FTD. *TAR DNA-Binding Protein* (*TARDBP*) encodes TDP-43, a ubiquitously expressed DNA and RNA binding protein containing a prion-like domain. It is involved in RNA and DNA processing: transcription, RNA splicing and transport, microRNA regulation, and stress granule formation (Ling, Polymenidou, & Cleveland, 2013). TDP-43 is usually localized to the nucleus, but it migrates to the cytoplasm under certain circumstances. In ALS, TDP-43 is found in cytoplasmic inclusions, cleaved, and hyperphosphorylated. ALS-causative mutations in TDP-43 pepper the glycine-rich domain of TDP-43 in the C-terminus involved in protein–protein interactions (ie, with hnRNP family proteins) that mediate RNA processing. *TARDBP* mutations may mediate their effect through loss of function owing to misfolding or mislocalization, or gain of function by sequestration of critical RNA and proteins associated with the cytoplasmic aggregates (Lee, Lee, & Trojanowski, 2012).

Mutations in *TARDBP* account for about 4% of cases of fALS cases and 1% of sALS (Renton et al., 2014; Sreedharan et al., 2008). Curiously, although TDP-43 pathology is common in FTD, *TARDBP* mutations are rarely reported in FTD (Borroni et al., 2009). *TARDBP*-mediated ALS is characterized by earlier and predominant upper limb onset as well as longer disease duration compared with sALS (Corcia et al., 2012). A few specific *TARDBP* mutations (eg, G298S1) have predictably aggressive phenotypes, but most do not (Corcia et al., 2012).

Patients with ALS associated with *TARDBP* mutations share neuropathology characteristics with those who have sALS, including neuronal loss, gliosis, and Bunina bodies in the anterior horns of the spinal cord as well as pallor in the corticospinal tracts (Al-Chalabi et al., 2012). With the exception of *SOD1* and *FUS* cases, TDP43 pathology appears to be one of the hallmarks of ALS pathology even in patients who are negative for *TARDBP* mutations. TDP-43–positive cytoplasmic inclusions are found in glial cells, motor neurons, and extramotor areas such as the frontal and temporal cortex (Van Deerlin et al., 2008).

TDP-43 pathology also links ALS with other neurodegenerative diseases; tau-negative FTLD with ubiquitin positive inclusions, Alzheimer disease (AD), Lewy body disease, and spinocerebellar ataxia 2 (SCA2) (Blokhuis et al., 2013).

Fused in Sarcoma/Translated in Liposarcoma Protein

FUS encodes an FET family RNA binding protein with structural similarities to TDP-43, including a low-complexity, prion-like domain and C-terminal nuclear localization domain. Like TDP-43, FUS is normally localized in the nucleus, whereas in the disease state it is mostly found in the cytoplasm. Unlike TDP-43, it is not ubiquinated, cleaved, or hyperphosphorylated in ALS. Most *FUS* mutations are missense and are inherited in a dominant manner with incomplete penetrance (Blair et al., 2010). However, patients with recessive inheritance and de novo mutations have been described. *FUS* mutations appear to affect mainly the protein C-terminal domain (nuclear localization domain), and p.Arg521Cys within the glycine-rich domain has been the most frequently observed among *FUS* mutations. Hypotheses of pathology mediated by *FUS* parallel TDP-43 (Ling et al., 2013), including loss of nuclear function (splicing, gene expression regulation, and microRNA biosynthesis) or toxic-gain-of-function in the cytoplasm (stress granules and aggregation). ALS is the most common phenotype associated with *FUS*, accounting for 4% of fALS and 1% of sALS. FUS-mediated ALS is associated with young onset, LMN phenotype (Hewitt et al., 2010), and, in some cases, rapidly progressive disease (Millecamps et al., 2010; Waibel, Neumann, Rabe, Meyer, & Ludolph, 2010). Like other ALS genes, mutations in *FUS* have been associated with ALS-FTD and FTD (Huey et al., 2012; Van Langenhove et al., 2010).

Pathologically, cases of *FUS* have been associated with severe loss of motor neurons in the spinal cord and relatively less in the brain stem and motor cortex, which is reflected in the LMN phenotype of these patients. Immunohistochemistry reveals normal *FUS* nuclear levels in many neurons and glial cells and dystrophic neurons with cytoplasmic FUS-positive inclusions. Surprisingly, patients with *FUS* fail to show other pathological hallmarks of ALS; there are no TDP-43–positive inclusions, and only rare ubiquitin- and p62-positive inclusions (Kwiatkowski et al., 2009; Vance et al., 2009).

FUS aggregates in all non-SOD1–mediated ALS, and in FTD, Huntington disease, and SCA (Blokhuis et al., 2013). That FUS aggregates may be composed of exclusively wild-type FUS highlights a role for FUS in neurodegeneration independent of genetic mutations.

Chromosome 9 Open Reading Frame 72 Hexanucleotide Repeat Expansion

In 2011, a hexanucleotide repeat expansion in the intronic region of the *Chromosome 9 Open Reading Frame 72* (*C9orf72*) gene was identified as the most frequent mutation associated with patients with ALS, FTD, and ALS-FTD (DeJesus-Hernandez et al., 2011; Renton et al., 2011). Approximately 23 to 30 repeats of the expansion (GGGGCC) are found in normal individuals. The minimum repeat length associated with disease is not fully known and somatic expansions vary within and between tissues (Rohrer et al., 2015), but in disease the repeats frequently expand to reach several hundred or even thousands. The function of the *C9or72* protein is uncertain, but it may have function as a guanine exchange factor activating Rab GTPases regulating membrane trafficking (Levine, Daniels, Gatta, Wong, & Hayes, 2013; Zhang, Iyer, He, & Aravind, 2012). The repeat expansion may act through either loss of function or toxic-gain-of-function.

Decreased numbers of *C9orf72* RNA transcripts in patients carrying the expansion (Mackenzie, Frick, & Neumann, 2014; Rohrer et al., 2015) support haploinsufficiency. In addition, the presence of intranuclear neuronal foci, presumably able to sequester RNA-binding proteins, could support gain-of-function (Lee et al., 2013; Mori et al., 2013). Specifically, *C9orf72* repeats form highly stable RNA G-quadruplex structures that could sequester transcription factors such as hnRNPA1 that are essential for DNA/RNA metabolism (Fratta et al., 2012; Zamiri, Reddy, Macgregor, & Pearson, 2014). An alternate theory of *C9orf72* gain-of-function focuses on dipeptides generated by repeat-associated non-ATG–dependent (RAN) translation of the repeats (Ash et al., 2013; Mori et al., 2013). These dipeptides colocalize with p62-positive inclusions in brains of *C9orf72* carriers (Gendron et al., 2013; Zu et al., 2013).

C9orf72 expansions are inherited in an autosomal dominant fashion with variable penetrance, and it is still unclear whether intergenerational anticipation occurs. *C9orf72* expansion is common, explaining up to 40% of fALS and 25% of familial FTD. In addition, mutations in *C9orf72* account for up to 7% of sALS and 6% of sporadic FTD (Majounie et al., 2012; Renton et al., 2014). Penetrance, age of onset, and disease course remain highly variable (Yokoyama, Ishiyama, Hasegawa, Uchihara, & Yagishita, 2014). More frequent bulbar onset was described in *C9orf72*-ALS. In addition, behavioral variant FTD is most commonly associated with *C9orf72* (Rohrer et al., 2015), and neuropsychiatric symptoms (psychosis, delusions, hallucinations, apathy, and disinhibition) are common. Patients with the *C9orf72* expansion have often a positive family history for neurodegenerative and psychiatric diseases (Boeve & Graff-Radford, 2012; Byrne, Elamin, Bede, Shatunov, et al., 2012).

C9orf72-ALS resembles sALS pathologically (Mackenzie et al., 2014), and *C9orf72*-FTD shows TDP-43 pathology involving predominantly the frontal and temporal neocortices and hippocampus, whereas *C9orf72*-ALS/FTD has features of both. Unique pathology in *C9orf72*-ALS includes the presence of aggregated and/or soluble RAN translation dipeptides, p62-positive/TDP43-negative neuronal cytoplasmic inclusions (NCIs), and intranuclear RNA inclusions in the cerebellum, hippocampus, frontotemporal neocortex, and pyramidal neurons (Bigio et al., 2013; Mackenzie et al., 2014). p62--Ubiquitin–positive but TDP-43–negative NCIs containing RAN translation dipeptides are often particularly numerous in the cerebellum (Mackenzie et al., 2014), even in typical patients with ALS. However, in rare cases, repeat expansions in

C9orf72 have also been associated with Parkinsonism (Lesage et al., 2013), Parkinson disease, dementia with Lewy bodies, multiple system atrophy, and corticobasal syndrome or hyperkinetic Huntington disease—like syndromes (Harms & Baloh, 2013) and cerebellar ataxia (Rohrer et al., 2015). The link between these clinical phenotypes, the underlying pathological findings, and the gene hexanucleotide expansion are an explosive area of research and may hold the keys to understanding ALS subtypes and/or neurodegeneration in general.

TATA-Binding Protein-Associated Factor 15

TATA-Binding Protein-Associated Factor 15 (*TAF15*) is a component of the RNA polymerase II. Similarly to TDP-43 and FUS, TAF-15 is an RNA-binding protein harboring a prion-like domain and prone to aggregation. Like FUS, it belongs to the FET family of proteins. Mislocalized cytoplasmic TAF15-positive punctae (negative for TDP-43) have been observed in spinal cord motor neurons in patients with sALS (Couthouis et al., 2011) and in FTLD-FUS (Mackenzie & Neumann, 2012).

Ewing Sarcoma Breakpoint Region 1

Ewing Sarcoma Breakpoint Region 1 (*EWSR1*) is an RNA-binding protein that also belongs to the FET family of proteins and appears to share properties similar to FUS and TAF15, such as the propensity to aggregate and a prion-like domain. Variants in *EWSR1* associated with ALS cases and EWSR1 cytoplasmic mislocalization were observed in spinal cord motor neurons of ALS (Couthouis et al., 2012) and in association with FTLD-FUS (Mackenzie & Neumann, 2012).

Angiogenin

Angiogenin (*ANG*) encodes angiogenin, which has a role in the vascular endothelial growth factor pathway, which has been implicated in ALS. Angiogenin is involved in inducing neovascularization and, as a member of the pancreatic ribonuclease A superfamily, it acts as a RNAase. Variants affecting its function may influence neuronal survival and may alter the formation of stress granules (Thiyagarajan, Ferguson, Subramanian, & Acharya, 2012). Some *ANG* variants were found with higher frequency in ALS and may increase the risk of developing ALS (van Es et al., 2011). Rare *ANG* mutations have been found in patients with fALS and sALS (Greenway et al., 2004, 2006). A higher frequency of *ANG* variants has also been observed in Parkinson disease (van Es et al., 2011; Rayaprolu et al., 2012).

Senataxin

Senataxin (*SETX*) encodes senataxin, a DNA/RNA helicase (Chen et al., 2004), which could be involved in RNA processing. In rare cases, mutations in *SETX* have been associated with young-onset, slowly progressive, LMN-predominant ALS (Avemaria et al., 2011; Chen et al., 2004), sometimes with ataxia (Hirano et al., 2011). *SETX* mutations have also been associated with ataxia-ocular apraxia type 2 (Asaka, Yokoji, Ito, Yamaguchi, & Matsushima, 2006).

Ataxin2

Ataxin2 (*ATXN2*) has a role in mRNA polyadenylation, stress granule formation, polyribosome assembly, and microRNA synthesis (Blokhuis et al., 2013). CAG trinucleotide expansion of the *ATXN2* gene encoding for polyglutamines (polyQ) causes SCA2. Intermediate size repeats, particularly in the 30—33 range, have been associated with increased risk for ALS (Neuenschwander, Thai, Figueroa, & Pulst, 2014) and ALS-FTD (Lattante et al., 2014) and predict shorter survival with more than 30 repeats (Chio, Calvo, et al., 2015). *ATXN2* intermediate-length polyQ repeats appear to mediate their effect by enhancing TDP-43 (Daoud et al., 2011; Elden et al., 2010; Lee et al., 2011) and FUS toxicity (Farg et al., 2013).

Heterogeneous Ribonucleoproteins

The heterogeneous ribonucleoproteins (hnRNPs) are proteins involved in pre-mRNA processing and mRNA export, localization, translation, and stability through interaction with TDP-43 (Buratti et al., 2005; Leblond, Kaneb, Dion, & Rouleau, 2014). Mutations associated with disease alter the low-complexity, prion-like domains leading to dysregulated protein polymerization and altered stress granule formation (Leblond et al., 2014). hnRNPA2B1 and hnRNPA1 mutations were observed in families with IBMPFD, FTD, and ALS, and in patients with ALS (Kim et al., 2013). However, additional studies have not confirmed these associations (Calini et al., 2013; Le Ber et al., 2014; Seelen et al., 2014).

Matrin 3

Matrin 3 (*MATR3*) encodes matrin 3 protein, a nuclear matrix protein that binds RNA and DNA and interacts with TDP-43. *MATR3* mutations have been associated with fALS (Johnson, Pioro, et al., 2014) and sALS (Lin et al., 2015). *MATR3* pathology has been observed in spinal cords of ALS patients independently from the presence of the mutation (Johnson, Pioro, et al., 2014).

Excitotoxicity

Excitotoxicity describes neuronal death from calcium influx resulting from overstimulation of glutamate receptors. Its role in the spread of ALS has long been theorized. Indeed, riluzole, the only approved therapy for ALS, is antiglutamatergic (Lacomblez, Bensimon, Leigh, Guillet, & Meininger, 1996). In addition, the astrocytic glutamate transporter excitatory amino acid transporter 2, which clears synaptic glutamate, has been shown to be downregulated or dysfunctional in ALS (Howland et al., 2002; Rothstein et al., 2005; Sattler & Rothstein, 2006; Turner, Bowser, et al., 2013). Finally, genetic data have begun to suggest a role for excitotoxicity in the initiation of ALS, as well.

D-Amino-Acid Oxidase (DAO)

D-Amino-Acid Oxidase (*DAO*) regulates the levels of D-serine, a co-agonist of the *N*-methyl-D-aspartate glutamate receptor (Paul & de Belleroche, 2012; Sasabe et al., 2012). A mutation in *DAO* (R199W) was reported to segregate with disease in a family with ALS (Mitchell, Paul, et al., 2010). In cell culture models, this mutation promoted autophagy, the formation of ubiquinated aggregates and neuronal apoptosis (Paul & de Belleroche, 2014). Excitotoxicity is likely a downstream effect of a cascade of events within motor neurons and may be more involved in disease spread than onset, yet it is intriguing that this mutation suggests at least plausible links between excitotoxity and disease causation.

Cytoskeleton Disruption and Cellular and Axonal Transport Dysfunction

Cytoskeletal architecture, cellular transport, axonal function, and synaptic vesicles release are vital functions of a neuron and have also been implicated in the pathogenesis of ALS (Schmidt, Pasterkamp, & van den Berg, 2009). More specifically, involvement of axonal transport in ALS pathogenesis is supported by the association between ALS and mutations in *neurofilament heavy chain* (*NEFH*), dynactin (*DCTN1*), *vesicle-associated membrane protein-associated protein B* (*VAPB*), and the disease modifier, kinesin-associated protein 3 (*KIFAP3*) (discussed subsequently in the Modifiers section). The relevance of cytoskeletal disruption in ALS is supported by the presence of ALS-associated mutations in genes encoding cytoskeletal proteins: peripherin (*PRPH*), spastin (*SPG4*), profilin 1 (*PFN1*), tubulin alpha 4A protein (*TUBA4*), and the disease-modifying gene, ephrin receptor (*EPHA4*) (discussed subsequently in the Modifiers section). Many of these genes encode proteins with overlapping function, even across the cytoskeletal and axonal transport pathways. *EPHA4* encodes components of the axonal repellent system, which may interact with Alsin and is modified by *VAPB* (Robberecht & Philips, 2013).

Neurofilament Heavy Chain

NEFH encodes neurofilament heavy chain, a key protein in axonal transport and maintenance. Changes in its cross-linking properties may contribute to the aberrant neurofilamentous accumulation found in the perikarya and proximal axons of ALS motor neurons directly, even in sporadic ALS (Figlewicz et al., 1994). In fact, *NEFH* variants have been associated mainly with sALS cases (Al-Chalabi et al., 1999; Figlewicz et al., 1994; Figlewicz, Rouleau, Krizus, & Julien, 1993).

Dynactin

Dynactin binds microtubules and is involved in cellular and axonal transport. *DCTN1* mutations have been observed in patients with sALS and fALS and in a family with ALS-FTD (Munch et al., 2005, 2004; Puls et al., 2003). *DCTN1* mutations were also associated with Perry syndrome (early-onset parkinsonism with hypoventilation, depression, and TDP-43 pathology) (Farrer et al., 2009). No mutations in *DCTN1* were found in a study of Parkinson disease, FTLD, and ALS (Vilarino-Guell et al., 2009). Thus, although its role is widely important, mutations seem to cause only limited clinical manifestations, a truism across most ALS-causative genes.

Vesicle-Associated Membrane Protein-Associated Protein B

VAPB is located in the endoplasmic reticulum (ER) and may be involved in ER stress, vesicle trafficking, Ca^{2+} homeostasis, and ephrin-induced signaling modulation (Morotz et al., 2012; Tsuda et al., 2008). Mutations in *VAPB* have been associated with ALS, but only uncommonly (Landers et al., 2008; Millecamps et al., 2010; Nishimura et al., 2004). *VAPB's* pathogenic role may be associated with the disruption of these functions (all of which overlap with other putative disease mechanisms, pathways, and/or causative mutations). Alternatively, its deleterious effect may be mediated by reduced anterograde axonal transport of mitochondria and disturbance of cellular Ca^{2+} homeostasis (Morotz et al., 2012). *VAPB* mRNA levels were decreased in the spinal cord of patients with ALS (Anagnostou et al., 2010).

Peripherin

PRPH is a cytoskeletal protein that maintains the neurofilament network. It is expressed in motor neurons and mutations increase its propensity to aggregate (Leung et al., 2004) in motor neurons. Variants of the gene have been found in a few patients with fALS and sALS cases (Leung et al., 2004; McKee et al., 2010; Steele & McGeer, 2008).

Spastin

Spastin (*SPG4*) encodes spastin, a microtubule-severing protein. Mutations in *SPG4* represent the most common cause of hereditary spastic paraplegia (HSP). *SPG4* mutations were also reported in a case of early-onset, slowly progressive ALS (Meyer et al., 2005) and in an adult-onset, rapidly progressive ALS case (Munch, Rolfs, & Meyer, 2008). Spastin mutations could alter microtubule dynamics, presumably through a toxic gain-of-function (Solowska et al., 2014).

Profilin 1

PFN1 encodes profilin 1, a G-actin—binding protein involved in the cytoskeletal dynamics and actin polymerization. Mutations have been associated with ALS (Wu et al., 2012), possibly owing to destabilization of the native conformation of PFN1 and subsequent propensity to aggregate (Boopathy et al., 2015), or through axonal outgrowth inhibition, integrity loss, and transport disruption (Leblond et al., 2014; Wu et al., 2012). Mutations in *PFN1* appear to account for less than 1% of all ALS (Ingre et al., 2013; Renton et al., 2014; Smith et al., 2015), whereas one variant, p.E117G, was found in a few patients with FTD (van Blitterswijk et al., 2013; Dillen et al., 2013). However, some studies have failed to demonstrate *PFN1* mutations causative of either ALS or FTD (Daoud et al., 2013; Tiloca et al., 2013).

Tubulin Alpha 4A Protein

TUBA4A encodes tubulin alpha 4A, a component of microtubules, that when mutated may disrupt microtubules dynamics and stability, affecting their ability to repolymerize (Smith et al., 2014). *TUBA4A* variants have been associated with fALS (Smith et al., 2014).

Uncertain Mechanism

A few genes have been associated with ALS, but their role in the disease pathogenesis appears particularly unclear.

Spatacsin

Spatacsin (*SPG11*) encodes spatacsin, a protein with unknown functions. It has been hypothesized that spatacsin may have a role in protein trafficking, gene expression, or axon maintenance, and variants of *SPG11* have been reported in cases of autosomic recessive juvenile-onset ALS (Daoud et al., 2012; Orlacchio et al., 2010) and HSP. The lack of data linking it to typical ALS is puzzling and may set this mutation apart from other ALS-causative mutations.

Alsin

ALS2 encodes Alsin, a guanine nucleotide exchange factor for GTPase, which may be involved in membrane organization, endocytosis as well as axon and dendrite development. *ALS2* mutations cause juvenile-onset ALS and infantile HSP (Eymard-Pierre et al., 2002; Hadano et al., 2001; Robberecht & Philips, 2013; Siddiqi et al., 2014; Yang et al., 2001). Alsin may mediate disease through a loss of function. Again, because it does not appear to cause typical ALS, its relationship to the disease and to other causative mutations remains uncertain.

Gene Modifiers

The influence of genetic factors in complex diseases such as ALS is not limited to disease-causative mutations. Disease phenotype may be influenced by multiple genetic variants with favorable or unfavorable effects on disease susceptibility, onset of symptoms, disease course, and/or duration of survival. This interplay of genetic factors modifying the phenotype may explain the challenging disease heterogeneity and mechanisms underlying both sALS and fALS.

Putative disease-modifying genes associated with ALS include *EPHA4*, *KIFAP3*, *UNC13A*, and *ELP3*. Additional possible modifiers and low-frequency variants have been explored but need further clarification (Marangi & Traynor, 2014).

Ephrin Receptor

EPHA4 is a tyrosin kinase receptor of the ephrin axonal repellent system which may participate in the formation of synapses and regulation of long-term synaptic plasticity and memory (Klein, 2009; Van Hoecke et al., 2012). Reduced *EPHA4* expression was associated with later ALS onset and longer survival (Van Hoecke et al., 2012), which suggests that its inhibition could serve as a potential therapeutic mechanism.

Kinesin-Associated Protein 3

Kinesin-Associated Protein 3 (*KIFAP3*) encodes a kinesin-associated protein that, in association with motor proteins, forms a complex involved in anterograde transport and chromosomal cytokinesis (Landers et al., 2009). Single-nucleotide polymorphisms reducing the expression of *KIFAP3* have been associated with increased survival in sALS (Landers et al., 2009). However, this was not replicated (van Doormaal et al., 2014; Traynor et al., 2010).

UNC13A

UNC13A encodes a member of the *UNC13* family, including presynaptic proteins of the central and neuromuscular synapses that is implicated in neurotransmitter secretion regulation and synaptic glutamate release (van Es et al., 2009). *UNC13A* was initially identified as a susceptibility locus for ALS (van Es et al., 2009) and subsequently associated with shorter survival in ALS (Chio et al., 2013; Diekstra et al., 2012).

RNA Elongator Acetyltransferase Complex Subunit 3

RNA Elongator Acetyltransferase Complex Subunit 3 (*ELP3*) is an RNA-processing protein that is a component of the RNA polymerase II involved in transcription elongation. *ELP3* variants were associated with patients with ALS (Simpson et al., 2009).

Possible Genetic Modifiers of C9orf72

Some gene variants have been identified as potential disease modifiers, specifically in carriers of the *C9orf72* repeat expansion (van Blitterswijk, Mullen, Wojtas, et al., 2014). These may help clarify disease heterogeneity within *C9orf72*-mediated ALS. For example, age at onset has been associated with variants in ubiquitin-associated protein 1, prion protein, and metallothionein 1E (*MT-Ie*), and survival with variants in granulin precursor, *MT-Ie*, *ELP3*, the epsilon 4 allele of apolipoproteins, *UNC13A*, and delta-aminolevulinate dehydratase (van Blitterswijk, Mullen, Wojtas, et al., 2014). In addition, variants in the transmembrane protein 106 B appeared to protect carriers against FTD (van Blitterswijk, Mullen, Nicholson, et al., 2014) and were associated with later age at onset and death in *C9orf72* expansion carriers with FTD (Gallagher et al., 2014). Furthermore, intermediate *ATXN2* repeats were found to be more frequently associated to motor neuron disease in patients with *C9orf72* expansions (van Blitterswijk, Mullen, Heckman, et al., 2014).

AMYOTROPHIC LATERAL SCLEROSIS MODELS

Genetic advances have led to the development of a panoply of preclinical models. The first and most widely used is a transgenic rodent model overexpressing mutant *SOD1 (G93A)* (Gurney et al., 1994). In many ways, it is an outstanding model of the disease, developing pathology and a clear phenotype owing to motor neuron loss. On the other hand, it develops little upper motor neuron pathology, progresses rapidly, and represents the pathology of only a single mutation. Substantial controversy surrounds the ability of this one model to reflect ALS pathology in humans or response to therapy.

Additional preclinical animal models include a host of other murine models (*C9orf72*, *TARDBP*, etc.), nematode, fruit fly, and zebrafish. Each has shown some use for exploring the disease. Induced pluripotent stem cell (iPSC)-derived motor neurons are increasingly coming into their own for understanding disease pathophysiology and for screening novel therapies in vitro (Turner, Bowser, et al., 2013).

DISCUSSION AND FUTURE DIRECTIONS

ALS is a clinically, pathologically, and genetically heterogeneous disease, and although it is typified by vulnerability of motor neurons, it is clear that the disease is far less cell type specific than once thought. The current sophisticated genetic, molecular, proteomic, advanced imaging, neuropathological technologies and computational capabilities offer an unprecedented opportunity to build new knowledge about ALS pathogenesis through a multimodal approach. Pathological, imaging, genetic, and clinical data are now demonstrating the involvement of a broad set of brain regions and cell types that are leading the research community to rethink long-held beliefs about the disease. In particular, the discovery of numerous ALS genes and ongoing elucidation of their biological role are leading us to clarify disease mechanisms and identify new potential therapeutic targets.

Pathways that appear to be of critical importance include mutant-protein and mutant-RNA toxicities causing disruption of cytoskeleton, axonal transport, mitochondrial function, and protein transport and degradation. Many of these deleterious effects may be carried out by disturbances in RNA and protein localization and resultant disruption of function. Neuroinflammation may be a direct result or a reactive process but appears to be critical to disease progression. The degree of interplay among these pathways within a given person remains unclear. Important subtypes of the disease may exist, distinguished not by clinical phenotype, but by the overriding pathophysiology. These subtypes may result from complex gene−gene (van Blitterswijk et al., 2012; Cady et al., 2015; Chio et al., 2012; Luigetti et al., 2011), gene−epigenetic (Paez-Colasante, Figueroa-Romero, Sakowski, Goutman, & Feldman, 2015), and gene−environment (Al-Chalabi & Hardiman, 2013) interactions. Because this is an uncommon disease, large-scale collaborative efforts to create large and detailed data sets of detailed clinical phenotype, and multiple -omics, perhaps from iPSC-derived motor neurons, may be required to answer these questions. An improved understanding of patient subsets by pathophysiology would lead to a paradigm shift in therapy development, with new efforts focused on specific disease mechanisms in carefully selected patient subsets. Antisense oligonucleotides against *SOD1*, for example, have been already tested in a phase 1, first-in-human trial showing their safety (Miller et al., 2013). Prognostication and patient counseling in the clinical setting could also improve dramatically.

While addressing causative disease pathology and patient subsets, studies of the genetics and processes of disease spread may lead to distinct therapeutic developments. Disease spread may be independent of mechanisms of onset. Much remains unknown about disease spread; it may occur between contiguous cells and/or through changes in neuronal networks (Eisen & Turner, 2013; Ravits, 2014; Talbot, 2014). The wide variability in rates of progression suggests that it may be possible to target mechanisms of spread and affect disease progression independent of cause.

In addition, the genetic and pathologic association of ALS with other neurodegenerative diseases ultimately may point to potential general common mechanisms of neurodegeneration, leading to the discovery of possible new shared biomarkers and therapeutic targets and approaches. Thus, understanding the complex mechanisms underlying ALS may benefit not only ALS but neurodegeneration in general.

"It is an old experience that through its mistakes Nature often offers us unexpected insights into its otherwise inscrutable secrets."

Loewy and Neuberg (1904).

REFERENCES

Aggarwal, A., & Nicholson, G. (2002). Detection of preclinical motor neurone loss in SOD1 mutation carriers using motor unit number estimation. *Journal of Neurology, Neurosurgery and Psychiatry, 73*(2), 199−201.

Agosta, F., Valsasina, P., Riva, N., Copetti, M., Messina, M. J., Prelle, A., ... Filippi, M. (2012). The cortical signature of amyotrophic lateral sclerosis. *PLoS One, 7*(8), e42816. http://dx.doi.org/10.1371/journal.pone.0042816. pii:PONE-D-12-11764.

Ajroud-Driss, S., & Siddique, T. (2014). Sporadic and hereditary amyotrophic lateral sclerosis (ALS). *Biochimica et Biophysica Acta*. http://dx.doi.org/10.1016/j.bbadis.2014.08.010. pii:S0925-4439(14)00263-4.

Al-Chalabi, A., Andersen, P. M., Nilsson, P., Chioza, B., Andersson, J. L., Russ, C., ... Leigh, P. N. (1999). Deletions of the heavy neurofilament subunit tail in amyotrophic lateral sclerosis. *Human Molecular Genetics, 8*(2), 157−164. pii:ddc036.

Al-Chalabi, A., & Hardiman, O. (2013). The epidemiology of ALS: a conspiracy of genes, environment and time. *Nature Reviews Neurology, 9*(11), 617–628. http://dx.doi.org/10.1038/nrneurol.2013.203.

Al-Chalabi, A., Jones, A., Troakes, C., King, A., Al-Sarraj, S., & van den Berg, L. H. (2012). The genetics and neuropathology of amyotrophic lateral sclerosis. *Acta Neuropathologica, 124*(3), 339–352. http://dx.doi.org/10.1007/s00401-012-1022-4.

Alami, N. H., Smith, R. B., Carrasco, M. A., Williams, L. A., Winborn, C. S., Han, S. S., … Taylor, J. P. (2014). Axonal transport of TDP-43 mRNA granules is impaired by ALS-causing mutations. *Neuron, 81*(3), 536–543. http://dx.doi.org/10.1016/j.neuron.2013.12.018. pii:S0896-6273(13)01179-3.

Albagha, O. M., Visconti, M. R., Alonso, N., Langston, A. L., Cundy, T., Dargie, R., … Ralston, S. H. (2010). Genome-wide association study identifies variants at CSF1, OPTN and TNFRSF11A as genetic risk factors for Paget's disease of bone. *Nature Genetics, 42*(6), 520–524. http://dx.doi.org/10.1038/ng.562. pii:ng.562.

Anagnostou, G., Akbar, M. T., Paul, P., Angelinetta, C., Steiner, T. J., & de Belleroche, J. (2010). Vesicle associated membrane protein B (VAPB) is decreased in ALS spinal cord. *Neurobiology of Aging, 31*(6), 969–985. http://dx.doi.org/10.1016/j.neurobiolaging.2008.07.005. pii:S0197-4580(08)00249-2.

Andersen, P. M., Forsgren, L., Binzer, M., Nilsson, P., Ala-Hurula, V., Keranen, M. L., … Marklund, S. L. (1996). Autosomal recessive adult-onset amyotrophic lateral sclerosis associated with homozygosity for Asp90Ala CuZn-superoxide dismutase mutation. A clinical and genealogical study of 36 patients. *Brain, 119*(Pt 4), 1153–1172.

Andersen, P. M., Nilsson, P., Ala-Hurula, V., Keranen, M. L., Tarvainen, I., Haltia, T., … Marklund, S. L. (1995). Amyotrophic lateral sclerosis associated with homozygosity for an Asp90Ala mutation in CuZn-superoxide dismutase. *Nature Genetics, 10*(1), 61–66. http://dx.doi.org/10.1038/ng0595-61.

Asaka, T., Yokoji, H., Ito, J., Yamaguchi, K., & Matsushima, A. (2006). Autosomal recessive ataxia with peripheral neuropathy and elevated AFP: novel mutations in SETX. *Neurology, 66*(10), 1580–1581. http://dx.doi.org/10.1212/01.wnl.0000216135.59699.9b. pii:66/10/1580.

Ash, P. E., Bieniek, K. F., Gendron, T. F., Caulfield, T., Lin, W. L., Dejesus-Hernandez, M., … Petrucelli, L. (2013). Unconventional translation of C9ORF72 GGGGCC expansion generates insoluble polypeptides specific to c9FTD/ALS. *Neuron, 77*(4), 639–646. http://dx.doi.org/10.1016/j.neuron.2013.02.004. pii:S0896-6273(13)00135-9.

Avemaria, F., Lunetta, C., Tarlarini, C., Mosca, L., Maestri, E., Marocchi, A., … Corbo, M. (2011). Mutation in the senataxin gene found in a patient affected by familial ALS with juvenile onset and slow progression. *Amyotrophic Lateral Sclerosis, 12*(3), 228–230. http://dx.doi.org/10.3109/17482968.2011.566930.

Bannwarth, S., Ait-El-Mkadem, S., Chaussenot, A., Genin, E. C., Lacas-Gervais, S., Fragaki, K., … Paquis-Flucklinger, V. (2014). A mitochondrial origin for frontotemporal dementia and amyotrophic lateral sclerosis through CHCHD10 involvement. *Brain, 137*(Pt 8), 2329–2345. http://dx.doi.org/10.1093/brain/awu138. pii:awu138.

Bartolome, F., Wu, H. C., Burchell, V. S., Preza, E., Wray, S., Mahoney, C. J., … Plun-Favreau, H. (2013). Pathogenic VCP mutations induce mitochondrial uncoupling and reduced ATP levels. *Neuron, 78*(1), 57–64. http://dx.doi.org/10.1016/j.neuron.2013.02.028. pii:S0896-6273(13)00186-4.

Bede, P., Bokde, A. L., Byrne, S., Elamin, M., McLaughlin, R. L., Kenna, K., … Hardiman, O. (2013). Multiparametric MRI study of ALS stratified for the C9orf72 genotype. *Neurology, 81*(4), 361–369. http://dx.doi.org/10.1212/WNL.0b013e31829c5eee. pii:WNL.0b013e31829c5eee.

Bede, P., Elamin, M., Byrne, S., McLaughlin, R. L., Kenna, K., Vajda, A., … Hardiman, O. (2013). Basal ganglia involvement in amyotrophic lateral sclerosis. *Neurology, 81*(24), 2107–2115. http://dx.doi.org/10.1212/01.wnl.0000437313.80913.2c. pii:01.wnl.0000437313.80913.2c.

Bede, P., Elamin, M., Byrne, S., McLaughlin, R. L., Kenna, K., Vajda, A., … Hardiman, O. (2014). Patterns of cerebral and cerebellar white matter degeneration in ALS. *Journal of Neurology, Neurosurgery and Psychiatry*. http://dx.doi.org/10.1136/jnnp-2014–308172. pii:jnnp-2014-308172.

Berry, J. D., & Cudkowicz, M. E. (2011). New considerations in the design of clinical trials for amyotrophic lateral sclerosis. *Clinical Investigation (London), 1*(10), 1375–1389. http://dx.doi.org/10.4155/cli.11.127.

Bigio, E. H., Weintraub, S., Rademakers, R., Baker, M., Ahmadian, S. S., Rademaker, A., … Mesulam, M. M. (2013). Frontotemporal lobar degeneration with TDP-43 proteinopathy and chromosome 9p repeat expansion in C9ORF72: clinicopathologic correlation. *Neuropathology, 33*(2), 122–133. http://dx.doi.org/10.1111/j.1440-1789.2012.01332.x.

Bjorkoy, G., Lamark, T., & Johansen, T. (2006). p62/SQSTM1: a missing link between protein aggregates and the autophagy machinery. *Autophagy, 2*(2), 138–139. pii:2405.

Blair, I. P., Williams, K. L., Warraich, S. T., Durnall, J. C., Thoeng, A. D., Manavis, J., … Nicholson, G. A. (2010). FUS mutations in amyotrophic lateral sclerosis: clinical, pathological, neurophysiological and genetic analysis. *Journal of Neurology, Neurosurgery and Psychiatry, 81*(6), 639–645. http://dx.doi.org/10.1136/jnnp.2009.194399. pii:jnnp.2009.194399.

van Blitterswijk, M., Baker, M. C., Bieniek, K. F., Knopman, D. S., Josephs, K. A., Boeve, B., … Rademakers, R. (2013). Profilin-1 mutations are rare in patients with amyotrophic lateral sclerosis and frontotemporal dementia. *Amyotrophic Lateral Sclerosis and Frontotemporal Degeneration, 14*(5–6), 463–469. http://dx.doi.org/10.3109/21678421.2013.787630.

van Blitterswijk, M., van Es, M. A., Hennekam, E. A., Dooijes, D., van Rheenen, W., Medic, J., … van den Berg, L. H. (2012). Evidence for an oligogenic basis of amyotrophic lateral sclerosis. *Human Molecular Genetics, 21*(17), 3776–3784. http://dx.doi.org/10.1093/hmg/dds199. pii:dds199.

van Blitterswijk, M., Mullen, B., Heckman, M. G., Baker, M. C., DeJesus-Hernandez, M., Brown, P. H., … Rademakers, R. (2014). Ataxin-2 as potential disease modifier in C9ORF72 expansion carriers. *Neurobiology of Aging, 35*(10), 2421.e13–2421.e17. http://dx.doi.org/10.1016/j.neurobiolaging.2014.04.016. pii:S0197-4580(14)00318-2.

van Blitterswijk, M., Mullen, B., Nicholson, A. M., Bieniek, K. F., Heckman, M. G., Baker, M. C., … Rademakers, R. (2014). TMEM106B protects C9ORF72 expansion carriers against frontotemporal dementia. *Acta Neuropathologica, 127*(3), 397–406. http://dx.doi.org/10.1007/s00401-013-1240-4.

van Blitterswijk, M., Mullen, B., Wojtas, A., Heckman, M. G., Diehl, N. N., Baker, M. C., ... Rademakers, R. (2014). Genetic modifiers in carriers of repeat expansions in the C9ORF72 gene. *Molecular Neurodegeneration, 9*, 38. http://dx.doi.org/10.1186/1750-1326-9-38. pii:1750-1326-9-38.

Blokhuis, A. M., Groen, E. J., Koppers, M., van den Berg, L. H., & Pasterkamp, R. J. (2013). Protein aggregation in amyotrophic lateral sclerosis. *Acta Neuropathologica, 125*(6), 777–794. http://dx.doi.org/10.1007/s00401-013-1125-6.

Boeve, B. F., & Graff-Radford, N. R. (2012). Cognitive and behavioral features of c9FTD/ALS. *Alzheimer's Research & Therapy, 4*(4), 29. http://dx.doi.org/10.1186/alzrt132. pii:alzrt132.

Boopathy, S., Silvas, T. V., Tischbein, M., Jansen, S., Shandilya, S. M., Zitzewitz, J. A., ... Bosco, D. A. (2015). Structural basis for mutation-induced destabilization of profilin 1 in ALS. *Proceedings of the National Academy of Sciences of the United States of America, 112*(26), 7984–7989. http://dx.doi.org/10.1073/pnas.1424108112. pii:1424108112.

Borroni, B., Bonvicini, C., Alberici, A., Buratti, E., Agosti, C., Archetti, S., ... Padovani, A. (2009). Mutation within TARDBP leads to frontotemporal dementia without motor neuron disease. *Human Mutation, 30*(11), E974–E983. http://dx.doi.org/10.1002/humu.21100.

Bosco, D. A., Lemay, N., Ko, H. K., Zhou, H., Burke, C., Kwiatkowski, T. J., Jr., ... Hayward, L. J. (2010). Mutant FUS proteins that cause amyotrophic lateral sclerosis incorporate into stress granules. *Human Molecular Genetics, 19*(21), 4160–4175. http://dx.doi.org/10.1093/hmg/ddq335. pii:ddq335.

Bosco, D. A., Morfini, G., Karabacak, N. M., Song, Y., Gros-Louis, F., Pasinelli, P., ... Brown, R. H., Jr. (2010). Wild-type and mutant SOD1 share an aberrant conformation and a common pathogenic pathway in ALS. *Nature Neuroscience, 13*(11), 1396–1403. http://dx.doi.org/10.1038/nn.2660. pii:nn.2660.

Brooks, B. R., Miller, R. G., Swash, M., & Munsat, T. L. (2000). El Escorial revisited: revised criteria for the diagnosis of amyotrophic lateral sclerosis. *Amyotrophic Lateral Sclerosis and Other Motor Neuron Disorders, 1*(5), 293–299.

Brotherton, T. E., Li, Y., & Glass, J. D. (2013). Cellular toxicity of mutant SOD1 protein is linked to an easily soluble, non-aggregated form in vitro. *Neurobiology of Disease, 49*, 49–56. http://dx.doi.org/10.1016/j.nbd.2012.08.010. pii:S0969-9961(12)00297-5.

Bruijn, L. I., Houseweart, M. K., Kato, S., Anderson, K. L., Anderson, S. D., Ohama, E., ... Cleveland, D. W. (1998). Aggregation and motor neuron toxicity of an ALS-linked SOD1 mutant independent from wild-type SOD1. *Science, 281*(5384), 1851–1854.

Buchan, J. R., Kolaitis, R. M., Taylor, J. P., & Parker, R. (2013). Eukaryotic stress granules are cleared by autophagy and Cdc48/VCP function. *Cell, 153*(7), 1461–1474. http://dx.doi.org/10.1016/j.cell.2013.05.037. pii:S0092-8674(13)00643-0.

Buratti, E., Brindisi, A., Giombi, M., Tisminetzky, S., Ayala, Y. M., & Baralle, F. E. (2005). TDP-43 binds heterogeneous nuclear ribonucleoprotein A/B through its C-terminal tail: an important region for the inhibition of cystic fibrosis transmembrane conductance regulator exon 9 splicing. *Journal of Biological Chemistry, 280*(45), 37572–37584. http://dx.doi.org/10.1074/jbc.M505557200. pii:M505557200.

Byrne, S., Elamin, M., Bede, P., & Hardiman, O. (2012). Absence of consensus in diagnostic criteria for familial neurodegenerative diseases. *Journal of Neurology, Neurosurgery and Psychiatry, 83*(4), 365–367. http://dx.doi.org/10.1136/jnnp-2011-301530. pii:jnnp-2011-301530.

Byrne, S., Elamin, M., Bede, P., Shatunov, A., Walsh, C., Corr, B., ... Hardiman, O. (2012). Cognitive and clinical characteristics of patients with amyotrophic lateral sclerosis carrying a C9orf72 repeat expansion: a population-based cohort study. *Lancet Neurology, 11*(3), 232–240. http://dx.doi.org/10.1016/S1474-4422(12)70014-5. pii:S1474-4422(12)70014-5.

Cady, J., Allred, P., Bali, T., Pestronk, A., Goate, A., Miller, T. M., ... Baloh, R. H. (2015). Amyotrophic lateral sclerosis onset is influenced by the burden of rare variants in known amyotrophic lateral sclerosis genes. *Annals of Neurology, 77*(1), 100–113. http://dx.doi.org/10.1002/ana.24306.

Calini, D., Corrado, L., Del Bo, R., Gagliardi, S., Pensato, V., Verde, F., ... Silani, V. (2013). Analysis of hnRNPA1, A2/B1, and A3 genes in patients with amyotrophic lateral sclerosis. *Neurobiology of Aging, 34*(11), 2695.e11–2695.e12. http://dx.doi.org/10.1016/j.neurobiolaging.2013.05.025. pii:S0197-4580(13)00238-8.

Chalasani, M. L., Kumari, A., Radha, V., & Swarup, G. (2014). E50K-OPTN-induced retinal cell death involves the Rab GTPase-activating protein, TBC1D17 mediated block in autophagy. *PLoS One, 9*(4), e95758. http://dx.doi.org/10.1371/journal.pone.0095758. pii:PONE-D-13-53356.

Charcot, J. M., & Joffory, A. (1869). Deux cas d'atrophie musculaire progressive avec lesions de la substance grise et des faisceaux antero-lateraux de la moelle epiniere. *Arch Physiol. Neurol. Pathol., 2*, 744–754.

Chaussenot, A., Le Ber, I., Ait-El-Mkadem, S., Camuzat, A., de Septenville, A., Bannwarth, S., ... Paquis-Flucklinger, V. (2014). Screening of CHCHD10 in a French cohort confirms the involvement of this gene in frontotemporal dementia with amyotrophic lateral sclerosis patients. *Neurobiology of Aging, 35*(12), 2884.e1–2884.e4. http://dx.doi.org/10.1016/j.neurobiolaging.2014.07.022. pii:S0197-4580(14)00491-6.

Chen, Y. Z., Bennett, C. L., Huynh, H. M., Blair, I. P., Puls, I., Irobi, J., ... Chance, P. F. (2004). DNA/RNA helicase gene mutations in a form of juvenile amyotrophic lateral sclerosis (ALS4). *American Journal of Human Genetics, 74*(6), 1128–1135. http://dx.doi.org/10.1086/421054. pii:S0002-9297(07)62840-7.

Chio, A., Calvo, A., Moglia, C., Canosa, A., Brunetti, M., Barberis, M., ... D'Alfonso, S. (2015). ATXN2 polyQ intermediate repeats are a modifier of ALS survival. *Neurology, 84*(3), 251–258. http://dx.doi.org/10.1212/WNL.0000000000001159. pii:WNL.0000000000001159.

Chio, A., Mora, G., Restagno, G., Brunetti, M., Ossola, I., Barberis, M., ... Calvo, A. (2013). UNC13A influences survival in Italian amyotrophic lateral sclerosis patients: a population-based study. *Neurobiology of Aging, 34*(1), 357.e1–357.e5. http://dx.doi.org/10.1016/j.neurobiolaging.2012.07.016. pii:S0197-4580(12)00406-X.

Chio, A., Mora, G., Sabatelli, M., Caponnetto, C., Traynor, B. J., Johnson, J. O., ... Lunetta, C. (2015). CHCH10 mutations in an Italian cohort of familial and sporadic amyotrophic lateral sclerosis patients. *Neurobiology of Aging, 36*(4), 1767.e3–1767.e6. http://dx.doi.org/10.1016/j.neurobiolaging.2015.01.017. pii:S0197-4580(15)00050-0.

Chio, A., Pagani, M., Agosta, F., Calvo, A., Cistaro, A., & Filippi, M. (2014). Neuroimaging in amyotrophic lateral sclerosis: insights into structural and functional changes. *Lancet Neurology, 13*(12), 1228–1240. http://dx.doi.org/10.1016/S1474-4422(14)70167-X. pii:S1474-4422(14)70167-X.

Chio, A., Restagno, G., Brunetti, M., Ossola, I., Calvo, A., Canosa, A., ... Consortium, S.. (2012). ALS/FTD phenotype in two Sardinian families carrying both C9ORF72 and TARDBP mutations. *Journal of Neurology, Neurosurgery and Psychiatry, 83*(7), 730–733. http://dx.doi.org/10.1136/jnnp-2012-302219.

Chow, C. Y., Landers, J. E., Bergren, S. K., Sapp, P. C., Grant, A. E., Jones, J. M., ... Meisler, M. H. (2009). Deleterious variants of FIG4, a phosphoinositide phosphatase, in patients with ALS. *American Journal of Human Genetics, 84*(1), 85–88. http://dx.doi.org/10.1016/j.ajhg.2008.12.010. pii:S0002-9297(08)00631-9.

Cirulli, E. T., Lasseigne, B. N., Petrovski, S., Sapp, P. C., Dion, P. A., Leblond, C. S., ... Goldstein, D. B. (2015). Exome sequencing in amyotrophic lateral sclerosis identifies risk genes and pathways. *Science, 347*(6229), 1436–1441. http://dx.doi.org/10.1126/science.aaa3650. pii:science.aaa3650.

Corcia, P., Valdmanis, P., Millecamps, S., Lionnet, C., Blasco, H., Mouzat, K., ... Camu, W. (2012). Phenotype and genotype analysis in amyotrophic lateral sclerosis with TARDBP gene mutations. *Neurology, 78*(19), 1519–1526. http://dx.doi.org/10.1212/WNL.0b013e3182553c88. pii:WNL.0b013e3182553c88.

Couthouis, J., Hart, M. P., Erion, R., King, O. D., Diaz, Z., Nakaya, T., ... Gitler, A. D. (2012). Evaluating the role of the FUS/TLS-related gene EWSR1 in amyotrophic lateral sclerosis. *Human Molecular Genetics, 21*(13), 2899–2911. http://dx.doi.org/10.1093/hmg/dds116. pii:dds116.

Couthouis, J., Hart, M. P., Shorter, J., DeJesus-Hernandez, M., Erion, R., Oristano, R., ... Gitler, A. D. (2011). A yeast functional screen predicts new candidate ALS disease genes. *Proceedings of the National Academy of Sciences of the United States of America, 108*(52), 20881–20890. http://dx.doi.org/10.1073/pnas.1109434108. pii:1109434108.

Cox, L. E., Ferraiuolo, L., Goodall, E. F., Heath, P. R., Higginbottom, A., Mortiboys, H., ... Shaw, P. J. (2010). Mutations in CHMP2B in lower motor neuron predominant amyotrophic lateral sclerosis (ALS). *PLoS One, 5*(3), e9872. http://dx.doi.org/10.1371/journal.pone.0009872.

Cozzolino, M., Rossi, S., Mirra, A., & Carri, M. T. (2015). Mitochondrial dynamism and the pathogenesis of amyotrophic lateral sclerosis. *Frontiers in Cellular Neuroscience, 9*, 31. http://dx.doi.org/10.3389/fncel.2015.00031.

Cudkowicz, M. E., McKenna-Yasek, D., Chen, C., Hedley-Whyte, E. T., & Brown, R. H., Jr. (1998). Limited corticospinal tract involvement in amyotrophic lateral sclerosis subjects with the A4V mutation in the copper/zinc superoxide dismutase gene. *Annals of Neurology, 43*(6), 703–710. http://dx.doi.org/10.1002/ana.410430604.

Cudkowicz, M. E., McKenna-Yasek, D., Sapp, P. E., Chin, W., Geller, B., Hayden, D. L., ... Brown, R. H. (1997). Epidemiology of mutations in superoxide dismutase in amyotrophic lateral sclerosis. *Annals of Neurology, 41*(2), 210–221. http://dx.doi.org/10.1002/ana.410410212.

Daoud, H., Belzil, V., Martins, S., Sabbagh, M., Provencher, P., Lacomblez, L., ... Rouleau, G. A. (2011). Association of long ATXN2 CAG repeat sizes with increased risk of amyotrophic lateral sclerosis. *Archives of Neurology, 68*(6), 739–742. http://dx.doi.org/10.1001/archneurol.2011.111. pii:68/6/739.

Daoud, H., Dobrzeniecka, S., Camu, W., Meininger, V., Dupre, N., Dion, P. A., & Rouleau, G. A. (2013). Mutation analysis of PFN1 in familial amyotrophic lateral sclerosis patients. *Neurobiology of Aging, 34*(4), 1311.e1–1311.e2. http://dx.doi.org/10.1016/j.neurobiolaging.2012.09.001. pii:S0197-4580(12)00457-5.

Daoud, H., Zhou, S., Noreau, A., Sabbagh, M., Belzil, V., Dionne-Laporte, A., ... Rouleau, G. A. (2012). Exome sequencing reveals SPG11 mutations causing juvenile ALS. *Neurobiology of Aging, 33*(4), 839.e5–839.e9. http://dx.doi.org/10.1016/j.neurobiolaging.2011.11.012. pii:S0197-4580(11)00483-0.

DeJesus-Hernandez, M., Mackenzie, I. R., Boeve, B. F., Boxer, A. L., Baker, M., Rutherford, N. J., ... Rademakers, R. (2011). Expanded GGGGCC hexanucleotide repeat in noncoding region of C9ORF72 causes chromosome 9p-linked FTD and ALS. *Neuron, 72*(2), 245–256. http://dx.doi.org/10.1016/j.neuron.2011.09.011. pii:S0896-6273(11)00828-2.

Deng, H. X., Chen, W., Hong, S. T., Boycott, K. M., Gorrie, G. H., Siddique, N., ... Siddique, T. (2011). Mutations in UBQLN2 cause dominant X-linked juvenile and adult-onset ALS and ALS/dementia. *Nature, 477*(7363), 211–215. http://dx.doi.org/10.1038/nature10353. pii:nature10353.

Diekstra, F. P., van Vught, P. W., van Rheenen, W., Koppers, M., Pasterkamp, R. J., van Es, M. A., ... Veldink, J. H. (2012). UNC13A is a modifier of survival in amyotrophic lateral sclerosis. *Neurobiology of Aging, 33*(3), 630.e3–630.e8. http://dx.doi.org/10.1016/j.neurobiolaging.2011.10.029. pii:S0197-4580(11)00460-X.

Dillen, L., Van Langenhove, T., Engelborghs, S., Vandenbulcke, M., Sarafov, S., Tournev, I., ... van der Zee, J. (2013). Explorative genetic study of UBQLN2 and PFN1 in an extended Flanders-Belgian cohort of frontotemporal lobar degeneration patients. *Neurobiology of Aging, 34*(6), 1711.e1–1711.e5. http://dx.doi.org/10.1016/j.neurobiolaging.2012.12.007. pii:S0197-4580(12)00638-0.

van Doormaal, P. T., Ticozzi, N., Gellera, C., Ratti, A., Taroni, F., Chio, A., ... Veldink, J. H. (2014). Analysis of the KIFAP3 gene in amyotrophic lateral sclerosis: a multicenter survival study. *Neurobiology of Aging, 35*(10), 2420.e13–2420.e14. http://dx.doi.org/10.1016/j.neurobiolaging.2014.04.014. pii:S0197-4580(14)00316-9.

Eisen, A., & Turner, M. R. (2013). Does variation in neurodegenerative disease susceptibility and phenotype reflect cerebral differences at the network level? *Amyotrophic Lateral Sclerosis and Frontotemporal Degeneration, 14*(7–8), 487–493. http://dx.doi.org/10.3109/21678421.2013.812660.

Elamin, M., Phukan, J., Bede, P., Jordan, N., Byrne, S., Pender, N., & Hardiman, O. (2011). Executive dysfunction is a negative prognostic indicator in patients with ALS without dementia. *Neurology, 76*(14), 1263–1269. http://dx.doi.org/10.1212/WNL.0b013e318214359f.

Elden, A. C., Kim, H. J., Hart, M. P., Chen-Plotkin, A. S., Johnson, B. S., Fang, X., ... Gitler, A. D. (2010). Ataxin-2 intermediate-length polyglutamine expansions are associated with increased risk for ALS. *Nature, 466*(7310), 1069–1075. http://dx.doi.org/10.1038/nature09320. pii:nature09320.

van Es, M. A., Schelhaas, H. J., van Vught, P. W., Ticozzi, N., Andersen, P. M., Groen, E. J., ... van den Berg, L. H. (2011). Angiogenin variants in Parkinson disease and amyotrophic lateral sclerosis. *Annals of Neurology, 70*(6), 964–973. http://dx.doi.org/10.1002/ana.22611.

van Es, M. A., Veldink, J. H., Saris, C. G., Blauw, H. M., van Vught, P. W., Birve, A., ... van den Berg, L. H. (2009). Genome-wide association study identifies 19p13.3 (UNC13A) and 9p21.2 as susceptibility loci for sporadic amyotrophic lateral sclerosis. *Nature Genetics, 41*(10), 1083–1087. http://dx.doi.org/10.1038/ng.442. pii:ng.442.

Eymard-Pierre, E., Lesca, G., Dollet, S., Santorelli, F. M., di Capua, M., Bertini, E., & Boespflug-Tanguy, O. (2002). Infantile-onset ascending hereditary spastic paralysis is associated with mutations in the alsin gene. *American Journal of Human Genetics, 71*(3), 518–527. http://dx.doi.org/10.1086/342359. pii:S0002-9297(07)60332-2.

Farg, M. A., Soo, K. Y., Warraich, S. T., Sundaramoorthy, V., Blair, I. P., & Atkin, J. D. (2013). Ataxin-2 interacts with FUS and intermediate-length polyglutamine expansions enhance FUS-related pathology in amyotrophic lateral sclerosis. *Human Molecular Genetics, 22*(4), 717–728. http://dx.doi.org/10.1093/hmg/dds479. pii:dds479.

Farrer, M. J., Hulihan, M. M., Kachergus, J. M., Dachsel, J. C., Stoessl, A. J., Grantier, L. L., ... Wszolek, Z. K. (2009). DCTN1 mutations in Perry syndrome. *Nature Genetics, 41*(2), 163–165. http://dx.doi.org/10.1038/ng.293. pii:ng.293.

Fecto, F., Yan, J., Vemula, S. P., Liu, E., Yang, Y., Chen, W., ... Siddique, T. (2011). SQSTM1 mutations in familial and sporadic amyotrophic lateral sclerosis. *Archives of Neurology, 68*(11), 1440–1446. http://dx.doi.org/10.1001/archneurol.2011.250. pii:68/11/1440.

Figlewicz, D. A., Krizus, A., Martinoli, M. G., Meininger, V., Dib, M., Rouleau, G. A., & Julien, J. P. (1994). Variants of the heavy neurofilament subunit are associated with the development of amyotrophic lateral sclerosis. *Human Molecular Genetics, 3*(10), 1757–1761.

Figlewicz, D. A., Rouleau, G. A., Krizus, A., & Julien, J. P. (1993). Polymorphism in the multi-phosphorylation domain of the human neurofilament heavy-subunit-encoding gene. *Gene, 132*(2), 297–300.

Filimonenko, M., Stuffers, S., Raiborg, C., Yamamoto, A., Malerod, L., Fisher, E. M., ... Simonsen, A. (2007). Functional multivesicular bodies are required for autophagic clearance of protein aggregates associated with neurodegenerative disease. *Journal of Cell Biology, 179*(3), 485–500. http://dx.doi.org/10.1083/jcb.200702115. pii:jcb.200702115.

Foerster, B. R., Welsh, R. C., & Feldman, E. L. (2013). 25 years of neuroimaging in amyotrophic lateral sclerosis. *Nature Reviews Neurology, 9*(9), 513–524. http://dx.doi.org/10.1038/nrneurol.2013.153. pii:nrneurol.2013.153.

Fratta, P., Mizielinska, S., Nicoll, A. J., Zloh, M., Fisher, E. M., Parkinson, G., & Isaacs, A. M. (2012). C9orf72 hexanucleotide repeat associated with amyotrophic lateral sclerosis and frontotemporal dementia forms RNA G-quadruplexes. *Scientific Reports, 2*, 1016. http://dx.doi.org/10.1038/srep01016.

Freischmidt, A., Wieland, T., Richter, B., Ruf, W., Schaeffer, V., Muller, K., ... Weishaupt, J. H. (2015). Haploinsufficiency of TBK1 causes familial ALS and fronto-temporal dementia. *Nature Neuroscience, 18*(5), 631–636. http://dx.doi.org/10.1038/nn.4000. pii:nn.4000.

Gallagher, M. D., Suh, E., Grossman, M., Elman, L., McCluskey, L., Van Swieten, J. C., ... Chen-Plotkin, A. S. (2014). TMEM106B is a genetic modifier of frontotemporal lobar degeneration with C9orf72 hexanucleotide repeat expansions. *Acta Neuropathologica, 127*(3), 407–418. http://dx.doi.org/10.1007/s00401-013-1239-x.

Gary, J. D., Sato, T. K., Stefan, C. J., Bonangelino, C. J., Weisman, L. S., & Emr, S. D. (2002). Regulation of Fab1 phosphatidylinositol 3-phosphate 5-kinase pathway by Vac7 protein and Fig4, a polyphosphoinositide phosphatase family member. *Molecular Biology of the Cell, 13*(4), 1238–1251. http://dx.doi.org/10.1091/mbc.01-10-0498.

Gendron, T. F., Bieniek, K. F., Zhang, Y. J., Jansen-West, K., Ash, P. E., Caulfield, T., ... Petrucelli, L. (2013). Antisense transcripts of the expanded C9ORF72 hexanucleotide repeat form nuclear RNA foci and undergo repeat-associated non-ATG translation in c9FTD/ALS. *Acta Neuropathologica, 126*(6), 829–844. http://dx.doi.org/10.1007/s00401-013-1192-8.

Greenway, M. J., Alexander, M. D., Ennis, S., Traynor, B. J., Corr, B., Frost, E., ... Hardiman, O. (2004). A novel candidate region for ALS on chromosome 14q11.2. *Neurology, 63*(10), 1936–1938. pii:63/10/1936.

Greenway, M. J., Andersen, P. M., Russ, C., Ennis, S., Cashman, S., Donaghy, C., ... Hardiman, O. (2006). ANG mutations segregate with familial and 'sporadic' amyotrophic lateral sclerosis. *Nature Genetics, 38*(4), 411–413. http://dx.doi.org/10.1038/ng1742. pii:ng1742.

Gurney, M. E., Pu, H., Chiu, A. Y., Dal Canto, M. C., Polchow, C. Y., Alexander, D. D., ... Deng, H. X. (1994). Motor neuron degeneration in mice that express a human Cu,Zn superoxide dismutase mutation. *Science, 264*(5166), 1772–1775.

Hadano, S., Hand, C. K., Osuga, H., Yanagisawa, Y., Otomo, A., Devon, R. S., ... Ikeda, J. E. (2001). A gene encoding a putative GTPase regulator is mutated in familial amyotrophic lateral sclerosis 2. *Nature Genetics, 29*(2), 166–173. http://dx.doi.org/10.1038/ng1001-166. pii:ng1001-166.

Harms, M. B., & Baloh, R. H. (2013). Clinical neurogenetics: amyotrophic lateral sclerosis. *Neurologic Clinics, 31*(4), 929–950. http://dx.doi.org/10.1016/j.ncl.2013.05.003. pii:S0733-8619(13)00060-1.

Hewitt, C., Kirby, J., Highley, J. R., Hartley, J. A., Hibberd, R., Hollinger, H. C., ... Shaw, P. J. (2010). Novel FUS/TLS mutations and pathology in familial and sporadic amyotrophic lateral sclerosis. *Archives of Neurology, 67*(4), 455–461. http://dx.doi.org/10.1001/archneurol.2010.52. pii:67/4/455.

Hirano, M., Quinzii, C. M., Mitsumoto, H., Hays, A. P., Roberts, J. K., Richard, P., & Rowland, L. P. (2011). Senataxin mutations and amyotrophic lateral sclerosis. *Amyotrophic Lateral Sclerosis, 12*(3), 223–227. http://dx.doi.org/10.3109/17482968.2010.545952.

Howland, D. S., Liu, J., She, Y., Goad, B., Maragakis, N. J., Kim, B., ... Rothstein, J. D. (2002). Focal loss of the glutamate transporter EAAT2 in a transgenic rat model of SOD1 mutant-mediated amyotrophic lateral sclerosis (ALS). *Proceedings of the National Academy of Sciences of the United States of America, 99*(3), 1604–1609. http://dx.doi.org/10.1073/pnas.032539299. pii:032539299.

Hu, W. T., Shelnutt, M., Wilson, A., Yarab, N., Kelly, C., Grossman, M., ... Glass, J. (2013). Behavior matters—cognitive predictors of survival in amyotrophic lateral sclerosis. *PLoS One, 8*(2), e57584. http://dx.doi.org/10.1371/journal.pone.0057584.

Huey, E. D., Ferrari, R., Moreno, J. H., Jensen, C., Morris, C. M., Potocnik, F., ... Momeni, P. (2012). FUS and TDP43 genetic variability in FTD and CBS. *Neurobiology of Aging, 33*(5), 1016.e9–1016.e17. http://dx.doi.org/10.1016/j.neurobiolaging.2011.08.004. pii:S0197-4580(11)00305-8.

Ilieva, H., Polymenidou, M., & Cleveland, D. W. (2009). Non-cell autonomous toxicity in neurodegenerative disorders: ALS and beyond. *Journal of Cell Biology, 187*(6), 761–772. http://dx.doi.org/10.1083/jcb.200908164. pii:jcb.200908164.

Ingre, C., Landers, J. E., Rizik, N., Volk, A. E., Akimoto, C., Birve, A., ... Weishaupt, J. H. (2013). A novel phosphorylation site mutation in profilin 1 revealed in a large screen of US, Nordic, and German amyotrophic lateral sclerosis/frontotemporal dementia cohorts. *Neurobiology of Aging, 34*(6), 1708.e1–1708.e6. http://dx.doi.org/10.1016/j.neurobiolaging.2012.10.009. pii:S0197-4580(12)00520-9.

Johnson, J. O., Glynn, S. M., Gibbs, J. R., Nalls, M. A., Sabatelli, M., Restagno, G., ... Traynor, B. J. (2014). Mutations in the CHCHD10 gene are a common cause of familial amyotrophic lateral sclerosis. *Brain, 137*(Pt 12), e311. http://dx.doi.org/10.1093/brain/awu265. pii:awu265.

Johnson, J. O., Mandrioli, J., Benatar, M., Abramzon, Y., Van Deerlin, V. M., Trojanowski, J. Q., ... Traynor, B. J. (2010). Exome sequencing reveals VCP mutations as a cause of familial ALS. *Neuron, 68*(5), 857–864.

Johnson, J. O., Pioro, E. P., Boehringer, A., Chia, R., Feit, H., Renton, A. E., ... Traynor, B. J. (2014). Mutations in the Matrin 3 gene cause familial amyotrophic lateral sclerosis. *Nature Neuroscience, 17*(5), 664–666. http://dx.doi.org/10.1038/nn.3688. pii:nn.3688.

Kachaner, D., Genin, P., Laplantine, E., & Weil, R. (2012). Toward an integrative view of Optineurin functions. *Cell Cycle, 11*(15), 2808–2818. http://dx.doi.org/10.4161/cc.20946. pii:20946.

Kim, H. J., Kim, N. C., Wang, Y. D., Scarborough, E. A., Moore, J., Diaz, Z., ... Taylor, J. P. (2013). Mutations in prion-like domains in hnRNPA2B1 and hnRNPA1 cause multisystem proteinopathy and ALS. *Nature, 495*(7442), 467–473. http://dx.doi.org/10.1038/nature11922. pii:nature11922.

Klein, R. (2009). Bidirectional modulation of synaptic functions by Eph/ephrin signaling. *Nature Neuroscience, 12*(1), 15–20. http://dx.doi.org/10.1038/nn.2231. pii:nn.2231.

Kwiatkowski, T. J., Jr., Bosco, D. A., Leclerc, A. L., Tamrazian, E., Vanderburg, C. R., Russ, C., ... Brown, R. H., Jr. (2009). Mutations in the FUS/TLS gene on chromosome 16 cause familial amyotrophic lateral sclerosis. *Science, 323*(5918), 1205–1208. http://dx.doi.org/10.1126/science.1166066. pii:323/5918/1205.

Lacomblez, L., Bensimon, G., Leigh, P. N., Guillet, P., & Meininger, V. (1996). Dose-ranging study of riluzole in amyotrophic lateral sclerosis. Amyotrophic lateral sclerosis/riluzole study group II. *Lancet, 347*(9013), 1425–1431.

Lagier-Tourenne, C., Polymenidou, M., & Cleveland, D. W. (2010). TDP-43 and FUS/TLS: emerging roles in RNA processing and neurodegeneration. *Human Molecular Genetics, 19*(R1), R46–R64. http://dx.doi.org/10.1093/hmg/ddq137. pii:ddq137.

Landers, J. E., Leclerc, A. L., Shi, L., Virkud, A., Cho, T., Maxwell, M. M., ... Brown, R. H., Jr. (2008). New VAPB deletion variant and exclusion of VAPB mutations in familial ALS. *Neurology, 70*(14), 1179–1185. http://dx.doi.org/10.1212/01.wnl.0000289760.85237.4e. pii:01.wnl.0000289760.85237.4e.

Landers, J. E., Melki, J., Meininger, V., Glass, J. D., van den Berg, L. H., van Es, M. A., ... Brown, R. H., Jr. (2009). Reduced expression of the Kinesin-associated protein 3 (KIFAP3) gene increases survival in sporadic amyotrophic lateral sclerosis. *Proceedings of the National Academy of Sciences of the United States of America, 106*(22), 9004–9009. http://dx.doi.org/10.1073/pnas.0812937106. pii:0812937106.

Lattante, S., Millecamps, S., Stevanin, G., Rivaud-Pechoux, S., Moigneu, C., Camuzat, A., ... Kabashi, E. (2014). Contribution of ATXN2 intermediary polyQ expansions in a spectrum of neurodegenerative disorders. *Neurology, 83*(11), 990–995. http://dx.doi.org/10.1212/WNL.0000000000000778. pii:WNL.0000000000000778.

Laurin, N., Brown, J. P., Morissette, J., & Raymond, V. (2002). Recurrent mutation of the gene encoding sequestosome 1 (SQSTM1/p62) in Paget disease of bone. *American Journal of Human Genetics, 70*(6), 1582–1588. http://dx.doi.org/10.1086/340731. pii:S0002-9297(07)60711-3.

Le Ber, I., Camuzat, A., Guerreiro, R., Bouya-Ahmed, K., Bras, J., Nicolas, G., ... Brice, A. (2013). SQSTM1 mutations in French patients with frontotemporal dementia or frontotemporal dementia with amyotrophic lateral sclerosis. *JAMA Neurology, 70*(11), 1403–1410. http://dx.doi.org/10.1001/jamaneurol.2013.3849. pii:1738444.

Le Ber, I., Van Bortel, I., Nicolas, G., Bouya-Ahmed, K., Camuzat, A., Wallon, D., ... Brice, A. (2014). hnRNPA2B1 and hnRNPA1 mutations are rare in patients with "multisystem proteinopathy" and frontotemporal lobar degeneration phenotypes. *Neurobiology of Aging, 35*(4), 934.e5–934.e6. http://dx.doi.org/10.1016/j.neurobiolaging.2013.09.016. pii:S0197-4580(13)00418-1.

Leblond, C. S., Kaneb, H. M., Dion, P. A., & Rouleau, G. A. (2014). Dissection of genetic factors associated with amyotrophic lateral sclerosis. *Experimental Neurology, 262*(PB), 91–101. http://dx.doi.org/10.1016/j.expneurol.2014.04.013. pii:S0014-4886(14)00115-0.

Lee, B. L., Lee, V. M.-Y., & Trojanowski, J. Q. (2012). Gains or losses: molecular mechanisms of TDP-43-mediated neurodegeneration. *Nature Reviews Neuroscience, 13*.

Lee, T., Li, Y. R., Ingre, C., Weber, M., Grehl, T., Gredal, O., ... Gitler, A. D. (2011). Ataxin-2 intermediate-length polyglutamine expansions in European ALS patients. *Human Molecular Genetics, 20*(9), 1697–1700. http://dx.doi.org/10.1093/hmg/ddr045. pii:ddr045.

Lee, Y. B., Chen, H. J., Peres, J. N., Gomez-Deza, J., Attig, J., Stalekar, M., ... Shaw, C. E. (2013). Hexanucleotide repeats in ALS/FTD form length-dependent RNA foci, sequester RNA binding proteins, and are neurotoxic. *Cell Reports, 5*(5), 1178–1186. http://dx.doi.org/10.1016/j.celrep.2013.10.049. pii:S2211-1247(13)00648-7.

Lesage, S., Le Ber, I., Condroyer, C., Broussolle, E., Gabelle, A., Thobois, S., ... Brice, A. (2013). C9orf72 repeat expansions are a rare genetic cause of parkinsonism. *Brain, 136*(Pt 2), 385–391. http://dx.doi.org/10.1093/brain/aws357. pii:aws357.

Leung, C. L., He, C. Z., Kaufmann, P., Chin, S. S., Naini, A., Liem, R. K., ... Hays, A. P. (2004). A pathogenic peripherin gene mutation in a patient with amyotrophic lateral sclerosis. *Brain Pathology, 14*(3), 290–296.

Levine, T. P., Daniels, R. D., Gatta, A. T., Wong, L. H., & Hayes, M. J. (2013). The product of C9orf72, a gene strongly implicated in neurodegeneration, is structurally related to DENN Rab-GEFs. *Bioinformatics, 29*(4), 499–503. http://dx.doi.org/10.1093/bioinformatics/bts725. pii:bts725.

Lillo, P., Mioshi, E., Zoing, M. C., Kiernan, M. C., & Hodges, J. R. (2011). How common are behavioural changes in amyotrophic lateral sclerosis? *Amyotrophic Lateral Sclerosis, 12*(1), 45–51. http://dx.doi.org/10.3109/17482968.2010.520718.

Lin, K. P., Tsai, P. C., Liao, Y. C., Chen, W. T., Tsai, C. P., Soong, B. W., & Lee, Y. C. (2015). Mutational analysis of MATR3 in Taiwanese patients with amyotrophic lateral sclerosis. *Neurobiology of Aging, 36*(5), 2005.e1–2005.e4. http://dx.doi.org/10.1016/j.neurobiolaging.2015.02.008. pii:S0197-4580(15)00108-6.

Ling, S. C., Polymenidou, M., & Cleveland, D. W. (2013). Converging mechanisms in ALS and FTD: disrupted RNA and protein homeostasis. *Neuron, 79*(3), 416–438. http://dx.doi.org/10.1016/j.neuron.2013.07.033. pii:S0896-6273(13)00657-0.

Loewy, A., & Neuberg, C. (1904). Uber cystinurie. *Hoppe-Seyler's Zeitschrift fur Physiologische Chemie, 43*, 338–354.

Luigetti, M., Lattante, S., Zollino, M., Conte, A., Marangi, G., Del Grande, A., & Sabatelli, M. (2011). SOD1 G93D sporadic amyotrophic lateral sclerosis (SALS) patient with rapid progression and concomitant novel ANG variant. *Neurobiology of Aging, 32*(10), 1924.e5–1924.e8. http://dx.doi.org/10.1016/j.neurobiolaging.2011.04.004. pii:S0197-4580(11)00119-9.

Mackenzie, I. R., Bigio, E. H., Ince, P. G., Geser, F., Neumann, M., Cairns, N. J., ... Trojanowski, J. Q. (2007). Pathological TDP-43 distinguishes sporadic amyotrophic lateral sclerosis from amyotrophic lateral sclerosis with SOD1 mutations. *Annals of Neurology, 61*(5), 427–434. http://dx.doi.org/10.1002/ana.21147.

Mackenzie, I. R., Frick, P., & Neumann, M. (2014). The neuropathology associated with repeat expansions in the C9ORF72 gene. *Acta Neuropathologica, 127*(3), 347–357. http://dx.doi.org/10.1007/s00401-013-1232-4.

Mackenzie, I. R., & Neumann, M. (2012). FET proteins in frontotemporal dementia and amyotrophic lateral sclerosis. *Brain Research, 1462*, 40–43. http://dx.doi.org/10.1016/j.brainres.2011.12.010. pii:S0006-8993(11)02186-X.

Mahoney, C. J., Beck, J., Rohrer, J. D., Lashley, T., Mok, K., Shakespeare, T., ... Warren, J. D. (2012). Frontotemporal dementia with the C9ORF72 hexanucleotide repeat expansion: clinical, neuroanatomical and neuropathological features. *Brain, 135*(Pt 3), 736–750. http://dx.doi.org/10.1093/brain/awr361. pii:awr361.

Majounie, E., Renton, A. E., Mok, K., Dopper, E. G., Waite, A., Rollinson, S., ... Traynor, B. J. (2012). Frequency of the C9orf72 hexanucleotide repeat expansion in patients with amyotrophic lateral sclerosis and frontotemporal dementia: a cross-sectional study. *Lancet Neurology, 11*(4), 323–330. http://dx.doi.org/10.1016/S1474-4422(12)70043-1. pii:S1474-4422(12)70043-1.

Makki, A. A., & Benatar, M. (2007). The electromyographic diagnosis of amyotrophic lateral sclerosis: does the evidence support the El Escorial criteria? *Muscle & Nerve, 35*(5), 614–619. http://dx.doi.org/10.1002/mus.20748.

Mancuso, R., Osta, R., & Navarro, X. (2014). Presymptomatic electrophysiological tests predict clinical onset and survival in SOD1(G93A) ALS mice. *Muscle & Nerve, 50*(6), 943–949. http://dx.doi.org/10.1002/mus.24237.

Marangi, G., & Traynor, B. J. (2014). Genetic causes of amyotrophic lateral sclerosis: new genetic analysis methodologies entailing new opportunities and challenges. *Brain Research, 1607*, 75–93. http://dx.doi.org/10.1016/j.brainres.2014.10.009. pii:S0006-8993(14)01361-4.

Maruyama, H., Morino, H., Ito, H., Izumi, Y., Kato, H., Watanabe, Y., ... Kawakami, H. (2010). Mutations of optineurin in amyotrophic lateral sclerosis. *Nature, 465*(7295), 223–226. http://dx.doi.org/10.1038/nature08971. pii:nature08971.

McKee, A. C., Gavett, B. E., Stern, R. A., Nowinski, C. J., Cantu, R. C., Kowall, N. W., ... Budson, A. E. (2010). TDP-43 proteinopathy and motor neuron disease in chronic traumatic encephalopathy. *Journal of Neuropathology & Experimental Neurology, 69*(9), 918–929. http://dx.doi.org/10.1097/NEN.0b013e3181ee7d85.

Meyer, H., Bug, M., & Bremer, S. (2012). Emerging functions of the VCP/p97 AAA-ATPase in the ubiquitin system. *Nature Cell Biology, 14*(2), 117–123. http://dx.doi.org/10.1038/ncb2407. pii:ncb2407.

Meyer, T., Schwan, A., Dullinger, J. S., Brocke, J., Hoffmann, K. T., Nolte, C. H., ... Linke, P. (2005). Early-onset ALS with long-term survival associated with spastin gene mutation. *Neurology, 65*(1), 141–143. http://dx.doi.org/10.1212/01.wnl.0000167130.31618.0a. pii:65/1/141.

Mezzapesa, D. M., D'Errico, E., Tortelli, R., Distaso, E., Cortese, R., Tursi, M., ... Simone, I. L. (2013). Cortical thinning and clinical heterogeneity in amyotrophic lateral sclerosis. *PLoS One, 8*(11), e80748. http://dx.doi.org/10.1371/journal.pone.0080748. pii:PONE-D-13-19897.

Millecamps, S., Salachas, F., Cazeneuve, C., Gordon, P., Bricka, B., Camuzat, A., ... Meininger, V. (2010). SOD1, ANG, VAPB, TARDBP, and FUS mutations in familial amyotrophic lateral sclerosis: genotype-phenotype correlations. *Journal of Medical Genetics, 47*(8), 554–560. http://dx.doi.org/10.1136/jmg.2010.077180. pii:jmg.2010.077180.

Miller, R. G., Mitchell, J. D., & Moore, D. H. (2012). Riluzole for amyotrophic lateral sclerosis (ALS)/motor neuron disease (MND). *Cochrane Database of Systematic Reviews, 3*, CD001447. http://dx.doi.org/10.1002/14651858.CD001447.pub3.

Miller, T. M., Pestronk, A., David, W., Rothstein, J., Simpson, E., Appel, S. H., ... Cudkowicz, M. E. (2013). An antisense oligonucleotide against SOD1 delivered intrathecally for patients with SOD1 familial amyotrophic lateral sclerosis: a phase 1, randomised, first-in-man study. *Lancet Neurology, 12*(5), 435–442. http://dx.doi.org/10.1016/S1474-4422(13)70061-9. pii:S1474-4422(13)70061-9.

Mioshi, E., Lillo, P., Yew, B., Hsieh, S., Savage, S., Hodges, J. R., ... Hornberger, M. (2013). Cortical atrophy in ALS is critically associated with neuropsychiatric and cognitive changes. *Neurology, 80*(12), 1117–1123. http://dx.doi.org/10.1212/WNL.0b013e31828869da. pii:WNL.0b013e31828869da.

Mitchell, J., Paul, P., Chen, H. J., Morris, A., Payling, M., Falchi, M., ... de Belleroche, J. (2010). Familial amyotrophic lateral sclerosis is associated with a mutation in D-amino acid oxidase. *Proceedings of the National Academy of Sciences of the United States of America, 107*(16), 7556–7561. http://dx.doi.org/10.1073/pnas.0914128107. pii:0914128107.

Mitchell, J. D., Callagher, P., Gardham, J., Mitchell, C., Dixon, M., Addison-Jones, R., ... O'Brien, M. R. (2010). Timelines in the diagnostic evaluation of people with suspected amyotrophic lateral sclerosis (ALS)/motor neuron disease (MND)—a 20-year review: can we do better? *Amyotrophic Lateral Sclerosis, 11*(6), 537–541. http://dx.doi.org/10.3109/17482968.2010.495158.

Mori, K., Weng, S. M., Arzberger, T., May, S., Rentzsch, K., Kremmer, E., ... Edbauer, D. (2013). The C9orf72 GGGGCC repeat is translated into aggregating dipeptide-repeat proteins in FTLD/ALS. *Science, 339*(6125), 1335–1338. http://dx.doi.org/10.1126/science.1232927. pii:science.1232927.

Morotz, G. M., De Vos, K. J., Vagnoni, A., Ackerley, S., Shaw, C. E., & Miller, C. C. (2012). Amyotrophic lateral sclerosis-associated mutant VAPBP56S perturbs calcium homeostasis to disrupt axonal transport of mitochondria. *Human Molecular Genetics, 21*(9), 1979–1988. http://dx.doi.org/10.1093/hmg/dds011. pii:dds011.

Muller, K., Andersen, P. M., Hubers, A., Marroquin, N., Volk, A. E., Danzer, K. M., ... Weishaupt, J. H. (2014). Two novel mutations in conserved codons indicate that CHCHD10 is a gene associated with motor neuron disease. *Brain, 137*(Pt 12), e309. http://dx.doi.org/10.1093/brain/awu227. pii:awu227.

Munch, C., Rolfs, A., & Meyer, T. (2008). Heterozygous S44L missense change of the spastin gene in amyotrophic lateral sclerosis. *Amyotrophic Lateral Sclerosis, 9*(4), 251–253. http://dx.doi.org/10.1080/17482960801900172. pii:791845771.

Munch, C., Rosenbohm, A., Sperfeld, A. D., Uttner, I., Reske, S., Krause, B. J., … Ludolph, A. C. (2005). Heterozygous R1101K mutation of the DCTN1 gene in a family with ALS and FTD. *Annals of Neurology, 58*(5), 777–780. http://dx.doi.org/10.1002/ana.20631.

Munch, C., Sedlmeier, R., Meyer, T., Homberg, V., Sperfeld, A. D., Kurt, A., … Ludolph, A. C. (2004). Point mutations of the p150 subunit of dynactin (DCTN1) gene in ALS. *Neurology, 63*(4), 724–726. pii:63/4/724.

Neuenschwander, A. G., Thai, K. K., Figueroa, K. P., & Pulst, S. M. (2014). Amyotrophic lateral sclerosis risk for spinocerebellar ataxia type 2 ATXN2 CAG repeat alleles: a meta-analysis. *JAMA Neurology, 71*(12), 1529–1534. http://dx.doi.org/10.1001/jamaneurol.2014.2082. pii:1910853.

Neumann, M., Sampathu, D. M., Kwong, L. K., Truax, A. C., Micsenyi, M. C., Chou, T. T., … Lee, V. M. (2006). Ubiquitinated TDP-43 in frontotemporal lobar degeneration and amyotrophic lateral sclerosis. *Science, 314*(5796), 130–133. http://dx.doi.org/10.1126/science.1134108. pii:314/5796/130.

Ngo, S. T., Steyn, F. J., & McCombe, P. A. (2014). Body mass index and dietary intervention: implications for prognosis of amyotrophic lateral sclerosis. *Journal of the Neurological Sciences, 340*(1–2), 5–12. http://dx.doi.org/10.1016/j.jns.2014.02.035. pii:S0022-510X(14)00140-3.

Nishimura, A. L., Mitne-Neto, M., Silva, H. C., Richieri-Costa, A., Middleton, S., Cascio, D., … Zatz, M. (2004). A mutation in the vesicle-trafficking protein VAPB causes late-onset spinal muscular atrophy and amyotrophic lateral sclerosis. *American Journal of Human Genetics, 75*(5), 822–831. http://dx.doi.org/10.1086/425287. pii:S0002-9297(07)63787-2.

Okamoto, K., Mizuno, Y., & Fujita, Y. (2008). Bunina bodies in amyotrophic lateral sclerosis. *Neuropathology, 28*(2), 109–115. http://dx.doi.org/10.1111/j.1440-1789.2007.00873.x. pii:NEU873.

Orlacchio, A., Babalini, C., Borreca, A., Patrono, C., Massa, R., Basaran, S., … Kawarai, T. (2010). SPATACSIN mutations cause autosomal recessive juvenile amyotrophic lateral sclerosis. *Brain, 133*(Pt 2), 591–598. http://dx.doi.org/10.1093/brain/awp325. pii:awp325.

Paez-Colasante, X., Figueroa-Romero, C., Sakowski, S. A., Goutman, S. A., & Feldman, E. L. (2015). Amyotrophic lateral sclerosis: mechanisms and therapeutics in the epigenomic era. *Nature Reviews Neurology, 11*(5), 266–279. http://dx.doi.org/10.1038/nrneurol.2015.57. pii:nrneurol.2015.57.

Paganoni, S., Macklin, E. A., Lee, A., Murphy, A., Chang, J., Zipf, A., … Atassi, N. (2014). Diagnostic timelines and delays in diagnosing amyotrophic lateral sclerosis (ALS). *Amyotrophic Lateral Sclerosis and Frontotemporal Degeneration, 15*(5–6), 453–456. http://dx.doi.org/10.3109/21678421.2014.903974.

Parkinson, N., Ince, P. G., Smith, M. O., Highley, R., Skibinski, G., Andersen, P. M., … Fisher, E. M. (2006). ALS phenotypes with mutations in CHMP2B (charged multivesicular body protein 2B). *Neurology, 67*(6), 1074–1077. http://dx.doi.org/10.1212/01.wnl.0000231510.89311.8b. pii:01.wnl.0000231510.89311.8b.

Paul, P., & de Belleroche, J. (2012). The role of D-amino acids in amyotrophic lateral sclerosis pathogenesis: a review. *Amino Acids, 43*(5), 1823–1831. http://dx.doi.org/10.1007/s00726-012-1385-9.

Paul, P., & de Belleroche, J. (2014). The role of D-serine and glycine as co-agonists of NMDA receptors in motor neuron degeneration and amyotrophic lateral sclerosis (ALS). *Frontiers in Synaptic Neuroscience, 6*, 10. http://dx.doi.org/10.3389/fnsyn.2014.00010.

Phukan, J., Elamin, M., Bede, P., Jordan, N., Gallagher, L., Byrne, S., … Hardiman, O. (2012). The syndrome of cognitive impairment in amyotrophic lateral sclerosis: a population-based study. *Journal of Neurology, Neurosurgery and Psychiatry, 83*(1), 102–108. http://dx.doi.org/10.1136/jnnp-2011-300188. pii:jnnp-2011-300188.

Phukan, J., Pender, N. P., & Hardiman, O. (2007). Cognitive impairment in amyotrophic lateral sclerosis. *Lancet Neurology, 6*(11), 994–1003. http://dx.doi.org/10.1016/S1474-4422(07)70265-X. pii:S1474-4422(07)70265-X.

Pottier, C., Bieniek, K. F., Finch, N., van de Vorst, M., Baker, M., Perkersen, R., … Rademakers, R. (2015). Whole-genome sequencing reveals important role for TBK1 and OPTN mutations in frontotemporal lobar degeneration without motor neuron disease. *Acta Neuropathologica, 130*(1), 77–92. http://dx.doi.org/10.1007/s00401-015-1436-x.

Prudencio, M., Hart, P. J., Borchelt, D. R., & Andersen, P. M. (2009). Variation in aggregation propensities among ALS-associated variants of SOD1: correlation to human disease. *Human Molecular Genetics, 18*(17), 3217–3226. http://dx.doi.org/10.1093/hmg/ddp260. pii:ddp260.

Puls, I., Jonnakuty, C., LaMonte, B. H., Holzbaur, E. L., Tokito, M., Mann, E., … Fischbeck, K. H. (2003). Mutant dynactin in motor neuron disease. *Nature Genetics, 33*(4), 455–456. http://dx.doi.org/10.1038/ng1123. pii:ng1123.

Ralston, S. H., & Albagha, O. M. (2011). Genetic determinants of Paget's disease of bone. *Annals of the New York Academy of Sciences, 1240*, 53–60. http://dx.doi.org/10.1111/j.1749-6632.2011.06228.x.

Ratti, E., Cudkowicz, M. E., & Berry, J. D. (2014). Motor neuron diseases. In E. G. Nabel (Ed.), *Scientific American medicine (online)*. Hamilton, ON: Decker Intellectual Properties. http://dx.doi.org/10.2310/7900.1266.

Ravits, J. (2014). Focality, stochasticity and neuroanatomic propagation in ALS pathogenesis. *Experimental Neurology, 262*(Pt B), 121–126. http://dx.doi.org/10.1016/j.expneurol.2014.07.021. pii:S0014-4886(14)00249-0.

Rayaprolu, S., Soto-Ortolaza, A., Rademakers, R., Uitti, R. J., Wszolek, Z. K., & Ross, O. A. (2012). Angiogenin variation and Parkinson disease. *Annals of Neurology, 71*(5), 725–727. http://dx.doi.org/10.1002/ana.23586. author reply 727-728.

Reich-Slotky, R., Andrews, J., Cheng, B., Buchsbaum, R., Levy, D., Kaufmann, P., & Thompson, J. L. (2013). Body mass index (BMI) as predictor of ALSFRS-R score decline in ALS patients. *Amyotrophic Lateral Sclerosis and Frontotemporal Degeneration, 14*(3), 212–216. http://dx.doi.org/10.3109/21678421.2013.770028.

Renton, A. E., Chio, A., & Traynor, B. J. (2014). State of play in amyotrophic lateral sclerosis genetics. *Nature Neuroscience, 17*(1), 17–23. http://dx.doi.org/10.1038/nn.3584.

Renton, A. E., Majounie, E., Waite, A., Simon-Sanchez, J., Rollinson, S., Gibbs, J. R., … Traynor, B. J. (2011). A hexanucleotide repeat expansion in C9ORF72 is the cause of chromosome 9p21-linked ALS-FTD. *Neuron, 72*(2), 257–268. http://dx.doi.org/10.1016/j.neuron.2011.09.010. pii:S0896-6273(11)00797-5.

Rezaie, T., Child, A., Hitchings, R., Brice, G., Miller, L., Coca-Prados, M., … Sarfarazi, M. (2002). Adult-onset primary open-angle glaucoma caused by mutations in optineurin. *Science, 295*(5557), 1077–1079. http://dx.doi.org/10.1126/science.1066901. pii:295/5557/1077.

Ringholz, G. M., Appel, S. H., Bradshaw, M., Cooke, N. A., Mosnik, D. M., & Schulz, P. E. (2005). Prevalence and patterns of cognitive impairment in sporadic ALS. *Neurology, 65*(4), 586–590. http://dx.doi.org/10.1212/01.wnl.0000172911.39167.b6. pii:65/4/586.

Robberecht, W., & Philips, T. (2013). The changing scene of amyotrophic lateral sclerosis. *Nature Reviews Neuroscience, 14*(4), 248–264. http://dx.doi.org/10.1038/nrn3430. pii:nrn3430.

Rohrer, J. D., Isaacs, A. M., Mizielinska, S., Mead, S., Lashley, T., Wray, S., … Warren, J. D. (2015). C9orf72 expansions in frontotemporal dementia and amyotrophic lateral sclerosis. *Lancet Neurology, 14*(3), 291–301. http://dx.doi.org/10.1016/S1474-4422(14)70233-9. pii:S1474-4422(14)70233-9.

Rosen, D. R. (1993). Mutations in Cu/Zn superoxide dismutase gene are associated with familial amyotrophic lateral sclerosis. *Nature, 364*(6435), 362. http://dx.doi.org/10.1038/364362c0.

Rothstein, J. D., Patel, S., Regan, M. R., Haenggeli, C., Huang, Y. H., Bergles, D. E., … Fisher, P. B. (2005). Beta-lactam antibiotics offer neuroprotection by increasing glutamate transporter expression. *Nature, 433*(7021), 73–77.

Rotunno, M. S., & Bosco, D. A. (2013). An emerging role for misfolded wild-type SOD1 in sporadic ALS pathogenesis. *Frontiers in Cellular Neuroscience, 7*, 253. http://dx.doi.org/10.3389/fncel.2013.00253.

Rubino, E., Rainero, I., Chio, A., Rogaeva, E., Galimberti, D., Fenoglio, P., … Pinessi, L. (2012). SQSTM1 mutations in frontotemporal lobar degeneration and amyotrophic lateral sclerosis. *Neurology, 79*(15), 1556–1562. http://dx.doi.org/10.1212/WNL.0b013e31826e25df. pii:WNL.0b013e31826e25df.

Sasabe, J., Miyoshi, Y., Suzuki, M., Mita, M., Konno, R., Matsuoka, M., … Aiso, S. (2012). D-amino acid oxidase controls motoneuron degeneration through D-serine. *Proceedings of the National Academy of Sciences of the United States of America, 109*(2), 627–632. http://dx.doi.org/10.1073/pnas.1114639109. pii:1114639109.

Sattler, R., & Rothstein, J. D. (2006). Regulation and dysregulation of glutamate transporters. *Handbook of Experimental Pharmacology, 175*, 277–303.

Schmidt, E. R., Pasterkamp, R. J., & van den Berg, L. H. (2009). Axon guidance proteins: novel therapeutic targets for ALS? *Progress in Neurobiology, 88*(4), 286–301. http://dx.doi.org/10.1016/j.pneurobio.2009.05.004. pii:S0301-0082(09)00077-X.

Schuster, C., Kasper, E., Dyrba, M., Machts, J., Bittner, D., Kaufmann, J., … Prudlo, J. (2014). Cortical thinning and its relation to cognition in amyotrophic lateral sclerosis. *Neurobiology of Aging, 35*(1), 240–246. http://dx.doi.org/10.1016/j.neurobiolaging.2013.07.020. pii:S0197 4580(13)00325-4.

Seelen, M., Visser, A. E., Overste, D. J., Kim, H. J., Palud, A., Wong, T. H., … van den Berg, L. H. (2014). No mutations in hnRNPA1 and hnRNPA2B1 in Dutch patients with amyotrophic lateral sclerosis, frontotemporal dementia, and inclusion body myopathy. *Neurobiology of Aging, 35*(8), 1956.e9–1956.e11. http://dx.doi.org/10.1016/j.neurobiolaging.2014.01.152. pii:S0197-4580(14)00177-8.

Seguin, S. J., Morelli, F. F., Vinet, J., Amore, D., De Biasi, S., Poletti, A., … Carra, S. (2014). Inhibition of autophagy, lysosome and VCP function impairs stress granule assembly. *Cell Death & Differentiation, 21*(12), 1838–1851. http://dx.doi.org/10.1038/cdd.2014.103. pii:cdd2014103.

Shimizu, T., Nagaoka, U., Nakayama, Y., Kawata, A., Kugimoto, C., Kuroiwa, Y., … Komori, T. (2012). Reduction rate of body mass index predicts prognosis for survival in amyotrophic lateral sclerosis: a multicenter study in Japan. *Amyotrophic Lateral Sclerosis, 13*(4), 363–366. http://dx.doi.org/10.3109/17482968.2012.678366.

Siddiqi, S., Foo, J. N., Vu, A., Azim, S., Silver, D. L., Mansoor, A., … Khor, C. C. (2014). A novel splice-site mutation in ALS2 establishes the diagnosis of juvenile amyotrophic lateral sclerosis in a family with early onset anarthria and generalized dystonias. *PLoS One, 9*(12), e113258. http://dx.doi.org/10.1371/journal.pone.0113258. pii:PONE-D-14-30213.

Simpson, C. L., Lemmens, R., Miskiewicz, K., Broom, W. J., Hansen, V. K., van Vught, P. W., … Al-Chalabi, A. (2009). Variants of the elongator protein 3 (ELP3) gene are associated with motor neuron degeneration. *Human Molecular Genetics, 18*(3), 472–481. http://dx.doi.org/10.1093/hmg/ddn375. pii:ddn375.

Skibinski, G., Parkinson, N. J., Brown, J. M., Chakrabarti, L., Lloyd, S. L., Hummerich, H., … Collinge, J. (2005). Mutations in the endosomal ESCRTIII-complex subunit CHMP2B in frontotemporal dementia. *Nature Genetics, 37*(8), 806–808. http://dx.doi.org/10.1038/ng1609. pii:ng1609.

Smith, B. N., Ticozzi, N., Fallini, C., Gkazi, A. S., Topp, S., Kenna, K. P., … Landers, J. E. (2014). Exome-wide rare variant analysis identifies TUBA4A mutations associated with familial ALS. *Neuron, 84*(2), 324–331. http://dx.doi.org/10.1016/j.neuron.2014.09.027. pii:S0896-6273(14)00847-2.

Smith, B. N., Vance, C., Scotter, E. L., Troakes, C., Wong, C. H., Topp, S., … Shaw, C. E. (2015). Novel mutations support a role for Profilin 1 in the pathogenesis of ALS. *Neurobiology of Aging, 36*(3), 1602.e17–1602.e27. http://dx.doi.org/10.1016/j.neurobiolaging.2014.10.032. pii:S0197-4580(14)00692-7.

Solowska, J. M., D'Rozario, M., Jean, D. C., Davidson, M. W., Marenda, D. R., & Baas, P. W. (2014). Pathogenic mutation of spastin has gain-of-function effects on microtubule dynamics. *Journal of Neuroscience, 34*(5), 1856–1867. http://dx.doi.org/10.1523/JNEUROSCI.3309-13.2014. pii:34/5/1856.

Sreedharan, J., Blair, I. P., Tripathi, V. B., Hu, X., Vance, C., Rogelj, B., … Shaw, C. E. (2008). TDP-43 mutations in familial and sporadic amyotrophic lateral sclerosis. *Science, 319*(5870), 1668–1672. http://dx.doi.org/10.1126/science.1154584. pii:1154584.

Steele, J. C., & McGeer, P. L. (2008). The ALS/PDC syndrome of Guam and the cycad hypothesis. *Neurology, 70*(21), 1984–1990. http://dx.doi.org/10.1212/01.wnl.0000312571.81091.26. pii:70/21/1984.

Strong, M. J., Grace, G. M., Freedman, M., Lomen-Hoerth, C., Woolley, S., Goldstein, L. H., … Figlewicz, D. (2009). Consensus criteria for the diagnosis of frontotemporal cognitive and behavioural syndromes in amyotrophic lateral sclerosis. *Amyotrophic Lateral Sclerosis, 10*(3), 131–146.

Swinnen, B., & Robberecht, W. (2014). The phenotypic variability of amyotrophic lateral sclerosis. *Nature Reviews Neurology, 10*(11), 661–670. http://dx.doi.org/10.1038/nrneurol.2014.184. pii:nrneurol.2014.184.

Talbot, K. (2014). Amyotrophic lateral sclerosis: cell vulnerability or system vulnerability? *Journal of Anatomy, 224*(1), 45–51. http://dx.doi.org/10.1111/joa.12107.

Tan, C. F., Eguchi, H., Tagawa, A., Onodera, O., Iwasaki, T., Tsujino, A., … Takahashi, H. (2007). TDP-43 immunoreactivity in neuronal inclusions in familial amyotrophic lateral sclerosis with or without SOD1 gene mutation. *Acta Neuropathologica, 113*(5), 535–542. http://dx.doi.org/10.1007/s00401-007-0206-9.

Teyssou, E., Takeda, T., Lebon, V., Boillee, S., Doukoure, B., Bataillon, G., … Millecamps, S. (2013). Mutations in SQSTM1 encoding p62 in amyotrophic lateral sclerosis: genetics and neuropathology. *Acta Neuropathologica, 125*(4), 511–522. http://dx.doi.org/10.1007/s00401-013-1090-0.

Thiyagarajan, N., Ferguson, R., Subramanian, V., & Acharya, K. R. (2012). Structural and molecular insights into the mechanism of action of human angiogenin-ALS variants in neurons. *Nature Communications, 3*, 1121. http://dx.doi.org/10.1038/ncomms2126. pii:ncomms2126.

Tiloca, C., Ticozzi, N., Pensato, V., Corrado, L., Del Bo, R., Bertolin, C., … Silani, V. (2013). Screening of the PFN1 gene in sporadic amyotrophic lateral sclerosis and in frontotemporal dementia. *Neurobiology of Aging, 34*(5), 1517.e9–1517.e10. http://dx.doi.org/10.1016/j.neurobiolaging.2012.09.016. pii:S0197-4580(12)00473-3.

Traynor, B. J., Nalls, M., Lai, S. L., Gibbs, R. J., Schymick, J. C., Arepalli, S., … Chio, A. (2010). Kinesin-associated protein 3 (KIFAP3) has no effect on survival in a population-based cohort of ALS patients. *Proceedings of the National Academy of Sciences of the United States of America, 107*(27), 12335–12338. http://dx.doi.org/10.1073/pnas.0914079107. pii:0914079107.

Tsuda, H., Han, S. M., Yang, Y., Tong, C., Lin, Y. Q., Mohan, K., … Bellen, H. J. (2008). The amyotrophic lateral sclerosis 8 protein VAPB is cleaved, secreted, and acts as a ligand for Eph receptors. *Cell, 133*(6), 963–977. http://dx.doi.org/10.1016/j.cell.2008.04.039. pii:S0092-8674(08)00613-2.

Turner, M. R., Bowser, R., Bruijn, L., Dupuis, L., Ludolph, A., McGrath, M., … Fischbeck, K. H. (2013). Mechanisms, models and biomarkers in amyotrophic lateral sclerosis. *Amyotrophic Lateral Sclerosis and Frontotemporal Degeneration, 14*(Suppl. 1), 19–32. http://dx.doi.org/10.3109/21678421.2013.778554.

Turner, M. R., Hardiman, O., Benatar, M., Brooks, B. R., Chio, A., de Carvalho, M., … Kiernan, M. C. (2013). Controversies and priorities in amyotrophic lateral sclerosis. *Lancet Neurology, 12*(3), 310–322. http://dx.doi.org/10.1016/S1474-4422(13)70036-X.

Turner, M. R., Parton, M. J., Shaw, C. E., Leigh, P. N., & Al-Chalabi, A. (2003). Prolonged survival in motor neuron disease: a descriptive study of the King's database 1990–2002. *Journal of Neurology, Neurosurgery and Psychiatry, 74*(7), 995–997.

Udan-Johns, M., Bengoechea, R., Bell, S., Shao, J., Diamond, M. I., True, H. L., … Baloh, R. H. (2014). Prion-like nuclear aggregation of TDP-43 during heat shock is regulated by HSP40/70 chaperones. *Human Molecular Genetics, 23*(1), 157–170. http://dx.doi.org/10.1093/hmg/ddt408. pii:ddt408.

Van Deerlin, V. M., Leverenz, J. B., Bekris, L. M., Bird, T. D., Yuan, W., Elman, L. B., … Yu, C. E. (2008). TARDBP mutations in amyotrophic lateral sclerosis with TDP-43 neuropathology: a genetic and histopathological analysis. *Lancet Neurology, 7*(5), 409–416. http://dx.doi.org/10.1016/S1474-4422(08)70071-1. pii:S1474-4422(08)70071-1.

Van Hoecke, A., Schoonaert, L., Lemmens, R., Timmers, M., Staats, K. A., Laird, A. S., … Robberecht, W. (2012). EPHA4 is a disease modifier of amyotrophic lateral sclerosis in animal models and in humans. *Nature Medicine, 18*(9), 1418–1422. http://dx.doi.org/10.1038/nm.2901. pii:nm.2901.

Van Langenhove, T., van der Zee, J., Sleegers, K., Engelborghs, S., Vandenberghe, R., Gijselinck, I., … Van Broeckhoven, C. (2010). Genetic contribution of FUS to frontotemporal lobar degeneration. *Neurology, 74*(5), 366–371. http://dx.doi.org/10.1212/WNL.0b013e3181ccc732. pii:74/5/366.

Vance, C., Rogelj, B., Hortobagyi, T., De Vos, K. J., Nishimura, A. L., Sreedharan, J., … Shaw, C. E. (2009). Mutations in FUS, an RNA processing protein, cause familial amyotrophic lateral sclerosis type 6. *Science, 323*(5918), 1208–1211. http://dx.doi.org/10.1126/science.1165942. pii:323/5918/1208.

Vilarino-Guell, C., Wider, C., Soto-Ortolaza, A. I., Cobb, S. A., Kachergus, J. M., Keeling, B. H., … Farrer, M. J. (2009). Characterization of DCTN1 genetic variability in neurodegeneration. *Neurology, 72*(23), 2024–2028. http://dx.doi.org/10.1212/WNL.0b013e3181a92c4c. pii:72/23/2024.

Waibel, S., Neumann, M., Rabe, M., Meyer, T., & Ludolph, A. C. (2010). Novel missense and truncating mutations in FUS/TLS in familial ALS. *Neurology, 75*(9), 815–817. http://dx.doi.org/10.1212/WNL.0b013e3181f07e26. pii:WNL.0b013e3181f07e26.

Walhout, R., Westeneng, H. J., Verstraete, E., Hendrikse, J., Veldink, J. H., van den Heuvel, M. P., & van den Berg, L. H. (2015). Cortical thickness in ALS: towards a marker for upper motor neuron involvement. *Journal of Neurology, Neurosurgery and Psychiatry, 86*(3), 288–294. http://dx.doi.org/10.1136/jnnp-2013-306839. pii:jnnp-2013-306839.

Watts, G. D., Wymer, J., Kovach, M. J., Mehta, S. G., Mumm, S., Darvish, D., … Kimonis, V. E. (2004). Inclusion body myopathy associated with Paget disease of bone and frontotemporal dementia is caused by mutant valosin-containing protein. *Nature Genetics, 36*(4), 377–381. http://dx.doi.org/10.1038/ng1332. pii:ng1332.

Weihl, C. C., Pestronk, A., & Kimonis, V. E. (2009). Valosin-containing protein disease: inclusion body myopathy with Paget's disease of the bone and fronto-temporal dementia. *Neuromuscular Disorders, 19*(5), 308–315. http://dx.doi.org/10.1016/j.nmd.2009.01.009. pii:S0960-8966(09)00009-1.

Whitwell, J. L., Weigand, S. D., Boeve, B. F., Senjem, M. L., Gunter, J. L., DeJesus-Hernandez, M., … Josephs, K. A. (2012). Neuroimaging signatures of frontotemporal dementia genetics: C9ORF72, tau, progranulin and sporadics. *Brain, 135*(Pt 3), 794–806. http://dx.doi.org/10.1093/brain/aws001. pii:aws001.

Wicks, P., Abrahams, S., Papps, B., Al-Chalabi, A., Shaw, C. E., Leigh, P. N., & Goldstein, L. H. (2009). SOD1 and cognitive dysfunction in familial amyotrophic lateral sclerosis. *Journal of Neurology, 256*(2), 234–241. http://dx.doi.org/10.1007/s00415-009-0078-0.

Wu, C. H., Fallini, C., Ticozzi, N., Keagle, P. J., Sapp, P. C., Piotrowska, K., … Landers, J. E. (2012). Mutations in the profilin 1 gene cause familial amyotrophic lateral sclerosis. *Nature, 488*(7412), 499–503. http://dx.doi.org/10.1038/nature11280. pii:nature11280.

Yang, Y., Hentati, A., Deng, H. X., Dabbagh, O., Sasaki, T., Hirano, M., ... Siddique, T. (2001). The gene encoding alsin, a protein with three guanine-nucleotide exchange factor domains, is mutated in a form of recessive amyotrophic lateral sclerosis. *Nature Genetics, 29*(2), 160–165. http://dx.doi.org/10.1038/ng1001-160. pii:ng1001-160.

Yokoyama, T., Ishiyama, M., Hasegawa, K., Uchihara, T., & Yagishita, S. (2014). Novel neuronal cytoplasmic inclusions in a patient carrying SCA8 expansion mutation. *Neuropathology, 34*(1), 27–31. http://dx.doi.org/10.1111/neup.12042.

Zamiri, B., Reddy, K., Macgregor, R. B., Jr., & Pearson, C. E. (2014). TMPyP4 porphyrin distorts RNA G-quadruplex structures of the disease-associated r(GGGGCC)n repeat of the C9orf72 gene and blocks interaction of RNA-binding proteins. *Journal of Biological Chemistry, 289*(8), 4653–4659. http://dx.doi.org/10.1074/jbc.C113.502336. pii:C113.502336.

van der Zee, J., Van Langenhove, T., Kovacs, G. G., Dillen, L., Deschamps, W., Engelborghs, S., ... Van Broeckhoven, C. (2014). Rare mutations in SQSTM1 modify susceptibility to frontotemporal lobar degeneration. *Acta Neuropathologica, 128*(3), 397–410. http://dx.doi.org/10.1007/s00401-014-1298-7.

Zhang, D., Iyer, L. M., He, F., & Aravind, L. (2012). Discovery of novel DENN proteins: implications for the evolution of eukaryotic intracellular membrane structures and human disease. *Frontiers in Genetics, 3*, 283. http://dx.doi.org/10.3389/fgene.2012.00283.

Zhu, G., Wu, C. J., Zhao, Y., & Ashwell, J. D. (2007). Optineurin negatively regulates TNFalpha- induced NF-kappaB activation by competing with NEMO for ubiquitinated RIP. *Current Biology, 17*(16), 1438–1443. http://dx.doi.org/10.1016/j.cub.2007.07.041. pii:S0960-9822(07)01721-6.

Zu, T., Liu, Y., Banez-Coronel, M., Reid, T., Pletnikova, O., Lewis, J., ... Ranum, L. P. (2013). RAN proteins and RNA foci from antisense transcripts in C9ORF72 ALS and frontotemporal dementia. *Proceedings of the National Academy of Sciences of the United States of America, 110*(51), E4968–E4977. http://dx.doi.org/10.1073/pnas.1315438110. pii:1315438110.

Chapter 43

Eating Disorders

Cynthia M. Bulik
University of North Carolina Center of Excellence for Eating Disorders, Chapel Hill, NC, United States; Karolinska Institutet, Stockholm, Sweden

Garret D. Stuber
University of North Carolina at Chapel Hill, Chapel Hill, NC, United States

The primary eating disorders, anorexia nervosa (AN), bulimia nervosa (BN), and binge-eating disorder (BED), remain some of the most misunderstood of neuropsychiatric disorders. Whereas other disorders have emerged from dark decades of fallacious attributions of causality, misconceptions about eating disorders still abound. Misattributions of cause and stigma persist even after an influential position paper by the Academy for Eating Disorders (AED) declared, "It is the position of the AED that anorexia nervosa and bulimia nervosa, along with their variants, are biologically based, serious mental illnesses (BBMI) that warrant the same level and breadth of health care coverage as conditions currently categorized in this way [e.g., schizophrenia, bipolar disorder, depression, obsessive-compulsive disorder (OCD)]. …we advocate this position unequivocally based on an emerging science that affirms with a reasonable degree of scientific and medical certainty that eating disorders are significantly heritable; influenced by alterations of brain function; significantly impair cognitive function, judgment, and emotional stability; and restrict the life activities of persons afflicted with these illnesses" (Klump, Bulik, Kaye, Treasure, & Tyson, 2009).

In line with this position paper, we take as a starting point that eating disorders, like other neuropsychiatric conditions, are complex traits influenced by an array of genetic and environmental factors and their co-action. Our incomplete knowledge about the underlying neurobiology of these illnesses reflects the fact that we have been overattentive to psychosocial and underattentive to biological causal factors for several decades. In many ways, the virtual explosion of neurobiological research on obesity has legitimized the study of the biology of eating disorders because many of the parameters under investigation such as appetite and weight dysregulation, impulse dyscontrol, and aberrant experience of reward are directly relevant to the core psychopathology of eating disorders.

Many of the parameters of dysregulation evident in eating disorders (ie, binge eating, increased physical activity, caloric restriction) are readily quantifiable and amenable to animal models. This enables us to parse out key behavioral traits and identify factors that contribute to their dysregulation. Critics claim that these animal models fail to capture the psychological complexity of the disorders as psychological features such as body dissatisfaction or drive for thinness that are considered to be pathognomonic of the illnesses. Others contend that these psychological traits might simply represent contemporary cultural epiphenomena of the disorders, and that the core features of the illnesses such as maintenance of extremely low body weight in AN, and binge eating in BN and BED have existed throughout time encased by other cultural explanatory models (eg, spirituality or saintliness in AN).

Nonetheless, animal models have been enormously helpful in drilling down into fundamental neurobiological pathways that contribute to processes that are wildly awry in eating disorders, such as appetite, weight regulation, initiation, and inhibition of eating. Such knowledge is critically valuable in understanding why these disorders emerge and why they persist, and it will serve as a concrete foundation for understanding how environmental risk factors may differentially affect individuals depending on their biological setting.

Genomics, Circuits, and Pathways in Clinical Neuropsychiatry. http://dx.doi.org/10.1016/B978-0-12-800105-9.00043-3

CLINICAL SYNDROMES

Anorexia Nervosa

AN is a perplexing psychiatric illness characterized by restriction of energy intake relative to requirements, leading to a significantly low body weight for the individual's age, sex, developmental trajectory, and physical health. Those who are still growing may fail to make expected increases in weight, height, and bone density. Despite low weight, individuals with AN experience an intense fear of gaining weight and strive for additional weight loss. They may see or experience their bodies as fat even though emaciated, engage in extreme weight-loss behaviors (eg, purging, dieting, excessive exercise, and fasting), place undue influence of body weight or shape on self-evaluation, or fail to recognize the seriousness of the low body weight. In the restricting subtype, individuals achieve low weight only through energy restriction and increased physical activity. In the binge-eating/purging type, individuals also engage in binge eating and inappropriate compensatory behaviors.

AN onsets most commonly in adolescence but can and do occur prepubertally and across the lifespan (Beck, Casper, & Andersen, 1996; Cooper, Watkins, Bryant-Waugh, & Lask, 2002; Nicholls, Lynn, & Viner, 2011). Both sexes develop AN, but it is more common in women (American Psychiatric Association, 2013). The estimated prevalence in the United States is 0.9% in women and 0.3% in men (Hudson, Hiripi, Pope, & Kessler, 2007).

Bulimia Nervosa

BN does not occur exclusively during the course of AN and is characterized by recurrent binge-eating episodes, defined as eating an amount of food definitely larger than what most individuals would eat in a similar period of time under similar circumstances while experiencing a sense of loss of control over eating. The symptom picture of BN includes recurrent inappropriate compensatory behaviors (eg, self-induced vomiting, laxative, diuretic, or other medication misuse; fasting; or excessive exercise). Self-evaluation is unduly influenced by body shape and weight. Binge eating and compensatory episodes occur on average once a week for at least 3 months.

BN onsets most frequently occurs in later adolescence or early adulthood, although onset can occur at any time. Although BN is more likely to be diagnosed in women (prevalence of 1.5% in women and 0.5% in men) (Hudson et al., 2007), the BN diagnostic criteria may be gender-biased, leading to underdetection in men. Bulimia may be expressed somewhat differently in males, with more common nonpurging compensatory behavior, such as excessive exercise (Anderson & Bulik, 2003; Lewinsohn, Seeley, Moerk, & Striegel-Moore, 2002) and more common reliance on supplements and steroids for weight and muscle control (Hildebrandt et al., 2011).

Binge-Eating Disorder

BED was first officially recognized as an eating disorder in the *Diagnostic and Statistical Manual of Mental Disorders*, Fifth Edition (DSM-5) (American Psychiatric Association, 2013). BED is marked by recurrent binge eating (at least weekly for 3 months, defined as in BN) coupled with the sense of lack of control over eating. However, it diverges from BN in that recurrent compensatory behaviors do not occur. Distress regarding the binge eating must be present and at least three other features such as the following must be present: eating much more rapidly than normal; eating until feeling uncomfortably full; eating large amounts of food when not feeling physically hungry; eating alone because of feeling embarrassed by how much one is eating; or feeling disgusted with oneself, depressed, or guilty afterward. Although commonly associated with overweight or obesity, BED can occur at any body weight and is diagnosed only if neither AN nor BN criteria are met.

The sex distribution in BED is more equal than in the other eating disorders (prevalence of DSM-IV BED 3.5% in women and 2% in men) (Hudson et al., 2007). Onset may be somewhat later than AN or BN, typically in late adolescence or the early twenties; however, children and adolescence exhibit loss of control eating, which may be a precursor to BED (Tanofsky-Kraff et al., 2011).

There is considerable diagnostic flux across the eating disorders. Individuals with AN commonly cross over to BN at some point during the course of illness, and around 30% of individuals presenting with BN report a history of AN (Allen, Byrne, Oddy, & Crosby, 2013; Tozzi et al., 2005). Migration from BN to BED can also occur. Transitions from BED to AN are less common, but do happen. Pure forms of restricting AN, BN, and BED exist (ie, maintaining a consistent diagnostic profile over time).

Subphenotypes That Cross-Cut the Eating Disorders

Conceptualizing eating disorders with a focus on the core features that cross-cut the diagnostic boundaries provides a convenient mechanism for identifying subphenotypes that do or do not lend themselves well to animal models. Core physiological or behavioral indices such as appetite dysregulation (eg, aberrant initiation, regulation, or inhibition of feeding), weight dysregulation (eg, underweight, overweight, uncoupling of predicted associations between energy intake and expenditure), reward responsivity, and hedonic response lend themselves well to animal models. The more psychological features such as dieting for weight control, body dissatisfaction, drive for thinness, and body image distortion represent complex psychological parameters associated with eating disorders that are harder if not impossible to model in animals.

Common Comorbid Conditions

Eating disorders are also highly comorbid; the most common co-occurring disorders are anxiety disorders, depressive disorders, and substance use disorders. Additional data are emerging on comorbid attention-deficit hyperactivity disorder (ADHD). These, too, lend themselves well to animal models. Cross-disorder genome-wide association studies (GWAS) have begun to identify genetic variants that are common to overlapping disorders with cross-cutting features, and are aiding in determining whether psychiatric disorders are etiologically distinct or share genetic causes (Cross-Disorder Group of the Psychiatric Genomics Consortium et al., 2013). The case for studying cross-disorder genetics in eating disorders is particularly strong with anxiety disorders, depressive disorders, substance use disorders, and ADHD, with some support for AN and autism.

Anxiety Disorders and Obsessive-Compulsive Disorder

Anxiety disorders are common among individuals with eating disorders, with as many as 71% of treatment-seeking patients presenting with at least one lifetime anxiety disorder (Godart et al., 2003). Between 55% and 83% of treatment-seeking individuals with AN report a lifetime history of at least one anxiety disorder (Godart, Flament, Lecrubier, & Jeammet, 2000; Kaye et al., 2004). Population-based studies estimate the prevalence of anxiety disorders to be between 24% and 48% in AN (Hudson et al., 2007; Swanson, Crow, Le Grange, Swendsen, & Merikangas, 2011). Anxiety disorders typically precede AN and onset during childhood (Kaye et al., 2004; Raney et al., 2008). OCD occurs in 15–29% of individuals with AN (Blinder, Cumella, & Sanathara, 2006; Halmi et al., 2003; Salbach-Andrae et al., 2008).

Two-thirds and three-quarters of treatment-seeking patients with BN report a lifetime anxiety disorder, often before the onset of BN (Godart, Flament, Perdereau, & Jeammet, 2002). In population-based studies, 65.2% of adults worldwide (Kessler et al., 2013), 80.6% of US adults (Hudson et al., 2007), and 66.2% of US adolescents (Swanson et al., 2011) with lifetime BN also meet criteria for a lifetime anxiety disorder. Lifetime-specific phobia, social phobia, and posttraumatic stress disorder are especially common (Hudson et al., 2007; Kessler et al., 2013). Anxiety disorders often have an onset before or contemporaneously with BN (Hudson et al., 2007).

Among treatment-seeking individuals with BED, around one-third have a lifetime anxiety disorder (Grilo, White, Barnes, & Masheb, 2013; Grilo, White, & Masheb, 2009; Wilfley et al., 2000), somewhat lower than in community-based samples; panic disorder is typically the most prevalent anxiety disorder (Grilo et al., 2009; White & Grilo, 2006; Wilfley et al., 2000). In population-based studies, 56.1% of adults worldwide (Kessler et al., 2013), 65.1% of US adults (Hudson et al., 2007), and 65.2% of US adolescents (Swanson et al., 2011) also meet criteria for a lifetime anxiety disorder, which is significantly higher than in individuals without lifetime BED (Hudson et al., 2007; Kessler et al., 2013), even after adjusting for body mass index (BMI) (Javaras, Pope, et al., 2008). Anxiety disorders typically have an onset before or contemporaneously with BED (Hudson et al., 2007); social phobia and specific phobia are the strongest predictors of subsequent BED (Kessler et al., 2013), although BED predicts the onset of panic disorder/agoraphobia (Kessler et al., 2013).

Depressive and Bipolar Disorders

Up to 60% of individuals with AN report a lifetime mood disorder episode (Salbach-Andrae et al., 2008); major depressive disorder is most common comorbid depressive or bipolar disorder and is present in up to 86% in patients with AN (O'Brien & Vincent, 2003). Estimates from population-based studies are somewhat lower, with the prevalence of depression estimated to be 9–39% (Hudson et al., 2007; Swanson et al., 2011). Epidemiologic data suggest a lower comorbidity between AN and bipolar disorder (BD) (about 2–3%), which is on par with the general population (Hudson et al., 2007; Swanson et al., 2011).

Among treatment-seeking individuals with BN, up to two-thirds meet criteria for lifetime major depressive disorder (MDD) (Brewerton et al., 1995). In population-based studies, 50.1% and 17.7% of US adults (Hudson et al., 2007) and 31.0% and 18.5% of US adolescents (Swanson et al., 2011) with lifetime BN also meet criteria for lifetime MDD and BD I/II, respectively. The lifetime prevalences of MDD and BD are significantly higher than analogous prevalences among control individuals without lifetime BN (Hudson et al., 2007; Swanson et al., 2011). Depressive disorders can occur contemporaneously with or before BN but are more likely to occur subsequent to BN (Hudson et al., 2007).

Among treatment-seeking individuals with BED, the prevalence of lifetime MDD is approximately 50% (Grilo et al., 2013, 2009; White & Grilo, 2006; Wilfley et al., 2000), even higher than in community-based samples, but the prevalence of current MDD is around 15–18% (Grilo et al., 2013, 2009; Wilfley et al., 2000), comparable to community-based samples (Pike, Dohm, Striegel-Moore, Wilfley, & Fairburn, 2001). In population-based studies, 32.3% and 12.5% of US adults (Hudson et al., 2007) and 35.4% and 9.0% of US adolescents (Swanson et al., 2011) with lifetime BED also meet criteria for lifetime MDD and BD I/II, respectively. The lifetime prevalences of MDD and BD are significantly higher than among control individuals without lifetime BED (Hudson et al., 2007; Swanson et al., 2011), even after adjustment for BMI (Javaras, Pope, et al., 2008). Based on retrospective report, about one-third of depressive disorder cases occur subsequent to BED, but they are more likely to occur contemporaneously with or before BED (Hudson et al., 2007).

Substance Use Disorders

As many as a quarter of patients with eating disorders experience alcohol use–related problems (Baker, Mitchell, Neale, & Kendler, 2010). In population-based studies, the prevalence of alcohol abuse/dependence are between 9% and 25% for AN and 14–46% for BN (Bulik et al., 2004; Dansky, Brewerton, & Kilpatrick, 2000; Hudson et al., 2007; Kessler et al., 2013; Root et al., 2010; Swanson et al., 2011). The prevalence is lower among individuals with AN-restricting (AN-R) type than the binge-eating/purging type (AN-BP) and in BN (Krug et al., 2009; Root et al., 2010; Salbach-Andrae et al., 2008). In a study conducted in female patients with eating disorders who were receiving inpatient treatment, alcohol abuse/dependence was present in only 3% of AN-R patients in contrast to 14% of AN-BP and 26% of BN patients (Blinder et al., 2006). BN may precede alcohol use disorder in most patients (Bulik et al., 2004).

Between 13% and 27% of individuals with AN and between 18% and 37% of individuals with BN meet criteria for substance abuse or dependence (Hudson et al., 2007; Kessler et al., 2013; Swanson et al., 2011). In treatment-seeking samples, substance abuse or dependence is significantly more common among those who engage in binge eating and purging (18% in BN and 10% in AN-BP) compared with AN-R (3%) (Blinder et al., 2006). Whereas substance abuse often precedes the onset of BN, the opposite pattern may be more common for AN (Baker et al., 2010, 2013).

In population-based studies, 21.7% and 14.4% of adults worldwide (Kessler et al., 2013), 21.4% and 19.4% of US adults (Hudson et al., 2007), and 13.9% and 22.5% of US adolescents (Swanson et al., 2011) with lifetime BED also meet criteria for lifetime alcohol abuse/dependence and drug abuse/dependence, respectively. Furthermore, substance use disorders tend to onset before or contemporaneously with BED, although one-quarter of lifetime substance disorders occur subsequent to BED (Hudson et al., 2007), with BED tending to predict drug abuse/dependence in particular (Kessler et al., 2013). Among treatment-seeking individuals with current BED, the prevalence of lifetime alcohol and drug abuse/dependence is similar (Grilo et al., 2009; Wilfley et al., 2000) or slightly lower (Grilo et al., 2013; White & Grilo, 2006) than in community-based samples, and men are more likely to have a lifetime substance use disorder than are women (Grilo et al., 2013, 2009). One study has suggested that there may be two subtypes of BED, with greater impulsivity present in the group with a history of substance abuse (Peterson, Miller, Crow, Thuras, & Mitchell, 2005). However, the presence of substance use disorders does not appear to affect treatment outcomes for BED (Wilfley et al., 2000).

Attention-Deficit Hyperactivity Disorder

Epidemiological studies estimate the prevalence of ADHD to be in the range of 2–16% for AN and 15–35% in BN (Hudson et al., 2007; Kessler et al., 2013; Swanson et al., 2011). Retrospective studies suggest that up to 24% of individuals with BN may have a childhood history of ADHD (Seitz et al., 2013; Yilmaz, Kaplan, Zai, Levitan, & Kennedy, 2011). Furthermore, women with ADHD are significantly more likely to report a history of AN or BN than are women without ADHD (Cumyn, French, & Hechtman, 2009).

In population-based studies, 10.2% of adults worldwide (Kessler et al., 2013), 19.8% of US adults (Hudson et al., 2007), and 12.6% of US adolescents (Swanson et al., 2011) with lifetime BED also meet criteria for lifetime ADHD.

In population-based studies, the lifetime prevalence of ADHD is significantly higher among individuals with lifetime BED than are individuals without lifetime BED (Hudson et al., 2007; Kessler et al., 2013). Although no study has examined whether the association between ADHD and BED remains after adjusting for BMI or obesity, which is itself associated with ADHD (Cortese et al., 2008), research based on nationally representative samples suggests that BED partially mediates the association between adult ADHD and overweight as well as obesity (Pagoto et al., 2009). In population-based studies, ADHD predicts subsequent BED (odds ratio [OR], 4.7 after demographic adjustments; OR, 1.9 after additional adjustment for other Axis I disorders comorbid with ADHD), but BED does not predict subsequent ADHD (OR, 0.0) (Kessler et al., 2013). Finally, there is limited evidence regarding the prevalence of ADHD among treatment-seeking individuals with BED, because studies of psychiatric comorbidity among those seeking treatment have not routinely assessed ADHD.

HUMAN GENETIC RESEARCH

Family and Twin Studies

The familiality and heritability of eating disorders, at least in European ancestry populations, have been clearly documented. Female relatives of individuals with AN are 11 times more likely to develop AN themselves than are the relatives of individuals without AN (Strober, Freeman, Lampert, Diamond, & Kaye, 2000). Consistent with the protean nature of eating disorder symptomatology, eating disorders do not breed true in families: Individuals who have a relative with AN or BN are at increased risk for developing either disorder (Lilenfeld et al., 1998; Strober et al., 2000). BED also aggregates in families independent of obesity (Hudson et al., 2006).

Twin studies estimate the heritability of AN to be somewhere between 0.28 and 0.74, which means that up to 74% of phenotypic variation can be explained by additive genetic factors (Bulik et al., 2006; Kipman, Gorwood, Mouren-Simeoni, & Ades, 1999; Klump, Miller, Keel, McGue, & Iacono, 2001; Kortegaard, Hoerder, Joergensen, Gillberg, & Kyvik, 2001). The broad range reflects varying definitions of illness and imprecise estimates owing to the relative rarity of the condition.

The heritability of BN has been estimated to be between 0.28 and 0.83 (Bulik, Sullivan, & Kendler, 1998; Bulik et al., 2010). The genetic correlation between AN and BN has been estimated to be 0.79 (Bulik et al., 2010), which may explain the high cross over rates and co-familiality between the two presentations.

The heritability of BED using varying definitions of illness has been estimated to be between 0.39 and 0.45 (Javaras, Laird, et al., 2008; Mitchell et al., 2010; Reichborn-Kjennerud, Bulik, Tambs, & Harris, 2004).

History of Genetic Studies in Eating Disorders

The history of the study of genetic factors contributing to eating disorders has followed the path of other psychiatric disorders. Linkage studies, which aimed to identify genomic regions that have an increased likelihood of containing genes associated with a disorder or trait, yielded a few positive findings (Bulik et al., 2003; Devlin et al., 2002; Grice et al., 2002) but did not lead to follow through on identifying specific contributing genes. A decade of candidate gene association studies ensued involving specific hypotheses based on biological function. However, no reliably replicated results emerged that were specific to the pathology of any of the eating disorders.

Current Status of Genetic/Genomic Research

GWAS have commenced in eating disorders, with the earliest efforts focused on AN. GWAS scan the entire genome in a hypothesis-free manner (Corvin, Craddock, & Sullivan, 2010).

The first genome-wide study of AN, carried out in 1033 patients with AN, reported a large and rare copy number variant on 13q12 present in two individuals but found no genome-wide significant loci in the case–control analysis (Wang et al., 2011). A GWAS collaboration on a twin sample also failed to find genome-wide significant susceptibility loci but provided some evidence for the possible involvement of eight loci with various eating disorder–related phenotypes such as drive for thinness, body dissatisfaction, bulimia, and weight fluctuations, as well as traits associated with OCD personality (Boraska et al., 2012). Similarly, another GWAS did not report genome-wide significant findings, but a number of genes were implicated in AN- and BN-spectrum disorder phenotypes (Wade et al., 2013).

The largest and most rigorous GWAS in AN yet conducted, part of the Wellcome Trust Case–Control Consortium 3, included 2907 patients with AN of European ancestry and 14,860 ancestry-matched female controls in the discovery meta-analysis (Boraska et al., 2014). Although there were no genome-wide significant findings in the discovery data set, 72 independent markers with the lowest *P* values were selected for replication, and in sign tests approximately 76% of these markers yielded results in the same direction as the discovery sample (Boraska et al., 2014). This observation suggests that the prioritized set of genomic variants likely contained true positive signals for AN; however, the sample size did, too (Boraska et al., 2014). Efforts are under way to dramatically increase sample size in AN. No GWAS for BN or BED have been conducted.

Emerging Epigenetic Research

Gene expression or methylation patterns may be altered in individuals with eating disorders. Although preliminary, initial epigenetic studies of eating disorders have focused on the promoter-specific methylation of the candidate genes that had been previously studied in eating disorders and primarily focused on AN. To date, no epigenetic studies in BED have been published. Studies of global methylation have revealed a pattern of significant global DNA hypomethylation in 22 patients with AN compared with 30 healthy control subjects, and a similar trend was observed in 24 patients with BN (Frieling et al., 2007). A modest reduction in whole-blood global DNA methylation was also reported in 32 adolescents with AN (Tremolizzo et al., 2014); however, this observation has not been universally replicated (Saffrey, Novakovic, & Wade, 2014).

Epigenetic research may ultimately make important contributions to our understanding of the role of elements outside of DNA coding sequence in susceptibility to eating disorders. In addition, eating disorders are at least partly brain disorders, and considering the tissue specificity of epigenetic modifications, use of blood and buccal cells as proxies to brain tissue is another limitation of the published studies (Yilmaz, Hardaway, & Bulik, 2014).

NEURAL CIRCUIT DISSECTION OF FEEDING BEHAVIOR RELATED TO EATING DISORDERS

Although eating disorders such as BED and BN (and perhaps AN) are uniquely human conditions, understanding the neural circuit maladaptation that contributes to these disorders can readily be studied at the basic science level using animal models. To accomplish this, many studies to date have used both wild-type and transgenic rodents to investigate the precise neural circuit components that regulate feeding behavior, as well as associated phenotypes related to reward and anxiety. These studies have now identified many important neural circuit contributors to these phenotypes. Because precise recordings and manipulations of genetically defined neural circuits are not possible in humans, these studies discussed subsequently highlight some of the critical mechanistic insights gained in relation to the neural circuitry that can regulate feeding.

Common Behavioral Strategies for Studying Adaptive and Maladaptive Feeding in Animal Models

Both adaptive and maladaptive aspects of feeding can readily be studied in animals. Most studies investigating the neural circuitry that regulated feeding have simply focused on performing perturbations that can increase or decrease free-feeding in animals. Although this approach does not model eating disorders per se, it can reliably determine whether precise neural circuit alterations can modulate normal food intake. These measurements are easy to perform for extended periods of time, and thus can also be coupled with measurements of metabolism and body weight/composition if needed. In addition, animal models for specific eating disorders such as BED and AN have been developed. Many distinct animal models for eating disorders have been previously reviewed (Kas & Adan, 2011). Whereas the precise details of many of the distinct models can vary, AN-associated phenotypes can be modeled in rodents typically by combining periods of food restriction and/or stress with increases in activity (activity-based anorexia) (Klenotich & Dulawa, 2012). In contrast, aspects of BED can also be modeled in rodents by providing limited, intermittent access to highly palatable foods (Avena, Bocarsly, & Hoebel, 2012; Avena, Rada, & Hoebel, 2009). Collectively, these animal models could be readily coupled with manipulations of functional neurocircuitry to establish the sufficiency and necessity of molecularly defined cell types in animal models of eating disorders.

Genetic Targeting and Optogenetic Actuation of Neuronal Subtypes to Regulate Feeding

To manipulate precise neural circuit components in the brain and then assay how these perturbations can change feeding and related phenotypes, it is first critical to establish that specific brain cell types can be selectively targeted for manipulation. Neuroscience has traditionally used lesion/ablation, electrical stimulation, and pharmacological activation and inactivation to study how neural function regulated feeding. However, these classical techniques have fundamental limitations that preclude their use to discern cell-type or pathway-specific functions related to feeding. Optogenetic manipulations directly address these limitations by enabling cell-type and circuit-specific investigation of neural function. A comprehensive overview of the entire field of optogenetics is beyond the scope of this chapter; however, many reports have thoroughly documented the history and development of the field (Bernstein & Boyden, 2011; Fenno, Yizhar, & Deisseroth, 2011; Tye & Deisseroth, 2012; Yizhar, Fenno, Davidson, Mogri, & Deisseroth, 2011; Zhang et al., 2010).

Genetic targeting strategies to introduce light-gated opsins into neuronal populations to manipulate them are the key advance of optogenetic approaches compared with classical techniques to study brain function. Neural tissue is composed of a heterogeneous mixture of phenotypically diverse cell types that vary with respect to their morphological, physiological, synaptic, and molecular properties. Because of this complexity, parsing the contribution of any neuronal population embedded within a network of heterogeneous cell types was formerly an insurmountable task. Optogenetic actuators of neural activity can readily be introduced in specific brain cell types, typically by using a combination of transgenic animal lines that permit cellular targeting (Gerfen, Paletzki, & Heintz, 2013; Witten et al., 2011) coupled with viral constructs (Atasoy, Aponte, Su, & Sternson, 2008) employed for transgene delivery to a particular anatomical location. In addition to cell-type specificity, optogenetics affords precise activation or inhibition events to study neural dynamics on a millisecond time scale, which is consistent with the temporal dynamics of endogenous neural activity. Although neural circuit function is temporally modulated over a range of time intervals, the moment-to-moment processing of information is mediated by rapid electrical and synaptic signals. Temporally precise optogenetic perturbations also allow for patterned activation of neural circuit elements time-locked to discrete environmental events or at key moments during behavior, such as the initiation of feeding.

Neural Circuits That Regulate Feeding and Related Phenotypes Relevant to Eating Disorders

A number of studies have incorporated optogenetics to study the neural circuitry that controls phenotypes related to eating disorders. Although this is not an exhaustive review of all studies that have incorporated optogenetics to regulate feeding, the studies discussed subsequently provide important examples by which optogenetic circuit perturbations have resulted in profound alterations in feeding behavior as well as providing new mechanistic insight into the neural circuits that control feeding.

Circuits for Promoting Excessive Feeding

A number of discoveries have uncovered important neural circuit components to promote excessive feeding, which may be relevant for over eating disorders such as BED. It was demonstrated that gamma aminobutyric acid-ergic (GABAergic) synaptic input from the bed nucleus of the stria terminalis (BNST) that project to and in nerve cells located in the lateral hypothalamus (LH), serves as a critical neuronal circuit node for regulating both feeding behavior as well as reward-related behaviors (Jennings, Rizzi, Stamatakis, Ung, & Stuber, 2013). Optogenetic stimulation of the pathway resulted in immediate and robust increases in feeding behavior. In addition, optogenetic inhibition of the BNST to the LH GABAergic pathway disrupted natural feeding behavior, which suggests that this pathway is required for adaptive food intake. Interestingly, these manipulations also increased reward-seeking behavior when food was not present. This suggests that the BNST to LH circuit may orchestrate hedonic aspects of feeding instead of promoting hunger. Further characterization of this circuit demonstrated that these inhibitory projections from the BNST to the LH preferentially synapse onto postsynaptic glutamatergic neurons to suppress their activity. Collectively, this work demonstrates the importance of circuit connectivity between the BNST and LH in regulating excessive feeding and reward.

The arcuate nucleus of the hypothalamus is another important neuronal substrate for the regulation of feeding behavior. Many important satiety-signaling molecules, such as leptin and ghrelin, are thought to regulate the activity of arcuate neuronal subpopulations, which in turn promote or suppress food intake (Bouret, Draper, & Simerly, 2004; Pinto et al., 2004). Studies have also demonstrated that direct optogenetic and chemogenetic stimulation of a neuronal subpopulation that produces the neuropeptide Agouti-related peptide (AgRP) readily increases feeding (Aponte, Atasoy, & Sternson, 2011; Krashes et al., 2011). In addition, arcuate AgRP neurons project to many areas that are important for aspects of feeding behavior, including but not limited to the BNST, LH, central amygdala (CeA), and periventricular hypothalamus (PVH) (Betley, Cao, Ritola, & Sternson, 2013). Interestingly, not all AgRP projections when activated optogenetically produce the same feeding-associated behavioral phenotypes. For example, a subset of arcuate nucleus AgRP neurons project to and inhibit PVH neurons that produce oxytocin, which in turn is sufficient to promote feeding (Atasoy, Betley, Su, & Sternson, 2012). Moreover, a subset of neurons within the PVH that produce thyrotropin-releasing hormone send excitatory projects back to the arcuate nucleus AgRP neurons to increase their activity and promote feeding (Krashes et al., 2014). Collectively, these emerging studies demonstrate the importance of arcuate nucleus AgRP neurons and their associated circuit connectivity in regulated feeding behavioral phenotypes that may have important relevance to eating disorders.

Critical Neuronal Circuits for the Suppression of Feeding

These studies highlight some of the important neural circuit contributors for promoting both adaptive and maladaptive feeding phenotypes. Additional work has identified key circuit components that can suppress feeding behavior, and thus may be relevant for unraveling the brain systems that underlie AN. For example, optogenetic stimulation of LH glutamatergic neurons can robustly suppress normal feeding behavior and produce aversion-related phenotypes (Jennings et al., 2013). These LH glutamatergic neurons project to other brain regions such as the parabrachial nucleus (PB) (see subsequent discussion) important for the suppression of feeding, and future studies will likely characterize which subset of LH glutamatergic neurons is critical for feeding suppression.

At the gross neuroanatomical level, the PB is an important neuronal substrate for regulating appetite (Becskei, Grabler, Edwards, Riediger, & Lutz, 2007; DiPatrizio & Simansky, 2008; Wu, Clark, & Palmiter, 2012), although the PB contains many genetically diverse cell types. It has been demonstrated that a subpopulation of PB neurons that express calcitonin gene-related peptide (CGRP) can negatively regulate feeding behavior. Optogenetic stimulation of PB CGRP neurons suppresses feeding, whereas genetic ablation of these neurons increases feeding (Carter, Soden, Zweifel, & Palmiter, 2013). These PB CGRP neurons send a strong glutamatergic projection to the CeA, and optogenetic stimulation of the PB CGRP projections directly suppresses feeding behavior. Consistent with this, a subset of CeA neurons that express the gene coding for protein kinase C-δ (PKC-δ) are activated by anorexigenic signals, and direct optogenetic stimulation of PKC-δ CeA neurons inhibits feeding (Cai, Haubensak, Anthony, & Anderson, 2014). Collectively these findings have identified a number of critical neurocircuit nodes within the LH, PB, and CeA that are important for the suppression of feeding, and thus may be dysregulated in AN and other eating disorders.

Summary and Future Directions

Eating disorders such as BED and AN exact a tremendous toll on society and those who have them. Nevertheless, additional research is needed to elucidate further the genetic and neurocircuit mechanisms that underlie these devastating disorders. As summarized here, both clinical studies from populations diagnosed with eating disorders and basic science studies on the neural circuitry that controls feeding and related behaviors are rapidly advancing our knowledge of the biological underpinnings of eating disorders. Further integration of these disparate fields of inquiry may provide a more holistic view of the neurobiological underpinnings of eating disorders. Next, we discuss experimental strategies to bridge the clinical and basic science further.

Applying Data Generated in the Clinic to Identify and Validate Neurocircuit Underpinnings of Eating Disorders

GWAS and brain functional imaging from clinical populations provide a wealth of data that could be potentially used to further test neural circuit function that regulates feeding in animals. For example, GWAS can identify single gene and gene network changes that may be important for the phenotypes associated with eating disorders. Animal models can then be generated with similar genetic alterations, and these animals could then be screened for maladaptive changes in feeding. This approach has been extremely successful in other fields, such as the study of autism and schizophrenia, but has yet to be applied to the study of eating disorders. In addition, neuroimaging data from patient populations with eating disorders could provide important novel neurocircuit targets to investigate further in animal models using optogenetics and related technologies. For example, if alterations in the functional connectivity between two or more brain regions are selectively observed in individuals with eating disorders relative to healthy controls, the relevance of these distinct brain regions could be explored experimentally by perturbing these circuit nodes in animals to assess whether direct alterations occur in feeding. Collectively, these two approaches could then leverage experimental data from clinical populations to investigate the neural circuitry that underlies eating disorders in animal models.

Transcriptional Profiling of Genetically Defined Neurons That Regulate Feeding

Although we have described strategies in which clinical data could be leveraged for more mechanistic basic science experiments related to eating disorders, it may also be possible to use genetic and cellular phenotype data generated in animal models to uncover novel therapeutic targets for the treatment of eating disorders. Because of the relative ease of cell type−specific targeting in animals models, it is possible to perform high-throughput transcriptional profiling from genetically defined neuronal populations based on their connectivity, gene expression, or activity dynamics (Ekstrand et al., 2014; Heiman et al., 2008; Knight et al., 2012). These approaches could be combined with animal models of eating disorders to assay how the transcriptional landscape of genetically defined neurons changes as a function of feeding phenotype. These data, in turn, could be mined for putative gene targets that could provide additional functional insight into how specific neuronal cell populations are altered. Collectively, this novel approach provides a provocative starting point for uncovering novel therapeutic targets for the development of new treatments for eating disorders.

CONCLUDING REMARKS

Although eating disorders are complex psychiatric conditions, future research should aim to render their neurobiological underpinnings decipherable. At their heart, eating disorders represent dysregulation in the most fundamental of human behaviors. Research into their biology ultimately has the potential to inform all disorders of weight and appetite dysregulation as well as an array of related neuropsychiatric conditions. A richer understanding of the neurobiological causes of specific eating disorders and also neurobiological mechanisms underlying successful pharmacologic and therapeutic interventions will require constant bridging of diverse fields including psychiatry, metabolomics, genetics, and neuroscience, and reciprocal information sharing between basic and clinical science.

ACKNOWLEDGMENTS

GDS was supported by funds from the Klarman Family Foundation, the Brain and Behavior Research Foundation, the Foundation of Hope, and the National Institute on Drug Abuse (DA032750 and DA038168). CMB was supported by the Anorexia Nervosa Genetics Initiative (ANGI), an initiative of the Klarman Family Foundation, and the National Institute of Mental Health (MH093615 and MH080065; and MH095860; Bardone-Cone). Dr Bulik is a grant recipient from Shire Pharmaceuticals. Dr. Bulik acknowledges funding from the Swedish Research Council (VR Dnr: 538-2013-8864).

REFERENCES

Allen, K. L., Byrne, S. M., Oddy, W. H., & Crosby, R. D. (2013). DSM-IV-TR and DSM-5 eating disorders in adolescents: prevalence, stability, and psychosocial correlates in a population-based sample of male and female adolescents. *Journal of Abnormal Psychology, 122*(3), 720−732. http://dx.doi.org/10.1037/a0034004.

American Psychiatric Association. (2013). *Diagnostic and statistical manual of mental disorders* (5th ed.). Washington, DC: American Psychiatric Association.

Anderson, C., & Bulik, C. M. (2003). Gender differences in compensatory behaviors, weight and shape salience, and drive for thinness. *Eating Behaviors, 5*, 1−11.

Aponte, Y., Atasoy, D., & Sternson, S. M. (2011). AGRP neurons are sufficient to orchestrate feeding behavior rapidly and without training. *Nature Neuroscience, 14*(3), 351−355. http://dx.doi.org/10.1038/nn.2739.

Atasoy, D., Aponte, Y., Su, H. H., & Sternson, S. M. (2008). A FLEX switch targets Channelrhodopsin-2 to multiple cell types for imaging and long-range circuit mapping. *The Journal of Neuroscience, 28*(28), 7025−7030. http://dx.doi.org/10.1523/JNEUROSCI.1954-08.2008.

Atasoy, D., Betley, J. N., Su, H. H., & Sternson, S. M. (2012). Deconstruction of a neural circuit for hunger. *Nature, 488*(7410), 172−177. http://dx.doi.org/10.1038/nature11270.

Avena, N. M., Bocarsly, M. E., & Hoebel, B. G. (2012). Animal models of sugar and fat bingeing: relationship to food addiction and increased body weight. *Methods in Molecular Biology, 829*, 351−365. http://dx.doi.org/10.1007/978-1-61779-458-2_23.

Avena, N. M., Rada, P., & Hoebel, B. G. (2009). Sugar and fat bingeing have notable differences in addictive-like behavior. *The Journal of Nutrition, 139*(3), 623−628. http://dx.doi.org/10.3945/jn.108.097584.

Baker, J. H., Mitchell, K. S., Neale, M. C., & Kendler, K. S. (2010). Eating disorder symptomatology and substance use disorders: prevalence and shared risk in a population based twin sample. *International Journal of Eating Disorders, 43*(7), 648−658. http://dx.doi.org/10.1002/eat.20856.

Baker, J. H., Thornton, L. M., Strober, M., Brandt, H., Crawford, S., Fichter, M. M., … Bulik, C. M. (2013). Temporal sequence of comorbid alcohol use disorder and anorexia nervosa. *Addictive Behaviors, 38*(3), 1704−1709. http://dx.doi.org/10.1016/j.addbeh.2012.10.005.

Beck, D., Casper, R., & Andersen, A. (1996). Truly late onset of eating disorders: a study of 11 cases averaging 60 years of age at presentation. *International Journal of Eating Disorders, 20*(4), 389−395.

Becskei, C., Grabler, V., Edwards, G. L., Riediger, T., & Lutz, T. A. (2007). Lesion of the lateral parabrachial nucleus attenuates the anorectic effect of peripheral amylin and CCK. *Brain Research, 1162*, 76−84. http://dx.doi.org/10.1016/j.brainres.2007.06.016.

Bernstein, J. G., & Boyden, E. S. (2011). Optogenetic tools for analyzing the neural circuits of behavior. *Trends in Cognitive Sciences, 15*(12), 592–600. http://dx.doi.org/10.1016/j.tics.2011.10.003.

Betley, J. N., Cao, Z. F., Ritola, K. D., & Sternson, S. M. (2013). Parallel, redundant circuit organization for homeostatic control of feeding behavior. *Cell, 155*(6), 1337–1350. http://dx.doi.org/10.1016/j.cell.2013.11.002.

Blinder, B., Cumella, E., & Sanathara, V. (2006). Psychiatric comorbidities of female inpatients with eating disorders. *Psychosomatic Medicine, 68*, 454–462.

Boraska, V., Davis, O. S., Cherkas, L. F., Helder, S. G., Harris, J., Krug, I., … Zeggini, E. (2012). Genome-wide association analysis of eating disorder-related symptoms, behaviors, and personality traits. *American Journal of Medical Genetics. Part B, Neuropsychiatric Genetics, 159B*(7), 803–811. http://dx.doi.org/10.1002/ajmg.b.32087.

Boraska, V., Franklin, C. S., Floyd, J. A., Thornton, L. M., Huckins, L. M., Southam, L., … Bulik, C. M. (2014). A genome-wide association study of anorexia nervosa. *Molecular Psychiatry, 19*(10), 1085–1094. http://dx.doi.org/10.1038/mp.2013.187.

Bouret, S. G., Draper, S. J., & Simerly, R. B. (2004). Formation of projection pathways from the arcuate nucleus of the hypothalamus to hypothalamic regions implicated in the neural control of feeding behavior in mice. *The Journal of Neuroscience, 24*(11), 2797–2805. http://dx.doi.org/10.1523/JNEUROSCI.5369-03.2004.

Brewerton, T., Lydiard, R., Herzog, D., Brotman, A., O'Neil, P., & Ballenger, J. (1995). Comorbidity of axis I psychiatric disorders in bulimia nervosa. *Journal of Clinical Psychiatry, 56*, 77–80.

Bulik, C. M., Devlin, B., Bacanu, S. A., Thornton, L., Klump, K. L., Fichter, M. M., … Kaye, W. H. (2003). Significant linkage on chromosome 10p in families with bulimia nervosa. *American Journal of Human Genetics, 72*(1), 200–207.

Bulik, C., Klump, K., Thornton, L., Kaplan, A., Devlin, B., Fichter, M., … Kaye, W. (2004). Alcohol use disorder comorbidity in eating disorders: a multicenter study. *Journal of Clinical Psychiatry, 65*, 1000–1006.

Bulik, C., Sullivan, P., & Kendler, K. (1998). Heritability of binge-eating and broadly defined bulimia nervosa. *Biological Psychiatry, 44*(12), 1210–1218.

Bulik, C., Sullivan, P., Tozzi, F., Furberg, H., Lichtenstein, P., & Pedersen, N. (2006). Prevalence, heritability and prospective risk factors for anorexia nervosa. *Archives of General Psychiatry, 63*, 305–312.

Bulik, C., Thornton, L., Root, T., Pisetsky, E., Lichtenstein, P., & Pedersen, N. (2010). Understanding the relation between anorexia nervosa and bulimia nervosa in a Swedish national twin sample. *Biological Psychiatry, 67*, 71–77.

Cai, H., Haubensak, W., Anthony, T. E., & Anderson, D. J. (2014). Central amygdala PKC-delta(+) neurons mediate the influence of multiple anorexigenic signals. *Nature Neuroscience, 17*(9), 1240–1248. http://dx.doi.org/10.1038/nn.3767.

Carter, M. E., Soden, M. E., Zweifel, L. S., & Palmiter, R. D. (2013). Genetic identification of a neural circuit that suppresses appetite. *Nature, 503*(7474), 111–114. http://dx.doi.org/10.1038/nature12596.

Cooper, P., Watkins, B., Bryant-Waugh, R., & Lask, B. (2002). The nosological status of early onset anorexia nervosa. *Psychological Medicine, 32*, 873–880.

Cortese, S., Angriman, M., Maffeis, C., Isnard, P., Konofal, E., Lecendreux, M., … Mouren, M.-C. (2008). Attention-deficit/hyperactivity disorder (ADHD) and obesity: a systematic review of the literature. *Critical Reviews in Food Science and Nutrition, 48*, 524–537.

Corvin, A., Craddock, N., & Sullivan, P. F. (2010). Genome-wide association studies: a primer. *Psychological Medicine, 40*(7), 1063–1077. http://dx.doi.org/10.1017/S0033291709991723.

Cross-Disorder Group of the Psychiatric Genomics Consortium, Lee, S. H., Ripke, S., Neale, B. M., Faraone, S. V., Purcell, S. M., … Wray, N. R. (2013). Genetic relationship between five psychiatric disorders estimated from genome-wide SNPs. *Nature Genetics, 45*(9), 984–994. http://dx.doi.org/10.1038/ng.2711.

Cumyn, L., French, L., & Hechtman, L. (2009). Comorbidity in adults with attention-deficit hyperactivity disorder. *Canadian Journal of Psychiatry, 54*(10), 673–683.

Dansky, B., Brewerton, T., & Kilpatrick, D. (2000). Comorbidity of bulimia nervosa and alcohol use disorders: results from the national women's study. *International Journal of Eating Disorders, 27*, 180–190.

Devlin, B., Bacanu, S., Klump, K., Bulik, C., Fichter, M., Halmi, K., … Kaye, W. H. (2002). Linkage analysis of anorexia nervosa incorporating behavioral covariates. *Human Molecular Genetics, 11*(6), 689–696.

DiPatrizio, N. V., & Simansky, K. J. (2008). Activating parabrachial cannabinoid CB1 receptors selectively stimulates feeding of palatable foods in rats. *The Journal of Neuroscience, 28*(39), 9702–9709. http://dx.doi.org/10.1523/JNEUROSCI.1171-08.2008.

Ekstrand, M. I., Nectow, A. R., Knight, Z. A., Latcha, K. N., Pomeranz, L. E., & Friedman, J. M. (2014). Molecular profiling of neurons based on connectivity. *Cell, 157*(5), 1230–1242. http://dx.doi.org/10.1016/j.cell.2014.03.059.

Fenno, L., Yizhar, O., & Deisseroth, K. (2011). The development and application of optogenetics. *Annalu Review of Neuroscience, 34*, 389–412. http://dx.doi.org/10.1146/annurev-neuro-061010-113817.

Frieling, H., Gozner, A., Romer, K. D., Lenz, B., Bonsch, D., Wilhelm, J., … Bleich, S. (2007). Global DNA hypomethylation and DNA hypermethylation of the alpha synuclein promoter in females with anorexia nervosa. *Molecular Psychiatry, 12*(3), 229–230. http://dx.doi.org/10.1038/sj.mp.4001931.

Gerfen, C. R., Paletzki, R., & Heintz, N. (2013). GENSAT BAC cre-recombinase driver lines to study the functional organization of cerebral cortical and basal ganglia circuits. *Neuron, 80*(6), 1368–1383. http://dx.doi.org/10.1016/j.neuron.2013.10.016.

Godart, N., Flament, M., Perdereau, F., & Jeammet, P. (2002). Comorbidity between eating disorders and anxiety disorders: a review. *International Journal of Eating Disorders, 32*, 253–270.

Godart, N. T., Flament, M. F., Curt, F., Perdereau, F., Lang, F., Venisse, J. L., … Hasler, G. (2003). Anxiety disorders in subjects seeking treatment for eating disorders: a DSM-IV controlled study. *Psychiatry Research, 117*(3), 245–258.

Godart, N. T., Flament, M. F., Lecrubier, Y., & Jeammet, P. (2000). Anxiety disorders in anorexia nervosa and bulimia nervosa: co-morbidity and chronology of appearance. *European Psychiatry, 15*(1), 38–45.

Grice, D. E., Halmi, K. A., Fichter, M. M., Strober, M., Woodside, D. B., Treasure, J. T., … Berrettini, W. H. (2002). Evidence for a susceptibility gene for anorexia nervosa on chromosome 1. *American Journal of Human Genetics, 70*(3), 787–792.

Grilo, C., White, M., Barnes, R., & Masheb, R. (2013). Psychiatric disorder co-morbidity and correlates in an ethnically diverse sample of obese patients with binge eating disorder in primary care settings. *Comprehensive Psychiatry, 54*, 209–216.

Grilo, C., White, M., & Masheb, R. (2009). DSM-IV psychiatric disorder comorbidity and its correlates in binge eating disorder. *International Journal of Eating Disorders, 42*, 228–234.

Halmi, K. A., Sunday, S. R., Klump, K. L., Strober, M., Leckman, J. F., Fichter, M., … Kaye, W. H. (2003). Obsessions and compulsions in anorexia nervosa subtypes. *International Journal of Eating Disorders, 33*(3), 308–319. http://dx.doi.org/10.1002/eat.10138.

Heiman, M., Schaefer, A., Gong, S., Peterson, J. D., Day, M., Ramsey, K. E., … Heintz, N. (2008). A translational profiling approach for the molecular characterization of CNS cell types. *Cell, 135*(4), 738–748. http://dx.doi.org/10.1016/j.cell.2008.10.028.

Hildebrandt, T., Lai, J. K., Langenbucher, J. W., Schneider, M., Yehuda, R., & Pfaff, D. W. (2011). The diagnostic dilemma of pathological appearance and performance enhancing drug use. *Drug and Alcohol Dependence, 114*(1), 1–11. http://dx.doi.org/10.1016/j.drugalcdep.2010.09.018.

Hudson, J., Lalonde, J., Pindyck, L., Bulik, C., Crow, S., McElroy, S., … Pope, H. (2006). Familial aggregation of binge-eating disorder. *Archives of General Psychiatry, 63*, 313–319.

Hudson, J. I., Hiripi, E., Pope, H. G., Jr., & Kessler, R. C. (2007). The prevalence and correlates of eating disorders in the national comorbidity survey replication. *Biological Psychiatry, 61*, 348–358.

Javaras, K. N., Laird, N. M., Reichborn-Kjennerud, T., Bulik, C. M., Pope, H. G., Jr., & Hudson, J. I. (2008). Familiality and heritability of binge eating disorder: results of a case-control family study and a twin study. *International Journal of Eating Disorders, 41*(2), 174–179.

Javaras, K. N., Pope, H. G., Lalonde, J. K., Roberts, J. L., Nillni, Y. I., Laird, N. M., … Hudson, J. I. (2008). Co-occurrence of binge eating disorder with psychiatric and medical disorders. *Journal of Clinical Psychiatry, 69*(2), 266–273.

Jennings, J. H., Rizzi, G., Stamatakis, A. M., Ung, R. L., & Stuber, G. D. (2013). The inhibitory circuit architecture of the lateral hypothalamus orchestrates feeding. *Science, 341*(6153), 1517–1521. http://dx.doi.org/10.1126/science.1241812.

Kas, M. J., & Adan, R. A. (2011). Animal models of eating disorder traits. *Current Topics in Behavioral Neurosciences, 6*, 209–227. http://dx.doi.org/10.1007/7854_2010_84.

Kaye, W., Bulik, C., Thornton, L., Barbarich, B. S., Masters, K., & Price Foundation Collaborative Group. (2004). Comorbidity of anxiety dsorders with anorexia and bulimia nervosa. *American Jounal of Psychiatry, 161*, 2215–2221.

Kessler, R. C., Berglund, P. A., Chiu, W. T., Deitz, A. C., Hudson, J. I., Shahly, V., … Xavier, M. (2013). The prevalence and correlates of binge eating disorder in the World Health Organization World Mental Health surveys. *Biological Psychiatry, 73*(9), 904–914. http://dx.doi.org/10.1016/j.biopsych.2012.11.020.

Kipman, A., Gorwood, P., Mouren-Simeoni, M. C., & Ades, J. (1999). Genetic factors in anorexia nervosa. *European Psychiatry, 14*(4), 189–198.

Klenotich, S. J., & Dulawa, S. C. (2012). The activity-based anorexia mouse model. *Methods in Molecular Biology, 829*, 377–393. http://dx.doi.org/10.1007/978-1-61779-458-2_25.

Klump, K. L., Bulik, C. M., Kaye, W. H., Treasure, J., & Tyson, E. (2009). Academy for eating disorders position paper: eating disorders are serious mental illnesses. *International Journal of Eating Disorders, 42*(2), 97–103.

Klump, K. L., Miller, K. B., Keel, P. K., McGue, M., & Iacono, W. G. (2001). Genetic and environmental influences on anorexia nervosa syndromes in a population-based twin sample. *Psychological Medicine, 31*(4), 737–740.

Knight, Z. A., Tan, K., Birsoy, K., Schmidt, S., Garrison, J. L., Wysocki, R. W., … Friedman, J. M. (2012). Molecular profiling of activated neurons by phosphorylated ribosome capture. *Cell, 151*(5), 1126–1137. http://dx.doi.org/10.1016/j.cell.2012.10.039.

Kortegaard, L. S., Hoerder, K., Joergensen, J., Gillberg, C., & Kyvik, K. O. (2001). A preliminary population-based twin study of self-reported eating disorder. *Psychological Medicine, 31*(2), 361–365.

Krashes, M. J., Koda, S., Ye, C., Rogan, S. C., Adams, A. C., Cusher, D. S., … Lowell, B. B. (2011). Rapid, reversible activation of AgRP neurons drives feeding behavior in mice. *Journal of Clinical Investigation, 121*(4), 1424–1428. http://dx.doi.org/10.1172/JCI46229.

Krashes, M. J., Shah, B. P., Madara, J. C., Olson, D. P., Strochlic, D. E., Garfield, A. S., … Lowell, B. B. (2014). An excitatory paraventricular nucleus to AgRP neuron circuit that drives hunger. *Nature, 507*(7491), 238–242. http://dx.doi.org/10.1038/nature12956.

Krug, I., Pinheiro, A. P., Bulik, C., Jimenez-Murcia, S., Granero, R., Penelo, E., … Fernandez-Aranda, F. (2009). Lifetime substance abuse, family history of alcohol abuse/dependence and novelty seeking in eating disorders: comparison study of eating disorder subgroups. *Psychiatry and Clinical Neurosciences, 63*(1), 82–87. http://dx.doi.org/10.1111/j.1440-1819.2008.01908.x.

Lewinsohn, P., Seeley, J., Moerk, K., & Striegel-Moore, R. (2002). Gender differences in eating disorder symptoms in young adults. *International Journal of Eating Disorders, 32*, 426–440.

Lilenfeld, L., Kaye, W., Greeno, C., Merikangas, K., Plotnikov, K., Pollice, C., … Nagy, L. (1998). A controlled family study of anorexia nervosa and bulimia nervosa: psychiatric disorders in first-degree relatives and effects of proband comorbidity. *Archives of General Psychiatry, 55*, 603–610.

Mitchell, K. S., Neale, M. C., Bulik, C. M., Aggen, S. H., Kendler, K. S., & Mazzeo, S. E. (2010). Binge eating disorder: a symptom-level investigation of genetic and environmental influences on liability. *Psychological Medicine, 40*(11), 1899–1906. http://dx.doi.org/10.1017/S0033291710000139.

Nicholls, D. E., Lynn, R., & Viner, R. M. (2011). Childhood eating disorders: British national surveillance study. *British Journal of Psychiatry, 198*(4), 295–301. http://dx.doi.org/10.1192/bjp.bp.110.081356.

O'Brien, K. M., & Vincent, N. K. (2003). Psychiatric comorbidity in anorexia and bulimia nervosa: nature, prevalence, and causal relationships. *Clinical Psychology Review, 23*(1), 57–74.

Pagoto, S. L., Curtin, C., Lemon, S. C., Bandini, L. G., Schneider, K. L., Bodenlos, J. S., & Ma, Y. (2009). Association between adult attention deficit/hyperactivity disorder and obesity in the US population. *Obesity, 17*, 539–544.

Peterson, C., Miller, K., Crow, S., Thuras, P., & Mitchell, J. (2005). Subtypes of binge eating disorder based on psychiatric history. *International Journal of Eating Disorders, 38*, 273–276.

Pike, K., Dohm, F., Striegel-Moore, R., Wilfley, D., & Fairburn, C. (2001). A comparison of black and white women with binge eating disorder. *American Journal of Psychiatry, 158*, 1455–1460.

Pinto, S., Roseberry, A. G., Liu, H., Diano, S., Shanabrough, M., Cai, X., ... Horvath, T. L. (2004). Rapid rewiring of arcuate nucleus feeding circuits by leptin. *Science, 304*(5667), 110–115. http://dx.doi.org/10.1126/science.1089459.

Raney, T. J., Thornton, L. M., Berrettini, W., Brandt, H., Crawford, S., Fichter, M. M., ... Bulik, C. M. (2008). Influence of overanxious disorder of childhood on the expression of anorexia nervosa. *International Journal of Eating Disorders, 41*(4), 326–332.

Reichborn-Kjennerud, T., Bulik, C., Tambs, K., & Harris, J. (2004). Genetic and environmental influences on binge eating in the absence of compensatory behaviours: a population-based twin study. *International Journal of Eating Disorders, 36*, 307–314.

Root, T. L., Pisetsky, E. M., Thornton, L., Lichtenstein, P., Pedersen, N. L., & Bulik, C. M. (2010). Patterns of co-morbidity of eating disorders and substance use in Swedish females. *Psychological Medicine, 40*, 105–115.

Saffrey, R., Novakovic, B., & Wade, T. D. (2014). Assessing global and gene specific DNA methylation in anorexia nervosa: a pilot study. *International Journal of Eating Disorders, 47*(2), 206–210. http://dx.doi.org/10.1002/eat.22200.

Salbach-Andrae, H., Lenz, K., Simmendinger, N., Klinkowski, N., Lehmkuhl, U., & Pfeiffer, E. (2008). Psychiatric comorbidities among female adolescents with anorexia nervosa. *Child Psychiatry and Human Development, 39*(3), 261–272.

Seitz, J., Kahraman-Lanzerath, B., Legenbauer, T., Sarrar, L., Herpertz, S., Salbach-Andrae, H., ... Herpertz-Dahlmann, B. (2013). The role of impulsivity, inattention and comorbid ADHD in patients with bulimia nervosa. *PLoS One, 8*(5), e63891. http://dx.doi.org/10.1371/journal.pone.0063891.

Strober, M., Freeman, R., Lampert, C., Diamond, J., & Kaye, W. (2000). Controlled family study of anorexia nervosa and bulimia nervosa: evidence of shared liability and transmission of partial syndromes. *American Journal of Psychiatry, 157*(3), 393–401.

Swanson, S. A., Crow, S. J., Le Grange, D., Swendsen, J., & Merikangas, K. R. (2011). Prevalence and correlates of eating disorders in adolescents. Results from the national comorbidity survey replication adolescent supplement. *Archives of General Psychiatry, 68*(7), 714–723. http://dx.doi.org/10.1001/archgenpsychiatry.2011.22.

Tanofsky-Kraff, M., Shomaker, L. B., Olsen, C., Roza, C. A., Wolkoff, L. E., Columbo, K. M., ... Yanovski, J. A. (2011). A prospective study of pediatric loss of control eating and psychological outcomes. *Journal of Abnormal Psychology, 120*(1), 108–118. http://dx.doi.org/10.1037/a0021406.

Tozzi, F., Thornton, L., Klump, K., Bulik, C., Fichter, M., Halmi, K., ... Kaye, W. (2005). Symptom fluctuation in eating disorders: correlates of diagnostic crossover. *American Jounal of Psychiatry, 162*, 732–740.

Tremolizzo, L., Conti, E., Bomba, M., Uccellini, O., Rossi, M. S., Marfone, M., ... Nacinovich, R. (2014). Decreased whole-blood global DNA methylation is related to serum hormones in anorexia nervosa adolescents. *World Journal of Biological Psychiatry, 15*(4), 327–333. http://dx.doi.org/10.3109/15622975.2013.860467.

Tye, K. M., & Deisseroth, K. (2012). Optogenetic investigation of neural circuits underlying brain disease in animal models. *Nature Reviews Neuroscience, 13*(4), 251–266. http://dx.doi.org/10.1038/nrn3171.

Wade, T. D., Gordon, S., Medland, S., Bulik, C. M., Heath, A. C., Montgomery, G. W., & Martin, N. G. (2013). Genetic variants associated with disordered eating. *International Journal of Eating Disorders, 46*(6), 594–608. http://dx.doi.org/10.1002/eat.22133.

Wang, K., Zhang, H., Bloss, C. S., Duvvuri, V., Kaye, W., Schork, N. J., ... Hakonarson, H. (2011). A genome-wide association study on common SNPs and rare CNVs in anorexia nervosa. *Molecular Psychiatry, 16*(9), 949–959. http://dx.doi.org/10.1038/mp.2010.107.

White, M., & Grilo, C. (2006). Psychiatric comorbidity in binge-eating disorder as a function of smoking history. *Journal of Clinical Psychiatry, 67*, 594–599.

Wilfley, D., Freidman, M., Dounchis, J., Stein, R., Welch, R., & Ball, S. (2000). Comorbid psychopathology in binge eating disorder: relation to eating disorder severity at baseline and following treatment. *Journal of Consulting and Clinical Psychology, 68*, 641–649.

Witten, I. B., Steinberg, E. E., Lee, S. Y., Davidson, T. J., Zalocusky, K. A., Brodsky, M., ... Deisseroth, K. (2011). Recombinase-driver rat lines: tools, techniques, and optogenetic application to dopamine-mediated reinforcement. *Neuron, 72*(5), 721–733. http://dx.doi.org/10.1016/j.neuron.2011.10.028.

Wu, Q., Clark, M. S., & Palmiter, R. D. (2012). Deciphering a neuronal circuit that mediates appetite. *Nature, 483*(7391), 594–597. http://dx.doi.org/10.1038/nature10899.

Yilmaz, Z., Hardaway, A., & Bulik, C. (2014). Genetics and epigenetics of eating disorders. *Advances in Genomics and Genetics, 5*, 131–150.

Yilmaz, Z., Kaplan, A. S., Zai, C. C., Levitan, R. D., & Kennedy, J. L. (2011). COMT Val158Met variant and functional haplotypes associated with childhood ADHD history in women with bulimia nervosa. *Progress in Neuropsychopharmacology and Biological Psychiatry, 35*(4), 948–952. http://dx.doi.org/10.1016/j.pnpbp.2011.01.012.

Yizhar, O., Fenno, L. E., Davidson, T. J., Mogri, M., & Deisseroth, K. (2011). Optogenetics in neural systems. *Neuron, 71*(1), 9–34. http://dx.doi.org/10.1016/j.neuron.2011.06.004.

Zhang, F., Gradinaru, V., Adamantidis, A. R., Durand, R., Airan, R. D., de Lecea, L., & Deisseroth, K. (2010). Optogenetic interrogation of neural circuits: technology for probing mammalian brain structures. *Nature Protocols, 5*(3), 439–456. http://dx.doi.org/10.1038/nprot.2009.226.

Section IV

Clinical Translation and Drug Development

Chapter 44

Psychiatric Pharmacogenomics: Translating Genomics

Roy H. Perlis
Harvard Medical School, Boston, MA, United States

DEFINITION AND OVERVIEW

For more than 100 medications approved by the US Food and Drug Administration (FDA), their label reflects the relevance of considering genetic variation to guide prescribing (Anonymous, 2014). At least 10% of those identified by the FDA are psychotropic medications, including some of the most widely prescribed. Perhaps unsurprisingly, both clinician (Hoop, Roberts, Green Hammond, & Cox, 2008) and patient (Herbild, Bech, & Gyrd-Hansen, 2009) surveys anticipate the usefulness of genomic testing for psychiatric prescribing. On the other hand, with the exception of a handful of hepatic enzymes known to have a role in drug metabolism, investigation of the role of genetic variation as it relates to psychotropics remains substantially less advanced than the investigation of neuropsychiatric disorder risk variants.

The term "pharmacogenomics" refers to the study of the impact of common or rare genetic variation on any aspect of drug response. This definition encompasses clinically apparent outcomes (treatment efficacy, tolerability, or safety) as well as elements of drug kinetics or dynamics. To date, the bulk of these efforts has focused on treatment efficacy and safety for antidepressants and antipsychotics, with more modest investigations of other treatments.

Whereas pharmacogenomics as originally described referred to phenotypes observable only in the presence of a particular drug or other perturbation, this assumption often is not tested, recognizing that some treatment response characteristics may be correlated with other clinically observable phenotypes. For example, as discussed subsequently, a common variant associated with antipsychotic-induced weight gain may also be associated with body mass index independent of treatment. On the other hand, the assumption is often made that variants influencing drug response may be under less selection pressure and thus more likely to exert large effects without being selected against; that is, the model posits that common variants may exert large effects.

CONNECTION TO GENOMICS IN PSYCHIATRY AND IMPLICATIONS FOR ARCHITECTURE

Progress in neuropsychiatric and neurodevelopmental genetics suggests a complex architecture in which the vast majority of risk variants exert modest individual effects. For psychotropic pharmacogenomics, the genetic architecture is entirely unknown. The most widely studied area of pharmacogenomics relates to the cytochrome P450 (CYP450) system. Here common and rare variants, including single-nucleotide polymorphisms (SNPs)/mutations as well as duplications and deletions, combine to produce substantial variation in the ability to metabolize diverse categories of medications and other environmental toxins.

Whereas the traditional first step in genetic investigation in human studies has been family and twin investigations, such data are more difficult to collect where treatment studies are concerned, particularly in light of changes in the management of depression and psychosis such that parental treatment exposure may differ from offspring. Small studies examined tricyclic antidepressant response among family members treated for depression, and were consistent with familiality (Pare, Rees, & Sainsbury, 1962). Notably, an early twin study also demonstrated the heritability of tricyclic antidepressant

Genomics, Circuits, and Pathways in Clinical Neuropsychiatry. http://dx.doi.org/10.1016/B978-0-12-800105-9.00044-5

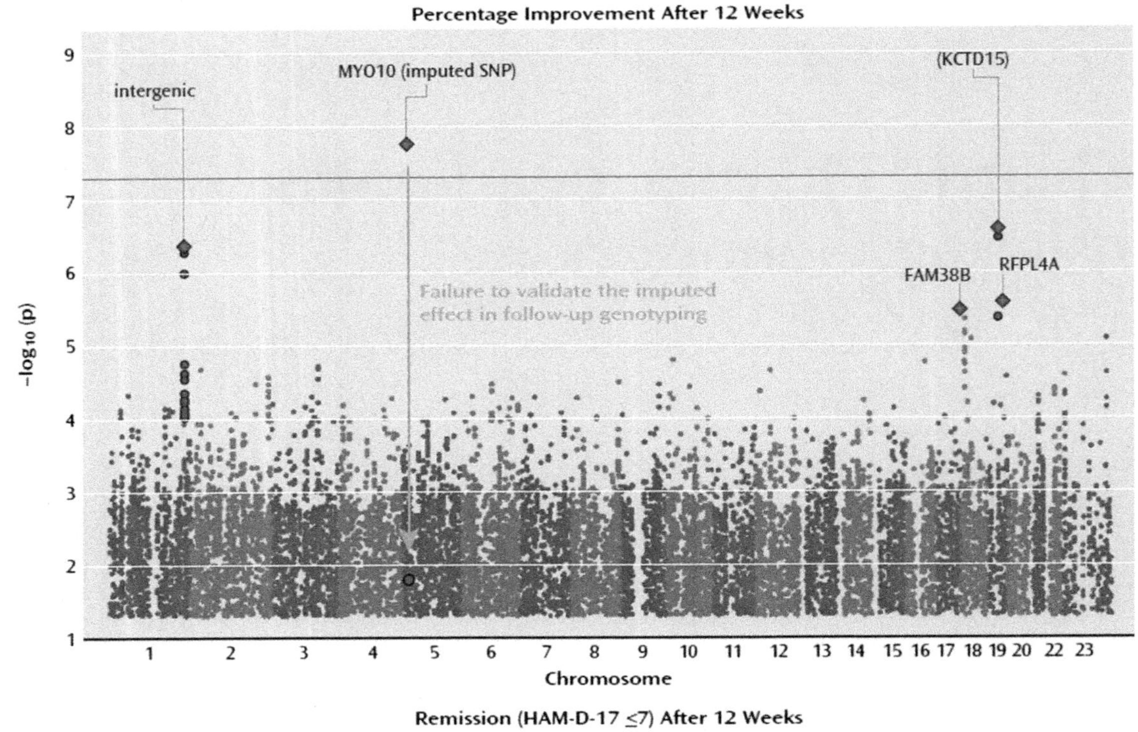

FIGURE 44.1 Meta-analysis of genome-wide association data for remission after antidepressant treatment, drawn from three antidepressant response cohorts. No single locus exceeded a genome-wide threshold for association. *Gendep, MARS, STARD investigators, AJP 2013; Available at https://www.broadinstitute.org/mpg/ricopili/.*

pharmacokinetics (Alexanderson, Evans, & Sjoqvist, 1969). One study reported a pedigree in which eight depressed family members responded only to monoamine oxidase inhibitor but not other antidepressants (O'Reilly, Bogue, & Singh, 1994). Among newer antidepressants, only fluvoxamine has been examined; a small study suggested that fluvoxamine response in a proband significantly increased the probability of response in a first-degree family member (Franchini, Serretti, Gasperini, & Smeraldi, 1998) (Fig. 44.1).

Only one study has examined the familiality of lithium response: In an early study, rates of lithium response were greater in first-degree family members of people who responded to lithium compared with those who did not (Grof et al., 2002). (More generally, other studies have suggested that lithium response may be associated with greater loading for affective illness, with increased mood disorder rates among first-degree family members of those who respond to lithium (Smeraldi et al., 1984).) Twin studies for antipsychotic response are limited to case series of monozygotic twins that suggest concordance in treatment outcomes (Arranz & de Leon, 2007). One early twin study suggested heritability of lithium uptake into red blood cells (Dorus, Pandey, & Davis, 1975), that is, greater correlation between monozygotic than dizygotic twins.

A convergent line of evidence suggesting the influence of genetic variation on treatment response comes from rodent studies of antidepressant response phenotypes. Although such models are often criticized as corresponding poorly to psychiatric illness per se, these studies at least suggest that psychotropic effects on behavior are influenced in a detectable fashion by strain-specific differences. For example, one study examined two mouse strains with differential response to citalopram on the tail-suspension test (Crowley, Brodkin, Blendy, Berrettini, & Lucki, 2006). This difference was mapped to a single peak, and subsequently to nonsynonymous SNPs in vesicular monoamine transporter 2 (slc18a2).

Similarly, differences between 27 inbred mouse strains in vacuous chewing movements after haloperidol exposure, which may represent the murine equivalent of antipsychotic-induced tardive dyskinesia (TD), have been used to map susceptibility loci (Crowley, Adkins, et al., 2012). A genome-wide association study (GWAS) identified multiple loci with strong evidence of association with this phenotype (Crowley, Kim, et al., 2012). Taken together, these studies suggest that even in the absence of strong human data for familiality, psychotropic-related phenotypes may be sufficiently heritable to allow identification of risk loci in human investigations.

CHALLENGES

The advantages of drug-response phenotypes for genetic study are largely hypothetical at this point, but they include the possibility of observing common variants that exert large effects (in the absence of strong selection pressure) and the possibility of a simpler mechanism of action (and thus fewer risk variants of larger effect) compared with disease etiology. Conversely, the major challenges in psychiatric pharmacogenomics are apparent and require careful consideration before examining the extant literature.

First, for most psychotropics, the drug–placebo differences observed in randomized, controlled trials are modest. For example, among antidepressants, the drug–placebo difference in a large meta-analysis was approximately 2–3 points on the Hamilton Depression Rating scale (Fountoulakis & Moller, 2011). Although drug-placebo differences are generally assumed to be less of a problem for antipsychotic and antimanic drugs, this is not necessarily the case. The problem is not a lack of efficacy per se, but rather that the population of apparent drug responders really encompasses true drug responders as well as placebo-like responders. (As a result, published GWAS of antidepressant response may also be interpreted as GWAS of placebo response, because there are insufficient data to determine specificity of response.) Whereas the genetics of placebo response would be interesting and potentially actionable in their own right (eg, as a means of enriching studies for people who likely do not respond to placebo), the phenotype is itself likely to be complex and multifactorial.

A related challenge is the paucity of studies incorporating an active comparator medication, which are necessary to determine whether an association is predictive of poor outcome in general or poor outcome with a particular treatment. The two alternative hypotheses have different implications, because one relates to the disease but not the intervention per se. Although arguably this is a downstream or second-order question, it is important for the potential clinical translation of any pharmacogenomics finding. For any controlled trial, a barrier to investigating treatment specificity has been the reliance on statistical tests of interaction. Formally, one examines the variant-by-treatment effect on the outcome of interest. The difficulty with this test is that it will have greatest power to detect association when effects are in the opposite direction: that is, drug A is superior to drug B in one group, and inferior to drug B in another. Whereas this assumption is biologically plausible in some circumstances, in others one might assume association only for one drug (eg, a CYP450 variant that affects the metabolism of only one of the two drugs). A possible alternative strategy which may be advantageous in some circumstances is discussed subsequently as the biomarker-stratified parallel design.

Yet another complication in pharmacogenomics studies is the need to consider treatment nonadherence, which can lead to diminished treatment effects and represent a form of misclassification. For antidepressants, nonadherence rates in clinical practice may exceed 50%, depending on the means of assessment and the study population (Cantrell, Eaddy, Shah, Regan, & Sokol, 2006; Stein, Cantrell, Sokol, Eaddy, & Shah, 2006; Warden et al., 2014). In one investigation, antidepressant levels were undetectable in about 20% of antidepressant-treated individuals (Roberson, Castro, Cagan, & Perlis, 2016). Although adherence in treatment studies may be greater because of the frequency of visits and common application of pill counts or other monitoring methods, modern antidepressant studies still yield high rates of nonadherence. Similar problems with adherence have been observed in antipsychotic trials (Julius, Novitsky, & Dubin, 2009). A computer simulation quantified these effects, illustrating the substantial decrement in statistical power attributable to nonadherence (Malhotra, Zhang, et al., 2012).

Yet another difficulty in interpreting pharmacogenomics results arises from the potential confounding effects between efficacy and tolerability. That is, whereas these two phenotypes are often treated as distinctly different, they have important impacts on one another. Patients may be more willing to continue a medication (a measure of tolerability) if they perceive benefit (efficacy); conversely, the apparent efficacy of a medication may be affected by its tolerability. Because it is difficult to distinguish the two sets of outcomes perfectly, any pharmacogenomic study of one outcome is also to some extent vulnerable to misclassification based on the other outcome (For an example of the complexity of this relationship as it relates to antidepressant response, see Keers et al. (2011)). The practical consequence of this correlation is the need to consider treatment discontinuation, and its impact, in any pharmacogenomic analysis. A sensitivity analysis including different models of study dropout can be valuable in demonstrating the robustness of a particular finding.

Another obstacle in pharmacogenomics studies is related to the underlying biology itself, rather than to patient factors. A limited understanding of mechanism of action remains for most psychotropic agents, and a lack of understanding of the key site of action for essentially all of them, because even the selective serotonin reuptake inhibitors (SSRIs) exert diverse and varying effects at other targets. As a result, unlike pharmacogenomics studies of, for example, antihypertensives or oral hypoglycemics, it has been difficult to set prior probabilities and thus to conduct successful candidate-based association studies, or even to prioritize loci from GWAS with suggestive evidence of association.

All of these challenges are dwarfed by the practical problems posed by the nature of pharmacogenomics phenotypes themselves. That is, to capture this phenotype reliably, most studies require longitudinal assessment with a well-defined medication start time and some sort of continued treatment over time. These phenotypes are thus far more difficult to capture than the cross-sectional phenotypes (even with some retrospective assessment) used in disease association studies. Thus, achieving adequate sample sizes to enable sufficient power to detect association can be challenging. Nearly all pharmacogenomics studies in psychiatry focus on short-term (8–12 weeks or less) trials; the exception is studies of lithium response, in which a year or more may be required to establish change in disease course attributable to medication.

For psychiatric disease phenotypes, the sample size problem has been overcome by meta-analysis of large cohorts drawn from multiple sources. Although reconciling phenotypes across studies has been a concern, to date there is no strong evidence that 4-hour diagnostic interviews yield cohorts with larger genetic effects than rapid phenotyping: for example, using disease screens or claims data (Ripke et al., 2013). For pharmacogenomics, studies rarely use sufficiently similar methodologies and interventions to allow them to be analyzed together. For example, a meta-analysis of antidepressant response combined three different cohorts: an outpatient effectiveness study, an outpatient randomized two-arm study, and an inpatient study, each of which used distinct pharmacotherapies. A further challenge is that many of the large DNA biobanks derived from clinical trials were derived from industry-supported studies, and have either not been genotyped or have not been made available for collaboration.

PRINCIPAL FINDINGS BY THERAPEUTIC AREA

Despite a large number of reports of pharmacogenomic analyses over the past decade, there are few consistent findings. Most of these studies examined a single small cohort at one or a few candidate genes. In general, the candidate-based approach is more readily justified in pharmacogenomics than in psychiatric disease genomics, inasmuch as the important metabolic enzymes and pharmacodynamic targets are often relatively well understood, certainly more so than the pathophysiology of neuropsychiatric disease. The following section focuses on positive results from GWAS or meta-analyses of candidate gene studies, organized by medication category and by type of outcome: efficacy or safety/tolerability. The small but growing number of rare variant studies is also included when they are available.

Antidepressants and Anxiolytics

Efficacy Studies

The variants most extensively studied for association with antidepressant and other psychotropic outcomes are those of the CYP450 system, particularly CYP2D6 and CYP2C19. The enzymes of the CYP450 system have a central role in hepatic phase I metabolism of numerous pharmacotherapies: One estimate indicates that 80% of medications in the pharmacopeia are metabolized by one or more of these enzymes, including many commonly prescribed antidepressant medications. The CYP450 genes, and most notably CYP2D6, present a primer in genomic variation, with SNPs, insertion/deletion polymorphisms, and copy number variants all demonstrated to have functional significance (see, eg, http://www.cypalleles.ki.se). These variants are generally analyzed in aggregate in terms of their impact on the activity of the enzyme: Individuals who are wild-type at all loci are referred to as extensive metabolizers, whereas those with diminished function are poor metabolizers (PMs) and those with enhanced function are ultrarapid metabolizers. Individuals who carry one diminished-function allele are labeled as intermediate metabolizers (IMs). The impact of most of these variants on metabolism is believed to be less substantial than for PMs. Of note, the frequencies of these groups and the specific alleles contributing to them vary widely across ethnicities.

For CYP450 variations to influence treatment response, two conditions must be satisfied. First, the variants examined must have meaningful effects on resulting blood levels (ie, pharmacokinetics) of the medication of interest. Whereas certain variants have strong evidence of their effects on metabolism, the impact of others (eg, IMs carrying one deleterious allele and one wild-type allele) is far from clear. Second, the blood level must be associated with treatment response. At the extremes, this relationship should be intuitive: Individuals with very low drug levels are unlikely to exhibit true drug effects (although, as noted previously, they may still exhibit placebo-like responses), whereas individuals with very high drug levels are probably at substantially elevated risk for adverse events.

The most extensively studied CYP450 substrate medications in psychiatry are the tricyclic antidepressants. Most of these are demethylated by CYP2C19 and hydroxylated to inactive compounds by CYP2D6. Unlike most other psychotropic medications, serum tricyclic antidepressant levels generally display a predictable association with treatment response (although the precise pattern varies from drug to drug). As a result, the association between CYP450 variation and tricyclic response is among the better established ones; guidelines have been published for adjusting tricyclic antidepressant dosing based on CYP450 status (Hicks et al., 2013).

For the far more widely prescribed SSRIs, evidence for the importance of CYP450 variation is best characterized as mixed. A 2007 review noted that most data were derived from small cohorts using heterogeneous designs (Berney, 2005). One major challenge in connecting CYP450 to clinically meaningful outcomes of antidepressant treatment is the apparent lack of association between SSRI levels and outcomes; only a few exceptions pertain to venlafaxine (Berney, 2005) or fluoxetine (Fava et al., 1994, 2002).

In an illustration of the limitations of genotyping in capturing functional status, a clever study examined 900 individuals with major depression disorder treated with the known CYP450 2D6 substrate venlafaxine, and compared genotyping results with apparent metabolic status based on the ratio of venlafaxine to its metabolite, desvenlafaxine. Only 4% of individuals were genotypic PMs, compared with 27% who were functionally PMs (presumably because of concomitant medications or other environmental effects, although these were not characterized directly).

Nonetheless, to date the strongest evidence for CYP450 genotyping effects on newer antidepressants comes from a pooled analysis of clinical trials of venlafaxine in major depressive disorder. Across four studies, more individuals reached remission among the wild-type metabolizers compared with the PMs (Lobello et al., 2010). Evidence for other SSRIs is even more limited. An investigation of the CYP2C19 substrate citalopram found that tolerability was poorer among non−wild-type metabolizers (Mrazek et al., 2011).

Another relevant protein in psychotropic response may be *p*-glycoprotein, which is responsible for transport of molecules across the blood−brain barrier and thus out of the central nervous system. The gene coding for this protein, ABCB1, was associated with the dose requirement for escitalopram, although the strongest effect was observed among heterozygous individuals, which suggests either interesting dose effects or type I error (Singh, Bousman, Ng, Byron, & Berk, 2012).

Most studies of antidepressant response genomics have examined monoaminergic candidate genes and provide an excellent illustration of the pitfalls of candidate-based studies, with limited understanding of mechanism leading to publication bias and type I error. Examination of candidate genes in larger cohorts has generally not supported associations with variants such as HTR2A (GENDEP Investigators, MARS Investigators, & STAR*D Investigators, 2013). Although far from their only target, the serotonin transporter SLC6A4, the proximal site of action of SSRIs, has been the target most often investigated, and in fact has been widely studied in nearly every psychiatric phenotype. A meta-analysis suggested an association with poorer treatment outcome for SSRIs, an effect observed among only Caucasian individuals. Importantly, the same meta-analysis also found evidence of publication bias (Porcelli, Fabbri, & Serretti, 2012).

Regardless of the utility of this association, it raises several relevant points for interpreting pharmacogenomics results. First, one small study suggested that the poorer outcome is restricted to SSRIs, but in general there are few data here, such that an equally valid hypothesis is that the risk variant predicts generally poorer outcome regardless of treatment. Second, two early studies suggested differential rates of adverse effects or discontinuation (Perlis et al., 2003; Popp, Leucht, Heres, & Steimer, 2006), another illustration of the importance of considering interactions between efficacy and tolerability. Finally, the effect has not been observed among Asian cohorts. In fact, some studies suggest effects in the opposite direction. This may reflect interesting gene-by-environment or epistatic effects, or more likely, the challenges of overcoming publication bias.

Three cohorts form the bulk of published GWAS. The first used data from the United Kingdom−based GENDEP study, in which outpatients were randomized to the selective serotonin reuptake inhibitor escitalopram or the norepinephrine reuptake inhibitor nortriptyline; the second used inpatient antidepressant response data from the German MARS study, whereas the third was drawn from the US effectiveness study STAR*D. STAR*D is notable for multiple treatment levels, beginning with the SSRI citalopram for all patients and then randomizing to next-step treatments after nonresponse (Ferreira et al., 2008; International Schizophrenia Consortium et al., 2009; Sklar et al., 2008).

In a meta-analysis, no single loci met a genome-wide threshold for significance (GENDEP Investigators et al., 2013). Multiple phenotypes, including percent change in symptoms (using mixed-effects models) and categorical remission (using last observation carried forward), were examined, an illustration of sensitivity analysis to try to distinguish efficacy from tolerability. In the context of emerging psychiatric disease genetics, the failure of this effort is less surprising, because the entire analysis spanned at most 2256 individuals. To encourage other groups to build on these analyses, the results were made public at www.ricopili.org; any individual locus can be queried, which may be helpful for future investigations examining nonoverlapping cohorts. A second meta-analysis, referred to as NEWMEDS, added cohorts, including one drawn from some industry studies, to the UK data (Tansey et al., 2012). This, too, failed to find any genome-wide associations.

Recognizing the substantial impact of placebo response on initial antidepressant treatment trials, a subsequent study investigated the more extreme phenotype of treatment resistance, that is, failure to reach remission despite multiple treatment trials. That study, which included a subset of the STAR*D cohort as well as a new treatment response cohort identified using electronic health records, failed to identify any genome-wide associations (O'Dushlaine et al., 2014). Other candidate-based association studies have identified suggestive but as yet unreplicated results (Kloiber et al., 2013; Perlis et al., 2008).

To date, three studies have examined the role of rare variants in antidepressant efficacy. The first examined copy number variation in two treatment cohorts: the STAR*D antidepressant effectiveness study, and a cohort ascertained using electronic health records (O'Dushlaine et al., 2014). The specific strategy employed represents a possible means of circumventing the limitations on collecting pharmacogenomics cohorts of adequate size, and is subsequently discussed in more detail in the section "Emerging Directions". In that study, no single locus reached a genome-wide threshold for association with efficacy. However, a common test of overall burden of copy number variants, that is, to what extent copy number variants' are more or less common in patients, suggested enrichment of a subset of duplications among individuals with treatment resistance.

A second study, using the NEWMEDS cohort, also examined deletions and duplications (Tansey et al., 2014). In that study, evidence of enrichment of duplications was not observed, although subtle methodologic differences make direct comparison with the prior study difficult.

Finally, a third, small study examined the role of rare variants using whole-exome sequencing. That study is most notable for examining a small cohort of Mexican American individuals randomized to one of two antidepressants (Wong et al., 2014). Employing a less stringent threshold for significance than the standard genome-wide $P < 5e{-}8$, one variant reached an exome-wide threshold, although the variant itself represents an SNP previously assessed in numerous GWASs. Surprisingly, no replication was attempted. However, a previous antidepressant meta-analysis of GWAS data found no evidence of association at that locus (www.ricopili.org).

Tolerability Studies

Beyond efficacy, a host of antidepressant tolerability phenotypes have been investigated. Weight gain with most antidepressants tends to be modest; a large naturalistic study suggested modest differences between interventions over 1 year (Blumenthal et al., 2014). Nonetheless, this phenotype has multiple advantages for pharmacogenomics investigation: It is readily and reliably quantified; it can be analyzed in terms of a continuous rather than dichotomous outcome (ie, a quantitative trait); and in contrast to efficacy, the underlying biology of obesity is better understood. In particular, obesity in general is known to be strongly heritable, with estimates ranging up to 70% depending on the specific phenotype examined (Allison et al., 1996). Short-term weight gain has been examined in multiple cohorts using a candidate gene approach, but no single locus has been replicated successfully (Keers & Aitchison, 2011).

Among the more common adverse effects observed with older as well as newer antidepressants are those related to sexual functioning, including difficulty achieving arousal and orgasm, as well as a decrease in libido. One study estimated that up to half of antidepressant-treated outpatients experienced such symptoms (Perlis, Laje, et al., 2009); they represent a major contributor to treatment nonadherence.

The only GWAS of sexual dysfunction associated with antidepressants examined a small Japanese cohort (n = 201) treated with SSRIs or serotonin-norepinephrine reuptake inhibitors (SNRIs); no loci were associated with sexual dysfunction at the genome-wide level (Kurose et al., 2012). Previously suggested involvement of glutamatergic genes (Perlis, Laje, et al., 2009) was not specifically examined, unfortunately.

Another antidepressant response phenotype examined in GWAS is treatment-associated (or treatment-emergent) suicidal ideation. More than 20 years ago, it was noted that a subset of individuals experience symptomatic worsening, and, in particular worsening or emergence of suicidal thoughts after initiation of antidepressants (Teicher, Glod, & Cole, 1993). Subsequent work also suggested that a subset of these cases were associated with activating effects of antidepressants (Perlis et al., 2007). Meta-analyses of FDA data suggest these effects are rare but occur more often than with placebo among younger patients (ie, those aged 24 or less) (Stone et al., 2009), which is consistent with prior reports implicating younger age as a risk factor (Perlis et al., 2007).

No candidate gene associations with treatment-associated suicidality have been replicated or supported by meta-analysis. The three antidepressant GWAS cohorts noted previously, STAR*D, GENDEP, and MARS, were individually examined for association with suicidality, with no loci achieving a genome-wide significance. Likewise, in meta-analysis, no genome-wide loci were identified (Ripke, unpublished data).

Antipsychotics

Although the familiality of antipsychotic treatment response has not been established, early understanding of antipsychotic pharmacokinetics and pharmacodynamics provided appealing candidates for study. Multiple antipsychotics are substrates for the CYP450 system already described, particularly CYP2D6; these include both typical antipsychotics (chlorpromazine, haloperidol, and perphenazine) and atypical antipsychotics (aripiprazole and risperidone). In addition, antagonism of dopamine D2 receptors and antipsychotic potency (Kapur & Mamo, 2003) focused attention on candidate gene studies of dopaminergic function. With the recognition that binding at serotonin receptors contributed to the "atypical" nature of second-generation antipsychotics (Kusumi, Boku, & Takahashi, 2014), subsequent studies also examined serotonergic variation. In light of the relatively large number of candidate-based studies and the tendency toward nonreplication, the next sections review primarily positive meta-analyses. For a review of individual candidates, see Zhang and Malhotra (2011).

Efficacy Studies

As with antidepressants, the potential relationship between CYP450 status and efficacy is intuitive but largely hypothetical. PMs might experience poorer apparent efficacy (because adverse effects lead to dropout), whereas ultrarapid metabolizers might experience poorer efficacy because standard doses yield insufficient blood levels. In reality, across multiple antipsychotic studies, CYP450 2D6 did not influence treatment outcomes (Zhang & Malhotra, 2011). This otherwise confusing finding might be explained by the observation that among risperidone-treated patients, PMs have higher blood levels, but these levels do not predict efficacy (and may even be inversely correlated with efficacy) (Riedel et al., 2005).

Not surprisingly, the most-studied pharmacodynamics gene in antipsychotic efficacy has been DRD2, in which several common variants were amenable to early small-scale genotyping efforts. A single base-pair insertion/deletion polymorphism was identified in the 5′ promoter, rs1799732 (sometimes referred to in older studies as −141C Ins/Del), although its functional implications are unclear. In a meta-analysis of clinical improvement among 687 antipsychotic-treated patients, individuals homozygous for the cytosine insertion were more likely to exhibit response (Zhang, Lencz, & Malhotra, 2010). Both first- and second-generation antipsychotics were examined and (unusual in psychiatric genomic studies) analyses included a large proportion of non-Caucasian subjects. Studies of other dopaminergic genes, including the D3 and D4 receptor, have yielded inconsistent results (Zhang et al., 2010).

Another commonly investigated candidate gene has been HTR2A, which codes for the serotonin 2A receptor, to which atypical antipsychotics bind with high affinity (Kusumi et al., 2014). A promoter SNP in this gene (rs6311) was associated with clinical improvement after atypical antipsychotic treatment (Chen, Shen, & Chen, 2009) across multiple studies.

Antipsychotic treatment outcomes have been investigated in GWAS. As with antidepressant studies, a substantial limitation has been the paucity of cohorts with DNA and prospectively defined treatment outcomes. The largest cohort with published GWAS data is drawn from the CATIE study, and reported 738 genotyped individuals who had been treated with predominantly atypical antipsychotics (McClay, Adkins, Aberg, Stroup, et al., 2011). One intergenic SNP (rs17390445) reached genome-wide significance for one phenotype, improvement in positive symptoms with ziprasidone. A second analysis in the same cohort examined associations with improvement in cognitive measures attributed to individual antipsychotics (McClay, Adkins, Aberg, Bukszar, et al., 2011), whereas several loci were considered suggestive based on a false discovery rate of 10%. None exceeded a genome-wide threshold for significance.

To date, one study has applied exome sequencing to examine rare variants for association with antipsychotic response. Among a small number of subjects ($n = 11$), one rare variant was identified with modest support among a larger patient subset, as well as a second cohort of patients with schizophrenia. The latter was notable for being drawn from the Xhosa population (Drogemoller et al., 2014).

Efforts are ongoing to align treatment response phenotypic data with the much larger cohorts involved in the Psychiatric Genomics Consortium (PGC) schizophrenia work group. A challenge in these data sets will be the absence of prospective treatment response data in most individuals, which necessitates reliance on retrospective assessment.

Tolerability Studies

Extrapyramidal symptoms represent a common adverse effect of first-generation and some second-generation antipsychotics, which contributes to nonadherence and treatment discontinuation. At the extreme, patients may develop involuntary choreoathetotic movements, most often perioral or involving the upper extremities, referred to as TD. Because individuals who metabolize antipsychotics poorly might be exposed to greater blood levels, a putative TD risk factor (Tenback, van Harten, & van Os, 2009), it follows that PMs could be at elevated risk. In fact, meta-analyses suggested that this is indeed the case (Patsopoulos, Ntzani, Zintzaras, & Ioannidis, 2005), with odds of TD development estimated to be about 40% greater among PMs. Multiple individual studies have also suggested increased risk for extrapyramidal symptoms in this subgroup.

As with antipsychotic efficacy, *DRD2* has been the most commonly studied gene for TD. Whereas an early meta-analysis suggested that a particular SNP (Taq1A, rs1800497) was associated with risk, with about a 50% increase in odds of TD among individuals homozygous for the risk allele, a more recent one failed to confirm this effect (Zhang et al., 2010). Moreover, the SNP actually lies in an adjacent ankyrin-repeat containing kinase gene, *ANKK1*, rather than DRD2 itself.

One cohort has been examined for extrapyramidal symptoms using genome-wide association. In the multicenter CATIE study, in which individuals received one of several antipsychotics, 738 individuals were examined using measures of parkinsonism, akathisia, and TD (Aberg et al., 2010). For the four phenotypes examined, only one (intergenic) SNP exceeded a traditional genome-wide threshold for statistical significance, on a measure of parkinsonism.

Numerous studies have examined candidate genes for antipsychotic-associated weight gain, motivated by the advantages of this phenotype in terms of reliability, power, and at least partial understanding of relevant biology (Lett et al., 2012). Although the magnitude of weight gain is greatest for a subset of the atypical antipsychotic people, even the typical antipsychotics may contribute to significant weight gain (Kahn et al., 2008). In general, dopaminergic and serotonergic genes have been a focus of study, yielding inconsistent results. The notable exception is a promoter SNP, rs3813829, in the serotonin 2C receptor (HTR2C), found in meta-analysis to be associated with differential weight gain with antipsychotic treatment (De Luca, Mueller, de Bartolomeis, & Kennedy, 2007). More recent studies have also tended to support this association, with a suggestion that other SNPs in HTR2C may be associated with emergence of metabolic syndrome as well (Ma et al., 2014). Investigations of mice with region-specific knockout of HTR2C support the relevance of the receptor to glucose metabolism (Berglund et al., 2013).

To date, two cohorts have been examined for genome-wide association with weight gain. The first examined the CATIE antipsychotic study (Adkins et al., 2011). That study examined 12 different metabolic adverse events and treated multiple antipsychotics individually; to address the problem of multiple comparisons, the analysis relied on a false discovery rate of 10% rather than a standard genome-wide significance threshold. With this more liberal threshold, multiple loci of interest were identified, although no evidence of replication has been presented.

The second study investigated multiple small cohorts of antipsychotic-treated children and adolescents, beginning with a cohort of 139 treatment-naïve patients. Although no loci exceeded a traditional genome-wide threshold, one intergenic SNP yielded $P < 5e{-}7$ and was replicated in three other small cohorts (Malhotra, Correll, et al., 2012). Of note, this SNP is about 150 kb from the melanocortin 4 receptor (MC4R), a gene in which rare variants are associated with obesity. Interestingly, MC4R also regulates HTR2C function, which raises the possibility that these two variants both point to a common biological pathway.

Another phenotype of substantial interest related to antipsychotic response has been clozapine-associated agranulocytosis. Although this outcome, in which neutrophil count drops rapidly, is exceedingly rare, it initially led clozapine to be withdrawn from the US market and now requires weekly or biweekly monitoring of blood levels. Moreover, despite the widely acknowledged efficacy advantage for clozapine in schizophrenia, the safety concerns resulted in FDA labeling indicating that it should not be considered a first-line treatment.

Both rare and common variants have also been investigated in 163 cases of clozapine-induced agranulocytosis, building on prior studies using GWAS and candidate-based approaches. Strong support ($P < 5e{-}14$ for one and 6e−10 for the other) was identified for two human leukocyte antigen (HLA)-associated changes: one in HLA-DQB1 and the other in HLA-B (Goldstein et al., 2014), consistent with prior candidate gene studies. In one case the rare allele is protective, whereas in the other it is risk-increasing.

Another clinically relevant antipsychotic adverse effect is prolongation of the cardiac repolarization or QT interval. The genome-wide study used a small ($n = 183$) cohort of patients with schizophrenia who were treated with iloperidone, who were of multiple ancestries, and who had no identified genome-wide loci (Volpi et al., 2009). A study of the individual antipsychotics included in CATIE revealed no individual genome-wide significant loci; the authors highlighted an SNP (rs4959235) in the ion transporter SLC22A23, associated with quetiapine QTc prolongation with $P < 2e{-}7$ (Aberg et al., 2012). Unfortunately, the latter study did not specifically examine loci implicated in the former, nor was replication attempted in another cohort.

Lithium, Anticonvulsants, and Opiates

Lithium preparations remain a reference standard treatment for bipolar disorder (Yatham et al., 2009) and are increasingly under investigation for other neurologic disorders as well. As a probe to understand the pathophysiology of bipolar disorder, one could hardly imagine a less appealing candidate. Lithium has an incredibly diverse range of actions at the cellular level, a long time to onset of clinical action, and a relatively narrow therapeutic index such that even identifying an in vitro equivalent to an in vivo serum level is challenging. Nonetheless, convergent evidence suggests one of two mechanisms for lithium action, related to inositol monophosphate signaling or the Wnt signaling pathway. Of these, the latter has been the most intense area of study (Can, Schulze, & Gould, 2014; Pan et al., 2011).

Because of the complex and interesting neurobiology of lithium response, it is not surprising that numerous candidate gene studies have investigated targets other than the usual monoaminergic suspects. These have included Wnt-related genes, glutamate-related genes, and genes related to circadian rhythms (the latter because of hypotheses about bipolar disorder rather than lithium per se). However, no single locus has demonstrated replicated associations with lithium response.

The first published GWAS examining lithium treatment response included 458 individuals drawn from the multicenter US STEP-BD study of bipolar disorder, as well as a replication data set drawn from 359 individuals in a UK bipolar disorder cohort (Perlis, Smoller, et al., 2009). It represents one of the only longer-term outcome studies in psychiatric pharmacogenomics, with subjects analyzed in terms of time to relapse or recur after initial recovery. Strongest association was identified for an intergenic SNP (rs10795189; $P < 6 \times 10[-7]$), but replication did not support this association.

In 2014, a second study reported GWAS results for a small cohort of Han Chinese individuals (n = 294), contrasting good and poor lithium-responsive patients. In that study, a single locus in glutamate decarboxylase-like protein 1 was identified with strong (rs17026651; $P < 3e{-}37$) evidence of association with response (Chen et al., 2014). The authors also included a 100-patient replication and a small longitudinal follow-up cohort (n = 24), in which they again observed an association. Unfortunately, the authors' decision not to pursue replication in a truly independent cohort prompted multiple response letters based on other treatment cohorts, none of which identified any evidence of association (Consortium on Lithium Genetics et al., 2014). Moreover, despite its name, the gene itself appears to be unrelated to glutamatergic neurotransmission. As a result, the effect must be assumed to represent either type I error, perhaps arising from stratification artifact, or an effect limited to a specific population group.

Ongoing efforts by a consortium of sites studying lithium, the Consortium on Lithium Genetics, based on the retrospective assignment of lithium response, may be helpful in addressing the major problem of limited power to detect association in published cohorts. Preliminary reports suggest that the clinical measure is only somewhat reliable across cohorts; however (with kappa values of about 0.54−0.66, and evidence of three overlapping response phenotypes), the impact on power remains to be determined (Manchia et al., 2013).

Anticonvulsants

Anticonvulsants have also been investigated in candidate gene studies in psychiatry, not in terms of efficacy but of adverse effects. In particular, carbamazepine and lamotrigine (as well as older anticonvulsants not generally used in psychopharmacology, such as phenytoin) may be associated with effects on bone marrow or severe rash, either of which has been suggested to represent an autoimmune phenomenon or direct toxicity, similar to hypotheses advanced about clozapine agranulocytosis. As a result, the focus of investigation has been HLA alleles, particularly because that they were among the first loci to be readily genotyped in the context of transplant biology. In the case of carbamazepine, multiple case−control association studies have identified a robust association with HLA-A and -B alleles, most notably with HLA-B*15:02, for which the odds ratio for a cutaneous adverse reaction was estimated to be 81 (95% confidence interval [CI], 46−143) in a meta-analysis (Grover & Kukreti, 2014). This allele is rare among Europeans; to date, no European individual with carbamazepine rash has been found to carry the risk allele (Amstutz et al., 2014). However, emerging data suggest association for other HLA alleles among Caucasian individuals as well (Amstutz et al., 2013; McCormack et al., 2011).

A number of small GWASs have examined carbamazepine adverse effects. Among European individuals, a small study (n = 65 cases) in 2011 associated HLA-A*31:01 with serious rash risk (McCormack et al., 2011); no other locus reached a genome-wide threshold.

Lamotrigine

Whereas the anticonvulsant lamotrigine, like carbamazepine, has been associated with severe cutaneous adverse effects, the association with HLA-B*15:02 is more modest than effects observed with some other anticonvulsants (odds ratio, 3.59; 95% CI, 1.15−11.22) (Cheung et al., 2013). Other HLA variants have also been associated with lamotrigine rashes (Li et al., 2013).

In addition to HLA, pharmacokinetic candidate genes have been investigated for anticonvulsants. Among these medications, beyond CYP450 effects already noted, phase II metabolism may be particularly relevant. The uridine

5′-diphosphate-glucuronosyltransferase (UGT) enzymes transform substrate drugs into more hydrophilic and thus more readily excreted compounds. In the case of lamotrigine, two phase II enzymes appear to be relevant: UGT1A4 and UGT2B7. Lamotrigine blood levels have primarily been shown to be influenced by UGT1A4 alleles. For example, in a Turkish cohort (n = 131), a functional variant (referred to as L48V) resulted in lamotrigine levels 52% lower than for wild-type alleles (Gulcebi et al., 2011). Similar effects have been observed in Han Chinese people (Chang, Yang, Zhang, & Liu, 2014). In a Thai population, other variants appear to influence lamotrigine levels, particularly UGT2B7 -161C > T (Singkham, Towanabut, Lertkachatarn, & Punyawudho, 2013).

Topiramate

Although the anticonvulsant topiramate proved to be ineffective for the treatment of bipolar disorder, it has found widespread off-label use for appetite suppression and alcohol use disorders. One candidate gene, GRIK1, which codes for the kainate-selective glutamate receptor GluR5, has been examined based on animal studies of topiramate binding (Kaminski, Banerjee, & Rogawski, 2004). In an initial small but placebo-controlled study, an intronic SNP in GRIK1 (rs2832407) was nominally associated with the severity of topiramate-induced side effects and with serum levels of topiramate (Ray et al., 2009). In a randomized trial of topiramate for alcohol misuse, the effect of topiramate on heavy drinking days was substantially and statistically significantly greater than that for placebo only in rs2832407 C-allele homozygotes (n = 122) (Kranzler, Covault, et al., 2014). Interestingly, this variant did not appear to influence changes in body mass index observed with topiramate in the same cohort (Kranzler, Armeli, et al., 2014).

A dose-limiting adverse effect with topiramate may be cognitive slowing. An intriguing GWAS examined cognitive effects topiramate at 100 mg among 158 healthy volunteers (Cirulli et al., 2012), although no genome-wide effects were identified even after joint analysis with another cohort of 290 topiramate-treated patients with epilepsy. Notably, the authors observed wide variation in topiramate plasma levels (more than 50-fold), which suggests that further investigation of pharmacokinetic variants may be warranted.

Opiates

Investigation into variation in opiate response has focused on the mu-opioid receptor OPRM1. A functional SNP, rs1799971 (sometimes referred to as A118G), changes asparagine to aspartic acid and has been well-studied in association with opiate requirement, particularly after surgery. A meta-analysis (Hwang et al., 2014) examined 18 studies involving 4607 participants. The mean opioid dose was significantly greater among G-allele carriers than AA homozygotes. Lending support to this association, multiple subgroup analyses were also nominally significant, including one examining only individuals of Asian ancestry.

EFFORTS AT CLINICAL TRANSLATION

Clinical Trials

Whereas many widely used diagnostic tools in medicine entered practice without randomized, controlled investigations, this approach adapted from pharmacologic study remains the reference standard for demonstrating the impact of pharmacogenomic testing. To date, one such study has compared testing with a panel including cyp450 2D6, 2C19, and 1A2 as well as serotonin transporter and one receptor. In that study, outpatients with major depression (n = 51) were randomized to testing or treatment as usual, with follow-up every 2 weeks out to 8 weeks (Furmaga NCDEU 2012). No significant difference in the magnitude of improvement was detected (31% in the test group versus 19% in the treatment as usual group; $P = 0.3$), although greater improvement in the test group was observed only at week 4. Two unblinded cohort studies with the same panel also suggested benefit. Among 44 patients, mean improvement was 31% versus 18% ($P = 0.08$). Finally, in a larger study (Hall-Flavin et al., 2012) of 165 patients, mean improvement of 47% versus 30% ($P < 0.0001$) was observed.

A key limitation of all three studies was the absence of data regarding the specific application of the test results and the individual loci that were informative. That is, in lieu of reporting results at individual loci, the publications describe a color-coded aggregate across pharmacokinetic and (presumably) pharmacodynamic loci. As a result, Although they can suggest the possible use of the approach, they do not help in developing future investigations. One possibility is that all the benefit accrued with the (more evidence-based) CYP450 aspect of the test. Moreover, the latter two studies were unblinded, which likely contributed to a substantial placebo effect. In a secondary analysis of these cohorts, more testing benefit was observed among patients receiving medications noted by the test to require greater caution, which might argue for a specific testing benefit. Conversely, some benefit was also observed in individuals not receiving such medications, which hints at the magnitude of a placebo-like effect. Taken together, these studies must be viewed as promising but far from convincing.

Cost-Effectiveness Studies

Genotyping costs are likely to continue to plummet, although albeit perhaps more slowly for clinically reported genotypes, for which additional administrative overhead and quality control are mandated by the US FDA. As such, estimating cost-effectiveness of pharmacogenomics testing is challenging because the cost is a moving target and should ultimately approach zero. In the intermediate term, however, this parameter has become critical for convincing payers to reimburse for testing; it is no longer sufficient simply to demonstrate efficacy.

In an effort to estimate the magnitude of effect for a single-locus test required for cost-effectiveness, we initially developed a model based on an association reported with efficacy in STAR*D (Perlis, Patrick, et al., 2009) (Fig. 44.2). Base-case assumptions suggested that even a moderate effect size would not be cost-effective, in part because simply using an alternative antidepressant in all cases dominates the benefit of test-driven prescribing. However, under circumstances in which an informative test is relatively common, and/or the effect size predicted by the test is large, even a relatively expensive test could be cost-effective. Moreover, as the test cost diminishes, even modest effect sizes are worth capturing through testing.

Even the most rigorous cost-effectiveness models must make numerous assumptions about costs, probabilities, and utilities. (In one of the few direct investigations of the usefulness patients assign to earlier antidepressant response, a Danish survey found that patients would pay about $100 to decrease the duration of antidepressant dosage adjustment by a month, and about $280 to avoid a single medication change (Herbild et al., 2009)). Cost-effectiveness analyses may thus be most useful simply to help understand the conditions under which genomic testing may be most useful: for example, helping to estimate the effect sizes to use for calculating power in future randomized trials. Another strategy for deciding on optimal effect sizes relies on UK guidance regarding clinically meaningful effects, and extrapolates to biomarker results. A calculator (available at http://www.depressiontools.org/onlinecalculator.html) allows entry of a single-locus effect and compares it with the NICE (National Institute for Health and Care Excellence) standard (Uher, Tansey, Malki, & Perlis, 2012).

A more direct alternative to understanding the cost–benefit relationship is to use health utilization data drawn from insurance claims databases or electronic medical records. One study examined the impact of CYP450 variants, aggregated with pharmacodynamics variants with widely varying degrees of support in the literature. Non–wild-type individuals who were receiving a CYP450 substrate medication, or another medication scored as "less desirable," based on an obfuscated algorithm, had greater health care costs in the prior year, including greater number of medical visits, medical absences from work, and disability claims (Winner, Allen, Altar, & Spahic-Mihajlovic, 2013).

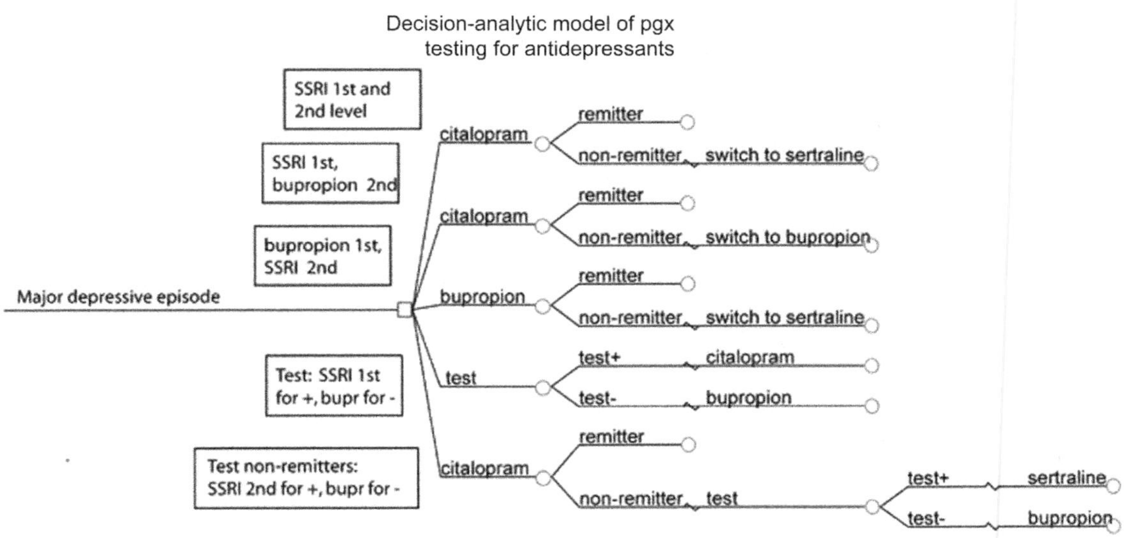

FIGURE 44.2 Illustration of a decision-analytic model of pharmacogenomic testing for antidepressant response, comparing alternate testing strategies. *Perlis, Neuropsychoparmacology 2009.*

Intriguing as this result may be, the risk for confounding by indication is high; that is, an alternative explanation would be that the use of CYP450 substrate medications is simply increased in sicker (and more treatment-resistant) populations.

Another application of health claims data directly examined outcomes among a cohort of 111 individuals who had received a commercial test combining CYP450 and pharmacodynamics variants, and compared them with a propensity score–matched cohort that did not receive testing (Fagerness et al., 2014). Propensity scores represent a means of attempting to match the treated group as closely as possible, to minimize confounding risk. In that study, outpatient treatment costs were 9.5% lower among tested patients. Notably, medication adherence was also significantly greater among the tested group, perhaps reflecting another benefit of testing: Patients may have more confidence in their assigned treatment. Unfortunately, as with other antidepressant pharmacogenomics studies to date, the presented data do not allow a determination of which aspects of the test are most informative and how clinician behavior is changed.

A question which has not been addressed directly to date, but is critical in considering clinical translation, is the optimal population in which to deploy pharmacogenomic testing. A useful analogy in psychiatry is screening for hypothyroidism: Although not especially useful or cost-effective in an unselected population, the rates of hypothyroidism may be substantially greater in individuals with treatment-resistant depression, such that testing could be justified. So, too, for example, an optimal population for studying CYP450 testing may be individuals treated with multiple medications, in whom risk for drug–drug interaction is likely to be greater (Manolopoulos, Ragia, & Alevizopoulos, 2012).

Just as diagnostic dissemination has correlates with phase 1–3 clinical studies of a novel therapeutic drug, there is also a postmarketing or phase 4 equivalent. Although diagnostic development for most psychotropics is still focused on earlier phases, outside the United States, testing for anticonvulsant adverse outcomes has begun to increase.

Perhaps the best example of this kind of investigation comes from an electronic health record study in Hong Kong, where testing in individuals receiving carbamazepine began in 2008 (Chen, Liew, & Kwan, 2014 #19142). Examining outcomes in more than 100,000 individuals, including more than 4000 tested for HLA alleles, is instructive. Most notably, the rates of Stevens–Johnson syndrome (SJS)/toxic epidermal necrolysis after carbamazepine decreased from 0.24% to 0% after initiation of testing; that is, the test appeared to "work" in the sense that high-risk individuals presumably did not receive treatment. On the other hand, another observation from this study is equally critical: The overall rate of SJS/10 remained unchanged and the rate of prescribing of carbamazepine decreased. This finding strongly suggests that the need to test dissuaded clinicians from prescribing carbamazepine and that they instead prescribed other (non–test-driven) anticonvulsants which also are associated with serious rash.

The case of carbamazepine poses additional conundrums likely to apply to other pharmacogenomics tests in psychiatry. First, the selection of a therapeutic agent is often time sensitive, as in the case of an individual presenting with new onset of seizures. Although rapid turnaround of HLA genotyping is possible, clinicians may be reluctant to wait for test results and may simply default to other strategies, as the Hong Kong example might indicate. Second, the optimal population for testing remains to be established, and may be influenced by race and ethnicity. The HLA-B*1502 allele is rare among Europeans, so despite its potential risk, the usefulness of testing non-Asians is debatable in the absence of data. Finally, the positive predictive value for HLA testing remains low; most individuals who are risk allele–positive will not develop a serious rash. As a result, although guidelines suggest avoiding carbamazepine as first-line treatment in these individuals (Amstutz et al., 2014), there is some risk that testing will lead to the use of less efficacious or safe interventions in many patients who could tolerate carbamazepine.

APPLICATIONS IN CLINICAL INVESTIGATION

The failure rate of new clinical entities (primarily drugs) in psychiatric clinical trials is among the highest in all of medicine. Fewer than 10% of drugs entering a phase 1 study successfully achieve approval by the US FDA. Although these failures are multifactorial, one likely contributor is the heterogeneity of psychiatric illness itself. That is, if one imagines that an intervention is highly effective in 30% of treated individuals, but that population is dispersed among the 70% in whom it is ineffective, the challenges in detecting benefit in that subgroup are apparent.

Outside psychiatry, examples have emerged of interventions "rescued" by secondary analysis of otherwise negative trials. For example, bapineuzumab phase II clinical trials were unsuccessful, but a subgroup of patients was identified who responded well and in whom the likelihood of success in subsequent study would be greater. Although the industry was slow to embrace this approach, reasoning that fragmenting the market for a medication would diminish returns, this decreased market is balanced against the greater success rate.

There are multiple possible designs for randomized intervention trials incorporating pharmacogenomics (Fig. 44.3A–D) (Perlis, 2011). The most standard is the post hoc or retrospective approach, in which DNA is simply collected during a trial

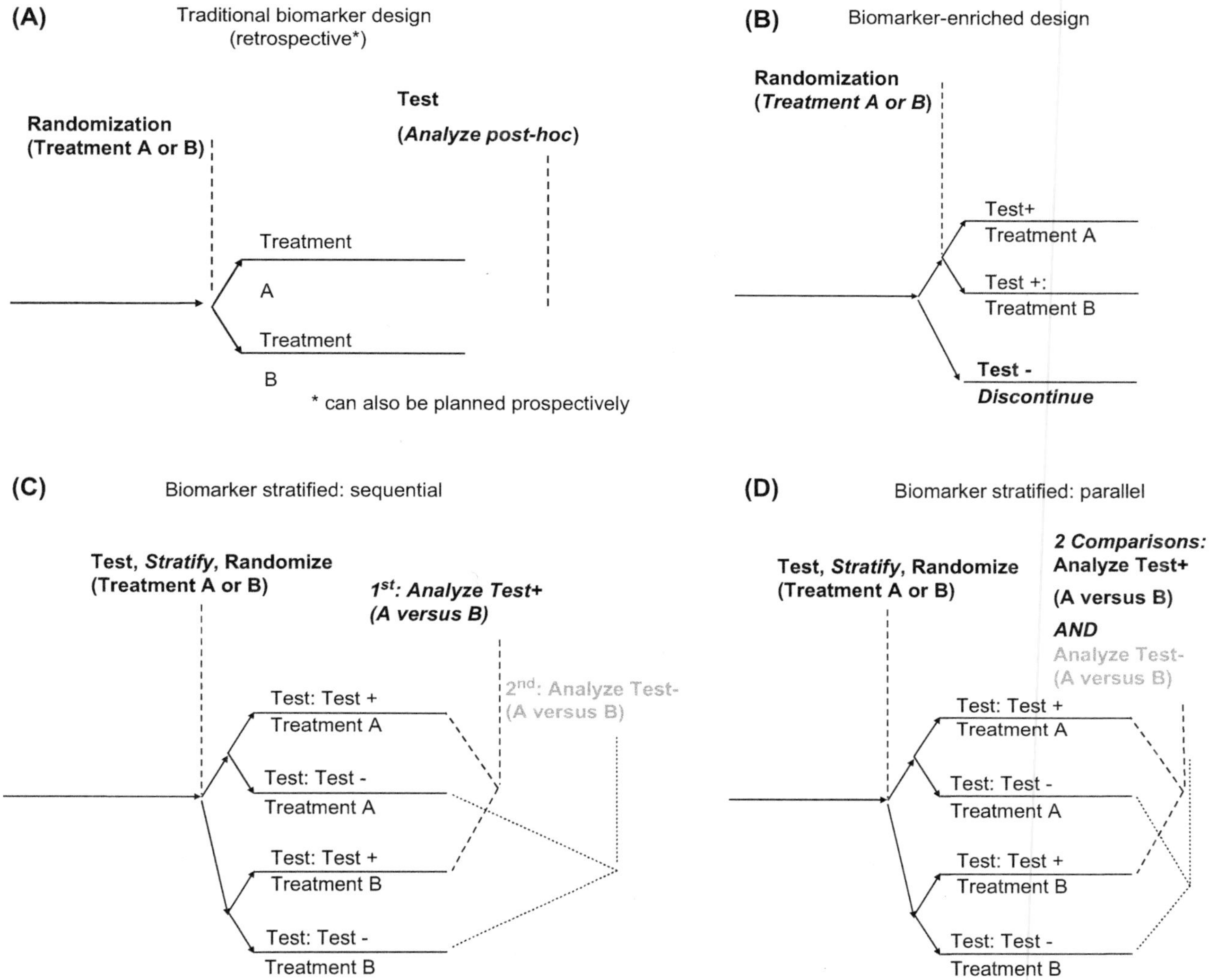

FIGURE 44.3 Example of study designs incorporating biomarkers. (A) Standard approach to a biomarker study, in which all markers are analyzed post hoc. (B) A biomarker-enriched design, in which only individuals with the risk variant of interest are randomized (as, eg, in a study targeting a rare causal variant.) (C, D) Two alternate approaches to the analysis of a biomarker-stratified design, in which individuals are randomized within biomarker-defined groups. The approaches differ in whether hypotheses are tested sequentially or in parallel, with important implications for statistical power. *Perlis Mol Psych 2011.*

and analyzed after study completion. This approach is most appropriate for studies in which the relevant genetics are unknown at time of study initiation, or are not necessarily important in treatment response.

An alternative design, referred to as biomarker-enriched, relies on genotyping (or sequencing) at time of study entry. Only those subjects carrying the allele of interest are randomized, whereas those without it do not receive treatment. This study is highly efficient provided the marker of interest is well-established. However, in the case where the relevance of a marker is less clear, it risks excluding an entire group of patients who might otherwise benefit. Apart from the obvious impact on feasibility and speed of recruitment, regulatory agencies might look askance at an intervention seeking approval for a group of patients when its safety has not been established.

A third class of study designs, referred to as biomarker-stratified, also genotypes at study entry but simply ensures that randomization is stratified by genotype, to maximize the ability to compare outcomes with or without the marker of interest. The two strategies differ in a subtle way. In stratified-parallel, the study is powered to examine two distinct comparisons: outcomes in the marker-positive and marker-negative groups. This allows for the possibility that direction of effect is unknown or incorrectly specified; that is, it is possible that the marker-negative rather than marker-positive group exhibits differential response. However, it requires a doubling of sample size (assuming experiment-wise type I error is to be controlled) because two independent tests are done. An alternative approach, sequential-parallel, assumes that the primary comparison is done only in the marker-positive group. If no effect is observed, no further analysis is done and the study is assumed to be negative. However, if an effect is observed, a test of the secondary hypothesis (ie, effect in the marker-negative group) is done to evaluate the specificity of effect. In principle, this does not require a greater sample size but allows some evaluation of outcomes in the marker-negative group if warranted.

POLICY AND REGULATORY CONSIDERATIONS

In other areas of medical genetics, the implications of providing test results to patients have been examined prospectively. In principle, one would expect all of the relevant literature regarding other kinds of imperfect tests (mammography, screening for colon cancer, prostate antigen screening, and so forth) to apply. In practice, genetic testing is often considered special because it is essentially unchanging and carries so much uncertainty. Perhaps the best analog is prenatal screening; still, the values and priorities involved may be different when one makes decisions about one's own health.

In Alzheimer disease, careful study suggests that receiving test results indicative of high risk for dementia does not substantially increase anxiety in the near term (Chao et al., 2008; Roberts, Cupples, Relkin, Whitehouse, & Green, 2005), but it predicts an increase in health-promoting behavior. Although very different than pharmacogenomic testing, this result at least suggests that test results suggestive of poorer outcomes may not exacerbate illness (although it will be important to understand their impact on individuals with preexisting psychopathology).

In addition to patient education, preliminary experience with genomic testing suggests the necessity of clinician education. The probabilistic results yielded from pharmacogenomics testing are qualitatively different from the typical "normal/abnormal" or "high/low" results from many other blood tests. Anecdotally, clinicians may interpret a finding that a patient is a non-wild-type metabolizer as indicating that drugs relying on that enzyme for metabolism cannot be used. This perception may be exacerbated by reports which list such substrates in red, or with a stop sign or similar icon. Notably, a tricyclic antidepressant guideline (Hicks et al., 2013) can be read as indicating that individuals who are PMs cannot be treated with this class of medications at all.

Of note, FDA approval of other diagnostic tools, ranging from melanoma screening to positron emission tomography for diagnosis or monitoring of Alzheimer's disease, included a mandate for clinician education. As such, the importance of education need not be a barrier to adoption of a test, provided there is parallel investment in developing and delivering educational materials.

A policy question of particular relevance to psychiatric pharmacogenomics is the role of regulation. Currently, pharmacogenomic tests (as with other diagnostics) may take one of two paths to commercial use. The first follows the traditional FDA approval process, with ongoing dialog culminating in a packet of evidence reviewed by an FDA committee. The second, far more efficient route recognizes that any laboratory-developed test may be marketed, provided the laboratory follows certain practices for assay design and validation. To date, commercial testing for psychiatric pharmacogenomics has exclusively followed the latter route. However, the likelihood that the FDA would eventually scrutinize these products more closely has probably led many commercial efforts to wait until greater clarity emerges.

In 2014, the US FDA indicated its intention to restrict the laboratory-developed test route, which is likely to entail substantially more evidence accrual before marketing future pharmacogenomics and other biomarker-based tests. Whether the increased certainty about FDA scrutiny has the effect of increasing investment in developing such tests, and outweighs the additional resources required to pursue FDA approval, remains to be seen.

EMERGING DIRECTIONS

It should be apparent from the review of published studies that the major barrier to progress in psychiatric pharmacogenomics is the availability of cohorts with reliable treatment response data to enable genetic association studies. Although efforts continue to develop analytic approaches to enhance power, for example, by using orthogonal data sets of transcriptomic or other data types, the lesson from disease genomics is that success may simply be a question of numbers.

Table 44.1 summarizes the strengths and limitations of various sources of DNA for pharmacogenomics study. Randomized controlled trials represent the recognized standard for establishing treatment efficacy, but at a substantial cost: in many cases, $10,000 per patient or more. Moreover, the vast majority of such studies are supported by industry, and achieving cross-industry collaboration has been challenging. In psychiatric pharmacogenomics, although individual companies have been willing to collaborate, no PGC-like effort has emerged.

TABLE 44.1 Comparison of Alternate Sources of DNA for Pharmacogenomics Studies

Source	Outcome Data	Treatment Data	DNA	Sample Size	Relative Cost
Clinical trial	+++	+++	+++	+	$$$
Disease registry	+	++	+++	++	$$
Claims data	+	+	Requires recontact	+++	$
EHR[a]	++	++	Requires recontact or biobank	+++	$

[a]*EHR, electronic health record.*

As an alternative, investigations have incorporated subjects drawn from biobanks, large DNA collections from subjects phenotyped broadly, or from electronic health records or national registries. Two notable examples of the latter have already been discussed. To identify a large cohort of individuals with major depression who are poorly responsive to standard antidepressants, one group developed machine-learning tools to classify drug response longitudinally, enabling a collection of more than 1000 antidepressant-response DNA's in less than 2 years (O'Dushlaine et al., 2014). Similarly, a cohort of clozapine-treated individuals was identified based on a monitoring database required to mitigate the agranulocytosis risk (Goldstein et al., 2014). In general, apart from the efficiency of collecting samples solely from individuals with informative phenotypes, such collections have the benefit of added face validity and generalizability, sometimes referred to as population-based pharmacogenomics (Simon & Perlis, 2010). On the other hand, a potential disadvantage is the absence of randomization and some lack of precision in defining treatment response. The latter may improve with the increasing incorporation of self-report or clinician-rated measures in psychiatric practice. Still, electronic health records or registry-based ascertainment represents a clear trade-off of efficiency for precision. This approach may be most powerful in cases where a relatively rare phenotype such as a serious adverse event is to be studied.

A related approach uses patient self-report as part of direct-to-consumer genomics initiatives. Although it is easy for researchers to doubt the seriousness of such efforts, they may provide large data sets capable of replicating associations identified through traditional approaches (Tung et al., 2011).

From an analytic perspective, new statistical methods or the application of existing approaches to longitudinal data may improve the efficiency of pharmacogenomic studies. Nearly all published studies rely on the method of the last observation carried forward to address dropouts in trials, when more data-efficient and less biased strategies such as mixed-effects models have become the norm for analyzing longitudinal data. Techniques to better account for the multiple correlated phenotypes typically captured in longitudinal studies are also required; for larger cohorts such as STAR*D and CATIE, there may be multiple publications, each examining overlapping outcomes. Correcting for all of these tests is likely far too conservative, whereas false-discovery—based approaches are too liberal (Adkins et al., 2011). In addition, strategies to use orthogonal data sets (eg, proteomics or transcriptomics data) to better estimate prior probabilities of association or refine associations may be especially relevant to pharmacogenomic studies. For example, all other things being equal, if common variation in a given gene is associated with an outcome and a drug is known to modulate expression of that gene's product, it might lead one to prioritize that gene for follow-up.

A final emerging area in psychiatric pharmacogenomics is integration of these data with other putative outcome predictors: neuroimaging, EEG, serum proteins, or simply clinical predictors. For example, C-reactive protein has been associated with differential response to SSRIs and SNRIs; integrating that marker with clinical predictors explains up to about 15% of variance in response (Uher et al., 2014). Likewise, another effort to predict antidepressant response integrated common genetic variation with body mass index (Papakostas et al., 2014). Whether either of these findings replicates, they illustrate the potential usefulness of integrating across independent modalities for prediction. In particular, the value of readily available clinical predictors is often overlooked. This strategy has become more feasible as large effectiveness studies and other large clinical data sets drawn from claims or electronic health records become available (Gallagher et al., 2012; Perlis et al., 2011). Although they may not provide strong discrimination in themselves, in aggregate they may at least be helpful in estimating risk. An example of this approach to prediction of treatment resistance may be found in Perlis (2013).

Because a theme of this text is moving across biological scale, the potential to use neuroimaging, EEG, or protein biomarkers to facilitate pharmacogenomics should also be considered. If these biomarkers could be identified, they might allow identification of common variants (alone or in aggregate) that are simply easier to measure. From this perspective, the more labor-intensive biological measure becomes a means of developing the eventual clinical diagnostic tool, rather than representing the tool itself.

REFERENCES

Aberg, K., Adkins, D. E., Bukszar, J., Webb, B. T., Caroff, S. N., Miller, D. D., ... van den Oord, E. J. (2010). Genomewide association study of movement-related adverse antipsychotic effects. *Biological Psychiatry, 67*(3), 279–282. http://dx.doi.org/10.1016/j.biopsych.2009.08.036. PubMed PMID:19875103; PubMed Central PMCID:PMC3388725.

Aberg, K., Adkins, D. E., Liu, Y., McClay, J. L., Bukszar, J., Jia, P., ... van den Oord, E. J. (2012). Genome-wide association study of antipsychotic-induced QTc interval prolongation. *The Pharmacogenomics Journal, 12*(2), 165–172. http://dx.doi.org/10.1038/tpj.2010.76. PubMed PMID:20921969; PubMed Central PMCID:PMC3388904.

Adkins, D. E., Aberg, K., McClay, J. L., Bukszar, J., Zhao, Z., Jia, P., ... van den Oord, E. J. (2011). Genomewide pharmacogenomic study of metabolic side effects to antipsychotic drugs. *Molecular Psychiatry, 16*(3), 321–332. http://dx.doi.org/10.1038/mp.2010.14. PubMed PMID:20195266; PubMed Central PMCID:PMC2891163.

Alexanderson, B., Evans, D. A., & Sjoqvist, F. (1969). Steady-state plasma levels of nortriptyline in twins: influence of genetic factors and drug therapy. *British Medical Journal, 4*(5686), 764–768. PubMed PMID:5391106; PubMed Central PMCID:PMC1630282.

Allison, D. B., Kaprio, J., Korkeila, M., Koskenvuo, M., Neale, M. C., & Hayakawa, K. (1996). The heritability of body mass index among an international sample of monozygotic twins reared apart. *International Journal of Obesity and Related Metabolic Disorders: Journal of the International Association for the Study of Obesity, 20*(6), 501–506. PubMed PMID:8782724.

Amstutz, U., Ross, C. J., Castro-Pastrana, L. I., Rieder, M. J., Shear, N. H., Hayden, M. R., ... CPNDs Consortium. (2013). HLA-A 31:01 and HLA-B 15:02 as genetic markers for carbamazepine hypersensitivity in children. *Clinical Pharmacology and Therapeutics, 94*(1), 142–149. http://dx.doi.org/10.1038/clpt.2013.55. PubMed PMID:23588310; PubMed Central PMCID:PMC3839910.

Amstutz, U., Shear, N. H., Rieder, M. J., Hwang, S., Fung, V., Nakamura, H., ... CPNDS Clinical Reccomendation Group. (2014). Recommendations for HLA-B*15:02 and HLA-A*31:01 genetic testing to reduce the risk of carbamazepine-induced hypersensitivity reactions. *Epilepsia, 55*(4), 496–506. http://dx.doi.org/10.1111/epi.12564. PubMed PMID:24597466.

Anonymous. (2014). *Table of pharmacogenomic biomarkers in drug labeling*. Available from: http://www.fda.gov/drugs/scienceresearch/researchareas/pharmacogenetics/ucm083378.htm.

Arranz, M. J., & de Leon, J. (2007). Pharmacogenetics and pharmacogenomics of schizophrenia: a review of last decade of research. *Molecular Psychiatry, 12*(8), 707–747. http://dx.doi.org/10.1038/sj.mp.4002009. PubMed PMID:17549063.

Berglund, E. D., Liu, C., Sohn, J. W., Liu, T., Kim, M. H., Lee, C. E., ... Elguist, J. K. (2013). Serotonin 2C receptors in pro-opiomelanocortin neurons regulate energy and glucose homeostasis. *The Journal of Clinical Investigation, 123*(12), 5061–5070. http://dx.doi.org/10.1172/JCI70338. PubMed PMID:24177424; PubMed Central PMCID:PMC3859401.

Berney, P. (2005). Dose-response relationship of recent antidepressants in the short-term treatment of depression. *Dialogues in Clinical Neuroscience, 7*(3), 249–262. Epub 2005/09/15. PubMed PMID:16156383; PubMed Central PMCID:PMC3181733.

Blumenthal, S. R., Castro, V. M., Clements, C. C., Rosenfield, H. R., Murphy, S. N., Fava, M., ... Perlis, R. H. (2014). An electronic health records study of long-term weight gain following antidepressant use. *JAMA Psychiatry, 71*(8), 889–896. http://dx.doi.org/10.1001/jamapsychiatry.2014.414. PubMed PMID:24898363.

Can, A., Schulze, T. G., & Gould, T. D. (2014). Molecular actions and clinical pharmacogenetics of lithium therapy. *Pharmacology, Biochemistry, and Behavior, 123*, 3–16. http://dx.doi.org/10.1016/j.pbb.2014.02.004. PubMed PMID:24534415; PubMed Central PMCID:PMC4220538.

Cantrell, C. R., Eaddy, M. T., Shah, M. B., Regan, T. S., & Sokol, M. C. (2006). Methods for evaluating patient adherence to antidepressant therapy: a real-world comparison of adherence and economic outcomes. *Medical Care, 44*(4), 300–303. http://dx.doi.org/10.1097/01.mlr.0000204287.82701.9b. PubMed PMID:16565629.

Chang, Y., Yang, L. Y., Zhang, M. C., & Liu, S. Y. (2014). Correlation of the UGT1A4 gene polymorphism with serum concentration and therapeutic efficacy of lamotrigine in Han Chinese of Northern China. *European Journal of Clinical Pharmacology, 70*(8), 941–946. http://dx.doi.org/10.1007/s00228-014-1690-1. PubMed PMID:24820767.

Chao, S., Roberts, J. S., Marteau, T. M., Silliman, R., Cupples, L. A., & Green, R. C. (2008). Health behavior changes after genetic risk assessment for Alzheimer disease: the REVEAL Study. Epub 2008/03/05 *Alzheimer Disease and Associated Disorders, 22*(1), 94–97. http://dx.doi.org/10.1097/WAD.0b013e31815a9dcc. PubMed PMID:18317253; PubMed Central PMCID:PMC2483341.

Chen, C. H., Lee, C. S., Lee, M. T., Ouyang, W. C., Chen, C. C., Chong, M. Y., ... Cheng, A. T.-A. (2014). Variant GADL1 and response to lithium therapy in bipolar I disorder. *The New England Journal of Medicine, 370*(2), 119–128. http://dx.doi.org/10.1056/NEJMoa1212444. PubMed PMID:24369049.

Chen, S. F., Shen, Y. C., & Chen, C. H. (2009). HTR2A A-1438G/T102C polymorphisms predict negative symptoms performance upon aripiprazole treatment in schizophrenic patients. *Psychopharmacology, 205*(2), 285–292. http://dx.doi.org/10.1007/s00213-009-1538-z. PubMed PMID:19387614.

Chen, Z., Liew, D., & Kwan, P. (November 25, 2014). Effects of a HLA-B*15:02 screening policy on antiepileptic drug use and severe skin reactions. *Neurology, 83*(22), 2077–2084. http://dx.doi.org/10.1212/WNL.0000000000001034. Epub 2014/10/29.

Cheung, Y. K., Cheng, S. H., Chan, E. J., Lo, S. V., Ng, M. H., & Kwan, P. (2013). HLA-B alleles associated with severe cutaneous reactions to antiepileptic drugs in Han Chinese. *Epilepsia, 54*(7), 1307–1314. http://dx.doi.org/10.1111/epi.12217. PubMed PMID:23692434.

Cirulli, E. T., Urban, T. J., Marino, S. E., Linney, K. N., Birnbaum, A. K., Depondt, C., ... Goldstein, D. B. (2012). Genetic and environmental correlates of topiramate-induced cognitive impairment. *Epilepsia, 53*(1), e5–8. http://dx.doi.org/10.1111/j.1528-1167.2011.03322.x. PubMed PMID:22091778; PubMed Central PMCID:PMC3253145.

Consortium on Lithium Genetics, Hou, L., Heilbronner, U., Rietschel, M., Kato, T., & Kuo, P. H. (2014). Variant GADL1 and response to lithium in bipolar I disorder. *The New England Journal of Medicine, 370*(19), 1857–1859. http://dx.doi.org/10.1056/NEJMc1401817#SA4. PubMed PMID:24806176.

Crowley, J. J., Adkins, D. E., Pratt, A. L., Quackenbush, C. R., van den Oord, E. J., Moy, S. S., ... Sullivan, P. F. (2012). Antipsychotic-induced vacuous chewing movements and extrapyramidal side effects are highly heritable in mice. *The Pharmacogenomics Journal, 12*(2), 147–155. http://dx.doi.org/10.1038/tpj.2010.82. PubMed PMID:21079646; PubMed Central PMCID:PMC3117923.

Crowley, J. J., Brodkin, E. S., Blendy, J. A., Berrettini, W. H., & Lucki, I. (2006). Pharmacogenomic evaluation of the antidepressant citalopram in the mouse tail suspension test. *Neuropsychopharmacology, 31*(11), 2433–2442. http://dx.doi.org/10.1038/sj.npp.1301065. PubMed PMID:16554742.

Crowley, J. J., Kim, Y., Szatkiewicz, J. P., Pratt, A. L., Quackenbush, C. R., Adkins, D. E., ... Sullivan, P. F. (2012). Genome-wide association mapping of loci for antipsychotic-induced extrapyramidal symptoms in mice. *Mammalian Genome: Official Journal of the International Mammalian Genome Society, 23*(5–6), 322–335. http://dx.doi.org/10.1007/s00335-011-9385-8. PubMed PMID:22207321; PubMed Central PMCID:PMC3356790.

De Luca, V., Mueller, D. J., de Bartolomeis, A., & Kennedy, J. L. (2007). Association of the HTR2C gene and antipsychotic induced weight gain: a meta-analysis. *International Journal of Neuropsychopharmacology, 10*(5), 697–704. http://dx.doi.org/10.1017/S1461145707007547. PubMed PMID:17291373.

Dorus, E., Pandey, G. N., & Davis, J. M. (1975). Genetic determinant of lithium ion distribution. An in vitro and in vivo monozygotic-dizygotic twin study. *Archives of General Psychiatry, 32*(9), 1097–1102. PubMed PMID:1237277.

Drogemoller, B. I., Niehaus, D. J., Chiliza, B., van der Merwe, L., Asmal, L., Malhotra, A. K., ... Warnich, L. (2014). Patterns of variation influencing antipsychotic treatment outcomes in South African first-episode schizophrenia patients. *Pharmacogenomics, 15*(2), 189–199. http://dx.doi.org/10.2217/pgs.13.218. PubMed PMID:24444409.

Fagerness, J., Fonseca, E., Hess, G. P., Scott, R., Gardner, K. R., Koffler, M., ... Lombard, J. (2014). Pharmacogenetic-guided psychiatric intervention associated with increased adherence and cost savings. *The American Journal of Managed Care, 20*(5), e146–e156. PubMed PMID:25326929.

Fava, M., Alpert, J., Nierenberg, A., Lagomasino, I., Sonawalla, S., Tedlow, J., ... Rosenbaum, J. F. (2002). Double-blind study of high-dose fluoxetine versus lithium or desipramine augmentation of fluoxetine in partial responders and nonresponders to fluoxetine. *Journal of Clinical Psychopharmacology, 22*(4), 379–387. PubMed PMID:12172337.

Fava, M., Rosenbaum, J. F., McGrath, P. J., Stewart, J. W., Amsterdam, J. D., & Quitkin, F. M. (1994). Lithium and tricyclic augmentation of fluoxetine treatment for resistant major depression: a double-blind, controlled study. *American Journal of Psychiatry, 151*(9), 1372–1374. PubMed PMID:8067495.

Ferreira, M. A., O'Donovan, M. C., Meng, Y. A., Jones, I. R., Ruderfer, D. M., Jones, L., ... Wellcome Trust Case Control Consortium. (2008). Collaborative genome-wide association analysis supports a role for ANK3 and CACNA1C in bipolar disorder. *Nature Genetics, 40*(9), 1056–1058. http://dx.doi.org/10.1038/ng.209. PubMed PMID:18711365; PubMed Central PMCID:PMC2703780.

Fountoulakis, K. N., & Moller, H. J. (2011). Efficacy of antidepressants: a re-analysis and re-interpretation of the Kirsch data. *International Journal of Neuropsychopharmacology, 14*(3), 405–412. http://dx.doi.org/10.1017/S1461145710000957. PubMed PMID:20800012.

Franchini, L., Serretti, A., Gasperini, M., & Smeraldi, E. (1998). Familial concordance of fluvoxamine response as a tool for differentiating mood disorder pedigrees. *Journal of Psychiatric Research, 32*(5), 255–259. http://dx.doi.org/10.1016/S0022-3956(98)00004-1. PubMed PMID:9789202.

Gallagher, P. J., Castro, V., Fava, M., Weilburg, J. B., Murphy, S. N., Gainer, V. S., ... Perlis, R. H. (2012). Antidepressant response in patients with major depression exposed to NSAIDs: a pharmacovigilance study. Epub 2012/10/04 *The American Journal of Psychiatry, 169*(10), 1065–1072. http://dx.doi.org/10.1176/appi.ajp.2012.11091325. PubMed PMID:23032386.

GENDEP Investigators, MARS Investigators, & STAR*D Investigators. (February 2013). Common genetic variation and antidepressant efficacy in major depressive disorder: a meta-analysis of three genome-wide pharmacogenetic studies. *American Journal of Psychiatry, 170*(2), 207–217. http://dx.doi.org/10.1176/appi.ajp.2012.12020237. Pub Med PMID:23377640.

Goldstein, J. I., Jarskog, L. F., Hilliard, C., Alfirevic, A., Duncan, L., Fourches, D., ... Sullivan, P. F. (2014). Clozapine-induced agranulocytosis is associated with rare HLA-DQB1 and HLA-B alleles. *Nature Communications, 5*, 4757. http://dx.doi.org/10.1038/ncomms5757. PubMed PMID:25187353; PubMed Central PMCID:PMC4155508.

Grof, P., Duffy, A., Cavazzoni, P., Grof, E., Garnham, J., MacDougall, M., ... Alda, M. (2002). Is response to prophylactic lithium a familial trait? *The Journal of Clinical Psychiatry, 63*(10), 942–947. PubMed PMID:12416605.

Grover, S., & Kukreti, R. (2014). HLA alleles and hypersensitivity to carbamazepine: an updated systematic review with meta-analysis. *Pharmacogenetics and Genomics, 24*(2), 94–112. http://dx.doi.org/10.1097/FPC.0000000000000021. PubMed PMID:24336023.

Gulcebi, M. I., Ozkaynakci, A., Goren, M. Z., Aker, R. G., Ozkara, C., & Onat, F. Y. (2011). The relationship between UGT1A4 polymorphism and serum concentration of lamotrigine in patients with epilepsy. *Epilepsy Research, 95*(1–2), 1–8. http://dx.doi.org/10.1016/j.eplepsyres.2011.01.016. PubMed PMID:21601426.

Hall-Flavin, D. K., Winner, J. G., Allen, J. D., Jordan, J. J., Nesheim, R. S., Snyder, K. A., ... Mrazek, D. A. (2012). Using a pharmacogenomic algorithm to guide the treatment of depression. *Translational Psychiatry, 2*, e172. http://dx.doi.org/10.1038/tp.2012.99. PubMed PMID:23047243; PubMed Central PMCID:PMC3565829.

Herbild, L., Bech, M., & Gyrd-Hansen, D. (2009). Estimating the Danish populations' preferences for pharmacogenetic testing using a discrete choice experiment. The case of treating depression. *Value Health, 12*(4), 560–567. http://dx.doi.org/10.1111/j.1524-4733.2008.00465.x. PubMed PMID:18980634.

Hicks, J. K., Swen, J. J., Thorn, C. F., Sangkuhl, K., Kharasch, E. D., Ellingrod, V. L., ... Clinical Pharmacogenetics Implementation Consortium. (2013). Clinical pharmacogenetics implementation consortium guideline for CYP2D6 and CYP2C19 genotypes and dosing of tricyclic antidepressants. *Clinical Pharmacology and Therapeutics, 93*(5), 402–408. http://dx.doi.org/10.1038/clpt.2013.2. PubMed PMID:23486447; PubMed Central PMCID:PMC3689226.

Hoop, J. G., Roberts, L. W., Green Hammond, K. A., & Cox, N. J. (2008). Psychiatrists' attitudes regarding genetic testing and patient safeguards: a preliminary study. *Genetic Testing, 12*(2), 245–252. http://dx.doi.org/10.1089/gte.2007.0097. PubMed PMID:18452395.

Hwang, I. C., Park, J. Y., Myung, S. K., Ahn, H. Y., Fukuda, K., & Liao, Q. (2014). OPRM1 A118G gene variant and postoperative opioid requirement: a systematic review and meta-analysis. *Anesthesiology, 121*(4), 825–834. http://dx.doi.org/10.1097/ALN.0000000000000405. PubMed PMID:25102313.

International Schizophrenia Consortium, Purcell, S. M., Wray, N. R., Stone, J. L., Visscher, P. M., O'Donovan, M. C., … Sklar, P. (2009). Common polygenic variation contributes to risk of schizophrenia and bipolar disorder. *Nature, 460*(7256), 748–752. http://dx.doi.org/10.1038/nature08185. PubMed PMID:19571811.

Julius, R. J., Novitsky, M. A., Jr., & Dubin, W. R. (2009). Medication adherence: a review of the literature and implications for clinical practice. *Journal of Psychiatric Practice, 15*(1), 34–44. http://dx.doi.org/10.1097/01.pra.0000344917.43780.77. PubMed PMID:19182563.

Kahn, R. S., Fleischhacker, W. W., Boter, H., Davidson, M., Vergouwe, Y., Keet, I. P., … EUFEST Study Group. (2008). Effectiveness of antipsychotic drugs in first-episode schizophrenia and schizophreniform disorder: an open randomised clinical trial. *Lancet, 371*(9618), 1085–1097. http://dx.doi.org/10.1016/S0140-6736(08)60486-9. PubMed PMID:18374841.

Kaminski, R. M., Banerjee, M., & Rogawski, M. A. (2004). Topiramate selectively protects against seizures induced by ATPA, a GluR5 kainate receptor agonist. *Neuropharmacology, 46*(8), 1097–1104. http://dx.doi.org/10.1016/j.neuropharm.2004.02.010. PubMed PMID:15111016.

Kapur, S., & Mamo, D. (2003). Half a century of antipsychotics and still a central role for dopamine D2 receptors. *Progress in Neuro-Psychopharmacology & Biological Psychiatry, 27*(7), 1081–1090. http://dx.doi.org/10.1016/j.pnpbp.2003.09.004. PubMed PMID:14642968.

Keers, R., & Aitchison, K. J. (2011). Pharmacogenetics of antidepressant response. *Expert Review of Neurotherapeutics, 11*(1), 101–125. http://dx.doi.org/10.1586/ern.10.186. PubMed PMID:21158559.

Keers, R., Bonvicini, C., Scassellati, C., Uher, R., Placentino, A., Giovannini, C., … Gennarelli, M. (2011). Variation in GNB3 predicts response and adverse reactions to antidepressants. *Journal of Psychopharmacology, 25*(7), 867–874. http://dx.doi.org/10.1177/0269881110376683. PubMed PMID:20826553.

Kloiber, S., Ripke, S., Kohli, M. A., Reppermund, S., Salyakina, D., Uher, R., … Lucae, S. (2013). Resistance to antidepressant treatment is associated with polymorphisms in the leptin gene, decreased leptin mRNA expression, and decreased leptin serum levels. *European Neuropsychopharmacology: The Journal of the European College of Neuropsychopharmacology, 23*(7), 653–662. http://dx.doi.org/10.1016/j.euroneuro.2012.08.010. PubMed PMID:23026132; PubMed Central PMCID:PMC4221661.

Kranzler, H. R., Armeli, S., Tennen, H., Gelernter, J., & Covault, J. (2014). GRIK1 genotype and daily expectations of alcohol's positive effects moderate the reduction of heavy drinking by topiramate. *Experimental and Clinical Psychopharmacology, 22*(6), 494–501. http://dx.doi.org/10.1037/a0038350. PubMed PMID:25436841; PubMed Central PMCID:PMC4251491.

Kranzler, H. R., Covault, J., Feinn, R., Armeli, S., Tennen, H., Arias, A. J., … Kampman, K. M. (2014). Topiramate treatment for heavy drinkers: moderation by a GRIK1 polymorphism. *The American Journal of Psychiatry, 171*(4), 445–452. http://dx.doi.org/10.1176/appi.ajp.2013.13081014. PubMed PMID:24525690; PubMed Central PMCID:PMC3997125.

Kurose, K., Hiratsuka, K., Ishiwata, K., Nishikawa, J., Nonen, S., Azuma, J., … Saito, Y. (2012). Genome-wide association study of SSRI/SNRI-induced sexual dysfunction in a Japanese cohort with major depression. *Psychiatry Research, 198*(3), 424–429. http://dx.doi.org/10.1016/j.psychres.2012.01.023. Epub 2012/03/27. PubMed PMID:22445761.

Kusumi, I., Boku, S., & Takahashi, Y. (2014). Psychopharmacology of atypical antipsychotic drugs: from the receptor binding profile to neuroprotection and neurogenesis. *Psychiatry and Clinical Neurosciences*. http://dx.doi.org/10.1111/pcn.12242. PubMed PMID:25296946.

Lett, T. A., Wallace, T. J., Chowdhury, N. I., Tiwari, A. K., Kennedy, J. L., & Muller, D. J. (2012). Pharmacogenetics of antipsychotic-induced weight gain: review and clinical implications. *Molecular Psychiatry, 17*(3), 242–266. http://dx.doi.org/10.1038/mp.2011.109. PubMed PMID:21894153.

Li, L. J., Hu, F. Y., Wu, X. T., An, D. M., Yan, B., & Zhou, D. (2013). Predictive markers for carbamazepine and lamotrigine-induced maculopapular exanthema in Han Chinese. *Epilepsy Research, 106*(1–2), 296–300. http://dx.doi.org/10.1016/j.eplepsyres.2013.05.004. PubMed PMID:23830818.

Lobello, K. W., Preskorn, S. H., Guico-Pabia, C. J., Jiang, Q., Paul, J., Nichols, A. I., … Ninan, P. T. (2010). Cytochrome P450 2D6 phenotype predicts antidepressant efficacy of venlafaxine: a secondary analysis of 4 studies in major depressive disorder. *Journal of Clinical Psychiatry*. http://dx.doi.org/10.4088/JCP.08m04773blu. Epub 2010/05/06, PubMed PMID:20441720.

Malhotra, A. K., Correll, C. U., Chowdhury, N. I., Muller, D. J., Gregersen, P. K., Lee, A. T., … Kennedy, J. L. (2012). Association between common variants near the melanocortin 4 receptor gene and severe antipsychotic drug-induced weight gain. *Archives of General Psychiatry, 69*(9), 904–912. http://dx.doi.org/10.1001/archgenpsychiatry.2012.191. PubMed PMID:22566560; PubMed Central PMCID:PMC4166499.

Malhotra, A. K., Zhang, J. P., & Lencz, T. (2012). Pharmacogenetics in psychiatry: translating research into clinical practice. *Molecular Psychiatry, 17*(8), 760–769. http://dx.doi.org/10.1038/mp.2011.146. PubMed PMID:22083729; PubMed Central PMCID:PMC3928114.

Ma, X., Maimaitirexiati, T., Zhang, R., Gui, X., Zhang, W., Xu, G., & Hu, G. (2014). HTR2C polymorphisms, olanzapine-induced weight gain and antipsychotic-induced metabolic syndrome in schizophrenia patients: a meta-analysis. *International Journal of Psychiatry in Clinical Practice, 18*(4), 229–242. http://dx.doi.org/10.3109/13651501.2014.957705. PubMed PMID:25152019.

Manchia, M., Adli, M., Akula, N., Ardau, R., Aubry, J. M., Backlund, L., … Alda, M. (2013). Assessment of response to lithium maintenance treatment in bipolar disorder: a consortium on lithium genetics (ConLiGen) report. *PLoS One, 8*(6), e65636. http://dx.doi.org/10.1371/journal.pone.0065636. PubMed PMID:23840348; PubMed Central PMCID:PMC3686769.

Manolopoulos, V. G., Ragia, G., & Alevizopoulos, G. (2012). Pharmacokinetic interactions of selective serotonin reuptake inhibitors with other commonly prescribed drugs in the era of pharmacogenomics. Epub 2012/06/22 *Drug Metabolism and Drug Interactions, 27*(1), 19–31. http://dx.doi.org/10.1515/dmdi-2011-0033. PubMed PMID:22718622.

McClay, J. L., Adkins, D. E., Aberg, K., Bukszar, J., Khachane, A. N., Keefe, R. S., … van den Oord, E. J. (2011). Genome-wide pharmacogenomic study of neurocognition as an indicator of antipsychotic treatment response in schizophrenia. *Neuropsychopharmacology, 36*(3), 616–626. http://dx.doi.org/10.1038/npp.2010.193. PubMed PMID:21107309; PubMed Central PMCID:PMC3055694.

McClay, J. L., Adkins, D. E., Aberg, K., Stroup, S., Perkins, D. O., Vladimirov, V. I., … van den Oord, E. J. (2011). Genome-wide pharmacogenomic analysis of response to treatment with antipsychotics. *Molecular Psychiatry, 16*(1), 76–85. http://dx.doi.org/10.1038/mp.2009.89. PubMed PMID:19721433; PubMed Central PMCID:PMC2888895.

McCormack, M., Alfirevic, A., Bourgeois, S., Farrell, J. J., Kasperaviciute, D., Carrington, M., … Pirmohamed, M. (2011). HLA-A*3101 and carbamazepine-induced hypersensitivity reactions in Europeans. *The New England Journal of Medicine, 364*(12), 1134–1143. http://dx.doi.org/10.1056/NEJMoa1013297. PubMed PMID:21428769; PubMed Central PMCID:PMC3113609.

Mrazek, D. A., Biernacka, J. M., O'Kane, D. J., Black, J. L., Cunningham, J. M., Drews, M. S., … Weinshilboum, R. M. (2011). CYP2C19 variation and citalopram response. *Pharmacogenetics and Genomics, 21*(1), 1–9. PubMed PMID:21192344; PubMed Central PMCID:PMC3090085.

O'Dushlaine, C., Ripke, S., Ruderfer, D. M., Hamilton, S. P., Fava, M., Iosifescu, D. V., … Perlis, R. H. (2014). Rare copy number variation in treatment-resistant major depressive disorder. *Biological Psychiatry, 76*(7), 536–541. http://dx.doi.org/10.1016/j.biopsych.2013.10.028. PubMed PMID:24529801; PubMed Central PMCID:PMC4104153.

O'Reilly, R. L., Bogue, L., & Singh, S. M. (1994). Pharmacogenetic response to antidepressants in a multicase family with affective disorder. *Biological Psychiatry, 36*(7), 467–471. PubMed PMID:7811843.

Pan, J. Q., Lewis, M. C., Ketterman, J. K., Clore, E. L., Riley, M., Richards, K. R., … Haggarty, S. J. (2011). AKT kinase activity is required for lithium to modulate mood-related behaviors in mice. *Neuropsychopharmacology, 36*(7), 1397–1411. http://dx.doi.org/10.1038/npp.2011.24. PubMed PMID:21389981; PubMed Central PMCID:PMC3096809.

Papakostas, G. I., Shelton, R. C., Zajecka, J. M., Bottiglieri, T., Roffman, J., Cassiello, C., … Fava, M. (2014). Effect of adjunctive L-methylfolate 15 mg among inadequate responders to SSRIs in depressed patients who were stratified by biomarker levels and genotype: results from a randomized clinical trial. *The Journal of Clinical Psychiatry, 75*(8), 855–863. http://dx.doi.org/10.4088/JCP.13m08947. PubMed PMID:24813065.

Pare, C. M., Rees, L., & Sainsbury, M. J. (1962). Differentiation of two genetically specific types of depression by the response to anti-depressants. *Lancet, 2*(7270), 1340–1343. PubMed PMID:13941389.

Patsopoulos, N. A., Ntzani, E. E., Zintzaras, E., & Ioannidis, J. P. (2005). CYP2D6 polymorphisms and the risk of tardive dyskinesia in schizophrenia: a meta-analysis. *Pharmacogenetics and Genomics, 15*(3), 151–158. PubMed PMID:15861039.

Perlis, R. H. (2011). Translating biomarkers to clinical practice. *Molecular Psychiatry, 16*(11), 1076–1087. http://dx.doi.org/10.1038/mp.2011.63. PubMed PMID:21709685; PubMed Central PMCID:PMC3752672.

Perlis, R. H. (2013). A clinical risk stratification tool for predicting treatment resistance in major depressive disorder. Epub 2013/02/06 *Biological Psychiatry, 74*(1), 7–14. http://dx.doi.org/10.1016/j.biopsych.2012.12.007. PubMed PMID:23380715; PubMed Central PMCID:PMC3690142.

Perlis, R. H., Beasley, C. M., Jr., Wines, J. D., Jr., Tamura, R. N., Cusin, C., Shear, D., … Fava, M. (2007). Treatment-associated suicidal ideation and adverse effects in an open, multicenter trial of fluoxetine for major depressive episodes. *Psychotherapy and Psychosomatics, 76*(1), 40–46. http://dx.doi.org/10.1159/000096363. PubMed PMID:17170562.

Perlis, R. H., Iosifescu, D. V., Castro, V. M., Murphy, S. N., Gainer, V. S., Minnier, J., … Smoller, J. W. (2011). Using electronic medical records to enable large-scale studies in psychiatry: treatment resistant depression as a model. Epub 2011/06/21 *Psychologie Medicale*, 1–10. http://dx.doi.org/10.1017/S0033291711000997. PubMed PMID:21682950.

Perlis, R. H., Laje, G., Smoller, J. W., Fava, M., Rush, A. J., & McMahon, F. J. (2009). Genetic and clinical predictors of sexual dysfunction in citalopram-treated depressed patients. *Neuropsychopharmacology, 34*(7), 1819–1828. http://dx.doi.org/10.1038/npp.2009.4. PubMed PMID:19295509.

Perlis, R. H., Mischoulon, D., Smoller, J. W., Wan, Y. J., Lamon-Fava, S., Lin, K. M., … Fava, M. (2003). Serotonin transporter polymorphisms and adverse effects with fluoxetine treatment. *Biological Psychiatry, 54*(9), 879–883. PubMed PMID:14573314.

Perlis, R. H., Moorjani, P., Fagerness, J., Purcell, S., Trivedi, M. H., Fava, M., … Smoller, J. W. (2008). Pharmacogenetic analysis of genes implicated in rodent models of antidepressant response: association of TREK1 and treatment resistance in the STAR(*)D study. *Neuropsychopharmacology, 33*(12), 2810–2819. http://dx.doi.org/10.1038/npp.2008.6. PubMed PMID:18288090.

Perlis, R. H., Patrick, A., Smoller, J. W., & Wang, P. S. (2009). When is pharmacogenetic testing for antidepressant response ready for the clinic? A cost-effectiveness analysis based on data from the STAR*D study. Epub 2009/06/06. doi: npp200950 [pii] *Neuropsychopharmacology, 34*(10), 2227–2236. http://dx.doi.org/10.1038/npp.2009.50. PubMed PMID:19494805.

Perlis, R. H., Smoller, J. W., Ferreira, M. A., McQuillin, A., Bass, N., Lawrence, J., … Purcell, S. (2009). A genomewide association study of response to lithium for prevention of recurrence in bipolar disorder. *The American Journal of Psychiatry, 166*(6), 718–725. http://dx.doi.org/10.1176/appi.ajp.2009.08111633. PubMed PMID:19448189; PubMed Central PMCID:PMC3908470.

Popp, J., Leucht, S., Heres, S., & Steimer, W. (2006). Serotonin transporter polymorphisms and side effects in antidepressant therapy—a pilot study. *Pharmacogenomics, 7*(2), 159–166. http://dx.doi.org/10.2217/14622416.7.2.159. PubMed PMID:16515395.

Porcelli, S., Fabbri, C., & Serretti, A. (2012). Meta-analysis of serotonin transporter gene promoter polymorphism (5-HTTLPR) association with anti-depressant efficacy. *European Neuropsychopharmacology, 22*(4), 239–258. http://dx.doi.org/10.1016/j.euroneuro.2011.10.003. PubMed PMID:22137564.

Ray, L. A., Miranda, R., Jr., MacKillop, J., McGeary, J., Tidey, J. W., Rohsenow, D. J., … Monti, P. M. (2009). A preliminary pharmacogenetic investigation of adverse events from topiramate in heavy drinkers. *Experimental and Clinical Psychopharmacology, 17*(2), 122–129. http://dx.doi.org/10.1037/a0015700. PubMed PMID:19331489; PubMed Central PMCID:PMC3682424.

Riedel, M., Schwarz, M. J., Strassnig, M., Spellmann, I., Muller-Arends, A., Weber, K., ... Moller, H. J. (2005). Risperidone plasma levels, clinical response and side-effects. *European Archives of Psychiatry and Clinical Neuroscience, 255*(4), 261–268. http://dx.doi.org/10.1007/s00406-004-0556-4. PubMed PMID:15565299.

Ripke, S., O'Dushlaine, C., Chambert, K., Moran, J. L., Kahler, A. K., Akterin, S., ... Sullivan, P. F. (2013). Genome-wide association analysis identifies 13 new risk loci for schizophrenia. *Nature Genetics, 45*(10), 1150–1159. http://dx.doi.org/10.1038/ng. 2742. PubMed PMID:23974872; PubMed Central PMCID:PMC3827979.

Roberson, A. M., Castro, V. M., Cagan, A., & Perlis, R. H. (March 2016). Antidepressant nonadherence in routine clinical settings determined from discarded blood samples. *Journal of Clinical Psychiatry, 77*(3), 359–362. http://dx.doi.org/10.4088/JCP.14m09612.

Roberts, J. S., Cupples, L. A., Relkin, N. R., Whitehouse, P. J., & Green, R. C. (2005). Genetic risk assessment for adult children of people with Alzheimer's disease: the risk evaluation and education for Alzheimer's disease (REVEAL) study. Epub 2005/11/25 *Journal of Geriatric Psychiatry and Neurology, 18*(4), 250–255. http://dx.doi.org/10.1177/0891988705281883. PubMed PMID:16306249.

Simon, G. E., & Perlis, R. H. (2010). Personalized medicine for depression: can we match patients with treatments? *The American Journal of Psychiatry, 167*(12), 1445–1455. http://dx.doi.org/10.1176/appi.ajp.2010.09111680. PubMed PMID:20843873; PubMed Central PMCID:PMC3723328.

Singh, A. B., Bousman, C. A., Ng, C. H., Byron, K., & Berk, M. (2012). ABCB1 polymorphism predicts escitalopram dose needed for remission in major depression. *Translational Psychiatry, 2*, e198. http://dx.doi.org/10.1038/tp.2012.115. PubMed PMID:23188198; PubMed Central PMCID:PMC3565756.

Singkham, N., Towanabut, S., Lertkachatarn, S., & Punyawudho, B. (2013). Influence of the UGT2B7 -161C>T polymorphism on the population pharmacokinetics of lamotrigine in Thai patients. *European Journal of Clinical Pharmacology, 69*(6), 1285–1291. http://dx.doi.org/10.1007/s00228-012-1449-5. PubMed PMID:23263737.

Sklar, P., Smoller, J. W., Fan, J., Ferreira, M. A., Perlis, R. H., Chambert, K., ... Purcell, S. M. (2008). Whole-genome association study of bipolar disorder. *Molecular Psychiatry, 13*(6), 558–569. PubMed PMID:18317468.

Smeraldi, E., Petroccione, A., Gasperini, M., Macciardi, F., Orsini, A., & Kidd, K. K. (1984). Outcomes on lithium treatment as a tool for genetic studies in affective disorders. *Journal of Affective Disorders, 6*(2), 139–151. PubMed PMID:6233346.

Stein, M. B., Cantrell, C. R., Sokol, M. C., Eaddy, M. T., & Shah, M. B. (2006). Antidepressant adherence and medical resource use among managed care patients with anxiety disorders. *Psychiatric Services, 57*(5), 673–680. http://dx.doi.org/10.1176/appi.ps.57.5.673. PubMed PMID:16675762.

Stone, M., Laughren, T., Jones, M. L., Levenson, M., Holland, P. C., Hughes, A., ... Rochester, G. (2009). Risk of suicidality in clinical trials of antidepressants in adults: analysis of proprietary data submitted to US Food and Drug Administration. *BMJ, 339*, b2880. http://dx.doi.org/10.1136/bmj.b2880. PubMed PMID:19671933; PubMed Central PMCID:PMC2725270.

Tansey, K. E., Guipponi, M., Perroud, N., Bondolfi, G., Domenici, E., Evans, D., ... Uher, R. (2012). Genetic predictors of response to serotonergic and noradrenergic antidepressants in major depressive disorder: a genome-wide analysis of individual-level data and a meta-analysis. *PLoS Medicine, 9*(10), e1001326. http://dx.doi.org/10.1371/journal.pmed.1001326. PubMed PMID:23091423; PubMed Central PMCID:PMC3472989.

Tansey, K. E., Rucker, J. J., Kavanagh, D. H., Guipponi, M., Perroud, N., Bondolfi, G., ... Uher, R. (2014). Copy number variants and therapeutic response to antidepressant medication in major depressive disorder. *The Pharmacogenomics Journal, 14*(4), 395–399. http://dx.doi.org/10.1038/tpj.2013.51. PubMed PMID:24445990.

Teicher, M. H., Glod, C. A., & Cole, J. O. (1993). Antidepressant drugs and the emergence of suicidal tendencies. *Drug Safety: An International Journal of Medical Toxicology and Drug Experience, 8*(3), 186–212. PubMed PMID:8452661.

Tenback, D. E., van Harten, P. N., & van Os, J. (2009). Non-therapeutic risk factors for onset of tardive dyskinesia in schizophrenia: a meta-analysis. *Movement Disorders: Official Journal of the Movement Disorder Society, 24*(16), 2309–2315. http://dx.doi.org/10.1002/mds.22707. PubMed PMID:19645070.

Tung, J. Y., Do, C. B., Hinds, D. A., Kiefer, A. K., Macpherson, J. M., Chowdry, A. B., ... Eriksson, N. (2011). Efficient replication of over 180 genetic associations with self-reported medical data. *PLoS One, 6*(8), e23473. http://dx.doi.org/10.1371/journal.pone.0023473. PubMed PMID:21858135; PubMed Central PMCID:PMC3157390.

Uher, R., Tansey, K. E., Malki, K., & Perlis, R. H. (2012). Biomarkers predicting treatment outcome in depression: what is clinically significant? *Pharmacogenomics, 13*(2), 233–240. http://dx.doi.org/10.2217/pgs.11.161. PubMed PMID:22256872; PubMed Central PMCID:PMC3566553.

Uher, R., Tansey, K. E., Dew, T., Maier, W., Mors, O., Hauser, J., ... McGuffin, P. (December 1, 2014). An inflammatory biomarker as a differential predictor of outcome of depression treatment with escitalopram and nortriptyline. *American Journal of Psychiatry, 171*(12), 1278–1286. http://dx.doi.org/10.1176/appi.ajp.2014.14010094. Epub 2014/10/31.

Volpi, S., Heaton, C., Mack, K., Hamilton, J. B., Lannan, R., Wolfgang, C. D., ... Lavedan, C. (2009). Whole genome association study identifies polymorphisms associated with QT prolongation during iloperidone treatment of schizophrenia. *Molecular Psychiatry, 14*(11), 1024–1031. http://dx.doi.org/10.1038/mp.2008.52. PubMed PMID:18521091.

Warden, D., Trivedi, M. H., Carmody, T., Toups, M., Zisook, S., Lesser, I., ... Rush, A. J. (2014). Adherence to antidepressant combinations and monotherapy for major depressive disorder: a CO-MED report of measurement-based care. *Journal of Psychiatric Practice, 20*(2), 118–132. http://dx.doi.org/10.1097/01.pra.0000445246.46424.fe. PubMed PMID:24638046.

Winner, J., Allen, J. D., Altar, C. A., & Spahic-Mihajlovic, A. (2013). Psychiatric pharmacogenomics predicts health resource utilization of outpatients with anxiety and depression. *Translational Psychiatry, 3*, e242. http://dx.doi.org/10.1038/tp.2013.2. Epub 2013/03/21, PubMed PMID:23511609; PubMed Central PMCID:PMC3625917.

Wong, M. L., Dong, C., Flores, D. L., Ehrhart-Bornstein, M., Bornstein, S., Arcos-Burgos, M., & Licinio, J. (December 1, 2014). Clinical outcomes and genome-wide association for a brain methylation site in an antidepressant pharmacogenetics study in Mexican Americans. *American Journal of Psychiatry, 171*(12), 1297–1309. http://dx.doi.org/10.1176/appi.ajp.2014.12091165. Epub 2014/10/31.

Yatham, L. N., Kennedy, S. H., Schaffer, A., Parikh, S. V., Beaulieu, S., O'Donovan, C., … Berk, M. (2009). Canadian Network for Mood and Anxiety Treatments (CANMAT) and International Society for Bipolar Disorders (ISBD) collaborative update of CANMAT guidelines for the management of patients with bipolar disorder: update 2009. *Bipolar Disorders, 11*(3), 225–255. http://dx.doi.org/10.1111/j.1399-5618.2009.00672.x. PubMed PMID:19419382.

Zhang, J. P., Lencz, T., & Malhotra, A. K. (2010). D2 receptor genetic variation and clinical response to antipsychotic drug treatment: a meta-analysis. *The American Journal of Psychiatry, 167*(7), 763–772. http://dx.doi.org/10.1176/appi.ajp.2009.09040598. PubMed PMID:20194480; PubMed Central PMCID:PMC2896457.

Zhang, J. P., & Malhotra, A. K. (2011). Pharmacogenetics and antipsychotics: therapeutic efficacy and side effects prediction. *Expert Opinion on Drug Metabolism & Toxicology, 7*(1), 9–37. http://dx.doi.org/10.1517/17425255.2011.532787. PubMed PMID:21162693; PubMed Central PMCID:PMC3057913.

Index

'*Note:* Page numbers followed by "f" indicate figures, "t" indicate tables and "b" indicate boxes.'

C

D

E

F

G

H

M

N

O

P

Q

R

S

S

T

U